Handbook of Superconducting Materials

Volume I: Superconductivity, Materials and Processes

www.superconductingmaterials.net

Handbook of Superconducting Materials

Volume I: Superconductivity, Materials and Processes

Edited by

David A Cardwell

University of Cambridge

and

David S Ginley

National Renewable Energy Laboratory

Institute of Physics Publishing
Bristol and Philadelphia

British Library Cataloguing-in-Publication Data

A catalogue record for this book is available from the British Library.

ISBN 0 7503 0432 4 (Vol. I)
0 7503 0897 4 (Vol. II)
0 7503 0898 2 (2 Vol. set)

Library of Congress Cataloging-in-Publication Data are available

Online version of encyclopedia at www.superconductingmaterials.net

Development Editor: David Morris
Production Editor: Simon Laurenson
Production Control: Sarah Plenty
Cover Design: Victoria Le Billon
Marketing: Nicola Newey and Verity Cooke

Published by Institute of Physics Publishing, wholly owned by The Institute of Physics, London

Institute of Physics Publishing, Dirac House, Temple Back, Bristol BS1 6BE, UK

US Office: Institute of Physics Publishing, The Public Ledger Building, Suite 929, 150 South Independence Mall West, Philadelphia, PA 19106, USA

Typeset in the UK by Alden Bookset
Printed in the UK by MPG Books Ltd, Bodmin, Cornwall

Contents

VOLUME II: CHARACTERIZATION, APPLICATIONS AND CRYOGENICS

Editors-in-Chief

Editorial Board

Scientific Advisory Board

List of contributors

J S Abell (B2.2.1–4)
School of Metallurgy and Materials,
University of Birmingham,
Birmingham,
UK

L A Abelson (E4.1)
TRW Space and Electronics,
Redondo Beach, CA,
USA

N McN Alford (B4.3)
Centre for Physical Electronics and Materials,
South Bank University,
London,
UK

M F Ashby (H3)
Department of Engineering,
University of Cambridge,
Cambridge,
UK

P A Beharrell (E2.2)
Magnetic Applications Group,
Department of Physics and Astronomy,
University of Southampton,
Southampton,
UK

K Behnia (D3.2.2)
Laboratoire de Physique Quantique,
UPR 5 – CNRS,
Ecole Supérieure de Physique et de Chimie Industrielles,
Paris,
France

E Bellingeri (B3.2.3, C3)
Département de Physique de la Matière Condensée,
Université de Genève,
Genève,
Switzerland

S J Bending (D2.8)
Department of Physics,
University of Bath,
Bath,
UK

A Bitterman (H1)
WestTech Market Information Services,
Los Angeles, CA,
USA

R D Blaugher (E1.3.1, E1.4.2, E1.3.4)
National Renewable Energy Laboratory,
Golden, CO,
USA

N E Booth (E5.1)
Department of Physics,
University of Oxford,
Oxford, UK

A I Braginski (B4.4.3)
Institute of Thin-Film and Ion Technology (ISI),
Research Center Juelich,
Juelich,
Germany (retired)

E H Brandt (A4.2)
Max-Planck-Institut für Metallforschung,
Stuttgart,
Germany

R G Buckley (D1.6)
Industrial Research,
Lower Hutt,
New Zealand

Yu V Bugoslavsky (D2.4)
General Physics Institute,
Moscow,
Russia

D P Butler (E5.2)
Department of Electrical Engineering,
Southern Methodist University,
Dallas, TX,
USA

T W Button (B2.2.1–4, B4.3)
School of Metallurgy and Materials,
University of Birmingham,
Birmingham,
UK

C Buzea (C5)
Research Institute of Electrical Communication
and New Industry Creation Hatchery Center,
Tohoku University,
Sendai,
Japan

A M Campbell (A3.1, D2.1, introduction to D3)
IRC in Superconductivity,
University of Cambridge,
Cambridge, UK

A D Caplin (introduction to D1 and D2)
Blackett Laboratory,
Imperial College,
London,
UK

P Caracino (E1.3.2)
Pirelli Cables and Systems,
Milan,
Italy

D A Cardwell (A1.3, introduction to A1, A2, A3, A4 and B4; B2.3.1)
IRC in Superconductivity,
University of Cambridge,
Cambridge,
UK

A Carrington (D3.2.1)
H H Wills Laboratory of Physics,
University of Bristol,
Bristol,
UK

J R Cave (introduction to E2)
Institut de Recherche d'Hydro-Québec,
Varennes,
Canada

H J Chaloupka (E3.3)
Department of Electrical Engineering,
University of Wuppertal,
Wuppertal,
Germany

C W Chu (G5)
Department of Physics and Texas Center for Superconductivity,
University of Houston,
Houston, TX,
USA

and

Lawrence Berkeley National Laboratory,
Berkeley, CA,
USA

and

Hong Kong University of Science and Technology,
Hong Kong

J R Clem (foreword)
Ames Laboratory,
Iowa State University,
Ames, IA, USA

R Cloots (B2.1)
SUPRAS,
University of Liège,
Chemistry Institute B6,
Liège,
Belgium

L Cooley (B3.3.2)
Applied Superconductivity Center,
University of Wisconsin – Madison,
Madison, WI,
USA

J R Cooper (D3.1)
IRC in Superconductivity and Physics Department,
University of Cambridge,
Cambridge,
UK

G W Crabtree (A4.1)
Argonne National Laboratory,
Argonne, IL,
USA

R Cywinski (D2.5)
Department of Physics and Astronomy,
University of Leeds,
Leeds,
UK

G Desgardin (B2.3.3)
Laboratoire CRISMAT,
ISMRA,
Caen,
France

M Dhallé, (D2.2)
Département de Physique de la Matière Condensée,
Université de Genève,
Genève,
Switzerland

P Diko (D1.4)
Institute of Experimental Physics,
Slovak Academy of Sciences,
Kosice,
Slovak Republic

R Dittmann (B4.4.3)
Institut für Elektrokeramische Materialien (EKM),
Institut für Festkörperforschung,
Forschungszentrum Jülich GmbH,
Jülich,
Germany

G B Donaldson (E4.4)
Department of Physics and Applied Physics,
University of Strathclyde,
Glasgow,
UK

J Donley (H2)
National Renewable Energy Laboratory,
Golden, CO,
USA

S X Dou (B3.1, C2)
Institute of Superconducting and Electronic Materials,
University of Wollongong,
Wollongong, NSW,
Australia

G Duperray (B3.2.1)
La Norville,
France (retired). Previously with Alcatel,
France

J Ekin (B5)
National Institute of Standards and Technology,
Boulder, CO
USA

D M Feldmann (D3.4)
Applied Superconductivity Center,
University of Wisconsin – Madison,
Madison, WI,
USA

D K Finnemore (D1.1.2)
Ames Laboratory,
USDOE and Department of Physics,
Iowa State University,
Ames, IA,
USA

J Flokstra (E4.2)
Low Temperature Division,
Faculty of Applied Physics,
University of Twente,
Enschede,
The Netherlands

R Flükiger (B1, B2.5, introduction to B3, B3.3.1, B3.3.6, C3)
Département de Physique de la Matière Condensée,
Université de Genève,
Genève,
Switzerland

T Forgan (D1.7.2)
School of Physics and Astronomy,
University of Birmingham,
Birmingham, UK

H C Freyhardt (B4.2.1)
Institut für Materialphysik,
Universität Göttingen,
Göttingen,
Germany

J Gallop (introduction to E3, E4 and E5; E5.3)
National Physical Laboratory,
Teddington,
UK

R F Giese (E1.3.3)
Naperville, IL,
USA (retired). Previously with Argonne National Laboratory,
Argonne, IL,
USA

D S Ginley (introduction to A1, A2, A3, A4 and B4; H2)
National Renewable Energy Laboratory,
Golden, CO,
USA

R Gladyshevskii (D1.1.1)
Department of Inorganic Chemistry,
L'viv National University,
L'viv,
Ukraine

W Goldacker (D3.3)
ITP, Forschungszentrum Karlsruhe,
Karlsruhe,
Germany

C E Gough (A2.6)
Department of Physics and Astronomy,
University of Birmingham,
Birmingham, UK

G Grasso (B3.2.2)
INFM-Research Unit of Genoa,
Physics Department,
University of Genoa,
Genoa,
Italy

C R M Grovenor (D1.3)
Department of Materials,
University of Oxford,
Oxford,
UK

D P Hampshire (D2.3, G1)
Department of Physics,
University of Durham,
Durham,
UK

T Hase (B3.3.3)
Kobe Steel Ltd,
Kobe,
Japan

H Hayakawa (B4.4.2)
Department of Quantum Engineering,
Nagoya University,
Nagoya-City,
Japan

P F Herrmann (B3.2.1)
Alcatel,
Optical Fiber Division,
Conflans,
France

A Hewat (D1.7.1)
Institut Laue-Langevin,
Grenoble,
France

M Hervieu (D1.2)
Laboratoire CRISMAT,
Université de Caen,
Caen,
France

P J Hirst (B4.4.1)
QinetiQ,
Malvern Technology Centre,
Malvern,
UK

B Holzapfel (B4.1)
Leibniz Institute for Solid State and Materials Research
Dresden (IFW Dresden),
Dresden,
Germany

R P Hübener (A2.2, A2.3)
Experimentalphysik II,
Universität Tübingen,
Tübingen,
Germany

J R Hull (E2.1)
Energy Technology Division,
Argonne National Laboratory,
Argonne, IL,
USA

R G Humphreys (B4.4.1)
QinetiQ,
Malvern Technology Centre,
Malvern,
UK

Y Iwasa (G4)
Japan Advanced Institute of Science and Technology,
Ishikawa,
Japan

T Izumi (B2.3.2)
Superconductivity Research Laboratory,
ISTEC,
Tokyo,
Japan

H Jones (E1.1, introduction to B5, E1 and F)
Clarendon Laboratory,
University of Oxford,
Oxford, UK

S Julian (G2)
Cavendish Laboratory,
University of Cambridge,
Cambridge,
UK

A M Kadin (E4.6)
HYPRES Inc.,
Elmsford, NY,
USA

D L Kaiser (B2.4)
Ceramics Division,
NIST,
Gaithersburg, MD,
USA

J Kellers (A1.3)
American Superconductor Europe,
Kaarst,
Germany

P H Kes (A4.3)
Kamerlingh Onnes Laboratory,
Leiden University,
Leiden,
The Netherlands

S A Keys (D2.3)
Department of Physics,
University of Durham,
Durham,
UK

S H Kilcoyne (D2.5)
Condensed Matter Group,
Department of Physics and Astronomy,
University of Leeds, Leeds,
UK

K Kitazawa (foreword)
Japan Science and Technology Corporation,
Kawaguchi City,
Japan

A Koblischka-Veneva (C1)
Superconductivity Research Laboratory,
ISTEC,
Tokyo,
Japan

H Koch (E4.3)
Physikalisch-Technische Bundesanstalt (PTB),
Berlin,
Germany

W K Kwok (A4.1)
Argonne National Laboratory,
Argonne, IL,
USA

D C Larbalestier (B3.3.2, D3.4)
Applied Superconductivity Center,
University of Wisconsin – Madison,
Madison, WI,
USA

P Lee (B3.3.2)
Applied Superconductivity Center,
University of Wisconsin – Madison,
Madison, WI,
USA

S Lee (D1.7.2, D2.8)
School of Physics and Astronomy,
University of St Andrews,
St Andrews,
UK

A J Leggett (A3.2)
Department of Physics,
University of Illinois at Urbana-Champaign,
Urbana, IL,
USA

D Lopez (A4.1)
Lucent Technologies,
Murray Hill, NJ,
USA

J L MacManus-Driscoll (B3.2.4)
Department of Materials,
Imperial College,
London,
UK

E Martínez (D2.6)
Universidad de Zaragoza,
Zaragoza,
Spain

L J Masur (A1.3)
American Superconductor Corporation,
Westborough, MA,
USA

B W McConnell (E1.3.5)
Engineering Science and Technology Division,
Oak Ridge National Laboratory,
Oak Ridge, TN,
USA

M McCulloch (E1.4.1, E1.4.3, E1.4.4)
Department of Engineering Science,
University of Oxford,
Oxford,
UK

C Meingast (D3.2.3)
IFP,
Forschungszentrum Karlsruhe,
Karlsruhe,
Germany

R Mele (E1.3.2)
Pirelli Cables and Systems,
Milan,
Italy

P N Mikheenko (C2)
Institute for Superconducting and Electronic Materials,
University of Wollongong,
Wollongong, NSW,
Australia

T Miyatake (B3.3.3)
Kobe Steel Ltd,
Kobe,
Japan

T Miyazaki (B3.3.3)
Electronics Research Laboratory,
Kobe Steel Ltd,
Kobe,
Japan

I Monot (B2.3.3)
Laboratoire CRISMAT,
ISMRA,
Caen,
France

J C Moore (D1.3)
Department of Materials,
University of Oxford,
Oxford,
UK

O A Mukhanov (E4.6)
HYPRES Inc.,
Elmsford, NY,
USA

M Murakami (B1, introduction to B2, B2.3.4, C1)
Superconductivity Research Laboratory,
ISTEC,
Tokyo,
Japan

J M Murduck (E4.1)
Northrop Grumman Science & Technology Center,
Baltimore, MD,
USA

M Nassi (E1.3.2)
Pirelli Cables and Systems,
Milan,
Italy

J Niemeyer (B4.4.2)
Quantum Electronics Department,
Physikalisch-Technische Bundesanstalt,
Braunschweig,
Germany

K Nomura (B2.3.2)
Superconductivity Research Laboratory,
ISTEC,
Tokyo,
Japan

X Obradors (B3.2.5)
Institut de Ciència de Materials de Barcelona,
Universitat Autònoma de Barcelona,
Bellaterra,
Spain

C M Pegrum (A1.3)
Department of Physics and Applied Physics,
University of Strathclyde,
Glasgow,
UK

S J Penn (B4.3)
Centre for Physical Electronics and Materials,
South Bank University,
London,
UK

B Pippard (A1.1)
Cavendish Laboratory,
University of Cambridge,
Cambridge, UK

A A Polyanskii (D3.4)
Applied Superconductivity Center,
University of Wisconsin – Madison,
Madison, WI,
USA

A Porch (A2.5, D2.7)
School of Engineering,
Cardiff University,
Cardiff, UK

R Radebaugh (F)
Cryogenic Technologies Group,
National Institute of Standards and Technology,
Boulder, CO,
USA

B Raveau (D1.2)
Laboratoire CRISMAT,
Université de Caen,
Caen,
France

D L Rayner (E2.3)
Magnex Scientific Limited,
Abingdon,
UK

D Ryan (E1.2)
Oxford Instruments,
Abingdon,
UK

G Saito (G3)
Department of Chemistry,
Graduate School of Science,
Kyoto University,
Kyoto,
Japan

N Sakai (C1)
Superconductivity Research Laboratory,
ISTEC,
Tokyo,
Japan

P V P S S Sastry (C4)
National High Magnetic Field Laboratory,
Florida State University,
Tallahassee, FL,
USA

J Satchell (E4.5)
QinetiQ,
Malvern Technology Centre,
Malvern,
UK

L F Schneemeyer (B2.4)
Bell Laboratories,
Lucent Technologies,
Murray Hill, NJ,
USA

J Schwartz (C4)
National High Magnetic Field Laboratory,
Florida State University,
Tallahassee, FL,
USA

B Seeber (B3.3.5)
Institute of Applied Physics – (GAP),
University of Geneva,
Geneva,
Switzerland

D Shaw (introduction to C and G)
New York State Institute on Superconductivity,
State Universtiy of New York at Buffalo,
Amherst, NY,
USA

Z-Y Shen (E3.1)
DuPont Superconductivity,
Wilmington, DE,
USA

H R Shercliff (H3)
Department of Engineering,
University of Cambridge,
Cambridge,
UK

J Shimoyama (B2.3.5)
Department of Superconductivity,
Graduate School of Engineering,
University of Tokyo,
Tokyo,
Japan

Y Shiohara (B2.3.2)
Superconductivity Research Laboratory,
ISTEC,
Tokyo,
Japan

A H Silver (E4.1)
Rancho Palos Verdes, CA,
USA

A Sin (B3.2.5)
Institut de Ciència de Materials de Barcelona,
Universitat Autònoma de Barcelona,
Bellaterra, Spain

R Sobolewski (E5.2)
Department of Electrical and Computer Engineering and
Laboratory for Laser Energetics,
University of Rochester,
Rochester,
NY, USA
and
Institute of Physics,
Polish Academy of Sciences,
Warszawa,
Poland

O Stadel (B4.2.2)
IOPW,
Technische Universität Braunschweig,
Braunschweig,
Germany

M Strasik (E2.4)
Superconductivity/Magnetics Group,
Boeing Phantom Works,
Seattle, WA,
USA

H L Suo (B3.3.6)
Département de Physique de la Matière Condensée,
Université de Genève,
Genève,
Switzerland

K Tachikawa (B2.2.6)
Faculty of Engineering,
Tokai University,
Japan

S Tajima (C1)
Superconductivity Research Laboratory,
ISTEC,
Tokyo, Japan

T Takeuchi (B3.3.4)
National Institute for Materials Science,
Tsukuba Magnet Laboratory,
Tsukuba, Japan

J Tallon (foreword)
Industrial Research Ltd and Victoria University of Wellington,
Lower Hutt,
New Zealand

E J Tarte (A2.7)
IRC in Superconductivity,
University of Cambridge,
Cambridge, UK

P Toulemonde (B3.3.6)
Département de Physique de la Matière Condensée,
Université de Genève,
Genève,
Switzerland

H J Trodahl (D1.6)
School of Chemical and Physical Sciences,
Victoria University of Wellington,
Wellington,
New Zealand

C Uher (A2.4)
Department of Physics,
University of Michigan,
Ann Arbor, MI,
USA

K K Uprety (C2)
Institute for Superconducting and Electronic Materials,
University of Wollongong,
Wollongong, NSW,
Australia

A Usoskin (B4.2.1)
Zentrum für Funktionswerkstoffe gGmbH,
Göttingen,
Germany

B Utz (B4.2.1)
Siemens AG,
Research Laboratories,
Erlangen,
Germany

O G Vendik (E3.2)
Electronics Department,
Electrotechnical University,
St Petersburg,
Russia

W F Vinen (A1.2)
Department of Physics and Astronomy,
University of Birmingham,
Birmingham, UK

G Wahl (B4.2.2)
IOPW,
Technische Universität Braunschweig,
Braunschweig,
Germany

J H P Watson (E2.2)
Magnetic Applications Group,
Department of Physics and Astronomy,
University of Southampton,
Southampton, UK

H W Weber (B2.6)
Atominstitut der Österreichischen Universitäten,
Vienna,
Austria

F Weiss (B4.2.2)
LMGP – ENSPG – INP Grenoble,
St Martin d'Hères,
France

D Welch (A2.1)
Brookhaven National Laboratory,
Upton, NY,
USA

U Welp (A4.1)
Argonne National Laboratory,
Argonne, IL,
USA

J Wiesmann (B4.1)
Luneburg,
Germany

D Winkler (E4.7)
Imego AB and Chalmers University of Technology and
Göteborg University,
Gothenburg, Sweden

R Wördenweber (B4.2.1)
Institut für Schichten und Grenzflächen,
Forschungszentrum Jülich,
Jülich, Germany

Y Yamada (B4.2.1)
SLR – ISTEC,
Nagoya,
Japan

Y Yamada (B2.2.6)
Faculty of Engineering,
Tokai University,
Japan

T Yamashita (C.5)
New Industry Creation Hatchery Center,
Tohoku University,
Sendai,
Japan
and
CREST Japan Science and Technology Corporation (JST)

Y Yang (D2.6)
Institute of Cryogenics and Energy Research,
University of Southampton,
Southampton, UK

M Yeadon (D1.5)
Institute of Materials Research and Engineering,
Singapore

I R Young (E2.3)
Robert Steiner Magnetic Resonance Unit,
Imperial College School of Medicine,
Hammersmith Hospital,
London, UK

Acknowledgments

The Editors-in-Chief would like to acknowledge the contributions of a number of people and institutions, who have played a major role in the development of the Handbook from concept to publication. The production of the *Handbook* would not have been possible without their input.

Four IOP Editors have been involved with the *Handbook*; Tony Wayte, Gillian Lindsey, Victoria Le Billon and David Morris. Each has worked tirelessly to extract articles and reports from authors and referees against ever-changing deadlines.

Archie Campbell, David Caplin and Masato Murakami have made particularly significant contributions to the *Handbook*, above and beyond their roles of Section Editors. They have advised on all aspects of the *Handbook* and frequently provided solutions to seemingly insurmountable problems that have occurred on a regular basis. They have done this with unflagging enthusiasm despite the logistical nightmares of such a long project.

We acknowledge the contributions of John Clem and David Larbalestier, particularly at the early stages of development of the *Handbook*. Their encouragement and guidance was a pivotal influence in the tutorial concept of the work.

Finally, we would like to acknowledge the support of our home institutions the National Renewable Energy Laboratory and the IRC in Superconductivity, Cambridge University. We would like to acknowledge the assistance of our administrative assistants without whom very little would get done; Carole Allman at NREL and Alicia Kelleher at the IRC. We would also like to acknowledge all those colleagues (you know who you are) who helped with the many aspects of the *Handbook* usually under the auspices of ASAP.

David A Cardwell
David S Ginley

The support of the following agencies is gratefully acknowledged.

B2.3.2: Parts of the studies presented in this paper were supported by New Energy and Industrial Technology Development Organization (NEDO) as a part of its Research and Development of Fundamental Technologies for Superconductor Applications Project under the New Sunshine Program administrated by the Agency of Industrial Science and Technologies MITI of Japan. B2.6: Financial support by the Austrian Science Foundation (Grant No. 11712) and by the European Union (TMR network SUPERCURRENT) is gratefully acknowledged. B3.1: Sincere thanks are given to the Australian Research Council and Metal Manufactures Ltd for financial support. B5: The author acknowledges with much gratitude support for the research and writing of this chapter from the National Institute of Standards and Technology, the US Department of Energy High Energy Physics and Energy Systems Programs, and the Department of Physics, University of Colorado, Boulder. Contribution of NIST, an agency of the US Government; not subject to copyright. Trade names, when used, are provided for the completeness of documentation and to help illustrate the techniques used; they neither constitute nor imply endorsement by NIST. D1.1.2: Work at Ames Laboratory was supported by the US Department of Energy (DOE), Office of Basic Energy Sciences and the Office of Energy Efficiency and Renewable Energy under Contract No. W-7405-ENG-82. D2.3: One of the authors (SAK) wishes to thank the UK Engineering and Physical Sciences Research Council (EPSRC) and Oxford Instruments plc for their support. D3.4: The work in Madison has been supported by EPRI, DOE, the Air Force Office of Scientific Research and the NSF-supported MRSEC at the University of Wisconsin. E2.1: The submitted manuscript has been created by the University of Chicago as Operator of Argonne National Laboratory ("Argonne") under Contract No. W-31-109-ENG-38 with the US Department of Energy. The US Government retains for itself, and others acting on its behalf, a paid-up, nonexclusive, irrevocable worldwide license in said article to reproduce, prepare derivative works, distribute copies to the public, and perform publicly and display publicly, by or on behalf of the Government. E5.2: This research was supported by the US Office of Naval Research grant N00014-02-1-0026 and the National Science Foundation grant DMR-0073366 (Rochester), and the National Science Foundation grant ECS-9800062 and Army Research Office grant ARO-38673PH (SMU). F: Contribution of NIST, notsubject to copyright in the US. G4: The author is supported in part by the JSPS 'Future Program' (RFTF96P00104) and also by Grant-in-Aid for the Priority Area 'Fullerenes and Nanotubes' from the Ministry of Education, Science, Sports, and Culture, Japan. G5: The work in Houston is supported in part by NSF Grant No. DMR-9804325, the T L L Temple Foundation, the John J and Rebecca Moores Endowment, and the State of Texas through the Texas Center for Superconductivity at the University of Houston; and at Lawrence Berkeley National Laboratory by the Director, Office of Science, Office of Basic Energy Sciences, Division of Materials Sciences and Engineering of the US Department of Energy under Contract No. DE-AC03-76SF00098.

Forewords

The *Handbook of Superconducting Materials* covers the full spectrum of properties, materials and applications of both low and high temperature superconductors in a coordinated series of articles written by researchers from all over the world. For such a breadth of introductory and state-of-the-art material it was considered appropriate to develop a foreword that properly reflects the nature of the work. As a result, the *Handbook of Superconducting Materials* is foreworded by three eminent scientists from different continents with distinct expertise in the fields of theory, materials and applications. In this way, we hope to give the reader a real flavour of both the overall scope of the *Handbook* and its contribution to the subject area.

David A Cardwell
David S Ginley

Theory

The field of superconductivity was born in 1911 when Onnes discovered this amazing phenomenon. For a number of years thereafter, only a handful of experimentalists with access to liquid helium could study the known superconductors, and newcomers to the field could read everything written about superconductivity. The field grew, with important milestones being the 1957 publication of the Bardeen–Cooper–Schrieffer (BCS) theory, the 1962 prediction and observation of Josephson effects, and the discovery in the late 1950s and early 1960s of new superconducting alloys that could be fabricated into composites capable of carrying high currents in high magnetic fields. After the discovery in the late 1980s of cuprates with transition temperatures above the boiling point of liquid nitrogen, superconductivity came to be observable in simple tabletop experiments. This generated renewed interest and an unprecedented amount of worldwide research on superconductors. As a consequence, superconductivity has grown into a huge field. About 10^5 scientific papers on this subject have been published just since Bednorz and Müller's 1986 pioneering paper on the cuprates. This poses a daunting challenge not only to graduate students beginning research but also to scientists with previous experience. They may ask, out of all these papers, where do I start? Which ones should I read? Where can I learn about the fundamental principles of how superconductivity works? How can I learn the vocabulary of superconductivity? How can I expand my knowledge about how superconductivity affects the physical properties of metals? How can I learn what kinds of experiments have been done? How can I avoid duplicating previous workers' research? The *Handbook of Superconducting Materials* is a good source for answers to these questions.

For experimentalists, the *Handbook* provides readable introductions to the basic physics of superconductivity without going into technical theoretical details. For theorists, the *Handbook* provides nice descriptions of the broad spectrum of experimental properties that well-educated theorists should know about. The *Handbook* presents theoretical concepts mostly at the phenomenological level, which both experimentalists and theorists should learn, and provides a sound basis for students interested in studying superconductivity theory at the microscopic level.

The section editors and authors of articles in the *Handbook* have had broad experience in the field of superconductivity. They have contributed their valuable time in writing for the *Handbook* chiefly because they felt an obligation to pass on some of their wisdom and distilled knowledge to the next generation of scientists. For them, producing the *Handbook of Superconducting Materials* has been a labour of love. Take full advantage of their gift to you.

John R Clem

Materials

Superconductivity is one of the most remarkable physical states yet discovered. Perhaps more than any other known effect superconductivity brings quantum mechanics to the scale of the everyday world where a single superconducting quantum state may extend over a distance of metres, or even kilometers—depending on the size of a coil or length of wire. There is something in this mysterious state akin to the starry wonders of the sky above that never fails to enchant its beholders: researcher, student or lay-person alike. And, like the motions of the planets, superconductivity resisted explanation for a very long time. The eventual breakthrough with the Bardeen–Cooper–Schrieffer theory, 46 years after their discovery, actually preceded, by about a decade, any significant commercial development of these remarkable materials. Today, half a century later still, the progress of science and technology is greatly accelerated. And yet a theory of the cuprate high temperature superconductors, discovered 16 years ago, still remains elusive. For these materials the tables are turned–a range of applications are now nudging their way onto the market even though we do not really understand them.

That of course overstates the matter. We know that the supercarriers are Cooper pairs and their symmetry in reciprocal space is predominantly d-wave. We quantitatively understand many properties of HTS materials such as the effect of impurities in suppressing superconductivity and the temperature dependence of the specific heat, thermal expansion and superfluid density. However, this description is primarily based on the observed d-wave symmetry. What we lack is a clear understanding of the mechanism that binds the pairs in the first place and a relationship between the magnitude of this interaction and the energy scale of the superconductivity set by the maximum d-wave gap amplitude. Of course the difficulty here is the strongly interacting electronic system and HTS cuprates, in this sense, are just part of a much wider problem of strongly correlated transition metal oxides that incorporates manganites, ruthenates, cuprates, vanadates and tungstates to name just a few, not to mention hybrid materials such as the ruthenocuprates in which magnetism and superconductivity appear to coexist. Such materials not only formally defy a suitable perturbation treatment but they exhibit many different types of ground-state correlation that compete with each other. Thus, in the HTS cuprates, we currently struggle with the issues of charge ordering, spin ordering and superconductivity and the question as to whether these are intimately linked or, in fact, compete.

The central approach to these issues is systematic measurement in high-quality materials. With the combined improvement in quality of single crystals as well as resolution in low-energy spectroscopic techniques much progress has been made in recent years leading to many surprising new results. By ‘systematic’ I mean variation in properties with carrier concentration, temperature, magnetic field and disorder. It is really only recently that the effect of small increments in carefully-controlled doping levels has been employed and these studies have uncovered abrupt changes in physical properties such as a ground-state metal insulator transition and the possibility of a quantum critical point driving the essential physics and phase behaviour.

If any relatively uncharted territory were to be identified it might be high pressure. Many pressure-dependent studies have been reported and, recently, elemental superconductivity has been discovered at high pressure in iron, sulphur, and lithium. Nonetheless, high pressures allow one to significantly modify the magnetic exchange interaction in oxides and much more could be done to explore the links between magnetism and other competing correlations through the combined investigation of pressure and doping dependent systematics.

Materials issues also lie at the heart of commercial application. Superconducting technologies such as magnets, motors, power cables, transformers, NMR, telecommunications and computing all push the present horizons of physical performance and their improvement, not to mention ultimate commercial success, is predicated in ongoing materials development.

So, while much has been done in this remarkable field, there is much yet to be done. These intellectual and engineering challenges form part of the ongoing *puzzle and promise* of superconductivity. This *Handbook* provides a snap-shot of our current knowledge of the field. It is necessarily in introductory form but its sheer scope illustrates just how broad and multidisciplinary this field is. More than that the wide range of authors underscores a further critical element of science in any field, namely the organic human element. Those of us who have spent a good few years in the field have collaborated, debated and interacted with many scientists around the globe and in the process formed lasting friendships that otherwise would never have eventuated. It is especially pleasing to see the names of so many such friends and collaborators as authors of the various sections of this *Handbook*. Collectively, we trust that these volumes will not only provide a comprehensive information base for the physics and phenomenology of superconductors, but will also communicate something of the puzzle and promise of superconductivity that continues to fascinate us as the years roll by.

Jeffery Tallon

Applications

It is a great pleasure to learn that the *Handbook of Superconducting Materials* has been completed and is now available to the superconductivity community. The science of superconductivity has seen a great change following the discovery of high temperature superconductors by Bednorz and Müller in 1986. The enormous research activity that followed this discovery has generated an unprecedented number of publications, which has tended to swamp individual researchers under a flood of disorganized information. As a result, it has been a particularly difficult task for researchers to re-formulate state-of-the-art knowledge in a systematic manner. The *Handbook of Superconducting Materials* succeeds in achieving exactly this via a comprehensive series of easy to follow articles written by many authors from various fields, including theories, physics, chemistry, materials science and applications. The tutorial style of the *Handbook* fills the gaps in our knowledge and ties together the various strands of superconductivity.

Superconductivity is becoming increasingly accepted as a potential technology for the future with which we can address a number of global environmental problems. One of the most attractive of these applications is the construction of a global superconducting electrical power network, which should be able to increase the cost effectiveness of renewable energy sources such as wind and solar power. The current technology is developed sufficiently to enable reliable estimates of the energy efficiency, including the cooling system and the total cost of the superconducting cable network system, to be made. The variations in demand for electricity between day and night and summer and winter will be averaged out with the development of such a system, underlining the potential contribution that applications of superconductivity can make to the world.

Another promising candidate for applications of superconductivity is the Maglev train, which runs at speeds of up to 500 km/h (i.e. half the speed of a jet propelled aircraft!). This speed can be more than doubled if the train is operated in a de-pressurized tube, providing us with the prospect of a new era of transporting passengers and goods at ultra high speeds. Because of its very low emissions, Maglev can contribute to an improved global environment and to the more efficient use of energy resources.

Another recent exciting development is the so-called single flux quantum device (SFQ), which offers the potential for faster, more energy efficient circuits than is achievable currently with semiconductor-based devices. Establishing the global superconducting communication network with the aid of the SFQ devices, the world with high-level information technology for the first time can become an energy saving society.

A combination of the three, superconductivity-based global networks described above will enable the world to exist solely on renewable energy. These examples are but a few of a considerable range described in the *Handbook*. I believe that superconductivity will be a key technology of the 21st century for the realization of a renewable energy based society and that the *Handbook of Superconducting Materials* will be a significant resource for facilitating the development of these technologies.

Koichi Kitazawa

Introduction

There are a number of reference works in the field of superconductivity in the form of handbooks, encyclopedias and compendia of a variety of conference and workshop proceedings. The relative abundance of such works, therefore, begs the questions 'why another' and 'how will this one be different'? These are probably best answered by considering what the *Handbook of Superconducting Materials* seeks to deliver.

The overriding aim of the *Handbook* is to provide a compendium of practical information in a highly accessible way so as to serve a broad readership with a focus on the materials science, measurement techniques and applications of high and low temperature superconductors. It will also provide an introductory approach to the history and background physics of superconductivity. This information is presented through a collection of correlated articles written in a distinctive tutorial style, targeted specifically at graduate students and scientists who are new to the field. In addition, the *Handbook* includes detail of state-of-the-art materials, techniques and devices, which it is hoped will also form an invaluable reference/bench-top document for more experienced workers in superconductivity. Individual articles, sections and chapters in the *Handbook* are cross-referenced to lead the reader carefully through their field of interest, in contrast to a collection of uncorrelated, highly technical articles which invariably form the basis of reference works in this field. The *Handbook of Superconducting Materials*, therefore, breaks the 'norm' for these reasons.

The *Handbook* is structured into four major sections: Introduction, Materials, Measurement Techniques and Applications. Each section is sub-divided further by materials class, measurement type and nature of application, as appropriate. A total of 12 Section Editors were employed from three continents to oversee the development and construction of the *Handbook* in this way. More than 120 authors were commissioned to produce articles under the direction of the Section Editors and Editors-in-Chief. Each article was peer-reviewed and scrutinized for content and written style by the Section Editors and Editors-in-Chief before it was accepted for publication. This was done with a focus of maintaining the perspective of a practical hands-on handbook rather than merely a series of review articles.

Our intention is that the *Handbook* is of practical use to its readers. We envisage it laid open on the laboratory bench with its pages well thumbed-through by researchers for direction on equipment design, data interpretation or materials processes. For the first time we hope also to have produced a reference work that is of particular use to supervisors in the support of their students and one which does not rely on the possession of detailed prior knowledge for entry to its articles. A second key element of the *Handbook* is its publication in electronic form on the Internet. By this media we intend to widen the readership of the *Handbook* and to increase its accessibility to a wider readership. We also hope to be able to maintain current databases which when accessed will complement the materials in the *Handbook*.

The Editors-in-Chief would like to acknowledge the effort of the many authors who contributed to the *Handbook* and, in particular, the contribution of the Section Editors, Dick Blaugher, Archie Campbell, David Caplin, Julian Cave, Rene Flükiger, Herbert Freyhardt, John Gallop, Harry Jones, Masato Murakami and David Shaw, who really made it happen.

David A Cardwell
David S Ginley

PART A

FUNDAMENTALS OF SUPERCONDUCTIVITY

A1
Introduction to section A1: History, mechanisms and materials

D S Ginley and D A Cardwell

Section A1 is the cornerstone of the handbook, acting not only as a general introduction to superconductivity, but providing significant insight into its overall organization and motivation. Chapter A1.1 is an excellent account of the history of the subject, followed in chapter A1.2 by a basic introduction into the nature of superconductivity. Finally, in chapter A1.3, there is a detailed discussion on the development of practical forms of high temperature superconductors, specifically for power, electronic and magnetic applications.

Chapter A1.1 delves into the history of superconductivity via a series of key events in the field that defined the subject. This begins with Kammerlingh–Onnes' preconceptions of the low temperature behaviour of the resistivity of metals and the shock of observing a sudden loss of resistance to the flow of DC current in mercury at 4.2 K. This led to a number of early theories based on conjected mechanisms for the zero resistance state and, notably, the failure to manufacture a high field superconducting solenoid. With the subsequent discovery of the Meissner effect, this led to the analysis of superconducting phenomena as a thermodynamic phase change and, eventually, to the London equations. These led, in turn, to the development of a two fluid model of charge flow and the concepts of magnetic penetration depth, coherence length and the existence of non-local effects. The early work culminated with the development of the Ginzburg–Landau equations, which resulted in the identification and initial understanding of a number of important superconducting phenomena, including surface energy at the S–N interface to classify type I and II superconductors, the Mendelssohn sponge, Abrikosov vortices, flux quantization and the role of impurities in flux pinning. A number of post Mendelssohn theories soon followed, and most notably those of Heisenberg and Fröhlich and a description of the superconducting energy gap. Finally, there came the watershed of the BCS theory, which seemingly laid to rest most of the then unanswered theoretical questions. The advent of the high T_c cuprates was subsequently to revitalize interest in superconducting theory more than 30 years after Bardeen, Cooper and Schreifer's Nobel prize-winning work.

Chapter A1.2 presents a detailed discussion of the nature of the superconducting state. This begins with the observation that superconductivity derives from an ordering transition and leads to the occurrence of zero resistance and the observation of the Meissner effect. The Ginzburg–Landau and BCS theories developed to describe the ordered state are presented. The chapter builds on the fundamental theory to discuss persistent current and trapped flux, which leads to a discussion of the mixed state and the nature of type I and II superconductors. The chapter concludes with a description of unconventional superconductors and discusses specifically the high T_c cuprates.

Chapter A1.3 provides a detailed insight into the motivation for the development of wire and tape, bulk and thin film forms of high temperature superconductors, specifically for practical applications. One of the most high profile areas for the application of superconducting wires and tapes is in the arena of high power applications including power cables, motors, generators and solenoid, high field magnets. Bulk materials, on the other hand, have considerable potential for the generation of high DC magnetic fields for permanent magnet type applications, whereas superconducting thin films can be used for a wide range of electronic, microwave and sensory devices. The desired materials properties required for superconductors to replace conventional technology at acceptable cost are very demanding. The chapter presents a balance of the historical development of HTS conductors in the context of the need to understand and control microstructure to achieve the desired properties in the various materials forms. The primary focus is on BSCCO for the manufacture of wires and tapes and on YBCO for bulk and thin film devices.

Overall section A1 provides a context for the history of the subject, the underlying physics of superconductivity and the form and nature of high temperature superconducting materials required to realize practical applications.

A1.1
Historical development of superconductivity

Brian Pippard

The story of superconductivity falls into rather well marked periods. From the discovery in 1911 until 1933 exploration was largely confined to Leiden, Toronto and Berlin, almost the only laboratories with liquid helium. A great change came with the discovery of the Meissner effect in 1933, which stimulated widespread interest in superconductivity at the same time as the exodus of Jews from Germany was causing a diffusion of low temperature physics. After 1945, the postwar reconstruction of European research also saw the emergence of the United States as a leading participant (greatly helped by the Collins liquefier and unprecedented government funding). With the revelation of the BCS theory in 1957 and the introduction of superconducting solenoids at about the same time a new era opened; we shall take this to define the end of our historical period.

Early in 1911, Kamerlingh Onnes measured the resistance of a thin thread of mercury and was not surprised that it fell to an imperceptibly low value as the temperature was reduced below 4.2 K [1] (see figure A1.1.1). He had, after all, chosen mercury precisely because it could be made pure enough to show clearly the steady fall of resistivity to zero that he expected in an ideal metal. Only when he and Holst repeated the measurement with greater sensitivity did they experience the shock of seeing an abrupt disappearance just below the normal boiling point of helium. Within months, superconductivity[†] had been found in lead and tin, and the now voluminous catalogue of superconducting elements, alloys and compounds began to grow. By observing the persistence for hours of a current induced in a lead ring Onnes convinced himself that below a well-defined transition temperature the resistivity was, in Casimir's words, 'the zeroest quantity we know' [2, 3]. By 1914, it had been found that a moderate magnetic field sufficed to restore the resistance, and that too large a current had the same effect (and could explode the wire); the connection between the two effects was noted in 1916 by Silsbee [4]. It was thereafter taken for granted that solenoids wound with superconducting wires were useless for strong fields, until about 1954 when Yntema achieved 7 kG with a niobium-wound coil [5]; to be sure, it had an iron yoke, but this probably marks the moment when hope was renewed. Within a few years high-field solenoids began to appear, as will be mentioned later.

Once superconductivity had been found in several metals it was an open season for the theorists, and in the next 20 years almost as many hopeful theories were proposed and forgotten. All started, naturally enough, from the premise of perfect conductivity for which some mechanism had to be found that would allow unopposed movement of the electrons; condensation into a rigid lattice was a favourite. Since during most of this period there was no quantum mechanics, let alone a coherent picture

[†]Onne's original name, whose form survives in the German Supraleitung, was gradually replaced by superconductivity as English became the preferred language during the 1930s.

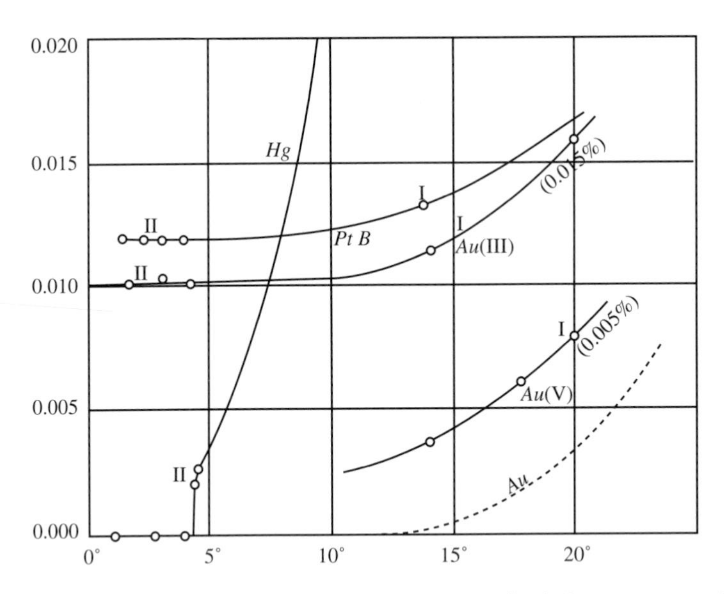

Figure A1.1.1. At the first Solvay conference, in November 1911, Kamerlingh Onnes presented an account of the extraordinary resistive behaviour of mercury at about 4 K. The resistance of gold and platinum wires drops steadily to a constant value as the temperature is lowered; in mercury there is a sudden drop to an imperceptibly low value.

of metallic conduction, there was unlimited scope for guesswork. With regret, we pass over this phase to take notice of the radical change that began in 1933 with the experiments of Meissner and Ochsenfeld, A2.2, A1.2 in Berlin, on the spontaneous expulsion of flux in the transition to superconductivity with a magnetic field present [6].

It should not be supposed that this important result made the truth immediately plain to all; it was not just a matter of cooling an ideal sample in a magnetic field and observing the complete expulsion of flux at the transition. The use of both solid and hollow cylinders confused the issue, and even with solid cylinders expulsion was far from complete. It is not surprising that behaviour so unexpected should take time to be grasped, and the now-accepted interpretation, that $B = 0$ in a superconductor came not from Berlin but from Gorter in Haarlem [7]. He had been impressed by the success of thermodynamics in accounting for the temperature variation of the critical magnetic field, and puzzled because the argument was invalid if the transition at H_c was not reversible. The new observation convinced him (though Meissner was doubtful at first) that indeed it was, and that superconductivity was something more than perfect conductivity. It was not long before repetitions of the experiment, with variations, set A1.2 doubts at rest, and in 1935 F and H London published their phenomenological theory that took Gorter's interpretation as the starting-point [8, 9]. They accepted the earlier idea that a supercurrent was accelerated by an electric field, $\Lambda\dot{\boldsymbol{J}} = \boldsymbol{E}$, but were dissatisfied because it implied too many solutions—it was not enough for them that the internal magnetic field should be unchangeable, it must be zero‡. This they achieved with the postulate $\Lambda\boldsymbol{J} + \boldsymbol{A} = 0$. Their first paper noted that such a relation would follow from the quantum-mechanical expression for $\boldsymbol{J}$ if only it could be assumed that a moderate magnetic field was powerless to perturb the electronic wave functions. This is not so in a normal metal, and they suggested that interactions between the electrons might confer the required rigidity. It is not clear how

‡They were not sure that Λ had to be interpreted as m/ne^2, as in earlier acceleration theories.

much notice was taken of this proposal at the time, but in the following years the latest round of tentative (and still abortive) molecular theories tended to seek an energy gap separating condensed and excited electron states. This might indeed make the condensed electron assembly unresponsive to a magnetic field, as in an insulator, but as yet no theories met the stiffer requirement of an energy gap that was compatible with a conduction current.

Meanwhile the new phenomenologies, thermodynamic (Gorter) and electromagnetic (London), encouraged attack from several quarters on the different problem of phase stability [10]. Penetration of magnetic field into a thin superconducting lamina allows it to survive in fields greater than H_c; why then is the transition of a pure metal at H_c so abrupt, yet in an alloy so gradual, with persistence of perfect conductivity even after complete field penetration? Mendelssohn was most concerned with alloys, and gave inhomogeneity as the reason—a network (sponge) of thin superconducting filaments of different composition from the rest could survive to higher fields. Gorter was of the opinion that in a pure metal there was a minimum size for a superconducting domain, so as to preclude a fine mixture of phases, stable in high fields. Heinz London postulated a surface energy at a phase boundary that would have the same discouraging effect. Gorter [11] admitted that a minimum size for normal domains was equally necessary, but any worries this may have caused him were small compared to what troubled both him and London—why should alloys be more tolerant than pure metals of finely divided phases? Little light A2.3 was cast until 1950, but the likely existence of an interface energy remained in the minds of those concerned with a related problem, the structure of the intermediate state. In essence, the problem is that the magnetic field can be uniform only at the surface of a long and thin superconductor. With any other shape some parts should reach H_c before others, and it was clear that then the superconducting body must become partially normal. Only phase division into thin laminae can prevent the field, H_c at each interface, being smaller somewhere in the normal phase. Landau in 1937 made the first theoretical attempt to resolve the difficulty, and there have been many delicate probings of the phase configuration at the surface, and progressively more sophisticated attempts to explain the observed patterns. But this is a study in itself, and we must be content to quote a source of references and return to the development of phenomenological models, especially those that came to be known, by analogy with similar ideas about superfluid helium, as two-fluid models [12, 13].

The idea of a two-fluid model, with coexistent 'superelectrons' and normal electrons, was probably not new when Heinz London wrote his doctoral thesis in 1933, but it seems that the first published mention was by Casimir and Gorter in 1934. London had searched unsuccessfully for resistive loss when the presumed normal electrons are stimulated by a high-frequency electric field in the penetration layer. His frequency of 40 MHz was far too low, but he succeeded in 1940 with 1500 MHz [14]; by then the two-fluid model was well established as a theoretical concept. In the explicit form of Casimir and Gorter it reproduced the thermodynamic behaviour, including the second-order critical behaviour in zero magnetic field—a recent and controversial invention by Ehrenfest which Laue and others (mistakenly) attacked as thermodynamically impossible. Casimir and Gorter postulated a steady diminution of the normal fraction, x, as the temperature fell below T_c and a corresponding increase of the superconducting fraction $(1-x)$. The superelectrons, carrying no entropy, contributed free energy proportional to $-(1-x)$, while the normal electrons contributed not the expected $-\gamma T^2 x$ but $-\gamma T^2 x^{1/2}$. When x took the value t^4, i.e. $(T/T_c)^4$, that minimized the free energy, the resulting critical magnetic field had the observed parabolic temperature variation. The model, an interesting artificial construction, received little attention until it proved relevant to measurements of the penetration depth [15, 16]. A1.2

It was inherent in the London phenomenological theory, as in earlier acceleration theories, that the currents screening magnetic fields from the interior should flow in a surface layer of effective thickness λ, i.e. $(\Lambda|\mu_0)^{1/2}$ in modern notation. The first direct studies of field penetration were made in 1940 by Shoenberg using pharmaceutical 'grey powder', finely ground mercury in chalk with droplets little more than 10^{-6} cm in diameter. This remains one of the few determinations of relative values, λ/λ_0, at different temperatures, and only in 1948 was it shown to agree well with the Casimir–Gorter model which predicted $\lambda/\lambda_0 = (1 - t^4)^{-1/2}$. Most measurements of penetration depth have used samples much larger than λ, and have determined $\lambda - \lambda_0$; on the assumption that the Casimir–Gorter model is generally valid it is possible to derive λ_0 [17, 18]. The few values obtained in the early post-war years were two or more times greater than expected from a naive acceleration theory. It also became clear that the high-frequency resistive losses were not readily explained, except in general terms, on the basis A3.1 of the two-fluid model, but these deficiencies of the phenomenological models were not seen as grave, and the proposals for modifying the London picture had other origins; they came in 1950 and shortly after from two independent sources, Cambridge (England) and Moscow. Since the political situation at the time practically excluded mutual discussion between the two countries[§] it is convenient to treat them separately.

The Cambridge side of the story began with measurements that showed little variation of λ as the magnetic field was increased to H_c, despite the thermodynamic requirement of an entropy increase which, if confined to the penetration layer, would imply a very significant change to the number of D2.1 superelectrons [19]. It seemed likely, then, that the entropy change and disturbance to the electron assembly were spread to a depth perhaps 20 times greater than λ, as if the superconducting phase could tolerate only gradual changes. This view was supported by the sharpness of the second-order transition and the absence of any foreshadowing of superconductivity above T_c, as if fluctuations in very small regions were not permitted. Gorter's 1934 suggestion of a minimum size had been forgotten, but the concept of coherence, as it came to be called, served to explain the origin of London's interface energy. If only gradual change was possible, the extended transition from superconducting to normal would exclude the magnetic field from a layer where the full energy of condensation was not available [20]. A little evidence that the penetration depth in an alloy was more susceptible to change by a magnetic field gave support to the view that the range of coherence was limited by electron scattering, so that the interface energy might be less in an alloy than in the pure metal, and might even become negative.

Substantial support for this interpretation came in 1953 when the penetration depth was found to be considerably larger in a tin–indium alloy than in pure tin, although the thermodynamic parameters were hardly affected. To incorporate the idea of coherence into the equation for the supercurrent, the local relation of the London theory was replaced by a non-local relation; J was now to be determined by an average of A over a range of about the coherence length, and therefore affected by impurity content. The integral equation, with solutions agreeing well with the limited data available, was inspired by the already-discarded Heisenberg–Koppe theory of superconductivity, but was soon more soundly A3.2 derived by Bardeen, even before the BCS theory gave it refined form [21, 22]. In 1950, to which we return for the Russian side of the question, such refinements did not concern Ginsburg and Landau; they adhered to the London local model until 1957 and the advent of BCS (or Bogoliubov who seized and developed the idea of Cooper pairs independently).

A3.1 The Ginsburg–Landau (GL) theory was a full-dress professional performance in comparison with the naive coherence idea [23]. It allowed arbitrary spatial variations of an order-parameter Ψ (which determined Λ) but at the cost of an extra energy of the form $|\text{grad}\,\Psi|^2$, with a multiplicative constant that was in no way arbitrary but fixed by the penetration depth and other parameters. When it was discovered that impurities could change λ without affecting the transition temperature they maintained this rigid connection at the price of credibility. The re-examination by Gorkov of the GL equation

---

[§]There were as yet no journals devoted to systematic translation of Russian papers, and the Russian publication, Journal of Physics, had been discontinued. Except for Cambridge, which enjoyed in David Shoenberg a fluent Russian-speaking physicist, there was great ignorance in the West of Russian work. Many Moscow physicists, on the other hand, spoke English well.

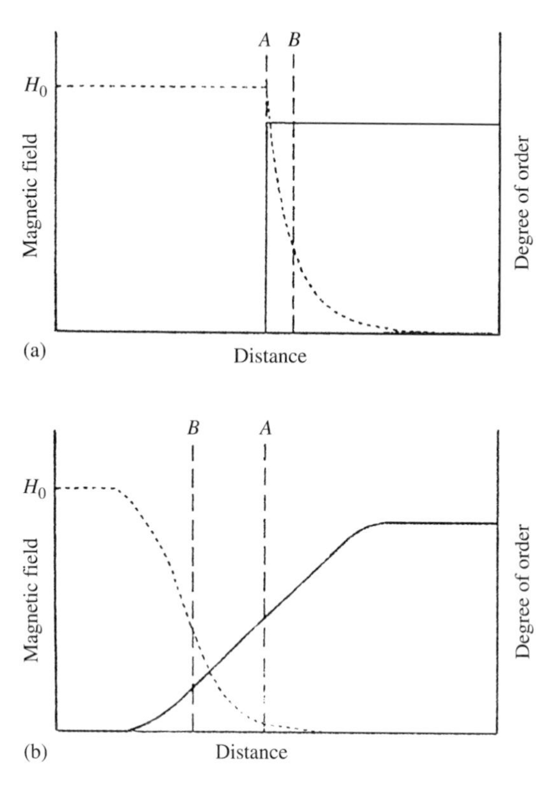

Figure A1.1.2. Diagrams to illustrate the origin of the interface surface energy [15]. In the upper diagram the interface is sharp (superconducting phase on the right) and field penetration puts the effective boundary of the magnetic field at B, within the superconductor; this leads to negative interface energy. In the lower diagram the transition is broadened and B lies outside the effective phase boundary A, to give positive surface energy.

in the light of BCS theory consolidated its status and in its revised form it remains a pillar of the theoretical structure [24]. In their original treatment, Ginsburg and Landau explained the interface energy in the same way (but in quantitative detail) as the coherence model. Strangely, they declined to consider the possibility that the parameters might take such values as would lead to a negative interface energy, and when Abrikosov took this step in 1952 they did their best to discourage him, so that he did not publish for some years [25]. He then pointed out that a negative interface energy would considerably change the shape of the magnetization curve [26]. Even in a long rod the field would begin to penetrate, before H_c was reached, to form a mixed phase which retained enough of superconductivity to conduct without resistance, and this could persist until the field was much greater than H_c. He drew attention to the pre-war experiments of Schubnikow and his co-workers who had taken trouble to homogenize alloys of indium and thallium, and measured magnetization curves like what he expected when the interface energy was negative. He distinguished between superconductors of type I (typically pure metals with positive interface energy) and those of type II (in ideal form homogeneous alloys with negative interface energy). Until then Schubnikow's work had been little regarded, partly because the Meissner effect had overshadowed it and partly because it was taken for granted that alloys were inhomogeneous and their magnetic behaviour explicable in terms of the Mendelssohn sponge. It was easy in those days to follow mandarins like Pauli in their scorn of 'dirt effects'.

A3.1
D2.1

Figure A1.1.3. Electron micrograph of the surface of a type II superconductor in the mixed state, with **B** normal to the surface. The flux lines, decorated with iron powder, have taken up a regular triangular lattice configuration [35].

The details of Abrikosov's analysis went much further than the categorization of different types of superconductor. In the Ginsburg–Landau theory the order parameter Ψ had the character of a wave function for the superelectrons, and was complex with a phase that depended on A. Abrikosov realized that continuity of Ψ meant that the magnetic flux through a normal region within a superconductor A2.6 could not take an arbitrary value, but only a multiple of a flux quantum h/e. This quantum had been foreseen for similar reasons by Fritz London in 1948 [27], but was not observed in a macroscopic ring until 1961 (and then with half London's value). In a superconductor with negative interface energy the smallest normal domain was what would become known as a flux line, with a core parallel to H on which Ψ dropped to zero, and around it a sheath where Ψ rose to its full value; one flux quantum was trapped in the sheath. The flux lines remained hypothetical, though not doubted, until 1966 when electron microscopy of a surface decorated with magnetic powder revealed a triangular array, hardly different from the square array Abrikosov had analysed [28]. The existence of flux lines justified a more precise picture of the migration of flux within the mixed state of superconductors, as well as the influence of inhomogeneities and defects in pinning flux lines, and the origin of electrical resistance accompanying flux flow.

The consequences were not realized until the first high-field superconducting solenoids became commercially available. They followed the initial exploration by Hulm and Matthias of new alloys and A1.2 compounds with high transition temperature [29] which, as extreme forms of type II superconductors such as Nb_3Sn and NbTi could retain flux lines in very strong fields and, being microscopically inhomogeneous, inhibited flux flow. It was only in 1962 that Goodman [30] pointed out the significance of Abrikosov's work and set the technological development on a sound basis. By this time the BCS theory had radically changed the physical understanding of superconductivity, and its applications were at long last being recognized as worthy of governmental and industrial support, such as has continued and enlarged with the discovery of cuprate superconductivity.

Here we may stop, except for brief remarks about the origin and immediate consequences of BCS A1.2 theory. The turning-point for theory was Fröhlich's appreciation (1950) that polaronic distortion of an ionic lattice could lead to an attractive force between electrons and create an energy gap at the Fermi surface—the conditions for realizing the Londons' dream of 1935 [31]. Simultaneously, it was found that the transition temperature depended on isotopic mass in the way predictcd by Fröhlich [32], so that his mechanism was immediately accepted even though his detailed theory of the ground state was seriously flawed. Several years elapsed before Bardeen, Cooper and Schrieffer overcame the enormous difficulties of a many-body theory in which electron pairs of opposite momentum and spin, Cooper's seminal inspiration, were coupled in a ground state [33]. The excited states had properties reminiscent of those postulated in the two-fluid model but, because of entirely novel symmetry properties, they were sufficiently different to succeed where the other had failed. For a while there was resistance from experienced theorists who were only well aware of the pitfalls besetting a path that they themselves had failed to discern. But experimenters received the theory enthusiastically and made such advances as convinced the most sceptical. Giaever's tunnelling that revealed the energy gap and the enhanced density of excited states above it; Josephson tunnelling with all its complexities and applications; the Knight shift in superconductors and the attenuation of ultrasonics; Andreev reflection and its consequences; the halving of the flux quantum to $h/2e$ as a direct consequence of electron pairing in the ground state; these are items in an unprecedented catalogue of success, and all within a year of two—witnesses to one of the scientific triumphs of the century.

A2.7

A2.6

Yet not, it seems the complete answer. There were critics, from the beginning, who agreed the theory described the phenomenon as no other had done, but were dissatisfied because it had little power to predict which metals would be superconductors, or what their transition temperature would be. With further theoretical development they had fallen silent before the events beginning in 1986, when the new class of cuprate superconductors was discovered and made clear that BCS might not be a universal theory [34]. Despite valiant efforts the high-temperature superconductors remain fundamentally mysterious. But those with transitions below about 25 K, and well described by BCS, have acquired and seem likely to retain the classical status of 'conventional superconductors'.

References

[1] Dahl P F 1992 *Superconductivity: Its Historical Roots and Development from Mercury to the Ceramic Oxides* (New York: American Institute of Physics) Chapter 3

[2] Dahl P F 1992 *Superconductivity: Its Historical Roots and Development from Mercury to the Ceramic Oxides* (New York: American Institute of Physics) p 83

[3] Casimir H B G 1983 *Haphazard Reality* (New York: Harper and Row) p 339

[4] Dahl P F 1992 *Superconductivity: Its Historical Roots and Development from Mercury to the Ceramic Oxides* (New York: American Institute of Physics) Chapter 4 and p 98

[5] Yntema G B 1995 *Phys. Rev.* **98** 1197

[6] Dahl P F 1992 *Superconductivity: Its Historical Roots and Development from Mercury to the Ceramic Oxides* (New York: American Institute of Physics) Chapter 9

[7] Dahl P F 1992 *Superconductivity: Its Historical Roots and Development from Mercury to the Ceramic Oxides* (New York: American Institute of Physics) Chapter 10

[8] Shoenberg D 1952 *Superconductivity* (Cambridge: Cambridge University Press) p 180

[9] Bardeen J 1956 *Encyclopedia of Physics* **Vol 15** ed S Fliigge (Berlin: Springer) p 284

[10] Dahl P F 1992 *Superconductivity: Its Historical Roots and Development from Mercury to the Ceramic Oxides* (New York: American Institute of Physics) Chapter 11

[11] Gorter C J 1935 *Physica* **2** 449

[12] Shoenberg D 1952 *Superconductivity* (Cambridge: Cambridge University Press) Chapter 4

[13] Faber T E 1958 *Proc. Roy. Soc.* A **248** 460

[14] London H 1940 *Proc. Roy. Soc.* A **176** 522

[15] Shoenberg D 1952 *Superconductivity* (Cambridge: Cambridge University Press) p 194

[16] Dahl P F 1992 *Superconductivity: Its Historical Roots and Development from Mercury to the Ceramic Oxides* (New York: American Institute of Physics) p 280

[17] Shoenberg D 1952 *Superconductivity* (Cambridge: Cambridge University Press) Chapter 5

[18] Bardeen J 1956 *Encyclopedia of Physics* **Vol 15** ed S Fliigge (Berlin: Springer) p 244

[19] Pippard A B 1950 *Proc. Roy. Soc.* A **203** 210
[20] Pippard A B 1951 *Proc. Camb. Phil. Soc.* **47** 617
[21] Dahl P F 1992 *Superconductivity: Its Historical Roots and Development from Mercury to the Ceramic Oxides* (New York: American Institute of Physics) p 245
[22] Bardeen J 1956 *Encyclopedia of Physics* **Vol 15** ed S Flügge (Berlin: Springer) p 299
[23] Bardeen J 1956 *Encyclopedia of Physics* **Vol 15** ed S Flügge (Berlin: Springer) p 324
[24] Waldram J R 1996 *Superconductivity of metals and cuprates* (Bristol: Institute of Physics Publishing) p 176
[25] Dahl P F 1992 *Superconductivity: Its Historical Roots and Development from Mercury to the Ceramic Oxides* (New York: American Institute of Physics) p 250
[26] Abrikosov A A 1957 *J. Phys. Chem. Solids* **2** 199
[27] London F 1948 *Phys. Rev.* **74** 562
[28] Waldram J R 1996 *Superconductivity of metals and cuprates* (Bristol: Institute of Physics Publishing) p 72
[29] Dahl P F 1992 *Superconductivity: Its Historical Roots and Development from Mercury to the Ceramic Oxides* (New York: American Institute of Physics) p 256
[30] Goodman B B 1962 *IBM J. Res. Dev.* **6** 63
[31] Bardeen J 1956 *Encyclopedia of Physics* **Vol 15** ed S Flügge (Berlin: Springer) p 359
[32] Dahl P F 1992 *Superconductivity: Its Historical Roots and Development from Mercury to the Ceramic Oxides* (New York: American Institute of Physics) p 252
[33] Bardeen J, Cooper L N and Schrieffer J R 1957 *Phys. Rev.* **108** 1175
[34] Bednorz J G and Müller K A 1986 *Z. Phys.* B **64** 189
[35] Träuble H and Essman U 1968 *J. Appl. Phys.* **39** 4052

A1.2
An introduction to superconductivity

W F Vinen

A1.2.1 Superconductivity as an ordering transition

Any material becomes more ordered as it cools, its entropy decreasing, provided it remains in thermodynamic equilibrium. As the material approaches the absolute zero of temperature, its entropy tends to vanish, so that it becomes completely ordered. Often the ordering proceeds through one or more discrete phase transitions. Perhaps the most remarkable of these phase transitions, and surely the most unexpected, are those leading to superfluid or superconducting phases, typically at temperatures of a few kelvin. At a superficial level these phases can be seen to exhibit frictionless flow, but they have other striking macroscopic quantum properties that are a reflection of the unique and special type of quantum A2.7 ordering that takes place within them.

In the case of superconducting metals, the phase transition involves the electron fluid that is responsible for electrical conductivity in the normal (high-temperature) phase. The frictionless flow is A2.1 seen as a loss of electrical resistivity, the material often showing a resistivity that is unmeasurably small. Electrical resistivity is due to the scattering of the conduction electrons by imperfections in the crystal lattice in which they are moving, so one might be tempted to think that in a superconductor these scattering processes are mysteriously turned off. As explained in Pippard's historical introduction, this A1.1 view is quite misleading. All superconductors exhibit the Meissner effect: in sufficiently small applied magnetic fields they behave as perfectly diamagnetic materials, in the sense that the magnetic flux is excluded in a reversible manner; they behave just like conventional diamagnetic materials, except that they have a much larger diamagnetic susceptibility. The required screening current is maintained as part A2.2 of the equilibrium state of the system just as it is in the diamagnetic screening current in a diamagnetic atom or molecule. Scattering processes, far from having disappeared, are helping to maintain this equilibrium, as we shall see a little later. A3.2

The superconducting state is a new thermodynamic phase of the metal, distinct from the normal phase. The form of the heat capacity of the metal in the neighbourhood of the transition is very similar to that found in ordering transitions of the type seen in, for example, a paramagnetic material when it transforms into a ferromagnet. In the superconducting phase the conduction electron fluid is in a more A2.4 strongly ordered state than it is in the normal phase, and the diamagnetism observed in the Meissner effect is an equilibrium property of this ordered state.

Distinct phases can be conveniently described in an appropriate phase diagram: that for most pure metals (type I) is shown schematically in figure A1.2.1, where the material has a shape with zero demagnetizing factor. The transition between the normal and superconducting phases in zero field is D2.4 second order (no latent heat); in a finite field it is first order. The transition from superconducting to normal phase, as the field is increased at fixed temperature, takes place when the free energy associated with the induced magnetic moment exceeds the free energy difference $(F_s - F_n)$ between the two phases

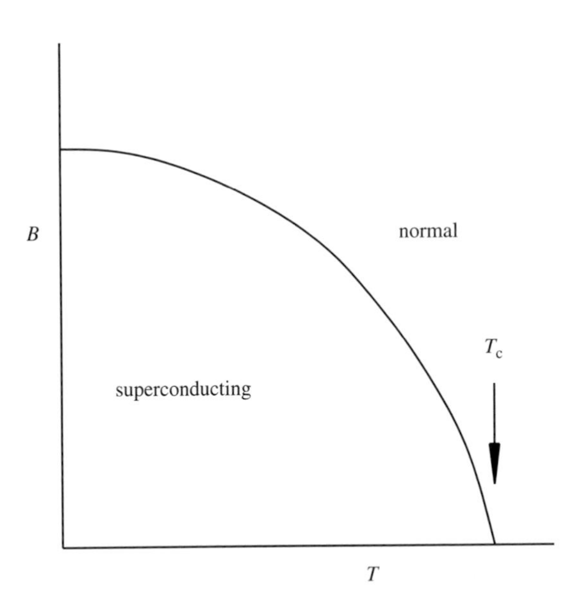

Figure A1.2.1. Schematic phase diagram for a type I superconductor: applied magnetic flux density (B) plotted against temperature (T).

in zero field, which leads to the following relationship for the critical field.

$$F_n - F_s = B_c^2/2\mu_0. \tag{A1.2.1}$$

Other and more complicated phase diagrams are also possible, as we shall see later.

A1.2.2 The Meissner effect and the nature of the ordering in a superconductor

The simplest form of diamagnetism in a single atom arises when, as is often the case, the atomic electron wavefunctions ψ are not significantly perturbed by the applied magnetic field. The current electric density in the atom, given in terms of the vector potential $\boldsymbol{A}$ for each electron acting independently by

A3.1

D2.1

$$\boldsymbol{J}(\boldsymbol{r}) = \frac{ie\hbar}{2m}(\psi^* \nabla\psi - \psi\nabla\psi^*) - \frac{e^2}{m}\psi\psi^* \boldsymbol{A}(\boldsymbol{r}) \tag{A1.2.2}$$

then reduces to

$$\boldsymbol{J}(\boldsymbol{r}) = -\frac{e^2}{m}\psi\psi^* \boldsymbol{A}(\boldsymbol{r}) \tag{A1.2.3}$$

since the unperturbed wavefunctions make no contribution. The resulting diamagnetic moment is very small, because the atom is very small. Fritz London noticed that, if equation (A1.2.3) was to apply to each conduction electron in a whole metal, the resulting diamagnetism would be very large, as in a superconductor, because the electron wavefunctions ψ extends over the very large volume of the whole metal. This does not happen in a normal metal because the conduction electron wavefunctions are not unperturbed; each electron wavefunction is strongly perturbed and describes a quantized cyclotron orbit. The two terms in equation (A1.2.2) almost cancel, so that there remains only the very weak 'Landau' diamagnetism. But if the conduction electron wavefunctions were not modified by the field, so that equation (A1.2.3) were to apply to a macroscopic number density (n_s) of electrons in the whole superconducting metal, the Meissner effect would be more or less correctly described. Equation (A1.2.3) would then imply that the total current in the superconductor would be given by A2.2

$$\boldsymbol{J} = -n_s \frac{e^2}{m} \boldsymbol{A} \qquad \text{(A1.2.4)}$$

from which we obtain, by taking the curl,

$$\text{curl } \boldsymbol{J} = -\frac{n_s e^2}{m} \boldsymbol{B} \qquad \text{(A1.2.5)}$$

where $\boldsymbol{B}$ is the magnetic induction. Combining equation (A1.2.5) with the Maxwell equation curl $\boldsymbol{B} = \mu_0 \boldsymbol{J}$, we obtain

$$\nabla^2 \boldsymbol{B} = \frac{1}{\lambda_L^2} \boldsymbol{B} \qquad \text{(A1.2.6)}$$

showing that the field tends to zero within the superconductor, with a penetration depth given by

$$\lambda_L^2 = \frac{m}{\mu_0 n_s e^2}. \qquad \text{(A1.2.7)}$$

If n_s is of the order of the total number of conduction electrons per unit volume in the metal, λ_L is very small (of order 100 nm) so that flux exclusion is almost complete, as observed. Careful experiments show that a finite penetration depth does exist, with the order of magnitude given by equation (A1.2.7), so we seem to have a good description of the superconducting behaviour.

The 'rigidity' in the wavefunction of the electrons in a superconductor that leads to the Meissner effect must presumably be a result of the ordering process that marks the onset of superconductivity. The nature of this ordering process became clear only when Bardeen, Cooper and Schrieffer had developed their theory, although the suggestion given much earlier by Fritz London that superfluidity in liquid helium and superconductivity in metals might have a common origin can now be seen to have pointed the way. Superfluidity is associated with Bose condensation: an ordering described by the accumulation of the helium atoms in a single quantum state. The BCS theory indicated, perhaps somewhat surprisingly, that a similar process is occurring in a superconductor. Since electrons are fermions, the individual electrons themselves cannot exhibit Bose condensation, but BCS told us that an attractive interaction between electrons, due to local distortion of the lattice (phonon exchange), leads to the formation of electron pairs (Cooper pairs), which can and do undergo a form of Bose condensation. A satisfactory description of this condensation is not straightforward, since each electron pair occupies a large volume (connected with the coherence length ξ_0 to which we refer later), so that there is massive overlap between the pairs. Bose condensation leads to long-range order in a particular type of correlation function, which, in the case of liquid helium, is the single particle density matrix. In the superconducting electrons it is the two-particle density matrix that exhibits the same long-range behaviour, as can be shown directly from the BCS wavefunction. A3.2 A2.3

It is the formation of the electron-pair condensate that gives rise to the rigidity of the superconducting wavefunction. Small perturbations to the wavefunction can be achieved only by mixing in states in which electron pairs have been removed from the condensate, and it turns out that this requires a minimum energy, Δ. This minimum energy is that required to produce thermal excitation of the system. In a normal metal the excited states of the system are obtained by taking an electron from below the Fermi surface and placing it above, thus creating an electron excitation and a hole excitation. In the superconductor the excitations involve the breaking of electron pairs, and they correspond to linear combinations of electron-like states and hole-like states in the normal metal. Each excitation has A2.3 A2.4 A2.1

an energy equal to $(\Delta^2 + \varepsilon_k^2)^{1/2}$, where ε_k is the energy of the corresponding excitation in the normal state. As the temperature of the superconductor is raised above absolute zero more and more excitations are produced and the energy gap Δ falls. Eventually at the critical temperature, T_c, Δ vanishes and the superconductor becomes a normal metal. In the BCS theory T_c is related to the energy gap at $T = 0$ by the relation $\Delta(0) = 1.76 k_B T_c$. However, it should be added that the existence of an energy gap is not essential for superconductivity; for example, superconductors containing magnetic impurities may be gapless; and superfluid helium is also gapless. The condensate in liquid helium owes its 'rigidity' to the form of the excitation spectrum, which is itself determined by the existence of the condensate.

Only the condensed electrons contribute to the Meissner effect, and the value of n_s in equation (A1.2.4) is the effective number of such electrons. The excitations behave to some extent like electrons in a normal metal. We are led therefore to a two-fluid model, in which n_s electrons behave as superconducting and the rest as normal. The falling value of n_s with increasing temperature is reflected in an increasing penetration depth ($\lambda_L \rightarrow \infty$ as $T \rightarrow T_c$).

A1.2.3 The Ginsburg–Landau equations; the Pippard and Ginsburg–Landau coherence lengths

A3.1 It is convenient in developing an understanding of superconductivity to introduce the 'condensate wavefunction' or Ginsburg–Landau wavefunction. This can be formally defined in terms of the correlation function to which we have already referred, but in essence it is a complex function, Ψ, the phase of which is the phase of the wavefunction of the condensed pairs and the amplitude of which is proportional to the local concentration of condensed pairs. The function Ψ obeys the Ginsburg–Landau equations when the temperature is near T_c; at lower temperatures these equations cease to be strictly correct, but they still provide a correct qualitative description. One of the G–L equations gives the supercurrent density

$$J(r) = \frac{ie\hbar}{2m}(\Psi^* \nabla \Psi - \Psi \nabla \Psi^*) - \frac{2e^2}{m} \Psi \Psi^* A(r) \qquad \text{(A1.2.8)}$$

(cf equation (A1.2.2)), while the other, which has the form of the nonlinear Schrodinger equation, describes in essence how the amplitude of Ψ varies with position, such as might occur near a normal-superconducting boundary. It is found that Ψ cannot change abruptly with position, but only gradually over a characteristic distance, ξ_{GL}, called the Ginsburg–Landau coherence length. We note the presence of the first term on the right-hand side of equation (A1.2.8), which implies that current-carrying states of the superconductor are possible independently of the magnetic field. We discuss such states in a moment. Since the modulus of Ψ is a measure of the density of condensed electrons it also a measure of the extent of the superconducting ordering. Indeed Ginsburg and Landau originally introduced Ψ as a (complex) A3.2 'order parameter,' and it is frequently referred to in this way. Often it is referred to in other ways: 'the gap parameter,' since it is proportional to the energy gap in a spatially homogeneous situation; or the 'pair potential,' since it appears as a self-consistent potential in some formulations of the theory of superconductivity.

The superconducting wavefunction is not completely rigid in its response to a perturbation. It is rigid only for perturbations that vary slowly with position. The relevant characteristic length scale is the size of a Cooper pair ξ_0, often called the Pippard coherence length. In practice ξ_0 is often larger than λ_L, so that our derivation of λ_L as the penetration depth, which relied on the assumption of a rigid wavefunction, breaks down. The relation (A1.2.4) between the current density and the vector potential has to be replaced by a nonlocal relation, involving a kernel with range ξ_0, and the expression for the penetration depth becomes more complicated ($\lambda \neq \lambda_L$). For a pure metal the coherence lengths ξ_{GL} and ξ_0 are equal at low temperatures and equal to $\pi\Delta(0)/\hbar v_F$, where v_F is the Fermi velocity in the normal metal, but at higher temperatures they behave differently, ξ_0 remaining constant but ξ_{GL} diverging to infinity at T_c.

A1.2.4 Persistent currents and the quantization of trapped flux

Suppose that the superconductor has the form of a long hollow cylinder, with radius R and wall thickness t. If t is large compared with the penetration depth, and if an external magnetic field is applied parallel to the axis of the cylinder, no magnetic flux penetrates to the inside of the cylinder. Suppose, however, that the cylinder is cooled into the superconducting phase while exposed to the external magnetic field. The state of lowest energy must surely then be one in which magnetic flux is trapped inside the cylinder. In such a state the Ginsburg–Landau wavefunction must, as it turns out, have the form D2.1

$$\Psi = \Psi_0 \exp[\mathrm{i}S(\boldsymbol{r})] \tag{A1.2.9}$$

where the Ψ_0 is the unperturbed wavefunction, and the phase S varies round the ring. The corresponding current density, obtained from (A1.2.8), is

$$\boldsymbol{J} = \frac{e\hbar\Psi_0^2}{m}\nabla S - \frac{2e^2\Psi_0^2}{m}\boldsymbol{A}. \tag{A1.2.10}$$

Taking the curl of this equation we recover equations (A1.2.5) and (A1.2.6), so that the magnetic field is still screened from the bulk of the superconductor. However, there is now a magnetic field within the cylinder, as we see by taking the line integral of equation (A1.2.10) round a circuit C enclosing the hole in the superconductor and lying at a distance much greater than the penetration depth from the surface (figure A1.2.2, where we have assumed that $\lambda \ll t$). $\boldsymbol{J} = 0$ on this circuit, and therefore, the flux within the circuit, equal to the line integral of $\boldsymbol{A}$ round the circuit, is given by

$$\Phi = \frac{\hbar}{2e}\oint \nabla S \cdot \mathrm{d}\boldsymbol{r}. \tag{A1.2.11}$$

The wavefunction Ψ must be single-valued, and therefore, the line integral in equation (A1.2.11) must be equal to $2\pi q$, where q is an integer. Therefore the trapped flux Φ can be nonzero, implying the existence of a magnetic field inside the cylinder, but the magnitude of the trapped flux is quantized in units of $\phi_0 = 2\pi\hbar/2e$. The factor $2e$ here has its origin in the fact that the condensate is composed of A2.6

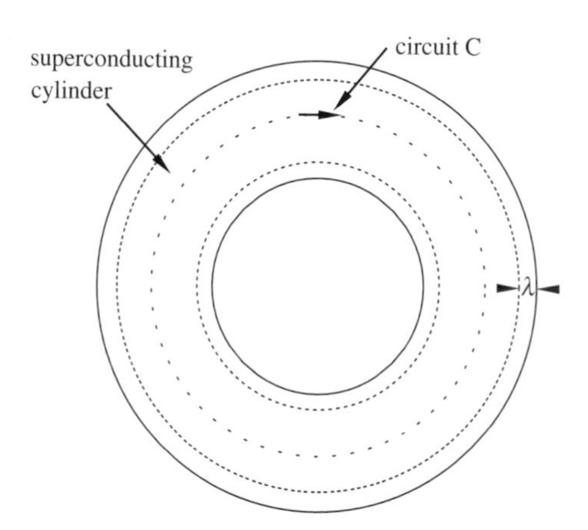

Figure A1.2.2. Illustrating the quantization of flux.

electron pairs. This quantization of trapped flux is rather directly related to the existence of the condensate wavefunction.

If the external field is removed from the superconducting ring the trapped flux remains. The superconductor is not then in a state of minimum possible free energy. It is, however, in metastable equilibrium, with an associated local minimum in the free energy. A transition to the state of absolutely minimum free energy would require that the flux passes through the superconductor and therefore penetrates it, at least transiently, by much more than the penetration depth, which is energetically very unfavourable (a consequence of the Meissner effect). More striking is the situation when the thickness, t, of the ring is much less than the penetration depth. A persistent current is still possible, although an extension of the argument in the preceding paragraph shows that the trapped flux is quantized in units less than $2\pi\hbar/2e$. The metastability of this persistent current does not follow simply from the Meissner effect. It is necessary to note that a sudden loss of the trapped flux, or equivalently loss of the persistent current, could occur only if all the Cooper pairs in the condensate were simultaneously to undergo a transition between two states (equation (A1.2.9)) with different $S(\boldsymbol{r})$, which has negligible probability. Alternatively, the persistent current would disappear if it exceeds a critical value, equal to roughly Δ/p_F, at which excitations are produced in such large numbers that superconductivity is suppressed (p_F is the Fermi momentum); below this critical current extra excitations may still be produced, the excitations being in a state of thermal equilibrium maintained by scattering, but the effect is only to reduce n_s without destroying it altogether. Under certain circumstances loss of a persistent current can occur with nonzero probability through the important process of gradual 'phase slippage', which we shall discuss later. The persistence of the current then depends on phase slippage having a sufficiently small probability. We emphasize that the persistent current is an equilibrium state of the system (a minimum in the free energy, maintained by collisions), subject only to the constraint that the phase of Ψ does not change.

A2.2

A3.2

A1.2.5 Flux lines, the mixed state, and type I and type II superconductors

We might ask whether a single quantum of flux, for example, could be trapped inside a bulk sample of superconductor. It can be easily shown from equation (A1.2.10) that in such a case the current would diverge to infinity along a line in the superconductor. In the presence of a very large current it is energetically favourable for the Cooper pairs to break up, so the material would become effectively normal along this line. The situation can be described by an appropriate solution of the Ginsburg–Landau equations. The amplitude of the Ginsburg–Landau wavefunction vanishes along the line, and it rises to its normal value over a distance from the line of order the Ginsburg–Landau coherence length ξ_{GL}. The phase $S(\boldsymbol{r})$ changes by 2π as the line is encircled, so that one flux quantum is associated with the line. The magnetic field penetrates from the line to a distance of order of the penetration depth λ (figure A1.2.3). The resulting structure is called a 'flux line' or 'vortex', the latter name originating from the fact that close to the line the electron velocity is similar to that in a hydrodynamic vortex (proportional to $1/r$). The ratio λ/ξ_{GL} is called κ. This ratio turns out to be approximately independent of temperature.

A3.1

Whether it is energetically favourable for such a structure to exist depends on its energy, ε, per unit length. Suppose that a superconductor in the form of a long thin solid cylinder (radius $\gg \lambda$) is exposed to a magnetic field, B, directed along its length. It can be shown that it is energetically favourable for a flux line to exist in the superconductor if B exceeds the value given by $B = B_{c1} = \varepsilon/\phi_0$. We recall that if B exceeds the value given by equation (A1.2.1) (the 'thermodynamic critical field') superconductivity is destroyed. Therefore, flux lines can be formed only if ε is sufficiently small, so that $B_{c1} < B_c$. It turns out that this condition is equivalent to the condition that $\kappa > 1/\sqrt{2}$.

If $\kappa < 1/\sqrt{2}$ the superconducting cylinder passes directly from the 'Meissner state' (no field penetration except in the penetration depth) to the normal state when B exceeds B_c. A superconductor of

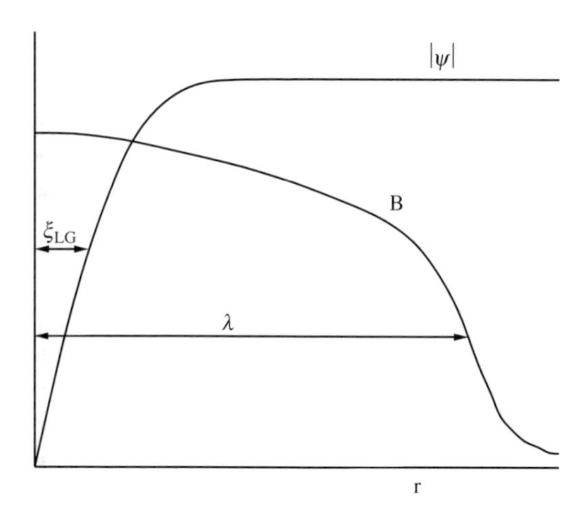

Figure A1.2.3. The modulus of the order parameter ($|\Psi|$) and the magnetic flux density (B) plotted against radial distance (r) from the centre of a flux line (schematic).

this type is called a type I superconductor; its phase diagram was shown in figure A1.2.1. If $\kappa > 1/\sqrt{2}$ flux lines can penetrate the superconducting cylinder at fields greater than B_{c1} ($< B_c$), the diamagnetic moment of the cylinder being reduced. In fact more and more flux lines will penetrate until the (repulsive) interaction between them causes it to be no longer energetically favourable for them to form. As the applied field is increased the density of flux lines increases, the diamagnetic moment falling, until the 'cores' of the flux lines (the regions of size ξ_{LG} where the superconductivity is suppressed) overlap, when the material becomes normal; the transition to the normal state occurs when $B >\sim \phi_0/\xi_{LG}^2$, which is equivalent to $B > B_{c2} = \sqrt{2}\kappa B_c$. A superconductor with $\kappa > 1/\sqrt{2}$ is called a 'type II superconductor'. When a type II superconductor contains an array of flux lines it is said to be in the mixed state. The magnetization curves for a type II superconductor is shown schematically in figure A1.2.4, and the phase diagram for a type II superconductor is shown in figure A1.2.5. (It is necessary to consider a long thin cylinder here because otherwise demagnetizing effects become important and result in a different type of D2.1

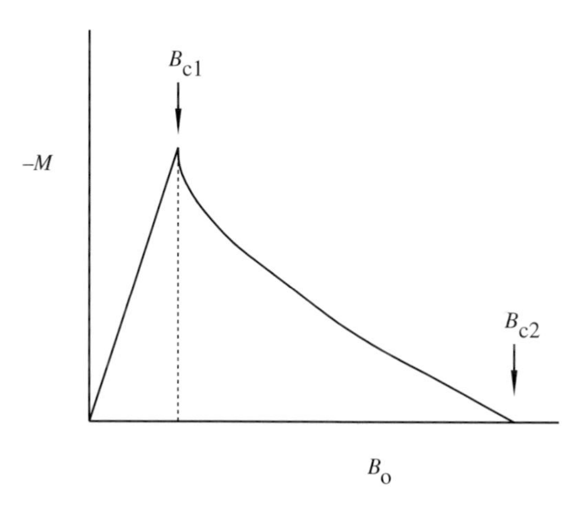

Figure A1.2.4. Magnetization curve for a type II superconductor. Magnetization (M) plotted against applied magnetic flux density (B_0) (schematic).

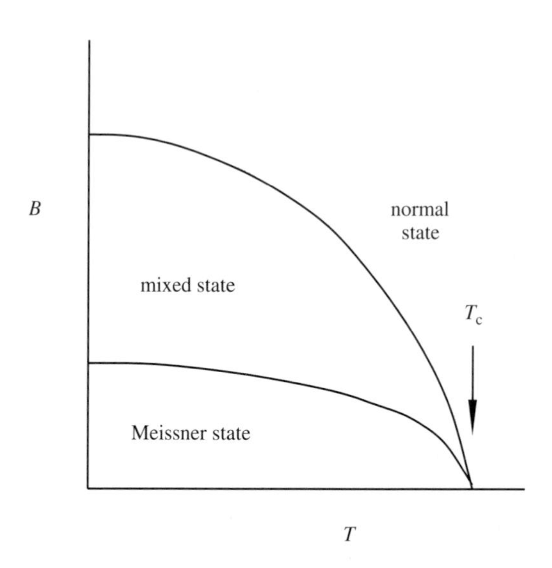

Figure A1.2.5. Schematic phase diagram for a type II superconductor: applied magnetic flux density (B) plotted against temperature (T).

behaviour. Even in type I superconductors flux penetration can then occur through the formation of the intermediate state in which relatively large areas of normal state are embedded in the superconductor.)

Type II superconductors are of great practical importance. Values of B_c are typically quite small (of order or less than 0.1 T), so that type I superconductors are useless in applications involving high magnetic fields. In contrast, values of B_{c2} can be very large, allowing, for example, the construction of superconducting coils for the generation of high magnetic fields. However, no pure single-element metal has a large value of κ, and indeed only niobium has a value of κ large enough (but only just) to make it type II. We are led therefore to look at other types of superconducting material, which we shall do in later sections.

A1.2.6 Phase slip

The existence of flux lines allows us to see how the 'persistent currents' we described earlier can decay by 'phase slippage'. If a single flux line were to pass through the walls of the cylinder described in section A1.2.4, the phase change round a circuit enclosing the hole in the cylinder would change by 2π. If the change is a decrease the effect is a decrease in the persistent current. This phase slippage, which involves only a localized perturbation to Ψ, can take place much more easily than any change involving simultaneously all the superconducting electrons. Creation of the flux line and its passage across the superconductor still generally encounters an energy barrier (some of it arising from an interaction between the flux line and the boundary of the superconductor). However, in the presence of a large enough current (the critical current) the barrier may be eliminated or reduced enough for thermally activated phase slippage to occur; the supercurrent can then decay in a relatively short time.

A1.2.7 Weak links and quantum interferometers

This type of phase slippage often occurs in narrow constrictions in the superconductor, and we then talk A2.7 of a weak link. A particularly simple type of weak link is the Josephson tunnel junction, formed by connecting two bulk volumes of superconductor through a thin insulating layer through which electrons can tunnel. The Cooper pairs can also tunnel, so that a supercurrent can pass across the junction. A simple analysis shows that for a weak junction the supercurrent is related to the difference in phase, ΔS, of the superconducting wavefunction across the junction by the relation

$$I = I_0 \sin \Delta S. \tag{A1.2.12}$$

The critical current is I_0, and we have described the dc Josephson effect. If I exceeds I_0 a potential difference, V, appears across the junction. A Cooper pair on one side of the junction then differs in energy by $2eV$ from one on the other side. Like an ordinary wavefunction the superconducting wavefunction, Ψ, has a time dependence $\exp(-\mathrm{i}Et/\hbar)$, so that the potential difference V gives rise to a continual slippage of the phase on one side of the junction relative to the other at a rate given by

$$\frac{\mathrm{d}\Delta S}{\mathrm{d}t} = \frac{2eV}{\hbar}. \tag{A1.2.13}$$

The junction therefore carries an oscillating supercurrent of angular frequency $2eV/\hbar$ (the ac Josephson effect). The phase slippage can be viewed as due to the steady flow of flux lines across the junction, each flux line causing a phase slip of 2π. The ac Josephson effect has been important in defining the volt in terms of frequency and the fundamental constants e and $\hbar$.

A2.7

Josephson junctions can be used to form quantum interferometers. Two junctions are arranged as shown in figure A1.2.6 to form a superconducting loop. Simple quantum mechanics shows that, if a magnetic flux Φ threads the loop, the differences in phase across the two junctions must differ by $2\pi\Phi/\phi_0$. The total critical current across the two junctions, I_0 (the maximum value of $I_{01}\sin\Delta S_1 + I_{02}\sin\Delta S_2$) must therefore be reduced by an amount that is periodic in Φ/ϕ_0. Since ϕ_0 is very small, we have the basis for a very sensitive magnetometer.

A1.2.8 The role of the normal electrons

We have seen that at a finite temperature there are condensed electrons in the superconductor and electronic excitations, which form, respectively, the superfluid and normal-fluid components in a two-fluid model. Any steady electric current in the superconductor is carried exclusively by the superconducting electrons; there is no electric field to drive the normal electrons which exhibit an

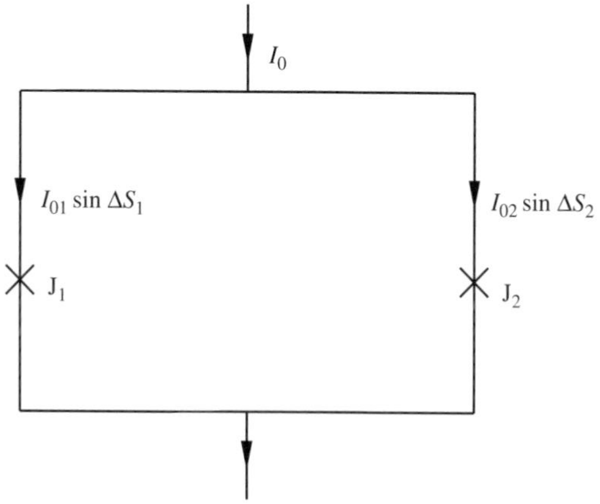

A2.1

Figure A1.2.6. Schematic quantum interferometer.

electrical resistivity due to the scattering of the excitations by lattice defects, as is the case for the electrons in an ordinary metal. At high frequencies, however, an electric field is required to move the superconducting electrons owing to their inertia, and this field will produce a response, and therefore A2.5 dissipation, in the normal electrons. The superconductor is therefore not free from resistance at high frequencies, especially at microwave frequencies. In calculating this resistance one must be careful to D2.6 remember that the normal electrons are really excitations with different properties from electrons in the normal metal. At still higher frequencies (usually in the infrared) the photon energy may be sufficient to break a Cooper pair, leading to greatly increased dissipation.

D2.4 The excitations play a role also in the thermal properties of the superconductor. Here, one must consider both the electronic excitations and the phonons. Both contribute to the heat capacity. Owing to the presence of the energy gap the electronic contribution to the heat capacity becomes very small for $T \ll T_c$, but the phonon contribution remains. The condensed electrons carry no entropy, so heat conduction is entirely due to the excitations. In normal pure metals the heat is carried largely by the electrons; the phonons carry little heat because they travel much more slowly than the electrons and because they are strongly absorbed by the electrons. In a superconductor the situation is very different, at least for $T \ll T_c$. The excitations disappear at low temperatures, so they themselves carry little heat. At the same time the phonons can interact only with the electronic excitations (they do not have enough energy to break Cooper pairs), so they are no longer strongly absorbed. Heat is therefore carried largely by the phonons, which can have a long mean free path as in a dielectric; the low-temperature phonon conductivity can therefore be high.

A1.2.9 Alloys

A3.2 So far we have confined our attention to pure single-element metals, which are described by the BCS theory. The BCS theory is valid only if the electron–phonon interaction that is responsible for the formation of Cooper pairs is weak. In some metals, for example lead, this is not true, and a development of the BCS theory is required to treat such strong-coupling superconductors quantitatively. However, the basic physics remains unchanged.

For the rest of this chapter we shall consider other types of superconducting material. As far as we know, superconductivity always involves the formation of electron pairs, but the pairs may be formed by a different (nonphonon) mechanism, and they may be formed in more complicated states than in a BCS superconductor, where the electrons in the pair have opposite spins and are in a state with no relative angular momentum (s-state).

We shall make a start, in this section, by considering a very important class of superconducting material formed from an alloy of two or more metals in which there is s-state pairing. In a pure metal at the low temperatures required for conventional superconductivity the electrons in the normal state have long mean free paths, much larger than the superconducting coherence length ξ_0. In an alloy this is generally no longer the case, owing to the scattering of electrons by the disorder in the alloy. At first sight one might expect that this would have a profound effect, since it would surely affect the way in which the pair wavefunction can be formed. Surprisingly, it has very little effect in the case of nonmagnetic components. The addition of a nonmagnetic alloying element to a pure metal has very little effect on the critical temperature. It turns out that this is because the pairing wavefunction can be set up on the basis of the real normal-state electronic wave functions, whatever they are, but still with the pairing of opposite spins; in the case of an alloy the wavefunctions include the effect of the scattering. The gap parameter as a function of temperature is essentially the same, as is the critical temperature. In many pure metals the electronic wavefunctions in the normal state exhibit isotropy associated with the details of the band structure, and this is reflected in anisotropy in the gap parameter, this parameter varying A2.1 round the Fermi surface in a way that is consistent with the crystal symmetry. Formation of Cooper pairs from wavefunctions that describe an electron that is being continually scattered removes this anisotropy, but otherwise there is little effect on the gap parameter.

In more detail, however, the properties of an alloy do differ in important ways from those of a pure metal, especially if the electron mean free path in the normal state, ℓ, is less than ξ_0 in the pure metal. The effective density of superconducting electrons is reduced by the scattering, by a factor of ℓ/ξ_0 in the limit $\ell \ll \xi_0$, so that the penetration depth is correspondingly increased. The coherence lengths are modified. The Pippard coherence length (governing the range of the nonlocal relation between $\boldsymbol{J}$ and $\boldsymbol{A}$) is reduced to ℓ in the same limit, while the Ginsburg–Landau coherence length at low temperatures becomes $(\ell\xi_0)^{1/2}$. We see therefore that the value of κ is increased. Alloys tend therefore to be type II superconductors. Suitably chosen alloys can have large values of κ, with a resulting large value of the upper critical field B_{c2}. Such alloys might therefore be used in applications in which the superconductor is exposed to a high magnetic field. A3.1

An important application is the production of high magnetic fields with a superconducting coil. However, a high κ is not sufficient for this purpose. In a high field the superconducting material will be in the mixed state. When it carries an electric current the flux lines will be subject to a force (roughly speaking a Lorenz force equal to $\phi_0 \times \boldsymbol{J}$ per unit length of line), which may cause the lines to move in the transverse direction, giving rise to phase slip and dissipation. To prevent such movement the flux lines must be pinned by suitable microstructure in the alloy. Much of the technology of magnet materials is concerned with the enhancement of this pinning. A1.3 A1.3

A1.2.10 Towards higher values of T_c and H_{c2}

Many practical applications would benefit from materials with higher values of both T_c and H_{c2}. Among conventional superconductors the best that can be done appears to be among intermetallic compounds with the cubic β-W (A15) structure and among the Chevrel phase compounds with composition $M_xMo_6X_8$, where M is a metal and X is either sulphur or selenium. Of these compounds the best known is probably the A15 compound Nb_3Sn, which has a critical temperature of about 18 K and an upper critical field at low temperatures exceeding 30 T, and which has been used extensively in superconducting magnets. These materials are of course strong-coupling superconductors, and the question arose whether stronger and stronger electron–phonon coupling can lead to higher and higher critical temperatures. Analysis that takes into account both the electron–phonon interaction and the Coulomb repulsion between electrons shows almost certainly that conventional mechanisms of superconductivity cannot lead to critical temperatures exceeding about 30 K, a view that seems to be confirmed by practical experience. G2 G1

A1.2.11 Unconventional superconductivity

So far this introduction has been concerned with ‘conventional’ superconductors; i.e. those in which superconductivity arises from the formation of a condensate of electron pairs, the attractive electron–electron interaction required to produce the pairs being due to a distortion of the lattice (phonon exchange). In the simplest form of the BCS theory, the electron pairs form in states with zero internal angular momentum (s-states) and therefore with antiparallel spins, and conventional superconductors are correctly described by this form of the theory. A3.2

During the past 10 years or so many superconducting materials have been discovered that appear not to be conventional. It appears that all involve electron pairing and, presumably, the formation of a condensate from the pairs, but they are unconventional in the sense that they involve pairing into states with nonzero angular momentum and/or a pairing mechanism that is not due to electron–phonon interaction.

A search for superconducting materials that do not depend on the electron–phonon interaction was based in part on the wish to find materials with higher critical temperatures. It was suggested by Little in 1964 that superconductivity might be found in solids composed of certain long-chain organic molecules, and that the electron–electron attraction might then arise through the electronic polarization of side-chain molecules. At that time even organic metals were hardly known, but over the years the search for, and study of, such materials has been intense, and it led eventually to the discovery of organic superconductors (although not in polymers). The normal-metallic properties of these materials involves much interesting and novel physics (and chemistry!), connected, for example, with their low dimensionality, and the study and understanding of these normal properties has been important. But the organic superconductors that have been discovered so far do not have high critical temperatures. The mechanism of superconductivity seems not to be settled, and it may be unconventional. As in other cases to which we shall shortly refer, the possibly unconventional superconductivity may be linked to unconventional normal behaviour.

Much interest has been shown in the possibility that non-s-state pairing might be found in some superconductors. We recall that atoms of the light isotope of helium, ^{3}He, are Fermions and have a nuclear spin of one half. The superfluid phase of liquid ^{3}He was discovered in 1972 at temperatures less than a few mK, and it was shown very quickly that it could be described by a development of BCS theory that incorporated p-state pairing and parallel nuclear spins ($L = 1$, $S = 1$), and which had been developed, at least in part, in earlier years. Pairing in s-states is not possible because of the strong short-range repulsion between two helium atoms; p-state pairing makes use of the long-range attraction. Owing to the p-state pairing, which gives 'structure' to a Cooper pair, superfluid ^{3}He is much more complicated than superfluid ^{4}He, and it exhibits a rich variety of phenomena that have no counterpart in ^{4}He. One aspect of this complication is the existence of (at least) two different superfluid phases of ^{3}He; there is p-state pairing in both phases (A and B), but the orbital and spin angular momenta are differently arranged around the Fermi surface in the two cases. The nontrivial structure of a Cooper pair means that the order parameter has more components than in the case of s-state pairing: for p-state pairing, for example, the order parameter has in principle 18 components (three possible spin orientations; three possible orientations of the orbital angular momentum, and each of the possible nine combinations has its own real and imaginary components of the order parameter); Ginsburg–Landau theory becomes correspondingly more complicated. It should also be mentioned that the study of normal liquid ^{3}He played an important role in the development of our understanding of normal Fermi liquids, a development that carried over into a much better understanding of the electrons in a normal metal than is possible in terms of a model in which the conduction electrons are regarded as a noninteracting Fermi gas. In the case of normal ^{3}He there are strong exchange interactions that cause the liquid to have a greatly enhanced Pauli susceptibility (due to the nuclear spins) and to be, in fact, almost a nuclear ferromagnet. As a result there are, within the liquid, strong local fluctuations in the spin magnetic moment (paramagnons). Exchange of paramagnons contributes in an important way to the interatomic interaction leading to pairing and superfluidity, and it also serves to allow types of superfluid phase to exist that would not otherwise be stable (e.g., the A phase).

A superconductor with non-s-state pairing can exhibit properties differing from those of a conventional superconductor, which can allow us to identify it. In some phases of a non-s-state superconductor the energy gap will vanish at certain points, or on certain lines, on the Fermi surface. This affects the temperature dependence of both the heat capacity and the normal fluid fraction (and hence the penetration depth); it gives rise to a power law dependence at the lowest temperatures instead of the exponential dependence that is characteristic of an energy gap which does not vanish in this way (it should be noted, however, that conventional superconductors containing magnetic impurities can also be gapless). However, this vanishing of the energy gap is not present in all phases of a non-s-state superfluid (e.g., the B-phase of superfluid ^{3}He), and a true vanishing of the energy gap may be difficult to distinguish experimentally from a very strong anisotropy in an s-state system. As we have already noted a non-s-state superconductor may exist in different superconducting phases, depending on the temperature and the applied magnetic field, as is the case in superfluid ^{3}He; different phases are seen in some superconductors that are based on heavy fermion metals, suggesting non-s-state pairing in these cases (again the heavy Fermion metals have interesting normal states, and similarities with liquid ^{3}He are likely, spin fluctuations being important in both the normal and superconducting phases). A fairly clear-cut indication of unconventional pairing comes from the effect of impurity scattering. We noted earlier that in a conventional superconductor such scattering has the effect of only smoothing out any anisotropy in the energy gap; in the case of unconventional pairing the scattering mixes states with different arrangements of the spin and/or orbital angular momenta, which destroys the superconductivity, such destruction occurring when the mean free path associated with the scattering is less than the superconducting coherence length. This effect is seen clearly in what is probably the best established case of p-state pairing in a superconductor: in Sr_2RuO_4. Another fairly clear-cut indication of non-s-state pairing can come from tunnelling studies. For example, d-state pairing ($L = 2, S = 0$) can lead to a situation where the order parameter has effectively a different sign on different parts of the Fermi surface; this can be observed in the behaviour of a quantum interferometer if the two junctions are formed with a second conventional superconductor on different sides of a single crystal of the unconventional superconductor. Finally, some types of unconventional pairing can involve a violation of time-reversal symmetry, so that the ground state can carry a nonzero (surface) current, even in the absence of an applied magnetic field.

G2

A1.2.12 The high T_c cuprates

The best known examples of what are almost certainly unconventional superconductors are the high T_c cuprates: $La_{2-x}(Ca,Sr)_x CuO_4$, with T_c in the range 30–40 K, discovered by Bednorz and Müller in 1978; $YBa_2Cu_3O_{7-x}$ (YBCO) with $T_c = 93$ K, discovered by Wu and co-workers in 1978; and all the others discovered over the subsequent years. Their potential applications have attracted enormous interest. But they are fascinating and important also in their basic physics. They differ from the classic conventional superconductor in at least four respects: their properties in the normal state are anomalous and not described by conventional Fermi liquid theory; they can have very high critical temperatures; they have layer structures, which lead to extreme anisotropy as between properties involving current flow within a layer and those involving current flow between layers; and they have very large (and anisotropic) values of κ, with very small (and anisotropic) coherence lengths, leading to inaccessibly high values of B_{c2}. It seems unlikely that such high values of T_c can result from a phonon mediated attraction between electrons, but there is as yet no agreement about the mechanism. Indeed there is no agreement about a theory of the normal state. The extreme anisotropy leads to interesting flux line structures, and this fact combined with the high T_c allows observation of new effects in the mixed state, such as the melting of the flux line lattice and the decoupling of the parts of a vortex in different layers (formation of pancake vortices). Finally, some at least of the cuprate superconductors are almost certainly not s-state superconductors: tunneling experiments of the type described in the preceding section show that YBCO probably involves d-state pairing. The physics of these materials continues to pose major challenges, as does the development of the practical applications of them.

C1

There was a feeling in about 1969 that superconductivity was fully understood and was a dead subject (see the volumes edited by Parks). How wrong that feeling turned out to be! The last two sections of this introduction provide an inadequate glimpse of the vast amount of interesting and important research on superconductivity that is still in progress.

Further Reading

In a survey of this kind it is neither possible nor desirable to refer to original papers. Even the number of books on superconductivity is too large. In any case detailed references can be found in subsequent chapters. It seems wise, therefore, to confine the references here to a few (somewhat arbitrarily) selected books and reviews, which can be used by a newcomer to the field to pursue further study.

London F 1950 *Superfluids* Vol 1 *Macroscopic Theory of Superconductivity* (New York: Wiley)

de Gennes P G 1966 *Superconductivity of Metals and Alloys* (New York: Benjamin)

Parks R D (ed) 1969 *Superconductivity* Vol 1 and 2 (New York: Dekker)

Tilley D R and Tilley J 1990 *Superfluidity and Superconductivity* 3rd edn (Bristol: Institute of Physics Publishing)

Tinkham M 1996 *Introduction to Superconductivity* 2nd edn (New York: McGraw-Hill)

Waldram J R 1996 *Superconductivity of Metals and Cuprates* (Bristol: Institute of Physics Publishing)

Ishiguro T, Yamaji K and Saito G 1997 *Organic Superconductors* (Berlin: Springer)

Ketterson J B and Song S N 1999 *Superconductivity* (Cambridge: Cambridge University Press)

Schofield A J 1999 Non-Fermi liquids *Contemporary Physics* **40** 95

Heffner R H and Norman M R 1999 Heavy Fermion Superconductivity *Comments in Condensed Matter Physics* (at press)

A1.3
Applied properties of superconducting materials

Lawrence J Masur, Jürgen Kellers, Colin M Pegrum and D A Cardwell

A1.3.1 Introduction

Traditionally, low temperature superconductors (LTS) have been used most commonly in applications that require large magnetic fields, such as MRI devices and maglev systems. The discovery of high temperature superconductors (HTS) in 1986, however, widened considerably the range of potential applications for superconducting materials in the electric power and electronics industries. The use of LTS in electric power applications, in particular, has been driven principally by their successful application in magnetic energy storage units. In these, LTS is a true enabling technology that allows the protection of sensitive electric machines or even an entire facility from the adverse effects of poor power quality. The more recent discovery of MgB_2 opens up potential for further application of superconductivity.

E2 E1.3.4 C5

In general HTS are available in three material forms: bulk, thin film and wires or tapes (LTS are not used practically in bulk form). Of these, bulk materials have the potential to generate large magnetic fields that are much greater than those achievable with conventional permanent magnets. Thin films have the potential for use in high-speed electronics, logic and microwave circuits where sharp bandwidth characteristics are required. Wires and tapes have the potential for use in transmission cables, electric rotating machinery and transformers as a more efficient and more compact replacement for copper technology.

E3 E1.3.5, E1.4.2

Bulk materials that can generate magnetic fields in excess of 2 T at 77 K and 13 T at 30 K are based entirely on the (RE)BCO family of HTS fabricated in the form of large grains by a melt processing technique. BSCCO-2223 fabricated as a multifilamenary composite, on the other hand, is accepted generally as being most appropriate for (pre-) commercial electrical conductor applications and is readily available in reasonably long lengths at average engineering critical current densities of 14 kA cm^{-2} (77 K, self-field, 1 μV cm^{-1}). Thin films for electronic, Josephson junction and microwave applications are usually fabricated from YBCO on a variety of single crystal substrates in different geometries by a number of deposition processes and have already been applied commercially in cell phone base stations and in fast digital logic devices. YBCO thin films deposited on long, flexible substrates are currently in development. These coated conductor composites can be manufactured as form-fit-function replacement of BSCCO-2223 wires and their potential for improved price performance ratio offers real optimism for an even wider use of HTS in electrical power applications.

B2.3.3 B3.2.2 B4.0 B3.2.4

While performance improvements in current carrying capability and uniformity are an important measure of the progress of HTS wire, other critical requirements must be met before HTS materials can be considered ready for general commercialization. The goals for all three forms of material are (a) technical viability, (b) reliability and (c) economic viability.

A1.3.2 Wires and tapes

A1.3.2.1 Long length manufacture of BSCCO-2223 wire

The industrial baseline for pilot manufacture and further expansion to commercial plant production is the multifilamentary composite (MFC) technology, based on the oxide powder in tube (OPIT) processing technique (see chapter B3.1). Superconducting oxide ceramic of nominal metal composition $Bi_{1.7}Pb_{0.3}Sr_2Ca_2Cu_3$ (BSCCO-2223) is used principally in this process. The starting oxide powder is prepared by pyrolysis of metal nitrates and then packed and sealed into cylindrical silver billets. The formation of mono-filamentary rods is carried out by conventional deformation processing of the billets using extrusion and drawing. Multi-filamentary composites are achieved by bundling the mono-filamentary rods together in a metallic tube that is deformed via axi-symmetric processing into round wire. Conversion of the round wire to tape geometry is achieved by rolling. Heat treatments are carried out within the stability range of the BSCCO-2223 phase. Multiple deform/sinter cycles result in a tape conductor with the desired dimensions and electrical properties. An overview on the manufacturing process and the resulting flexible multifilamentary composite conductor is given in figures A1.3.1 and A1.3.2, respectively.

B3.1

Conductor performance and process development

The challenge for wire manufacturers is to transfer results of R&D into long length wire manufacturing. A useful measure of a successful process development is to manufacture multiple wire length under nominally identical conditions and to observe the resulting process variability.

Masur *et al* [1] have reported an average long length performance of 118 A (77 K, self-field, 1 μV cm^{-1}) with a standard deviation of approximately 6% over a population of roughly 29 km of wire. A more recent production run, having a population of roughly 60 km of wire, shows an average performance of 130 A (77 K, self-field, 1 μV/cm^{-1}) with a standard deviation of approximately 6%.

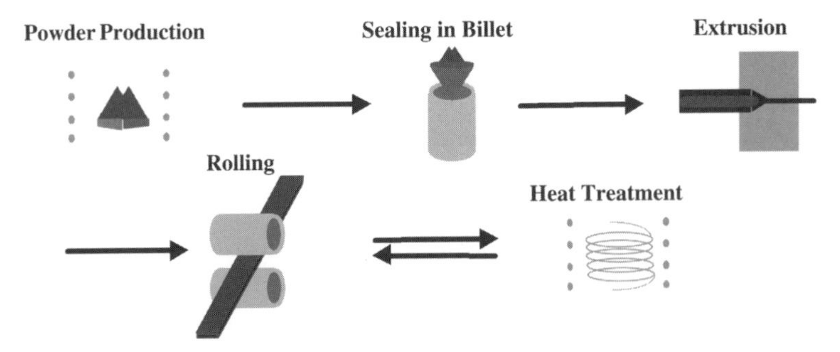

Figure A1.3.1. Schematic diagram of the manufacture of flexible multifilamentary composite conductor via the OPIT route.

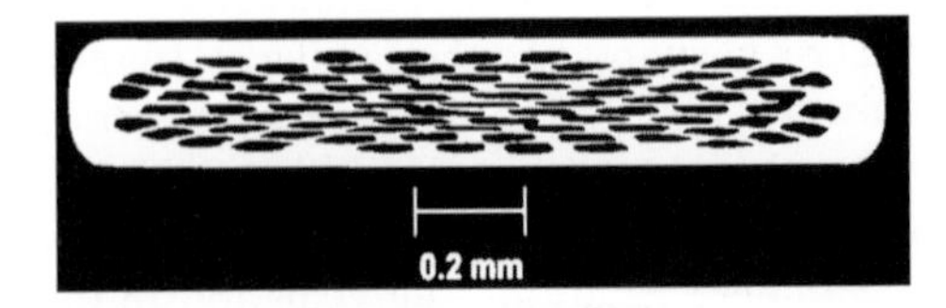

Figure A1.3.2. Microscopic transverse cross-section of a 85 filament conductor.

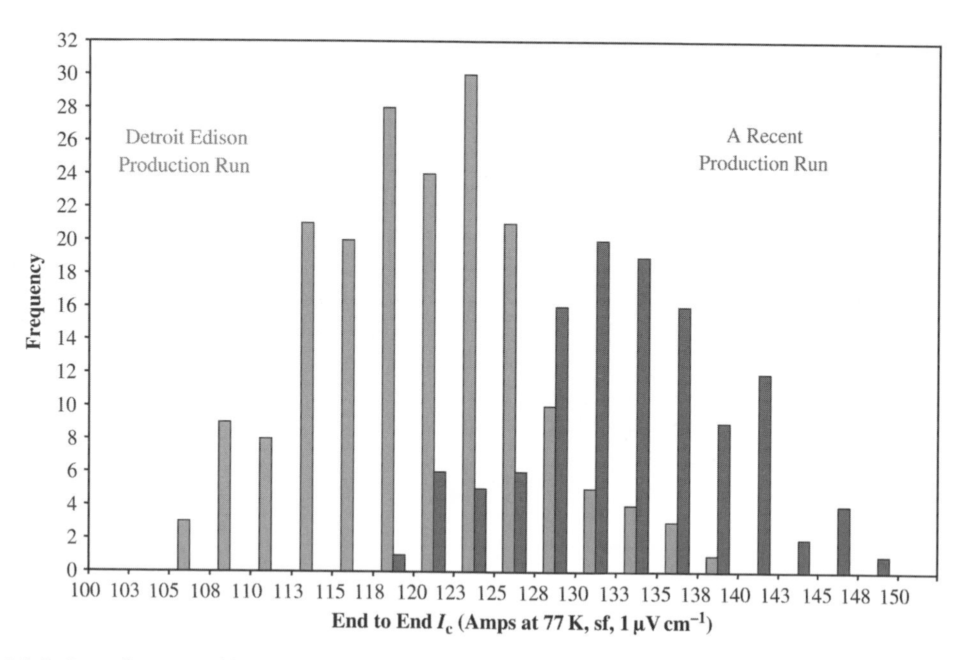

Figure A1.3.3. I_c performance histogram comparing two productions runs that are approximately 1 year apart. An increase of approximately 10% is noted in the two population averages.

Figure A1.3.3 shows a comparison of these two Bi-2223 wire populations, illustrating an average performance improvement of 10% over 1 year.

As long length Bi-2223 from manufacturing runs continues to improve, so do results from experimental short lengths that have dimensions and process elements that are compatible with straightforward introduction into the manufacturing process. A recent paper by Huang *et al* [2] indicates short length performance results of up to 170 A (77 K, self-field, 1 μV cm^{-1}), which is an additional 13% improvement over the best long length result of 150 A. This demonstration of high performance Bi-2223 wire gives confidence in a constant improvement in the properties of this form of material. Electrical performance is already appropriate for use in near-term applications such as motors, generators and power cables, in particular.

For applications of HTS that require operating conditions significantly different from 77 K and self-field, the values of critical current need to be scaled. Appropriate scaling factors have been published by Rodenbush *et al* [3], as shown in figure A1.3.4.

A1.3.2.2 Material design constraints

The current carrying capability of BSCCO-2223-MFC wires has reached the levels required for commercial applications and more recently the robustness and reliability of the wire have been demonstrated in successful prototypes. As a means to improve the mechanical properties of BSCCO-2223 conductors, many organizations have developed oxide-dispersion-strengthened silver matrices [4]. The strengthened silver significantly improves the tolerance to both tensile and compressive forces over conventional pure silver. Buczek *et al* [5] and Bray *et al* [6] have discussed this phenomenon for tensile and compressive forces, respectively. These data are illustrated in figures A1.3.5 and A1.3.6, respectively.

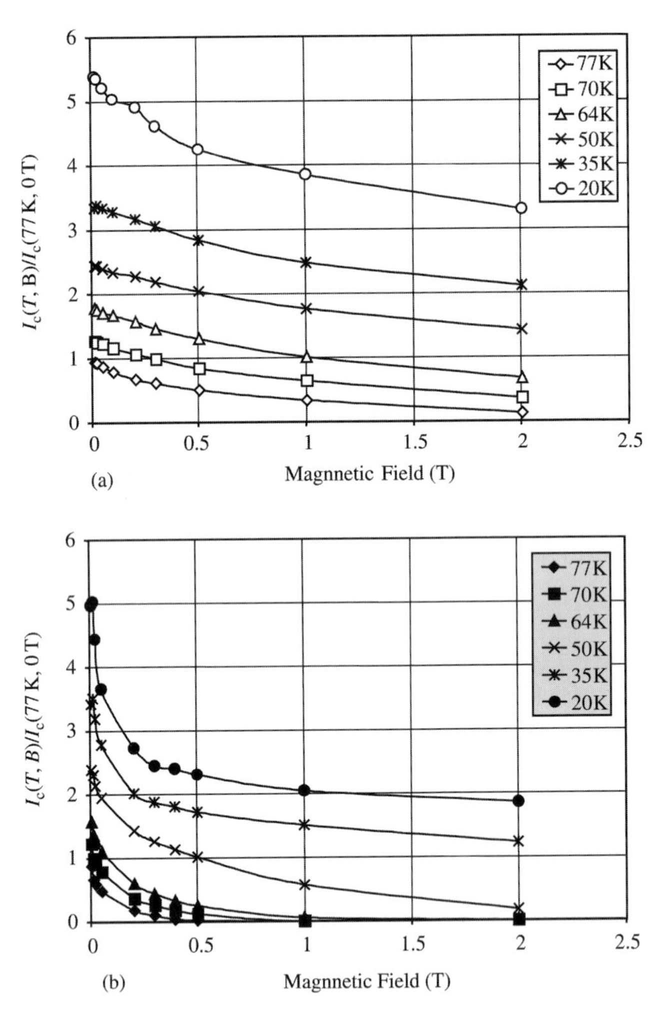

Figure A1.3.4. HTS performance at different temperatures and fields compared to the performance at 77 K. (*a*) Field parallel and (*b*) perpendicular to the conductor face. Data taken from [3].

The robustness of HTS conductor may be increased further by adding a thin (35 μm) reinforcing layer of stainless steel to both sides of the HTS tape [7]. Such a process has been developed by American Superconductor to yield the J_c versus tensile stress–strain performance shown in figure A1.3.7. While the addition of reinforcing material results in a net decrease in engineering critical current density by 33%, reinforced BSCCO-2223 tapes are able to withstand 265 MPa tensile stress and 0.4% tensile strain at 77 K. Thus, reinforced BSCCO-2223 tapes provide a mechanically robust and reliable product from which to make high performance prototypes. Reliability over 4 years of operation has been demonstrated in an HTS-based ion-beam steering magnet used in a tandem accelerator facility in the Institute for Geological and Nuclear Sciences in New Zealand. HTS-based power cables have been operating for 2 years in the Southwire manufacturing facility in Carrollton, Georgia, and for almost a year in a substation in Copenhagen, based on Southwire and NKT HTS cable technology, respectively.

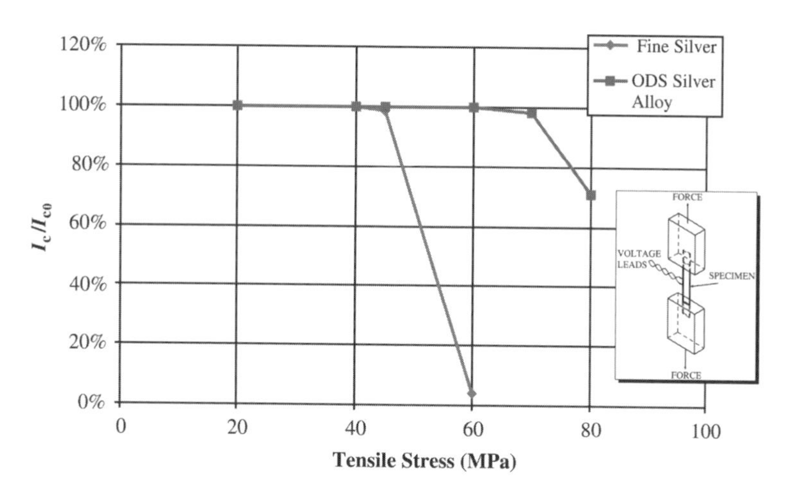

Figure A1.3.5. Comparison of critical current degradation under tensile loads for BSCCO-2223 wires using pure silver and ODS silver matrices. Data taken from [5].

A1.3.2.3 Economics of BSCCO-2223 wire

As HTS technology has matured beyond the levels of feasibility demonstrations, aspects relevant to commercial viability become essential. Lowest cost, highest performance is key to commercialization of HTS wire. This translates directly to: (a) an increase in the manufacturing capacity to ensure availability to end users; (b) a reduction in the manufacturing cost to commercialize broader markets; and (c) an improvement in process consistency by moving to a fully stabilized industrial manufacturing mode. These aspects are addressed simultaneously by a new large-scale manufacturing facility put into operation in early 2002 in Devens, USA by American Superconductor. This 33 000 m^2 manufacturing plant is dedicated solely to the production of Bi-2223-MFC and has a full capacity of 20 000 km per year. First wire production from this facility for the external market is expected at the end of 2002. This facility will enable a price/performance ratio in the range of \$50/(kAm). Broad market acceptance of HTS wire solutions depends on continuing to drive the price/performance of HTS wire to this range or lower.

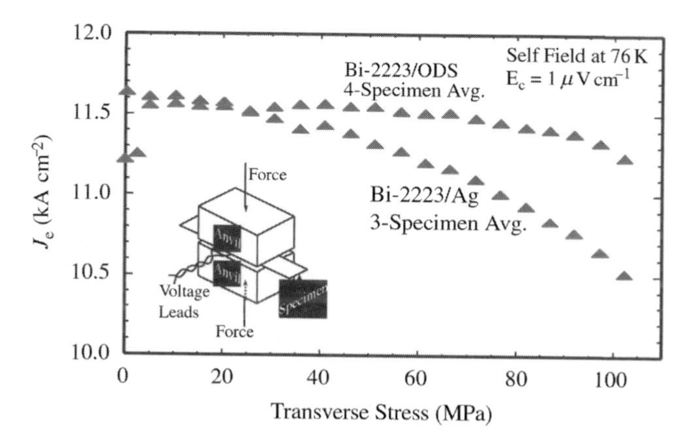

Figure A1.3.6. Comparison of critical current degradation under transverse compressive loads for BSCCO-2223 wires using pure silver and ODS silver matrices. Data taken from [6].

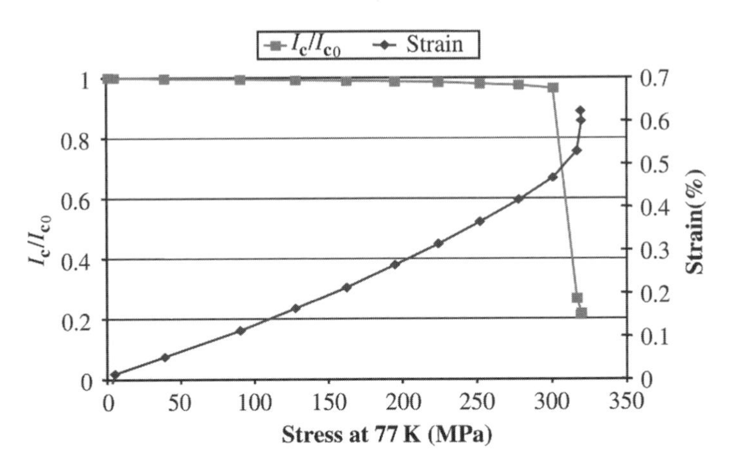

Figure A1.3.7. Critical current–stress–strain plot under tensile loads for BSCCO-2223 wires using stainless steel reinforcement on both sides of the tape.

A1.3.2.4 BSCCO-2212 conductors

B3.2.1 B3.2.2 Despite their chemical similarity, the two bismuth based HTS compounds $(BiPb)_2Sr_2Ca_2Cu_3O_x$ (BSCCO-2223) and $Bi_2Sr_2Ca_1Cu_2O_x$ (BSCCO-2212) offer quite different properties. BSCCO-2223 is generally preferred for electrical power applications, because of its higher critical temperature ($T_c = 110$ K). This offers a capability at 77 K, which is not achievable with BSCCO-2212 ($T_c = 80$ K). For applications operating well below 20 K, such as high-energy physics or ultra-high-field magnets, BSCCO-2212 can be the material of choice. BSCCO-2212 can be made in the form of a round wire because the required intergranular connectivity does not need to be achieved by deformation-induced texturing to a flat tape. Showa Electric Wire & Cable Co. Ltd., in collaboration with Chubu Electric Power Co. Inc., IGC-Advanced Superconductors and Lawrence Berkeley Laboratories, have developed a 20-strand Rutherford cable with $I_c > 4.5$ kA at 4.2 K in self-field using BSCCO-2212 round wire, as shown in figure A1.3.8. A design study of another Rutherford type cable consisting of 10 sub-cables, wherein each sub-cable has a 1×6 configuration fabricated by cabling six BSCCO-2212 wires around an Ag–Mg–Sb alloy core has been reported by Hasegawa *et al* [8]. This cable carries almost 10 kA at 4.2 K in self-field. Other high field tests conducted at NIMS in Tsukuba, Japan, and NHMFL in Tallahassee, Florida USA, have used tape geometry BSCCO-2212 conductors, such as those manufactured by Hitachi and Oxford Superconducting Technology.

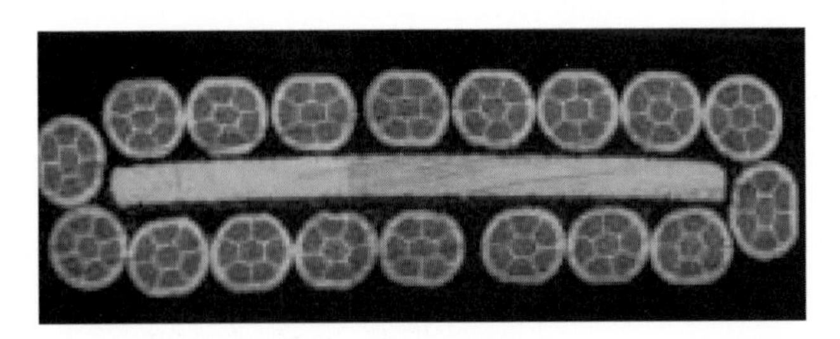

Figure A1.3.8. Example of a Rutherford type cable made of BSCCO-2212 round wire. Data provided by Hasegawa [8].

A1.3.2.5 YBCO coated conductor composites

Y–Ba–Cu–O (YBCO) coated conductor composite (CCC) wires utilize both YBCO HTS and metallic substrates. This technology presents an opportunity to reduce the price/performance of HTS wire below that of BSCCO MFC wire. Another potential advantage is that YBCO and also related compounds such as Tl–Ba–Ca–Cu–O (Tl-1223) or Hg–Ba–Ca–Cu–O (Hg-123), offer better performance in applied field, especially in the 77 K regime. This is because the so-called irreversibility line, $H_{irr}(T/T_c)$ is much higher than for the BSCCO superconductors [9], as discussed in further detail in section A1.3.4.2. One restriction on YBCO is the superconducting transition temperature of only 92 K, which could be an issue for applications where an operational temperature of 80 K and higher is required [10].

B3.2.4 B3.2.3 B3.2.5

While the above physical properties are well established and define the advantages and disadvantages with respect to BSCCO, there remain significant hurdles to overcome in the manufacturability of YBCO conductors (see chapter B4.1). The micaceous properties of BSCCO provide an easy path for texturing the compound and hence to the electrical coupling of adjacent crystallites. Due to a lack of easy cleavage, however, this is not possible with YBCO-123 so that biaxial texture needs to be induced by a substrate and transferred to a YBCO film by epitaxial growth. Textured substrates have been prepared with different processes including a ion-beam assisted deposition (IBAD) [11], inclined substrate deposition (ISD) [12], or by deformation texturing of the substrate [13, 14]. For any of these routes, the technology is still in the early stages with high performance (100 A cm^{-1} width) length of fully processed wires achieved at around the meter scale in 2002. Recently Fujikura has announced performance of 40 A cm^{-1} width over 30 m, using an IBAD texturing technique.

B4.1 B4.1

If YBCO is to substitute current BSCCO-2223 technology as the second-generation wire for electrical power applications, the importance of low cost processes needs to be emphasized. For example, running industrial quantities of conductor through a laser ablation process may lead to prohibitive cost with no advantage over BSCCO-2223-MFC. A lower cost approach, however, is offered by solution-based (sol–gel) coating techniques, which are under development by a number of research groups [15, 16]. An architecture appropriate for such a low cost process is illustrated in figure A1.3.9. This is similar to that produced by vacuum-based techniques but offers significant cost and time savings for the deposition of the HTS and buffer layers.

B4.2.2 B4.3

The approach of solution based metal-organic deposition has been validated generally by the deposition of YBCO-MOD films on a ceria-coated monocrystalline YSZ substrate to yield a critical current of 1 MA cm^{-2} over a thickness of 2 μm and on short textured nickel substrates with a thickness of 1 μm. The single crystal substrate result corresponds to 200 A cm^{-1} tape width, compared with 100 A cm^{-1} width for the textured substrate. In short, YBCO-CCC lies further out in the development cycle, but has the potential to improve price/performance over BSCCO-2223-MFC in the long term.

B4.2.2

HTS Layer: Solution-based, metal-organic deposition (MOD) process
Buffer Layers: Thin buffer layers deposited by e.g., solution-based deposition processes
Substrate: Deformation-textured metal or alloy

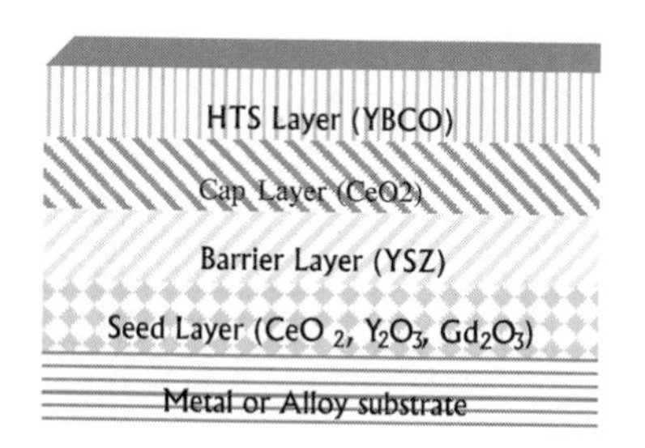

Figure A1.3.9. Schematic architecture of a flexible YBCO-123 coated conductor.

A1.3.2.6 Prototype commercial applications

Long length BSCCO-2223 wire has enabled a number of prototype devices to be developed, such as a cryogen-free HTS magnet system that generates 7.25 T at 21 K [17]. Various organizations are designing and testing HTS synchronous motors for use in industrial processing as well as power transmission cables. These prototype activities are an important step along the path to product commercialization.

HTS cables for power transmission

Given the strain on today's power grid and the difficulty of siting new overhead lines, high current density power cables represent a considerable market opportunity for HTS. Building on several successful demonstrations at shorter lengths, a number of significant pre-commercial prototype projects are running all over the world. HTS cables are already operational in Carrolton, Copenhagen and Tokyo. A summary of HTS cable projects to date, both past and present, is given by Malozemoff [18].

Pirelli Cables and Systems and the electric utility Detroit Edison are currently involved in a project which consists of replacing nine copper cables with three 125 m long HTS cables installed in existing conduits under a Detroit Edison substation to supply power to industrial and residential customers in the urban center of Detroit. The three HTS cables are designed to carry the same 100 MW load as the nine copper cables they are replacing in the original system. The project will demonstrate the feasibility of using existing infrastructure to relieve electrical distribution congestion in large metropolitan areas without costly and disruptive excavation and allowing vacated conduits to be used for higher-value purposes (e.g. carrying fiber optics, high-speed internet service etc). Further details on AC and DC power transmission can be found in chapter E1.3.2.

E1.3.2

Industrial HTS motors and generators

The overall goal of the industrial HTS motors and generators program is the commercialization of high efficiency, large-scale devices (see chapter E1.4.1 and E1.4.2). HTS rotating machinery is characteristically synchronous in nature with copper stators and HTS rotor coils in a rotor cold-mass. HTS motors and generators will significantly reduce the size, weight and electrical losses compared to conventional designs. Because of their smaller size, HTS machines can have lower synchronous reactance, which is helpful for voltage regulation and stability. In ship applications, the smaller size can facilitate significant noise reduction, space savings and streamlined ship design. Widespread interest in this technology is demonstrated by a string of successful prototypes of various sizes up to 5000 hp [19–21]. Finally, designs for marine propulsion motors and generators have been reported [22].

E1.4.1, E1.4.2

A1.3.3 Thin films

A1.3.3.1 Choice of superconductor, substrate and fabrication methods

Low temperature superconductors

Most LTS thin-film applications at 4.2 K in liquid helium are based on Nb, which has a transition temperature of just over 9 K. Almost without exception Nb films are made by DC magnetron sputtering in Ar. This well-established process [23] needs low levels of residual impurity gases (especially O_2) to grow high-quality films (see chapter B4.2.1). For these reasons the best films are grown in ultra-high vacuum deposition systems, with a sample load-lock and clean pumping by turbo-molecular or cryo-pumps. Figure A1.3.10 shows one such system. Films are usually 200–300 nm thick and are grown at

B4.2.1

Figure A1.3.10. An ultra-high vacuum cryo-pumped sputter deposition system, with Nb, Al, Mo and SiO_2 sources and sample load lock (University of Strathclyde, photograph courtesy of R G Weston).

typically 100 nm min^{-1}. Substrates can be as large as the sputter source size and substrate motion permit, though typically they are 50 or 75 mm in diameter. For compatibility with processing equipment designed for silicon-based semiconductor applications, the substrates are frequently silicon wafers, usually with a thermal oxide surface layer for enhanced adhesion, but other materials can be used for specialized requirements, such as sapphire, quartz or low thermal expansion glass. Films are normally grown at ambient temperature and are amorphous, but films with some epitaxial alignment can be grown on heated sapphire and are used for high-quality Josephson junctions with reduced sub-gap conductance [24]. Films of NbN (or variants such as $Nb_{1-x}Ti_xN$) are also important, primarily for digital applications, because their higher T_c of over 14 K makes operation at 10 K possible, which is within the range of closed-cycle cryo-coolers. These films, which are grown by reactive sputtering in an Ar/N_2 atmosphere [25, 26], are particularly promising for terahertz SIS mixers because of their higher superconducting energy gap [27].

Various non-superconducting materials are needed for multi-layer LTS devices. SiO_2 is commonly used for insulating layers, made either by RF sputtering or by low-temperature electron cyclotron resonance plasma-enhanced chemical vapour deposition (ECR PECVD). Mo, Pd, Au, Au alloys and other metals are used for resistors. Al is also needed for the most common Nb Josephson junction fabrication process (see section A1.3.3.3 on Josephson junctions). Multi-layer processing systems generally also include an Ar ion beam source for surface cleaning, to minimize contact resistance between different metal layers.

High temperature superconductors

The choices for HTS, substrate and deposition technique are much wider and the whole process of film deposition and processing is more challenging than for LTS. There are four families of cuprate materials, based on YBaCuO, BiSrCaCuO, TlBaCaCuO and HgBaCaCuO. For HTS devices that use Josephson junctions $YBa_2Cu_3O_{7-\delta}$ is the most commonly-used material, but $Tl_2Ba_2CaCu_2O_{8+\delta}$ is also attractive for microwave devices. Good BiSrCaCuO films can be made, but with some difficulty and so are less widely used; HgBaCaCuO films remain largely at the development stage. TlBaCaCuO and HgBaCaCuO have the added problems that Tl and Hg are highly toxic, which limits their uses. For each

C1, C2
C3, C4

family further variants can be made by partial or complete substitution of the rare earth element, for example $NdBa_2Cu_3O_{7-\delta}$, can be made to grow with very smooth surfaces for multi-layer structures [28].

Thin-film deposition of the metallic HTS cuprates must achieve the correct stoichiometry and oxygen content. Growth processes invariably require a high substrate temperature (750–800°C) and usually an O_2 atmosphere. In addition, they must be grown epitaxially and in the correct crystallographic orientation, to preserve the layered structure that is the key to their superconducting properties. This calls for specialised substrates with as close as possible lattice match between the single-crystal substrate and film. The substrate must also be chemically stable at the deposition temperature and there must be a good match of the coefficients of thermal expansion between substrate and film. For microwave devices the dielectric constant and dielectric loss factor (tan δ) of the substrate are also important issues [29]. For non-microwave use, $SrTiO_3$ is the best choice for most devices using $YBa_2Cu_3O_{7-\delta}$ films. $SrTiO_3$, however, exhibits a highly temperature-dependent dielectric constant and significant dielectric loss at the frequencies of interest for HTS microwave applications and so alternative substrates must be used, such as MgO, $LaAlO_3$, $GaNdO_3$, CeO buffered sapphire or yttria-stabilized ZrO_2 (YSZ) on Si.

There are several favoured deposition methods. Pulsed laser deposition (PLD) uses a laser to ablate B4.2.1 material in an O_2 atmosphere onto a substrate at typically 780°C (see chapter B4.2.1). It has the major advantage that for many HTS materials the stoichiometry of the target is preserved in the film, but the coverage area is relatively small (typically 1 cm^2); this is acceptable for some Josephson devices, but not for many microwave filter applications. RF sputtering from a composite target can cover larger areas for microwave devices, but an off-stoichiometric target is needed to achieve the correct stoichiometry in the film. Thermal or electron-beam evaporation from separate metallic sources produces excellent large-area films, but requires sophisticated deposition rate control for each source. Metal organic chemical vapour deposition (MOCVD) is potentially very attractive for commercial production over large areas, but has yet to emerge as a commonly-used technique. Molecular beam epitaxy (MBE) is ideal in principle for layered materials like HTS, but is little used, owing to its complexity and expense. All film growth processes require accurate control of the substrate temperature during deposition and a B4.2.1 controlled cool-down and anneal sequence in an O_2 atmosphere (see chapter B4.2.1 for further details of these film deposition techniques). TlBaCaCuO and HgBaCaCuO films require special growth procedures, due to both the volatility of Tl and Hg and their toxicity. For example, TlBaCaCuO can be grown initially without the Tl by PLD from a BaCaCu target and the Tl then added by reaction with Tl vapour in a sealed tube at about 800°C [30]. A similar process may be used to grow HgBaCaCuO films.

A1.3.3.2 Patterning and processing

Thin film structures in HTS and LTS films are patterned in spun-on resist by optical or electron-beam B4.4.1 lithography (see chapter B4.4.1). The former can routinely achieve feature sizes of 2 μm and less than 1 μm with specialized techniques. Electron-beam lithography reaches <0.1 μm and is used, in particular, to make ultra-small low-capacitance Josephson junctions with areas less than 10^{-2} μm^2. For LTS, two methods may be used to process the film. The lift-off technique uses a resist mask that precedes film deposition, and unwanted material deposited on top of areas of resist is then lifted off by immersion in acetone, which dissolves the underlying resist and leaves a patterned film. This process is possible because sputter-deposition is done with the substrate at ambient temperature and so there is no thermal damage to the resist. Alternatively, films may be etched by reactive ion etching (RIE) in a plasma, for example, using a CF_4/O_2 mixture for Nb, or CHF_3 for SiO_2; for RIE the resist layer follows film deposition and is used as an etch mask. For HTS the high deposition temperature means that lift-off is not feasible, and because the by-products of RIE are non-volatile, that method cannot be used either. The methods used instead are Ar ion beam etching (using typically 500 eV ions with a beam current density of $2\,\mathrm{mA\,cm^{-2}}$ and a rotating cooled substrate holder), or, less preferably, wet chemical etching. Special techniques are required for contacts to HTS films, usually by the deposition of Au pads that are subsequently annealed to achieve low contact resistance. Conventional ultrasonic bonds may then be made to these pads.

A1.3.3.3 Josephson junctions

Low temperature superconductors

For Nb and NbN devices the universally-preferred method of junction fabrication is the whole wafer technique originated by Kroger *et al* [31] and Nb technology advanced significantly with the development by Gurvitch *et al* [32] of the process to use a tri-layer sandwich of Nb–Al/Al_2O_3–Nb deposited across the entire substrate, to form one continuous superconductor–insulator–superconductor junction. A thin ($\approx$ 10 nm) Al layer is sputtered on top of the Nb base layer and the Al is then partially oxidized in O_2 to form the tunnel barrier. The thickness of Al_2O_3 is controlled by the O_2 pressure and oxidation time and determines the junction critical current density J_c, which may be in the range $1\,\mathrm{A\,cm^{-2}}$ to $2\times10^5\,\mathrm{A\,cm^{-2}}$. The top layer of Nb is then deposited to complete the tri-layer. The method produces high-quality repeatable junctions, for several reasons. The junction can be grown on a bare substrate, uncontaminated by any earlier resist processing. The deposition sequence is done without breaking vacuum, which avoids interface contamination. Al wets and covers Nb very effectively and the residual un-oxidized Al has little effect on tunnelling and the junction critical current, due to the proximity effect. The actual thickness of the Al deposited is therefore relatively uncritical. The oxidation process provides a high degree of control over the Al_2O_3 barrier thickness, which, because it is so thin (a few nm), could not be achieved by the deposition of the barrier as Al_2O_3 or any other material.

The tri-layer is subsequently etched down to the base layer to leave islands or mesas where junctions are required and the base layer patterned to form parts of the device's superconducting structure. A subsequent deposited and patterned layer of Nb provides a wiring layer for interconnections. For digital applications, there may be another continuous layer of superconductor and insulator under the tri-layer to form a ground-plane. There are several variations on this basic process, for example, to allow exposed edges of the junction mesas to be anodised to prevent contact to the wiring layers. Figure A1.3.11 shows the structure produced by one such process. For NbN a tri-layer technique may also be used, but with a deposited barrier, such as sputtered MgO [33] or AlN.

B4.4.1

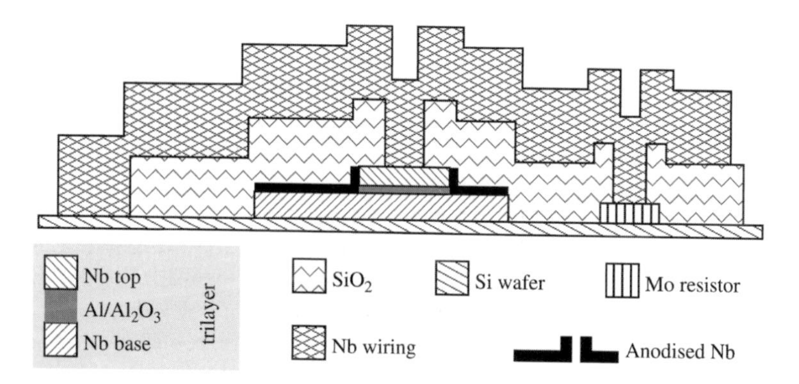

Figure A1.3.11. Cross-sectional schematic view for a typical LTS fabrication process.

High temperature superconductors

For these materials there is no equivalent of LTS Nb–Al/Al_2O_3–Nb junctions (due to fabrication constraints and to the intrinsic properties of HTS materials). Instead, most HTS junctions rely on the weak contact between grain boundaries [34] and such grain boundary junctions (GBJs) may be induced in the film where needed by several common techniques. Since supercurrents flow in the *a* or *b* crystallographic directions which are usually parallel to the substrate in HTS thin films, the GBJs generally need to be perpendicular to the substrate. Bicrystal GBJs (figure A1.3.12(*a*)) are formed on a substrate made from two halves fused together with a misalignment angle. A film on such a substrate forms a GBJ along the underlying substrate boundary. The J_c of this junction can be varied widely by choice of misalignment angle and, to a lesser extent, by film thickness and junction width [34], as shown in figure A1.3.13. All GBJs must lie on a straight line, which is a severe fabrication constraint. Step edge junctions (figure A1.3.12(*b*)) are formed by ion beam etching steps or pits into a single-crystal substrate, and GBJ forms where the film passes over the edge [35]. Step height, film thickness and junction width determine J_c and junctions can be placed almost anywhere, as in the double-loop SQUID illustrated in figure A1.3.14. Ramp-edge junctions are a more complex alternative, and can have a higher I_cR_n product than bicrystal or step-edge junctions, which is attractive for applications. In their original form (figure A1.3.12(*c*)) an angled edge is formed on an initial base layer of superconductor that is capped with an insulating layer [36]. The exposed edge is then coated with a thin layer of insulator (e.g. $PrBa_2Cu_3O_{7-\delta}$, $PrBa_2Cu_3O_{7-\delta}$ doped with Ga or Co, or Co-doped $YBa_2Cu_3O_{7-\delta}$) to form a junction barrier and a second HTS layer is then deposited. More recently most work on ramp-edge junctions has turned to the interface-engineered type. In this case, the barrier is formed by processing the base layer (e.g. by etching, annealing or chemical treatment), rather than by depositing a barrier layer [37]. Other junction techniques are well-developed, including *c*-axis microbridges [38] where the junction current flows perpendicular to the film surface and much work continues to develop an equivalent to the tri-layer technique used for LTS devices.

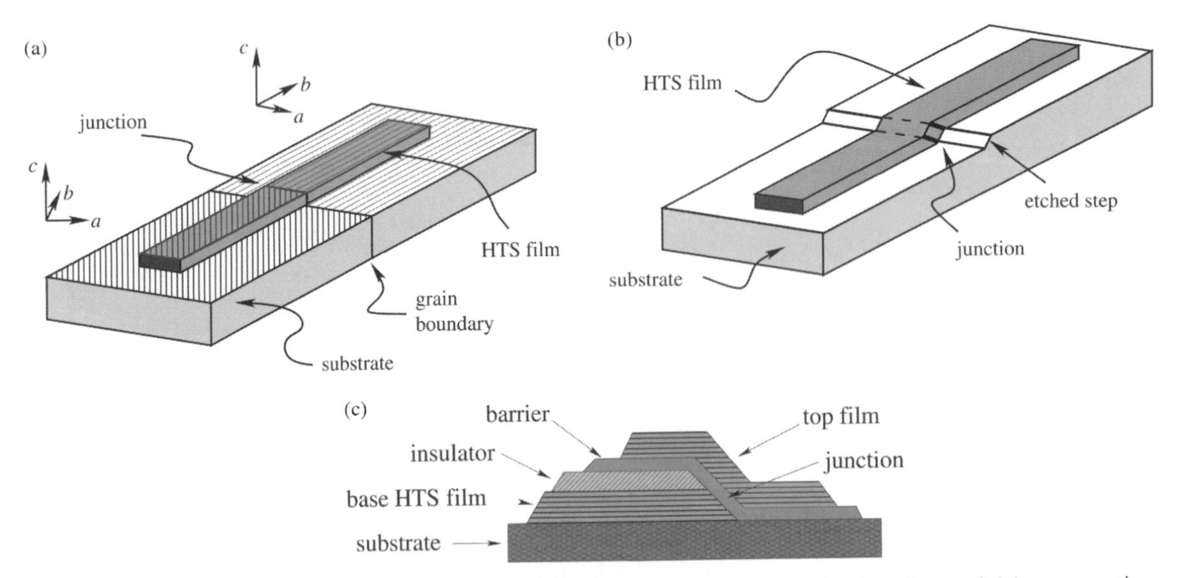

Figure A1.3.12. Three types of HTS junctions (*a*) bicrystal junction (*b*) step-edge junction and (*c*) cross-section through a ramp junction. Film thicknesses in (*c*) are not to scale.

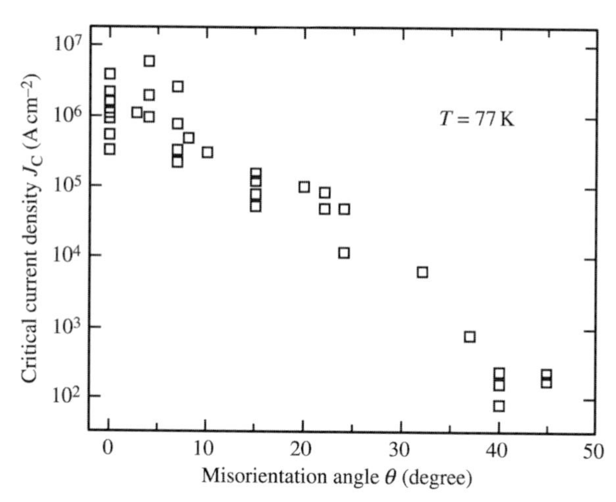

Figure A1.3.13. Variation of the critical current density J_c with misalignment angle θ for a range of different bicrystal junctions. (Based on data in [34].)

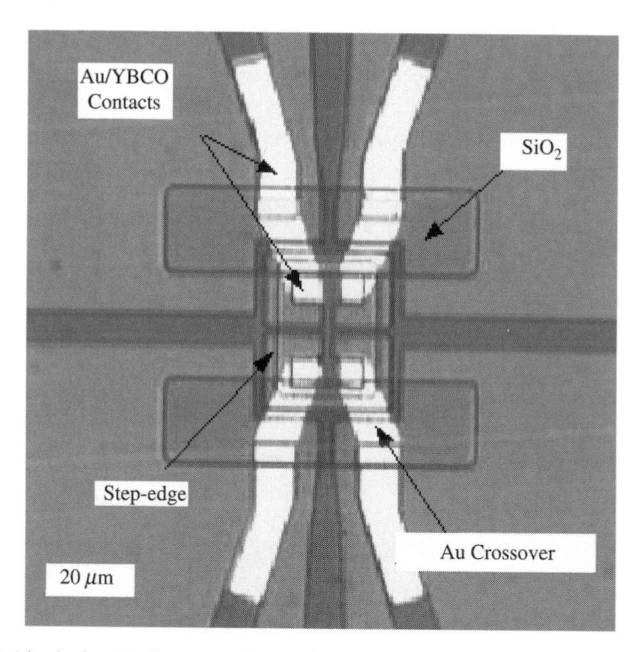

Figure A1.3.14. A double-loop gradiometric SQUID based on step-edge junctions [35].

A1.3.3.4 The current status of MgB_2 films

MgB_2 has the highest T_c (39 K) of the metal diboride series of superconductors (others all have $T_c < 10$ K) [39]. It appears to be largely BCS-like, with little anisotropy in its superconducting properties and, unlike the cuprates, has strong, non-Josephson contact between grains [40]. These properties should make it ideal for thin-films and devices. However, the high volatility of Mg and its sensitivity to oxidation are overriding issues in their fabrication. Thin films with the highest T_c ($\approx$ 39 K) and sharpest

C5

transitions are currently grown by an *ex situ* process. These techniques use PLD (or electron-beam evaporation) to grow a pure boron film, which is then annealed typically at 900°C in Mg vapour [41]. Current work aims to remove the need for the high-temperature *ex situ* anneal and find a lower temperature *in situ* process more suited to the fabrication of devices with multi-layers and Josephson junctions. Such *in situ* processes use PLD from an Mg-rich target or sputtering from multiple targets on to a heated substrate. This is usually followed by an anneal in an inert atmosphere at up to 600°C, for example [42], though Saito *et al* [43] have an as-grown process that does not require the post-deposition anneal. In-situ films have $T_c \approx 25-30$ K and generally exhibit broader transitions. Substrates include STO, sapphire, MgO and Si. Since there is no Josephson coupling between grains, grain-boundary junctions are not feasible and the high temperature deposition is not amenable to the fabrication of sandwich-style SIS or SNS junctions. As a consequence, Josephson devices fabricated to date with MgB_2 have used weak links [44], although structures with excellent Josephson properties have also been made by the focused ion beam (FIB) technique [45].

A1.3.3.5 Thin film devices and applications

E3, E4, E5

Sections E3–E5 of this handbook cover in detail both LTS and HTS devices and their many applications and only some highlights are mentioned here. In very broad terms there are two classes of device: those that use Josephson junctions, and passive devices, which do not. The superconducting quantum interference device or SQUID uses just two junctions and is the most sensitive detector possible for magnetic flux and has many applications, such as in biomagnetism or non-destructive evaluation; see [46] for an overview. Figures A1.3.15 and A1.3.16 show two examples of SQUIDs. LTS digital devices use up to $\approx 10^4$ junctions to make very fast digital logic systems based on the single flux quantum (SFQ) mode of operation [47] (see chapter E4.5). Arrays of $> 10^4$ single LTS junctions are routinely used in laboratories worldwide as voltage standards [48]. Single LTS junctions (or small-number arrays) are used as X-ray and phonon detectors [49] and single LTS junctions have unsurpassed performance as mixers and detectors for radio astronomy at up to ≈ 1 THz [50]. All of these LTS devices are possible in principle with HTS, but the main area of current interest is only in HTS SQUIDs (digital devices have been demonstrated, but only have a few 10s of junctions). The most promising application area for HTS films is in passive microwave filters, especially for cellular radio base-stations [51]. Such filters have lower insertion loss and more sharply-defined passbands than their normal-metal counterparts. The consequent increase in the number of signals that can be handled makes them commercially highly attractive, especially as highly-reliable closed-cycle coolers are now available.

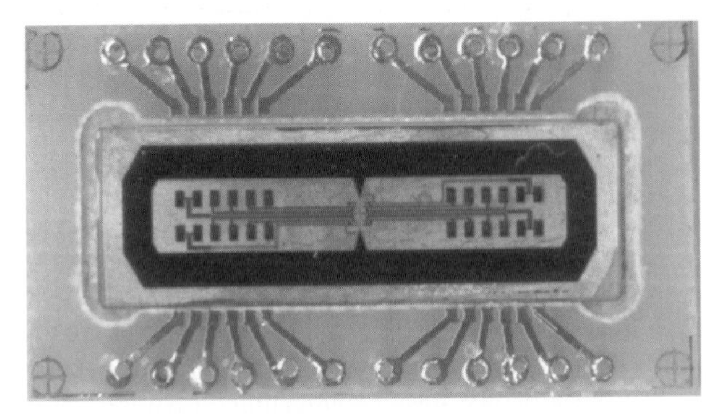

Figure A1.3.15. A single-layer HTS gradiometer made from YBCO on a 1 cm × 3 cm $SrTiO_3$ substrate. There are four SQUIDS at the center, directly coupled to the double-loop gradiometer [81].

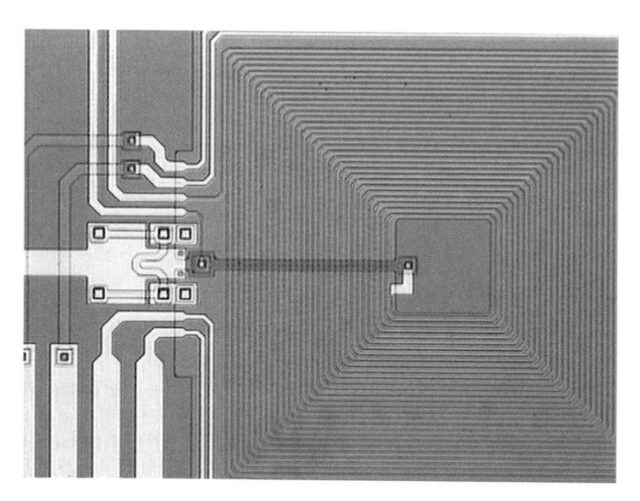

Figure A1.3.16. Part of a Nb multi-layer SQUID, showing the Josephson junctions, shunt resistors and part of a spiral signal input coil that has tracks 3 μm wide (University of Strathclyde).

HTS films are also attractive for use in NMR or MRI detector coils [52], where the lower losses increase the signal-to-noise ratio and decrease acquisition times.

A1.3.4 Bulk materials

A1.3.4.1 Introduction to bulk materials

Applications of bulk HTS are generally based on their ability to support a large persistent eddy current, which is usually induced in the material by exposure to a changing magnetic field. The energy density associated with so-called magnetized bulk HTS at 77 K, in particular, is typically greater than that of wires and tapes but less than that achievable in thin films. By virtue of their greater volume, however, bulk HTS have significant potential for high field 'permanent magnet' type engineering applications, which require large stored energies to generate sufficiently large magnetic fields. Such applications include magnetic bearings, fault current limiters, magnetic clamps and flywheel energy storage systems [53, 54]. In general the field generating capacity of bulk HTS is determined by the magnitude and homogeneity of the critical current density, J_c, and the length scale over which it flows. The magnetic moment of such a current loop is given simply by the product of its magnitude and area, and may be integrated over the sample volume to yield the net moment of the material. Significantly, fields generated by bulk materials at 77 K may exceed significantly the maximum practical field produced by a permanent magnet of around 1.5 T, which is essential for enhanced performance of HTS compared with conventional materials. It is, however, important to understand that the mechanism of field generation by permanent magnets and bulk superconductors is entirely different. The former is based on alignment of atomic spins, whereas macroscopic currents flowing within the samples are responsible for the latter, in a similar manner to which field is generated by a solenoid. The definition of magnetic moment per unit volume, or volume magnetization, $\boldsymbol{M}$, in permanent magnets and bulk superconductors is another common source of confusion (see chapter C2.1). $\boldsymbol{M}$ is constant in homogeneous permanent magnets and therefore generally easy to define. The two-dimensional nature of the magnetic moment of a macroscopic current loop, however, is not easy to interpret in a three-dimensional volume. Simply dividing the total magnetic moment of a bulk superconductor by the sample volume, as is frequently done, yields an effective volume magnetization, which increases with sample volume (i.e. as the current

C2.1

loop area increases). Hence, the respective definitions of $\boldsymbol{M}$ for permanent magnets and bulk superconductors are not entirely analogous (see chapter C2.1).

C2.1

A1.3.4.2 Irreversibility and bulk sample size

In general any variation of magnetic flux within a conductor is accompanied by the generation of an electric field normal to the direction of flux motion, which leads to energy dissipation. In order to support an induced current without energy loss, therefore, the bulk superconductor must necessarily be able to resist the motion of magnetic flux. Type II superconductors are unique in that magnetic field within their interior exists in the form of quantised filaments, or flux lines, each containing one quantum of magnetic flux ($= 2.07 \times 10^{-15}$ Wb) (see chapter A2.6). The ability of the superconductor to pin these flux lines under the influence of the Lorentz force per unit volume ($= B \times J_c$, where B is the applied field) is a direct measure of the critical current density of the material. Magnetic flux pinning in a bulk superconductor is usually achieved by inhomogeneities in the microstructure of the specimen. The strength of flux pinning in bulk HTS is therefore limited by the applied magnetic field, assuming constant J_c. The magnitude of field at which the Lorentz force exceeds the pinning force is called the irreversibility field and generally increases with decreasing temperature. The irreversibility field at a given temperature represents the field at which the magnetic behaviour of the material becomes reversible (i.e. still superconducting but unable to support a useful current density) and hence that at which flux pinning sites become ineffective. Bulk HTS are therefore characterized generally by the variation of irreversibility field with temperature, or so-called irreversibility line, which defines the field/temperature phase space above which the material becomes useless for current carrying applications. The irreversibility line represents the upper limit to the field trapping ability of the sample. The irreversibility lines of common bulk HTS are shown in figure A1.3.17. It can be seen that the $(RE)Ba_2Cu_3O_{7-\delta}$ [(RE)BCO where RE = Y, Nd, Sm, Eu, Gd etc)] class of supercondutors exhibits the highest irreversibility field over a wide temperature range and is therefore the most promising for application at high currents and magnetic fields. Unfortunately the ability of (RE)BCO to carry current, and hence to generate magnetic field, is limited severely by the presence of grain boundaries in the

A2.6

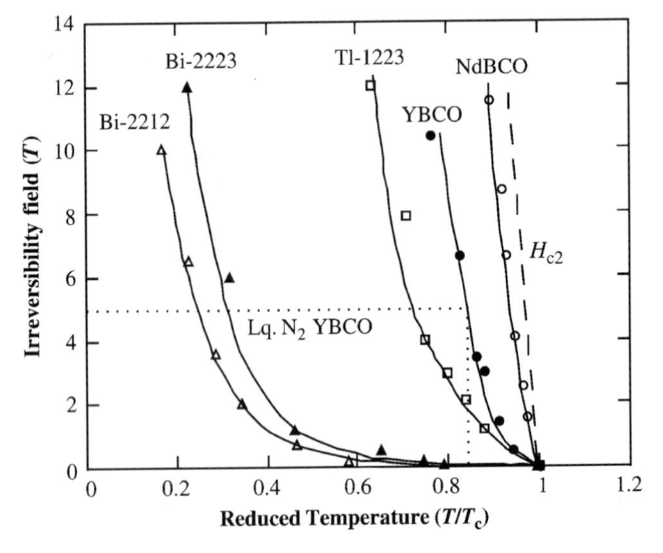

Figure A1.3.17. Irreversibility data for HTS compounds.

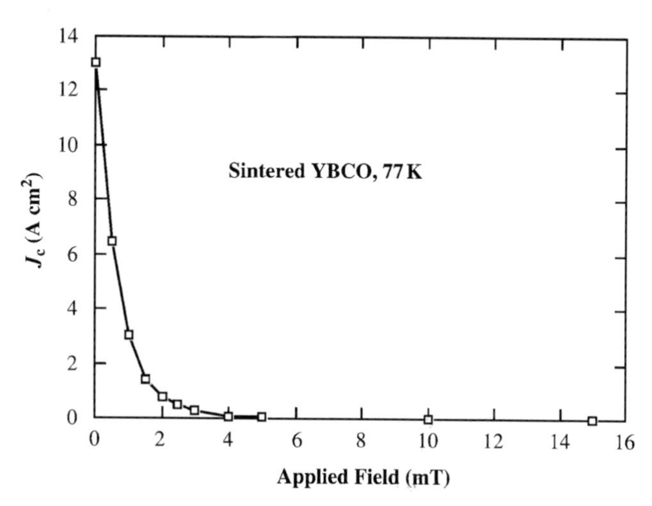

Figure A1.3.18. The field dependence of J_c for sintered YBCO at 77 K.

sample microstructure (the problem of granularity does not appear to limit the transport J_c in BSCCO, for example). This is manifest as a rapid decrease in J_c with applied field, as shown in figure A1.3.18, with a field of only a few mT reducing J_c to zero [55]. It is necessary, therefore, to process (RE)BCO in the form of large, single grains if the grain boundary problem is to be avoided.

This section describes an appropriate processing technique for the fabrication of large grain light rare earth (RE)BCO materials for high trapped field applications. The understanding and control of this process is essential for the realization of a variety of practical applications of bulk HTS which, for the reasons described above, are anticipated ultimately to form the basis of practical, medium scale industrial production processes of these materials.

B2.3.3
B2.3.4

A1.3.4.3 Melt processing of (RE)BCO

A variety of melt processing techniques have been developed for the fabrication of large grain (RE)BCO bulk superconductors. These are all based on a peritectic reaction, in which a solid decomposes to form a second solid and a liquid phase that occurs in these systems at around 1000°C, depending on the rare earth element, or combinations of elements. The target $(RE)Ba_2Cu_3O_{7-\delta}$ (RE-123) phase is formed from the peritectically decomposed, or semi-molten state from solid $(RE)_2BaCuO_5$ (RE-211), a Ba–Cu–O based liquid phase (L) and oxygen gas (G) according to the following reaction:

B2.3.3

$$\underset{\text{(RE-211)}}{(RE)_2BaCuO_5} + \underset{\text{(L)}}{Ba_3Cu_5O_{6.72}} + \underset{\text{(G)}}{0.42\ O_2} \rightarrow \underset{\text{(RE-123)}}{2(RE)Ba_2Cu_3O_{6.28}}.$$

The non-superconducting RE-211 phase (referred to as Nd-422 for neodymium to reflect the composition of one unit cell) and the liquid in this reaction can be produced by rapidly heating a pre-sintered green body of the desired composition to a temperature well above the peritectic temperature, T_p. In principle, formation of the required RE-123 phase can be achieved by cooling the peritectically molten (RE)BCO sample slowly through the peritectic temperature — hence the name of the process. In practice, however, the loss of liquid from the sample at elevated temperature limits severely the process times for stoichiometric RE-123 and the starting composition is usually enriched with up to 40 mol % excess RE-211 phase to reduce this effect (this increases the effective viscosity of the peritectically decomposed sample which consequently retains its structural integrity). This contrasts with the melt process used to fabricate Bi–Sr–Ca–Cu–O, which is characterized by a relatively complex phase diagram. Individual phases tend to melt incongruently in this system, which is therefore generally much more difficult to control than (RE)BCO (see chapter B3.2.2).

B2.3.5
B3.2.2

Seeded melt growth

The most common melt process technique for the fabrication of (RE)BCO is that based on Top Seeded Melt Growth (TSMG) in which a small single crystal with lattice parameters similar to that of the target RE-123 material is placed on the top surface of the presintered pellet to provide a homogeneous grain nucleation site, often under a thermal gradient [56] (see chapter B2.3.3). Fortunately, the peritectic decomposition temperature of different RE-123 compounds varies between around 950° and 1080°C (see chapter B2.3), as shown in figure A1.3.19 (taken from the data given in table B2.3.1, chapter B2.3). Since all the compositions illustrated in figure A1.3.19 are superconducting with similar lattice parameters, it is possible to use a RE-123 compound to seed the growth of a RE-123 superconductor with a lower pertitectic temperature, provided the difference in T_p is sufficiently large (i.e. >40°C) and that the RE-123 composition is available in the form of a small single crystal. For example, Y-123 ($T_p = 1000$°C), which is the easiest bulk (RE)BCO superconductor to fabricate by melt processing, is usually grown using either a Sm ($T_p = 1060$°C) or a Nd ($T_p = 1070$°C) seed. These seed compounds yield a processing window of approximately 60 and 70°C, respectively, for the growth of Y-123. This is sufficiently large to decompose the bulk YBCO composition without melting the seed. A typical large single YBCO grain grown by TSMG using a Nd-123 seed is illustrated in figure A1.3.20. Large, individual grains of YBCO have been grown up to 10 cm in diameter by the Dowa Mining Company, Japan.

B2.3.3
B2.3

The processing conditions for the fabrication of NdBCO and SmBCO are considerably more complex than those for YBCO. The properties of these materials are particularly sensitive to the processing atmosphere, for example, and typically require low oxygen partial pressures of 0.1% during melt growth. As a result a so-called oxygen controlled melt growth (OCMG) process has been developed for NdBCO and SmBCO which involves a similar melt process to that described above but under accurately controlled oxygen partial pressure down to 10^3 Pa [57, 58]. Samples of these materials melt processed under ambient oxygen pressure exhibit low transition temperatures that are typically broad in temperature and hence unsuitable for engineering applications at 77 K. In addition, it is generally not

B2.3.4

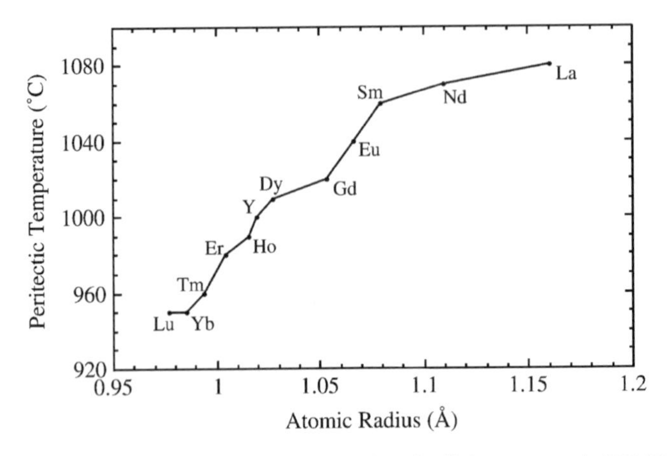

Figure A1.3.19. The variation of T_p with atomic radius for the light rare earth RE-123 HTS bulk phases.

Figure A1.3.20. Large grain YBCO fabricated by top seeded melt growth using a SmBCO seed (IRC in Superconductivity, University of Cambridge).

possible to grow RE-123 compounds with larger atomic radii, such as La-123, Nd-123 and Sm-123, by TSMG using other RE-123 single crystal seeds due to their relatively high T_p and consequent narrow thermal processing window available. As a result it is necessary to use other metal oxide single crystals such as MgO, $SrTiO_3$ or Al_2O_3. Of these MgO is more 'wettable' for the RE-123 phase and tends to be the most common choice of seed for TSMG of NdBCO, in particular. MgO does, however, have around a 22% mis-match in lattice parameter with Nd-123, which leads to difficulties in controlling grain orientation, depending on the exact growth conditions [59]. Alternatively, NdBCO may be grown by hot self-seeding, in which a Nd-123 seed crystal is added to the surface of the peritectically decomposed NdBCO bulk at a temperature close to T_p. This process generally yields larger grains with controlled orientations although it does have significant practical limits for medium scale production of bulk superconductors [60].

B2.3.4

Melt processed microstructure and properties

The rate of peritectic solidification during melt growth is determined by a number of compositional, kinetic and thermodynamic influences, as discussed in chapter B2.3.1. Of these, however, the solubility of the rare earth ion in the peritectically decomposed Ba–Cu–O liquid at the growth front is particularly significant and generally limits the extent to which RE-211 can dissolve during growth. As a result RE-211 inclusions remain trapped within the bulk RE-123 phase matrix to form a distribution of unconnected inclusions, as shown in figure A1.3.21. The RE-211 inclusion density correlates with an increase in effective flux pinning sites, which enhance J_c of the fully processed sample, as illustrated for YBCO by figure A1.3.22 [61–63]. The underlying pinning mechanism is associated with features of the RE-211 inclusions that have the length scale of the coherence length ($\sim 10^{-8}$ m), such as the RE-123/RE-211 interface, dislocations and twin boundaries. Whatever the mechanism, the result of increased flux pinning

B2.3.1 B2.3.4 A4.3

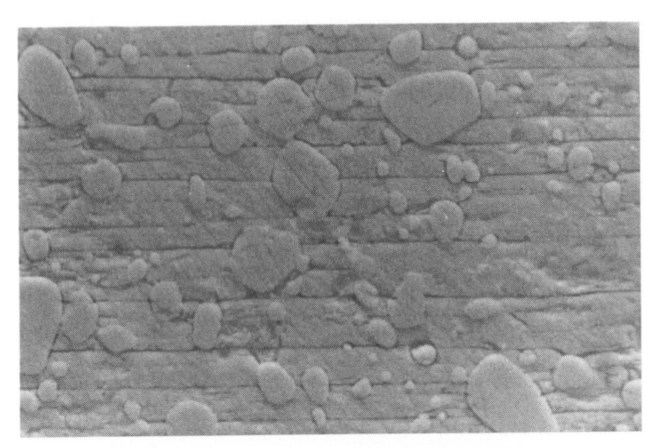

Figure A1.3.21. Y-211 inclusions in a melt processed Y-123 matrix (IRC in Superconductivity, University of Cambridge).

is higher trapped fields. Hence, a general processing aim is to generate a fine distribution of the second phase RE-211 inclusions in a large, homogeneous superconducting RE-123 matrix. The latter may be achieved by the addition of a variety of dopants to the precursor materials such as Pt or CeO_2, which have proved particularly effective in refining the size and distribution of the RE-211 phase.

The field trapped by homogeneous, large grain TSMG superconductors is typically inverse parabolic in shape, as shown in figure A1.3.23 for YBCO at 77 K. The peak field of this particular sample is around 0.5 T, although fields in excess of 2 T at 77 K and 13 T at 33 K have been reported by groups at the University of Houston and IFW, Dresden. Unfortunately, TSMG (RE)BCO materials tend to be brittle and therefore unable to resist the relatively large tensile forces associated with large trapped magnetic fields, which increase with B^2 (see chapter E2.1). As a result, bulk (RE)BCO materials require reinforcing by epoxy or CFRP encapsulation or by steel banding if they are to be used in high field applications.

E2 E2.1 B2.3.4

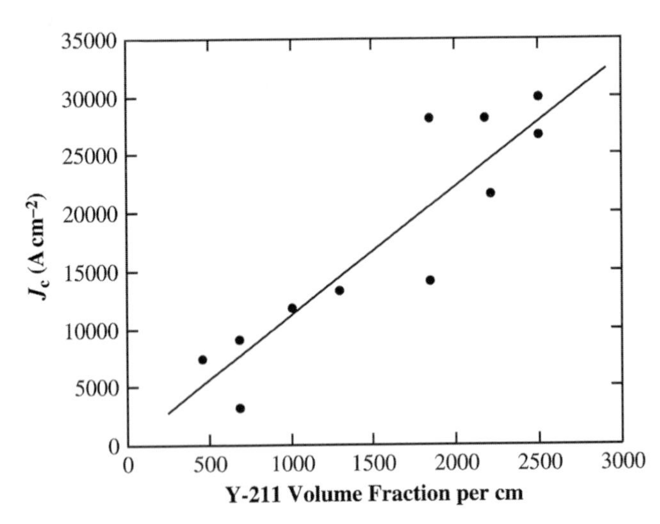

Figure A1.3.22. Correlation between critical current density and Y-211 particle size in melt processed YBCO (taken from [60]).

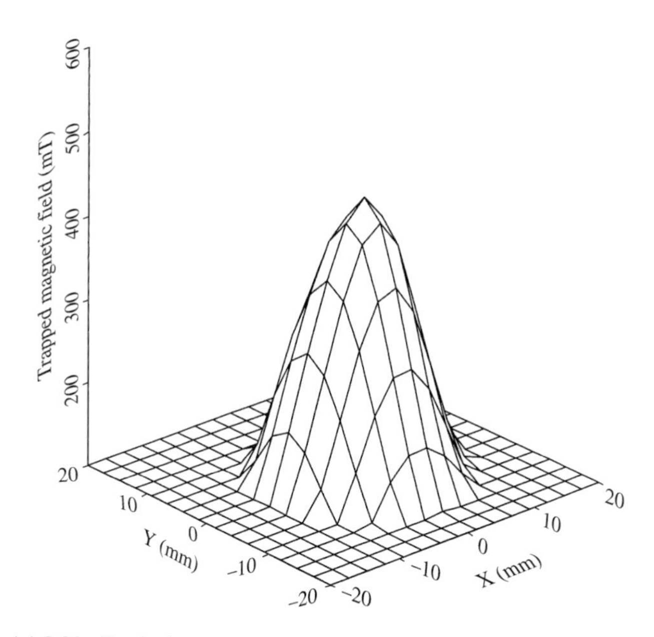

Figure A1.3.23. Typical trapped flux profile of magnetized bulk YBCO at 77 K.

Finally, a number of joining processes are being developed to yield high current carrying joins between individual melt processed grains to enable the generation of larger and more uniform trapped fields. The more practical of these are based generally on the use of a flux of lower peritectic decomposition temperature than the YBCO matrix [64], the generation of self-flux from liquid phase trapped within the melt processed microstructure [65] and the use of a silver-doped YBCO interfacial layer at the position of the join [66]. Encouraging results have been reported for small area samples and joins that can support significant current densities at 77 K in fields up to 5 T have been fabricated [67]. Continued development of this technology, however, is essential if large-scale applications of bulk melt processed material are to be realized and practical fabrication processes developed.

A1.3.4.4 Bulk devices and applications

There are four general categories of engineering device to which bulk, large grain HTS may be applied. These are: (a) motors and generators; (b) levitation devices, such as magnetic bearings and flywheels; (c) magnetic resonance imaging (MRI); and (d) trapped flux devices, such as dent pullers, magnetic clamps and magnetic separators. Each application is based on the field trapping properties of bulk HTS, which generally give rise to higher magnetic fields than can be achieved with permanent magnets and hence higher forces (trapped flux and levitation devices, motors and generators), or greater mechanical stability (flywheels). The principles of operation and the key properties of bulk materials required for each class of device are outlined in this section.

E2
E1.4.3

Motors and generators

The operation of rotating electrical machines is based on the torque produced by the interaction between magnetic fields and currents. The magnitude of this torque is limited generally by the saturation magnetic flux density of iron. The increased current, and hence power, density and higher trapped fields

E1.4.3

associated with bulk HTS materials, therefore, has the potential to enhance the performance of permanent magnet and reluctance electrical machines by increasing the torque available and by reducing the amount of iron required to contain the magnetic flux (see chapter E1.4.3). This translates to smaller, lighter, more efficient machine designs [68]. A number of prototype devices have been built to date with bulk HTS on the rotor, including hysteresis [69], linear [70] and reluctance [71] machines. The output power of these machines has increased from a few watts to in excess of 20 kW although they remain limited by the saturation flux density of iron. The next generation of trapped, rather than induced flux devices, has the potential to increase the performance of permanent magnet machines by up to 500%, providing practical techniques to magnetize the bulk HTS and to cool the rotor during service can be developed [68].

Levitation devices

E2.1 A range of passive levitation devices using bulk HTS have emerged over recent years based on the relatively high levitation force associated with field gradients generated by these materials and on the stability of such arrangements (see chapter E2.1) [72]. Superconducting magnetic bearings are an obvious application of bulk levitators and may be based on the forces between magnetized bulk superconductors, magnetized superconductors and permanent magnets, or hybrid magnetization induced superconductors in permanent magnet systems. In general, these can exceed the forces achievable in designs based on permanent magnets only and offer a significant reduction in the amount of energy dissipation and hence wear, associated with contact bearings. Hybrid bearings can generate magnetic pressures up to 0.3 MPa although they tend to be limited by their relatively low stiffness [73, 74]. Even so, they may find application in a variety of 'softer' journal or thrust bearing levitation systems used in flywheel energy storage systems, transport systems, cryo-pumps and electrical machines [75]. Extending superconducting bearing design to trapped field superconductor—superconductor systems has the potential to generate substantially enhanced forces and associated load supporting capability. Such systems, however, are currently at an early stage of development and are susceptible to problems associated with re-magnetization if they lose their charge (similarly, high trapped fields in bulk HTS can demagnetize permanent magnets).

The lateral mechanical stability associated with bulk HTS has the potential to simplify conventional permanent magnet bearing design. This translates ultimately to simpler bearing control systems, which is particularly important in flywheels for energy storage applications, which have to operate uninterrupted for long periods of time.

MRI

E2.3 In many regards MRI is extremely well-suited to the application of bulk HTS (see chapter E2.3). Here the requirement for a large background field and superimposed field gradient, which determines the frequency bandwidth of the device, can potentially be provided by bulk HTS. Furthermore, the availability of liquid helium in conventional LTS based systems reduces the cooling cost implications for the incorporation of HTS materials. Although existing MRI systems require a magnetic field stable to better than one part per million, a second generation of machines are under development for in-vitro, non-imaging applications which may be more tolerant to less stable magnetic fields (i.e. those associated with flux creep in bulk HTS). Even so, bulk HTS may be used to 'pre-polarize' specimens for MRI evaluation with a potential seven-fold improvement in signal to noise ratio. To date MRI applications based on bulk HTS are at an early stage of development, although their potential is considerable.

Trapped flux devices

Trapped flux devices have potential for use in the aerospace industry for dent pullers and magnetic clamps (see chapter E2.4) and for magnetic separators (see chapter E2.2). The operation of a dent puller is based on the application of a magnetic field trapped by a bulk superconductor to a sheet conductor at a relatively low rate to minimize forces on the sheet associated with eddy currents generated by the changing field (Faraday–Lenz law). The applied field of between 1 and 2 T is then removed rapidly, specifically to generate large eddy currents and associated forces on the dented area. These forces result in the removal of the dent from the sheet conductor. The biggest challenge with this device is the controlled rapid removal of the field from the region of the dent. This can be achieved by either removing the magnetized superconductor mechanically or by heating the sample rapidly to above its T_c. The use of dent pullers to repair aircraft can reduce down-time by the order of days which leads, in turn, to cost savings as high as $20 000 per repair.

E2.4
E2.2

Magnetic clamps are rather simpler in their operation, although they are limited to work pieces that are ferromagnetic in nature. Here, large grain bulk HTS are used to clamp the work piece with a force proportional to the product of the trapped field and the saturation flux density of the ferromagnet (assuming that the trapped field exceeds the latter). Such arrangements can exert a clamping pressure in excess of 0.3 MPa to an iron-based work piece, which exceeds significantly that achievable with a permanent magnet. In addition, trapped flux clamps can be controlled by thermally cycling and re-magnetizing the bulk HTS, which is a fundamental limitation of permanent magnet devices and are much safer to operate.

E2.4

Magnetic separators use the force applied to a ferromagnetic contaminant by a field gradient to purify a liquid charge for various ceramic and effluent applications. Flux tubes constructed from stacks of magnetized large grain bulk HTS have been proposed for this purpose [76] and demonstrated to offer significant cost and performance advantages over systems based on permanent magnets. Recently, a simple retrofit of magnetized bulk SmBCO for Nd–B–Fe in an effluent purification unit has yielded a higher quality product with a minimum of process modifications, which underlines the potential of bulk HTS for this application [77].

E2.2

The continued development of bulk materials will lead to an increased range of applications, within both existing and niche devices. There are, however, a number of technological issues that still require addressing, such as the ability to magnetize and re-magnetize bulk samples and to cool them reliably and

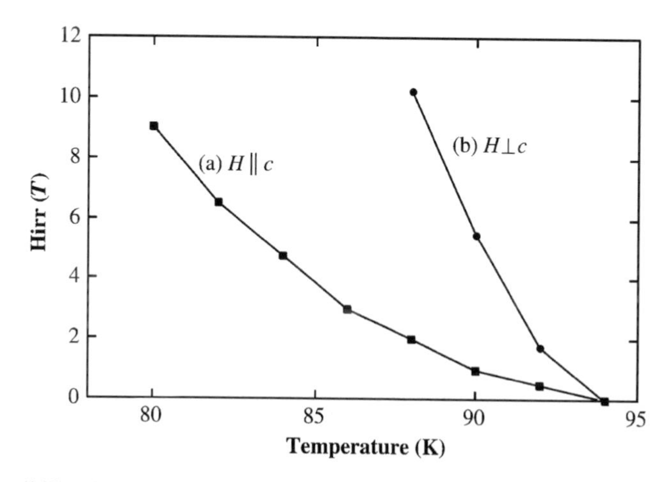

Figure A1.3.24. Irreversibility lines for NdBCO for field applied (*a*) parallel and (*b*) perpendicular to the *c*-axis.

economically before these will find wide scale application. Even so, bulk materials appear to offer real potential for medium term applications of HTS.

A1.3.4.5 Emerging bulk materials

B2.3.4 Recently, TSMG processes for other (RE)BCO systems such as GdBCO and (Nd–Eu–Gd)BCO have been developed [78]. These processes are still performed under a reduced oxygen atmosphere although the flux generating properties of the large grain samples appear to be significantly better than YBCO, B2.3 particularly in an applied magnetic field (see chapter B2.3). Despite the processing difficulties associated with NdBCO, this material exhibits significantly greater irreversibility fields than YBCO for fields applied parallel to the major crystallographic axes, as illustrated in figure A1.3.24 (from [79]). The large irreversibility fields of NdBCO at 77 K of around 12 T for $H \| c$ and greater than 40 T (extrapolated) for $H \perp c$ underline the considerable potential of this material for trapped field applications.

A further and more recent development is the fabrication of TSMG foams of YBCO [80]. These foams have the potential to generate large magnetic fields but offer the significant advantages of reduced weight, rapid cooling (i.e. an order of magnitude better that conventional TSMG bulks) and large surface area for heat extraction. In addition, bulk foams have the potential to be reinforced effectively by epoxy encapsulation, which has implications for their use in a variety of high field devices such as motors and generators, energy storage flywheels and faulty current limiters.

A1.3.5 Summary

There is an enormous range of potential application areas for LTS and HTS. In general, there are three main forms of superconducting material that are suitable for application in engineering devices. Of these wires and tapes are appropriate for the fabrication of transmission lines and wire-wound devices such as motors and generators and solenoidal magnets. Thin film superconductors are used extensively in high frequency microwave devices, including, resonators, filters, mixers and Josephson junction based devices such as SQUIDs. Finally, bulk materials are able to support relatively large induced currents and, therefore, can be used to generate large magnetic field for permanent magnet type applications, such as flywheel energy storage, magnetic bearings and clamps.

Materials development is established for LTS materials but is on-going for all forms of HTS. There are a number of exciting prospects for these materials including the development of long length, high current conductors, controllable Josephson junctions which operate at 77 K and the generation of magnetic fields which contain an order of magnitude more energy than those achievable with permanent magnets. The advent of such devices depends critically on the continued development of fabrication processes, cryogen-free cooling systems and on reducing the costs of materials and device production. These issues, therefore, will be central to the application of HTS materials for a number of years to come.

References

[1] Masur L *et al* 2001 *IEEE Trans. Appl. Supercond.* **11** 3256

[2] Huang Y B, Riley G N Jr, Yu D, Teplitsky M, Otto A, Fleshler S, Parrella R D, Cai X Y and Larbalestier D 2001 *Progress in Bi-2223 tape performance* (Madison: Int. Cryogenic Materials Conference)

[3] Rodenbush A J, Aized D and Gamble B B 1999 Conduction and vapor cooled HTS power leads for large scale applications *IEEE Trans. Appl. Supercond.* **9** 1233

[4] Ullmann B, Gabler A, Quilitz M and Goldacker W 1997 Transport critical currents of Bi(2223) tapes at 77 K under mechanical stress *IEEE Trans. Appl. Supercond.* **7** 2042

[5] Buczek D *et al* 1997 Manufacturing of HTS composite wire for a superconducting power transmission cable demonstration *IEEE Trans. Appl. Supercond.* **7** 2196

[6] Bray S L, Ekin J W, Clickner C C and Masur L 2002 Transverse compressive stress effects on the critical current of Bi-2223/Ag tapes reinforced with pure Ag and oxide-dispersion-strengthened Ag *J. Appl. Phys.* submitted

[7] King C, Herd K, Laskaris T and Mantone A 1996 Evaluation of a Strengthening and Insulation System for High Temperature BSCCO-2223 Superconducting Tape *Adv. Cryog. Eng.* **42** 855

[8] Hasegawa T, Koizumi T, Hikichi Y, Nakatsu T, Scanlan R M, Hirano N and Nagaya S, HTS conductors for magnets *Proc. Conf. on Magnet Technology* Geneva, September, 2001 submitted

[9] Suenaga M, Welch D O and Budhani R 1992 Magnetically measured irreversibility temperatures in superconducting oxides and alloys *Supercond. Sci. Technol.* **5** S1

[10] Malozemoff A P *et al* 1999 HTS wire at commercial performance levels *IEEE Trans. Appl. Supercond.* **9** 2469

[11] Iijima Y, Onabe K, Futaki N, Sadataka N, Kohno O and Ikeno Y 1993 *IEEE Trans. Appl. Supercond.* **3** 1510

[12] Hasegawa K *et al* 1998 *Advances in Superconductivity X* (Tokyo: Springer) p 607

[13] Yoshino H, Yamazaki M, Fuke H, Thanh T D, Kudo Y, Ando K and Oshima S 1998 *Advances in Superconducitvity X* (Tokyo: Springer) p 759

[14] Goyal A *et al* 1996 *Appl. Phys Lett.* **69** 1795

[15] Smith J A *et al* 1999 *IEEE Trans. Appl. Supercond.* **9** 1531

[16] Araki T, Yuasa T, Kurosaki H, Yamada Y, Hirabayashi I, Kato T, Hirayama T, Ijima Y and Saito T 2002 *Supercond. Sci. Technol.* **15** L1

[17] Snitchler G, Kalsi S S, Manleif M, Schwall R E, Sidi-Yekhlef A, Ige S and Medeiros R 1999 High-field warm-bore HTS conduction cooled magnet *IEEE Trans. Appl. Supercond.* **9** 553

[18] Malozemoff A P, Maguire J, Gamble B and Kalsi S *Proc. Conf. on Magnet Technology* Geneva, September, 2001 at press

[19] Driscoll D, Dombrovski V and Zhang B 2000 *IEEE Power Eng. Rev.* **20** 16

[20] Nick W *et al* 2001 *Proc. EUCAS Conf.* Copenhagen, August, 2001

[21] American Superconductor press release, July, 2001

[22] Kalsi S S *IEEE PES Meeting* New York, January 2002, IEEE CDCat#02CH37309C

[23] Wasa K and Hayakawa S 1992 *Handbook of Sputter Deposition Technology* (Norwich, NY: William Andrew)

[24] Blamire M G, Somekh R E, Lumley J M, Morris G W, Barber Z H and Evetts J E 1987 Effects of fabrication conditions on the properties of SIS tunnel-junctions *J. Phys. D: Appl. Phys.* **20** 1159

[25] Myoren H, Shimizu T, Iizuka T and Takada S 2001 Properties of NbTiN thin films prepared by reactive dc magnetron sputtering *IEEE Trans. Appl. Supercond.* **11** 3828

[26] Nakamura K, Akaike H, Ninomiya Y, Tate Y, Fujimaki A and Hayakawa H 2001 NbN/Nb/AlO_x/Nb/NbN junctions fabricated using an ultra-high vacuum dc-magnetron sputtering system *Supercond. Sci. Technol.* **14** 1144

[27] Iosad N N, Roddatis V V, Polyakov S N, Varlashkin A V, Jackson B D, Dmitriev P N, Gao J R and Klapwijk T M 2001 Superconducting transition metal nitride films for THz SIS mixers *IEEE Trans. Appl. Supercond.* **11** 3832

[28] Eulenburg A, Romans E J, Fan Y C and Pegrum C M 1999 Pulsed laser deposition of $YBa_2Cu_3O_{7-\delta}$ and $YBa_2Cu_3O_{7-\delta}$ thin films: a comparative study *Physica* C **312** 91

[29] Moeckly B H and Zhang Y 2001 Strontium titanate thin films for tunable $YBa_2Cu_3O_7$ microwave filters *Appl. Supercond.* **11** 450

[30] Bramley A P, O'Connor J D and Grovenor C R M 1999 Thallium-based HTS thin films, processing, properties and applications *Supercond. Sci. Technol.* **12** R57

[31] Kroger H, Smith L N and Jillie D W 1981 Selective niobium anodization process for fabricating Josephson tunnel-junctions *Appl. Phys. Lett.* **39** 280

[32] Gurvitch M, Washington M A and Huggins H A 1983 High-quality refractory Josephson tunnel-junctions utilizing thin aluminum layers *Appl. Phys. Lett.* **42** 472

[33] Kawakami A, Wang Z and Miki S 2001 Low-loss epitaxial NbN/MgO/NbN trilayers for THz applications *IEEE Trans. Appl. Supercond.* **11** 80

[34] Hilgencamp H and Mannhart J 2002 *Rev. Mod. Phys.* **74** 485

[35] Millar A J, Romans E J, Carr C, Eulenburg A, Donaldson G B and Pegrum C M 2001 Step-edge Josephson junctions and their use in HTS single-layer gradiometers *IEEE Trans. Appl. Supercond.* **11** 1351

[36] Antognazza L, Moeckly B H, Geballe T H and Char K 1995 Properties of high-T_c Josephson-junctions with $Y_{0.7}Ca_{0.3}Ba_2Cu_3O_{7-\delta}$ barrier layers *Phys. Rev.* B **52** 4559

[37] Heinsohn J K, Dittmann R, Contreras J R, Scherbel J, Klushin A, Siegel M, Jia C L, Golubov S and Kupryanov M Y 2001 Current transport in ramp-type junctions with engineered interface *J. Appl. Phys.* **89** 3852

[38] Hirst P J, Humphreys R G, Satchell J S, Wooliscroft M J, Reeves C L, Williams G, Pidduck A J and Willis H 2001 The role of interfaces in c-axis microbridges *IEEE Trans. Appl. Supercond.* **11** 143

[39] Buzea C and Yamashita T 2001 Review of the superconducting properties of MgB_2 *Supercond. Sci. Technol.* **14** R115

[40] Kambara M, Hari Babu N, Sadki E S, Cooper J R, Minami H, Cardwell D A, Campbell A M and Inoue I H 2001 High intergranular critical currents in metallic MgB_2 superconductor *Supercond. Sci. Technol.* **14** L5–L7

[41] Kang W N, Kim H J, Choi E M, Jung C U and Lee S L 2001 MgB_2 superconducting thin films with a transition temperature of 39 K *Science* **292** 1521

[42] Blank D H A, Hilgenkamp H, Brinkman A, Mijatovic D, Rijnders G and Rogalla H 2001 Superconducting Mg-B films by pulsed-laser deposition in an in situ two-step process using multicomponent targets *Appl. Phys. Lett.* **79** 394
[43] Saito A, Kawakami A, Shimakage H and Wang Z 2002 As-grown deposition of superconducting MgB_2 thin films by multiple-target sputtering system *Japan. J. Appl. Phys. Part 2 — Lett.* **41** L127
[44] Brinkman A, Veldhuis D, Mijatovic D, Rijnders G, Blank D H A, Hilgenkamp H and Rogalla H 2001 Quantum interference device based on MgB_2 nanobridges *Appl. Phys. Lett.* **79** 2420
[45] Burnell G, Kang D J, Lee H N, Moon S H, Oh B and Blamire M G 2001 Planar superconductor-normal-superconductor Josephson junctions in MgB_2 *Appl. Phys. Lett.* **79** 3464
[46] Weinstock H (ed) 1996 *SQUID Sensors: Fundamentals, Fabrication and Applications* NATO ASI Series (Dordrecht: Kluwer)
[47] Tahara S, Yorozu S, Kameda Y, Hashimoto Y, Numata H, Satoh T, Hattori W and Hidaka M 2001 Superconducting digital electronics *IEEE Trans. Appl. Supercond.* **11** 463
[48] Hamilton C A 2000 Josephson voltage standards *Rev. Sci. Instrum.* **71** 3611
[49] Kurakado M 1999 Progress in superconducting tunnel junction detectors *X-Ray Spectrom.* **28** 388
[50] Andreone D, Brunetti L, Lacquaniti V, Steni R and Thorpe J R 2001 SIS receivers for millimeter and submillimeter-wave detection *IEEE Trans. Appl. Supercond.* **11** 820
[51] Willemsen B A 2001 HTS filter subsystems for wireless telecommunications *IEEE Trans. Appl. Supercond.* **11** 60
[52] Bracanovic D, Esmail A A, Penn S J, Webb S J, Button T W and Alford N M 2001 Surface $YBa_2Cu_3O_7$ receive coils for low field MRI *IEEE Trans. Appl. Supercond.* **11** 2422
[53] Cardwell D A 1998 Processing and properties of large grain (RE)BCO *J. Mat. Sci. Eng.* B **B53** 1
[54] Campbell A M and Cardwell D A 1997 Bulk high temperature superconductors for magnet applications *Cryogenics* **37** 567
[55] Jones A R, Doyle R A, Blunt F J and Campbell A M 1992 *Physica* C **196** 63
[56] Cardwell D A, Lo W, Thorpe H D E and Roberts A 1995 A controllable temperature gradient furnace for the fabrication of large grain YBCO ceramics *J. Mater. Sci. Lett.* **14** 1444
[57] Yoo S I, Sakai N, Takaichi H, Higuchi T and Murakami M 1994 *Appl. Phys. Lett.* **65** 633
[58] Murakami M, Yoo S I, Higuchi T, Sakai N, Weltz J, Koshizuka N and Tanaka S 1994 *Japan J. Appl. Phys.* **33** L715
[59] Hari Babu N, Lo W, Cardwell D A and Shi Y H 1999 Fabrication and microstructure of large grain Nd–Ba–Cu–O Applied Superconductivity, ed X Obradors, F Sandiumenge and J Fontcuberta *Inst. Phys. Conf. Ser.* **167** p 41
[60] Kambara M, Hari Babu N, Shi Y H and Cardwell D A 2001 Growth of melt-textured Nd-123 by hot-seeding under reduced oxygen partial pressure *J. Mat. Res.* **16** 1163
[61] Murakami M 1992 *Supercond. Sci. Technol.* **5** 185
[62] Plain J, Puig T, Sandiumenge F, Obradors X and Rabier J 2002 *Phys. Rev.* B **65** 104526
[63] Martínez B, Obradors X, Gou A, Gomis V, Piñol S and Fontcuberta J 1996 *Phys. Rev.* B **53** 2797
[64] Noudem J G, Reddy E S, Tarka M, Noe M and Schmitz G J 2001 *Supercond. Sci. Technol.* **14** 363
[65] Lo W, Cardwell D A, Bradley A D, Doyle R A, Shi Y H and Lloyd S 1999 *IEEE Trans. Appl. Supercond.* **9** 2042
[66] Puig T, Ridrigues T Jr, Carrillo A E, Obradors X, Zheng H, Welp U, Chen L, Claus H, Veal B W and Crabtree G W 2001 *Physica* C **363** 75
[67] Cardwell D A, Bradley A D, Hari Babu N, Kambara M and Lo W 2002 Processing, microstructure and characterisation of artificial joins in top seeded melt grown Y–Ba–Cu–O *Supercond. Sci. Technol.* **15** 639
[68] McCulloch M D 2002 Future trends in the application of (RE)BCO to electrical machines *Supercond. Sci. Technol.* **15** 826
[69] Kovalev L K *et al* 1998 *Proc. ICEC* **17** 379
[70] Cruise R J *et al* 2000 *Physica* C **341–348** 2627
[71] Kovalev L K *et al* 1998 *Proc. ICEC* **17** 527
[72] Moon F C 1994 *Superconducting Levitation* (New York: Wiley)
[73] Basinger S A, Hull J R and Mulcahy T M 1990 *Appl. Phys. Lett.* **57** 2942
[74] Hull J R, Mulcahy T M, Salama K, Selvamanickam V, Weinberger B R and Lynds L 1992 *J. Appl. Phys.* **72** 2089
[75] Hull J R 2000 *Supercond. Sci. Technol.* **13** R1
[76] Watson J H P 1998 *Handbook of Applied Superconductivity* ed B Seeber (Institute of Physics Publishing) p 1371
[77] Saho N and Mizumori T 2002 HiTc bulk superconductor-based membrane magnetic separation for water purification *Proc. Am. Ceram. Soc. Meeting* (St. Louis, May, 2002) submitted
[78] Muralidhar M, Koblishka M R, Saitoh T and Murakami M 2000 *Advances in Superconductivity XII* (Tokyo: Springer) p 497
[79] Lo W, Hari Babu N, Cardwell D A, Shi Y H and Astill D M 1999 Preparation and Properties of Large Grain Peritectically Processed Nd–Ba–Cu–O Advances in Superconductivity XI **2** ed N Koshizuka and S Tajima p 721
[80] Reddy E S and Schmitz G J 2002 Superconducting foams *Supercond. Sci. Technol.* **15** p L21
[81] Eulenbury A, Romans E J, Carr C, Millar A J, Donaldson E B and Pegrum C M 1999 Highly balanced long-baseline single-layer high-T_c superconducting quantum interference device gradiometer *Appl. Phys. Lett.* **75** 2301–3

A2
Introduction to section A2: Fundamental properties

D S Ginley and D A Cardwell

A suite of properties characterizes the superconducting state of materials, including the loss of electrical resistivity to the flow of dc current, the Meissner effect and an anomaly in the thermal conductivity and specific heat. A particularly interesting feature of the superconducting state, compared to other phase changes, is that remnants of the normal state remain even when superconductors are cooled below their critical temperature, T_c. In addition, while the BCS theory explains well the behaviour of classical type I and type II superconductors, it is not generally applicable to the high temperature superconducting materials. Section A2 provides an introduction to superconducting phenomena and their relationship to the normal state. Six sections describe the properties of the normal state compared with that of the superconducting state, the Meissner–Ochsenfeld (M–O) effect, the loss of superconductivity due to external influences, such as temperature and applied magnetic field, thermal and high frequency electromagnetic properties of superconductors, flux quantization and finally, Josephson effects.

The nature of the superconducting state is introduced in chapter A2.1 by first discussing the nature of the normal state. The historical evolution of the key concepts required to describe the normal state are presented initially, followed by a deseription of the current theoretical understanding of the phenomenon. This leads to the description of the superconducting state in terms of the classical BCS approach. Chapter A2.2 presents a primarily phenomenological view of the M–O effect. The section builds on the theme of superconductivity presented in chapter A2.1 to demonstrate the M–O effect, which follows as a natural consequence. The concept of the magnetic penetration depth is then introduced and its implications with respect to superconductivity are discussed. The influence of magnetic field on the superconducting state is presented in chapter A2.3, which represents a critical area with respect to nearly all applications of both low and high temperature superconductors. The concept of coherence length is then presented and used to classify type I and type II superconducting materials. The key concepts of upper, lower and thermodynamic critical fields, H_{c2}, H_{c1} and H_c, are outlined as an introduction to some of the critical parameters that determine the practical application of superconductors in an external or self-generated magnetic field.

Chapter A2.4 develops the concept of the energy gap as a fundamental parameter for understanding the mechanism of superconductivity. The energy gap plays a crucial role in most superconducting properties and is central to the operation of superconducting tunnelling structures, which is one of the most promising area for technological devices. The section outlines the physical principles that underlie the origin of the energy gap and then describes the properties of type I and type II materials. Chapter A2.5 extends the magnetic properties discussed in chapters A2.2 and A2.3 to describe the high frequency electromagnetic behaviour of superconducting materials between 100 MHz and 500 GHz. The chapter introduces the concept of the surface impedance, Z_s, and its relationship to the surface resistance, R_s, which are critical parameters for superconducting devices operating in the microwave regime. The chapter presents a conventional description of high frequency conductivity (Mattis – Bardeen) and how HTS materials deviate from the model. The chapter concludes by discussing the effects of high magnetic field on the surface impedance.

Chapter A2.6 presents a detailed discussion of flux quantization and uses this to describe superconductivity as an example of a macroscopic quantum phenomenon. The chapter then discusses the detailed mechanism of the origin of flux quantization and presents striking experimental data to show the existence of flux lines in both conventional and HTS type II materials. Finally, many of the concepts from the previous chapters are integrated into a discussion of Josephson effects in chapter A2.7. The critical nature of the superconducting state is employed in many analogue and digital devices, such as SQUIDS and digital logic elements. A very thorough discussion of the Josephson effect is presented, starting with single particle tunnelling and then building to a variety of junction structures. The effects of field and junction geometry are subsequently discussed in detail.

Overall, section A2 presents an overview of the inter-relationship of the key superconducting parameters with each other, as well as the effects of external influences, such as field and temperature. This information is presented throughout the chapter within the context of the underlying physics.

A2.1
Normal state versus superconductor

David Welch

A2.1.1 Introduction

In recent years, there has been a tremendous expansion in our knowledge of the classes of materials which exhibit what is generally called 'metallic behaviour' and in the subclasses of these which exhibit superconductivity. In fact, it is not a trivial matter to define what constitutes a 'metal,' given that a number of organic compounds with structures which contain various low-dimension features, doped C_{60} fullerenes, a wide variety of intermetallic compounds, many ceramic oxides and even suitably-doped liquid ammonia, are now considered to be metallic in character and even normally gaseous elements such as iodine and hydrogen have been made to become metallic at high pressures. A working definition of 'metallic behaviour' is that the electrical resistivity (or at least one element of the resistivity tensor in anisotropic materials) remains finite, or falls to zero in the case of superconductors, as the temperature approaches absolute zero, as shown in figure A2.1.1.

G2 G4

At the turn of the 20th century, the behaviour of 'good metals' such as Cu, Ag, Al etc, began to be explained in terms of the behaviour of independent, non-interacting 'free' electrons by Drude and Lorentz, although these pioneering efforts were hampered by the necessity of using classical Boltzmann statistics. In the 1930s, the then-new quantum mechanics, and their attendant Fermic–Dirac statistics, were used with great success by Sommerfeld, Bloch, Wilson and Brillouin to shed great illumination on what constituted semiconductors, metals and insulators. This state of understanding was well-described in the classic work of Mott and Jones, *The Theory of the Properties of Metals and Alloys*, originally published in 1936, and still in print [1], and is still well worth consulting even today for a cogent introduction to the normal-state properties of simple metals, transition metals and their alloys.

In these early quantum theories, the electrons were considered to act in an independent manner, although this approximation was usually justified only by its success. The notion of independent electron behaviour received theoretical justification in Landau's elegant theory of the Fermi liquid [2, 3], which showed that under certain circumstances, not 'bare' electron, but 'quasiparticles' (i.e. electrons together with their accompanying regions of perturbed electron density, lattice distortions, etc) acted in an independent manner, and in many circumstances the effects of electron–electron interactions on electron dynamics and thermodynamics could be accounted for by means of an effective mass. Large effective masses arising from electron–electron interaction seem to be vital in understanding the properties of so-called 'heavy fermion' superconductors [4].

G2

Despite such advances in the understanding of metallic properties such as the low-temperature heat capacity and the temperature-dependent electrical resistivity in normal metals, certain experimental facts pointed to the vital importance of correlations between electronic states and dynamics, namely the existence in the metallic state of ferromagnetism and antiferromagnetism (spin correlations) and superconductivity (momentum correlations). Furthermore, the importance of correlated electron behaviour

A3.2

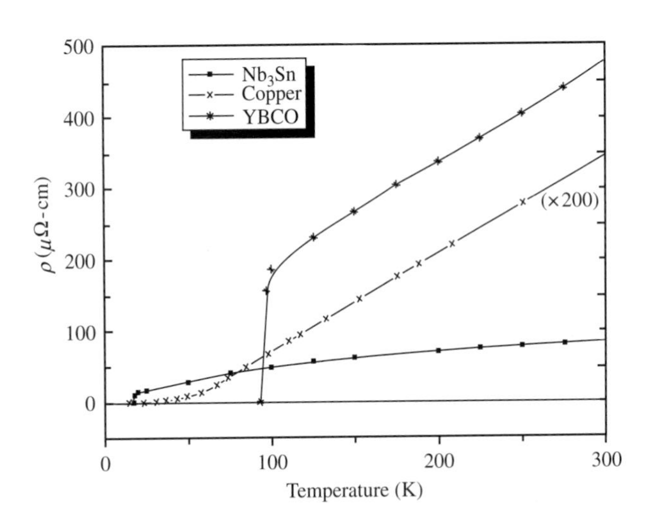

Figure A2.1.1. The temperature dependence of the resistivity of a normal metal (Cu), a conventional superconductor (Nb_3Sn), and a high-temperature superconductor ($YBa_2Cu_3O_7$). [Note: The resistivity of Cu is multiplied by 200 for the sake of visibility.] Data are from: (1) Cu; [13] (2) Nb_3Sn; [14] (3) $YBa_2Cu_3O_7$; [15].

was recognized by Wigner and Seitz in 1933 to be crucial in describing cohesion, even in the simplest monovalent metals such as Na, Li, etc [1, 2]. This important point was further developed by Mott in 1949 when he showed that below a certain volume density monovalent atoms such as Na, which should always be metallic, according to independent-electron models, cannot sustain delocalized, metallic behaviour and that electron–electron correlations and screening are required to understand the onset of metallic behaviour at a certain critical density; this approach was developed much more thoroughly by Hubbard in the 1960s (see discussion in [2, 3]). The basic ideas of Mott and Hubbard are of great importance as a basis for understanding the metallic, insulating and superconducting states in cuprate high-temperature superconductors [5].

As the brief discussion above indicates, the characterization of the electronic states and properties of metals, including superconductors, utilizes concepts ranging from the behaviour of 'nearly-free,' nearly-independent electrons (quasiparticles) to the correlations in their dynamics and spatial arrangements which result from the interaction between quasiparticles and which lead to ferromagnetism, antiferromagnetism and superconductivity. In the latter case, the crucial correlation effect that underlies the formation of the superconducting state is the formation of paired electronic states, either in momentum space (Cooper pairs) or in 'real' space (e.g. bipolarons). This pairing results in composite particles which obey Bose–Einstein statistics, which then permits Bose condensation of the paired electrons into a quantum superfluid, i.e., superconductivity occurs.

In the following two sections we will give a brief sketch of some characteristics of the electronic states and behaviour in the normal metallic state and in the superconducting state.

A2.1.2 Normal metallic state characteristics

In the normal state of metals it is a matter of practical necessity, as well as being a rather good approximation, to describe the electronic states as one-electron Bloch functions, which for a given state labelled by $\boldsymbol{k}$, are of the form:

$$\psi(\boldsymbol{r}) = u_k(\boldsymbol{r}) \exp(\boldsymbol{k}\cdot\boldsymbol{r}) \tag{A2.1.1}$$

where $\boldsymbol{k}$ is the wave vector (which describes the momentum associated with the state), $\boldsymbol{r}$ is the position coordinate of the electron and u is a function which reflects the effect of the periodic potential field experienced by the electron and includes the effect of the ion cores of the atoms and other electrons (in an averaged way). Each such state can accommodate two electrons of opposite spin. For a constant potential field (free electrons), the function u is a constant and the various wave functions are plane waves. In this case, the energy of an electron in state k is given simply by a parabola:

$$E(k) = \hbar^2 k^2/2m \tag{A2.1.2}$$

where $\hbar$ is Planck's constant divided by 2π and m is the electronic mass.

In the more general case of periodic potentials due to interaction with other electrons and ions, the energy versus k relation is not a simple parabola. Such non-constant potentials give rise to bands of electron energies periodic in the wave vector $\boldsymbol{k}$, and these bands of electrons may be separated by insulating energy gaps between them. For partly filled bands, the effects of the periodic lattice potential and electron–electron interactions can be approximated by the use of an effective mass tensor given by [2]:

$$m_{ij}^{*} = \frac{\hbar^2}{(\mathrm{d}^2E/\mathrm{d}k_i\mathrm{d}k_j)}. \tag{A2.1.3}$$

The one-electron states are filled with the available electrons according to the Pauli principle for fermions until all available electrons are exhausted; the degree of filling of the bands determines whether the system under consideration is a metal, semiconductor or insulator [1–3]. In the present case, we assume the system to be metallic, which means that at least one of the available bands is only partially filled and that many nearby empty states exist with energies very close to the maximum energy of filled states, E_F, the so-called 'Fermi level'. The boundary in the space of wave vectors $\boldsymbol{k}$ (k-space) between filled and empty states is called the Fermi surface. For a free-electron system with a parabolic energy relation (equation (A2.1.2)) the Fermi surface is a sphere with radius k_F, which is a simple function of the electron density, as discussed in numerous solid-state physics texts [6]. In less-idealized materials, including A15 compounds, heavy fermion superconductors and cuprate superconductors, the Fermi surface does not have a simple spherical geometry in k-space [7]. This is shown in figure A2.1.2 for a hypothetical metal with a simple cubic structure. However, in many metals and as is assumed in the Bardeen–Cooper–Schrieffer (BCS) theory of superconductivity [5, 8], the Fermi surface can be assumed to be approximately spherical and to be characterized by a Fermi energy E_F or, equivalently a Fermi wavevector $\boldsymbol{k}_F$ or an equivalent Fermi velocity v_F given by:

A3.2

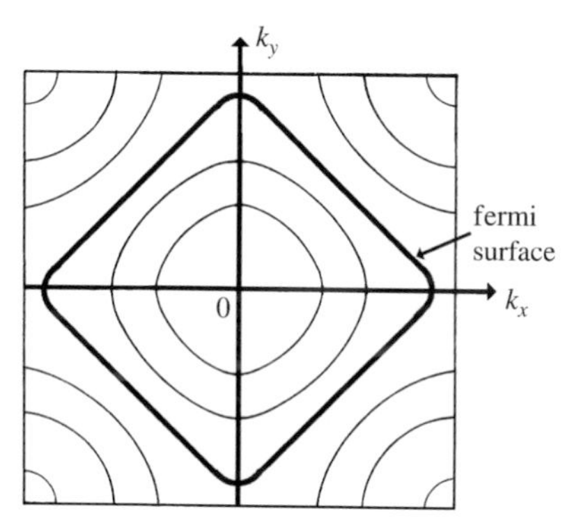

Figure A2.1.2. A cross-section through surfaces of constant energy, at equal energy intervals, in k-space for a hypothetical metal in a simple cubic structure. Note that the Fermi surface is not spherical.

$$v_F = \frac{\hbar k_F}{m^*}. \tag{A2.1.4}$$

[This Fermi velocity enters into expressions for the BCS superconducting coherence length (see equations (A2.1.8) and (A2.3.2) of chapter A2.3).]

Another characteristic parameter which describes the electronic structure and which appears in the BCS theory (and other theories) of superconductivity is the density of states in energy, evaluated at the Fermi surface, $N(E_F)$. The density of states $N(E)dE$ is defined as the number of states per atom which have energies between E and $E + dE$. [Sometimes the density of states at the Fermi level appearing in superconductivity literature is written $N(0)$ when the zero of energy is taken as the Fermi level and usually the density of states per atom and per spin is used.]

A3.2

The parameters above [$N(E_F)$, m^*, v_F, etc] can be evaluated from electronic structure theory (either *ab initio* or semi-empirical) [7] or by the interpretation of experimental data, such as paramagnetic susceptibilities and low-temperature heat capacities, as discussed in chapter A2.4. Such parameters are used as input in descriptions of the superconducting state, such as the BCS theory, etc.

A2.4

Another important characteristic of the normal state is that deviations from crystalline perfection, such as impurity atoms or structural defects, as well as atomic displacements due to thermal vibration, cause electrons to be scattered from filled to empty states in the vicinity of the Fermi surface. Such scattering gives rise to electrical resistance, and measurements of the electrical resistivity in the normal state, ρ_N, can be used to deduce valuable information about electron dynamics [9]. One important characteristic parameter of this scattering is the mean-free-path, ℓ, between scattering events caused by impurities and defects for electrons near the Fermi surface. This is an important parameter in determining certain superconducting properties such as the coherence length.

A1.2

A2.1.3 The superconducting state

It is implicit in the description of the normal metallic state discussed above that the only interaction between the electrons is the long-range Coulomb repulsion. Although Landau showed that the quasiparticle concept allows for correlations in the electron positions, which reduce the range of the repulsions by means of screening, this does not qualitatively change the picture. However, in the early 1950s Leon Cooper made a major advance by demonstrating that if there is any mechanism to cause an attractive interaction, no matter how small, between electrons, then the independent quasiparticle becomes unstable with respect to the formation of bound pairs of quasiparticles, now called Cooper pairs, in momentum space (k-space); a succinct description of this can be found in [2] and [8]. Bardeen, Cooper and Schrieffer showed how the electron–phonon interaction can give rise to the necessary attraction and developed the first successful, and now standard, theory of superconductivity. In this theory, even though the creation of a phonon by an electron scattering event costs an energy of $\hbar\omega$, the subsequent interaction of the phonon with a second electron can result in a reduction in energy, the magnitude of which we denote by V, and that for a free electron system with a spherical Fermi surface, the energy of the resulting bound pair of electrons at the Fermi surface is reduced by a binding energy Δ_0 given by

A3.2

$$\Delta_0 = \frac{2\hbar\omega}{\exp(1/N(0)V) - 1}. \tag{A2.1.5}$$

Note that this binding energy depends not only on the attractive energy V but also on the magnitude of the density of electronic states at the Fermi surface, $N(0)$.

These bound Cooper pairs of electrons formed from one-electron states at the Fermi surface are bosons and thus can undergo Bose condensation into a single coherent quantum superfluid state for which the wavefunction

$$\psi = \psi_0 \exp(\mathrm{i}\theta) \qquad \text{(A2.1.6)}$$

is characterized by an amplitude ψ_0, where $|\psi_0|^2$ is the density of Cooper pairs, and a phase θ. This quantum state is macroscopic in nature, with a size dictated by the size of the superconducting body. At zero temperature the ground state of the electronic system in this superconducting state can be described as a coherent mixture of occupied states within a somewhat diffuse Fermi surface (in k-space), separated from excited states which include unpaired quasiparticles by an energy gap given by 2Δ, where, at zero temperature, Δ is given by equation (A2.1.5). The difference, produced by the gap, between the energy of excitations in a normal metal and in a superconductor is shown in figure A2.1.3. The coherent character of the condensed state of paired electrons within a spread of k values at the Fermi surface and the exclusion of nearby available empty states because of the energy gap means that scattering of electrons by impurities, defects, phonons, etc does not occur. The gap is attached to the Fermi surface, and still exists even when the net momentum is not zero, i.e. even when the centre of the Fermi surface is shifted so that a net current of electrons is flowing: this state has no resistance; it exhibits superconductivity (see figure A2.1.4). A consideration of the thermodynamics of this state (including the effects of temperature), as discussed in chapter A2.2, reveals an even more remarkable property: the superconducting state is perfectly diamagnetic (the so-called 'Meissner effect').

A2.7

A1.2

A2.2

Including the effects of a non-zero temperature in the formation and properties of the condensed superconducting state shows that the magnitude of the energy gap 2Δ, as well as the density of paired electrons, diminishes continuously with increasing temperature and vanishes in a second-order transition (at zero magnetic field) at a critical value of the temperature T_c. (See [8] for details.) The value of T_c is determined by the size of the electron pair binding energy Δ_0:

$$k_B T_c = \alpha \Delta_0 \qquad \text{(A2.1.7)}$$

where k_B is the Boltzmann constant and α is a number of order unity; the BCS theory yields a value for α of 0.568. The value of the energy gap Δ_0 can be measured by a variety of experimental methods [10], including heat capacity measurements, shown in figure A2.1.5, as discussed in chapter A2.4. This permits evaluation of the constant α and is thus one test of the validity of particular theories, e.g. BCS theory, for particular classes of superconducting materials, e.g. cuprates.

A2.4

D3.2.1

The critical temperature T_c is one important thermodynamic characteristic of the superconducting state. Another is the critical magnetic field H_c, and its variants H_{c1} and H_{c2}, as discussed in chapter A2.3. There are also two important length scales which characterize the superconducting state. The first is the coherence length, ξ, first introduced by Pippard in 1953, which characterizes the distance which

A2.3

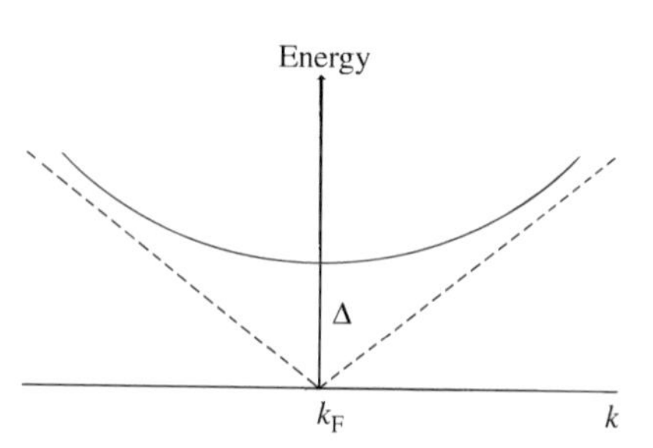

Figure A2.1.3. The excitation spectrum for quasi-particles in a BCS superconductor (solid line) compared with that for the normal state (dashed lines) [5].

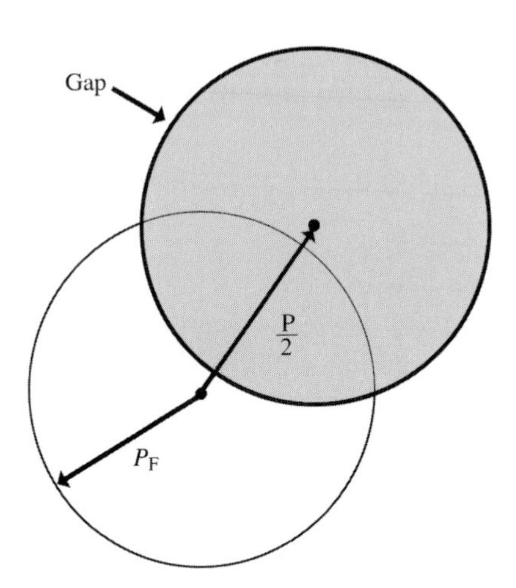

Figure A2.1.4. The momentum distribution in a current-carrying superconductor (shaded circle) compared to that in the normal state (open circle). P is the total momentum of a Cooper pair. In both cases the momentum vectors are uniformly distributed in spheres of radius p_F, the surfaces of which are the Fermi surface. Note that the energy gap in the superconducting state is carried by the Fermi surface.

A1.2 is required for the density of superconducting electrons n_s to change appreciably in an inhomogeneous superconductor [5, 8, 10]. In pure and defect-free superconductors, this length, ξ_0, is a characteristic of the material and depends on T_c and the Fermi velocity v_F (equation (A2.1.4)); the BCS theory result is:

$$\xi_0 = 0.18\frac{\hbar v_F}{k_B T_c} \tag{A2.1.8}$$

Electron scattering by impurities, defects, etc can reduce the coherence length, and in 'dirty' superconductors ξ is given approximately by $(\xi_0 \ell)^{1/2}$, where ℓ is the mean-free-path for electron scattering in the normal state; thus alloying or plastic deformation can be used to reduce the coherence length when desired.

The second characteristic length associated with the superconducting state is the so-called 'penetration depth,' λ. This length describes the distance required for magnetic fields to decay in going A1.2 from a region of normal material into the perfectly diamagnetic superconductor [8]. This length arose from the pioneering work by F. London and H. London on the electrodynamics of superconductors

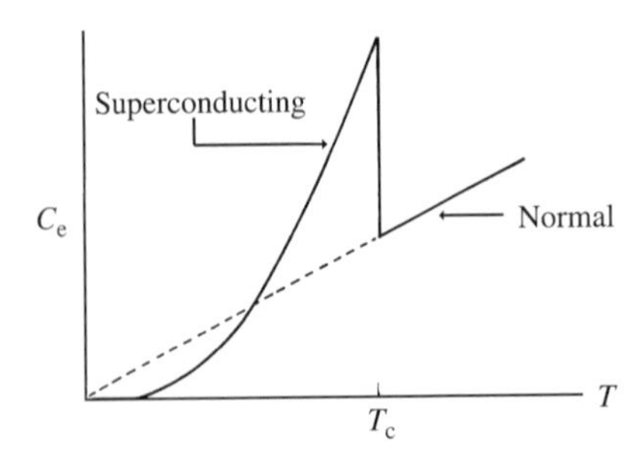

Figure A2.1.5. The temperature dependence of the electronic contribution to the specific heat of a conventional superconductor.

and is sometimes called the London penetration depth. For essentially pure superconductors, the London penetration depth is controlled by the density of superconducting electrons n_s and is given by:

$$\lambda_L = \left(\frac{mc^2}{4\pi n_s e^2}\right)^{1/2} \quad \text{(A2.1.9)}$$

where c is the speed of light in vacuum and, e and m are the electronic charge and mass. Where the electron mean-free-path path ℓ is limited by scattering of impurities or defects, the penetration depth is given approximately by $\lambda_L(\xi_0/\lambda)^{1/2}$. The nature of the behaviour (type I or type II) of superconductors in magnetic fields is determined in a very important way by the value of the ratio of the penetration depth to the coherence length. This is discussed in detail in chapter A2.3.

A2.3

The present state of understanding of the nature of the superconducting state is in an active state of development. It is not yet clear whether or not a 'BCS-like' theory of superconductivity will suffice to describe high-T_c superconductors. A review [11] of the present state of the theoretical development has recently appeared which discusses some of the issues. Other aspects of the situation are described in [5, 10]. Not only are the superconducting-state properties of high-T_c cuprate superconductors different in many respects from those of conventional superconductors, but also there are important differences between the normal states of these two classes of superconductors. For example, there is now considerable evidence that the symmetry of the pair state in cuprate superconductors is not s-wave, as in conventional BCS superconductors, but rather is d-wave in nature [12]. This symmetry is reflected in the character of the superconducting gap, for example, resulting in lines of nodes (zeros of the gap) on the Fermi surface, and this, in turn, is reflected in many properties which depend on the excitation spectrum [5]. Regarding differences between the normal states, one of the most important is the existence of a gap or at least a substantial reduction in the density of states, a 'pseudogap,' near the Fermi surface in the normal state of underdoped cuprates. The pseudogap is manifested in anomalies in the heat capacity and the paramagnetic susceptibility of underdoped cuprates [5].

References

[1] Mott N F and Jones H 1936 *The Theory of the Properties of Metals and Alloys* (New York: Dover)
[2] Cottrell A H 1988 *Introduction to the Modern Theory of Metals* (London: Institute of Materials)
[3] Mott N F 1990 *Metal–Insulator Transitions* 2nd edn (London: Taylor and Francis)
[4] Fisk Z, Hess D W, Pethick C J, Pines D, Smith J L, Thompson J D and Willis J O 1988 Heavy-electron metals: new highly correlated states of matter *Science* **239** 33
[5] Waldram J R 1996 *Superconductivity of Metals and Cuprates* (Bristol: Institute of Physics Publishing)
[6] Quéré Y 1998 *Physics of Materials* (Amsterdam: Gordon and Breach)
[7] Harrison W A 1999 *Elementary Electronic Structure* (Singapore: World Scientific)
[8] de Gennes P G 1966 *Superconductivity of Metals and Alloys* (Redwood City, CA: Addison-Wesley)
[9] Rossiter P G 1987 *The Electrical Resistivity of Metals and Alloys* (Cambridge: Cambridge University Press)
[10] Kresin V G, Morantz H and Wolf S A 1993 *Mechanisms of Conventional and High T_c Superconductivity* (Oxford: Oxford University Press)
[11] Ruvalds J 1996 Theoretical prospects for high-temperature superconductors—topical review *Supercond. Sci. Technol.* **9** 905
[12] Tsuei C C and Kirtley J R 2000 Pairing symmetry in cuprate superconductors *Rev. Mod. Phys.* **72** 969
[13] Bass, J and Fischer K H eds 1982 *Landolt–Börnstein Numerical Data and Functional Relationships in Science and Technology, vol 15a Metals: Electronic Transport Phenomena* (Berlin: Springer)
[14] Woodward D W and Cody G D 1964 *Phys. Rev.* **136** 166
[15] Iye Y, Tamegai T, Sakakibara T, Goto T, Miura N, Takeya H and Takei H 1988 *Physica* C **153–155** 26

A2.2
The Meissner–Ochsenfeld effect

Rudolf P Huebener

Following Kamerlingh Onnes discovery of superconductivity, in 1933 Walther Meissner and his collaborator. Ochsenfeld at the Physikalisch Technische Reichsanstalt in Berlin discovered the most fundamental property of a superconductor, namely its ability to expel magnetic flux from its interior [1]. The perfect diamagnetism associated with this Meissner–Ochsenfeld effect results from electric shielding currents flowing without resistance near the surface of the superconductor. In figure A2.2.1, we show schematically how the superposition of the applied magnetic field and the field generated by the shielding current result in zero magnetic flux density B inside the superconductor. We define $B(r)$ in terms of the local field $H(r)$ produced by the superposition of the external field H, produced by external currents (e.g. the field by a long solenoid), and the field generated by currents flowing within the superconductor. From Ampere's theorem, we can see that for $B(r) = \mu_0 H(r) = 0$, the total current I per unit length flowing around the outer surface of a superconducting cylinder in a parallel field is $I = -H$. Since the supercurrent flows without resistance is a necessary consequence of the existence of the Meissner–Ochsenfeld effect, whereas the inverse conclusion does not hold, the Meissner–Ochsenfeld effect is clearly more fundamental than just the disappearance of the electric resistance (although only the latter phenomenon is suggested by the name 'superconductivity'). D2.1

Supercurrent flow without resistance is a necessary consequence of the Meissner–Ochsenfeld effect, whereas a transition to zero resistance does not imply flux exclusion. Indeed, a transition to zero resistance would simply trap within the superconductor any field previously present; the final state would then depend on the magnetic and thermal history of the sample. This is shown in figure A2.2.2 where the critical magnetic field $H_c(T)$, separating the superconducting state from the normal state, is plotted versus temperature. We consider two different ways to pass from point (1) in the normal state to point (4) in the superconducting state. First, we assume only infinite electric conductivity and the absence of the Meissner–Ochsenfeld effect in the superconductor. Then, along the path 1–2–4 at point (4) we have $B = 0$ in the superconductor. In contrast, on path 1–3–4 at the transition to zero resistance, the applied field would be trapped inside the superconductor, so that the magnetic induction would remain constant with $B = \mu_0 H$. The exclusion of magnetic flux associated with the Meissner effect ensures that the final state is in practice independent of the thermodynamic path. This means that superconductivity is a thermodynamic state of the system, so that equilibrium thermodynamics can be applied to the superconducting phase transition. D2.1

The magnetic energy per unit volume required for achieving magnetic flux expulsion is given by

$$-\mu_0 \int_0^H M(H) \mathrm{d}H = \frac{\mu_0}{2} H^2 \qquad \text{(A2.2.1)}$$

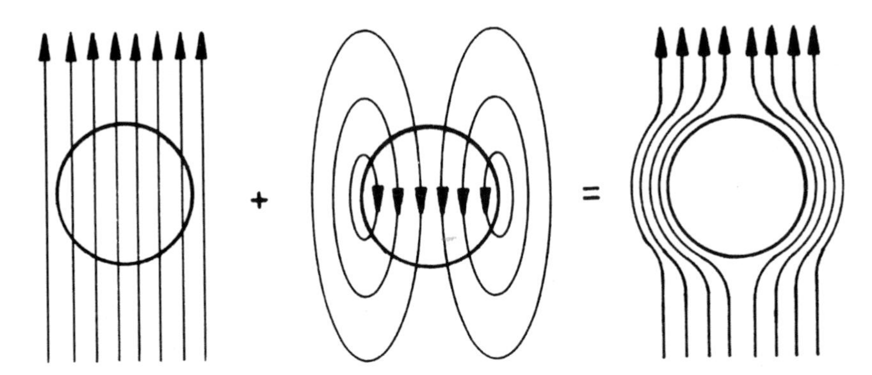

Figure A2.2.1. The superposition of the applied magnetic field and the magnetic field generated by the shielding supercurrent results in zero magnetic flux density inside the superconductor.

since for perfect diamagnetism we have $M = -H$. Here, μ_0 is the vacuum permeability and $M(H)$ the magnetization. The magnetic energy of equation (A2.2.1) must be overcompensated by the gain in free energy density for the superconducting state to be energetically favorable. At the thermodynamic critical magnetic field $H_c(T)$, the energy gain of the superconducting state vanishes. Denoting the free energy densities in zero magnetic field in the normal and in the superconducting state by $f_n(T)$ and $f_s(T)$, respectively, we obtain for their difference [2]

A1.2

$$f_n(T) - f_s(T) = \frac{\mu_0 H_c^2(T)}{2}. \tag{A2.2.2}$$

For a typical value of $H_c = 10^4\,\mathrm{A\,m^{-1}}$, we have $f_n - f_s = 40\,\mathrm{J\,m^{-3}}$.

Since the density of the Meissner shielding current cannot become infinite, the shielding currents are spread over a distinct distance from the surface. As a consequence, the magnetic field drops from its value $\mu_0 H$ to zero only within a characteristic length scale given by the London penetration depth λ_L. The first theoretical discussion of this finite magnetic field penetration was given in the London theory, yielding the exponential decay of the local magnetic flux density $H(x)$ within the superconductor with increasing distance x from the surface

A1.2

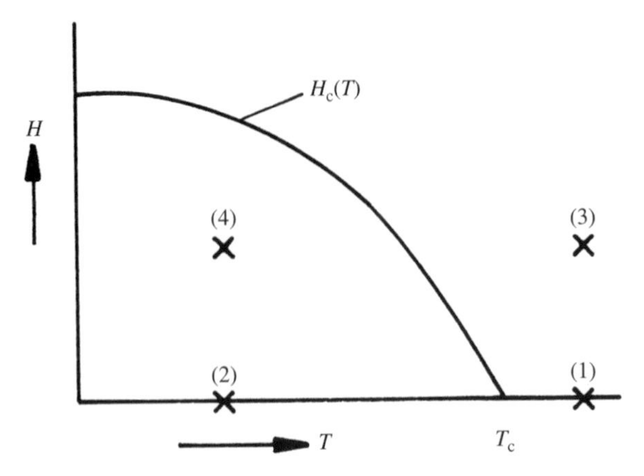

Figure A2.2.2. Critical magnetic field H_c versus temperature. From point (1) one can reach point (4) via point (2) or point (3).

$$H(x) = H(0)\exp\left(-\frac{x}{\lambda_L}\right). \tag{A2.2.3}$$

The London penetration depth λ_L for electrons flowing without dissipation is given by [3]

$$\lambda_L = \left(\frac{m}{\mu_0 e^2 n_s}\right)^{1/2} \tag{A2.2.4}$$

where m, e, and n_s are the electron mass, charge, and number density, respectively. For superconductors, both the effective mass and the charge are doubled; the mass is the effective mass involved in electrical transport and $n_s(T)$ is the number density of the superconducting electrons, which increases from zero at T_c to the normal state density at low temperatures. Because of the appearance of the density n_s in the denominator of equation (A2.2.4), λ_L is temperature dependent and diverges for $T \rightarrow T_c$. The experimental values of the temperature dependent penetration depth λ can be well fitted by the empirical relation

$$\lambda(T) = \lambda(0)\left[1 - \left(\frac{T}{T_c}\right)^4\right]^{-1/2}. \tag{A2.2.5}$$

A similar relation has been predicted in an early two-fluid model by Gorter. For pure metallic superconductors, such as Pb, Sn, In, Al, and Hg, the value of $\lambda(0)$ is typically around 50 nm. From this we see that the perfect diamagnetism from the Meissner–Ochsenfeld effect is well established only if both sample dimensions perpendicular to the magnetic field are much larger than this value of $\lambda(0)$. From Maxwell's equation, $j = (1/\mu_0)\,\mathrm{curl}\,H$, and from the fact that H drops almost to zero within the distance λ from the surface, we find the approximate expression for the maximum density of the shielding supercurrent $j_c = H_c/\lambda$.

A1.1

From equation (A2.2.4), we note that $\lambda^{-2}(T)$ is proportional to the density of superconducting electrons $n_s = n_n - n_{qp}$, where n_{qp} is the density of quasi-particles thermally excited across the gap in a superconductor. Measurements of $\lambda(T)$ are therefore very important in testing microscopic models for superconductivity. For conventional BCS superconductors, the existence of a near isotropic, s-wave energy gap gives rise to a vanishingly small exponential variation in $\lambda(T)$ at low temperatures. In contrast, for the recently discovered cuprate superconductors, one finds approximately

A3.2

$$\lambda^{-2}(T) = \lambda^{-2}(0)[1 - aT] \tag{A2.2.6}$$

over a very wide range of temperatures, for screening currents flowing in the cuprate planes. This is consistent with the now firmly established d-wave symmetry of the cuprate superconductors, which has nodes in the energy gap along certain directions. At low temperatures, the presence of such nodes leads to power law dependences of many of the thermal and transport properties instead of the exponential dependences expected for conventional BCS superconductors with an isotropic energy gap in all directions (s-wave). However, as shown recently by Schopohl and Dolgov, in the low-temperature limit, the temperature dependence of λ must vanish in order to remain consistent with the third law of thermodynamics.

The experimental values $\lambda(0)$ of the penetration depth at zero temperature are up to five times larger than the quantity $\lambda_L(0)$ in equation (A2.2.4). This deviation has been explained by Pippard in terms of the superconducting coherence length ξ_0 and by extending the London theory accordingly. According to Pippard, in dirty superconductors, in which the electron mean free path $l \ll \xi_0$, the magnetic

A1.2

A2.3 A4.2
penetration depth is increased by a factor $\sim(\xi_0/l)^{1/2}$. For discussion of the length ξ_0, see chapters A2.3 and A4.2.

A2.3
The Meissner–Ochsenfeld effect and the associated screening currents flowing within the penetration depth of the surface are the intrinsic properties of all the superconductors at sufficiently small fields. In chapter A2.3, we will introduce the concept of two types of superconductors: type I in which, on application of a sufficiently strong field, a direct transition is made from the Meissner state with $B = 0$ to the normal state with $B = \mu_0 H$, and type II, in which partial flux penetration is nucleated at a smaller field H_{cl} in the form of quantized flux lines, so that $0 < B < \mu_0 H$.

D2.1 D2.4 A2.3
In the previous discussion, we have essentially been assuming superconductors in the form of long rods or cylinders aligned parallel to the applied field, where the maximum field at the surface is simply the applied field H. However, for an arbitrary shaped sample, flux exclusion implies a concentration of flux around the surface and a higher local surface field $H/(1-D)$, where D is known as the demagnetization factor. The description of demagnetization in terms of the single quantity D is only possible if the sample geometry consists of a rotational ellipsoid with the rotational axis oriented parallel to the magnetic field. For a cylinder in a field transverse to its length, $D = 1/2$, while for a sphere $D = 1/3$. When the surface field exceeds $H_c(1 - D)$ or $H_{c1}(1 - D)$ for a type I or a type II superconductor, respectively, flux will begin to penetrate and we no longer have a complete Meissner effect (see further discussion in chapter A2.3).

C
Because of the fundamental importance of the Meissner–Ochsenfeld effect and since magnetization measurements can be performed relatively easily, more recently the detection of this effect has played an important role in the discovery and confirmation of new superconductors and, in particular, in the discovery of high-temperature superconductivity by Bednorz and Müller [4].

References

[1] Meissner W and Ochsenfeld R 1933 *Naturwiss.* **21** 787
[2] Gorter C J and Casimir H B G 1934 *Physica* **1** 306
[3] London F and London H 1935 *Proc. R. Soc.* A **149** 71
[4] Bednorz J G, Takashige M and Müller K A 1987 *Europhys. Lett.* **3** 379

Further Reading

Huebener R P 2001 *Magnetic Flux Structures in Superconductors 2nd edition* (Berlin: Springer)

Shoenberg D 1965 *Superconductivity* (Cambridge University Press: Cambridge)

Tinkham M 1996 *Introduction to Superconductivity* (New York: McGraw Hill)

Waldram J R 1996 *Superconductivity of Metals and Cuprates* (Bristol: Institute of Physics Publishing)

Tilley D R and Tilley J *Superfluidity and Superconductivity* (New York: Van Nostrand Reinhold)

A2.3
Loss of superconductivity in magnetic fields

Rudolf P Huebener

A2.3.1 The superconducting coherence length

In addition to the magnetic penetration depth λ discussed in chapter A2.2, there is another important length scale: the superconducting coherence length ξ_0 first introduced by Pippard [1]. The length ξ_0 is a measure of the spatial extent of the wave function describing Cooper pairs and is therefore the minimum distance over which the density n_s of superconducting electrons can change significantly. It indicates the spatial rigidity of the superconducting wave function. An estimate based on the Heisenberg uncertainty principle yields

A2.2 A1.2

$$\xi_0 \approx \frac{\hbar v_F}{\Delta} \approx \frac{\hbar v_F}{k_B T_c} \tag{A2.3.1}$$

where $\hbar$ is Planck's constant divided by 2π, v_F is the Fermi velocity, Δ is the superconducting energy gap, and k_B is Boltzmann's constant. A more accurate result can be obtained from the BCS theory [2]

A3.2

$$\xi_0 = \frac{\hbar v_F}{\pi\Delta} = 0.18\frac{\hbar v_F}{k_B T_c} \tag{A2.3.2}$$

where we have substituted the BCS relation $\Delta = 1.76 k_B T_c$. For pure metallic superconductors, such as Pb, Sn, In, and Hg, the experimental values of ξ_0 range around 100–300 nm. For Al, a value as large as $\xi_0 = 1.4\,\mu$m has been reported.

These experimental values given are for the clean limit with the electron mean free path $l \gg \xi_0$. In the presence of strong impurity scattering and in most alloys, $l \ll \xi_0$. Pippard [3] introduced an effective coherence length to account for electron scattering of the form

$$\frac{1}{\xi} = \frac{1}{\xi_0} + \frac{1}{l}. \tag{A2.3.3}$$

Many of the newly discovered cuprate superconductors are believed to be in the clean limit with mean free paths within the CuO planes much greater than the coherence length of typically 2–3 nm. The out-of-plane coherence length is believed to be less than ~ 0.1 nm, reflecting the highly localized nature of the electronic wave functions largely confined to the CuO planes, which are responsible for the metallic and superconducting properties.

C1

Although the coherence length describes the extent of the superconducting wave function of a Cooper pair, the co-operative nature of the superconducting transition involves interactions between the pairs which lead to strongly temperature dependent coherence and penetration lengths. This can most conveniently be considered using the phenomenological Ginzburg–Landau (GL) theory [4], which was formally justified from BCS theory near T_c by Gorkov, but is believed to be more widely applicable for

A3.1

all temperatures and fields. In the clean limit ($l \gg \xi_0$), it can be shown that

$$\xi(T) = 0.74\xi_0 \left(\frac{1}{1-t}\right)^{1/2} \tag{A2.3.4}$$

$$\lambda(T) = \frac{\lambda_L(0)}{\sqrt{2}} \left(\frac{1}{1-t}\right)^{1/2}. \tag{A2.3.5}$$

In the dirty limit ($l \ll \xi_0$), scattering modifies the prefactors with $\xi_{eff}(0) \sim l$ and $\lambda_{eff}(0) \sim \lambda_L(0) \times (\xi_0/l)^{1/2}$ leaving the temperature factors unchanged.

The dimensionless GL parameter,

$$\kappa = \frac{\lambda(T)}{\xi(T)} \tag{A2.3.6}$$

is therefore independent of temperature but is strongly dependent on scattering. In sections A2.3.2 and A2.3.3, we will show how the GL parameter differentiates between type I (when $\kappa < 1\sqrt{2}$) and type II (when $\kappa > 1\sqrt{2}$) superconductors (see article A4.2).

A4.2

A2.3.2 Type I superconductors

With the exception of Nb, V and Tc, all elemental superconductors are type I, exhibiting a full Meissner flux expulsion before making a transition to the normal state at $H_c(T)$. However, as indicated in chapter A2.2, this is only true for samples with a negligible demagnetizing factor.

A2.2

For a type I superconductor with a finite demagnetizing factor, the critical field at the surface will reach the thermodynamic transition field when the external field is $H_c(1 - D)$. Normal regions will be nucleated at the surface and flux will enter generating a domain structure of normal regions in which the field is equal to H_c and superconducting regions excluding flux. A schematic example of this 'intermediate state' for a thin superconducting slab in a field perpendicular to its surface is shown in figure A2.3.1. The intermediate state is established in the magnetic field regime $H_c(1 - D) < H < H_c$.

D2.1

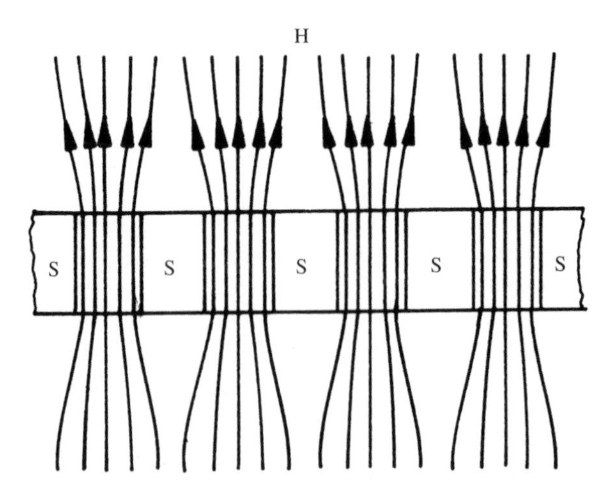

Figure A2.3.1. Intermediate state of a type I superconductor. The normal domains carry the flux density $\mu_0 H_c$ and are separated from each other by the superconducting phase with zero flux density.

An important feature of the domain configuration of figure A2.3.1 is the wall energy associated with the interface between the normal and superconducting phases. This wall energy is somewhat analogous to the Bloch wall energy separating regions of opposite magnetization in a ferromagnet. It can be obtained in the following way. At the interface, the transition from the superconducting to the normal state can take place only over the distance $\xi(T)$ because of the rigidity of the superconducting wave function. $\xi(T)$ is the temperature-dependent coherence length, discussed in section A2.3.1. Over this distance $\xi(T)$, the superconducting condensation energy is lost, yielding the contribution $[\mu_0 H_c^2(T)/2]\xi(T)$ per unit area (see equation (A2.2.2)). However, because of the magnetic flux penetration into the superconducting domain over the distance $\lambda(T)$, no gain and also no loss of condensation energy occur. Therefore, by subtracting this part we find the wall energy α per unit area

A1.1

A1.2

$$\alpha = \frac{\mu_0 H_c^2(T)}{2}[\xi(T) - \lambda(T)]. \tag{A2.3.7}$$

The difference $\delta \equiv \xi(T) - \lambda(T)$ is referred to as the wall energy parameter. In type I superconductors we have $\xi(T) > \lambda(T)$ and, hence, the quantities δ and α are positive. More careful calculation shows that the cross-over between positive and negative wall energies occurs when $\lambda(T)/\xi(T) = 1/\sqrt{2}$.

The first theoretical treatment of the domain structure in the intermediate state of a type I superconductor was given by Landau [5]. He minimized the total free energy taking into account the interface wall energy, the magnetic energy within and outside the superconductor, and the energy associated with changes in domain structure as the field leaves the sample. The Landau domain theory yields the periodicity length a of the domain configuration. The length a is the sum of the lengths a_n and a_s of the normal and superconducting domains, respectively: $a = a_n + a_s$. For the length a one finds

$$a = \left[\frac{\delta d}{f(H/H_c)}\right]^{1/2} \tag{A2.3.8}$$

where δ is the wall energy parameter, $f(H/H_c)$ is a numerical function of the normalized field H/H_c and d is the thickness of the plate measured along the magnetic field direction (see figure A2.3.1). Similar concepts are also used for calculating the domain size in other cases, such as a ferromagnet.

In the discussion leading to equation (A2.3.8), a structure of long laminar domains is assumed, such that the spatial variations can be restricted to two dimensions, as is often observed at intermediate fields. This is shown in figure A2.3.2 for a type I superconducting lead film in a perpendicular field. However, in addition, normal domains with nearly circular cross-section ('flux tubes') can also be found, in particular at low magnetic fields. In the intermediate state the volume fraction filled by the normal domains is equal to H/H_c, since the normal domains carry the field value H_c. For the laminar domain pattern this yields the relation $a_n/a = H/H_c$. From equation (A.2.3.8), we see that the length scale a of the laminar domain structure varies proportional to $d^{1/2}$, decreasing with decreasing thickness of the superconductor. Similarly, with decreasing d the diameter of the flux tubes also becomes smaller.

If the magnetic field is oriented exactly perpendicular to the surface of a large flat plate or film, the long, laminar domains are statistically oriented in the plane of the superconductor, as shown in figure A2.3.2. However, if the magnetic field is inclined at an angle, the domains become oriented parallel to this field component in the plane. This effect was first demonstrated by Sharvin. This Sharvin geometry is advantageous for accurately measuring the periodicity length a of the magnetic domain structure.

Superconductivity can also be destroyed, even in zero applied field, if the current flowing through a superconductor creates a field at its surface in excess of H_c. For a cylindrical wire of radius r, this leads to a critical current $I_c = H_c r/2$, known as Silsbee's rule.

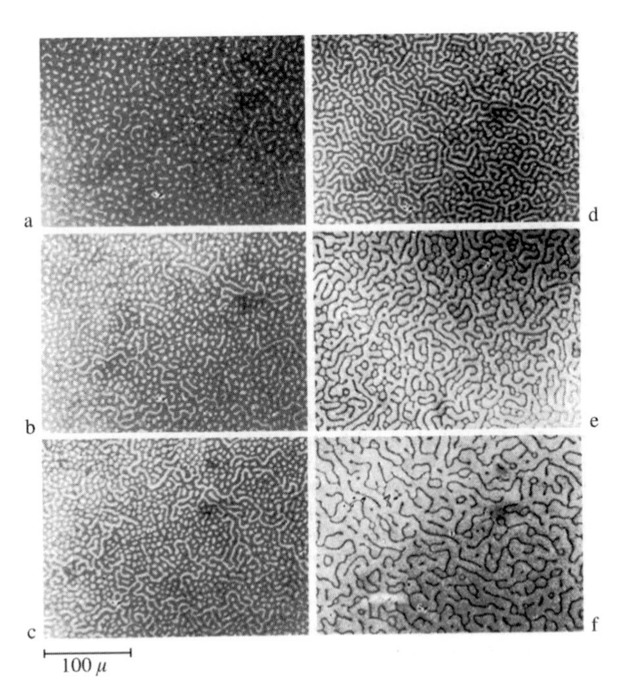

Figure A2.3.2. Intermediate state structure of a lead film in a perpendicular magnetic field observed magneto-optically for the following field values: (*a*) 7.6 kA m^{-1}; (*b*) 10.5 kA m^{-1}; (*c*) 14.2 kA m^{-1}; (*d*) 17.4 kA m^{-1}; (*e*) 27.8 kA m^{-1}; (*f*) 32.7 kA m^{-1}. The superconducting phase is dark, $T = 4.2$ K, film thickness = 9.3 μm.

A2.3.3 Type II superconductors

Type I superconductivity, discussed so far, is established in metals of high purity where the electron scattering in the normal state is relatively weak. However, an increasing amount of evidence has accumulated, indicating the existence of another type of superconductors with various 'unusual' properties. It was Shubnikov in the early 1930s who provided this evidence. The 'unusual' behaviour such as the appearance of electric resistance at magnetic fields below the critical field $H_c(T)$ or the observation of an incomplete Meissner effect was found in metals with a large impurity concentration and in alloys, where the electron scattering is relatively strong. It was not until Abrikosov's 1955 derivation of the magnetic properties of superconductors with κ values $> 1/\sqrt{2}$ that such properties were recognized as intrinsic and characteristic of what are now known as type II superconductors [6].

A2.2

Type II superconductors are distinguished by the fact that the penetration depth $\lambda(T)$ is larger than the coherence length $\xi(T)$. In this case, the wall energy α from equation (A2.3.7) becomes negative and the magnetic domains within the superconductor are reduced to the smallest possible unit of magnetic flux, namely individual magnetic flux quanta Φ_0 (see chapter A2.6).

A1.2

A2.6

In type II superconductors, the Meissner–Ochsenfeld effect is established only up to what is known as the lower critical magnetic field $H_{c1} < H_c$ (see chapter A2.2). It then becomes energetically favourable for the sample to undergo a transition in which flux tubes (flux lines) carrying a single quantum of flux Φ_0 are nucleated at the surface, forming a flux line lattice in the bulk. This is known as the 'mixed state', representing a mixture between the superconducting and normal phases. In this state, the spatially averaged magnetic flux density B is

A2.2

$$B = n \cdot \Phi_0 \tag{A2.3.9}$$

where n is the areal density of the flux lines. The magnetic flux quantum is

$$\Phi_0 = \frac{h}{2e} = 2.07 \times 10^{-15}\,\mathrm{T\,m^2}. \qquad \text{(A2.3.10)}$$

The mixed state extends to the upper critical magnetic field H_{c2} above which the normal state is formed. The thermodynamic critical field H_c defined from the difference between the free energy densities of the normal and superconducting phases according to equation (A2.2.2) lies between H_{c1} and H_{c2} : $H_{c1} < H_c < H_{c2}$. As the field is increased above H_{c1} the density of flux lines increases until their cores with radius $\sim\xi(T)$ start overlap leading to suppression of superconductivity and a second-order transition to the normal state at a field $H_{c2}(T) = \Phi_0/2\pi\xi(T)^2$. Figure A2.3.3($a$) shows magnetic flux density versus magnetic field and figure A2.3.3(b) shows magnetization versus magnetic field for a type II superconductor. If the sample geometry causes demagnetization effects, the mixed state of a type II superconductor is established in the magnetic field range $H_{c1}(1 - D) \leq H \leq H_{c2}$. D2.1

A phenomenological understanding of type II superconductivity has been provided by the GL theory [4]. Here, the spatial and temporal properties of the superconductor are described by a macroscopic complex wave function,

$$\psi = |\psi| \exp(\mathrm{i}\varphi) \qquad \text{(A2.3.11)}$$

acting as an order parameter. The squared amplitude $|\psi|^2$ is identified as the Cooper pair density n_s or the superconducting energy gap. Both the absolute value $|\psi|$ and the phase φ can be space and time dependent. Strictly speaking, the GL theory is only applicable close to the second-order phase transition at $H_{c2}(T)$, though in practice it appears to provide an excellent description of the superconducting state over a large part of the magnetic phase diagram. Within this theory, magnetic flux quantization results from the requirement that the wave function ψ must be single-valued at any point in the superconductor and, hence, that the phase φ can change only by multiples of 2π following a complete closed path within the superconductor. Therefore, flux quantization is the macroscopic analog of the Bohr–Sommerfeld quantum condition in atomic physics and of the quantization of hydrodynamic circulation or vorticity in superfluid helium, first observed by Vinen in 1958, $\oint v_s\,\mathrm{d}l = nh/m$. The formation of a regular lattice of magnetic flux quanta was first predicted theoretically by Abrikosov from an ingenious solution of the GL equations. Therefore, this lattice is often referred to as the Abrikosov vortex lattice. A3.1 A1.1

In a type II superconductor, flux penetrates the bulk in the form of flux lines with line cores along which ψ drops to zero. This allows solutions of the form $|\psi(r)|\mathrm{e}^{in\theta}$, as in a multiply connected superconducting ring. These vortex-like solutions involve currents circulating around the flux core with associated magnetic flux $n\Phi_0$. Abrikosov showed that singly quantized flux lines would be nucleated in A2.7

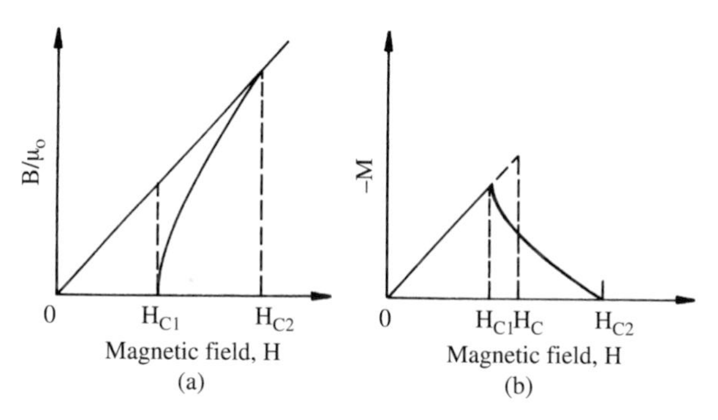

Figure A2.3.3. (a) Magnetic flux density B and (b) magnetization M versus magnetic field for a type II superconductor with a demagnetization factor $D = 0$.

a type II superconductor at a field H_{c1}. The structure of an isolated magnetic flux line carrying a single flux quantum and representing the building block of the mixed state is shown schematically in figure A2.3.4. At the centre of the flux line, $|\psi(r)|$ rises to its equilibrium bulk value over a distance $\sim\xi(T)$. The local flux density $h(r)$ is a maximum at the centre of the core and falls to zero as $\sim e^{-r/\lambda(t)}$ at large distances. The circulating current peaks at a distance $\sim\xi(T)$ and falls off with distance $\sim e^{-r/\lambda(t)}/r$. Crudely speaking, the core of a flux line represents a tube of normal phase with radius ξ imbedded in the superconducting phase. At least for the classical superconductors this is a reasonable picture, since these materials reside in the dirty limit where the electron mean free path l is small compared to the radius ξ of the vortex core. However, such a picture does not hold any more for the cuprate superconductors, since they are in the clean limit with $l \gg \xi$. In the latter materials, the electronic vortex structure is strongly affected by the Andreev bound states in the core. For clarification of this area, further experimental and theoretical work is needed at present. The role of the d-wave symmetry of the order parameter [7] in the electronic vortex structure of many cuprate superconductors represents another intriguing question which has not yet been answered completely.

For fields in excess of H_{c1}, Abrikosov showed that flux lines penetrate to form a lattice, held apart by their mutual repulsion. On increasing the field, the magnetic fields of the flux lines overlap leading to

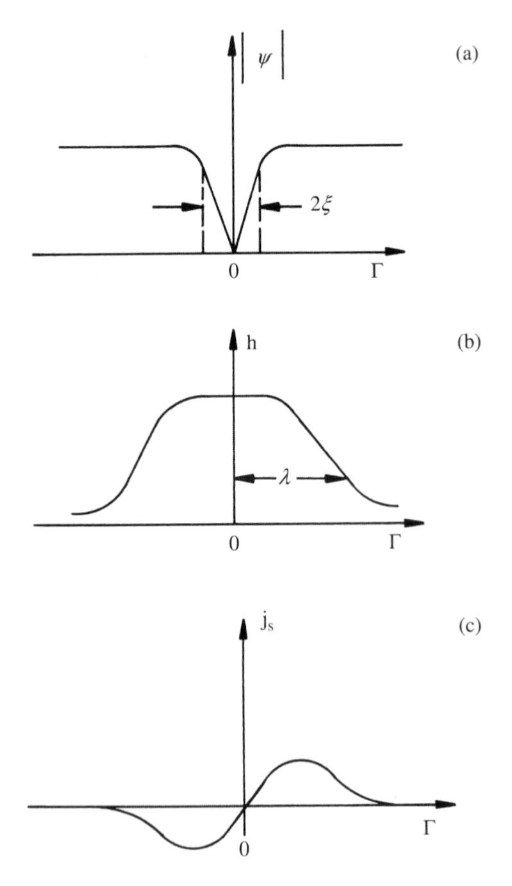

Figure A2.3.4. Structure of an isolated magnetic flux line carrying a single flux quantum. (*a*) Amplitude $|\psi|$ of the superconducting wave function, (*b*) local magnetic flux density h, and (*c*) supercurrent density j_s versus the distance from the vortex axis.

weaker variations in the relative values of the internal fields, though the absolute magnitude of the variations in internal field remains approximately constant. In materials with κ close to unity, the interactions between flux lines can become positive, giving rise to an instability in which flux lines clump together in regions at a constant flux density within a matrix of superconducting material free of flux-lines. This is known as the 'intermediate mixed' state.

The following are useful relations for the lower and upper critical fields in a type II superconductor:

$$H_{c1} \approx \frac{\Phi_0}{4\pi\mu_0\lambda^2}\ln\kappa \approx \frac{H_c}{\sqrt{2}\kappa}\ln\kappa \qquad \text{(A2.3.12)}$$

$$H_{c2} = \frac{\Phi_0}{2\pi\mu_0\xi^2} = \sqrt{2}\kappa H_c. \qquad \text{(A2.3.13)}$$

An electric transport current passing through a superconductor residing in the mixed or intermediate state always generates a Lorentz force acting on the magnetic flux structure. An induced motion of the flux structure results in an emf and dissipation. The superconducting properties of a type II superconductor can only be maintained if the flux structure is purposely pinned. Investigating the pinning of flux lines and the development of metallurgical microstructures to produce efficient pinning is essential for the successful application of both conventional and cuprate superconductors for magnets and other power engineering applications.

A4.2

A4.3

References

[1] Pippard A B 1950 *Proc. R. Soc. A* **203** 210
[2] Bardeen J, Cooper L N and Schrieffer J R 1957 *Phys. Rev.* **108** 1175
[3] Pippard A B 1953 *Proc. R. Soc. A* **216** 547
[4] Ginzburg V L and Landau L D 1950 *Zh. Eksp. Teor. Fiz.* **20** 1064
[5] Landau L D 1937 *Zh. Eksp. Teor. Fiz.* **7** 371
[6] Abrikosov A A 1957 *Zh. Eksp. Teor. Fiz.* **32** 1442
[7] Tsuei C C and Kirtley J R 2000 *Rev. Mod. Phys.* **73** 969

Further Reading

Huebener R P 2001 *Magnetic Flux Structures in Superconductors 2nd edition* (Berlin: Springer)

Saint-James D, Sarma G, Thomas E J 1969 *Type II Superconductivity* (New York: Pergamon)

Shoenberg D 1965 *Superconductivity* (Cambridge: Cambridge University Press)

Tinkham M 1996 *Introduction to Superconductivity* (New York: McGraw Hill)

Waldram J R 1996 *Superconductivity of Metals and Cuprates* (Bristol: Institute of Physics Publishing)

A2.4
Thermal properties of superconductors

Ctirad Uher

A2.4.1 Introduction

In general, under thermal properties of superconductors one usually includes physical phenomena in which temperature and thermal energy represent the dominant thermodynamic variables. One is typically interested in how the system responds to the applied thermal gradient or how the internal degrees of freedom absorb thermal energy. The archetypal examples of thermal properties are heat conduction and specific heat. The former is a transport coefficient revealing how efficiently a solid carries heat, and the latter is a static parameter containing information about the internal energy and the excitation levels of the system. Both the thermal conductivity and specific heat are important material parameters that have considerable influence on the choice and selection of materials for various technological applications. In superconductors, the two thermal properties gain further importance because the entry into the superconducting domain leaves a clear and unmistakable signature on their magnitude and temperature dependence. In fact, their unique behaviour as the sample is cooled through the transition temperature or when one monitors sample properties at very low temperatures has provided invaluable insight into the remarkable environment of the superconducting state. The gap in the energy spectrum that arises as a consequence of conduction electrons forming bound, two-electron states called Cooper pairs, has a dominant influence on the behaviour of the thermal conductivity and specific heat at temperatures below the superconducting transition temperature. Because the energy gap plays a central role in essentially all physical properties of superconductors, including their thermal properties, some of the fundamental issues pertaining to the emergence of the energy gap are first outlined. Then the specific heat and the input it provides concerning the excitation spectrum of superconductors are discussed. This is followed by a presentation of the behaviour of thermal conductivity in both conventional and high-temperature superconductors (HTS), and, finally, the trend in thermal expansion is briefly described.

D3.2.1
A2.1

A2.4.2 Energy gap

The concept of an energy gap is of paramount importance to the understanding of the mechanism of superconductivity. This energy gap plays a crucial role in most of the superconducting properties and it also stands at the core of one of the most promising technological devices—superconducting tunnelling structures. It is thus instructive to outline physical principles that give rise to the superconducting energy gap.

A2.1 A2.7 E4.1

The full many-body treatment of the energy gap appeared for the first time in the BCS theory [1]. However, the existence of the energy gap was clearly forecasted in several earlier experiments on superconductors—most notably in an exponential dependence of the low temperature specific heat

A3.2

and in far-infrared absorption studies, where a certain minimum frequency (threshold) was required in order to absorb electromagnetic radiation.

Development of the BCS theory was preceded by two key theoretical findings: the work of Fröhlich [2], who showed that phonons are essential to establish an effective attractive interaction between two electrons, and the work of Cooper [3] who considered a scenario of two electrons with opposite spins and momenta forming a bound state of the type $(k\uparrow, -k\downarrow)$ called Cooper pair. The task left to Bardeen, Cooper and Schrieffer was to extend the pairing interaction to all electrons of the system and to describe what the ground state of such an assembly would look like and what would be the spectrum of excitations from the ground state. The result of their work is the celebrated BCS theory.

A3.2

A complete quantum mechanical treatment, often presented using a formalism of second quantization, can be found in most of the texts on superconductivity. A particularly illuminating treatment is given by Leggett [4]. Our overview will focus on how the energy gap arises in the BCS theory and how it relates to the thermal properties of superconductors. We also present a brief discussion of the density of states of the quasiparticles (the broken Cooper pairs), as this quantity is of fundamental importance to both equilibrium and non-equilibrium phenomena.

According to Cooper, formation of a two-electron bound state—a Cooper pair—lowers the energy of the system. As a result, even more Cooper pairs are formed. However, each subsequent Cooper pair does not lower the energy by the same amount as the first Cooper pair, i.e. the total energy reduction is not obtained by multiplying a contribution due to a single Cooper pair by the number of pairs. The reason is that with more and more electrons forming Cooper pairs, there are progressively fewer empty states to scatter into. With fewer scattering processes available, a point will be reached when the cost of raising the electrons above the Fermi level (kinetic energy cost) exceeds the gain arising from a reduction in the potential energy upon pair formation. To find the ground state, one must make a variational calculation to search for the minimum total energy of the entire system for all possible pair configurations.

A2.1

Making simplifying assumptions such as considering spherical Fermi surfaces and by taking the electron–phonon interaction as independent of the electron wavevectors, i.e. as a constant V, the BCS theory leads to an expression for the energy gap at $T = 0$ of the form

$$\Delta(0) = \frac{\hbar\omega_D}{\sinh[1/N(E_F)V]}. \tag{A2.4.1}$$

Here ω_D is the Debye frequency and $N(E_F)$ is the density of states at the Fermi energy. Since energies are measured with respect to the Fermi energy, $N(E_F)$ is often written as $N(0)$. The product $N(0)V$ reflects the strength of the electron–phonon coupling and for the usual case of weak-coupling superconductors (most elemental superconductors) $N(0)V \ll 1$. Under such conditions, equation (A2.4.1) becomes

A2.1

$$\Delta(0) = 2\hbar\omega_D \exp\left[-\frac{1}{N(0)V}\right]. \tag{A2.4.2}$$

The expression for the superconducting transition temperature T_c has, apart from a numerical factor, a form similar to the right-hand side (RHS) of equation (A2.4.2), and this allows us to relate the zero-temperature gap to the transition temperature as

$$2\Delta(0) = 3.52\, k_B T_c. \tag{A2.4.3}$$

Typical values of $\Delta(0)$ are of the order of 1 meV. This is to be contrasted with the energy gap in semiconductors that is of the order of 1 eV, larger by a factor of a thousand. It is also useful to keep in mind that the gap is much smaller than the typical Fermi energy, $\Delta(0) \approx 10^{-4} E_F$. In spite of its small size, the superconducting energy gap is responsible for all the striking properties that one associates with the superconducting state.

Unlike the essentially temperature independent gap in semiconductors, the superconducting energy gap is a strong function of temperature. While it initially changes a little from its maximum value at $T = 0$, it decreases rapidly above about $T_c/2$ and vanishes at $T = T_c$. The temperature dependence follows from an implicit relation for the temperature dependent gap $\Delta(T)$ given by

$$\frac{1}{N(0)V} = \int_{-\hbar\omega_D}^{\hbar\omega_D} \frac{\tanh\left\{\frac{[\varepsilon^2+\Delta^2(T)]^{1/2}}{2k_B T}\right\}}{2[\varepsilon^2 + \Delta^2(T)]^{1/2}} \, d\varepsilon \qquad \text{(A2.4.4)}$$

and is shown in figure A2.4.1 as a plot of reduced energy gap versus reduced temperature.

Excitations from the superconducting ground state, i.e. the broken pairs, are called quasiparticles. The presence of the energy gap modifies the excitation spectrum of quasiparticles in a significant way and the possible states that are accessible to the quasiparticles are given by

$$E_k = (\epsilon_k^2 + \Delta^2)^{1/2}. \qquad \text{(A2.4.5)}$$

At high energies, $\epsilon_k \gg \Delta$, the quasiparticle excitations are nothing else but normal electron excitations, $E_k \approx \epsilon_k = \hbar^2 k^2/2m - E_F^o$, measured with respect to the Fermi energy. However, equation (A2.4.5) also implies that there are no excitation states available with the energy below Δ. While in the normal state excitations have no lower energy limit, in the superconducting state the energy gap imposes a finite threshold for allowed excitations. This is illustrated in figure A2.4.2, where electron excitations in the normal metal (figure A2.4.2(*a*)) and in a superconductor (figure A2.4.2(*b*)) are sketched.

The question arises of what happens to the states within the gap region? Since the total number of states is not changed by the interaction, the states in the gap are simply 'pushed out' by the interaction

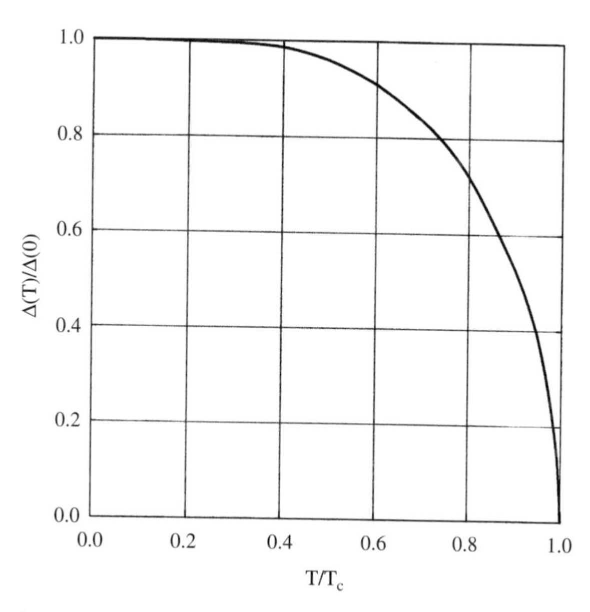

Figure A2.4.1. Temperature dependence of the energy gap $\Delta(T)$, relative to its zero temperature value $\Delta(0)$, plotted as a function of the reduced temperature T/T_c.

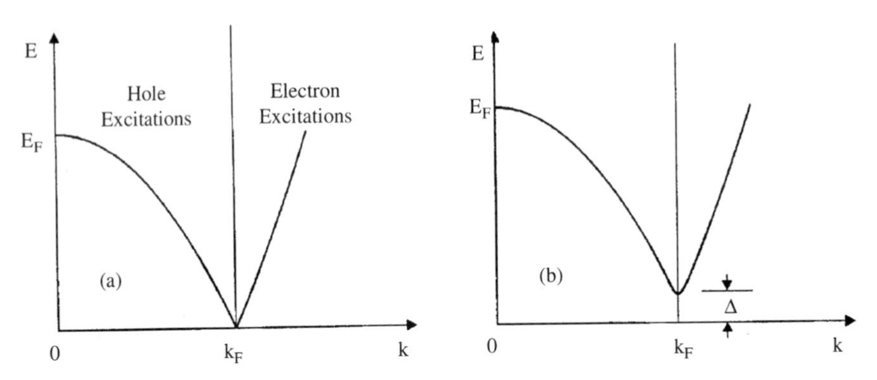

Figure A2.4.2. (*a*) Excitation energies in a normal metal. (*b*) Excitation energies in a superconductor. Note a gap Δ in the excitation spectrum that, for the reasons of clarity, is shown to be much exaggerated.

and pile up in the energy region just above and below the energy gap. Formally, this is expressed with the aid of the density of excitation states in the condensate $N_s(E)$, which has the form

$$N_S(E) = \frac{N_n(0)E}{(E^2 - \Delta^2)^{1/2}}. \tag{A2.4.6}$$

Here $N_n(0)$ is the density of normal electron states at the Fermi energy. For large excitation energies, the quasiparticle density of states $N_s(E)$ essentially coincides with the density of states of the normal electrons. The largest difference between the two is near the edges of the gap where the function $N_s(E \rightarrow \Delta)$ has a pole. It is often advantageous to plot the density of states so that it explicitly shows both the quasielectron and quasihole branches of excitations, figure A2.4.3. The excitation energy is measured here as positive outward in both directions from the centre.

We have noted the relevance of the energy gap to the low temperature specific heat of superconductors, where it leads to an exponentially varying temperature dependence. We have also pointed out the existence of a threshold for optical and microwave absorption that reflects the minimum energy needed in order to break a Cooper pair. The energy threshold is also revealed in the I–V curves of tunnel junctions, where at very low temperatures ($T \rightarrow 0$) the gap prevents any single-particle flow until the bias voltage satisfies $eV = 2\Delta(0)$. At higher tempertures the bias is lowered and the onset of current is less sharp reflecting the temperature dependence of the energy gap $\Delta(T)$. Such tunnelling of quasiparticles was first observed by Giaever [5] in 1960 and it led to him winning the Nobel Prize in 1973. Quasiparticle tunnelling has matured into an important field of physics with many practical applications. It is also the most convenient and accurate technique for determining the value of the energy gap. Readers interested in this topic, and in superconducting devices in general, are referred to an excellent monograph by Van Duzer and Turner [6].

A2.1
A2.7
E4.7

The author has tried to highlight the main physical arguments that explain the origin of the energy gap in superconductors and some of the physical properties where the presence of the gap has an overwhelming influence were noted. In spite of its very small magnitude, the energy gap plays a pivotal role in the physical properties of superconductors.

A2.4.3 Specific heat

Among the property measurements, studies of specific heat are one of the most rewarding in terms of the wealth of information one gains about a particular material. Specific heat is not a probe that yields sharp, spectral-like resolution of the excitation processes taking place in a material. Rather, its power

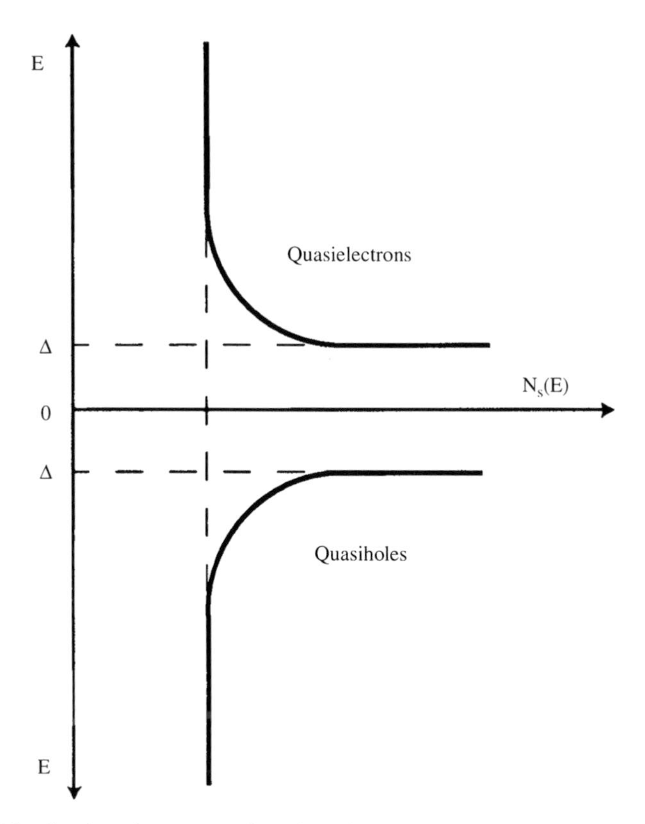

Figure A2.4.3. A plot of the density of states as a function of the excitation energy displaying both the quasielectron and quasihole branches.

rests in providing an integral assessment of the thermodynamic state of a material and in pinpointing the temperatures where the system undergoes dramatic changes in its internal energy.

To alter the temperature of a substance, one must add or take away energy in the form of heat. The amount of heat depends on how 'large' the object is and on its composition. Normalizing this heat per unit mass (frequently per mole of a substance or per gram-atom) and per unit increase in temperature, one arrives at an intrinsic material parameter called specific heat. The heat supplied to an object can be absorbed by its crystal lattice—we say that heat excites quantized entities called phonons—or by conduction electrons. Thus, specific heat data shed light on both the vibrational and electronic properties of a material as well as on the elementary excitations and phase transitions that may constitute the realm of a substance.

In superconductors, specific heat has proved to be an exceptionally important physical parameter. First of all, its unusual exponential temperature dependence at very low temperatures ($T \ll T_c$) supplied an all-important clue (together with the far-infrared and microwave absorption studies) to the existence of an energy gap in the excitation spectrum of superconductors. Furthermore, the jump in specific heat at the superconducting transition temperature has been an invaluable and most reliable probe of the truly bulk nature of superconductivity. To appreciate the unique character of specific heat in superconductors, some fundamental points regarding its behaviour in the normal state are first outlined.

As already mentioned, heat can be absorbed, and thus can excite, both the electron and phonon distributions. Excitations lead to an increase in the internal energy of the system. Internal energy is conceptually a very useful physical quantity but is experimentally inaccessible. One thus works with the next best thing—its temperature derivative—which is the specific heat. Actually, it is a partial derivative of the internal energy with respect to the temperature calculated under the assumption that the sample volume is constant, $C_V = (\partial U/\partial T)_V$. In practice, the specific heat is measured at constant pressure rather than at constant volume, i.e. $C_P = (\partial U/\partial T)_P$. The difference between C_P and C_V is small, especially at low temperatures, and will thus be neglected here.

A full mathematical treatment of the specific heat can be found in most of the texts on solid state physics and will not be repeated here. Rather, we will only highlight the fundamental points on which the theory is based and give results that will be useful in discussions of the main topic—specific heat of superconductors.

Vibrations of the crystal lattice are present in solids regardless of whether they are metals or insulators. Lattice vibration modes* are modelled as harmonic oscillators, the energy of which is quantized in units of $\hbar\omega$ called phonons. Here ω is the frequency and $\hbar = h/2\pi$ where h is the Planck constant. Phonons have spin zero and thus obey the Bose–Einstein statistics with the distribution function $\epsilon(\omega,T)$. The number of phonon modes within the frequency range between ω and $\omega + \mathrm{d}\omega$ is $\Delta(\omega)\mathrm{d}\omega$ where $\Delta(\omega)$ is the phonon density of states. The form of $\Delta(\omega)$ depends on the assumptions made regarding the dispersion relation (how frequency relates to the wavevector) and what criterion is used to determine the highest vibrational frequency that can be excited in the system. Clearly, there must be some kind of high frequency cut-off, otherwise the energy of the system would be infinite. In the Debye model†, the lattice specific heat is written as

$$C_\mathrm{L} = \left[\frac{\partial U(T)}{\partial T}\right]_V = \int \Delta(\omega)\frac{\partial \varepsilon(\omega,T)}{\partial T}\hbar\omega\mathrm{d}\omega = 9Nk_\mathrm{B}(T/\theta_\mathrm{D})^3\int_0^{\frac{\theta_\mathrm{D}}{T}}\frac{x^4\mathrm{e}^x}{\mathrm{e}^x-1}\mathrm{d}x \tag{A2.4.7}$$

where θ_D is the Debye temperature, $\theta_\mathrm{D} = \hbar\omega_\mathrm{D}/k_\mathrm{B}$, and $x = \hbar\omega/k_\mathrm{B}T$. At high temperatures, $k_\mathrm{B}T > \hbar\omega_\mathrm{D}$ and equation (A2.4.7) returns the well known Dulong–Petit law, i.e. the temperature independent $C_\mathrm{L} = 3Nk_\mathrm{B} \approx 25\,\mathrm{J\,K^{-1}\,mole^{-1}}$. We are more interested in the low temperature behaviour, $T \ll \theta_\mathrm{D}$, in which case equation (A2.4.7) becomes

$$C_\mathrm{L} = \frac{12\pi^4}{5}Nk_\mathrm{B}\left(\frac{T}{\theta_\mathrm{D}}\right)^3 \quad (\mathrm{J\,mole^{-1}\,K^{-1}}). \tag{A2.4.8}$$

Equation (A2.4.8) is the well-known Debye T^3 law. It states that for all solids, the lattice specific heat is a universal function that scales with a single parameter—the Debye temperature θ_D. In general, equation (A2.4.8) gives an excellent fit to the low temperature lattice specific heat data.

The finite density of conduction electrons also contributes to the internal energy of a system and gives rise to the electronic specific heat. The starting point in this case is an assumption that the conduction electrons behave as a gas of weakly interacting particles. The mathematical description that acknowledges electrons as particles with a half-integer spin (fermions) and recognizes the fact that an electron is not permitted to enter a state already occupied by an identical electron is the Fermi–Dirac distribution function, $f(E,T)$. The number of states an electron can occupy in an energy interval $\mathrm{d}E$ between energies E and $E + \mathrm{d}E$ is given by $N(E)\mathrm{d}E$ where $N(E)$ is the density of states. The highest occupied energy level

*A crystal consisting of N lattice points (not the same as N atoms unless the structure is monatomic) can vibrate in $3N$ acoustic modes, the factor of three implying three polarization branches, one longitudinal and two transverse. The optical modes are neglected as their energy is too high to be excited, especially at low temperatures.

†The Debye model assumes an elastic isotropic medium with a linear dispersion for each branch of the phonon spectrum and the same phonon velocities for all three polarizations. The high frequency cut-off—Debye frequency—is determined by the requirement that the total number of vibrational modes is $3N$.

at $T = 0\,\mathrm{K}$ is called the zero temperature Fermi energy, E_F^0, and a characteristic temperature of the electron gas, $T_F = E_F^0/k_B$, is called the Fermi temperature (typical values for metals $\sim 5\times 10^4\,\mathrm{K}$). By measuring the energy with respect to the Fermi energy E_F, the electronic specific heat becomes

$$C_e(T) = \left(\frac{\partial U}{\partial T}\right)_V = \int (E - E_F)\frac{\partial f(E,T)}{\partial T}N(E)\mathrm{d}E = \frac{\pi^2 k_B^2 N(0)T}{3} \equiv \gamma T. \qquad \text{(A2.4.9)}$$

Here $N(0)$ is the density of states at the Fermi energy and γ is a coefficient specific to a given metal and known as the Sommerfeld constant. There are two essential points to note. The first is the linear temperature dependence of the electronic specific heat, which assures that at sufficiently low temperatures the electronic specific heat will exceed the rapidly falling (T^3-dependence) lattice specific heat. The other point concerns the magnitude of the electronic specific heat at room temperatures, $C_e(300\,\mathrm{K}) \sim 0.1\,\mathrm{J\,K^{-1}\,mole^{-1}}$. This value is quite insignificant in comparison to the lattice specific heat, $C_L(300\,\mathrm{K}) \sim 3R = 25\,\mathrm{J\,K^{-1}\,mole^{-1}}$. Until the development of quantum mechanics, it was a major puzzle why *experimental* values of the specific heat of metals and insulators come out to be comparable at ambient temperatures. After all, the classical equipartition law predicted for metals a specific heat about 50% higher on account of the term $3R/2\,\mathrm{J\,K^{-1}\,mole^{-1}}$ representing the specific heat of free electrons. Providing an explanation for the 'missing' electronic degrees of freedom in the room temperature specific heat of metals was one of the early major triumphs of the quantum-statistical theory. Unlike lattice vibrations, where all $3N$ acoustic phonon modes contribute to the lattice specific heat, only a small fraction, $\sim k_BT/E_F \leq 0.01$, of the N electrons—those within the energy range k_BT near the Fermi energy—can move into the free states in their neighbourhood, i.e. to contribute to the specific heat. Electrons 'deeper' in the distribution do not have empty states in their vicinity and the available energy of k_BT per electron is not high enough to elevate them to the Fermi level.

D3.2.1

Summarizing, we write for the low temperature specific heat of metallic systems (including superconductors in their normal state)

$$C(T) = C_e(T) + C_L(T) = \gamma T + \beta T^3. \qquad \text{(A2.4.10)}$$

Plotting the experimental data as $C(T)/T$ versus T^2 offers a convenient way to extract the Sommerfeld constant (from the intercept) and the coefficient β (from the slope). Occasionally such a plot indicates an upturn at the lowest temperatures that signals the Schottky contribution due to nuclear hyperfine splitting. This can be accounted for by adding a AT^{-2} term to equation (A2.4.10).

Turning our attention to superconducting materials, one of the most interesting and important manifestations of the superconducting state is the unusual behaviour of the specific heat. Figure A2.4.4 shows the specific heat data for aluminium in the superconducting state contrasted with the normal state behaviour obtained by suppressing superconductivity with the aid of an external magnetic field. The reader will note the following characteristic features:

(a) a sharp jump in the specific heat at $T = T_c$,

(b) excess specific heat in the superconducting state near T_c,

(c) a crossover followed by lower values of C_{es} in comparison to C_{en},

(d) exponential rather than power law behaviour at the lowest temperatures.

It should be emphasized that all the above features are the result of changes in the electrodynamic properties of the conduction electrons. The phonon system is substantially oblivious of the superconducting transition and the lattice specific heat undergoes no change at T_c.

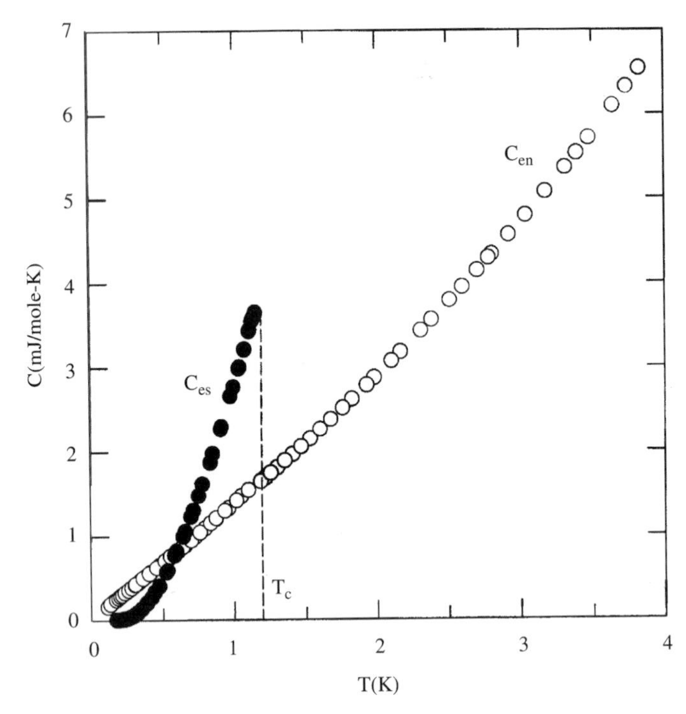

Figure A2.4.4. Specific heat of aluminium in the superconducting state C_{es}, and in the normal state, C_{en}. The normal state data below T_c were obtained using a magnetic field of 300 Oe, which is sufficient to suppress superconductivity. Adapted from the data of Phillips [7].

The behaviour depicted in figure A2.4.4 can be understood qualitatively using thermodynamic arguments pertaining to phase transitions—in this case a transition between the normal and superconducting phases. Again, providing only an outline of the argument, we note that the condition of equilibrium between the normal and superconducting phases at a given temperature T and magnetic field H is expressed as the equality of the free energies of the two phases. From this follows a Clausius–Clapeyron-like equation for the difference in entropies of the normal and superconducting states,

$$S_n - S_s = -\frac{V}{4\pi} H_c \frac{dH_c}{dT}. \tag{A2.4.11}$$

Here H_c is the critical field and dH_c/dT is the slope of the transition temperature curve. As noted elsewhere in this handbook, $dH_c/dT \leq 0$ and thus the entropy of the superconducting state is lower, i.e. the superconducting state is more ordered than the normal state. The entropy difference implies the existence of the latent heat, $Q = T(S_n - S_s) > 0$, which must be absorbed by the superconductor in the transformation from the superconducting to the normal state. This situation of course refers to a transition in the presence of an external magnetic field which is the first-order phase transition. When the transition takes place in zero external field—it does so at $T = T_c$ and $H_c = 0$—there is no latent heat and the transition is of the second-order. Differentiating equation (A2.4.11) with respect to temperature and multiplying by T leads to the difference between specific heats in the superconducting and normal states,

A2.1

$$C_s - C_n = \frac{T}{4\pi}\left[H_c \frac{d^2 H_c}{dT^2} + \left(\frac{dH_c}{dT}\right)^2\right]. \tag{A2.4.12}$$

It follows that at $T = T_c$, i.e. for $H_c = 0$, $C_s - C_n = (T_c/4\pi)\,(dH_c/dT)^2 > 0$, as is indeed observed experimentally. At lower temperatures, we expect a reversal in the above inequality because the first term in equation (A2.4.12) is negative and will dominate as the H_c versus T curve flattens at lower temperatures. To go beyond the qualitative description, one requires a microscopic theory of superconductivity, i.e. one needs to address the physical origin of the superconducting state. Before we take this step, it is important to mention that careful fits of the specific heat data at the lowest temperatures often suggested the temperature behaviour of the form $C_s/\gamma T_c = a\exp(-bT_c/T)$. The exponentially decreasing specific heat hinted at the existence of some kind of energy gap over which the superconducting electrons had to be excited in order to be able to absorb energy (heat). The notion of an energy gap turned out to be one of the key experimental inputs towards the development of a microscopic theory of superconductivity.

It is important to recognize that the superconducting condensate—a state of perfect order at $T = 0$—has no entropy. The entities that carry entropy are the quasiparticles. At $T > 0$, there is always a finite number of quasiparticles around and their density increases with increasing temperature. In this sense it is useful to picture the superconducting state as a two-fluid mixture, comprising a fraction x of the paired electrons (superconducting fluid) and a fraction $(1 - x)$ of quasiparticles (normal fluid). At $T = 0$, all electrons are paired, $x = 1$, while at $T = T_c$ all pairs have been broken, $x = 0$. At very low temperatures, $T \ll T_c$, it can be shown that the specific heat follows an exponential dependence of the form

$$\frac{C_{es}}{\gamma T_c} = 1.34\left[\frac{\Delta(0)}{k_B T}\right]^{3/2} \exp\left[\frac{-\Delta(0)}{k_B T}\right]. \tag{A2.4.13}$$

The specific heat in equation (A2.4.13) is normalized to the normal state specific heat at T_c assumed to be of purely electronic origin, $C_L \ll C_{en}$. In the BCS model, the exponential dependence of the specific heat at the lowest temperatures has an immediate explanation: at these temperatures a vast majority of all electrons are paired and separated from the single-electron states (quasiparticle excitations) by a gap of size Δ ($\sim$1 meV) that is almost temperature independent. The uptake of energy by the condensate then requires the break-up of Cooper pairs, i.e. excitations over the gap. As the temperature decreases, it becomes exponentially more and more difficult to supply the necessary energy. A3.2

At the transition temperature T_c, there is a discontinuity in the specific heat that, for the BSC model, can be expressed as

$$\frac{\Delta C}{C_{en}} = \frac{C_{es} - C_{en}}{C_{en}} = 1.43, \qquad \text{at } T = T_c. \tag{A2.4.14}$$

The jump at T_c arises from the difference in the electronic specific heats in the superconducting and normal states. Remember, the phonon spectrum is unchanged at T_c. In itself, a significant jump at T_c provides strong experimental evidence for the bulk nature of superconductivity. This is to be contrasted with the effect of filamentary superconducting shorts or the minority phase that may exist in connected loops throughout the sample. Both such morphologies may yield zero resistance and the latter may even mimic a complete diamagnetic signal. Yet, they would hardly be representative of real bulk superconductivity. Only specific heat measurements can make such a claim. Table A2.4.1 gives experimental values for the size of the jump for several conventional superconductors. The agreement with the BCS model is quite good except for lead and mercury, the two strongly-coupled superconductors for which the BCS model is inadequate. The appropriate starting point for describing the properties of strongly-coupled superconductors is the Eliashberg equations.

Table A2.4.1. Experimental values of the ratio $(C_{es} - C_{en})/C_{en}$ at T_c

Superconductor	Ratio
Aluminium	1.60
Tantalum	1.58
Thallium	1.15
Tin	1.60
Vanadium	1.57
Zinc	1.25
Lead	2.65
Mercury	2.18

To obtain the temperature dependence of the specific heat in the entire range from $T = 0$ to $T = T_c$, it is convenient to use tables available in the literature, e.g. Muhlschlegel [8].

Except for complications arising with the strongly-coupled superconductors, the BCS theory provides quite a good description of the specific heat of conventional superconductors. The success has a lot to do with the simple *s*-wave nature of the pairing mechanism, low T_c values and, last but not least, the high homogeneity, structural integrity, and one- or at most two-component nature of the metallic structures we call conventional superconductors. In the high-temperature perovskite superconductors we encounter a very different and far more complicated situation.

As fascinating as HTS are, they are not the easiest materials to work with. They have a complicated oxide structure; they are sensitive to preparation conditions; they frequently contain minute amounts of secondary phases; and they are often inhomogeneous and structurally unstable. Their properties depend critically on the stoichiometry and on the oxygen content in particular, and they all have rather low carrier densities. The remarkably high transition temperature—the signature property that attracts attention—is actually a considerable drawback in the measurements of the specific heat. With the $T_c \sim 100$ K and low carrier densities, the electronic specific heat amounts to not more than 1–2% of the total specific heat. Thus, the jump at T_c, if any, is very difficult to resolve against the large background of the lattice specific heat. An extremely short coherence length ($\xi \sim 20$ Å) results in very large critical fields, well beyond the capability of the laboratory magnets. Thus, one cannot rely on a magnetic field to suppress the superconducting state over more than a narrow range of temperatures below T_c. Consequently, in spite of great efforts on the part of researchers, the collected data sets lack the richness and clarity of information and must be subjected to *tour de force* analysis that injects assumptions concerning the behaviour of the lattice specific heat. Because of the overwhelming dominance of the lattice specific heat term, even relatively small anharmonic contributions and dilatation corrections can easily account for as much as 50% of the entire electronic specific heat. The main task is therefore a proper assessment of the lattice vibrational spectrum—not a trivial problem for a system with many atoms in the unit cell and a temperature range where anharmonicity cannot be neglected.

One further complication must be mentioned even though it potentially provides an interesting input relevant to the nature of the superconducting state—fluctuation contribution to the specific heat. In conventional superconductors, except for very thin films or filamentary structures where the reduced dimensionality tends to enhance fluctuations, one can usually neglect fluctuation contributions to the specific heat. In perovskite superconductors, on account of their high transition temperature, short coherence length, and the distinctly layered character of their structure, fluctuations in the order parameter are an inherent feature of the physical properties in the neighbourhood of T_c. For instance, fluctuations round resistivity curves above T_c and extend the resistive state below T_c. They also have a direct influence on the behaviour of the specific heat. An important parameter in assessing the nature of fluctuations is the critical reduced temperature, t_c, given by

$$t_c = \left[\frac{8\pi k_B T_c}{H_c^2(0)\xi^3(0)}\right]^2 \qquad \text{(A2.4.15)}$$

where $H_c(0)$ is the critical field at $T = 0$ and $\xi(0)$ is the coherence length. The temperature t_c delineates two distinct regimes of fluctuations. For $t \equiv |T - T_c|/T_c > t_c$, the fluctuations are weak, do not interact, and can be described in terms of the mean field Ginzburg–Landau theory as the Gaussian fluctuations. For $t < t_c$, interactions between fluctuations cannot be neglected, the fluctuation specific heat becomes the leading term, and the system finds itself in the so-called critical fluctuation regime. Specific functional forms have been derived for different regimes of fluctuations and these include dimensional parameters to reflect the appropriate dimensionality of the superconducting system. Relevant formulae can be found, for instance, in a review article by Triscone and Junod [9]. A3.1

The number of papers addressing experimental aspects of specific heat in HTS is very large. The reader should make use of review articles that describe in detail the various aspects of specific heat in several families of perovskite superconductors [10–13]. For our purposes, we only illustrate the general trend in the specific heat, notably the limiting low temperature behaviour and the temperature range near T_c.

One of the surprises of the early specific heat studies on sintered samples of $La_{2-x}Sr_xCuO_4$ (LSCO) and $YBa_2Cu_3O_{7-\delta}$ (YBCO) was the observation of a rather large linear term (γ-term) at low temperatures (a positive intercept on the C/T versus T^2 curves of 5–20 mJ K^{-2} $mole^{-1}$). This was, of course, quite an unexpected result because at $T \ll T_c$ the electronic specific heat should be negligible. Detection of such a large linear term led to speculations as to whether it is intrinsic to HTS (perhaps due to nodes or line of zeros on the anisotropic gap, spinons of the resonance valence band (RVB) model, or even a group of normal carriers that for some reason did not condense), or whether the linear term is of extrinsic origin (a minute amount of impurity phase with a large γ-term, or oxygen vacancies forming two-level tunnelling systems). It was also noted that an unexpected T-linear term was observed in the low temperature thermal conductivity, see section A.2.4.4. As single crystals became available, and with their continuous improvement in quality, the measured values of the linear term came down to a level of a few mJ K^{-2} $mole^{-1}$. However, as hard as the researchers tried, a complete suppression of the linear term in YBCO has not yet been achieved. Whether the values of the order of 2–3 mJ K^{-2} $mole^{-1}$ are intrinsic to YBCO, or whether further improvements in the quality of the samples will eventually lead to a complete disappearance of the linear term, remains to be seen. It is important to point out that in the family of Bi-based HTS, $Bi_2Sr_2Ca_{n-1}Cu_nO_{2n+4}$ with $n = 2$ (BSCCO-2212) and $n = 3$ (BSCCO-2223), the measured linear term has always been smaller than in YBCO, and a large number of reports conclude that it is, in fact, zero. Tl-based compounds, $Tl_mBa_2Ca_{n-1}Cu_nO_{2(n+1)+m}$, have not been explored as thoroughly but the existing data suggest a small linear term in the low temperature specific heat. However, these compounds contain Ba, and trace amounts of the $BaCuO_2$ impurity phase—the same impurity believed to be responsible for the early large linear terms in YBCO—might interfere with the intrinsic behaviour of the specific heat. Small linear terms are also found in the Hg-based perovskites, the materials with the highest reproducible T_c. In contrast, a cubic and copperless superconductor $(Ba_{0.6}K_{0.4})BiO_3$ seems to have no linear term in its low temperature specific heat. C2 C3

At high temperatures, the most striking aspect of specific heat is the lack of a prominent discontinuity (jump) at T_c. Even in high quality crystals with very narrow transition widths and a near complete Meissner effect, it is difficult to ascertain the presence of the jump. Within a few degrees of T_c the specific heat data are strongly influenced by fluctuations and whether the jump is real or not often depends on how the fluctuation contribution is modelled. There seems to be a consensus that in YBCO

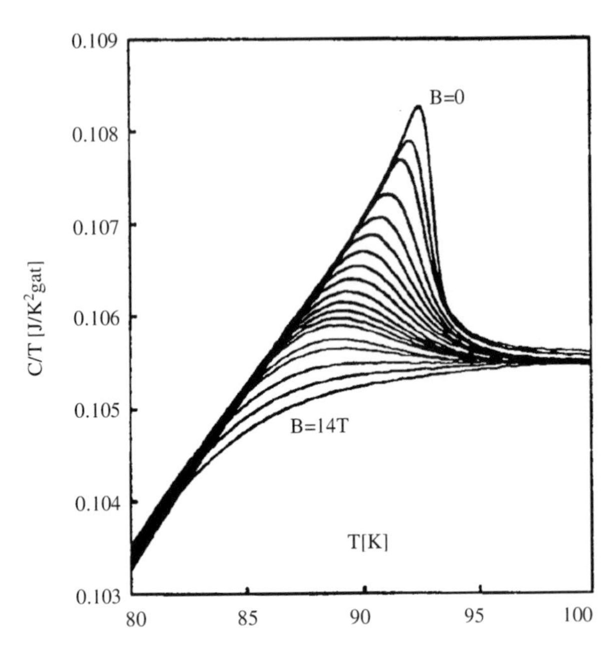

Figure A2.4.5. Specific heat plotted as C/T versus T for a range of magnetic fields oriented parallel to the c-axis of a $YBa_2Cu_3O_{6.9}$ crystal. From top to bottom the fields are: $B = 0, 0.25, 0.5, 1, 1.5, 2, 2.5, 3, 3.5, 4, 4.5, 5, 5.5, 6, 7, 8, 10$, 12 and 14 T. From [14].

a small jump exists (figure A2.4.5) (possibly associated with the long range three-dimensional nature of superconductivity), but that it is missing in much more anisotropic BSCCO.

Very close to the transition temperature, critical fluctuations dominate and they lead to a near logarithmic divergence of the specific heat. Further from the critical temperature the fluctuations are weaker and the Gaussian approximation is valid. Moreover, in highly anisotropic structures such as BSCCO, the fluctuations may undergo a dimensional crossover from two-dimensional (far from T_c) to three-dimensional (nearer T_c). Appropriate forms of two- and three-dimensional Gaussian fluctuations can describe and model this crossover.

There are three principal methods to measure specific heat: adiabatic calorimetry, modulation methods and relaxation calorimetry. A detailed description of the various techniques of measuring specific heat of solids can be found in [15–17]. Experimental issues pertaining to measurements of heat capacity of superconductors are discussed in chapter D3.2.1.

D3.2.1

A2.4.4 Thermal conductivity

One of the important properties of solids is their ability to conduct heat. How well a given material conducts heat is clearly of great technological interest. Beyond its technological relevance, knowledge of the nature of the entities that facilitate heat transport and how they depend on temperature and on any deliberate modification of the crystal structure are important issues that provide insight into the electronic and vibrational properties of materials. The physical parameter that indicates how efficiently a material conducts heat is called thermal conductivity, κ. According to Fourier's law, the thermal gradient ∇T imposed across an isotropic sample of a cross-sectional area A results in a heat flow $\dot{Q}$ given by

$$\dot{Q} = -\kappa \; A \; \nabla T. \tag{A2.4.16}$$

The negative sign implies that the heat flows down the thermal gradient, i.e. from the warmer to the colder end of the sample. For small temperature gradients, one can replace ∇T by $-\Delta T/L$, where $\Delta T = T_h - T_c$ is the temperature difference between two points separated by a distance L along the sample. Equation (A2.4.16) can then be written as

$$\kappa = \frac{\dot{Q}L}{A\Delta T} \tag{A2.4.17}$$

which is a convenient form for obtaining experimental values of thermal conductivity based on the longitudinal steady-state method—perhaps the most frequently used technique for measuring thermal conductivity illustrated in figure A2.4.6. The heat flow $\dot{Q}$, in this case, is generated by dissipating electrical power in a small heater attached to the free end of the sample.

D3.2.1

Heat in a solid is carried by two principal entities: free carriers responsible for the so-called electronic contribution κ_e, and lattice vibrations (phonons) that contribute the term κ_p. To a first approximation, the two contributions act as independent heat-conducting channels and the total thermal conductivity is given as

$$\kappa = \kappa_e + \kappa_p. \tag{A2.4.18}$$

While the phonon contribution, κ_p, is present in all solids, the electronic contribution varies depending on the type of material under consideration. In insulators there are no free carriers and κ_e is zero. On the other hand, free carriers abound in metals and κ_e dominates their heat transport. In semiconductors the carrier density depends on doping and so does the relative importance of the electronic term κ_e. Conventional superconductors are metals or alloys and, above their transition temperature, T_c, heat transport is dominated by the electronic term κ_e. Phonons represent only a small contribution, which increases as the metal becomes less pure and more disordered. HTS have

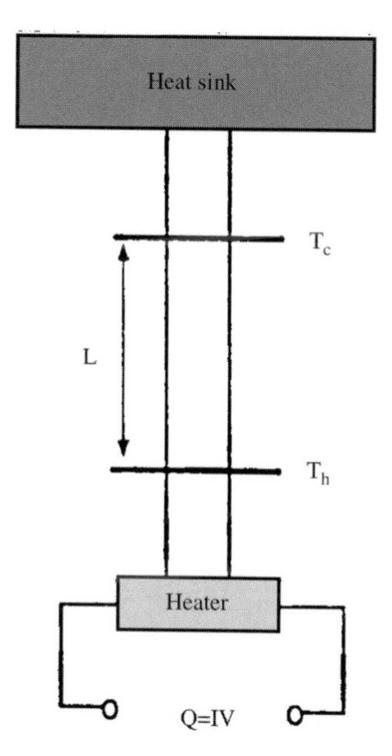

Figure A2.4.6. Experimental set-up to measure thermal conductivity using the longitudinal steady-state method.

a significantly reduced carrier density in comparison to typical metals and, as a consequence, phonons become the more important heat conducting channel. For instance, in bulk sintered cuprates, phonons account for 90–95% of the total thermal conductivity and, even in the best single crystals available, phonons carry at least half of all the heat in the normal state.

Each heat conducting channel—free carriers and phonons—is subject to relaxation mechanisms (scattering) which ensure the stationary nature of the heat conducting process. For instance, charge carriers are scattered by phonons, yielding the thermal resistivity contribution W_{e-p}, by defects, giving rise to the thermal resistivity W_{e-d} and by other charge carriers, resulting in the thermal resistivity W_{e-e}. Phonons can scatter on free carriers, W_{p-e}, on defects, W_{p-d} and in interactions with other phonons, W_{p-p}. According to Mathiessens's rule, the scattering processes within each heat conducting channel are additive, i.e.

$$W_e = 1/\kappa_e = W_{e-p} + W_{e-d} + W_{e-e} \tag{A2.4.19}$$

$$W_p = 1/\kappa_p = W_{p-e} + W_{p-d} + W_{p-p}. \tag{A2.4.20}$$

Depending on the carrier density, the density of defects and the temperature range, the magnitude and the temperature dependence of each term in equations (A2.4.19) and (A2.4.20) can be evaluated. For instance, a normal metal ($\kappa \approx \kappa_e$) can be modelled by the expression

$$W_e \equiv 1/\kappa_e = a\ T^2 + b/T \tag{A2.4.21}$$

where the first term on the RHS stands for the carrier scattering by phonons and the second term represents the interaction of carriers with static lattice defects. The carrier–carrier interaction is usually small and is neglected in equation (A2.4.21).

Since the charge carriers transport not only the charge but also heat (the excess of the thermal energy given by the Fermi distribution function), it is not surprising that the electronic thermal conductivity, κ_e, is related to the electrical conductivity, σ, or to its inverse, the electrical resistivity, ρ. This interdependence is known as the Wiedemann–Franz law,

$$\kappa_e = \sigma L_o T = L_o T/\rho. \tag{A2.4.22}$$

Here L_o is the Lorenz number equal to $2.45 \times 10^{-8}\ V^2\ K^{-2}$. The Wiedemann–Franz law is valid provided the non-equilibrium carrier distributions generated by the electric field and by the thermal gradient relax to the state of thermal equilibrium at the same rate. This happens when the carrier scattering is elastic, i.e. when the carriers neither lose nor gain energy while being scattered. In that case, equation (A2.4.22) offers a convenient way to determine the electronic thermal conductivity from a (much simpler) measurement of the electrical resistivity.

A2.1 In the normal state, superconductors do not display any unusual features in their heat transport behaviour and, in general, conform to the description appropriate for metals (in the case of conventional superconductors) or for more-or-less insulating solids (in the case of HTS). As the temperature is lowered below T_c, the formation of the Cooper condensate gives rise to a drastic modification of the heat A3.2 flow pattern. Three properties of the condensate provide the overriding influence. (a) Cooper pairs carry no entropy and therefore the usual electronic thermal conductivity should vanish rapidly below T_c. (b) Cooper pairs do not scatter phonons, which means that the phonon mean free path may increase as the sample is cooled below T_c. (c) Electrons may be excited from the condensate into quasiparticle states, and this 'normal gas' of particles, together with phonons, can carry heat below T_c. Which one of the three properties dominates depends on the type of superconductor. On account of their large charge carrier density, a vast majority of conventional superconductors display a sharp drop in their thermal conductivity below T_c. In fact, in pure elemental superconductors such as Sn or Zn, the thermal conductivity below T_c may decrease by three to four orders of magnitude in comparison to its normal state value, as illustrated in figure A2.4.7. Such pure superconductors can be used as heat switches. The normal state thermal conductivity below T_c is achieved with the aid of a small magnetic field just large enough to destroy superconductivity. Less pure conventional superconductors show a smaller suppression of the thermal conductivity, and in the extreme but rare case of certain alloys, e.g. Pb–Bi, one may even observe a small rise below T_c. This is due to a large contribution associated with the phonon thermal conductivity that is further enhanced below T_c as the condensing electrons can no longer limit the phonon mean free path. Theoretical treatments developed by Bardeen, Rickayzen and Tewordt, the so-called BRT-theory [19], and by Geilikman and Kresin [20] are appropriate for the description of the thermal conductivity of conventional superconductors below T_c.

A2.3

Thermal conductivity studies on HTS have proved to be an exciting endeavour and, even more than fifteen years after the discovery of HTS, the measurements continue to generate fascinating new results. Because of HTS' superconducting transition temperatures in excess of 100 K, the study of the normal state properties—the essential component of a complete picture of the physical world of these materials—poses a challenge. The usual transport properties such as the electrical resistivity, Hall effect, and thermopower are ineffective below a T_c, and in HTS this represents a wide temperature range. In contrast, no limitations are imposed on thermal conductivity measurements.

D3.1

While the normal state thermal conductivity of HTS is low (somewhat less than the value for stainless steel) and only weakly temperature dependent, the behaviour of thermal conductivity in the superconducting state is striking, as shown in figure A2.4.8. A dramatic enhancement seen below T_c leads to a prominent peak near $T_c/2$ and a subsequent rapid decrease at the lowest temperatures. These features have initially been analysed and explained in terms of the rising phonon mean free path, i.e. in the context of the BRT-theory. As more insight has been gained into the exotic world of HTS, it has become clear that the relaxation time of quasiparticles is dramatically enhanced below T_c. The unusually long quasiparticle lifetime then provides an alternative explanation for the rising thermal conductivity

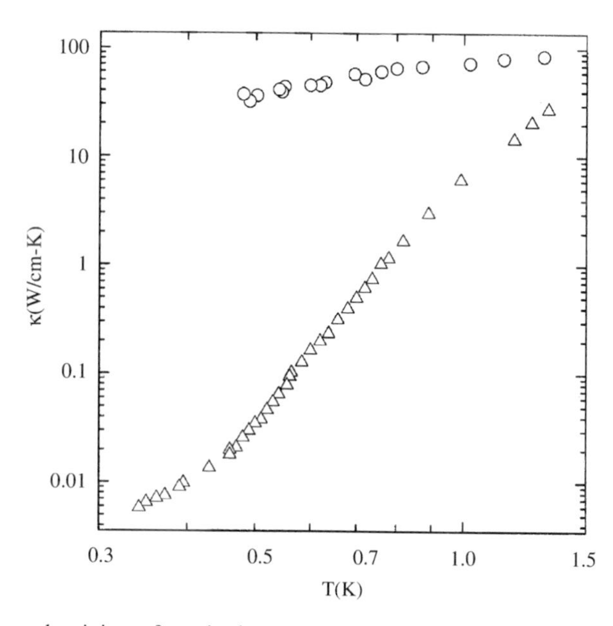

Figure A2.4.7. Thermal conductivity of a single crystal of Sn in the superconducting state (triangles) and in alongitudinal magnetic field of 500 Oe that is strong enough to drive the crystal into the normal state. Adapted from [18].

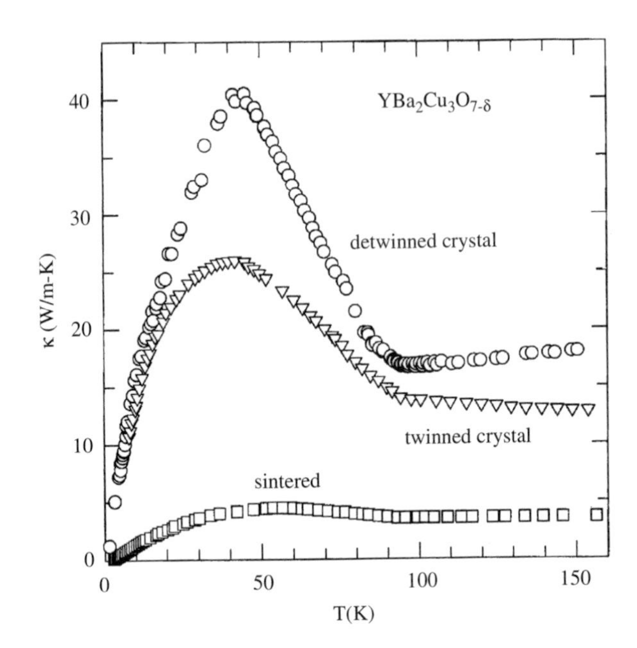

Figure A2.4.8. Thermal conductivity of $YBa_2Cu_3O_{7-\delta}$. Circles are the data for a detwinned crystal, triangles represent the data of a twinned crystal, and squares are the typical data for a sintered material. Adapted from [21].

below T_c, [22]. The two competing viewpoints—one favouring phonons and the other quasiparticles—have both yielded convincing fits to the experimental data, and both seem equally plausible. The difficulty in making an unambiguous choice between the two competing interpretations rests on the fact that the charge carriers and phonons both make significant contributions to the heat transport—in single crystals the contribution of the charge carriers in the normal state may be as high as 30% of the total thermal conductivity. Furthermore, below T_c, both phonons and quasiparticles have rather similar behaviours, leading to an eventual peak in the thermal conductivity near $T_c/2$. An attempt to resolve the issue was made by Krishana *et al.* [23]. In this case the thermal conductivity was measured in both the longitudinal (the usual configuration for measuring thermal conductivity) and the transverse (the Righi-Leduc configuration) directions in the presence of a mutually perpendicular magnetic field. Magnetic vortices introduce asymmetry (handedness) in the scattering of quasiparticles while phonons scatter symmetrically. Consequently, the transverse thermal conductivity κ_{xy} (the Righi-Leduc term) is given entirely by quasiparticles without any phonon background. From the ratio of the transverse and longitudinal thermal conductivities it is possible to estimate individual contributions due to quasiparticles and phonons. The experiment indicated that about one half of the heat is carried by quasiparticles and the other half by phonons.

At very low temperatures, $T \ll T_c$, the quasiparticle excitations should cease and the only mode of heat transport should be via phonons with the thermal conductivity $\kappa \propto T^3$ on account of crystal boundary scattering. While this is the case with insulating cuprates such as $YBa_2Cu_3O_6$, the superconducting forms of the structure, in particular $YBa_2Cu_3O_{7-\delta}$ and $La_{2-x}Sr_xCuO_4$, have consistently displayed a T-linear limiting dependence, figure A2.4.9. This suggests the functional form

$$\kappa = aT + bT^3 \tag{A2.4.23}$$

is representative of the behaviour at the lowest temperatures. There seems to be a correlation here with the T-linear term (the γ-term) often seen in the specific heat of HTS, as discussed in section A2.4.3.

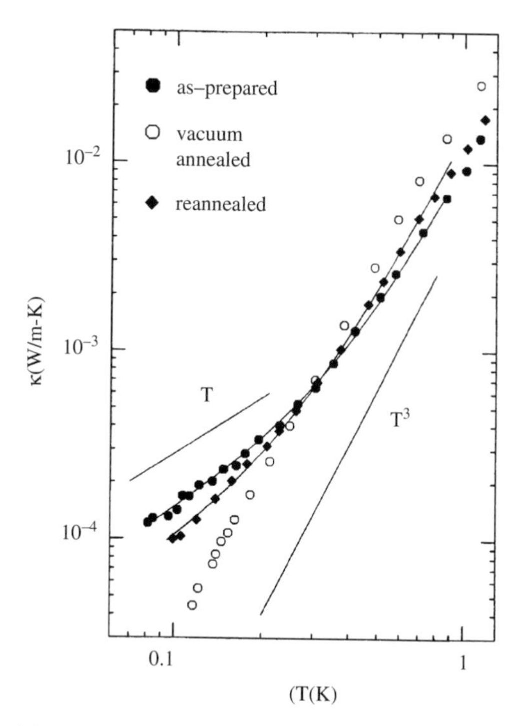

Figure A2.4.9. Thermal conductivity of sintered $YBa_2Cu_3O_{7-\delta}$ in the as-prepared state (superconductor); following a vacuum anneal that reduces the amount of oxygen and makes the sample insulating; and after reannealing in oxygen to return the sample to its superconducting state. The lines through the data are fits to equation (A2.4.23). From [24].

It should be noted, however, that the T-linear term is not present in all HTS; for instance, Bi-based cuprates display a robust T^2-dependence below 2 K regardless of their structural form.

On account of their substantially layered structure, thermal transport in HTS is highly anisotropic, with the in-plane thermal conductivity exceeding the c-axis thermal conductivity by an order of magnitude. Moreover, high accuracy measurements on detwinned single crystals of $YBa_2Cu_3O_{7-\delta}$ show a substantial in-plane anisotropy in the superconducting state, wherein the thermal conductivity measured along the b-axis is almost 30% higher than the thermal conductivity measured in the a-direction. Higher values of κ_b indicate a contribution of the Cu-O chains towards the thermal transport.

Thermal conductivity measurements on HTS have yielded an unusually rich set of data and continue to provide an important insight into the physical mechanisms responsible for the striking properties of these materials. A detailed account of thermal conductivity studies on a variety of families of HTS, including effects of magnetic fields, is given by Uher [25]. Experimental techniques employed to measure the thermal conductivity of superconducting samples are described in chapter D3.2.2.

D3.2.2

A2.4.5. Thermal expansion

Although often relegated to the status of a footnote in the text on solid-state theory, thermal expansion is of enormous practical importance in the design of instruments and tools expected to operate in an environment subjected to large temperature variations. For instance, a special near-zero thermal expansion glass is very desirable for large scale optical instruments such as primary telescope mirrors.

D3.2.3

On the other hand, a picture of buckled train rails is perhaps the most vivid example of what can happen when one ignores the fact that solids do change their dimensions when exposed to large temperature swings. What causes thermal expansion?

Larger amplitude of vibrations, i.e. greater displacement of atoms from their equilibrium positions as the solid absorbs thermal energy, is commonly viewed as the origin of thermal expansion. While substantially correct, this statement should be sharpened in the following sense. In the elementary treatment of lattice vibrations, one assumes that each atom or ion is confined in a potential well that has the shape of a parabola. In other words, the potential energy is a quadratic function of atomic displacements from the equilibrium position. Such a description is referred to as the harmonic approximation. In this model, the motion of atoms or ions is symmetrical with respect to their equilibrium positions, albeit with increasingly larger displacements as the temperature increases. Thus, although the atoms vibrate more vigorously, the time average of their motion does not deviate from their equilibrium positions. Hence, within the strictly harmonic approximation, there is no thermal expansion. What give rise to the thermal expansion of a lattice are the anharmonic interactions that are described by higher order terms (e.g. cubic) in the potential function. Atoms vibrating in such an anharmonic (i.e., asymmetric) potential well spend on average more time at greater separations and thus the lattice expands.

To quantify the effect of thermal expansion, one introduces the volume coefficient of thermal expansion β, defined as

$$\beta = \frac{1}{V}\left(\frac{\partial V}{\partial T}\right)_P \equiv \left(\frac{\partial \ln V}{\partial T}\right)_P. \tag{A2.4.24}$$

Here V is the volume, T the absolute temperature and the subscript P reminds us that the derivative is taken at constant pressure. For cubic and isotropic solids, all three dimensions L_i, $i = 1-3$ expand equally and one can write

$$\beta = 3\alpha = 3\left(\frac{\partial \ln L}{\partial T}\right)_P \tag{A2.4.25}$$

where α is the coefficient of linear expansion, and L is the length. Typical values of α at ambient temperatures are of the order of $10^{-5}\,\mathrm{K}^{-1}$.

Expansion of crystals with the axial symmetry (hexagonal, tetragonal and rhombohedral) is specified by two principal coefficients of linear expansion:
$\alpha_\perp = (\partial \ln a/\partial T)_P$ and $\alpha_\parallel = (\partial \ln c/\partial T)_P$, where a and c are lengths perpendicular and parallel to the main crystal axis. The volume coefficient of thermal expansion for this class of solids is then

$$\beta = 2\alpha_\perp + \alpha_\parallel. \tag{A2.4.26}$$

The expansion of solids with still lower symmetry is described by three principal coefficients of linear expansion.

The dimensional change a given solid undergoes reflects the fact that the system is trying to minimize its free energy. A thermodynamic description of thermal expansion is most conveniently done in terms of the Helmholtz free energy $F(V,T)$ which is an explicit function of volume and temperature. Using standard thermodynamic transformations,

D2.1

$$\beta = \left(\frac{\partial \ln V}{\partial T}\right)_P = -\left(\frac{\partial \ln V}{\partial P}\right)_T\left(\frac{\partial P}{\partial T}\right)_V = \chi_T\left(\frac{\partial P}{\partial T}\right)_V = -\chi_T\frac{\partial^2 F}{\partial V \partial T} = \chi_T\left(\frac{\partial S}{\partial V}\right)_T \tag{A2.4.27}$$

one can relate the volume expansion coefficient β to the isothermal compressibility χ_T and the volume derivative of the entropy S. As a rule of thumb, soft materials, i.e. materials with large compressibility, have a high thermal expansion coefficient. Since χ_T is invariably positive, the sign of β, i.e. whether the solid expands or contracts, is determined by the functional dependence of entropy on volume. For most solids, the entropy (the degree of disorder) increases with the increasing volume and β is positive, i.e. the lattice expands. Occasionally, the volume derivative of the entropy is negative, leading to lattice contraction. This may happen when transverse rather than longitudinal modes dominate the vibration spectrum. The situation is vaguely analogous to the behaviour of a guitar string [26] which, when stretched, exerts an attractive force between the two points where it is attached, and thus tends to bring them closer together. Such 'cross-contraction' effects [27] influence low temperature thermal expansion of many diamond-structure crystals that contract on warming over at least some range of temperatures.

During the first two decades of the 20th century, as the data on the specific heat and thermal expansion of solids became more widely available, Grüneisen noticed that the thermal expansion coefficient at ambient and at very low temperatures mimics the behaviour of the specific heat. More precisely, both tend to be constant at ambient and higher temperatures while their low temperature behaviour is proportional to T^3. The dimensionless form of the ratio of the coefficients of expansion and specific heat,

$$\gamma^G = \frac{\beta V}{C_P \chi_s} = \frac{\beta V}{C_V \chi_T} \approx \text{constant} \tag{A2.4.28}$$

has since been known as the Grüneisen parameter γ^G. The superscript G is used here to distinguish the Grüneisen parameter from the coefficient of specific heat introduced in section A2.4.3. C_P and C_V appearing in equation (A2.4.28) are the molar specific heats at constant pressure and constant volume, respectively, χ_S is the adiabatic compressibility, and V is the molar volume. Experiments by Grüneisen and others have shown that γ^G is typically between one and three, and is nearly temperature independent for a given solid.

So far, we have considered only lattice vibrations as a contributing factor towards the thermal expansion of solids. What happens in the presence of free charge carriers, i.e. how do metals expand? Since the thermal expansion coefficient depends on the entropy, any entity that contributes to the entropy of a solid is likely to exert influence, via equation (A2.4.27), on its thermal expansion.

In section A2.4.3 we noted that, due to the Fermi–Dirac statistics, the electronic contribution to the specific heat at ambient temperatures is negligible. We also pointed out that at low temperatures the electronic specific heat is a linear function of temperature and thus, at low enough temperatures, it will dominate the cubic term arising from the lattice specific heat. Likewise, the coefficient of thermal expansion of non-magnetic** metals can be written as consisting of two terms, one due to the lattice and the other due to electrons,

$$\beta = \beta_e + \beta_L = \chi_T \left(\frac{C_e \gamma_e^G + C_L \gamma_L^G}{V} \right) = aT + bT^3 \qquad (T \ll \theta_D). \tag{A2.4.29}$$

Evidently, the electronic contribution to thermal expansion is linear in temperature and should be the dominant term at low enough temperatures. In analogy with equation (A2.4.28), the electronic Grüneisen parameter γ_e^G is defined as $\gamma_e^G = \beta_e V / \chi_T C_e$, and it can be shown that it reflects the volume derivative of the density of states at the Fermi energy,

**Expansion in magnetic metals is frequently affected by a large magnetostriction contribution that arises from the strain dependence of sample magnetization. Furthermore, elementary excitations such as spin waves (and other excitations that contribute to the entropy) may influence the thermal expansion of magnetic solids.

$$\gamma_{\mathrm{e}}^{\mathrm{G}} = \left[\frac{\partial \ln N(E)}{\partial \ln V}\right]_{T,E=E_{\mathrm{F}}}. \tag{A2.4.30}$$

For free electrons, on account of the density of states $N(E_{\mathrm{F}})$ being proportional to $V^{2/3}$, the electronic Grüneisen parameter $\gamma_{\mathrm{e}}^{\mathrm{G}} = 2/3$. In real metals, complications due to specific band structure effects result in deviations from this value.

Considering thermal expansion in superconductors, we recall from section A2.4.3 that at the superconducting transition temperature T_{c}, the conduction electrons form pairs and undergo condensation while the lattice properties are left essentially intact. Since at T_{c} one observes a small jump in the magnitude of the specific heat, one also expects a small change in the thermal expansion coefficient as the sample is cooled through the transition temperature. In the presence of a magnetic field, the transition is classified as a first-order transition, and there will be differences in volume between the normal (n) and superconducting (s) phases. According to Shoenberg [28], the volume difference can be written as

$$V_{\mathrm{n}} - V_{\mathrm{s}}(0) = V_{\mathrm{s}}\frac{H_{\mathrm{c}}}{4\pi}\left(\frac{\partial H_{\mathrm{c}}}{\partial P}\right)_{T}. \tag{A2.4.31}$$

By differentiating equation (A2.4.31) with respect to temperature and considering the result at $H_{\mathrm{c}} = 0$, one obtains the difference between the expansion coefficients:

$$\beta_{\mathrm{n}} - \beta_{\mathrm{s}} = \frac{1}{4\pi}\left(\frac{\partial H_{\mathrm{c}}}{\partial T}\right)_{P}\left(\frac{\partial H_{\mathrm{c}}}{\partial P}\right)_{T}. \tag{A2.4.32}$$

Setting also $T = T_{\mathrm{c}}$ (i.e. considering a second-order transition), one can derive, after some manipulations, the Ehrenfest equation

$$\beta_{\mathrm{n}} - \beta_{\mathrm{s}} = \frac{C_{\mathrm{n}} - C_{\mathrm{s}}}{V}\frac{\mathrm{d}\ln T_{\mathrm{c}}}{\mathrm{d}P}. \tag{A2.4.33}$$

Equation (A2.4.33) can be written as

$$\frac{\mathrm{d}T_{\mathrm{c}}}{\mathrm{d}P} = VT_{\mathrm{c}}\frac{\beta_{\mathrm{n}} - \beta_{\mathrm{s}}}{C_{\mathrm{n}} - C_{\mathrm{s}}} \tag{A2.4.34}$$

where V is the molar volume. Equation (A2.4.34) provides an alternative route to direct pressure measurements of a very important superconducting parameter—the pressure dependence of the transition temperature.

The above formulae apply strictly to type-I superconductors. For type-II materials, there is no discontinuity in length at either the lower or upper critical fields, but only a change in the slope of $L(H)$, for details see [29]. As an example of thermal expansion in conventional superconductors, figure A2.4.10 shows the data for niobium, tantalum and vanadium as each material is cooled through its respective transition temperature. For niobium and tantalum $\Delta\beta \equiv \beta_{\mathrm{n}} - \beta_{\mathrm{s}}$ is positive, while for vanadium and many type-II superconductors, $\Delta\beta$ is negative. The behaviour of conventional superconductors that are used in high-field magnet applications (multifilamentary wires of Nb–Ti, Nb_3Sn and V_3Ga) is described by Clark *et al.* [31].

E1.1

Thermal expansion in high-T_{c} superconductors has attracted much interest. The early studies focussed mostly on sintered samples and suggested that thermal expansion of high-T_{c} perovskites is similar to those of non-superconducting ceramics. These early results were reviewed by White [32].

Availability of single crystals, particularly the untwinned forms of $YBa_2Cu_3O_{7-\delta}$, has opened the road for the study of structural anisotropy and provided insight into the interplay between the structural

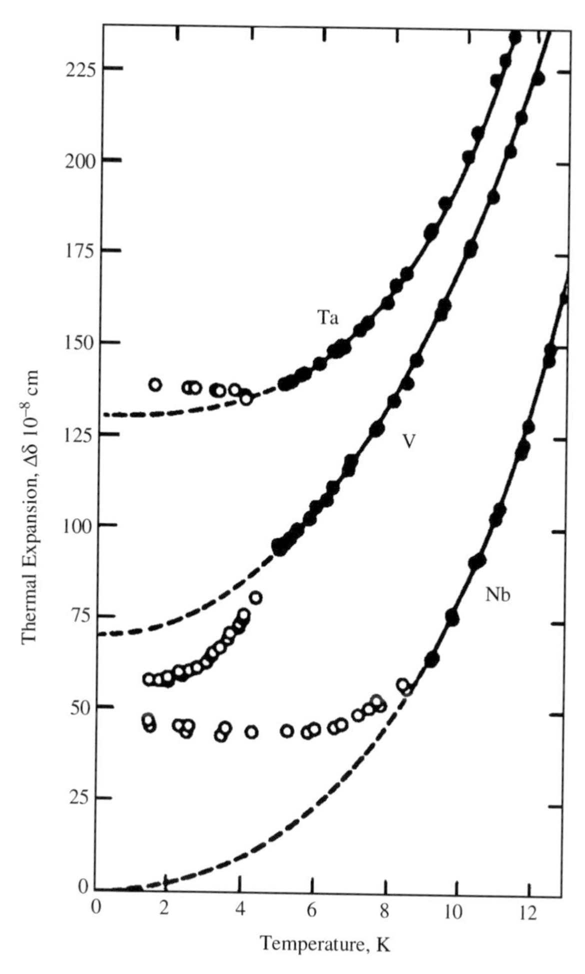

Figure A2.4.10. Linear thermal expansivities in pure Nb, Ta and V rods in normal (filled circles) and superconducting (open circles) states. Broken lines represent the normal state below T_c. From the data of White [30].

and superconducting parameters. Due to their layered structure, crystals of high T_c perovskites are rather delicate and it is a distinct advantage to have available a thermodynamic assessment of the pressure dependence (or uniaxial stress dependence) of T_c, via the Ehrenfest equation, rather than risk sample damage by directly applying external pressure. Since $YBa_2Cu_3O_{7-\delta}$ has an orthorhombic structure, one needs to measure linear coefficients of thermal expansion along all three crystallographic axes: the in-plane *a*- and *b*-axes, and in the *c*-direction perpendicular to the CuO_2 planes. An example of such measurement is the data of Meingast *et al.* [33] shown in figure A2.4.11. It is clear from figure A2.4.11(*a*) that there is considerable anisotropy in the thermal expansion of detwinned $YBa_2Cu_3O_{7-\delta}$ with the largest coefficient of thermal expansion being along the *c*-axis and the smallest along the *b*-axis—in the direction of the CuO chains. Figure A2.4.11(*b*) shows an expanded view of the change in the expansion coefficient $\Delta\alpha$, defined here as $\Delta\alpha = \alpha_s - \alpha_n$, near the transition temperature for each axis. While no jump is observed in the *c*-direction, the jumps along the *a*-axis and along the *b*-axis are comparable but have opposite signs, $\Delta\alpha^a \approx -\Delta\alpha^b$. Figure A2.4.11(*b*) also depicts significant deviations from the mean-field behaviour (solid lines) that are likely to be due to a contribution associated with

C1

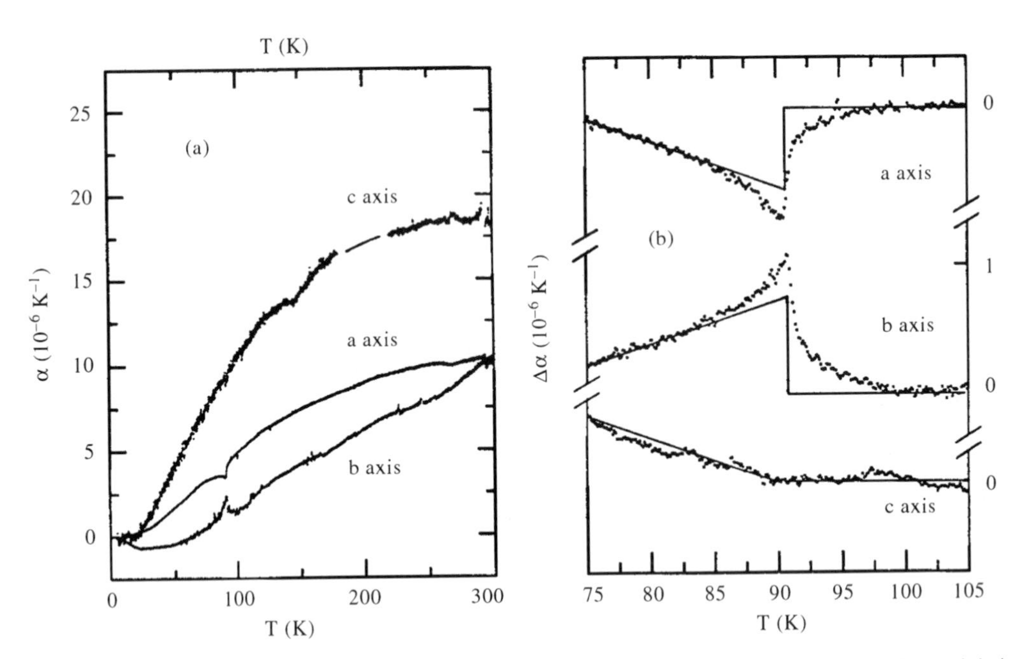

Figure A2.4.11. Linear coefficients of thermal expansion for a detwinned crystal of $YBa_2Cu_3O_{7-\delta}$. (*a*) An overall behaviour along the three principal axes. (*b*) An expanded view showing the behaviour of $\Delta\alpha$ near T_c. Notice the opposite signs of $\Delta\alpha$ for the *a*-axis and *b*-axis, and no jump at T_c for the *c*-axis. The data of Meingast *et al.* [33].

superconducting fluctuations. By combining the expansivity jumps with the jumps in specific heat, and using the Ehrenfest equation, the authors of [33] obtained the following values for the uniaxial pressure derivatives of the transition temperature in the detwinned sample of $YBa_2Cu_3O_{7-\delta}$:

D1.4

$$\frac{dT_c}{dP_a} = -1.9\,\mathrm{K}\,(\mathrm{GPa})^{-1}$$

$$\frac{dT_c}{dP_b} = +2.2\,\mathrm{K\,GPa}^{-1}$$

$$\frac{dT_c}{dP_c} \approx 0\,\mathrm{K\,GPa}^{1}.$$

The hydrostatic pressure dependence of the transition temperature follows from

$$\frac{dT_c}{dP} = \frac{dT_c}{dP_a} + \frac{dT_c}{dP_b} + \frac{dT_c}{dP_c} \approx 0.3\,\mathrm{K\,GPa}^{-1}.$$

This value is in good agreement with direct hydrostatic pressure measurements that yield $0 \leq dT_c/dP \leq 0.7\,\mathrm{K}\,(\mathrm{GPa})^{-1}$. A rather small value of the pressure derivative of T_c for the detwinned crystals of $YBa_2Cu_3O_{7-\delta}$ is the consequence of near-cancellation of two large terms with opposite signs that represent the *a*-axis and *b*-axis expansion.

Depending on the desired level of precision, several techniques are available for measuring the dimensional changes of solids. The detailed description of various techniques of measuring thermal expansion is given, e.g., by Yates [34], and a recently available book by Taylor [35] is an excellent reference source for those seeking to learn more about thermal expansion in solids. High resolution techniques of thermal expansion appropriate for the study of superconducting materials (including the effect of magnetic field) are discussed in chapter D3.2.3.

D3.2.3

References

[1] Bardeen J, Cooper L N and Schrieffer J R 1957 Theory of superconductivity *Phys. Rev.* **108** 1175

[2] Fröhlich H 1950 Theory of the superconducting state: I. The ground state at the absolute zero of temperature *Phys. Rev.* **79** 845

[3] Cooper L N 1956 Bound electron pairs in a degenerate Fermi gas *Phys. Rev.* **104** 1189

[4] Leggett A 1975 A theoretical description of new phases of liquid ^{3}He *Rev. Mod. Phys.* **47** 331

[5] Giaever I 1960 Energy gap in superconductors measured by electron tunneling *Phys. Rev. Lett.* **5** 147

[6] Van Duzer T and Turner C W 1981 *Principles of Superconducting Devices and Circuits* (New York: Elsevier) p 36–91

[7] Phillips N E 1959 Heat capacity of aluminum between 0.1 K and 4 K *Phys. Rev.* **114** 676

[8] Muhlschlegel B 1959 Die thermodynamischen funktionen des supraleiters *Z. Phys.* **155** 313

[9] Triscone G and Junod A 1996 Thermal and magnetic properties *Bismuth-Based High-Temperature Superconductors*, ed H Maeda and K Tagano (New York: Marcel Dekker) p 33–74

[10] Junod A 1989 Specific heat of high temperature superconduction; a review *Physical Properties of High Temperature Superconductors* **vol. 2**, ed D M Ginsberg (Singapore: World Scientific) p 13–120

[11] Phillips N E, Fisher R A and Gordon J E 1992 The specific heat of high-T_c superconductors *Prog. Low-Temp. Phys.* **13** 267

[12] Srinivasan R 1995 Specific heat studies of superconductivity *Superconductivity*, ed P N Butcher and Y Lu (Singapore: World Scientific) p 241–278

[13] Junod A 1996 Specific heat of high temperature superconductors in high magnetic fields *Studies of High-Temperature Superconductors* **19**, ed V A Narlikar (New York: Nova Science) p 1–68

[14] Roulin M, Junod A and Walker E 1988 Observation of second-order transitions below T_c in the specific heat of $YBa_2Cu_3O_x$ *Physica C* **296** 137

[15] Barron T H K and White G K 1999 *Heat Capacity and Thermal Expansion at Low Temperatures* (New York: Kluwer Academic/Plenum)

[16] Gmelin E 1987 Low temperature calorimeter: a particular branch of thermal analysis *Thermochimica Acta* **110** 183

[17] Ho C Y and Cezairliyan A 1988 *Specific Heat of Solids* (New York: Hemisphere)

[18] Peshkov V P and Parshin A Y a 1965 The efficiency of superconducting thermal switches *Proc. 9th Int. Conf. Low Temperature Physics* (London: Plenum) p 517–520

[19] Bardeen J, Rickayzen G and Tewordt L 1959 Theory of the thermal conductivity of superconductors *Phys. Rev.* **113** 982

[20] Geilikman B T and Kresin V Z 1958 Phonon thermal conductivity of superconductors *Sov. Phys. – Dokl.* **3** 116

[21] Uher C, Liu Y and Whitaker J F 1994 The peak in the thermal conductivity of Cu-O superconductors: electronic or phononic origin? *Can. J. Superconductivity* **7** 323

[22] Yu R C, Salamon M B, Lu J P and Lee W C 1992 Thermal conductivity of an untwinned $YBa_2Cu_3O_{7-\delta}$ single crystal and a new interpretation of the superconducting state thermal transport *Phys. Rev. Lett.* **69** 1431

[23] Krishana K, Harris J M and Ong N P 1995 Quasiparticle mean-free path in $YBa_2Cu_3O_7$ measured by thermal Hall conductivity *Phys. Rev. Lett.* **75** 3529

[24] Cohn J L, Peacor S D and Uher C 1988 Thermal conductivity of $YBa_2Cu_3O_{7-\delta}$ below 1 K: evidence for normal-carrier transport well below T_c *Phys. Rev. B* **38** 2892

[25] Uher C 1992 Thermal conductivity of high-temperature superconductors *Physical Properties of High Temperature Superconductors* **vol. 3**, ed D M Ginsberg (Singapore: World Scientific) pp 159–284

[26] Barron T H K, Collins J G and White G K 1980 Thermal expansion of solids at low temperatures *Adv. Phys.* **29** 609

[27] White G K 1998 Thermal expansion *Handbook of Applied Superconductivity* **vol. 1**, ed B Seeber (Bristol: Institute of Physics Publishing) p 1107–1119

[28] Shoenberg D 1952 *Superconductivity* Ch. 3. (Cambridge: Cambridge University Press)

[29] Hake R R 1969 Thermodynamics of type-I and type-II superconductors *J. Appl. Phys.* **40** 5148

[30] White G K 1962 Thermal expansion of vanadium, niobium and tantalum at low temperatures *Cryogenics* **2** 292

[31] Clark A F, Fujii G and Ranney M A 1981 The thermal expansion of several materials for superconducting magnets *IEEE Trans. Magn.* **MAG-17** 2316

[32] White G K 1993 Thermal expansion and Grüneisen parameters of high T_c superconductors *Studies of High Temperature Superconductors* **vol. 9**, ed A Narlikar (New York: Nova) p 121–147

[33] Meingast C, Kraut O, Wolf T, Wuhl H, Erb A and Muller-Vogt G 1991 Large a–b anisotropy of the expansivity at T_c in untwinned $YBa_2Cu_3O_{7-\delta}$ *Phys. Rev. Lett.* **67** 1634

[34] Yates B 1972 *Thermal Expansion* (New York: Plenum) p 1–121

[35] Taylor R E 1998 *Thermal Expansion of Solids* (Materials Park, Ohio: ASM International)

A2.5
High frequency electromagnetic properties

Adrian Porch

A2.5.1 Introduction

Measurements of the electromagnetic response of superconductors in the microwave spectrum (for the purposes of this chapter defined to be in the range from 100 MHz to 500 GHz) and in the infrared spectrum (>500 GHz) allow the simultaneous investigation of the dynamics and energy states of both paired carriers and unpaired carriers (quasiparticles) in a superconductor. This allows information to be deduced regarding the nature of the superconducting pairing state, the energy gap, the dynamics of quasiparticles and the effects of defects and impurities [1–3]. High frequency measurements of conventional (i.e. low T_c) superconductors [4] provided the springboard for the BCS theory, and later confirmation of many of the theory's predictions [1–3]. Measurements in static applied magnetic fields [5] probe the dynamics of flux lines, yielding flux line viscosities, pinning strengths and information regarding quasiparticle states within the flux line cores.

A2.4 A3.2 D2.2, D2.4

While the electrodynamics of conventional superconductors are well understood, there are many open issues regarding HTS materials that warrant further investigation using high frequency techniques. Examples of these are the origins of the nonlinear effects at high microwave field levels (discussed in detail in chapter D2.7), of direct relevance to potential HTS microwave device applications (see section E3), and fundamental issues regarding the symmetry and mechanism of the pairing in HTS materials.

D2.7, E3 A1.2

A2.5.2 Surface impedance, complex conductivity and the two fluid model

The response of a superconducting sample at microwave frequencies is characterized by its *surface impedance*, which is a function of the electrical conductivity, and can be measured by a number of standard microwave cavity techniques (see chapter D2.7). At infrared frequencies, the conductivity is extracted directly from measurements of the power transmitted through very thin film samples or reflected from thicker samples [6, 7].

D2.7

When a high frequency electromagnetic wave falls on the surface of a good conductor, the electromagnetic fields are confined within a shallow *skin depth*, which for a superconductor is the magnetic penetration depth λ. If the conductor is flat on the scale of the skin depth, the *surface impedance* Z_s of the conductor is defined to be

$$Z_s = \frac{E}{H} = R_s + iX_s \qquad \text{(A2.5.1)}$$

where E and H are the magnitudes of the tangential electric and magnetic fields, respectively, at the surface. Physically, the *surface resistance* of the conductor R_s quantifies the rate of energy dissipation within the skin depth, whilst the *surface reactance* X_s quantifies the peak electromagnetic energy stored within the skin depth. Consequently, the measurement of these parameters is relevant to high frequency applications (see section E3).

E3

The general theory of the surface impedance of superconductors in low amplitude ac fields was developed by Mattis and Bardeen [8, 9] using the BCS theory. The full theory is complicated and beyond the scope of this handbook. However, in certain limits, simple results emerge depending on the relative sizes of the superconducting coherence length ξ, the quasiparticle mean free path l and λ; typical values of ξ and λ are listed in table A2.5.1.

A3.2 A1.2, A2.3

Of most relevance are the *local limits* (i.e. $l \ll \lambda$) of the Mattis–Bardeen theory, of which there are two. The *clean local limit* (often called the *London limit*) is attained when $\lambda \gg l \gg \xi$. In this limit, the supercurrent density $\mathbf{J}_s$ (i.e. that part of the current associated with paired carriers) at some point in the superconductor is related to the electric field $\mathbf{E}$ *at the same point* by the first London equation

A1.2

$$\frac{\partial \mathbf{J}_s}{\partial t} = \frac{1}{\mu_0 \lambda_L^2} \mathbf{E} \qquad \text{(A2.5.2)}$$

where the subscript 'L' is used to denote the London limit. From table 1 it can be seen that metals like Pb and Nb fall outside this limit and require a rigorous solution of the Mattis–Bardeen theory. However, HTS materials with *ab*-plane currents fall within the London limit and exhibit local electrodynamics; the same is also true for A-15 superconductors like Nb_3Sn. Consequently, the high frequency electrodynamics of both the main classes of superconducting materials for microwave applications (i.e. HTS and A-15) can be modelled quite simply using the first London equation. The other local limit is the *dirty local limit*, attained when $\lambda \gg l \approx \xi$; equation (A2.5.2) still applies, but with λ_L^2 replaced by $\lambda_{\text{dirty}}^2 = \lambda_L^2(1 + \xi/l)$. This limit is appropriate for alloy superconductors.

A2.3

The effect of the electric field within the penetration depth is to accelerate both quasiparticles and pairs, so that in either local limit the current density is the sum of the quasiparticle current and supercurrent densities (denoted $\mathbf{J}_n$ and $\mathbf{J}_s$, respectively). Applying equation (A2.5.2) for an incident electromagnetic wave of angular frequency ω gives

$$\mathbf{J} = \mathbf{J}_n + \mathbf{J}_s = (\sigma_1 - i\sigma_2)\mathbf{E} \qquad \text{(A2.5.3)}$$

Table A2.5.1. Typical values of the material parameters of a selection of superconducting materials. Mean free paths vary significantly with sample quality and are not quoted, but a large value of the ratio λ/ξ is usually sufficient to push the material into the local limit. The list is ordered so that those materials at the bottom of the table are furthest into the local limit

Material	$\lambda(0)$ (nm)	$\xi(0)$ (nm)	$2\Delta(0)/kT_c$	T_c (K)
Al	16	1500	3.4	1.2
In	25	400	3.5	3.3
Sn	28	300	3.6	3.7
Pb	28	110	4.1	7.2
Nb	32	39	3.7	9.2
Nb_3Sn	50	5	4.4	18
$YBa_2Cu_3O_7$ (*ab*-plane)	140	2	Unclear	93

where the conductivity of the superconductor is a complex, frequency and temperature dependent quantity

$$\sigma(\omega, T) = \sigma_1(\omega, T) - \mathrm{i}\sigma_2(\omega, T). \tag{A2.5.4}$$

The real part of (A2.5.4), σ_1, is associated with the dissipative response of the quasiparticles, and the imaginary part $\sigma_2 = 1/\omega\mu_0\lambda^2$ with the nondissipative response of the supercurrent. This description is called the *two fluid model* and is appropriate for local electrodynamics ($l < \lambda$).

A1.2

The surface impedance of any good conductor with a local conductivity σ can be derived using the *classical skin effect* theory

$$Z_s = R_s + \mathrm{i}X_s = \left(\frac{\mathrm{i}\omega\mu_0}{\sigma}\right)^{1/2}. \tag{A2.5.5}$$

Unsurprisingly, for a superconductor at microwave frequencies not too close to T_c, it is found that $\sigma_1 \ll \sigma_2$, in which case, expanding equation (A2.5.5) to the first order in σ_1/σ_2 yields the following results for the surface resistance and surface reactance, respectively,

$$R_s \approx \left(\frac{\mu_0\omega}{\sigma_2}\right)^{1/2}\frac{\sigma_1}{2\sigma_2} \equiv \frac{1}{2}\omega^2\mu_0^2\lambda^3\sigma_1, \quad X_s \approx \left(\frac{\mu_0\omega}{\sigma_2}\right)^{1/2} \equiv \omega\mu_0\lambda. \tag{A2.5.6}$$

Unlike a normal metal where $R_s \propto \omega^{1/2}$, for a superconductor $R_s \propto \omega^2$; this result has considerable importance for microwave applications, as discussed in section E3. Equation (A2.5.6) also states that a superconductor is only truly lossless (i.e. $R_s = 0$) at non-zero frequencies when the temperature is reduced to absolute zero, since then $\sigma_1 = 0$.

A2.5.3 The conventional (Mattis–Bardeen) picture of the high frequency conductivity

The conductivity $\sigma(\omega, T)$ is most easily calculated in the dirty local limit [8, 9], and some numerical results for σ_n/σ are shown in figure A2.5.1, where σ_n is the corresponding (real) conductivity of the normal metal at the same frequency and temperature.

The dirty limit conductivity ratios of figure A2.5.1 can be qualitatively understood in terms of the energy gap Δ in the quasiparticle density of states. The gap frequency is defined as $\omega_g = 2\Delta/\hbar$, of the order of 1 THz for conventional superconductors at low temperatures. Referring to figure A2.5.1(*a*), $\sigma_1 = 0$ when $T = 0$ for frequencies $\omega < \omega_g$ since there are no quasiparticles present under these conditions. However, when $\omega > \omega_g$, σ_1 increases rapidly even at $T = 0$, corresponding to an energy loss mechanism as a result of the incident infrared photons creating pairs of quasiparticles from the ground state. At very high frequencies, σ_1 approaches σ_n.

At low temperatures when $\omega < \omega_g$, the presence of the energy gap ensures that $\sigma_1 \propto \exp(-\Delta/kT)$ (figure A2.5.1(*b*)), and σ_1 exhibits a broad peak just below T_c as a result of two characteristics of the BCS theory: the enhanced density of states at the gap edge and the *coherence factors* associated with electromagnetic absorption; the latter feature is often called the *coherence peak*.

For frequencies well below ω_g, it is found that $\sigma_2 \propto 1/\omega$ for all limits of the Mattis–Bardeen theory, so that λ is approximately independent of frequency. In the normal state $\sigma_2 = 0$ and there is a rapid rise in σ_2 on entering the superconducting state which ensures that $\sigma_2 \gg \sigma_1$, apart from temperatures very close to T_c (figure A2.5.1(*c*)), and which causes a precipitous drop in surface resistance. The quantity $\sigma_2(0) - \sigma_2(T)$ is called the quasiparticle *backflow*, where $\sigma_2(0) - \sigma_2(T) \propto \exp(-\Delta/kT)$ at low temperatures, thus giving rise to $\lambda_L(T) - \lambda_L(0) \propto \exp(-\Delta/kT)$; this is closely linked to the low T form of $\sigma_1(T)$, when also $\sigma_1(T) \propto \exp(-\Delta/kT)$.

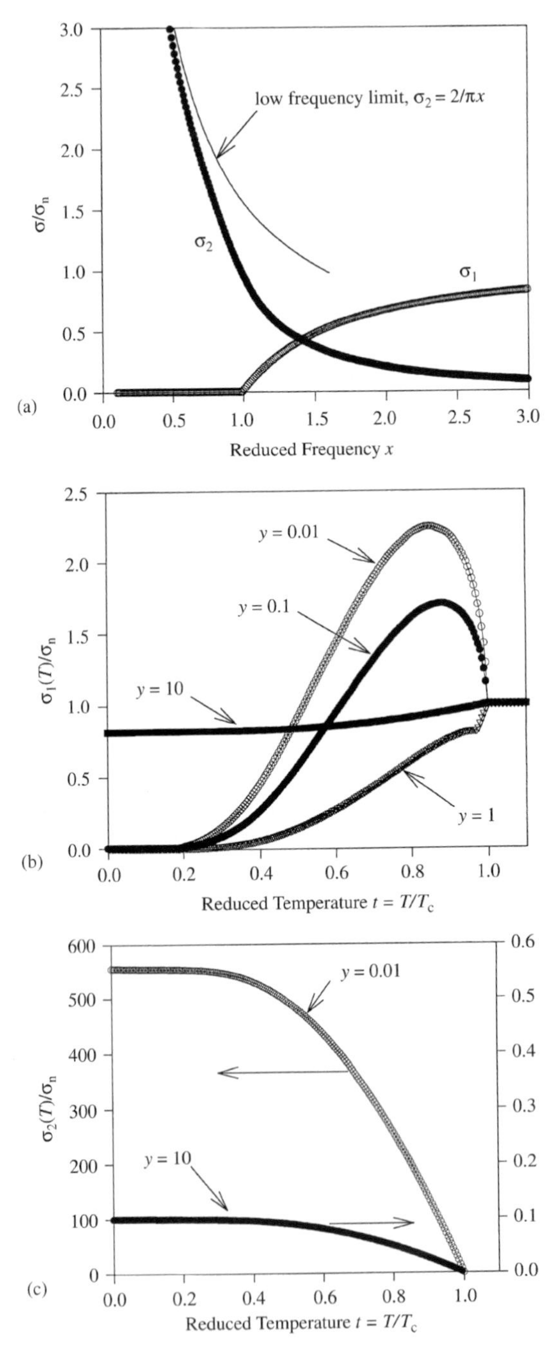

Figure A2.5.1. (*a*) The dirty limit conductivity ratios at $T = 0$ as a function of reduced frequency $x = \hbar\omega/2\Delta$. Not shown is the δ-function in σ_1 at zero frequency, which arises as a result of the presence of the energy gap Δ in the quasiparticle density of states, and which ensures that $\sigma_2 \propto 1/\omega$ at low frequencies. The solid line is the low frequency BCS prediction for σ_2. (*b*) The temperature dependence of σ_1 in the dirty limit for various frequencies $y = \hbar\omega/kT_c$ above and below the gap frequency (which has a BCS value of $y = 3.52$ at low T). (*c*) The temperature dependence of σ_2 in the dirty limit for various frequencies $y = \hbar\omega/kT_c$.

The conductivity in the London limit can be expressed in terms of the Mattis–Bardeen dirty limit conductivity ratios

$$\left(\frac{\sigma}{\sigma_{\rm n}}\right)_{\rm London} = \left(\frac{\sigma_1}{\sigma_{\rm n}}\right)_{\rm dirty} - {\rm i}\frac{\xi}{l}\left(\frac{\sigma_2}{\sigma_{\rm n}}\right)_{\rm dirty}. \qquad \text{(A2.5.7)}$$

Since $\sigma_{\rm n} \propto l$ in the London limit, σ_2 is independent of l, as we expect from equation (A2.5.2). The effect of increasing l (thus pushing the material from the dirty to the clean limit) is to suppress the coherence peak in σ_1, of relevance to HTS materials. However, the characteristic low temperature features of both σ_1 and σ_2 in the dirty limit are retained in the London limit, as are the characteristic frequency dependences.

In the *Pippard* limit ($l, \xi >> \lambda$) the electrodynamics are nonlocal (see chapter A2.3). The limiting result of the Mattis–Bardeen theory for the surface impedance is that calculated using the *anomalous skin effect* theory

A2.3

$$\frac{Z_{\rm s}}{Z_{\rm n}} = \left(\frac{\sigma}{\sigma_{\rm n}}\right)^{-1/3}_{\rm dirty} \qquad \text{(A2.5.8)}$$

where $Z_{\rm n} \propto ({\rm i}\omega\mu_0)^{2/3}(\sigma_{\rm n}/l)^{-1/3}$ is the surface impedance of the normal metal in the extreme anomalous limit (i.e. when l is much greater than the skin depth). A number of pure superconductors with very low $T_{\rm c}$ (e.g. Al) are in this limit.

The Mattis–Bardeen predictions for $\sigma(\omega,T)$ have been successfully verified by experiments at microwave and infrared frequencies for conventional superconductors under a wide range of conditions [1–3]. The temperature dependence of the surface resistance of a Nb sample is shown in figure A2.5.2, where at low temperatures $R_{\rm s} \propto \sigma_1\omega^2 \propto \omega^2 \exp(-\Delta/kT)$. In practice, $R_{\rm s}$ reaches a limiting value $R_{\rm s,res}$ at very low temperatures, which is highly dependent on sample purity and microstructure, in addition to frequency and applied magnetic field. In high quality Nb samples, it is possible to reduce $R_{\rm s,res}$ to a few nΩ at 10 GHz.

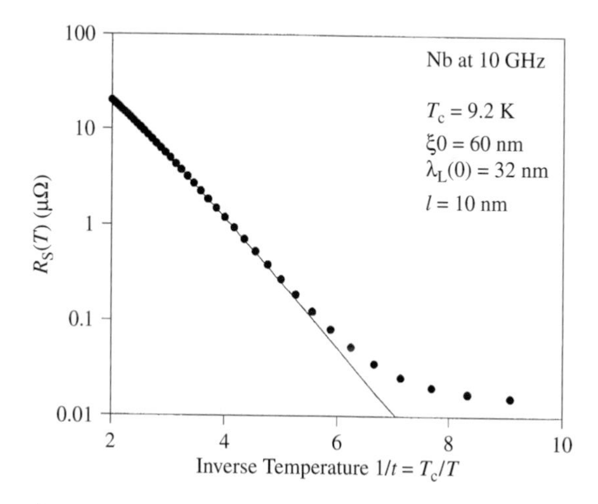

Figure A2.5.2. The surface resistance at 10 GHz for a pure Nb sample as a function of $T_{\rm c}/T$ at low temperatures. The residual resistance of this sample is 12 nΩ. The solid line is the fit to $\exp(-\Delta/kT)$, with $2\Delta(0) = 3.7\,kT_{\rm c}$.

A2.5.4 The high frequency conductivity of HTS materials

Without an accepted pairing mechanism to describe the superconductivity in HTS materials, it is impossible to predict the precise form of the high frequency conductivity $\sigma(\omega,T)$. However, since these materials are in the London limit, it is straightforward to determine $\sigma(\omega,T)$ experimentally from measurements of the surface impedance, followed by inversion of equation (A2.5.6).

There are a number of reasons why the conductivity of HTS materials should differ from that predicted by Mattis–Bardeen theory, even if the pairing mechanism is BCS-like (i.e. phonon mediated with s-wave pairing). Any anisotropy of the energy gap would cause rounding of the quasiparticle density of states at the gap edge, thus suppressing the coherence peak in σ_1 just below T_c; similar effects occur as a result of strong phonon coupling, and also due to the fact that l is strongly temperature dependent, increasing very rapidly below T_c and pushing HTS materials well into the London limit.

A2.1 A2.4

There is conclusive experimental evidence to suggest that hole doped HTS materials exhibit an order parameter with $d_{x^2-y^2}$ symmetry ('d-wave pairing') [10, 11]. This opens up an interesting new set of phenomena at low temperatures owing to the presence of line nodes in the energy gap on the Fermi surface. Such a symmetry predicts a limiting value of σ_1 at absolute zero [12], independent of the quasiparticle scattering rate

$$\sigma_{00} \approx \frac{ne^2\hbar}{m\pi\Delta(0)} \approx \frac{\hbar}{2\pi\mu_0\lambda_L^2(0)kT_c}. \tag{A2.5.9}$$

This is of the order of $10^5\,\Omega^{-1}\,m^{-1}$ for $YBa_2Cu_3O_7$. Correspondingly, there is a fundamental limit to the low temperature value of R_s of around $1\,\mu\Omega$ at 10 GHz; however, this is difficult to distinguish from the sample dependent residual resistance $R_{s,res}$, which is present even in conventional materials. Within the London limit at low temperatures, for *ab*-plane currents the quasiparticle backflow for d-wave symmetry is proportional to T, resulting in $\sigma_1(T) \approx \sigma_{00} + \alpha T$ and $\sigma_2(T) \approx \sigma_2(0) - \beta T$, where α and β are constants; since $\sigma_2 \propto 1/\lambda^2$, the latter of these results in $\lambda_L(T) - \lambda_L(0) \propto T$ [13, 14].

The most appropriate phenomenological description of the *ab*-plane conductivity of HTS materials is a two fluid model (consistent with the London limit) within the context of d-wave pairing symmetry [15, 16]. The quasiparticle dynamics can be described using the simple Drude model, including a temperature dependent quasiparticle scattering rate $\tau(T) = l(T)/v_F$, where v_F is the Fermi velocity. The temperature dependences of the pair and quasiparticle densities can be introduced in a nonrigorous manner using the dimensionless parameters $x_s(T)$ and $x_n(T)$, respectively, which can be calculated from $\lambda_L(T)$ since $x_s = \sigma_2(T)/\sigma_2(0) \equiv [\lambda_L(0)/\lambda_L(T)]^2$ and $x_n(T) = 1 - x_s(T)$; the two fluid conductivity is then [15, 16]

A2.1

$$\sigma(\omega, T) = \frac{1}{\mu_0\lambda_L^2(0)}\left[\frac{x_n(T)\tau(T)}{1 - i\omega\tau(T)} - i\frac{x_s(T)}{\omega}\right] + \sigma_{00}. \tag{A2.5.10}$$

Simulated results for the conductivity of $YBa_2Cu_3O_7$ single crystals and thin films for *ab*-plane microwave currents are plotted in figure A2.5.3, assuming that $\sigma_1(0)$ is very small. The broad peak in $\sigma_1(T)$ is *not* a coherence peak, but is due to the competition between the rapidly increasing $l(T)$ and the rapidly decreasing $x_n(T)$ as the temperature is reduced, since from equation (A2.5.10) it is found that approximately $\sigma_1(T) \propto x_n(T)l(T)$. For thin films, $l(T)$ is reduced owing to the increased quasiparticle scattering rate due to the greater density of defects compared to high quality crystals, although it is not possible to associate this behaviour to any particular type of defect. This is sufficient to suppress the peak in $\sigma_1(T)$ and shift it to higher temperatures, in addition to changing the quasiparticle backflow term from $\propto T$ to $\propto T^2$ within the d-wave framework [15, 16].

The corresponding *ab*-plane surface impedance is obtained by substituting the conductivity of equation (A2.5.10) in equation (A2.5.6), which is plotted in figure A2.5.4, having first subtracted

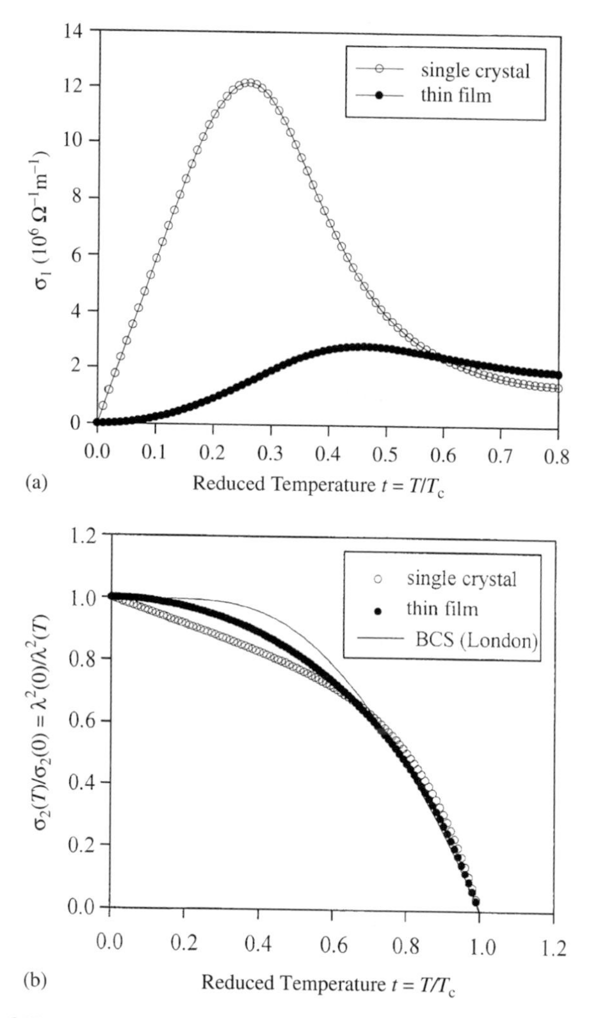

Figure A2.5.3. (*a*) $\sigma_1(T)$ of $YBa_2Cu_3O_7$ single crystals and thin films modelled using equation (A2.5.10), ignoring the effects of any residual conductivity at $T = 0$. (*b*) The corresponding $\sigma_2(T)$. The quasiparticle backflow at low T is $\propto T$ for crystals, but $\propto T^2$ for thin films. Also shown are the results expected from the Mattis–Bardeen theory in the London limit.

the residual resistance at low temperatures. The form of $R_s(T)$ reflects that of $\sigma_1(T)$, where the increased amount of scattering present in thin films can lead to lower values of $R_s(T)$ than those of the best quality crystals at low temperatures, a significant result for thin film microwave applications.

A2.5.5 The surface impedance in large dc magnetic fields

Static flux lines are nucleated within a type II superconductor by applying a dc magnetic field whose magnitude exceeds the lower critical field B_{c1}. A low amplitude, high frequency applied field will then cause these flux lines to oscillate reversibly about their pinning centres, leading to a redistribution of magnetic energy and additional energy dissipation associated with their motion; both effects can be studied by measuring the surface impedance.

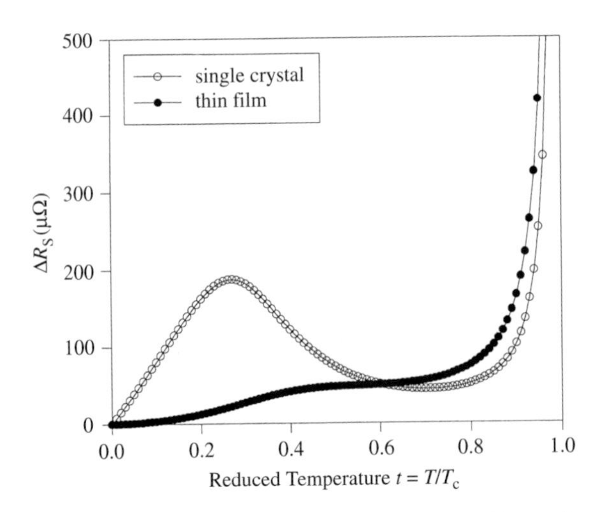

Figure A2.5.4. $\Delta R_s(T)$ at 10 GHz for $YBa_2Cu_3O_7$ single crystals and thin films using the conductivities of figure A2.5.3 after subtracting the residual surface resistance. The peak in $\sigma_1(T)$ is reflected in the surface resistance of the single crystal. Below $t \approx 0.5$, high quality thin films have lower surface resistance than high quality single crystals owing to the reduced quasiparticle mean free path l. Note that the surface resistance values are much higher than those for Nb (figure A2.5.2) at the same reduced temperature.

Unified modelling of the dc flux line response to high frequency fields was performed by Coffey and Clem [17], and also by Brandt [18], including the effects of thermally activated flux creep. To obtain a physical picture of the processes occurring it is useful to resort to the simple model of Gittleman and Rosenblum [19], which works best for large applied fields. For the case of parallel dc and ac magnetic fields, they proposed an equation of motion for individual flux lines of the form $\mathbf{F} = \eta\mathbf{v} + \kappa_p\mathbf{x}$, where $\mathbf{F}$ is the force on the flux line, $\mathbf{v}$ and $\mathbf{x}$ are the flux line velocity and displacement, respectively, η is the flux line viscosity and κ_p is the pinning force constant. The effective high frequency conductivity in a static magnetic field B is then

$$\sigma_f = \frac{\eta}{B\Phi_0}\left(1 - i\frac{\omega_p}{\omega}\right) \quad \text{(A2.5.11)}$$

A2.6 where Φ_0 is the flux quantum and $\omega_p \equiv \kappa_p/\eta$ is called the *depinning frequency*. This conductivity can be inserted into the local expression for surface impedance $Z_s = R_s + iX_s = \sqrt{i\omega\mu_0/\sigma_f}$ to quantify the high frequency response of the flux lines.

There are two important limiting regimes. At large B and very high frequencies ($\omega >> \omega_p$) one enters the *flux flow regime*, which is highly dissipative. The effective conductivity is $\sigma_f \approx \eta/B\Phi_0$ (i.e. it is almost real), and the surface resistance and reactance are approximately equal

$$R_s \approx X_s \approx \sqrt{\frac{\omega\mu_0 B\Phi_0}{2\eta}} \quad \text{(A2.5.12)}$$

with both R_s and X_s large, and both proportional to $\sqrt{\omega B}$. Surface impedance measurements performed in this flux flow regime are a useful means of determining η.

At low frequencies ($\omega \ll \omega_p$) one enters the *pinned regime*, in which case the flux line impedance is almost purely reactive and $\sigma_f \approx -i\kappa_p/\omega B\Phi_0 \equiv -i/\omega\mu_0\lambda_c^2$, where $\lambda_c = \sqrt{B\Phi_0/\mu_0\kappa_p}$ is the *Campbell*

D2.5 *penetration depth*. If B is large $X_s \approx \omega\mu_0\lambda_c$, but for smaller values of B (close to B_{c1}) one has to resort to the Coffey–Clem model [17]. The surface reactance is then $X_s \approx \omega\mu_0(\lambda^2 + \lambda_c^2)^{1/2}$ and the surface resistance differs little from its zero field value $R_s(0) \approx \sigma_1\omega^2\mu_0^2\lambda^3/2$. The small surface resistance change $\Delta R_s(B) = R_s(B) - R_s(0)$ on applying a small field B is approximately proportional to $\omega^2 B$; furthermore, in the small field limit it can be shown that the ratio $\Delta R_s(B)/\Delta X_s(B) \approx \omega/\omega_p$. Therefore, single frequency measurements of the surface impedance in the pinned regime can be used to determine the depinning frequency $\omega_p = \kappa_p/\eta$. From this η can be found, since κ_p can be determined directly from $X_s(B)$. In practice, it is found that κ_p is itself a function of B at large fields owing to the effects of collective flux pinning [5].

Some simulated results based on the simple model of Gittleman and Rosenblum for the frequency dependence of the surface impedance of $YBa_2Cu_3O_7$ in an applied magnetic field of 5 T are shown in figure A2.5.5. The measured surface impedance in the pinned regime can be influenced by thermally activated flux creep at high temperatures and low frequencies [18], affecting these characteristic frequency dependences at low fields. Depinning frequencies for conventional superconductors are typically < 100 MHz [19], whilst those for HTS materials are typically in the range 10–100 GHz due to the large values of κ_p.

Measurements of the flux line viscosity are of fundamental importance since they probe the quasiparticle dynamics within the flux line cores. The simplest model for flux line dissipation is that of Bardeen and Stephen [5], which predicts that the normal state core conductivity is $\sigma_n = \eta/B_{c2}\Phi_0$. This is valid for a continuum of core quasiparticle energy states, and measurements of η and σ_n for conventional superconductors agree well with this model. However, measurements of η in HTS materials with fields applied parallel to the c-axis imply that at low temperatures η exceeds 10^{-6} Nsm^{-2} putting them in the so-called *superclean limit*; here the quasiparticle scattering time τ within the cores is long enough for the core energy states to become discrete, and the Bardeen–Stephen model becomes

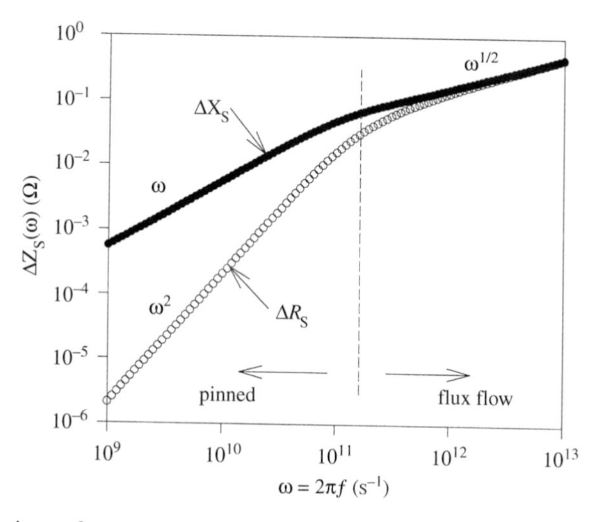

Figure A2.5.5. The change in surface resistance ΔR_s and surface reactance ΔX_s as a function of frequency on applying a static magnetic field of magnitude 5 T calculated using the model of Gittleman and Rosenblum [19]; here $\kappa_p = 4 \times 10^4$ Nm^{-2} and $\eta = 3 \times 10^{-7}$ Nsm^{-2}, appropriate for $YBa_2Cu_3O_7$ thin films around 60 K [5], yielding a depinning frequency $\omega_p = 1.3 \times 10^{11}$ s^{-1} $\equiv$ 21 GHz. Qualitatively similar results are obtained for conventional superconductors, but with lower depinning frequencies [19]. The characteristic frequency dependences above and below the depinning frequency are labelled, where the low frequency results can be affected significantly by flux creep.

invalid (since this assumes a continuum of core states, as in the bulk metal). The quasiparticle dynamics are further complicated by the possibility of d-wave pairing and the necessity to include a Hall effect term in the flux line equation of motion. Needless to say, a complete model of the high frequency dynamics of flux lines in HTS materials at low temperatures is yet to be established.

References

[1] Waldram J R 1964 Surface impedance of superconductors *Adv. Phys.* **13** 1
[2] Halbritter J 1974 On surface resistance of superconductors *Z. Phys.* **238** 466
[3] Klein O, Nicol E J, Holczer K and Grüner G 1994 Conductivity coherence factors in the conventional superconductors Nb and Pb *Phys. Rev.* B **50** 6307
[4] Pippard A B 1953 An experimental and theoretical study of the relation between magnetic field and current in a superconductor *Proc. R. Soc.* A **216** 547
[5] Golosovsky M, Tsindlekht M and Davidov D 1996 High frequency vortex dynamics in $YBa_2Cu_3O_7$ *Supercond. Sci. Technol.* **9** 1
[6] Palmer L H and Tinkham M 1968 Far infrared absorption in thin superconducting lead films *Phys. Rev.* **165** 588
[7] Holmes C C, Kamal S, Bonn D A, Liang R, Hardy W N and Clayman B P 1998 Determination of the condensate from optical techniques in unconventional superconductors *Physica* C **230** 230–240
[8] Mattis D C and Bardeen J 1958 The theory of the anomalous skin effect in normal and superconducting metals *Phys. Rev.* **111** 412
[9] Turneaure J P, Halbritter J and Schwettman H A 1991 The surface impedance of superconductors and normal metals: the Mattis–Bardeen theory *J. Supercond.* **4** 341
[10] Kirtley J R, Tsuei C C, Sun J Z, Chi C C, Yujahnes L S, Gupta A, Rupp M and Ketchen M B 1995 Symmetry of the order parameter in the high T_c superconductor $YBa_2Cu_3O_7$ *Nature* **373** 6511
[11] Tsuei C C, Kirtley J R, Ren Z F, Wang J H, Raffy H and Li Z Z 1997 Pure $d_{x^2-y^2}$ order parameter symmetry in the tetragonal superconductor $Tl_2Ba_2CuO_6$ *Nature* **387** 6632
[12] Lee P A 1993 Localised states in a d-wave superconductor *Phys. Rev. Lett.* **71** 1887
[13] Kamal S, Ruixing Liang, Hosseini A, Bonn D A and Hardy W N 1998 Magnetic penetration depth and surface resistance in ultrahigh purity crystals *Phys. Rev.* B **58** R8933
[14] Panagopoulos C, Cooper J R, Xiang T, Peacock G B, Gameson I and Edwards P P 1997 Probing the order parameter and the *c*-axis coupling of high-T_c cuprates by penetration depth measurements *Phys. Rev. Lett.* **79** 2320
[15] Bonn D A, Kamal S, Kuan Zhang, Ruixing Liang, Baar D J, Klein E and Hardy W N 1994 Comparison of the influence of Ni and Zn impurities on the electromagnetic properties of $YBa_2Cu_3O_7$ *Phys. Rev.* B **50** 4051
[16] Fink H J 1998 Residual and intrinsic surface resistance of $YBa_2Cu_3O_7$ *Phys. Rev.* B **58** 9415
[17] Coffey M W and Clem J R 1991 Unified theory of effects of vortex pinning and flux creep upon the rf surface impedance of type II superconductors *Phys. Rev. Lett.* **67** 386
[18] Brandt E H 1991 Penetration of ac fields into type II superconductors *Phys. Rev. Lett.* **67** 2219
[19] Gittleman J I and Rosenblum B 1966 Radio frequency resistance in the mixed state for subcritical currents *Phys. Rev. Lett.* **16** 734

Further Reading

Waldram J R 1996 *Superconductivity of Metals and Cuprates* (Bristol and Philadelphia: Institute of Physics Publishing)

An excellent, topical account of all aspects of superconductivity, with much discussion of the high frequency properties. Contains particularly good accounts of both the theoretical and experimental results of conventional materials, together with comparisons with HTS.

Portis A M 1993 *Lecture Notes in Physics—Vol. 48: Electrodynamics of High-Temperature Superconductors* (Singapore: World Scientific)

Discusses all aspects of microwave studies of HTS. Comprehensive discussions of nonlinear effects at high field amplitudes.

Tinkham M 1996 *Introduction to Superconductivity* 2nd edn, (New York: McGraw-Hill)

The classic educational text on superconductivity, recently updated to include HTS and more recent developments in conventional superconductivity. Thorough discussions of BCS theory and dirty limit conductivity ratios.

Bonn D A *et al.* 1996 Surface Impedance Studies of YBCO *Proc. 21st Conf.* on *Low Temperature Physics, Czech. J. Phys.* **46** *(S6)* 3195–3202

A summary of the pioneering surface impedance studies of HTS crystals undertaken at the University of British Columbia, including the effects of impurities.

A2.6
Flux quantization

Colin E Gough

A.2.6.1 Introduction

The quantization of magnetic flux is a direct manifestation of the macroscopic quantum description of the superconducting state. Flux quantization is a defining property of any superconductor, as is the closely related Meissner effect (see chapter A2.2).

A2.2

Almost as an afterthought, while discussing the quantum-mechanical, wave-like properties of superconductors, London (1950), in his monograph on Superfluids [1], added a footnote predicting that flux in a superconducting ring would be quantized in units of h/q, where q was the charge of the superconducting electrons. London assumed that q was equal to the single electron charge. The subsequent BCS theory [2] showed that superconductivity involved paired electron states (Cooper pairs). Flux is therefore quantized in units of

$$\phi_0 = h/2e \ (2.07 \times 10^{-15}\,\mathrm{Tm}^2 \text{ or in equivalent units of Vs}). \tag{A2.6.1}$$

The flux Φ within a thick superconducting ring or cylinder is therefore given by

$$\Phi = \int_{\mathrm{area}} B\,\mathrm{d}S = n\phi_0 \tag{A2.6.2}$$

where n is an integer.

Although London assumed that the quantization of flux was of purely academic interest, worthy only of a footnote, this property now underpins many of the most important device applications of superconductors. These include:

(a) superconducting quantum interference devices (SQUIDs), where the response to a magnetic field is periodic in the flux quantum (see chapter E4.2);

A4.2

(b) rapid single flux quantum logic (RSFQ), where the quantum of flux in a superconducting ring is used as the elementary bit of information (see chapter E4.5); and

A4.5

(c) the primary voltage standard, where the volt is now defined in terms of a frequency, $f = V/\phi_0$, equivalent to the number of flux quanta per second passing through a Josephson junction, where V is the voltage across the junction. This is known as the ac Josephson relation [3], where the frequency $f = 484\ \mathrm{MHz}\,\mu\mathrm{V}^{-1}$. Using an array of superconducting junctions in series, the frequency and, hence, the voltage can be measured to very high precision (see chapter A2.7).

A2.7

The observation of quantized flux in units of $h/2e$ provided the first convincing evidence for the pairing of electrons in conventional superconductors [4, 5]. A similar experiment using an early ceramic $YBa_2Cu_3O_8$ sample [6] confirmed electron-pairing in the cuprate high temperature superconductors. Because such superconductors are believed to have d-wave symmetry, superconducting circuits that result in quantized flux trapping with $\Phi = (n + 1/2)\phi_0$ can be devised. The observation of trapped states involving half a flux quantum provides the most convincing demonstration to date for d-wave superconductivity in the cuprate high temperature superconductors [7–9].

In 1955, Abrikosov showed that the magnetic state of type II superconductors involved singly quantized flux lines penetrating the bulk of the material [10]. At the centre of the flux line the superconducting order parameter is reduced to zero, so that states involving a circulating vortex of current can exist within the bulk, the flux line core essentially acting as line singularity around which the quantized current can flow. The flux associated with the circulating current is ϕ_0, so that their number density is $B/\phi_0\,\mathrm{m}^{-2}$. In many cuprate high temperature superconductors, the coupling between the superconducting CuO_2 planes is so weak that they are essentially two-dimensional. Flux penetrates such layers in the form of two-dimensional vortices, or flux pancakes, again involving a single flux quantum ϕ_0. As the interlayer coupling is increased, the flux pancakes align along the field direction to form the equivalent of a conventional flux line in a three-dimensional, but strongly A4.3 anisotropic type II superconductor (see chapter A4.3).

A2.6.2 Theory

A2.6.2.1 Quantization of flux in a ring or cylinder

Flux quantization is a direct consequence of a macroscopic wave-mechanical description of the superconducting state. This was first suggested by London and was further developed by Abrikosov A3.1 [10] within the context of the Ginsburg–Landau (GL) theory [11], based on Landau's earlier model of second order phase transitions. In such a model, the superconducting state is described by a macroscopic wave function $\Psi(r)\,e^{i\theta(r)}$, where $|\Psi|^2 = n_s$ is the number density of superconducting electrons and $\theta(r)$ is a position-dependent phase. By analogy with the conventional wave-mechanical treatment of particles in a magnetic field, the supercurrent density $\mathbf{J}_s$ can be written as

$$\mathbf{J}_s = \frac{e^*}{2m^*}[\psi^*(-i\hbar\nabla - e^*A)\psi + \text{complex conjugate}] \qquad \text{(A2.6.3)}$$

D2.1 where the term involving the vector potential $\mathbf{A}$ arises from the generalized momentum, $p \rightarrow p - e^*\mathbf{A}$, of a particle with effective mass m^* and charge e^* in a magnetic field $\mathbf{B} = \mathrm{curl}\,\mathbf{A}$. The above expression can be written more simply as

$$\mathbf{J}_s = \frac{e^* n_s}{m^*}[\hbar\nabla\theta(r) - e^*\mathbf{A}]. \qquad \text{(A2.6.4)}$$

We now consider a superconducting ring or cylinder and evaluate the line integral of the above expression well inside the superconductor, where the screening currents associated with the Meissner A2.2 effect are effectively zero, as shown in figure A2.6.1(a) (see chapter A2.2). Along this path $\mathbf{J}_s = 0$, so that

$$\mathbf{A} = \frac{\hbar}{e^*}\nabla\theta(r). \qquad \text{(A2.6.5)}$$

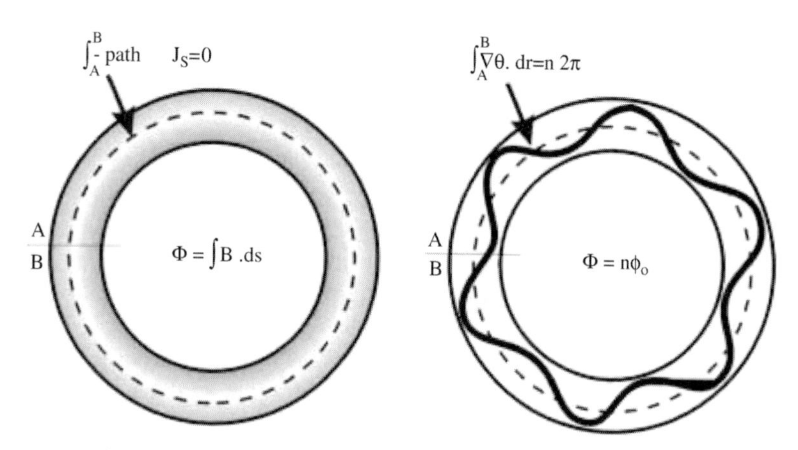

Figure A2.6.1. A superconducting ring or cylinder in an external field **B**, illustrating the path of integration well inside the superconductor along which $\mathbf{J}_s = 0$; (*b*) an allowed wave solution with $n = 6$ waves around the loop, so that the enclosed flux is $6\phi_0$.

On integrating around the path, we have

$$\oint_{\text{path}} \mathbf{A}\,\mathrm{d}r = \frac{\hbar}{e^*}\oint_{\text{path}} \nabla\theta(r)\mathrm{d}r. \tag{A2.6.6}$$

Now,

$$\oint_{\text{path}} \nabla\theta(r)\mathrm{d}r = \oint_{\text{path}} \frac{\partial\,\theta(r)}{\partial r}\mathrm{d}r = [\theta(r)]_{\mathbf{A}}^{\mathbf{B}} = n2\pi \tag{A2.6.7}$$

since $\Psi(r)$ must be single valued, as shown by a typical wave solution with $n = 6$ in figure A2.6.1(*b*). Furthermore,

$$\oint_{\text{path}} A\,\mathrm{d}r = \int_{\text{area}} \text{curl}\,\mathbf{A}\,\mathrm{d}S = \int_{\text{area}} \mathbf{B}\,\mathrm{d}S \tag{A2.6.8}$$

is simply the magnetic flux Φ enclosed within the integration path. Combining these two results, we have the principal result that

$$\Phi = \frac{\hbar n 2\pi}{e^*} = n\frac{h}{2e} \tag{A2.6.9}$$

where we have assumed Cooper pairing with $e^* = 2e$. The experimental confirmation of flux quantization in units of $h/2e$ therefore provides a direct confirmation of electron pairing in the superconducting state.

The Meissner state with $\Phi = 0$ is a special case of the above relationship with $n = 0$. This is the only allowed solution for a 'singly connected' superconductor—a superconductor with no hole or line singularity, such as a flux line core, passing through it.

If the thickness of the superconducting ring or cylinder is less than the penetration depth, or if we take an integration path close to the inner surface, the screening current will not be zero. It is then not the total flux that is quantized but what is termed the fluxoid. In this more general case we can write

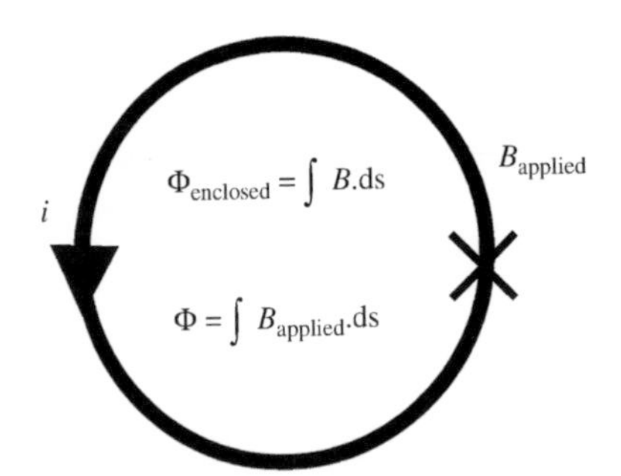

Figure A2.6.2. A superconducting ring intersected by a Josephson junction.

$$\int_{area} B\,dS + \mu_0\lambda^2 \oint_{path} j\,dr = n\phi_0 \tag{A2.6.10}$$

where we have substituted the relationship $m^*/n_s e^{*2} = \mu_0\lambda_L^2$, where λ_L is the London penetration depth over which the surface screening currents decay. E4.2

Fundamental to the operation of any SQUID is the periodicity of its properties with the flux in units of the flux quantum ϕ_0. This may easily be shown by considering the energy of a superconducting ring intersected by a single Josephson junction, as shown in figure A2.6.2.

The stored energy E in a ring with inductance L can be written as

$$E = (\Phi - \Phi_{enclosed})^2/2L - E_J\cos(\theta_A - \theta_B). \tag{A2.6.11}$$

The first term is the stored magnetic energy $(1/2)Li^2 = (1/2)(\Phi - \Phi_{enclosed})^2/L$, where $(\Phi - \Phi_{enclosed})$ is the difference between the applied flux Φ ($\mathbf{B}_{applied} \times$ area of ring) and the flux enclosed within the superconducting ring, while the second term is the Josephson energy $-E_J\cos(\theta_A - \theta_B)$, where $\theta_A - \theta_B$ is the difference in phase of the superconducting order parameter across the Josephson junction and $E_J = I_c\phi_0/2\pi$ (see chapter A2.7). If we again consider a line integral well within the superconducting ring, where $J_s = 0$, we obtain $\theta_A - \theta_B = 2\pi\Phi_{enclosed}/\phi_0$. A2.7

The total energy E can therefore be written as

$$E = (\Phi - \Phi_{enclosed})^2/2L - E_J\cos[(2\pi/\phi_0)\Phi_{enclosed}]. \tag{A2.6.12}$$

Minimizing the energy with respect to $\Phi_{enclosed}$, we obtain

$$\Phi - \Phi_{enclosed} = LI_c\sin[(2\pi/\phi_0)\Phi_{enclosed}]. \tag{A2.6.13}$$

The supercurrent flowing around the ring, $(\Phi - \Phi_{enclosed})/L$, is therefore periodic in $\Phi_{enclosed}$ in units of the flux quantum ϕ_0. This is the basis of operation of both the rf and dc SQUID. For a dc SQUID the critical current is periodic in the applied flux Φ such that

$$I_c = 2I_0|\cos\pi\Phi/\phi_0| \tag{A2.6.14}$$

as discussed in more detail in the section on SQUIDs (chapter E4.2). E4.2

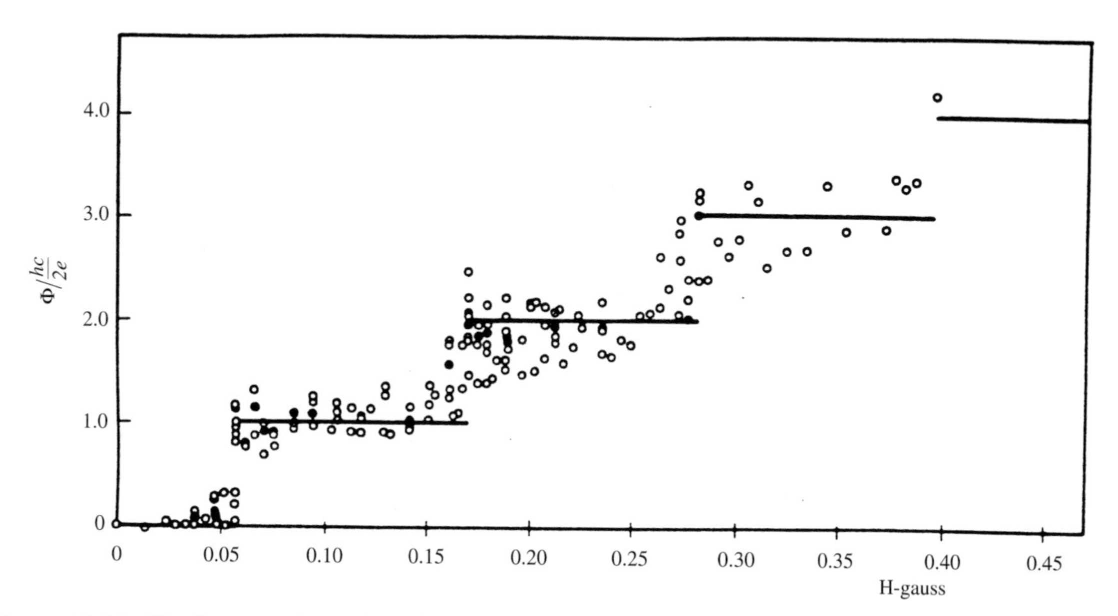

Figure A2.6.3. The first experimental confirmation of flux quantization in units of $h/2e$, demonstrated by the discrete values of flux trapped in a small tin cylinder when cooled in a small axial magnetic field [4].

A2.6.3 Experimental

A2.6.3.1 Flux quantization in conventional superconductors

The confirmation of flux quantization in conventional superconductors was published independently by two groups in 1961 [4, 5]. The same journal issue also included a number of related theoretical papers. Both experiments involved the use of very small diameter superconducting cylinders coated on the surface of a non-magnetic core.

In the measurements of Deaver and Fairbank [4], a thin cylinder of tin was deposited on a 1 cm length of 13 μm diameter copper wire. This was cooled in a small axial field and the magnetic moment trapped was deduced from the voltage induced when the sample was vibrated inside a pick-up coil. Many successive measurements were made on cooling in different fields. The trapped flux was shown to be a stepped function of applied field consistent with flux quantization in units of $h/2e$, as shown in figure A2.6.3. Doll and Näbauer [5] performed a slightly different experiment, trapping field in a lead cylinder coated on the surface of a 10 mm quartz fibre. This was supported on a very light torsion fibre and the trapped magnetic moment deduced from the torsional oscillations of the sample in the applied field. To their surprise, they observed the flux trapped in units of about half the value predicted by London, but consistent with later ideas of electron pairing and the quantization of flux in units of $h/2e$.

A2.6.3.2 Flux quantization in the cuprate high temperature superconductors

Many exotic theories were initially proposed to account for the very high transition temperatures of the cuprate superconductors, including superconducting states involving 2, 4, 8 or even 16 electrons. At the same time, several experienced researchers remained highly sceptical of the 'superconducting' transition, believing it could simply be a transition to a very much lower resistance state. Early measurements [6] on a multiphase ceramic ring of Y-Ba-Cu-O demonstrated the existence of very long-lived, trapped flux

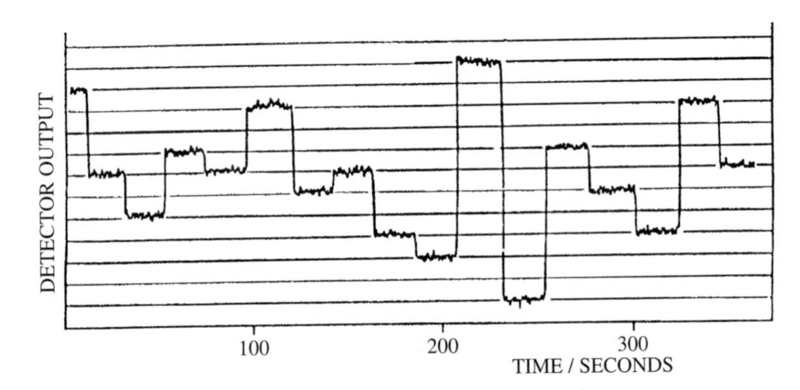

Figure A2.6.4. Demonstration of flux quantization in units of $h/2e$ in a ceramic YBCO ring [6]. The lines are spaced exactly $h/2e$ apart and the regular transitions between the quantum states were induced by short bursts of electromagnetic radiation.

states differing in flux by $h/2e$, as shown in figure A2.6.4. These measurements not only confirmed the existence of a truly superconducting, zero resistance, quantum mechanical state, but also confirmed electron pairing, just as in conventional superconductors. However, such measurements cannot reveal anything about the nature of the microscopic mechanisms leading to such pairing.

A2.6.3.3 1/2 ϕ_0 flux trapping in the cuprate HTc superconductors

It is now firmly established that the cuprate HTc superconductors have a predominantly d-wave symmetry, with nodes in the superconducting wave function at 45° to the in-plane CuO bonds. The sign of the wave function therefore reserves on changing between the a- and b-directions, as illustrated schematically in figure A2.6.5(*a*). Several experiments have confirmed this sign change, consistent with d-wave symmetry of the HTS wave-function [7–9].

In one such experiment, Wellstood and co-workers used a conventional s-wave superconductor to make a superconducting loop connecting adjacent edges of a square HTS crystal. Because of the symmetry of the d-wave function, the weak-link Josephson junctions formed at the boundaries between the d-wave and s-wave superconductors have to accommodate the change in sign with direction of the superconducting d-wave function. This implies that the flux trapped in such a loop is given by

A2.7

$$\Phi_{\text{enclosed}} = (n + 1/2)\phi_0. \tag{A2.6.15}$$

The smallest flux that can be trapped is, therefore, half a flux quantum and all trapped flux states are offset from the origin by half a flux quantum, as Mathai *et al.* [9] confirmed in a series of careful measurements.

Half quantum flux trapping was also observed by Kirtley *et al.* [8] in a series of measurements in which thin film HTc superconductors were deposited on a tricrystal substrate (three crystals cut in specific crystalline orientations and fused together along their carefully cut edges). As shown in figures A2.6.5(*a*) and (*b*), a series of thin film rings were lithographically patterned, one enclosing the intersection of all three interfaces, two rings crossing a single interface and another ring well away from the interface areas. The tricrystal was carefully designed with crystalline orientations chosen so that, over the three junctions crossing the central ring, symmetry dictates that there has to be a π phase

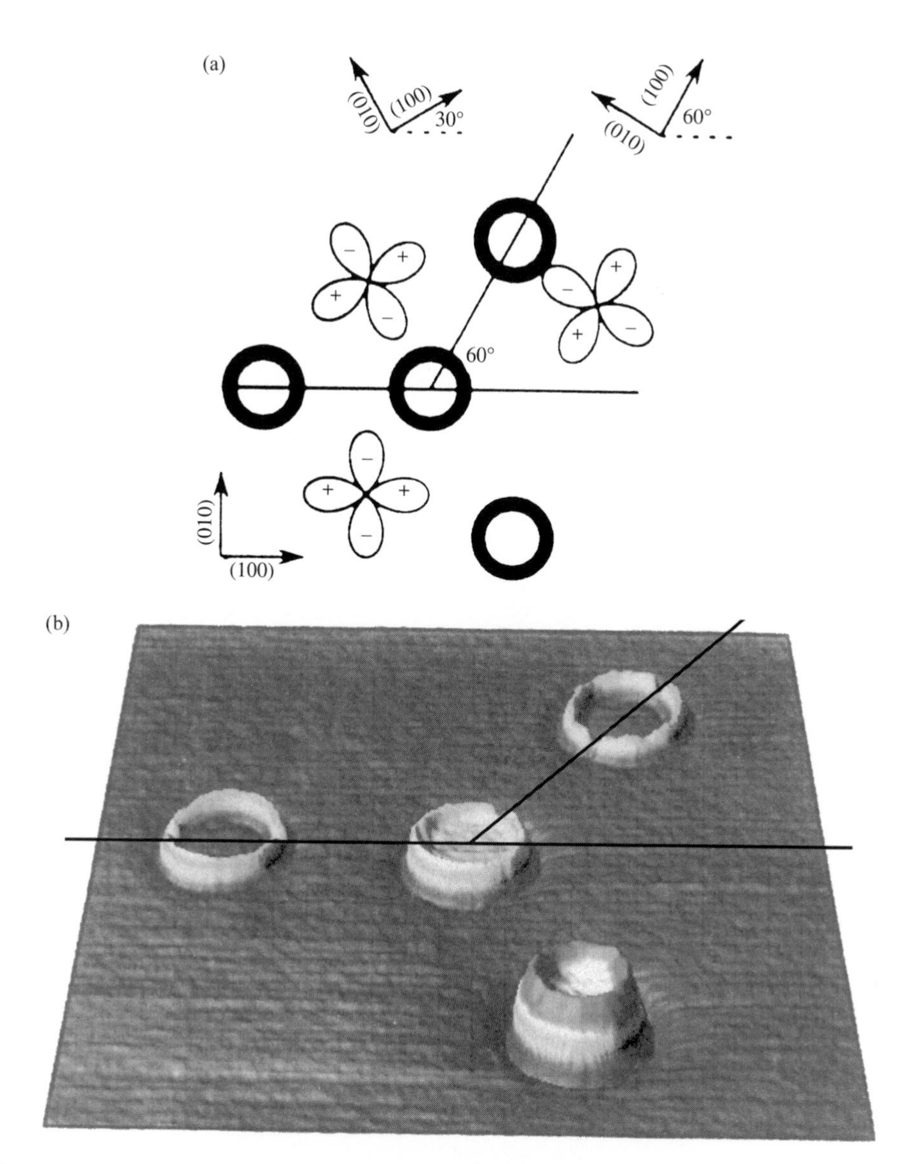

Figure A2.6.5. Scanning flux microscope measurements illustrating the d-wave symmetry of cuprate superconductors by Tsuei and co-workers. (*a*) the orientations of the tricrystal and epitaxially grown HTc superconductors and the position of the lithographically patterned rings, which are intersected by weak-link, Josephson junctions, on crossing the substrate grain boundaries. (*b*) The trapping of half a flux quantum in the central ring crossing three grain boundaries, with no flux quanta trapped in the two rings crossing single grain boundaries, and one flux quantum (though it could have been any integer) in the isolated ring [courtesy of Tsuei].

change in the superconducting phase. The lowest flux that such a ring can trap is therefore $\pm 1/2\phi_0$, as shown in figure A2.6.5(*b*), taken from the cover of Science (Vol **271**, 1995).

Experiments demonstrating 1/2 quantum flux trapping provide the strongest evidence for the d-wave symmetry of the cuprate superconductors. Such trapping has been confirmed for all cuprate superconductors investigated to date, but has never been observed for conventional superconductors.

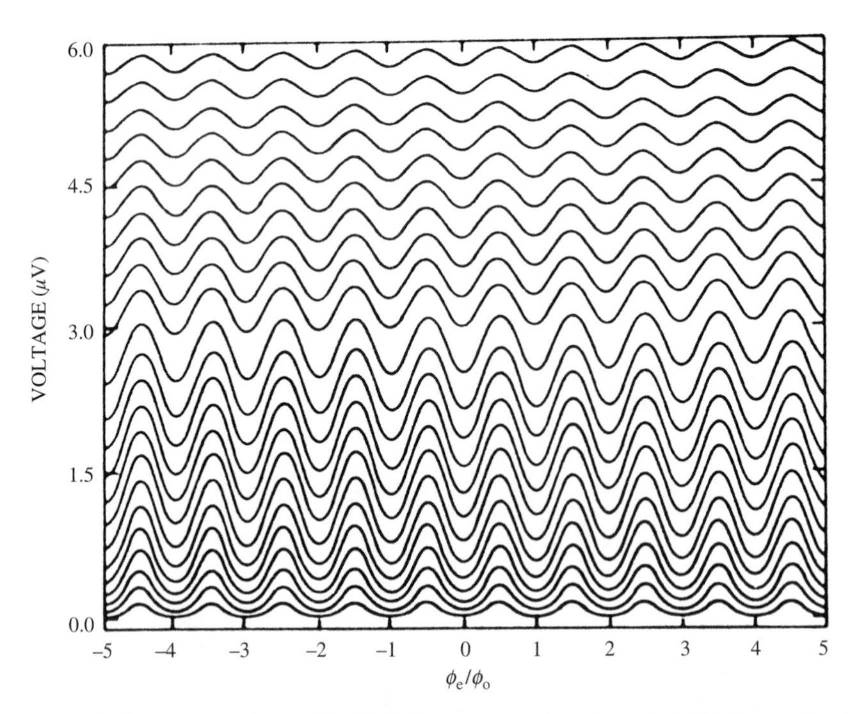

Figure A2.6.6. The periodic response of an rf HTS SQUID as a function of applied field for various levels of *rf* bias. The response can be shown to be periodic in ϕ_0 over many thousands of flux quanta [courtesy of Gross].

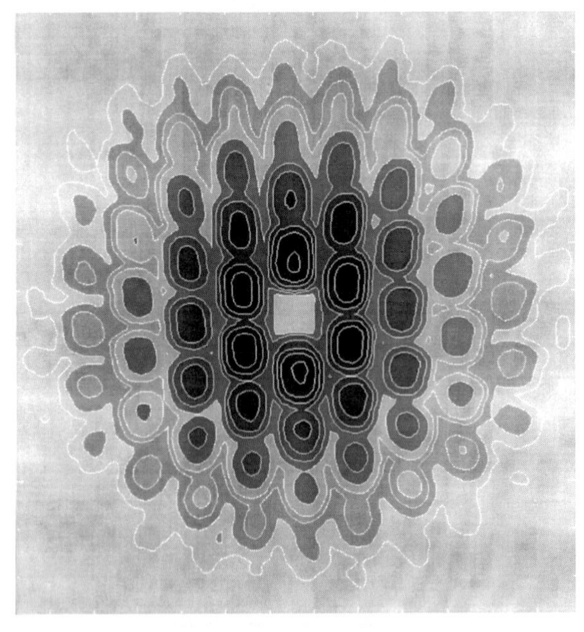

Figure A2.6.7. Small angle neutron diffraction from the lattice of quantized flux lines in a niobium single crystal [ILL Grenoble, by courtesy Forgan].

A2.6.3.4 *The SQUID and flux quantization*

As indicated in the previous section, the flux enclosed by a superconducting ring containing one or more Josephson junctions is periodic in the flux quantum ϕ_0. SQUIDs (described in chapter E4.2) can measure changes in applied flux ($B \times$ area of SQUID) to an accuracy approaching 10^{-6} $\phi_0\,\mathrm{H/z}^{-1/2}$, making a SQUID the most sensitive of any magnetic sensor.

Figure A2.6.6 illustrates the voltage response of a HTS rf SQUID as a function of the externally applied flux ϕ_e, illustrating the response periodic in ϕ_0 [12]. The curves are obtained for different values of rf bias.

E4.2

A2.6.4 Flux quanta and the mixed state of type 2 superconductors

Although quantized flux lines in the bulk of a superconductor cannot be visualized directly, their existence can be inferred from small angle neutron diffraction. In such measurements the magnetic moment of the neutron probes the spatial variations of the internal magnetic field associated with the circulating currents around each flux line. Such measurements by Cribier *et al.* [13] provided the first direct evidence for the flux line lattice in the mixed state of type II superconductors.

D1.7.2

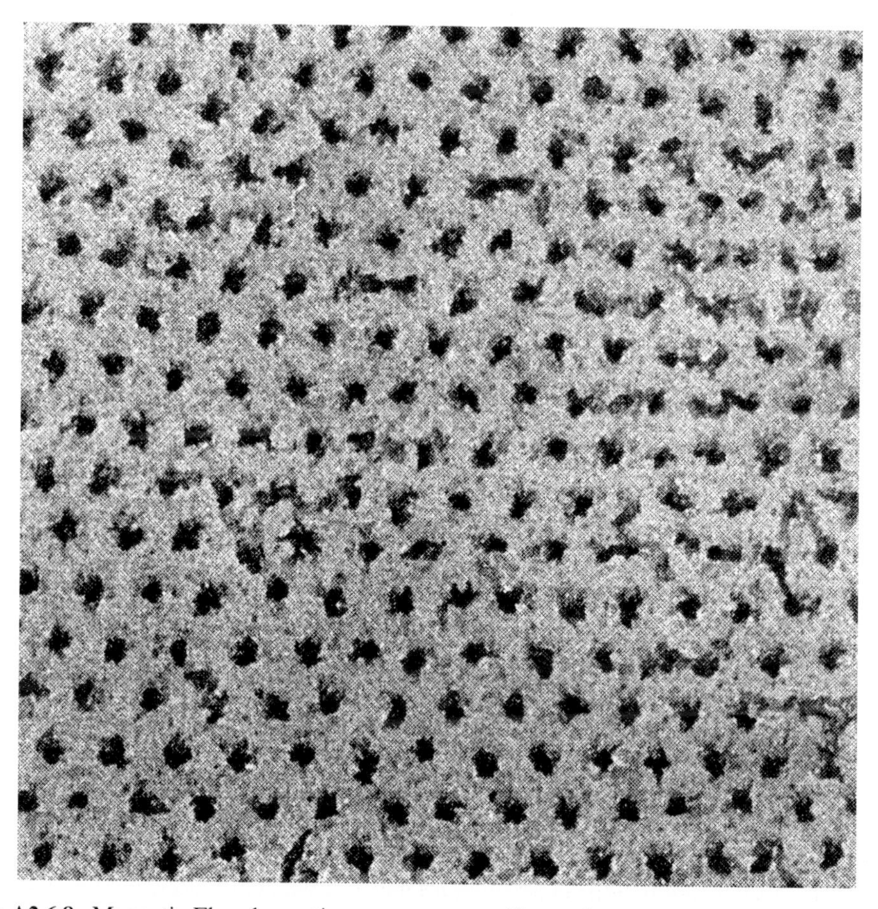

Figure A2.6.8. Magnetic Flux decoration measurements illustrating a near perfect triangular lattice.

Figure A2.6.7 shows a more recent example of a very high resolution neutron diffraction pattern obtained from a triangular flux lattice in the mixed state of a large single crystal of niobium. Note the similarity with an X-ray diffraction pattern. Neutron scattering provides detailed information on the spatial variations of the local flux density $B(x)$, whereas X-rays probe the variation in electron density $\rho(x)$.

D1.7.1

A2.6.5 Flux decoration measurements

It is possible to decorate individual flux lines emerging from the surface of a superconductor with very fine magnetic powder, which can then be visualized using a SEM. The magnetic grains tend to segregate in the regions where flux lines leave the surface of a superconductor. This technique was pioneered by Essmann and Trauble [14] in the 1970s. Figure A2.6.8 shows an example of such a measurement, which illustrates a near perfect triangular lattice of singly quantized flux lines emerging from the surface of a superconductor. Occasional defects, such as dislocations in the flux line lattice, can be identified.

D1.7.2

A2.6.6 Scanning SQUID and Hall probe microscopes

More recently, microscopes using dc SQUIDS with small pick up coils (< 10 μm) as magnetic pick up coils have been used to map out the fields produced by quantized flux lines emerging from the surface of thin films and single crystals. Figure A2.6.5(*b*) shows the use of such a microscope to measure the 1/2 flux quantum of flux in a HTS thin film ring on a tricrystal substrates (see figure A2.6.5(*a*)). A small Hall probe can also be used to probe flux line structures with almost the same sensitivity [15].

A2.6.7 Direct measurement of quantized flux by electron holography

D1.2

Another exciting modern advance has been the development by Tonomura's group of transmission electron microscopy using interfering coherent electron beams to image quantum vortices in thin cross-sections of superconducting materials [16]. An example of such an image in a thin section of a Nb superconductor is shown in figure A2.6.9, where every 'dimple' represents an individual flux line. Video image capture techniques enable the dynamics of the individual flux quanta to be investigated as a function of time, temperature and applied fields (for examples, visit, www.aaas.org/science/matsuda.htm). The thermal activation of individual flux lines between pinning sites is often observed to nucleate instabilities involving the correlated motion of a large number of vortices. It is important to be able to visualize such events because such processes are almost certainly always important but are often ignored in theoretical models for thermally activated and quantum creep.

A2.6.8

Very recently, researchers from the University of Oslo and Bell Laboratories have succeeded in observing flux lines using an optical microscope for the first time [17]. Flux lines in $NbSe_2$ were imaged with ~ 1 μm resolution making use of the magneto-optic Faraday effect, involving rotation of the plane of polarisation of light passing through a ferrite garnet film placed in close proximity to the surface.

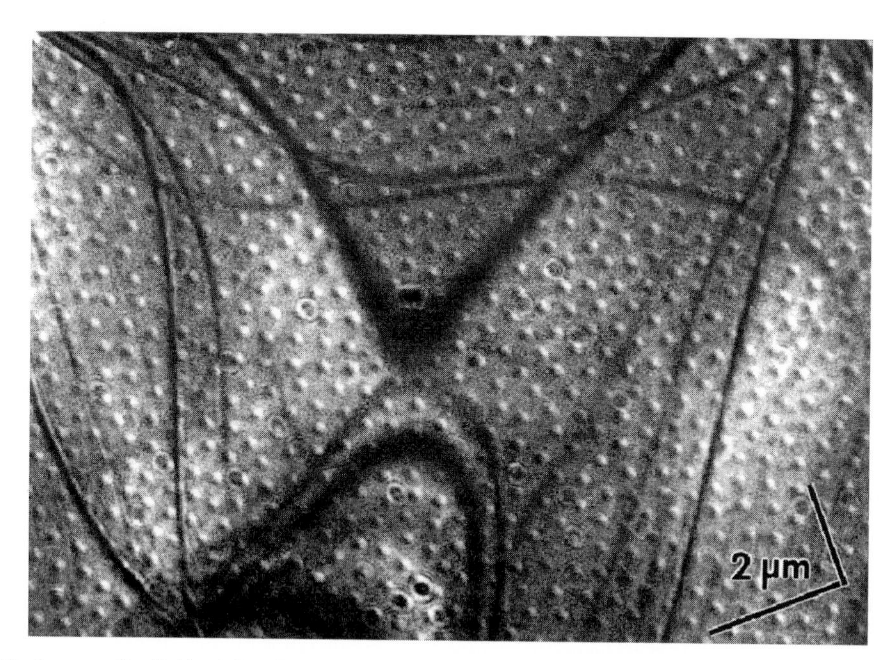

Figure A2.6.9. Image of individual flux lines in a thinned single crystal of niobium. Individual vortices are images as 'dimples' with one side darker than the other. The larger streaks crossing the image are interference effects associated with crystal bending and differences in thickness [courtesy of Tonomura].

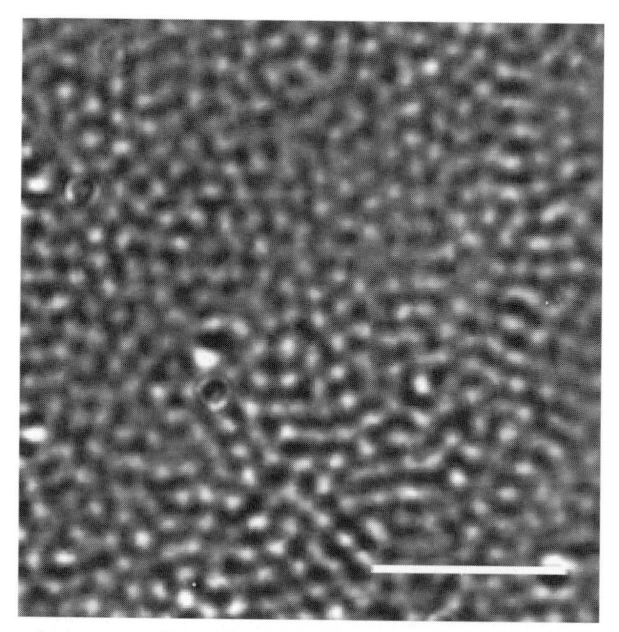

Figure A2.6.10. Magneto-optic images of vortices in $NbSe_2$ at 4.2 K in a field of 8 gauss. The white marker represents 10 μm.

References

[1] London F 1950 *Superfluids* **Vol 1**, (New York: Wiley) p 152
[2] Bardeen J, Cooper L N and Schrieffer J R 1957 *Phys. Rev.* **108** 1175
[3] Josephson B D 1962 *Phys. Lett.* **1** 251
[4] Deaver B S and Fairbank W M 1961 *Phys. Rev. Lett.* **7** 43
[5] Doll R and Näbauer M 1961 *Phys. Rev. Lett.* **7** 51
[6] Gough C E, Colclough M S, Forgan E M, Jordan R G, Keene M, Muirhead C M, Rae A I M, Thomas N, Abell J S and Sutton S 1987 *Nature* **326** 855
[7] Wollman D A, Van Harlingen D J, Giapintzakis J and Ginsberg D M 1995 *Phys. Rev. Lett.* **74** 797
[8] Kirtley J R, Tsuei C C, Rupp M, Sun J Z, Yu-Jahnes L -S, Gupta A, Ketchen M B, Moler K A and Bhushan M 1996 *Phys. Rev. Lett.* **76** 1336
[9] Mathai A, Gim Y, Black R C, Amar A and Wellstood F C 1995 *Phys. Rev. Lett.* **74** 4523
[10] Abrikosov A A 1957 *Sov. Phys. JETP* **5** 1174
[11] Ginsburg V L and Landau L D 1950 *Zh. Eksperim. I. Teor. Fiz.* **20** 1064 (in Russian)
[12] Gross R, Chadhari P, Kawasaki M, Ketchen M B and Gupta A 1990 *Appl. Phys. Lett.* **57** 727
[13] Cribier D, Jacort B, Rao L M and Farnoux B 1964 *Phys. Lett.* **9** 106
[14] Essmann U and Träuble H 1967 *Phys. Lett.* **A24** 526
[15] Oral A, Bending S J, Humphreys R G and Heneni M 1996 *J. Low Temp. Phys.* **105** 1135
[16] Harada K, Matsuda T, Bonevich J, Igarisho M, Kopndo S, Pozzi G, Kawabe U and Tomura A 1992 *Nature* **360** 51
[17] Goa P E, Haughlin H, Bazijevich M, Il'yashenko E, Gammel P and Johansen T H 2001 *Supercond. Sci. Technol.* **14** 729

Further Reading

Rose-Innes A C and Rhoderick E H *Introduction to Superconductivity* (1978 Oxford: Pergamon)

Tinkham M *Introduction to Superconductivity* (2nd Ed) (1996 Singapore: McGraw Hill)

Waldram JR *Superconductivity of Metals and Cuprates* (1990 Bristol: IOP Publishing)

Standard textbooks providing detailed accounts of the wave mechanical description of superconductors and further theoretical and experimental details on quantised flux, flux lines and the magnetic states of superconductors involving quantised flux lines.

A2.7
Josephson effects

E J Tarte

A2.7.1 Introduction

The Josephson effects [1] are very important to many areas of modern superconductivity research. Their most obvious and direct relevance is to the study and development of active superconducting electronic devices. Here, Josephson junctions form the basic elements of superconducting quantum interference devices (SQUIDs) [2], single flux quantum (SFQ) logic [3], junction arrays for the voltage standard [4] and for high frequency radiation sources [5]. However, for most high temperature superconductors (HTS), they govern the transport of supercurrent in a significant number of situations. For example, a number of HTS materials appear to be Josephson coupled perpendicular to the copper oxide planes [6]. In addition, the Josephson effect often limits the flow of current in poorly aligned polycrystalline material [7].

E4.2, E4.5

A wide variety of structures exhibit Josephson effects, but in general they can be classified into three groups: systems in which superconducting electrodes are separated by insulating layers; conducting (but not superconducting) layers; or a region where the superconductivity is weakened because it has an extremely narrow cross-section. In this chapter, tunnel junctions are concentrated upon because these form the basis of most of the established applications of Josephson effects. To this end, the properties of superconducting tunnel junctions will also be described first. Following this some of the most basic features of Josephson effects will be described.

A2.7.2 Single particle tunnelling

Tunnel junctions usually consist of electrode layers of superconductor and/or normal metal separated by an insulating barrier, although it is important to note that other structures such as grain boundaries may exhibit tunnelling phenomena [8]. The insulator may be another material sandwiched between the electrodes or may be a vacuum. The transmission of electrons through the barrier is determined by a transmission coefficient D, which according to elementary quantum mechanics should be of the form $\exp[-l\sqrt{(2m\Gamma/\hbar^2)}]$ where Γ is the barrier height, l the barrier width, m the electron mass and $\hbar$ Planck's constant. If the insulator is simply a vacuum layer, the barrier height Γ is simply the work function of the metal, whereas, if it consists of a dielectric layer, Γ is fixed by the bandgap in the barrier's electronic structure. A number of factors may complicate this simple picture, but the overall features are generally preserved and as a rule of thumb we can take D^2 to be approximately given by $\exp(-l)$ if l is measured in Å [9].

A very important feature of superconducting tunnel junctions is that their current–voltage characteristics can be used to determine the energy gap Δ and to explore the density of single particle excitations of one spin in the superconductor $N(E)$. Somewhat surprisingly, it turns out that the structure

E4.7

can be treated using a relatively simple 'semiconductor model', and that the tunnelling current I as a function of voltage V is given by equation (A2.7.1) where $f(E)$ is the Fermi function, E the excitation energy referred to the Fermi level and e the charge on the electron. The R and L subscripts refer to the right or left side of the barrier.

$$I = \frac{8\pi D^2 e}{\hbar} \int_{-\infty}^{\infty} [f_R(E - eV) - f_L(E)] N_R(E - eV) N_L(E) \mathrm{d}E. \tag{A2.7.1}$$

The simplicity of this equation is surprising because it totally ignores the presence, in the superconductors, of the Cooper pairs to which the single particle excitations are intimately connected and which might have been expected to affect the tunnelling rates. The explanation for this is rather subtle and can be found in the review article by Waldram [9].

E4.7 The three simplest examples of tunnelling systems are shown in figure A2.7.1: normal–insulator–normal (NIN), superconductor–insulator–normal (SIN) and superconductor–insulator–superconductor (SIS) tunnelling. Using the semiconductor model, the current–voltage characteristic is constructed by shifting the equilibrium density of states by an energy equal to eV and comparing the occupation of states on either side of the barrier. Experimental data for an SIN tunnel junction and an B4.4.2 SIS tunnel junction are shown in figure A2.7.2.

For NIN tunnelling, the density of states can be taken to be constant (at least for small voltages) in the vicinity of the Fermi level, where $f_R(E - eV) - f_L(E)$ is finite and equation (A2.7.1) reduces to Ohm's law $I = V/R$ where R is the tunnel junction resistance.

For SIN tunnelling, the current is very small until the voltage reaches the energy gap voltage Δ/e in the superconductor and then rises very rapidly. The amount of current at voltages $< \Delta/e$ is very sensitive to temperature since, in equilibrium, the occupation of the levels above the gap is determined by the thermal tail of the Fermi function. For an SIN junction, it is easy to show

$$\frac{\mathrm{d}I}{\mathrm{d}V} = \frac{8\pi D^2 e N_N(0)}{\hbar} \int_{-\infty}^{\infty} \frac{\mathrm{d}f_R(E - eV)}{\mathrm{d}E} N_S(E) \mathrm{d}E \tag{A2.7.2}$$

A2.1 where $N_N(0)$ is the density of states in the normal metal at the Fermi level. At very low temperatures, the derivative of the Fermi function becomes a delta function and then it is clear that the voltage dependence of the differential conductance gives the form of the superconducting density of states. As the density of states at the edge of the tunnel barrier on the superconductor side is the quantity measured, tunnelling can be used to study the anisotropy of the gap in single crystals [10] or to compare tunnelling from the superconducting and normal parts of an SN bilayer. SIN tunnelling is A3.2 important in verifying the prediction of the BCS theory for the form of the superconducting density of states.

In figure A2.7.1, tunnelling between dissimilar superconductors, known as SIS tunnelling, is shown. The current–voltage characteristic contains two important features. As the voltage across the junction is increased, the current rises because the states become available in the high gap superconductor S for excitations in the low gap superconductor S′ to tunnel into. At a voltage where the peaks in the densities of states are aligned in energy $(\Delta_S - \Delta_{S'})/e$, the current reaches a maximum and then, as the voltage is increased, declines again. When the voltage reaches $(\Delta_S + \Delta_{S'})/e$, the heavily occupied peak in the density of states of S′ is aligned with the upper sparsely occupied peak in S and the current rises very rapidly. For a symmetric SIS tunnel junction, the difference gap peak at $(\Delta_S - \Delta_{S'})/e$ is obviously absent. Sufficiently below the superconducting transition temperature, the current flowing at voltages below $2\Delta_S/e$ is very small and the differential conductance at this voltage is E4.7 extremely small. The latter feature is very useful in creating quasiparticle mixer devices for mm-wave signals.

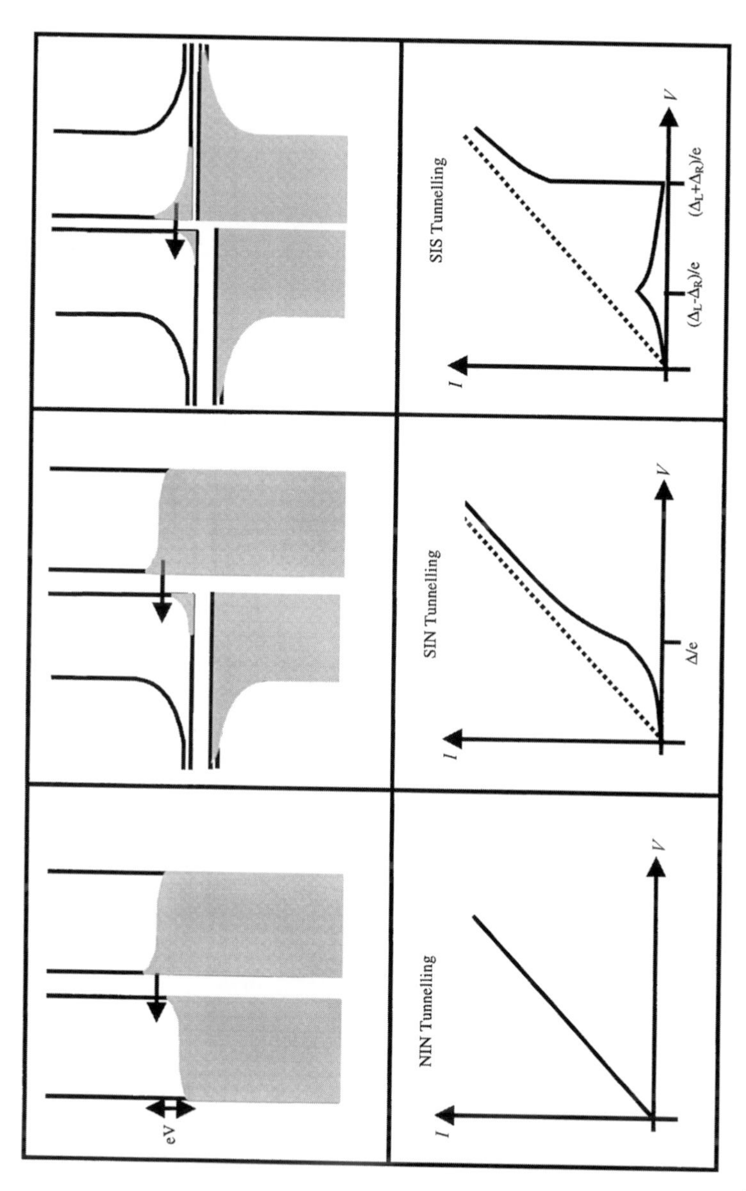

Figure A2.7.1. Single particle tunnelling showing the densities of states and occupation at finite temperature with the corresponding current–voltage characteristic.

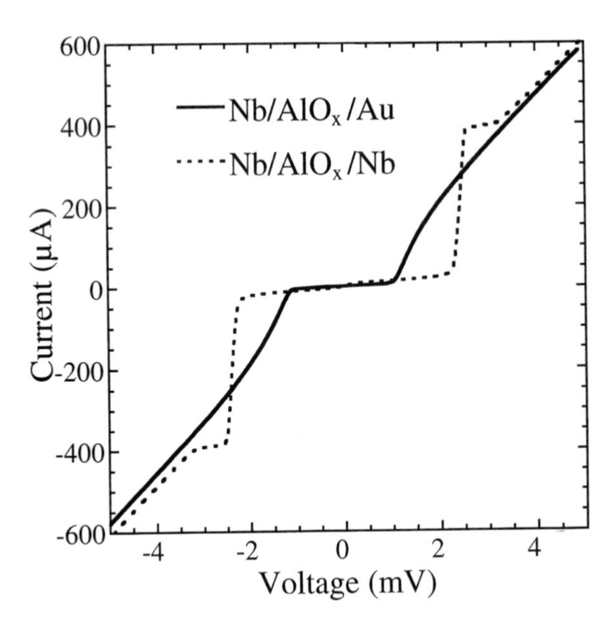

Figure A2.7.2. Experimental I–V characteristics for a Nb/AlO$_x$/Nb SIS tunnel junction and a Nb/AlO$_x$/Au SIN tunnel junction (courtesy of Dr Gavin Burnell and Dr Mark Blamire).

A2.7.3 The dc Josephson effect and quantum interference

The Josephson effects are consequences of the macroscopic quantum nature of the superconducting state and provide some of the most striking demonstrations of this. We can define a single wavefunction $\boldsymbol{\Psi}(r)$ for the superfluid of pairs, of the form $|\boldsymbol{\Psi}(r)|\exp \mathrm{i}\theta(r)$ where $\theta(r)$ is the phase of the wavefunction. When two superconductors are weakly coupled in one of the ways described before, a supercurrent whose density depends on the difference in the phases on the two sides of the barrier ϕ flows between them. In the absence of an external magnetic field, ϕ is $(\theta_2 - \theta_1)$.

Often we can assume that the coupling is mediated by the weak penetration of the wavefunction from one superconducting electrode to the other side of the barrier and vice versa. In this case we can consider the total wavefunction $\boldsymbol{\Psi}_{\mathrm{tot}}$ to be a linear superposition of the two contributions, then we can use $\boldsymbol{\Psi}_{\mathrm{tot}}$ to calculate the current as follows. The quantum mechanical equation for the current is

$$\boldsymbol{J} = \frac{\mathrm{i}e\hbar}{2m}(\boldsymbol{\Psi}^* \nabla \boldsymbol{\Psi} - \boldsymbol{\Psi}\nabla \boldsymbol{\Psi}^*) - \frac{2e^2}{m}\boldsymbol{\Psi}\boldsymbol{\Psi}^* \boldsymbol{A} \tag{A2.7.3}$$

but, in zero applied field, we can take $\boldsymbol{A} = 0$ and the two contributions to $\boldsymbol{\Psi}_{\mathrm{tot}}$, $\boldsymbol{\Psi}_1$ and $\boldsymbol{\Psi}_2$ have phases which do not vary individually with position. Associating $\boldsymbol{\Psi}_2$ with the left (negative x) side of the barrier and $\boldsymbol{\Psi}_1$ with the right (positive x) side, if we write $\boldsymbol{\Psi}_{\mathrm{tot}} = \boldsymbol{\Psi}_1 + \boldsymbol{\Psi}_2\exp(\mathrm{i}\phi)$ and calculate the current at a point in one of the electrodes then we obtain the simple Josephson current–phase relationship:

$$J_{\mathrm{s}} = J_0 \sin\phi. \tag{A2.7.4}$$

Note that ϕ is defined to be *minus* the phase difference measured across the junction in the direction of current flow, so that J_0 is positive. The critical supercurrent density is J_0 and the junction critical current I_0 is its integral over the junction area. Thus, a Josephson junction can support a dc supercurrent I_{s} with $-I_0 < I_{\mathrm{s}} < I_0$, dependent on the phase difference across the junction.

Equation (A2.7.4) holds for a wide variety of weak links. However, the current–phase relationship can deviate from this in the microbridge case [11]. It is this relationship between the supercurrent density and the quantum mechanical phase difference which gives rise to quantum interference effects. These can be observed when a magnetic field is applied to a single isolated junction or to a circuit containing Josephson junctions and superconducting loops. The latter forms the basis for the SQUID. The response of a single junction to an external magnetic field encompasses a wide range of behaviour determined by its geometry as described below.

E4.2

A2.7.3.1 Short junctions in a magnetic field

When an external magnetic field is applied to the superconductor, the critical current of the junction can be modified because the field makes the phase difference across the junction vary with position. This can be understood by considering figure A2.7.3. When the magnetic field is applied in the y direction, a screening current is set up in the surface of the superconductor, flowing within the magnetic penetration depths on either side of the barrier. We can relate this to the phase of the wavefunction and the magnetic vector potential using the quantum mechanical equation for the current (A2.7.3) which, on substituting $\boldsymbol{\Psi} = |\boldsymbol{\Psi}(r)|\exp[\mathrm{i}\theta(r)]$, gives

$$\nabla\theta = -\frac{2\pi}{\Phi_0}\left(\frac{m}{|\boldsymbol{\Psi}|^2 e^2}\boldsymbol{J} + \boldsymbol{A}\right) \qquad \text{(A2.7.5)}$$

where Φ_0 is the magnetic flux quantum. However, we must be careful of what we mean by the phase difference ϕ, because we cannot simply write it as the difference of the phases of the two wavefunctions $(\theta_2 - \theta_1)$, since this is not gauge invariant. In fact, according to equation (A2.7.5), the only way to ensure $\boldsymbol{J}$ is invariant under the gauge transformation $\boldsymbol{A} \rightarrow \boldsymbol{A} + \nabla\chi$ is if the corresponding transformation for θ is $\theta \rightarrow \theta - (2\pi/\Phi_0)\chi$. This suggests that we should use the following definition for the gauge invariant phase difference:

$$\phi = \theta_2 - \theta_1 - \frac{2\pi}{\Phi_0}\int_2^1 \boldsymbol{A}\cdot \mathrm{d}\boldsymbol{l} \qquad \text{(A2.7.6)}$$

where 1 and 2 refer to points on opposite sides of the junction barrier, e.g. A_1 and D_2 or D_1 and A_2. Remember that ϕ is defined to be positive in the opposite direction to the Josephson current flow. Hence, in figure A2.7.3, between points at x and $x + \mathrm{d}x$, the change in the gauge invariant phase difference is

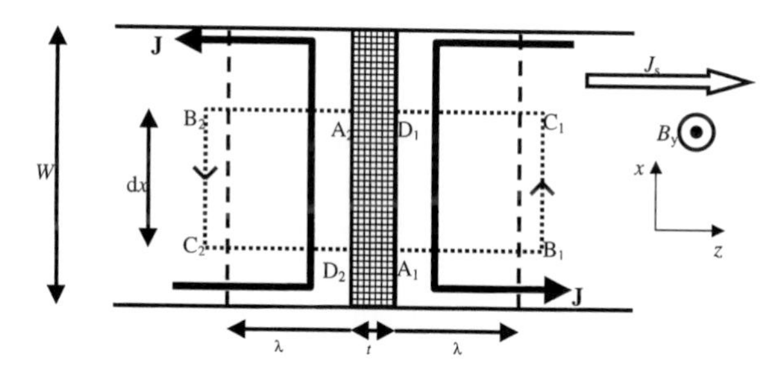

Figure A2.7.3. Contours of integration used to derive the magnetic field dependence of the gauge invariant phase difference. The screening current $\boldsymbol{J}$ is also indicated.

$$\phi(x+\mathrm{d}x)-\phi(x)=(\theta_{A_2}-\theta_{D_1})-(\theta_{D_2}-\theta_{A_1})-\frac{2\pi}{\Phi_0}\left(\int_{A_2}^{D_1}\boldsymbol{A}\cdot\mathrm{d}\boldsymbol{l}-\int_{D_2}^{A_1}\boldsymbol{A}\cdot\mathrm{d}\boldsymbol{l}\right). \quad \text{(A2.7.7)}$$

Using equation (A2.7.5) we can write the change in phase between the points A_1 and D_1 caused by the magnetic field as

$$\theta_{D_1}-\theta_{A_1}=-\frac{2\pi}{\Phi_0}\int_{ABCD_1}\left(\boldsymbol{A}+\frac{m}{|\Psi|^2e^2}\boldsymbol{J}\right)\cdot\mathrm{d}\boldsymbol{l} \quad \text{(A2.7.8)}$$

and a similar expression for the contour $(ABCD)_2$. Here $\boldsymbol{J}$ is the screening current flowing in the electrodes as a response to the applied magnetic field. In carrying out this integration, we make sure that the contours $(ABCD)_1$ and $(ABCD)_2$ extend a distance into the electrodes beyond the penetration depth where $\boldsymbol{J}$ is equal to zero. Then $\boldsymbol{J}$ makes no contribution at all to the integral because inside the penetration regions, it is perpendicular to the contour. Thus we can rewrite equation (A2.7.7) as

$$\phi(x+\mathrm{d}x)-\phi(x)=\frac{2\pi}{\Phi_0}\oint\boldsymbol{A}\cdot\mathrm{d}\boldsymbol{l}. \quad \text{(A2.7.9)}$$

The integral represents the flux enclosed by the contours and the paths across the junction which is equal to $B(2\lambda+t)\mathrm{d}x$, so the phase difference across the junction can be obtained from

$$\frac{\mathrm{d}\phi(x)}{\mathrm{d}x}=2\pi\frac{B(2\lambda+t)}{\Phi_0}. \quad \text{(A2.7.10)}$$

If equation (A2.7.10) is integrated and $\phi(x)$ is substituted in equation (A2.7.4), then we can integrate J_s over the junction area to obtain the current I and maximize I with respect to the constant of integration ϕ_0 to determine the critical current. For a uniform critical current density J_0, the junction critical current is

$$I_c(\Phi)=I_0\left|\frac{\sin\left(\frac{\pi\Phi}{\Phi_0}\right)}{\frac{\pi\Phi}{\Phi_0}}\right| \quad \text{(A2.7.11)}$$

B4.4.3 where $\Phi=BLd$ and $d=(2\lambda+t)$ is the magnetic width of the barrier. This function is plotted in figure A2.7.4(a)for a range of applied flux values. An important feature to note is that the central maximum of this curve is twice as wide as the secondary maxima. This curve is often called the 'Fraunhofer' pattern due to the analogy between the quantum interference effects which define its shape and Fraunhofer diffraction from a single slit. We can also define a magnetic area for the junction equal to Ld. In general, for an arbitrary critical current density distribution $J_0(x,y)$, the critical current as a function of magnetic field is related to the Fourier transform of a function $j(x)$ as follows:

$$I_c(k)=\left|\int\int_{\text{area}}\mathrm{d}x\,\mathrm{d}y\,J_0(x,y)\exp ikx\right|=\left|\int_{-L/2}^{L/2}\mathrm{d}x j(x)\exp(ikx)\right| \quad \text{(A2.7.12)}$$

where $k=2\pi Bd/\Phi_0$. Thus, the variation of the junction critical current with applied magnetic field is a very important measure of their quality since it gives information about the uniformity of $J_0(x,y)$ over the junction area. Deviations from a uniform critical current density distribution are easily detected by comparing the observed Fraunhofer pattern with the ideal one.

An example of the variation of critical current with applied field is shown in figure A2.7.4(b), for a 6 μm wide 24° misoriented grain boundary junction in a 100 nm thick $YBa_2Cu_3O_{7-\delta}$ film. At first sight,

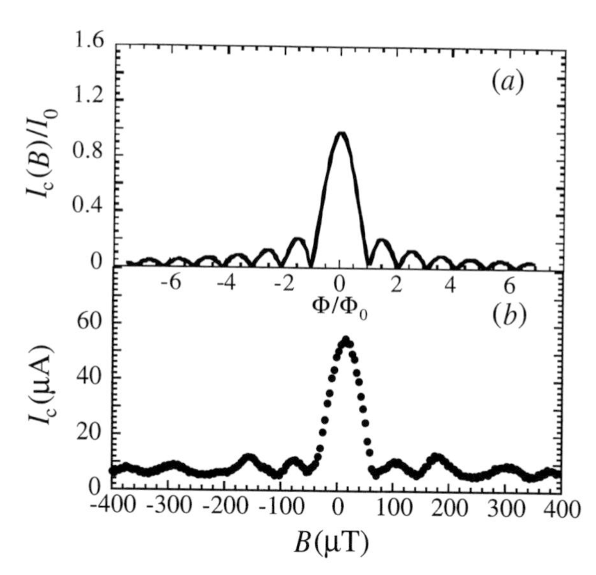

Figure A2.7.4. (*a*) The ideal critical current versus applied flux curve (Fraunhofer pattern) for a uniform junction. (*b*) Critical current versus field, for a 6 μm wide 24° misoriented grain boundary junction in a $YBa_2Cu_3O_{7-\delta}$ film.

the agreement between these data and the theoretical curve above it is reasonable, but the experiment and theory disagree in two important ways. Firstly, the critical current never goes to zero at the minima and this is a common feature of grain boundary junctions in high temperature superconductors. In contrast, complete suppression may be observed with low T_c tunnel junctions [9] or high T_c junctions produced by focussed electron beam irradiation [12]. The offset seen here is of the type predicted by Yanson [13] for a junction whose barrier is extremely non-uniform on a very fine scale. It has recently been shown by Hilgenkamp *et al.* [14] that the behaviour of these devices can be understood in this way, due to the $d_{x^2-y^2}$ pairing state of the high temperature superconductors and the facetting of the grain boundary.

Secondly, using the theory described before, the magnetic field required to reach the first minimum should be equal to $B_{min} = \Phi_0/L(t + 2\lambda) = 1$ mT, assuming a penetration depth of 150 nm, but the actual value is much smaller, 50 μT. This discrepancy arises because this junction has its dimension parallel to the field direction less than the penetration depth. In this case the screening currents set up in response to the applied field penetrate much further into the electrodes and it is not possible to draw the contour $(ABCD)_1(ABCD)_2$ so that $\boldsymbol{J}$ makes no contribution to the phase difference in equation (A2.7.8). Hence, the magnetic thickness of the barrier is no longer $d = 2\lambda + t$ and the position of the minima is then more difficult to estimate. This problem has been solved for a very thin film where the first minimum occurs at $B_{min} = 1.8\Phi_0/L^2$ [15], which in this case would give a value close to 100 μT.

A2.7.3.2 *Junctions in superconducting loops*

The above argument can easily be extended to treat the behaviour of one or more Josephson junctions in superconducting loops. This is important because such structures form the basis of SQUIDs and SFQ electronics which are dealt with in chapters E4.2 and E4.5. However, at this point we will only give the results for the simplest cases.

E4.2, E4.5

For a single junction in a loop the gauge invariant phase difference is given by

$$\phi = -2\pi\frac{\Phi}{\Phi_0} \tag{A2.7.13}$$

where Φ is the flux inside the loop. If the loop has inductance L_s, then $\Phi = \Phi_{ext} + L_s I_s$ where Φ_{ext} is the externally applied flux and I_s is the screening current set up in the loop equal to $I_0 \sin\phi$. The relationship between Φ and Φ_{ext} is

$$\phi + \frac{2\pi L_s I_0}{\Phi_0}\sin\phi + \frac{2\pi\Phi_{ext}}{\Phi_0} = 0 \tag{A2.7.14}$$

and this equation governs the behaviour of rf SQUIDs.

The result equivalent to equation (A2.7.13) for a loop containing two junctions is

$$\phi_1 - \phi_2 = 2\pi\frac{\Phi}{\Phi_0} \tag{A2.7.15}$$

and this can be used to determine the magnetic field dependence of the critical current if the junctions are connected electrically in parallel. However, if the loop has a finite inductance, this cannot be done analytically. For two junctions in parallel with a very low inductance loop, the critical current is given by

$$I_c(\Phi) = 2I_0\left|\cos\frac{\pi\Phi}{\Phi_0}\right| \tag{A2.7.16}$$

as may be shown using equation (A2.7.12).

A2.7.3.3 Long junctions

When one or both of the lateral dimensions (width and length) of a Josephson junction become large then the simple quantum interference effects described above are modified by the screening of magnetic fields out of the junction region. This applies to both externally applied fields and the self fields associated with the current flowing in the leads for the junction. The detailed behaviour depends on the geometry of its electrical connections, which determine the magnitude and direction of self fields. The five important classes are shown in figure A2.7.5. Junctions (a)–(d) can be fabricated using thin film tunnel junction technologies, whereas the slab geometry corresponds to grain boundaries in bulk material. For the in-line, slab and cross geometries, the leads generate appreciable magnetic fields in the plane of the junction barrier. Grain boundaries in linear strips of HTS film have been shown to behave like overlap junctions [16] where the self field is negligible.

Since the magnetic field is screened out of the junction region, we cannot assume that it is uniform within the barrier, as we did earlier. Hence, we use the Maxwell equation $dB_y/dx = \mu_0 J_z$ and equation (A2.7.10) to obtain an equation for the spatial variation of the phase along the junction:

$$\frac{d^2\phi}{dx^2} = \frac{2\pi\mu_0 d}{\Phi_0}J_z = \frac{2\pi\mu_0 d}{\Phi_0}J_0\sin\phi = \frac{\sin\phi}{\lambda_J^2} \tag{A2.7.17}$$

where $\lambda_J = [\Phi_0/(2\pi\mu_0 dJ_0)]^{1/2}$ is the Josephson penetration depth. When the junction's largest dimension is significantly longer than the Josephson penetration depth, the device is said to be in the long limit. It should be noted that this expression is not strictly valid for grain boundaries in HTS films because of the problem of defining d. However, the behaviour of long HTS grain boundary junctions can be understood at least qualitatively using the solutions of equation (A2.7.17).

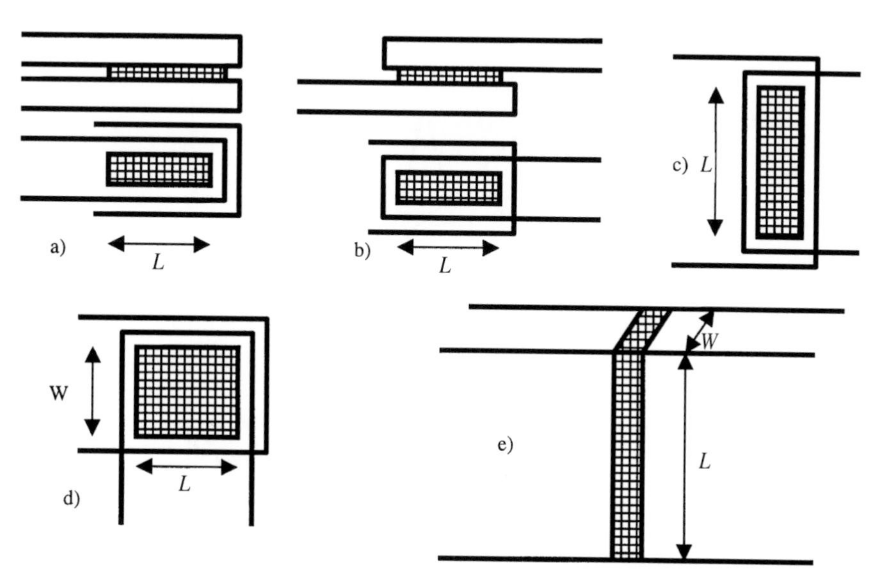

Figure A2.7.5. Josephson junction geometries: (*a*) symmetric in-line, (*b*) asymmetric in-line, (*c*) overlap, (*d*) cross and (*e*) slab. The upper parts of (*a*) and (*b*) are side views, whilst the lower parts are top views. The shaded area represents the active area of the junction barrier.

The critical current as a function of applied field for two types of long junction is shown in figure A2.7.6. Both are grain boundary junctions in YBCO films but, in one case the film is patterned to form an asymmetric in-line device. In the other, the film is patterned into a linear strip crossing the grain boundary. Both curves have a triangular form for low fields with a number of small triangular features appearing for larger fields. The shape for small fields is associated with the screening of the applied field from the junction and the curve extrapolates to a critical field value of $4\mu_0 J_0 \lambda_J$ where one Josephson vortex enters the junction. For the asymmetric in-line junction, the self field is reflected in its critical current versus field curve and the maximum critical current no longer occurs at zero field. The smaller triangular features represent states with one or more Josephson vortices in the junction and several of these states can occur at the same field value with the $I_c(B)$ curve exhibiting hysteresis.

B4.4.3

The form of the $I_c(B)$ curve for all types of junction listed is similar to those shown in figure A2.7.6. However, if we plot the critical current in zero field $I_c(0)$ versus length L for the devices then major differences appear, as shown in figure A2.7.7. Initially, all of the junctions are in the small limit and the critical current increases in proportion to length. When the length of the asymmetric in-line junction becomes comparable to $2\lambda_J$, the self field begins to be excluded from the interior of the junction and the rate of increase of $I_c(0)$ with L decreases eventually saturating at $2WJ_0\lambda_J$. This also happens for the symmetric in-line and the slab geometry, but not until L becomes comparable to $4\lambda_J$ and $I_c(0)$ saturates at $4WJ_0\lambda_J$. In contrast, although the overlap junction shows long junction behaviour for the same range of L values as the other devices, its $I_c(0)$ value never saturates because its self field is so small.

HTS grain boundaries provide examples of both overlap and slab geometry long junctions. For grain boundaries in linear strips of thin HTS film [16], W is typically < 300 nm whilst L is $\geq 2\,\mu$m. These dimensions result in a negligible self field and it has been shown experimentally that $I_c(0)/2WJ_0\lambda_J$ does not saturate with increasing L/λ_J. However, single grain boundaries in bulk material having dimensions of order $100\,\mu\text{m} \times 100\,\mu\text{m}$ are clearly slab-like and their critical currents are of order $4WJ_0\lambda_J$, assuming the same critical current densities as equivalent thin films' grain boundaries [17].

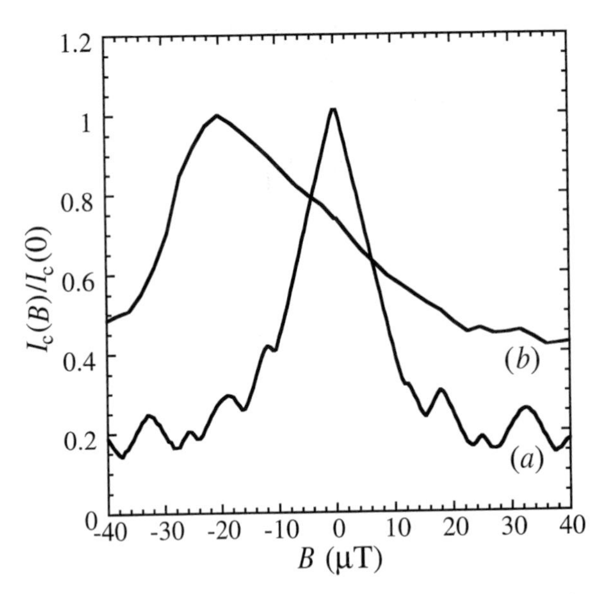

Figure A2.7.6. Critical current versus field curves for YBCO thin film grain boundary along Josephson junctions. Curve (*a*) is for a junction formed in a strip of film whilst curve (*b*) is for an asymmetric in-line junction (courtesy of Dr Stephen Isaac and Dr Mark Blamire).

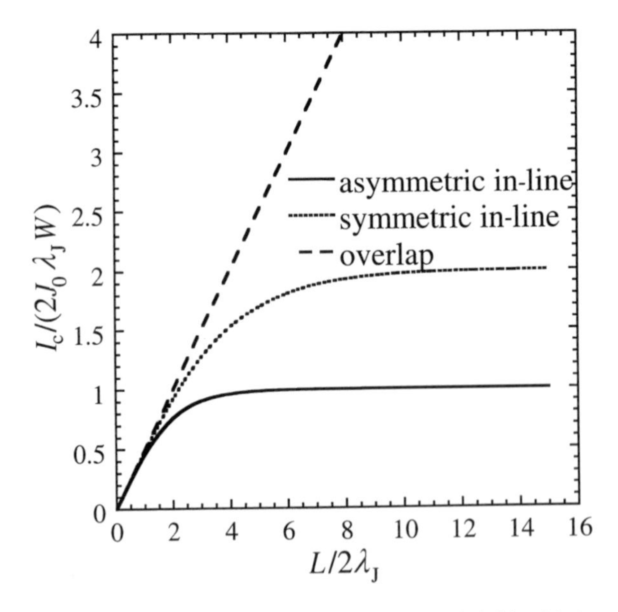

Figure A2.7.7. Variation of maximum critical current in zero external field with junction length for (*a*) overlap geometry, (*b*) symmetric in-line and slab geometry and (*c*) asymmetric in-line geometry.

A2.7.4 AC Josephson effects

In order to understand what happens when a Josephson junction is in the finite voltage state, we must look at the energy of the superfluid of pairs. Just as we can apply the quantum mechanical equation for the current to the pair wavefunction, we can use the quantum mechanical energy operator ($\hat{E} = i\hbar\partial/\partial t$)

to calculate its energy eigenvalue. If we assume that the superconductor is in equilibrium with a reservoir of particles with an electrochemical potential of μ, then when a pair of electrons enters the superfluid, they must do so with an energy of $2\,\mu$. Thus, in a steady state of the form $|\boldsymbol{\Psi}(\boldsymbol{r})|\exp[\mathrm{i}\theta(r)]$, we have $\hbar\partial\theta/\partial t = -2\mu$. If a junction is sufficiently weakly coupled, we can apply this result to both its electrodes independently and since the voltage across the junction $V = (\mu_1 - \mu_2)/e$, we can write:

$$\hbar\frac{\partial\phi}{\partial t} = 2eV. \tag{A2.7.18}$$

A2.7.4.1 Voltage biased Josephson junctions and the ac Josephson effect

When a constant finite voltage is applied across a junction, the contribution of the supercurrent to the total current can be found by integrating equation (A2.7.18) and substituting for the phase in equation (A2.7.4) to obtain

$$I_{\mathrm{s}} = I_0 \sin\left(\frac{2eV}{\hbar}t + \phi_0\right) = I_0\sin(\omega_{\mathrm{J}}t + \phi_0). \tag{A2.7.19}$$

Thus, the supercurrent flowing through the junction at a finite voltage oscillates at a frequency $f_{\mathrm{J}} = \omega_{\mathrm{J}}/2\pi = V/\Phi_0$ which is 0.48 GHz/μV. This is known as the ac Josephson effect.

The normal current I_{n} that flows in addition to the supercurrent depends on the nature of the barrier. As we have seen in section A2.7.1, for a tunnel junction, the normal current is a very strong function of the voltage. However, the essential physics can be understood by taking the junction resistance to be ohmic, so that $I_{\mathrm{n}} = V/R$. This gives the resistively shunted junction model which can also be applied to other types of junction such as microbridges and normal barrier devices.

Figure A2.7.8(a) shows a schematic diagram of the current–voltage characteristic of a voltage biased junction based on this model. The time averaged dc current at each voltage value is simply the normal current flowing in the resistance, but the presence of the ac component suggests that the device could be used as a tunable oscillator. However, each junction can only provide power of order RI_0^2, which is typically less than 1 μW, so usually an array of such devices is used [5].

If we apply a combination of dc and ac voltage to the device of the form $V = V_0 + V_{\mathrm{rf}}\cos(\omega_{\mathrm{rf}}t)$, using a microwave source for example, then the phase itself oscillates:

$$\phi = \frac{2eV}{\hbar}t + \frac{2eV_{\mathrm{rf}}}{\hbar\omega_{\mathrm{rf}}}\sin\omega_{\mathrm{rf}}t + \phi_0. \tag{A2.7.20}$$

This generates ac supercurrents at the Josephson frequency ω_{J} and the side frequencies $\omega_{\mathrm{J}} \pm n\omega_{\mathrm{rf}}$, so that:

$$I_{\mathrm{s}} = I_0\sum_n J_{\mathrm{n}}\left(\frac{2eV_{\mathrm{rf}}}{\hbar\omega_{\mathrm{rf}}}\right)\sin(\omega_{\mathrm{J}}t \pm n\omega_{\mathrm{rf}}t + \phi_0) \tag{A2.7.21}$$

where the amplitude functions J_n are Bessel functions. The ac supercurrents have no effect at most voltages but, when $2eV = n\hbar\omega_{rf}$, one of the side frequencies will be zero and there is a corresponding vertical spike in the I–V characteristic. This is shown schematically in figure A2.7.8(b). The appearance of these Shapiro spikes in the I–V characteristic is known as the inverse ac Josephson effect. Arrays of Josephson junctions biased on the same spike are currently used in standard laboratories to maintain the International Standard Value of the volt [4].

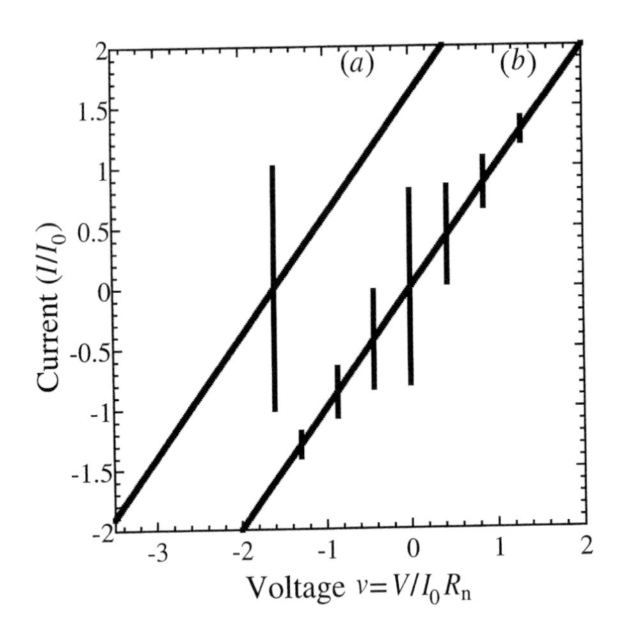

Figure A2.7.8. Schematic diagram of the I–V curve for (a) a voltage biased Josephson junction (shifted) and (b) a voltage biased Josephson junction under microwave irradiation.

A2.7.4.2 Current biased Josephson junctions

Although in the previous section, we have assumed that we could apply a constant dc voltage to the junction, in practice this is rather difficult. This is because Josephson junctions typically have resistances of a few ohms, whereas the transmission lines used to connect the device to a source have an impedance of 50 Ω. Therefore, it is much more realistic to treat the junction as biased by a constant dc current. We are then interested in the voltage across the junction; this is determined by the Josephson element and the components of its impedance in parallel (capacitance and resistance). Both the current flowing in each of these components and the voltage are then time dependent, oscillating at the Josephson frequency corresponding to the average dc voltage, as will be shown later.

Equation (A2.7.18) enables us to generate a model for the junction in the finite voltage state with a constant current bias: the resistively and capacitively shunted junction model (RCSJ). Here, we treat the Josephson current, quasiparticle current and displacement current as parallel contributions to the total current through the device. This gives

$$I = I_0 \sin\phi + \frac{V}{R} + C\frac{\mathrm{d}V}{\mathrm{d}t}. \qquad \text{(A2.7.22)}$$

We can write this in normalized form in terms of the phase as:

$$i = \sin\phi + \frac{\mathrm{d}\phi}{\mathrm{d}\tau} + \beta_c\frac{\mathrm{d}^2\phi}{\mathrm{d}\tau^2} \qquad \text{(A2.7.23)}$$

where $i = I/I_0$, $\tau = (2\pi I_0 R/\Phi_0)t = \omega_c t$ and $\beta_c = \omega_c RC$. The normalized voltage is $v(\tau) = V(t)/I_0R = \mathrm{d}\phi/\mathrm{d}\tau$.

We can now use equation (A2.7.18) to calculate the voltage for a particular bias current from the phase difference $\phi(t)$. This gives a voltage which is a periodic function of time, but the dc voltage across the junction is proportional to the time average of $\mathrm{d}\phi/\mathrm{d}t$ so that

$$V_{\mathrm{dc}} = \frac{\Phi_0}{2\pi}\left\langle\frac{\mathrm{d}\phi}{\mathrm{d}t}\right\rangle = \frac{\Phi_0}{2\pi T}\int_0^{T_J}\frac{\mathrm{d}\phi}{\mathrm{d}t}\,\mathrm{d}t = \frac{\Phi_0}{T_J} = \Phi_0 f_J \qquad \text{(A2.7.24)}$$

where T_J is the period of the Josephson oscillation and f_J is its frequency. Figure A2.7.9 shows the time dependence of the voltage across the junction for different values of bias current. For large biases the voltage reduces to a sinusoidal waveform with a constant voltage equal to V_{dc} superimposed upon it. As the bias current is decreased, the voltage waveform changes to a series of narrow pulses which contain a large number of harmonics, but whose fundamental frequency decreases in proportion to V_{dc}. However, it is apparent from equation (A2.7.24) that the area under each pulse is equal to Φ_0. This is also true for other Josephson circuits where the waveform is more complicated, e.g. a SQUID, and the pulses are therefore often known as SFQ pulses (see chapter E4.5).

E4.2
E4.5

A2.7.4.3 Current–voltage characteristics of current biased junctions

Using equations (A2.7.14) and (A2.7.15), we can obtain the current–voltage (I–V) characteristics of the junction which depend on the value of the junction capacitance and hence β_c. For a junction with no capacitance we have an analytic solution:

$$I = \sqrt{I_0^2 + \frac{V^2}{R^2}}. \qquad \text{(A2.7.25)}$$

However for a finite β_c value, the detailed form of the I–V characteristic must be calculated numerically, which is complicated by the appearance of hysteresis for $\beta_c \geq 1$. Figure A2.7.10 shows examples of I–V characteristics for junctions with β_c values of 0, 1 and ∞. The curves for $\beta_c = 0$ and ∞ are theoretical, whereas the curve for $\beta_c = 1$ is experimental data. For the latter, hysteresis can be readily observed with a well-defined discontinuity in the voltage at $I = I_0$ and an increase in voltage

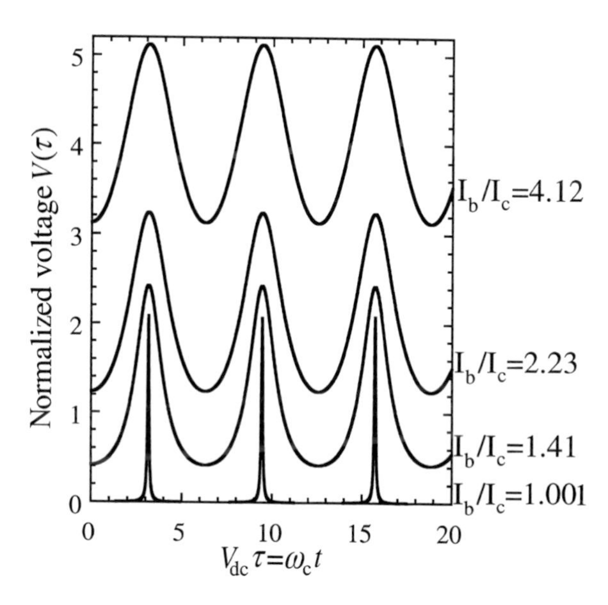

Figure A2.7.9. The time dependence of the voltage across a Josephson junction for three different bias currents.

as the bias current increases which is slower than the $\beta_c = 0$ case. As the bias current decreases through I_0, the voltage does not return to zero, but persists until a smaller value called the return or retrapping current I_r which is a non-linear function of β_c. In the limit of $\beta_c = \infty$, $I_r = 0$ which means that the I–V characteristic reduces to a supercurrent branch and an ohmic branch, as shown.

Here the resistance is ohmic, which it is clearly not in the case of a Josephson tunnel junction and to a first approximation we can usually replace the ohmic branch by the quasiparticle tunnelling characteristics described in section A2.7.1. In general, LTS Josephson tunnel junctions always show hysteresis, whereas HTS Josephson junctions are only hysteretic at low temperatures (with the exception of intrinsic Josephson junctions between copper oxide planes [6]). This is mainly due to the difference in their resistance-area products, which are typically smaller than $1\,\Omega\,\mu\mathrm{m}^2$ for HTS junctions as compared to $100\,\Omega\,\mu\mathrm{m}^2$ above the gap voltage for LTS tunnel junctions.

Just as is the case with finite capacitance, many other properties of current biased Josephson junctions are difficult to derive analytically. Important cases which have been treated numerically and using other techniques are the I–V characteristics of SQUIDs [18], the inverse ac Josephson effect and the effect of thermal noise. Under microwave irradiation, the Shapiro spikes which appear in the voltage biased case are replaced by Shapiro steps at the same voltages, but the step heights are only equal to that of the spikes when $\hbar\omega_{\mathrm{rf}} > 2eI_0R$. An example is shown in figure A2.7.11. Following Ambegaokar and Halperin [19] the effect of thermal noise is quantified by $\gamma = \hbar I_0/ekT$. This causes the rounding of the current–voltage characteristic in the vicinity of the critical current. This is illustrated by figure A2.7.12 where the I–V characteristic of a grain boundary Josephson junction at 77 K can be compared to the $\beta_c = 0$ case in figure A2.7.10. For high temperature superconductors, thermal noise is a serious problem because of the much higher operation temperature.

E4.2

B4.4.3

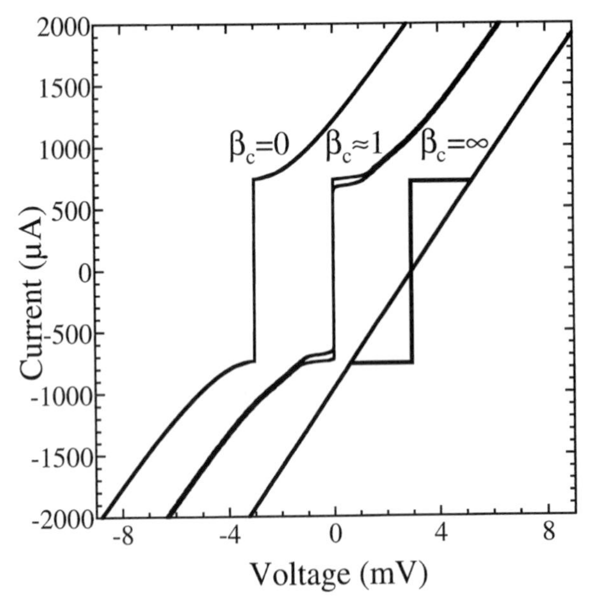

Figure A2.7.10. I–V characteristics of current biased Josephson junctions with different values of β_c. The curve for $\beta_c \approx 1$ is experimental data measured using a bicrystal grain boundary Josephson junction. The curves for $\beta_c = 0$ and $\beta_c = \infty$ have been calculated using the resistance and critical current of the real junction.

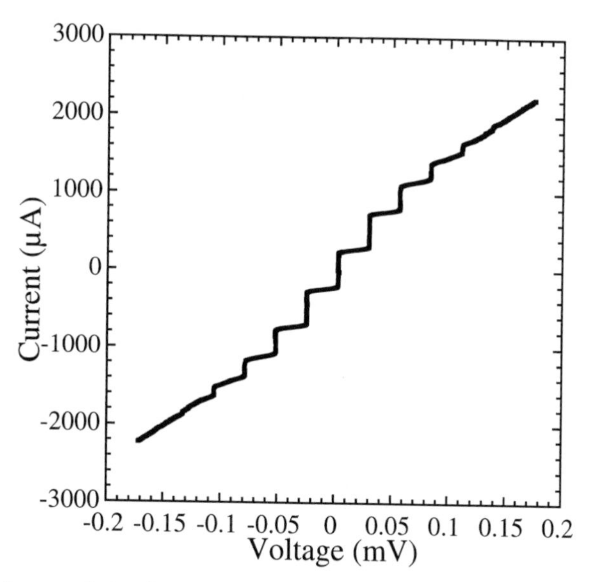

Figure A2.7.11. The I–V characteristic of a Nb–Au–Nb junction under microwave irradiation showing Shapiro steps (courtesy of Richard Moseley and Dr Mark Blamire).

A2.7.5 The magnitude of the critical current

This subject has been left until last because, unlike the other topics covered earlier, it is very model dependent and can only be determined using microscopic theory. We will give the results for two important cases that may be said to represent the extremes: classic tunnel junctions and superconductor–normal metal–superconductor (SNS) junctions. There is a range of behaviour in between these examples which is very sensitive to the device structure. E4.7

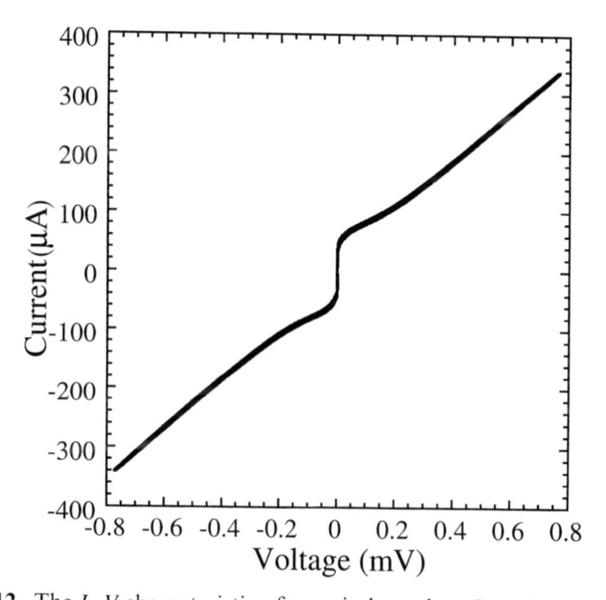

Figure A2.7.12. The I–V characteristic of a grain boundary Josephson junction at 77 K.

Tunnel junctions were the case originally treated by Josephson, but the form of $I_0(T)$ was derived by Ambegoakar and Baratoff [20]:

$$I_0R = \frac{\pi\Delta}{2e}\tanh\left(\frac{\Delta}{2kT}\right) \tag{A2.7.26}$$

which near T_c can be approximated by

$$I_0R = \frac{2.34\pi k}{e}(T_c - T) \tag{A2.7.27}$$

The temperature dependence of I_0 is largely determined by that of the gap $\Delta(T)$ and the behaviour near T_c is determined by the fact that in this region $I_0 \propto \Delta^2(T)$. If the junction had electrodes with different transition temperatures, we would have $I_0 \propto \sqrt{(T_c - T)}$. At $T = 0$, $I_0R = \pi\Delta(0)/2e$, which sets an upper limit on ω_c and, hence, on the range of frequencies over which the ac Josephson effects can be observed. Close to T_c, for microbridges, I_0R is given by equation (A2.7.27), but at lower temperatures rises above the value for a tunnel junction and has a form which depends on the mean free path in the superconductor [11]. At $T = 0$, $I_0R = \pi\Delta(0)/e$ in the clean limit and $I_0R = 1.3\pi\Delta(0)/2e$ in the dirty limit.

For SNS junctions, a simple expression can only be written down for special cases in the dirty limit, we show here two limiting cases. Close to T_c the temperature dependence of I_0R is given by [21]

$$I_0R \approx \frac{\pi\Delta(0)}{e}\frac{\rho_n^2\xi_n l}{\rho_s^2\xi_s^2}\left(1 - \frac{T}{T_c}\right)^2 \exp\left(\frac{-l}{\xi_n}\right) \tag{A2.7.28}$$

for junctions whose barrier length l is longer than the normal metal coherence length ξ_n, where ξ_s is the superconductor's coherence length and the barrier and superconductor have normal state resistivities ρ_n and ρ_s, respectively. The exponential dependence of I_0R upon the barrier length clearly makes its value typically much smaller than that of a tunnel junction or a microbridge near T_c. However, equation (A2.7.28) becomes invalid when l is shorter than ξ_n and, for very small barrier lengths, an SNS junction approaches the microbridge limit [11]. At very low temperatures we have

$$I_0R = \frac{29\pi\Delta(0)}{e}\left[\frac{\xi_n(T_c)}{l}\right]^2. \tag{A2.7.29}$$

The detailed behaviour of a particular device is determined by its structure and the materials used in its fabrication, which give a wide variety of behaviour even with conventional superconducting systems, but all of the results given in this section have been verified for systems which are close to the models based on which they have been derived. An excellent survey of the experimental situation for high temperature superconductors is given in Ref. [22].

References

[1] Josephson B D 1962 Possible new effects in superconductive tunnelling *Phys. Lett.* **1** 251
[2] Ryhänen T, Seppä H, Ilmoniemi R and Knuutila J 1989 SQUID magnetometers for low-frequency applications *J. Low Temp. Phys.* **76** 287
[3] Likharev K K 1996 Ultrafast superconductor digital electronics: RSFQ technology roadmap *Czech. J. Phys.* **46** 3331
[4] Kohlmann J, Muller F, Gutmann P, Popel R, Grimm L, Dunschede F W, Meier W and Niemeyer J 1997 Improved 1 V and 10 V Josephson voltage standard arrays *IEEE Trans. Appl. Supercond.* **7** 3411
[5] Wan K, Jain A K and Lukens J E 1989 Submillimeter wave generation using Josephson junction arrays *Appl. Phys. Lett.* **54** 1805
[6] Kleiner R and Muller P 1994 Intrinsic Josephson effects in high T_c superconductors *Phys. Rev.* B **49** 1327

[7] Chaudhari P, Mannhart J, Dimos D, Tsuei C C, Chi J, Oprysko M M and Scheuermann M 1988 Direct measurement of the superconducting properties of single grain-boundaries in $YBa_2Cu_3O_{7-\delta}$ *Phys. Rev. Lett.* **60** 1653
[8] Gross R and Mayer B 1991 Transport processes and noise in $YBa_2Cu_3O_{7-\delta}$ grain–boundary junctions *Physica* C **180** 235
[9] Waldram J R 1976 The Josephson effects in weakly coupled superconductors *Rep. Prog. Phys.* **39** 751
[10] Wolf E L 1996 Scanning tunneling spectrometry of a superlattice superconductor: $Bi_2Sr_2CaCu_2O_8$ *Superlattice Microstruct.* **19** 305
[11] Likharev K K 1979 Superconducting weak links *Rev. Mod. Phys.* **51** 101
[12] Pauza A J, Campbell A M, Moore D F, Somekh R E and Broers A N 1994 Josephson-junctions in $YBa_2Cu_3O_{7-\delta}$ by electron-beam irradiation *Physica* B **194** 119
[13] Yanson I K 1970 Effect of fluctuations on the dependence of the Josephson current on the magnetic field *Sov. Phys.–JETP* **31** 800
[14] Hilgenkamp H, Mannhart J and Mayer B 1996 Implications of $d_{x^2-y^2}$ symmetry and faceting for the transport properties of grain boundaries in high-T_c superconductors *Phys. Rev.* B **53** 14586
[15] Rosenthal P A, Beaseley M R, Char K, Colclough M S and Zaharchuk G 1991 Flux focusing effects in planar thin-film grain-boundary Josephson-junctions *Appl. Phys. Lett.* **59** 3482
[16] Mayer B, Schuster S, Beck A, Alff L and Gross R 1993 Magnetic-field dependence of the critical current in $YBa_2Cu_3O_{7-\delta}$ bicrystal grain-boundary junctions *Appl. Phys. Lett.* **62** 783
[17] Gray K E, Field M B and Miller D J 1998 Explanation of low critical currents in flat, bulk versus meandering, thin-film [001] tilt bicrystal grain boundaries in $YBa_2Cu_3O_{7-\delta}$ *Phys. Rev.* B **58** 9543
[18] Tesche C D and Clarke J 1977 dc SQUID: Noise and Optimisation *J. Low Temp. Phys.* **29** 301
[19] Ambegaokar V and Halperin B I 1969 Voltage due to thermal noise in the dc Josephson effect *Phys. Rev. Lett.* **22** 1364
[20] Ambegaokar V and Baratoff A 1963 Tunneling between superconductors *Phys. Rev. Lett.* **10** 486
[21] de Gennes P G 1964 Boundary effects in superconductors *Rev. Mod. Phys.* **36** 225
[22] Delin K A and Kleinsasser A W 1996 Stationary properties of high-critical-temperature proximity effect Josephson junctions *Supercond. Sci. Technol.* **9** 227

A3
Introduction to section A3: Elementary theory

D A Cardwell and D S Ginley

Introduction

Superconductivity was discovered by Kammerlingh-Onnes in 1911. The remarkable and unpredictable nature of this solid state phenomenon is highlighted by the extended period that followed, during which no reasonable explanation for its occurrence could be found. In fact, it was not until 1935 that the London brothers made significant progress in formulating a classical phenomenological theory (i.e. one in which *ad hoc* assumptions are made to describe experimental observations correctly) for the magnetic properties of superconductors and the electrodynamics of the supercurrent. The resulting London equations, which relate current density to electric and magnetic field, were based on the experimental observation of the Meissner state and assumed that two types of charge carriers were present in the superconducting state (the so-called two fluid model). These equations were subsequently able to predict the penetration and exponential decay of magnetic field within the surface of the superconductor.

Another period of low theoretical activity followed the Londons' work. In 1950, however, Frölich proposed that electrons could couple, or interact, via the exchange of a phonon. At about the same time Ginzburg and Landau developed a quantum mechanical, phenomenological theory to predict the effect of magnetic field on superconductors. The Ginzburg–Landau theory was based on the assumption that the behaviour of superconducting electrons can be described by an order parameter, which, in turn, is derived from an effective wave function. This theory was able to predict subtle differences in the behaviour of superconductors in the presence of a magnetic field, compared with the London theory. In addition, it established the criteria for formation of type I and type II superconductivity, based on the coherence length and penetration depth of the material.

Despite the success of the London and Ginzburg–Landau approaches, is was not until 1956 that Cooper proposed the existence of Cooper pairs in an extension of Frölich's electron coupling hypothesis. This proved to be a major development in the formulation of a microsopic theory of superconductivity. One year later in 1957 Bardeen, Cooper and Schreiffer proposed the formation of a Bose condensate of Cooper pairs in a superconducting ground state and the energy gap associated with this condensation. The so-called BCS theory is able to predict the transition temperature of most low T_c materials from the Debye phonon frequency and density of electron states close to the Fermi energy.

It was not until the discovery of high temperature superconductors in 1986 that the success of the BCS theory was questioned (BCS predicts a maximum theoretical transition temperature of below 30 K). HTS materials have subsequently provided theoreticians with considerable new challenges over the past decade and a consensus of the mechanism of superconductivity above 30 K has so far proved ellusive.

The two papers in this section describe in detail the key elements of both phenomenological and microscopic theories of superconductivity, although the division between the two is somewhat ill-defined. They account for the development of the theories described above and outline their successes and limitations. One can not help feeling a sense of history repeating itself, however, with the so far unexplained behaviour of HTS materials.

A3.1
Phenomenological theories

A M Campbell

A3.1.1 Introduction

Phenomenological theories take some experimental results, make a simple assumption and calculate the consequences. If they agree with experiment, the phenomenological theory is considered a success. With this definition all theories in physics, including BCS, are phenomenological theories, but conventionally A3.2 the term is used in superconductivity to describe the London theory [1] and Ginzburg–Landau theory (G–L) [2] in which the assumptions do not involve the interactions between electrons. An early phenomenological theory is the 'two fluid model' [3] which splits the electrons into two separate populations, the normal electrons and the superconducting electrons. Like many other concepts in superconductivity this model originated in work on liquid helium and, in spite of being contrary to the spirit of the quantum mechanical picture of electrons as identical particles, remains an important part of the language of superconductivity.

A3.1.2 The London theory

Both London and Ginzburg–Landau based their ideas on a macroscopic wave function (see chapter A1.2), but it is instructive to see how the London equations arise from a simple picture of free A1.2 electrons, although strictly such a system should be treated in the extreme non-local limit.

For free electrons $e\mathbf{E} = m\dot{\mathbf{v}}$ and $\mathbf{J} = n_s e\mathbf{v}$ where n_s is the number of superconducting electrons. Here $\mathbf{J}$ refers to the supercurrent. There may also be a normal current which behaves conventionally and independently of the supercurrent. (This can occur at high frequencies and in temperature gradients [4].) In this section n_s is assumed constant, in contrast to the G–L theory where it can vary in space and time for a number of reasons. The effect of variation of n_s due to composition variations is treated in [5]. Another important situation in which n_s varies is when a normal current is injected into a superconductor and the normal electrons gradually convert to super-electrons. This is the proximity effect.

Ignoring these complications it follows that:

$$\mathbf{E} = (m/n_s e^2)\dot{\mathbf{J}} \tag{A3.1.1}$$

and

$$\nabla \times ((m/n_s e^2)\dot{\mathbf{J}}) = -\dot{\mathbf{B}}\,. \tag{A3.1.2}$$

We lump the material parameters into a single parameter $\Lambda = (m/n_s e^2)$ and integrate equation (A3.1.2) to get

$$\mathbf{E} = \Lambda \dot{\mathbf{J}} \tag{A3.1.3}$$

and

$$\nabla \times (\Lambda \mathbf{J}) = -\mathbf{B} + f(\mathbf{r}) \tag{A3.1.4}$$

where Λ is a material parameter.

Equation (A3.1.4) allows an arbitrary time independent magnetic field to be present, the trapped field we would expect in a perfect conductor. However, the discovery of the Meissner effect suggested that in equilibrium there was no flux in the superconductor and London made the hypothesis that $f(\mathbf{r}) = 0$ so that the local field was always uniquely related to the local current density. This removed the multiple equilibrium states possible if the superconductor were merely a perfect conductor and was therefore much more consistent with classical thermodynamics. The London equations are

A2.2

$$\mathbf{E} = \Lambda \dot{\mathbf{J}} \tag{A3.1.5}$$

and

$$\nabla \times (\Lambda \mathbf{J}) = -\mathbf{B}\,. \tag{A3.1.6}$$

Taking the curl of equation (A3.1.6) leads immediately to the exponential decay of current and field with distance from the surface of the superconductor.

$$\nabla^2 \mathbf{J} = \mathbf{J}/\lambda^2 \tag{A3.1.7}$$

where

$$\lambda = (\Lambda/\mu_{0,})^{1/2} = (m/\mu_0 n_s e^2)^{1/2} \tag{A3.1.8}$$

λ is the London penetration depth.

Since the field is not completely excluded, the effective susceptibility, χ_{eff}, of a slab of half width d, parallel to the external field is then:

$$\chi_{\mathrm{eff}} = (\lambda/d) \tanh\,(d/\lambda) - 1\,. \tag{A3.1.9}$$

Here, χ_{eff} is defined as the total magnetic moment per unit volume divided by the external field. The connection between the effective susceptibility of a body and the susceptibility of the material of which it is composed is discussed in chapter D2.1, see equation (A3.1.31). This is only relevant to a uniform magnetization, which is not the case here, except in the limit of zero penetration depth. In this limit χ_{eff} is -1 for a slab, -2 for a transverse cylinder and -1.5 for a sphere.

D2.1

For a cylinder of radius r, parallel to the field, the equivalent expression is

$$\chi_{\mathrm{eff}} = (\lambda/r)\left(\frac{I_1(r/\lambda)}{I_0(r/\lambda)}\right) - 1 \tag{A3.1.10}$$

where I_1 and I_0 are Bessel functions of the second kind.

For spheres:

$$\chi_{\mathrm{eff}} = 1.5\left[(3\lambda/r)\coth\,(r/\lambda) - 3(\lambda/r)^2\right] - 1. \tag{A3.1.11}$$

A3.1.2.1 Multiply connected systems

Multiply connected samples require care. The term means samples with holes containing trapped flux, or solid samples containing vortices. Suppose we have a hole in a superconductor and apply Faraday's law and equation (A3.1.5) to a loop round it within the superconductor

$$\oint \dot{\mathbf{B}}\ \mathbf{dS} = -\oint \mathbf{E}\ \mathbf{dl} = -\oint \mu_0\lambda^2\dot{\mathbf{J}}\ \mathbf{dl} \tag{A3.1.12}$$

Hence the integral $\oint(\mathbf{B}\ \mathbf{dS} + \mu_0\lambda^2\mathbf{J}\ \mathbf{dl})$ is constant with time. This integral is the 'fluxoid', which reduces to the flux in macroscopic samples when the contour can be taken in a region deep in the superconductor where $\mathbf{J} = 0$. In small samples, or around vortices in type II superconductors, the flux is less but this is compensated for by the fact that there will be a current density $\mathbf{J}$. For a loop which does not contain the hole the fluxoid is zero.

A2.6

An extension to the London equation (A3.1.6) is the equation if there is a vortex present at a point $\mathbf{r}_0$. A vortex has a singularity at its centre so the equation is

$$\nabla\times(\mu_0\lambda^2\mathbf{J}) + \mathbf{B} = \phi_0\delta(\mathbf{r}-\mathbf{r}_0)\,. \tag{A3.1.13}$$

It can be seen that this is equation (A3.1.6) at all points except the vortex core. However, by integrating over a surface spanning $\mathbf{r}_0$, we see that it also fulfills the condition that the fluxoid of the vortex is equal to a constant ϕ_0, shown below to be quantized in units of $h/2e$ (see section A3.1.3.3).

The single vortex solution is

$$B(r) = \phi_0 K_0(r/\lambda)/(2\pi\lambda^2)\,. \tag{A3.1.14}$$

A vortex lattice is made up by the superposition of such vortices at various values of $\mathbf{r}_0$ and, since in high T_c superconductors, the core is extremely small the London limit is an excellent approximation for calculating the properties of the vortex lattice (see chapter A4.2).

A4.2

Flux in a thin walled cylinder

To illustrate the London equations in a multiply connected system we consider a hollow cylinder of radius a with a wall thickness much less than the penetration depth. We start with zero field and current and increase the external field H_0 parallel to the axis. We assume that the wall is sufficiently thin for the current density to be treated as uniform and the field to be assumed equal to the external field as a first approximation.

From Faraday's law, at the wall, $2\pi aE = \pi a^2\mu_0\dot{H}_0$ so $E = a\mu_0\dot{H}_0/2$. Substituting this in equation (A3.1.5) and integrating over time gives

$$J = aH_0/(2\lambda^2)\,. \tag{A3.1.15}$$

The degree of screening, i.e. the drop in field across the wall, is J times the wall thickness, d,

$$\delta H/H_0 = ad/(2\lambda)^2\,. \tag{A3.1.16}$$

Significant flux is excluded when λ^2 becomes comparable to ad. The penetration depth is then intermediate in size between the wall thickness and radius, a thickness comparable to λ is not needed for nearly complete screening. (This is a rather general result for thin sheets, another example is a film perpendicular to an applied field. See chapters D2.1 and D2.2)

D2.1, D2.2

This situation cannot continue indefinitely, there comes a point at which the current density is too great, and we now find this value. We can argue that the material will go normal when the change in free energy, $\mu_0 H_c^2/2$ per unit volume, which in this case is $\pi a d \mu_0 H_c^2$, is equal to the work done by the solenoid which is $\mu_0 H_0 \pi a^2 J d$. Substituting for H_0 from equation (A3.1.12) gives

$$J = H_c/\lambda\,. \tag{A3.1.17}$$

This is the maximum current density the material can carry, and is equal to the surface current density when the surface field in a bulk samples reaches H_c. It is an overestimate, as the high current density reduces the free energy from its value at zero field and current. (Also the thermodynamics is slightly suspect as the transition is irreversible, the cylinder will immediately become superconducting again once the current has died away so it is only valid in the limit of thin walls.) A better figure, although only differing by a numerical constant, comes from the G–L theory below (section A3.1.3.2). The analysis here brings out the point that, although this geometry is topologically multiply connected, provided we have not exceeded the critical current, there is no trapped flux and it is simply connected as far as the London equations are concerned. Trapped flux is considered in section A3.1.3.4.2.

D2.6

A3.1.2.2 The gauge of **A**

D2.1

We can express the London equations in terms of the vector potential defined by $\nabla \times \mathbf{A} = \mathbf{B}$. With this definition we can add the gradient of a scalar to **A** without changing **B** and this is called a gauge transformation. It must not change any observable quantity. To define **A** more precisely we often define $\nabla \cdot \mathbf{A} = 0$, although equation (A3.1.30) shows that this gauge may not always be consistent with the G–L theory. **A** is still not defined uniquely until we define the boundary conditions, and this is not trivial. There are many different boundary conditions for the definition of a unique electrostatic potential and, since the vector potential has three components there are three times as many possible boundary conditions to define a unique value for **A**.

However, if we define $\nabla \cdot \mathbf{A} = 0$, it follows from equation (A3.1.6) that $\mathbf{J} + \mathbf{A}/(\mu_0 \lambda^2) = \nabla \phi$ where ϕ is any scalar function which satisfies $\nabla^2 \phi = 0$. To define **A** uniquely in a simply connected body London imposed the boundary condition that $\mu_0 \lambda^2 \mathbf{J}_n + \mathbf{A}_n = 0$ where subscript n denotes the normal component, i.e the normal component of $\nabla \phi$ is zero. This is the boundary condition for the London gauge. The electrostatic analogy is a body with no charge density (since $\nabla^2 \phi = 0$) and no normal component of the electric field at any point on the surface. In this case the only solution for the electrostatic potential is a constant. Hence in the London gauge we can assume $\phi = 0$ since the addition of a constant to **A** cannot affect the magnetic field. In multiply connected samples this boundary condition does not define **A** uniquely, see section A3.1.3.4.

Now equations (A3.1.5) and (A3.1.6) become

$$\mathbf{E} = -\dot{\mathbf{A}} \tag{A3.1.18}$$

and

$$\mu_0 \lambda^2 \mathbf{J} = -\mathbf{A}\,. \tag{A3.1.19}$$

Equation (A3.1.19) can be interpreted as saying that the current density at any point is proportional to the flux that has crossed that point, in contrast to a normal material where it is proportional to the rate at which flux crosses the point, i.e. the electric field. Note that we have now defined **A** uniquely; we can no longer add even a constant without affecting the current density.

The importance of the vector potential brings out the involvement of wave mechanics in superconductivity, something that London realized from the outset since he postulated a Bose–Einstein condensation as the cause of superconductivity. Equation (A3.1.19) can be put in the form $\mathbf{p} = 0$ where **p**, the momentum vector in wave mechanics of a particle of charge e and mass m, is given by:

$$\mathbf{p} = m\mathbf{v} + e\mathbf{A}\,. \tag{A3.1.20}$$

For superconductivity to occur this momentum vector must remain zero when a magnetic field is applied, and this is what is meant by the 'rigidity' of the superconducting state.

Since the phase of a wave-function θ is related to the momentum by

$$\mathbf{p} = \hbar\mathbf{k} = \hbar\nabla\theta\,. \tag{A3.1.21}$$

It follows that

$$m\mathbf{v} + e\mathbf{A} = \hbar\nabla\theta\,. \tag{A3.1.22}$$

Integrating round a closed loop

$$\oint (m\mathbf{v} + e\mathbf{A})/\hbar \quad \mathbf{dl} = \Delta\theta\,. \tag{A3.1.23}$$

The quantity in the integral is proportional to the fluxoid defined above, and London pointed out that this produced a natural unit for the fluxoid, $\phi_0 = h/e$. Since we would expect the change in phase round a closed loop $\Delta\theta$ to be an integer multiple of 2π the equation also implies the quantisation of flux, but this does not seem to have been stated explicitly until it was derived from the Ginzburg–Landau theory.

A3.1.3 Ginzburg–Landau theory

The Ginzburg–Landau theory of superconductors has been an astonishingly rich source of new physics. It has been much more widely productive than even the original authors expected since one sentence in their paper stated that the solutions of the Ginzburg–Landau equations for the limit κ tend to infinity, offer no intrinsic interest and would not be discussed. (However it should be added that they did derive the maximum field for a solution to exist, which we now know as B_{c2}, and recognised that an instability occurs if $\kappa > 1/\sqrt{2}$. κ is defined in equation (A3.1.38).) Although proposed for temperatures near T_c the theory can be derived from the microscopic theory at absolute zero, and so can be used at all temperatures, at least semiquantitatively. Different authors use different solutions of these equations to make their points and therefore different simplifications. Only an outline can be given here, with some examples which are chosen because they do not appear in the standard texts, but more details can be found in the books by Tinkham [6] and de Gennes [7]. The book by St. James *et al* [8] probably has the widest range of solutions, particularly those involving vortices, and van Duzer and Turner [9] is the most relevant to thin films and devices. A recent book by Waldram [4] explains in detail the connection between Ginzburg–Landau and the microscopic theory, and includes the applications to cuprate superconductors.

The G–L theory is an extension of their general theory of second order phase transitions, which is based on a free energy expanded in a power series of some kind of order parameter. For example, in a ferromagnetic transition it is the number of aligned electron spins. In a superconductor they made the plausible assumption that, since the energy is lowered when the electrons become superconducting, we can write the free energy near the critical temperature as a power series in the number of superconducting electrons, n_s. (The fact that fluctuation effects mean that near T_c the expansion as a power series is invalid has in no way diminished the predictive value of the G–L theory.) However this assumption is not sufficient to describe a superconductor in a magnetic field, since in this situation the lowest energy would be obtained by splitting the sample into a series of laminae small compared with the penetration depth, so that we gain the energy of the superconducting electrons without the disadvantage of having to exclude the field. It was therefore postulated that there was an energy term dependent on the gradient of the order parameter (squared since energies must be positive), as well as its magnitude. This will produce a surface energy which prevents the subdivision lowering the energy.

The London theory had been based on the idea of the superconducting state as a macroscopic quantum state, so it seems natural (with hindsight) to postulate a macroscopic wave function ψ such that $n_s = |\psi|^2$ is the order parameter. Now a wave function has both magnitude and phase and can be written as $\psi = f\exp(i\theta)$, where f and θ are real functions of position. f must be single valued, θ can be multivalued, but only by a multiple of 2π. This has important consequences. The phase θ is related to the momentum **p** by $\mathbf{p} = \hbar\mathbf{k} = \hbar\nabla\theta$. We suppose the current carriers have mass m^* and charge e^*. (Note that some authors use e as the electronic charge while others use $-e$ which causes differing signs in the G–L equations if expressed in terms of the electronic charge.) In a magnetic field $\mathbf{p} = m^*\mathbf{v} + e^*\mathbf{A}$ and the observable quantity, the velocity **v**, cannot depend on the gauge of **A**. (The phase is not directly observable, only the gauge independent phase difference, see section A3.1.3.4.3.) Hence if we change **A** by adding an arbitrary $\nabla\phi$ then we must add $e^*\phi/\hbar$ to the phase of ψ. This means that in the expression for the free energy, which must also be gauge invariant, we cannot have a simple $(\nabla f)^2$ or ∇f^2 but, instead, the gauge invariant expression $(1/2m^*)|(i\hbar\nabla + e^*\mathbf{A})\psi|^2$. (Although ∇f^2 might seem the simplest answer it is not compatible with the idea of a macroscopic wave function.) The physical significance of this term in the free energy is brought out by substituting $f\,\exp(i\theta)$ for ψ and rewriting it as $1/2m^* n_s v^2 + (1/2m^*)(\hbar\cdot\nabla f + e^* fA)^2$. This shows that as well as the gradient and field term, we have automatically included the kinetic energy of the electrons, which is an important term in a superconductor.

To these terms we add the magnetic energy of the field in the superconductor, **b**, giving the following expression for the free energy in the superconducting state F_s

$$F_s(b) = F_n(0) + \alpha|\psi|^2 + \frac{1}{2}\beta|\psi|^4 + (1/2m^*)|(i\hbar\nabla + e^*\mathbf{A})\psi|^2 + \mathbf{b}^2/2\mu_0 \qquad \text{(A3.1.24)}$$

α and β are the only material parameters, which will be temperature dependent. In order to get a transition as we go through T_c, α must be positive above T_c and negative below it. To ensure an energy minimum as opposed to a maximum, β must be positive at all temperatures. From the case where $\psi = 0$ it follows that $F_n(b) = F_n(0) + b^2/2\mu_0$. As pointed out in chapter C2.2 this means that if we want to add the work done on an isolated sample by an external solenoid to make a Gibbs function, or an availability, we must use **H·dB** integrated over all space, not **BdH** or **HdM**.

D2.1

G–L then used the calculus of variations to turn the minimum energy integrated over the sample volume into two differential equations. Refer to [4] and [8] for more details. Although it is normally assumed that the material is uniform, we get the same equations if α and β are functions of position so we can include pinning centres and the proximity effect. This allows us to calculate the critical current of a region of reduced gap, such as a grain boundary [10]. Anisotropy can be added by making m^* an effective mass tensor instead of a scalar (although strictly tensors can only be used in linear systems of equations, and the equations are not linear). These equations are the famous Ginzburg–Landau equations which we will just quote here. The first is

$$-\mathbf{j} = (ie^*\hbar/2m^*)(\psi^*\nabla\psi - \psi\nabla\psi^*) + (e^{*2}/m^*)|\psi|^2\mathbf{A} \qquad \text{(A3.1.25)}$$

where **j** is the local current density. This is identical to the general equation for the current density in wave mechanics. The second contains the material parameters and is

$$\alpha\psi + \beta|\psi|^2\psi + (1/2m^*)(i\hbar\nabla + e^*\mathbf{A})^2\psi = 0\,. \qquad \text{(A3.1.26)}$$

The surface contribution to the energy gives the G–L boundary condition on ψ [4]:

$$(\mathrm{i}\hbar\nabla_{\mathrm{n}} + e^*\mathbf{A}_{\mathrm{n}})\psi = 0 \qquad \text{(A3.1.27)}$$

where subscript n means the normal component.

If the boundary is a vacuum there can be no current across it, which implies that $\mathrm{d}\psi/\mathrm{d}x = 0$ at the boundary (equation (A3.1.25). If the boundary is with a normal metal the boundary condition must be derived from the microscopic theory, see [7] and [4] which discuss the boundary conditions in some detail. A3.2

Since the units of ψ are arbitrary we can chose m^* as an arbitrary parameter before fixing the material parameters α and β. m^* is usually chosen as twice the electron mass, which means that ψ^2 is a number per unit volume. The value of e^* is not arbitrary since it determines the size of the flux quantum and, from experiment (and the microscopic theory), is twice the electronic charge (negative for electrons and positive for holes). This means that the properties of a superconductor at any given temperature are determined by only two material parameters α and β. We usually replace these with two directly measurable parameters, for example the critical field and penetration depth, which are derived below in terms of α and β. Further simplifications are to normalize the variables with these parameters to give a dimensionless equation with only one parameter, κ, and to use a gauge transformation to remove the phase of the order parameter where possible. This is done below. A1.2

Although Maxwell's equations for the magnetic field might appear to be an independent set of equations, the equations for **b** can be derived, like the G–L equations, by minimizing the magnetic energy $b^2/2\mu_0$, which is included in the G–L free energy, so the G–L equations are automatically consistent with the field equations. The free energy is a local expression for the energy density on the scale of several atoms. By convention lower case letters are usually used for local fields. An exception is **A**, the vector potential, which is also a local field. Upper case letters are usually used for averages, or integrals over the sample.

The general question of equilibrium conditions is covered in detail in chapter D2.1. If we are interested in the internal structure of the superconductor it is easiest to minimize the Helmholtz free energy $F = U - TS$, with no external work done, i.e. constant total flux. This can be done both for wires or films carrying a current, or an isolated sample in a magnetic field. However, if we want to know the external field, $\mathbf{H}_0$, in equilibrium with the resulting structure in the latter case we must vary **B** and put the change in F equal to the work done by an external solenoid. The free energy F includes the free space magnetic energy, so for long cylinders parallel to the external field we must use a work term $\delta w = \mathbf{H}_0\boldsymbol{\delta\phi}$, where $\boldsymbol{\phi}$ is the flux entering the sample, in order to be consistent with the change in F in the normal state. If we have a sample of unit cross-sectional area this can be written $\mathbf{H}_0\delta\mathbf{B}$ where **B** is the average of **b** over the sample cross-section. We therefore need to minimize $F - \mathbf{H}_0\mathbf{B}$ at constant T and $\mathbf{H}_0$. If the sample is in thermodynamic equilibrium we can then, if we wish, define a local $\mathbf{H} = \mathbf{H}_0$ and minimize a local Gibbs function $F - \mathbf{H}\cdot\mathbf{B}$. Note however that in this expression **A** and **b** are fields on a scale between the atomic and the coherence length, while **H** is uniform across the sample and **B** is an average over the sample cross-section. If the sample has a finite demagnetising factor the calculation of the magnetic energy requires the integration of b^2 over all space. This is very intractable in most cases so normally minimizing a Gibbs function is only done for a sample with no demagnetizing factor. The other extreme for thin small samples perpendicular to the field is, however, also soluble. Extensive examples of numerical solutions can be found in [11] and [12] and references therein. D2.1

Although in principle we can find the equilibrium state of a superconductor by minimizing the free energy numerically with respect to all variables at all points, this requires too much computing power for macroscopic samples and to obtain analytic solutions we usually do the minimization in several steps with increasing length scales. We illustrate this with the steps used to derive the Abrikosov vortex lattice A1.1

[19]. First, we minimize F with respect to $\mathbf{A}$ and ψ. This leads to the standard G–L equations above, equations (A3.1.25), (A3.1.26), and ensures that $\mathbf{b}$ and ψ are in equilibrium on the scale of many atoms. There are many solutions to these equations, of which one is an array of vortices in arbitrary positions. We now try a periodic solution with vortices on a regular lattice, and minimize F by moving a fixed number of vortices around. This is equivalent to surrounding the sample containing the vortices with a Type I material, to keep the total flux fixed, and allowing the vortices to settle into their equilibrium structure. It tells us the structure for a given flux density, and does not involve anything outside the superconductor, so avoiding complications with $\mathbf{H}$ or applied fields. It is the equivalent to a constant volume experiment in gases.

Having determined the structure of the vortex state for a given vortex density, we then change the vortex density $\mathbf{B}$, keeping the same structural symmetry (for example a hexagonal array), and allow the external field to do work. Then putting $\delta F = \mathbf{H}_0 \delta \mathbf{B}$ gives the equilibrium flux density with an external field $\mathbf{H}_0$ which is then the local relation between $\mathbf{B}$ and $\mathbf{H}$ in equilibrium on the scale of many flux lines. In gases this gives the relation between the volume and the external pressure.

Finally, if we have typical high T_c crystals with a large demagnetising factor, we use the relation between $\mathbf{B}$ and $\mathbf{H}$ we have derived, combined with classical electromagnetism, to find the magnetic properties of the crystal, just as if it were a ferromagnetic crystal.

A3.1.3.1 Material properties

It is necessary to relate the empirical material parameters α and β to measurable quantities, which is done by considering a long sample parallel to the external field (no demagnetizing effects). To do this we separate the amplitude and phase of the order parameter by a gauge transformation. We put

$$\psi = f \exp(\mathrm{i}\theta) \tag{A3.1.28}$$

with f and θ real, and

$$\mathbf{A} = \mathbf{A}' + (\hbar/e^*)\nabla\theta . \tag{A3.1.29}$$

Since in what follows we will use this new vector potential we now drop the prime after $\mathbf{A}$. Equation (A3.1.25) becomes

$$-\mathbf{j} = (e^{*2} f^2/m^*)\mathbf{A} . \tag{A3.1.30}$$

With a good deal of algebra, (including the use of equation (A3.1.30) to show that $\nabla(f^2\mathbf{A}) = 0$) equation (A3.1.26) becomes

$$(\hbar^2/2m^*)\nabla^2 f = (\alpha + \frac{1}{2}\beta f^2 + (e^*2/2m^*)\mathbf{A}^2)f . \tag{A3.1.31}$$

The free energy is equation (A3.1.24) with f written for ψ

$$F_{\mathrm{s}}(b) = F_{\mathrm{n}}(0) + \alpha f^2 + \frac{1}{2}\beta f^4 + (1/2m^*)|(\mathrm{i}\hbar\nabla + e^*\mathbf{A})f|^2 + \mathbf{b}^2/2\mu_0 . \tag{A3.1.32}$$

We now derive a number of measurable parameters from these equations.

Inside a uniform bulk material in zero field $\mathbf{b} = 0$, $\mathbf{A} = 0$ and there are no gradients, so the order parameter is found by putting $\mathrm{d}F/\mathrm{d}\psi = 0$ and is given by $f^2 = \psi_0^2 = -\alpha/\beta$. It is common to normalize ψ by this factor.

In low magnetic fields the order parameter will not be very different from this value, so putting $\psi = \psi_0$ in equation (A3.1.25) and putting

$$\mu_0 \mathbf{j} = \nabla \times \mathbf{B} = \nabla \times \nabla \times \mathbf{A} = -\nabla^2 \mathbf{A}\,. \tag{A3.1.33}$$

Then

$$-\nabla^2 \mathbf{A} = (\mu_0 e^{*2} \alpha / \beta m^*)\mathbf{A}\,. \tag{A3.1.34}$$

This is the same as the London equation and the penetration depth is given by

$$\lambda^2 = -m^* \beta / (\mu_0 e^{*2} \alpha) \tag{A3.1.35}$$

(note that α is negative below T_c).

In zero field if we ignore the second order in $|\psi|^2$, equation (A3.1.31) becomes

$$(\hbar^2/2m^*)\nabla^2 f = \alpha f\,. \tag{A3.1.36}$$

This shows that above T_c (where α is positive and ψ small) a fluctuation in the order parameter varies over a characteristic length ξ, the coherence length, where

$$\xi^2 = |\hbar^2/2\alpha m^*|\,. \tag{A3.1.37}$$

Small changes in ψ decay exponentially over this distance. This is the temperature dependent, or G–L, coherence length which describes the length scale over which fluctuations in the order parameter occur. At absolute zero it corresponds to the size of a Cooper pair, but tends to infinity at T_c while the Cooper pair size remains constant. (Below T_c α changes sign and the solutions are more complex but ξ is still the characteristic length for changes in order parameter, see [4, 6].)

We now define the G–L dimensionless parameter κ

$$\kappa = \lambda/\xi = (2m^{*2}\beta/(\mu_0 e^{*2}\hbar^2))^{1/2}\,. \tag{A3.1.38}$$

To find the critical field we put the sample in a solenoid with a field B_c, and allow the material to change reversibly from the superconducting state to the normal state. In the superconducting state $B = 0$ in most of the sample so

$$F_s(B) = F_n(0) + \alpha\psi_0^2 + \frac{1}{2}\beta\psi_0^4\,. \tag{A3.1.39}$$

In the normal state

A2.1

$$F_n(B) = F_n(0) + B_c^2/2\mu_0\,. \tag{A3.1.40}$$

The work done by the solenoid as the flux enters the sample is (B_c^2/μ_0)

Putting the change in F equal to the work done gives:

$$B_c^2 = \mu_0 \alpha^2/\beta\,. \tag{A3.1.41}$$

Since there are only two independent material parameters (initially α and β) the parameters we have derived can be expressed in terms of any two of them. For solution the equations are usually put in dimensionless form by dividing distances by λ, fields by $\sqrt{2}B_c$ and the order parameter by ψ_0. In this form the equations depend only on κ.

The secondary dimensionless variables are then

$$\text{vector potential } \mathbf{a} = \mathbf{A}/\sqrt{2}B_c\lambda \tag{A3.1.42}$$

$$\text{flux} \quad \Phi = (2\pi/\kappa)\Phi/\phi_0 \tag{A3.1.43}$$

$$\text{current density} \quad \mathbf{j} = (\lambda\mu_0/\sqrt{2}B_c)\mathbf{J} \tag{A3.1.44}$$

$$\text{free energy difference} \quad \Delta E = \Delta F/2\Delta F_0 \tag{A3.1.45}$$

ΔF_0 is the free energy difference in zero field.

Equations (A3.1.25) and (A3.1.26) become:

$$\mathbf{j} = i/(2^* \kappa)(\psi^* \nabla\psi - \psi\nabla\psi^*) - \mathbf{a}\psi^* \psi \tag{A3.1.46}$$

$$(\kappa^{-1}\nabla + \mathbf{a})^2\psi + (\psi^* \psi - 1)\psi = 0 \tag{A3.1.47}$$

and equations (A3.1.30) and (A3.1.31) become

$$\mathbf{j} = -\mathbf{a}f^2 \tag{A3.1.48}$$

$$-\kappa^{-2}\nabla^2 f + (|\mathbf{a}|^2 + f^2 - 1)f = 0. \tag{A3.1.49}$$

The dimensionless free energy is

$$\Delta E = -f^2 + \frac{1}{2}f^4 + k^{-2}|\nabla f^2| + |\mathbf{a}^2|f^2 + \mathbf{b}^2. \tag{A3.1.50}$$

The fact that the G–L equations can be written in dimensionless form with only one parameter, κ, means that if we vary temperatures and materials we can plot the curves on top of each other by suitable scaling. Thus, all materials with the same value of κ will give the same magnetization curve at all temperatures if scaled. If we now add pinning centres the energies involved are those of the G–L theory so most simple theories lead to another dimensionless parameter and the hysteretic magnetization curves will also scale, but now with two adjustable parameters. This has been found to be true for many systems, (e.g. [13]) but is not universal. Scaling of experimental results is an important technique, but as the number of free parameters increases the physical significance decreases. Most experimental curves can be fitted to any reasonable power law with two or three adjustable parameters.

A3.1.3.2 The maximum current density

Equation (A3.1.25) shows that if we have a current density there is a gradient in ψ and a corresponding energy term attributable to the kinetic energy of the electrons. Therefore, the kinetic energy of the electrons at high current densities has a similar effect on the order parameter to that of a magnetic field. High current densities reduce ψ and if high enough cause a transition to the normal state. We assume a wire with a thickness much less than λ so that the current density j is uniform and gradient terms are zero. We substitute the expression for A in equation (A3.1.25) into equation (A3.1.26). From equations (A3.1.35) and (A3.1.41), $(H_c/\lambda)^2 = -(e^{*2}\alpha^3)/(m^*\beta^2)$. If we also put $s = (\psi/\psi_0)^2 = -\psi^2\beta/\alpha$ then equation (A3.1.26) becomes

$$s^2 - s^3 = -j^2\lambda^2/2H_c^2. \tag{A3.1.51}$$

We need a solution for s between 0 and 1. For zero j the low energy solution is $s = 1$ and the solution for s decreases as j increases. The largest value of j for which a solution exists is when the solution is at the maximum of $s^2 - s^4$ which is when $s^2 = 2/3$.

Then

$$j = \sqrt{(8/27)}H_c/\lambda \tag{A3.1.52}$$

This is the maximum possible current density and is similar to the previous result, equation (A3.1.14).

It is worth pointing out that this procedure is rather different from the way we derive critical fields. These are derived from the G–L equations assuming reversibility and putting the change in F equal to the work done by a solenoid. This is a conventional phase transition. A transport current is not a macroscopic equilibrium state and so cannot be part of a conventional phase transition. It is not possible to equate a change in free energy to the work done by the battery since the battery works continuously in the normal state. What this derivation does is to minimize energy on a local scale, i.e. start with the G–L equations, but then show that these have no solutions if the current density is larger than a limiting value.

This maximum current is close to the depairing current of the microscopic theory and is an absolute maximum critical current density. It can be interpreted as the current at which the kinetic energy of the superelectrons reaches the condensation energy, so that it becomes favourable to revert to the normal state. An ideal pinning array consisting of a hole down the core of each vortex can cause currents of this size. If we take an energy of $1/2\mu_0 H_c^2$ over a volume $\pi\xi^2$ per unit length this will produce a force of $1/2\pi\mu_0 H_c^2\xi$ per unit length. Putting this equal to the Lorentz force on a vortex, $J_c\phi_0$, and using the relations in section A3.1.4.1, gives a critical current density of $1.1H_c/\lambda$ which is in fact above the depairing current. Hence, although at first sight the maximum critical current due to pinning centres appears to be quite different from the depairing current, an ideal pinning array should be able to produce critical currents comparable with the depairing current. This can be of the order of 10^7–10^8 A cm^{-2} in both low and high T_c superconductors and values within a factor of ten of the depairing current have been achieved. Many high T_c superconductors can carry extremely high current densities, it is only the grain boundaries that limit the current for practical applications.

A4.2

A4.3

A3.1.3.3 Flux quantization

In the London theory the current density **j** is proportional to **A**, but since **A** well away from the hole is proportional to the total flux in a hole, while **j** tends to zero, this expression cannot hold in a multiply connected system with trapped flux, nor if there are vortices present.

D2.1

If we put $\psi = f\ \exp(i\theta)$ (but do not change the gauge) equation (A3.1.25) becomes

$$\mathbf{j} = (e^*\hbar/m^*)f^2\nabla\theta - (e^{*2}/m^*)f^2\mathbf{A}. \tag{A3.1.53}$$

Putting $f^2 = -\alpha/\beta$ and $\lambda^2 = -m^*\beta/(\mu_0 e^{*2}\alpha)$ this can be written

$$\mathbf{A} + \mu_0\lambda^2\mathbf{J} = \hbar\nabla\theta/e^*. \tag{A3.1.54}$$

If observables are to be single valued, then integrating round a closed loop the integral of $\nabla\theta$ is $2n\pi$, where n is an integer, so

$$\int \dot{\mathbf{B}}\ \mathrm{d}\mathbf{S} + \oint \mu_0\lambda^2\mathbf{j}\ \ \mathrm{d}\mathbf{l} = n\phi_0 \tag{A3.1.55}$$

where $\phi_0 = h/e^*$.

Hence the G–L theory leads to the conclusion that the fluxoid defined by London is quantized in units of h/e^*. The flux in any macroscopic hole must be an integer number of the flux quantum $\phi_0 = h/e^*$. In small systems the flux is less but the quantum is made up by the term in **J**. Since this result is also derivable from the London theory it only requires that the properties depend on the electrons being described by a macroscopic wave function. It is not dependent on the power law expansion of the G–L theory but is a more general result. The BCS theory and experiment are consistent in showing that e^* is $2e$.

A3.1.3.4 The phase of the order parameter and the gauge of the vector potential

The question of gauge invariance in the Ginzburg–Landau equations is a difficult one, and needs the study of a number of treatments for a full understanding. In [8] the problem is largely avoided by taking the curl of the equations and working in terms of the magnetic field. Clem [14] recommends using the superfluid current density, which is gauge invariant, rather than the phase and vector potential. (This is essentially the gauge invariant phase difference of equation (A3.1.59) below.) However, most texts use the phase and vector potential, so some remarks on the subject are appropriate.

Let us return to the transformation of the G–L equations by putting $\psi = f\exp(\mathrm{i}\theta)$ and $\mathbf{A} = \mathbf{A}' + (\hbar/e^*)\nabla\theta$ (equations (A3.1.28) and (A3.1.29).) One consequence is that since $\nabla\cdot\mathbf{j} = 0$, equation (A3.1.30) shows that $\nabla(f^2\mathbf{A}) = 0$. Therefore if the order parameter is not uniform we cannot use a gauge in which $\nabla\cdot\mathbf{A} = 0$. However in most cases f is a constant so we can put $\nabla\cdot\mathbf{A} = 0$.

More important is the fact that we have removed the phase from the G–L equations, which suggests that we can always use a gauge with a real order parameter. Unfortunately this is not true, because in making the substitution we have implicitly assumed that the phase is single valued.

If the phase is single valued the substitution in equations (A3.1.26) and (A3.1.27) to obtain equations (A3.1.30)–(A3.1.32) is straightforward. From section A3.1.3.3 there is no trapped flux, and there are no vortices. Furthermore, since the phase no longer appears in the equations we can ignore it and assume it to be zero. (This is probably a better definition of the London gauge, i.e. a real order parameter but see ‘A single vortex’ and ‘A thin walled cylinder’.)

The other possibility is a multivalued phase of the order parameter, which is a physically different situation. In general the potential θ used in the gauge transformation of equations (A3.1.24)–(A3.1.26) must be single valued. If this is not the case different values lead to a different integral of A round a circuit and, hence, a change in the flux enclosed, which is observable and must not be changed by a gauge transformation. Hence we cannot use this substitution to get rid of the phase.

In spite of this, theoreticians have been able to simplify derivations by using a real order parameter in multiply connected systems. This can be done for some purposes if the phases at a point differ by an integer times 2π. Taking the curl removes the constant and we get single valued currents and fields. Thus using equations (A3.1.30) and (A3.1.31) when there are vortices present gives correct values for most of the variables of interest.

However, we can expect problems with anything that depends on the flux enclosed, such as the electric field of moving vortices, since this will depend on the integral of $\mathbf{A}$ round a contour loop, rather than its differential. Waldram [15] has described the technique in physical terms as the insertion down the centre of a vortex of a very small solenoid containing one flux quantum in the opposite sense to that of the vortex. Since this solenoid has no external field it has no influence on the currents and fields outside the vortex. Therefore to find local currents and fields it may be permissible to use a real order parameter if vortices are present. However this is clearly a different physical situation from the original problem since if the vortex moves there is no net flux transfer and no voltage induced, so any results which depend on the flux of the vortex will be wrong. These problems should be made clearer by the analysis of the two geometries which follow.

A single vortex

A single vortex at the centre of a cylinder has been analysed in two gauges by Tinkham [3]. A single vortex has cylindrical symmetry so it is physically reasonable to assume an order parameter in which the phase is equal to the azimuthal angle. This is multivalued in differences of 2π. Since the currents are all in the theta direction this must also be true of the vector potential. A_θ is a function of r and other components are zero. Near the centre of the vortex:

$$A_\theta(r) = \frac{1}{2}b(0)r\,. \tag{A3.1.56}$$

This describes a uniform field $b(0)$.
At large distances

$$A_\theta(r) = \phi_0/(2\pi r)\,. \tag{A3.1.57}$$

This gives the correct flux enclosed, ϕ_0.

The solution satisfies the magnetic conditions for the London gauge since $\nabla\cdot\mathbf{A} = 0$ and the normal component of the superfluid velocity at the surface of the cylinder is zero. However the order parameter is not real, and a real order parameter is the best definition of the London gauge. As pointed out above the normal boundary condition is not sufficient to define the gauge uniquely.

If on the other hand we impose the condition that the order parameter is real, which is the London gauge, then $\mathbf{j}$ is proportional to $\mathbf{A}$ so that $\mathbf{A}$ vanishes exponentially at large radii. Now the new vector potential $\mathbf{A}'$ is given by

$$\mathbf{A}'_\theta(\mathbf{r}) = \mathbf{A}_\theta(\mathbf{r}) - \phi_0/(2\pi r)\,. \tag{A3.1.58}$$

In this case we still get the same currents and magnetic fields. However if we integrate $\mathbf{A}$ round a circle enclosing the vortex core we conclude there is no flux enclosed. This is because the physical situation requires a multivalued phase of the order parameter so the gauge transformation involves a potential which is not single valued. The fields and currents in the vortex are correct, but the flux is not, so we do not have the same physical situation as before the transformation in all respects. The G–L equations have many solutions and it is always necessary to check their consistency with all the boundary conditions, which means more than just matching fields at material boundaries.

A thin walled cylinder

As a second illustration, we consider a field B_0 applied parallel to the axis of a hollow circular cylinder of radius a, initially uniform, but later with a weak link to make the connection to the chapter on SQUIDs. We assume a thin walled cylinder so that the current density is uniform and the field inside is approximately equal to that outside. The London treatment of this is in section A3.1.2.1 and we now consider the G–L treatment (which is very similar) and in particular the phase. To do this we rewrite equation (A3.1.54) as

$$\mu_0\lambda^2 e^*\mathbf{J}/\hbar = \nabla\theta - e^*\mathbf{A}/\hbar\,. \tag{A3.1.59}$$

$\int(\nabla\theta - e^*\mathbf{A}/\hbar)\,\mathrm{d}\mathbf{l}$ is defined as the gauge invariant phase difference between two points, $\Delta\theta_g$. Locally this phase is related to the superfluid current density, $\mathbf{J}$ (see chapter A2.7). A2.7

If we first look at a solution with θ zero, i.e. a real order parameter, then integrating equation (A3.1.59) round the cylinder:

$$\mu_0\lambda^2 2\pi a J = -\pi a^2 B_0\,. \tag{A3.1.60}$$

This is the situation if we start with no trapped flux in the cylinder; the fluxoid is zero.

In this case the Coulomb and London gauges coincide and, although there is flux within the cylinder, it is not quantized and will disappear if the external field is removed. The system is a simply connected system since the fluxoid is zero. The current density is

$$J = -aB_0/(2\mu_0\lambda^2). \qquad (A3.1.61)$$

This will apply as the external field is increased until the critical current is reached, when flux will enter the cylinder and be trapped.

If the order parameter is not single valued, the fluxoid is not zero and there is trapped flux in the A2.6 cylinder. We try a phase which differs by $2n\pi$ going round the cylinder. Now integrating equation (A3.1.59) gives

$$2\mu_0\pi a\lambda^2 J = n\phi_0 - \pi a^2 B_0. \qquad (A3.1.62)$$

This means that if in the normal state the applied field puts an integer number of flux quanta, n, into the circle defined by the cylinder there will be no screening current on cooling to the superconducting state. If this is not the case the current will be positive or negative depending on the sign of the difference between the applied number of fluxoids, and the fluxoid in the sample. If B_0 is increased the screening current density increases until the critical value is reached. There will then be a sudden change in n and the trapped flux changes by a whole number of quanta.

For a circular cylinder the London and Coulomb gauges coincide, since there are no electrostatic charges. If the cylinder is a different shape the derivation above is similar in the London gauge. For the initial application of the field the current density is $-B_0A/(\mu_0\lambda^2 L)$ where A/L is the ratio of the area to circumference. However, there will now be electrostatic charges on the surface of the cylinder to direct the current round any corners so that, in the conventional Coulomb gauge, we must include an electrostatic potential term.

Weak links

A2.7 The language used to describe Josephson junctions is very similar to that of the G–L equations but was originally derived from the tunnelling of Cooper pairs. However, it soon became clear that the Josephson equations did not only apply to tunnel junctions, but to any small weak section of a superconductor. There is therefore a continuum starting with Josephson tunnel junctions, extended to other weak links, which are small pinning centres for Abrikosov vortices, and therefore also to the bulk pinning of vortices in type II superconductors. For example [17] links the RSJ model of a Josephson junction with an isolated pinning centre; in this section we make the connection between the phase of the order parameter in the G–L theory and the phase difference across a weak link.

We introduce a weak link into the cylinder of the previous section as a thin line parallel to the axis, E4.2 to make a SQUID, and revert to a real order parameter with $\theta = 0$, i.e. the initial application of a field. (Real SQUIDS have a very different geometry but this does not alter the principles of the argument.) Equation (A3.1.59) becomes

$$\mu_0\lambda^2\mathbf{J} = -\mathbf{A}. \qquad (A3.1.63)$$

If the weakness is in the material then λ will be increased at this point since λ is proportional to $n_s^{-1/2}$. Alternatively, if there is a constriction, $\mathbf{J}$ is increased. In either case the effect is the same, that is from equation (A3.1.59) the gradient of the gauge invariant phase difference, $\nabla\theta_g = \mu_0\lambda^2\mathbf{J}e^*/\hbar$, will be greatly increased at the weak link compared with the rest of the circuit. This is the phase that appears in the Josephson equations, not the phase of the order parameter θ which is zero everywhere until a quantum enters the ring. (A close analogy is if we were to introduce a thin but very resistive line down the cylinder. If a current is induced the voltage drop will take place almost entirely across the resistive strip).

To take a specific example, suppose the subscript s refers to the superconductor and w to the weak link, and that these are of length l_s and l_w respectively. We make a weak link consisting of a thin strip of reduced order parameter and so increased λ. Then integrating equation (A3.1.63) round the ring:

$$\mu_0 J(\lambda_s^2 l_s + \lambda_w^2 l_w) = \pi a^2 B_0. \tag{A3.1.64}$$

The change in gauge independent phase in the two regions is

$$\Delta\theta_{gs} = \lambda_\sigma^2 l_s e^* / \hbar(\lambda_s^2 l_s + \lambda_w^2 l_w) \text{ and } \Delta\theta_{gw} = \lambda_\omega^2 l_w e^* / \hbar((\lambda_s^2 l_s + \lambda_w^2 l_w). \tag{A3.1.65}$$

It can be seen that if $\lambda_w^2 l_w$ is much greater than $\lambda_s^2 l_s$, as will be the case for a weak link, nearly all the phase difference occurs across the weak link. Similarly, if we impose an alternating current, the voltage drop will appear almost entirely at the junction since either λ or $\mathbf{J}$ is increased at this point. Across the link

$$\Delta V = E l_w = -\dot{A} l_w = \mu_0 \lambda^2 \dot{J} l_w = (\hbar/e^*)\mathrm{d}(\Delta\theta_{gw})/\mathrm{d}t. \tag{A3.1.66}$$

It can be seen that the electric field is proportional to the rate of change of θ_g, as deduced by Josephson.

A2.5

Equation (A3.1.66) leads to the 'kinetic inductance'. The origin of the term can be seen by considering a junction of area S and thickness d. In terms of the voltage V and current I, equation (A3.1.66) becomes

$$V/d = \mu_0 \lambda^2 \dot{I}/S. \tag{A3.1.67}$$

This implies an inductance $\mu_0 \lambda^2 d/S$ associated with the junction (or indeed any section of superconducting material).

The energy in the inductance is $1/2LI^2$ which can then be written $1/2n_s m v^2$ per unit volume. This is the kinetic energy of the electrons, hence the name kinetic inductance.

Weak links made from narrow sections (microbridges) and thin sections of reduced order parameter (SS′S or SNS junctions) are the easiest type of Josephson junction to understand within the framework of the GL theory since they are essentially pinning centres which operate on single vortices. From the point of view of this section the kinetic inductance is due to the reversible motion of a vortex in the potential well of the junction.

A2.7

It is perhaps unfortunate for the understanding of the subject that the brilliant intuition of Josephson led to his equations being derived for the most sophisticated type of junction, a tunnel junction. Many physicists working on the much easier route of small type II superconductors would have reached the same conclusions later and less elegantly, but the subject might have been more easy to understand. For example, the famous relation between voltage and frequency follows at once from Faraday's law of induction. If f vortices enter per second then $V = f\phi_0 = fh/e^*$. Josephson derived this relation from tunnelling theory as $Ve^* = hf$, [16], which can be interpreted as the Planck relation between frequency and energy.

Differences in V–I characteristics between junctions and bulk type II superconductors arise from two sources. One is that a short junction is a single pinning centre acting on one vortex at a time, while in a bulk sample the large number of vortices and pinning centres leads to collective effects. This causes the difference in curvature of the V–I curve at the critical current. There are two kinds of long Josephson junctions. One is an ideal smooth junction when the length is greater than the Josephson penetration depth, but a single quantum is pinned. This only requires a more complex solution of the Josephson equations. However, in grain boundaries in high T_c superconductors there are likely to be several Josephson vortices in the boundary and the critical current depends on the pinning of this linear vortex

A4.3

array by inhomogeneities in the boundary. In this case the V–I characteristics approach the A2.7 characteristics of type II superconductors due to the collective pinning of Josephson vortices along the junction.

More fundamental is the hysteresis in the V–I characteristic of tunnel junctions. This arises because the Josephson vortex has no normal core in which dissipation takes place so there is almost no damping, in contrast to an Abrikosov vortex which is very heavily damped. Therefore the mass term in the equation of motion dominates the viscous term and inertial effects cause hysteresis. This is the only measurement that can show that the coupling between layers is due to Josephson tunnelling. (type II superconductors can also show hysteretic V–I curves, but for quite different reasons.) For the case of microbridges, most Josephson effects appear if the bridge is smaller than a penetration depth, since this forces vortices to enter one at a time. However, to show the hysteretic behaviour in the V–I characteristic found in tunnel junctions the microbridge must be smaller than the coherence length. But having pointed out that the parallel between junctions and pinning centres is useful in understanding the effects qualitatively, it must be added that many quantitative results can only be obtained using Josephson's theory.

A3.1.4 Type II superconductors

A1.2 The G–L theory introduced the idea of a surface energy to explain the Meissner effect, which requires the surface energy to be positive. G–L recognised that if $\kappa > 1/\sqrt{2}$ an instability of some kind would occur. Pippard introduced the idea of a coherence length, a distance over which the order parameter could vary. He pointed out that, if this was less than the penetration depth, a boundary could gain the magnetic energy by allowing flux penetration while costing little in condensation energy because the coherence length was short [18]. The surface energy would then be negative and lead to a fine subdivision of the normal and superconducting phases. (An analogy is the surface energy between sugar and water which is also negative and causes a similar subdivision called solution.) A laminar model was developed from this idea but, clearly, if a laminar structure lowers the energy it can be further A1.1 subdivided by perpendicular boundaries to produce more surface and the result is essentially the vortex lattice arrived at earlier by Abrikosov [19].

A3.1.4.1 The Abrikosov theory

The most celebrated solution of the Ginzburg–Landau equations is the vortex lattice derived by Abrikosov [19] This is covered in more detail elsewhere. For completeness we quote some of his results here. These can be derived approximately by physical arguments which are given here, while the equations quoted are the more accurate versions from the Abrikosov theory.

Firstly, for a hexagonal lattice (figure A3.1.1) each unit cell contains flux ϕ_0. Hence if the spacing of the vortices is a then

$$Ba^2 = \phi_0(2/\sqrt{3}). \tag{A3.1.68}$$

For example if we cool a film in the Earth's field there will be trapped vortices every six microns. This can have a major impact on phenomena such as microwave losses.

Since the vortex needs energy to enter the superconductor an external field B_{c1} is needed to push them in. Once it is strong enough and a vortex enters it will migrate to the centre of a macroscopic specimen so a large number can enter until the internal field is comparable with the external field to equalise the magnetic pressure. Since they stop entering when their circulating currents overlap, a flux density of B_{c1} corresponds to a vortex spacing of about λ.

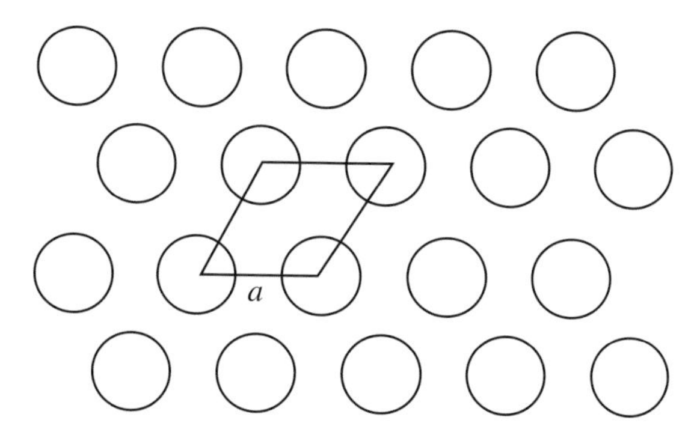

Figure A3.1.1. A hexagonal array with vortex spacing a. The unit cell area is $a^2\sqrt{3}/2$.

$$B_{c1}\lambda^2 = \phi_0((\ln\kappa + 0.8)/4\pi). \tag{A3.1.69}$$

As the external field increases the vortices are pushed closer together until at the upper critical field B_{c2} the cores overlap and the spacing is about ξ. Since this is a second order transition there is no sudden change between the superconducting state and the normal state at this point.

$$B_{c2}\xi^2 = \phi_0(1/2\pi). \tag{A3.1.70}$$

From the vortex spacing at B_{c1} and B_{c2}, it follows that B_{c1}/B_{c2} is approximately $(\xi/\lambda)^2 = 1/\kappa^2$.

$$B_{c1}/B_{c2} = \ln(\kappa + 0.08)/(2\kappa^2). \tag{A3.1.71}$$

The resulting magnetization curve is roughly a straight line between B_{c1} and B_{c2} and the area under it is $B_c^2/2\mu_0$ (figure A3.1.2.) A4.2

Hence

$$B_{c1}B_{c2} = B_{c2}\ln(\kappa + 0.08). \tag{A3.1.72}$$

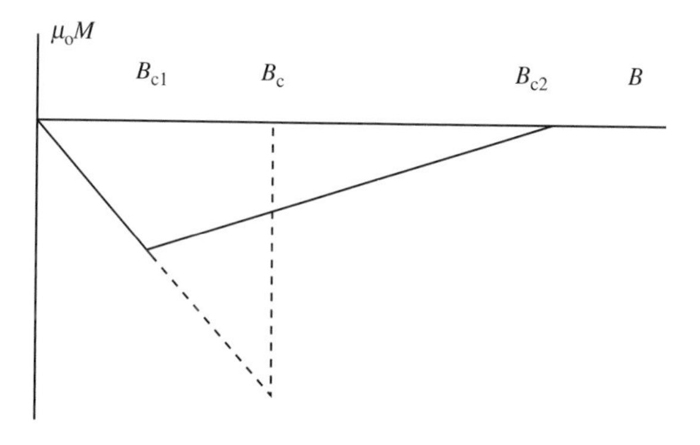

Figure A3.1.2. The magnetization curve based on the qualitative vortex picture. The area under the magnetization curve must equal that of the triangle based on B_c. This means that many of the results of Abrikosov can be found approximately from Euclidean geometry and the vortex spacing at B_{c1} (about λ) and B_{c2} (about ξ).

The gradient of the magnetization curve at high fields is approximately $-B_{c1}/(B_{c2}-B_{c1})$, so

$$\mu_0 M = (B_{c2} - B)/(1 + 1.16(2\kappa^2 - 1)). \quad \text{(A3.1.73)}$$

In the London approximation (i.e. small coherence length) the magnetization at intermediate fields is given by [7]

$$\mu_0 M = -(\phi_0/8\pi\lambda^2)\ln(0.368B_{c2}/B). \quad \text{(A3.1.74)}$$

From these results

$$B_{c1} = B_c \ln(\kappa + 0.08)/\sqrt{2}\kappa \quad \text{(A3.1.75)}$$

and

$$B_{c2} = \sqrt{2}\kappa B_c. \quad \text{(A3.1.76)}$$

There are many combinations of these parameters but there are only two independent ones, ultimately related to α and β in the G–L free energy. It can be seen that if any one parameter behaves strangely it has implications for all the others. For example some underdoped oxide superconductors appear to have a B_{c2} which increases exponentially at very low temperatures [20]. If this is true, and if B_c stays fairly constant, it follows that B_{c1} must decrease exponentially and the penetration depth must increase similarly. Alternatively, if the penetration depth behaves normally, then B_c must increase exponentially. Neither scenario seems likely, which forces us to look for explanations outside the conventional Abrikosov theory, such as fluctuation effects.

Acknowledgments

This article has concentrated on aspects of the Ginzburg–Landau theory more relevant to applications, and I am grateful to a number of people who have helped me to link these results with the more fundamental theory, which they understand much better than I do. In particular I would like to thank Dr J.R.Waldram and Dr E.H. Brandt who have put me right on a number of occasions.

References

[1] London F 1961 *Superfluids* (New York: Dover)
[2] Ginzburg V L and Landau L D 1950 On the theory of superconductivity *Zh. Eksperim. Teor. Fiz.* **20** 1064 English translation in collected papers of L D Landau, D Ter Haar ed, (Oxford: Pergamon) 1965
[3] Gorter C J and Casimir H B G 1934 *Phys.* Z **35** 963; Z. *Techn. Phys.* **15** 539
[4] Waldram J R 1997 *Superconductivity* (Cambridge: Cambridge University Press)
[5] Cave J R and Evetts J E 1986 Critical temperature profile determination using a modified London equation for inhomogeneous superconductors *J. Low Temp. Phys.* **63** 35–55
[6] Tinkham M 1996 *Introduction to superconductivity* (New York: McGraw-Hill)
[7] de Gennes P G 1966 *Superconductivity of metals and alloys* (New York: W A Benjamin)
[8] Saint James D, Thomas E J and Sarma G 1969 *Type II Superconductors* (Oxford: Pergamon)
[9] van Duzer T and Turner C W 1981 *Principles of superconductive devices and circuits* (New York: Elsevier)
[10] Campbell A M 1989 Ginzburg–Landau calculations of the critical current densities of grain boundaries *Physica* C **162–164** 1609–1610
[11] Schweigert V A and Peeters F M 2000 Transitions between different superconducting states in mesoscopic disks *Physica* C **144** 266–271
[12] Chibotaru L F, Ceulemans A, Bruyndoncx V and Moshchalkov V V 2001 Vortex entry and nucleation of anti-vortices in a mesoscopic superconducting triangle *Phys. Rev. Lett* **86** 1323–1326
[13] Coote R I, Campbell A M and Evetts J E 1972 Vortex pinning by large precipitates *Can. J. Phys.* **50** 421–427

[14] Clem J R Lecture notes, unpublished
[15] Waldram J R Private communication
[16] Josephson B D 1962 Possible new effects in superconducting tunnelling *Phys. Lett.* **1** 251–253
[17] AM Campbell 1987 Pinning and critical currents in type II superconductors, Proc. 18th Int. Conf. Low Temp. Phys. Kyoto. *Japanese J. of Appl. Phys.*, **26** suppl. 26-3, 2053–2058
[18] Pippard A B 1955 Trapped flux in superconductors *Phil. Trans. R. Soc.* **248** 97–129
[19] Abrikosov A 1957 On the magnetic properties of superconductors of the second group *Sov. Phys. JETP* **5** 1174–1182
[20] Mackenzie A P, Julian S R, Lonzarich G G, Carrington A, Hughes S D, Liu R S and Sinclair D 1994 Resistive upper critical field of $Tl_2Ba_2CuO_6$ *J. Supercond.* **7** 271–277

A3.2
Microscopic theory

A J Leggett

The essentials of the microscopic theory of superconductivity presented by BCS [1] in their classic 1957 paper, and generally believed to describe at least a large subclass (referred to hereafter as 'classic') of the currently known superconductors will be outlined in this chapter. Exactly which aspects of it, if any, are relevant to the cuprates is, at present, an open question.

A3.2.1 Normal state: Cooper instability

By the time a metal has been cooled down to the critical temperature T_c for the onset of superconductivity, the electrons are already highly degenerate, that is, their distribution is profoundly affected by the Pauli principle, which states that no more than one electron can occupy a given single-particle state. Under these conditions the normal state of the 'classic' superconductors is usually fairly well described by the standard 'textbook' model of a metal. In the very simplest version of this model, usually associated with the name of Sommerfield, the conduction electrons move freely in a constant potential; thus they occupy plane-wave states with wave vector **k**, momentum $\hbar\mathbf{k}$ and energy $\epsilon(\mathbf{k}) = \hbar^2\mathbf{k}^2/2m$. At the next (Bloch) level one takes into account the effect of the periodic potential of the static ionic lattice and, as a result, the relevant electron states are no longer plane waves but 'Bloch waves'; the energy of the state in general depends on the direction as well as the magnitude of the wave vector, giving the usual band structure. [Also, the electrons can be scattered by small vibrations of the ionic lattice (phonons) as well as by impurities]. Finally, at the most sophisticated (Landau–Silin) level, it is recognized that the interactions between conduction electrons, up to now neglected, are actually very important and mean that the true energy eigenstates correspond to the occupation of a Bloch-wave state not by a single electron but by a 'quasiparticle', that is, an electron surrounded by a 'screening cloud' of other electrons. The resulting picture, while not quite in one–one correspondence with the Bloch scheme, is close enough to it that in the present context we may treat a quasiparticle as effectively equivalent to an actual electron.

There is one fundamental property which persists throughout this increasingly sophisticated description: imagine that we could cool the normal phase to zero temperature without it becoming superconducting. Then the basic 'fermionic' entities, be they real electrons, Bloch waves or quasiparticles, would fill up the available states one by one up to the Fermi surface, that is the locus in **k**-space of states with the maximum (Fermi) energy. At finite temperatures of the order of T_c most of this 'Fermi sea' is inert and, for any phenomenon which involves only relatively weak excitation of the system, all the action comes from electrons in a narrow shell of states, of width $\sim k_B T$ in energy, close to the Fermi surface. Although superconductivity is not quite of this type, it is still true that the states mainly involved are indeed close to the Fermi surface and the bulk of the Fermi sea can be ignored.

A2.1

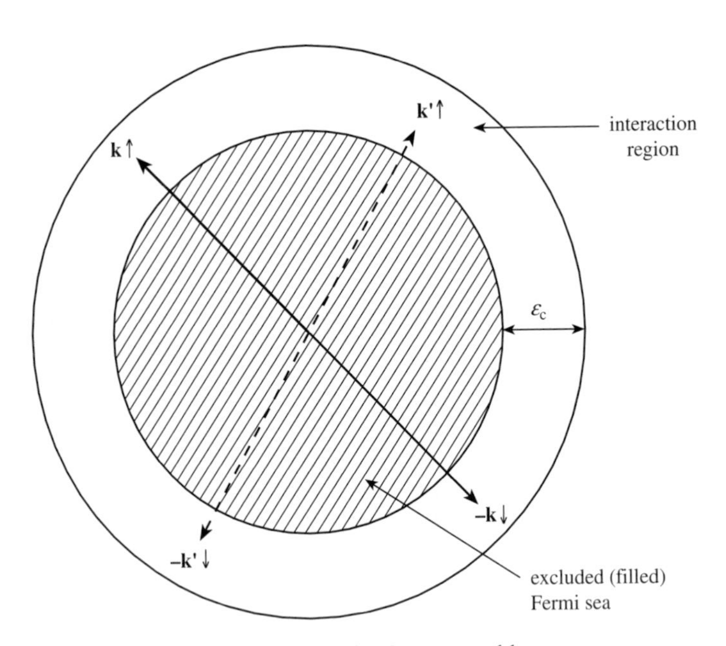

Figure A3.2.1. The Cooper problem.

The historical jumping-off point for BCS theory was the following observation by Cooper [2]: consider a gas of $N-2$ free electrons at $T=0$, so that all states below the Fermi energy ϵ_F are filled and all those above empty, as illustrated in figure A3.2.1. Then imagine we introduce two more electrons, with opposite spin and total momentum zero, which, to maintain the Pauli principle, are allowed to occupy only the vacant states and which are subject to a weak mutual interaction whose matrix elements for scattering from $(\mathbf{k}\uparrow:-\mathbf{k}\downarrow)$ to $(\mathbf{k}'\uparrow:-\mathbf{k}'\downarrow)$ are constant whenever the states $\mathbf{k}$ and $\mathbf{k}'$ both have energy (relative to the Fermi energy) less than ϵ_c and zero otherwise. If we denote this constant by $-V_0$, then for $V_0<0$ (repulsive interaction) nothing interesting happens. If, however, $V_0>0$ (attraction), we find that the Schrödinger equation for the two added electrons always has a solution which lies *below* the minimum energy $2\epsilon_F$ which they would have if non-interacting, by an amount E given for small V_0 by

$$E = 2\epsilon_c \exp - (2/N(0)V_0) \tag{A3.2.1}$$

where $N(0)$ is the density of single-particle states of one spin per unit energy near the Fermi surface (for a free gas this is $3N/4\epsilon_F$). Moreover, the wave function corresponding to this negative-energy state indeed corresponds to a 'bound' state; the two-electron wave function is constant as a function of the center-of-mass coordinate, but falls off fast as a function of their *relative* coordinate $\mathbf{r}$, so that the result is a sort of 'di-electronic molecule' with a radius $\xi' \sim \hbar v_F/E$ (where $v_F = (2m\epsilon_F)^{1/2}$ is the Fermi velocity). This bound state is called a *Cooper pair*; note that it exists for any attraction, however weak.

A3.2.2 Effective interaction

If Cooper's observation is to have any relevance to real metals, it is necessary that the effective interaction between electrons in states close to the Fermi surface be attractive. At first sight this seems unlikely, since the 'bare' Coulomb interaction $e^2/(4\pi\epsilon_0\mathbf{r})$ between any two electrons is certainly repulsive. However, it turns out that in a metal this bare interaction is strongly screened by the collective

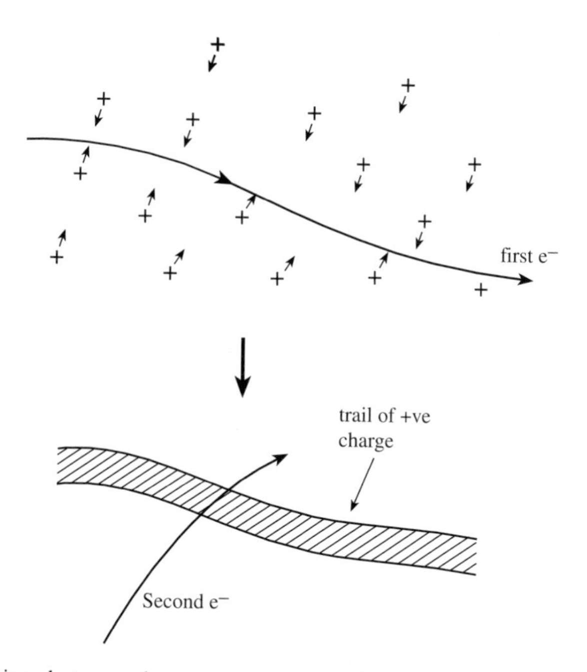

Figure A3.2.2. The effective electron–electron interaction induced by polarization of the ionic background.

effects of the other electrons, and the resulting 'effective' interaction falls off exponentially with a characteristic length which can be of the order of the mean distance between electrons.

Equally important, the interaction with phonons (lattice vibrations) can actually generate an effective *attraction* between electrons. One way of seeing this is that the first electron attracts the positive ions towards its path and thereby leaves behind a trail of (slowly relaxing) positive charge, which subsequently attracts a second electron; thus we generate an effective electron–electron attraction, as illustrated in figure A3.2.2. An alternative point of view is that the exchange of virtual phonons leads to an attraction between electrons, in the same way as in the Yukawa theory of nuclear forces exchange of virtual mesons leads to one between nucleons. In any event, detailed calculation shows that this phonon-generated attraction, when added to the screened Coulomb repulsion, can generate a net interaction the relevant matrix elements of which are attractive. This is not invariably so—if it were, all metals should be superconductors at low enough temperature—and it is actually not at all trivial to determine the sign of the net interaction from first principles.

The actual form of the interaction as a function of the wave vectors $\mathbf{k}$ and $\mathbf{k}'$ involved in the scattering process is quite complicated, and this is taken into account in more sophisticated calculations; however, it turns out to be a surprisingly good approximation for many purposes to replace it, as BCS did in their original work, by something close to the simple form used above for the Cooper problem, with the 'cutoff' energy ϵ_c taken to be of the order of the characteristic (Debye) phonon energy.

A3.2.3 Nature of the superconducting groundstate

While Cooper's simple calculation gives much insight into the basic mechanism of superconductivity (at least in the classic superconductors), it is obvious that it is internally inconsistent in treating the last two 'special' electrons which form the Cooper pair on a different footing from the remaining $N - 2$, whose only function is to fill up the Fermi sea and thereby exclude the paired electrons from it. A complete calculation which treats all N electrons on the same footing was given by BCS in [1]. A somewhat intuitive interpretation of the principal BCS results will now be presented.

For most purposes, and in particular when one wishes to calculate the *changes* that various physical properties undergo on passing from the normal to the superconducting state, it is adequate to visualize the superconducting groundstate as having all the N electrons bound into di-electronic 'molecules' (Cooper pairs) which are all described by *the same* two-particle wave function. This fundamental property, that all Cooper pairs have to occupy the same pair state, is characteristic not only of the groundstate but of all low-energy states in which a finite fraction of electrons are paired; it is tempting to think of it as a kind of 'Bose condensation' (the pairs have total spin zero and thus are indeed bosons!) but, irrespective of the terminology, it is the essential key to understanding the abnormal properties of superconductors such as the Meissner effect (see chapter A2.2).

A2.2

What is the basis of the assertion that all Cooper pairs must occupy the same two-particle state? Why, for example, could we not put half of them into a pair state with centre-of-mass momentum $\mathbf{K}$ equal to zero and the other half into one with finite $\mathbf{K}$? The answer is that we could, but this would lose a large fraction of the gain in potential energy obtained by the 100% 'condensation' into a single state described above, a loss which is not adequately compensated by entropic or other factors. (This is because to enjoy the maximum benefit of the attractive interaction, all pairs of electrons must be able to scatter into the same set of states, something that is only possible if they have a common value of $\mathbf{K}$. For details see [3].)

Given, then, that all Cooper pairs must be described by the same two-particle wave function F, what does this function look like? For the classic superconductors, at least, it is believed to be a product of space and spin functions, with the latter corresponding to the singlet state:

$$F = F(\mathbf{r}_1, \mathbf{r}_2)\psi_{\text{spin}}(S = 0). \tag{A3.2.2}$$

If, furthermore, we restrict our attention for the moment to the groundstate (in zero magnetic field) then the spatial wave function $F(\mathbf{r}_1, \mathbf{r}_2)$ should correspond to the center of mass (COM) of the pair being at rest, i.e. it should be independent of the COM coordinate $\mathbf{R} \equiv (1/2)(\mathbf{r}_1 + \mathbf{r}_2)$ and a function only of the *relative* coordinate $\mathbf{r}_1 - \mathbf{r}_2 = \boldsymbol{\rho}$. Moreover, for the classic superconductors it is believed that the net orbital angular momentum of the pairs is zero, which means that $F(\boldsymbol{\rho})$ is independent of the direction of $\boldsymbol{\rho}$ and a function only of the relative distance $|\boldsymbol{\rho}| \equiv \rho$. An approximate representation of $F(\rho)$, valid for most distances of interest, is

$$F(\rho) \sim \text{const}[\sin(2k_F\rho)/2k_F\rho]\exp[-(\rho/\xi_p)] \tag{A3.2.3}$$

where $\xi_p \gg 1/k_F$ (see below). Thus, at 'short' distances ($\rho \ll \xi_p$) the relative wave function is indistinguishable from that of two free electrons each with magnitude of momentum k_F, with centre of mass at rest and in a relative s-state (the term in square brackets), but if one goes out to longer distances one sees that the pair is indeed bound, forming a 'molecule' with radius $\sim\xi_p$ illustrated in figure A3.2.3. This 'pair radius' is of the same order as the more familiar 'Pippard coherence length' ξ_0, which is traditionally used to describe the electrodynamics (see chapters A4.2 and A2.3), i.e. of order $\hbar v_F/\Delta(0)$, where the quantity $\Delta(0)$, whose physical significance will be explained in the next section, is given in the simple model used by BCS by the formula

A2.3
A4.2, A2.3

$$\Delta(0) = 2\epsilon_c \exp - \{1/[N(0)V_0]\}. \tag{A3.2.4}$$

Thus, for weak attraction V_0 the quantity ξ_p can be very large, in practice as much as $\sim 10^{-6}$ m; the different 'molecules' (Cooper pairs) therefore overlap one another very substantially, to the extent that within the volume of a single pair one may find 10^9–10^{12} electrons belonging to 'other' pairs (though the fundamental indistinguishability of electrons makes this form of words of dubious meaning!). That ξ_p and ξ_0 are of the same order is to be expected, since the non-local effects described by the Pippard

A2.3

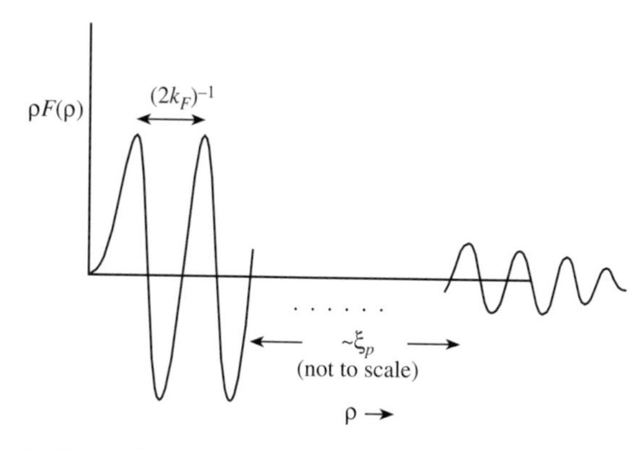

Figure A3.2.3. Qualitative behaviour of the Cooper-pair wave function. Note that the factor $(2k_{\mathrm{F}}\rho)^{-1}$ has been extracted for clarity.

coherence length arise in some sense from the fact that the two electrons of a pair cannot be treated as independent.

As yet, the constant in equation (A3.2.4) has not been specified, and this raises a rather delicate point about the 'number' of Cooper pairs. It turns out that the *changes* in 'two-particle' properties such as the interaction energy induced by pair formation may be correctly calculated by treating the 'pair wave function' (equation (A3.2.3)) (or a slightly more accurate version) exactly like an ordinary molecular wave function, that is, by multiplying the two-particle quantity in question [e.g. $V(\rho)$] by the square of $F(\rho)$ and integrating over ρ. However, it is then necessary to know the overall constant; we cannot assume *a priori* that F is normalized to one! In fact, it turns out that the correct choice is such that the integral of F^2 is of order $N(0)\Delta(0)$; since $N(0)$ is of order N/ϵ_{F}, this means that in some sense, although the pair wave function (equation (A3.2.3)) characterizes all N particles, the 'number of Cooper pairs' N_{c} is only a fraction of order $\Delta(0)/\epsilon_{\mathrm{F}} \sim 10^{-4}$ of N. Further, since, as we shall see, the quantity $\Delta(0)$ is a measure of the binding energy of a pair, the total energy gained by condensation into the superconducting state is of order $N(0)[\Delta(0)]^2$—typically a fraction of order 10^{-8} of the total energy of the normal groundstate.

A3.2.4 BCS theory at finite temperature

Having established the nature of the superconducting groundstate in BCS theory, we will now investigate the excited states. As already mentioned, the states obtained by relaxing the condition of 'uniqueness' of the condensate (i.e. of the pair wave function) have energies far too large to be thermodynamically relevant. However, there is another way of obtaining excited states, as follows: formation of a 'complete' pair state requires partial occupation of each pair of plane-wave states $(\mathbf{k}, -\mathbf{k})$ by a *pair* of electrons with opposite spins. If a given state $\mathbf{k}$ is occupied by an electron which does not have an opposite-spin 'partner' in the state $-\mathbf{k}$, it turns out that this costs, relative to the ground-state, an energy $\{\epsilon_k^2 + [\Delta(0)]^2\}^{1/2}$, where ϵ_k is the normal-state energy relative to the Fermi energy. Thus, the minimum energy for such excitation is $\Delta(0)$, as a result of which, this quantity is called the (zero-temperature) 'energy gap'. The electron states which are pushed out from the energy region below $\Delta(0)$ accumulate at energies slightly greater than $\Delta(0)$, giving a larger density of states there than in the normal state. A2.1

At finite temperature entropy considerations make it thermodynamically favourable for a finite number of such 'broken pairs' to exist in the system. As a result, while the condensate wave function is still unique, the effective number N_{c} of Cooper pairs is reduced, and this in turn decreases the energy gap Δ, which is thus temperature-dependent. When the temperature reaches a 'critical' value T_{c} given by $\Delta(0)/1.76$, the gap tends to zero, as $(T_{\mathrm{c}} - T)^{1/2}$; illustrated in figure A3.2.4. It should be emphasized that

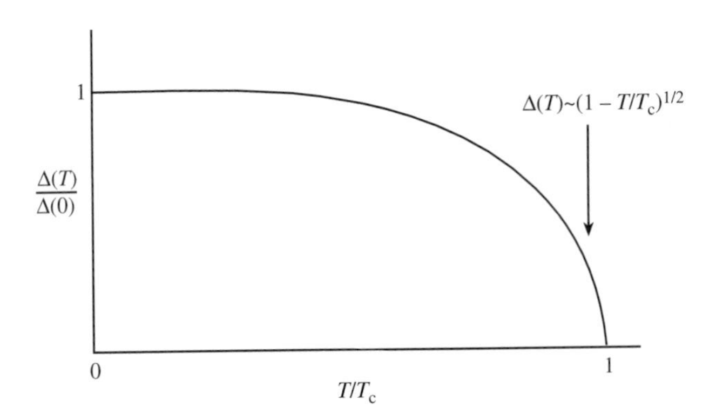

Figure A3.2.4. Temperature-dependence of the energy gap $\Delta(T)$.

the 'pair radius' which characterizes the fall-off of the Cooper-pair wave function F is, like the Pippard coherence length which is of the same order, only rather weakly temperature-dependent and does *not* diverge as T approaches T_c from below. (Neither of these lengths should be confused with the 'Ginzburg–Landau (GL) correlation length' (healing length) $\xi\,(T)$ which does diverge as $(T_c - T)^{-1/2}$; see equation (A2.3.9) and chapter A4.2.)

At finite T the condensate is still 'inert' just as at $T = 0$ (see next section) but the broken pairs form a 'normal component' which behaves qualitatively like the electrons in a normal metal (although their dynamics and scattering rates are somewhat different). Thus, for example, quantities like the specific heat or the Pauli spin susceptibility receive contributions only from the normal component (the latter because the Cooper pairs, having total spin zero, cannot be polarized by an external magnetic field).

A3.2.5 Meissner effect: relation of BCS and GL descriptions

The most fundamental property of superconductors is the Meissner effect, that is, the property of excluding a weak magnetic field: see chapter A2.2. For simplicity, let us consider an 'extreme type II' superconductor (see A2.3.3), for which the electrodynamics is local and well described by the original London theory (see chapter A3.1). In this case the fundamental equation, which when coupled with the standard Maxwell equations leads to exclusion of a weak field, can be written, in the simplest case, as a relation between the local electric current density $\mathbf{j}(\mathbf{r})$ and the electromagnetic vector potential $\mathbf{A}(\mathbf{r})$:

A2.2 A3.1 D2.1

$$\mathbf{j}(\mathbf{r}) = -\Lambda(T)\mathbf{A}(\mathbf{r}) \tag{A3.2.5}$$

Equations (A3.2.1 and A3.2.2) follow from equation (A3.2.5) on taking the time derivative and the curl, respectively. London's expression for the coefficient $\Lambda(T)$ was $n_s e^2/m$, where the 'superfluid density' $n_s(T)$ tends to the total electron density as T tends to zero and to zero as T tends to T_c. As we shall see, equation (A3.2.5) is a direct consequence of the uniqueness and single-valuedness of the pair wave function F.

The argument proceeds by analogy with one which may be familiar in the context of atomic diamagnetism. Consider a single particle of charge e in the presence of a weak magnetic vector potential $\mathbf{A}(\mathbf{r})$ at $T = 0$. The standard gauge-invariant expression for the electric current density $\mathbf{j}(\mathbf{r})$ is

$$\mathbf{j}(\mathbf{r}) = \mathrm{Im}[(e/m)(-i\hbar\psi\nabla\psi)] - (e^2/m)|\psi(\mathbf{r})|^2\mathbf{A}(\mathbf{r}) \tag{A3.2.6}$$

where $\psi(\mathbf{r})$ is the Schrödinger wave function. Suppose that, as is normally the case, the wave function is real for $\mathbf{A} = 0$ and thus $\mathbf{j}(\mathbf{r}) = 0$. If, on the application of a weak potential $\mathbf{A}(\mathbf{r})$, the wave function does not change its form, then the gradient terms still do not contribute and we find

D2.1

$$\mathbf{j}(\mathbf{r}) = -(e^2/m)\rho(\mathbf{r})\mathbf{A}(\mathbf{r}) \tag{A3.2.7}$$

where $\rho(\mathbf{r}) \equiv |\psi(\mathbf{r})|^2$ is the probability density. An exactly similar result follows for a many-electron system, interacting or not, provided $\rho(\mathbf{r})$ is now interpreted as the total electron density at point $\mathbf{r}$ (this is the result used in deriving the standard formula for the diamagnetism of rare gas atoms). At finite temperature T a similar result will follow, provided, (a) the thermal average of the current density is zero for $\mathbf{A} = 0$ and, (b) a finite but weak value of $\mathbf{A}$ produces no change in either the wave functions or in the occupation factors for the different states; however, $\rho(\mathbf{r})$ is now the density only of the condensed electrons (see below).

We thus see that at $T = 0$ we recover equation (A3.2.5), with London's value of the constant $\Lambda(T)$, *provided* that the groundstate many-body wave function is 'rigid' (inert) against application of a weak vector potential. But why should the groundstate possess this kind of rigidity? For definiteness consider a very thin ring of radius R, so that application of a flux ϕ through the ring leads to a circumferential vector potential $\mathbf{A}_\theta = \phi/2\pi R$. For $\mathbf{A} = 0$, it is obvious from the symmetry that the pair wave function $F(\mathbf{r}_1, \mathbf{r}_2)$ is constant as a function of the centre-of-mass coordinate $\mathbf{R} = (1/2)(\mathbf{r}_1 + \mathbf{r}_2)$. Now, a weak but finite vector potential acts symmetrically on the two electrons of the Cooper pair and thus cannot change the dependence of F on the relative coordinate; the only possibility is that it changes its dependence on the COM coordinate and, from the symmetry of the problem, the only possibility is to multiply the latter by a phase factor $\exp(i\alpha\theta)$, where θ is the angular component of the vector $\mathbf{R}$ in cylindrical polar coordinates. But the crucial point, now, is that since the wave function must return to its original value when the two electrons of a pair are carried together once around the ring (the 'single-valuedness' condition), the only possible values of α are integers n (including zero). Since a non-zero value of n costs a large extra kinetic energy (actually of order $N\hbar^2/2mR^2$, although to discuss why the factor is N and not N_c would take more space than I can afford here), this state cannot be energetically favourable for small $\mathbf{A}$ and the wave function must remain unchanged from its $\mathbf{A} = 0$ value, precisely the 'rigidity' necessary to justify equation (A3.2.5). At finite temperatures the same argument goes through for those electrons which are still condensed, but not all are, and the quantity $\Lambda(T)$ is therefore proportional, crudely speaking, to the density of condensed electrons ('superfluid density'). A2.6

Why does a similar argument not go through for a normal metal? It turns out that the simplest case to analyse is that of low but not ultra-low temperature, so that $k_B T \gg \hbar^2/mR^2$. The point is that although the single-particle wave functions, like those of the Cooper-pair COM, must of course be single-valued and thus of the form $\exp(in\theta)$ with n integral, even for $\mathbf{A} = 0$ there is a considerable population of finite-n states and it is only the thermal average of $\mathbf{j}(\mathbf{r})$ which is zero. A finite value of $\mathbf{A}$ now can (and does) induce shifts in the relative populations of these states in just such a way as to preserve zero average current. It is essential to appreciate that this argument would work equally for an uncondensed system of bosons, and that the only reason it does not work for the Cooper pairs is that to distribute the latter between many different pair states (corresponding to different n-values) would be prohibitively costly in energy, i.e. that even at finite temperature the pairs must still be 'Bose-condensed'. (Actually, the phenomenon of superfluidity in liquid ^{4}He, which is the analogue of superconductivity for a neutral system, is usually attributed to the onset of Bose condensation in that system, see e.g. [4]).

It is appropriate, finally, to indicate the connection between the microscopic BCS theory and the phenomenological GL one, which historically preceded it. Actually, the complex scalar order parameter $\Psi(\mathbf{r})$ of GL theory turns out to be nothing but the pair wave function $F(\mathbf{r}_1, \mathbf{r}_2) = F(\mathbf{R}, \boldsymbol{\rho})$ of BCS theory evaluated, at $\boldsymbol{\rho} = 0$, as a function of $\mathbf{R} \equiv \mathbf{r}$, i.e. it is nothing but *the COM wave function of the Cooper pairs*. This statement is true up to a normalization factor which in GL theory is arbitrary; moreover, while the correspondence can be made formally for arbitrary conditions, it tends not to be very useful unless the variation of Ψ with $\mathbf{r}$ is slow on the scale of the pair radius—a condition which is usually A3.1

assumed in applications of GL theory. Once this is realized, one can obtain the free energy functional $F\{\Psi(\mathbf{r})\}$ of GL theory from BCS theory by allowing $\Psi(\mathbf{r})$ to have arbitrary (slow) variations in amplitude and/or phase but requiring that, subject to this constraint, the system be in thermal equilibrium. The resulting expression is in general extremely messy, but for temperatures close to critical takes the simple form (A3.1.1), where the coefficients α, β and γ can be calculated from BCS theory.

References

[1] Bardeen J, Cooper L N and Schrieffer J R 1957 *Phys. Rev.* **108** 1175

[2] Cooper L N 1956 *Phys. Rev.* **104** 1189

[3] Leggett A J 1997 *Electron*, ed M Springford (Cambridge, UK: Cambridge University Press) pp 148–181 see especially pp 159–162

[4] Leggett A J 1995 *Twentieth Century Physics* vol. 2, ed L M Brown, A Pais and B Pippard (Bristol, UK: IOP Publishing and AIP Press) pp 913–966

Further Reading

Tinkham M 1996 *Introduction to Superconductivity* (New York: McGraw-Hill)

de Gennes P-G 1966 *Superconductivity of Metals and Alloys* trans. Pincus P A (New York: Benjamin)

A4
Introduction to section A4: Critical currents of type II superconductors

D A Cardwell and D S Ginley

Introduction

The current carrying properties of type II superconductors are of considerable importance for both fundamental and applied studies of these materials. The critical current density, J_c, in particular, is regarded by engineers as the key parameter to the practical application of low and high temperature superconductors in a diversity of engineering applications. As a result, its optimization for different materials has attracted enormous attention over a number of decades. J_c is essentially a characteristic of the mixed state in type II materials, in which normal and superconducting phases co-exist. Magnetic flux enters type II superconductors in the form of lines of magnetic flux quanta, equivalent to Faraday's lines of force. These thread current vortices which circle predominantly normal regions of the material and give rise to some fascinating magnetic behaviour.

The critical current is a direct measure of the resistance to motion of flux quanta in type II materials. In general a wide diversity of microscopic features such as dislocations, lattice strains and second phases, form effective barriers, or pins, to flux motion under a variety of conditions. The basic physics of the effects of flux pinning are straightforward. Individual flux lines will remain pinned as long as the Lorentz force they experience, given by $B \times J$, remains below the temperature and field-dependent potential energy of the pinning site. The current density at which this potential is exceeded at a given field, therefore, corresponds directly to J_c. The subsequent motion of these flux lines on a microscopic scale will generate an associated electric field and cause dissipation in the material. The nature and properties of the various flux pinning centres and how individual vortices interact with one another, on the other hand, is generally complex and influence the resultant J_c in a variety of ways. This is the subject of chapter 4.

The terminology used to describe the behaviour of flux lines in type II superconductors can be the source of some confusion. The magnetization, or M–H, curve of defect-free single crystals, for example, is often described as ideal. Clearly, such materials are unable to pin magnetic flux and consequently exhibit zero J_c, which makes them entirely unsuitable and far from ideal for most engineering applications. A recent and more useful trend, therefore, has been to describe the magnetization of type II materials in terms of their magnetic reversibility, or hysteresis. J_c is directly proportional to the latter and may be extracted using a simple model from the M–H loop.

This section contains three very different but entirely complementary articles which address the mixed state, vortices and their interaction and flux pinning in general. There is necessarily some overlap between these articles which has been retained given that it differs slightly in emphasis in each case. Each article provides an effective illustration of a particular aspect of the complexity of critical current in type II high and low temperature superconductors.

A4.1
The mixed state

G Crabtree, W K Kwok, D Lopez and U Welp

A4.1.1 Introduction

The mixed state of high temperature superconductors (HTS) contains a rich variety of vortex solid and liquid phases whose properties depend on magnetic field, temperature and disorder. In clean, nearly defect-free crystals at low temperature, an Abrikosov vortex, or flux, lattice [1] is expected. A1.1 In the presence of weak random point defects, a vortex glass state [2, 3] has been proposed where long range translational and orientational symmetry of the vortex lattice is broken. A Bose glass state [4], has been predicted for vortices disordered by one-dimensional columnar defects, such as the amorphous tracks induced by heavy ion irradiation or for two-dimensional planar twin boundaries. These various solid phases are shown in figure A4.1.1, which depicts the generic magnetic phase diagram of the high temperature superconductor $YBa_2Cu_3O_{7-\delta}$ (YBCO). C1

The nature of the solid phase is determined by the competition between the vortex interaction energy, which favours a lattice, and the vortex–pin site interaction, which favours a glass. Both of these energies are overwhelmed by the thermal energy at higher temperature, and the vortex solid melts to form a liquid. Indeed, the Ginzburg number, which characterizes the strength of thermal fluctuations, is about four orders of magnitude larger in HTS compared to conventional low temperature systems. In general, fluctuations in HTS are enhanced by their small coherence length, large anisotropy and high A2.3 transition temperatures. In these materials, therefore, the vortex liquid occupies a large portion of the magnetic phase diagram. Compared to the familiar vortex solid, the vortex liquid is a novel phase where the mechanical properties of the lattice are described by a shear viscosity, rather than a shear modulus. The layered structure of HTS gives rise to two kinds of vortices; two-dimensional pancake vortices [5], which move nearly independently in adjacent layers, and conventional three-dimensional line vortices where the coupling between layers is strong. Transitions between these two limits can be induced by temperature or field [6]. High anisotropy in the liquid state, leads to novel 'entangled' behaviour [7], where neighbouring flexible flux lines approach each other. If these cannot pass through each other because the vortex 'cutting energy' is too high, their viscosity may be enhanced. At low temperatures, the entangled vortex liquid may freeze into a polymer glass-like state via a Volger–Fulcher type transition [7]. A myriad of new vortex states in high temperature superconductors may be studied which are not A4.2 observable in conventional superconductors where the temperature range is restricted and the vortex melting line is almost coincident with the upper critical field line.

The vortex phases in HTS are highly accessible because all the relevant experimental parameters can be controlled readily in the laboratory. The vortex density, thermal energy and Lorentz driving force, for example, can be established by tuning the magnetic field, temperature and measuring current, respectively. Vortex lattice disorder can be tuned by adjusting the defect type and density through controlled proton or electron irradiation for point defects, and heavy ion irradiation for line defects.

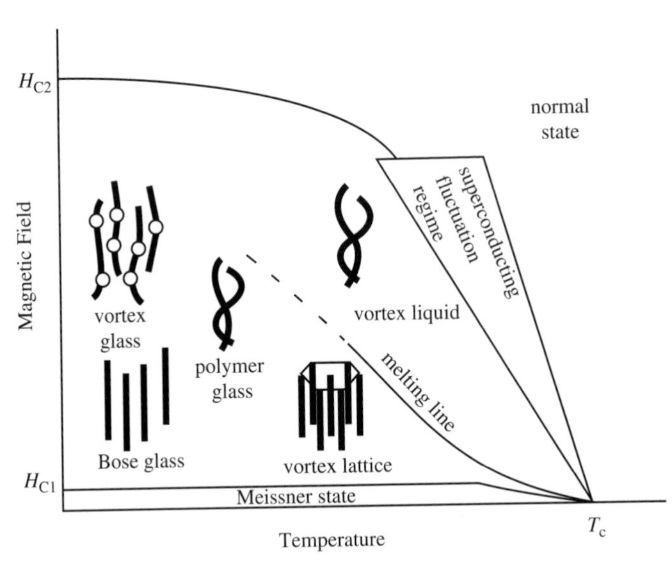

Figure A4.1.1. Generic magnetic phase diagram of YBCO indicating the various vortex liquid and solid phases.

Control of these parameters provides experimental access to the various liquid, lattice and glassy vortex phases and enables systematic studies to be made of their various thermodynamic states and non-linear dynamics.

In this chapter, we review some of the basic characteristic behaviour of the vortex phases in high temperature superconducting YBCO as identified in specific experiments. We begin with the phase diagram for a clean, nearly defect-free single crystal and discuss the implications of introducing various types of defects into this system in terms of their effect on the phase diagram. These pinning centers B2.6 include point defects generated via electron and proton irradiation, columnar defects via heavy ion irradiation and naturally occurring planar twin boundary defects. We discuss the effect of these defects on the vortex melting and irreversibility lines of the YBCO system and the anisotropic pinning associated, in particular, with correlated defects.

A4.1.2 Vortex lattice melting

In clean untwinned single crystals of YBCO, a first order phase transition separates the vortex solid from the vortex liquid phase as shown by the circles in the experimental magnetic field-temperature phase A4.2 diagram of figure A4.1.2. The vortex melting line was first observed in transport measurements as a sharp 'kink' in the temperature dependence of the resistivity [8–10]. The first order nature of the transition was strongly suggested by the sharpness of the kink and by the hysteretic resistive behaviour at the transition [11, 12]. However, while transport measurements are very effective at locating the melting line, they are intrinsically non-equilibrium in nature and therefore cannot provide thermodynamic information about the transition itself. Later magnetic measurements [13–16] of the vortex melting line confirmed its thermodynamic nature. The lower inset of figure A4.1.2 shows simultaneous magnetization and transport measurements of the vortex melting transition in a field of 4 T parallel to the crystallographic *c*-axis of a clean untwinned YBCO crystal [15]. The onset of the sharp 'kink' in the resistivity corresponds directly to the onset of a discontinuous jump in the magnetization, indicated by a jump in the SQUID magnetometer voltage—($U_{sq} - U_{sqs}$). This measurement confirms that the kink in the resistivity measurement indeed corresponds to the first order vortex lattice to liquid melting transition. Direct

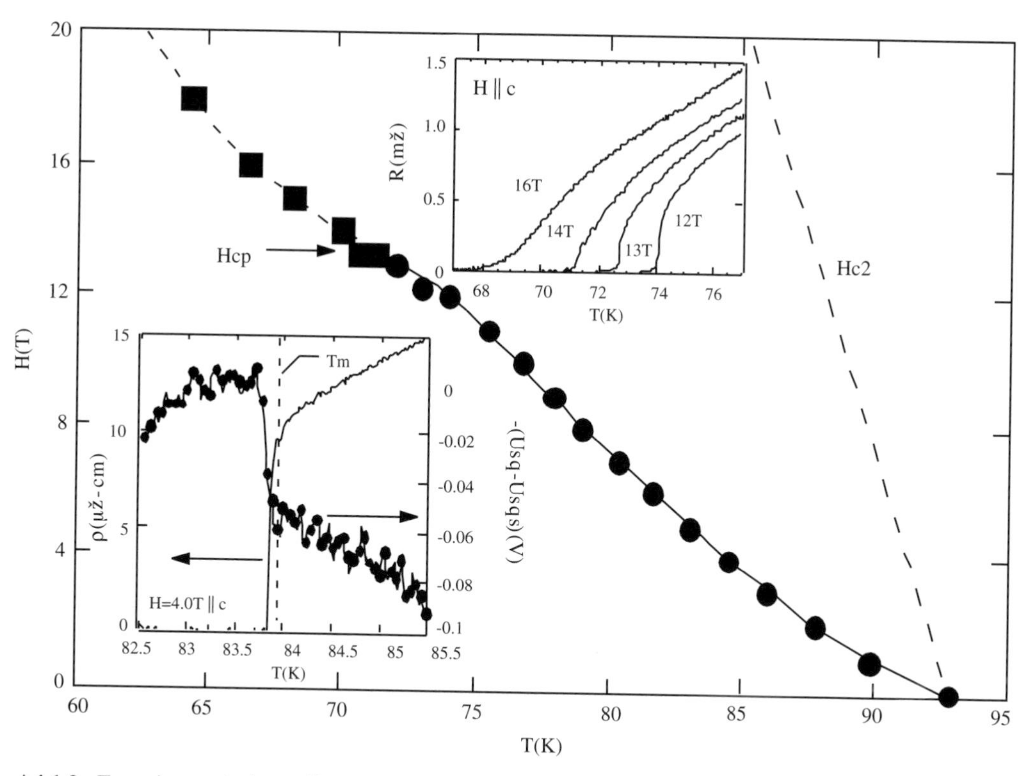

Figure A4.1.2. Experimental phase diagram of a clean untwinned YBCO single crystal showing the vortex lattice melting line determined from transport measurements. Lower inset: simultaneous measurement of the temperature dependence of the magnetization and resistivity, demonstrating the thermodynamic first order nature of the transition associated with the kink in the resistivity. The variation in the magnetization is shown as a jump in the SQUID voltage, U_{sq}, with respect to a linear extrapolation of the low temperature variation, U_{sqs}. Upper inset: tail of the resistive transition at high magnetic fields near the upper critical end point.

determination of the entropy change associated with this transition by calorimetry [17–19] agrees quantitatively with the value derived from magnetization and proves thermodynamic consistency. The measured value of the latent heat of this transition is unusually large because the configurational entropy is dominated by a contribution from internal degrees of freedom [20, 21] in this system.

The first order melting line obtained from transport measurements terminates at a critical point at high field [9, 22, 23] and is labelled in the main panel of figure A4.1.2 as H_{cp}. This critical point is identified as the magnetic field where the discontinuity in the resistivity is replaced by a smooth monotonic resistive transition as shown in the upper inset of figure A4.1.2. This 'kink' is still quite prominent at 12 T but diminishes rapidly at higher fields and disappears completely above 14 T. This critical point has been observed in thermodynamic specific heat experiments [24, 25]. It is sensitive to point disorder and may be lowered systematically along the first order melting line by controlled introduction of point disorder using electron irradiation [26, 27].

A4.1.3 Vortex pinning by disorder: points, lines and planes

Many applications of high temperature superconductivity depend on a high critical current at high temperature, which requires effective pinning by defects. Pinning in HTS is severely degraded in the liquid A4.3

phase, which occupies much of the high temperature portion of the phase diagram. Various methods have been proposed to pin the vortex liquid or reduce its extent in H–T space by raising the melting line of the glass phase. The main panel of figure A4.1.3 shows typical examples of the resistive transition to the superconducting state in a magnetic field of 4 T parallel to the crystallographic *c*-axis for (i) a nearly defect-free untwinned YBCO crystal, (ii) a crystal with dense twin boundaries perpendicular to the direction of vortex motion, (iii) a crystal irradiated with 9 MeV protons, (iv) a crystal irradiated with 1 MeV electrons and (v) a crystal irradiated with 1.4 GeV uranium ions along the *c*-axis. The vortex liquid phase above the 'kink', or melting temperature T_m, for the untwinned crystal, displays ohmic behaviour in contrast to non-ohmic behaviour below T_m. The introduction of defects suppresses the first order melting transition as evidenced by the absence of a discontinuity in the resistivity of the defect-containing crystals in figure 4.1.3. Ohmic behaviour is prevalent in the latter throughout the entire resistive region down to the sensitivity-limited zero-resistivity temperature.

A4.3

The addition of defects to YBCO reduces dissipation in the vortex liquid regime compared to the untwinned nearly defect-free crystal. The largest reduction is seen in the crystal irradiated with heavy ions, where zero resistivity is attained near $t = T/T_{co} \sim 0.93$. Here, the defect sites consist of linear amorphous tracks about 100 Å in diameter [28]. The second largest reduction in dissipation is seen in

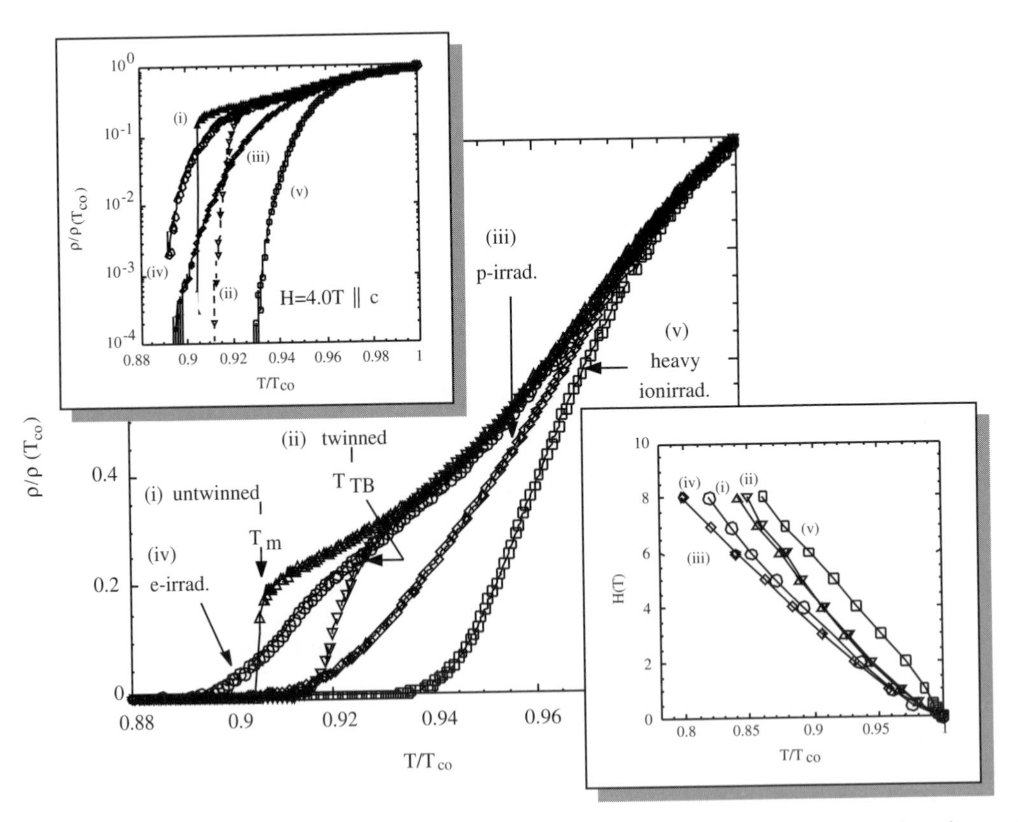

Figure A4.1.3. Main panel: resistive transitions at $H = 4\,\mathrm{T} \| c$ for (i) an untwinned crystal, (ii) a twinned crystal, (iii) 9 MeV $3 \times 10^{16}\,\mathrm{p}^+\,\mathrm{cm}^{-2}$ proton-irradiated untwinned crystal, (iv) 1 MeV $1 \times 10^{19}\,\mathrm{e}^-\,\mathrm{cm}^2$ electron-irradiated untwinned crystal, and (v) 1.4 GeV $^{238}\mathrm{U}^{64+}$ ion-irradiated untwinned crystal with $B_\Phi = 1.0\,\mathrm{T}$. Upper inset: semi-log plot showing the tail of the resistive transition. Lower inset: melting line for the untwinned crystal and the irreversibility lines for the defected crystals.

the densely twinned crystal where the resistivity goes to zero near $t \sim 0.91$. Twin boundary pinning, on the other hand, does not become significant until $T_{TB} < 0.927$, as evidenced by the sharp down-turn in resistivity in figure A4.1.3. Indeed, above T_{TB}, the proton-irradiated crystal shows a larger reduction in resistivity than does the twinned crystal. The zero resistance point for the proton- and electron-irradiated crystals actually lies below that of the defect-free crystal, as shown by the semi-log plot in the upper inset of figure A4.1.3. This is quite surprising at first sight, since the introduction of point defects would appear naturally to assist in the pinning of vortices. However, if an entangled vortex liquid state is created by the introduction of point defects, we can imagine that the entangled state frustrates the formation of a vortex lattice at the freezing temperature of the clean system. The entangled vortex liquid may freeze subsequently into a *disordered* solid at a lower temperature by a second or higher order transition.

The lower inset of figure A4.1.3 shows the irreversibility lines for the same crystals. Correlated defects, such as planar twin boundaries and columnar amorphous tracks induced by heavy ion irradiation shift the irreversibility line to higher temperatures. This is in striking contrast to the effect of point defects induced by electron and proton irradiation which shift the irreversibility line to lower temperatures [26, 29]. All the curves display a concave upward curvature, except for the crystal irradiated with heavy ions (v) which displays a change in slope from concave up to linear behaviour at $H = 1$ T. This crystal was irradiated with 1.4 GeV uranium ions to a dose matching field of $B_\Phi = 1$ T, where the density of columnar defects matches that of the vortices. The change in the irreversibility line at $H = 1$ T for this crystal, therefore, reflects the change in pinning behaviour when relatively free interstitial vortices begin to appear.

A1.3

A4.2

A4.1.4 Pinning anisotropy

Correlated defects exhibit preferential pinning directions, unlike point defects, which are isotropic in nature. The anisotropic vortex pinning behaviour in the liquid state due to columnar defects and twin boundaries is shown in figure A4.1.4, together with that for an untwinned crystal. The largest reduction in dissipation is observed for magnetic field applied parallel to the columnar defects (i.e. along the c-axis irradiated direction) and parallel to the twin planes at $H \| c(\theta = 0°)$. Any tilt of the magnetic field away from the c-axis decreases the pinning strength and increases the dissipation due to misalignment of the vortices with the relevant pinning structure. A depinning angle, θ_p, can be identified where the dissipation is at its maximum, with the superconducting anisotropy determining the resistive anisotropy for greater field misalignment. The depinning angle is approximately 25° for twin boundary pinning about the c-axis in a densely twinned crystal and exceeds 50° for crystals irradiated with U-ions. The pinning anisotropy in the ion-irradiated crystal is large enough to reverse the underlying intrinsic superconducting anisotropy. The untwinned crystal reveals the intrinsic anisotropy for YBCO, which is a smooth decrease in resistivity from a maximum for $H \| c$ to a minimum for $H \| ab$. In contrast, the ion-irradiated crystal exhibits a maximum at $H \| ab$ and a minimum at $H \| c$. It may be desirable for technological application to splay uniformly the correlated defects in all directions to obtain strong isotropic pinning. Experiments based on internal fissioning of heavy ions by high energy protons [30] to produce random splayed defects have recently shown some promise in this direction.

B2.6

Anisotropic pinning due to correlated defects leads to a Bose glass state at low temperatures [4, 31, 32], in contrast to an isotropic vortex glass state for untwinned crystals with point defects induced by proton irradiation [29]. Figure A.4.1.5 shows the angular dependence of the vortex glass transition temperature for a proton-irradiated crystal and the Bose glass transition temperature for a densely-twinned crystal. The data were obtained from fits of the resistivity just above the transition to $\rho = \rho_0(T - T_g)^s$, where s is a field-independent constant and T_g is the second-order vortex glass or Bose glass transition temperature (for a review of glass scaling see [33]). The two glasses display very different

A4.3

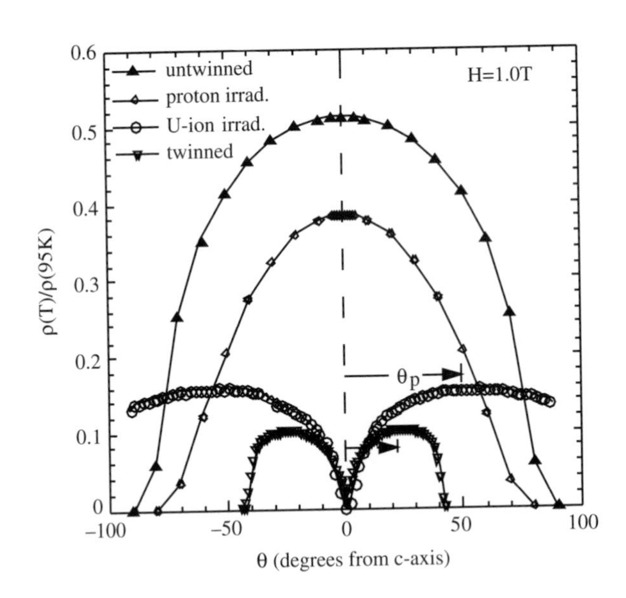

Figure A4.1.4. Angular dependence of the resistivity for (i) an untwinned crystal, (ii) a twinned crystal, (iii) an untwinned crystal with columnar defects and (iv) a proton-irradiated untwinned crystal at various temperatures in the liquid state.

angular dependences of T_g. The Bose glass transition depends strongly on the pinning strength of the correlated defects responsible for the glassy disorder. The associated angular dependence of the pinning causes a cusp in the glass transition temperature for fields applied along the high pinning direction, which is seen as a sharp maximum along the *c* direction in figure A4.1.5 where the vortices are aligned with the twin boundaries. In contrast, the point defects responsible for pinning in the vortex glass have

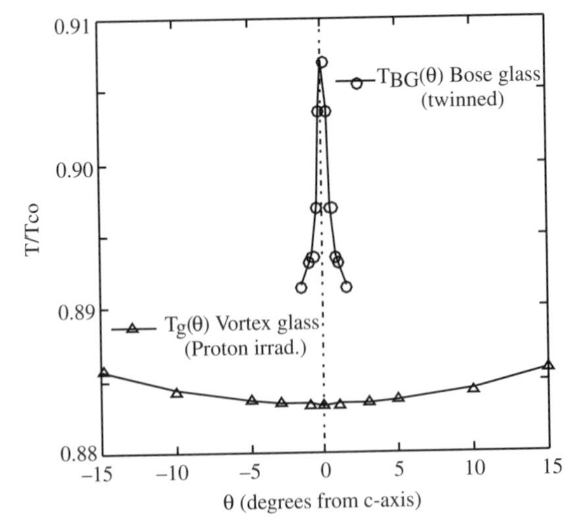

Figure A4.1.5. Angular dependence of the vortex glass and Bose glass transition temperatures for the untwinned proton-irradiated crystal and for the twinned crystal (the data for negative angles are a mirror image of the positive angle data).

isotropic pinning strength. The transition temperature in these systems has no special angular signature but instead follows the intrinsic anisotropy of the layered structure, with a broad minimum for $H \| c$ and a maximum for $H \| ab$, which is qualitatively similar to the first order melting temperature in the clean YBCO system. This natural angular dependence of the glass transition temperature is shown in figure A4.1.5 for the proton-irradiated crystal.

A4.1.5 Conclusion

The vortex phases and phase transitions in high temperature superconductors display remarkable diversity arising from the competition of interaction energy, pinning energy and thermal energy. The variety of thermodynamic behaviour in vortex matter rivals that of ordinary atomic matter and offers an accessible vehicle for systematic studies of phase equilibria in clean and disordered media. Disorder in the form of randomly placed pinning defects has a strong effect on melting and the transition from solid to liquid states. Point disorder drives the lattice to a vortex glass phase which melts at a *lower* temperature than the lattice. Correlated disorder in the form of one-dimensional heavy ion tracks and two-dimensional twin boundary planes drives the lattice to a Bose glass that melts at a *higher* temperature than the lattice. The pinning strength and upward shift in the melting temperature due to correlated defects can be dramatic, raising the irreversibility line to nearly 93% of the transition temperature in the case of U-ion generated tracks. This large effect is due to the compatible geometries of the vortex line and pinning defect, which allows strong pinning when the vortex is aligned with the defect structure. The extended defect geometry is likewise responsible for highly anisotropic pinning. For heavy ion tracks along the c direction, the pinning anisotropy is so strong that it reverses the intrinsic anisotropy of the defect-free crystal, lowering the resistivity along c to less than that along ab. Strong pinning of correlated defects is advantageous for applications. The associated strong anisotropy, however, precludes high performance in all but a single field direction, and is potentially a disadvantage.

This survey has outlined some of the basic features of the superconducting vortex phases. They offer many interesting opportunities for research on fundamental problems in superconductivity, phase equilibria in disordered media and non-linear electrodynamics. There are equally interesting technological challenges in controlling vortex behaviour by selection of pinning defects and adjustment of their density and orientation. Understanding vortex behaviour in disordered environments is a key element for continued development of the scientific and technological potential of superconductivity.

References

[1] Abrikosov A A 1957 On the magnetic properties of superconductors of the second group *Sov. Phys.– JETP* **5** 1174

[2] Fisher M P A 1989 Vortex-glass superconductivity: a possible new phase in bulk high- T_c oxides *Phys. Rev. Lett.* **62** 1415

[3] Fisher D S, Fisher M P A and Huse D A 1991 Thermal fluctuations, quenched disorder, phase transitions and transport in type-II suprconductors *Phys. Rev.* B **43** 130

[4] Nelson D R and Vinokur V M 1993 Boson localization and correlated pinning of superconducting vortex arrays *Phys. Rev.* B **48** 13060

[5] Clem J R 1991 Two-dimensional vortices in a stack of thin superconducting films: a model for high-temperature superconducting multilayers *Phys. Rev.* B **43** 7837

[6] Glazman L I and Koshelev I E 1991 Thermal fluctuations and phase transitions in the vortex state of a layered superconductor *Phys. Rev.* B **43** 2835

[7] Nelson D R and Seung H S 1989 Theory of melted flux liquids *Phys. Rev.* B **39** 9153

[8] Safar H, Gammel P L, Huse D A, Bishop D J, Rice J P and Ginsberg D M 1992 Experimental evidence for a first order vortex-lattice-melting transition in untwinned, single crystal $YBa_2Cu_3O_7$ *Phys. Rev. Lett.* **69** 824

[9] Kwok W K, Fleshler S, Welp U, Vinokur V M, Downey J and Crabtree G W 1992 Vortex lattice melting in untwinned and twinned single crystals of $YBa_2Cu_3O_{7-\delta}$ *Phys. Rev. Lett.* **69** 3370

[10] Charalambous M, Chaussy J and Lejay P 1992 Evidence from resistivity measurements along the *c* axis for transition within the vortex state for $H \| ab$ in single-crystal $YBa_2Cu_3O_7$ *Phys. Rev.* B **45** 5091

[11] Kwok W K, Fendrich J, Fleshler S, Welp U, Downey J and Crabtree G W 1994 Vortex liquid disorder and the first order melting transition in $YBa_2Cu_3O_{7-\delta}$ *Phys. Rev. Lett.* **72** 1092

[12] Crabtree G W, Kwok W K, Welp U, Fendrich J A and Veal B W 1996 Static and dynamic vortex transitions in clean $YBa_2Cu_3O_7$ *J. Low Temp. Phys.* **105** 1073

[13] Liang R, Bonn D A and Hardy W N 1996 Discontinuity of reversible magnetization in untwinned YBCO single crystals at the first order vortex melting transition *Phys. Rev. Lett.* **76** 835

[14] Welp U, Fendrich J A, Kwok W K, Crabtree G W and Veal B W 1996 Thermodynamic evidence for a flux lattice melting transition in $YBa_2Cu_3O_{7-\delta}$ *Phys. Rev. Lett.* **76** 4809

[15] Fendrich J A, Welp U, Kwok W K, Koshelev A E, Crabtree G W and Veal B W 1996 Static and dynamic vortex phases in $YBa_2Cu_3O_{7-\delta}$ *Phys. Rev. Lett.* **77** 2073

[16] Willemin M, Schilling A, Keller H, Rossel C, Hoffer J, Welp U, Kwok W K, Olsson R J and Crabtree G W 1998 First order vortex-lattice melting transition in $YBa_2Cu_3O_{7-\delta}$ near the critical temperature detected by magnetic torque *Phys. Rev. Lett.* **81** 4236

[17] Schilling A, Fisher R A, Phillips N E, Welp U, Dasgupta D, Kwok W K and Crabtree G W 1996 Calorimetric observation of a latent heat at the vortex lattice melting transition of untwinned $YBa_2Cu_3O_{7-\delta}$ *Nature* **382** 791

[18] Schilling A, Fisher R A, Phillips N E, Welp U, Kwok W K and Crabtree G W 1997 Anisotropic latent heat of vortex-lattice melting in untwinned $YBa_2Cu_3O_{7-\delta}$ *Phys. Rev. Lett.* **78** 4833

[19] Roulin M, Junod A, Erb A and Walker E 1998 Calorimetric transitions in the melting line of the vortex system as a function of oxygen deficiency in high purity YBCO *Phys. Rev. Lett.* **80** 1722

[20] Koshelev A E 1997 Point-like and line-like melting of the vortex lattice in the universal phase diagram of layered superconductors *Phys. Rev.* B **56** 11201

[21] Dodgson M J W, Geshkenbein V B, Nordborg H and Blatter G 1998 Characteristics of first order vortex lattice melting: jumps in entropy and magnetization *Phys. Rev. Lett.* **80** 837

[22] Safar H, Gammel P L, Huse D A, Bishop D J, Lee W C, Giapintzakis J and Ginsberg D M 1993 Experimental evidence for a multicritical point in the magnetic phase diagram for the mixed state of clean, untwinned $YBa_2Cu_3O_7$ *Phys. Rev. Lett.* **70** 3800

[23] Crabtree G W and Nelson D R 1997 Vortex physics in high temperature superconductors *Phys. Today* **50** 38

[24] Roulin M, Revaz B, Junod A, Erb A and Walker E 1999 High resolution specific heat experiments on the vortex melting line in $MBa_2Cu_3O_x$ (M = Y, Dy, Eu) crystals: observation of first and second order transitions up to 16 T *Proc. NATO Advanced Study Institute on the Physic and Materials Science of Vortex States, Flux Pinning and Dynamics (Kuşadasi, Turkey, July 26 – August 8, 1998)* ed S Bose and R Kossowski (Dordrecht, The Netherlands: Kluwer Academic Publishers) p 489

[25] Bouquet F, Marcenat C, Calemczuk R, Erb A, Junod A, Roulin M, Welp U, Kwok W K, Crabtree G W, Phillips N E, Fisher R A and Schilling A 1999 Calorimetric study of the transitions between the different vortex states in $YBa_2Cu_3O_7$ *Proc. NATO Advanced Study Institute on the Physic and Materials Science of Vortex States, Flux Pinning and Dynamics (Kuşadasi, Turkey, July 26 – August 8, 1998)* ed S Bose and R Kossowski (Dordrecht: Kluwer Academic Publishers) p 743

[26] Fendrich J A, Kwok W K, Giapintzakis J, van der Beek C J, Fleshler S, Welp U, Viswanathan H K, Downey J and Crabtree G W 1995 Vortex liquid state in an electron irradiated untwinned $YBa_2Cu_3O_{7-\delta}$ crystal *Phys. Rev. Lett.* **74** 1210

[27] Crabtree G W, Kwok W K, Welp U, Lopez D and Fendrich J A 1999 Vortex melting and the liquid state in $YBa_2Cu_3O_x$ *Proc. NATO Advanced Study Institute on the Physic and Materials Science of Vortex States, Flux Pinning and Dynamics (Kuşadasi, Turkey, July 26 – August 8, 1998)* ed S Bose and R Kossowski (Dordrecht: Kluwer Academic Publishers) p 357

[28] Paulius L M, Fendrich J A, Kwok W K, Koshelev A E, Vinokur V M and Crabtree G W 1997 Effects of 1-GeV uranium ion irradiation on vortex pinning in single crystals of the high temperature superconductor $YBa_2Cu_3O_{7-\delta}$ *Phys. Rev. B* **56** 913

[29] Petrean A M, Paulius L M, Kwok W K, Fendrich J A and Crabtree G W 2000 Experimental evidence for the vortex glass phase in untwinned proton irradiated $YBa_2Cu_3O_{7-\delta}$ *Phys. Rev. Lett.* **84** 5852.

[30] Krusin-Elbaum L, Thompson J R, Wheeler R, Marwick A D, Li C, Patel S, Shaw D T, Lisowski P and Ullmann J 1994 Enhancement of persistent currents in $Bi_2Sr_2CaCu_2O_8$ tapes with splayed columnar defects induced with 0.8 GeV protons *Appl. Phys. Lett.* **64** 3331

[31] Kwok W K, Fendrich J A, Fleshler S, Welp U and Crabtree G W 1994 Effect of twin boundary density on the vortex melting transition in $YBa_2Cu_3O_{7-\delta}$ single crystals *Critical Currents in Superconductors*, H W Weber (Singapore: World Scientific) pp 15–20

[32] Grigera S A, Morre E, Osquiguil E, Balseiro C, Nieva G and de la Cruz F 1998 Bose-glass phase in twinned $YBa_2Cu_3O_{7-\delta}$ *Phys. Rev. Lett.* **81** 2348

[33] Blatter G, Feigel'man M V, Geshkenbein V B, Larkin A I and Vinokur V M 1994 Vortices in high temperature superconductors *Rev. Mod. Phys.* **66** 1125

A4.2
Vortices and their interaction

E H Brandt

A4.2.1 Introduction

The existence of vortices in type II superconductors was predicted first by Alexei A. Abrikosov when he discovered a two-dimensional (2D) periodic solution of the Ginzburg–Landau (GL) equations. Abrikosov correctly interpreted this solution as a periodic arrangement of flux lines, the so-called flux line lattice (FLL). Each flux line (or, alternatively, fluxon or vortex line) carries one quantum of magnetic flux $\Phi_0 = h/2e = 2.07 \times 10^{-15}\,\mathrm{Tm}^2$, generated by a vortex of circulating supercurrents (see chapter A2.6). Consequently, the magnetic field peaks at the vortex positions. The vortex core is a tube in which the superconductivity is weakened with its centre defined by the line at which the superconducting order parameter vanishes. For well separated or isolated vortices, the radius of the tube of magnetic flux is the magnetic penetration depth, λ, and the core radius approximates to the superconducting coherence length, [1, 2] (see chapter A2.3). The spacing a of the vortices decreases with increasing applied magnetic field and the average flux density, $\bar{B} = 2\Phi_0/(\sqrt{3}a^2)$, increases for the triangular FLL (see figure A4.2.1 and chapter A2.6). The flux tubes begin to overlap with further increase in $\bar{B}$ such that the periodic induction $B(x,y)$ is nearly constant, with only a small relative modulation around its average value. Eventually the vortex cores begin to overlap such that the amplitude of the order parameter decreases until it vanishes when $\bar{B}$ reaches the upper critical field, $B_{c2} = \Phi_0/(2\pi\xi^2)$, where the superconductivity disappears. Figure A4.2.2 shows profiles of the induction $B(x,y)$ and of the superconducting order parameter $|\psi(x,y)|^2$ for two values of $\bar{B}$ corresponding to flux-line spacings $a = 4\lambda$ and $a = 2\lambda$.

A3.1 A1.1 A2.6 A3.1 A2.6

The periodic solution which describes the FLL exists when the GL parameter $\kappa = \lambda/\xi$ exceeds the value $1/\sqrt{2}$ (this condition defines type II superconductors) and when the applied magnetic induction B_a ranges between the lower and upper critical fields $B_{c1} \approx \Phi_0 \ln\kappa/(4\pi\lambda^2)$ (where $\bar{B} = 0$) and B_{c2} (where $\bar{B} = B_a = B_{c2}$). The superconductor is in the Meissner state for $|B_a| < B_{c1}$, which expells all magnetic flux, forcing $B \equiv 0$ inside the superconductor (see article A2.2). With increasing $B_a > B_{c1}$ the inner induction $\bar{B}$ increases monotonically. The volume magnetization for long cylinders with field applied parallel to their axes (i.e. for which demagnetising or shape effects are negligible) is defined as $M = (\bar{B} - B_a)/\mu_0$. In this case the negative magnetization $-M$ initially increases linearly, $M = -H_a$ for $B_a \leq B_{c1}$; $-M$ decreases sharply at $B_a = B_{c1}$, however, as flux lines start to penetrate and then approximately linearly at higher B_a until it vanishes at B_{c2}, see also section A4.2.5 and chapter A4.3 by P H Kes. The area under the magnetization curve $-M(H_a)$ is $B_c^2/2\mu_0$ where $B_c = \Phi_0/(\sqrt{8}\pi\xi\lambda) = B_{c2}/\kappa$ is the thermodynamic critical field. The three critical fields coincide (i.e. $B_{c1} = B_c = B_{c2}$) in superconductors with $\kappa = 1/\sqrt{2}$ (this is almost realized exactly in pure niobium).

A1.2 A2.2 D2.4 A4.3 A1.2

The properties of the vortex lattice may be calculated from London theory at low inductions $\bar{B} \ll B_{c2}$ and large $\kappa \gg 1$, to which the GL theory reduces when the magnitude of the order parameter is

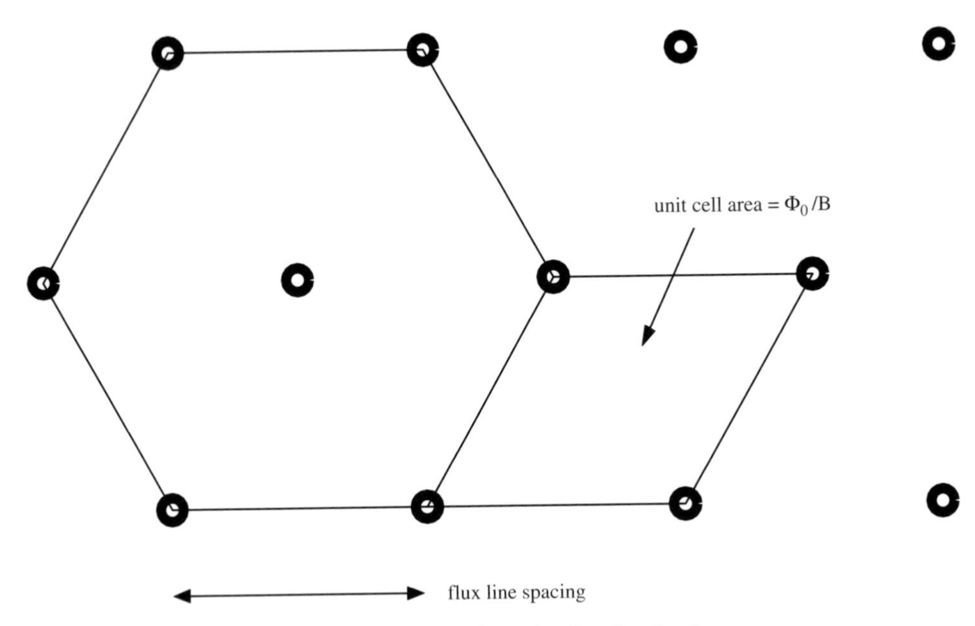

Figure A4.2.1. The triangular flux line lattice.

nearly constant. In the London limit, $B(x,y)$ is the linear superposition of the fields of isolated vortices in which case the London expressions for $B(x,y)$ and for the energy also apply to a non-periodic arrangements of vortices. The London theory was further extended to describe curved vortices and to anisotropic superconductors. The GL theory, however, has to be used at larger $\bar{B} > 0.25B_{c2}$ or smaller $\kappa < 2$, although analytical solutions are available only for the periodic FLL near B_{c2}. This theory cannot be applied to a distorted FLL or at lower $\bar{B}$. A3.1

The elastic moduli of the vortex lattice may be obtained by expanding the energy of the superconductor with respect to small displacements of the vortices from their ideal lattice positions. The elasticity of the vortex lattice is *non-local* in contrast to the local elasticity of atomic lattices, i.e., the energy of compressional and tilt deformations of the FLL is strongly reduced when the wave vector k of the strain field is large (i.e. $k > \lambda^{-1}$) [3].

A4.2.2 Results from London theory

A1.2 The London theory may be formulated by minimizing the sum F of the potential energy of the magnetic field $\mathbf{B}(\mathbf{r})$ [$E_p = \mathbf{B}^2(\mathbf{r})/2\mu_0$ per unit volume] and the kinetic energy of the supercurrent density $\mathbf{J}(\mathbf{r}) = \mu_0 \nabla \times \mathbf{B}(\mathbf{r})$ [$E_k = \lambda^2(\nabla \times \mathbf{B})^2/2\mu_0$ per unit volume]

$$F = \frac{1}{2\mu_0}\int_V [\mathbf{B}^2 + \lambda^2(\nabla \times \mathbf{B})^2]\,\mathrm{d}^3 r \tag{A4.2.1}$$

with respect to $\mathbf{B} = \nabla \times \mathbf{A}$ where $\mathbf{A}$ is the vector potential. This yields the homogeneous London equation $\mathrm{B} - \lambda^2\nabla^2\mathbf{B} = 0$ or $\mathbf{J} = -\mu_0^{-1}\lambda^{-2}\mathbf{A}$, assuming the Maxwell equations $\nabla\cdot\mathbf{B} = 0$ and $\nabla \times \mathbf{B} = \mu_0\mathbf{J}$.

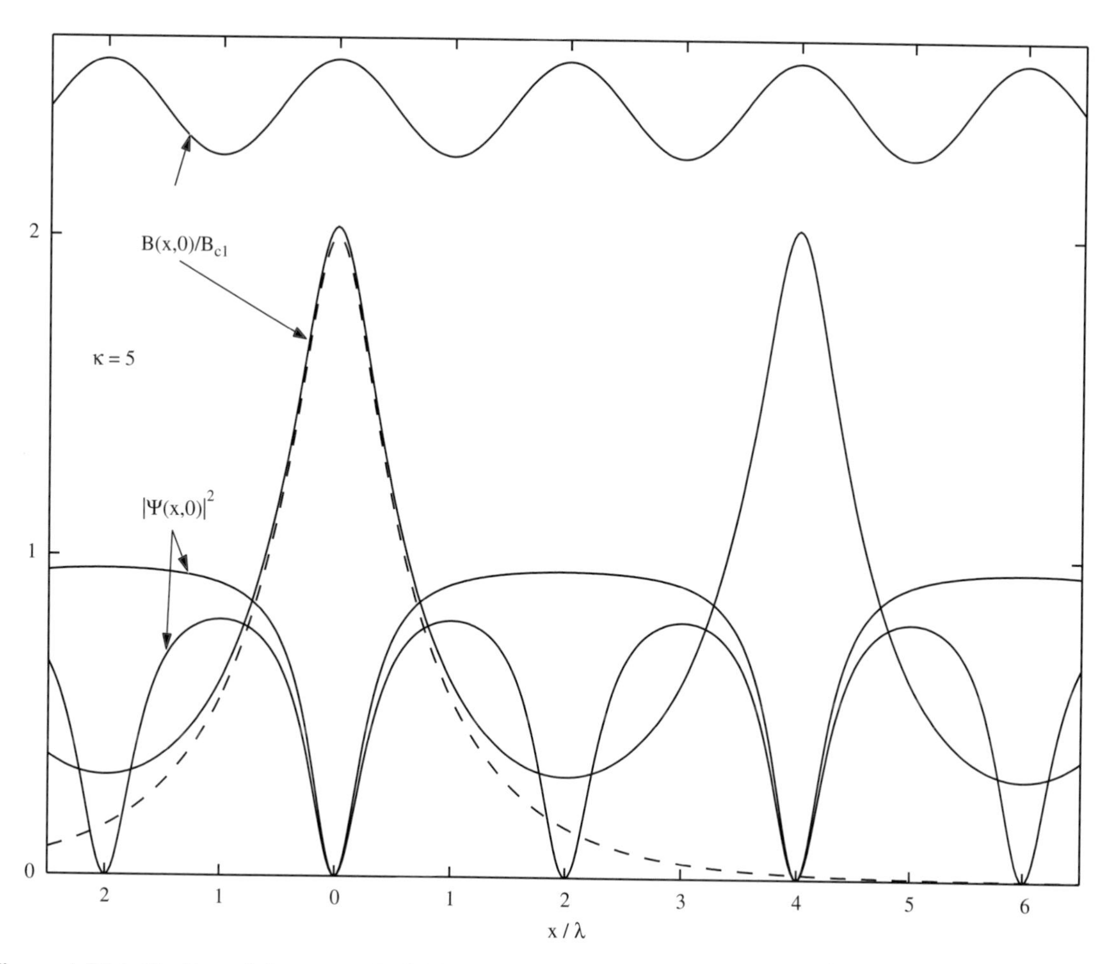

Figure A4.2.2. Profiles of the magnetic field $B(x,y)$ and order parameter $|\Psi(x,y)|^2$ along the x-axis (a nearest neighbour direction) for two flux-line lattices with lattice spacing $a = 4\lambda$ (solid lines) and $a = 2\lambda$ (thin lines). The dashed line shows the magnetic field of an isolated flux line. The data is derived from the Ginzburg–Landau theory for $\kappa = 5$.

A4.2.2.1 Parallel vortices

In the presence of vortices, one has to add singularities to the formulation which describe the vortex core. The result is the modified London equation for straight, parallel vortex lines aligned along $\hat{\mathbf{z}}$

$$\mathbf{B}(\mathbf{r}) - \lambda^2\nabla^2\mathbf{B}(\mathbf{r}) = \hat{\mathbf{z}}\Phi_0\sum_\nu \delta_2(\mathbf{r} - \mathbf{r}_\nu). \tag{A4.2.2}$$

Here $\mathbf{r}_\nu = (x_\nu, y_\nu)$ are the two-dimensional (2D) vortex positions and $\delta_2(\mathbf{r}) = \delta(x)\delta(y)$ is the 2D delta function. This linear equation may be solved by Fourier transform using $\int \exp(i\mathbf{kr})\mathrm{d}^2k = 4\pi^2\delta_2(\mathbf{r})$ and $\int \exp(i\mathbf{kr})(k^2 + \lambda^{-2})^{-1}\mathrm{d}^2k = 2\pi K_0(|\mathbf{r}|/\lambda)$. Here $K_0(x)$ is a modified Bessel function with the limits $K_0(x) \approx -\ln(x)$ for $x \ll 1$ and $K_0(x) \approx (\pi/2x)^{1/2}\exp(-x)$ for $x \gg 1$. The resulting magnetic field of any arrangement of parallel vortices is then the sum of individual vortex fields centred at the positions $\mathbf{r}_\nu$,

$$\mathbf{B}(\mathbf{r}) = \hat{\mathbf{z}}\frac{\Phi_0}{2\pi\lambda^2}\sum_\nu K_0\left(\frac{|\mathbf{r}-\mathbf{r}_\nu|}{\lambda}\right). \tag{A4.2.3}$$

The energy F_{2D} of this 2D arrangement of vortex lines with length L is obtained by inserting equation (A4.2.2) into equation (A4.2.1). The London energy is determined by the magnetic field values at the vortex positions and is obtained by integrating over the delta function

$$\begin{aligned} F_{2D} &= L\frac{\Phi_0}{2\mu_0}\sum_\mu B(\mathbf{r}_\mu) \\ &= L\frac{\Phi_0^2}{4\pi\mu_0\lambda^2}\sum_\mu\sum_\nu K_0\left(\frac{|\mathbf{r}_\mu-\mathbf{r}_\nu|}{\lambda}\right). \end{aligned} \tag{A4.2.4}$$

This expression shows that the energy is composed of the self-energy of the vortices (i.e. terms $\mu = \nu$) and a pairwise interaction energy (i.e. terms $\mu \neq \nu$). To avoid the divergence of the self-energy, the logarithmic infinity of B has to be terminated at the vortex centers $\mathbf{r}_\nu$ by introducing a finite radius of the vortex core of order ξ, the coherence length of the GL theory. This cut-off may be achieved by replacing the distance $r_{\mu\nu} = |\mathbf{r}_\mu - \mathbf{r}_\nu|$ in equation (A.4.2.4) by $\tilde{r}_{\mu\nu} = (r_{\mu\nu}^2 + 2\xi^2)^{1/2}$ and multiplying by a normalization factor ≈ 1 to conserve the flux Φ_0 of the vortex. This analytical expression suggested by Clem [4] for a single vortex, and later generalized to the vortex lattice [5], is an excellent approximation as was shown numerically [6] by solving the GL equation for the periodic FLL in the entire range of $\bar{B}$ and κ, $0 \leq \bar{B} \leq B_{c2}$, $\kappa \geq 1/\sqrt{2}$.

A2.3

A4.2.2.2 Curved vortices

Arbitrary three-dimensional (3D) arrangements of curved vortices at positions $\mathbf{r}_\nu(z) = [x_\nu(z), y_\nu(z), z]$ satisfy the following 3D London equation [3]

$$\mathbf{B}(\mathbf{r}) - \lambda^2\nabla^2\mathbf{B}(\mathbf{r}) = \Phi_0\sum_\nu\int d\mathbf{r}_\nu\delta_3(\mathbf{r}-\mathbf{r}_\nu). \tag{A4.2.5}$$

Here the integral is along the vortex lines and $\delta_3(\mathbf{r}) = \delta(x)\delta(y)\delta(z)$. The resulting magnetic field and energy are, with $\tilde{r}_{\mu\nu} = [|\mathbf{r}_\mu(z) - \mathbf{r}_\nu(z)|^2 + 2\xi^2]^{1/2}$,

$$\mathbf{B}(\mathbf{r}) = \frac{\Phi_0}{4\pi\lambda^2}\sum_\nu\int d\mathbf{r}_\nu\frac{\exp(-\tilde{r}_{\mu\nu}/\lambda)}{\tilde{r}_{\mu\nu}},$$

$$\begin{aligned} F_{3D} &= \frac{\Phi_0}{2\mu_0}\sum_\mu\int d\mathbf{r}_\mu B(\mathbf{r}_\mu) \\ &= \frac{\Phi_0^2}{8\pi\mu_0\lambda^2}\sum_\mu\sum_\nu\int d\mathbf{r}_\mu\int d\mathbf{r}_\nu\frac{\exp(-\tilde{r}_{\mu\nu}/\lambda)}{\tilde{r}_{\mu\nu}}. \end{aligned} \tag{A4.2.6}$$

This indicates that all vortex segments interact with each other in a similar manner to the interaction between magnetic dipoles or tiny current loops, but with their long-range magnetic interaction ($\propto 1/r$) screened by a factor $\exp(-\tilde{r}_{\mu\nu}/\lambda)$. The 3D interaction between curved vortices is illustrated schematically in figure A4.2.3.

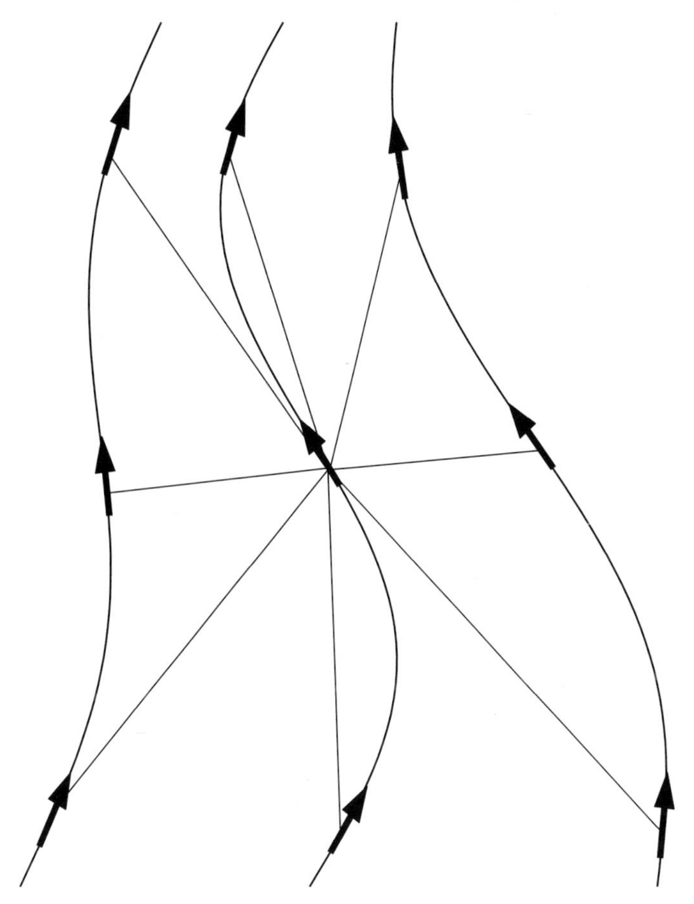

Figure A4.2.3. Visualization of the pairwise interaction between the line elements (arrows) of curved flux lines within London theory.

A4.2.2.3 Vortices near a surface

The solutions of equations (A4.2.6) apply to vortices in the bulk and have to be modified near the surface of the superconductor. In simple geometries, such as for superconductors with one or two planar surfaces surrounded by vacuum, the magnetic field and energy of a given vortex arrangement is obtained by adding the field of appropriate images (in order to satisfy the boundary condition that no current leaves the surface) and a magnetic stray field generated by a fictitious surface layer of magnetic monopoles to ensure continuity of the total magnetic field across the surface [7]. The magnetic field and interaction of straight vortices oriented perpendicular to a superconducting film of arbitrary thickness is calculated in [8].

A4.2.2.4 Thin films and layered superconductors

The self and interaction energies are modified near the surface of a superconductor. In films of thickness $d \ll \lambda$, the short 2D vortices interact mainly via their magnetic stray field outside the superconductor over an effective penetration depth $\Lambda = 2\lambda^2/d$. At short distances $r \ll \Lambda$ this interaction is logarithmic as in the bulk case, and decreases as $\exp(-r/\Lambda)$ at large $r \gg \Lambda$. The Fourier transform of the 2D vortex interaction $V(r) = \int(\mathrm{d}^2k/4\pi^2)\tilde{V}(k)\,\exp(i\mathbf{kr})$ changes with decreasing thickness d, from $\tilde{V}(k) = E_0(k^2+\lambda^{-2})^{-1}$ $(d \gg \lambda)$ to $\tilde{V}(k) = E_0(k^2+k\Lambda^{-1})^{-1}$ $(d \ll \lambda)$ where $E_0 = d\Phi_0^2/(\mu_0\lambda^2)$. A similar (but 3D) magnetic interaction exists between the 2D pancake vortices in the superconducting CuO layers of high temperature superconductors (HTS), $\tilde{V}(\mathbf{k}) = E_0 dk_3^2 k_2^{-2}(\lambda^{-2}+k_3^2)^{-1}$, where d is now the distance between the layers, $k_2 = k_x^2 + k_y^2$, $k_3^2 = k_2^2 + k_z^2$ and $\lambda = \lambda_{ab}$ is the penetration depth for currents flowing within them [3].

A4.2.2.5 Anisotropic superconductors

For many purposes HTS may be considered as uniaxially anisotropic materials, may be characterized by two penetration depths $\lambda_a \approx \lambda_b \approx \lambda_{ab}$ (for currents in the *ab* plane) and λ_c (for currents along the *c*-axis) within the London theory. The anisotropy ratio $\Gamma = \lambda_c/\lambda_{ab} = \xi_{ab}/\xi_c \geq 1$ describes also the anisotropy of the GL coherence lengths, ξ_{ab} and ξ_c, which define the inner cut-off lengths in the anisotropic London theory. The general solution for arbitrarily arranged straight or curved vortex lines is [3]

$$B_\alpha(\mathbf{r}) = \Phi_0 \sum_\mu \int \mathrm{d}\mathbf{r}_\mu^\beta f_{\alpha\beta}(\mathbf{r}-\mathbf{r}_\mu) \tag{A4.2.7}$$

$$F_{3D} = \frac{\Phi_0^2}{2\mu_0}\sum_\mu\sum_\nu \int \mathrm{d}\mathbf{r}_\mu^\alpha \int \mathrm{d}\mathbf{r}_\nu^\beta f_{\alpha\beta}(\mathbf{r}_\mu - \mathbf{r}_\nu)$$

with the tensorial interaction $(\alpha, \beta = x, y, z)$

$$f_{\alpha\beta}(\mathbf{r}) = \int \frac{\mathrm{d}^3k}{8\pi^3}\,\exp(i\mathbf{kr})\,f_{\alpha\beta}(\mathbf{k}) \tag{A4.2.8}$$

$$f_{\alpha\beta}(\mathbf{k}) = \frac{\exp[-2g(k,q)]}{1+\Lambda_1 k^2}\left(\delta_{\alpha\beta} - \frac{q_\alpha\, q_\beta\, \Lambda_2}{1+\Lambda_1 k^2+\Lambda_2 q^2}\right).$$

Here $g(k,q) = \xi_{ab}^2 q^2 + \xi_c^2(k^2-q^2) = (\Lambda_1 k^2 + \Lambda_2 q^2)\xi_c^2/\lambda_{ab}^2$ enters the cut-off factor $\exp(-2g)$; $\mathbf{q} = \mathbf{k}\times\hat{\mathbf{c}}$, $\hat{\mathbf{c}}$ is the unit vector along the *c*-axis, $\Lambda_1 = \lambda_{ab}^2$, $\Lambda_2 = \lambda_c^2 - \lambda_{ab}^2 \geq 0$, and the sums and integrals are over the μth and νth vortex line. The contribution to $\mathbf{B}(\mathbf{r})$ of the segment $\mathrm{d}\mathbf{r}_\mu$ is now in general not parallel to $\mathrm{d}\mathbf{r}_\mu$ due to the tensorial character of $f_{\alpha\beta}(\mathbf{r})$.

A4.2.3 Ginzburg–Landau, Pippard and BCS theories

A1.2 A3.1 A3.2 The London theory was extended in two ways, both of which introduce a second length ξ. These are the measures of Ginzburg–Landau (GL), which is non-linear and of Pippard, which is non-local. All three were shown later to follow from the microscopic BCS theory in limiting cases.

A4.2.3.1 Ginzburg–Landau theory

A3.1 The GL theory of 1950 introduces a complex order parameter $\psi(\mathbf{r})$ in addition to the magnetic field $\mathbf{B}(\mathbf{r}) = \nabla\times\mathbf{A}(\mathbf{r})$. The GL function $\psi(\mathbf{r})$ is proportional to the BCS energy-gap function $\Delta(\mathbf{r})$, and its square $|\psi(\mathbf{r})|^2$ to the density of Cooper pairs. The superconducting coherence length ξ gives the scale over which $\psi(\mathbf{r})$ can vary, while the penetration depth λ governs the variation of the magnetic field as in

London theory. Both λ and ξ diverge at the superconducting transition temperature T_c according to $\lambda \propto \xi \propto (T_c - T)^{-1/2}$, but their ratio, the GL parameter $\kappa = \lambda/\xi$, is nearly independent of the temperature T. The GL theory reduces to the London theory (which is valid down to $T = 0$) in the limit $\xi \ll \lambda$, which means that the magnitude of $|\psi(\mathbf{r})|$ is constant, except in the vortex cores, where it vanishes. The GL equations are obtained by minimizing a free energy functional $F\{\psi, \mathbf{A}\}$ with respect to the GL function $\psi(\mathbf{r})$ and the vector potential $\mathbf{A}(\mathbf{r})$. With the length unit λ and magnetic field unit $\sqrt{2}B_c$ the GL functional becomes

$$F\{\psi, \mathbf{A}\} = \frac{B_c^2}{\mu_0} \int \left[-|\psi|^2 + \frac{1}{2}|\psi|^4 + \left| \left(-\frac{\mathrm{i}\nabla}{\kappa} - \mathbf{A} \right) \psi \right|^2 + (\nabla \times \mathbf{A})^2 \right] \mathrm{d}^3 r. \qquad \text{(A4.2.9)}$$

F$\{\psi, A\}$ and the resulting GL equations may be expressed in terms of the real function $|\psi|$ and the gauge-invariant supervelocity $\nabla\varphi/\kappa - \mathbf{A}$, where $\varphi(\mathbf{r})$ is the phase of $\psi = |\psi|\exp(\mathrm{i}\varphi)$. The supercurrent density is $\mathbf{J} = \mu_0^{-1}\lambda^2(\nabla\varphi/\kappa - \mathbf{A})|\psi|^2$. In an external field $\mathbf{H}$ it is necessary to minimize $G = F - \mathbf{BH}$, rather than F. Which yields the equilibrium field $\mathbf{H} = \partial F/\partial \mathbf{B}$. The reversible magnetization curves $B(H)$ of pin-free superconductors are calculated in this way from GL theory [6], as were the field profiles shown in figure A4.2.2.

The GL theory modifies the London interaction between vortices in two ways firstly the range of the magnetic repulsion at large inductions $\bar{B}$ beomes larger, $\lambda' = \lambda/(1 - \bar{B}/B_{c2})^{-1/2}$ and, secondly, a weak attraction of range $\xi' = \xi/(2 - 2\bar{B}/B_{c2})^{-1/2}$ is added, caused by the condensation energy gained by the overlap of the vortex cores. Parallel vortex lines then interact by an effective potential $V(r) \propto K_0(r/\lambda') - K_0(r/\xi')$ which no longer diverges at zero distance r [9]. A3.1

A4.2.3.2 Pippard theory

Inspired by Chamber's non-local generalization of Ohm's law, Pippard introduced a superconductor coherence length ξ in 1953 by generalizing the London equation $\mu_0\mathbf{J} = -\lambda_L^2\mathbf{A}$ to a non-local relationship [2] (see also chapter A2.3) A2.3 A1.2

$$\mu_0\mathbf{J}(\mathbf{r}) = -\lambda_P^{-2}\frac{3}{4\pi\xi^2}\int \frac{\mathbf{r}'(\mathbf{r}'\mathbf{A}(\mathbf{r} - \mathbf{r}'))}{r'3} \mathrm{e}^{-r'/\xi}\mathrm{d}^3 r'. \qquad \text{(A4.2.10)}$$

In the presence of electron scattering with mean free path l, the Pippard penetration depth $\lambda_P = (\lambda_L^2\xi_0/\xi)^{1/2}$ exceeds the London penetration depth λ_L of a pure material with coherence length ξ_0, since the effective coherence length ξ is reduced by scattering, $\xi^{-1} \approx \xi_0^{-1} + l^{-1}$ [2]. In the limit of small $\xi \ll \lambda_P$, equation (A4.2.10) reduces to the local relation $\mu_0\mathbf{J}(\mathbf{r}) = -\lambda_P^{-2}\mathbf{A}(\mathbf{r})$. Pippard's equation (A4.2.10) in Fourier space becomes $\mu_0\mathbf{J}(\mathbf{k}) = -Q_P(k)\mathbf{A}(\mathbf{k})$ with

$$Q_P(k) = \lambda_P^2 h(k\xi), \quad h(x) = \frac{3}{2x^3}\left[(1 + x^2)\mathrm{atan}\, x - x\right], \quad h(0) = 1. \qquad \text{(A4.2.11)}$$

A4.2.3.3 BCS theory

The microscopic BCS theory (in the Green function formulation of Gor'kov) for weak magnetic fields yields a similar non-local relation $\mu_0\mathbf{J}(\mathbf{k}) = -Q(k)\mathbf{A}(\mathbf{k})$ as suggested by Pippard, in which the Pippard kernel $Q_P(k)$ is replaced by the BCS kernel [10] A3.2

$$Q_{\rm BCS}(k) = \lambda^{-2}(T)\sum_{n=1}^{\infty}\frac{h[k\xi_{\rm K}/(2n+1)]}{1.0518\ (2n+1)^3}. \tag{A4.2.12}$$

Here $h(x)$ is defined in equatioin (A4.2.11), $\lambda(T) = Q_{\rm BCS}(0)^{-1/2} \approx \lambda(0)(1 - T^4/T_c^4)^{-1/2}$ is the temperature dependent magnetic penetration depth, and $\xi_{\rm K} = \hbar v_{\rm F}/(2\pi k_{\rm B}T) \approx 0.844\lambda(T)T_{\rm c}/(\kappa T)$ ($v_{\rm F}$ = Fermi velocity, κ = GL parameter). The range of the BCS Gor'kov kernel is of the order of the BCS coherence length $\xi_0 = \hbar v_{\rm F}/(\pi\Delta_0)$ where Δ_0 is the BCS energy gap at $T = 0$.

With the non-local relation $\mu_0\mathbf{J}(\mathbf{k}) = -Q(k)\mathbf{A}(\mathbf{k})$ equation (A4.2.2) for a vortex line at $\mathbf{r}_\nu = 0$ now becomes $[1 + Q(k)^{-1}k^2]\tilde{B}(k) = \Phi_0$ with the solution

$$\tilde{B}(k) = \frac{\Phi_0 Q(k)}{Q(k)+k^2}, \quad B(r) = \frac{\Phi_0}{2\pi}\int_0^\infty \frac{Q(k)}{Q(k)+k^2}J_0(kr)\,k\,{\rm d}k \tag{A4.2.13}$$

where $J_0(x)$ is a Bessel function. The Pippard or BCS field $B(r)$ equation (A4.2.13) of an isolated vortex line is no longer monotonic as compared with the London field, [cf equation (A4.2.3)], but exhibits a field reversal with a negative minimum at large distances $r \gg \lambda_{\rm P}$ from the vortex core. This effect should be observable if $\xi \approx \lambda$, i.e., for clean superconductors with small GL parameter κ at low temperatures.

The field reversal of the vortex field is partly responsible for the attractive interaction between flux lines at large distances, which has been observed in clean niobium at temperatures significantly below $T_{\rm c}$ and which follows from BCS theory at $T < T_{\rm c}$ for pure superconductors with GL parameter κ close to $1/\sqrt{2}$. This attraction leads to abrupt jumps in the magnetization curve and to an agglomeration of flux lines that can be observed in superconductors with demagnetization factor $N \neq 0$ as FLL islands surrounded by Meissner state, or Meissner islands surrounded by FLL [11]. For the definition of N see chapter A3.1 by A. M. Campbell and section A4.2.5.

A3.1

Another BCS effect which differs from the GL result is that in clean superconductors at low temperatures, the periodic magnetic field of the FLL near B_{c2} is not a smooth spatial function as in GL theory, but has sharp conical maxima and minima such that the profile $B(x, 0)$ along a nearest neighbour direction has zig–zag shape [3].

A4.2.4 Elasticity of the vortex lattice

The flux-line displacements caused by pinning forces and by thermal fluctuations may be calculated using the elasticity theory of the FLL. Figure A4.2.4 illustrates the three basic distortions of the triangular FLL, namely shear, uniaxial compression and tilt. The linear elastic energy $F_{\rm elast}$ of the FLL is obtained by expanding its free energy F with respect to small displacements $\mathbf{u}_\nu(z) = \mathbf{r}_\nu(z) - \mathbf{R}_\nu = (u_{\nu x}, u_{\nu y})$ of the flux lines from their ideal parallel lattice positions $\mathbf{R}_\nu$ and keeping only the quadratic terms. This yields [3]

$$F_{\rm elast} = \frac{1}{2}\int_{\rm BZ}\frac{{\rm d}^3k}{8\pi^3}u_\alpha(\mathbf{k})\Phi_{\alpha\beta}(\mathbf{k})u_\beta^*(\mathbf{k}) \tag{A4.2.14}$$

where $\mathbf{u}(\mathbf{k})$ is the Fourier transform of the displacement field $\mathbf{u}_\nu(z)$, now $(\alpha, \beta) = (x, y)$ and $\mathbf{k} = (k_x, k_y, k_z)$. The k-integral in equation (A4.2.14) is over the first Brillouin zone (BZ) of the FLL, since the 'elastic matrix' $\Phi_{\alpha\beta}(\mathbf{k})$ is periodic in the k_x, k_y plane. The finite vortex core radius restricts the k_z integration to $|k_z| \leq \xi^{-1}$. For an elastic medium with uniaxial symmetry the elastic matrix becomes

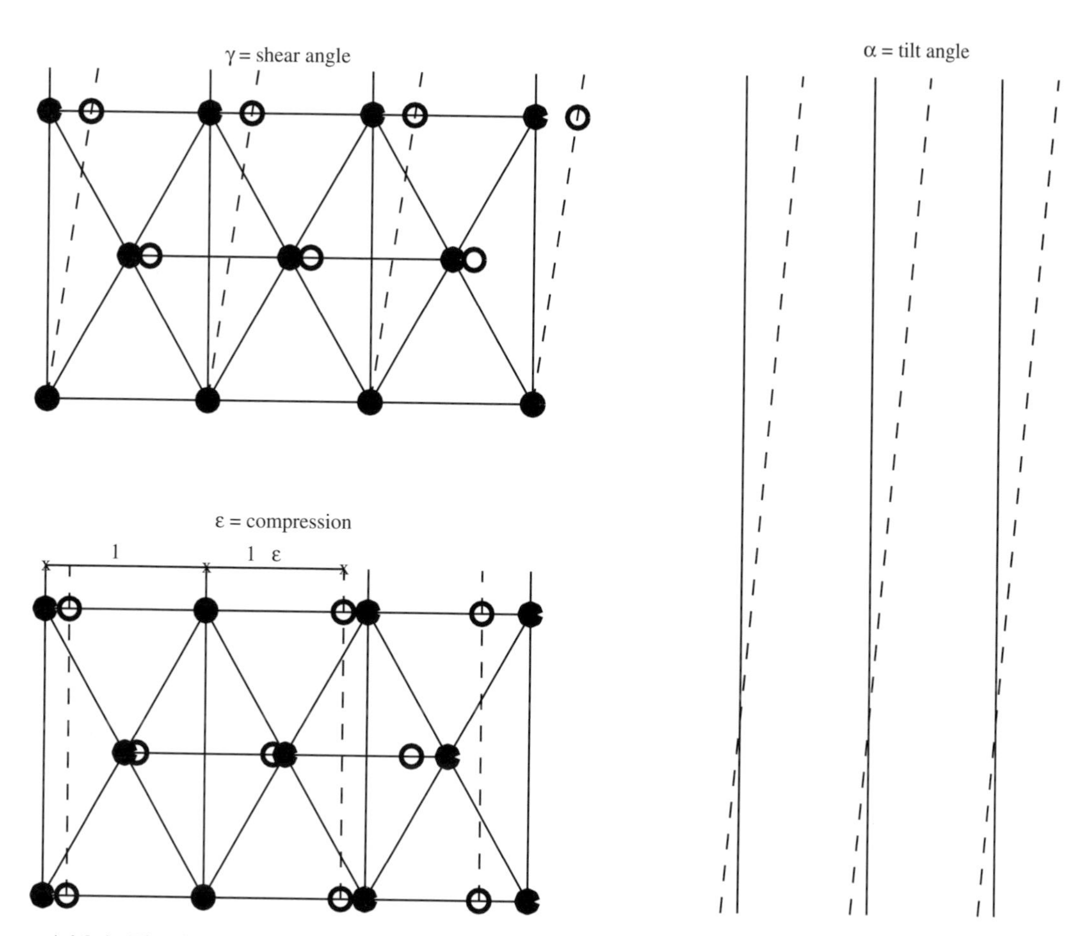

Figure A4.2.4. The three basic homogeneous elastic distortions of the triangular flux-line lattice. The full dots and solid lines mark the ideal lattice and the hollow dots and dashed lines the distorted lattice.

$$\Phi_{\alpha\beta}(\mathbf{k}) = (c_{11} - c_{66})k_\alpha k_\beta + \delta_{\alpha\beta}[(k_x^2 + k_y^2)c_{66} + k_z^2 c_{44}]. \tag{A4.2.15}$$

The coefficients c_{11}, c_{66}, and c_{44} are the elastic moduli of uniaxial compression, shear and tilt, respectively. $\Phi_{\alpha\beta}(\mathbf{k})$ has been calculated from GL and London theories [3] for the FLL. The result, a sum over reciprocal lattice vectors, should coincide with expression A4.2.15 in the continuum limit, i.e., for small $|\mathbf{k}| \ll k_{\mathrm{BZ}}$, where $k_{\mathrm{BZ}} = (4\pi B/\Phi_0)^{1/2}$ is the radius of the circularized (actually hexagonal) Brillouin zone of the triangular vortex lattice with area πk_{BZ}^2. In the London limit the elastic moduli for isotropic superconductors, becomes

$$c_{11}(k) \approx \frac{B^2/\mu_0}{1 + k^2\lambda^2}, \quad c_{66} \approx \frac{B\Phi_0/\mu_0}{16\pi\lambda^2}, \quad c_{44}(\mathbf{k}) \approx c_{11}(k) + 2c_{66}\ln\frac{\kappa^2}{1 + k_z^2\lambda^2}. \tag{A4.2.16}$$

The GL theory yields an additional factor $(1 - B/B_{c2})^2$ in c_{66} [i.e. $c_{66} \propto B(B - B_{c2})^2$]. The $\mathbf{k}$ dependence (dispersion) of the compression and tilt moduli $c_{11}(k)$ and $c_{44}(\mathbf{k})$ means that the elasticity of the vortex lattice is *non-local*, i.e., strains with short wavelengths $2\pi/k \ll 2\pi\lambda$ have a much lower elastic energy than a homogeneous compression or tilt (corresponding to $\mathbf{k} \to 0$). This elastic non-locality comes from the fact that the magnetic interaction between the flux lines typically has a range λ much longer than the flux-line spacing a. Each flux line, therefore, interacts with many other flux lines.

Both the compressional modulus c_{11} and the typically much smaller shear modulus $c_{66} \ll c_{11} \approx c_{44}$ originate from the flux-line interaction. The last term in the tilt modulus c_{44} (A4.2.16), however, originates from the line tension of isolated flux lines, defined by $P = \lim_{B \to 0}(c_{44}\Phi_0/B)$. In isotropic (or cubic) superconductors like Nb and its alloys, the line tension coincides with the self energy of a flux line, $P = F_s$, $F_s = \Phi_0 B_{c1}/\mu_0 \approx (\Phi_0^2/4\pi\mu_0\lambda^2)(\ln\kappa + 0.5)$ for $\kappa \gg 1$. In anisotropic materials, the line tension and line energy of flux lines in general are different and depend on the angle θ of the vortex line with respect to the c-axis, with $P(\theta) = F_s(\theta) + \partial^2 F_s/\partial\theta^2$. Using $F_s(\theta) = F_s(0)(\cos^2\theta + \Gamma^{-2}\sin^2\theta)^{1/2}$ with $\Gamma = \lambda_c/\lambda_{ab} \geq 1$, one obtains $P(0) = F_s(0)/\Gamma^2$ and $P(\pi/2) = F_s(\pi/2)\Gamma^2 = P(0)/\Gamma^3$ [3]. In isotropic superconductors the uniaxial symmetry of the ideal vortex lattice (i.e., the appearance of a preferred axis) is induced by the applied magnetic field. This induced anisotropy leads to a small difference between the compressional and tilt moduli c_{11} and c_{44}, but not to a difference between line energy and line tension.

As a consequence of non-local elasticity, the flux-line displacements $\mathbf{u}_\nu(z)$ caused by local pinning forces, and also the space and time averaged thermal fluctuations $\langle \mathbf{u}_\nu(z)^2 \rangle$, are much larger than they would be if $c_{44}(\mathbf{k})$ had no dispersion, i.e., if it were replaced by $c_{44}(0) \approx B^2/2\mu_0$. The maximum displacement $u(0) \propto f$ caused at $\mathbf{r} = 0$ by a point force of density $f\delta_3(\mathbf{r})$, and the thermal fluctuations $\langle u^2 \rangle \propto k_B T$, are given by similar expressions [3],

$$\frac{2u(0)}{f} \approx \frac{\langle u^2 \rangle}{k_B T} \approx \int_{BZ} \frac{d^3k}{8\pi^3} \frac{1}{(k_x^2 + k_y^2)c_{66} + k_z^2 c_{44}(\mathbf{k})} \approx \frac{k_{BZ}^2 \lambda}{8\pi[c_{66}c_{44}(0)]^{1/2}}. \qquad \text{(A4.2.17)}$$

In this result, a large factor $[c_{44}(0)/c_{44}(k_{BZ})]^{1/2} \approx k_{BZ}\lambda \approx \pi\lambda/a \gg 1$ originates from the elastic non-locality. In anisotropic superconductors with $B \| c$, the length λ in the numerator of equation (A4.2.17) is replaced by the larger length λ_c which effectively enhances the thermal fluctuations by an additional factor $\Gamma = \lambda_c/\lambda_{ab}$.

A4.2.5 Continuum description of the vortex state

A continuum description of the vortex state may be used to calculate the distributions of magnetic field and current in superconductors of arbitrary shape for length scales large compared with the vortex spacing a and the penetration depth λ. Two different algorithms have been proposed [12, 13] which, in principle, allow computation of the electromagnetic behaviour of superconductors of arbitrary shape with and without vortex pinning.

Apart from the Maxwell equations, a continuum description requires the constitutive laws of the superconductor. These may be obtained, for example, from the London equation by taking the limits $a \to 0$, $\lambda \to 0$, and from appropriate models of vortex dynamics. One constitutive law is the reversible magnetization curve of a pin-free, or ideal, superconductor, $M(B_a)$ or $B(B_a)$. In the simplest case, this may read $\mathbf{M} = 0$ or $\mathbf{B} = \mu_0\mathbf{H}$, which is valid if everywhere B is larger than several times the lower critical field B_{c1}. In general, however, the reversible $M(B_a)$ computed from London or GL theories should be used.

D2.1

The correct $M(B_a)$ for finite B_{c1} and general geometry will lead to an irreversible magnetization loop, even in complete absence of vortex pinning. This irreversibility is caused by a geometric barrier for the penetration of magnetic flux, which is absent only if the superconductor has the shape of an ellipsoid or is a cone with a sharp cusp or edge where flux lines can penetrate easily. In superconductor cylinders or strips with rectangular cross section $2a \times 2b$ ($2a$ = diameter or width, $2b$ = height) in increasing B_a for example, the magnetic flux penetrates first reversibly at the four corners in the form of nearly straight flux lines, as illustrated in figure A4.2.5. When the field of first flux entry B_{en} [13] is reached, these flux

D2.1

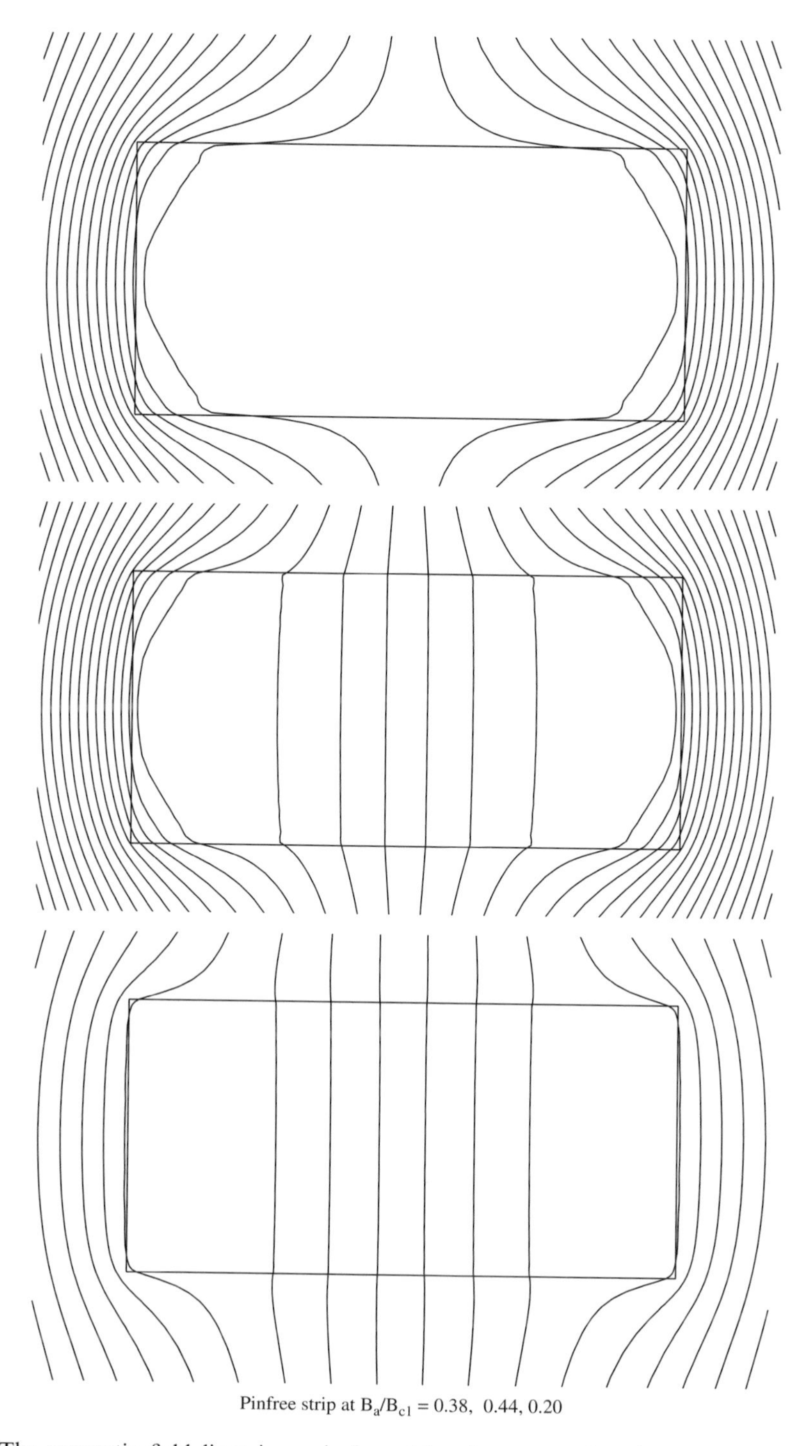

Pinfree strip at B$_a$/B$_{c1}$ = 0.38, 0.44, 0.20

Figure A4.2.5. The magnetic field lines in a pin-free strip of aspect ratio $b/a = 0.5$ in an increasing applied perpendicular field at two field values $B_a/B_{c1} = 0.38$ (top) and 0.44 (middle) just below and above the entry field $B_{en} = 0.40B_{c1}$, and in decreasing field at $B_a/B_{c1} = 0.20$ (bottom). Calculation by the method [13].

lines join at the equator and jump to the specimen centre, from where they gradually fill the entire superconductor. On decreasing B_a, some flux initially exits reversibly, until a reversibility field $B_{rev} > B_{en}$ is reached at which the magnetization loop opens since the barrier for flux exit is weaker than the barrier for flux entry. In the limit of thin films there is no barrier for flux exit. For arbitrary aspect ratio b/a the entry field is given by

$$B_{en} \approx B_{c1} \tanh \sqrt{cb/a} \tag{A4.2.18}$$

where $c = 0.36$ for strips and $c = 0.67$ for discs or cylinders. This geometric barrier should not be confused with the Bean–Livingstone barrier for the penetration of a straight vortex line into the planar surface of a superconductor, which would lead to a similar asymmetric magnetization loop (see chapter A4.3). The geometric barrier is caused by the line tension of the vortices penetrating at sharp [12, 13] or rounded [14] corners. This line tension (equal to $\Phi_0 B_{c1}/\mu_0$ in isotropic superconductors [3]) is balanced by the Lorentz force exerted on the vortex ends by the surface screening currents and directed towards the specimen centre.

The calculated irreversible magnetization loops of pin-free cylindrical superconductors with various aspect ratios $b/a = 0.08$ (thin disc) to $b/a = \infty$ (long cylinder) in an axial field B_a are illustrated

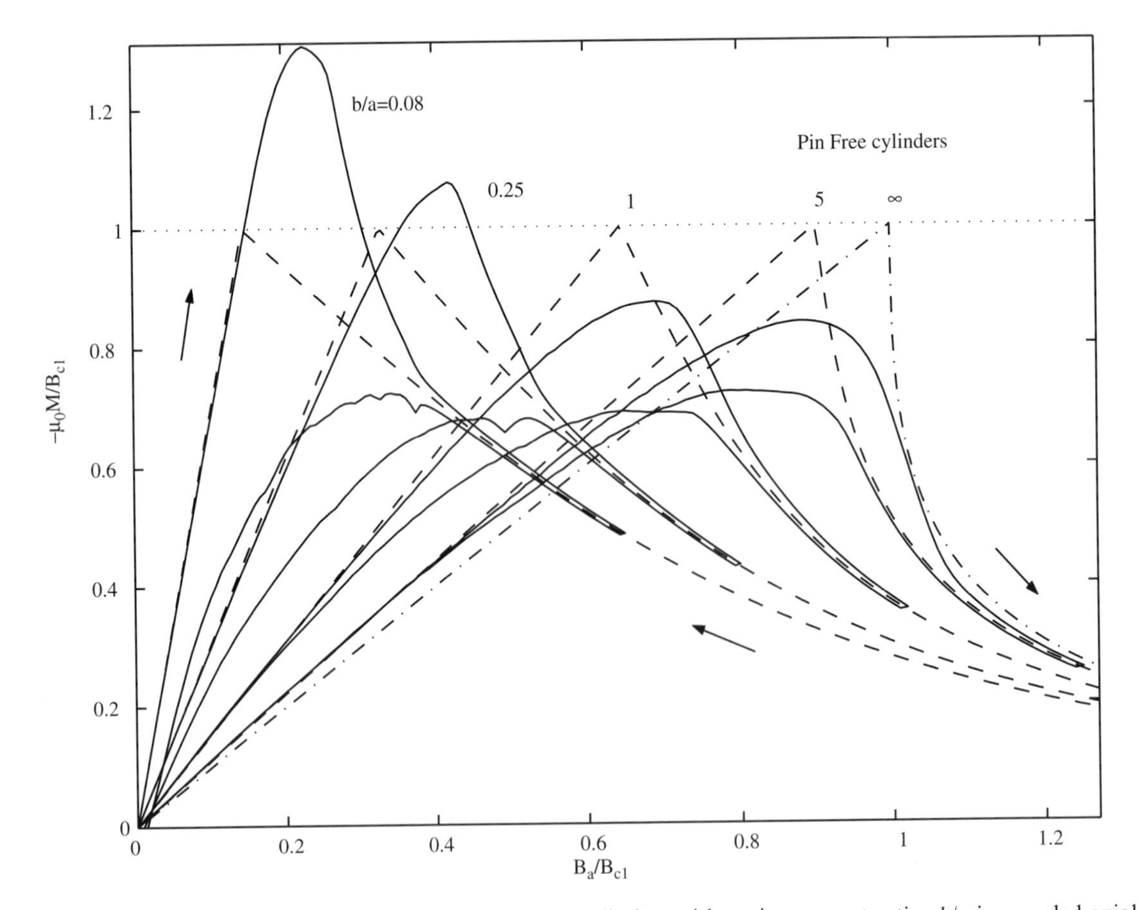

Figure A4.2.6. The irreversible magnetization of pin-free cylinders with various aspect ratios b/a in a cycled axial magnetic field B_a (solid lines). The dashed lines show the reversible magnetization of ellipsoids with the same initial slope as the cylinder.

in figure A4.2.6, together with the corresponding reversible magnetization curves of ellipsoids which have the same initial (Meissner state) slope as the cylinders. The magnetization loops of these cylinders (like those of other non-ellipsoids) have a maximum at $B_a \approx B_{en}$, and are reversible at $B_a > B_{rev}$ at which field they coincide with the magnetization of the corresponding ellipsoid. All pin-free magnetization loops are symmetric, $M(-B_a) = -M(B_a)$ with $M(0) = 0$, i.e., no remanent flux can remain at $B_a = 0$ since bulk

A2.2

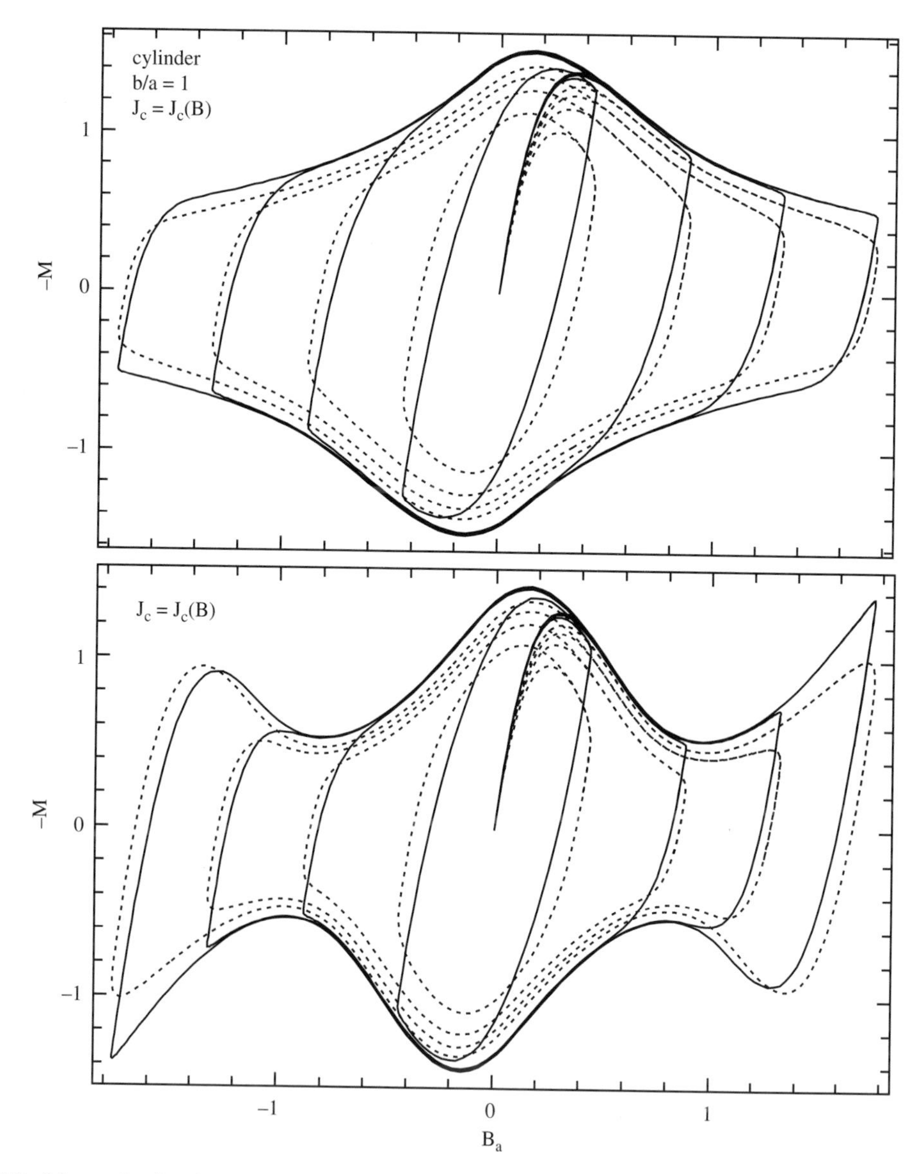

Figure A4.2.7. Magnetization loops of cylinders with aspect ratio $b/a = 1$, creep exponents $n = 51$ (bold lines) and $n = 5$ (dashed lines), with $B_{c1} = 0$, in cycled applied field $B_{\bar{a}}$ for two induction dependent critical current densities $J_c(B) = J_{c0}/(1 + 3\beta)$ (top, Kim model) and $J_c(B) = J_{c0}(1 - 3\beta + 3\beta^2)$ (bottom, a 'fish-tail' model), where $\beta = B_a/(\mu_0 J_{c0} a)$. The magnetization M is in units $J_{c0}a/(2\pi b)$ and B_a in units $\mu_0 J_c a$ where a, b are the radius and half height of the cylinder, respectively.

pinning is absent and there is no barrier for flux exit when $B_a = 0$. The small finite $M(0)$ in figure A4.2.6 is caused by the finite ramp rate used in these calculations, leading to a viscous drag force on the vortices.

The reversible magnetization curves of pin-free ellipsoids $M(B_a, N)$ follow from the magnetization curve $M(B_a, N = 0)$ of long cylinders in parallel field B_a by the concept of a demagnetization factor N. N takes the values $0 \leq N \leq 1$, $N = 0$ for parallel geometry, $N = 1/2$ for infinite cylinders in perpendicular field, $N = 1/3$ for spheres, and $N = 1$ for thin films in perpendicular field. The implicit equation for an effective internal field B_i has to be solved for ellipsoids with $N \neq 0$

D2.4

$$B_i = B_a - N\mu_0 M(B_i, N = 0) \qquad \text{(A4.2.19)}$$

to obtain the reversible magnetization $M(B_a, N) = M(B_i, N = 0)$ (dashed lines in figure A4.2.6).

A second constitutive law is based on the local electric field **E**(**J**,**B**) which is generated by moving vortices and which in a compact way can desribe flux flow and Hall effect, vortex pinning and thermally activated depinning. A simple but still quite general isotropic model is $\mathbf{E} = \rho(J, B)\mathbf{J}$ with $\rho = \text{const}\cdot B \cdot |J/J_c|^{n-1}$, where J_c is the critical current density and n the creep exponent [3, 15]. Within this realistic model, flux flow is described by $n = 1$, flux creep (relaxation with approximately logarithmic time law) by $n \gg 1$, and the critical state model of vortex pinning by the limit $n \to \infty$ (see also chapters A4.3 and A3.1). In general, both $J_c(B,T)$ and $n(B,T)$ may depend on the local induction B and on the temperature T. Figure A4.2.7 shows some magnetization loops of a short cylinder calculated in [15] for two different $J_c(B)$ models and two creep exponents. The finite London penetration depth λ can be accounted for by adding to $E(J, B)$ a term $E = \mu_0\lambda^2\dot{J}$, where $\dot{J}$ is the time derivative of the current density. This generalization is described in [16].

References

[1] DeGennes P G 1966 *Superconductivity of Metals and Alloys* (New York: Benjamin)
[2] Tinkham M 1975 *Introduction to Superconductivity* (New York: McGraw-Hill)
[3] Brandt E H 1995 The flux-line lattice in superconductors *Rep. Prog. Phys.* **58** 1465–1594
[4] Clem J R 1975 Simple model for the vortex core in type II superconductors *J. Low Temp. Phys.* **18** 427–434
[5] Hao Z, Clem J R, Mc Elfresh M W, Civale L, Malozemov A P and Holtzberg F 1991 Model for the reversible magnetization of high-κ type II superconductors: Application to high-T_c superconductors *Phys. Rev.* B **43** 2844–2852
[6] Brandt E H 1997 Precision Ginzburg–Landau solution of the flux-line lattice with arbitrary induction and symmetry *Phys. Rev. Lett.* **78** 2208–2211
[7] Brandt E H 1981 Properties of the distorted flux-line lattice near a planar surface *J. Low Temp. Phys.* **42** 557–584
[8] Jung-Chun Wei and Tzong-Jer Yang 1996 Current distribution and vortex–vortex interaction in a superconducting film of finite thickness *Jpn. J. Appl. Phys.* **35** 5696–5700
[9] Brandt E H 1986 Elastic and plastic properties of the flux-line lattice in type II superconductors *Phys. Rev.* B **34** 6514–6517
[10] Abrikosov A A, Gorkov L P and Dzyaloshinski I E 1963 *Methods of Quantum Field Theory in Statistical Physics* (Englewood Cliffs: Prentice Hall)
[11] Brandt E H and Essmann U 1987 The flux-line lattice in type II superconductors *Phys. Stat. Solidi* b **144** 13–38
[12] Labusch R and Doyle T B 1997 Macroscopic equations for the description of the quasi-static magnetic behaviour of a type II superconductor of arbitrary shape *Physica* C **290** 143–160
[13] Brandt E H 1999 Geometric barrier and current string in type II superconductors obtained from continuum electrodynamics *Phys. Rev.* B **59** 3369–3372
[14] Benkraouda M and Clem J R 1998 Critical current from surface barriers in type II superconducting strips *Phys. Rev.* B **58** 15103–15107
[15] Brandt E H 1998 Superconductor disks and cylinders in an axial magnetic field. I. Flux penetration and magnetization curves *Phys. Rev.* B **58** 6506–6522
[16] Brandt E H 2001 Theory of type II superconductors with finite London penetration depth *Phys. Rev.* B **64** 024505

A4.3
Flux pinning

P H Kes

A4.3.1 Introduction

Type II superconductors with high upper critical fields H_{c2} have considerable potential for a variety of practical applications, such as high magnetic field solenoids, permanent magnet and energy storage devices. However, the interplay between currents and flux lines in superconducting materials results in a driving force on the flux lines that puts them into motion. According to the mechanism first discussed by Bardeen and Stephen, such motion leads to dissipation of energy, manifested as an electric potential within the material, which therefore can no longer be considered to be superconducting, i.e. its resistance becomes non-zero. Prevention of flux line motion, or flux pinning, up to a high critical current density J_c, is therefore essential for practical applications of superconducting materials. Fortunately, flux pinning is a quite general phenomenon related to the presence of lattice defects in commonly produced materials. In the following sections an overview is given of this interesting phenomenon and some related issues.

A4.3.2 Hysteresis—the Bean model

A4.3.2.1 Experimental techniques

Experimentally, flux pinning manifests itself in many ways which provide us with a variety of techniques to measure the critical current density J_c. Most direct of these techniques is to measure the current–voltage (*IV*) characteristic of a superconducting wire or thin film. Ideally, this should be constant (i.e. $V = 0$) up to $\boldsymbol{J} = \boldsymbol{J}_c$ and then rise almost linearly with a slope slightly greater than the flux flow resistivity ρ_f for $\boldsymbol{J} > \boldsymbol{J}_c$, (see below and figure A4.3.3). A reasonable approximation for ρ_f is $\rho_f \approx \rho_n B/B_{c2}$, where ρ_n is the resistivity of the normal state, suggesting that the dissipation takes place mainly in the core of the moving flux lines. Non-uniformities give rise to rounding of the *IV* curves and hence it is common practice to define $\boldsymbol{J}_c$ (rather arbitrarily) by a voltage criterion of 1 μV dropped over 1 cm. For technical superconductors the *IV* curves can often be fitted successfully to an (empirical) power law, $V \propto (J/J_c)^n$, to characterize the quality of a wire. A large value of n generally signifies a high material uniformity. In strong pinning materials $\boldsymbol{J}_c$ can be as large as a few tenths of the depairing current density, $\boldsymbol{J}_0 \approx H_c/\lambda$, which is typically $10^{12}\ \mathrm{A\,m^{-2}}$. To achieve such large current densities in transport is difficult due to heating effects and therefore inductive probes are frequently used.

D2.2

A4.3

In the absence of pinning centres, surface or geometrical effects, the flux distribution inside a superconductor in a magnetic field $H > H_{c1}$ is uniform with density B. Screening currents in a surface layer of thickness λ produce a magnetic moment m and an associated volume magnetization M which is in equilibrium with H, as described by the reversible magnetization curve of Abrikosov (see chapter A4.2), i.e. $M = M_{rev}(H)$ and $B_{rev}(H) = \mu_0(H + M_{rev})$. When pinning centres are present the flux lines will be

A4.2

A4.2

D2.1

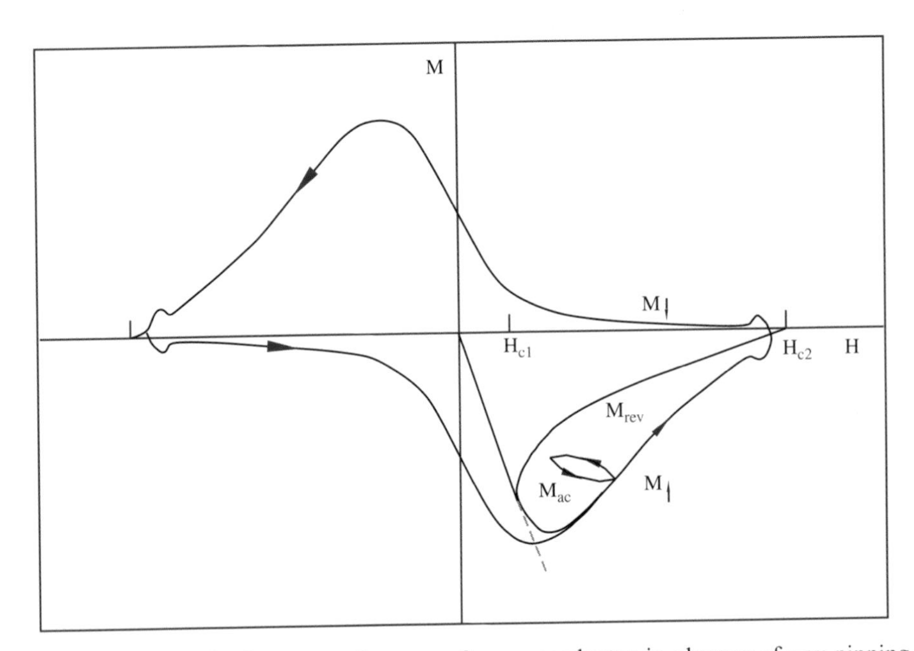

Figure A4.3.1. Typical magnetization curves for a $\kappa \approx 2$ superconductor in absence of any pinning mechanism ($M_{\rm rev}$) and in presence of bulk pinning only ($M_\uparrow$, increasing field, $M_\downarrow$, decreasing field). Also shown is the magnetization loop traversed when a small ac-field of moderate frequency is superimposed on the dc-field. The little 'humps' near H_{c2} illustrate the peak effect.

trapped upon entering or leaving the superconductor leading to flux density gradients and an irreversible magnetization loop, as illustrated in figure A4.3.1.The irreversibility is a direct measure of $\boldsymbol{J}_c$, as will be discussed below. More spatially resolved information about the flux density profiles can be obtained from ac techniques in which a ripple field of varying amplitude is superimposed onto a much larger dc field. The induced voltage signal in this case is a measure of the depth x to where the ac field has penetrated the superconductor, providing information about $\boldsymbol{J}_c(x)$. For meaningful interpretation, uniform material properties in the direction perpendicular to the direction of field penetration have to be assumed. Very often, materials are much less uniform and flux penetration occurs along material defects, such as twin or grain boundaries in $YBa_2Cu_3O_7$, for example. Local magnetic probes can be used to visualize the flux density distribution. The Faraday effect is exploited in magneto-optics, in which the polarization of light is rotated in presence of a magnetic field to identify flux concentration gradients. Alternatively arrays of micron-sized Hall probes or scanning Hall probes may be used. For this use, if the Hall probes are small enough and $B \leq \mu_0 H_{c1}$ so that the field profiles do not overlap, the positions of individual vortices may be detected. The latter can also be achieved by Bitter decoration where small clusters of magnetic atoms are deposited on the surface of a superconductor. The magnetic forces attract the clusters to the sites where the flux lines leave the surface. Very detailed information about flux line distribution and flux line lattice defects can be obtained with this technique (see chapter A2.6).

D2.4, A1.3
C1
D3.4
D1.7.2
A2.6

A4.3.2.2 The Bean model

Magnetic experiments in sample configurations for which demagnetization effects are negligible are usually interpreted in terms of the critical state equation, first introduced in 1962 by Bean [1].

$$(\partial B/\partial H)^{-1}_{\text{rev}}|\vec{B}\times\nabla\times\vec{B}| = BJ_c(B) = F_p \qquad \text{(A4.3.1)}$$

This equation expresses the fact that the magnetic pressure caused by the gradient in flux density, B, is compensated by the pinning force per unit volume F_p. For fields well above H_{c1} and for a Ginzburg–Landau parameter $\kappa = \lambda/\xi \gg 1$, the prefactor $(\partial B/\partial\mu)^{-1}_{\text{rev}}$ (which follows from the slope of the reversible magnetization curve) can be replaced by μ_0^{-1}. It turns out that under the same conditions $\boldsymbol{J}_c$ can be assumed to be constant throughout the sample. This model can be applied to irreversible magnetization curves to determine the deviation Δm from the reversible magnetization using

A3.1

$$\Delta M = m/V = (1/2V)\int |\boldsymbol{J}_c \times r|\mathrm{d}V \qquad \text{(A4.3.2)}$$

where m is the measured magnetic moment, r is the length scale over which the current flows and V is the sample volume. For a cylindrical sample with radius a, for example, equation A4.3.2 predicts $\Delta M = |M_\uparrow - M_{\text{rev}}| = M_\downarrow - M_{\text{rev}} = \boldsymbol{J}_c a/3$. It follows that the reversible magnetization lies midway between the increasing and decreasing field branches of the hysteresis loop (indicated in figure A4.3.1 by $M_\uparrow$ and $M_\downarrow$), provided the Bean model is applicable.

A minor magnetization loop may be traversed repeatedly by superimposing a small ac field of amplitude h_0 on the background dc field, as indicated in figure A4.3.1 by M_{ac}. Using the same procedure, but varying h_0, it is possible to determine $\boldsymbol{J}_c(r)$ from the shape of the minor loops and to separate surface barrier effects from bulk pinning [2]. As was first realized by Campbell [2], this method also offers the possibility to measure the range and the curvature (described by the Labusch parameter) of the flux pinning potential. The first harmonic of the in- and out-of-phase component of the susceptibility χ_1' and χ_1'' is usually measured with a lock-in technique. These quantities approximately represent the average slope and the area of the minor loop, respectively. From a slightly different point of view, ac experiments on superconductors can be considered as a measurement of the electromagnetic skin depth, characteristic of metals. The theoretical analysis is totally equivalent to that for metals except that Ohm's law has to be replaced by the expression describing the flux flow resistivity, $E = \rho(J)J$, which is clearly non-linear in J. This non-linearity is a characteristic of irreversible superconductors and can be detected most directly by measuring the third harmonic of the ac susceptibility in order to locate the irreversibility line (see section A4.3.4.).

D1.2

A1.3

A4.3.2.3 Bean–Livingstone barrier

It is well known that parallel flux lines repel each other, which raises question about the existence and stability of a flux lattice. This is due to an inward force on the flux lines provided by diamagnetic screening currents in the surface layer of thickness λ, which effectively form a magnetic container. This force decays exponentially, i.e. $\propto \exp(-x\lambda)$, where x is the distance between a flux line and the surface of the sample. Supercurrents around the vortex core on the other hand, will be distorted close to the surface when $x < \lambda$. It is necessary, therefore, to superimpose a mirror image anti-vortex at $-x$ to obtain the appropriate boundary conditions. The attractive interaction between the vortex and its image will produce an outward force that decays approximately as $\exp(-2x/\lambda)$. The result is that an entering flux line has to overcome a net outward force, or equivalently, a surface energy barrier, in contrast to an outgoing vortex which experiences no barrier. The resulting magnetization is shown in figure A4.3.2; here, the field at which flux first enters $H_s \simeq H_c$, with $M_\uparrow \simeq -H_s^2/2H$ and $M_\downarrow \simeq -H_{c1}/(\pi\ln\kappa) \approx 0$ [3], with the latter being the hallmark of the Bean–Livingstone barrier. This barrier can be suppressed by appropriate surface treatment to yield a surface layer thicker than λ in which T_c changes gradually from zero to its bulk value, such as an oxygen diffusion layer at the surface of Nb or V.

A4.2

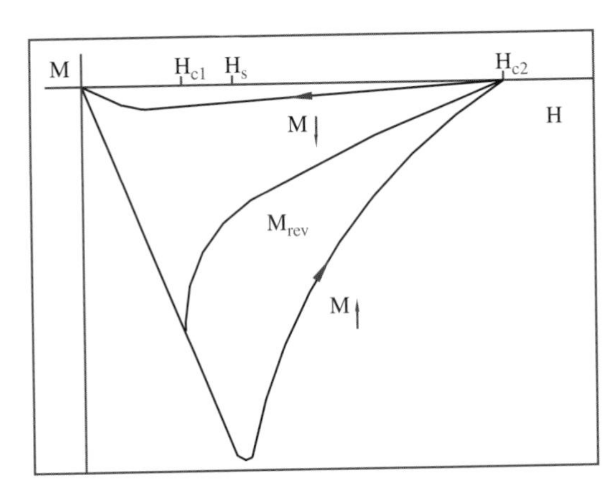

Figure A4.3.2. The effect of a Bean–Livingstone surface barrier on magnetization in the absence of bulk pinning or geometrical effects. Note the characteristics: the first penetration of flux occurs at $H_s \approx H_c$, the magnetization is zero at $H = 0$, the peaks in $M(H)$ are more or less symmetric around H_{c1}.

A4.3.2.4 Geometrical barrier

A4.2 Experiments on single crystals of high temperature superconductors are often carried out with the field aligned parallel to the c-axis. In this direction the physical size of actual crystals is much smaller than in the ab plane due to anisotropy in the growth rate along the different crystallographic axes so that large demagnetization effects occur. The external field H_i at the surface of an ellipsoid shaped sample in the absence of bulk pinning is enhanced according to the expression $H_i = H_a - n_x M_{rev}(H_i)$. In this case the applied field H_a is parallel to one of the principal axes (x) and n_x is the corresponding demagnetization D2.4 coefficient. This would reduce the effective H_{c1} by a factor $(1-n_x)$ which, for a slab of thickness t and width $2w$, is about t/w. In practice, however, flux lines penetrate through the sharp corners at the edge of the slab, as discussed in detail in chapter A4.2. The increasing line energy creates a counteracting force up to the field where the flux lines have penetrated a distance of order $t/2$, above which they can enter freely. This field of first penetration $H_p \simeq H_{c1}(t/w)^{1/2}$ is much larger than $(t/w)H_{c1}$. The line energy thus creates a geometrical barrier for increasing field, which does not exist for decreasing field and therefore gives rise to strong magnetic hysteresis, at low fields only, for $H_a \leq H_{c1}$ [4].

A4.3.3 Pinning

A4.3.3.1 Elementary pinning mechanisms

Crystal defects locally alter material properties and consequently the superconducting parameters in their environment. These local changes may couple to the periodic variations of both the order A4.2 parameter and the local field, which are characteristic of the mixed state (see chapter A4.2). In principle the interaction should follow by solving the Ginzburg–Landau equations with the appropriate boundary conditions imposed by the defects. Depending on specific conditions such as flux density, size and character of the defect, etc., it is possible to classify the elementary interactions in terms of the predominant coupling mechanism as either magnetic or core.

Examples of the magnetic interaction are the effect of surfaces parallel to the applied magnetic field (this might be the external surface as well as some large precipitate interface within the sample bulk) and thickness variations of thin films for fields normal to the film. The Bean–Livingstone barrier, which is the first case, has been discussed above. In the case of thickness variations, the vortices are trapped at the sites of least thickness where the line energy of the vortex is minimum. Evidently, the typical length scale related to the magnetic interaction is λ. In materials with a large κ, therefore, this kind of interaction is small and generally disappears with increasing magnetic field.

The coupling to the variation of the order parameter (related to the density of Cooper pairs) is the origin of flux pinning for almost all defects, including dislocations, point defects, voids, grain boundaries and precipitates. Defects deviate typically from the surrounding material by differences in density, elasticity, electron-phonon coupling or electron mean free path. The first three properties give rise to a local change in T_c, whereas the latter leads predominantly to a variation in the elastic scattering length ℓ of the electrons. Consequently, one may distinguish between δT_c- and $\delta\ell$-pinning. The typical length scale r_f of the core interaction depends on the spatial variations of the order parameter. Therefore, $r_f \approx \xi$ for low ($B < 0.2B_{c2}$) and $r_f \approx a_0/2$ for high flux densities respectively, where a_0 is the vortex lattice parameter ($\frac{1}{2}a_0^2\sqrt{3} = \Phi_0/B$). The elementary pinning force $f(r)$ has the same properties, with its maximum value denoted generally by f_p.

A2.3
A3.1

A4.3.3.2 Summation of pinning forces and vortex lattice disorder

It is necessary to consider random distributions of small point defects (dimensions $\ll r_f$) to elucidate the summation concept. The net effect of all pinning centres in a volume V_c of the vortex lattice is given by a collective force F_c, with a distribution related to the different distributions of point defects. The average of F_c is zero with its associated fluctuation $\delta F_c = (n_p V_c \langle f^2 \rangle)^{1/2}$, where n_p is the concentration of pins and $\langle f^2 \rangle \approx \frac{1}{2} f_p^2$ the mean square of $f(r)$ averaged over a primitive cell of the vortex lattice with area Φ_0/B. A pinning strength W can be defined by $W \equiv n_p \langle f^2 \rangle \approx \frac{1}{2} n_p f_p^2$ in order to express the volume pinning force as

$$F_p = \delta F_c / V_c = (W/V_c)^{\frac{1}{2}} \qquad \text{(A4.3.3)}$$

The task remains to compute V_c. In a seminal paper Larkin and Ovchinnikov [5] (an extensive review is given in [6]) showed that V_c follows from the balance between the elastic deformation energy (see section A4.2.4) and the work done by the pinning centres. The physical picture is that the collective action of the pinning centres breaks up the vortex lattice in elastically independent domains of size L_c parallel to the flux lines and R_c in the transverse direction so that

$$R_c^2 L_c = V_c \qquad \text{(A4.3.4)}$$

The pinning correlation lengths R_c and L_c can be determined from

$$\frac{1}{2} c_{44} \left(k_{xy} \approx \frac{\pi}{R_c} \right) \left(\frac{r_f}{L_c} \right)^2 = \frac{1}{2} c_{66} \left(\frac{r_f}{R_c} \right)^2 = \left(\frac{W}{R_c^2 L_c} \right)^{\frac{1}{2}} r_f \qquad \text{(A4.3.5)}$$

It should be noted that the dispersion of the tilt modulus c_{44} plays an important role. This leads to

$$L_c = [c_{44}(0)/c_{66}]^{\frac{1}{2}} R_c, \quad \lambda_h \ll R_c \ll L_c \qquad \text{(A4.3.6a)}$$

$$L_c = [c_{44}(0)/c_{66}]^{\frac{1}{2}} R_c^2 / \pi \lambda_h, \quad R_c \ll L_c \ll \lambda_h \qquad \text{(A4.3.6b)}$$

for local and non-local elasticity respectively. Here $\lambda_h = \lambda/(1-b)^{\frac{1}{2}}$, with $b \equiv B/B_{c2}$. In principle, R_c, L_c and W can be computed from experimental F_p data, but it turns out that the disorder of the vortex lattice is often dominated by the plasticity and the occurrence of dislocations [7], rather than elastic deformations as is assumed theoretically [5, 6]. The ultimate limit of disorder (the amorphous limit) is reached when the pin energy in equation (A4.3.5) becomes larger than the shear energy, which can be expressed by putting $R_c = a_0$. The independently pinned objects are now single flux line segments of length L_c given by

$$L_c \simeq \left\{ \frac{c_{44}(0) a_0^3 r_f}{\lambda_h^2 W^{\frac{1}{2}}} \right\}^{\frac{2}{3}} \quad \text{(A4.3.7)}$$

In strong pinning materials this situation of 'single vortex pinning' occurs frequently. This gives rise to dome-shaped F_p versus B curves, which can be well described by a scaling relation $F_p \propto B_{c2}^n(T) b^p (1-b)$ with $n \approx 2.5$ and $p = 7/6$ for δT_c pinning and $p = 15/6$ for $\delta \ell$ pinning [8].

L_c is very large in very weak pinning materials, such as amorphous thin films of thickness t. When the field is perpendicular to the film, $L_c \gg t$, which means that the flux lines are straight across the film and the disorder only develops because of shear deformations. This is the case in two-dimensional collective pinning (2DCP) for which R_c follows by substituting $L_c = t$ in equation (A4.3.5) giving

$$R_c \approx 2 r_f c_{66} [2\pi t / W \ln(w/R_c)]^{\frac{1}{2}} \quad \text{(A4.3.8)}$$

Here w is the width of the film. Good agreement is obtained [7, 8] between this expression and experiment.

A4.3.3.3 *Thermal effects, vortex lattice melting and thermal depinning*

In the considerations above, the effect of temperature is ignored except for a trivial temperature dependence entering via $H_c(T)$, $\xi(T)$ and $\lambda(T)$. A more subtle effect is that of thermal motion of flux lines around their equilibrium positions (in the absence of bulk pinning) or around their metastable positions in the random pinning potential. The vortex lattice melts when the thermal displacements become larger than a fraction $c_L a_0$ of the inter-vortex spacing. Here c_L is the Lindemann constant, $c_L \approx 0.2$. This criterion gives rise to a melting line in the field-temperature phase diagram which, for conventional superconductors, is located near the upper critical field $H_{c2}(T)$. The liquid phase is characterized by a linear resistivity ($E = \rho J$), the low temperature phase in the presence of pinning by a non-linear resistivity ($\rho = \rho(J)$ and true superconductivity, i.e. $\rho \to 0$ for $J \to 0$ [6], see below (and chapter A4.1). When the thermal displacements exceed the range of the elementary pinning force r_f the pinning energy will decrease by thermal smearing and eventually lead to thermal depinning at the depinning line in the phase diagram. This line is marked by a sudden decrease of the irreversibility. By comparing length scales it is clear that at low fields (i.e. $B < 0.2 B_{c2}$) the depinning line should lie below the melting line, whereas the opposite is true at high fields. Finally, for a lattice composed of straight vortices the melting criterion follows from the theory of Kosterlitz-Thouless for 2D melting. The shear modulus at the melting line should fall to zero, although in practice disorder smears out the phase transition.

A4.1
A4.2

A4.3.3.4 *The peak effect*

A sudden increase is often apparent in plots of $\boldsymbol{J}_c$ versus H or T just before $\boldsymbol{J}_c$ goes to zero. The points where $\boldsymbol{J}_c$ becomes zero form a line in the HT phase diagram which may be identified as the irreversibility line or sometimes as the melting line. In conventional superconductors this usually lies very near the upper critical field $H_{c2}(T)$. In dc resistance measurements the peak in $\boldsymbol{J}_c$ causes a dip in $R(T)$ or $R(H)$, while in ac-susceptibility experiments a peak in χ'' occurs each time the condition $\boldsymbol{J}_c = \alpha h_0 / a$ is fulfilled, where α is a geometry-dependent constant. The peak occurs when the field or temperature dependence of the mechanisms determining the balance between deformation and pinning energies changes upon entering a different elastic regime. Examples of such cross-over regimes are from elastic to plastic deformations or from shear dominated to non-local tilt dominated behaviour. The observed increase of

J_c originates generally from the softening of the vortex lattice, which enables it to find configurations of stronger pinning.

A4.3.4 Flux creep

Thermal fluctuations cause the vortex lattice to jump between different metastable configurations of almost equal energy. In the presence of a driving force this process leads to a unidirectional creep of the flux line lattice, accompanied by the reappearance of resistance. In an early description of this problem [9] the jumping units are elastically independent bundles of flux lines with volume V_b that move over distances of order r_f. The work done by the driving force per bundle, therefore, is JBV_br_f, with the height of the energy barrier between metastable states characterized by $J_cBV_br_f$. Taking into account both forward and backward bundle jumps leads to a net average velocity of the flux line lattice given by

$$v = r_f\nu_0[\mathrm{e}^{-\varepsilon_c(1-J/J_c)} - \mathrm{e}^{-\varepsilon_c(1+J/J_c)}] = 2r_f\nu_0\mathrm{e}^{-\varepsilon_c}\sinh(\varepsilon_c J/J_c) \qquad \text{(A4.3.9)}$$

where the abbreviation $\varepsilon_c = U_c/k_BT = J_cBV_br_f/k_BT$ has been introduced, k_B is the Boltzmann constant and ν_0 a typical attempt frequency. The creep resistivity follows from $\rho_{creep} = Bv/J$. When $\varepsilon_c \gg 1$, a measurable effect of creep is seen in the proximity of J_c. In this case, only the contribution of the forward term in equation (A4.3.9) is relevant and ρ_{creep} depends strongly on J, as illustrated schematically in figure A4.3.3. Linear resistivity is encountered for relatively low thermal energy $\varepsilon_c \leq 10$) for small J in the regime of thermally assisted flux flow (TAFF) [10] with

$$\rho_{TAFF} \approx \rho_f\exp(-\varepsilon_c) \qquad \text{(A4.3.10)}$$

Here the prefactor has been replaced by the flux flow resistivity which yields the right limit for $\varepsilon_c < 1$.

It can be shown easily that the flux distribution in the TAFF regime should obey a diffusion equation for B, with $D = \rho_{TAFF}/\mu_0$ as the diffusion constant. Conditions for the irreversibility line can be derived from the solutions of this equation. For ac susceptibility experiments it follows, for instance,

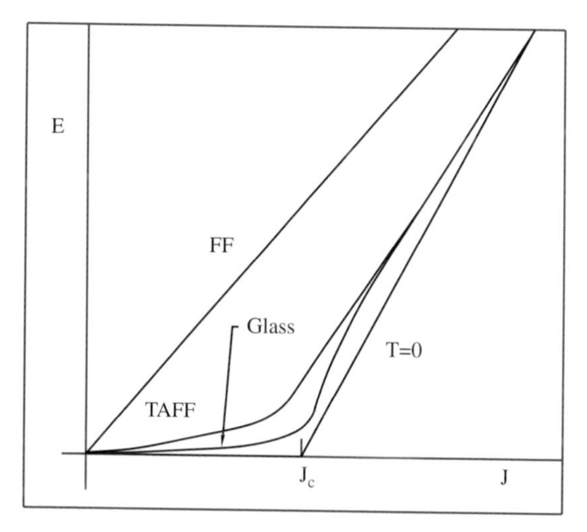

Figure A4.3.3. Typical $E(J)$ characteristics in the mixed state in the absence of pinning (FF), with bulk pinning at zero temperature ($T = 0$) and at finite temperature for finite energy barriers (TAFF) and infinite barriers (glass) for $J \to 0$. Note that the differential resistivity above the critical current density J_c is larger than or equal to the flux flow (FF) resistivity in the absence of pinning. This effect is due to velocity fluctuations and is usually ignored in textbooks.

that the peak in χ'' occurs when the penetration depth of the ac field equals the sample size, i.e. $\lambda_{ac} \approx a$ and $\lambda_{ac} \approx (D/\omega)^{1/2}$ where ω is the angular frequency of the ac field. The irreversibility line is thus obtained from $U_c(B,T) \approx k_B T \ln(\rho_f/\mu_0 a^2 \omega)$, which depends on frequency as is observed experimentally.

Although ρ_{TAFF}/ρ_f may be extremely small, theoretically TAFF describes a normal metal with Ohmic resistance for small $\boldsymbol{J}$ and not a true superconductor with zero resistance. The reason is that the energy barriers between metastable states are assumed to be finite and independent of $\boldsymbol{J}$. For an elastic medium in a random potential Feigelman and Vinokur [11] (see also [6]) showed, however, that both the size of the flux bundle and the hopping distance should increase when $\boldsymbol{J}$ decreases in order to maintain the balance between deformation energy and the work done by the driving force. The energy barriers in this model of collective creep grow correspondingly according to $U(\boldsymbol{J}) \sim U_c(\boldsymbol{J}_c/\boldsymbol{J})^{\mu}$, where the exponent μ depends on the dimensionality of the elastic medium and the random nature of the environment. This ranges from 1/7 for single vortices to 16/9 in the large bundle limit. The important issue is that the barrier becomes infinite for decreasing $\boldsymbol{J}$, which leads to true zero resistivity $\rho_{cc} \sim \rho_f e^{-c(\boldsymbol{J}_c/\boldsymbol{J})^{\mu}}$. The corresponding I–V characteristic is illustrated in figure A4.3.3.

A similar result has been obtained for the vortex glass phase introduced by M.P.A. Fisher [12] to describe the continuous phase transition from a vortex liquid to a disordered vortex solid at the vortex glass transition line $T_g(B)$. The vortex glass theory predicts a specific behaviour for the resistivity in the cirtical regime near the phase transition in which case all data for different temperatures and fields collapse on two curves, F_+ for $T > T_g$ and F_- for $T < T_g$ [13] in plots of $(E/\boldsymbol{J}).|T - T_g|^{-\nu(z-1)}$ versus $(\boldsymbol{J}/T).|T - T_g|^{-2\nu}$. The critical exponent ν describes the vortex glass phase correlation length ξ_g which diverges at T_g according to $\xi_g \propto |T - T_g|^{-\nu}$, while z is the dynamic exponent describing critical slowing down. A typical scaling plot is shown in figure A4.3.4. The exact functional dependence of $F_\pm$ is unknown $F_+ \rightarrow 1$ for small denoting linear resistivity, and $F_{(x)} \rightarrow \exp(-1/x^{\mu})$, describing true superconductivity. At T_g E should depend on $\boldsymbol{J}$ according to the power law $E \propto \boldsymbol{J}^{(z+1)/2}$.

Flux creep is usually studied by measuring the time decay of the magnetic moment in constant applied field. The related expression in the collective creep regime for the time dependence of the diamagnetic screening currents is

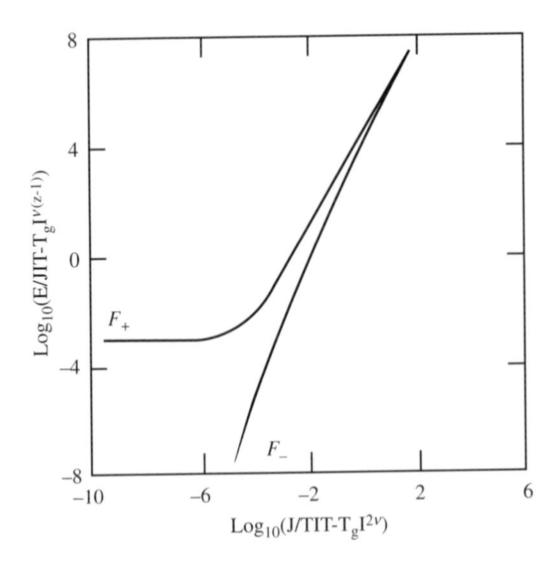

Figure A4.3.4. Collapse of over 100 I–V curves in a vortex-glass scaling plot for $YBa_2Cu_3O_7$ in a field of 4 T for the parameter values $T_g = 74.5$ K, $z = 4.8$ and $\nu = 1.7$ [13].

$$J(t) \approx J_c \left[1 + \frac{\mu}{\varepsilon_c} \ln\left(1 + \frac{t}{t_0} \right) \right]^{-1/\mu} \tag{A4.3.11}$$

where t_0 is a typical microscopic timescale of the system. This formula interpolates between the famous logarithmic time decay originally proposed by Anderson [9] for systems with large ε_c and the collective creep prediction. Information about the energy barrier $U(\boldsymbol{J})$ from magnetic decay experiments is obtained by plotting $k_B T \ln(c - |\partial M/\partial t|)$ versus $\Delta M \propto \boldsymbol{J}$. Repeating the experiment at slightly different temperatures yields a series of curves over finite $\boldsymbol{J}$ intervals which can be overlapped to enable determination of the constant c [14]. In this way the $U(\boldsymbol{J})$ dependence is determinable over three decades in $\boldsymbol{J}$. In principle, this can also be achieved by waiting long enough ($\sim 10^{40}$ years) for the current to decay! The true J_c is obtained by extrapolating to $U = 0$. It is interesting to note that the $U(\boldsymbol{J})/k_B T$ ratio always falls within the range 10–30. This remarkable observation can be understood in terms of the self-organizing mechanism of creep and the experimental time window of the experiments.

A4.3.5 Layered superconductors

The consequences of a layered structure and the related anisotropy has received much attention following the discovery of HTS. Both organic superconductors and artificial multilayers, however, also belong to this interesting class of materials. The mixed state properties of $YBa_2Cu_3O_7$ are presented in chapter A4.1. This material has a moderate anisotropy with a value for the anisotropy parameter of $\gamma = 7$. A significantly greater anisotropy is observed in $Bi_2Sr_2CaCu_2O_8$ (Bi-2212) on the other hand, in which $\gamma \approx 200$–300. This material can be considered to consist of a stack of superconducting (CuO_2) and insulating (BiO) layers. Superconducting screening currents, or pancakes vortices, are located in the CuO_2 planes. Rather than Abrikosov line vortices, therefore, pancakes vortices stack to form the flux structure in the material [15]. Above a field $B_{2D} \approx \Phi_0/(\gamma s)^2$, where s is the distance between the CuO_2 double layers, the intralayer interaction between pancake vortices is stronger than the inter-layer interaction, which makes the system quasi two-dimensional. Individual pancake vortices are pinned by oxygen vacancies [16], that account for J_c values at low temperatures as large as 0.1 $J_0 (\approx 4 \times 10^{10}\ \mathrm{Am^{-2}})$, but with very small pinning energies ($U_c \approx 40$ K), subsequently leading to very strong flux creep effects and a low-lying irreversibility line. At low fields this line actually coincides with the first order phase transition line, at which the vortex lattice sublimates to a gas of decoupled pancake vortices [17]. This first order transition is marked by a small jump in the reversible magnetization. As can be seen in figure A4.3.5(a), the flux density decreases upon cooling through the phase transition, so that the transition is driven by the gain in entropy of the pancake gas phase. At low temperatures the jump disappears and is replaced by a sharp peak effect caused by a sudden reappearance of bulk pinning, see figure A4.3.5b. Small angle neutron scattering and μ-SR experiments show that the disorder in the vortex lattice at the peak field increases dramatically, suggesting a disorder driven (phase) transition from a weakly disordered line lattice to a strongly disordered entangled vortex configuration. A phase diagram is shown in figure A4.3.6 [18].

When the magnetic field is rotated away from the c-axis the vortex properties scale with the field component along the c direction, nicely illustrating the quasi-2D nature. Only when the angle with the CuO_2 planes becomes smaller than γ^{-1} Josephson vortices develop between the pancake vortices in adjacent CuO_2 planes. The Josephson vortices can be easily depinned when the driving force is directed along the ab planes. In contrast, a driving force in the c direction is opposed by a very strong intrinsic pinning force due to the layeredness. Another consequence of the quasi-2D nature is the very

C
A4.1
C2
A1.1
A2.7

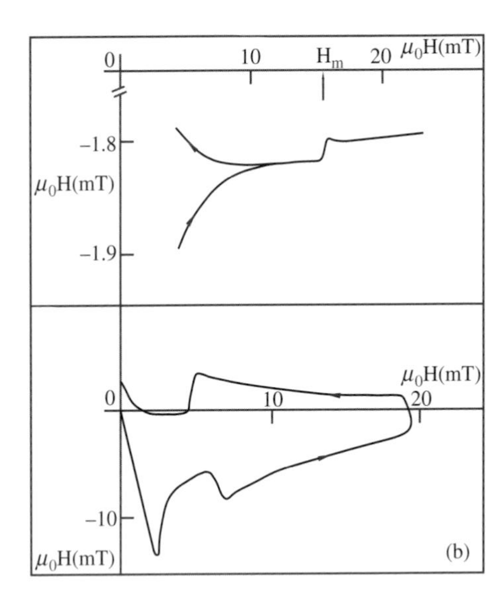

Figure A4.3.5. (*a*) Typical magnetization curve of a Bi-2212 single crystal (as-grown, 64 K) going through the first order transition line marked by the jump in magnetization at the ('melting') field H_m. (*b*) Low temperature, M versus H curve (same crystal, 25 K) showing a disorder induced 'second' peak effect at about $B_{sp} = 6\,\mathrm{mT}$.

non-uniform current distribution in electric transport experiments when current contacts are placed on the top surface of the sample. The driving force on pancake vortices in the top layers is thus much larger than in the bottom layers, which leads to phase slips and a concomitant voltage between the top and bottom surfaces.

Interesting effects also occur when Bi-2212 is irradiated with high energy heavy ions which form amorphous tracks of radius $c_0 \approx 3.5\,\mathrm{nm}$. These columnar defects form ideal, strong, linear pinning centres. The core interaction dominates at temperatures above 0.4 T_c where the vortex core size $\xi(T)\sqrt{2}$ exceeds c_0. The strong linear correlation in the direction of the columnar defects effectively suppresses the deteriorating effect of the large anisotropy on the flux pinning which results in an upward shift of the irreversibility line. The new position of this line is determined by the concentration of columnar defects expressed in terms of a dose equivalent field B_ϕ (at B_ϕ the number of vortices equals the number of columnar defects). The low temperature phase is now called a Bose glass (see chapter A4.1). In figure A4.3.6 the irreversibility line for $B_\phi = 2$ T is indicated on a logarithmic scale. For a regular array of columnar defects a Mott insulator phase is predicted at $B = B_\phi$, although in practice the distribution of amorphous tracks is entirely random and the competition between pinning energy and intervortex interaction suppresses a true phase transition. The properties of the high-temperature phase are very rich and depend strongly on whether the flux density is larger or smaller than the density of columnar defects. At present the situation is not yet fully resolved, although it has been shown [19] that the pinning energy of a columnar defect can be determined from the reversible magnetization curve in the high-temperature phase. This curve has an s-shape, associated with the energy costs of adding one vortex to the system. This vortex will profit from the pinning energy well below B_ϕ whereas this is prohibited for $B \gg B_\phi$ because all columnar defects are occupied. It shows once more that the mixed state of type II superconductors with strong pinning centres not only is a very practical phenomenon, but that it also forms a rich playground for solid state and statistical physics.

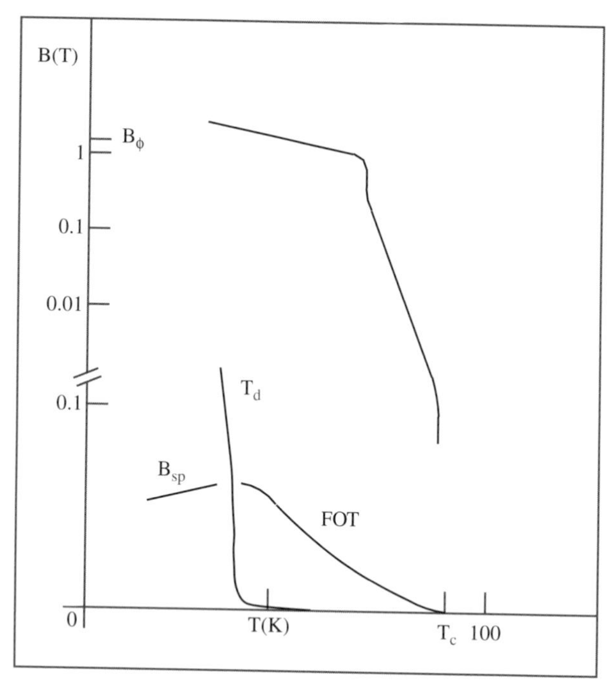

Figure A4.3.6. Phase diagrams for defect-free and irradiated ($B_\phi = 2$ T) single crystals of Bi-2212. The low field phase diagram for defect-free Bi-2212 shows the first order transition (FOT) line separating a weakly pinned ordered solid from a gas of pancake vortices, the second peak line B_{sp} separating a pinned ordered solid at low fields from a pinned entangled solid above B_{sp}. T_d marks the depinning crossover line. The line for the irradiation crystal separates the low temperature Bose glass phase from a high-temperature Bose liquid. The shape of the line is determined by different dynamic processes which are still a subject of research.

References

[1] Bean C P 1962 Magnetization of hard superconductors *Phys. Rev. Lett.* **8** 250

[2] Campbell A M and Evetts J E 1972 Flux vortices and transport currents in type II superconductors *Adv. Phys.* **21** 199

[3] Clem J, 1974 A model for flux pinning in superconductors *Low Temperature Physics LT-13* **3**, eds K D Timmerhaus, W J O'Sullivan and E F Hammel (Plenum: New York) p 102

[4] Zeldov E, Larkin A I, Geshkenbein V B, Konczykowski M, Majer D, Khaykovich B, Vinokur V M and Shtrikman H 1994 Geometrical barriers in high temperature superconductors *Phys. Rev. Lett.* **73** 1428

[5] Larkin A I and Ovchinnikov Y u N 1979 Pinning in type II superconductors *J. Low Temp. Phys.* **34** 409

[6] Blatter G, Feigelman M V, Geshkenbein V B, Larkin A I and Vinokur V M 1994 Vortices in high-temperature superconductors *Rev. Mod. Phys.* **66** 1125

[7] Wördenweber R, Kes P H and Tsuei C C 1986 Peak and history effects in two-dimensional collective flux pinning *Phys. Rev.* B **33** 3172

[8] Kes P H 1992 Flux pinning and the summation of pinning forces *Concise Encyclopedia of Magnetic & Superconducting Materials* ed J E Evetts (Oxford: Pergamon) p 163

[9] Anderson P W 1962 Theory of flux creep in hard superconductors *Phys. Rev. Lett.* **9** 309

[10] Kes P H, Aarts J, Van den Berg J, Van der Beek C J and Mydosh J A 1989 Thermally assisted flux flow at small driving forces *Supercond Sci. Technol.* **1** 242

[11] Feigelman M V and Vinokur V M 1990 Thermal fluctuations of vortex lines, pinning and creep in high- T_c superconductors *Phys. Rev.* B **41** 8986

[12] Fisher D S, Fisher M P A and Huse D 1991 Thermal fluctuations, quenched disorder, phase transitions and transport in type II superconductors *Phys. Rev.* B **43** 130

[13] Koch R H, Foglietti V, Gallagher W J, Koren G, Gupta A and Fisher M P A 1989 Experimental evidence for vortex-glass superconductivity in Y–Ba–Cu–O *Phys. Rev. Lett.* **63** 1115

Koch R H, Foglietti V, and Fisher M P A 1990 *Reply to Comment on Experimental evidence for vortex- glass superconductivity in Y–Ba–Cu–O* eds S N Coppersimith, M Inui and P B Littlewood *Phys. Lett.* **64** 2586

[14] Van der Beek C J, Kes P H, Maley M P, Menken M J V and Menovsky A A 1992 Flux pinning and creep in the vortex-glass phase in $Bi_2Sr_2CaCu_2O_{8+\delta}$ single crystals *Physica* C **195** 307

[15] Clem J R 1991 Two-dimensional vortices in a stack of thin superconducting films: a model for high-temperature superconducting multilayers *Phys. Rev.* B **43** 7837

[16] Li T W, Menovsky A A, Franse J J M and Kes P H 1996 Flux pinning in Bi-2212 single crystals with various oxygen contents *Physica* C **257** 179

[17] Zeldov E, Majer D, Konczykowski M, Geshkenbein V B, Vinokur V M and Shtrikman H 1995 Thermodynamic observation of first-order vortex-lattice melting in $Bi_2Sr_2CaCu_2O_8$ *Nature* **375** 373

[18] Fuchs D T, Zeldov E, Tamegai T, Ooi S, Rappaport M and Shtrikman H 1998 Possible new vortex matter in $Bi_2Sr_2CaCu_2O_8$ *Phys. Rev. Lett.* **80** 4971

[19] Van der Beek C J, Konczykowski M, Li T W, Kes P H and Benoit W 1996 Large effect of columnar defects on the thermodynamic properties of $Bi_2Sr_2CaCu_2O_8$ single crystals *Phys. Rev.* B **54** R792

PART B

PROCESSING

B1
Introduction to processing methods

Masato Murakami and Réne Flükiger

This chapter presents a comprehensive review of the various processing techniques for high and low T_c superconducting materials. Processing is the key to industrial applications of all engineering materials. As a result, the development of appropriate processing techniques to yield desirable properties and shapes of superconductors is fundamental to achieving technological applications of these materials.

Superconductors can only function under restricted conditions below the critical planes consisting of H_c (critical field), J_c (critical current density) and T_c (critical temperature). Of these critical parameters, H_c and T_c represent intrinsic properties of superconducting materials, although inappropriate processing can degrade these values. In contrast, J_c is a structure-sensitive parameter and therefore strongly dependent on microstructure and hence processing route. Most applications of superconductors require large J_c values of order $10^5\,\mathrm{A\,cm^{-2}}$ so that one must develop processing techniques that can produce superconducting materials with sufficiently large J_c in the form of practical final product geometries, such as long wires.

B1.1 Critical current density

By definition, there are two different values of J_c. One is the de-pairing current, at which the kinetic energy of Cooper pairs exceeds the condensation energy, or the difference in the free energy between the superconducting state and the normal conducting state. The de-pairing current is intrinsic to materials and between 10^2 and 10^4 times larger than engineering J_c values of practical superconductors (i.e. the total current flowing divided by the total cross-section of the conductor, including and sheathing and channels for cryogens). With the exception of electronic devices, most applications of superconductors require large values of engineering J_c. As shown schematically in figure B1.1, J_c values depend on the number and distribution of defects in the microstructure, such as normal conducting particles that prevent the flux motion in the mixed state. Such defects are called pinning centres, and processing methods to distribute these homogeneously and in a controlled way throughout the superconductor must be developed for the manufacture of materials for practical, current carrying applications. Of the many low T_c materials, only NbTi and Nb_3Sn have been used for electric power applications simply due to the fact that these materials can be made into wires with sufficient flux pinning ability.

A2.1 A1.3 A4.3 A4.3

In contrast to power applications, the defect density or the number of pinning centres should be as low as possible in thin films for electronic device and sensor applications. Pinning defects trap fluxoids and hence form a source of noise for these applications. Therefore, pure, defect-free single element metal superconductors such as Pb and Nb have been used mainly for the development of superconducting devices or sensors, such as superconducting quantum interference devices (SQUIDS). It is interesting that a large number of on-going superconductivity projects are targeting applications of high

E4.2

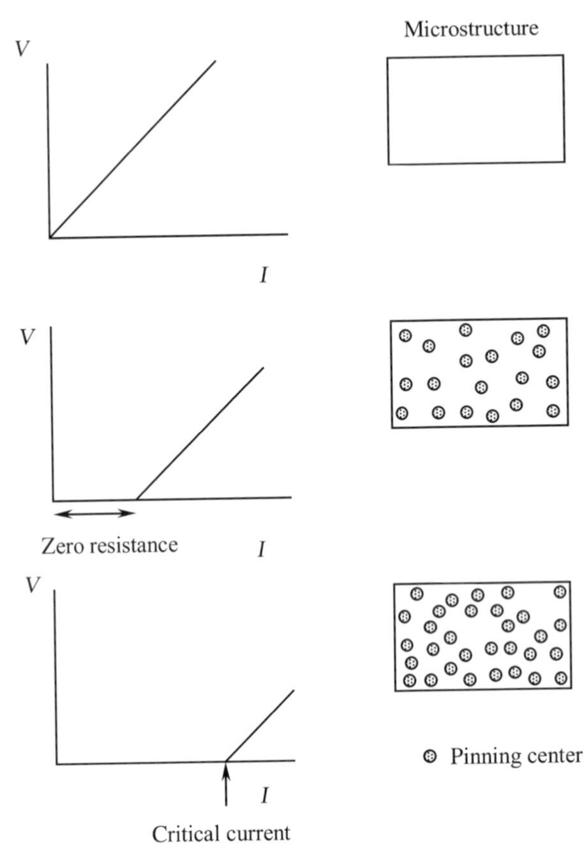

Figure B1.1. V–I characteristics of type II superconductors in magnetic fields. The J_c value is zero in a clean type II superconductor.

temperature superconductors (HTS) for the areas of both power and electronics. However, it should be borne in mind that the microstructural properties appropriate for these applications are very different, with defect structures required for electric power devices and nearly perfect, defect-free structures required for electronic devices.

B1.2 Single crystal growth

Microstructural control for achieving large values of J_c requires a thorough understanding of the basic properties of superconducting materials. For example, construction and understanding the phase diagram of a particular superconducting material is essential for optimizing its processing route. H3

B2.4 It is also important to grow single crystals for characterizing materials properties. In particular, single crystal growth of HTS, which are very complex with a large anisotropy, is essential for understanding their basic properties. As a result, the fabrication of HTS single crystals became a worldwide challenge after their discovery in 1986. The quality of HTS single crystals has consequently improved dramatically over the interim period, with an associated increase in the basic understanding of their properties. The techniques for growing single crystals of superconductors are similar to those commonly employed for other materials. The easiest method is flux growth, which involves growth from B2.4 a solution at high temperature. Silicon, for example, is grown from solution by a top-seeded technique,

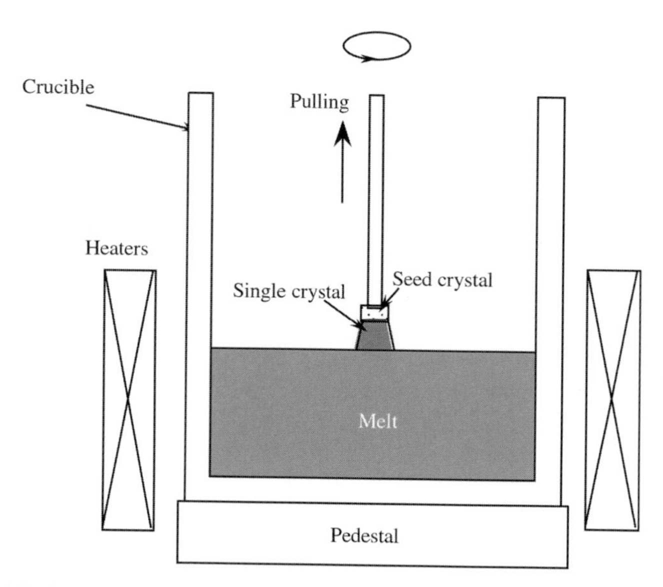

Figure B1.2. Typical experimental arrangement for the top-seeded growth method.

which produces large diameter wafers for semiconductor technologies. Figure B1.2 shows a schematic illustration of the experimental arrangement for the top-seeded solution growth process. Similar techniques have been applied successfully to the growth of Y–Ba–Cu–O single crystals. For most HTS materials, however, floating zone growth methods are used more commonly to grow single crystals, since this technique avoids the use of a crucible. Sample contamination by crucibles is the most serious problem for the growth of HTS materials. The experimental arrangement for the floating zone technique is shown schematically in figure B1.3. This chapter describes the details of single crystal growth methods appropriate for both high and low T_c superconducting materials.

B2.3.3

B1.3 Sintering

In the early stage of their development, most HTS compounds were synthesized by a sintering technique, which is a common ceramic processing route and generally easy to apply. As a result, sintering techniques are often employed in the search for and development of new compounds.

Figure B1.4 shows a flow chart for 'standard' sintering of HTS. The first step of this process is to prepare good quality starting, or precursor, powders. HTS powder for any specified composition can be made by a variety of routes and the characteristics of the powder often reflect the preparation technique employed. Solid state reaction, for example, involves heating the raw precursor powders to a temperature typically within 20% of the melting point of the target composition (this process is called calcination). The contaminants, size distribution, porosity and other factors can lead to wide variations in sample properties. With the evolution of more thorough and precise preparation techniques, however, good-quality HTS powders can now be fabricated routinely. Such powder preparation techniques are reviewed in this chapter.

B2.2.2

B2.1

The second step in the sintering process is to transform a compacted pellet of the calcined precursor powders to a solid ceramic body by heating the powder compact to typically within 5% of the melting point of the target composition. The important parameters in this process are the temperature, sintering time and environment. The optimum sintering conditions are determined empirically rather than

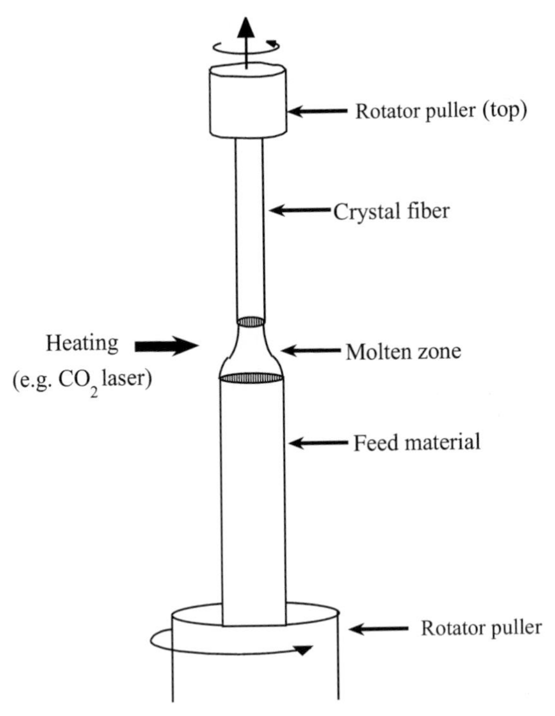

Figure B1.3. Typical experimental arrangement for the floating zone method.

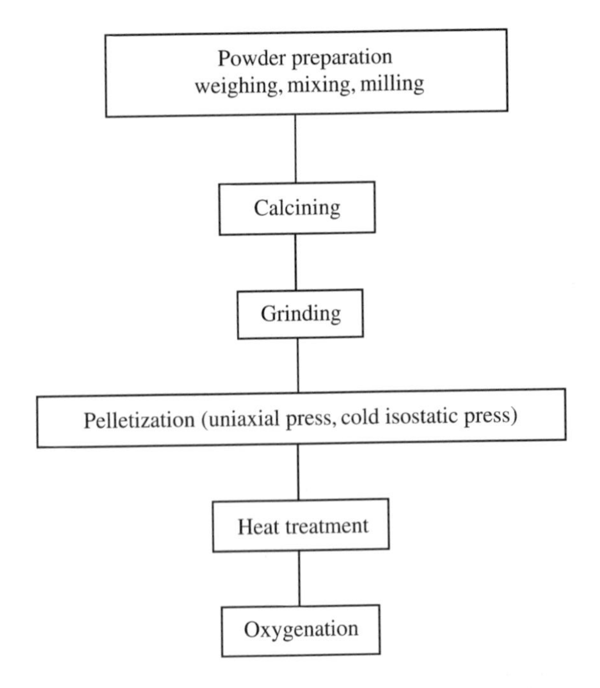

Figure B1.4. Flow chart of a standard sintering process for oxide superconductors.

theoretically. For some HTS materials such as $YBa_2Cu_3O_y$, post-sintering oxygen annealing is necessary to achieve superconducting properties in the bulk sintered ceramic. B2.3.3

From the viewpoint of engineering applications, sintering is an important technology for producing Bi–Sr–Ca–Cu–O bars that may be used as current leads to supply power to a cryo-cooled superconducting solenoid. Sintering techniques are also employed for manufacturing large tubes to shield external fields for medical diagnosis of the human brain or heart. Bi–Sr–Ca–Cu–O plates fabricated by a sintering-based diffusion process have been particularly successful for this application. In addition to industrial applications, sintering is an important process for optimizing processing conditions in general. B2.2.3

B1.4 Wire technology

One of the most attractive features of superconductors for electric power applications is the generation of high fields with small energy consumption, associated mainly with the requirement to cool the superconductor. The output of the power devices is simply given by the Lorentz force relation: $F = J \times B$, where F is the electromagnetic force per unit volume on the conductor, J is the current density and B the magnetic induction. This simple equation shows that a larger F is achieved with increased B at constant J, which is the basis of high power devices. Another advantage of large B is that the same F may be achieved with smaller J, which translates directly to more compact power devices with the same output. For a normal conducting magnet, however, an increase in B is restricted by heat generation. As a result, the net gain by increasing B is easily out-weighed by the increase in energy consumption used to generate this increase. A superconducting magnet running in a persistent current mode, however, can be used to generate an extremely large B (>10 T) with extremely low energy consumption. This is the most important feature of superconducting power devices. E1.1

B1.4.1 Low T_c materials

Most practical magnetic fields in excess of 2 T are generated by superconducting solenoids. For this purpose, it is necessary to manufacture long superconducting wires (i.e. in km lengths) with large values of J_c. In low T_c materials, the thermal instability associated with an extremely low specific heat proved to be the most difficult problem to overcome. The development of multi-filamentary structures for long conductors was a key breakthrough in the evolution of a process technology for these materials. This involves embedding small-diameter filamentary superconducting wires in a metal matrix with good thermal conductivity, which avoids local heat generation and thus makes the wire thermally stable. E1.1 A1.3

The NbTi and Nb_3Sn low T_c superconducting systems are already suitable for the industrial production of long lengths of wires. The current preferred wire fabrication method for these materials starts with a ductile metallic rod (e.g. Cu, Cu–Sn bronze, Nb, NbTi, etc), which can be deformed with an area reduction ratio of $>10^6$. Indeed, a Nb rod with an initial diameter of 10 mm, will have at the end of the process a diameter of only a few micrometers. Generally, the production process for LTS superconducting wires begins with an extrusion step with the area reduction ratio varying between typically 36 and100, depending on the material. B3.3.1, B3.3.2

The size reduction steps for NbTi may be performed without the need to anneal the material intermediately, which is important in reducing the costs of fabrication. The flux pinning properties of the wire at the end of the fabrication process can be improved significantly by the so-called thermo-mechanical treatment, which consists of a combination of heat treatment and drawing steps. The heat treatments are performed at approximately 400°C to precipitate out normal conducting phases of micrometer dimensions. These precipitates are elongated (so-called 'α-bands') during final drawing to a thickness comparable to the coherence length in NbTi of around 6 nm. Large current carrying

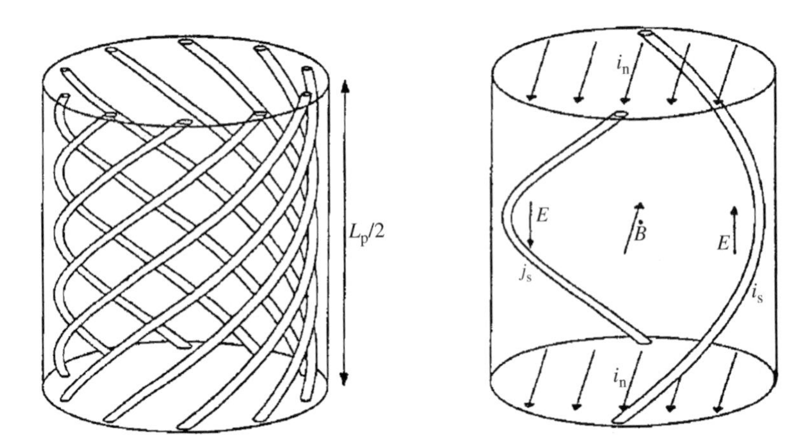

Figure B1.5. Tubes of Cu–Sn alloys (bronzes) with up to 15.6 wt.% Sn which contain a Nb core.

enhancements have been produced in recent years by this technique, with J_cs in excess of 5000 A mm^{-2} achieved in multifilamentary wires of NbTi at 4.2 K and 5 T.

The superconductor used for producing fields up to 22 T is Nb_3Sn. Because of the brittleness of the A15 phase, however, it is not possible to apply the usual deformation process to this material. A number of alternative techniques have been developed for processing Nb_3Sn wires, however, but only two have reached an industrial level: 'the bronze route' and the 'internal Sn diffusion process', which are described schematically in figure B1.5. In both cases, the Nb_3Sn wires are fabricated by deforming ductile components to the final configuration, which consists of more than 10 000 Nb filaments with diameters ranging between 4 and 6 μm. The brittle superconducting A15 phase is then formed by reaction heat treatment at temperatures between 670 and 700°C. Sn diffuses out the Cu–Sn alloy during the reaction, called the *bronze diffusion process*, and reacts with the Nb to form the Nb_3Sn phase. Once the A15 phase is formed the irreversible strain limit of the wire to maintain its critical current density is of the order of $\varepsilon = 0.6\%$, and the wire must be treated with great care.

B2.5

The bronze route

B3.3.1

The bronze route begins with tubes of Cu–Sn alloys (bronzes) with up to 15.6 wt.% Sn, which contain a Nb core, as indicated in the flow diagram in figure B1.5. This composition is the highest solubility limit for Sn in Cu at equilibrium. These rods are drawn to a hexagonal shape, bundled and inserted into a bronze tube, which is closed at both ends under vacuum to form a billet. The arrangement is then extruded hydrostatically at around 650°C. The extruded rods are again drawn and bundled with rods of pure Cu (for thermal stabilization), and protected from contamination by Sn by a Ta barrier. Finally, these rods are surrounded by an external bronze tube to form a second billet, and again extruded. Three extrusion steps are usually necessary to yield a wire containing 10 000 filaments or more.

The bronze route produces long lengths of homogeneous wire but has the disadvantage that the Cu–Sn bronze undergoes considerable cold working during deformation. This problem is addressed by applying so-called recovery heat treatments of 0.5 h at intermediate temperatures, which are low enough to minimize formation of the unwanted A15 type phase. This recovery treatment, however, has to be applied after each 50% reduction of area, with between 15 and 20 anneals typically required per wire, which is a costly and time-consuming process. Indeed, a full production run using the 'bronze route' ordinarily takes 4–6 months. The advantage of the method, however, resides in the very high homogeneity of the product wire and that it is particularly well suited for the construction of high field magnets operating in 'persistent mode'.

The internal Sn diffusion process

The internal Sn diffusion process avoids the need for recovery heat treatments, which makes the process particularly attractive. A further advantage is that this process is not limited by the solubility of Sn in Cu, as it is the case for the 'bronze route': the Sn content in the core can be raised to between 18 and 20 wt.%, in principle yielding a higher total area of reacted A15 phase and higher values of critical current density. There are two modifications of the internal Sn diffusion process (see figure B1.5). B3.3.1

(a) Starting with Nb rods inside Cu tubes, which are bundled around a central core consisting of a Cu tube filled with Sn.

(b) Replacing Nb rods by expanded Nb foils (with rhombic holes in order to form a Nb network). The Nb foils are rolled along with Cu foils around Cu tubes, which are filled with Sn.

An inherent disadvantage of the internal Sn diffusion process is, however, the contact between filaments, which cannot be avoided and leads to considerably higher ac losses when working in an alternating current regime. In addition, the overall homogeneity of wires produced by this technique is less pronounced compared to wires processed by the 'bronze route'. A further disadvantage is that the Ta barrier is difficult to remove, limiting its potential for use in persistent mode operation.

Wire twisting

Ordinarily, multifilamentary wires are delivered in a twisted state, with twist pitches of the order of between 15 and 25 mm. When using Nb_3Sn wires in alternating current regimes, the ac losses induced by coupling currents between the filaments in a multifilamentary wire have to be taken into account. These inductively generated currents exist due to the changes in the local magnetic field between the filaments (see chapter C2.6). The existence of these currents can be understood from figure B1.6, which shows schematically the filaments in a multifilamentary wire. Here the induced current flows along a filament, A1.3

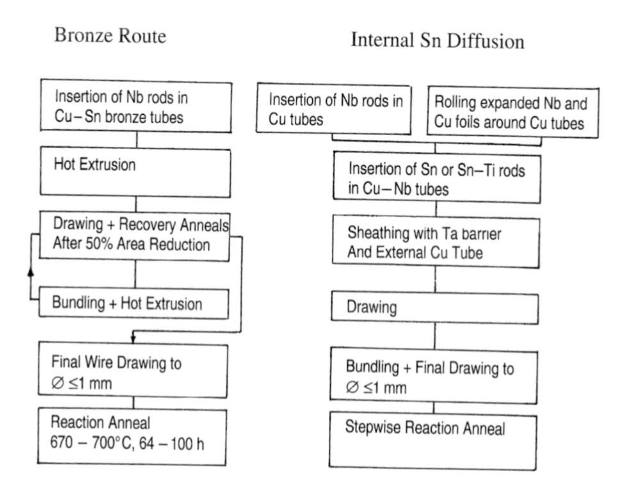

Figure B1.6. Filaments in a multifilamentary wire.

across the matrix, back into a second filament, back across the matrix and, finally, into the original filament. The coupling currents cause mainly ohmic heating as they pass the matrix, which leads to the coupling current losses. In order to reduce the coupling currents, the filaments in the wire are twisted with a twist pitch as short as possible to limit the extent of the inductive loops to half the twist pitch.

D2.6

B1.4.2 High T_c materials

HTS compounds generally form fragile ceramic materials, and it was believed originally that the fabrication of long wires would be difficult. Fortunately, a powder-in-tube (PIT) technique was developed to fabricate long wires of Bi–Sr–Ca–Cu–O, which exhibit reasonably high values of J_c (albeit at relatively low fields). Figure B1.7 shows a schematic drawing of the PIT processing technique. In this method, raw powders are packed in a silver or silver alloy sheath and drawn or rolled into long tapes. The rolled tapes are then annealed under a controlled atmosphere to form the superconducting phase. Due to an extremely large anisotropy in the crystal structure of Bi–Sr–Ca–Cu–O superconductors, long tapes with a highly aligned structure can be synthesized by simple mechanical rolling of these compounds. A large anisotropy, on the other hand, translates to weak pinning due to a small interaction distance along the flux lines for fields applied parallel to the c-axis. As a result, the power applications of Bi compound wires and tapes are restricted to below 30 K.

B3.2.1, B3.2.2

RE–Ba–Cu–O (where RE represents a rare earth element) HTS materials contain strong pinning centres, even at 77 K. Unfortunately, however, the PIT method is not suitable for the manufacture of (RE)BCO compound wires and tapes due to the lower anisotropy and increased granularity of these materials. At present, a so-called 'next generation' wire development fabrication process is being developed worldwide. This process is based in principle on thin film fabrication technology, which

C1

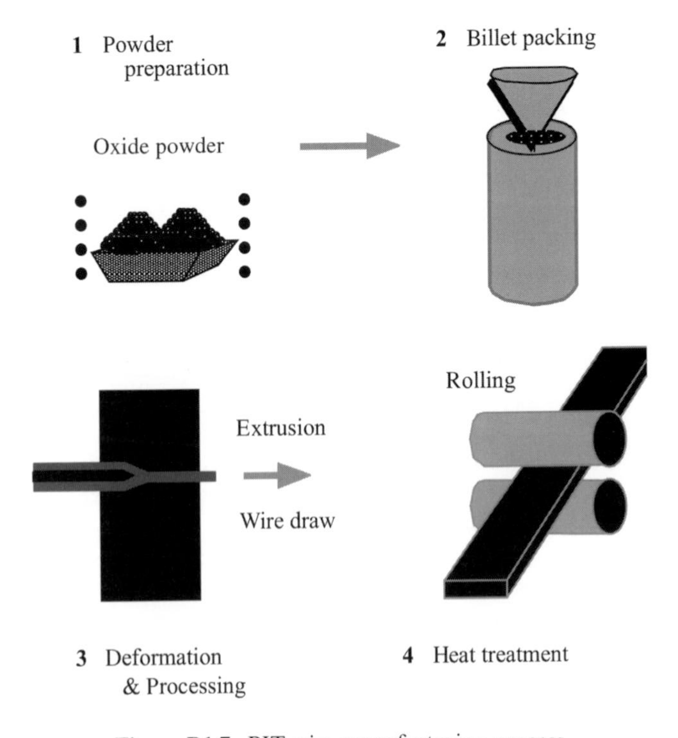

Figure B1.7. PIT wire manufacturing process.

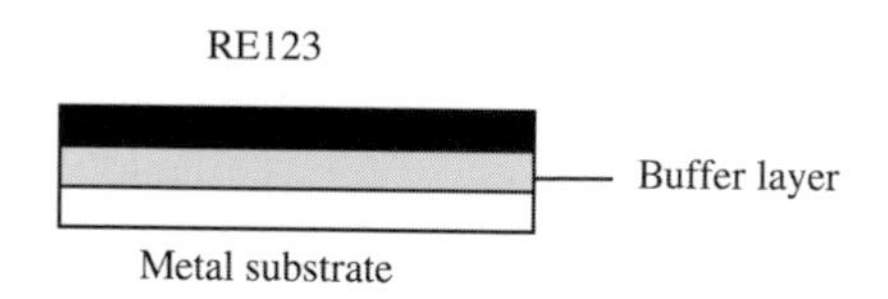

Figure B1.8. The layer structure of a coated conductor.

involves depositing RE–Ba–Cu–O onto textured metal tapes with certain buffer layers, as shown schematically in figure B1.8. The production cost of this process is considerable and, together with problems of stabilization, will be difficult barriers to overcome. Wire fabrication techniques for both high and low T_c materials are reviewed in this chapter. B4.2.1

Zero resistivity is another particularly useful property of superconductors. Significant energy, and hence cost, savings are potentially possible, for example, in superconducting transmission power cables. Until recently, however, such applications were considered impractical simply due to the problems of liquid helium based refrigeration. The discovery of high T_c materials has led to the development of more practical superconducting power cable designs using liquid nitrogen as a cryogen, which have real potential for application in the utility network. Superconducting cables with lengths of over 100 m, for example, have already been constructed using Bi–Sr–Ca–Cu–O tapes. The low flux pinning of this compound is not a serious problem in this application since the wires are subjected only to relatively small (self) magnetic fields.

B1.5 Electronic device applications

Most electronic device applications are based on thin film fabrication technology. The deposition of a good quality thin film onto a single crystal substrate is usually the first step in this process. There are a number of techniques available used to grow thin films on a variety of substrates, and these are described in this chapter. One such method, based on ion beam sputtering, is shown in figure B1.9 and this can be used generally to illustrate thin film physical deposition processes. Here a target, usually prepared by a sintering method, is used as a source of material. The target is vaporized, or sputtered, in an inert gas atmosphere or under vacuum by an incident ion beam and deposited onto the surface of the substrate. Sputtering is a standard procedure for the physical deposition of thin films although a number of other, related techniques exist, including electron beam evaporation, laser ablation and chemical vapour deposition (CVD). Chemical, solution-based techniques such as sol–gel and dip coating can also be used to fabricate thin films. Physical deposition methods, however, are more common for the growth of HTS thin film structures. B4.2.1

It is relatively straightforward to grow a good quality thin film for single element low T_c superconductors such as Pb and Nd for device applications. However, for HTS compounds consisting of four elements or more, it has been difficult to deposit homogeneous, good-quality films particularly over large areas. Unwanted impurity phases are likely to form within the film unless the free energy of the superconducting phase structure is significantly lower than other phases with similar structure.

In the early stage of HTS development, high J_c values of almost two orders of magnitude greater than bulk materials were achieved in thin films. However, it is important to realize that high J_c in the presence of magnetic field reflects a high defect density, so that such films are generally not suitable for device applications. The synthesis of a good quality HTS thin films, therefore, still requires further development.

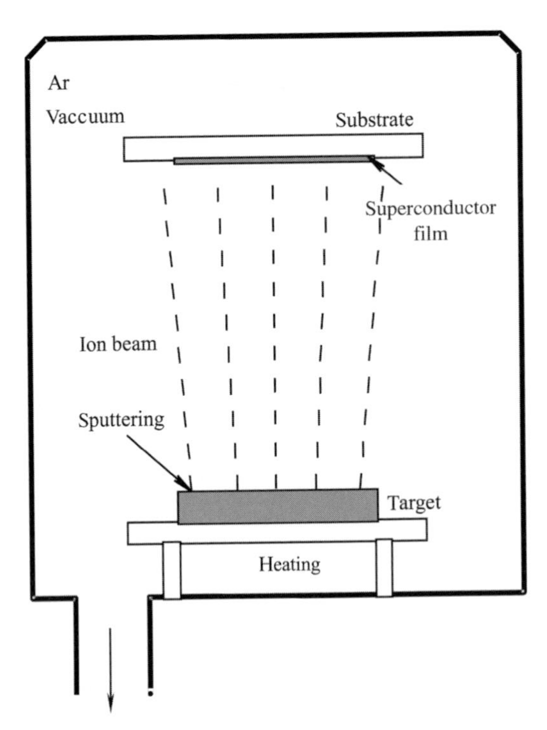

Figure B1.9. Physical method of thin film fabrication.

The final step for the processing of thin films for devices is the construction of a multi-layer B4.4.1 structure, which is even more difficult, since the interface between layers must be clean and sharp. Hence, any chemical reaction or diffusion at the interface should be carefully controlled. Recent progress in A2.7 multi-layer thin film processing technology has enabled the production of ramp-edge type Josephson E4.2 junctions with reasonable reproducibility. SQUID sensors based on HTS materials, in particular, have already been commercialized by a number of companies.

HTS thin films have significant potential for application in the field of wireless networks and communications. Since superconductors are nearly lossless at microwave frequencies, they can act as E3.1 almost perfect RF filters. This performance is what is required for wireless base station receivers, in particular.

B1.6 Bulk fabrication technology

There are no practical applications of low T_c superconductors of large dimensions in bulk form, simply due to the thermal instability of such structures. Local heat generation within the bulk materials leads easily to an abrupt quenching of the superconducting state for bulk forms of low T_c superconductors.

In contrast to low T_c materials, bulk HTS are thermally stable even in large sample sizes due to their relatively large specific heat. As a result, engineering applications of bulk HTS, such as sintered Bi–Sr–Ca–Cu–O current leads, are possible. Most bulk applications, however, are based on RE–Ba–Cu–O A1.3 [(RE)BCO], which exhibits strong pinning in the presence of a magnetic field at 77 K (see chapter A1.3). A top seeded melt process is commonly used to fabricate bulk (RE)BCO materials in the form of large B2.3.3 grains to overcome the severe weak-link problem associated with grain boundaries. Figure B1.10 shows such a melt-growth process. A seed crystal is placed on top of the sintered precursor body, which is

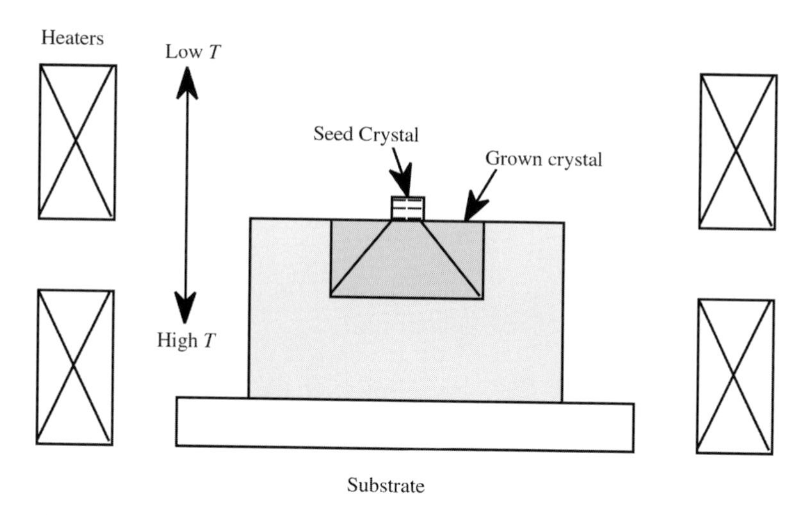

Figure B1.10. Schematic drawing of the top-seeded melt-growth process.

heated to its peritectic decomposition temperature and melt-grown during slow cooling under a temperature gradient. With this process, a single grain of (RE)BCO up to 10 cm in diameter can be grown. It is also usual to enrich the starting powder composition with RE_2BaCuO_5 (or $RE_4Ba_2Cu_2O_{10}$) to stabilize the bulk shape during the partial melting process and to enhance flux pinning.

Large grain bulk (RE)BCO superconductors have significant potential for various industrial applications. A heavy object can be suspended stably using the interaction between bulk superconductors and permanent magnets, which has been used for non-contact bearings and flywheel energy storage systems. Bulk superconductors can also trap large fields (>10 T) at reduced temperatures (<50 K). Various prototype electric power devices have been developed based on trapped-field bulk superconductor magnets such as electric motors, magnetic separation systems and laboratory magnets.

E1, E2

E1.4.3, E2.2

B1.7 Contact techniques

Although not related directly to materials processing, this chapter also describes techniques to make effective electrical contacts to HTS materials. Unlike metallic superconductors, making a low resistivity contact is difficult for these materials but never the less important for both materials characterization, such a measurement of J_c, and for industrial applications such as current leads.

B5.1

In conclusion, the ability to process materials in a controlled and effective way is fundamental to both basic studies and industrial applications. Without understanding the basic properties of materials, one cannot propose and develop industrial applications that require good-quality materials with reproducible properties. Furthermore, the future use of superconducting technology relies primarily on the development of production methods that are both economically feasible and easy to scale up. In this regard, superconducting materials are well on their way to form the basis of wide scale practical applications.

B2
Introduction to section B2: Bulk materials

Masato Murakami

Low temperature superconductors in bulk form are thermally unstable due to their extremely low specific heat at liquid helium temperature. Thermal stability against quenching has only been achieved in the form of multifilamentary structures: low T_c wires of small diameter embedded in copper or aluminum with high thermal conductivity. There were several trials attempting to use bulk low T_c materials for power applications. Trapped-field magnets were once made with sintered Nb_3Sn bulk materials that were broken into pieces after quenching. Hence, bulk low temperature superconducting materials had only been used for basic studies until the discovery of high temperature superconductors in 1987.

Due to their large specific heats at higher temperatures, bulk high temperature superconductors are thermally stable and hence can be used for various engineering applications. Sintered Bi2223 bars were the first commercial products applied to the power leads or the electrical connectors for superconducting magnets. Compared with conventional Cu leads, the thermal load and thus the consumption of liquid helium were greatly reduced. The employment of high T_c current leads made it possible to run the superconducting magnet refrigerated by cryocoolers. The market for cryogen-free superconducting magnets is rapidly growing. However, Bi2223 materials still have a problem of low flux pinning, and therefore they can be used for high field applications only at temperatures below 30 K.

Large grain RE123 (RE: rare earth elements) pellets are now widely used for a variety of engineering applications even at 77 K due to their large flux pinning. At an early stage of research, however, sintered RE123 bulk materials suffered from a weak link problem at grain boundaries due to a combination of large anisotropy and an intrinsically small coherence length. This problem was overcome by employing the melt-textured process, in which RE123 materials are solidified from the partially molten state in a temperature gradient, resulting in a highly aligned structure with strongly coupled boundaries. Recent progress in melt processing combined with the top seeding technique enabled the production of large-grain RE123 based pellets of >5 cm diameter with controlled crystal orientation. Such a pellet can trap a field >2 T at 77 K and >10 T at temperatures below 50 K, and therefore they are attractive for many industrial applications.

The synthesis of single crystals of both low and high T_c superconductors is still important, because understanding of the basic properties of materials is the key to successful industrial applications. In particular, for highly anisotropic materials like high T_c superconductors, thermal, mechanical and superconducting properties must be clarified for the design of engineering devices.

Sintered bulk materials of thallium and mercury based cuprates are still mainly used for basic and applied research, since single crystals are difficult to grow in these systems. Some engineering applications of these series of materials are still under consideration, because they have an advantage of higher T_c over RE123.

In this section, standard production methods for bulk superconducting materials including sintering, melt processing and single crystal growth techniques are presented. The fundamental aspects of bulk fabrication techniques are also reviewed.

B2.1
Powder processing techniques

Rudi Cloots

B2.1.1 Introduction

The ultimate objectives in the chemical processing of high-T_c superconductors are to control the physical and chemical aspects and properties of these materials. In fact, the physical properties of high-T_c superconducting materials are strongly dependent on their chemical processing due to the fact that the grain connectivity plays a significant role in the expression of the practical parameters to be controlled for the intended application (see section B2: Bulk materials). Some emphasis can be put also on molecular-level control like that of the nanosized secondary phase distributions or on controlled surface compositional gradients in order to achieve desired properties.

B2

Nevertheless the intrinsic values of the physical parameters to be considered for characterizing a material, and in particular its anisotropy, are only directly accessible from single crystals (see chapter B2.4). Due to the difficulties generally encountered in the preparation of sizeable appropriate single crystals, scientists have to deal with compressed powder-like systems. It is in fact possible to extract quite valuable data for the fundamental physical parameters like the coherence length, the London penetration depth or the intragranular critical current density when merely operating on powders. It is also necessary to point out that powders provide an alternative to single crystals, when these are not available for obtaining the structural parameters of the superconductors like the lattice constants or the cationic site distributions as a function of doping. Last, but not least, the preparation of advanced high-T_c ceramic materials such as thin films, tapes or wires, and quasi-single crystals, requires that the properties of the end-product be controlled and reproduced starting from powder products. In view of such considerations a precise control of powder properties by a chemical approach seems to be an essential condition for the manufacturing of reproducible high-T_c components.

B2.4 A2.3 A1.2

The microstructure and physical properties of a high-T_c ceramic material are strongly dependent on the physical and chemical characteristics of the starting powders. Moreover, a good knowledge of powder characteristics is strongly recommended in order to achieve some reproducibility in the performances of the end-product.

In this chapter we address the most popular methods for characterizing the precursor powders. A description of the most valuable parameters is given. The type of data to be obtained and measurement limitations are also presented.

B2.1.1.1 Physical properties

The major physical properties of high-T_c or other advanced ceramic precursor powders are (i) the size distribution of the particles, (ii) the degree of aggregation and (iii) the morphology of the grains.

The size distribution of the particles may be determined by a variety of techniques like laser diffusion, optical or electronic microscopy, gravity sedimentation, etc. In laser diffusion, scattered light produced by the laser irradiation of the particles in suspension in a solvent can be analysed in the framework of the Raleigh equation. This gives an average value of the particle size distribution, in the range of 0.03–900 μm, with the restrictive condition that the particles are assumed to be spherical. The average values are expected to differ from one batch to another because of the particle morphology and the powder degree of aggregation.

D1.3, D1.4

For the particle morphology and its degree of aggregation investigations, depending on the size range and level of resolution, optical-, scanning electron- or transmission electron-microscopy techniques are to be preferred (see chapter D1: Structure/microstructure). The main difficulty remains in the sample preparation due to the fact that grains tend to aggregate when deposited onto the most commonly used sample holders. Notice that automated image analysis softwares are now commercialy available in order to determine precisely the particle size distribution.

D1

The principle of the gravity sedimentation technique is based on the relationship between the sedimentation velocity of the particles suspended in a fluid (which is directly related to the particle size and degree of aggregation) and the viscosity of the medium. Local density changes evaluated by light or X-ray absorption techniques may be used to determine the particle size distribution in the suspension. Again clusters have to be removed from the suspension by using appropriate surfactant molecules dissolved in the liquid medium.

B2.1.1.2 Chemical composition

A large number of techniques is available for determining the chemical phase composition of the powder depending on the concentration of secondary phases or impurities. Some techniques require the dissolution of the powder in an acidic medium. It is not possible to mention all techniques commonly used in a laboratory. Only a few are selected here below because of their large field of applicability.

Inductively coupled plasma (*ICP*): This method requires the atomization following plasma injection of the powder constituents dissolved in an aqueous solution. The elements are identified and their proportion quantified on the basis of their electronic emission spectra

X-ray fluorescence spectrometry and electron-probe micro-analysis (*EPMA*): The powder dispersed in a matrix (or dissolved in a solution) is submitted to an X-ray or electron beam producing some specific X-ray fluorescence at different energies characteristic of the constitutive elements. The fluorescence intensities can be analysed with a quantitative goal. In order to do so, standards are required allowing the conversion of intensities to absolute concentrations.

X-ray powder diffraction (*XRD*): The principle is based on the analysis of the diffraction lines produced by the interaction between a monochromatic X-ray beam and the powder. The position of the different diffraction lines is a signature of the crystal structure, while their scattered intensities can be directly related to the atomic composition of the measuring crystals. (For more information, see chapter D1.1.2)

D1.1.2

In order to produce bulk superconducting materials from a powder, depending on the specified requirements, it is crucial to carefully control the as-produced powder. For example, the preparation of a long-length conductor, for which a huge value of the transport critical current density is intended, requires the optimisation of (i) the alignment of the crystallites during the process and (ii) the density of the end-product (see section B3: Wires and tapes). The precursor powder is generally packed into a metallic tube, submitted to successive rolling and sintering steps. The density of the starting material is

B3

crucial in order to overcome the formation of voids and cracks during the mechanical and thermal treatments. Fine grain size distribution and grain uniform morphology are thus key requirements for obtaining high quality long length conductors. Sinterability of the precursor powder is also crucial. These desired properties must be balanced with respect to acceptable costs.

For a large number of ceramic materials including oxide superconductors, one conventional technique is most often used for the preparation of powders: the solid state reaction route (SSRR) [1]. However, the major drawback of solid state reactions is that reactants are not well mixed on the atomic scale. Therefore, while the solid state reaction route has been applied in order to produce batches of powder for basic physical and structural studies, the requirement for better control of chemical homogeneity and specific particle size distribution and morphology, has prompted the development of non-conventional processes often referred to as 'Chimie Douce' or 'soft chemistry' methods like (i) the co-precipitation method, (ii) the sol–gel method, and (iii) the vapour-phase transport method.

Here we review some of these main methods used to produce high-T_c superconducting powders. The main objective is merely to obtain pure monophasic materials with acceptable superconducting properties. Parameters like the cation distributions, the oxygen stoichiometry, the charge carriers concentration have to be optimized. That needs developing new synthetic strategies to control the chemistry of the ceramic materials.

B2.1.2 The solid-state reaction route for raw powders

The so-called solid-state route involves mixing together powdered reactants in the appropriate ratio. Such reactants are usually oxide or carbonate precursor compounds commercially available. The key point for choosing the appropriate precursors is that the reactants have to be of high purity, of controlled chemistry, and chemically stable at room temperature resulting in an accurate control of the stoichiometry throughout the mixture of the starting powders. If one of the reactant evaporates or is sensitive to the atmosphere, the process has to be conducted in a sealed evacuated tube. The mixed reactants may have perhaps to be pressed into pellets and calcined at relatively high temperature for prolonged periods in an electrical furnace in order to produce the intended target composition.

The solid-state reaction between reactants takes place typically for 24–48 h at around 900°C in either air or controlled atmosphere. The powder is usually milled during the process and re-calcined to ensure that the reaction is complete. A knowledge of the phase diagram is obviously useful in defining the appropriate experimental conditions.

This not too sophisticated technique is, however, very attractive for producing new compounds with variable starting chemical composition. Nevertheless the solid-state reaction route presents several drawbacks. First, chemical impurities are sometimes introduced into the powdered reactants from the grinding media and/or the crucible used for the synthesis, often made of silica, alumina, zirconia, stainless steel, etc. Second, the reactants are mixed at the level of individual particles; thus it is difficult to achieve chemical homogeneity at the molecular level in the end-product. Third, since powders made from solid-state reactions are generally aggregated, their particle size and morphology are difficult to control. Finally, the solid-state reaction route usually requires high temperatures (carbonates typically have high decomposition temperatures) and thus needs much energy consumption. The driving force for the synthesis of $YBa_2Cu_3O_7$ superconductor, for example, can be evaluated on the basis of the following equation

B2.2.2, A1.3

$$Y_2O_3 + 4BaO + 6CuO \Leftrightarrow 2YBa_2Cu_3O_{6.5}$$

which requires of the order of 40 kcal mol^{-1}. For comparison, any processing techniques involving the vapour phase requires a huge amount of energy to make the vapour precursors from the solid ones (oxides or carbonates). This energy is released into the atmosphere when the target compound is formed from the precursors.

On the other hand, soft solution processing (following sections) consumes very little energy due to the fact that the lattice energy of the solid precursors can be compensated by the solution of the ions.

In the case of the solid-state reaction route, the reaction between the reactants is thus only limited by the diffusion of the reactive species through the solid or via a dissolution–reprecipitation process when a liquid phase is present. Based on this process, consider the different stages involved in the solid-state reaction route. At the interfaces between the solid particles, the nucleation of the target compound takes place. Subsequent crystal growth is much more difficult because the reactants are no longer in contact, but are separated by a rather impenetrable solid layer of the target compound. A solid-state diffusion process is often required for completion of the solid-state reaction (the rate of which could be increased by compaction). Intermediate grindings are often required to break up reactant/product interfaces and to bring fresh reactant surfaces into contact. Another possibility is when a liquid-phase assisted transport of matter can occur. However, residual liquid phase is generally segregated at the grain boundaries of the bulk material limiting their transport properties.

B2.2.5

A1.3

The degree of homogeneity in the end-product is also very limited by the concentrated areas of porosity and density gradients generally developed during the firing process owing to an intrinsically poor sinterability of the components. This is particularly true when one or more of the starting materials evaporates during the heat treatment. As a consequence, improvements in the sinterability of the compact powder is needed, and can be provided by paying attention to synthetic procedures, as for example by looking for ultrafine, high purity, homogeneous, and highly chemically reactive starting powders. The need for a reduced level of contamination generally introduced by conventional mixing and grinding tools impose also to operate with great care during the preparation process.

In the case of superconducting materials, the solid-state reaction route involves grinding and mixing the reactants (oxides or carbonates), then heating the mixed product, generally pressed into a pellet at a sufficiently high temperature. Factors like the nature of the reactants, the degree of homogeneity in the mixed product, the heating cycle have to be carefully controlled.

B2.1.2.1 Y–Ba–Cu–O ('123') system

The solid-state reaction route is currently much used for the preparation of $YBa_2Cu_3O_7$, thus involving repeated grinding and sintering of the parent oxides (Y_2O_3 and CuO) and carbonates ($BaCO_3$), followed by a calcination at high temperature (of the order of 950°C) [2]. Further annealing under oxygen atmosphere from 950 to 400°C with a 24 h plateau thermal treatment at 400°C is needed in order to produce the desired orthorhombic superconducting $YBa_2Cu_3O_7$ phase [3]. The $YBa_2Cu_3O_7$ powder so obtained is typically contaminated by secondary phases such as the Y_2BaCuO_5 (211), the $BaCuO_2$ (011) and the CuO (001) phases [4]. The contamination results from the non-homogeneous dispersion of the reactants in the powder before heat treatment. It is clear from the phase diagram reported for the Y_2O_3–BaO–CuO system (figure B2.1.1) [4] that a deviation from the 123 stoichiometry leads to the formation of other relatively stable phases, most of them are relatively inert, non-reactive solids or liquids which in the latter case tend to segregate at the $YBa_2Cu_3O_7$ grain boundaries. In fact, use of this contaminated powder in the manufacturing of bulk materials has resulted in poor critical current density properties due to the presence of structural defects, in particular chemical inhomogeneities at grain boundaries. Although the problems associated with powder synthesis by the solid-state reaction have been clearly set out, $YBa_2Cu_3O_7$ synthesis from the reaction

A1.3

$$Y_2O_3 + 4BaO + 6CuO + 1/2O_2 \rightarrow 2YBa_2Cu_3O_7$$

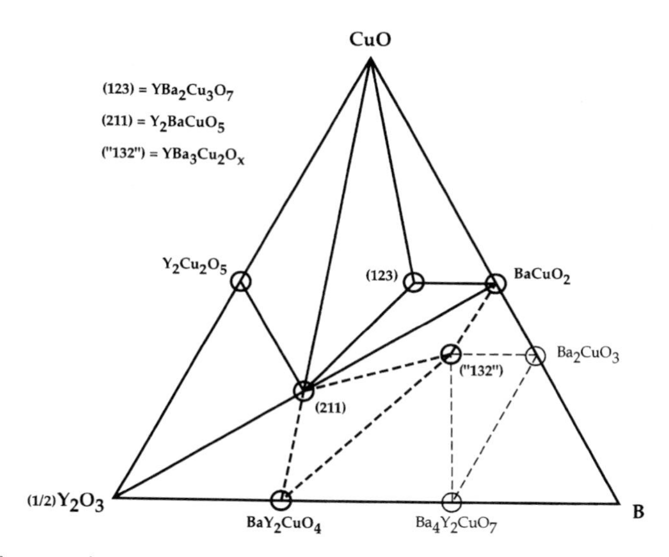

Figure B2.1.1. Ternary phase diagram of the Y_2O_3–BaO–CuO system at 1220 K (reproduced from [4]).

introduces some specific problems associated with the nature of the starting reactants. The barium source, BaO, readily picks up CO_2 from the atmosphere to form $BaCO_3$ which is very difficult to decompose [5]. BaO_2 can be used instead: it has a lower decomposition temperature than $BaCO_3$, and it can act as an internal source of oxygen so the duration of the thermal treatment can be reduced [6]. CuO is very reactive towards commonly used crucibles at high temperature leading to contaminated or off-stoichiometric $YBa_2Cu_3O_7$ phases. Foreign atoms can also be introduced into the structure modifying the superconducting properties.

Other rare-earths, substituted for yttrium, in 123 compounds can be produced by the same procedure, all with a superconducting critical temperature around 90 K [7]. Specific care is needed when large rare-earth (Nd or Sm) 123 compounds are produced due to the fact that neodymium or samarium ion partially replaces barium at their crystallographic sites leading to degraded superconducting properties. The initial thermal treatment has to be conducted under oxygen partial pressure in order to get the appropriate 123 stoichiometry. Calcium substituted for yttrium 123 compounds have also been produced by the solid-state reaction route [8]. For a 20% level of doping, it is interesting to note that the optimized charge carriers concentration can be achieved for a reduced oxygen content of the order of 6.8, so the duration of annealing in the oxygen atmopshere is reduced to a considerable extent.

B3.2.4

B2.1.2.2 Bi–Sr–Ca–Cu–O systems

The preparation of the $Bi_2Sr_2Ca_{n-1}Cu_nO_{2n+4}$ ($n = 1, 2$ and 3) superconductors is essentially conducted by the solid-state reaction route. The procedure is exactly the same as for the $YBa_2Cu_3O_7$ superconductor except for the calcination temperature. However, it is very difficult to obtain pure monophasic products due to thermodynamic and kinetic reasons [9, 10]. The $n = 3$ so-called '2223' phase has one of the highest T_c (110 K). In spite of the numerous investigations on the preparation process found in the literature, the 2223 phase, which is thermodynamically metastable and not accessible directly from a liquid phase, is very difficult to obtain as a single phase material. The coexistence of the three members $n = 1, 2$ and 3 in the form of polytypic intergrowths in a single grain is known also to be very detrimental for obtaining good superconducting properties [11] (see chapter D1.2).

B2.2.3

The standard method of sample preparation is to fire intimate mixtures of the constituent binary oxides and carbonates. Samples produced by such a powder processing have low density and suffer from lack of liquid phase sintering. The plate-like morphology of the grains and the very low packing density suggest that powder methods which require liquid or solid diffusion may be very inefficient as a synthetic route. Moreover, it has been argued that the solid-state reaction route, involving calcination and sintering of the component oxides, is not suitable to obtain single phase Bi-2223 product primarily because of the content loss due to evaporation of the more volatile element, i.e., Bi_2O_3, while other components are relatively sluggish during reaction, hence leading to heterogeneous phases [12]. The reaction temperature is limited by the presence of low melting components and the reaction leaves unreacted oxides and impurity phases behind. Because of the similarities in structures, lattice dimensions though different in the c-axis parameter, and formation kinetics, all three $Bi_2Sr_2Ca_{n-1}Cu_nO_{2n+4}$ phases can be present to some degree in any prepared sample. Only a careful control of the processing conditions would permit us to obtain any one of these phases in a pure single phase form [13].

For understanding the relationship between composition, structure, and properties, and also for searching optimum synthesis conditions we need to know phase diagrams. Let us recall that in the following, our discussion is limited to the compounds which may exist in the quaternary Bi_2O_3–SrO–CaO–CuO system in the ranges of interest for this review, i.e., the regions resulting in superconducting single phases. We will thus concentrate on this part of the phase diagram. Strobel *et al* investigated the phase diagram of the lead-substituted quaternary Bi_2O_3–SrO–CaO–CuO system in the temperature range 825–1100°C, and in air along the line $Bi_{1.6}Pb_{0.4}Sr_2Ca_{n-1}Cu_nO_{2n+4+y}$ [14, 15]. Strobel *et al* have examined several compositions lying on a suitable selected line defined by the index x in the formula $(Bi_{1.6}Pb_{0.4}Sr_2CuO_6)_{1-x}$–$(CaCuO_2)_x$ $(0 < x < 1)$, and specified the compositions by the index n, varying between $n = 1$ and $n = \infty$ $[n = 1/(1-x)]$, in the formula $Bi_{1.6}Pb_{0.4}Sr_2Ca_{n-1}Cu_nO_{2n+4+y}$. This line thus includes the nominal compositions of the stable superconducting phases. Strobel *et al* observed the occurence of nine distinct solid phases and two liquid ones. The resulting phase diagram is shown in figure B2.1.2. The transformation and melting lines of the superconducting $n = 1, 2$, and 3 phases are located below 900°C. Below the solidus (855°C), the sequence of stable $[n]$ phases is [1], [1] + [2], [2], [2] + [3], and finally [3]. The $n = 3$ Bi-2223 phase exists as the dominant phase between 835 and 875°C

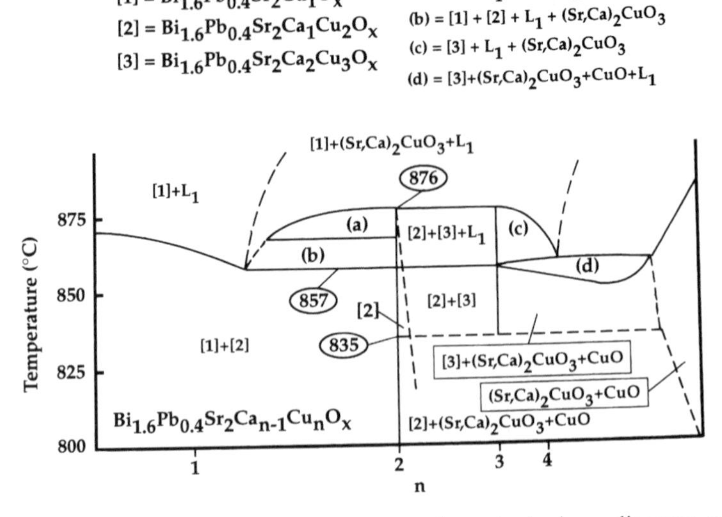

Figure B2.1.2. Part of interest (see text) of the temperature-dependent phase diagram reported only for the $Bi_{1.6}Pb_{0.4}Sr_2Ca_{n-1}Cu_nO_{2n+4+y}$ chemical composition (reproduced from [13]).

only for $3 \leq n < 4$. Let us recall also that the $n = 1$ and $n = 2$ phases present a large solubility range, i.e., the substitution of calcium for strontium [16]. This solubility range becomes narrower with increasing temperature, and shifts to lower calcium content. It was also noticed that T_c of the 2212 phase decreased with increasing calcium content. An appreciable degree of Bi/Sr substitution was also reported [17].

Partial substitution of bismuth with lead was found to favor the development of the $n = 3$ 2223 phase [18]. Nevertheless, the superconducting materials produced by this method with respect to the phase diagram reported here above often contain several phases. Reduced superconducting properties are found in such samples.

Therefore, we intend now to discuss the ways for enhancing the 2223 phase formation by processing via a solid-state reaction route.

Adjustment of the sintering temperature and time

It has been shown that precise control over the sintering temperature is very relevant in order to get a $T_{c\ zero}$ higher than 100 K. Many authors reported that the 2223 phase was formed at temperatures in the range 827°C < T < 856°C [19–21]. The 2223 phase was found to be thermodynamically unstable at temperature above 856°C. According to the same authors, a 50–100 h time heat treatment is necessary to develop significant proportions of the high-T_c phase. Longer sintering times might be detrimental due to a large de-densification owing to an evaporation of bismuth and lead. A repetitive grinding to break up the grain boundaries, pressing and annealing procedure was later on developed in order to obtain the desired 2223 single phase material [20, 22, 23].

B2.2.3

Modification of the starting nominal composition

From the large number of experimental studies reported in the literature concerning the preparation of the Bi-2223 phase, it was demonstrated that the best starting stoichiometric compositions to obtain 2223 phase rich samples were in the region close to $Bi_{1.6}Pb_{0.4}Sr_2Ca_2Cu_3O_y$ with a little excess of calcium and copper [24]. An excess of calcium and copper was effectively reported to greatly enhance the kinetics of formation of the 2223 phase [25]. As a consequence, owing to the lowering of the melting temperature, the formation of the 2223 phase at 850°C was very substantially accelerated. T_c was insignificantly affected by these compositional changes. It seemed that a liquid–solid reaction played an important role in the growth of the high-T_c phase. Morgan *et al* [26, 27] effectively showed that the 2223 phase was slowly formed from precursors by small liquid droplets migrating over growing platelets of the 2212 phase. The role of this transient liquid phase, by dissolving calcium and copper, is to provide a reservoir and a pathway for calcium and copper ion diffusion. As the material is epitaxially deposited upon the 2212, it is almost inevitable that some 2212 will become encapsulated and remain as syntactic intergrowths. It is thus of common belief that an increase of the calcium and copper content in the starting stoichiometry is likely to enhance the formation of the 2223 phase, in particular when lead doping was envisaged, in this latter case through an increase of the Ca_2PbO_4 phase quantity. A large amount of Ca_2PbO_4 would result in much liquid phase that enhances the diffusion rate and thus accelerates the formation of the intended 2223 phase [19, 28–30]. Thermal differential analyses reported in reference [31] on Pb-free and Pb-doped Bi-2223 initial chemical composition systems clearly demonstrate that the reaction producing the 2223 phase should not be an exothermic process. The driving force of the formation of the thermodynamically less stable 2223 phase, in the presence of lead, is a great increase in entropy related to the redistribution of calcium, and lead during sintering. One crucial role of lead in the formation of the 2223 phase is thus to enable the transfer of low-reactivity species (such as calcium ions) to a more highly reactive phase (Ca_2PbO_4). Therefore, it seems clear that the formation and development of the microstructure of the ceramic superconductor are regulated mainly by diffusion

B2.2.3
B2.3.5

processes. Mechanisms such as intergrowth, disproportion, and dissolution-precipitation have commonly been advanced.

Among the different routes proposed in the literature in order to enhance 2223 phase formation, the use of precursors is expected to give better results than those obtained when the conventional solid-state route is followed [32, 33]. The technique has been reported to produce nearly single-phase 2223 materials via a classical solid-state reaction between two pre-synthesized crystalline phases. In this method sluggish reactants are treated at high temperature to give precursors which will be mixed with a low-melting one, the so-called two-powder process. Obviously, a two-compound reaction should give better results than a multiple one. By reducing the number of reactants we can expect better products. Moreover the technique requires short reaction times and provides 2223-rich materials. This precursor method was optimized by choosing the appropriate precursors. Beltran *et al* [34] for example used Bi_2CuO_4 as one of the precursors for its topology, open structure and low melting point. This makes Bi_2CuO_4 an excellent solid matrix for the topochemical insertion of alkaline earth ions in the preparation of $Bi_2Sr_2Ca_{n-1}Cu_nO_{2n+4}$ phases. This precursor matrix method applied to the preparation of Bi-2223 phase was reported in several papers. Among these reports, we would like to mention the paper of Dorris *et al* [35]. In this paper, an original experimental procedure was proposed based on a two-powder process between $Sr_xCa_{1-x}CuO_2$ and $Bi_{1.8}Pb_{0.4}Sr_{2-x}Ca_{1+x}Cu_2O_8$, producing highly pure Bi-2223 materials.

Beside the two-powder process, the glass route was also used for the preparation of the Bi-based superconductors [36, 37]. Due to their ability to form glass, the Bi-based superconductors prepared by the glass-to-ceramic route offer several advantages:

(a) preparation of superconducting ceramics free of porosity,

(b) ease of fabrication of specific shapes of superconductors, and

B2.4

(c) control of the crystallization process.

This method involves a rapid solidification technique by which samples are melted first and then splat quenched to form an amorphous structure. The crystallization of the phases is then performed through subsequent annealing of quenched glasses. [For appropriate reading see [38]].

The glass-to-ceramic method is very attractive due to the fact that the mixing of the components occurs at a molecular level. Chemical homogeneity is therefore accessible. Nevertheless, although theoretically the formation of the 2223 phase should be easily controlled by the glass-to-ceramic route, the formation of the 2223 phase by annealing the amorphous phase seems to follow the same sequence as the 2223 phase synthesis by the solid-state reaction route described above. This is essentially due to the existence of different glass forming regions in the quaternary phase diagram of the system which leads to multiphasic systems with different chemical compositions in the glassy material. A 2223 initial composition leads during splat quenching to a two-phase system made of crystalline phase with a chemical composition close to $(Sr,Ca)_3Cu_5O_8$ dispersed in a fully dense glassy matrix of copper-rich 2212 chemical composition. The presence of the crystalline phase has clearly some strong influence on the following crystallization process as for the so-called two-powder process [39]. From such a consideration, the glass-to-ceramic route can be considered as an alternative to the two-powder process.

B2.1.2.3 Tl–Ba–Ca–Cu–O systems

C3
The solid-state reaction route as described above is no longer applicable to the synthesis of the thallium cuprate superconductors owing to the high toxicity and volatility of thallium oxide.

Essentially two types of thallium-based superconductors can be synthesized: the $Tl_2Ca_{n-1}Ba_2Cu_nO_{2n+4}$ and $TlCa_{n-1}Ba_2Cu_nO_{2n+3}$ families with $n = 1, 2$ and 3. Special care must be paid since Tl evaporates during the heat treatment. Synthesis has to be conducted in a sealed evacuated tube, in such a way that better control of stoichiometry can be achieved. The mixed reactants, pressed into a pellet are wrapped in a gold foil and sealed in an evacuated quartz ampoule [40]. In so doing, the Tl(2223) phase with $T_c = 122$ K can be prepared from the corresponding mixture of the parent oxide or carbonate compounds, by a calcination step at 850°C.

The use of precursors is often employed for the synthesis of thallium-based superconductors. Tl can be incorporated into the final structure by a reaction at high temperature in a sealed evacuated tube between thallium oxide and a Ba–Ca–Cu–O oxide precursor [41, 42]. The precursor was synthesized by a classical solid-state route in air at 900°C from a mixed reactant powder. This method is preferred when carbonates are used as the starting compounds. Problems related to the existence of structural analogies between the different members of the thallium-based family compounds have to be solved, as it is the case for the Bi-based superconductors, in order to prepare pure monophasic products. Thallium-excess starting composition could be preferred in order to compensate the possible loss of thallium during the process [43]. Nevertheless, it is clear that an excess of thallium in the starting chemical composition, corresponding to the ideal $Tl_{1(+x)}Ca_1Ba_2Cu_2O_y$ (1122) system, leads to a mixture of Tl-1122 and Tl-2122 phases [44]. It has been demonstrated that a thallium-deficient composition gives better control of the stoichiometry in the final product resulting in the formation of pure monophasic 1122 phase [45]. It is also important to note that the charge carriers concentration is directly related to the thallium content in the end-product. It has been established from experimental evidences that the formation of a pure Tl-2223 material requires the use of a calcium-rich thallium-deficient starting chemical composition, close to $Tl_1Ca_3Ba_2Cu_3O_y$[44, 46, 47]. It can be noticed that the thallium cuprates belong to the family of metastable superconducting phases as in the case of bismuth cuprates. If the thermal treatment is prolonged for a long period, the Tl-2223 phase decomposes into Tl-1223 probably due to the evaporation of thallium during the process. On the other hand, at temperatures higher than 870°C, Tl-2223 decomposes into Tl-2212 and $CaCuO_2$. Nevertheless, despite the difficulties encountered in the synthesis of pure monophasic materials, the Tl-2223 phase is much more easily formed than the Bi-2223, even if the final composition of the end-product does not reflect the composition of the starting mixture.

B2.1.2.4 Hg–Ba–Ca–Cu–O systems

Only the $HgBa_2Ca_{n-1}Cu_nO_{2n+4}$ family compounds has been discovered up to date. As is the case for the thallium-based compounds, owing to the high toxicity and volatility of HgO, it is not possible to incorporate directly mercury into the precursor of the high-T_c 1212 and 1223 superconductors. The solid-state process requires two steps: a precursor of nominal composition $Ba_2Ca_2Cu_3O_x$ is synthesized by a classical solid-state route. Mercury is then incorporated into the structure by mixing the precursor with HgO and by calcining the mixture in a sealed evacuated tube at 850°C during 12 h [48]. Rhenium is sometimes partially substituted for mercury in order to increase the thermodynamic stability of the Hg-1223 phase [49]. Rhenium also participates in the self-optimization of the charge carrier concentration which is strongly dependent on the oxygen content. The highest T_c for these systems can be obtained without post-annealing of the Hg-1223 phase.

C4

B2.1.2.5 Other cuprates

Other materials like lead cuprates, electron-doped superconductors, infinite-layer cuprates can be also prepared following the traditional solid-state reaction route. Most of them require non-conventional experimental conditions owing to their particular behaviour, as for example in the case of lead cuprates where the mixing and calcination of the parent oxides or carbonates yields the formation of the stable $SrPbO_3$ perovskite-like compound. A modified two-powder process is required for the preparation of nearly pure $Pb_2Sr_2(Y,Ca)Cu_3O_{8+x}$ with good superconducting properties [50]. In the case of electron-doped superconductors, annealing under reduced or inert atmosphere is often required for getting optimized superconducting properties, making sure that the cerium(4+) doping does not increase the oxygen concentration which could result in a 'dissimulated' hole-doped superconductor. Hydrostatic pressures are also sometimes required in order to stabilize the desired crystallographic structure as in the case of infinite-layer cuprates [51, 52].

D2, D3

B2.1.2.6 Superconducting physical properties (see sections D2 and D3)

In order to evaluate the superconducting properties of the material, the powder has to be pressed and re-annealed at high temperature for a prolonged period. Transport properties are strongly dependent on the presence or not of weak links at the grain boundaries. A percolation path can be established at any temperature below T_c through the entire sample volume if a strong coupling between individual grains is initiated. A zero critical temperature (T_{c0}) can be defined as the temperature at which the resistivity goes to zero, indicating the establishment of a supercurrent through the sample volume. The presence of a foot structure in the electrical resistance as a function of temperature [$R(T)$] transition is clearly connected to the difficulty for the supercurrent to pass through individual grains, whence strongly indicative of the bad quality of the intergranular matrix. AC susceptibility measurements are also very helpful in the determination of the superconducting properties of the green body. AC magnetic susceptibility is intimately linked to resistivity. The susceptibility onset occurs at the superconducting critical temperature T_c. In between T_c and T_{c0}, the grains are completely superconducting but decoupled from each other. This temperature range coincides exactly with a first drop in the real part of the susceptibility, caused by intragranular shielding currents. As the temperature is decreased below T_{c0}, lossless currents can flow across intergranular junctions and thus participate to a further decay of the real part of the susceptibility. It is also clear for such polycrystalline materials that the intragranular shielding currents are reversible (no loss signal, in the alternative part of the susceptibility is observed), suggesting that no vortex is penetrating the grains (i.e. $H < H_{c1}$). Since H_{c1} goes to zero near T_c, another reason for explaining the absence of an intragranular peak has to be considered, like the fact that the London penetration depth near T_c is much larger than the small grain diameter, so that there is no vortex in the grains. Figure B2.1.3 gives the electrical resistivity and the AC magnetic susceptibility as a function of temperature measured for several AC magnetic fields for a Bi-2212 ceramic sample prepared by the solid-state reaction route.

D3.1

D2.5

We emphasize that the major drawback of the solid-state reaction route is that reactants are not mixed on an atomic scale. For the reaction to proceed reasonably, high mobility of the reactants and maximum contact between the reacting particles are needed. Powders have thus to be finely crushed and heated at high temperatures so that ion diffusion can easily proceed. All alternative methods described in the next sections address this problem by achieving atomic scale mixing of the reactants. In most of the following methods, precursors are mixed in solution and then transformed into a solid by precipitation or gelation. Heating the resulting product at relatively low temperature leads to the formation of a finely divided powder precursor. Such a precursor can be transformed into the desired material by firing next at high temperature even for a short time. Long-range diffusion of the constitutive ions is thereby no longer required.

The following methods which are reviewed pertain to the solution chemistry science and involve condensation, co-precipitation and sol–gel process.

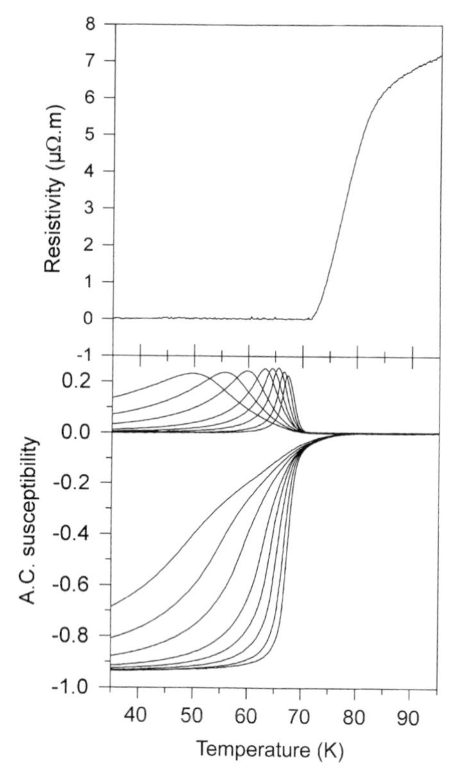

Figure B2.1.3. Electrical resistivity and AC magnetic susceptibility as a function of temperature measured for several AC magnetic fields on a Bi-2212 ceramic material. From right to left: μ0HAC = 0.14, 0.28, 0.70, 1.40, 2.80, 7.00, 14.00, and 28.00 G.

B2.1.3 The coprecipitation route

Coprecipitation involves the formation of a crystalline or an amorphous solid phase with the appropriate stoichiometry by precipitating the reactant ions from a solution. The major difficulty along this route in getting a solid precursor for the synthesis of the high-T_c superconductors resides in the control of the stoichiometry. In order to do so, the precipitating agent must be a multivalent organic compound coordinated to more than one metal ion, preferentially coordinated to all reactant ions present in solution, and in the appropriate ratio. To separate the solid precursor from the solution whatever the initial concentration of the reactants, it is also necessary that the precipitate is strongly insoluble in the solution.

The main precipitating agents used for producing the precursors of high-T_c superconductors are citrates, oxalates, triethylamine or hydroxydes.

The solid precursor when isolated from the solution is dried, then calcined at a relatively high temperature in a suitable atmosphere, in order to obtain the intended superconducting material. The main advantage of coprecipitation over the solid-state reaction route is that the reactant ions are mixed at a molecular level. This results in a more homogeneous distribution of the reactants in the precursor of the superconducting materials. A decrease in the reaction temperature and in the duration of annealing follow because the diffusion barriers are significantly reduced compared to those of conventional solid-state synthesis. Owing to the easy thermal decomposition of the precursor in a controlled manner, a smaller particle size and a lesser degree of aggregation are expected resulting in a higher density of the end-product.

C1 *B2.1.3.1 Y–Ba–Cu–O (123) system*

When $YBa_2Cu_3O_7$ is prepared by the oxalate route, Oxalic acid is added to an aqueous solution of yttrium, barium and copper nitrates and the pH is adjusted to 7.5 by using ammonia [53–56]. A precipitate is formed. After filtration and drying it can be converted into the orthorhombic $YBa_2Cu_3O_7$ phase by heating in air at 870°C for a few days followed by annealing under an oxygen atmosphere at 400°C. The powder so-obtained is characterized by a small particle size and a very low degree of aggregation. Nevertheless secondary phases like Y-211 and CuO are generally formed owing to the moderate solubility of barium oxalate as compared to yttrium and copper. Such a barium-deficiency needs to be overcome in order to produce $YBa_2Cu_3O_7$ with good superconducting properties. To reduce the solubility of barium oxalate, the coprecipitation can be conducted in an alcoholic medium by using triethylammonium oxalate as the precipitating agent added to an aqueous ethanol solution [57–59]. An alternative to the coprecipitation of oxalates in aqueous solution consists in adding both oxalic acid and urea into the solution of nitrates [60]. The pH of the solution can be adjusted in situ by the thermal decomposition of urea into CO_2 and NH_3. Very fine particulates of $YBa_2Cu_3O_7$ can be obtained. Generally speaking, for all these methods, great care is needed in order to avoid the formation of barium carbonate during the drying step. This difficulty can be overcome by first drying the precipitate under vacuum. Owing to its very high decomposition temperature, barium carbonate induces some chemical inhomogeneity and barium deficiency in the end-product leading to an impure $YBa_2Cu_3O_7$ material.

C2 *B2.1.3.2 Bismuth cuprates*

Coprecipitation is not easily applicable to the formation of powder precursors of the bismuth cuprates owing to the difficulty of finding compounds of the metal ions which could be all soluble in a common solvent. A precise control of the cationic stoichiometry is thus very difficult to achieve. Last, but not least, bismuth nitrate, which is probably the best choice in getting a homogeneous solution with all the other metal ions, is not stable in water and precipitates into a hydroxynitrate following the reaction:

$$Bi(NO_3)_3(aq) + 2H_2O \rightarrow Bi(OH)_2NO_3(s) + 2H^+.$$

Nitric acid has thus to be added to the solution in order to overcome the formation of the hydroxynitrate phase, which is not very compatible with the basic conditions generally required for the coprecipitation of multidentate organic compounds [61]. From such considerations, it is obvious that the control of stoichiometry is only possible by experimenting with various compositional sequences of all the metal ions present in the initial solution.

Another possibility to avoid the presence of the bismuth hydroxynitrate salt in the precursor powder is by starting with acetates of all the metal ions in acetic acid, and then by precipitating the oxalates by adding excess oxalic acid to the solution [62]. Again pH has to be adjusted by adding ammonia to the solution. Triethylamine can also be used [63]. The advantage of triethylamine over ammonia is that it has a higher basicity and a lower complexing ability towards copper(2+). The oxalates are converted into the intended bismuth cuprates by heating the dried precipitate at 850°C in air for a few days.

B2.1.4 The sol–gel route

The sol–gel process involves heating a gel formed from a colloidal suspension in order to get the appropriate solid phase. Since the reactants are mixed on an atomic scale into the gel, the reaction generally goes to completion in a short time and can be conducted at a lower temperature as compared to the classical solid-state reaction route. The starting solution consists of a mixture of metal-containing compounds (such as metal alkoxides, acetylacetonates, soluble inorganic compounds like nitrates, acetates,...), water as the hydrolysis agent or the solvent, acid or base as the catalyst, and the organic solvent. In non-aqueous-based sol–gel process [64–66], the metal compounds undergo hydrolysis and polycondensation at room temperature resulting in a 'sol'. The term 'sol' designates a suspension of nanometer-sized particles dispersed into the solution without any precipitation. As the hydrolysis reaction goes further, the particles tend to aggregate to each other forming a 'wet' gel which still contains water and solvent. Evaporation of water and solvent gives rise to a 'dry' gel. Heating the dry gel to relatively high temperature produces the ceramic material by thermal decomposition of the remaining organics. The formation of the ceramic material proceeds at relatively low temperature as compared to the conventional solid-state reaction route owing to the fact that the reactants in the sol–gel process are mixed on an atomic scale into the solution. For comparison with non-aqueous-based sol–gel process, in aqueous-based sol–gel process [67–70], water is used as the solvent and a gelation agent like ethanolamine is added to the solution when the metal-counterions have no tendency to gel in water by themselves as it is the case by using nitrate salts. Water is then eliminated from the solution, the resulting medium being in a rather viscous state: the so-called gel. Thermal heat treatment of the gel at relatively high temperature gives rise to a solid ceramic.

Another possibility is by using a mixed nitrate solution with ethylene glycol as solvent. Ethylene-diamine-tetra-acetic acid (EDTA) is then used as the complexing agent (the molar amount of EDTA to be added is equal to that of the total molar amount of cations). The nitrate solution is added dropwise to the EDTA solution and the pH is adjusted. Wet gel are obtained by heating the solution at 80°C for 2 – 3 days. Further evaporation of the solvent under reduced pressure leads to a dry gel. Calcination is performed at 800°C to obtain an agglomerate powder.

The sol–gel process is also very effective for producing particles with controlled morphology and degree of aggregation, especially when modifying parameters like the pH of the solution, the firing temperature,.... The feasibility of such a process has been demonstrated to result in controlled porosity and mechanical properties of the end-product.

The Sol–gel process offers also an attractive method for producing thick or thin films of superconducting materials owing to the possibility to fabricate the material in one step with a high degree of control of the stoichiometry of the end-product (see chapters B4.2 and B4.3).

B4.2, B4.3

B2.1.4.1 $YBa_2Cu_3O_7$ system

C1

In the synthesis of $YBa_2Cu_3O_7$ by the sol–gel process, the alkoxide precursors are both very expensive and difficult to produce. Therefore an aqueous-based sol–gel method is preferred. In this method, yttrium, barium and copper acetates are dissolved in acetic acid [71]. The pH of the solution is adjusted to 3.6 by using ammonia and the solution is maintained at 80°C for one day until a transparent gel is formed. The adjustment of the pH is necessary for obtaining homogeneous solutions during ageing and for enhancing the gelation process. Then the gel is dried until a black solid is produced. Further annealing at relatively high temperature (of the order of 920°C) gives rise to the $YBa_2Cu_3O_7$ phase. If nitrates are used as the starting reactants, ethanolamine is added to the solution and plays both the role of base and gelation agent [72]. In this case, a minimum of water is necessary to guarantee the gelation of the solution within one or two days. The temperature of the transformation of the gel into the solid phase can be determined by thermal differential analysis (see figure B2.1.4), and strongly depends on the precursor type.

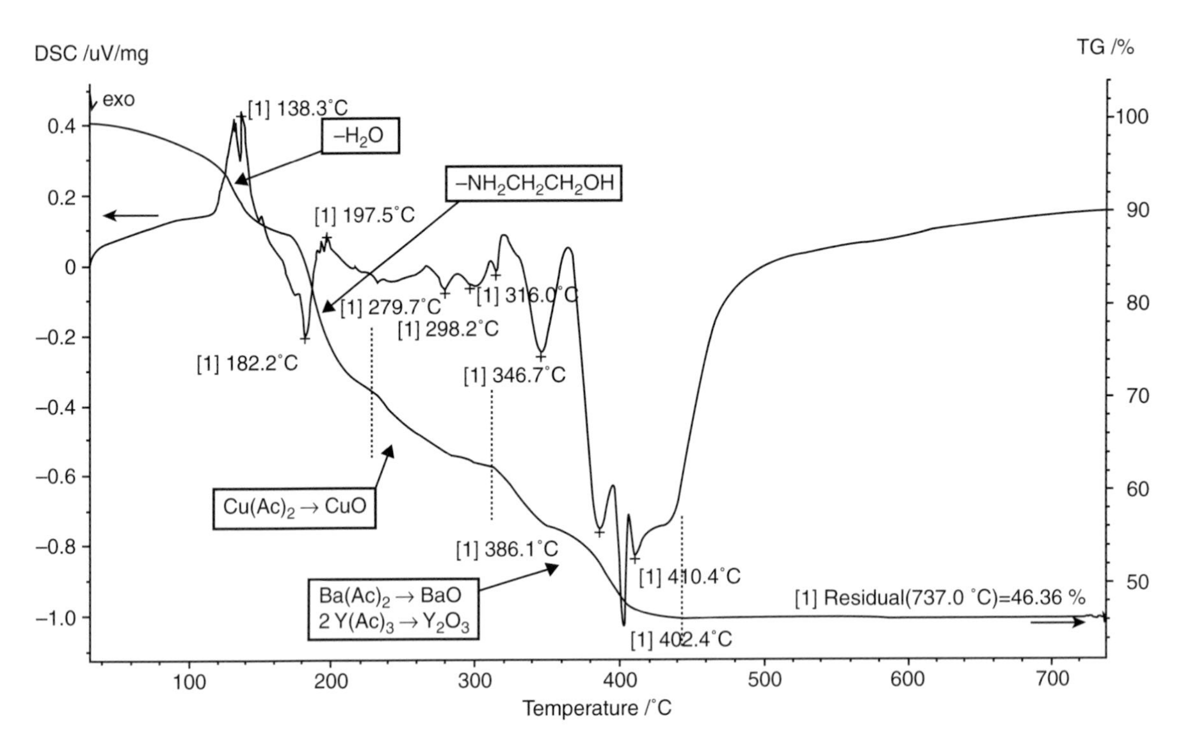

Figure B2.1.4. Differential thermal analysis of the YBCO acetate sol–gel precursor carried out in air between room temperature and 850°C at a heating rate of $10°C\,min^{-1}$, using a thermoanalyser Netzsch. Its decomposition follows a three step process. Weight losses during heating are also reported and revealed the evaporation of the rest of the solvents (water and ammonia) first, followed by the successive decompositions of the acetates, copper, barium and yttrium, from 200 to 450°C. The film consists then of BaO, CuO and Y_2O_3, converted next into the $YBa_2Cu_3O_7$ phase by a reaction-growth process conducted at 850°C.

C2 *B2.1.4.2 Bismuth cuprates*

There are only a few reports related to the preparation of bismuth cuprates by the alkoxide sol–gel process, essentially due to the fact that the alkoxide of the metals, especially for bismuth and lead, are not easily available [73]. It is also nearly impossible to find alkoxides soluble in a common solvent of all the metals involved in the preparation of Bi-based superconductors. An aqueous-based sol–gel process has been developed based on the use of nitrate salts [74]. The nitrates have to be generated *in situ* into the solution from carbonates or oxides initially dissolved in nitric acid. In fact, most nitrates are hygroscopic, thereby making it very difficult to control precisely the stoichiometry of the end-product by using solid nitrates. It is also necessary to mix the components as separate solutions owing to their tendency to form hydroxides when strontium, calcium and barium carbonates are present together into the solution. High pH values are generated in the vicinity of the solid particles leading to the formation of very stable hydroxydes. Ethylene-diamine-tetra-acetic acid (EDTA) is used as the gelation agent. After mixing the solutions together, EDTA first and then ammonia are added to the final solution in order to adjust the pH to 5. In such conditions, EDTA is transformed into its complexing un-protonated subunit stabilizing all the cations in the solution. After removal of water, a gel is formed which can be transformed into the superconducting Bi-based material by subsequent annealing at relatively low temperature, depending on the stoichiometry of the end-product. It is important to note that carbonates are produced first by the thermal decomposition of the complexant EDTA. Nevertheless, carbonates of the constitutive cations are very easily transformed into corresponding oxides at temperatures far below the temperatures needed when the carbonates are processed via a classical solid-state route, as demonstrated by differential thermal analysis (see figure B2.1.5). Since EDTA also forms a stable complex with thallium, such a method can be used for producing Tl-based superconductors via a sol–gel process. C3

It should be emphasized that soft solution chemistry processes have been developed in order to control precisely the superconducting properties of the precursor powder. Optimum microstructure is required in the end-product, i.e., a fine grain size to avoid microcracking during the manufacturing of the bulk superconducting materials. For optimum transport and magnetic properties, the highest J_c is also required whatever the initial chemical composition. The question to a chemist is: how to improve the critical current density by chemical routes? The best solution is by using a reactive precursor powder prepared by a chemical process. For example, the sol–gel process gives rise to a molecular-level homogeneity in the starting powder. In fact the starting powder has been shown to greatly affect J_c. The manufacture of long length conductors for example needs Bi-based powder with high homogeneity, ultrafine particles and low carbon content.

Nevertheless, a significant problem remains when a large quantity of powder is needed. Other techniques are thus necessary at an industrial level in order to produce an appropriate amount of powder.

B2.1.5 The spray pyrolysis route

In order to produce very rapidly and in a reproducible way large quantities of powder with controlled stoichiometry and morphology, the spray pyrolysis technique was developed. A solution containing

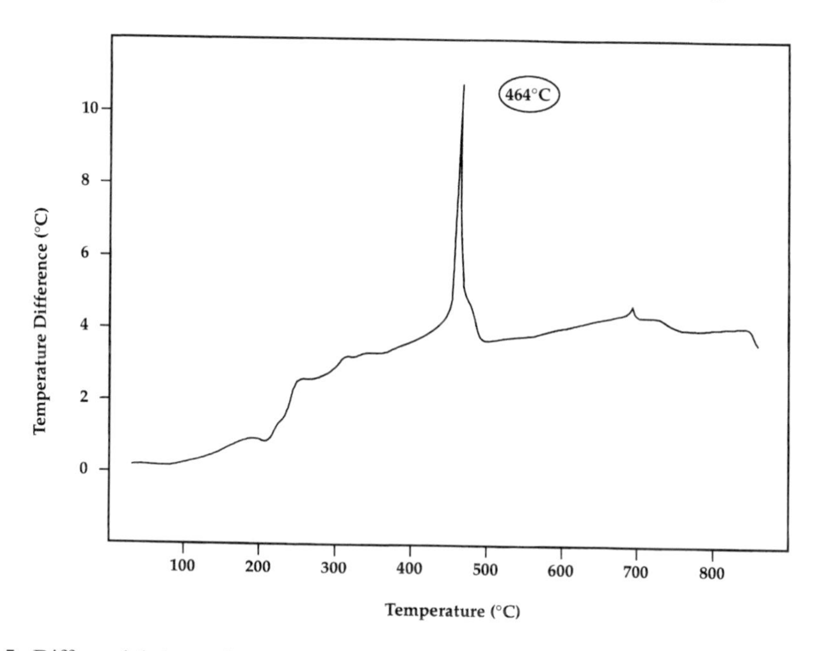

Figure B2.1.5. Differential thermal analysis of the decomposition of the Bi-2212 oxide precursor produced by a sol–gel process using EDTA as the gelation agent. The DTA peak reported at 464°C occurs when the as-produced carbonates dissociate to form the oxides. This shows that the formation of oxides from the carbonates is complete at a temperature far below the temperature needed when processing Bi-2212 ceramic by solid-state reaction from carbonates reagents (reproduced from [72]).

the appropriate cations is used and droplets of this solution are generated by an aerosol generator. Different (4) modes of powder production can be considered [75–77]: first the droplets are splash deposited onto a heated substrate where the solvent evaporates and the thermal decomposition of the precursor powder is carried out. In the second mode, dry precipitates of the precursor powder are deposited onto the substrate where the thermal decomposition is initiated. In a third mode, the precipitates are thermally decomposed into gaseous products which can be deposited on the top of a substrate. In the fourth mode, the gaseous products are transformed into solid ones in such a way that a full thermal decomposition takes place far from the substrate. In both latter cases, not only the temperature of the substrate is a key parameter, but a precise control of the surroundings is needed in order to garantuee the high thermal decomposition of the precursor solution. The last mode is very similar to the so-called spray drying process which can be described in the framework of the spray pyrolysis technique [78–80]. The last mode is also very useful for the preparation of large quantities of fine powders.

B4.1

The size of aerosol droplets is around 30 μm or even larger when the aerosol is generated pneumatically (like in the spray drying process) [81]. On the other hand, when ultrasonic generation is used, the droplets may be smaller by one order of magnitude. Depending on the excitations frequency, it is possible to precisely control their size, from a small diameter size when high frequencies are used (up to 3 MHz) to a large diameter size when low frequencies are used (lower than 1 MHz)[81].

Nitrate solutions are usually used as the solution precursor. A similar procedure used for the preparation of nitrate solutions in co-precipitated and sol–gel processes has to be considered in order to avoid the non-stoichiometry in the end-product whatever the intended chemical composition of the superconducting phase.

A precise knowledge of the thermal behavior of the precursor solution is needed in order to fix the experimental conditions when one mode has been selected. Differential thermal analyses (DTA) have to be performed prior to start of the production. It is clear from DTA data that the individual components of the solution are degraded following totally different mechanisms (see figure B2.1.6), inducing some deviation from stoichiometry usually when the substrate is heated up to 200°C, i.e., above the thermal decomposition of copper nitrate. By adjusting the chemical composition of the starting solution it is possible to overcome this problem, in particular when a heating mode of the substrate is preferred.

C1, C4

$YBa_2Cu_3O_7$, $Bi_2Sr_2Ca_2Cu_3O_{10}$, Tl-2212 and Tl-2223 and mercury-based bulk superconductors have been produced by this technique. The process needs a thermal treatment of the precursor. By controlling precisely the thermal parameters, it is possible to produce a powder with particles characterized by a well-defined spherical shape with typical size of the order of a few microns, depending on the method used to generate the aerosol [75].

This technique offers also the possibility to produce very rapidly one micron thick films of the superconductors which permit to determine physical characteristics like the London penetration depth by measuring the microwave surface resistance as a function of the initial chemical composition and the microstructure of the material. Measurements have been done on YBCO films and give a London penetration depth of the order of 1.3 μm [75].

The spray pyrolysis is thus very appropriate when high speed and low cost production of high-T_c superconducting powders or tapes are required, in spite of the polycrystallinity of the films prepared when tapes are needed.

B2.1.6 Freeze drying

Soft solution chemical processes like sol–gel or spray-drying methods offer the possibility to control carefully the distribution of the components on a molecular level. However, even if the process of solvent evaporation to give the precursor is very fast, components can segregate into the solution leading to undesirable inhomogeneities in the end-product. When high level of homogeneity is required, especially

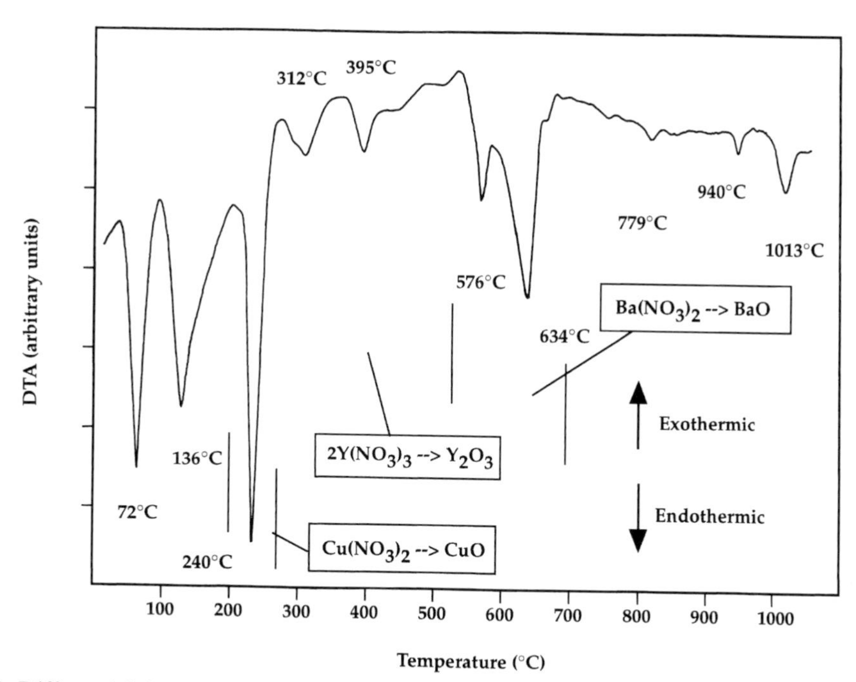

Figure B2.1.6. Differential thermal analysis of the as-spray dried YBCO powder from a nitrate solution precursor. After the initial loss of water, the as-spray dried Y-123 powder exhibits three distinct regimes of thermal decomposition: the first major DTA peak occuring at 240°C is referred to the copper nitrate decomposition into CuO, directly followed by the decomposition of yttrium nitrate into Y_2O_3. The last major DTA peak occuring at 634°C corresponds to the decomposition of $Ba(NO_3)_2$. After dissociation the material consists of oxides, the synthesis of YBCO being initiated at around 780°C.

when mercury-based superconductors are produced, the freeze-drying technique gives the best results [82]. In this method a freeze-dried mixture of metal nitrates is used to prepare the appropriate precursor. The solvent, often water, is sublimated in the frozen state in a such a way that the components cannot migrate during the preparation step preventing inhomogeneities in the end-product. This method offers also the possibility to work in a carbon-free environment preventing degradation of the superconducting properties often due to the presence of residual carbonates segregated at the grain boundaries [83]. C4

In order to control the microstructural parameters of the precursor powder, it is also possible to initiate a spherical shape of the particles by creating droplets of the nitrate solution to be sprayed into liquid nitrogen [84]. Again typical particle size of the order of a few microns are produced by this technique. Special care is needed when the metal nitrates tend to absorb water from the environment. X-ray diffraction analysis shows that the frozen droplets are mainly constituted by nitrate of the metals with different degrees of hydration. As a consequence, due to the presence of this residual water, the mixture tends to melt during the heat treatment. A pre-heating stage at low temperature is often required to eliminate the residual water. It is also very convenient to reduce copper losses during the thermal treatment, especially when nitrates are decomposed. During the pre-heating stage, the copper nitrate is transformed into a hydroxy-nitrate salt which gives rise to copper oxide by further thermal decomposition without any loss of copper. A flash decomposition process, consisting of a rapid heating process above the temperature where all the components decompose, is also often required in order to avoid segregation of the components [84]. This temperature can be determined by thermal gravimetric analysis coupled with differential thermal analysis. D1.1.2

B2.1.7 Conclusion

Different procedures have been described in this report in order to control precisely the characteristics of the precursor powder of a superconducting material. This implies:

(a) to improve the processibility; in fact pure monophasic materials are often desired

(b) to enhance the current carrying capabilities of the superconducting materials prepared from the powder, through the improvement of the chemical homogeneity and then the manufacturing of dense materials with enhanced mechanical properties

(c) to develop new materials and concepts

(d) to investigate the influence of impurities on the superconducting properties.

In this regard, soft chemical processing using solutions have been attracting increased interest.

Acknowledgment

I would like to address my warmest thanks to all the people from the LCIS and SUPRAS groups for either participating to part of this work and for helpful discussions.

References

[1] Rao C N R and Gopalakrishnan J 1989 *New Directions in Solid State Chemistry* (Cambridge: Cambridge University Press)
[2] Cava R J, Batlogg B, Vandover R B, Murphy D W, Sunshine S, Siegrist T, Rameika J P, Rietman E A, Zahurak S M and Espinosa G P 1987 *Phys. Rev. Lett.* **58** 1676
[3] Cava R J, Batlogg B, Chen C H, Rietman E A, Zahurak S M and Werder D 1987 *Phys. Rev.* B **36** 5179
[4] Clarke D R 1987 *Int. J. Mod. Phys.* B **1** 170
[5] Clarke D R, Shaw T M and Dimos D 1989 *J. Am. Ceram. Soc.* **72** 1103
[6] Rao C N R 1988 *J. Solid State Chem.* **74** 147
[7] Tarascon J M, McKinnon W R, Greene L H, Hull L W and Vogel E M 1987 *Phys. Rev.* B **36** 226
[8] Manthiram A, Lee S J and Goodenough J B 1988 *J. Solid State Chem.* **73** 278
[9] Michel C, Hervieu M, Borel M M, Grandin A, Deslandes F, Provost J and Raveau B 1987 *Z. Phys.* B **68** 421
[10] Maeda H, Tanaka Y, Fukutomi M and Asano T 1988 *Jpn. J. Appl. Phys.* **27** L209
[11] Tanaka A, Kamehara N and Niwa K 1989 *Appl. Phys. Lett.* **55** 1252
[12] Chen Y L and Stevens R 1992 *J. Am. Ceram. Soc.* **75** 1160
[13] Stassen S, PhD Thesis University of Liège (Belgium)
[14] Strobel P, Toledano J C, Morin D, Schneck J, Vacquier G, Monnereau O, Primot J and Fournier T 1992 *Physica* C **201** 27
[15] Toledano J C, Strobel P, Morin D, Schneck J, Vacquier G, Monnereau O, Barmole V, Primot J and Fournier T 1993 *Appl. Supercond.* **1** 581
[16] Müller R, Schweizer T, Bohac P, Suzuki R O and Gauckler L J 1992 *Physica* C **203** 299
[17] Hong B and Mason T O 1991 *J. Am. Ceram. Soc.* **74** 1045
[18] Sunshine S A, Siegrist T, Schneemeyer L F, Murphy D W, Cava R J, Batlogg B, Van Dover R B, Fleming R M, Glarum S H, Nakahara S, Farrow R, Krajewski J J, Zahurak S M, Waszczak J V, Marshall J H, Marsh P, Rupp L W and Peck W F 1988 *Phys. Rev.* B **38** 893
[19] Chen Y L and Stevens R 1992 *J. Am. Ceram. Soc.* **75** 1150 and references therein
[20] Mei Y, Green M, Jiang C and Luo H L 1989 *J. Appl. Phys.* **66** 1777
[21] Button T W, McN Alford N, Birchall J D, Wellhofer F, Gough C E and O'Connor D A 1989 *Supercond. Sci. Technol.* **2** 224
[22] Ullrich M, Schaper W and Freyhardt H C 1990 *Supercond. Sci. Technol.* **3** 602
[23] Kusano Y, Nauba T, Takada J, Ikeda Y and Takano M 1994 *Physica* C **235–240** 477
[24] Koyama S, Endo U and Kawai T 1988 *Jpn. J. Appl. Phys.* **27** L1861
[25] Toledano J C, Morin D, Schneck J, Faqir H, Monnereau O, Vacquier G, Strobel P and Barmole V 1995 *Physica* C **253** 53
[26] Morgan P E D, Housley R M, Porter J R and Ratto J J 1991 *Physica* C **176** 279
[27] Morgan P E D, Piche J D and Housley R M 1992 *Physica* C **191** 179
[28] Huang Y T, Liu R G, Lu S W, Wu P T and Wang W N 1990 *Appl. Phys. Lett.* **56** 779

[29] Shi D, Boley M, Chen J G, Xu M, Vandervoort K, Liao Y X, Zangvil A, Akujieze J and Segre C 1989 *Appl. Phys. Lett.* **55** 699
[30] Huang Y T, Shei C Y, Wang W N, Chiang C K and Lee W H 1990 *Physica* C **169** 76
[31] Kim S H, Kim Y Y, Lee S H and Kim K H 1992 *Physica* C **196** 27
[32] Rao C N R, Ganapathi L, Vijayaraghavan R, Ranga Rao G, Kumari M and Mohan Ram R A 1988 *Physica* C **156** 827
[33] Sastry P V P S S, Gopalakrishnan I K, Sequeira A, Rajagopal H, Gangadharam K, Phatak G M and Iyer R M 1988 *Physica* C **156** 230
[34] Beltran D, Caldes M T, Ibanez R, Martinez E, Escriva E and Beltran A 1989 *J. Less-Common Met.* **150** 247
[35] Dorris S E, Prorok B C, Lanagan M T, Sinha S and Poeppel R B 1993 *Physica* C **212** 66
[36] Hinks D G, Soderholm L, Capone D W II, Dabrowski B, Mitchell A W and Shi D 1988 *Appl. Phys. Lett.* **53** 423
[37] Howard P J 1989 *Supercond. Sci. Technol.* **2** 216
[38] Abe Y (ed) 1997 *Superconducting Glass Ceramics in BSSCO: Fabrication and Applications* (Singapore: World Scientific)
[39] Cloots R, Stassen S, Rulmont A, Godelaine P A, Diko P, Duvigneaud P H and Ausloos M 1994 *J. Cryst. Growth* **135** 496
[40] Parkin S S P, Lee V Y, Nazzal A I, Savoy R, Huang T C, Gorman G and Beyers R 1988 *Phys. Rev. B* **38** 6531
[41] Vijayaraghavan R, Rangavittal N, Kulkarni G U, Grantscharova E, Guru Row T N and Rao C N R 1991 *Physica* C **179** 183
[42] Gopalakrishnan I K, Sastry P V P S S, Gangadharan K, Phatak G M, Yakhmi J V and Iyer R M 1988 *Appl. Phys. Lett.* **53** 414
[43] Barry J C, Iqbal Z, Rama Krishna B L, Sharma R, Eckhardt H and Reidinger F 1989 *J. Appl. Phys.* **65** 5207
[44] Ganguli A K, Nanjundaswamy K S, Subbanna G N, Rajumon M K, Sarma D D and Rao C N R 1988 *Mod. Phys. Lett. B* **2** 1169
[45] Vijayaraghavan R, Gopalakrishnan J and Rao C N R 1992 *J. Mater. Chem.* **2** 237
[46] Goretta K C, Chen J G, Chen N, Hash M O and Shi D 1990 *Mater. Res. Bull.* **25** 791
[47] Parkin S S P, Lee V Y, Engler E M, Nazzal A I, Huang T C, Gorman G, Savoy R and Beyers R 1988 *Phys. Rev. Lett.* **60** 2539
[48] Antipov E V, Loureiro S M, Chaillout C, Capponi C C, Tholence J L, Putilin S N and Marezio M M 1993 *Physica* C **215** 1
[49] Yamasaki H, Nakagawa Y, Mawatari Y and Cao B 1997 *Physica* C **273** 213
[50] Cava R J 1988 *Nature* **336** 211
[51] Takano M, Azuma M, Hiroi Z, Bando Y and Takeda Y 1991 *Physica* C **176** 441
[52] Azuma M, Hiroi Z, Takano M, Bando Y and Takeda Y 1992 *Nature* **356** 775
[53] Manthiram A and Goodenough J B 1987 *Nature* **329** 701
[54] Wang X Z, Henry M, Livage J and Rosenmann I 1987 *Solid-state Commun.* **64** 881
[55] Clark R J, Harrison L L, Skirius S A and Wallace W J 1990 *Mol. Cryst. Liq. Cryst.* **184** 377
[56] Vos A, Carleer R, Mullens J, Yperman J, Vanhees J and Van Poucke L C 1991 *Eur. J. Solid State Inorg. Chem.* **28** 657
[57] Vilminot S, El Hadigui S and Desory A 1988 *Mater. Res. Bull.* **23** 521
[58] Kellner K, Wang X Z, Gritzner C and Bauvale D 1991 *Physica* C **173** 208
[59] Praminik P, Biswas S, Singh C, Bhattacharya D, Dey T K, Sen D, Ghatak S K and Chopra K L 1988 *Mater. Res. Bull.* **23** 1693
[60] Liu R S, Chang C T and Wu P T 1989 *Inorg. Chem.* **28** 154
[61] Rao C N R, Nagarajan R and Vijiyaraghavan R 1993 *Supercond. Sci. Technol.* **6** 1
[62] Das Santos D I, Balachandran U, Guttschow R A and Poeppel R B 1990 *J. Non-Cryst. Solids* **121** 441
[63] Shei C Y, Liu R S, Chang C T and Wu P T 1991 *Inorg. Chem.* **29** 3117
[64] Moore G, Kramer S and Kordas G 1989 *Mater. Lett.* **7** 415
[65] Kramer S, Kordas G, McMillan J, Hilton G and Van Harligen D 1988 *Appl. Phys. Lett.* **53** 156
[66] Monde T, Kozuka H and Sakka S 1988 *Chem. Lett.* **2** 287
[67] Khan S A, Bagley B G, Barboux P and Torres F E 1989 *J. Non-Cryst. Solids* **110** 142
[68] Kozuka H, Umeda T, Jin J and Sakka S 1988 *Adv. Ceram. Mater.* **3** 520
[69] Barboux P, Tarascon J, Green L H, Hull G and Bagley B 1988 *J. Appl. Phys.* **63** 2725
[70] Chen F and Tseng T 1990 *J. Am. Ceram. Soc.* **73** 889
[71] Kullberg M L, Lanagan M T, Wu W and Poepell R B 1991 *Supercond. Sci. Technol.* **4** 337
[72] Fujiki M, Hikita M and Sukegawa K 1987 *Jpn. J. Appl. Phys.* **26** L1159
[73] Dhalle M, Van Haesendonck C, Bruynseraede Y, Kwarciak J and Van der Biest O 1990 *J. Less-Common Metals* **164–165** 663
[74] Fransaer J, Roos J R, Delaey L, Van der Biest O, Arkens O and Cellis J P 1989 *J. Appl. Phys.* **65** 3277
[75] Jergel M 1995 *Supercond. Sci. Technol.* **8** 67
[76] Viguié J C and Spitz J 1975 *J. Electrochem. Soc.* **122** 585
[77] Blandenet G, Court M and Lagarde Y 1981 *Thin Solid Films* **77** 81
[78] Kourtakis K, Robbins M and Gallagher P K 1989 *J. Solid State Chem.* **82** 290
[79] Tomizawa T, Matsunaga H, Fujishiro M and Kakegawa H 1990 *J. Solid State Chem.* **89** 212
[80] Tripathi R B and Johnson D W Jr 1991 *J. Am. Ceram. Soc.* **74** 247
[81] Langlet M, Senet E, Deschanvres J L, Delabouglise G, Weiss F and Joubert J C 1989 *Thin Solid Films* **174** 263
[82] Ichinose N, Ozaki Y and Kashu S 1992 *Superfine Particle Technology* (Berlin: Springer) p 153
[83] Coppa N V, Myer G H, Salomon R E, Bura A, O'Reilly J W and Crow J E 1992 *J. Mater. Res.* **7** 2017
[84] Lee S, Shlyakhtin O A, Mun M-O, Bae M-K and Lee S-I 1995 *Supercond. Sci. Technol.* **8** 60

B2.2 Introduction to bulk firing techniques

J S Abell and T W Button

Superconducting oxides in bulk form have considerable potential for technological exploitation in certain key niche areas of power engineering. These include current leads, fault current limiters and magnetic shielding. The low thermal conductivity of the superconducting oxides coupled with the lack of joule heating below T_c can reduce much of the heat leakage to conventional superconducting magnets when compared with a conventional metallic copper lead. These components are usually in the form of an assembly of wires or tapes or large diameter rods or tubes. The two higher members of the BSCCO series and YBCO are candidates for this application. YBCO and Bi-2223 are prepared by bulk sintering routes described in this section, whereas Bi-2212 is fabricated by a melt casting process covered in chapter B2.3.

E1.2

B2.3.5

A high critical current density (J_c) is an essential requirement of most applications of this type and the usual microstructural requirements apply: crystallographic alignment, good grain connectivity, optimized flux pinning features, although the relative importance of these factors varies. Thus, the BSCCO phases are more anisotropic than YBCO and so alignment is crucial to realising optimum J_c. However, J_c in BSCCO is not grain boundary limited like it is in YBCO, but requires development of intra-grain flux pinning centres. Intergrowths of lower member phases in the BSCCO system and in the multilayer thallium and mercury cuprates are a common feature of their microstructures and require careful processing in order to control their distribution. The quality and degree of misorientation of grain boundaries is crucial in YBCO, the control of which is equally dependent on thermal treatment. BSCCO phases are more tolerant of grain misorientations, particularly with respect to *c*-axis twist boundaries.

A1.3

YBCO in sintered form is usually prepared from stoichiometric mixtures of powders in a one step process, the sintering taking place as close to the peritectic temperature as possible. Sintering aids can be added, particularly those which can provide liquid phase sintering, thus accelerating the diffusion process and enhancing intergrain contact. In contrast, the BSCCO compounds are often prepared by a two step process involving the initial preparation of intermediate precursor powders which promote the evolution of the required phase, probably by the formation of liquid phases.

B2.2.2

B2.2.3

Bulk sintered YBCO is usually prepared in disc, ring, rod or wire form, and the routes employed will be discussed in chapter B2.2.3. Bi-2212 in bulk form is normally in the form of rods or tubes and the best properties are realized by a melt processing route providing enhanced alignment and grain interconnectivity. This will be dealt with in chapter B2.3. Bi-2223 is not readily prepared from the melt and so most artefacts are prepared by a sintering route, which is the subject of chapter B2.2.3.

B2.2.3

The chapter B2.2.4 on sintering of other superconducting oxides will concentrate on the two most studied systems, namely the thallium and the mercury series of cuprates. Most of the sintering work

B2.2.4

performed on these superconducting oxides has been primarily designed to produce single phase material rather than to create a particular artefact, and so the routes employed are not application-driven to produce a particular shape or geometry, but rather the main consideration is to aid sintering by compaction. Both one-step and two-step approaches have been reported.

B2.2.1
Fundamentals of sintering techniques

J S Abell and T W Button

B2.2.1.1 Theoretical considerations

The starting point for the fabrication of any bulk ceramic artefact is powder of the appropriate stoichiometry which then has to be shaped, compacted and sintered to produce a dense product having the structural and/or functional properties required for the application. Inherent in this fabrication route are changes in the bulk density and volume of the article. Loose powder has a volume fraction of 0.2–0.3, increasing to 0.4–0.6 after compaction into a green body. Sintering then leads to densification to a final volume fraction of 0.85–1.0 (100% theoretical density). In order to ensure good sinterability a ceramic powder must be fine. This concept is embodied in Herring's Law [1], which notes that as the particle size of the powder reduces the time taken to sinter to a given density also reduces. Hence it is very important to use fine powders and furthermore it is important that the powders be free from agglomerates as these will cause differential sintering, produce inhomogeneities and trap internal stresses [2]. As a corollary, if the powder is coarse and badly agglomerated, full density will never be achieved and the resulting defects will produce a weak ceramic.

The driving force for sintering is the reduction of surface energy by the reduction of surface area and this is achieved by reaction of the particles at high temperature. The reduction of surface area is achieved by material transport. If we imagine two spherical superconducting particles of radii r, then the contact between the particles will be characterized by a contact diameter x. In the initial stages of sintering the surface curvature at the contact between the two particles is directly related to the vapour pressure, which is far higher in areas of high surface curvature. The initial stage of sintering is by evaporation and condensation, and this is depicted schematically in figure B2.2.1.1. The difference in free energy or chemical potential between the neck area and the surface of the particle provides a driving force which causes transfer of material by the fastest means possible. If the vapour pressure is low there are other solid state processes by which material transport can occur. For example, matter can move from the particle surface, from the particle bulk, or from the grain boundaries between particle by surface, lattice or grain boundary diffusion as shown in figure B2.2.1.2 and table B2.2.1.1 [3]. The relative rates of these processes in a particular system determine which are dominant in the sintering process. It should be noted, however, that not all of these processes result in a decrease in porosity. Only transfer of matter from the particle volume or from the grain boundary between the particles causes shrinkage and pore elimination.

Two main processes can be distinguished in the sintering of HTS materials. The first is solid state diffusion in which vacancy migration is usually the rate-determining step and this is described in equation (B2.2.1.1) [4],

B2.2.5

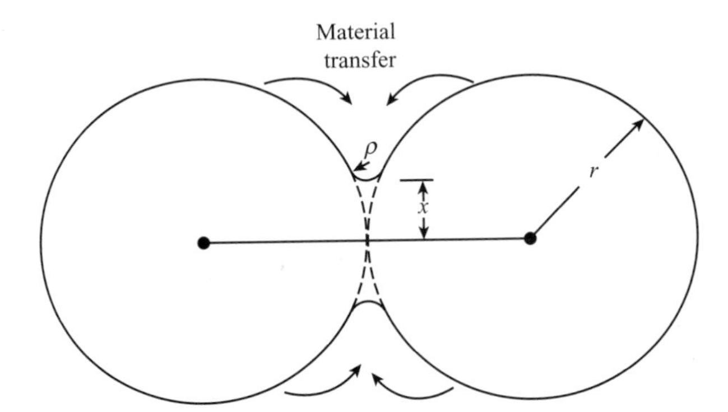

Figure B2.2.1.1. Schematic representation of the initial stages of sintering by evaporation–condensation (after Kingery *et al* [3], figure 10.18, p 471).

$$\frac{x}{r} = \left(\frac{40\gamma a^3 D^* t}{kT}\right)^{\frac{1}{5}} r^{\frac{3}{5}} \tag{B2.2.1.1}$$

where γ is the surface energy, a^3 is the volume of the lattice vacancy, D^* is the diffusion coefficient and t is time. It should be noted here that all HTS materials are highly anisotropic with regard to diffusion and essentially all mass transport takes place in the a–b plane. The model presented would therefore need to be modified accordingly before rigorous use. In addition, other types of diffusion processes may also need to be considered as, for example, diffusion of oxygen in BSCCO is thought to be by interstitials rather than vacancies [5]. The second important process is that of liquid phase sintering, where the presence of a liquid phase promotes diffusion leading to rapid material transport and densification. The liquid phase, which becomes liquid at temperatures at or below those used for sintering, can arise through thermodynamically driven phase changes as the material is heated, or can be added as an extra component to the powder compact prior to sintering. In the YBCO system, for example, the presence of the peritectic decomposition (around 1030°C in oxygen and 1000°C in air) can be used to advantage, and this has led onto the so-called melt processing techniques for bulk and thick film materials, which are described in more detail in chapters B2.3 and B4.3, respectively. Alternatively, non-stoichiometric compositions can be used so that liquid rich in Cu–O (for YBCO) is present at the sintering temperature. The situation for the BSCCO materials is more complex and transient liquid phases which can influence both the crystallization and sintering behaviour can arise by a number of different routes. This is discussed in more detail in chapter B2.2.3. It should be noted, however, that due to the strong influence of the structure and morphology of the grain boundaries on the properties of HTS materials, the presence of additional grain boundary phases in the sintered product is usually avoided. When sintering in the presence of a liquid phase equation (B2.2.1.2) [6, 7] relating to viscous flow applies;

B2.3, B4.3

B2.2.3

$$\frac{x}{r} = \left(\frac{3\gamma t}{2\nu r}\right)^{\frac{1}{2}} \tag{B2.2.1.2}$$

where ν is the viscosity of the liquid phase. Under these conditions rapid grain growth is also observed. This can be beneficial for some materials and material systems but, for reasons discussed in more detail in the following section, the limitation of grain size in some sintered YBCO materials is crucial. An isothermal grain growth model has described the kinetics of grain growth where the grain size l_g is related to a power law in time [8].

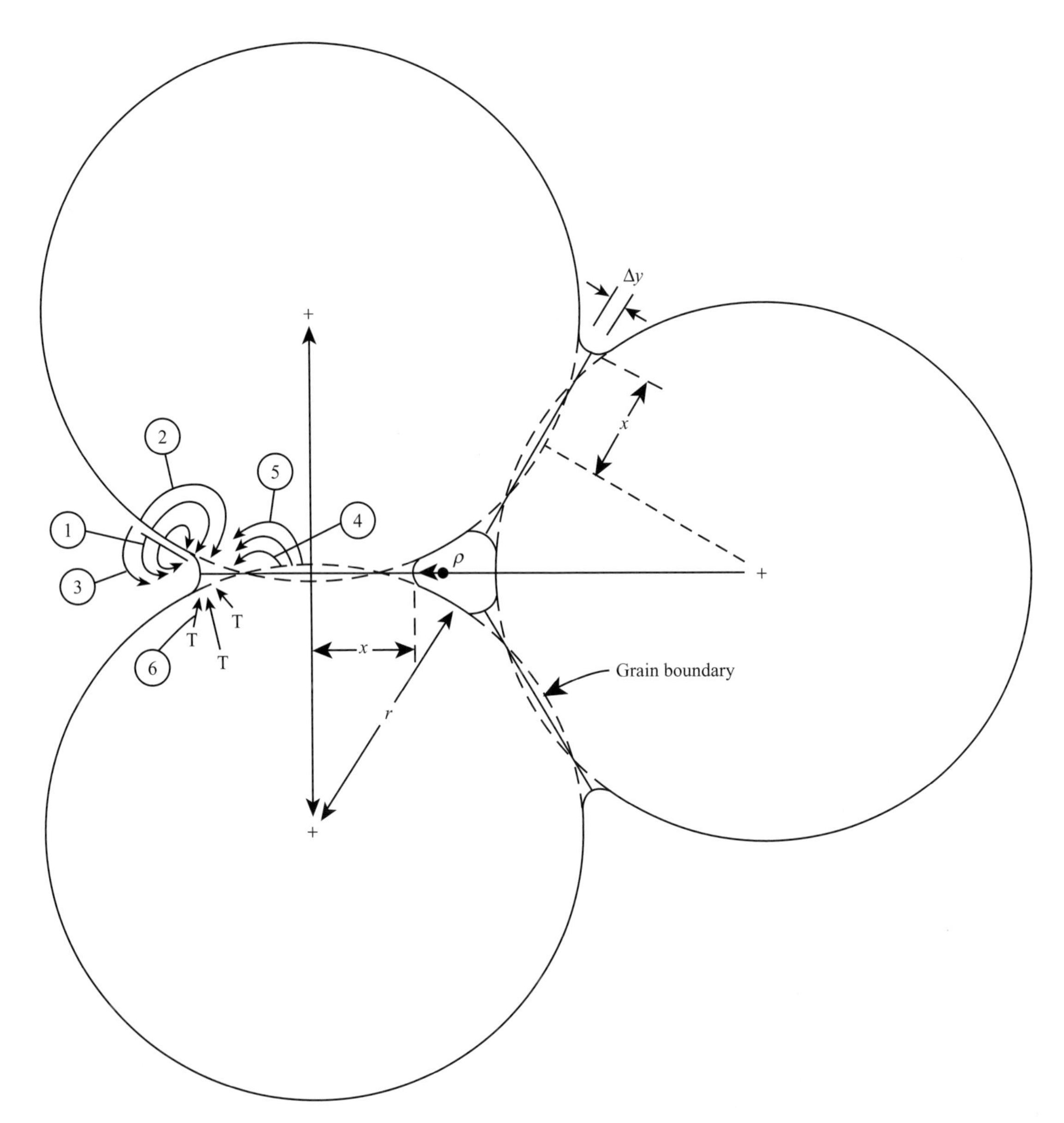

Figure B2.2.1.2. Alternative paths for material transport during the initial stages of sintering. See also table B2.2.1.1 (after Kingery *et al* [3], figure 10.21, p 475).

$$l_g = At^n \tag{B2.2.1.3}$$

where l_g is the grain size and A is a constant with an Arrhenius type temperature dependence. The application of these types of equations is invariably an exercise in approximation and a wide range of values can be found. As an example, however, the activation energy for YBCO has been cited as 125 kJ mol^{-1} [8], with the exponent n having a value of 1/5 for a sintering temperature of e.g. 930°C with no liquid phase, and of 1/3 when sintered above the peritectic temperature in the presence of a liquid phase (equations (B2.2.1.1) and (B2.2.1.2)).

Table B2.2.1.1. Alternative paths for material transport during the initial stages of sintering[a]

Mechanism number	Transport path	Source of matter	Sink of matter
1	Surface diffusion	Surface	Neck
2	Lattice diffusion	Surface	Neck
3	Vapour transport	Surface	Neck
4	Boundary diffusion	Grain boundary	Neck
5	Lattice diffusion	Grain boundary	Neck
6	Lattice diffusion	Dislocations	Neck

[a] See also figure B2.2.1.2 (after Kingery *et al* [3]).

In HTS materials of all three main materials systems the presence of a liquid phase in the early stages of sintering is generally beneficial, but great care must be taken in the time temperature profile of the sintering step. In BSCCO 2223, for example, although a very small amount of liquid is beneficial, if the sintering temperature is too high the 2223 phase will be lost and BSCCO 2212 is the resultant phase. In fact, BSCCO 2212 is the preferred phase for some manufacturers of HTS wire because there is a wider processing window and because the properties ($J_c(H)$) of BSCCO 2212 at liquid helium temperatures (and up to about 30 K) are preferred for current carrying applications.

B2.2.3

B2.2.1.2 Practical considerations

In general, ceramic materials require sintering processes that are carried out at temperatures above ambient and therefore some form of heating chamber is required. For binder burnout operations and the sintering of the bulk BSCCO materials, heating is usually carried out in air using a muffle furnace. The maximum temperatures required for sintering all the HTSC materials are $<1050°C$ and so furnaces with resistive wire heating elements and fibreboard insulation materials are suitable. Most furnaces are now provided with quite sophisticated temperature controllers that can be programmed to carry out a series of segments each with a defined rate of heating or cooling or isothermal dwell time. However, care has to be taken over the measurement of temperature. The indicated controller temperature should not be taken as the sample temperature and variations of $\pm 10°C$ compared to the actual temperature of the sample are not uncommon. It is good practice, therefore, to position a monitoring thermocouple close to the sample, and to calibrate the actual temperature profile of the furnace throughout the heating sequence. This is particularly important in the BSCCO system where temperature control to $\pm 1°C$ can be required during sintering in order to ensure the correct phase is formed. Additional precautions may need to be taken if this degree of temperature uniformity is required over the whole furnace volume.

B2.2.3

YBCO materials are conventionally sintered in an oxygen atmosphere, and this is most conveniently achieved using a tube furnace. Seals at the tube ends provide inlet and outlet points for the gas. The remarks above concerning temperature measurement are equally valid in the case of tube furnaces, as the controlling thermocouples are usually positioned outside the tube, whereas the monitoring thermocouple needs to be positioned inside the tube close to the sample. The choice of tube material is usually dictated by cost and availability. Quartz, non-porous mullite and alumina furnace tubes have all been used with success for sintering YBCO. Quartz has an upper usable temperature limit of around 1000°C, above which devitrification is observed after prolonged use. Mullite and alumina can be used to higher temperatures of 1500 and 1600°C, respectively, without any problems, although for

B2.2.3

tubes with larger diameters and lengths the cost of these materials can be prohibitive compared to quartz.

The final consideration is that of the material used to support the sample during sintering. As noted above, sintering is accompanied by large dimensional changes and care has to be taken to ensure that these can be accommodated without subjecting the sample to excessive stresses which can lead to cracking. Conventional practice is to place the sample on a bed of its own powder, or that of an inert material, although this may be impractical if any liquid phase is formed. Alternatively, high grade alumina substrates or porous supports have been used effectively, provided there is little or no liquid phase present during the process. However, it should be noted that aluminium could degrade the properties of HTS materials. Alternative materials are zirconia and barium zirconate which are less prone to detrimental reactions and are used successfully for some melt processing and single crystal growth techniques.

B2.2.1.3 Characterization of sintered materials

Characterization techniques for superconducting materials are described in detail in chapter C and the reader should refer to the relevant section for more information where appropriate. However, there are a number of techniques that are particularly useful for the characterization and quality assessment of bulk sintered materials and these will be briefly outlined here. Standard XRD, SEM and optical microscopy

D1.1.2, D1.3, D1.4

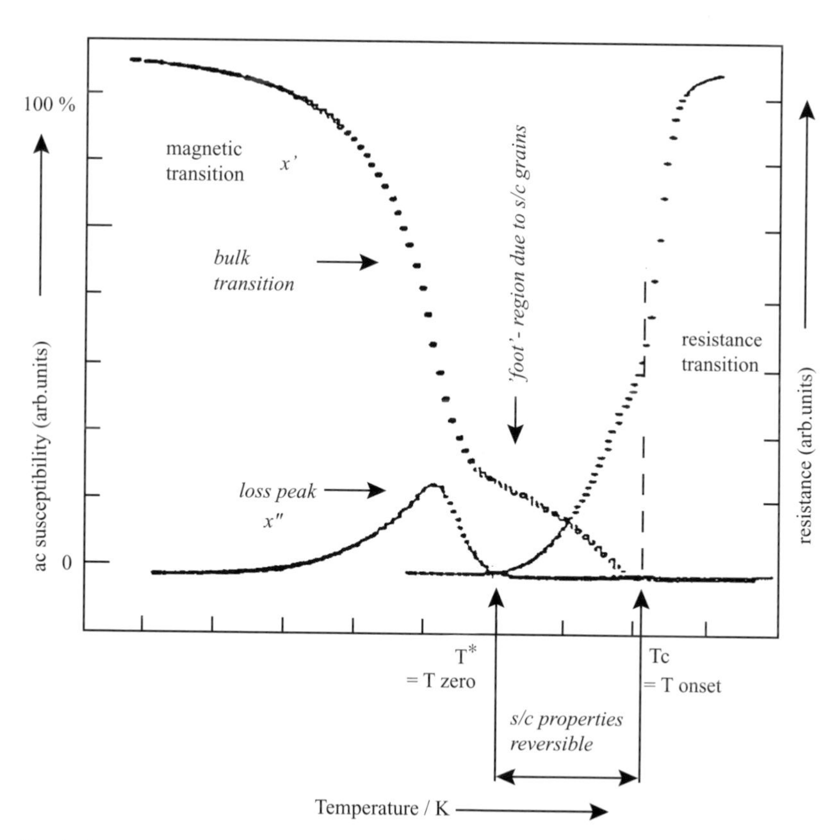

Figure B2.2.1.3. Typical temperature dependence of resistance and ac-susceptibility measured on a sintered sample of polycrystalline $YBa_2Cu_3O_x$.

techniques are invaluable for assessing the phase purity and microstructural features (impurity phases, grain size, porosity, etc), as are standard T_c measurements (resistive and inductive), transport J_c (four-terminal techniques) and magnetization for assessing the superconducting properties. Of particular importance in determining the quality and degree of the grain connectivity is the measurement of the low field ac susceptibility at fixed frequency. By applying an additional dc magnetic field at low temperature even small differences between samples are made much more pronounced, thus assisting in process optimization. Additional advantages of this technique are the ability to assess powders and sintered materials and the contactless nature of the measurement. A typical plot of the temperature dependence of resistance and ac susceptibility of a sintered polycrystalline YBa_2Cu_3O is shown in figure B2.2.1.3, where the contributions of the intergranular and intragranular currents to the real and imaginary parts of the susceptibility are noted.

D2.2
D2.4

References

[1] Herring C 1950 *J. Appl. Phys.* **1** 301
[2] Kendall K 1998 Agglomerate strength *Powder Metall.* **31** 28
[3] Kingery W D, Bowen H K and Uhlmann D R 1976 *Introduction to Ceramics* 2nd edn (New York: Wiley)
[4] Kuczynski G C 1949 *J. Met.* **1** 169
[5] Routbort J L and Rotherman S J 1994 *J. Appl. Phys.* **76** 5615
[6] Frenkel J 1945 *J. Phys.* **9** 385
[7] Exner H E and Petzow G 1973 *Sintering and Catalysis* ed G C Kuczynski (New York: Plenum) p 279
[8] Shin M W, Hare T M, Kingon A I and Koch C C J 1991 *J. Mater. Res.* **6** 2026

B2.2.2
Sintering techniques for YBCO

J S Abell and T W Button

Much work has been carried out on sintering YBCO materials and $YBa_2Cu_3O_x$ in particular, both for the requirement to produce sintered shapes with the necessary properties for specific applications and in order to produce samples for assessment, characterization and scientific study. The main fabrication routes involve fairly standard ceramics processing procedures and, where these are mentioned below, only brief details are given. However, particular attention is drawn to the more unusual aspects of the fabrication procedures, and those specific to YBCO. It is recommended that the reader should consult one of the many standard texts on ceramics processing [1] should they be unfamiliar with any of the standard techniques. The description below will concentrate on the sintering of $YBa_2Cu_3O_x$ (YBCO) in the solid state i.e. at temperatures below the peritectic temperature. A1.3

B2.2.2.1 Powder preparation/calcination

Appropriate powders can be prepared by a number of different techniques as described in chapter B2.1. By far the most common technique is to use a solid state processing route in which $BaCO_3$, CuO and Y_2O_3 are mixed in stoichiometric proportions to yield $YBa_2Cu_3O_x$. A typical production route would be to mix the constituent powders together in a ball mill or vibro-energy mill in a non-aqueous medium for several hours. The slurry is then dried, and the powder is calcined in air at temperatures between 900 and 960°C for 12–24 h. Calcination is sometimes carried out in two stages with an intermediate grinding/mixing stage in order to improve homogeneity. The calcined material is then ground further using zirconia grinding media in a non-aqueous fluid until the desired particle size and/or surface area is achieved. Typical values for a YBCO powder would be a mean particle size of 1–5 μm and a surface area of $> 3\,m^2\,g^{-1}$. It is usual practice to optimize the calcination conditions (temperature, time and number of intermediate mixes) in order to achieve a phase-pure powder. However, as noted in the previous section, these multi-cation systems have complex phase relationships, and there is scope to tailor the calcination conditions so that some lower melting point phases are present in the powder. These can promote and enhance the sintering process. B2.1

Alternative powder preparation routes include co-precipitation, sol–gel and spray pyrolysis, all of which have been used to take advantage of the more homogeneous mixing that can be achieved by these solution techniques. The reader should also be aware of the availability of commercial sources of suitable YBCO powders from, for example, Rhone Poulenc (USA) and Praxair Specialty Ceramics (Seattle, USA). B2.1

B2.2.2.2 Pressing–shaping techniques

Powders can be shaped using a number of different techniques. Shapes with simple geometries (e.g. discs, rods, tubes, bars) are usually formed by conventional powder pressing techniques, where powder is placed in a suitable die and pressure is applied in order to consolidate the powder. The green density of the powder compact is influenced by the particle size and size distribution of the powder, and by the pressure applied. Typical pressing pressures are ~100 MPa; the use of high pressures can lead to the generation of cracks and defects as the pressure is released. Small amounts (~1% by weight) of polymeric binders (e.g. PVA, PEG or PVB) can be added to the powder during the final stages of preparation in order to improve the strength and handleability of the compact. Variations in density throughout the compact due to the friction of the powder on the walls of the die can arise from uniaxial pressing techniques. Density variations in the green state can lead to non-homogeneous sintering and distortion of the sample. A more homogeneous density distribution can result from isostatic pressing.

More complex shapes can be formed by combining the powder with organic binders and solvents to form pastes which can be shaped using plastic processing techniques such as extrusion and calendering. The use of high shear mixing techniques such as viscous processing [2] can aid the breakdown of agglomerates in the powder resulting in more homogeneous green bodies leading to smaller defect sizes in the sintered sample and improved mechanical properties. Bulk YBCO artefacts made by this technique have the highest reported mechanical strength [2].

D3.3

B2.2.2.3 Burnout and sintering

The organic components in the green body need to be carefully removed prior to any sintering operation. Most binder systems are designed to decompose completely to gaseous products when heated in air at temperatures below 500°C. The precise heating profile will depend on the binder type, the percentage of binder in the component, and the size and shape of the component, but heating rates of 0.5–1°C min^{-1} in the temperature range RT–500°C are typical. A flow of air over the samples and isotherms at temperatures where the decomposition rate is highest can assist the removal of the decomposition products and help prevent any deleterious reaction between decomposition products and the YBCO, which has been observed in the burnout of large components. Once complete removal of the binder has been achieved, or for samples containing no organic components, a faster heating rate to the sintering temperature can be used.

Factors which affect the sintering rate of YBCO include the particle size of the powder used, the sintering temperature and time, and the atmosphere. Figure B2.2.2.1 shows the effect of atmosphere on the densification of calcined YBCO powder, with a faster sintering rate and larger overall shrinkage being observed for materials sintered in air compared to oxygen. Herring's Law [3] (see chapter B2.2.1) is exemplified in the series of microstructures shown in figure B2.2.2.2. In this experiment [3] samples from a calcined YBCO powder were ground for periods from 0–336 h, the longer grinding times resulting in powders with a smaller average particle size. Samples produced from these powders were all subjected to the same sintering conditions (960°C for 12 h in a flowing oxygen atmosphere). Samples produced from the coarser powders (short grinding times) are highly porous and comprise large particles which are poorly sintered, whereas the finer particles resulting from the longer grinding times produce a fine-grained, highly dense product.

B2.2.1

Sintering is normally carried out in a flowing oxygen atmosphere at ambient pressures (typical flow rates 0.1–1.0 l min^{-1}), and oxygen annealing can be carried out as part of the cooling segment. Typical YBCO sintering temperatures are 900–960°C for times of 2–24 h. Although YBCO can be sintered to 100% theoretical density, the diffusion rate of oxygen in dense YBCO is slow (see chapter B2.2.5). It is

B2.2.5

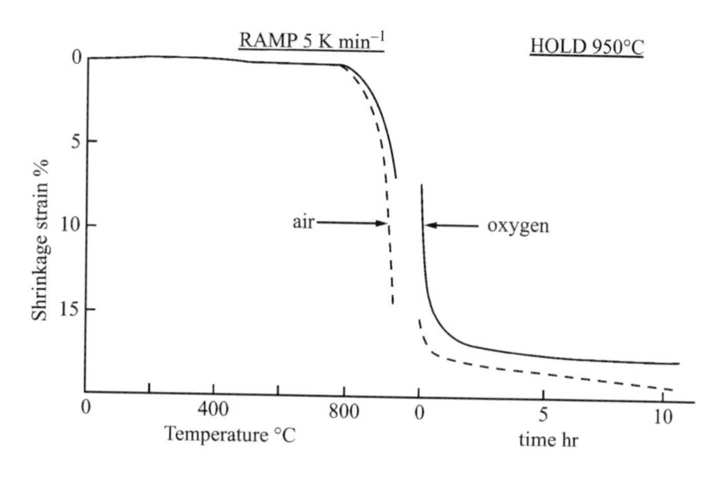

Figure B2.2.2.1. The effect of atmosphere on the sintering of $YBa_2Cu_3O_x$.

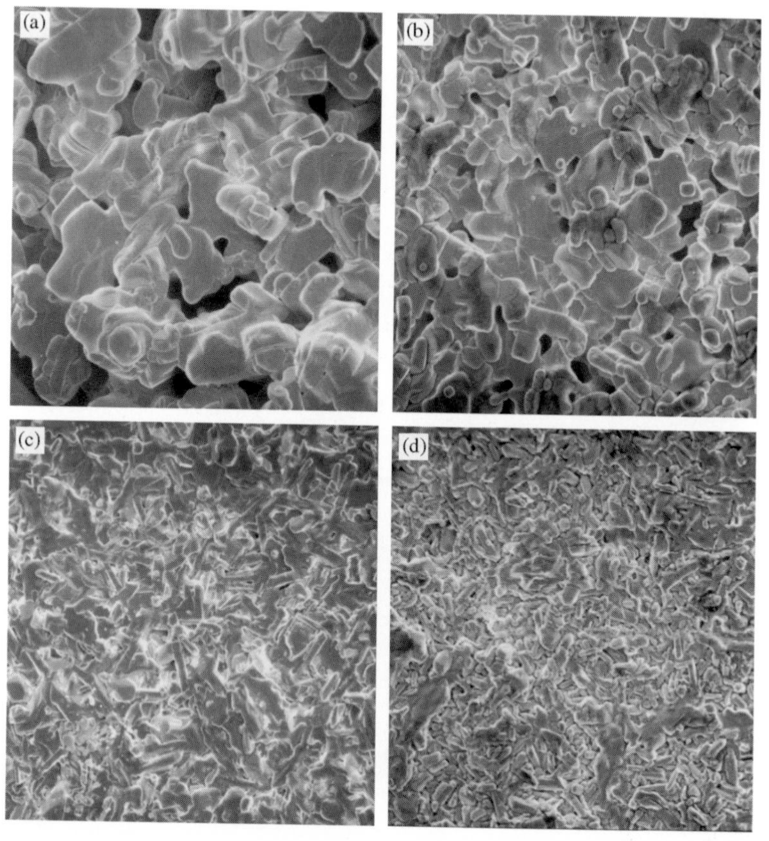

Figure B2.2.2.2. SEM micrographs of sintered $YBa_2Cu_3O_x$ rods showing the effect of powder particle size on the sintered microstructure. Starting powders were calcined and then ground for (*a*) 0, (*b*) 16, (*c*) 168 and (*d*) 336 h respectively. The sintering conditions for all samples were identical: 960°C in oxygen for 12 h (after Alford *et al* [4], figure 5, P 1513).

therefore important to retain an open pore network at the end of the sintering stage in order to ensure that a homogeneous and optimum oxygen content is obtained in the annealing stage. Therefore, sintered densities are usually kept $\leq 90\%$ of theoretical density. This is discussed in more detail below. A typical thermal profile for a complete burnout, sintering and annealing cycle of a YBCO component is shown in figure B2.2.2.3, and a typical microstructure of sintered YBCO is shown in figure B2.2.2.4.

Alternative process routes could be used in which the sintering step is carried out in air. However, the importance of achieving the correct oxygen stoichiometry in the sintered body is paramount, and a subsequent annealing step in an oxygen atmosphere, along the lines outlined above, would be required in

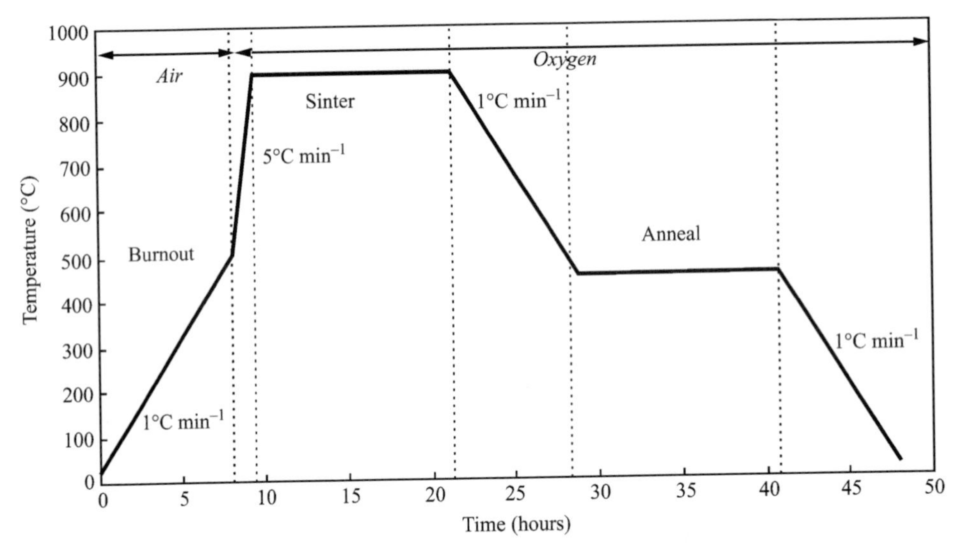

Figure B2.2.2.3. Schematic thermal profile for the burnout, sintering and annealing of bulk YBCO.

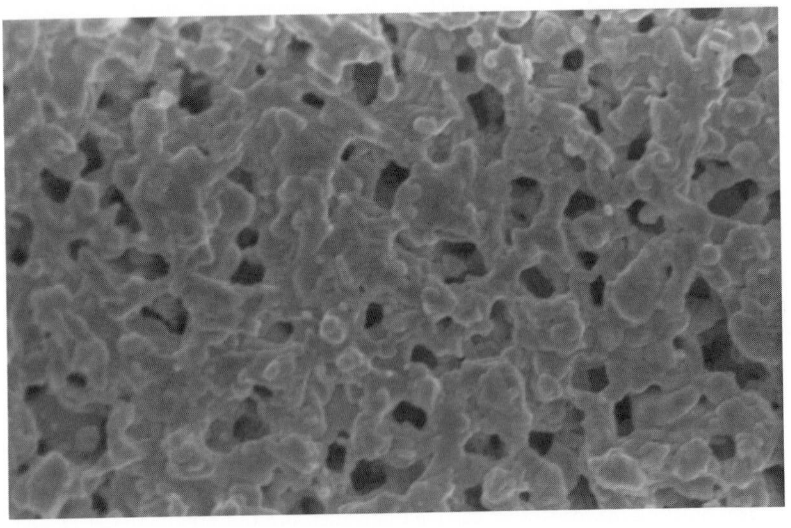

Figure B2.2.2.4. SEM micrograph showing typical microstructure of polycrystalline $YBa_2Cu_3O_x$ sintered to 85% theoretical density.

order to optimize the superconducting properties. Further details regarding oxygen annealing are dealt with in chapter B2.2.5.

B2.2.5

B2.2.2.4 Properties of sintered YBCO

The physical and mechanical properties of sintered $YBa_2Cu_3O_x$ are strongly dependent on the overall fabrication route used and the resulting microstructure which has been achieved. However, table B2.2.2.1 lists some typical properties for both sintered $YBa_2Cu_3O_x$ and Pb-doped BSCCO (2223) materials that will give a guide as to the values which should be attainable using the techniques outlined above. The reader should refer to chapter B2.2.3 for details regarding the fabrication of the BSCCO materials.

B2.2.3

The remainder of this section will outline the effects of some of the key process related variables on the properties of YBCO. These have been exemplified in a series of papers by Alford *et al* [2, 4–7]. Experiments to deliberately vary the density of YBCO rods have revealed the influence of Young's modulus on porosity (see figure B2.2.2.5) [6]. From these data the value of Young's modulus for a sample with zero porosity was estimated as being between 180 and 200 GPa. The high scatter of the data observed for samples with high density was attributed to microcracking arising due to the different thermal expansion coefficients of the three crystallographic axes. This effect can be minimized by preventing grain growth above a critical value which was calculated as 4 μm. In addition, cracking is less likely if the stiffness is less, so for these reasons the optimum strength results are obtained on materials approximately 0.85 of the theoretical density.

The influence of the density (or porosity) on the oxygen stoichiometry and J_c has also been examined in some detail [7]. It was found that the J_c is low at high porosities due to the poor connectivity

Table B2.2.2.1. Typical properties of sintered $YBa_2Cu_3O_x$ and Pb-doped BSCCO (2223) [3, 4]

		$YBa_2Cu_3O_x$	BSCCO
Flexural strength (MPa)		216 ± 16	174 ± 17
Young's modulus (GPa)		141.8 ($\rho = 87\%$) 180–200 ($\rho = 100\%$)	150–216
Fracture toughness, $K1_c$ (Mpam$^{0.5}$)		1.07 ± 0.18	1.3 ± 0.17
Critical flaw size (μm)		15	30
Thermal expansion 50–450°C ($\times 10^{-6}$ K^{-1})		11.5	11.7
Specific heat at 82% theoretical density (J kg^{-1} K^{-1})		431	390
Thermal conductivity (W m^{-1} K^{-1})	300 K	2.67 ($\rho = 85\%$)	0.89 ($\rho = 80\%$)
	77 K	1 ($\rho = 85\%$)	
Resistivity ($\mu\Omega$ cm)	300 K	670	
	100 K	220	
Critical temperature, T_c (K)		92 (width 1 K)	107
Critical current density, J_c, at 77 K, zero field (A cm^{-2})		500–1000 ($\rho = 85\%$)	300 ($\rho = 80\%$)

ρ = Bulk density.

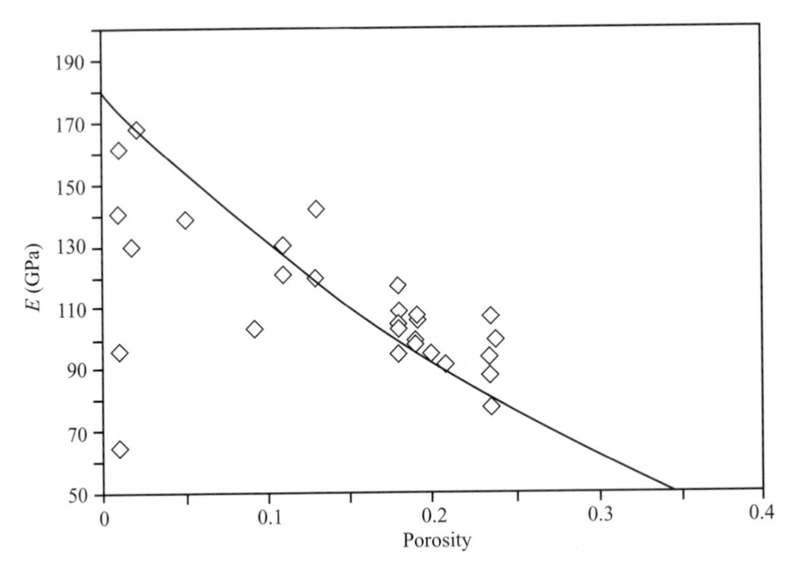

Figure B2.2.2.5. Effect of porosity on the Young's modulus of polycrystalline $YBa_2Cu_3O_x$ rods. Curve represents equation $E = E_0(1 - P)^3$, where P is the pore volume fraction and $E_0 = 180$ GPa (after Alford *et al* [6], figure 5, P 765).

in the samples. Figure B2.2.2.6 shows that, as the porosity decreases, the J_c rises reaching a maximum value of approximately 10^3 A cm^{-2} in 1 mm diameter rods at densities of about 0.9 of theoretical density. Densities in excess of this resulted in a marked decrease in the critical current density and this was attributed to the existence of closed, not interconnected porosity at the higher densities. The result was a reduction in the oxygen content of the denser samples and a corresponding reduction in the J_c was also

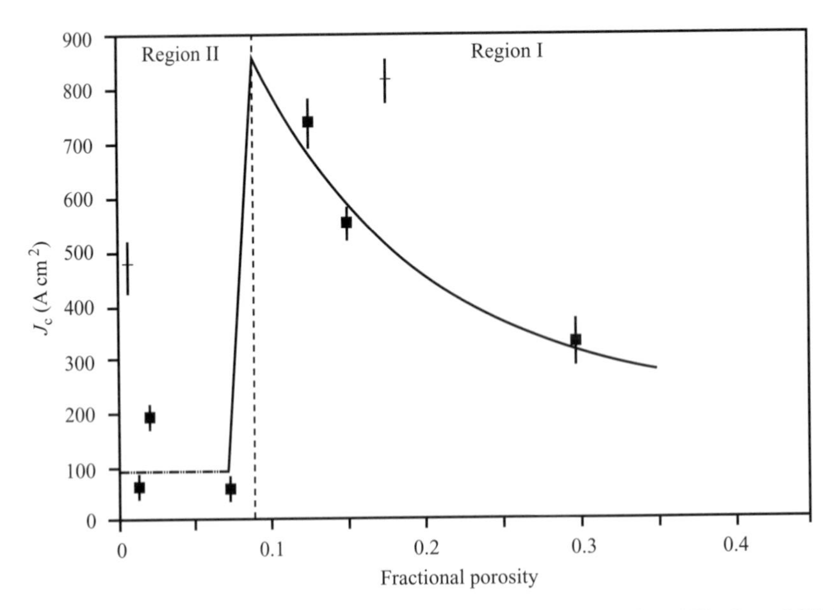

Figure B2.2.2.6. Variation in J_c with porosity for sintered $YBa_2Cu_3O_x$ rods (after Alford *et al* [7], figure 1, P 58).

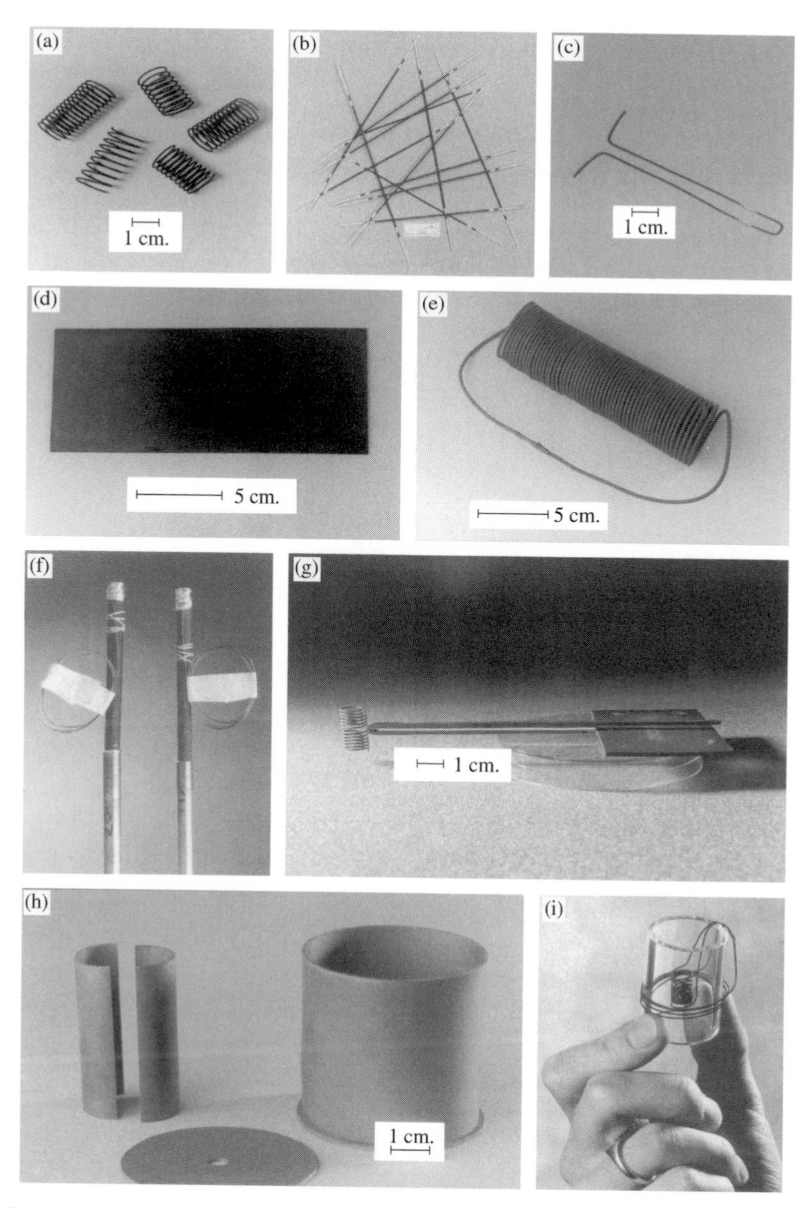

Figure B2.2.2.7. Examples of prototype bulk sintered YBCO artefacts fabricated by ICI Superconductors* 1988–1994. (*a*) helical coils, (*b*) 1 mm diameter rods, (*c*) dipole antenna, (*d*) substrate, (*e*) persistent mode coil, (*f*) current leads, (*g*) helical antenna, (*h*) maser cavity and (*i*) fluxtransformer. *now incorporated within the Functional Materials Group, IRC in Materials, University of Birmingham, UK.

found to be linearly related to the oxygen content. Thus, both from mechanical and superconducting considerations, an optimum density for sintered bulk YBCO materials is around 0.85–0.90 of theoretical density.

A similar study of the factors influencing the radio frequency and microwave properties of sintered bulk YBCO materials has also been carried out [4]. Although thick and thin film materials

B4 have now largely superseded the use of sintered materials in these application areas, a brief discussion of the findings is included here for completeness. In contrast to metallic materials, the surface resistance R_s in sintered YBCO was found not to be directly related to the surface roughness, and was determined more by the microstructure. Experiments were conducted to evaluate the effect of pore volume, grain size, starting powder size and surface area. The influence of T_c, J_c and normal state B2.7 conductivity was also examined. The lowest R_s (77 K) in bulk sintered ceramics was obtained in rods with grain sizes averaging 6–8 μm where R_s was approximately 0.4 mΩ at 500 MHz. Interestingly, even in fully dense rods which displayed immeasurably small J_c values, the R_s was lower than wires which displayed higher J_c but which contained porosity. Thus a fully oxygenated outer skin is sufficient for good rf properties as would be expected, but complete oxygenation throughout the bulk of the material is required for good J_c properties. The reader should refer to the original texts for more detailed information on all these issues. Examples of a range of bulk sintered YBCO artefacts are shown in figure B2.2.2.7.

References

[1] Terpstra R A, Pex P P A C and de Vries A H 1995 *Ceramic Processing* (London: Chapman and Hall)

[2] Alford N McN, Button T W and Birchall J D 1990 Processing, properties and devices in high T_c superconductors *Supercond. Sci. Technol.* **3** 1

[3] Herring C 1950 *J. Appl. Phys.* **1** 301

[4] Alford N McN, Button T W, Peterson G E, Smith P A, Davis L E, Penn S J, Lancaster M J, Wu Z and Gallop J C 1991 Surface resistance of bulk and thick film $YBa_2Cu_3O_x$ *IEEE Magn.* **27** 1510–1518

[5] Alford N McN, Penn S J and Button T W 1997 High temperature superconducting thick films and bulk materials *Functional Materials in New Millennium Systems*, ed M J Kelly (London: The Institute of Materials) p 105

[6] Alford N McN, Birchall J D, Clegg W J, Harmer M A, Kendall K and Jones D H 1988 The physical and mechanical properties of $YBa_2Cu_3O_{9-y}$ superconductors *J. Mater. Sci.* **23** 761

[7] Alford N McN, Clegg W J, Harmer M A, Birchall J D, Kendall K and Jones D H 1988 The effect of density on critical current and oxygen stoichiometry of $YBa_2Cu_3O_x$ superconductors *Nature* **332** 58

B2.2.3
Sintering techniques for BSCCO

J S Abell and T W Button

There are three members of the BSCCO series of compounds of general formula $Bi_2Sr_2Ca_{n-1}Cu_nO_x$ with $n = 1, 2, 3$ ($T_c = 6$, 85 and 110 K, respectively). Much of the technological development of the BSCCO compounds 2212 and 2223 has been concentrated on the fabrication of wires and tapes usually sheathed in silver or silver alloys (see section B3). However, bulk BSCCO materials are of considerable interest in a number of niche potential applications such as current leads, magnetic shielding, fault current limiters. Examples of Bi(Pb)-2223 current leads in tubular form are shown in figure B2.2.3.1.

B3

E1.2, E1.3.3

The choice of compound in the BSCCO system depends on the requirements of the particular application area, but phase stability considerations determine the appropriate processing route for the two high temperature superconducting phases [1]. High density grain aligned artefacts of the 2212 phase are best produced by a melt casting and recrystallization route, whereas bulk 2223 materials are best achieved by a more conventional sintering approach, although it should be recognized that a transient liquid phase is usually involved during the sintering of Bi-2223. This paper will thus confine itself primarily to techniques and issues concerning the sintering of bulk materials of the 2223 phase.

B3.2.1

B2.2.3.1 Precursor powder preparation

Precursor powders of an appropriate composition can be prepared by a number of different routes, as described in chapter B2.1. Both solid state reaction and wet chemical routes can be employed to produce high quality calcined powders of both phases. This is particularly significant for the 2223 phase, where the nature of the precursor powder particles determines the sintering characteristics of the bulk material; it is less so for the 2212 phase where partial melting or melt casting and recrystallization is the preferred route, rendering the powder characteristics less important.

B2.1

The conventional and most convenient method is the ceramic solid state reaction route. This involves multiple mixing and grinding and heating of powders of metal oxides or carbonates. Prolonged heat treatment is necessary to enable solid state diffusion to come to equilibrium. To promote better mixing and enhanced reaction rates, many chemical routes to prepare precursors have been tried, including co-precipitation, sol–gel and spray pyrolysis, to take advantage of the better mixing at the molecular level to produce more homogeneous final material in shorter times and lower temperatures.

C2

B2.1

B2.2.3.2 Pressing techniques/shaping

Several consolidation and shape forming techniques have been employed in the fabrication of 2223 bulk material. These separate into two groups, room temperature-based techniques such as uniaxial pressing, cold isostatic pressing (CIP) and those undertaken at elevated temperatures like hot pressing

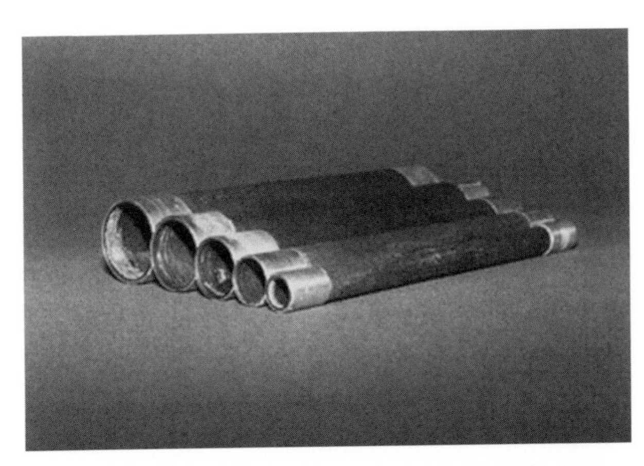

Figure B2.2.3.1. Bi(Pb)-2223 bulk tubes fabricated by cold isostatic pressing and sintering (courtesy of Y. Yamada—figure 13.1 in Bismuth-based high temperature superconductors, ed H Maeda and K Togano, Marcel Dekker Inc. 1996. p 278).

and sinter forging, which will be dealt with separately. Uniaxial compaction is performed in a die with a fixed geometry and therefore provides the essential ingredient of a net shape forming process. The shape and size of the product is fixed in two dimensions, the third being determined by the volume of material in the die and the compaction pressure along the chosen direction. Isostatic pressing provides equal compaction pressure in all directions and is usually achieved by placing the material in a flexible enclosure (rubber, polymer or soft metal bag) and then subjecting this to uniform pressure in a high pressure cell, where the pressure transmitting medium is either a liquid for CIP or a gas (typically argon or nitrogen) for hot isostatic pressing (HIP). Extrusion of a paste containing precursor powders offers an alternative shaping technique, with particular relevance to wires and rods for current leads.

B2.2.3.3 Sintering of the 2201 and 2212 phase

There are few reports of the conventional sintering of the first two members of the BSCCO series of compounds. As is the case with the higher members of the series, the preparation of phase pure 2201 is not straightforward. The phase purity is improved by substitution of Sr by La, which also leads to a composition dependent increase in T_c. This has led to investigation of replacing Sr by other trivalent rare earth ions and increases of T_c up to 33 K have been reported [2]. The optimum sintering temperature varies with rare earth ion but lies in the range 800–850°C.

B3.2.1

The standard fabrication route for this phase is by solid state reaction from high purity Bi_2O_3, $SrCO_3$ and CuO. Appropriate amounts of the starting powders are weighed, mixed, ground and calcined at 750°C for 10 h in air. The resulting powders are reground, pressed and sintered between 800 and 850°C for 35 h followed by slow cooling or quenching.

A very similar approach has been adopted for the preparation of bulk sintered 2212 where this has been reported. Appropriate quantities of the constituent powders are mixed thoroughly and calcined at 780°C for 24 h followed by a further 24 h at 820°C. The ground powder is then pressed into discs and sintered at 840–855°C for extended periods (up to 200 h). Some degree of texture can be introduced into the final product by such a thermo-mechanical treatment, although the best texture and superconducting properties are realized by partial melt processing as discussed in section B2.3.

There is evidence that T_c is enhanced by quenching, with values of 93–97 K for quenched material compared with 50–80 K for slowly cooled samples. This enhancement has been attributed to a decrease in oxygen content with associated reduction in hole concentration to an optimal level for the best superconducting properties, whereas slow cooling results in over-doping. However, this may depend on whether the phase is prepared Pb free or Pb-doped [3]. Pb is found to enhance homogeneity, stability and T_c by partially substituting trivalent Bi ions with divalent Pb ions, thus creating oxygen vacancies and lowering the doping. Typical Pb doping levels give a nominal composition of $Bi_{1.6}Pb_{0.4}CaCu_2O_8$. No significant change in phase purity was observed in these experiments but changes in T_c, degree of texture and microstructure were reported. In the Pb-free samples T_c was always greater in the quenched material and independent of sintering time, whereas for Pb-doped samples the behaviour was time dependent, being identical to the Pb free material for long sinter times (48–144 h), but for short sinters (12–36 h) T_c was higher in the slow cooled samples.

B2.2.3.4 Sintering of the 2223 phase

The kinetics of formation and the formation route of the 2223 phase have been the subject of intense research, particularly in the context of the development of 2223 tape technology; similar arguments apply to the fabrication of bulk materials. Among the important variables are composition and phase assemblage of precursors, calcination conditions, particle size, firing temperature and time, atmosphere, heating/cooling rates and intermediate grinding. It is now generally accepted that partial substitution of Bi by Pb greatly enhances the kinetics of formation of the 2223 phase. The phase relationships in the Bi–Pb–Sr–Ca–Cu–O system are rather complicated due to the large number of component elements. No general agreement about the formation mechanism of the Bi(Pb)-2223 phase has been reached. Several mechanisms have been proposed: (a) an intercalation process in which the pre-existing Bi(Pb)-2212 phase transforms directly into Bi(Pb)-2223 platelets through the insertion of additional Ca–O and Cu–O layers into the structure; (b) a disproportionation of the Bi(Pb)-2212 phase into Bi-2201 and Bi(Pb)-2223; (c) a nucleation and growth process in which the Bi(Pb)-2223 phase is formed by precipitation from a so-called partially melted phase [4–6]. Which mechanism is appropriate seems to depend on the starting powder composition and phase assemblage of the precursor. All the proposed formation mechanisms, except for intercalation, appear to depend on the presence of a liquid phase, but even the intercalation process could be facilitated by faster cation diffusion in the liquid. In Pb-doped material it is generally believed that the calcium plumbate phase plays an important role in the liquid phase, but in Pb-free Bi-2223 the liquid phase is thought to arise from the incongruent melting of Bi-2212 creating a Bi-rich liquid phase from which the Bi-2223 precipitates [7].

B3.2.2

B3.2.2

The role of liquid phases in the phase evolution remains obscure. The 2223 phase can equilibrate with intermediate phases Ca_2CuO_3, $(Sr, Ca)_{14}Cu_{24}O_{41}$, CuO, a Pb-rich phase, 2212 and the liquid. One of the keys to preparing samples with high critical current density is to control the size and distribution of the non-superconducting phases remaining in the superconducting core matrix after thermo-mechanical processing. Manipulation and control of the liquid phase can be crucial in promoting the growth of the 2223 phase and healing cracks caused by the intermediate grinding and consolidation.

Most routes involve the 2212 phase either as a pre-prepared component of a particular phase assemblage, or as an intermediate phase occurring in the course of a prolonged set of heat treatments. A mixture of the oxides and carbonates of the individual cations is usually the starting point for the solid state reaction method. In an extensive systematic investigation of the influence of starting composition, particularly the Bi:Pb ratio, Bernik [7] has identified a single phase region for the 2223 phase as shown in figure B2.2.3.2. Thus, powders of Bi_2O_3, PbO, $SrCO_3$, $CaCO_3$ and CuO in the cationic proportions Bi : Pb : Sr : Ca : Cu = 1.84 : 0.34 : 1.91 : 2.03 : 3.06 have been found to be the most effective initial composition. The Pb substitution has been found necessary to maximize the proportion of the 2223 phase and optimize the superconducting properties. The starting powders are dried at 150°C to constant

B3.2.1

B2.1

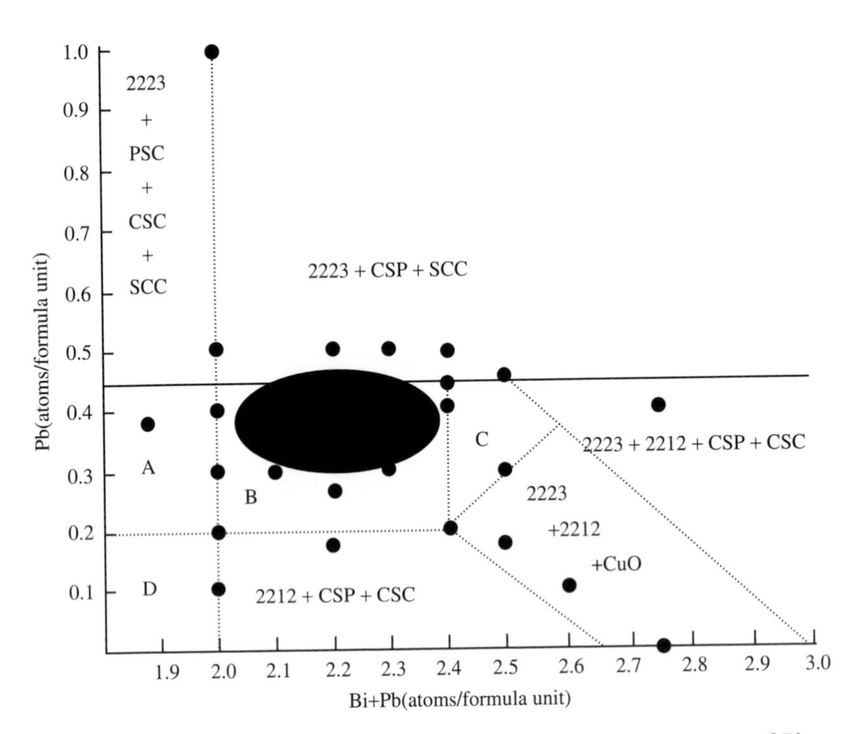

Figure B2.2.3.2. Extent of the 2223 phase field (elliptical shaded region) defined in terms of Pb content and Bi:Pb ratio (after Bernik [7] figure 6).

weight, mixed thoroughly under isopropanol and dried at 40°C. The powders are then pelletized in a stainless steel die under uniaxial pressure of 200 MPa. The pellets are then heat treated in stationary air in a tube furnace in an alumina dish, placing the samples on sacrificial pellets of the same composition to prevent contamination. The sequential heat treatments are performed three times at 800°C for 20 h each time, then at 850–855°C twice for 20 h each time and finally 850–855°C once for 60 h. The samples are air quenched at each stage and re-ground and pelletized. The primary purpose of the particular heat treatment regime adopted in this work was in order to follow the development of the phase formation from the starting components through to the final product. However, the accumulated time involved (160 h) is not unlike the time period reported by many groups, which also employ intermediate grindings to enhance phase conversion. Similar lengthy and multi-stage heat treatments are also reported for Bi-2223/Ag tapes (see chapter B3).

Some groups report the addition of excess PbO to compensate for evaporation losses during these typical extended heat treatments. Such losses may be dependent on sample geometry, indicating that evaporation from the sample surface provides the main contribution to the losses and not diffusion through the solid. The temperature of 2223 phase formation decreases and the range is widened as the oxygen partial pressure decreases. Thus, processing under reduced partial pressure of oxygen (typically $Ar : O_2 = 10 : 1$) can be beneficial to phase purity [8].

B2.1 Appropriate solutions are formed for the chemical routes. Variations on wet chemical routes, including spray pyrolysis and sol–gel have been devised in attempts to reduce the overall sintering time involved in the preparation of phase pure 2223 bulk material. Some of these involve the initial preparation of 2212 and then mixing with chemically prepared powders containing the other cations [9], while others employ a single powder process, which results in different phase assemblages depending on the prevailing atmosphere [10]. Variations on this theme are probably too numerous to mention and the situation is changing all the time as groups develop alternative approaches.

Although the extended heat treatments are necessary for full phase development of the 2223 phase, problems can arise from the associated de-densification which inevitably accompanies these long anneals [11]. Typical changes in the microstructure with sintering time are shown in figure B2.2.3.3. After 2 h at 850°C a fairly dense material is obtained (approximately 82% of theoretical) as seen in figure B.2.2.3.3(*a*). The microstructure is characterized by a random orientation of small platelets ($5–10 \times 0.5\,\mu$m). Sintering for 50 h (figure B2.2.3.3(*b*)) results in a dramatic change to much larger platelets ($15–25\,\mu$m). No significant increase in platelet thickness is observed and so the microstructure is much more open and less dense. The platelets continue to grow with longer heat treatments (figure B.2.2.3.3(*c*)) and the structure remains very open with little evidence of sintering in the conventional sense (i.e. a reduction in surface area and densification) is observed. These observed density changes can be quantified and figure B2.2.3.4 shows the variation in the bulk density with sintering time. The de-densification is due partly to a decrease in mass (loss of Pb by evaporation) but mainly to the substantial increase in volume as the 2223 phase evolves and develops. The corresponding effect this behaviour has on the superconducting properties is depicted in figure B2.2.3.5. For low sinter times the material will not support a transport current. Although the density is optimized, the proportion of the superconducting phase and its intergrain connectivity is not sufficiently well established at this stage. Current density is optimized after 50 h sinter, but longer times lead to reduced values and greater scatter as de-densification develops.

The importance of intermediate pressing at regular intervals during these prolonged heat treatments is clearly demonstrated by these observations. The density is improved, the degree of texture is enhanced by alignment of the platelet grains under pressure and this leads to improved superconducting behaviour through more effective intergrain connectivity. The change in microstructure between a conventionally sintered sample and one subject to intermediate uniaxial pressure is clearly seen in the micrographs in figure B2.2.3.6.

An alternative approach involves the use of organic additives to facilitate extrusion of rods and wires, particularly with reference to the production of current leads [12]. The extrudable paste is prepared by mixing precursor powders with binder, plasticiser, lubricant, dispersant and solvent in appropriate quantities. The extruded rod is dried to remove the solvent and then heated slowly in oxygen to burn out the remaining organics. The rod is then sintered at 830–850°C for over 100 h, usually in

E1.2

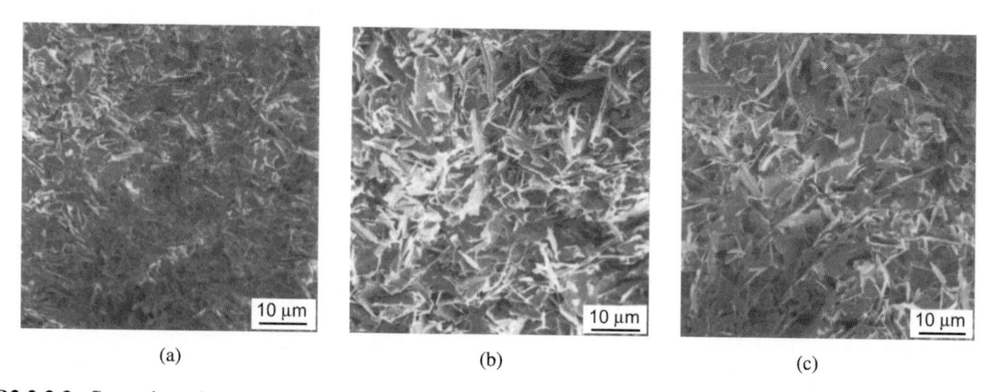

Figure B2.2.3.3. Scanning electron micrographs of the fracture surfaces of Bi(Pb)-2223 sintered in air at 850°C for (*a*) 2 h, (*b*) 59 h and (*c*) 116 h (after Button *et al* [11] figure 5).

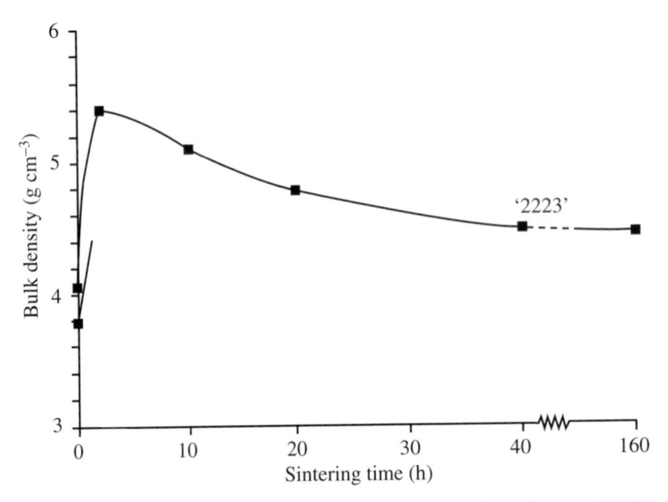

Figure B2.2.3.4. Variation in bulk density with sintering time at 850°C for Bi(Pb)-2223.

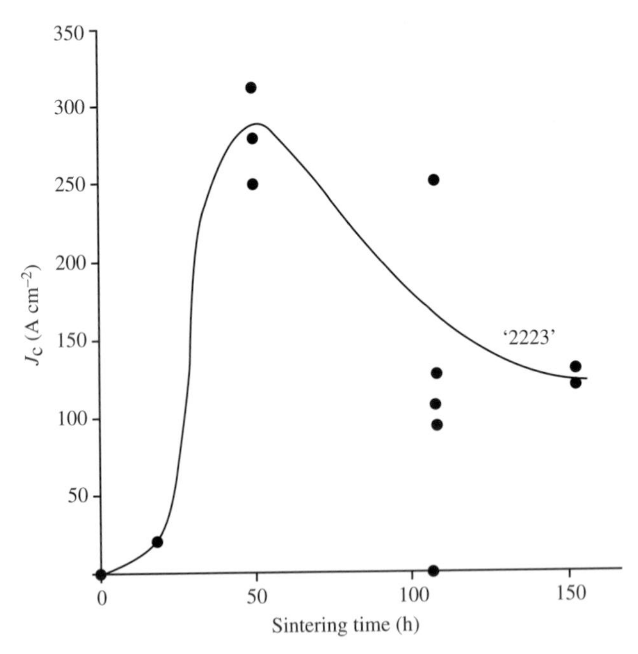

Figure B2.2.3.5. Variation in critical current density, J_c, with sintering time at 850°C for Bi(Pb)-2223.

a reduced partial pressure of oxygen to form the 2223 phase. The associated de-densification in this case can be remedied by cold isostatic pressing of the low density rod, a process that enhances the mechanical alignment of the 2223 platelet grains. The pressed rod is subsequently sintered at ~840°C for 72 h in the same partial pressure of oxygen to give the final product. Critical current densities up to 10^3 A cm^{-2} have been obtained using this method [12].

C2

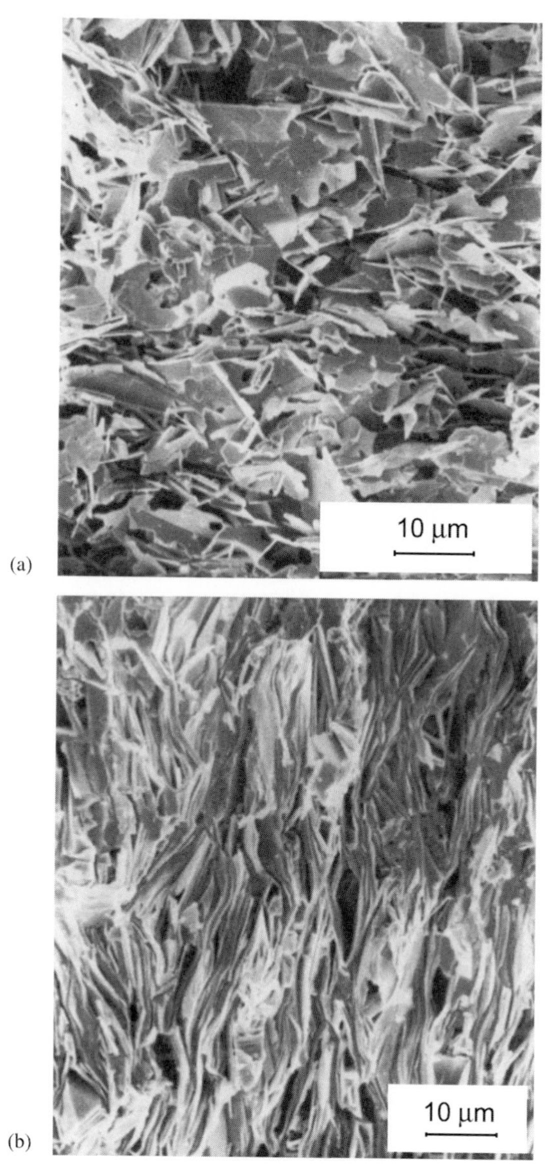

Figure B2.2.3.6. SEM micrographs of typical fracture surfaces of (*a*) sintered and (*b*) pressed and sintered Bi(Pb)-2223 (after Wellhofer *et al* [12] figure 4).

B2.2.3.5 Hot pressing/sinter forging

As already discussed, most of the conventional sintering routes involve an extended period (>100 h) of heat treatment, either continuous or accumulative. Such lengthy anneals are not industrially attractive. One way of shortening the heat treatment time is to undertake the compaction at elevated temperatures by hot pressing or sinter forging. These methods also allow stoichiometric charges to be used which eliminates the problem of the slow reaction kinetics for the formation of the 2223 phase from a mixed assemblage of intermediate compounds. The distinction of sinter forging is that the uniaxial pressure is

B3.2.2

applied at a controlled rate and without lateral constraint allowing some movement and rotation of the grains. Some mechanical texturing is the usual bonus of this technique. Again recipes vary but the following is a typical experimental treatment [13].

Powders of nominal composition given above were prepared by solid state reaction of commercial powders of Bi_2O_3, PbO, $SrCO_3$, $CaCO_3$ and CuO powders. The powders are thoroughly mixed or ground in alcohol, dried and calcined in air in steps with intermediate grinding. A typical calcining schedule would be 750°C for 5 h, 800°C for 5 h, 850°C for 20 h and 850°C for 50 h. The fraction of 2223 at this stage is ~95%. This powder is compacted into bars by uniaxial pressing in a steel die resulting in 50% dense bars. The bars are then placed between MgO or Al_2O_3 plates (with or without silver foil to reduce reaction) and sinter forged in air at 830–860°C up to 2–10 MPa for 5–20 h. Final bars are 85–95% dense and highly textured with *c*-axes parallel to the forging direction. $\boldsymbol{J}_c$s of $\sim 10^4\,\mathrm{A\,cm^{-2}}$ have been measured on such samples. Some reaction can occur between the sample and the plates but can be polished off.

B2.2.3.6 Microstructure/superconducting properties

In the superconducting state the electrical and magnetic properties depend heavily on the sample microstructure and therefore on the efficacy of the thermo-mechanical treatment, particularly with respect to the quantity and distribution of non-superconducting second phases but also in relation to the control of intergrowths of the $n = 2$ phase and the quality of the grain interconnections. Measurement of the dc transport current will provide a definitive critical current density but useful information on the performance of bulk sintered material can be obtained from ac susceptibility measurements. The real and imaginary parts of the diamagnetic signal can be interpreted in terms of inter and intra-granular current flow and their dependence on magnetic field provides essential data on the quality of inter grain connectivity and the efficacy of the thermo-mechanical treatment (see figure B2.2.1.3 of chapter B2.2.1). Some typical properties of sintered BSCCO 2223 are given in table B2.2.2.1 of chapter B2.2.2.

B2.2.1
B2.2.2

References

[1] Majewski P J 1996 *Bismuth Based High Temperature Superconductors*, eds H Maeda and K Togano (New York: Dekker) p 129

[2] Lan Y C, Che G C, Jia S L, Wu F, Dong C, Chen H and Zhao Z X 1996 The effects of composition, synthesis conditions, oxygen content and F doping on superconductivity and structure for R-substituted Bi-2201 *Supercond. Sci. Technol.* **9** 297

[3] Padam G K, Ekbote S N, Suri D K, Gogia B, Ravat K B and Das B K 1997 A comparative study of rapid and slow furnace cooling effects on the superconducting properties of Pb-free and Pb-doped Bi-2212 HTSC *Physica* C **277** 43

[4] Hu Q Y, Liu H K and Dou S X 1995 *Physica* C **250** 7

[5] Grivel J-C and Flukiger R 1996 *Supercond. Sci. Technol.* **9** 555
Grivel J-C and Flukiger R 1998 *Supercond. Sci. Technol.* **11** 288

[6] Morgan P E D, Housley R M, Porter J R and Ratto J J 1991 *Physica* C **176** 279

[7] Bernik S 1997 Synthesis and phase relations of the 110 K superconducting 2223 phase in the Bi–Sr–Ca–Cu–O system *Supercond. Sci. Technol.* **10** 671

[8] Hudakova N, Plechacek V, Dordor P, Plachbart K, Knizek K, Kovac J and Reiffers M 1995 *Supercond. Sci. Technol.* **8** 324

[9] Kashimura T, Isobe T, Senna M, Itoh M and Koizumi T 1996 Reactive sintering for oriented and connected Bi-2223 superconductive oxides from a mixture of delaminated Bi2212 and fine powdered supplemental ingredients *Physica* C **270** 297

[10] Mao C, Zhou L, Wu X and Sun X 1998 Rapid one-powder process to synthesise phase assemblage composed of Bi-2212, Ca2CuO3 and CuO *Physica* C **303** 28

[11] Button T W, Alford N Mc N, Birchall D, Wellhofer F, Gough C E and O'Connor D A 1989 A comparison of Pb-doped BSCCO superconductors at the 2223 and 2234 compositions *Supercond. Sci. Technol.* **2** 224

[12] Wellhofer F, Gough C E, O'Connor D A, Button T W and Alford N Mc N 1990 The effect of Pb-doping and grain alignment on BSCCO superconductors *Supercond. Sci. Technol.* **3** 611

[13] Murayama N and Shin W 1999 Decomposition of Bi2223 phase during sinter forging *Physica* C **312** 255

B2.2.4
Sintering techniques for other HTC compounds

J S Abell and T W Button

There are many series of superconducting oxides, which could be discussed in this chapter. However, we shall confine our attention to the thallium and mercury based cuprates, because over and above reports of their successful synthesis, considerable research has been devoted to the fabrication of particular device configurations of these compounds. Their enhanced critical temperatures provide the principal motivation for their technological exploitation. Both these series are characterized by volatility at elevated temperatures and the associated toxicity of the Tl and Hg bearing constituents, so care should be exercised in the synthesis of these compounds, which should only be undertaken with the appropriate equipment and safeguards.

C3, C4

B2.2.4.1 Fabrication of the mercurocuprates

There are seven members of the mercurocuprate family of general formula $HgBa_2Ca_{n-1}Cu_nO_{2n+2+\partial}$. The first three ($n = 1, 2, 3$) have received the most attention since their discovery in 1993, largely because they can be produced by ambient pressure routes and not least due to their high critical temperatures (94, 128 and 134 K, respectively). The phases with $n > 3$ require high pressures for their synthesis, with the exception of one report of the production of the $n = 4$ phase by encapsulation. The highest T_c 1223 phase exhibits very respectable critical current densities, even above 77 K, but perhaps of more importance for applications are the significant irreversibility fields (>0.5 T) which both this phase and the $n = 2$ phase show even above 100 K. The fabrication of the mercurocuprates in technologically useful forms remains the main challenge to their successful exploitation.

C4

The synthesis and preparation of bulk material of this series of compounds Hg-12($n-1$)n is extremely complex and still poorly understood. It is therefore difficult to provide a general recipe for each phase; each research group which has attempted synthesis has developed its own strategy. The formation of the superconducting phases depends critically on the synthesis conditions, which include the Hg vapour pressure, the partial pressure of oxygen, the reaction temperature and the atmospheric environment. Much of the work described in the literature is primarily designed to achieve single phase material and not necessarily to produce high density bulk material for particular applications requirements. The practical synthesis of the mercurocuprate compounds is dominated by their toxic nature, particularly at elevated temperatures. One of the key issues in the synthesis of the Hg-12($n-1$)n series is whether the phases form by a solid/solid diffusion or a vapour/solid reaction. The decomposition of mercuric oxide at temperatures (~500°C) below those generally employed for solid state reaction of the mercurocuprate phases suggests a vapour/solid reaction, but in any case necessitates complete containment of the reaction mixtures.

This is usually achieved by one of two routes, the application of high pressure to reduce decomposition or the encapsulation of the reaction mixtures within a sealed silica reaction tube. The high pressure route is used almost exclusively for the preparation of the $n > 3$ members of the family, whereas the encapsulation method is generally used for the phases with $n < 3$. The standard encapsulation route inevitably results in a somewhat porous microstructure due to the generation of Hg vapour and gaseous oxygen from the dissociation of HgO within the material. Modifications to this approach have been devised to alleviate this disadvantage. High pressure routes promote densification but have other drawbacks, mainly related to the non-equilibrium nature of the process.

The choice of starting reaction mixtures for both low and high pressure routes presents two different, broad approaches. One involves the preparation of a precursor compound followed by the mercuration of the precursor in a two-step route, and the other is a single step reaction of the constituent monoxides in suitable stoichiometric proportions [1]. In the precursor route, it is vitally important to avoid or minimize subsequent exposure to air due to the high reactivity of the precursors to carbon dioxide and water vapour. All operations should therefore be carried out in a glove box environment. A precursor of general formula $Ba_2Ca_{n-1}Cu_nO_x$ can be prepared by calcination of a mixture of oxides or nitrates at temperatures in excess of 920°C in flowing oxygen for several days. The precursor is then mixed in a glove box with an appropriate amount of mercuric oxide and pelletized. The pellet is wrapped in gold foil to prevent reaction with the walls of the silica tube in which it is then sealed, usually under vacuum. The silica ampoule is then sealed in a steel container by welding or screwed end caps, designed to contain any explosion which may occur. The whole assembly is then heated in a furnace to between 650 and 900°C (depending on the end product) for 1–20 h. Heating and cooling rates can also be important variables in obtaining phase purity.

B2.2.4.2 Fabrication of Hg-1201

B3.2.5 Preparation of the first member of the series is generally regarded as more straightforward than the higher, calcium-containing members. Although there are reports of high pressure synthesis of this phase, the precursor method is most widely used, the precursor compound in question being $Ba_2CuO_{3+\partial}$ [1]. This phase has an affinity for CO_2 and so the mixing with mercuric oxide should be undertaken in a glove box in inert atmosphere. In the first report of the Hg-1201 phase [2], the powder mixture was heated to 800°C and slowly cooled to give a T_c of 94 K. A similar route [3] gave a T_c of 84 K, which was improved to 95 K by an oxygen anneal at 500°C for 24 h. Most reports recommend a firing time in excess of 5 h. A single step route from the monoxides by firing at 700°C for 1 h was also successful in producing the correct phase, but subsequent argon anneal was necessary to reduce the overdoping and optimize T_c [4]. Reports that improvements in phase purity can be achieved by rapid heating to the reaction temperature and by quenching [1] should be noted. It is thought that such action avoids the preferential formation temperature range of the major impurity phase $BaHgO_2$.

Other impurity phases observed to occur during the preparation of the 1201 phase are the various BaO–CuO phases, $BaCuO_2$, $Ba_2CuO_{3+\partial}$ and $Ba_2Cu_3O_{5+\partial}$. The relative stabilities of these phases are dependent on the partial pressure of oxygen during the reaction. It is claimed that single phase 1201 can only be obtained with a partial pressure of oxygen between 0.012 and 0.15 bar [5]. Control of the prevailing environment is clearly of supreme importance in achieving single phase material.

A typical two-step method would be as follows. The precursor Ba_2CuO_3 is prepared by calcining appropriate amounts of intimately mixed BaO and CuO powders in air or oxygen (flowing or static) at 930°C for 24 h with intermediate grindings. Some groups use $BaNO_3$ or a sol–gel route to the precursor. The precursor should be handled in a glove box environment and then mixed stoichiometrically with HgO, ground and pressed into pellets. The pellets are sealed in quartz ampoules under vacuum ($\sim 10^{-2}$ to 10^{-6} Torr), placed in a preheated furnace at 800–845°C and fired for 8–15 h, followed by a quench. Rapid heating and cooling appear to aid phase purity [6]. Although most groups use evacuated ampoules, rapid synthesis at ambient pressure in an open system has been reported [7]. The precursor was ground with HgO, pelletized and placed in a silica tube with one end sealed and introduced into a preheated furnace between 800–900°C and heated for 45–90 seconds and quenched. The material was 'nearly phase pure' with a small trace of $BaHgO_2$.

B2.2.4.3 Fabrication of Hg1212 and Hg1223

The Ca bearing members of the family appear to pose more preparation problems than the first family member. Phase formation has been found to depend critically on several key reaction parameters. These two members are considered together because similar approaches are adopted for the two, and experience indicates that they can often form together and co-exist in bulk materials. In particular, 1212 can readily appear as an impurity phase in partially successful attempts to produce single phase 1223.

While high pressure routes have shown considerable promise for these compounds, conventional encapsulation routes have also been successful. Considerable effort has been invested in the preparation of the most suitable precursor for the successful synthesis of these compounds. The use of nitrates to form the precursor has proved more successful than oxides and carbonates in the standard solid state reaction route, as demonstrated by the efficacy of the final mercuration synthesis. The intimate mixing achieved by solution-based and sol–gel methods has also proved beneficial.

A typical two step processing route for 1212 and 1223 would be as follows. A precursor material with a nominal composition of $Ba_2CaCu_2O_5$ (for $n = 2$) or $Ba_2Ca_2Cu_3O_x$ (for $n = 3$) is prepared by forming an intimate mixture of appropriate quantities of $BaCO_3$, CuO and CaO, pressing the powder into pellets and calcining in oxygen for 24–48 h at 900–920°C to decompose the $BaCO_3$. Sometimes this is repeated to promote homogeneity. Pelletisation is used to reduce the surface to volume ratio of the precursor to minimize subsequent degradation on exposure to air between synthesis steps. Greater phase purity and homogeneity have been achieved by using nitrates as starting constituents. Appropriate amounts of cation nitrates are dissolved in de-ionised water and heated on a hot plate with continuous magnetic stirring. The solution is slowly dried, with the evolution of brown NO_2 fumes, to a blue powder. The powder is ground and heated in an alumina crucible to over 500°C (M.Pt ~540°C) for 30 min, and to over 600°C (B.Pt ~630°C) for 1 h until all the NO_2 is removed. The resulting black mixture is then ground, compacted and sintered in flowing oxygen at 900°C for 24–48 h.

B2.1

The precursor is then mixed (preferably in a glove box to minimize exposure to air) with HgO powder and compacted into pellets, which are sealed in an evacuated silica ampoule. This is sealed into a stainless steel tube, primarily for safety reasons. This assembly is placed into a tube furnace and heated slowly at 150–250°C h^{-1} to 800–850°C, held for 6 h and slow cooled. An alternative single step route to Hg 1223 has been described [8]. The monoxides are ground in appropriate molar amounts in a glove box and pressed into pellets, wrapped in silver foil and sealed in air into silica tubes. The tubes were placed in steel bombs and heated at 700°C h^{-1} up to 750–800°C, held for 5–10 h and furnace cooled. A subsequent post anneal at 300°C in oxygen for 15 h was necessary to optimize T_c.

High pressure routes have been successfully applied to the preparation of high phase purity samples of these two phases [1]. The use of high temperatures (in excess of 900°C) and short reaction times (~1 h) are the main advantages of this approach. High pressure regimes can act to suppress decomposition, particularly of Hg containing constituents, and to promote densification and grain connectivity. Typical pressures are 10–50 kbar. Conditions in the pressure cell seem to favour the formation and stability of the mercurocuprates at the expense of impurity phases. Most reports of high pressure experiments have used precursor compounds as opposed to a single step monoxide approach. Although successful, high pressure techniques do not lend themselves to the fabrication of bulk materials for technological exploitation. However, this approach confirms the importance of the partial pressure of Hg in the reaction vessel to the production of single phase material. This has led to modification of the encapsulation route designed to reduce the volume within the vessel by space filling and, thereby, effectively increasing the prevailing pressure during the reaction.

One of the problems of the two step encapsulation route is that the rapid release of Hg from HgO at ~500°C results in the formation of $CaHgO_2$ before the reaction to form Hg-12($n-1$)n occurs at ~800°C. An alternative modification of the encapsulation technique is variously described as the controlled vapour–solid reaction technique [9], or the Hg diffusion process [10], or the internal versus external Hg source approach [11]. Here, the decomposition of HgO is actively employed as the mercuration route, with the mercury vapour supplied by another source diffusing into the precursor to produce the correct phase. Two samples, one a mercury-free precursor and the other the precursor plus an excess of HgO are sealed together in the silica tube. Using this route the 1212 can be formed, at temperatures around 700°C, and 1223 at the higher temperature of ~800°C. This approach has the distinct advantage of overcoming the problem of porosity generated in the solid state reaction route arising from the decomposition of the HgO to produce Hg and O_2 within the sample during the reaction.

A typical example of the diffusion route from an external source for Hg-1223 is given. The Hg-free precursor and the Hg-containing reactant material are prepared from the same precursor powder. Starting powders of BaO, CaO and CuO are mixed in the molar ratio 2:2:3 and ground and calcined in an alumina crucible in air at 900°C for 26 h. To prepare the precursor, the porous calcined material is ground and pressed into pellets and sintered at 900°C for 26 h in air. These pellets can then be further ground and hydrostatically pressed (~150 MPa) into bars. The reactant bars are also prepared from the calcined material which is ground, pressed into pellets and sintered at 900°C for 26 h in air. The pellets are then reground immediately and mixed with HgO powder according to the composition $Hg_{1.4}Ba_2Ca_2Cu_3O_{8+\partial}$. The excess Hg provides sufficient Hg vapour to diffuse into the precursor bar. The reactant and the precursor (mass ratio 3:1) are sealed in an evacuated silica tube and then in a stainless steel tube. The assembly is placed in a tube furnace, heated to 910°C in 50 min, held for 3 h and furnace cooled. The samples are subsequently oxygenated at 300°C in flowing oxygen for 10 h to remove any residual $HgCaO_2$ impurity phase.

Control of the Hg vapour pressure appears crucial for the successful formation of the Hg12($n-1$) phases. Various ways of addressing this issue have been reported. Reactant pellets of varying mass can be sealed in quartz tubes of fixed volume or, conversely, reactant pellets of fixed mass within tubes of varying volume. Alternatively, precursor pellets and reactant pellets with varying mass ratio (typically 1:3) can be sealed in a fixed volume silica tube. A more recent variation of these methods have been the reports that additives to the reactant can serve to control the release of and therefore the vapour pressure of Hg in the ampoule. For example, the addition of ~1 wt% Se to the reactant pellet can affect the Hg vapour pressure and thereby promote the formation of both 1212 and 1223 [10].

B2.2.4.4 Fabrication of HgBCCO for $n > 3$

High pressure routes are used almost exclusively for the higher members of the series. However, these methods involve essentially non-equilibrium processes and can result in intergrowths and phase mixing. Few laboratories have access to such equipment and the means of applying the pressure can vary considerably. A general description of a route is therefore not appropriate and the reader is referred to the particular literature of these techniques [12–14]. Hot isostatic pressing has been used to prepare the 1223 phase using either nitrates or carbonates to form the precursor. The HIP reaction conditions (pressures, temperatures, times) are significantly different from those used in the quartz ampoule method. Pressures in the range 100–160 MPa at temperatures up to 900°C for times of ~30 min can give satisfactory results [15]. A piston cylinder geometry has also been applied to prepare samples for $n = 3, 4$ and 5. Here a pressure of 2 GPa was typically employed with a synthesis time of 2–4 h. By varying the composition and temperature, all the members of the family could be prepared, from 800–860°C for $n = 1$ up to 1000–1040°C for $n = 6$.

The exceptions to the use of high pressure for the synthesis of these higher mercurocuprates are few. The addition of Re, which assists the formation of the $n = 3$ compound, has a similar effect on the $n = 4$ phase allowing preparation under low pressure conditions [1]. There is one report [16] of the fabrication of the unsubstituted $n = 4$ phase by encapsulation and the controlled vapour–solid reaction route. Prolonged firing (100 h) at 800°C of pellets of nominal cation compositions 0223 and 1223 in an evacuated silica ampoule resulted in a domination of the $n = 4$ phase.

B2.2.4.5 Thallium-based cuprates

Thallium cuprates comprise two homologous series based on a single Tl–O layer and a double Tl–O layer. The single layer family are of general formula $TlBa_2Ca_{(n-1)}Cu_nO_x$ or $12(n-1)n$ and the double layer series are $Tl_2Ba_2Ca_{(n-1)}Cu_nO_x$ or $22(n-1)n$. The maximum T_c is found for the single layer compound with $n = 3$ at 128 K. The problems encountered in the synthesis of the thallium cuprates are almost identical to those of the mercurocuprates, particularly in terms of toxicity and volatility at elevated temperatures, and this is reflected in the similarity of approaches adopted to synthesize these compounds. Indeed, substitution of Tl in the mercurocuprates and vice versa demonstrates the similarities in synthesis procedures for the two families [17]. Thus, the majority of methods are based on the preparation of a Tl-free precursor followed by an exposure to Tl vapour at elevated temperatures and ambient pressures to allow diffusion of the Tl into the precursor to produce the desired compound. Alternatively, one-step high pressure routes have also been successfully employed, although these seem to be aimed primarily at the two-layer family.

B3.2.3

Much of the literature is concerned with the use of substitutions designed to stabilize the individual phases in order to achieve single phase materials. Thus, both Pb and Bi have been substituted in this manner, particularly for the members of the single layer series 1212 and 1223 [18, 19]. The likelihood of a change in hole carrier density as a result of substitution by a different valence cation should be taken into account when adopting this approach. This may be adjusted subsequently by appropriate post annealing.

The most common synthesis route used is very similar to that employed for the mercurocuprates. A Tl-free precursor is prepared by solid state reaction from appropriate amounts of BaO_2 ($BaCO_3$), CaO($CaCO_3$) and CuO. This mixture is usually calcined at ~850–950°C under 1 bar oxygen for several hours. The calcined precursor is ground and mixed with Tl_2O_3 and Pb_2O_3, pelletized and sealed under oxygen in a silica tube, usually wrapped in Au foil to minimize reaction with the walls of the silica ampoule. The whole assembly is then heat treated at ~960°C for a few hours. Typical oxygen pressure would be such that the pressure at 960°C was 1 bar. A similar method, but using Ni foil, has been used for the 2223 compound [20]. Pelletizing and pressing at each stage is beneficial and can introduce some degree of texturing in the final product [21].

The high pressure route [22] employed for the double layer series typically involves a sintering stage at 1050–1100°C for 1 h or so under 5–6 GPa, and thus has two advantages: a single step route and a short time scale, but requires special equipment not available in most laboratories.

B2.2.4.6 Cutting and shaping

The brittle nature of the superconducting oxides at room temperature means that care must be taken when cutting and shaping sintered products. Net shape forming is the preferred option where possible. A variant of this approach is green shaping, where the bulk material is cut or shaped in the green state into the shape required when the artefact is relatively soft. Due allowance for shrinkage on subsequent sintering has to be made, a process which is not always easy to control reproducibly.

D3.3

Where post-production shaping is required the use of ultrasonic drilling, slow speed diamond wheels or diamond impregnated wire slicing are the best methods. An alternative approach is grinding, but the same precautions apply. Centreless grinding of misshapen rods could be required and some limited use of a lathe to perform internal grinding of artefacts may be necessary.

Many of the superconducting oxides are reactive to water leading to corrosion and degradation of the bulk artefact and deterioration of superconducting properties. Thus, non-aqueous lubricants are preferred in any cutting or shaping process. Light paraffin oils or high molecular weight alcohols are recommended. There are also many proprietary preparations which can be used. These lubricate and remove heat from the cutting and can subsequently be evaporated off or removed by solvents such as acetone.

Spark erosion can be useful in producing more complicated shapes which would be difficult or impossible by mechanical means. The spark machine tool can be fabricated in the shape required from brass or stainless steel.

References

[1] Peacock G B, Zhou W and Edwards P P 1997 Crystal chemistry of the mercurocuprates *Studies of High Temperature Superconductors* vol 25 ed A Narlikar (New York: Nova) p 185

[2] Putulin S N, Antipov E V, Chmaissem O and Marezio M 1993 *Nature* **362** 226

[3] Wagner J L, Radaelli P D, Hinks D G, Jorgensen J D, Mitchell J F, Dabrowski B, Knapp G S and Bend M A 1993 *Physica* C **210** 447

[4] Itoh M, Tokiwa-Yamamoto A, Adachi S and Yamauchi H 1993 *Physica* C **212** 271

[5] Alyoshin V A, Mikhailova D A and Antipov E V 1995 *Physica* C **255** 173

[6] Asab A, Gameson I and Edwards P P 1995 *Physica* C **255** 180

[7] Sheng Z Z, Li Y F and Pederson D O 1995 *Solid State Comm.* **95** 277

[8] Peacock G B, Gameson I, Slaski M, Zhou W, Cooper J R and Edwards P P 1995 *Adv. Mater.* **7** 925

[9] Antipov E V, Loureiro S M, Chaillout C, Capponi J J, Bordet P, Tholence J L, Putilin S N and Marezio M 1993 *Physica* C **215** 1

[10] Li J Q, Lan C C, Hung K C and Shen L J 1998 *Physica* C **304** 133

[11] Amm K M, Sastry P V P S S, Knoll D C, Peterson S C and Schwarz J 1998 *Supercond. Sci. Technol.* **11** 793

[12] Scott B A, Suard E Y, Tsuei C C, Mitzi D B, McGuire T R, Chen B-H and Walker D 1994 *Physica* C **230** 239

[13] van Tendeloo G, Chaillout C, Capponi J J, Marezio M and Antipov E V 1994 *Physica* C **223** 19

[14] Louriero S M, Antipov E Y, Alexandre E T, Kopnin E, Gorius M F, Souletie B, Perroux M, Argoud R, Gheorghe O, Tholence J G and Capponi J J 1994 *Physica* C **235–240** 905

[15] Lechter W, Toth L, Osofsky M, Skelton E, Soulen R J, Qadri S, Schwartz J, Kessler J and Wolters C 1995 *Physica* C **249** 213

[16] Usami R, Adachi S, Itoh M, Tatsuki T, Tokiwa-Yamamoto A and Tanabe K 1996 *Physica* C **262** 21

[17] Wu X J, Tokiwa-Yamamoto A, Tatsuki T, Adachi S and Tanabe K 1999 *Physica* C **314** 219

[18] Lebbou K, Trosset S, Abraham R, Cohen-Adad M, Ciszek M and Liang W Y 1998 *Physica* C **304** 21

[19] Konig W and Gritzner G 1998 *Physica* C **294** 225

[20] Piskunov Y V, Mikhalev K N, Zhdanov Y I, Gerashenko A P, Verkovskii S V, Okulova K A, Medvedev U Y, Yakubovskii A Y, Shustov L D, Bellot P V and Trokiner A 1998 *Physica* C **300** 225

[21] Konig W T and Gritzner G 1998 *Physica* C **294** 225

[22] Wu X J, Tatsuki T, Adachi S and Tanabe K 1998 *Physica* C **301** 39

B2.2.5
Diffusion process for the synthesis of high-T_c superconductors

Kyoji Tachikawa and Yutaka Yamada

B2.2.5.1 Introduction

Originally, the diffusion process was adapted to the successful fabrication of A15 superconducting compounds, such as Nb_3Sn and V_3Ga [1]. It usually proceeds in a composite, composed of a high melting point component and a low melting point component, which results in the formation of new phases stable at the reaction temperature. The high-T_c superconducting oxides with denser and more homogeneous structure can be synthesized by the diffusion process in a shorter reaction time than by a conventional sintering process. The authors have applied the diffusion process for the synthesis of $YBa_2Cu_3O_{7-x}$(Y-123), $Bi_2Sr_2CaCu_2O_{8+x}$(Bi-2212) and $TlBa_2Ca_2Cu_3O_y$(Tl-1223) high-T_c superconductors. Moreover, this process has been applied for the preparation of Y-123 magnetic shields on large diameter metal tubes, Bi-2212 current leads for superconducting magnets and Tl-1223 tapes with improved I_c–B performance at 77 K.

C1, C2, C3

B2.2.5.2 YBCO and RE(Ho, Dy, Gd, Sm, Nd)BCO systems

B2.2.5.2.1 YBCO system

The principle for preparing diffusion composites is schematically illustrated in figure B2.2.6.1. In the Y–Ba–Cu–O system, the $YBa_2Cu_3O_{7-x}$(Y-123) superconducting layer, about 100 μm in thickness, is synthesized between the Y_2BaCuO_5(Y-211) substrate and the $Ba_3Cu_5O_8$(Y-035) coating layer after the reaction at 930°C for 10 h [2]. The Y-123 layer formed by the reaction is composed of columnar and random crystal grains.

C1

A1.3

The diffusion process was applied for the coating of the thick Y-123 layer on a metal tube. A mixed oxide of Y-211 and Y-123 with a ratio of 1:1 was coated on a metal tube as the substrate layer. The mixing of Y-123 to Y-211 improves the adherence of the substrate layer against the metal tube. Then, the Y-035 layer was coated on the substrate – the thickness of the substrate and the coating layer being 150 and 50 μm, respectively. These coatings were provided by a low-pressure plasma spray onto the metal tube [3]. This spray facilitates the deposition of thick and dense oxide layer over a large area in a short operation time. The Y-035 oxide diffuses into the substrate and acts as a flux which enhances the grain growth of Y-123 in the diffusion layer. The excess flux simultaneously reacts with Y-211 to form the Y-123 phase. The addition of 10 wt% Ag to the Y-035 coating layer reduces the reaction temperature to 880°C [4].

Y-123 magnetic shielding tubes were prepared based on the above mentioned technique, where AISI 316L tubes were used as the base metal. Figure B2.2.5.2 is the appearance of the Y-123 magnetic shielding

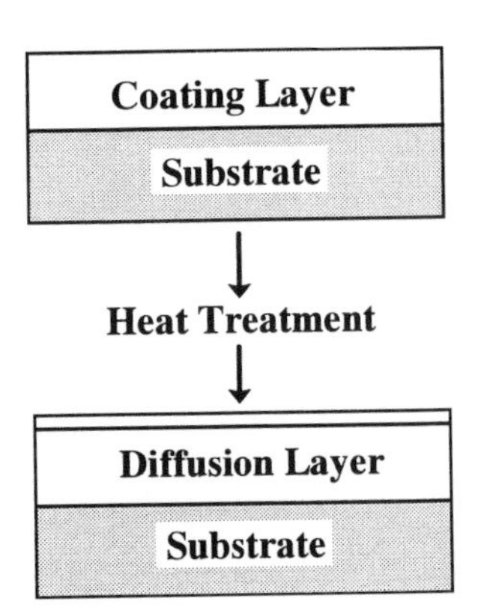

Figure B2.2.5.1. Schematic diagram of the diffusion process.

tube [5]. The magnetic field inside the superconducting shield was measured in a shielding room surrounded by a permalloy and aluminium laminated plate. The magnetic field inside the superconducting shield becomes lower as the measuring point approaches the centre of the tube, as shown in figure B2.2.5.3 [5]. The magnetic field at the centre, which is 140 mm in depth, is below the sensitivity level of 10^{-13} T $Hz^{-0.5}$ for the SQUID (superconducting quantum interference device) sensor over almost all frequency ranges.

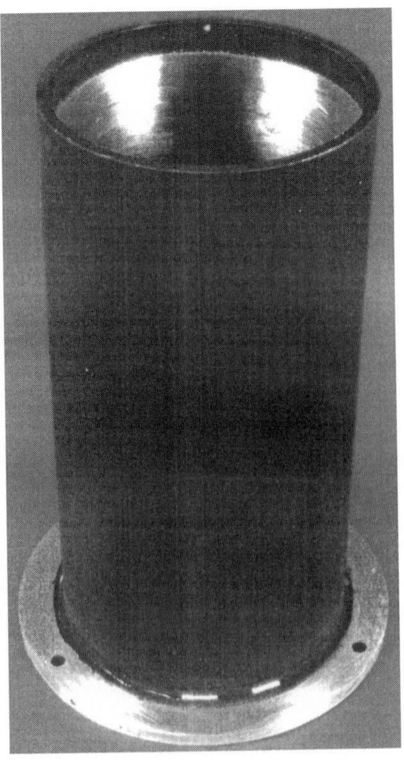

Figure B2.2.5.2. Appearance of the superconducting magnetic shield with Y-123 coated layer on AISI 316L tube.

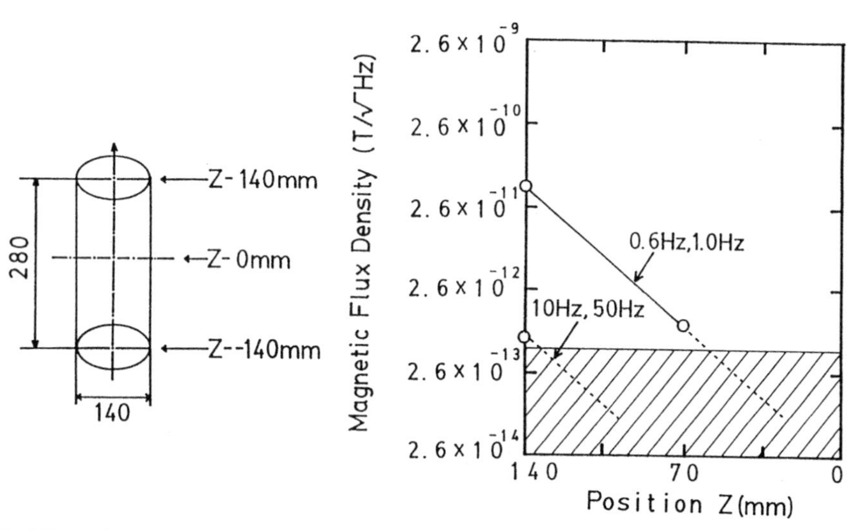

Figure B2.2.5.3. Magnetic flux density in the superconducting shield – 280 mm in length and 140 mm in diameter – measured by the SQUID sensor. The hatched area in the figure is below the sensitivity of the SQUID sensor.

B2.2.5.2.2 Rare earth(Ho, Dy, Gd, Sm, Nd)BCO systems

The thick and uniform RE-123 layer is easily prepared through a diffusion composite composed of RE-211 substrate and RE-035 coating layer. Figures B2.2.5.4 (*a*) and (*b*) show the EPMA line scanning charts taken on the cross-section of RE-035/211(RE: Gd and Nd) composites reacted at 950°C for 10 h in 20 vol% O_2/Ar gas atmosphere. The thick and uniform diffusion layers, a few hundreds of μm in thickness, are formed in all specimens. XRD peaks taken from the surface of reacted composites are assigned to those of the RE-123 phase. The thickness of the Gd-123 layer is substantially more than that

B2.3.4

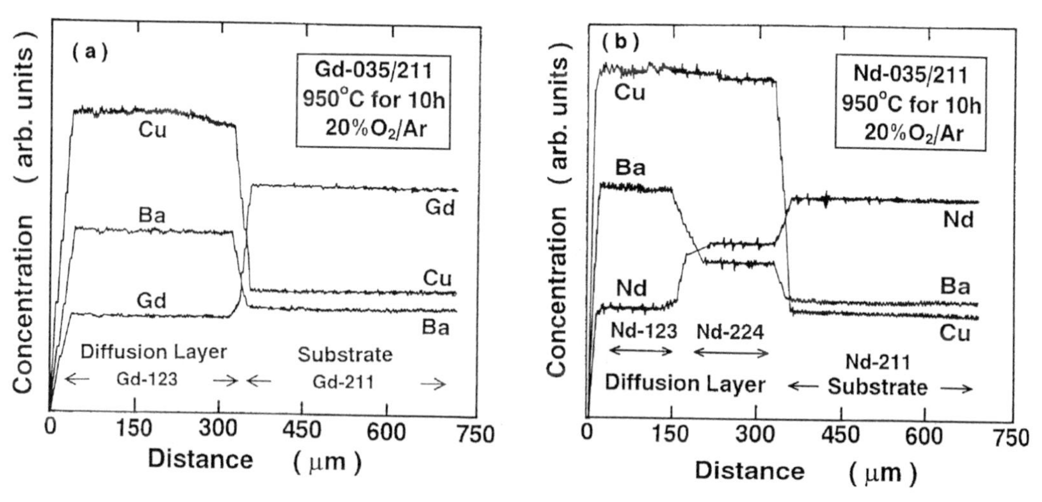

Figure B2.2.5.4. EPMA line scanning charts of the cross-section of (*a*) Gd-based diffusion composite and (*b*) Nd-based diffusion composite; both composites were reacted at 950°C for 12 h in 20% O_2/Ar mixed gas atmosphere.

of the Y-123 layer, suggesting that the Gd-123 phase is easier to form by the diffusion rection than the Y-123 phase. Moreover, the RE-224 layer is formed between the substrate and the RE-123 layer in Sm and Nd-based composites [6].

The total thickness of the diffusion layer in RE composites depends on the O_2 partial pressure in O_2/Ar mixed gas atmosphere, and reaches a maximum at 20% O_2 partial pressure. The reaction at 950°C for 10 h in 100% O_2 gas atmosphere suppresses the formation of the Sm-224 phase in Sm-based composites, and also decreases the fraction of the Nd-224 layer in the diffusion layer of the Nd-based composites. The offset T_c of all RE-123 layers formed by the diffusion reaction exceeds 90 K.

B2.2.5.3 BSCCO systems

C2 In the Bi–Sr–Ca–Cu–O system, the $Bi_2Sr_2CaCu_2O_{8+x}$ (Bi-2212) superconducting layer with a dense and oriented structure is synthesized by the diffusion reaction between a high melting point Sr–Ca–Cu oxide and low melting point Bi–Cu oxide coating layer [7]. The substrate is composed of Bi-free Sr–Ca–Cu oxide with the Sr:Ca:Cu composition ratio of 2:1:2 (Bi-0212). The calcined Bi-0212 oxide powder is formed into cylindrical rods 3 mm in diameter, and tubes with an outside diameter of 20 mm and an B2.2.3 inside diameter of 16 mm by cold isostatic pressing, and then sintered. The coating layer is composed of Bi–Cu oxide with the Bi:Cu composition ratio of 2:1 (Bi-2001). The calcined Bi-2001 oxide powder with 30 wt% Ag_2O addition is coated around the substrate rod and tube. The heat treatment was performed at 820–860°C for 10–40 h in open air to produce the superconducting Bi-2212 diffusion layer. Then some of the specimens were post-annealed at 400–750°C for 6 h in Ar gas atmosphere.

Figure B2.2.5.5 shows the EPMA composition mappings taken on the cross-section of the tube specimen reacted at 850°C for 20 h. A Bi-2212 diffusion layer about 150 μm in thickness is synthesized uniformly around the cylindrical tube. Most of the Ag added to the coating layer precipitates on the surface of the specimen after the reaction. The addition of Ag to the coating layer apparently enhances the diffusion reaction [4]. Some Cu oxide remains near the surface, maybe due to the extra Cu concentration in the diffusion composites (Bi-2001/Bi-0212).

The SEM micrograph taken on the fractured cross-section of the rod specimen reacted at 850°C for 20 h is shown in figure B2.2.5.6. The diffusion layer is composed of plate-like grains grown in the diffusion direction, i.e. in the radial direction of the cylindrical rod [8]. Figure B2.2.5.7 presents the XRD pattern taken on the surface of the cylindrical tube after removing the Ag precipitation. The Bi-2212 XRD patterns of both the outside and inside of the tube specimen indicate an

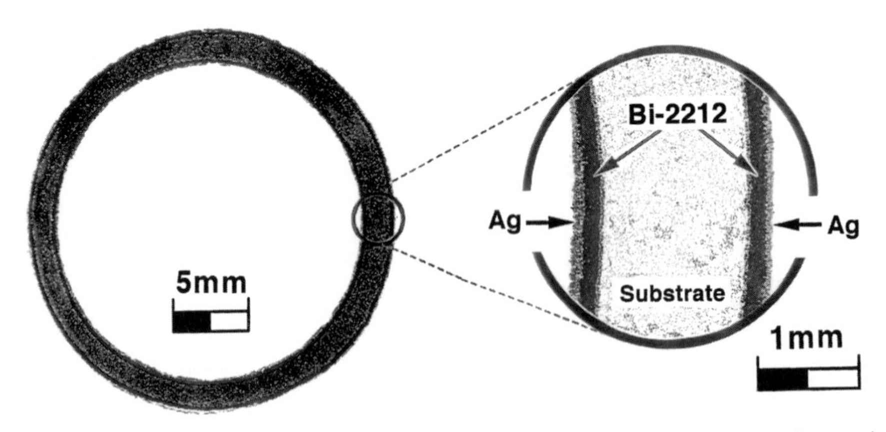

Figure B2.2.5.5. EPMA composition mappings taken of the cross-section of the Bi-2212 tube specimen after the reaction at 850°C for 20 h.

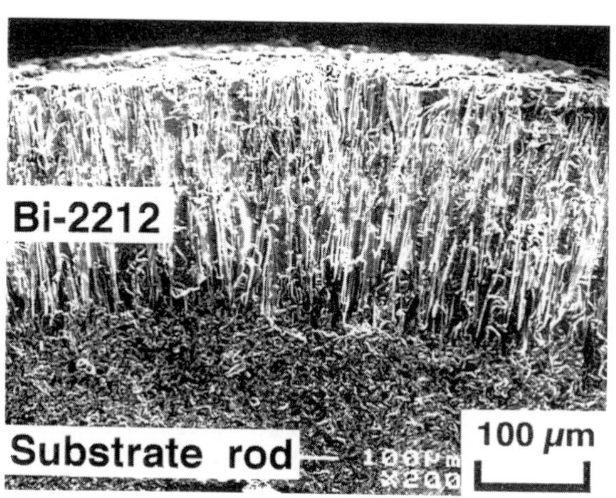

Figure B2.2.5.6. SEM micrograph taken of the fractured cross-section of the Bi-2212 rod specimen after the reaction at 850°C for 20 h.

extremely strong (200) peak in comparison with that of the random powder diffraction pattern. Thus, plate-like grains grown in the diffusion layer are mainly composed of ab-plane grains.

Figure B2.2.5.8 shows the temperature dependence of the transport critical current I_c and critical current density J_c in magnetic fields up to 3 T for the rod specimen reacted at 850°C for 20 h, and that for the specimen post-annealed at 500°C for 6 h in Ar gas atmosphere after the diffusion reaction. The I_c

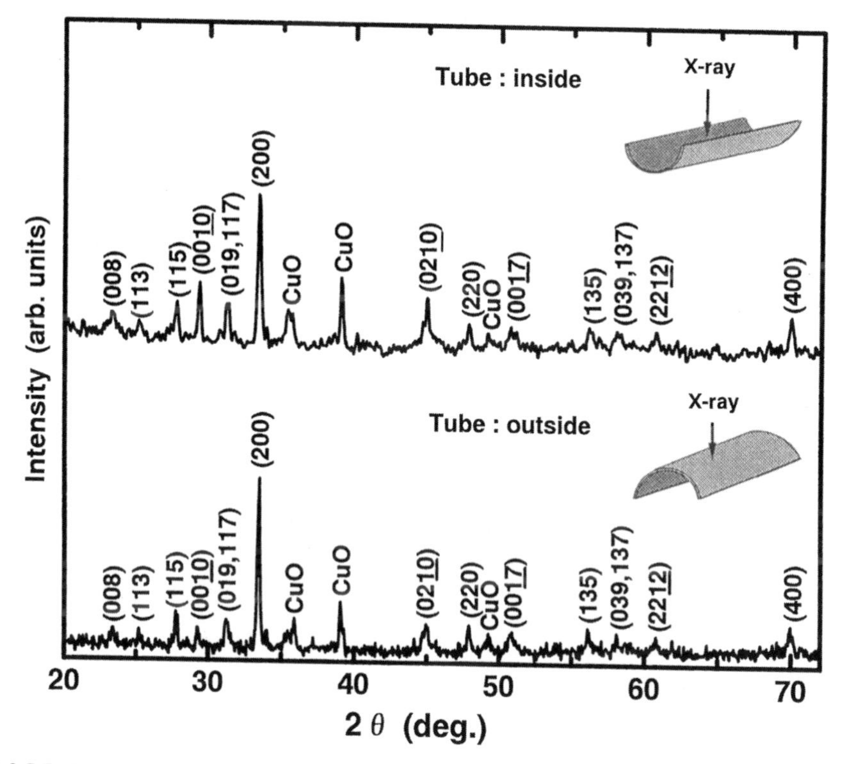

Figure B2.2.5.7. XRD pattern taken of the surface of the Bi-2212 tube specimen reacted at 850°C for 20 h.

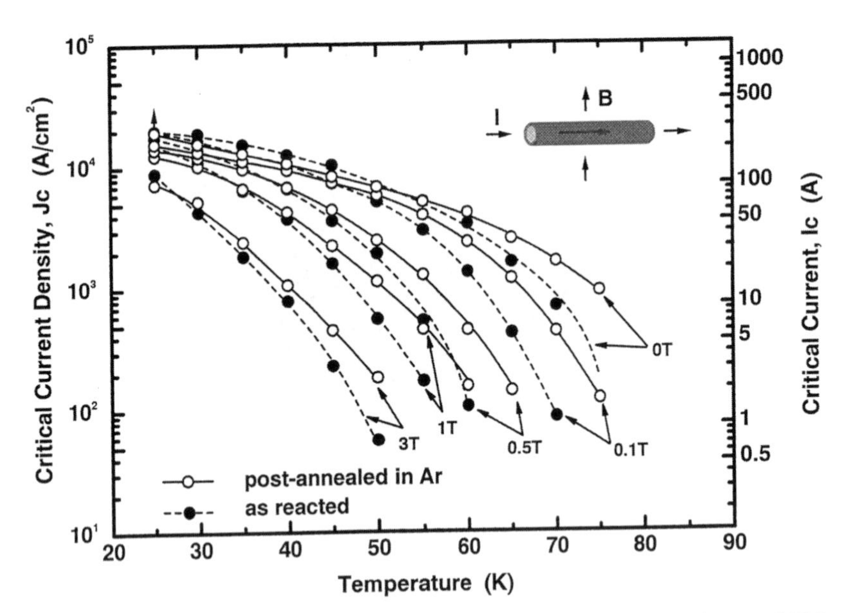

Figure B2.2.5.8. Temperature dependence of transport I_c and J_c in the magnetic field up to 3 T for the Bi-2212 rod specimen reacted at 850°C for 20 h and post-annealed at 500°C for 6 h in Ar gas atmosphere.

for the reacted specimen exceeds 300 A at 4.2 K, and the I_c and J_c are about 300 A and 20 000 A cm^{-2} at 25 K under self-field, respectively. The J_c decreases with increasing temperature and magnetic field, and is about 10 000 A cm^{-2} at 30 K under 1 T, or at 45 K under 0 T [9]. The J_c performance of the post-annealed specimen against temperature and magnetic field is appreciably higher than that of the reacted specimen above 50 K. The improvement in J_c at higher temperatures and magnetic fields may result from the increase of T_c caused by the annealing in Ar gas atmosphere. In the diffusion processed specimens, a fairly large current can be transmitted in the axial direction, while the Bi-2212 layer has an a-axis oriented structure against the substrate.

Figures B2.2.5.9 (*a*) and (*b*) demonstrate the Bi-2212 cylindrical tube, 20/16 mm in outside/inside diameter and 55 mm in length. (*a*) is the reacted specimen covered with the precipitated Ag, and (*b*) is the specimen with Ag contact formed on both ends. Precipitated Ag on the surface is removed except for both the ends using an etching reagent. The I_c of the diffusion specimen hardly falls by removing the Ag precipitation on the surface of the specimen. The transport performance of the tube specimen has been evaluated at 4.2 K and self-field. The transport current up to 6250 A was successfully passed at a ramp rate of 50 A/s with no induced voltage on the HTS specimen. The voltages of both joint increased with the transport current and remained constant for 30s after reaching 6250 A. The transport current density of the Bi-2212 diffusion layer corresponds to 35000 A/cm^2. The transport current was shut down at 6350 A due to the sharp rise of the negative joint voltage [10]. The thermal conductivity through the diffusion specimen is fairly small by the removal of the precipitated Ag from the surface of the specimen. Therefore, Bi-2212 oxide cylinders prepared by the diffusion process with large transport I_c and J_c as well as low thermal conductivity may be attractive as a current lead for superconducting magnets [11]. Meanwhile, the Pb substitution for Bi in the coating layer facilitates the formation of a thick Bi-2223 layer with no crystal orientation [7].

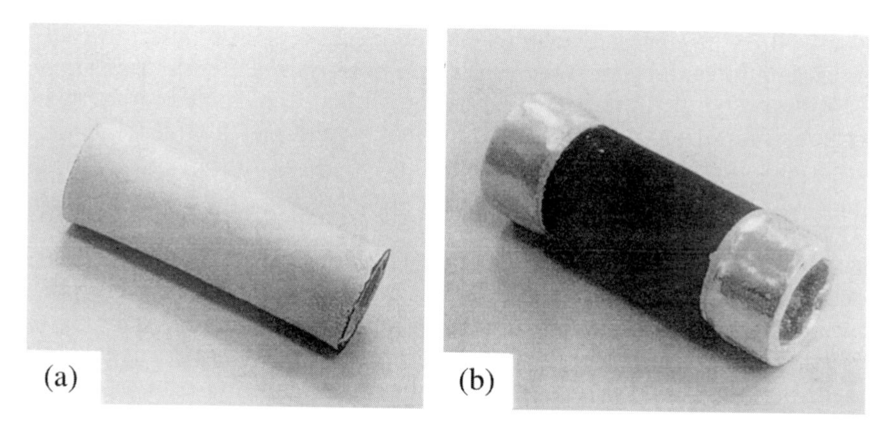

Figure B2.2.5.9. Bi-2212 cylindrical tubes: (*a*) as reacted and (*b*) with Ag contact on both ends. Precipitated Ag on the surface shown in (*a*) is removed except for both ends using an etching reagent.

B2.2.5.4 TBCCO systems

In the Tl–Ba–Ca–Cu–O system, the $TlBa_2Ca_2Cu_3O_y$ (Tl-1223) phase with T_c of 110 K and relatively low I_c degradation under magnetic field at 77 K can be synthesized by the reaction between Ba–Ca–Cu oxide(Tl-0122) substrate and Tl–Ba oxide(T1-2100) coating layer. The TlF substitution for Tl_2O_3 in the coating layer plays a key role in the synthesis of the Tl-1223 phase [12–14]. Figure B2.2.5.10 shows the XRD patterns for the Tl-2(F)100/0122 diffusion layer reacted at 830°C for different reaction times. Here, Tl-2(F)100 is the Tl:Ba:Ca:Cu composition ratio of the coating layer in which the F is added using TlF. Tl-2212 is predominant in the specimen reacted for 10 min. The Tl-2212 phase transforms to the Tl-2223 phase, and subsequently to the Tl-1223 phase. A small amount of Tl-1223 phase appears already in

C3
D1.1.2

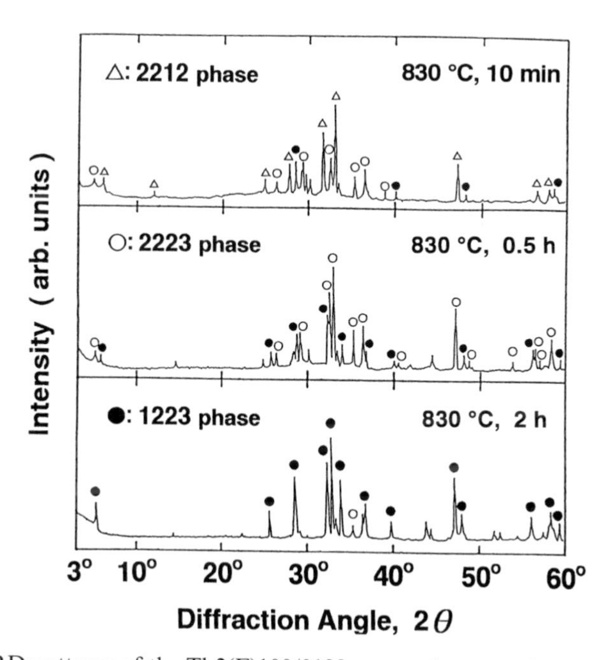

Figure B2.2.5.10. XRD patterns of the Tl-2(F)100/0122 composite reacted at 830°C for different times.

the specimen reacted for 10 min and becomes predominant in the specimens reacted for 1–2 h [12]. When Tl_2O_3 is used in the coating layer, the Tl-2223 phase is formed after the reaction at 830°C for 2 h, and the Tl-1223 phase is hardly formed. The formation of the Tl-1223 phase is reported to be enhanced by the Pb substitution for Tl and Sr substitution for Ba in the TBCCO system [15]. The F addition to the TBCCO diffusion composite promotes the phase transformation to the T1-1223 phase without any element substitution.

D2.2 Figure B2.2.5.11 shows the transport I_c at 77 K versus the applied field (B) curves for different specimens after the reaction at 830°C for 2 h. The F-free composite with T1-2223 layer shows a double-step degradation in I_c with increasing applied magnetic field. The first drop in I_c below 0.1 T, followed by a small plateau, and the subsequent second drop at about 0.5 T are considered to be associated with the weak link and the vortex glass–liquid transition, respectively [13]. The I_c of the specimen with F substitution using TlF, Tl-2(F)100/0122, does not show the second drop in the I_c–B curve at 77 K, and the specimen still retains a relatively large I_c at 1.5 T and 77 K. The slow cooling between 830 and 800°C results in the further improvement of I_c–B performance of the specimen. The I_c at 77 K and 0 T is 112 A, the corresponding J_c being about 1.5×10^4 A cm^{-2}. The amount of first drop in the I_c–B curve of the specimen is somewhat reduced by the slow cooling. No anisotropy in I_c at 77 K is observed for all specimens shown in figure B2.2.5.11 with respect to the direction of the applied field. The thick Tl-1223 layer prepared by this process shows a fairly large J_c in spite of its random grain structure.

The zero T_cs of the Tl-2(F)100/0122 and Tl-2100/0122 specimens reacted at 830°C for 2 h are plotted on a magnetic field versus temperature map as shown in figure B2.2.5.12. Specimen currents are

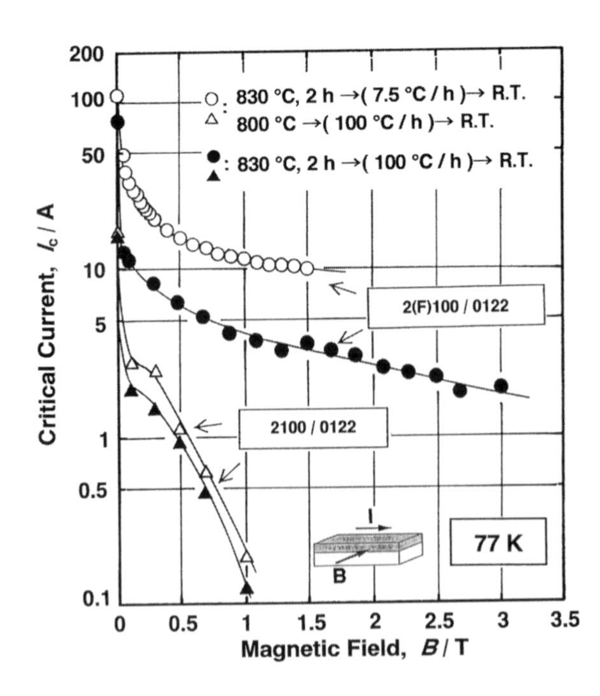

Figure B2.2.5.11. I_c at 77 K versus applied field curves of Tl-2 (F) 100/0122 and Tl-2100/0122 composites reacted at 830°C for 2 h. The open and closed circles indicate the I_c–B performance of the specimens with a cooling rate of 7.5 and 100°C h^{-1} between 830 and 800°C, respectively.

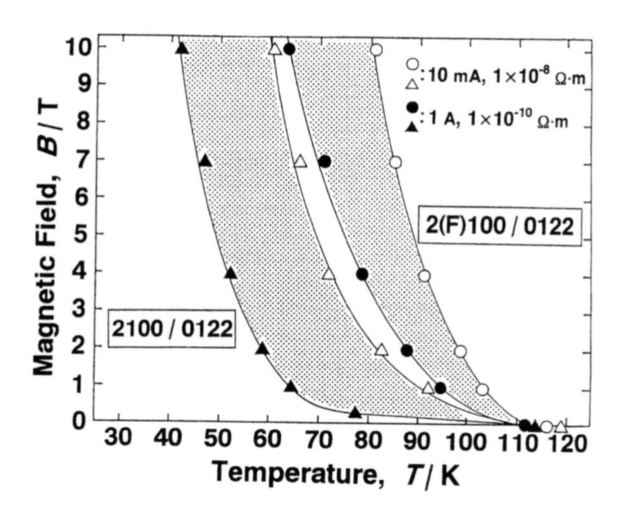

Figure B2.2.5.12. Zero resistive transition temperature of Tl-2(F)100/0122 and Tl-2100/0122 composites under magnetic field. Specimen currents were 10 mA and 1 A for open and closed symbols, respectively.

10 mA and 1 A for open and closed symbols, respectively, the corresponding resistivity criteria being about 1×10^{-8} and 10^{-10} Ωm. Transition points are significantly shifted to high temperatures by the F substitution. The 1 A transition point at 77 K reaches about 4.5 T in the F-substituted specimens, which is one order higher than that in the F-free specimen.

A thick layer of Tl-1223 is formed on the Ni and Ag tape [14]. When the Ag tape is used as a base metal, a substantial amount of Ag migrates into the diffusion layer after the reaction. However, migration of Ni is not observed in the case of Ni tape. A Tl-1223 layer about 300 μm thick was formed on the Ni tape. The bonding between the diffusion layer and the Ni tape is sufficiently good. Thus, the Ni tape is found to be quite favourable as the base metal for the thick Tl-1223 layer formed by the diffusion reaction. The formation of the Ni–Cu–O layer, about 10 μm in thickness, is identified by the EPMA analysis, which may improve the bonding between the Ni tape and the diffusion layer. The 5 mm wide Tl-1223 coated tape carries 75 A at 77 K and ambient field [14].

B2.2.5.5 Summary

The diffusion process facilitates the synthesis of thick and homogeneous high-T_c oxide layers on the substrate in a relatively short reaction time. The thickest diffusion layer is formed in the Gd-123 layer among different Y- and RE-based high-T_c oxides, although no crystal grain alignment is found. A highly oriented grain structure is formed in the Bi-2212 layer around rod and tube substrates, by which a large transport current is obtained in the axial direction. A thick Tl-1223 layer with favourable J_c–B performance at 77 K is synthesized by the diffusion reaction with the TlF substitution for Tl_2O_3 in the coating layer. The diffusion process has been applied for the preparation of Y-123 magnetic shielding tubes, Bi-2212 current leads and Tl-1223 coated tape conductors.

References

[1] Tachikawa K 1980 Development of A15 filamentary composite superconductors *Filamentary A15 Superconductors* (New York: Plenum) p 1

[2] Tachikawa K, Sadakata N, Sugimoto M and Kohno O 1988 Fabrication of superconducting Y–Ba–Cu oxide through an improved diffusion process *Japan. J. Appl. Phys.* **27** L1501

[3] Tachikawa K, Shimbo Y, Ono M, Kabasawa M and Kosuge S 1989 High-Tc superconducting films of Y–Ba–Cu oxides prepared by a low-pressure plasma spraying *11th Int. Conf. on Magnet Tech.* (London, New York: Elsevier Applied Science), p 1494

[4] Tachikawa K, Inoue T, Zama K and Hikichi Y 1992 Effects of some additional elements on the reaction diffusion in high-Tc oxide composites *Supercond. Sci. Technol.* **5** 386

[5] Shimbo Y, Niki K, Kabasawa M and Tachikawa K 1994 High-Tc magnetic shields prepared by a low-pressure plasma spray *Adv. Cryo. Engng.* **40** 253

[6] Tachikawa K, Kohchi N, Matsumoto H and Minemoto T 1999 *Advances in Superconductivity* vol 11, (Tokyo: Springer-Verlag) p 797

[7] Tachikawa K, Watanabe T, Inoue T and Shirasu K 1991 Bi–Sr–Ca–Cu–O superconducting oxides synthesized from different diffusion couples *Japan. J. Appl. Phys.* **30** 639

[8] Tachikawa K, Yamada Y, Satoh M and Hishinuma Y 1995 Structures and superconducting properties of oriented Bi-2212 oxide layer synthesized by a diffusion process *Proc. Conf. Topical Int. Cryo. Mat. (1994)* (Singapore: World Scientific) p 307

[9] Yamada Y, Yamashita F, Wada K and Tachikawa K 1998 Structure and superconducting properties of Bi-2212 cylinders prepared by diffusion process *Adv. Cryo. Engng.* **44** 547

[10] Yamada Y, Takiguchi M, Suzuki O, Tachikawa K, Tamura H and Mito T 2001 Transport performance of Bi-2212 current leads prepared by a diffusion process. *IEEE Trans. Appl. Superconductivity* **11** 2555

[11] Yamada Y, Sakuraba J, Hasebe T, Hata F, Chong C K, Ishihara M and Watanabe K 1994 High-Tc oxide current leads and superconducting magnet using no liquid helium *Adv. Cryo. Engng.* **40** 281

[12] Kikuchi A, Kinoshita T, Nishikawa N, Komiya S and Tachikawa K 1995 Enhanced critical current in Tl–Ba–Ca–Cu–O superconductors prepared by diffusion process with fluorine addition *Japan. J. Appl. Phys.* **34** L167

[13] Tachikawa K, Kikuchi A, Kinoshita T, Nakamura T and Komiya S 1996 Critical current in Tl-base high-Tc oxides prepared by a diffusion process *Advances in Superconductivity* **vol 8**, (Tokyo: Springer-Verlag) p 915

[14] Tachikawa K, Kikuchi A, Nakamura T and Komiya S 1998 Synthesis of thick Tl-based high-Tc oxide layer through a diffusion process *J. Supercond.* **11** 147

[15] Aihara K, Doi T, Soeta A, Takeuchi S, Yuasa T, Seido M, Kamo T and Matsuda S 1992 Flux pinning in Tl-(1223) superconductor *Cryogenics* **32** 936

B2.3.1
Introduction to bulk melt processing techniques

D A Cardwell

Bulk high temperature superconductors (HTS) have significant potential for a variety of high field 'permanent magnet' applications such as magnetic bearings, fault current limiters, magnetic clamps and flywheel energy storage systems. In general the field generating capacity of HTS materials, which may exceed significantly the maximum practical field produced by a permanent magnet (1.5 T), is determined by the magnitude and homogeneity of the critical current density (J_c) of the material and the length scale over which it flows. J_c, in turn, is determined by magnetic flux pinning under the influence of the Lorentz force, which is usually achieved by inhomogeneities in the microstructure of the specimen. The so-called irreversibility field represents the field at which the magnetic behaviour of the material becomes reversible at a given temperature and, hence, that at which flux pinning sites become ineffective. HTS are characterised in general by an irreversibility line (i.e. the variation of irreversibility field with temperature) which defines the field/temperature phase space above which the material becomes useless for current carrying applications (see figure B2.3.1.1). The $(RE)Ba_2Cu_3O_{7-\delta}$ [(RE)BCO where RE = Y, Nd, Sm, Eu, Gd etc)] class of superconductors exhibits the highest irreversibility fields over a wide temperature range and these are therefore the most promising materials for application at high currents and magnetic fields. Unfortunately, the ability of (RE)BCO to carry current and, hence, to generate magnetic field, is severely limited by the presence of grain boundaries in the sample microstructure (the problem of granularity does not appear to limit the transport J_c in BSCCO, for example), which is manifested as a rapid decrease in J_c with applied field (see figure B2.3.1.2). It is necessary, therefore, to process (RE)BCO in the form of large, single grains if the grain boundary problem is to be avoided.

E2.1, E1.3.3
E1.3.4
A4.3
A1.3
A1.3

A variety of melt processing techniques has been developed for the fabrication of large grain (RE)BCO. These are all based on a peritectic reaction, occuring in these systems between 1000 and 1080°C (i.e. the peritectic temperature, T_p), in which the $(RE)Ba_2Cu_3O_{7-\delta}$ (RE-123) phase is formed from solid $(RE)_2BaCuO_5$ (RE-211), a Ba–Cu–O based liquid phase (L) and oxygen gas (G), i.e.:

B2.3.2

$$\underset{(RE-211)}{(RE)_2BaCuO_{10}} + \underset{(L)}{Ba_3Cu_5O_{6.72}} + \underset{(G)}{0.42\,O_2} \rightarrow \underset{(RE-123)}{2(RE)Ba_2Cu_3O_{6.28}}.$$

The non-superconducting RE-211 phase (referred to as Nd-422 for neodymium to reflect the composition of one unit cell) and the liquid in this reaction can be produced by rapidly heating a pre-sintered green body of the desired composition to a temperature well above T_p. In principle, formation of the required RE-123 phase is then achieved by cooling the peritectically molten (RE)BCO sample slowly through the peritectic temperature—hence the name of the process. In practice, however, the loss of liquid from the sample at elevated temperature limits severely the process times for stoichiometric RE-123 and the starting composition is usually enriched with up to 40 mol% excess RE-211 phase

B2.3.3
B2.3.4

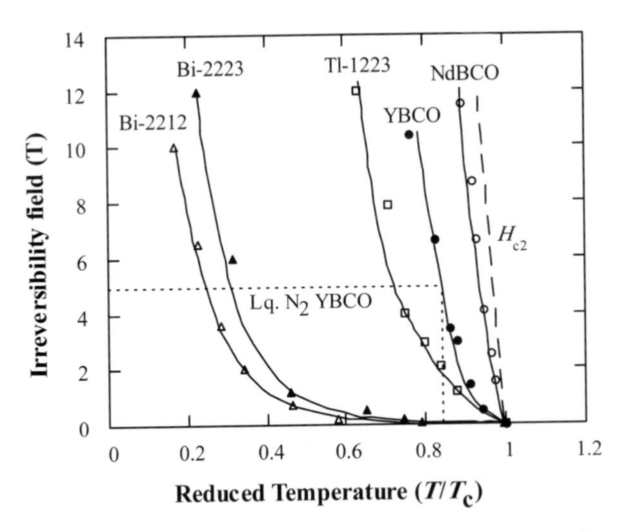

Figure B2.3.1.1. Irreversibility data for HTS compounds.

B2.3.5 to reduce this effect. This contrasts with the melt process used to fabricate Bi–Sr–Ca–Cu–O which is characterised by a relatively complex phase diagram. Individual phases tend to melt incongruently in this system which is therefore generally much more difficult to control than (RE)BCO.

The most common melt process technique for the fabrication of (RE)BCO is based on top seeded melt growth (TSMG), in which a small single crystal with lattice parameters similar to that of the target RE-123 material is placed on the top surface of the presintered pellet to provide a heterogeneous grain nucleation site, often under a thermal gradient. A typical large single YBCO grain grown by this technique is shown in figure B2.3.1.3. In this case, the seed crystal must necessarily have a higher peritectic temperature than the bulk (RE)BCO sample if it is to remain undecomposed during the melt process. Sm-123 or Nd-123 crystals provide suitable seeds for the growth of YBCO, for example,

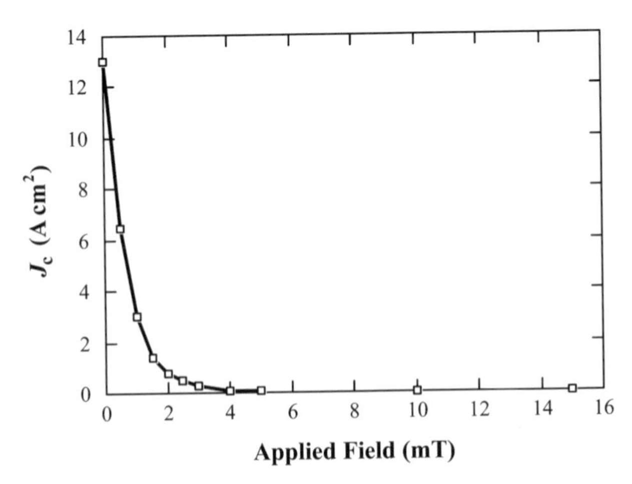

Figure B2.3.1.2. The field dependence of J_c for sintered YBCO at 77 K.

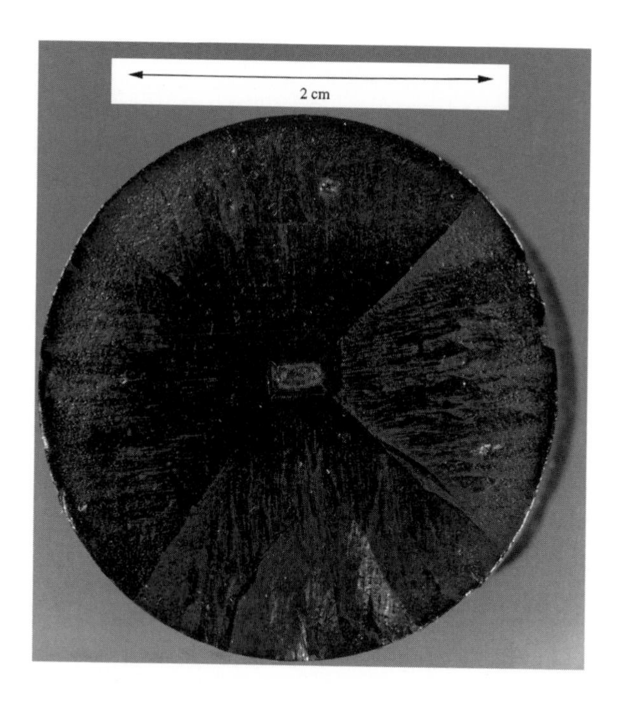

Figure B2.3.1.3. Large grain YBCO fabricated by seeded melt growth using a SmBCO seed.

whereas MgO is frequently used as a seed in other (RE)BCO materials. In general, the processing conditions for the fabrication of NdBCO and SmBCO are more complex than those for YBCO. The properties of these materials are particularly sensitive to the processing atmosphere, for example, and typically require low oxygen partial pressures of 0.1% during melt growth. As a result an oxygen controlled melt growth (OCMG) process has been developed for NdBCO and SmBCO, involving a similar melt process to that described above but under accurately controlled oxygen partial pressure down to 10^3 Pa. Samples of these materials melt processed under ambient oxygen pressure exhibit low transition temperatures which are typically broad in temperature and hence unsuitable for engineering applications at 77 K.

The rate of peritectic solidification during melt growth is determined by a number of compositional, kinetic and thermodynamic influences which are addressed in detail in this paper. Of these, however, the solubility of the rare earth ion in the peritectically decomposed Ba–Cu–O liquid at the growth front is particularly significant. For the case of YBCO, for example, the solubility of Y is limited to around 0.6 mol% which limits, in turn, the extent to which Y-211 can dissolve. As a result Y-211 inclusions remain trapped within the bulk Y-123 phase matrix to form a distribution of unconnected inclusions, as shown in figure B2.3.1.4(a). Enrichment of the starting stoichiometry with RE-211 has two main beneficial effects for the sample processing and properties. Firstly, the excess Y-211 sustains peritectic solidification over greater sample sizes so that larger grain specimens may be produced by the melt process technique. Secondly, the RE-211 inclusion density correlates with an increase in effective flux pinning sites, which enhance J_c of the fully processed sample and yield higher trapped fields. Hence, a general processing aim is to generate a fine distribution of the second phase RE-211 inclusions in a large, homogeneous superconducting RE-123 matrix. The latter may be achieved by the addition of a variety of dopants to the precursor materials, such as Pt or CeO_2, which have proved particularly effective in refining the size and distribution of the RE-211 phase (figure B2.3.1.4(b)).

B2.3.2
B2.3.3
B2.3.4

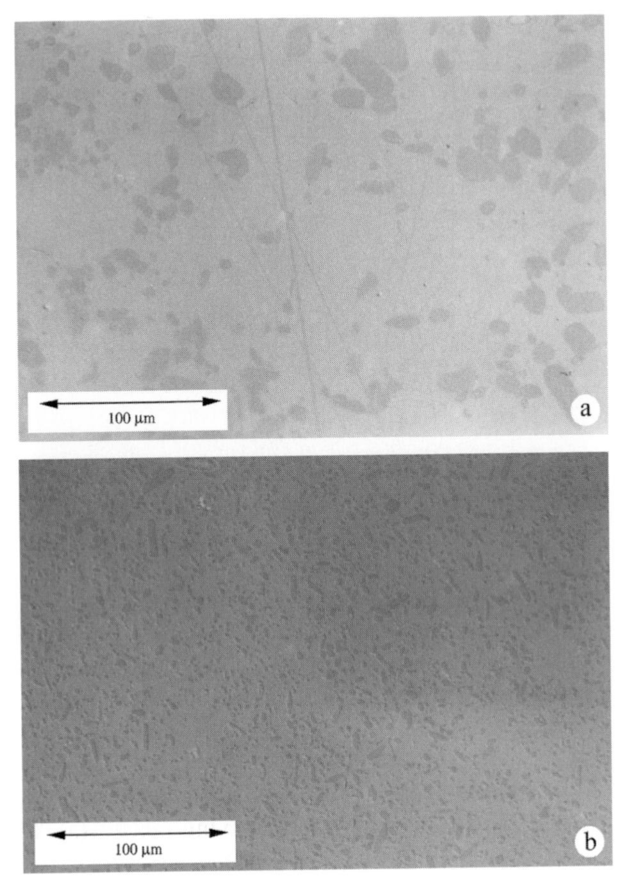

Figure B2.3.1.4. (*a*) Y-211 inclusions in a melt processed Y-123 matrix. The sample in (*b*) has been doped with Pt and exhibits a significantly finer distribution of Y-211.

This chapter describes the fundamentals of the melt processing technique in general and specifically for YBCO, BSCCO and other light (RE)BCO materials. The understanding and control of these processes is essential for the realization of a variety of practical applications using HTS, which are anticipated ultimately to form the basis of practical, medium scale industrial production processes.

B2.3.2
Melt processing techniques: fundamentals of melt process

Yuh Shiohara, Katsumi Nomura and Teruo Izumi

B2.3.2.1 Introduction

The critical current density (J_c) is one of the important superconductivity properties for applications such as wires, cables and magnets etc, and depends strongly on the macro and microstructures, which are influenced by the fabricating process. In order to obtain high J_c superconducting materials, the following factors should be attained: (a) high density, (b) elimination of weak links, (c) highly orientated structure, (d) introduction of effective pinning centres and (e) optimization of oxygen content. Appropriate annealing after growth can solve factor (e). Melt processing techniques including directional solidification processes help in achieving factors (a)–(c). Regarding factor (d), addition of several other elements, such as Pt or Ce, has been recently identified to be effective for generating a fine dispersion of non-superconducting Y_2BaCuO_5 (Y211) particles as pinning centres for magnetic flux lines in the superconducting $YBa_2Cu_3O_y$ (Y123) matrix through the peritectic reaction. Consequently, melt processing is recognized as the most effective and important technique for the development of high quality bulk oxide superconductors. Furthermore, a basic understanding of the fundamentals of crystal growth and processing is essential to control the melt process. In this chapter, the important aspects of melt processing are reviewed, mainly for the YBCO superconductors, since this system has been extensively studied from engineering, material science and crystal growth points of view, which are all necessary for the understanding of the phenomenon.

A4.0

A4.3

B2.3.2.2 Phase diagram

The phase diagram is instructive for understanding the growth mechanism and also for optimizing the process parameters. Most of the investigations for phase diagram research, however, have concentrated on the solid state [1–15] and there have been few reports for the high temperature range, which is the process temperature range for melt growth techniques. In this section, the characteristics of the phase diagram in the high temperature range of the Y–Ba–Cu–O system are described.

B2.3.3

Figures B2.3.2.1(*a*) and (*b*) show the calculated isothermal sections of the Y_2O_3–BaO–CuO ternary system at 1223 and 1273 K, respectively, at 0.21 atm oxygen pressure [16, 17]. The phase diagrams are obtained by fitting thermodynamic calculations to experimental results [10, 18]. The composition region of the two-phase equilibrium between Y123 and liquid reduces rapidly with increasing

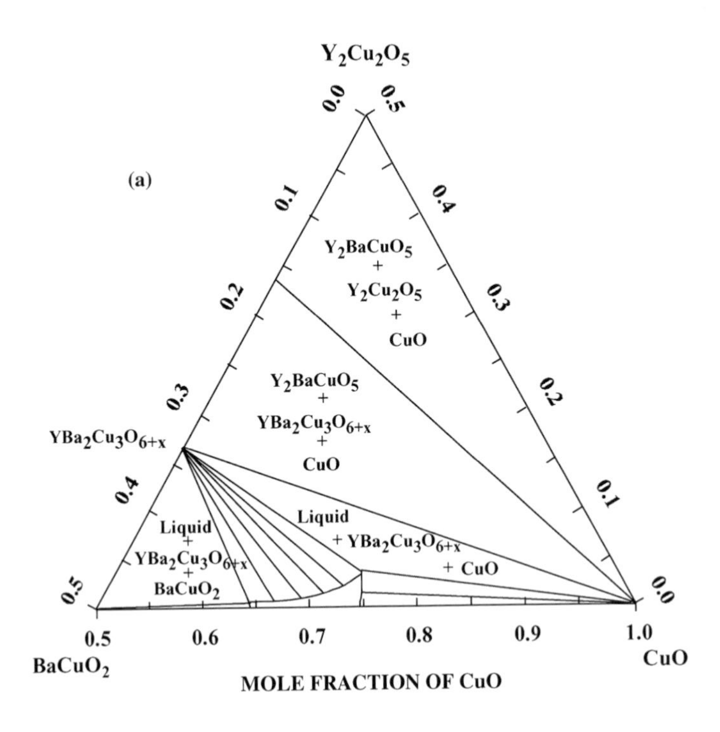

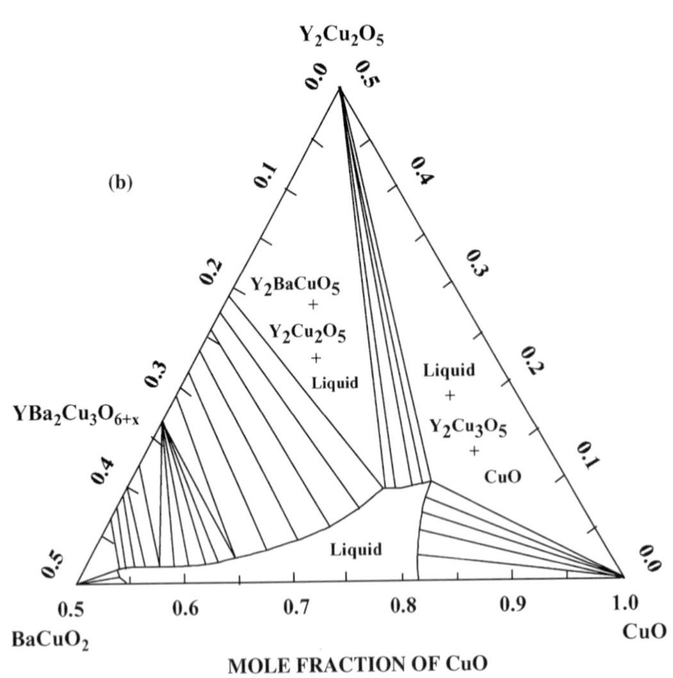

Figure B2.3.2.1. Calculated isothermal sections of the Y_2O_3–BaO–CuO ternary system at 1223 K (*a*) and 1273 K (*b*) under 0.21 atm oxygen pressure [16].

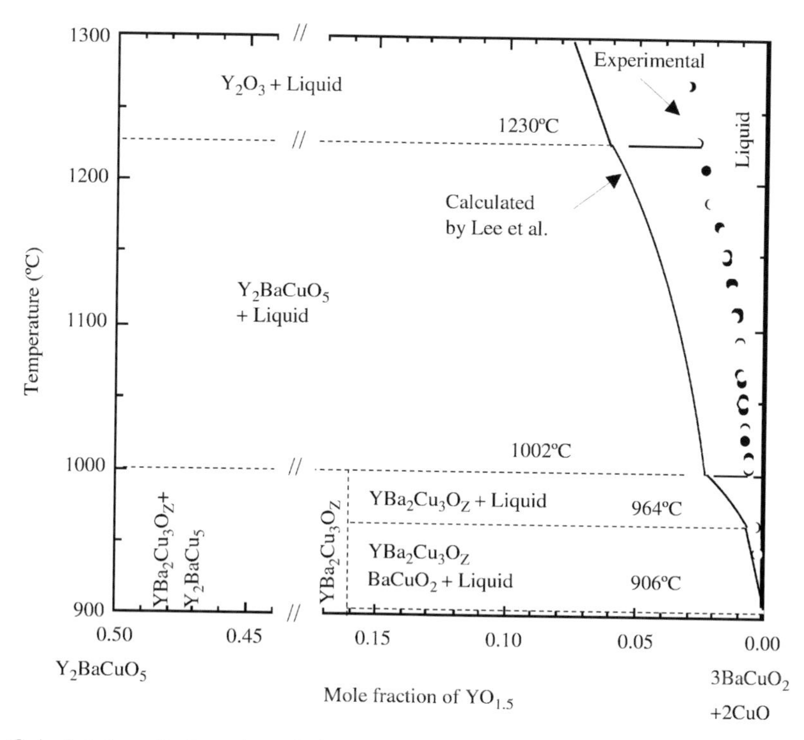

Figure B2.3.2.2. Calculated vertical section of the Y_2O_3–BaO–CuO system under 0.21 atm oxygen pressure [16] and experimental results of the liquidus line [20].

temperature because of the evolution of Y211 even in the low Y_2O_3 region. In order to indicate the temperature dependences, a vertical section on the (Y123)–(Y211)–($3BaCuO_2$ and 2CuO) tie line is shown in figure B2.3.2.2. In this figure, the much more reliable liquidus compositions of Y measured experimentally are superimposed. According to this figure, it was found that the Y solubility is very low (0.6 mol%) and the Y123 phase is formed by the peritectic reaction between Y211 and liquid. Therefore it is impossible to grow Y123 crystals from a congruent melt. Additionally, the liquidus slope near the Y123 peritectic reactions is very high, which suggests that it is difficult to obtain a higher growth rate even if large undercoolings can be applied.

The phase diagram depends also on the oxygen partial pressure [19] and type of rare earth (RE) element [20–27]. In general, the peritectic temperature decreases with decreasing oxygen partial pressure. Additionally, RE123 has a solid solution for the relatively large and light ionic RE element-systems, such as Nd and Sm. In these cases, the liquid composition can be changed during growth which leads to non-steady state growth. Therefore, the process parameters should be selected carefully with a consideration of the phase diagram. B2.3.4

B2.3.2.3 Solidification processing for HTSC materials

Solidification methods for high T_c superconducting materials are classified into two groups—one is a 'growth from semi-solid' method for high J_c applications, and the other is a 'growth from solution' method for high crystallinity single crystals. The dominant phenomena considered to affect the microstructure in the two categories are illustrated schematically in figures B2.3.2.3 and B2.3.2.4. In the growth from semi-solid system, Y211 particles, which is the high temperature stable phase, exist E1 B2.4

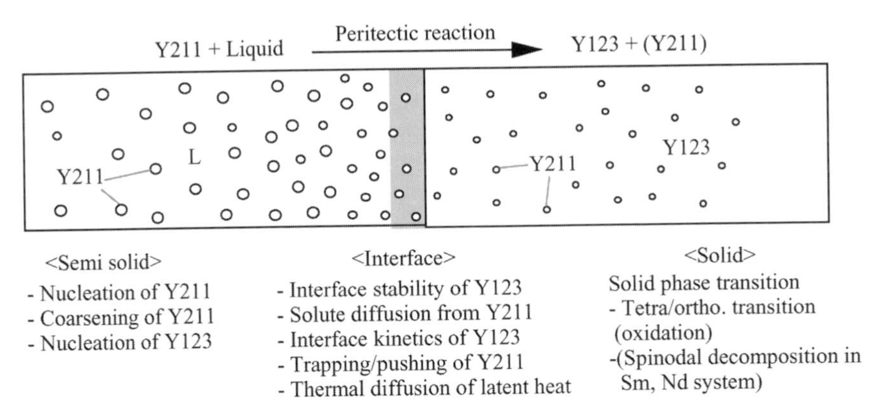

Figure B2.3.2.3. Dominant phenomena affecting microstructure of the Y–Ba–Cu–O system in melt processing.

relatively near the growing interface of the Y123 phase. The necessary solute for the growth is supplied from the Y211 through a concentration gradient in the liquid. Some of the Y211 particles can be entrapped into the growing Y123 solid phase and they can act consequently as a pinning centre. On the other hand, a liquid diffusion zone without Y211 particles exists in front of the growth interface, and the solute is transported to the zone near the interface by bulk convection in the case of solution growth. Therefore, solution growth is a suitable method for producing high quality crystals without inclusions, although it is difficult to obtain a high growth rate. Both categories are further divided into two groups, which are free growth and constrained directional solidification. The relationships between the methods are summarized in figure B2.3.2.5. As single crystal growth, which is one of the solution growth methods, is described in another part of this handbook (chapter B2.4), only the typical phenomena and model analysis for melt processing are reviewed here, in the following section.

A1.3
B2.4

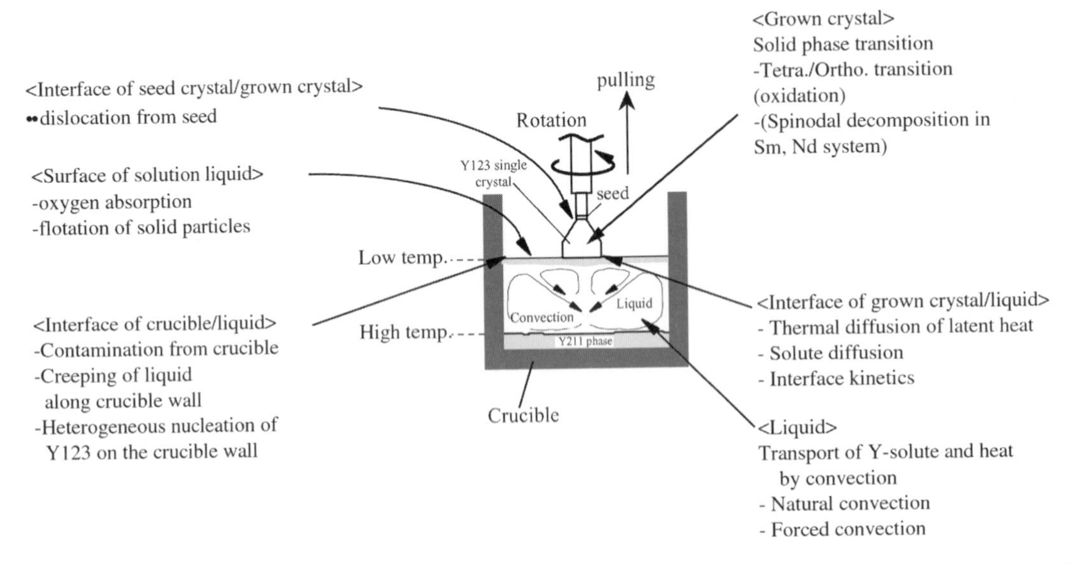

Figure B2.3.2.4. Dominant phenomena affecting size and crystallinity of the Y123 single crystal in the TSSG method (e.g. SRL-CP method).

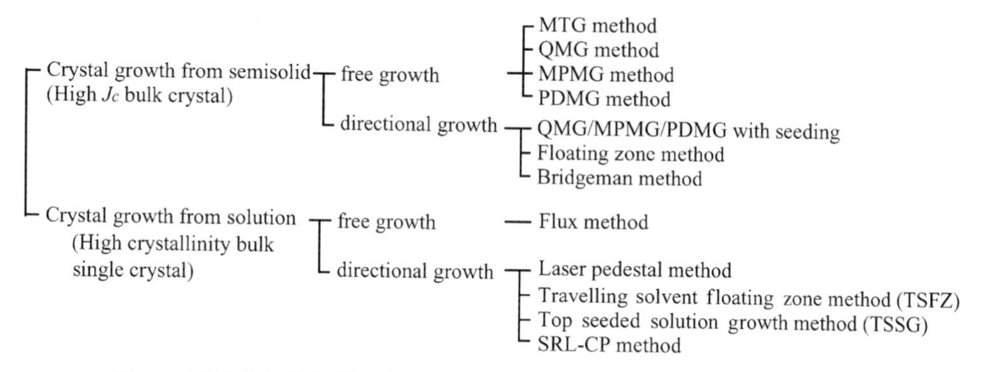

Figure B2.3.2.5. Solidification methods for producing bulk HTSC crystals.

B2.3.2.4 Solute diffusion through liquid for peritectic reaction

B2.3.2.4.1 Experimental results

As mentioned earlier, the Y123 crystal is formed by a peritectic reaction between Y211 and liquid during solidification processing. In general, solidification from the melt is limited by interface kinetics and/or mass transport. In particular, it has been recognized that the effect of interface kinetics is important in the crystal growth, which exhibits faceted growth. However, mass transport is also important in the case of the low Y_2O_3 solubility in the liquid. In this section, a diffusion-limiting process is shown to understand the special peritectic reaction through the liquid.

B2.2.5

Figure B2.3.2.6 shows the interface morphology of the directionally solidified sample [28]. The necessary solute for the peritectic reaction in the metallic alloy system is normally supplied through solid diffusion, as shown in figure B2.3.2.7(*a*). However, the drastic change in the Y211 volume fraction from the liquid to the Y123 crystal and a sharp interface are observed in figure B2.3.2.6. This phenomenon cannot be explained by the peritectic reaction with diffusion in the solid but should be considered as diffusion in the liquid as shown in figure B2.3.2.7(*b*), which is similar to some unique metallic alloy systems such as Al–Mn and Cu–Sn. Almost simultaneously, three different research groups proposed

B2.2.5

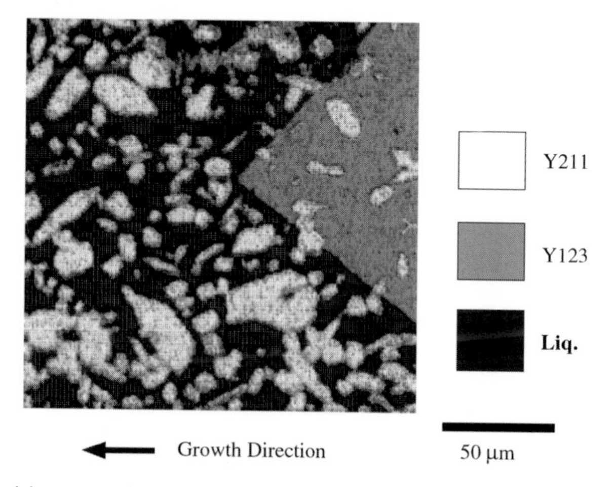

Figure B2.3.2.6. Y composition mapping image around the Y123 growth interface grown at $1\,\mathrm{mm\,h^{-1}}$. White particles are the Y211 phase, grey region is the Y123 phase and black matrix is the liquid phase.

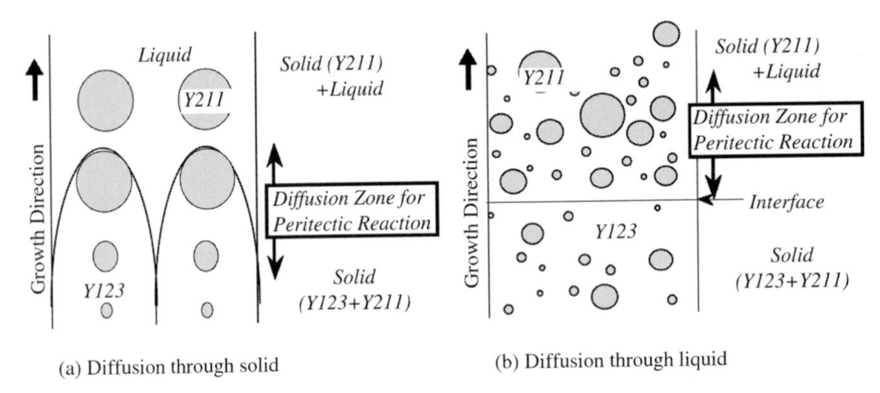

Figure B2.3.2.7. Expected structures around the interface in two different systems of peritectic reactions: (*a*) necessary solutes are transported through solid; (*b*) necessary solutes are provided through liquid.

similar diffusion-limiting models with diffusion in the liquid phase [28–31]. One of the solidification models is described briefly here using the model proposed by Izumi *et al* [28].

B2.3.2.4.2 Model analysis

B2.3.3 Figure B2.3.2.8 shows the principle of the solute diffusion-limiting model. The yttrium diffusion flux for Y123 phase growth from Y211 particles dispersed in the liquid ahead of the interface of the growing Y123 crystals was estimated using the local equilibrium assumption. This flux was considered from

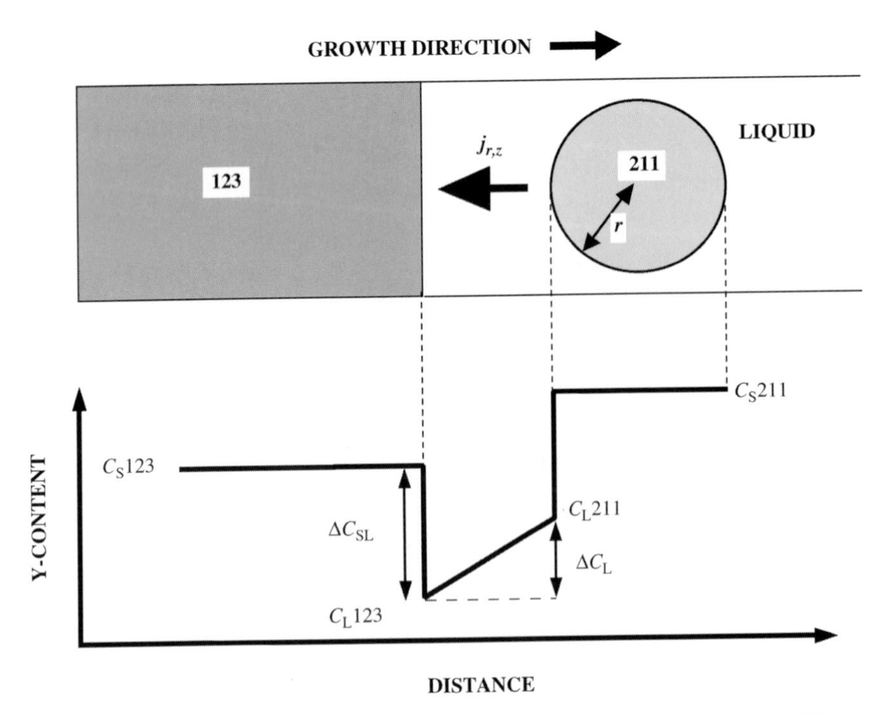

Figure B2.3.2.8. Sketch of the principle of the proposed solidification model. Composition difference, as a driving force providing the necessary solute for growth, is indicated in the composition profile.

the composition difference in the liquid at the two different interfaces, liquid/Y211 and liquid/Y123. This composition difference has three origins in the case of directional solidification (ΔC_1: due to the curvature of Y211 particles dispersed in the liquid, ΔC_2: due to undercooling and ΔC_3: due to the temperature gradient as a function of the distance z between the interface and each Y211 particle), which are indicated on the schematic phase diagram in figure B2.3.2.9.

The diffusion flux from one Y211 particle, $j_{r,z}$, with a radius of r at a certain distance, z, from the Y123 interface can be calculated using Fick's first law:

$$j_{r,z} = D_L \frac{\Delta C_L}{z} = D_L \frac{(\Delta C_1 + \Delta C_2 + \Delta C_3)}{z} = \frac{D_L}{z}\left[\frac{1}{m_L^{211}}\left(\frac{2\Gamma^{211}}{r} + G_z\right) + \left(\frac{1}{m_L^{123}} - \frac{1}{m_L^{211}}\right)\Delta T\right] \quad \text{(B2.3.2.1)}$$

where D_L is the diffusivity, m_L^{211} is the liquidus slope of the Y211 phase, Γ^{211} is the Gibbs–Thomson coefficient ($\sigma^{211}/\Delta S_f$), σ^{211} is the Y211/liquid interfacial energy, ΔS_f is the volumetric entropy of fusion, r is the radius of the Y211 particle, G is the temperature gradient, z is the distance from the Y123 interface, m_L^{123} is the liquidus slope of the Y123 phase and ΔT is the undercooling at the Y123 interface.

The total Y flux, j_{211}, from the Y211 particles can be estimated by the integration of the flux from individual Y211 particles with respect to z and r within the diffusion zone, δ_c, based on the concept of maximum flux ($j_{max} = 4\pi r^2 k_{max}$)

$$\begin{aligned} j_{211} &= N \int_0^\infty \int_0^{\delta c} j_{r,z} \Psi[r_0(r,z)] \mathrm{d}z\, \mathrm{d}r \\ &= N \int_0^\infty \pi r^2 \left\{4k_{max}^{211} \int_0^{\delta *} \Psi[r_0(r,z)] \mathrm{d}z + D_L \int_{\delta *}^{\delta_c} \frac{\Psi[r_0(r,z)]}{z} \Delta C_L \, \mathrm{d}z\right\} \mathrm{d}r \end{aligned} \quad \text{(B2.3.2.2)}$$

where $\psi[r_0(r,z)]$ is the size distribution function of the Y211 particles in the diffusion zone, N is the number of Y211 particles per unit volume and k_{max} is the maximum decomposition rate.

On the other hand, the diffusion flux per unit area, j_{123}, required for the crystal growth is calculated using the composition difference, ΔC_{SL}, between the yttrium concentration of the Y123 solid, C_S^{123},

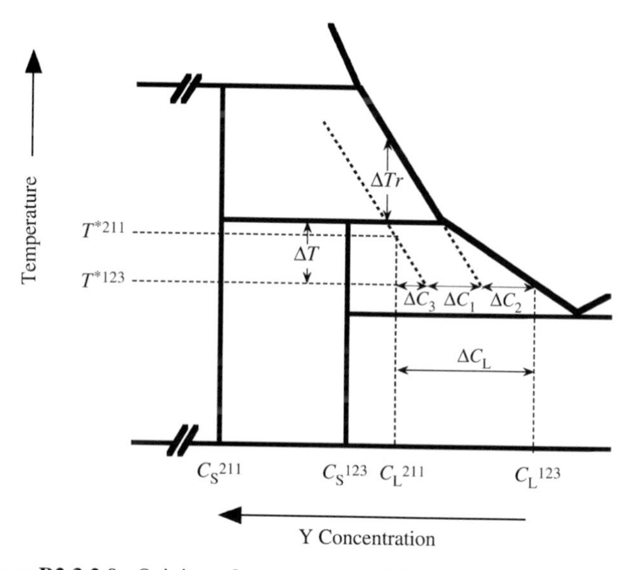

Figure B2.3.2.9. Origins of composition difference used in the model.

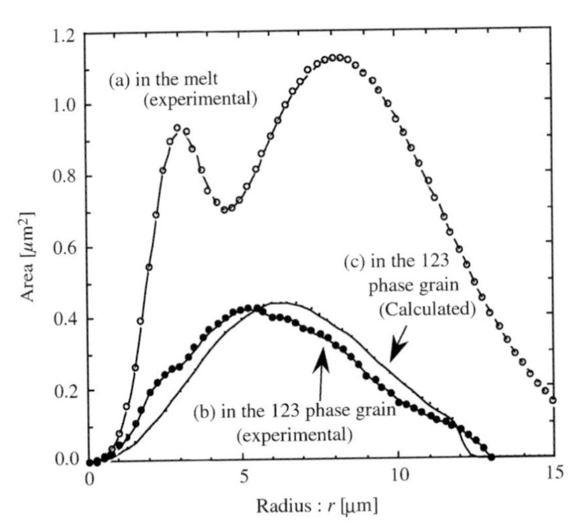

Figure B2.3.2.10. Change in size distributions of Y211 particles before and after the growth of Y123 crystal at 1 mm h^{-1}. Curves include experimentally measured distributions (before and after solidification) and the calculated distribution in the Y123 phase grain using the initial experimentally measured distribution.

and that of the liquid at the liquid/Y123 interface, C_L^{123}

$$j_{123} = \Delta C_{SL} R = (C_S^{123} - C_L^{123})R = \left(C_S^{123} - C_L^e + \frac{\Delta T}{m_L^{123}}\right)R \qquad \text{(B2.3.2.3)}$$

where C_L^e is the yttrium concentration of the liquid at the peritectic temperature, m_L^{123} is the liquidus slope of the Y123 phase and R is the growth rate of the Y123 crystal.

In order to achieve steady state growth of the Y123 crystal at the interface, the total flux from the Y211 particles described in equation (B2.3.2.3) should be equated with the necessary flux given by equation (B2.3.2.4). Then the condition for steady state growth is given as follows

$$R = \frac{N\int_0^\infty \pi r^2 \{4k_{\max}^{211}\int_0^{\delta^*} \Psi[r_0(r,z)]\mathrm{d}z + D_L\int_{\delta^*}^{\delta_c} \frac{\Psi[r_0(r,z)]}{z}\Delta C_L\,\mathrm{d}z\}\mathrm{d}r}{C_S^{123} - C_L^e + \Delta T/m_L^{123}}. \qquad \text{(B2.3.2.4)}$$

This model predicted that higher balanced velocity for the continuous steady state growth of the Y123 crystal could be attained with larger undercoolings and/or smaller average particle size in the initial size distribution of Y211 particles. The order of magnitude of the calculated growth rates is several millimetres per hour, which is in good agreement with the experimentally confirmed growth rate for continuous growth. This model can be also utilized to estimate the change in size distribution of Y211 particles from the liquid to Y123 crystals as shown in figure B2.3.2.10.

B2.3.2.5 Pushing/trapping of 211 particles

Varanasi *et al* [32, 33] have observed Y211 particles to be present in an inhomogeneous manner delineating distinguishable patterns in textured Y123 crystals with Pt additions. Kim *et al* [34] also reported that fine particles were pushed out of the advancing Y123/liquid interface towards the liquid phase in the Y–Ba–Cu–O system with the metal oxide (CeO) addition. In both cases, the observed patterns of the Y211 segregation seem to exist along boundaries between different growth directions.

Cima *et al* [35] have observed, for the first time, the occurrence of significant anisotropy of the Y211 particle macrosegregation during seeded growth of melt textured Y123 crystals. Investigations into segregation of the Y211 particles have been systematically carried out by Endo *et al* [36, 37], which show that segregation depends on the growth rate (R) as a function of undercooling (ΔT) as well as on the growth direction. B2.3.3

From the quantitative analysis of the Y211 particles in the four areas with different undercoolings (ΔT) and different growth directions for the bulk sample grown by the seeding method with different ΔT, the number of Y211 particles in the region of $\Delta T = 30\,\mathrm{K}$ is clearly much larger than that in the $\Delta T = 10\,\mathrm{K}$ region in a whole range of diameters, as shown in figure B2.3.2.11. In particular, Y211 particles with a diameter under 0.5 μm were rarely observed in the growth region of $\Delta T = 10\,\mathrm{K}$. Also it is noticed that the total volume fractions (V_{f211}) of the Y211 particles in $\Delta T = 30\,\mathrm{K}$ for both directions, obtained by integrating the volume fraction in the range of all diameters in figure B2.3.2.11, are in good agreement with the theoretical value (V_{f211} cal. $= 0.23$), which is estimated using the lever rule from the nominal composition (Y123 + 0.4Y211). On the other hand, the V_{f211} in $\Delta T = 10\,\mathrm{K}$ regions ($V_{f211} = 0.13$ and 0.081 for the *a*- and *c*-direction, respectively) are much smaller than the theoretical values. These results indicate that the excess Y211 particles could not enter the Y123 grain completely; in other words, control of the excess amount of Y211 particles in the Y123 could not be achieved by simple precursor composition control.

The pushing/trapping phenomenon of Y211 particles at the Y123 interface depends on the growth rate, the growth orientation and the radius of the particle. Smaller particles are pushed by the growing Y123 interface. The critical condition may be introduced as follows:

$$R^* \propto \frac{\Delta\sigma_0}{\eta r^*} \tag{B2.3.2.5}$$

where R^* is the critical growth rate, η is the melt viscosity, r^* is the critical radius of the Y211 particle and $\Delta\sigma_0$ is the interfacial energy difference obtained from the following formula; pushing phenomena can occur when the following condition is satisfied:

$$\Delta\sigma_0 = \sigma_{\mathrm{SP}} - \sigma_{\mathrm{LP}} - \sigma_{\mathrm{SL}} > 0 \tag{B2.3.2.6}$$

where σ_{SP}, σ_{LP} and σ_{SL} are the solid-particle, liquid-particle and solid–liquid interfacial energies, respectively. This phenomenon could be a reason for the prevention of steady state growth. Nakamura

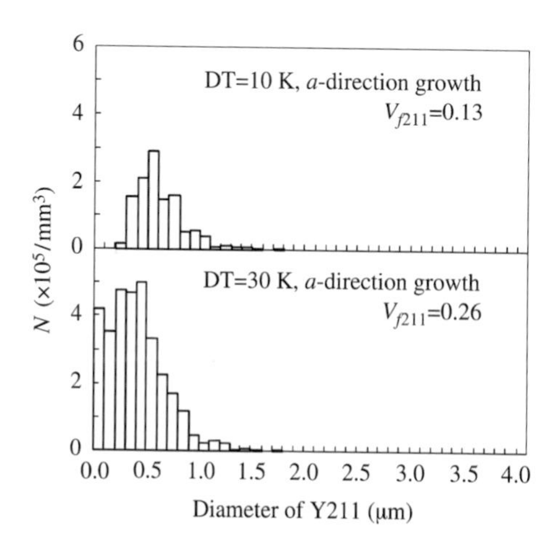

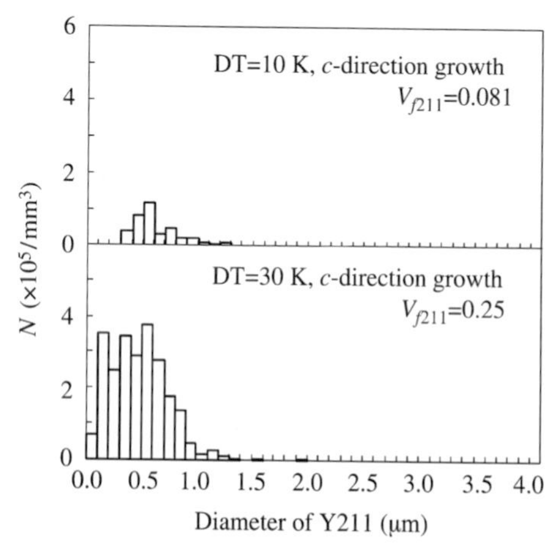

Figure B2.3.2.11. Histograms of the Y211 distribution for each area of different growth direction under different ΔT.

and Kambara *et al* [38] reported that growth of the Y123 crystal is forced to terminate when the volume fraction of the Y211 phase particles in the liquid reaches around 0.6 in front of the Y123 growing interface.

Considering the relationship between R and ΔT, we can at least explain qualitatively the macrosegregation of Y211 as a function of R and growth direction. In the case of $\Delta T = 10$ K, the critical radius (r_a*) of the Y211 particles for the a-direction growth is smaller than r_c* because of the anisotropy of R and the assumption of $\Delta\sigma_0$. r_a* and r_c* at $\Delta T = 10$ K are considered to be relatively large in the size distribution of Y211 dispersed in front of the interface because the total volume fractions of the Y211 particles are much less than the calculated value for both directions. When ΔT is changed from 10 to 30 K the growth rates for both directions increased, leading to a significant decrease of both r_a* and r_c*. Accordingly, even smaller Y211 particles, which were pushed at $\Delta T = 10$ K, are now trapped by the Y123 crystals. From the consistency between the experimental and the calculated total volume fractions, r_a* and r_c* for $\Delta T = 30$ K could be expected to be smaller than the smallest radius of the Y211 particles in front of the interface. Fedorov [39] observed an anisotropic dependence of R* on r* in their investigations of the pushing/trapping of the Ni metal particles by different faces, (110) and (111), in a salol crystal, which is a faceted organic material.

One should be careful when the pushing/trapping theory is being applied to the Y123/Y211 system because this theory was proposed for explaining phenomena between inclusions, which are inert for both liquid and solid, and non-faceted material. On the other hand, the Y211 particles are reactive inclusions since they supply the Y solute to Y123, giving rise to faceted growth of the Y123 crystals.

B2.3.2.6 Doping effect for coarsening of 211 particles

The control of the Y211 particle size is one of the important factors for achieving high J_c because Y211 particles can act as pinning centres. The size increases in the semi-solid state of Y211 and liquid by a coarsening mechanism. Addition of several other elements, such as Pt or Ce, has been determined to be effective for fine dispersion of non-superconductive Y211 particles [40, 41]. Figure B2.3.2.12 shows the

A4.3
B2.3.3

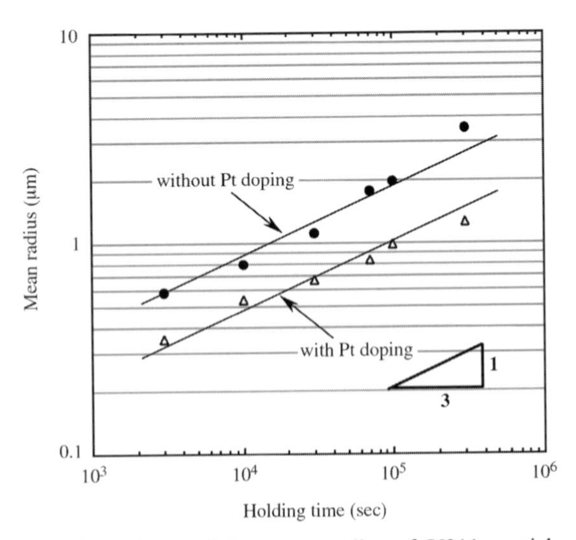

Figure B2.3.2.12. Holding time dependence of the mean radius of Y211 particles with and without Pt doping; logarithmic plot.

holding time dependence of the mean radius of the Y211 particles with and without Pt doping on a logarithmic plot. From this, the time dependence may be explained using coarsening theory [42], which is indicated by the following equation:

$$r_{\mathrm{m}} = \alpha(D_{\mathrm{L}}\Gamma t)^{1/3} \quad \text{(B2.3.2.7)}$$

where r_{m} is the mean radius of the Y211 particles, α is a constant, D_{L} is the diffusivity and Γ is the Gibbs–Thompson coefficient given earlier. The fine Y211 particles in the liquid are obtained by the suppression of particle coarsening which is caused by the change in the product $D_{\mathrm{L}}\ \Gamma$ in equation (B2.3.2.7).

B2.3.2.7 Joining of superconductors

From a practical point of view, joining of bulk superconductors is a very important for producing artefacts which are larger and/or longer scale than laboratory scale specimens. Furthermore, it is desired that the superconducting properties at the join are of the same quality as elsewhere in the superconductor. As mentioned earlier, the superconducting properties, especially J_{c}, are influenced strongly by the effective current path that is related to both the weak links and the orientation of the crystals, and the melt process is an advantageous technique for addressing these issues.

Several efforts to join superconductors in the RE–Ba–Cu–O system have been reported [43–45]. Kimura *et al* proposed a soldering method [44]. In this the $YbBa_2Cu_3O_y$ (Yb123) superconducting powder was sandwiched as the solder between the Y123 superconducting bulks. The peritectic temperature of Yb123 is lower than that of Y123 [20, 46]. They succeeded in melting only the Yb123 solder without the decomposition of the Y123 matrix. Maeda *et al* studied this soldering method in more detail [47]. They reported that the Yb_2BaCuO_x (Yb211) particles and pores (gas bubbles) were easily segregated at the final solidified region after the soldering, and these phenomena were explained by the *U-C-J* pushing–trapping model [48, 49] as mentioned earlier. Segregation of the non-superconducting phases reduced the effective current path, which degraded the superconducting properties. Maeda *et al* modified the soldering method from the sandwich type to the bridge type [50]. Figure B2.3.2.13 shows schematic illustrations and a photograph of both soldering methods: (*a*) the sandwich type soldering and (*b*) the bridge type soldering. Segregation of the non-superconducting phases at the final solidified region does not prevent the current path on the bridge type soldering method. They succeeded in obtaining the suitable joining structure without inhibiting significantly the current path by the non-superconducting phases as shown in figure B2.3.2.13(*b*).

B2.3.2.8 Other important phenomena and materials

A unique technique for the fabrication of superconducting layers using semi-solid and diffusion processes has been reported by Tachikawa *et al*. This is based on a diffusion reaction process to fabricate a superconducting layer that is composed of a high melting point substrate and a low melting point coating layer which react with each other and transform to the superconducting phase [51–56]. They applied the diffusion process to synthesize RE, Bi, and Tl based oxide superconductors. It was of interest to note that the *a*-axis oriented structures were easily obtained by the diffusion process in the Bi based system [53, 54]. For RE–Ba–Cu–O superconductor, $Ba_3Cu_5O_x$ powder with a low melting point was coated on the high melting point RE_2BaCuO_x (RE211) substrate. The RE123 thick layer (over 100 μm) could be formed at a relatively lower temperature (920–980 °C) and shorter heating duration (10 h) than those of other semi-solid melting processes [52].

B2.2.5

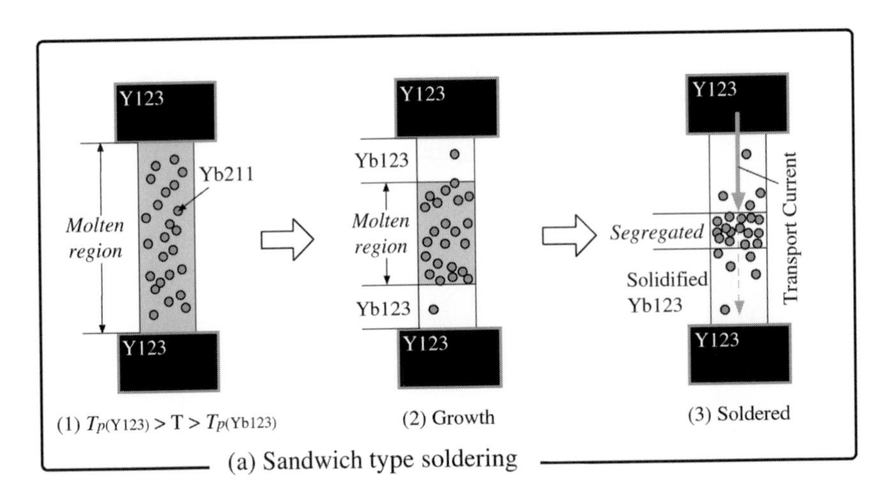

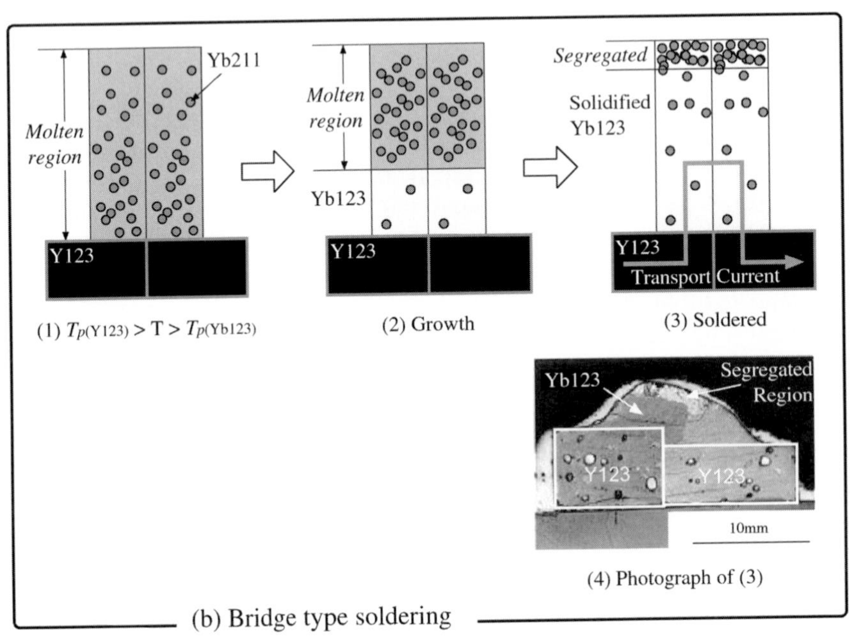

Figure B2.3.2.13. Schematic illustrations and photograph of the soldering methods: (*a*) sandwich type soldering and (*b*) bridge type soldering.

C2 B3.2.1 The $Bi_2Sr_2CaCu_2O_x$ (Bi-2212) superconductor is well known as one of the candidates for practical superconducting wires. The partial melting and slow cooling process [57] which is a kind of the melt processing realizes a well-oriented microstructure, good connectivity between superconducting grains and excellent critical current density below 30 K for the superconducting wires. The powder-in-tube (PIT) method [58–62] or the coating method [63–68] using an Ag or Ag alloy sheath or tape has been usually selected for the fabrication method of the Bi-2212 wires and tapes. Ag lowers the melting point of Bi-2212 and promotes the formation of a highly textured structure [69, 70]. Many investigations B3.2.1 of phase changes during the heat treatment [71–75] have been carried out and the results have been helpful for controlling the microstructure, thus improving the superconducting properties. Bi-2212

Table B2.3.2.1. Substituted values for the estimation of growth rate, estimated results and experimental growth rates for the Y, Sm, Nd systems (in air)

	Y	Sm	Nd
Temperature at the bottom T_b (K)	1288	1335	1362
Temperature at the surface T_s (K)	1273	1330	1357
Equilibrium concentration at T_b C_L^{211} (at%)	0.64	1.85	3.30
Equilibrium concentration at T_s C_i (at%)	0.55	1.70	3.00
Concentration in RE123 C_S^{123} (at%)	16.7	16.7	16.7
Diffusion coefficient of the solute D_L ($cm^2\,s^{-1}$)	1×10^{-5}	1×10^{-5}	1×10^{-5}
Coefficient of kinetic viscosity ν ($cm^2\,s^{-1}$)	1×10^{-2}	1×10^{-2}	1×10^{-2}
Angular velocity of the crystal ω (s^{-1})	4π	4π	4π
Estimated maximum growth rate R_{max} ($cm\,s^{-1}$)	1.5×10^{-5}	2.2×10^{-5}	4.8×10^{-5}
Experimental growth rate R_{ex} ($cm\,s^{-1}$)	2.7×10^{-6}	3.8×10^{-6}	1.5×10^{-5}

decomposes to the $(SrCa)CuO_2$, the $Bi_2(SrCa)_4O_7$, and the liquid around 1153 K in air. The $(SrCa)CuO_2$ phase transforms to the $(SrCa)_2CuO_3$ phase, then transforms to the (SrCa)O phase with increasing temperature. In the cooling process, the reverse reaction occurs; however, $(SrCa)CuO_2$ particles easily grow and remain as the $(SrCa)CuO_2$ phase, which leads to not only the incompletion of the reaction for the Bi-2212 phase but also the prevention of the effective current path [71]. Zhang *et al* reported that control of the oxygen partial pressure during the heat treatment was effective to control the melting temperature of the Bi-2212 and the solid phases after the melting [72]. There were several studies to show that a control of the composition of the superconducting powders was effective in reducing the impurity phases after the melting process and improving the superconducting properties [76–78]. Basically, the Bi based superconductor is a very complex and unsolved system because the Bi based system consists of many elements compared with the Y system. For further details in the Bi based system, see chapter B2.3.5.

B3.2.1

B2.3.5

As mentioned earlier, an understanding of the melt processing for high T_c oxide superconducting materials has certainly been progressing; however, there are a lot of unsolved phenomena. Further progress of basic understanding of the solidification and crystal growth fields is required not only for process development but also for materials science, because the solidification of oxides has not been well investigated.

References

[1] Hinks D G, Soderholm L, Capone D W II, Jorgensen J D, Schuller I K, Segre C U, Zhang K and Grace J D 1987 *Appl. Phys. Lett.* **50** 1688
[2] Aselage T and Keefer K 1988 *J. Mater. Res.* **3** 1279
[3] Osamura K, Zhang W, Yamashita T, Ochiai S and Predel B 1988 *Z. Metallkde.* **79** 693
[4] Oka K, Nakane K, Ito M, Saito M and Unoki H 1988 *Japan. J. Appl. Phys.* **27** L1065
[5] Maeda M, Kadoi M and Ikeda T 1989 *Japan. J. Appl. Phys.* **28** 1417
[6] Ullman J E, McCallum R W and Verhoeven J D 1989 *J. Mater. Res.* **4** 752
[7] Dembinski D, Gervais M, Odier P and Coutures J P 1990 *J. Less-Common. Met.* **164–165** 177
[8] Ahn B T, Lee V Y and Beyers R 1990 *Physica* C **167** 529
[9] Kawabata S, Hoshizaki H, Kawahara N, Enami H, Shinohara T and Imura T 1990 *Japan. J. Appl. Phys.* **29** L1490
[10] Lay K W and Renlund G M 1990 *J. Am. Ceram. Soc.* **73** 1208
[11] Kimura S, Shimpo R and Nakamura Y 1992 *J. Jpn. Inst. Metals* **56** 1145
[12] Karen P, Braaten O and Kjekshus A 1992 *Acta Chem. Scandinavica* **52** 805
[13] Zhou Z and Navrotsky A 1992 *J. Mater. Res.* **7** 2920
[14] Erb A, Biernath T and Muller-Vogt G 1993 *J. Cryst. Growth* **132** 389
[15] Ng W W, Cook L P, Paretzkin B, Hill M D and Stalick J K 1994 *J. Am. Ceram. Soc.* **77** 2354
[16] Lee B J and Lee D N 1991 *J. Am. Ceram. Soc.* **74** 78
[17] Lee B J and Lee D N 1989 *J. Am. Ceram. Soc.* **72** 314
[18] Lindemer T B, Hunley J F, Gates J E, Sutton A L Jr, Brynestad J, Hubbard C R and Gallagher P K 1989 *J. Am. Ceram. Soc.* **72** 1775
[19] Nakamura M, Krauns Ch, Yamada Y and Shiohara Y 1996 *Japan. J. Mater. Res* **11** 1076–81
[20] Krauns Ch, Sumida M, Tagami M, Yamada Y and Shiohara Y 1994 *Z. Phys.* B **96** 207
[21] Iwata T, Hikita M and Tsurumi S 1989 *Advances in Superconductivity*, ed K Kitazawa and T Ishiguro (Tokyo: Springer-Verlag) p 197
[22] Yoo S I and McCallum R W 1993 *Physica* C **210** 147
[23] Wada T, Suzuki N, Maeda T, Maeda A, Uchida S, Uchinokura K and Tanaka S 1988 *Appl. Phys. Lett.* **52** 1989
[24] Daeumling M, Seuntjens J M and Larbalestier D C 1990 *Nature* **346** 332
[25] Kambara M, Nakamura M, Shiohara Y and Umeda T 1997 *Physica* C **275** 127
[26] Sumida M, Tagami T, Krauns Ch, Shiohara Y and Umeda T 1995 *Physica* C **249** 47
[27] Murakami M, Yoo S I, Higuchi T, Sakai N, Weltz J, Koshizuka N and Tanaka S 1994 *Japan. J. Appl. Phys.* **33** L715
[28] Izumi T, Nakamura Y and Shiohara Y 1993 *J. Cryst. Growth* **128** 757–761
[29] Cima M J, Flemings M C, Figuredo A M, Nakade M, Ishii H, Brody H D and Haggerty J S 1992 *J. Appl. Phys.* **72** 179
[30] Mori N, Hata H and Ogi K 1992 *J. Jpn. Inst. Metals* **6** 648
[31] Izumi T, Nakamura Y and Shiohara Y 1992 *J. Mater. Res.* **7** 1621
[32] Varanasi C and McGinn P J 1993 *Physica* C **207** 79
[33] Varanasi C, Black M A and McGinn P J 1996 *J. Mater. Res.* **11** 565
[34] Kim C J, Kim K B, Won D Y, Moon H C, Suhr D S, Lai S H and McGinn P J 1994 *J. Mater. Res.* **9** 1952
[35] Cima M J, Rigby K, Flemings M C, Haggerty J S, Honjo S, Shen H and Sung T M 1995 *Proc. Inter. Workshop on Superconductivity* (Hawaii: Maui) p 55
[36] Endo A, Chauhan H S, Nakamura Y and Shiohara Y 1995 *Proc. Int. Workshop on Superconductivity* (Hawaii: Maui) p 59
[37] Endo A, Chauhan H S, Egi T and Shiohara Y 1996 *J. Mater. Res.* **11** 795
[38] Nakamura Y, Kambara M, Izumi T, Shiohara Y, *Physica* C at press
[39] Fedorov O P 1992 *Growth of Crystals* (New York: Consultant Bureau) p 169
[40] Murakami M, Morita M, Doi K and Miyamoto K 1989 *Japan. J. Appl. Phys.* **28** 1189
[41] Fujimoto H, Murakami M, Gotoh S, Shiohara Y, Koshizuka N and Tanaka S 1990 *Advances in Superconductivity* **vol 2**, ed T Ishiguro and K Kajimura (Tokyo: Springer-Verlag) p 285
[42] Izumi T, Nakamura Y and Shiohara Y 1993 *J. Mater. Res.* **8** 1240
[43] Salama K and Selvamanickam V 1992 *Appl. Phys. Lett.* **60** 898
[44] Kimura K, Miyamoto K and Hashimoto M 1995 *Advances in Superconductivity* **vol 7**, ed K Yamafuji and T Morishita (Tokyo: Springer-Verlag) p 681
[45] Sakai N, Matthews D N, Hedderich R, Takaichi H and Murakami M 1994 *Advances in Superconductivity* **vol 6**, ed T Fujita and Y Shiohara (Tokyo: Springer-Verlag) p 803
[46] Morita M, Takebayashi S, Tanaka M, Kimura K, Miyamoto K and Sawano K 1991 *Advances in Superconductivity* **vol 3**, ed K Kajimura and H Hayakawa (Tokyo: Springer-Verlag) p 733
[47] Maeda J, Seiki S, Kambara M, Nakamura Y, Izumi T and Shiohara Y 2000 *Advances in Superconductivity* **vol 12**, ed T Yamashita and K Tanabe (Tokyo: Springer-Verlag) p 449
[48] Endo A, Watanabe Y, Miyake K, Umeda T, Murata K and Shiohara Y 1997 *J. Jpn. Inst. Met.* **61** 963
[49] Uhlmann D R, Chalmers B and Jackson K A 1964 *J. Appl. Phys.* **35** 2986

[50] Maeda J, Kawase T, Nakamura Y, Izumi Y, Murata K and Shiohara Y, unpublished.
[51] Tachikawa K, Sadakata N, Sugimoto M and Kohno O 1988 *Japan. J. Appl. Phys.* **27** L1501
[52] Tachikawa K, Kohchi N, Matsumoto H and Minemoto T 1999 *Advances in Superconductivity* **vol 11**, ed N Koshizuka and S Tajima (Tokyo: Springer-Verlag) p 797
[53] Tachikawa K, Hikichi Y, Zama K, Moriyasu T and Suzuki T 1993 *IEEE Trans. Appl. Supercond.* **3** 1174
[54] Tachikawa K, Watanabe T, Inoue T and Shirasu K 1991 *Japan J. Appl. Phys.* **30** 639
[55] Tachikawa K, Zama K and Kikuchi A 1993 *Japan. J. Appl. Phys.* **32** L654
[56] Kikuchi A, Kinoshita T, Nishikawa N, Komiya S and Tachikawa K 1995 *Japan J. Appl. Phys.* **27** L167
[57] Kase J, Togano K, Kumakura H, Dietderich D R, Irisawa N, Morimoto T and Maeda H 1990 *Japan J. Appl. Phys.* **29** L1096
[58] Heine K, Tenbrink J and Thoner M 1989 *Appl. Phys. Lett.* **55** 2441
[59] Halder P, Hoehn J G Jr, Rice J A and Motowidlo L R 1992 *Appl. Phys. Lett.* **60** 495
[60] Nomura K, Seido M, Kitaguchi H, Kumakura H, Togano K and Maeda H 1993 *Appl. Phys. Lett.* **62** 2131
[61] Okada M, Tanaka K, Fukushima K, Sato J, Kitaguchi H, Kumakura H, Kiyoshi T, Inoue K and Togano K 1996 *Japan J. Appl. Phys.* **35** 63
[62] Hase T, Shibutani K, Hayashi S, Ogawa R and Kawate Y 1995 *Cryogenics* **35** 127
[63] Shimoyama J, Morimoto T, Kitaguchi H, Kumakura H, Togano K, Maeda H, Nomura K and Seido M 1992 *Japan J. Appl. Phys.* **31** L163
[64] Nomura K, Sasaoka T, Sato J, Kuma S, Kumakura H, Togano K and Tomita N 1994 *Appl. Phys. Lett.* **64** 112
[65] Nomura K, Sato J, Kuma S, Kumakura H, Togano K and Tomita N 1994 *Appl. Phys. Lett.* **64** 912
[66] Tomita N, *et al*, 1995 *IEEE Trans. Appl. Supercond.* **5** 520
[67] Walker M S, *et al*, 1997 *IEEE Trans. Appl. Supercond.* **7** 664
[68] Hasegawa T, Hikichi Y, Koizumi T, Imai A, Kumakura H, Kitaguchi H and Togano K 1997 *IEEE Trans. Appl. Supercond.* **7** 1703
[69] Dietderich D R, Ullmann B, Freyhardt H C, Kase J, Kumakura H, Togano K and Maeda H 1990 *Japan J. Appl. Phys.* **29** L1100
[70] Kase J, Morimoto T, Togano K, Kumakura H, Dietderich D R and Maeda H 1991 *IEEE Trans. Mag.* **27** 1254
[71] Hasegawa T, Kobayashi H, Kumakura H, Kitaguchi H and Togano K 1994 *Supercond. Sci. Technol.* **7** 579
[72] Zhang W and Hellstrom E E 1995 *Supercond. Sci. Technol.* **8** 430
[73] Hasegawa T, Kobayashi H, Kumakura H and Togano K 1995 *Advances in Superconductivity* **vol 7**, ed K Yamafuji and T Morishita (Tokyo: Springer-Verlag) p 719
[74] Kumakura H, Kitaguchi H, Togano K and Sugiyama N 1996 *J. Appl. Phys.* **80** 5162
[75] Fujii H, Kumakura H, Kitaguchi H, Togano K, Zhang W, Feng Y and Hellstrom E E 1997 *IEEE Trans. Appl. Supercond.* **7** 1707
[76] Shimoyama J, Tomita N, Morimoto T, Kitaguchi H, Kumakura H, Togano K, Maeda H, Nomura K and Seido M 1992 *Japan J. Appl. Phys.* **31** 1328
[77] Hasegawa T, Hikichi Y, Kumakura H, Kitaguchi H and Togano K 1997 *Proc. 1997 Int. Workshop on Superconductivity* p 64
[78] Zhang W, Pupysheva O V, Ma Y, Polak M, Hellstrom E E and Larbalestier D C 1997 *IEEE Trans. Appl. Supercond.* **7** 1544

B2.3.3
Melt processing techniques: melt processing of YBCO

Isabelle Monot and Gilbert Desgardin

B2.3.3.1 Introduction

During recent years, there have been spectacular demonstrations of the capability of high temperature superconductors for the electrical power grid; prototype power cables, transformers, fault current limiters, energy storage devices and motors have been constructed. Cryomagnetic applications of superconductors require the fabrication of bulk ceramic with different shapes and sizes and high critical current densities (J_c). The melt processing techniques are successful in overcoming the J_c limitations associated with grain boundaries' weak links and increasing the flux pinning force through the introduction of pinning centres in the microstucture. The YBCO system has been extensively studied in this way. While reported J_c values of melt processed YBCO superconductors have already surpassed the lower limits for practical applications, J_c enhancement above 10^5 A cm^{-2} and less severe degradation in high fields at 77 K are still required for better performances and improved safety margin. The fabrication of textured YBCO samples with high critical current values has been accomplished by several groups with various melt processing methods. The main characteristic of these processings is to decompose the material peritectically above the decomposition temperature of $YBa_2Cu_3O_{7-\delta}$ (Y123) (1020°C in air) and then to cool it down very slowly. They may also use an additional driving force such as temperature gradient, magnetic field or seeds, for the growth of highly textured YBCO ceramics. The non superconducting Y_2BaCuO_5 (Y211) phase always remains in the final microstructure in the form of inclusions, entrapped in the Y123 matrix, due to the incomplete peritectic recombination. These Y211 inclusions, like several other types of defects such as twin boundaries, dislocations and stacking faults have been considered as possible flux pinning centres in melt textured YBCO. Numerous different processing routes were developed in order to optimize the properties, the microstructure and the material shape required for applications. All electro-magnetic properties of bulk superconductors depend on the current load, the product of current density and the size of the current loop. The maximum size of the current loop, fixed by the material size and shape, will be defined by the application design. Once the material shape is given, only an increase in the critical current density will lead to an improvement of the performances (see sections E1 and E2). In order to meet these requirements, complete investigations of the defects formation mechanisms in relationship with the processing methodology and with the critical currents or levitation forces of these materials have been carried out.

E1 B2.3.2 A1.3 A1.3 B2.3.2 E1, E2

Among all the melt processings which have been developed to date, one can observe a large difference between the original melt texture growth (MTG) process, firstly developed by Jin [1], and the melt powder melt growth (MPMG) process, introduced more recently by Murakami [2]. The first one uses the stoichiometric Y123 sintered precursor while the latter involves powder oxide precursors

B2.2.2

with 40 mol% molar excess of Y_2BaCuO_5 (Y211) phase and 0.3–0.5 wt% of Pt. After the texture formation, their microstructures are drastically different, mainly due to the size and the dispersion of the Y211 inclusions. Very high magnetic J_c can be achieved with MPMG processed samples, especially between 0 and 1 T [3]. Besides these two processings, several modified procedures have been developed. The most characteristic ones will be briefly described in this chapter. Various compositions, Y211 excess, or dopants have been used by different groups leading to a rather large variety of microstructures and properties. Among those various results, it is sometimes difficult to determine which parameter plays a significant role in the melt textured 123 microstructure, the growth mechanisms and the superconducting properties. Furthermore, phase diagrams are not fully known in these new complex systems, which makes it difficult to choose suitable conditions for sample preparation. However, after complete investigations of several groups on the nucleation and growth mechanisms of Y123 during melt processings [4–9], it is now commonly accepted that the Y211 excess and all the factors which might refine and homogeneously distribute the Y211 inclusions and increase their volume fraction are beneficial to the microstructure and to the superconducting properties of the textured Y123 [10–12].

B2.3.2

The fracture toughness and thermal shock resistance have also be taken into account from a functional point of view [13]. Thermal shock resistance of the melt textured composite Y123/Y211 has been shown to be better in the melt textured ceramics than in the sintered ones due to the laminated microstructure which undergoes a very small [14] temperature gradient.

D3.3

After all the research on the YBCO melt processings, the recent goal that has to be achieved is to produce samples suitable for industrial applications. This implies the synthesis of samples with J_c close to 10^5 A cm^{-2} at 77 K and in self field, produced with a melt processing as simple as possible, in a reproducible manner with the lowest amount of wasted material and the best homogeneity of pieces.

A1.3

In recent years, many efforts have been made to increase the textured domain size, and the actual developments are to the process itself and to the driving force of the texture formation, all the results indicating that the only use of slow cooling is that it does not yield extended domains. Three main driving forces will be described in this chapter: temperature or composition gradients, magnetic field and top seeding. The fundamental and technological aspects of the different melt processings will also be developed, taking into account that the precursor, composition and processing conditions are constantly linked and never independent. The major aspects of the microstructural characteristics will be presented for each case. Finally the most representative superconducting perfomances will be extracted from the results to illustrate the evolution and progress of melt processings.

B2.3.3.2 First developments of the melt processings

To overcome the J_c limitations, due to weak link properties of the grain boundaries in sintered YBCO ceramics, the anisotropic growth of Y123 crystal favoured in the a,b direction has been used to develop texture formation processings. Weak link free current path in the texture direction is thus created due to the limited number of grain boundaries.

A1.3

In order to obtain highly oriented samples Jin *et al* in 1988 developed a melt process called MTG [1, 15]. In this process, a sintered YBCO sample was melted and slowly cooled in a thermal gradient of 20–50°C cm^{-1}. The shape of the sample can be preserved, thanks to the peritectic decomposition of the Y123 phase, above 1010°C in air.

$$2YBa_2Cu_3O_{7-x} \rightarrow Y_2BaCuO_5 + (3BaCuO_2 + 2CuO)\text{liquid, for } 1010°C < T < 1050°C. \quad \text{(B2.3.3.1)}$$

The decomposition of the Y123 phase, controlled by temperature and time, at high temperature leads to a highly viscous melt which limits liquid losses.

The following slow cooling allows the preferred orientation of the grain growth along the ab planes to take place, and thus grains are aligned along this direction. The growth in the c direction is intrinsically three to five times lower than in the ab directions, thereby the final microstructure results in a stacking of 123 plates. The Y123 phase matrix includes Y211 particles, randomly distributed throughout the oriented grains, due to the liquid losses and the incomplete peritectic recombination. The external thermal gradient imposed in this process provides a supplementary driving force which helps a planar Y123 growth front to take place and progress. Several observations have concluded that the growth of Y123 does not proceed through the classical peritectic reaction [3], which consists of the formation of a 123 envelope around individual Y211 particles. The investigators [6, 16–18] proposed a growth mechanism based on the dissolution of 211 particles, which can be summarized as follows: the Y123 phase nucleates in the melt where the composition is favourable and rapidly grows along the ab planes. The Y211 particles dissolve ahead of the growth front to provide the Y supply. Their dissolution creates a concentration gradient between 211/liquid and liquid/123 interfaces which consist here in the intrinsic driving force for the diffusion of Y through the liquid phase to sustain the growth. Once the Y211 particles are entrapped by the growth front, they will not contribute to the Y supply anymore. For further details see chapter B2.3.2. The continuous growth of Y123 requires a steady supply of Y which is only provided by the dissolution of Y211 particles in the liquid ahead of the advancing front. Since Y is only slightly soluble in the liquid [19] (< 2 mol%) it is likely that its rate of transport to the interface will be the limiting factor of growth.

B2.3.2

B2.3.2

Transport J_c values of the MTG samples obtained by Jin *et al* in 1988 exceeded 10^4 A cm^{-2} in zero field, at the liquid nitrogen temperature, 77 K. This result indicated that grain alignment may have extensively reduced the weak links and lead to a great material improvement.

The typical microstructure is shown in figure B2.3.3.1 and consists of large needle shaped 123 grains, occasionally including secondary phases. However, J_c values are below 4000 A cm^{-2} under an applied magnetic field of 1 T, at 77 K, which are still very low for practical applications.

In order to improve J_c values in magnetic fields, effective pinning centres must be introduced. This requires a careful control of the microstructure, as explained hereafter. Growth mechanism studies have revealed that an abundance of small Y211 particles, ahead of the growth front in the molten part of the sample, will enhance the Y concentration gradient in this region and increase the maximum growth rate of 123 [20–23]. Realizing the need for small 211 inclusion size and for the homogeneous distribution of these inclusions particles in the melt, Murakami *et al* developed a process termed MPMG [3] very similar to the QMG process developed by Morita [24–27] at the same time. In this method a sintered Y rich YBCO powder is rapidly heated to approximately 1400°C in a Pt crucible, where the sample decomposes into Y_2O_3 + Liquid and incorporates 0.3–0.5 wt% Pt. The sample is then quenched to room temperature and grown into powder. At this stage the sample consists of Y_2O_3 particles and the solidified liquid phase (a mixture of barium cuprates and amorphous phases). The powder is subsequently pressed and rapidly heated to 1100°C in air and held at this temperature during an appropriate holding time for partial melting. During this stage the homogeneously dispersed Y_2O_3 particles are converted to evenly distributed small 211 precipitates. The sample is then rapidly cooled to 1000°C just above the peritectic temperature followed by slow cooling at a rate of 1–5°C h^{-1} to approximately 950°C. The sample is finally annealed in flowing oxygen to incorporate oxygen and to obtain optimized superconducting properties.

A1.3, A4.3

This process leads to improved domain size reaching cubic centimetre easily and microstructural observations reveal that the size of the 211 phase is much finer and its distribution is much more uniform than in classical MTG samples. Figure B2.3.3.2 shows the microstructure of a MPMG YBCO sample with the composition of Y : Ba : Cu = 1.4 : 2.2 : 3.2. It is notable that fine 211 inclusions are dispersed in the 123 matrix. The number of 211 inclusions dispersed in the matrix can be controlled by changing the starting composition.

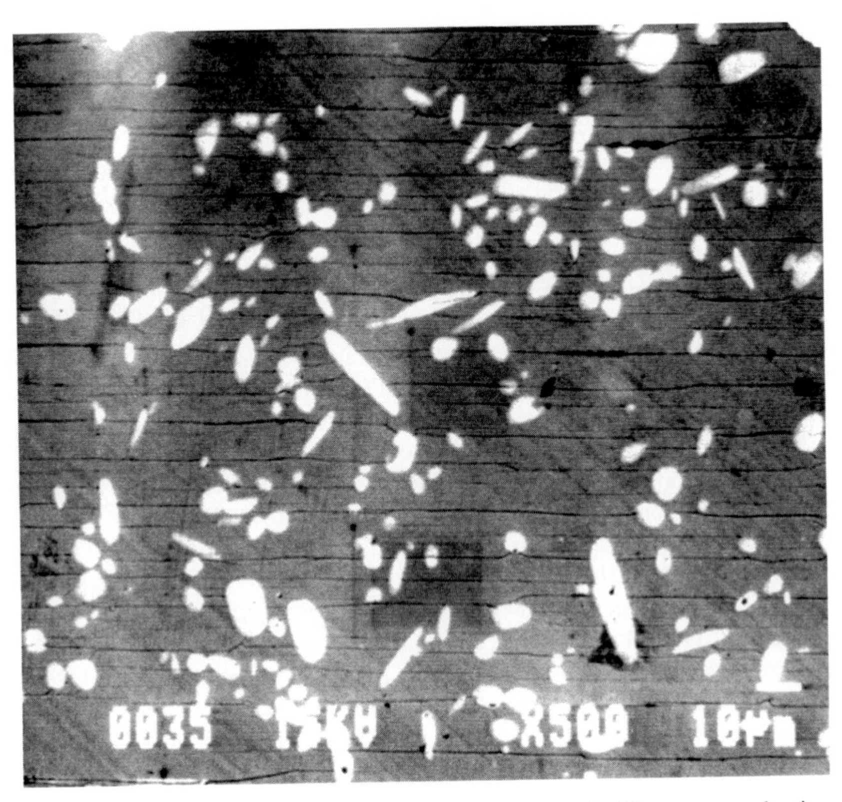

Figure B2.3.3.1. Typical microstructure of an MTG $Yba_2Cu_3O_{7-d}$ sample. The superconducting matrix includes Y_2BaCuO_5 particles (10–50 μm) randomly distributed.

This process leads to improved J_c values especially under a magnetic field of 1 T. $J_c(B)$ variations calculated from magnetization measurements on MPMG processed samples exhibit $5-6 \times 10^4$ A cm^{-2} in self field and over 2×10^4 A cm^{-2} under 1 T. Murakami *et al* [28–30] have shown the improvement of J_c with increasing 211 content reducing the 211 particle size which proves the role of the microstructure in the pinning properties of these materials. However, the MPMG process has shown the possibility of improving the superconducting properties of the YBCO system drastically but it remains very complex and difficult to handle. This explains why, in between these two characteristic melt processings, many investigations were devoted to the development of an effective simple route to produce YBCO melt textured samples. These developments can be classified into two groups—the first one, devoted to the study of starting composition and precursor optimization, and the second one, focused on the processings themselves with different configuration and driving forces for texture formation.

B2.3.3.3 Precursors and composition developments

Besides the use of the simple stoichiometric Y123 sintered sample on melt quench Y rich powder, several modified procedures have been developed with different precursors in order to try to control the growth conditions and the microstructure better and to simplify the precursor preparation and costs. Thus the powder melt process (PMP) starts from a mixture of Y_2BaCuO_5 (Y211) [31, 32], $BaCuO_2$ and CuO pressed precursor, the solid liquid melt growth process (SLMG) uses

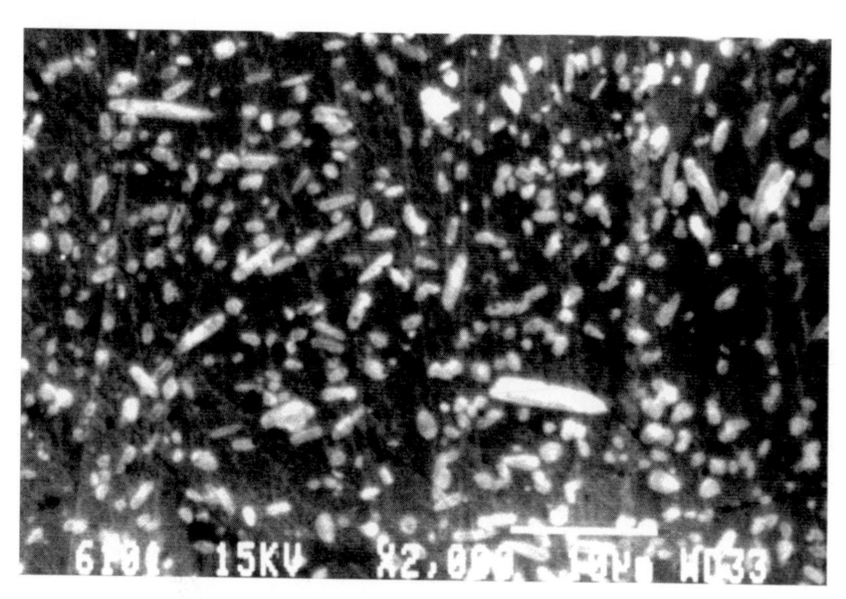

Figure B2.3.3.2. Microstructure obtained with the MPMG process from the $Y_{1.4}Ba_{2.2}Cu_{3.2}O_x$ composition. Y_2BaCuO_5 are micron size and very homogeneously dispersed in the Y123 matrix.

$Y_2O_3/BaCuO_2/CuO$ mixed powders [33–35] and the powder melt growth (PMG) starts from Y123/Y211 pressed pellets.

The essential difference between these processes lies in the combination of the different starting powders prior to melt texturing. Another difference lies in the initial stoichiometry, which means that the Y211 excess content can be largely varied, typically between 0 and 40 mol% (nominal compositions up to 60 mol% have even been used).

These variations result in microstructures which essentially differ by the size and distribution of the Y211 inclusions as shown in figure B2.3.3.3 with several examples. However, melt textured Y123 samples fabricated by these derived methods are found to consist of multiple oriented domains which are randomly directed with respect to their neighbours. The boundaries separating these domains are determined to be high angle grain boundaries and might include secondary phases [36, 37].

As further improvements, realizing from the MPMG process results the need for small Y211 particle size and their homogeneous distribution in the melt, other investigators have studied the influence of dopants such as Pt, Ce and Sn (for the most effective ones) on the texture formation and growth of Y123 [38–41]. The actions of the dopants are not yet completely understood but their benefit is essentially attributed to their action on the Y211 particles, especially for Ce or Pt dopants. The addition of a few weight per cent of dopant in the nominal composition seems to modify the viscosity of the melt and thus the interfacial energy between Y211 grains and the melt. These modifications prevent the Y211 grain coarsening in the molten state and induce some drastic changes in the 211 morphology as shown in figure B2.3.3.4 for the Pt doped samples. The Y211 grain morphology is modified above the peritectic decomposition temperature from round shapes in undoped samples to long aciculate shapes in the MPMG processed YBCO. In this latter case, 0.3 wt% of Pt is incorporated from the Pt crucible during the high temperature decomposition at 1400°C.

B2.3.2

In the case of Pt doping, usually in the form of metallic Pt or PtO_2 powders, numerous studies have confirmed its action; if only prereacted Y211 are employed as additive to Y123 in the nominal composition, Pt definitely plays the role of an effective growth inhibitor in the partially molten state.

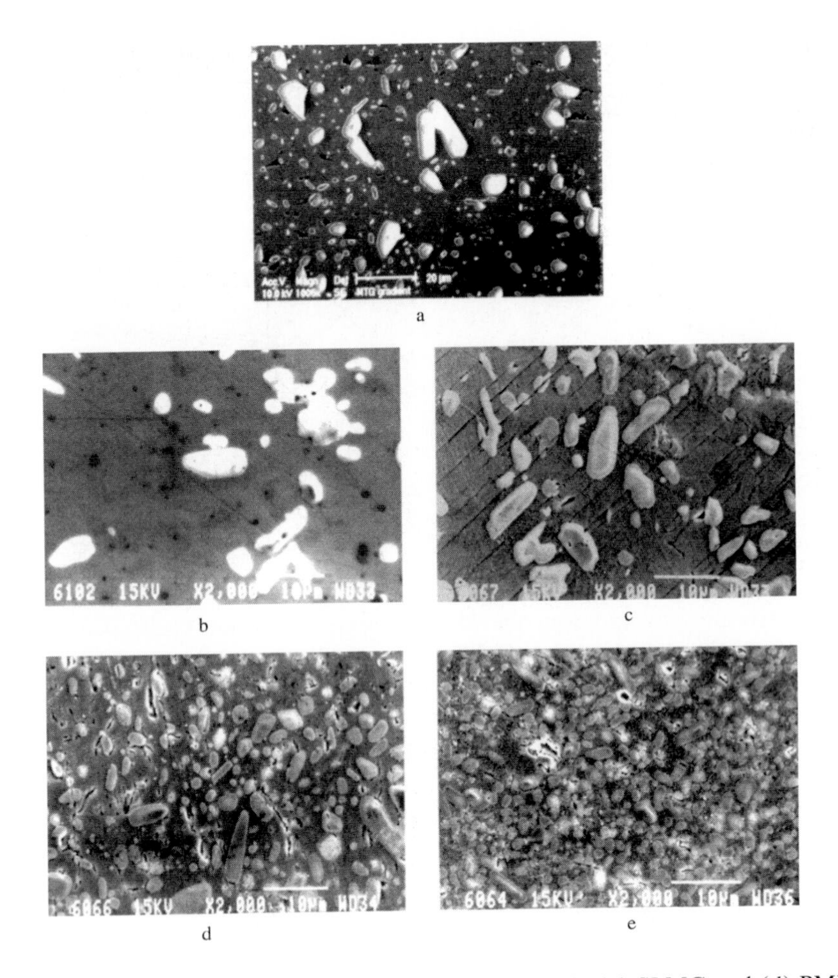

Figure B2.3.3.3. Examples of microstructures obtained by (*a*) MTG, (*b*) SLMG and (*d*) PMP processings with 20 mol% excess of Y211 and (*c*) SLMG and (*d*) PMP processings with 40 mol% Y211 in excess in the nominal compositions.

If Y211 nucleate and grow from Y_2O_3 and liquid (like in the SLMG or MPMG process) or from the decomposition of Y123, Pt also plays a role in the effective nucleation site for the Y211 refinement. Many authors have reported on the effect of Pt [42–47], Ce [48–52], or Sn [53–60], or their combination on the Y211 morphology and further Y123 grain growth.

In any case if the 211 particles are refined in the melt and prevented from grain growth coarsening or Ostwald ripening, the resulting microstructures exhibit finely dispersed Y211 inclusions. An example is given, for doped and undoped samples synthesized by the MTG process, in figure B2.3.3.5. In the case of Ce or Sn doping, usually in the form of CeO_2 and SnO_2 additions from 0.5 to 2 wt%, the 211 morphology is not drastically modified but the particle size is reduced. These dopants act mainly as grain growth inhibitors by modification of the viscosity of the melt and the interfacial energies. It is worth pointing out the importance of these doping effects on the Y211 particle size, morphology and distribution at high temperature, because these controls will ensure the control of the melting stage which includes liquid formation, phase homogenization and liquid losses.

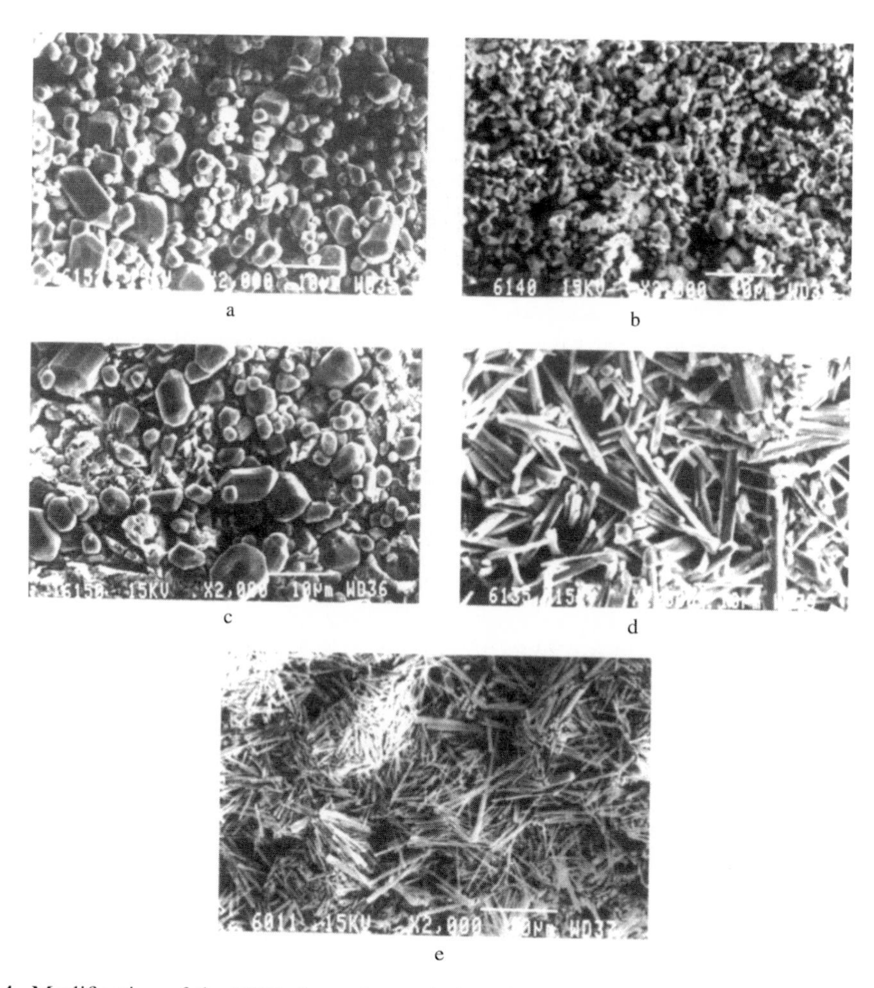

Figure B2.3.3.4. Modification of the Y211 size and morphology in the melt due to Pt doping in the MPM process (*a*) without Pt, (*b*) with 0.5 wt% PtO_2 in the SLMG process, (*c*) without Pt, (*d*) with Pt and (*e*) in the MPMG process.

As it is now commonly accepted that Y supply is the limiting factor of Y123 grain growth, the growth conditions are closely linked to the microstructure of the melt. The shift of the composition towards Y rich composition and modification of the radius curvature of the Y211 particles will modify the phase diagram and thus the solidification conditions. Further fundamental studies have revealed the importance of the Y211 radius curvature in the dissolution of these particles [5, 61, 62]. They have predicted that when Y is no more the limiting factor of the Y123 growth, the interface kinetics will mainly govern the growth rate. The main result is that when an abundance of small 211 particles, either due to the use of Y rich starting composition or due to Y211 refinement, are in the melt, the concentration gradient of Y ahead of the growth front will be enhanced and the maximum 123 growth rate will be increased. As a consequence the conditions for a planar continuous growth front will be maintained and larger single oriented domains will be produced.

Several other dopants such as Zr, Al_2O_3 [63–66] have been tried but Ce, Pt and Sn remain the most effective ones for microstructural and domain size improvements. The last point that should be

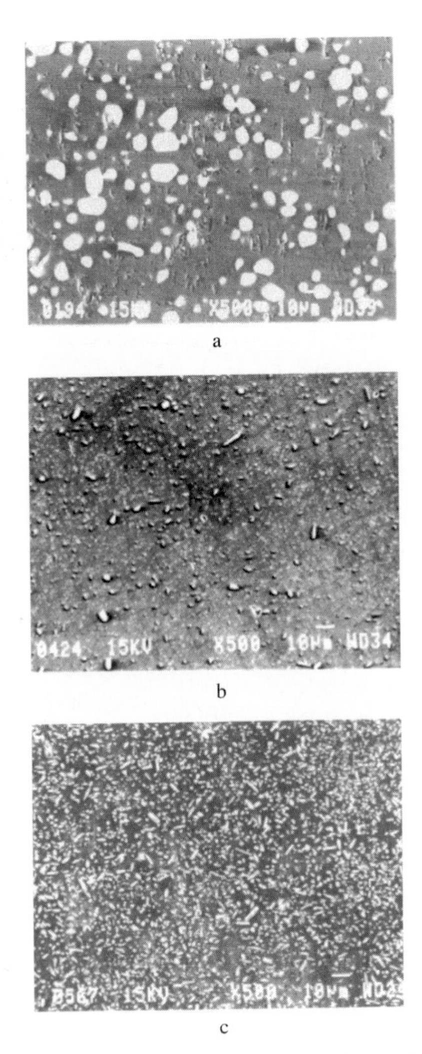

Figure B2.3.3.5. SEM micrograph of $Y_{1.4}Ba_{2.2}Cu_{3.2}O_x$ MTG processed samples: (*a*) undoped, (*b*) 0.5 wt% Pt doped and (*c*) 2 wt% Ce doped.

emphasized is the optimization of the composition of melt processed YBCO by silver addition. Indeed, melt textured YBCO samples exhibit poor toughness and thermal shock resistance. Their behaviour is typically brittle. This could be a real problem for their practical use in superconducting devices, especially for large pieces. Mechanical studies have shown the effective role of silver addition, in the proportion of 10–15 wt%, in increasing the toughness and the thermal shock resistance of highly textured Y123/Y211 composites as shown in figure B2.3.3.6 [14, 67–70]. In this new composite system, the bridging of cracks by plastically deformed silver inclusion, shown in figure B2.3.3.7, prevents the microcrack propagation and improves the mechanical resistances of this highly anisotropic material. Moreover, for silver addition below 15 wt% the 123 growth conditions are not largely modified and the resulting superconducting properties are not affected. The second part of the development of Y123 single domains has been devoted to the process itself. The process is generally determined to produce a necessary shape (bars, rod shaped samples, pellets of small or large diameters etc.) and also to produce

D3.3

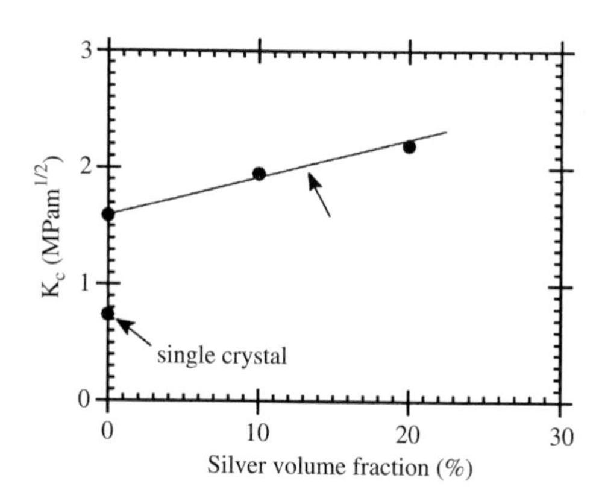

Figure B2.3.3.6. Improvement of the fracture toughness with Ag content in Y123/Y211/Ag melt textured composites. The value for a single crystal is mentioned for comparison.

a supplementary driving force for the texture formation while the processing conditions have to be determined precisely for an accurate knowledge and control of the growth conditions. A brief review of the different processes successful for the fabrication of highly textured YBCO samples will be proposed hereafter.

B2.3.3.4 Developments of the processing conditions

Taking into account the possible improvements of growth conditions with variations of the composition in order to get a better control of the microstructure, the developments of the melt processings of YBCO all over the world have revealed the importance of the processing conditions on the Y123 grain growth. This means that the precursor density (sintered or pressed), the temperature and time for the peritectic decomposition, the cooling or moving rate, the external thermal gradient submitted to the sample are all important parameters to know and control precisely. B2.2.2

A classification can be proposed from the additional driving force of the Y123 growth provided by these different melt processings. Four groups of melt processings can thus be distinguished for the texture formation: horizontal Bridgman, floating zone (with a laser zone melting or in a microwave cavity), high magnetic field and top seeding (with single crystal seed or cold finger). B2.4

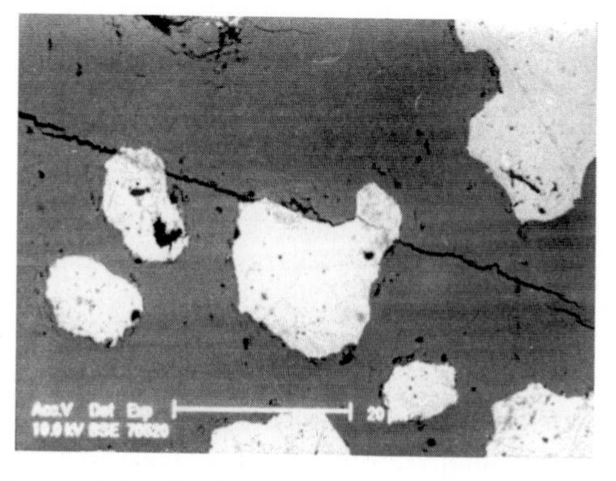

Figure B2.3.3.7. Bridging of the crack by plastically deformed silver particles in Y123/Y211/Ag composites.

Horizontal Bridgman: MTG

B2.4 In the early developments of the melt texturing techniques for YBCO, many research groups have studied the horizontal Bridgman technique [71]. In this method the whole sample lying on a support is decomposed above the peritectic temperature and recrystallized by slow cooling of the furnace or low rate translation of the sample or the furnace. An additional driving force for the texture formation is provided by the thermal gradient which can be submitted to the sample during its recrystallization. Theoretical studies by Cima *et al* [72, 73] have shown that a planar growth front and thus a continuous growth can be obtained with a compromise between the ratio of G and R (G – thermal gradient and R – the processing rate) and the product (G × R) (expressed in °C h^{-1}) representing the interfacial recrystallization rate which will determine the grain size, the latter varying inversely with the product (G × R).

Enhanced domain size has been obtained from these considerations with the horizontal Bridgman method. However, this method does not allow the whole sample to be submitted to the same thermal treatment. The growth front progress, from one extremity of the sample to the other, imposes a different holding time above the recrystallization temperature all along the sample. This leads to variation of the composition of the melt of the liquid losses and of the size of the Y211 inclusions along the sample.

Moreover, the necessary presence of a support below the sample provides reactivity, nucleation sites and unexpected radial gradients, making the formation of a single domain difficult, in a reproducible manner. In this regard, the use of Y211 sintered bars [74] has been found to be one of the best candidates for Y123 support. Their similar components avoid the pollution phenomena and their relatively close thermal expansion coefficients prevent cracks formation in Y123, initiated by the contact between the sample and the support.

A variation of the thermal gradient melt growth method is the composition gradient melt growth technique which utilizes the variation in formation temperatures of the $REBa_2Cu_3O_{7-d}$ class of compounds to control the direction of the grain growth during the melt process [75]. This is achieved by shaping a sample of gradually varying composition through its thickness so that the compound at the upper surface of the specimen forms at a higher temperature than that at the bottom. The low temperature formation of $YbBa_2Cu_3O_{7-d}$ (900°C) compared with that of $YBa_2Cu_3O_{7-d}$ (1000°C) makes this compound suitable for this purpose. The samples are usually made of $Y_{1-x}Yb_xBa_2Cu_3O_{7-d}$ and $Y_{2-x}Yb_xBaCuO_5$ with increasing concentration of Yb from the top to the bottom of the sample. The resultant sample is presintered and submitted to melt processing. The variation in composition causes the sample to solidify directionally from the top ($x = 0$) downwards on cooling which simulates an external applied thermal gradient. Limited success has been achieved in the fabrication of large grain, although the optimum processing conditions are not fully developed. Grain sizes below 1 cm have been achieved which exhibit moderated critical current densities at 77 K. The main advantage of this method is that no temperature gradient is required, but the actual poor superconducting results and the complex sample preparation, which is both difficult and time consuming, lead us to conclude that the full potential of this technique has yet to be established.

Zone melting: vertical MTG

B2.4 In order to avoid all the problems associated with the contact between the support and the Y123 samples, vertical melt processings have been developed since the early stage of YBCO melt processings [76–78]. In this method long bar samples are held and translated through a narrow hot zone where the sample is ‘zone melted’. The high temperature gradient generated on both sides of the hot zone associated with the pulling rate provides the driving force for the texture formation. With this technique, the whole sample is submitted to the same thermal treatment, the problems associated with the support do not exist anymore and the process is continuous, allowing the growth of long sample rods.

The only difficulty to overcome is to avoid the formation of a neck in the molten zone and the sample fall. This problem can be partially solved by adjusting the Y211/Y123/ dopants composition in order to generate a high viscosity of the melt and by also adjusting the sample weight that means the sample length. Very good results have been obtained with the longitudinal gradient of 100–150°C cm^{-1} [79]. Variations of this vertical zone melted process consist of limiting the extent of the molten zone by modification of the heating apparatus. Instead of the classical furnace, laser radiations have been focused in one point providing gradients over 1000°C cm^{-1} [80, 81]. Another original system involves the use of a TE102 microwave cavity where the sample, correctly located in a maximum of the magnetic field, will be locally heated by induction [82]. This process will be hereafter called microwave melt textured growth (MMTG). This heating mode, based on the interaction between radiation and matter, needs an accurate and constant temperature control by adjusting the delivered power necessary to obtain an homogeneously treated sample. An improvement of this process is to place a tubular shaped conductor in the microwave cavity. The latter interacts with the generated magnetic field and is heated itself by induction. The heating zone in the middle of the cavity is totally correlated with the magnetic field profile $H(y)$ and is not sample dependent anymore. The microwave power is then fixed and the resonance conditions maintained constant. In this configuration thermal instabilities do not occur anymore and sample radiation is limited, avoiding this rapid cooling of the sample surface and multi-nucleation during the recrystallization. The thermal gradient obtained with this configuration is about 250°C cm^{-1} (intermediate between the classical furnace and laser zone melting apparatus) [83]. Very long nearly single grains have been synthesized with this technique with a pulling rate close to 2 mm h^{-1} as shown in figure B2.3.3.8. One of the beneficial points of this method is that only longitudinal grain boundaries are formed during this process. These longitudinal grain boundaries will not be so detrimental to the current flow as the transverse grain boundaries often produced in classical zone melting processes.

In all these zone melting processes, the high temperature gradients generated should have allowed relatively high growth rates according to the theory of constitutional supercooling (see chapter B2.3.2). All the results have shown that the lower the growth (or pulling) rate, the longer the single domain. The best results for pulling rates were obtained below 2 mm h^{-1}, indicating that the supplementary driving force provided by the thermal gradient is not sufficient for the Y supply to the growth front. A second common result is the control of the direction of the (ab) planes which is really difficult to align with the sample rod axis. In spite of these limitations the transport critical current density of the textured rods exceeds several 10^4 A cm^2 at 77 K in zero field, although the direction of the growth of the (a,b) planes is not yet optimized.

These developments are an important step with regard towards applications such as current leads (see also chapter E1.2).

E1.2

Melt texturing under magnetic field: FMTG

In the research of a supplementary driving force for the continuous growth of large single domains, some successful attempts have been made with the help of the driving force of an external magnetic field [84–86]. This technique is based on the fact that Y123 crystals possess an anisotropic paramagnetic susceptibility ($\chi//c > \chi\perp c$). When placed in a magnetic field this anisotropy leads to an alignment of the c-axis of each crystal parallel to the direction of the applied field. Both calculations and experiments have shown that at temperatures sufficiently high to melt Y123, this magnetic force is large enough to overcome misorienting forces arising from thermal vibrations or shape anisotropy. In the melting stage and during the recrystallization of Y123, the crystallites nucleated in the melt rotate under the influence of the magnetic field to align their c-axis parallel to the applied field. In this way, one can carefully

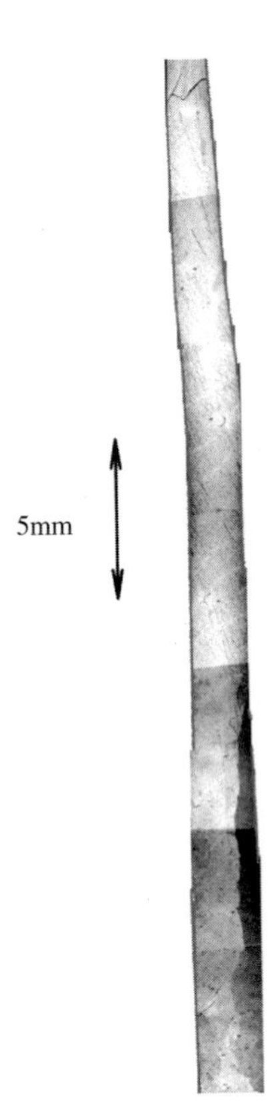

Figure B2.3.3.8. Polished section of a MMTG processed at 1.9 mm h^{-1}. A nearly single domain appears over more than 5 cm in length and only longitudinal grain boundaries can be observed.

control the direction of the induced orientation. This step was performed in a vertical tubular furnace place in the room temperature bore of a superconducting magnet. Magnetic fields up to 7 T may be applied parallel to the long axis of the furnace. It has been confirmed that it is the magnetic field, and not, for example, the presence of an unexpected temperature gradient within the furnace, which is responsible for the induced orientation. Slow cooling rates are employed to encourage growth of large crystal grains and result in a material with a bulk textured structure. Figure B2.3.3.9 shows a scanning electron microscope micrograph of a field melt textured sample. Thin Y123 platelets with their c-axis aligned parallel to the annealing field and 211 precipitate 10–30 μm are evidenced. The influence of the starting material composition of the thermal cycle and the field strength and direction on the superconducting properties have been intensively investigated. This technique allows a good control of the sample orientation but needs a relatively heavy apparatus.

D2.2 High transport critical current densities have been obtained in these field textured samples, especially under high magnetic fields. For an external field applied parallel to both a–b planes and the direction of the current flow, transport critical current densities of 16 200 A cm^2 in self field and 11 000 A cm^{-2} under 1 T were reproducibly obtained [87].

Top seeding: TSMTG

A1.3 The extensive technological development of melt textured bulk monoliths is possible only if the sample synthesis is optimized and suitable for massive production. This means that the sample preparation

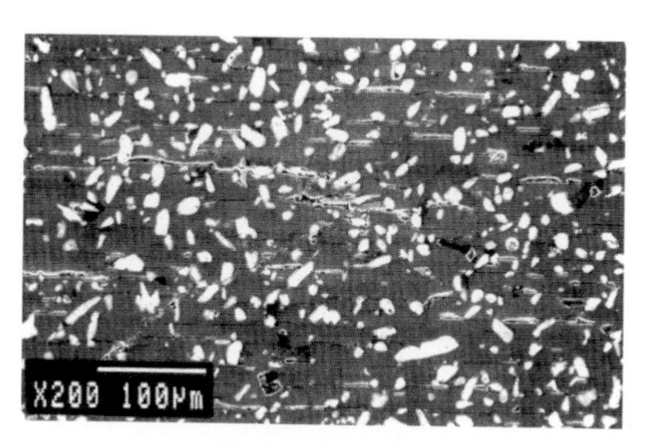

Figure B2.3.3.9. Reproduced with the permission of Tounier (Matformag, France). SEM of a Y123 FMTG YBCO sample.

process should be as simple as possible, with high reproducibility and a minimum of wasted material. Taking these considerations into account the top seeding method, suggested in the early stage of the developments of melt processings [89, 88, 3,], is extensively studied in the recent investigations [90, 91]. This process seems to be actually the most promising for melt texture progress towards industrialization [92].

A single crystal of Sm123 and, in a later context, Nd123 and MgO crystals have been shown to be effective seeds for Y123 superconductors [93–95]. Here, a seed is defined as a nucleation centre where the crystal orientation of the growth matches that of the nucleation source. Due to the higher melting temperature of the seed, the latter remains solid when Y123 is heated above its peritectic decomposition temperature. On subsequent cooling, epitaxial growth of Y123 occurs at the seed (Liquid + Y211) interface due to the similar lattice parameters of the two 123 compounds. Seeded growth from the Sm123 single crystal or the melt textured cleaved surface has been demonstrated as one of the most effective ones. Figure B2.3.3.10 shows the result of the growth from a Sm123 seed placed on the top surface of a Y123/Y211 composite pellet. This method has been successfully used by several groups with different precursor powders, and some variations in the process, showing thus the promising results of this method. The combination of seeding with pre-existing methods such as horizontal Bridgman [96] or vertical zone melting [97] is actually under investigation to achieve highly reproducible results, with an accurate growth control.

B2.3.2

However, the success of this method which combined several driving forces for the Y123 control growth is conditioned by an accurate control of the growth conditions combined with the composition [98]. Besides, the use of a seed severely modifies the phase diagram, especially the undercooling for the 123 nucleation which is considerably reduced by the presence of a seed [99, 100]. On the other hand, in order to grow the whole sample from the seed and by further sympathetic nucleation, homogeneous nucleation in the melt should be totally avoided. These two restrictive conditions define a solidification window, that means a temperature range defined by the nucleation temperature from the seed for the upper one and the homogeneous nucleation in the melt for the lower one, in which the whole sample has to grow. Moreover experimental results have shown that this solidification window from the phase diagram is also composition dependant [101–103]. Once the solidification window is determined for a given process and composition, the cooling rate has to be adjusted in order to grow the desired domain size.

In the most recent developments of the top seeding method, batches of pellets have been successfully produced in one run, in a homogeneous temperature profile, that means without supplementary driving force for the growth, thus proving the high efficiency of this technique for mass production of

Figure B2.3.3.10. Micrograph of the top surface of a Y123 single domain grown from a cleaved melt textured Sm123 seed.

A1.3 melt textured monoliths [104]. The only difficulty to overcome is to have high quality seeds, extracted from melt textured Sm123. The low margin between the peritectic decomposition temperature of Y123 (1020°C) and that of Sm123 (1060°C) needs a high purity of the Sm123 phase. Furthermore, B2.3.4 independently of the phase purity, this system might easily form a solid solution between the Sm and the Ba sites $Sm_{1+x}Ba_{2-x}CuO_y$ especially when synthesized in air or oxygen. The substituted phases exhibit a lower decomposition temperature than stoichiometric Sm123, thus reducing the safety margin of the decomposition temperature between the two materials. Recent investigations on REBaCuO systems have shown the possibility of growing high quality large grains of $SmBa_2Cu_3O_{7-x}$, thus producing high quality seeds [105]. Investigations are still under way to find other candidates appropriate for seeding, in particular MgO and Nd123 crystals are suggested.

A1.3 The actual challenge is to grow very large diameter pellets exhibiting large levitation forces for application. The combination of seed-composition, thermal gradients and thermal treatment has to be accurately adjusted and the best results concern actually up to 10 cm diameter pellets grown in Japan [106, 107] and in France [108].

One of the actual limitations of this method is the limited superconducting properties (especially the levitation force) resulting from the difficulty to reoxygenate these single domains properly and completely due to their large size and high density [109, 110].

Variations of the top seeding melt texturing process were to bring the seed after the high temperature stage [111]. This procedure allows the decomposition of the YBCO material at a higher temperature (above 1100°C) to obtain a better homogeneity of the melt in order to reduce the number of nuclei and thus to limit the growth of undesirable grains afterwards. However, the control of seed deposition is difficult to handle and to reproduce; moreover this procedure is not suitable for batch production. Another way to control the nucleation and the following growth of 123 is to introduce a cold finger formed with a small tube in which circulating air constantly cools the finger placed very close from the top surface of the pellet. This cold point gives the nucleation point for the beginning of the growth and also generates a small radial thermal gradient of a few degrees per centimetre which helps for the orientation of the growth as a supplementary driving force. This method is not widely used for the same reasons as that of the previous one which means its reproducibility is not evidenced and is difficult to optimize for the synthesis of batches.

Other processings considerations

All the YBCO samples produced by the various melt processing techniques require oxygen uptake before they exhibit optimum superconducting properties (T_c, J_c, levitation forces). This treatment involves heating the sample above its orthorhombic to tetragonal structural phase transition, which occurs typically between 450 and 600°C under atmospheric pressure, followed by cooling to room temperature under flowing oxygen.

Complete oxygenation is difficult to achieve, particularly in large, dense and highly textured samples. Typical oxygenation times depend on sample porosity, homogeneity and the size of specimen and may vary from several hours to several weeks.

This point might originate some large variations of the superconducting properties in samples synthesized with a similar melt texturing process but with a different annealing procedure. The post-fabrication annealing of bulk materials will be developed in chapter B2.5.

The industrial production of Y123 melt textured material needs a perfect control of the grain growth and for the future achievement of these specimens, a reliable synthesis method has to be found in order to adjust together the precursor (purity, granulometry ...), the composition (Y211 additions, dopants and doping level ...), the employed driving force (isothermal conditions, thermal gradient, top seeding, cold finger, sample pulling ...), the apparatus configuration, the sample geometry and the thermal procedure. This combination should yield high quality materials with optimized microstructure and superconducting properties.

Many contributors have tried to correlate the composition and the processing conditions to the resulting microstructure and the superconducting performances. Among all those interesting and useful correlations, it is, however, undoubtedly difficult to extract the best parameters to choose and combine to get a good result. From a practical point of view, some attempts have been made to try to find out the suitable process for a given composition. Two procedures, both developed by Tournier's group in France, will be described here.

The first one measures the magnetic susceptibility of the sample all along the melt processing [112]. This characteristic, with an example given in figure B2.3.3.11, reveals all the critical points of the process: first, the susceptibility decreases during the sample heating because of oxygen losses, then it increases rapidly at the decomposition temperature due to the higher magnetic susceptibility of the melt (Ba and Cu). The following drop is due to oxygen losses which induce copper reduction of Cu^{2+} in Cu^{+}. The recrystallization is connected with a decrease of the magnetic susceptibility in a narrow temperature range. After solidification, the material still incorporates oxygen and c increases again. The oxygen exchanges have been confirmed by the thermogravimetry recording during the susceptibility measurement; further details concerning these measurements are given in [110] and [111].

D2.5

In this way it is possible to correlate immediately the thermal treatment with the sample response in order to determine the optimum temperature and time necessary to get an homogeneous melt and to fit the cooling rate to the sample growth, mainly related to its composition and size. These experiments have underlined the importance of the oxygen losses and uptake in the decomposition and solidification temperature and rates of this material.

The second procedure has been developed by this same group to optimize the melt texture formation of large size pellets rapidly and efficiently [108, 111]. It is based on the direct observation of a top seeded sample during the heat treatment with a video camera placed above a vertical tube furnace. The recording of the sample surface evolution evidences the solidification window in which a single domain grows from the seed before the nucleation of new Y123 grains in the remaining melt. Moreover, a frame-by-frame analysis allows the calculation of (a,b) growth rates. Finally, this in-live growth observation helps to determine the best thermal treatment fitted to a given sample size and composition.

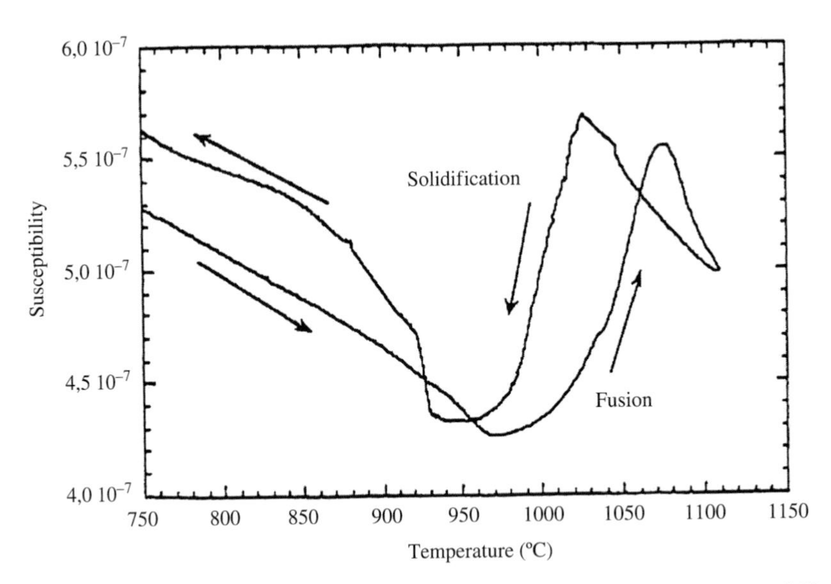

Figure B2.3.3.11. Reproduced with the permission of Tournier (Matformag, France). Susceptibility measurements of YBCO during a melt texturing process.

Once this thermal cycle is determined, it can be transferred reliably to a batch process for example, if the thermal environment (gradients, ramp rates etc.) of the sample is similar.

B2.3.3.5 Superconducting characteristics in melt textured YBCO

A4.0 Besides the critical temperature(T_c), critical current densities (J_c) are the most important controlling factors for the potential utilization of HTS in practical applications. Since virtually no grain boundaries exist along the current path for current flowing in the (a,b) planes, melt textured Y123 has been found to be able to carry J_c at least three orders of magnitude higher than those of sintered samples. The achieved critical current densities are already within the requirements of some superconducting devices (see E chapter E). Another important superconducting characteristic for the application of large size melt textured YBCO samples is the levitation force; however, it is actually difficult to compare the results obtained with the different processings because of the various measurement methods and procedures, especially due to the nature and geometry of the permanent magnet compared to the sample dimensions.

Typically, J_c of melt textured high T_c superconductors can be determined by either magnetization or transport methods. From the magnetization measurements by SQUID or the vibrating sample D2.4 magnetometer (VSM), the modified Bean model [113] is used to calculate J_c. With the external field applied parallel to the sample c axis, J_c is relatively sensitive to the magnetic field at 77 K. However, when the temperature is reduced to 60 K and below, J_c is only weakly dependant on the field strength. The limitation and disadvantage of this method is that if the magnetic field is directed away from the c direction, the screening current will flow along the (a,b) planes, the c direction and the grain boundaries, where the extraction of the real J_c value is quite complicated. In addition, errors in the J_c results might also arise from the uncertainty in the grain dimension and the demagnetization factors. However, the magnetization measurements are comparatively easy to perform.

D2.2 When J_c is determined by transport measurements, a variety of current–field configurations can be realized as shown in figure B2.3.3.12(a), which represents the J_c variations obtained from a field textured sample in different configurations of current flow and applied field direction. Another example is given in figure B2.3.3.12(*b*), which describes the angular dependence of the transport critical current densities with the applied field (fixed at 8 T in this example). The disadvantages of the transport method lie mainly in the Joule heating generated at the current contacts and the relatively high voltage criterion used in the measurement (1–5 mV cm^{-1}). Moreover, the high critical current densities need to prepare very small sample sections, thus adding some difficulties in the sample preparation.

Examples of the variations in J_c with the magnetic field will be given hereafter to illustrate the effects of composition and melt processings on the superconducting performances of these materials. The developments of melt processings have been mainly devoted to improve the microstructure homogeneity and yield extended single domains with low material losses. However, obtaining high superconducting performances by the introduction of effective pinning centres in the microstructure has been, and remains, the constant goal of all the studies.

Effect of Y_2BaCuO_5 (Y211) additions

The superconducting properties of the samples prepared by melt processed methods with varying concentrations and sizes of Y_2BaCuO_5 added to the $YBa_2Cu_3O_x$ precursor powder have been largely investigated. Murakami [3, 114] has clearly shown the improvement of J_c values with Y211 excess in MPMG YBCO samples. Figure B2.3.3.13(*a*) shows the magnetic field dependence of J_c at 77 K for three MPMG samples with different Y211 contents. The J_c values can be significantly enhanced by increasing the Y211 content. This result supports the idea that Y211 inclusions are effective in flux pinning. Figure B2.3.3.13(*b*), which plots J_c at 77 K and under 1 T external field for various volume fraction ratios of Y211 over inclusion size, evidences the direct proportionality between these two variables. This confirms the efficiency of controlling the microstructure of melt textured samples especially for what concerns the content, distribution and size of the Y211 secondary phase. TEM and HREM studies, shown in figure B2.3.3.14, have tried to correlate the microstructure and the super-

A1.3

D1.2

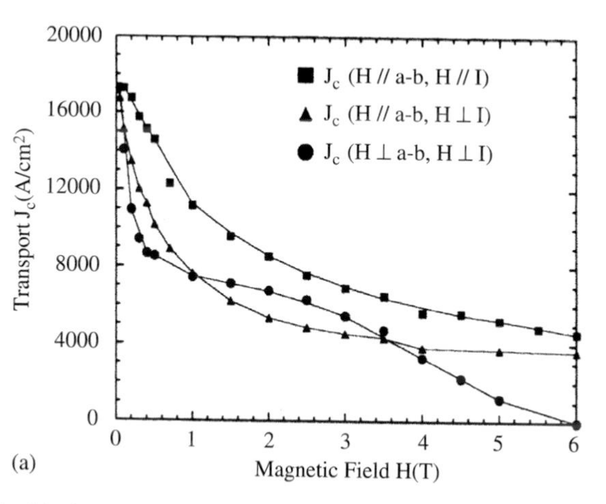

Figure B2.3.3.12. Reproduced with the permission of Tournier (Matformag, France). (*a*) Transport measurements of the critical current densities in a FMTG sample with different configurations of the flowing current and the external field at 77 K. (*b*) Angular dependence of the transport J_c at 77 K and under an applied field of 8 T with the current flowing along the (a,b) planes and zero angle corresponding to the field applied parallel to the ab planes and still perpendicular to the current flow in a TSMTG sample.

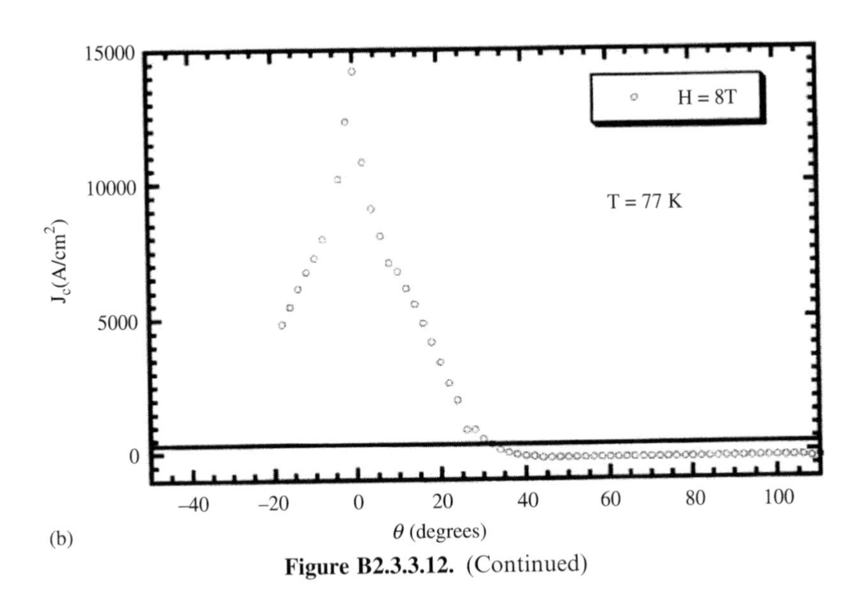

Figure B2.3.3.12. (Continued)

conducting properties in melt processed YBCO samples. Interfacial pinning between the superconducting matrix and the normal inclusions undoubtedly originates the enhancement of the properties, but other indirect effects have been also suggested such as the increase of the dislocation density in the vicinity of the Y211 particles [115, 116] or the modification of the twin structure and density [117, 118].

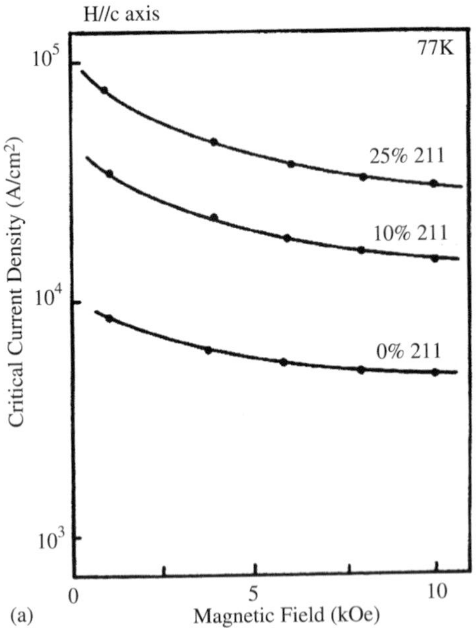

Figure B2.3.3.13. Reproduced with the permission of Murakami (ISTEC, Japan). (*a*) Magnetic field dependence of magnetic J_c at 77 K, with H parallel to c for three MPMG samples with different Y211 contents. (*b*) A plot of J_c (77 K, $B//c$, 1 T) versus V_f/d of the Y211 inclusions for various melt processed YBCO with different Y211 contents and sizes. V_f/d corresponds to the effective surface area of phase boundary for flux pinning.

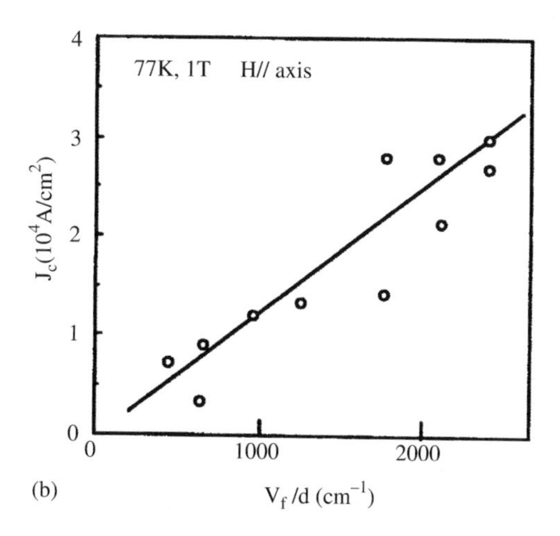

Figure B2.3.3.13. (Continued)

These considerations are still under debate and investigation, to determine the most effective contribution of the Y211 particles in the pinning properties under low and high magnetic field.

Effects of the dopants

The effect of a dopant such as Pt, Ce or Sn is mainly related to the Y123 growth conditions by the modification of the molten state (Y211 morphology and dissolution, interfacial energies and kinetics, viscosity of the melt) and the modification of the growth rates, in the (a,b) planes and in the c direction,

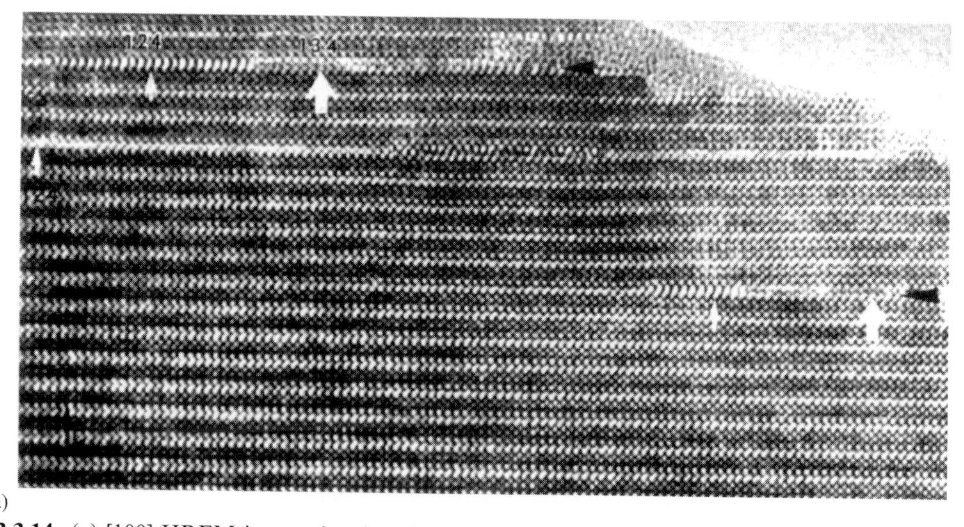

Figure B2.3.3.14. (*a*) [100] HREM image showing the stacking defects of the type 124 (small white arrows) and more complex 224 defects (big white arrows). (*b*) small torn like twins near the 211/123 interface. (*c*) [001] bright field image showing the twin structure near $BaCeO_3$ inclusion, and partial dislocations running along the [100] or [010] directions.

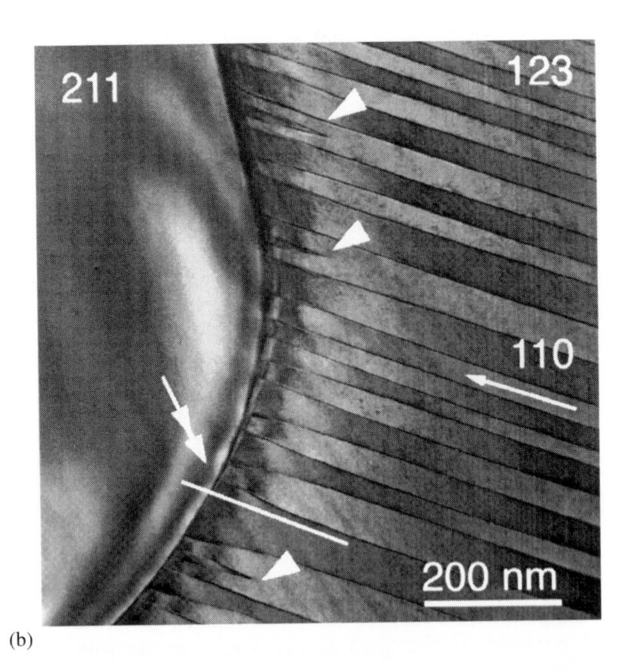

(b)

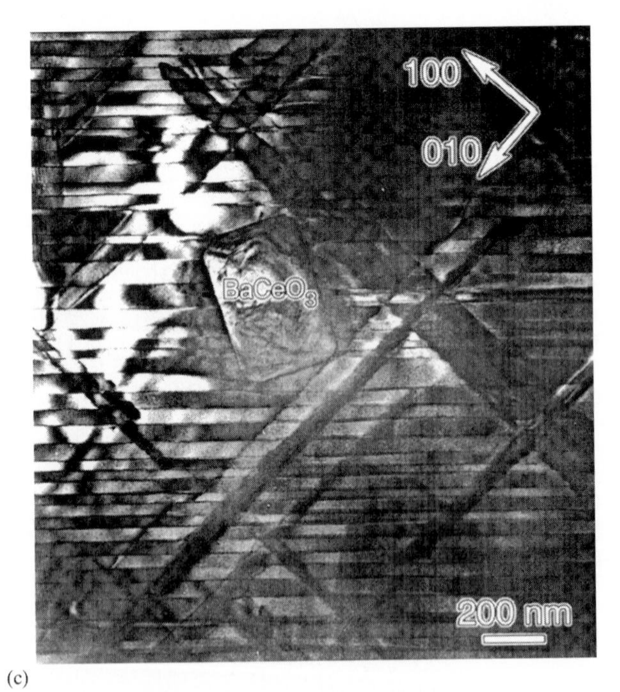

(c)

Figure B2.3.3.14. (Continued)

leading to an improved microstructure. With the introduction of such dopants in the nominal composition some secondary phases may be formed or the superconducting Y123 matrix might incorporate these foreign elements. As a consequence the superconducting transition temperature might be affected. Figure B2.3.3.15 shows that the T_c values might be slightly depressed by 2 wt% Ce doping while 0.5 wt% Pt preserves a high T_c value, above 92.5 K for the T_c onset; the combination of Ce and Pt gives an intermediate result with a T_c onset at 92 K. In these three cases, the transition is similar and quite broad (about 3 K) [50].

The effect of these same dopants on the superconducting properties of MTG samples grown in a thermal gradient with the horizontal Bridgman method are shown in figure B2.3.3.16. Pt is effective in maintaining the J_c values under magnetic fields up to 1 T, while Ce greatly increases the J_c values in self field and very low field. As expected, the combination of these two dopants (0.5 wt% CeO_2 and 0.5 wt% PtO_2) leads to very interesting J_c values at 77 K, exceeding 6×10^4 A cm^{-2} in self field and still $4 \times 5\ 10^4$ A cm^{-2} under 1 T).

B2.4

The second point to be noted is that the doping level should be adjusted for each composition and process in order to optimize its effectiveness and to find out the best compromise between T_c and J_c. An example is given in figure B2.3.3.17 which shows that in a MPMG processed sample doped with Sn, the best J_c variation is obtained for 10 mol% of $BaSnO_3$ addition, while 5 mol% tends to depress the superconducting properties. Sn doping also tends to increase the superconducting transition temperature [56].

Effects of the melt processing

Y211 addition and some dopants have been proved to be effective in enhancing the microstructure and consequently the superconducting properties. However, the optimum level of addition and doping is closely related to the thermal conditions, the diffusivity and the mass transport; thus with the same precursors, large variations can be observed as shown in figure B2.3.3.18. These results obtained in TSMTG samples differ clearly from those obtained in MTG processed samples with the same compositions and are presented in figure B2.3.3.16[119, 120]. In the case of TSMTG lower variations are observed with the different dopant tested and in this case, Pt doping seems to be the most effective. A last illustration of the role of the process is given with the comparison of Ce doped samples in three different melt processings – MMTG, MTG and TSMTG. Ce doping does not seem to be effective in enhancing MMTG sample properties while large low field J_c values may be obtained in MTG and TSMTG processed samples. However, Ce has been shown to be very effective in the vertical zone melting process by other authors [81].

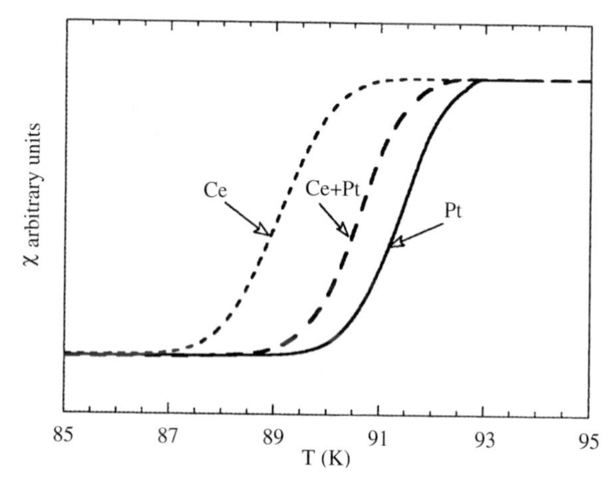

Figure B2.3.3.15. Superconducting transition of three $Y_{1.4}Ba_{2.2}Cu_{3.2}O_x$ MTG processed samples doped with 0.5 wt% PtO_2, or 2 wt% CeO_2 or both of them.

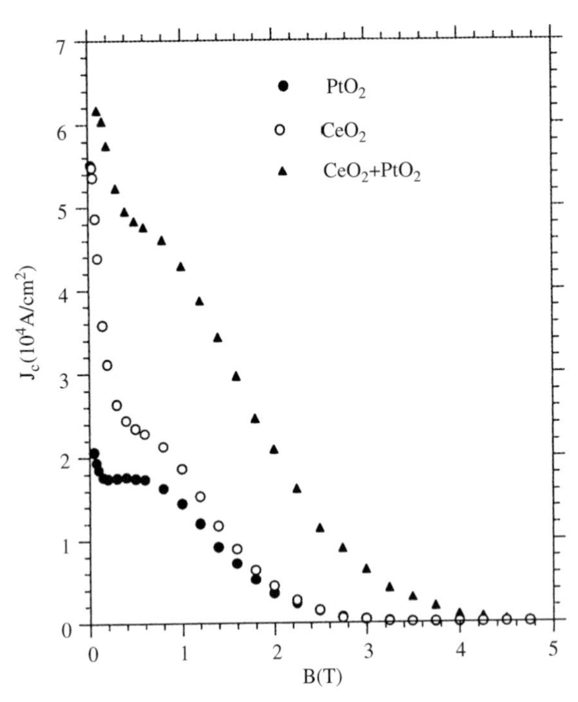

Figure B2.3.3.16. Magnetic critical current densities J_c of $Y_{1.4}Ba_{2.2}Cu_{3.2}O_x$ MTG processed samples as a function of the applied field at 77 K and with Ce, Pt and Ce + Pt doping; the magnetic field is applied parallel to the sample c axis.

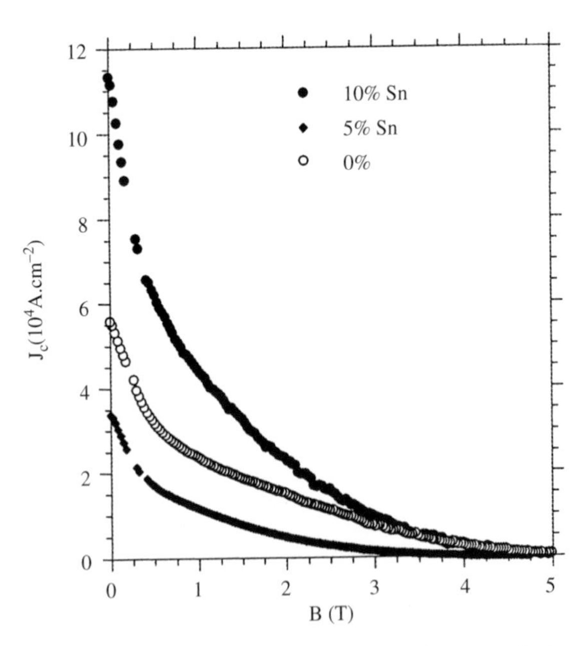

Figure B2.3.3.17. Influence of the doping level on $J_c(B)$ behaviour in the case of 0, 5 and 10 mol% $BaSnO_3$ doping in $Y_{1.8}Ba_{2.4}Cu_{3.4}O_x$ MPMG processed samples (77 K, $B//c$).

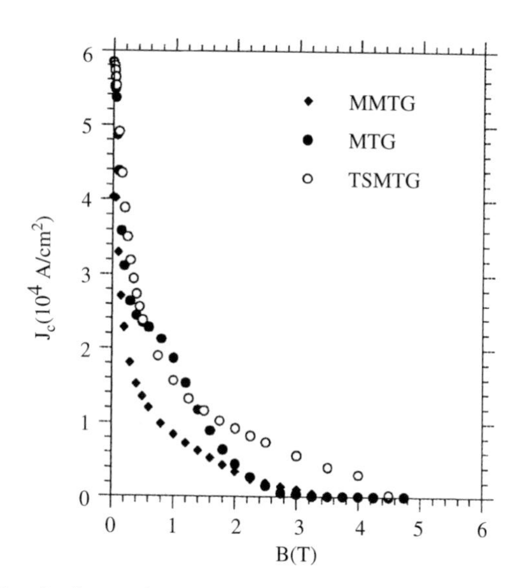

Figure B2.3.3.18. Influence of the doping and process; J_c (B) behaviour in the case of $Y_{1.4}Ba_{2.2}Cu_{3.2}O_x$ TSMTG processed samples for Pt, Ce and Ce + Pt doped compositions (77 K, $B//c$).

These examples cannot be a representative of all the results obtained with the different melt processes, but they simply provide evidence for the close relation between precursor, composition and processing, which cannot be independently considered in the search for an optimized and reproducible way of production of high performances of large quantities of YBCO superconductors.

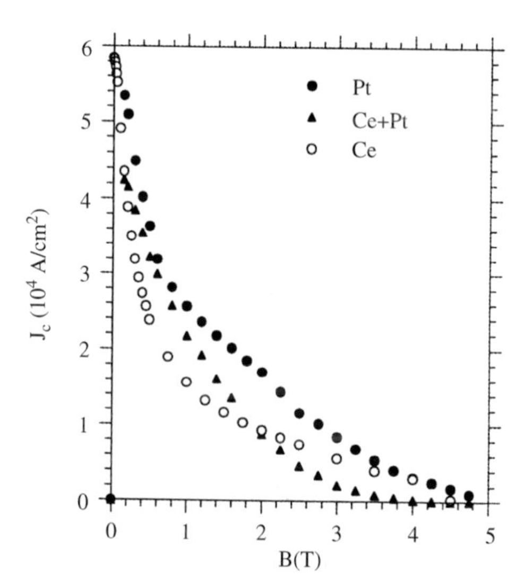

Figure B2.3.3.19. Influence of the process; J_c (B) variations for optimized Ce doped compositions in MTG, MMTG and TSMTG processed samples (77 K, $B//c$).

B2.3.3.6 Conclusion

Over the past 10 years, different melt texturing methods have been developed, based on the peritectic reaction which occurs at temperatures between 1000 and 1050°C in the Y–Ba–Cu–O system. Significant progress has been made, principally due to the efforts which have been made to modify the microstructure and to increase the size as well as the alignment of the oriented domains.

After the submission to melt processing, YBCO material typically contains platelet boundaries and characteristic Y211 second phase inclusions in their microstructure. The size and distribution of these inclusions, which have been successfully refined by effective dopants like Pt or Ce, have been correlated with J_c improvement.

In all cases, the melting temperature, the peritectic solidification temperature and the optimum cooling rate are key process parameters which can vary significantly with sample composition and melt processing configuration. Since the continuous growth of Y123 is believed to be sustained by Y diffusion from the dissolving Y211–liquid interface to the Y123–liquid growth front, the maximum allowable growth rate remains small. Consequently, a slow cooling rate and/or a slow travelling rate have to be utilized to maintain the stability of the planar Y123 growth front. This slow growth rate, however, can be marginally improved by decreasing the inter-particle spacing of the Y211 precipitates. This has been achieved by the addition of an excess of Y211 in the nominal compositions and by the action of dopants as Y211 grain growth inhibitors.

In non directional melt texturing, the domain size is limited to multiple Y123 nucleation sites and further impingement of the domains. The first developments of the MTG process have lead to improved superconducting properties but on multi domain samples. On the other hand, when an additional driving force is imposed, or when the nucleation site is controlled, in so-called directional solidification processings, much extended domains can be obtained.

Thermal gradients or the magnetic field used as a supplementary driving force yield large well oriented domain size, but the sample length or diameter is still limited by the apparatus configuration. With the vertical Bridgman configuration with zone melting by conventional or laser or modified microwave heating, very long samples having a diameter of up to 3–4 mm, with grain alignment relatively parallel to the sample axis, have been fabricated. The restriction of this method is due to the self weight of the sample which limits the possible length and diameter. Moreover for diameters larger than 5 mm, the control of the orientation is more difficult and the number of domains is increased. For samples with larger cross-section, seeded directional solidification was shown to be the most effective and actually the most promising for applications.

Due to grain alignment in melt textured Y123, the critical currents densities J_c obtained from magnetic or transport measurements, with the current flowing in the highly superconductive (a,b) planes, can reach 10^5 A cm^{-2} at 77 K. Even though these J_c values are relatively large and promising, a strong J_c anisotropy is observed when the direction of the applied field is varied and these values are limited by depinning. Improvement in J_c has been achieved by introducing additional flux pinning centres to the microstructure with an accurate control of the latter. Recent melt texturing methods have been developed to accomplish this task, especially with Y211 and the addition of dopants which results in an increasing number of small Y211 precipitates with size less than 2 μm. Several studies have shown that the interfaces of small high surface curvature are responsible for the J_c enhancement.

The main idea arising from the summary of all the studies concerning the developments of the melt processing techniques is that the composition, the process parameters and the heating configuration are always closely linked and determine the growth conditions in each melt texturing method. After 10 years of development, with the advances in melt texturing methods and J_c improvements, use of bulk high T_c superconductors in simple applications should be possible shortly.

Acknowledgments

The authors thank their collaborators C Leblond, S Marinel, M P Delamare and J Wang for extensive experimental and analytical work and discussions. They also thank B Raveau for leading this work.

References

[1] Jin S, Tiefel T H, Sherwood R C, Van Dover R B, Davis M E, Kammlott G W and Fastnacht R A 1988 Melt textured growth of polycrystalline $Yba_2Cu_3O_{7-\delta}$ with high transport J_c at 77 K *Phys. Rev.* B **37** 7850

[2] Murakami M, Morita M, Doi K and Miyamoto K 1989 A new process with the promise of High J_c in oxide superconductors *Japan. J. Appl. Phys.* **28** 1189

[3] Murakami M 1992 *High Temperature Supeconductor* (Singapore: World Scientific)

[4] St John D H 1990 The peritectic reaction *Acta. Metall. Mater.* **38** 631

[5] Cima M J, Flemings M C, Figueredo A M, Nakade M, Ishii H, Brody H D and Hagerty J S 1992 Semi solid solidification of high temprature superconducting oxides *J. Appl. Phys.* **72** 1868

[6] Izumi T, Nakamura Y and Shiohara Y 1992 Crystal growth mechanism of $Yba_2Cu_3O_{7-\delta}$ superconductors with peritectic reaction *J. Mater. Res.* **7** 1621

[7] Izumi T, Nakamura Y and Shiohara Y 1992 Diffusion solidification model on Y-system superconductors *J. Mater. Res.* **7** 395

[8] Izumi T and Shiohora Y 1992 Growth mechanism of $YBa_2Cu_3O_y$ superconductors prepared by the horizontal Bridgman method *J. Mater. Res.* **7** 16

[9] Wolf T 1996 Crystal growth mechanism and $Yba_2Cu_3O_{7-\delta}$ growth anisotropy of crystals *J. Cryst. Growth* 4397

[10] Chen B J, Rodriguez M A, Mitsure S T and Snyder R L 1993 Effect of undercooling temperature on the solidification kinetics and morphology of Y–Ba–Cu–O during melt texturing *Physica* C **217** 367

[11] Jin S, Kammlott G W, Tiefel T H, Kodas T T, Ward T L and Kroeger P M 1991 Microstructure and properties of the Y–Ba–Cu–O superconductor with submicron 211 dispersions *Physica* C **181** 57

[12] McGinn P N, Zhu N, Chen W, Sengupta S and Li T 1991 Microstructure and critical current density of zone melt textured $Yba_2Cu_3O_{7-\delta}$ with Y_2BaCuO_5 additions *Physica* C **176** 203

[13] Kim C J, Kim K B, Chang I S, Won D Y, Moon H C and Suhn D S 1993 The effect of Y_2BaCuO_5 on the microstructure and formation of cracks in the partially melted YBaCuO oxides *J. Mater. Res.* **8** (4)

[14] Tancret F *PhD Thesis* University of Caen, France

[15] Jin S, Tiefel T H, Sherwood R C, Davis M E, Van Dover R B, Kammlott G W, Fashnacht R A and Kuth H D 1988 High critical currents in YBaCuO superconductors *Appl. Phys. Lett.* **52** 2074

[16] Rodriguez M A, Chen B J and Snyder R L 1992 The formation mechanism of textured $YBa_2Cu_3O_{7-\delta}$ *Physica* C **195** 185

[17] Jin S, Kammlott G W, Tiefel J H and Chen S K 1992 Formation of layered microstructure in the Y–Ba–Cu–O and Bi–Sr–Ca–Cu–O superconductors *Physica* C **198** 333

[18] Athur S P, Selvamanickam V, Balachandran U and Salama K 1996 Study of growth kinetics in melt-textured $YBa_2Cu_3O_{7-x}$ *J. Mater. Res.* **11** 2976

[19] Krauns C h, Sumida M, Tagami M, Yamada Y and Shiohara Y 1994 Solubility of RE elements into BaCuO melts and the enthalpy of dissolution *Z. Phys.* B **96** 207

[20] Pellerin N, Odier P, Simon P and Chateigner D 1994 Nucleation and growth mechanisms of textured YBaCuO and the influence of Y_2BaCuO_5 *Physica* C **222** 133

[21] Griffith M L, Huffman R T and Halloran J W 1994 Formation and coarsening behavior of Y_2BaCuO_5 from peritectic decomposition of $Yba_2Cu_3O_{7-x}$ *J. Mater. Res.* **9** 1633

[22] Sakai N, Yoo S I and Murakami M 1995 Control Y_2BaCuO_5 size and morphology in melt-processed $YBa_2Cu_3O_{7-\delta}$ superconductor *J. Mater. Res.* **10** 1611

[23] Frangi F, Higuchi T, Deguchi M and Murakami M 1995 Optimization of Y_2BaCuO_5 phase morphology for the growth of large bulk YBCO grains *J. Mater. Res.* **10** 2241

[24] Kimura K, Tanaka M, Horiuchi H, Morita M, Tanaka M, Matsuo M, Morikawa H and Sawano K 1991 A new domain structure in $YBa_2Cu_3O_{7-x}$ prepared by the quench and melt growth method *Physica* C **174** 263

[25] Morita M, Tanaka M, Sasaki T, Hashimoto H and Sawano K 1991 Magnets made of QMG crystals *Proc. ISS IV*

[26] Kimura K, Morita M, Tanaka M, Takebayashi S, Muyamoto K and Sawano K 1991 Critical current density of a $YBa_2Cu_3O_x$ by melt (QMG) process *Physica* C **185–189** 2467

[27] Morita M, Sawamura M, Takebayashi S, Kimura K, Teshima H, Kiyamoto K and Hashimoto M 1994 Processing and properties of QMG materials *Physica* C **235–240** 209

[28] Murakami M, Gotoh S, Fujimoto H, Yamaguchi K, Koshizuka N and Tanaka S 1991 Flux pinning and critical currents in melt processed YBCO *Supercond. Sci. Technol.* **4** S43

[29] Yamagushi K, Murakami, Fujimoto H, Gotoh S, Oyama T, Shiohara Y, Koshizuka N and Tanaka S 1991 Microstructure of the melt powder melt growth processed YBCO *J. Mater. Res.* **6** 1404

[30] Ni B, Kobayashi M, Fumaki K, Yamafuji K and Matsushita T 1991 Effect of $YBaCuO_5$ particles on pinning characteristics of YBaCuO prepared by Melt Powder Melt Growth Method *Japan. J. Appl. Phys.* **30** L1861

[31] Lian Z, Pingxiang Z, Oing J, Keguang W, Jingrong W and Xiaoto W 1990 The properties of YBCO superconductors prepared by a new approach: the powder melting process *Supercond. Sci. Technol.* **3** 490

[32] No K, Chung D S and Kim J M 1990 Fabrication of textured $YBa_2Cu_3O_x$ superconductor using unidirectional growth *J. Mater. Res.* **5** 2610

[33] Pavate V, Williams L B, Kvam E P, Vanarasi C and Mc Ginn P J 1994 Effects of platinum and oxygenation on microstructure in $Yba_2Cu_3O_{7-\delta}/YBaCuO_5$ bulk materials *J. Electron. Mater.* **23**

[34] Varanasi C, McGinn P J, Pavate V and Kvam E P 1994 Critical current density and microstructure of melt processed $YBa_2Cu_3O_x$ with PtO_2 additions *Physica* C **221** 46

[35] Shi D, Sengupta S and Luo J S 1993 Extremely fine precipitation and flux pinning in melt-processed $YBa_2Cu_3O_x$ *Physica* C **213** 179

[36] Ogawa J and Yamashita T 1995 Effect of the misorientation angle on the magnetic properties of YBCO grain boundary Josephson Junctions *IEEE Trans. Appl. Supercond* **5** 2204

[37] Müller D and Freyhardt H C 1996 Twin-boundary characteristics of melt-textured $YBa_2Cu_3O_7$ *Phil. Mag. Lett.* **73** 63

[38] Delamare M P, Monot I, Wang J, Provost J and Desgardin G 1996 Influence of CeO_2 BaCeO3 or PtO_2 additions on the microstructure and the critical current density of melt processed YBCO samples *Supercond. Sci. Technol.* **9** 534

[39] Monot I, Wang J, Delamare M P, Marinel S, Hervieu M, Provost J and Desgardin G 1997 Influence of the precursor and dopants on the chemistry and the texture formation of melt processed $Yba_2Cu_3O_{7-\delta}$ superconductors *Physica* C **282–287** 507

[40] Varanasi C, Black M A and McGinn P J 1994 A comparison of the effects of PtO_2 and $BaSnO_3$ additions on the refinement of Y_2BaCuO_5 and magnetization of textured $Yba_2Cu_3O_{6+x}$ *Supercond. Sci. Technol.* **7** 10

[41] Izumi T, Nakamura Y and Shiohara Y 1993 Doping effects on coarsening of Y_2BaCuO_5 phase in liquid *J. Mater. Res.* **8** 467

[42] Ogawa N, Hirabayashi I and Tanaka S 1991 Preparation of high J_c YBCO bulk superconductor by platinum doped Melt Growth Method *Physica* C **177** 101

[43] Varanasi C, McGinn P J, Pavate V and Kvam E P 1994 Critical current density and microstructure of melt-processed $YBa_2Cu_3O_x$ with PtO_2 additions *Physica* C **221** 46

[44] Morita M, Tanaka M, Takebayashi S, Kimura K, Miyamoto K and Sawano K 1991 Effect of Pt addition on Melt-processed YBaCuO superconductors *Japan. J. Appl. Phys.* **30** L813

[45] Kim W, Shim G, Jang D, Suh C, Shin W and Kwangsoono K 1994 Effects of Pt doping on microstructure of $YBa_2Cu_3O_x$ superconductor prepared by directional solidification *Japan. J. Appl. Phys.* **33** 999

[46] Wegmann M R and Lewis J A 1995 The role of platinum in partial melt textured growth of bulk YBCO *IEEE Trans. Appl. Supercond.* **5**

[47] Fagan J F, Partis D A, Richmond-Hope I A and Amarakoon V R W 1994 Influence of excess yttrium and platinum on the growth behavior of $YBa_2Cu_3O_{7-\delta}$ and Y_2BaCuO_5 *J. Electron. Mater.* **23**

[48] Pinol S, Sandiumenge F, Martinez B, Gomis B, Foncuberta J, Obradors X, Snoeck E and Roucau C H 1994 Enhanced critical currents by CeO_2 additions in directional by solidified $YBa_2Cu_3O_7$ *Appl. Phys. Lett.* **65** 1448

[49] Pinol S, Sandiumenge F, Martinez B, Vitalta N, Granados X, Gomis V, Galante F, Fontcuberta J and Obradors X 1995 Modified growth mechanism in directionnally solidified $YBa_2Cu_3O_7$ *IEEE Trans. Appl. Supercond.* **5** 1459

[50] Monot I, Verbist K, Hervieu M, Laffez P, Delamare M P, Wang J, Desgardin G and Van Tendeloo G 1997 Microstructure and flux pinning properties of melt textured grown doped $Yba_2Cu_3O_{7-\delta}$ *Physica* C **274** 523

[51] Ogawa N and Yoshida H 1992 Cerium oxide doped YBCO superconductor by melt growth method Proc. *Advances in Superconductivity IV (Tokyo)* p 455

[52] Monot I, Wang J, Delamare M P, Marinel S, Hervieu M, Provost J and Desgardin G 1997 Influence of the precursor and dopants on the chemistry and the texture formation of melt processed $Yba_2Cu_3O_{7-\delta}$ superconductors *Physica* C **282–287** 507

[53] Marinel S, Monot I, Provost J and Desgardin G 1998 Effect of SnO_2 and CeO_2 doping on the microstructure and superconducting properties of $Yba_2Cu_3O_{7-\delta}$ textured melted zone sample *Supercond. Sci. Technol.* **11** 563

[54] McGinn P, Chen W, Zhu N, Tan L, Varanasi C and Sengupta S 1991 Microstructure and critical current density of zone melted textured $Yba_2Cu_3O_{6+x}/YBaCuO_5$ with $BaSnO_3$ additions *Appl. Phys. Lett.* **59** 120

[55] Lepropre M *et al* 1994 Critical currents up to 71000 A cm^{-2} at 77 K in melt textured YBCO doped with $BaSnO_3$ *Cryogenics* **34** 63

[56] Monot I, Higuchi T, Sakai N and Murakami M 1994 Effect of $BaSnO_3$ additions in MPMG-processed YBCO *Physica* C **233** 155

[57] Shimoyama J I, Kase J, Kondoh S, Yanagagisawa E, Matsubara T, Suzuki M and Morimoto T 1990 Addition of new pinning center to unidirectionnaly melt solidified Y–Ba–Cu–O superconductor *Japan. J. Appl. Phys.* **29** L1999

[58] Varanasi C, Balkin D and McGinn P 1992 The chemical stability of $BaSnO_3$ in the melt $Yba_2Cu_3O_{6-x}$ during solidification *Mater. Lett.* **13** 363

[59] Song Y and Gaines J R 1993 Addition of $BaSnO_3$ to melt-textured $Yba_2Cu_3O_{7-\delta}$ *Supercond. Sci. Technol.* **6** 761

[60] Osamura K, Matsukura N, Kusumoto Y, Ochiai S, Ni B and Matsushita T 1990 Improvement of critical current density in $Yba_2Cu_3O_{6+x}$ superconductor by Sn addition *Japan. J. Appl. Phys.* **29** L1621

[61] Schmitz G J, Laakmann J, Wolters C h, Rex S, Gawalek W, Habisreuter T, Bruchlos G and Gornert P 1996 Influence of Y_2BaCuO_5 particles on the growth morphology of pentectically solidified $Yba_2Cu_3O_{7-\delta}$ *J. Mater. Res.* reprint

[62] Nakamura Y and Shiohara Y 1996 Peritectic solidification model for Y-system superconducting oxides *J. Mater. Res.* **11** 2450

[63] Juang J Y, Wu C L, Wang S J, Chu M L, Wu K H, Uen T M, Gou Y S, Chang H L, Wang C and Tsai M J 1994 Effects of ZrO_2 on the texturing and properties of melt processed $Y_1Ba_2Cu_3O_{7-\delta}$ *Appl. Phys. Lett.* **64**

[64] Chakrapani V, Balkin D and McGinn P 1993 The effect of second phase additions (SiC, $BaZrO_3$, $BaSnO_3$) on the microstructure and superconducting properties of zone melt textured $Yba_2Cu_3O_{7-\delta}$ *Appl. Supercond.* **1** 71

[65] Chen Y L, Zhang L, Chan H M and Harmer M P 1993 Controlled heterogeneous nucleation of melt textured $Yba_2Cu_3O_{7-\delta}$ by addition of Al_2O_3 particles *J. Mater. Res.* **8** 2128

[66] Cloots R, Robertz B, Auguste F, Rulmont A, Bougrine H, Vandewalle N and Ausloos M 1998 Effect Of $BaZrO_3$ additions on the microstructure and physical prperties of melt-textured Y123 superconducting materials *Mater. Sci. Eng.* **B53** 154

[67] Tiefel T H, Jin S, Sherwood R C, Medavis M, Kammlott G, Gallagher P K and Johnson D W 1989 Grain growth enhancement in $YBa_2Cu_3O_{7-\delta}$ superconductor by silver oxide doping *Mater. Lett.* **7** 363

[68] Mironova M, Lee D F and Salama K 1993 TEM and critical current density studies of melt-textured $YBa_2Cu_3O_x$ with silver and Y_2BaCuO_5 additions *Physica* C **211** 188

[69] Yun J, Marner M P and Chou Y C T 1994 Effect of silver addition on the microstructure of $YBa_2Cu_3O_{7-x}$ *J. Mater. Res.* **9**

[70] Tancret F, Monot I and Osterstock F 1997 Toughness and thermal shock resistance of melt textured YBCO ceramic superconductors *Euroceramic V, Key Engineering Materials* **132–136** 611

[71] Brice J C 1965 *The Growth of Crystals from the Melt* (Amsterdam: North-Holland) p 125

[72] Cima M J, Flemming M C, Figueredo A M, Nakade M, Ishii H, Brody H D and Haggerty J S 1992 Semi solid solidification of high temperature superconducting oxides *J. Appl. Phys.* **72** 179

[73] Cima M J, Jiang X P, Chow H M, Haggerty J S, Fleming M C, Brody H D, Laudise R A and Johnson D W 1990 Influence of growth parameters on the microstructure of directionally solidified $Bi_2Sr_2CaCu_2O_y$ *J. Mater. Res.* **5** 1834

[74] Monot I *PhD Thesis* University of Caen, France

[75] Morita M, Takebayashi S, Tanaka M, Kimura K, Miyamoto K and Sawano K 1990 *Processing Advances in Superconductivity* 3rd edn (Sendai) p 733

[76] McGinn P J, Chen W, Zhu N, Balachandran U and Lanagan M T 1990 Texture processing of extruded $YBa_2Cu_3O_{6+x}$ wires by melting zone *Physica* C **165** 48

[77] Imagawa Y and Shiohara Y 1996 Orientation control of Pt added YBaCuO by the directional solidification method *Physica* C **262** 243

[78] Van Tol H 1996 Critical current and pinning mechanisms in directionally solidified $Yba_2Cu_3O_{7-d}/Y_2BaCuO_5$ composites *Phys. Rev.* B **53** 2797

[79] Brand M, Gross C, Elschner S, Gauss S and Assmus W 1993 Transport critical current density of melt textured $Yba_2Cu_3O_{7-x}$ rods prepared by zone melting using a high temperature gradient *Proc. Eucas* (Göttingen) p 369

[80] Pellerin N, Odier P and Gervais M 1993 Texturation of YBaCuO by laser zone melting *J. Cryst. Growth* **129** 21

[81] Sandiumenge F, Martinez B and Obradors X 1997 Tailoring of the microstructure and critical currents in directionally solidified $Yba_2Cu_3O_{7-x}$ *Supercond. Sci. Technol.* **10** A93

[82] Marinel S, Desgardin G, Provost J and Raveau B 1998 A microwave melt texture growth process of $YBa_2Cu_3O_{7-\delta}$ *Mater. Sci. Eng.* B **52** 47

[83] Marinel S and Desgardin G 1998 A new inductive furnace based on microwave irradiation for growing long $Yba_2Cu_3O_{7-\delta}$ single-domain bars *Adv. Mater.* **10** 1448

[84] De Rango P, Lees M R, Lejay P, Sulpice A, Ingold M, Germi P and Pernet M 1991 Texturing of magnetic materials at high temperature by solidification in a magnetic field *Nature* **349** 770

[85] Bourgault D, De Rango P, Barbut J M, Braithwaite D, Lees M R, Lejay P, Sulpice A and Tournier R 1992 Transport properties of magnetically melt textured $YBa_2Cu_3O_{7-\delta}$ *Physica* C **194** 171

[86] Lees M R, Bourgault D, De Rango P, Lejay P, Sulpice A and Tournier R 1992 A study of the use of a magnetic field to control the microstructure of the high-temperature superconducting oxide *Phil. Mag.* B **65** 1395

[87] Lees M R, Bourgault D, Braithwaite D, de Rango P, Lejay P, Sulpice A and Tournier R 1992 Transport properties of magnetically textured $Yba_2Cu_3O_{7-\delta}$ *Physica* C **191** 414

[88] Morita M, Sawamura M, Takebayashi S, Kimura K, Teshima H, Kiyamoto K and Hashimoto M 1994 Processing and properties of QMG materials *Physica* C **235–240** 209

[89] Meng R L, Gao L, Gautier-Picard P, Ramirez D, Sun Y Y and Chu C W 1994 Growth an possible size limitation of quality single-grain $YBa_2Cu_3O_7$ *Physica* C **232** 337

[90] Cardwell D A 1998 Processing and properties of large grain RE(BaCuO) *Mater. Sci. Eng.* B **53** 1

[91] Sengupta S, Corpus J, Gaines J R, Todt V R, Zhang X F, Miller D J, Varanais C and McGinn P J 1997 Fabrication and characterization of melt processed YBCO *IEEE Trans. Appl. Supercond.* **7** 1723

[92] Sengupta S, Corpus J, Agarwal M and Gaines J R 1998 Feasibility of manufacturing large domain YBaCuO levitators by using melt processing techniques *Mater. Sci. Eng.* B **53** 62

[93] Marinel S, Wang J, Monot I, Provost J and Desgardin G 1997 Top seeding melt texture growth mechanisms of superconducting YBCO pellets *Supercond. Sci. Technol.* **10** 147

[94] Wang J, Monot I, Chaud X, Erraud A, Marinel S, Provost J and Desgardin G 1998 Fabrication and characterization of large grain $Yba_2Cu_3O_{7-\delta}$ superconductors by seeded melt texturing *Physica* C **304** at press

[95] Müller D and Freyhardt H C 1995 Growth model for melt textured $Yba_2Cu_3O_{7-\delta}$ *Physica* C **242** 283

[96] Lee D F, Partsinevelos C S, Presswood R G and Salama K 1994 Melt texturing of preferentially aligned $YBa_2Cu_3O_x$ superconductors by a seeded directional solidification method *J. Appl. Phys.* **76** 603

[97] Marinel S and Desgardin G unpublished.

[98] Chow J C L, Lo W, Dewhurst C D, Leung H T, Cardwell D A and Shi Y H 1997 The influence of process parameters on the growth morphology of large grain Pt doped YBCO fabricated by seeded peritectic solidification *Supercond. Sci. Technol.* **10** 435

[99] Nakamura Y, Endo A and Shiohara Y 1996 The relation between the growth rate of superconductive oxide *J. Mater. Res.* **11** 1094

[100] Wang J, Monot I, Delamare M P, Marinel S, Provost J and Desgardin G 1997 Processing of superconductive YBCO single grain monoliths by top-seeding melt texturing *Physica* C **282–287** 501

[101] Lo W, Cardwell D A, Dewhurst C D, Leung H T, Chow J C L and Shi Y H 1997 Controlled processing and properties of large Pt-doped YBaCuO pseudo-crystals for electromagnetic applications *J. Mater. Res.* **12** 2889

[102] Wang J, Monot I, Marinel S and Desgardin G 1997 Controlled growth of single grain $Yba_2Cu_3O_{7-\delta}$ by top seeding melt texturing *Mater. Lett.* **33** 215

[103] Leblond C, Monot I, Provost J and Desgardin G 1998 Optimisation of the texture and characterization of large size top-seeded-melt-grown YBCO pellets *Physica* C **311** at press

[104] Litzkendorf D, Habisreuther T, Wu M, Strasser T, Zeisberger M, Gawalek W, Helbig M and Görnert P 1998 Batch-processing of melt textured YBCO for motor applications *Mater. Sci. Eng.* B **53** 75

[105] Hayashi N, Diko P, Nagashima K, Yoo S I, Sakai N and Murakami M 1998 Fabrication of large single domain Sm-123 supercondcutors by OCMG method *Mater. Sci. Eng.* B **53** 104

[106] Teshima H, Morita M and Hashimoto M 1996 Comparison of the levitation forces of melt-processed YBaCuO supoerconductors for different magnets *Physica* C **269** 182

[107] Nagaya Shigeo 1997 *ISTEC Journal* **10** 31

[108] Gautier-Picard P, Chaud X, Beaugnon E, Erraud A and Tournier R 1998 Growth of YBaCuO single domains up to seven centimetres *Mater. Sci. Eng.* B **53** 66

[109] Hyun O B, Yoshida M, Kitamura T and Hirabayashi I 1996 Peak effect, oxygen deficiency ans scaling behavior of magnetization for $Yba_2Cu_3O_{7-\delta}$ crystals prepared by top seeded solution growth *Physica* C **258** 365

[110] Shi D, Qu D and Tent B A 1997 Effect of oxygenation on levitation force in seeded melt grown single-domain $Yba_2Cu_3O_{7-d}$ *Physica* C **291** 181

[111] Gautier-Picard P *PhD Thesis* University Joseph Fourier, Grenoble (France)

[112] Chaud X, Beaugnon E and Tournier R 1992 Magnetic susceptibility during the peritectic recombination of YBaCuO *Physica* C **282–287** 525

[113] Bean C P 1962 Magnetization of hard superconductors *Phys. Rev. Lett.* **8** 250

[114] Murakami M, Fujimoto H, Gotoh S, Yamaguchi K, Koshizukha N and Tanaka S 1991 Flux pinning due to nonsuperconducting particles in melt processed YBaCuO superconductors *Physica* C **185–189** 321

[115] Sadiumenge F, Pinol S, Obradors X, Snoeck E and Roucau C 1994 Microstructure of directionally solidified high critical current $Yba_2Cu_3O_7$–Y_2BaCuO_5 composites *Phys. Rev.* B **50** 7032

[116] Obradors X, Martinez M, Sadiumenge F, Pinol S, Fontcuberta J, Gomis V, Granados X, Vitalta N and Mora J 1995 Directional solidification of $YBa_2Cu_3O_7$: defects and pinning mechanisms *Fourth Euro Ceramics* **6** 173

[117] Daumling M, Erb A, Walker E, Genoud J Y and Flükiger R 1996 Monotonic dependence of J_c on magnetic field in twinned crystals of $YBa_2Cu_3O_{7-d}$ and $ErBa_2Cu_3O_{7-d}$ *Physica* C **257** 371

[118] Muller D and Freyhardt H C 1996 Twin-boundary characteristics of melt-textured $YBa_2Cu_3O_{7-d}$ *Phil. Mag. Lett.* **73** 63

[119] Marinel S, Monot I, Provost J and Desgardin G 1997 Effect of SnO_2 and CeO_2 doping on the microstructure and superconducting properties of $Yba_2Cu_3O_{7-d}$ textured melted zone sample *Supercond. Sci. Technol.* **11** 563

[120] Delamare M P, Monot I, Wang J, Hervieu M, Desgardin G, Verbist K and Van Tendeloo G 1996 Combination of CeO_2 and PtO_2 doping for strong enhancement of J_c under magnetic field in melt textured YBCO superconductor *Physica* C **262** 381

B2.3.4
Melt processing techniques: melt processing for RE–Ba–Cu–O

Masato Murakami

B2.3.4.1 Introduction

Rare earth (RE) elements generally form a $REBa_2Cu_3O_y$ (RE123) phase which exhibits superconductivity above 90 K. Notable exceptions to this are Ce and Tb for which the RE123 phase does not exist and there are no data available for radioactive Pm. Although Pr forms the 123 phase, it is still controversial whether Pr123 is superconducting or not. B2.3.3

Table B2.3.4.1 lists the RE elements which form a superconducting RE123 phase, along with their ionic radii and the peritectic decomposition temperatures (T_p) [1]. Superconducting RE123 compounds are classified into two groups depending on the ionic radius. Only a stoichiometric RE123 compound is formed for RE elements with radius smaller than that of Dy, while RE ions larger than Gd form a $RE_{1+x}Ba_{2-x}Cu_3O_y$ type solid solution [2]. A1.3

Figure B2.3.4.1(*a*) shows the variation of T_c of $Nd_{1+x}Ba_{2-x}Cu_3O_y$ (Nd123ss) sintered in air as a function of x [3]. The T_c of Nd123ss gradually decreases with increasing x and is accompanied by an orthorhombic to tetragonal phase transition at $x = 0.18$ as shown in figure B2.3.4.1(*b*). It can be seen that superconductivity is finally lost at around $x = 0.4$. It is commonly observed in all RE123ss that T_c is depressed with increasing amount of RE^{3+} substitution on the Ba^{2+} site. Such a depression in T_c can be explained partly in terms of a decrease in the carrier concentration or the hole density as trivalent RE replaces bivalent Ba. The structural change is associated with disorder in the CuO chain site caused by extra oxygen ions at anti-chain (O5) site to maintain charge balance [4].

Figures B2.3.4.2(*a*)–(*c*) show sub-solidus phase diagrams for the Y–Ba–Cu–O, Nd–Ba–Cu–O, and Sm–Ba–Cu–O systems [5]. The phase diagrams of Eu and Gd systems are similar to that of the Sm system except for a narrower range of RE–Ba solid solution. In the Y–Ba–Cu–O system, both Y123 and Y211 are point compounds, while for La, Nd, Sm, Eu and Gd systems, RE123 is not a stoichiometric compound but forms a RE–Ba solid solution. The second phase is a stoichiometric RE211 for Sm, Eu and Gd, while it is $RE_{4-2x}Ba_{2+2x}Cu_{2-x}O_z$ (RE422ss) with a RE–Ba solid solution for La and Nd. It should also be born in mind that the phase relations are strongly dependent on the temperature and oxygen partial pressure (pO_2), in addition to the species of RE element. B2.3.2

RE123ss superconductors exhibit low onset T_cs with a broad superconducting transition when melt-processed in air. This is ascribed to the fact that melt-grown $RE_{1+x}Ba_{2-x}Cu_3O_y$ has a wide range of x values when solidified in air or under relatively high pO_2 [6], because the solidification temperature is almost independent of the chemical composition of RE123ss.

In contrast, the substitution of RE on Ba site is largely suppressed when RE–Ba–Cu–O superconductors are melt-processed in a reduced oxygen atmosphere. This is due to the fact that

Table B2.3.4.1. Rare earth elements which form superconducting RE123, ionic radius and T_p of their RE123 phase in air

Rare earth	La	Nd	Sm	Eu	Gd	Dy	Y	Ho	Er	Tm	Yb	Lu
Ionic radius (A)	1.160	1.109	1.079	1.066	1.053	1.027	1.019	1.015	1.004	0.994	0.985	0.997
T_p of RE123 (C)	1080	1070	1060	1040	1020	1010	1000	990	980	960	950	950

x in $RE_{1+x}Ba_{2-x}Cu_3O_y$.
T_p: peritectic decomposition temperature.

the decomposition or solidification temperatures of RE123ss are depressed with increasing RE content under low pO_2 [7], which allows the preferential formation of near stoichiometric RE123ss.The melt-process for RE123ss under controlled pO_2 is known as the oxygen-controlled melt-growth (OCMG) process [7]. In the following section, melt processing of RE–Ba–Cu–O and its superconducting properties are summarized.

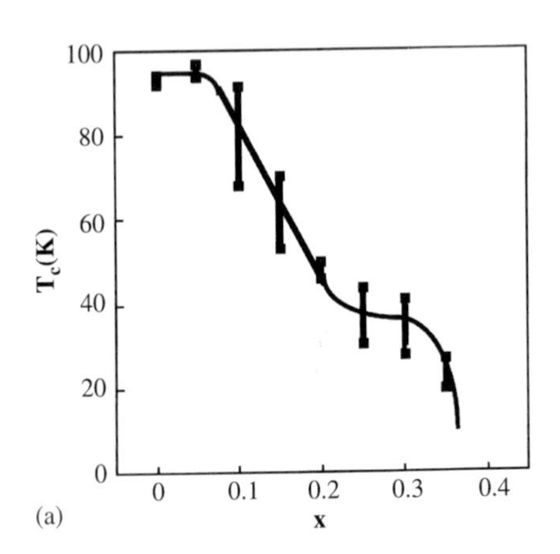

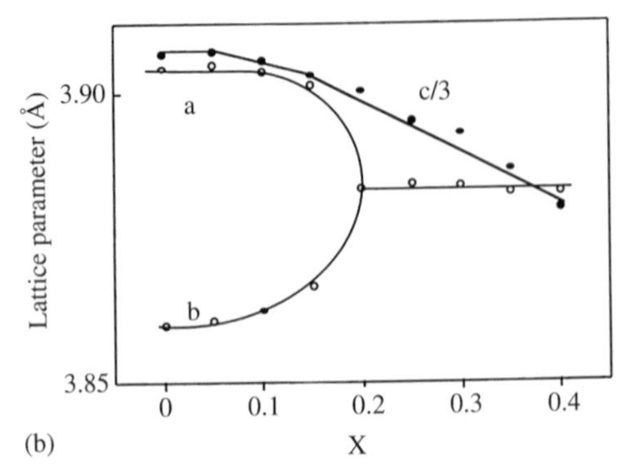

Figure B2.3.4.1. Plots of (a) T_c and (b) the lattice constants as a function of x in $Nd_{1+x}Ba_{2-x}Cu_3O_y$ sintered in air. T_c is depressed with increasing x accompanied by a decrease in orthorhombicity.

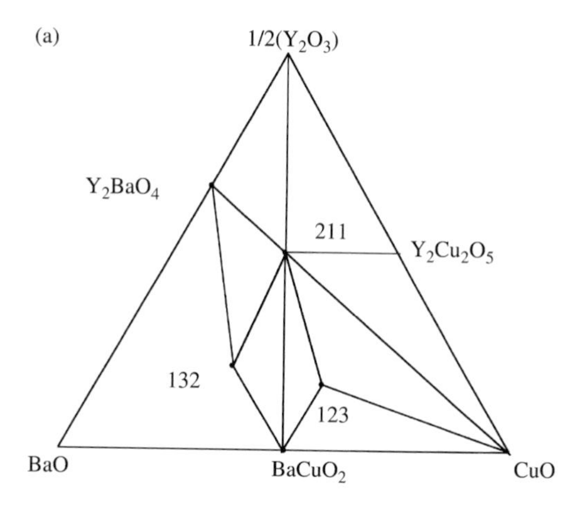

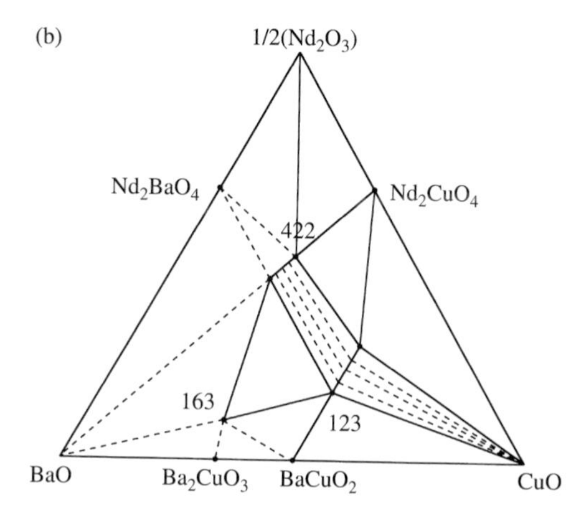

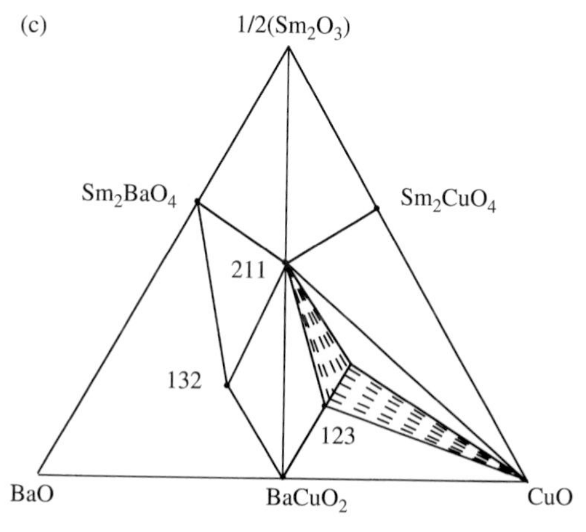

Figure B2.3.4.2. Quasi-ternary sub-solidus phase diagram for (*a*) Y, (*b*) Nd and (*c*) Sm–Ba–Cu–O systems.

B2.3.4.2 Melt processing

B2.3.4.2.1 Effect of pO_2 *on* T_c

Control of pO_2 during solidification is the most critical parameter affecting the T_c of RE–Ba–Cu–O [7–11] compounds. This is illustrated in figure B2.3.4.3, which shows the effect of pO_2 on the superconducting transition of melt-processed Nd–Ba–Cu–O [7]. It is clear that lowering pO_2 is effective in increasing the T_c of Nd–Ba–Cu–O superconductors and that the sample melt-processed under controlled pO_2 of 0.1 atm exhibits the highest onset T_c of 96.5 K with a sharp transition, which is higher than the typical T_c values observed in Y123 (90–92 K). The effect of pO_2 on T_c and the superconducting transition for other RE–Ba–Cu–O superconductors (RE: Sm, Eu, Gd) is similar to the Nd–Ba–Cu–O, but varies less dramatically as the ionic radius decreases, as shown in figure B2.3.4.4 [12]. The fabrication of high T_c La–Ba–Cu–O by the OCMG process is difficult since this system exhibits the widest range of La–Ba solid solution of all RE systems, in addition to the added complication of the La_2BaCuO_4 phase forming during melt processing [13]. Both T_c and the superconducting transition of this compound have been greatly improved with subsequent compositional control including the addition of Ag, employment of Ba-rich La422ss and pO_2 control at the 0.1% level [14].

OCMG-processed RE–Ba–Cu–O superconductors have significant potential for industrial applications, since they exhibit high T_c values of 94–96 K with a sharp transition and, moreover, larger J_c values than Y–Ba–Cu–O. The OCMG process is basically identical to other melt-processes, including the top-seeded melt-growth (TSMG) process for the Y–Ba–Cu–O system, except for pO_2 control at the melt-growth stage where the solidification of the RE123 phase takes place. In general, thermal profiles for the melt-growth need to be defined in accordance with the peritectic decomposition temperature of the particular RE123 phase being processed [15–18].

B2.3.3

The source of high T_c in the OCMG processed samples in the RE123 family is not clearly understood at present. It was once proposed that RE123 with large RE ions exhibit intrinsically high T_c due to the larger lattice constants [19]. It has also been suggested that the composition of a high T_c sample may lie in the region $x < 0$ for $Nd_{1+x}Ba_{2-x}Cu_3O_y$ (i.e. Ba^{2+} substitution for Nd^{3+} site) [20]. This type of substitution can increase hole carrier concentration, as in the case of substituting ca^{2+} for Y^{3+} in

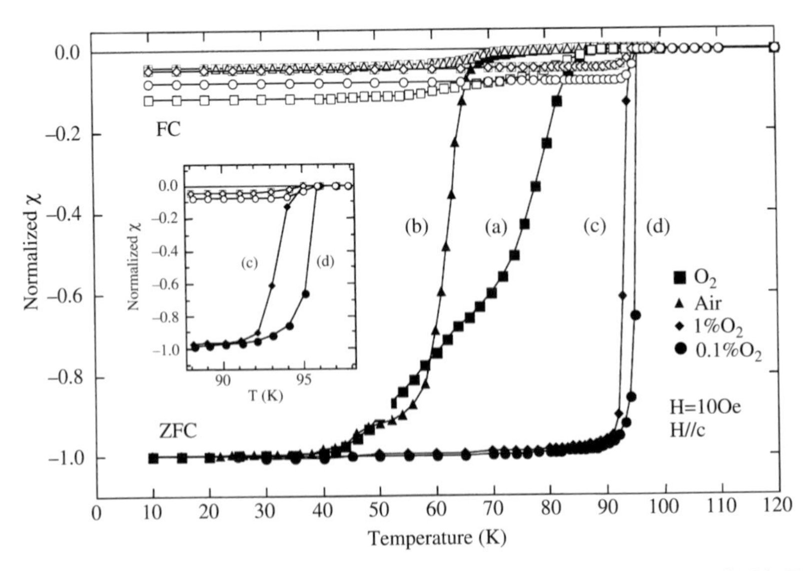

Figure B2.3.4.3. Effect of pO_2 on the superconducting transition of melt-textured Nd–Ba–Cu–O.

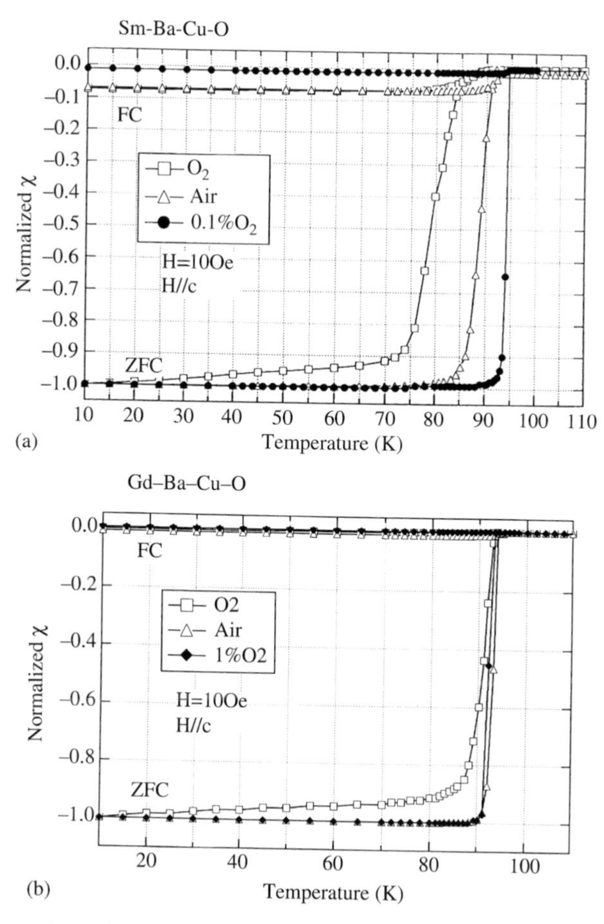

Figure B2.3.4.4. Temperature dependence of magnetic susceptibility for (*a*) Sm–Ba–Cu–O, (*b*) Gd–Ba–Cu–O melt-processed under different pO_2.

$YBa_2Cu_4O_8$ [21–23], which will lead to an increase in T_c. Furthermore, satellite peaks in the 3d core signal of the Ba^{2+} ion have been observed in x-ray photoelectron spectroscopy of OCMG-processed Nd–Ba–Cu–O samples, which reflects the different Ba environment where it substitutes onto the Nd site [24]. Recent powder x-ray diffraction analyses also confirm the presence of $Nd_{1+x}Ba_{2-x}Cu_3O_y$ with $x = -0.1$ within the RE123 phase matrix [25]. D1.1.2

The melting point of RE123 is depressed by either decreasing pO_2 or RE ionic radius as shown in figure B2.3.4.5 [26]. For RE123 with an ionic RE radius smaller than Gd, the melting point of RE123 approaches that of $BaCu_2O_2$ when pO_2 is lower than 0.01 atm, and thus the preferential formation of RE123 is prohibited. This results in the deterioration of the quality of RE123 crystals subsequently grown [26]. It is thus obvious that pO_2 is a common critical processing parameter for the successful melt growth of RE123 superconductors.

B2.3.4.2.2 Melt-processing of RE–Ba–Cu–O in air

Although the OCMG-processed RE–Ba–Cu–O has a great potential for the applications of the RE123 family, the control of oxygen partial pressure makes the scale-up of this process difficult and raises

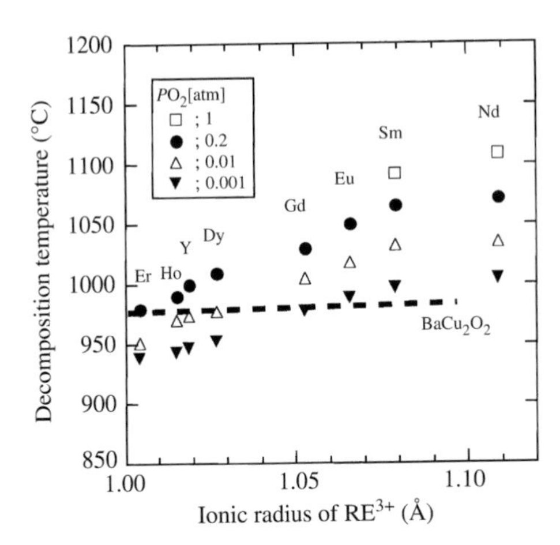

Figure B2.3.4.5. Plots of the peritectic decomposition temperature (T_p) versus ionic radius as a parameter of pO_2. $BaCuO_2$ is not stable, in a low $pO_2 < 0.01$ atm, and $BaCu_2O_2$ plays a prominent role in phase formation. The melting point of this phase (plotted as a broken line in the figure under pO_2 of 0.001 atm) is higher than that of RE123 for RE ions smaller than Dy, which inhibits the stable peritectic growth.

the cost of the product material. This has motivated the development of an air-based melt-process technique for high performance RE–Ba–Cu–O.

According to the $NdO_{1.5}$–BaO–CuO ternary phase diagram [see figure B2.3.4.2(*b*)], stoichiometric Nd123 has a tie line to a Ba-rich melt [27]. Therefore, if one shifts the precursor composition towards the Ba-rich direction, nearly stoichiometric Nd123 can be formed preferentially. In fact T_c could be improved in the sample melt-processed in air by using Ba rich composition [28]. Ba, however, generally migrates to the substrate during the partial melting stage, and thus high T_c is achieved only at the early stage of the grain growth. Hu *et al* [29] succeeded in raising the T_c of air-processed Nd–Ba–Cu–O to 95 K by subsequent post-process annealing in a reduced oxygen atmosphere. However, trapped fields in the large grain sample did not recover completely with such a simple annealing process.

The Nd–Ba–Cu–O system forms a different kind of second phase, compared to Y–Ba–Cu–O, with the chemical formula of $Nd_{4-2x}Ba_{2+x}Cu_{2-x}O_z$ (Nd422ss), which has a solid solution towards the Ba-rich direction. Kojo *et al* [30] succeeded in growing high T_c Nd123 by using Ba-rich $Nd_{3.6}Ba_{2.4}Cu_{1.8}O_y$ precursor, which has a tie line to stoichiometric Nd123. In contrast to a Ba-rich melt, the beneficial effect of a Ba-rich composition could be preserved for a longer time if one uses a Ba-rich Nd422ss solid.

Kohayashi *et al* [31] succeeded in producing high T_c Sm–Ba–Cu–O in air by adding Ag. The beneficial effect of Ag has also been reported in Gd–Ba–Cu–O [32]. Molten Ag can absorb oxygen, and hence it is probable that internal oxygen content at the growth front may naturally be controlled at a low pO_2 level when Ag is present during melt-processing, even in air.

B2.3.4.3 J_c–B properties of OCMG-processed RE–Ba–Cu–O bulks

B2.3.4.3.1 Secondary peak effect

D2.4 Figure B2.3.4.6 shows J_c–B curves of OCMG-processed Nd–Ba–Cu–O and Sm–Ba–Cu–O samples together with melt-processed Y–Ba–Cu–O for comparison. Both Nd–Ba–Cu–O exhibits a secondary peak effect of fishtail shape, which contributes to enhanced J_c values at intermediate fields compared B2.3.3 to melt-processed Y–Ba–Cu–O. In the Y-system, the secondary peak effect has also been observed in

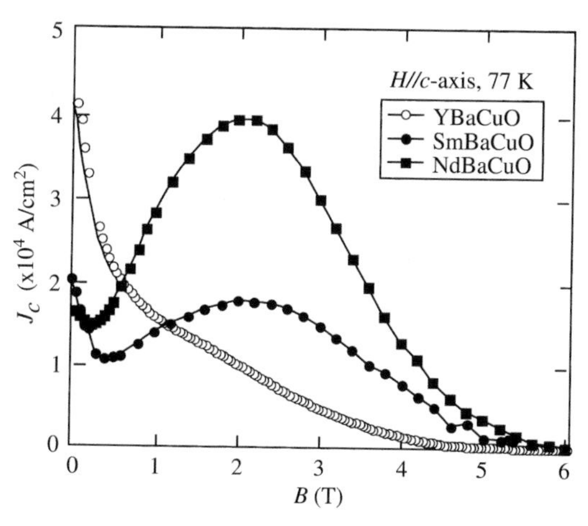

Figure B2.3.4.6. Field dependence of J_c (77 K, $H//c$) for OCMG-processed Nd–Ba–Cu–O and Sm–Ba–Cu–O and also for Y–Ba–Cu–O melt-processed in air. Note that J_c decreases monotonically with field for Y–Ba–Cu–O, while a well-developed secondary peak effect is observed in Nd and Sm–Ba–Cu–O.

flux-grown single crystals [33] and even in some melt-processed bulks [34–36]. Commonly, however, this effect is not observed in sintered or thin film materials. The peak effect is believed to originate from local oxygen-deficient regions [33] when YBCO contains insignificant impurities since it disappears after a full oxygenation for this material. Some reports [37] for Y123 single crystals which contradict this interpretation are probably ascribable to contamination by impurities since a non-negligible depression in the onset T_c of the sample is usually observed, even after full oxygenation. In this case, the regions with locally depressed T_c caused by impurities are responsible for the field-induced flux pinning, and thus the secondary peak effect.

B2.2.1, B4.0

B2.4

The peak effect in OCMG-processed RE–Ba–Cu–O superconductors is also attributed to field-induced pinning, although its origin is different from that in melt-processed Y–Ba–Cu–O superconductors. RE–Ba–Cu–O forms a $RE_{1+x}Ba_{2-x}Cu_3O_y$ type solid solution, in which T_c is depressed with increasing RE content. In OCMG-processed Nd–Ba–Cu–O, scanning tunnelling microscopy (STM) [38] and TEM [39] observations revealed that Nd-rich regions about 10–50 nm in size with composition close to 123 (see figure B2.3.4.7) are distributed in the superconducting matrix. Such Nd rich regions with depressed T_c are believed to be responsible for the observed field-induced flux pinning and thus for the secondary peak effect. Even if OCMG-processed RE–Ba–Cu–O superconductors are fully oxygenated, the oxygen site disorder on the Cu–O chain in the RE-substituted Ba regions is unavoidable since the trivalent RE^{3+} ion in the bivalent Ba^{2+} site requires extra oxygen at the anti-chain site to satisfy charge neutrality. Rietveld refinement of neutron diffraction data for $Nd_{1+x}Ba_{2-x}Cu_3O_y$ has revealed that the oxygen site disorder at the Cu–O chain increases as x increases for the orthorhombic structure ($x < 0.25$) [40]. Therefore, as long as this type local inhomogeneity is present, the peak effect is observed even after full oxygenation.

Recently, Chikumoto *et al* [41] found that compositional fluctuations in Nd–Ba–Cu–O disappear after high temperature annealing. They also confirmed that the secondary peak effect is not observed in such samples, which strongly supports the fact that the secondary peak effect is caused by fluctuation in the Nd:Ba local stoichiometry.

However, the formation mechanism of compositional fluctuation is still controversial. When Nd substitutes on the Ba site, extra oxygen must be added to the anti-chain site to maintain charge balance. Since oxygen ion has a -2 valence, it attracts another Nd ion to the neighbour, and thus the Nd ion on

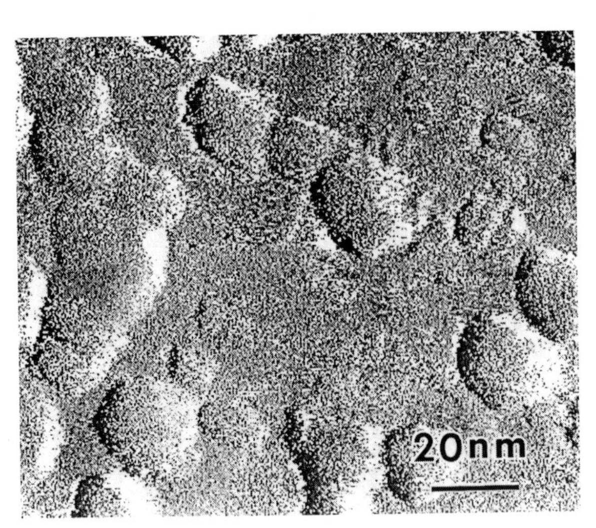

Figure B2.3.4.7. Scanning tunnelling micrograph for single crystalline Nd123 flux-grown under controlled pO_2 of 1%. Clusters with different contrast about 10–50 nm in size are distributed in the matrix.

a Ba site intrinsically has a tendency to form a pair [4, 40]. However, the cluster formation cannot be simply explained with this observation.

It has been proposed that chemical variation occurs as a result of spinodal decomposition, giving rise to a periodical variation in Nd concentration [42]. So far, however, no direct evidence has been presented to support this hypothesis other than the presence of Nd/Ba fluctuation [43]. It has also been proposed that perturbation at the growth front may cause compositional fluctuation, which subsequently leads to the formation of Nd-rich clusters as a result of re-distribution of Nd solute [1, 10]. Further studies are clearly necessary to draw a definite conclusion.

B2.3.4.3.2 Second phase particles

A1.3 Like Y211, particles in Y–Ba–Cu–O, RE211 particles can be dispersed in the RE123 matrix and contribute significantly to J_c enhancement. In almost all the RE–Ba–Cu–O systems, J_c enhancement through RE211 dispersion has already been established [44]. Pt and CeO_2 addition, for example, are found to be effective in refining the size of RE211 in YBCO and thus in increasing J_c [45]. As a result, achieving a fine dispersion of RE211 is a common technique for J_c enhancement in all RE123 systems, both with and without a RE–Ba solid solution.

Here one should note that La and Nd have a different kind of second phase particle which forms a $RE_{4-2x}Ba_{2+2x}Cu_{2-x}O_y$ type solid solution (RE422ss). As already mentioned, due to the presence of RE–Ba solid solution, Ba-rich Nd422ss can be used for the preferential growth of nearly stoichiometric Nd123, even in air. Like RE211, it has been confirmed that Nd422ss particles can be dispersed in the Nd123 matrix, and can function as pinning centres, contributing to J_c enhancement. However, Pt addition is not so efficient in the size refinement of Nd422ss as it is in the case of YBCO [46]. CeO_2 addition is more efficient, on the other hand, but much reduced, compared with its effect on B2.3.3 RE211 [47]. Recently, it has been found that a combined addition of Pt and CeO_2 leads to refinement of Nd422ss down to $< 1\ \mu m$, which can enhance J_c values up to the level of $10^5\ A\ cm^{-2}$ at 77 K in self-field [48].

B2.3.4.3.3 J_c improvement through mixing different RE elements

Partial substitution of the Y site by other RE elements in Y123 has been studied extensively, mainly with the aim of increasing J_c [49, 50]. However, no dramatic increase in J_c has been observed except in some isolated reports [51]. Moreover, partial substitution of the Y site with RE elements which form a RE–Ba solid solution have led to a depression of T_c, which is ascribed to the partial substitution of RE on the Ba site.

With the advent of the OCMG process, which inhibits the substitution of RE on the Ba site, one can explore the various RE123 systems by corresponding RE site with different RE elements without decreasing T_c [52]. Two main interesting results have been obtained from such studies. Firstly, field-induced pinning is greatly affected by the ratio and kind of RE element, which is understandable by considering the fact that this determines the range of RE–Ba solid solution [52]. Indeed, the peak J_c values are improved greatly by compounding the RE site with Nd, Eu, and Gd [52]. It was found that a change in their ratio also strongly influences the peak effect [53]. Secondly, refinement of RE211 particles down to the level of $< 0.1\,\mu m$ was achieved in (Nd, Eu, Gd) composites, as shown in figure B2.3.4.8. It is interesting to note that fine 211 particles smaller than 0.1 μm contain only Gd, despite the fact that three RE elements were present in the system [54]. Furthermore, fine Eu211 particles can be selectively dispersed for other combinations of RE elements [55]. At present, the mechanism of selective refinement of RE211 particles by a certain RE element is not understood clearly. According to the pushing theory of foreign particles at the solid/liquid interface [56], particles smaller than a critical size are pushed away from the growth front, which is known as the source of macroscopic segregation of Y211 in large, single-grain Y–Ba–Cu–O [57]. It is probable that a difference between the solubility and solidification temperatures may cause extreme refinement of RE211 particles. Recent microstructural observation revealed no macroscopic segregation of RE211 particles in (Nd, Eu, Gd) composites [58], which is beneficial for the production of large-grain bulk samples.

D1.2
B2.3.2
A1.3

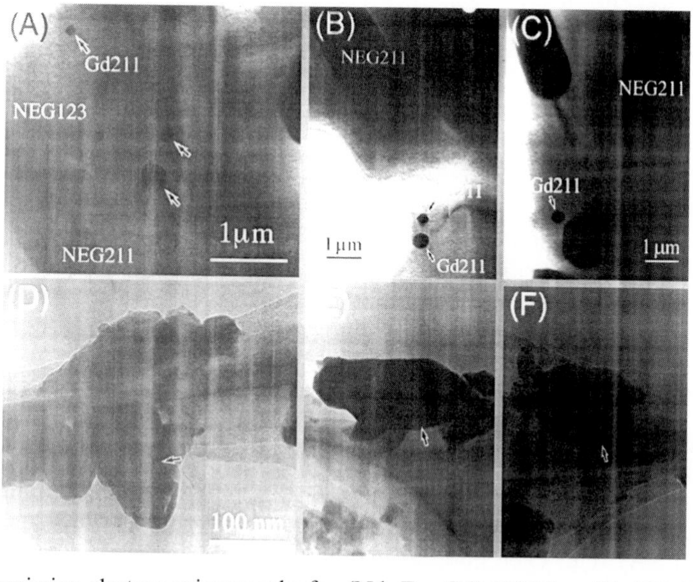

Figure B2.3.4.8. Transmission electron micrographs for (Nd, Eu, Gd) 123/40 mol% (Nd, Eu, Gd)211 composite with 0.5 wt% Pt addition prepared by the OCMG process in 0.1% O_2. Extremely fine 211 inclusions are observed along with relatively large 211. The former consists of Gd alone, while the latter contains Nd, Eu and Gd in equal concentrations.

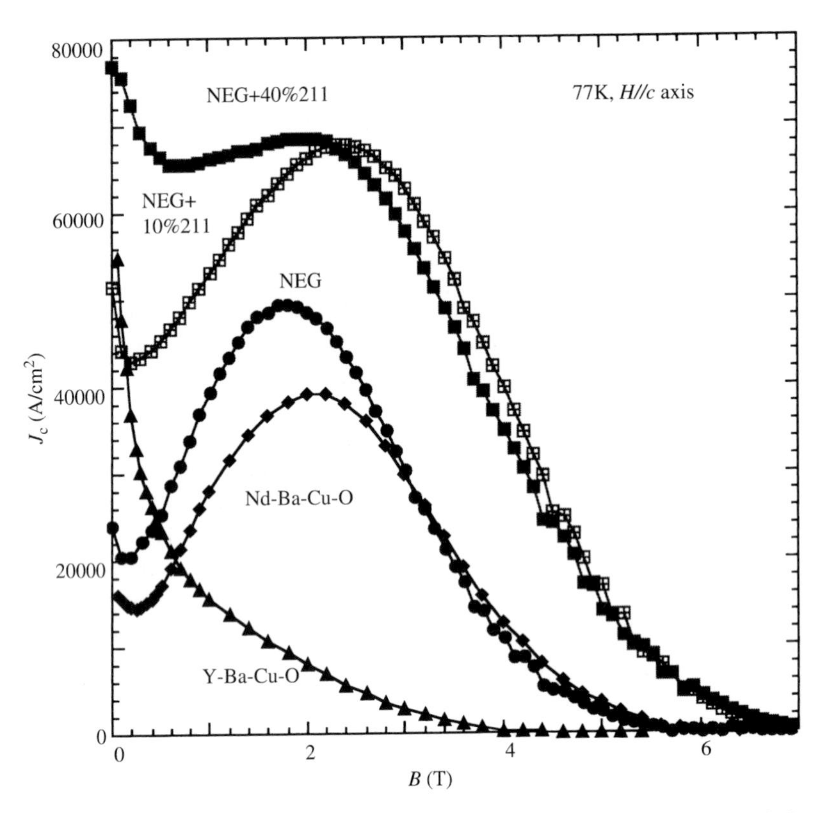

Figure B2.3.4.9. Summary of J_c–B properties (77 K, $B//c$) for various RE–Ba–Cu–O. Nd–Ba–Cu–O exhibits higher J_c values than Y–Ba–Cu–O at intermediate fields due to the secondary peak effect. Compounding the RE site with Nd, Eu, and Gd elements can enhance the peak J_c values. Overall J_c values can be further improved by adding RE211 inclusions.

D2.4 Figure B2.3.4.9 summarizes the J_c–B properties in melt-processed RE–Ba–Cu–O and demonstrates clearly how J_c values have been improved. Due to the presence of the secondary peak effect, J_c is improved significantly at intermediate fields for Nd123 compared to Y123. Field-induced pinning is enhanced further by compounding the RE site with several RE ions. An addition of RE211 particles could further improve J_c without reducing the field-induced pinning. As a result, extremely high J_c values have been achieved in (Nd, Eu, Gd)–Ba–Cu–O even at 77 K.

B2.3.4.4 Grain enlargement

B2.3.3 The TSMG process developed for growing large grain Y–Ba–Cu–O can be applied to the RE–Ba–Cu–O without major modification [59]. A seed crystal of RE123 with a higher melting point is positioned at the centre of the precursor pellet and solidified either by slow-cooling or isothermal heat treatment under controlled pO_2. Large grain samples have been successfully synthesized by such a melt-process in various RE–Ba–Cu–O compounds [60].

Figure B2.3.4.10 shows an optical micrograph of a Sm–Ba–Cu–O pellet about 36 mm in diameter melt-grown with a Nd123 seed crystal placed on its top surface at room temperature. The sample was solidified by slow-cooling from 1050 to 980 °C with a rate of $\approx 0.5\,\mathrm{K\,h^{-1}}$ in a vertical temperature gradient of $5\,\mathrm{K\,cm^{-1}}$ under controlled pO_2 of 1%. It is clear that the Sm123 crystal grows in a large

Figure B2.3.4.10. Photograph of the top surface of a single-grain Sm–Ba–Cu–O pellet 36 mm in diameter melt-grown with the TSMG process under controlled pO_2 of 1%. A Nd123 seed is positioned at the centre.

single domain. It has been shown that single grain materials can be grown in almost all the RE–Ba–Cu–O systems [60] and even in systems compounded with several RE elements [61].

Figure B2.3.4.11 campares the trapped field distribution for the 36 mm diameter single-grain Sm–Ba–Cu–O sample with that of a 46 mm diameter YBCO grain. Both samples were cooled in a magnetic field of 3 T down to 77 K and field mapped using Hall sensors. Although the Sm–Ba–Cu–O sample size is small, it is significant that it can trap a field exceeding 1 T, compared to that of the larger size Y–Ba–Cu–O, which is about 0.6 T. Recently, a trapped field over 2 T was recorded in a single grain of Sm–Ba–Cu–O [62] and Gd–Ba–Cu–O [63] at 77 K. E2

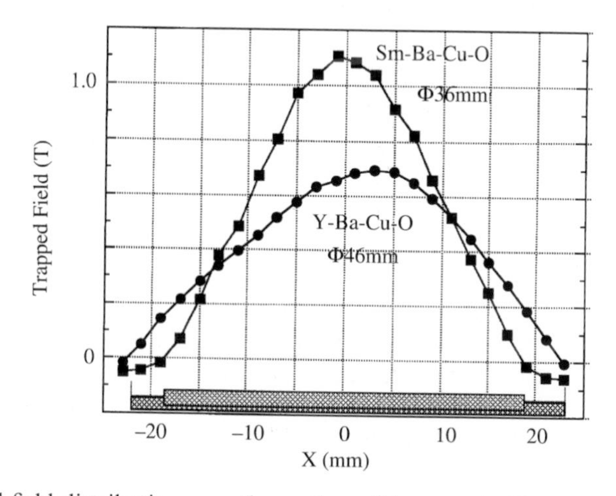

Figure B2.3.4.11. Trapped field distribution over the surface of large-grain Y–Ba–Cu–O (46 mm) and Sm–Ba–Cu–O (36 mm) at 77 K. The samples were cooled in the presence of 3 T from room temperature to 77 K and the external field was removed.

B2.3.4.5 Improvement of mechanical properties

D3.3 Compared to melt-textured Y–Ba–Cu–O, most reports suggest that the mechanical properties of OCMG-processed RE–Ba–Cu–O superconductors were poor [64–66]. In particular, cracking is observed even in the as-grown state for Nd–Ba–Cu–O and Sm–Ba–Cu–O [64]. Ag particles dispersed in the RE123 matrix can improve mechanical properties significantly, however, and Ag addition is essential for the melt-growth of large-grain RE–Ba–Cu–O for applications. The source of the poor mechanical properties in these materials is not yet clear. It is possible that RE ions with larger ionic radii will form RE123 with larger lattice constants, which leads to intrinsically poor mechanical properties [65]. In addition, oxygen gas is released during peritectic decomposition of the RE123 phase, which causes the formation of gas bubbles, that tend to remain in the sample under low pO_2 [65]. It is possible to reduce the amount of residual pores by melt-processing under pO_2 of 1 atm, although this degrades the superconducting properties of the sample.

Resin impregnation has been found to be very efficient in improving mechanical properties [67]. When bulk RE–Ba–Cu–O sample are immersed in molten resin under an exhausted environment gas, resin can penetrate, or back-fill, into the bulk interior through the surface cracks, and even fill the open pores connected to the surface cracks, as shown in figure B2.3.4.12. Resin can also spread into the interior region of large-grain samples through holes drilled artificially in the centre regions of the bulk sample. It has been confirmed that the mechanical properties are improved dramatically with resin impregnation [67]. It should also be noted that corrosion resistance is also greatly improved with this treatment, which again is critically important for industrial applications [67].

B2.3.4.6 Summary

RE–Ba–Cu–O (RE: Nd, Sm, Eu, Gd) superconductors with a RE–Ba solid solution exhibit low T_c when melt-processed in air. In contrast they show high $T_c > 95$ K and better flux pinning than the Y–Ba–Cu–O system when melt-processed in a reduced oxygen atmosphere. The characteristic strong

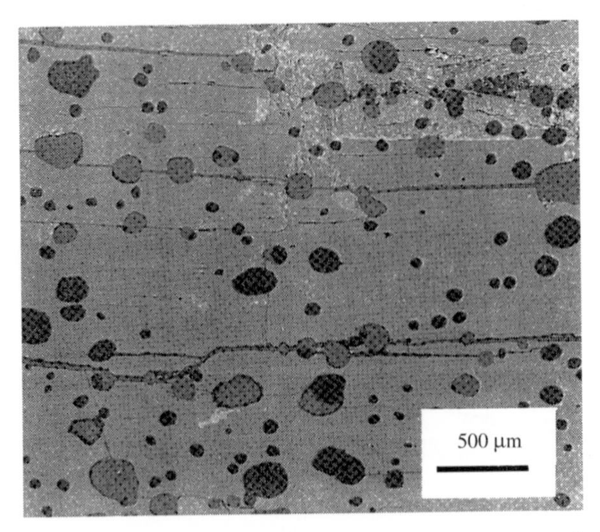

Figure B2.3.4.12. Polarized optical micrograph of (100) cross section of large-grain Sm–Ba–Cu–O with resin impregnation. Resin penetrates into the sample through the surface cracks and fills the open pores connected to the surface cracks.

flux pinning in RE–Ba–Cu–O superconductors originates from chemical modulation of the RE123 matrix by local RE-substitution on the Ba site. A fine dispersion of the second phase (RE422 for La, Nd; RE211 for Sm, Eu, Gd) is also critically important for many aspects of the processing, including the minimization of liquid loss during melt-processing, reduction of cracks, improvement of mechanical properties and further enhancement in J_c. Large grain RE–Ba–Cu–O has already been synthesized by a top-seeded melt-growth process under controlled pO_2. Large grain RE–Ba–Cu–O shows larger trapped fields than Y–Ba–Cu–O due to enhanced flux pinning.

References

[1] Murakami M, Sakai N, Higuchi T and Yoo S I 1997 *Supercond. Sci. Technol* **9** 1015
[2] Wong-Ng W, Paretzkin B and Fuller E R 1990 *J. Solid State Chem.* **85** 117
[3] Takita K, Akinaga H, Katoh H, Uchino T, Ishigaki T and Asano H 1987 *Japan. J. Appl. Phys.* **26** L1323
[4] McCallum R W, Kramer M J, Dennes K W, Park M, Wu H and Hofer R 1995 *J. Electron. Mater.* **24** 1931
[5] Yoo S I and McCallum R W 1993 *Physica* C **210** 147
[6] Wu H, Dennis K W, Kramer M J and McCallum R W 1998 *Appl. Supercond.* **6** 87
[7] Murakami M, Yoo S I, Higuchi T, Sakai N, Weltz J, Koshizuka N and Tanaka S 1994 *Japan. J. Appl. Phys.* **33** L715
[8] Yoo S I, Murakami M, Sakai N, Higuchi T and Tanaka S 1994 *Japan. J. Appl. Phys.* **33** L1000
[9] Murakami M, Yoo S I, Higuchi T, Sakai N, Watahiki M, Koshizuka N and Tanaka S 1994 *Physica* C **235–240** 2781
[10] Yoo S I, Sakai N, Higuchi T and Murakami M 1995 *IEEE Trans. Appl. Supercond.* **5** 1568
[11] Saitoh T, Segawa K, Kamada K, Sakai N, Yoo SI and Murakami M 1995 *The 1995 International Workshop on Superconductivity* (Hawaii, June 1995) Extended Abstracts, p 330
[12] Murakami M, Yoo S I, Higuchi T, Oyama T and Sakai N 1995 *Critical State Superconductors: Proc 1994 Topical International Cryogenic Materials Conf.* (Hawai, October 1994), (Singapore: World Scientific 1997) p 52
[13] Sakai N, Yoo S I, Goshima S and Murakami M 1997 *Advances in Superconductivity* vol 9 (Tokyo: Springer) p 709
[14] Sakai N, Yoo S I, Watahiki M and Murakami M 1998 *Mater. Sci. Eng.* B **53** 109
[15] Sakai N, Yoo S I, Goshima S and Murakami M 1997 *Advances in Superconductivity* vol 9 (Tokyo: Springer) p 709
[16] Appelboom H M, Matijasevic V C, Mathu F, Rietveld G, Anczykowski B, Peterse W J A M, Tuinstra F, Mooij J E, Sloof W G, Rijken H A, Klein S S and Van Ijzendoorn L J 1993 *Physica* C **214** 323
[17] Lindemer T B, Specht E D, MacDougall C S, Taylor G M and Pye S L 1993 *Physica* C **216** 99
[18] Yoo S I, Sakai N, Takaichi H, Higuchi T and Murakami M 1994 *Appl. Phys. Lett.* **65** 633
[19] Tallon J L and Flower N E 1993 *Physica* C **204** 237
[20] Lindemer T B, Chakoumakos B C, Spechet E D, Williams R K and Chen Y J 1994 *Physica* C **231** 80
[21] Miyatake T, Gotoh S, Koshizuka N and Tanaka S 1989 *Nature* **341** 41
[22] Wada T, Sakurai T, Suzuki N, Koriyama S, Yamauchi H and Tanaka S 1990 *Phys. Rev.* B **41** 11209
[23] Fischer P, Kaldis E, Karpinski J, Rusiecki S, Jilek E, Trounov V and Hewat A W 1993 *Physica* C **205** 259
[24] Murakami M, Yoo S I, Sakai N, Takaichi H, Higuchi T and Tanaka S 1998 *US Patent* 5849667.
[25] Osabe G, Takizawa T, Yoo S I, Sakai N, Higuchi T and Murakami M 1999 *Mater. Sci* B **65** 11
[26] Takahashi M, Sakai N, Yoo S I and Murakami M 1997 Advances in Superconductivity **vol 9**, ed S Nakajima and M Murakami (Tokyo: Springer) p 713
[27] Osamura K and Zhang W 1993 *Z. Metallkd.* **84** 523
[28] Yao Y and Shiohara Y 1998 *Mater. Sci. Eng.* **B53** 11
[29] Hu A M, Jia S L, Chen H and Zhao Z X 1996 *Physica* C **272** 297
[30] Kojo H, Yoo S I and Murakami M 1997 *Physica* C **289** 85
[31] Kohayashi S, Miyairi H, Yoshizawa S, Haseyama S, Nagaya S, Satoh M and Nakane H 1998 *Advances in Superconductivity* **vol 10**, (Tokyo: Springer) p 693
[32] Hinai H, Nariki S, Sakai N, Murakami M and Otsuka M 2000 *Supercond. Sci. Technol.* **13** 676
[33] Daeumling M, Seuntjens J M and Larbalestier D C 1990 *Nature* **346** 332
[34] Nakamura N, Murakami M, Fujimoto H and Koshizuka N 1992 *Cryogenics* **32** 949
[35] Groot P, Beduz C, Zhu-An Y, Yanru R and Smith S 1991 *Physica* C **185–189** 2471
[36] Ullrich M, Müller D, Heinemann K, Niel L and Freyhardt H C 1993 *Appl. Phys. Lett.* **63** 406
[37] Werner M, Sauerzopf F M, Weber H W, Veal B D, Licci F, Winzer K and Koblischka M R 1994 *Physica* C **235–240** 2833
[38] Ting Wu, Egi T, Itti R, Kuroda K and Koshizuka N 1996 *Advances in Superconductivity* **8** 481
[39] Egi T, Wen J G, Kuroda K, Unoki H and Koshizuka N 1995 *Appl. Phys. Lett.* **67** 2406
[40] Kramer M J, Yoo S I, McCallum R W, Yelon W B, Xie H and Allenspach P 1994 *Physica* C **219** 145
[41] Chikumoto N, Ozawa S, Yoo S I, Hayashi N and Murakami M 1997 *Physica* C **278** 187
Chikumoto N, Yoshioka J and Murakami M 1997 *Physica* C, **291** (1997) 79

[42] Nakamura M, Yamada Y, Hirayama T, Ikuhara Y, Shiohara Y and Tanaka S 1996 *Physica* C **259** 295
[43] Hirayama T, Ikuhara Y, Nakamura M, Yamada Y and Shiohara Y 1997 *J. Mater. Res.* **12** 293
[44] Mase A, Ikeda S, Yoshikawa M, Yanagai Y, Itoh Y, Oka T, Ikuta H and Mizutani U 1998 Advances in Superconductivity vol 10 (Tokyo: Springer) p 737
[45] Kim C J, Park H W, Kim K B, Lee K W, Kuk I H and Hong G W 1996 *Mater. Lett.* **29** 7
[46] Kojo H, Yoo S I, Sakai N and Murakami M 1997 *Superlattices and Microstructures* **21** 83
[47] Frangi F, Yoo S I, Sakai N and Murakami M 1997 *J. Mater. Res.* **12** 1990
[48] Chauhan H S and Murakami M 1999 Mater. Sci *Eng.* **B65** 48
[49] Matthews D N, Cochrane J W and Russell G J 1995 *Physica* C **249** 255
[50] Mahmoud A S and Russell G J 1998 *Supercond. Sci. Technol.* **11** 1036
[51] Mahmoud A S and Russell G J 1999 *Physica* C **322** 193
[52] Muralidhar M, Koblischka M R, Saitoh T and Murakami M 1998 *Supercond. Sci. Technol.* **11** 1349
[53] Muralidhar M, Koblischka M R, Das A, Sakai N and Murakami M 2000 *Advances in Superconductivity* vol 12 (Tokyo: Springer) p 497
[54] Muralidhar M, Saitoh T, Segawa K and Murakami M 1998 *Appl. Supercond.* **6** 139
[55] Muralidhar M, Koblischka M R and Murakami M 2000 *Advances in Superconductivity* **vol 12** (Tokyo: Springer) p 494
[56] Uhlmann D R, Chalmers B and Jackson K A 1964 *J. Appl. Phys.* **35** 2986
[57] Dewhurst C D, Wai Lo, Shi Y H and Cardwell D A 1998 Mater. Sci. Eng. B **53** 169
[58] Diko P, Muralidhar M, Koblischhka M R and Murakami M 2000 *Advances in Superconductivity* **vol 12** (Tokyo: Springer) p 488
[59] Hayashi N, Diko P, Nagashima K, Yoo S I, Sakai N and Murakami M 1998 *Mater. Sci. Eng.* B **53** 104
[60] Mizutani U, Mase A, Ikuta H, Yanagi Y, Yoshikawa M, Itoh Y and Oka T 1999 *Mater. Sci. Eng.* B **65** 66
[61] Muralidhar M, Segawa K and Murakami M 1999 *Mater. Sci. Eng.* B **65** 42
[62] Ikuta H, Mase A, Yanagi Y, Yoshikawa M, Itoh Y, Oka T and Mizutani U 1998 *Supercond. Sci. Technol.* **11** 1345
[63] Nariki S and Murakami M 2000 *Supercond. Sci. Technol.*, submitted
[64] Miyamoto T, Katagiri J, Nagashima K and Murakami M 1999 *IEEE Trans. Appl. Supercond.* **9** 2066
[65] Sakai N, Seo S J, Inoue K, Miyamoto T and Murakami M 1999 *Advances in Superconductivity* **vol 11** (Tokyo: Springer) p 685
[66] Ikuta H, Mase A, Hosokawa T, Yanagi Y, Yoshikawa M, Itoh Y, Oka T and Mizutani U 1999 *Advances in Superconductivity* vol 11 (Tokyo: Springer) p 657
[67] Tomita M and Murakami M *Supercond. Sci. Technol.* **13** (2000) submitted

B2.3.5
Melt processing techniques: melt process for BSCCO

J Shimoyama

Among the Bi-based superconductors, the melt-solidification process has been extensively applied mainly for Bi2212 compound in order to obtain single crystals as well as the melt-solidified tapes and wires. In the case of Bi2223 phase, the melt process for fabrication of tapes is quite difficult to apply, since this phase does not directly form from the partially molten state. However, its single crystals were recently grown by the floating zone method with very slow growth rate. Melt-solidification process applied for Bi2201 phase is quite similar to Bi2212, however, it has been performed only for single crystal growth due to its extremely low T_c for its practical application as tapes, etc. In this section, various kinds of melt process for Bi2212 including their superconducting properties are discussed.

C2 B3.2.1 B3.2.2 A1.3

B2.3.5.1 High temperature phase diagram of Bi2212

The chemical formula of Bi2212 is generally described as $Bi_2Sr_2CaCu_2O_{8+\delta}$, however, this compound has large cation nonstoichiometry and the stoichiometric $Bi_2Sr_2CaCu_2O_{8+\delta}$ hardly forms through the synthesis under an ambient air. The stable composition in air is always slightly Bi-rich, 5–15% excess from the stoichiometric ratio. In addition, two alkali-earth elements, Sr and Ca, have a wide solid-solution range of Sr/Ca = 0.67–2.75 and the total amount of these elements in the chemical formula can take smaller values than 3 down to approximately 2.6. These cation nonstoichiometries strongly affect the high temperature phase diagram of the Bi2212. The composition dependent high temperature phase diagrams of the Bi2212 are well summarized by Majewski [1]. High temperature phase diagrams of $Bi_xSr_2CaCu_2O_{8+\delta}$ and $Bi_{2.18}Sr_{3-y}Ca_yCu_2O_{8+\delta}$ are displayed in figures B2.3.5.1 and B2.3.5.2, respectively. Partial melting temperature of the Bi2212 is influenced both by the chemical composition and the partial pressure of oxygen, P_{O_2}. The composition which shows the highest partial melting temperature, 1163 K in air, is $Bi_{2.18}Sr_2CaCu_2O_{8+\delta}$ and the solidification temperature range is systematically broadened with leaving this composition, down to approximately 1103 K in air for Ca-rich composition. The effect of P_{O_2} in the atmosphere on partial melting point is roughly described as 5 K/log(P_{O_2}/Pa) when $P_{O_2} > \sim 30$ Pa. The partial melting reaction from the Bi2212 phase does not occur under further reducing atmospheres, where a decomposition of Bi2212 to Cu_2O and $Bi_2(Sr,Ca)_3O_z$ takes place in the solid state. Although the cation composition and P_{O_2} strongly affect the partial melting point of Bi2212, most of the studies and developments on the melt-solidification of Bi2212 have been done for nearly stoichiometric composition, such as $Bi_{2.1}Sr_{1.95}CaCu_2O_{8+\delta}$, because the T_c of the Bi2212 decreases with the cation nonstoichiometry. The atmosphere usually adopted is ambient air or flowing oxygen. Therefore, the partial melting point of Bi2212 is usually thought to be approximately 1163 K. On the other hand, in the case of Bi2201, the representative partial melting point is 1183 K, which is higher than that of Bi2212. Some papers reported that the quenched Bi2212 samples from approximately 1173 K

B3.2.1 B3.2.1 C2

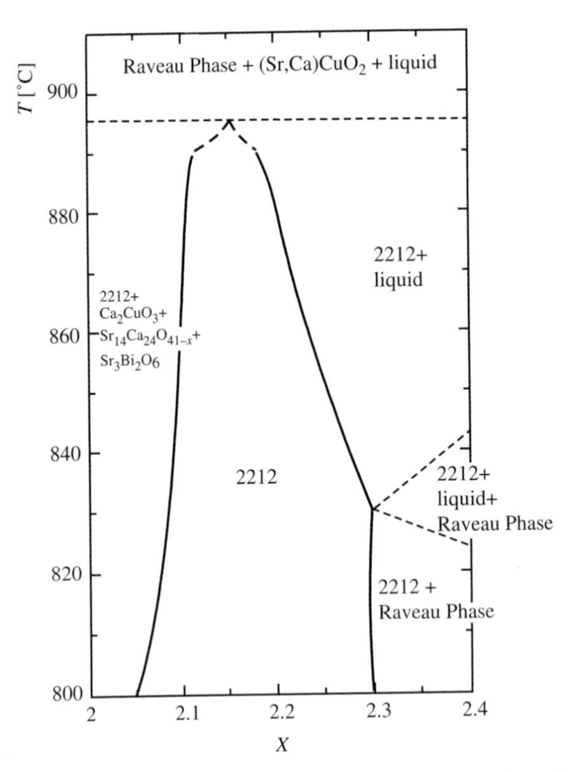

Figure B2.3.5.1. High temperature phase diagram of $Bi_xSr_2CaCu_2O_{8+\delta}$ (redrawn from ref. 1).

contain Bi2201 as a dominant phase with very little or no trace of Bi2212 phase. However, it should be noted that this does not mean the Bi2212 phase forms via Bi2201 from partial melting state as far as the nominal composition, Bi:(Sr + Ca):Cu, is nearly 2:3:2.

In the partial melting state, constituent phases change with temperature and P_{O_2}. One of the available review papers on this subject was written by Hellstrom and Zhang [2]. With increasing temperature over the partial melting point of Bi2212, Bi-free and Cu-free phases form besides the liquid phase. In the pure oxygen atmosphere, $(Sr,Ca)_{14}Cu_{24}O_z$ [$z \sim 41$: Bi-free] and $Bi_2(Sr,Ca)_4O_z$ [Cu-free] phases appear just above the partial melting point, while the $(Sr,Ca)CuO_2$ phase appears instead of $(Sr,Ca)_{14}Cu_{24}O_x$ in air. Further increases in temperature change these $(Sr,Ca)_{14}Cu_{24}O_z$ and $(Sr,Ca)CuO_2$ phases to $(Sr,Ca)_2CuO_3$ in both atmospheres and, at temperatures approximately 40 K higher than the partial melting point, the (Sr,Ca)O phase forms and the Cu-free phase disappears. Under the reductive atmospheres, such as in $P_{O_2} = 1000$ Pa, $Bi_2(Sr,Ca)_3O_z$ phase appears as a Cu-free phase, while variation of Bi-free phases with temperature is almost similar to the case of air. These phase relations in the partial melting state are well established by many studies.

B2.3.5.2 Synthesis of Bi2212 single crystals

B2.4 Synthesis of the single crystalline Bi2212 was started by the self-flux method at an early stage after its discovery. The alumina crucible is available for the crystal growth of this system. By this method, plate-like crystals up to $\sim 5 \times 5\ mm^2$ (*ab*-plane) have been obtained, however, each grown crystal has slightly different cation compositions even when they are grown at the same time.

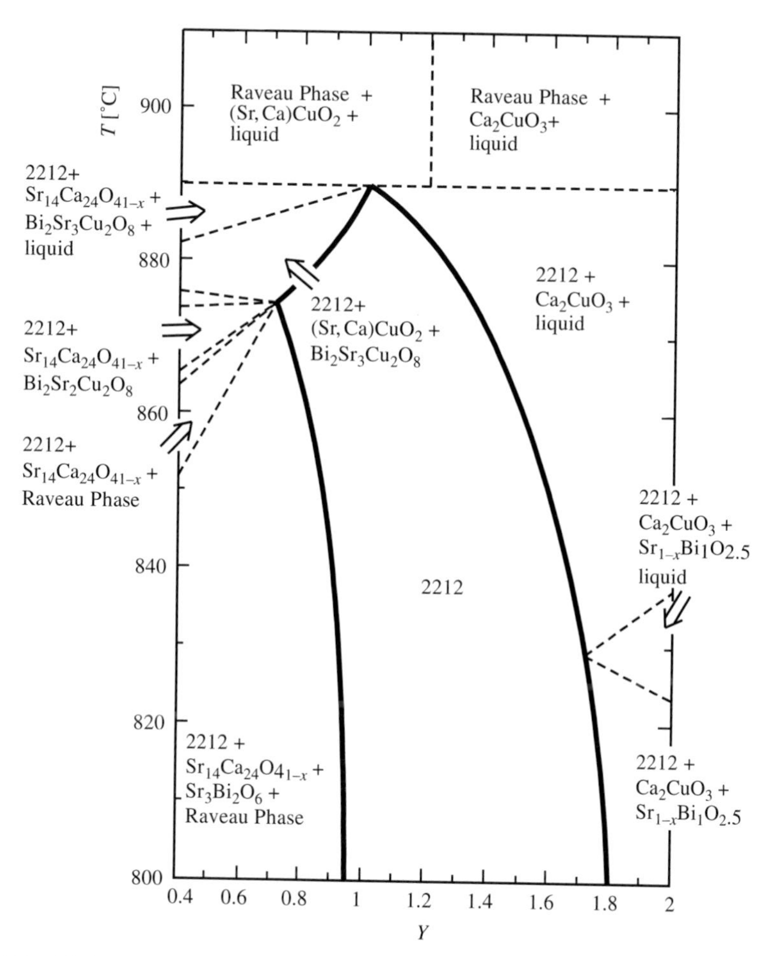

Figure B2.3.5.2. High temperature phase diagram of $Bi_{2.18}Sr_{3-y}Ca_yCu_2O_{8+\delta}$ (redrawn from ref. 1).

The floating zone (FZ) method has been also applied for the crystal growth of Bi2212. The Bi-rich composition is sometimes used for seed rod as a solvent (travelling solvent floating zone; TSFZ). Figure B2.3.5.3 shows a schematic illustration of crystal growth procedure by the FZ method. In both methods, FZ and TSFZ, the grown boule is a single domain which contains large number of Bi2212 single crystals with their *a*-axes almost parallel to the growth direction. The advantage of these methods is the homogeneous cation composition in the single crystals, which is almost uniform except in the initial part of the boule. In order to obtain high quality crystals, with good crystallinity and large single domain, extremely slow growth rate is required, usually less than 0.3 mm/h. The largest single crystal is over 30 mm in length along the *a*-axis with a width (along the *b*-axis) up to the diameter of the boule ($\sim$6 mm). Thickness of the crystal (along the *c*-axis) is typically 100 μm. Figure B2.3.5.4 shows a typical surface x-ray diffraction pattern of the Bi2212 single crystal. Only sharp 00*l* reflections can be seen. It should be noted that the as grown single crystals seldom show fine superconducting properties, such as sharp superconducting transition, because the distribution of cation and oxygen is inhomogeneous in the crystal. These inhomogeneities and thermal distortion in the as grown crystal can be removed by post-annealing under moderate conditions for a long time until reaching the equilibrium between crystal and atmosphere and quenching.

B2.4
D1.1.2

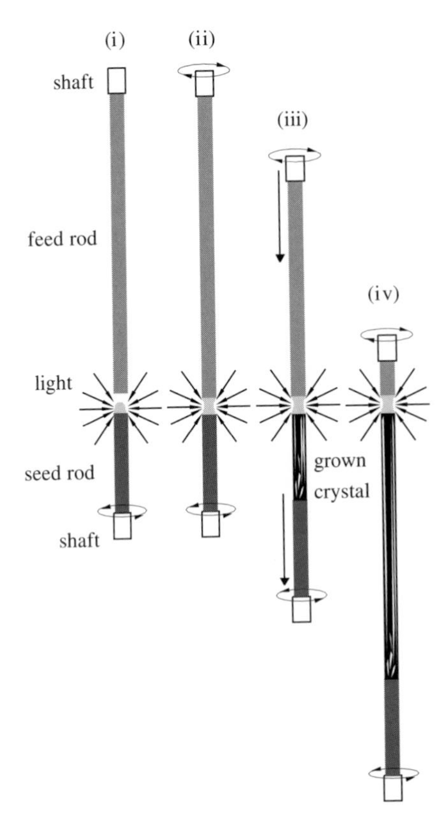

Figure B2.3.5.3. Schematic drawing of crystal growth procedure by the FZ or TSFZ method. (i) Top of the seed rod is melted. (ii) Feed rod is connected with melt. (iii) Both rods are slowly moved downwards less than 0.3 mm/h. (iv) After a few weeks, long crystal boule containing Bi2212 single crystals is obtained.

B2.3.5.3 Synthesis and development of Bi2212/Ag melt-solidified tapes and wires

The first successful study on the synthesis of Bi2212/Ag melt-solidified tape was reported by Kase *et al* in 1991 [3]. They applied the partial-melt and slow-cooling temperature schedule for Bi2212 thick film on silver foil, ~50 μm in thickness, and found a dense and strongly grain aligned Bi2212 thin layer of ~20 μm formed with *c*-axis normal to the foil surface. Such tape structure is quite desirable for carrying large superconducting current and, in fact, higher J_c than Nb–Ti and Nb_3Sn wires at 4.2 K in high magnetic field >10 T was attained. In their study, the initial Bi2212 thick film was prepared by the doctor–blade–cast method using slurry containing Bi2212 calcined powder, organic binders, dispersant and solvents.

B3.2.1
B4.3

The partial melting point of Bi2212 is slightly lowered by contacting silver [2]. For example, it is 1154 K in air. The optimized temperature schedule for obtaining high J_c tapes is shown in figure B2.3.5.5. This schedule can be used for the synthesis under wide P_{O_2} range, higher than 10^2 Pa. In the cooling process from 1110 K to room temperature, fast cooling is preferable, because the Bi2212 phase is unstable at middle temperature range, 800–1000 K, which causes the partial decomposition of Bi2212 to Bi2201 phase [4]. Figure B2.3.5.6 shows typical fractured cross section of Bi2212/Ag tape.

B3.2.1

After the report by Kase *et al*, various methods for preparing Bi2212/Ag tape have been developed. The dip-coating technique and painting method were found to be suitable processes for preparing long melt-textured tapes by applying a similar temperature pattern. These long tapes led to trial production

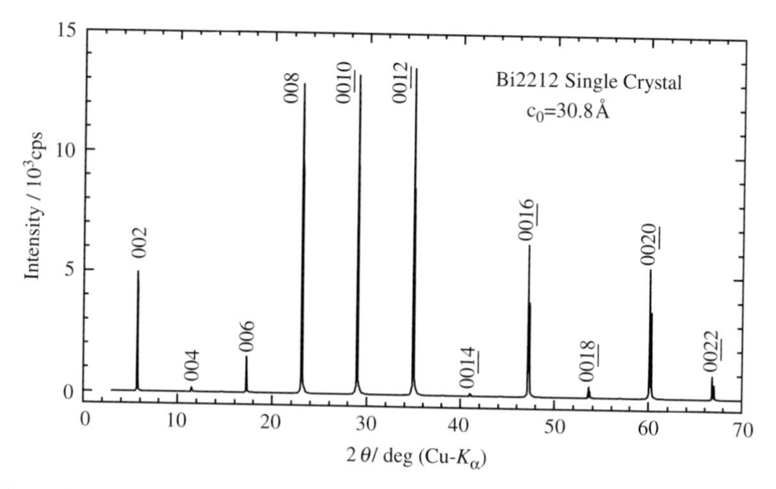

Figure B2.3.5.4. Typical x-ray diffraction pattern of Bi2212 single crystal surface.

of small pancake coils in the early stage. Instead of decreasing temperature, the melt-solidification can be performed by increasing P_{O_2} at an appropriate temperature. Funahashi *et al* applied the isothermal partial melting method for the synthesis of Bi2212/Ag tape [5]. They changed atmosphere from nitrogen flow (partial melting state) to $P_{O_2} = 2 \times 10^4$ Pa (solid) at a constant temperature at 1138 K and found J_c increases with an increase of annealing time at 1138 K after solidification, which reduces the size and the amount of secondary phases.

B3.2.1

The continuous melt-solidification method for long tape fabrication up to approximately 100 m was successfully established at Showa Electric Co. using a long tube furnace [6], which provides the special temperature gradient taking the desirable partial-melt and slow-cooling procedure for moving tapes into account. Very recent progress on the Bi2212/Ag tape is the pre-annealing and intermediate rolling (PAIR) process developed at National Research Institute for Metals, Japan [7]. The rolling process for

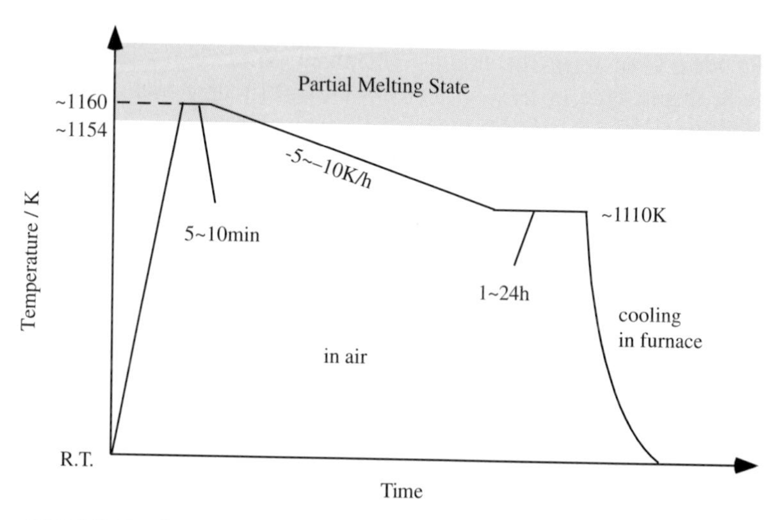

Figure B2.3.5.5. Typical temperature pattern for melt-solidification of Bi2212/Ag tape.

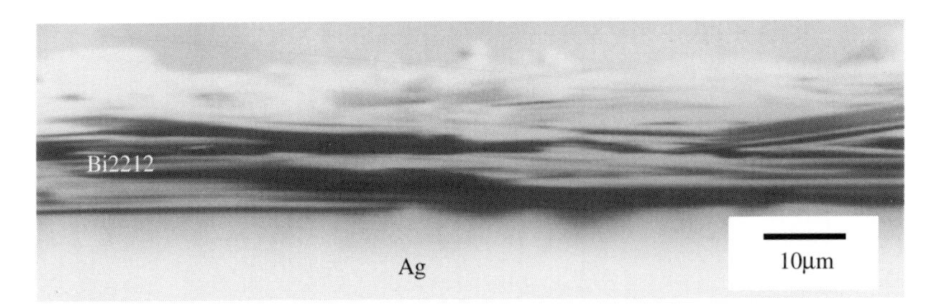

Figure B2.3.5.6. Secondary electron image of fractured cross section of Bi2212/Ag tape.

pre-annealed tape is effective for densifying the Bi2212 layer, resulting in a dense and homogeneous textured layer without the large voids formed after the conventional melt-solidification heat process. Reflecting such ideal microstructure, the PAIR processed Bi2212/Ag tape can carry record-high J_c at liquid helium temperature in high magnetic field, for example, $J_c = 5 \times 10^5\ \mathrm{A/cm^2}$ at 4.2 K in 10 T. By combining the continuous melt-solidification method for long tape, the PAIR processed tape up to 100 m in length has been fabricated.

B3.2.1 The melt-solidification method has been also applied in the synthesis of Bi2212 silver sheathed tapes and wires. Typical temperature schedule is almost as same as for Bi2212/Ag tape, because the silver sheath is essentially transparent for oxygen diffusion at high temperatures. However, in the synthesis of silver sheathed tapes and wires, the residual carbon must be eliminated prior to the melt-solidification, because it makes large voids in the superconducting core, resulting in non-uniform and grain misoriented, low J_c tapes and wires. In order to increase J_c by increasing interface area between silver and Bi2212, most of the recently developed Bi2212 silver sheathed tapes and wires are of multi-filament or jelly-roll type. The multi-filamentary tape and wires have already been manufactured at an industrial scale in Japan, USA and Europe mainly for superconducting magnet operation at low temperatures. Special target of Bi2212 superconducting magnet is high static field generation for high resolution nuclear magnetic resonance system; for example, static bias field of 23.5 T corresponding to the 1 GHz nuclear magnetic resonance system. At present, record-high magnetic field of 22.8 T only by superconducting magnet has been achieved by a Bi2212/Nb_3Sn/Nb-Ti hybrid magnet [8].

In all the high-T_c tapes and wires, Bi2212 melt-solidified materials have great advantage for fabricating the persistent current circuit, because one can make a superconducting joint with high I_c by melt-solidifying the local area even for multi-filamentary tapes and wires [9]. Recently developed rotation symmetric arranged tape in tube wire (ROSAT wire) [10], which is a melt-processed, multi-filamentary wire with threefold symmetry in the cross section, allowed us to make a high J_c solenoid coil, since its angular dependence of J_c with magnetic field direction is quite small. This wire is quite promising for fabricating practical superconducting magnets with homogeneous field distribution.

The most serious problem for Bi2212 melt-solidified tapes and wires for practical application is their weak flux pinning nature at high temperatures above 30 K, while the T_c is higher than 80 K. This A4.3 problem originates from the substantially poor flux pinning properties of Bi2212 crystals, as shown in figure B2.3.5.7. Therefore, tremendous efforts have been made for the improvement of intrinsic flux pinning properties of Bi2212 by various methods. The heavily Pb-doped Bi2212 single crystals [$\mathrm{Pb/(Bi+Pb)} > 20\%$], which can be prepared both by self-flux and FZ methods, show excellent J_c properties up to high temperature region when they are in carrier overdoped state [11]. In spite of such promising properties, it is quite difficult to synthesize Bi(Pb)2212 melt-solidified tapes with high J_c at the present stage due to broadened solidification temperatures and accompanied compositional fluctuation in a large scale. Other approaches for improving melt-solidified Bi2212 tapes and wires have not been successful or not applicable for industrial production.

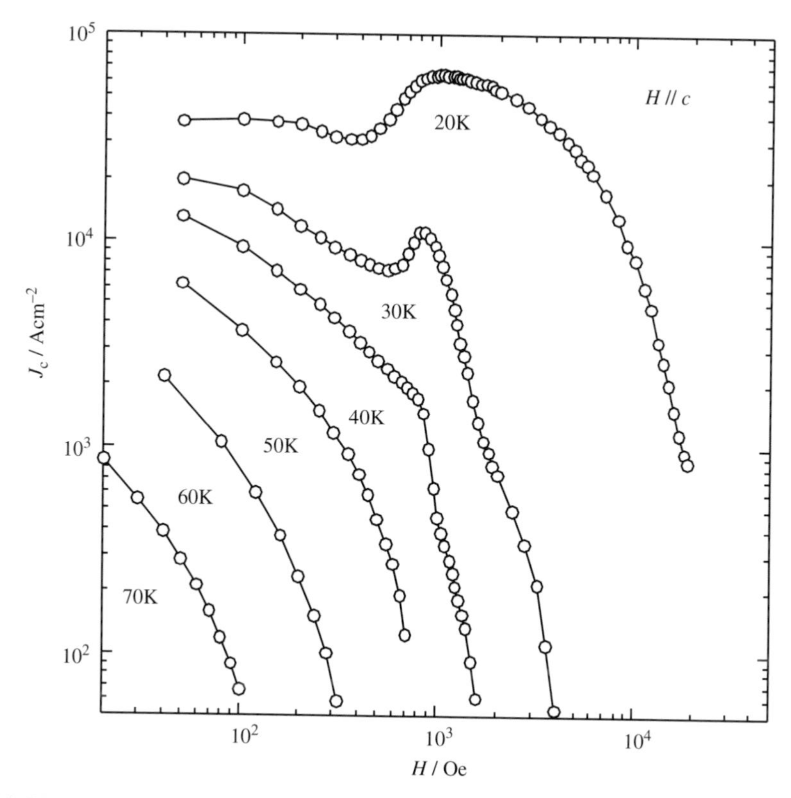

Figure B2.3.5.7. Typical J_c–H properties of Bi2212 single crystal ($T_c = 82$ K, carrier slightly overdoped).

B2.3.5.4 Development of Bi2212 melt-solidified rods

The Bi2212 melt-solidified rods have been developed for their practical application as superconducting current leads. In the case of the conventional FZ method, the growth rate must be quite slow, which is less than 1 mm/h, in order to control crystal alignment to be crystallographic *ab*-plane parallel to the growth direction. Such a slow rate is not realistic from an industrial point of view. The laser pedestal method was discovered as an effective method for fast growth of Bi2212 melt-solidified rods with preferential grain orientation by Sumitomo Electric Industries [12]. In this method, CO_2 laser is used for making local partial melting zone with high temperature gradient and, therefore, a high growth rate of several cm per hour is achievable. The rod diameter is variable by changing rates both of rod feed and crystal pulling. Another approach is the addition of silver (~ 10 wt%) to Bi2212 rod [13]. In this method, both Bi2212 crystals and included silver filaments grow parallel to the growth direction, however the growth rate is rather slow, a few mm per hour.

B2.3.5.5 Single crystal growth of Bi2223

The Bi2223 is the first material recorded as having a T_c higher than 100 K [14]. In the early stage after its discovery, however, synthesis of single-phased Bi2223 sample was found to be quite difficult and then, almost all of the research and development was performed for lead-doped Bi2223. The role of doped lead is in promoting phase formation of Bi2223 via local partial melting state. High temperature phase diagram of Bi(Pb)2223 had been studied eagerly prior to 1990. Unfortunately, the conditions for direct

B2.4

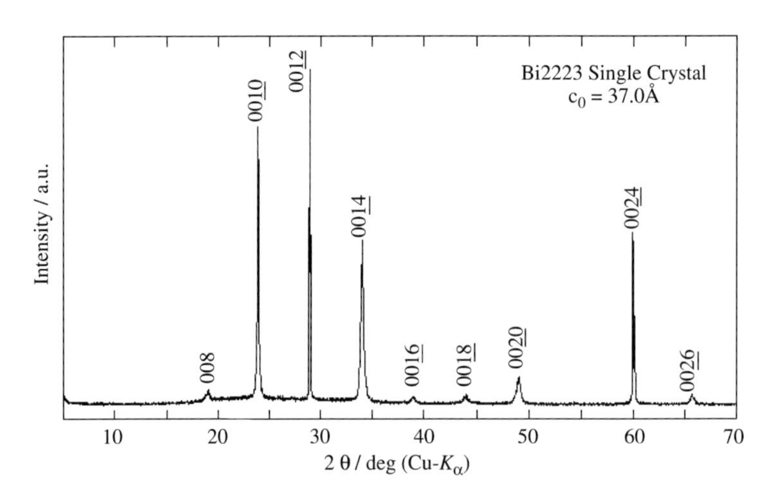

Figure B2.3.5.8. Surface x-ray diffraction pattern of Bi2223 single crystal.

crystal growth of Bi(Pb)2223 from a partial melter state did not emerge from these studies. Therefore, single crystal growth of Bi(Pb)2223 has been attempted by the flux method. By this method, Bi(Pb)2223 single crystals do not grow much less than 1 mm^2 in size [15]. For pure Bi2223, its high temperature phase diagram has not been clarified yet and very few attempts were made to obtain its single crystals. However, quite recently, large Bi2223 single crystals having similar size to Bi2212 were found to grow by the FZ method [16]. In this synthesis, very slow growth rate of less than 0.05 mm/h is required to obtain Bi2223 crystals. Typical x-ray diffraction pattern of the Bi2223 single crystal surface is shown in figure B2.3.5.8. The highest T_c of the Bi2223 single crystal is 110 K. Critical current performance of Bi2223 phase is quite similar to that of Bi2212 phase by considering the difference in T_c [17].

References

[1] Majewski P J 1996 *Bi-based Superconductors*, ed K Togano and H Maeda (New York: Dekker) pp 129–151
[2] Hellstrom E E and Zhang W 1996 *Bi-based Superconductors*, ed K Togano and H Maeda (New York: Dekker) pp 427–449
[3] Kase J, Togano K, Kumakura H, Dietderich D R, Irisawa N, Morimoto T and Maeda H 1990 *Japan. J. Appl. Phys.* **29** L1096
[4] Shimoyama J, Kase J, Morimoto T, Mizusaki J and Tagawa H 1991 *Physica* C **185–189** 931
[5] Funahashi R, Matsubara I, Ogura T, Ueno K and Ishikawa H 1997 *Physica* C **273** 337
[6] Hasegawa T, Hikichi Y, Kumakura H, Kitaguchi H and Togano K 1995 *Japan. J. Appl. Phys.* **34** L1638
[7] Miao H, Kitaguchi H, Kumakura H, Togano K and Hasegawa T 1998 *Physica* C **301** 116
Miao H, Kitaguchi H, Kumakura H, Togano K, Hasegawa T and Koizumi T 1998 *Physica* C **303** 65
[8] Okada M, Tanaka K, Matsuda S, Sato K, Sato J, Kitaguchi H, Kiyoshi T, Kumakura H, Togano K and Wada H 1998 *Advances in Superconductivity; Proc. 10th Int. Symp. Superconductivity (ISS '97) (Oct. 27–30, 1997, Gifu)* vol 10 pp 1381
[9] Fukushima K, Okada M, Sato J, Kiyoshi T, Kumakura H, Togano K and Wada H 1997 *Japan. J. Appl. Phys.* **36** L1433
[10] Okada M, Tanaka K, Wakuda T, Ohata K, Sato J, Kumakura H, Kiyoshi T, Kitaguchi H, Togano K and Wada H 1999 *IEEE Trans. Appl. Supercond.* **9** 920
[11] Chong I, Hiroi Z, Izumi M, Shimoyama J, Nakayama Y, Kishio K, Terashima T, Bando Y and Takano M 1997 *Science* **276** 770
Shimoyama J, Nakayama Y, Kishio K, Hiroi Z, Chong I and Takano M 1997 *Physica* C **281** 69
[12] For example, Kasuu O, Nonoyama H, Takahashi K, Sato K and Hayashi K 1994 *Advances in Superconductivity: Proc. 6th ISS'93, (Oct. 26–29, Hiroshima)* vol 6 pp 731
[13] Michishita K, Shimizu N, Sugawara Y, Ito W, Sasaki Y, Kubo Y, Nagaya S and Inoue T 1993 *Japan. J. Appl. Phys.* **32** L572
[14] Maeda H, Tanaka Y, Fukutomi M and Asano T 1988 *Japan. J. Appl. Phys.* **27** L209
[15] For example, Lee S, Yamamoto A and Tajima S 2001 *Advances in Superconductivity: Proceedings 13th ISS'00 (Oct. 14–16, Tokyo)* vol 13 p 341
[16] Fujii T, Watanabe T and Matsuda A 2001 *J. Cryst. Growth* **223** 175
[17] Shimizu K, Okabe T, Horii S, Otzschi K, Shimoyama J and Kishio K *Proc. MRS2001 Fall Meeting* at press

B2.4
Growth of superconducting single crystals

Debra L Kaiser and Lynn F Schneemeyer

B2.4.1 Introduction

Single crystals are ordered, three-dimensional arrangements of atoms that can possess an extremely high degree of crystalline perfection. As such, crystals play an important role in research studies of intrinsic properties and crystallographic structure as well as in technological applications. Many of the superconducting compounds are highly anisotropic, so single crystals have been essential for determining the directionality of key properties such as lower and upper critical fields, H_{c1} and H_{c2}, critical current density J_c, coherence length and penetration length. Intrinsic measurements of high quality crystals have provided the basis for designing a variety of electronic and magnetic devices and products for diverse applications including electrical power transmission, high-speed signal propagation on striplines, and magnetic imaging and storage. Single crystals have also permitted full determinations of the fundamental structures of pure superconducting phases, thereby paving the way to understanding the mechanism for superconductivity and predicting the existence of undiscovered superconducting compounds. There is significant overlap between the factors that influence the growth of crystals of the superconducting cuprate phases from eutectic melts and processes which involve a small amount of similar liquid compositions present at the growth interface. Thus, investigations of single crystal growth have provided guidelines for defining processing conditions in some bulk and thin film growth techniques, such as melt texturing and liquid phase epitaxy.

A1.3
E3.2
E2.3
B3.2.1, B2.3.2
B2.3.2
B4.3

The first section of this chapter discusses fundamentals of crystal growth with emphasis on special considerations for superconducting compounds. The following section describes the various techniques that have been employed to grow single crystals of both high and low temperature superconductors. Crystal chemistry and state-of-the-art crystal growth approaches for major classes of high temperature oxide superconductors are reviewed in the next section. The compounds considered include rare earth-doped lanthanum cuprates, $YBa_2Cu_3O_7$ and its rare earth isomorphs, the Bi–Sr–Ca–Cu–O phases, and the Tl and Hg-containing cuprates, as well as other copper oxide and non-cuprate (i.e., $LiTi_2O_4$, $Ba_{1-x}K_xBiO_3$ and $BaPb_{1-x}Bi_xO_3$) superconductors. The last section focuses on crystal growth of important classes of low temperature superconductors.

C1, C2, C3, C4
B2.5

B.2.4.2 Fundamentals of single crystal growth

There are numerous textbooks which provide excellent tutorials on the fundamentals of single crystal growth, e.g. [1, 2]. Crystal growth involves two stages, nucleation and growth. Crystals of the superconducting compounds are typically grown from liquid solutions or melts. When the solution becomes supersaturated with respect to the solute, nuclei will form. Nucleation can occur spontaneously in the solution, on the container walls, or on a single crystal or polycrystalline seed of the solute material or a different material of similar crystal structure. Nuclei that exceed the critical size grow into crystals. In unseeded growth processes there are many nucleation events, resulting in aggregates of intergrown crystals in addition to individual crystals. The crystals typically have highly faceted morphologies, with the largest dimensions lying along the fastest growing crystallographic directions, such as [100]. The facets are usually low index planes like (100) or (111).

In general, crystal growth of complex oxides like the high temperature superconductors requires the development of processing conditions unique to the chemistry of each compound. All of the known high T_c superconductors are incongruently-melting solids, so crystals are usually prepared by solution or flux growth methods which use a solvent or melt of a different composition than the compound of interest. Further, the superconducting oxides are complex phases composed of four or more elements and thus have very complex chemistries. The processing variables of importance in the most widely used growth techniques are solution composition, solutionizing temperature, atmosphere and crucible material. Phase equilibria studies can guide the selection of the solution composition, temperature and atmosphere. However, given the complexity of most of the oxide superconductors, a full determination of the primary phase field for the desired compound is a massive undertaking and empirical conditions for crystal growth were first formulated with little or no phase equilibria data. As the detailed phase relations become better understood, refinements of the crystal growth processes to produce larger crystals of higher perfection should be possible. Oxide melts are highly reactive, and interactions of the crucible with the melt frequently result in the formation of undesired phases or contamination of the crystal product. Crucible compatibility studies were often an early step in the determination of crystal growth conditions. In several of the superconducting oxides, highly volatile and/or toxic elements have further complicated crystal growth efforts. Finally, oxygen content has proven to be critical in many of the oxide phases, requiring careful control of the growth atmosphere or the inclusion of a post-annealing process to optimize superconducting properties.

B2.4.3 Growth techniques

B2.5
This section gives an overview of the techniques which have been employed to grow single crystals of high and low temperature superconductors. More comprehensive descriptions of the various crystal growth techniques can be found in any of the numerous texts on crystal growth, e.g. [2–4]. Since the oxide superconductors melt incongruently, Bridgman and Czochralski growth techniques which are used for congruently melting compounds will not be considered here. Most of the techniques described below are categorized as solution or flux growth processes.

B2.4.3.1 Sintering and grain growth

B2.2.1
Sintering is the agglomeration of compacted powders by the application of heat below the melting point. During sintering, large grains typically grow at the expense of finer grains to form crystallites. Frequently, growth is assisted by the presence of a small amount of lower melting composition material, often localized at the grain boundaries. The driving force for grain growth is a reduction in total energy resulting from a decrease in the grain boundary area and the total grain boundary energy. Individual crystallites extracted from a sintered mass can be large enough for x-ray diffraction studies as well as for some physical property measurements. For example, large grains obtained from sintered $YBa_2Cu_3O_7$ [5–7] and $HgBa_2CuO_4$ [8, 9] samples have been used for structure determinations.

D1.1.2 B2.3.3, C4

B2.4.3.2 Flux growth

Flux growth is the growth of crystals from high temperature solutions of different composition to the desired phase. In flux growth, a charge composed of the flux solvent and the desired phase A is initially heated in a crucible to a temperature above which the flux melts and A is dissolved in the melt. The solubility of the A is then reduced, typically by cooling the melt, resulting in the formation of crystals of A. This is illustrated schematically in the simple binary phase diagram in figure B2.4.1, where A is the desired phase, B is the flux material, and X is the initial charge composition. Crystals are removed from the solidified melt by mechanical separation or by chemical dissolution of the flux material.

The composition of the flux may be different from that of the desired phase; for example, KCl has been used for the growth of Bi–Sr–Ca–Cu–O [10, 11] and $BaPb_{1-x}Bi_xO_3$ [12] crystals. Alternatively, the flux may be composed of one or more of the elements of the desired phase, as in the case of the $BaCuO_2$–CuO mixtures used in $YBa_2Cu_3O_7$ crystal growth [13, 14]. This latter situation is often referred to as eutectic or non-stoichiometric melt growth. Many different fluxes with varying composition, degree of covalency, redox behaviour and acid–base characteristics have been employed to grow oxide crystals. The choice of a suitable flux for a particular compound is largely empirical, although chemical reasoning and specific needs regarding temperature stability, melting point and reactivity can guide the selection process. Many other processing parameters have a strong effect on crystal growth, including the composition of the charge material (80 wt% flux is a typical starting point), the choice of the crucible material, the growth atmosphere, the maximum hold temperature (must be below the decomposition temperature of the desired phase) and the cooling rate. These processing parameters are also selected empirically and are optimized through iterative experimentation.

C2
C1

Single crystals of most of the high temperature superconductors have been grown by flux methods. The crystals are frequently doped with foreign elements present in the flux material or the crucible, as for Hg–Ba–Ca–Cu–O crystals grown from Bi_2O_3-containing fluxes [15], and $YBa_2Cu_3O_7$ crystals grown in gold crucibles [14].

C4

B2.4.3.3 Top-seeded solution growth

Top-seeded solution growth (TSSG) is a variation of flux growth. As the name implies, a seed crystal is brought into contact with a melt as illustrated in figure B2.4.2. The furnace is arranged to produce a well-defined temperature gradient, and the temperature in the vicinity of the melt surface is maintained near the temperature for the onset of solidification. The seed is attached to a support rod which acts as a heat leak making the seed slightly cooler than the nearby liquid; growth initiates on the seed. To promote additional growth, the temperature is slowly lowered and, simultaneously, the growing crystal slowly raised. At the end of the growth experiment, the sample is raised out of the melt and the adhering

B2.3.3,
B2.3.4

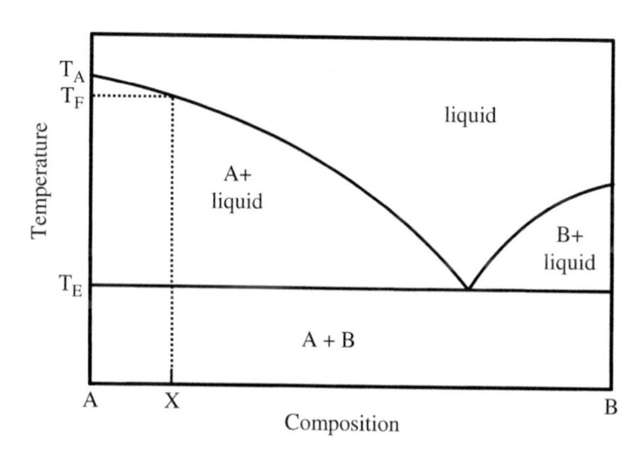

Figure B2.4.1. Schematic phase diagram for a simple binary mixture.

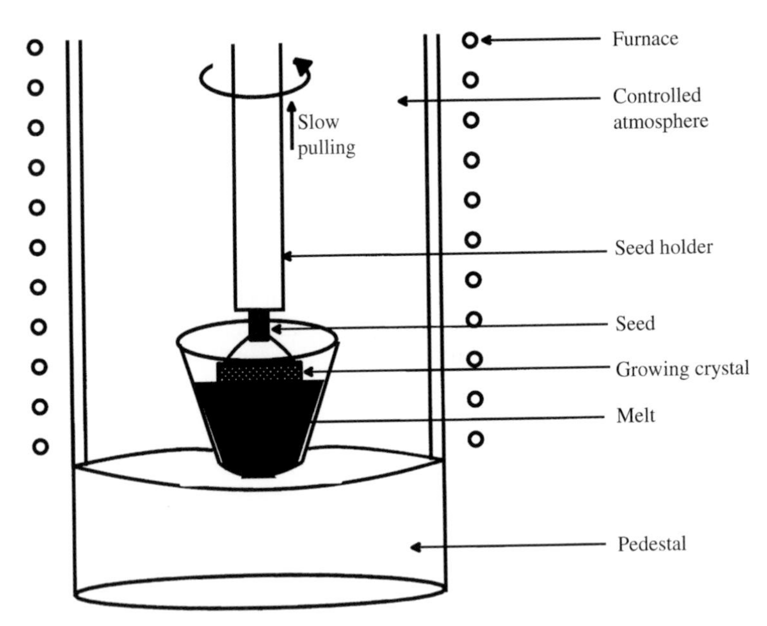

Figure B2.4.2. Schematic diagram of the experimental arrangement for TSSG.

melt is spun off, typically leaving the newly-grown crystal largely flux-free. As in flux growth, contamination of the crystal by elements leached from the crucible is a concern.

Top-seeding has been used to grow large crystals of La_2CuO_4 and related materials from CuO-rich [16, 17] and $Li_4B_2O_5$ [18, 19] melts. Single crystals of $YBa_2Cu_3O_7$ [20, 21] and Bi–Sr–Ca–Cu–O [22, 23] have also been grown by top-seeding approaches.

C1, C2

B2.4.3.4 Floating zone

The floating zone technique uses a heat source to melt locally a portion of a feed rod of a congruently melting compound. This method involves no crucible, so sample contamination by interaction with the crucible is not an issue. Typical heat sources include lasers, focused radiant lamps, radio frequency generators and resistive heaters. The feed rod can be fabricated from a single crystal or polycrystalline sintered ceramic. A schematic diagram of the experimental setup is shown in figure B2.4.3. During an experiment, the molten zone is translated slowly along the length of the feed rod, either by moving the feed rod through the stationary hot zone or by scanning the zone along the fixed feed rod. The two ends of the feed rod are rotated in opposite directions. Surface tension forces act to retain the shape of the zone along the vertical axis of the feed rod. The molten zone may be seeded at the trailing edge with a single crystal to promote crystal growth. The molten material solidifies readily into a single crystal or oriented sample.

A modification of the floating zone method, where the feed cylinder has a much larger diameter than the pulled crystal rod, is known as the pedestal technique [24]. In this method, the surface of the feed cylinder is melted by means of an arc image furnace [24] or a laser [25]; the seed is lowered into the molten pool and then extracted slowly. A miniaturized version of the pedestal method that has been used to grow single crystal fibres of many different materials by laser heating is referred to as the laser-heated pedestal growth (LHPG) technique [25–27]. The seed may be a single crystal or polycrystalline rod of the desired fibre material, or an inert refractory metal such as Pt or Ir.

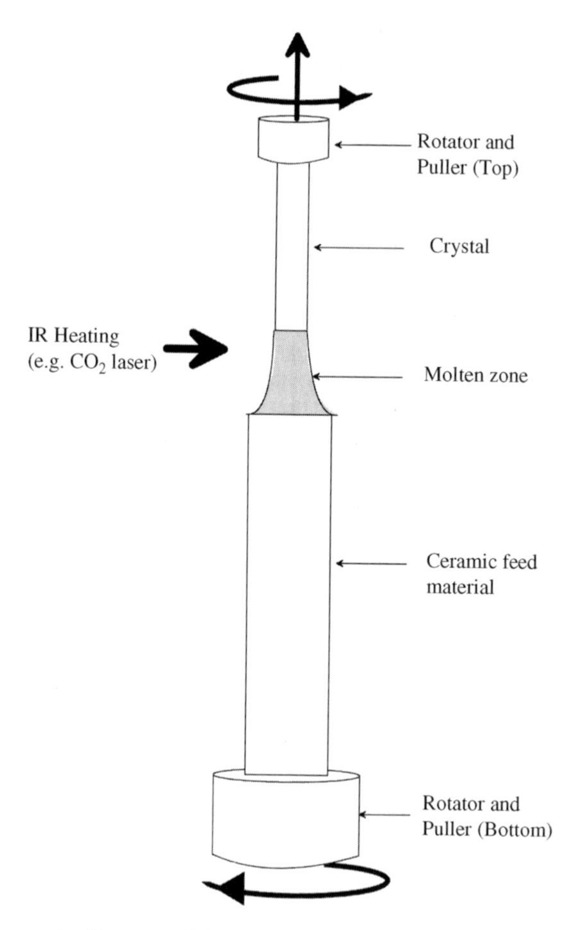

Figure B2.4.3. Schematic diagram of the experimental arrangement for floating zone growth.

The application of floating zone techniques to high T_c single crystal growth has recently been reviewed by Revcolevschi and Jegoudez [28]. The LHPG technique has been used to grow highly-textured polycrystalline $Bi_2CaSrCu_2O_8$ fibres composed of single crystal grains up to 3 mm long [29, 30].

B2.4.3.5 Travelling solvent method

The travelling solvent floating zone (TSFZ) method, a combination of the floating zone and flux growth techniques, is used for incongruently melting compounds. This method uses no crucible and thus has the same advantages as the floating zone approach. In the TSFZ technique, a seed rod of the desired compound A is connected to a feed rod of the same composition A by a plug of solvent material with composition B. The solvent plug is melted by a heat source and the molten zone is moved slowly through the feed rod, with the crystallization front initiating at the seed end. The feed rod melts and the cations move through the molten zone to the growth front at the seed end, supplying material for the growing crystal while maintaining the composition of the molten zone. Ideally, compound A is the only product to crystallize from the liquid solvent B. The seed, either a single crystal or an epitaxial thin film, provides a template for the growing crystal.

The TSFZ technique has been applied to the growth of $La_{2-x}Sr_xCuO_4$ [31] and (RE)$Ba_2Cu_3O_7$ (RE = rare earth cations) [32, 33] crystals using CuO-rich solvents. In the case of $La_{2-x}Sr_xCuO_4$, large, high quality single crystal boules were prepared; the (RE)$Ba_2Cu_3O_7$ boules contained a number of millimetre-sized grains with good alignment along the growth direction.

B2.4.3.6 Electrocrystallization

Electrocrystallization involves the deposition of a crystalline material by the reduction or oxidation of ionic species in a solution at an electrode. The process can be run at very low temperatures and is thus advantageous for compounds that have poor thermal stability. A schematic diagram of a simple one-cell apparatus [34] is presented in figure B2.4.4. In the growth of a compound A from a flux solvent B, the solvent is first melted in a crucible. After adding compounds that contain the various elements in A, the electrodes are lowered into the melt and electrolysis is initiated. Crystals form at the working electrode, which is typically an inert conductor such as Pt, Au or Ag. The process is run isothermally and potentiostatically (at a constant voltage) or galvanostatically (at a constant current). At the end of the deposition run, the electrodes are withdrawn from the hot melt, cooled and rinsed with an appropriate solution to remove residual solvent. Experimental variables include the crucible and electrode materials, flux solvent composition, temperature and fixed voltage or current.

Millimetre-sized crystals of $Ba_{1-x}K_xBiO_3$ [35] and $LiTi_2O_4$ [36] have been grown by electrocrystallization. Small crystallites of superconducting $La_{2-x}Na_xCuO_4$ [37] and $EuBa_2Cu_3O_7$ [38] have also been grown from NaOH fluxes by an electrochemical approach.

B2.4.3.7 Hydrothermal growth

Hydrothermal growth is the growth of crystals from aqueous solutions at elevated temperatures and pressures. This is an important crystal growth approach which has received particular attention from the mineralogy community because naturally occurring crystals are often produced under hydrothermal-like conditions. The starting materials are dissolved in water under controlled pH conditions, heated

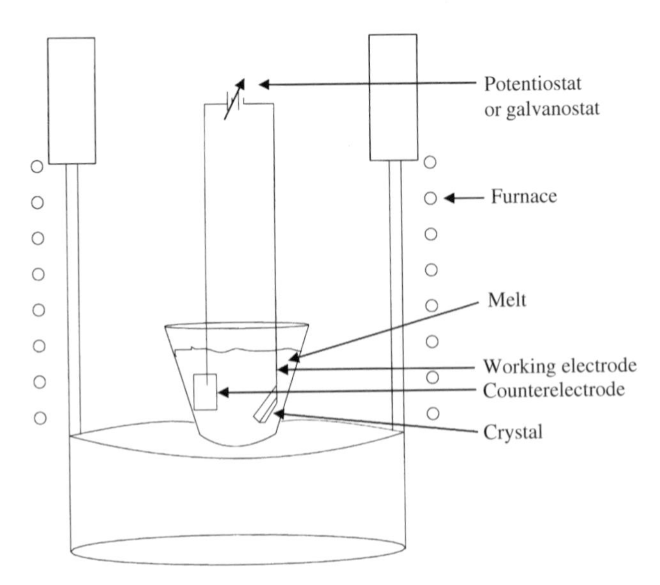

Figure B2.4.4. Schematic diagram of an electrochemical crystallization apparatus. Both anodic ($A^- \rightarrow A_{XL} + e^-$) and cathodic ($A^+ + e^- \rightarrow A_{XL}$) crystal growth processes are possible.

under pressure to produce a solution and then slow-cooled to allow precipitation and crystal growth of the phase of interest. The utility of the approach depends on the enhanced solvent properties and mobilities of supercritical water ($T > 373°C$). Other solvents besides water have also been used for this purpose, a growth approach termed solvento-thermal growth. The general experimental approach for hydrothermal growth has been reviewed by Byrappa [39].

High quality $BaPb_{1-x}Bi_xO_3$ crystals have been obtained hydrothermally using KCl [40] and NaOH [41] solutions. Attempts to establish hydrothermal growth conditions or identify a suitable non-aqueous solvent for the other cuprate superconductors have been unsuccessful thus far.

B2.4.4 Growth of high temperature superconducting compounds (HTSC) single crystals

In this section we describe the crystal chemistry of the different classes of high temperature superconducting compounds, including cuprate and non-cuprate phases, and discuss the techniques that have been used to grow crystals of the various compounds. Where relevant, post-annealing processes are also discussed briefly. For some compounds, it is virtually impossible to include all of the crystal growth references, given the tremendous amount of information in the literature. We have endeavoured to include the major early references on each different technique, with emphasis on new developments published within the last five years. In preparing this paper, we referred to a number of excellent review papers devoted to the crystal growth of high T_c cuprates [42–45].

B2.4.4.1 La_2CuO_4 and $La_{2-x}A_xCuO_4$

The material which initiated the explosion of interest in cuprate superconductors was La_2CuO_4. Bednorz and Muller first reported indications of superconductivity at high temperatures in a mixed phase sample in the La–Ba–Cu–O phase system in 1986 [46]. Investigators worldwide subsequently showed that the perovskite-related K_2NiF_4 phase La_2CuO_4 (figure B2.4.5) with lanthanum partially substituted by barium was a bulk superconductor at a then record setting temperature of 30 K. Explorations of chemically related materials such as $La_{2-x}Sr_xCuO_4$ soon raised the T_c to 40 K for the optimal composition $x = 0.15$ [47, 48]. Researchers quickly found approaches to the crystal growth of these materials for detailed investigations of the underlying physics.

Large crystals of $La_{2-x}Ba_xCuO_4$ and $La_{2-x}Sr_xCuO_4$ with volumes approaching 1 cm^3 were grown from CuO fluxes in Pt crucibles by slow-cooling from soak temperatures of 1100 to 1350°C [17, 49–53]. The resulting as-grown crystals all had broad superconducting transitions at onset temperatures below 25 K, presumably due to Pt contamination and non-ideal or inhomogeneous composition (x and oxygen stoichiometry). T_c was increased by post-annealing in flowing oxygen, but the transition width was still large [49].

Since the La_2CuO_4–CuO system forms a stable melt at about 1050°C in air, TSSG approaches can be employed. Millimetre-sized crystals of $La_{2-x}Sr_xCuO_4$ and $La_{2-x}Ba_xCuO_4$ have been grown from CuO [16, 17, 54–56] or $Li_4B_2O_5$ [18, 19] rich melts in Pt crucibles. Growth temperatures ranged from 1150 to 1350°C, and seed rods of Pt, $LaSrFeO_4$ and La_2CuO_4 were used. Crystals grown by TSSG were either not superconducting, even after oxygen annealing, or exhibited T_c values of 10–20 K after annealing. The poor superconducting properties may be attributed to Pt contamination, Li contamination (for the $Li_4B_2O_5$ fluxes), and inhomogeneous or non-ideal oxygen or alkaline earth concentration.

The largest and highest quality $La_{2-x}Sr_xCuO_4$ ($0 \leq x \leq 0.3$) and $La_{2-x}Ba_xCuO_4$ ($0.06 \leq x \leq 0.09$) crystals have been grown by the crucibleless TSFZ approach [31, 57–63]. The feed and seed rods are stoichiometric or slightly CuO-rich sintered, polycrystalline $La_{2-x}A_xCuO_4$ (A = Sr or Ba) and the solvent rods are composed of 55–85 mole % CuO with a balance of $La_{2-x}A_xCuO_4$. Growth runs were typically performed under an oxygen pressure of 0.2–1 MPa (2–10 atm). A modified TSFZ method

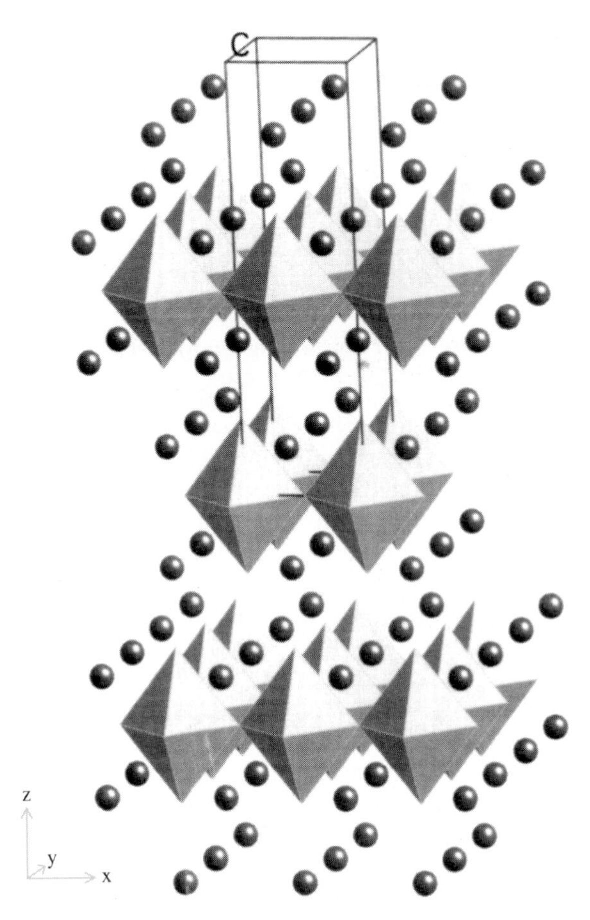

Figure B2.4.5. Extended lattice view of the K_2NiF_4-type structure of La_2CuO_4.

which uses no solvent rod has also been reported [62]. Typical single crystal boule dimensions are 5–7 mm in diameter and 10–15 mm in length. Sharp superconducting transitions at $T_c = 33$–38 K for $La_{1.85}Sr_{0.15}CuO_4$ and $T_c = 27$–29 K for $La_{2-x}Ba_xCuO_4$ ($0.065 \leq x \leq 0.09$) were obtained, mostly for the as-grown state. These transitions are higher than those obtained in crystals prepared by flux growth or TSSG methods where it was likely that contamination by the crucible caused reduced T_c values. The TSFZ method has also been used to grow crystals of $La_{2-x}Ca_xCuO_4$ ($0.075 \leq x \leq 0.09$) with broad transitions at $T_c = 15$ K [63] and $(La,Sm,Sr)_2CuO_4$ with broad transitions at $T_c = 17$ K after oxygen annealing [64].

B2.4.4.2 $Nd_{2-x}Ce_xCuO_4$ and related phases

The compounds $Ln_{2-x}Ce_xCuO_4$ (LnCCO, Ln = Nd, Pr, Sm, Eu) [65, 66] and $Ln_{2-x}Th_xCuO_4$ (Ln = Nd, Pr) [66, 67] were the first cuprate high T_c materials in which the carriers of the superconducting current were electrons (n-type) rather than holes (p-type). In the $Ln_{2-x}Ce_xCuO_4$ family, only the compositions $0.14 \leq x \leq 0.17$ are superconducting, with T_c up to 24 K [68, 69]. These compounds have the Nd_2CuO_4, or T′, structure with square planar coordination of the Cu-O layers as illustrated in figure B2.4.6.

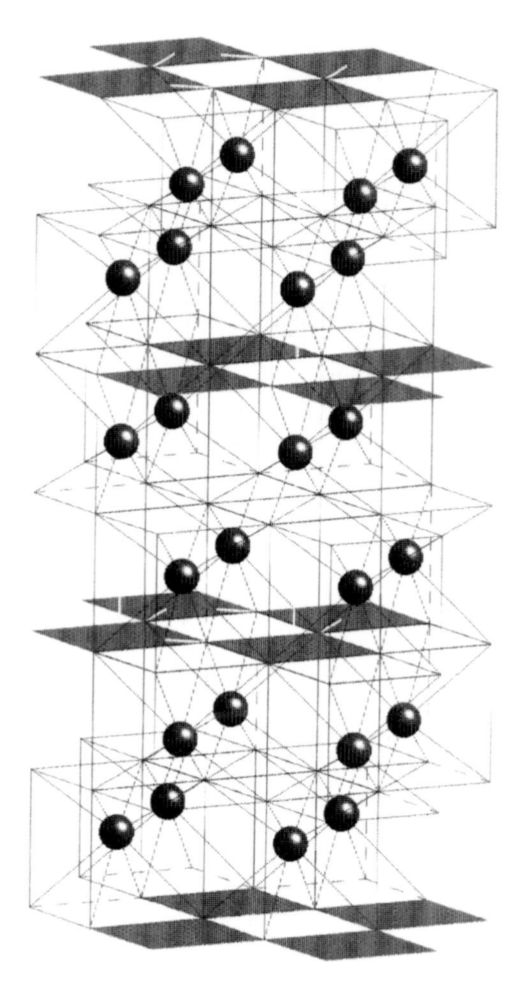

Figure B2.4.6. Extended lattice view of the T′-type structure of Nd_2CuO_4.

All reported methods for LnCCO crystal growth involve a CuO-rich self-flux or solvent that promotes melting at temperatures as low as 1050°C. NdCCO [69–74], SmCCO [73, 74] and PrCCO [74–76] crystals with dimensions up to $10 \times 10 \times 0.3\,mm^3$ have been grown by a slow-cooling flux growth approach in alumina or Pt crucibles. A recent study of crucible/melt compatibility demonstrated that MgO was the optimal crucible material for PrCCO crystal growth [77]. A modified flux growth method in which a NdCCO pellet is reacted with Cu shot yielded millimetre-sized platelet crystals [78]. NdCCO crystals have also been grown from CuO-rich melts in Pt crucibles by a TSSG approach [79, 80]. Finally, TSFZ methods have been used to grow NdCCO crystal boules up to 7 mm in diameter by 50 mm in length [71, 81–83].

In contrast to some of the other oxide superconductors, which must be oxygen-annealed to attain high T_c values, as-grown $Ln_{2-x}Ce_xCuO_4$ crystals are oxygen-rich and must be annealed under reducing conditions at temperatures above 850°C to optimize T_c. Large, high quality $Ln_{2-x}Ce_xCuO_4$ crystals with sharp superconducting transitions are difficult to prepare due to inhomogeneities in both oxygen and Ce content. A two-step process involving a high temperature (850°C) anneal in an inert gas followed by a lower temperature (500°C) anneal in oxygen has been used to achieve sharp transitions at 27 K in large NdCCO boules [83].

B2.4.4.3 $YBa_2Cu_3O_7$ and its lanthanoid isomorphs

$YBa_2Cu_3O_7$, referred to as YBCO or 213, was the first superconductor to have a T_c above the boiling point of liquid nitrogen [84, 85]. This compound, and many of the isomorphous rare earth analogues $(RE)Ba_2Cu_3O_7$ (RE = rare earth cations = La, Nd, Sm, Eu, Gd, Dy, Ho, Er, Tm, Lu), have T_c values near or slightly above 90 K [85–88]. These $(RE)Ba_2Cu_3O_7$ ((RE)BCO) phases are orthorhombically-distorted, oxygen-deficient perovskites with a unit cell structure as shown in figure B2.4.7. The Ba and R cations on the A-site are ordered, leading to a tripling of the standard ABO_3 perovskite cell along the *c*-axis.

B2.2.2 Small crystallites of YBCO formed by sintering Y–Ba–Cu–O powder compacts at temperatures near 970°C were used for some early structural and magnetic measurements [5–7]. YBCO decomposes peritectically near 1120°C, precluding crystal growth from stoichiometric melts. However, there are B2.3.2 some compositions in the pseudoternary BaO–CuO–$YO_{1.5}$ phase diagram that partially melt at 975–1000°C in air [89]; the first well-formed single crystals of suitable size and quality for many research studies were grown from charges with these starting compositions [13, 14]. Many groups have explored the use of various crucible materials for crystal growth from CuO–BaO self-fluxes, including Al_2O_3 [13, 14, 90–94], Au [14], Pt [13, 14, 49, 90, 95, 96], ZrO_2 [13, 94, 97], yttria-stabilized zirconia [98], MgO [13, 14, 98], ThO_2 [13, 98] and $BaZrO_3$ [99]. Corrosion of the crucible leading to doping of the YBCO crystals with the crucible cations occurs to some extent for all of these crucible materials, with the exception of $BaZrO_3$ [99]. The crucible materials which routinely give good quality crystals with T_c values exceeding 90 K after oxygen annealing are Au, ZrO_2 and $BaZrO_3$. Self-flux growth processes have yielded small crystals with nearly perfect euhedral morphology as shown in figure B2.4.8, as well as larger crystals with dimensions up to $5 \times 5 \times 2\,mm^3$ [93, 94, 98].

Watanabe [101] did an extensive survey of other high temperature fluxes, including carbonates, borates, oxides, fluorides and hydroxides, and concluded that CuO–BaO self-fluxes were superior. However, two groups have reported on the growth of large YBCO crystals (up to $8 \times 7 \times 2\,mm^3$) by

Figure B2.4.7. Polyhedral representation of the fully oxidized, perovskite-related structure of $YBa_2Cu_3O_7$.

Figure B2.4.8. Scanning electron micrograph of a $YBa_2Cu_3O_7$ single crystal grown by a self-flux method [100].

slow-cooling NaCl–KCl fluxes from temperatures in the range 980–1140°C [102, 103]. CuO–BaO fluxes containing up to 65% NaCl–KCl–RbCl were used to grow large ($10 \times 5 \times 0.05\,mm^3$) crystal plates [104]. Small (30 μm × 130 μm), plate-like (RE)BCO (RE = Nd, Sm, Eu, Gd) crystallites were precipitated directly from molten NaOH–KOH solutions at 450°C [105]. Moderate sized crystals ($0.5 \times 0.5 \times 0.5\,mm^3$) were grown from dry KOH fluxes at 750°C [106]. Crystals grown from In_2O_3 fluxes in alumina crucibles [107] had a range of T_c values, all below 78 K, presumably due to Al doping. A K_2CO_3 flux was used to grow millimetre-sized YBCO crystals in alumina crucibles by slow-cooling from 1000°C [108].

TSSG approaches using seed rods to pull crystals from CuO–BaO fluxes held at 1000–1080°C have been reported by several groups. Clusters of millimetre-sized YBCO crystals have been grown from Pt [20] and alumina [21] crucibles using pull rods of the same material as the crucible; T_c values below 70 K were reported after oxygen annealing, due to oxygen deficiency or contamination. Shiohara and co-workers [109, 110] have used modified TSSG methods to grow large single crystals (up to $2 \times 2 \times 2\,cm^3$) with T_c values near 90 K. YBCO crystals were pulled onto textured BSmCO rods from yttria crucibles containing CuO–BaO flux on top of a Y_2BaCuO_5 solute layer, used to maintain Y supersaturation of the melt [109]. BNdCO crystals were grown on BNdCO epitaxial thin film seeds from CuO–BaO melts in Nd_2O_3 crucibles [110]. The melting temperature of a $BaCuO_2$ flux was decreased to 935°C by adding a small amount of KCl, permitting lower temperature growth of YBCO crystal clusters [111]. B2.3.3

TSFZ techniques have been used to grow large crystalline boules of BRCO. High-density, polycrystalline BRCO seed and feed rods are connected via a plug of CuO/BaO-rich solvent; the plug is melted and moved through the feed rod. (RE)BCO boules with diameters of 3–6 mm and lengths of several cm have been grown by this technique [32, 33]. The crystalline boules were composed of millimetre-sized grains of nearly identical orientation with respect to the growth axis and had T_c values of 91–94 K. A modified horizontal Bridgman-like method has been used to grow clusters of YBCO single crystals from molten CuO/BaO-rich charges held in alumina boats in a temperature gradient [112].

Most as-grown crystals are oxygen deficient and must be annealed to attain superconductivity near 90 K. The effect of annealing temperature and oxygen partial pressure on the oxygen stoichiometry is illustrated in figure B2.4.9 [113]. Twinning occurs during cooling through the tetragonal–orthorhombic phase transition, so superconducting (RE)BCO crystals usually contain microtwins in the a–b plane. Twins can be removed by thermomechanical treatment [114–117], resulting in a single-domain, 'true' single crystal. A purely thermal treatment for producing large untwinned regions in crystals by quenching from the tetragonal state and subsequently annealing in oxygen has also been reported [118]. B2.3.4

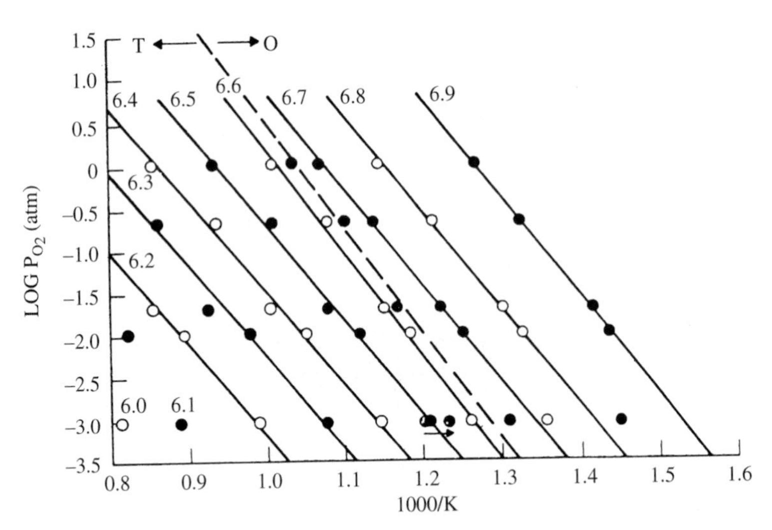

Figure B2.4.9. Lines of constant oxygen composition x in $Ba_2YCu_3O_x$ single crystals plotted in oxygen potential versus reciprocal temperature space. Reproduced from an article by Gallagher [113] (copyright American Ceramic Society).

B2.4.4.4 $Ba_4Y_2Cu_7O_{15}$ *and* $Ba_4Y_2Cu_8O_8$ *and their lanthanoid isomorphs*

The Ba–Y–Cu–O pseudoternary system contains a family of superconducting compounds of the general formula, $Ba_4Y_2Cu_{6+n}O_{14+n}$, all of which contain planar Cu–O layers separated by copper oxide chain structures. The first three members of the series ($n = 0$–2) are superconducting: the widely studied $Ba_2YCu_3O_7$ or 213 phase discussed in the previous section; the $Ba_4Y_2Cu_7O_{15}$ or 247 phase [119]; and the $Ba_4Y_2Cu_8O_{16}$ or 248 phase [120]. The optimal T_c values for the 213, 247 and 248 compounds are 92, 70 [121] and 81 K [120], respectively. All of the phases contain infinite two-dimensional sheets formed from five-coordinate square-pyramidal Cu–O units which share oxygen vortices. The Y ions occupy sites between these layers. In 213, infinite one-dimensional Cu–O single chains link the two-dimensional sheets, while in 248 the sheets are linked by infinite Cu–O double chains (see figure B2.4.10). The 247 phase contains alternating single and double chains, as shown in figure B2.4.11. The double chains in these structures have fixed stoichiometry while the single chains have variable oxygen stoichiometry. Thus, the 213 and 247 phases exhibit considerable non-stoichiometry and T_c is strongly dependent upon the oxygen content. In contrast, the 248 phase exists over a relatively narrow range of oxygen concentration; T_c may be increased to 90 K by substituting an alkaline earth element such as calcium for Y [122]. Most of the rare earth analogues of the undoped and Ca-doped 248 phase form; T_c is inversely related to the rare earth ion size [123–125]. In contrast to the 213 phase which exhibits extensive twinning, the 247 and 248 phases do not contain twins in the a/b plane.

Single crystal growth of the 247 and 248 phases requires high oxygen pressures and high temperatures [121]. The specialized equipment necessary to carry out such studies is limited to a few laboratories responsible for most of the work in this area. Since the phases melt incongruently, both 248 and 247 crystals have been grown by a flux method using CuO/BaO-rich self-fluxes. Thin (<0.1 mm), millimetre-sized Y-248 crystal platelets were grown from melts in alumina [121, 126–129], yttria-stabilized zirconia [127–130], yttria [131] or $BaZrO_3$ [132] crucibles under oxygen partial pressures of 20–280 MPa. Crystals with sharp transitions at T_c above 80 K were obtained using all crucible materials except alumina. Crystals of 248 grown in alumina had T_c values below 75 K, most likely due to Al incorporation [131]. Other rare earth 248 crystals [130, 133] as well as Ca [130], Zn [129] and Ni [129] substituted crystals were also grown from a self-flux. Millimetre-sized platelet crystals of Y-247 have

Figure B2.4.10. Polyhedral representation of the fully-oxidized structure of $Ba_4Y_2Cu_8O_{16}$ [248]. Note that the structure contains only double Cu–O chains.

also been grown from self-fluxes at high oxygen partial pressures (5–160 MPa) in alumina [121,126] and $BaZrO_3$ [132] crucibles. T_c values in the range of 20–70 K were obtained for the as-grown crystals; the low values are attributed to Al contamination and oxygen non-stoichiometry.

B2.4.4.5 $Bi_2Sr_2CaCu_2O_8$ and its homologues

The discovery of superconductivity at 84 K in a Bi–Sr–Ca–Cu–O composition in 1988 [134] demonstrated the existence of a cuprate phase other than $YBa_2Cu_3O_7$ with a T_c above the boiling point of liquid nitrogen. Three distinct superconducting phases in the homologous series of approximate formula

C2
C1

Figure B2.4.11. Polyhedral representation of the fully-oxidized structure of $Ba_4Y_2Cu_7O_{15}$ [247]. Note that the structure contains alternating single and double Cu–O chains.

$Bi_2Sr_2Ca_{n-1}Cu_nO_{4+2n}$ (BSCCO) have been identified: $n = 1$ or Bi-2201 [135]; $n = 2$ or Bi-2212 [136]; and $n = 3$ or Bi-2223 [136]. Like $YBa_2Cu_3O_7$, the Bi-22($n-1$)n phases contain infinite two-dimensional CuO planes. T_c increases with n, the number of CuO planes: $T_c = 10$, 85 and 110 K for $n = 1$, 2 and 3, respectively.

C2 Structures for the Bi-2201, Bi-2212 and Bi-2223 phases are illustrated schematically in figures B2.4.12, B2.4.13 and B2.4.14, respectively. The compounds are layered phases with rock-salt bismuthate double-layer structural units alternating with perovskite-type cuprate structural units. Because these structural units have slightly different sizes, the Bi – 22(n – 1)n phases have incommensurate structures. This incommensurate character, as well as anti-site disorder and stacking faults, cause the actual compositions to differ slightly from the ideal formulae.

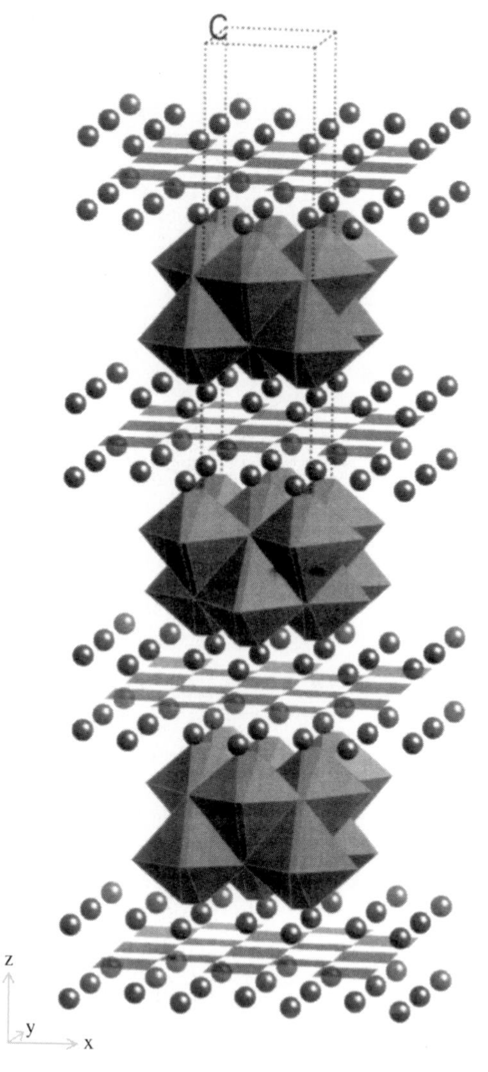

Figure B2.4.12. Schematic drawing of the $Bi_2Sr_2CuO_6$ (Bi-2201) structure, which contains only infinite one-dimensional Cu–O planes.

There are a few reports on Bi-2201 single crystal growth; this phase is of lesser interest because of its low T_c. There have been many papers on the growth of Bi-2212 crystals by a variety of different techniques. T_c values for as-grown Bi-2212 crystals are in the range of 80–93 K unless stated otherwise. It is much more difficult to grow high quality crystals of the highest T_c phase, Bi-2223. Since substitutional doping with Pb appears to stabilize Bi-2223 in the phase pure form [137–139], there have been numerous studies on the growth of (Bi,Pb)-2223 as well as (Bi,Pb)-2212 crystals.

B2.4.4.6 Bi-2201

Flux growth approaches have been used to grow thin platelet crystals of Bi-2201 with $T_c = 9$ K from self-fluxes contained in Au [140] or alumina [141, 142] crucibles. Needle-shaped crystals 7–15 mm long

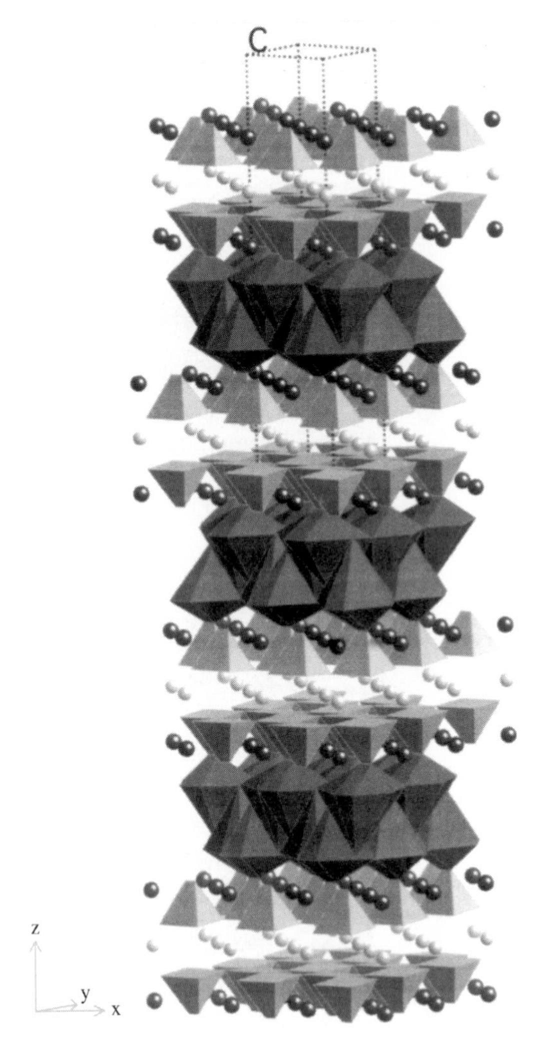

Figure B2.4.13. Schematic drawing of the $Bi_2Sr_2CaCu_2O_8$ (Bi-2212) structure, which contains pairs of Cu–O planes.

[141] and (Bi,Pb)-2201 crystals with $T_c = 7$ K [143] were also prepared by self-flux methods. Thicker Bi-2201 crystals, with dimensions up to $6 \times 4 \times 0.2$ mm^3, were obtained by slow-cooling KCl–KF fluxes in Pt crucibles [144].

B2.4.4.7 Bi-2212

B3.2.1 B2.3.5 Thin (50–200 μm), millimetre-sized platelet crystals of Bi-2212 have been grown from near stoichiometric melts [141, 145–147] or self-fluxes [141, 148–154] in alumina crucibles by slow-cooling from temperatures of 950 to 1200°C. Crystals grown by slow-cooling from self-flux [155] and near stoichiometric [156] melts in Pt crucibles have also been reported. Directional solidification approaches involving large temperature gradients have been used to grow large platelet crystals from stoichiometric and off-stoichiometric melts in alumina [157–159] and MgO [160] crucibles. A stoichiometric charge melted by inductive heating in a cold crucible yielded crystals with $T_c = 81$ K after oxygen annealing [161].

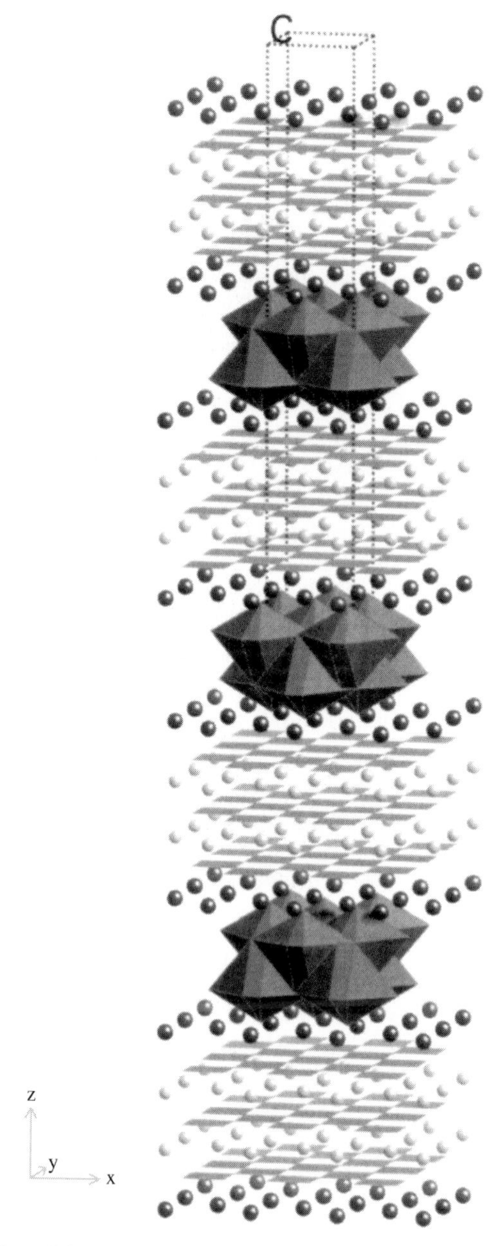

Figure B2.4.14. Schematic drawing of the $Bi_2Sr_2Ca_2Cu_3O_{10}$ (Bi-2223) structure, which contains triple Cu–O plane structures.

A number of groups have reported the growth of (Bi,Pb)-2212 crystals by self-flux approaches [143, 157, 162–164]. Y [157, 165–168], Gd [167], Fe [168] and Al [154] doped Bi-2212 crystals have also been grown from Bi-rich self-fluxes or near stoichiometric melts. Al contamination was found to decrease the T_c of Bi-2212 crystals by up to 10 K [154]; thus, alumina may not be the optimal crucible material. Extremely sharp transitions have been reported for crystals grown in MgO [160].

Alkali chloride fluxes, mainly KCl, in Pt crucibles have been used to grow very thin (0.4–50 μm) Bi-2212 crystals with areas approaching 1 cm^2 [10, 11, 169–173]. KBr has also been used as a flux [174]. The advantages of using alkali halide fluxes are ease of separation of the crystals by dissolution of the flux in water and stability of Pt crucibles to the melt. Thicker crystals have been grown from K_2CO_3 fluxes in Pt crucibles [175].

C3 TSSG approaches have been used to grow Bi-2212 crystals from self-fluxes in Pt or alumina crucibles using $YAlO_3$ [23], $Bi_4Ge_3O_{12}$ [23] or Bi-2212 [176] seed crystals. Crystal aggregates [22, 23] as well as single crystal boules 10 mm in diameter by 6 mm in length [176] have been reported.

Floating zone approaches have been applied to grow aggregates of Bi-2212 crystals using stoichiometric feed and seed rods [177, 178]. A modified floating zone approach [179] yielded crystals up to $4 \times 3 \times 1$ mm^3. TSFZ methods using Bi-2212 feed and seed rods and a Bi_2O_3/CuO-rich solvent plug have also been reported [180–182]. Boules are composed of single crystal plates up to $50 \times 5 \times 2$ mm^3 [178, 182]. Y [183] and Ni [180] doped Bi-2212 crystals have also been grown by TSFZ approaches.

Bi-2212 fibers composed of highly aligned crystal plates have been fabricated by LHPG, a miniaturized version of the float zone process [29, 30, 184, 185]. Fibers up to 1 mm in diameter and several centimetres in length having high critical current densities were grown from stoichiometric source rods using polycrystalline Bi-2212 seeds.

A thin Bi-2212 crystal with an area of 4×10 mm^2 and a T_c of 90 K was grown by electrocrystallization from a Bi–Sr–Ca–Cu–O melt [186]. A novel process in which the oxygen partial pressure above a sintered (Bi,Pb)-2212 compact was progressively increased at a constant temperature (875°C) below the incongruent melting point was used to grow millimetre-sized plates of (Bi,Pb)-2212 [187]. Vertical Bridgman approaches, where Bi-2212 charges are loaded into alumina ampoules or Au capsules, melted and then slowly drawn through a temperature gradient, have also been reported [188, 189]. Plate-like crystals with areas up to 15×7 mm^2 were obtained for pull-down rates of 0.5–1.0 mm h^{-1}.

Whisker crystals having a blade-like shape are of great interest because they can have a very high degree of crystalline perfection with a low density of defects. The first Bi-2212 whiskers were grown on the surfaces of Bi–Sr–Ca–Cu–O pellets sintered in air at 825°C [190]. These whiskers were a few micrometers thick, 10–30 μm wide, and a few millimeters long. Much larger (Bi,Pb)-2212 whiskers were obtained by heating 1 mm thick melt-quenched Bi(Pb)–Sr–Ca–Cu–O glass samples in a stream of oxygen gas for 120 h [191]. The whiskers, which grew perpendicular to the glassy plate, were each composed of a stack of plate-like crystals and had dimensions up to 10 μm $\times$ 0.3 mm $\times$ 15 mm. This method has also been used to grow Pb-free Bi-2212 whiskers with cross-sectional areas down to 0.1 μm^2 [192]. Slow-cooling a stoichiometric Bi-2212 melt in an alumina crucible also yielded whiskers [193, 194]. Finally, La-doped Bi-2212 whiskers have been grown by slow-cooling Bi-rich melts in alumina crucibles [195].

Oxygen concentration, first reported to have a large influence on T_c in bulk Bi-2212 [196], also has a large effect on the T_c in single crystals [160, 197, 198]. The transition onset temperature is reported to shift reversibly from 72 to 86 K for Bi-2212 crystals [160], and from 76 to 90 K in (Bi,Pb)-2212 crystals [197] when the annealing conditions are varied from oxidizing to reducing. There appears to be a structural distortion unrelated to the oxygen concentration which also causes a decrease in T_c with increasing oxygen partial pressure above 13 Pa (0.1 Torr) [198].

B2.4.4.8 Bi-2223

B3.2.2 Platelet crystals of (Bi,Pb)-2223 have been grown from KCl [172, 199] or KCl/KNO_3 [200] fluxes in Pt or alumina crucibles by slow-cooling from about 825°C. The crystals had T_c values of 105–110 K [172, 200] and were as large as $4 \times 3 \times 0.01$ mm^3 [199]. Both whiskers [201] and very thin single crystals [201, 202] of (Bi,Pb)-2223 with $T_c = 107$ K have been grown by annealing Bi-2212 whiskers or crystals in $Bi_2Sr_2Ca_4Cu_6Pb_{0.5}O_x$ powder.

B2.4.4.9 $Tl_2Ba_2CaCu_2O_8$ and its homologues

Some of the highest superconducting transition temperatures, up to 125 K, have been found in the Tl–Ba–Ca–Cu–O system [203]. The superconducting compounds in this homologous series of nominal composition $Tl_2Ba_2Ca_{n-1}Cu_nO_{6+2n}$, referred to as Tl-22($n-1$)$n$, have T_c values of 90 K for $n = 1$ [204], 110 K for $n = 2$ [205], 125 K for $n = 3$ [205] and 115 K for $n = 4$ [206]. These phases are structurally similar to the BSCCO family of compounds (see figures B2.4.12–B2.4.14), with Tl replacing Bi and Ba replacing Sr. All of the Tl superconductors are layered phases with rock-salt thallate double-layer structural units alternating with perovskite-type cuprate structural units. Unlike their Bi analogues, the Tl-containing phases appear to have essentially commensurate structures. C3

Single crystal growth of the various Tl-containing cuprates is complicated due to the high volatility, chemical reactivity and toxicity of Tl_2O_3 at high temperatures. Thus, crystal growth temperatures should be minimized, and appropriate precautions observed to contain the reagents. Another difficulty in preparing single crystals is the strong tendency to form syntactic polycrystals, consisting of two or more phases grown epitaxially on one another along the c-axis direction. These intergrowth structures are revealed by high resolution transmission electron microscopy studies [207, 208]. Despite these challenges, single crystals of the $n = 1 - 4$ phases have been obtained by flux growth methods, generally with sealed or covered crucibles to contain the Tl–O vapour. The Tl crystals all have a platelet growth morphology, but are not micaceous like the BSCCO crystals.

Small crystals of Tl-2212 [209], Tl-2223 [210] and Tl-2201 [204] suitable for x-ray structural studies and some physical property measurements were first grown in sealed Au tubes by slow-cooling CuO/CaO or CuO-rich melts. Other investigators have used the same self-fluxes to grow larger (up to $2 \times 2 \times 0.2$ mm^3) crystals of Tl-2212 [206, 211–215], Tl-2223 [206, 208, 211, 214–216] or Tl-2201 [206, 214] in sealed Pt [211, 213] or Au [206, 208, 216] crucibles, or covered alumina crucibles [212, 214, 215]. One of these investigators [208] used a step-cooling approach, which is commonly used in liquid phase epitaxy growth of compound semiconductors. Crystals of Tl-2234 with dimensions up to $0.8 \times 0.8 \times 0.2$ mm^3 have been grown from CuO/CaO-rich melts in Au tubes [206] or alumina crucibles [207] or from stoichiometric charges in alumina crucibles [217].

Tl–O/CuO or Tl–O/CaO/CuO self-fluxes have also been used to grow crystals of Tl-2212 [218–220], Tl-2223 [218, 219] and Tl-2201 [220]. Two modified flux growth approaches, one in which the chemicals were handled and loaded into a sealed Pt crucible in a dry box atmosphere [219] and another involving two heating/cooling cycles [220], have permitted the growth of crystals of the $n = 1–3$ phases with dimensions up to $10 \times 10 \times 1$ mm^3. A novel approach, in which near millimetre-sized Tl-2223 crystals were grown from a solid/liquid mixture held at 900°C for 3 h, slowly cooled to 870°C and then quenched to room temperature, has also been reported [221]. Finally, mixed Tl-2201/Tl-2212 and Tl-2212/Tl-2223 crystals were grown by heating a charge near the peritectic temperature (907°C) to a solid/liquid state, where grain growth occurs readily [222]. These latter two studies were guided by the results of phase equilibria investigations.

Rb-doped single crystals of Tl-2212 containing Tl-1223 intergrowths were obtained by slow-cooling fluxes having Rb/Tl ratios up to 30/70 in covered Pt crucibles under an oxygen atmosphere [223]. The Rb-doped crystals were larger and thicker than undoped Tl-2212 crystals grown under similar conditions and had more intergrowths. Small ($0.5 \times 0.5 \times 0.05$ mm^3) crystals of $(Tl,Hg)_2Sr_2CaCu_2O_z$ were prepared by slow-cooling a stoichiometric oxide charge from 1140°C in an open barium zirconate crucible under an Ar pressure of 0.95 kbar [224].

Like many of the other superconducting cuprate phases, the properties of Tl-based cuprate superconductors are improved by post-growth annealing [211, 225]. Morosin and co-workers [225] found that the T_c values of Tl-2223 crystals increased upon annealing in oxygen, vacuum or forming gas above 300°C, suggesting that the crystals were not equilibrated during the growth process. These annealing effects may result from alterations of the Tl-O layers, which can oxidize to Tl^{3+} during mild anneals.

B2.4.4.10 $TlBa_2CaCu_2O_8$ and its homologues

C3 A homologous family of Tl–O cuprate superconductors with monolayer Tl–O sheets separating the Cu perovskite-like units was discovered [226] shortly after reports of the double Tl–O layer compounds Tl-22($n-1$)n discussed in the previous section. Four phases with differing numbers n of Cu–O infinite planes have been reported in the $TlBa_2Ca_{n-1}Cu_nO_{2n+3}$ [Tl-12($n-1$)n] family: $n=1$ which is not superconducting [227]; $n=2$ with $T_c = 65$–85 K [227]; $n=3$ with $T_c = 110$ K [226]; and $n=4$ with $T_c = 122$ K [228]. Structures for the Tl-12($n-1$)n family are identical to the structures of the Tl-22($n-1$)n compounds except that there is only a single Tl–O layer separating the Cu–Ba–Ca–O perovskite blocks rather than a double layer. A schematic drawing of the Tl-2212 structure is presented in figure B2.4.15. Two related compounds where Tl is partially substituted with Pb and Ba is replaced by Sr, $(Tl_{0.5}Pb_{0.5})Sr_2CaCu_2O_7$ with $T_c = 85$ K and $(Tl_{0.5}Pb_{0.5})Sr_2Ca_2Cu_3O_9$ with $T_c = 120$ K, have also been reported [229].

Crystal growth considerations are similar to those discussed previously for the double Tl–O layer cuprate superconductors. Difficulties associated with the use of Tl_2O_3 as well as the formation of undesirable intergrowth layers of differing composition [227] are issues for the single Tl–O layer compounds. There are only a few reports on the growth of Tl-1212 and Tl-1223 crystals by self-flux techniques.

Microcrystals of Tl-1212 [211] and Tl-1223 [230] were first grown from a CuO-rich flux in a sealed Pt crucible by slow-cooling from 950°C. These experiments were conducted under 1 atm of oxygen to suppress Tl loss. Small Tl-1212 crystal platelets were also grown from a stoichiometric melt in an alumina capsule [231]. Larger crystals, with dimensions up to $1 \times 1 \times 0.1\ mm^3$ and $T_c = 90$ K, were obtained by slow-cooling CuO/CaO-rich melts in covered alumina crucibles from 950°C [215]. Tl-O/CuO-rich fluxes have been used to grow crystals of Tl-1212 and Tl-1223 from oxide charges handled and loaded into sealed Pt crucibles under an Ar atmosphere in a dry box [219]. Very large Tl-1223 crystals, up to $10 \times 10 \times 1\ mm^3$, were grown by this method. The broad transitions observed for these large crystals were attributed to stoichiometry variations due to cation site substitution [219]. Rb-doped Tl-1223 and Tl-1212 crystals containing intergrowths of other Tl superconducting compounds were prepared by a flux growth method using covered Pt crucibles and an oxygen atmosphere [223]. Single crystals of Sr-rich $Tl(Ba,Sr)_2Ca_{n-1}Cu_nO_{2n+3}$, where $n = 2$ and 3, have been grown from a charge of nominal composition $TlBaSrCa_2Cu_3O_9$ melted at 1000°C in an alumina crucible sealed in an evacuated quartz tube [232].

B2.4.4.11 $HgBa_2CaCu_2O_8$ and its homologues

C4 Superconductivity in the Hg-based cuprate family having the generic formula $HgBa_2Ca_{n-1}Cu_nO_{2n+2+}$ [Hg-12($n-1$)n] was first reported in 1993 [233] for the $n = 1$ compound (Hg-1201). Shortly thereafter, a record high T_c of 133 K was reported for the $n = 3$ compound (Hg-1223) under ambient conditions [234]. Subsequently, it was found that T_c values in excess of 150 K could be induced in Hg-1223 by the application of high pressure [235]. Within the next year, compounds covering the full composition range $n = 1$ to 8 were reported [236–239].

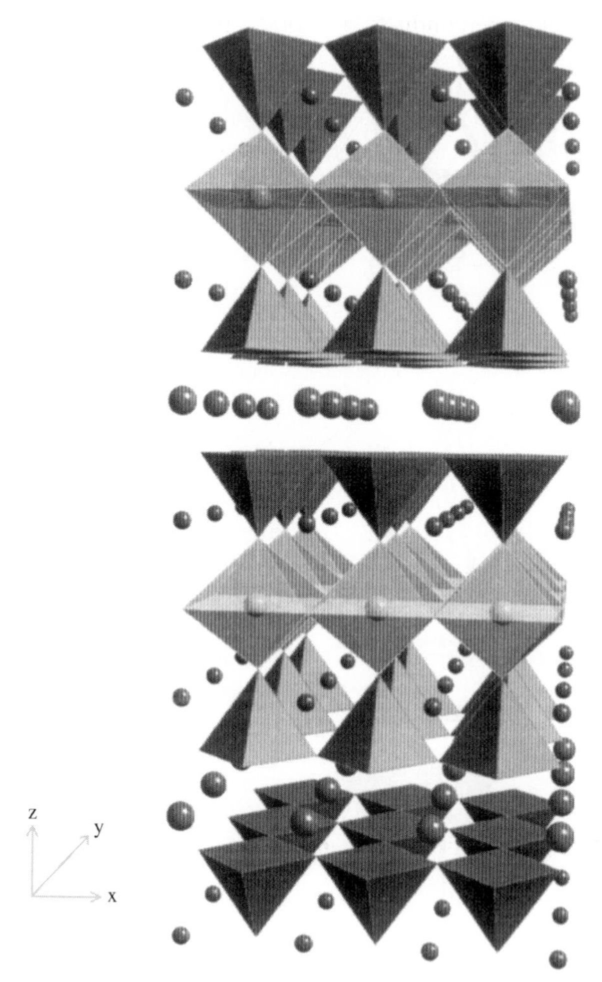

Figure B2.4.15. Schematic drawing of the $TlBa_2CaCu_2O_8$ (Tl-1212) structure, which contains single Tl–O layers.

All of the compounds in the homologous series have tetragonal structures with P4/mmm symmetry [233, 238, 239]. These structures can be represented as rock-salt-like slabs, $BaO/HgO_\delta/BaO$, alternating with perovskite-like slabs composed of n CuO_2 layers separated by $(n-1)$ Ca^{2+} cations. The structures of the first four members of the series are shown in figure B2.4.16 [239]. The Hg compounds are structurally similar to the Tl–O single-layer cuprates $TlBa_2Ca_{n-1}Cu_nO_{2n+3}$; the principle difference is a lower oxygen occupancy of the Hg layer as compared to the Tl layer. Approximate T_c values for the $n = 1 - 8$ compounds in the series at ambient pressure and optimal oxygen concentration are presented in table B2.4.1.

Single crystal growth of the Hg phases is complicated because the compounds decompose at temperatures below the peritectic melting points at ambient pressure and the resulting Hg-containing species volatilize. Crystal growth experiments are conducted under high pressure to suppress Hg vaporization, or in a sealed evacuated tube to contain the Hg vapour species and control their partial pressures. In addition, temperature must be carefully controlled during growth due to the small differences in crystallization temperatures for the various Hg-12$(n-1)n$ compounds.

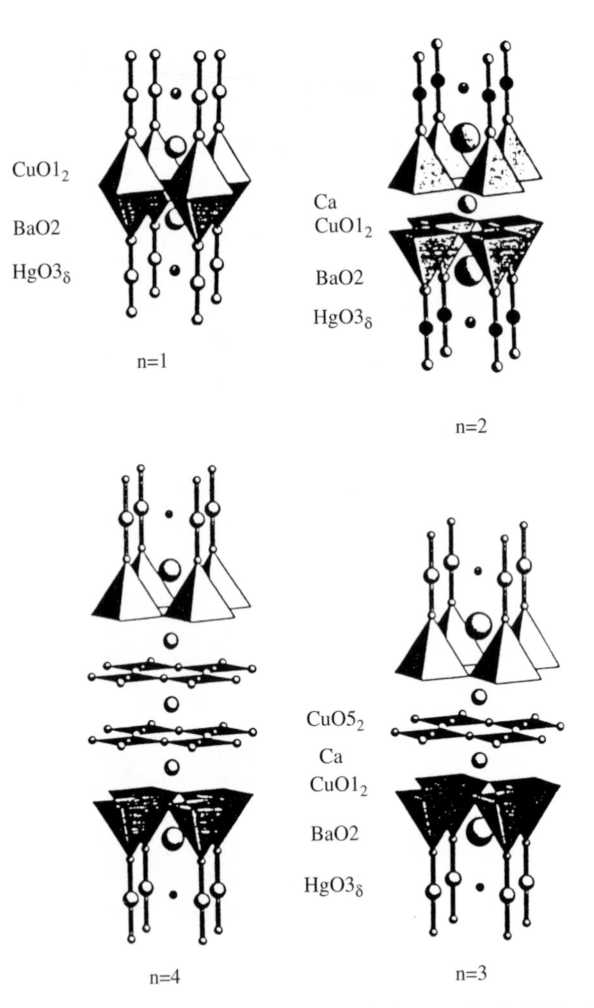

Figure B2.4.16. Schematic representation of the structures of the $n = 1$–4 $HgBa_2Ca_{n-1}Cu_nO_{2n+2+\delta}$ phases. Reproduced from an article by Marezio *et al* [239] (copyright Elsevier).

Table B2.4.1. T_c values for the $HgBa_2Ca_{n-1}Cu_nO_{2n+2+\delta}$ compounds

n	Compound	T_c (K)	Reference
1	Hg-1201	95–97	[9, 240–242]
2	Hg-1212	126	[243]
3	Hg-1223	133–135	[8, 234, 244]
4	Hg-1234	129	[245]
5	Hg-1245	112	[238]
6	Hg-1256	92	[238]
7	Hg-1267	88	[238]
8	Hg-1278	< 90	[238]

Despite these barriers, three different techniques have been developed to grow crystals of the first five members of the series. The first method, derived from the solid-state synthesis technique of Meng *et al* [237], involved heating three pellets wrapped in a Au foil in an evacuated sealed tube to 800°C [8]. Two of the oxide pellets had the composition Hg : Ba : Ca : Cu = 1 : 2 : $(n-1)$: n and the third had the composition Ba : Ca : Cu = 2 : $(n-1)$: n. Growth resulted from a reaction between the gaseous Hg-containing species and the solid pellet. This technique has been used to grow crystals of the $n = 3$ [8] and 1 [241, 242] phases.

The second technique was a flux growth method performed under Ar pressure ($P_{Ar} = 0.1$ kbar) using charges composed of Hg-12$(n-1)n$ and a Ba–Cu–O rich flux [245]. Crystals of the $n = 2$ [243] and 3–5 [243, 245] phases have been grown by this method. In addition, crystals with $n = 6$ were obtained as a byproduct of a $n = 4$ growth run [243].

Finally, crystals have been obtained by sintering or grain growth of an oxide powder charge of nominal composition Hg-1201 contained in an alumina crucible in an evacuated, sealed quartz tube [9]. Only crystals of the $n = 1$ phase have been grown by this approach.

Superconducting crystals of the general formula $Hg_{1-x}A_xBa_2Ca_{n-1}Cu_nO_{2n+2+\delta}$, where A = Pb, Bi or Re have also been reported. Pb-substituted crystals of the $n = 1$ [246], 2 [246, 247] and 3–5 [246–248] phases were grown by the high pressure, flux growth technique described above using PbO as a flux. A sintering approach with a Bi_2O_3-containing charge was used to grow Bi-substituted crystals with $n = 1$ [249], 2 [15] and 3 [15]. Similarly, Re-doped crystals of the $n = 2$ and 3 phases were grown by sintering Hg–Re–Ba–Cu–Ca–O charges in sealed, evacuated quartz ampoules [250]. Finally, a single crystal of a Re-substituted Hg-1223/Hg-1234 intergrowth phase (HgRe-2457) was grown by a high pressure technique using a Ba–Cu–Ag–O flux [251].

Oxygen content has a strong effect on the T_c values of the Hg-12$(n-1)n$ compounds, and both oxygen underdoping and overdoping decrease T_c from the optimal values [238–240]. There have been a number of annealing studies on single crystals of the Hg phases [8, 241–243].

Stacking faults are a common defect in the Hg-12$(n-1)n$ compounds with $n \geq 2$. These defects have been identified as intergrowth layers of other members of the homologous series by high resolution electron microscopy [234, 238, 239, 248, 252], electron diffraction [236] and single crystal x-ray diffraction studies [8].

D1.2

D1.1.2

B2.4.4.12 $Pb_2Sr_2M_{1-x}Ca_xCu_3O_8$ and related phases

Compounds of the formula $Pb_2Sr_2M_{1-x}Ca_xCu_3O_{8+\delta}$, or PSACO-2213, where A = $M_{1-x}Ca_x$ and M is a lanthanoid element, form a class of cuprate superconductors with T_c values as high as 80 K [253, 254]. The structure has double-layers of infinite planes of corner-shared CuO_5 pyramids interleaved by a lanthanoid/alkaline earth solid solution, as shown in figure B2.4.17. Separating the double layers are the Pb–O/Cu/Pb–O planes containing linearly coordinated Cu(I) atoms. The stoichiometric end-member of the series, $Pb_2Sr_2MCu_3O_8$ ($x = 0$), is not superconducting. Partial substitution of the lanthanoid constituent by an alkaline earth element (Ca or Sr) introduces holes into the conducting Cu-O planes giving them metallic-like superconducting character. Like $YBa_2Cu_3O_7$, the PSACO-2213 phase has a rich substitutional chemistry. For example, Ag can partially substitute for the linearly coordinated Cu(I) atoms [255], and the phase can be formed for the majority of the lanthanoids including Ce [256]. Relatively reducing, low oxygen partial pressure conditions are necessary for the preparation of these materials as might be expected from the presence of Cu(I) in the structure.

Millimetre-sized crystals of PSACO-2213 containing various lanthanoid elements were first grown by a flux method under low oxygen pressure (1% O_2) from PbO-based melts in alumina crucibles [256]. The T_c values of the resulting crystals were variable, due to differences in both x and the oxygen stoichiometry. Improved crystal growth results were obtained by adding PbF_2 [257, 258],

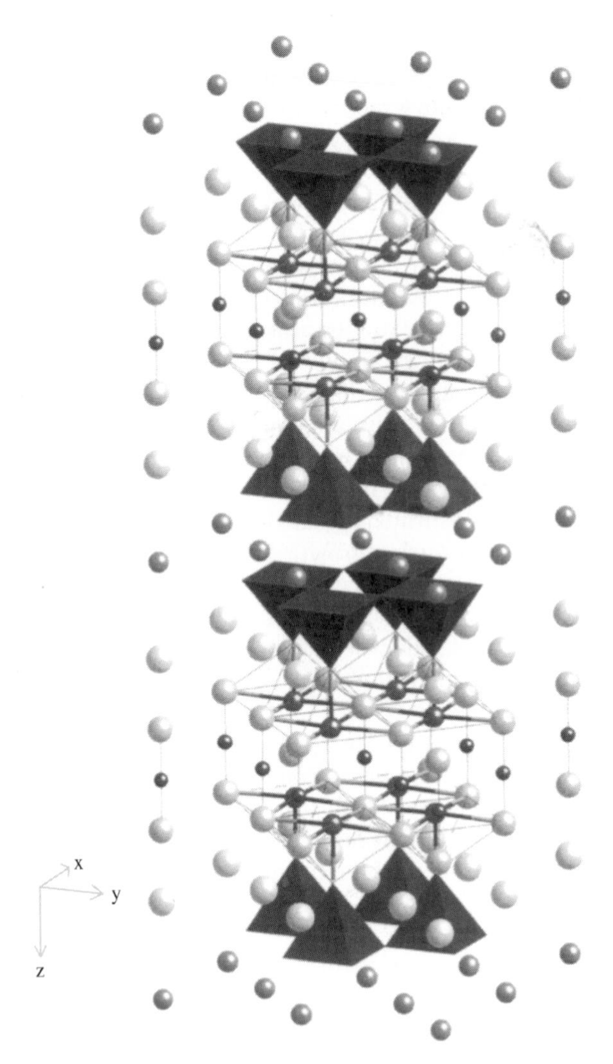

Figure B2.4.17. Polyhedral representation of the idealized structure of $Pb_2Sr_2YCu_3O_8$, PSACO-2213.

NaCl [259–261] or KCl [261] to the PbO flux and using Pt crucibles. Crystals with dimensions up to $3 \times 3 \times 2\,mm^3$ and T_c values up to 84 K have been reported for the modified PbO fluxes.

The PSACO-2213 structure can accommodate a wide range of oxygen stoichiometries, up to $\delta = 1.6$ for the $x = 0$ compound [255]. Further, the nature of the cations in PSACO-2213 strongly affects the stability of the compound under different oxygen partial pressures. Since the T_c is sensitive to both δ and x, it is difficult to achieve optimal superconducting properties in PSACO-2213 crystals.

A related phase, $Pb_{0.5}Sr_{2.5}Y_{1-x}Ca_xCu_2O_{7-\delta}$, referred to as PSACO-1212, was found to exist over the composition range $0 \leq x \leq 0.6$ [262, 263]. This tetragonal oxide has a structure consisting of double rock-salt type layers containing Pb and Sr intergrown with double oxygen-deficient perovskite layers. The compound was found to be superconducting for $0.5 \leq x \leq 0.6$, with the T_c ranging from 50 to 75 K.

Millimetre-sized crystals of PSACO-1212 with broad superconducting transitions at $T_c = 40–65$ K were grown from a self-flux containing excess PbO, CuO and SrO [264]. Other attempts to grow crystals from a partially molten stoichiometric charge [265] or a CuO flux [266] resulted in large crystals (up to

$5 \times 4 \times 3\,mm^3$) that were not superconducting, even after extended oxygen anneals. The addition of 10 wt% $AgNO_3$ to a CuO flux permitted the growth of millimetre-sized PSACO-1212 crystals with $T_c = 65\,K$ and $\Delta T_c = 15\,K$ in the as-grown state [267].

B2.4.4.13 Other superconducting cuprates

The emerging structural picture for superconducting cuprates shows that all have infinite two-dimensional Cu–O planes with various spacing layers separating the planes. The chemistry of the spacing layers permits tuning of the band filling in order to attain a superconducting condition. Examples of spacing layers include Cu–O single and double chains for the YBCO phases and rock salt-type Bi–O layers for the BSCCO phases. Many other cuprate superconductors which contain a variety of different spacing layers have been discovered; some are listed in table B2.4.2. This is not, however, a comprehensive list. These systems are chemically complicated, requiring extreme synthesis and growth conditions and post-annealing to attain superconductivity. Most of the compounds listed in table B2.4.2 have not yet been grown in single crystal form.

The first superconducting cuprate phase with a different spacing layer, Ga–$O_{4/4}$, was $Sr_2(Ln,Ca)$-$GaCu_2O_7$ (Ln = rare earth element) with a T_c of 40 K [268, 269]. Thin crystal plates ($0.3 \times 0.2 \times 0.02\,mm^3$) of

Table B2.4.2. Other cuprate superconductorss

General formula	M	n	T_c (K)	Reference
$Sr_2(Ln,Ca)_{n-1}MCu_nO_{2n+3}$	Ga	2–4	30–107	[268, 269, 286, 287]
	Al	3–5	78–110	[288, 289]
	Ge/Cu	3–6	78–90	[290, 291]
	Cu/V	3–7	up to 107	[292]
$Sr_2(Ln,Ce)_2Mcu_2O_9$	Nb or Ta		28	[271]
	Ga		12	[293]
	Ga/Cu		28	[294]
	Ru		30	[295]
	Ru/Nb		25–32	[296]
	Cu/Ce		20–50	[297]
	Pb/Cu		25	[298]
$Sr_2(Ln,Ca)(Cu,M)_3O_x$	Fe		20	[299]
	Ti		26	[300]
$(Ba_x Sr_{1-x})_2Cu_{1+y}O_{2.2+2y+\delta}(CO_3)_{1-y}$			40	[272]
$(Y,Ca)_{0.95}Sr_{2.05}Cu_{2.4}(CO_3)_{0.6}O_y$			63	[301]
$(Cu_{0.5}C_{0.5})_2Ba_3Ca_2Cu_3O_{11}$			91	[302]
$CCa_2CaCu_2O_{7+\delta}$			47	[303]
$(B,Cu)(Sr,Ba)_2YCu_2O_7$			51	[304]
$Sr_2(Ca,Sr)_{n-1}Cu_n(CO_3)_{1-x}(BO_3)_xO_y$		1–3	50–115	[305]
$(Ca,Na)_2CuO_2Cl_2$			26	[277]
$(Sr,Ca)_3Cu_2O_{4+\delta}Cl_{2-y}$			80	[278]
$Sr_2CuO_2F_{2+\delta}$			46	[279]
$(Sr,Ba)_2CuO_2F_{2+\delta}$			64	[280]
$(Sr,Nd)CuO_2$			40	[306]
$(Ca,Sr)_{1-y}CuO_2$			up to 110	[307]
$Sr_2CuO_{3.1}$			70	[308]
$Sr_3Cu_2O_{5+\delta}$			100	[308]

$HoSr_2GaCu_2O_7$ were grown from a $HoSr_4GaCu_{10}O_x$ melt in an alumina crucible by slow-cooling from 1040°C [268]. Very small (10 μm) crystals of $YSr_2GaCu_2O_7$ were obtained by raising a stoichiometric charge just above its decomposition temperature for a short time [270]. Neither of these Ga cuprate crystals was superconducting. The compound $Sr_2Nd_{1.5}Ce_{0.5}NbCu_2O_{10-\delta}$ with NbO_6 octahedral spacing layers was found to have a T_c of 28 K [271]. This material is stable only in a very narrow temperature window, between 1100 and 1125°C, in flowing oxygen.

A cuprate superconductor containing carbonate slabs, $(Ba_xSr_{1-x})_2Cu_{1+y}O_{2.2+2y+\delta}(CO_3)_{1-y}$, was first reported by Kinoskita [272]. The structure of the pure Sr form of this compound ($x = 0$ and $y = 0$) is shown in figure B2.4.18 [273]. Other carbonate and carbonate/borate superconducting phases are listed in table B2.4.2. Millimetre-sized crystals of $(Ba_xSr_{1-x})_2Cu_{1.1}O_{2.2+\delta}(CO_3)_{0.9}$ with $T_c = 32$ K have been grown from stoichiometric pellets by sintered grain growth under an oxygen partial pressure of 50 atm [274]. Although single crystal growth of other superconducting carbonate phases has not been reported, single crystals of non-superconducting $A_4CuM(CO_3)_2O_4$ (A = Sr, Ba; M = Li, Na, Ca) [275] and $Ba_2M_xCu_{2-x-y}(CO_3)_yO_{2+\delta}$ (M = Cu, Cd, Ca) [276] have been grown by a sintering approach under atmospheric conditions. The synthetic challenge in the carbonate phases is that they are unstable above their formation temperature, releasing carbon as CO_2 gas.

Superconducting phases have also been identified in the copper oxychlorides $(Ca,Na)_2CuO_2Cl_2$ [277] and $(Sr,Ca)_3Cu_2O_{4+\delta}Cl_{2-y}$ [278] and the copper oxyfluorides $A_2CuO_2F_{2+\delta}$ (A = Sr, Sr/Ba, Sr/Ca) [279, 280]. Crystals of non-superconducting $Sr_2CuO_2Cl_2$ with volumes approaching 0.5 cm^3 have been grown by slow-cooling stoichiometric melts in alumina boats from 1135°C in air [281], and by a floating zone technique [282]. Many of these materials, particularly the bromides and chlorides, are water sensitive.

One of the more interesting cuprate systems to display high temperature superconductivity is the infinite layer family. These materials, of the general formula $MCuO_2$, have infinite two-dimensional CuO_2 sheets separated by M ions. Crystals of the first material of this type, $Ca_{0.86}Sr_{0.14}CuO_2$, grown from stoichiometric melts at 1150°C, were insulating [283]. A number of related superconducting compounds with T_c up to 110 K have been synthesized under high pressure (see table B2.4.2 and [284]). Millimetre-sized single crystals of $CaCuO_2$ were grown from a $BaCuO_2$/CuO flux in yttria or alumina crucibles at an Ar gas pressure of 1000 MPa [248]. The crystals had broad transitions at superconducting onset temperatures of 70–100 K.

Other phases that are structurally related to the high T_c cuprates but which have not been successfully doped into a metallic regime suitable for superconductivity have been reported. For example, $LaSrCuAlO_5$ crystallizes in an oxygen-deficient perovskite structure [285], and platelet crystals of this material have been grown from a CuO rich melt. We will not cover these non-superconducting materials in this paper.

B2.4.4.14 Non-copper superconductors

$BaPb_{1-x}Bi_xO_3$

The discovery of superconductivity at critical temperatures up to 13 K in the perovskite oxide $BaPb_{1-x}Bi_xO_3$ in 1975 [309] was the first important milestone in the history of oxide superconductors. This phase is a three-dimensional material, exhibits superconductivity in a specific composition range ($0.05 \le x \le 0.3$), and has a low density of states.

Millimetre-sized $BaPb_{1-x}Bi_xO_3$ crystals with T_c near 11 K were first grown by high temperature (1050°C) flux methods from KCl [12] and Ba–Pb–Bi–O [310] melts in Pt crucibles. Hydrothermal techniques were used to grow millimetre-sized $BaPb_{1-x}Bi_xO_3$ crystals with $x \le 0.3$ from KCl [40] and NaOH [41] solutions at much lower temperatures, 400–450°C; crystals with $x \approx 0.25$ had good superconducting properties, with T_c near 12 K and ΔT_c less than 1.8 K. Flux growth techniques were

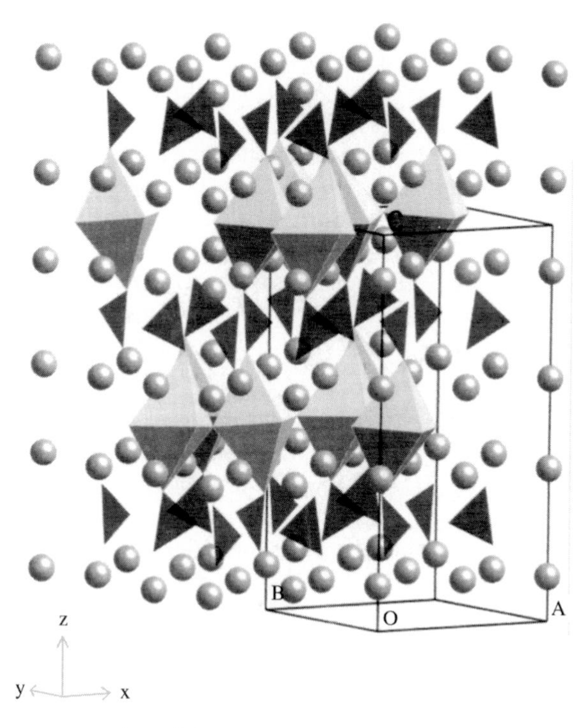

Figure B2.4.18. Schematic representation of the carbonate-containing cuprate superconductor $Sr_2CuO_2(CO_3)$.

employed to grow Pb-rich $BaPb_{1-x}Bi_xO_3$ ($x < 0.3$) crystals from PbO/Bi_2O_3-rich melts and Bi-rich crystals ($x > 0.6$) from $Bi_2O_3/PbO/Ba(NO_3)_2$ melts in covered Pt crucibles [311].

$Ba_{1-x}K_xBiO_3$

The discovery of superconductivity above 20 K in $Ba_{1-x}K_xBiO_3$ (BKBO) [312, 313] was viewed as remarkable; unlike the other high T_c phases, BKBO contains no copper and has an ideal cubic perovskite structure rather than a layered structure. BKBO is superconducting for $0.35 < x < 0.5$; the T_c depends upon x [314] and has a maximum value slightly above 30 K for $x = 0.4$.

Submillimetre-sized crystals with $x \approx 0.4$ and $T_c \approx 30$ K were first grown from KOH fluxes by air oxidation [315], an approach similar to that used for the growth of lower K content, non-superconducting BKBO crystals [316]. The most widely used technique for BKBO crystal growth is electrocrystallization, first reported by Norton [35]. Clusters of crystals with each crystal up to 7 mm on an edge were deposited on inert anodes (Pt, Ag or Au) from molten fluxes composed of KOH, Bi_2O_3 and $Ba(OH)_2{\cdot}8H_2O$ at temperatures of 150–350°C [317–320]. Seeding the working electrode with BKBO or $BaBiO_3$ and replacing $Ba(OH)_2{\cdot}8H_2O$ with BaO permitted the growth of single crystals with volumes up to 2 cm^3 [321, 322]. Crucibles are teflon, graphite or Pt; cathodes are graphite, Pt or Bi; and reference electrodes are Bi and BKBO.

$LiTi_2O_4$

The superconducting spinel $LiTi_2O_4$, first identified in 1978, has a relatively high T_c of 13 K [323]. Comparisons between the properties of this compound and the other high T_c oxides may provide interesting insights into mechanisms of superconductivity or new classes of high temperature superconductors. Like the cuprates, $LiTi_2O_4$ has a relatively low density of states combined with a relatively high T_c. In contrast to the cuprates, $LiTi_2O_4$ is a three-dimensional superconductor.

Millimetre-sized octahedral crystals have been grown on a Ti wire cathode by electrochemical reduction of a molten $NaBO_2$–$LiBO_2$–NaF mixture containing TiO_2 [36]. The reaction of Li metal and TiO_2 at 1100°C in a sealed iron vessel has also been used to obtain $LiTi_2O_4$ crystals [324] with a T_c of 12 K. Smaller cubic $LiTi_2O_4$ crystals (up to 0.1 mm on an edge) and large crystal clusters were grown on a Ti electrode from a $LiBO_2$ flux containing TiO_2 by electrocrystallization [325]. In a direct reduction process, octahedral $LiTi_2O_4$ crystals were deposited onto a Zr plate immersed in a $LiBO_2/TiO_2$ melt covered with a NaCl protective layer [325].

B2.4.5 Summary

Single crystals have played a key role in fundamental studies of high T_c superconductors, particularly chemical and structural investigations. The availability of large, high quality crystals together with a better understanding of the type and distribution of defects in the crystals has enabled critical experiments on the underlying physics of these remarkable materials. Single crystal growth methods developed for high T_c cuprates have been applied to other complex oxides of increasing technological importance, such as colossal magnetoresistive and ferroelectric materials. Commercial applications of high temperature superconductors are emerging in a vast array of technology areas including cellular communications and medical diagnostics. Crystal growers continue to have an important role in advancing the knowledge underlying both the development and application of these extraordinary materials.

References

[1] Hurle D T J, ed 1993 *Handbook of Crystal Growth: Fundamentals* **Vol. 1** (London: North Holland)

[2] Laudise R A 1970 *The Growth of Single Crystals* (Englewood Cliffs, NJ: Prentice-Hall)

[3] Elwell D and Scheel H J 1975 *Crystal Growth from High Temperature Solution* (London: Academic Press)

[4] Hurle D T J, ed 1994 *Handbook of Crystal Growth: Bulk Crystal Growth, Basic Techniques* **vol. 2a** (London: North Holland)

[5] Oda Y, Kohara T, Nakada I, Fujita H, Kaneko T, Toyoda H, Sakagami E and Asayama K 1987 The Meissner effect of the small single crystals of $Ba_2YCu_{2.89}O_{6.80}$ *Japan. J. Appl. Phys.* **26** L809

[6] Dinger T R, Worthington T K, Gallagher W J and Sandstrom R L 1987 Direct observation of electronic anisotropy in single-crystal $Y_1Ba_2Cu_3O_{7-x}$ *Phys. Rev. Lett.* **58** 2687

[7] Liu J Z, Crabtree G W, Umezawa A and Zongquan L i 1987 Superconductivity and structure of single crystal $YBa_2Cu_3O_x$ *Phys. Lett.* A **121** 305

[8] Colson D, Bertinotti A, Hammann J, Marucco J F and Pinatel A 1994 Synthesis and characterization of superconducting single crystals of $HgBa_2Ca_2Cu_3O_{8+\delta}$ *Physica* C **233** 231

[9] Pelloquin D, Hardy V, Maignan A and Raveau B 1997 Single crystals of the 96 K superconductor (Hg, Cu) $Ba_2CuO_{4+\delta}$: growth, structure and magnetism *Physica* C **273** 205

[10] Schneemeyer L F, van Dover R B, Glarum S H, Sunshine S A, Fleming R M, Batlogg B, Siegrist T, Marshall J H, Waszczak J V and Rupp L W 1988 Growth of superconducting single crystals in the Bi–Sr–Ca–Cu–O system from alkali chloride fluxes *Nature* **332** 422

[11] Katsui A 1988 Crystal growth of superconducting Bi–Sr–Ca–Cu–O compounds from KCl solution *Japan. J. Appl. Phys.* **27** L844

[12] Katsui A and Suzuki M 1982 Single crystal growth of Ba(Pb, Bi)O_3 from molten KCl solvent *Japan. J. Appl. Phys.* **21** L157

[13] Schneemeyer L F, Waszczak J V, Siegrist T, van Dover R B, Rupp L W, Batlogg B, Cava R J and Murphy D W 1987 Superconductivity in $YBa_2Cu_3O_7$ single crystals *Nature* **328** 601

[14] Kaiser D L, Holtzberg F, Scott B A and McGuire T R 1987 Growth of $YBa_2Cu_3O_x$ single crystals *Appl. Phys. Lett.* **51** 1040

[15] Pelloquin D, Hardy V and Maignan A 1996 Synthesis and characterization of single crystals of the superconductors $Hg_{0.8}Bi_{0.2}Ba_2Ca_{n-1}Cu_nO_{2n+2+\delta}$ ($n = 2, 3$) *Phys. Rev.* B **54** 16246

[16] Inoue T, Hayashi S, Komatsu H and Shimizu M 1987 Growth of $(La_{1-x}Sr_x)_2CuO_{4-\delta}$ crystals from high temperature solution *Japan. J. Appl. Phys.* **26** L732

[17] Oka K and Unoki H 1987 Phase diagram of the La_2O_3–CuO system and crystal growth of $(LaBa)_2CuO_4$ *Japan. J. Appl. Phys.* **26** L1590

[18] Birgeneau R J, Chen C Y, Gabbe D R, Jenssen H P, Kastner M A, Peters C J, Picone P J, Thio T, Thurston T R and Tuller H L 1987 Soft-phonon behavior and transport in single-crystal La_2CuO_4 *Phys. Rev. Lett.* **59** 1329
[19] Picone P J, Jenssen H P and Gabbe D R 1987 Top seeded solution growth of La_2CuO_4 *J. Cryst. Growth* **85** 576
[20] Oka K, Saito M, Ito M, Nakane K, Murata K, Nishihara Y and Unoki H 1989 Phase diagram and crystal growth of $NdBa_2Cu_3O_{7-y}$ *Japan. J. Appl. Phys.* **28** L219
[21] Rao S M, Loo B H, Wang N P and Kelley R J 1991 A new technique for the growth of superconducting $YBa_2Cu_3O_{6+\delta}$ crystals completely separated from flux *J. Cryst. Growth* **110** 989
[22] Shigematsu K, Takei H, Higashi I, Hoshino K, Takahara H and Aono M 1990 Growth of single crystals of Bi–Sr–Ca–Cu–O *J. Cryst. Growth* **100** 661
[23] Wang Y, Bennema P, Schreurs L W M, van Bentum P J M, van Kempen H, van de Leemput L E C, Wnuk J and van der Linden P 1990 Seeded growth, morphology and surface topology of superconducting Bi–Sr–Ca–Cu–O single crystals *J. Cryst. Growth* **99** 933
[24] Poplawsky R P and Thomas J E Jr 1960 Floating zone crystals using an arc image furnace *Rev. Sci. Instrum.* **31** 1303
[25] Burrus C A and Stone J 1975 Single-crystal fiber optical devices: a Nd:YAG fiber laser *Appl. Phys. Lett.* **26** 318
[26] Fejer M M, Nightingale J L, Magel G A and Byer R L 1984 Laser-heated miniature pedestal growth apparatus for single-crystal optical fibers *Rev. Sci. Instrum.* **55** 1791
[27] Feigelson R S 1986 Pulling optical fibers *J. Cryst. Growth* **79** 669
[28] Revcolevschi A and Jegoudez J 1997 Growth of large high-T_c single crystals by the floating zone method: a review *Progress in Materials Science* **Vol 42**, (London: Pergamon) pp 321–39
[29] Feigelson R S, Gazit D, Fork D K and Geballe T H 1988 Superconducting Bi–Ca–Sr–Cu–O fibers grown by the laser-heated pedestal growth method *Science* **240** 1642
[30] Brody H D, Haggerty J S, Cima M J, Flemings M C, Barns R L, Gyorgy E M, Johnson D W, Rhodes W W, Sunder W A and Laudise R A 1989 Highly textured and single crystal $Bi_2CaSr_2Cu_2O_x$ prepared by laser heated float zone crystallization *J. Cryst. Growth* **96** 225
[31] Tanaka I, Yamane K and Kojima H 1989 Single crystal growth of superconducting $La_{2-x}Sr_xCuO_4$ by the TSFZ method *J. Cryst. Growth* **96** 711
[32] Oka K and Ito T 1994 Crystal growth of $REBa_2Cu_3O_{7-y}$ (RE = Y, La, Pr, Nd and Sm) by the travelling-solvent floating-zone method *Physica* C **227** 77
[33] Kuroda K, Choi I H, Egi T, Unoki H and Koshizuka N 1997 $NdBa_2Cu_3O_{7-\delta}$ Single crystal growth by the traveling-solvent floating-zone method *Proc. Conf. Sixteenth International Cryogenic Engineering/International Cryogenic Materials* **vol. 3**, (Oxford: Elsevier) pp 1467–70
[34] Shanks H R 1972 Growth of tungsten bronze crystals by fused salt electrolysis *J. Cryst. Growth* **13/14** 433
[35] Norton M L 1989 Electrodeposition of $Ba_{0.6}K_{0.4}BiO_3$ *Mater. Res. Bull.* **24** 1391
[36] Durmeyer O, Kappler J P, Derory A, Drillon M and Capponi J J 1990 Magnetic superconducting properties of $LiTi_2O_4$ single crystal *Solid State Commun.* **74** 621
[37] Tang H Y, Lee C S and Wu M K 1994 Crystallization of $La_{2-x}Na_xCuO_4$ superconductor by low-temperature electrochemical deposition *Physica* C **231** 325
[38] Tang H Y, Hshu H Y, Lee C S, Yang J L and Wu M K 1997 Electrochemical synthesis of superconducting $EuBa_2Cu_3O_{7-x}$ from low-temperature molten salt *J. Electrochem. Soc.* **144** 16
[39] Byrappa K 1994 Hydrothermal growth of crystals *Handbook of Crystal Growth* **vol. 2a**, ed D T J Hurle (London: North-Holland) pp 465–562
[40] Hirano S and Takahashi S 1986 Hydrothermal crystal growth of $BaPb_{1-x}Bi_xO_3$ ($0 < x < 0.30$) *J. Cryst. Growth* **78** 408
[41] Hirano S and Takahashi S 1987 NaOH solution hydrothermal growth and superconducting properties of $BaPb_{1-x}Bi_xO_3$ single crystals *J. Cryst. Growth* **85** 602
[42] Scheel H J and Licci F 1991 Phase diagrams and crystal growth of oxide superconductors *Thermochimica Acta* **174** 115
[43] Schneemeyer L F 1993 Growth of single crystals of various high-T_c superconductors *Processing and Properties of High-T_c Superconductors*, ed S Jin (London: World Scientific) pp 45–86
[44] Assmus W and Schmidbauer W 1993 Crystal growth of HTSC materials *Supercond. Sci. Technol.* **6** 555
[45] Karpinski J, *et al* 1999 High-pressure synthesis, crystal growth, phase diagrams, structural and magnetic properties of $Y_2Ba_4Cu_nO_{2n+x}$, $HgBa_2Ca_{n-1}Cu_nO_{2n+2+\delta}$ and quasi-one-dimensional cuprates *Supercond. Sci. Technol.* **12** R153
[46] Bednorz J G and Muller K A 1986 Possible high T_c superconductivity in the Ba–La–Cu–O system *Z. Phys. B – Condens. Matter.* **64** 189
[47] Tarascon J M, Greene L H, McKinnon W R, Hull G W and Geballe T H 1987 Superconductivity at 40 K in the oxygen-defect perovskites $La_{2-x}Sr_xCuO_{4-y}$ *Science* **235** 1373
[48] Cava R J, van Dover R B, Batlogg B and Rietman E A 1987 Bulk superconductivity at 36 K in $La_{1.8}Sr_{0.2}CuO_4$ *Phys. Rev. Lett.* **58** 408
[49] Hidaka Y, Enomoto Y, Suzuki M, Oda M and Murakami T 1987 Single crystal growth of $(La_{1-x}A_x)_2CuO_4$ (A = Ba or Sr) and $Ba_2YCu_3O_{7-y}$ *J. Cryst. Growth* **85** 581
[50] Shamoto S, Hosoya S and Sato M 1988 Single crystal growth of high-T_c superconductors *Solid State Commun.* **66** 195

[51] Chen C, Watts B E, Wanklyn B M, Thomas P A and Haycock P W 1988 Phase diagram and single crystal growth of $(La, Sr)_2CuO_4$ from CuO solution *J. Cryst. Growth* **91** 659
[52] Veselago V G, Gamajunov K V, Zorya V I, Ivanov A L, Osiko V V, Tatarintsev V M, Fradkov V A, Chernikov M A and Chernov A I 1990 Strontium content of $La_{2-x}Sr_xCuO_{4-\delta}$ single crystals grown from CuO flux *Supercond. Sci. Technol.* **3** 121
[53] Ito T, Takagi H, Ishibashi S, Ido T and Uchida S 1991 Normal-state conductivity between CuO_2 planes in copper oxide superconductors *Nature* **350** 596
[54] Rytz D, Wechsler B A, Nelson C C and Kirby K W 1990 Top-seeded solution growth of $BaTiO_3$, $KNbO_3$, $SrTiO_3$, $Bi1_2TiO_{20}$ and $La_{2-x}Ba_xCuO_4$ *J. Cryst. Growth* **99** 864
[55] Cassanho A, Keimer B and Greven M 1993 Growth of large pure, doped and co-doped La_2CuO_4 single crystals *J. Cryst. Growth* **128** 813
[56] Maljuk A N, Zhokhov A A, Emel'chenko G A, Zver'kova I I, Turanov A N and Shekhatman V Sh 1993 Cu-deficiency in $La_{2-x}Sr_xCu_{1-y}O_{4-\delta}$ single crystals and how it affects superconducting properties *Physica* C **214** 93
[57] Yu J, Yanagida Y, Takashima H, Inaguma Y, Itoh M and Nakamura T 1993 Single crystal growth of superconducting $La_{2-x}Ba_xCuO_4$ by TSFZ method *Physica* C **209** 442
[58] Oka K, Menken M J V, Tarnawski Z, Menovsky A A, Moe A M, Han T S, Unoki H, Ito T and Ohashi Y 1994 Crystal growth of of $La_{2-x}Sr_xCuO_{4-\delta}$ by the travelling-solvent floating-zone method *J. Cryst. Growth* **137** 479
[59] Ito T and Oka K 1994 Growth and transport properties of single-crystalline $La_{2-x}Ba_xCuO_4$ *Physica* C **235–40** 549
[60] Duijn V H M, Hien N T, Menovsky A A and Franse J J M 1994 Growth and characterization of $La_{2-x}Sr_xCuO_{4-\delta}$ bulk single crystals *Physica* C **235–40** 559
[61] Hosoya S, Lee C H, Wakimoto S, Yamada K and Endoh Y 1994 Single crystal growth of $La_{2-x}Sr_xCuO_4$ with improved lamp-image floating-zone furnace *Physica* C **235–40** 547
[62] Zhang K, Mogilevsky R, Hinks D G, Mitchell J, Schultz A J, Wang Y and Dravid V 1996 Crystal growth of $(La,Sr)_2CuO_4$ by float zone melting *J. Cryst. Growth* **169** 73
[63] Kojima H, Yamamoto J, Mori Y, Khan M K R, Tanabe H and Tanaka I 1997 Single crystal growth of superconducting $La_{2-x}M_xCuO_4$ (M = Ca, Sr, Ba) by the TSFZ method *Physica* C **293** 14
[64] Oka K, Unoki H, Hayashi K, Nishihara Y and Takeda Y 1990 Single-crystal growth of $(LaSmSr)_2CuO_4$ *Japan. J. Appl. Phys.* **29** L1807
[65] Tokura Y, Takagi H and Uchida S 1989 A superconducting copper oxide compound with electrons as the charge carriers *Nature* **337** 345
[66] Markert J T, Early E A, Bjornholm T, Ghamaty S, Lee B W, Neumeier J J, Price R D, Seaman C L and Maple M B 1989 Two new electron cuprate superconductors $Pr_{1.85}Th_{0.15}CuO_{4-y}$ and $Eu_{1.85}Ce_{0.15}CuO_{4-y}$, and properties of $Nd_{2-x}Ce_xO_{4-y}$ *Physica* C **158** 178
[67] Markert J T and Maple M B 1989 High temperature superconductivity in Th-doped Nd_2CuO_{4-y} *Solid State Commun.* **70** 145
[68] Takagi H, Uchida S and Tokura Y 1989 Superconductivity produced by electron doping in CuO_2-layered compounds *Phys. Rev. Lett.* **62** 1197
[69] Tarascon J M, *et al* 1989 Growth, structural, and physical properties of superconducting $Nd_{2-x}Ce_xCuO_4$ crystals *Phys. Rev.* B **40** 4494
[70] Hidaka Y and Suzuki M 1989 Growth and anisotropic superconducting properties of $Nd_{2-x}Ce_xCuO_{4-y}$ single crystals *Nature* **338** 635
[71] Oka K and Unoki H 1989 Phase diagram and crystal growth of superconductive $(NdCe)_2CuO_4$ *Japan. J. Appl. Phys.* **28** L937
[72] Pinol S, Fontcuberta J, Miravitlles C and Paul D McK 1990 Crystal growth and phase diagrams for the Nd_2O_3-CeO_2-CuO system *Physica* C **165** 265
[73] Peng J L, Li Z Y and Greene R L 1991 Growth and characterization of high quality single crystals of $R_{2-x}Ce_xCuO_{4-y}$ (R = Nd, Sm) *Physica* C **177** 79
[74] Dalichaouch Y, de Andrade M C and Maple M B 1993 Synthesis, transport, and magnetic properties of $Ln_{2-x}Ce_xCuO_{4-y}$ single crystals (Ln = Nd, Pr, Sm) *Physica* C **218** 309
[75] Matsuda M, Endoh Y and Hidaka Y 1991 Crystal growth and characterization of $Pr_{2-x}Ce_xCuO_4$ *Physica* C **179** 347
[76] Brinkmann M, Rex T, Bach H and Westerholt K 1996 Crystal growth of high-T_c superconductors $Pr_{2-x}Ce_xCuO_{4+\delta}$ with substitutions of Ni and Co for Cu *J. Cryst. Growth* **163** 369
[77] Kaneko N, Hidaka Y, Hosoya S, Yamada K, Endoh Y, Takekawa S and Kitamura K 1999 Optimum crucible material for growth of $Pr_{2-x}Ce_xCuO_4$ crystal *J. Cryst. Growth* **197** 818
[78] Markl J, Strobel J P, Klauda M and Saemann-Ischenko G 1991 Preparation of $Ln_{2-x}Ce_xCu_1O_{4-\delta}$ single crystals (Ln = Nd, Sm) by a modified flux flow method *J. Cryst. Growth* **113** 395
[79] Cassanho A, Gabbe D R and Jenssen H P 1989 Growth of single crystals of pure and Ce-doped Nd_2CuO_4 *J. Cryst. Growth* **96** 999
[80] Zhigunov D I, Shiryaev S V, Kurnevich L A, Kalanda N A, Kurochkin L A, Barilo S N, Vashuk V V and Smakhtin L A 1999 Growth and chemical analysis of bulk $Nd_{2-x}Ce_xCuO_4$ single crystals *J. Cryst. Growth* **199** 605
[81] Tanaka I, Watanabe T, Komai N and Kojima H 1991 Growth and superconductivity of $Nd_{2-x}Ce_xCuO_4$ single crystals *Physica* C **185–9** 437
[82] Gamayunov K, Tanaka I and Kojima H 1994 Single-crystal growth of $Nd_{2-x}Ce_xCuO_4$ at low oxygen pressure in ambient atmosphere *Physica* C **228** 58

[83] Balbashev A M, Shulyatev D A, Panova G K h, Khlopkin M N, Chernoplekov N A, Shikov A A and Suetin A V 1996 The floating zone growth and superconductive properties of $La_{1.85}Sr_{0.15}CuO_4$ and $Nd_{1.85}Ce_{0.15}CuO_4$ single crystals *Physica* C **256** 371
[84] Wu M K, Ashburn J R, Torng C J, Hor P H, Meng R L, Gao L, Huang Z J, Wang Y Q and Chu C W 1987 Superconductivity at 93 K in a new mixed-phase Y–Ba–Cu–O compound system at ambient pressure *Phys. Rev. Lett.* **58** 908
[85] Cava R J, Batlogg B, van Dover R B, Murphy D W, Sunshine S, Siegrist T, Remeika J P, Rietman E A, Zahurak S and Espinosa G P 1987 Bulk superconductivity at 91 K in single-phase oxygen-deficient perovskite $Ba_2YCu_3O_{9-\delta}$ *Phys. Rev. Lett.* **58** 1676
[86] Hor P H, Meng R L, Wang Y Q, Gao L, Huang Z J, Bechtold J, Forster K and Chu C W 1987 Superconductivity above 90 K in the square-planar compound system $ABa_2Cu_3O_{6+x}$ with A = Y, La, Nd, Sm, Eu, Gd, Ho, Er, and Lu *Phys. Rev. Lett.* **58** 1891
[87] Schneemeyer L F, Waszczak J V, Zahorak S M, van Dover R B and Siegrist T 1987 Superconductivity in rare earth cuprate perovskites *Mater Res. Bull.* **22** 1467
[88] Hulliger F and Ott H R 1987 Superconducting and magnetic properties of $Ba_2LnCu_3O_{7-x}$ compounds (Ln = Lanthanides) *Z. Phys.* B – *Condens. Matter.* **67** 291
[89] Roth R S, Davis K L and Dennis J R 1987 Phase equilibria and crystal chemistry in the system Ba–Y–Cu–O *Adv. Ceram. Mater.* **2** 303
[90] Das B N, Toth L E, Singh A K, Bender B, Osofsky M, Pande C S, Koon N C and Wolf S 1987 Growth of single crystals of $YBa_2Cu_3O_7$ *J. Cryst. Growth* **85** 588
[91] Scheel H J and Licci F 1987 Crystal growth of $YBa_2Cu_3O_{7-x}$ *J. Cryst. Growth* **85** 607
[92] Katayama-Yoshida H, Okabe Y, Takahashi T, Sasaki T, Hirooka T, Suzuki T, Ciszek T and Deb S K 1987 Growth of $YBa_2Cu_3O_{7-\delta}$ single crystals *Japan. J. Appl. Phys.* **26** L2007
[93] Wolf T h, Goldacker W and Obst B 1989 Growth of thick $YBa_2Cu_3O_{7-x}$ single crystals from Al_2O_3 crucibles *J. Cryst. Growth* **96** 1010
[94] Sadowski W and Scheel H J 1989 Reproducible growth of large free crystals of $YBa_2Cu_3O_{7-x}$ *J. Less-Common Met.* **150** 219
[95] Balestrino G, Barbanera S and Paroli P 1987 Growth of single crystals of the high-temperature superconductor $YBa_2Cu_3O_{7-x}$ *J. Cryst. Growth* **85** 585
[96] Taylor K N R, Cook P S, Puzzer T, Matthews D N, Russell G J and Goodman P 1988 Surface morphology of flux-grown single crystals of $YBa_2Cu_3O_{7-\delta}$ *J. Cryst. Growth* **88** 541
[97] Vanderah T A, Lowe-Ma C K, Bliss D E, Decker M W, Osofsky M S, Skelton E F and Miller M M 1992 Growth of near-free-standing $YBa_2Cu_3O_7$-type crystals using a self-decanting flux method *J. Cryst. Growth* **118** 385
[98] Liang R X, Dosanjh P, Bonn D A, Baar D J, Carolan J F and Hardy W N 1992 Growth and properties of superconducting YBCO single crystals *Physica* C **195** 51
[99] Erb A, Walker E and Flukiger R 1995 $BaZrO_3$: the solution for the crucible corrosion problem during the single crystal growth of high-T_c superconductors $REBa_2Cu_3O_{7-\delta}$; RE = Y, Pr *Physica* C **245** 245
[100] Kaiser D L, Holtzberg F, Chisholm M F and Worthington T K 1987 Growth and microstructure of superconducting $YBa_2Cu_3O_x$ single crystals *J. Cryst. Growth* **85** 593
[101] Watanabe K 1991 An approach to the growth of $YBa_2Cu_3O_{7-x}$ single crystals by the flux method II *J. Cryst. Growth* **114** 269
[102] Bosi S, Puzzer T, Russell G J, Town S L and Taylor K N R 1989 Large single crystals of $YBa_2Cu_3O_{7-\delta}$ superconductors from chloride fluxes *J. Mater. Sci. Lett.* **8** 497
[103] Gencer F and Abell J S 1991 The growth of $YBa_2Cu_3O_{7-\delta}$ single crystals with the aid of a NaCl–KCl flux *J. Cryst. Growth* **112** 337
[104] Nakamura N and Shimotomai M 1991 Growth of $YBa_2Cu_3O_x$ single crystals by a self-flux method with alkali chlorides as additives *Physica* C **185–9** 439
[105] Marquez L N, Keller S W and Stacy A M 1993 Synthesis of twin-free, orthorhombic $EuBa_2Cu_3O_{7-\delta}$ superconductors at 450°C by direct precipitation from molten NaOH and KOH *Chem. Mater.* **5** 761
[106] Sunshine S A, Siegrist T and Schneemeyer L F 1997 Single crystal growth of cuprates from hydroxide fluxes *J. Mater. Res.* **12** 1210
[107] Ono A, Nozaki H and Ishizawa Y 1988 Preparation and properties of $Ba_2YCu_3O_{7-y}$ single crystals grown using an indium oxide flux *Japan. J. Appl. Phys.* **27** L340
[108] Murugaraj P, Maier J and Rabenau A 1989 Preparation of large crystals of Y–Ba–Cu–O superconductor *Solid State Commun.* **71** 167
[109] Yamada Y and Shiohara Y 1993 Continuous crystal growth of $YBa_2Cu_3O_{7-x}$ by the modified top-seeded crystal pulling method *Physica* C **217** 182
[110] Nakamura M, Kutami H and Shiohara Y 1996 Fabrication of $NdBa_2Cu_3O_{7-\delta}$ single crystals by the top-seeded solution-growth method in 1%, 21%, and 100% oxygen partial pressure atmosphere *Physica* C **260** 297
[111] Rao S M, Chang R H, Law K S, Wu M K and Khattak C P 1996 Influence of KCl on the growth of $YBa_2Cu_3O_{6+\delta}$ from high temperature solutions *J. Cryst. Growth* **162** 48
[112] Shibata S, Unoki H, Kuroda K and Koshizuka N 1997 Crystal growth of $NdBa_2Cu_3O_{7-y}$ and $Ba_2Cu_3O_{5+\delta}$ single crystals by a horizontal Bridgman like method *J. Mater. Sci. Lett.* **16** 1295
[113] Gallagher P K 1987 Characterization of $Ba_2YCu_3O_x$ as a function of oxygen partial pressure, Part I: thermoanalytical measurements *Adv. Ceram. Mater.* **2** 632

[114] Kaiser D L, Gayle F W, Roth R S and Swartzendruber L J 1989 Thermomechanical detwinning of superconducting $YBa_2Cu_3O_{7-x}$ single crystals *J. Mater. Res.* **4** 745

[115] Schmid H, Burkhardt E, Sun B N and Rivera J P 1989 Uniaxial stress induced ferroelastic detwinning of $YBa_2Cu_3O_{7-\delta}$ *Physica* C **157** 555

[116] Hatanaka T and Sawada A 1989 Ferroelastic domain switching in $YBa_2Cu_3O_x$ single crystals by external stress *Japan. J. Appl. Phys.* **28** L794

[117] Giapintzakis J, Ginsberg D M and Han P D 1989 A method for obtaining single domain superconducting $YBa_2Cu_3O_{7-x}$ single crystals *J. Low Temp. Phys.* **77** 155

[118] Rice J P and Ginsberg D M 1991 A method for producing untwinned $YBa_2Cu_3O_{7-\delta}$ crystals without subjecting them to stress *J. Cryst. Growth* **109** 432

[119] Bordet P, Chaillout C, Chenavas J, Hodeau J L, Marezio M, Karpinski J and Kaldis E 1988 Structure determination of the new high-temperature superconductor $Y_2Ba_4Cu_7O_{14+x}$ *Nature* **334** 596

[120] Karpinski J, Kaldis E, Jilek E, Rusiecki S and Bucher B 1988 Bulk synthesis of the 81 K superconductor $YBa_2Cu_4O_8$ at high oxygen pressure *Nature* **336** 660

[121] Karpinski J, Rusiecki S, Kaldis E, Bucher B and Jilek E 1989 Phase diagrams of $YBa_2Cu_4O_8$ and $YBa_2Cu_{3.5}O_{7.5}$ in the pressure range 1 bar $\leq P_{O2}$ $\geq$ 3000 bar *Physica* C **160** 449

[122] Miyatake T, Gotoh S, Koshizuka N and Tanaka S 1989 T_c increased to 90 K in $YBa_2Cu_4O_8$ by Ca doping *Nature* **341** 41

[123] Morris D E, *et al* 1989 Eight new high-temperature superconductors with the 1:2:4 structure *Phys. Rev.* B **39** 7347

[124] Adachi S, Adachi H, Setsune K and Wasa K 1991 Synthesis of $LnBa_2Cu_4O_8$ (Ln = rare earth elements) ceramics at one atmosphere oxygen pressure *Physica* C **175** 523

[125] Liu H B, Morris D E and Sinha A P B 1992 T_c enhancement versus rare earth size in $R_{0.9}Ca_{0.1}Ba_2Cu_4O_8$ (R = Sm, Eu, Gd, Dy, Ho, and Er) *Phys. Rev.* B **45** 2438

[126] Miyatake T, Takata T, Yamaguchi K, Takamuku K, Koshizuka N, Tanaka S, Shibutani K, Hayashi S, Ogawa R and Kawate Y 1992 Crystal growth of $YBa_2Cu_4O_8$ and $Y_2Ba_4Cu_7O_{15}$ under high pressure *J. Mater. Res.* **7** 5

[127] Dabrowski B, Zhang K, Pluth J J, Wagner J L and Hinks D G 1992 Single-crystal growth and characterization of $YBa_2Cu_4O_8$ with $T_c \sim 80$ K *Physica* C **202** 271

[128] Sengupta S S, Han P D and Payne D A 1994 Self-flux growth of $YBa_2Cu_4O_8$ crystals in a hot isostatic press. An improved method for accelerated crystal growth *Physica* C **232** 283

[129] Dabrowski B 1998 Single-crystal growth and properties of substituted $YBa_2Cu_4O_8$ *Supercond. Sci. Technol.* **11** 54

[130] Hijar C A, Stern C L, Poeppelmeier K R, Rogacki K, Chen Z and Dabrowski B 1995 *Physica* C **252** 13

[131] Schwer H, Karpinski J, Kaldis E, Meijer G I, Rossell C and Mali M 1996 Evidence for Al doping in the CuO_2 planes of $YBa_2Cu_4O_8$ single crystals *Physica* C **267** 113

[132] Genoud J Y, Erb A, Revaz B and Junod A 1997 Growth of high purity $YBa_2Cu_4O_8$ and $Y_2Ba_4Cu_7O_{15-\delta}$ single crystals in $BaZrO_3$ crucibles under high oxygen pressure, and absence of magnetic 'fishtail' effect *Physica* C **282–287** 457

[133] Schneemeyer L F, Waszczak J V, Herzog T, Glarum S H and Siegrist T 1992 Growth of single crystals of $Ba_4Er_2Cu_7O_{15}$ *The International Workshop on Superconductivity* (Pittsburgh: International Superconductivity Technology Center: Toyko and Materials Research Society) pp 49–52

[134] Maeda H, Tanaka Y, Fukutomi M and Asano T 1988 A new high-T_c oxide superconductor without a rare earth element *Japan. J. Appl. Phys.* **27** L209

[135] Michel C, Hervieu M, Borel M M, Grandin A, Deslandes F, Provost J and Raveau B 1987 Superconductivity in the Bi–Sr–Cu–O system *Z. Phys.* B – *Condens. Matter.* **68** 421

[136] Tallon J L, Buckley R G, Gilberd P W, Presland M R, Brown I W M, Bowden M E, Christian L A and Goguel R 1988 High-T_c superconducting phases in the series $Bi_{2.1}(Ca,Sr)_{n+1}Cu_nO_{2n+4+\delta}$ *Nature* **333** 153

[137] Sunshine S A, *et al* 1988 Structure and physical properties of single crystals of the 84-K superconductor $Bi_{2.2}Sr_2Ca_{0.8}Cu_2O_{8+\delta}$ *Phys. Rev.* B **38** 893

[138] Takano M, Takada J, Oda K, Kitaguchi H, Miura Y, Ikeda Y, Tomii Y and Mazaki H 1988 High-T_c phase promoted and stabilized in the Bi,Pb–Sr–Ca–Cu–O system *Japan. J. Appl. Phys.* **27** L1041

[139] Endo U, Koyama S and Kawai T 1988 Preparation of the high-T_c phase of Bi–Sr–Ca–Cu–O superconductor *Japan. J. Appl. Phys.* **27** L1476

[140] Strobel P, Kelleher K, Holtzberg F and Worthington T 1988 Crystal growth and characterization of the superconducting phase in the Bi–Sr–Cu–O system *Physica* C **156** 434

[141] Han P D and Payne D A 1990 Crystal growth of high T_c superconductors in the system Bi–Ca–Sr–Cu–O *J. Cryst. Growth* **104** 201

[142] Remschnig K, Tarascon J M, Ramesh R and Hull G W 1991 Growth and properties of large area $Bi_{2+x}Sr_{2-x}CuO_{6+y}$ single crystals *Physica* C **175** 261

[143] Kishida S, Tokutaka H, Nakanishi S, Fujimoto H, Nishimori K, Ishihara N, Watanabe Y and Futo W 1990 *J. Cryst. Growth* **99** 937

[144] Changkang C, Wanklyn B M, Smith D T and Wondre F R 1991 Crystal growth of $Bi_2Sr_2CuO_6$ from flux system KCl–KF and its superconductivity *J. Mater. Sci.* **26** 5323

[145] Ren Z, Yan X, Jiang Y and Guan W 1988 Growth of superconducting single crystals of composition Bi : Sr : Ca : Cu = 2 : 2 : 1 : 2 in the Bi–Sr–Ca–Cu–O system *J. Cryst. Growth* **92** 677

[146] Campa J A, Gutierrez-Puebla E, Monge M A, Rasines I and Ruiz-Valero C 1992 Single-crystal growth of superconducting $Bi_2Sr_2CaCu_2O_8$ using rotary crucibles *J. Cryst. Growth* **125** 17

[147] Jayavel R, Sekar C, Murugakoothan P, Venkateswara C R, Subramanian C and Ramasamy P 1993 Growth of large size single crystals and whiskers of $Bi_2Sr_2CaCu_2O_8$ by step-cooling method *J. Cryst. Growth* **131** 105

[148] Takagi H, Eisaki H, Uchida S, Maeda A, Tajima S, Uchinokura K and Tanaka S 1988 Transport and optical studies of single crystals of the 80-K Bi–Sr–Ca–Cu–O superconductor *Nature* **332** 236

[149] Nomura S, Yamashita T, Yoshino H and Ando K 1988 Single-crystal growth of $Bi_2(Sr,Ca)_{3-d}Cu_2O_x$ and its superconductivity *Japan. J. Appl. Phys.* **27** L1251

[150] Guo S J and Easterling K E 1990 Improved single crystal growth in the Bi–Sr–Ca–Cu–O system using a sealed cavity technique *J. Cryst. Growth* **100** 303

[151] Fujii T, Nagano Y and Shirafuji J 1991 Growth by self-flux method and post-annealing effect of $Bi_2Sr_2Ca_1Cu_2O_x$ single crystals *J. Cryst. Growth* **110** 994

[152] Ciszek T F and Evans C D 1991 Single-crystal growth and low-field AC magnetic susceptometry of $YBa_2Cu_3O_{7-\delta}$, $ErBa_2Cu_3O_{7-\delta}$, and $Bi_2Sr_2Ca_{0.8}Cu_2O_8$ superconductors *J. Cryst. Growth* **109** 418

[153] Kishida S, Hosokawa E and Liang W Y 1999 Growth of $Bi_2Sr_2CaCu_2O_y$ single crystals by a self-flux method using precursors *J. Cryst. Growth* **205** 284

[154] He Y, Zhou F and Zhao Z X 1999 Contamination by alumina crucible and the effect of Al on T_c for Bi-2212 single crystals *Physica* C **328** 207

[155] Liu J Z, Crabtree G W, Rehn L E, Geiser Urs, Young D A, Kwok W K, Baldo P M, Williams J M and Lam D J 1988 Crystal growth and superconductivity in the Bi–Ca–Sr–Cu–O system *Phys. Lett.* A **127** 444

[156] Chowdhury A J S, Wanklyn B M R, Wondre F R, Hodby J W, Volkozub A V and de Groot P A J 1994 Growth of high-quality 2212 BSCCO crystals in Pt crucibles and characterisation *Physica* C **225** 388

[157] Mitzi D B, Lombardo L W, Kapitulnik A, Laderman S S and Jacowitz R D 1990 Growth and properties of oxygen- and ion-doped $Bi_2Sr_2CaCu_2O_{8+\delta}$ single crystals *Phys. Rev.* B **41** 6564

[158] Wu W B, Li F Q, Jia Y B, Zhou G N, Qian Y T, Qin Q N and Zhang Y H 1993 Growth of $Bi_2Sr_2CaCu_2O_8$ single crystals from Bi-rich melts *Physica* C **213** 133

[159] Sun X F, Wu W B, Zhu J S, Zhao X R, Jia Y B, Zhou G N, Li X -G and Zhang Y H 1996 A growth and annealing experiment concerning large crystals of $Bi_2Sr_2CaCu_2O_y$ with only a single domain *Supercond. Sci. Technol.* **9** 750

[160] Lombardo L W and Kapitulnik A 1992 Growth of $Bi_2Sr_2CaCu_2O_8$ single crystals using MgO crucibles *J. Cryst. Growth* **118** 483

[161] Steinberg A N, Raduchev V A, Denisevich V V, Lysikov S V, Laukhin V N, Homenko A G, Zvarykina A V, Buravov L I and Topnickov V N 1992 High-T_c single-crystal growth in the Bi–Sr–Ca–Cu–O system by inductive melting in a cold crucible *Supercond. Sci. Technol.* **5** 327

[162] Balestrino G, Gambardella U, Liu Y L, Marinelli M, Paoletti A, Paroli P and Paterno G 1988 Growth of thick single crystals of the high T_c superconductor $Bi_2Sr_2CaCu_2O_{8+x}$ *J. Cryst. Growth* **92** 674

[163] Zhang Lu, Liu J Z, Lan M D, Klavins P and Shelton R N 1993 Crystal growth and superconductivity of $Bi_{1.7}Pb_{0.3}Sr_2CaCu_2O_8$ *J. Cryst. Growth* **128** 734

[164] Yang G, Abell J S, Shang P, Jones I P and Gough C E 1996 Growth and microstructure in $Bi_2Sr_2CaCu_2O_y$ single crystals *J. Cryst. Growth* **166** 820

[165] Jayavel R, Thamizhavel A, Murugakoothan P, Subramanian C and Ramasamy P 1993 Growth, twin and domain structure studies of superconducting $Bi_2Sr_2Ca_{1-x}Y_xCu_2O_{8+\delta}$ single crystals *Physica* C **215** 429

[166] Chowdhury A J S, Wanklyn B M R, Volkozub A V and Hodby J W 1996 Growth of doped and undoped BSCCO 2212 crystals in platinum crucibles by repeated remelting and recrystallization *J. Cryst. Growth* **166** 863

[167] Chowdhury A J S, Charnley N R, Wondre F R, Volkozub A V, de Groot P A J, Wanklyn B M R and Hodby J W 1996 Growth of BSCCO 2212 crystals in flat dish-shaped crucibles *J. Cryst. Growth* **169** 405

[168] Villard G, Pelloquin D, Maignan A and Wahl A 1997 Growth and superconductivity of $Bi_2Sr_2Ca_{1-x}Y_xCu_2O_{8+\delta}$ single crystals in the T_c optimum region *Physica* C **278** 11

[169] Chen C and Wanklyn B 1989 Evaporation kinetics of a halide flux system for the growth of Bi–Sr–Ca–Cu–O superconducting crystals *J. Cryst. Growth* **96** 547

[170] Keszei B, Szabo G Y, Vandlik J, Pogany L and Oszlanyi G 1989 Growth of BCSCO single crystals by a slow-cooling flux method *J. Less-Common Met.* **155** 229

[171] Shishido T, Shindo D, Ukei K, Sasaki T, Toyota N and Fukuda T 1989 Growth of single crystals in the Bi–Sr–Ca–Cu–O system using KCl as a flux *Japan. J. Appl. Phys.* **28** L791

[172] Balestrino G, Milani E, Paoletti A, Tebano A, Wang Y H, Ruosi A, Vaglio R, Valentino M and Paroli P 1994 Fast growth of $Bi_2Sr_2Ca_2Cu_3O_{10+x}$ and $Bi_2Sr_2CaCu_2O_{8+x}$ thin crystals at the surface of KCl fluxes *Appl. Phys. Lett.* **64** 1735

[173] Wang X L, Horvat J, Liu H K and Dou S X 1997 Spiral growth of $Bi_2Sr_2CaCu_2O_y$ single crystals using KCl flux technique *J. Cryst. Growth* **173** 380

[174] Shishido T, Toyota N, Shindo D and Fukuda T 1990 Growth and characterization of $Bi_2(Sr_{1-x}Ca_x)_3Cu_2O_y$ single crystals extracted from KBr flux *Japan. J. Appl. Phys.* **29** 2413

[175] Jayavel R, Murugakoothan P, Venkateswara Rao C R, Subramanian C and Ramasamy P 1993 Growth of superconducting $Bi_2Sr_2CaCu_2O_8$ single crystals using K_2CO_3 flux *Supercond. Sci. Technol.* **6** 349

[176] Oka K, Han T -S, Ha D -H, Iga F and Unoki H 1993 Crystal growth of $Bi_2Sr_2CaCu_2O_8$ by the top-seeded solution-growth method *Physica* C **215** 407

[177] Takekawa S, Nozaki H, Umezono A, Kosuda K and Kobayashi M 1988 Single crystal growth of the superconductor $Bi_{2.0}(Bi_{0.2}Sr_{1.8}Ca_{1.0})Cu_{2.0}O_8$ *J. Cryst. Growth* **92** 687

[178] Shigaki I, Kitahama K, Shibutani K, Hayashi S, Ogawa R, Kawate Y, Kawai T, Kawai S, Matsumoto M and Shirafuji J 1990 Optimization of single crystal preparation of $Bi_2Sr_2CaCu_2O_x$ superconductor by the travelling solvent floating zone method *Japan. J. Appl. Phys.* **29** L2013

[179] Pandey R K, Hannan M and Raina K K 1994 Single crystal growth of $Bi_2CaSr_2Cu_2O_{8+x}$ superconductor by traveling zone method *J. Cryst. Growth* **137** 268

[180] Menken M J V, Winkelman A J M and Menovsky A A 1991 Crystal growth of Bi-based (85 K) high-T_c superconducting compounds with a mirror furnace *J. Cryst. Growth* **113** 9

[181] Emmen J H P M, Lenczowski S K J, Dalderop J H J and Brabers V A M 1992 Crystal growth and annealing experiments of the high T_c superconductor $Bi_2Sr_2CaCu_2O_{8+\delta}$ *J. Cryst. Growth* **118** 477

[182] Gu G D, Takamuku K, Koshizuka N and Tanaka S 1993 Large single crystal Bi-2212 along the *c*-axis prepared by the floating zone method *J. Cryst. Growth* **130** 325

[183] Ha D H, Kim I S, Park Y K, Oka K and Nishihara Y 1994 Crystal growth of Ba–Sr–Ca–Y–Cu–O by the traveling solvent floating zone method *Physica* C **222** 252

[184] Gazit D, Peszkin P N, Moulton L V and Feigelson R S 1989 Influence of growth rate on the structure and composition of float zone grown $Bi_2Sr_2CaCu_2O_8$ superconducting fibers *J. Cryst. Growth* **98** 545

[185] Lu Z, Moulton L V, Feigelson R S, Raymakers R J and Peszkin P N 1990 Factors affecting the growth of single crystal fibers of the superconducting $Bi_2Sr_2CaCu_2O_8$ *J. Cryst. Growth* **106** 732

[186] Zhang X J, Zeng X B, Fang M H, Xu Z A, Zhang Q R, Wang Y W and Tang X M 1993 An approach to the growth of $Bi_2Sr_2CaCu_2O_y$ single crystals by electrocrystallization *High-Temperature Superconductivity (BHTSC '92)* ed Z Z Gan, S S Xie and Z X Zhao (London: World Scientific) pp 203–4

[187] Mansori M, Faqir H, Satre P, Bendriss A, Syono Y and Sebaoun A 1999 A new single crystal growth method of $(Bi,Pb)_2Sr_2CaCu_2O_x$ superconductor *J. Cryst. Growth* **197** 141

[188] Ono A, Sueno S and Okamura F P 1988 Preparation and properties of single crystals of the high-T_c oxide superconductor in the Bi–Sr–Ca–Cu–O system *Japan. J. Appl. Phys.* **27** L786

[189] Kishida S and Hosokawa E 1998 Optimum growth condition of $Bi_2Sr_2CaCu_2O_y$ single crystals in a vertical Bridgman method *J. Cryst. Growth* **192** 136

[190] Jung J, Franck J P, Mitchell D F and Claus H 1988 Whisker growth of superconducting Bi–Sr–Ca–Cu oxide *Physica* C **156** 494

[191] Matsubara I, Kageyama H, Tanigawa H, Ogura T, Yamashita H and Kawai T 1989 Preparation of fibrous Bi(Pb)–Sr–Ca–Cu–O crystals and their superconducting properties in a bending state *Japan. J. Appl. Phys.* **28** L1121

[192] Latyshev Yu I, Gorlova I G, Nikitina A M, Antokhina V U, Zybtsev S G, Kukhta N P and Timofeev V N 1993 Growth and study of single-phase 2212 BSCCO whiskers of submicron cross-sectional area *Physica* C **216** 471

[193] Krapf A, Lacayo G, Kastner G, Kraak W, Pruss N, Thiele H, Dwelk H and Herrmann R 1991 Flux growth of Bi–Sr–Ca–Cu–O whiskers *Supercond. Sci. Technol.* **4** 237

[194] Kraak W, Krapf A, Pruss N, Neubert G, Rogaschewski S, Hahnert I, Wilde W and Thiele P 1996 Growth, characterization and physical properties of Bi–Sr–Ca–Cu–O superconducting whiskers *Phys. Status Solidi* A **158** 183

[195] Jin H, Skwirblies S and Kotzler J 1999 Growth and characterization of BiSCCO 2212 whiskers from melts containing various La contents *J. Cryst. Growth* **207** 154

[196] Morris D E, Hultgren C T, Markelz A M, Wei J Y T, Asmar N G and Nickel J H 1989 Oxygen concentration effect on T_c of the Bi–Ca–Sr–Cu–O superconductor *Phys Rev.* B **39** 6612

[197] Fournier P, Kapitulnik A and Marshall A F 1996 Growth and annealing of $(Bi_{1-x}Pb_x)_2Sr_2CaCu_2O_{8+\delta}$ single crystals *Physica* C **257** 291

[198] Sun X F, Wu W B, Zheng L, Zhao X R, Shi L, Zhou G N, Li X G and Zhang Y H 1997 Effects of oxygen doping and structural distortion on the superconductivity of $Bi_2Sr_2CaCu_2O_y$ single crystals *J. Phys.: Condens. Matter.* **9** 6391

[199] Prabhakaran D, Subramanian C and Ramasamy P 1999 Synthesis of Bi-2223 phase using different precursors through different routes and their characterisation *Mat. Sci. Eng. B-Solid* **58** 199

[200] Chu S and McHenry M E 1998 Growth and characterization of $(Bi,Pb)_2Sr_2Ca_2Cu_3O_x$ single crystals *J. Mater. Res.* **13** 589

[201] Matsubara I, Tanigawa H, Ogura T, Yamashita H, Kinoshita M and Kawai T 1991 Superconducting whiskers and crystals of the high T_c $Bi_2Sr_2Ca_2Cu_3O_{10}$ phase *Appl. Phys. Lett.* **58** 409

[202] Inoue T, Hayashi S, Miyashita S, Shimizu M, Nishimura Y and Komatsu H 1992 Conversion of superconducting Bi-system single crystals from 2212 to 2223 by the annealing method *J. Cryst. Growth* **123** 615

[203] Sheng Z Z and Hermann A M 1988 Bulk superconductivity at 120 K in the Tl–Ca/Ba–Cu–O system *Nature* **332** 138

[204] Torardi C C, Subramanian M A, Calabrese J C, Gopalakrishnan J, McCarron E M, Morrissey K J, Askew T R, Flippen R B, Chowdhry U and Sleight A W 1988 Structures of the superconducting oxides $Tl_2Ba_2CuO_6$ and $Bi_2Sr_2CuO_6$ *Phys. Rev.* B **38** 225

[205] Hazen R M *et al* 1988 100-K superconducting phases in the Tl–Ca–Ba–Cu–O system *Phys. Rev. Lett.* **60** 1657

[206] Kotani T, Kaneko T, Takei H and Tada K 1989 Crystal growth and superconductivity of Tl–Ca–Ba–Cu–O system *Japan. J. Appl. Phys.* **28** L1378

[207] Kotani T, Nishikawa T, Takei H and Tada K 1990 Structural study of $Tl_2Ca_3Ba_2Cu_4O_x$ single crystal by transmission electron microscopy *Japan. J. Appl. Phys.* **29** L902

[208] Liu R S, Zhou W, Cooper J R, Bennett M J, Janes R, Zheng D N and Edwards P P 1990 Preparation of Tl–Ca–Ba–Cu–O single crystals with T_c up to 118 K by the step-cooling method *Supercond. Sci. Technol.* **3** 568

[209] Subramanian M A, Calabrese J C, Torardi C C, Gopalakrishnan J, Askew T R, Flippen R B, Morrissey K J, Chowdhry U and Sleight A W 1988 Crystal structure of the high-temperature superconductor $Tl_2Ba_2CaCu_2O_8$ *Nature* **332** 420

[210] Torardi C C, Subramanian M A, Calabrese J C, Gopalakrishnan J, Morrissey K J, Askew T R, Flippen R B, Chowdhry U and Sleight A W 1988 Crystal structure of $Tl_2Ba_2Ca_2Cu_3O_{10}$, a 125 K superconductor *Science* **240** 631

[211] Ginley D S, Morosin B, Baughman R J, Venturini E L, Schirber J E and Kwak J F 1988 Growth of crystals and effects of oxygen annealing in the Bi–Ca–Sr–Cu–O and Tl–Ca–Ba–Cu–O superconductor systems *J. Cryst. Growth* **91** 456

[212] Onoda M, Kondoh S, Fukuda K and Sato M 1988 Structural study of superconducting Tl–Ba–Ca–Cu–O system *Japan. J. Appl. Phys.* **27** L1234

[213] Bukowski E D and Ginsberg D M 1989 Growth and oxygenation of single-crystal BiSrCaCuO and TlBaCaCuO superconductors *J. Low Temp. Phys.* **77** 285

[214] Kawaguchi K and Nakao M 1990 Growth and magnetic properties of Tl–Ca–Ba–Cu–O single crystals *J. Cryst. Growth* **99** 942

[215] Ren Z F, Naughton M J, Lee P and Wang J H 1991 Growth of superconducting single crystals of $Tl_mBa_2Ca_{n-1}Cu_nO_y$ in a convenient way *J. Cryst. Growth* **112** 587

[216] Kajitani T, Hiraga K, Nakajima S, Kikuchi M, Syono Y and Kabuto C 1989 X-ray diffraction analysis on 2223 Tl oxide single crystals *Physica* C **161** 483

[217] Togano K, Kumakura H, Mukaida H, Kawaguchi K and Nakao M 1989 Resistive transition in magnetic fields for a single crystal of a Tl–Ba–Ca–Cu–O superconductor *Japan. J. Appl. Phys.* **28** L907

[218] Liu J Z, Jia Y X, Klavins P, Shelton R N, Downey J and Lam D J 1991 Preparation and magnetic measurements of single crystal $Tl_2Ba_2CaCu_2O_8$ *J. Cryst. Growth* **109** 436

[219] Venturini E L, Tigges C P, Baughman R J, Morosin B, Barbour J C, Mitchell M A and Ginley D S 1991 Stoichiometry and irradiation effects in melt grown Tl–Ca–Ba–Cu–O single crystals *J. Cryst. Growth* **109** 441

[220] Duan H M, Kaplan T S, Dlugosch B, Hermann A M, Swope J, Drexler J, Boni P and Smyth J 1992 Single crystal growth of Tl-based superconductors *Physica* C **203** 257

[221] Matsushita Y, Hasegawa M, Sakai F and Takei H 1995 Crystal growth of $Tl_2Ba_2Ca_2Cu_3O_{10}$ (Tl-2223) superconductors *Japan. J. Appl. Phys.* **34** L1263

[222] Takei H, Sakai F, Hasegawa M, Nakajima S and Kikuchi M 1993 Crystal growth of Tl-based cuprate superconductors *Japan. J. Appl. Phys.* **32** L1403

[223] Morosin B, Venturini E L, Dunn R G, Newcomer Provencio P, Missert N and Padilla R R 1998 Structural and compositional characterization of rubidium-containing crystals of the Tl–Ba–Ca–Cu–O superconductors *Physica* C **302** 119

[224] Valldor M, Bryntse I and Morawski A 1999 Synthesis and x-ray single-crystal analysis of a 2212-type superconductor in the Tl–Hg–Sr–Ca–Cu–O system *Physica* C **314** 27

[225] Morosin B, Venturini E L and Ginley D S 1991 Annealing study on single crystal Tl-2223 superconductors *Physica* C **175** 241

[226] Parkin S S P, Lee V Y, Nazzal A I, Savoy R, Beyers R and La Placa S J 1988 $Tl_1Ca_{n-1}Ba_2Cu_nO_{2n+3}$ ($n = 1, 2, 3$): a new class of crystal structures exhibiting volume superconductivity at up to ≈ 110 K *Phys. Rev. Lett.* **61** 750

[227] Parkin S S P, Lee V Y, Nazzal A I, Savoy R, Huang T C, Gorman G and Beyers R 1988 Model family of high-temperature superconductors: $Tl_mCa_{n-1}Ba_2Cu_nO_{2(n+1)+m}$ ($m = 1, 2$; $n = 1, 2, 3$) *Phys. Rev.* B **38** 6531

[228] Ihara H, Sugise R, Hirabayashi M, Terada N, Jo M, Hayashi K, Negishi A, Tokumoto M, Kimura Y and Shimomura T 1988 A new high-T_c $TlBa_2Ca_3Cu_4O_{11}$ superconductor with $T_c > 120$ K *Nature* **334** 510

[229] Subramanian M A, Torardi C C, Gopalakrishnan J, Gai P L, Calabrese J C, Askew T R, Flippen R B and Sleight A W 1988 Bulk superconductivity up to 122 K in the Tl–Pb–Sr–Ca–Cu–O system *Science* **242** 249

[230] Morosin B, Ginley D S, Schirber J E and Venturini E L 1988 Crystal structure of $TlCa_2Ba_2Cu_3O_9$ *Physica* C **156** 587

[231] Kolesnikov N N, Korotkov V E, Kulakov M P, Lagvenov G A, Molchanov V N, Muradyan L A, Simonov V I, Tamazyan R A, Shibaeva R P and Shchegolev I F 1989 Structure of superconducting single crystals of $TlBa_2(Ca_{0.87}Tl_{0.13})Cu_2O_7$, $T_c = 80$ K *Physica* C **162–4** 1663

[232] Martin C, Maignan A, Huve M, Labbe P h, Ledesert M, Leligny H and Raveau B 1993 A Sr-rich 1223 cuprate, $Tl_{1+x}Ba_{2/3}Sr_{4/3}Ca_{2-x}Cu_3O_9$, with a T_c of 110 K *Physica* C **217** 106

[233] Putilin S N, Antipov E V, Chmaissem O and Marezio M 1993 Superconductivity at 94 K in $HgBa_2CuO_{4+\delta}$ *Nature* **362** 226

[234] Schilling A, Cantoni M, Guo J D and Ott H R 1993 Superconductivity above 130 K in the Hg–Ba–Ca–Cu–O system *Nature* **363** 56

[235] Chu C W, Gao L, Chen F, Huang Z J, Meng R L and Xue Y Y 1993 Superconductivity above 150 K in $HgBa_2Ca_2Cu_3O_{8+\delta}$ at high pressures *Nature* **365** 323

[236] Antipov E V, Loureiro S M, Chaillout C, Capponi J J, Bordet P, Tholence J L, Putilin S N and Marezio M 1993 The synthesis and characterization of the $HgBa_2Ca_2Cu_3O_{8+\delta}$ and $HgBa_2Ca_2Cu_3O_{8+\delta}$ phases *Physica* C **215** 1

[237] Meng R L, Beauvais L, Zhang X N, Huang Z J, Sun Y Y, Xue Y Y and Chu C W 1993 Synthesis of the high-temperature superconductors $HgBa_2CaCu_2O_{6+\delta}$ and $HgBa_2Ca_2Cu_3O_{8+\delta}$ *Physica* C **216** 21

[238] Scott B A, Suard E Y, Tsuei C C, Mitzi D B, McGuire T R, Chen B H and Walker D 1994 Layer dependence of the superconducting transition temperature of $HgBa_2Ca_{n-1}Cu_nO_{2n+2+\delta}$ *Physica* C **230** 239

[239] Marezio M, Antipov E V, Capponi J J, Chaillout C, Loureiro S, Putilin S N, Santoro A and Tholence J L 1994 The superconducting $HgBa_2Ca_{n-1}Cu_nO_{2n+2+\delta}$ homologous series *Physica* B **197** 570

[240] Xiong Q, Xue Y Y, Cao Y, Chen F, Sun Y Y, Gibson J, Chu C W, Liu L M and Jacobson A 1994 Unusual hole dependence of T_c in $HgBa_2CuO_{4+\delta}$ *Phys. Rev.* B **50** 10346

[241] Bertinotti A, Viallet V, Colson D, Marucco J -F, Hammann J, Le Bras G and Forget A 1996 Synthesis, crystal structure and magnetic properties of superconducting single crystals of $HgBa_2CuO_{4+\delta}$ *Physica* C **268** 257

[242] Pissas M, Billon B, Charalambous M, Chaussy J, LeFloch S, Bordet P and Capponi J J 1997 Single-crystal growth and characterization of the superconductor $HgBa_2CuO_{4+\delta}$ *Supercond. Sci. Technol.* **10** 598

[243] Morawski A, Lada T, Paszewin A and Przybylski K 1998 High gas pressure for HTS single crystals and thin layer technology *Supercond. Sci. Technol.* **11** 193

[244] Maignan A, Putilin S N, Hardy V, Simon C H and Raveau B 1996 Magnetic study of $HgBa_2Ca_2Cu_3O_{8+\delta}$ single crystals: effect of doping on irreversibility and fishtail lines *Physica* C **266** 173

[245] Karpinski J, Schwer H, Mangelschots I, Conder K, Morawski A, Lada T and Paszewin A 1994 Crystals of Hg superconductors *Nature* **371** 661

[246] Schwer H, Conder K, Kopnin E, Meijer G I, Molinski R and Karpinski J 1997 X-ray single crystal structure analysis of $Sr_{0.73}CuO_2$ and $Hg_{1-x}Pb_xBa_2Ca_{n-1}Cu_nO2_{n+2+\delta}$ compounds (n = 1 – 5, x = 0–0.5) *Physica* C **282–7** 901

[247] Karpinski J, Conder K, Schwer H, Lohle J, Lesne L, Rossel C, Morawski A, Paszewin A and Lada T 1995 Single crystals of Hg-12($n-1$)n and infinite-layer $CaCuO_2$ obtained at gas pressure P = 10 kbar *J. Supercond.* **8** 515

[248] Karpinski J, Schwer H, Mangelschots I, Conder K, Morawski A, Lada T and Paszewin A 1994 Single crystals of $Hg_{1-x}Pb_xBa_2Ca_{n-1}Cu_nO_{2n+2+\delta}$ and infinite-layer $CaCuO_2$ – synthesis at gas-pressure 10 kbar, properties and structure *Physica* C **234** 10

[249] Pelloquin D, Maignan A, Guesdon A, Hardy V and Raveau B 1996 Single crystal study of the '1201' superconductor $Hg_{0.8}Bi_{0.2}Ba_2CuO_{4+\delta}$ *Physica* C **265** 5

[250] Lin C T, Yan Y, Peters K, Schonherr E and Cardona M 1998 Flux growth $Hg_{1-x}Re_xBa_2Ca_{n-1}Cu_nO_{2n+2+\delta}$ single crystals by self-atmosphere *Physica* C **300** 141

[251] Schwer H, Molinski R, Kopnin E M, Angst M and Karpinski J 1999 Single crystal of the 1223/1234 intergrowth phase $Hg_{1.44}Re_{0.5}Ba_4Ca_5Cu_7O_{20}$: structure and properties *Physica* C **311** 49

[252] Yao Y S, Liu W, Su Y J, Xiong Y F, Ma W J, Cao G H, Zou G T and Zhao Z X 1997 New phase of Hg-2435 in the Hg–Ba–Ca–Cu–O oxide system *Physica* C **282–7** 897

[253] Cava R J, *et al* 1988 Superconductivity near 70 K in a new family of layered copper oxides *Nature* **336** 211

[254] Subramanian M A, Gopalakrishnan J, Torardi C C, Gai P L, Boyes E D, Askew T R, Flippen R B, Farneth W E and Sleight A W 1989 Superconductivity near liquid nitrogen temperature in the Pb–Sr–R–Ca–Cu–O system (R = Y or rare earth) *Physica* C **157** 124

[255] Gallagher P K, O'Bryan H M, Cava R J, James A C W P, Murphy D W, Rhodes W W, Krajewski J J, Peck W F and Waszczak J V 1989 Oxidation–reduction of $Pb_2Sr_2Ln_{1-x}M_xCu_{3-y}AgyO_{8+\delta}$ *Chem. Mater.* **1** 277

[256] Schneemeyer L F, *et al* 1989 Crystal growth and substitutional chemistry of $Pb_2Sr_2MCu_3O_8$ *Chem. Mater.* **1** 548

[257] Capponi J J, Chmaissem O, Fournier T, Gorius M F, Korczak S, Mareizo M and Tholence J L 1990 Preparation and characterization of 80 K superconducting $Pb_2Sr_2Y_{1-x}Ca_xCu_3O_{8+\delta}$ single crystals *J. Less-Common Met.* **164, 165** 808

[258] Korczak Z, Korczak W, Kolesnik S, Skoskiewicz T and Igalson J 1990 Preparation and characterisation of PbSr(YCa)CuO superconducting crystals *Supercond. Sci. Technol.* **3** 370

[259] Xue J S, Reedyk M, Lin Y P, Stager C V and Greedan J E 1990 Synthesis, characterization and superconducting properties of $Pb_2Sr_2(Y/Ca)Cu_3O_{9-\delta}$ single crystals *Physica* C **166** 29

[260] Changkang C, Wanklyn B M, Dieguez E, Cook A J, Hodby J W, Schwartzbrod A, Dabkowski A and Dabkowska H 1992 Phase diagram and crystal growth of Pb_2Sr_2 $(Y_xCa_{1-x})Cu_3O_{8+\delta}$ *J. Cryst. Growth* **118** 101

[261] Kriebel C l, Banazadeh M, Winkel G, Knauf N, Roden B, Braden M and Freitag B 1994 Crystal growth, separation and characterization of superconducting $Pb_2Sr_2(Y_{1-x}Ca_x)Cu_3O_{8+\delta}$ *J. Cryst. Growth* **141** 124

[262] Rouillon T, Provost J, Hervieu M, Groult D, Michel C and Raveau B 1989 Superconductivity up to 100 K in lead cuprates: a new superconductor $Pb_{0.5}Sr_{2.5}Y_{0.5}Ca_{0.5}Cu_2O_{7-\delta}$ *Physica* C **159** 201

[263] Rouillon T, Provost J, Hervieu M, Groult D, Michel C and Raveau B 1990 The solid solution $Pb_{0.5}Sr_{2.5}Y_{1-x}Ca_xCu_2O_{7-\delta}$: superconductivity and structure *J. Solid State Chem.* **84** 375

[264] Dabkowski A, Dabkowska H, Greedan J E, Xue J S and Stager C V 1993 Growth and characterization of superconducting single crystals of layered 1212 PbSrYCaCu oxide *J. Cryst. Growth* **126** 471

[265] Jin H, Chen Z J, Cao L Z, Zhou G E and Mao Z W 1995 Growth of 1212 phase Pb–Sr–Y–Ca–Cu–O single crystals *J. Cryst. Growth* **148** 106

[266] Jin H, Wang N L, Chong Y, Deng M, Cao L Z, Chen Z J, Zhou G E and Mao Z W 1995 Single-crystal growth and characterization of the $Pb_{0.5}Sr_{2.5}Y_{1-x}Ca_xCu_2O_y$ system *J. Mater. Res.* **10** 2211

[267] Aswal D K, Narang S N, Gupta S K, Mudher K D S, Purandare S C and Sabharwal S C 1997 Growth of superconducting $Pb_{0.5}Sr_{2.5}Y_{1-x}Ca_xCu_2O_y$ single crystals from charges containing silver *J. Cryst. Growth* **179** 665

[268] Vaughey J T, Theil J P, Hasty E F, Groenke D A, Stern C L, Poeppelmeier K R, Dabrowski B, Hinks D G and Mitchell A W 1991 Synthesis and structure of a new family of cuprate superconductors: $LnSr_2Cu_2GaO_7$ *Chem. Mater.* **3** 935

[269] Cava R J, van Dover R B, Batlogg B, Krajewski J J, Schneemeyer L F, Siegrist T, Hessen B, Chen S H, Peck W F Jr. and Rupp Jr. L W 1991 Superconductivity in multiple phase $Sr_2Ln_{1-x}Ca_xGaCu_2O_7$ and characterization of $La_{2-x}Sr_xCaCu_2O_{6+\delta}$ *Physica* C **185–9** 180

[270] Roth G, Adelmann P, Heger G, Knitter R and Wolf T h 1991 The crystal structure of $RESr_2GaCu_2O_7$ *J. Phys. I* **1** 721

[271] Cava R J, Krajewski J J, Takagi H, Zandbergen H W, van Dover R B, Peck W F Jr. and Hessen B 1992 Superconductivity at 28 K in a cuprate with a niobium oxide intermediary layer *Physica* C **191** 237

[272] Kinoshita K and Yamada T 1992 A new copper oxide superconductor containing carbon *Nature* **357** 313

[273] Miyazaki Y, Yamane H, Kajitani T, Oku T, Hiraga K, Morii Y, Fuchizaki K, Funahashi S and Hirai T 1992 Preparation and crystal structure of $Sr_2CuO_2(CO_3)$ *Physica* C **191** 434

[274] Shibata H, Kinoshita K and Yamada T 1994 Single crystal growth of $(Ba_{1-x}Sr_x)_2Cu_{1.1}O_{2.2+\delta}(CO_3)_{0.9}$ by solid state recrystallization *Physica* C **232** 181

[275] Calestani G, Ganguly P, Matacotta F C, Nozar P, Migliori A, Thomas K A and Tomasi A 1995 Synthesis, single crystal growth and structural determination of new copper oxycarbonates $A_4CuM(CO_3)_2O_4$, A = Sr, Ba, M = Li, Na, Ca *Physica* C **247** 359

[276] Calestani G, Matacotta F C, Migliori A, Nozar P, Righi L and Thomas K A 1996 Synthesis, crystal growth and structural characterisation of barium–copper oxycarbonates $Ba_2M_xCu_{2-x-y}(CO_3)_yO_{2+\delta}$ (M = Cu, Cd, Ca; $0.05 < x < 0.25$) *Physica* C **261** 38

[277] Hiroi Z, Kobayashi N and Takano M 1994 Probable hole-doped superconductivity without apical oxygens in $(Ca,Na)_2CuO_2Cl_2$ *Nature* **371** 139

[278] Jin C-Q, Wu X-J, Laffez P, Tatsuki T, Tamura T, Adachi S, Yamauchi H, Koshizuka N and Tanaka S 1995 Superconductivity at 80 K in $(Sr,Ca)_3Cu_2O_{4+\delta}Cl_{2-y}$ induced by apical oxygen doping *Nature* **375** 301

[279] Al-Mamouri M, Edwards P P, Greaves C and Slaski M 1994 Synthesis and superconducting properties of the strontium copper oxy-fluoride $Sr_2CuO_2F_{2+\delta}$ *Nature* **369** 382

[280] Slater P R, Hodges J P, Francesconi M G, Edwards P P, Greaves C, Gameson I and Slaski M 1995 An improved route to the synthesis of superconducting copper oxyfluorides $Sr_{2-x}A_xCuO_2F_{2+\delta}$ (A = Ca, Ba) using transition metal difluorides as fluorinating reagents *Physica* C **253** 16

[281] Miller L L, Wang X L, Wang S X, Stassis C, Johnston D C, Faber J Jr and Loong C-K 1990 Synthesis, structure and properties of $Sr_2CuO_2Cl_2$ *Phys. Rev.* B **41** 1921

[282] Hien N T, Franse J J M, Pothuizen J J M, Li T W and Menovsky A A 1997 Growth and characterisation of bulk $Sr_2CuO_2Cl_2$ single crystals *J. Cryst. Growth* **171** 102

[283] Siegrist T, Zahurak S M, Murphy D W and Roth R S 1988 The parent structure of the layered high-temperature superconductors *Nature* **334** 231

[284] Lagues M, *et al* 1996 The infinite layer family of superconducting cuprates *Coherence in High Temperature Superconductors*, eds G Deutscher and A Revcolevschi (Singapore: World Scientific) pp 70–98

[285] Wiley J B, Sabat M, Hwu S -J and Poeppelmeier K R 1990 $LaSrCuAlO_5$: a new oxygen-deficient perovskite structure *J. Solid State Chem.* **88** 250

[286] Isobe M, Matsui Y and Takayama-Muromachi E 1994 High-pressure synthesis of $Y_{1-x}Ca_xSr_2GaCu_2O_{7+\delta}$ ($0 \leq x \leq 1.0$) *Physica* C **222** 310

[287] Takayama-Muromachi E and Isobe M 1994 New series of high T_c superconductors $GaSr_2Ca_{n-1}Cu_nO_{2n+3}$ ($n = 3$; $T_c = 70$ K, $n = 4$; $T_c = 107$ K) prepared at high pressure *Japan. J. Appl. Phys.* **33** L1399

[288] Matveev A T and Takayama-Muromachi E 1995 High-pressure synthesis of a new superconductor $AlSr_2Ca_{2-x}Y_xCu_3O_9$. The $n = 3$ member of the $AlSr_2Ca_{n-1}Cu_nO_{2n+3}$ family *Physica* C **254** 26

[289] Isobe M, Kawashima T, Kosuda K, Matsui Y and Takayama-Muromachi E 1994 A new series of high-T_c superconductors $AlSr_2Ca_{n-1}Cu_nO_{2n+3}$ ($n = 4$, $T_c = 110$ K; $n = 5$, $T_c = 83$ K) prepared at high pressure *Physica* C **234** 120

[290] Matveev A T, Ramirez-Castellanos J, Matsui Y and Takayama-Muromachi E 1996 New high-T_c superconductor $(Ge_zCu_{1-z})Sr_2Ca_{2-x}Y_xCu_3O_y$ ((Ge,Cu)-1223) prepared under high pressure *Physica* C **262** 279

[291] Matveev A T, Ramirez-Castellanos J, Matsui Y and Takayama-Muromachi E 1997 New high-T_c superconductors $(Ge_zCu_{1-z})Sr_2Ca_{n-1-x}Y_xCu_nO_y$ ($n = 4$, 6) prepared at high pressure *Physica* C **274** 48

[292] Zhigadlo N D, Anan Y, Asaka T, Ishida Y, Matsui Y and Takayama-Muromachi E 1999 High-pressure synthesis and characterization of a new series of V-based superconductors $(Cu_{0.5}V_{0.5})Sr_2Ca_{n-1}Cu_nO_y$ *Chem. Mater.* **11** 2185

[293] Li R K, Kremer R K and Maier J 1992 A new layered superconductor: $GaSr_2(Y, Ce)_2Cu_2O_{9-\delta}$ *Physica* C **200** 344

[294] Adachi S, Nara A and Yamauchi H 1992 Synthesis of $(Ga/Cu)(Sr/Eu)_2(Eu/Ce)_2Cu_2O_y$ superconductor *Physica* C **201** 403

[295] Kaiban T, Yitai Q, Yadun Z, Li Y, Zuyao C and Yuheng Z 1996 Synthesis and characterization of a new layered superconducting cuprate: $RuSr_2(Gd,Ce)_2Cu_2O_z$ *Physica* C **259** 168

[296] Ono A 1995 Preparation of new superconducting cuprate $(Ru,Nb)Sr_2(Sm,Ce)_2Cu_2O_z$ *Japan. J. Appl. Phys.* **34** L1121

[297] Ono A and Horiuchi S 1993 Synthesis of a new superconductor $(Cu_{0.6}Ce_{0.4})Sr_2Y_{1.2}Ce_{0.8}Cu_2O_z$ under high oxygen pressures *Physica* C **216** 165

[298] Maeda T, Sakuyama K, Koriyama S, Ichinose A, Yamauchi H and Tanaka S 1990 New superconducting cuprates $(Pb,Cu)(Eu,Ce)_2(Sr,Eu)_2Cu_2O_z$ *Physica* C **169** 133

[299] Smith M G, Chan J and Goodenough J B 1993 High-temperature superconductivity in $Y_{1-z}Ca_zSr_2Cu_{2.5}Fe_{0.5}O_{6+x}$ *Physica* C **208** 412

[300] Palacin M R, Fuertes A, Casan-Pastor N and Gomez-Romero P 1995 Synthesis structure and superconductivity in all-perovskite-layered titanium cuprates *J. Solid State Chem.* **119** 224

[301] Akimitsu J, Uehara M, Ogawa M, Nakata H, Tomimoto K, Miyazaki Y, Yamane H, Hirai T, Kinoshita K and Matsui Y 1992 Superconductivity in the new compound $(Y_{1-x}Ca_x)_{0.95}Sr_{2.05}Cu_{2.4}(CO_3)_{0.6}O_y$ *Physica* C **201** 320

[302] Kawashima T, Matsui Y and Takayama-Muromachi E 1994 A new oxycarbonate superconductor $(Cu_{0.5}C_{0.5})_2$-$Ba_3Ca_2Cu_3O_{11}$ ($T_c = 91$ K) prepared at high pressure *Physica* C **233** 143

[303] Matveev A T, Matsui Y, Yamaoka S and Takayama-Muromachi E 1997 High-pressure synthesis of a new oxycarbonate superconductor $CCa_3Cu_2O_{7+\delta}$ *Physica* C **288** 185

[304] Zhu W J, Yue J J, Huang Y Z and Zhao Z X 1993 $(B,Cu)Sr_2YCu_2O_7$, a new layered copper-oxide based on the boron–oxygen group *Physica* C **205** 118

[305] Uehara M, Uoshima M, Ishiyama S, Nakata H, Akimitsu J, Matsui Y, Arima T, Tokura Y and Mori N 1994 A new homologous series of oxycarbonate superconductors $Sr_2(Ca,Sr)_{n-1}Cu_n(CO_3)_{1-x}(BO_3)_xO_y$ ($n = 1, 2$ and 3) *Physica* C **229** 310

[306] Smith M G, Manthirum A, Zhou J, Goodenough J B and Markert J T 1991 Electron-doped superconductivity at 40 K in the infinite-layer compound $Sr_{1-y}Nd_yCuO_2$ *Nature* **351** 549

[307] Azuma M, Hiroi Z, Takano M, Bando Y and Takeda Y 1992 Superconductivity at 110 K in the infinite-layer compound $(Sr_{1-x}Ca_x)_{1-y}CuO_2$ *Nature* **356** 775

[308] Hiroi Z, Tanako M, Azuma M and Takeda Y 1993 A new family of copper oxide superconductors $Sr_{n+1}Cu_nO_{2n+1+\delta}$ stabilized at high pressure *Nature* **364** 315

[309] Sleight A W, Gillson J L and Bierstedt P E 1975 High-temperature superconductivity in the $BaPb_{1-x}Bi_xO_3$ system *Solid State Commun.* **17** 27

[310] Katsui A 1982 Single crystal growth of superconducting $BaPb_{1-x}Bi_xO_3$ from PbO_2–Bi_2O_3–$BaPbO_3$ solution *Japan. J. Appl. Phys.* **21** L553

[311] Batlogg B 1984 Superconductivity in Ba(Pb, Bi)O_3 *Physica* B **126** 275

[312] Mattheiss L F, Gyorgy E M and Johnson D W Jr 1988 Superconductivity above 20 K in the Ba–K–Bi–O system *Phys. Rev.* B **37** 3745

[313] Cava R J, Batlogg B, Krajewski J J, Farrow R, Rupp L W Jr, White A E, Short K, Peck W F and Kometani T 1988 Superconductivity near 30 K without copper: the $Ba_{0.6}K_{0.4}BiO_3$ perovskite *Nature* **332** 814

[314] Shiyou P, Jorgensen J D, Dabrowski B, Hinks D G, Richards D R, Mitchell A W, Newsam J M, Sinha S K, Vaknin D and Jacobson A J 1990 Structural phase diagram of the $Ba_{1-x}K_xBiO_3$ system *Phys. Rev.* B **41** 4126

[315] Schneemeyer L F, Thomas J K, Siegrist T, Batlogg B, Rupp L W, Opila R L, Cava R J and Murphy D W 1988 Growth and structural characterization of superconducting $Ba_{1-x}K_xBiO_3$ single crystals *Nature* **335** 421

[316] Wignacourt J P, Swinnea J S, Steinfink H and Goodenough J B 1988 Oxygen atom thermal vibration anisotropy in $Ba_{0.87}K_{0.13}BiO_3$ *Appl. Phys. Lett.* **53** 1753

[317] Norton M L and Tang H -Y 1991 Superconductivity at 32 K in electrocrystallized Ba–K–Bi–O *Chem. Mater.* **3** 431

[318] Rosamilia J M, Glarum S H, Cava R J, Batlogg B and Miller B 1991 Anodic synthesis and characterization of millimeter crystals of $Ba_{0.6}K_{0.4}BiO_3$ *Physica* C **182** 285

[319] Mosley W D, Liu J Z, Matsushita A, Lee Y P, Klavins P and Shelton R N 1993 Preparation and superconductivity of $Ba_{1-x}K_xBiO_3$ single crystals *J. Cryst. Growth* **128** 804

[320] Uchida T, Nakamura S, Suzuki N, Nagata Y, Mosley W D, Lan M D, Klavins P and Shelton R N 1993 Effect of growth conditions on the superconductivity of $Ba_{1-x}K_xBiO_3$ crystals *Physica* C **215** 350

[321] Han P D, Chang L and Payne D A 1993 Top-seeded growth of superconducting $(Ba_{1-x}K_x)BiO_3$ crystals by an electrochemical method *J. Cryst. Growth* **128** 798

[322] Barilo S N, *et al* 1999 A new method for growing $Ba_{1-x}K_xBiO_3$ single crystals and investigation of their properties *J. Cryst. Growth* **198/199** 636

[323] Johnston D C 1976 Superconducting and normal state properties of $Li_{1+x}Ti_{2-x}O_4$ spinel compounds I. Preparation, crystallography, superconducting properties, electrical resistivity, dielectric behavior and magnetic susceptibility *J. Low Temp. Phys.* **25** 145

[324] Akimoto J, Gotoh Y, Kawaguchi K and Oosawa Y 1992 Preparation of $LiTi_2O_4$ single crystals with the spinel structure *J. Solid State Chem.* **96** 446

[325] Campa J A, Velez M, Cascales C, Gutierrez Puebla E, Monge M A, Rasines I and Ruiz-Valero C 1994 Crystal growth of superconducting $LiTi_2O_4$ *J. Cryst. Growth* **142** 87

B2.5
Growth of A15 type single crystals and polycrystals and their physical properties

René Flükiger

B2.5.1 Introduction

B2.5.1.1 Electronic properties of A15 type compounds

Before the discovery of the high T_c superconductors, the highest values of T_c were observed in compounds crystallizing in the A15 type structure (figure B2.5.1) and more than 5000 works describing their properties have been published. Actually, 64 A15 type compounds with the general formula $A_{1-\beta}B_{\beta}$, are known, where A stays for a transition metal: Ti, V, Cr, Zr, Nb, Mo, Hf, Ta, W, while B represents either transition elements or elements or elements at the right side of the periodic system. The compounds Nb_3Ge, Nb_3Ga, Nb_3Al, Nb_3Sn, V_3Si and V_3Ga exhibit the highest T_c values, in the range between 23.0 and 15.4 K. Stoichiometric A15 type compounds are characterized by the formula A_3B ($\beta = 0.25$), where A is a transition element (A = Nb, V, Ta, Cr, Mo, W, Ti, Zr), while the B element can be either a transition or a non-transition element. The A15 phase field often extends over a certain range of compositions and β may comprise compositions at both sides of stoichiometry: in the very most cases, β lies within the range $0.18 \le \beta \le 0.30$. For example, the phase field in the system V_3Si ranges from $\beta = 0.20$ to 0.27, while for Nb_3Sn, it extends from $\beta = 0.18$ to 0.255. There are compounds, e.g. the binaries Nb_3Ga and Nb_3Al or the pseudobinary $Nb_3Al_{0.80}Ge_{0.20}$, where the stoichiometric composition is stable at very high temperatures only. In these cases, the stoichiometric composition can only be stabilized by ultrarapid quenching or by non-equilibrium formation processes, e.g., sputtering or chemical vapour deposition. The values for the superconducting transition temperature, T_c, the electronic specific heat, γ and the upper critical field, $B_{c2}(0)$ for a series of A15 type compounds are listed in table B2.5.1, using the data of [1–6].

Effect of composition and ordering on the electronic properties of A15 type compounds

Since their discovery, A15 compounds have encountered a particular interest, in particular because of their high critical magnetic fields, B_{c2}. The most important industrial application is related to superconducting coils producing magnetic fields, reaching up to 21 T at 4.2 K, for nuclear magnetic resonance (NMR) and for laboratory magnets. From the fundamental point of view, another property was at the center of interest, the electronic density of states, N(0), since A15 type compounds exhibit the highest values among all known type II superconductors, as can be seen in table B2.5.1, where the values of γ, the linear term of the electronic specific heat, are listed.

E1.1
E2.3
A2.4

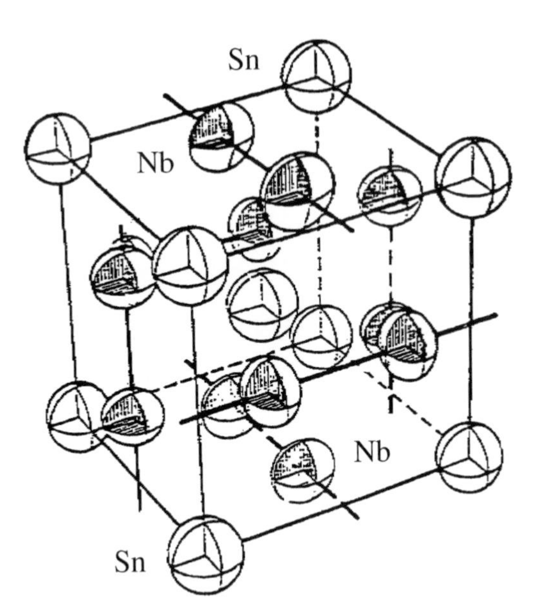

Figure B2.5.1. The A15 type structure. Example: Nb_3Sn.

There is a strong correlation between the long-range Bragg–Williams atomic order parameter, S and the electronic properties in A15 type compounds, e.g. T_c, γ, N(0), B_{c2} and ρ_0. This correlation has been described in a large number of papers: a very complete review on the effects of atomic ordering on the electronic properties of A15 type compounds has been compiled by the author [5, 6]. It follows that the values of the electronic properties of A15 type compounds can exhibit strong variations with the order parameter, S, as well as with composition. The variation of the atomic order parameter and its effect on the electronic properties has been studied either after fast quenching (Nb_3Al [7], V_3Ga [8], V_3Au [9]) or after high energy irradiation (V_3Si [10], Nb_3Sn [11]).

From the data reported in [1–11], it follows that the variation of the electronic properties is very important both at the vicinity of the stoichiometric composition, $\beta = 0.25$, and for order parameters close to perfect ordering, characterized by $S = 1.00$ (see figures B2.5.2 and B2.5.3). The variation of the electrical resistivity, ρ_0, is a very useful tool for the characterization of A15 type single crystals, this property showing the largest variation with atomic ordering. As an example, figure B2.5.1 shows the variation of ρ_0 in Nb_3Sn as a function of composition, using the single crystal data of Hanak *et al* [12]

Table B2.5.1. T_c, $B_{c2}(0)$ and γ for a series of A15 type compounds [1–4]

Compound	T_c [K]	$B_{c2}(0)$ [T]	γ [mJ K^{-2} gram atom]
Nb_3Ge	23.0	38	7.6
$Nb_3Al_{0.7}Ge_{0.3}$	21.0	43	8.7
Nb_3Ga	20.7	38	11.5
Nb_3Al	19.1	34	10.8
Nb_3Sn	18.0	32	2.8
$Nb_3Au_{0.7}Pt_{0.3}$	13.0	30	9.0
V_3Si	17.0	24	13.9
V_3Ga	15.9	22	24.2

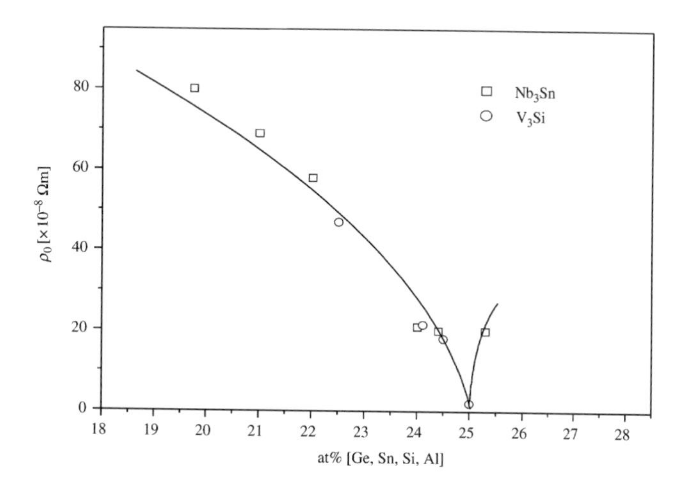

Figure B2.5.2. Variation of ρ_0 in Nb_3Sn and V_3Si with composition χ: Nb_3Sn, after Hanak *et al* [12] and Devantay *et al* [13], O: V_3Si, after Flükiger *et al* [5, 18].

and the polycrystal data of Devantay *et al* [13]. It follows that the ρ_0 between 24.4 and 25 at.% Sn decreases by a factor of 10 (from 20 to 2 $\mu\Omega$ m), while the factor between the much larger domain between 20 and 24.5 at.% Sn is only 4. It will be shown later in this review that this behaviour is directly responsible for the particular variation of the upper critical field, B_{c2}, at compositions close to stoichiometry.

The corresponding variation of ρ_0 with the atomic order parameter S for Nb_3Sn films after irradiation with 20 MeV ^{32}S ions by Nölscher *et al* [11] is shown in figure B2.5.2. It is seen that ρ_0 increases by a factor of 20 for the very small variation in S from 1.00 to 0.98, further decreases in S being much less effective.

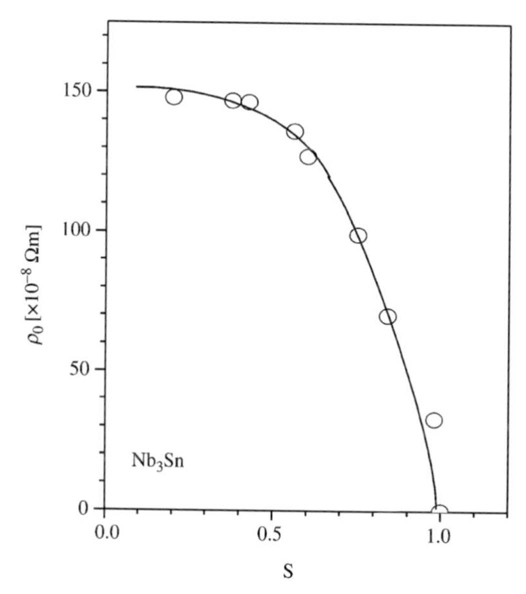

Figure B2.5.3. Variation of ρ_0 in Nb_3Sn as a function of the order parameter S, after irradiation at various fluences with 20 MeV ^{32}S ions (after Nölscher *et al* [11]).

Table B2.5.2. Single crystal formation in A15 type compounds. T_c indicates the highest reported value at the stoichiometric composition [5]

Compound	Phase formation	Composition β in $A_{1-\beta}B_\beta$	T_c (K)	Single crystals
V_3Si	Congruent	0.20–0.27	17.0	yes
V_3Ge	Congruent	0.24	6.0	yes
V_3Ir	Congruent	0.20–0.30	0.1	yes
Nb_3Ir	Congruent	0.22–0.28	1.7	yes
Nb_3Sn	Peritectic	0.18–0.255	18.2	yes
Nb_3Sb	Peritectic	0.22–0.24	0.8	yes
Cr_3Si	Congruent	0.23–0.28	< 0.01	yes
Ti_3Au	Congruent	0.25	1.0	yes

B2.5.2 Single crystal growth techniques for A15 type compounds

The accurate study of fundamental physical properties of superconducting compounds requires to it be performed on single crystals. The availability of single crystals depends in a decisive way on the details of the high temperature phase diagram, in particular about the phase formation. It must be noted that it is not always possible to obtain single crystals in a sufficiently large size for the measurement of physical properties. The only A15 type compounds where data on single crystals have been reported are listed in table B2.5.2.

In the following, the most current techniques for the single crystal growth of A15 type compounds are briefly discussed. These techniques are: zone melting, recrystallization, chemical vapour transport and top seeded solution growth. Table B2.5.2 gives a list of all A15 type compounds which were formed and characterized as single crystals. The six compounds V_3Si, V_3Ge, Nb_3Ir, V_3Ir, Cr_3Si and Ti_3Au form by a congruent reaction, which is most appropriate for crystal growth. Single crystals have also been reported for the two compounds Nb_3Sn and Nb_3Sb, which are formed by a peritectic reaction, as most of the A15 type compounds. Four A15 type compounds (V_3Au, V_3Ga, Nb_3Au, Mo_3Pt) form by a peritectoidic reaction, which renders the formation of single crystals particularly difficult: so far, none of these systems has been grown as a single crystal, but it is thought that chemical vapour transport methods could be successful.

B2.4

In the following, various methods for growing single crystals with the A15 type structure will be presented. Some selected physical properties will also be discussed.

B2.5.2.1 Zone melting of congruently melting compounds: The system V_3Si

Single crystals of all six congruently forming compounds can be grown by zone melting procedures. Since most of the work has been undertaken in V_3Si, which has the highest T_c value among the V based A15 compounds, 17.0 K, this compound will be described in detail in the following. Most authors described the phase V_3Si as forming peritectically, but Jorda and Muller [14] proved that this phase forms congruently. They performed a very detailed metallurgical study and also presented a series of micrographs for a wide range of compositions, showing clearly two eutectics at both sides of the stoichiometric composition, an irrefutable proof for a congruent phase formation. They showed that the reason for the apparent peritectic phase formation was due to the reactivity of Si with various crucible materials at temperatures above 1600°C. They found that the phase field of the phase V_3Si ranges from $\beta = 0.20$ to 0.27 and thus contains the stoichiometric composition, $\beta = 0.25$.

The experimental procedure for V_3Si single crystal growth by zone melting is described by Seeber and Nickl [15]. A similar method was used by Jurisch *et al* [16] and Pulver [17]. Note that all these authors believed, erroneously, that it formed a peritectic phase (see the above remarks), but this does not influence their results and their conclusions. A two stage process was adopted [15]: in the first stage, V metal of a purity of 99.8% and Si (semiconductor quality) were melted to buttons by electron beam melting in a water cooled Cu hearth in a vacuum of 10^{-4} Pa. The weight loss during melting was negligible. To ensure uniformity of the composition, the buttons were turned and remelted several times. To get polycrystalline bars, the buttons were crushed and remelted in a water cooled Cu boat by r.f. melting in a pure Ar atmosphere. These polycrystalline bars had a diameter of 7 mm and a length of 70 mm. The zone travelling rate for the first pass was 9 cm h^{-1}, while the lower part of the rod was rotating at 40 rpm. Several passes were performed, the last one at 3 cm h^{-1}.

Single crystals of approximately 0.5 cm in diameter and 4 cm length were prepared. The analysis was performed by Laue X-ray diffraction and by transmission X-ray topography. A micrographic investigation showed the absence of any secondary phases. The density of dislocations, determined by an etch pit technique, was 4×10^5 pits cm^{-2}. The direction of growth was near to [115]. The measurement of rocking curves yielded a half width of 20 inch, which indicates a very good crystal quality. This particular V_3Si single crystal was slightly non-stoichiometric, at $\beta \approx 0.235$, which is indicated by quantitative composition determination and by the value of the lattice parameter: $a = 0.47265$ nm. The deviation from stoichiometry is due to Si losses during the zone melting process. An additional indication for a deviation from stoichiometry in the single crystal produced in [15] is the absence of martensitic lattice transformation down to 10 K. Indeed, the low temperature cubic–tetragonal transformation occurs only in the composition range between $\beta = 0.245$ and 0.25 in the systems V_3Si and Nb_3Sn.

B2.5.2.2 Recrystallization

V_3Si

It is well known that single crystals up to a certain size can be grown by recrystallization, provided that the annealing temperature is sufficiently close to the melting point. This holds for peritectically forming phases, but is particularly efficient for congruently melting phases. The case of the congruently forming V_3Si, where the A15 phase field reaches from 20 to 27 at.% Si [14] is well suited for illustrating the efficiency of recrystallization. Flükiger *et al* [18], prepared a series of polycrystals between 22 and 25.5 at.% Si by arc melting and submitted them to a heat treatment of 10 days at 1800°C in a high vacuum W furnace. The result was striking: for all compositions, single crystals of more than 1 mm size were obtained, the maximum size, more than 1 cm^3, occurring for a composition close to 25 at.% Si. Recrystallization has not been used very often for growing large single A15 phase single crystals, which is mainly due to the fact that not many high vacuum furnaces are available to perform anneals of such long periods at such high temperatures.

Nb_3Sn

Recrystallization has also been successfully used for the formation of large, high quality Nb_3Sn samples with crystallites of $>100\,\mu m$ size. In order to avoid Sn losses by evaporation, Goldacker *et al* [19] and Kapoor and Wright [20] used a powder metallurgical high pressure approach, starting with Sn and Nb powders of 35–45 μm size. Goldacker *et al* [19] performed a heat treatment of 24 h at 1100°C at 100 MPa argon pressure, using a hot isostatic pressure (HIP) device. The containment consisted of a stainless steel cylinder, which was sealed by electron bombardment. The reaction between Nb_3Sn and steel was avoided by thick Ta foils, which covered the whole inside steel surface. Very large samples, with

B3.3.3

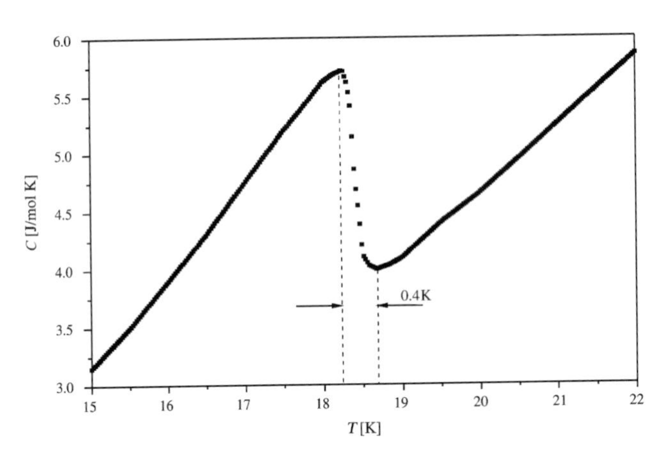

Figure B2.5.4. Specific heat curve showing a very narrow transition for Nb_3Sn polycrystals prepared by high pressure sintering (after Goldacker *et al* [19]).

masses up to 700 g have been prepared by this process and much larger samples could be easily reacted. The quality of the samples was remarkable, the superconducting transition width obtained by specific heat measurements being only 0.4 K (figure B2.5.4), corresponding to a variation of the Sn content of less than ± 0.1 at.% over the macroscopic sample size.

The nominal sample composition was 25.4 at.% Sn and the lattice parameter after the HIP reaction was 0.5292 nm. After 48 h at 1050°C, this value decreased to 0.5292 nm, which is explained as being caused by the release of internal stresses. The value of the lattice parameter is slightly higher than that reported by Devantay *et al* [13] for the stoichiometric composition, thus confirming a value of β above 0.25. The overall density of samples close to the stoichiometric composition was 98.5% and no cracks or voids were observed. In contrast to the already mentioned systems Nb_3Al and Nb_3Ga, the stoichiometric composition in Nb_3Sn is not stable at high temperatures. As found by Charlesworth *et al* Sn [21], the Nb_3Sn phase forms at 2130°C around 22 at.% Sn, and the Sn rich limit of the A15 phase field varies from $\approx$ 23 at.% Sn $>$1600 to 25 at.% at 930°C. At the reaction temperature of 1100°C chosen by Goldacker *et al* [19], it is clear that the participation of a liquid phase close to stoichiometric compositions largely contributes to the densification during the HIP synthesis. This is in contrast to samples with lower Sn contents, where recrystallization must occur at much higher temperatures, as follows from the equilibrium phase diagram [21].

B2.2.1 The Nb_3Sn samples prepared by HIP process [19] had a T_c value of 18.0 K and an extrapolated B_{c2}(4.2 K) value of 23 T, which is comparable to other literature values. The present samples had a residual resistivity value of 22 $\mu\Omega$ cm, which corresponds well to the value reported by Devantay *et al* [13]. Because of the reaction at 1100°C, the grain sizes reached up to 100 μm, with the consequence of a low pinning centre density. As expected, the critical current density values, J_c, are very low, with 4×10^3 A cm^{-2} at 10 T, which are two orders of magnitude smaller than the J_c values in Nb_3Sn wires. It should be recalled here that the average grain size in Nb_3Sn wires is of the order of 0.1 μm.

B2.5.2.3 Crystal growth by chemical vapour deposition

The system Nb_3Sn

B3.3.3 Nb_3Sn is a peritectic compound [21] with a peritectic temperature of 2130°C. At this temperature, the compound forms highly non-stoichiometric, at a composition of $\approx$ 22 at.% Sn. Unlike V_3Si,

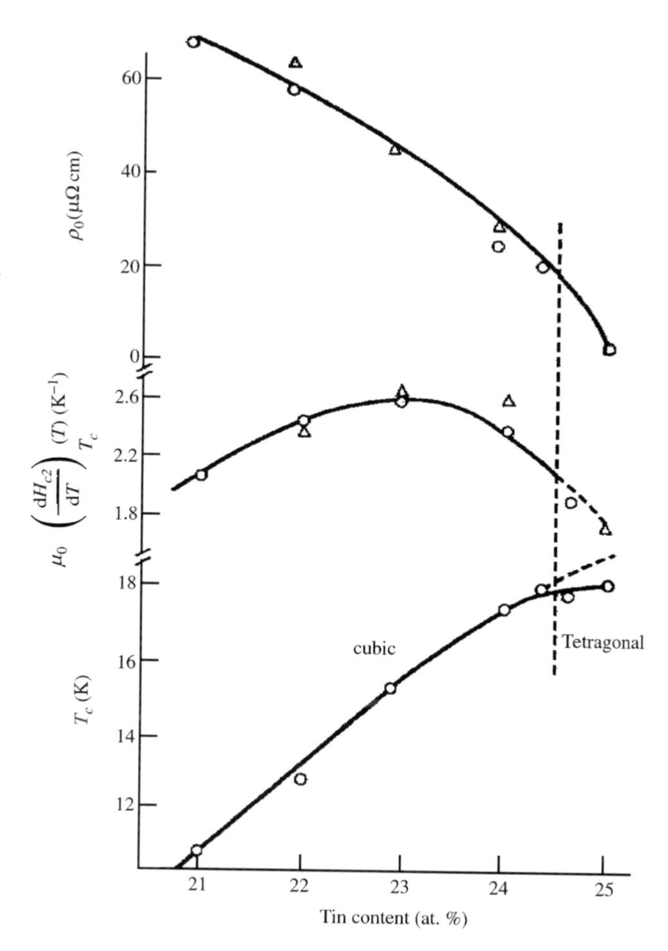

Figure B2.5.5. Variation of T_c, ρ_0 and the initial slope $(dH_{c2}/dT)_{T=T_c}$ *versus* Sn content in Nb_3Sn. At low temperatures, binary Nb_3Sn transforms into the tetragonal phase at $\beta > 0.245$. After Flükiger *et al* [5, 6, 29].

stoichiometric single crystals thus cannot be formed by zone melting. In addition, high temperature recrystallization would yield crystallites with compositions well below $\beta = 0.25$, with T_c values of the order of 10–12 K (see figure B2.5.5), i.e. much lower than the value of 18 K for the stoichiometric composition. Thus, single crystal growth of stoichiometric Nb_3Sn crystals with an appreciable size must be performed by alternative methods. Most used is chemical vapour transport (or chemical vapour deposition, CVD).

B4.2.1
B4.2.2

The first single crystals of Nb_3Sn have been grown by Hanak *et al* [12, 22]. The synthesis of this compound by the gas phase reaction appeared to be feasible because both Nb and Sn can be obtained by the hydrogen reduction of their gaseous chlorides at temperatures well below 1000°C. A simultaneous reduction of a mixture of these chlorides yields crystalline Nb_3Sn without the intermediate formation of the free metals. The overall reaction for the production of Nb_3Sn deposits is

$$3NbCl_4\,(g) + SnCl_2\,(g) + 7\,H_2\,(g) \rightarrow Nb_3Sn\,(s) + 14\,HCl\,(g).$$

The equilibrium constant K_p for this CVD reaction has a value of about 5×10^{-3} at 950°C; it increases with temperature. From this result, it follows that this reaction is reversible, because at equilibrium sufficient amounts of all the reacting gases are present, and it is possible to transport Nb_3Sn from lower temperatures (T_1) to higher temperatures (T_2) by means of HCl gas.

Transport of Nb_3Sn by HCl

B3.3.3 B2.2.1
As reported by Hanak and Berman [22], the starting material at the first transport experiments was sintered Nb_3Sn powder, which was sealed in quartz ampoules under a pressure of about 0.75 bar HCl. Crystals of millimetre size were obtained after 5 days at temperatures $T_1 = 800°C$ and $T_2 = 900°C$. The transported Nb_3Sn was considerably purer than the starting sintered material, but still contained several thousand ppm impurities, originating from the starting material as well as from the quartz ampoule. From the standpoint of microstructure, it was found that the best crystals were grown from stoichiometric Nb_3Sn which contained little or no additional phases. A 5% excess of free Sn in the starting material gave rise to free Sn in the transported sample, causing a decrease in the transport rate and yielding a polycrystalline deposit. When higher temperatures were used, Si contamination occurred, and the compound Nb_3Si having the Cu_3Au structure was formed at $T_2 \approx 1150°C$. Prior to the successful attempts to grow larger single crystals of Nb_3Sn it was established that continued single-crystalline growth is critically limited by the growth rate. Continued single-crystalline growth took place when the average growth rate was approximately 1.8 nm s^{-1} or less for $T_1 \approx 850°C$ and $T_2 \approx 900°C$. At a growth rate of 14.4 nm s^{-1}, which was achieved by higher temperature gradients ($T_1 = 800°C$ and $T_2 = 950$ to $1000°C$), only polycrystalline deposits were obtained.

In order to minimize the sources of impurities, the starting material was Nb_3Sn which was obtained by CVD, while the transport ampoules were made of arc-cast Mo instead of quartz. The reason of Mo as ampoule material is explained by the fact that Mo is not transported by HCl. As shown schematically in figure B2.5.6, the Mo ampoules were positioned vertically in the furnace, in order that the transport was carried out by means of gaseous diffusion alone. A massive Mo block (figure B2.5.6) was used to transfer heat directly to the seed crystal to confine nucleation and growth to the desired region containing the crystal seed. The temperature T_2 was held at 870–930°C and T_1 at 830–870°C. The hydrogen flow was maintained around the sealed ampoule to compensate for Sn losses due to diffusion.

After 3–4 months, crystals with a mass between 1 and 8 g ($\approx 1\ cm^3$) were grown, the rate of transport being between 2 and 6 mg h^{-1}. The single crystals grew on the seed crystals and there was only a minor incidence of additional nucleation. The lattice parameter of the single crystals was 0.5295 nm, i.e. slightly higher than the value for the stoichiometric composition, 0.5290 nm. Within an experimental error of $\pm$ 0.5 at.%, the measured macroscopic density was the same as the calculated one, based on the lattice parameters.

The Nb_3Sn single crystals grown by Hanak *et al* [12, 22] were used for the measurement of a series of physical properties. As mentioned above, the martensitic transformation to the low temperature tetragonal phase was determined by X-rays by Vieland *et al* [23] and Mailfert *et al* [24], who found $T_m = 43$ K. After a heat treatment in vacuum of 100 h at 1000°C, Rehwald *et al* [25] measured the sound velocities at 40 MHz by the pulse-echo method, and measured the temperature variation of the elastic moduli c_{ij} of these single crystals and confirmed the lattice instability at 43 K of the compound Nb_3Sn (see figure B2.5.7).

Transport of Nb_3Sn by HBr

B3.3.3 B4.2.2
It is interesting that transport of Nb_3Sn can also be accomplished by using another transport gas, HBr, under similar experimental conditions [22].

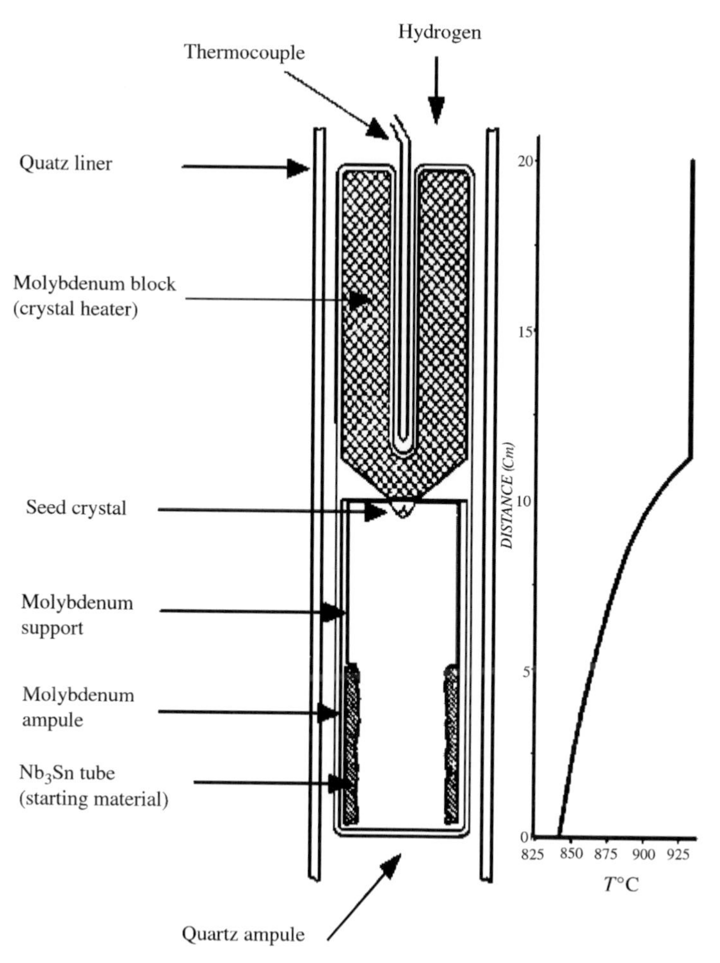

Figure B2.5.6. Experimental arrangement for the growth of Nb_3Sn single crystals by HCl transport (after Hanak *et al* [12, 22]).

B2.5.2.4 *Top seeded solution growth*

Another crystal growth technique for the system Nb_3Sn that starts from Sn–Nb solutions is the technique of 'top seeded solution growth'. This technique was used by Watanabe *et al* [26] on binary Nb_3Sn and on pseudobinary $(Nb_{1-x}Ti_x)_3Sn$ compounds. The work was motivated by the observation of Bussière *et al* [27] and Kumakura *et al* [28], who performed Young's modulus and internal friction measurements and found that the addition of Ti to Nb_3Sn decreases the martensitic transformation temperature T_m. As will be mentioned later in this review, Ti is an important additive to Nb_3Sn since it enhances the upper critical field B_{c2} by 3 to 4 T, leading to the development of superconducting solenoids producing higher magnetic fields.

B2.4

Watanabe *et al* [26] grew three crystals $(Nb_{1-x}Ti_x)_3Sn$ of the composition $x = 0, 0.2$ and 1.5 at.% Ti, of dimensions close to $0.1 \times 0.15 \times 0.25\,mm^3$, with lattice parameters of 0.52905, 0.52909 and 0.52903 nm, respectively. The mosaic spread, defined by the full width at half maximum (FWHM) of the reflection peak of the rocking curve was measured in the range of 0.03–0.06°, thus indicating a high

D1.1.2

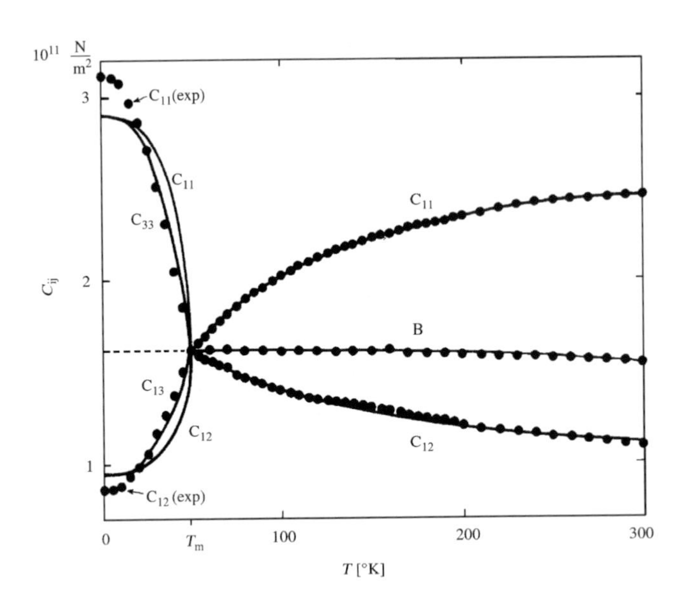

Figure B2.5.7. Elastic moduli c_{11}, c_{12} and B *versus* temperature for Nb_3Sn single crystals. Circles represent measured values, full lines result from theory (after Rehwald [25]).

single crystal quality. From the intensity profiles shown in figure B2.5.8, the martensitic transition temperature T_m for the compositions $x = 0$ and 0.2 was determined to be 36.5 and 33.5 K. For the sample with 1.5 at.% Ti, no transition was observed down to temperatures of 16 K. At the same time, the ratio a/c of the low temperature tetragonal lattice decreased from 1.0047 for the binary Nb_3Sn to 1.0012 for 0.2 at.% Ti. These results confirm the decrease of T_m with Ti content observed by other authors. It has to be noted that a spread is observed in the values of T_m reported by various authors: 45 K [23, 24], 43 K [30], 38.5 K [31], 37 K [32] and 37.5 K [26]. The reasons for this spread are essentially due to slight deviations from stoichiometry, but also to the different measuring techniques.

It was found that the variation of the martensitic low temperature phase transition temperature in single crystalline and polycrystalline Nb_3Sn decreases when substituting Nb (by Ta or Ti) or Sn (by Sb or Ga). As shown in figure B2.5.9, no transition above 10 K is observed for Ti contents above $\approx$ 1 at.% [33] or Ta contents above 1.5 at.% [34].

B2.5.3 Characterization of A15 type single crystal quality by electrical resistivity measurements

D3.1 The compound V_3Si is one of the rare A15 type compounds that exhibits a perfect atomic ordering (characterized by the long-range atomic Bragg–Williams order parameter $S = 1$) and is stable at the stoichiometric composition ($\beta = 0.25$). Besides V_3Si, there is only one A15 type compound with high electronic density of states exhibiting perfect atomic ordering at the stoichiometric composition: Nb_3Sn. As discussed by Toyota *et al* [35] and Flükiger [5, 6, 29], it is no coincidence that these two compounds are the only ones exhibiting a martensitic transformation at low temperature: V_3Si transforms at 21 K [36], while Nb_3Sn transforms at 43 K [24–26]. The quality of single crystals can be characterized quite easily by measuring the electrical resistivity ρ_0, taken just above the transition temperature T_c. In the case of V_3Si, the variation of the electrical resistivity was on a series of single crystals as a function of the Si content [18]. Since the single crystals had a quite irregular shape, these authors [18] reported the value of $RRR = \rho_{300\,K}/\rho_0$ rather than the value of ρ_0. The results are shown in figure B2.5.10 and show a very

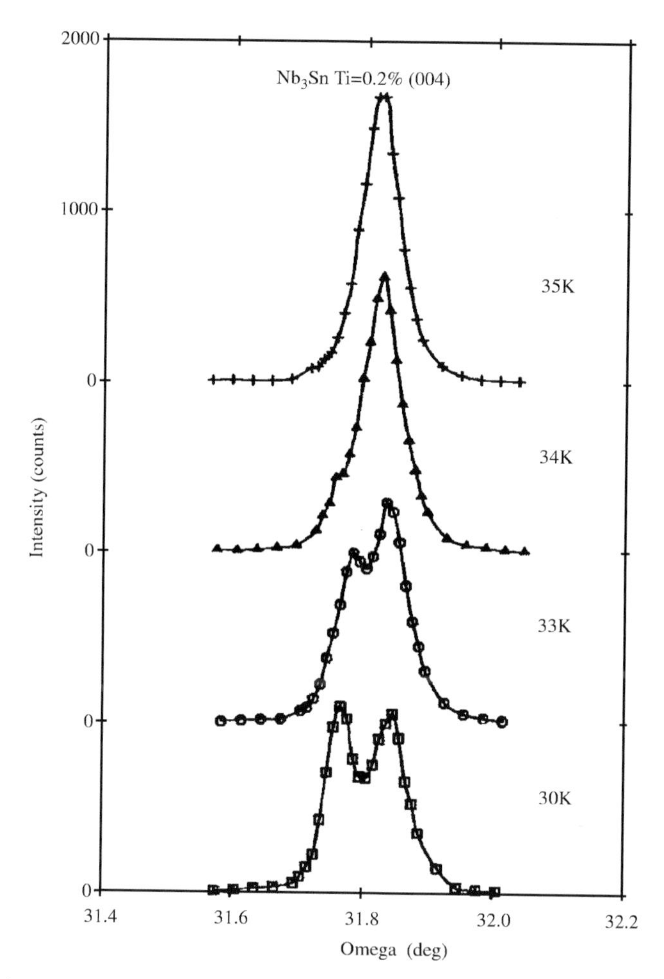

Figure B2.5.8. Intensity profiles of the (004) reflection at various temperatures for the composition $x = 0.2$ in $(Nb_{1-x}Ti_x)_3Sn$ (after Watanabe *et al* [26]).

sharp variation of the RRR ratio in the proximity of the stoichiometric composition. Between 22 and 24.5 at.% Si, RRR increases from 1 to 6, while the addition of only 0.5 at.% Si from 24.5 to 25 at.% Si causes an increase up to RRR above 80! The variation of RRR seems to be symmetrical on both sides of stoichiometry. Figure B2.5.10 shows that a very small deviation from stoichiometry, leading to a partial replacement of Si atoms on the 2a sites by V atoms, has a considerable effect on the value of ρ_0.

The behaviour of the RRR ratio as a function of composition in the V_3Si system showing a sharp discontinuity at stoichiometry as shown in figure B2.5.10 is certainly the most impressive way to demonstrate that this system is perfectly ordered [5, 6]. The change of ρ_0 with composition is strongest when approaching the stoichiometric composition. At the vicinity of 25 at.%Si, very small perturbations, e.g. deviations from stoichiometry, impurities, small amounts of ternary additives cause a strong increase of ρ_0. Once the perturbation is high enough, e.g. at compositions sufficiently far away from stoichiometry, e.g. < 24.5 at.% Si, it is seen that further changes of the state of the system lead to smaller changes in the electrical resistivity ρ_0. Indeed, between 0.22 and 0.245, RRR varies only from 2 to 5.

D3.1

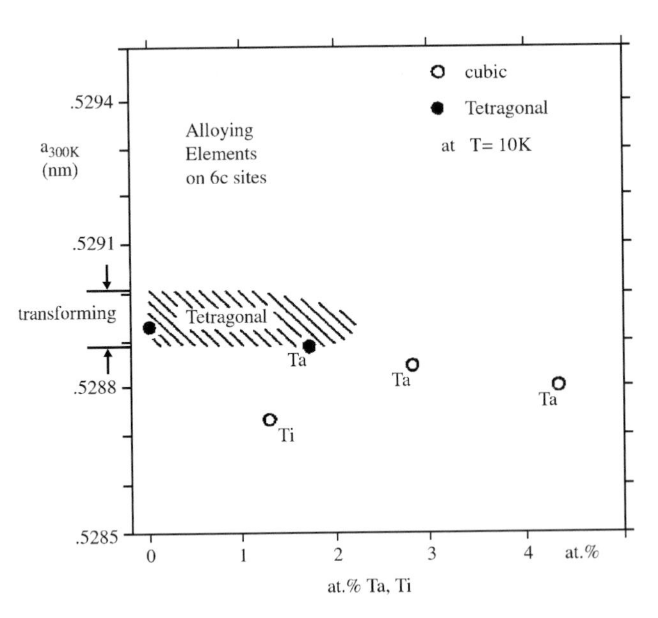

Figure B2.5.9. Stability range of the low temperature tetragonal phase in Nb_3Sn as a function of Ta additions (Tachikawa *et al* [33]) and Ti additions (Tafto *et al* [34]).

With a value of $\rho(300\,\mathrm{K}) = 75 \times 10^{-8}\,\Omega\mathrm{m}$, a resistivity ratio RRR = 83.7 corresponds to $\rho_0 = 0.9 \times 10^{-8}\,\Omega\mathrm{m}$, which is the lowest value ever reported for A15 type compounds [18]. This shows that for congruently melting systems, recrystallization can be a valuable method for growing single crystals. As mentioned above, the combination of extremely careful arc melting (with melting losses below 0.1%) and of excessively long recrystallization times at temperatures close to the solidus (14 days at 1850°C) yields very homogeneous polycrystalline samples, with large crystallites having sizes from $> 1\,\mathrm{mm}^3$ to $> 1\,\mathrm{cm}^3$ [6, 18]. The sharp variation of the RRR ratio close to stoichiometry in figure B2.5.10 suggests that the value of ρ_0 could even be substantially lower than the value of $0.9 \times 10^{-8}\,\Omega\mathrm{m}$ reported in [16]. At this high quality level, it may even be expected that ρ_0 of V_3Si could be further lowered by using Vanadium of a higher purity (the V in [18] had a purity of 99.7%).

B2.2.1

If follows that the lowest reported ρ_0 values for A15 type compounds at the stoichiometric composition is in general seriously affected by composition gradients. Indeed, even small inhomogeneities, of the order of < 0.5 at.% Si reduce the RRR ratio by a factor of 2 and more! As can be seen from figure B2.5.10, the effect of the compositional gradient is considerably reduced if the average concentration is more than 0.5 at.% Si away from stoichiometry.

B2.5.4 Effect of additions on the electronic properties of A15 type compounds

Following Labbé and Friedel [41] the observed low temperature structural instability in V_3Si and Nb_3Sn is correlated to an electronic instability arising from the very high electronic density of states in these compounds. From the present analysis, it follows that an additional necessary condition for the structural instability is the simultaneous availability of stoichiometry and perfect ordering. The compound V_3Ga, for example, has an even higher electronic density of states than V_3Si (see table B2.5.1), but its Bragg–Williams order parameter reaches only the value of $S = 0.98$ [8], thus rendering impossible a low temperature phase transformation.

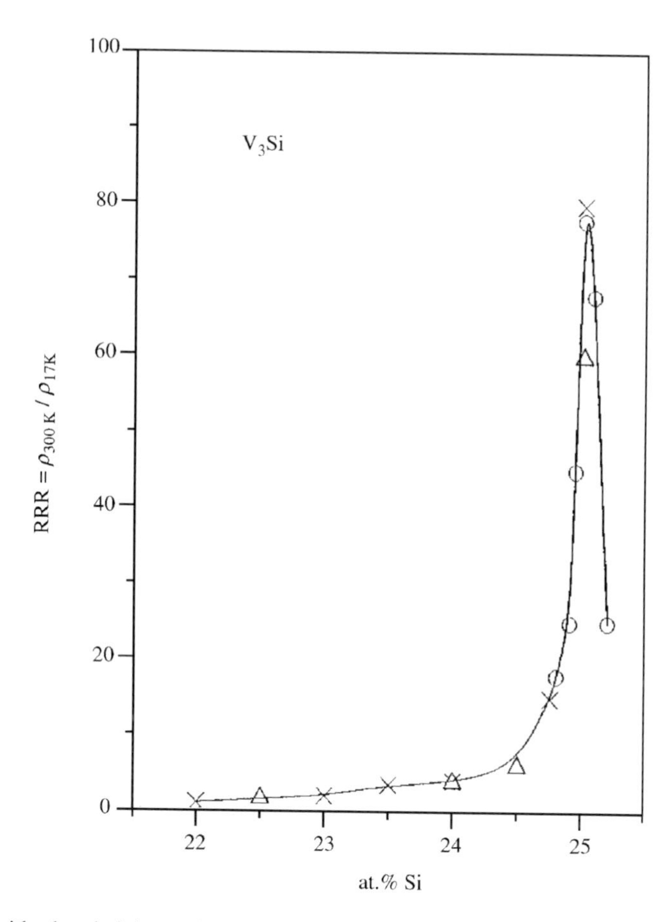

Figure B2.5.10. Residual resistivity ratio RRR *versus* Si content in the system V–Si (after Flükiger *et al* [18]).

The martensitic cubic–tetragonal phase transition leads in both compounds V_3Si and Nb_3Sn to a slight decrease of the superconducting transition temperature. Devantay *et al* [13] have shown that stoichiometric, cubic Nb_3Sn would exhibit a transition temperature T_c above 18.5 K, i.e. 0.5 K above the reported value for tetragonal Nb_3Sn: this follows from the extrapolation carried out in figure B2.5.5, which clearly shows a flattening of the T_c *versus* β relationship above 24.5 at.% Sn. This is confirmed by the fact that both, Ta or Ti additions to Nb_3Sn lead to a slight enhancement of T_c to values around 18.3 K. However, this slight effect on T_c is not sufficient to explain the observed strong decrease of the electrical resistivity and of the upper critical fields when approaching the stoichiometric composition. From all available data, it can be followed that the martensitic transformation is only a consequence of the simultaneous occurrence of high electronic density of states, stoichiometry and perfect ordering, and can thus not be at the origin of the observed strong electronic effects.

D3.1

The real origin of this strong decrease is essentially due to an atomic ordering effect: as mentioned above, Nb_3Sn (and V_3Si) are indeed the only A15 type compounds which are stable at the stoichiometric composition ($\beta = 0.25$) and in addition exhibit perfect ordering ($S = 1$). It is the combined effect of stoichiometry and atomic ordering, shown in figures B2.5.2 and B2.5.3, which is the real cause for the decrease of ρ by more than one order of magnitude between 24.5 and 25 at.% Sn.

As mentioned earlier in this review, the effect of substituting Ta and Ti for Nb in Nb_3Sn is to lower the martensitic transition temperature and thus to stabilize the cubic phase [33, 34]. However, the description would not be complete without mentioning the variation of the electrical resistivity, ρ_0, as a function of the Ti content. A systematical investigation [5, 6] of the variation of ρ_0 with various additives to Nb_3Sn showed that the addition of Ti, Ta, Ga, Ni, and H to Nb_3Sn leads in all cases to the enhancement of ρ_0 [29]. For Ti additions, these authors found a considerable enhancement of the value of ρ_0, from $\approx 1\,\mu\Omega$ cm for binary Nb_3Sn to $40\,\mu\Omega$ cm for ≈ 2 at.% Ti.

This result is important for the industrial application of Nb_3Sn wires at very high fields. Indeed, the enhancement of ρ_0 is directly correlated to the enhancement of the upper critical field, B_{c2}, through the relationship $B_{c2} \sim T_c\,\gamma\,\rho_0$. The variation of the upper critical field as a function of the Sn content in binary Nb_3Sn has been plotted by the author [37] and is shown in figure B2.5.11. It is based on the data of Orlando *et al* [38], Arko *et al* [39] and Foner *et al* [40] and shows a maximum of ρ_0 around 24 at.% Sn, followed by a sharp decrease in the vicinity of the stoichiometric composition. This sharp decrease is caused by the very strong decrease of ρ_0 in the same composition range (see figure B2.5.5), reflecting the dominant effect of atomic ordering on the electrical resistivity. At 25 at.% Sn, the value of $\rho_0 \approx 1\,\mu\Omega$ cm (clean limit) corresponds to upper critical field values around 20 T. The substitution of various elements in the A15 lattice leads to local disorder, which causes a shift towards the dirty limit: the raise of ρ_0 to $40\,\mu\Omega$ cm is correlated to the observed strong increase in B_{c2} from 20 to more than 30 T!

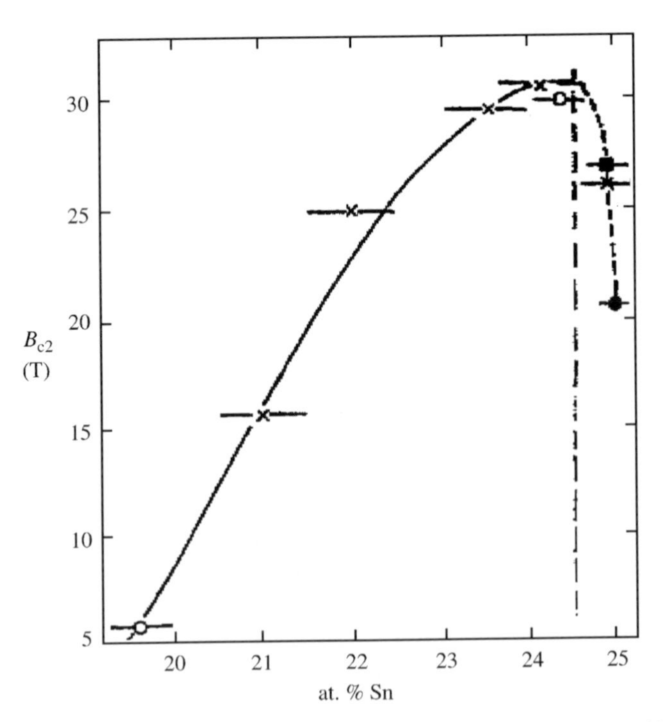

Figure B2.5.11. Variation of B_{c2} *versus* Sn content for Nb_3Sn [37], using the data of [38–40]. The decrease of B_{c2} above 24.5 at.% Sn is not due to the tetragonal transformation, but to the decrease of ρ_0 close to the stoichiometric composition.

B2.5.5 Conclusions

Among the 64 different binary A15 type compounds, eight have so far been grown as single crystals. The various methods for single crystal growth of A15 type compounds have been reviewed and are the following ones, depending on the phase individual A15 phase formation in each compound:

(a) Zone melting, for the congruently melting compounds V_3Si, V_3Ge, Nb_3Ir and Ti_3Ir. B2.4

(b) Recrystallization near the melting point for very long time periods, for V_3Si and Nb_3Pt. In the case of Nb_3Sn, high pressures were necessary during recrystallization in order to avoid excessive evaporation. B2.4

(c) Chemical vapour deposition (Nb_3Sn, Nb_3Sb) and B4.2.2

(d) Top seeded solution growth [Nb_3Sn, $(Nb_{1-x}Ti_x)_3Sn$]. B2.4

The formation of single crystals in the systems Nb_3Sn and V_3Si has been analysed with particular care. These two compounds exhibit (with the exception of V_3Ga, which forms congruently from the solid) the highest electronic density of states among the A15 type compounds, but have another important property: of all the A15 type compounds with T_c values above 10 K, they are the only ones for which two essential conditions are fulfilled: (a) the stoichiometric composition, $\beta = 0.25$, is comprised in the equilibrium phase field, and (b) they crystallize in the perfectly ordered state, characterized by the Bragg–Williams long-range atomic order parameter $S = 1.00$, as follows from the systematical investigation by the author [5], who determined the following Bragg–Williams order parameters on all A15 type compounds.

It can be said that the occurrence of high quality A15 single crystals and polycrystals at various compositions has been the necessary basis for the understanding of the exciting variation of the electronic properties of this class of compounds at compositions close to stoichiometry. The importance of the present analysis is not only restricted to fundamental considerations, but has a direct impact on industrial applications. It has indeed been possible to explain the strong enhancement of B_{c2} in Nb_3Sn wires after various substitutions, and thus to optimize the performances of these wires at magnetic fields exceeding 20 T.

The present analysis can also be applied to the high field behaviour of the recently discovered, perfectly ordered superconductor MgB_2, where an enhancement of B_{c2} from 14 to > 24 T has been observed after irradiation: this strong enhancement is comparable to the observed enhancement in Nb_3Sn shown in figure B2.5.11. C5, B3.3.6

References

[1] Muller J 1980 A15 type superconductors *Rep. Prog. Phys.* **43** 643
[2] Junod A 1982 *J. Less-Common Metals*
[3] Junod A 1982 *Superconductivity in d- and f- Band Metals* ed W Buckel and W Weber (Karlsruhe: Academic Press) p 89
[4] Foner S and McNiff E J 1978 *Appl. Phys. Lett.* **32** 122
Foner S and McNiff E J 1976 *Phys. Lett.*, **58A** 318
[5] Flükiger R 1987 *Atomic Ordering, Phase Stability and Superconductivity in Bulk and Filamentary A15 Type Compounds* **1** May, Karlsruhe, Germany: Kern-Forschungszentrum p 306
[6] Flükiger R 1992 *Encyclopedia of Materials Science & Engineering* ed J Evetts (Oxford: Pergamon Press) pp 1–15
[7] Flükiger R, Jorda J L, Junod A and Fischer P 1981 *Appl. Phys. Commun.* **1** 9
[8] Flükiger R, Staudenmann J L and Fischer P 1976 *J. Less-Common Metals* **50** 253
[9] Junod A, Bellon P, Flükiger R, Heiniger F and Muller J 1972 *Phys. Kond. Materie* **15** 133
[10] Viswanathan R and Caton R 1978 *Phys. Rev.* B **18** 15
[11] Nölscher C and Saemann-Ischenko G 1985 *Phys. Rev.* **32B** 1519
[12] Hanak J J, Strater K and Cullen G W 1964 *RCA Rev.* **25** 342

[13] Devantay H, Jorda J L, Decroux M, Muller J and Flükiger R 1991 *J. Mater. Sci.* **16** 2145
[14] Jorda J L and Muller J 1982 *J. Less-Common Metals* **84** 39
[15] Seeber B and Nickl J 1973 *Phys. Stat. Sol. (a)* **15** 73
[16] Jurisch M, Berthel K H and Ullrich H J 1977 *Phys. Stat. Sol. (a)* **44** 277
[17] Pulver M 1972 *Z Physik* **257** 22
[18] Flükiger R, Küpfer H, Jorda J L and Muller J 1987 *IEEE Trans. Magn.* **23**
[19] Goldacker W, Ahrens R, Nindel M, Obst B and Meingast C 1993 *IEEE Trans. Appl. Supercond.* **3** 1322
[20] Kapoor D and Wright R N 1980 *Metall. Trans.* **11A** 685
[21] Charlesworth J P, MacPhail I and Madsen P E 1980 *J. Mater. Sci.* **5** 580
[22] Hanak J J and Berman H S 1967 *J. Phys. Chem. Solids* **C7** 249
[23] Vieland L J and Wicklund A W 1968 *Phys. Rev.* **166** 424
[24] Mailfert R, Batterman B W and Hanak J J 1967 *Phys. Lett.* **24A** 315
[25] Rehwald W 1968 *Phys. Lett.* **27A** 287
Rehwald W, Rayl M, Cohen R W and Cody G D 1972 *Phys. Rev.* B **6** 363
[26] Watanabe Y, Toyota N, Inoue T, Komatsu H and Iwasaki H 1988 *Japan. J. Appl. Phys.* **27** 2218
[27] Bussière J F, Faucher B, Snead C L and Suenaga M 1982 *Adv. Cryo. Eng. Mat.* **28** 453
[28] Kumakura H, Tachikawa K, Snead CL and Suenaga M, Nippon Kinzoku Gakkaishi
[29] Goldacker W and Flükiger R 1985 *Physica* B **135** 359
[30] Axe J D and Shirane G 1973 *Phys. Rev.* B **8** 1965
[31] Kamigaki K, Sakashita H, Tearuchi H, Maeda H and Toyota N 1986 *Proc. Int. Conf. Martensitic Transformations* Nara (Sendai: The Japan Institute of Metals) p 138
[32] Fuji Y, Hastings J B, Kaplan M and Shirane G 1982 *Phys. Rev.* **25** 364
[33] Tachikawa K, Asano T and Takeuchi T 1981 *Appl. Phys. Lett.* **39** 766
[34] Tafto J, Suenaga M and Welch D O 1984 *J. Appl. Phys.* **55** 4330
[35] Toyota N, Kobayashi T, Kataoka M, Watanabe H F Y, Fukase T, Muto Y and Takei F 1988 *Phys. Soc. Japan* **57** 3089
[36] Batterman B W and Barrett C S 1964 *Phys. Rev. Lett.* **13** 390
[37] Flükiger R, Schauer W and Goldacker W 1982 *Superconductivity in d- and f- Band Metals* eds W Buckel and W Weber (Karlsruhe: Academic Press) p 41
[38] Orlando T P, Alexander J A, Bending S J, Kwo J, Poon S J, Hammond R H, Beasley M R, McNiff E J Jr and Foner S 1981 *IEEE Trans. Magn.* **17** 368
[39] Arko A J, Lowndes D H, Müller F A, Roeland L W, Wolfrat J, Van Kessel A T and Webb G W 1978 *Phys. Rev. Lett.* **40** 1590
[40] Foner S and McNiff E J Jr 1981 *Solid State Commun.* **39** 959
[41] Labbé J and Friedel J 1966 *J. Phys.* **27** 153

B2.6 Irradiation

Harald W Weber

B2.6.1 Introduction

Almost every conceivable kind of radiation has been applied to high temperature superconducting materials, especially in the early months (years) of euphoria following their discovery by Bednorz and Müller in 1986. It soon turned out that their properties degraded in most cases, a fact that was related to the displacement of oxygen atoms and the corresponding reduction of the transition temperature T_c. A few types of radiation, in particular fast neutrons, high energy protons and, later on, very high energy heavy ions were found to improve significantly the irreversible magnetic properties of these superconductors, although some (smaller) changes of T_c still prevailed. This is related to the introduction of flux pinning centres, i.e. of extended defects with sizes comparable to the superconducting coherence A4.3 length, into the superconducting matrix and still represents the most successful way of improving the material performance with respect to the critical current densities J_c, the trapped magnetic fields in large bulk superconductors and the location of the irreversibility lines in the (H,T)-phase diagram as per today. A1.3

Since we are concerned in this chapter with bulk materials, i.e. with superconductors of 'macroscopic' dimensions (with a thickness in the range from a few mm to a few cm), another property of the irradiating particles becomes of paramount importance, namely the penetration range in a particular material. This range is generally very small (of the order of a few 10–100 μm) for charged particles, unless we consider very light ions (e.g. protons) with an energy of several hundred MeV. As a consequence, only neutrons fulfil both of the required criteria, i.e. they are able to produce sufficiently large defects for optimal flux pinning and to penetrate matter over distances of several 10 cm. The interaction of radiation with matter and the defects resulting from this interaction will, therefore, be discussed in some detail in chapters B2.6.2 and B2.6.3. Based on this knowledge, we will proceed in chapter B2.6.4 with the role these defects play in improving the critical current densities, the trapped fields and the irreversibility lines in bulk RE-123 superconductors. Chapter B2.6.5 will comment on A1.3 recent developments in another melt processed material, namely Bi-2223 tapes (see section B3). Chapter B3 B2.6.6 will summarise our conclusions on the subject.

B2.6.2 Interaction of radiation with matter

The interaction of electromagnetic radiation (γ, x-rays) with high temperature superconductors has occasionally been mentioned in the literature. Consider photons with energies above 1.02 MeV. In this case, a non-zero cross section for pair production exists, i.e. the creation of two electrons with an energy of 510 keV each becomes feasible. These electrons would just be able to transfer sufficient energy to a constituent of the superconductor to remove it from its lattice site (threshold energy [1, 2]), i.e. to create a Frenkel pair. Since such individual pairs of a vacancy and an interstitial atom do not play a significant role for flux pinning, we will not consider this kind of radiation any further.

Particles interact with matter through nuclear reactions, initially transferring energy to the 'primary knock-on atom' (PKA) through elastic, at higher energies also through inelastic, collisions. The further development of the radiation-induced defect depends entirely on the energy transferred to this PKA. The evolution of such defect structures formed by the interaction of ions and matter can be readily assessed on the basis of the TRIM code [3]. In the case of uncharged particles such as neutrons, these two interaction channels exhaust the number of possibilities, the inelastic collisions being quite insignificant, if the neutron spectrum of a fission reactor is used as the radiation source, because the number of neutrons with energies above ~ 10 MeV (onset of non-negligible inelastic cross sections) is very small. If we move up in energy and particularly consider heavier particles, an additional interaction mechanism starts to set in, namely an electronic interaction between the charged particles and matter leading to very high ionisation energy losses per unit length of the material, which is characterised by the electronic stopping power S_e. The evolution of this interaction mechanism, which dominates the nuclear interactions at these particle energies by orders of magnitude, with particle energy and mass is shown schematically in figure B2.6.1 [4].

Beginning from the right-hand side, where the mass and the energy are highest, the energy transfer leads to local melting of the material and to the formation of a continuous amorphous zone along the particle trajectory ('columnar track'). At its end (i.e. after travelling about 100 μm through the superconductor), the incident particle has lost all energy. In the case of Y-123, an electronic stopping power exceeding 25 keV nm^{-1} is needed to form such a continuous cylindrical track [5]. A decrease of mass and energy leads to a deterioration of this ideal defect structure, i.e. we start with almost cylindrical defects ($S_e > 15$ keV nm^{-1}, 'string of beads'), move on to aligned elongated areas of amorphous material and aligned collision cascades ($S_e > 10$ keV nm^{-1}) and end up with clusters of point defects at low energies and mass.

In view of the above mentioned short penetration range, a direct production of such defects in superconductors of practical interest is not feasible, not even with, e.g., 5 GeV lead ions. However, extended elongated defects can be created by a trick, i.e. by doping the superconductor, e.g., with uranium prior to processing and by exposing the final product to a beam of thermal neutrons [6, 7]. In this way, fission of the ^{235}U nuclei is induced. The corresponding two fission products are of intermediate mass number and carry energies of approximately 100 MeV each. They are thus able to create columnar defects in the form of the 'strings of beads' shown in figure B2.6.1 with a total length

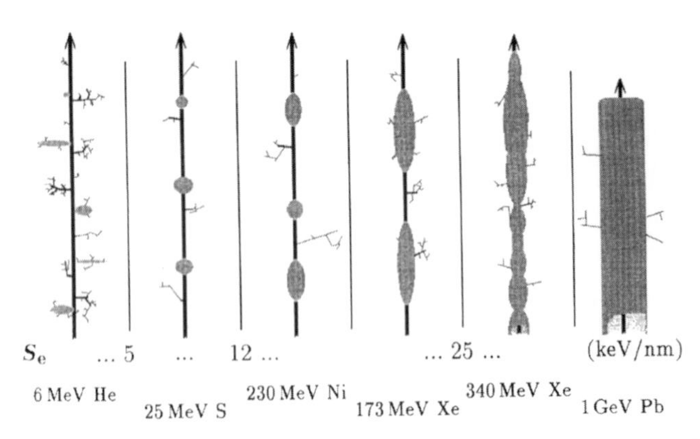

Figure B2.6.1. Defect evolution with the electronic stopping power S_e. Typical ions and ion energies needed for certain stopping powers are listed in the bottom line.

of about 10 μm. The orientation of these fission tracks is totally random and is always originating from the position of the individual uranium atoms. Clearly, limitations resulting from penetration range considerations are completely removed, since thermal neutrons have a very large penetration range, i.e. they can reach the localised uranium atoms anywhere in the superconducting matrix. A similar arrangement of fission tracks can be produced in selected HTS by another radiation technique. Two nuclei, ^{209}Bi and ^{200}Hg, have a reasonable fission cross section for high energy (e.g. 800 MeV) protons. Exposure of these HTS to such a beam, therefore, again produces randomly oriented fission tracks of the same nature as discussed above [8].

B2.6.3 Defects

This section will briefly summarise the experimental evidence (by TEM) that we have on the defects produced by the radiation techniques discussed in the previous sections.

Beginning with the neutrons, it is necessary to consider the energy distribution of neutrons prevailing in fission reactors, since they are used almost exclusively for the purpose of neutron irradiation. Figure B2.6.2 shows that a very broad energy spectrum, spanning almost 10 orders of magnitude, will interact with matter [9].

Fast neutrons, i.e. those with energies above 0.1 MeV, are of primary interest, since they will be able to transfer energies of the order of 1 keV or more to the PKA, which will in turn transfer energy to the neighbouring atoms and thereby produce a so called *collision cascade*. TEM investigations of Y-123 and Bi-2212 [10, 11] showed that the resulting defects are amorphous regions with a diameter of about 2.5 nm, which are surrounded by an inwardly directed strain field of approximately the same size (figure B2.6.3(*a*)). Their flux pinning capability is depicted schematically in figure B2.6.3(*b*) (note the almost ideal match with the flux line core at elevated temperatures). Their spatial distribution is completely random, since the incident particles come from all directions and interact with matter in a completely random way.

The density of collision cascades scales linearly with neutron fluence, a property that also makes such irradiation experiments very useful from a fundamental point of view, since the concentration of pinning-active defects can be varied at will.

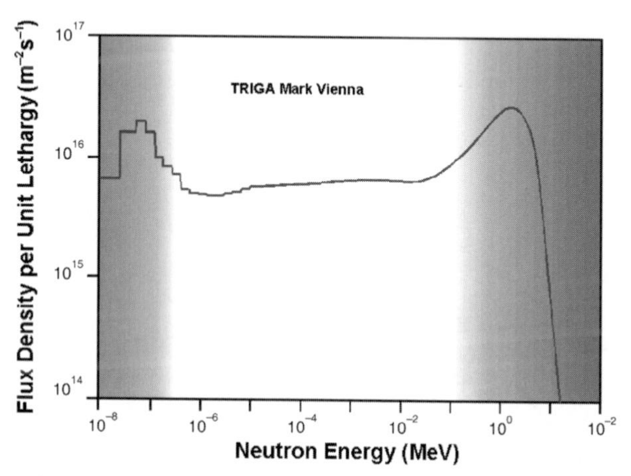

Figure B2.6.2. Distribution of neutron energies in the core of a fission reactor (Triga reactor, Vienna). The shaded area on the left hand side marks the energy range of the thermal neutrons, the shaded area on the right hand side refers to the fast neutrons.

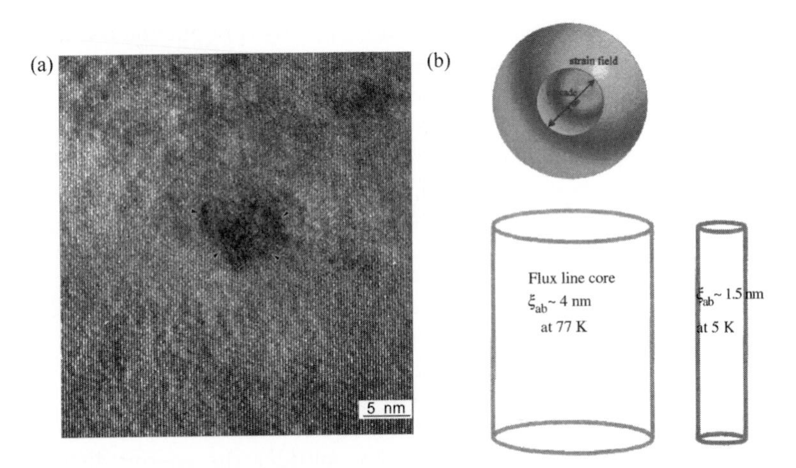

Figure B2.6.3. (*a*) TEM picture of a collision cascade (dark area surrounded by arrows); (*b*) schematic view of the defect including the strain field and comparison with typical flux line dimensions in Y-123 at 77 and 5 K.

Considering figure B2.6.2 again, we will then have a range of neutron energies (between ~1 and 100 keV) that is difficult to assess with regard to the resulting defects. We know that a neutron energy of ~1 keV represents the lower limit for displacing individual atoms from their lattice positions, i.e. all neutrons with energies *below* this limit do not contribute to a change of the superconductor at all (under normal circumstances, see below). *Above* ~1 keV, they are certainly able to displace atoms and to form point defects, with increasing energy predominantly point defect *clusters*, but their size is too small for detection in TEM (i.e. smaller than 1 nm). They will contribute to flux pinning to some extent, but will mainly lead to displacements of certain constituents of the superconductor from their regular lattice site, in particular of the lightest atom, i.e. oxygen. This will affect the superconducting transition temperature to a certain extent [12].

The second type of defects, the fission tracks, are produced via the thermal neutron spectrum (figure B2.6.2, left-hand side) of the reactor because of their high fission cross section for ^{235}U. Two kinds of defect configurations have been observed, which are shown schematically in figure B2.6.4.

In the first case (figure B2.6.4(*a*)) the starting points of all tracks are concentrated in small clusters, since uranium forms a chemical compound (precipitate) in 123 superconductors [13] with the composition

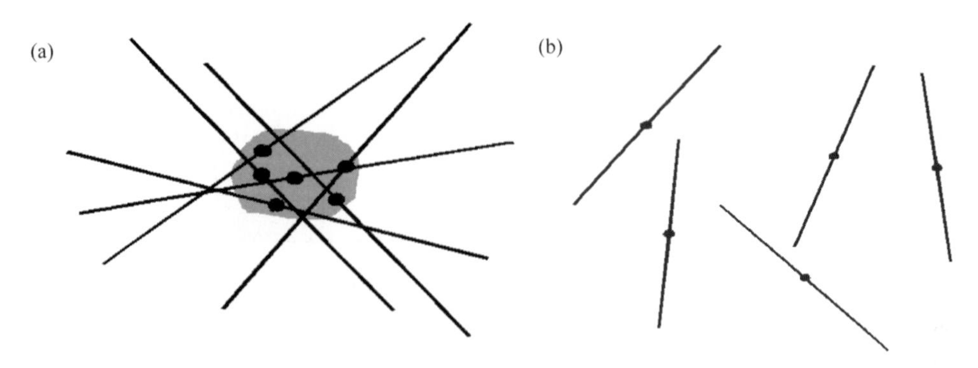

Figure B2.6.4. Schematic view of the defect configurations following fission of ^{235}U in Y-123 (*a*) and in Bi-2223 (*b*).

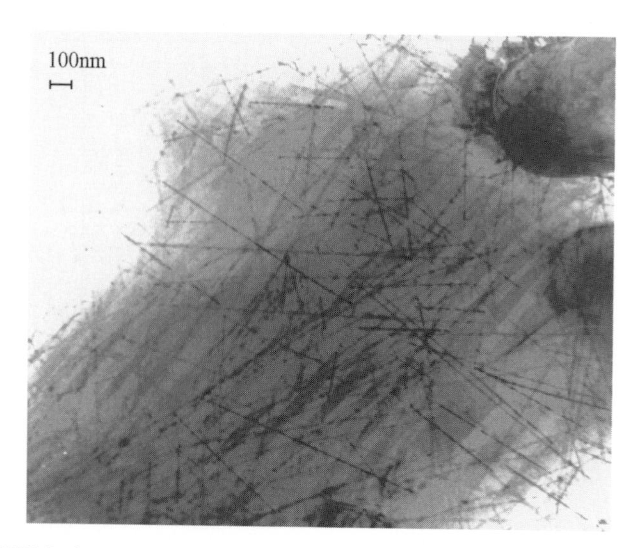

Figure B2.6.5. TEM picture of fission tracks in Y-123 following exposure to thermal neutrons.

$(U_{.6}Pt_{.4})YBa_2O_6$ (fcc, $a = 0.85$ nm). The formation of these clusters with a diameter of approximately 300 nm leads to a very inhomogeneous defect distribution, the consequences of which will be discussed in the following section. A TEM picture of the tracks [13] originating from one of these clusters is shown in figure B2.6.5.

The 'string of beads' character corresponding to an initial ion energy of $\sim$100 MeV can clearly be seen there. The diameter of the tracks is $\sim$10 nm, their total length (two tracks starting from one U atom) is $\sim 7\,\mu$m. The second possible configuration (figure B2.6.4(b)) occurs in Bi-2223 tapes, where no precipitate formation could be observed so far, at least at low uranium concentrations (below 0.6 wt%) [14, 15]. Consequently, the final defect configuration consists of a completely random array of fission tracks with roughly the same dimensions as mentioned above for Y-123. Proton-induced fission of the Bi nuclei results in exactly the same defect configuration.

B2.6.4 Results

B2.6.4.1 Critical current densities and irreversibility lines

It is not completely straightforward, of course, to measure the critical current densities in bulk melt textured HTS. Among the various options (see e.g. [16]), we will concentrate in the following on data assessed on the basis of the 'flux profile' technique [17], a magnetic method, where a small ac ripple field (of varying amplitude at fixed low frequency) is superimposed onto a dc field and the amplitude dependence of the 'pick-up' voltage analysed in terms of the Bean model [18]. Two sets of recent experiments will be summarised. D2.5

In the first, various RE-123 bulk samples are analysed. The most remarkable development during recent years is based on the substitution of Y by various other rare earth elements and mixtures of them, e.g. [19]. Data on Y-, Nd-, and SmGd-123 [20] are presented in figure B2.6.6(a) (for $H \| c$, i.e. for the current flowing in the basal plane). Due to the pronounced 'fishtail' effect [21] in the substituted materials, the critical current densities remain almost constant up to very high fields (of the order of 8 T) and the irreversibility lines reach unprecedented high values (figure B2.6.6(b)). B2.3.4

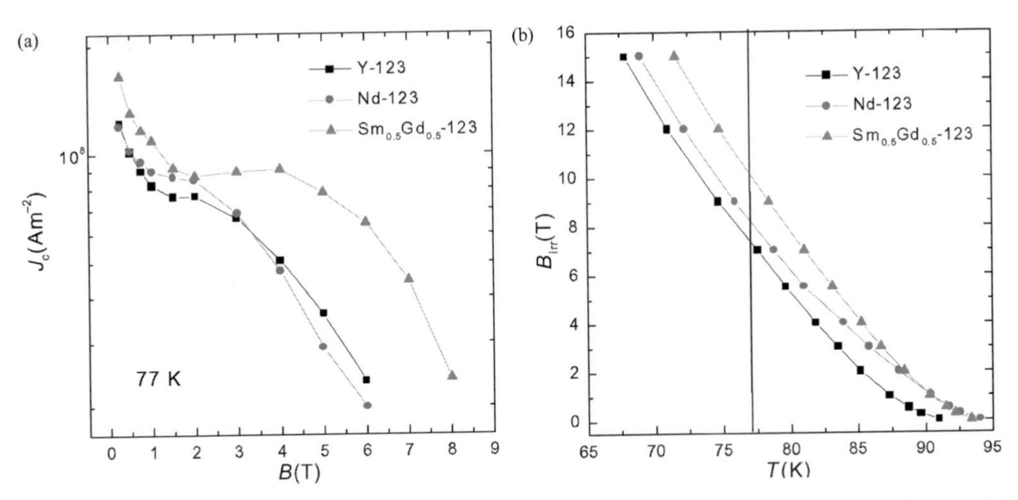

Figure B2.6.6. Comparison of critical current densities at 77 K (*a*) and of irreversibility lines (*b*) in several RE-123 superconductors.

Fast neutron irradiation, i.e. the addition of spherical collision cascades, leads to enhancements of J_c by typically one order of magnitude at 77 K and to systematically smaller enhancements at lower temperatures (figure B2.6.7).

This is consistent with the expected flux pinning behaviour (figure B2.6.3(*b*)) based on core pinning considerations. Systematic investigations of the fluence dependence of J_c (figure B2.6.8) consistently show a flat maximum at a fast neutron fluence of $4 \times 10^{21}\,m^{-2}$ (E > 0.1 MeV), i.e. at a defect concentration of $2 \times 10^{22}\,m^{-3}$, followed by a slow decrease of J_c, which is attributed to the accompanying decrease of the transition temperature due to oxygen displacements by the lower energy neutrons.

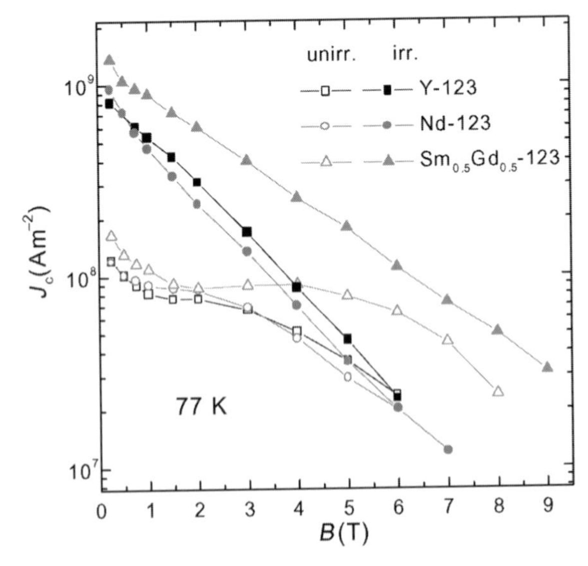

Figure B2.6.7. Enhancement of J_c in several RE-123 superconductors by fast neutron irradiation.

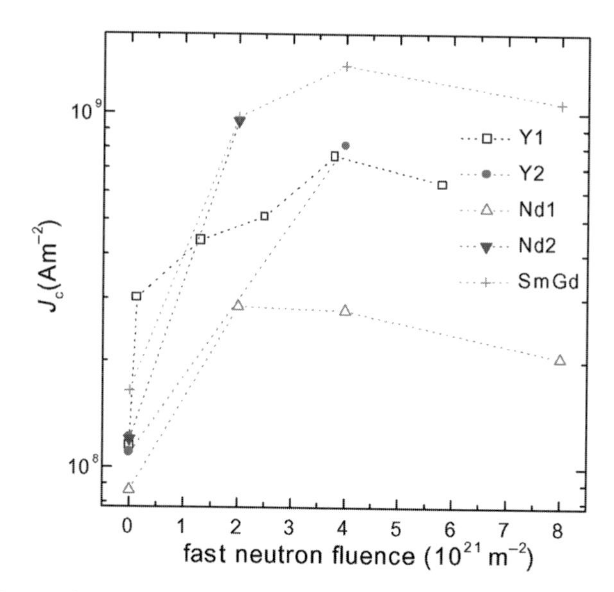

Figure B2.6.8. Dependence of J_c (77 K, 0.25 T) on fast neutron fluence in several RE-123 superconductors.

The irreversibility lines remain almost unchanged, thus indicating that the nature of flux pinning does not change, but the number of strongly pinning defects is optimised. A1.3

Very similar results were obtained on uranium doped Y-123 bulk materials. This may, at first glance, seem surprising in view of the completely different nature of the radiation-induced defects, but can be understood in terms of a simple model [22] developed for the strongly inhomogeneous defect configuration discussed in section B2.6.3. Consider the volume displayed in figure B2.6.5 ($2 \times 1.5 \times 0.01\ \mu\mathrm{m}^3$). The projection of the associated three-dimensional track configuration onto the xy-plane is shown in figure B2.6.9 for two cases, (*a*) a cluster being located in the left-hand corner of the upper panel, and (*b*) no such cluster occurs in this particular plane.

The defect distributions are very inhomogeneous in both cases. Suppose, furthermore, that a flux line is oriented parallel to the y-direction. You will note immediately that the flux pinning along the entire length of such a track does not occur practically and that the flux lines will, under most circumstances, just cross the tracks under a certain angle, thus reducing the pinning effective track length considerably. This can be evaluated further using simple arguments, the final result of figure B2.6.9(*c*) indicates that nearly 90% of the tracks have a 'pinning effective' defect length of between 10 and 20 nm. Hence, the radiation-induced defect structures are quite similar under both conditions: the fast neutron induced collision cascades have a size of about 6 nm and are homogeneously and randomly distributed, the thermal neutron induced fission tracks have effective sizes of 10–20 nm and are very inhomogeneously, but still randomly, distributed. Again, the irreversibility lines remain almost unchanged, slight shifts (from ~6.6 to ~7.6 T) were observed in Y-123 at 77 K at a thermal neutron fluence of $8.8 \times 10^{20}\ \mathrm{m}^{-2}$ [23]. Uranium doping of other RE-123 superconductors has not been reported so far. A4.3 A1.3

B2.6.4.2 Trapped fields

Based on the results on radiation-induced J_c enhancements, it is natural to expect that the trapped magnetization ('trapped field distribution', 'maximum trapped field') should increase by the same factor,

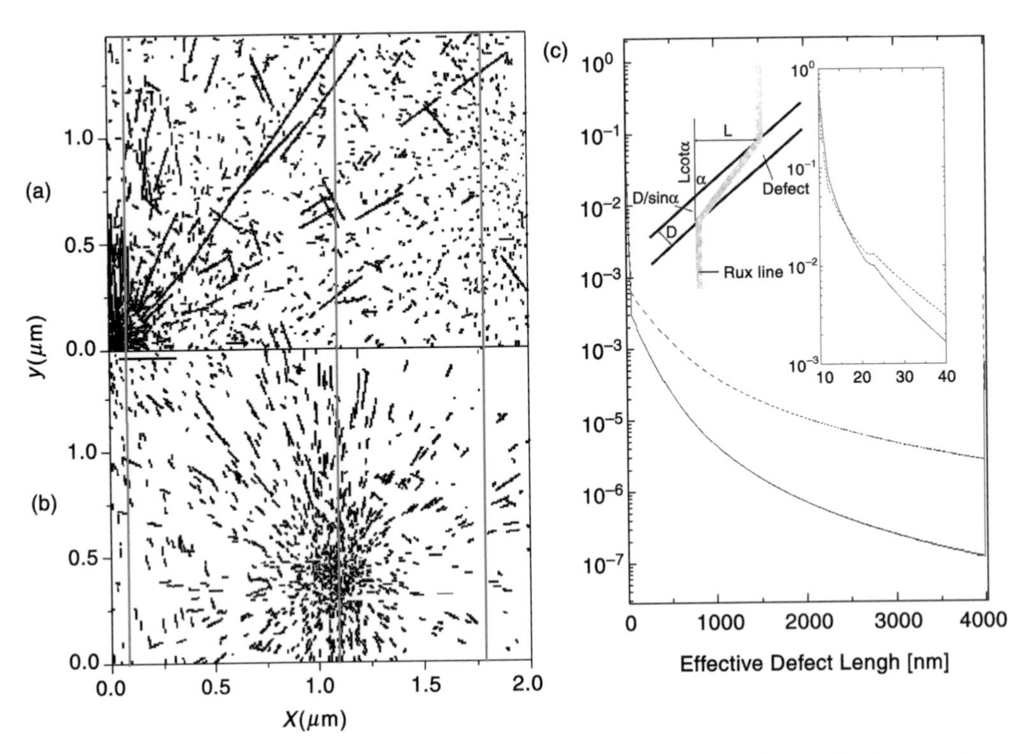

Figure B2.6.9. Projection of the three-dimensional fission track distribution of figure B2.6.5. onto the *xy*-plane. (*a*) An uranium containing precipitate is situated in the left side corner; (*b*) no deposit is found in this material section; (*c*) effective length of the tracks for flux pinning.

since both are related by the sample dimension only (better: the dimension of unimpeded supercurrent flow) according to the Bean model or its extensions. However, some care must be taken here. All data on J_c are assessed from small samples (typically $3 \times 3 \times 3\,\mathrm{mm}^3$), while results on the trapped fields are of primary interest for large pellets (say $30 \times 30 \times 10\,\mathrm{mm}^3$) and usually obtained by Hall probe scanning, i.e. by measuring the magnetic field distribution in the remnant state following a field cycle into the fully penetrated critical state ('activation'). Consequently, any microstructural differences between 'small' and 'large' samples would immediately manifest themselves in such comparisons.

D1.5

Many irradiation experiments on large pellets confirmed, in principle, that very significant enhancements of the trapped fields can indeed be achieved. A typical example is shown in figure B2.6.10, where the as-grown and the irradiated state (fast neutrons, $2 \times 10^{21}\,\mathrm{m}^{-2}$) are compared in the left and centre panels. Trapped fields well above 2 T (at 77 K), in some cases even above 3 T, were recorded. However, upon simply repeating the activation and the measuring process, clear degradations of the maximum trapped fields are usually observed (figure B2.6.10, right panel). Although the field distributions do not show any unexpected kinks or dips, which could be related to the formation of macroscopic cracks (this is also observed occasionally), we believe that 'micro-cracking' occurs under the tremendously enhanced Lorentz forces, thus reducing the supercurrent flow systematically.

A1.3

This view is supported by the fact that particles of various sizes always splinter off the pellets during activation. Microscopic investigations of the samples and attempts to reinforce the pellets are currently under way to get a better view of these essential problems which are impeding a further optimization of the materials for applications.

A1.3

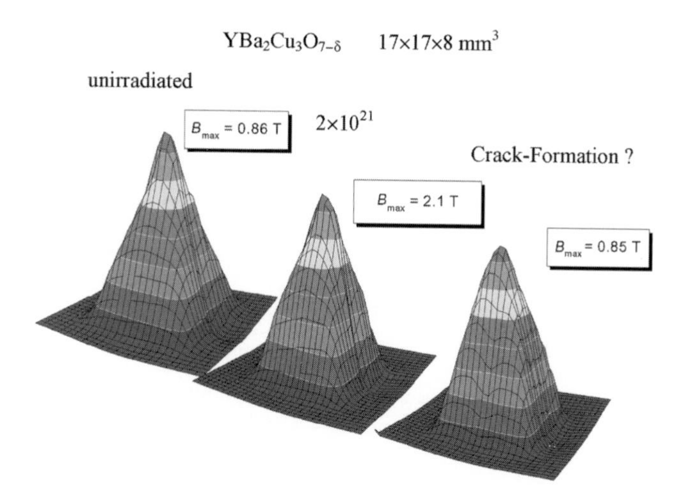

Figure B2.6.10. Hall probe scanning results of the trapped field distribution in a large Y-123 pellet. Left: as-grown, middle: following fast neutron irradiation to a fluence of 2×10^{21} m^{-2}, right: same as in the middle, but after two activations.

B2.6.5 Results on Bi-2223 tapes

An overview of radiation effects would be incomplete if recent results on the property changes of melt-processed Bi-2223 tapes were not at least mentioned. Fast neutron irradiation has been explored in some detail, e.g. [24], and has led to the following results. The critical current densities (now assessed from transport experiments) are enhanced, typically by factors of 2–5, particularly for magnetic fields applied perpendicular to the tape surface ($H \| c$ in the following). At the same time, the low field transport currents, which are determined to a major extent by the 'weak link' structure of the tape, generally decrease at a typical fast neutron fluence of 2–4×10^{21} m^{-2} by 20–40%. This is explained by the damaging of the weak links through the energy transfer by the fast neutrons, as observed previously in transport experiments on ceramics.

B3.2.2

Dramatic improvements in this system were found by introducing fission tracks (either through high energy proton induced direct fission of the Bi nuclei [8] or by the addition of uranium and the subsequent exposure to a beam of thermal neutrons [14, 15, 25]). Results on a tape containing 0.15 wt% of 'pure' ^{235}U are shown in figure B2.6.11.

The irradiation was made in a highly 'thermalised' neutron spectrum (thermal neutron flux density: 2.7×10^{15} m^{-2} s^{-1}, fast neutrons: 1.4×10^{14} m^{-2}s^{-1}) to a thermal fluence of 4×10^{19} m^{-2}. The density of tracks amounts to 6×10^{19} m^{-3} at this fluence (we count the two tracks originating from one U nucleus as *one* pinning centre), the contribution of fast neutron induced collision cascades is totally negligible under these conditions. The critical current densities (figure B2.6.11(*a*)) show drastic enhancements for both major field orientations, but in particular for $H \| c$. At the same time, the low field properties of the tapes remain completely unchanged, i.e. the above mentioned damaging of the weak/strong link structure does not occur due to the absence of the fast neutron component. This radical change of the in-field behaviour is emphasized most clearly by studying the angular dependence of J_c at various fields (figure B2.6.11(*b*)). The usual flux pinning anisotropy, i.e. the ratio of critical current densities for $H \| a, b$ versus $H \| c$, is reduced from ~ 37 in the unirradiated state to ~ 2 after irradiation. Furthermore, the irreversibility fields at 77 K (figure B2.6.12) are more than doubled for both major field orientations. The latter two observations clearly represent a major breakthrough considering the application potential of this superconductor.

C2

A1.3

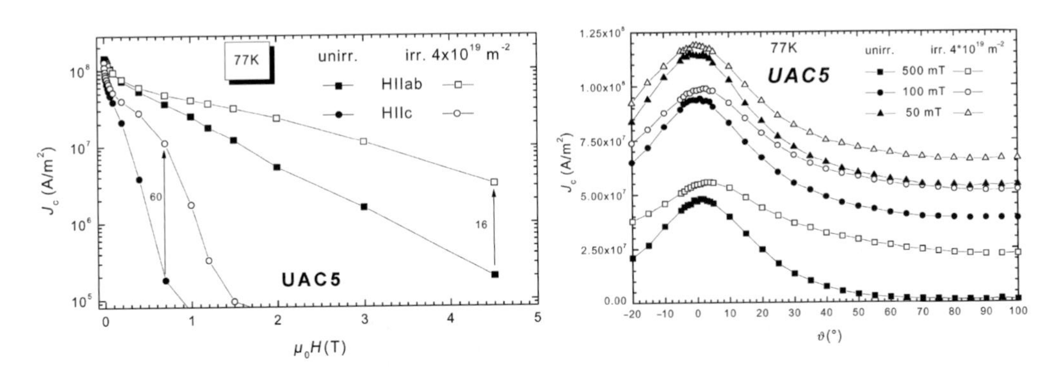

Figure B2.6.11. Field (a) and angular (b) dependence of J_c in an uranium doped Bi-2223 monofilamentary tape at 77 K following thermal neutron irradiation to a fluence of $4 \times 10^{19}\ \mathrm{m}^{-2}$.

A rather straightforward explanation of these fundamental property changes is based on the pancake nature [26] of the magnetic microstructure in this two-dimensional material and depicted schematically in figure B2.6.13. Consider the defect introduced by fast neutrons (spherical collision cascade) first. Due to the limited size of this defect and the missing (strong) correlation among the pancakes, the cascade will be able to pin effectively only two pancakes, see e.g. [27]. An array of randomly oriented extended defects, such as the fission tracks in Bi-2223, will, however, be able to pin a large number of pancakes for any orientation. The optimum will occur for $H \| c$, but for all other orientations we expect scaling with the orientation angle, in agreement with recent results for various splay angles introduced into Bi-2212 single crystals by heavy ion irradiation [28].

A4.3

B2.4

Of course, there is also another 'side of the coin'. All kinds of radiation techniques generally lead to induced radioactivity in the sample. In the case of Ag-sheathed PIT tapes, ^{110}Ag is significantly activated by thermal neutrons. However, by optimising the content of ^{235}U in the tapes, the exposure time to the neutron beam can be significantly reduced (while keeping the number of fission tracks the same), thus alleviating the problem considerably. The overall radioactivity is much smaller than in the case of proton

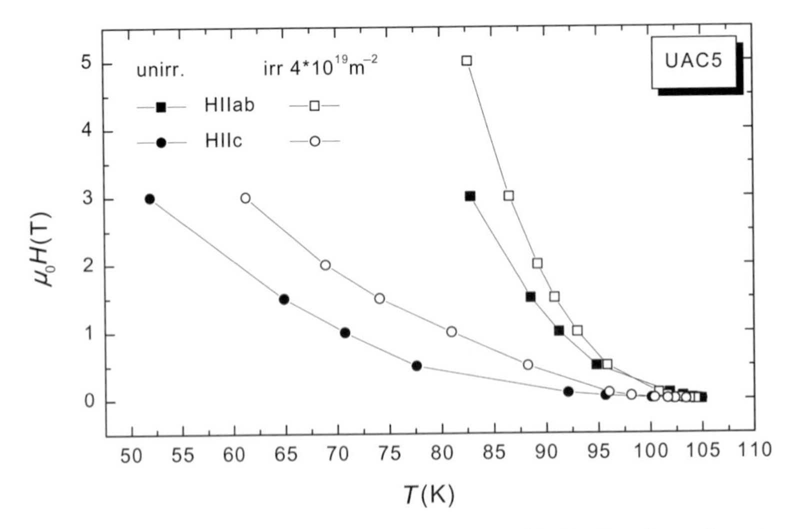

Figure B2.6.12. Irreversibility lines of the same tape prior to and following thermal neutron irradiation.

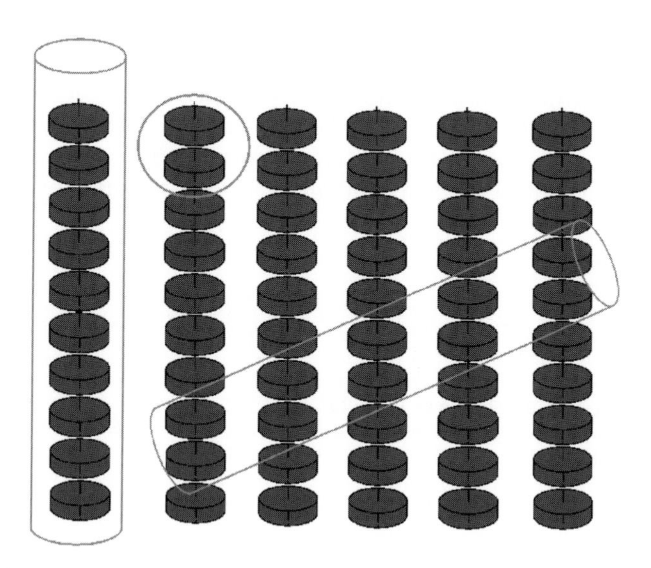

Figure B2.6.13. Schematic view of various defect configurations and the pancake structure in two-dimensional superconductors.

induced fission of the Bi nuclei in the tapes. This and the much easier availability of neutron beams (even over large volumes) certainly make the addition of uranium a highly attractive technique.

B2.6.6 Conclusions

The applicability of radiation techniques to bulk HTS materials was reviewed. The most stringent requirement for the selection of a certain kind of radiation is the penetration range, which needs to be of the order of several cm. Only neutrons and high energy light ions, such as protons, meet this criterion. The resulting defects were discussed in section B2.6.3. They can either be spherical amorphous regions with a diameter of ~ 6 nm (fast neutrons) or cylindrical extended tracks of amorphous material (fission) with diameters of 7–10 nm and lengths of the order of 10 μm. Both act as strong pinning centres in the superconductors.

Under optimal irradiation conditions (optimal non-overlapping defect density versus acceptable T_c degradations due to oxygen displacements), the critical current densities and the trapped fields in RE-123 bulk materials are improved by approximately one order of magnitude at 77 K. The location of the irreversibility lines in the (H,T)-phase diagram is hardly affected as the materials show strong pinning characteristics and a more or less three-dimensional behaviour of their magnetic microstructure anyway. As a consequence of the formation of uranium precipitates and the resulting strongly inhomogeneous defect configuration, the flux pinning action of both kinds of defects is comparable. Melt-textured Bi-2223 tapes, however, can be pushed to new limits that are of considerable interest for applications. In this case, the randomly oriented extended defects (individual unclustered fission tracks) optimally interact with the pancake microstructure, thus enhancing J_c by about two orders of magnitude, removing the J_c anisotropy almost completely and doubling the irreversibility fields at 77 K.

A1.3

B3.2.2

It is, therefore, worth exploring radiation effects in any HTS material, irrespective of whether or not irradiated materials will be directly suitable for commercial exploitation, since they certainly provide us with benchmark results on limitations imposed by their intrinsic properties.

Acknowledgments

I am pleased to acknowledge the continuous and most valuable cooperation of many group members, in particular F.M. Sauerzopf, M. Eisterer, S. Tönies, and H. Niedermaier. Close contacts with M. Murakami (ISTEC, Tokyo), R. Weinstein (University of Houston), S.X. Dou (University of Wollongong), and H.C. Freyhardt (University of Göttingen) were essential for many aspects of the work described in the text.

References

[1] Kirk M A, Baker M C, Liu J Z, Lam D J and Weber H W 1988 *MRS Symp. Proc.* **99** 209

[2] Kirsanov V V and Musin N N 1991 *Phys. Lett.* A **153** 493

[3] Biersack J P and Haggmark L G 1980 *Nucl. Instrum. Methods* **174** 257

[4] Kraus W, Leghissa M and Saemann-Ischenko G 1994 *Phys. Bl.* **50** 333

[5] Hardy V, Groult D, Hervieu M, Provost J, Raveau B and Bouffard S 1991 *Nucl. Instrum. Methods* B **54** 472

[6] Fleischer R L, Hart J R Jr, Lay K W and Luborsky F E 1989 *Phys. Rev.* B **40** 2163

[7] Weinstein R, Ren Y, Liu J, Chen I G, Sawh R, Foster C and Obot V 1993 *Proc. Int. Symp. Supercond.* (Tokyo: Springer) p 855

[8] Krusin-Elbaum L, Thompson J R, Wheeler R, Marwick A D, Li C, Patel S, Shaw T D, Lisowski P and Ullmann J 1994 *Appl. Phys. Lett.* **64** 3331

[9] Weber H W, Böck H, Unfried E and Greenwood L R 1986 *J. Nucl. Mat.* **137** 236

[10] Frischherz M C, Kirk M A, Farmer J, Greenwood L R and Weber H W 1994 *Physica* C **232** 309

[11] Aleksa M, Pongratz P, Eibl O, Sauerzopf F M, Weber H W, Li T W and Kes P H 1998 *Physica* C **297** 171

[12] Sauerzopf F M 1998 *Phys. Rev.* B **57** 10959

[13] Weinstein R, Sawh R, Ren Y, Eisterer M and Weber H W 1998 *Supercond. Sci. Technol.* **11** 959

[14] Schulz G W, Klein C, Weber H W, Moss S, Zeng R, Dou S X, Sawh R, Ren Y and Weinstein R 1998 *Appl. Phys. Lett.* **73** 3935

[15] Dou S X, Guo Y C, Marinaro D, Boldeman J W, Horvat J, Weinstein R, Sawh R, Ren Y 2000 *Adv. Cryog. Eng.* **46** 761

[16] Weber H W 2001 *Handbook on the Physics and Chemistry of Rare Earths* eds K A Gschneidner Jr, L Eyring and M B Maple (Amsterdam: Elsevier) **31** ch 196 pp 187–250

[17] Campbell A M 1969 *J. Phys.* C **2** 1492

[18] Bean C P 1962 *Phys. Rev. Lett.* **8** 250

[19] Muralidhar M, Chauhan H S, Saitoh T, Kamada K, Segawa K and Murakami M 1997 *Supercond. Sci. Technol.* **10** 663

[20] Eisterer M, Weber H W, Schätzle P, Krabbes G, Chikumoto N, Muralidhar M, Murakami M, Weinstein R, Sawh R and Ren Y 2000 *Adv. Cryog. Eng.* **46** 655

[21] Däumling M, Seuntjens J M and Larbalestier D C 1990 *Nature* **346** 332

[22] Eisterer M, Tönies S, Novak W, Weber H W, Weinstein R and Sawh R 1998 *Supercond. Sci. Technol.* **11** 1101

[23] Eisterer M 1999 *PhD Thesis* TU Wien.

[24] Hu Q Y, Weber H W, Sauerzopf F M, Schulz G W, Schalk R M, Neumüller H W and Dou S X 1994 *Appl. Phys. Lett.* **65** 3008

[25] Tönies S, Klein C, Weber H W, Zeimetz B, Guo Y C, Dou S X, Sawh R, Ren Y and Weinstein R 2000 *Adv. Cryog. Eng.* **46** 755

[26] Clem J R 1991 *Phys. Rev.* B **43** 7837

[27] Brandstätter G, Sauerzopf F M and Weber H W 1997 *Phys. Rev.* B **55** 11693

[28] Drost R J, van der Beek C J, Zandbergen H W, Konczykowski M, Menovsky A A and Kes P H 1999 *Phys. Rev.* B **59** 13612

Further Reading

Thompson M W 1969 *Defects and Radiation Damage in Metals* (Cambridge: Cambridge University Press)-text book

Trushin Yu V 1996 *Theory of Radiation Processes in Metal Solid Solutions* (Commack: Nova)-text book

Kirk M A and Weber H W 1992 Electron microscopy investigations of irradiation defect structures in the high T_c superconductor $YBa_2Cu_3O_{7-x}$ *Studies of High Temperature Superconductors*, vol 10 ed A V Narlikar (New York: Nova)

B3
Introduction to section B3: Processing of wires and tapes

René Flükiger

Superconducting wires or tapes must fulfill strong technical and economical requirements before being considered reliable and competitive for industrial applications. The technical criteria are: sufficiently high critical current density at the operation conditions, availability in kilometer lengths, thermal stability, low energy losses in the alternative current regime and mechanical strength with respect to uniaxial stresses. Wires or tapes satisfying these conditions are produced, in two different kinds of configurations: filamentary and layered. The choice of the configuration depends on the anisotropy of both crystal structure and electronic properties of a given compound, which influence the conditions at the grain boundaries and thus the transport critical current density, J_c.

The low T_c superconductors NbTi, Nb_3Sn, Nb_3Al and $PbMo_6S_8$ can be fabricated in a simple multifilamentary wire configuration: their J_c value can be optimized without any texturing of the grains. On the other hand, *c* axis textured grains are required in the filaments of the high T_c compounds Bi(2212) and Bi,Pb(2223) and of the newly discovered MgB_2, thus leading to a tape configuration with flat filaments. For compounds like Y(123), Tl(1223) and Hg(1223), however, *c* axis texturing is not sufficient anymore and a biaxial texturing is required, thus leading to a coated conductor configuration. In this section, the discussion is restricted to coated conductors with *thick* superconducting layers ($>1\ \mu m$), produced by *non-vacuum* processing routes, coated conductors with thin layers ($<1\ \mu m$) are described in this Handbook.

A great variety of production processes have been developed, each superconducting material requiring a particular, individual processing route. In some cases, several routes are known for producing wires and tapes of the same material. The final choice of a specific route will depend not only on the performances but also on the production costs. The main production routes for superconducting wires and tapes are described, the main interest residing in the scientific and technical aspect of the processing routes for each material, while little attention has been given to the economical aspect. The present section describes all known processing routes for wires and tapes based on both, high T_c and low T_c compounds. All aspects of fabrication and microstructure characterization at a submicron scale are treated, as well as those of the measurement of various superconducting properties, in particular J_c.

The fabrication of the tapes based on the Bi based compounds Bi(2212) and Bi,Pb(2223) by the so-called powder-in-tube process is described in this section. The aspects of powder characterization and deformation are discussed, with a particular emphasis on the final thermomechanical treatment of the tapes, which leads to optimized values of J_c. Bi based tapes are actually produced in industrial lengths, for cables and solenoids.

The three systems Y(123), Tl(1223) and Hg(1223) are so far available at a laboratory scale only. The interest in these compounds resides in the techniques used for the preparation of thick film coatings, e.g. liquid phase epitaxy, spray pyrolysis and electrodeposition.

NbTi is the most widely used superconductor for magnets producing fields up to 9 T. NbTi wire fabrication involves a thermomechanical process for the control of the microstructure at a nanometric scale. Nb_3Sn wires are used for producing magnetic fields up to 21 T, the main processing routes being the internal Sn diffusion and the bronze route. Both these materials are produced at an industrial scale, in lengths up to 10 km. The last two compounds, Nb_3Al and $PbMo_6S_8$ wires have particularly high upper critical magnetic field values. The production of thermally stabilized Nb_3Al wires in long lengths by a continuous quenching process as well as the J_c measurements up to 27 T is described.

B3.1
High T_c conductor processing techniques

S X Dou

B3.1.1 Introduction

The discovery of ceramic-type materials, which exhibit superconductivity at liquid nitrogen temperature (77 K), has led to extensive research into materials formulation, characterization and methods of fabrication. The field has developed to the stage where a range of high-T_c superconductors (HTS) materials have been identified and characterized and many of the problems of fabrication into long lengths or thin films have been overcome. More than 30 companies around the world have set out to commercialize HTS and related products during the last 10 years. The emerging HTS industry is well on its way to fulfilling its initial expectations. The technical performance of long-length, state-of-the-art, powder-in-tube Bi-2223/Ag wires produced by a number of manufacturing companies has reached the level required for the demonstration of possible large-scale applications. For example, several consortia have produced demonstration models of the power transmission cable, fault current limiter, transformer and motors of various types. For several reasons full commercialization of such HTS devices has not yet been achieved: the levels of electrical performance remain just below those required for technical and commercial success, and cost needs to be substantially reduced.

B3.2.2

A1.3

Despite enormous advances in high transition temperature (T_c) superconductors through intensive research efforts, practical applications of these materials have been hampered by two major difficulties: the low critical current density (J_c) and high cost of precious metal matrix materials. These problems are a consequence of the characteristics of ceramics and, in superconductors, exist as a result of their high critical transition temperature itself. Extensive research work has revealed that the known copper-oxide superconductors all have the following features in common: (a) layer structures, which lead to two dimensionality and strongly anisotropic properties; (b) mixed valencies, which are the origin of the charge carriers, which have a very low density compared to conventional superconductors [3]; and (c) extremely short coherence lengths and large penetration depths [1].

C2

A2.3, A1.2

The consequences of these inherent properties are the following two problems. The first is weak links at grain boundaries, which lead to very low intergrain current densities since the boundaries act as easy paths for Josephson vortices. So the critical current density will be determined by the pinning of these vortices. Another problem is that, in HTS, the magnetic flux line creep is unusually rapid. As the 'giant flux creep' proceeds, the critical current density declines. The flux creep in effect controls the size of the critical current density. Especially, the giant flux creep dominates the magnetic behaviour of the Bi–Sr–Ca–Cu–O (BSCCO) even more strongly than $YBa_2Cu_3O_{7-x}$ (YBCO) since the BSCCO systems are more anisotropic than YBCO. BSCCO exhibited a wide reversible regime and low melting temperature of the Abrikosov Lattice [2]. To give rise to an appreciable pinning strength, it is necessary to introduce defects the size of the dimension of a coherence length. Otherwise, these defects may well act

A2.7

C1, C2

A1.1

in decoupling and weak link roles. Thus, it is a great challenge to introduce effective pinning without destroying the superconductivity.

D3.3 B2.2.4 A1.3
In addition, high-T_c materials, as with all ceramics, are very brittle and difficult to shape and handle. For large-scale application, superconducting wires have to be reasonably flexible so that they can be wound into coils for handling, storage and installation. Thus, it is important to overcome the brittleness problem of high-T_c materials before practical applications can be realized. Silver has been successfully employed to fabricate metal/superconductor composites, such as metal-clad wires, tapes and multifilaments, to compensate for the ceramic's brittleness. Furthermore, the metal provides a means of thermal dissipation, thus stabilizing the superconductor environment.

B3.2.1 B3.2.2
The powder-in-tube (PIT) technique, among various methods to produce metal/superconductor composites, has proved to be an attractive route for producing superconducting wires. Although this technique has been applied to all major HTS materials for fabrication of wires and tapes, little progress has been made in these materials except for Bi-based HTS. This is because the Bi-based HTS has a unique feature: micaceous morphology, which allows for the formation of a continuous core of HTS during mechanical deformation. All other HTS materials suffer serious grain boundary weak link problems and are undesirable for making wires and tapes using the PIT technique. For these reasons HTS wire and tapes have been focussed on Bi-based HTS. This will also be the central topic of this chapter.

A significant improvement in the critical current density (J_c) and flexibility has been achieved in the Ag-clad $Bi_2Sr_2CaCu_2O_{8+x}$ [3] and Ag-clad $(Bi,Pb)_2Sr_2Ca_2Cu_3O_{10+x}$ [4–6]. Progress in understanding the weak link problem and pinning behaviour has been made. Critical current densities of 6×10^4 A cm^{-2} for Ag-clad $(Bi,Pb)_2Sr_2Ca_2Cu_3O_{10+y}$ tape [7] at 77 K and zero field has been achieved. It is more encouraging that the J_c of Ag/BiSrCaCuO wires at 4.2 K is already comparable to conventional metallic superconductors and even superior at fields above 20 T [8]. Recently, J_c over 20 000 A cm^{-2} at 77 K has been achieved for long length tapes over 1000 meters by a number of research groups.

It has been well established that the Bi-based superconductors have three compositions $Bi_2Sr_2CuO_{6+y}$ (2201, $T_c = 7$ K), $Bi_2Sr_2CaCu_2O_{8+y}$ (2212, $T_c = 85$ K) and $(Bi,Pb)_2Sr_2Ca_2Cu_3O_{10+y}$ (2223, $T_c = 110$ K) corresponding to one-, two- and three-CuO layer compounds, respectively [9–10].

C2
The details of these compounds can be found in chapter C2. The latter two compounds are the candidates for fabrication of conductors. In this chapter, the conductor processing techniques for producing Ag/Bi-based superconducting wires by PIT method will be emphasized. These include starting composition, powder processing, various mechanical deformation techniques and matrix materials. Electrical, magnetic and mechanical properties are presented and the weak link behaviour and flux pinning properties will be discussed.

B3.1.2 Powder processing

B3.1.2.1 Chemical composition of precursor powder

B2.1 B3.2.1 B2.3.5
As a first step of PIT process the quality and reactivity of the precursors is of vital importance to the final properties of superconducting wires. It has been realized that it is difficult to produce single-phase $Bi_2Sr_2Ca_2Cu_3O_{10+y}$ (2223), despite the use of variable starting compositions and heat-treatment conditions. Partial Pb substitution for Bi promotes the formation of Bi-2223 and stabilizes the 110 K phase [11, 12]. Thus, Pb-substituted Bi-2223 is commonly used for fabrication of HTS wires and tapes. For the formation of Bi-2212, Pb substitution is not required, so the majority of groups uses Bi-2212 without Pb. However, Pb substitution into Bi-2212 has recently been found to have a beneficial effect on reduction of anisotropy and introduction of pinning centres, hence improvement of flux pinning in single crystals. Some groups have tried to use Pb-substituted Bi-2212 powder for PIT process.

Ideally, the nominal composition of precursor powder should be in the vicinity of stoichiometric 2212 and 2223 in order to form single phase in the end products. However, because these compounds all have a wide range of solid solution and inter-substitution between elements, the exact starting stoichiometric composition does not necessarily ensure the single phase in the end products. Furthermore, the large number of compounds that coexist in the system and complicated thermo-mechanical processes used for fabrication of the wires and tapes make it more difficult to achieve the single phase, in particular for Bi-2223. Phase relation and phase diagram of these two systems can be used as an essential guide for determining the processing parameters. A large number of systems have been investigated and reported in the literature, for example, the system Bi_2O_3–SrO–CaO–CuO [13]. The starting chemical composition used by various groups is largely based on experience and is closely related to the processing routes used. Table 3.1.1 lists some starting compositions for Bi-2212 and Bi-2223 [13].

B3.2.1
B2.3.5

It is noted that various cation stoichiometries in the starting materials have been used for 2223 wire fabrication. The main feature for Bi-2223 powder is a slight excess in Bi and Ca because Bi forms the major part of the liquid phase which is essential for the formation of the Bi-2223 phase while the excessive Ca will ensure the correct stoichiometry of Bi-2223 phase, as incorporation of Ca is a slow process. It has been suggested that the finely dispersed calcium- and copper-rich precipitates (Ca_2CuO_3) act as pinning centres in this material [6, 14]. However, the Ca_2CuO_3 precipitates are non-superconducting phase, so

Table B3.1.1. Nominal compositions used for (*a*) 2212 and (*b*) Pb-2223 conductors [13]

(*a*)

Organization	Bi	Sr	Ca	Cu	Conductor Form/comments
Kobe Steel	2.1	2	1	1.9	Tape, powder contains 0.1 Ag
Mitsubishi Cable	2	2	0.64	1.64	Tape
NRIM	2	2	0.95	2	Film
Showa Electric	2.05	2	1	1.95	Tape
Sumitomo Metals	2	2.3	0.85	2	Tape, 1% O_2
Vacuumschmelze	2	2	1	2	Round wire

(*b*)

Organization	Bi	Pb	Sr	Ca	Cu
American Superconductor Corp.	1.8	0.3	1.9	2.0	3.1
Intermagnetics General Corp.	1.8	0.4	2.0	2.2	3.0
Sumitomo Electric	1.8	0.4	2.0	2.2	3.0
Siemens	1.8	0.4	2.0	2.1	3.0
Australian Superconductors	1.84	0.35	1.95	2.05	3.05
Toshiba	1.72	0.34	1.83	1.97	3.13

the amount and configuration of the impurity phases must be well controlled. It appears that the starting composition with higher Ca and Cu content causes accelerated formation of the 2223 phase, but a significant amount of non-superconducting phases present in the samples reduces the volume fraction of superconducting phase and hence transport critical current density.

B3.1.2.2 Powder processing methods

B2.1 C2 Nearly all of the conventional powder processing techniques have been used for preparation of HTS powders. General requirements for HTS powder include:

(a) homogeneity in chemical and phase compositions;

(b) small particle size and narrow range of size distribution;

(c) high reactivity;

(d) free carbon contamination; and

(e) desirable phase assemblage.

Methods that can meet the above requirements are briefly described below:

Solid-state reaction technique

B2.1 Bi-2212 and Bi2223 powders can be prepared from mixtures of Bi_2O_3, PbO, $SrCO_3$, $CaCO_3$ and CuO by normal powder metallurgy procedures: mixing, calcining at 830°C for 12 h and 840°C for 12 h, pressing into pellets and sintering at 870–910°C for (BSCCO) and 840–850°C for (BPSCCO) in 0.01–1.00 atm B2.2.1 oxygen for 3–240 h. The conventional ceramic-processing techniques are simple, inexpensive and very convenient to use. With good milling facilities such as attrition milling and high efficiency ball milling, high quality powder can be obtained as demonstrated by the excellent results of tape performance by Sumitomo Ltd, using the solid state reaction procedure [4]. The disadvantages are the possible contamination with carbon and other impurities from starting materials milling process. In particular, when $CaCO_3$ and $SrCO_3$ are used as starting materials, large segregated particles of Sr–Ca–Cu–O, B2.2.3 Ca–Cu–O and Cu–O phases are commonly found in the sintered materials.

Co-precipitation technique

B2.1 All the oxide or nitrate is dissolved in nitric acid solution. The solution is co-precipitated by using oxalic acid at a designated pH value. The precipitates are washed, filtered and calcined at high temperatures. The advantages of this process are the high reactivity and atomic level mixing. Furthermore, the use of carbonates can be avoided so that carbon content may be better controlled. The disadvantages are the difficulty to control the stoichiometry since the solubility for each cation is different no matter how the pH value is adjusted. Furthermore, oxalic acid used in the co-precipitation process may still end up with formation of carbonates, such as $SrCO_3$ and $CaCO_3$. This process is not widely used in the HTS community.

A technique to precipitate simultaneously all metal ions as hydroxide-carbonates using a pyrolyzable precipitating agent was developed [15]. In this method, a mixture of tetramethylammonium hydroxide and tetramethylammonium carbonate is used as the precipitating agent as the precipitates can be easily decomposed and removed from the final powder. Furthermore, the high pH of these agents also offers flexibility in pH control. In preparation of Bi-based powder, the precipitation was found to be affected by the carbonization time. A careful monitoring of the carbonization time is required to fully precipitate all metal ions in a desired stoichiometry [16].

Co-decomposition technique

A mixture of powder is prepared by drying aqueous solutions of Bi, Sr, Ca, Cu and Pb nitrates, calcining at 800°C for 12 h and 840°C for 24 h, and pressing into pellets and sintering at 850°C for 20–50 h [12]. Ball milling or attrition milling is required to achieve homogeneity since the precipitation for each cation does not take place simultaneously. This procedure is simple and easy to use but the original mixing is not at the atomic level and, hence, phase segregation may occur during calcining.

B2.1 B2.2.3

Sol–gel procedure

Very fine powders have been prepared by a sol–gel technique. Sol is defined as a colloidal suspension of fine particles. Upon the destabilization of the sol, aggregation takes place, and a rigid network is formed that is called a gel [17]. Appropriate amounts of citric acid are added to solutions of Bi, Sr, Ca, Cu and Pb nitrates. The viscous solutions are evaporated under vacuum at 50–60°C for 24 h, dried in an oven for 12 h, and fired at 650°C for 10 h. The fine powders were pressed into pellets and sintered at 850°C for 10–120 h. This procedure has been used to produce ultrafine powders with uniform particle-size distribution. The advantages are the atomic level mixing of the cations, which results in high reactivity and fine grain size in the final products. However, the dissociation of citric acid used in this process may also yield carbonates. It is not suitable for large quantity production and is expensive.

B2.1

Freeze-drying technique

This technique has been used to produce high-quality powders of $YBa_2Cu_3O_{7-x}$ [18]. This technique involved the freeze drying of solutions. A solution of Bi, Sr, Ca, Cu, and Pb nitrates is flash frozen by spraying into liquid nitrogen, freeze dried at −30° and 10^{-2} Torr for 48 h and calcined at a temperature range of 780 to 840°C for some period of time, depending on the requirements of phase assemblage. Freeze-drying is an advantageous technique in that a solution is instantaneously frozen to prevent segregation, thus giving mixing at the atomic level without recourse to the use of precipitating agents or the risk of carbonate formation. This procedure is suitable for large scale production with good reproducibility. However, the calcining procedure is critical or demixing could occur at temperatures below 200°C.

B2.1 B2.2.1

Spray-drying technique

Aqueous solution of cation nitrates is sprayed into a hot chamber set at a temperature of about 200°C to evaporate water. The dried powder is collected and calcined at elevated temperatures to yield desirable powder with atomic level mixing. There is no carbon involved during this process so the carbon content can be better controlled. One of the problems is that particle size may not be desirable and require additional milling.

B2.1

Aerosol or spray prolysis technique

This procedure is similar to spray-drying process except that the spray chamber is a high temperature furnace. An aqueous solution of nitrate salts is generally used as a starting solution to minimize carbon contamination from the solvent or the precursor. An aerosol generator is used to make droplets about 1 μm in size. The fine mist of aqueous solution passing through the furnace is decomposed to a mixture of oxide components. The mixture is calcined at high temperatures (900°–1000°C) to yield desirable phase composition by controlling the temperatures and time. The quality of the powder depends on the reactor parameters such as residence time and the processing temperature. Using this technique, good

B2.1

quality powder with particles having a similar size and homogeneous composition can be produced. This process is suitable for large batch production and has been used by some commercial companies. One of the disadvantages is the possibility of losing stoichiometry since some elements such as Pb may be evaporated at high temperature. The prolysis temperature must be properly controlled.

The reactivity of the powders produced by the above procedures has a general tendency that solution routes are better than solid state reaction routes. For example, annealing at 840°C for 50 h is sufficient to form a nearly pure 2223 phase for the freeze dried powder, whereas more than 100 h annealing is required in order to form a nearly pure 2223 phase when the dry mixing powder is used. To explain the acceleration of the (2223) phase formation in the samples made from the freeze dried nitrate powders, the effect of carbonates on the sintering should be considered. It is easier to form SrO and CaO from the dissociation of the nitrates rather than the carbonates owing to the lower decomposition temperatures of the former. The powders made through the freeze drying and co-decomposition procedures were carbon-free. In contrast, the powders made through the sol–gel or solid-state routes contained carbonates. The dissociation rates of $SrCO_3$ and $CaCO_3$ at calcining temperatures 800–820°C are slow. Furthermore, Bi_2O_3 and PbO easily form a liquid phase at these temperatures. Consequently, the undecomposed $CaCO_3$ and $SrCO_3$ remain isolated phases, which are free to react with CuO to form $SrCaCu_4O_6$ and $SrCaCu_2O_4$, which are often detected by energy dispersive spectrometry (EDS) analysis. These impurity phases have high melting points ($>$ 1000°C), suggesting that further phase formation B2.2.5 and reaction take place through solid-state diffusion with concomitantly slow kinetics. It is evident that the solution route provides highly reactive, homogeneous and carbonate-free powders from which the 110 K phase (2223) can be formed effectively.

B3.1.2.3 Effect of phase assemblage in precursor powder

Incompletely reacted Bi-2223 powder

B3.3.2 In the case of Bi-2212, the starting and ending phase composition are the same. Thus, the starting phase assemblage is not so critical as long as it is in the vicinity of 2212 stoichiometry. However, the phase assemblage in the precursor powder of Bi-2223 is vitally important. The predominant phase being 2223 or 2212 plus unreacted components in the starting powders have been used and compared [19]. It was found that the predominant phase being 2223 in the starting powder resulted in lower J_c levels, probably B2.3.5 due to the loss of powder reactivity during sintering prior to loading of the samples into the Ag tubes. On the other hand, the starting powder consisting of a mixture of the major phase being 2212 plus unreacted components showed superior J_c–H behaviour, probably due to the subsequent reactions and a liquid phase sintering within the wires. Liquid phase sintering or partial melt process have been found to improve the grain alignment and connectivity [5]. The liquid phase sintering has been proposed to explain the positive effect of PbO addition for promoting the formation of 2223. It was suggested that eutectic melt reaction between Ca_2PbO_4 and the 2201 phase enhanced the formation of the high T_c phase. This accounts for the fact that predominant 2223 packing powder gives low J_c while the incompletely reacted powder packing in the Ag tube results in better J_c [20, 21]. For the 2223 single phase there is no liquid phase formation below 830°C while the eutectic reaction in the incompletely reacted multiphase takes place at temperatures from 800 to 840°C during the thermomechanical treatment of the Ag/Bi-based composites.

Pb distribution in the precursor powder

Phase transformation plays an important role in the powder processing stage of the fabrication of Ag/Bi-2223 tapes. It is commonly believed that the rate of formation of 2223 is an important factor in determining the processing time, microstructure and critical current density of these tapes [1]. The important role of Ca_2PbO_4 present in precursor powder has been well recognised in terms of benefit for formation of Bi-2223. However, it has been realized that the temperature for liquid phase formation in the presence of Ca_2PbO_4 is below 800°C which is not in the single Bi-2223 regime. Thus, Pb should be transformed to 2212 phase in order to raise the liquid phase formation temperature. In recent work reported by several groups, the Pb distribution in the precursor powder has been found to have a significant effect on sintering temperature, time and critical current density in the final Ag/Bi-2223 composite [22–23]. For a tape made from precursor powder containing Pb-free 2212 as the major phase, i.e. Pb is the form of Ca_2PbO_4, a low sintering temperature and long sintering time over 200 h are required, whereas a higher sintering temperature and shorter sintering time can be used for tapes made from a precursor powder with Pb-2212 as the major phase. By incorporating Pb into Pb-2212 phase in the precursor powder, not only was the total sintering reduced to 100 h but also J_c increased significantly.

B2.2.3

Table B3.1.2 gives the details of correlation between Pb distribution and sintering time and temperature. It is evident that sintering temperature increases and sintering time decreases with increasing fraction of Pb-2212 in the precursor powder. Figure B3.1.1 shows the effect of Pb distribution in the powder on the J_c. The term of matured powder means the full incorporation of Pb to form Pb-2212 phase.

C2

B3.1.2.4 The effect of carbon in the starting materials

The presence of carbon and carbon contamination during powder processing has a serious consequence as it results in blistering and bubbling in subsequent thermal processes. Although there are a number of origins causing the blistering and bubbling in Bi-based HTS tapes it is commonly believed that carbon is the major source of these problems. Systematic studies on the mechanism and prevention of bubble formation in the processing of Bi-2212 has been carried out by Helstrom's group [25] and Ag/Bi-2223 by Flukiger *et al.* [26]. They prepared the Bi-based precursors using carbonates and co-precipitation routes with varying calcining times, which allow the powders to have different carbon content before filling into the Ag tube. It was found that the J_c was drastically affected by the presence of carbon. The presence of carbon, perhaps in carbonate form, results in poor grain connectivity and high porosity. To prevent this problem, carbon-free powder processing routes such as a solution route should be used and powder handling should be in a carbon free environment.

Table 3.1.2. Effect of phase composition in precursor powder on annealing times and temperature, and final phase assemblage of Ag/Bi-2223 tapes [24]

Tape no	Calcination procedure	Main secondary phases in precursor powder	y-Pb moles in 2212[a]	Annealing time and temperature
1	A = 800°C/24 h	SrCaCuO, Ca_2PbO_4, Bi_6Ca_7O,CuO,CaO	0	250 h, 830–832°C
2	B = A+810°C/12 h	SrCaCuO, Ca_2PbO_4, CaCuO,CuO,BiSrCuO	0.13	200 h, 832–834°C
3	C = B+820°C/12 h	Ca_2PbO_4, SrCaCuO	0.2	150 h, 836–838°C
4	D = B+820°C/24 h	SrCaCuO, Ca_2PbO_4	0.27	100–125 h, 838–840°C
5	E = B+820°C/36 h	SrCaCuO	0.34	80–100 h, 840–843°C

[a] Semi-quantitative y was defined by the areas with Ca_2PbO_4 peak at about 17.5° over [115] peak of the 2212 when there is no Pb in 2212 using sample 1 as a standard.

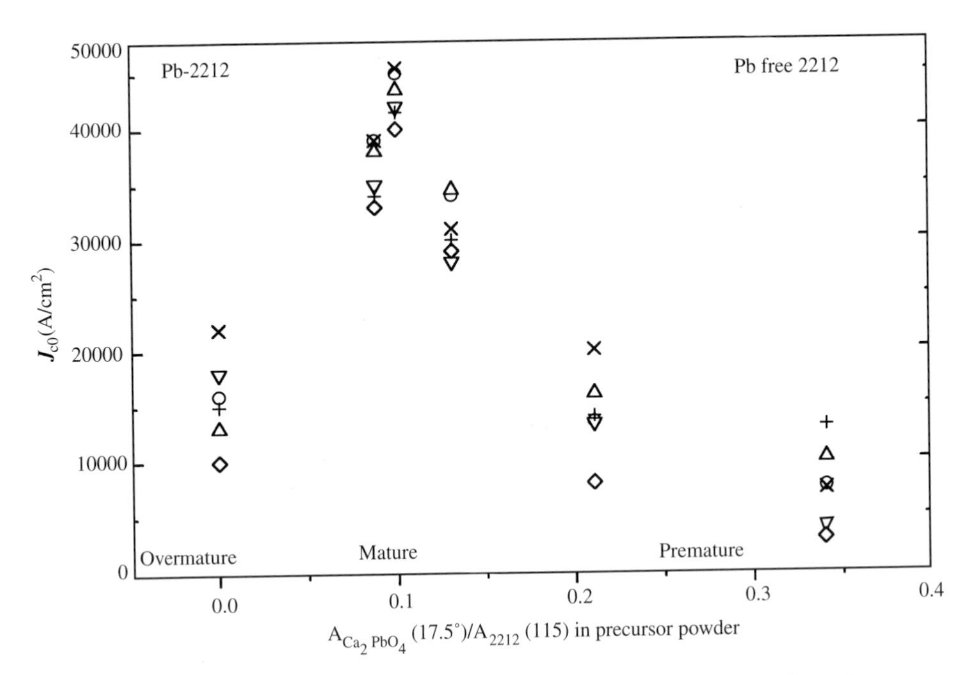

Figure B3.1.1. J_c versus fraction of Ca_2PbO_4 over total 2212 in various precursor powders; six tape samples were used for each powder composition.

B3.1.3 Mechanical deformation

B3.1.3.1 Powder-in-tube (PIT) technique

B3.2.1 The PIT technique has been used to process Ag-sheathed Bi-2212 and Bi-2223 wires. The PIT process consists of powder packing, swaging, drawing, pressing or rolling and thermal treatment. Alternatives B3.2.2 to the PIT process have also been developed to overcome the shortcomings of PIT. A number of sophisticated PIT routes have been developed to optimize the performance of Ag/Bi-HTS tapes in the past. A typical PIT procedure is described as follows. The precalcined powders are pressed into round bars of designed dimensions. The bars are loaded into silver tubes, and the composites are swaged and drawn to a final diameter of 0.7–1.0 mm. The wires are rolled into tapes of overall thickness ~ 0.1–0.2 mm and width ~ 3 mm. The resultant tapes are heat treated at 830–850°C for varying times up to 60 h in air or a mixture of oxygen and nitrogen at $Po_2 = 5.0 \times 10^3$ Pa. The tapes are then uniaxially pressed at 1 GPa and heat-treated under the same conditions for 120 h. This process is repeated once or twice. Figure B3.1.2 illustrates this process [27].

The effect of packing procedure has been studied by Yamada *et al.* [5]. They found that low density packing by filling the loose powder in the Ag tube leads to an irregular shape of the oxide core and hence a low J_c, whereas high density packing with cylindrical rods prepared by swaging showed a smooth interface between Ag and the oxide core, resulting in high J_c. Mechanical deformation is a critical step in the PIT process that plays a major role in achieving grain alignment and grain connectivity. Several mechanical deformation processes have been developed for fabricating HTS materials. In particular, hot deformation including hot-pressing, isostatic hot-pressing, hot forging and hot rolling has been intensively investigated in the past 10 years.

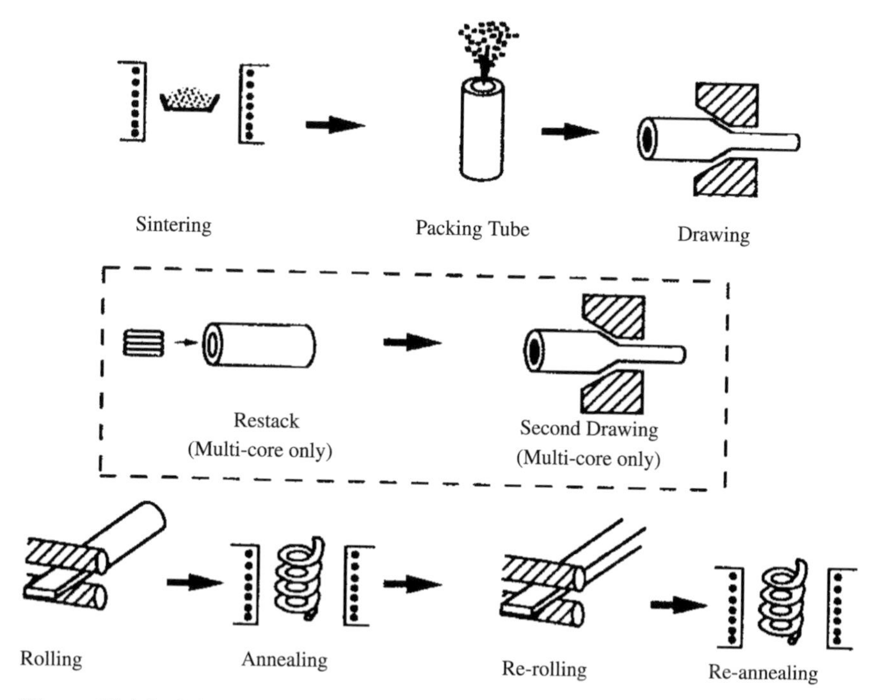

Figure B3.1.2. Schematic diagram of the PIT method to make wires and tapes [27].

The mechanical deformation process of Ag-sheathed HTS composites is inhomogeneous because the mechanical properties of ceramic powder are incompatible with those of silver sheath. The common problems of mechanical deformation are the formation of so-called sausaging and cracks, which severely limit the overall current of the conductors. A detailed analysis of the mechanical deformation process of PIT Ag-sheathed composites was reported by Han *et al.* [28].

Owing to the large separation and week bonding between the two Bi–O layers (0.3 nm), the Bi-based materials exhibit a micaceous morphology and are easily cleaved. Mechanical deformation, in particular during the rolling and pressing process takes advantage of the plate-like morphology for grain alignment in the a–b plane direction. The degree of alignment has been considered a major contributing factor for the high J_c in the Ag-sheathed Bi-based tapes. It has been realized that repetitive deformation and annealing is necessary to achieve a high degree of grain alignment and hence high J_c [1]. The degree of texturing is dependent on the oxide core thickness. It was found that the degree of texturing decreased from the Ag/superconductor interface towards the centre of the oxide layer. Thus, the high J_c values observed in the thinner tapes are attributed to the improvement of grain alignment. The J_c dependence on the thickness may also be attributable to the effect of self-generated fields. The relative importance of these two factors depends on the processing procedure used. C2

The degree of grain alignment of the tapes can be demonstrated by X-ray diffraction (XRD) patterns. However, XRD patterns only give the degree of alignment in a very thin layer on the surface and do not reflect the alignment situation in the entire thickness of sample since the texturing degree may vary from surface to the centre of the tapes. By rotating a tape sample in a magnetic field, a peak was observed in the plot of J_c versus θ which is the angle between the rotating tape and the applied field. The J_c shows a maximum when the applied field (H) is parallel to the tape surface and a minimum when H is perpendicular to the tape surface. The ratio of $\alpha = J_c(H \perp \mathrm{c})/J_c(H \| \mathrm{c})$ gives a good indication of the degree D1.1.2 A1.3

of grain alignment in the entire thickness of the tape since the magnetic field penetrates the tape [29]. Thus, this can be used to compare the degree of texturing in various samples.

B3.1.3.2 Cold deformation process

Single core tape

Conventional cold drawing, pressing or rolling is commonly used for fabrication of single core (monofilamentary) tapes. A typical example is given below. The powders are filled into pure Ag tubes and compacted using a pressure of about 2 kbar, reaching a density of about 5 g cm^{-3}. The tubes are properly sealed with plugs and then deformed, initially by swaging and then by drawing into an outer diameter of about 1.0–1.5 mm, the cross-sectional reduction for each step being about 10%. At this level, the powder density inside the wire reaches 6 g cm^{-3}. Finally the wires are cold rolled to reduce the total thickness to typically 0.1–0.2 mm. The oxide core area usually represents about 30–35% of the total tape cross-section, and the powder density can be higher than 90% of the theoretical density for the Bi(2223) phase. Particular attention has to be given to the deformation steps, in order to avoid the formation of sausaging in the oxide longitudinal section. In general, a reduction of $\approx$ 5–10% between two consecutive rolling steps is used, in order to minimize the longitudinal fluctuations.

Multifilamentary tapes

A1.3
E1.4.2
The fabrication of multifilamentary Bi,Pb(2223) tapes is not essentially different from the process mentioned above for monofilamentary tapes, except for an intermediate bundling step. In the following, we describe the fabrication technique used by several groups and manufacturers. The deformation of the monofilamentary rod is performed up to a diameter of several millimetres (1–3 mm), then it is drawn to a hexagonal shape. Several hexagonal bars (generally 37 or 55) are bundled and stacked into a second Ag tube of a diameter ranging up to 20 mm, which is drawn down to the final diameter (1–2 mm). The composite wire is rolled to tape thickness of 0.2–0.3 mm, with width varying between 2.5 and 3 mm. The deformation reduction rate for drawing and rolling is around 5–15% between two consecutive rolling steps. The optimized deformation speed depends upon several parameters, e.g. the reduction rate, the diameter of the rolls and the state of the surface of the rolls. The process described here concerns the production of tapes at a laboratory scale. For the fabrication of tapes at industrial lengths ($>$1 km), the initial diameter of the rod after bundling must be much larger: manufacturers may introduce an extrusion step, but no details have been published so far.

B3.1.3.3 Sandwich rolling

In the flat rolling process, the sausaging morphology in the oxide core and the undulating interface between the silver sheath and core is commonly observed. The J_c is limited by the worst segment of the tapes. In order to improve J_c over the whole length of long tapes, the position sensitivity must be reduced and transverse sausaging in the tape must be prevented. A new deformation process called 'sandwich' rolling has been developed to prevent the formation of sausaging and cracks in longitudinal direction [30]. In this 'sandwich' rolling, the tape is deformed between two steel sheets. The stress–strain state of the tape in 'sandwich' rolling is the same as that of uniaxial pressed tape because the deformation of steel sheets is negligible in comparison to that of Ag-clad Bi-2223 tape. The stress–strain distribution of superconducting oxide core can be controlled by the changing the curvature of the steel sheets. The critical current density of the sandwich rolled tape is improved in comparison with the flat rolled tape.

Figure B3.1.3 shows longitudinal section micrographs of tapes obtained by flat rolling, pressing and 'sandwich' rolling. It is evident that the Ag/oxide core interface obtained by pressing and 'sandwich' rolling

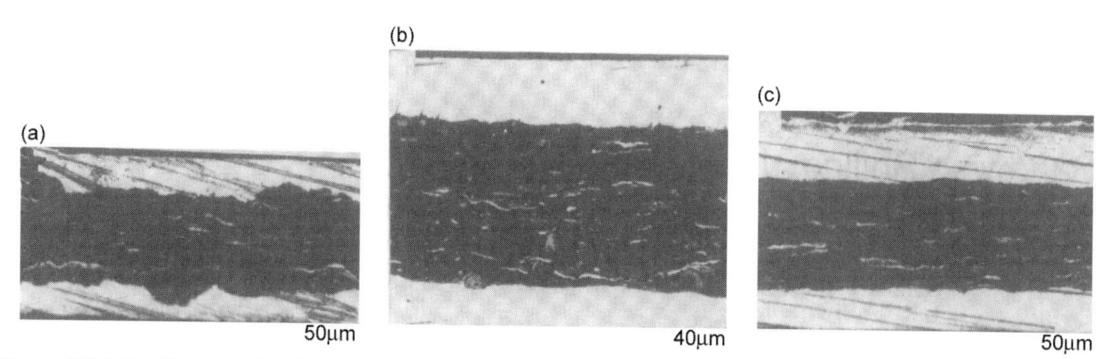

Figure B3.1.3. Micrographs of longitudinal section of tapes produced with (*a*) flat rolling, (*b*) pressing, and (*c*) 'sandwich' rolling.

is much smoother and flatter than that by flat rolling. Figure B3.1.4 shows the different longitudinal surface morphologies of tapes fabricated by different deformation routes. The sausaging direction is the same in the pressed and 'sandwich' rolled tapes, parallel to the length of the tape. However, the sausaging in the flat-rolled tape is perpendicular to the length of tape. The interface instability affects the crystal growth along the a–b plane and therefore decreases the degree of texture. What is more serious is that the transverse sausaging introduces transverse microcracks, which block the current flow in the entire filament.

'Sandwich' rolling with flat steel sheets gives a similar stress–strain state to pressing, except that the strain in the longitudinal direction is larger in 'Sandwich' rolling than that in pressing due to the curvature and slight deformation of the steel sheets. The transverse direction expansion is constrained by the sheets, because the silver tape is impressed into the steel sheets. This kind of stress–strain state constrains more serious longitudinal sausaging as shown in figure B3.1.4. Through the control of the stress–strain state, long multifilament tapes with high J_c and improved stability and reproducibility may be achieved by the 'sandwich' rolling process.

B3.1.3.4 Four axis rolling

Recently, a new deformation technique for multifilamentary tapes was introduced, based on a motor-driven four-roll machine, that permits the simultaneous reduction of the thickness and width of the samples [31]. A schematic representation of the four-roll machine is given in figure B3.1.5. The main difference between this machine and the common 'turk's head' resides in the fact that the rolling force is much larger in the motor-driven case. As will be shown below, the rolling force on the tape is also more homogeneous than when using a conventional two-roll machine. The role of the rolling force was first

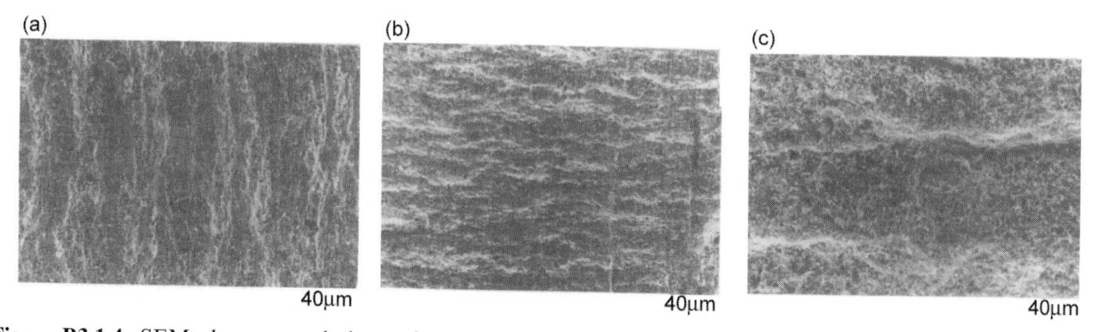

Figure B3.1.4. SEM planar morphology of core surface of tapes with (*a*) rolling, (*b*) pressing, and (*c*) 'sandwich' rolling.

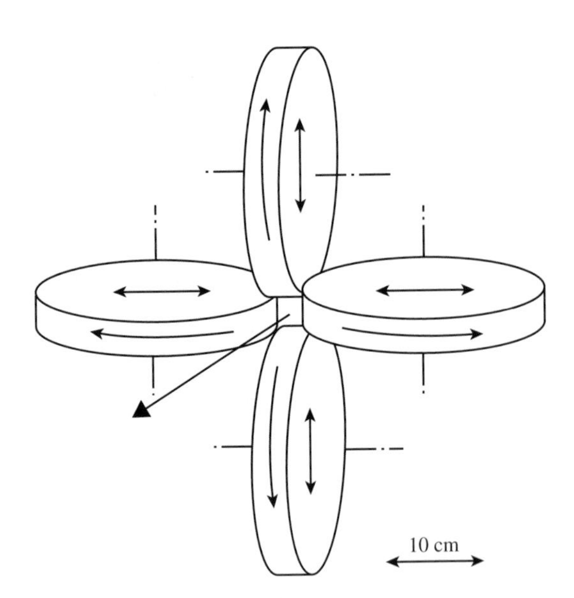

Figure B3.1.5. A schematic representation of the four roll machine, with four rolls simultaneously reducing the thickness and width of the tape.

emphasized by Grasso *et al.* [32], who found that there is for each tape configuration an optimum rolling force leading to optimized critical current density values.

B3.2.2 Square monofilamentary wires of about 1–1.5 mm width, prepared by using the four-roll machine, are stacked into a square Ag tube, which is again deformed using the same machine, the final dimensions being the same as for the usual tapes, ie. 250 μm thickness and 3 mm width. The advantages of the square symmetry are evident when comparing the cross-sections of the multifilamentary tapes. The cross-section of figure B3.1.6 represents a four-rolled tape with 100 filaments and a superconducting fraction of about 30%. The filaments near the centre in standard multifilamentary tapes are more compressed than those at the sides, whereas four-rolled tapes exhibit a more homogeneous density. In addition, the distances between the single filaments show higher fluctuations for the standard tape than for the four-rolled tape.

B3.1.3.5 Cryogenic deformation

Hot deformation has been found to be superior to cold deformation in achieving high texture and grain connectivity, resulting in improvements in J_c and J_c-field performance [33–35]. In the case of Ag/Bi-HTS wires, the silver sheath becomes very soft at high temperature while the oxide core remains hard. This incompatibility between silver sheath and oxide core results in non-uniformity throughout the core and at the metallic interface during deformation. Thus, the degree of densification and texture inside the oxide core, is restricted by the deformation pressure that can be applied during mechanical process.

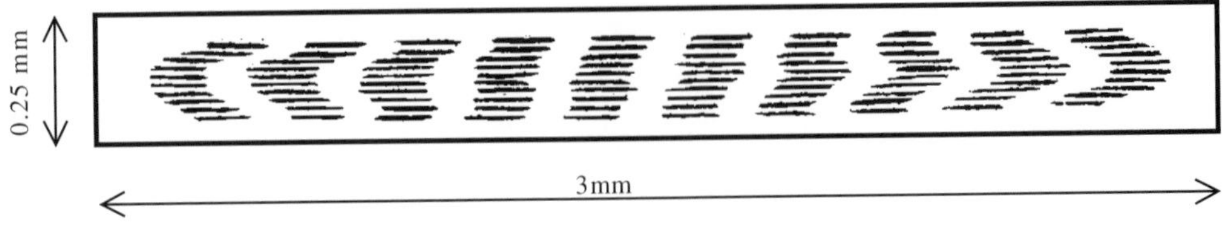

Figure B3.1.6. Transverse section of multifilamentary four-rolled Ag/Bi-2223 tape with 100 filaments.

In order to overcome this problem, a cryogenic deformation process has been developed and investigated for use in the processing of Ag/Bi–Pb–Sr–Ca–Cu–O superconducting wires [36, 37]. A typical cryogenic deformation process comprises rolling or pressing the wires into tapes at 77 K and then heat treating the tapes at 838–842°C for a period of 40–60 h. It was found that the cryogenic deformation improved the density, grain alignment, Ag/oxide core interface and critical current density. Critical current densities for cryogenically rolled and pressed Ag-clad mono- and multi-filamentary tapes showed a 10–20% increase on those observed in normally processed tapes. In comparison with hot deformation processed tapes, the cryogenic processed tapes showed improved flux pinning capability whereas the hot deformation was found to only improve weak links. During cryogenic deformation, the hardness of the silver sheath was significantly increased, which allowed a greater pressure to be applied during the deformation process. This large pressure was found to be the principle cause of the enhanced densification, texturing and increased dislocation densities.

C2

Cryogenically rolled Ag-clad Bi-2223 mono-filament tapes: the PIT composites were drawn to a final diameter of 1.2–1.6 mm. The resultant wires were then rolled to 0.15 mm thick in liquid nitrogen at 77 K. The tapes were annealed at 840°C for 50 h, pressed and further sintered at 840°C for 30 h and at 825°C for 30 h [36]. The average I_c for cryogenically rolled tapes was 15–20% higher than that for normally rolled tapes. SEM images show that the cryogenically rolled tape had a higher density and better texture than that of normal rolled tape. SEM images of normally pressed and cryogenically pressed tapes, both having a similar reduction rate, show that filaments of the cryogenically pressed tape have higher density, better alignment, less impurities and smoother interface than the normally pressed tapes. For normally pressed tapes, a large load resulted in the formation of undulating filaments and the 'sausaging effect' while, for the same applied pressure, there was no sausaging effect in cryogenically pressed tape, owing to the increase of hardness of silver at low temperatures. This suggests that the improvement in J_c using cryogenic deformation may be attributable to the improvement in grain connectivity and alignment. The grain connectivity can be improved by increasing deformation pressure. However, because of the softness of silver sheath, large deformation pressures will lead to a poor interface between silver sheath and oxide core.

D1.3

The effect of mechanical deformation rate was investigated with tapes pressed at varying pressures in liquid nitrogen and normal conditions. Figure B3.1.7 shows the I_c versus deformation rate for both the cryogenically pressed and normally pressed tapes. It is clear that at higher deformation rates (greater than 15%) the I_c declines with increasing deformation rate for normally pressed tapes, while it increases with increasing deformation rate to the maximum I_c at a reduction rate of about 25% for the cryogenically processed tape compared to 15% for a normally processed tape. Figure B3.1.8 shows the dependence of J_c on magnetic fields at 77 K with field applied perpendicular to the tape surface. At low magnetic fields, the rapid drop in J_c was attributed to Josephson weak links at the grain boundaries. J_c showed a large drop at low magnetic fields in a rolled sample. The J_c of cryogenically pressed tapes was also lower than for hot-pressed tapes in low fields up to 100 mT. In high fields, however, the J_c of the cryogenically pressed tape crossed over with hot-pressed tapes and then dropped more slowly with increasing field above 150 mT than the hot-pressed and rolled tapes. These results indicate that the hot-pressing improves the grain connectivity and, hence, the weak link behaviour in low fields, whereas the cryogenic pressing induces more defects, resulting in an improvement in flux pinning.

A2.7 A4.3

B3.1.3.6 Hot deformation

Hot-pressing

It has been reported that the critical current density (J_c) of Ag sheathed $Bi_{1.6}Pb_{0.4}Ca_{2.0}Cu_{3.0}O_{10+y}$ is enhanced through the use of hot isostatic pressing (HIP) [39, 40]. A relative density up to 95% was

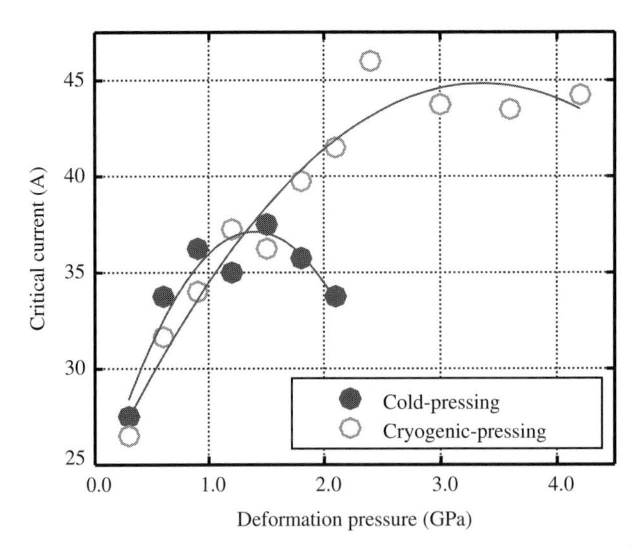

Figure B3.1.7. Critical current versus deformation pressure for Ag/Bi-2223 tapes pressed at room temperature and liquid nitrogen temperature [38].

achieved through HIPing at 650°C for 2 h under 200 MPa argon. Under these conditions, J_c was four times that without HIPing. T_0 was unaffected by HIPing for samples encapsulated with Pyrex glass while T_0 was suppressed from 103 to 86 K for samples encapsulated with stainless steel.

It should be pointed out that HIPing at high pressures and temperatures below 810°C causes the decomposition of the 2223 phase, resulting in a new Pb-rich compound $(Bi,Pb)_3Sr_2Ca_2CuO_y$ (3221). Electron diffraction and X-ray diffraction data revealed that the 3221 phase has unit cell parameters of $a = 0.9919$ nm and $c = 0.3471$ nm with a hexagonal symmetry [41], which is in contrast to the commonly found tetragonal or orthorhombic symmetry in the Bi–Pb–Sr–Ca–Cu–O systems. Since the 3221 phase is non-superconductive the amount and distribution have to be well controlled.

D1.1.2
D1.7.1

Uniaxial hot-pressing has also been used to improve the relative density and preferred grain orientation of the nearly pure (2223) phase in the Bi–Pb–Sr–Ca–Cu–O system. Owing to the plate-like

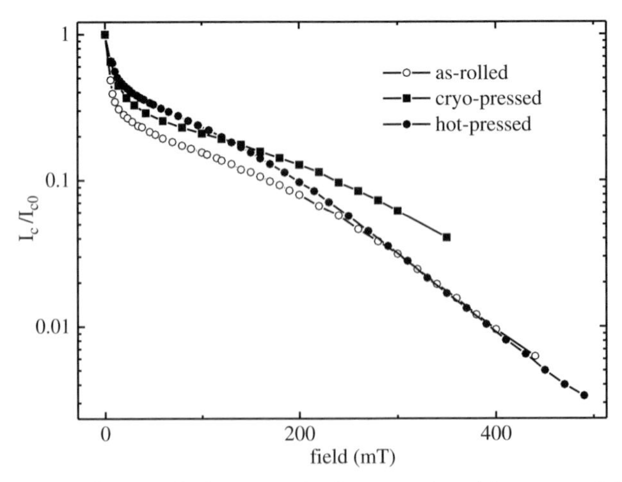

Figure B3.1.8. J_c dependence for as-rolled, cryogenically-pressed and hot-pressed Ag/Bi-2223 tapes [37].

morphology of the 2223 phase, uniaxial hot-pressing is more effective than HIPing for promoting grain alignment and improving the density. Hot-pressing has been carried out at 650–855°C and 30–200 MPa. Density up to 99% of theoretical [42] and J_c as high as 10 000 A cm^{-2} at 77 K and zero field [43] have been achieved in the hot-pressed bulk 2223 sample. However at high pressure, e.g. 200 MPa, the 2223 phase was partially destroyed during hot-pressing, resulting in the formation of a new phase 3221 similar to the case of HIPing, due to the weak bonding between the Bi–O layers. Although the 2223 phase can be recovered by post-annealing at 840°C for 50 h, density and hence the J_c was lowered.

In order to study grain connectivity in Ag-sheathed Bi-2223 tapes, hot-pressing technique was also used to prepare Ag-sheathed multifilamentary Bi-2223 tape. A typical example is given below. Ag-sheathed Bi-2223 samples were prepared using the oxide-powder-in-tube (OPIT) method. Precursor powder with a nominal chemical composition of $Bi_{1.8}Pb_{0.35}Sr_{1.91}Ca_{2.05}Cu_{3.06}O_x$ with Bi-2212 as the major phase was used to fabricate Ag/Bi-2223 tapes. Multifilamentary tapes were fabricated using a single re-stacking procedure, sintered for 100 h and slowly cooled with intermediate pressing. Some of these tapes were sandwiched between two pieces of ceramic sheets and hot-pressed under a pressure of 15 MPa at 780–825°C for 10–360 min. After hot-pressing, samples were slowly cooled. B2.2.3

The self-field critical current density (J_c) of tapes after hot pressing was significantly enhanced, the maximum increase was more than double. For example, hot-pressing raised the J_c ($B = 0$) from 22 000 A cm^{-2} ($I_c = 18$ A) to 56 800 A cm^{-2} ($I_c = 33$ A) for a tape [44]. XRD, SEM and TEM analysis showed that hot-pressing technique chiefly improved grain connectivity and reduced the weak link by increasing the density of the oxide core, but at the same time, it reduced impurities to some degree. D1.1.2 D1.5 D1.2

B3.1.3.7 Eccentric rolling and sequential pressing

As discussed in the previous section, flat rolling will induce transverse cracks, which are more damaging to the current path because their occurrence definitely eliminates the transport current for such a material. These cracks are responsible for the large difference in J_c values of press-sintered short samples and roll-sintered long tapes. In addition to the sandwich-rolling discussed above, attempts to use uniaxial pressing for manufacturing of long tapes have been made utilizing 'sequential pressing' [45] and by 'semi-continual pressing' [46]. However, sequential and semi-continual pressings have some limitations in tape manufacturing, e.g. the existence of overlapped zones and low speed of deformation. A1.3

A new rolling technique, so called 'eccentric rolling', has been developed for texture and density improvement in Bi(2223)/Ag tapes made by PIT [47]. It consists of rolling deformation between the outer surface of the inner roller and inner surface of the outer hollow cylinder. This technique allows deformation of the BSCCO/Ag composite in a way that is similar to flat rolling using rollers of very large radius or to a continual pressing, but without any overlapping parts. In comparison to classical rolling with roller of a similar radius, a higher core density and widening of tape have been obtained. The transport critical current densities for the eccentric rolled tape have been noticeably improved compared to the flat rolled tape.

B3.1.4 Thermal treatment

B3.1.4.1 Phase relation and phase diagram of (Bi,Pb)-2223

In PIT process, heat treatment is a vitally important step to achieve high quality HTS tapes. Heat treatment is responsible for transformation of Bi-2212 and other phases to Bi-2223, improvement of grain alignment, grain growth and densification. It is generally believed that attainment of pure Bi-2223 phase in Ag-sheathed Bi-2223 tapes is critical in improving the critical current density (J_c). The phase B3.2.1 B2.3.5

relationship and phase diagram of the BPSCCO systems have been extensively studied in the past decade. Some phase diagrams are presented in chapter D1.2. Furthermore, the phase relationships and formation conditions in the $(BiPb)_2Sr_2Ca_2Cu_3O_x$ system are complicated due to the unresolved influences of stoichiometry [48]. Many groups [49–54] have studied the 2223 phase transformation and used this information to propose some new processing to improve the properties. A van't Hoff diagram has shown the stability limits of the already formed 2223 phase at certain temperatures and oxygen pressures [55], and other workers have reported thermodynamic phase diagrams [56]. Recently, Moon *et al.* have constructed a stability diagram for (Bi,Pb)-2223 compiled from studies [56–60] on preformed (Bi,Pb)-2223 materials [61], as shown in figure B3.1.9.

D1.2 B3.2.1 B2.3.5

The stability diagram is composed of five regions: (Bi,Pb)-2223 stable region, solid-state decomposition regions A and B and incongruent melting regions A and B. In the PIT process, silver sheath is primarily used in forming composite wires and tapes. Studies have demonstrated that silver decreased the partial melting temperature and hence the sintering temperature up to 10–15°C, depending on the amount of silver addition and the filling factor of wires and tapes [56, 62]. At a higher temperature range, 885–905°C, silver dissolved into and diffused through the (Bi,Pb)-2223 incongruent melt liquid facilitates further incongruent melting [61]. However, in the case of Ag/Bi-2223 processing, the important temperature region is between 700 and 850°C. It is noticed in figure B3.1.9 that the stability region of Bi-2223 increases with decreasing oxygen partial pressure. It is advantageous to have a wide temperature window for heat treatment of Ag/Bi-2223 tapes, in particular, for large-scale production. Thus, a low oxygen partial pressure around 7% is commonly used for heat treatment instead of air.

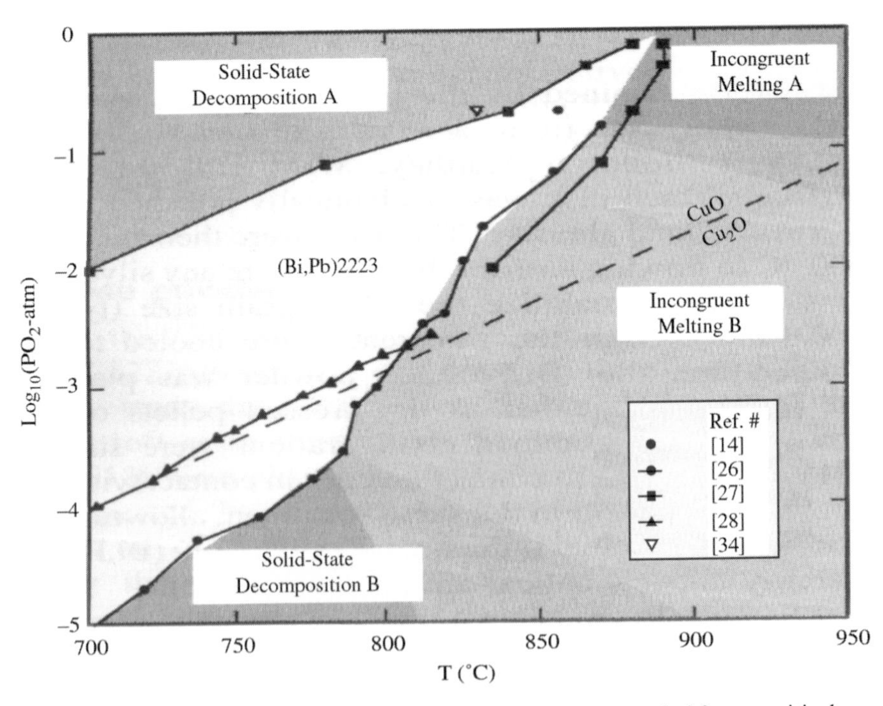

Figure B3.1.9. The PO_2-temperature (Bi,Pb)2223 phase stability diagram compiled from a critical assessment of the literature and the data of the individual studies used to construct it [60].

B3.1.4.2 *Phase evolution and formation mechanism during processing*

The phase evolution and Bi-2223 phase formation during heat treatment of Ag/Bi-2223 have been a central research topic in the last several years. Factors affecting the phase evolution and transformation include the nominal composition and phase composition of the precursor powder, heat treatment procedure, sintering temperature and duration, heating and cooling rate as well as oxide core/silver ratio. Because of the multiple variables simultaneously contributing to the phase evolution process the phase conversion or Bi-2223 phase formation mechanism remains controversial, although numerous studies have been performed in the past.

It is commonly believed that the presence of liquid phase is essential to the formation and grain alignment of Bi-2223 [63]. There are different mechanisms for the formation and alignment of Ag/Bi-2223 tapes. Based on the observation of droplets in the interior cavities of the sintered powder mixture, Morgan *et al.* [64] suggested that Bi-2223 formed by depositing dissolved cations from liquid onto ledges of Bi-2212 or existing Bi-2223 platelets. Some authors have used Avrami relationship between the fraction of Bi-2223 and the sintering time to fit various kinetic models. These have led to the conclusion that the conversion process is a nucleation and growth mechanism [65]. This was further confirmed by careful and consecutive observations on the surface of a sintered pellet during heat treatment [66]. Using synchrotron XRD, *in situ* studies have also come to the same conclusion that nucleation and growth is the predominant mechanism for the Bi-2223 formation [67]. On the other hand, using transmission electron microscopy (TEM) and *ex situ* transmission XRD techniques, an intercalation of the Ca and Cu layers into the existing Bi-2212 platelets was proved to be the primary phase conversion mechanism [68, 69]. However, the mechanism of Bi-2223 formation depends on the amount of liquid present and the powder processing routes. The intercalation process is the primary mechanism when very little liquid is present in the system, while, if a large amount of liquid is present, the nucleation and growth of blocks of Bi-2223 appears to be the dominant mechanism [69].

B2.2.3
D1.1.2
D1.2

B3.1.4.3 *Phase transformation during cooling process*

As discussed in the previous section the presence of liquid is essential for a number of reasons. However, the question is how we deal with the liquid at the end of the thermal cycle. Recent work using *ex situ* static quenching [70], *in situ* synchrotron XRD [67, 71] and *in situ* neutron diffraction [72] reveal that Ag/Bi-2223 tapes have a substantial amount of liquid or amorphous phase during processing. The conversion of the liquid phase on cooling has a significant effect on the phase composition in the final tape and hence J_c, and depends upon the cooling procedure [73]. Slow cooling and two step annealing procedures at the end of the final thermal cycle have shown clear advantages in critical current densities and flux pinning over conventional processing [74, 75].

The cooling rate at the end of the final thermal cycle of the PIT process was manipulated to examine the effect on J_c of multifilament Bi-2223 tapes [75, 76]. It was found that the cooling rate influences both the intergranular connectivity and intragranular flux pinning strength of the polycrystalline filaments. As the cooling rate from 825 to 730°C in 7.5% O_2 was decreased over a range of 5–0.005°C min^{-1}, J_c (77 K, 0 T) increased from similar to 8 to similar to 24 kA cm^{-1}(2), and the irreversibility field increased from similar to 120 to similar to 200 mT, The J_c (4.2 K, 0 T) increased in a similar fashion. Cooling slowly also sharpened the critical temperature transition and increased the critical onset temperature from 107 to 109 K. These improvements in the superconducting properties occurred despite partial decomposition of the $(Bi, Pb)_2Sr_2Ca_2Cu_3O_x$ phase into non-superconducting impurity phases during the slow cooling.

A two-step sintering procedure used in the final thermal cycle of the PIT process is shown in figure B3.1.10 [71, 74]. In the second step, the annealing process was carried out at temperatures in the range 700–845°C. At the end of the final annealing step, the tapes were either quenched or

B3.2.1
D1.1.2
B3.2.2
B2.2.3

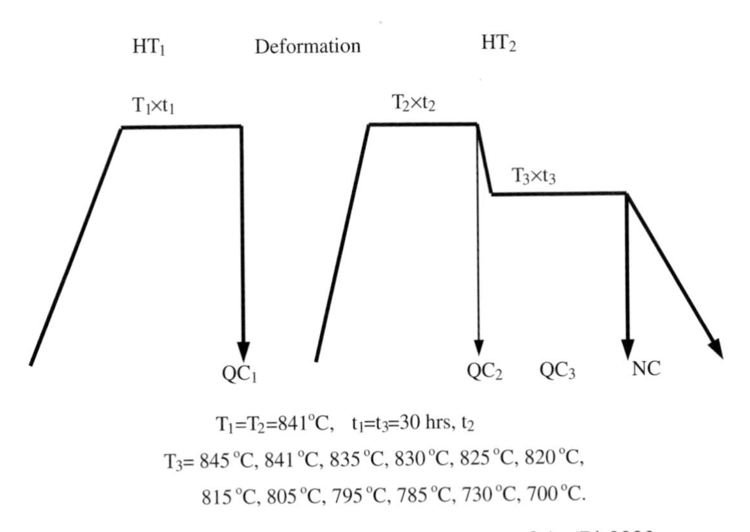

Figure B3.1.10. Schedule for thermal treatment of Ag/Bi-2223 tapes.

cooled at 0.5°C min^{-1}. The details of phase assemblage of the tape versus annealing temperatures are shown in figure B3.1.11. A striking feature of all the curves is the special annealing temperature around 825°C where the 2223 volume fraction shows a maximum while all other phases show a minimum. Above 825°C, liquid phase coexists with 2223 and Ca–Sr–Cu–O, and its amount increases with increasing temperature. When the annealing temperature is set at 845°C, there is a large fraction of liquid present. On quenching, the liquid phase is converted to amorphous phase. If the annealing temperature is set at 825°C, the lowest temperature at which liquid forms and which leads to the formation of 2223, the liquid phase is largely converted to 2223 through reaction with Ca–Sr–Cu–O, until all the liquid phase is exhausted. This results in the 2223 maximum at this temperature. Below 825°C (which is out of the 2223 stability regime), the amorphous phase (and liquid phase, if any) is converted to 2212, leading

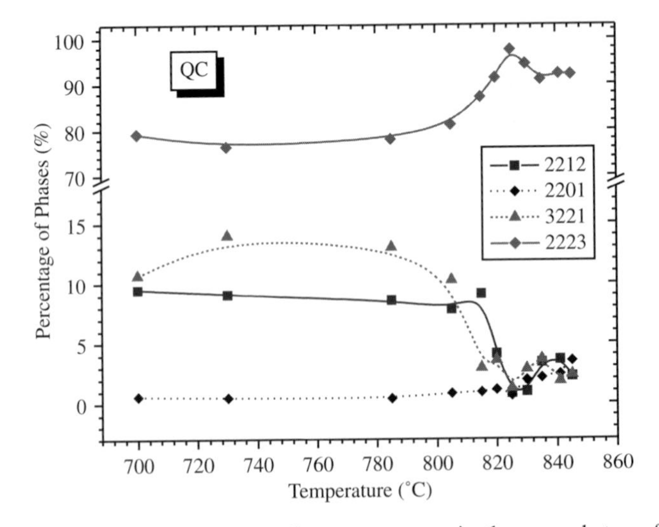

Figure B3.1.11. Various phase fraction versus annealing temperature in the second step of annealing for quenched tapes.

to a rapid increase in 2212. These results suggest that 825°C is an important temperature at which the liquid phase is largely converted to 2223 with minimum 2201 and 2212. These results are consistent with *in situ* observations using synchrotron XRD and neutron diffraction [67, 72].

2201 increases with increasing annealing temperature, suggesting that liquid phase coexists with 2223 above 825°C. This is consistent with direct observation using high temperature neutron diffraction *in situ* technique [72]. However, 2201 may be also partly formed from the liquid phase during rapid cooling since the conversion from liquid to 2201 is a very fast process. This is evidenced by *in situ* synchrotron XRD observation [67], which shows that 2201 appeared only on cooling. 3221 increases rapidly below 815°C, either due to conversion from the amorphous phase or the decomposition of 2223. In a separate work, when the finished tapes were annealed at 780°C for 10 h with sampling every 2 h the 3221 increased and 2223 decreased monotonically while 2212 remained nearly unchanged [77]. The fraction of 3221 was up to 17% at the end of the annealing. The 2223 decomposition at this temperature regime was also confirmed by using synchrotron XRD *in situ* study [67].

Figure B3.1.12 shows the J_c of tapes both quenched and slow cooled (2°C min^{-1}). It is interesting to note the same maximum at about 825°C. It appears that the higher 2223 fraction is the first and obvious reason for this maximum. At higher temperatures, the liquid phase converted to amorphous phase. The amount of liquid phase increased with increasing annealing temperature. The large fraction of liquid phase preserved as amorphous phase at colony boundaries after quenching form networks, which act as weak links to depress J_c. At annealing temperatures below 825°C, the lower J_c is attributable to the fact that in this regime, the amorphous phase was only converted to 2212 since it is outside the 2223 stability regime. The best strategy to improve J_c in tapes is to take advantage of both thermodynamic equilibrium and kinetically metastable conditions to maximize the conversion of liquid to 2223, reduce 2201, 2212 and 3221 formation on cooling and ensure a full oxygenation of the tapes.

B3.2.1 B2.3.5

B3.1.4.4 Reduction of the processing time of Ag/Bi-2223 tapes

Fabrication of Ag/Bi-2223 tapes usually entails multi-step cycles often requiring a total annealing time of up to several hundred hours. This tedious process is very undesirable for large scale production. The rate

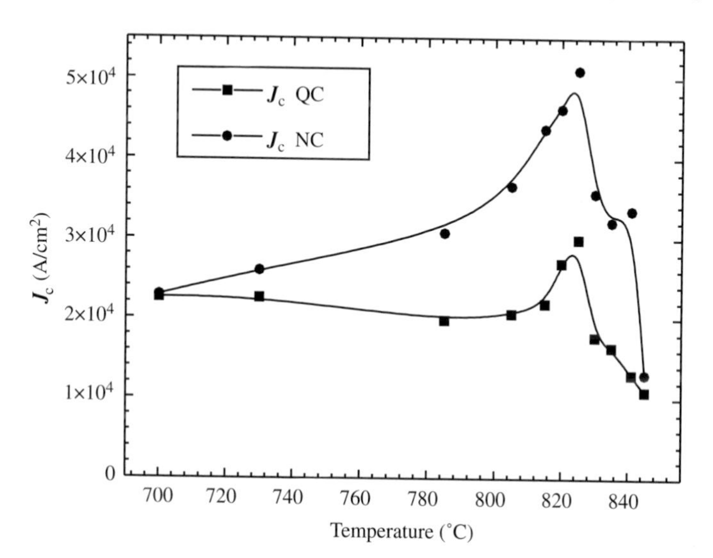

Figure B3.1.12. Critical current density of the tapes treated by quenching and slow cooling (2°C min^{-1}) at two steps after final heat treatment at different temperatures.

of formation of Bi-2223 is an important factor determining the annealing time, and this is also related to the phase assemblage and nominal stoichiometry in the precursor powder, e.g. Pb distribution in the precursor powder has been found to have a significant effect on sintering temperature and time in the final Ag/Bi-2223 composite [22, 24, 25]. For a tape made from precursor powder containing Pb-free Bi-2212 as the major phase, a low sintering temperature and long sintering time is required, whereas a higher sintering temperature and shorter sintering time can be used for tapes made from a precursor powder B2.2.3 with Pb-2212 as the major phase. By incorporating Pb into Bi-2212 phase in the precursor powder, the total sintering was reduced to 100 h.

To further reduce the sintering time, knowledge and information on the phase transformation during heating and cooling are required. In recent work, we showed that a tape quenched at 840°C at the end of the second thermal cycle contained a higher Bi-2223 fraction, along with small quantities of Bi-2201, but no Bi-2212 [78]. These results indicate that a high volume fraction of Bi-2223 may have already formed during the first thermal cycle and that it may have decomposed during the multi-stage heating and cooling. On the basis of this understanding, a new processing schedule was proposed in which the tape was quenched, mechanically deformed and then quickly re-heated to the sintering temperature after the first thermal cycle [79]. In this process any unnecessary phase decomposition and recovery stages were avoided. An intermediate pressing or rolling stage is required to improve the ceramic density and D1.1.2 microstructural texture. XRD examination revealed that the volume fraction of Bi-2223 phase reached 91 and 94% after a total sintering time of 30 h in the multifilament tapes. This indicates that 30 h sintering time is more than enough to form Bi-2223 in the multifilament tapes if the quenching procedure C2 is used.

B3.1.5 Continuous tube filling and forming process

One of the problems associated with the PIT procedure is an unavoidable variation in the initial powder packing density along the tube length, which adversely affects the uniformity of the final tape. Another disadvantage of the PIT technique is that the length of superconducting wire is limited by the length and diameter of the initial silver tube. Very long wire and tape can only be fabricated by starting with a long or large silver tube. Increasing the length of tube makes the packing process more difficult, consequently resulting in an even greater variation in the initial packing density. Increasing the diameter of the tube means increasing the extent of requisite mechanical deformation in order to obtain fine wire and tape. It is well established that extensive mechanical deformation on this type of metal/ceramic composite (ceramic is very hard and silver is very soft and ductile) degrades the uniformity of the superconductor core inside the wires and tapes. The sausaging effect is often observed in PIT wires and tapes that have undergone extensive mechanical deformation.

A1.3 A new wire manufacturing method, called the 'continuous tube forming/filling (CTFF)' technique has been developed, entailing a continuous feeding of loose superconductor powder onto long silver strip by a mass control unit; the silver then being formed into a tube encasing the powder, forming wire. Figure B3.1.13 illustrates the CTFF process schematically [80]. To fabricate multifilament wire, a bundle of monofilamentary wires from a multiple wire feeding unit are continuously fed to a long silver strip and the strip is wrapped into tube to encase the monofilamentary wires, forming CTFF multifilamentary wire. Because the length of silver strip is virtually unlimited, extremely long lengths of CTFF wire can be easily manufactured. Moreover, the quantity of superconductor powder is fed very accurately by machine, making the powder packing density very uniform. During CTFF processing, the superconductor powder and sheath material is directly converted into fine silver-sheathed wire. The extent of mechanical deformation is, thus, largely reduced compared to PIT processing, and the sausaging effect is minimized. Most importantly, the CTFF procedure is a simple, fast and continuous wire processing technique. A1.3 It is more suitable for large industrial scale fabrication of HTSC superconducting wires and tapes.

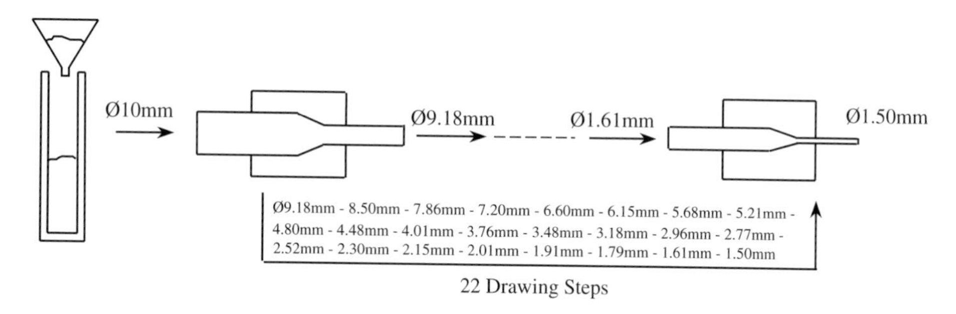

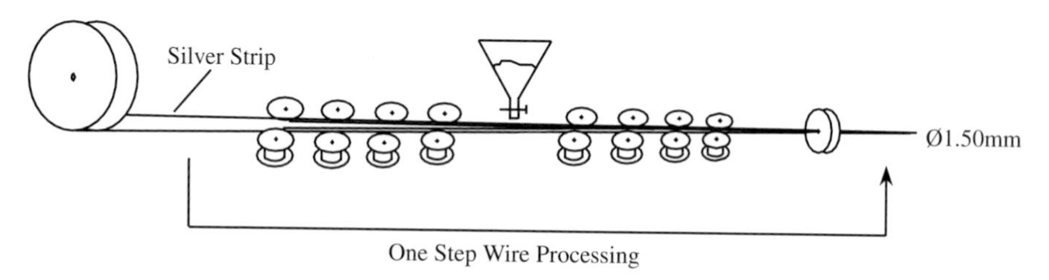

Figure B3.1.13. Comparison of PIT process and CTFF method [35].

By incorporating CTFF processing, processing-induced inhomogeneities including the rough silver/core interface, cracks and density variation along the lengths of wires and tapes associated with PIT processing can be largely eliminated.

An example for CTFF processing is given below. Bi2223 wires have been fabricated using normal multiphase precursor powder and a pure silver strip. The silver strip used is about 0.15 mm thick and 6 mm wide and the resultant wire is approximately 1.5 mm in diameter. The uniformity of the CTFF wires have been examined and found that the interface of silver/core is extremely smooth and the variation of packing density along the length of wire is $< 5\%$. The CTFF wires are converted into flat tapes and heat-treated by a thermomechanical process consisting of either rolling or pressing and sintering. An improved superconductor core uniformity and substantially reduced sausaging and cracking is found in tapes produced from CTFF wires. This is believed to be due to the uniform powder packing density and reduced required mechanical deformation compared to PIT tapes. A powder leaking problem encountered with CTFF wires during tape processing can be circumvented by pre-annealing at temperatures greater than 750°C before flat rolling. A parallel comparison of CTFF tapes and the conventional PIT tapes has been performed by heat treating samples from both techniques under identical conditions. The results indicate that the CTFF tapes are superior in both J_c and I_c behaviour in magnetic field. The CTFF process will be a superior superconductor wire production technique over the conventional PIT process.

B3.1.6 Doctor-blade or dip-coating technique

Because of the volatility of the Pb and decomposition of Bi-2223 phase upon melt, the Ag/Bi-2223 tapes are processed by a combination of rolling and sintering heat treatment. For Bi-2212 wires and tapes, a number of techniques have been used for processing, including doctor-blade, dip-coating, B4.3

jelly-roll, etc. Melting process is normally incorporated in the process as it can achieve better densification and texture without decomposition of Bi-2212 phase.

B3.1.6.1 Doctor-blade technique

B4.3 The doctor-blade method was developed to process Bi-2212 tapes [81]. Bi-2212 powder is prepared by using a conventional calcination and milling procedure. The powder is mixed with an organic formulation consisting of solvent (trichloroethylene), binder (polyvinyl butyral) and dispersant (sorditantrioleate), and milled for 2 days to obtain slurry. The slurry is cast, using a doctor-blade, into a typically 125 mm-wide and 50 μm-thick green sheet. Short tapes 50 mm in length and 3–5 mm in width are cut from this green sheet, mounted on Ag foils 50 mm thick and then heat-treated. Since the microstructure and J_c of the Bi-2212 are very sensitive to the heat-treatment condition, a melt-solidification method has been developed to obtain grain-oriented microstructure of Bi-2212. The composite tape is heated at 500°C for 2 h, in order to remove the organic materials, and the temperature is increased to 885–890°C, slightly higher than the melting point of the Bi-2212. The tape is then slowly cooled to 830–840°C at a rate of 10°C h^{-1} and, finally, the tape is furnace cooled to room temperature. B2.2.3

B3.1.6.2 Dip-coating technique

B4.3 Bi2212/Ag short tapes are easily obtained by the doctor-blade method as mentioned earlier. For fabrication of longer tapes and coils, a dip-coating method is desirable. Bi-2212 pancake coils are fabricated by this method as shown in figure B3.1.14 [82]. Slurry consisting of Bi-2212 powder and organic materials is prepared in a similar manner as slurry for the doctor-blade process except for the larger volume fraction of solvent (trichloroethylene). An Ag tape 10–30 mm in width and 50 μm in thickness is continuously coated with slurry on both sides by passing the tape into the slurry. In order to obtain a uniform layer of the slurry on the tape, it is very important to properly adjust the viscosity of the slurry. Viscosity of the slurry can be controlled by changing the volume fraction of the Bi-2212 in

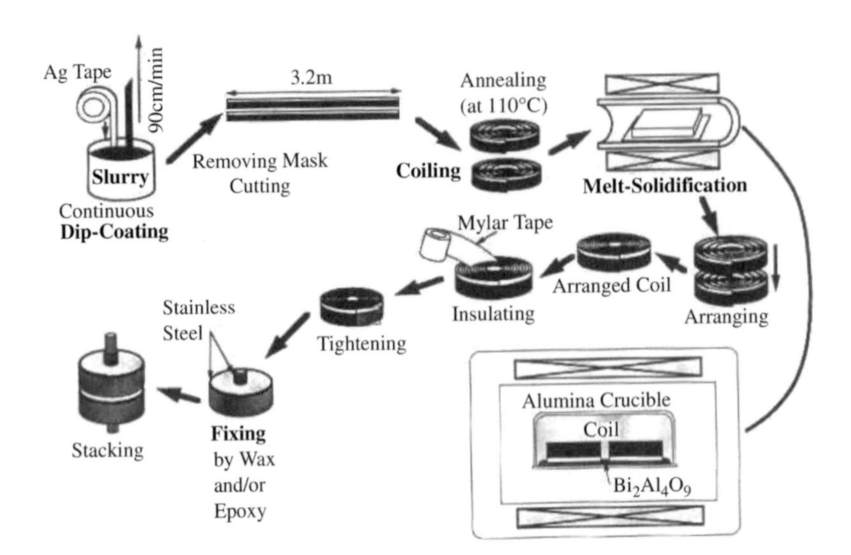

Figure B3.1.14. Schematic illustration of Bi-2212 tape and coil made using a dip-coating technique [82].

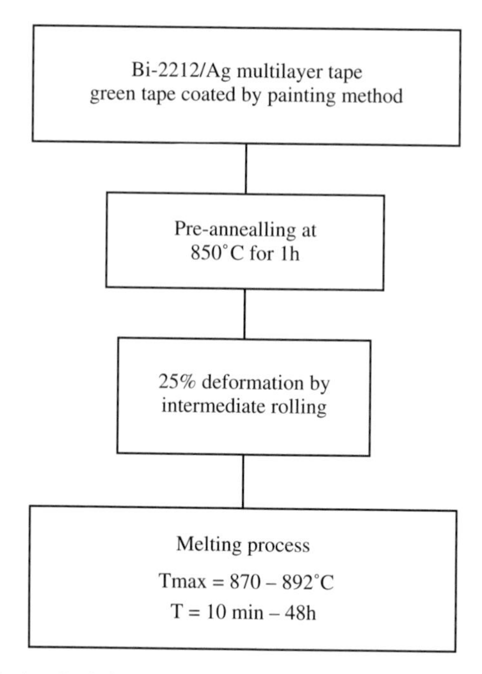

Figure B3.1.15. A schedule of a PAIR procedure for processing Bi-2212 tape.

the slurry. After the solvent in the slurry is completely evaporated, the coated tape is wound loosely into a coil, with a small gap between each turn, and then heat-treated. The heat-treatment condition is similar to that of the doctor-blade process, consisting of partial melting and subsequent slow cooling. After the heat treatment, two or three coils are coaxially combined together by inserting coils to a gap of another coil. Mylar tapes of 50 μm thickness are also inserted into each gap of the coils for insulation.

A process consisting of pre-annealing and intermediate rolling has been developed to treat the doctor-blade or dip-coated Bi-2212 tapes [83]. This is called a PAIR process. In this method, several dip-coated tapes are stacked together and wrapped with 50 μm thick Ag–Mg alloy foil. The green multilayer tapes are processed in a schedule as shown in figure B3.1.15. The PAIR process has significantly improved the microstructure; in particular, the texture and grain connectivity. A record high J_c of 4×10^5 A cm^{-2} in 10 T at 4.2 K has been achieved by this process [84].

B3.1.7 Sheath matrix materials

Sheath materials form an integral part of the PIT wires and tapes. The sheath material should fulfil the following requirements:

(a) chemical compatible with BPSCCO oxide, that is, it must not react with the HTS core;

(b) it must be stable in the ambient atmosphere and does not oxidize in oxygen;

(c) it should have the required mechanical properties such as tensile strength and hardness;

(d) it should have the same thermal expansion coefficient as the HTS core; and

(e) it should be abundant and relatively cheap.

While silver is widely used as the sheath material for superconductor wires and tapes, its mechanical properties are inadequate to withstand the stresses developed during fabrication and service. Stresses developed in the material could lead to the degradation of transport properties; in large and/or high-field magnets, the electromagnetic/magnetic loop stresses developed as a consequence of Lorentz forces could even reach the ultimate strength of the material. Such forces are likely in devices like power cables, motors, transformers and coils, and are especially true for high field magnet insert coils, which will be subject to large hoop stresses on the windings. Another problem with the pure silver sheath is the sausaging effect along the length of tapes. The sausaging core of tapes reduces the effective superconducting cross-sectional area and disrupts the grain alignment near the silver/core interface, which is believed to carry most of the current passing through the tape. More seriously, microcracks are often found to accompany sausaging, which block the current path. The sausaging is due to the mismatch of the mechanical properties of the oxide core and the silver sheath. In order to minimize sausaging, the extent of mechanical deformation of wire processing should be reduced and the mechanical strength of the sheath should be enhanced.

A1.3

B2.3.5

D3.3

To improve the mechanical characteristics of tapes, thereby creating a more robust final product with minimal sausaging, alternative sheath materials are required. The choice of sheath material will be dictated by its mechanical strength, chemical inertness with the superconducting core, ductility, electrical conductivity, coefficient of thermal expansion, permeability to oxygen and cost. A series of silver alloys including AgSb, AgAu, AgAl, AgCu, AgMg, AgNi and AgNiMg [85, 86] have been used as sheath materials. However, the J_c values seen in PIT tapes fabricated from these alloys were lower than that of pure Ag sheathed tapes, stressing the importance of investigating the effect of such sheath alloys on the phase formation of Bi-2223 in PIT tapes. A small quantity of doping element such as Zr, Ti or Hf has been doped to the Ag–Cu alloy sheath for fabrication of Bi-2223 tapes. Improved J_c and J_c behaviour in magnetic field has been reported for these alloy sheathed Bi-2223 tapes [85]. A transport J_c of 2.2×10^4 A cm^{-2} at 77 K and self-field has been achieved in Ag–10 at% Cu–0.1 at%Hf alloy sheathed tape.

Silver alloy containing small amounts of Ni and Mg was first used by Tenbrink *et al.* [86] to fabricate Bi-2212 wire. The Ni and Mg provide solid solution strengthening for the alloy sheath. When the composite is heat treated in an oxidizing atmosphere, these alloying elements oxidize internally, forming oxide precipitates, which strengthen the alloy sheath. Ag–Al alloys used as sheath materials have the same strengthening mechanisms [87]. The strength of fully processed tapes and wires made with these alloy sheaths is comparable with conventional Cu-sheathed Nb_3Sn conductors, as shown in figure B3.1.16 [13]. Most work has concentrated on the commercially available Mg–Ag alloys. Ag–Mg–Ni alloy sheathed Bi-2223 tapes show an increase in mechanical strength versus pure silver, the improvement being ascribed to a segregation of MgO and NiO oxide at the Ag grain boundaries during processing. The effect of such an internal oxidation reaction on the formation mechanism of superconducting materials should be considered as a competitive mechanism. Ag–Mg alloy sheath has a significant effect on the phase formation kinetics. Ag–Mn and Ag–Pt alloys react extensively with Bi-2223 precursor powder.

B2.2.3

A survey of yield points and electrical resistivity of more than 25 Ag alloys shows that AgMn alloys are most appropriate for strengthening the Bi-2223 tapes [88]. Ag sheath containing 0.06–0.4 wt% Mn, 0.75–1.5 wt% Pd and 0.25–0.5 wt% Au used for fabrication of Bi-2223 tapes have been studied by Fisher *et al.* [89]. The results of tensile test show that the yield point of the Ag–Mn sheath (approaching 100 MPa) is double that of pure Ag sheath.

E1.2

For the application of current leads, it is important to use sheath materials with low thermal conductivity in order to reduce the heat leakage. Ag–Au alloy sheath has been studied for this purpose. A 100 cm long current lead of 1000 A with 0.2 W kA^{-1} heat leakage was designed using Ag–11 at% Au alloy sheath [90]. The overall J_c was about 1700 A cm^{-2} at 77 K and self-field with superconductor core cross-section ratio being 0.65. The Ag–Au sheathed Bi-2223 tapes were found to be suitable for the application of current leads.

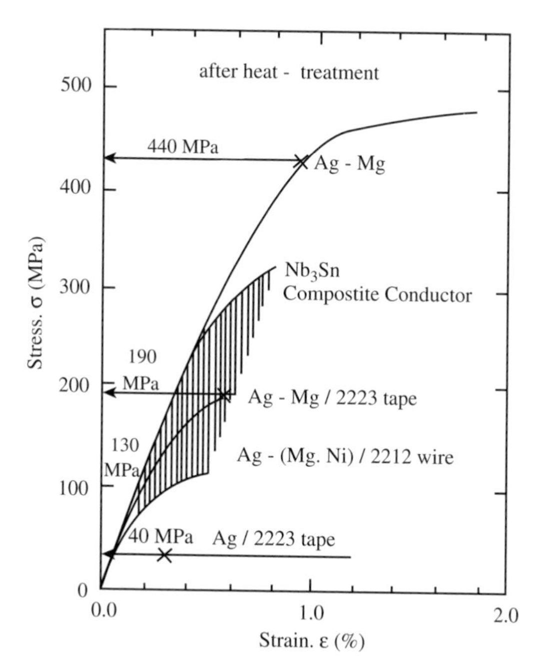

Figure B3.1.16. Strength of fully processed Ag-sheathed BSCCO conductors: 2223 tapes with pure Ag and Ag–Mg sheaths and Ag-sheathed 2212 wire. The strength of a Nb_3Sn composite conductor is shown for comparison. The data for 2212 wire were plotted using a 0.1% offset from the modulus for Ag–Mg [13].

It was found previously that Ag addition in the Bi-2223 tapes lowers the annealing temperature and accelerates the formation of 2223 phase. However, the undesirable particle shape causes grain misorientation and hence leads to a decrease in J_c. In contrast, Ag addition with large particle size shows a beneficial effect on the grain alignment and grain connectivity. During the drawing and rolling processes Ag acts as soft medium or 'lubricant' to assist the powder flow, resulting in smoother interface between the Ag layers and oxide layers [91]. SEM image of Ag-doped and undoped tapes shows that oxide layers adjacent to the Ag lamina are dense and well aligned in comparison with the undoped tape. A noticeable improvement in I_c – B behaviour of the Ag-doped tape was observed in comparison with undoped tape. The improvement in J_c is evident in the low field region while the J_c field dependence is about the same for Ag-doped and undoped tapes in higher fields. This suggests that Ag addition mainly improves weak links in the tape.

D1.3

B3.1.8 Resistive interfilamentary barriers for reducing ac loss

As reported previously the Ag/Bi-2223 multifilamentary tape with untwisted filaments behaves as if its filaments were fully coupled, resulting in substantial loss. By twisting the filaments with a small pitch, the loss is reduced. However, there is still significant coupling loss component in this case, in particular when the field is applied perpendicular to the tape surface. Another procedure to reduce the ac loss is to increase the resistivity of the sheath materials, as discussed in previous section. But most Ag alloys are still rather conductive and the choice of high resistive Ag alloys is restricted due to the requirement of chemical compatibility with superconductor oxides. Moreover, both oxide core and alloy sheath are brittle, limiting the magnitude of twisting pitch.

D2.6

A technique to incorporate the resistive oxide barrier layers in the Ag/Bi-2223 multifilamentary tapes has been developed in which the ac loss has been substantially reduced [92]. The oxide barriers used for this purpose include $BaZrO_3$, $SrZrO_3$ [92] and $SrCO_3$ [93]. The barrier layers are reasonably uniform. The transverse resistivity was increased in the twisted multifilamentary Bi-2223 tapes by a factor of 10, leading to a substantial coupling loss. A similar approach having a resistive $SrCO_3$ barrier layer between the filaments has been used [93]. These barriers are cheap, commercially available and compatible with Ag and oxide core. J_c over 20 kA/cm^2 at 77 K has been achieved.

Acknowledgments

The author gratefully acknowledge Drs Y.C. Guo, J. Horvat, R. Zeng, W.G. Wang, B. Zeimetz, Q.Y. Hu, M Ionescu and Prof H.K. Liu for their contribution to this work. Sincere thanks are given to Prof T.Beales and Dr M. Apperley for their help and support.

References

[1] Dou S X and Liu H K 1996 *Supercond. Sci. Technol.* 6 297 and Ref. therein
[2] Gammel P L, Schneemeyer L F, Waszczak J V and Bishop D J 1988 *Phys. Rev. Lett.* **61** 1666
[3] Heine K, Tenbrink J and Thoner M 1989 *Appl. Phys. Lett.* **55** 2441
[4] Sato K, Hikata T and Iwasa Y 1990 *Appl. Phys. Lett.* **57** 1928
[5] Yamada Y, Oberst B and Flükiger R 1991 *Supercond. Sci. Technol.* **4** 165
[6] Dou S X, Liu H K and Guo Y C 1992 *Physica* C **194** 343
[7] Li Q, Brodersen K, Hjuler H A and Freltoft T 1993 *Physica* C **217** 360
[8] Tenbrink, J, Wilhelm, M, Heine, K and Krauth, H *Applied Superconductivity Conf.*, (Snowmass Village, CO, September 1990)
[9] Michel C, Hervieu M, Borel M M, Grandin F, Deslandes F, Provost J and Raveau B 1987 *Z. Phys.* B **68** 421
[10] Maeda H, Tanaka Y, Fukutomi M and Asano T 1988 *Japan. J. Appl. Phys. Lett.* **27** L209
[11] Dou S X, Liu H K, Bourdillon A J, Kviz M, Tan N X and Sorrell C C 1989 *Phys. Rev.* B **40** 5266
[12] Liu H K, Dou S X, Savvides N, Zhou J P, Tan N X, Bourdillon A J and Sorrell C C 1989 *Physica* C **157** 93
[13] Hellstrom E E 1995 *High-Temperature Superconducting Materials Science and Engineering, New Concepts and Technology*, ed D L Shi (Pergamon: Elsevier) p 383
[14] Shi D, Chen J G, Welp U, Boley M S and Zanggvil A 1989 *Appl. Phys. Lett.* **55** 1354
[15] Bunker B, Lamppa D L, and Voight J A. US Patent 4839339 (1989)
[16] Sengupta S 1998 *JOM* **50** 19
[17] Ring T 1996 *Fundamentals of Ceramic Powder Processing and Synthesis* (New York: Academic) p 179
[18] Johnson S M, Gusman M I, Rowcliffe D J, Geballe T H and Sun J Z 1987 *Adv. Ceram. Mater.* **2** 337
[19] Dou S X, Guo Y C and Liu H K 1992 *Physica* C **194** 343
[20] Dou S X, Guo Y C and Liu H K 1992 *Mater. Res. Soc. Symp. Proc.* **275** 227
[21] Hatano T, Aota K, Ikeda S, Nakamura K and Ogawa K 1988 *Japan. J. Appl. Phys.* **27** L2055
[22] Grasso G, Jeremie A and Flukiger R 1995 *Supercond. Sci. Technol.* **8** 827
[23] Maroni V A, *Extended Abstract of The 1997 International Workshop on Superconductivity*, cosponsored by ISTEC and MRS, (Hawaii, USA, June 1997) p 167.
[24] Wang W G, Horvat J, Liu H K and Dou S X 1996 *Supercond. Sci. Technol.* 875
[25] Guo Y C, Wang W G, Liu H K and Dou S X *Proc. 16th ICMC/ICEC*, p 1393 ed T Huruyama, T Mitsuii and K Yamafuji
[26] Flukiger, R, Jeremie, A, Hensel, B, Seibt, E, Xu, J Q and Yamada, Y *ICMC* (Huntsville, USA, June 1991)
[27] Motowidlo L R, Haldar P, Jin S and Spencer N D 1993 *IEEE Trans. Appl. Supercond.* **3** 942
[28] Han Z, Skov-Hansen P and Freltoft T 1997 *Supercond. Sci. Technol.* **10** 371
[29] Hu Q Y, Liu H K and Dou S X 1992 *Cryogenics* **32** 1038
[30] Dou S X, Ionescu M, Wang W G, Liu H K, Babic E and Kusevic I 1995 *J Electronic Mater.* **24** 1801
[31] Grasso G and Flukiger R 1997 *Advances in Superconductivity* **vol 9**, ed S Nakajima and M Murakami (Tokyo: Springer) p 835
[32] Grasso G, Jeremie A and Flukiger R 1995 *Supercond. Sci. Technol.* **8** 827
[33] Dou S X, Liu H K, Apperley M, Song K H and Sorrell C C 1990 *Physica* C **167** 525
[34] Balachandran U, Langan M T, Einloth M C, Singh J P, Goretta K C and Poeppel R B 1990 *Ceramic Trans.* **13** 621
[35] Däumling M, Grasso G, Grindatto D P and Flükiger R 1995 *Physica* C **250** 30
[36] Hu Q Y, Liu H K and Dou S X 1997 *Physica* C **274** 204
[37] Dou S X, Hu Q Y, Guo Y C, Horvat J and Liu H K 1998 *Supercond. Sci. Technol.* **11** 781
[38] Guo Y C, Liu H K, Liao X Z and Dou S X 1998 *Physica* C **301** 199

[39] Liu H K, Guo Y C and Dou S X 1992 *Supercond. Sci. Technol.* **5** 591
[40] Dou S X, Liu H K, Wang J and Bian W M 1991 *Supercond. Sci. Technol.* **4** 21
[41] Dou S X, Liu H K, Zhang Y L and Bian W M 1991 *Supercond. Sci. Technol.* **4** 203
[42] Song K H, Sorrell C C, Dou S X and Liu H K 1991 *J. Am. Ceram. Soc.* **74**
[43] Rouessac V, Desgardin G and Gomina M 1997 *Physica* C **282–287** 2573
[44] Zeng R, Ye B, Horvat J, Guo Y C, Zeimetz Z, Yang X F, Beales T B, Liu H K and Dou S X 1998 *Physica* C **307** 29
[45] James M P, Ashworth S P, Glowacki B A, Garrre R and Conti S 1995 EUCAS *Inst. Phys. Conf. Ser.* **148** 343
[46] Tomsic M 1997 *Supercond. Industry* **10** 18
[47] Kopera L, Kovac P and Husek I 1999 *Supercond. Sci. and Technol.* **11** 433
[48] Majewski P 1997 *Supercond. Sci. Technol.* **10** 453
[49] Takada Y, Kanno R, Tanigawa F, Yamamoto O, Ikeda Y and Takanno M 1989 *Physica* C **159** 247
[50] Fisher K, Rojek A, Thierfeldt S, Lippert H and Arons R R 1989 *Physica* C **160** 466
[51] Triscone G, Genoud J Y, Graf T, Junod T and Muller J 1991 *Physica* C **176** 247
[52] Wang J, Wakata M, Kaneko T, Takano S and Yamauchi H 1993 *Physica* C **208** 323
[53] Guo Y C, Liu H K and Dou S X 1992 *Physica* C **200** 147
[54] Daumling M, Maad R, Jeremie R and Flukiger R J 1997 *Mater. Res.* 6 **12** 1445
[55] Rubin L M, Orlando T P, Vander Sande J B, Gorman G, Sacoy R, Swope R and Beyers R 1993 *Physica* C **217** 217
[56] MacManus-Driscoll J, Bravman J, Savoy R J, Gorman G and Beyers R B 1994 *J. Am. Ceram. Soc.* **77** 2305
[57] Chen Y L and Steven R 1992 *J. Amer. Cerc. Soc.* **75** 1150
[58] Zhu W and Nicholson P S 1993 *J. Appl. Phys.* **73** 8423
[59] Tetenbaum M, Hash M, Tani B S, Luo J S and Maroni V A 1995 *Physica* C **249** 396
[60] Majewski P, Kaesche S and Aldinger F 1997 *J. Am. Ceram. Soc.* **80** 1174
[61] Moon R J, Trumble P and Bowman K J 1999 *J. Mater. Res.* **14** 653
[62] Guo Y C, Liu H K and Dou S X 1993 *J. Mater. Res.* **8** 2187
[63] Morgan P E D, Housley R M, Porter J R and Ratto J J 1991 *Physica* C **176** 279
[64] Morgan P E D, Piche J D and Housley R M 1992 *Physica* C **191** 179
[65] Hu Q Y, Liu H K and Dou S X 1995 *Physica* C **250** 7
[66] Grivel J C and Flukiger R 1996 *Supercond. Sci Technol.* **9** 555
[67] Poulsen H F, Frello T, Andersen N H, Bentzon M D and von Zimmermann M 1998 *Physica* C **298** 265
[68] Bian W, Zhu Y, Wang Y L and Suenaga M 1995 *Physica* C **248** 119
[69] Cai Z X and Zhu Y 1997 *Mater. Sci Eng.* **A238** 210
[70] Zeng R, Fu S K, Liu H K, Beales T and Dou S X 1999 *IEEE Appl. Supercond.* **9** 2742
[71] Thurston T R, Wildgrouber U, Jirawi N, Haldar P, Seenaga M and Wang Y L 1996 *J. Appl. Phys.* **79** 3122
[72] Giannimi E, Bellingeri E, Passerini R and Flukiger R 1999 *Physica* C **315** 185
[73] Dou S X, Zeng S X, Hoevat J, Beales T and Liu H K 1999 *IEEE Appl. Supercond.* **9** 2436
[74] Wang W G, Horvat J, Liu H K and Dou S X 1997 *Physica* C **291** 1
[75] Parrell J A, Larbalestier D C, Riley G N, Li Q, Carter W L, Parrella R D and Teplitsky M 1997 *J. Mater. Res.* **12** 2997
[76] Parrell J A, Larbalestier D C, Riley G N, Li Q, Carter W L, Parrella R D and Teplitsky M 1996 *Appl. Phys. Lett.* **69** 2915
[77] Wang W G, Horvat J, Zeimetz B, Liu H K and Dou S X 1998 *Physica* C **297** 1
[78] Guo Y C, Horvat J, Liu H K and Dou S X 1998 *Physica* C **300** 38
[79] Dou S X, Zeng R, Beales T and Liu H K 1998 *Physica* C **303** 21
[80] Guo Y C, Liu H K, Dou S X and Collings E W 1997 *Physica* C **282–287** 2597
[81] Kase J, Irisawa N, Togano K, Kumakura H, Dietderich D R and Maeda H 1991 *IEEE Trans. Magn.* **17** 1254
[82] Shimoyama J, Kadowaki K, Kitaguchi H, Kumakura H, Togano K, Maead H, Nomura K and Seido M 1993 *Appl. Supercond.* **1** 43
[83] Miao H, Kitaguchi H, Kumakura H, Togano K and Hasegawa T 1998 *Physica* C **301** 116
[84] Miao H, Kitaguchi H, Kumakura H, Togano K, Hasegawa T and Koizumi T 1999 *Physica* C **320** 77
[85] Ishizuka M, Tanaka Y and Maeda H 1995 *Physica* C **252** 339
[86] Tenbrink J, Wilhelm M, Heine K and Krauth H 1993 *IEEE Trans. Magn.* **3** 1123
[87] Motowidlo L R, Galingski G, and Haldar P *The Materials Research Society Spring Meeting*, SF (1993)
[88] Hutten A, Schubert M, Rodig C, Schlafer U, Verges P and Fisher K 1997 *Appl. Supercond.* **2** 1255
[89] Fisher K, Fahr T, Hutten A, Schlafer U, Schubert M, Rodig C and Trinks H P 1998 *Supercond. Sci. Technol.* **11** 995
[90] Sasaoka T, Normura K, Sato J, Kuma S, Fujishiro H, Ikebe M and Noto K 1994 *Appl. Phys. Lett.* **64** 1304
[91] Zeimetz B, Dou S X and Liu H K 1996 *Supercond. Sci. Technol.* **9** 888
[92] Huang Y B, Grasso G, Marti F, Dhalle M, Witz G, Clerc S, Kwasnitza K and Flukiger R 1997 EUCAS *Inst. Phys. Conf. Ser.* **158** 1385
[93] Goldacker W, Quilitz M, Obst B and Eckelmann H 1998 *Physica* C **310** 182

B3.2.1
Processing of high T_c conductors: the compound Bi(2212)

Gérard Duperray and Peter F Herrmann

B3.2.1.1 Introduction

It has been shown rapidly after the discovery of the BiSrCaCuO superconductor family that flexible conductors can be made by using silver as the sheath material [1]. During the first half of the 1990s much efforts were devoted to the development of Bi(2223) powder in tube (PIT) conductors [2, 3]. In spite of its lower T_c, up to 93 K for the Bi(2212) phase and 110 K for the Bi(2223) phase, Bi(2212) presents various advantages. Simple processing conditions allow the use of comparably cost effective Bi(2212) powders made by the melt cast process (MCP) [4] and a cost effective conductor fabrication that does not require a multi-step thermo-mechanical heat treatment of the conductor. High connectivity between Bi(2212) grains can be achieved by melt growth processing using a simple single-step heat treatment. In the intermediate temperature range (20–60 K), the current densities of Bi(2212) conductors of comparable degree of development are significantly above the current density of Bi(2223) conductors. Over the past few years, interest in Bi(2212) has increased considerably resulting in various conductor processing technologies.

C2 A1.3 B2.3.5 B3.1

The aim of this chapter is to provide the reader with a background sufficient to understand most of the publications dealing with the constant progress of bulk and multifilamentary Bi(2212) based conductors. This contribution is organized so that a novice in this subject could collect a set of the most papers and most of the information on the leading teams in the field. He will also be informed of the main 'hard points' to be solved as a priority for bringing such superconductors to a real economical level.

A1.3 B3.1

Knowledge of the phase diagram contents and the kinetics of the reactions is necessary to define the major role of the thermodynamic parameters during elaboration of the conductors and to reach the final phase purity needed for performance of the conductors. So the chemistry of the Bi(2212) is detailed from the basic Bi–Sr–Ca–Cu–O phase diagram down to particular effects of each one of numerous parameters: starting composition, smoothness of the interfaces, O_2 partial pressure, treatment temperature, heating, cooling rates etc, by means of the best papers dealing with this complicated chemical system. After a review as exhaustive as possible of the multiple phases identified, attention is paid to the remarkably large stability domain of most of these phases as a function of stoichiometry.

B2.3.5

Common difficulties like sausaging and a more specific defect, the 'bubbling phenomenon' affecting sheathed Bi(2212), are carefully examined.

A wide panel of all the thermal cycles proposed in the literature is reviewed with an attempt to check the effect of each specific step on the final quality of the corresponding conductors.

Dip coated and PIT conductors are developed [5], a round and isotropic conductor containing round filaments is developed [6], recorded current densities are reached by the PAIR process [7] and a new method for assembling anisotropic tapes to form a round conductor ROSAT with almost isotropic behaviour has been achieved [8]. The development of rectangular PIT tapes [9] is strongly oriented towards reproducible fabrication of long length conductors for 'Wind and React' (W&R) or 'React and Wind' (R&W) applications. Some transverse sections of these conductor concepts are shown in figure B3.2.1.1 [6–9]. The common feature of these conductors is that the Bi(2212) precursor powder is protected by a silver envelope. At high temperatures this envelope is permeable to oxygen and allows the adjustment of the oxygen content in the superconductor during the thermal treatment. A second important fact is that the alignment of the superconducting phase is initiated at the interface between the silver and the superconducting filaments. In conductors with flat filaments, this results in a high degree of texture near the flat interface with the crystallographic a–b planes parallel to the conductor.

A1.3 B2.1

Today, PIT conductors using either Bi(2223) or Bi(2212) can be realized in kilometre lengths which are used in prototype magnet applications and in various ac demonstration devices. This increase in interest for the Bi(2212) phase compared to the Bi(2223) phase is also related to economic aspects. Simple economic considerations [10, 11] show that optimum dc application conditions under moderate fields will be in the 20 K temperature range. This conclusion weakly depends only on the conductor cost and remains valid even if the well known 10 $/kAm cost target is reached and places the Bi(2212) phase in a favourable position for these application conditions. However, for self field ac applications the optimum conditions are in the liquid nitrogen temperature range where the Bi(2223) phase is currently the best choice. Today's performance of Bi(2212) and Bi(2223) do not allow their use in high field applications at liquid nitrogen temperatures. For this application domain the best conductor candidate might be the Y(123) based coated conductor (see chapter B4.3), which so far can only be made in short lengths.

A1.3 A1.3 B4.3 B4.2.1

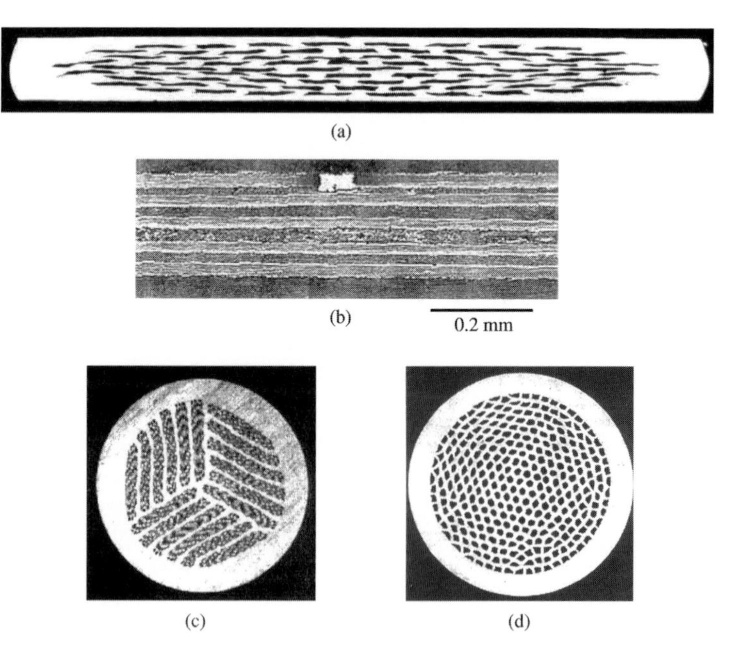

Figure B3.2.1.1. Different Bi(2212) conductor geometries: Alcatel: Rectangular PIT tape, Ag–Pd matrix, Showa Electric: PAIR-Conductor, Courtesy of H. Kitaguchi from NRIM, Hitachi: ROSAT-Structure, Courtesy of M. Okada from HRL, IGC: Round conductor, Courtesy of E.W. Collings from IGC.

This chapter is divided into six sections. In section B3.2.1.2, the Bi–Sr–Ca–Cu–O phase diagram is reviewed. The Bi(2212) melting step is analysed in section B3.2.1.3 with a special focus on *fusion-decomposition of the Bi(2212) phase*. Parameters that influence the *solidification and reformation* are reviewed in section B3.2.1.4. The reaction heat treatment to optimize J_c is discussed in section B3.2.1.5. Different conductor fabrication technologies are briefly reviewed in section B3.2.1.6. The last section, B3.2.1.7, presents an exhaustive review of the Bi(2212) conductors manufacturing processes and sheath composition linked to their respective performances and specific advantages like tolerance to high magnetic field, mechanical stress, feasibility, and cost effectiveness.

B2.5.3

B3.2.1.2 The phase diagram

The Bi–Sr–Ca–Cu–O phase diagram has been intensively studied since the discovery of the superconducting Bi phases. These chemical data have been used to optimize the thermal cycles to produce pure and well-textured Bi(2212). Figure B3.2.1.2 shows the phase diagram provided by Majewski [12]. The figure shows the most common phases in the Bi–Sr–Ca–Cu–O system. The representation has been simplified to a tetrahedron by using the four respective oxides at each corner instead of five elements. Two categories of species can be distinguished:

B2.5.3

- The Bi(2212) and Bi(2223) superconducting phases are located inside the tetrahedron because they contain all the four oxides. These two phases having a range of stoichiometry are not point compounds.

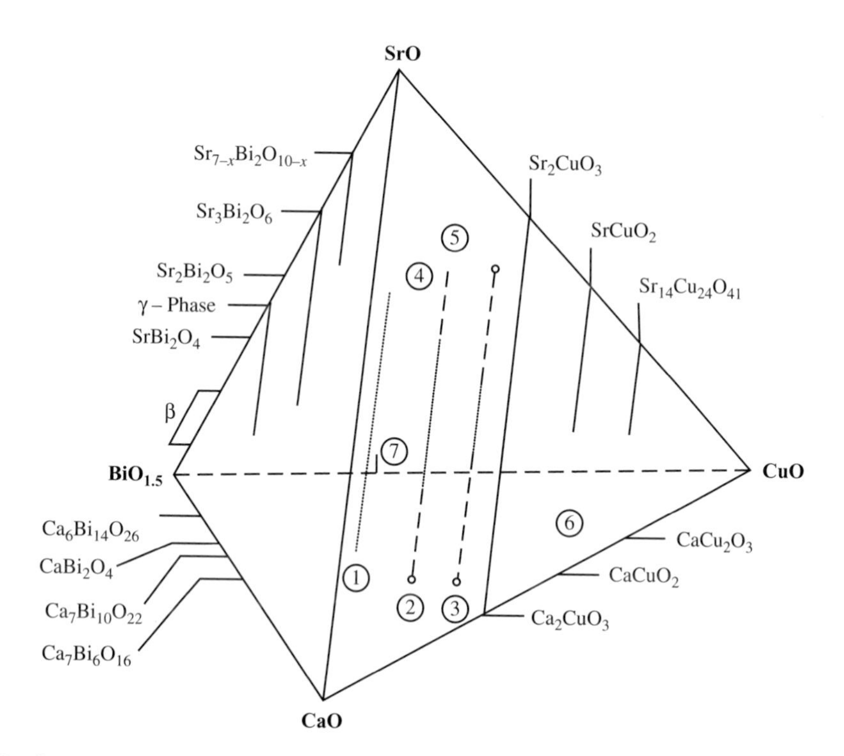

Figure B3.2.1.2. Compounds in the quaternary system Bi_2O_3–SrO–CaO–CuO at 850°C in air © 1996 ed. Marcel Dekker [12].

Table B3.2.1.1. Identified phases formed during the decomposition of the Bi(2212)

Superconducting phases	
$Bi_2Sr_2CuO_y$ or $Bi_2(Sr,Ca)_2CuO_y$	Bi(2201)
$Bi_2Sr_2CaCu_2O_y$ or $Bi_2(Sr,Ca)_3Cu_2O_y$	Bi(2212)
Alkaline earth cuprates	Cu free phases
$(Sr,Ca)CuO_2$	$Bi_2(Sr,Ca)_3O_y$
$(Sr,Ca)_2CuO_3$	$Bi_9Sr_{11}Ca_5O_y$
$(Sr,Ca)_{14}Cu_{24}O_x$ (generally $Sr_8Ca_6Cu_{24}O_x$)	$Bi_2(Sr,Ca)_4O_y$
$(Sr,Ca)Cu_2O_2$	$Bi_3(Sr,Ca)_7O_y$
$(Sr,Ca)_3Cu_5O_8$	

- The other phases that contain only three oxides are located on the faces of the tetrahedron. Most of these phases are generated during the solid or liquid decomposition of the superconducting phases and will be studied in section B3.2.1.3.

  * The alkaline earth cuprates (AEC) located in the SrO–CaO–CuO plane are characterized by their hardness and high melting temperatures. Commonly designed as 'refractory residual phases' these phases show a large Sr–Ca substitution which can be complete for $(Sr,Ca)_2CuO_3$. They are detailed in table B3.2.1.1.
  * The 'Cu-free phases' are in the Bi_2O_3–SrO–CaO plane. These phases also have Sr–Ca substitution seen in the AEC.
  * An important phase in the Bi_2O_3–SrO–CaO plane is Bi(2201). This phase has a range of stoichiometry and exists within the tetrahedron because of Sr–Ca substitution with a representative formula $Bi_2(Sr,Ca)_2CuO_y$.
  * No other interesting phases are found in the Bi_2O_3–CaO–CuO plane.

Special attention must be paid to the relatively large stability domain of the Bi(2212) as a function of stoichiometry variations. Majewski [12] provided exhaustive data on the Bi(2212)'s stability as a function of the stoichiometric index y of Ca in $Bi_xSr_{3-y}Ca_yCu_2O_8$ and the stoichiometric index x of Bi in $Bi_xSr_2Ca_2Cu_2O_8$. The quantitative representation of these variations is difficult in the three-dimensional diagram, since it cannot provide temperature information. Figure B3.2.3 for Sr–Ca substitution shows more clearly the stability domain of Bi(2212) as a function of temperature for stoichiometric indexes varying from $Bi_2Sr_{2.6}Ca_{0.4}Cu_2O_y$ to $Bi_2SrCa_2Cu_2O_y$. Figure B3.2.4 shows the stability domain of Bi(2212) with excess Bi as a function of temperature for stoichiometric indexes varying from $Bi_2Sr_2CaCu_2O_y$ to $Bi_{2.4}SrCa_2Cu_2O_y$. In both cases the phase relations surrounding the stability domain are indicated.

Another important point explained in [12] is the oxygen index. This parameter is an indicator of the oxidation level of the copper between Cu_2O and CuO_{1+d} that directly influences the superconducting properties of Bi(2212). The oxygen index was found to depend on Ca content, Bi content and temperature:

- The oxygen index increases from 8.15 in $Bi_{2.18}Sr_{2.2}Ca_{0.8}Cu_2O_{8.15}$ up to 8.3 in $Bi_{2.18}Sr_{1.3}Ca_{1.7}Cu_2O_{8.3}$ and from 8.12 in $Bi_{2.1}Sr_2CaCu_2O_{8.12}$ up to 8.37 in $Bi_{2.3}Sr_2CaCu_2O_{8.37}$.

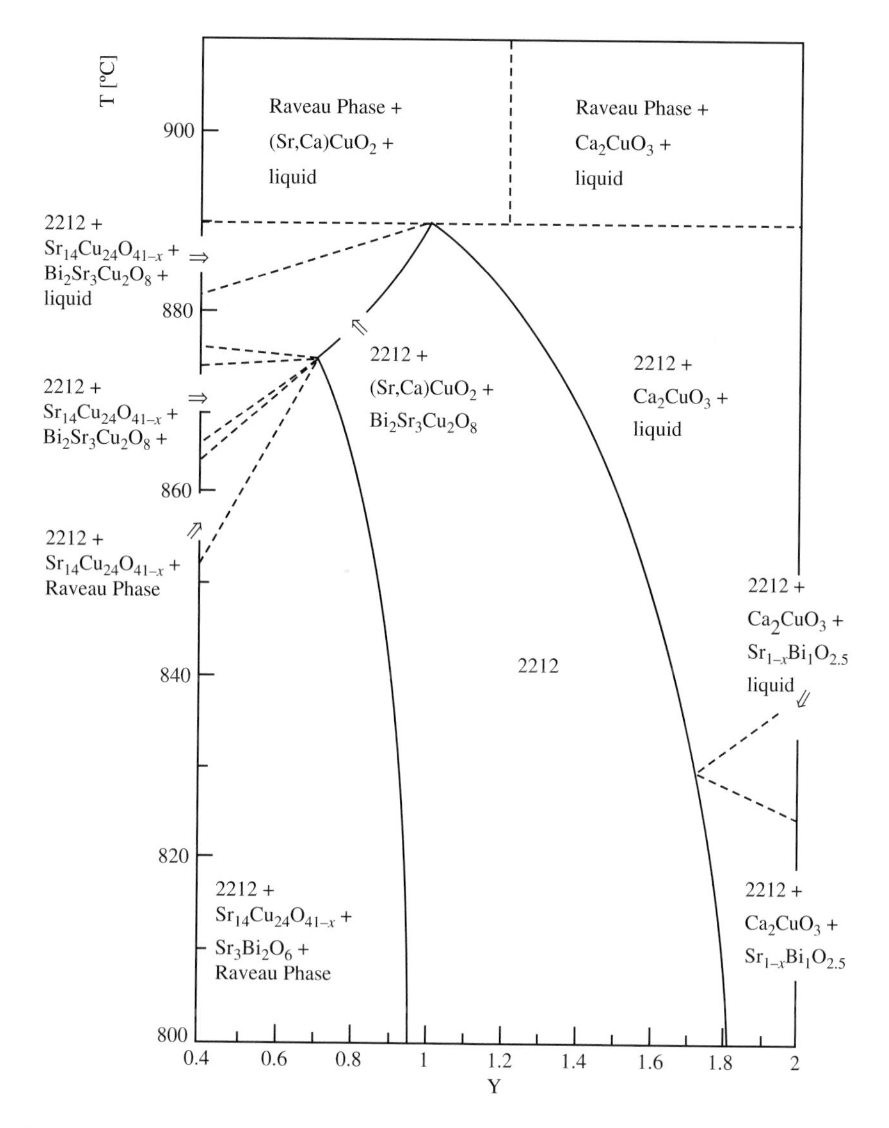

Figure B3.2.1.3. Stability range of $Bi_{2.18}Sr_{3-y}Ca_yCu_2O_{8+\delta}$ as a function of temperature and Ca content © 1996 ed. Marcel Dekker [12].

- The oxygen content increases with decreasing temperature and this effect is more pronounced in Ca rich compositions. Thermogravimetric measurements showed a maximum oxygen increase when the temperature of the $Bi_{2.18}Sr_{1.3}Ca_{1.7}Cu_2O_y$ composition was lowered from 820 to 700°C.

A discontinuity at 800°C accompanied by a sudden mass variation between an oxygen depleted Bi(2212) α and a Bi(2212) β phase with a constant oxygen index of 8.33 is described in [13]. This oxygen variation mechanism occurring in the solid state should not be confused with the oxygen evolving during melting, which will be examined in the next section.

For a clearer and better understanding of the reversibility, the melting and reformation of Bi(2212) are separated.

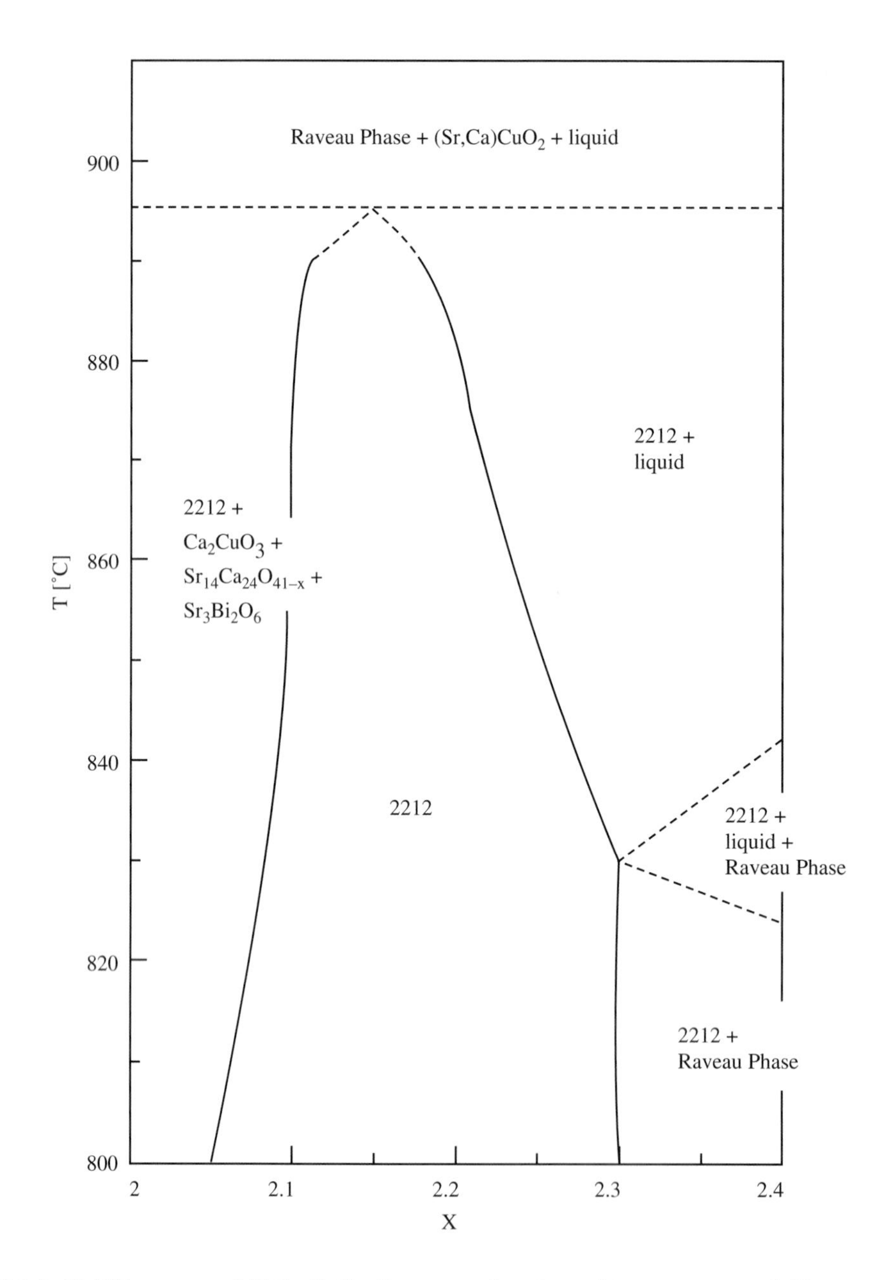

Figure B3.2.1.4. Stability range of $Bi_xSr_2CaCu_2O_{8+\delta}$ as a function of temperature and Bi content © 1996 ed. Marcel Dekker [12].

B3.2.1.3 Fusion-decomposition of the Bi(2212) phase

Many studies have examined the stability and decomposition of Bi(2212). Direct observation of melting in air was studied by hot stage microscopy with simultaneous thermal analysis and final SEM/EDS characterization by Morgan *et al* [14]. Table B3.2.1.1 lists the phases formed during the decomposition of Bi(2212) as a function of temperature and oxygen partial pressure.

The peritectic fusion is observed at 880°C in air with thin needles of (Sr,Ca)CuO_2, $(Sr,Ca)_2CuO_3$ and small 'Cu free phase' crystals floating on the surface of the liquid [14]. After 6 min at 882°C, the surface is entirely covered by thin needles with their size increasing as a function of time and temperature. These needles do not dissolve during cooling. The melting of Bi(2212) corresponds to a reduction: Bi(2212) → Solid(s) + Liquid + O_2 and its temperature depends on the pO_2. The variation of enthalpy and entropy for melting was deduced in [15] from the variation of the melting temperature as a function of pO_2:

$$\Delta H = 731\,\mathrm{kJ\,mol^{-1}} \quad \Delta S = 624\,\mathrm{J\,mol^{-1}\,K^{-1}}.$$

An exhaustive study of the stability of Bi(2212) as a function of temperature and pO_2 was reported in [16–19], leading to the frequently provided phase diagram (figure B3.2.1.5), which shows a phase boundary quite similar to that of CuO/Cu_2O with a change in slope at 1060 K (790°C). Five domains are delimited, one stability domain at low temperature and high pO_2 (upper right part of the diagram) and four decomposition domains with O_2 elimination.

At low pO_2, the lower part of the diagram below the curve is the subsolidus where the Bi(2212) phase decomposes in a solid state reaction:

$$\text{(A domain)}\quad Bi_2Sr_2CaCu_2O_{8+\delta} \rightarrow Bi_2Sr_2CaO_{6+\delta} + Cu_2O + 1/2\,O_2.$$

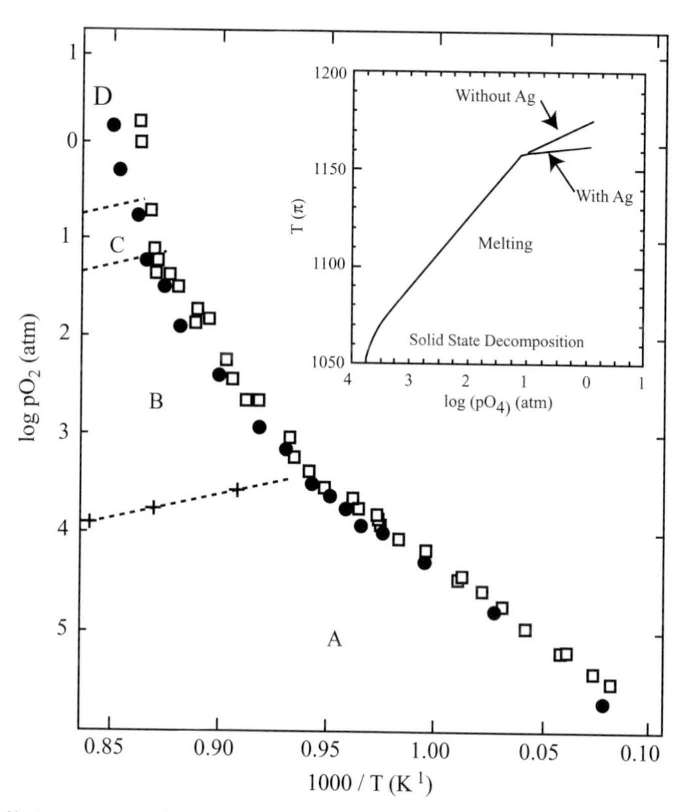

Figure B3.2.1.5. Van't-Hoff plot showing the stability limits of Bi(2212) (□) with Ag and (●) without Ag © 1996 ed. Marcel Dekker [22, 18].

At higher pO_2 (upper left part of the diagram below the curve) the Bi(2212) phase decomposes into one liquid phase and one or more solid phases:

$$\text{(B domain)}\quad \text{Bi(2212)} \rightarrow \text{liquid} + \text{(Sr, Ca)}_2\text{CuO}_3$$

$$\text{(C domain)}\quad \text{Bi(2212)} \rightarrow \text{liquid} + \text{Bi}_2\text{(Sr, Ca)}_3\text{O}_y + \text{(Sr, Ca)CuO}_2$$

$$\text{(D domain)}\quad \text{Bi(2212)} \rightarrow \text{liquid} + \text{Bi}_2\text{(Sr, Ca)}_3\text{O}_y + \text{(Sr, Ca)}_{14}\text{Cu}_{24}\text{O}_x.$$

This diagram also shows the influence of Ag which lowers the melting temperature at high pO_2 values.

The liquid in domain D has a composition between Bi(2212) and Bi(2201) with 0.28 dissolved Ag. Below 0.1 atm O_2 and above 900°C $(Sr,Ca)_{14}Cu_{24}O_x$ decomposes into $(Sr,Ca)CuO_2 + CuO$.

Another study [20] was made in the domain 540°C $< T <$ 800°C, 7.4×10^{-3} atm $< pO_2 <$ 15 atm. Above 8 atm O_2 the decomposition begins at 700°C and corresponds to an oxidation:

$Bi(2212) + 1/2O_2 \rightarrow Bi_9Sr_{11}Ca_5O_y + Bi_2(Sr,Ca)_2CuO_y + CuO$. In air the decomposition begins at 695°C and is very slow. Above 0.1 atm O_2 the upper stability line depends on the redox couple Bi^{3+}/Bi^{5+} and below 0.1 atm O_2 the lower stability line depends on the redox couple Cu^{2+}/Cu^{+}. Gannon and Sandhage [21] studied the stability of Bi(2212) in PIT and thick films up to 600 atm O_2. When the temperature is increased under 600 atm O_2 there is an initial solid state decomposition domain limited by the melting of Ag:

$$\text{Bi(2212)} + 1/2\text{O}_2 \rightarrow \text{Bi}_2\text{(Sr, Ca)}_2\text{CuO}_y + \text{Bi}_9\text{Sr}_{11}\text{Ca}_5\text{O}_y + \text{CuO}$$

then the melting occurs beyond 910°C even under pO_2 as high as 660 atm. In [22] the lines corresponding to the phase boundaries of the four decomposition domains of temperature and pO_2 are expressed as equations and the chemical species present in the system are examined. In any case the phase rule predicting a maximum of three solid phases in the liquid is verified:

$$\text{From 0.5 to 1 atm O}_2\quad \text{liquid} + \text{(Sr, Ca)}_{14}\text{Cu}_{24}\text{O}_x + \text{Bi}_2\text{(Sr, Ca)}_4\text{O}_y$$

$$0.2\ \text{atm O}_2\quad \text{liquid} + \text{(Sr, Ca)CuO}_2 + \text{Bi}_2\text{(Sr, Ca)}_4\text{O}_y$$

$$0.075\ \text{atm O}_2\quad \text{liquid} + \text{(Sr, Ca)CuO}_2 + \text{(Sr, Ca)}_2\text{CuO}_3 + \text{Bi}_2\text{(Sr, Ca)}_4\text{O}_y$$

$$0.01\ \text{atm O}_2\quad \text{liquid} + \text{(Sr, Ca)CuO}_2 + \text{(Sr, Ca)}_2\text{CuO}_3 + \text{Bi}_2\text{(Sr, Ca)}_3\text{O}_y$$

$$1 \times 10^{-3}\ \text{atm O}_2\quad \text{liquid} + \text{(Sr, Ca)}_2\text{CuO}_3 + \text{Bi}_2\text{(Sr, Ca)}_3\text{O}_y + \text{(Sr, Ca)O}.$$

An important aspect is the size and the reactivity of these solid compounds. The same study [22] determined that between 0.5 and 1 atm O_2 the grains of $(Sr,Ca)_{14}Cu_{24}O_x$ and $Bi_2(Sr,Ca)_4O_y$ were small, while in air $(Sr,Ca)CuO_2$ grains grow fast. Between 0.1 and 0.075 atm O_2 the proportion of $(Sr,Ca)CuO_2$ is low and the $(Sr,Ca)_2CuO_3$ grains are small and reactive. In 0.01 atm O_2 the proportion of $(Sr,Ca)CuO_2$ is low but $Bi_2(Sr,Ca)_3O_y$ grains grow quickly leading to a large quantity of residual phases after synthesis. In 1×10^{-3} atm O_2 the $(Sr,Ca)CuO_2$ is replaced by large grains of $Bi_2(Sr,Ca)_3O_y$ which is impossible to recombine.

The starting composition, especially the Sr/Ca ratio, is an important parameter influencing the composition of the melt state and further recombination. Depending on the starting composition, Schartman *et al* [23] distinguished four composition domains in the melt state. Reeves *et al* [24] studied 1.3, 2 and 2.75 Sr/Ca ratios and Yoshida *et al* [25, 26] studied 1.14, 1.5, 2 and 3.28

Table B3.2.1.2. Data from [27] in 1996

Organization	Bi	Sr	Ca	Cu	Conductor form/comment
Kobe steel	2.1	2	1	1.9	Tape/precursors with 0.1 Ag
Mitsubishi Cable	2	2	0.64	1.64	Tape
NRIM	2	2	0.95	2	Film
Showa Electric	2.05	2	1	1.95	Tape
Sumitomo Metals	2	2.3	0.85	2	Tape/treatment under 1% O_2
Vacuumschmelze	2	2	1	2	Round wire

Sr/Ca ratios. Indeed, for production of Bi(2212) conductors each company uses its own composition as indicated in table B3.2.1.2.

This composition's influence cannot be dissociated from the recombination study examined in the next section.

B3.2.1.4 Study of the cooling-reformation of the Bi(2212) phase

The performance of the conductor depends on this crucial step. The performance is bound very tightly to three parameters.

(a) Phase purity, e.g. a phase recombination as complete as possible with a low proportion of small residual phase grains.

(b) Alignment of the Bi(2212) platelets in the direction of the flow of current. The alignment generally begins along the silver interface.

(c) Density of the superconducting material, e.g. with absence of voids.

B3.2.1.4.1 Phase purity

Work has been performed to identify the influence of various parameters on the final phase purity. Reeves *et al* [24] and Yoshida [26] studied the influence of the Sr/Ca ratio, Hellstrom and Zhang [22] and Yoshida [26] studied the influence of pO_2, and Kumakura [28] and Hayashi *et al* [29] studied the influence of the cooling rate.

Influence of Sr/Ca ratio

Different Sr/Ca ratios have been studied to generate various compositions of the partial melt state to allow a more complete synthesis of Bi(2212), for example with less Cu free phase and smaller grains of $(Sr,Ca)_{14}Cu_{24}O_x$. Table B3.2.1.3 in [26] and figure B3.2.1.6 in [30] summarize the different collected data.

Reactions during cooling:

For Sr/Ca = 2.75 the $Bi_2(Sr,Ca)_4O_y$ and Bi(2201) grains become very large and 200 h are needed to convert Bi(2201) into Bi(2212).

The melting temperature is minimum at 865°C for Sr/Ca = 2. Bi(2201) is not formed and the synthesis of Bi(2212) is fast with a low residual phase content.

For Sr/Ca = 1.3 the $(Sr,Ca)_{14}Cu_{24}O_x$ grains, which convert into $(Sr,Ca)_2CuO_3$ during melting, reappear and grow, leading to a poor reformation of Bi(2212) with a heterogeneous distribution of the residual phases.

Table B3.2.1.3. Data collected from [26]

Sr/Ca ratio	Composition of the melt in air	Composition of the melt in $pO_2 = 0.01$ atm
1.14		Bi(2212) → liquid ($Bi_{2.44}Sr_{1.69}Ca_{0.70}CuO_y$) + $Sr_{0.21}Ca_{2.04}CuO_3$
1.5		Bi(2212) → liquid ($Bi_{2.38}Sr_{1.83}Ca_{0.67}CuO_y$) + $Bi_{2.1}Sr_{1.90}Ca_{0.98}O_6$ + $Sr_{0.40}Ca_{1.83}CuO_3$
2	Bi(2212) → liquid ($Bi_{2.75}Sr_{1.78}Ca_{0.95}CuO_y$) + $Bi_2Sr_{2.20}Ca_{1.30}O_7$ + $Sr_{0.63}Ca_{0.53}CuO_2$	Bi(2212) → liquid ($Bi_{2.04}Sr_{1.61}Ca_{0.55}CuO_y$) + $Bi_2Sr_{1.93}Ca_{0.99}O_6$ + $Sr_{0.64}Ca_{0.51}CuO_2$ + $Sr_{0.45}Ca_{1.88}CuO_3$
3.28	Bi(2212) → liquid ($Bi_{2.79}Sr_{2.03}Ca_{0.68}CuO_y$) + $Bi_2Sr_{2.35}Ca_{1.21}O_7$ + $Sr_{0.78}Ca_{0.37}CuO_2$	Bi(2212) → liquid ($Bi_{2.08}Sr_{1.63}Ca_{0.54}CuO_y$) + $Bi_2Sr_{2.03}Ca_{0.82}O_6$ + $Sr_{0.77}Ca_{0.31}CuO_2$

B4.3 Films from electrophoretically deposited precursors with a Cu–Ca or Sr deficient stoichiometry were studied by Balmer *et al* [31]. Very small changes in stoichiometry were found to affect the melting behaviour and the performance of the conductor. The highest T_c (90 K) and J_c at 77 K (80 A mm^{-2}) were observed for both Ca and Cu deficient films.

Influence of the pO_2

B3.1 Examining the type and size of grains in the melt state at various thermodynamic conditions, Hellstrom and Zhang [22] and Polak *et al* [32] recommended the use of a pO_2 between 0.5 and 1 atm where

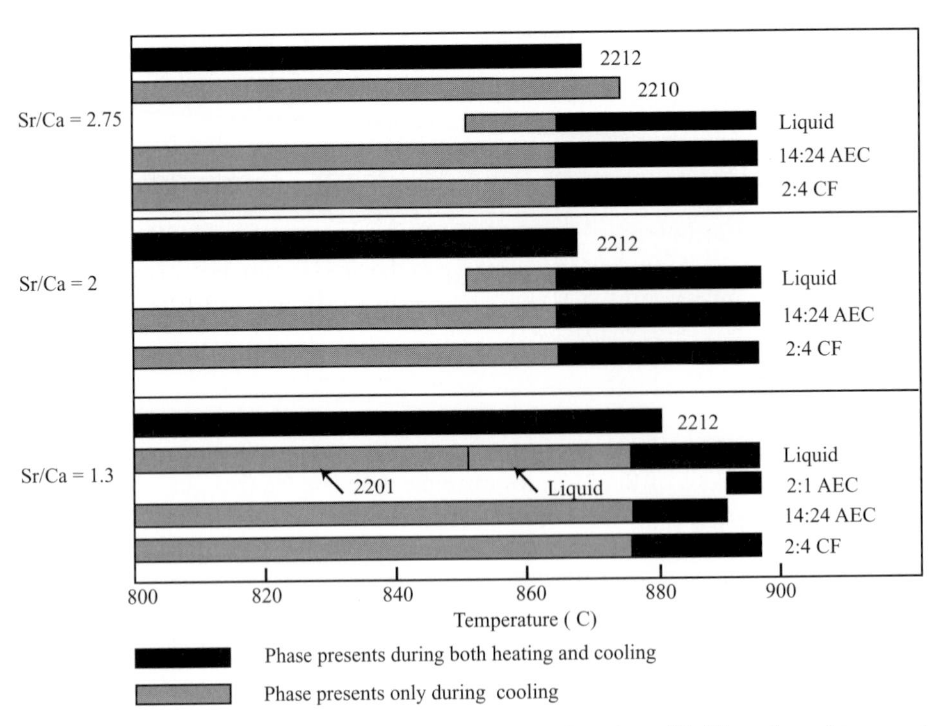

Figure B3.2.1.6. Phases present in a Bi(2212) melt in pure O_2 as a function of Sr/Ca ratio and temperature © 1997 IEEE [30].

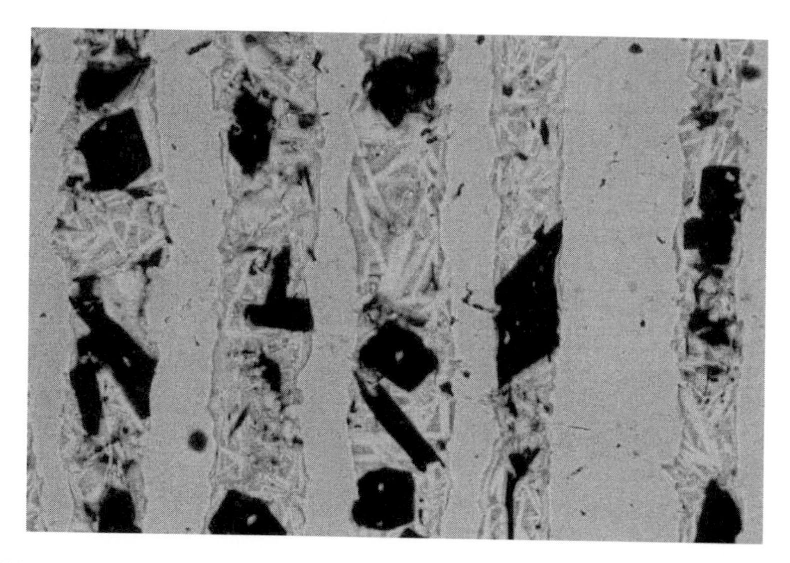

Figure B3.2.1.7. Large-grained structure obtained after melting under $pO_2 = 0.03$ atm at 870°C.

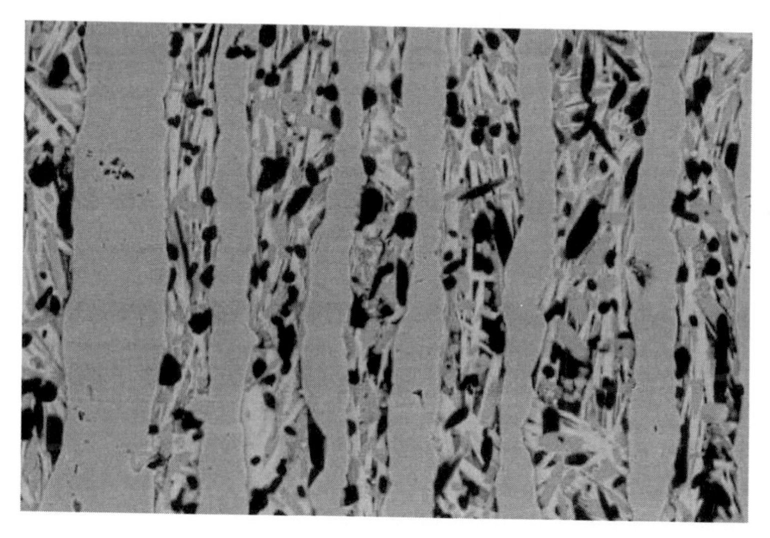

Figure B3.2.1.8. Fine-grained structure obtained after melting under $pO_2 = 1$ atm at 900°C.

the $(Sr,Ca)_{14}Cu_{24}O_x$ and $Bi_2(Sr,Ca)_4O_y$ grains are small in size, thereby allowing the best chances of recombination into pure Bi(2212) (figures B3.2.1.7 and B3.2.1.8).

Applying the lever rule to the ternary phase diagram $BiO_{1.5}$–(Sr,Ca)O–CuO, Yoshida [26] tried to deduce the main residual phases from the composition of the liquid and that of the formed Bi(2212): for example in $pO_2 = 0.01$ atm at 850°C the formed Bi(2212) phase has the composition $Bi_{2.2}Sr_{1.8}CaCu_2O_y$ very close to the line between the liquid and $Bi_2(Sr,Ca)_3O_6$ leading to an excess of this Cu free phase after complete solidification. Another example is proposed in air at 890°C: the formed Bi(2212) phase has the composition $Bi_{2.2}Sr_{1.8}CaCu_2O_y$ very close to the line between the liquid and $(Sr,Ca)CuO_2$ leading to an excess of this AEC.

In [33], the influence of pO_2 on the kinetics of the Bi(2212) reformation in bulk melt cast precursors at 1000°C is studied. After many TGA/DTA experiments on bulk or ground precursors treated under air or pure N_2 and XRD analysis it was deduced that O_2 acts as an inhibitor for the synthesis of Bi(2212), the kinetics being much faster in deoxygenated precursors.

B2.1 D1.1.2

Influence of the cooling rate

A slow cooling rate in the partial melt state not only promotes the formation of large Bi(2212) platelets, but also the growth of grains of the non superconducting phases. The J_c in dip coated conductors increased by a factor of 2 when the cooling rate was decreased from 30 to 2°C min^{-1} [34]. Hayashi *et al* [29] obtained large, well aligned grains by applying a fast cooling, 50°C/min^{-1}, just after the melting step down to 875°C before slow cooling at 2°C min^{-1} to 837°C in air. These parameters will be discussed further in the thermal cycle section.

B4.3

B3.2.1.4.2 Alignment of the Bi(2212) platelets

Several studies tried to understand the process leading to alignment of the Bi(2212) platelets. Inoescu *et al* [35] consider that the nucleation occurs in the liquid but at the contact with silver or with the $(Sr,Ca)_{14}Cu_{24}O_x$ grains, which are themselves oriented along the silver interface. These grains are slowly dissolved, leaving voids surrounded by precipitated silver formerly dissolved in the melt. Lang *et al* [36] found that, in air, nucleation begins at 875°C inside the liquid independent of AEC grains and that no Bi(2201) is formed. The O_2 flow seems to play an important role in the alignment: Hellstrom [27] observed that, in dip coated films on MgO substrate, the initial nucleation and alignment occurred at the atmosphere interface and [37] at the silver interface of PIT conductors. In both cases these interfaces provide the O_2. The presence of more textured Bi(2212) at the silver interface has frequently been proved by crystallographic, magneto-optic or J_c measurements as a function of distance from the interface [38, 39].

B4.3

Ag is considered to be an important agent in the alignment process. Hellstrom [27] and Lang *et al* [36] measured its solubility to be between 4 and 5% in the liquid and zero in Bi(2212). Kumakura [28] indicates that Ag lowers the melting temperature of Bi(2212) by dissolution in the liquid with the formation of Ag–CuO eutectic. The ΔT_f was found to be greatest for 2% Ag [40], it is also a function of pO_2: with ΔT_f being 25, 20 and 10°C, in 1, 20 and 1×10^{-3} atm O_2, respectively. MacManus-Driscoll *et al* [19] mention that Ag lowers the viscosity of the liquid above the solidus at 798°C. The smoothness of the Ag interface is also of great importance and it was assumed that the addition of alloying elements to Ag degrades the Bi(2212) grain alignment and J_c [41].

B3.2.2

It is observed that thin (< 25 μm) filaments with smooth interfaces promote the alignment of Bi(2212) platelets by a steric action. Thin filaments are difficult to obtain by rolling due to the hardness difference between the ceramic core and the Ag sheath. Below a critical thickness the well known sausaging phenomenon occurs, leading in extreme cases to filaments fracturing with propagation angle at 45° (figure B3.2.1.9). Three methods are generally used to lower this defect: (a) using special fine-grained powder [42, 43]; (b) hardening the sheath by adding small amounts of Ni, Mn, Zr, Mg, Cu, Au, Pd, etc [6, 28], which can poison the superconductivity, the performances being reduced by 30–40% for doping beyond a few 0.1 at.%; and (c) optimizing the rolling sequence. Figure B3.2.1.10 shows a recently developed rolling process applied to a PIT conductor with 76 filaments at Alcatel.

A1.3

B3.2.1.4.3 Density of the core

The presence of residual voids and bubbles constitute a major difficulty for reaching high J_c in Ag sheathed Bi(2212) conductors. Voids appear during the Bi(2212) reformation step when the platelets are

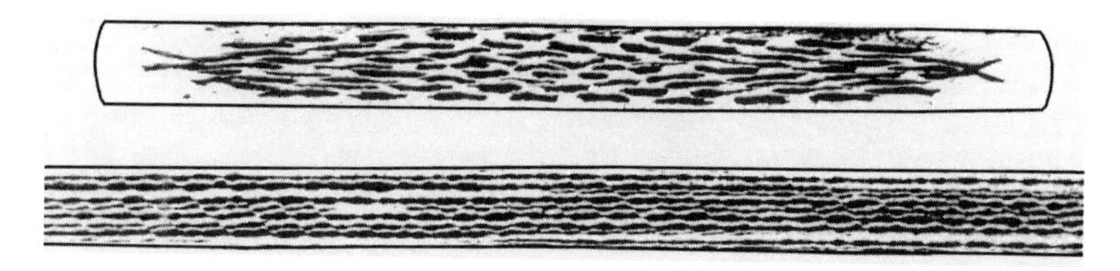

Figure B3.2.1.9. 76 filament PIT conductor sausaged after rolling.

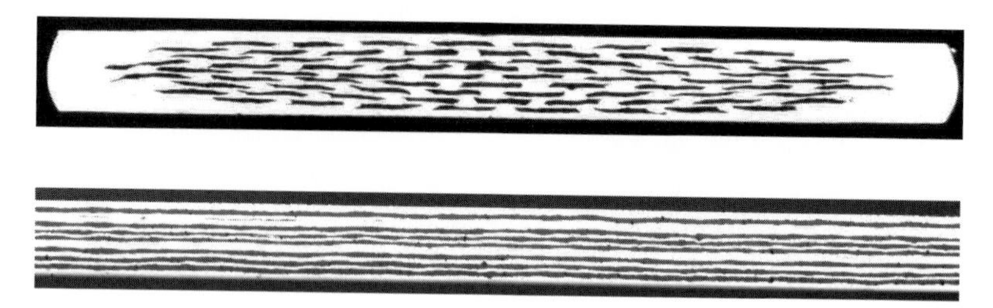

Figure B3.2.1.10. Same conductor as in figure B3.2.1.9 processed according to an optimized rolling sequence.

slowly formed and the AEC grains are consumed (figure B3.2.1.11). Because some of them have reached a size in the melt that is of the same order as the filaments' thickness, voids are left in the middle of the filaments when the AEC is consumed. In the worst case, some of the AEC or 'Cu free' grains remain in the void: it appears as if at some point in the reformation no liquid phase could access the grain to fully 'digest' it. Another reason for void formation is the coalescence of the residual N_2 from air between the grains of precursor powder that was not completely evacuated from the first stage billet. These voids filled with N_2 degrade the performance of the conductor but are not responsible for the catastrophic 'bubble' defect. The bubbling phenomenon affects Ag sheathed PIT Bi(2212) conductors, starting from the inside of the filaments (figure B3.2.1.12). B3.1

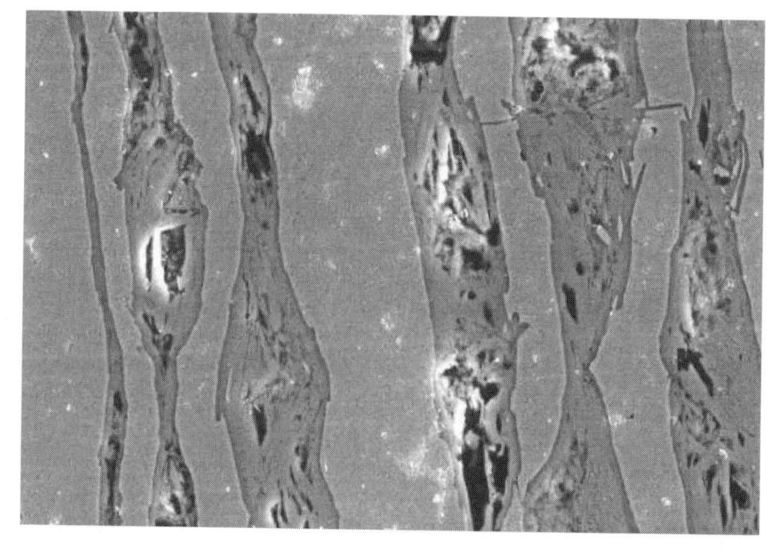

Figure B3.2.1.11. Voids left by the dissolution of AEC grains during the synthesis.

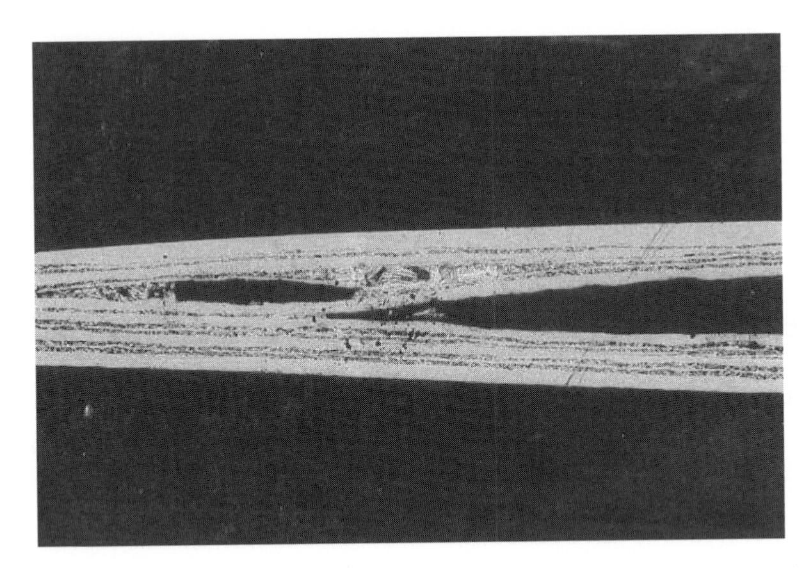

Figure B3.2.1.12. Bubbles starting from the inside of the filaments in a PIT conductor.

Bubbling has been intensively studied in order to understand the mechanism and to propose solutions. Hellstrom [27] and Fujii *et al* [44] identified four possible agents responsible for bubbling during the final treatment:

(a) adsorbed H_2O making the Ag sheath swell between 400 and 600°C;

(b) Decomposition of $Sr(OH)_2$ and $Ca(OH)_2$ with the release of H_2O occurring between 650 and 800°C;

(c) $SrCO_3$ evolving CO_2 above 800°C until fusion;

(d) O_2 evolved during the peritectic melting at the rate of 1/2 O_2 per mole of Bi(2212).

The same authors suggest a pre-treatment of the material at 800°C under O_2 flow or 500–600°C under vacuum and proposed the following mechanism [45]: during final heat treatment the pO_2 inside/outside the gradient is low, so every gaseous species (N_2, H_2O or CO_2) acts as a bubbling agent. They concluded that a counter pressure is needed in the furnace atmosphere. For example, when treating under 1% O_2 at atmospheric pressure, a 0.99 atm counter pressure is applied. In contrast, treating under pure O_2 allows no counter pressure. We have observed bubbling for the same conductor when treated with pure O_2 and not with 1% O_2. It must be taken into account that the applied temperatures during the melting step have to be adjusted: ~780°C for 1% O_2 and ~850°C for pure O_2 and the heating rate to the melting step must not be too fast to allow equalization of the inside and outside pO_2. If treatments with $pO_2 = 1$ atm are used to get a fine-grained structure in the melt state, an artificial counter pressure must be implemented, as done by Fujii *et al* [44], where they observed the suppression of bubbling under 1 atm $O_2 + 7$ atm N_2. As an additional advantage they observed a decrease in the residual voids compared to conductors treated in low pO_2 atmosphere. Microstructures with smaller and more uniformly distributed pores were obtained by Reeves *et al* [46] in thin walled multifilamentary conductors after pre-annealing under vacuum and annealing at 835°C in 100% O_2 followed by a final treatment in air at 5 atm ($pO_2 = 1$ atm) yielding $J_c = 820$ A mm^{-2} at 4.2 K, 0 T.

B3.1

A1.3

Hase *et al* [47] found that bubble formation is related to the sausaging phenomenon that occurs when rolling the tape. Two preventative treatments were proposed.

- Heating at a rate of less than 10°C h^{-1} with a pressure of less than 4×10^{-7} bar to eliminate bubbling due to H_2O.
- Heating at a rate of less than 10°C h^{-1} with a pressure of less than 2.7×10^{-5} bar up to 750°C then keeping it for 20 h at 835°C under pure O_2 to eliminate CO_2 bubbling.

Ray *et al* [48] tried a capillary transport of the liquid phase called the 'wicking process' to eliminate bubbles.

Another detrimental effect of the residual carbon, besides forming CO_2 induced bubbles, is to form $SrCO_3$ and possibly $Bi_2Sr_4Cu_2CO_3O_y$ [27]. These Sr rich phases lower the melting temperature from 865° to 850°C in air with the increase in carbon content from 100 to 1600 ppm. A detailed study was made by [49] concerning the Bi(2212) phase diagram with 1×10^{-4} bar $< pCO_2 < 0.1$ bar. They found a complete decomposition of Bi(2212) when $pCO_2 > 0.01$ bar.

The final purity problems to be cited are the detrimental effects of adding Pb to the B-2212 composition [50, 51] and the loss of Bi_2O_3 from dip coated films during heat treatment. The loss of Bi_2O_3 was compensated by heating the samples in the presence of $Bi_2Al_4O_9 + Al_2O_3$.

B3.2.1.5 Reaction heat treatment for J_c optimization

B3.2.1.5.1 Generic thermal cycle

Figure B3.2.1.13, showing a generic 'melt texturing' cycle as studied step by step by Hellstrom [27], distinguishes 4 regions: B2.3.5

- In the first region Bi(2212) melts evolve 1/2 O_2 and dissolve $\approx$ 5% at. Ag in the liquid. The refractory phases grow, specially $(Sr,Ca)CuO_2$ in PIT conductors.
- In the second region the Bi(2212) nucleates and grows from the reaction between the liquid and the refractory phases in the melt. If the non-superconducting grains are small, the Bi(2212) platelets can relatively grow without any hindrance, with growth being faster along the *a,b* axes than along *c*. B2.3.5
- In the third region the Bi(2212) platelets' purity and alignment continue to increase, but even after 100 h some $(Sr,Ca)CuO_2$ grains and Bi(2201) intergrowths still remain.

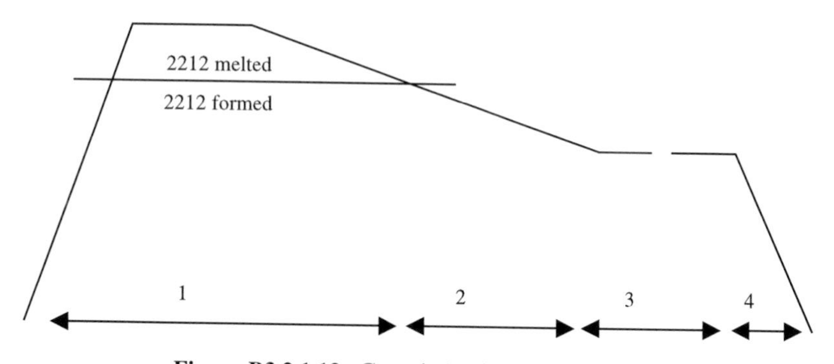

Figure B3.2.1.13. Generic 'melt texturing' cycle.

- The final cooling occurs in the fourth region. For thermodynamic reasons, quenching is the best way to cool Bi(2212) but mechanical aspects have to be taken into account, so that the cooling rate is a compromise between the decomposition kinetics of Bi(2212) and the prevention of cracks formed by differential ceramic/Ag shrinkage. Practically, a cooling rate around 300°C h^{-1} is used. This rate cannot be achieved easily when large Wind&React coils are to be cooled.
- An additional region is needed when Bi(2212) is to be used at 68–77 K. It consists of a 4–10 h step between 450 and 600°C in pure N_2 [12, 13, 52]. After 15 h at 500°C [53] T_c increased to 93 K, which is due to creation of holes by the elimination of O from the Bi(2212) lattice. It is observed that only the intragranular J_c was enhanced with the grain connectivity damage leading to increase in J_c by $\approx$ 15% at 77 K and decrease in J_c by $\approx$ 15% at 4.2 K.

C2

B3.2.1.5.2 Specific thermal cycles

Different groups working on Bi(2212) are optimizing thermal cycles with the aim to decrease the residual phase content and to achieve a large-grained, well-textured microstructure. Ray [54] replaced the continuous melt texturing cooling by a set of successive steps at decreasing temperatures (figure B3.2.1.14) to create some nuclei during the falling edge followed by subsequent growth of the nuclei and platelets during the constant temperature portion of the step.

As discussed in section B3.2.1.4 the phase assemblage in the system after the crucial partial melting step depends on the pO_2 in the furnace atmosphere. Because the Bi(2212) reformation kinetics and completeness of the Bi(2212) formation depends on the reactivity of the species generated during partial melting, Zhang and Hellstrom [55] examined samples of PIT conductors quenched at different steps in the thermal treatment of pO_2 varying from 0.001 to 1 atm. This study, illustrated by numerous micrographic SEM examinations of the processed tapes, concludes that between 0.001 and 0.01 atm O_2 the key phase in the assemblage is $Bi_2(Sr,Ca)_3O_x$, from 0.075 to 0.30 atm O_2 it is $(Sr,Ca)CuO_2$, and from 0.40 to 1 atm O_2 it is $(Sr,Ca)_{14}Cu_{24}O_x$. In this last pO_2 range the homogeneity of the microstructure of fully processed tape is optimum and J_c values are the highest (1200 A mm^{-2} at 4.2 K, 0 T).

A1.3
D1.3

Hauck *et al* [56] tried five melting processes and determined an optimum process to melt a deoxygenated composition of Bi(2212) at 891°C under 0.5–1 atm O_2. The deoxygenation is obtained by prereaction with $pO_2 = 10^{-3}$ atm at $798 < T < 865$°C.

A different heat treatment to form Bi(2212), called 'isothermal melt growth' [37, 57–59] consists of:

B2.3.5

- preparing vitreous precursors by grinding quenched Bi(2212) melt;
- shaping the powder into a conductor form, then melting in Ar at 760–860°C for 10 to 60 min;
- without changing the temperature, changing 10–100% O_2 atmosphere.

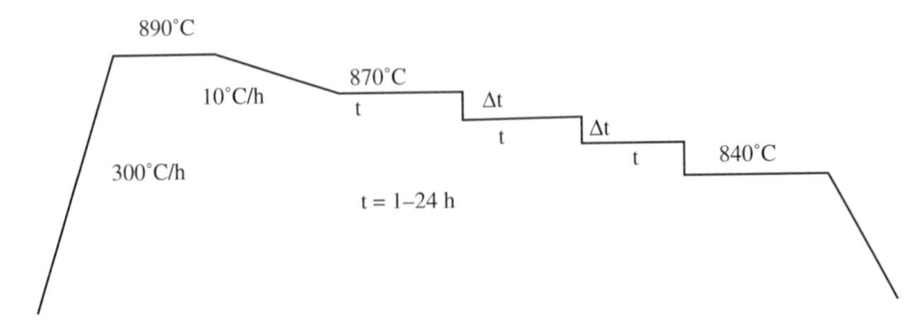

Figure B3.2.1.14. Melt texturing cycle with nuclei formation control.

Experiments on dip-coated conductors showed five temperature domains.

- At $T < 770$°C the viscosity of the liquid is too high to wet the substrate and the structure is not textured, with small, randomly oriented Bi(2212) grains and residual Ag, CaO, CuO and Bi2201 grains.
- Between 770 and 790°C, Ag is wetted by the liquid phase, alignment occurs along the Ag interface with small quantities of $(Sr,Ca)CuO_2$ and $(Sr,Ca)_{14}Cu_{24}O_x$ residual phase.
- Between 790 and 840°C, large grains of Cu_2O and $Bi_2(Sr,Ca)_3O_y$ appear along with $(Sr,Ca)CuO_2$ growing from CaO nucleation sites.
- Between 840 and 860°C, the growth of Cu_2O and $Bi_2(Sr,Ca)_3O_y$ grains is stationary and $(Sr,Ca)CuO_2$ needles keep growing.
- Above 860°C the $(Sr,Ca)CuO_2$ needles reach $\approx 500\ \mu$m. Large grains of $Bi_2(Sr,Ca)_4O_y$ and Bi(2201) are also present.

Only domains with 770°C $< T <$ 790°C and 840°C $< T <$ 860°C result in a reasonable J_c of 1200 and 800 A mm^{-2}, respectively. C2

A continuous process was studied by Burgoyne *et al* [60] and Morgan *et al* [61] for treating dip-coated conductors (Ag/0.25%Ni/0.25%Mg) in a furnace having seven temperature zones. The melting step occurred at 874–876°C and annealing after melting occurred at 860–830°C. Processing speeds of 0.55, 1.1 and 1.6 m h^{-1} were studied. B2.3.5

- At 0.55 m h^{-1} 20–30 μm agglomerates of Bi(2201) were found.
- At 1.1 m h^{-1} many $(Sr,Ca)_2CuO_3$ needles had grown.
- At 1.6 m h^{-1} reduction of the melting time led to fine (and reactive) $(Sr,Ca)CuO_2$ and $(Sr,Ca)_2CuO_3$ grains.

They combined melting at 1.6 m h^{-1} and annealing at 0.55 m h^{-1} to obtain the best microstructure. New studies were made in a furnace with five temperature zones and a melting temperature of 885°C: after a preliminary decarburization treatment of 1 h at 400°C plus 16 h at 800°C under O_2, a J_c of 1200 A mm^{-2} (I_c = 100 A) was achieved.

Continuous processing was also experimented on with Ag/Mg sheathed PIT conductors in a 3 zone quartz furnace by Hu *et al* [62]. For an optimized melting temperature (886°C) and pulling speed (15 m min^{-1}) J_c = 140 A mm^2 at 4.2 K, 0 T was found. A1.3 C2

For its own PIT conductors Alcatel developed a special thermal cycle in 3% O_2. This heat treatment, based on the good wetting between Ag and Bi(2201) based liquid (figure B3.2.1.15), consists of three steps (figure B3.2.1.16):

(a) melting at 860–880°C for 10–30 min;

(b) quenching to complete solidification at 750°C;

(c) annealing for 50–100 h at 780–830°C to form the Bi(2212).

This cycle was used on 76–115 filament Ag sheathed PIT conductors made with Bi(2212) precursors provided by Alcatel High Temperature Superconductors at Cologne. The time of the melt and the rate of cooling to 750°C were optimized to achieve small grained refractory phases and the rate A1.3

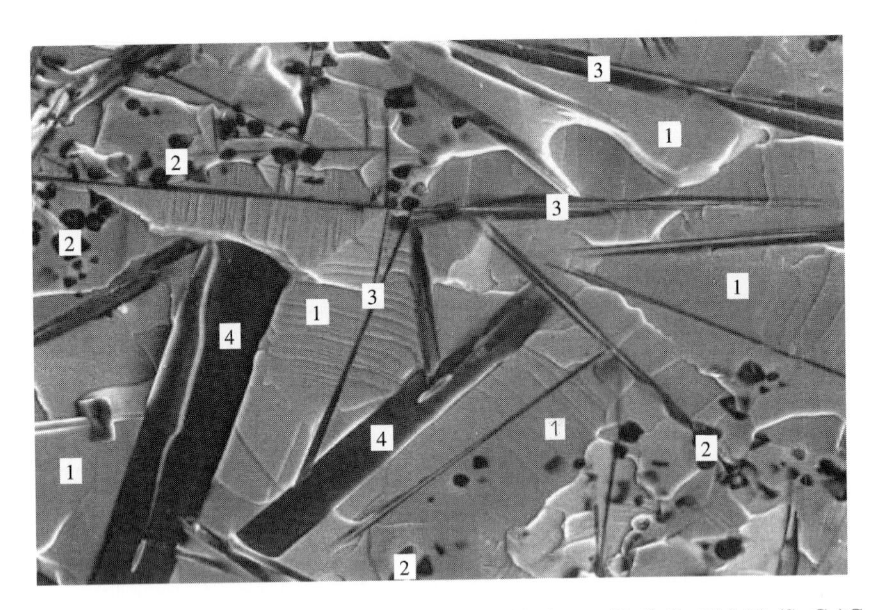

Figure B3.2.1.15. Melted Bi(2212) on Ag foil. (1) liquid phase, (2) CaO, (3)&(4) $(Sr,Ca)CuO_2$.

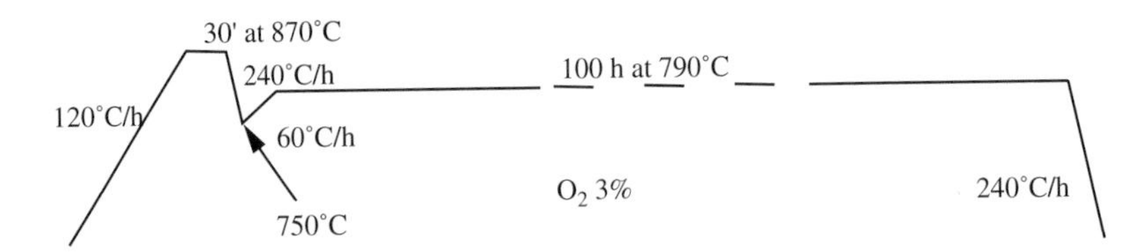

Figure B3.2.1.16. Melt texturing cycle with typical parameters developed at Alcatel.

of the ascending slope to the synthesis temperature was optimized to generate only a small quantity of liquid phase. We presume that during the 50–100 h step the Bi(2212) is formed by a quasi-solid synthesis mechanism from the large, flat, textured Bi(2201) grains that are present at 750°C. Due to the synthesis in a quasi-solid state, the kinetics are slow, the residual phase content is rather high, and the structure is heterogeneous. Figure B3.2.1.17 and B3.2.1.18 show, respectively, an area in the same conductor with a nearly perfect structure and another with poor texturing, voids and residual phases. Nevertheless such conductors have a high J_c as indicated in section B3.2.1.6 due to the presence of Bi(2212) platelets at the Ag interface in the areas with a poor structure, that act as shunts between the heterogeneous and more ideal regions.

B2.3.5

B3.2.1.6 Conductors fabrication routes

B4.3

Most of the studies on the phase evolution of the Bi(2212) system were made on open structures obtained by dip-coating or doctor blading where observation is easier than in silver sheathed conductors. A major inconvenience of this route is the lack of protection against the environment. The conductors shown in figure B3.2.1.1(a)–(d) are all closed structures and are made either by dip-coating or by the PIT process. The latter are single-stack multifilamentary assemblies except for the ROSAT conductor where multifilamentary wires are assembled from multifilamentary tapes. PIT conductors can be classified as flat, round or rectangular PIT by their geometry and by the width/thickness ratio,

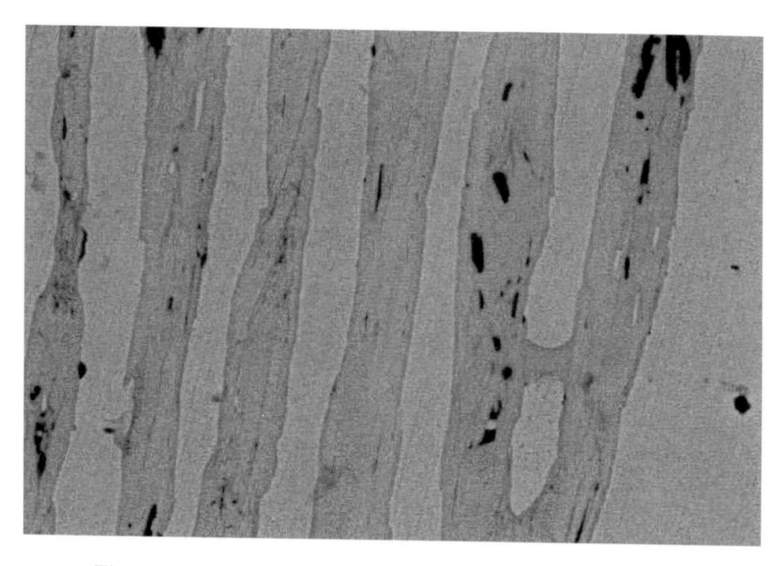

Figure B3.2.1.17. Filament area with a nearly perfect structure.

or aspect ratio of the final conductor. In the following, a conductor shall be called wire for an aspect ratio less than 3 and shall be called a tape for an aspect ratio greater than 3.

B3.2.1.6.1 Dip coating technique

The continuous dip-coating method is reviewed by Ilyushechkin [63]. It consists of depositing an organic slurry on a silver or alloy tape that is dried in a furnace. The thickness of the superconducting layer depends on the oxide content and viscosity of the slurry, and on the drawing speed. Generally a suspension of finely milled Bi(2212) powder in trichlorethylene-alcohol is used. To increase the oxide B4.3

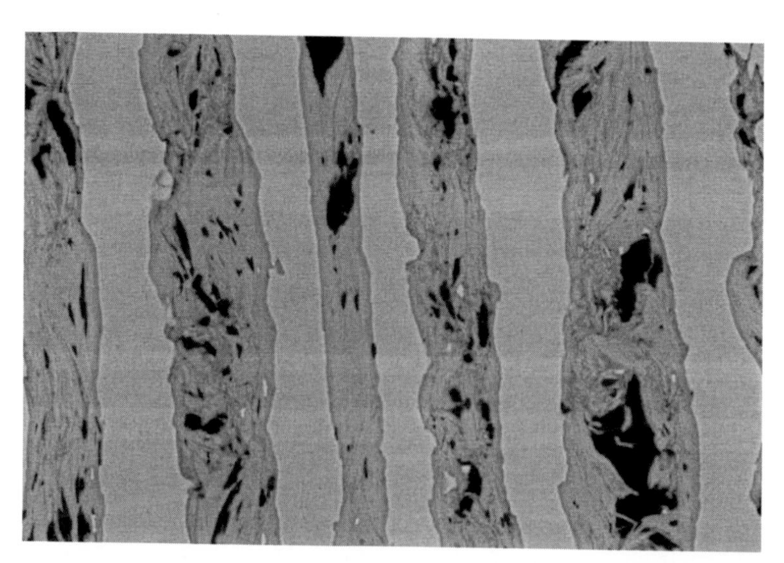

Figure B3.2.1.18. Poor structure with voids and much misalignment except a few Bi(2212) platelets along the Ag interface.

C2

content and to reduce the viscosity, a dispersing agent can be used. High molecular weight esters or partially oxidized fish oils like 'menhaden oil', used for paints, are very efficient. A widely used temporary binder for such composition is polyvinyl formal. After drying, an additional rolling step can increase the density, which lowers the shrinkage and risk of cracks during the thermal treatment. The conditions used by Morgan *et al* [14] and Marken *et al* [64] covered a 50 μm Ag substrate on both sides with a green layer 50–200 μm thick. After burning the binders the layer thickness was reduced to 10–50 μm resulting in a superconductor/total sections ratio (fill factor) of 30–66%. Lengths up to 100 m are produced on a rig described in [63] with $J_e = 106\,\mathrm{A\,mm^{-2}}$ at 77 K, 0 T. Dip coated conductors made on pure silver or on AgMg alloys, result in improved mechanical strength. For practical use, the dip-coated tapes can be continuously sheathed between two thin protecting silver tapes. This protects the superconducting layer from humidity or carbon uptake but reduces the fill factor.

B3.2.1.6.2 The PAIR tape

B4.3 C2

The pre-annealing and intermediate rolling (PAIR) process (figure B3.2.1.1(*b*)) consists of stacking and wrapping dip-coated ribbons in a silver foil, then applying a heat treatment followed by an intermediate rolling step before the final thermal treatment. The dip coated conductor is improved to make multilayer conductor [52, 65, 66]. The smoothness of the silver surface is very important for the performance, making the surface preparation of the silver an important part of the work. A PAIR conductor was made with three pure silver tapes 25 μm thick, 4 mm wide and 100 m long with a 25 μm coating layer stacked and wrapped in a Ag/Mg alloy foil. Samples are pre-annealed in pure O_2 before cold rolling and then melt textured also in pure O_2. After optimizing the experimental conditions, this process led to the highest J_c known in short samples: $J_c = 5000\,\mathrm{A\,mm^{-2}}$ at 4.2 K, 10 T. The optimization study and the corresponding microstructures are described in [67]. An optimum was determined for the following parameters: pre-annealing at 840°C for 1 h, intermediate rolling with a 25% reduction factor, melting 10 min at 888°C max, melt-texturing at 5°C min to 840°C and rapid cooling to room temperature. In fully processed PAIR process tape a typical fill factor of 15–20% is achieved.

B3.2.1.6.3 Standard PIT tape

B3.1 A1.3

Standard PIT process is widely used for the fabrication of Bi(2223) conductors, e.g. round drawing followed by rolling (see chapter B3.1). The difference is that the heat treatment is adapted to the Bi(2212) phase.

B3.2.1.6.4 Rectangular PIT tape

B3.1

The conductor shown in figure B3.2.1.1(*a*) is made by the rectangular PIT process which differs from the standard PIT process (see chapter B3.1.) in that the multifilamentary billet begins as a rectangular tube that is packed with an arrangement of rectangular monofilament conductors that are deformed using passive four-roll-drawing-dies (also called Turk-heads) to a final dimension of about $3.5 \times 0.25\,\mathrm{mm^2}$. In this single stage assembled conductor a typical fill factor of 25–30% is achieved. 76, 110 or 115-filament conductors with green tape lengths up to 1 km (figure B3.2.1.19) and 400 m lengths of reacted conductor are made. Disadvantages of rectangular PIT tape are the slightly increased cost of the silver tubes and the powder rods. An advantage of the rectangular conductor is a flatter silver-BSCCO-filament interface that is favourable for highly oriented interface texturing that reduces the sensitivity to the magnetic field if it is applied parallel to the tape. A second advantage, related to the four-roll-drawing-die, is that it provides a low friction deformation of the wire, reducing both heating during deformation and mechanical fatigue of the materials. Finally, the internal arrangement and position of each filament in the final tape conductor is fixed when the multifilamentary billet is finished. The orientation of the wire is

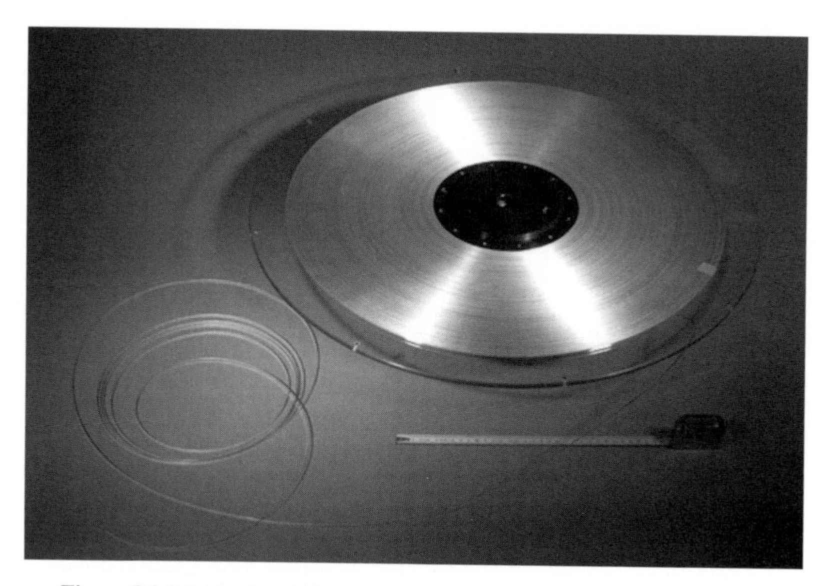

Figure B3.2.1.19. One kilometre long rectangular PIT tape conductor.

controlled during the whole deformation process avoiding spontaneous twisting that results in the filaments rotating along the conductor axis. The process results in a DC-tape conductor where the position of each filament is maintained over kilometre lengths.

B3.2.1.6.5 ROSAT wire

ROSAT wire (see figure B3.2.1.1(*c*)) is a variation of the PIT process with a diamond stacking of the monofilamentary bundles. It is assembled from 18 flat PIT tape conductors in three batches rotated by 120° which are introduced into a round silver tube. The conductor which contains 990 filaments reaches a critical current of 900–1180 A at 4.2 K which is almost independent of field orientation due to its weak anisotropy (± 10%). An advantage of this second stage assembly is that filaments with high aspect ratio are obtained in a round conductor. This is important for magnet design where non-consideration of the orientation of the magnetic field inside the windings is an advantage. A disadvantage is that most filaments 'see' a transverse field component so this conductor is more sensitive to magnetic field than a standard tape conductor in parallel field. In this second stage assembled conductor a typical fill factor of 22% is achieved.

B3.2.1.6.6 The round PIT wire

The IGC round conductor fabrication process shown in figure B3.2.1.1(*d*) is similar to the process that is used to make low temperature conductors. With this technology [68] single stage round cross-section conductors with 240–430 round filaments are assembled. Round PIT wires have also been shaped into conductors having rectangular cross section with an aspect ratio of up to 2. These latter conductors have the highest current densities. Second stage assemblies of 7 × 61 filament conductors led to 427 filament conductors. The current carrying capacity of this conductor is also insensitive to the orientation of the magnetic field. Twenty 427 filament wires have been used to fabricate an 80 m length of Rutherford Cable.

A1.3

B3.2.1.7 Current densities in Bi(2212) conductors

C2
Two types of current densities are commonly used for assessing the quality of superconductors. First, the critical current density J_c, which is the critical current most commonly determined by the 1 μV cm^{-1} (100 μV m^{-1}) criterion, divided by the superconducting cross-section in the wire. The superconducting cross-section is determined by image analysis. With careful calibration, a precision of $\pm$ 10% can be achieved. J_c assesses the quality of the superconducting phase and is of interest for scientists who are interested in the performance of the superconducting material. Second, the engineering current density J_e is obtained by dividing the critical current by the entire cross section of the wire. J_e qualifies the usefulness of a conductor for applications, and, when possible, J_e is used in this section.
A1.3

B3.2.1.7.1 Critical currents and current densities in self field

C2
In the following, the available results (table B3.2.1.4) for different Bi(2212) conductors are discussed. The cross section of these conductors is slightly below 1 mm^2 except for the second stage assembly ROSAT where the section is 2.5 mm^2. It is this conductor which achieves the large critical current of 1180 A with a filament number of 990. Almost the same value (1150 A) is achieved in the PAIR conductor with only four superconducting layers. The filament thickness in all conductors are in the range 10–20 μm which appears to represent an optimum for actual processing conditions independent of the processing route. The short sample results in self field at 4 K of the PAIR, rectangular PIT tape and round PIT wire are best values which have been achieved on different samples of length 50–100 mm. The I_c value of the ROSAT conductor is reported for a 20 m length. The long sample results of PIT conductors are about 25–30% below the short sample values except for the PAIR conductor where a reduction of between 50 and 60% is found.

The data for the PAIR conductor are extrapolated from 10 tesla measurements using the scaling factor of 1.9. The J_c value of 9500 A mm^2 for the PAIR conductor is the highest value achieved so far in Bi(2212) conductors. The potential to increase this J_c is of course unknown today. However, this value

Table B3.2.1.4. Bi(2212) conductor data: geometry and self field properties at 4 K

	PAIR [7, 65] Showa-electric, NRIM	Rect. PIT tape [9, 69] Alcatel	ROSAT [8, 70] Hitachi	Round PIT wire [68] and [71] IGC
Cross-section	0.15–0.2 × 4–4.8 mm^2	0.2–0.28 × 3–4 mm^2	Ø = 1.78 mm	Ø = 0.81 mm
No of filaments or layers	4	76	990	427
Diameter or thickness of filaments	10 μm	20 μm	~15 μm	15 μm
Superconductor fill-factor σ_s	17%	28%	19–22%	18%
I_c(short[a],4 K,0T)	1150 A[b]	600 A	1180 A (20 m)	315 A
J_c(short[a],4 K,0T)	9500 A mm^2 [b]	2800 A mm^2	2500 A mm^2 (20 m)	3400 A mm^2
J_e(short[a],4 K,0T)	1500 A mm^2 [b]	775 A mm^2	475 A mm^2 (20 m)	611 A mm^2
I_c(long,4 K,0T)	550 A (100 m)	450 A (100 m)	900 A (400 m)	220 A (~150 m)
J_c(long,4 K,0T)	4120 A mm^2	1800 A mm^2	1660 A mm^2	2370 A mm^2
J_e(long,4 K,0T)	700 A mm^2	500 A mm^2	370 A mm^2	390 A mm^2

[a] short sample: ~5 cm except for ROSAT: 20 m!.
[b] extrapolated from 10 tesla measurements.

can be used to estimate the potential of more classical PIT conductors. The PAIR result is roughly a factor of 4 above the J_c values in the other conductors. This, together with possibly increasing the fill factor to a maximum of about 50%, leads to an estimate of a factor of about 10 increase in the current carrying capacity of non-PAIR conductors.

B3.2.1.7.2 Critical currents and current densities in high magnetic field

Bi(2212) conductors are most promising for high field application above 20 T. Available results for four conductor options are shown in figure B3.2.1.20. All conductors show a relatively strong decrease in the critical current at low fields. This is probably still related to the presence of weak grain boundaries that are progressively switched off by the magnetic field. For PIT tape above 20 T, the critical current becomes constant which indicates that no more grain boundaries are affected and that the bulk critical field is much higher than the field of 30 T. C2

The stronger field dependence in ROSAT and round PIT wires is related to the presence of the perpendicular field component in the a–b plane of the superconductor. However, it is not clear from geometrical reasons why the field dependence should be improved by ROSAT. Although the difference clearly exists, this difference remains small in the high field domain. The values of ROSAT are measured at above 18 T and the low field extrapolation is possibly somewhat optimistic.

It is easily understood that the PAIR and rectangular PIT tape conductors are less sensitive to the magnetic field, which is applied in the favourable direction, parallel to the tape. The relative field dependence of these tapes should be somehow related to the flatness of the interface with the silver. For reasons discussed earlier, we expect the interface to be flatter in the PAIR than in the rectangular PIT conductor. A lower field dependence would be expected for the PAIR conductor but the data are not consistent with this expectation. The stronger field dependence of the PAIR conductor is possibly related to the much higher value of the critical current density or to a transverse field component when carrying out the measurements.

To compare the field dependence, the available I_c/I_{co} values at 10 and 30 T are indicated in table B3.2.1.5. At 10 T the I_c reduction in different conductors is in the range of a factor of 2; 10% less in A1.3

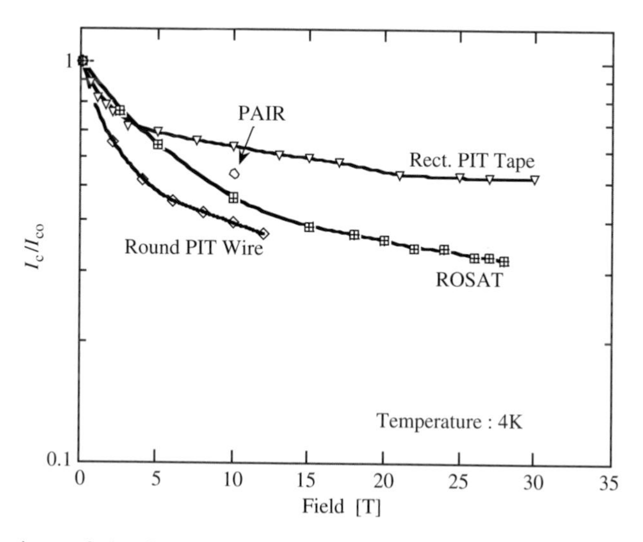

Figure B3.2.1.20. Comparison of the field sensitivity of the critical currents of different Bi(2212) conductors. Courtesy of H. Kitaguchi (PAIR), M. Okada ROSAT and L. Motowidlo Round PIT Wire.

Table B3.2.1.5. Bi(2212) conductor high field properties at 4 K; the results are not necessarily measured on the same conductor sample as in table B3.2.1.4

	PAIR [65, 7] Showa-electric, NRIM	Rect. PIT tape [11] Alcatel	ROSAT [70, 8] Hitachi	Round PIT wire[a] [71] IGC
I_c/I_{co} (10 T)	0.53 (10 T)	0.62 (10 T)	0.46 (10 T)	0.40 (10 T)
I_c/I_{co} (max. Field)	NA	0.51 (30 T)	0.33 (28 T)	NA
I_c(short,4 K,)	625 A (10 T)	200 A (30 T)	340 A (28 T)	100 A (12 T)
J_c(short,4 K) At a field of:	5000 A mm^2 (10 T)	790 A mm^2 (30 T)	800–1000 A mm^2 (28 T)	1047 A mm^2 (12 T)
J_e(short,4 K) At a field of:	800 A mm^2 (10 T)	220 A mm^2 (30 T)	190 A mm^2 (28 T)	260 A mm^2 (12 T)

[a] Round processed PIT wire with a rectangular final shaping.

tapes, 10–20% more in wires. In the 30 T range, PIT tapes still carry about 1/2 of the zero field critical current while in the ROSAT conductor this value is reduced to 1/3. The critical currents of 200–340 A in the 30 T field range show the interesting potential of these conductors for very high field applications.

E1

B3.2.1.7.3 Temperature dependence of critical currents

C2

The temperature and field dependence of the critical current in a representative sample from a 1000 m length of a rectangular PIT tape for parallel field is shown in figure B3.2.1.21. It shows a relatively flat field behaviour for temperatures lower than 30 K. At 20 K the J_c drop at 5 T is about a factor of 2 and a transport current of 100 A is still obtained. At higher temperatures, I_c drops faster until, at 77 K, only applications in low field can be considered.

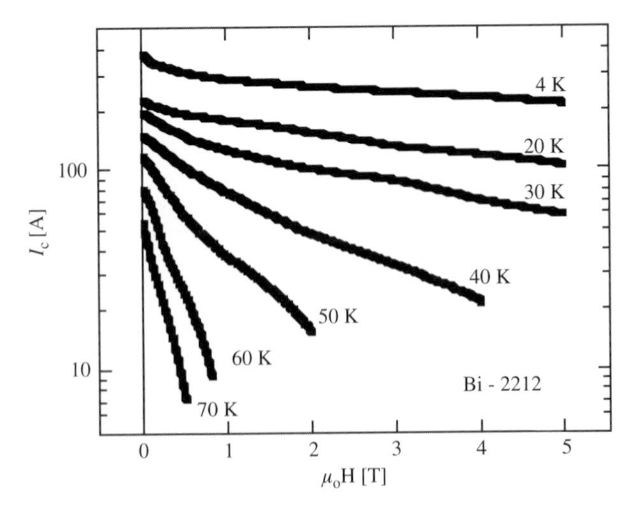

Figure B3.2.1.21. Field and temperature dependence of the critical current in Bi-2212 conductors. The self-field current density at 4 K is 400 A mm^2.

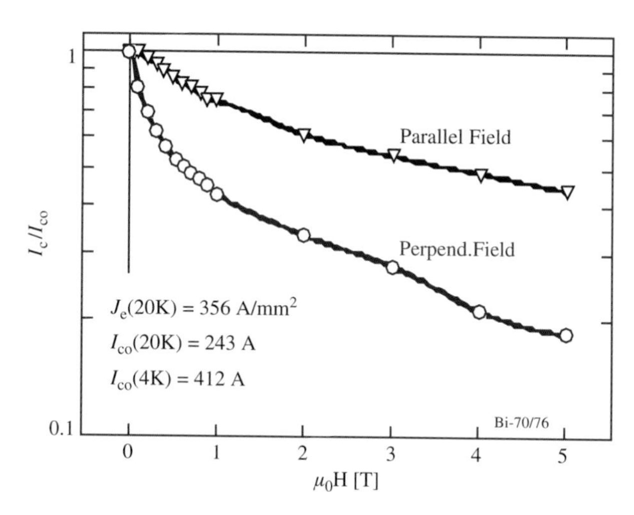

Figure B3.2.1.22. Anisotropy of critical current in Bi(2212) conductors at 20 K.

B3.2.1.7.4 Anisotropy of critical currents

The field dependence and the anisotropy of the critical current measured at 20 K are shown in figure B3.2.1.22. At 2 T, the critical current in perpendicular field is 55% of the parallel field value indicating that this conductor shows a comparatively low anisotropy. In the PAIR conductor [65] at 2 T and at 20 K, this value is in the range of 25–35% indicating the higher degree of texture achieved in this conductor. Additional measurements of anisotropy in rectangular PIT tape conductors are given in [72].

B3.2.1.7.5 Mechanical strength of silver and silver alloy sheathed Bi(2212) conductors

The Bi(2212) conductors are usually made with pure silver as the matrix material. It is soft material that may not protect the superconducting filaments against damage during standard coil winding processes and at operation at high field where very high magnetic pressures occur. D3.3

A detailed axial strain study on Bi(2212) tape conductors was made by Haken *et al* [73]. The samples were soldered to an U shaped sample holder allowing compressive and tensile strain measurements. Three domains of strain ranges have been identified.

(a) A linear reduction of the critical current is observed when a compressive strain up to −0.6% is applied. This reduction is the intrinsic response of the superconducting material under compressive strain conditions. A1.3

(b) At moderate tensile stress a plateau with constant critical current is observed. In this domain, the superconducting filaments that are under axial pre-compression at low temperature are progressively unloaded. A1.3

(c) At a critical strain of 0.4%, a drastic reduction of the critical current is observed. This corresponds to the appearance of the first net axial strain on the filaments, which causes irreversible crack formation and a dramatic reduction of I_c.

The plateau is therefore not an intrinsic property of Bi(2212) itself but rather the result of the interplay between the silver matrix and superconductor. This critical behaviour was observed many

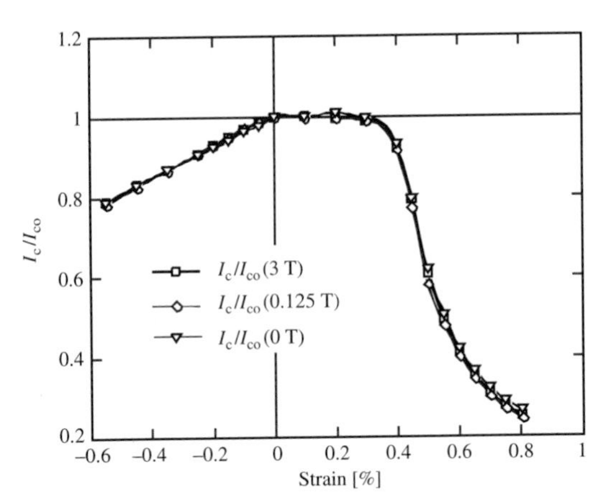

Figure B3.2.1.23. Critical current measurement of a rectangular PIT tape under axial tensile strain. Measurements courtesy of B. ten Haken. I_c is 345 A (self field), 273 A (125 mT), and 134 A (3 T) all at 4 K.

times and is illustrated in [74] on a rectangular PIT tape conductor. Figure B3.2.1.23 shows the strain behaviour for different magnetic fields. These curves are not reversible, once the damage has occurred the critical current is permanently degraded. This and the fact that the curves for different magnetic field superimpose perfectly (see inset) is compatible with crack formation as the underlying mechanism.

Similar behaviour was found in strain measurements [75] on PAIR samples. For tensile stress, constant J_c is observed until the critical stress is reached at a strain of 0.5%, where a dramatic decrease in I_c is observed. For compressive strain a classical linear I_c reduction to 0.8 I_{co} is observed for a strain value of −0.6%.

A1.3 The general strain behaviour described above occurs for silver and silver alloys. However, the critical strain value is reached at very different values of mechanical stress in silver and silver alloys. For example, the critical mechanical stress is about 50 MPa in pure silver, about 80 MPa in AgPd, and 250 MPa in AgMg or AgMgSb. The typical stress in a winding machine is about 80 Mpa, so silver is too soft and AgPd is still not strong enough. Several alloys have been proposed but the best known candidate for a high strength conductor is the silver magnesium alloy [76]. However, even this conductor must be further strengthened to resist magnetic pressures and Lorentz forces which exceed 250 MPa at high fields.

Acknowledgments

We would like to thank Dr H. Kitaguchi from NRIM, Dr L. R. Motowidlo from IGC and Dr M. Okada from Hitachi for fruitful discussions and up to date information on PAIR, round PIT wires, and ROSAT wires. Further, we would like to thank Dr J. Hascicek from NHMFL for high field measurement and Dr Bennie ten Haken from Univ. Twente for strain measurements. Finally we also want to thank Drs J. Bock and M. Bäcker from Alcatel High Temperature Superconductors for supplying experimental Bi(2212) powders and valuable discussions.

References

[1] Heine K, Tenbrink J and Thoner M 1989 High-field critical current densities in $Bi_2Sr_2CaCu_2O_{8+x}$/Ag wires *Appl. Phys. Lett.* **55** 2441

[2] Sato K, Hikata T, Mukai H, Ueyama M, Shibuta N, Kato T, Masuda T, Nagata M, Iwata K and Mitui T 1991 High-J_c silver-sheathed Bi-based wires *IEEE. Trans. Magn.* 1231

[3] Otto A, Masur L G, Gannon J, Podtburg E, Daly D, Yurek G J and Malozemoff A P 1993 Multifilamentary Bi-2223 composite tapes made by a metallic precursor route *IEEE. Trans. Appl. Supercond.* **3** 915

[4] Bock J, Elschner S and Preisler E 1991 The impact of oxygen on melt processing of Bi-HT superconductors *Adv. Supercond. (Proc ISS'90)* **3** 797

Bock J, Bestgen H, Elschner S and Preisler E 1993 Large shaped parts of melt cast BSCCO for applications in electrical engineering *IEEE. Trans. Appl. Supercond.* **3** 1653

[5] Dai W, Marken K R, Hong S, Cowey L, Timms K and McDougall I 1995 Fabrication of HT_c coils from Bi(2212) PIT and dip-coated tape *IEEE. Trans. Appl. Supercond.* **5** 516

[6] Collings E W, Sumption M D, Scanlan R M, Dietderich D R, Motowildo L R, Sokolowski R S, Aoki Y and Hasegawa T 1999 Bi(2212)/Ag-based Rutherford cables: production, processing and properties *Supercond. Sci Technol.* **12** 87

[7] Kitaguchi H, Kumakura H, Togano K, Miao H, Hasegawa T and Koizumi T 1999 Bi(2212)/Ag multilayer tapes with $J_c > 500\,000$ A/cm^2 at 42 K and 10 T by using pair process *IEEE. Trans. Appl. Supercond.* **9** 1794

[8] Okada M, Sato J, Kumakura H, Kitaguchi H, Togano K and Wada H 1998 A new symmetrical arrangement of tape-shaped multifilaments for Bi(2212) round-shaped wire *IEEE. Trans. Appl. Supercond.* **9** 1904

[9] Herrmann P F, Béghin E, Bock J, Duperray G, Grivon F, Legat D, Leriche A, Marlin P, Parasie Y and Tavergnier J P 1998 Pre-industrial PIT conductor and coil development at Alcatel *IEEE. Trans. Appl. Supercond.* **9** 2738

[10] Herrmann P F, Allais A, Bock J, Cottevieille C, Duperray G, Legat D, Leriche A, Melin J, Ryan D, Tavergnier J P, Tessier C, Verhaege T and Parasie Y 2000 BSCCO based superconductors for magnet applications *Adv. in Supercond.* **vol 12**, (ISS'99) (Tokyo: Springer-Verlag) p 730

[11] Herrmann P F, Bock J, Bruzek C E, Cottevieille C, Duperray G, Hascicek J, Legat D, Leriche A, Verhaege T and Parasie Y 2000 Long length PIT conductors realized by rectangular deformation route *Supercond. Sci. Technol.* **13** 477 (*Proc. EUCAS 99*)

[12] Majewski P 1996 Phase Equilibria and crystal chemistry of the high-temperature superconducting compounds of the system Bi_2O_3–SrO–CaO–CuO *Bismuth-Based High-Temperature Superconductors*, ed H Maeda and K Togano (New York: Dekker) p 129

[13] Idemoto Y and Fueki K 1990 Oxygen nonstoichiometry and valences of Bi and Cu in $Bi_2Sr_{1.88}CaCu_{2.14}O_y$ *Physica* C **168** 167

[14] Morgan C G, Priestnall M, Hyatt N C and Grovenor C R 1995 Characterisation of the partial melt processing of Bi(2212) *Proc. EUCAS*, ed D D Hughes (Edinburgh: Institute of Physics) p 331

[15] Idemoto Y and Fueki K 1990 Melting point of superconducting oxides as a function of oxygen partial pressure *Japan. J. Appl. Phys.* **29** 2729

[16] Rubin L M, Orlando T P, Vander Sande J B, Gorman G, Savoy R J, Swope R and Beyers R B 1992 Phase stability limits of Bi(2212) and Bi(2223) *Appl. Phys. Lett.* **61** 1977

[17] Rubin L M, Orlando T P, Vander Sande J B, Gorman G, Savoy R J, Swope R and Beyers R B 1993 Phase stability limits and solid-state decomposition of Bi(2212) and Bi(2223) in reduced oxygen pressures *Physica* C **217** 227

[18] MacManus-Driscoll J L, Wang P C, Bravman J C and Beyers R B 1994 Phase equilibria and melt processing of Bi(2212) tapes at reduced oxygen partial pressures *Appl. Phys Lett.* **65** 2872

[19] MacManus-Driscoll J L, Bravman J C, Savoy R J, Gorman G and Beyers R B 1994 Effects of Ag and Pb on the phase stability of Bi(2212) and Bi(2223) above and below the solidus temperature *J. Am. Ceram. Soc.* **77** 2305

[20] MacManus-Driscoll J L, Li Y H and Yi Z 1997 Upper phase stability of Bi(2212) *J. Am. Ceram. Soc.* **80** 807

[21] Gannon J J and Sandhage K H 1997 Solid-state high-oxygen-fugacity processing of Bi(2212) superconductors *IEEE. Trans. Appl. Supercond.* **7** 1533

[22] Hellstrom E E and Zhang W 1996 Melt processing Bi(2212) conductors, the influence of oxygen on phase relations in the melt *Bismuth-Based High-Temperature Superconductors*, eds H Maeda and K Togano (New York: Dekker) p 427

[23] Schartman R R, Sakidja R and Hellstrom E 1993 Supersolidus phase investigation of the Bi–Sr–Ca–Cu oxide system in silver tape *J. Am. Ceram. Soc.* **76** 724

[24] Reeves J L, Polak M, Zhang W, Hellstrom E E, Babcock S E, Larbalestier D C, Inoue N and Okada M 1997 Overpressure processing of Ag-sheathed Bi(2212) tapes *IEEE Trans. Appl. Supercond.* **7** 1541

[25] Yoshida M and Endo A 1993 Improvement of J_c of Ag-sheathed Bi(2212) tapes using melt-growth technique under reduced partial pressure *Japan. J Appl. Phys.* **32** L1509 Part 2

[26] Yoshida M 1995 Melt state of $Bi_2Sr_xCa_{3-x}Cu_2O_z$ with $15 < x < 23$ in 001 atm oxygen partial pressure *Japan J Appl. Phys* **34** 98 Part 1

[27] Hellstrom E E 1995 Processing Bi-based high-T_c superconducting tapes, wires and thick films for conductor applications *HTSC Materials Science and Engineering new Concepts and Technology*, ed L Shi (New York: Pergamon Press) p 383

[28] Kumakura H 1996 Bi(2212)/Ag composites tapes processed by doctor blade or dip-coating process *Bismuth-Based High-Temperature Superconductors*, eds H Maeda and K Togano (New York: Dekker) p 451

[29] Hayashi S, Shibutani K, Hase T, Ogawa R and Kawate Y 1996 Properties of Bi(2212) Ag sheathed tape and its application to magnets *Bismuth-Based High-Temperature Superconductors*, eds H Maeda and K Togano (New York: Dekker) p 411

[30] Zhang W, Pupysheva O V, Ma Y, Polak M, Hellstrom E E and Larbalestier D C 1997 Study of the effect of Sr/Ca ratio on the microstructure and critical current density of Bi(2212) Ag-sheathed tapes *IEEE. Trans. Appl. Supercond.* **7** 1544

[31] Balmer B R, Grovenor C R M and Riddle R 1999 Stoichiometric variations of Bi(2212) electrophoretically deposited thick films *IEEE. Trans. Appl. Supercond.* **9** 1888

[32] Polak M, Zhang W, Polyanskii A, Pashitski A, Hellstrom E E and Larbalestier D C 1997 The effect of the maximum processing temperature on the microstructure and electrical properties of melt processed Ag-sheathed Bi(2212) tape *IEEE. Trans. Appl. Supercond.* **7** 1537
[33] Bock J and Preisler E 1989 Preparation of single phase Bi(2212) by melt processing *Solid State Commun.* **72** 453
[34] Muroga T, Sato J, Kitaguchi H, Kumakura H, Togano K and Okada M 1998 Enhancement of critical current density for Bi(2212)/Ag tape conductors through microstructure control *Physica* C **309** 236
[35] Ionescu M, Dou S X, Apperley M and Collings E W 1998 Phase and texture formation in Bi(2212)/Ag tapes processed in oxygen *Superconduct. Sci. Technol.* **11** 1095
[36] Lang T h, Buhl D, Cantoni M and Gauckler L J 1995 Decomposition and reformation of Bi(2212) during the partial melt processing in oxygen *Proc. EUCAS*, ed D D Hughes (Edinburgh: Insitute of Physics) p 195
[37] Holesinger T G, Miller D J, Viswanathan H K, Dennis K W, Chumbley L S, Winandy P W and Youngdahl A C 1993 Directional isothermal growth of highly textured Bi(2212) *Appl. Phys. Lett.* **63** 982
[38] Hishinuma Y, Kitaguchi H, Kumakura H, Togano K, Miao H and Chenevier B 1998 Local J_c distribution in superconducting oxide layer of Bi(2212)/Ag tapes *IEEE. Trans. Appl. Supercond.* **9** 1908
[39] Hishinuma Y, Kitaguchi H, Miao H, Kumakura H, Itoh K and Togano K 1998 Critical current density distribution in the superconducting oxide layer of pre-annealing and intermediate rolling processed Bi(2212)/Ag composite tapes *Supercond. Sci. Technol.* **11** 1237
[40] Lang T h, Buhl D, Cantoni M, Wu Z and Gauckler L J 1995 Melt processing of Bi(2212) thick films and bulk components *Proc. EUCAS*, ed D D Hughes (Edinburgh: Institute of Phyiscs) p 203
[41] Kumakura H, Kitaguchi H, Togano K, Muroga T, Sato J and Okada M 1999 Influence of Ag substrate on grain alignment and critical current density of Bi(2212) tape conductors *IEEE. Trans. Appl. Supercond.* **9** 1804
[42] Motowildo L R, Galinsky G, Ozeryansky G, Zhang W and Hellstrom E E 1994 Dependance of critical current density on filament diameter in round multifilament Ag-sheathed Bi(2212) wires processed in O_2 *Appl. Phys. Lett.* **65** 2731
[43] Sengupta S, Caprino E, Kird K, Gaines J R, Motowildo L R, Sokolowski R S, Garcia R R and Mukhopadhyay S 1998 Synthesis of Bi–Sr–Ca–Cu oxide powders for Ag composite wires with uniform micron sized filaments *IEEE. Trans. Appl. Supercond.* **9** 2601
[44] Fujii H, Kamakura H, Kitaguchi H, Togano K, Zhang W, Fen Y g and Hellstrom E E 1997 The effect of oxygen partial pressure during heat treatment on the microstructure of dip-coated Bi(2212)/Ag and Ag alloy tapes *IEEE. Trans. Appl. Supercond.* **7** 1707
[45] Hellstrom E E and Zhang W 1995 Formation and prevention of bubbles when melt processing Ag-sheathed Bi(2212) conductors *Supercond. Sci Technol.* **8** 317
[46] Reeves J L, Hellstrom E E, Irizarry V and Lehndorff B 1999 Effects of overpressure processing on porosity in Ag-sheathed Bi(2212) multifilamentary tapes with various geometries *IEEE. Trans. Appl. Supercond.* **9** 1836
[47] Hase T, *et al* 1996 Summary (translated from Japanese) of the paper Mechanism and control of bubbling in Ag sheathed Bi(2212) superconducting tapes *J. Japan Inst. Met.* **60** 1020
[48] Ray R D II, Smith P A and Olsen E A 1995 Synthesis of Ag sheathed, Bi(2212) tapes by a novel liquid wicking method *Physica* C **251** 1
[49] Majewski P, Nast R and Aldinger F 1999 The influence of CO_2 on the phase stability of Bi(2212) *Supercond. Sci Technol.* **12** 249
[50] Miao H, Kitaguchi H, Kumakura H and Togano K 1998 Microstructure and superconducting properties of Pb-substitued Bi(2212)/Ag dip-coated tapes *Physica* C **298** 312
[51] Crossley Y H, Caplin A D and MacManus-Driscoll J L 1999 The influence of high Pb doping on flux pinning and phase formation in bulk and tapes of $Bi_{2.2-x}Pb_xSr_{1.8}CaCu_2O_{8+\delta}$ *IEEE. Trans. Appl. Supercond.* **9** 1832
[52] Miao H, Kitaguchi H, Kumakura H, Togano K, Hasegawa T and Koizumi T 1998 Bi(2212)/Ag multilayer tapes with $J_c > 500\,000$ A/cm^2 at 4.2 K and 10 T by using pre-annealing and intermediate rolling process *Physica* C **303** 81
[53] Fukumoto Y, Moodenbaugh A R, Suenaga M, Fisher D A, Shibutani K, Hase T and Hayashi S J 1996 Effect of oxygen partial pressure during post heat treatment on Bi(2212)/Ag tapes *J. Appl. Phys.* **80** 331
[54] Ray II R D 1993 *PhD Thesis* University of Wisconsin-Madison.
[55] Zhang W and Hellstrom E E 1995 The effects of oxygen on melt-processing Ag-sheathed Bi(2212) conductors *Supercond. Sci. Technol.* **8** 430
[56] Hauck J, Bickmann K, Chernyaev S and Mika K 1995 Critical current densities in Bi(2212) thick films *Proc. EUCAS*, ed D Dew Hughes (Edinburg: Institute of Physics) p 187
[57] Holesinger T G, Phillips D S, Willis J O and Peterson D E 1995 Relationship between processing temperature and microstructure in isothermal melt processed Bi(2212) thick film *IEEE. Trans. Appl. Supercond.* **5** 1939
[58] Willis J O, Holesinger T G, Coulter J Y and Maley M P 1997 Magnetic field orientation dependance of J_c in Bi(2212) round wire *IEEE. Trans. Appl. Supercond.* **7** 2022
[59] Holesinger T G, Baldonado P S, Van Vo N, Dai W, Marken K R and Hong S 1999 Isothermal melt processing of Bi(2212) tapes *IEEE. Trans. Appl. Supercond.* **9** 1800
[60] Burgoyne J W, Eastell C J, Morgan C G, East D, Jenkins R G, Storey R, Yang M, Dew-Hughes D, Jones H, Grovenor C R and Goringe M G 1995 A novel continuous process for the production of long lengths of Bi(2212)/Ag dip-coated tape *Proc. of EUCAS*, ed D D Hughes (Edinburg: Institute of Physics) p 335

[61] Morgan C G, Henry B M, Eastell C J, Goringe M G, Grovenor C R, Burgoyne J W, Dew-Hughes D, Priestnall M, Storey R and Jones H 1997 Continuous melt processing of Bi(2212)/Ag dip coated tapes *IEEE. Trans. Appl. Supercond.* **7** 1711

[62] Hu Q Y, Viouchkof Y, Weijers H W and Schwarz J 1999 Continuous processing of AgMg-sheathed Bi(2212) tapes *IEEE Trans. Appl. Supercond.* **9** 1808

[63] Ilyushechkin Y, Williams B, Lo F, Yamashita T and Talbot P 1999 Continuous production of Bi(2212) thick films on silver tapes *IEEE. Trans. Appl. Supercond.* **9** 1912

[64] Marken K R, Dai W, Cowey L, Ting S and Hong S 1997 Progress in Bi(2212)/Ag composite tape conductors *IEEE. Trans. Appl. Supercond.* **7** 2211

[65] Kitaguchi H, Miao H, Kumakura H and Togano K 2000 Relationship between $Bi_2Sr_2CaCu_2O_x$ layer thickness and J_c enhancement by PAIR process *Physica* C at press Kitaguchi H, Itoh K, Takeuchi T, Kumakura H, Miao H, Wada H, Togano K, Hesegawa T and Koizumi T Performance at 10–50 K of Bi(2212)/Ag multilayer tape fabricated by using PAIR process *Physica* C at press Kitaguchi H Private communication

[66] Hasegawa T, Koizumi T, Aoki Y, Kitaguchi H, Miao H, Kumakura H and Togano K 1999 Reaction mechanism and microstructure of PAIR processed Bi(2212)/Ag tapes *IEEE. Trans. Appl. Supercond.* **9** 1884

[67] Miao H, Kitaguchi H, Kumakura H, Togano K, Hasegawa T and Koizumi T 1999 Optimization of melt-processing temperature and period to improve critical current density of Bi(2212)/Ag multilayer tapes *Physica* C **320** 77

[68] Motowidlo L R, Sokolowski R S, Hasegawa T, Aoki Y, Koizumi T, Ohtani N, Scanlan R, Deitderich D and Nagaya S, M^2S-HTSC-VI Conference to be published

[69] Alcatel unpublished results 1999

[70] Okada M Private communication 1999

[71] Motowidlo L R, Sokolowski R S, Hasegawa T, Aoki Y, Koizumi T, Ohtani N, Scanlan R, Deitderich D and Nagaya S, M^2S-HTSC-VI Conference to be published

[72] Ryan D, Wilson M N, van Beersum J, Forestier Y, Herrmann P and Marken K 1999 A variable temperature test facility with variable field orientation *Supercond. Sci. Technol.* **13** 1259 Proc. EUCAS

[73] Haken B ten, Godeke A, Schuver H S and Kate H H J ten 1995 A descriptive model for the critical current as a function of axial strain in Bi(2212)/Ag wires *Inst. Phys. Conf. Ser.* **148** p 73 (Bristol: Institute of Physics Publishing)

[74] Haken B ten, Private communication

[75] Kitaguchi H, Takeuchi T, Itoh K, Kumakura H, Togano K, Hasegawa T and Koizumi T 1999 Strain Effect in Bi(2212)/Ag PAIR Processed Tapes *Advances in Supercond.* **vol 12**, (ISS 99) (Tokyo: Springer) p 730

[76] Goldacker W, Eckelmann H, Quilitz M and Ullmann B 1997 Effect of twisting on the filaments of multifilamentary BSCCO(2223)/Ag and /AgMg tapes *IEEE. Trans. Appl. Supercond.* **7** 1670

B3.2.2
Processing of high T_c conductors: the compound Bi,Pb(2223)

Giovanni Grasso

B3.2.2.1 Introduction

The Bi-based high T_c superconductor with the highest potential for industrial applications is the compound $Bi_2Sr_2Ca_2Cu_3O_{10}$, commonly abbreviated to Bi(2223), which undergoes a superconducting transition at 110 K [1]. This phase is commonly stabilized by the addition of Pb which substitutes about 10–20% of the Bi sites, and is then denominated Bi,Pb(2223). In spite of the complexity of the crystalline structure of this system, several industrial manufacturers have already succeeded in fabricating multifilamentary conductors surrounded by a silver sheath in batch lengths exceeding 1 km. The current carrying capacity of these tapes and their mechanical properties have been constantly improved in recent years; meanwhile, their behaviour in the presence of alternating currents and/or fields has been also considered. A significant advantage of Ag sheathed Bi,Pb(2223) tapes is that they can find an application over a wide temperature range, which typically extends between the boiling temperatures of liquid helium and liquid nitrogen. They can withstand large magnetic fields of the order of several tesla at temperatures below 40 K, while above this temperature they can be still employed in lower magnetic field applications, as power cables: this is still a major advantage with respect to a second Bi-based superconductor, the Bi(2212) phase, which actually shows a lower T_c value and therefore worse field dependence at high temperatures.

C2
A1.3

In this chapter, a brief discussion of the known facts concerning the Bi,Pb(2223) compound and its phase diagram are initially presented. In the following, the thermo-mechanical treatment of Ag sheathed Bi,Pb(2223) tapes will be described, starting from the monofilamentary configuration. It has to be said that this tape structure is not perfectly appropriate for an industrial scope, but it allows for a simpler microstructural and physical characterization of the conductor than for multifilamentary tapes. In particular, the monofilamentary configuration has been employed by Grivel *et al* [2] for studying the complex mechanisms of the Bi,Pb(2223) phase formation. Another experiment that will be cited regards the correlation between the initial texturing of the precursor Bi(2212) phase after tape deformation (before reaction) and the final texturing of the current carrying Bi,Pb(2223) grains after reaction. Monocore tapes have been also used for studying the effect of uniaxial pressing and of cold rolling, thus giving an answer to questions related to the observed differences in the transport properties of tapes deformed by these alternative techniques.

B3.2.1
B2.3.5

The process of mechanical deformation by the usual rolling procedure with two cylinders and by a modified four roll technique will be described, and it will be shown that the achieved local critical current distribution is more homogeneous for the tapes deformed by four rolls, where the pressure exerted on the filaments is more controlled, which is also reflected by a more regular filament dimension and density. The typical variation of the critical current density of a Bi,Pb(2223) tape as a function of the applied magnetic field and of its orientation are reported for temperature of 77 K and 4.2 K. Results

B3.1

will be qualitatively discussed by taking into account the mechanisms for current transfer between neighbouring grains.

A1.3 An important part of this chapter will be devoted as well to the fabrication and characterization of multifilamentary tapes in view of their use in ac applications. The effect of filament twisting will be D2.6 presented and discussed. Nevertheless, a substantial reduction of the ac losses in multifilamentary Bi,Pb(2223) tapes has been mainly obtained by the concept of the 'oxide barrier', which surrounds each filament and leads to an enhancement of the transversal resistivity, thus causing a drastic decrease of the coupling losses.

B3.2.1 In view of any practical application, it is also important to consider that a high critical current density is not a sufficient criterion for the description of high quality Bi,Pb(2223) tapes, but that this functional conductor should also withstand the mechanical stress applied during the fabrication of a practical device and in operating conditions without considerable damage. The reinforcement of the Ag matrix is usually carried out by suitable additions combined to a process of internal oxidation. A very high level of mechanical strength has been achieved nowadays on such Ag-sheathed conductors.

B3.2.2.2 Details of the Bi,Pb(2223) phase diagram

B3.2.1 The system in which the high-temperature superconducting phase Bi,Pb(2223) exists comprises a large amount of components and, therefore, its description at varying temperatures and constant pressure requires a six-dimensional space that cannot be handled. This is the most important problem one has to face when phase diagrams are required for the optimisation of the processing of Bi-based bulk materials. The most advanced works concerning phase diagram studies in the Bi–Pb–Sr–Ca–Cu–O–Ag system have been carried out by Majevski [3].

In the recent past, various simplified conditions have been studied in order to reduce the number of components and achieve a reasonable representation of the phase diagram of this system. Typically, the metaloxides Bi_2O_3, PbO, SrO, CaO, and CuO have been considered as components instead of the single elements, and a two-dimensional representation of the system at different temperatures was achieved while keeping constant four of the many independent parameters. Both the Pb-free and the Pb-doped systems have been studied in this simplified way.

B3.2.2.2.1 Phase diagram of the Pb-free Bi,Pb(2223) phase

B3.2.1 Concerning the Pb-free phase, it has been found that it is stable only over a very narrow temperature range and exhibits phase equilibria with a few compounds existing within the system. In contrast to the Bi(2212) phase, the Bi(2223) phase exhibits only a very narrow variation of the Sr- and Ca-content within the limits Sr:Ca$\sim$1.9:2.1–2.2. The Pb-free Bi(2223) phase also exhibits an excess of the Bi content of about 2.5. The known four-phase equilibria of the Bi(2223) phase as determined by Majevski [4] and Schulze *et al* [5] are shown in table B3.2.2.1 for a temperature of 850°C in air atmosphere.

Whole sections of the phase diagram of the Pb-free phase studied so far are dominated by phase equilibria of the superconducting phases with alkaline earth cuprates and bismuthates, copper oxide, a liquid phase and equilibria of the superconducting phases themselves. Small variations in concentration or temperature can result in quite different phase compositions and in a significant change of the volume content of the Bi(2223) phase. This behaviour is of extreme importance for the preparation of Bi(2223) ceramics, as any local inhomogeneity due to an improper powder processing may cause a significant reduction of the phase purity.

Table B3.2.2.1. Four-phase equilibria of the Pb-free 2223 phase at 850°C in air

1	2223–(Ca, Sr)$_2$CuO$_3$–CuO–liquid
2	2223–2212– CuO–liquid
3	2223–2212–(Ca, Sr)$_2$CuO$_3$–liquid
4	2223–(Ca, Sr)$_2$CuO$_3$–(Sr, Ca)$_{14}$Cu$_{24}$O$_{41-x}$–CuO
5	2223–2212–(Sr, Ca)$_{14}$Cu$_{24}$O$_{41-x}$–CuO
6	2223–2212–(Ca, Sr)$_2$CuO$_3$–(Sr, Ca)$_{14}$Cu$_{24}$O$_{41-x}$

B3.2.2.2.2 Phase diagram of the Bi,Pb(2223) phase

The study of the Pb-doped system is of much greater interest than the Pb-free phase for practical applications because of the discovery done by several authors [6–10] that the formation of the Bi,Pb(2223) phase is significantly promoted by the partial substitution of Bi by Pb. A detailed study of the Pb solubility as a function of the temperature and cation ratio has been carried out in order to give some hints for optimising the processing route of Bi,Pb(2223) ceramics. The Pb solubility of the Bi,Pb(2223) phase has been observed to be significantly temperature dependent [11]. A temperature versus concentration diagram including the single-phase region and surrounding four- and five-phase regions is represented in figure B3.2.2.1.

B2.3.5

B3.2.1

The most essential fact regarding the Bi and Pb content dependence on temperature is the maximum in Pb solubility at 850°C in air. At about 750°C the phase contains almost no Pb, but nevertheless it is still stable. Even at lower temperatures, the decomposition of the Bi,Pb(2223) phase occurs at a very slow rate, and it is hardly detectable.

The varying Pb solubility of the Bi,Pb(2223) phase is a limiting parameter for the processing of a single phase material. While slowly cooling Bi,Pb(2223) samples after synthesis at 850°C, their composition moves away from the single phase domain and turns into multi-phase systems consisting of the Bi,Pb(2223) phase with a lower Pb content and secondary phases. This process is often used at the end of the heat treatment of Ag-sheathed tapes, leading to a higher T_c and improved magnetic field dependence of the critical current density.

In figure B3.2.2.2, a schematic section through the single-phase region at 850°C is represented, including determined multi-phase regions surrounding the Bi,Pb(2223) phase.

All the known four- and five-phase regions taking place with the Pb-doped phase identified by Majevski [3] are listed in table B3.2.2.2. However, the actual number of possible phase regions around the Bi,Pb(2223) phase is probably much larger.

B3.2.2.2.3 The effect of Ag to the phase diagram of the Bi,Pb(2223)

Finally, taking Ag into account in sheathed Bi,Pb(2223) conductors introduces an additional dimension into the problem. At temperatures below the peritectic melting of the Bi,Pb(2223) phase, the phase relations appear not to be largely affected by the addition of silver, even if it is known that the optimal temperature for the formation of the phase is shifted to about 835° in air atmosphere. From the material processing it is indeed known that the Bi,Pb(2223) phase as well as all phases known to be in equilibrium with it are in thermodynamic equilibrium with silver.

B3.2.1

However, when the Bi,Pb(2223) phase starts to decompose at temperatures above about 845°C, Ag appears to significantly affect the phase relations. In particular, Ag appears to reduce the decomposition temperature from about 890°C down to about 860°C at Ag concentration of about 50 mol%.

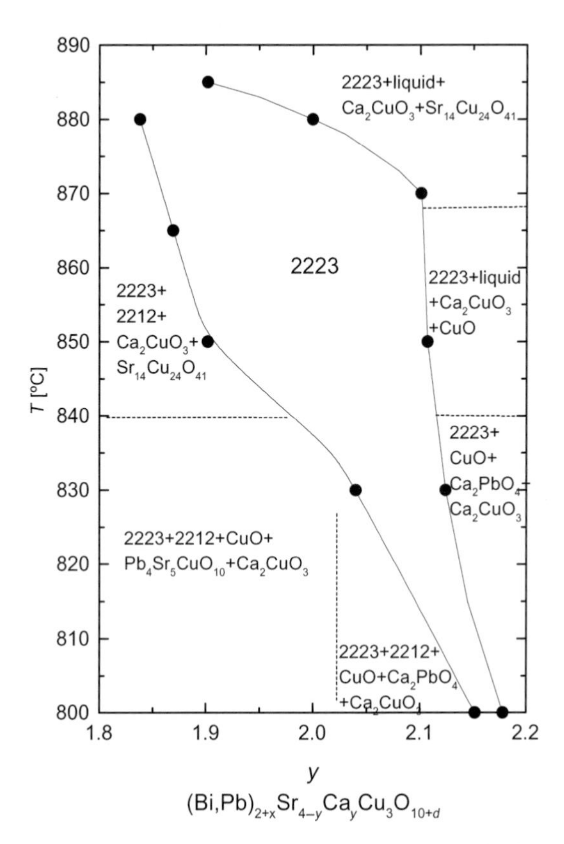

Figure B3.2.2.1. Temperature *versus* concentration diagram including the single-phase region and surrounding four- and five-phase regions surrounding the pure Bi,Pb(2223) domain.

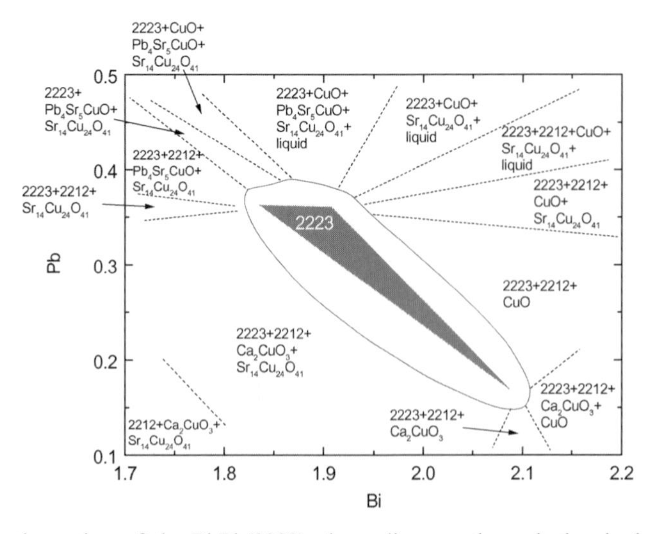

Figure B3.2.2.2. Schematic section of the Bi,Pb(2223) phase diagram through the single-phase region at 850°C, including determined multi-phase regions surrounding the Bi,Pb(2223) phase as determined by Majevski [3].

Table B3.2.2.2. All the known four- and five-phase regions of the Bi,Pb(2223) phase

	4-phase regions	5-phase regions
1	2223–2212–(Ca, Sr)$_2$CuO$_3$–(Sr, Ca)$_{14}$Cu$_{24}$O$_{41-x}$	2223–(Sr, Ca)$_{14}$Cu$_{24}$O$_{41-x}$–Pb$_4$Sr$_5$CuO$_{10}$–CuO–liquid
2	2223–2212–(Ca, Sr)$_2$CuO$_3$–CuO	2223–(Sr, Ca)$_{14}$Cu$_{24}$O$_{41-x}$–Ca$_2$PbO$_4$–CuO–liquid
3	2223–2212–(Sr, Ca)$_{14}$Cu$_{24}$O$_{41-x}$–CuO	2223–2212–(Sr, Ca)$_{14}$Cu$_{24}$O$_{41-x}$–Pb$_4$Sr$_5$CuO$_{10}$–CuO
4	2223–2212–(Sr, Ca)$_{14}$Cu$_{24}$O$_{41-x}$–Pb$_4$Sr$_5$CuO$_{10}$	2223–2212–(Sr, Ca)$_{14}$Cu$_{24}$O$_{41-x}$–Ca$_2$PbO$_4$–CuO
5	2223–(Sr, Ca)$_{14}$Cu$_{24}$O$_{41-x}$–CuO–liquid	2223–2212–(Ca, Sr)$_2$CuO$_3$–Pb$_4$Sr$_5$CuO$_{10}$–CuO
6	2223–(Ca, Sr)$_2$CuO$_3$–(Sr, Ca)$_{14}$Cu$_{24}$O$_{41-x}$–liquid	2223–2212–(Ca, Sr)$_2$CuO$_3$–Ca$_2$PbO$_4$–CuO
7	2223–(Ca, Sr)$_2$CuO$_3$–CuO–liquid	2223–2212–(Ca, Sr)$_2$CuO$_3$–Pb$_4$Sr$_5$CuO$_{10}$–Ca$_2$PbO$_4$
8	2223–(Ca, Sr)$_2$CuO$_3$–CuO–Ca$_2$PbO$_4$	2223–2212–CuO–Pb$_4$Sr$_5$CuO$_{10}$–Ca$_2$PbO$_4$
9	2223–(Ca, Sr)$_2$CuO$_3$–Pb$_4$Sr$_5$CuO$_{10}$–Ca$_2$PbO$_4$	
10	2223–CuO–Pb$_4$Sr$_5$CuO$_{10}$–Ca$_2$PbO$_4$	
11	2223–2212–Pb$_4$Sr$_5$CuO$_{10}$–Ca$_2$PbO$_4$	

The temperature of the maximum Pb solubility decreases from 850°C to about 835°C and the maximum content of Pb dissolved in the Bi,Pb(2223) phase appears to be reduced by Ag.

B3.2.2.3 Thermomechanical processing

B3.2.2.3.1 Considerations about the Bi,Pb(2223) phase formation process

Bearing in mind the difficulties appearing with the synthesis procedure of a pure Bi,Pb(2223) sample and its relevance in view of a technological application, a large number of laboratories have addressed the problem of the phase formation mechanism of this compound starting from a given precursor. It has to be said that no general agreement about the different reaction routes involved in the Bi,Pb(2223) phase formation process has been achieved so far. A1.3

While the nature of the starting precursors can undoubtedly have an influence on the phase formation mechanism, the presented results will mainly concern the study of the formation process occurring from a heterogeneous powder obtained through the calcination of the constituents at a given temperature, which is actually slightly lower than that of the Bi,Pb(2223) phase formation. This is the most common method used for the preparation of bulk superconducting samples. B2.1

As first proposed by Ikeda *et al* [12] and confirmed by Jeremie *et al* [13] in powder mixtures and by Grivel *et al* [14] in Bi,Pb(2223) tapes, the reaction from the phase Bi(2212) to Bi,Pb(2223) passes through the intermediate phase Bi,Pb(2212), which temporarily forms during the temperature ramp. The dissolution of a certain amount of Pb in the Bi(2212) phase was evidenced both by direct EDX measurements on single Bi(2212) grains, and by DTA measurements, when comparing the melting temperatures in calcined powders of otherwise identical composition, the only difference being the presence or absence of Pb. A direct micrographic observation of the morphological transformations occurring in the precursor powders during the reaction is not possible in this case. The entire phase formation has been more recently studied by Grivel *et al* [2], who have employed both scanning and transmission electron microscopy on pressed pellets and Ag-sheathed tapes to elucidate the formation process. In figure B3.2.2.3, a selection of SEM photographs of a fixed location of a Pb-containing precursor powder pellet after various total sintering periods at 852°C are shown. It appears that the Bi,Pb(2212) plate-like grains present on the surface after the densification step are still visible after a short sintering time. However, as the sample is further heat treated at high temperature, the Bi,Pb(2212) B2.1 D1.5 D1.3

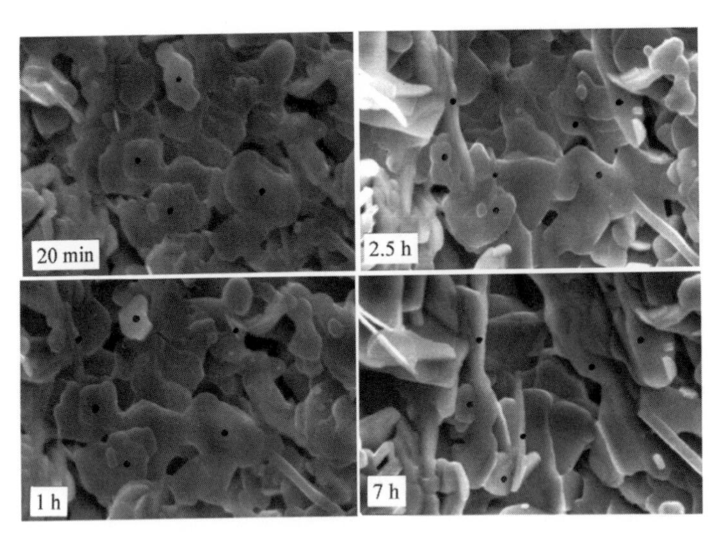

Figure B3.2.2.3. SEM photographs of the same location on the surface of a Pb-containing pellet after pressing and after various total sintering times at 852°C. The same magnification was used throughout.

grains progressively disappear, whereas new crystallites grow on the sample surface. Compositional analysis performed by EDX on several Bi,Pb(2212) grains during the first stages of the reaction have revealed the continuous incorporation of Pb into the phase before its decomposition.

A slight decrease of the Sr:Ca ratio suggests that some Ca also enters this phase at the beginning of the reaction. On the other hand, no time variation of the Cu:Sr atomic ratio was detected. Although the newly formed crystallites appear to have a lower degree of texture than the initial Bi,Pb(2212) grains, it was possible to determine their composition, that was found to correspond to that of the Bi,Pb(2223) phase. In particular, these grains do not seem to originate directly from the pre-existing Bi,Pb(2212) grains, as would be the case if the transformation proceeded through a Ca and Cu–O layer-by-layer intercalation mechanism.

The larger in-plane dimensions of the Bi,Pb(2223) grains as compared with those of the initial Bi,Pb(2212) grains also provide evidence for a growth phenomenon.

Investigations on Ag-sheathed tapes were performed on samples sintered at 838°C for various times and quenched. The composition of several grains on an atomic level was analysed by TEM after each sintering time and after peeling off the Ag sheath. For three different sintering times, 9, 17 and 25 h, the stacking of the $(Bi,Pb)_2Sr_2Ca_{n-1}Cu_nO_{2n+4}$ structure type layers along the c-axis of the grains was studied. For each sintering time, 35 grains resulting in a total of around 3000 layers were analysed. From these data, the proportion of Bi,Pb(2212) layers in each grain was determined.

D1.2

The raw results are plotted in the histograms presented in figure B3.2.2.4. It appears that only Bi,Pb(2212) grains with very few Bi,Pb(2223) intergrowths and Bi,Pb(2223) grains with less than 30% of Bi,Pb(2212) layers are present. In the case of a layer-by-layer insertion in the Bi,Pb(2212) phase, one would instead expect to find grains with an arbitrary proportion of layers during the transformation. In Ag-sheathed tapes as well as in pellets, the Bi,Pb(2223) phase forms independently of the initial Bi,Pb(2212) crystallites present in the sample. This formation process can be well described by a classical nucleation and growth process.

It has to be noted, however, that the investigations performed on the Ag-sheathed tapes were restricted to samples which were subjected to a single heat treatment, without any densification step (pressing or rolling). The corresponding results therefore describe the processes occurring during

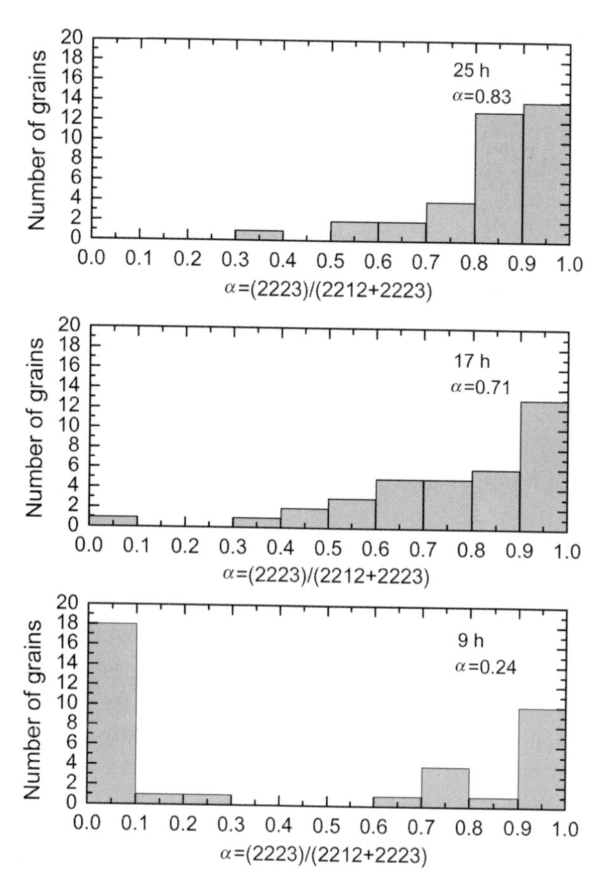

Figure B3.2.2.4. Histograms of the number of Bi,Pb(2212) and Bi,Pb(2223) grains as determined by TEM analysis of various grains extracted from Ag-sheathed tapes reacted for different sintering times.

the initial stages of the thermo-mechanical treatment usually performed on tapes. No definitive answers can be therefore given at present on the complete transformation process from Bi(2212) to Bi,Pb(2223) on a microscopic scale.

B3.2.2.3.2 Thermo-mechanical treatment for optimizing J_c of monofilamentary Bi,Pb(2223) tapes

The critical current density is the most important property of high T_c and low T_c superconductors that are intended for power applications. In polycrystalline high-T_c superconductors, a critical current density J_c flowing through the entire sample is strongly influenced both by microscopic and macroscopic factors. It is well known from the study of bicrystal junctions that in-plane and out-of-plane textures of the superconducting grains can improve the transport properties at the grain boundaries by several orders of magnitude. At the same time, macroscopic factors like cracks, presence of secondary phases at the grain boundaries interrupting the current paths, intergrowth, voids, etc, can also strongly affect the J_c value. A1.3

The powder-in-tube (PIT) method is largely employed for the fabrication of long Bi-based conductors showing the relevant superconducting properties at the liquid nitrogen temperature [15]. Within this process, mono- and multifilamentary Ag-sheathed Bi,Pb(2223) superconducting tapes can B3.1

be prepared by cold deformation and subsequent heat treatment of pure and alloyed silver tubes, previously filled with suitably pre-reacted powders. Optimized powders should present an overall stoichiometry near to Bi,Pb(2223), but they must be only partly reacted in such a way that they would contain a large amount of Bi(2212) phase (up to 80%) with the addition of other secondary phases such as CuO, Ca_2CuO_3 and Ca_2PbO_4, each of them in a proportion typically below 10%.

As silver is a very ductile metal, standard deformation techniques of metallurgy can be employed. Hot extrusion, swaging, wire drawing, flat- and four-rolling are the most common processes which are usually employed for the fabrication of long Bi,Pb(2223) conductors [16]. Hot extrusion, swaging and drawing are used in the first part of the tape fabrication, because they all allow for the cold deformation of round wires but with different strain components. Cold rolling is instead the finishing deformation process, as it is essential in order to achieve a tape-like shape for the conductor. After cold rolling, optimized monocore tapes present a thickness of about 0.1–0.3 mm, and a width of 2–4 mm. The entire deformation process is essential in order to facilitate the *c*-axis texturing of the superconducting grains needed after completing the fabrication route [17].

B3.1

In figure B3.2.2.5, a typical transverse cross-section of a monofilamentary tape is shown. The silver sheath surrounds completely the inner superconducting core. Due to the particular deformation process employed for the tape preparation, the shape of the filament is ellipsoidal-like, showing a regular narrowing moving towards the tape sides.

Considering the whole fabrication process, many parameters can have a relevant influence on the final tape transport properties. The most important effects of the cold working procedure are related to the type of deformation, its speed and reduction rate. In a more general way, the die or roll material and size, or the lubricant used during the cold working process can represent delicate choices as well. These factors strongly affect both the 'sausaging' of the filament cross section, the precursor powder density, and the pre-texturing of the Bi(2212) grains, which are all in turn correlated to the tape transport properties. A direct analysis carried by SEM together with the oxide density data are highly required to estimate the validity of a given choice of cold deformation parameters. Furthermore, it has to be noted that, after the cold deformation process, the precursor powders enclosed into the silver sheath present a very small grain size ($<1\ \mu m$) and become almost amorphous, mostly due to the high stresses applied to the tape during the fabrication.

D1.3

After the cold deformation process, it is still necessary to react the Ag-sheathed conductors at high temperature in oxygen partial pressure in order to form the Bi,Pb(2223) phase. The high temperature treatment turns out to be a crucial process for achieving high critical current densities, particularly on very long samples: Bi,Pb(2223) phase purity, grain size and grain boundary properties are indeed seriously influenced by the selected temperature as well as by the total duration of the process, also including the cooling phase as well. One or more deformation steps are also required at different stages of the heat treatment in order to further increase the oxide powder density, which is a dominant factor to achieve much larger J_c values.

C2

A differential thermal analysis (DTA) measurement can be of great help for sorting out optimal heat treatment parameters. DTA measurements performed on a typical unreacted tape as well as on the precursor powders are shown in figure B3.2.2.6. According to the DTA measurements, the peak related

B2.1

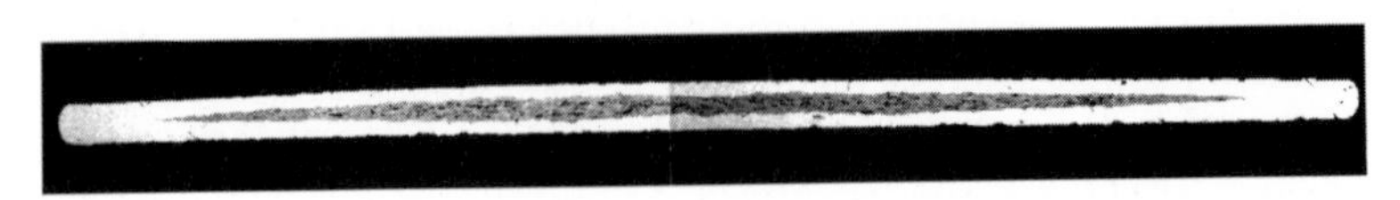

Figure B3.2.2.5. Typical transversal cross section of a monofilamentary tape. The overall tape thickness is about 90 μm, the tape width being 2.5 mm.

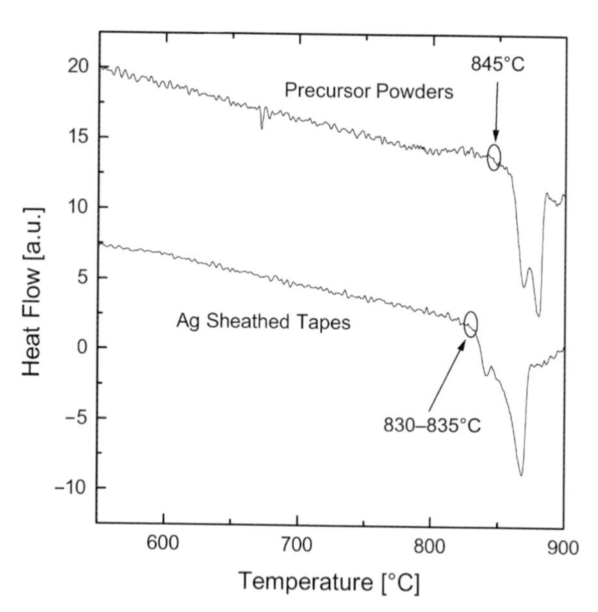

Figure B3.2.2.6. DTA measurements as a function of the temperature for an Ag-sheathed tape and for calcined powders. The onset of the reaction into Bi,Pb(2223) is evidenced.

to the formation of the Bi,Pb(2223) phase is shifted to lower temperatures by about 10–15°C in the Ag sheathed tape.

The lowering of the formation temperature of the Bi,Pb(2223) phase has essentially three origins: (a) the presence of Ag in direct contact with the phase, as shown by Grivel *et al* [18]; (b) the reduced grain size in the tapes, which accelerates the kinetics of the phase formation; and (c) the presence of the sheath which is partly permeable to oxygen, but inevitably changes the environmental conditions in which the reaction into Bi,Pb(2223) takes place. The DTA measurement is a very powerful tool to determine straightforwardly the optimal temperature for the formation of pure Bi,Pb(2223) phase. Indeed, the optimal temperature for the formation of the Bi,Pb(2223) phase can vary within a temperature range of 10°C just if the precursor powders are treated by slightly different procedures. Hopefully, a direct correlation can be established between the onset temperature of the DTA peaks and the optimal heat treatment of Bi,Pb(2223) tapes, as determined by X-ray and SEM analysis performed on samples treated at different temperatures.

B2.1

D1.1.2

D1.3

From the DTA measurements of figure B3.2.2.6, it appears that no other evident reactions are taking place at temperatures lower than the Bi,Pb(2223) formation temperature. However, by studying the X-ray diffraction patterns of quenched tapes at intermediate temperatures, it is possible to show that more complex mechanisms are taking place. In figure B3.2.2.7, the XRD patterns of the superconducting filament surface are shown for temperatures between 500°C and 837°C. At relatively low temperature ($\leq$ 700°C), the diffraction peaks are almost unchanged.

D1.1.2

Starting from 750°C, the Bi(2212) peaks begin to raise, which is a sign of the re-crystallization and growth of these grains. At this stage of the heating ramp, the growth of the Bi(2212) grains allows for the partial release of the air which has filled the interstitials of the oxide filament. Moreover, this gas tends to expand due to its increased pressure at high temperature, and a bubbling effect of the silver sheath at the tape surface can occur, especially if the temperature is raised too fast. Above 800°C the peak of the Ca_2PbO_4 phase reduces in intensity, and in parallel the (200)–(020) peaks of the Bi(2212) phase are splitting. The Pb has therefore started to dope the Bi(2212) phase.

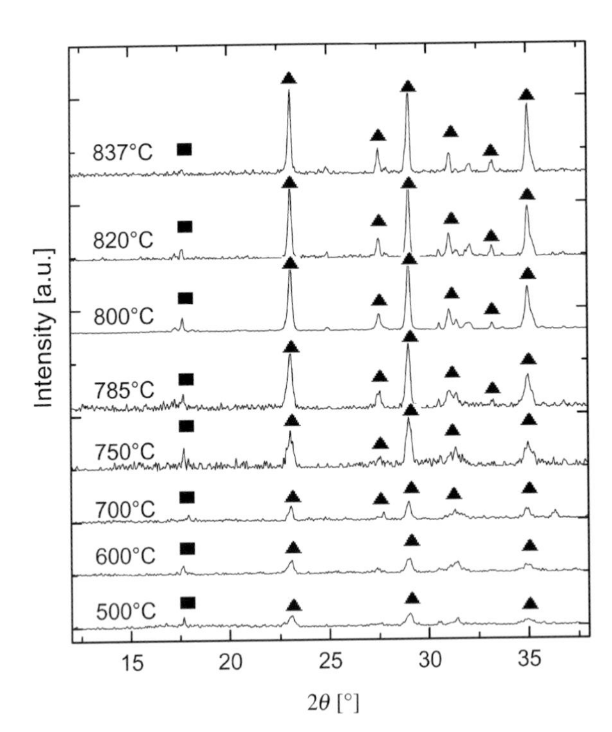

Figure B3.2.2.7. X-ray diffraction patterns of quenched tapes at temperatures between 500 and 837°C. The identified phases are: ■: Ca_2PbO_4 and ▲: Bi(2212).

At a given heat treatment temperature of 837°C, the Ca_2PbO_4 phase is completely decomposed, but no traces of the Bi,Pb(2223) phase are initially detected. At this stage of the reaction, the Bi(2212) grain size can reach up to 10–20 μm in size. Longer heat treatment stages at high temperature are required in order to form the Bi,Pb(2223) phase. In figure B3.2.2.8, the evolution of the Bi,Pb(2223) phase formation is shown as a function of the heat treatment time. The first detectable peaks of the Bi,Pb(2223) phase appear only after several hours of heat treatment.

During this period of time (0–6 h), it is hardly possible to observe any variation in the X-ray diffraction pattern or by SEM analysis. However, it is believed that the composition of the Pb-doped Bi(2212) phase is slowly but continuously changing during the first 5–10 h of treatment, and that it is going towards a critical instability point in which the formation of the Bi,Pb(2223) phase suddenly starts. The formation process of the Bi,Pb(2223) phase inside the silver sheath has been monitored in its entirety, the behaviour being reported in figure B3.2.2.9. The transformation from Bi(2212) into Bi,Pb(2223) clearly presents the expected step-like behaviour.

D1.3

C2

The reaching of a high Bi,Pb(2223) phase purity inside the silver sheath does not imply at all that such a conductor would be also able to carry a large transport current. The optimization of the critical current density is indeed a multi-parameter problem, and the solution to it giving the best results is often a compromise. In order to simplify the problem, it is often arbitrarily assumed that there is no direct relation between the powder preparation technique and the cold deformation process, that is, an optimized deformation process should always give the best-possible results achievable with any particular treatment employed for the powder preparation. The final heat treatment parameters, instead, have been constantly tuned, because they turned out to be extremely sensitive to any variation of the precursor powders and of the deformation process.

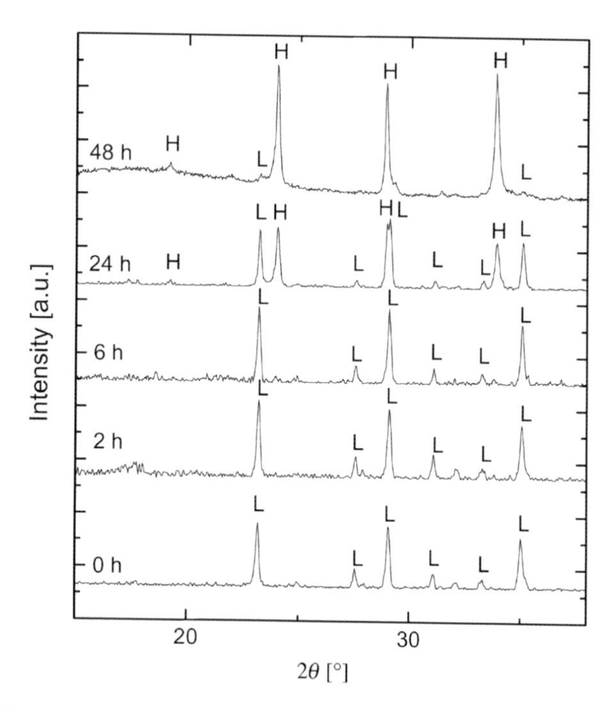

Figure B3.2.2.8. X-ray diffraction patterns of tapes treated at 837°C for 0, 2, 6, 24 and 48 h. L indicates the Bi(2212) peaks, H the Bi,Pb(2223) ones.

The influence of the preparation process of the precursor powders has been therefore studied on monofilamentary tapes which have been prepared by a standard deformation process and treated in air. Even if a lower oxygen partial pressure atmosphere is often used as well, it does not introduce substantial changes in the tape properties and will not be discussed. Different calcined powders of nominal composition $Bi_{1.72}Pb_{0.34}Sr_{1.83}Ca_{1.97}Cu_{3.13}O_x$ have been filled inside pure Ag tubes, with a packing density of about 65% of the theoretical density (which corresponds to 4.3 g cm^{-3}), and they have been deformed by swaging and drawing down to an outer diameter of 1.0 mm. Afterwards, these wires have been cold rolled in steps of about 10% of thickness reduction down to a thickness of 100 μm. B2.1

Short straight pieces (up to 30 cm long) of tapes have been heat treated inside tubular furnaces at temperatures between 830 and 840°C for up to 300 h, with several (up to three) intermediate re-densification steps. The precursor powders have been first prepared by calcination of coprecipitated powders at temperatures ranging from 780 to 830°C, for a duration of 12 h. B3.1

The best result of critical current density of 30 kA cm^{-2} at the liquid nitrogen temperature achieved within this batch of samples has been reached with precursor powders calcined at 820°C, provided that the tapes are exclusively treated at 838°C. The plot of figure B3.2.2.9 gives other useful information. First of all, the critical current density appears to be extremely sensitive to the heat treatment temperature of the tapes: a difference of ±2°C of the temperature can reduce J_c by a factor as high as two. C2

This result confirms that particular care of furnace homogeneity should be taken. From the results reported in figure B3.2.2.9, we can also deduce that there is a clear correlation between the optimal temperature for the heat treatment of the tapes and the temperature at which the precursor powders have been calcined.

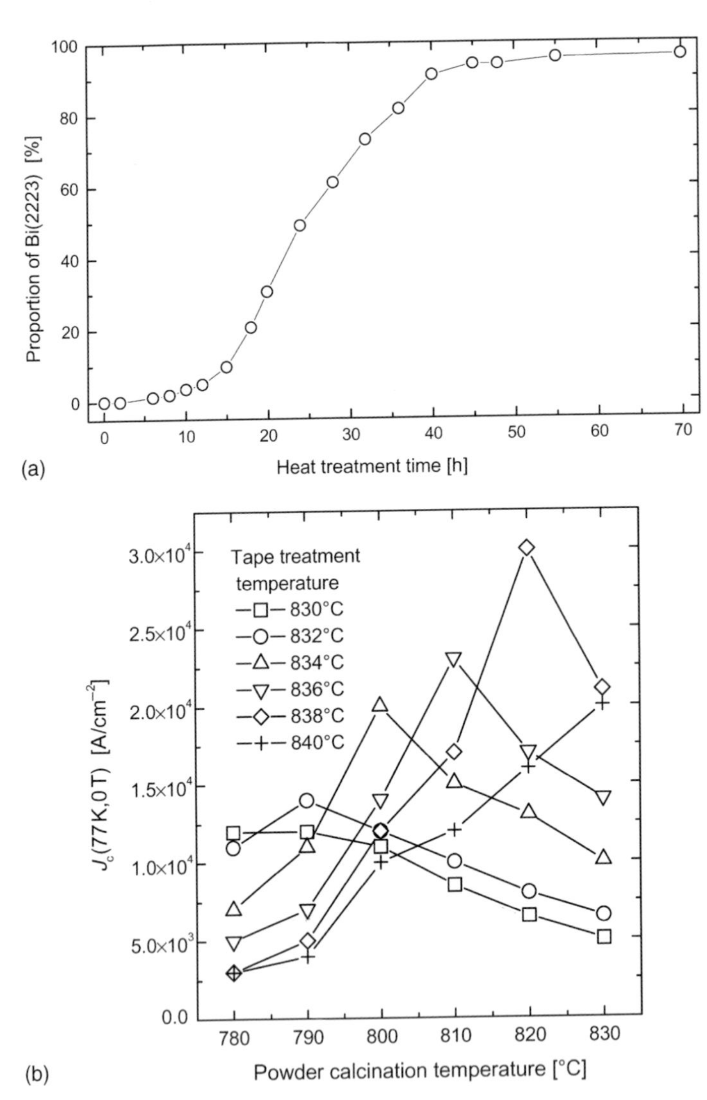

Figure B3.2.2.9. (*a*) Proportion of Bi,Pb(2223) phase in a tape as a function of time for a temperature of 837°C. (*b*) J_c as a function of the powder calcination temperature, and of heat treatment temperature of the silver sheathed tapes.

In a very schematic way, it is possible to say that in the powders calcined at lower temperatures, a lower content of Bi(2212) phase is present, and therefore the secondary phases, which are present in larger amount, will produce a larger amount of liquid phase during the reaction into Bi,Pb(2223). The higher amount of liquid phase will speed up the formation of the Bi,Pb(2223) phase, which will also take place at a lower temperature. This is essentially the reason for the required reduction of the heat treatment temperature of the tapes.

C2 On the other hand, the maximum critical current density is reducing for calcination temperatures below 820°C, essentially because the final Bi,Pb(2223) phase purity we observed was not comparable to that of the best tapes. On the contrary, the powders calcined at 830°C have led to worse results because they cannot supply enough liquid phase for the formation of the Bi,Pb(2223) phase, particularly for the construction of high quality grain junctions. The duration of the calcination process has a marginal influence on the critical current density, provided that it is long enough to avoid the carbon contamination at the Bi,Pb(2223) grain boundaries.

As already mentioned, the tape deformation must be carefully controlled in order to (a) improve the oxide powder density and (b) obtain a smooth silver-filament interface. Experimental correlation between critical current density and both oxide density and sausaging can be deduced. In a first experiment, several tapes have been prepared with the same precursor powders, filled inside pure Ag tubes with different initial packing density, from 40% (hand packing), up to 65% of the theoretical density. Even higher densities have not been studied because they would lead to a strong sausaging effect of the filament.

The tubes have been deformed in a standard way by swaging, drawing and rolling, and the resulting tapes have been reacted at 837.5°C for about 200 h. The oxide density has been deduced from simple geometrical considerations. The results are shown in figure B3.2.2.10.

A relative smooth variation of the oxide density leads to strong variations of J_c. This effect is not related to a reduced Bi,Pb(2223) phase purity as verified by X-ray analysis of the various samples. It is more likely probable that the formation of the strong links between the grains requires the highest possible density during the reaction. Another coexistent explanation is that the higher oxide density forces the Bi,Pb(2223) grains to grow in a more aligned way in the tape direction, improving then the degree of texture of the grains.

A good compromise between high critical current density, limited filament sausaging, and sufficient mechanical strength of monofilamentary tapes has been found for conductors fabricated by cold rolling of round wires of outer diameter in the range between 1 and 2 mm, that produce a tape of about 2.5–4 mm in width. Keeping the wire diameter fixed to 1 mm, the influence of the tape thickness on J_c has been analysed. As can be seen in figure B3.2.2.11, the critical current shows a rather narrow peak centred at about 30–40 μm of filament thickness, that is, for about 100 μm of overall tape thickness.

B3.1
A1.3

For both lower and higher thickness, the critical current density strongly decreases. The drop for lower thickness has been imput to the increased sausaging of the filament, as the oxide powders have already reached the saturation value of the density.

The main reason for the drop of J_c in thicker tapes is essentially related to the lower degree of texture of the Bi,Pb(2223) grains.

C2

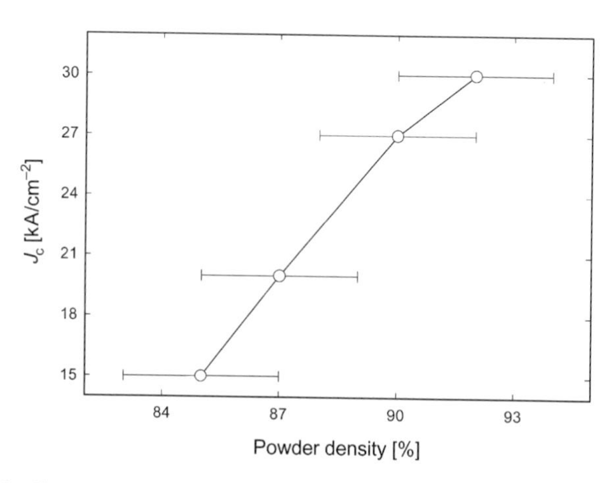

Figure B3.2.2.10. Correlation between the critical current density and the oxide powder density.

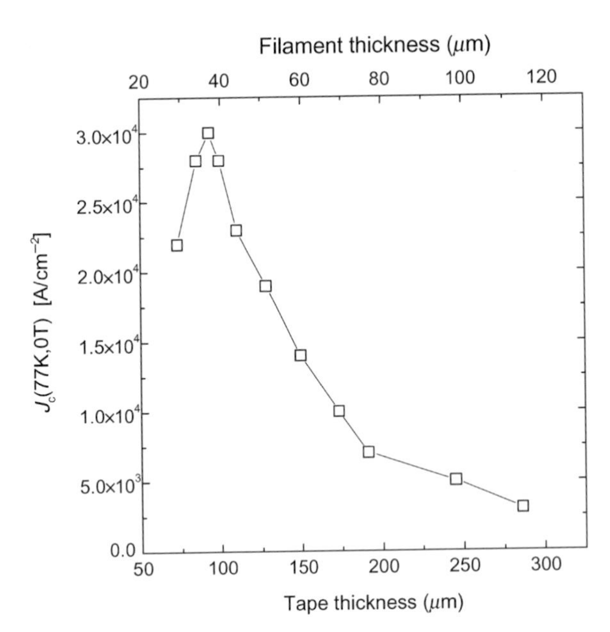

Figure B3.2.2.11. Tape thickness dependence of the critical current density.

This lack of texture is the consequence of the lower pre-texturing of the Bi(2212) grains achieved during the cold deformation process prior to the heat treatment. X-ray diffraction analysis has been used to reveal the mechanical texturing of the precursor grains. The X-ray diffraction pattern measured on randomly oriented calcined precursor powders (figure B3.2.2.12, left) can be compared with the pattern obtained on an unreacted monofilamentary tape of 245 μm in thickness after the removal of the silver sheath (figure B3.2.2.12, right). The latter shows a relevant difference in the intensity of the Bi(2212) peaks. The difference (115) – (008) in the peak intensity is explained by the presence of a c-axis texture.

D1.1.2

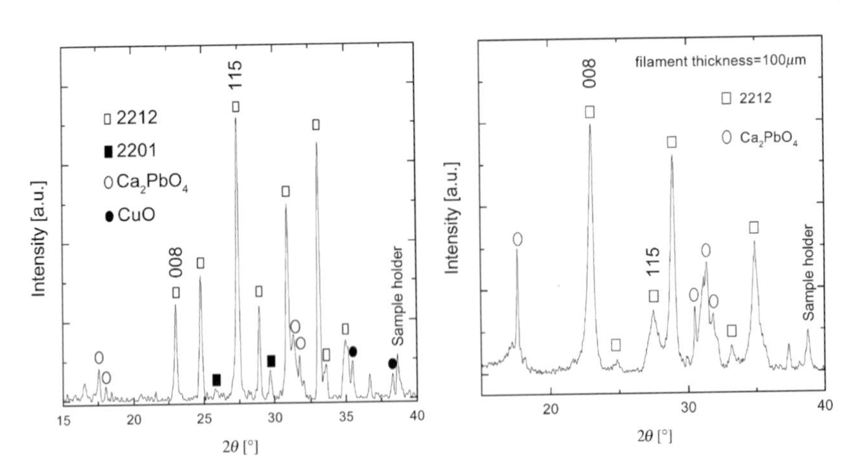

Figure B3.2.2.12. X-ray diffraction pattern of the pre-reacted powders (left), and of the unreacted tape surface (right) after mechanical remotion of the Ag sheath.

In order to extract quantitative information about the pre-texturing of the Bi(2212) phase, it is possible to define a texture parameter t as

$$t = \frac{L(115)}{L(115) + L(008)}$$

where L(115) and L(008) are the integrals of the (115) and (008) peaks of the Bi(2212), respectively. For the isotropic powders, the t value lies typically between 2 and 4, while for the green tape of figure B3.2.2.12 we have found $t = 0.23$.

The X-ray analysis was carried out on green tapes of different filament thickness between 30 and 115 μm (overall thickness between 75 and 286 μm), in tapes where the filament thickness represents about 40% of the overall thickness. For each tape the texture parameter t has been evaluated and the complete set of data is presented in figure B3.2.2.13 left, where it results that the texture generally increases if the tape thickness is reduced. D1.1.2

The higher texture of the Bi(2212) grains is reflected in a direct way on the final Bi,Pb(2223) texture. This effective correlation can be verified by comparing the graphs of figure B3.2.2.13: on the right of this figure, the misalignment angle of the Bi,Pb(2223) grains in reacted tapes of various thickness is presented. D1.1.1

The heat treatment of Bi,Pb(2223) tapes is generally performed at temperatures between 830 and 840°C, depending on the particular powder preparation process which is employed. However, the achievement of high critical current densities with an appropriate heat treatment process is a much more complex task than simply reacting the Bi,Pb(2223) phase. The Bi,Pb(2223) phase can indeed be achieved with a phase purity of about 85–90% after 48 h of heat treatment, but much longer times, as said before, are required in order to reach appreciably high critical currents.Unfortunately, with the duration of the heat treatment alone, it is hardly possible to raise the critical current density above 5–10 kA cm^{-2} at the liquid nitrogen temperature. Intermediate densification steps are strictly required in order to further increase the critical current density. In fact, during the first heat treatment, the thickness of the tape can increase by about 10–20%. As silver cannot change its thickness, this means that the oxide filament has increased its thickness by about 40%, or as well that the density has reduced to about 50%. The straightforward way to again pack the superconducting grains is to uniaxially press the tape several times B3.1 C2

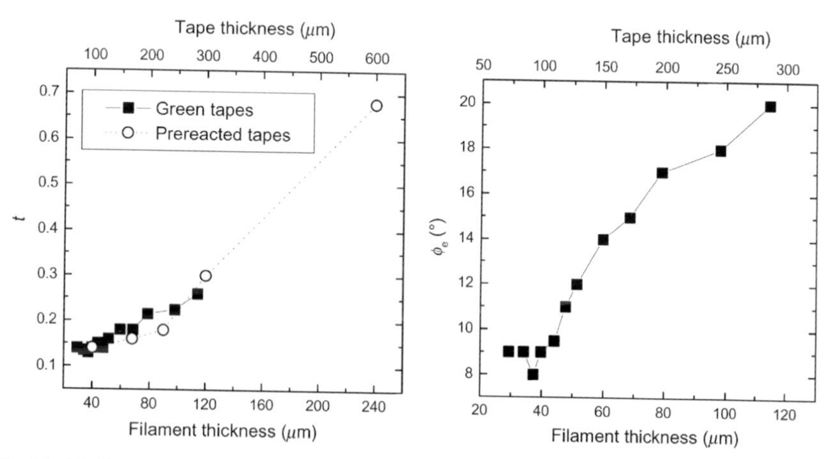

Figure B3.2.2.13. Bi(2212) texture parameter t of unreacted and pre-reacted tapes as a function of the tape thickness (left), and tape thickness dependence of the effective transport mean misalignment angle ϕ_e (right).

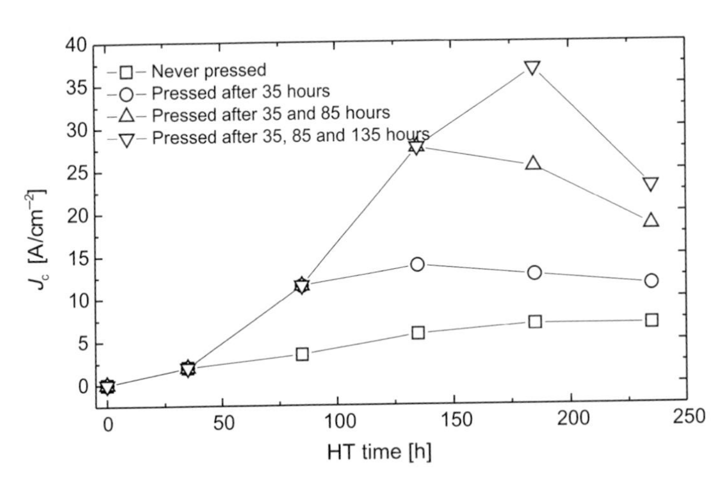

Figure B3.2.2.14. Critical current density as a function of the heat treatment time, for tapes which have been uniaxially pressed 0, 1, 2 and 3 times with a pressure of 2 GPa.

during the heat treatment. The critical current density as a function of the heat treatment time and of the number of intermediate pressing stages is shown in figure B3.2.2.14.

The re-densification of tapes by uniaxial pressure has a clear positive effect on the critical current density, that can be increased up to 37 kA cm^{-2} at 77 K, that is about six times higher than without any pressing stage. Possible explanations for this very clear improvement can be different, but two of them seem to be more appropriate. The first reason is that a re-densification with such a high pressure will push the Bi,Pb(2223) grains to be in contact again, and the formation of strong grain links can be facilitated. The second reason is based on X-ray diffraction pattern analysis performed on pressed and not-pressed samples.

D1.1.2

As can be seen in figure B3.2.2.15, where the X-ray diffraction patterns of both samples are shown, the pressed ones reaches a clearly higher Bi,Pb(2223) purity, and no traces of Bi(2212) are found in it. Both tapes have been heat treated for 185 h. The uniaxial pressing is therefore pushing together the secondary phases which are still present inside the filament, forcing them to complete their transformation into Bi,Pb(2223). However, the uniaxial pressure cannot be applied at an arbitrary moment of the heat treatment process. In particular, the first densification step has to be introduced at a well-defined condition.

In figure B3.2.2.16, the critical current of a standard tape is shown as a function of the first heat treatment duration.

C2

The critical current shows a step-like behaviour, which seems to follow the similar curve which has been determined for the formation of the Bi,Pb(2223) phase. Tapes have been pressed after 15, 24, 35, 40, and 48 h of first heat treatment, and then every 50 h, in order to complete a total heat treatment time of about 200 h for all the tapes. The critical current density is maximum for the tape pressed after 35 h of first heat treatment, when the Bi,Pb(2223) phase content in the tape is about 75%. All the other tapes show much lower J_c values.

The reasons for this behaviour are not yet completely clear. It is possible, however, that the uniaxial pressure has to be applied first when the Bi,Pb(2223) grains are already formed, but at the same time when there are still enough secondary phases that can in turn transform into Bi,Pb(2223), and act as a glue for the further formation of strongly linked current paths.

The uniaxial pressing is a very useful densification technique, but unfortunately it can be used only for very short samples (typically 2.5 cm long) which are interesting only from a fundamental point of

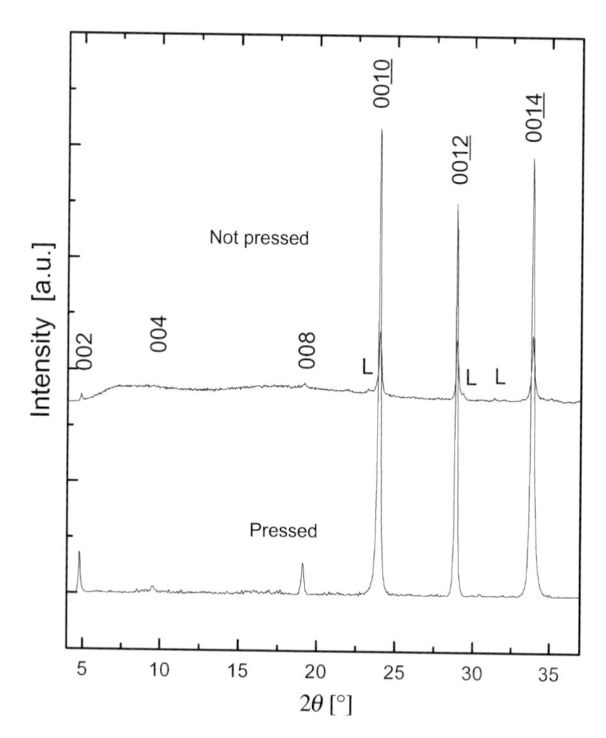

Figure B3.2.2.15. X-ray diffraction of pressed and not pressed tapes after 185 h of heat treatment at 837.5°C. The (00l) peaks of the Bi,Pb(2223) have been indexed. The residual Bi(2212) peaks in the not pressed tape have been indicated by L.

view. In view of practical applications, it is necessary to employ a densification technique that can be also applied for the preparation of much longer samples. In this case, cold rolling has to be used instead of uniaxial pressing [19]. By cold rolling, however, it is much more difficult to apply very strong pressures compressing the tapes without elongating them beyond the point where cracks of the filaments in transversal direction start to dominate. Force captors have therefore been installed on the roll axis in order to monitor carefully the force applied to the tapes during rolling. While rolling a tape, the section S over which the rolling force is exerted can be theoretically estimated and depends mainly on the tape thickness reduction and on the tape width. According to Thomas [20], a rough estimation of the average section S is given by

$$S = lR\sqrt{\frac{\Delta t}{R}\left(1 - \frac{\Delta t}{4R}\right)}$$

where l is the tape width, R the cylinder radius, and Δt the thickness reduction. In a typical case, section values S are of the order of 1–2 mm^2 (with $l = 2$–3 mm, $R = 50$ mm, $\Delta t = 10$–20 μm). Tapes have been densified using different pressures during rolling, and the results are shown in figure B3.2.2.17.

A sharp peak of J_c has been found for a rolling pressure of about 0.6 GPa, while a fast drop has been detected for higher forces. Moreover, it has been found that the J_c dependence on the pressure during rolling has remarkable similarities with that on the uniaxial pressure on short tapes prepared for comparison (figure B3.2.2.17). For pressures up to 0.6 GPa the two curves are almost identical. However, for uniaxially pressed tapes, the critical current density increases monotonically even when the pressure is

C2

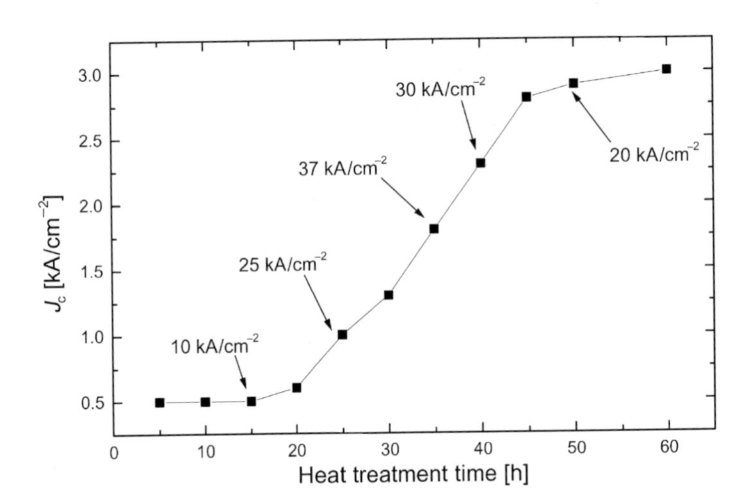

Figure B3.2.2.16. Evolution of the critical current density of a standard tape during the first heat treatment at 837.5°C. The final critical current density values of tapes uniaxially pressed after 15, 20, 35, 40, and 48 h are also indicated.

increased over 0.6 GPa, up to 2 GPa. The drop of J_c in cold rolled tapes for high pressures can be explained by the formation of cracks in transversal direction with respect to the tape orientation, due to the higher elongation during rolling.

The intrinsically different deformation of the tapes induced by uniaxial pressing and cold rolling will also lead to a clearly different tape microstructure. In figure B3.2.2.18, the cross section of a uniaxially pressed tape and a cold rolled tape are shown, respectively. The tape pressed with 2 GPa

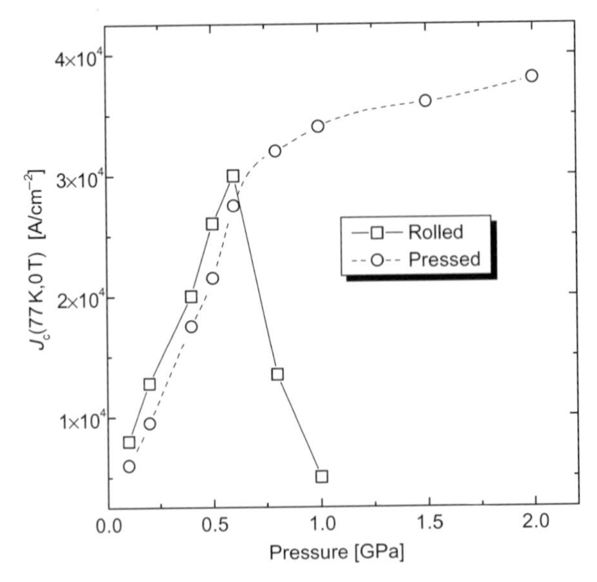

Figure B3.2.2.17. J_c value as a function of the pressure exerted on the tape during both uniaxial pressing and cold rolling applied between the heat treatments.

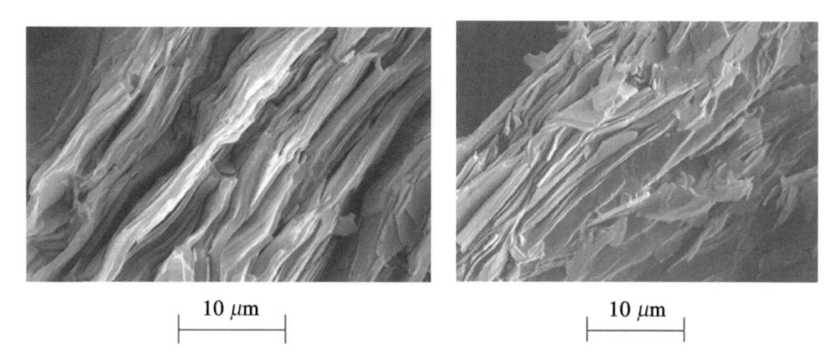

Figure B3.2.2.18. Typical cross section of a uniaxially pressed tape (left) and of a cold rolled tape (right) after the heat treatment at 837.5°C for 200 h.

presents a very dense stacking of Bi,Pb(2223) grains, which seem to follow the filament sausaging. The high pressure has probably deformed the grains themselves, which are very often not straight anymore. On the contrary, in the cold rolled tape, with a pressure of about 0.6 GPa, the grains are still straight, and they does not appear to be as dense as in the pressed samples.

B3.2.2.3.3 Heat treatment of multifilamentary Bi,Pb(2223) tapes

As described before, monofilamentary Ag-sheathed Bi,Pb(2223) tapes are ideal samples for studying the Bi,Pb(2223) phase formation and the superconducting properties of highly textured materials. B3.1

They are also relatively straightforward to be prepared by the PIT method; furthermore they do not require any intermediate heat treatment during the cold deformation. However, they also present several critical disadvantages, which discourage the practical application of such single core tapes. Indeed, the main disadvantages of the single filament tapes are (a) the extremely poor mechanical strength [21]; and (b) the relevant hysteretic losses occurring in presence of alternating currents [22]. The fabrication of very long tapes, as well as practical applications as high field magnets, power cables and transformers cannot be seriously considered without a substantial improvement of the tape mechanical strength and a reduction of the energy losses with ac currents. Moreover, the poor mechanical strength of the Bi,Pb(2223) tapes is a property inherent to the ceramic nature of the high T_c superconductors, and the silver sheath surrounding the filament is not strong enough to compensate for its brittleness.

Multifilamentary tapes can be prepared as well as the single core ones through the PIT method. Precursor powders calcined with the same procedure than for monofilamentary tapes, are filled inside pure silver tubes, which are deformed by swaging and drawing into round wires. At this stage of the preparation, the cold rolling deformation is replaced by a final drawing step with a hexagonal shaped die of suitable dimension. The drawn wire is then cut into several pieces, which are packed again inside a second silver tube. The stacking operation is crucial for the achievement of a homogeneous distribution of the different filaments. After a bundling heat treatment needed to glue the silver walls together, the tubes are deformed in the usual way as for monofilamentary tapes, that is, by swaging and/or drawing, again into round wires typically of 1.25–2 mm of diameter. The wires are then cold rolled in several steps of about 10% of thickness reduction each, down to a thickness that can vary between 150 and 350 μm. A1.3

At this stage of the preparation process, the multifilamentary tapes still need to be reacted at high temperature in order to form the Bi,Pb(2223) phase. As the powders have been calcined in the same way then for monofilamentary tapes, it is expected that the heat treatment parameters have not dramatically changed. A DTA measurement performed on a 37-filament tape prepared with powders of composition $Bi_{1.72}Pb_{0.34}Sr_{1.83}Ca_{1.97}Cu_{3.13}O_x$ is shown in figure B3.2.2.19. B2.1

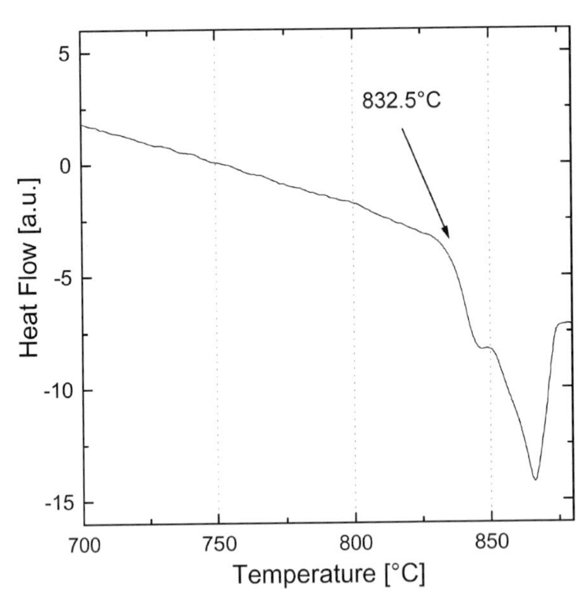

Figure B3.2.2.19. DTA measurement performed on a silver-sheathed 37-filament tape with calcined powders of composition $Bi_{1.72}Pb_{0.34}Sr_{1.83}Ca_{1.97}Cu_{3.13}O_x$.

B2.1 The shape of the DTA curve is very similar to that of monofilamentary tapes, except for a slight shift of the entire curve to lower temperatures, by about 1–2°C. This means that heat treatment temperatures in the range 835–838°C should still be appropriate for an optimal reaction into Bi,Pb(2223). However, the DTA measurement does not give very precise information about the kinetics of the phase formation, which is, however, clearly modified in multifilamentary tapes.

The phase formation and therefore the critical current density reach their optimal values in a time that can be about 50% shorter than for monofilamentary tapes, independently from the starting powder composition. In figure B3.2.2.20 the critical current density has been reported as a function of the heat treatment time, and for three different compositions of the precursor powders. From this graph it is evident that the powder stoichiometry has a critical effect on the tape transport properties.

C2 The choice of the composition A brings to a faster reaction compared to the compositions B and C, but the highest critical current density value of 28 kA cm^{-2} has been achieved within this batch of samples only with the slowest reacting composition.

The heat treatment temperature has been fixed to 837°C for all the tapes, as suggested by the DTA measurements, while the duration has been optimised for each particular composition. An intermediate cold rolling densification has been applied after 20, 30, and 40 h of heat treatment for the compositions A, B and C, respectively.

D1.1.2 The X-ray analysis of the multifilamentary tapes is much more problematic than for the single core ones, where silver can be easily peeled off. The best solution is to chemically etch the outer silver sheath, and to perform the X-ray diffraction analysis on the filaments that appear at the surface. It will not give, of course, any global information on the composition of the tape, but just of a few filaments, as the penetration depth of the X-rays in the Bi,Pb(2223) phase is of the order of 5–10 μm. The X-ray diffraction pattern measured at the surface of a 55-filament tape after reaction into Bi,Pb(2223) is shown in figure B3.2.2.21. The critical current density value of this tape is of about 25 kA cm^{-2} at 77 K.

D1.1.1 The Bi,Pb(2223) phase is highly textured, because only the (00*l*) peaks can be indexed. However, it is also possible to individuate the most intense peaks of the Bi(2212) phase, which are not observable, in

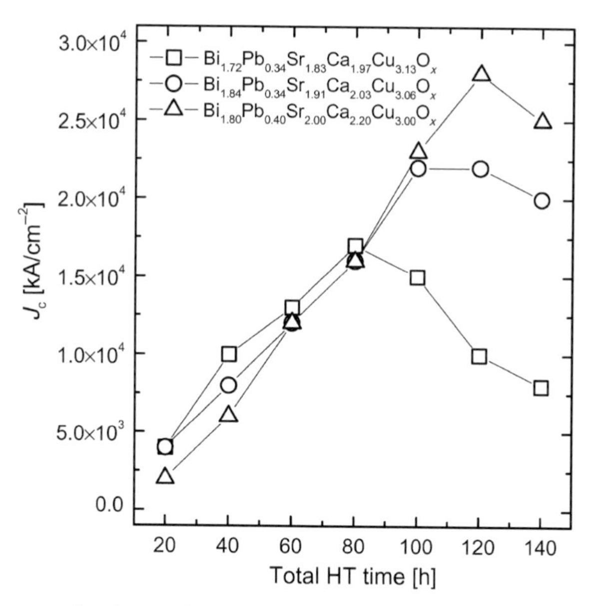

Figure B3.2.2.20. Critical current density as a function of the heat treatment time for 37-filament tapes prepared with three different compositions: (A) $Bi_{1.72}Pb_{0.34}Sr_{1.83}Ca_{1.97}Cu_{3.13}O_{10+x}$ [15], (B) $Bi_{1.84}Pb_{0.34}Sr_{1.91}Ca_{2.03}Cu_{3.06}O_{10+x}$ [23], and (C) $Bi_{1.80}Pb_{0.40}Sr_{2.00}Ca_{2.20}Cu_{3.00}O_{10+x}$ [24].

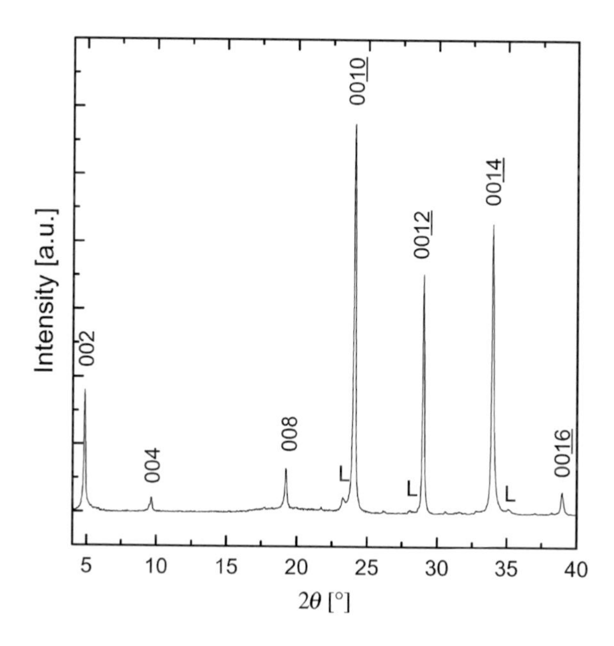

Figure B3.2.2.21. X-ray diffraction pattern of the etched surface of a 55-filament tape after complete reaction of 140 h at 837°C. L indicates the peaks relative to the Bi(2212) phase.

general, in reacted monofilamentary tapes. This means that the reaction from Bi(2212) into Bi,Pb(2223) is not perfectly completed; the reasons are however, still obscure.

The oxygen diffusion through the silver sheath is often considered as the origin of the lower purity in multifilamentary tapes. It is well known that during the reaction into Bi,Pb(2223) an exchange of oxygen with the atmosphere surrounding the tape is required, and this process should be undoubtedly slower in the thicker multifilamentary samples. However, as said before, the X-rays reach only the outer shell of filaments, which are separated by a thickness of a few tenths of microns of silver from the atmosphere, exactly the same as in monofilamentary tapes. The explanation of the lower Bi,Pb(2223) purity should therefore be found elsewhere. In the thin filaments, an imperfect and fluctuating local composition of the phase mixture prior to reaction into Bi,Pb(2223) seems to be a more appropriate explanation of the experimental observations.

A1.3

C2

A more correct way to study the Bi,Pb(2223) phase purity and texture in multifilamentary tapes is to perform the measurements on isolated single filaments. After several unfortunate attempts, the filaments have been successfully isolated through complete chemical etching of the silver sheath from a rather thick 7-filament tape. The filaments measure, in average, 500 μm in width, and 20 μm in thickness. Unfortunately, they are too brittle to be handled, and therefore the transport properties have not yet been measured on them in a systematic way. On the contrary, the X-ray diffraction measurements on isolated filaments have given some complementary information to that observed before on the entire tape. In figure B3.2.2.22, the X-ray diffraction pattern of an isolated filament is compared to that of a standard monofilamentary tape.

At first sight, the two patterns show similar features, with very intense (*00l*) peaks of the Bi,Pb(2223) phase; however, when looking at the peaks which are just above the limit of the noise, some differences will finally appear. First, the isolated filament still shows the presence of the Bi(2212) phase, which is not the case for the monofilamentary tape. It is very difficult to estimate the residual amount of Bi(2212), but this should be anyway greater than 10%.

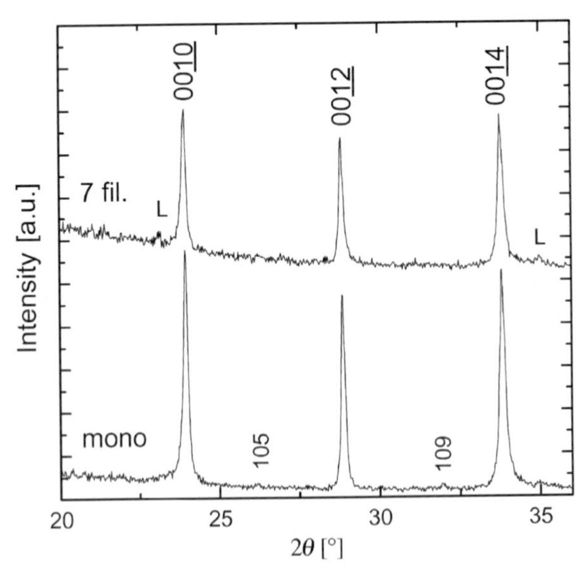

Figure B3.2.2.22. X-ray diffraction patterns of a monofilamentary tape surface, and of a single filament extracted from a 7-filament tape. The Bi,Pb(2223) peaks have been indexed, as well as the Bi(2212) peaks visible in the extracted filament (indicated by L).

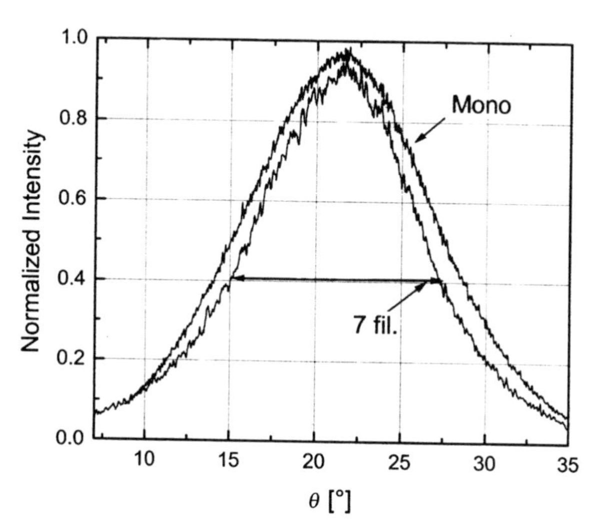

Figure B3.2.2.23. Rocking curves of a monofilamentary tape and of an isolated filament, both measured on the (0010) peak.

The second difference concerns the presence of detectable (*10l*) peaks of the Bi,Pb(2223) phase in the monofilamentary tape. The intensity of these peaks is highly sensitive to the *c*-axis texture of the Bi,Pb(2223) grains, and they are used, when observed, to estimate the degree of orientation of this phase. The fact that they are absent in the isolated filament would suggest that the Bi,Pb(2223) grains are better textured in the multi-core tapes.

This hypothesis can be eventually confirmed by performing rocking curve measurements around the (*00l*) peaks. For this measurement, the analysis of the isolated filament is preferable to that of the multifilamentary tape. In fact, the particular ellipsoidal-like distribution of the filaments inside the silver sheath would inevitably affect the FWHM of the rocking curve. The rocking curves measured for the isolated filament and for the reference monofilamentary tape are shown in figure B3.2.2.23.

D1.1.2

The FWHM of the rocking curve, the rocking angle, is about 12° for the isolated filament, while it is about 14° for the monofilamentary tape. In general, an X-ray mean misalignment angle of the grains with respect to the tape normal is defined as 1/2 FWHM; this means that in the isolated filament the grains are better textured than in a monofilamentary tape.

By adjusting the heat treatment process, very high critical current densities can be also achieved on relatively thick and wide tapes (0.25×3.7 mm), which have the advantage of a higher mechanical strength, especially compared to the monofilamentary samples. The mechanical strength of a 37-filament tape to an applied bending strain has been measured, in order to verify that they perform better than monofilamentary tapes. The measurement is shown in figure B3.2.2.24.

The critical bending strain reaches about 0.5–0.6%, which is clearly higher than in monofilamentary tapes, in which it hardly reaches 0.2%; they can withstand a bending diameter of about 25 mm without a serious reduction of the critical current density. Moreover, the fabrication process is easily scalable to very long lengths, which makes multifilamentary tapes very promising for industrial applications.

A1.3

The four-roll deformation applied to the fabrication of Bi,Pb(2223) tapes

The four-roll machine (figure B3.2.2.25) has the clear advantage, compared to standard deformation processes, of being able to deform rectangular shaped wires of, in principle, indefinite length. New procedures for the preparation of multifilamentary Bi,Pb(2223) tapes can be therefore introduced.

B3.1

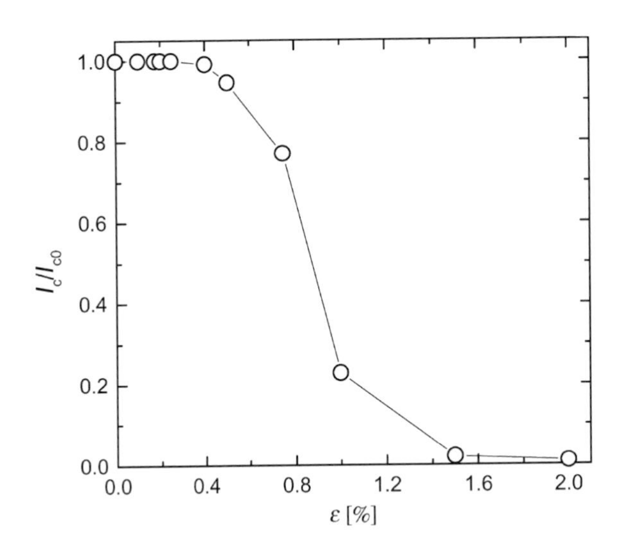

Figure B3.2.2.24. Critical current density as a function of the bending strain in a 37-filament Bi,Pb(2223) tape.

B2.1 Calcined precursor powders have been filled inside round silver tubes that are deformed by usual swaging and drawing down to wires of a diameter of about 1.5 mm.

B3.1 They are subsequently deformed by the four-roll machine into square shaped wires of about 1 × 1 mm. Several pieces of these wires are stacked into square-shaped tubes of various sizes (up to 12 × 12 mm), and with wall thickness of about 1 mm. A schematic representation of the stacking of the wires and of the deformation process for the standard and four-roll tape is shown in figure B3.2.2.26.

After the usual bundling treatments at 650°C, the tubes are directly deformed by the four-roll machine, the thickness and the width of the tubes being reduced at the same time by about 50 μm per step. When the wires have reached a dimension of typically 3 × 3 mm, the width is kept fixed, while the

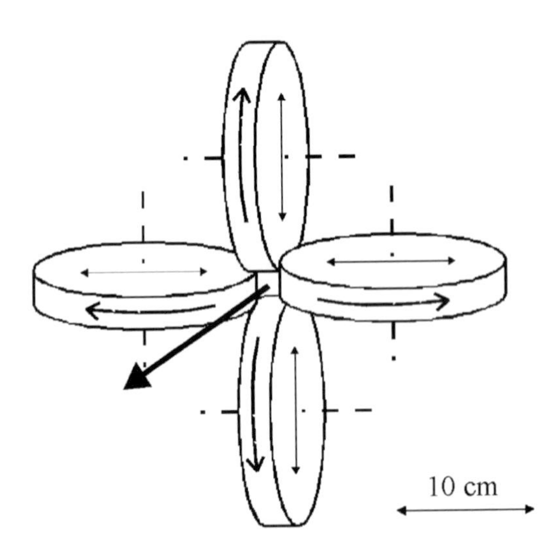

Figure B3.2.2.25. Schematic representation of the four-roll machine. The motor-driven rolls can reduce at the same time thickness and width of a rectangular shaped wire.

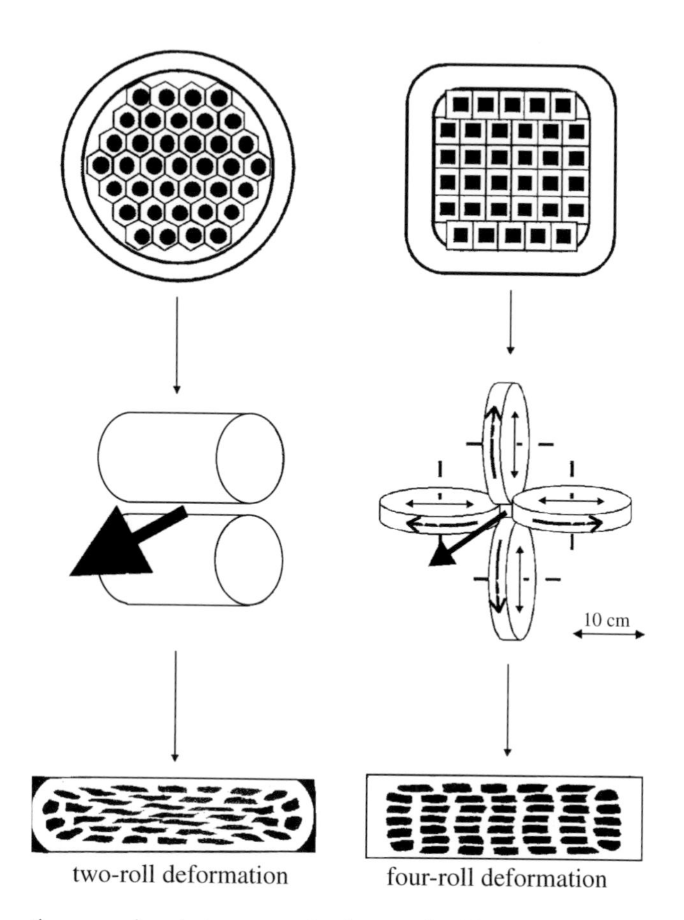

Figure B3.2.2.26. Schematic comparison between standard two-roll rolling (left), and four-roll rolling of Bi,Pb(2223) tapes. The four-roll rolling leads to a clearly improved tape homogeneity.

thickness is further reduced, generally below 0.5 mm, in steps of 30–50 μm. The perfect control of the tape cross section during the fabrication with the four-roll machine leads to clearly improved filament configuration in the final sample.

The filaments near the tape centre as well as those at the sides underwent similar deformation stresses, due to the additional pressure applied on the tape by the horizontal rolls. Moreover, by four-rolling a much higher superconducting fraction can be reached in the tape with respect to the standard PIT process, without any formation of cracks over long lengths (10 m).

In figure B3.2.2.27, the cross section of a 100 filament tape is shown.

The filaments have been stacked in a 10×10 configuration, and their thickness in the final tape is of the order of 5–10 μm. The four-rolled tapes have been reacted in pieces up to 10 m in length, at a temperature

A1.3

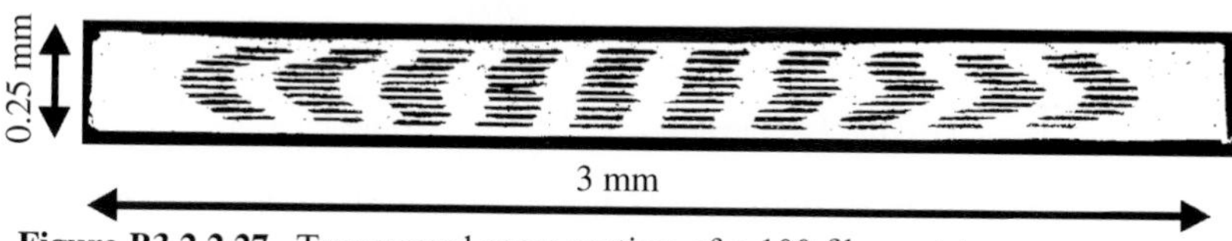

Figure B3.2.2.27. Transversal cross section of a 100-filament tape.

Table B3.2.2.3. Critical current density values of four-roll tapes with number of filaments up to 100

Number of filaments	Supercond. fraction	Tape size (mm^2)	Heat treatment process at 837°C	J_c ($kA\,cm^{-2}$)
18	20%	3 × 0.20	50 h + 100 h	25
34	25%	3 × 0.20	40 h + 100 h	26
34	25%	3 × 0.30	50 h + 100 h	26
45	35%	3 × 0.20	40 h + 100 h	23
45	35%	3 × 0.30	50 h + 100 h	23
45	35%	3 × 0.40	50 h + 100 h	21
100	20%	3 × 0.20	35 h + 80 h	15

of 837°C, for up to 200 h with an intermediate rolling step. In table B3.2.2.3, the critical current densities achieved on the four-roll tapes are summarized.

C2 Critical current density values above 20 $kA\,cm^{-2}$ have been reproducibly achieved on tapes with up to 45 filaments, and for a tape thickness up to 0.4 mm. Only the 100 filament tapes have reached a lower critical current, probably due to the appearing of sausaging when the filament thickness goes below 10 μm. Further optimization of the deformation process is therefore required for tapes with high number of filaments.

Periodic pressing of Bi,Pb(2223) tapes

Many authors have already demonstrated the advantage of pressing in the evolution of the transport C2 properties. The highest reported J_c values obtained for monofilament tapes deformed by pressing and rolling are 69 and 40 $kA\,cm^{-2}$ respectively (77 K, 0 T) [23, 25]. In this paragraph a new deformation process called 'periodic pressing' (PP) is introduced, by this method, it is possible to achieve high J_c values on long lengths of Bi,Pb(2223) multifilamentary tapes while retaining all the advantages of pressing. A prototype deformation machine, called the periodic pressing machine (see figure B3.2.2.28),

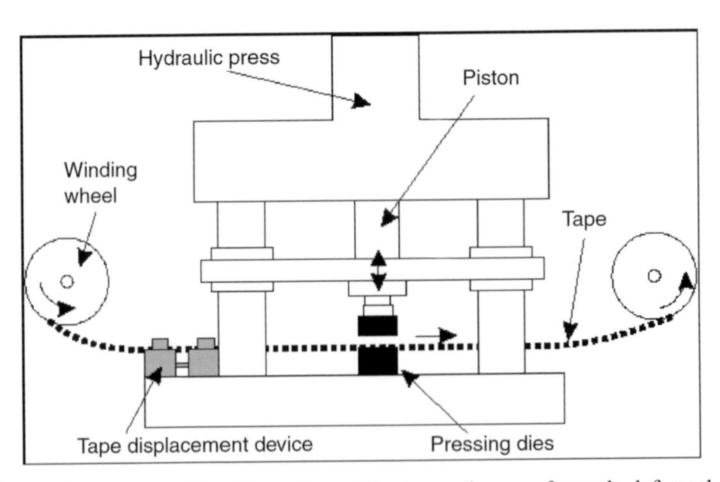

Figure B3.2.2.28. Schematic drawing of the PP machine. The tape advances from the left to the right by the use of a tape displacement device (which also controls the overlap length).

was used for all the uniaxial pressing steps. It allows pressing with a force up to 40 tf, while the tape advances step by step by means of a device which is synchronized with the hydraulic system.

The step length can be controlled between 0 and 5 cm, while the step frequency is typically 1 Hz. Therefore the average speed of the deformation is comparable with that obtained in the rolling process (1–2 cm s^{-1}). Moreover, the speed could be easily increased by further enhancing the step frequency of the hydraulic system. In order to avoid irretrievable damage to the ceramic filaments, dies with special shapes were designed. These shapes facilitate the transition from the at and pressed region to the rounded low-pressure region, thus avoiding the creation of damaged zones by the appropriate choice of the overlapping lengths.

Figure B3.2.2.29 shows the improvement of the I_c value obtained by the new route (PP) compared with the standard one (rolling). C2

After the first heat treatment (50 h), the I_c value is lower; this is because the higher density after pressing causes slower reaction kinetics in the pressed tape. For longer annealing times, however, the advantage of the PP process becomes evident. The I_c value reaches 59 A for the pressed tape instead of 38 A for the rolled one. This corresponds to J_c values of 34.5 and 24.8 kA cm^{-2} (77 K, 0 T), respectively. Figure B3.2.2.30 presents longitudinal sections of tapes deformed using standard dies (figure B3.2.2.30(*a*)) or with specially shaped dies figure (B3.2.2.30(*b*)). In the first case, the pressing process creates a sharp change in the tape thickness over a region of about 100 μm in length for a reduction of 50 μm in thickness. It can be reasonably assumed that the Bi,Pb(2223) filaments suffer a strong and detrimental bending in this zone. A1.3

In the second case, however, a similar change of the thickness is distributed over a much longer length (the horizontal and vertical scales are different, and only part of the transition region is shown), and a smooth overlap with the pressed zone can be obtained, which drastically limits the possible damage.

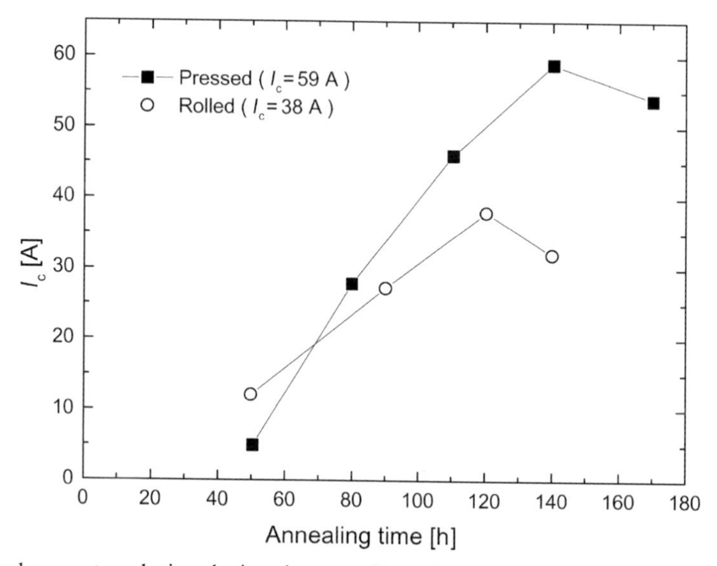

Figure B3.2.2.29. Critical current evolution during the annealing of 37-filament tapes deformed by the pressing and standard rolling processes. The pressed tape is successively submitted to four pressing steps, instead of just a single deformation step for the rolled one.

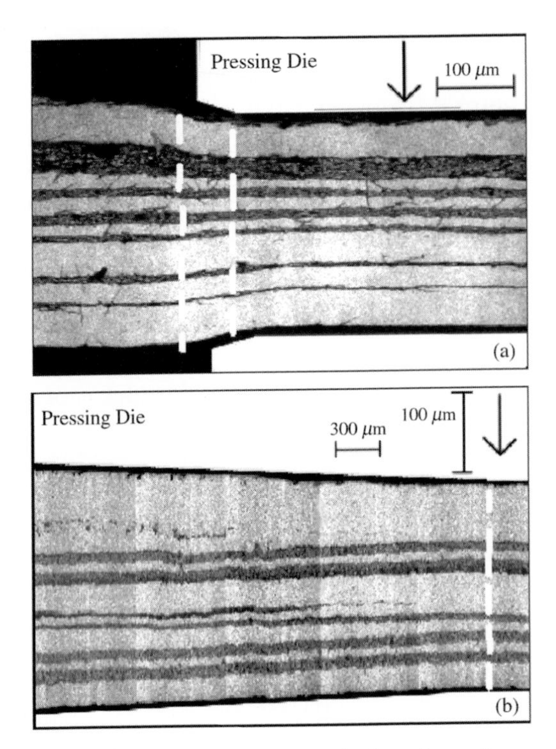

Figure B3.2.2.30. SEM micrographs showing detailed longitudinal sections of tapes pressed (*a*) with standard dies (with sharp angle) and (*b*) with new dies (small angle to avoid cutting off the surface of the tape). It should be noted that, to show clearly the shape of the dies (white figures) in the second picture, the x and y scales are not similar.

B3.2.2.4 Critical current density of Bi,Pb(2223) tapes

C2 In optimized Ag sheathed Bi,Pb(2223) tapes, critical current density values well over 60 kA cm^{-2} have been reached at the liquid nitrogen temperature by several research groups. The flow of such relevant transport currents over large distances (compared to the typical grain dimension) in these polycrystalline superconductors is made possible by the high quality of the grain boundary microstructure, as shown by D1.2 Grindatto *et al* [26] by means of a detailed HRTEM analysis.

In order to further develop Bi,Pb(2223) tapes for industrial applications, it is, however, extremely important to be aware of the influence of the magnetic field on the transport J_c value, as this exists virtually in any practical device under study at present. Moreover, the knowledge of the magnetic and temperature dependence of the critical current density is fundamental for the understanding of the A1.3 current transport mechanisms as well, which has in turn helped in identifying the most important current limiting factor occurring in these conductors.

B3.2.2.4.1 Temperature and magnetic field dependence of the critical current

C2 The critical current determined from the measure of a voltage–current (V–I) characteristics is influenced partly by intrinsic factors, as the pinning energy and critical current distribution within the D2.2 superconducting core, and partly by extrinsic factors, as the presence of the highly conductive Ag sheath in parallel to the Bi,Pb(2223) core. In particular, knowing the numerous differences between mono- and multifilamentary Bi,Pb(2223) tapes, it is expected to observe substantial variations both in the

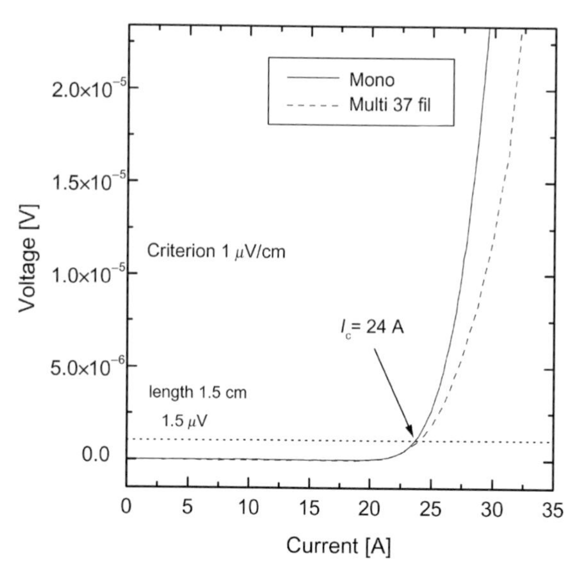

Figure B3.2.2.31. *V–I* characteristics at 77 K in self-field of a mono- and a multifilamentary tape with similar critical current.

V–I characteristics and in the magnetic field dependence of the critical current density. The *V–I* characteristics of typical mono- and multifilamentary tapes are shown in figure B3.2.2.31. Both samples show a critical current value of 24 A corresponding to a J_c value of 28 kA cm^{-2}, considering that the monofilamentary tape presents an overall cross section of $2.7 \times 0.09\,mm^2$, with a superconducting fraction of 35%, while the 37-filament tape presents a cross section of $2.7 \times 0.16\,mm^2$, with a superconducting fraction of about 20%. In spite of the similar current density, the self-field *V–I* characteristics is much steeper in the monofilamentary tape, in which the current begins to be shared with the silver sheath well above I_c. D2.2 C2

The higher proportion of silver in the tape cross section (80% instead of 65%), and the facilitated redistribution of the current within the different filaments are responsible for the smoothness of the transition curve of figure B3.2.2.31.

For many applications, as in the case of current limiters, a smoother transition can be even advantageous, because it can improve the stability of the device in case of flowing currents I higher than the critical value I_c. For other applications, as in the case of high field magnets and nuclear magnetic resonance devices that are fabricated with low temperature superconducting cables, a steeper transition is strictly required, because it also implies a lower flux creep relaxation of non-dissipative currents that allows its use in a persistent mode. E1.3.3 A4.3

The typical temperature dependence of the critical current density of a Bi,Pb(2223) tape is shown in figure B3.2.2.32. Besides the self-field curve, measurements in various external magnetic field are shown for a field orientation parallel to the tape plane (often indicated as $\theta = 90°$, where θ is the angle between the tape normal and the magnetic field). A linear variation of J_c with the temperature is observed, and the effect of the external magnetic field is to shift the curves in the direction of the origin, without qualitative changes of the temperature dependence. The common slope of the data is about 1.8 kA (cm·K)$^{-1}$. Only very near to the irreversibility temperature, a continuously decreasing slope is observed. C2

From the measurements of $J_c(T)$ alone, it is not possible to conclude whether residual Bi(2212) is incorporated to the Bi,Pb(2223) current path or not. However, it cannot be excluded that Bi(2212)

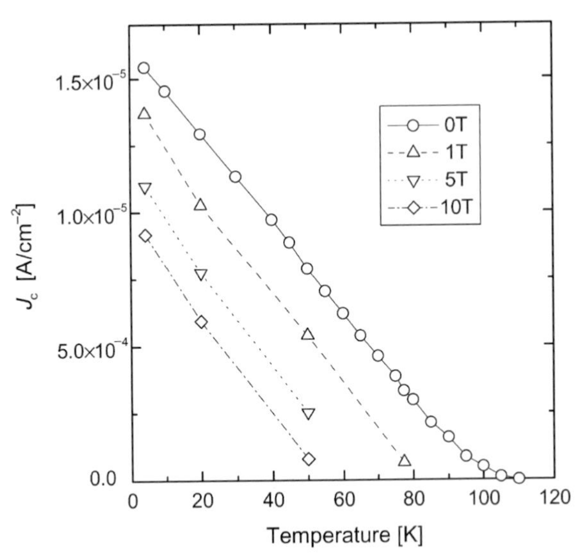

Figure B3.2.2.32. Temperature dependence of the critical current density in various applied magnetic field up to 10 T. The field is oriented parallel to the tape plane.

intergrowths are even present in the grains of high-current tapes, although with strongly reduced probability, as they are found in almost all grains of Bi,Pb(2223) [Hen95].

C2 Besides these considerations it can be stated that the temperature dependence $J_c(T)$ (with and without external magnetic field) of Bi,Pb(2223) tapes with $J_c(T = 77\,\mathrm{K},\ B = 0\,\mathrm{T}) > 30\,\mathrm{kA\,cm^{-2}}$ resembles the one that is characteristics of high-quality thin films, and gives no hints for transport phenomena dominated by the presence of weak links in the current path. This does not mean that weak links are absent in the samples as the current transport measurements test only the best parts of them.

By extrapolating the linear dependence of the $J_c(T)$ curve to the zero critical current value, it is possible to define an irreversibility line in the H–T plane, above which the tape is not carrying any appreciable transport current. This line is plotted in figure B3.2.2.33, together with the irreversibility lines deduced by the AC susceptibility, transport resistivity, and magnetization measurements, and for a field orientation parallel to the tape normal.

A1.3 D2.5, D2.4, D3.1

A1.3 All the irreversibility lines (IL) fall on the same curve, except for the resistive one, which is, however, strongly dependent from the chosen criterion. The equivalence of all the IL strongly supports the hypothesis that the limiting mechanism for the macroscopic current transport is essentially the intragrain current density, at least in the temperature and field intervals analysed in figure B3.2.2.33.

B3.2.2.4.2 The 'Railway-Switch' model for the current transport mechanism

The peculiar behaviour of the transport properties of Bi,Pb(2223) tapes cannot be explained with the standard model for the current flow in polycrystalline superconductors. A first attempt to interpret such novel behaviour has been carried out by Bulaevskii *et al* [27], with the so-called 'Brick-Wall' model. In this model, the microstructure of the Bi,Pb(2223) filament has been compared to a brick-wall, as shown in figure B3.2.2.34. The grains are the bricks, which therefore show a large contact area along the a–b planes. In this model, a sudden interruption of a current path (a 'weak-link') can be overcome by flowing the current in c-direction between adjacent grains, and their large contact area minimizes the weak-link effect. A hypothetical current path according to the brick-wall model is also represented in figure B3.2.2.34.

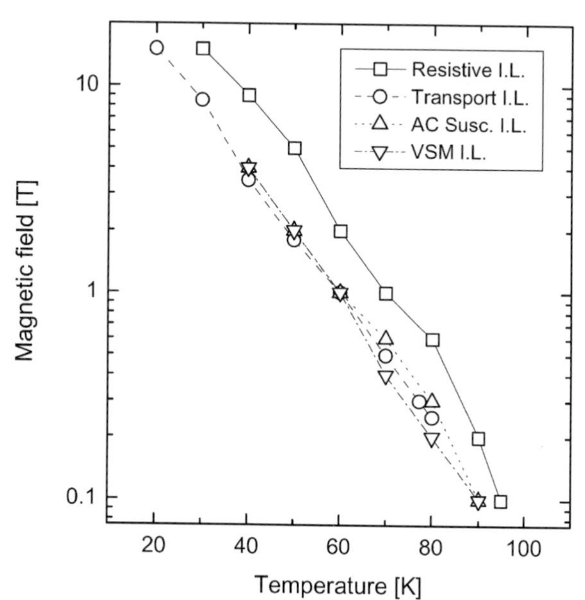

Figure B3.2.2.33. Irreversibility lines (I.L.) of a monofilamentary Bi,Pb(2223) tape measured with various techniques.

However, the observation of the tape microstructure has revealed a completely different arrangement of the Bi,Pb(2223) grains, that does not necessarily show wide surface contact areas. A typical cross section of a real Bi,Pb(2223) tape is instead shown in figure B3.2.2.35.

Hensel *et al* [28], Owing to a complex structure of the Bi,Pb(2223) grains, have suggested an alternative approach to the current transport mechanism modeling, in which the macroscopic current is flowing through grains connected by low angle *c*-axis grain boundaries, exactly as a train does in an elaborate railway network. The most common low angle *c*-axis grain boundaries which are encountered in high-quality Bi,Pb(2223) tapes are of two types: (*a*) edge-on *c*-axis tilt grain boundaries or ECTILT, and (*b*) small-angle *c*-axis tilt grain boundaries or SCTILT.

D1.1.1

The main assumption of the Railway-Switch model is that these small-angle grain boundaries constitute strong superconducting links between adjacent grains, and therefore that they are the origin of the high critical current density. A typical example of a real SCTILT grain boundary between Bi,Pb(2223) grains is shown in figure B3.2.2.36.

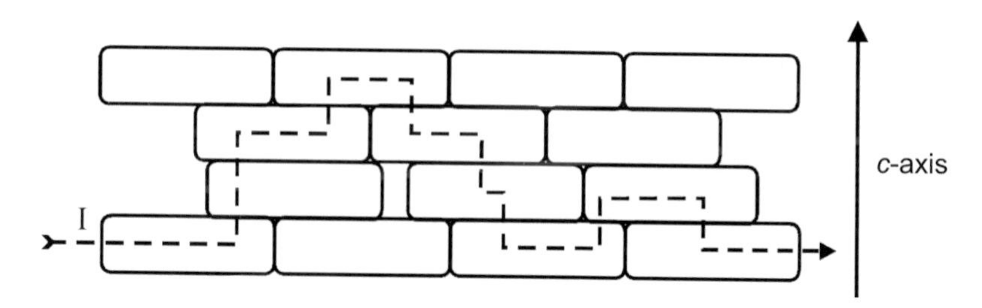

Figure B3.2.2.34. Schematic representation of the tape microstructure and of the current flow as predicted by the Brick-Wall model. The current preferably flows between grains in the *c*-axis direction. The grain aspect ratio does not reflect the real situation, the grain thickness being typically 10–50 times smaller than the grain width.

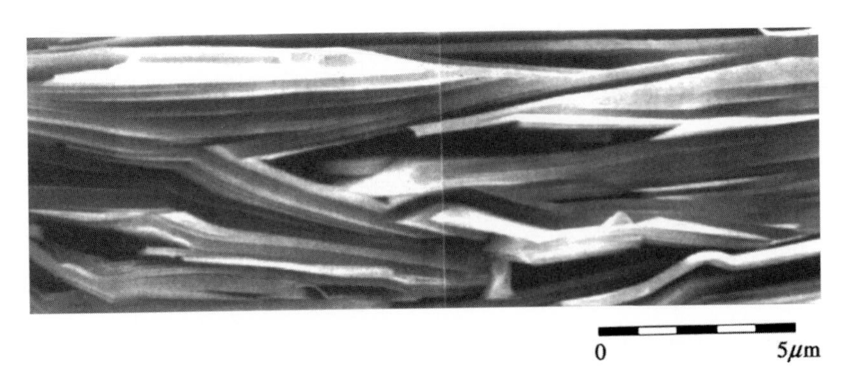

Figure B3.2.2.35. Typical longitudinal cross section of a monofilamentary Bi,Pb(2223) tape showing the Railway-Switch like structure of the grains.

At first sight, this assumption can appear to be in contrast with previous results of Dimos *et al* [29], and Amrein *et al* [30]. These authors have measured the transport properties of bi-crystalline thin films as a function of the well-defined angle of the single grain boundary present in the sample. They have observed an empirical relation between the critical current density and the misalignment angle of the grain boundary. This relation, that appears to be valid for YBCO as well as Bi(2212), is characterized by a critical angle θ_c, that is of the order of 5°:

C1, C2

$$J_c^{gb} = J_c \exp\left(\frac{-\theta}{\theta_c^-}\right)$$

where θ is the angle between the two grains, while J_c and J_c^g are the intra- and intergrain critical current densities, respectively. Hensel *et al* [28] have found, however, that for Railway-Switch like grain boundaries this equation is no longer valid. Indeed, a correction factor taking into account the effective cross section of the grain boundary has to be introduced, and therefore it is possible to write:

Figure B3.2.2.36. SCTILT grain boundary in Bi,Pb(2223) tapes.

$$\frac{I_c^{gb}}{I_c} = \exp\left(-\frac{\theta}{\theta_c}\right)\frac{1}{\sin\theta}$$

where I_c and I_c^{gb} are the critical current of the grain and of the grain boundary, respectively. For an angle θ up to 13.4°, the critical current of the grain boundaries results to be higher than that of the grains themselves, and therefore within this model they cannot be considered as a limiting factor for the macroscopic transport critical current density.

B3.2.2.4.3 Interpretation of the $J_c(B,T,\theta)$ behaviour of Bi,Pb(2223) tapes

The analysis of the field and angular dependence of $J_c(T)$ of monofilamentary tapes gives a further support to the validity of the Railway-Switch model for the current transport. The critical current density of a standard tape with J_c(77 K, 0 T) of 34 kA cm^{-2} is shown in figure B3.2.2.37, as a function of the magnetic field, and for $\theta = 0$ and 90°. A1.3

If weak-links are present in the current path, they necessarily lead to a strong decrease of J_c in magnetic fields, but a strong decrease alone does not necessarily mean that weak-links must be involved. It has been shown by Bulaevskii *et al* [27] that the field dependence of the critical current, including the initial drop at low fields can be remarkably well described when the imperfect texture is taken into account. A lack of pinning due to high crystalline perfection with only few pinning centres together with an imperfect texture can therefore lead to a weak-link like behaviour of the critical current. The Railway Switch interpretation of the $J_c(B,T)$ curves relies just on these assumptions and does not need any weak links in the current path. B4.2.1 A4.3

Looking at the tape microstructure and at the transport properties, Kaneko *et al* [31] have introduced the parallel-pipe model, that is essentially an extension of the Railway-Switch model to the macroscopic current flow. According to these authors, the main factor limiting the critical current in Bi,Pb(2223) tapes is the density of parallel paths of strongly linked Bi,Pb(2223) grains that allows the current flow along the superconductor. A schematic representation of the model is presented in figure B3.2.2.38.

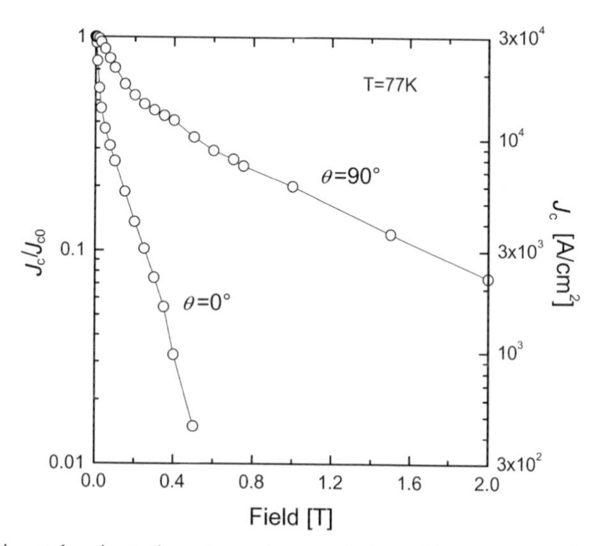

Figure B3.2.2.37. Magnetic and orientation dependence of the critical current density of a monofilamentary Bi,Pb(2223) tape with self-field J_c of 30 kA cm^{-2}.

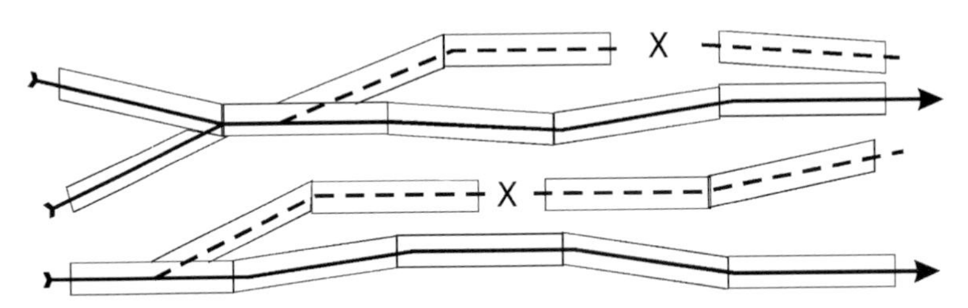

Figure B3.2.2.38. Schematic representation according to the parallel-pipe model of the current flow in a Bi,Pb(2223) tape showing strongly linked grains. The current flows only through the continuous current paths.

The current can flow only through continuous current paths, while those which are interrupted by secondary phases, or more generally by weak-links, are completely ineffective for the current transport. In this model, the critical current density is simply limited by the number of continuous 'pipes' which are able to carry the current over a distance of many grain lengths.

In the framework of the Railway-Switch model, it is possible to extract important information about the degree of texture of the Bi,Pb(2223) grains from the angular dependence of the critical current density. From measurements performed on Bi-based superconducting films and single crystals, it has been possible to determine that J_c is almost insensitive to the magnetic field, if the latter is applied along the a–b planes, while an exponential drop is observed if it is applied along the c-axis. For all the other intermediate orientations, the drop of the critical current is dependent just from the component of the field along the c-axis, that is, $B\cos\theta$. It is therefore possible to write:

$$J_c(B) = J_{c0}\exp\left(-\frac{B\cos\theta}{B_t}\right)$$

where J_{c0} is the zero-field critical current, and B_t is a characteristic field that depends on the temperature, and on the properties intrinsic to the particular sample.

C2 In our case, the Bi,Pb(2223) grains are not perfectly textured with their c-axis along the tape normal. When the field is approximately oriented along the tape normal ($\theta \approx 0°$), the critical current is only marginally sensitive to the angular distribution of the grain orientation: the $J_c(\theta)$ measurements performed with $\theta \approx 0°$ are mostly dependent from the intrinsic properties of the Bi,Pb(2223) grains. On the contrary, when $\theta \approx 90°$, the component of B along the tape normal varies considerably with θ, and therefore $J_c(\theta)$ is much more dependent from the distribution of the grain orientation.

A typical angular dependence of the critical current density of a monofilamentary tape is shown in figure B3.2.2.39, for various field amplitudes between 0.01 and 0.5 T.

When $\theta \approx 0°$, a moderate variation of the critical current density with the angle is effectively observed, while when $\theta \approx 90°$ a relatively sharp maximum is found for all the field amplitudes we have investigated. The shape of the peak at $\theta \approx 90°$ is obviously affected by the angular distribution of the grain orientations.

It is interesting to re-plot the critical current density as a function of $B\cos\theta$, which is the component of the applied magnetic field oriented along the tape normal. This plot is shown in figure B3.2.2.40, for a fixed field amplitude of 0.5 T. At low angles, a plateau-like behaviour of the critical current density is observed, after which the critical current gradually drops to zero with a substantially linear dependence in the logarithmic scale.

According to the interpretation of Hensel *et al* [32], the plateau-like behaviour is given by the angular distribution of the grains inside the tape. The information about the average degree of texture of the grains are therefore given by the characteristic field B^*, defined by the linear interpolation shown in figure B3.2.2.40.

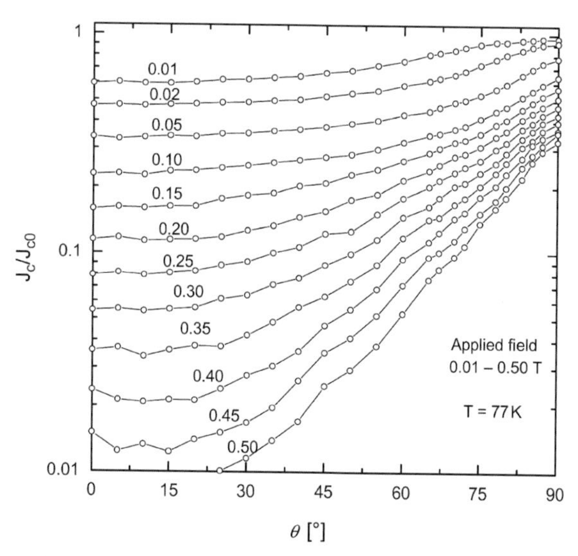

Figure B3.2.2.39. Angular dependence of the critical current density at 77 K for a monofilamentary Bi,Pb(2223) tape for various magnetic field amplitudes between 0.01 and 0.5 T.

A characteristic angle θ^* can be therefore defined as

$$B^* = B\cos(90° - \theta^*) \text{ or } \theta^* = \sin^{-1}\left(\frac{B^*}{B}\right).$$

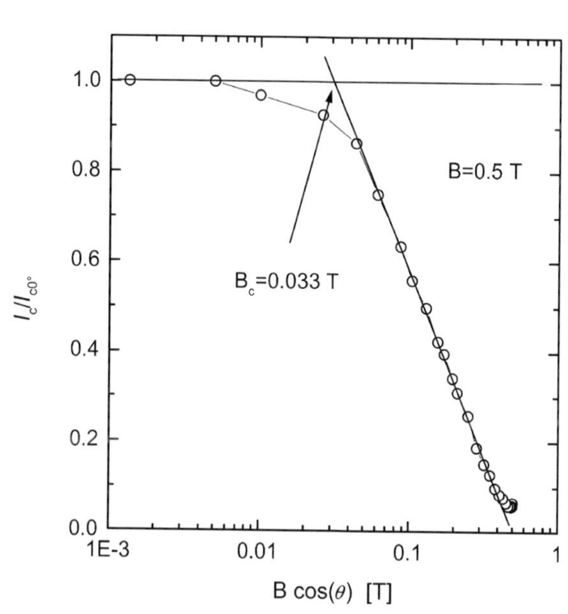

Figure B3.2.2.40. Angular dependence of the critical current density with a fixed field amplitude of 0.5 T. A characteristic field value $B*$ of 0.033 T has been determined for this sample.

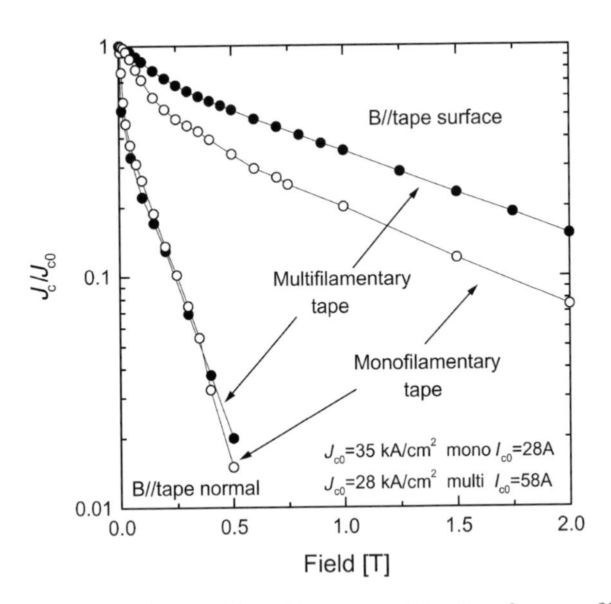

Figure B3.2.2.41. Magnetic field dependence of the critical current density of a monofilamentary tape at 77 K, for two different orientations of the tape with respect to the field. The measurements are compared with those performed on a monofilamentary tape.

The angle θ^* is therefore considered as the mean misalignment angle or the Bi,Pb(2223) grains responsible for the current transport. This angle is *a priori* a function of the magnetic field and temperature, and it can be different from the rocking angle deduced by X-ray measurements.

D1.1.2

The magnetic field dependence and the anisotropy of the critical current density of multifilamentary tapes can be largely different from those of monocore tapes, as a consequence of their largely different microstructure. The critical current of a 37-filament tape with $J_c(77\,\mathrm{K}, 0\,\mathrm{T}) = 28\,\mathrm{kA\,cm^{-2}}$ ($I_{c0} = 58\,\mathrm{A}$) has been measured as a function of the tape orientation at the liquid nitrogen temperature. The measurements are shown in figure B3.2.2.41, and compared with those performed on a reference monofilamentary tape with $J_c(77\,\mathrm{K}, 0\,\mathrm{T}) = 35\,\mathrm{kA\,cm^{-2}}$ ($I_{c0} = 28\,\mathrm{A}$).

C2

A substantial difference between the magnetic field dependence of mono- and multifilamentary tapes is clearly observed when the field is applied parallel to the tape plane. For an external field of 1 T applied in this direction, the critical current density of the 37-filament tape reduces by a factor of 2.5–3.0, while the usual reduction factor monofilamentary tapes is between 5 and 6. As already shown before, in this orientation the field dependence of J_c is strongly correlated to the angular distribution of the Bi,Pb(2223) grains inside the tape, in such a way that a reduced field dependence corresponds to a higher degree of texture.

A1.3

By means of the technique derived from the Railway-Switch model for the calculation of the mean misalignment angle of the grains, it has been possible to determine the influence of the number of filaments on the degree of texture of the Bi,Pb(2223) phase. The results have been summarized in figure B3.2.2.42.

It is clear from this graph that the misalignment angle of standard PIT tapes gradually reduces for increasing number of filaments in the tape, the mean misalignment angle of 55- and 61-filament tapes being of the order of 5°, that is, 2–3° less than in monofilamentary tapes. As all the multifilamentary tapes measured for the graph of figure B3.2.2.42 have similar overall dimension ($\approx 3 \times 0.2$ mm), the number of filaments should directly influence the filament thickness itself. In figure B3.2.2.43, the average filament thickness of all these tapes has been reported.

A1.3

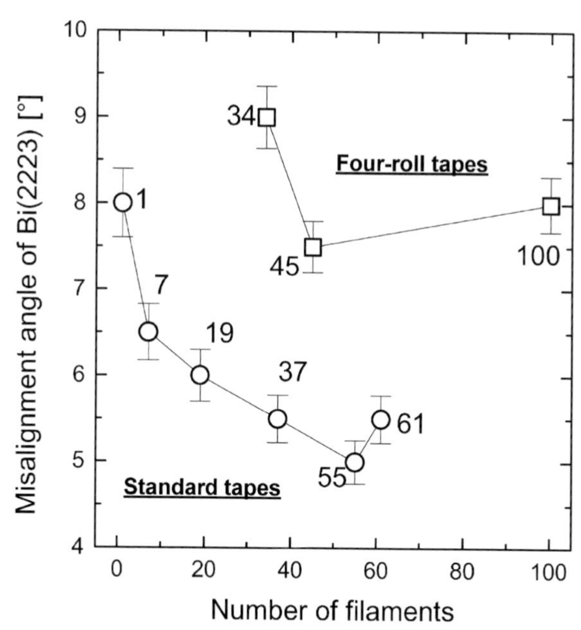

Figure B3.2.2.42. Mean misalignment angle of the Bi,Pb(2223) grains of multifilamentary tapes determined by transport measurements. Standard tapes have been deformed with the usual technique, that is, by swaging, drawing and rolling.

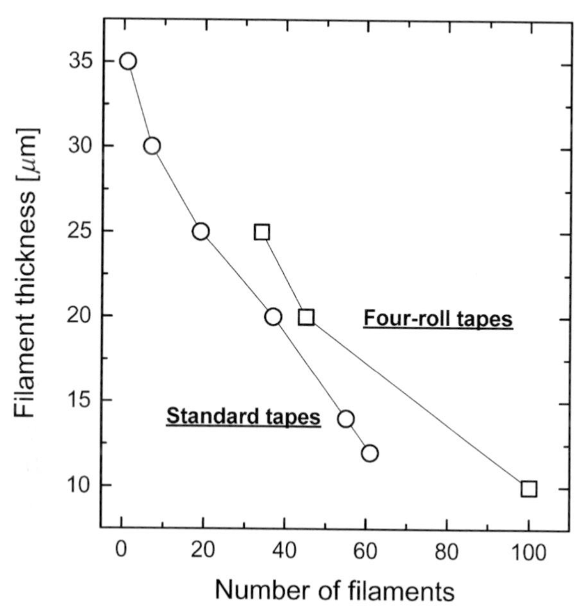

Figure B3.2.2.43. Filament thickness plotted as a function of the number of filaments and of the preparation technique that has been employed.

While the results achieved on the standard deformed tapes would suggest us about a direct correlation between filament thickness and misalignment angle of the grains, this does not seem to be the case anymore if we look at the results concerning the four-roll deformed tapes. In fact, the 100-filament tape shows an average filament thickness of 10 μm, clearly lower than for all the others, but with a misalignment angle of the Bi,Pb(2223) grains still comparable to that of monofilamentary tapes, of about 8°. It is therefore clear that the degree of texture of the Bi,Pb(2223) grains is not only influenced by the filament thickness, but also more in general by the whole deformation process employed for the fabrication of the tape.

A1.3 In figure B3.2.2.41 it has been shown that (in a normalized scale) the multifilamentary tape can in principle carry twice more current than a monofilamentary tape at fields of 1 T and above. This results cannot be entirely explained by the slightly higher degree of texture of the Bi,Pb(2223) grains, that C2 mainly governs the high field slope of the $J_c(B)$ curves. Instead, it can be easily seen from figure B3.2.2.41 that the main difference lies in the first initial drop of the critical current at low fields.

There is still an open controversy about the origin of this drop in Bi,Pb(2223) tapes; it can be easily demonstrated, however, that the relevant suppression of this drop in multifilamentary tapes is essentially due to the presence of non-negligible components of the self field oriented along the tape normal. In fact, as indicated in figure B3.2.2.41, the self-field critical current of the 37-filament tape was 58 A, that is, more than twice the critical current of the single-core tape.

B3.2.2.4.4 History dependence of the critical current density at low temperatures

It has been shown that the Railway-Switch model can adequately describe the transport properties of Bi,Pb(2223) tapes, in all the investigated temperature and field range. However, in the low temperature region, a peculiar behaviour of the transport critical current, not predicted by the model of Hensel *et al* [32], can be observed.

C2 In fact, a history dependence of the critical current with respect to the applied magnetic field distinctly appears, as shown in figure B3.2.2.44. In this figure, the critical current of a monofilamentary tape is shown at different temperatures between 4.2 and 40 K, and both for increasing and decreasing

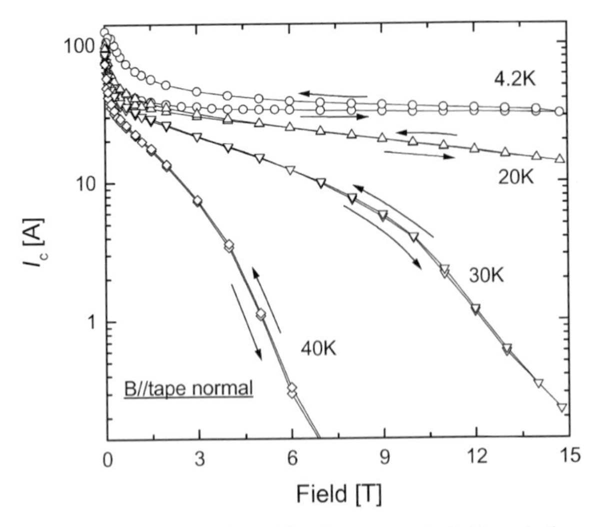

Figure B3.2.2.44. History dependence of J_c induced by the magnetic field variation at various temperatures between 4.2 and 40 K. The field has been applied parallel to the tape normal.

external magnetic field. During these measurements, the magnetic field direction was always parallel to the tape normal.

As can be seen from figure B3.2.2.44, the critical current measured after having increased the magnetic field is generally lower than in the opposite case, especially at 4.2 K, where the effect is still evident at a field of 14 T.

The explanation of this phenomenon is not yet clear, even if a very similar behaviour has been already observed in other polycrystalline superconductors with not perfect grain connectivity; so far, the only theoretical interpretation of this mechanism has been proposed by D'yachenko [33], for YBCO melt textured samples.

C1

B2.2.3

The observation of such hysteretic behaviour seems to contradict the conclusion that in standard tapes the transport critical current density is limited just by the intragrain properties of the Bi,Pb(2223) phase. However, the hysteretic behaviour is confined in a limited domain of fields and temperatures, as shown in the H–T plane of figure B3.2.2.45: at 30 K, the irreversibility field is already larger than the history field by about an order of magnitude.

A further investigation of the mechanism leading to the history effect is therefore required before coming to more precise conclusions.

B3.2.2.4.5 Lateral distribution of the critical current density

One of the questions that are still open about the transport properties of Bi,Pb(2223) tapes is where the current preferentially flows inside the superconducting filament. Generally, impurities, voids, defects, secondary phases and weak links are opposing to an homogeneous current distribution inside the filament. The investigation of the local lateral current distribution inside the Bi,Pb(2223) filament of Ag sheathed tapes has been carried by a simple and effective direct transport measurement.

C2

The measurements were first performed on high J_c monofilamentary tapes (J_c(77 K, 0 T) of 23 kA cm^{-2}) prepared by rolling. In order to measure the lateral J_c distribution inside the oxide core, the

A1.3

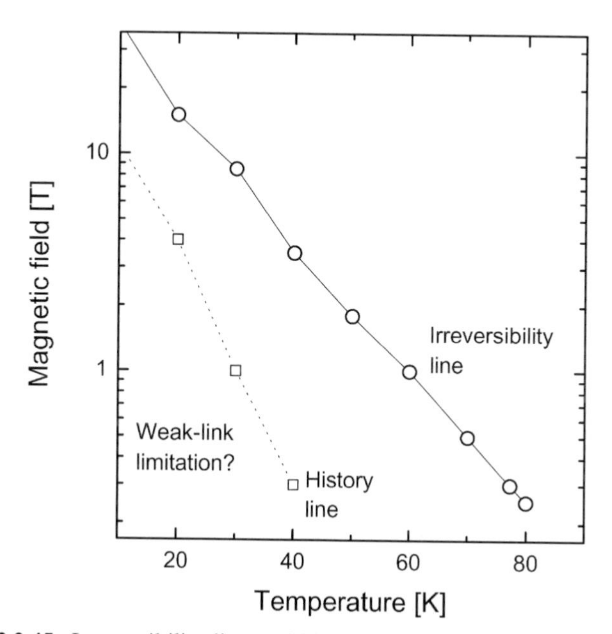

Figure B3.2.2.45. Irreversibility line and history line plotted in the H–T plane.

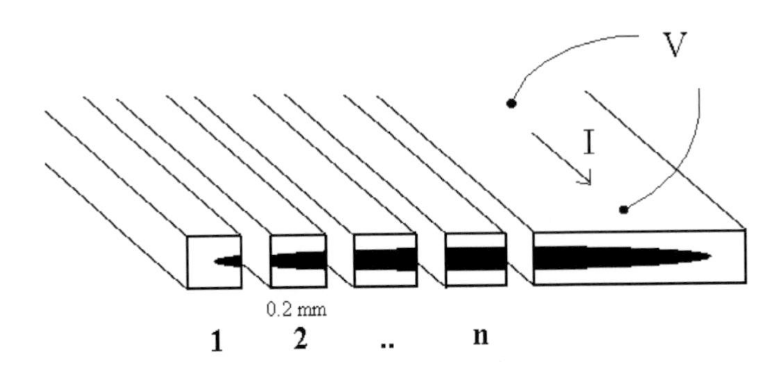

Figure B3.2.2.46. Schematic description of the technique used to determine the local J_c distribution.

tapes have been mechanically cut starting from one side and in longitudinal direction by means of a razor blade, without taking off the silver sheath. Typically, strips of about 0.2 mm in width have been successively cut away without damaging the superconducting properties of the filaments. The J_c values of the samples have been measured as usual by the standard four probe technique (criterion of $1\,\mu\mathrm{V\,cm^{-1}}$), with six different voltage contacts placed at about 2 mm from each other. With the razor blade, narrow strips of the sample of about 0.2 mm in width have been cut starting from one lateral side, as shown in figure B3.2.2.46. After each cut, I_c of the residual tape has been measured.

C2 In figure B3.2.2.47 the decrease of I_c has been plotted versus the number of strips which have already been cut. The tape cross section S has been evaluated after each cut (figure B3.2.2.47) by means of a standard optical technique. The local J_c has been simply determined as follows: after the cut n, the I_c^n and S^n of the residual tape are measured, and so we calculate

$$J_c^n = \frac{I_c^n - I_c^{n-1}}{S^n - S^{n-1}}.$$

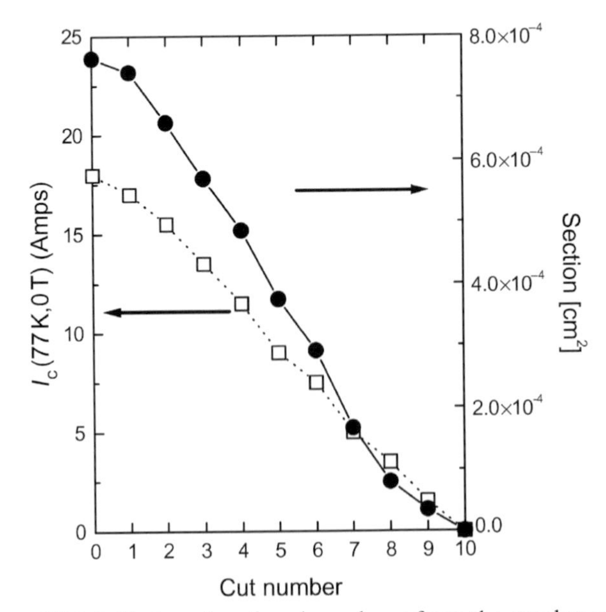

Figure B3.2.2.47. I_c and section dependence from the number of cuts.

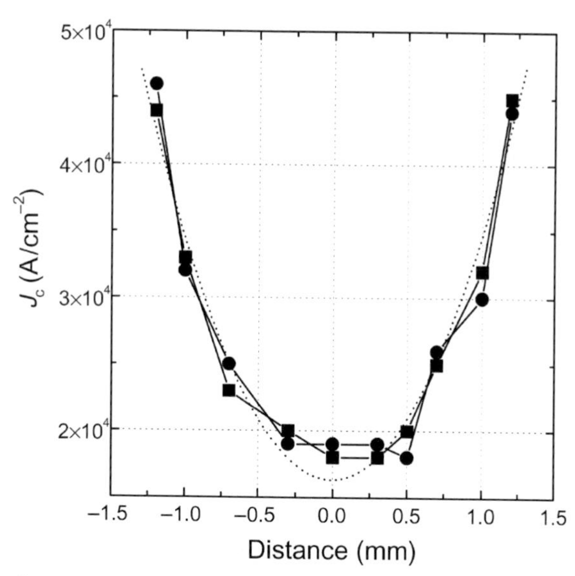

Figure B3.2.2.48. Local J_c distribution for two different tapes with respect to the distance from the tape centre.

The local J_c has been plotted in figure B3.2.2.47. The same operation has been repeated on several samples in order to minimize the fluctuation errors. C2

The J_c distribution is shown in figure B3.2.2.48 for two different samples of the same batch. A parabolic behaviour seems to describe quite well the J_c distribution. It is interesting to note that the central part of the tape has a significantly lower J_c (about 18 kA cm^{-2} at 77 K, 0 T) with respect to the average measured on the whole sample (about 23 kA cm^{-2}).

The maximum J_c value (about 46 kA cm^{-2} at 77 K, 0 T) is reached at the external strips of the tape and is three times higher than the value found in the centre.

As transport measurements have shown that no remarkable differences in the texture of Bi,Pb(2223) grains are present, SEM investigation has been extensively used in order to study in detail the local microstructure of the tapes. A large number of polished transversal sections of cold rolled tapes have been analysed. One of these sections is shown in figure B3.2.2.49. By analysing the section with a high magnification, a clear difference has been found between the microstructure in the external strips and in the centre of the filament. As shown in figure B3.2.2.50, secondary phases [mainly $(Sr,Ca)_{14}Cu_{24}O_{41}$] tend to agglomerate in the centre, while at the sides very pure Bi,Pb(2223) regions are found. D1.3

For multifilamentary tapes, the knowledge of the lateral critical current distribution is also very important, because it can help for a better understanding of the complex ac loss mechanisms. However, the origins of a possible presence of a lateral distribution of J_c should be at least partially different than for monofilamentary tapes. In multifilamentary tapes, the lateral critical current distribution should mainly arise as a result of an inhomogeneous deformation process, while in monofilamentary samples the reaction conditions inside the silver sheath can also play an important role. A1.3 D2.6

It is unfortunately not possible to repeat the same technique used before on multifilamentary tapes, essentially because (a) they are thicker and the razor blade is not sharp enough to cut them properly, and (b) a slight misalignment of the cut with respect to the tape direction can completely interrupt several filaments, which will not contribute anymore to the current transport, leading then to an improper result. The lateral distribution has been therefore roughly investigated by cutting 3 mm wide, 250 μm thick tapes into three longitudinal slices, and measuring independently their transport properties.

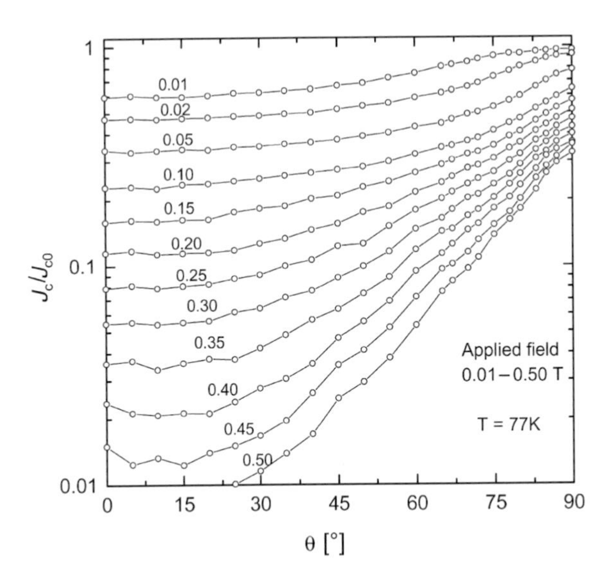

Figure B3.2.2.49. Typical transversal section of a rolled tape after the heat treatment. The white circles indicate the positions where the two pictures of figure B3.2.2.50 have been taken.

The experiment has been performed both on standard and four-rolled deformed multifilamentary tapes with 37 and 45 filaments, respectively. The fill factor of the standard tape is about 20%, while the superconducting fraction for the four-roll tape is about 35% of the overall cross section.

The cross sections of both tapes are shown in figure B3.2.2.51

As can be seen from these pictures, the filaments near the tape centre of the standard tape are much more compressed than those at the tape sides, while the four-roll deformed tape looks much more homogeneous.

The critical current densities of the different slides are reported in figure B3.2.2.52.

C2 For the standard tape with average J_c at 77 K of 28 kA cm^{-2}, the critical current density is clearly higher in the central slice (35 kA cm^{-2}) than at the sides (22 kA cm^{-2}), which is a totally opposite behaviour compared to what found in monofilamentary tapes. The amplitude of the local J_c variation

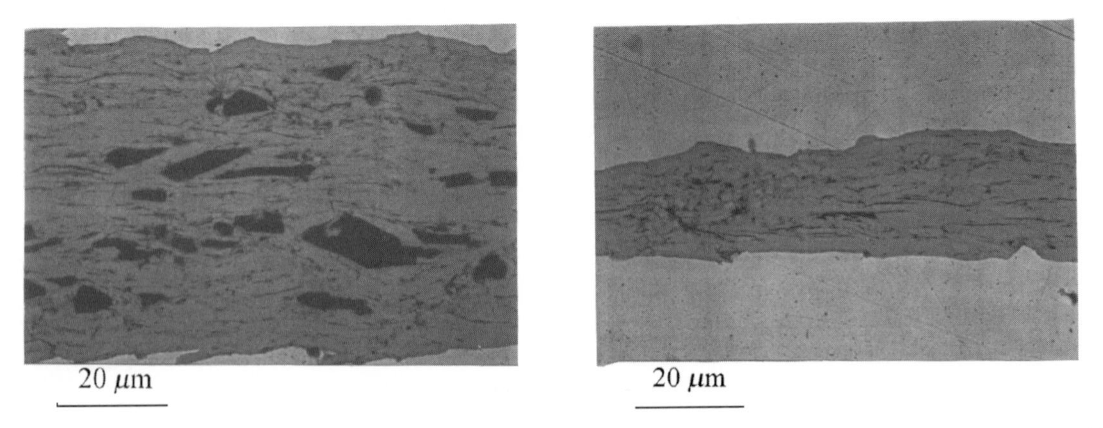

Figure B3.2.2.50. Details of the longitudinal section of a rolled tape at the tape centre (left) and at one tape side (right).

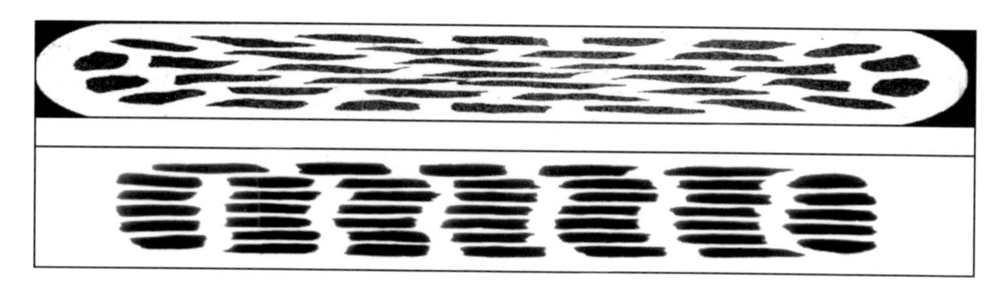

Figure B3.2.2.51. Transversal cross section of the multifilamentary tapes which have been investigated for the lateral current distribution study. The upper cross section is relative to a standard-deformed tape, while the lower is a cross section of a four-roll deformed tape. The tape width is about 3 mm, and the thickness 250 μm.

seems to be slightly smaller than for monofilamentary tapes. The four-roll deformed tape (average J_c value of 26 kA cm^{-2}) shows a much lower fluctuation of J_c, of the order of 10% of the average value, between the centre and the tape sides.

The innovative four-roll deformation is therefore decisive in homogenizing the current distribution throughout all the filaments, which is clearly not the case for the standard PIT tapes. The explanation of this effect has to be found in the different compression of the filaments due to the deformation process in the standard and four-roll deformed tapes. A confirmation of the higher homogeneity of the four-roll deformed tapes has been obtained by Vickers microhardness measurements performed on single filaments of both standard and four-rolls deformed tapes. The measurements are summarized in figure B3.2.2.53. The Vickers microhardness of single filaments has been plotted as a function of the lateral distance from the filament centre to the tape centre. B3.1

Every filament has been measured in three different points, in order to have a higher accuracy. For the standard tapes, a significant variation of the microhardness has been observed between the filaments which are near the tape centre and those which are near to the sides. Typically, the filaments near to the centre present a Vickers microhardness value of about 130–140 Hv, while near to the sides it decreases to about 90 Hv. For the four-roll deformed tapes, the Vickers microhardness is much less position dependent, going from about 145 Hv at the centre to 125 Hv near to the tape sides. Moreover, the B3.1

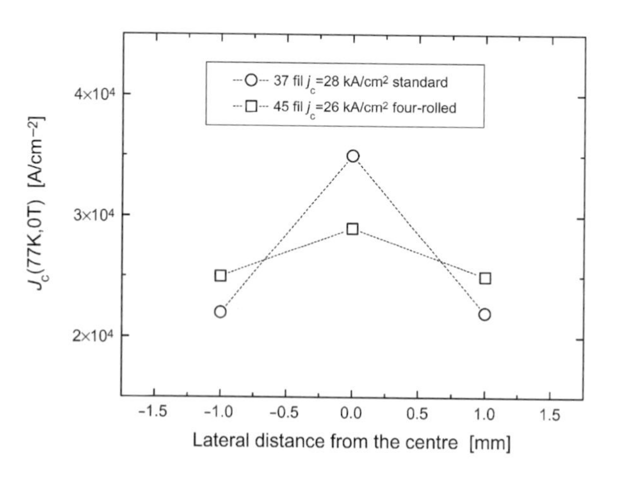

Figure B3.2.2.52. Lateral distribution of the critical current for multifilamentary tapes deformed with both standard and four-roll technique.

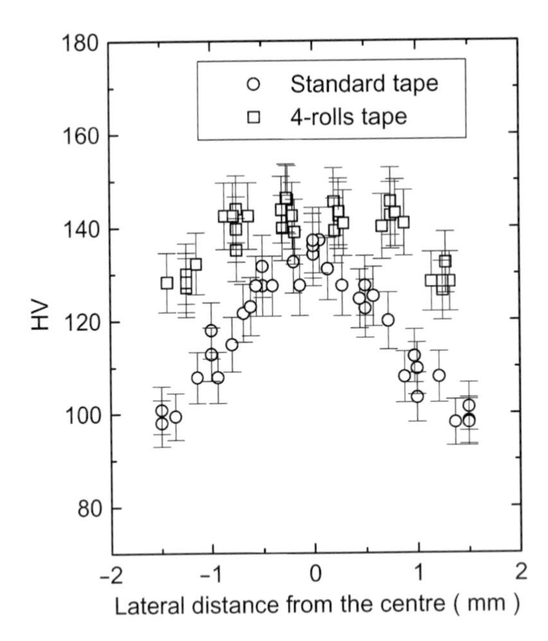

Figure B3.2.2.53. Variation of the Vickers microhardness for a standard and four-roll deformed tape as a function of the distance of the filament from the tape centre.

Vickers microhardness of the four-roll deformed tapes is reproducibly higher than that of the standard deformed tape.

B3.2.2.5 Mechanical properties of Bi,Pb(2223) tapes

D3.3 Pure silver is chosen as a base material for the sheath for multiple reasons: it is very ductile, it is permeable to oxygen, and it is not an obstacle for the Bi,Pb(2223) phase formation. The very low electrical resistivity and high thermal conductivity of silver can either be positive or negative aspects A1.3 depending on the practical application under study. In fact, for current lead applications, a low electrical resistivity, low thermal conductivity sheath would be ideal, while for ac applications like power cables E1.3.5 and transformers, high thermal conductivity and high electrical resistivity are preferable. But in all cases, E1.3.2 a much harder and stronger sheath is required instead of pure silver, in order to make their handling much more comfortable, and to compensate the intrinsic brittleness of the cuprate high T_c superconductors. For high field magnets, high mechanical strength is also required to sustain the strong Lorentz forces acting on the coil. The most common and easiest way to increase the hardness of the sheath is to replace the pure silver with a harder silver alloy [21]. Many of the silver alloy compounds are very well known for their very high hardness, as Ag–Cu, for example, but it should not be neglected that the replacing sheath should have a similar behaviour to pure silver with respect to the Bi,Pb(2223) phase formation.

Conventional additions chosen for the reinforcement of the Ag sheath are Mg, Ti and Mn. They have been selected for their limited tendency to react with the Bi,Pb(2223) phase [34]. These alloys are usually prepared for the purpose of fabricating tapes with an addition level up to 4 at %. Tubes made from these B3.1 alloys can be filled with calcined powders, and treated by the usual PIT process. The single core tape is the simplest and most suitable configuration for the study of the interaction between the alloyed sheath and

Table B3.2.2.4. Vickers microhardness of the sheath of Bi,Pb(2223) tapes prepared with different silver alloys. The proportion of the additive element is given in atomic units

Sheath material	Vickers microhardness (HV)
Pure Ag	90
Ag + 0.2%Ti	95
Ag + 2%Ti	110
Ag + 0.2%Mn	100
Ag + 2%Mn	140
Ag + 1%Mg	100
Ag + 2%Mg	140
Ag + 4%Mg	160

the Bi,Pb(2223) core. The Vickers microhardness of the sheath has been measured for all the different sheath material at the end of the deformation process. The results are shown in table B3.2.2.4.

The Vickers microhardness of the sheath increases from about 90 HV for the reference sample, made with pure silver, to 160 HV for the 4 at % Mg alloyed sheath. The addition of a very small amount of Ti and Mn does not seem to be very effective for the improvement of the tape strength though Mn and Mg additions have a similar, positive influence on the tape mechanical strength, while Ti is only marginally improving the hardness of the sheath at the actual level of doping.

DTA analysis of the various tapes have been performed in order to determine the optimal temperature for the reaction into Bi,Pb(2223) and the achievement of the highest critical current density. The DTA measurements performed on a standard tape as well as on a 4% Mg added Ag sheathed tape are shown in figure B3.2.2.54. It is clear that the AgMg alloy has a reaction with the oxide filament, as an

B2.1

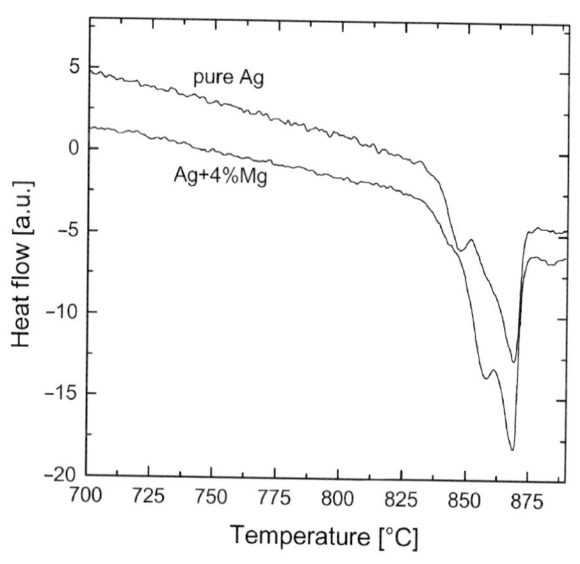

Figure B3.2.2.54. DTA measurements performed on a standard tape and a 4% Mg added Ag sheathed Bi,Pb(2223) tape.

Table B3.2.2.5. Critical current density and Vickers microhardness values of the reinforced monofilamentary Bi,Pb(2223) tapes

Sheath material	Optimal reaction temperature (°C)	J_c at 77 K (A cm^2)	Vickers microhardness (HV)
Pure Ag	837.5	30	50
Ag+0.2%Ti	835	10	60
Ag+2%Ti	832	5	70
Ag+0.2%Mn	837.5	25	60
Ag+2%Mn	835	15	100
Ag+1%Mg	837.5	25	90
Ag+2%Mg	835	20	110
Ag+4%Mg	835	10	120

B2.1 additional peak in the DTA has appeared at a temperature of about 855°C. However, the onset temperature of the formation of the Bi,Pb(2223) phase has not substantially changed from 832 to 834°C.

The monofilamentary tapes have been heat treated at the temperatures individuated by the DTA measurements, and the critical current density values, as well as the microhardness results are shown in table B3.2.2.5. The Vickers microhardness of the reacted tape sheaths have reduced by about 30–50% with respect to the values measured just after the cold deformation (see table B3.2.2.4).

This is essentially due to the important sheath grain growth, which is facilitated by the prolonged high temperature heat treatment, but also to the partial internal oxidation of Mg, Mn, and Ti. The most deceiving result has been observed with the Ti-addition, which leads to a considerable reduction of J_c without a clear improvement of the sheath strength. The Mn- and Mg additions show more promising results, as the Vickers microhardness has been increased above 100 HV, without depressing J_c below

C2 20 kA cm^{-2}. In particular, the Ag+2 at. % Mg allow seems to be the better compromise between high enough J_c and improved mechanical strength [35].

It has to be noted, however, that the reduction of the critical current density is not observed if a protective layer of pure silver is put in between the superconducting core and the alloyed silver sheath. Critical current density values of 30 kA cm^{-2} at 77 K have been achieved on monofilamentary tapes where a 10–20 μm thick protective layer of pure silver is surrounding the oxide core. In this way, the outer alloy sheaths that have been successfully employed are AgMg, AgMn, and AgCu as well.

D2.2 The transport critical current density of the AgMg tapes has been extensively measured as a function of the temperature and of the applied field, and compared to that of a standard reference tape with pure Ag sheath. In figure B3.2.2.55 and B3.2.2.56, the transport measurements performed at 77 and 4.2 K are shown, respectively.

The AgMg-sheathed tapes show a clearly faster drop of the critical current density both at 77 and 4.2 K. At 77 K and a field of 1.5 T (applied parallel to the tape plane), the Ag sheathed tape carries a twice higher current density compared to the AgMg-sheathed one.

The faster drop of J_c is probably due to the presence of a higher residual amount of Bi(2212) phase in the current paths of AgMg-sheathed tapes. A further improvement of the heat treatment parameters could be useful in improving the Bi,Pb(2223) phase purity and therefore the tape transport properties too.

D3.2.2 The thermal conductivity of Ag and Ag–Mg sheathed Bi,Pb(2223) tapes has been measured at temperatures between 4.2 and 300 K. In figure B3.2.2.57 the comparison between the thermal conductivity of Ag and Ag–Mg sheathed tapes is presented from 20 to 200 K. At low temperatures ($T \cong 20$ K), the thermal conductivity of the Ag sheathed tapes shows the presence of a peak below 20 K

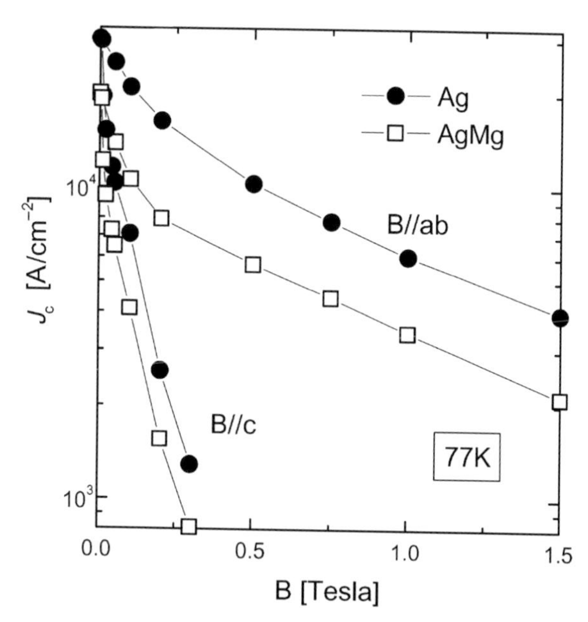

Figure B3.2.2.55. Critical current density at 77 K as a function of the applied magnetic field for Ag- and AgMg-sheathed Bi,Pb(2223) tapes. The measurements have been performed for B//tape normal and B//tape plane.

which is about seven times higher than Ag–Mg sheathed tapes. It is noticeable that the peak is completely destroyed by the small amount of Mg solute. At intermediate temperatures both thermal conductivity converges to a nearly constant value of about 2 W(cm K)$^{-1}$. D3.2.2

Measurements of thermal and electrical conductivity at 4.2 K have been performed separately both on the Ag and Ag–Mg sheaths. It has been found that the electrical RRR ($R^{300\,K}/R^{4.2\,K}$) is about 100 and

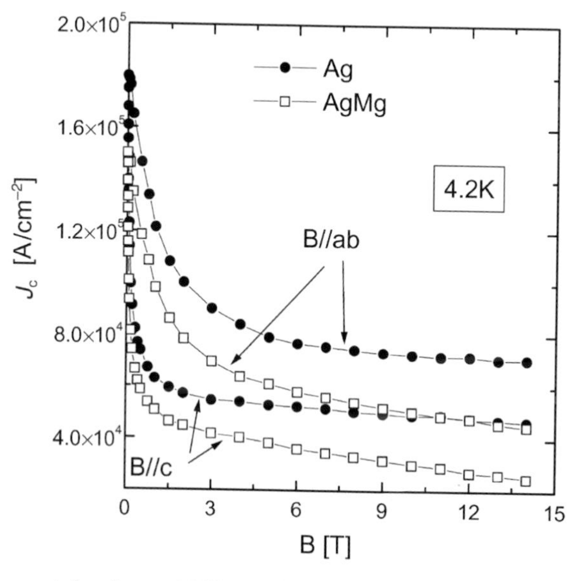

Figure B3.2.2.56. Critical current density at 4.2 K as a function of the applied magnetic field for Ag- and AgMg-sheathed Bi,Pb(2223) tapes. The measurements have been performed for B//tape normal and B//tape plane.

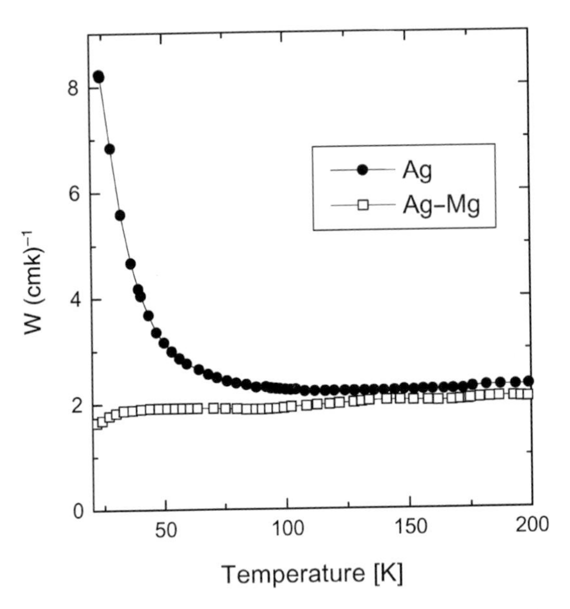

Figure B3.2.2.57. Thermal conductivity of Ag and Ag–Mg sheathed tapes.

17 for the Ag and the Ag–Mg sheaths, respectively, and in particular $RRR_{Ag}/RRR_{Ag-Mg} \cong 6$. It is interesting to note that the ratio between the residual thermal conductivity of the two types of tapes has a similar value [$K_{Ag}(4.2\,K) \cong 3.5\,W\,(cm\,K)^{-1}$, and $K_{Ag-Mg} \cong 0.6\,W\,(cm\,K)^{-1}$, that is, $K)_{Ag}/K_{Ag-Mg} \cong 6$].

This result is in agreement with the Wiedemann–Franz law:

$$K\rho = TL_0 \text{ with } L_0 = 2.45\,W\,\Omega\,K^{-2} \Rightarrow \frac{K_{Ag}(4.2\,K)}{K_{Ag}(300\,K)}\frac{300\,K}{4.2\,K} = RRR_{Ag}.$$

A similar behaviour has been found for the Ag–Mg sheathed tapes.

A1.3 The improvement of the strength of the silver sheath can be also carried out in multifilamentary tapes. Knowing that the alloyed silver can be safely employed if it is not in direct contact with the superconducting oxide, the schematic configuration of the filaments shown in figure B3.2.2.58 has been considered.

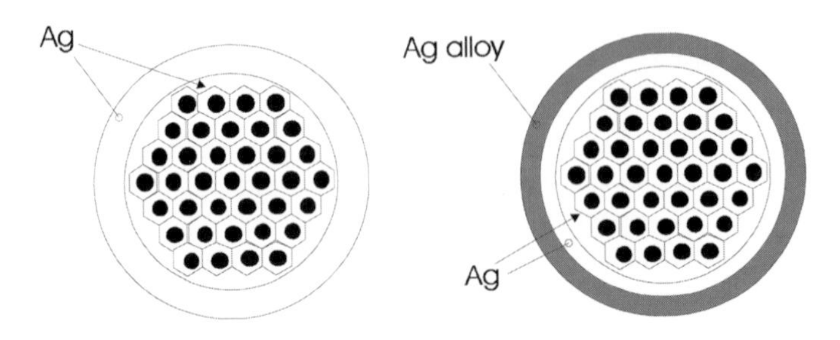

Figure B3.2.2.58. Schematic representation of the configuration of the silver sheath and of the filaments in standard and reinforced multifilamentary tapes.

Table B3.2.2.6. Critical current, critical strain, and critical stress of both Ag and AgMg sheathed multifilamentary Bi,Pb(2223) tapes

Sample	I_c (A)	ϵ_{cr} ($\times 10^{-3}$)	σ_{cr} (Mpa)
Ag, 19 fil	18.8	2.5	46
Ag, 37 fil	30.2	2.5	63
Ag, 37 fil	37	1.8	43
Ag, 55 fil	37.8	2.2	63
Ag, Mg 37 fil	31.7	2.5	105
Ag, Mg 37 fil	29.5	2.6	116

The reinforced silver sheath is placed only as an outer layer, while pure silver is employed both for the hexagonal wires and for the inner walls of the surrounding tube. In this way, the bundling between the various filaments and the outer tube is still governed by the silver diffusion properties in the pure metal.

The reinforced wires have been deformed as usual by swaging, drawing and rolling, and they have been heat treated in the same conditions than the standard tapes, that is, at a temperature of 837°C for a total time of about 150 h, with an intermediate cold rolling deformation step. B3.1

No appreciable variation of the critical current density has been observed between standard and reinforced tapes, typical values obtained on AgMg-sheathed 37 filament tapes being in the range 25–28 kA cm^{-2} at 77 K.

The improvement in mechanical properties of the alloyed sheath tapes has been verified by performing tensile strain experiments, using a specially conceived set-up that allows stress–strain and critical current versus stress or strain measurements, both at room temperature and in liquid nitrogen. A load cell is used to measure the applied force, while the applied deformation is measured by cryogenic strain gauges glued to the tapes.

The results are summarized in table B3.2.2.6. A criterion of 10% of critical current reduction has been employed for the determination of the critical strain ϵ_{cr}, and of the critical stress (σ_{cr}).

B3.2.2.6 AC losses in Bi,Pb(2223) tapes

B3.2.2.6.1 General remarks

Nearly all the possible industrial applications of Bi,Pb(2223) tapes require the transport of alternating currents through the superconductor. It is therefore extremely important to optimize the tape configuration in order to reduce the ac losses as much as possible. This is not a simple task, because there are several mechanisms contributing to the energy dissipation [22]. In single core tapes, the main origin of energy dissipation in the presence of alternating currents and/or magnetic fields is the hysteretic behaviour of the magnetization, typical of type II superconductors. This irreversible behaviour of the magnetization leads to the so-called hysteresis loss Q_h per cycle. E1.3.2 D2.6 A1.3

A very simple estimation of the Q_h losses can be derived from the critical state model, developed by bean. In the simplest case of an indefinitely extended slab of thickness d and constant J_c with an applied magnetic induction B_a oriented parallel to the slab, Bean has calculated the losses as a function of the applied field amplitude ΔB_a.

If $\Delta B_a > 2\,B_p = \mu_0 J_c\,d$, then

$$Q_h = \frac{B_p \Delta B_a}{\mu_0}\left(1 - \frac{4}{3}\frac{B_p}{\Delta B_a}\right) = \frac{J_c d \Delta B_a}{2}\left(1 - \frac{2}{3}\frac{\mu_0 J_c d}{\Delta B_a}\right)$$

where B_p is the field where the applied field and the screening current have reached the centre of the slab.

As, in general, $\Delta B_a >> 2\,B_p$, then

$$Q_h = \frac{1}{2}\Delta B_a J_c d.$$

The hysteretic losses are therefore directly proportional to the slab thickness d.

It is possible to demonstrate that this is still the case when the flux creep of the vortices in real superconductors is taken into account. In fact:

$$\text{in flux flow regime, and } \Delta B_a >> 2\,B_p: \quad \frac{Q_h}{V} = \frac{1}{2}\Delta B_a d\left(J_c + \frac{a\dot{B}}{\rho_f}\right), \text{ while,}$$

$$\text{in flux creep regime, and } \Delta B_a >> 2\,B_p: \quad \frac{Q_h}{V} = \frac{1}{2}\Delta B_a d J_c\left[1 + \frac{kT}{U_0}(\ln a\dot{B} - \ln E_0)\right],$$

where $a = d/4$, V is the sample volume, $\dot{B}$ is the field sweep rate, ρ_f is the flux flow resistivity, E_0 is a constant with the dimension of an electric field, and U_0 is the pinning energy.

D2.6 According to these equations, the easiest way to reduce the ac losses is to reduce the filament size. The multifilamentary tapes should therefore present much lower hysteretic losses than the single-core ones. In figure B3.2.2.59, the loss voltage measurements performed on mono- and multifilamentary tapes with various filament number are presented. The alternate current frequency is 59 Hz.

From these measurements it appears already clear that ac losses are not significantly reduced by changing from a simple mono- to a more complex multifilamentary tape structure. Just the 19-filament tape presents somewhat lower loss voltages than all the other samples. The origin for this unexpected D2.6 result is the appearance of an additional dissipation mechanism in ac conditions, namely the coupling losses Q_c. Coupling current losses in multifilamentary conductors are caused by a magnetic coupling of

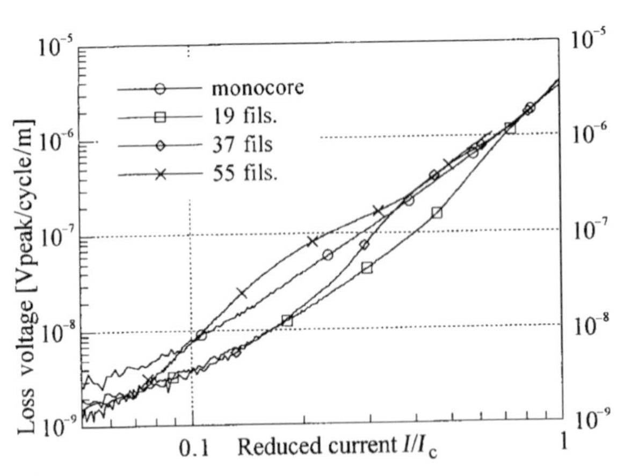

Figure B3.2.2.59. Loss voltage as a function of the reduced current for mono- and multifilamentary tapes at 77 K, self field, and for a frequency of 59 Hz.

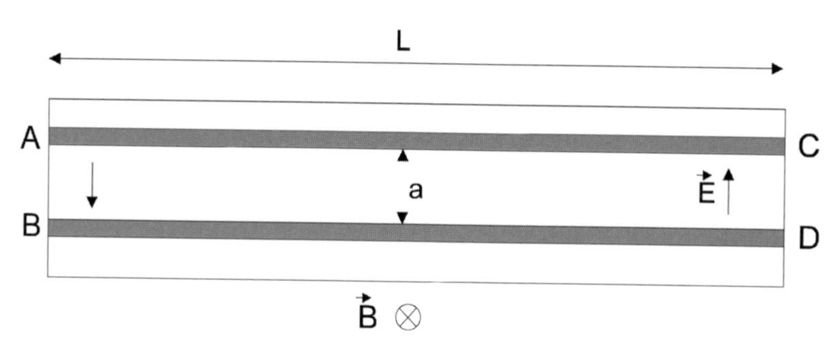

Figure B3.2.2.60. Schematic representation of a multifilamentary tape of length L, immersed in a time dependent magnetic field parallel to the tape normal.

the individual filaments in the presence of a time varying magnetic field B oriented parallel to the tape normal.

In a given tape of length L (figure B3.2.2.60), two parallel filaments at a distance a experience a time dependent magnetic field B. If $\dot{B}$ is constant, then a voltage V will be induced in the loop ABCD

$$V = -\dot{B}La.$$

In the superconducting paths AD and BC, no voltage drop should instead appear. Therefore, the electrical field in the silver matrix can be written as

$$E = V/a = -\dot{B}L.$$

The corresponding power loss density ($P = EJ$) between AB can be written as

$$P \propto E^2/\rho = \dot{B}^2 L^2/\rho$$

where ρ is the matrix electrical resistivity. Therefore, in multifilamentary tapes, an energy loss term proportional to the square of the tape length L appears in the balance of the various loss mechanisms.

B3.2.2.6.2 *Twisting of multifilamentary Bi,Pb(2223) tapes*

A possible solution to the coupling loss problem can be found by twisting the filaments. This technique has been already studied for round and flat cables, and some analytical results have been presented by Campbell [36] and Kwasnitza and Bruzzone [37]. In figure B3.2.2.61, a schematic representation of two twisted filaments inside a multifilamentary tape is shown.

D2.6
B3.1

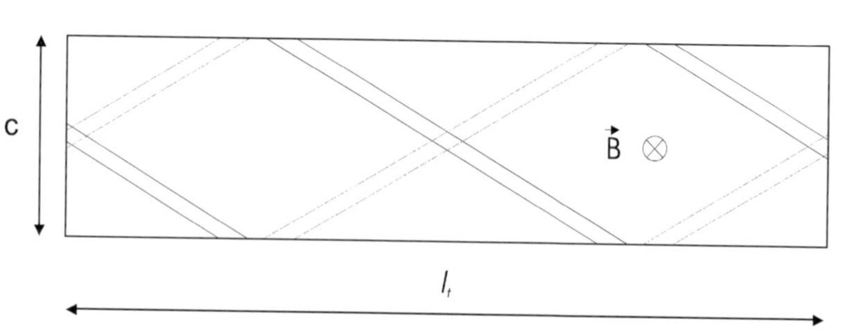

Figure B3.2.2.61. Schematic representation of a tape with twisted filaments.

In a round conductor, Campbell has found that

$$\frac{P_c}{V} = n\tau\frac{\dot{B}^2}{\mu_0}$$

where P_c is the power loss term, n is a form factor ($n = 1/(1 - N)$, where N is the demagnetization factor), and τ is the coupling current decay time constant. Again according to Campbell, it is possible to write

$$\tau = \frac{1}{24\pi^2}\frac{\mu_0}{\rho_e}\,l_t^2$$

where l_t is the twist pitch, and ρ_e is the effective transverse electrical resistivity between two filaments in the normal conducting matrix.

Moreover, Kwasnitza and Bruzzone [37] have found that, in a flat conductor of thickness d and width c ($d << c$), it is possible to write

$$\frac{P_c}{V} = \frac{1}{144}\left(\frac{c}{d}\right)^2\frac{l_t^2\dot{B}^2}{\rho_e}.$$

In twisted tapes, the power loss is proportional to the square of the twist pitch, and inversely proportional to the effective transverse electrical resistivity between two filaments. It is therefore important to fabricate Bi,Pb(2223) tapes with the shortest possible twist pitch. Twisted multifilamentary tapes can be easily prepared by slightly modifying the standard deformation process of the PIT method. The easiest way to introduce a twist of the filaments is just before the final cold rolling process, when the conductor still presents a round shape.

A quick heat treatment at low temperature ($\approx 200°C$) would further help for twisting the round wires; a final twist pitch as short as 1 mm can therefore be achieved. Afterwards, the wires are rolled as usual into flat tapes, where the twist pitch previously introduced becomes generally 2–4 times longer, due to the obvious elongation of the tape. The surface of a twisted 7-filament tape is shown in figure B3.2.2.62, after partial etching of the outer silver sheath. In this case, the twist pitch of the filaments is about 12 mm.

The twisted tapes are generally heat treated in exactly the same conditions than the untwisted ones. In figure B3.2.2.63, the typical behaviour of the critical current density of a series of 37-filament tapes prepared with various twist pitches is shown.

C2

The critical current density, plotted as a function of $1/l_t$, shows a plateau-like behaviour until when the twist pitch reaches a critical value of about 10 mm. For pitches shorter than 10 mm, J_c inevitably drops, due to the heavily distorted filament configuration. However, for a relatively short twist pitch of 5 mm, the critical current density is depressed by about 10–20% only, which is in many cases an

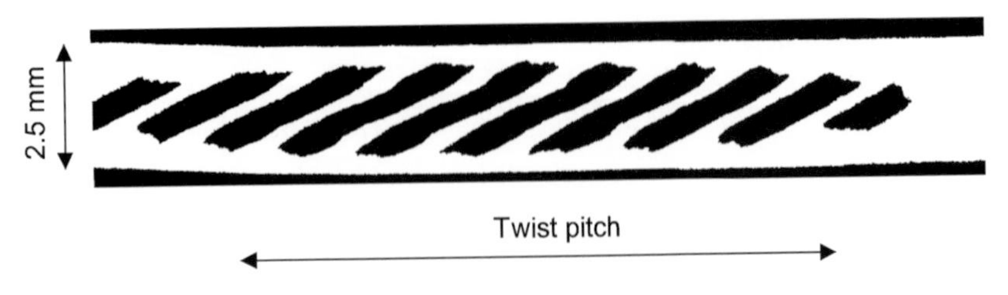

Figure B3.2.2.62. Etched surface of a 7-filament Bi,Pb(2223) tape with a twist pitch of about 12 mm. The tape thickness is 180 μm, and the width is 2.5 mm.

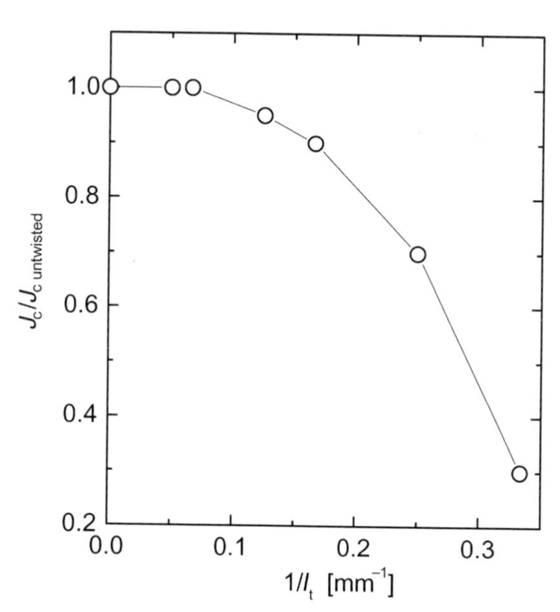

Figure B3.2.2.63. Critical current density as a function of $1/l_t$ for 37-filament tapes. The J_c value of the reference untwisted tape was 26 kA cm^{-2} at 77 K, self field.

acceptable compromise. A similar behaviour has been also reported for tapes with different number of filaments and/or superconductor fill factor.

The transport ac losses have been measured at 77 K in self-field for twisted and untwisted multifilamentary tapes with 7, 19, 37, and 55 filaments, and for various twist pitches. Unfortunately, the result is not as positive as expected, when the matrix is composed by pure silver. Only for the 7- and 19- filament tapes a slight reduction of the ac losses has been observed, while with higher number of filaments the differences are hardly detectable. The measurements performed on the 19-filaments are shown in figure B3.2.2.64.

D2.6
A1.3

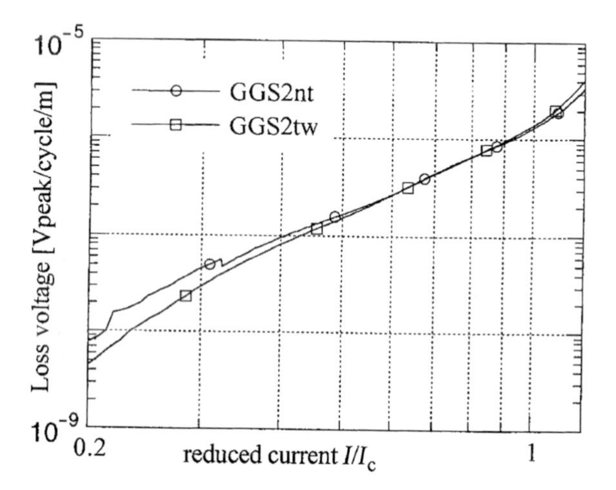

Figure B3.2.2.64. Loss voltage as a function of the reduced current for twisted and untwisted 19-filament tapes. The twist pitch is 10 mm. Square symbols are relative to the twisted tape measurement.

D2.6 These results can be hardly described by the ac loss theory, the possible explanation should be instead found in the peculiar tape microstructure. In fact, if some superconducting (or with very low electrical resistance) links between adjacent filaments are present, then the effect of twisting on the ac losses will be largely reduced.

The easiest way to determine whether the filaments in twisted tapes are electrically joined is to measure the transport properties of twisted conductors cut exactly in two halves in the longitudinal direction. If no superconducting leakage between filaments is present, then the two halves should present a markedly resistive behaviour; on the contrary, if some superconducting links are present, the I–V characteristics will show the typical V–I characteristics of a superconducting tape.

D2.2 In figure B3.2.2.65, the I–V characteristics of each half of the tapes are shown, for increasing number of filaments. The superconducting fraction is about 20% for all the tapes, as well as the critical current value, which was about 30 A for all the tapes (before cutting). As can be seen from the figure, apparent differences are observed between the samples. For the 7- and 19-filament tapes, a clearly resistive behaviour is observed, which is an indication that the conductor does not present any relevant superconducting bridging between neighbouring filaments.

However, the slope of the I–V characteristics, which is related to the transverse electrical resistance between the filaments, gradually reduces for increasing number of filaments. For the 37- and 55- filament tapes, indeed, the transverse electrical resistivity is almost negligible in the case of low currents. This behaviour should therefore explain the disappointing results of the ac loss measurements performed on the twisted samples. D2.6

C2 Finally, it is also interesting to note that all the samples show a clear change in the I–V curvature when the current has reached 25% of the critical current value of the entire tape.

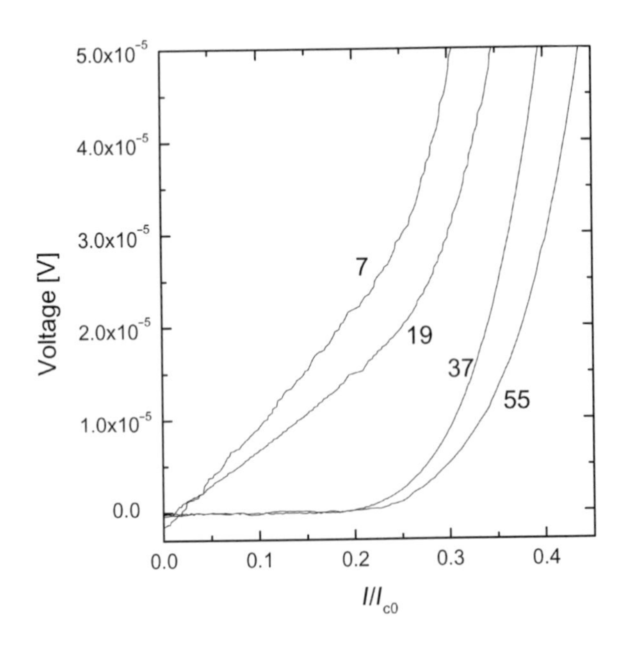

Figure B3.2.2.65. I–V characteristics for various half-tapes with number of filaments varying from 7 to 55. The distance between the voltage contacts is 45 mm. The current is normalized to the critical current value of the entire tapes before the cut.

B3.2.2.7 Reduction of AC losses by oxide barriers

Recently, Huang and Flükiger [38] have successfully addressed the problem of reducing coupling losses by introducing highly resistive oxide layers as current barriers between each filament of Bi 2223 tapes. The advantages of this method are: (a) the introduced material is not in direct contact with the core, making it easier to avoid additional reactions; and (b) the transverse resistivity can be increased significantly without changing the longitudinal one which results in a strong reduction of the coupling. However, due to the presence of these oxide barriers, the achievement of a high J_c value faces with two new problems, increased brittleness and reduced oxygen diffusion, that require a re-optimization of the preparation parameters in order to achieve satisfying J_c values. D2.6

A variety of materials, CuO, NiO, TiO, TaO, MgO, MnO, Bi(2212) and BaZrO, was initially tested, inserted either directly as oxide powder or as metallic layers which were subsequently oxidized *in situ*. Ideally, such a material should have a much higher resistivity than Ag, and should be easily introduced into the tape structure without seriously affecting the Bi,Pb(2223) phase formation. Two possible routes have been explored by these authors: insertion and deformation of metallic layers which were subsequently oxidized *in situ*, and direct insertion and deformation of oxide powders.

Initially, the use of pure metals, such as Cu, Ni, and Ti was investigated as barrier materials. These metals present the advantage that they are highly ductile and transform easily into oxides which have almost no solubility in Ag. A typical cross section of an Ag–Cu sheathed Bi,Pb(2223) tape is shown in Figure B3.2.2.66(*a*) after deformation and before the heat treatment. Although Cu has a higher yield stress than Ag, uniform 1–5 μm thick Cu layers around every filament can be obtained in a reproducible way after cold working.

However, after reaction at 837°C for 150 h the Cu layers are clearly destroyed, as shown in figure B3.2.2.66(*b*).

By means of SEM and EDX analysis, they found that, although the Cu layers were already completely oxidized at temperatures below 650°C, they started to diffuse quickly towards the Bi,Pb(2223) filaments with further increase of temperature even though CuO has a negligible solubility in Ag. On the other hand, cold working of Ag–Ni and Ag–Ti sheathed tapes turned out to be more difficult than for Ag–Cu sheathed tapes due to the higher yield stress and hardness of these metals compared to pure Ag. Cracks were observed in both the Ni and Ti barriers, indicative of the fact that the deformation stresses are mainly concentrated in these metal layers. Ni barriers can be fully oxidized at temperatures similar to Cu. Using NiO barriers, however, two main problems were observed: (a) the very thin 0.3–2 mm NiO layers start to agglomerate during heat treatment, and (b) some NiO diffuses through the Ag, reacts with the cores and promotes low T phase formation. Critical current values of 8–10 kA cm^{-2} 77 K, 0 T have been achieved by Huang and Flükiger [38] on these tapes and the transverse resistivity was increased by a factor of 3. Ti was found to react even more with Bi,Pb(2223) D1.3, D1.5

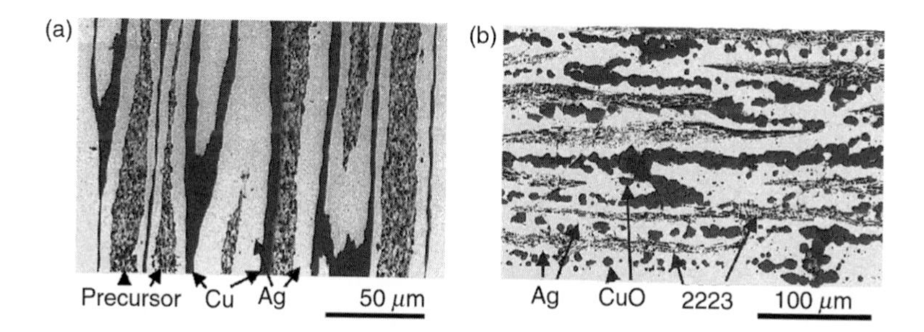

Figure B3.2.2.66. Details of the cross section of a 37-filament Bi,Pb(2223) tape with CuO barriers (*a*) after deformation and (*b*) after heat treatment.

phase. In conclusion, none of the three metals (Cu, Ni and Ti) examined was found to be suitable in the preparation of oxide barriers.

The authors have therefore investigated an alternative process of directly inserting oxide powder as a barrier material. A range of oxides was tested for its suitability in producing uniform and stable barriers: TaO, Bi(2212), MgO, MnO and BaZrO. Bi(2212) was chosen as a candidate for the following reasons: (a) it is very similar to Bi,Pb(2223) from both physical and chemical point of view, which should make the deformation and phase formation easier; (b) its grains have a plate-like shape and agglomeration of the thin barrier layer can therefore be reduced; (c) the Bi(2212) superconductor will transfer into a poor conductor at 77 K under a very small external field.

From all the oxides examined so far, $BaZrO_3$ turned out to be best suited in producing highly resistive barriers, either from the point of view of deformation or of Bi,Pb(2223) phase formation. The results obtained with the other oxides can be summarized as follows: TaO can be easily formed into a very thin barrier layer, but a strong agglomeration appears during the heat treatment. On the other side, commercially available MgO powders are not so easy to introduce due to their particle size of 10 μm in average and shape. Finally, MnO reacts with the Bi,Pb(2223) phase, while Bi(2212) layers are less stable than $BaZrO_3$ due to the decomposition effects occurring during the reaction heat treatments.

$BaZrO_3$ is a very suitable compound to fabricate inert crucibles for melting high T materials. In a preliminary study, Huang and Flükiger [38] succeeded in obtaining phase pure Bi,Pb(2223) phase even after mixing 20 wt. % of the $BaZrO_3$ compound to the precursor powders. Also in the Ag/$BaZrO_3$/Ag sheathed Bi,Pb(2223) tapes, no trace of relevant reaction was found, neither by DTA/TG nor by SEM/EDX analysis.

B3.1 Using the conventional PIT route, the authors succeeded in obtaining Ag/$BaZrO_3$ sheathed Bi,Pb(2223) multifilamentary tapes with different number of filaments (7, 19 and 37) and barriers with uniform thickness in the range of 0.3–2 μm. Figure B.3.2.2.67 presents cross-sectional views of such a 19-filament Bi 2223 wire after the final heat treatment.

The final barrier layer thickness is 1 μm. In this picture, it can be seen that each filament is well encapsulated by the $BaZrO_3$ barrier layers. Only very few holes appear in the resistive layers, indicating that the barrier is stable whilst treating the tapes at high temperature. It was possible to twist these tapes with a pitch length of 2 cm, which resulted in a further enhancement of the decoupling effect. J_c values of this tape, both under self- or external fields, are very similar to those without twisting.

C2 The value of T_c for these tapes is always close to 110 K, regardless of the thickness of the barrier layers. However, this type of tapes yields a wide range of J_c values, from 5 to 13 kA cm^{-2} at 77 K, 0 T, depending on the barrier thickness. Thicker barriers generally lead to lower J_c values, e.g. for thickness over 1 μm, J_c is consistently lower than 7 kA cm^{-2} at 77 K, 0 T. As shown in figure B3.2.2.67, where fluctuations in filament size are clearly noticeable, a strong sausaging of the core has been observed in the tapes with thick barriers. Moreover, some filaments of the tape are broken, as indicated by the arrows. Moreover, from measurements of J_c as a function of the angle between applied field and tape normal, it was found that due to the strong sausaging the misalignment angle of Bi,Pb(2223) grains of

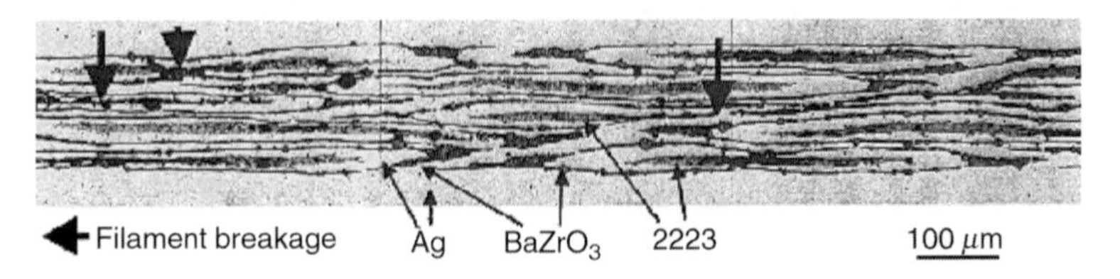

Figure B3.2.2.67. Cross section of a fully heat treated 19-filament tape with 1 μm thick $BaZrO_3$ barriers made by the normal PIT method.

these tapes is as high as 8°, compared to about 5° for tapes sheathed by pure Ag. The sausaging appears mainly during the wire drawing and laminating processes.

The investigation of a suitable material to be used as a resistive barrier has been also carried out by other groups. Goldacker *et al* [39] have successfully introduced a resistive barrier based on $SrCO_3$, reaching again critical current densities in excess of 10^4 A cm^{-2} at 77 K. Zhang *et al* [40] have extended the idea of using a resistive barrier based on a low-T_c superconducting phase by employing the Bi-2201 phase. With this barrier material, they achieved a critical current density of about 15 kA cm^{-2}, together with a reduction of the ac losses by about 70%. C2

Since the hard $BaZrO_3$ particles within the tapes are difficult to deform with the soft Ag and the mica-like BSCCO powder, stresses concentrate at these particles and introduce cracking during deformation. Therefore, the presence of high hardness $BaZrO_3$ barriers not only increases the deformation force, but also reduces the strength of sheath. When a round wire is laminated into a flat tape the strongest sausaging can be found in the centre part of tape since the filaments in this part are deformed much more heavily than those on the edge. Therefore, some modification of the conventional PIT method is necessary for the further optimization of this type of tapes.

Figure B3.2.2.68 is a cross-sectional view of a 30-filament tape with a barrier layer of 1 μm thickness made by the four-roll deforming technique. Figure B3.2.2.68(*a*) shows an intermediate stage in the rolling process, while figure B3.2.2.68(*b*) presents a micrograph of the final tape. The filaments have a regular size and are neatly stacked with continuous barrier layers between them. The advantage of this method is that no tensile force is applied during wire forming and every filament is deformed equally while laminating. Consequently, the sausaging problems are diminished and J_c values are increased, up to 15 kA cm^{-2} at 77 K, 0 T, compared to those of tapes made by the normal PIT method with barrier thickness exceeding 1 μm.

The effectiveness of $BaZrO_3$ barriers in reducing filament coupling has been clearly demonstrated by different experiments. In the double Hall sensor magnetization measurement, the tapes are mounted on a Hall sensor in an ac field generating coil ($H//c$). A second sensor is placed in the same coil at a sufficient distance from the sample. The set-up thus measures simultaneously H and B (the external field and magnetic induction at the tape surface), which allows a direct calculation of the loss per cycle Q. The loss contribution due to coupling currents can be characterized using the decay time constant $\tau \propto l_t^2/\rho_e$, where l_t is the filament twist pitch length and ρ_e is the transverse resistivity. For similar untwisted samples, $l_t = 2 \times$ sample length. The frequency dependence of the coupling loss Q_c is expected to display the following features: (a) at low frequencies $Q_c \propto \Delta B^2 f \tau$, with ΔB the ac field amplitude and f the frequency, (b) the occurrence of a maximum in Q_c at $2\pi f \tau = 1$ and (c) for large ΔB values saturation at $f > f_c$, with $f_c \propto \rho_e J_c d/(\Delta B l_t^2)$ and d the total thickness of superconducting material in the direction of

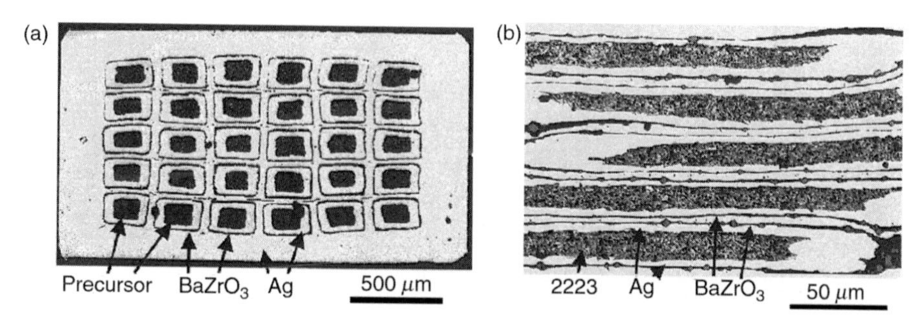

Figure B3.2.2.68. SEM picture of the cross-section of a 30-filament Ag/$BaZrO_3$ sheathed Bi,Pb(2223) (*a*) wire and (*b*) tape made by a modified PIT method.

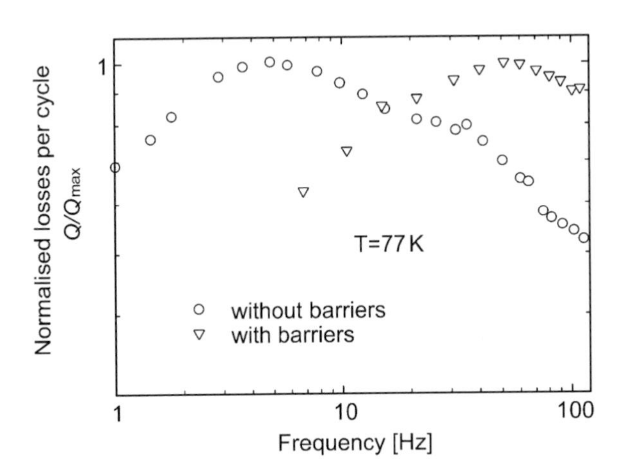

Figure B3.2.2.69. Frequency dependence of normalized total losses per cycle for two 19-filament tapes twisted with a pitch length of 2 cm and sheathed by either pure Ag (circles) or Ag/$BaZrO_3$ (triangle).

D2.6 magnetic field penetration. To get low coupling current losses at 50 Hz, ρ_e must be as large as possible. Increasing ρ_e causes a corresponding decrease of the time constant τ, shifting the maximum coupling losses to higher frequencies and bringing the 50 Hz working range into the linear, low loss part of the curve. Figure B3.2.2.69 shows the loss per cycle Q (normalized by its maximum value Q_{max}) *versus* frequency for two samples of identical length, with and without oxide barrier, twisted with a pitch length of 2 cm.

It clearly shows the efficiency of the 1 μm thick $BaZrO_3$ barriers in increasing the transverse electrical resistivity ρ_e of the matrix and shifting up the frequency of the loss maximum. The total losses per cycle of twisted 19-filament tapes with and without the $BaZrO_3$ barriers are compared in a ΔB range where the coupling losses dominate. For the tape without barriers and peak-to-peak amplitude $\Delta B = 9.3$ mT, $Q_{max} = 7.4$ J m^{-3} while for the tape with barriers and $\Delta B = 4.6$ mT, $Q_{max} = 1.7$ J m^{-3}. Due to the presence of the barriers the frequency at the loss maximum, which is proportional to the transverse resistivity, increases approximately by a factor of 10.

References

[1] Maeda H, Tanaka Y, Fukutomi M and Asano T 1988 *Japan. J. Appl. Phys.* **27** L209
[2] Grivel J C, Grindatto D P, Grasso G and Flükiger R 1998 *Supercond. Sci. Technol.* **11** 110
[3] Majevski P 1997 *Supercond. Sci. Technol.* **10** 453
[4] Majevski P 1994 *Adv. Mater.* **6** 460
[5] Schulze K, Majevski P, Hettich B and Petzow G 1990 *Z. Metallkde* **81** 836
[6] Sastry P and West A R 1994 *J. Mater. Chem.* **4** 647
[7] Endo U, Koyama S and Kawai T 1989 *Japan. J. Appl. Phys.* **28** L190
[8] Sato K, Shibuta N, Mukai H, Hikata T, Ueyama M and Kato T J 1991 *Appl. Phys.* **70** 6484
[9] Dou S X, Liu H K, Zhang Y L and Blan W H 1991 *Supercond. Sci. Technol.* **41** 203
[10] Cava R J *et al* 1988 *Physica* C **153–155** 560
[11] Kaesche S, Majevski P and Aldinger F 1996 *Z. Metallkde* **87** 587
[12] Ikeda S, Ichinose A, Kimura T, Matsumoto T, Maeda H, Ishida Y and Ogawa K 1988 *Jpn. J. Appl. Phys.* **27** L999
[13] Jeremie A, Alami-Yadri K, Grivel J C and Flükiger R 1993 *Supercond. Sci. Technol.* **6** 730
[14] Grivel J C, Jeremie A, Hensel B and Flükiger R 1993 *Supercond. Sci. Technol.* **6** 725
[15] Yamada Y, Obst B and Flükiger R 1991 *Supercond. Sci. Technol.* **4** 165
[16] Grasso G, Perin A, Hensel B and Flükiger R 1993 *Physica* C **217** 335
[17] Grasso G, Perin A and Flükiger R 1995 *Physica* C **250** 43
[18] Grivel J C and Flükiger R 1994 *Physics* C **229** 177

[19] Grasso G, Jeremie A and Flükiger R 1995 *Supercond. Sci. Technol.* **8** 827
[20] Thomas, G G 1970 *Production Technology* (Oxford: Oxford University Press) p 63
[21] Kessler J, Blüm S, Wildgruber U and Goldacker W 1996 *J. Alloys. Comp.* **195** 511
[22] Kwasnitza K and Clerc S 1994 *Adv. Cryogen. Eng.* **40** 53
[23] Li Q, Brodersen K, Hjuler H A and Freltoft T 1993 *Physica* C **217** 360
[24] Sato K, Hikata T, Mukai H, Ueyama U, Shibuta N, Kato T, Masuda T, Nagata M, Iwata K and Mitsui T 1991 *IEEE Trans. Magn.* **27** 1231
[25] Sato K, Ohkura K, Hayashi K, Ueyama M, Fujikami J and Kato T 1996 *Proc. Int. Workshop on Advances in High Magnetic Fields, Physica* B **216** 258
[26] Grindatto D P, Hensel B, Grasso G, Nissen H U and Flükiger R 1996 *Physica* C **271** 155
[27] Bulaevskii L N, Clem J R, Glazman L I and Malozemoff A P 1992 *Phys. Rev.* B **45** 2545
[28] Hensel B, Grasso G and Flükiger R 1995 *Phys. Rev.* B **21** 15456
[29] Dimos D, Chaudari P and Mannhart J 1990 *Phys. Rev.* B **41** 4038
[30] Amrein T, Schultz L, Kabius B and Urban K 1995 *Phys. Rev.* B **51** 6792
[31] Kaneko T, Kobayashi S, Hayashi K and Sato K 1997 *Advances in Superconductivity IX* **vol 2** p 907
[32] Hensel B, Grivel J C, Jeremie A, Perin A, Pollini A and Flükiger R 1993 *Physica* C **205** 329
[33] D'yachenko A I 1994 *Physica* C **213** 167
[34] Grivel J C and Flükiger R 1996 *J. Alloys Compd* **241** 127
[35] Grasso G, Marti F, Castellazzi S, Ferdeghini C, Cimberle M R, Putti M, Siri A S, Reimann N, Dutoit B and Flükiger R *Advances in Superconductivity VIII* **vol. 2** p 851
[36] Campbell A M 1982 *Cryogenics* **22** 3
[37] Kwasnitza K and Bruzzone P 1986 *Proc. ICEC* **11** 741
[38] Huang Y B and Flükiger R 1998 *Physica* C **294** 71
[39] Goldacker W, Quilitz M, Obst B and Eckelmann H 1999 *IEEE Trans. Appl. Supercond.* **9** 2155
[40] Zhang P X, Inada R, Uno K, Oota A and Zhou L 2000 *Supercond. Sci. Technol.* **13** 1505

B3.2.3
Processing of high T_c conductors: the compound Tl(1223)

Emilio Bellingeri

Introduction

The Tl(1223) compound is one of the most promising materials for the preparation of superconducting cables in view of its high critical temperature (120 K) while its irreversibility field is also very high. When Tl is partially substituted by Cu this material presents the highest irreversibility line reported up to now [1].

C3 A1.3

However, many difficulties prevent the application of this material. Some of them are due to the Tl volatility and toxicity that makes the synthesis of this material relatively complicated.

Specific synthesis paths have been developed to overcome this problem. One of the most effective is the use of a moderate gas pressure (50 bar) during the high temperature step of the phase formation. This pressure was shown to be enough to prevent significant Tl losses up to more than 1000°C, but does not require extremely complicated or expensive techniques that can prevent eventual industrial application.

Since an exhaustive discussion on usual preparation method is given in chapter C3, this chapter will be mainly focused on high pressure synthesis and on fluorine substitutions, that, similar to more conventional cation substitution, enlarge the phase stability domain and thus have a positive effect during the phase preparation.

C3

B3.2.3.1 Details of the phase diagram and phase formation

One of the more serious difficulties in processing Tl based superconductors is the high volatility of Tl and Tl oxides at the temperatures required for synthesis and sintering of the superconducting phases. Furthermore, Tl is highly toxic which adds a significant safety problem.

B2.2.4

There are basically three ways of synthesis of Tl based superconductor [2]: in a sealed (or well closed) crucible (e.g. [3]); in a two zone furnace (e.g. [4]; figure B3.2.3.1), providing a Tl oxide source or in a (moderately) high pressure furnace preventing the evaporation of Tl oxides (e.g. [5]). For a comprehensive discussion on the Tl based superconductors phase diagrams refer to chapter C3 of this handbook.

C3

The diagram in figure B3.2.3.2 [6] shows the stability of different Tl containing phases as a function of $p(O_2)$ and p(TlO) partial pressures. The composition is fixed at Ba:Ca:Cu = 2:2:3 and the diagram is drawn for temperatures a few degrees lower than the melting point. The melting temperatures depend on $p(O_2)$ so the diagram does not represent a section at $T =$ constant.

The single TlO layer phase Tl(1223) transforms in Tl(2223) with increasing p(TlO). The region of stability of this latter phase is fairly narrow. For a further increase in p(TlO) Tl(2212) is found. All these transformations are reversible; and thus at thermodynamical equilibrium—but with a remarkable difference in the kinetics of formation—the phases with three CuO_2 layers [Tl(1223) and Tl(2223)] have slower growth kinetics than the double CuO_2 layer compounds [Tl(1212) and Tl(2212)].

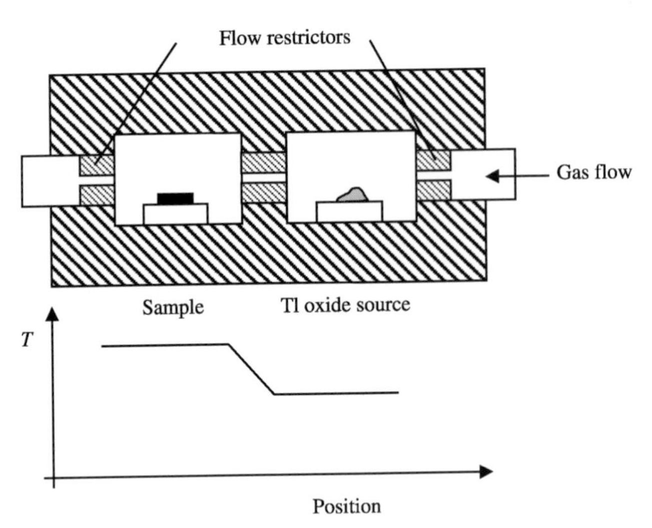

Figure B3.2.3.1. Two zone thallination of precursor samples by the use of a flowing carrier gas. The schematic temperature profile is also shown.

In the case of the Sr homologues, the diagrams are different; the double TlO layer members of the family are absent. In this case, the phase with the larger stability domain is Tl(1212) so it is necessary to pass through this phase for the formation of Tl(1223).

C3 The thermodynamic stability of the Tl(1223) phase can be improved by appropriate cationic and anionic substitutions. The compounds $Tl_{0.5}Pb_{0.5}Sr_{1.6}Ba_{0.4}Ca_2Cu_3O_x$, $Tl_{0.7}Pb_{0.2}Bi_{0.2}Sr_{1.6}Ba_{0.4}Ca_2Cu_3O_x$ (also with $Sr_{1.8}Ba_{0.2}$) and $Tl_{0.78}Bi_{0.22}Sr_{1.6}Ba_{0.4}Ca_2Cu_3O_x$ have a large stability domain and show a very high superconducting irreversibility field: for these reasons they are widely used in the preparation of tapes [7–9].

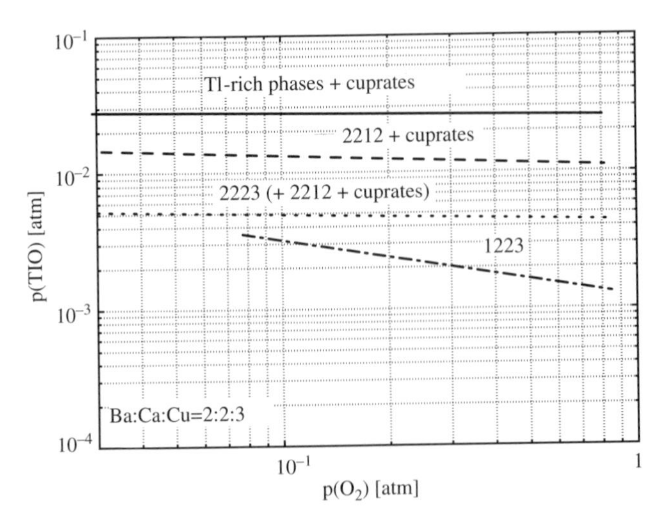

Figure B3.2.3.2. Phase equilibrium diagram for superconducting Tl based compounds at a temperature just below melting (the diagram is not a section at constant temperature) in the $p(O_2)$,p(TlO) plane, for a precursor composition Ba:Ca:Cu = 2:2:3. Non-superconducting cuprates are not indicated [6].

During synthesis by means of a two zone furnace, p(TlO) is a parameter that can be controlled by setting the temperature of the thallium oxide source (figure B3.2.3.1), while in a sealed crucible it is determined (at fixed reaction temperature) by the ratio between the volume of the Tl_2O_3 and the volume of the crucible.

B2.2.1

If the synthesis is performed in open or quasi-closed systems, the starting composition usually has an excess of thallium oxide to compensate for the Tl losses during the treatment. During the reaction one moves down, with time, in the diagram in figure B3.2.3.2 through the different phases and the final product can be controlled by accurate setting of the duration of the thermal treatment.

The synthesis path is completely different if the reaction is performed under high isostatic pressure. Experiments show that a moderately high pressure of 50 atm is already sufficient to prevent significant Tl losses up to 1100°C. In this case, the phase formation is mainly realised through a solid state diffusion reaction [10]. In figure B3.2.3.3, the stability region of the Tl(1223) phase is represented in a temperature –oxygen partial pressure diagram ($P_{He} = 50$ bar). Tl(1212) is the more stable phase at temperatures both lower and higher than the Tl(1223) stability region. Since the formation kinetics of the double CuO layer phase is very fast, under normal conditions, Tl(1223) is formed from Tl(1212). In some cases, with a premelting step at very high temperature (1100°C), Tl(1223) can be formed following another path, avoiding the formation of Tl(1212) [10].

In order to improve the grain connectivity, particular attention is usually paid to the preparation of Tl(1223) at a temperature close to the melting point.

Differential thermal analysis (DTA) of $Tl_{0.7}Pb_{0.2}Bi_{0.2}Sr_{1.6}Ba_{0.4}Ca_2Cu_3O_x$ reveals three endothermic peaks at 945, 963 and 981°C. It is known that above 945°C decomposition of Tl(1223) takes place. Above 945°C the amount of secondary phases, especially (Ca,Sr)-cuprates and plumbates, increases with increasing temperature. At 963°C clear indications of Tl(1212) appear, in agreement with the observation that (Ca,Sr)- and Cu-rich phases have precipitated from the original Tl(1223) phase. The X-ray diffraction pattern at 981°C indicates that the main phase is Tl(1212), and evidence for the occurrence of melting is observed.

B2.1

Basic information about a material's microstructural development, texture and densification can be obtained by the analysis of the grain size with respect to temperature and time. The grain growth

D1.7.1

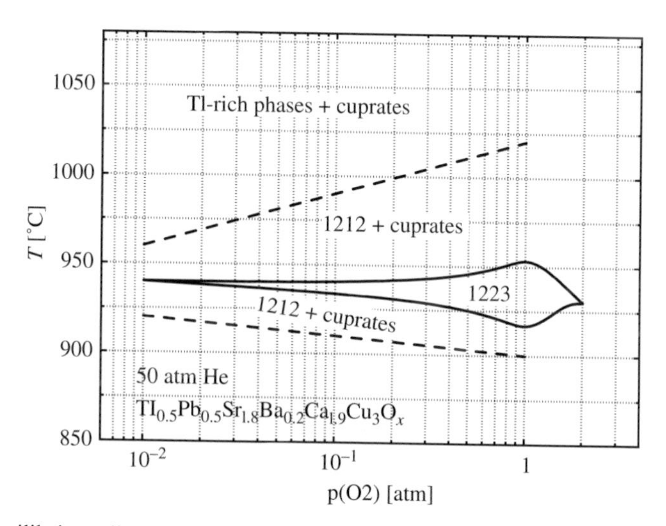

Figure B3.2.3.3. Phase equilibrium diagram for the TlSrCCO system where the double Tl layer compounds are absent. The diagram in the $p(O_2)$,T plane is drawn for a pressure of 50 bar of He.

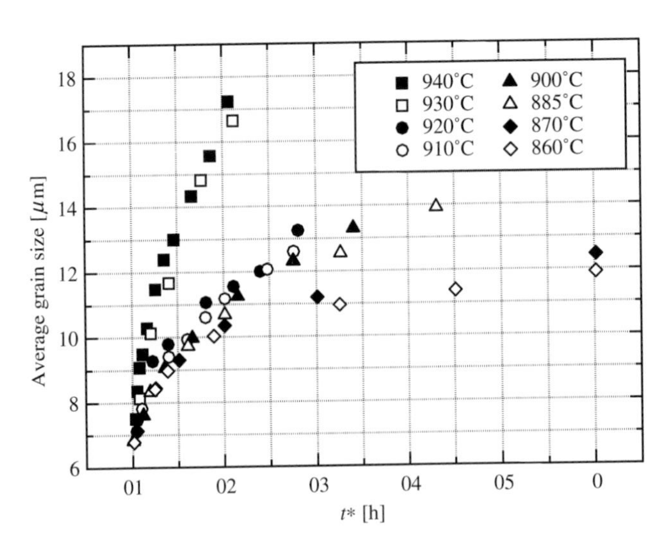

Figure B3.2.3.4. Average grain size versus annealing time at different temperatures for the composition $(Tl_{0.6}Pb_{0.2}Bi_{0.2})(Sr_{1.8}Ba_{0.2})Ca_2Cu_3O_x$ [11].

B2.2.4 D1.3 kinetics during the thermal treatment of polycrystalline Tl(1223) in flowing oxygen was examined by measuring grain sizes by SEM at various times and temperatures between 860 and 940°C [11]. Figure B3.2.3.4 indicates that the average grain size increases with temperature as well as with time. For the interpretation of the growth kinetics, the isothermal grain-growth-model equation, which has the form

$$\sigma = At^n \tag{B3.2.3.1}$$

was used, where σ denotes the average grain size, A a constant that exhibits Arrhenius temperature dependence, t the time and n the grain growth exponent. With respect to the initial grain size σ_0 the following relations were applied

$$\sigma^m - \sigma_0^m = at^* \tag{B3.2.3.2}$$

and

$$\sigma = (at^* + \sigma_0^m)^n \tag{B3.2.3.3}$$

where $m = n^{-1}$ and σ_0 denotes the grain size at $t^* = 0$. Considering equations (B3.2.3.1) and (B3.2.3.3) the grain-growth-model equation can be written as

$$\sigma = A(t^* + t_0)^n \tag{B3.2.3.4}$$

where t_0 corresponds to the time needed to grow grains of size σ_0, and $t = (t^* + t_0)$. By defining $\sigma_0^m = A^m t_0$ and $a = A^m$, the parameter t_0 can be calculated from the initial grain size with the relation $t_0 = (\sigma_0^m)/a$ by a fit of equation (B3.2.3.2). Taking the logarithm of equation (B3.2.3.4), the slope yields the exponent n and the intersection with the ordinate axis the rate constant A.

In most ceramic systems, the grain growth exponent, $n < 0.5$ [12, 13] due to the existence of pores, inclusions of secondary phases, grain segregation and non-stoichiometry, which increase the energy necessary to move grain boundaries and inhibit grain growth. Another factor that may highly influence grain growth is the presence of a liquid phase. It has been observed for other HTSC-ceramics [14, 15], B2.2.1

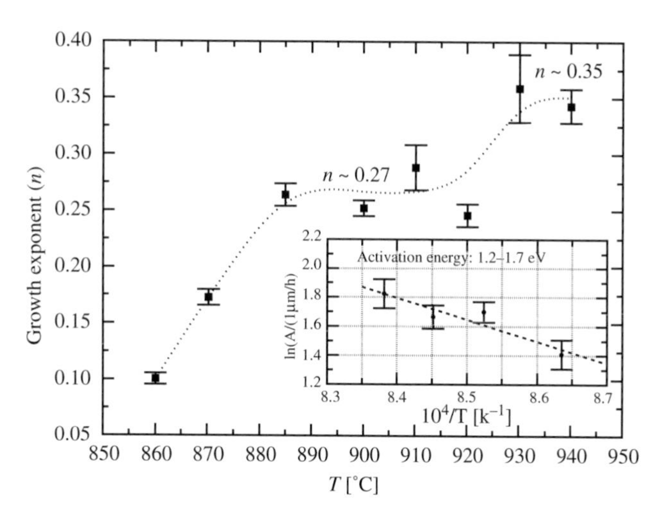

Figure B3.2.3.5. Temperature dependence of the grain growth exponent. The inset presents Arrhenius temperature dependence of the rate constant A, corresponding to an activation energy for the grain boundary motion of 1.2–1.7 eV [11].

that grain growth follows a cubic law ($n = 1/m = 1/3$), which corresponds to the coexistence with the liquid. Figure B3.2.3.5 shows the dependence of the exponent n on the annealing temperature. With increasing temperature the exponent n increases rapidly to a value of about 0.27 at 885°C, remains nearly constant over a wide temperature range and increases again to 0.35 at 930°C. Above 885°C, n is between the ideal value of 1/5, which corresponds to a non-liquid process and 1/3, which indicates a growth from a liquid. This indicates a mixed process, probably with a very small amount of liquid involved. The fact that the value is lower than 1/3 might also be attributed to the presence of highly anisotropic grain boundary energies as discussed by Grest *et al* [16], which would favour the formation of plate-like grains during sintering. The second shift of n at about 930°C is related to the onset of peritectic/eutectic liquid, which might correspond to a very small endothermic peak observed at 923°C by DTA. B2.1

The incomplete formation of Tl(1223), as evaluated from X-ray diffraction measurements, and the existence of an eutectic between $BaCuO_2$ and CuO at approximately 890°C may explain the low formation temperature of a liquid phase which appears during annealing. Since no decomposition behaviour could be detected by DTA measurements at this temperature, one may conclude that a very small amount of liquid between particles is sufficient to favour a nearly cubic dependence of the grain growth. This study is in agreement with other evidence of melting [17–19]. D1.1.2

From the Arrhenius temperature dependence of the rate constant A [$A = A_0 \exp(-Q/KT)$], the activation energy Q for grain boundary motion between 885 and 920°C, corresponding to $n \approx 0.27$, was determined to be in the range of 1.2–1.7 eV, which is considerably lower than values reported for Y-123 (8–11 eV [14]) and Bi-2223 (5–8 eV [20]). Since grain growth simulations suggest that the grain growth exponent is related to the grain boundary energy it can be concluded that the growth rate and hence the grain boundary motion of Tl(1223) are much faster than for Y-123 and Bi-2223.

B3.2.3.2 The influence of fluorine on the phase stability [21]

Several groups have succeeded in preparing nearly single-phase Tl(1223) materials by partly substituting cations in $TlBa_2Ca_2Cu_3O_{9-\delta}$ (e.g. Pb and Bi on the Tl site, Sr on the Ba site) [8, 22]. Precise studies on C3 the influence of substitution are facilitated by a synthesis under high isostatic gas pressure, which avoids losses of volatile components [23, 24].

Partial substitution of oxygen by fluorine has been shown to have positive effects on some superconducting systems, e.g. by making it possible to prepare superconducting $Sr_2CuO_2F_{2+\delta}$ at ambient pressure [25]. Studies on Hg-based superconductors doped by fluorine have also led to interesting results [26] and new superconducting phases were obtained in the BiSSCO system [27]. Fluorine may be introduced by annealing under fluorine atmosphere or from solid fluorides. Fluoride precursors were also used for the preparation of different Tl-based superconductors [28, 29].

Three preparation routes are considered:

Path I: Already reacted Tl,Pb(1223) is annealed at low temperature (max 300°C) in the presence of a fluorine gas. This gas can be produced for example by the decomposition of an ammonium salt (e.g. NH_4HF_2). This method has proven to be very efficient to partially substitute oxygen by fluorine, but the process is difficult to control.

Path II: Different precursors are prepared from commercial simple oxides (Tl_2O_3, PbO, SrO, CaO and CuO), replacing each time one oxide (exceptionally two oxides) by the corresponding fluoride [TlF (and PbF_2), SrF_2, CaF_2 or CuF_2]. The cation ratio in the nominal composition was fixed at $Tl_{0.6}Pb_{0.5}Sr_2Ca_{1.9}Cu_3O_{9-x-\delta}F_x$. The mixtures of powders were pelletized and heated in a high-pressure furnace at 940°C (3 h, 50 bar He and 1 bar O_2). When starting with SrF_2 the reaction is unsatisfactory since the majority of the fluoride is found unreacted in the final product. Using CuF_2, CaF_2 or TlF (and PbF_2), the Tl(1223) phase forms, and the cell parameters are found to be slightly smaller for the F-containing sample ($a = 3.812$, $c = 15.245$ Å) than for the F-free sample ($a = 3.813$, $c = 15.263$ Å). Starting from TlF and PbF_2, or from CuF_2, partial melting took place at the reaction temperature (940°C) and relatively large Tl(1223) crystals (up to 100 μm) are observed in the sample.

B2.1 *Path III*: The third reaction path consists of two steps. In the first step a Tl- and Pb-free precursor is prepared by calcination of a mixture of high-purity Sr, Ba and Ca carbonates and Cu oxide at 900°C (flowing oxygen) and 980°C (air) for a total of 48 h. In the second step appropriate amounts of Tl oxide and -fluoride and Pb oxide are added to obtain the nominal composition $Tl_{0.6}Pb_{0.5}Sr_{1.8}Ba_{0.2}Ca_{1.9}Cu_3O_{9-x-\delta}F_x$. Powders are prepared by heating the pelletized mixtures at high isostatic pressure (900–970°C, 3 h, 50 bar He and 1 bar O_2). Varying the amount of TlF in the precursor mixture, the highest phase purity is observed for a $TlO_{1.5}$:TlF mole ratio 1:5. For this ratio the purity is comparable to that observed for the corresponding F-free samples, whereas starting not only from a higher, but also from a lower TlF content, larger amounts of impurity phases are formed. Studies on the reaction conditions showed that, as expected, the presence of fluorine decreases the formation temperature of Tl(1223). Already at 900°C (3 h, 50 bar), 90% pure samples can be produced, the amount of SrF_2 being significantly reduced with respect to that observed in samples prepared by path II. The upper temperature limit for high-purity samples is surprisingly high. As can be seen from figure B3.2.3.6, where D1.1.2 X-ray diffraction diagrams of F-containing Tl(1223) powders prepared at different temperatures are compared, even powders reacted at 960°C contained a large amount of the Tl(1223) phase. The samples started to melt at 955°C, which is approximately 10–15°C lower than for F-free Tl(1223) (970°C). To obtain high-purity powders of the latter it is necessary to decrease the reaction time to a few minutes, which has a negative effect on the grain size. The presence of fluorine was found to extend the temperature range favourable for Tl(1223) grain growth, i.e. the region where Tl(1223) is in equilibrium with a liquid phase. As for F-free samples, the reaction time also had an influence on the phase purity. Reaction times longer than 3 h at 955°C produced a progressive decomposition of the Tl(1223) phase and we conclude that, with longer reaction times, Tl(1223) is no longer in equilibrium with a liquid phase.

B2.2.1 Particular attention was paid to reactions at temperatures close to the melting point of the superconducting oxide. Contrary to what was observed for F-free samples, the Tl(1223) phase was still present in F-containing samples even when substantial melting had taken place. Tl(1212) was detected

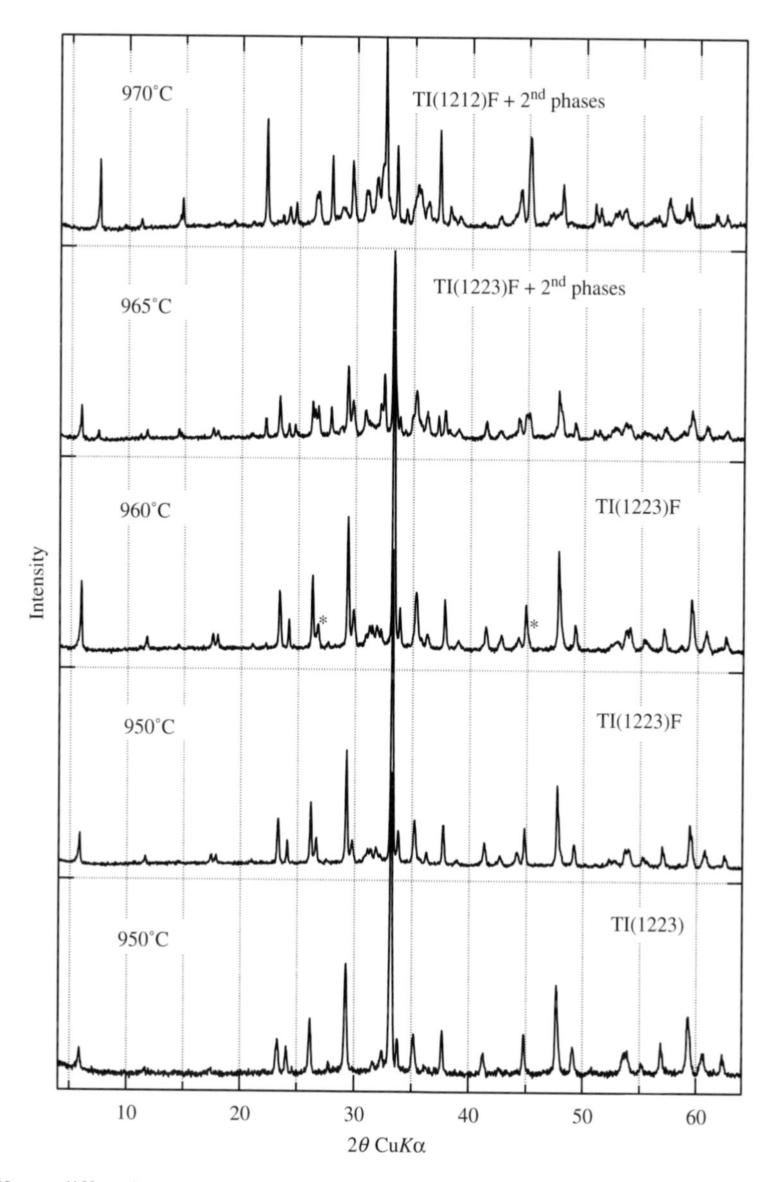

Figure B3.2.3.6. X-ray diffraction spectra of F-containing Tl(1223) powders prepared at different temperatures (3 h, 50 bar He/1 bar O_2). Asterisks indicate peaks of SrF_2 [21].

both when the reaction was incomplete (low temperature) and when Tl(1223) was partly decomposed (high temperature). The narrow temperature region between the starting point of melting and decomposition of the Tl(1223) phase observed for F-free samples was found to be enlarged for F-containing ones, allowing to obtain high-purity samples with larger grains. However, the colonies of large plate-like grains were systematically separated by secondary phases and/or the samples contained cracks. These results are summarised in figure B3.2.3.7, where the Tl(1223) 'single' phase regions are plotted in a temperature, time plane. The dashed lines surround the region where the F-free Tl(1223) phase can be formed with purity greater then 90%, whereas the same region for F doped Tl(1223) is enclosed in the solid lines. The samples in which a partial melting is observed are framed. Figure B3.2.3.8(*a*) and (*b*) show SEM

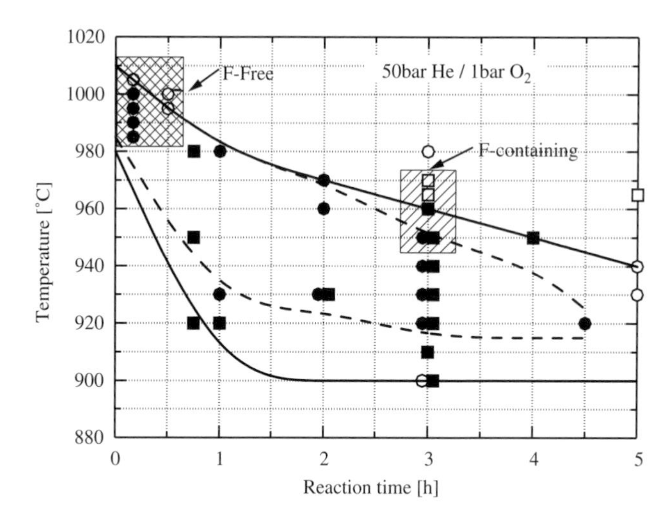

Figure B3.2.3.7. Purity of Tl,Pb,Bi(1223) (● > 90%, ○ < 90%) and Tl,Pb(1223)F (■ > 90%, □ < 90%) samples prepared at different reaction temperatures versus the reaction time. Regions where melting took place are framed [21].

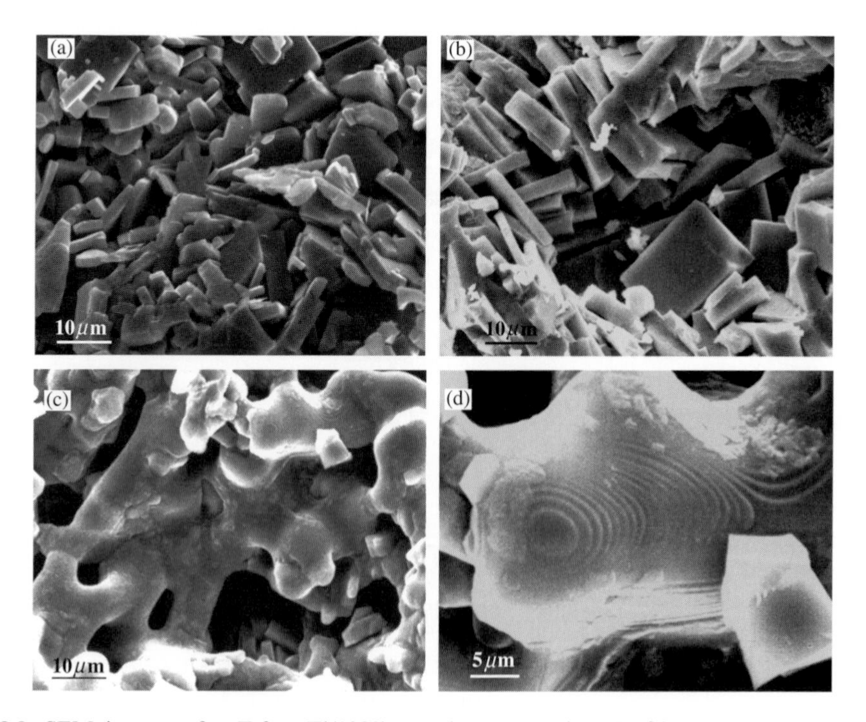

Figure B3.2.3.8. SEM images of a F-free Tl(1223) powder prepared at 940°C (*a*) and F-containing Tl(1223) powders prepared at 940°C (*b*) and 965°C (*c*, *d*) (3 h, 50 bar He / 1 bar O_2) [21].

images of Tl(1223) samples produced at 940°C. The grains of F-free Tl(1223) samples (910–940°C) are small and of irregular shape, whereas those of F-doped samples prepared at 900–955°C were larger ($7 \times 7 \times 2\ \mu m^3$) and had a pronounced plate-like shape with well defined edges. As can be seen from figure B3.2.3.8(*c*) and (*d*), reaction temperatures of 960–970°C produced grains of irregular shape with clear traces of melting.

Chemical analysis confirms that F is present in the sample at the end of the thermal treatments, giving an overall composition $Tl_{0.57}Pb_{0.50}Sr_{1.67}Ba_{0.20}Ca_{1.95}Cu_3O_yF_{0.47}$ that corresponds well to the starting stoichiometry. EPMA measurements on Tl(1223) grains show that fluorine is effectively D1.5 incorporated in the structure (0.4 fluorine atoms per formula unit).

Structural refinements carried out on X-ray diffraction data from both F-free and F-containing D1.1.2 Tl(1223) confirm previously reported positions and occupations of the cation sites [22]. Small amounts of Tl were refined on the Ca site, whereas the Tl site was found to be displaced from the ideal position on the four-fold rotation axis along the short cell vectors (~ 0.3 Å). Refinements on neutron diffraction data from F-free samples showed that the O site in the TlO layer is off-centered and located at 0.51 Å from the ideal position (see also [30]). No extra anion sites were found in the structure of F-doped Tl(1223), however, the anion site in the TlO layer is displaced only by 0.22 Å, increasing the Tl-anion in-layer distances. This leads us to believe that the fluorine atoms partly occupy the anion site in the TlO layer. The F:O ratio on this site is assumed to be the same as the Tl:Pb ratio (1:1) on the cation site in the same layer. Complete crystallographic parameters of the two phases are listed in table B3.2.3.1.

The superconducting transition temperatures, determined from AC susceptibility measurements, D2.5 did not differ for the two kinds of sample and no systematic trends in the slight variations of T_c were

Table B3.2.3.1. Crystallographic parameters for $Tl_{0.6}Pb_{0.5}Sr_{1.8}Ba_{0.2}Ca_{1.9}Cu_3O_{9-\delta}$ (regular) and $Tl_{0.6}Pb_{0.5}Sr_{1.8}Ba_{0.2}Ca_{1.9}Cu_3O_{8.5-\delta}F_{0.5}$ (bold) from neutron diffraction data. Space group *P*4/*mmm*, $a = 3.8220(1)$, $c = 15.3690(2)$ Å and $a = 3.8243(1)$, $c = 15.3560(4)$ Å, respectively

Site	*X*	*Y*	*Z*	*B* (Å^2)
0.5Tl + 0.5Pb[a] in 4(*m*)	0.068(4)	0	1/2	1.0
	0.081(5)	**0**	**1/2**	**1.0**
1.42Sr + 0.38Ca + 0.20Ba[b] in 2(*h*)	1/2	1/2	0.3284(5)	0.6(2)
	1/2	**1/2**	**0.3221(6)**	**0.7(3)**
2Cu in 2(*g*)	0	0	0.2118(5)	0.7(1)
	0	**0**	**0.2159(6)**	**0.7(1)**
1.92Ca + 0.08Tl[b] in 2(*h*)	1/2	1/2	0.1083(8)	0.6(2)
	1/2	**1/2**	**0.1105(8)**	**0.4(3)**
1Cu in 1(*a*)	0	0	0	0.7
	0	**0**	**0**	**0.7**
1O(F) in 4(*o*)	0.634(7)	1/2	1/2	0.5
	0.558(6)	**1/2**	**1/2**	**0.5**
2O in 2(*g*)	0	0	0.3716(7)	0.4(1)
	0	**0**	**0.3645(9)**	**0.7(2)**
4O in 4(*i*)	0	1/2	0.2081(5)	0.4
	0	**1/2**	**0.2085(5)**	**0.7**
2O in 2(*f*)	0	1/2	0	0.4
	0	**1/2**	**0**	**0.7**

[a] Tl/Pb ratio from nominal composition.
[b] Sr/Ca/Ba and Ca/Tl ratios from X-ray diffraction data on F-free sample.

observed with respect to the fluorine content or the reaction temperature. Superconductivity was observed even for the highest amount of fluorine introduced (about one atom per formula unit). The two samples for which crystallographic data are given here showed superconducting onset temperatures of 116 (F-free) and 114.5 K (F-containing). These values suggest that the doping state of the CuO_2 superconducting layers, and consequently the environment of the Cu atoms, has not been changed by the incorporation of fluorine. This observation further supports the idea that the fluorine atoms are located in the TlO layer, the partial substitution of O^{2-} by F^- being equilibrated by a partial reduction of thallium from Tl^{3+} to Tl^+.

B3.2.3.3 Reaction heat treatments for optimising J_c

The preparation methods of Tl(1223) conductor can be divided into two basic approaches: the closed and the open one. The closed-system process is typically the powder in tube (PIT) method, where the ceramic powder is enclosed by a metallic sheath and sealed. In this way, loss of Tl during processing is reduced. In the open-system process, the Tl is generally introduced into the precursor—deposited on a substrate — in a separate step, by inducing an appropriate partial pressure of Tl vapour using an outside source. The different preparation approaches are schematically shown in figure B3.2.3.9.

B3.1
A1.3

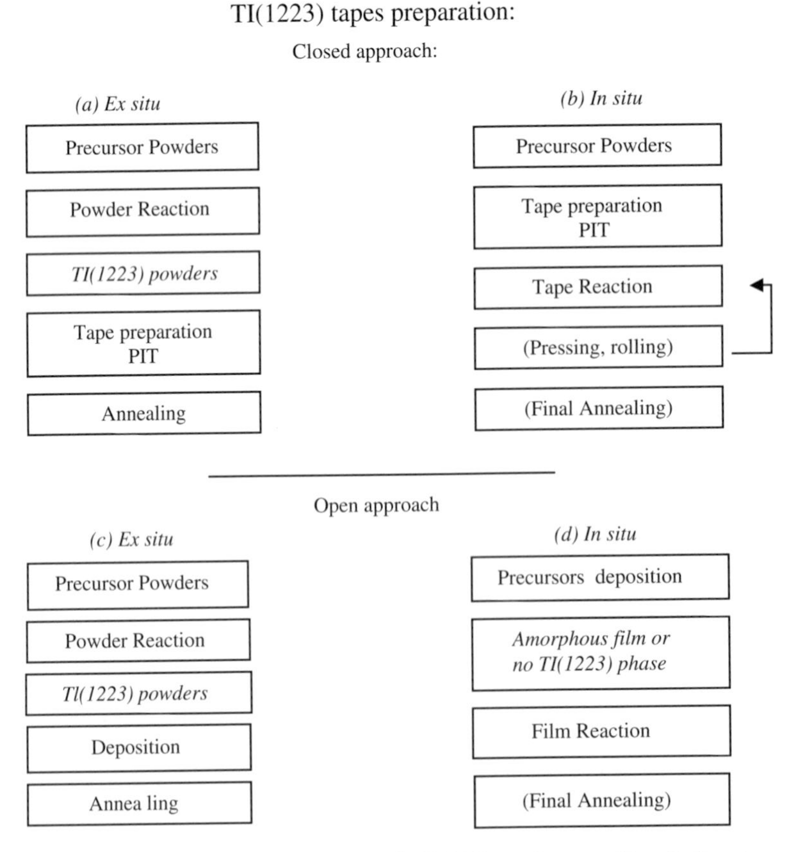

Figure B3.2.3.9. Different approaches for the synthesis of Tl(1223) conductors. Facultative steps are in parenthesis.

Figure B3.2.3.10. Transverse cross-sections of Ag-sheathed Tl,Pb,Bi(1223) wires and tapes: 7- (diameter 1.25 mm) and 259-filament (∅1.5 mm) wire, a 37-filament (thickness 140 μm) tape, all prepared by the PIT method, and a 3-layer (120 μm) tape prepared by electrophoretic deposition.

Cross-sections of different of kinds of tapes and wires are shown in figure B3.2.3.10

B3.2.3.3.1 Closed approach

This approach is based on the PIT method. In brief, a Ag (or Ag alloy) tube is filled with the ceramic powders and then swaged, drawn and rolled to the final tape form. If the reaction that forms the Tl (1223) phase is performed before the tape preparation, the method is referred to as *ex situ* (figure B3.2.3.11); if the phase is formed inside the tape it is referred to as *in situ* (figure B3.2.3.12). B3.1

In both cases a Tl, Pb and Bi free precursor is generally prepared by calcination (~ 1000°C 20 h, O_2 or air) of Ba-, Ca- and Sr carbonates with Cu oxide.

The following three compounds are usually found in different proportions in the precursor powders after calcination: $(Sr,Ca)CuO_2$, $(Ca,Sr)_2CuO_3$ and '$BaCuO_2$'; the carbonates are fully decomposed. C3

The compositions: $Tl_{0.7}Pb_{0.2}Bi_{0.2}Sr_{1.8}Ba_{0.2}Ca_{1.9}Cu_3O_x$, $Tl_{0.6}Pb_{0.5}Sr_{1.8}Ba_{0.2}Ca_{1.9}Cu_3O_x$ and $Tl_{0.78}Bi_{0.22}Sr_{1.6}Ba_{0.4}Ca_2Cu_3O_x$ are commonly employed.

Different reaction paths are used to treat these powders and to form the superconducting phase. The reaction of the pelletised powders under ~ 50 bar of gas pressure (50 bar He/1 bar O_2) at a temperature of 900–940°C for a few hours is shown to be one of the the most suitable methods for this purpose, allowing one to start with the desired composition without Tl excess.

The reacted powders are used to prepare *ex situ* Ag-sheathed tapes. Unreacted powders of the same composition can be used for the *in situ* preparation of Ag-alloyed sheathed tapes. Under 1 bar of oxygen pressure, Ag melts at 931°C, i.e. below the temperature typically used to synthesize the Tl(1223) phase. To obtain sheaths that are more resistant to high temperature treatments, Ag is alloyed with Au or Pd. As an example the addition of 20 wt% Au or 6 wt% Pd raises the melting temperature of the alloys to 985 and 980°C, respectively. B2.1

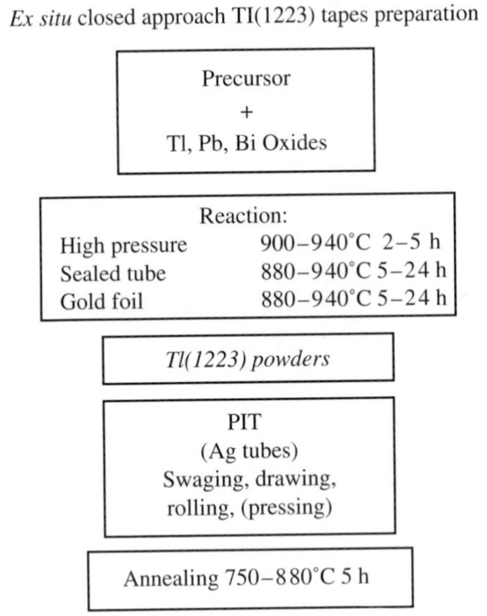

Ex situ closed approach Tl(1223) tapes preparation

Precursor + Tl, Pb, Bi Oxides	
Reaction:	
High pressure	900–940°C 2–5 h
Sealed tube	880–940°C 5–24 h
Gold foil	880–940°C 5–24 h
Tl(1223) powders	
PIT (Ag tubes) Swaging, drawing, rolling, (pressing)	
Annealing 750–880°C 5 h	

Figure B3.2.3.11. Schematic diagram of *ex situ* Tl(1223) powder reaction and tape manufacturing in the closed approach.

The metallic tubes are filled with the powders and then mechanically deformed by swaging and drawing to a diameter of 1.25–1.31 mm. Tapes are obtained by rolling to a thickness of 80–200 μm, the final width of the tapes being 2.8–3.2 mm. Multifilamentary tapes are produced by the restacking technique.

The tapes that contain unreacted powders are heated for 3–6 h at 900–960°C to produce the Tl(1223) phase, either in flowing oxygen or at a pressure up to 2 kbar [$p(O_2)$ up to 4 bar] [24].

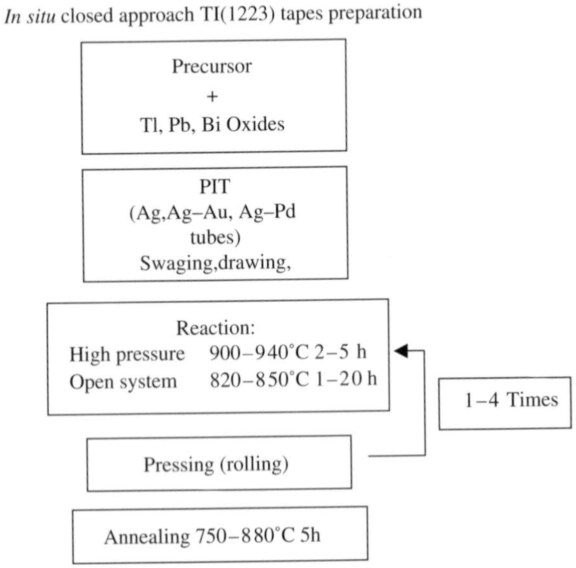

In situ closed approach Tl(1223) tapes preparation

Precursor + Tl, Pb, Bi Oxides	
PIT (Ag,Ag–Au, Ag–Pd tubes) Swaging,drawing,	
Reaction:	
High pressure	900–940°C 2–5 h
Open system	820–850°C 1–20 h
Pressing (rolling)	(repeated 1–4 Times, back to Reaction)
Annealing 750–880°C 5h	

Figure B3.2.3.12. Schematic diagram of *in situ* tape manufacturing and Tl(1223) phase formation reaction in the closed approach.

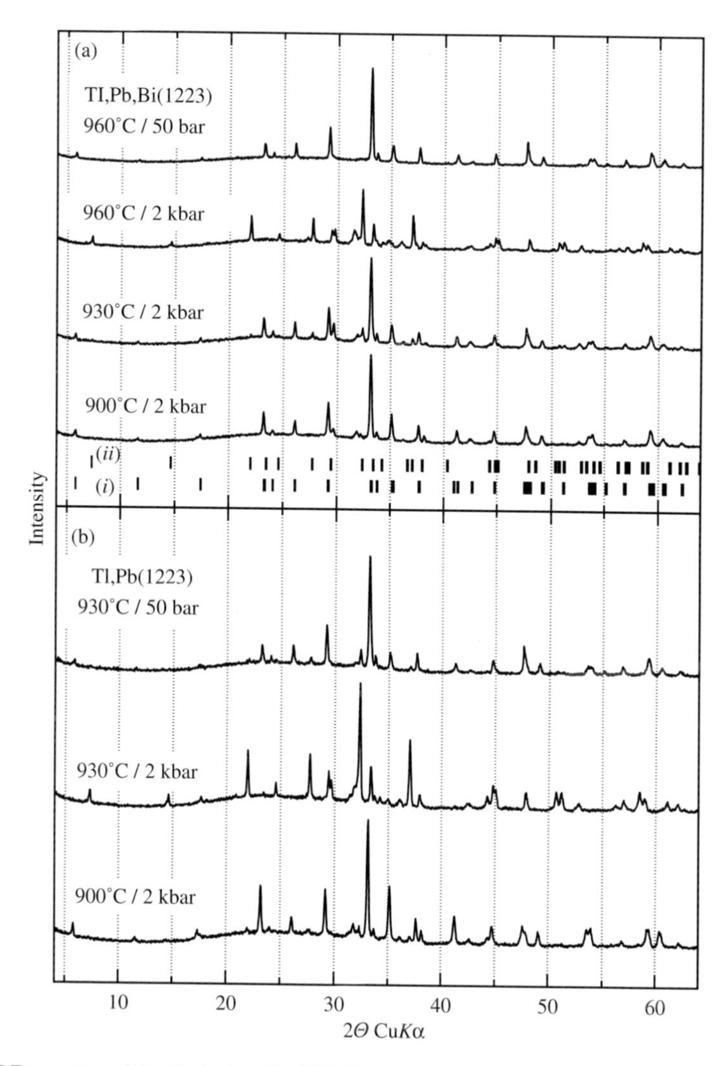

Figure B3.2.3.13. XRD spectra of Ag(Au)-sheathed Tl,Pb,Bi(1223) (*a*) and Tl,Pb(1223) (*b*) tapes. Bars indicate the peak positions of Tl(1223) (i) and Tl(1212) (ii) [24].

Both *in situ* and *ex situ* tapes are usually subjected to a further annealing step at a temperature between 750 and 850°C in flowing oxygen for a few hours. This annealing adjusts the oxygen content and is an essential requirement for the sintering of the grains in the *ex situ* reacted tapes.

The XRD diagrams of four Ag(Au)-sheathed Tl,Pb,Bi(1223) tapes, prepared at different temperatures and pressures, are compared in figure B3.2.3.13(*a*). It can be seen that most of the tapes contain a high amount of the tetragonal Tl(1223) phase. Its weight fraction in the sample prepared at 960°C and 50 bar exceeds 95%, the main impurity phase being $(Ca,Sr)_2CuO_3$ ($\sim$ 1%). More than 90% of the Tl(1223) phase was also obtained for other samples prepared at 50 bar and temperatures up to 980°C, as long as the reaction times were shorter than 3 h. In samples prepared at temperatures above 980°C, or at a higher pressure, the Tl(1212) phase appeared, accompanied by other impurities such as $(Ca,Sr)_2CuO_3$, $(Sr,Ca)CuO_2$, Ca_2PbO_4 and $BaPbO_3$, indicating a partial decomposition of Tl(1223) under these conditions.

D1.1.2
C3

D1.1.2 As can be seen from the XRD diagrams of Tl,Pb(1223) tapes shown in figure B3.2.3.13(*b*), similar features were observed for samples without Bi. The Bi-free Tl(1223) phase, however, starts to decompose at lower temperatures, in agreement with the common observation that Bi additions stabilise the Tl(1223) phase.

The chemical composition of the samples was checked by structure refinements based on XRD data, D1.5 by EDX and by wet chemical analysis. The cation ratios are in good agreement with the nominal compositions of the powders, e.g. the composition $Tl_{0.58}Pb_{0.50}Sr_{1.80}Ba_{0.20}Ca_{1.91}Cu_3O_{8.48}$ was found by chemical analysis immediately after the high-pressure synthesis for the nominal $Tl_{0.6}Pb_{0.5}Sr_{1.8}Ba_{0.2}Ca_{1.9}Cu_3O_x$ powder. The low oxygen content of the reaction product confirms the need for postannealing in oxygen.

Figure B3.2.3.14 shows SEM images of cross-sections of fractured Tl,Pb,Bi(1223) tapes subjected to different treatments. In all tapes square, brick shaped grains can be seen. The typical grain size was $\sim 5\,\mu m$ for reaction temperatures of 900–960°C and $\sim 10\,\mu m$ for higher temperatures. Increasing the reaction time from 3 to 6 h has a positive effect on the grain size, but the Tl(1223) phase starts to decompose. The grains in the Bi-free samples are slightly smaller and of irregular shape. For both compositions and all reaction temperatures, tapes prepared at higher pressures show a higher density. The texture of samples treated at 2 kbar is slightly improved with respect to the tapes prepared at 50 bar or at ambient pressure, but they also contain more impurity phases. Moreover, cracks are observed in these tapes, possibly due to the partial melting of the oxide at higher temperatures and the different dilatation coefficient of Ag and the ceramic.

Samples prepared at temperatures above 980°C contain even larger grains (25–50 μm), but the decomposition of Tl(1223) into Tl(1212) and byproducts becomes very significant.

No reaction between the powder and the sheath could be detected under the conditions applied here. Tapes prepared at 50 bar resisted the high-pressure treatment well. Holes in the sheath appeared only for tape thicknesses below 130 μm and for reaction temperatures above 970°C. However, holes were present on the surface of all tapes (including 180 μm thick ones) prepared at 2 kbar. The fact that

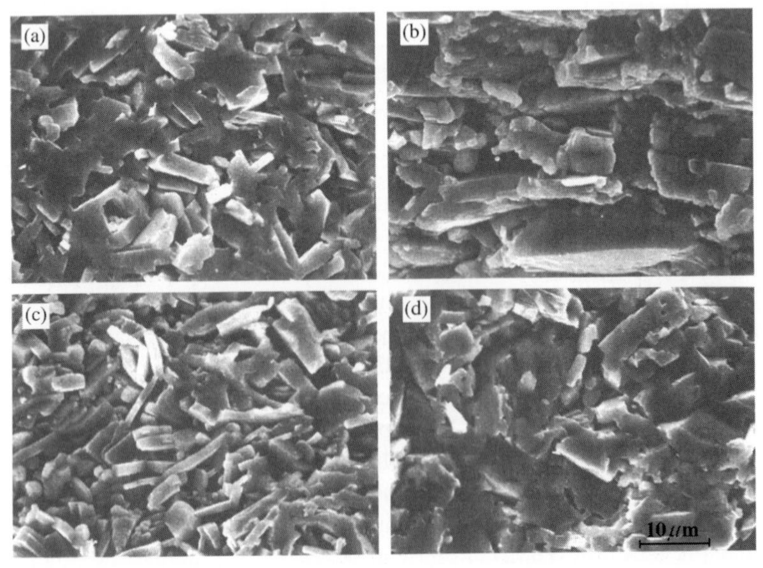

Figure B3.2.3.14. SEM images of the ceramic inside Ag(Au)-sheathed Tl,Pb,Bi(1223) tapes treated at 960°C / 50 bar for 3 h (*a*); 960°C / 50 bar for 6 h (*b*); 930°C / 50 bar for 3 h (*c*); 930°C / 2 kbar for 3 h (*d*) [24].

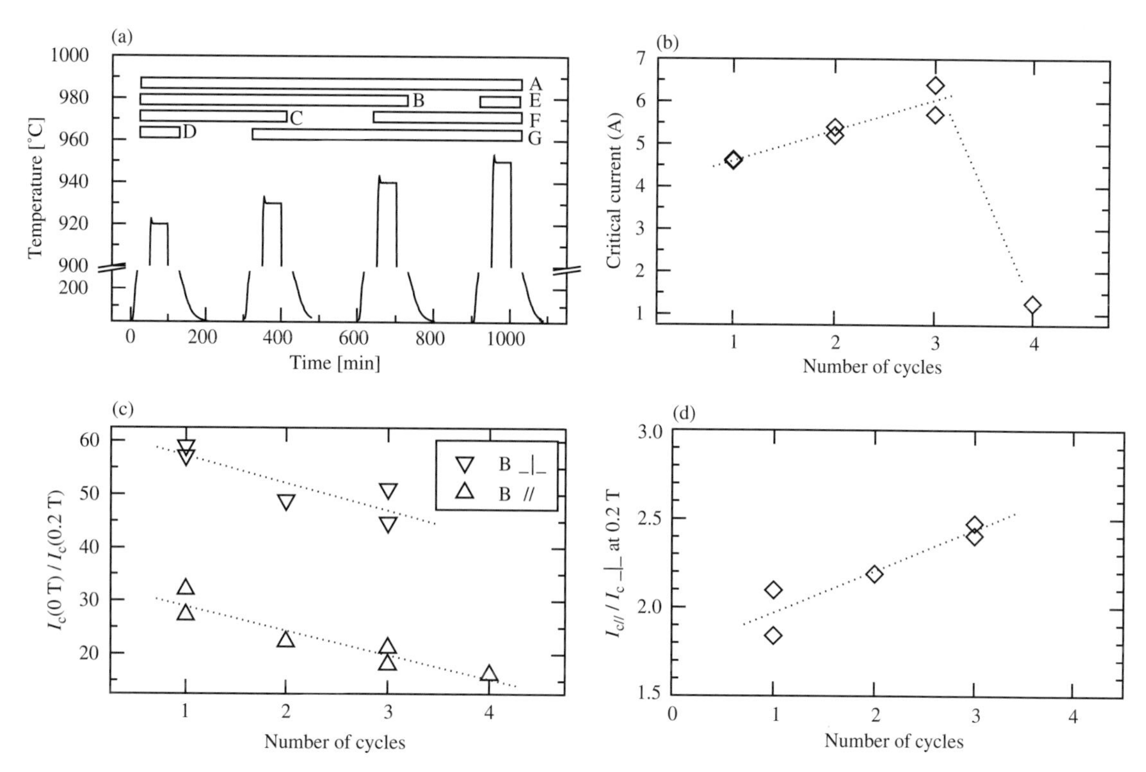

Figure B3.2.3.15. Four temperature profiles applied to 7 Tl,Pb(1223)F tapes (e.g. sample A was treated four times) (*a*) critical current at 0 T, 77 K (*b*) ratio of the critical currents with the magnetic field (0.2 T) parallel and perpendicular to the tape surface (*c*) and ratio of the critical currents at 0 and 0.2 T (*d*) versus the number of cycles [cycles 1, 2, 3 and 4 correspond to sample E(D), C(F), B(G) and A in (*a*) respectively].

holes appear in tapes containing Tl(1223), as well as in tapes where the Tl(1223) phase has partly decomposed into Tl(1212) and other phases, confirms the presence of a liquid phase during the reaction.

Ag(Au)-sheathed Tl,Pb(1223)F tapes were submitted to different heat- and deformation treatments which are summarised in figure B3.2.3.15(*a*)[31]. Up to four cycles of pressing with 1 GPa (10 kbar) followed by an *in situ* reaction at a temperature between 920 and 950°C were performed. As can be seen from figure B3.2.3.15(*b*), the critical current (77 K, 0 T) increased during the first three cycles, reaching a critical current density of 10 000 A cm^{-2}, but then dropped down. The drastic fall can be explained by the disappearance of the liquid phase, needed to 'heal' the cracks formed during the pressing, after a certain number of thermomechanical treatments. The drop of the critical current at 0.2 T with respect to its value in zero field (figure B3.2.3.15(*c*)) is the smallest (a factor of 16) after four cycles. The ratio of the critical current with the magnetic field applied parallel and perpendicular to the tape surface reaches a value of 2.5 after the third cycle (figure B3.2.3.15(*d*)), indicating that some anisotropy has been introduced during the treatments.

B3.1

B3.2.3.3.2 Open approach

Also the open approach can be divided into *in situ* and *ex situ* processes. In the *ex situ* case (figure B3.2.3.16) an already reacted (see above) Tl(1223) powder is deposited on a substrate by different methods, e.g. electrophoresis, spin coating etc. Also in this case, as for *ex situ* reacted tapes, a further

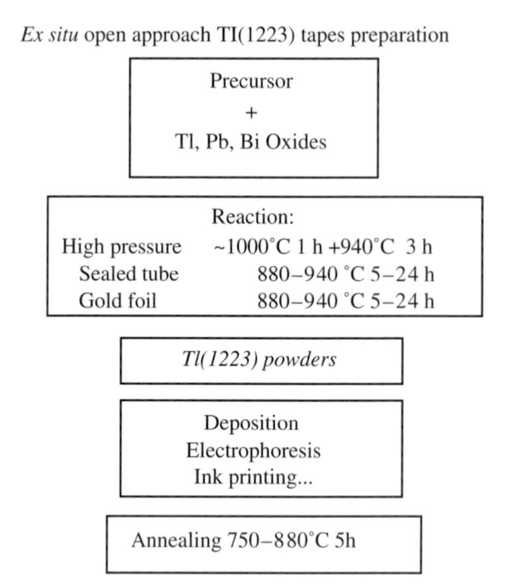

Figure B3.2.3.16. Schematic diagram of *ex situ* Tl(1223) phase formation reaction and film deposition in the open approach.

annealing step (typically 700–850°C in flowing O_2) is required to obtain the sintering of the ceramic grains. If the grains have a platelike morphology it is possible to introduce texture in the deposit e.g. by applying a uniaxial pressure. Tl(1223) powders produced by a premelting step at very high temperature (1080°C) and then reacted at the usual temperature (940°C), both under 50 bar He/1 bar O_2, shows a pronounced platelike shape as shown in figure B3.2.3.17. Tapes prepared by an electrophoretic technique, using such a powder, have a texture comparable to the one commonly obtained in Bi(2223) tapes (figure B3.2.3.18) [32].

Figure B3.2.3.17. SEM image of $(Tl,Pb,Bi)(Sr,Ba)_2Ca_2Cu_3O_{9-\delta}$ powders heated at 1080°C, ground and then reacted at 940°C (both treatments at 50 bar He / 1 bar O_2) [10].

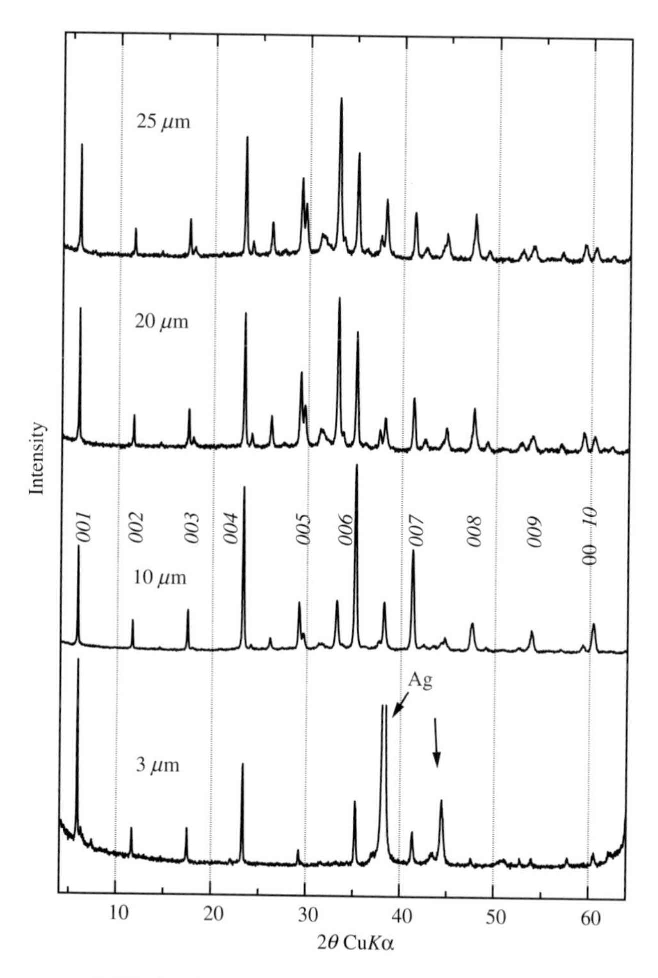

Figure B3.2.3.18. XRD spectra of $(Tl,Pb,Bi)(Sr,Ba)_2Ca_2Cu_3O_{9-\delta}$ electrophoretically deposited tapes of different thickness. The 0 0 l reflections are labelled [32].

In the *in situ* case (figure B3.2.3.19) a precursor is deposited on the substrate and the reaction is performed afterwards. Very few attempts to form the phase directly during the deposition are reported.

Usually the deposited precursor does not yet contain Tl. It is supplied by performing the synthesis reaction in a Tl and O atmosphere (two-zone process or sealed tube with Tl oxide source). For the deposition of the Tl-free precursor all the commonly adopted methods to produce thick layers can be used (spray pyrolysis, sputtering, MBE, etc). For the deposition of a Tl-containing precursor, less 'dispersive' techniques are usually employed. Electrochemical deposition is the most suitable technique for this kind of preparation because no vapours of toxic Tl are produced [33]. The deposition systems need to allow good control and reproducibility of the quality and of the stoichiometry of the film. However, fluctuation of the stoichiometry within 10% is tolerable, these variations do not significantly affect the phase formation and the superconducting properties of the Tl(1223) phase, making this material particularly suitable to be prepared by simple and cheap coating techniques.

B4.2.1, B4.2.2

The synthesis reaction of these kind of samples is critical because they are very sensitive to Tl losses and thus it is necessary to react them in the presence of Tl vapour (as for the Tl free precursor) or under

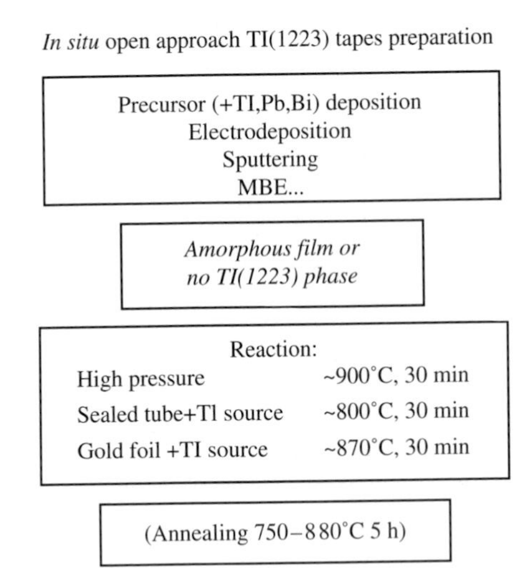

Figure B3.2.3.19. Schematic diagram of *in situ* film deposition and Tl(1223) phase formation reaction in the open approach.

high pressure. On the other hand, the reaction temperature and time can be drastically reduced with respect to those for powders (down to 780°C for few minutes), especially when the deposited precursors are in an amorphous form.

The main advantage of performing the reaction on a substrate is the texturing potential of the latter. C3 As discussed in the previous session, a good level of *c*-axis texture can be obtained by the *ex situ* technique with an open approach, but the transport properties of Tl(1223) tapes prepared by such methods are still weak link dominated. This suggests that, to overcome the weak links problem and reach high values of transport critical current in magnetic field, an almost perfect in-plane alignment is also necessary. This argument seems to be valid for all the high temperature superconductors different from the Bi-based ones.

A biaxial alignment and consequently high values of critical current under magnetic field has been achieved up to now only by deposition of the superconducting ceramics on texturing substrates such as B4.2.1 buffered Ni RABiT or IBAD tapes. These techniques, however, require complicated, long and expensive procedures to prepare suitable substrates for HTSC deposition. Also, the deposition of the superconducting material by vacuum techniques (laser ablation, sputtering, etc) is a slow and expensive process. In order to simplify tape manufacture it is possible to develop Ag textured tapes as substrates [34]: this material can be used without any buffer layer due to its weak chemical interaction with the HTCS. The surfaces (001) and (110) of silver present Ag–Ag distances of 4.09 Å (square) and 4.09 Å, 2.89 Å (rectangular), respectively. The matching of these parameters with those of the HTSC (~ 3.85 Å) was found to be good enough to obtain epitaxial growth on Ag [35]. Samples prepared on silver foil show a pronounced *c*-axis texture but random in plane orientation is observed. In contrast, the deposition on the textured substrates yields also strong in-plane alignment. Tl(1223) grows epitaxially on Ag, and {100} [36] {110} [37] orientation as is clearly shown in the SEM image in figure B3.2.3.20.

Proof that this kind of process is able to produce well textured structures, capable of reaching very high values of critical current densities, is given by the fact that Tl(1223) deposited on $LaAlO_3$ [38] or $SrTiO_3$ [37] grows epitaxially on the whole substrate as shown in the ϕ-scan reported in figure B3.2.3.21.

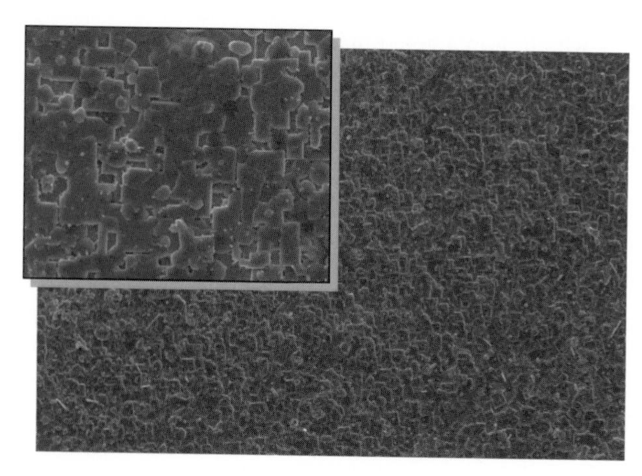

Figure B3.2.3.20. SEM image of the surface of 3 mm thick film of Tl(1223) deposited on the (110) face of a Ag single crystal (magnification in the inset) [37].

B3.2.3.4 Critical current densities of Tl(1223)

In this section transport critical currents, measured by the four probe technique and with a criterion of $1\,\mu V\,cm^{-1}$ are discussed. D2.2

B3.2.3.4.1 Closed approach

Critical current densities of about $25000\,A\,cm^{-2}$ at 77 K and 0 T were achieved for 3 cm long monofilamentary tapes [39]. For 15 cm long unpressed monofilamentary tapes J_c decreased to $12000\,A\,cm^{-2}$ and for 200 cm long (long-length) tapes to $10000\,A\,cm^{-2}$ in part because of localized mechanical damage induced during the manufacturing. This problem should not be present for D2.2

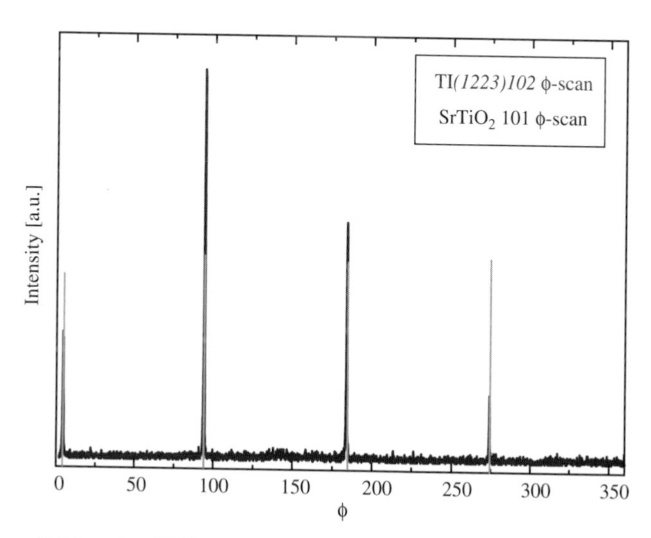

Figure B3.2.3.21. ϕ-scan of 102 peak of Tl(1223) deposited on $SrTiO_3$ (single crystal) showing epitaxial growth [37].

A1.3 B3.1 multifilamentary tapes, which show a better mechanical resistance (it is, for instance, possible to bend such a tape to a radius of about 1.5 cm without significant reduction of the critical current). However, lower J_c values of about 6500 A cm^{-2} were achieved for multifilamentary tapes (37 filaments) [40]. The decrease of the critical current in multifilamentary tapes with respect to monofilamentary tapes can be attributed to two main factors:

(a) the reduced thickness of the superconducting filaments results in filament sausaging, a problem which is aggravated by the presence of grain agglomerates in the starting powder;

(b) the larger quantity of silver reduces the oxygen diffusion, making it difficult to reach optimal doping.

Critical current densities versus applied magnetic field at 77 K for short-length mono- and 37-filamentary tapes are shown in figure B3.2.3.22. In both cases there is a strong magnetic field

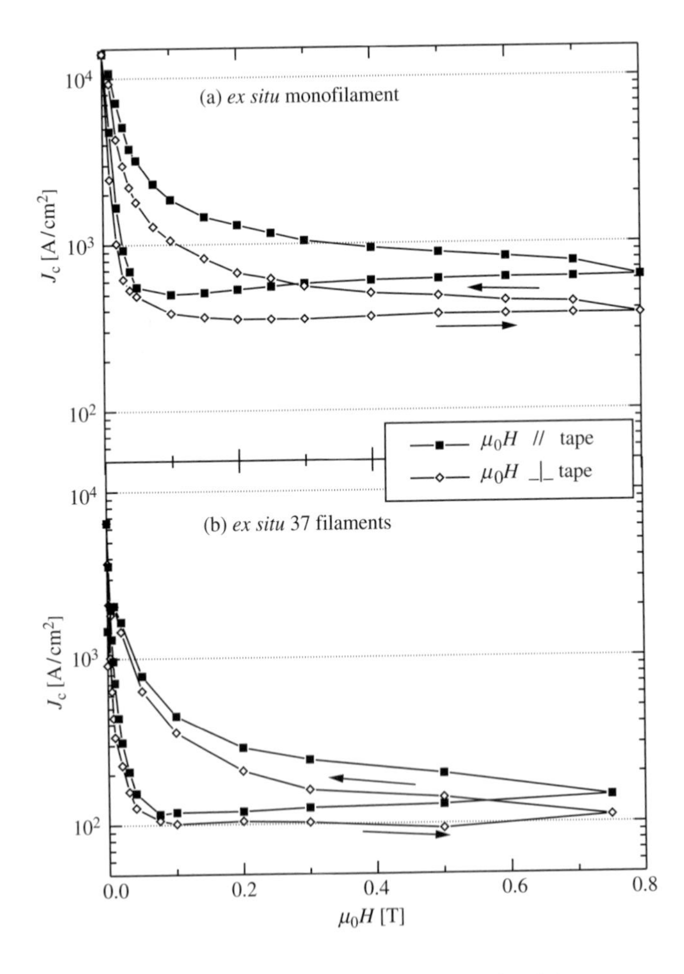

Figure B3.2.3.22. Critical current density (77 K) versus magnetic field for (*a*) a mono- and (*b*) a 37-filamentary Tl(1223) tape [40].

dependence, with a decrease in J_c by a factor of 23 and 54, respectively, when the magnetic field is increased from 0 to 0.2 T. This sharp drop and the strong hysteretic behaviour clearly indicate that the current is limited by weak links between the grains. This is confirmed by the temperature dependence of the critical current. The persistence of a non-zero critical current at high magnetic field (up to 10 T at 4.2 K) can, however, not be explained in terms of weak links, but indicates the existence of a few strongly coupled current paths.

The presence of weak links is mainly due to the low degree of texture induced during the tape manufacturing. figure B3.2.3.22 shows only a very small anisotropy, the ratio of the critical current densities with the magnetic field parallel and perpendicular to the tape surface being less than 2. XRD measurements performed on stripped tapes also confirm the lack of significant texture.

D1.1.2

Figure B3.2.3.23 shows J_c as a function of the tape thickness for multifilamentary tapes. The critical current densities are comparable for the 7- and 37-filamentary tapes, but lower for the tape with 259 filaments. The highest values of J_c correspond to a tape thickness of $\sim 140\,\mu m$ for 7 and 37 filaments, and $\sim 200\,\mu m$ for 259 filaments. The less good value observed for the latter tape can be explained by the fact that the individual filaments become very thin ($\sim 5\,\mu m$). In the inset of figure B3.2.3.23, the same data are plotted versus filament thickness, estimated from cross-sectional SEM images. The critical current density is independent of the filament thickness down to $\sim 10\,\mu m$ but drops when the filament thickness becomes comparable to the grain size.

D1.3

Measurements of the critical current performed on a monofilamentary tape after cutting it into $\sim 250\,\mu m$ wide strips showed that the current is uniformly distributed inside the core (see figure B3.2.3.24) [40, 41]. In contrast to Bi(2223) tapes, where the regions close to the silver sheath show a higher critical current density [42], in Tl(1223) tapes the silver seems to have no significant effect on the texture or on the grain connectivity.

Critical current densities of 10 000 A cm^{-2} (77 K, 0 T) were observed for Ag-sheathed F-containing Tl(1223) tapes with *ex situ* reacted powder. As can be seen from figure B3.2.3.25, the critical current density of such a tape showed a less pronounced field dependence than that of the corresponding F-free tape. The ratio of the critical currents with the magnetic field parallel and perpendicular to the tape surface was higher and the hysteresis width smaller for the F-containing sample. Similarly, F-containing

C3

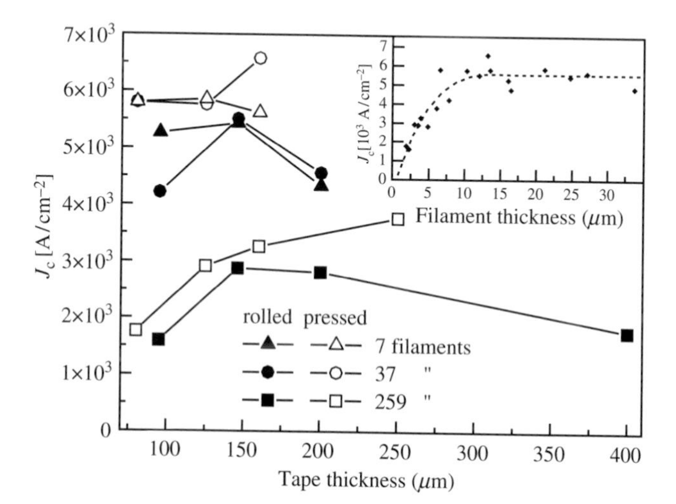

Figure B3.2.3.23. Critical current density (77 K,0 T) versus tape thickness for multifilamentary Tl(1223) tapes. Inset: critical current density versus filament thickness [40].

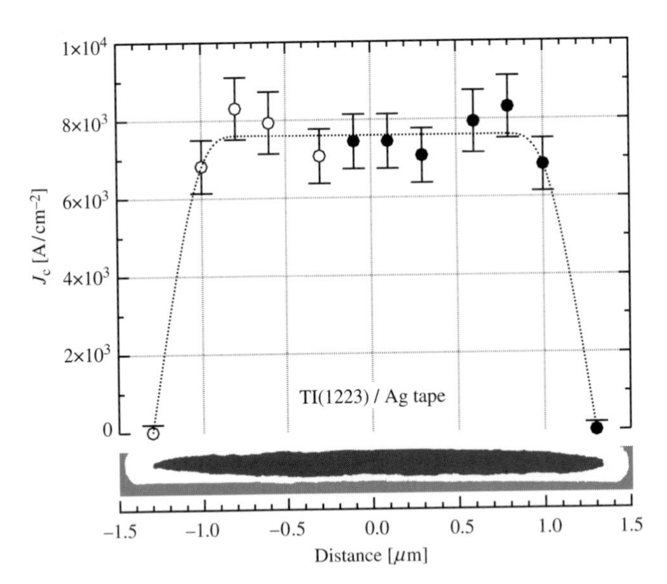

Figure B3.2.3.24. Distribution of the critical current density inside a monofilamentary Tl(1223) tape [40].

Tl(1223) tapes with *in situ* reacted powders showed better transport properties (figure B3.2.3.25) than the corresponding samples without fluorine.

B3.2.3.4.2 Open approach

D2.4 Magnetisation measurements showed that electrophoretically deposited samples exhibit superconducting behaviour also before the annealing, while the transport properties can be measured only after the final thermal treatment. A value of 11 000 A cm^{-2} (4.5 A) at 77 K and 0 T was reached for this kind of tape (figure B3.2.3.26). Despite the remarkable value of the zero field current, the transport properties exhibited a strong magnetic field dependence, and the critical current density was decreased by a factor of 60 in a magnetic field of 0.2 T. This can be explained by the fact that tapes with deposited layers are more sensitive to heat treatments than bulk materials or thick tapes prepared by the PIT technique. The transport properties, despite the high degree of *c*-axis texture, are still dominated by weak links.

B4.3 Aerosol deposition from a solution [43, 44] and electro-deposition [33] on single crystal substrates, followed by an *in situ* reaction, showed very promising values of critical currents: $J_c >$ B4.1 10^5 A cm^{-2} at 77 K/0 T, with current densities higher then 10^4 A cm^{-2} persisting up to high magnetic field (5 T) at LN_2 temperature. The reason for such a tremendous improvement over the previously discussed methods lies in the essential biaxial grain alignment of the poly-crystalline films, which was found to be a fundamental condition to obtain high J_c values in magnetic fields. The grains should be both *c*- and *a*-axis aligned in such a way that the microstructure approaches that of a single crystal. In E1 this respect important progress has been made demonstrating that biaxially textured poly-crystalline Ag substrates allows alignment of the superconducting grains. For power applications, high-I_c values B4.0 are also required, which means one needs to grow HTCS films with thickness of the order of 10 μm or more. From this point of view, techniques such as spray deposition from an ink, sol–gel and screen-printing are of considerable interest. The great challenge associated with the problem of thick film B4.3 growing involves developing and sustaining the substrate texture up to the very surface of the thick film.

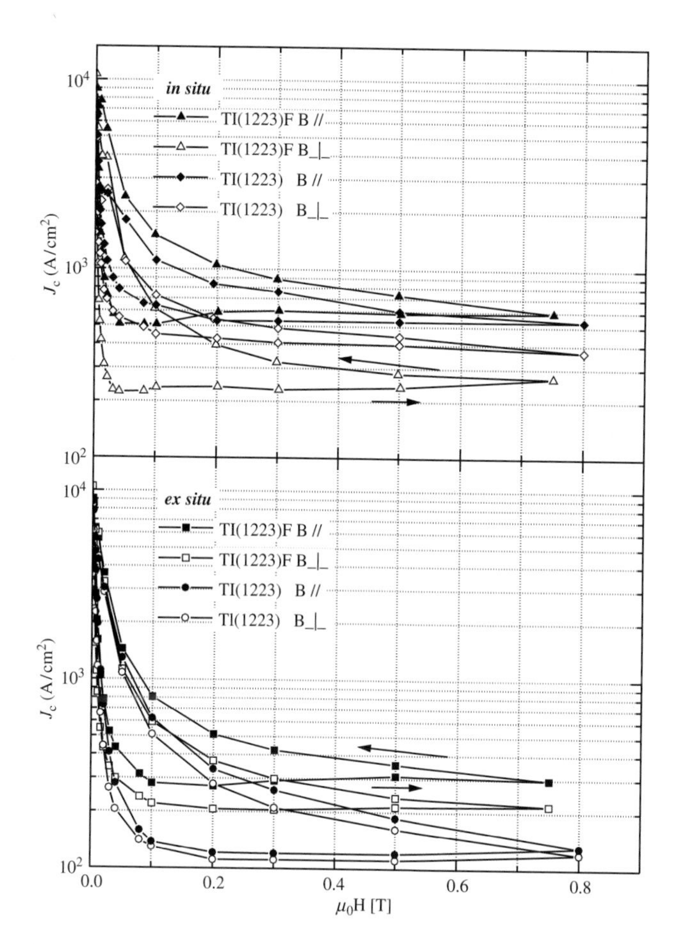

Figure B3.2.3.25. Critical current densities (77 K) versus magnetic field for F-containing Tl(1223) tapes with *ex situ* and *in situ* reacted powders [21].

Critical current densities for different kind of tapes are summarised in table B3.2.3.2 and figure B3.2.3.27.

B3.2.3.5 Conclusion

In this paper, the current situation in the developmentof Tl(1223) tapes and thick films was reviewed. This high temperature superconductor is a promising material for future applications in power transmission line and various magnet system. These types of application requires J_c values in the range of 104–105 A cm^{-2} which are sustainable in magnetic field of 3–5 T and operating temperature close to 77 K. C3

Two basic processing procedures of Tl(1223) conductor were discussed here: the so called closed approach represented by the PIT method and the open approach represented by various deposition or coating methods such as aerosol deposition, electrodeposition, sol–gel screening etc.

The closed approach process is able to produce conductors with good properties in the absence of magnetic field but J_c decreases dramatically in very low magnetic field of only 0.1 T due to the well known but always present weak link problem.

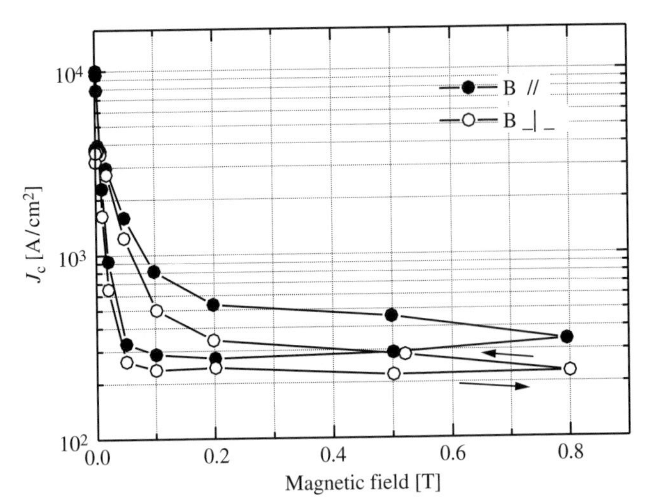

Figure B3.2.3.26. Critical current density (77 K) versus magnetic field applied parallel and perpendicular to the surface, for an electrophoretically deposited tape [32].

C2 C1 B3.2.1, B3.2.2

Texturing is an important issue for tape preparation but plays a very different role in the Bi-2223 and RE-123 / Tl-2223 cases, resulting in the fundamental problem in the latter. It is nearly impossible to obtain by mechanical methods the necessary grain alignment to overcome this problem, as successfully happens in the BiSCCO based conductors.

Table B3.2.3.2. Reported critical current densities of substituted Tl(1223) tapes prepared by different methods

Powder/Sheath Preparation, Length	J_c [A cm^{-2}] at 77 K,0 T	I_c [A] at 77 K,0 T	I_c (0 T)/I_c(0.5 T) at 77 K	Ref
Tl,Pb,Bi(1223)/Ag *ex situ*, PIT, mono, 3 cm	25 000	21	45	[39]
Tl,Pb,Bi(1223)/Ag *ex situ*, PIT, mono, 2 m	10 000	7	–	[40]
Tl,Pb,Bi(1223)/Ag *ex situ*, PIT, 37 filaments, 3 cm	6500	6	54	[40]
Tl,Pb(1223)F/Ag *ex situ*, PIT, mono, 3 cm	10 000	8	24	[7]
Tl,Pb,Bi(1223)/Ag,Au PIT, *in situ*, mono, 3 cm	11 000	10	16	[22]
Tl,Pb(1223)/Ag,Au PIT, *in situ*, mono, 3 cm	6000	6	12	[24]
Tl,Pb(1223)F/Ag,Au PIT, *in situ*, mono, 3 cm	10 000	9	16	[21]
Tl,Pb,Bi(1223)/Ag *ex situ*, deposition, mono, 3 cm	9000	1	29	[32]
Tl,Pb,Bi(1223)/Ag *ex situ*, deposition, 3 layers, 3 cm	11 000	4.5	59	[32]
Tl (1223)/textured Ag *in situ*, spray, ?	92 000	?	3.5	[44]

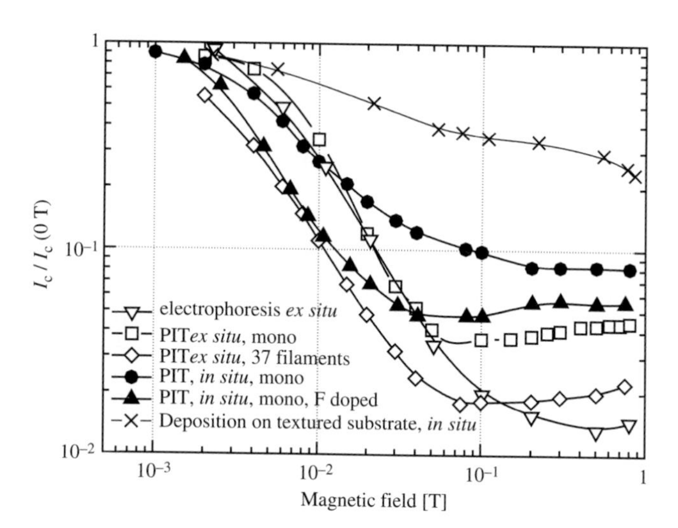

Figure B3.2.3.27. Normalized critical current (77 K) versus magnetic field for Tl(1223) tapes prepared by different methods.

In the open approach the possibility of using a textured substrate to induce the correct grain orientation during the superconducting phase formation appears very promising. As an alternative to the RABITs and IBAD coated conductor techniques, it is possible to produce high quality textured Ag ribbons and to utilise non vacuum, fast and cheap deposition techniques for the Tl(1223) phase. The Tl(1223) superconductor, in view of its large phase stability and tolerance for out stoichiometry and substitution is the optimum candidate for this kind of preparation techniques.

B4.2.1

Long textured Ag ribbons can be produced by rolling and annealing. Ag is therefore an interesting alternative for coated tape preparation. The tapes produced by this technique shows very promising structural characteristics but still some difficulties should be overcome in particular on the surface quality and texturing of the substrate and on the homogeneity of the deposition.

References

[1] Ihara H, Sekita Y, Khan N A, Ishida K, Harashima E, Kojima T, Yamamoto H, Tanaka K, Tanaka K, Terada N and Obara H 1999 *IEEE Trans. Appl. Supercond.* **9** 1551
[2] Holstein W L 1994 *Appl. Supercond.* **2** 345
[3] Ruckenstein E and Wu N L 1994 *Thallium -Based High-Temperature Superconductors*, ed A M Hermann and J V Yakhmi (New York: Dekker) p 119
[4] DeLuca J A, Garbauskas M F, Bolon R B, McMullin J G, Balz W E and Karas P L 1991 *J. Mater. Res.* **6** 1415
[5] Flükiger R, Gladyshevkii R E and Bellingeri E 1998 *J. Supercond.* **11** 23
[6] Aselage T L, Venturini E L and van Deusen S B 1993 *J. Appl. Phys.* **75** 1023
[7] Doi T, Okada M, Soeta A, Yuasa T, Aihara K, Kamo T and Matsuda S-P 1991 *Physica* C **183** 67
[8] Liu R S, Wu S F, Shy D S, Hu S F and Jefferson D A 1994 *Physica* C **222** 278
[9] Ren Z F, Wang J H, Miller D J and Goretta K C 1995 *Physica* C **247** 163
[10] Gladyshevkii R E, Bellingeri E and Flükiger R 1998 *J. Supercond.* **11** 109
[11] Heede S, Ullrich M, Freyhardt H C, Gladyshevkii R E, Bellingeri E and Flükiger R 1998 *J. Supercond.* **11** 109
[12] Rhines F R and Craig K R 1974 *Metall. Trans.* **5** 413
[13] Bolling G F and Winegard W C 1958 *Acta Metall* **6** 283
[14] Chu C T and Dunn B 1990 *J. Mater. Res.* **5** 1819
[15] Richards L E, Hoff H A and Aggarwal P K 1993 *J. Electron Mater.* **22** 1233
[16] Grest G S, Srolovitz D J and Anderson M P 1985 *Acta Metall.* **33** 520
[17] Lanagan M T, Hu J, Foley M, Hagen M R, Goretta K C, Kostic P and Miller D J 1996 *Physica* C **256** 387

[18] Morgan P E D, Doi T J and Housley R M 1993 *Physica* C **213** 438
[19] Selvamanickam V, Finkle T, Pfaffenbach K, Haldar P, Peterson E J, Salazaar K V, Roth E P and Tkaczyk J E 1996 *Physica* C **260** 313
[20] Grivel J-C and Flükiger R 1996 *J. Alloys Comp.* **235** 53
[21] Bellingeri E, Gladyshevskii R, Marti F, Dhallè M and Flükiger R 1998 *Supercond. Sci. Technol.* **11** 810
[22] Gladyshevskii R E, Galez P h, Lebbou K, Allemand J, Abraham R, Couach M, Flükiger R, Jorda J-L and Cohen-Adad M Th 1996 *Physica* C **267** 93
[23] Opagiste C, Couach M, Khoder A F, Abraham R, Jondo T K, Jorda J-L, Cohen-Adad M Th, Junod A, Triscone G and Muller J 1993 *J. Alloys Comp.* **195** 47
[24] Gladyshevskii R E, Bellingeri E, Perin A and Flükiger R 1996 *High Temperature Superconductors: Synthesis, Processing, and Large-Scale Applications*, eds U Balachandran, P J McGinn and J S Abell (TMS Warrendale: USA) 321
[25] Al-Mamouri M, Edwards P P, Greaves C and Slaski M 1994 *Nature* **369** 382
[26] Wang Y T and Hermann A M 1995 *Physica* C **254** 1
[27] Bellingeri E, Grasso G, Gladyshevskii R E, Dhallé M and Flükiger R 2000 *Physica* C **329** 267
[28] Tachikawa K, Kikuchi A, Kinoschita T and Komiya S 1995 *IEEE Trans. Appl. Supercond.* **5** 2019
[29] Hamdan N M, Ziq Kh A and Al-Harti A S 1999 *Physica* C **314** 125
[30] Gladyshevskii R E, Galez Ph, Lebbou K, Bellingeri E, Couach M, Flükiger R, Jorda J-L and Cohen-Adad M Th 1996 *Czech. J. Phys.* **46** 1415
[31] Bellingeri E, Gladyshevskii R E and Flükiger R 1997 *Proc. EUCAS'97 Inst. Phys. Conf. Ser.* **158** (Bristol: Institute of Physics Publishing)
[32] Bellingeri E, Gladyshevkii R E and Flükiger R 1998 *J. Supercond.* **11** 77
[33] Bhattacharya R N, Blaugher R D, Ren Z F, Li W, Wang J H, Paranthaman M, Verebelyi D T and Christen D K 1998 *Physica* C **304** 55
[34] Suo H L, Genoud J-Y, Schindl M, Walker E, Tybell T, Cleton F, Zhou M and Flükiger R 2000 *Supercond. Science Technol.* **13** 912
[35] Schindl M, Koller E, Genoud J-Y, Suo H-L, Walker E, Fischer Ø and Flükiger R *Proc. EUCAS '99 Inst. Phys. Conf. Ser.* (Bristol: Institute of Physics Publishing)
[36] Doi T J, Yuasa T, Ozawa T and Higashiyama K 1994 *Japan. J. Appl. Phys.* **33** 5692
[37] Bellingeri E, Suo H L, Genoud J-Y, Schindl M, Walker E and Flükiger R 2001 *IEEE Trans. Appl. Supercond. Proc. ASC 2000* **11** 3122
[38] Bhattacharyaa R N, Wu H L, Wang Y-T, Blaugher R D, Yang S X, Wang D Z, Ren Z F, Tu Y, Verebelyi D T and Christen D K 2000 *Physica* C **333** 59
[39] Jeong D Y, Kim H K and Kim Y C 1999 *Physica* C **314** 139
[40] Bellingeri E, Gladyshevskii R E and Flükiger R 1997 *Il Nuovo Cimento* **19** 1117
[41] Fox S, Moore J C, Jenkins R, Grovenor C R M, Boffa V, Bruzzese R and Jones H 1996 *Physica* C **257** 332
[42] Grasso G, Hensel B, Jeremie A and Flükiger R 1995 *Physica* C **241** 45
[43] Li W, Wang D Z, Lao J Y, Ren Z F, Wang J H, Paranthaman M, Verebelyi D T and Christen D K 1999 *Supercond. Sci. Technol.* **12** L1
[44] Doi T J, Sugiyama N, Yuasa T, Ozawa T, Higashiyama K, Kikuchi S and Osamura K 1995 *Adv. Supercond.* **8** 903

B3.2.4
Processing of high T_c conductors: the compound YBCO

J L MacManus-Driscoll

B3.2.4.1 Introduction

Of the various superconductor compounds that can be utilized for conductor applications, the rare earth barium cuprate phase, $REBa_2Cu_3O_{7-x}$ (REBCO) has the greatest potential for achieving high current densities in field. However, even after 13 years, the great promise of this material has not yet been realized. The problems are inherent to the crystallographic nature of REBCO which, in bulk, forms insufficiently well aligned grains to give strong grain boundary coupling. This is in contrast to the Bi-based superconductors, which exhibit strongly coupled grains, although with much weaker intragrain pinning. Thus, the conventional powder metallurgical tape processing routes utilized for the Bi-based superconductors are inappropriate for REBCO. C1 A1.3 B3.1

In the early 1990s, renewed insight into the problem of REBCO conductor brought forth two novel processing routes, so called 'IBAD' and 'RABiTS', for fabricating suitable substrates onto which textured REBCO would grow. Utilizing thin film technologies for the growth of the REBCO layer has, so far, yielded metre length conductors with excellent properties. The shift of research attention back to REBCO has meant that progress has been rapid and a whole array of alternative thin and thick film deposition technologies are once again being revisited. In this review, the current status and future of the REBCO conductor processing routes will be discussed, with a strong emphasis on the materials issues underlying the conductor properties. B4.2.1 B4.1

B3.2.4.2 REBCO microstructural issues

The Bi–Sr–Ca–Cu–O (BSCCO) superconductors, $Bi_2Sr_2Ca_{n+1}Cu_{n+2}O_x$ $(n = 0, 1)$, have weak intragrain pinning, which means that above 40 K, in the presence of a magnetic field, the critical current density, J_c, vanishes to zero. The origin of the weak pinning is structural in nature and can only be overcome by increasing superconductive coupling between the layers in the very anisotropic unit cell, or through chemical or physical introduction of defects to pin the flux lattice. While there are many research programs directed at engineering flux pinning centres into BSCCO, there has been little success in finding an effective and practical route to increased pinning. Therefore, at the present time, superconductors with stronger intrinsic pinning are required for applications at 77 K, in fields of >1 T. Possible candidate materials include $TlBa_2Ca_2Cu_3O_9$ (Tl-1223, $T_c \sim 120$ K), $HgBa_2Ca_2Cu_3O_8$ (Hg-1223, $T_c \sim 135$ K) and REBCO ($T_c \sim 91$–96 K). Considering the toxic nature of Tl and Hg, combined with the fact that the phase chemistry of Tl-1223 and Hg-1223 are very complex and not well understood, REBCO is the material of choice for conductor applications at 77 K. In REBCO, while it is possible to process the material so that there is strong pinning within the grains, it is also important to consider C2 A4.3 C4 A1.3

the influence of the grain boundaries on current transport, since a conductor fabricated from the material will necessarily be polycrystalline (containing tens of thousands of grain boundaries). B2.2.2

All the high temperature superconductors have anisotropic crystal structures with weak interlayer coupling in the *c* direction. This is in contrast to strong coupling in the *a–b* planes. Consequently, the coherence length, ξ, along *c* is shorter than along *a–b* (e.g. in $YBa_2Cu_3O_{7-x}$ (YBCO) at 77 K, in zero field, $\xi_c \sim 0.5$ nm and $\xi_{ab} \sim 1.5$ nm). Since superconducting charge transport occurs in the *a–b* planes, for conductor applications it is necessary to align the grains of the material such that *a–b* is along the conductor length and *c* is perpendicular to the conductor length. This gives rise to *c*-axis aligned grain boundaries. In practice, it is impossible to perfectly align the boundaries along *c*, which means that the grain boundary plane will have components of the *a–b* and *c* directions (the coherence length contains components of ξ_c and ξ_{ab}). The problem of imperfectly aligned boundaries is that they are of finite width due to the presence of atomic disorder in the boundary region. In order for superconducting electrons to couple across the boundary, its width must be less than ξ. This means that only low angle grain boundaries, with the least disorder, can transmit the superconductor current. A2.3

From the the early Dimos [1] bicrystal studies of *c*-axis tilt grain boundaries (rotation of grains in the *a–b* plane, around the *c*-axis) in YBCO thin films, it was clear that high angle grain boundaries were very detrimental to current transport. As a rough rule of thumb, above a grain boundary angle of ~5–10°, the intergranular J_c was found to decrease by one order of magnitude for every 10° increase in grain boundary angle. Subsequent, more detailed studies have shown that for most boundaries there is a linear increase in boundary width with boundary angle [2]. The increase in boundary width is related to the rearrangement of atoms near the dislocation cores at the boundaries, and the associated strain, charge distribution and disorder [3]. On the other hand, the presence of low angle boundaries may, in fact, be beneficial to REBCO conductors, since in many studies it has been found that J_c does not degrade very much, if at all, at $<5°$, for fields applied along *c* [4]. It is believed that insulating dislocation cores, which act like *c* oriented columnar defects, may be responsible for this behaviour. However, strong pinning in the vicinity of low angle grain boundaries is not a universal feature of YBCO boundaries fabricated by all methods (even boundaries fabricated by different thin film methods behave differently). Therefore, the precise nature of the boundary on the atomic scale needs to be understood. In practical terms, engineering of dislocation pinning centres into the boundaries will require very precise control of the growth conditions. In short, grain boundary science and engineering is not yet advanced enough to yield significant improvement of REBCO conductor properties. B4.2.1 B4.1

B3.2.4.3 Tape architecture

A typical tape architecture is composed of up to four layers:

(a) substrate B4.1

(b) buffer layer B4.2.1

(c) REBCO layer

(d) passivation/insulation layer.

The optimum substrate is non-reactive with REBCO, eliminating the need for an additional buffer layer. The additional functional requirements of the substrate include good thermal expansion matching, good flexibility, good texturability and non-oxidizability. The substrate must also be deformable down to $<100\,\mu$m and stable at this thickness. Such thin substrates are required to optimize engineering critical current (current/total cross sectional area of conductor + substrate).

There is no ideal substrate material which meets all the above criteria, and for this reason, most conductor fabrication routes now include an oxide buffer layer between substrate and REBCO conductor layer. The buffer layer provides a chemical barrier between substrate and REBCO, thus widening the substrate selection criteria to include 'reactive' materials.

B4.1 B4.2.1

To prevent the formation of high angle grain boundaries, the REBCO layer must be very sharply textured both in-the-plane and perpendicular-to-the-plane of the substrate. Therefore, the buffer layer must also be sharply textured to allow for hetero-epitaxial growth of the REBCO layer. Ni-based metallic alloys have proven to be suitable substrate materials, with their particularly good thermal expansion match, reasonable oxidation resistance and excellent texturability. The buffer layers which have proven to be suitable are the same as those used in YBCO thin film devices grown on silicon, namely yttria stabilized zirconia (YSZ), MgO and CeO_2.

The passivation/insulation layers are required to protect the tape from the ambient as well as to prevent current leakage between conductor filaments once they are multilayered. These final layers do not have the strict functional materials requirements of the layers beneath and they can be fabricated at low temperatures. Further work is required in materials selection studies of the passivation/insulation layer, once the materials challenges of the initial layers are resolved.

B3.2.4.4 Characterization of tape texture

The transport properties of REBCO tapes are much more strongly dependent on grain alignment than those of BSCCO tapes. Not only does the level of out-of-plane (perpendicular-to-substrate) texture strongly influence intergranular J_c, but so does the in-plane texture. For this reason, conventional rocking curves which give information on out-of-plane texture do not suffice to give a complete texture picture. Instead, three-dimensional studies of grain orientation of REBCO, buffer and substrate layers are required.

B3.1

X-ray pole figure analysis allows full spatial information on grain orientations to be acquired (as discussed in more detail in chapter B.3.1). A pole figure shows the statistical distribution of a selected family of atomic planes in relation to the substrate normal. In the case of a rolled metal substrate, the rolling direction is also included in the pole figure. With orientation information about one family of planes, the orientation of the entire crystal structure is probed (since each set of planes within the crystal are uniquely related to one another). A pole figure is determined with a fixed, monochromatic x-ray beam and fixed detector position. The incident x-ray angle is set to the Bragg angle for the chosen family of planes to be analysed. The scanning is brought about by rotating the sample in a spiral configuration to account for all possible grain orientations in three-dimensional space. The x-ray intensity recorded at any position in space is related to the number of planes at that position, and thus is a measure of the pole density. Typically, sample areas of 5×20 mm, and depths of $10-50\ \mu$m can be analysed in one scan.

D1.1.2

If a sample is well textured, the positions of high intensity in the pole figure will reveal the crystal symmetry of the material. A typical pole figure [Bragg angle set to $\sim 32°$ for the (103) family of planes] for a textured YBCO conductor grown by spray pyrolysis on a (110) Ag single crystal is shown in figure B3.2.4.1 [5]. The breadth of the peaks in the crystallographic directions, ϕ (circumferential direction) and χ (radial direction), reveal the misalignment of the grains in the in-plane and out-of-plane sample directions, respectively.

D1.1.2 B4.3

A radial slice of a few degrees through one of the peaks (at a constant ϕ) would provide equivalent information to a conventional rocking curve. A circumferential section of 360°through all the peaks (at a constant χ) would be equivalent to a ϕ scan, in which the in-plane grain alignment is probed. Often, the level of texture in a sample is reported by showing only a ϕ scan, e.g. as in figure B3.2.4.2 [6] which is a ϕ scan of the (103) peak in the same YBCO sample grown on (110) Ag, of figure B3.2.4.1. Clearly, the whole texture information is not represented here but only the in-plane spread of the grains which are

D1.1.2

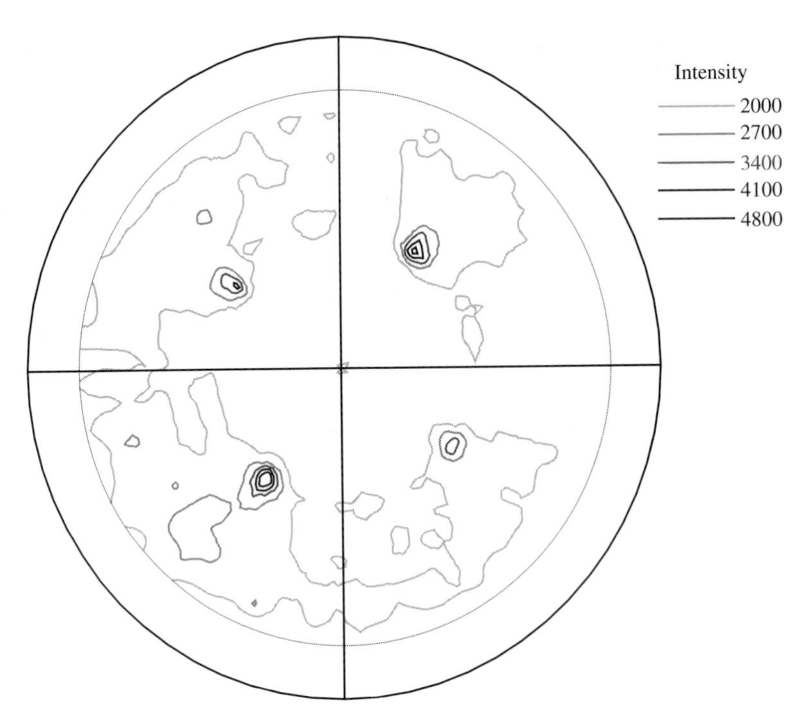

Figure B3.2.4.1. (103) pole figure of textured YBCO grown by spray pyrolysis on (110) Ag single crystal. Four main poles are evident, indicative of epitaxial growth of the film. After Wells and MacManus-Driscoll [5].

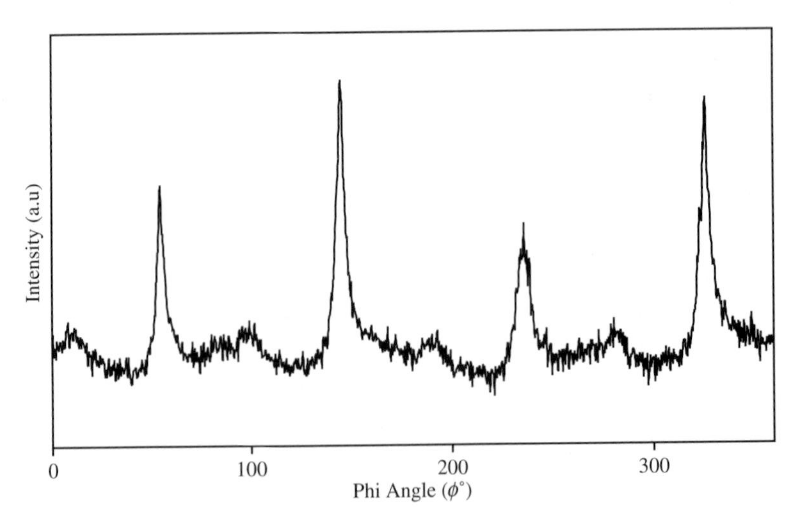

Figure B3.2.4.2. ϕ scan of (103) peak in textured YBCO grown by spray pyrolysis on (110) Ag single crystal (same as for figure B3.2.4.1), showing four main peaks spaced by 90° in ϕ and four minor peaks, 45° rotated from the main peaks, indicative of a minor in-plane rotated component. After Wells *et al* [6].

well aligned in the plane of the substrate. ϕ scans are, in fact, more useful when it is known that the overall texture in a sample is very high (such as in epitaxial thin films), although they are often used incorrectly to assess texture in less well textured REBCO tapes.

B4.2.1

Another way to measure sample texture is using electron backscattered patterns (EBSP) in the scanning electron microscope (SEM). In the EBSP technique, the SEM beam impinges on the sample, and Kikuchi lines are formed by diffraction of diffusely scattered electrons. The configuration of the lines depends upon the orientation of the crystallite from which the electrons are scattered, and, therefore, with knowledge of the crystal structure, and utilizing a beam size smaller than the grain size, it is possible to determine the orientation of individual grains within a sample. On a similar time scale to the generation of an x-ray pole figure, thousands of grains can be analysed by probing the beam in a grid pattern across the sample. An example EBSP image for partially textured silver foil is shown in figure B3.2.4.3. Three different shades of grey are used to indicate different boundary angles. It is observed that most of the grain boundaries are misoriented by $<40^\circ$. However, a large number of twin boundaries are also present whose misorientation angles with respect to the parent grains are >40.

D1.3

The advantage of EBSP over x-ray pole figure analysis is the ability to probe micron sized regions and to obtain statistics on the boundary orientations. In addition, surface defects can be imaged and one can obtain a pictorial view of possible current percolation paths. The clear advantage of the x-ray technique is that large areas can be probed. Both techniques are used widely to characterize REBCO conductors, and the underlying buffer layer and substrate materials.

D1.1.2

B4.2.1, B4.

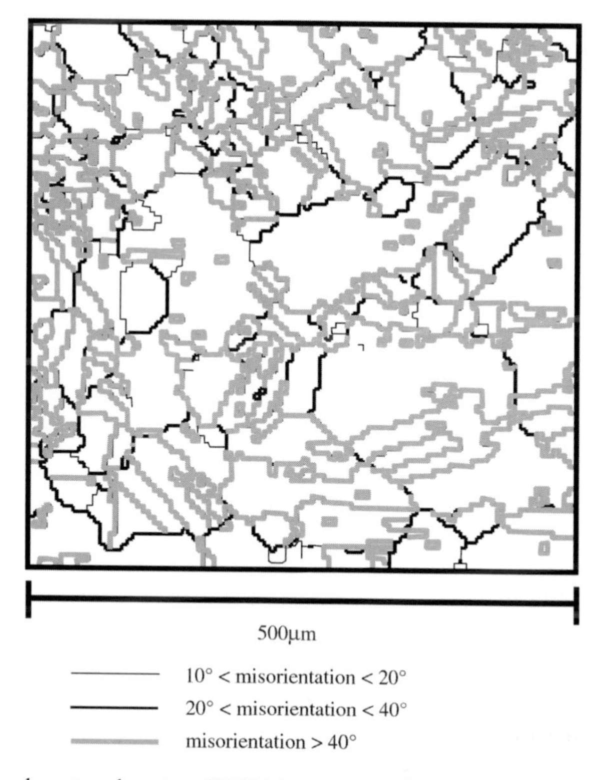

Figure B3.2.4.3. Electron backscattered pattern (EPSP) image of partially textured silver foil. Three different shades of grey are used to indicate different misorientation angles, θ. Fine dark grey lines: $10^\circ < \theta < 20^\circ$. Thick dark grey lines: $20^\circ < \theta < 40^\circ$. Thick light grey lines: $\theta > 40^\circ$ (twin boundaries, mainly). After Wells and MacManus-Driscoll [5].

B3.2.4.5 Routes to REBCO conductor fabrication

C1 B2.3 B2.1

It was realized very soon after the discovery of YBCO that the bulk, polycrystalline ceramic could carry no more current than copper because of the poorly coupled grain boundaries. Later, it was shown that by melt texturing of the ceramic, more strongly coupled, low angle grain boundaries are formed. In melt texturing, the control of domain nucleation and growth is critical to the formation of a stable, planar growth front and to the formation of biaxially oriented grains. However, the very slow growth rates (of the order of microns per second) and the need to control temperature to a precise degree mean that the standard melt texturing routes are not practical for formation of metre length conductors.

A4.:

The alternative to 'bulk' powder routes for conductor synthesis is to use thin or thick film routes. The advantages are two-fold: texture control through epitaxial growth and, in the case of vapour deposited films, there is the added advantage of enhanced flux pinning, which is specific to the nature of thin film processing and yields around an order of magnitude enhancement in J_c. In the following subsections, the substrate and buffer production technologies which have allowed epitaxial growth of buffer and/or REBCO layers are reviewed. In addition, the promising vapour phase and non-vapour phase methods for buffer layer and REBCO conductor fabrication are discussed.

B3.2.4.5.1 Textured metallic substrate fabrication

B4.2 B4.

A method for producing highly textured nickel or nickel alloy metallic substrates was developed at Oak Ridge National Lab (ORNL) in 1994, and is termed rolling-assisted-biaxially textured-substrates (RABiTs). The development of RABiTs was spurred on by promising results which showed YBCO thin films could be grown, with biaxial alignment, directly onto a metallic (silver single crystal) surface. Ni-based alloys are the currently preferred substrates to be produced by RABiTs because of the ease and sharpness of texture (unlike, for example, Ag which produces twin orientations). Ni also has a good thermal expansion match with REBCO.

In the RABiTs process, the metal is rolled using highly polished rolls, and then recrystallized to give cube textured Ni. Currently, FWHM values in ϕ of $\sim 6°$ are now being achieved by several groups worldwide, in foils of thickness $< 100\ \mu m$. The root mean square surface roughness of the grain surfaces is only ~ 10 nm [7]. The current challenges facing RABiTs processing are: (a) Ni has a tendency to undergo grain boundary grooving upon recrystallization, giving groove depths of several tens of nm and widths of up to $1\ \mu m$; (b) outgrowths of NiO precipitates form on the surface of the Ni foil before or during deposition of the buffer layer and these can lead to cracking in the buffer layer; (c) Ni is ferromagnetic which leads to enhanced hysteretic losses of conductors used in ac applications. Alloying additions with other transition metals can eliminate the problem of losses, but their influence on the formability of the alloy still require further investigation.

B3.2.4.5.2 Buffer layer fabrication

Buffer layers on RABiTs substrates

B4.: B4.

The most reliable route to buffer layer deposition on RABiTs substrates is by conventional physical vapour deposition (PVD) such as laser ablation, sputtering and electron beam evaporation (as discussed in chapter B.4.1). However, unlike deposition of REBCO, where relatively high oxygen activities are required during deposition to ensure phase stability, the oxygen activity at the Ni surface must be maintained at a very low level to minimize formation of NiO. Various buffer layer configurations have been, and continue to be, investigated. The buffer layer sequences which include YSZ and/or CeO_2 on Ni appear to be optimum for achieving a well textured, dense, lattice-matched and thermal-expansion-matched surface on which to grow the surface REBCO layer. There are several microstructural issues of

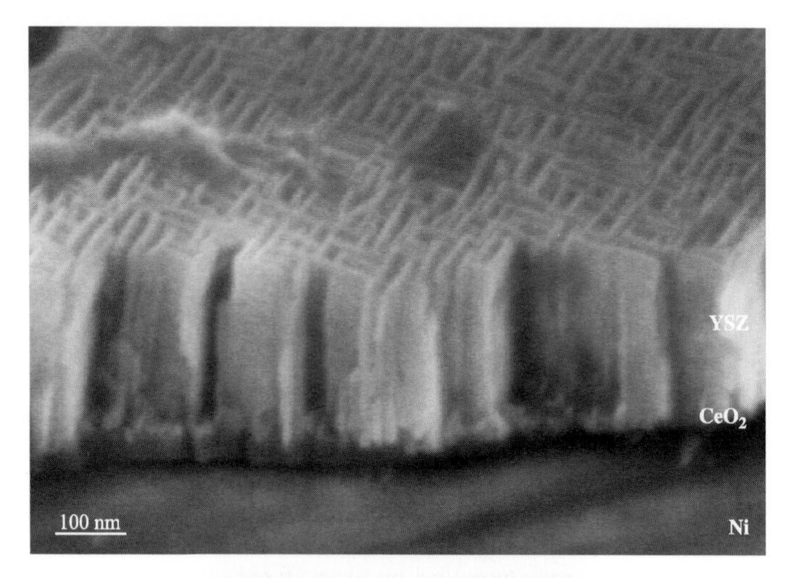

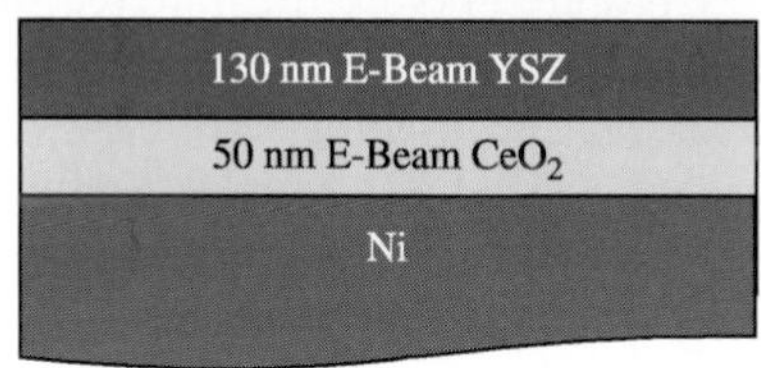

Figure B3.2.4.4. High resolution scanning electron micrograph of a fracture surface of a YSZ film grown on CeO_2 on RABiTs Ni, after Yang *et al* [8].

buffer layers on RABiTs substrates which need to be studied further. Typically, YSZ and CeO_2 layers grown by PVD routes have finger-like, columnar morphologies which can be relatively rough and sometimes porous. The columnar morphology and porosity is carried through into the surface REBCO film. Figure B3.2.4.4 shows a high resolution scanning electron micrograph of a fracture surface of a YSZ film grown on CeO_2 on RABiTs Ni [8]. The columnar grain structure of approximate dimensions $50 \times 10 \times 150$ nm is evident in the YSZ. It is speculated that a defective, columnar microstructure could be beneficial to intergranular pinning. However, it seems likely that columnar growth is also detrimental to mechanical properties. A further uncertainty to be resolved is how the Ni surface roughness and oxidation influences the buffer layer morphology and subsequent REBCO morphology and properties.

B4.2.1
B4.1

Buffer layer growth by IBAD

An alternative 'substrate + buffer' template onto which REBCO can be grown is an untextured Ni-based metallic substrate with a biaxially textured ceramic buffer layer formed by ion beam assisted deposition (IBAD). A group led by Iijima at the Fujikura Company in Japan first demonstrated the formation of texture in a ceramic thin film on an untextured substrate by utilizing IBAD [9]. A dual ion beam sputtering technique was used, with an Ar ion sputtering source directed at a YSZ buffer target, and a second, assisting, Ar + O ion source directed at the hastealloy substrate. Several years on, the technique has been further developed by several groups, worldwide, to demonstrate metre length buffered substrates.

B4.2.1

We shall not discuss the theory of texture evolution by ion bombardment, except to point out that there are two basic models which invoke theories of ion channelling [10] and differential sputtering [11] to explain texture formation. Of importance here are the predictions of the models in terms of practical implementation of IBAD: film growth is slow (0.05–$0.4\,\mathrm{nm\,s^{-1}}$) and the quality of the overall film texture is strongly dependent on ion beam incident angle. The ion beam divergence should be kept as small as possible in order to give minimum deviation from the channelling angle and, thus, sharp $\Delta\phi$ values. This is particularly important for process scale-up where long, linear ion sources will be required for buffer coating of long sections of hastealloy tape.

YSZ and CeO_2 may not be the ideal buffers since they grow in the form of columnar grains, whereby the texture improves with thickness. A minimum layer thickness for adequate in-plane texture (of $\sim 10°$ FWHM in ϕ) requires a buffer layer thickness of $\sim 1\,\mu m$ and processing time $\sim 3\,h$. A more suitable buffer layer might be MgO, which exhibits discontinuous growth involving stacking of three-dimensional islands [12]. This type of growth process means that the texture does not improve with film thickness. Indeed, high quality films (in-plane texture, $\Delta\phi$, $\sim 7°$) are observed for MgO film thicknesses of only $\sim 10\,nm$. The challenges to be overcome for MgO buffers are (a) since the optimum MgO layer is very thin, a very smooth substrate is required otherwise the film surface is rough, (b) the MgO film is highly defective as a result of the ion beam interaction and is therefore unstable to the atmosphere (this means that further capping layers are required before the buffered substrate can be translated from the IBAD chamber to the YBCO deposition chamber).

D1.1.2

The ultimate success of IBAD for producing long conductor lengths is the cost and scaleability of the tape fabrication. A commonly cited cost target to make superconductors attractive for large scale application is \$10/kA-m. The IBAD and RABiTs routes to substrate/buffer production are roughly two orders of magnitude more costly than the target number. Nevertheless, several groups are now demonstrating the feasibility of IBAD scale-up with metre length conductors being produced using continuous, reel-to-reel coating of the buffer and subsequent REBCO layer.

B4.2.1 B4.1

An alternative to IBAD for formation of a biaxially textured buffer on polycrystalline substrate is using a technique called 'inclined substrate deposition' (ISD). The method was first demonstrated at the Sumitomo Osaka Research Center. YSZ was deposited on a polycrystalline hastealloy substrate which was inclined with respect to a laser plume [13, 14]. While the method is relatively cheap and straightforward compared to IBAD, the in-plane texture has, so far, not been as good.

B4.2.1

Buffer layer growth by direct oxidation of metallic substrate

The prospect of using an unbuffered metallic substrate for REBCO conductors is very attractive. The benign nature of Ag makes it the only possible candidate material, but ways to form a sharp Ag texture have not yet been found. The next best alternative to an unbuffered metal is a surface oxidized metal. A Japanese group at ISTEC have been the first to demonstrate a surface oxidized Ni substrate [15]. A NiO(001) layer was formed on a Ni(001) substrate. YBCO has been shown to grow epitaxially on NiO(001). Many groups are now researching into control of the surface oxidation of Ni with the aim of forming the correct NiO(001) texture in a dense, smooth film. It appears that the oxidation is very sensitive to annealing conditions.

B3.2.4.5.3 REBCO thin and thick film fabrication routes and properties

PVD growth of REBCO

For the RABiTs, IBAD, and ISD methods of substrate/buffer production discussed in the previous section, thin film routes based on PVD have been used to produce the buffer and, in most cases, the REBCO layer. Typical current carrying properties for YBCO conductors grown by 'all thin film' routes

B4.2.1 B4.2.2

at 77 K and 0 T field over 1 m lengths, are $\sim 7 \times 10^5\ \mathrm{A\,cm^{-2}}$ for $>1\ \mu\mathrm{m}$ thick YBCO on RABiTs or IBAD substrates [16, 17] and $\sim 2 \times 10^5\ \mathrm{A\,cm^{-2}}$ on ISD substrates. For short conductor lengths ($\sim$1 cm) on RABiTs or IBAD substrates of a few μm thickness, J_cs are in excess of $1\ \mathrm{MA\,cm^{-2}}$ and the field dependence of the current is excellent, with J_c degradation of only one order of magnitude for fields in-the-plane of the substrate of ~ 10 T.

While the use of PVD ensures the production of high quality superconductor films with strong intrinsic flux pinning, the growth rates are slow ($<0.5\ \mathrm{nm\,s^{-1}}$) and there appears to be a thickness limitation of $<10\ \mu\mathrm{m}$, before epitaxial growth of the film breaks down. Many alternative, faster rate, deposition routes are being explored both for REBCO and oxide buffer growth on RABiTs substrates. Below, the most promising of these non-PVD routes are outlined, with emphasis on REBCO film growth rather than buffer layer growth, although the same basic principles apply to the buffer growth. B4.2.1

MOCVD and CVD growth of REBCO

Metal organic chemical vapour deposition (MOCVD) is used in several industries to produce uniform coatings of various materials on large area substrates. In MOCVD of REBCO, metal organics of the rare earth, barium, and copper are individually evaporated, transported in an inert carrier gas, and then mixed with oxidants such as O_2 and N_2O. The mixture is decomposed on a heated substrate to form an REBCO film at a rate of up to $1\ \mu\mathrm{m\,min^{-1}}$. So far, J_cs in excess of $10^5\ \mathrm{A\,cm^{-2}}$ at 77 K have been achieved on short, ceramic-buffered, RABiTs substrates. The main drawbacks of MOCVD are (a) film composition is very sensitive to substrate temperature; (b) the surface areas of the conventional solid state sources change during deposition leading to poor stoichiometry control; and (c) the available Ba sources are unstable and show poor volatility. B4.2.2 B4.1

Chemical vapour deposition is similar to MOCVD except that commonly available salts of the constituent cations of REBCO, such as the halides, are used as precursors instead of metal organics. The off-gases from both MOCVD and CVD are toxic and require special recovery procedures.

BaF_2 process for growth of REBCO

In the BaF_2 process, first pioneered at Bell laboratories in 1987, the RE, Cu and BaF_2 constituents are evaporated at atmospheric pressure onto a substrate at low temperatures ($\sim$100°C), at rates of up to $100\ \mathrm{nm\,s^{-1}}$. BaF_2 is used instead of Ba because of the instability of Ba to the processing atmosphere. The amorphous film is annealed, *ex situ*, at $\sim$700°C in a moisture-rich, reducing atmosphere, in order to crystallize the REBCO phase. The method has been successfully demonstrated on RABiTs substrates, yielding J_cs (77 K, 0 T) of $\sim 5 \times 10^{-5}\ \mathrm{A\,cm^{-2}}$ in films of $2.5\ \mu\mathrm{m}$ thickness, deposited at a rate of $0.3\ \mathrm{nm\,s^{-1}}$ [18]. The technique has considerable promise for coated conductor fabrication, although there are still uncertainties remaining to be solved, namely determination of (a) the mechanism of the REBCO crystallization and epitaxy formation; (b) the upper thickness limit on epitaxial film formation; and (c) the requirements of surface finish and texture in the underlying substrate. B4.2.1

LPE

Liquid phase epitaxy (LPE) is a common route for growth of semiconductor thin films [19]. The method is now being widely investigated for growth of REBCO thick films for conductor applications. A molten 'self-flux' of composition on the primary crystallization surface for REBCO is melted in a crucible, and the REBCO is grown, typically by top seeded solution growth, from this flux. The primary crystallization surface is indicated on a section through the Y–Ba–Cu–O phase diagram (between $YBa_2Cu_3O_{7-x}$ and $Ba_3Cu_5O_8$) in figure B3.2.4.5 [20]. A closely lattice-matched substrate is rotated near the top of the crucible and REBCO forms on this substrate 'seed'. Rotation at $\sim$100 rpm is necessary to B4.2.2

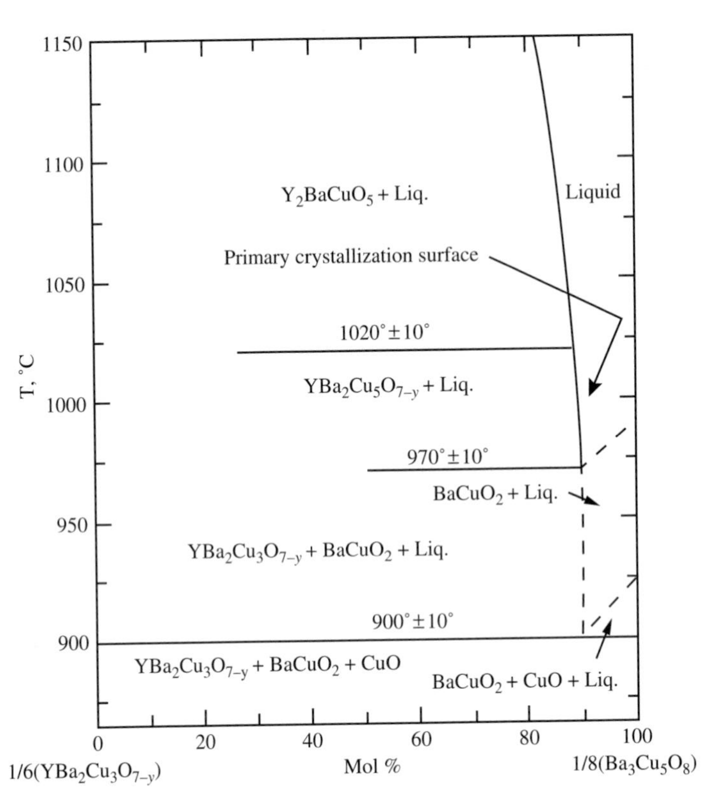

Figure B3.2.4.5. Section through the Y–Ba–Cu–O phase diagram (between $YBa_2Cu_3O_{7-y}$ and $Ba_3Cu_5O_8$) showing the primary crystallization surface which is the flux composition for LPE growth of YBCO, after Oka and Unoki [20].

ensure surface fluid flow from the centre of the crucible to the walls and so to prevent any floating crystals of YBCO from attaching to the substrate. In LPE, it is important to control supersaturation in order to control the growth mode of the film. Supersaturation, σ, is defined as $(n-n_e)/n_e$, where n = actual composition of the solute (REBCO) in the solvent (the primary crystallization liquid surface), and n_e = equilibrium concentration of the solute in the solvent. It is desirable to have a high supersaturation (by undercooling and addition of excess RE to the flux) in order to have a strong thermodynamic driving force for growth and two-dimensional nucleation of grains.

B4.2.2 A major advantage of LPE growth is that growth rates are rapid. Indeed, films can be grown at rates of up to 10 μm min^{-1}, although the atomic order in the most rapidly grown films may be rather low. YBCO (10 μm thick) has been grown on single crystal substrates with J_cs approaching 1 MA cm^{-2} at 77 K [21].

Owing to the high growth temperatures, substrate reactivity is a serious problem for the formation of LPE films on metallic substrates. However, initial results of YBCO grown on MgO buffered Ag C1 appear promising, with J_cs of up to 10^4 A cm^{-2} at 77 K [22]. Ways of reducing the flux temperature have been, and are continuing to be, investigated. While fluorine and silver additions can reduce the flux temperature by up to 100°C (to ~900°C) [22], other chemical means of reducing the flux temperature need to be sought to reduce this temperature further.

Sol–gel

B2.1 B4.3 The sol–gel technique is commonly used in the formation of electronic oxide thin films, such as ferroelectrics. For REBCO, organic solutions of the Ba- and RE- alkoxides are mixed in 2-methoxy ethanol, and copper oxide is mixed in pyridine. The mixture is vacuum distilled followed by hydrolysis to form a gel. The substrate is either dip-coated into the gel or the gel is spin coated onto the substrate. The layer is pyrolysed at ~200°C to vapourize the organic solvent, followed by application of another layer. The pyrolysis/layering sequence is repeated until a film of the desired thickness is obtained. Finally, crystallization of the film is carried out at ~800°C to yield the REBCO phase. While sol–gel has the advantage of being a non-vacuum technique and yields high film deposition rates, crystallization of the REBCO film requires very precise annealing conditions, and the upper limit on film thickness for epitaxial film formation is uncertain. The presence of organics also poses problems for waste gas disposal and there could be carbon contamination in the film from the organics.

Spray pyrolysis

Spray pyrolysis involves the dispensing of an aerosol solution of nitrates or carboxylate precursor solutions cation constituents onto a heated substrate. The carrier gas can be inert or partially oxidizing and ultrasonic evaporation of the solution may also be used to produce a mist. Film composition is controlled empirically by adjusting cation ratios in the spray solution, flow rates and substrate temperature. Spray pyrolysed films are typically rough and porous because they are formed from $>10\ \mu$m-sized particle droplets. Further, post-heat treatment is required to improve crystallinity in the films and to optimize the microstructure.

B2.1 B4.3

Results of spray pyrolysed YBCO films show J_cs of up to 2×10^4 A cm^{-2} in $\sim3\ \mu$m thick films deposited at a rate of $\sim0.1\ \mu$m min^{-1} on partially textured silver substrates [5]. The films are very dense with surface roughnesses of ~1000 Å. The level of biaxial texturing in the films is found to be equivalent to that in the substrate.

MOD

Metal organics decomposition (MOD) has similarities both to sol–gel and spray pyrolysis. Trifluoroacetates of RE, Ba and Cu are used in an organic solvent and the precursor solution is applied to the substrate by dip or spin coating. The reaction temperatures and resultant film morphology are similar to films grown by spray pyrolysis. There have been no reports, so far, of the use of the technique for making thick film YBCO on textured metallic substrates for potential conductors.

B2.1 B4.3

References

[1] Dimos D, Chaudhari P, Mannhart J and LeGoues F 1988 Orientation dependence of grain-boundary critical currents in $YBa_2Cu_3O_{7-\delta}$ bicrystals *Phys. Rev. Lett.* **61** 219–222

[2] Browning N D, Buban J P, Nellist P D, Norton D P, Chisholm M F and Pennycook S J 1998 The atomic origins of reduced critical currents at [001] tilt grain boundaries in $YBa_2Cu_3O_{7-\delta}$ thin films *Physica* C **294** 183–193

[3] Heinig N F, Redwing R D, I-Fei-Tsu, Gurevich A, Nordman J E, Babcock S E and Larbalestier D C 1996 Evidence for channel conduction in low misorientation angle [001] tilt $YBa_2Cu_3O_{7-x}$ bicrystal films *Appl. Phys. Lett.* **69** 577–579

[4] Safar H, Coulter J Y, Maley M P, Foltyn S, Arendt P, Wu X D and Willis J O 1995 Anisotropy and Lorentz-force dependence of the critical currents in $YBa_2Cu_3O_{7-\delta}$ thick films deposited on nickel-alloy substrates *Phys. Rev.* B **52** R9875–R9878

[5] Wells J J, and MacManus-Driscoll J L 1999 High rate deposition $YBa_2Cu_3O_{7-x}$ thick films on oxide and silver single crystal substrates by spray pyrolysis *unpublished*

[6] Wells J J, Crossley A L and MacManus-Driscoll J L 1999 In-plane aligned YBCO thick films on {110} rolled and single crystal silver by ultrasonic mist pyrolysis *IEEE. Trans. Supercond.* submitted (ASC 1998 Conference, Palm Desert, CA, Sept. 1998)

[7] Goyal A, Norton D P, Budai J D, Paranthaman M, Specht E D, Kroeger D M, Christen D K, He Q, Saffian B, List F A, Lee D F, Marton P M, Klabunde D E, Hartfiel E and Sikka V K 1996 High critical current density superconducting tapes by epitaxial deposition of $YBa_2Cu_3O_x$ thick films on biaxially textured metals *Appl. Phys. Lett.* **69** 1795–1797

[8] Yang C-Y, Babcock S E, Goyal A, Paranthaman M, List F A, Norton D P, Kroeger D M and Ichonse A 1998 Microstructure of electron-beam-evaporated epitaxial yttria-stabilized zirconia/CeO_2/bilayers on biaxially textured Ni tape *Physica* C **307** 87–98

[9] Iijima Y, Tanabe N and Kohno O 1992 Biaxially aligned YSZ buffer layer on polycrystalline substrates *Advances in Superconductivity IV: Proc. 4th Intl. Symp. Supercond. (ISS)* (Tokyo: Springer) pp 679–682

[10] Dobrev D 1982 Ion beam assisted texture formation in vacuum-condensed thin metal films *Thin Solid Films* **92** 41–53

[11] Bradley R M, Harper J M E and Smith D A 1986 Theory of thin film orientation by ion bombardment during deposition *J. Appl. Phys.* **60** 4160–4163

[12] Wang C P, Do K B, Beasley M R, Geballe T H and Hammond R H 1997 Deposition of in-plane textured MgO on amorphous Si_3N_4 substrates by ion-beam-assisted deposition and comparisons with ion-beam-assisted deposited yttria-stabilized-zirconia *Appl. Phys. Lett.* **71** 2955–2957

[13] TEPCO, 1996 Sumitomo Report 1 Meter YBCO Tape with J_c of 1.5×10^5 A cm^{-2} *Supercond. Week.* **10**

[14] Hasegawa K 1997 *In-plane Aligned YBCO Thin Film Tape Fabricated by All Pulsed Laser Deposition: Spring MRS Meeting* (*San Francisco*)

[15] Matsumoto K, Kim S B, Wen J G, Hirabayashi I, Watanabe T, Uno N and Ikeda M 1999 Fabrication of in-plane aligned YBCO films on polycrystalline Ni tapes buffered with surface-oxidized NiO layers *IEEE Trans. Appl. Superconductivity* **9** 1539–1542

[16] Goyal A 1997 *Status of the RABiTs Approach to Fabricate Biaxially Aligned, High J_c Superconductors: Spring MRS Meeting* (*San Francisco*)

[17] LANL superconductivity group *private communication* 1999

[18] Solovyov V F, Wiesmann H J, Suenaga M and Feenstra R 1998 Thick $YBa_2Cu_3O_7$ films by post-annealing of the precursor by high rate e-beam deposition on $SrTiO_3$ substrates *Physica* C **309** 269–274

[19] Small B M, Giess E A and Ghez R 1994 *Handbook of Crystal Growth* **vol. 3**, ed D J Hurle (Amsterdam: Elsevier) pp 223–253

[20] Oka K and Unoki H 1990 Primary crystallization fields and crystal growth of $YBa_2Cu_3O_{7-y}$ *J. Crys. Growth* **99** part-2 922–924

[21] Miura S, Hashimoto K, Wang F, Enomoto Y and Morishita T 1997 Structural and electrical properties of liquid phase epitaxially grown $YBa_2Cu_3O_x$ films *Physica* C **278** 201–206

[22] Niiori Y, Yamada Y and Hirabayashi I 1998 Low temperature LPE growth of $YBa_2Cu_3O_{6+x}$ thick film on silver substrate using silver saturated Ba–Cu–O–F flux *Physica* C **296** 65–68

Further Reading

MacManus-Driscoll J L 1998 Recent advances in processing of high H_{irr} conductors *Annu. Rev. Mater. Sci.* **28** 421–462.

Goyal A *et al* 1997 Conductors with controlled grain boundaries: an approach to the next generation, high temperature superconducting wire *J. Mater. Res.* **12** 29240.

Klemenz C and Scheel H J 1993 Liquid phase epitaxy of high-T_c superconductors *J. Crys. Growth* **129** 421–428.

Gurevich A, Babcock SE, Cai XY, Heinig N, Larbalestier DC, Pashitskii EA, I Fei-Tsu 1997 Mechanisms of current transport through grain boundaries in high-temperature superconductors. *In Workshop on Superconductivity (The 3rd Joint ISTEC/MRS Workshop). Program and Extended Abstracts.* (*ISTEC-Int. Supercond.* Tokyo, Japan: Technol. Center) pp. 116–119.

Arendt P, Foltyn S, Xin-Di Wu, Townsend J, Adams C, Hawley M, Tiwari P, Maley M, Willis J, Moseley D, Coulter Y 1995 Fabrication of biaxially oriented YBCO on (001) biaxially oriented yttria-stabilized-zirconia on polycrystalline substrates *Epitaxial Oxide Thin Films and Heterostructures: Proc. Fall Materials Research Society Meeting* (Pittsburgh, PA, USA: Mater. Res. Soc.) pp. 209–214.

B3.2.5
Processing of high T_c conductors: the compound Hg(1223)

X Obradors and A Sin

B3.2.5.1 Introduction

The superconductor family $HgBa_2Ca_{n-1}Cu_nO_y$ ($n = 1, 2, 3, \ldots$) has been intensively studied since its discovery in 1993 [1]. The $n = 3$ member of this Hg-superconductor series has the record T_c of 135 K at ambient pressure [2] and under high pressures a T_c of 160 K may be reached (around 30 GPa) [3]. Such material may display current densities as high as 10^6 A cm^{-2} at 77 K and 10^5 A cm^{-2} at 110 K and 1 T [4]. Irreversibility fields above 5 T at 100 K can be also achieved [5]. These characteristics demonstrate the high potential that these superconductors have. The properties and future potential of mercury superconductors are fully described in chapter C1. The synthesis process of mercury superconductors is relatively complex because of the high vapour pressure of Hg which requires preparation of the materials in confined vessels avoiding the loss of Hg vapours. The ceramic materials are usually prepared either at high pressures (0.1–7.5 GPa) or by means of the sealed quartz tube technique ($<$ 5 MPa). C4

The preparation of films (thin and thick films) usually consists of two steps: (a) deposition of the precursor phase and (b) a mercuration process using a source of Hg vapour within the same closed tube. Epitaxial thin films on single crystalline substrates can be achieved with this technique similarly to Tl superconductors. B4.2.1 B4.3

Interest in tape synthesis has been very limited up to now due to the difficulty to find non-reactive metallic substrates. In this chapter we discuss the fundamental issues which must be considered when processing Hg superconductors, including the preparation of ceramics, thin films and tapes.

B3.2.5.2 Ceramic synthesis

The main difficulty in the synthesis of ceramic mercurocuprates is the decomposition at low temperatures of mercury compounds. The HgO is used as a source of mercury in all the synthesis of mercurocuprates. This oxide decomposes into metallic Hg vapour and oxygen at 1 bar and 460°C. Thus, to avoid the loss of stoichiometry the synthesis of these compounds has to be carried out in closed vessels or in high pressure chambers. Basically, two different processing routes may be followed to prepare the raw materials for the superconducting synthesis: (a) the direct method [6] and (b) the precursor method [7]. The direct method consists of the preparation of a green pellet containing a mixture of the stoichiometric simple oxides required for the superconducting phase. In the precursor method, on the other hand, a multiphase compound corresponding to the stoichiometry $Ba_2Ca_{(n-1)}Cu_nO_x$ is first prepared. The synthesis of this precursor phase can be done starting from nitrates, carbonates or oxides. This first step of the synthesis of Hg superconductors is very important to control because it has a strong influence on the formation B2.2.4 B2.1

of the superconducting phase [8, 9]. Then, the precursor multiphase powder is mixed in stoichiometric form with the required HgO and pelletized. Moreover, the binary precursors used for the synthesis are very sensitive to atmospheric CO_2 and moisture. It is therefore necessary to handle the materials in controlled atmosphere to avoid exposure to air. In addition, it is essential to pay attention to and maintain safety procedures due to the toxicity of mercury and the mercury compounds.

C4 High pressure synthesis of Hg superconductors allows production of optimal quality materials for all the members of the series $HgBa_2Ca_{n-1}Cu_nO_y$. This type of synthesis involves the use of external high pressures (0.1–7.5 GPa) to suppress the decomposition of HgO and prevent mercury losses. The external pressure can be generated mainly by two techniques: (a) the belt-type system [1] and (b) hot isostatic pressing (HIP) [10, 11]. The belt-type apparatus apply a mechanical pressure on the samples and allows generation of pressures of around 7.5 GPa. However, their applicability is limited because the sample size is very small. The HIP technique allows increase in the quantity of material and uses lower pressures (0.2 GPa) than belt-type systems. This technique may be useful to process materials for applications, since the sintered powders may reach high density and the ceramics can be shaped into specific forms by using stiff metal covers.

The first three members of the series $HgBa_2Ca_{n-1}Cu_nO_y$ can also be synthesized by means of the sealed quartz tube [6, 9], while the higher members require high pressure synthesis. With this technique it turns out to be very beneficial to carry out a partial substitution of mercury by higher valence cations, such as Re, because the precursors become chemically stable against CO_2 and moisture.

The synthesis of the mercurocuprates may be modelled as a solid–gas reaction [12, 13] (figure B3.2.5.1(*a*)). Therefore, one of the most important synthesis parameters to stabilize the superconductor phase is the total pressure produced by $Hg_{(g)}$ and $O_{2(g)}$ inside the quartz tube. The pressure influences the competition between the formation of Hg12(n − 1)n cuprates and the binary compound $HgCaO_2$ [13, 14]. This last binary oxide turns out to be the hardest impurity to eliminate, remaining frequently in the final product. In figure B3.2.5.1(*b*) a typical thermal treatment profile for this type of synthesis is shown. It is important to use a high heating ramp to avoid the presence of $HgCaO_2$ and a moderate cooling ramp rate to allow the gas absorption by the solid phase.

C4 The development of an *in situ* pressure monitoring system within the quartz tube (thermobaric analyzer [13]) has allowed detailed analysis of the reaction mechanism of the mercurocuprates and determination of the optimized synthesis parameters, such as filling factor, heating and cooling ramps, etc [9, 13] (figure B3.2.5.2). Additionally, this system allows one to determine the thermodynamic A4 stability conditions of the different phases, an essential step to control the crystalline grain growth and the sintering process (figure B3.2.5.3). The critical currents and the magnetic irreversibility line of A1.3 polycrystalline Hg-based superconductors have been investigated by several authors and it has been found that at high temperatures the magnetic irreversibility is mainly associated with surface A4.2 barriers [15–18]. Bulk pinning of vortices is evidenced only at lower temperatures through the

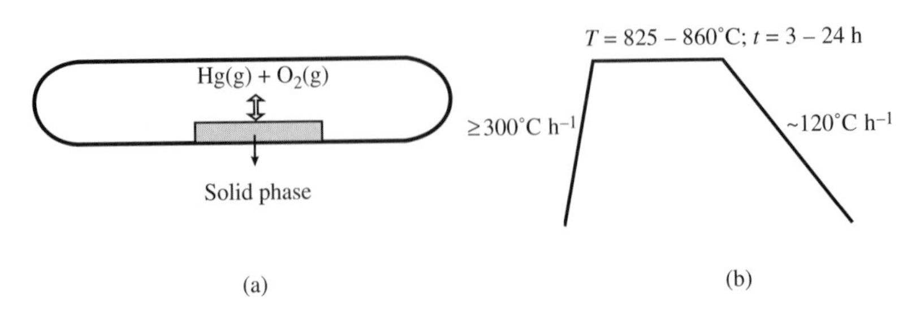

Figure B3.2.5.1. (*a*) Scheme of the gas–solid reaction inside the quartz tube. (*b*) Diagram of a general thermal treatment to prepare (Hg, M)-1223 within a quartz tube.

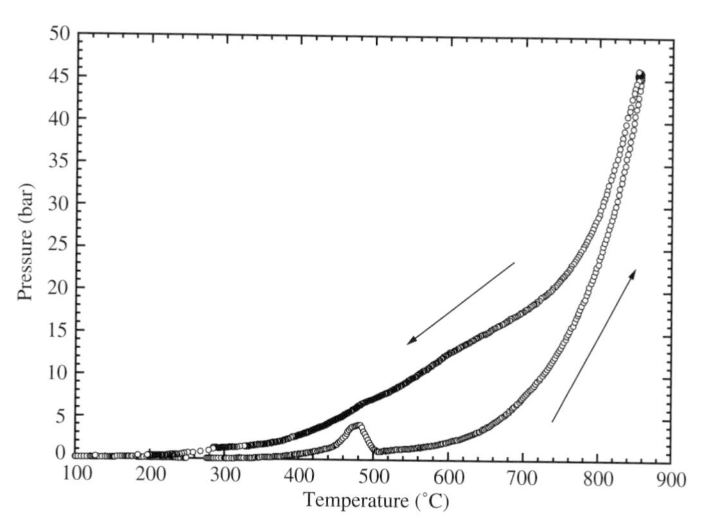

Figure B3.2.5.2. Pressure versus temperature curve of a typical (Hg,Re)-1223 synthesis. The anomaly observed at $T \approx 480°C$ corresponds to the formation of the binary compound $HgCaO_2$.

Figure B3.2.5.3. SEM image of a $(Hg,Re)Ba_2Ca_2Cu_3O_{8+\delta}$ ceramic sample where the platelet-like morphology of the crystals may be observed.

development of symmetric hysteresis loops. Enhanced bulk pinning was suggested to appear with partial Re doping which could indicate a decrease of the intrinsic anisotropy [16, 19].

B3.2.5.3 Thin film synthesis

B4.2.1 E4 The deposition of epitaxial thin films of mercury-based superconductors in single crystalline substrates has its own interest due to the possible electronic applications of these materials. Additionally, however, it must be considered as a necessary step towards the development of coated conductors over metallic substrates, as required for practical use in electrotechnical applications.

The activity developed in this field may be considered as moderate due to the difficulties in mastering the solid–gas reaction. The most common method used to perform the synthesis is based on two steps. The first one consists of preparing a deposition of the precursor multiphase $Ba_2Ca_{(n-1)}Cu_nO_x$ mixture in a substrate. This deposition can be carried out by different techniques such as laser ablation, rf sputtering, spray pyrolysis, sol–gel, etc [4, 20–26] The second step involves a mercurization process similar to that described for ceramic materials.

B4.1 The choice of single crystalline substrates to grow mercury-based superconductors has been large, with reasonably good quality in the following cases: MgO, $SrTiO_3$, $LaAlO_3$ and YSZ. The lattice mismatch appears to be minimal in the case of $SrTiO_3$, where we have $a = 3.89$ and 3.82 Å for the substrate and the Hg-1223 phase.

B4.2.1 The deposition process defines the final thickness of the film, which is limited to about 0.1–0.2 μm in the case of laser ablation or sputtering [4, 20–23]. In these cases several authors have also prepared the samples through an intercalation at room temperature of alternate layers of $Ba_2Ca_{(n-1)}Cu_nO_x$ and HgO precursors, leaving a HgO layer on top to avoid the carbonation of the precursor multiphase deposited material.

B4.3 Other techniques may be used to prepare thicker films. Spray pyrolysis uses a precursor nitrate solution to generate an aerosol through a pneumatic nebulizer or by means of piezoelectric ultrasonic excitation [24, 25]. The aerosol is carried by an inert gas (Ar) to a preheated substrate where the deposition occurs and the solvent is vapourized. This technique allows production of films with thickness in the range 1–10 μm. After the deposition a higher temperature anneal is necessary to remove all the traces of nitrates and to adjust the oxygen content of the precursor film.

B4.2.2 The most common procedure to carry out the vapour reaction of the deposited films with Hg is by means of the quartz tube technique, similar to the reaction leading to ceramic materials. The thermodynamic conditions where the superconducting phase may be formed are very similar to those found in ceramic materials, thus the temperatures and pressures which must be used to prepare single phase materials do not differ too much either.

C4 The critical currents achieved in thin films prepared by the different techniques described so far appear to be very high ($>10^6$ A cm^{-2} at 77 K), and at low fields they still remain above 10^5 A cm^{-2} at 110 K [21, 24]. Very recently, it has been found that thin films or single crystals with partial substitution of Hg by Re or Pb ions led to enhanced superconducting properties, besides the chemical stability effect already mentioned before [26]. The origin of this phenomenon is still controversial and it has been ascribed either to an enhanced metallization of the charge transfer block $(Hg,Re)O_\delta$ separating the superconducting CuO_2 planes or to an overdoping effect [19, 23, 27, 28].

B3.2.5.4 Tape preparation

B3.1 The reports of successful preparation of mercurocuprate superconducting tapes are very limited up to now. The first difficulty with the well known technique of powder in tube is that of the solubility of Ag in Hg, which can modify the inner tube Hg pressure during the high temperature annealing heat treatments. This loss of Hg stoichiometry may be compensated for by introducing small amounts of excess Hg which then improve the final phase purity. This process was used by Peacock *et al* [29] to prepare $HgBa_2CuO_{4+\delta}$ tapes and allowed them to reach nearly single-phased samples. A similar procedure has been followed in the preparation of Tl-based superconducting tapes [30]. However, in the Hg superconductors any uniaxial texture was induced and the phase purity at the grain boundaries was unknown. As a consequence the transport critical currents of the $HgBa_2CuO_{4+\delta}$ tapes prepared by powder in tube were very low. Further effort on the preparation of the higher T_c phases should be carried out to ascertain if the powder in tube technique is suitable for these materials.

A1.3 B3.1

More successful results have been achieved when preparing coated tapes consisting of cold rolled Ni flexible substrates with buffer layers of Cr or Cr+Ag having a thickness between 5 and 50 nm deposited by sputtering [31, 32]. The low solubility of Ni in Hg avoided any noticeable degradation of the substrate and the intermediate metallic layers were introduced to improve the adherence of the multiphase $Ba_2Ca_{(n-1)}Cu_nO_x$ precursor. The deposition of the precursor may be carried out by spraying to obtain a thickness in the range 10–40 μm while a cold rolling process of the substrate enhanced the compacity and the uniaxial texture of the final tape. Finally, a closed tube annealing with Hg vapour similar to that described for thin film preparation must be performed to grow textured tapes. The microstructure of the crystalline coatings prepared with this procedure appears to be onion-skin like, with the platelets oriented parallel to the metallic substrate. Some additives (HgX_2, X = Cl, I, F) were claimed to promote the grain growth of Hg-1223 while the use of a topotactic reaction to grow the Hg-1223 phase from small Hg-1212 grains previously oriented mechanically by cold rolling also seems to lead to an improvement in the quality of the coated tapes [32].

B4.2.1

Transport critical current densities as high as 7×10^4 A cm^{-2} at 77 K were reported for these tapes, when a uniaxial texture is achieved [32]. Attempts to prepare thick films deposited directly on metallic Ag substrates have been only partially successful, probably due to the formation of an amalgam at the interface with the superconductor [33, 34]. Further enhancement of J_c should be expected either through improvements of the uniaxial texture or if an additional in-plane texture is achieved by using biaxially textured substrates, similar to those prepared by IBAD or RABIT, which have led to $YBa_2Cu_3O_7$ second generation tapes with very high critical currents at 77 K. An additional advantage of second generation Hg-1223 tapes with biaxial texture should be the improved capability to generate fields at higher temperatures, near 100 K thus minimizing the cryogenic costs. It appears then that Ni-based coated conductors of Hg-1223 seem to be very promising candidates for high current applications.

D2.2 B4.2.1 F

B3.2.5.5 Summary

The synthesis of Hg-based superconductors is a complex issue due to the need to control the Hg vapour pressure at high temperature where a gas–solid reaction occurs. The progress in understanding the equilibrium properties of this reaction and the discovery of the stabilization effect of Re doping have promoted a great progress in the achievement of reliable processing techniques of Hg-based superconducting materials. It is particularly encouraging that several thin film deposition techniques have been successfully used to grow high quality materials. Unfortunately, there is an important hindrance in the development of Hg-based tapes by means of the powder in tube technique due to the formation of Ag–Hg amalgams which modify the Hg vapour pressure and, hence, make the process very unreliable. On the other hand, the use of metallic coatings chemically compatible with Hg seems to be a promising way to achieve Hg-based coated conductors.

C4

Troubleshooting in the synthesis of Hg-12($n-1$)n superconductors

Symptom	Remedy
Presence of $HgCaO_2$	Increase heating ramp ($>300°C\,h^{-1}$) Increase the filling factor ($>0.7\,g\,cm^{-3}$)*
Presence of low T_c members ($n = 1, 2$)	Moderate the cooling ramp ($\sim 120°C\,h^{-1}$) Increase the filling factor ($>0.7\,g\,cm^{-3}$)* $pO_2 \leq 0.2$ bar in precursor annealing Increase the synthesis temperature
Low T_c	$pO_2 \leq 0.2$ bar during precursor annealing ($n \geq 3$) $pO_2 \geq 0.2$ bar during precursor annealing ($n < 3$)
Low superconducting volume fraction	Improve homogeneity of precursors: • Additional thermal treatments • Sol–gel precursor synthesis Increase the synthesis time ≥ 3 h* Addition of excess of Hg(l)

* Only quartz tube synthesis.

Hazards

- Explosion of quartz ampoule due to an excessive pressure generation. Decrease the maximum temperature or protect the sample to avoid the chemical reaction of the sample with the quartz tube.
- All the high temperature anneals must be carried out in special, isolated and well aired laboratory rooms to avoid any human contact with Hg vapours in the case that a quartz ampoule explodes.

References

[1] Putilin S N, Antipov E V, Chmaissen O and Marezio M 1993 Superconductivity at 94 K in $HgBa_2CuO_{4+\delta}$ *Nature* **362** 226

[2] Schilling A, Cantoni M, Guo J D and Ott H R 1993 Superconductivity above 130 K in the Hg–Ba–Ca–CuO system *Nature* **363** 56

[3] Chu C W, Gao L, Chen F, Huang Z H, Meng R L, Xue Y Y, Chu C W, Gao L, Chen F, Huang Z J, Meng R L and Xue Y Y 1993 Superconductivity above 150 K in $HgBa_2Ca_2Cu_3O_{8+x}$ at high pressures *Nature* **356** 323

[4] Krusin-Elbaum L, Tsuei C C and Gupta A 1995 High current densities above 100 K in the high-temperature superconductor $HgBa_2CaCu_2O_{6+\delta}$ *Nature* **373** 679

[5] Krusin-Elbaum L, Blatter G, Thomson J R, Petrov D K, Wheeler R, Ullmann J and Chu C W 1998 Anisotropic rescaling of a splayed pinning landscape in Hg-cuprates: strong vortex pinning and recovery of variable range hopping *Phys. Rev. Lett.* **81** 3948

[6] Meng R L, Beauvais L, Zhang X N, Huang Z J, Sun Y Y, Xue Y Y and Chu C W 1993 Synthesis of the high temperature superconductors $HgBa_2CaCu_2O_{6+x}$ and $HgBa_2Ca_2Cu_3O_{8+\delta}$ *Physica* C **216** 21

[7] Fukuoka A, Tokiwa-Yamamoto A, Itoh M, Usami R, Adachi S and Tanabe K 1997 Dependence of T_c and transport properties on the Cu valence in $HgBa_2Ca_{n-1}Cu_nO_{2(n+1)+\delta}$ ($n = 2, 3$) *Phys. Rev.* B **55** 6612

[8] Loureiro S M, Stott C, Philip L, Gorius M F, Perroux M, Le Floch S, Capponi J J, Xenikos D, Toulemonde P and Tholence J L 1996 The importance of the precursors in high pressure synthesis of Hg-based superconductor *Physica* C **272** 94

[9] Sin A, Cunha A G, Calleja A, Orlando M T D, Emmerich F G, Baggio Saitovich E, Segarra M, Piñol S and Obradors X 1999 Influence of precursor oxygen stoichiometry in the formation of Hg, Re-1223 superconductors *Supercond. Sci. Technol.* **12** 120

[10] Lechter W, Toth L, Osofky M, Skelton E, Soulen R J Jr, Qadri S, Schwartz J, Kessler J and Wolters C 1995 One-step reaction and consolidation of Hg based high-temperature superconductors by hot isostatic pressing *Physica* C **249** 213

[11] Tampieri A, Calestani G, Celotti G, Micheletti C and Rinaldi D 1998 Preparation of Hg-1201 superconductor by hot isostatic pressing *Physica* C **298** 10

[12] Alyoshin V A, Mikhailova D A and Antipov E V 1995 Synthesis of $HgBa_2CuO_{4+\delta}$ under controlled mercury and oxygen pressures *Physica* C **255** 173

[13] Sin A, Cunha A G, Calleja A, Orlando M T D, Emmerich F G, Baggio-Saitovich E, Piñol S, Segarra M and Obradors X 1998 Pressure-controlled synthesis of the $Hg_{0.82}Re_{0.18}Ba_2Ca_2Cu_3O_{8+\delta}$ superconductor *Adv. Mater.* **10** 1126

[14] Sin A, Cunha A G, Calleja A, Orlando M T D, Emmerich F G, Baggio-Saitovich E, Piñol S, Chimenos J M and Obradors X 1998 Formation and stability of $HgCaO_2$, a competing phase in the synthesis of $Hg_{1-x}Re_xBa_2Ca_2Cu_3O_{8+\delta}$ superconductor *Physica* C **306** 34

[15] Kim Y C, Yhompson J R, Christen D K, Sun Y R, Paranthaman M and Specht E D 1995 Surface barriers, irreversibility line, and pancake vortices in an aligned $HgBa_2Ca_2Cu_3O_{8+\delta}$ superconductor *Phys. Rev.* B **52** 4438

[16] Fábrega L, MartÍnez B, Fontcuberta J, Sin A, Piñol S and Obradors X 1998 The Re-doped high T_c superconductor $HgBa_2Ca_2Cu_3O_x$: magnetic irreversibility versus anisotropy *J. Appl. Phys.* **83** 7309

[17] Fábrega L, MartÍnez B, Fontcuberta J, Sin A, Piñol S and Obradors X 1998 Surface barriers and magnetic irreversibility in grain-aligned Re-doped Hg-1223 *Physica* C **296** 29

[18] Krelaus J, Reder M, Hoffman J and Freyhardt H C 1999 Magnetization and relaxation in Hg-1223: bulk vs. surface irreversibility, anisotropy and the influence of Re-doping *Physica* C **314** 81

[19] Shimoyama J I 2000 Control of pinning strength in Hg- and Bi-based superconductors *Supercond. Sci. Technol.* **13** 43

[20] Yun S H, Wu J Z, Kang B W, Ray A N, Gadup A, Yang Y, Farr R, Sun G F, Yoo S H, Xin Y and He W S 1995 Fabrication of c-oriented $HgBa_2Ca_2Cu_3O_{8+\delta}$ superconducting films *Appl. Phys. Lett.* **67** 2866

[21] Yun S H, Wu J Z, Tidrow S C and Eckart D W 1996 Growth of $HgBa_2Ca_2Cu_3O_{8+\delta}$ thin films on $LaAlO_3$ substrates using fast temperature ramping Hg-vapour annealing *Appl. Phys. Lett.* **67** 2866

[22] Yun S H and Wu J Z 1996 Superconductivity above 130 K in high-quality mercury-based cuprate thin films *Appl. Phys. Lett.* **68** 862

[23] Yun S H, Pedarnig J D, Rössler R, Bäuerle D and Obradors X 2000 In-plane and out-of-plane resistivities of vicinal Hg-1212 thin films *Appl. Phys. Lett.* **77** 1369

[24] Moriwaki Y, Sugano T, Gasser C, Fukuoka A, Nakanishi K, Adachi S and Tanabe K 1996 Epitaxial $HgBa_2Ca_2Cu_3O_y$ films on $SrTiO_3$ substrates prepared by spray pyrolysis technique *Appl.Phys. Lett.* **69** 3423

[25] Sin A, Supardi Z, Sulpice A, Odier P, Weiss F, Ortega L and Núñez-Regueiro M 2001 Synthesis by aerosol process of superconductor films and buffer layer materials *IEEE Trans. Appl. Supercond.* **11** 2877

[26] Moriwaki Y, Sugano T, Tsukamoto A, Gasser C, Nakanishi K, Adachi S and Tanabe K 1998 Fabrication and properties of *c*-axis Hg-1223 superconducting thin films *Physica* C **303** 65

[27] Tallon J L, Bernhard C, Niedermayer C h, Shimoyama J, Hahakura S, Yamaura K, Hiroi Z, Takano M and Kishio K 1996 A new approach to the design of high-T_c superconductors: metallised interlayers *J. Low Temp. Phys.* **105** 1379

[28] Fábrega L, Fontcuberta J, Calleja A, Sin A, Piñol S and Obradors X 1999 Muon spin relaxation in Re substituted $HgA_2Ca_{n-1}Cu_nO_{2n+2+\delta}$ ($A = Sr, Ba$; $n = 2, 3$) superconductors *Phys. Rev.* B **60** 7579

[29] Peacock G B, Gameson I, Edwards P P, Khaliq M, Yang G, Shields T C and Abell J S 1997 Fabrication of high-temperature superconducting $HgBa_2CuO_{4+\delta}$ within silver-sheathed tapes *Physica* C **273** 193

[30] Jeong D Y, Kim H K and Kim Y C 1999 Much enhanced J_c by intermediate rolling in just-rolled Tl-1223/Ag tapes *Physica* C **314** 139

[31] Meng R L, Hickey B, Wang Y Q, Sun Y Y, Gao L, Xue Y Y and Chu C W 1996 Processing of highly oriented $(Hg_{1-x}Re_x)Ba_2Ca_2Cu_3O_{8+\delta}$ tape with $x \approx 0.1$ *Appl. Phys. Lett.* **68** 3177

[32] Meng R L, Wang Y Q, Lewis K, Garcia C, Gao L, Xue Y Y and Chu C W 1997 Fabrication and microstructure of Hg-1223 *Physica* C **282–287** 2553

[33] Sastry P V P S S, Su J, Atwell S L, Durbin S M and Schwartz J 2001 Fabrication and characterization of $(HgRe)Ba_2CaCu_2O_y$ thin films *IEEE Trans. Appl. Supercond.* **11** 3098

[34] Su J, Sastry P V P S S and Schwartz J 2001 Growth of $Hg_{0.8}Pb_{0.2}Ba_2Ca_2Cu_3O_{8+\delta}$ thick films on Ag using a modified process route *IEEE Trans. Appl. Supercond.* **11** 3118

B3.3.1
Overview of low T_c materials for conductor applications

René Flükiger

B3.3.1.1 Introduction

B3.3.1.1.1 Literature about low T_c superconductors

For more than 20 years, the highest reported superconducting transition temperature was $T_c = 23$ K, for the A15 type compound Nb_3Ge [1] and most of the known superconductors were either metallic or intermetallic compounds. Superconductivity had already been found in oxide systems, the highest transition temperature being reported for $LiTi_2O_4$, with $T_c = 13$ K [2]. After the discovery of a new class of perovskites with much higher T_c values by Müller and Bednorz [3] in 1986, the definitions 'Low T_c' and 'High T_c' superconductors were introduced for a better distinction. In the meantime, several classes of superconductors other than Cu-based perovskites, with T_c values well above that of Nb_3Ge were discovered. Some well known superconductors found after 1986 are listed in table B3.3.1.1. The first one discovered was the fullerene Rb_3C_{60} (or $RbCs_2C_{60}$), with $T_c = 28$ K [4], then the compound $Ba_{1-x}K_xBiO_3$ with $T_c = 32$ K [5], followed by the compounds $YPd_5B_3C_{0.3}$ [6] and finally, MgB_2 with $T_c = 39$ K [7]. C

The term 'Low T_c' superconductor is somewhat arbitrary, as there is no clear criterion used to make a clear distinction from other classes of superconductors. There is a common point between the known metallic and intermetallic superconductors: it is the large coherence length, of the order of 2.5 nm and more, in contrast to 'High T_c' superconductors, where this quantity does not exceed 1.5 nm. Since MgB_2 has a coherence length of $\xi_0 = 5.0$ nm, it is now customary to classify this material as a 'Low T_c' superconductor, in spite of its peculiar properties (two gaps, deviation from BCS theory). C5 B3.3.6

The superconducting materials already used for industrial applications, e.g. NbTi and Nb_3Sn, are already discussed in detail in chapter B3.3, as well as the systems Nb_3Al, $M_xMo_6S_8$ (Chevrel phases) and MgB_2. The scope of the present review is to discuss other 'Low T_c' superconducting materials, which never made their way to the market, but are interesting from the fundamental point of view, e.g. V_3Ga, Nb_3Ge and NbN. In addition, some additional information is given about the ordering effects on the superconducting properties of A15 type compounds. A particular interest is given to the properties of these materials at high fields, in view of their use in magnets producing fields above 20 T. The production of magnets with constantly increasing magnetic fields in view of their application for NMR spectrometry is an important issue for the field of superconductivity: the progress achieved in the last 20 years is shown in figure B3.3.1.1. B3.3 E1.1

Table B3.3.1.1. Binary and pseudobinary A15 type compounds with $T_c > 10$ K

Compound	T_c	Compound	T_c
Nb_3Ge(f)	23.0	V_3Si	17.0
$Nb_3(Al_{0.80}Ge_{0.20})$	21.0	V_3Ga	15.9
Nb_3Ga(q)	20.7	V_3Al (f)	12.0
Nb_3Al(q)	19.1		
Nb_3Sn	18.2	$Mo_{0.65}Re_{0.35}$ (f)	15
Nb_3Si(f)	> 17	$Mo_{0.40}Tc_{0.60}$	13.4
$Nb_3(Au_{0.75}Pt_{0.25})$	13.0	Mo_3Os	12.7
Nb_3Au	11.5	$W_{0.60}Re_{0.40}$ (f)	11
Nb_3Pt	11.1		

(f) compounds obtained by thin film deposition, (q) composition $\beta = 0.25$ obtained by quenching from high temperatures.

B3.3.1.1.2 Historical remarks on low T_c superconductors

A complete overview of all known 'Low T_c' superconductors would be far beyond the scope of the present Handbook. Data compilations are available where the formulas of the reported superconductors (elements and compounds) are listed in an alphabetical order. The first compilation was published in 1976 by Roberts [8] and contains the values of T_c, crystal structures and references. A much more detailed compilation, containing all known data from the discovery of superconductivity in 1911–1987, was published by the author in the Landolt–Börnstein Series (Springer) in six sub

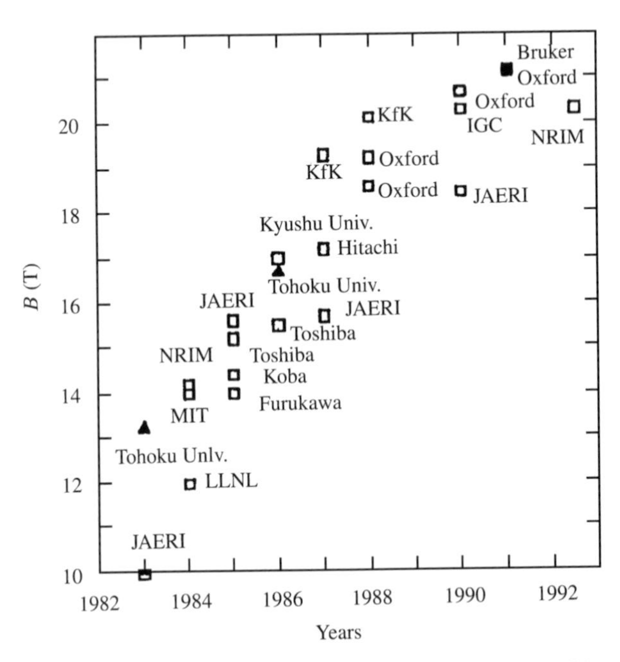

Figure B3.3.1.1. Progress in high magnetic fields in the last 20 years.

volumes [9]. It contains not only the T_c values of all known reported materials up to 1987 (more than 30 000 references), but also information about the material preparation and low temperature properties, in particular the electronic specific heat, γ, and the critical fields $\mu_0 H_{c1}$ and $\mu_0 H_{c2}$. For all these references, it is also indicated whether other normal state low temperature properties have been measured, e.g. electrical resistivity, Seebeck coefficient etc.

After 1986, most laboratories initiated intense research work programs on 'High T_c' superconductors, and the interest in the so-called 'Low T_c' superconductors decreased markedly. However, at present, more than 90% of the world market in the superconductivity field is still centred on the 'Low T_c' compounds, NbTi and Nb_3Sn. There is no doubt that further progresses in the materials Bi,Pb(2223), Bi(2212), Y(123) and now MgB_2 will gradually increase their market share. The newly discovered MgB_2 compound ($T_c = 39$ K) shows that surprises with other new 'Low T_c' superconductors cannot be excluded. Indeed, the structure of this class of materials with the formula XB_2 (where X is a metal) has been known for nearly 40 years, but no T_c values above 8 K were reported. It is important to say that with the present state of knowledge, there were no reasons to foresee this discovery.

B3.3.1 B3.3.2 C5 B3.3.6

In the last decades, most articles dealing with superconductors were centred on a relatively small number of materials, where the study of the electronic properties or the superconducting parameters was of particular interest. One must go back in history to find really general papers investigating the occurrence of superconductivity and possible correlations or mechanisms, for a large number of materials. The first work establishing a correlation between structural and material properties and superconductivity was published by Matthias [10] in 1963. He discussed the occurrence as well as the systematic variation of T_c for materials crystallizing in various crystal structures. From his overview of metallic and intermetallic compounds of over 32 crystal structures, he derived an empirical rule between the number N of valence electrons per atom, e/a (electrons per atom, counting all electrons outside a filled shell) and the observed maximum of T_c for binary materials. He found that T_c reaches maxima close to values of $N = 5$ and 7, as shown in figure B3.3.1.2, where the variation of T_c versus e/a for compounds crystallizing in the A15 phase is shown. After the discovery of a great number of new material classes, however, it was found that the validity of this rule is very much restricted: it can only be applied to the elements of the periodic systems and to some binary metallic and intermetallic compounds, e.g. A15, Laves and σ phases.

When classifying binary Low T_c compounds, the number of electrons is still useful, as can be shown by studying the systematic variation of the homogeneity ranges of compounds crystallizing in their respective crystal structures. For binary compound series based on transition elements only, e.g. Nb–W, Nb–Os, Nb–Ir, Nb–Pt,... or Mo–W, Mo–Os, Mo–Ir, Mo–Pt,..., the phase fields of the various intermetallic compounds are stable at certain values of e/a, as shown for the Mo–X series in figure B3.3.1.3 (Brewer plots). This does not only hold for the superconducting A15 phase, but also for the adjacent intermetallic phases, the sequence for the most important ones being $A2 \rightarrow A15 \rightarrow \sigma \rightarrow \chi \rightarrow A3 \rightarrow A1$. However, this simple rule is not valid when the second element (X in Nb–X or V–X) is a nontransition element, e.g. X = Sn, Ga, Al, Ge: phase stability is now influenced by stoichiometric considerations rather than by electron numbers. At present, no simple rules have been found for ternary or quaternary compounds, neither for the Chevrel phases, nor for the layered perovskite or organic compounds. It has to be noted that even the binary MgB_2 compound constitutes an exception.

C5

The BCS correlation between T_c and the electronic density of states has motivated a large number of low temperature specific heat investigations. Early general works treating various physical properties for a large number of materials with different crystal structures have been published, and will be mentioned here. Inspired by the work of Matthias [11], an overview of the linear term of the electronic specific heat, γ, for a great variety of material classes was published by Heiniger *et al* [11]. A review of topologically close-packed structures of transition metal alloys has been given by Sinha [12]. Theoretical works which had a large impact on the understanding of 'Low T_c' materials by strong coupling arguments are due to

A3.2

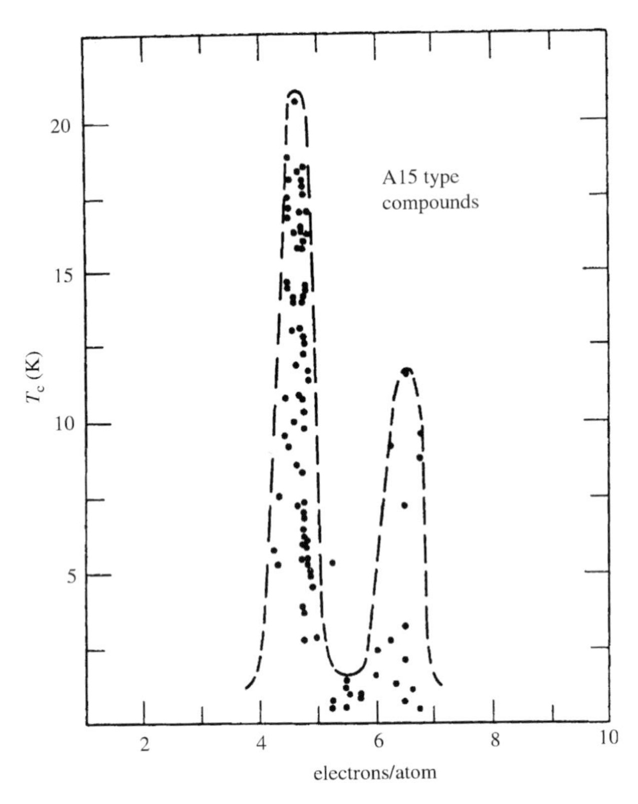

Figure B3.3.1.2. Superconducting transition temperature *versus* electron concentration for binary compounds crystallizing in the A15 type structure (Matthias plot).

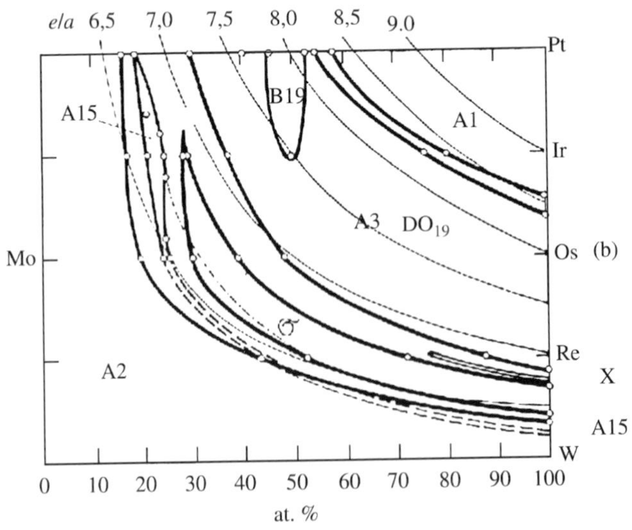

Figure B3.3.1.3. The Brewer plots for Cr–X binary systems, where X is a transition element. The phase fields follow the number of electrons/atom [20].

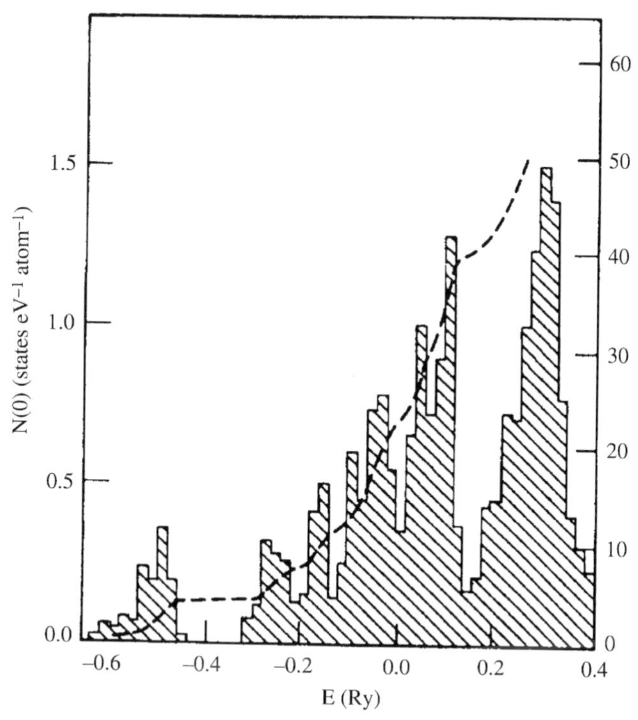

Figure B3.3.1.4. Theoretical band-structure density of states *versus* energy for A15 type compounds. The dashed curve represents the number of electrons per unit cell [13].

Mattheiss [13], MacMillan [14] and later, Allen and Dynes [15]. The electronic density of states of A15 type compounds as a function of energy calculated by Mattheiss [13] is shown in figure B3.3.1.4. It is interesting to note that these theories apply within a certain limit to a large number of 'Low T_c' superconductors, but do not describe the situation for 'High T_c' compounds. For the latter, it was found that the BCS theory does not apply: the search for new theories describing HTS compounds is an area of intense research. It appears that new superconductor classes may be discovered where the current theories must be modified. For example, the compound MgB_2 does not follow the BCS model at all temperatures and fields as shown by Junod *et al* [16], who showed the existence of two superconducting gaps. The behaviour at high magnetic fields of superconductors was also described very early for LTS, by Hake [17] and Rainer and Bergmann [18]. It was found that the Ginzburg–Landau–Abrikosov–Gorkov (GLAG) formalism used for describing the pinning behaviour of LTS superconductors can also be qualitatively applied for HTS as well as for MgB_2.

A3.2
C5
A3.1

In most LTS it was found that there is a correlation between the electron–phonon interaction parameter λ and the value of T_c, with $T_c \sim \omega_D \exp[-(1+\lambda)/(1-\mu^*)]$, where μ^* is the effective Coulomb interaction parameter, of the order of 0.1 in an ordinary metal. The parameter λ increases for higher density of states $N(0)$ at the Fermi surface. For most superconductors, there is a correlation between $N(0) = 3/2\gamma/(\pi k_B^2)$ and T_c, as shown in figure B3.3.1.5.

B3.3.1.2 Properties of low T_c compounds

In section B3.3.1.2.1, a brief summary of the properties of various 'low T_c' compounds will be given. This section is meant to complement the following paragraphs B3.3.1.2–B3.3.1.6, where the current carrying

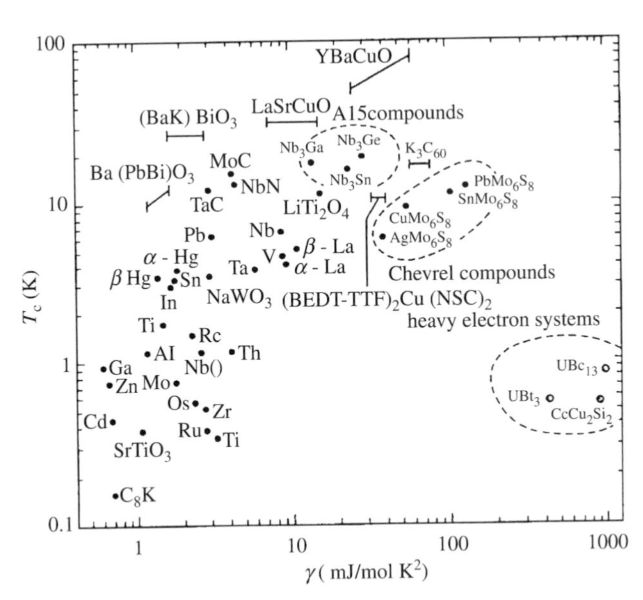

Figure B3.3.1.5. The relation between the Sommerfeld constant γ and T_c for a variety of superconducting systems.

properties of all these systems have been described in detail. Additional details of metallurgy and crystal structure, as well as of the physical properties of a series of compounds are given.

B3.3.1.2.1 The NbTi compound

B3.3.2 The system NbTi, described in chapter B3.3.2, crystallizes in the cubic centred structure. The interest in this compound resides in the fact that it is the only ductile superconducting material with an upper critical field $B_{c2}(0)$ above 10 T. This allows a straightforward deformation of NbTi to industrial multifilamentary wires, without the final reaction heat treatment required for all other high field superconductors. The pinning properties in NbTi can be strongly enhanced by performing a complex thermomechanical treatment at the end of the deformation process (see chapter B3.3.2); since this additional treatment occurs at quite low temperatures, e.g. $\leq 400°C$, it does not substantially increase the price of this material: the production costs of NbTi are thus the lowest among all high field superconductors. At present, it can be foreseen that only MgB_2 conductors, described in chapter B3.3.6, will be available at comparable costs. C5, B3.3.6

From these remarks, it is not surprising that NbTi is the most common industrial superconductor, and is used for different purposes. NbTi wires are used in magnetic resonance imaging (MRI) devices, which are a current diagnostic tool in hospitals and medical centres. E2.3 The magnetic fields in today's MRI devices vary between 1.5 and 2 T for full-body installations. With progress in medical research and in the treatment of large quantities of data, there is a clear development towards higher magnetic fields for MRI magnets, and an increasing number of MRI magnets producing 3 T are actually being installed. A prototype full body MRI magnet producing 7 T for research purposes has already been built and tested. For research purposes, e.g. on animals, smaller MRI magnets producing up to 11 T at 1.8 K are planned.

B3.3.2 In order to profit from the low costs of NbTi wires, the latter are used to produce the background field around interior high field Nb_3Sn coils. Such magnets used for research and for performing nuclear magnetic resonance (NMR), can reach up to 21 T. At a much larger scale, NbTi wires are used for the construction of dipoles in accelerators, such as the Large Hadron Collider at CERN. The flux pinning properties of NbTi have been described in detail by Cooley *et al* in chapter B3.3.2. A4.3

B3.3.1.2.2 The A15 type compounds

Of the 69 presently known binary A15 type compounds, at least 54 are superconducting, and about 15 of them exhibit a critical temperature above 10 K. The latter are summarized in table B3.3.1.1, together with some pseudobinary compounds, e.g. $Nb_3(Al–Ge)$, $Nb_3(Au–Pt)$. Note that the stoichiometric composition is not stable at equilibrium in all A15 type compounds, e.g. Nb_3Ga and Nb_3Al. The A15 phase field of two compounds, shown in figure B3.3.1.6, illustrates that the composition $\beta = 0.25$ is only stable at high temperatures. In other compounds, e.g. Nb_3Ge, Nb_3Si or V_3Al, the stoichiometric composition can only be obtained by thin film deposition, i.e. by nonequilibrium methods. Review articles about these compounds and their properties have been published by Muller [9], who described the physical properties, and by Flükiger [20], who studied in detail the phase diagrams of a large number of binary A15 type compounds.

The A15 type structure

The A15 type structure Cr_3Si occurs generally, but not always, at the stoichiometry A_3B. It belongs to the space group O_h^3–$Pm3n$ and has a primitive unit cell containing eight atoms. Assuming the origin to be at the centre, the following atomic positions are obtained:

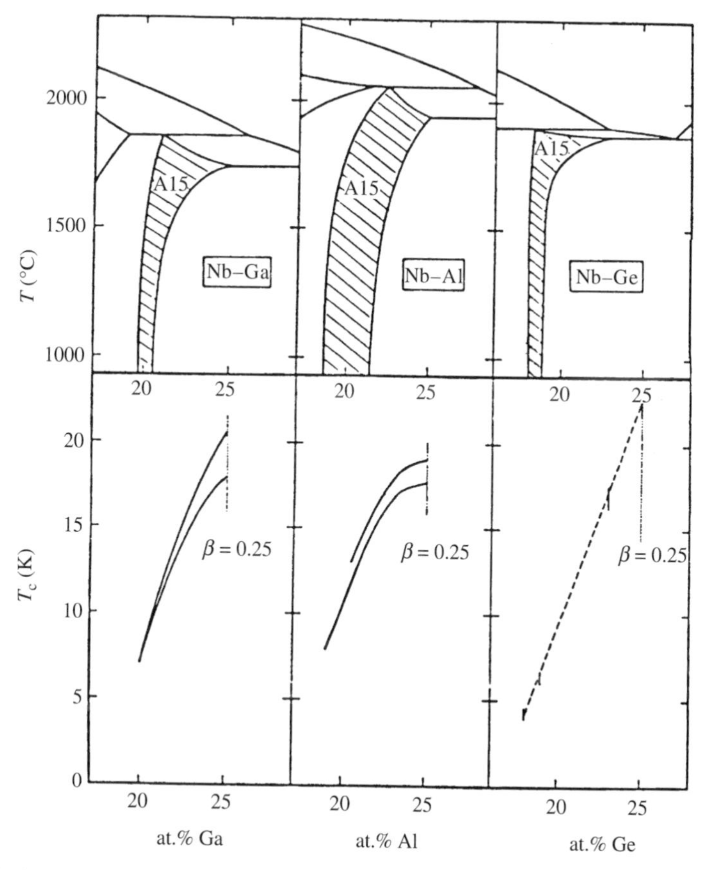

Figure B3.3.1.6. A15 phase fields of the systems Nb–Ga, Nb–Ge and Nb–Ge. The Nb–Ge phase field does not include the stoichiometric composition [20, 25].

2 B atoms in 2(a):	0, 0, 0	1/2, 1/2, 1/2
6A atoms in 6(c):	1/4, 0, 1/2 0, 1/2, 1/4 1/2, 3/4, 0	1/2, 1/4, 0 3/4, 0, 1/2 0, 1/2, 3/4

The cubic unit cell of the A15 type structure is shown in figure B3.3.1.7. The B atoms have 12 neighbors (A atoms) at a distance $(\sqrt{5}/4)\,(r_A + r_B)$, and these are arranged in the form of a distorted icosahedron. The A atoms themselves have a larger co-ordination number of 14; the CN14 polyhedron around each A atom consists of two A atoms at a distance $a/2$, four B atoms at $(\sqrt{5}/4)\,a$, and eight A atoms at a distance $(\sqrt{6}/4)\,a\,(= 2\,r_A)$.

The long-range atomic order parameter

As found by Van Reuth *et al* [21], there is a strong correlation between the T_c values of A15 type compounds and the Bragg–Williams long range order parameter, S (table B3.3.1.2). The values of the electrical resistivity, ρ, and of the upper critical field, B_{c2}, are also affected by changes of the order parameter, and so the critical current density of a superconducting wire is also affected. Thus, the transport properties of superconducting wires based on A15 type compounds at high fields can only be understood by taking into account the respective values of S. This is particularly true for the binary compound Nb_3Sn, where J_c can be strongly enhanced by adding additional elements, e.g. Ta, Ti or Ta + Ti.

A3.1
B3.3.3

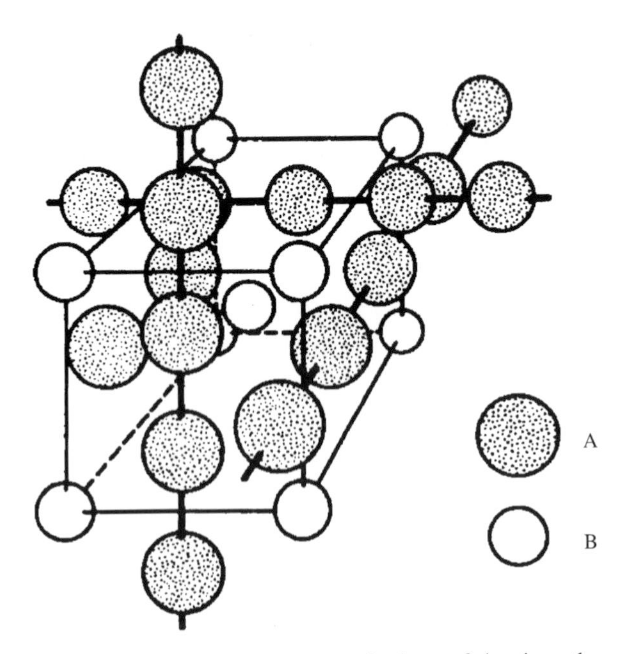

Figure B3.3.1.7. The A15 type structure A_3B. For clarity, only three of the six orthogonal chains of A atoms are shown.

Table B3.3.1.2. Values of the Bragg–Williams order parameter, S, for A15 type compounds after various heat treatments and the corresponding values of the superconducting transition temperature

Compound	Order parameter S	Change in T_c	Degree of ordering
V_3Si	1.0	–	Perfectly ordered
Nb_3Sn	1.0	–	Perfectly ordered
Nb_3Al	0.95–0.98	18.1–20.1	Varies with quenching rate
V_3Ga	0.94–0.98	13.6–15.9	Varies with quenching rate
V_3Au	0.93–0.98	< 0.1–3.2	Varies with quenching rate

The degree of atomic ordering in an A15 type compound of the general formula $A_{1-\beta}B_{\beta}$ can be described by the values S_a and S_b for each lattice site, defined by

$$S_a = [r_a - (1-\beta)]/\beta \quad \text{and} \quad S_b = (r_b - \beta)/(1-\beta)$$

where r_a is the fraction of a sites occupied by A atoms and r_b the fraction of b sites occupied by B atoms. At the composition β, the degree of atomic ordering for a given sample is thus sufficiently defined by one parameter only, either S_a or S_b. At the stoichiometric composition, $\beta = 0.25$, there is $S_a = S_b = S$.

Atomic ordering is a collective phenomenon and is understood as a thermodynamic equilibrium distribution of atoms over the lattice points of the crystal. The long range atomic order parameter thus depends on temperature: $S = S(T)$. Theoretically, the order parameter value of A15 type compounds at low temperature would be 1. However, a certain amount of disorder can be 'frozen in' since the cooling process from the melt or from the annealing temperature, T_A, occurs over finite time. Thus, the order parameter for a given A15 type compound will, in general, be $S \leq 1$, the deviation from unity depending on the thermal history of the measured sample. Therefore all order parameter measurements performed at room temperature describe in reality a nonequilibrium state. The equilibrium order parameter can only be measured at temperatures above the diffusion limit, i.e. above $\sim 0.6\,T_F$, where T_f is the melting temperature. The change of the order parameter by quenching procedures is limited by the cooling rates, which are usually of the order of 10^4–10^{5}°C s^{-1}, and changes of only $\Delta S \leq 0.05$ have been reported. Ordinary quenching rates ($\leq 10^{5}$°C s^{-1}) are not sufficient to lower its atomic order parameter and high energy irradiation is necessary. Much higher local quenching rates can be obtained by high energy irradiation, either with neutrons ($E > 1$ MeV) or heavy ions [22], thus yielding atomic order parameters S well below 0.5.

B2.2.5

B2.6

The effect of atomic ordering on T_c can be demonstrated for the compound V_3Ga. The value of T_c of V_3Ga can be varied between 13.5 and 15.3 K by changing the thermal history, the highest value being achieved after a prolonged ordering heat treatment at 560°C and corresponding to the highest degree of atomic ordering, $S = 0.98$ (see figure B3.3.1.8). This is in contrast to Nb_3Sn, which has much faster diffusion kinetics and is thus always perfectly ordered.

The atomic order parameters of a large number of A15 type compounds have been determined by Van Reuth *et al* [21], Blaugher *et al* [23] and later by Flükiger *et al* [25, 26]. From these data, it is possible to extract a general correlation between the normalized electronic specific heat γ/γ_0 and the order parameter S/S_0 represented in figure B3.3.1.9. This picture shows the direct influence of the degree of ordering on the electronic density of states.

Both parameters, the atomic composition and the atomic order parameter have a strong influence on the electrical resistivity ρ_0 (just above T_c) of A15 type compounds [24]. This can be illustrated in figure B3.3.1.10, which shows that the perfectly ordered compounds V_3Si, Nb_3Sn and Nb_3Ge ($S = 1$)

D3.1

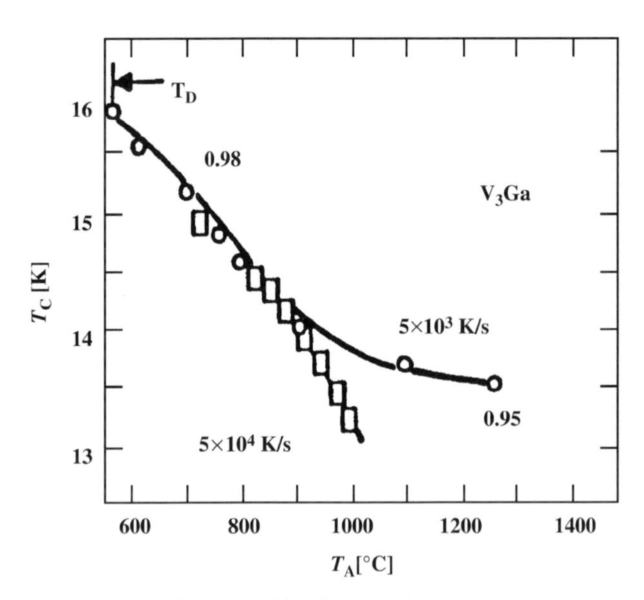

Figure B3.3.1.8. Variation of T_c of V_3Ga after quenching from various temperatures T_A at two cooling rates: 5×10^3 and 5×10^4°C/s [26].

compounds have a very similar behaviour [25]. The fact that ρ_0 decreases by more than one order of magnitude between 24.5 and 25 at.% Si or Sn, from 19 to 2 $\mu\Omega$ cm reflects the very high degree of ordering in these compounds. It has to be noted that the variation ρ_0 vs β for these compounds shown in figure B3.3.1.10 cannot be influenced by ordinary heat and quench cycles. Only high energy irradiation will cause a strong enhancement of ρ_0, even at the stoichiometric composition [22].

B3.3.4

The situation is quite different for the compound Nb_3Al, which exhibits a certain amount of disorder, with a maximum value of $S = 0.98$ after long term anneals at temperatures below 850°C [25].

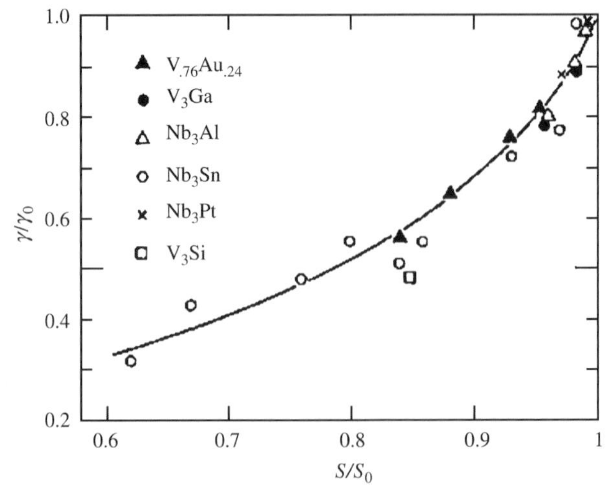

Figure B3.3.1.9. Normalized linear term of the specific heat γ/γ_0 (representing the electronic density of states) *versus* the atomic order parameter S/S_0 for various A15 type compounds after quenching or irradiation.

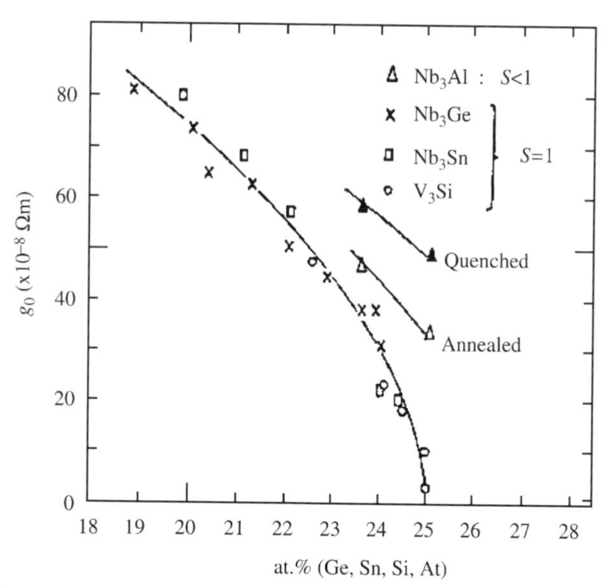

Figure B3.3.1.10. Electrical resistivity ρ_0 *versus* atomic composition β in the A15 type systems $Nb_{1-\beta}Sn_\beta$, $V_{1-\beta}Si_\beta$, $Nb_{1-\beta}Ge_\beta$ and $Nb_{1-\beta}Al_\beta$. At the stoichiometric composition ($\beta = 0.25$), ρ_0 has always a minimum [26].

The difference in ρ_0 between the quenched and the annealed state in figure B3.3.1.10 reflects the change in atomic ordering. The situation of Nb_3Al is analogous to the one encountered in V_3Ga, illustrated in figure B3.3.1.9.

The system V_3Si is particularly suited for studying the effect of composition on perfectly ordered A15 type compounds, since a series of single crystals can be prepared between 22 and 25.2 at.% Si. At the stoichiometric composition, the value of $\rho_0 = 0.9\ \mu\Omega$ cm for V_3Si is the lowest reported for A15 type compounds. Figure B3.3.1.11 shows the variation of the RRR value as a function of the Si composition: the sharp increase by a factor 80 at the vicinity of 25 at.% Si illustrates the behaviour of a perfectly ordered system [27, 28].

The upper critical field $\mathbf{B_{c2}}$

The research on A15 type phases has been strongly motivated by the fact that several compounds exhibit upper critical field values, B_{c2}, in the range between 20 and 45 T. The variation of B_{c2} as a function of temperature is shown in figure B3.3.1.12 for the compounds Nb_3Al [29], Nb_3Ge [30], $Nb_3Al_{0.7}Ge_{0.3}$ [31], $Nb_3Au_{0.7}Pt_{0.3}$ [32], Nb_3Sn [33] and Mo_3Os [34]. The highest value has been measured for the pseudobinary compound, $Nb_3Al_{0.7}Ge_{0.3}$. It has to be noted that the value $B_{c2}(0) = 24$ T for Nb_3Sn corresponds to the 'clean' limit of polycrystalline samples, where $\rho_0 < 10\ \mu\Omega$ cm. In industrial Nb_3Sn wires, the addition of Ta or Ti to Nb_3Sn leads to an increase of the resistivity ρ_0 to $> 30\ \mu\Omega$ cm ('dirty' limit), and thus to $B_{c2}(0)$ to values exceeding 30 T.

A3.1

B3.3.3

For comparison, the upper critical fields of V based A15 type compounds have been plotted in figure B3.3.1.13. A representation of the critical current densities of multifilamentary '*in situ*' V_3Ga wires [35] is given in figure B3.3.1.14. Magnetic fields of 16 T have been achieved in magnets based on V_3Ga wires. However, due to the relatively low value of $B_{c2}(0)$ for the system V_3Ga, the fabrication of this compound had to be abandoned in favour of Nb_3Sn, the other compound being formed by the bronze process.

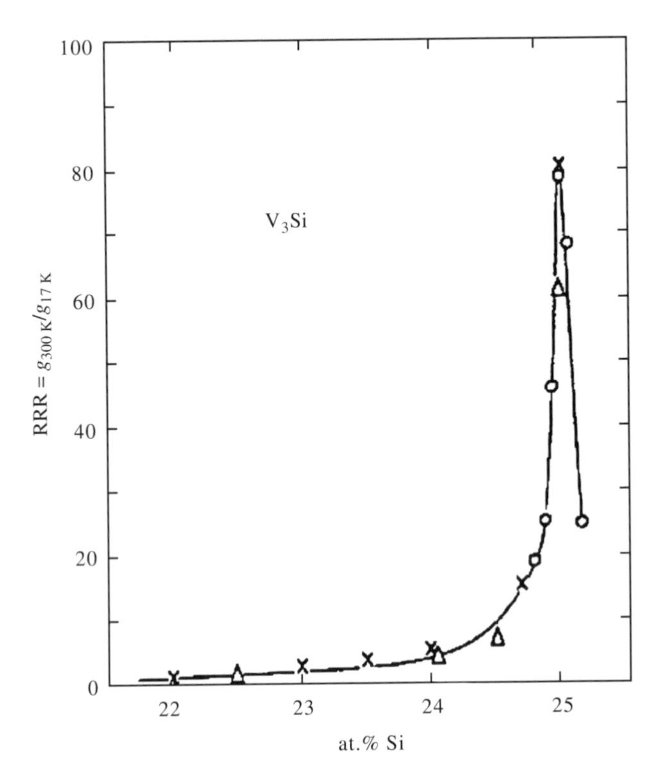

Figure B3.3.1.11. Variation of the RRR ratio in V_3Si single crystals as a function of the Si content ○: Berthel and Pietrass [27], ×,△ Flükiger *et al* [28].

High field properties of V_3Ga multifilamentary wires

A3.1 E1.1 The A15 type compound V_3Ga exhibits a T_c value of 15.9 K. It was developed in view of high field magnets, but its upper critical field of $B_{c2} = 24$ T markedly lower than that of Nb_3Sn, which is at present the only industrial material for fields above 18 T. In spite of these characteristic differences, there is

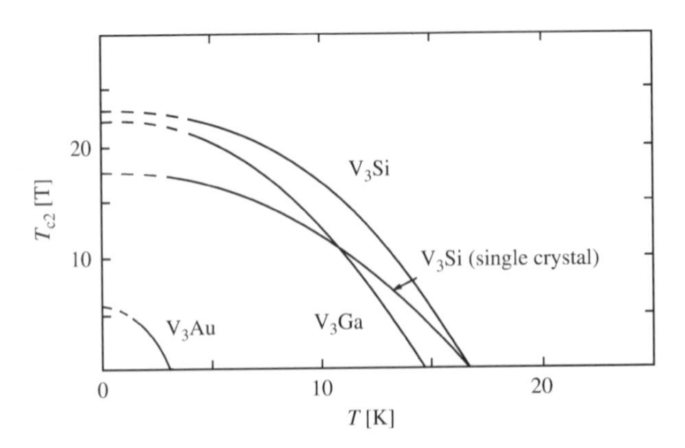

Figure B3.3.1.12. B_{c2} *versus* T for the compounds V_3Ga, V_3Si and V_3Au. The B_{c2} value of the 'clean' V_3Si single crystal is markedly lower than the value of V_3Si polycrystals [19].

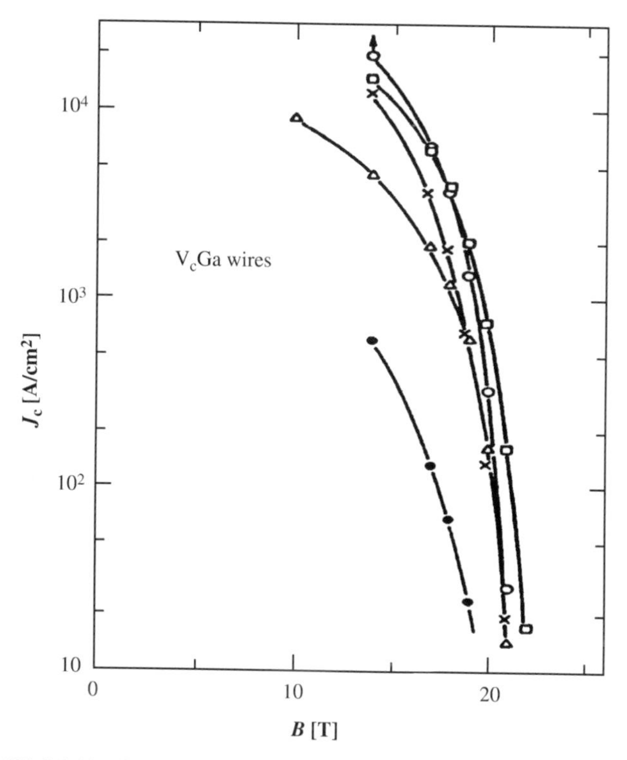

Figure B3.3.1.13. Critical current densities of V_3Ga multifilamentary wires [35].

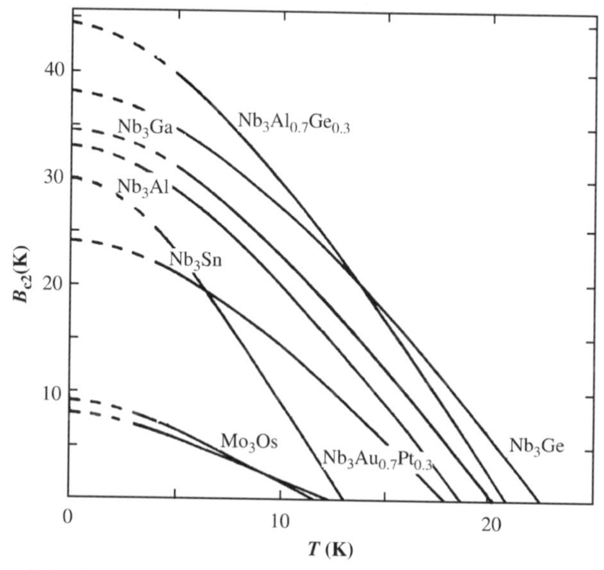

Figure B3.3.1.14. B_{c2} *versus* T for the compounds $Nb_3Al_{0.7}Ge_{0.3}$, $Nb_3Au_{0.7}Pt_{0.3}$, Nb_3Sn (clean limit), Nb_3Ga, Nb_3Al and Mo_3Os [19].

a strong link between these two systems in view of their use in high field superconducting wires: the bronze diffusion reaction. Indeed, the bronze diffusion process described for Nb_3Sn in chapter B3.3.3 can also be applied to the V_3Ga system. This analogy still holds for the so-called *in situ* wire production process. As reported by Noto *et al* [35], Cu-25 at.% V can be induction melted in CaO crucibles, followed by rapid solidification in a cooled mold. After deformation to the final size, the wire was Ga plated and heat treated at 450–550°C for two to six days. The J_c characteristics of these wires after several heat treatment conditions are shown in figure B3.3.1.13. The wire heat treated at 550°C for two days showed the highest J_c values, with 10^4 A cm^{-2} at fields as high as 16.2 T [35].

B3.3.3

The upper critical fields of the V based A15 type compounds V_3Si [36] and V_3Ga [37] have been plotted in figure B3.3.1.12. A representation of the critical current densities of multifilamentary V_3Ga wires is given in figure B3.3.1.13. Magnetic fields of 16 T have been achieved in magnets based on V_3Ga wires. However, due to the relatively low value of $B_{c2}(0)$ for the system V_3Ga, the fabrication of this compound had to be abandoned in favour of Nb_3Sn, the other compound formed by the bronze process.

A3.1 B3.3.3

High field properties of Nb_3Ge thin films

At the stoichiometric composition, the compound Nb_3Ge exhibits with 23 K the highest T_c value among A15 type compounds (see table B3.3.1.1). However, this composition is not stable at thermal equilibrium, as shown in figure B3.3.1.6, and nonequilibrium deposition methods have to be used. Nb_3Ge films were prepared by dc sputtering [1] and by chemical vapour deposition [38, 39].

In the case of sputtering, a single target technique was adopted [1]. Targets were made by arc casting the elements in various ratios and the effects of target composition on the growth of A15 phase were investigated. For the preparation of CVD films, $NbCl_5$ powder and liquid $GeCl_4$ were used as starting materials. The chloride vapour was mixed with H_2 gas and then reduced for 30 min in a quartz reactor held at a deposition temperature T_d between 750 and 900°C. Most Nb_3Ge films prepared by sputtering normally exhibit T_c values exceeding 20 K, as found by resistive measurements. The stoichiometric composition was confirmed by X-ray diffraction analysis, showing a single phase pattern at the ratio Nb : Ge = 3 : 1. With decreasing Nb/Ge ratio of targets, an increase in the amount of the hexagonal Nb_5Ge_3 phase was observed. (Note that the equilibrium phase Nb_5Ge_3 is tetragonal.) SEM pictures showed a columnar grain structure, and the critical current densities were very high, as shown in figure B3.3.1.15.

B4.2.2

The T_c values of Nb_3Ge thin films prepared by CVD were somewhat lower, and did not exceed 20.7 K [38], and their microstructure was considerably influenced by the deposition temperature T_d. The size of A15 grains increased markedly when raising T_d from 800 to 900°C, grain sizes up to 1 μm being observed at 900°C [38]. From electron microprobe analysis, it was found that CVD films have ratios Nb:Ge closer to 2.5, and contain a small amount of second phase. From the results of sputtered and CVD processed Nb_3Ge thin films shown in figure B3.3.1.15, it follows that at temperatures above 10 K, high J_c values are obtained at 15 T, 6×10^4 A cm^{-2}.

The upper critical fields B_{c2}(4.2 K) of sputtered and CVD processed Nb_3Ge thin films were determined by a Kramer plot and reached 31.6 and 33.4 T, respectively [38]. The highest B_{c2} value for Nb_3Ge, 37 T, was obtained by Foner *et al* [23] on a sputtered film. It is clear that without the discovery of Chevrel phases, or later HTS compounds, Nb_3Ge would have been a promising candidate for very high field magnets. It is remarkable that the world first superconducting solenoid producing fields above 20 T was based on a CVD processed Nb_3Ge tape. It produced a field of 0.5 T in a background field of 20 T [39].

B4.2.2 E1.1

B3.3.1.2.3 The Chevrel phases $M_xMo_6S_8$

G1

A very detailed review on the physical properties of Chevrel phase compounds has been published by Fischer [40]. In the present Handbook, the Mo chalcogenides or Chevrel phase compounds [41] is

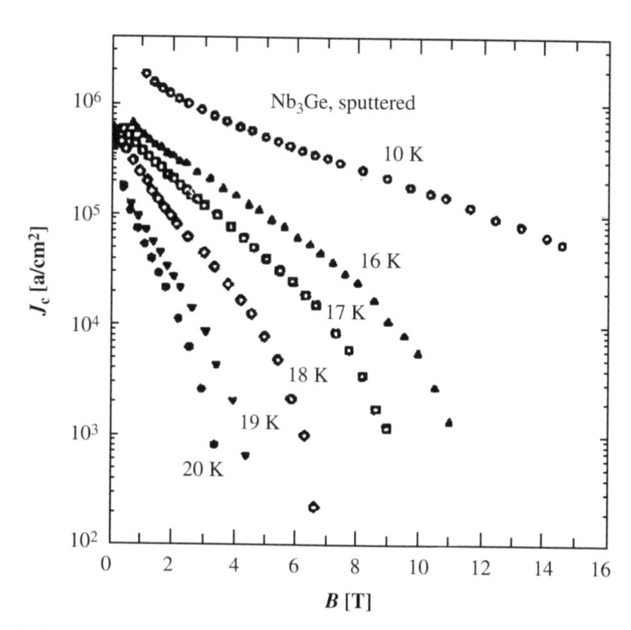

Figure B3.3.1.15. Critical current density *versus* applied field for sputtered Nb_3Ge thin films [38].

described by Seeber in chapter B3.3.5. These compounds are characterized by the general formula $M_xMo_6X_8$ (M = metal, X = Chalcogen: S, Se, Tl, $0 \leq x \leq 4$). The Chevrel phase compounds exhibit the highest upper critical field, B_{c2}, of all 'Low T_c' superconductors: in spite of their relatively low T_c value of 15 K, the compound $PbMo_6S_8$ exhibits B_{c2} values exceeding 50 T. These compounds exhibit remarkable physical properties, which arise from the unique configuration of Mo_6X_8 building blocks, which are held together by the insertion of a weakly bonded third component M. This particular arrangement can accomodate very different ions, which can either be magnetic or nonmagnetic. Depending upon the nature of the M element (transition or nontransition element, rare earth), the physical properties of the corresponding compound are very different. Little is known about the metallurgy of ternary Mo chalcogenides. The high vapour pressure of the chalcogen and of some of the M elements, e.g. Pb, Sn, ... render the preparation of homogeneous, single phase samples very difficult.

The most common method used to form the rhombohedral compounds $M_xMo_6X_8$ consists of reacting appropriate amounts of the three elements (or binary combinations of them) in quartz tubes at temperatures up to 1200°C. Dense samples allowing microscopic analysis can be obtained by hot pressing or by melting under high argon pressure. The higher pressures cause a drastic decrease of the diffusion rate of the evaporated atoms in the argon atmosphere at the vicinity of the surface, resulting in a lower evaporation rate. The generally used crucible material is Al_2O_3, but BeO and BN have also been used. Chevrel phase wires are produced by filling prereacted $Pb_{1.2-x}Sn_x$ Mo_6S_8 powders in a Nb or Ta tube, which is placed in a stabilizing Cu tube and surrounded by a stainless steel tube (see figure in chapter B3.3.5). The steel jacket has a double function: it acts as a mechanical reinforcement, and in addition, it must produce a compressive prestress on the $Pb_{1.2-x}Sn_x$ Mo_6S_8 core. The wires produced were cold worked by drawing from 14 to 1 mm diameter, before a final anneal at temperatures between 900 and 1000°C under HIP conditions (1 kbar Argon pressure).

For the composition $Pb_{0.96}Sn_{0.24}Mo_6S_8$, J_c values of 2.0×10^4 and 4.0×10^4 A cm^{-2} were obtained at 4.2 and 2.0 K for the superconducting core [42, 43]. It is clear that these values are superior to those already published for Nb_3Sn wires, thus rendering Chevrel phase compounds the ideal candidate for

B3.3.3

E1.1 magnets exceeding 20 T. The reason for these high values resides in the extraordinarily high upper critical field B_{c2}, which reaches 40 T at the composition yielding the maximum J_c. However, in spite of these very promising values, Chevrel phase wires have never reached the industrial level. This can be explained by the following reasons:

A2.3 (a) The coherence length of $Pb_{1.2}Mo_6S_8$ is 2.2 nm and lies between that of Nb_3Sn (4.0 nm) and of HTS superconductors. It follows that the conditions at the grain boundaries are very important in view of the current carrying properties. The question arises whether highly textured Chevrel phase grains would lead to higher J_c values. However, it is a particularity of Chevrel phases that no texturing is observed after cold deformation, which is due to the particular mechanical properties of the grains.

(b) In order to compensate the differential thermal expansion, a significant amount of stainless steel is needed. Rimikis *et al* [42] have shown that the cross-section of the wires should contain $> 50\%$ of stainless steel. If one takes into account that $> 25\%$ of Cu are necessary for thermal stabilization, it follows that the cross-section of the superconducting core cannot exceed 25%, thus reducing the overall J_c value by a factor of 4.

(c) The necessity of a stainless steel sheath involves the use of long stainless tubes of less than 2 mm inner diameter. The availability of these thin tubes is restricted to lengths of the order of 50 m, which limits the total length of the superconducting wire to less than 100 m. It should be mentioned here that the industrial requirement is of the order of 1000 m.

D3.2.1 (d) The requirement of thermal stability makes it necessary to reduce the filament diameter to less than 10 μm. Due to the difficult deformability of this material, little work has been done in the preparation of multifilamentary wires.

G1 These remarks explain why Chevrel phase wires have so far not been used, in spite of their very promising high field properties.

B3.3.1.2.4 The compound NbN

The binary compound NbN has the cubic structure B1 and has been studied very thoroughly in the literature: in contrast to all other known superconductors, NbN can be produced at remarkably low temperatures, the lowest reported one being of the order of 100°C. For this reason, NbN is particularly suited for electronic devices, where complex architectures are considerably easier to produce at lower temperatures.

In view of the production of NbN films with high J_c values, the compound was produced by rf sputtering [44] on substrates of quartz or sapphire, from a Nb disc target in an Ar/N_2 atmosphere. Substrates were heated at temperatures between < 300 and 700°C, and the partial pressures of Ar and N_2 were varied independently. The T_c values of NbN films are strongly influenced by the nitrogen partial pressure, and values exceeding 16 K were obtained at $p_{N_2} = 4 \times 10^{-6}$ atm. A further increase in p_{N_2} provides a slight decrease in T_c and causes an appreciable increase in the normal state resistivity ρ_0, the structure being still B1 as determined by X-ray analysis. For films deposited on quartz, SEM observations showed columnar grains of 20–50 nm in diameter with the (111) plane parallel to the film surface. Films on sapphire contained an initial growth layer of 150 nm thickness with coarser grains [44]. Sputtered NbN films exhibit a considerable anisotropy: $B_{c2}^{\perp}$ and $B_{c2}^{\parallel}$ are similar for films with low ρ_0 values, but differ quite strongly above 200 $\mu\Omega$ cm, where B_{c2} starts to increase rapidly. The critical current density of thin NbN films are plotted in figure B3.3.1.16, showing that NbN tapes would be

D1.1.2, D1.3

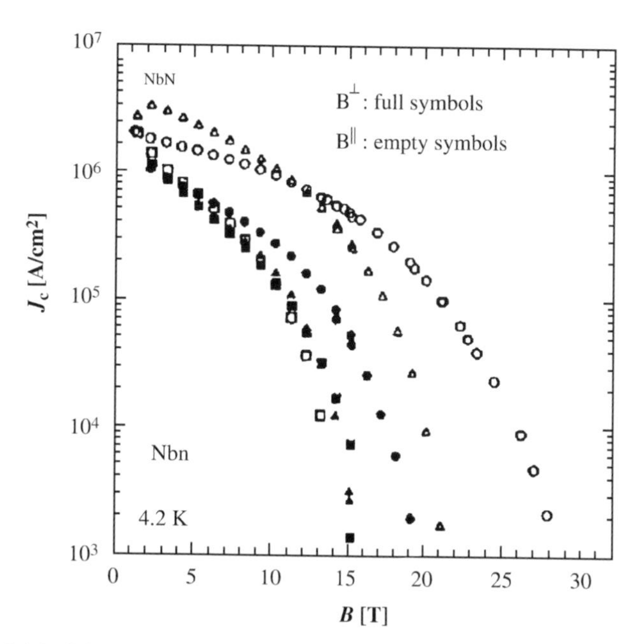

Figure B3.3.1.16. Critical current density *versus* applied field for thin NbN films [39].

a very competitive system for high field magnets at 4.2 K: the field anisotropy, however, is quite strong: a J_c value of 1×10^5 A cm^{-2} is found for 21 T with fields parallel to the surface, but only at 14 T for normal fields. E1.1

The strain dependence of 700 nm thick NbN films deposited on Hastelloy B substrates show irreversible strain values of $\varepsilon_{irr} = 1.3\%$. For thinner films (300 nm), the value of ε_{irr} even increases to 1.4%. The effect of the substrate seems to be determinant, NbN films sputtered on stainless steel substrates showing considerably lower ε_{irr} values.

B3.3.1.2.5 The compound MgB_2

A very detailed description of the fabrication of multifilamentary tapes and wires based on MgB_2 is given in this Handbook by Suo *et al* (chapter B3.3.6). This compound has been discovered only recently and is certainly promising in view of MRI magnets and possibly current limiters. C5 B3.3.6 E2.3, E1.3.3

References

[1] Gavaler J R 1973 *Appl. Phys. Lett.* **23** 480
[2] Johnston D C, Prakash H, Zachariasen W H and Viswanathan R 1973 *Mater. Res. Bull.* **8** 777
[3] Bednorz J G and Müller K A 1986 *Z. Phys. B* **64** 189
[4] Lüders K 1994 *Phys. Bl* **50** 166
[5] Cava R J, BaPbO
[6] Cava R J, Takagi H, Batlogg B, Zandbergen H W, Krajewski J J, Peck W F Jr, van Dover R B, Felder T, Siegtrist T, Mizuhashi K, Lee J O, Eisaki H, Carter S A and Uchida S 1994 *Nature* **367** 146
[7] Nagamatsu J, Nakagawa N, Muranka T, Zenitani Y and Akimitsu J 2001 *Nature* **410** 63
[8] Roberts B W 1976 *J. Phys. Chem. Ref. Data* **5** 581
[9] Flükiger R and Klose W 1990 Landolt & Börnstein New Series *Superconductors, Transition Temperatures and Characterization of Elements* **21:a**, Flükiger R and Klose W 2002 Landolt & Börnstein New Series, '*Superconductors, Transition Temperatures and Characterization of Elements, Alloys and Compounds*', (Berlin, Heidelberg, New York:Springer-Verlag), 21:f. (Berlin: Springer)

[10] Matthias B T, Geballe T H and Compton V B 1963 *Rev. Mod. Phys.* **35** 1
[11] Heiniger F, Bucher E and Muller J 1966 *Phys. Kondens. Materie* **5** 243
[12] Sinha A K 1972 *Progress in Material Science* **vol. 15** p 184
[13] Mattheiss L F 1965 *Phys. Rev.* **138** A112
[14] McMillan W L 1968 *Phys. Rev.* **167** 166
[15] Allen P B and Dynes R C 1975 *Phys. Rev.* **12** 905
[16] Junod A, Wang Y, Bouquet F, Sheikin I, Toulemonde P, Eskildsen MR, Eisterer M, Weber HW, Lee S and Tajima S *Low Temperature Conference, LT 23*, Yokohama, August 2002, Proceedings submitted
[17] Hake R R 1967 *Phys. Rev.* **158** 356
[18] Rainer D and Bergmann G 1974 *J. Low Temp. Phys.* **14** 501
[19] Muller J 1980 A15 type superconductors *Rep. Prog. Phys.* **43** 643
[20] Flükiger R 1981 *Superconducting Materials* ed S Foner and B B Schwartz Plenum p 511
[21] Van Reuth E C and Waterstrat R M 1968 *Acta Crystallogr. B* **24** 186
[22] Sekula S T 1978 *J. Nucl. Mater.* **72** 91
[23] Blaugher R D, Hein R E, Cox J E and Waterstrat R M 1969 *J. Low Temp. Phys.* **1** 539
[24] Flükiger R 1982 *Advances in Cryogenic Engineering Materials* **vol 28** ed T Reed and A Clark p 399
[25] Flükiger R 1992 *Encyclopedia of Advanced Materials* ed D Bloor, M C Flemings, R J Brook and S Mahajan Pergamon p 1
[26] Flükiger R 1987 Atomic Ordering, Phase Stability and Superconductivity in Bulk and Filamentary A15 Type Compounds, *Report KfK 4204* Institut für Technische Physik, Karlsruhe, Germany
[27] Berthel K H and Pietrass B 1978 *J. Phys.* **39** C6
[28] Flükiger R, Küpfer H, Jorda J L and Müller J 1987 *IEEE Trans. Magn.* **23** 980
[29] Foner S, McNiff E J Jr, Geballe T H, Willens R H and Buehler E 1971 *Physics* **55** 534
[30] Foner S, McNiff E J Jr, Gavaler J R and Janocko M A 1974 *Phys. Lett.* **47A** 485
[31] Foner S, McNiff E J Jr, Matthias B T, Geballe T H, Willens R H and Corenzwit E 1970 *Phys. Lett.* **31A** 349
[32] Flükiger R, Foner S, McNiff E J Jr and Fischer Ø 1979 *Solid State Commun.* **30** 723
[33] Foner S and McNiff E J Jr 1976 *Phys. Lett.* **58A** 318
[34] Bongi G, Fischer Ø, Jones H, Flükiger R and Treyvaud A 1972 *Helv. Phys. Acta* **45** 13
[35] Noto K, Matsukawa M and Watanabe K 1992 Sci. Reports of the Research Institutes, Tohoku University, Series A, vol **37** p 7
[36] Foner S and MacNiff E J 1978 *Appl. Phys. Lett.* **32** 122
[37] Dew-Hughes D 1978 *Cryogenics* **15** 435
[38] Suzuki M, Ouchi H and Anayama T 1982 *Proc. Int. Conf. on Cryogenic Materials* ed K Tachikawa and A Clark Butterworths, Guildford p 242
[39] Weiss F 1978 *PhD thesis* Grenoble
[40] Fischer O 1978 *Appl. Phys.* **16** 1
[41] Chevrel R, Sergent M and Prigent J 1971 *J. Solid. State Chem.* **3** 515
[42] Rimikis G, Goldacker W, Specking W and Flükiger R 1991 *IEEE Trans. Magn.* **27** 1116
[43] Yamasaki H, Umeda M, Kimura Y and Kosaka S 1991 *IEEE Trans. Magn.* **27** 1112
[44] Suzuki M, Kiboshi T, Anayama T and Nagata A 1982 Proc. MRS Meeting on Advanced Materials **6** 77

B3.3.2
Processing of low T_c conductors: the alloy Nb–Ti

Lance Cooley, Peter Lee and David C Larbalestier

B3.3.2.1 Overview

Niobium–titanium alloys have been widely used in superconducting applications since the early 1960s. The success of Nb–Ti has been due to its combination of excellent strength and ductility with high current-carrying capacity at magnetic fields sufficient for most applications. Moreover, these advantages are obtained with raw material and fabrication costs that are significantly lower than other technological superconductors for magnetic fields in the 2–8 T range. A significant factor driving down the cost is the widespread use of Nb–Ti in magnetic resonance imaging (MRI) magnets, which amounts to a consumption of approximately 1000 tons per year of finished Cu and Nb–Ti, strand. Another unique advantage of Nb–Ti superconductors is the fact that heat treatments that form flux-pinning centres can be applied prior to cabling, winding and other magnet assembly steps. This is made possible by the lack of any strong dependence of the superconducting properties on strain and by the mechanical toughness of Nb–Ti strands. High yield strength, comparable to that of steels [1], further relaxes constraints on the support structure. These excellent properties will ensure continued widespread use of Nb–Ti alloys for a long time to come. Primary applications of Nb–Ti alloy superconductors include magnets for MRI, nuclear magnetic resonance (NMR), laboratory apparatus, particle accelerators, electric power conditioning, minesweeping, ore separation, levitated trains and superconducting magnetic energy storage (SMES).

B3.3.1 E2.3 E2.3 E2.2, E1.3.4

This chapter reviews the state of the art of Nb–Ti superconductors. Nb–Ti alloys belong to a larger class of transition-metal alloys, whose members consist of a Group IVa element (Ti, Zr, Hf) and either a Group Va (V, Nb, Ta) or a Group VIa element (Cr, Mo, W). An important property of this class is that Group IVa elements undergo an allotropic transformation from a hexagonal close-packed (hcp) structure to body-centered cubic (bcc) structure at high temperature; for pure Ti this transition occurs at 882°C [2]. Since Group Va and VIa elements are also bcc metals, bcc solid-solution alloys are favoured at high temperature (the β phase), and can be retained by quenching to room temperature. Many β alloys are good superconductors [3], as would be expected from the high transition temperatures (among elemental superconductors) of V, Nb and Ta. However, there is low solubility of Group Va and/or Group VIa elements in the low-temperature hcp α phase, less than about 2.5 at% Nb in Ti, for example. As discussed in section B3.3.2.2, the competition between these phases and the incipient phase transition of a quenched β alloy to $\alpha+\beta$ is the origin of many observed physical properties of Nb–Ti alloy.

B3.3.1

Strong flux pinning is crucial, because good overall current carrying capacity up to fields of ~8 T is the most important technological attribute of Nb–Ti. As outlined in chapter A4, magnetic flux lines penetrate all high field superconductors at field ranges of interest. These lines will move under the

A4

A1.3 Lorentz force of an applied electrical current. However, defects and other areas of weakened A4.3 superconductivity (the pinning sites, or pins) exert a strong attractive force (the pinning force) on the flux lines. The balance of Lorentz force and pinning force is broken at the maximum useful current, the critical current. Thus, the higher the pinning force is, the higher the critical current will be. This fact is quite important: magnet conductor cost is inversely related to flux pinning strength, because the overall amount of conductor needed to generate a given magnetic field is inversely proportional to the current-A1.3 carrying capacity. Often, the critical current density J_c, i.e. critical current divided by conductor cross-sectional area, is used as an index of conductor performance. Figure B3.3.2.1 shows typical J_c values for state-of-the-art Nb–Ti composites as a function of applied field H.

One key to obtaining high J_c is the fact that the α phase is not superconducting at typical operating temperatures, the boiling point of liquid helium, 4.2 K at standard atmospheric pressure, or in the working superfluid helium-4, 1.8–2.2 K range. This makes it possible to produce many flux-pinning sites by precipitating α-Ti during moderate heat treatments. This is the central goal of the so-called 'conventional process', which is used to make almost all commercial strands. An alternative is to remove any heat treatments and, instead, hand-assemble the superconducting and non-superconducting phases, called the 'artificial pinning centre' (APC) process. APC strands now exhibit the best performance below 6 T field, but they have not yet been implemented beyond laboratory studies. Both fabrication processes produce ribbon-like nanostructures, as shown in figure B3.3.2.1. The ribbons result from the development of fibre texture in the bcc phase, which in turn creates folding and curling of the pinning centres.

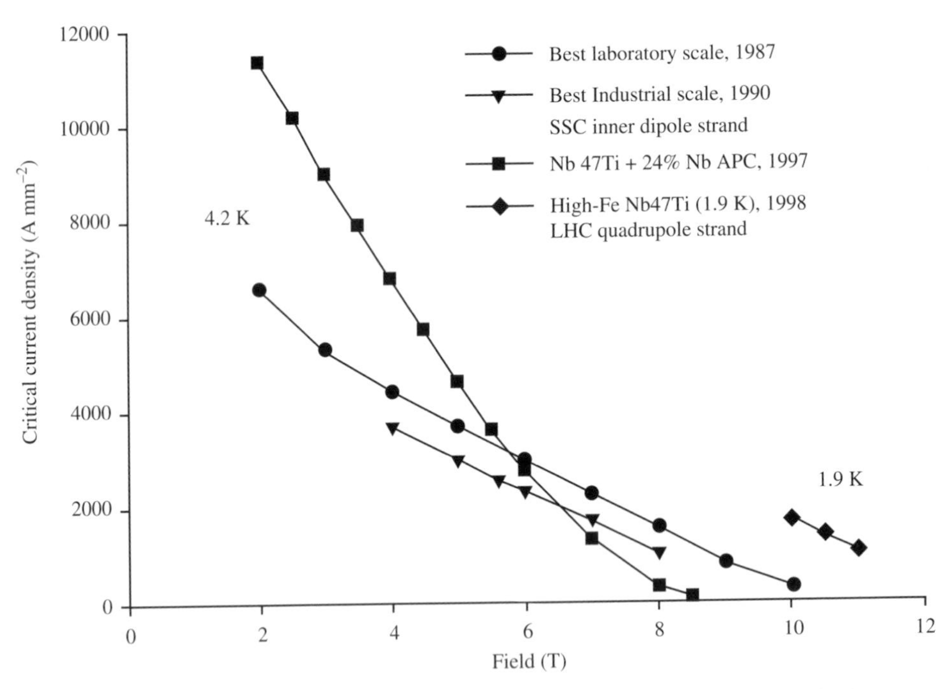

Figure B3.3.2.1. Plot of the critical current density as a function of field for various Nb–Ti wire composites that represent the present state of the art. The strongest driver of research and development of conventionally processed, binary alloy wire has been high-energy physics magnet programs, such as the Superconducting Super-collider (SSC—cancelled in 1993), and the Large Hadron Collider (LHC—scheduled for operation in 2005). Also shown is an artificial pinning-centre (APC) composite, which performs significantly better than conventional strands at low field but so far has not been implemented beyond laboratory studies. (These data were obtained from [4–7].)

This underlying metallurgical fact means that, although the precipitates start out with an ellipsoidal shape at ~200 nm size, they deform into ribbons with 1–4 nm thickness, 4–10 nm separation and large (> 50) aspect ratio by the time flux pinning is optimum. Seeing the true nanostructure was an essential component of optimizing the conventional process [2, 8], which was largely empirical prior to electron microscopy work.

A4.3

The development of a two-phase laminar nanostructure is also crucial because every flux line encounters a pinning site. This point is illustrated by the equilibrium flux line spacing in figure B3.3.2.2. Direct summation then applies; that is, the total pinning force is directly proportional to the number of precipitates or artificial pins. In support of this point, the plot in figure B3.3.2.3 shows the expected linear J_c increase with volume fraction of α-Ti precipitates. The presently attainable limit is about 25% of precipitates in the conventional process, while the pin fraction can be higher (~35%) for the APC process. Flux pinning by other defects, such as grain boundaries, contributes a weak background of about 700 A mm^{-2} at 5 T. Direct summation is also important because knowledge about the pinning interactions of specific defects (precipitates, grain boundaries, etc.) can be extracted from measurements of bulk samples and from observations of the nanostructure.

A3.1

This chapter is organized as follows: an overview of the Nb–Ti alloy system is given in section B3.3.2.2. Fabrication of composites by the conventional process is discussed in section B3.3.2.3, with the important principles of nanostructure control discussed next in section B3.3.2.4. Fabrication of composites by APC techniques is then summarized in section B3.3.2.5. Underlying issues for optimizing

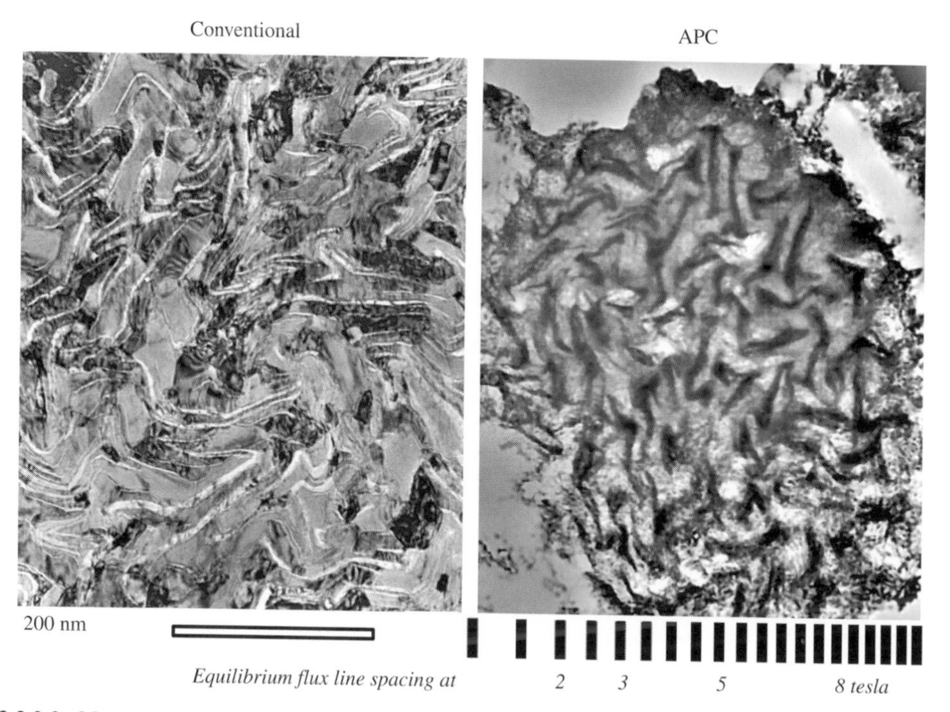

Figure B3.3.2.2. Nanostructures of optimized Nb–Ti composites made by the conventional process (left half) and by an APC process (right half). In these TEM images, the α-Ti precipitates are lighter and the Nb artificial pinning sites are darker in contrast than the gray continuous regions representing the Nb–Ti and (in the right half) copper matrices. At the lower right is a diagram of the magnetic flux lines at equilibrium, indicating their diameter at a temperature of 4.2 K and their separation as a function of magnetic field.

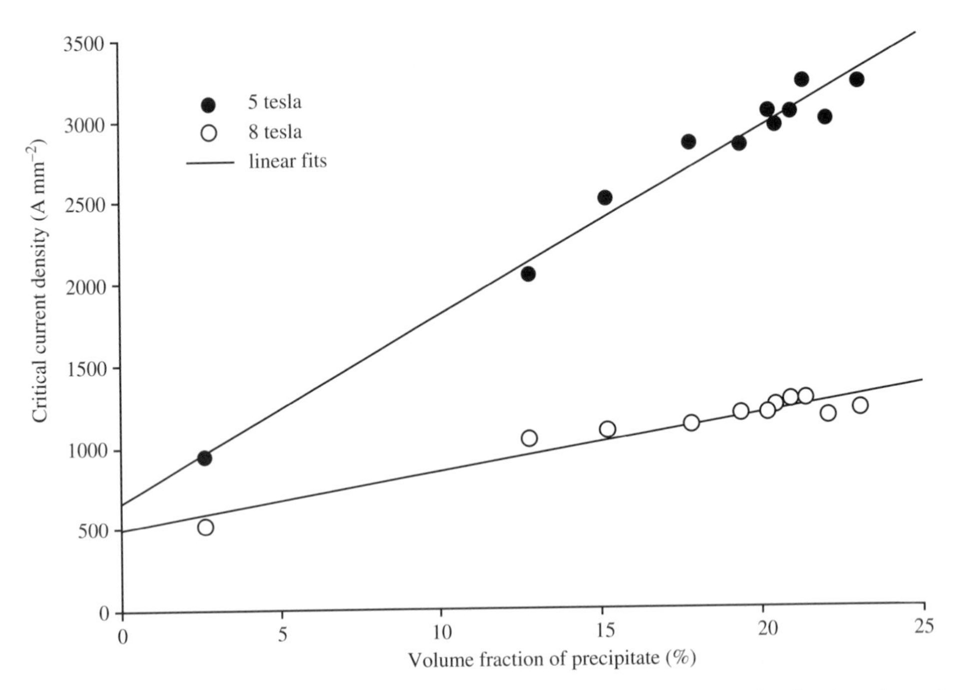

Figure B3.3.2.3. The critical current density for a number of different Nb–Ti composites is plotted as a function of the observed volume fraction of α-Ti precipitate. Results are shown for 5 and 8 T field, at 4.2 K. The data were taken from [4] and [9].

A4.3 the superconducting and especially the flux pinning properties follows in section B3.3.2.6. The chapter concludes with a look toward future topics. Extensive reviews of Nb–Ti alloy superconductors have been recently published by Lee [2], Kreilick [10] and Lee and Larbalestier [11]. These works update earlier reviews by Larbalestier [1], Collings [3] and McInturff [12]. Access to many pertinent documents can be gained over the internet at www.asc.wisc.edu.

B3.3.2.2 The Nb–Ti alloy system

B3.3.2.2.1 Phase diagram of the Nb–Ti system

H3 The goal of fabrication processes is to destroy the homogeneous structure of annealed, single phase alloys and replace it with a two-phase nanostructure through integrated deformation and heat treatment steps. The various techniques that can be employed are limited by the basic thermodynamic relationships of the alloy system. In keeping with the industry standard, compositions herein will be designated by mass fraction, e.g. Nb47Ti represents the standard commercial alloy that is 47% Ti by mass. This convention is one of convenience because it directly reflects alloy cost. Some data, such as that regarding physical properties, are more conveniently discussed in terms of atomic fractions, where e.g. Nb47Ti is equivalent to Ti37 at. %Nb, or electron-to-atom ratio (e/a), e.g. 4.37. A good estimate of the Ti atomic fraction of commercial alloys (Nb44–62Ti) can be obtained by adding 15 to the Ti mass fraction, e.g. Nb47Ti is about Nb62 at. %Ti.

H3 An important property of the Nb–Ti phase diagram, shown in figure B3.3.2.4, is that the β phase starts to decompose only well below the melting temperature. This means that diffusion is sluggish, so the formation of α precipitates due to $\beta \rightarrow \alpha + \beta$ does not occur even for moderate rates of cooling from

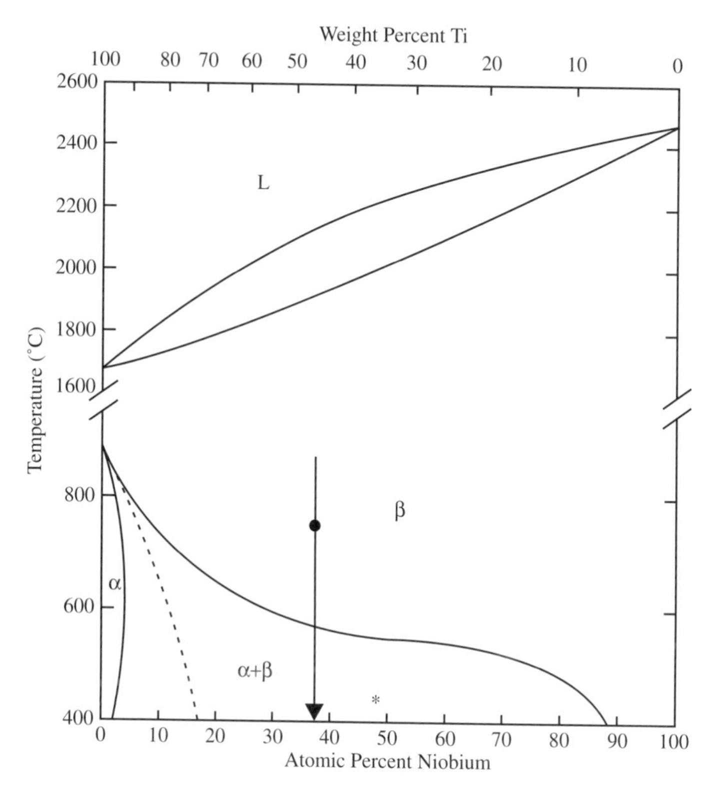

Figure B3.3.2.4. Estimated equilibrium phase diagram of the binary Ti–Nb alloy system. The high temperature phase boundaries are adapted from Hansen *et al* [13], while the low temperature phase boundaries are based on calculations by Kaufman and Bernstein [14]. The dashed line represents the martensitic transformation inferred by Moffatt and Larbalestier [15]. The standard Nb47Ti composition is indicated by the arrow, and a typical recrystallization annealing temperature is indicated by the black dot. The asterisk represents the composition of the β matrix after extensive heat treatments have been applied to form 20–25% of precipitate.

the ~800°C annealing temperature and for fairly high concentrations of Ti. Quenching retains the β phase for Ti-rich alloys, although this may result in martensite formation (α') [15]. The standard composition used in industry, Nb47Ti, is outside the composition where martensite is of concern [15], as indicated in figure B3.3.2.4. Thus, at room temperature and below, the standard alloy consists of the metastable β phase, and any phase transition is latent.

Several metastable phases can also form [16, 17], both upon quenching the β phase and during ageing heat treatments. The hexagonal ω phase can form during ageing heat treatments between 100 and 500°C, and is of concern to the conventional process when the amount of prior cold work is low. The orthorhombic martensite α'' is a transitional phase between α' and β.

B3.3.2.2.2 Physical properties of the Nb–Ti system

An overview of the basic physical properties was given by Larbalestier [1]. The lattice parameter of Nb–Ti alloy changes by less than 2% across the entire β phase region [1, 3], which indicates that many physical properties of Nb–Ti alloy exhibit a gradual variation as Ti is progressively substituted for Nb. Generally speaking, properties related to the bcc structure, such as specific heat, paramagnetic susceptibility, critical

B3.3.1
A2.4

D3.2.3 temperature T_c and thermal expansion, obey this rule. In competition with this trend is the incipient decomposition of the β phase, which becomes more prevalent as the Ti content increases. The incipient D3.1 phase transition strongly affects scattering-dependent properties, such as resistivity ρ and thermal conductivity, which have a generally faster than linear increase with Ti concentration. The coefficient of D3.2.1 electronic specific heat γ of Nb–Ti is about 1000 J m^{-3} K^{-2}. The Debye temperature lies between 250 and 300 K [1].

The variations of T_c, ρ and the upper critical field H_{c2} as a function of alloy composition are plotted in figure B3.3.2.5. The critical temperature shows a mild variation between pure Nb (9.23 K) and Nb50Ti (8.5 K), with a weak peak at about Nb30Ti (9.8 K). Addition of Ti is more potent at reducing T_c for alloys with Ti content above 50 mass%. The resistivity of Group IVa and Group Va alloys was studied long ago by Berlincourt and Hake [19], and the data shown in figure B3.3.2.5 is representative. Resistivity increases at a rate of about 1 $\mu\Omega$ cm per at. % for atoms with closely matched size, such as Ti and Nb, and at a higher rate of about 1.4 $\mu\Omega$ cm per at. % for atoms with different size, such as Zr and Nb. The rate of increase is concave upward, tending towards the Mott localization limit (>100 $\mu\Omega$ cm) for more than 70 at % Ti. Recently thin-films were used to show that these trends continue for higher Ti content [20], where bulk samples are difficult to make. Except near the endpoints (pure Nb or pure Ti) where higher resistivity ratios can be obtained, residual resistivity ratios of these alloys are close to 1 [21]. D3.1 The resistivity of commercial alloys is 55–65 $\mu\Omega$ cm, compared to 12.5 $\mu\Omega$ cm at room temperature for pure Nb and ~40 $\mu\Omega$ cm for pure Ti.

A3.1 The upper critical field at 4.2 K exhibits a broad dome-like curve with a maximum of about 11.6 T at a composition of Nb44Ti and a value above 11 T for a composition range from about Nb40Ti to about Nb52Ti. The peak results from a balance of the trends above, where the zero-temperature value can be predicted by

$$\mu_0 H_{c2}(0) = 3.11 \times 10^3 \rho\gamma T_c \tag{B3.3.2.1}$$

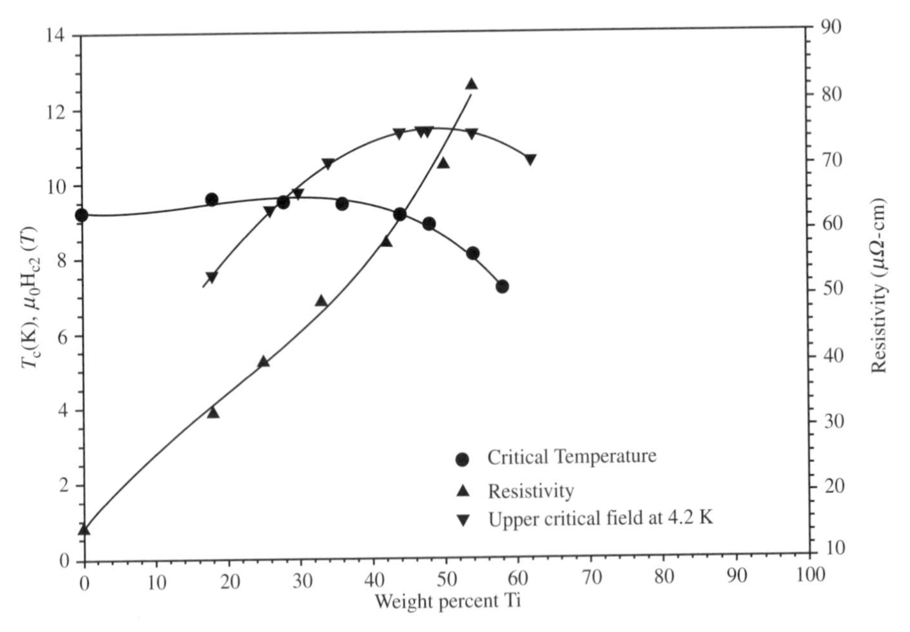

Figure B3.3.2.5. The critical temperature, upper critical field at 4.2 K, and resistivity at 293 K are plotted as a function of the mass fraction of Ti across the binary Nb–Ti alloy system. The data were adapted from [2] and [18].

with all quantities being in SI units. The above equation is an extension of the Ginzburg–Landau theory [22]; in more familiar terms, the peak in H_{c2} arises from a weak decrease in the critical field H_c and a large increase in the Ginzburg–Landau parameter κ with increasing Ti content, where $H_{c2} = \sqrt{2}\kappa H_c$. The value of H_c is linked to T_c via the condensation energy of the superconducting state, while the value of κ is linked by scattering theories to ρ. Naus *et al* [23] and Heussner *et al* [24] recently investigated how the value of H_{c2} depends somewhat on the measurement technique, using modern equipment. Their investigations also showed that the irreversibility field (see chapter A4), is about 1 T lower than H_{c2}, similar to earlier results obtained by Suenaga *et al* [25].

A3.1

B3.3.2.2.3 Reduction of paramagnetic limiting in ternary Nb–Ti–Ta and Nb–Ti–Hf alloys

An important limitation of binary Nb–Ti alloys is the strong paramagnetic moment of Nb, where the Pauli paramagnetic susceptibility is about 220 ppm. This is not changed much by alloying with Ti [1]. Heussner also reported susceptibilities of 250–280 ppm for Nb47Ti and Nb62Ti mixed with Nb artificial pins measured well above the upper critical field, where signals due to the superconducting state are absent, in agreement with this range [26]. The fairly strong paramagnetism contributes to the energy of the normal state, which opposes the superconducting state. This limits the value of $\mu_0 H_{c2}(0)$ to about 16.7 T, based on the Clogston formula $\mu_0 H_{c2}(0) = 1.84 T_c$ T [27]. By contrast, the value expected [28] from the slope of H_{c2}(T) at T_c is $H_{c2}(0) = 0.69 T_c (dH_{c2}/dT)|_{T_c} \approx 17.8$ T, using the values 2.85 T K^{-1} and 9.1 K found by Meingast *et al* for Nb48Ti wire [18]. The paramagnetic limit reduces the experimental upper critical field for alloys with ≥ 40 mass% Ti [21], where measured values at 1.2 K are all less than 15 T.

B3.3.1

An important reduction of the normal-state energy is provided by spin-orbit scattering. Since the scattering rate is proportional to Z^4, elements with high atomic number Z are especially strong scatterers. The incorporation of moderate amounts of Ta and Hf into ternary and quaternary alloys based on Nb–Ti thus extends the limit of superconductivity to somewhat higher fields. Ta is the most common alloying element because past work has shown it to be more potent [10, 29, 30]. Ta is also present in most Nb-bearing ores and is difficult to separate out due to its very high melting temperature.

Adding Ta or Hf reduces the critical temperature by 1–2 K but increases the resistivity and the slope of H_{c2}(T) at T_c. This results in comparable H_{c2} values at 4.2 K for ternary and binary alloys, about 11.3 T. At 1.8 K, however, H_{c2} is 0.3–1.5 T higher for the ternary alloys. Figure B3.3.2.6 shows contours of different H_{c2} values at 2 K as a function of composition, based on data of Hawksworth and Larbalestier [21, 31]. The heavy line in the plot denotes a constant e/a ratio of 4.37, which is the standard binary Nb–Ti composition. Moving upward along this line represents increasing Ta content, where it can be seen that additions of 15–25 mass% of Ta increase the upper critical field from 14.2 to 15.5 T. The greater difference between 2 and 4.2 K H_{c2} values, 3.5–4 T for the ternary alloy versus 3 T for the binary alloy, means that properties such as the critical current can be shifted farther upward in field by cooling for ternary alloys.

B3.3.2.3 Fabrication of conventional Nb–Ti wires

The standard fabrication techniques for making fine-filament Nb–Ti strands were largely researched in conjunction with accelerator magnet programs during the 1980s and early 1990s. The state of the art in this regard is represented by magnet strands currently in production for the Large Hadron Collider. In parallel to these programs, improvements in and increased demand for MRI magnets have contributed towards improving process efficiency and reducing variability, while also identifying areas where the strict tolerances of accelerator-magnet grade strand could be relaxed. The fabrication process described in this section is representative of strands produced for accelerator magnets.

B3.3.1

E1.1

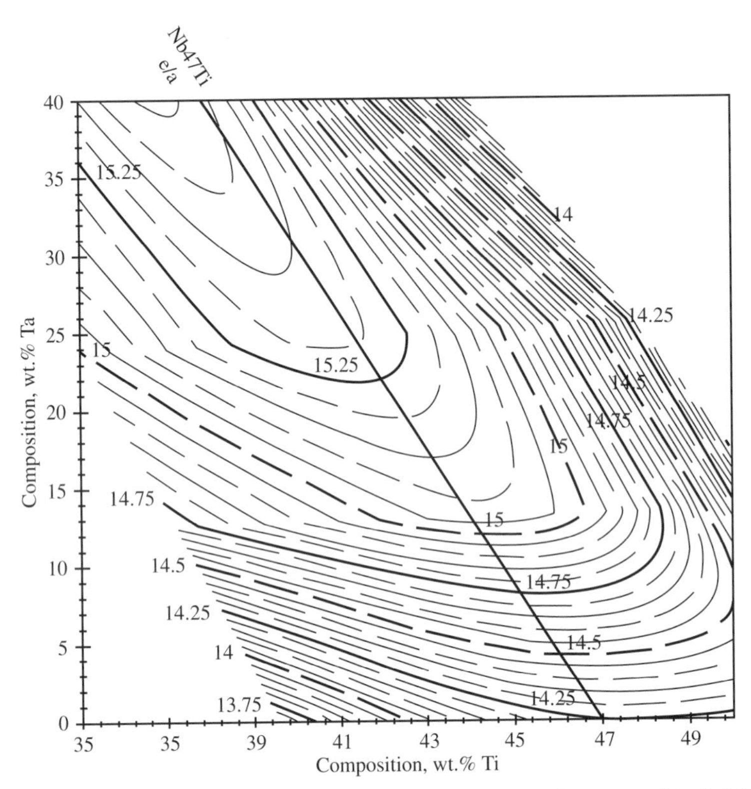

Figure B3.3.2.6. Contours of constant values of the resistivity determined upper critical field (in T) at 2 K are plotted as a function of ternary alloy composition. The x-axis of the plot represents binary Nb–Ti alloys, for which the standard composition corresponds to an e/a ratio of 4.37. The heavy line running through the contours (light lines) represents a constant value of e/a = 4.37 as Ta is added. This line passes near to the region with highest H_{c2}, based on data of Hawksworth and Larbalestier [21, 31].

An overview of the conventional fabrication process is given in figure B3.3.2.7. Each component represented in this figure is discussed in detail later in this section. Note that the Nb–Ti diameter ranges over almost 5 orders of magnitude, from the starting size of the recrystallized, wrought alloy to the final filament diameter. There are essentially three stages of this process, labelled with respect to the drawing strain relative to the precipitation heat treatments used to form α-Ti precipitates. The prestrain regime encompasses extrusion and rod and large wire drawing. Multiple precipitation heat treatments are given next, usually separated by strain increments of 1–1.25. The pinning centres formed during the heat treatments are finally refined over a large final strain of 4–5, during which the flux-pinning properties become optimum. At the top right of figure B3.3.2.7, light microscopy and digital image analysis provide information about filament uniformity and integrity of the barrier (the bright ring around the filaments), which are of concern from extrusion through the final heat treatment. At right centre and right bottom, scanning electron microscopy in electron backscatter detection mode provides a convenient assessment of the nanostructure during the intermediate and final strain regimes.

B3.3.2.3.1 Metallurgy of Nb–Ti cast alloys

The optimum composition range of Nb–Ti alloys is between 46–50 mass% Ti due to a combination of practical considerations and processing limitations. Higher Ti amounts produce α-Ti precipitates at

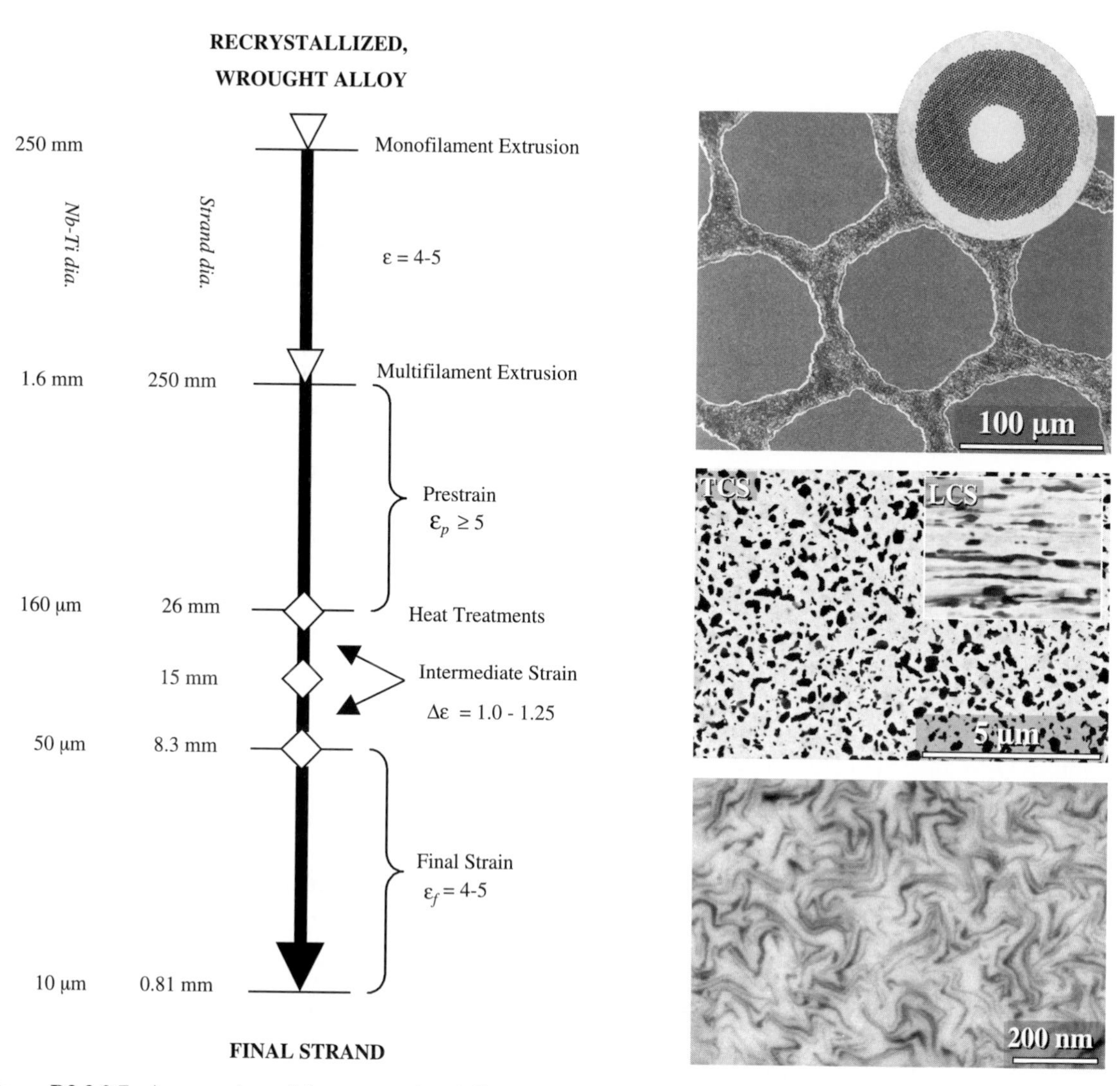

Figure B3.3.2.7. An overview of the conventional filament Nb–Ti process, as implemented during the fabrication of high-Fe Nb–Ti strand for LHC quadrupole magnets, is shown. The flow chart on the left side indicates characteristic diameters of a Nb–Ti filament and the multifilament strand, along with the different strain regimes (prestrain, intermediate strain and final strain). The images on the right side show various magnifications of the strand transverse and longitudinal cross-sections and address important aspects at each level of the fabrication process (filament uniformity and barrier integrity, precipitate size and volume fraction, and nanostructure optimization).

a faster rate and in greater volume, but require more area reduction prior to the first heat treatment (prestrain) than is typically available in order to suppress undesirable precipitate morphologies [32]. Lower Ti amounts produce insufficient precipitate for obtaining high critical current density. In terms of physical properties, the composition range above is close to the composition with highest H_{c2}, Nb44Ti, and somewhat below the range where T_c begins to fall off sharply with increasing Ti content, above Nb50Ti, as shown in figure B3.3.2.5.

A4

High-homogeneity alloys are essential for developing consistent, optimum electrical and mechanical properties of Nb–Ti strands. These are achieved by consumable-electrode arc melting, electron beam melting, or plasma arc melting of pure metals [33]. At least one remelting step is common so as to minimize coring due to the large separation between the liquidus and the solidus boundaries and the high melting point of pure Nb. Without these steps, cast alloys tend to contain unacceptable (± 5 mass %Ti) variations in local and overall chemistry and, possibly, gross defects such as unmelted Nb inclusions. An example of this problem is shown in figure B3.3.2.8. With remelting steps, composition variations can be held to within 1.5 mass %Ti across the entire billet, and Nb inclusions are rare. Further improvements can be achieved by additional annealing, but at the expense of increased grain size. Gross inhomogeneities are cost-effectively observed by using flash-radiographs of billet cross-sectional slices. Specifying a lack of Nb inclusions or a lack of Ti-rich 'freckles' in such slices ensures that the billet has undergone a well-controlled melting and subsequent solidification. Typical specifications call for less than 200 ppm of most impurities, except for oxygen (200–1000 ppm) and tantalum (500–2000 ppm). Recent work shows that relaxed specifications for iron, up to 600 ppm or higher, can be beneficial [6, 34]. Etching and analysis techniques that are used to characterize inhomogeneities are discussed in [2].

The cost of wrought Nb–Ti alloys depends primarily on the alloy composition and the cost of raw materials. Generally speaking, alloys with compositions other than the standard Nb47Ti are not carried in suppliers' inventories and, therefore, such alloys must be specially prepared. Manufacturers of MRI devices consume the majority of Nb–Ti alloy produced annually, of the order of 1000 tons. Raw alloys must have high purity and are therefore rather expensive, where the approximate cost of Ti is $25 per kg and Nb is $150 per kg in the purity specification ranges above. Wrought alloy can be obtained for about $140 per kg, although a significant increase from these prices can be expected for non standard alloy compositions and small ingot sizes.

E2.3

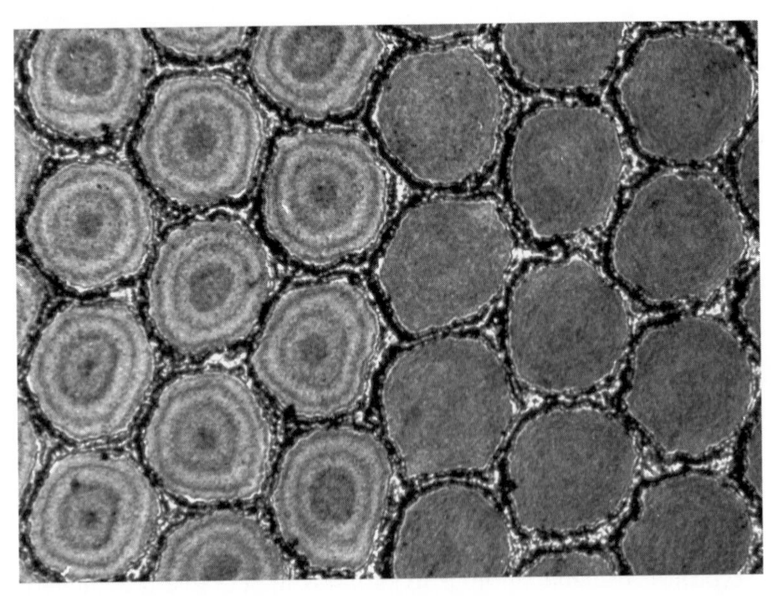

Figure B3.3.2.8. This billet was inadvertently stacked with Nb–Ti rods from two different raw material sources. On the left, the characteristic 'tree ring' pattern indicates that these rods were not properly remelted, and have variations in local chemistry due to the large separation between the liquidus and the solidus. On the right, much more uniform chemistry is indicated by the uniform contrast of the Nb–Ti rods.

The need for high homogeneity stems both from mechanical and performance demands. Cast ingots are typically 200–600 mm in diameter and are hot forged to a diameter of about 50–100 mm before being annealed. Starting from a recrystallized diameter of 150 mm, the total strain,

$$\varepsilon = 2\ln d_0/d_f \qquad \text{(B3.3.2.1)}$$

required to reduce the filament diameter below 50 μm is 16. The tremendous amount of area reduction (a factor of 10 million) is unique among metallurgical systems, with possible exceptions being piano wire and light-bulb filaments. Nb–Ti rods and filaments must therefore have excellent ductility, low hardness and a predictable hardening rate, be free from hard inclusions and have a uniform grain size. In addition, high critical current density depends on having a high number density of uniformly sized and distributed α-Ti precipitates. This cannot be achieved unless there is a fine initial grain size (ASTM 6 or less) and a uniform grain distribution. Homogeneous distribution of Ti throughout the entire alloy ingot is required because the precipitate morphology and the precipitation rate are very sensitive to composition [32]. These initial quality concerns are of the utmost importance to the later stages of the conventional process, as discussed in more detail in section B3.3.2.4.

B3.3.2.3.2 Stabilizer materials

Except in special applications, practical Nb–Ti superconductors are multifilamentary composites. This means that hundreds to thousands of Nb–Ti filaments are embedded in a stabilizing matrix, often oxygen-free high conductivity (OFHC) copper. Some applications may use single filaments with relatively large diameter, for instance when low-loss superconducting joints are needed. However, flux-jump instabilities generally result when the filament diameter is above 30 μm (see sections A2 and A4). Due to the magnetization of the superconducting filaments themselves, high-homogeneity magnets also require filament diameters smaller, by a factor of 1000 or more, than the magnet bore diameter.

B3.3.1

In the event of a quench (i.e. an uncontrollable phase change from the superconducting to the normal state, due to heat generated by friction, movements, etc.), sufficient stabilizer must be included in the composite design to both conduct heat to the coolant bath and carry the electric current until the magnet can be shut down. Candidate stabilizer materials are therefore good electrical and thermal conductors, such as high-purity aluminium and copper. The area ratio of the stabilizer to that of the superconductor in an accelerator magnet strand ranges from 1.5:1 for small magnets to as much as 10:1 for very large stored energies. Low ratios can be specified when magnets are small enough to be self-protecting, while much higher ratios (>10) provide protection against any perturbation [35].

The residual resistivity ratio (RRR) is an index of stabilizer quality that is often used. The RRR value is defined by the resistivity at 293 K divided by the saturation value at low temperature (typically 4.2 K). Saturation of the resistivity occurs when electron scattering by thermal vibrations, which depends on temperature, falls below scattering by impurities and defects, which does not depend on temperature. RRR values of 30–150 are called for by typical magnet specifications, which can be attained by annealing, e.g. at ~ 250°C for 1 h for standard oxygen-free high conductivity (OFHC) copper. Since the resistivity of copper is a strong function of field, there is little advantage to increase RRR above 100. Collings noted that too high a value of RRR can produce unacceptable coupling between filaments [36], discussed below. Aluminium is another option. Unlike Cu, its resistivity saturates quickly with increasing field [10], and very high RRR values can be taken advantage of.

Another important role of the stabilizer is the prevention of circulating currents between filaments in ac applications or when the rate of field change is high. From Maxwell's equations, there will be an electric field applied along the circumference of a current loop when the flux density inside the loop changes with time. When neighbouring superconducting filaments are joined by a good conductor, closure of such current loops can be achieved when a current path crosses the intervening stabilizer.

D2.1

To minimize the area enclosed by such current loops, most practical strands are twisted, while strands in B2.6 cables are also fully transposed. However, it is sometimes desirable to further reduce the induced losses by changing the stabilizer material to one with higher resistivity. Adequate protection and thermal conductivity can then be provided by incorporating a region of high-purity stabilizer far away from the superconducting filaments in the strand cross-section.

Coupling losses can be exacerbated by the proximity effect, a phenomenon in which superconductivity decays into a neighbouring normal metal over a finite distance. Pronounced effects of coupling were noted in [37] when the interfilament separation became of the order of this decay length, 0.1–1 μm. A simple way to shorten the proximity length and suppress coupling currents is to increase the resistivity of the matrix. Cu–Ni alloys have been used in this application, with between 5 and 30% Ni being substituted for Cu. Fickett showed that the rate of resistivity increase was about 0.8 μΩ cm per % Ni [38]. Cu–Ni alloys form a solid solution with a common face-centred cubic structure, so the alloying of Ni does not strongly affect the ductility. In fact, Cu–Ni stabilizer actually provides a better match of strength and hardness to the Nb–Ti filaments, allowing requirements on filament separation to be relaxed somewhat, as discussed later. Another possibility to suppress coupling is spin-flip scattering. In this case, a magnetic atom such as Mn, Cr or Fe that does not form a solid-solution with Cu can be incorporated into the matrix. The large magnetic moments of these ions break the superconducting electron pairs, providing a more direct suppression of the proximity effect than can be achieved by shortening the proximity length. Mn is very potent in this regard. Collings found that 0.9 mass% of Mn provides a similar degree of decoupling as 30% of Ni [36]. Additions of both Mn and Ni can be used when the magnetization of the stabilizer is not a major concern [39].

B3.3.2.3.3 Strand geometry

The overall geometry of a Nb–Ti strand reflects several different requirements of the application. Many magnets require a Nb–Ti filament diameter d_f less than about 30 μm, with a tendency to permit larger d_f values at higher fields due to their lower J_c. Some composites have 2–10 μm filaments, however, to minimize hysteretic losses and unwanted magnetic field harmonies in magnets, both of which are caused by the magnetization of the filaments themselves. The filaments must be uniform over a very long length, typically 1–20 km or longer, in order to produce cables and wind a large solenoid with few joints. Each A1.3 multifilamentary strand must also carry 100–1000 A, or each cable 1–20 kA, to reach typical designed magnetic field strengths. This final requirement means that the superconductor must have a total Nb–Ti cross-sectional area of about 0.04–0.4 mm^2, assuming its current-carrying capacity is $\sim$ 2500 A mm^{-2} at 5 T. The current-carrying capacity is defined by the critical current density J_c and a suitable safety margin, although often magnets are operated very close to J_c in the high-field regions of the coil. Together with the other requirements on d_f above, a typical composite strand thus contains 40–16 000 filaments. For instance, the outer conductor design for the LHC contains about 6400 filaments, each with a diameter of 6 μm, in a strand of diameter 0.83 mm.

The stabilizer should make intimate contact with each filament to ensure good heat flow and current transfer. This is commonly achieved by embedding the filaments within a continuous matrix of the stabilizer. However, this also means that soft metals like Cu and Al must be co-deformed with hard Nb–Ti alloy. Gregory *et al* [40] noted that poor piece length and filament sausaging (i.e. variations of the cross-sectional area along the length of a filament, like a string of sausages) was a common problem in Nb–Ti/Cu composites with widely separated filaments. By contrast, good piece length and uniform filaments were obtained in Nb–Ti/Cu composites with closely spaced filaments. The optimum ratio of the edge-to-edge separation s and the filament diameter is $s/d_f \approx 0.15$ for Nb–Ti/Cu composites. Since this requirement fixes the local stabilizer to superconductor area ratio, it is generally necessary to incorporate an additional block of pure copper into the composite cross-section. It should be noted that with an s/d_f ratio of 0.15 and a final filament diameter of $< 5\ \mu m$, the separation is of the order of the proximity length of pure Cu. For example, Collings showed that the proximity length at 4.2 K and zero field is about 300 nm for copper with an RRR of 30, but this increased to 2 μm for an RRR of 200 [36]. This means that alloyed Cu stabilizers must be used when the desired filament diameter is small. Alloyed Cu surrounding the filaments can be combined with pure copper strand cores to provide a combination of stable deformation, quench protection and filament decoupling.

B3.3.2.3.4 Diffusion barriers

Hard intermetallic phases, formed at the Cu/Nb–Ti interface during hot extrusion and precipitation heat treatments, must be avoided. The chief intermetallics are those of Ti and Cu, especially $TiCu_4$ [41–43]. Intermetallic particles do not co-deform with the strands or stabilizer matrix; instead, they distort the filaments around them, which results in gross variation of the filament cross section and sausaging. As a result, locally reduced filament cross-sectional areas pinch off the current, forcing it to transfer around the damaged area and reducing the overall J_c value (also see section B3.3.2.6). Agglomerated particles ultimately lead to strand failure. B3.1

Nb is the most widely used barrier material. Nb and Cu have very low mutual solubility, and pure Nb has excellent ductility and makes good mechanical bonds with both Nb–Ti and Cu. The Nb barrier does not prevent diffusion of Ti or Cu across it, but interdiffusion of either atom can be kept low enough to prevent the $TiCu_4$ phase from forming. The standard practice is to wrap several layers of Nb foil around the Nb–Ti with a small overlap. The barrier cannot be too thick because it consumes conductor cross-section normally used to carry current. However, since the barrier thickness is reduced during wire drawing, there must be sufficient material to retain isolation of Cu and Ti during the final heat treatment. Faase *et al* [43] showed that the required barrier thickness at this stage is 0.6 μm. Working backward from a fixed final strain of 4–5, this means that the required barrier area fraction *increases* inversely with smaller target filament diameter of the application. For final filament diameters of 6–9 μm, a barrier representing 4% of the strand cross-section is typically used.

Faase *et al* [43] observed that the primary location for intermetallic formation was between the Nb–Ti filament and the Nb barrier. Evidently, the Nb grain boundaries provide short-circuit pathways for Cu diffusion. They also observed that there was a region of reduced precipitate diameter extending into the Nb–Ti, shown in figure B3.3.2.9, which they reasoned was due to a poisoning effect of diffused copper. A small grain size of the Nb–Ti also helps preserve the barrier integrity. Heussner *et al* [44] showed that plane-strain deformation leads to the development of a ragged interface between the Nb–Ti and the barrier. The interface roughness was found to be proportional to the Nb–Ti grain size. Heussner *et al* surmised that as the wire diameter is reduced, intercurling of grains necessary to maintain grain continuity eventually thins the barrier or, in rarer cases, causes a complete breach. B3.3.1

B3.3.2.3.5 Extrusion and rod drawing

The present state of the art of composite wire fabrication uses extrusion to facilite strong bonds between the Nb–Ti rods, the Nb barrier and the stabilizer. Strong bonds are essential for obtaining good piece length and high yield, so it is important to establish good bonds right from the beginning of the conventional process. Preheating of the extrusion billet and adiabatic heating during extrusion both promote good interface bonding. Extrusion billets are stacked from large pieces with low surface-area to volume ratio. This improves the effectiveness of etching and cleaning procedures to remove oxides present on the surfaces of composite elements. Oxides, grease and other impurities are the primary enemies of bond formation. B3.3.1

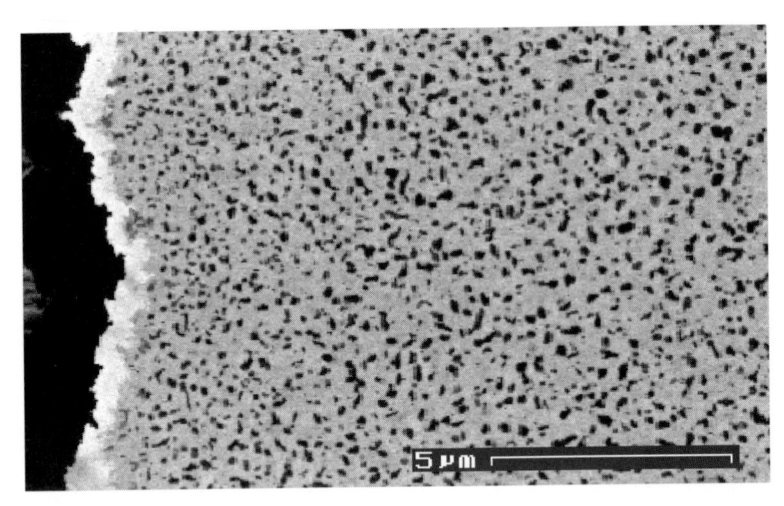

Figure B3.3.2.9. This digitally-enhanced scanning electron microscopy image shows the variation of the precipitate size and number density near the Nb diffusion barrier. The α-Ti precipitates appear as black dots in the gray Nb–Ti matrix, while the bright region running along the left of the image is the Nb barrier. Although the precipitates have smaller size near the barrier, the number of precipitates is greater, giving overall an approximately constant volume fraction of precipitate across the filament cross-section. The refinement of precipitate size near the barrier is thought to be due to having small amounts of copper in the Nb–Ti, which crossed the barrier along its grain boundaries.

A1.3 E2.3

Multifilamentary extrusion is the initial fabrication step for applications that require a relatively low number (100) of filaments, such as MRI magnet strands. A typical billet is assembled by inserting Nb–Ti rods with their optional diffusion barriers into holes that have been gun-drilled into a copper cylinder. The Nb–Ti rods can be 5–25 mm in diameter, while the overall diameter of the copper cylinder is limited by the capacity of the extrusion press (250–400 mm).

Two extrusion steps are used to fabricate strands with many filaments (10 000), such as LHC magnet strands, as shown in figure B3.3.2.7. The first step is monofilamentary, using a billet made from a single core of Nb–Ti alloy with a diffusion barrier wrap that is inserted in a copper sheath. The overall dimensions of the billet depend on the total piece length needed for the subsequent restack. After monofilamentary extrusion, cold rod drawing is used to reduce the overall diameter to 1–3 mm. The second extrusion step is multifilamentary, using a billet made from similar lengths of the monofilament strands stacked together in a copper can. All of the outer surfaces of pieces employed in this step are the same material (Cu), which helps good bond formation. The monofilament strands are usually drawn through a hexagonal die ('hexing') before being cut into similar lengths to help eliminate void space from the billet. Special pieces, such as bars, strips or grooved elements, and hot isostatic pressing (HIPing) can also be used to eliminate void space. A central copper core is usually used because it is thought to reduce the tendency for cavitation or centre bursting during extrusion [45], in addition to the stabilization advantages described earlier. After extrusion, cold rod drawing is again used to reduce the diameter to 25–40 mm, depending on the maximum capability of heat-treatment facilities. Additional restacking and multifilamentary extrusion steps can be used to attain very large ($>10^6$) numbers of filaments or very small (<0.1 μm) filament diameters. However, these tend to reduce yield and degrade the filament uniformity. The ability to manufacture strands using a single stack of Cu-clad filaments was a requirement of successful accelerator magnet strand production, and was a major technological advance.

Extrusion procedures are similar in all the above cases. After the billet is assembled, it is welded, evacuated, isostatically pressed, machined to fit the chamber and preheated to between 500 and 650°C.

It is desirable to limit the recovery of cold work already applied between the first and second extrusion by careful control of extrusion conditions. It is also important to monitor the rod drawing and later wire drawing processes to maintain quality control. Parrell *et al* [46] established the usefulness of micro-hardness measurements to monitor the hardening rate of filaments and the overall composite. These measurements can also reveal lost cold work due to excessive heating during the multifilamentary extrusion. High *et al* [47] applied digital image analysis to monitor filament shape uniformity and overall robustness of the filament assembly. These measurements showed that filaments with a high degree of uniformity gave the best performance.

B3.3.2.3.6 Thermomechanical cycle

As described in more detail in section B3.3.2.4, flux-pinning centres are produced by a series of heat treatment and wire drawing steps at an intermediate stage. This procedure generally applies three heat treatments with intervening wire drawing in a cycle. The heat treatments can be for constant times at constant temperature, for example 3 × 80 h at 420°C, or can be widely varying in time and temperature, depending on the desired precipitate morphology. A4.3

Work mostly supported by US Department of Energy, High-Energy Physics programs between 1980 and 1989 led to an understanding of desirable times, temperatures, wire die spacings and other conditions for the thermomechanical cycle. An empirical optimization came before the materials science was fully worked out. Aggressive heat treatments, 10–160 h at 405–435 °C, gave dramatic improvement in J_c over the standard industry heat treatment, 40 h at 375°C [48]. The optimum combination was found to be three heat treatments of 80 h duration at 420°C. The heat treatments were separated by a strain increment of about 1.1, corresponding to five die steps at an area reduction of 20% per die. The effects of various heat treatments and the change of J_c with final drawing strain are shown in figure B3.3.2.10. A vital prerequisite for success was the availability of starting Nb–Ti ingots with low levels of chemical variability [33]. As stressed earlier, the precipitation rate and morphology are quite sensitive to alloy composition, and it was impossible to determine the best heat-treatment parameters until studies of high-homogeneity alloy became available.

After the development of high-homogeneity alloys, substantial progress in understanding the interrelationship between precipitate amount and morphology was made. Increasing the heat treatment temperature increases the precipitation rate but also results in larger precipitates [4]. Similar results could also be obtained by increasing the heat treatment time at constant temperature. However, both routes tended to exacerbate barrier breakdown and promote mechanical instabilities. Increasing the number of heat treatments could be used to obtain similar results, as found by Li and Larbalestier [48] in figure B3.3.2.10. For instance, [9] showed that short heat treatments (1–10 h) that are close together in strain ($\Delta\epsilon \approx 0.6$) nucleate a large number density of precipitates. These later grow when a more aggressive heat treatment is applied at the final increment. The results of a wide variety of heat treatment parameters have been summarized in [2]. B3.3.1

B3.3.2.3.7 Final wire drawing

Wire drawing is used to reduce the filament diameter to design specifications, and also to develop and refine the nanostructure to produce strong flux pinning. A strain after the last precipitation heat treatment ϵ_f of 3–5 is generally needed to attain optimum J_c. Working backward from a final d_f of 5 μm, the final wire diameter d_w can be calculated from

$$d_f = \frac{d_w}{\sqrt{N(1+R)}} \qquad \text{(B3.3.2.3)}$$

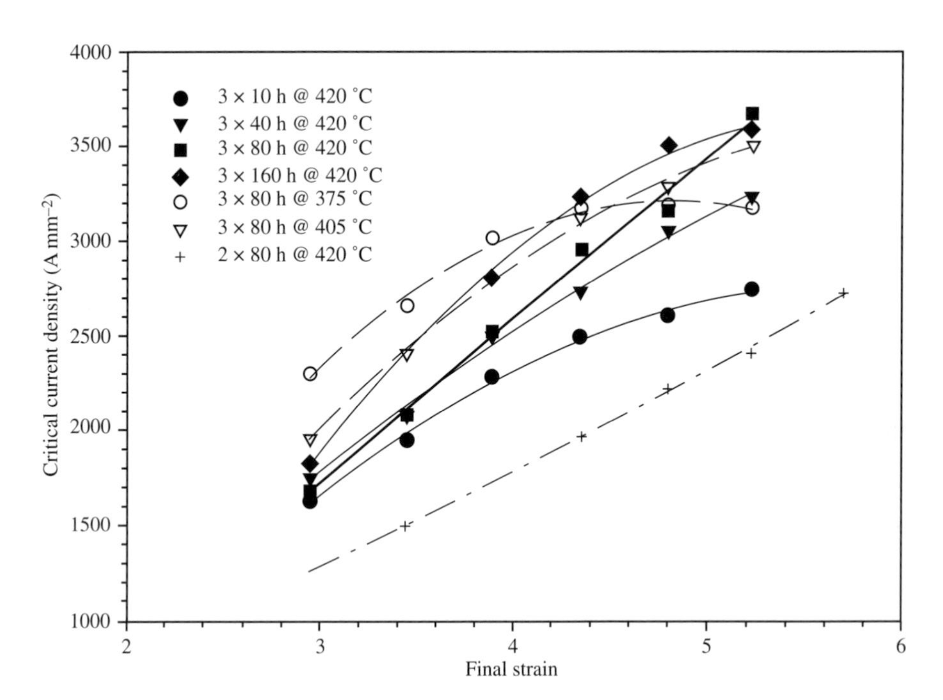

Figure B3.3.2.10. The critical current density at 5 T, 4.2 K is plotted as a function of the final drawing strain for a variety of heat treatments applied to Nb46.5Ti alloy monofilament strands. The data were taken from [48].

where N is the number of filaments and R is the volume (area) ratio of stabilizer to superconductor. Taking $N = 10\,000$ and $R = 1.3$ (typical of an accelerator magnet strand), this gives $d_w = 0.76$ mm. Applying the value of ϵ_f above, this also means that the final heat treatment is given at a wire diameter of about 7 mm.

Multiple die machines with 15–25% area reduction per die are commonly used during this process. Just above the final diameter, the composite is twisted to reduce filament coupling currents. The twist pitch is typically 10–20 times the wire diameter, with generally tighter pitches being given to composites used in applications with fast rates of field change.

B3.3.2.3.8 Cabling

A1.3 The current-carrying capacity required for magnets can be many times the capability of a strand. For this reason, cabling is an important step in the manufacture of superconducting magnets. Cables may combine any or all of the following: the superconducting composite strands, insulation, additional stabilizer, support elements, coolant channels and an outlet conduit or armoured jacket. Cabling techniques are similar to those used elsewhere in the electric power industry.

Rutherford cables provide a high packing fraction and full transposition of the strands. They are assembled by twisting 20–30 multifilament strands together, then compacting them into a rectangular or trapezoidal ('keystoned') cross-section using a Turk's head. The flattened cable has two layers, so that many filaments run parallel to each other on either cable face. Sharp bends at the cable edges degrade the current-carrying capacity by a few per cent, because the strand shape is heavily deformed.

Collings and Sumption [49] analysed the effect of various coatings and native oxide layers on the A1.3 coupling currents of Rutherford cables. Although similar in origin to eddy-current and proximity losses within individual strands, losses in cables can be very high due to the large cross-sectional area of the cable face that can be enclosed by a coupling current path. Collings and Sumption noted that these face-on losses were significantly reduced by the native copper oxide layer on the surface of composite strands. The effectiveness of the oxide layer was reduced when strong compaction was applied to the cable and, in that case, other coatings, insulating layers or a flat core can be added between the layers to reduce face-on losses. In opposition to the need to limit face-on losses is the need to establish good connectivity between adjacent strands on a tape face. This allows current sharing as the cable nears its critical current, and offers protection during a quench. D2.6

B3.3.2.4 Details of nanostructural control

The conventional process relies on the fact that α-Ti precipitates are the primary flux-pinning centres. Therefore, *all processing steps must serve the development of the heterogeneous nanostructure at the final wire diameter*. If this is properly done, the final precipitate arrangement satisfies three fundamental requirements for obtaining high J_c:

(a) The number density of flux-pinning interactions should be as high as possible. A4.3

(b) Each pinning interaction should be strong.

(c) There should not be a strong variation in pinning center size or distribution.

Since the number of pinning sites and, therefore, the number of pinning interactions, increases with the precipitate volume fraction at constant final precipitate thickness, J_c increases with the percentage of α-Ti precipitate [2, 4, 9], as shown in figure B3.3.2.3. Thus, item (a) can be restated as achieving as high a volume fraction of precipitate as possible.

Intergranular α-Ti precipitates that are located at the grain-boundary triple points are desired. This precipitate mode produces consistent hardness, workability and critical current properties if care is taken to minimize chemical variations across the starting billet. The key understanding of the conventional process has come from correlating the effects of the different processing steps on the formation and refinement of the triple-point precipitates. As emphasized earlier, seeing the true nanostructure by electron microscopy was a pivotal advance in forming the correlation above. In particular, obtaining images of transverse cross-sections of strands was crucial. Almost all of the initial nanostructural characterization was done by transmission electron microscopy (TEM), for which sample preparation is difficult and sample turnaround is very slow. More recently Faase *et al* [50] showed that scanning electron microscopy (SEM) could be used to rapidly analyse the precipitate distribution at heat-treatment size when used in electron backscatter detection mode (EBSD). Since backscattered electron number density is proportional to atomic number, contrast directly reflects chemical variations in the sample. Thus, in figure B3.3.2.9, the precipitates are clearly seen as black dots in the gray Nb–Ti matrix, while the pure Nb barrier is almost white. The development of high-resolution field-emission SEM instruments has provided spatial resolution for analysis of Nb–Ti specimens down to nearly final size, as indicated in figure B3.3.2.7. SEM–EBSD now provides very rapid characterization and a wealth of nanostructural information. D1.2 D1.3

The cold work given to the composite is the primary means of controlling the production and refinement of the precipitates. Cold work can be divided into three regimens: prestrain ϵ_p given between the multifilamentary extrusion (if warm) and the first heat treatment, incremental strain $\Delta\epsilon$ given between heat treatments, and the final strain ϵ_f given after the last heat treatment. Lee [2] noted that cold work serves seven functions:

Prestrain serves to

(a) promote the triple-point morphology

(b) improve the chemical homogeneity on a local scale

(c) increase the density of precipitate nucleation sites

(d) increase the grain boundary density and Ti diffusion rate, and

(e) reduce the average diffusion distance to the precipitate nucleation site.

Incremental strain also serves functions (c)–(e), as well as

(f) increasing the volume of precipitate.

Final strain plays the important role of

(g) reducing the precipitate dimensions from 300–100 nm to 4–1 nm thickness and 10–4 nm separation.

These steps must be combined to fit within the typical 'strain space' of ~12 determined by the ratio of extrusion billet diameter to final wire diameter. The elevated temperatures (~600°C) used for multifilament extrusion usually prevent the full retention of the cold work applied during monocore rod drawing, thus effectively limiting the available prestrain from extrusion to the first heat treatment diameter to ~5. However, because multifilamentary extrusion is 'warm', cold work applied during monofilament rod drawing is at least partially retained, therefore adding to the starting prestrain in the multifilamentary billet. Parrell *et al* [46] showed that warmer extrusion increases the recovery of monofilament cold work.

The importance of giving an appropriate prestrain can be examined by noting that the α-Ti precipitation temperature lies well below the melting point. This means that large amounts of stored energy must be present to drive the nucleation of triple-point precipitate. Work by Buckett and Labalestier [17] found that undesirable precipitate modes, notably ω and Widmanstätten α, occurred for Nb47Ti when $\epsilon_p < 5$. The intragranular precipitates are not desirable because they can be irregularly distributed and they reduce the workability of the composite. Intergranular precipitates or grain-boundary films (a precursor of the triple-point precipitate) were the only precipitation modes above $\epsilon_p = 5$. Later work by Lee *et al* [32] extended this investigation to alloys with different composition. Desirable precipitate modes for a 420°C heat treatment can generally be found when

$$\varepsilon_p > 0.8(W - 42) \tag{B3.3.4}$$

where W is the Ti mass fraction. Thus, $\epsilon_p = 5$ is sufficient for Nb47Ti, but ϵ_p must exceed 9 to obtain a desirable precipitate morphology in Nb54Ti.

Once an appropriate amount of prestrain has been provided, precipitate formation depends on supplying Ti first to the grain boundaries and then to the triple points via grain boundary diffusion. West and Larbalestier [51, 52] noted that short heat treatments, such as 3 h at 300°C, produced a grain boundary film of Ti but not triple-point precipitates. However, Lee *et al* [9] later showed that the film precipitates produced by multiple heat treatments, e.g. at 300°C for 3 h, could be converted with one 80 h, 420°C heat treatment into a volume fraction of triple-point precipitates close to that formed by the standard 3 × 80 h at 420°C schedule. The importance of the incremental strain, therefore, is to relocate sources of Ti close to the grain boundaries, since grain boundary diffusion is much faster than bulk diffusion in this material. In this manner, the standard 3-HT schedule produces about 20% of precipitate. Reduction of $\Delta\epsilon$ and/or lengthening of the heat treatment time can produce more precipitate [53]. However, this must be balanced against barrier breakdown and filament sausaging. Many industrial heat treatment schedules use 4 or even 6 HT at lower temperature ($\sim$375°C) successfully with different prescriptions for the incremental strain.

In addition to strain, the amount of impurity atoms, such as iron, can produce dramatic differences in the microstructure. For constant overall precipitate volume fraction, a higher number of smaller precipitates is desirable because the amount of final strain needed to optimize flux pinning is reduced. The reclaimed strain can be added to the prestrain or used to add more heat treatments. Increased levels of iron impurities (600 vs 200 ppm) are a key difference in strands made for low-beta quadrupoles for the LHC, which represents the present state of the art. In addition to providing increased nucleation of precipitates, higher levels of impurity atoms may also reduce the tendency to form very large precipitates during aggressive heat treatments. This is shown in figure B3.3.2.11, which compares SEM backscatter images of post heat-treatment microstructures for a SSC-vintage high-homogeneity composite, a high-iron LHC quadrupole strand and a high-iron R&D strand heat treated for a total of 4000 h. Images (*a*) and (*b*) compare the standard 3 HT for 80 h at 420°C, where the refinement of the microstructure is evident in (*b*). Image (*c*) shows that the coarsening of the precipitate after 4000 h of heat treatment at 420°C is no worse than that of the high-homogeneity strand in (*a*). However, the composite in

D3.3
D1.3

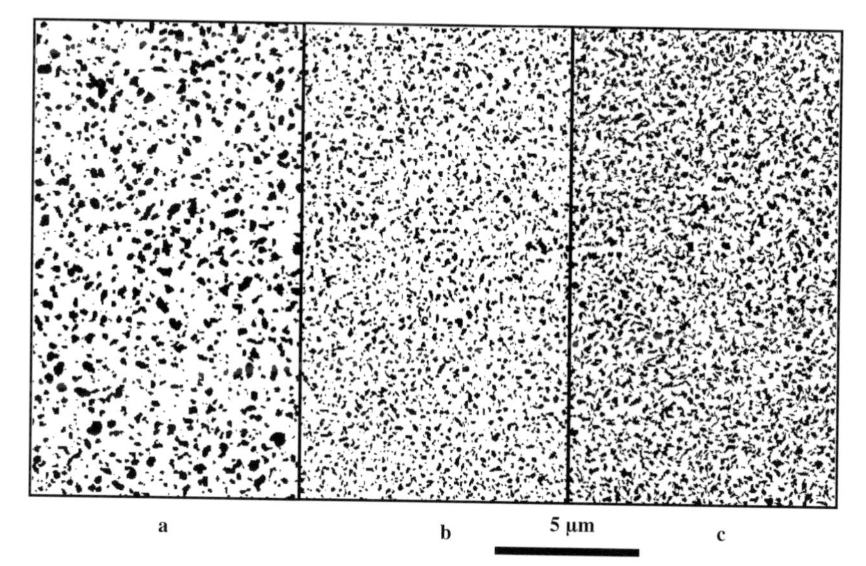

Figure B3.3.2.11. These digitally-processed SEM images show the effect of relaxing tolerances on iron content from high-homogeneity alloy specifications. In (*a*), about 20% of precipitates (the black regions) is produced after three heat treatments of 80 h each at 420°C are given to a high-homogeneity Nb–Ti composite. After the same heat treatments are given to a composite with higher iron content (600 ppm versus less than 200 ppm for the high-homogeneity composite) a similar volume fraction of precipitates is formed. However, as indicated in (*b*), the precipitate size is significantly smaller and the number of precipitates seen in this cross-section is significantly higher, suggesting that the higher iron content produces some refinement of the microstructure. In (*c*), more extensive heat treatment of a high-iron Nb49Ti strand, up to 4000 total h at 350°C, results in a higher volume fraction of precipitates, about 28%, but with a maximum precipitate size similar to that seen in (*a*). The composite shown in (c) is the first conventionally-processed strand to break the 'barrier' of 4000 A mm^{-2} at 5 T, 4.2 K [54, 55].

(*c*) contains about 28% of precipitate and attains the present record J_c of > 4000 A mm^{-2} at 5 T, 4.2 K [54, 55].

Rapid refinement of the microstructure is achieved during the final drawing strain. This sequence is depicted in the images on the right side of figure B3.3.2.7. Meingast *et al* [18] traced the evolution of precipitate shape with increasing ϵ_f. As a result of the different deformation of the hcp precipitates and the bcc matrix, the precipitates develop high cross-sectional aspect ratios while they elongate along the wire axis. This results in a transformation from a more or less equiaxed shape with 100–200 nm diameter after heat treatment, to a fine ribbon-like shape with 1–4 nm thickness, 50–100 nm width and 1–10 μm length. Plane-strain deformation of the β matrix produces heavy folding and intercurling of the ribbons for $\epsilon_f > 3$, similar to the process by which Nb diffusion barriers roughen. Meingast *et al* found that the precipitate thickness, perimeter and separation were reduced at a rate proportional to $d_w^{1.6}$, which is almost as fast as the rate of area reduction. This result suggests that J_c increases approximately linearly with final drawing strain, which is a common observation for wire composites that have good filament uniformity. This is shown in figure B3.3.2.10.

A1.3

B3.3.2.5 Artificial pinning-centre wires

In 1985, Dorofejev *et al* proposed an alternative method of introducing flux-pinning centers into Nb–Ti wire composites [56]. Their idea was to use restacking techniques to hand-assemble the superconductor and the pins at a large diameter. No heat treatment is thus required to form the flux-pinning structure; in fact, high temperatures are avoided to limit interdiffusion of the superconductor and the pins. In principle, this approach confers several advantages over the conventional process:

A4.3 B3.3.1

- (a) Pin materials not native to the Nb–Ti system can be used, which might have a stronger pinning force than α-Ti.
- (b) The pins can be arranged in special geometries, which might be more efficient at matching the flux-line lattice.
- (c) The pins can have identical size and shape at the beginning of the process, potentially leading to a more uniform distribution of pinning centres at final wire size.
- (d) The pinning centre material can improve other superconducting properties of the final wire, such as its critical temperature.
- (e) The volume fraction of pins can be much higher than 20% without significant changes in the composite design.

A4.3

Because of these potential advantages, APC composites received widespread research and development attention in the late 1980s and early 1990s. The status of APC composite development has recently been reviewed in [57].

In practice, the best success has been found for APC composites that incorporate Nb pins with volume fractions that are comparable to those made by precipitation heat treatment (20–25% [7, 58, 59]). The limit of the APC process is not known; as discussed later, multilayer results suggest that the critical current density might exceed 5000 A mm^{-2} at 5 T, 4.2 K.

B3.3.2.5.1 Fabrication techniques

In contrast to the conventional process, where the starting phase is a single component of homogeneous alloy, APC composites begin with a mixture of superconducting Nb–Ti alloy and a metal that is not

B3.3.1

superconducting at the target field and temperature. Excessive heat is usually avoided, since this would tend to diffuse the different phases together. An alternative method used to make some APC composites is to transform precursor materials (e.g. pure Nb and pure Ti) into two solid-solution alloys (e.g. a Nb-rich and a Ti-rich Nb–Ti alloy) by heat treatment at a micrometre scale [60].

Two approaches have evolved to make APC composites; these are shown schematically in figure B3.3.2.12. Hard-assembled composites give the freedom to position the pinning centres in arrays, such as in a hexagonal lattice. Although this arrangement is intended to match the spacing of flux-line rows when reduced to tens of nanometers, it is not possible to maintain the array below about 100 nm separation, as described later. The overall billet design is similar to a conventional multifilamentary billet. Designs include inserting rods of the pin material into a gun-drilled Nb–Ti billet, inserting rods of one material into tubes of the other and stacking together many such elements, and assembling together rods of both materials. Hand-assembled composites ensure that all pins begin with the same size and

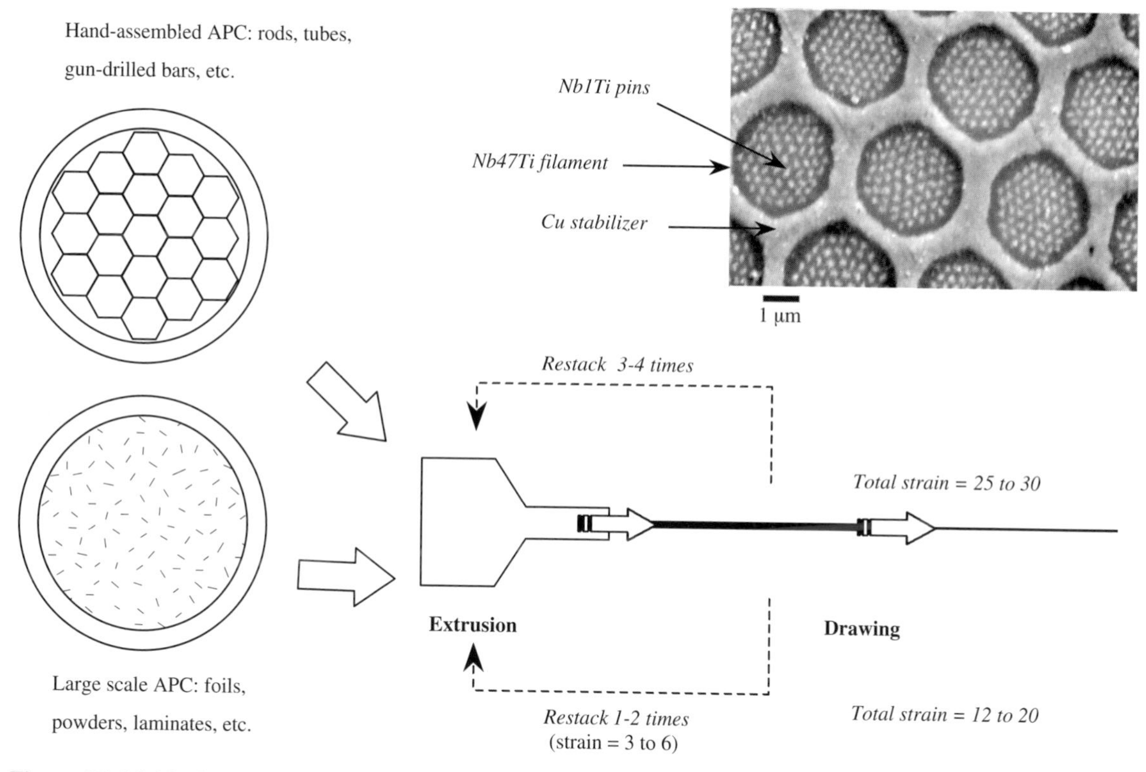

Figure B3.3.2.12. An overview of APC fabrication processes is illustrated in this figure. Hand-assembled APC composites tend to be small in scale, due to the requirement that pieces be handled and are of the order of millimetres in size. Hand-assembled elements include rods, tubes, gun-drilled cores, and other shapes. To reduce the elements to 10 nm size, a total strain of 25–30 is required, and 3–4 restacking and extrusion steps are needed. Nanostructural analyses show that uniform pin shape and arrangement can be maintained down to approximately 100 nm in effective diameter. An example of this kind of nanostructural control is shown in the photograph. Large-scale APC composites make use of powders, foils, jelly-rolls, chips of laminated foils, pellets or other constituents that are typically micrometres in size. Due to the smaller starting size, the total strain required to reach the desired nanostructure is 12–20, much less than for hand-assembled routes. However, control over the arrangement of the artificial pins is lost.

spacing. The disadvantage of hand-assembled techniques is the large strain required, more than double that of the conventional proces, because material starting sizes must be at least a few millimeters, while the final size is about 10 nm. Equation (B3.3.2.2) gives a minimum strain of 23 for these parameters. Thus, multiple extrusions (3–4) are necessary, and material workability poses greater challenges than in the conventional process.

Large-scale APC composites use powder-metallurgy, laminated foils, sprayed coatings and other techniques to reduce the starting component size to about 10 μm, thus reducing the required strain to a value that is comparable to the conventional process. These techniques sacrifice precise control over the starting position, shape and arrangement of the pinning centres, although control over certain aspects, such as foil thickness, can be retained. Control over the distribution of starting material sizes is vital to the optimization of large-scale APC composites, in analogy to controlling the chemistry, grain size and hardness of the wrought alloy in the conventional process. If there are significant variations in the initial component sizes, these can be carried through to the final wire size, causing a weak optimization of J_c and a lower overall maximum J_c value. For example, a broad distribution in starting powder size was observed to extend to the final wire size in several powder-metallurgy APC composites in [61, 62], leading to J_c much lower than expected.

B3.3.2.5.2 Nanostructural development in APC composites

Plane-strain deformation of the Nb–Ti matrix controls the nanostructural development of APC composite wires as it does for conventional Nb–Ti composite wires. Studies in [58], [63] and [64] showed that the onset of distortion of round pins occurred at about 100 nm diameter, despite these composites being hand-assembled and showing highly symmetrical arrays of the pins at larger diameters. The onset was found in [63] to depend somewhat on the grain sizes of the Nb pins and the Nb47Ti matrix, as well as on the degree to which warm extrusion was used during processing. Since grain size determines the relative scale of intercurling at the Nb/Nb–Ti interface, much like it does at the interface of the Nb–Ti matrix and the Nb diffusion barrier in conventional composites, fine grained materials increase the pin shape stability and help retain uniformity of the pin array to lower equivalent pin diameter d_p. The continued deformation of the pins below $d_p = 30$ nm eventually results in the development of large aspect ratios in cross section. The increasing aspect ratio of the pins finally produces a distortion of their overall arrangement, so that, by the time flux pinning is optimum, an overall nanostructure very similar to that produced by the conventional process is found, as shown in figure B3.3.2.2. At maximum pinning force, Nb pins are somewhat thicker than α-Ti precipitates are, 10–15 nm versus 1–4 nm, and they have an aspect ratio of about 10 [64]. Nanostructural studies of large-scale composites show that an intercurled, ribbonlike nanostructure is formed somewhat earlier [62, 65], and generally the nanostructure development obeys the properties of hand-assembled composites above.

A4.3 Artificial pins made from face-centered cubic (fcc) metals may produce different optimum nanostructures. Scanning tunnelling microscopy investigations reported by Levi *et al* noted that Cu–Ni pins did not have a high aspect ratio at 20 nm scale [66]. The composite investigated was one made by Rizzo *et al* [67], which incorporated pinning centres made by combining Ni rods in a Cu sheath.

B3.3.2.5.3 Choice of the pinning centre materials

A4.3 Early APC composites were designed from the viewpoint of attaining the highest possible pinning force by simply increasing the volume fraction of pins. Success was obtained after optimizing Nb47Ti APC composites with Nb pins [7, 59]. These were the first Nb–Ti composites to break the technological 'barrier' of 4000 A mm^{-2} at 5 T, 4.2 K. The present best value is about 4600 A mm^{-2} [7]. Nb was chosen as the pinning centre for two reasons: since bulk Nb is not superconducting at high field and its critical temperature is the same as that of the superconductor, Nb pins were thought to provide strong flux pinning without depressing the critical temperature due to the proximity effect. Second, Nb was practical for metallurgical reasons, due to its excellent ductility and identical crystal structure and atomic size as compared to Nb47Ti.

However, APC composites with Nb pins also exhibited undesirable flux-pinning behaviour and superconducting properties, as compared to conventional composites. The reasons for their different properties are discussed in section B3.3.2.6. The most significant variation is a depression of H_{c2} by almost 2 T, making even the best APC composites inferior to conventional wires above 6 T field, as shown in figure B3.3.2.1. The problem stems from having the wrong overall composition of the APC composite, since the final-size properties are close to those of a homogeneous alloy with the same composition. Thus, properties of a Nb30–35Ti alloy were found in early APC composites, leading to the steeply reduced H_{c2} values (see figure B3.3.2.5). Another significant variation is the fact that the optimized nanostructure was somewhat larger in scale than in conventional composites, as shown in figure B3.3.2.2. This larger nanostructural scale matched flux lines better at low field, producing a peak of the bulk pinning force curve at 2–3 T instead of at the 5–7 T field regime desired for most magnets. Flux pinning at 2–3 T was *double* that of conventional composites, with J_c values reaching 12–15 kA mm^{-2}.

Later APC composites investigated pinning centre materials that were thought to give more desirable superconducting properties while still providing a strong pinning force. One solution to the problem of H_{c2} depression is to carefully choose both the composition of the Nb–Ti alloy superconductor and the volume fraction and composition of the pin to give an overall composition that is close to the one with maximum H_{c2}, Nb44Ti. Preliminary experiments have yielded mixed results: Matsumoto *et al* [68] obtained a US Patent based on this principle, in which high-field J_c values that were comparable to those obtained by the conventional process were claimed. Morris [65] measured H_{c2} values close to 11 T at 4.2 K, much higher than the above Nb-pin composites, when a Nb31Ti matrix was combined with pure Ti pins. However, H_{c2} was only 7 T at 4.2 K for a composite made from Nb62Ti and Nb rods by Heussner [26]. Another view [64] suggests that alloyed Nb or Ti pins may be beneficial for increasing J_c at high fields. Heussner *et al* [69] explored this possibility by comparing Nb artificial pins to Nb7.5Ta and Nb10W pins, but found inconclusive results.

Perhaps the ultimate manifestation of the artificial pinning centre concept is embodied in composites that use ferromagnetic pins. Because magnetism evolves from the interaction of electrons with aligned spins, it destroys the pairing of electrons with opposite spins associated with superconductivity. Therefore, superconductivity is suppressed completely within about 1 nm of the superconductor/ferromagnet interface. The flux-pinning force of a magnetic pin is thought to be close to the maximum possible pinning force, so very high critical current densities might be achieved. However, the pins must be widely separated to ensure that there are regions where superconductivity is unaffected by the pins, and this requirement limits the field range where ferromagnetic pins have maximum benefit. Several APC composites having Ni or Fe pins have been made [67, 70], and the best of these attains the highest bulk pinning force achieved at 4.2 K, 30 GN m^{-3}, at 2 T field [57].

A4.3

B3.3.2.6 Optimization of superconducting properties

The individual steps that make up the conventional process emphasize only the optimization of the flux-pinning properties; optimization of the critical temperature and upper critical field are secondary because Nb47Ti has T_c and H_{c2} (4.2 K) values close to the maximum values available for the alloy system. This is not the situation in APC composites, which can encompass vastly different combinations of starting alloy compositions while still attaining comparable flux-pinning properties. The optimization of APC composites thus involves a more complicated matrix of parameters, including but not limited to H_{c2}, T_c

and the pinning force. To address these issues, it is important to understand how the superconducting properties change as a composite wire processed by either route approaches its final size.

B3.3.2.6.1 Upper critical field and critical temperature

A3.1 A2.3 de Gennes [71] showed that superconductivity does not end abruptly at the interface between a superconducting and a non-superconducting metal; rather it decays over characteristic lengths, the coherence length ξ in the superconductor and the proximity length ξ_N in the non-superconducting metal. When superconducting and non-superconducting metals are in close proximity, as they are in Nb–Ti composites, the proximity effect changes superconducting properties, especially the critical temperature. It was discussed earlier how coupling currents flowing between neighbouring Nb–Ti filaments can be enhanced by the proximity effect when the filament separation approaches ξ_N, approximately 300 nm in pure copper. The predicted decrease of the critical temperature is commonly observed in Nb–Ti strands with filament diameter and separation less than 0.1 μm [37, 72].

Detailed investigations of an optimized Nb48Ti composite by Meingast *et al* [18, 73] showed that the superconducting properties changed with final wire drawing in a manner consistent with de Gennes' model above. The critical temperature decreased from 9.5 K immediately after the final heat treatment to A4.3 9.0 K when flux pinning became optimized; H_{c2} and ρ also exhibited a change with ϵ_f. In this case, however, proximity effect with copper could be ruled out because the composite was monofilamentary. Meingast *et al* reasoned that the proximity effect instead produced a 'homogenization' of the nanostructure, by 'mixing' properties over the 5 nm scale of the coherence length as final wire drawing progressed. That is, properties that were characteristic of the Ti-depleted matrix after the final heat treatment, when microstructural dimensions were much larger than ξ, changed to properties characteristic of the starting wrought alloy composition at the final wire diameter. This occurred even though the two-D1.2 phase nanostructure was clearly visible by TEM. Apparently the optimized nanostructure functioned as a homogeneous medium because the precipitate thickness and separation were comparable to, or even less than, the 5–10 nm scale of superconductivity.

This work showed that an important side effect of optimizing the flux-pinning properties is to simultaneously reduce variations of superconductivity across the nanostructure. The appearance of superconducting properties characteristic of a homogeneous single-phase alloy was quite unexpected, since flux-pinning mechanisms were thought to require heterogeneous properties. The mixing hypothesis of Meingast *et al* has other important consequences. Since many APC composites mix together Nb47Ti with 15–27% Nb pins, the equivalent single-phase alloy with the same overall composition is in the range between Nb25Ti and Nb35Ti. Thus, there is a significant reduction of the upper critical field, by as much as 2–3 T at 4.2 K, when the nanostructure is fully refined. The need for a more Ti-rich matrix alloy is an important reason why APC composites have not been implemented widely. Adjusting the overall composition of the composite has had some beneficial results, as mentioned earlier. This idea has been explored more extensively in [65] and [68]. In addition, the wide range of mutual solubility of the β B3.3.1 phase, discussed in section B3.3.2.2, allows ternary Nb–Ti–Ta alloys to be prepared with many different Ta compositions. The mixing hypothesis suggests that these alloys should be chosen to lie close to the line given by e/a = 4.37, indicated in figure B3.3.2.6.

It is more difficult to extend the Meingast mixing hypothesis to incorporate APC composites with metals outside of the binary Nb–Ti system because an equivalent homogeneous alloy would be made of Nb, Ti and other elements, a combination that probably cannot be made as a bulk alloy. Work on ultra fine-filament composites with copper or copper alloy stabilizers mentioned earlier shows that the long proximity length of copper produces deleterious proximity coupling well before the filament diameter approaches the Nb–Ti matrix thickness found in optimized, conventionally processed strands. The large body of work on B4.1 superconducting thin-film multilayers may also be instructive in this case, since the laminar nanostructure of Nb–Ti composites resembles a folded and curled-up multilayer. In many cases (see [74] for a review), the proximity effect produces properties consistent with the averaged solid-state properties (i.e. resistivity, specific heat, etc.) of the constituent layers when the layer thickness is less than ξ or ξ_N.

An interesting property of multilayers is a change from three-dimensional to two-dimensional behaviour when the layer thickness becomes comparable to ξ or ξ_N, at length scales just above those of the mixing effect. This change produces steep changes in the upper critical field in multilayers with pure metals as the superconductor and the pin layers, with parallel-field values becoming strongly *enhanced* [74]. An interesting question is to what extent a dimensional change affects the upper critical field of Nb–Ti alloy, where the electron mean free path is less than ξ. A dimensional crossover in Nb–Ti/Cu-alloy multilayers was explored recently in [75], which showed that the upper critical field was dependent on the field angle, being highest for parallel field alignment. The angular dependence was found to vary with layer thickness and the pin material used. This result suggests that the value of the upper critical field might also vary across the nanostructure of an optimized Nb–Ti composite, depending on the local alignment of pins to the applied field. Since the random nanostructure of round strands can be effectively aligned by rolling them into tapes [76], the presence of a dimensional crossover can be tested for. These experiments were performed very recently [77], and showed that the parallel upper critical field is about 1 T higher than the value in perpendicular field.

A2.3

A3.1

B3.3.2.6.2 Critical current density and flux pinning

As stated in section B3.3.2.1, high critical current density is the single most important parameter that defines the technological value of Nb–Ti alloy. The benchmark value at 5 T field and 4.2 K temperature has increased steadily with time [78], and very recently the 'barrier' of 4000 A mm^{-2} has been broken in a conventionally processed composite [54, 55]. The best APC composite attains an even higher value of 4600 A mm^{-2} [7]. Much improvement has come from understanding and removing extrinsic limitations, first by studying monofilament composites in the laboratory, and later by understanding how to control filament instabilities in multifilament composites.

A1.3

No other superconductor that can be made in kilometre lengths reaches as high a fraction of its theoretical current-density limit. For Nb–Ti the depairing current limit is about 10^{11} Am^{-2}, so practical conductors reach about 4% of this value at 5 T, 4.2 K. An important advantage of Nb–Ti over other superconductors is the fact that the number density of pins is larger than that of magnetic flux lines. Each individual pinning interaction then contributes fully to the overall pinning force, and differences between what is observed in real conductors and what is possible theoretically can be traced directly to features in the nanostructure. Part of the difference between measured J_c values for real conductors and theoretical limits comes in the inefficiency of the random orientation of the precipitates, since roughly half of them lie perpendicular to the flux lines and therefore produce no change in the flux-line energy for a transverse Lorentz force. This was demonstrated in [76], where by flattening a round wire into a tape conductor an alignment of the precipitates within 10° of the tape face and a factor of almost 2 increase in J_c could be achieved. The statistical distribution of precipitate thickness makes up another part of the difference, since the elementary pinning force f_p (the pinning force exerted by a single precipitate) is proportional to the thickness. In multilayer thin films, which have nearly identical pin layer thickness, the highest J_c values exceed 20% of the theoretical limit [79]. Since the statistical and orientational limitations are inevitable, the two-phase nanostructure created in round Nb–Ti strands is nearly optimum for attaining the highest possible current density available to this superconducting system.

The combination of nanostructural information with measurements of the flux-pinning properties provides the information needed to optimize a Nb–Ti composite. Flux-pinning measurements also provide quality control. Vital information is obtained by measuring J_c because of the force balance

A4.3

$$\mu_0 J_c H = F_p \tag{B3.3.2.5}$$

A1.3 where the left side represents the applied Lorentz force and the right side the total pinning force due to the nanostructure. The total pinning force can be broken down in terms of the number density of pins n_p and the elementary pinning force f_p per pin if the vector sum of pinning forces is known. Improving the elementary pinning force, such as by changing the nature of the pinning centre, thus produces an A2.6 increase in the fundamental limit to performance. In the limit of very high n_p, comparable to the flux-line density $\mu_0 H/\phi_0 \approx 10^{14}–10^{15}\ \mathrm{m}^{-2}$, the approximation

$$F_p = n_p f_p \tag{B3.3.2.6}$$

is valid, which is the case for the optimized nanostructure. Thus, experimental values for f_p can be obtained from measurements of the critical current density and inverting the summation equation,

$$f_p = Q = \mu_0 J_c H / n_p.$$

A1.3 Often the term ‘specific pinning force’ Q is used to distinguish observed values from theoretical estimates of f_p (see e.g. [51]). It is important to take into consideration that critical *current* I_c is the quantity that is actually measured, whereas critical current *density* is inferred by dividing I_c by the overall superconductor area. This means that variations in the cross-sectional area, and other similar ‘extrinsic’ limitations, combine with natural variations of the intrinsic pinning force in a manner that is not easy to deconvolute. Fortunately, variations of the intrinsic pinning force for Nb–Ti wires are thought to be small, which is often not the case for other superconductors. Extrinsic limitations are thus the dominant source of degradation of I_c, and it is possible to both study the fundamental properties and to achieve performance that is as close to ideal as possible by exercising care during wire fabrication and eliminating extrinsic problems. Monofilament wires in particular have been widely studied for this reason.

The primary extrinsic limitation of Nb–Ti multifilamentary conductors is the variation in filament cross-sectional area (sausaging). An assessment of the degree of sausaging can be determined from the A4 current–voltage transition $V(I)$ (see chapter A4). A baseline voltage is found for $I \ll I_c$, which is caused by system noise and thermal offsets. This abruptly rises when the current reaches I_c and, in this regime, D2.2 the transition can be characterized by a power law,

$$V = (I - I_c)^n.$$

A4.3 It is important to keep in mind that the actual value of I_c is determined by an arbitrary criterion of electric field, e.g. $0.1\ \mu\mathrm{V\ cm}^{-1}$, or resistivity, e.g. $10^{-14}\ \Omega\ \mathrm{m}$, which is chosen to reasonably separate the transition region from the baseline voltage. Assuming the critical current density is constant, the voltage rise for regions with small cross-sectional area will occur at lower current than for regions with large cross-sectional area. This allows the degree of sausaging to be quantified by the steepness of the $V(I)$ transition. Warnes and Larbalestier [80] related n to the observed variation of filament sausaging due to intermetallic formation. This showed that low n values were associated with poor filament quality, while n values higher than 30 were found for uniform filaments at 5 T field. Warnes [81] also noted that sausaging was comparatively easier to detect in monofilament than in multifilament composites because sharing of current among several filaments did not occur. In particular, it was thought that voltages associated with current sharing might fall below easy detection. This meant that multifilament composites with $n = 60$ would have equivalent filament area variations of a monofilament composite with $n = 30$. Further developments of these ideas were summarized by Edelman and Larbalestier [82], relating statistical distributions of filament area to n and the $V(I)$ transition. High *et al* [83] found a correlation between the statistical coefficient of variation (COV—standard deviation expressed as a percentage of the mean) of filament cross-sectional area with n, based on digital image analysis of SSC conductor cross-sections. That work showed that a COV of 6% was associated with 'good' composites with $n > 30$, while a COV of 10–12% was associated with $n < 30$ and filaments damaged by intermetallic particles. High quality industrial composites now have COV values of 5% or less. Thus, measurement of n is an important quality control parameter.

In good monofilament wires that have a high n value, area variations are thought to have a weaker influence on $V(I)$ than the intrinsic behaviour of the flux-pinning mechanism. In this case it is possible to infer the basic properties if information about the nanostructure is also obtained. The intrinsic behaviour as a function of ϵ_f can be represented by a family of bulk pinning-force curves $F_p(H)$, as shown in figure B3.3.2.13(a). A general shift of the curve peak towards higher field is seen as it increases in magnitude. This can be correlated with the development of the nanostructure shown in figure B3.3.2.7. As the pin separation becomes smaller and n_p becomes higher, a good match of pin density to flux density applies at progressively higher fields. At the same time, the elementary pinning force is at its maximum value for $\epsilon_f \leq 4$ and begins to fall only near the final wire diameter. A similar family of curves for a typical APC composite with Nb pins is shown in figure B3.3.2.13(b). The primary difference between APC and conventional composites is the strong peak at low fields in the former. The maximum pinning force at 2 T field, which can attain 30 GN m^{-3} [7], is almost double that of conventional composites. Thus, APC composites with Nb pins are particularly well suited for low-field applications.

D2.1

It is worth examining figure B3.3.2.13 more closely to understand the implications for optimizing the flux-pinning mechanism. As the pinning-center thickness decreases in either a conventional or an APC composite, the peak of the $F_p(H)$ curve moves upward and shifts toward higher field. This indicates that the primary contribution to the increase of F_p is the increase in n_p. Meingast and Larbalestier [5] determined that thick precipitates have a specific pinning force that is about the same as that of Nb–Ti grain boundaries, 300–400 N m^{-2} [51]. Although this value drops off to about 120 N m^{-2} when the precipitates become very thin near the final wire size, the number density of precipitates grows much

A4.3

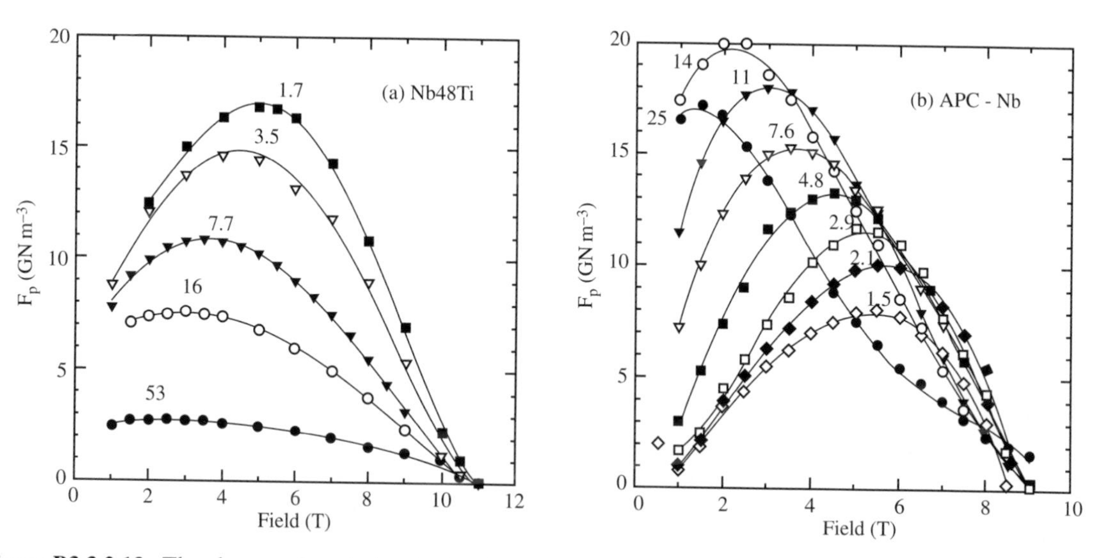

Figure B3.3.2.13. The change of the bulk pinning force curve at 4.2 K with decreasing pinning-centre thickness is illustrated in these plots. In (a), bulk pinning force curves are presented for a conventional Nb48Ti composite with α-Ti precipitates (from [5]) and, in (b), similar curves are shown for an Nb47Ti APC composite with 15% Nb1Ti pins (from [64]). The curves are labelled by the value of the pinning-centre thickness (nm) in both plots, which decreases together with the average pinning-centre separation and the overall filament diameter.

faster than that of grain boundaries, so precipitates provide the primary contribution to the bulk pinning A4.3 force as the final wire diameter is approached. Note that flux-pinning theories are usually discussed in terms of the flux-line tension, which is of order 10^{-5} to 10^{-4} N m^{-1}. In systems such as the present one, where the pinning centres are planar, it is convenient to multiply this by the number density of flux lines along the axial length of the pin, about 10^7–10^8 per meter, giving a range of 100–1000 N m^{-2} for the maximum theoretical specific pinning force.

B3.3.1 As the final wire diameter is approached, proximity coupling of the pinning centres to the surrounding Nb–Ti matrix increases. In conventional composites, plot (*a*) of figure B3.3.2.13, proximity coupling weakly affects the overall superconducting properties with decreasing pinning-centre thickness, as discussed earlier, and a high number density of pins promotes direct summation of the elementary pinning forces. The optimum stage of the conventional Nb48Ti composite is thus obtained at the end of the fabrication process, when precipitates are 1–4 nm thick and 5–10 nm apart, less than half the average separation of flux lines at 5 T, 22 nm. This is why the optimum $F_p(H)$ curve has a peak near 5 T field, an important consideration for high-field magnets because this is about half of the irreversibility field where $F_p(H) \rightarrow 0$. By contrast, in some APC composites, such as the Nb47Ti + 15% Nb1Ti pins in plot (*b*) of figure B3.3.2.13, proximity coupling has a rather significant effect on the superconducting properties as the pinning-centre thickness decreases, which weakens the elementary pinning force and offsets the increasing pinning-centre number density. The optimum curve for the APC composite occurs well before the smallest pinning-centre thickness is reached, with a peak near 2 T. This corresponds to a much larger pinning-centre thickness of 14 nm and a pinning-centre separation of about 30 nm, somewhat less than the separation of flux lines at high field. Further reduction of the APC strand diameter results in a continued shift of the peak of $F_p(H)$ to higher field, ultimately to a field two-thirds that of the irreversibility field. However, strong proximity coupling at A4.3 this stage has depleted the elementary pinning mechanism and circumvented a strong bulk pinning force at high field.

The different behaviour indicated in plots (*a*) and (*b*) are thought to reflect different aspects of the elementary pinning mechanism for Nb pins and α-Ti precipitates. This problem has been widely debated. Before APC composites became available, Meingast and Larbalestier [5] discussed the pinning mechanism of α-Ti precipitates for a carefully processed and optimized Nb48Ti composite. They found that the bulk pinning force curve shape was consistent with a core-pinning mechanism at 4.2 K. Since their extensive microstructural analysis [18] showed that precipitates were the dominant pinning centre, core pinning was plausible. However, deeper analysis showed that the variations of the critical field, the central A3.1 parameter of core pinning, were somewhat less than variations of the Ginzburg–Landau parameter kappa (κ), suggesting that both core pinning and delta-kappa pinning should contribute to the shape of the pinning force curve. At high temperature, variations of the critical field were clearly dominant but, in this case, the shape of the bulk pinning force curve did not correspond to the expected field dependence of core pinning. Meingast and Larbalestier speculated that stronger depression of superconductivity in precipitate clusters might affect the bulk pinning force at high temperature. Similar variations of the bulk pinning force curve were explored by Matsushita and Küpfer [84], although no microstructural study was performed.

However, the core-pinning ideas above could not account for the shapes of the pinning-force curves seen for APC composites, and more recent attention has focused on magnetic pinning mechanisms. One model centres on the fact that both optimized conventional and APC composites have laminar nanostructures. Since the structure of a single quantized flux line consists of a cylindrical wall of current flowing around a non-superconducting core, planar defects can distort the current wall and thereby provide a magnetic pinning force. Early experimental work on one of the first Nb47Ti APC composites with Nb pins suggested that indeed such variations might be related to the elementary pinning force, based on the shape of the $F_p(H)$ curve [85]. A later experiment on Nb47Ti thin films showed that flux pinning was stronger in Nb47Ti thin films due to the magnetic interactions with image vortices across the film surfaces [86]. Theoretical work by Gurevich [87] showed that a similar image concept could be used to solve the magnetic distortions of fluxons near planar defects, such as precipitates or artificial pins. Even weak perturbations in this case are enough to disrupt the current wall and provide a strong elementary pinning force [88], which suggests that strong proximity coupling might not be incompatible with a strong pinning force. This idea was applied to the specific case of Nb–Ti composites later in [64], where a modification was made to assess the pin thickness in terms of the proximity length of the pin. That work predicted that the optimum pin thickness should be $\xi_N/3$; the optimum thickness thus is not related to the coherence length. Since this corresponds to ≈ 3 nm for α-Ti precipitates and ≈ 10 nm for Nb artificial pins, the difference in the observed optimum pin thickness in figure B3.3.2.13 could be accounted for. Another important result of magnetic pinning models is that the shape of the bulk pinning force curve need not be fixed for a given elementary pinning mechanism, as had been widely believed based on the pioneering work by Fietz and Webb [89].

A4.3

A second idea was proposed by Matsushita *et al* [90, 91]. In this theory, it is supposed that a fraction of the available condensation energy of the superconductor is lost due to the proximity coupling of the pins. The rationale for this supposition is the assumption that if the pins were replaced by voids, the full condensation energy (as indicated by the value of T_c, for example) would be restored. When flux lines occupy the proximity-coupled pins, the energy lost due to coupling is restored because the non-superconducting cores locally suppress superconductivity. This energy difference makes up the pinning energy and the elementary pinning force. Since the primary contribution to the energy difference came from the supercurrent flowing through the pin, this energy is primarily magnetic in origin. The different shapes of the pinning force curves in figure B3.3.2.13 might then be due to the special situation of having Nb pins: since pure Nb actually has a *higher* condensation energy than strong Nb–Ti alloys, Nb pins can function as repulsive barriers [91], which would be most efficient when their separation was somewhat larger than the separation of flux lines.

A3.2 A2.6

B3.3.2.7 Other topics

B3.3.2.7.1 Alloys with Ti content outside the standard alloy range

There are many superconducting applications that do not emphasize high magnetic fields, such as MRI magnets, electric power conditioning, SMES and ore separation. Nb–Ti alloys with Ti compositions above the standard Nb47–50Ti alloy range might provide a simpler fabrication process in these cases. McKinnell *et al* [92] explored Nb52–62Ti for electric power applications. Because of the higher concentration of Ti, 20% of precipitate could be produced with a single heat treatment. This benefit was offset somewhat by the higher prestrain needed to get the triple-point morphology. As indicated by equation (B3.3.2.4), prestrains of 7–12 are needed to obtain the necessary amount of stored cold work, as opposed to 5 for Nb47Ti. The added prestrain pushes up the amount of wire that must be accommodated in heat-treatment facilities.

E2.3 E1.3.4, E2.2 A1.3

Since Ti is cheaper than Nb, high-Ti alloys provide the additional benefit of lower cost when scaled to a production process. The performance of high-Ti alloys relative to the overall cost may then be competitive with APC composites, which attain 10 000–15 000 A mm^{-2} but are more difficult to produce and, at present, incorporate Nb as pinning centres.

Tachikawa [93] examined Nb25Ti alloys and boron-doped variants. This work showed that intermetallic formation could be avoided by limiting the amount of Ti available. No diffusion barrier would then be necessary. The target applications for such composites might be ultrafine filament composites ($d_f \leq 0.1$ μm) for ac uses, described below. Due to the small filament diameter requirement, Nb barriers cannot be used since they would be < 50 nm thick at heat-treatment diameter, well below the

0.5 μm limit discussed in section B3.3.2.3. An interesting result uncovered by Tachikawa was an enhancement of H_{c2} due to an increase in ρ by the boron addition. This effect should apply to Nb47–50Ti alloys as well, although the general effect of high interstitial content ($>$1000 ppm) is largely unexplored.

B3.3.2.7.2 Ultrafine multifilamentary composites for ac applications

D2.6 Applications of Nb–Ti superconductors at 50–60 Hz generally require that the hysteretic loss should be small. Since these losses are proportional to the filament diameter, composites fabricated for ac use often have multiple restacking and extrusion steps and filament diameters less than 100 nm. Hlasnik *et al* [72] showed that the interface between the Nb–Ti filaments and the stabilizer provides extremely strong flux-pinning, and ample current-carrying capacity can be achieved without any heat treatment. The flux-pinning mechanism is due to the surface barrier, described long ago by Clem [94]. The peak critical current density can exceed 10 000 A mm^{-2} at 2 T, 4.2 K, for composites with $d_f \approx 50$ nm. Solenoids operating at 50 Hz with a peak field of 4 T have been constructed from such composites [95].

APC composites with Nb pins are also especially useful in this regard. Matsushita [91] noted that Nb pins may be repulsive at low field, which means that they can provide surface-like barriers in the interior of the Nb–Ti filaments. This property was applied by Miura *et al* [96], who constructed a 60 Hz magnet with a peak field of 2.5 T using APC wire [97].

B3.3.2.8 Future work

The Large Hadron Collider (LHC), for which dipole magnets will operate at 1.8–2.0 K and attain 8.6 T, will be perhaps the ultimate test of the capability of Nb–Ti. However, the LHC may represent the end of the Nb–Ti era as the mainstay of particle accelerator magnets. This is significant, because the high-energy physics community has provided the strongest push for higher performance in Nb–Ti. By comparison, the MRI industry, which consumes approximately 1 Tevatron (at Fermi National Accelerator Laboratory) worth of Nb–Ti conductor each year, is now market driven, and the need for very advanced conductors is low. In 1998, the Gilman subpanel [98] recommended that new high-energy physics activities be directed at the energy frontier, about 100 TeV in the centre of mass. Several scenarios exist with different siting restrictions, and many of these foresee the eventual use of 10–20 T magnets, a field range B3.3.3 where Nb–Ti loses out to Nb_3Sn and other conductors. Still, some cost estimates for a very high energy collider find that a somewhat larger machine with Nb–Ti magnets operating at 6–8 T may be the cheapest option. As a testament to the design advantages of Nb–Ti, production magnets built by Northrup-Grumman for the Relativistic Heavy Ion Collider at Brookhaven National Laboratory cost about \$109 000 each [99], which is comparable to the cost of modern laboratory magnet systems.

Nonetheless, several open questions remain. The finding of two-dimensional properties in Nb–Ti multilayers and tapes indicates that there must be two-dimensional superconductivity locally within round wires, in regions surrounding precipitates or artificial pins lying parallel to each other. To some degree, two-dimensional properties of multilayers must be effectively integrated in the random laminar nanostructure found in round strands. It is not yet known whether the large number of discontinuous precipitates in round wires, in contrast to continuous layers of a multilayer or an APC composite, creates differences in the flux pinning behaviour. It is also not known whether the onset of two-dimensional properties in multilayers and more recently conventionally-processed Nb–Ti tapes is linked to the development of strong flux pinning. Recent experiments [75, 77] show that two-dimensional properties are established at nanostructural dimensions much larger than those associated with A4.3 optimum flux pinning. The possibility of local two-dimensional properties also opens up new possibilities for understanding and perhaps optimizing the values of the critical temperature and the upper critical field of APC composites.

The results above are also important in understanding the limit of flux pinning in Nb–Ti. Because optimum flux pinning is always accompanied by strong proximity coupling, the properties of an overall fully coupled nanostructure must be considered. In the case of conventionally processed alloys, there does not appear to be a significant deleterious effect on overall superconducting properties as the precipitates are effectively mixed back into the superconducting matrix by the proximity effect. However, this is not the case for APC composites, and the overall composition of the composite must be balanced carefully. Recent multilayer work has extended this idea to elements that are not part of the Nb–Ti alloy system. In principle, it is possible to predict what the properties of a generic APC composite will be when proximity coupling becomes strong, and the optimization of such a design may push the flux pinning limit attained so far. Moreover, although the various elementary pinning mechanisms predict different behaviour of the bulk pinning force when viewed in detail, to first order adequate pinning is supplied by having a large number density of any second phase. Thus, in addition to optimizing the composite design for the proximity effect, it is also necessary to do this for as fine a laminar nanostructure as can be made.

The effect of additional elements, such as iron and boron, is another area that is largely unexplored in modern high-homogeneity alloys. An increase in both T_c and H_{c2} was noted earlier for Nb25Ti with 1% boron. However, the effect of boron and other interstitial additions has not been explored on alloys in the standard range, which have much higher resistivity. Interstitial elements, especially oxygen, may also provide stabilization of the α-Ti phase and increased precipitation rates. As was found for relaxed specifications for iron, new advances may be seen for other modifications of the interstitial atom concentration of the starting alloy. Since the iron atoms appeared to aid nucleation and retard grain growth, it would be desirable to distribute uniformly larger numbers of iron impurities. Such 'homogeneously inhomogeneous' composites might provide new insight into the mechanisms of precipitate formation. Also, since little work has been performed over the past 20 years on basic alloy properties, new basic studies could benefit greatly from the availability of modern preparation and characterization tools.

Continued improvement in ternary Nb–Ti–Ta alloys has been reported very recently [100], but it is still difficult to obtain consistent results. As also reported in [101], alloys with 12–37 mass% Ta appear to be more difficult to process. The proper balance of cold work and heat treatment parameters is not known, and there has not been the level of nanostructural control that is exercised in conventionally processed binary composites. Achieving sufficient homogeneity of the ternary alloy has been identified as a central need. Unusual precipitation rates have also been noted and, based on these results, a new ternary alloy phase diagram has been recently proposed [102].

References

[1] Larbalestier D C 1981 Niobium-titanium superconducting materials *Superconductor Materials Science*, eds S Foner and B B Schwartz (New York: Plenum) pp 133–199

[2] Lee P J 1999 Abridged metallurgy of ductile alloy superconductors *Wiley Encyclopedia of Electrical and Electronics Engineering* (New York: Wiley)

[3] Collings E W 1983 *A Sourcebook of Titanium Alloy Superconductivity* (New York: Plenum)

[4] Lee P J and Larbalestier D C 1983 An examination of the properties of SSC Phase II R&D strands *IEEE Trans Appl. Supercond.* **3** 833–841

[5] Meingast C and Larbalestier D C 1989 Quantititative description of a very-high critical current density Nb–Ti superconductor during its final optimization strain: II. Flux pinning mechanisms *J. Appl. Phys.* **66** 5971–5983

[6] Lee P J, Fischer C M, Gabr-Rayan W, Larbalestier D C, Naus M T, Squitieri A A, Starch W L, Barzi E Z A, Limon P J, Sabbi G, Zlobin A, Kanithi H, Hong S, McKinnell J C and Neff D 1999 Development of high performance Nb–Ti(Fe) multifilamentary superconducting strand for the LHC high gradient quadrupole conducter *IEEE Trans. Appl. Supercond.* **9** 1559–1562

[7] Heussner R W, Marquardt J D, Lee P J and Larbalestier D C 1997 Increased critical current density in Nb–Ti wires having Nb artificial pinning centers *Appl. Phys. Lett.* **70** 901–903

[8] Lee P J and Larbalestier D C 1987 Development of nanometer scale structures in composites of Nb–Ti and their effect on the superconducting critical current density *Acta. Metall.* **35** 2526–2536
[9] Lee P J, McKinnell J C and Larbalestier D C 1990 Restricted novel heat treatments for obtaining high J_c in Nb–46.5 wt%Ti *Adv. Cryo. Eng. (Materials)* **36** 287–294
[10] Kreilick, T S 1990 Niobium–Titanium Superconductors *Metals Handbook*, 10th edn, vol 2, Properties and Selection: Nonferrous Alloys and Special-Purpose Materials (Metals Park, OH: ASM International) pp 1043
[11] Lee P J and Larbalestier D C 1994 Fabrication methods—1. BCC Alloys *Composite Superconductors*, eds T Matsushita, P J Lee and S Ochiai (New York: Marcel Dekker) pp 237–258
[12] McInturff A D 1980 *The Metallurgy of Superconducting Materials*, eds T Luhman and D Dew-Hughes (New York: Plenum) Chapter 3
[13] Hansen M, Kamen E L, Kessler H D and McPerson D J 1951 Systems titanium–molybdenum and the titanium–columbium *J. Metals* **3** 881–888
[14] Kaufman L and Bernstein H 1970 *Computer calculations of phase diagrams* (New York: Academic)
[15] Moffatt D L and Larbalestier D C 1988 The competition between martensite and omega in quenched Ti–Nb alloys *Metall. Trans. A* **19** 1677–1686
[16] Moffatt D L and Larbalestier D C 1988 The competition between the alpha and omega phases in aged Ti–Nb alloys *Metall. Trans. A* **19** 1687–1694
[17] Buckett M I and Larbalestier D C 1987 Precipitation at low strains in Nb46.5wt%Ti *IEEE Trans. Magn.* **23** 1638–1641
[18] Meingast C M, Lee P J and Larbalestier D C 1989 Quantitative description of a very high critical current density Nb–Ti superconductor during its final optimization strain I. Microstructure, T_c, H_{c2} and resistivity *J. Appl. Phys.* **66** 5962–5970
[19] Berlincourt T G and Hake R R 1963 *Phys. Rev.* **131** 140
[20] Mani A, Vaidhyanathan L S, Hariharan Y and Radhakrishnan T S 1966 Structural and superconducting properties of Nb–Ti alloy thin films *Cryogenics* **36** 937–941
[21] Hawksworth, D G 1981 The upper critical field and high field critical current density of niobium–titanium and niobium–titanium based alloys PhD thesis University of Wisconsin
[22] Goodman B B 1966 *Rep. Prog. Phys.* **49** 445
[23] Naus M T, Heussner R W, Squitieri A A and Larbalestier D C 1997 High field flux pinning and the upper critical field of Nb–Ti superconductors *IEEE Trans. Appl. Supercond.* **7** 1122–1125
[24] Heussner R W, Bormio Nunes C, Lee P J, Larbalestier D C and Jablonski P D 1996 Properties of niobium–titanium superconducting wires Nb artificial pinning centers *J. Appl. Phys.* **80** 1640–1646
[25] Suenaga M, Ghosh A K, Xu Y and Welch D O 1991 Irreversibility temperatures of Nb_3Sn and Nb–Ti *Phys. Rev. Lett.* **66** 1777–1780
[26] Heussner, R W 1998 Flux pinning in superconducting Nb–Ti wires with Nb artificial pinning centres *PhD thesis* University of Wisconsin
[27] Clogston A M 1962 Upper limit for the critical field in hard superconductors *Phys. Rev. Lett.* **9** 266–267
[28] Werthamer N R, Helfand E and Hohenberg P C 1966 *Phys. Rev.* **147** 295
[29] Gregory E, Kreilick T S, von Goeler F S and Wong J 1988 Preliminary results on properties of ductile superconducting alloys of operation to 10 tesla and above *Proc. 12th Int. Cryogenic Engineering Conf.* eds R G Scurlock and C A Bailey (Butterworth) p 874
[30] McInturff A D, Carson J, Larbalestier D C, Lee P J, McKinnell J, Kanithi H, McDonald W K and O'Larey P M 1990 Ternary superconductor NbTiTa for high field superfluid magnets *IEEE Trans. Magn.* **26** 1450–1452
[31] Hawksworth D G and Larbalestier D C 1980 Enhanced values of H_{c2} in Nb–Ti ternary and quaternary alloys *Adv. Cryo. Eng.* **26** 479–486
[32] Lee P J, McKinnell J C and Larbalestier D C 1989 Microstructure control in high Ti NbTi alloys *IEEE Trans. Magn.* **25** 1918–1921
[33] Larbalestier D C, et al 1985 High critical current densities in industrial scale composites made from high homogeneity Nb46.5Ti, *IEEE Trans. Magn.* **21** 269–272
[34] Smathers D B, Leonard D A, Kanithi H C, Hong S, Warnes W H and Lee P J 1996 Improved niobium 47 weight % titanium composition by iron addition *Mater. Trans. Japan. Inst. Met.* **37** 519–526
[35] Wilson M N 1983 *Superconducting Magnets* (Oxford: Clarendon)
[36] Collings E W 1988 Stabilizer design considerations in fine-filament Cu/Nb–Ti composites *Adv. Cryo. Eng. (Materials)* **34** 867–878
[37] Matsumoto K, Akita S, Tanaka Y and Tsukamoto O 1990 Proximity coupling effect in NbTi fine-multifilamentary superconducting composites Appl *Phys. Lett.* **57** 816–818
[38] Fickett F R 1982 Electric and magnetic properties of CuSn and CuNi alloys at 4 K *Cryogenics* **22** 135–137
[39] Sumption M D and Collings E W 1994 Influence of Ni additions on the low temperature magnetic properties of a Cu–1% Mn alloy *J. Appl. Phys.* **76** 7461–7467
[40] Gregory E, Kreilick T S, Wong J, Ghosh A K and Sampson W B 1987 Importance of spacing in the development of high current densities in multifilamentary superconductors *Cryogenics* **27** 178–182

[41] Garber M, Suenaga M, Sampson W B and Sabatini R L 1985 Effect of Cu_4Ti compound formation on the characteristics of NbTi accelerator magnet wire *IEEE Trans. Nucl. Sci.* **32** 3681–3683

[42] Larbalestier D C, Lee P J and Samuel R W 1986 The growth of intermetallic compounds at a copper–niobium–titanium interface *Adv. Cryo. Eng. (Materials)* **32** 715–722

[43] Faase K J, Lee P J, McKinnell J C and Larbalestier D C 1992 Diffusional reaction rates through the Nb wrap in SSC and other advanced multilfilamentary Nb–46.5wt%Ti composites *Adv. Cryo. Eng. (Materials)* **38** 723–730

[44] Heussner R W, Lee P and Larbalestier D 1993 Non-uniform deformation of the niobium diffusion barriers in niobium–titanium wire *IEEE Trans. Appl. Supercond.* **3** 757–760

[45] Avitzur B A 1983 *Handbook of Metal-Forming Processes* (Wiley: New York)

[46] Parrell J, Lee P and Larbalestier D 1993 Cold work loss during heat treatment and extrusion of Nb–46.5wt%Ti composites as measured by microhardness *IEEE Trans. Appl. Supercond.* **3** 734–737

[47] High Y, Lee P, McKinnell J and Larbalestier D 1992 Quantitative analysis of sausaging in Nb barrier clad filaments of Nb–46.5wt%Ti as a function of filament diameter and heat treatments *Adv. Cryo. Eng. (Materials)* **38** 647–652

[48] Chengren L and Larbalestier D C 1987 Development of high critical current densities in niobium 46.5 wt% titanium *Cryogenics* **27** 171–177

[49] Collings E W, Sumption M D, Kim S W, Wake M, Shintomi T, Nijhuis A, ten Kate H H J and Scanlan R M 1997 Suppression and control of coupling currents in Stabrite-coated Rutherford cable with cores of various materials and thicknesses *IEEE Trans. Appl. Supercond.* **7** 962–966

[50] Faase K J, Warnes W H, Lee P J and Larbalestier D C 1995 Microstructural and compositional gradients in the filament-matrix region of Nb–Ti wire composites *IEEE Trans. Appl. Supercond.* **5** 1197–1200

[51] Larbalestier D C and West A W 1984 New perspective on flux pinning in niobium-titanium composite superconductors *Acta Metall.* **32** 1871–1881

[52] West A W and Larbalestier D C 1984 Microstructural changes produced in a multifilamentary Nb–Ti composite by cold work and heat treatment *Metall. Trans.* A **15** 843–852

[53] Chernyi O V, Tikhinskij G F, Storozhilov G E, Lazareva M B, Kornienko L A, Andrievskaya N F, Slezo V V, Sagalovich V V, Starodubov Ya D and Savchenko S I 1991 Nb–Ti superconductors of a high current-carrying capacity *Supercond. Sci. Technol.* **4** 318–323

[54] Chernyi O V, Storozhilov G E, Tikhinskij G F, Gogulya V F, Mette V L, Gulyakin Yu A, Tsoraev A K S, Bogdanova L D and Belozerov Yu A 1992 Production of Nb–Ti superconductors of a high critical current density *Cryogenics* **32** 601–604

[55] Chernyi O V, Andrievskaya N F, Ilicheva V O, Storozhilov G E, Lee P J and Squitieri A A 2002 The microstructure and critical current density of Nb-48 Wt.%Ti superconductor with very high alpha-Ti precipitate volume and very high critical current, paper I-11 A-04 presented at ICMC-CEC 2001, accepted for publication in *Advances in Cryogenic Engineering* **48**

[56] Dorofejev G L, Klimenko E Y and Frolov S V 1985 Current carrying capacity of superconductors with artifical pinning centers *Proc. 9th Int. Conf. Magnet Technology*, eds C Marinucci and P Weymuth (Villigen: Swiss Inst. for Nucl. Research) pp 564–566

[57] Cooley L D and Motowidlo L R 1999 Advances in high field superconducting composites by addition of artificial pinning centers to niobium–titanium *Supercond. Sci. Technol.* **12** R135–R151

[58] Cooley L D, Lee P J, Larbalestier D C and O'Larey P M 1994 Periodic pin array at the fluxon lattice scale in a high-field superconducting wire *Appl. Phys. Lett.* **64** 1298–1300

[59] Matsumoto K, Takewaki H, Tanaka Y, Miura O, Yamafuji K, Funaki K, Iwakuma M and Matsushita T 1994 Enhanced J_c properties in superconducting NbTi composites by introducing Nb artificial pins with a layered structure *Appl. Phys. Lett.* **64** 115–117

[60] Seuntjens J M, Rudziak M K, Renaud C, Wong T and Wong J 1995 High energy physics conductor scale-up progress of he Supercon artificial pinning center process *IEEE Trans. Appl. Supercond.* **5** 1185–1188

[61] Lee P J, Larbalestier D C and Jablonski P D 1995 Quantification of pinning center thickness in conventionally processed and powder processed artificial pinning center microstructures *IEEE Trans. Appl. Supercond.* **5** 1701–1704

[62] Jablonski P D, Lee P J and Larbalestier D C 1994 Artificial two-phase Nb–Ti nanostructures using powder metallurgy techniques *Appl. Phys. Lett.* **65** 767–769

[63] Heussner R W, Bormio Nunes C, Lee P J, Larbalestier D C and Jablonski P D 1996 Properties of niobium–titanium superconducting wires with Nb artificial pinning centers *J. Appl. Phys.* **80** 1640–1646

[64] Cooley L D, Lee P J and Larbalestier D C 1996 Flux-pinning mechanism of proximity-coupled planar defects in conventional superconductors: evidence that magnetic pinning is the dominant pinning mechanism in niobium–titanium alloy *Phys. Rev.* B **53** 6638–6652

[65] Morris, P F Artificial pinning centers in Nb–Ti superconductor utilizing expanded metal in a jelly-roll conductor *MS Thesis* The University of Alabama in Huntsville

[66] Levi Y, Millo O, Rizzo N D, Prober D E and Motowidlo L R 1998 *Appl. Phys. Lett.* **72** 480–482

[67] Wang J Q, Rizzo N D, Prober D E, Motowidlo L R and Zeitlin B A 1997 Flux pinning in multifilamentary superconducting wires with ferromagnetic artificial pinning centers *IEEE Trans. Appl. Supercond.* **7** 1130–1133

[68] Matsumoto K, Tanaka Y, Yamada K, and Miura O 1994 Nb–Ti alloy type superconducting wire Patent no. 5,374,320

[69] Heussner R W, Bormio Nunes C, Cooley L D and Larbalestier D C 1997 Artificial pinning center Nb–Ti superconductors with alloyed Nb pins *IEEE Trans. Appl. Supercond.* **7** 1142–1145
[70] Cooley L D, Jablonaki P D, Heussner R W and Larbalestier D C 1996 Nb–Ti composite wires with artificial ferromagnetic pins *Adv. Cryo. Eng. Materials* **42** 1095–1102
[71] de Gennes P G 1989 *Superconductivity of Metals and Alloys* (Addison-Wesley: Redwood City)
[72] Hlasnik I, Takacs S, Burjak V P, Majoros M, Krajcik J, Krempansky L, Polak M, Jergel M, Korneeva T A, Mironova O N and Ivan I 1985 Properties of superconducting NbTi superfine filament composites with diameters ≤ 0.1 μm *Cryogenics* **25** 558–565
[73] Meingast C, Daeumling M, Lee P J and Larbalestier D C 1987 Proximity effect depression of the critical temperature in two-phase Nb–Ti superconductors *Appl. Phys. Lett.* **51** 688–689
[74] Jin B Y and Ketterson J B 1989 Artificially metallic superlattices *Adv. Phys.* **38** 189–366
[75] Cooley L D and Hawes C D 1999 Effects of a dimensional crossover on the upper critical field of practical Nb–Ti alloy superconductors *J. Appl. Phys.* **86** 5696–5704
[76] Cooley L D, Jablonski P D, Lee P J and Larbalestier D C 1991 Strongly enhanced critical current density in Nb 47 wt% Ti having a highly aligned microstructure *Appl. Phys. Lett.* **58** 2984–2986
[77] Patel A A and Cooley L D 2001 Anisotropy of the upper critical field in Nb–Ti tapes *IEEE Trans. Appl. Supercond.* submitted
[78] See plots and graphs at http://www.asc.wisc.edu
[79] McCambridge J D, Rizzo N D, Hess S T, Wang J Q, Ling X S, Prober D E, Motowidlo L R and Zeitlin B A 1997 Pinning and vortex lattice structure in NbTi alloy multilayers *IEEE Trans. Appl. Supercond.* **7** 1134–1137
[80] Warnes W H and Larbalestier D C 1986 Critical current distributions in superconducting composites *Cryogenics* **26** 643–653
[81] Warnes W H 1988 A model for the resistive critical current transtion in composite superconductors *J. Appl. Phys.* **63** 1651–1662
[82] Edelman H S and Larbalestier D C 1993 Resistive transistions and the origin of the *n* value in superconductors with a Gaussian critical-current distribution *J. Appl. Phys.* **74** 3312–3315
[83] High Y E, Lee P J, McKinnell J C and Larbalestier D C 1992 Quantitative analysis of sausaging in Nb barrier clad filaments of Nb–46.5wt%Ti composites *Adv. Cryo. Eng. (Materials)* **38** 647–652
[84] Matsushita T and Küpfer H 1988 Enhancement of the superconducting critical current from saturation in Nb–Ti wire I *J. Appl. Phys.* **63** 5048–5057 Matsushita T and Küpfer H 1988 Enhancement of the superconducting critical current from saturation in Nb–Ti wire II *J. Appl. Phys.* **63** 5060–5065
[85] Cooley L D, Lee P J and Larbalestier D C 1991 Is magnetic pinning a dominant mechanism in Nb–Ti? *IEEE Trans. Magn.* **27** 1120–1124
[86] Stejic G, Gurevich A, Kadyrov E, Christen D, Joynt R and Larbalestier D C 1994 Effect of geometry on the critical currents of thin films *Phys. Rev.* B **49** 1274–1288
[87] Gurevich A 1992 Nonlinear Josephson dynamics and pinning in superconductors *Phys. Rev.* B **46** 3187–3190
[88] Gurevich A and Cooley L D 1994 Anisotropic flux pinning in a network of planar defects *Phys. Rev.* B **50** 13563–13576
[89] Fietz W A and Webb W W 1969 Hysteresis in superconducting alloys— temperature and field dependence of dislocation pinning in niobium alloys *Phys. Rev.* **178** 657–667
[90] Otabe E S and Matsushita T 1993 Critical current density in superconducting Nb–Ti under the proximity effect *Cryogenics* **33** 531–540
[91] Matsushita T, Iwakuma M, Funaki K, Yamafuji K, Matsumoto K, Miura O and Tanaka Y 1996 Flux pinning property of artificially introduced Nb in Nb–Ti superconductors *Adv. Cryo. Eng. (Materials)* **42** 1103–1108
[92] McKinnell J C, Lee P J and Larbalestier D C 1989 The effect of titanium content on the pinning force in Nb44wt%Ti to Nb62wt%Ti *IEEE Trans. Magn.* **25** 1930–1933
McKinnell J C, Lee P J, Remsbottom R, Larbalestier D C, O'Larey P M and McDonald W K 1988 High titanium Nb–Ti alloys—initial high critical current density properties. *Adv. Cryo. Eng. (Materials)* **34** 1001–1007
[93] Tachikawa K 1999 Some properties of Nb25Ti alloy *IEEE Trans. Appl. Supercond.* **9** 1563–1566
[94] Clem J R 1974 A model for flux pinning in superconductors *Low Temperature Physics - Lt 13* eds K D Timmerhaus, W J O'Sullivan and B F Hammel (New York: Plenum) pp 102–106
[95] Polák M, Pitel J, Majoroš M, Kokavec J, Suchoň D, Kedrová M, Kvitkovič J, Fikis H and Kirchmayr H 1995 superconducting DC/AC magnetic system for loss and magnetization experiments operating up to 50/60 Hz *IEEE Trans. Appl. Supercond.* **5** 717–720
[96] Miura O, Matsumoto K, Tanaka Y, Yamafuji K, Harada N, Iwakuma M, Funaki K and Matsushita T 1992 Pinning characteristics in multifilamentary Nb–Ti superconducting wires with submicrometre filaments introduced artificial pinning centres *Cryogenics* **32** 315–322
[97] Miura O, Inoue I, Suzuki T, Matsumoto K, Tankaa Y, Funaki K, Iwakuma M, Yamafuji K and Matsushita T 1993 The development of a 2.5 T/100 kV A AC superconducting magnet using a high- J_c NbTi superconducting wire having Nb artificial pins *Supercond. Sci. Technol.* **6** 748–754
[98] Gilman F J 1998 *HEPAP Subpanel Report on planning for the future of U.S. high-energy physics* U.S. Department of Energy, Office of Energy Research, Division of High Energy Physics

[99] Willen, E 1998 Superconducting magnets *Hadron Colliders at the Highest Energy and Luminosity: Proc. 34th Workshop of the INFN Eloisatron Project* p 141

[100] Chernyi O V, Storozhilov G E, Ilicheva V O, Starodubov Y D, Lazareva M B and Andrievskaya N F 2001 Current characteristics and microstructure in a multifilamentary Nb-37Ti-22Ta superconductor *IEEE Trans. Appl. Supercond.* **11** 3796–99

[101] Lee P J, Fischer C M, Larbalestier D C, Naus M T, Squitieri A A, Starch W L, Werner R J, Limon P J, Sabbi G, Zlobin A and Gregory E 1999 Development of high performance Nb–Ti–Ta multifilamentary superconducting strand for the LHC high gradient quadrupole conductor *IEEE Trans. Appl. Supercond.* **9** 1571–1574

[102] Na L and Warnes W H 2001 Thermodynamics of the Nb–Ti–Ta ternary superconducting system *IEEE Trans. Appl. Supercond.* **11** 3800–03

B3.3.3
Processing of low T_c conductors: the compound Nb_3Sn

Takayoshi Miyazaki, Takashi Hase and Takayuki Miyatake

B3.3.3.1 Introduction

Among the numerous known superconducting materials, only NbTi alloy and Nb_3Sn compound are produced on a mass production commercial scale at present. Nb_3Sn multifilament wire is used in magnets to generate high magnetic fields over 8 T. Magnets with such field generating capability are required in an increasing number of R&D fields. In the field of life science research, for example nuclear magnetic resonance (NMR) analysis requires magnetic fields over 21 T. High field magnetic resonance imaging (MRI) on the other hand yields very high resolution biological images. In high energy physics research, the Very Large Hadron Collider (VLHC) project is expected to realize high field dipole magnets over 10 T. The International Thermonuclear Experimental Reactor (ITER) fusion project plans further to use magnetic fields of about 12 T to trap high temperature plasmas. In other R&D fields, higher T_c materials than NbT, such as Nb_3Sn, are sometimes used in liquid helium free magnets for a fast ramping rate even in generating magnetic fields less than 10 T. Nb_3Sn wire is often used for such applications.

B3.3.2 E1.1 E2.3

In early stages, tape shape Nb_3Sn conductor was employed in magnetic devices. Tachikawa and Kaufman in 1969–1971 proposed 'the bronze process' using Cu to decrease the reaction temperature. Moreover, third element addition to Nb_3Sn, most typically Ti or Ta, was found to be very effective in increasing the critical current density in high fields over 12 T in the early 80s. These techniques accelerated the development of scale up to industrial production from the laboratory record level.

B1

Nb_3Sn is an intermetallic compound, and is therefore very brittle. As a result many processing techniques are applied to the development of commercial wires having a long continuous length. In this review, several basic properties of Nb_3Sn composite wire are introduced. At first, thermodynamic phase diagrams of both binary, Nb–Sn and ternary, Nb–Sn–Cu, systems are presented. Then, basic physical properties of Nb_3Sn such as critical temperature and critical field are mentioned briefly. Various kinds of Nb_3Sn multifilament wire fabrication techniques are introduced. Finally critical current density and mechanical properties, which are the most important and fundamental parameters for applications, are described.

A1.3

B3.3.3.2 Phase diagram of the Nb–Sn system

B3.3.3.2.1 High temperature region

The phase diagram of the Sn and Nb alloy system is shown in figure B3.3.3.1 [1]. Nb_3Sn has an A15 type crystal structure. The A15 phase starts to form below a temperature of 2130°C with an Sn content of about 18 at.%. Below 1800°C, the range of the A15 phase is from ~18 to 25.1 at.%Sn including the stoichiometric composition. This relatively broad composition range means that the Nb–Sn system

B2.5

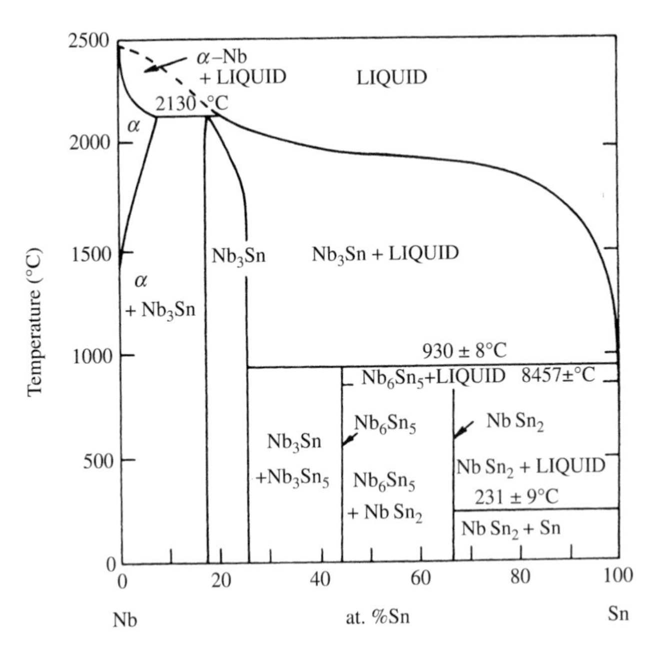

Figure B3.3.3.1. Phase diagram for Nb–Sn binary system in high temperature region [1].

needs no unequilibrium process, such as quenching, to form the stoichiometric A15 phase. Figure B3.3.3.1 shows that the A15 phase is the only stable phase above ~930°C in the binary system. Below ~930°C, two other phases, Nb_6Sn_5 and $NbSn_2$ can be formed. These three phases can be present during heat treatment at a temperature of below 930°C. Therefore, we have to apply temperatures above ~930°C in order to create a Nb_3Sn single phase in the binary system. This temperature is rather high in view of applying Nb_3Sn wire to commercial products, most of which are prepared via the wind and react technique because materials other than Nb_3Sn, i.e. Cu stabilizer, insulator, coil bobbin etc, degrade at such high temperatures.

B3.1

However, in addition of Cu to the Nb–Sn binary system is known to decrease the temperature to form an A15 phase below 700°C. Figure B3.3.3.2 shows the Nb–Sn–Cu ternary phase diagram at 700°C [2]. This figure illustrates the formation of the single A15 phase. As mentioned below, most wire processing techniques employ phases present in the Cu–Sn–Nb ternary system.

B3.3.3.2.2 Low temperature region

B2.5

A phase diagram of Nb–Sn alloy in the low temperature region is shown in figure B3.3.3.3 [3]. Nb_3Sn, which has stoichiometric composition, shows the martensitic phase transformation at a temperature of T_m. The crystal structure of A15 phase is cubic above T_m and becomes tetragonal below T_m with $a/c = 1.0062$ at 4 K [4] and 1.008 at 10 K [3]. Here, numerous authors, based on mechanical measurements [5, 6], x-ray diffraction [3] and neutron scattering [4], report that $40 < T_m < 50$ K. In addition, Flukiger *et al* [3] point out that under mechanical stress tetragonality can be partially present, even at room temperature. For the composite Nb_3Sn wire, the prestress arising from large thermal contraction between the Cu based alloy based matrix and Nb_3Sn on cooling from the reaction temperature to 4.2 K, the measuring temperature, plays a significant role as a driving force for the formation of the tetragonal phase.

D1.1.2

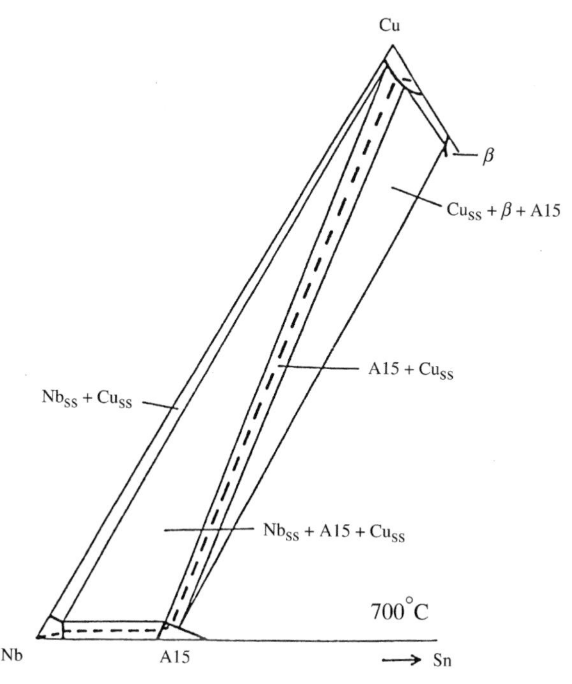

Figure B3.3.3.2. Phase diagram for Nb–Sn–Cu ternary system at 700°C [2].

B3.3.3.3 Physical properties

B3.3.3.3.1 Binary Nb_3Sn

Figure B3.3.3.4 indicates the upper critical field, B_{c2}, as a function of temperature T. The solid line shows the calculated results by the Werthamer–Helfand–Hohenberg (WHH) theory [7]. For Nb_3Sn, the experimental results (indicated by black dots in figure B3.3.3.4) show larger B_{c2} than the value predicted by the conventional WHH theory in the low temperature region. The main reason for the enhancement of B_{c2} is strong coupling between electrons and phonons [7, 8]. The result of calculations A3.1

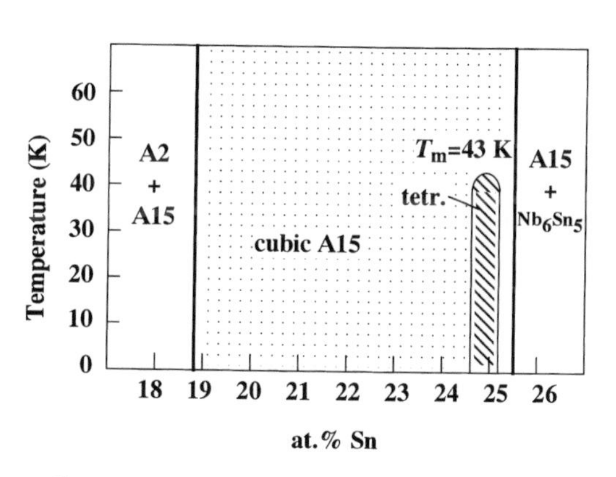

Figure B3.3.3.3. Phase diagram for Nb–Sn binary system in the low temperature region [3].

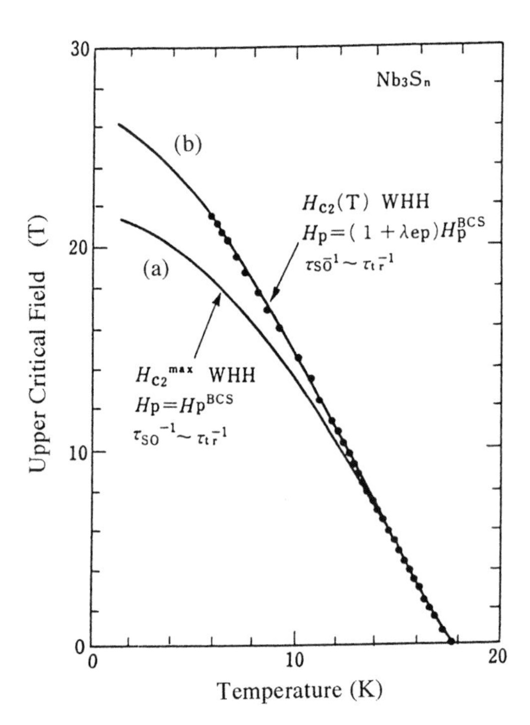

Figure B3.3.3.4. Dependence of upper critical field on temperature: comparison of experimental data with results of calculations [7]. Solid lines are results of calculation by (*a*) the conventional WHH theory and (*b*) corrected WHH theory with the electron–phonon coupling.

based on the WHH theory corrected by the electron–phonon coupling shows good agreement with measured data.

B2.5 On the other hand, B_{c2} of tetragonal Nb_3Sn is suppressed in the low temperature region as shown in figure B3.3.3.5 [9]. The B_{c2} of untransformed Nb_3Sn with a cubic crystal structure is about 29 T, though Nb_3Sn shows a B_{c2} of only about 24.2 T at 0 K. Figure B3.3.3.6 shows the relationship between B_{c2} and the Sn content in the A15 phase. Below 24.5 at.% Sn, B_{c2} increases monotonically with Sn content, though B_{c2} decreases very rapidly over 24.5 at.% Sn [10]. Figure B3.3.3.7 shows the dependence of critical temperature, T_c, on Sn content [10]. In this figure, T_c increases monotonically with Sn content. In the vicinity of 24.5 at.%, the T_c versus Sn content curve shows a discontinuity and is shifted to lower value.

As mentioned above, the crystal system of Nb_3Sn with the Sn content over 24.5 at.% should be transformed from cubic to tetragonal. This is thought to be one reason for the very rapid decrease of B_{c2} and T_c shift to a lower level over 24.5 at.% Therefore, it is possible that B_{c2} of Nb_3Sn increases with the Sn content in the A15 phase if the martensitic phase transformation does not occur.

To obtain high performance in B_{c2} and T_c, it is important to prevent the martensitic transformation by introducing appropriate disorder, such as that associated with a Sn content just below 24.5 at.% or by the addition of other elements, as discussed below.

B3.3.3.3.2 Ternary addition to Nb_3Sn

Several kinds of additive elements for Nb_3Sn are proposed, such as Zr, Hf, Mg, Ge, Ti and Ta. Of these elements, Ti and Ta are employed widely in practical conductors [11]. Figure B3.3.3.8 shows

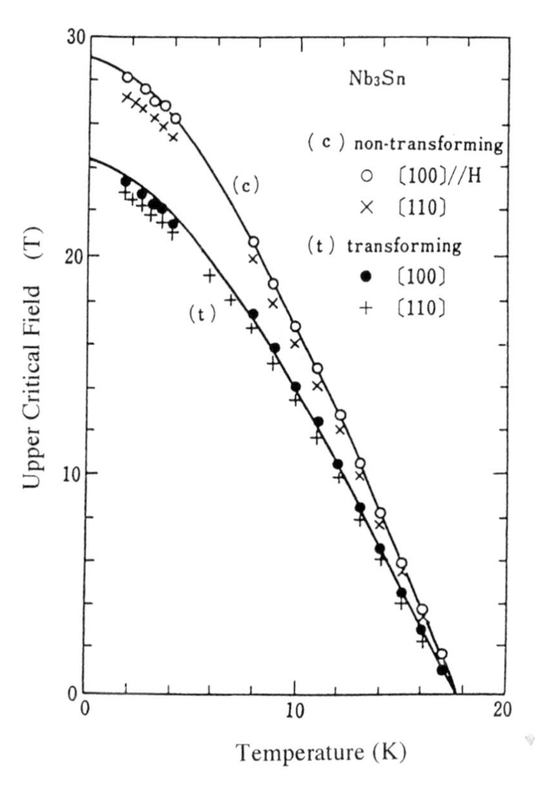

Figure B3.3.3.5. Comparison of $H_{c2}(T)$ curves of transformed and non-transformed Nb_3Sn single crystal [9].

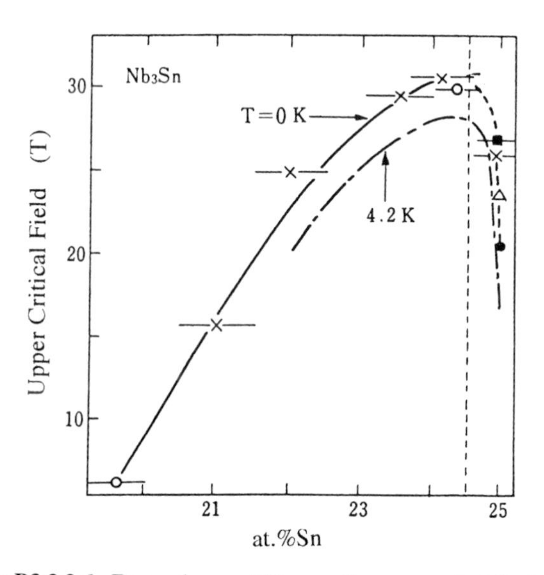

Figure B3.3.3.6. Dependence of H_{c2} on Sn content in Nb_3Sn [10].

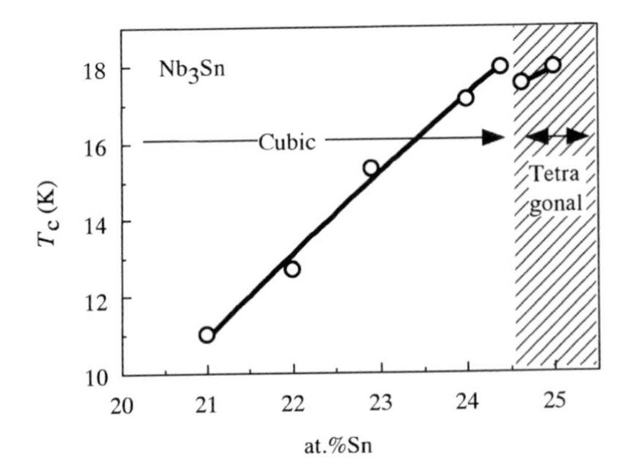

Figure B3.3.3.7. Dependence of T_c on Sn content in Nb_3Sn [10].

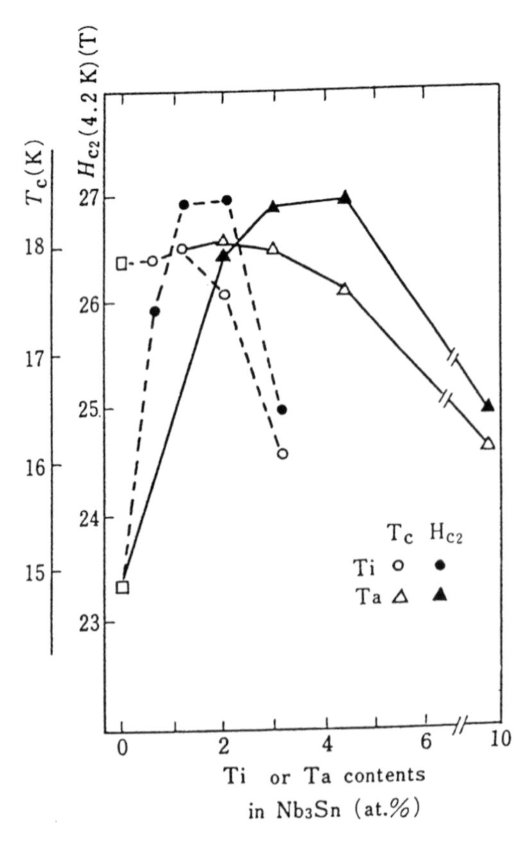

Figure B3.3.3.8. Dependence of T_c and H_{c2} on Ti and Ta content [12–15].

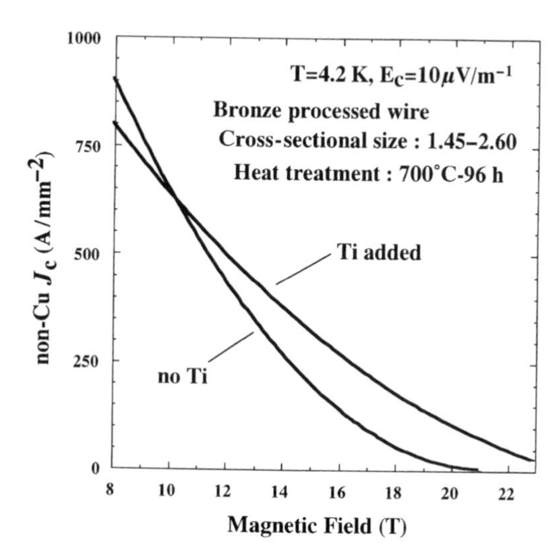

Figure B3.3.3.9. Comparison of relationship between J_c and magnetic field for Ti added Nb_3Sn and non-Ti added Nb_3Sn.

the dependence of T_c and B_{c2} on the Ti content [12–14] or Ta addition [15]. The dependence of B_{c2} indicates a peak of about 27 T for both Ti and Ta though the undoped system indicates a B_{c2} of about 23 T. Lower concentrations of Ti yield a different peak value of B_{c2} compared with Ta addition. A4

High critical current density at high field is expected from these improvement in B_{c2}. The dependence of critical current density on magnetic field for binary and ternary Nb_3Sn are shown in figure B3.3.3.9. It is clear that the J_c of Ti added Nb_3Sn is higher than that of non-Ti added Nb_3Sn in a magnetic field higher than 10 T.

Recently, both Ti and Ta containing conductors have been developed to improve J_c in the high field region [16–19]. Moreover, it is reported that addition of Ta to the filament prevents sausaging and it is useful for suppression of the ac loss [20, 21]. B3.1 D2.6

B3.3.3.4 Nb_3Sn multifilamentary wire fabrication

Multifilamentary Nb_3Sn wire can be fabricated by several processes. Figure B3.3.3.10 shows schematic cross sections of Nb_3Sn wire fabricated by some of these processes. Roughly speaking, fabrication methods can be categorized into three types in terms of the method of Sn supply. In the first type, Sn is supplied from Cu–Sn alloy matrix. This process is the so-called bronze route process. The second type is, where Sn is supplied from solid Sn located inside the wire. This process is called the internal tin diffusion process. Finally, in the third type Sn is supplied from plated Sn on the surface of the wire. This is called the external tin diffusion process containing the powder metallurgy (P/M) process, the ECN process, infiltration process and the *in situ* process. Details of each process will be explained below. A1.3 B1

B3.3.3.4.1 Bronze route

The bronze route process was developed in the early 1970s [22, 23]. In this process, bronze (Cu–Sn alloy) is used as a source of Sn, with Nb_3Sn produced at a lower reaction temperature [24]. Since this process needs no solid Sn rods or foils, it is possible to process at high temperature i.e. hot extrusion. This means B1

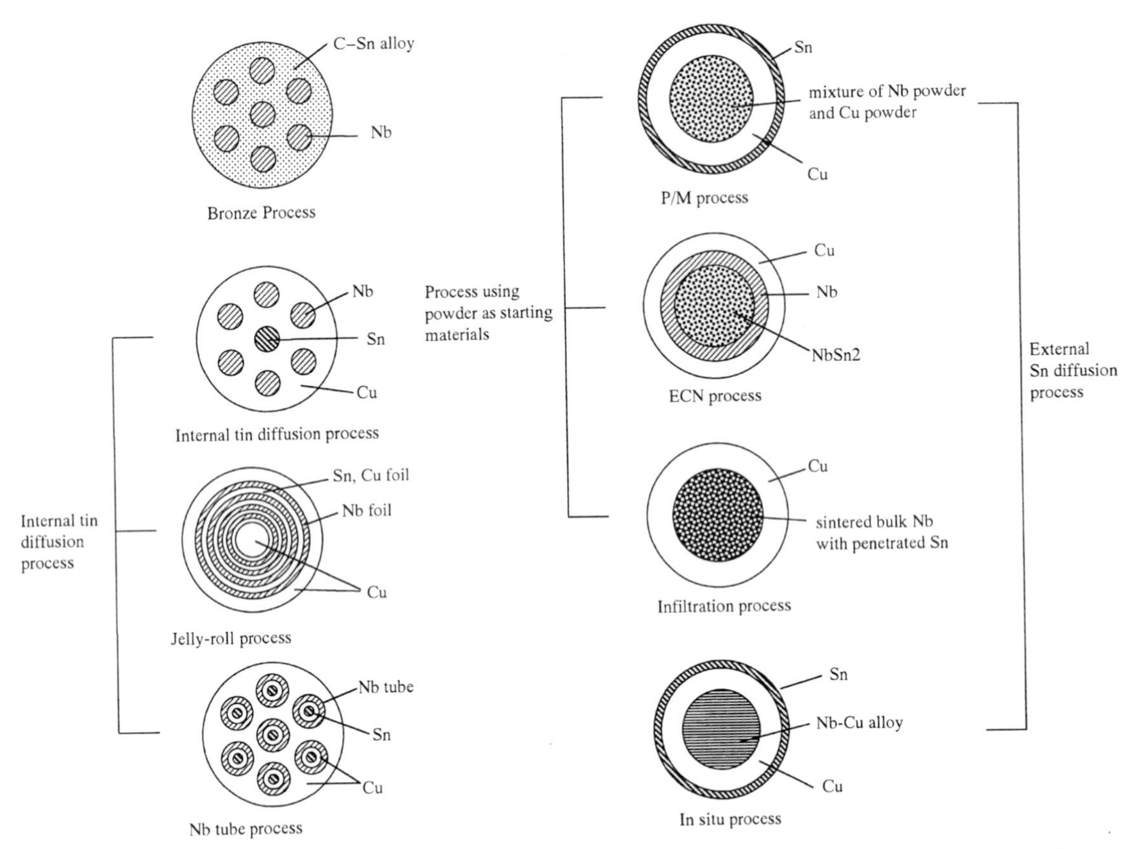

Figure B3.3.3.10. Schematic cross sections of Nb_3Sn wires processed by each fabrication method.

B3.1 that a uniform wire containing fewer sausaging filaments can be easily obtained. Additionally, a large reduction of cross sectional area during extrusion should be possible. Great care must be taken, however, not to form Nb_3Sn while processing at high temperature.

B1 A schematic flow diagram of the bronze process is shown in figure B3.3.3.11. A billet is assembled by inserting Nb rods into drilled bronze and its diameter reduced by extrusion and drawing. In the wire making process, intermediate annealing has to be performed several times because bronze exhibits a marked tendency for work hardening. Next the billet is assembled by stacking the primary Nb–bronze composite, Cu stabilizer and Nb or Ta barrier to prevent Sn pollution in Cu during heat treatment. The diameter of the secondary billet is reduced to a wire with an appropriate diameter by extrusion, drawing and annealing.

Characteristic cross sections of the wires manufactured by this process at Kobe Steel Ltd are shown in figure B3.3.3.12. In these figures, (*a*) and (*b*) are termed internally stabilized and externally stabilized Nb_3Sn wire, respectively.

B2.2.1 Bronze processed wire is heated at temperatures from 500 to 750°C to form Nb_3Sn by solid state diffusion and reaction. To obtain high J_c in a high field region, the wire is heated to a relatively high temperature of 700–750°C to form a stoichiometric composition of Nb_3Sn. To prepare a wire for application in a lower or an intermediate field region, heat treatment at a lower temperature of around 650°C is adopted to generate fine grains. Details of the relationship between J_c and heat treatment

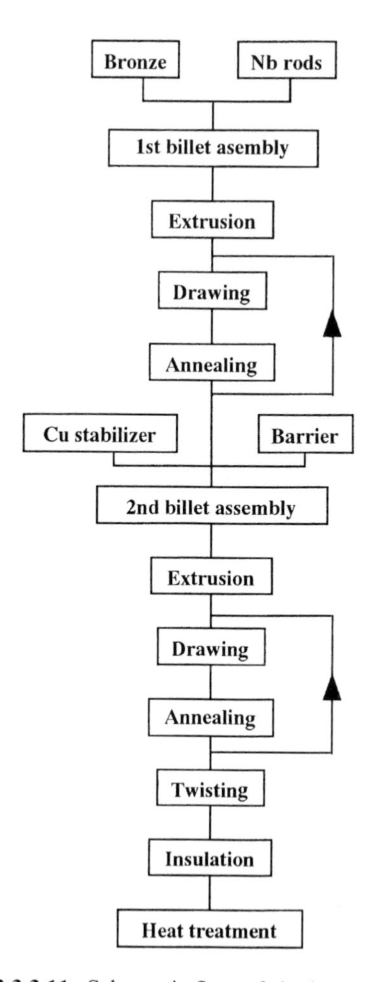

Figure B3.3.3.11. Schematic flow of the bronze-route process.

conditions will be discussed in the following section. The wire for ac use is heated at a lower temperature of about 600°C to reduce ac loss.

D2.6

Improvement in J_c by enhancing the Sn content in the matrix for the bronze-processed Nb_3Sn superconducting wire has been reported [25, 26]. However, it is difficult to fabricate single phase high Sn bronze ingots without inter-metallic precipitations. This is because the solubility limit of Sn to Cu is 15.8 wt.% on the equilibrium phase diagram of Cu–Sn alloy [27]. These inter-metallic precipitations lead to the breakage of the wire in the process of drawing. Recently, a Nb_3Sn superconductor with a bronze matrix of Cu–15 wt%Sn–0.3 wt%Ti has been manufactured successfully [25, 26, 28].

The cross sectional area ratio of bronze to Nb is called the bronze ratio. The bronze ratio is defined as a trade off between possible superconducting areas and the Sn content. The bronze ratio of ~3 is commonly adopted for dc use. For ac use, a larger ratio such as ~20 should be adopted to reduce coupling ac loss between filaments.

D2.6

The filament diameter is another important parameter in wire design. Fine filaments need a shorter heat treatment duration than thicker ones and uniform fine grains can be formed in the fine filaments. That is why it is possible to realize high J_c and low ac loss for fine filaments. Actually, a

D2.6

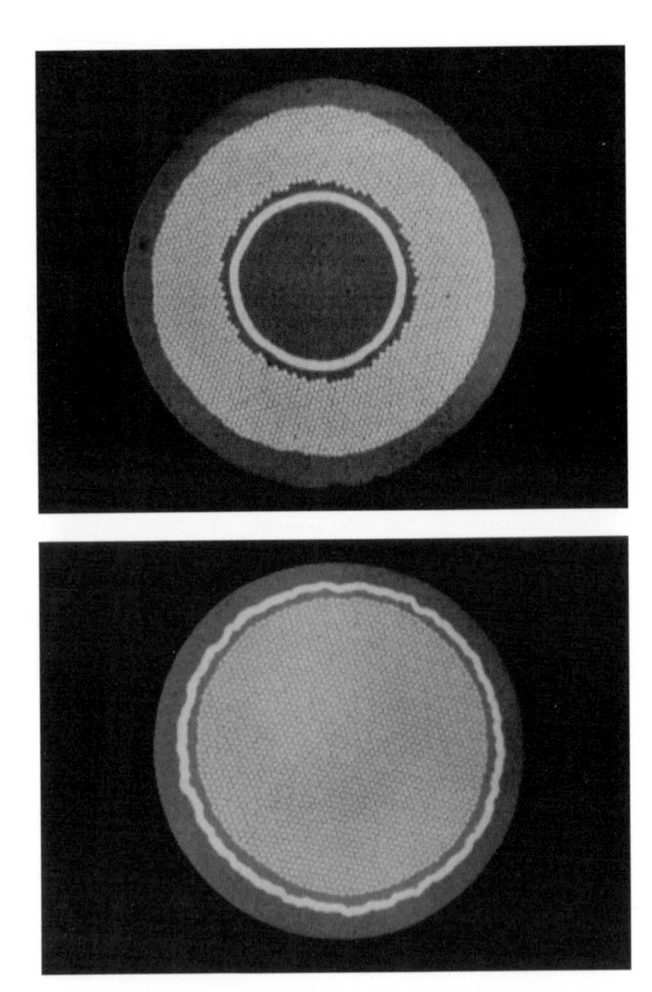

Figure B3.3.3.12. Typical cross sections of the bronze-route Nb_3Sn wires for high field NMR magnets (by Kobe Steel Ltd). (*a*) Internally stabilized and (*b*) externally stabilized Nb_3Sn wire.

practical wire diameter for ac use is about 0.2 μm, and for pulse or dc use a diameter from 2 to ~8 μm is satisfactory.

B3.3.3.4.2 Internal Sn diffusion processing

B1 In the internal Sn process, solid pure or a slightly impure Sn is used as a source of Sn [29]. A cross section of an internal Sn processed Nb_3Sn wire manufactured by Mitsubishi Electric Co. for ITER is shown in figure B3.3.3.13. In this process, there is basically no limit to the amount of Sn, though the bronze processed wire has to use bronze with the Sn content of up to 15.8%. Therefore, high J_c can be expected in this process by considering only the method of Sn supply. However, the Sn area should be a dead area in the conductor, containing no superconducting filaments. Eventually, these two factors, the amount of Sn and the superconducting area, have to be balanced to realize high J_c in this process.

In general, alloys that exhibit marked work hardening are not used to reduce the amount of intermediate annealing. This means that the cost and delivery time of commercial products can be

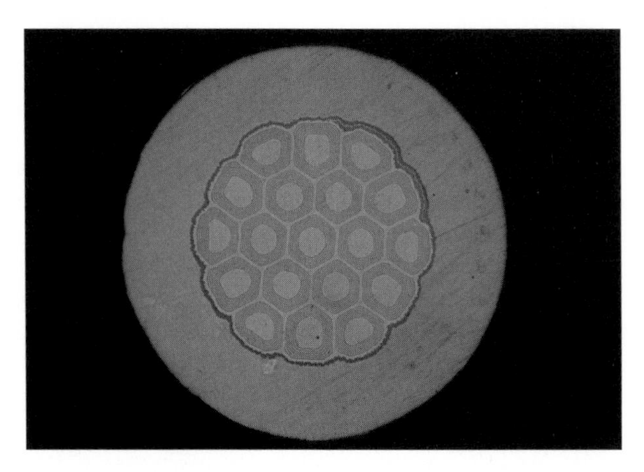

Figure B3.3.3.13. A typical cross section of internal tin diffusion processed Nb_3Sn wire for the ITER project (by Mitsubishi Electric Co.).

reduced compared with the bronze processed wire. A typical process flow of the internal Sn process is shown in figure B3.3.3.14 [30, 31]. In this process, the insert procedure of Sn rods is very important because the melting temperature and the hardness of Sn are very low compared with other materials. The Sn rod is assembled after extrusion to avoid unexpected deformation of Sn by work-heat during the extrusion. Two different processing techniques are adopted as shown in figure B3.3.3.14. In the first method, the Sn rod is inserted after drilling the extruded billet as shown in figure B3.3.3.14(*a*) [30]. In the second method, the Sn rod is inserted after pipe extrusion of billet as shown in figure B3.3.3.14(*b*) [31]. The three basic elements, Cu, Nb and Sn, have levels of different workability and melting temperature. Therefore, it would be more difficult to process uniform wire even by a process at room temperature compared with the bronze-route processed Nb_3Sn wire. Poor uniformity of the filament would be an origin of low n-value of the wire processed in this route.

In the fabrication process using solid Sn, pre-heating to form bronze is generally needed. After the pre-heating, the heat treatment for Nb–Sn reaction is performed.

Finally, the composition of the matrix in a cross section is not uniform because of the existence of hard intermetallic phases. These intermetallic phases lead to degradation of mechanical properties of the internal Sn processed Nb_3Sn wire after heat treatment.

The Tin-tube-source (TTS) method was developed in the IGC/Advanced Superconductors, Inc. [32]. In this process, Nb filaments are surrounded by Sn tube located inside the barrier as shown in figure B3.3.3.15. Other fabrication methods, such as the Nb tube method [33] and the modified jelly roll (MJR) method [34], were proposed which can be considered as variations of the internal Sn diffusion method.

B3.3.3.4.3 P/M processing

In a broad sense of the powder metallurgy P/M process, several kinds of powders are employed as starting raw materials. In this sense, the P/M process can be classified into various categories as listed in table B3.3.3.1.

In table B3.3.3.1, procedures 1 to 4 can be referred to as the P/M process. These processes have different arrangements of Cu and Sn, though Nb or Nb alloy powder is commonly used as shown in table B3.3.3.1. Procedures no. 5 through 7 are known as the ECN method, details of which will be described in the following section. Procedure no. 8 is an infiltration process using sintered bulk consisting of Nb powder. Sn penetration into the bulk composite is achieved by dipping it into molten Sn bath. This method is described after the following section.

Micrographs of Nb_3Sn wire fabricated by procedure no. 1 are shown in figure B3.3.3.16. Basically, these wires have the advantage that the stacking process to obtain the multifilamentary wire is not needed. As shown in figure 3.3.3.16(*a*) and (*b*), fine Nb filaments are dispersed in the Cu matrix. The wires manufactured by procedures No.1 through 4 have basically discontinuous filaments in the longitudinal direction as shown in figure B3.3.3.16(*c*).

A1.3

If Nb powder with good workability is applied, intermediate annealing is hardly required during the manufacturing process. Since the workability of Nb is degraded rapidly by only slight pollution by gas elements such as oxygen and hydrogen, it is very important to suppress the adsorption of the impurity

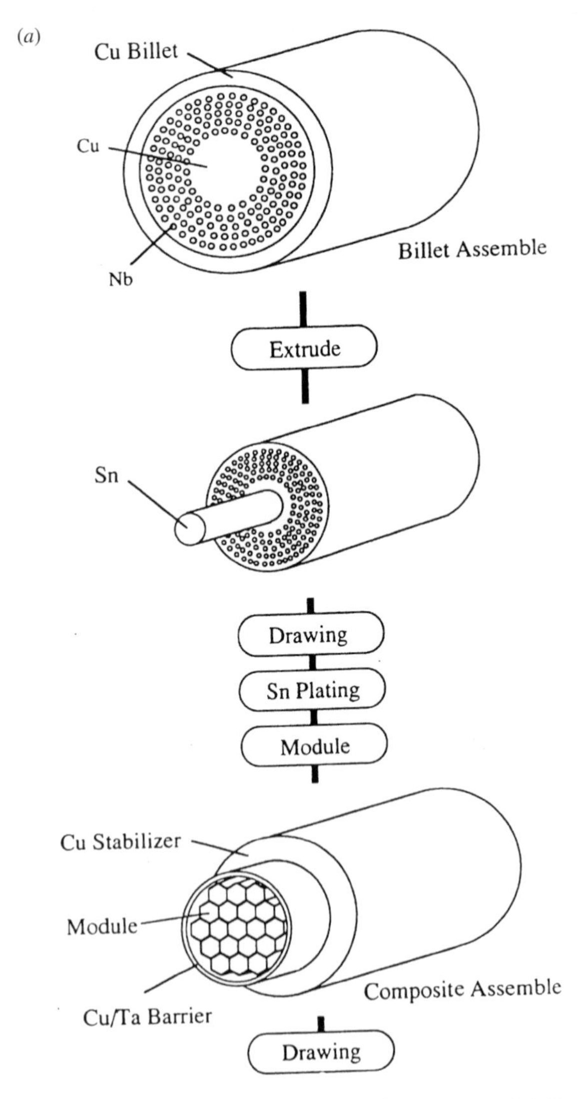

Figure B3.3.3.14. Schematic flow diagram of internal tin diffusion process. (*a*) Sn rod is inserted into a drilled extruded Cu–Nb billet [30]. (*b*) Sn rod is inserted into a pipe extruded Cu–Nb billet [31].

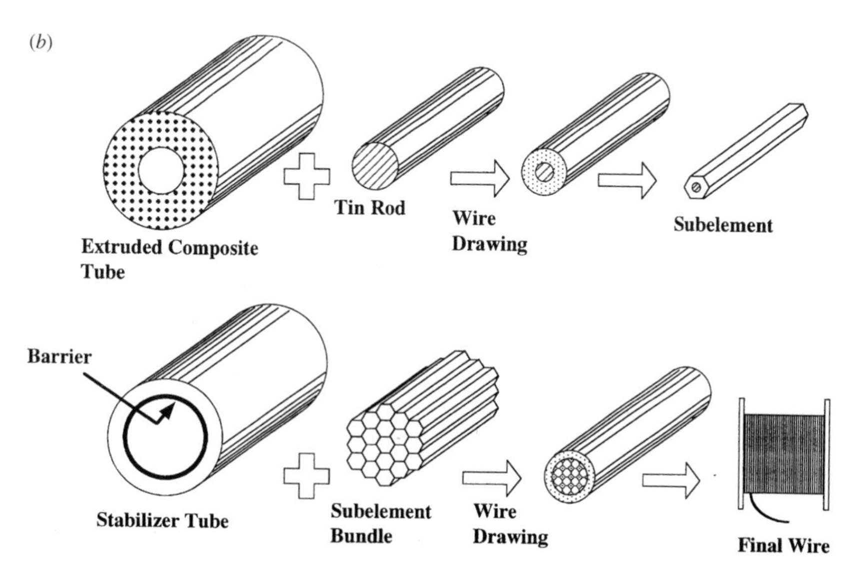

Figure B3.3.3.14. Continued.

gas on the surface of Nb powder to a minimum level, in addition to maintaining the purity of the raw material itself.

Procedure no. 1

In this process, Sn is supplied from externally plated Sn on the surface of the wire. The process is shown schematically in figure B3.3.3.17 [35]. In this case, Nb and Cu powders are prepared by the plasma rotating electrode process (PREP). Figure B3.3.3.18 is a typical photograph of magnified Nb powders made by the PREP [35]. In this process, both Nb and Cu powders are mixed and pressed into a bulk mixture. Alloy powders like Nb–Ta and Cu–Mn can be used depending on the application. Alloy powder can also be applied to procedures no. 2 and 3. The pressed bulk mixture is inserted into the Cu jacket to build up a billet. The billet is extruded and drawn into wire with an appropriate diameter. After Sn is plated onto the surface of the wire, heat treatment for Sn diffusion and reaction of Sn and Nb is performed.

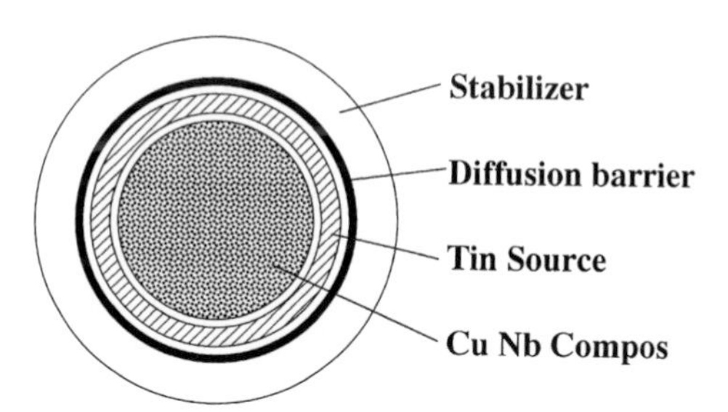

Figure B3.3.3.15. A schematic cross section of wire produced by the tin-tube-source (TTS) method [32].

Table B.3.3.3.1. Summary of the various kinds of Nb_3Sn process using powder metals as starting materials

No.	Powder	Sn source	Process
1	Nb*, Cu*	Sn plating	Composite powder is placed in Cu jacket, drawn to wire, plated Sn and heated.
2	Nb*, Cu*	Sn rod	Composite powder is placed in Cu jacket with Sn rods, drawn to wire and heated.
3	Nb*, Sn*	Sn powder	Composite powder is placed in Cu jacket, drawn to wire and heated.
4	NbTa	Brozne matrix	Composite powder is placed in bronze jacket, drawn to wire and heated.
5	($NbSn_2$, Nb) or (Nb_6Sn_5, Sn)	$NbSn_2$	Composite powder is placed in Nb tube, drawn to wire and heated.
6	Nb_6Sn_5, Cu	Nb_6Sn_5	Composite powder is placed in Nb tube, drawn to wire and heated.
7	Nb_6Sn_5, Nb	Nb_6Sn_5	Composite powder is placed in Ta tube, drawn to wire and heated.
8	Nb	infiltrated Sn	Composite powder is sintered, infiltrated Sn, placed in Nb or Cu jacket, drawn to wire and heated.

* Occasionally elementary addition is examined.

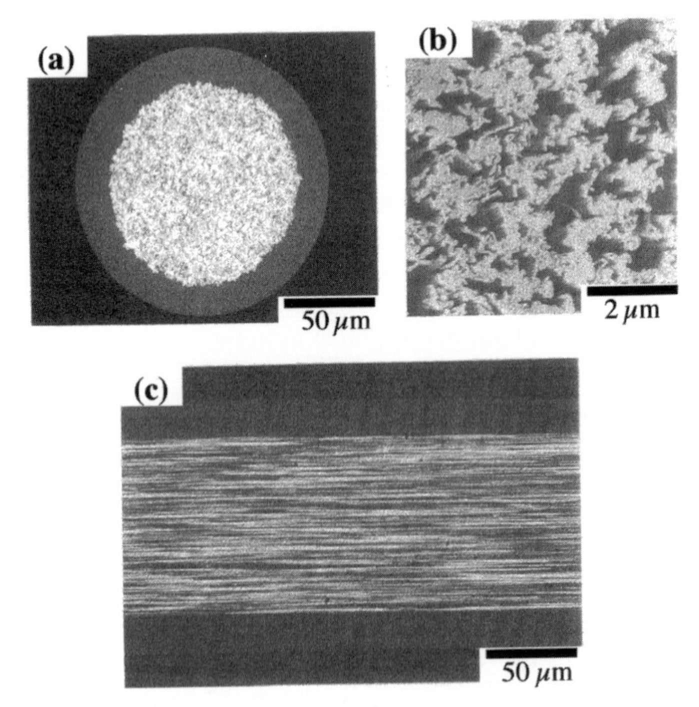

Figure B3.3.3.16. Micrographs of typical P/M processed Nb_3Sn wire. (*a*) An overall cross section, (*b*) a magnified micrograph of the area in the powder composite and (*c*) a longitudinal cross section.

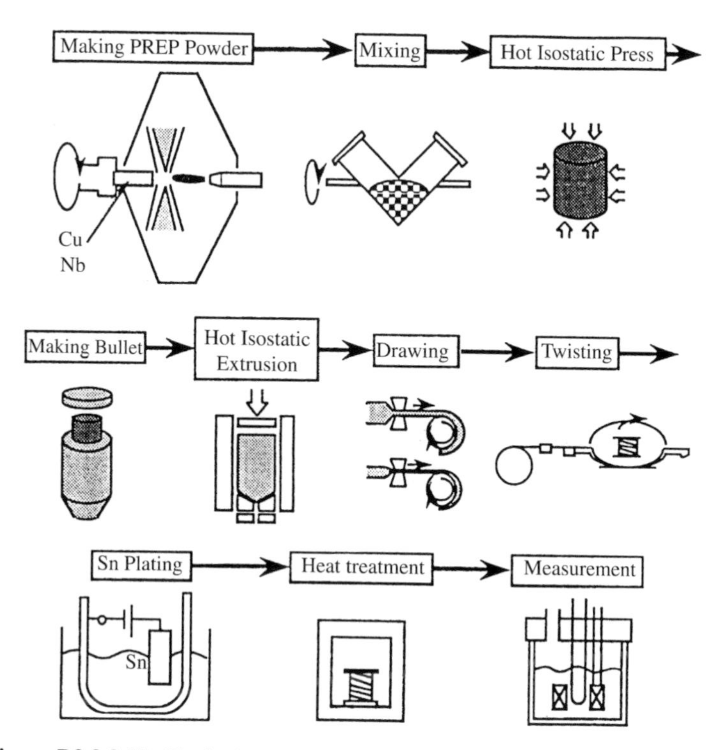

Figure B3.3.3.17. Typical manufacturing process flow of the P/M method.

Procedure no. 2

In this process, Sn is supplied by internal diffusion from the solid Sn rod. A pressed mixture of Nb and Sn powders is drilled in order to accomodate Sn rods. After coating with Cu, this is fabricated into wire by extrusion and drawing. This process does not require plating of Sn on the surface of the wire but processing is difficult since the workability of Sn and the bulk mixture of powders are different. Cu or Ga may be added to Sn to enhance its workability to that of Nb and to improve the overall workability of the wire [36].

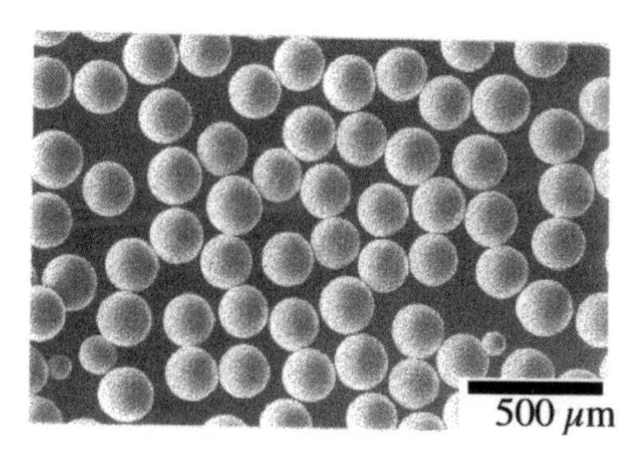

Figure B3.3.3.18. Typical photograph of Nb powder made by the PREP method.

Procedure no. 3

Nb and Sn powder mixture is pressed into a bulk in this process, which is coated with Cu. After that, extrusion and drawing are performed to obtain a wire. Element additions may also be incorporated in this process [37].

Procedure no. 4

A mixture of Nb and Ta powders is inserted into a bronze tube, which is extruded and drawn into a wire.
A4.3 In this process, improvement in J_c by introducing artificial pinning centers in filaments is achieved [38].

B3.3.3.4.4 ECN technique

The ECN method was proposed by the Energieonderzoek Centrum Nederland (the Netherlands Energy Research Foundation, ECN) [39]. In this process, powder of the intermetallic compound $NbSn_2$ is adopted as shown in procedure no. 6 in table B3.3.3.1. In the process, $NbSn_2$ is transformed to Nb_3Sn via the Nb_6Sn_5 compound. $NbSn_2$ powder and a small amount of Cu powder to promote the reaction are introduced into a Nb tube. After stacking Nb tubes filled with powders into the Cu, extrusion and drawing are carried out to fabricate a wire.

Two processes to form Nb_3Sn from Nb_6Sn_5 are proposed [40]. One is the creation of Nb_3Sn directly from Nb_6Sn_5, whereas the other involves reacting Nb and Sn directly to form Nb_6Sn_5. The grain size of Nb_3Sn formed in the first process is larger, i.e. from 5 to 15 μm, than that in the second process, i.e. less than 0.1 μm. The latter process is thought to be preferable to achieve higher J_c.

Procedure no. 6

To improve J_c, only Nb_6Sn_5 powder is used (rather than $NbSn_2$ powder) as a starting material [40]. In this process, since only fine Nb_3Sn grains formed by the reaction of Sn from Nb_6Sn_5 and Nb exist, higher J_c can be expected.

Procedure no. 7

This process was proposed by Tachikawa *et al* [41]. Nb_6Sn_5 and Nb powder are used as raw materials. By using Nb powder, the physical distance between reacting elements could be reduced. This procedure makes it possible to produce fine Nb_3Sn grains by the heat treatment. Recently, a new method, in which reacted Ta–Sn powder is encased in a Nb or a Nb–Ta tube, has been proposed [42].

B3.3.3.4.5 Infiltration technique

Procedure no. 8

This process is considered as a powder process but using a solid phase (Nb) – liquid phase (Sn) reaction. A schematic diagram of this process is shown in figure B3.3.3.19 [43]. Nb powder is made into a porous
B2.2.1 bulk by pressing and sintering. The bulk Nb is dipped into a molten Sn bath to introduce Sn to porosity in the bulk.

In an early stage of the development, the tape formed conductor was processed by rolling since the workability of Nb is unsatisfactory. Hydro–dehydro Nb powder with a diameter of order of 10 μm is used as an initial material in the following stage. The Nb bulk is heated under a pressure of 10^{-5} Torr at 2250°C for 3 min after rolling. The resulting Nb bulk contains some voids at this point in the process.

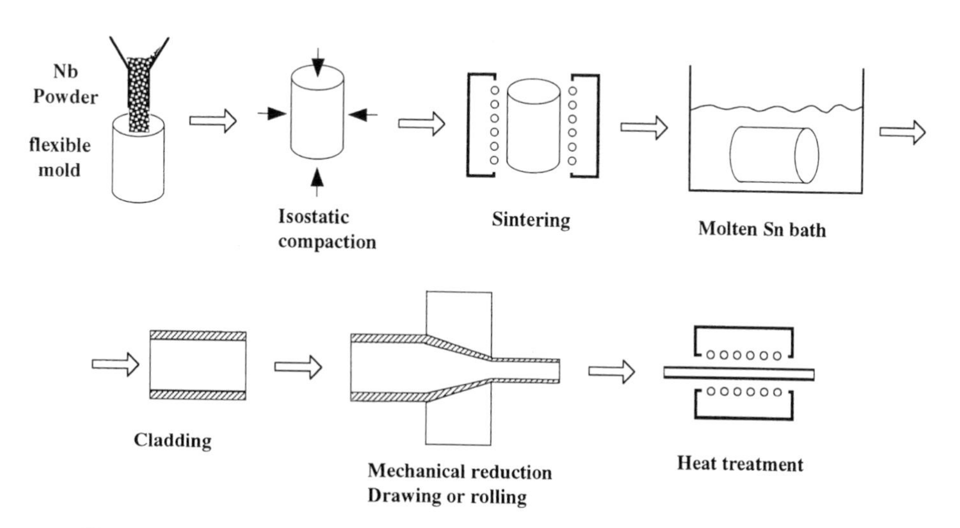

Figure B3.3.3.19. Schematic process diagram of the infiltration technique [43].

The Nb bulk is rolled into the tape and then the arrangement is dipped into a molten Sn bath to penetrate Sn into the Nb bulk. The Nb bulk with penetrated Sn is rolled and heated [43].

After the discovery of the deoxidization technique for Nb, wires may be produced by drawing using Nb powder with good workability.

B3.3.3.4.6 In situ technique

This process uses Nb–Cu alloy as a starting material [44–46]. Since Cu and Nb do not form a solid solution, Nb dendrites in a Cu matrix are formed in an ingot. Such Nb dendrites after etching the Cu matrix are shown in figure B3.3.3.20. By reducing the diameter of the Nb–Cu alloyed ingot, Nb dendrites are processed into fine filaments dispersed in Cu as shown in figure B3.3.3.21. Therefore, this process does not require a process, similar to the P/M process.

In the actual process, the Nb–Cu ingot is covered with a Cu jacket and then the billet is drawn into the wire. Sn is supplied by Sn plated on the surface of the wire or by Sn rod embedded in the Nb–Cu alloy.

The activation energy of Nb for oxidation is very low, while the workability of Nb is degraded by oxidation. Therefore, Ti, Al, Hf and Zr are commonly added to the Nb–Cu alloy to absorb oxygen [47].

B3.3.3.5 Critical current density of Nb_3Sn wire

The critical current density (J_c) is a fundamental parameter for practical applications such as superconducting magnet design. In Nb_3Sn multifilamentary wires, three kinds of J_c have been defined as follows:

A4, A1.3, E1.1

$$[\text{overall } J_c] = I_c/S_{\text{all}} = [\text{non}-\text{Cu } J_c]/(1+R_{\text{cu}}) \quad \text{(B3.3.3.1)}$$

$$[\text{non}-\text{Cu } J_c] = I_c/S_{\text{non-Cu}} = [\text{overall } J_c](1+R_{\text{Cu}}) \quad \text{(B3.3.3.2)}$$

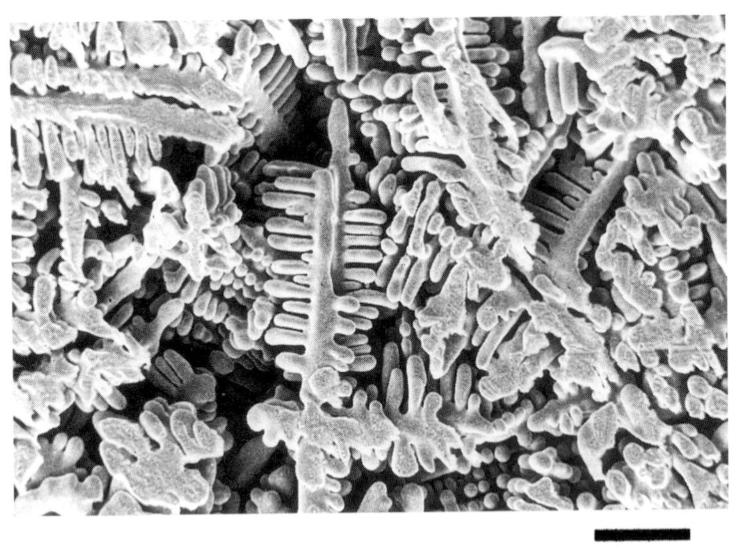

Figure B3.3.3.20. Dendrite structure of Nb in etched Cu–Nb alloy (by Fujikura Ltd.).

$$[\text{core } J_c] = I_c/S_{\text{core}}. \quad (B3.3.3.3)$$

Here I_c is the critical current, determined from I–V characteristics using appropriate voltage criteria. Electric field criteria of 100 or $10\,\mu\text{V m}^{-1}$ or resistivity criteria of 10^{-13} or 10^{-14} Ωm are

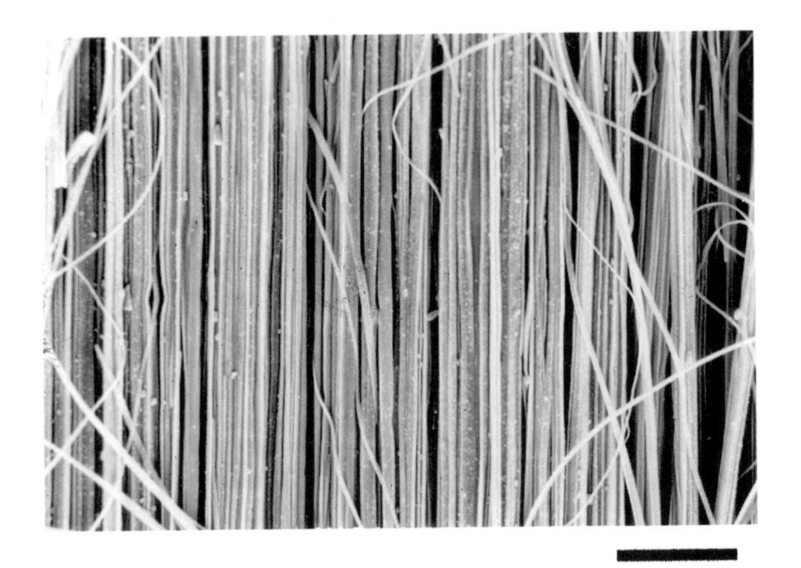

Figure B3.3.3.21. Nb filaments of the in-situ processed wire (by Fujikura Ltd.).

adopted. S_{all} and $S_{non\text{-}Cu}$ are the cross sections of the wire and non-Cu area, respectively. R_{cu} is the so-called Cu ratio which is the cross sectional ratio of Cu to non-Cu. S_{core} expresses the cross section of Nb_3Sn or the composite of Nb_3Sn and non-superconducting matrix for multifilamentary wires.

The overall J_c can be regarded as the most important engineering parameter. The product of the overall J_c and packing factor of the superconducting magnet leads directly to the critical current density per winding area. The overall J_c should be changed by the copper ratio which is determined from examination of the thermo-electric stability in the wire design. The precise Cu ratio has to be measured to evaluate the non-Cu J_c.

The core J_c basically shows the critical current capacity per superconducting area. However, since estimation of the cross section of the total reacted layer is very difficult for multifilamentary wire, practically, the core J_c expresses the critical current per composite area estimated by subtracting the Cu and diffusion barrier area from the total cross section of the wire. In this sense, the core J_c can be said to have the same meaning as the non-Cu J_c.

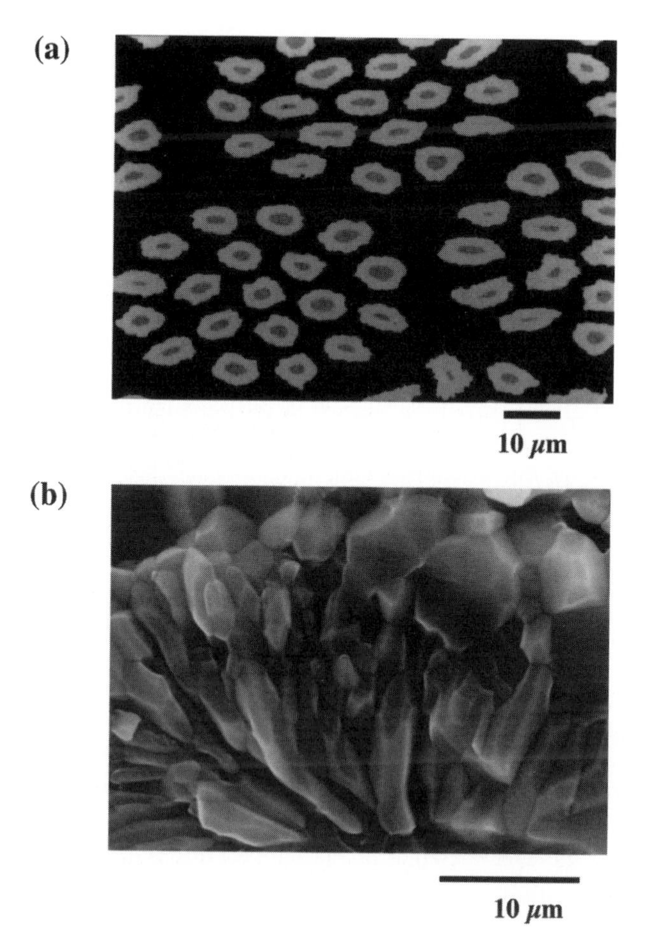

Figure B3.3.3.22. (*a*) Typical cross section of Nb_3Sn filaments in the bronze processed wire observed by SEM using refracted electrons, (*b*) micrograph of the cross section of a fractured filament by high resolution field emission (FE) SEM.

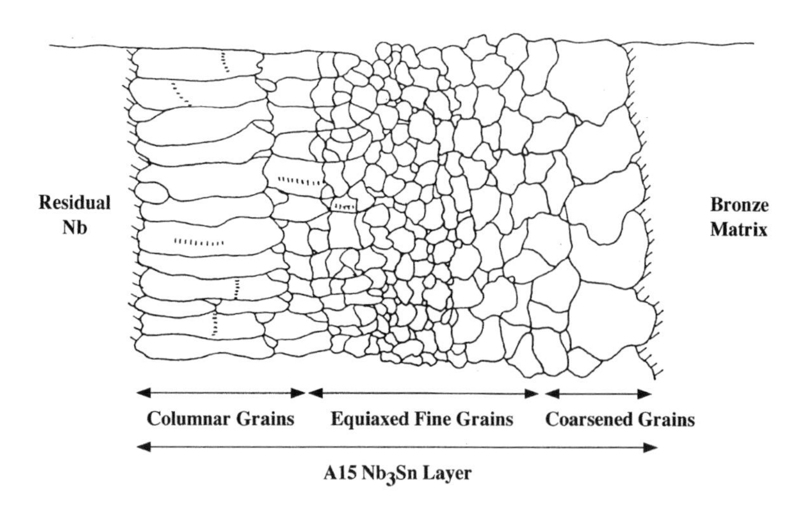

Figure B3.3.3.23. Schematic illustration of a typical microstructure for Nb_3Sn layer in the practical multifilamentary wire [48].

Figure B3.3.3.22(*a*) shows a magnified cross section of filaments in the bronze route processed multifilamentary wire after the reaction to form Nb_3Sn. In this figure, the brightest circle areas indicate Nb_3Sn and residual Nb can be seen in the centre of each filament.

Figure B3.3.3.22(*b*) is a microphotograph of the cross section of a broken filament. In this figure, the morphology of grains in the Nb_3Sn reacted layer can be observed. A schematic illustration of a typical microstructure for the Nb_3Sn layer in the practical multifilamentary wire is shown in figure B3.3.3.23. After reaction to form Nb_3Sn, columnar grains exist in the vicinity of the boundary of Nb and coarsened grains are formed in the vicinity of the border of matrix from which Sn is diffused to Nb_3Sn layer. Equiaxed fine grains are generated between columnar and coarse grains [48].

Grain boundaries form effective pinning centres in Nb_3Sn superconductors. Several researchers report the dependence of J_c or maximum pinning force F_{pmax} on grain size [49–51]. The finer grains make it possible to realize higher J_c in an intermediate magnetic field because F_{pmax} is increased with the inverse of the grain size or the density of the grain as shown in figure B3.3.3.24 [50, 51]. A4.3

Distributions of the composition in the reacted layer and in the grain have been investigated [48, 52]. The Sn content at the boundary of Nb and the reacted layer is between ~18 to ~20 at.% and at the boundary of bronze, is between ~28 to ~30 at.%. Heat treatment at high temperature can be used to generate a well distributed stoichiometric Nb_3Sn layer rather than heat treatment for a long duration at low temperature. In addition, a plateau of the profile of Sn content of about 25 at.% appear during this process.

The J_c of Nb_3Sn wire depends on heat treatment conditions which determine its microstructure and chemical composition. Figure B3.3.3.25 shows micrographs of the reacted layer in the bronze processed Nb_3Sn wire heated at (*a*) 650°C and (*b*) 720°C for 150 h [25]. Generally, heat treatment at low temperature generates fine equiaxed grains, while at high temperature it generates large columnar grains. Therefore, higher core J_c can be expected for the wire heated at lower temperature. On the other hand, the Nb_3Sn layer thickness of the wire heated at higher temperature is greater than that heated to lower temperature for the same duration. Therefore, the balance of grain size and layer thickness by optimizing the heat treatment condition is important to realize a high value of the non-Cu J_c.

It is known that the Nb_3Sn conductor saturates in the F_p versus b curves in the high field region near B_{c2}, where b is the magnetic field normalized by B_{c2} [53]. In figure B3.3.3.26, an example of the saturated characteristic for the practical bronze processed wire heated in several conditions and with different Sn A4.2

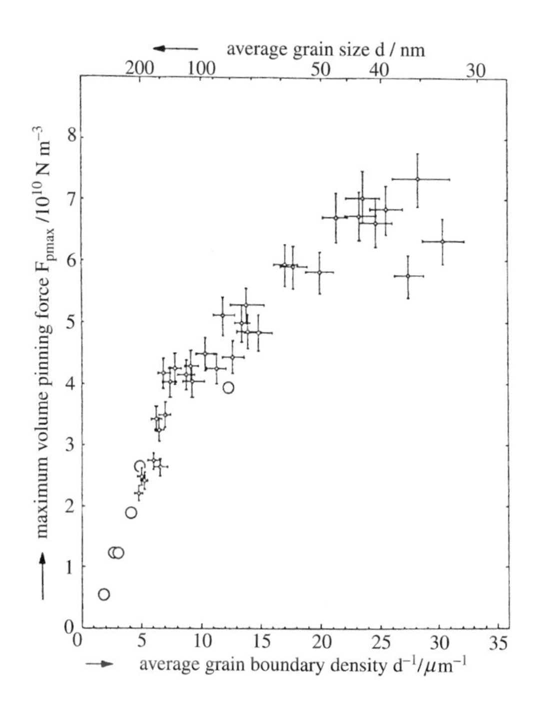

Figure B3.3.3.24. Dependence of F_{pmax} on inverse of the grain size. Data with error bars are from [50] and other open circles are from [51].

content in bronze is shown [54]. In the region of b over ~0.8, each F_p versus b curves overlap. The reason for these saturation properties can be explained by flux lattice shearing [55]. A4.3

Figure B3.3.3.27 is a schematic illustration of the relation between (*a*) pinning force and the reduced magnetic field and (*b*) the J_c and the magnetic field, for Nb_3Sn. In low or intermediate fields, J_c increases with decreasing grain size as shown in figure B3.3.3.27(*a*). On the other hand, in a high field region, F_p versus b curves show saturated properties and fine grains do not always correspond to high J_c values. In this region, J_c is increased with B_{c2} as shown in figure B3.3.3.27(*b*). B_{c2} is increased with the Sn content in the Nb_3Sn layer if the martensitic transformation is suppressed by a small amount of element addition or other appropriate atomic disorder as mentioned earlier. Therefore, to improve the J_c in a

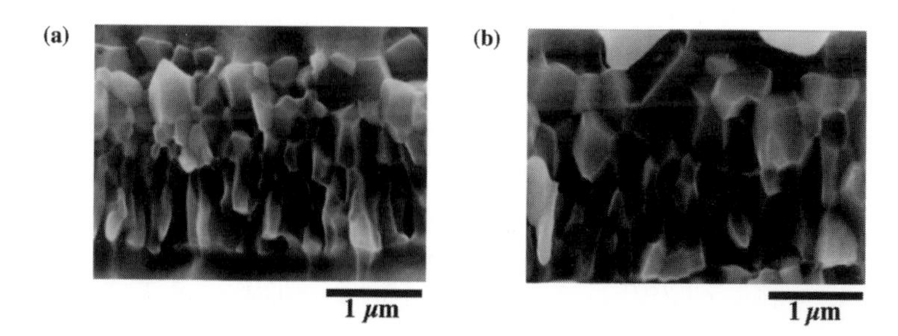

Figure B3.3.3.25. Micrographs of the reacted layer in the bronze processed Nb_3Sn wire heated at (*a*) 650°C and (*b*) 720°C for 150 h [52].

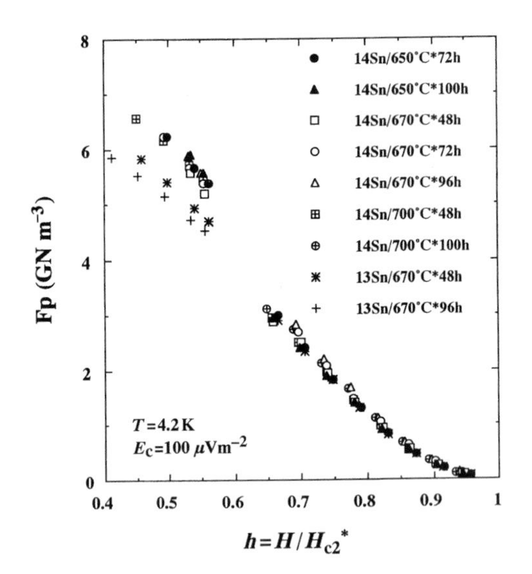

Figure B3.3.3.26. An example of the saturated characteristics for the bronze processed wire heated in different conditions and with different Sn content in bronze matrix [54].

high field region near B_{c2}, heat treatments with longer duration or at higher temperature are preferred for wire with three or four added elements.

The dependence of J_c on filament diameter in various magnetic fields is shown in figure B3.3.3.28. Though a fine filament could realize a high J_c in a low field, J_c increases with the filament diameter at high fields. The observed dependence of J_c on filament diameter is due either to high magnetic force or to less sausaging in the filament. More investigation is needed to clarify the cause of dependence of J_c on filament diameter at each field region.

B3.1

In many cases, the size or the scale of the wire affects very seriously its workability when fabricating continuous and uniform conductors. Even if very high J_c is reported for experimental wire, there is no guarantee that the same value will be achieved for the wire at industrial lengths. The piece length or weight of the wire has to be checked continuously.

Generally, each wire has a different Cu ratio so thermal pre-stress on the Nb_3Sn should be different and will bring about differences in J_c as described in detail in the next section.

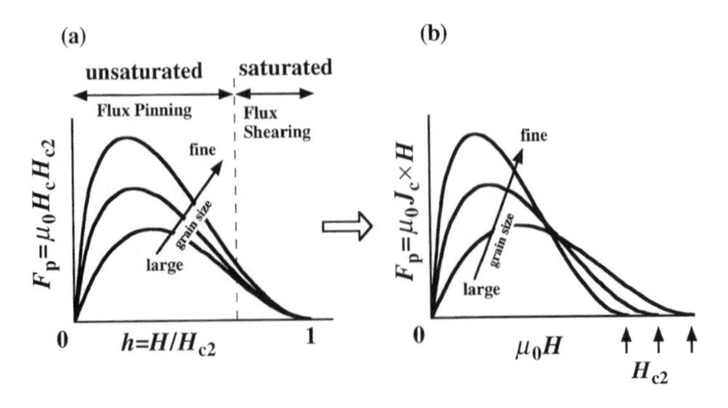

Figure B3.3.3.27. Schematic figures explaining the relation between pinning properties and J_c dependence on the magnetic field.

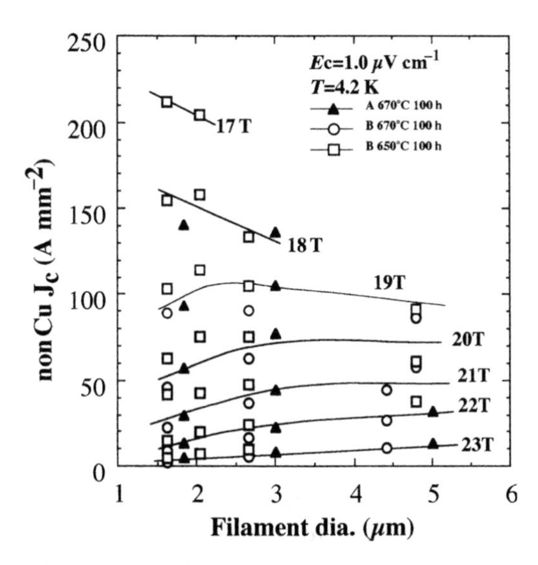

Figure B3.3.3.28. Dependence of J_c on filament diameter in applied magnetic field.

B3.3.3.5.1 Typical J_cs in Nb_3Sn wires prepared by several fabrication routes

Typical dependence of J_c on the magnetic field for Nb_3Sn wires processed in several ways are introduced for reference as follows:

Bronze route

An improvement of J_c in the bronze route produced Nb_3Sn wire by using 15 wt.% Sn bronze has been reported [25, 56]. Figure B3.3.3.29 shows typical J_c–B curves of developed wire compared to wire with

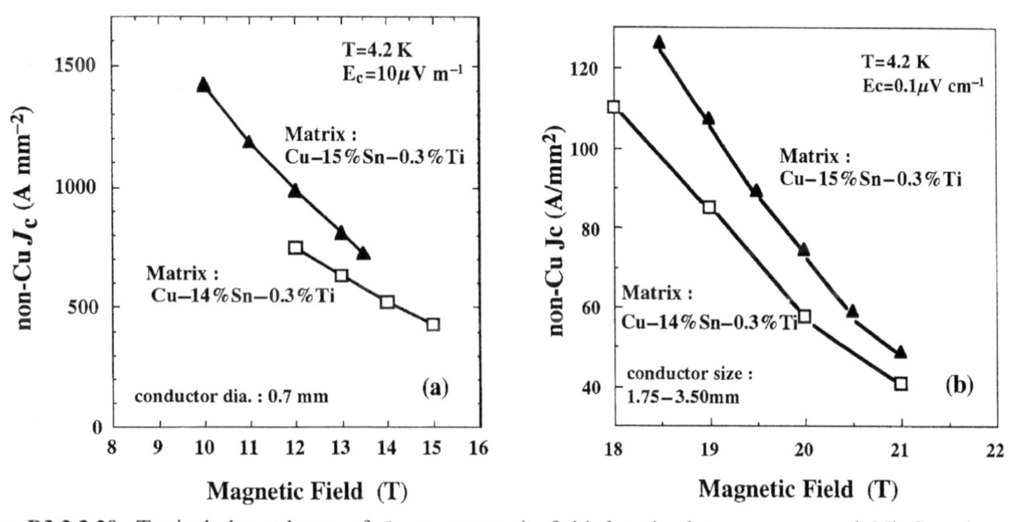

Figure B3.3.3.29. Typical dependence of J_c on magnetic field for the bronze processed Nb_3Sn wires: (*a*) in intermediate magnetic field region [56], (*b*) in high field region.

14 wt.% Sn bronze. Improvement in J_c of about 30% and 980 A mm^{-2} at 12 T has been achieved in the wire with a diameter of 0.7 mm as shown in figure B3.3.3.29(*a*). Improvement in J_c in the high field region has also been achieved in wire with a rectangular cross section of $1.75 \times 3.50\,mm^2$ as shown in figure B3.3.3.29(*b*).

The main reason for the improvement in J_c, by increasing the Sn content in the bronze matrix, is thought to be the decrease of the Nb_3Sn grain size [56].

Internal Sn diffusion processing

Figure B3.3.3.30 shows typical dependencies of J_c on magnetic field for TTS and MJR processed Nb_3Sn conductors [32, 57]. The J_c of over 2000 A mm^{-2} at 12 T is achieved for the MJR processed Nb_3Sn wires. MJR wire shows a J_c of almost twice large as that for the TTS wire.

P/M processing

The J_c of the P/M processed wire increases with the volume fraction of Nb powder up to 50%. However, J_c is decreased as the volume fraction increases to over 50% as shown in figure B3.3.3.31 [58]. Foner *et al* pointed out that the cause of the dependence of J_c on Nb volume fraction over 50% is that excess Nb prevents Sn diffusion [58].

ECN technique

J_c of about 1700 A mm^{-2} for ECN processed wire using $NbSn_2$ powder as a starting material has been reported [59]. Tachikawa *et al* [42] reported on a new process using Ta–Sn powder in a Nb or Ta added Nb tube. J_c–B characteristics for the sample processed by this technique in a high magnetic field is shown in figure B3.3.3.32. J_c of the Nb_3Sn core reaches about 500 A mm^{-2} at 23 T and 4.2 K in the specimen with a Nb–3.3 at.%Ta sheath. They reported improvement in J_c by the addition of Ta to a Nb tube as shown in figure B3.3.3.32. However, in this report, the specimen contains no Cu stabilizer, so

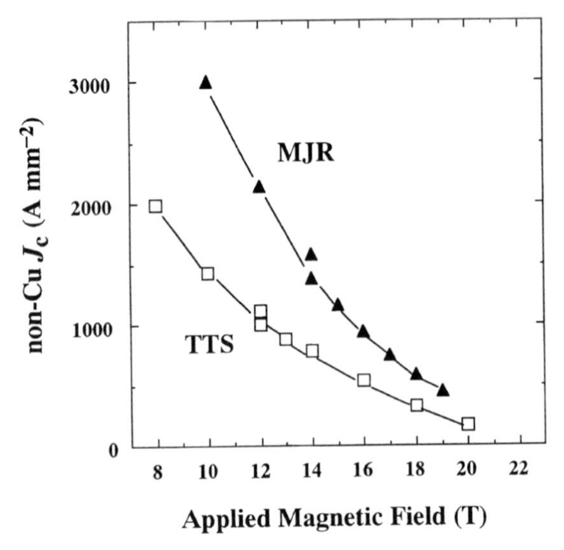

Figure B3.3.3.30. Typical dependence of J_c on the magnetic field for the Nb_3Sn wires by the internal tin process [32, 57].

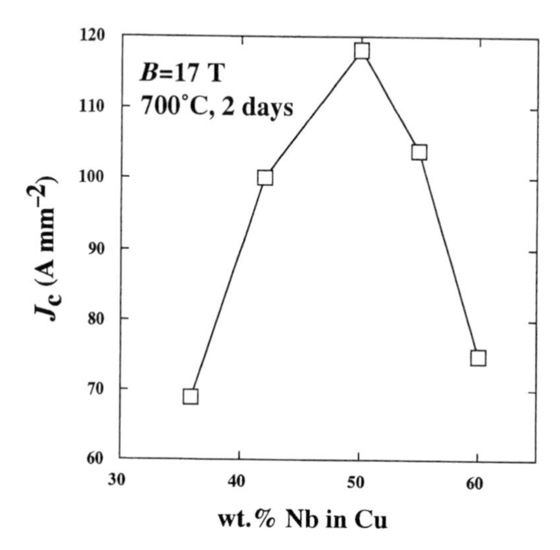

Figure B3.3.3.31. Dependence of J_c on Nb content for P/M processed Nb_3Sn wire [58].

the effect of prestress and decrease of heat treatment temperature, i.e. 925°C at present, have to be considered for future applications.

Besides these data, the reader can refer to J_c–B characteristics for other superconducting materials given at the website in [60].

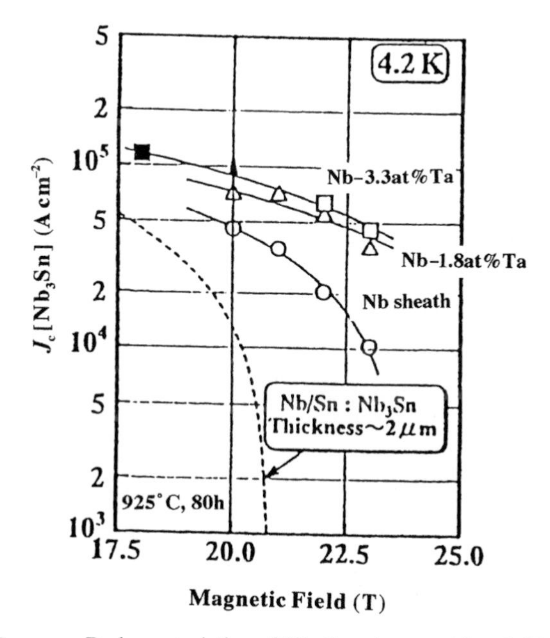

Figure B3.3.3.32. J_c versus B characteristics of Nb_3Sn wire considered (Tachikawa *et al* [42]).

B3.3.3.6 Mechanical properties of Nb_3Sn wire

In many cases, superconducting wires are applied under magnetic stress. Nb_3Sn is an intermetallic compound and known to be a very brittle material. Therefore, it is important to improve its tensile properties for high stress applications. In addition, J_c of Nb_3Sn wires is known to depend on the strain of the wire. This dependence has to be taken into consideration for the optimum design of magnets.

E1.1

B3.3.3.6.1 Tensile properties

To improve the mechanical properties of composite Nb_3Sn wire, several internal reinforcements have been investigated by using Al_2O_3 dispersed in copper [61], Cu–Nb alloy [62] and Ta [28]. When reinforcement by materials with good mechanical properties is considered, the following points should be considered.

(a) good workability as a composite wire consisting of a reinforcer and superconducting components;

(b) decrease in overall J_c due to incorporation of the non-superconducting reinforcer into the wire;

(c) degradation of mechanical properties of the wire due to the heat treatment for Nb_3Sn formation; and

(d) no shearing between reinforcer and other matrix materials under stress.

Recently, Ta reinforced Nb_3Sn wire has been developed for a 1 GHz (23.5 T) nuclear magnetic resonance magnet [28]. Figure B3.3.3.33 shows a cross section of a Ta reinforced Nb_3Sn wire. In this photograph, the Ta reinforcer is located in the center of the cross section. It is obvious that the round deformation of the Ta suggests good workability in the composite. The volume fraction of Ta is 11% and the other parameters can be found in [28].

Figure B3.3.3.34 shows stress versus strain curves for non-reinforced and Ta reinforced wires at 4.2 K [28]. In this figure, a marked improvement of the tensile strength of the wire is observed by reinforcement. The 0.2% yield strength at 4.2 K for Ta reinforced and non-reinforced Nb_3Sn wires are

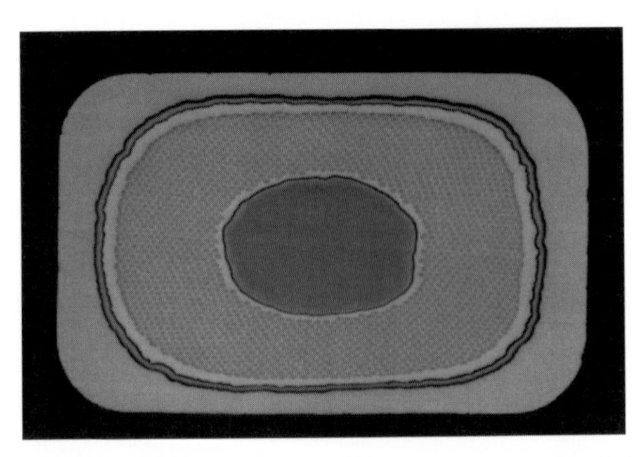

Figure B3.3.3.33. A cross sectional view of Ta reinforced Nb_3Sn wire (by Kobe Steel Ltd.) [28].

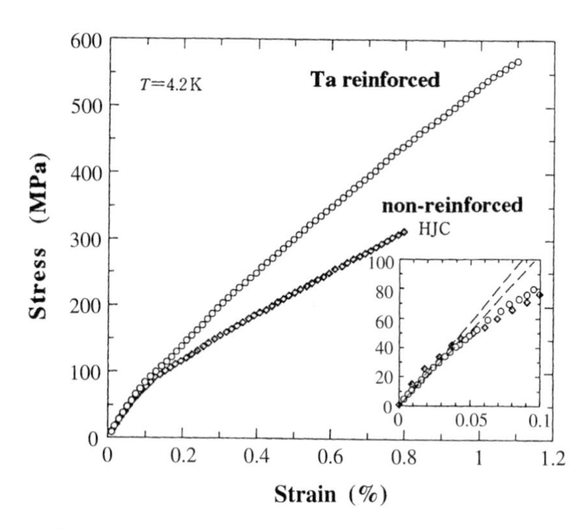

Figure B3.3.3.34. Comparison of stress–strain curves for Ta reinforced Nb_3Sn wire and non-Ta reinforced wire [28].

305 and 170 MPa, respectively. Because of the Ta melting point of 3000°C, softening by heat treatment for Nb_3Sn formation at about 700°C is very slight.

The rule of mixtures expressed as follows was investigated to confirm whether the shearing between reinforcer and matrix occurs.

$$\sigma(\varepsilon) = V_{Ta}\sigma_{Ta}(\varepsilon) + (1 - V_{Ta})\sigma_{non\text{-}Ta}(\varepsilon) \qquad \text{(B3.3.3.4)}$$

where V_{Ta} is volume fraction of Ta in the wire. $\sigma_{Ta}(\varepsilon)$ and $\sigma_{non\text{-}Ta}(\varepsilon)$ represent the dependence of stress on strain for Ta and a non-Ta matrix, respectively. $\sigma(\varepsilon)$ is a stress–strain curve for the Ta reinforced wire as a whole. Calculated results using equation (B3.3.3.4) are plotted in figure B3.3.3.35. In this figure, data of the stress–strain curve for a Ta core were measured using the Ta extracted from the wire by

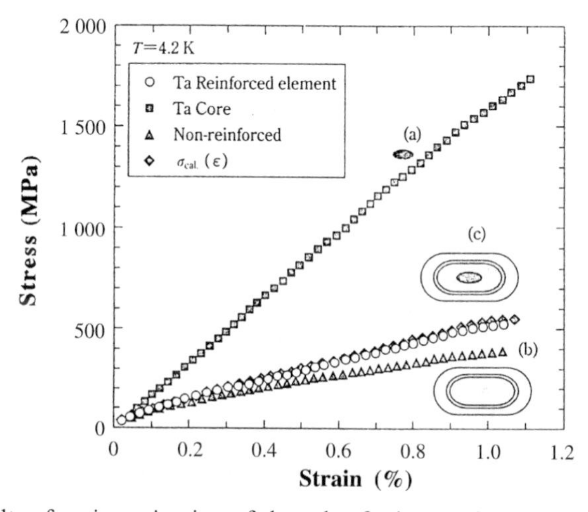

Figure B3.3.3.35. Results of an investigation of the rule of mixtures for Ta reinforced Nb_3Sn wire.

etching. The data are for a non-reinforced conductor acquired from a Nb_3Sn wire having almost the same Cu to non-Cu ratio. It is obvious that the stress–strain curve for reinforced wire is in good agreement with the result of calculation using equation (B3.3.3.4). This means that no shearing occurred between the Ta reinforcer and the matrix.

B3.3.3.6.2 Irreversible strain

Generally, J_c of Nb_3Sn wire depends on strain as described in the following section. The effect of strain on J_c is reversible in the low strain region, though there is a limit in strain for this reversibility. This means that cyclic energizing of the superconducting magnet must remain below the strain limit. This limitation in strain is called irreversible strain and is expressed by ε_{irr}. Irreversible strain is thus one of the most important factors for practical application at high fields. When the wire experiences strain over ε_{irr}, superconducting filaments are damaged intrinsically [63].

E1.1

A1.3

Typical values of ε_{irr} are given in table B3.3.3.2 for several types of Nb_3Sn wire [64]. In this table, intrinsic irreversible strain $\varepsilon_{0,irr}$ is described by

$$\varepsilon_{0,irr} = \varepsilon_{irr} - \varepsilon_m. \tag{B3.3.3.5}$$

Here, ε_m is the strain at which J_c shows a maximum value. For solid-filament bronze processed wire, the value of $\varepsilon_{0,irr}$ is constant at about 0.5%, regardless of the value of ε_m. Tubular-filament bronze processed wire and tape conductors are more susceptible to tensile fracture than solid filaments as evidenced by their significantly low value of $\varepsilon_{0,irr}$ as shown in table B3.3.3.2. In contrast, wire with finer filaments such as powder and in situ processed wires show greater intrinsic irreversible limits than those of tubular or solid filament wires. This increase in ε_{irr} is presumably a result of the fine filament size and increase in yield strength of the matrix located in interfilaments. From table B3.3.3.2, it is also obvious that cabling is effective in increasing the intrinsic reversible strain. In this case, overall stress on the cable can be relieved by shearing between each elementary wire.

A1.3

B3.3.3.6.3 Axial-strain effect on critical current

J_c of Nb_3Sn wire is very sensitive to strain. The dependence of J_c on both strain and magnetic field is described by the scaling law as follows [65].

Table B.3.3.3.2. ε_m and $\varepsilon_{0,irr}$ for several Nb_3Sn wires [64]

	Filament diameter (μm)	ε_m (%)	$\varepsilon_{0,irr}$ (%)
Tape	2–10	0.4–0.6	$\leq$0.2
Monofilament (>5 μm reaction layer)	>100	–	$\leq$0.2
Tubular filaments			
Internal bronze	11–26	0.2–0.4	0.2–0.25
Internal $NbSn_2$ powder	57	0.06	0.44
Bronze process, solid filaments	5–50	0.2–0.4	0.4–0.5
Powder process, infiltrated	0.8	0.13	0.9
In situ	0.4	0.5–0.6	0.9
Cable	3.5	0.24	0.9

$$\frac{I_c}{I_{cm}} = \left[\frac{B^*_{c2}(\varepsilon)}{(B^*_{c2m})}\right]^{n-p} \left[\frac{1 - B/B^*_{c2}(\varepsilon)}{(1 - B/B^*_{c2m})}\right]^q . \tag{B3.3.3.6}$$

Here $B^*_{c2}(\varepsilon)$ is the average upper critical field which depends on strain ε. The dependence of $B^*_{c2}(\varepsilon)$ on strain is expressed by:

$$B^*_{c2}(\varepsilon) = B^*_{c2m}(1 - a|\varepsilon_0|^u) \tag{B3.3.3.7}$$

where ε_0 is the intrinsic strain expressed by $\varepsilon_0 = \varepsilon - \varepsilon_m$.

The dependence of calculated I_c/I_{cm} on the intrinsic strain at each magnetic field is plotted in figure B3.3.3.36 [64]. For the Nb_3Sn conductor, the values of parameter a in equation (B3.3.3.6) are 900 for compressive strain ($\varepsilon_0 < 0$) and 1250 for tensile strain ($\varepsilon_0 > 0$). Other parameters for Nb_3Sn are indicated in figure B3.3.3.36. The effects of strain shown in figure B3.3.3.36 are much more pronounced at higher fields, leading to extreme strain dependence as B approaches B^*_{c2}. Typical dependencies of B^*_{c2} on the strain are shown in figure B3.3.3.37 [65]. It is pointed out that the dependence of the upper critical field or critical temperature on the elastic strain is an intrinsic property of the superconducting materials. Figure B3.3.3.37 shows a typical result of fitting the experimental data to equation (B3.3.3.7) [66]. Good agreement between the results of calculation and experiment, such as shown in figure B3.3.3.37, are reported not only for Nb_3Sn with ternary addition but also for other A15 type superconducting materials [66].

The origin of such dependence of I_c/I_{cm} on strain in compound wires is thought to be prestress on Nb_3Sn arising from the difference between thermal contraction of Nb_3Sn and other matrices such as bronze, copper and barrier layers [67–69]. In many cases, the reaction temperature is about 600–750°C, so that the wire experiences a temperature difference of about 1000 K when it operates at 4.2 K. For example, the difference between the relative thermal contraction of Nb_3Sn and bronze is about 1.0% [70]. It can be expressed as ε_m as a strain to relieve the prestress on Nb_3Sn by the difference of thermal contraction. Easton *et al* [68] explained this situation clearly and predicted ε_m by considering the elastic–plastic behavior of the stress–strain curves of each element in the wire, i.e. Nb_3Sn, Nb, Ta, copper and bronze.

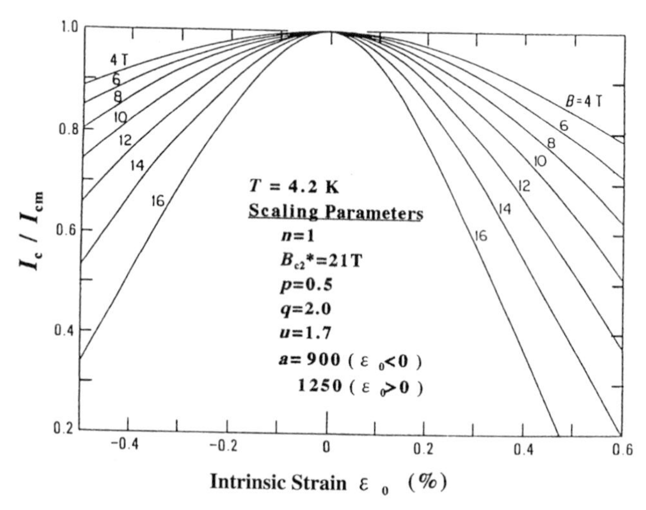

Figure B3.3.3.36. Dependence of normalized critical current on intrinsic strain calculated by the scaling law [64].

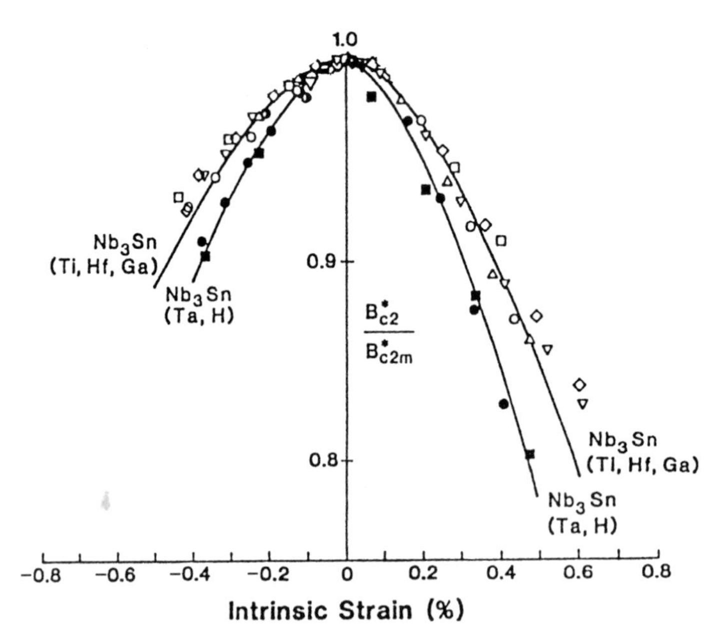

Figure B3.3.3.37. Dependence of normalized upper critical field on intrinsic strain [66].

B3.3.3.6.4 Transverse stress effect

Another important mechanical property is the dependence of J_c on transverse stress [66]. Figure B3.3.3.38 shows a typical relationship between J_c and compressive stress applied transversely to a bronze-processed Nb_3Sn conductor [71]. In a discussion for mechanical properties in the transverse direction, data are plotted by stress rather than strain because of the difficulty in measuring the transverse displacement over a small dimension of the filament diameter. In figure B3.3.3.38, J_c decreases

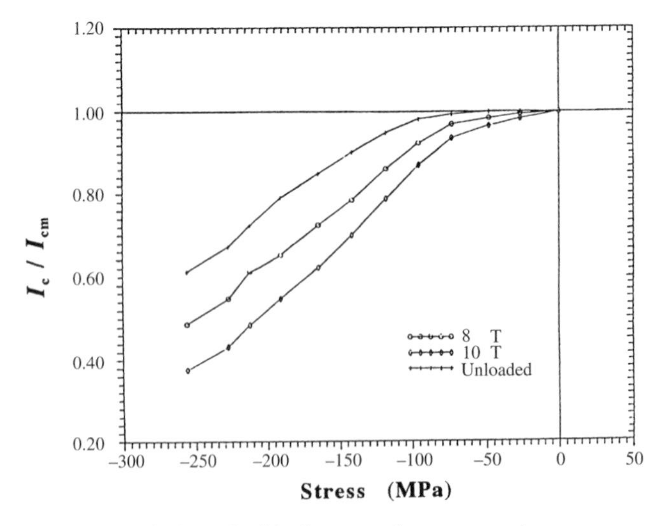

Figure B3.3.3.38. Degradation of critical current by compressive transverse stress [71].

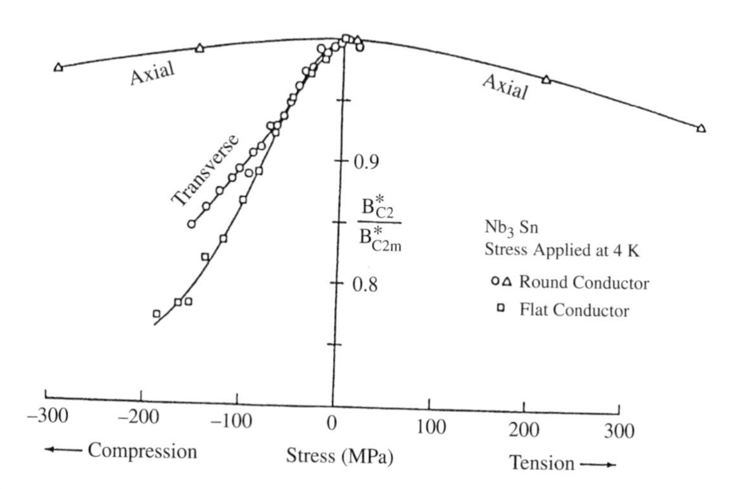

Figure B3.3.3.39. Comparison of dependence of upper critical field on axial and transverse stress [63].

monotonically with compressive transverse stress and degradation of critical current by transverse stress is steeper than that by axial stress. Generally, it is pointed out that the level of stress which causes a given amount of I_c degradation at 10 T is about seven times lower for transverse stress than that for axial stress. I_c degradation by transverse stress is reversible in low stress regions with a maximum between 50 and 100 MPa. Even in this low stress region, I_c recovery is thought to be incomplete because several filaments are still compressed by the plastically deformed matrix.

Ekin *et al* [70] pointed out that the transverse stress effect changes J_c mainly through changes in the upper critical field estimated by the Kramer approximation, B^*_{c2}, similar to the axial stress effect. Figure B3.3.3.39 shows a comparison of the axial stress and transverse on B^*_{c2}. It is shown that the effect of transverse stress on B^*_{c2} is much greater than that of axial stress. For a given B^*_{c2} degradation, the transverse stress is generally about 10 times less than that of the axial stress. The origin of greater sensitivity in transverse stress than axial stress is still not clear at present [63]. A3.1

B3.3.3.7 Summary

In this review, very basic properties such as phase diagrams, manufacturing techniques, T_c, H_{c2}, J_c and mechanical properties for the compound Nb_3Sn, were introduced. The Nb–Sn binary system has an advantage in the phase diagram because its A15 phase can be created under equilibrium conditions. Cu addition decreases the heat treatment temperature from over 930°C to under 700°C for the creation of single phase material. Ternary element addition, e.g. Ta or Ti is very effective to improve the superconducting properties of Nb_3Sn, H_{c2}, J_c in high field. Several kinds of manufacturing processes were introduced. In many of these processes, the above mentioned techniques using Cu and ternary addition are adapted. To improve non-Cu J_c, an increase in thickness of reacted layer and decrease in grain size are required. In the high field region, the influence of H_{c2} has to be considered. Improvements in mechanical strength of Nb_3Sn multi-filamentary wire by using high strength materials were introduced. The wire using Ta as a reinforcing material showed a yield strength of over 300 MPa at 4.2 K. Finally, the dependence of J_c on transverse and radial strain was reviewed together with a scaling law. When Nb_3Sn wire is used in magnets, the basic properties described here have to be considered in addition to processing technology to produce good conductors appropriate to each application. E1.1

Acknowledgment

We thank Dr. Y. Kubo, Mitsubishi Electric Co. for the photograph of the cross section of the internal tin processed Nb_3Sn wire for the ITER project in figure B3.3.3.13 and also Dr. K. Goto, Fujikura Ltd., for photographs of the in-situ processed wire in figures B3.3.3.20 and B3.3.3.21. Figures B3.3.3.16 and B3.3.3.18 are parts of results of the work in Kobe Steel Ltd. under the New Sunshine Program of AIST, MITI being consigned by NEDO.

References

[1] Suenaga M 1980 Metallurgy of continuous filamentary A15 superconductors *Superconductor Materials and Science: Metallurgy, Fabrication and Applications*, ed S Forner and B B Schwartz (New York: Plenum Press) p 213
[2] Hopkins R H, Roland G W and Daniel M R 1977 Phase relations and diffusion layer formation in the system Cu–Nb–Sn and Cu–Nb–Ge *J. Metall. Trans.* **8A** 91
[3] Flukiger R, Schauer W, Specking W, Schmit B and Springer E 1981 Low temperature phase transformation in Nb_3Sn multifilamentary wires and the strain dependence of their critical current density *IEEE Trans. Magn.* **17** 2285
[4] Shirane G and Axe J D 1971 Neutron scattering study of the lattice-dynamical phase transition in Nb_3Sn *Phys. Rev.* B **4** 2957
[5] Snead C L Jr., Kumakura H and Suenaga M 1983 Effect of disorder on the martensitic phase transformation in Nb_3Sn *Appl. Phys. Lett.* **43** 311
[6] LeHuy H, Bussiere J F and Berry B S 1983 Effect of interstitial hydrogen on Young's modulus and the martensitic transformation of Nb_3Sn *IEEE Trans. Magn.* **19** 893
[7] Orlando T P, McNiff E J Jr., Foner S and Beasley M R 1979 Critical fields, Pauli parametric limiting, and material parameters of Nb_3Sn and V_3Si *Phys. Rev.* B **19** 4545
[8] Beasley M R 1982 New perspectives on the physics of high-field superconductors *Adv. Cryog. Eng. Mater.* **28** 345
[9] Foner S and McNiff E J Jr. 1981 Upper critical fields of cubic and tetragonal single crystal and polycrystalline Nb_3Sn in DC fields to 30 tesla *Solid State Commun.* **39** 959
[10] Flukiger R, Schauer W and Goldacker W 1982 *Proc. 4th. Conf. on Superconductivity in d- and f- band metals* **41**
[11] Toyota N 1984 H_{c2} in A15 superconductors *Japan. J. Cryo. Eng.* **19** 319 (originaly from Suenaga M 1983 *US-Japan workshop for high field superconductors*)
[12] Sekine H, Iijima Y, Itch K, Tachikawa K, Tanaka Y and Furuto Y 1983 Improvements in current-carrying capacities of Nb_3Sn composites in high fields through titanium addition to the matrix *IEEE Trans. Magn.* **19** 1429
[13] Suenaga M, Tsuchiya K and Higuchi N 1984 Superconducting critical current density of bronze processed pure and alloyed Nb_3Sn at very high magnetic field (up to 24 T) *Appl. Phys. Lett.* **44** 919
[14] Tachikawa K, Sekine H and Iijima Y 1982 Composite-processed Nb_3Sn with titanium addition to the matrix *J. Appl. Phys.* **53** 5354
[15] Suenaga M, Aihara K, Kaiho K and Luhman T S 1979 Superconducting properties of $(Nb, Ta)_3Sn$ wires fabricated by the bronze process *Adv. Cryog. Eng.* **26** 442
[16] Miyazaki T, Fukumoto Y, Matsukura N, Miyatake T, Shimada M and Kobayashi N 1996 Improvement of critical current density of bronze-processed Nb_3Sn superconducting wire *Adv. Cryog. Eng.* **42** 1385
[17] Hirose R, Kamikado T, Ozaki O, Yoshikawa M, Hase T, Shimada M, Kawate Y, Takabatake K, Kosuge M, Kiyoshi T, Inoue K and Wada H 1998 21.7 T Superconducting magnet using $(Nb, Ti)_3Sn$ conductor with 14%-Sn bronze *Proc. 15th Int. Conf. Magnet Technology* 874
[18] Iwaki G, Sakai S, Kamata K, Sasaki K, Inaba S, Moriai H and Yoshida K 1993 Development of bronze-processed $(Nb Ti)_3Sn$ superconducting wires for central solenoid model coil of ITER *IEEE Trans. Appl. Supercond.* **3** 998
[19] Sakamoto H, Yamada K, Yamada N, Tanaka Y and Ando T 1993 Properties of bronze-processed multifilamentary Nb_3Sn wires for fusion experimental reactor *IEEE Trans. Appl. Supercond.* **3** 994
[20] Matsukura N, Fukumoto Y, Miyazaki T, Inoue Y, Miyatake T, Shimada M, Ogawa R and Kurahashi H 1996 Effect of tantalum addition on hysteresis losses and critical current densities of powder-metallurgy processed Nb_3Sn superconducting wires *Adv. Cryog. Eng.* **42** 1345
[21] Sakai S, Miyashita K, Kamata K, Endoh K, Tachikawa K and Tanaka H 1993 Critical current densities and magnetic hysteresis losses in submicron filament bronze-processed Nb_3Sn wires *IEEE Trans. Appl. Supercond.* **3** 990
[22] Tachikawa K 1970 *Proc. Int. Conf. on Cryog. Eng. Berlin* Iliffe Science and Technical Publications p 339
[23] Kaufman A R and Pickett J J 1970 *Bull. Am. Soc.* **15** 883
[24] Dew-Hughes D 1979 Physical metallurgy of A15 compounds *Treaties on Materials Science and Technology* **vol. 14** ed T S Luhman and D Dew-Hughes (New York: Academic) p 137
[25] Miyazaki T, Matsukura N, Miyatake T, Shmiada M, Takabatake K, Itoh K, Kiyoshi T, Sato A, Inoue K and Wada H 1998 Improvement of critical current density in the bronze-processed Nb_3Sn superconductor *Adv. Cryog. Eng.* **44** 943

[26] Miyazaki T, Murakami Y, Hase T, Shimada M, Itoh K, Kiyoshi T, Takeuchi T, Inoue K and Wada H 1999 Development of Nb_3Sn superconductors for a 1 GHz NMR magnet — Dependence of high-field characteristics on tin content in bronze matrix *IEEE Trans. Appl. Supercond.* **9** 2500

[27] Hansen P M 1958 *Constitution of binary alloys* (New York: McGraw-Hill) p 633

[28] Miyazaki T, Matsukura N, Miyatake T, Shimada M, Takabatake K, Itoh K, Kiyoshi T, Sato A, Inoue K and Wada H 1998 Development of bronze-processed Nb_3Sn superconductors for 1 GHz NMR magnets *Adv. Cryog. Eng.* **44** 935

[29] Hashimoto Y, Yoshizaki K and Tanaka M 1974 Processing and properties of superconducting Nb_3Sn filamentary wire *Proc. ICEC-5* p 332

[30] Egawa K, Kubo Y, Nagai T, Wakata M, Uchikawa F, Taguchi O, Wakamoto K, Morita M, Isono T, Nunoya Y, Yoshida K, Nishi M and Tsuji H 1997 Mass production of Nb_3Sn strands by internal tin diffusion process for the ITER project *Japan. J. Cryo. Eng.* **32** 173 (in Japanese)

[31] Zeitlin B A, Ozeryansky G M and Hamachalam K 1985 An overview of the IGC internal tin Nb_3Sn conductor *IEEE Trans. Magn.* **21** 293

[32] Ozeryansky G M and Gregory E 1991 A new internal tin Nb_3Sn conductor made by a novel manufacturing process *IEEE Trans. Magn.* **27** 1755

[33] Murase S, Koizumi O, Shiraki H, Koike Y, Suzuki E, Ichihara M, Nakane F and Aoki N 1979 Multifilamentary niobium-tin conductors *IEEE Trans. Magn.* **15** 83

[34] McDonald W K, Curtis C W, Scanlan R M, Larbalestier D C, Marken K and Smathers D B 1983 Manufacture and evaluation of Nb_3Sn conductors fabricated by the MJR method *IEEE Trans. Magn.* **19** 1124

[35] Matsukura N, Miyazaki T, Miyatake T, Shimada M and Chiba M 1997 Effect of niobium concentration on the critical current density and hysteresis loss of powder-metallurgy processed Nb_3Sn superconducting wires *J. Japan. Inst. Metal.* **61** 807 (in Japanese)

[36] Thieme C L H and Foner S 1993 Improved superconducting Nb_3Sn wire using Nb (Ti), Sn (Ga), Cu, and Ag powders *IEEE Trans. Appl. Supercond.* **3** 1326

[37] Pourrahimi S, Thieme C L H and Forner S 1987 21 Tesla powder metallurgy processed Nb_3Sn (Ti) using (Nb–1.2 wt%Ti) powders *IEEE Trans. Magn.* **23** 661

[38] Gauss S and Flukiger R 1988 PM preparation of Nb–Ta composite rods for use in $(NbTa)_3Sn$ multifilamentary superconducting wire *Met. Powder Rep.* **43** 41

[39] van Beijnen C M A and Elen J D 1979 Multifilament Nb_3Sn superconductors produced by the E.C.N. technique *IEEE Trans. Mag.* **15** 87

[40] Neijmeijer W L and Kolster B H 1990 Characteristics of a production route for filamentary Nb_3Sn superconductor based on a reaction between Nb and Nb_6Sn_5 *J. Less-Common Met.* **160** 161

[41] Tachikawa K, Natsuume M, Tomori H and Kuroda Y 1997 High-field superconductors prepared through a new route *Adv. Cryog. Eng.* **42** 1359

[42] Tachikawa K, Yamamoto T, Yokoyama T and Kato T 1999 New high-field Nb_3Sn conductors prepared from Ta–Sn compound powder *IEEE Trans. Appl. Supercond.* **9** 2500

[43] Pickus M R, Holthuis J T and Rosen M 1980 A15 multifilamentary superconductors by the infiltration process *Filamentary A15 Superconductors*, ed M Suenaga and A F Clark (New York: Plenum) p 331

[44] Roberge R and Foner S 1980 In situ and powder metallurgy multifilament superconductors: fabrication and properties *Filamentary A15 Superconductors*, ed M Suenaga and A F Clark (New York and London: Plenum) p 241

[45] Finnemore D K, Verhoeven J D, Gibson E D and Ostenson J E 1980 Preparation and properties of in situ prepared filamentary Nb_3Sn–Cu superconducting wire *Filamentary A15 Superconductors*, ed M Suenaga and A F Clark (New York and London: Plenum) p 259

[46] Bevk J and Tinkham M 1980 Superconducting properties and coupling mechanisms in in situ filamentary composites *Filamentary A15 Superconductors*, ed M Suenaga and A F Clark (New York and London: Plenum) p 271

[47] Quincey P G and Dew-Hughes D 1987 Effect of titanium additions to in-situ Nb_3Sn wire *IEEE Trans. Magn.* **23** 633

[48] Wu I W, Dietderich D R, Holthuis J T, Hong M, Hassenzahl W V and Morris J W Jr. 1983 The microstructure and critical current characteristics of a bronze-processed multifilamentary Nb_3Sn superconducting wire *J. Appl. Phys.* **54** 7139

[49] Livingston J D 1977 Grain size in A-15 reaction layers *Phys. Stat. Sol.* **44** 295

[50] Schauer W and Schelb W 1981 Improvement of Nb_3Sn high field critical current by Nb_3Sn two-stage reaction *IEEE Trans. Magn.* **17** 374

[51] Scanlan R M, Fietz W A and Koch E F 1975 Flux pinning centers in superconducting Nb_3Sn *J. Appl. Phys.* **46** 2244

[52] Suenaga M and Jansen W 1983 Chemical compositions at and near the grain boundaries in bronze-processed superconducting Nb_3Sn *Appl. Phys. Lett.* **43** 791

[53] Kramer E J 1975 Microstructure–critical current relationships in hard superconductors *J. Electron. Mater.* **4** 839

[54] Miyazaki T, Matsukura N, Miyatake T, Shimada M, Kurahashi H, Tatara I and Kobayashi N 1997 *Annual report of Institute for Materials Research in Tohoku Univ.* 120 (in Japanese).

[55] Kramer E J 1973 Scaling laws for flux pinning in hard superconductors *J. Appl. Phys.* **44** 1360

[56] Zhang Y, McKinnell J C, Hentges R W and Hong S 1999 Recent development of niobium–tin superconducting wire at OST *IEEE Trans. Appl. Supercond.* **9** 1444

[57] Pourrahimi S, Thieme C L H, Schwartz B B and Foner S 1985 Powder metallurgy processed Nb_3Sn employing extrusion and varying Nb content *IEEE Trans. Magn.* **21** 764
[58] Hornsveld E M and Elen J D 1990 Development of niobium–tin conductors at ECN *Adv. Cryog. Eng.* **36** 157
[59] Lee P J http: //www.engr.edu/centers/asc/ascimages/plots/index.html
[60] Murase S, Nakayama S, Koyanagi K, Masegi T, Nomura S, Urata M, Shimamura K, Amano K, Shiga N, Watanabe K and Kobayashi N 1997 Development of the Alumina-Copper Reinforced Nb_3Sn Wire for Coil Fabrication *Proc. ICEC 16/ICMC* 1707.
[61] Iwasaki S, Goto K, Sadakata N, Saito T, Kohno O, Awaji S and Watanabe K 1997 Mechanical and Superconducting Properties of Multifilamentary Nb_3Sn Wire with CuNb Reinforced Stabilizer *Proc. ICEC 16/ICMC* 1723.
[62] Matsushita T and Ekin J W 1994 Superconducting properties *Composite superconductors*, ed K Osamura (New York: Dekker) p 93
[63] Ekin J W 1983 Strain effects in superconducting compounds *Adv. Cryog. Eng.* **30** 823
[64] Ekin J W 1980 Strain scaling law for flux pinning in practical superconductors. Part 1: Basic relationship and application to Nb_3Sn conductors *Cryogenics* **20** 611
[65] Matsushita T and Ekin J W 1994 Superconducting properties *Composite superconductors*, ed K Osamura (New York: Dekker) p 95
[66] Rupp G 1978 Effect of bronze on the compression of Nb_3Sn in multifilamentary conductors *Cryogenics* **18** 663
[67] Easton D S, Kroeger D M, Specking W and Koch C C 1980 A prediction of the stress state in Nb_3Sn superconducting composites *J. Appl. Phys.* **51** 2748
[68] Ekin J W 1980 Mechanical properties and strain effects in superconductors *Superconductor Materials and Science: Metallurgy, fabrication and Applications*, ed S Forner and B B Schwartz (New York: Plenum) p 455
[69] Rupp G 1980 The importance of being prestressed *Filamentary A15 Superconductors*, ed M Suenaga and A F Clark (New York: Plenum Press) p 155
[70] Ekin J W, Bray S L, Danielson P, Smathers D, Sabatini R L and Suenaga M 1989 Transverse stress effect on the critical current of internal tin and bronze processed Nb_3Sn superconductors *Proc. 6th Japan/U.S. Workshop on High Field Superconductors* p 50
[71] Ekin J W 1987 Effect of transverse compressive stress on the critical current and upper critical field on Nb_3Sn *J. Appl. Phys.* **62** 4829

B3.3.4 Processing of low T_c conductors: the compound Nb_3Al

Takao Takeuchi

B3.3.4.1 Introduction

Nb_3Al may be considered as an alternative to Nb_3Sn for high field and large scale magnet applications, such as nuclear fusion, high energy particle accelerator and GHz class nuclear magnetic resonance analysis, since it demonstrates extremely high critical current densities J_c in high fields and has an excellent tolerance to mechanical strains. However, a commercially available fabrication process has not been established yet, unlike the bronze process for Nb_3Sn. The formation of unwanted ternary compounds including Cu prevents application of the bronze process to Nb_3Al conductors. Furthermore, the deviation of Nb_3Al phase compositions from A15 stoichiometry at temperatures lower than 1900°C leads to difficulties in developing the fabrication process to yield excellent high field properties. The processes are roughly classified into three groups by the reaction temperature. (a) A great deal of effort has been devoted to low temperature processes [jelly-roll (JR), powder metallurgy (PM), clad chip extrusion (CCE), rod-in-tube (RIT)], in which the microscale assembly of the elemental constituents Nb/Al is directly diffusion reacted at temperatures less than 1000°C to suppress Nb_3Al grain coarsening. An Nb_3Al cable-in-conduit (CIC) conductor with a current capacity of 60 kA at 12.5 T has been fabricated using the JR process for nuclear fusion magnets. Even if the diffusion spacing between Nb and Al of less than 0.1 μm is employed to complete the reaction in a short time, the Nb_3Al phase composition still remains nonstoichiometric and, thus, J_c rapidly decreases in high fields. (b) Straightforward diffusion reaction of Nb/Al composites at high temperature (>1800°C), produced by laser beam irradiation, improves the high field properties, but results at the same time in low J_c at low fields due to grain coarsening of the Nb_3Al phase. (c) Rapid quenching from high temperature and subsequent phase transformation from bcc supersaturated solid solution to the A15 phase, however, solves the above difficulty. Stoichiometric Nb_3Al can be formed with a fine grain structure and, hence, high J_c over a wide magnetic field range, comprising high fields, with $B > 20$ T. The excellent strain tolerance of Nb_3Al conductors with respect to Nb_3Sn and the improved high field performance are very promising properties of Nb_3Al wires. The properties of Nb and Nb alloy superconductors are described in chapter B3.3.

B3.3.3
E1.1
B2.0
B3.3.3
B2.5
B3.3.3

B3.3.4.2 Phase diagram

Nb_3Al belongs to a group of superconductors that have the A15 structure. The highest reported values of critical temperature T_c and upper critical magnetic field $\mu_0 H_{c2}$ (B_{c2}) at 4.2 K are 18.6 K and 32 T, respectively, both being higher than those of Nb_3Sn. This is why Nb_3Al is promising as an alternative to Nb_3Sn for high field applications. The phase diagram of the Nb–Al system [figure B3.3.4.1(*a*)][1]

B2.5
B3.3.3

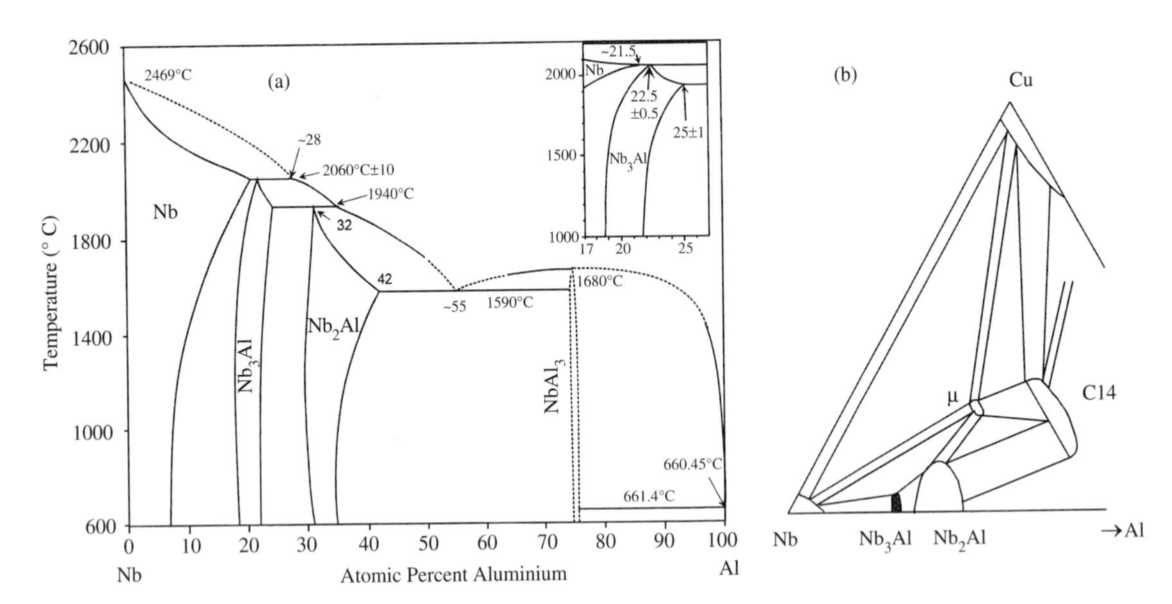

Figure B3.3.4.1. (*a*) Nb–Al binary phase diagram [1], (*b*) part of a section of the Nb–Cu–Al ternary phase diagram at 1000°C [2].

B3.3.3 has a common feature with the Nb–Sn system: the A15 phase is formed by a peritectic reaction, bcc + liquid → A15, at 2060°C. The bcc solid solution phase $Nb(Al)_{ss}$ is extensive: the solubility limit of Al is 21.5 at% at 2060°C and 9 at% up to 1000°C. The phase limit of the A15 phase strongly depends on temperature. For the Nb-rich side the Al content varies from ≈ 19 at% at 1500°C to 22.5 A1.3 at% at the peritectic temperature. The Al-rich limit of the A15 phase includes the stoichiometric composition at the second peritectic temperature (1940°C). However, the Al-rich phase limit is shifted to lower Al concentrations with decreasing temperature, down to ~ 21.5 at% at 1000°C. This deviation of Nb_3Al phase compositions from A15 stoichiometry at low temperatures leads to difficulties when fabricating Nb_3Al conductors with good superconducting properties. It has been reported that the critical temperature T_c of a bulk sample with nominal composition $Nb_{0.75}Al_{0.25}$ after long annealing at temperatures below 1000°C is of the order of 12 K, corresponding to the T_c value obtained for the Al-rich limit of ~ 21 at% Al.

The ternary phase diagram of Nb–Al–Cu [figure B3.3.4.1(*b*)] [2] is, furthermore, different from B3.3.3 that of Nb–Sn–Cu. Thus, the 'bronze process', the established production route for fully stabilized Nb_3Sn multifilamentary wires (Nb filaments are drawn in Cu–Sn bronze matrix and reacted to produce B2 Nb_3Sn), is not available for Nb_3Al. Apart from stable Nb_2Al and $NbAl_3$ phases, additional phases appear in the centre region of the diagram: the μ and Laves (C14) phases. These unwanted ternary compounds block the tie-line routes between the Cu–Al bronze and the Nb solid solution via the A15 (Nb_3Al) phase. This is in contrast to the Cu–Sn–Nb phase diagram in which the A15 (Nb_3Sn) phase is the only stable relevant stable phase, other than the terminal Nb and Cu–Sn bronze. Replacement of Cu–Al bronze with Ag–Al solid solution renders these ternary compounds unstable and enables instead the formation of Nb_3Al together with Nb_2Al by the solid state diffusion reaction between Nb filaments and the Ag–Al solid solution matrix [3]. However, the resultant Nb_3Al shows an optimized T_c of 13.9 K, much less than that of stoichiometric Nb_3Al.

Thus, the achievement of stoichiometry is essential for producing high T_c and, hence, high B_{c2} A3.1 Nb_3Al conductors. A heat treatment at elevated temperatures above 1800°C with a commonly used vacuum furnace enables the stoichiometric Nb_3Al to form, while it trades off the high critical current density characteristics at low magnetic fields: the grain coarsening of Nb_3Al dramatically degrades J_c.

B3.3.4.3 Approach to stoichiometry

There are two conceptual routes to achieve the stoichiometry of Nb_3Al without grain coarsening, and numerous methods have been proposed, as listed in table B3.3.4.1. A great deal of effort has been devoted to the low temperature process (JR [4, 5], RIT [6, 7], CCE [8], PM [9, 10]), in which the microscale assembly of the elemental constituents Nb/Al (overall composition: Nb–25 at%Al) is directly reacted with diffusion at temperatures less than 1000°C to suppress the grain growth of Nb_3Al. As the diffusion distance between Nb and Al becomes small ($\sim$ 100 nm), the free energy of Nb_2Al increases faster than that of Nb_3Al because the surface/volume ratio is much larger for Nb_2Al layer [figure B3.3.4.2(*a*)]. Thus, the composition of the A15 phase in equilibrium with Nb_2Al phase shifts towards the stoichiometry, suppressing the formation of unwanted Nb_2Al [5, 7].

The phase formation sequence for the reaction of Nb and Al has been investigated with multilayer film samples [11]. As shown in figure B3.3.4.3, the first phase to form is $NbAl_3$, and then both Nb_2Al and Nb_3Al phases start to grow when all the Al has reacted to form $NbAl_3$. For the overall composition 25 at%Al the multilayer periodicity (diffusion distance) plays a role. In films of large periodicity, the phases present after long annealing treatments are Nb_2Al and Nb_3Al, as predicted by the equilibrium

Table B3.3.4.1. Fabrication process for Nb_3Al conductors

Group	Technique	Nb/Al spacing	React temp.	Formation of A15-type Nb_3Al	Grain size of Nb_3Al	T_c (K)	J_c (4.2 K, 10 T) A mm^{-2}	J_c (4.2 K, 21 T) A mm^{-2}	Length achievement m
I	PM	$< 0.1\ \mu m$	low $< 1000°C$	Straight-forward diffusion reaction	fine	$\cong 15.5$	1200	0	(no description)
	JR								4600
	CCE								(no description)
	RIT								30
II	Laser/electron beam heating	$< 10\ \mu m$	high $> 1800°C$		large	18.3	500	480	50
III	Liquid quench	–	$> 1900°C$ + rapid quench + $< 1000°C$	Transformation from bcc super-saturated solid solution $Nb(Al)_{ss}$	fine	17.8	> 3000	340	1
	RHQT	$< 1\ \mu m$							300

J_c is defined as the critical current divided by the Nb_3Al compound cross-section.

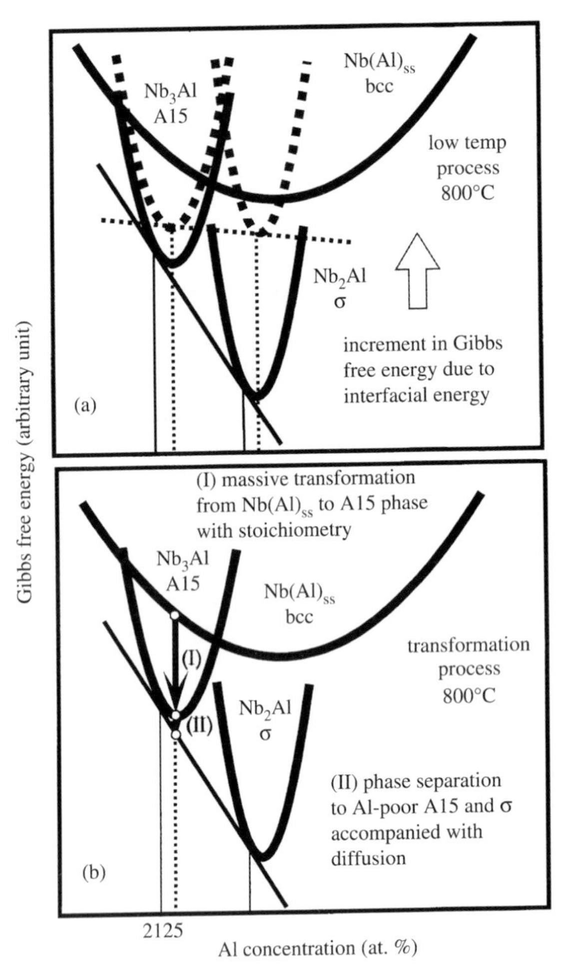

Figure B3.3.4.2. The concept of possible ways to the stoichiometric Nb_3Al formation [14]. (*a*) Direct diffusion reaction in enhanced interfacial energy state, (*b*) via bcc supersaturated solid solution route.

phase diagram. Only in the multilayers of small periodicity (143 nm) is the final product fully A15, supporting the shift of the A15 composition towards stoichiometry. For much thinner periodicity (40 nm), Bormann proposed a different phase formation sequence: the intermediate Nb/$NbAl_3$ was transformed to a bcc solid solution of Al in Nb by isothermal transformation at 700°C, with a subsequent extensive transformation to the A15 phase at higher temperatures [12].

The second route occurs via a supersaturated bcc solid solution $Nb(Al)_{ss}$, which is formed by rapid cooling ($>10^4$°C s^{-1}) from temperatures above 1900°C [13]. For an overall composition Nb–25 at%Al, a two-phase structure is finely distributed so as to minimize the free energy [figure B3.3.4.2(*b*)] [14]. This microstructure consists of Nb_2Al particles dispersed in an off-stoichiometric A15 phase matrix. Since the A15 and Nb_2Al phases have different compositions, the transition $Nb(Al)_{ss} \rightarrow A15 + Nb_2Al$ requires the redistribution of elements by long range diffusion. According to the step rule, the phase transition from metastable systems commonly keeps the free energy hierarchy. Thus, the $Nb(Al)_{ss} \rightarrow A15$ (stoichiometry) transformation could occur by a massive transformation, which requires only a change in crystal structure and not in local composition. The two-phase separation should lower the free energy

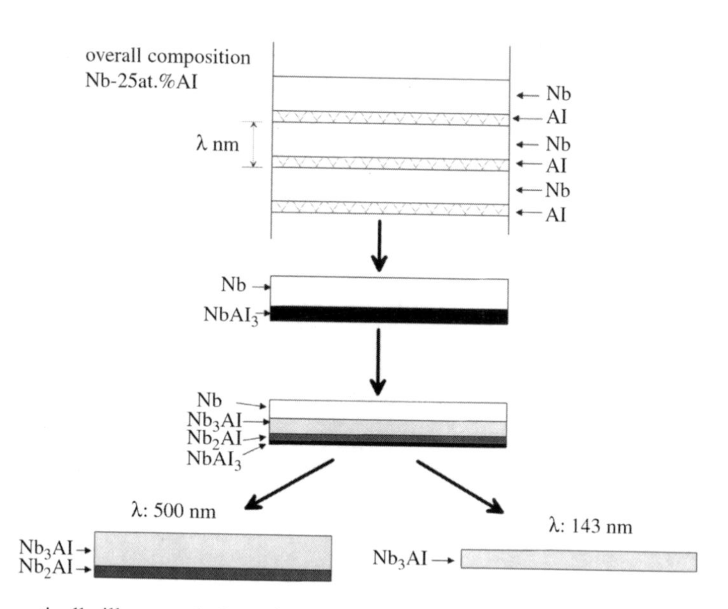

Figure B3.3.4.3. Schematically illustrated phase formation sequence for the reaction of Nb/Al multilayer thin films at less than 1000°C [11].

more than the massive transformation, but requires long range diffusion, which takes more time. Although the massive transformation decreases the free energy to a smaller extent, it will occur faster since no long range diffusion is required.

Straightforward diffusion reaction of Nb/Al composites at high temperature (> 1800°C) with laser beam irradiation [15] is a direct route to achieve stoichiometry, consistently with the equilibrium phase diagram. Higher cooling rates suppress the considerable grain growth.

B3.3.4.4 Fabrication process

The processing of A15 Nb_3Al superconducting strands is a two-step operation. The first step consists of making the final size strand, with the constituents in a fine state of subdivision, and the second step is the heat treatment that converts them into the desired A15 phase. The processes are roughly classified into three groups by the reaction temperature.

B3.3.4.4.1 Low temperature process (I)

In the manufacture of Nb_3Al, in contrast to Nb_3Sn, diffusion considerations require that the elemental constituents be assembled so that Al dimensions are less than 100 nm in the finished wire. As shown in figure B3.3.4.4, the JR, RIT, CCE and PM techniques enable the microscale assembly of the elemental constituents Nb/Al in order to complete the diffusion reaction in a short time and suppress the grain growth of A15 Nb_3Al. In the JR process, alternate foils of Nb and Al are wound, like a jelly roll, onto a Cu rod and inserted into holes drilled in a Cu matrix. The total reduction in area R_0 to achieve an Al layer thickness of ~100 nm is ~10^5. In the RIT process, an Al-alloy rod (core) is inserted into a Nb tube and drawn down. A triple stacking operation results in a final core diameter of ~100 nm. Since the RIT process requires much larger R_0 (~10^{10}), the relative hardness of the core and the matrix must be adjusted by alloying Al with additives Mg, Ag, Cu etc to improve the workability of the Nb/Al composite. In the CEC process, the three-layered Al/Nb/Al clad foil is cut into square chips, and placed

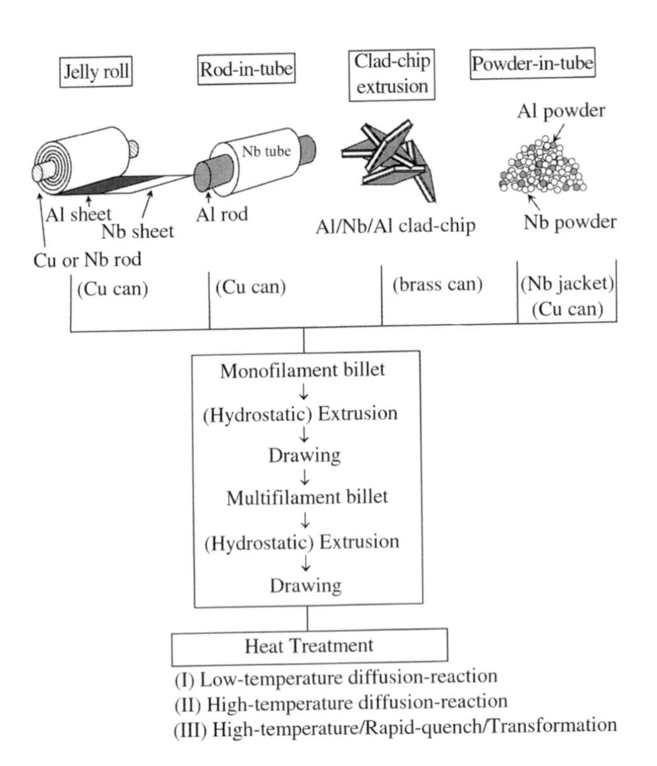

Figure B3.3.4.4. Various fabrication processes for Nb_3Al multifilamentary conductors.

randomly into a brass cannister to be extruded. In the PM process, a mixture of hydride-dehydride Nb powder ($< 40\,\mu m$) and Al powder ($< 9\,\mu m$) is put into a Cu (or CuBe) can. Then, the resulting composites are extruded and drawn down into a wire as the monofilament. A bundle of these packed into a Cu can and reduced to a wire gives an Al layer thickness (Al core diameter) of 100 nm. The resultant short diffusion distance between Nb and Al is one of the key parameters controlling the quality of the A15 phase to be formed at low temperatures less than 1000°C.

A4.3 Figure B3.3.4.5 shows T_c, B_{c2}, the maximum pinning force density F_{pmax}, lattice parameter, and the phase formation determined by x-ray diffraction as a function of Al core size, which is a parameter representing the diffusion spacing for the RIT Nb_3Al conductor. With decreasing Al core diameters in the range from 2.3 μm to about 50 nm, B_{c2} (4.2 K) and T_c increase and reach 21 T and 15.8 K, values that are much higher than those of the bulk Nb_3Al sample. This enhancement is accompanied by the development of A15 phase and a decrease in the lattice parameter of A15 phase, indicating the formation of the metastable A15 phase with near-stoichiometric composition. The peak F_{pmax} occurred at a smaller Al size than the peak in T_c and B_{c2} (4.2 K), since these latter two parameters are limited by proximity effect degradation, presumably due to the presence of some unreacted Nb. This in turn may be causing an increase in F_p by 'interface pinning' [16]. As demonstrated in figure B3.3.4.6, the low temperature processed JR Nb_3Al conductor shows much higher tolerance to axial strain and transverse compressive stress than the Nb_3Sn conductor [17, 18]. This is the main reason why the JR Nb_3Al conductor is promising as the candidate material for the International Thermonuclear Experimental Reactor (ITER) magnets, despite having a J_c value lower than that of Nb_3Sn (figure B3.3.4.7).

B2.5

B3.3.3

The Japan Atomic Energy Research Institute has developed the Cu-stabilized Nb_3Al CIC conductor with a current capacity of 60 kA at 12.5 T, using the JR process, for the Nb_3Al insert to be tested in the ITER central solenoid model coil [19]. A cable of 1152 ($3 \times 4 \times 4 \times 4 \times 6$) strands is encased in a stainless

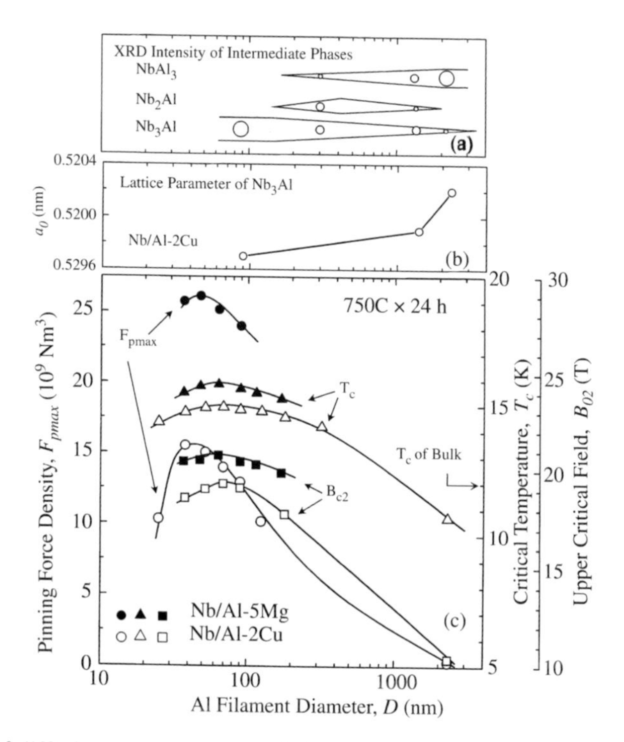

Figure B3.3.4.5. Effects of diffusion spacing between Nb and Al on microstructure and critical parameters of the RIT processed Nb_3Al [7].

steel (JN1HR) conduit as shown in figure B3.3.4.8. The strand has a J_c above 600 A mm^{-2} at 12 T under the bending strain of 0.4%, a hysteresis loss density lower than 410 mJ cc^{-1} in the alternative magnetic field of $\pm$ 3 T parallel to the wire axial direction and a residual resistivity lower than 1.6×10^{-10} Ωm. A large multifilament billet to be volumetrically drawn into a 16 km strand is available. The yield of 90% and 70% is obtained for the strand length > 0.5 and > 1.5 km, respectively [20]. A1.3

B3.3.4.4.2 High temperature process (II)

Even if the diffusion spacing between Nb and Al is made less than 0.1 μm to complete the reaction in a short time, off-stoichiometry of the Nb_3Al phase composition still remains in the processed wire and thus J_c rapidly decreases in high fields. Therefore, an attempt was made to carry out the straightforward diffusion reaction of Nb/Al composites at high temperature (>1800°C) with laser [15] and electron [21, 22] beam irradiation techniques. The Nb/Al composites for this technique have been prepared by PM and RIT processes. High power density beam irradiation on the long length Nb/Al composite, which moves from reel to reel at high speed in a vacuum chamber, enables a short time high-temperature reaction. This improves the high field properties as shown in figure B3.3.4.7. However, this trades off oppositely high J_c in low fields due to the unavoidable excess grain coarsening of the Nb_3Al phase.

B3.3.4.4.3 Transformation process (III)

For the transformation process, preparation of the long length bcc supersaturated solid solution $Nb(Al)_{ss}$ is essential. In the liquid quench process [23, 24], the Nb–25 at%Al molten alloy is ejected

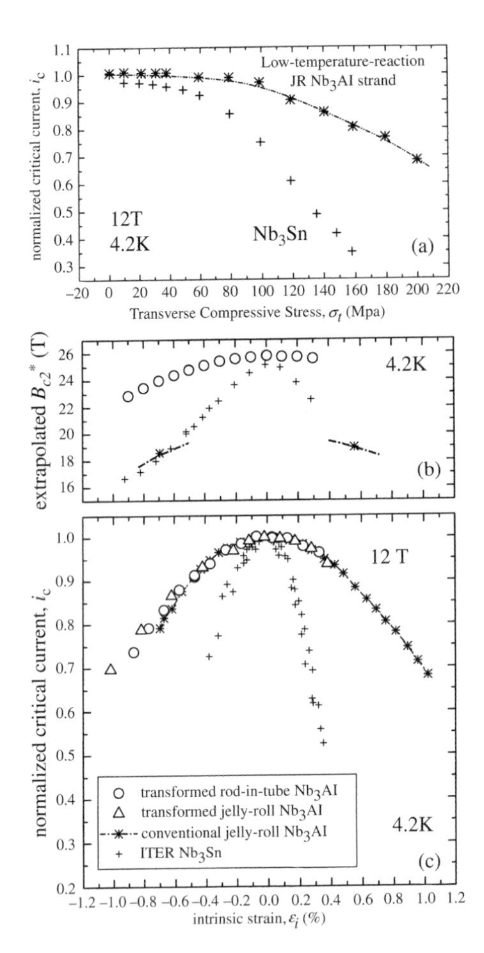

Figure B3.3.4.6. Degradation of critical parameters with (*a*) transverse compressive stress [18], (*b*) and (*c*) uniaxial intrinsic strain [17, 29].

through a nozzle on a continuously reeled-out Cu-substrate tape, which is held at 500°C to enhance the wettability and hence the cooling rate. The undesirable freezing of molten alloy at the nozzle precludes making the $Nb(Al)_{ss}$ longer than 1 m [24].

In the rapid heating, quenching and transformation (RHQT) process [25, 26], the JR and RIT processed Nb/Al multifilamentary wires, 300 m in length, have been reel-to-reel ohmically heated to a very high temperature (~ 1900°C) and quenched into a molten gallium bath (~ 50°C) to form metastable $Nb(Al)_{ss}$ filaments embedded in a Nb matrix (figure B3.3.4.9). The resulting $Nb/Nb(Al)_{ss}$ composites were annealed at 800°C to transform massively the $Nb(Al)_{ss}$ filaments to Nb_3Al with high stoichiometry. Many stacking faults form in the transformed A15 phase [27], and thus T_c and B_{c2} (4.2 K) of RHQT Nb_3Al conductors are 17.8 K and 26 T, which are slightly lower than those of high temperature processed (electron beam irradiated) Nb_3Al conductors. However, the Nb_3Al phase transformed from the metastable $Nb(Al)_{ss}$ has fine grains so that the J_c of RHQT JR Nb_3Al conductor is somewhat higher than that of the conventional JR Nb_3Al and the modified JR Nb_3Sn conductors over the whole range of magnetic fields examined (figure B3.3.4.7).

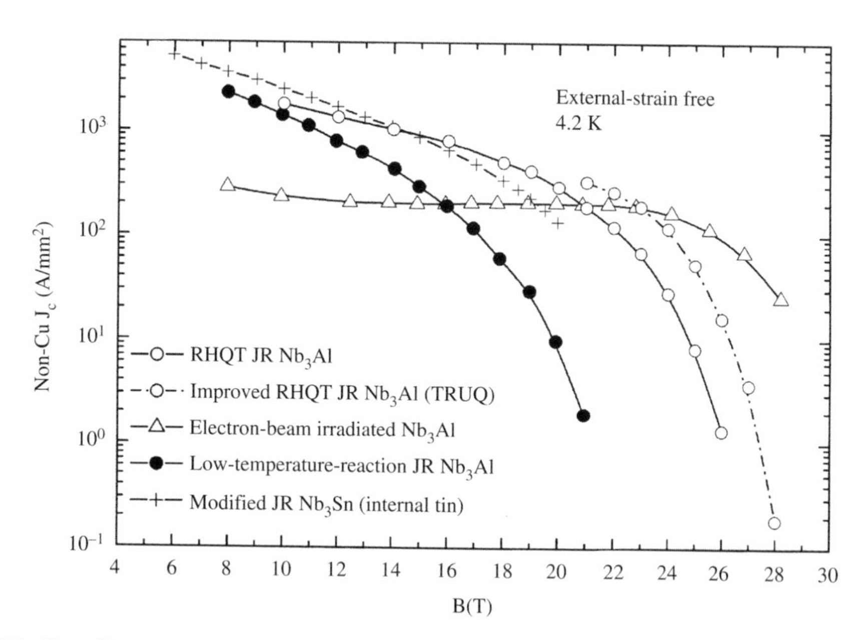

Figure B3.3.4.7. Non-Cu J_c versus magnetic field curves of the low temperature reaction JR Nb_3Al [5], electron beam irradiated PM Nb_3Al [15], RHQT JR Nb_3Al [14], improved RHQT JR Nb_3Al [32] and modified JR Nb_3Sn (internal Sn). For the Nb_3Al conductors, non-Cu J_c is defined as the J_c divided by (1 + Nb/Nb_3Al-ratio).

The strain sensitivity of J_c of A15 phases was explained to increase dramatically with long range atomic order parameter [28]. If this is the case, the resulting high stoichiometry would balance against the excellent strain tolerance. In line with the prediction, the high stoichiometry achieved in the present transformed Nb_3Al conductors slightly increased the strain sensitivity of the B_{c2} [figure B3.3.4.6(*b*)]. However, B_{c2} of the transformed Nb_3Al conductor was larger by about 5 T than that of the conventionally heat

A3.1

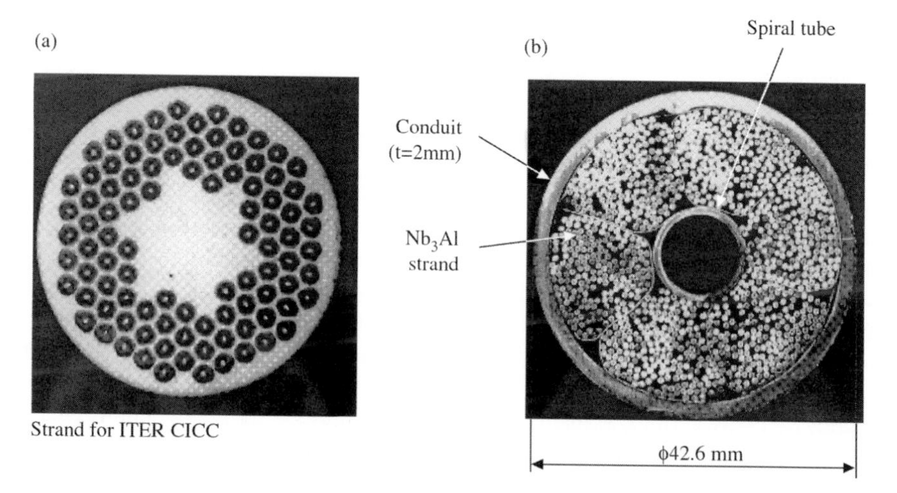

Figure B3.3.4.8. Cross-sectional images of (*a*) the strand and (*b*) the CIC conductor for the low temperature processed JR Nb_3Al conductors [19].

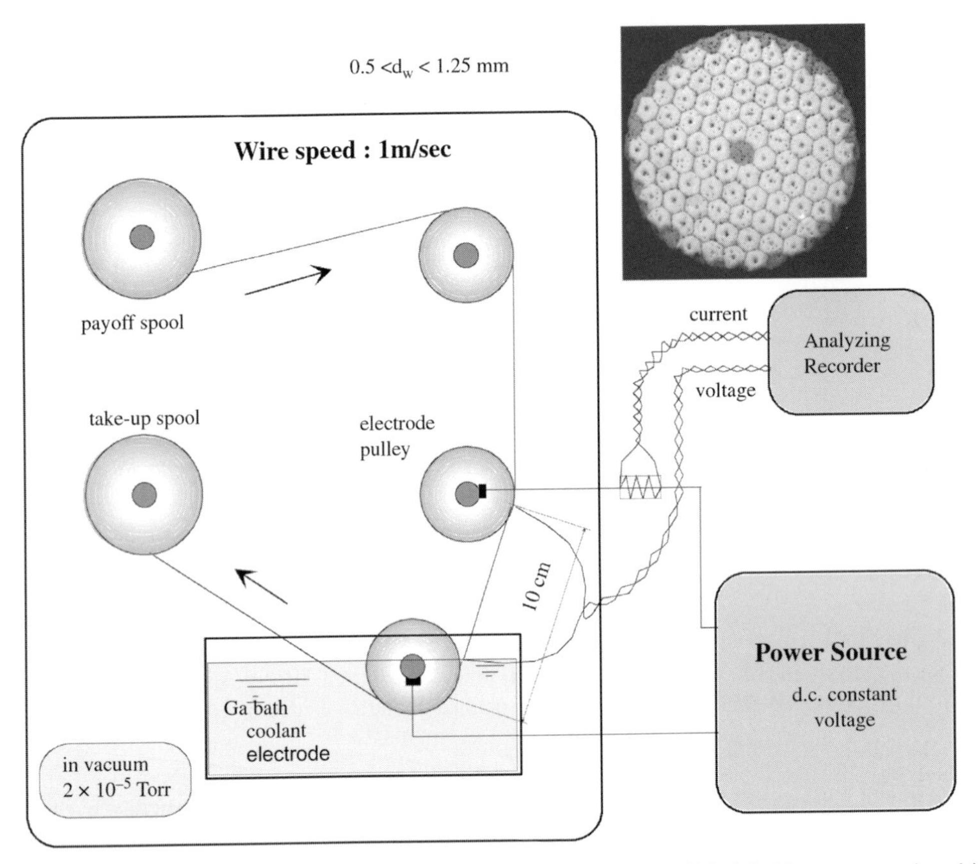

Figure B3.3.4.9. Reel-to-reel ohmic-heating and rapid quenching apparatus [25, 26]. The cross-sectional image of as-quenched JR Nb/Nb(Al)$_{ss}$ composite [14] is given in the figure.

treated Nb_3Al. Since J_c strongly depends on B_{c2} itself as well, J_c at a given magnetic field (12 T) looks less sensitive to strain if B_{c2} is larger. The J_c degradation with -0.7% intrinsic strain was only 20% for both the transformed RIT and JR Nb_3Al conductors, almost the same in magnitude as that for the conventional JR Nb_3Al conductor that has a B_{c2} lower by 5 T [figure B3.3.4.6(*c*)]. Consequently, the excellent strain tolerance is compatible with substantially improved high field performance for the transformed Nb_3Al conductors [29].

In contrast to the conventional JR conductor, the starting subelement is made by rolling Nb and Al foils around a Nb rod (not a Cu rod), because the Cu reacts unavoidably with Nb/Al to form the unwanted ternary Nb–Al–Cu compounds during the ohmic heating. A bundle of such rods was extruded in a Cu sheath (figure B3.3.4.4). The Cu sheaths assembled at extrusion must be removed by etching and must be usually unavailable as a stabilizer, since the RHQT process includes a short heat treatment around 1900°C, which is much higher than the melting point of Cu. Based on the ductile nature of the as-quenched Nb/Nb(Al)$_{ss}$ composite at room temperature, the Cu stabilizer is therefore incorporated, after quenching, by mechanical cladding [30]. Such deformation causes no serious degradation of J_c even at 50–70% reduction in cross-sectional area, rather an increase in J_c at the beginning of deformation. Another stabilization technique has been also developed: internally including the Ag stabilizer as a basic constituent of the strand where the Ag is separated from the JR Nb/Al filaments by a Nb diffusion barrier [30]. The resultant Cu-clad RHQT JR Nb_3Al conductor was wound into a coil, of which the inner and outer diameters and height are 19.7, 40.8 and 49.7 mm, respectively, transformed at 800°C for 10 h, and impregnated with beeswax [31]. The coil, while carrying a current of 179 A at 2.1 K in a superconducting back-up field of 21.2 T, generated an additional 1.3 T. The quenched current of the coil showed quite good agreement with the I_c of the short sample, indicating the characteristic uniformity in the longitudinal direction. The resulting total magnetic field of 22.5 T is the highest record ever for a metallic superconducting coil. Attempts are intensively underway to improve drastically the high field properties of RHQT Nb_3Al conductors by transforming the $Nb/Nb(Al)_{ss}$ composite at temperatures above 1000°C and suppressing the stacking faults: transformation heat based up quenching (TRUQ) method and others (figure B3.3.4.7) [32, 33]. The J_c (4.2 K) at 21 T is enhanced to 590 from 340 A mm^{-2}.

References

[1] Jorda J L, Flükiger R, Junod A and Muller J 1981 Metallurgy and superconductivity in Nb–Al *IEEE Trans. Magn.* **17** 557
[2] Hunt C R and Raman A 1968 Alloy chemistry of $\sigma(\beta U)$-related phases *Z. Metallkde.* **59** 701
[3] Takeuchi T, Kosuge M, Iijima Y and Inoue K 1993 Superconductivity of Nb_3Al formed by solid state reaction of Nb with Ag-based alloy *IEEE Trans. Appl. Supercond.* **3** 1014
[4] Ceresara S, Ricci M V, Sacchetti N and Sacerdoti G 1975 Nb_3Al formation at temperatures lower than 1000°C *IEEE Trans. Magn.* **11** 263
[5] Yamada Y, Ayai N, Takahashi K, Sato K, Sugimoto M, Ando T, Takahashi Y and Nishi M 1994 Development of Nb_3Al/Cu multifilamentary superconductors *Adv. Cryog. Eng.* **40** 907
[6] Takeuchi T, Iijima Y, Kosuge M, Kuroda T, Yuyama M and Inoue K 1989 Effects of additive elements on continuous ultra-fine Nb_3Al MF superconductor *IEEE Trans. Magn.* **25** 2068
[7] Takeuchi T, Kuroda T, Itoh K, Kosuge M, Iijima Y, Kiyoshi T, Matsumoto F and Inoue K 1992 Development of Nb tube processed Nb_3Al multifilamentary superconductor *J. Fusion Energy* **11** 7
[8] Saito S, Ikeda K, Ikeda S, Nagata A and Noto K 1989 Nb_3Al superconducting wires fabricated by the clad-chip extrusion method *Proc. 11th Int. Conf. Magnet Technology* (Tsukuba: Elsevier) p 974
[9] Akihama R, Murphy R J and Foner S 1980 Nb–Al multifilamentary superconducting composites produced by powder processing *Appl. Phys. Lett.* **37** 1107
[10] Thieme C L H, Pourrahimi S, Schwartz B B and Foner S 1984 Improved high performance of Nb–Al powder metallurgy processed superconducting wires *Appl. Phys. Lett.* **44** 260
[11] Barmak K, Coffey K R, Rudman D A and Foner S 1990 Phase formation sequence for the reaction of multilayer thin films of Nb/Al *J. Appl. Phys.* **67** 7313
[12] Bormann R, Krebs H U and Kent A D 1986 The formation of the metastable phase Nb_3Al by a solid state reaction *Adv. Cryog. Eng.* **32** 1041
[13] Webb G W 1978 Cold working Nb_3Al in the bcc structure and then converting to the A-15 structure *Appl. Phys. Lett.* **32** 773
[14] Takeuchi T *et al* 1999 Enhanced current capacity of jelly-roll processed and transformed Nb_3Al multifilamentary conductors *IEEE Trans. Appl. Supercond.* **9** 2682
[15] Kumakura H, Togano K, Tachikawa K, Tsukamoto S and Irie H 1986 Fabrication of Nb_3Al and $Nb_3(Al,Ge)$ superconducting composite tapes by electron beam irradiation *Appl. Phys. Lett.* **49** 46
[16] Takeuchi T, Iijima Y, Kosuge M, Inoue K, Watanabe K and Noto K 1988 Pinning mechanism in a continuous ultrafine Nb_3Al multifilamentary superconductor *Appl. Phys. Lett.* **53** 2444
[17] Specking W, Kiesel H, Nakajima H, Ando T, Tsuji H, Yamada Y and Nagata M 1993 First results of strain effects on I_c of Nb_3Al cable in conduit fusion superconductors *IEEE Trans. Appl. Supercond.* **3** 1342
[18] Zeritis D, Iwasa Y, Ando T, Takahashi Y, Nishi M, Nakajima H and Shimamoto S 1991 The transverse stress effect on the critical current of jelly-roll multifilamentary Nb_3Al wires *IEEE Trans. Magn.* **27** 1829
[19] Koizumi N *et al* 1996 Design of the Nb_3Al insert to be tested in ITER central solenoid model coil *IEEE Trans. Magn.* **32** 2236
[20] Yamada Y *et al* 1999 Development of Nb_3Al superconductors for international thermonuclear experimental reactor (ITER) *Cryogenics* **39** 115
[21] Kumakura H, Togano K, Tachikawa K, Yamada Y, Murase S, Nakamura E and Sasaki M 1986 Synthesis of Nb_3Ga and Nb_3Al superconducting composites by laser beam irradiation *Appl. Phys. Lett.* **48** 601
[22] Kosuge M, Iijima Y, Takeuchi T, Inoue K, Kiyoshi T and Irie H 1993 Multifilamentary Nb_3Al wires reacted at high temperature for short time *IEEE Trans. Appl. Supercond.* **3** 1010
[23] Togano K, Takeuchi T and Tachikawa K 1982 A15 $Nb_3(AlGe)$ superconductors prepared by transformation from liquid quenched body-centered cubic phase *Appl. Phys. Lett.* **41** 199
[24] Takeuchi T, Togano K and Tachikawa K 1987 Nb_3Al and its ternary A15 compound conductors prepared by a continuous liquid quenching technique *IEEE Trans. Magn.* **23** 956

[25] Takeuchi T 2000 Nb_3Al conductors for high field applications *Supercond. Sci. Technol.* **13** R101
[26] Iijima Y, Kosuge M, Takeuchi T and Inoue K 1994 Nb_3Al multifilamentary wires continuously fabricated by rapid-quenching *Adv. Cryog. Eng.* **40** 899
[27] Kikuchi A, Iijima Y and Inoue K 2001 Microstructures of rapidly-heated/quenched and transformed Nb_3Al multifilamentary superconducting wires *IEEE Trans. Appl. Supercond.* **11** 3615
[28] Flükiger R, Isernhagen R, Goldacker W and Specking W 1984 Long range atomic order, crystallographical changes and strain sensitivity of J_c in wires based on Nb_3Sn and other A15 type compounds *Adv. Cryog. Eng.* **30** 851
[29] Takeuchi T, Iijima Y, Inoue K, Wada H, ten Haken B, ten Kate H J, Fukuda K, Iwaki G, Sakai S and Moriai H 1997 Strain effects in Nb_3Al multifilamentary conductors prepared by phase transformation from bcc supersaturated-solid solution *Appl. Phys. Lett.* **71** 122
[30] Takeuchi T 2000 Nb_3Al conductors—rapid heating, quenching and transformation process *IEEE Trans. Appl. Supercond.* **10** 1016
[31] Takeuchi T *et al* 2001 Stabilization and coil performance of RHQT-processed Nb_3Al conductors *IEEE Trans. Appl. Supercond.* **11** 3972
[32] Takeuchi T, Banno N, Fukuzaki T and Wada H 2000 Large improvement in high field critical current densities of Nb_3Al conductors by the transformation-heat-based up-quenching method *Supercond. Sci. Technol.* **13** L11
[33] Kikuchi A, Iijima Y and Inoue K 2001 Nb_3Al conductor fabricated by DRHQ (double rapidly-heating-quenching) process *IEEE Trans. Appl. Supercond.* **11** 3968

B3.3.5
Processing of low T_c conductors: the compounds $PbMo_6S_8$ and $SnMo_6S_8$

B Seeber

B3.3.5.1 Introduction

With respect to the critical temperature and the upper critical field, $PbMo_6S_8$ (PMS) and $SnMo_6S_8$ (SMS) are the most interesting compounds of the family of Chevrel phases (CP). Although the critical temperatures of both of them are typically between 14 and 15 K, the critical field B_{c2} at 4.2 K is extremely high: ~51 T and ~30 T for PMS and SMS, respectively [1, 2]. In comparison, the commercial superconductor Nb_3Sn has a B_{c2} at 4.2 K between 23 and 29 T, depending on the exact chemical composition [3]. CP-conductors are good candidates for the next generation of superconducting wires allowing fields clearly above 21 T, the present limit of the low T_c technology. It is interesting to note that the lead in PMS can be substituted by tin and the phase exists in the whole concentration range from $x = 0$ to 1 in $Pb_{1-x}Sn_xMo_6S_8$ (PSMS) [4]. This opens the possibility to adjust physical properties in order to achieve the best performance. For instance, it was found that the critical current shows a maximum between $x = 0.2$ and 0.3 with only a slightly decreased B_{c2} [4, 5]. Many groups have developed techniques for the fabrication of monofilamentary and multifilamentary PMS and SMS wires [6–28, 65, 70, 72]. Unfortunately R&D work in this field was substantially reduced after the discovery of high-temperature superconductivity in 1986.

G1

A3.1

B3.3.5.2 Physical properties

B3.3.5.2.1 Critical temperature

The critical temperature of well prepared PMS and SMS is between 14 and 15 K. In spite of issues related to the synthesis of the material, the above given T_c range comes also from different definitions of the superconducting transition in the literature (e.g. onset or midpoint of the transition, resistive or inductive measurement). Since the coherence length of PMS and SMS is rather short (2.5–3.2 nm at 4.2 K), they may show a granular behaviour. This means that physical properties like T_c, B_{c2} and J_c are different inside grains and at grain boundaries, which must be taken into account when data from literature are compared. It is important to note that the granular behaviour can be suppressed by appropriate preparation techniques like hot isostatic pressing (HIPing). As discussed later, this is easier to achieve in the SMS system than in the PMS system. The critical temperature of $Pb_{1-x}Sn_xMo_6S_8$ as a function of x is shown in figure B3.3.5.1. With increasing substitution of Pb by Sn, first T_c decreases. If the initial T_c of the PMS is high, the T_c goes up again for x-values near unity [29]. For low T_c PMS

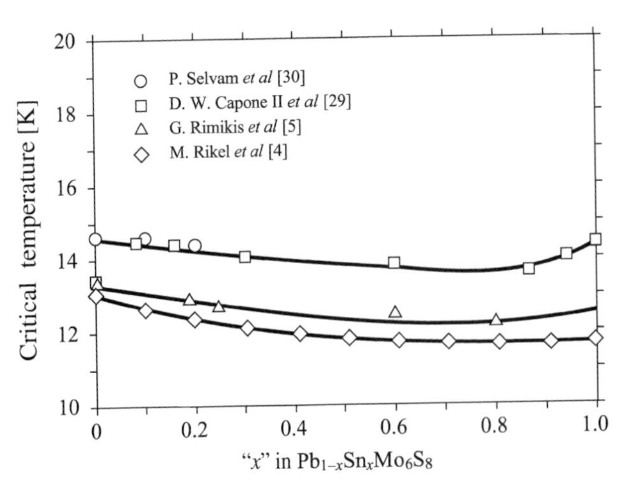

Figure B3.3.5.1. The critical temperature of the pseudo-binary $Pb_{1-x}Sn_xMo_6S_8$ system. Data were taken from Selvam *et al* [30], Capone [29], Rimikis [5] and Rikel [4].

the behaviour is qualitatively the same up to $x \sim 0.8$ but there is no increase for SMS [4]. For a more detailed discussion regarding the critical temperature of CP see chapter G1 in this handbook.

G1

B3.3.5.2.2 Upper critical field

A3.1

The upper critical field of PMS and SMS is more intriguing. For B_{c2} measurements, from the experimental point of view, steady state magnetic fields of up to ~35 T are available only in a few laboratories around the world and B_{c2}s above this limit have to be extrapolated. Fortunately the existing theories allow such an estimate [31, 32] and the result can eventually be cross-checked in pulsed magnetic fields. The important spread of B_{c2} values in the literature is not only due to sample preparation, but also due to imprecise distinction between B_{c2} of grains and of grain boundaries. For instance, the resistive transition from the normal to the superconducting state as a function of field probes only the best material along the percolation path. In contrast, measurements where critical currents are involved (e.g. screening currents, transport currents), are dominated by the properties of grain boundaries, in particular in short coherence length superconductors. The thus obtained B_{c2} is identical with the irreversibility field defined by the closure of the magnetization curve. A more detailed discussion can be found in [33]. In order to obtain the intrinsic B_{c2}, presumably the best method is to measure the specific heat as a function of temperature and field. This is a true bulk measurement and the width of the specific heat jump at the superconductor transition is a measure of the homogeneity of the material. Due to such measurements up to 25 T, there is no doubt about the B_{c2} of $PbMo_6S_8$. In figure B3.3.5.2 the behaviour of B_{c2} of PMS is shown as a function of the critical temperature. B_{c2} and T_c (isoentropic transition) have been obtained from specific heat measurements in a magnetic field [34–36]. Note that the upper critical field goes through a maximum at a $T_c \sim 13$ K. This is unusual, but can be explained by a cross over between a dirty limit (<13 K) and a clean limit behaviour (>13 K) [35, 37]. Note the very important increase of B_{c2} when the temperature is reduced from 4.2 to 1.8 K. No such detailed study is available for $SnMo_6S_8$, but B_{c2} values at 4.2 K converge in the literature to around 30 T.

A2.3

D3.2.1

A2.4

The anisotropy of B_{c2} has been measured for $PbMo_6S_8$, $PbMo_6Se_8$, $SnMo_6Se_8$, and Mo_6Se_8 single crystals [38]. It was found that there is no measurable anisotropy in the plane perpendicular to the ternary symmetry axis of the crystal structure. However, there is a different B_{c2} for a field in the direction of this axis. For $PbMo_6S_8$, $PbMo_6Se_8$ and $SnMo_6Se_8$, the B_{c2} anisotropy is about ~15% with $B_{c2}^{\perp} > B_{c2}^{\parallel}$.

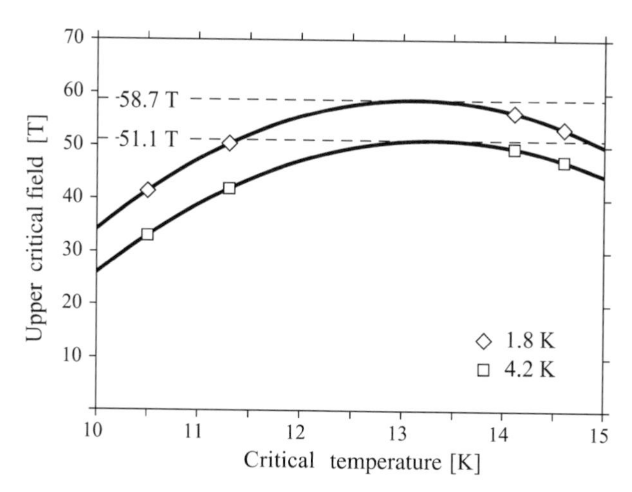

Figure B3.3.5.2. The upper critical field B_{c2} (orbital critical field in the dirty limit approximation) versus the critical temperature (isoentropic transition) of $PbMo_6S_8$. Data have been obtained by specific heat measurements in magnetic fields up to 25 T [34–36]. Solid lines are a quadratic fit to the data (reproduced by permission of Institute of Physics Publishing, Bristol, UK).

In the case of Mo_6Se_8 the difference is $\sim 10\%$ but with $B_{c2}^{\perp} < B_{c2}^{\parallel}$. It seems that the anisotropy is related to the rhombohedral angle α_R of the crystal structure [38]. For angles $\alpha_R < 90°$ $B_{c2}^{\perp} > B_{c2}^{\parallel}$ and the B_{c2} anisotropy grows with increasing deviation from 90°. On the contrary, for angles $\alpha_R > 90°$, as observed in Mo_6Se_8, $B_{c2}^{\perp} < B_{c2}^{\parallel}$. No data are available for $SnMo_6S_8$, but the situation is very similar to that of $PbMo_6S_8$. Because α_R is closer to 90°, a smaller B_{c2} anisotropy can be expected.

When lead is substituted by tin in $Pb_{1-x}Sn_xMo_6S_8$, the upper critical field changes as outlined in figure B3.3.5.3. The represented data are from the same samples as in figure B3.3.5.1. In order to obtain B_{c2} at

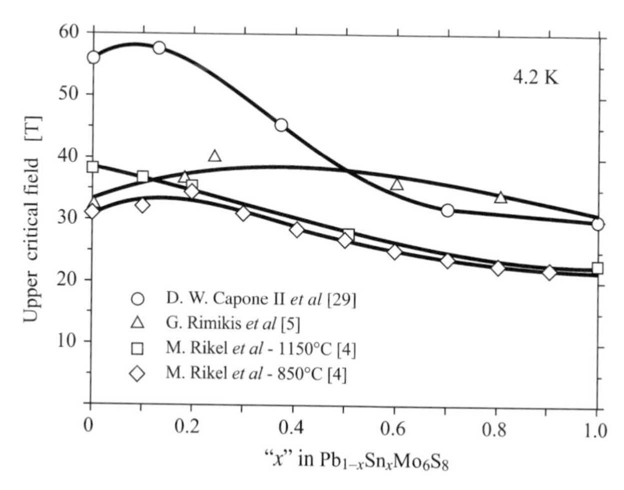

Figure B3.3.5.3. The effective upper critical field at 4.2 K as a function of "x" in the $Pb_{1-x}Sn_xMo_6S_8$ system. The data of Capone *et al* have been obtained by a critical current measurement up to 23 T and extrapolating the pinning force to zero [29]. Rimikis has obtained the B_{c2} data by a Kramer extrapolation [5]. Because experimental details are lacking, the data of Rikel *et al* [4] have to be considered as qualitative.

4.2 K from the work of Rikel *et al* [4], the given initial slopes of the critical field $|dH_{c2}/dT|_{T_c}$ have been used for the B_{c2} calculation. In general, B_{c2} is reduced for increasing x. However, there is some dispersion for x near zero, which probably comes from the conditions of sample preparation. For instance, annealing at 850°C gives a much lower B_{c2} than that at 1150°C [4]. According to the T_c and figure B3.3.5.2 the B_{c2} of the pure PMS sample from Rikel *et al,* annealed at 850°C, should be ~ 51 T (instead of the observed 30 T). This discrepancy may be explained by the fact that 51 T is the B_{c2} of the bulk PMS and the measured 30 T is the effective B_{c2} of grain boundaries. Additional uncertainties come from the measurement of T_c and of $|dH_{c2}/dT|_{T_c}$, which have been carried out by an inductive method in fields up to 8 T and it is not known how the superconducting transition was defined (onset, midpoint or others). Furthermore, the amplitude of the excitation field, which can considerably shift the transition on the temperature scale, is unknown. The same comparison for the Capone II *et al* [48] pure PMS sample gives, according to the B_{c2} versus T_c relation of figure B3.3.5.2, ~ 51 T and the difference to the reported 55 T is smaller than in the former case. Here the T_c was obtained by a resistive measurement and the B_{c2} was determined by a critical current measurement up to 23 T, followed by a linearization of the pinning force versus field and extrapolation to zero pinning force [29]. Another example, how B_{c2} of PMS may change according to the preparation condition is discussed in [39]. A hot isostatically pressed (HIPed) PMS bulk sample (800°C–8 h–200 MPa) shows a B_{c2} (4.2 K) of ~ 40 T, an improvement by a factor of almost two with respect to a sample without such a treatment. Because of all these uncertainties, figure B3.3.5.3 should be considered as a qualitative illustration how B_{c2} varies by the substitution of Pb with Sn.

B3.3.5.2.3 Structural phase transition

There are various indications that PMS and SMS may undergo a structural phase transition upon cooling. By analogy to other CP, like $BaMo_6S_8$ and $EuMo_6S_8$, this should be a transition from the rhombehedral R3 structure to a non-superconducting triclinic one with P1 space group [40]. D1.1.2 Investigations by high-resolution neutron powder diffraction [41] and by synchrotron and Guinier x-ray powder diffraction [42] show line broadening during cool down, but no splitting of lines or appearance of new lines. The broadening can be interpreted as a small (< 1%) triclinic distortion. According to Jorgensen *et al* [41], SMS shows a smaller distortion with respect to PMS. In PMS data are also available on the elastic constants as a function of temperature, obtained by sound velocity measurements [43, 44]. With decreasing temperature, there is a softening of elastic constants, but without any clear sign of a lattice transformation. In general, lattice softening can be considered as a precursor for a structural phase transition, although the latter need not necessarily take place. As an example, one can mention B3.3.3 V_3Si and Nb_3Sn which show lattice softening with subsequent phase transition from a cubic to a tetragonal structure. In cases where the lattice transformation temperature $T_m < T_c$, the superconducting transition stabilizes the lattice and no lattice transformation takes place [45].

Another possibility for the detection of a lattice transformation is the measurement of the thermal expansion as a function of temperature. Meingast *et al* [46] carried out such an investigation in the $Pb_xSn_yMo_6S_8$ system with the help of a high-resolution capacitive dilatometer. Their results are reproduced in figure B3.3.5.4. Note that, for a pure PMS bulk sample, there is an important jump starting at T_c ~ 14 K and extending to lower temperatures. From the thermodynamic point of view this jump is due to the onset of superconductivity and can be calculated according to the Ehrenfest relationship $\Delta\alpha = (\Delta C_p/V_m T_c)(dT_c/dp)$, where ΔC_p is the jump in the specific heat, V_m is the molar volume and dT_c/dp is the pressure dependence of T_c. Since $dT_c/dp < 0$, $\Delta\alpha$ must be negative [47, 48]. This is obviously in contradiction with what has been observed. In addition, the magnitude of $\Delta\alpha$ is too high for a superconducting transition. However, after powdering the PMS bulk sample, the $\Delta\alpha$ at T_c had the right sign and magnitude [46]. This means that the jump in the PMS bulk sample must have a second origin, which has something to do with the preparation. A comparison with Nb_3Sn may help to explain

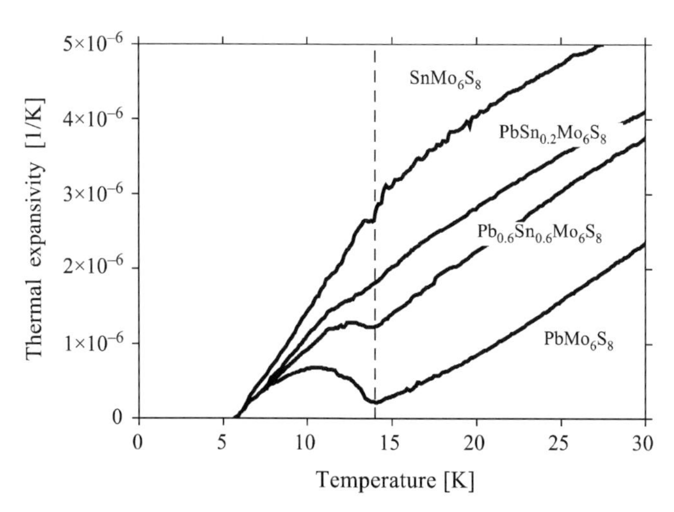

Figure B3.3.5.4. The coefficient of thermal expansion as a function of temperature for the $Pb_xSn_yMo_6S_8$ system (reproduced by permission of Meingast [46]).

the situation in PMS. In Nb_3Sn there is a distinct phase transition at $T_m = 40-50$ K, which can easily be seen in a thermal expansion measurement. With decreasing temperature, the lattice parameters of the new phase change continuously but are arrested by the onset of superconductivity at $\sim$18 K. The observed $\Delta\alpha$ is about 10 times larger as expected from the Ehrenfest relation. So the arrest of the ongoing phase transition is the main reason for the magnitude of $\Delta\alpha$. The sign of $\Delta\alpha$ depends on the pressure dependence of the phase transition temperature T_m [46]. The behaviour of PMS can be explained by the same arguments. In contrast to Nb_3Sn, the thermal expansion measurement of PMS cannot see the smoothly increasing distortion of the lattice during cool down. At the onset of superconductivity the distortion is abruptly stopped, resulting in a jump of the thermal expansion coefficient. The height of this jump is now a measure of the preceding lattice distortion. Since PMS powder has the expected Ehrenfest $\Delta\alpha$ at T_c, it is concluded that the preparation of bulk samples introduces stress (inhomogeneous and anisotropic) which is at the origin of a lattice distortion. In a recent paper Ingle *et al* [49] have shown that the $\Delta\alpha$ in HIPed $Pb_{0.8}Sn_{0.2}Mo_6S_8$ bulk samples do not depend on the temperature between 800 and 950°C, nor on the applied argon pressure between 14 and 250 MPa [49]. The used PSMS powder was from the same batch and the only difference, after HIPing, was the onset of superconductivity, namely 14.2 and 12.7 K for the 800°C/14 MPa and the 950°C/250 MPa treatment, respectively. They concluded that due to the same magnitude of the thermal expansion anomaly at T_c, the change of T_c does not come from the lattice distortion but most probably from oxygen contamination at higher temperature and pressure, manifested by a reduced c/a lattice constant ratio. As shown in figure B3.3.5.4, addition or substitution by tin reduces the α-anomaly at T_c and contributes substantially to the stabilization of the lattice structure. This is nicely confirmed by the work of Ingle *et al* at least in the above-mentioned temperature and pressure range.

D3.2.3

D3.2.3

B3.3.5.2.4 Critical current density and pinning of bulk samples

The critical current density, J_c, at low and medium fields is strongly influenced by metallurgical aspects, which are at the origin of the pinning of flux lines. At higher fields, the density of flux lines increases and the elastic or plastic properties of the flux line lattice, which is intrinsic to the material, must also be taken into account. The situation has recently been reviewed by Rikel [50]. In both PMS and SMS, the critical current density at low fields (<15 T) increases with decreasing grain size, which indicates that pinning occurs at grain boundaries. The dependence of J_c in SMS is exactly $J_c \propto 1/d$, where d is the mean grain size. In PMS J_c varies less with $1/d$, so grain boundaries act differently in these compounds. In addition, in PMS a saturation of J_c at grain sizes smaller than 0.1 μm has been observed [51]. At higher fields (>15 T), J_c in PMS and SMS no longer depends on the grain size.

The situation can be described in more detail by analysing the pinning force $F_p = J_c B$. If one pinning mechanism is working over the whole temperature and field range, the latter can be described by a scaling law: $F_p = J_c B = C B_{c2}^m b^p (1 - b)^q$ where C is a constant characterizing the microstructure, B_{c2} is the effective upper critical field (transport critical current density is zero) and $b = B/B_{c2}$, the reduced field [52]. The exponents m, p and q can be determined from experiment. It has been shown that in high quality PMS the m-exponent, determining the temperature dependence of the pinning force, is $m = 2.4 \pm 0.2$ [53]. By high quality is meant a width of the calorimetrically measured superconducting transition of 0.6 K and which does not show broadening in an applied field [36]. If the width of this transition increases, which means a more inhomogeneous material, the m-exponent also increases [53]. The field dependence of the pinning force is given by the second term $b^p(1 - b)^q$.

B3.3.5.3 Bulk materials

B3.3.5.3.1 $PbMo_6S_8$

In many cases of PMS a scaling law is observed for $p = 0.5$ and $q = 2$. This is just like in Nb_3Sn where J_c is determined by grain boundary pinning [54]. Here, the maximum pinning force is observed at a reduced field of $b = 0.2$. If the mean grain size of PMS increases, the maximum of F_p is shifted towards a higher reduced field. The situation is illustrated in figures B3.3.5.5 and B3.3.5.6 where data from Karasik *et al* [55] are summarized. For a grain size of 0.7 ± 0.2 μm there is a nice scaling (figure B3.3.5.5) but for a grain size of 1.3 ± 0.4 μm the reduced pinning force for different temperatures is no longer on a universal curve (figure B3.3.5.6). This means that more than one pinning mechanism acts and the individual contributions change with temperature. Note the continuous shift of the maximum of the pinning force

A4.3

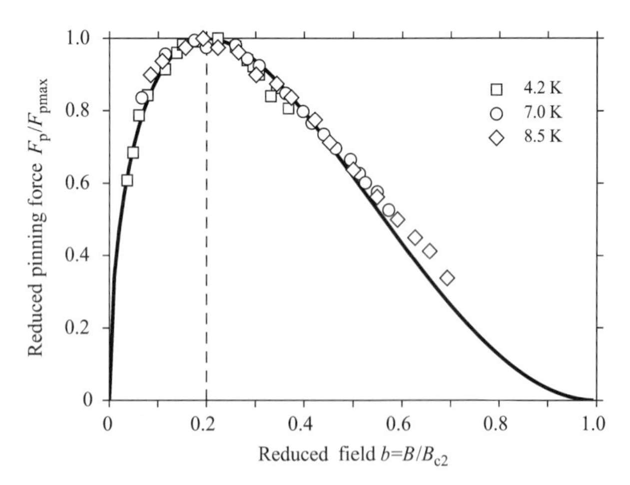

Figure B3.3.5.5. Reduced pinning force of PMS bulk material with a mean grain size of 0.7 ± 0.2 μm as a function of reduced field [55]. The drawn line corresponds to the Kramer scaling law with $F_p \propto b^{0.5}(1 - b)^2$ (reproduced by permission of American Institute of Physics, College Park, USA).

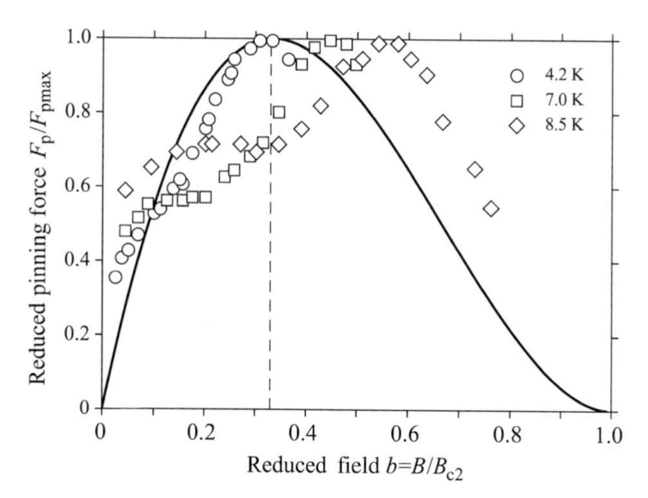

Figure B3.3.5.6. Reduced pinning force of PMS bulk material with a mean grain size of $1.3 \pm 0.4\,\mu$m as a function of reduced field [55]. Note that there is no scaling law observed. The drawn line is a guide to the eyes and corresponds to $F_p \propto b(1-b)^2$ (reproduced by permission of American Institute of Physics, College Park, USA).

towards higher reduced fields for increasing temperatures. The situation resembles that of NbTi where the position of the maximum pinning force can be adjusted on the b-scale by appropriate cold work [56].

B3.3.5.3.2 $SnMo_6S_8$

Pinning in the SMS system has recently been studied extensively by Bonney *et al* [57]. Bulk samples have been prepared by HIPing at 800°C for 8 h as a last step in order to consolidate the connectivity of individual grains. High resolution TEM revealed perfect grain boundaries, free of second phase precipitation. As well as intragranular defects a multitude of planar defects, like twins and stacking

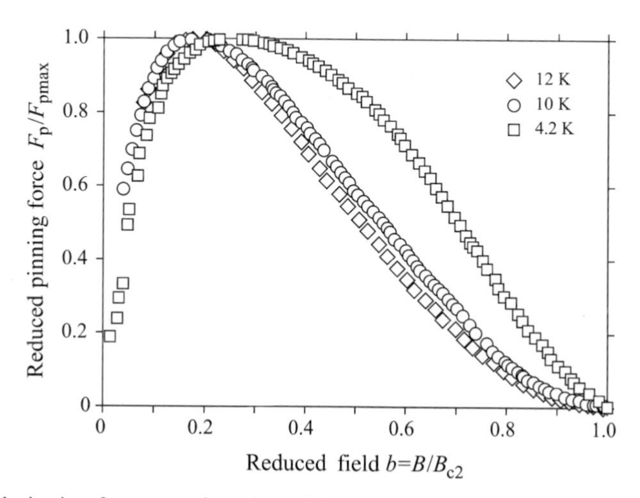

Figure B3.3.5.7. Reduced pinning force as a function of the reduced field for a $SnMo_6S_8$ bulk sample with a mean grain size of $0.26 \pm 0.04\,\mu$m (grain boundary length per unit area $L_{gb} = 4.0 \pm 0.6\,\mu$m) [57] (reproduced by permission of American Institute of Physics, College Park, USA).

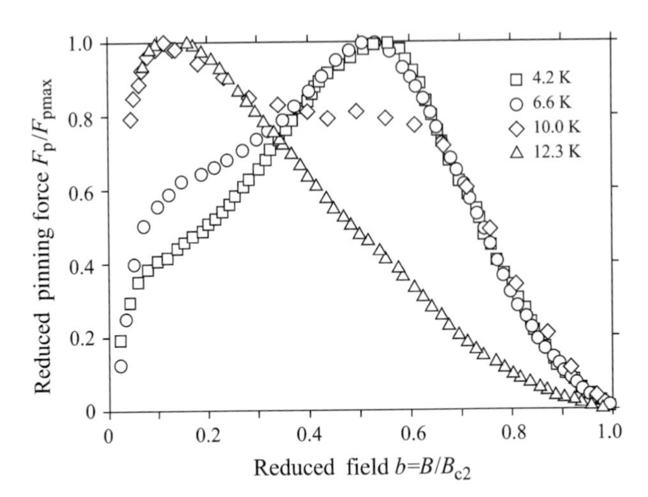

Figure B3.3.5.8. Reduced pinning force as a function of the reduced field for a $SnMo_6S_8$ bulk sample with a mean grain size of $0.48 \pm 0.04\,\mu m$ (grain boundary length per unit area $L_{gb} = 2.1 \pm 0.2\,\mu m$) [57] (reproduced by permission of American Institute of Physics, College Park, USA).

faults, have been observed. Post-HIP heat treatments yielded to an increase of the mean grain size, but not to a change in the density of defects inside grains. The reduced pinning force of SMS as a function of reduced field is shown in figures B3.3.5.7 and B3.3.5.8 with a small grain size ($d = 0.26 \pm 0.04\,\mu m$, grain boundary length per unit area $L_{gb} = 4.0 \pm 0.6\,\mu m^{-1}$) and after a re-crystallisation heat treatment at 1150°C for 250 h ($d = 0.48 \pm 0.04\,\mu m$, $L_{gb} = 2.1 \pm 0.2\,\mu m^{-1}$), respectively. At temperatures above 4.2 K, the peak of the pinning force in the sample with small grain size is situated at $b = 0.2$ (figure B3.3.5.7). By decreasing the temperature to 4.2 K, a high-field bulge appears. In the re-crystallized sample the bulge develops a peak of the pinning force at $b \sim 0.5$ (figure B3.3.5.8). Obviously in both cases no scaling is observed and the acting pinning mechanisms are strongly temperature dependent. For similar mean grain sizes, the observed behaviour is consistent with earlier work of Karisik *et al* [55] and a more recent one of Gupta *et al* [58]. However, for mean grain sizes $> 1\,\mu m$, the situation is similar to that of PMS described above (continuous shift of the maximum pinning force towards higher b for an increasing temperature) [55].

A position of the maximum pinning force above $b = 0.2$ suggests that intra-grain pinning dominates. Studies regarding this kind of pinning in CPs are rare. The above mentioned planar defects in SMS have also been observed in PMS thin films, but with a rather low density [59]. Rikel pointed out that the oxygen content in the CPs may introduce oxygen containing precipitates like MoO_2 or $Pb(Sn)Mo_4O_6$, which can contribute to pinning [50]. Another possibility is a substitution of sulphur by oxygen. In particular the S(2) atoms lying on the ternary axis (figure D1.4.3 of chapter D1.4) can be substituted by oxygen, leading to well defined point defects [60]. Finally it should be mentioned that Pb and Sn, due to their volatile character, can easily lead to under-stoichiometric PMS and SMS [61]. Whether the structural instability of PMS and, to a lesser extent, SMS plays a role regarding pinning is not clear. The coexistence of the R3 and the non-superconducting P1 phase in bulk or wire samples may reduce J_c for geometrical reasons (i.e. less superconducting cross section). However, the reduction of the superconducting phase, due to the presence of the non-superconducting one, should show up e.g. in specific heat measurements, which is not the case [37]. So the lattice instability must act on the critical current in a more subtle way.

D3.2.1

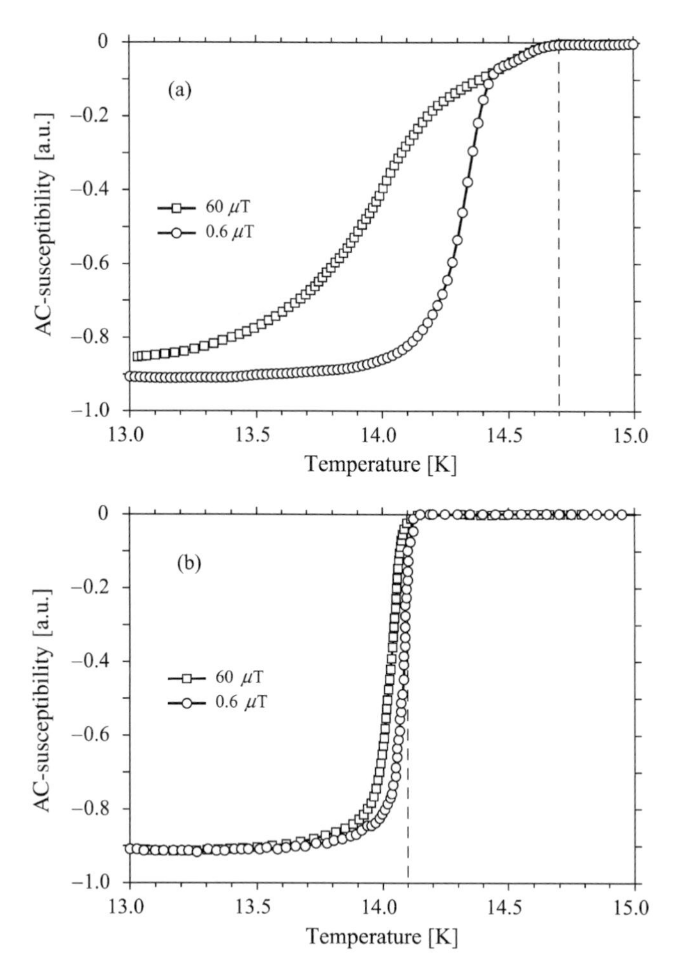

Figure B3.3.5.9. The normalized real part of the ac-susceptibility with 0.6 and 60 μT as amplitudes of the excitation field of (*a*) a hot pressed PMS bulk sample (1200°C, 1 h) and the same sample but with Sn addition according to the nominal composition $PbSn_{0.3}Mo_6S_8$ [53].

B3.3.5.3.3 Substitution of Pb by Sn

The substitution of Pb by Sn or the addition of Sn to PMS results in an improved critical current density [4, 5]. The optimum Sn concentration in $Pb_{1-x}Sn_xMo_6S_8$ is at $x = 0.2$–0.3. Although the physical reasons are not completely understood, there are several indications that the quality of grain boundaries is improved. In figure B3.3.5.9 the normalized superconducting transition of the ac-susceptibility of hot pressed PMS is compared with that of Sn doped PMS according to the nominal composition $PbSn_{0.3}Mo_6S_8$ [53]. The transition is shown for two distinct amplitudes of the excitation field, namely 0.6 and 60 μT. Note that the transition in figure B3.3.5.9(*a*) is typical for a granular system $((J_c^{\mathrm{grain}} > J_c^{\mathrm{grain-boundary}})$. The first transition (at high temperature) is due to PMS grains and the second one (at lower temperature) corresponds to the T_c of grain boundaries. The lower the critical current at the grain boundaries, the more the transition is enlarged towards lower temperatures when the excitation field is increased. Therefore, this information can be used to characterize the quality of grain boundaries.

In the case of $PbSn_{0.3}Mo_6S_8$ (figure B3.3.5.9(*b*)) there is no granular behaviour and only one transition for a given excitation field is seen. It has been shown in an accompanying specific heat measurement, that the addition of Sn shifts the onset of superconductivity from 14.7 to 14.1 K, as is also seen in figure B3.3.5.9(*b*). This indicates that Pb has been partially substituted by Sn. The increase of the excitation field over two decades has had only little influence on the transition. Further arguments relating to why the critical current increases in Sn substituted or doped PMS, were given by Selvam *et al* [30]. For instance D1.3 microstructural analysis shows a more uniform material and less porosity, Auger electron spectrometry indicates a reduction of grain boundary phases and the densification during hot pressing is much faster.

B3.3.5.4 Wires

B3.3.5.4.1 Wire manufacturing

B3.1 Because PMS and SMS cannot be synthesized directly from the melt (incongruent melting), wires and also bulk materials, have to be manufactured by powder metallurgical methods. Regarding wires, either pre-reacted PMS, SMS or PSMS powder is used, or so called precursor powders (e.g. PbS, SnS, MoS_2 and Mo). The selection of the kind of powder depends on the drawing process. As an example it should be mentioned that precursor powder, or a mixture of it with a pre-reacted one, is suited for deformation at ambient temperature. In a hot deformation process the use of a pre-reacted powder is convenient (no segregation of different kind of powders, very good homogeneity over the length of the wire).

It is worth mentioning that, from a thermodynamic point of view, PMS is an extremely stable phase, which can be synthesized at high quality in a wide temperature range, typically between 450 and 1650°C [37]. The low temperature synthesis allows to control the mean grain size down to $\sim 0.2\,\mu m$ [67]. The high temperature synthesis at 1650°C yields crystals up to $\sim 100\,\mu m$ in size. Recently, new methods for the synthesis of ultrafine precursor powders like PbS, MoS_2 and Mo (0.05–0.5 μm), as well as Mo_6S_8 (~ 0.3 to $0.5\,\mu m$) became available [68, 69]. Much less investigations are available regarding the synthesis of SMS, but the situation should be similar.

The starting powder is then cold isostatically pressed (CIPed), machined and inserted in a metallic tube like Ta, Nb or Mo, which is sealed under vacuum. These barrier materials have to be selected with respect to their compatibility with the superconducting phase (little or no chemical reaction) and must be suited for the wire drawing process. Ta and Nb can be deformed to large extents at ambient temperature, mostly without any intermediate heat treatment. In contrast, with Mo, hot deformation techniques have to be applied. Taking into account that optimal powder metallurgical deformation techniques require elevated temperatures anyway, this should be considered as an advantage. It has to be noted that hot wire drawing is state of the art of industry working in the field of powder metallurgy and therefore is commensurately more expensive than the cold drawing techniques that can be used with Nb or Ta. The barrier material must also withstand relatively high temperatures, which are required for a consolidation heat treatment after wire drawing. In the case of precursor powders, a heat treatment is necessary to form the superconducting phase. Regarding inertness a Mo barrier is the best choice. If measures can be taken to keep heat treatment temperatures as low as possible, the chemical reaction between PMS/SMS and Nb or Ta can be reduced. This will be discussed in more detail later. As a next step the billet is inserted in a copper tube, which serves as a stabilizer, and then in a stainless steel tube. The latter allows the adjustment of the thermal expansion of the whole matrix, together with the barrier, to that of the PMS [62, 63]. This is very important because PMS contracts much more upon cooling than Ta, Nb or Mo so the superconductor comes under tensile stress at 4.2 K. Since stainless steel contracts more than PMS, it can be used for thermal stress compensation. Wires with Mo matrix do not need a copper stabilizer because the electrical properties of Mo are similar to that of Cu [66]. This allows a lower matrix/superconductor ratio, α, and the overall critical current density is improved. Values of $\alpha = 1.5$

Table 3.3.5.1. Chronological compilation (non-exhaustive) of work done in the field of PMS, SMS and PSMS wires

Initial Powder	Barrier	Outer Matrix	o.d. [mm]	Annealing [°C]	[h]	HIP [MPa]	Ref.
$PbMo_6S_8$	Mo	ss	0.3–0.7	750–950	0.3–250	–	[10, 11]
$Pb–MoS_2–Mo$	Ta	Cu	1.05	850–1050	1–3	100–200	[17]
$PbMo_6S_8$	Nb	Cu	0.26–0.82	725–850	2–120	–	[22]
$SnS–MoS_2–Mo$	Nb	Cu	0.30–0.96	750–900	4–44	–	[24]
$PbMo_6S_8$	Nb or Ta	Cu + ss	0.25–1.00	750	90–120	–	[25]
$PbS–MoS_2–Mo$	Mo	ss	1.00	875–950	5–20	–	[19]
Pb–Mo–S	Nb	ss	1.50	1030–1100	0.5–2	–	[15]
$Pb–MoS_2–Mo_2S_3–Mo$	Ta or Nb	Cu or CuNi	0.84–1.10	975–1075	0.17–5	–	[18]
$Pb_{1.2-x}Sn_xMo_6S_8$	Ta or Nb	Cu + ss	1.00	1000	0.5–1	100	[5]
$Sn_{1.2}Mo_6S_8$	Nb or Ta	Cu + ss	1.00	800–1000	0.5–30	100	[72]
$PbS–MoS_2–Mo$	Mo	ss	1.00	950–1200	12	200	[20, 21]
Pb–Sn–Mo–S	Nb	ss	~ 0.8	1100	0.75	200	[16]
$PbMo_6S_8$ + 10w% (PbS,MoS_2,Mo)	Mo	ss	0.4	1225	4	110	[70]
$Pb_{1-x}Sn_xMo_6S_8$	Nb or V	CuNi	2.5	800	12	200	[65]
$Pb_{0.6}Sn_{0.4}Mo_6S_8$ + $Sn_{0.2}$ + 10w% (Pb,MoS_2,Mo)	Nb	CuNi + ss	0.4	900	0.5	190	[27]

(40% PMS, 60% matrix) have already been achieved [66]. After the outer tube has been sealed under vacuum, the assembly is swaged or extruded, followed by wire drawing. The feasibility of multifilamentary wires has been shown [64, 65]. There are many variable regarding the barrier and the outer matrix materials, as well as the deformation process. A non-exhaustive summary of work in the field of PMS, SMS and PSMS wires is given in chronological order in Table B3.3.5.1. Most of these wires have been developed on a laboratory scale resulting in typical lengths of 100–200 m. The only exception is the PMS wire development with a Mo barrier which has been carried out on standard industrial production machinery [10, 11, 19–21]. Production lengths of up to 1 km (o.d. = 0.4 mm), without breakage, have been reported [10, 11]. A1.3

B3.3.5.4.2 Critical current densities

The critical current density in the superconducting filament of different CP wires at 4.2 K is summarized in figure B3.3.5.10. For clarity, only wires with the highest J_c are considered. It is interesting to note that all curves converge at higher fields: $1.75 \pm 0.4 \times 10^8$ A m^{-2} at 20 T and $0.76 \pm 0.25 \times 10^8$ A m^{-2} at 25 T. This happens independently of the initial powders used, the applied barrier/matrix materials and the annealing conditions. The higher critical current densities at lower fields are achieved by smaller grains [27], but mainly with the addition of tin to PMS [5, 16, 27]. For instance, Cheggour *et al* used a tin containing powder [$Pb_{0.6}Sn_{0.4}Mo_6S_8$ + $Sn_{0.2}$ + 10w% (Pb,MoS_2,Mo)], as was the case for Rimikis *et al* [5] ($Pb_{0.96}Sn_{0.24}Mo_6S_8$). In contrast, Seeber *et al* worked without any tin in the initial powder ($PbMo_6S_8$ + 10w%(PbS,MoS_2,Mo)). In figure B3.3.5.11 available data of the critical current density at 1.8 K are presented. Cooling in superfluid helium at 1.8 K is becoming more and more common for high performance superconducting magnets. As an example of large scale applications of this cooling G1

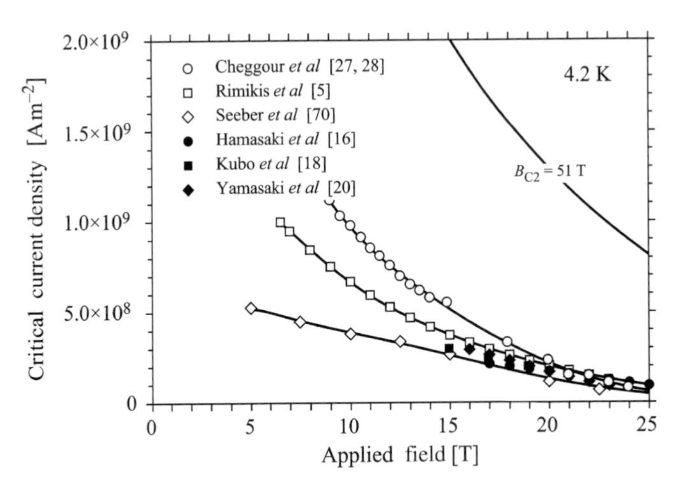

Figure B3.3.5.10. The critical current density with respect to the filament cross section of CP-wires at 4.2 K. The 51 T line corresponds to a calculation assuming the effective $B_{c2} = 51$ T and scaling the data of Cheggour.

technology, one can mention Tore Supra in Chadarache (France) and the large hadron collider, LHC, at CERN (Switzerland). Regarding figure B3.3.5.11, note the increase of J_c which is at 25 T from $0.76 \pm 0.25 \times 10^8$ (at 4.2 K) to $2.26 \pm 0.33 \times 10^8$ A m^{-2} (at 1.8 K). Since the microstructure does not change, the strong increase can be explained by the temperature term of the pinning force $F_p \propto B_{c2}^m$ with $m \cong 2.4$ and the high slope of the critical field $|dB_{c2}/dT|4.2$ K (see also figure B3.3.5.2). Again, different CP-wires converge at higher fields.

A4.3

B3.3.5.4.3 Upper critical field

A3.1

In figure B3.3.5.12 an estimate of the effective upper critical field for different CP-wires is shown. Values for B_{c2} have been obtained by plotting $B^{0.25}J_c^{0.5}$ versus B. According to Kramer, such a plot is a straight

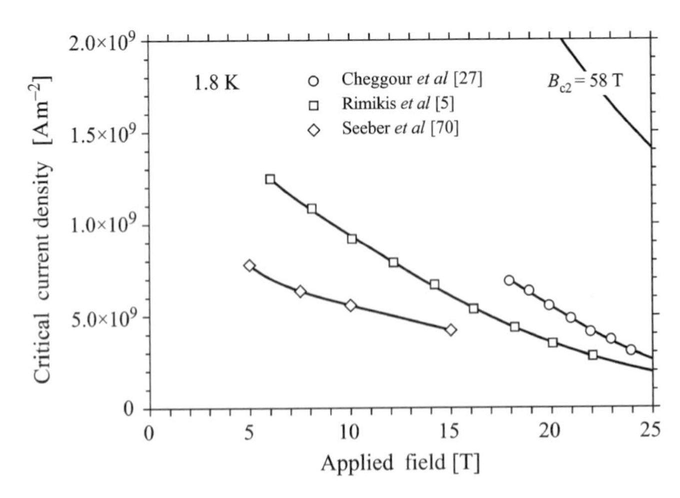

Figure B3.3.5.11. The critical current density with respect to the filament cross section of CP-wires at 1.8 K. The 58 T line corresponds to a calculation assuming the effective $B_{c2} = 58$ T and scaling the data of Cheggour.

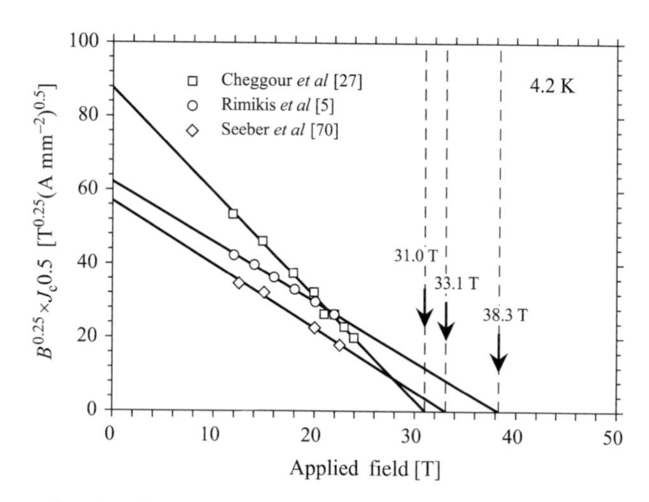

Figure B3.3.5.12. Kramer plots for the determination of the effective upper critical field of different CP-wires.

line and the intersection with the B-axis gives B_{c2} ($J_c = 0$) [71]. The so-called Kramer plot is correct in superconductors where pinning at grain boundaries is dominant (e.g. Nb_3Sn, most of CP-wires). In other words, the pinning force can be described by $F_p \propto B_{c2}^m b^{0.5}(1-b)^2$ and its maximum is at $b = 0.2$. However, in situations where other types of pinning are operating, such a plot is no longer linear. In CP-wires at 4.2 K and in high fields (> 10 T) $B^{0.25} J_c^{0.5}$ versus B is linear and can be used for comparison of B_{c2}s. For completeness it must be said that in CP-wires such a linear behaviour is not always observed, in particular at higher temperatures or lower fields (due to the shift of the maximum pinning force away from $b = 0.2$). The extrapolated B_{c2}s in figure B3.3.5.12 varies between 31.0 and 38.3 T, which is much below the expected bulk value of ~ 51 T. For clarity, in figure B3.3.5.12 the B_{c2}s of [16, 18, 20] are not shown, which are 37.8, 43.8 and 34.1 T, respectively. Because all these B_{c2}s are extrapolated for $J_c = 0$, they represent the B_{c2} at grain boundaries, or an effective upper critical field. If the pinning force $F_p = J_c B$ is plotted versus the reduced field $b = B/B_{c2}$, one finds the maximum of F_p at $b \sim 0.2$ for PMS-wires with Nb-barrier (Cheggour, Rimikis), but for wires with Mo-barrier (Seeber) the maximum is at $b \sim 0.33$. This means that in the latter case pinning is dominated by intra-grain defects and no longer by grain boundaries. As in bulk samples, it is easier to shift the maximum pinning force to higher reduced fields in the SMS-system. For instance, a maximum pinning force at $b \sim 0.4$ has been reported in SMS-wires with a Ta-barrier. However, the critical current at high fields is about half the values shown in figure B3.3.5.10 and the effective upper critical field is 27 T [72].

Supposing the effective B_{c2} is equal to the bulk value, it is interesting to calculate a projection of the achievable J_c. This has been carried out by taking the data of Cheggour *et al*, to fit them by $F_p = CB_{c2}^{2.4} b^{0.5}(1-b)^2$ and to recalculate F_p for an increase of B_{c2} to 51 T at 4.2 K and 58 T at 1.8 K. The results are shown in figure B3.3.5.10 and B3.3.5.11. Without any change of the microstructure, just by improving the effective B_{c2}, current densities in the range of 1×10^9 A m^{-2} can be expected at 25 T.

B3.3.5.4.4 Limits to critical current densities

Regardless of the microstructure and the connectivity of grains, the most important limitation of the critical current density at high fields is the effective upper critical field. One may ask the question why, up to now, was it not possible to increase B_{c2} to the bulk value? In CP-wires with Nb or Ta barrier there is a chemical reaction between the CP monofilament and the barrier. Because the reactivity of Nb with

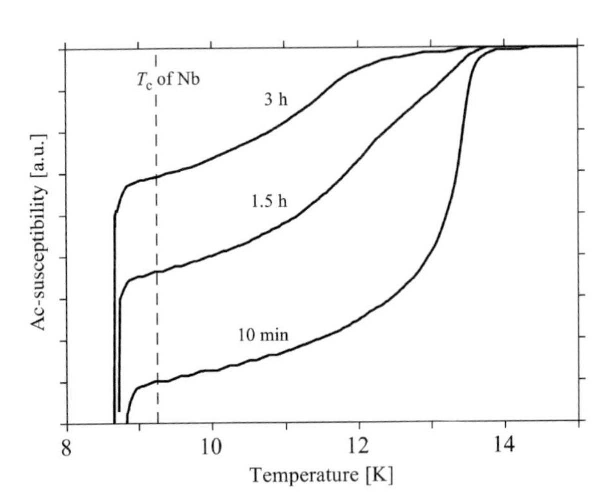

Figure B3.3.5.13. The inductive transition of a $PbSn_{0.4}Mo_6S_8$ wire with a Nb-barrier, which has been annealed at 1000°C for different times [75].

respect to Ta is less, Nb is a frequently used barrier material and the interface with the CP has been studied in detail [73, 74]. In the case of a PSMS wire with precursor powder (PbS, Sn, MoS_2 and Mo), corresponding to the composition $PbSn_{0.4}Mo_6S_8$, heat treatments between 750 and 1050°C have been carried out. As an example, the annealing at 1000°C is described. At the Nb–PSMS interface a layer with a lamellar structure is growing in the direction of the PSMS core. Microprobe analysis suggests a phase with a composition near to $Pb_{0.15}Nb_2S_3$. The square of the thickness of this layer is linear in time and stops growth when no further sulphur is available from the PSMS. At the end there is no superconducting phase left. The growth rate of the layer does not follow Arrhenius' law, suggesting that more than one diffusion mechanism is active. It has also been found that the density of the powder has an influence on the growth rate. The higher the density, the lower the growth rate. Regarding superconducting properties, the situation can be illustrated by an inductive T_c measurement as shown in figure B3.3.5.13 [75]. The transition at higher temperatures corresponds to PSMS. Further, the transition around 9 K is that of the niobium barrier. After an annealing of 10 min at 1000°C, the onset of superconductivity is at ~14 K and no distinction between grains and grain boundaries can be made. After 1.5 h at 1000°C the onset slightly decreases and there is a kink at ~12.2 K. The latter can be interpreted by the intersection of two transitions, as in a granular system. Note that the screened volume has been substantially reduced. After 3 h the degradation of PSMS continues and for annealing times > 4 h the transition of PSMS disappears. The observed behaviour shows clearly the degradation of the superconductor, in particular of its grain boundaries. Although the extraction of the T_c profile in the wire sample requires the deconvolution of the data (due to the temperature dependent penetration depth of an inhomogeneous superconductor), they give a qualitative overview of how T_c varies at grain boundaries. The larger the inductive transition, the larger the B_{c2} distribution (see figure B3.3.5.2) and the observed effective B_{c2} corresponds to a mean value. CP wires with a Mo barrier do not show any degradation of grain boundaries with increasing aggressive heat treatment. This is illustrated in figure B3.3.5.14, where the T_c of a PMS wires is shown [76]. A 4 h isochronal annealing at 990°C and ambient pressure is compared with HIP heat treatment at 990 and 1225°C (110 MPa). Under HIP conditions the long tail at low temperatures can be reduced and with increasing annealing temperature the onset is shifted to higher temperatures. The hatched area indicates the T_c inhomogeneity along the wire. There are parts with an onset at 14 K and a width of < 1 K, but there are other parts with a less good

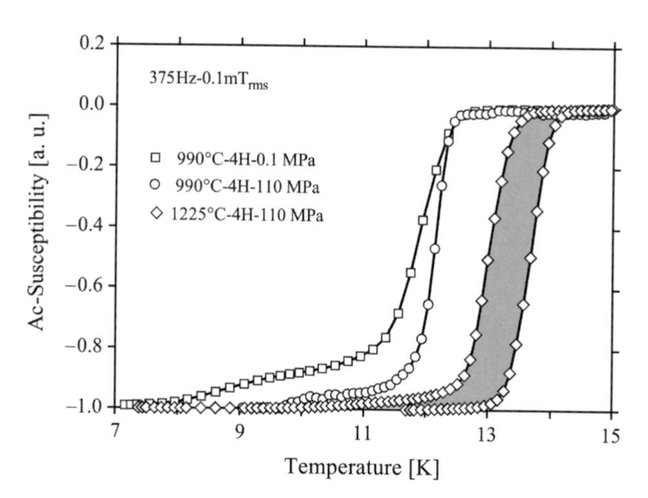

Figure B3.3.5.14. The inductive transition of a $PbMo_6S_8$ wire with a Mo barrier which has been subject to isochronal annealing at 990 and 1225°C. The hatched area indicates the variation of T_c along the wire length. Note the beneficial influence of HIPing with an argon pressure of 100 MPa [76] (reproduced by permission of IEEE, Piscataway, USA, © 1995 IEEE).

transition. From critical current measurements in fields up to 22.5 T, the effective critical field was estimated to 33.1 T for both the 990°C/110 MPa and 1225°C/110 MPa treatment (see also figures B3.3.5.10 and B3.3.5.12, Seeber *et al*). Presumably due to the T_c inhomogeneity along the length of the wire, the higher temperature did not improve the effective B_{c2}. Another measurement supports the tight correlation between high J_c in high fields with the T_c distribution and the corresponding mean effective upper critical field. Because of its short sample length, the 1225°C/110 MPa wire showed degradation after several weeks due to the release of thermally induced compressive prestress over the ends [70]. As a consequence, J_c decreased at high fields and the mean effective B_{c2} was reduced to 30 T. In addition, a low

A3.1

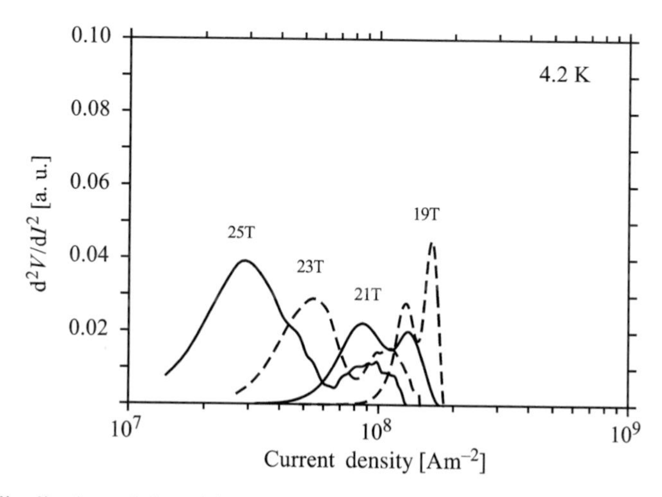

Figure B3.3.5.15. The distribution of the critical current density in a degraded $PbMo_6S_8$ wire [70] (reproduced by permission of Institute of Physics Publishing, Bristol, UK).

temperature tail appeared in the inductively measured transition. The potential of this wire became obvious when the V–I curves of the critical current measurement were analysed in detail. In figure B3.3.5.15 the second derivative d^2V/dI^2 of this wire is shown. As it was pointed out by Baixeras *et al*, and later by Warnes *et al*, the d^2V/dI^2 yields important information regarding the distribution of J_c [77, 78] (see Chapter C2.3). Although the J_c distribution before degradation was quite sharp, it is now very large and shows two peaks. Note that the field dependence of the right peak is much less than that of the left one. If the peak positions are used for a Kramer plot, two critical fields are obtained, namely 31.4 and 46.4 T, which may be interpreted as the effective B_{c2} at grain boundaries and the B_{c2} of grains, respectively. Due to the important difference between effective B_{c2} and bulk B_{c2} the heat treatment of this wire was far from optimal.

A3.1 In summary the effective B_{c2} in CP-wires, which is representative for the B_{c2} at grain boundaries, and which determines the J_c at high fields, cannot be increased easily to the bulk B_{c2} in wires with a reactive barrier like Nb or Ta. As has been shown, the diffusion of sulphur and lead from the CP-core towards the barrier degrades the grain boundaries. If the temperature for the phase formation and consolidation can be decreased, this problem should be less severe. This was the main reason to develop methods for low temperature synthesis of CP-powders or precursors as was carried out by the group at the University of Rennes, France [67–69]. At present it is not clear whether low temperature annealing, which involves automatically powders with very small grain size, can overcome the B_{c2} problem. A very small grain size increases the grain boundary area and its sensitivity for degradation. Apart from the handling, wire drawing at ambient temperature is difficult with fine-grained powders because they densify very fast to 100% of the theoretical density and further deformation is prevented due to the non-ductile behaviour of the powder. In cases where higher temperatures are necessary, short annealing times for reducing the degradation are not a good solution to the problem because in a real magnet this is impossible to achieve due to its important thermal mass and unallowed temperature gradients. Wires with an inert Mo barrier are much better suited and the main reason for the observed reduced effective B_{c2}, up to now, is insufficiently homogenous PMS powder (too large T_c and B_{c2} distribution) and a not yet optimized heat treatment. That the B_{c2} problem can be solved has been shown by Bonney *et al* [57] in the SMS system. By appropriate preparation (wrapping the bulk sample in a Mo foil, vacuum sealing in a stainless steel can and HIPing) magnetization measurements up to 30 T indicate that the effective B_{c2} at grain boundaries, in the paper nominated as irreversibility field (loop closure) is only slightly below the bulk B_{c2} (slope change in the magnetization curve).

B3.3.5.4.5 Behaviour under uniaxial stress

E1.1 Because in superconducting magnets there are winding sections, which must sustain particular high magnetic forces, the mechanical behaviour of the wire used must be known. Data regarding CP wires are rare and the so far known stress–strain curves are summarized in figure B3.3.5.16. The work of Katagiri *et al* [80] has been carried out on a PMS wire with a Nb barrier and a Cu–30Ni matrix (16% PMS, 16% Nb, 68% Cu–30Ni) [80]. Due to the relatively soft Cu–30Ni matrix, the $\sigma_{0.2}$ yield strength is only ~200 MPa. There is a remarkable strengthening by the addition of stainless steel, as is the case in the two other studies. Rimikis measured a SMS wire with a Nb barrier and a Cu–ss matrix (6% SMS, 9% Nb, 15% Cu and 70% ss) [79] and the data from Seeber *et al* comes from a PMS wire with Mo barrier D3.3 and a ss-matrix (30.5% PMS, 18.3% Mo and 51.2% ss) [66]. Note the very high $\sigma_{0.2}$ which is between 700 and ~850 MPa. For comparison, typical $\sigma_{0.2}$ values for Nb_3Sn are 100–250 MPa.

As a next step the influence of mechanical stress on the critical current density has to be known. Because the most important stress in the windings of a magnet is the hoop stress, it is common to measure the J_c versus uniaxially applied strain. It is interesting to note that CP wires with a stainless steel matrix can be tailored so that the irreversible strain, ε_{irr}, above which the critical current does not

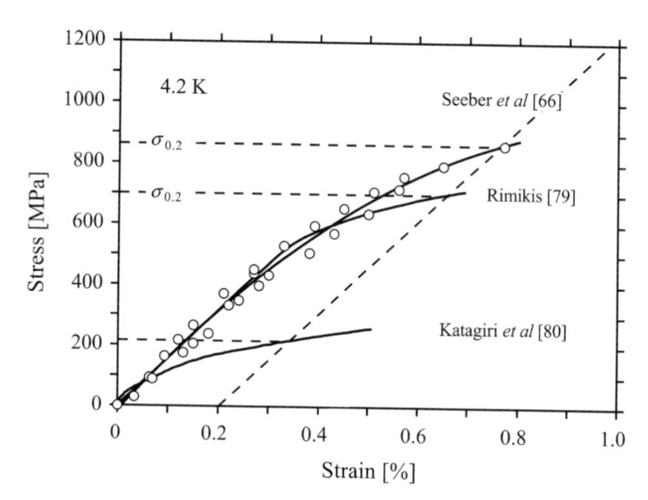

Figure B3.3.5.16. Stress–strain curves for different CP wires. Katagiri *et al*: 16% PMS, 16% Nb, 68% Cu-30Ni [80], Rimikis: 6% SMS, 9% Nb, 15% Cu and 70% ss [79] and Seeber *et al*: 30.5% PMS, 18.3% Mo and 51.2% ss [66].

recover upon unloading, can be adjusted to the requirement. This has been demonstrated by Goldacker *et al* who measured the same PMS wire with an increasing content of stainless steel. For instance 51% of ss yields an ε_{irr} of 0.35% which goes up to 0.50% and 0.85% for 64% and 73% of ss, respectively [81]. Seeber *et al* have seen a similar behaviour for PMS wires with a Mo barrier. By increasing the ss/Mo ratio from 2.8 to 3.8, ε_{irr} goes from 0.15 to 0.30% [66]. This is shown in figure B3.3.5.17 together with measurements of Goldacker and of Katagiri [80]. The increase of the critical current with strain in the CP wire with the highest ε_{irr} is due to a compressive prestress in the superconductor, like in Nb_3Sn. B3.3.3

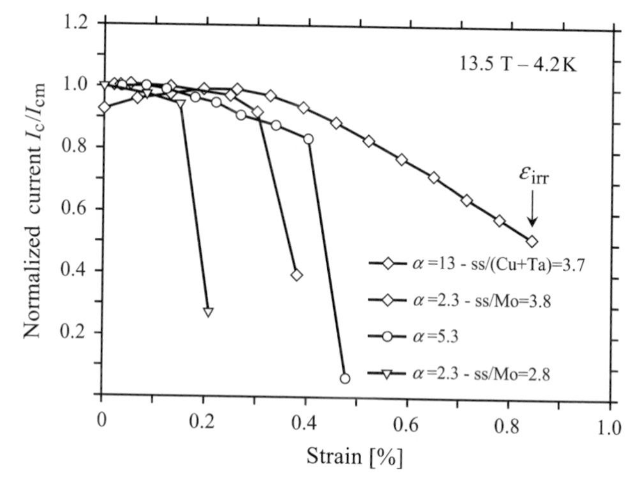

Figure B3.3.5.17. The normalized critical current with respect to its maximum value I_{cm} as a function of uniaxial strain for different CP wires. The wire with $\alpha = 13$ (α is the matrix/superconductor fraction) has the highest irreversible strain ε_{irr} (7% PMS, 14% Ta, 6% Cu and 73% ss)[81]. The PMS wire with $\alpha = 5.3$ has a Nb barrier and a Cu–30Ni matrix (16% PMS, 16% Nb, 68% Cu–30Ni) and has been measured at 14.4 T [80]. Finally both the wires with $\alpha = 2.3$ are distinguished by that different ss/Mo fraction (ss/Mo = 2.8 and 3.8) [66].

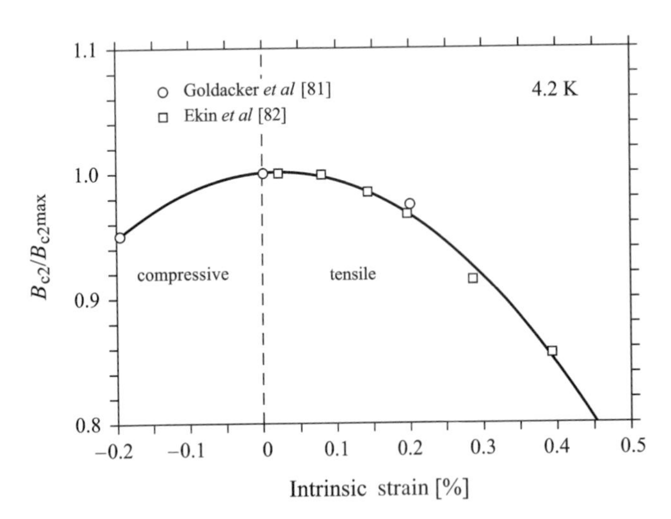

Figure B3.3.5.18. Normalized effective critical field at 4.2 K as a function of intrinsic strain (reprinted from reference [81], © 1989, with permission from Elsevier Science).

The prestress comes from the differential contraction of the components in the wire and can be adjusted by the amount of stainless steel. So, by applying an uniaxial stress, the superconductor comes first to a state with minimum stress (maximum J_c) and then under tensile stress where the J_c is reduced. A similar situation can be achieved in the other CP wires by appropriate design of the barrier and the matrix. The strain at which J_c has its maximum is considered as the state with minimum stress in the superconductor and the intrinsic strain is defined to be zero. Regarding the field dependence of J_c versus ε, Ekin has shown that there is virtually no influence between 8 and 24 T, which is an important advantage with respect to Nb_3Sn [82]. It should be mentioned that the intrinsic strain dependence of J_c is different from wire to wire and the reason for this is unknown. Finally, it is known that Nb_3Sn is rather sensitive to transverse compressive stress [83, 84]. For this reason it was also further studied in PMS and SMS wires [81]. For a transverse compressive stress of 100 MPa, which is an upper limit in many magnet designs, the J_c reduction is ~12% for the studied PMS wire and negligible for a SMS wire in fields of 20 and 18 T, respectively.

It has been found that in strain experiments (uniaxial and transverse), the effective upper critical field is a function of strain. B_{c2} is more sensitive with respect to transverse compressive stress in SMS than in PMS, but up to 100 MPa changes are negligible (a few%) [81]. Under axial stress, B_{c2} obtained by a Kramer extrapolation [81] and a fit to the pinning force [82], goes through a maximum when plotted as a function of intrinsic strain (figure B3.3.5.18). For instance, a compressive prestrain of 0.2% reduces the B_{c2} to ~95% of the value without prestrain. Consequently the prestress state of a wire, together with electromagnetic forces, can be at the origin of a reduced effective upper critical field and reduced J_cs at high fields. Again, by appropriate wire design it can be managed that the B_{c2} is near the optimum. In addition, if one takes into account that under stress primarily the T_c is changed [48] and any B_{c2} variation comes from this (see figure B3.3.5.2), there is another possibility to overcome the B_{c2} problem by the right choice of the initial CP powder.

B3.3.5.5 Concluding remarks

Due to the extraordinary phase stability of PMS and SMS, there is a good chance of optimizing parameters important for the critical current density. With respect to the upper critical field, PMS wires are obviously more important than SMS wires, although substitution or addition of tin to PMS can be of interest. Depending on the method of synthesis, the mean grain size of the starting powders can be varied in a wide range (~ 0.1 to $\sim 100\,\mu$m). This allows an improved connectivity of powder particles which is also of importance at high fields. In addition, due to the particular crystal structure, one has the possibility of introducing intra-grain pinning without significant T_c or B_{c2} reduction. Due to new routes for CP powders or precursors, the T_c and B_{c2} distribution can be kept small, which is the first step to an improvement of the effective B_{c2} at grain boundaries. At present it is unclear whether grain boundary degradation can be suppressed in CP wires with a reactive barrier like Nb or Ta, which is an important issue when dealing with small filament diameters. However, one has the possibility to overcome this difficulty by a further development of the Mo barrier technology. This implies an optimization of the hot deformation process and the availability of powders, within a wide range of particle sizes, is extremely helpful. For instance, the optimum deformation temperature for the CP powder can be adjusted to that of the barrier and the matrix. Then, appropriate hot deformation techniques improve the connection between grains and, maybe, a final heat treatment is no longer required. A scenario for a multifilament wire configuration could be a stainless steel matrix with CP filaments surrounded by a low resistivity Mo barrier (high residual resistivity ratio). Such a design allows a low matrix to superconductor ratio, α, and therefore an improved overall critical current density. The specific heat of stainless steel is about 20 times higher than that of copper, which may be helpful in the event of a quench. Further, the high electrical resistivity of the stainless steel matrix reduces coupling currents between filaments, which is of interest for time varying fields. Apart from the high critical field, probably the most important property of PMS wires is the mechanical strength due to the presence of stainless steel. Supposing an allowed tensile strain of 0.3%, and depending on the low temperature stainless steel used, tensile stress of up to 600 MPa can be applied. By an appropriate ss–Mo fraction the reduction of the effective upper critical field is negligible ($\sim$99% of its original value). Finally, the projected critical current density, in the case where the effective upper critical field is equal to the bulk value, is in the range of 1×10^9 A m^2 at 25 T.

References

[1] Fischer Ø, Jones H, Bongi G, Sergent M and Chevrel R 1974 Measurement of critical fields up to 500 kG in the ternary molybdenum sulphides *J. Phys.* C **7** L450

[2] Decroux M and Fischer Ø 1982 Critical fields of ternary molybdenum chalcogenides *Superconductivity in Ternary Compounds II*, eds M P Maple and Ø Fischer (Berlin: Springer) p 57

[3] Krauth H 1998 Commercially available superconducting wires: conductors for d.c. applications *Handbook of Applied Superconductivity*, ed B Seeber (Bristol: Institute of Physics Publishing) p 397

[4] Rikel M O, Togonidze T G and Tsebro V I 1986 Current-carrying capacity and other superconducting properties of $Pb_{1-x}Sn_xMo_6S_8$ solid solutions *Sov. Phys. Solid State* **28** 1496

[5] Rimikis G, Goldacker W, Specking W and Flükiger R 1991 Critical currents in $Pb_{1.2-x}Sn_xMo_6S_8$ wires *IEEE Trans. Mag.* **27** 1116

[6] Alekseevski N E, Glinski M, Dubrovol'skii N M and Tsebro V I 1976 Critical currents of certain superconducting molybdenum sulphides *JETP Lett.* **23** 412

[7] Decroux M, Fischer Ø and Chevrel R 1977 Superconducting wires of $PbMo_6S_8$ *Cryogenics* **17** 291

[8] Luhman T and Dew-Hughes D 1978 Superconducting wires of $PbMo_6S_8$ by powder technique *J. Appl. Phys.* **49** 936

[9] Rossel C, Seeber B and Fischer Ø 1980 Critical current densities in powder processed $PbMo_6S_8$ wires *Phys. Stat. Sol.* a **59** K43

[10] Seeber B, Rossel C and Fischer Ø 1981 $PbMo_6S_8$: a new generation of superconducting wires? *Ternary Superconductors*, eds G K Shenoy, B D Dunlap and F Y Fradin (New York, Amsterdam and Oxford: North Holland) p 119

[11] Seeber B, Rossel C, Fischer Ø and Glätzle W 1983 Investigation of the properties of $PbMo_6S_8$ powder processed wires *IEEE Trans. Mag.* **19** 402

[12] Kimura Y 1982 Superconducting properties of the $PbMo_6S_8$ conductor produced by solid state reaction *Phys. Stat. Sol.* a **69** K189

[13] Seeber B, Rossel C, Fischer Ø and Glätzle W 1983 Investigation of the properties of $PbMo_6S_8$ powder processed wires *IEEE Trans. Mag.* **19** 402

[14] Hamasaki K, Yamashita T, Komata T, Noto K and Watanabe K 1984 Superconducting wires of Pb–Mo–S by electroplating technique *Adv. Cryog. Eng.* **30** 715

[15] Hamasaki K, Noto K, Watanabe K, Yamashita T and Komata T 1989 Critical current densities and pinning mechanism of Chevrel-phase superconducting wires *Proc. MRS Int. meeting on Advanced materials, (Ikebukuro, Tokyo)* **6** 115

[16] Hamasaki K and Watanabe K 1992 (Pb,Sn)Mo_6S_8 monofilamentary wires produced by HIP technique *Science Report Res. Inst. Tohoku Univ. Serie A* **37** 51

[17] Kubo Y, Yoshizaki K, Fujiwara F and Hashimoto Y 1986 Superconducting properties of Chevrel-phase $PbMo_6S_8$ wires by an improved powder process *Adv. Cryog. Eng.* **32** 1085

[18] Kubo Y, Yoshizaki K, Fujiwara F, Noto K and Watanabe K 1989 Small coil tests of Chevrel-phase $PbMo_6S_8$ wires *Proc. MRS Int. meeting on Advanced materials, (Ikebukuro, Tokyo)* **6** 95

[19] Yamasaki H and Kimura Y 1988 Investigation of the fabrication process of hot-worked stainless steel and Mo sheathed $PbMo_6S_8$ wires *J. Appl. Phys.* **64** 766

[20] Yamasaki H, Umeda M, Kimura Y and Kosaka S 1991 Current carrying properties of the HIP treated Mo-sheath $PbMo_6S_8$ wires *IEEE Trans. Mag.* **27** 1112

[21] Yamasaki H, Umeda M, Kosaka S, Kimura Y, Willis T C and Larbalestier D C 1991 Poor intergrain connectivity of $PbMo_6S_8$ in sintered Mo-sheathed wires and the beneficial effect of hot-isostatic-pressing treatments on the transport critical current density *J. Appl. Phys.* **70** 1606

[22] Hirrien M, Chevrel R, Sergent M, Dubots P and Genevey P 1987 Cold powder metallurgy processed Chevrel phase superconducting wires. I. Prereacted $PbMo_6S_8$ powder with niobium-copper as sheath material *Mat. Lett.* **5** 173

[23] Yamashita T, Hamasaki K, Noto K, Watanabe K and Komata T 1988 High field performance of $PbMo_6S_8$ superconductors *Adv. Cryog. Eng.* **34** 669

[24] Sergent M, Hirrien M, Pena O, Padiou J, Chevrel R, Couach M, Genevey P and Dubots P 1988 First Chevrel phase $SnMo_6S_8$ wires using cold powder metallurgy process prepared from tin-doped in-situ reacted powders *Adv. Cryog. Eng.* **34** 663

[25] Goldacker W, Miraglia S, Hariharan Y, Wolf T and Flükiger R 1988 $PbMo_6S_8$ wires from HIP prereacted material *Adv. Cryog. Eng.* **34** 655

[26] Cheggour N, Gupta A, Decroux M, Flükiger R, Fischer Ø, Massat H, Langlois P, Bouquet V, Chevrel R and Sergent M 1996 Promising critical current density in the Chevrel phase superconducting wires *Physica* C **258** 21

[27] Cheggour N, Decroux M, Gupta A, Fischer Ø, Perenboom J A A J, Bouquet V, Sergent M and Chevrel R 1997 Enhancement of the critical current density in Chevrel phase superconducting wires *J. Appl. Phys.* **81** 6277

[28] Cheggour N, Decroux M, Fischer Ø and Hampshire D P 1998 Irreversibility line and granularity in Chevrel phase superconducting wires *J. Appl. Phys.* **84** 2181

[29] Capone II D W, Hinks D G and Brewe D L 1990 Improved high-field performance of $Pb_{1-x}Sn_xMo_6S_8$ ribbons *J. Appl. Phys.* **67** 3043

[30] Selvam P, Cattani D, Cors J, Decroux M, Niedermann P, Fischer Ø, Chevrel R and Pech T 1993 The role of Sn addition on the improvement of J_c in $PbMo_6S_8$ *IEEE Trans. Appl. Supercond.* **3** 1575

[31] Helfand E and Werthamer N R 1966 Temperature and impurity dependence of the superconductivity critical field H_{c2} *Phys. Rev.* **147** 288

[32] Werthamer N R, Helfand E and Hohenberg P C 1966 Temperature and purity dependence of the superconducting critical field H_{c2}. III Electron spin and spin orbital effects *Phys. Rev.* **147** 295

[33] Zheng D N, Ramsbottom H D and Hampshire D P 1995 Reversible and irreversible magnetization of the Chevrel-phase superconductor $PbMo_6S_8$ *Phys. Rev.* B **52** 12931

[34] Cors J Propriétés supraconductrices sous champ magnétique du composé $PbMo_6S_8$ étudiées par chaleur spécifique *PhD thesis* University of Geneva

[35] Cors J, Cattani D, Decroux M, Stettler A and Fischer Ø 1990 The critical field of $PbMo_6S_8$ measured by specific heat up to 14 T *Physica* B **165&166** 1521

[36] Van der Meulen H P, Peerenboom J A A J, Berendschot T T J M, Cors J, Decroux M and Fischer Ø 1995 Specific heat of $PbMo_6S_8$ in high magnetic fields *Physica* B **211** 269

[37] Decroux M, Selvam P, Cors J, Seeber B, Fischer Ø, Chevrel R, Rabiller P and Sergent M 1993 Overview on the recent progress on Chevrel phases and the impact on the development of $PbMo_6S_8$ wires *IEEE Trans. Appl. Supercond.* **3** 1502

[38] Decroux M, Fischer Ø, Flükiger R, Seeber B, Delesclefs R and Sergent M 1978 Anisotropy of H_{c2} in the Chevrel phases *Solid State Commun.* **25** 393

[39] Ramsbottom H D and Hampshire D P 1997 Improved critical current density and irreversibility line in HIPed Chevrel phase superconductor $PbMo_6S_8$ *Physica* C **274** 295

[40] Baillif R, Dunand A, Müller J and Yvon K 1981 Structural phase transformation in the cluster chalcogenides $EuMo_6S_8$ and $BaMo_6S_8$ *Phys. Rev. Lett.* **47** 672

[41] Jorgensen J D, Hinks D G and Felcher G P 1987 Lattice instability and superconductivity in the Pb, Sn and Ba Chevrel phases *Phys. Rev.* B **35** 5365

[42] François M, Yvon K, Cattani D, Decroux M, Chevrel R, Sergent M, Boudjada S and Wroblewski T h 1994 Synchrotron powder diffraction of the low-temperature lattice distortion of $PbMo_6S_8$ *J. Appl. Phys.* **75** 423

[43] Balankin A S, Bychkov Y u F and Kharchenkov A M 1985 Anomalies of the elastic properties of Chevrel phases *Sov. Tech. Phys. Lett.* **11** 28

[44] Wolf B, Molter J, Bruls G, Lüthi B and Jansen L 1996 Elastic properties of superconducting Chevrel-phase compounds *Phys. Rev.* B **54** 348

[45] Testardi L R 1973 Elastic behaviour and structural instability of high-temperature A-15 structure superconductors *Physical Acoustics*, eds W P Mason and R N Thurston (New York: Academic) p 194

[46] Meingast C, Goldacker W and Wildgruber U 1991 Thermal expansion anomalies at T_c in $Pb_xSn_yMo_6S_8$ *Proc. Int. Workshop on Chevrel Phase Superconductor* (Chavannes-de-Bogis) p 7

[47] Shelton R N, Lawson A C and Johnston D C 1975 Pressure dependence of the superconducting transition temperature for ternary molybdenum sulphides *Mat. Res. Bull.* **10** 297

[48] Capone I I D W, Guertin R P, Foner S, Hinks D G and Li H C 1984 Effect of pressure and oxygen defects in divalent Chevrel-phase superconductors *Phys. Rev.* B **29** 6375

[49] Ingle N J C, Willis T C, Larbalestier D C and Meingast C 1998 Effects of hot isostatic pressing on the lattice parameters and the transition temperature of $Pb_{0.8}Sn_{0.2}Mo_6S_8$ *Physica* C **308** 191

[50] Rikel M O 1999 Critical current and pinning mechanisms in Chevrel-phase superconductors: material science issues *IEEE Trans. Appl. Supercond.* **9** 1735

[51] Karasik V R, Rikel M O, Togonidze T G and Tsebro V I 1985 Investigation of current-carrying capacity of bulk single-phase $PbMo_6S_8$ samples with grains of $\sim 0.1\ \mu m$ size *Sov. Phys. Solid State* **27** 1889

[52] Fietz W A and Webb W W 1969 Hysteresis in superconducting alloys: temperature and field dependence of dislocation pinning in niobium alloys *Phys. Rev.* **178** 657

[53] Cattani, D Etude des densités de courant critique dans le composé $PbMo_6S_8$ *PhD thesis* University of Geneva

[54] Scanlan R M, Fietz W A and Koch E F 1975 Flux pinning centres in superconducting Nb_3Sn *J. Appl. Phys.* **46** 2244

[55] Karasik V R, Karyaev E V, Zakosarenko V M, Rikel M O and Zsebro V I 1984 Vortex-lattice pinning in bulky single-phase $PbMo_6S_8$ and $SnMo_6S_8$ samples with various grain sizes *Sov. Phys. JETP* **60** 1221

[56] Wada H, Itoh K, Tachikawa K, Yamada Y and Murase S 1985 Enhanced high-field current carrying capacities and pinning behaviour of NbTi-based superconducting alloys *J. Appl. Phys.* **57** 4415

[57] Bonney L A, Willis T C and Larbalestier D C 1995 Dependence of critical current density on microstructure in the $SnMo_6S_8$ Chevrel phase superconductor *J. Appl. Phys.* **77** 6377

[58] Gupta A, Decroux M, Selvam P, Cattani D, Willis T and Fischer Ø 1994 Critical currents and pinning powder metallurgical processed Chevrel phase bulk superconducting samples *Physica* C **234** 219

[59] Vasiliev A L, Uvarov O H, Gribeluk M A, Kiselev N A, Rickel M O, Ilyichev E A and Tsebro V I 1988 HREM of thin film $PbMo_6S_8$ superconducting compound *Ultramicroscopy* **25** 23

[60] Hinks D G, Jorgensen J D and Li H C 1983 Structure of the oxygen point defect in $SnMo_6S_8$ and Pb Mo_6S_8 *Phys. Rev. Lett* **51** 1911

[61] Le-Lay L, Powell D R and Willis T C 1992 Structure of $Sn_{0.854}Mo_6S_8$ *Acta Cryst.* **C48** 1179

[62] Seeber B, Glätzle W, Cattani D, Baillif R and Fischer Ø 1987 Thermally induced pre-stress and critical current density of $PbMo_6S_8$ wires *IEEE Trans. Mag.* **23** 1740

[63] Miraglia S, Goldacker W, Flükiger R, Seeber B and Fischer Ø 1987 Thermal expansion studies in the range 10 K–1200 K in $PbMo_6S_8$ by means of x-ray diffraction *Mat. Res. Bull.* **22** 795

[64] Grill R, Kny E and Seeber B 1989 Anwendung von Refraktärmetallen in Keramik Supraleiter *Proc. 12th Plansee Seminar* (Reutte) eds H Bildstein and R Eck p 989

[65] Willis T C, Jablonski P D and Larbalestier D 1995 Hot isostatic pressing of Chevrel-phase bulk and hydrostatically extruded wire samples *IEEE Trans. Appl. Supercond.* **5** 1209

[66] Seeber B 1998 New superconducting wires: Chevrel phases *Handbook of Applied Superconductivity*, ed B Seeber (Bristol: Institute of Physics) p 429

[67] Rabiller P, Rabiller-Baudry M, Even-Boudjada S, Burel L, Chevrel R, Sergent M, Decroux M, Cors J and Maufras J L 1994 Recent progress in Chevrel-phase synthesis: a new low temperature synthesis of the superconducting $PbMo_6S_8$ compound *Mat. Res. Bull.* **29** 567

[68] Even-Boudjada S, Burel L, Chevrel R and Sergent M 1998 New synthesis route of $PbMo_6S_8$ superconducting Chevrel-phase from ultrafine precursor mixtures: I. PbS, MoS_2 and Mo powders *Mat. Res. Bull.* **33** 237

[69] Even-Boudjada S, Burel L, Chevrel R and Sergent M 1998 New synthesis route of $PbMo_6S_8$ superconducting Chevrel-phase from ultrafine precursor mixtures: II. PbS, Mo_6S_8 and Mo powders *Mat. Res. Bull.* **33** 419

[70] Seeber B, Cheggour N, Perenboom J A A J and Grill R 1994 Critical current distribution of hot isostatically pressed $PbMo_6S_8$ wires *Physica* C **234** 343

[71] Kramer E J 1973 Scaling laws for flux pinning in hard superconductors *J Appl. Phys.* **44** 1360

[72] Goldacker W, Rimikis G, Seibt E and Flükiger R 1991 Improved $Sn_{1.2}Mo_6S_8$ wire preparation technique *IEEE Trans. Mag.* **27** 1779

[73] Rabiller P, Chevrel R, Sergent M, Ansel D and Bohn M 1992 Niobium antidiffusion barrier reactivity in tin-doped in situ $PbMo_6S_8$-based wires *J. Alloys Comp.* **178** 447

[74] Eastell, C Microstructure and properties of high temperature superconducting wires *PhD thesis* University of Oxford, UK

[75] Rabiller, P Étude et optimisation des propriétés supraconductrices de filaments à base de phase de Chevrel au plomb *PhD thesis* University of Rennes France

[76] Seeber B, Erbuke L, Schröter V, Perenboom J A A J and Grill R 1995 Critical current limiting factors of hot isostatically pressed (HIPed) $PbMo_6S_8$ wires *IEEE Trans. Appl. Supercond.* **5** 1205
[77] Baixeras J and Fournet G 1967 Pertes par déplacement de vortex dans un supraconducteur de type-II non idéal *J. Phys. Chem. Solids* **28** 1541
[78] Warnes W H and Larbalestier D 1986 Critical current distributions in superconducting composites *Cryogenics* **26** 643
[79] Rimikis, G Einflussgrössen und Methoden zur Optimierung der supraleitenden und mechanischen Eigenschaften von Chevrelphasendrährten *PhD thesis* University of Karlsruhe, Germany
[80] Katagiri K, Seo K, Okada T, Kubo Y, Fujiwara F, Noto K, Morii Y, Ishii M and Watanabe K 1992 Strain effects in $PbMo_6S_8$ wires *Adv. Cryog. Eng.* **38** 843
[81] Goldacker W, Specking W, Weiss F, Rimikis G and Flükiger R 1989 Influence of transverse compressive and axial tensile stress on the superconductivity of $PbMo_6S_8$ and $SnMo_6S_8$ wires *Cryogenics* **29** 955
[82] Ekin J W, Yamashita T and Hamasaki K 1985 Effect of uniaxial strain on the critical current and critical field of Chevrel phase $PbMo_6S_8$ superconductors *IEEE Trans. Mag.* **21** 474
[83] Ekin J W 1988 Transverse stress effect on multifilamentary Nb_3Sn superconductors *Adv. Cryog. Eng.* **34** 547
[84] Specking W, Goldacker W and Flükiger R 1988 Effect of transverse compression on I_c of Nb_3Sn multifilamentary wires *Adv. Cryog. Eng.* **34** 569

B3.3.6
Processing of low T_c conductors: the compound MgB_2

Hongli Suo, Pierre Toulemonde and René Flükiger

B3.3.6.1 Introduction

The discovery [1] of MgB_2 as a superconductor with a transition temperature near 40 K has promoted considerable interest in the area of basic and applied research on superconducting materials. High transport critical current densities have been reported in MgB_2 bulk samples [2–4]. Because of its intermediate transition temperature, weak link free grain boundaries [5], low weight and very low material cost, this material is interesting for potential applications such as magnetic resonance imaging (MRI) and transformers in an operating temperature in the range of 20–30 K. Canfield *et al* [6] were the first to report on MgB_2 wires (obtained by diffusion of Mg vapour into boron fibers reinforced by a W core) with inductive self-field J_c values of $10^5\,A\,cm^{-2}$ at 4.2 K. Large critical current densities of $10^6\,A\,cm^{-2}$ and significant enhancements of the irreversibility field, B_{irr}, and the upper critical field, B_{c2}, have later been reported in MgB_2 films grown on single crystalline substrates [7, 8]. In view of eventual industrial applications, several research groups have tried to demonstrate the feasibility of fabricating MgB_2 tapes with high critical current density, using the powder-in-tube (PIT) method, either with a heat treatment after deformation [9–16] or in the as-deformed state [17–19]. The transport critical current densities of these tapes at 4.2 K and in zero external field are in the range of $\sim 10^4$–$10^6\,A\,cm^{-2}$. It was found that both the grain size of starting powder [10] and the appropriate deformation, followed by post-annealing at 850–950°C, were determinant for obtaining higher J_c values.

C5 E2.3 A1.3 B3.1

Iron was found to be an excellent sheath material for MgB_2 tapes [9, 10, 12, 15]. So far, multifilamentary MgB_2 wires have been produced using sheathing tubes consisting of Cu–Ni [18], Cu/NbZr [20] or Fe [21, 22]. The highest J_c values were obtained after two-axial rolling, followed by annealing under Ar flow. The J_c value in these Fe/MgB_2 square wires after annealing at 950°C was $1.1\times10^5\,A\,cm^{-2}$ at 4.2 K/2 T. The self field value of J_c at 4.2 K of MgB_2 tapes with seven filaments [21] was estimated to $4\times10^5\,A\,cm^{-2}$, while at 1 T, $5\times10^4\,A\,cm^{-2}$ was obtained.

In view of applications, a careful study of the transport behavior and the physical properties of MgB_2 tapes is essential in order to get an insight into the flux pinning mechanism and thus to further improve their transport currents. Reports on the irreversibility fields, electronic anisotropy, transport critical current densities and n factors in monofilamentary MgB_2 tapes have appeared recently [23]. A significant effect of both sheath material and initial MgB_2 powder grain size on transport J_c values and upper critical fields was reported. More recently, Komori *et al* [24] reported a new approach for the fabrication of MgB_2 superconducting tapes with large in-field transport critical current density. The tape was prepared by depositing MgB_2 film on a Hastelloy tape buffered with an YSZ layer (PLD process).

B4.2.1

B3.3.2 The J_c values of these tapes exceeded $10^5\,A\,cm^{-2}$ at 4.2 K and 10 T, which is superior to those of conventional Nb–Ti wires.

This review summarizes the preparation and the most recent results and developments in fabricating MgB_2 monofilamentary tapes. We will mainly focus on Fe/MgB_2 mono- and multifilamentary tapes [9, 10, 23], and will also discuss the magnetic field dependence of the critical current A1.3 density as well as the temperature dependence of B_{c2} and B_{irr}. The anisotropy effects on the critical A4.3 current density are discussed and the first results on exponential n factor are presented.

B3.3.6.2 Preparation and microstructure of Fe/MgB_2 tapes

B3.1 There are so far three methods to fabricate MgB_2 wires and tapes: Mg diffusion into B whiskers, deposition of MgB_2 films on metallic ribbons and powder-in-tube (PIT) methods. Diffusion of Mg into B wires has been found to rapidly convert commercially available B fibres of 200 μm diameter into superconducting MgB_2 wires [6]. The Mg diffusion method was also successful for the fabrication of A1.3 tapes [25]. The PIT process is actually the only known method for preparing multifilamentary wires and tapes. One of the main problems encountered when producing industrial MgB_2 wires or tapes by this process is the hardness and brittleness of MgB_2. In particular, drawing or rolling into wires or tapes with very fine filaments constitutes a particular problem. Two different kinds of precursor powders can be used in the PIT process: pre-reacted MgB_2 powder (*ex situ*) or a mixture of Mg and B powders at the stoichiometric composition Mg : B = 1 : 2, with an *in situ* formation of the phase MgB_2. In both cases, powders were packed in the tube of sheath material, followed by swaging, drawing and rolling. In the following, the discussion will be restricted to the *ex situ* method with pre-reacted MgB_2 powders.

As an alternative to MgB_2 tapes, Komori *et al* [24] prepared MgB_2 tapes by depositing MgB_2 B4.1 film on a Hastelloy tape buffered with an YSZ layer. The substrate of Hastelloy (C-276) tape, $4 \times 30 \times 0.3\,mm^3$ was pre-coated with a YSZ (yttria-stabilized zirconia) buffer layer of 1 μm thickness. The films, of a final thickness of 400 nm, were deposited on a buffered substrate using a KrF excimer laser, and then heated for 20–30 min in the temperature range of 550–660°C (see chapter B3.2.5).

B3.3.6.2.1 Pre-reacted MgB_2 powders (ex situ process)

Tapes produced using pre-reacted MgB_2 powder are already superconducting after deformation, even before any treatment [17]. Nevertheless, it has been proven [9, 10] that appropriate heat treatments are determinant for getting higher J_c values. Usually, the wires and tapes were treated at 900–1000°C under a Ar atmosphere after deformation [9–14, 18–20]. The pre-reacted powder used for *ex situ* PIT process was a standard commercial MgB_2 powder (e.g. from Alfa-Aesar with a purity of 98%) after ball milling. At the University of Geneva, Suo *et al* [9, 10] and Flükiger *et al* [23] have demonstrated A1.3 the influence of the initial MgB_2 grain size on J_c and on B_{irr}. A typical transverse and longitudinal cross-section of deformed Fe/MgB_2 tapes is shown in figures B3.3.6.1(a) and B3.3.6.1(b), showing very uniform transverse and longitudinal sections. SEM micrographs of polished cross-sections of the tapes (figure B3.3.6.1(c)) show that the MgB_2 core consists of a tightly packed powder with grain sizes in the order of 0.5–2 μm, i.e. markedly smaller than the initial grain size, as a consequence of grain refinement during deformation. After heat treatment, the cross-sections of the tapes show a highly dense MgB_2 core (figure B3.3.6.1(d)), due to grain recrystallization, leading to a strongly improved intergranular connectivity, as can be seen from the almost flat and dense aspect of this image. B2.2.1 The powder x-ray diffraction patterns of the crushed MgB_2 core revealed ~5% of MgO as the only impurity phase after sintering.

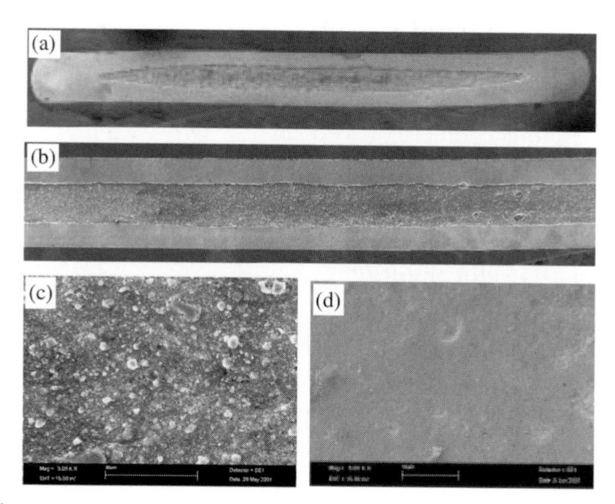

Figure B3.3.6.1. SEM microstructures of the Fe/MgB_2 tapes: (*a*) transverse cross-section of as-rolled Fe/MgB_2 tape; (*b*) longitudinal cross-section of as-rolled Fe/MgB_2 tapes; (*c*) MgB_2 core in as-rolled Fe/MgB_2 tapes; (*d*) Fe/MgB_2 core after annealing [9, 10, 22].

B3.3.6.3 Results and discussion

B3.3.6.3.1 Superconducting transitions in as-rolled and annealed Fe/MgB_2 tapes

Several authors [9–11] have reported ac susceptibility measurements of as-rolled MgB_2 tapes showing two transitions, related to intergranular and intragranular currents. The intergranular transition can be identified as a signature of weak links.

A4.3

Typical ac and dc susceptibility data for the MgB_2 cores in Fe/MgB_2 tapes before and after annealing and also for the starting powder are shown in figure B3.3.6.2 [9, 10]. As-rolled and annealed tapes showed an onset of the superconducting transition at 37 K, i.e. markedly lower than the starting powder (39 K). This decrease of T_c is explained by deformation induced stresses, which are not completely released after annealing, as confirmed by x-ray diffraction pattern refinement. Note that the *in situ* measurements of the pressure dependence of T_c have also revealed a decrease of T_c upon compressive strain (0.8–2 K GPa^{-1}) [26, 27], which is only partially recovered when the pressure is released [28]. Moreover, the deformation dramatically affects the grain connectivity as shown by the marked weak link behavior of the as-rolled tapes, indicated by a dissipative peak in χ'', and confirmed by SQUID magnetometry [29][†]. After annealing, the connectivity is recovered and the tapes present a unique and sharp superconducting transition.

D2.5 D1.1.2 D2.4

B3.3.6.3.2 The critical current density J_c (H,T) of Fe/MgB_2 tapes

Effect of the heat treatment

A J_c value of 10^5 A cm^{-2} at 4.2 K/0 T on as-deformed Ni–sheathed MgB_2 tapes was first reported by Grasso *et al* [17]. The critical current density (both inductively and resistively) in the monofilamentary

D2.2, D2.4

[†]We note that the annealed Fe/MgB_2 tape showed a small paramagnetic Meissner effect. The same behaviour was also measured in Nb samples with strong pinning.

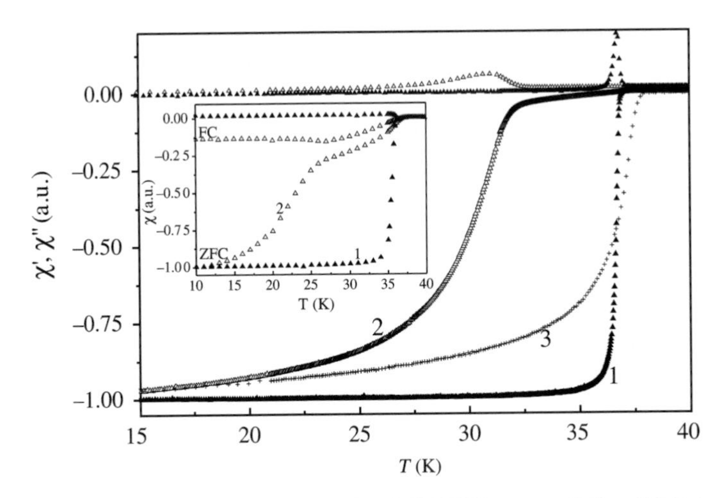

Figure B3.3.6.2. AC susceptibility (0.1 Oe, 8 KHz) and DC SQUID susceptibility (20 Oe) *versus* T. (1): Fe/MgB_2 tapes, annealed; (2) Fe/MgB_2 tapes, as-rolled; (3) MgB_2 starting powder. Note the dissipative peak in χ'' at ~32 K for as-rolled tapes. [9, 10].

MgB_2 tapes and wires obtained by PIT method, followed by annealing, is in the range of $\sim 10^4$–10^6 A cm^{-2} at 4.2 K and in zero field (actually, J_c values at 0 T can only be determined by extrapolation, due to quenching effects). The appropriate annealing of the tapes increases the core density and sharpens the superconducting transition, strongly raising J_c, as shown in figure B3.3.6.3 [9, 10]. In as-rolled Ni and Fe sheathed tapes, a self-field transport critical current density around 10^4 A cm^{-2} at 4.2 K/2 T was obtained. For the Ni/MgB_2 tape after annealing, the transport J_c values increase by more than a factor of 10, reaching 2.3×10^5 A cm^{-2} at 1.5 T (corresponding to a I_c value of 300 A). For the Fe/MgB_2 tape, a value of 10^4 A cm^{-2} at 6.5 T was found, corresponding to a I_c value

D2.2

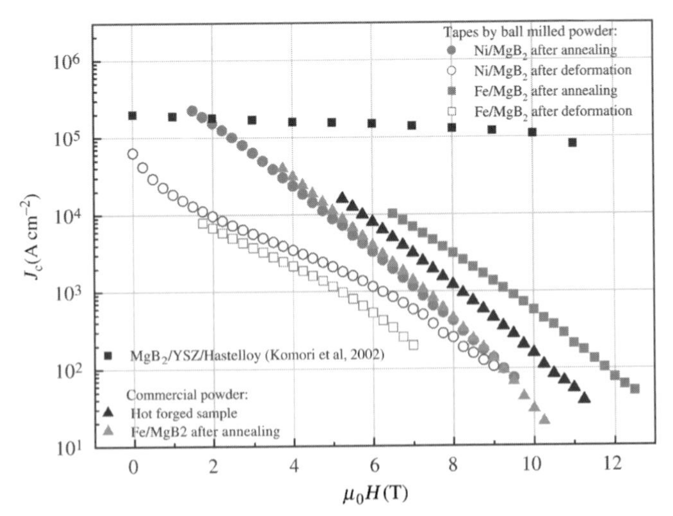

Figure B3.3.6.3. Field dependence of the transport J_c values at $T = 4.2$ K in both as-rolled and annealed Fe/MgB_2 tapes [9, 10], and for deposited MgB_2 film [24]. For comparison, the transport J_c curve of a dense hot forged bulk sample is also shown [4].

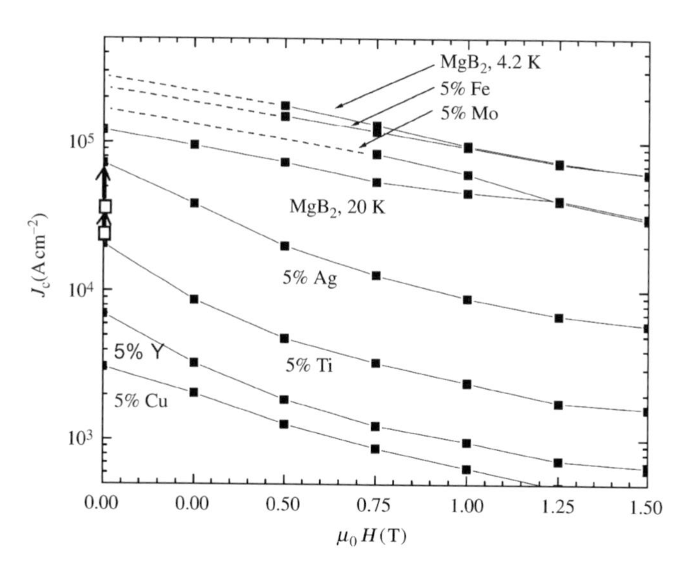

Figure B3.3.6.4. Inductive J_c *versus* B for various MgB_2 samples reported by Jin *et al* [12].

of 40 A. Because of insufficient thermal stabilization, sample quenching limits the measurements above a certain current in the tapes. Extrapolating the field dependence of transport J_c in Fe sheathed tapes yielded self-field values close to 10^6 A cm^{-2}. Such a high J_c value would correspond to critical currents well above 1000 A, which can only be measured after improving the thermal stability, for example by forming multifilamentary tapes with filament diameters of a few microns. A1.3

Jin *et al* [12] also reported a higher transport J_c ($>3.6 \times 10^4$ A cm^{-2}) at 20 K/0 T after a heat treatment at 900°C/30 min in Ar in their Fe/MgB_2 tapes (figure B3.3.6.4). Wang *et al* [30] studied the effect of sintering time on the critical current density of MgB_2 wires prepared by PIT method using Mg : B = 1 : 2 powder mixtures. The latter found the remarkable result that a heat treatment of a few minutes yields the same J_c values as those for much longer sintering times. At 15 K, these authors achieved J_c values of 4.5×10^5 and 1.0×10^5 A cm^{-2} at 0 and 2 T, respectively.

Effect of the sheath material

For the fabrication of MgB_2 tapes by PIT process it is important to find a suitable sheath material, which is not only flexible but also inert with respect to MgB_2. This sheath has to play the role of a diffusion barrier for the volatile and reactive Mg. Since Mg and MgB_2 tend to react with many metals, such as Cu and Ag, forming solid solutions or intermetallics with low melting points and rendering the metal cladding useless during sintering of MgB_2 at 900–1000°C. There is only a small number of metals which are not soluble or do not form intermetallic compounds with Mg [12]. These are Fe, Mo, Nb, V, Ta, Hf and W. So far, various sheath materials have been tried to fabricate metal clad MgB_2 wires/tapes, such as Ni [9, 10, 17], Cu–Ni [18], Nb–Ta/Cu/stainless steel [11], stainless steel [18, 19], Cu [11], Ag [11], Ag/stainless steel [11], Fe/stainless steel [14] and Fe [9, 10, 12, 13]. It was found that Fe is the most appropriate material as a practical cladding metal or diffusion barrier for MgB_2 wire and tape fabrication.

Suo *et al* [9, 10] observed a $\sim 15\,\mu$m thick reaction layer between the Ni sheath and the MgB_2 filament after annealing. Energy-dispersive x-ray (EDX) microanalysis showed these layers to consist of the phase $MgBNi_2$ (also reported by Grasso *et al* [17]). This reaction consumes some of the Mg within the filament, leading to an increase in the porosity, thus degrading the transport J_c value as described D1.3

below. On the other hand, only a faint reaction was found between the Fe sheath and the MgB_2. After annealing, the Ni/MgB_2 tape had, at 1.5 T, a J_c value of 2.3×10^5 A cm^{-2}. In a field of 6.5 T its value was B2.1 five times lower than that of Fe/MgB_2 tapes ($\sim 10^4$ A cm^{-2}) prepared using the same ball milled powder.

Effect of additive elements

An improvement of the B_{c2} or the flux pinning potential of MgB_2 materials can be realized through structural or microstructural modifications. The effect of grain boundaries can be improved by decreasing the grain size (see next paragraph). However, other methods can be used, e.g. chemical doping, introduction of precipitates, dispersion of nano-sized particles, all of which may lead to defects at the atomic scale, e.g. vacancies or dislocations.

The field and temperature dependence of J_c after introducing either additive elements or substitutions on the Mg site [31–43] or on the B site [44–46] has been studied in bulk MgB_2 samples, with the elements Li, Na, Ca, Ag, Cu, Al, Zn, Zr, Ti, Mn, Fe, Co and C. However, only Al has been found to really substitute Mg, thus yielding the solid solution $(Mg_{1-x}Al_x)B_2$ with T_c decreasing gradually.

An increase of J_c, observed after doping with Ti [47, 48] and Zr [49], was interpreted as being due to nano-sized defects in the MgB_2 grains. Ti does not occupy an atomic site in the MgB_2 structure but forms a thin TiB_2 layer with a thickness of about one unit cell at the MgB_2 grain boundaries, forming a strongly coupled nano-particle structure, which may be responsible for the enhancement of J_c. Oxygen additions have also been reported as a possible reason for the enhancement of J_c in both bulk materials and in thin films [7]. Recently, it has been shown by Liao *et al* [50] that coherent Mg(B,O) with nanometre size, observed inside MgB_2 grains may be correlated to the increase of flux pinning in such materials.

The preparation of doped $(Mg_{1-x}Al_x)B_2$ or $Mg(B_{1-x}C_x)_2$ has mainly been realized on bulk samples, and little data are known about wires and tapes. Jin *et al* [12] have reported on the effect of additive elements (Ti, Fe, Ag, Cu, Mo or Y) on the critical current densities of wires prepared by the PIT process. These authors have shown that alloying MgB_2 with 5 mol% of Ti, Ag, Cu, Mo or Y drastically reduces J_c, as shown in figure B3.3.6.4. This strong decrease of J_c is related to the reaction between MgB_2 and the additive elements. In the case of Fe, a very small decrease of J_c was observed. Cu additions caused a B3.1 decrease by two to three orders of magnitude, which is related to a reaction with MgB_2. Fe is thus particularly suitable for the wire preparation, either as sheath material or for diffusion barriers.

The effect of the powder grain size

In order to increase J_c in wires and tapes, the fabrication process must be optimized by using finer B2.1 starting powders or by incorporating chemically inert nano-scale particles that would inhibit the grain growth. Eom *et al* [7] obtained J_c values of 10^4 A cm^{-2} at fields as high as 14 T after doping of thin films with oxygen. Nano-inclusions of precipitates Mg(B,O) have also been effective in enhancing J_c [50]. These high current densities, exceeding 10^6 A cm^{-2} in thin films [7], thus demonstrating the potential for further improving the current carrying capacities of tapes using powder of the size of 1 μm and below.

The effect of the initial powder grain size on the transport critical current density was first studied by Suo *et al* [9, 10], who reduced the average grain size by ball milling. As shown in figure B3.3.6.5, the as-purchased powder contains a large number of agglomerated grains, with a wide size distribution centred at around 60 μm. After 2 h of ball milling in a glove box, 35% of the grains had diameters around 3 μm, while 60% were still centred around 30 μm. The transport critical current densities D2.2 for annealed Fe/MgB_2 tapes prepared by both as-purchased and ball-milled powders are also shown in figure B3.3.6.3. At 6.5 T, the tape with as-purchased powder had a J_c value of 2.4×10^3 A cm^{-2}, i.e. four times the value of 10^4 A cm^{-2} for the tape with ball-milled powder.

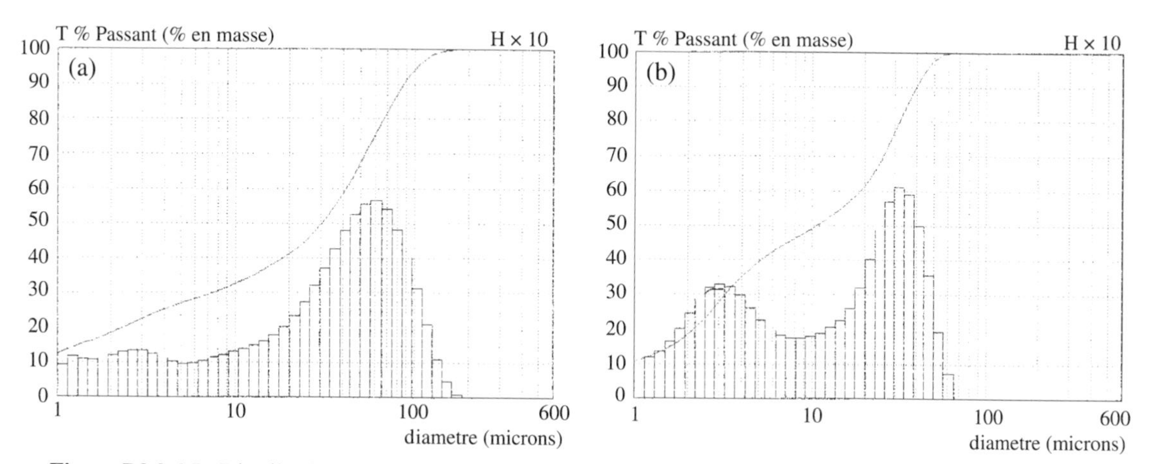

Figure B3.3.6.5. Distributions of powder grain size: (*a*) as-purchased powder; (*b*) ball-milled powder [21].

High B_{c2} values in deposited MgB_2 films

As mentioned earlier, Komori *et al* [24] have prepared tapes prepared by depositing thin MgB_2 films on a Hastelloy tape buffered with an YSZ layer. In this tape, very large J_c values of more than $10^4\,A\,cm^{-2}$ at fields as high as 12 T were observed, which is about two orders of magnitude larger than the typical value of PIT processed tapes (figure B3.3.6.3). This result shows that such tapes already have better magnetic field characteristics than conventional Nb–Ti wires. It follows that the small magnetic field dependence of the J_c values is correlated with the very high value of $B_{c2}(0) \approx 39$ T, which is twice as high as the value obtained for the actually best PIT Fe/MgB_2 tapes (figure B3.3.6.6). In these MgB_2 films, the MgB_2 phase had grain sizes of ≤ 10 nm, i.e. much smaller than the typical MgB_2 grain size of PIT processed tape (between 100 and 1000 nm [9, 18]). A strong grain boundary pinning force may thus be correlated with

B3.1
A4.3

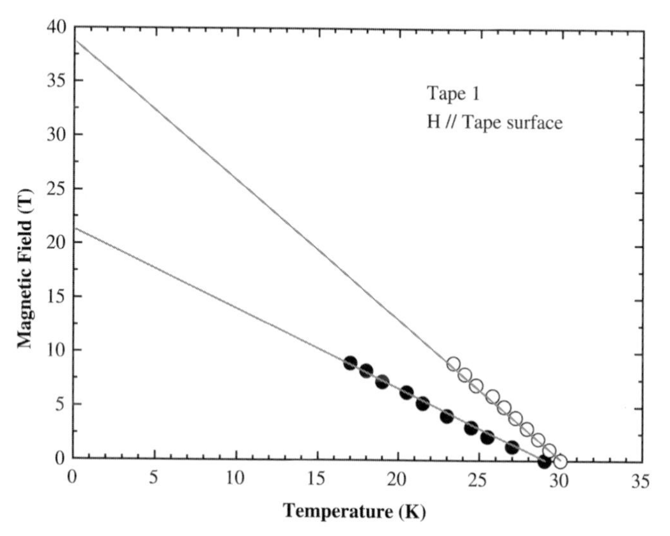

Figure B3.3.6.6. Temperature dependence of H_{c2} and H_{irr} as determined by the onset and offset temperatures of resistive transitions at magnetic fields up to 9 T for MgB_2 films deposited on a Hastelloy ribbons buffered with an YSZ layer (after Komori *et al* [24]).

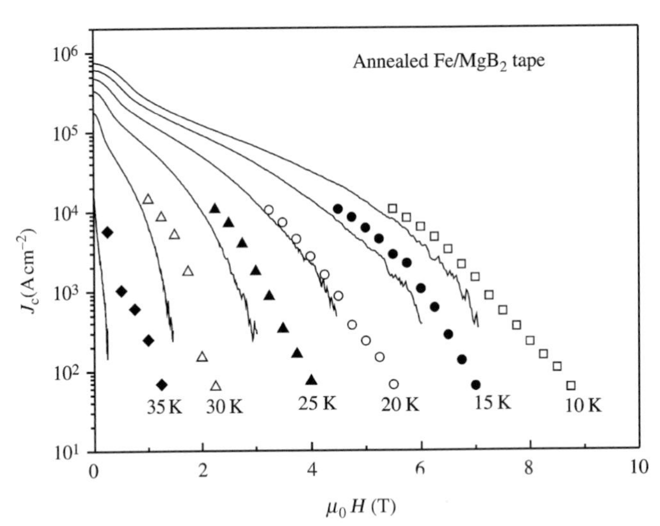

Figure B3.3.6.7. Field dependence of J_c for annealed Fe/MgB_2 tapes, measured either inductively (solid lines) or resistively (symbols) [23].

B2.5 the very small grains of these thin layers. This can be concluded by analogy with the case of A15 type superconductors, where a strong correlation between J_c and the grain size has been reported, the pinning force being inversely proportional to the grain size [51].

The combined effect of higher B_{c2} values and of smaller powder grain sizes on J_c is the key for further improvement of J_c in PIT processed tapes. B3.1

Critical current density at different temperatures

Several groups have reported the transport J_c values as a function of temperature [12, 13, 15, 16, 23]. Figure B3.3.6.7 shows both inductive and transport J_c values for one of the Fe/MgB_2 tapes in [23], measured every 5 K between 10 and 35 K and plotted against the magnetic field. The inductive $J_c(H,T)$ values were calculated from the magnetic hysteresis loops of the MgB_2 cores using the Bean critical state model. Above $J_c \approx 10^4$ A cm^{-2}, the transport measurements are affected by sample quenches due to the insufficient thermal stability, the electrical resistivity of Fe being too high. The temperature of the sample was monitored during the measurement, showing a temperature rise of approximately 0.5 K at high currents. The transport J_c values are consistent with those obtained from magnetization measurements. The field dependence of J_c is essentially the same for transport and magnetization and, well above the self field regime, J_c is found to decay almost exponentially, for inductive and transport data as well. The transport J_c values are higher than those of the magnetic ones at fixed field and temperature due to the different criteria fixed for the two independent experiments. At $T = 25$ and 30 K, transport J_c well above 10^4 A cm^{-2} were obtained at fields of 2.25 and 1.0 T, respectively. D2.2 D2.4

B3.3.6.3.3 Critical field

Upper critical field

A3.1 As reported by Flükiger *et al* [23], the main effect of the Ni sheath after annealing at 950°C is to lower the upper critical field of the Ni/MgB_2 tape, due to the reaction with the MgB_2 phase. This has a negative

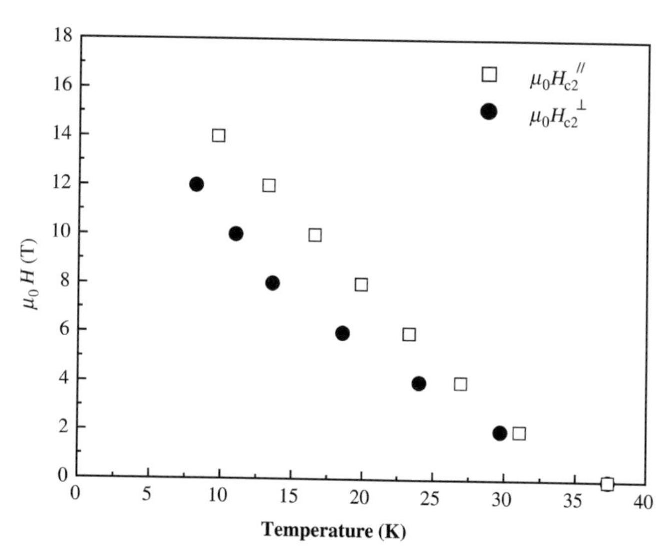

Figure B3.3.6.8. B_{c2} versus T of annealed Fe/MgB_2 tapes, for fields perpendicular and parallel to the sample [23].

effect on the core, thus causing a decrease of the transport critical current and upper critical field, in contrast to the Fe sheath, which is almost inert to MgB_2.

Eom *et al* [7] and Komori *et al* [24] have reported a strong enhancement of the upper critical field in their MgB_2 films after deposition. The latter determined the magnetic parameters H_{c2} and H_{irr} by taking the onset and offset temperatures of the resistive transition, respectively, with fields applied parallel to the tape surface. Figure B3.3.6.6 shows the plot of B_{c2} and B_{irr} for their films as a function of temperature. Linear extrapolation gives the B_{c2}(4.2 K) and B_{irr}(4.2 K) values of about 33 and 18 T, respectively. These very high values are comparable to those of Eom *et al* [7] for thin films.

A3.1
A1.3

Electronic anisotropy

Since MgB_2 consists of a layered structure of Mg and B hexagonal sheets, electronic anisotropy is expected. Because of very different sample shape and fabrication process, the reported values of $\gamma = \mu_0 H''_{c2}(0\,\mathrm{K})/\mu_0 H^{\perp}_{c2}(0\,\mathrm{K})$ show a considerable scattering, with γ values ranging between 1 [52] and 13 [53–56]. Experiments performed on single crystals [57–62] have given γ values between 2.6 and 3. A somewhat smaller anisotropy of 1.8–2.5 has also been observed in *c*-axis oriented thin films [63–65].

C5

Since MgB_2 is a highly promising candidate for technological applications, it is important to resolve the effect of anisotropy in Fe/MgB_2 tapes. Beneduce *et al* [23] measured the superconducting transition at magnetic fields up to 14 T. In figure B3.3.6.8, they report B_{c2} as a function of temperature for both orientations: the upper critical field is considerably higher when the field is perpendicular to the *c*-axis (i.e. parallel to the tape surface). As in other reports [6], the $B_{c2}(T)$ curve shows a positive curvature near T_c. The slopes dB_{c2}/dT of the $\mu_0 H_{c2}\,(T)$ curves for fields between 2 and 14 T were evaluated as -0.45 and -0.57 T/K for the perpendicular and the parallel configuration, respectively [23]. By using the dirty limit extrapolation, they found $\mu_0\, B''_{c2} = 15.1$ T and $\mu_0\, B^{\top}_{c2} = 11.9$ T, thus yielding an anisotropy factor of $\gamma = 1.3$, which shows only small variations with T. The same value for the anisotropic factor was found in both Ta/MgB_2 wires [16] and hot deformed bulk MgB_2 samples [66]. This value is smaller than that of the corresponding values for the *c*-axis oriented thin films [63–65] and for MgB_2 single crystals [57–62], which are $\gamma = 1.8$–2.2 and 2.6, respectively. This result suggests a partial alignment of the grains along a crystallographic axis, i.e. a preferential orientation of the core grains, especially at the interface with the iron sheath. Structural Rietveld refinements on XRD patterns of the MgB_2 core at the Fe sheath interface suggest a preferential texture [23]. The rocking curve around the (0 0 2) reflection gives a FWHM of ~18°, which indicates a slight preferential orientation along the c-axis of the MgB_2 grains near the Fe sheath. Further optimization of the deformation process is expected to give higher anisotropy factors.

Irreversibility field

A1.3 In view of MgB_2 applications, it is important to know the field values below which J_C is non-zero. The definition of 'irreversibility field', H_{irr} is currently used for this limit in MgB_2. In some papers, the expression H^* (which is in reality a depinning field) is used, but the same effect is described: a separation between pinned and unpinned vortices. Buzea *et al* [67] have summarized the reported irreversibility fields for a series of samples, showing very different results, depending on the microstructure. The highest values are reported for thin films [7], while bulk samples and tapes exhibit lower similar values. Note that the definitions for $\mu_0 H_{irr}$ are not always the same ones: certain authors define this value as the point of zero resistance under a given field, while others use a more realistic zero J_C criterion. For Fe/MgB_2 tapes prepared using prereacted powders, at 5 K Wang *et al* [30] have reported B_{irr} values of 6 T, while our tapes [23] showed somewhat higher values (8 T at the same temperature). The temperature dependence of B_{c2} for our tapes based on commercial and ball milled powders (size distributions in figure B3.3.6.5) is shown in figure B3.3.6.9. Ball milling does not affect B_{c2} i.e., the crushing procedure does not affect the grain cores. The present data do not yet allow us to give an unambiguous answer about the ball milling effect on B_{irr}.

The exponential n factor

A1.3 The upper critical field and the irreversibility field for MgB_2 are well above those of NbTi. As several groups are making rapid progress in producing MgB_2 based wires or tapes, the prospect of low-cost

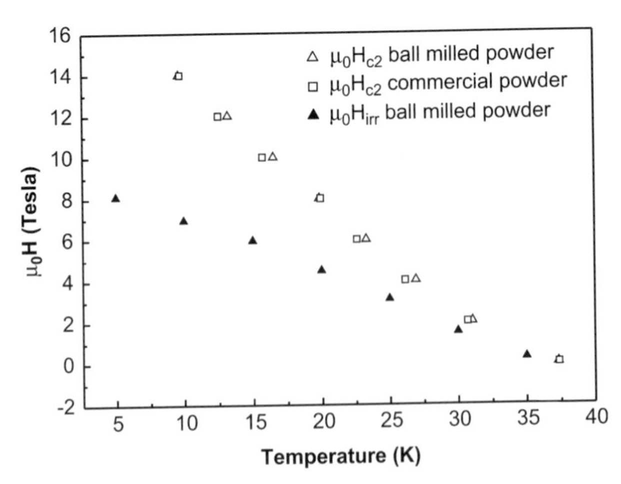

Figure B3.3.6.9. Temperature dependence of B_{c2} for the Fe sheathed tapes based on commercial and ball milled MgB_2 powders [23]. For the latter, B_{irr} is also given.

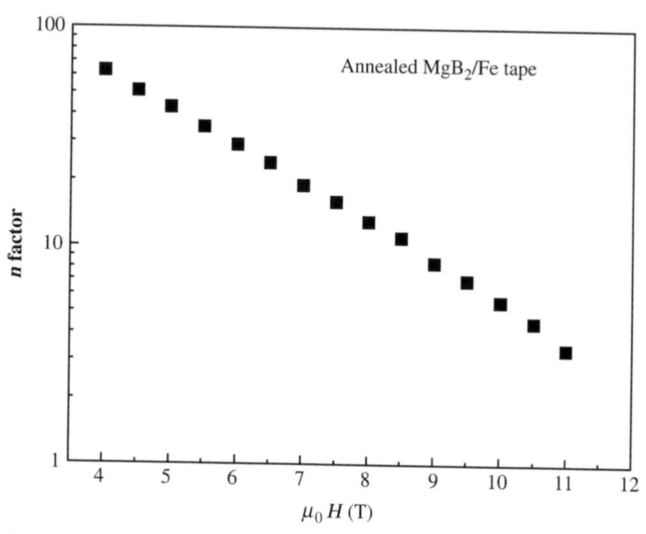

Figure B3.3.6.10. Field dependence of the exponential n factor in Fe/MgB_2 tapes [23].

superconducting solenoids [69] based on this compound becomes ever more realistic. To use such magnets in the persistent mode, the n factor of the superconducting wire should have a value greater than 30 at the operating field and temperature [70, 71]. The variation of the n factor with field in the annealed Fe/MgB_2 tapes reported by Flükiger *et al* [23] is shown in figure B3.3.6.10. The factor was determined to be 60 at 4 T and ~30 at 6 T, and decreases exponentially to 10 at 8.5 T. The high value of the n factors at intermediate magnetic field has an important practical consequence, since it can be advanced that such magnets could be operated in the persistent mode at fields smaller than 6 T at 4.2 K.

B3.3.6.4 Conclusion

The most used method for fabrication of MgB_2 wires and tapes is the PIT method. Among all metals, which have been tried as possible sheath materials, Fe is so far the only one being almost chemically inert with respect to MgB_2. The recrystallization during annealing leads to a densification and to a strong improvement of J_c values, which are more than a factor of ~10 higher than those of the as-deformed tapes. The study on the effect of the initial MgB_2 powder grain size shows that both the critical current density and the irreversibility field in Fe/MgB_2 tapes are enhanced after reducing the initial powder to sizes of the order of $\leq 3\ \mu$m under Ar atmosphere. Ball milling has no effect on the upper critical field, B_{c2}, showing that this process does not affect the grain cores. Transport J_c values above 10^4 A cm^{-2} are obtained in Fe/MgB_2 tapes at 4.2 K/6.5 T, 25 K/2.25 T and 30 K/1 T. Extrapolation to zero field at 4.2 K yields J_c values around 10^6 A cm^{-2}, but this value will only be reached after an improvement of the thermal stability of MgB_2 tapes. Anisotropy measurements indicate an upper critical field anisotropy ratio of 1.3: it is expected that this value will be enhanced by appropriate deformation processes. The Fe/MgB_2 tapes exhibit a very high n factor, which opens possibilities for a persistent mode operation at lower fields.

B3.1

A1.3

It appears that Fe/MgB_2 tapes have the potential to be produced industrially. However, thermal stability and the mechanical properties of the wires and tapes have to be improved, thus constituting a major challenge for the next few years.

A2.4 D3.3

References

[1] Nagamatsu J, Nakagawa N, Muranka T, Zenitanim Y and Akimitsu J 2001 *Nature* **410** 63
[2] Kambara M, Babu N H, Sadki E S, Cooper J R, Minami H, Cardwell D A, Campbell A M and Inoue I H 2001 *Supercond. Sci. Technol.* **14** L5
[3] Takano Y, Takeya H, Fujii H, Kumakura H, Hatano T and Togano K 2001 *Appl. Phys. Lett.* **78** 2914
[4] Dhallé M, Toulemonde P, Beneduce C, Musolino N, Decroux M and Flükiger R 2001 *Physica* C **363** 155
[5] Larbalestier D C, et al, 2001 *Nature* **410** 186
[6] Canfield P C, Finnemore D K, Budko S L, Ostenson J E, Lapertot G, Cunningham C E and Petrovic C 2001 *Phys. Rev. Lett.* **86** 2423
[7] Eom C B, Lee M K, Choi J H, Belenky L, Song X, Cooley L D, Naus M T, Patnaik S, Jiang J, Rikel M, Polyanskii A, Gurevich A, Cai X Y, Bu S D, Babcock S E, Hellstrom E E, Labalestier D C, Rogado N, Regan K A, Hayward M A, He T, Slusky J S, Inumaru K, Haas M K and Cava R J 2001 *Nature* **411** 558
[8] Paranthaman M, Cantoni C, Zhai H Y, Christen H M, Aytug T, Sathyamurthy S, Specht E D, Thompson J R, Lowndes D H, Kerchner H R and Christen D K 2001 *Appl. Phys. Lett.* **78** 3669
[9] Suo H L, Beneduce C, Dhallé M, Musolino N, Genoud J Y and Flükiger R 2001 *Appl. Phys. Lett.* **79** 3116
[10] Suo H L, Beneduce C, Dhallé M, Musolino N, Su X D, Walker E and Flükiger R 2001 *Adv. Cryog. Eng.* **48** 872
[11] Glowacki B A, Majoros M, Vickers M, Evetts J E, Shi Y and Mcdougall I 2001 *Supercond. Sci. Technol.* **14** 193
[12] Jin S, Mavoori H, Bower C and van Dover R B 2001 *Nature* **411** 563
[13] Soltanian S, Wang X L, Kusevic I, Babic E, Li A H, Liu H K, Collings E W and Dou S X 2001 *Physica* C **361** 84
[14] Nast R, Schlachter S I, Zimmer S, Reiner H and Goldacker W. *Physica* C, in press.
[15] Goldacker W, Schlachter S I, Zimmer S and Reiner H 2002 *Supercond. Sci. Technol.* **14** 787
[16] Pradhan A K, Feng Y, Zhao Y and Koshizuka N 2001 *Appl. Phys. Lett.* **79** 1649
[17] Grasso G, Malagoli A, Ferdeghini C, Roncallo S, Braccini V, Cimberle M R and Siri A S 2001 *Appl. Phys. Lett.* **79** 230
[18] Kumakura H, Matsumoto A, Fujii H and Togano K 2001 *Appl. Phys. Lett.* **79** 2435 cond-mat/0106002
[19] Song K J, Lee N J, Jang H M, Ha H S, Ha D W, Oh S S, Sohn M H, Kwon Y K and Ryu K S 2001 cond-mat/0106124.
[20] Liu C F, Du S J, Yan G, Feng Y, Wu X, Wang J R, Liu X H, Zhang P X, Wu X Z, Zhou L, Cao L Z, Ruan K Q, Wang C Y, Li X G, Zhou G E Zhang Y H 2001 cond-mat/0106061
[21] Suo H L, Beneduce C, Suo X D and Flükiger R 2002 *Supercond. Sci. Technol.* **15** 1058
[22] Suo H L, Beneduce C, Toulemonde P and Flükiger R 2001 *IEEE Trans. Appl. Supercond.* MT-17, (Geneva, September 2001)
[23] Flükiger R, Lezza P, Beneduce C and Suo H L, Presented at Boromag, Genova, June 2002, to be published in Proceedings
[24] Komori K, Kawagishi K, Takano Y, Arisawa S, Kumakura H, Fukutomi M and Togano K 2002 cond-mat/0203113.
[25] Che G C, Li S L, Ren Z A, Li L, Jia S L, Ni Y M, Chen H, Dong C, Li J Q, Wen H H and Zhao Z X, cond-mat/0105215
[26] Monteverde M, Núñez-Regueiro M, Rogado N, Regan K A, Hayward M A, He T, Loureiro S M and Cava R J 2001 *Science* **292** 75–77
[27] Tissen V G, Nefedova M V, Kolesnikov N N and Kulakov M P, cond-mat/0105475 (2001).
[28] Schlachter S I, Fietz W H, Grube K and Goldacker W, cond-mat/0107205 (2001).
[29] Kostíc P, Paulikas A P, Welp U, Todt V R, Gu C, Geiser U, Williams J M, Carlson K D and Klemm R A 1996 *Phys. Rev.* B **53** 791
[30] Wang X L, Soltanian S, Horvat J, Qin M J, Liu H K and Dou S X, cond-mat/0106148 and cond-mat/0107478.
[31] Slusky J S, Rogado N, Regan K A, Hayward M A, Khalifah P, He T, Inumaru K, Loureiro S M, Haas M K, Zandbergen H W and Cava R J 2001 *Nature* **410** 343
[32] Kazakov S M, Angst M, Karpinski J, Fita I M and Puzniak R 2001 *Solid State Commun.* **119** 1
[33] Mehl M J, Papaconstantopoulos D A and Singh D J 2001 *Phy. Rev.* B **64** 140509
[34] Paranthaman M, Thompson J R and Christen D K 2001 *Physica* C **355** 1
[35] Zhao Y, Feng Y, Cheng C H, Zhou L, Wu Y, Machi T, Fudamoto Y, Koshizuka N and Murakami M 2001 *Appl. Phys. Lett.* **79** 1154
[36] Mickelson W, Cumings J, Han W Q and Zettl A 2001 *Phys. Rev.* B **65** 052505
[37] Zhao Y G, Zhang X P, Qiao P T, Zhang H T, Jia S L, Cao B S, Zhu M H, Han Z H, Wang X L and Gu B L 2001 *Physica* C **361** 91
[38] Li S Y, Xiong Y M, Mo W Q, Fan R, Wang C H, Luo X G, Sun Z, Zhang H T, Li L, Cao L Z and Chen X H 2001 *Physica* C **363** 219
[39] Feng Y, Zhao Y, Sun Y P, Liu F C, Fu B Q, Zhou L, Cheng C H, Koshizuka N and Murakami M 2001 *Appl. Phys. Lett.* **79** 3983
[40] Tampieri A, Celotti G, Sprio S, Rinaldi D, Barucca G and Caciuffo R 2002 *Solid State Commun.* **121** 497
[41] Kühberger M and Gritzner G 2002 *Physica* C **370** 39
[42] Bharathi A, Balaselvi S J, Kalavathi S, Reddy G L N, Sastry V S, Hariharan Y and Radhakrishnan T S 2002 *Physica* C **370** 211
[43] Cimberle M R, Novak M, Manfrinetti P and Palenzona A 2002 *Supercond. Sci. Technol.* **15** 43
[44] Takenobu T, Ito T, Chi DamHieu, Prassides K and Iwasa Y 2001 *Phys. Rev. B* **64** 134513

[45] Paranthaman M, Thompson J R and Christen D K 2001 *Physica* C **355** 1
[46] Mickelson W, Cumings J, Han W Q and Zettl A 2001 *Phys. Rev. B* **65** 052505
[47] Zhao Y, Feng Y, Cheng C H, Zhou L, Wu Y, Machi T, Fudamoto Y, Koshizuka N and Murakami M 2002 *Appl. Phys. Lett.* **79** 1154
[48] Zhao Y, Huang D X, Feng Y, Cheng C H, Machi T, Koshizuka N and Murakami M 2002 *Appl. Phys. Lett.* **80** 1640
[49] Feng Y, Zhao Y, Sun Y P, Liu F C, Fu B Q, Zhou L, Cheng C H, Koshizuka N and Murakami M 2002 *Appl. Phys. Lett.* **79** 3983
[50] Liao X Z, Serquis A C, Zhu Y T, Huang J Y, Peterson D E, Mueller F M and Xu H F 2002 *Appl. Phys. Lett.* **80** 4398
[51] Matsumoto K, Miyake T, Osamura K and Hirabayashi I 2000 *Proc IWCCA-HTS 2000*, ed T T Matsushita p 33
[52] Chen X H, Xue Y Y, Meng R L and Chu C W 2001 *Phys. Rev.* B **64** 172501
[53] Shinde S R, Ogale S B, Biswas A, Greene R L and Venkatesan T 2001 cond-mat 0110541
[54] Simon F, Jánossy A, Fehér T and Murányi F 2001 *Phys. Rev. Lett.* **87** 047002
[55] Bud'ko S L, Kogan V G and Canfield P C 2001 *Phys. Rev.* B **64** 180506
[56] de Lima O F, Ribeiro R A, Avila M A, Cardoso C A and Coelho A A 2001 *Phys. Rev. Lett.* **87** 5974
[57] Lee S, Mori H, Masui T, Eltsev Y, Yamamoto A and Tajima S 2001 *J. Phys. Soc. Japan* **70** 2255
[58] Pradhan A K, Shi Z X, Tokunaga M, Tamegai T, Takano Y, Togano K, Kito H and Ihara H 2001 *Phys. Rev. B* **64** 212509
[59] Kim K H P, Choi J H, Jung C U, Chowdhury P, Lee H S, Park M S, Kim H J, Kim J Y, Du Z, Choi E M, Kim M S, Kang W N, Lee S I, Sung G Y and Lee J Y 2001 *Phys. Rev.* B **65** 100510
[60] Xu M, Kitazawa H, Takano Y, Ye J, Nishida K, Abe H, Matsushita A, Tsujii N and Kido G 2001 *Appl. Phys. Lett.* **79** 2779
[61] Eltsev Y u, Lee S, Nakao K, Chikumoto N, Tajima S, Koshizuka M and Murakami M 2002 *Phys. Rev.* B **65** 140501R
[62] Angst M, Puzniak R, Wisniewski A, Jun J, Kazakov S M, Karpinski J, Roos J and Keller H 2002 *Phys. Rev. Lett.* **88** 167004
[63] Ferdeghini C, Ferrando V, Grassano G, Ramandan W, Bellingeri E, Braccini V, Marré D, Manfrinetti P, Palenzona A, Borgatti F, Felici R and Lee T L 2001 *Supercond. Sci. Technol.* **14** 952
[64] Patnaik S, Cooley L D, Gurevich A, Polyanskii A A, Jiang J, Cai X Y, Squitieri A A, Naus M T, Lee M K, Choi J H, Belenky L, Bu S D, Letteri J, Song X, Schlom D G, Babcock S E, Eom C B, Hellstrom E E and Larbalestier D C 2001 *Supercond. Sci. Technol.* **14** 315
[65] Vaglio R, Maglione M G and Di Capua R, cond-mat/0203322
[66] Handstein A, Hinz D, Fuchs G, Muller K H, Nenkov K, Gutfleisch O, Narozhnyi V N and Schultz L 2001 *J. Alloys Compd.* **329** 285
[67] Buzea C and Yamashita T 2001 *Supercond. Sci. Technol.* **14** R115
[68] Gümbel A, Eckert J, Fuchs G, Nenkov K, Müller K H and Schultz L 2002 *Appl. Phys. Lett.* **80** 2725
[69] Soltanian S, Horvat J, Wang X L, Tomsic M and Dou S X 2002 cond-mat/0205406
[70] Seeber B 1998 *Handbook of Applied Superconductivity*, ed B Seeber (Bristol: Institute of Physics Publishing) p 307
[71] Tschopp W H and Laukien D D 1998 *Handbook of Applied Superconductivity*, ed B Seeber (Bristol: Institute of Physics Publishing) p 1191

B4
Introduction to section B4: Thick and thin films

D S Ginley and D A Cardwell

The primary focus of this section is on the development of thick, thin and HTS films for superconducting applications. Because of their ceramic nature and associated pronounced weak link behaviour, there has been an extended search for methods to grow very high quality or biaxially textured films for high current devices. This effort has led to the practical application of techniques such as pulsed laser ablation (PLA) and ion beam assisted deposition (IBAD), as well as to the development of buffer layer stack technology, which has impacted broadly outside the HTS field. In this area the superconductivity community has led the way with the growth of oxide and heterostructure films, which have revolutionized the way that people think about the practical applications of oxides.

Chapter B4 is concerned with state of the art in film growth. Chapter B4.1 presents a comprehensive discussion of the nature of the substrates required for the growth of HTS films, initially within the context of a range of applications from electronic (high and low frequency) to high power for coated conductors. This leads directly to a description of the key properties required for high quality substrates of chemical compatibility, minimal thermal expansion mismatch and minimal lattice mismatch. A discussion of the critical area of substrate preparation is also presented. Finally, the chapter discusses the growth of films both on polished, single crystal substrates, such as $LaAlO_3$, and on textured multigrain substrates, such as rolled nickel tape, for high current applications.

Chapter B4.2 presents the various deposition techniques for the growth of thin films. The chapter is sub-divided into two sections, which address the methods of physical deposition (section B4.2.1) and chemical deposition (section B4.2.2). Section B4.2.1 includes sputtering, pulsed laser deposition, reactive thermal co-evaporation and liquid phase epitaxy (LPE). The different constraints on film deposition associated with these techniques are presented and their relative strengths and weaknesses described, including a discussion of the appropriate diagnostics for each approach. Section B4.2.2 describes the chemical deposition technique as of one of two types: a two step process, such as screen printing or metal organic decomposition, in which the initial precursor composition is further reacted to generate the required superconductor, or a one step process, such as MOCVD, where the precursor is decomposed to yield directly the target compound. The section pays particular attention to the nature of the precursors used in both chemical deposition techniques and to the kinetics of the decomposition pathways.

Chapter B4.3 describes the preparation of YBCO, BiSCCO, TlBCCO and HgBCCO thick films. A wide variety of preparative approaches is discussed, including; preparation of a powder based thick film, electrophoretic deposition, sol–gel, plasma spraying, spray pyrolysis and LPE. The chapter relates directly to B4.1 by references to the appropriate substrate technology and discusses the areas of application for these films.

Chapter B4.4 describes the nature of and processes used for producing circuit elements from thin film superconductors. Section B4.4.1 discusses the processes required to produce HTS circuits, dealing specifically with photolithography, etching by a variety of techniques including chemical, plasma, e-beam and ion beam and the nature of insulating layers and the requirements for device planarization. Sections B4.4.2 and B4.4.3 present a detailed discussion of the development and processing of Josephson junctions for HTS and LTS materials, respectively. Such devices are the fundamental building blocks of SQUIDS, voltage standards and digital circuits based on single flux quanta. The articles in these sections describe the complex multilayer topology required to construct Josephson junction elements from both LTS and HTS, including the limitations and strengths of the different systems.

B4.1
Substrates and functional buffer layers

B Holzapfel and J Wiesmann

B4.1.1 Introduction

For the deposition of HTS thin films the choice of the substrate material is more crucial than for many other thin film processes and applications. HTS films are ceramics which have to be deposited or *ex situ* annealed at high temperatures with a sufficient partial pressure of oxygen. Additionally, because of the anisotropy of the mostly perovskitic structures, only a very high degree of texture in all three dimensions induces the high performance of HTS, therefore, the film deposition is focused on epitaxy. Furthermore, applications of HTS films are operating at low temperatures of 150 down to 4 K. Last but not least, not only physical, but also economical aspects have to be taken into account, if thin film techniques should lead to technological products.

B4.2.1 B4.2.2 A1.3

In this chapter a general overview on the very large and complex topic of substrate selection for HTS films will be given. In many cases the substrate does not consist of only one material, but of the main substrate coated with one or several different buffer layers. These buffer layers should enable the use of substrates which alone do not meet all necessary features for a successful HTS film deposition. Therefore, deposition techniques and special architectures of functional buffer layers will also be addressed in the following.

The first part of the chapter contains an overview of the various substrate requirements. Thereafter the spotlight is directed towards substrate selection from the viewpoint of HTS film applications. This section is divided into substrates for electronic, microwave and high-current applications. For the last, the development of an optimum combination of substrate and buffer layers is most complicated. Therefore, the two main important techniques for coated conductors, ion beam assisted deposition (IBAD) and rolling assisted biaxially textured substrates (RABiTS), will be presented in more detail.

B4.2.1 B3.2.4

Different aspects which have to be taken into account for the right substrate selection for superconducting films, are discussed in section B4.1.2. Features like the chemical compatibility between substrate and film, match of thermal expansion and lattice match for epitaxy are discussed. Even if suitable substrate materials are found, several pre-treatments of the substrate itself have to be made in order to achieve the surface quality and the cleanliness necessary for a successful deposition. They are discussed in section B4.1.3. In section B4.1.4 the problem of choosing the right substrate for special applications will be overviewed. Herein two different aspects are of importance. Firstly, what kind of substrate material or composition should be used not only to make a thin film with good superconducting properties but also for realizing an application using this film with a good performance? For example for high-frequency (HF) applications only substrate materials with a low HF-resistance will be competitive. Secondly, which economic features of the chosen substrate could be

D3.2.3 D2.7

important both for the technical and economic feasibility? This viewpoint leads to further aspects, namely the costs of the whole substrate and the possibility of up-scaling.

B4.1.2 Substrate requirements

B4.1.2.1 Chemical compatibility

First of all for the suitability of a substrate material it is very important that no interdiffusion with the HTS film occurs. A chemical reaction between them would strongly affect the superconducting properties of the HTS film. The substrate must be widely inert at the typical deposition conditions of an oxygen-rich ambient atmosphere and high temperatures in the range of 650 up to 850°C, depending on the deposition method. Therefore, oxide substrates like $SrTiO_3$, MgO, YSZ (yttria stabilized zirconia: ZrO_2 with 8–10 mol% Y_2O_3) or $LaAlO_3$ are, in principle, more suitable than metal substrates. Other typical substrates for thin film growth like silicon and sapphire show a diffusion of Si and Al into the HTS film. Nevertheless, a direct deposition of HTS films both on metal substrates and single-crystalline silicon and sapphire substrates was tried, but without remarkable success. For YBCO films deposited directly on metals J_c reached only a hundredth of the best values. AES analyses of YBCO on Ni substrates showed a Ni diffusion into the HTS films resulting in poor superconducting properties. The formation of metal oxides like Cr_2O_3 or NiO_2 between a hastelloy (CrNi alloy) substrate and the HTS film was also observed, leading to poor HTS properties [1]. For the deposition on Si the main problem is the high deposition temperature. Si diffusion is substantial at temperatures above 600°C [2]. On pure sapphire only poor J_c values can be achieved because of Al diffusion into the YBCO film and the development of a disturbing $BaAl_2O_4$ layer [3, 4]. Thus, today, the deposition of HTS films directly on metal surfaces Ag (apart from) and silicon or sapphire substrates plays no role.

Interdiffusion can be avoided if buffer layers, which do not react with the HTS films, are deposited on the bare substrate. The microstructure of these buffer layers becomes very important because only dense layers without diffusion paths e.g. through pores or open grain boundaries can work as a buffer. Using oxide buffers like CeO_2, YSZ, MgO, Y_2O_3 or $SrTiO_3$, HTS films can also be deposited on metals, silicon and sapphire with high J_c values above 1 MA cm^{-2} (see section 4.1.4). The best reported J_c value was achieved on a $SrTiO_3$ substrate [5]. From the viewpoint of chemical compatibility the use of suitable buffer layers enables the deposition of HTS films on nearly every substrate material.

B4.1.2.2 Thermal expansion

D3.2.3 Because of both the high deposition temperature and the low superconducting transition temperature the difference in thermal expansion across this temperature range between the substrate, the HTS film and, eventually, additional intermediate buffer layers plays a significant role for the choice of the D3.3 substrate–film combination. HTS films are ceramic and, therefore, very brittle materials so crack formation can easily occur. If the films are strained during cooling from the deposition temperature down to room temperature or from room temperature down to the application temperature compressive or tensile stresses in the film will be induced depending on the difference between the thermal expansion of the film and the substrate.

If the thermal expansion coefficient of the film exceeds that of the substrate, the film will be in tension, in the opposite case it will be in compression. The tensile stress is more critical for ceramic D4.2.1 materials. As expected, the results of bending of YBCO films on metal substrates show that the critical current is reduced by 20% at very small values of tensile stress (0.4%), whereas the same J_c reduction A1.3 occurs at a higher value (0.8%) for compressive stress [6].

Because of the stress accumulation with increasing film thickness there exists a critical thickness at which the film starts cracking. For YBCO on Si this value is in the range of 50–70 nm, where a tensile stress in the YBCO film is observed [7]. Only low temperature deposition methods like thermal co-evaporation introduced by Kinder and co-workers have enabled YBCO films to be deposited on Si with relatively high thicknesses (above 120 nm) without cracking, but only with buffer layers to prevent interdiffusion [8]. On pure sapphire substrates without a special treatment the limit is about 300 nm, because the difference between the thermal expansion of the substrate and the HTS film is slightly lower. Using buffer layers the critical thickness can be increased. But the buffers have only a small effect. Due to the large thickness the substrate dominates the origin of thermal stresses in the films on top during some temperature changes. Thus, materials with a large thermal expansion difference to the HTS films are eliminated as substrate candidates. For YBCO films the thermal expansion coefficient α_{YBCO} is (10–13) $\times 10^{-6}$ K^{-1}. Substrates with $\alpha < 0.9 \times \alpha_{YBCO}$ are not suitable because the critical thickness falls below 300 nm. The values of α for many important substrate materials are summarized in table B4.1.1.

D3.2.3

B4.1.2.3 Lattice mismatch

For the deposition of HTS films with good superconducting properties, mainly with high critical current densities J_c, the films have to exhibit a high degree of texture both in and out of plane. Very early, Dimos *et al* [9] and also other groups [10, 11] realized that the intergranular J_c drops very fast with an increasing grain boundary angle, e.g. for 5° boundaries the J_c value reaches only 10% of the maximum value. The best properties can only be reached if the texture of the HTS film is as high as possible. Therefore, an epitaxial growth on single-crystalline or very highly textured substrates is desired. The crystallographic structure of the substrate surface or the surface of a buffer layer on a substrate plays the most important role. The surface atoms should match with the atoms of the first HTS layer. This can be described with the near-coincident-site-lattice (NCSL) theory [12], an extension of the coincident-site-lattice (CSL) theory described by Ballufi *et al* [13]. The coincident sites are atomic positions with preferably the same or similar atomic sizes and valences that coincide on both sides of the interface. For a good epitaxy the number of coincidence sites should be as high as possible. From this viewpoint the most suitable substrate materials exhibit a similar crystal structure to the HTS films. For good superconducting properties HTS films should grow c-oriented which means that the substrate surface is oriented parallel to the a–b plane of the orthorhombic HTS structure. If a film grows in the undesired a-orientation large needles can often be seen in SEM micrographs. One degree of freedom enables the rotation of this rectangular plane—for YBCO with $a = 0.382$ nm and $b = 0.388$ nm around the normal. The ideal substrate surface should have typical atomic distances of coincidence sites with the same values for a and b or for a rotation of 45° with values for a and b of about 0.544 nm. In the following the lattice mismatch is calculated between the substrate material and the a–b plane of a YBCO crystal. Other HTS materials have a similar lattice parameter and mismatch, respectively. For HTS film growth the best results were achieved on oxides with perovskite crystal structure on (001) surfaces. These oxides exhibit in many cases a very small misfit to the YBCO a–b plane. Most results exist for (001) $SrTiO_3$ ($a = 0.3905$ nm) because of its easy availability. For large area deposition, $LaAlO_3$ (pseudocubic cell with $a = 0.5377$ nm)

A1.3

Table B4.1.1. Thermal expansion coefficients of the common substrate materials

	$SrTiO_3$	$LaAlO_3$	$NdGaO_3$	YSZ	MgO	α-Al_2O_3 (sapphire)
α ($\times 10^{-6}$ K^{-1})	9.4	10–13	9–11	11.4	14	9.4

B2.4 is the most favoured material. Its crystals can be grown by the Czochralski technique and can be scaled up to diameters of about 20 cm.

The second group of substrates contains the non-perovskitic oxides. They do not match perfectly to the HTS films, but because of the existence of an orientation with many coincidence sites, epitaxial growth becomes possible. Many HTS films were successfully grown on the easily available cubic MgO ($a = 0.4211$ nm), which has a lattice mismatch of 9%. Even such a high misfit enables an epitaxial deposition of HTS films. The best results were achieved on MgO substrates, which were annealed in oxygen above 1000°C before the HTS deposition to improve the surface crystallinity [14]. Oxide substrates with the CaF_2 structure such as YSZ ($a = 0.512$–0.515 nm) are often used because they can serve as buffer layers on substrates that show interdiffusion with HTS films. For YSZ an additional problem had to be solved. Although it exhibits a low misfit to HTS films for the 45° rotation between the [100] planes, other different in plane orientation relations can also occur. Because of the high lattice mismatch three epitaxial relations between (100) YSZ and *c*-axis oriented YBCO films were observed: besides the normal obtained 45° relation ([110]YBCO||[100]YSZ) also, at high deposition temperatures above 760°C, a 0° relation [15], i.e. [100]YBCO||[100]YSZ, and, more rarely, a 9° relation [16]. Optimized films with high J_c values show only the 45° relation. The orientation depends strongly on the deposition conditions. On very smooth substrates mainly the 0° epitaxy was observed. A widely used material with CaF_2 structure is cerium dioxide ($a = 0.543$ nm). CeO_2 has a smaller misfit than YSZ, but it can not be produced as a single crystal suitable for film deposition. Thus, it is only used as a buffer material. Another important substrate material is sapphire, which can be produced in large diameters and with a high surface quality. The (0001) oriented sapphire has no lattice match to the HTS materials, but for the r-cut sapphire, i.e. a crystal cut with the (1102) orientation as normal direction, the mismatch is only 6%. Many other oxide materials with different crystal structures were examined but without a breakthrough for applications. Substrates with the K_2NiF_4 structures (e.g. $CaNdAlO_4$ [17], $LaSrGaO_4$ [18]), spinel structure ($MgAl_2O_4$ [19]) or hexagonal lattice (e.g. $LiNbO_3$ [20], $MgLaAl_{11}O_{19}$ [21]) also exhibit a small lattice mismatch and so they enabled a more or less successful deposition of HTS films, but without technical relevance. An overview of these rare substrate types can be found within the review article by Phillips [22].

B4.2.1 Of the metal substrates for thick film applications of YBCO only silver or silver alloys were used successfully without additional buffer layers because of their chemical tolerance to YBCO. Due to a low lattice mismatch an epitaxial deposition is possible.

On semiconducting substrates such as GaAs [23] or Si [7, 8], which may be used for connecting HTS and microelectronic technology (e.g. in integrated circuits), an epitaxial deposition of buffer layers and HTS films is possible. However, differences in thermal conductivity and a bad chemical compatibility prevent good results.

The relevant properties of the common types of substrates for HTS film deposition are summarized in table B4.1.2. Besides the lattice mismatch in the three lattice directions *a*, *b* and *c* with respect to YBCO also an eventually occurring phase transformation temperature T_{Ph}, which may affect the film properties, and the melting points T_m are illustrated.

The misfit of the most common substrates is illustrated in figure B4.1.1. For $SrTiO_3$ and $LaAlO_3$ it is nearly invisible.

B4.1.3 Substrate pre-treatment

In every thin film deposition process the substrate treatment prior to film growth is an important first step to reach the desired film properties. The surface quality of the substrate is especially important because, in most cases, film growth is very sensitive to the initial film–substrate interaction. The cleaner, more homogeneous and more well-defined the substrate surface is prior to film deposition, the more

Table B4.1.2. Properties of the common substrate materials [22]

	$\Delta a/a$ (%)	$\Delta b/b$ (%)	$\Delta c/c$ (%)	T_{Ph} (K)	T_m (K)
$SrTiO_3$	+2.0	+0.7	+0.1	110	2353
$LaGaO_3$	+1.5	+0.7	−0.5	420	2023
$LaAlO_3$	−0.9	−2.2	−3.0	800	2453
$NdGaO_3$	+0.3	+0.3	−1.3	> 1300	1873
$NdAlO_3$	+2.2	+4.0	+4.4	1820	2363
$YAlO_3$	−4.5	−5.4	−4.0	> 1573	2148
MgO	−9.0	−6.7	−7.4	–	3100
YSZ	+3.6	+6.3	+5.8	–	3000
CeO_2	0.0	+1.3	+1.9	–	2900

reproducible is the resulting film growth, especially if epitaxial film growth is desired. Surface defects that have to be removed, or at least minimized, range from macroscopic defects like cracks, scratches or chemical contamination to submicron or atomic scale defects like micropores or ledges on cleavage planes. Careful mechanical or chemical polishing (or a combination of both) removes macroscopic defects and creates a smooth substrate surface. Additional substrate cleaning is typically performed by an ultrasonic assisted chemical treatment in various solvents. The specific cleaning steps that are necessary to remove all chemical contaminations may vary for different substrate materials or for different substrate suppliers and their optimization typically requires an empirical approach. The resulting surface quality is normally good enough for non-epitaxial film growth. Well controlled epitaxial film growth, however, requires further substrate treatment, which is further described for HTS film growth on $SrTiO_3$ as a typical example. An annealing heat treatment in pure oxygen above 800°C recrystallizes the substrate surface and, at 950°C, straight surface steps with single-unit cell height are formed due to the apparent miscut angle of the substrate. The required annealing time depends on this miscut angle. Standard commercial substrates with miscut angles of 0.1–0.3° show good surface morphology at an annealing time of 30 min, whereas a 0.05° miscut requires 10 h [25]. At higher

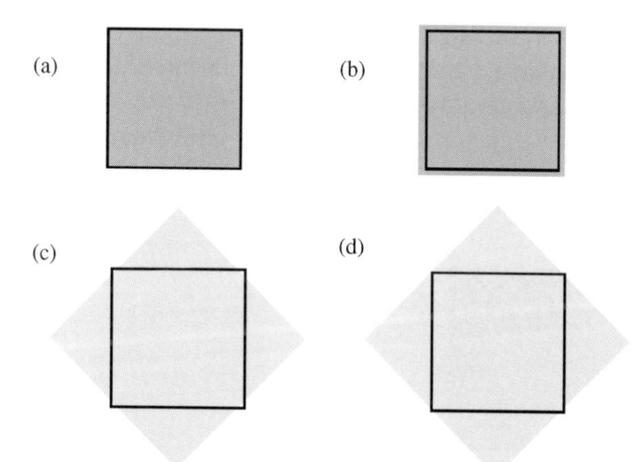

Figure B4.1.1. Scaled illustration of the misfit between YBCO (black bordered rectangle) and typical substrates (grey squares): (*a*) $SrTiO_3$, (*b*) MgO, (*c*) YSZ and (*d*) $LaAlO_3$. The squares represent an area of $a \times a$ (*a*: lattice constant).

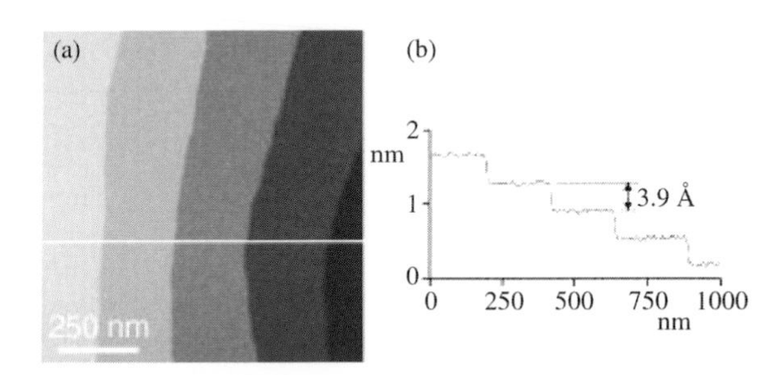

Figure B4.1.2. (*a*) AFM micrograph of a nearly perfect and atomically flat surface, obtained after soaking in water, followed by etching in a BHF solution and annealing, (*b*) cross-section, averaged over three scan lines along line in (*a*) [26].

annealing temperatures Sr segregations and off-stoichiometries can occur, worsening the substrate surface properties.

For oxide electronic applications, atomically smooth and well-defined surfaces are necessary. For perovskites like $SrTiO_3$, two different surface terminations (SrO or TiO) are possible. This may cause growth difficulties due to different chemical interactions of each termination with a growing thin film – e.g. growth of TiO on $SrTiO_3$ shows different morphologies due to different wetting properties of the two terminations.

However, these differences in the chemical behaviour have been used to create singly-terminated substrate surfaces by an additional chemical etching step. Kawasaki *et al* [24] used NH_4F buffered HF (BHF) with a pH value of 4.5 to selectively etch the SrO atomic plane, resulting in a pure TiO-terminated substrate surface [24]. Additional annealing at 850°C removes carbon contamination and results in straight, atomically flat terraces. But the required pH value is quite sensitive to the substrate surface conditions and a wrong choice can lead to uncontrolled etching and the formation of deep etch pits [25]. This etching approach can be simplified by first using water as a chemical reagent to transform the SrO terminated surface area to $Sr(OH)_2$ that is easily dissolved in acidic solutions. Using this two step chemical process (water + BHF-etching) and an additional annealing at 950°C, a nearly perfect single TiO terminated surface with very straight terrace ledges can reproducibly prepared as shown in figure B4.1.2 [26]. The quality of this surface treatment was proved by quasi-ideal layer by layer homoepitaxial growth monitored by reflection high energy electron diffraction (RHEED) (figure B4.1.3). Using non-oxide substrates like metals or semiconductors for the epitaxial growth of oxide films careful precautions have to be taken to avoid the uncontrolled formation of a self-oxide layer or to remove an already existing one. For Si substrates the native oxide layer can be removed by pre-etching the substrate in HF or NH_4F [27]. This results in an H termination of the Si surface that avoids

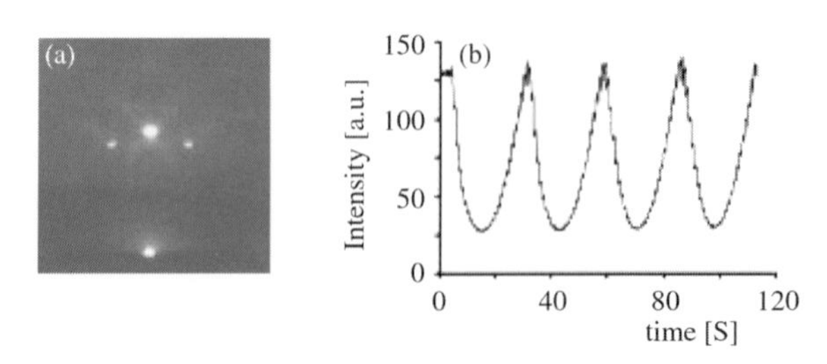

Figure B4.1.3. (*a*) RHEED pattern of the quasi-ideal $SrTiO_3$ surface shown in figure B4.1.1(*a*), (*b*) intensity of the specular reflection recorded during deposition of 4 monolayers of $SrTiO_3$ by PLD [26].

the formation of a fresh native oxide layer and enables the handling of the Si substrate in air. The H termination can be removed by heating the substrate immediately before the deposition, resulting in a clean, oxide-free surface.

B4.1.4 Bare substrates

B4.1.4.1 HF applications

One large field for HTS film applications could be in passive microwave (MW) devices for communication systems. The most important quality factor in devices such as filters, resonators and oscillators is the lowest possible MW surface resistance R_s of the HTS film. R_s can be extremely small in HTS films with very high surface quality. Substrates are required which allow not only perfect HTS film growth on mostly large areas, but also possess good properties for the desired MW application. Besides properties such as high crystalline perfection and mechanical strength, the dielectric constant ϵ and the losses $\tan\delta$ should be as low as possible. For the mainly used substrates the dielectric properties are summarized in table B4.1.3. Additionally, a deposition of HTS films on both sides of the substrates is often desired. Therefore two wafers are bonded together at their incompletely polished surface side. Because of the availability of both high quality surfaces and large areas, sapphire and $LaAlO_3$ are the best substrate candidates for the commercialization of HTS–MW devices. Also $NdGaO_3$ and MgO were often used with good results.

E3.1, E3.3 D2.7

On $LaAlO_3$ excellent YBCO films could be prepared with low R_s values. The main problem with this substrate is the high probability of twinning. The 45° grain boundary would strongly reduce the J_c in the HTS film. Additionally, a growth of undesired *a*-axis grains was often observed. To avoid this, a buffer layer is deposited before the HTS film deposition. The best results were achieved with CeO_2 buffer layers, which show a very low chemical reactivity and a small misfit to the HTS materials. The low misfit of the $LaAlO_3$ substrate together with very good chemical compatibility and relatively low dielectric losses give it a high potential for MW applications. On large area $LaAlO_3$ substrates 8 inches in diameter YBCO films were prepared [29]. Also double-sided wafers up to 4 inches in diameter could be successfully coated [30].

A better mechanical strength together with a low $\tan\delta$ loss favours sapphire as a substrate for MW applications. The poor chemical compatibility and the low lattice match demand an additional buffer layer. As for $LaAlO_3$, CeO_2 is the better choice than YSZ or MgO for the same reasons. But the large

D3.3

Table B4.1.3. Dielectric properties of substrates suitable for MW applications [28]

	ϵ	$\tan\delta$ at 10 GHz, 77 K
$SrTiO_3$	300	2×10^{-2}
YSZ	27–33	$>6\times10^{-4}$
MgO	9.6–10	6.2×10^{-6}
Sapphire	9.4–11.6	10^{-8}
YSZ buffered sapphire	see YSZ	see YSZ
CeO_2 buffered sapphire	21.2 for CeO_2	
$LaAlO_3$	20.5–27	7.6×10^{-6}–3×10^{-4}
$NdGaO_3$	23	4×10^{-4}

difference in the thermal expansivity between sapphire and HTS materials precludes higher thicknesses of crack-free HTS film. However, the thickness of crack-free YBCO films could be extended up to 700 nm on sapphire buffered with a sputtered CeO_2 layer by deliberately introducing defects into the YBCO film [31]. Large sapphire substrates with diameters of 22 cm could successfully be coated with B4.2.1 YBCO films with $J_c > 1\,\mathrm{MA\,cm^{-2}}$. Double-sided sapphire wafers up to 3 inches in diameter could also be successfully coated [32].

MgO substrates were buffered by a thin $SrTiO_3$ layer because of the large lattice misfit. Good MW properties could be achieved in YBCO films on these buffered MgO substrates [33]. A material which does not need any buffer layer is $NdGaO_3$. It has a very low lattice misfit to YBCO, shows no twinning and exhibits appropriate dielectric properties. Only the mechanical strength could be better and the thermal expansivity matches are not as good as for other perovskite materials. Because of the very low misfit, undesired *a*-axis growth of TBCCO [34] or YBCO [22] was also observed. Nevertheless, YBCO D1.1.2 films with very good crystallinity—namely with very small rocking curves and smooth surfaces—could be deposited [35]. This problem together with the lack of availability of low-price and large area substrates limits the technical feasibility.

B4.1.4.2 Electronic applications

A2.7 Since most electronic applications are based on Josephson junctions (JJs), the most important requirement for the substrates used is related to the reliable preparation of the JJs. For low T_c B4.4.2 superconductors this is not a critical issue, since the JJs are based on tunnel junctions prepared by low B4.4.3 temperature deposition of polycrystalline or amorphous films and it is sufficient to use a well-cleaned standard Si wafer. For HTS the situation is different since the JJs are based on intentionally created grain boundaries or the epitaxial deposition of oxide multilayers. A reproducible and well defined preparation of these basic structures is essential for all the applications. A common way to create reproducible HTS grain boundaries that can be used for JJs is the creation of steps in the bare substrate surface by chemical or ion beam etching or by the use of bicrystal substrates.

Since the properties of JJs are strongly related to the specific way a JJ is prepared, details of the preparation procedures will be described later in the chapter on JJ.

E4 Buffer layers play an important role in the realization of electronic applications. First, they are necessary to prepare high quality HTS films on classic semiconducting substrates like Si and, second, using buffer layers one can easily create grain boundaries in YBCO films due to different epitaxial growth relationships. A typical buffer layer for the purpose of using semiconducting substrates for HTS films is YSZ, which can be grown epitaxially on Si. Due to the large thermal expansion mismatch between Si and most oxides, the total layer thickness of the buffer/HTS multilayer has to be less than about 50–100 nm to avoid cracking. This limits drastically the possible buffer layer thickness and, for future all-oxide electronic applications, the buffer layers have to show perfect properties at a film thickness of only a few unit cells, or other substrates like silicon on sapphire (SOS) have to be used. An important step towards using standard Si substrates for oxide electronic applications was made recently by McKee *et al* [36], who demonstrated the reproducible growth of single unit cell oxides on Si. Artificial grain boundaries can be created using special buffer layer architectures since there exist different in-plane epitaxial growth relationships among the oxides used in HTS film growth technology. For example, cube-on-cube growth takes place for YBCO on MgO and for CeO_2 on MgO. The growth of YBCO on CeO_2, however, shows a 45° in-plane rotation. Therefore, using a patterned CeO_2 seed layer on a MgO substrate (or another MgO buffer layer) for the YBCO film growth, one can create a 45° artificial grain boundary, a so-called biepitaxial junction [37]. The creation of other grain boundaries is also possible. Using MgO as a seed layer on a (110)-oriented STO substrate one can obtain 45° *a*-axis twist or tilt grain boundaries [38].

B4.1.4.3 *Substrates for high current applications*

Low angle grain boundaries on technical substrates are very important for high current applications of HTS films. Simultaneously, large areas have to be coated with an HTS film. Actually two different substrate geometries are proposed: either long tapes (in one dimension) or large area substrates (in two dimensions). B4.2.1

HTS thin film applications as current limiters can be realized with large single-crystalline materials, for instance with sapphire substrates [39], but for cheaper, scaled up production, economical substrate materials with large dimensions (10×10 up to $20 \times 50\,\mathrm{cm}^2$) are also required. Depending on the application the substrates should be ceramics or metals. On polycrystalline substrates, the J_c value is limited because of the grain boundaries in the HTS film. The best reported value on a polycrystalline zirconia layer on a polycrystalline metal substrate for YBCO is a J_c of only $0.1\,\mathrm{MA\,cm^{-2}}$ [40]. Thus, for a film growth of HTS with very large J_c values the substrate surface has to be biaxially oriented, i.e. also in the lateral directions. Therefore two different approaches are used: the so-called IBAD and the RABiTS technique. The IBAD technique allows a coating of an in-plane textured thin oxide layer independently of the substrate type and orientation. This IBAD layer can serve as a template for an epitaxial deposition of the HTS film and, if necessary, it can prevent interdiffusion between substrate and HTS film. As ceramic substrates partially stabilized zirconia or alumina were used. The common metal substrates are stainless steels, Ni based alloys (hastelloy or Inconell) or pure nickel. For the deposition of thin films the substrates have to be polished to a very high quality. The surface roughness should be below 20 nm. Both electrochemical and mechanical polishing were established. An additional important demand is that the metal substrate should not recrystallize at the high temperatures required for the deposition of the HTS film. The RABiTS technique is a manufacturing method for long thin metal tapes up to several 10 m with a highly biaxial texture. This tapes can be coated epitaxially with buffer layers for avoiding interdiffusion between metal tape and HTS film. Both techniques allow the growth of HTS films with J_c values above $1\,\mathrm{MA\,cm^{-2}}$ without the use of a single crystalline substrate. E1.3.3 C1 B4.2.1 B4.2.1 B3.2.4

IBAD

The IBAD technique was first introduced for the deposition of metals by Yu *et al* [41] in 1985. They found that a bombardment of the growing Nb film during the deposition by argon atoms with energies of several hundred eV causes an in-plane orientation. Iijima *et al* [42] demonstrated in 1991 that this technique could also be used for an in-plane alignment of the ceramic material YSZ. They achieved an (001) in-plane orientation in the YSZ film by focussing the assisting beam with an angle of 55° with respect to the substrate normal. The argon beam aligns parallel to a (111) direction in the YSZ. On this IBAD layer an epitaxial deposition of YBCO is possible. Because the J_c value in HTS films strongly correlates with the degree of in-plane orientation of the YSZ layer, the optimization of the biaxial texture was first to the fore. The main quality parameter for the IBAD layers is the FWHM (full width at half maximum) $\Delta\phi$ of the four peaks in (111)-pole figures or phi-scans [42]. For reaching J_c values above $1\,\mathrm{MA\,cm^{-2}}$ in YBCO films the $\Delta\phi$ in the YSZ should be below 10° at the substrate surface [6]. This texture can be reached in optimized IBAD YSZ films at thicknesses above 600–800 nm, because $\Delta\phi$ drops only slowly with increasing film thickness (figure B4.1.4). Because of the divergence of the assisting beam a different saturation of $\Delta\phi$ of 7.0° [43], 8.9° [44] and 48.4° [45] was observed. In the case of a higher divergence of the assisting ion beam the saturation gets worse. Several groups in the world can deposit high quality YSZ films with the required degree of in-plane texture. The growth mechanism of the IBAD YSZ films was investigated by a few research groups, but up to now no widely accepted growth model exists. For more details see [43, 46]. The assisting ion beam was combined with different B4.2.1 B3.2.4 D1.1.2

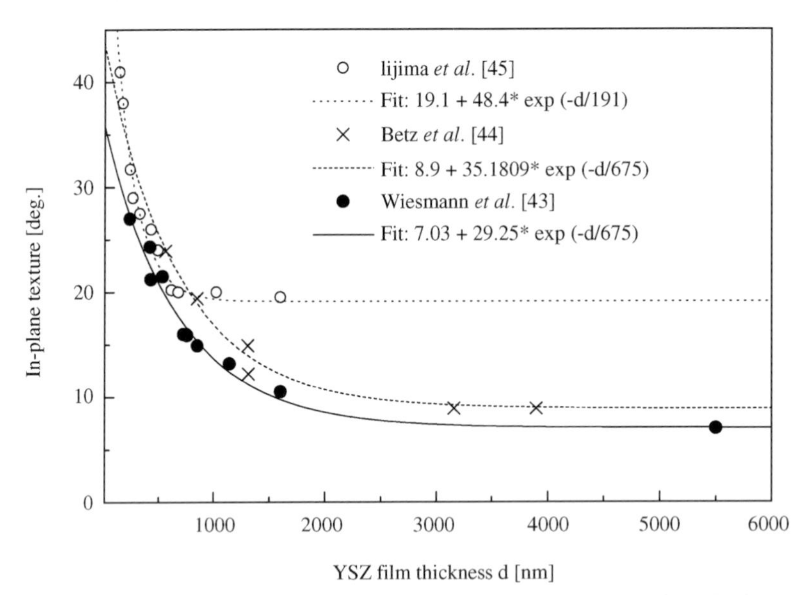

Figure B4.1.4. Dependence of the in-plane texture [$\Delta\phi$ of (111)-Phi-Scans] on the film thickness of the ion beam assisted deposited YSZ film. The three curves are fits through the data points of three different research groups. They show an exponential decay with a saturation depending on the quality of the assisting ion beam.

deposition methods, such as e-beam evaporation [47, 48], ion beam sputtering [42, 49–51] and pulsed laser deposition [44, 52]. The combination of the assisting ion beam with the deposition method of 'pulsed laser deposition' is also called IBALD (ion beam assisted laser deposition). It is the only method where the material flux on the substrate is discontinuous. But an effect of this on the quality of the layers could not be observed. The mechanism of in-plane texturing seems to be the same as in the case of continuous deposition methods. Ion beam sputtering and IBALD produced the best results with regard to the in-plane texture quality. Although YSZ contains structural oxygen vacancies, it serves as a good buffer layer for metallic substrates in order to avoid diffusion into the HTS films [51]. IBAD films were even better diffusion barriers than polycrystalline films because of the dense film structure caused by the ion impingement of the assisting beam. For economically and technically competitive applications of HTS films on IBAD substrates the most important point is now the development of deposition techniques on large areas with high volume deposition rates. Ceramic substrates of $20 \times 20\,cm^2$ can be deposited with IBAD YSZ layers with volume deposition rates above $10\,nm\,m^2\,min^{-1}$ using an assisting ion gun with a diameter of 10–20 cm [53]. Large metallic tapes could be coated up to a length of 0.5–0.9 m both with YSZ buffer layers and YBCO films [45, 54, 55]. Even curved substrates can be coated with an in-plane aligned YSZ film [56].

B4.2.1

B3.2.4

An alternative deposition method for IBAD and IBALD layers is the inclined substrate deposition (ISD). The atoms from the target, which themselves build up the ISD layer, serve as the assisting beam by setting an angle of 55° between the substrate normal and the particle flux. Some groups have used YSZ for ISD with sputtering [57] or PLD [58, 59] as deposition methods, but with respect to IBAD only Hasegawa *et al* have achieved a comparable performance with high deposition rates using laser ablation [58].

As an illustrative summary, the different deposition equipments for IBAD and ISD are schematically shown in figure B4.1.5. In the case of e-beam evaporation the e-beam source is located at the position of the target. For a successful epitaxial deposition of YBCO films on IBAD YSZ an additional epitaxial thin layer, for instance Y_2O_3 [60] or CeO_2 [48], is recommended because firstly YSZ

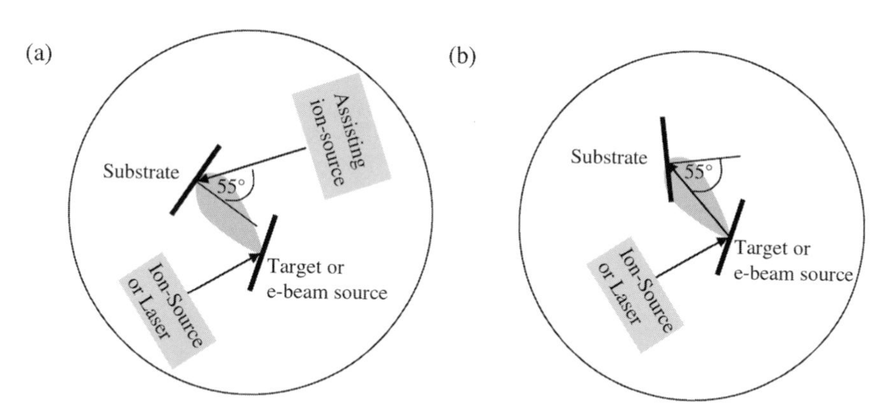

Figure B4.1.5. Typical deposition chamber for IBAD (*a*) and ISD (*b*).

and Ba can react together forming an amorphous, polycrystalline or poorly textured $BaZrO_4$ (BZO) layer, which can destroy or diminish the good in-plane texture of the YSZ, and secondly the lattice mismatch of BZO is a little bit higher than on a pure YSZ surface. This BZO layer is not a problem on single-crystalline YSZ substrates where it can be built epitaxially [61].

Because of the disturbing BZO interlayer and the slow texture development of the YSZ resulting in large IBAD film thicknesses several other alternative oxides, such as CeO_2 [62–64], $CeGdO_2$ [53] or Pr_6O_{11} [65], were examined. They could be textured with similar results using IBAD. CeO_2, which does not show a disturbing reaction with components of the HTS film, would theoretically be the best material choice for IBAD, but up to now the best in-plane texture was much lower, compared to YSZ layers, due to a worse evolution of the in-plane aligned grains.

B4.2.1
B3.2.4

In 1997 one alternative to YSZ was found with IBAD MgO layers [66]. A very good in-plane texture with $\Delta\phi = 7°$ could be achieved in very thin films at a thickness of only 10 nm on amorphous Si_3N_4. On metal substrates with such an IBAD MgO layer an additional epitaxial MgO layer is necessary in order to prevent interdiffusion into the YBCO film. On nickel based alloy substrates coated with an IBAD MgO layer and 600 nm epitaxial MgO a YBCO film has reached a J_c of 0.46 MA cm^{-2} [67]. This technique is now extended to a deposition of long tapes. It seems not to be limited to deposition onto an amorphous layer like Si_3N_4 [67]. Also with the ISD method MgO could be successfully in-plane aligned. With the deposition method of co-evaporation using an angle of 45° between particle flux and substrate normal thick MgO layers could be prepared with an in-plane degree $\Delta\phi$ of 8° using high deposition rates of up to 250 nm min^{-1} [68]. First results of YBCO films with J_c of 0.79 MA cm^{-2} with such a MgO layer deposited by ISD co-evaporation demonstrate the potential of this relatively new technique, which probably can be upscaled very economically.

C1

RABiTS

Another approach to realize high current carrying coated conductors is the RABiTS technique first introduced by the Oak Ridge National Laboratory in 1996 [69]. In contrast to IBAD biaxial aligned buffer layers are grown by epitaxial growth on already highly biaxially oriented metal tapes. Biaxially aligned metal tapes can be obtained by a strong cold-rolling deformation of fcc-metals like Ni or Cu and a subsequent recrystallization heat treatment, as shown schematically in figure B4.1.6. Using appropriate deformation and annealing conditions this can result in a sharp cube texture with the (001) plane parallel to the tape surface and a [100] direction parallel to the rolling direction, a well known metallurgical recrystallization effect known since the 1930s [70]. The challenge of this approach is to find

B4.2.1
B3.2.4

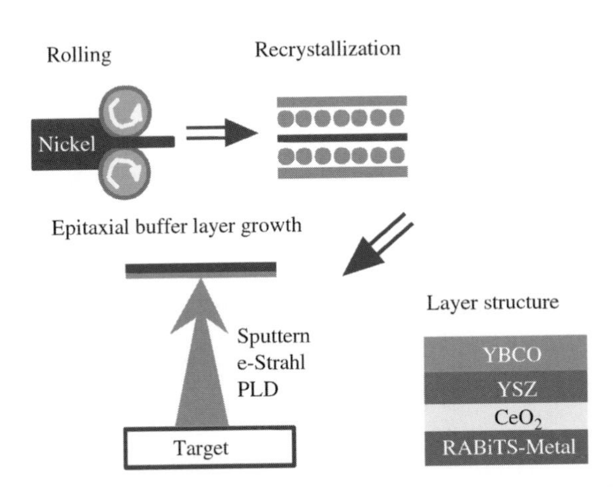

Figure B4.1.6. Schematic illustration of the Y123 tape preparation route by the RABiTS process.

a pure metal or an alloy that develops a sharp cube texture, matches the lattice constant and the thermal expansion coefficient of an YBCO-compatible oxide buffer layer and meets the other technical requirements of a coated conductor substrate such as strength and costs. Since epitaxial oxide film growth usually requires high temperatures and the presence of an oxide atmosphere during growth, there is also a serious danger of uncontrolled oxidation of the metal surface prior to the oxide film deposition, preventing the desired growth orientation of the buffer layer. Up to now Ni and Ni-alloys seem to be the best choice to meet all these quite different requirements. For producing long tapes with an optimized cube texture a typical preparation process is as follows [71]. Ni or Ni-alloys are prepared by melting elements with a purity of about 99.98% in an induction furnace. After a homogenization treatment to establish a suitable microstructure, the resulting square-shaped rod is cold rolled with several passes down to the final thickness of 80 μm, equivalent to a reduction of $\Delta d/d = 99.5\%$ or a true strain of $\epsilon = 5.52$. Finally, the samples are recrystallised at temperatures between 300 and 1100°C either in a mixed atmosphere of 10% H_2 in N_2 or in an atmosphere of pure H_2. Well-prepared Ni-tapes show a typical grain-to-grain misalignment of less than 10° and a grain size of about 50–100 μm as shown in figure B4.1.7 by electron backscattering diffraction pattern (EBSD) measurements [72]. From the locally resolved texture information of the EBSD pattern one can quantitatively extract the grain boundary network (GBN) of the Ni tape. This is exactly the relevant information for the current carrying capability of the resulting coated conductors, because the GBN is transferred to the epitaxially grown YBCO layer and determines there the upper limit for the critical current density.

To overcome the problem of uncontrolled (111) NiO formation during the first epitaxial buffer layer growth step either the metal oxidation is prevented before the first monolayers of the buffer film are deposited by choosing appropriate deposition conditions and buffer layer architectures or the metal surface oxidation is controlled in such a way that (100)-oriented NiO is formed. One way to avoid the metal oxidation is by growing first in high vacuum a noble metal layer like Pd, which is resident to unwanted oxidation during the following deposition steps in an oxygen atmosphere [69]. Such an epitaxially grown buffer layer architecture (Ni/Pd/CeO) prepared by PLD was one of the first chosen by Goyal *et al* to obtain a high YBCO J_c of 0.2 MA cm^{-2}. However, there are problems connected to the strong diffusion of Pd at the deposition temperatures of YBCO, limiting the attainable J_c to values below 1 MA cm^{-2}. A second, more successful, route is to choose carefully an oxide buffer layer system and the deposition conditions in such a way that the substrate metal oxide is thermodynamically unstable under the deposition conditions, whereas the oxide buffer is stable [73, 74].

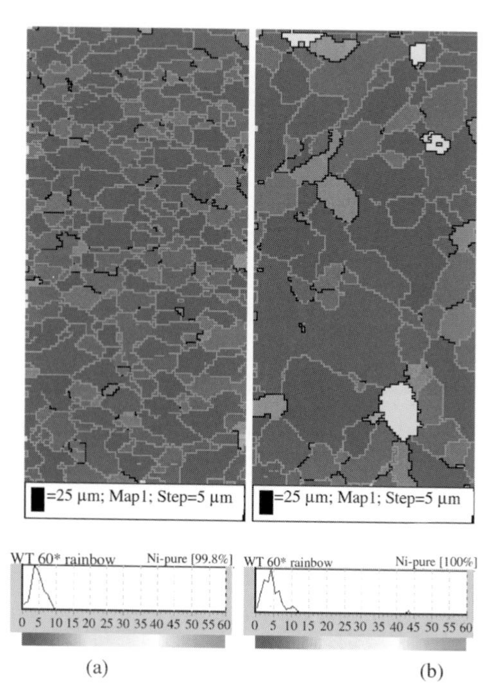

Figure B4.1.7. EBSP mappings of (*a*) nickel +5.0 at.% tungsten annealed over 45 min at 1000°C and (*b*) nickel +0.1 at.% tungsten annealed at 850°C for 30 min.

In this context Ni is a very promising substrate, since there are a few rare earth/rare earth oxides, which meet this thermodynamic requirement with respect to Ni/NiO. The best results were obtained with the deposition of CeO_2 on Ni using a reducing deposition atmosphere of H_2/Ar, which also removes the native NiO layer from the tape substrate. Nearly perfect epitaxial growth Ni(001)[100]||CeO_2(001)[110] can be obtained, but the CeO thickness is limited to about 50 nm due to crack formation [74]. To reach the necessary buffer thickness of several hundred nm that prevents poisoning of YBCO by Ni diffusion, a second YSZ oxide buffer layer is added. Since the lattice mismatch between YSZ and YBCO is quite large, YBCO deposition directly on YSZ often results in two epitaxial relationships YSZ(001)[100]||YBCO(001)[110] and YSZ(001)[100]||YBCO(001)[100], creating disastrous 45° grain boundaries in the YBCO film. To prevent this, a last thin CeO_2 film is added to the buffer layer architecture, reducing the lattice mismatch to YBCO that gives only one epitaxial growth relation CeO_2(001)[100]||YBCO(001)[110]. The epitaxial growth of such a multilayer structure is shown in figure B4.1.8. D1.1.2

Using this three layer buffer architecture, critical current densities in zero and applied magnetic field comparable to YBCO films on single-crystalline substrates were achieved by several different deposition techniques like PLD or the BaF_2-method [75, 76]. To simplify the deposition process, a single buffer layer design would be preferable. Recent results show that YSZ can also be epitaxially grown directly on Ni, without the CeO_2 seed-layer, resulting in J_c values of 1 MA cm^{-2} [77], and that Y_2O_3 could be used as a single buffer layer system [78].

Another way to overcome the pre-deposition oxidation problem was introduced by Matsumoto *et al* [79]. They used a controlled high temperature anneal of the metal tape to form an (100)-oriented epitaxially grown NiO-layer. This self-oxidation epitaxy (SOE) process offers a simple way to facilitate the formerly complex and sensitive buffer layer, deposition process. Using this technique in

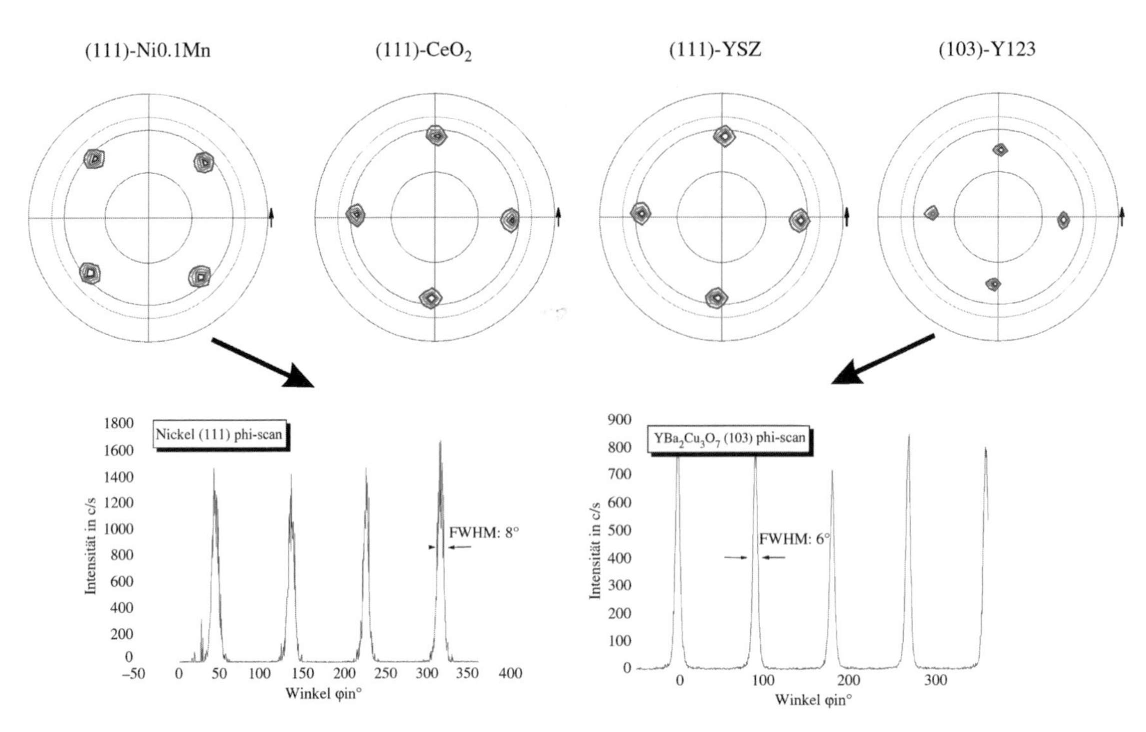

Figure B4.1.8. X-ray pole figure diagrams and corresponding phi-scans of a Y123/YSZ/CeO_2/Ni-tape heterostructure.

combination with an additional MgO buffer layer, YBCO films with a J_c up to 0.5 MA cm^{-2} could be prepared [80].

Most of the buffer layer architectures studied up to now consist of insulating buffer layers. However, conductive buffer layers show advantages for coated conductors, as they provide electrical stabilization of the superconducting layer in case of a transition to the normal conducting regime. One studied buffer layer architecture is the $SrRuO_3$/$LaNiO_3$ combination. Both oxides are conducting with room temperature resistivities between 300 and 600 μΩ cm. $LaNiO_3$ films show the better epitaxial growth on Ni, whereas high quality YBCO with current densities above 1 MA cm^{-2} can be grown epitaxially on $SrRuO_3$-buffered single crystals. The combination of both oxides for the deposition of YBCO on Ni resulted in an all conductive heterostructure with a critical current density of YBCO of 1.25 MA cm^{-2} [81]. One remaining problem is the formation of a thin insulating NiO-layer at the interface between Ni and $LaNiO_3$, reducing the electrical contact between the Ni-tape and the superconductor.

In general, nickel also has some disadvantageous properties: Abnormal grain growth may occur in the substrate temperature range, in which the buffer and $YBa_2Cu_3O_{7-\delta}$ film is deposited, destroying the cube texture [82]. The ferromagnetism of the nickel leads to magnetization losses, while its low tensile strength limits the possibility of producing and coating the very thin tapes that are necessary to achieve a high engineering current density (i.e. the current in the HTS divided by the cross section of the whole ribbon). On the other hand, the capability of forming the cube texture is also dependent on the alloying content. It is well known that alloying of Ni generally lowers the stacking fault energy (SFE), which leads to a change in the rolling texture and therefore a change in the recrystallisation texture so that the cube texture component disappears [A, B]. Also, solution hardening due to alloying without a significant change in the SFE affects the recrystallisation texture and prevents the formation of a sharp cube texture

[C]. As long as solid solution hardening is not strong enough to suppress the cube texture, it is beneficial for the tape and allows higher engineering current densities.

Despite the success in obtaining high critical current densities on pure Ni tapes there are still some substrate issues that need to be solved for coated conductors:

(a) Long length deposition

(b) Non-magnetic Ni-alloy tape substrates

(c) High strength tapes

The first issue is mainly connected to the difficulties of a continuous deposition process and therefore will be discussed in the corresponding chapter on film deposition. The availability of high quality biaxially aligned substrates is not a length issue due to the inherent preparation process. Since YBCO coated conductors are favoured for high temperature applications in magnetic fields, a non-magnetic Ni-alloy substrate would be preferable. Practical Ni-alloys, which are both non-magnetic and show a strong cube texture, are NiCr and NiV [71, 83]. In both alloys already a small alloying content of around 10% (9% for V and 13% for Cr) suppresses ferromagnetic ordering at 77 K. But the substrate surface oxidation prior to deposition is much more serious than for pure Ni, since both V and Cr form very stable oxides that make the epitaxial buffer layer growth more difficult. Nevertheless critical current densities up to 0.2 MA cm^{-2} can be obtained for NiCr alloys and, surprisingly, the SOE method works quite well with NiV [84]. Using Ni-alloys also the 0.2% yield strength of the tape can be increased due to solution hardening by a factor of about 3 [71] compared to the recrystallized soft pure Ni. Nevertheless, further improvement of the mechanical strength is needed to reduce the usable tape thickness, increasing the engineering critical current density.

C1
A1.3

References

[1] Simon, T 1994 Augerelektronentiefenprofilanalysen von hochtemperatursupraleitenden YBaCuO-Filmen auf metallischen und oxidischen Substraten *PhD Thesis* (in German) University of Göttingen

[2] Mogro-Campero A 1990 A review of high-temperature superconducting films on silicon *Supercond. Sci. Technol.* **3** 155

[3] Gao J, Klopman B B G, Aarnink W A M, Reitsma A E, Geritsma G J and Rogalla H 1992 Epitaxial $YBa_2Cu_3O_x$ thin films on sapphire with a $PrBa_2Cu_3O_x$ buffer layer *J. Appl. Phys.* **71** 2333

[4] Dovidenko K, Oktyabrsky S, Tokarchuk D, Michaltsov A and Ivanov A 1992 The influence of the native $BaAl_2O_4$ boundary layer on microstructure and properties of $YBa_2Cu_3O_{7-x}$ thin films grown on sapphire *Mater. Sci. Eng.* B **15** 25

[5] Kromann R, Bilde-Sørensen J B, de Reus R, Andersen N H, Vase P and Freltoft T 1992 Relation between critical current densities and epitaxy of $YBa_2Cu_3O_7$ thin films on MgO (100) and $SrTiO_3$ (100) *J. Appl. Phys.* **71** 3419

[6] Freyhardt H C, Hoffmann J, Wiesmann J, Dzick J, Heinemann K, Isaev A, Usoskin A and García-Moreno F 1998 Y-123 films on technical substrates *Appl. Supercond.* **4** 435

[7] Fork D K, Fenner D B, Barton R W, Phillips J M, Connell G A N, Boyce J B and Gaballe T H 1990 High critical currents in strained epitaxial $YBa_2Cu_3O_{7-\delta}$ *Appl. Phys. Lett.* **57** 1161

[8] Prusseit W, Corsepius S, Zwerger M, Berberich P, Kinder H, Eibl O, Jaekel C, Breuer U and Kurz H 1992 Epitaxial $YBa_2Cu_3O_{7-d}$ films on silicon using combined YSZ/Y_2O_3 buffer layers *Physica* C **201** 249

[9] Dimos D, Chaudari P and Mannhardt J 1990 Superconducting transport properties of grain boundaries in $YBa_2Cu_3O_7$ bicrystals *Phys. Rev.* B **41** 4038

[10] Ivanov Z G, Nilsson P A, Winkler D, Alarco J A, Claeson T, Stepantsov E A and Tzalenchuk A 1991 Weak links and DC squids on artificial nonsymmetric grain-boundaries in $YBa_2Cu_3O_{7-\delta}$ *Appl. Phys. Lett.* **59** 3030

[11] Hilgenkamp H and Mannhardt J 1998 Superconducting and normal-state properties of $YBa_2Cu_3O_{7-\delta}$-bicrystal grain boundary junctions in thin films *Appl. Phys. Lett.* **73** 265

[12] Hwang D, Ravi T S, Ramesh R, Chan S-W, Chen C Y, Nazar L, Wu X D, Inam A and Venkatesan T 1990 Application of a near coincidence site lattice theory to the orientations of $YBa_2Cu_3O_{7-x}$ grains on (001) MgO substrates *Appl. Phys. Lett.* **57** 1690

[13] Balluffi R W, Brokman A and King A H 1982 CSL DSC lattice model for general crystal–crystal boundaries and their line defects *Acta Metall.* **30** 1453
[14] Minamikawa T, Suzuki T, Yonezawa Y, Segawa K, Morimoto A and Shimizu T 1995 Annealing temperature-dependence of MgO substrates on the quality of $YBa_2Cu_3O_x$ films prepared by pulsed laser deposition *Japan. J. Appl. Phys.* **34** 4038
[15] Alarco J A, Brorsson G, Ivanov Z G, Nilsson P A, Olsson E and Lofgren M 1992 Effects of substrate temperature on the microstructure of $YBa_2Cu_3O_{7-\delta}$ films grown on (001) Y-ZrO_2 substrates *Appl. Phys. Lett.* **61** 723
[16] Fork D K, Garrison S M, Hawley M and Geballe T H 1992 Effects of homoepitaxial surfaces and interface compounds on the inplane epitaxy of YBCO films on yttria stabilized zirconia *J. Mater. Res.* **7** 1641
[17] Berkowski M, Pajaczskowska A, Gierlowski P, Lewandowski S J, Sobolewski R, Gorshunov B P, Kozlov G V, Lyudmirsky D B, Sirotinsky O I, Saltykov P A, Soltner H, Poppe U, Buchal C and Lubig A C 1990 C and AlO_4 perovskite substrate for microwave and far-infrared applications of epitaxial high Tc-superconducting thin-films *Appl. Phys. Lett.* **57** 632
[18] McConnell A W, Hughes R A, Dabkowski A, Dabkowska H A, Preston J S, Greedan J E and Timusk T 1994 Evaluation of $LaSrGaO_4$ as a substrate for $YBa_2Cu_3O_{7-\delta}$ *Physica* C **225** 7
[19] Chan S W, Chopra M, Chi C C, Frey T and Tsuei C C 1993 Growth of superconducting Y-Ba-Cu-O films on spinel and garnet *Appl. Phys. Lett.* **63** 2964
[20] Hohler A, Guggi D, Neeb H and Heiden C 1989 Fully textured growth of $Y_1Ba_2Cu_3O_{7-\delta}$ films by sputtering on $LiNbO_3$ substrates *Appl. Phys. Lett.* **54** 1066
[21] Xiong G C, Lian G J, Zhu X, Li J, Gan Z Z, Jing D, Shao K and Guo H Z 1993 $YBa_2Cu_3O_{7-\delta}$ thin films and microstrip resonators on $MgLaAl_{11}O_{19}$ substrates *J. Appl. Phys.* **74** 2983
[22] Phillips J M 1996 Substrate selection for high-temperature superconducting thin films *J. Appl. Phys.* **79** 1829
[23] Fork D K, Nashimoto K and Geballe T H 1992 Epitaxial $YBa_2Cu_3O_{7-\delta}$ on GaAs (001) using buffer layers *Appl. Phys. Lett.* **60** 1621
[24] Kawasaki M, Ohtomo A, Arakane T, Takahashi K, Yoshimoto M and Koinuma H 1996 Atomic control of $SrTiO_3$ surface for perfect epitaxy of perovskite oxides *Appl. Surf. Sci.* **107** 102
[25] Koster G, Kropman B L, Rijnders G J H M, Blank D H A and Rogalla H 1998 Influence of the surface treatment on the homoepitaxial growth of $SrTiO_3$ *Mater. Sci. Eng.* B **56** 209
[26] Koster G, Kropman B L, Rijnders G J H M, Blank D H A and Rogalla H 1998 Quasi-ideal strontium titanate crystal surfaces through formation of Sr-hydroxide *Appl. Phys. Lett.* **73** 2920
[27] Thanh V L, Bouchier D and Hincelin G 2000 Low-temperature formation of Si(001) 2×1 surfaces from wet chemical cleaning in NH_4F solution *J. Appl. Phys.* **87** 3700
[28] Wördenweber R 1999 Growth of high-T_c thin films *Supercond. Sci. Technol.* **12** R86
[29] Rao R A, Gan Q, Eom C B, Suzuki Y, McDaniel A A and Hsu J W P 1996 Uniform deposition of $YBa_2Cu_3O_7$ thin films over an 8 inch diameter area by a 90 degrees off-axis sputtering technique *Appl. Phys. Lett.* **69** 3911
[30] Berberich P, Utz B, Prusseit W and Kinder H 1994 Homogeneous high quality YBCO films on 3″ and 4″ substrates *Physica* C **219** 497
[31] Zaitsev A G, Ockenfuss G, Guggi D, Wördenweber R and Krüger U 1997 Structural perfection of (001) CeO_2 thin films on (1102) sapphire *J. Appl. Phys.* **81** 3069
[32] Lorenz M, Hochmuth H, Natusch D, Borner H and Tharigen T 1997 Large-area and double-sided pulsed laser deposition of Y-Ba-Cu-O thin films applied to HTSC microwave devices *IEEE Trans. Supercond.* **7** 1240
[33] Rao M R 1996 Influence of the microstructure of a $SrTiO_3$ buffer layer on the microwave properties of $Y_1Ba_2Cu_3O_{7-x}$ on MgO *Appl. Phys. Lett.* **69** 1957
[34] Holstein W L, Parisi L A, Flippen R B and Swartsfeger D G 1993 Effect of single-crystal substrates on the growth and properties of superconducting $Tl_2Ba_2Ca_2Cu_2O_8$ films *J. Mater. Res.* **8** 962
[35] Chin C C, Takahashi H, Morishita T and Sugimoto T 1993 Substrate dependence of the crystallinity and mosaic texture of $YBa_2Cu_3O_{7-\delta}$ ultrathin films deposited by laser ablation *J. Mater. Res.* **8** 951
[36] McKee R A, Walker F J and Chisholm M F 1998 Crystalline oxides on silicon—the first five mono layers *Phys. Rev. Lett.* **81** 3014
[37] Char K, Colclough M S, Garrison S M, Newman N and Zaharchuk G 1991 Bi-epitaxial grain-boundary junctions in $YBa_2Cu_3O_7$ *Appl. Phys. Lett.* **59** 733
[38] Dichiara A, Lombardi F, Granozio F M, Pepe G, Diuccio U S, Tafuri F and Valentino M 1996 A new type of biepitaxial *c*-axis tilted YBCO Josephson-junction *J. Supercond.* **9** 237
[39] Gromoll B, Krämer H-P, Ries G, Schmidt W, Heismann B, Neumüller H-W, Volkmar R R and Fischer S 2000 Development of resistive current limiters with YBCO films *Appl. Superconductivity (Inst. Phys. Conf. Ser.)* **167**
[40] Matsuno S, Umenura T, Uchikawa F and Ikeda B 1995 $YBa_2Cu_3O_x$ Thin films with yttria stabilized zirconia buffer layer on metal substrate by liquid source chemical vapor deposition using tetrahydrofuran solution of β-diketonates *Jpn. J. Appl. Phys.* **34** 2293
[41] Yu L S, Harper J M E, Cuomo J J and Smith D A 1985 Alignment of thin films by glancing angle ion bombardment during deposition *Appl. Phys. Lett.* **47** 932
[42] Iijima Y, Tanabe N, Ikeno Y and Kohno O 1991 Biaxially aligned $YBa_2Cu_3O_{7-x}$ thin film tapes *Physica* C **185** 1959

[43] Wiesmann J, Dzick J, Hoffmann J, Heinemann K and Freyhardt H C 1998 Growth mechanism of biaxially textured YSZ films deposited by IBAD *J. Mater. Res.* **13** 3149

[44] Betz V, Holzapfel B and Schultz L 1997 In situ reflection high energy electron bombardment analysis of biaxially oriented yttria-stabilized zirconia thin film growth on amorphous substrates *Thin Solid Films* **301** 28

[45] Iijima Y, Hosaka M, Tanabe N, Sadakata N, Saitoh T, Kohno O and Takeda K 1998 Processing and transport characteristics of YBCO tape conductors formed by IBAD method *Appl. Supercond.* **4**

[46] Dzick, J 2000 Mechanismen der ionenstrahlunterstützten texturbildung in yttrium-stabilisierten zirkondioxid-filmen PhD Thesis (in German)

[47] Sonnenberg N, Longo A S, Cima M J, Chang B P, Ressler K G, McIntyre P C and Liu Y P 1993 Preparation of biaxially aligned cubic zirconia films on pyrex glass substrates using ion-beam assisted deposition *J. Appl. Phys.* **74** 1027

[48] Maul M, Eckhardt H, Adrian H, Steinborn T and Fuess H 1993 $YBa_2Cu_3O_{7-x}$ Thin films on metallic substrates with biaxially aligned buffer layers made by ion beam assisted deposition *Appl. Supercond.* **1** 521

[49] Wu X D, Foltyn S R, Arendt P, Townsend J, Adams C, Campbell I H, Tiwari P, Coulter Y and Peterson D E 1994 High current $YBa_2Cu_3O_{7-x}$ thick films on flexible nickel substrates with textured buffer layers *Appl. Phys. Lett.* **65** 1961

[50] Wiesmann J, Heinemann K, Damaske M, Usoskin A, Freyhardt H C, Simon T and Neuhaus W 1993 Preparation and AES depth profile analysis of sputtered YSZ and CeO_2 buffer layers on technical substrates for HTSC applications *Appl. Supercond.* **1** 627

[51] Knierim A, Auer R, Geerk J, Linker G, Meyer O, Reiner H and Schneider R 1997 High critical current densities of $YBa_2Cu_3O_{7-x}$ thin films on buffered technical substrates *Appl. Phys. Lett.* **70** 661

[52] Reade R P, Berdahl P, Russo R E and Garrison S M 1992 Laser deposition of biaxially textured yttria stabilized zirconia buffer layers on polycrystalline metallic alloys for high critical current Y-Ba-Cu-O thin films *Appl. Phys. Lett.* **61** 2231

[53] Wiesmann, J 1998 Wachstum von biaxial texturierten YSZ-Dünnfilmen mittels ionenstrahlunterstützter Deposition PhD Thesis (in German)

[54] Foltyn S R, Arendt P N, Dowden P C, DePaula R F, Groves J R, Coulter J Y, Jia Q X, Maley M P and Peterson D E 1999 High T_c coated conductors—performance of meter-long YBCO/IBAD flexible tapes *IEEE Trans. Appl. Supercond.* **9** 1519

[55] Dzick J, Sievers S, Hoffmann J, Thiele K, García-Moreno F, Usoskin A, Jooss C and Freyhardt H C 2000 Biaxially textured buffer layers on large-area polycrystalline substrates *Appl. Superconductivity (Inst. Phys. Conf. Ser.)* **167**

[56] Hoffmann J, Dzick J, Wiesmann J, Heinemann K, García-Moreno F and Freyhardt H C 1997 Biaxially textured yttria stabilized zirconia buffer layers on rotating cylindrical surfaces *J. Mater. Res.* **12** 593

[57] Fukutomi M, Kumagai S and Maeda H 1997 Fabrication of $YBa_2Cu_3O_y$ Thin films on textured buffer layers grown by plasma beam assisted deposition *Aust. J. Phys.* **50** 381

[58] Hasegawa K, Yoshida N, Fujino K, Mukai H, Hayashi K, Sato K, Ohkuma T, Honjyo S, Ishii H and Hara T 1996 In-plane aligned YBCO thin film tape fabricated by pulsed laser deposition *Proc. ISS 96*

[59] Quinton W A J and Baudenbacher F 1997 Biaxial alignment of YSZ buffer layers on inclined technical substrates for $YBa_2Cu_3O_{7-x}$ tapes *Appl. Supercond. (Inst. Phys. Conf. Ser.)* **158** 1089

[60] Bauer M, TU Munich, Private communication

[61] Skofronick G L, Carim A H, Foltyn S R and Muenchausen R E 1994 Orientation of $YBa_2Cu_3O_{7-x}$ films on unbuffered and CeO_2-buffered yttria-stabilized zirconia substrates *J. Appl. Phys.* **76** 4753

[62] Zhu S, Lowndes D H, Budai J D and Norton D P 1994 In-plane aligned CeO_2 films grown on amorphous SiO_2 substrates by ion-beam assisted pulsed laser deposition *Appl. Phys. Lett.* **65** 2012

[63] Wiesmann J, Hoffmann J, Usoskin A, García-Moreno F, Heinemann K and Freyhardt H C 1995 Biaxially textured YSZ and CeO_2 buffer layers on technical substrates for large-current HTS applications *Appl. Supercond (Inst. Phys. Conf. Ser.)* **148** 503

[64] Gnanarajan S, Katsaros A and Savvides N 1997 Biaxially aligned buffer layers of cerium oxide, yttria stabilized zirconia and their bilayers *Appl. Phys. Lett.* **70** 2816

[65] Betz V, Holzapfel B and Schultz L 1997 In-plane aligned Pr_6O_{11} buffer layers by ion-beam assisted pulsed laser deposition on metal substrates *Appl. Phys. Lett.* **71** 2952

[66] Wang C P, Do K B, Beasley M R, Gaballe T H and Hammond R H 1997 Deposition of in-plane textured MgO on amorphous Si_3N_4 substrates by ion-beam assisted deposition and comparisons with ion-beam assisted deposited yttria-stabilized-zirconia *Appl. Phys. Lett.* **71** 2955

[67] Groves J R, Arendt P N, Foltyn S R, DePaula R F, Wang C P and Hammond R H 1999 Ion-beam assisted deposition of biaxially aligned MgO template films for YBCO coated conductors *IEEE Trans. Supercond.* **9** 1964

[68] Bauer M, Semerad R and Kinder H 1999 YBCO films on metal substrates with biaxially aligned MgO buffer layers *IEEE Trans. Appl. Supercond.* **9** 1502

[69] Goyal A, Norton D P, Budai J D, Paranthaman M, Specht E D, Kroeger D M, Christen D K, He Q, Saffian B, List F A, Lee D F, Martin P M, Klabunde C E, Hatfield E and Sikka V K 1996 High critical current density superconducting tapes by epitaxial deposition of $YBa_2Cu_3O_x$ thick films on biaxially textured metals *Appl. Phys. Lett.* **69** 1795

[70] Müller H G 1939 *Z. Metallk.* **31** 161

[71] de Boer B, Eickemeyer J, Reger N, Fernandez L, Richter J, Holzapfel B, Schultz L, Prusseit W and Berberich P 2001 Cube textured nickel alloy tapes as substrates for YBCO coated conductors *Acta Materialica* **49** 2021

[72] Eickemeyer J, Selbmann D, Opitz R, de Boer B, Holzapfel B, Schultz L and Miller U 2001 Nickel-refractory metal substrate tapes with high cube texture stability *Supercond. Sci. Technol.* **14** 152

[73] Jackson T J, Glowacki B A and Evetts J E 1998 Oxidation thermodynamics of metal substrates during the deposition of buffer layer oxides *Physica* C **296** 215

[74] Norton D P, Goyal A, Budai J D, Christen D K, Kroeger D M, Specht E D, He Q, Saffian B, Paranthaman M, Klabunde C E, Lee D F, Sales B C and List F A 1996 Epitaxial $YBa_2Cu_3O_7$ on biaxially textured nickel (001): an approach to superconducting tapes with high critical current density *Science* **274** 755

[75] Mathis J E, Goyal A, Lee D F, List F A, Paranthaman M, Christen D K, Specht E D, Kroeger D M and Martin P M 1998 Biaxially textured $YBa_2Cu_3O_x$ conductors on rolling assisted biaxially textured substrates with critical current densities of 2–3 MA cm^2 *Japan. J. Appl. Phys.* **37** L137

[76] Feenstra R, Cristen D K, Budai J D, Pennycook S J, Norton D P, Lowndes D H, Klabunde C E and Galloway M D 1992 Properties of low temperature, low oxygen pressure post-annealed YBaCuO thin films *High-T_c Superconductor Thin Film*, L Correra (North-Holland: Amsterdam) p 331

[77] Park C, Norton D P, Verebelyi D T, Christen D K, Budai J D, Lee D F and Goyal D F 2000 Nucleation of epitaxial yttria-stabilized zirconia on biaxially textured (001) Ni for deposited conductors *Appl. Phys. Lett.* **76** 2427

[78] Ichinose A, Kikuchi A, Tachikawa K and Akita S 1998 Deposition of Y_2O_3 buffer layers on biaxially-textured metal substrates *Physica* C **302** 51

[79] Matsumoto K, Kim S B, Wen J G, Hirabayashi I, Watanabe T, Uno N and Ikeda M 1999 Fabrication of in-plane aligned YBCO films on polycrystalline Ni tapes with surface-oxidized NiO layers *IEEE Trans. Appl. Supercond.* **9** 1539

[80] Matsumoto K, Kim S, Hirabayashi I, Watanabe T, Uno N and Ikeda M 2000 High critical current density $YBa_2Cu_3O_7$ tapes prepared by the surface-oxidation epitaxy method *Physica* C **330** 150

[81] Aytug T, Wu J Z, Cantoni C, Verebelyi D T, Specht E D, Paranthaman M, Norton D P, Christen D K, Ericson R E and Thomas C L 2000 Growth and superconducting properties of YBCO films on conductive SrRuO3 and LaNiO3 multilayers for coated conductor applications *Appl. Phys. Lett.* **76** 760

[82] de Boer B, Reger N, Opitz R, Eickemeyer J, Holzapfel B and Schultz L 2000 *Proc. ICOTOM*

[83] Petrisor T, Boffa V, Celentano G, Ciontea L, Fabbri F, Gambardella U, Ceresara S and Scardi P 1999 Development of biaxially aligned buffer layers on Ni and Ni-based alloy substrates for YBCO tapes fabrication *IEEE Trans. Appl. Supercond.* **9** 2256

[84] Boffa V, Petrisor T, Annino C, Fabbri F, Bettinelli D, Celentano G, Ciontea L, Gambardella U, Grimaldi G, Mancini A and Galluzzi V 2000 High J_c YBCO thick films on biaxially textured Ni-V substrates with CeO_2/NiO intermediate layers *Appl. Superconductivity (Inst. Phys. Conf. Ser.)* **167** 427

B4.2.1
Physical vapour thin film deposition techniques

H C Freyhardt, R Wördenweber, B Utz, A Usoskin and Y Yamada

B4.2.1.1 Introduction

Substantial progress in the understanding of high T_c superconducting (HTS) materials as well as in their application has been made due to the development of physical film-deposition techniques. In this section some of the important methods are considered, such as sputter deposition (SD), pulsed laser deposition (PLD), reactive thermal coevaporation (TCE) and liquid phase epitaxy (LPE), which at present are successfully employed in HTS film manufacturing. SD and LPE methods were widely known for a long time and, historically, PLD techniques were used narrowly, for instance for x-ray mirror manufacturing. On the other hand, TCE incorporating periodic post-oxygenation treatments was specifically developed for high T_c film deposition. SD and TCE methods represent typical vacuum deposition techniques, while PLD is a method that can be classified either as a vacuum or gas-flow-deposition technique. LPE represents a melt-growth technique. A1.3

The primary purpose of the above techniques is to provide a stoichiometric growth of multicomponent oxide-based films of the superconducting material. This becomes a challenge because of the generally large difference in the partial pressures (i.e. in bonding energies) of the different atomic components. This is especially relevant with respect to the oxygen content which commonly undergoes drastic variations with changes of deposition temperature and ambient pressure, but would also apply to other volatile constituents of the HTS compounds, e.g. Tl. The allowed oxygen pressures during a vacuum deposition cannot exceed 0.01 mbar, and should be smaller than 1 mbar in the case of PLD. On the other hand, the pressures needed to reach the optimal oxygen stoichiometry of the HTS material should exceed 100 mbar. The incompatibility of these requirements leads in many cases to a compromise where the film is grown under oxygen pressures which are optimal for a given deposition method and a subsequent post-oxygen-loading is performed by a relevant pressure increase at temperature lower than the growth temperature in order to establish the required stoichiometry. C3

Obviously, the proper composition is not sufficient for the manufacturing of HTS films with a high critical current density J_c. To yield high transport currents, an epitaxial, well textured, film is required. This means that the microcrystallites formed on the substrate surface must possess a good in-plane alignment, because grain boundaries, particularly with large misorientation angles between the grains, impair the current transport. For suitable single-crystalline substrates, such as $SrTiO_3$ or MgO, the epitaxial HTS film growth can be provided by an adjustment of the growth conditions, i.e. of temperature, deposition rate and ambient pressure. In the case of polycrystalline ('technical') substrates, e.g. ceramic or metallic ribbons or sheets, an artificially induced alignment by e.g. ion-beam assisted deposition (IBAD) [1, 2], inclined substrate deposition (ISD) [3] of the buffer layers or rolling assisted biaxial texturing of metallic substrate (RABiTS) [4] must be exploited. The degree of the in-plane crystalline perfection of HTS films determines the J_c values that can be achieved (see figure B4.2.1.1). B4.1 B3.2.4 D1.1.2

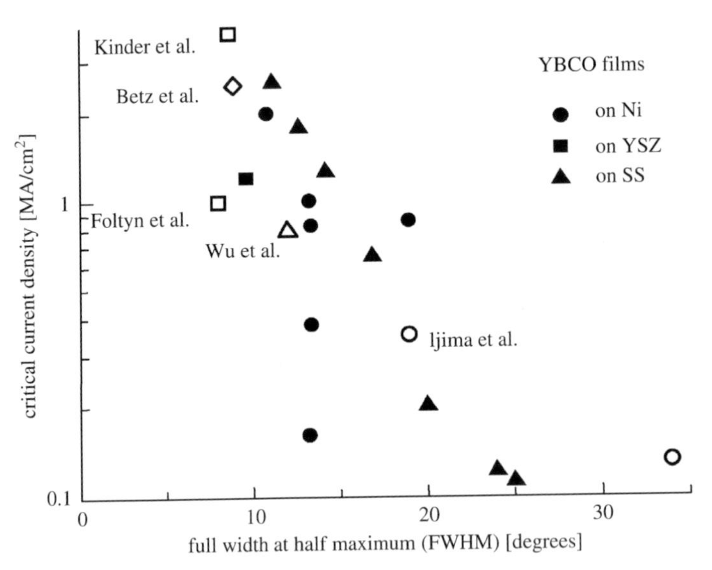

Figure B4.2.1.1. Critical current densities versus full width at half maximum (FWHM) of a Φ scan, which characterizes the quality of the in-plane texture, for YBCO films on different substrates [5 and references therein].

Apparently, the epitaxial microstructure should also be accompanied by a high homogeneity and density of film material to allow an improved current transport.

Whereas initially the general feasibility of the deposition techniques were considered, one is now in a position to coat lab-size substrates, tapes up to $\sim$ 100 cm in length and $20 \times 20\,\text{cm}^2$ in area. The next step, which is within reach, will lead to the processing of long-length respectively larger area HTS conductors suited e.g. for practical applications in fault current limiters, superconducting current leads or cables.

E1.3.3, E1.2, E1.3.2

B4.2.1.2 Sputter deposition

B4.2.1.2.1 Introduction of sputter deposition techniques

Sputtering is one of the most commonly used methods for deposition of thin films and multilayers. Its advantage stems from the simplicity of the physical processes involved, its versatility and flexibility for altering and customization. It is widely used by thin film suppliers for the semiconductor, photovoltaic, sensor and recording market. High melting point materials like ceramics and refractory metals which are hard to deposit by evaporation techniques are easily deposited by sputtering. The technology allows an upscaling and process parameters can be controlled very easily.

The classical sputter technologies range from simple dc glow discharge sputtering, which is limited to sputtering of conducting target material, through pulsed dc sputtering (typically at 10–200 kHz) preferentially used for reactive sputter deposition (i.e. SD with an additional reactive gas component), to rf (13.56 MHz) sputtering where any target regardless of its conductivity can be sputtered. Various cathode concepts and forms can be used: planar or hollow cathodes, cylindrical and rectangular cathodes, triodes or even ion beam guns. A further classification of the different technologies into on-axis and off-axis sputter deposition is defined by the geometrical arrangement of the cathodes with respect to the substrates. Finally, sputter yields (which represent one bottleneck for standard HTS SD) can strongly be enhanced by the use of magnetron assisted targets.

The very nature of the sputtering process can be utilized for tailoring the chemistry or structure of films. Bias sputtering (e.g. self bias) and ion-assisted ion beam sputtering (IBS and dual IBS) make use of the bombardment of the growing film during deposition. This leads to various effects like intermixing and enhanced ad-atom mobility. Similar effects are nowadays used in plasma assisted or plasma enhanced evaporation, chemical vapour deposition (CVD) or MBE technologies. B4.2.2

Disadvantages of sputtering include low target material utilization depending on the target configuration. For instance, in magnetron sputtering the plasma is restricted to a small surface area of the target, the so called 'race track'. Typically, in conventional magnetrons one expects to sputter only 25–35% of the total target material. This yield can be increased by the use of grounded or floating shields, rotating magnets, which keep the magnetron track moving [6–8] and/or flattening of the magnetic field lines parallel to the target surface [9, 10].

Whereas evaporation processes are typically carried out in high vacuum, which is partly due to the low energy and thus limited mean free path of the evaporated flux, sputtering is typically carried out in a gas atmosphere at typical gas pressures ranging between 10^{-3} and 5×10^{-2} mbar. Only nowadays lower and, especially, higher pressure ranges are utilized due to the special demands of epitaxial HTS thin films. The considerable gas pressure usually leads to intended or unintended incorporation of sputtering gas in the film. This effect can be increased by applying a negative bias voltage to the substrate holder which leads to ion bombardment at the substrate. The bombardment with ions can also lead to the formation of defects in crystalline films, which can be utilized for an improvement of the critical properties [11] or a release of internal stresses in the film [12]. Unlike CVD, where uniform deposition on a non-flat surface is possible, uniform sputter deposition is mainly restricted to planar substrates. Coating of non-planar surfaces can be improved by the application of a bias voltage to the substrate, which directs the sputter flux towards the substrate [13].

There is a tremendous amount of literature available on different aspects of sputter techniques. In this paper, major aspects of HTS thin film deposition via sputtering will be described. Before going into the details a brief introduction into the main principles and techniques of sputtering will be given. For further information refer to the literature [14]. Then the aspect of high-pressure reactive sputtering of HTS material will be covered.

B4.2.1.2.2 Principle of glow discharge sputtering

DC glow discharge is the simplest sputtering technique. A schematic sketch of the essential components is depicted in the inset of figure B4.2.1.2. The target material is attached to the water cooled planar cathode facing the substrates which are mounted to the substrate holder. The substrate holder can be grounded, biased or floating. Furthermore, for a number of applications (e.g. for deposition of HTS material) the substrate temperature plays an important role. For these applications, specially designed substrate coolers or heaters are mounted together with the substrate holder. During deposition the recipient is filled with the sputter gas, which for non-reactive sputtering consists of the inert gas Ar or Xe. In the case of reactive processes reactive components (e.g. O_2 or N_2) are added to the inert gas. If a dc power is applied to the target, initially a very small current flows through the system due to the limited number of charge carriers in the system (see figure B4.2.1.3). With increasing target voltage the density of charge carriers increases partly due to secondary electrons that are emitted from the target and partly due to impact ionization. The voltage rises and a 'Townsend discharge' is created. At this point an avalanche is started, which leads to a steady state where the numbers of electrons and ions

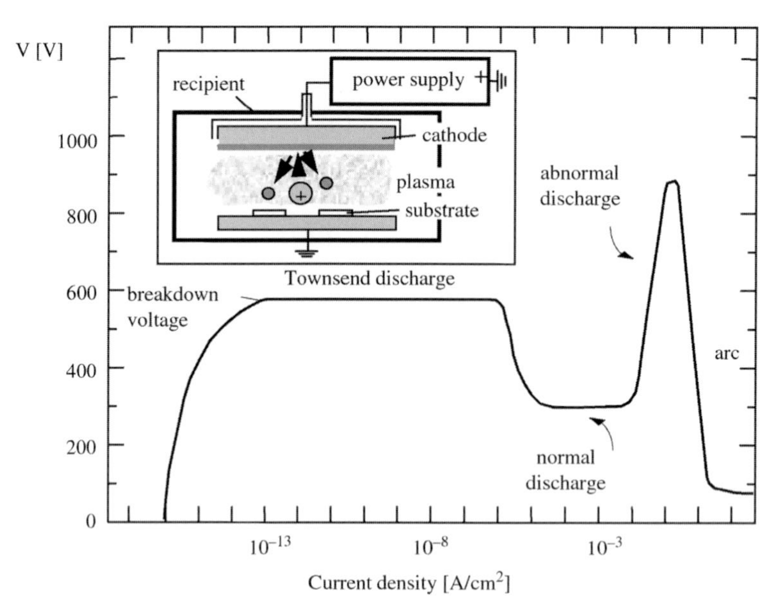

Figure B4.2.1.2. Schematic sketch of a voltage–current characteristic of a dc glow discharge and (inset) of a standard dc sputter recipient.

become the same and the plasma becomes self-sustaining. This regime is called the 'normal glow' state. Increasing the dc power leads to a rise in the target voltage (abnormal discharge) and finally to development of arcs. The deposition usually takes place in the normal or at the beginning of the abnormal sputter regime.

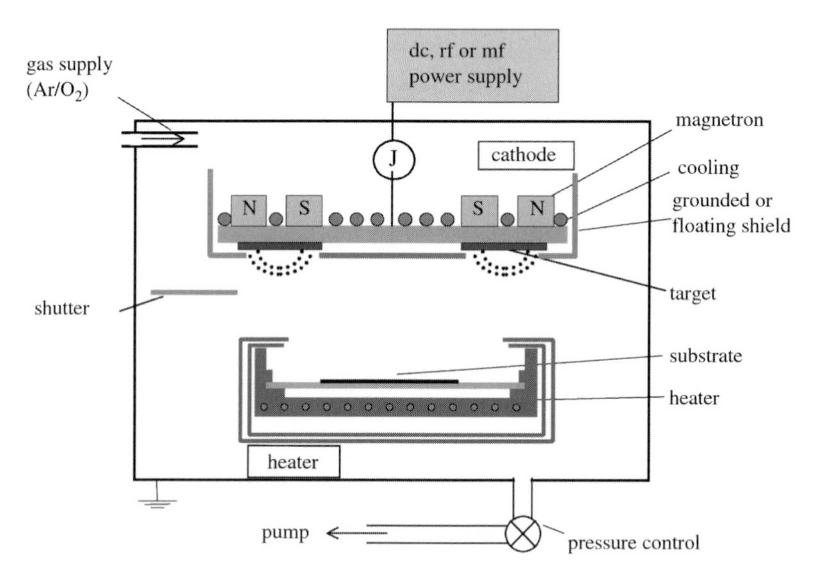

Figure B4.2.1.3. Schematic sketch of an on-axis magnetron sputter device for high pressure dc, mf and rf deposition of HTS and compatible materials [21].

B4.2.1.2.3 Sputter deposition of HTS-material

SD has the deserved reputation as being the technique for preparing thin films of alloys and complex materials. It was the natural method to try when HTS materials were first discovered. However, the problem of resputtering (due to oxygen ions generated in the plasma) was recognized in this class of materials some 18 years before the discovery of YBCO for $Ba(PbBi)O_3$ deposition [15] and has been thoroughly investigated later [16]. A method to avoid this problem is to thermalize the energetic species either by working at very high pressure (which also provides a particularly oxygen-rich environment) or by 'off-axis' sputtering at a modestly high sputtering pressure which forces all atoms emanating from the target to undergo a few energy-reducing collisions before approaching the substrate.

C1

Several different combinations and techniques using fully oxygenated stoichiometric targets are established, i.e. off-axis techniques using a cylindrical magnetron [17], facing magnetrons [18] or a single magnetron [19], and on-axis techniques using non-magnetron sputtering at very high pressure [20] or magnetron sputtering at high pressure [21, 22] (see figure B4.2.1.3). Furthermore, the cathodes can be charged with dc, rf (16.56 MHz) or mf (medium frequency ranging from 20 to 200 kHz) power or combination of these [23].

The sputter deposition process of HTS material is usually divided into three different parts. In the first part (presputtering), the target is slowly energized, the shutter between target and substrate is closed, and the substrate is slowly heated up to the deposition temperature. When substrate temperature and power at the cathode has reached the deposition parameters (see e.g. table B4.2.1.1) the second part, i.e., the actual deposition, is started by opening the shutter. The exact choice of the preparation parameters strongly depends upon the sputtering device (size, geometrical arrangement and sputtering mode). The process is terminated by the third part, i.e., the cooling down of the deposited HTS film. During this part of the process the oxygen content of the superconducting film is usually established *in situ*. This can be achieved by applying oxygen at high pressure (standard process, at 1 bar) or alternatively by offering activated oxygen at low pressure [24]. Although the standard process is certainly easier to handle, the alternative process has advantages for sputter deposition (activated oxygen can be generated by the sputter source) and large area deposition (avoidance of local overheating of the film), it is very reproducible, can be applied to complex high quality multilayers with top- or interlayers, through which oxygen transport is difficult (e.g. dense $NdGaO_3$ top-layers), and it can be

B4.1

Table B4.2.1.1. Typical sets of parameters for dc and mf sputtering of YBCO using 6″ planar magnetron cathodes for on-axis [23, 25]

Deposition technique	dc	Pulsed dc
Process pressure [Pa]	60	15–25
Ar/O_2 ratio	2:1	4:1
Heater temperature [°C]	820–860	880–900
Cathode power [W]	210	750
Frequency [kHz]	–	135–200
Target-substrate distance [mm]	80	100
Growth rate [nm min^{-1}]	1–5	5–10

integrated more easily into complex or automated preparation processes since it is a low pressure procedure.

Growth of cuprate films has remained a complicated matter, at best, due to its rather complex multi-component crystal structures of the layered cuprates, which are prone to a variety of defects and growth morphologies, and the extreme deposition parameters (high process pressure and temperature), which furthermore have to be controlled very accurately. Nevertheless, a number of typical parameters can be given, which are characteristic for high quality, sputter-deposited HTS films. Characteristic features of sputtered YBCO thin films are typically: surface roughness $r_{\text{peak-to-peak}} \approx 10-20\,\text{nm}$, FWHM of the rocking curve of the (005) reflex $\Delta\omega(005) \approx 0.1-0.3°$, a RBS minimum channelling yield of $\chi_{\min} \approx 2-3\%$, resistivity at 300 K of $\rho(300\,\text{K}) \approx 350-400\,\mu\Omega$ cm, a resistance ratio of $\rho(300\,\text{K})/\rho(100) \approx 3-3.3$, critical temperature (offset, inductive measurement) of $T_{c,\text{off}} \approx 87-90\,\text{K}$, transition width (inductive measurement) of $\Delta T_c \approx 0.5-0.8\,\text{K}$ and a microwave surface resistance of $R_s(77\,\text{K}, 10\,\text{GHz}) \approx 0.3-0.5\,\text{m}\Omega$. Films with different morphologies can be fabricated (see figure B4.2.1.4) depending upon the choice of sputtering technique and growth parameters.

C1 D1.1.2 D2.7

In conclusion, the different sputter techniques are usually highly reproducible. Advantages of these typically single target deposition techniques are that they are compatible with high pressures and the use of reactive gas components, and that both calibration and rate control are readily obtainable. The presence of a plasma can be a great advantage for the process. For instance oxygen can be activated which can make the oxygenation of the HTS thin films very easy. Homogeneous composition can be obtained over a rather large area (typically 1–2 inches for off-axis sputtering and 4 inches for on-axis sputtering). Recently a large area deposition was tested with 70 cm long planar cathodes [27]. However, it seems to be inherently slow (typically 20–300 nm h^{-1}) due to either off-axis geometries or high levels of thermalization. Higher rates can be achieved by modifications of the magnetron (dual magnetrons) or pulsed-dc deposition [23] by which the rate could be increased to 1.2 μm h^{-1}. Nevertheless, sputtering has been widely used for the deposition of compounds (e.g. nitrides and oxides) comparable to HTS material as the workhorse technique because it is reliable, cheap and relatively easy to run automatically or semiautomatically and can be scaled up to areas larger than 1–3 m. Furthermore, although the presence of a plasma and high deposition pressures hamper the use of most *in situ* characterization such as low energy electron diffraction (LEED) or reflection high energy diffraction (RHEED), superlattices and multilayer HTS films have been grown with various sputter techniques [28–31] as well as multilayer heterostructures involving e.g. HTS and magneto-resistive compounds [32].

B4.2.1.3 Pulsed laser deposition

B4.2.1.3.1 Physical principles of pulsed laser deposition

From the very beginning of HTS discovery, the pulsed laser deposition (PLD) represents one of the most successful and reliable techniques for HTS-film fabrication. The major part of outstanding experimental results in the past and now is achieved on HTS films prepared by PLD with which J_cs of (2–4) MA cm^2 can easily be reached for the films on single crystalline substrates.

The principle of PLD is based on the evaporation of the target material subjected to laser beam pulses. Each of these pulses causes a *complete* vaporization of a small volume of material in the area of the beam spot on the target surface. The duration of the laser pulse must be short enough (conventionally less than 50 ns) to provide, at least partly, adiabatic conditions for each evaporation event. This is absolutely necessary for suppressing target destoichiometrization originating from a macro-scale diffusion in the target. In reality, it is not imaginable to avoid completely the heat exchange between the laser heated volume and the rest of the target. As a result, a certain degree of destoichiometrization at the target surface increases in the beginning of ablation and stabilizes after ~ 100 laser pulses [33, 34]. Nevertheless,

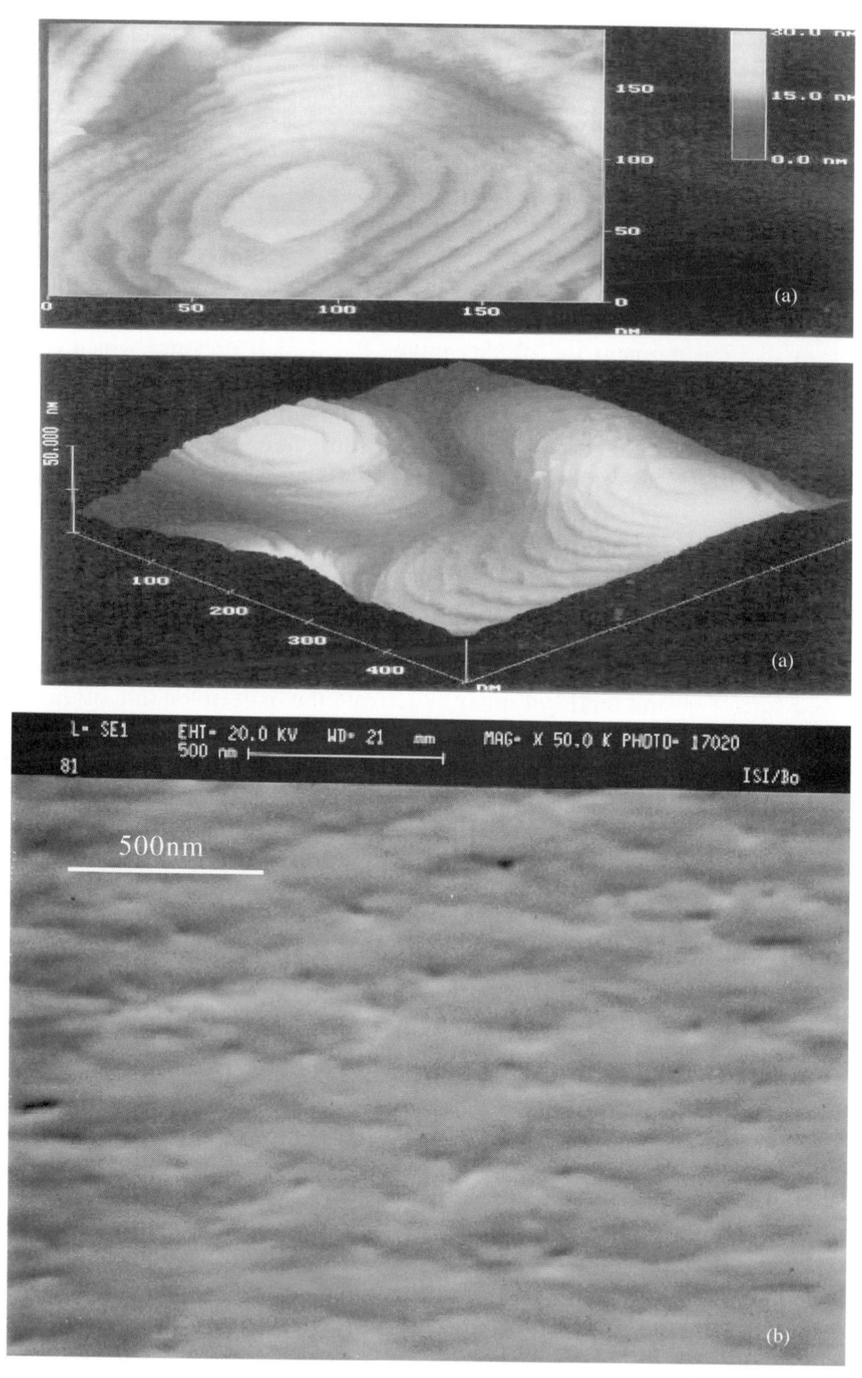

Figure B4.2.1.4. (*a*) STM images of a high-pressure DC sputtered YBCO film on $SrTiO_3$ [26] showing simultaneously spiral growth and two-dimensional nucleation and growth between the spirals and (*b*) SEM images of the surface of a 700 nm thick YBCO film on 2 × sapphire without spiral growth revealing a peak-to-peak roughness of ~ 10 nm [12].

if the ablation remains stable, such a surface deterioration does not cause an incongruity in the vaporized material due to self-balancing of the diffusion/evaporation processes similar to well known effects in the multicomponent sputtering [35].

The energy of photons together with the surface power density of the laser beam is sufficient to provide not only evaporation but also an emission of positive ions, monoxide ions [36] and electrons from the target surface [37]. The material evaporated from the hot target is then further heated up due to absorption of the laser radiation. In the case of excimer-laser ablation of YBaCuO targets the temperature of the plasma has been evaluated at 7000–20 000 K, although the evaporation temperature of the target material is much lower. At the end of the laser pulse a dense plasma cloud with a thickness of 0.01–0.1 mm forms near to the ablated area. As the excitation time provided by the laser pulse is relatively short, at the first moment the plasma plume has a high pressure which causes its further adiabatic expansion, preferably in the direction normal to the target surface. Depending on the ambient gas pressure in the vacuum chamber (which is typically in the range of 0.1–0.6 mbar) the length of the plume varies in between 20 and 100 mm. The lifetime of the plume amounts to $\sim$ 5–10 μs [37]. The energies of the particles in the laser plume forms a broad and complicated spectrum typically from 1 to 200 eV [36].

C1 B2.2.2

If a substrate is exposed to such a plume, a stoichiometric deposition of multicomponent films can be provided. A typical PLD scheme is depicted in figure B4.2.1.5(*a*). An important requirement is to deposit less than 0.1 nm of the film thickness per deposition pulse. In this case atoms have a much higher probability of finding their correct sites in the crystalline lattice in the course of surface migration [38]. For the oxide-based compositions, used e.g. for HTS films, it is necessary to maintain in addition a partial oxygen pressure of several tenths of mbar to activate the oxygen loading of the growing film as well as an oxygen post-loading. The latter one is conventionally performed after film deposition in an oxygen atmosphere of 300–1000 mbar and at film temperatures of 350–500°C.

In general, the deposition technique seems to be rather simple because neither thermal or high voltage (discharge) deposition sources, nor special means for a deposition control are required. The main advantage is that the film stoichiometry is pre-determined by target composition. Nevertheless, aside from this simplicity there are some specific problems which one has to overcome to establish an efficient and highly reproducible method for HTS film deposition.

The first point is an ejection of macro-particulates which hit the substrate during PLD [39–43]. This results in numerous particulates (droplets) with a characteristic size of 1 μm embedded in the film. In principle, two mechanisms can be responsible for this phenomenon: (a) a condensation of the particulates

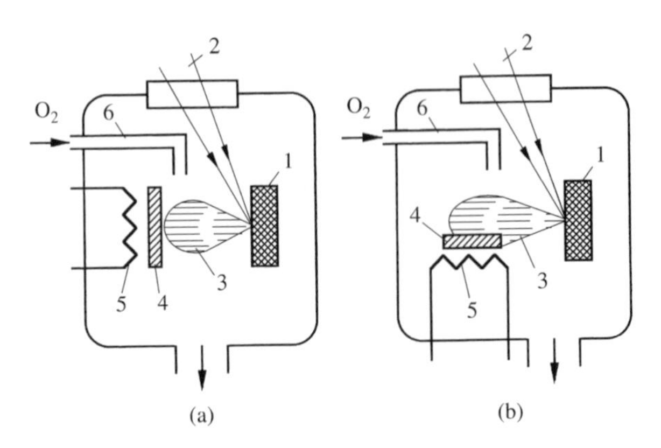

Figure B4.2.1.5. On-axis (*a*) and off-axis (*b*) PLD schemes. 1 target, 2 laser beam, 3 laser plume, 4 substrate, 5 substrate heater, 6 oxygen line.

directly in the plasma plume due to its overcooling in the course of plume expansion, and (b) an ejection of molten particulates from the target caused by ablation of loosely bonded target flakes [40] or by splashing of molten target material as a result of subsurface boiling [43, 44]. The experiments on off-axis film deposition where the substrate is oriented in parallel to the axis of plasma plume (see figure B4.2.1.5(*b*) [41] have shown a significant reduction in the number of particulates. This, at least indirectly, confirms a dominating role of the mechanism (b). Unfortunately, the off-axis deposition geometry also leads to a drastic reduction of the deposition rate [41] and has, in general, only a narrow field of applications. Nevertheless, the off-axis deposition was successfully employed for the deposition of YBaCuO films when it was combined with substrate rotation [45]. Several particulate-removal schemes for the on-axis deposition geometry have been reported. They are exploiting mechanical velocity filters [46], special pre-treatments of target surface [42, 47], double-beam laser ablation [39] and supplementary laser-assisted heating of the plume in the area near to the target [40, 48]. The two latter techniques allow production of almost particulate-free films. Nevertheless, a similar result can be achieved by ordinary PLD employing UV (193–308 nm) excimer lasers if the pulse energy density is just above the ablation threshold [39] which, e.g. for $YBa_2Cu_3O_{7-\delta}$, is estimated to be 1–2 J cm^{-2} depending on the laser wavelength and the target properties [49, 50].

B4.2.1.3.2 Alteration of target surface: influence on film deposition rate

The phenomenon of particulates ejection is closely linked to the generation of a particular relief of the target surface which undergoes great alterations during the ablation process [33, 34]. The initially ground or polished surface of the YBaCuO target after 20–100 pulses of 2–5 J cm^{-2} evolves into a ridge-cone morphology interspersed with elevated pedestals. After 100–2000 laser shots the target surface is filled with closely packed columnar elements of $\sim$ 10 μm in diameter (figure B4.2.1.6(*a*)). The cone axes are approximately aligned with the direction of the laser beam. This columnar alignment suggests that erosion accompanied by shadowing (due to segregated yttrium islands which are much more resistant to laser vaporization) is the primary cone formation mechanism [33, 34], as opposed to earlier suggested melting/resolidification model [44]. But even this reasonable explanation seems to be not sufficient for the understanding the regularity of the cone structure. On the other hand the Y 'protecting caps' in many cases cannot be observed at the top of the cones. A satisfactory description of the cone phenomenon seems to be given by recently developed angular selective ablation model, taking into account the second-order light scattering of the laser beam at the target surface [51].

The development of the cone structure is especially undesirable when the ablation process is exploited for film deposition because of: (a) a drastic reduction of the deposition rate [33, 34, 52, 53], (b) deflections of the laser plume from the normal direction [33, 34] and (c) an increasing number of droplets ejected from the target [42]. All these effects are dependent on time (i.e. on the number of ablation pulses) and, therefore, cause unstable conditions for film growth. For instance, a strong reduction of film deposition rate is observed not only at the beginning of the ablation, but also after 10^3 ablation pulses (see figure B4.2.1.7). The total decrease of the deposition rate corresponds to > 90 %. To maintain the same efficiency of PLD an increase of the beam energy would be needed, which in its turn further deepens the existing relief and leads to an irreversible deterioration of the target surface.

To solve the problem a lot of attempts have been made, and a variable azimuth ablation (VAA) method has been developed [52, 54]. The principle of VAA could be understood from the following simplified consideration. As mentioned above, at a constant angle, φ, of beam incidence, instead of the initial surface with a smaller roughness (schematically depicted in the figure B4.2.1.8(*a*))a ridge/cone relief forms (figure B4.2.1.8(*b*). If the angle of incidence is changed to the opposite inclination ($-\varphi$), a similar structure forms in the 'opposite' direction (see figure B4.2.1.8(*c*). Periodic variations of the beam direction (i.e. VAA) at a constant value of $|\varphi|$, i.e. relative variations of the beam azimuth ψ

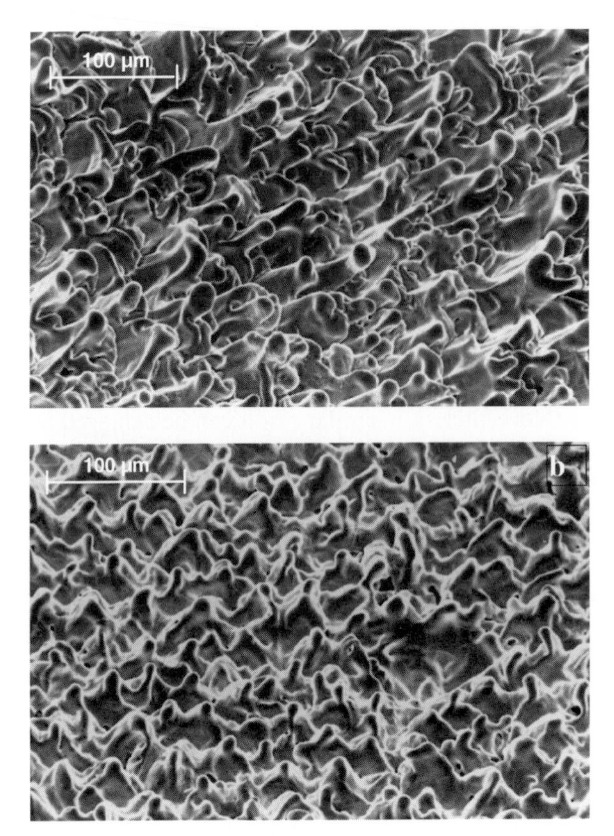

Figure B4.2.1.6. SEM micrographs of the ceramic $YBa_2Cu_3O_{7-\delta}$ target after laser ablation with an oblique beam incidence without (*a*) and with (*b*) VAA. Number of ablation pulses $N_a = 100$, angle of incidence $\varphi = 45°$ (d).

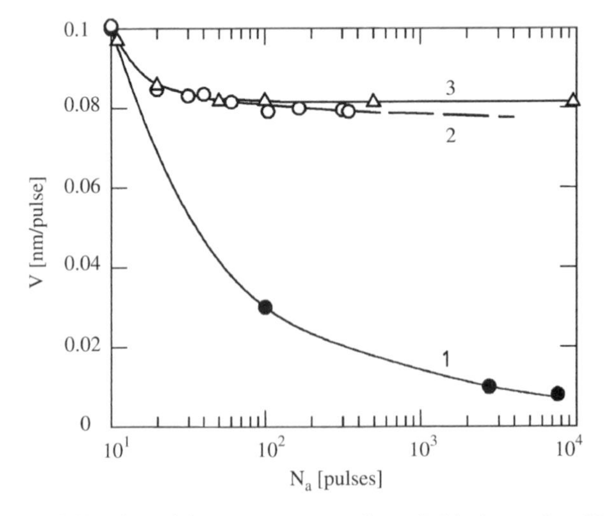

Figure B4.2.1.7. Dependences of film deposition rate V on number of ablation pulses N_a for conventional PLD (1), VAA with spiral scan (2), and VAA with meander scan (3). Energy density of the laser beam (308 nm) on the target is 2.5 J cm^{-2} [52].

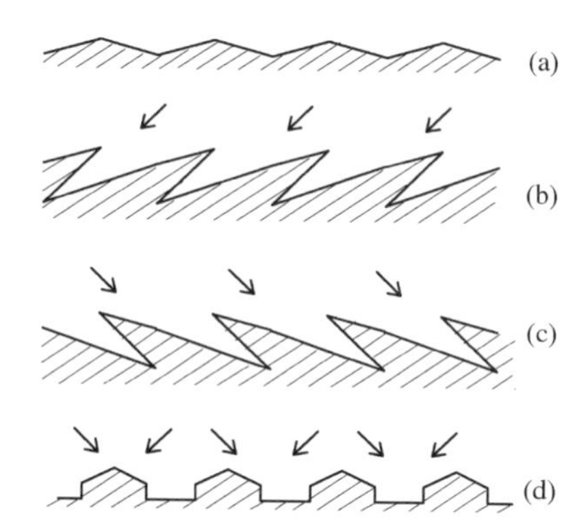

Figure B4.2.1.8. Schematic views of the initial target surface relief (*a*) and the surface relief after laser ablation with incident angle $\varphi = 45°$ for azimuth angles $\psi = 0°$ (*b*), $\psi = 180°$ (*c*), and with periodic variations of ψ between 0° and 180° (d).

against the target, avoid the development of a deep surface relief because of the competition of two differently tending processes shown in figures B4.2.1.8(*b*) and (*c*). The resulting surface topography corresponds to figure B4.2.1.8(*d*). A SEM-micrograph of the target surface after VAA (figure B4.2.1.6(*b*)) exhibits a decreased depth of the relief with the preferable orientation of ridges/cones in the normal direction. The efficiency of PLD in this case saturates quickly after ~ 30 pulses keeping a level of 0.08 nm/pulse (figure B4.2.1.7). The precision of such a 'stabilization' depends on the way of target scanning.

D1.3

At first, the VAA-effect was observed on a permanently rotating disc-like target which simultaneously performs in-plane oscillations of its rotation centre (denoted as r-scan) relatively to the beam spot (figure B4.2.1.9(*a*)) [55, 56]. The track of the beam on the target resembles a spiral. Apparently, a moment when the spot crosses the rotation centre corresponds to a 'jump' of the azimuth of the incoming beam according to the VAA conditions. Because of an inhomogeneous ablation 'load', an observed tendency of the initially flat target surface is to acquire a concave shape when a periodic

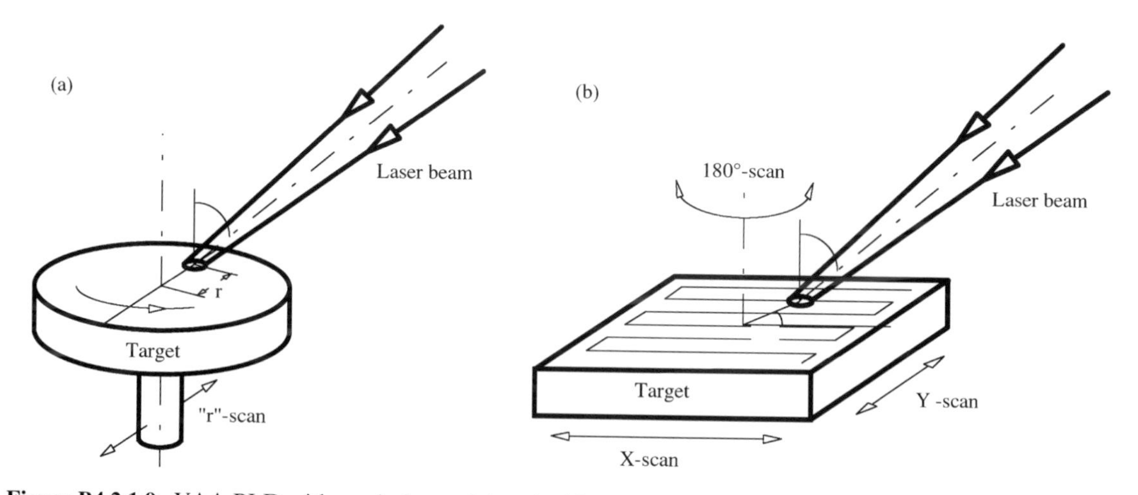

Figure B4.2.1.9. VAA PLD with a spiral scan (*a*) and with meander scan incooperating periodic 180° azimuth (ψ) variations (*b*). In both cases the incident angle φ is constant.

function $r(\tau)$ with a constant $dr/d\tau$ is used. The concave shape leads to undesirable angular deviations of the laser plume. To keep the surface in a flat state, a linear velocity $dr/d\tau$ of the oscillations must be dependent on r as $dr/d\tau = C/r$, where r is a variable distance between the rotating centre and the laser spot, C is a constant which depends on laser repetition rate and on the relation of the spot dimension to the dimension of the target surface. Obviously, the linear velocity $dr/d\tau$ becomes infinite when $r = 0$. It is not possible to meet this requirement, even approximately in the experiment. This is one of the main disadvantages of VAA with a continuously rotating target which results in a considerable instability of the laser plume and, therefore, in a destabilization of the film deposition rate by $\sim 6\%$ (figure B4.2.1.7). Particular conditions at the centre of rotation where the surface is exposed to the beam with random numbers of different azimuth angles cause unexpected peculiarities around this point [52].

A schematic principle of VAA with a meander scan is shown in figure B4.2.1.9(*b*). The target motion provides (a) a meander scan of its surface with regard to the laser beam and (b) periodic turns by 180° which result in changes of the beam azimuth ψ while the incident angle φ is kept constant. The periodic variations of ψ lead to a suppression of the cone structures if the 180° turns of the target are performed after not more than 3–5 complete X–Y scan cycles. Constant velocities of the X–Y scans ensure the homogeneity of the ablation load and, therefore, the maintenance of the surface flatness. The stability of the deposition rate ($< 3\ \%$) was found to be better than the one for the spiral scan (figure B4.2.1.7).

B4.2.1.3.3 High-rate PLD for large-area film deposition

A large area deposition based on VAA in many cases requires a scan of the substrate with a laser plume, as it improves the homogeneity of the thickness and stoichiometry of the grown films.

In the geometry of the spiral scan, the motion of the plume can only be provided by synchronically moving both the beam and the rotating/oscillating target. This seems to be extremely intricate. It is much easier and more efficient to employ the meander scan for this aim. In this case, instead of one or both of X–Y target motions an equivalent beam scan is introduced, so that the beam scanning area corresponds to the desirable dimensions of deposition area—e.g. a one-dimensional beam scan provided by an oscillating mirror was employed in our experiments. Due to that, it was possible to scale up the deposition area to $35 \times 200\ \mathrm{mm}^2$ [52]. Together with rotating/linear motions of the substrate it allows a deposition of HTS films on tubes and tapes of more than 5 m in length.

C1 It is known, particularly for YBaCuO that PLD must provide the deposition of < 0.1 nm of film thickness per deposition pulse to ensure good film stoichiometry. Because of a relatively small divergence of the plasma plume, only a small substrate area of 1–2 cm^2 is coated with the film during a single deposition pulse. On the other hand, an increase of the pulse repetition rate, f, is limited by a drop of J_c, which is found to start from $f = 10$–20 Hz. This results in low-rate film processing with a volume deposition rate, R, of only 0.5–2 nm m^2 h, which implies that a processing time of > 5 h is required to produce a $1000\ \mathrm{m} \times 10\ \mathrm{mm}$ CC tape with a 1 μm-thick film. On the other hand, the low-rate processing results in more film contamination, which degrades the film quality and the J_c.

The aim of high-rate PLD [57] is to allow a considerable increase in the *integral deposition rate* together with a large area deposition. The developed technique is based on scanning a large area target with the laser beam. The laser plume produced in the course of such scans causes film deposition in a random sequence of small areas of a large area substrate which is also smoothly moved during the film deposition. The target and substrate scans are performed independently in time, and, therefore, ensure a homogeneous film thickness and stoichiometry.

Under such conditions, the repetition rate of laser pulses can be increased by a factor of 10^3 while the local repetition rate of the deposition pulses at any particular substrate area is kept at the low level of < 20 Hz, i.e., in the optimal range for film growth. In principle, laser pulses with $f > 10$ kHz could be employed for such processing, but relevant excimer lasers with such frequencies are not available yet.

YBaCuO films with a J_c of up to 2.3 MA cm^{-2} were deposited on long stainless steel substrates using the high-rate PLD with a deposition rate of 35 nm m^2 h [57], i.e. an increase of the film deposition rate by at least a factor of 30 compared to conventional PLD could be utilized. C1

B4.2.1.3.4 Maintenance of the deposition temperature

During high T_c film deposition it is important to maintain the optimal temperature of the substrate. This optimal temperature can vary from 720 to 800°C depending on substrate and film compositions, PLD regimes, etc, while the tolerance for the optimum deposition temperature is only $\pm$ (2–4)°C. At these temperatures, high radiation losses of the substrate surface are influenced by the film growth which is accompanied by a pronounced change of the IR emissivity [58]. This influence becomes incredibly high, especially in the case when the substrate is not bonded directly to the heater (which allows a large area deposition due to the substrate translation). A principal solution for a precise temperature control is a quasi-equilibrium heating (QEH) technique [59]. QEH is based on the heating of the substrate in a high-temperature 'black cavity' combined with a short periodic exposure of the substrate surface to the synchronized laser plume. QEH is capable of maintaining a precise ($\pm$ 1°C) substrate temperature because the radiation losses from the substrate/film surface are reduced, by at least a factor of 10. Different constructions of QEH systems aimed at large-area PLD deposition on flat and tubular substrates have been developed [59, 61]. One of them, employed for PLD deposition of high T_c superconducting films on tubular substrates, is depicted schematically in figure B4.2.1.10. A tube substrate (1) is heated by a cylindrical heater (2) around which a chopper (3) rotates quickly. In both of them there are windows (4) enabling film deposition at the moment of their coincidence. At exactly this moment the laser triggered by a synchronizing electronics produces a beam pulse (5) hitting a target (6). For the rest of the time the deposition window (4) is closed by the chopper. To obtain a homogeneous film the tube substrate rotates and is also slowly pulled through the QEH system as shown in figure B4.2.1.10. It was found, that low heating powers (of only 100 and 800 W for 10 and 70 mm-diameter tubes, respectively) are required to reach a substrate temperature of 800°C, and to maintain it during PLD very precisely.

The tubular QEH system was also successfully employed for YBaCuO deposition on long (to 5.5 m) stainless steel tapes. In this case the tape was wound in a form of helical spiral around tube 1 shown in figure B4.2.1.10 which played the role of tape holder [57].

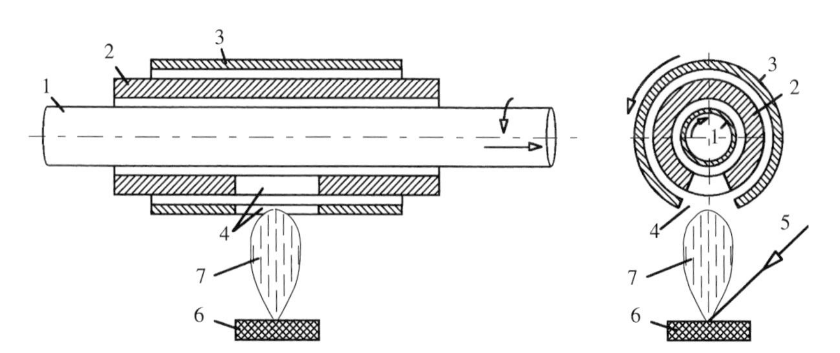

Figure B4.2.1.10. Schematic view of QEH-system for PLD on tubular substrates. 1 tubular substrate, 2 cylindrical heater, 3 chopper, 4 deposition window, 5 laser beam, 6 target, 7 laser plume scanning the target.

B4.2.1.3.5 Conclusion

C1 The recent success in PLD development resulting (a) in manufacturing YBCO films with $J_c >$ 1 MA cm^{-2} on buffered technical (metallic and ceramic) substrates, (b) in solving the problem of large area film deposition and (c) in the possibility of an extraordinary increase of the integral deposition rate A1.3 up to 35 nm m^2 h^{-1} [57] reveals real prospects of establishing new PLD technologies for HTS film industrial production in the near future.

B4.2.1.4 Thermal coevaporation

A1.3 A number of promising superconducting applications explored today rely on the availability of nearly single crystalline YBCO thin films. There is a demand especially for films with both large extensions and a thickness in the micron range that have to be manufacturable with high speed at modest costs in large numbers. To date the most prominent technologies based on such material are passive microwave E3, E1.3.3 devices (e.g. for mobile communication), resistive fault current limiters and coated conductors. Thus it is straightforward to consider thermal evaporation techniques which are clearly cheap, almost unlimited in deposition rate and film size, and well established in industrial production (e.g. aluminium coatings) for decades.

B2.2.2, B2.1 In a first approach one might try to simply evaporate YBCO powder or sintered material as used for plasma coating techniques. Electron beam evaporation of simple oxides like MgO or CeO_2 is common practice in the optical industry. Unfortunately, while these materials are chemically stable in vacuum at evaporation temperatures, YBCO decomposes rapidly at modest temperatures resulting in a mixture of materials that considerably differ in vapour pressure and thus evaporation rate. A stoichiometric evaporation can only be achieved by flash evaporation resulting in rates that are much too high to allow for epitaxial growth. No YBCO film growth has ever been reported for this evaporation approach.

As a consequence, the stoichiometry of YBCO thin films can only be controlled if the components Y, Ba and Cu are evaporated separately at the same time. In an single step process, the metals Y, Ba and Cu are deposited in a reactive atmosphere forming YBCO films *in situ* on heated substrates. In this section we focus on this technology called reactive TCE.

B4.2.1.4.1 A basic setup for YBCO thermal coevaporation

The evaporation materials Y, Ba and Cu are supplied in the form of granules or pieces of wire. They can be evaporated by electron beam the evaporators which direct an electron beam on the material placed in a cooled crucible heating it to evaporation temperature. This is a very robust and proven commerical technology. Nevertheless, most of the YBCO deposition systems in operation today heat the materials by placing them on strips of high melting point metal ('boats') made of Ta, W, or Mo and electrically contacted by clamps. The metals Y, Ba and Cu evaporate on heating the boats resistively by high currents.

The vapour metal atoms spread and move to the substrate as long as they do not suffer collisions with gas molecules scattering them back or at least off their path. Such 'ballistic' transport is only achieved if the mean free path for scattering is much larger than the distance from the metal sources to the substrate. Typical separations of 30 cm require the pressure of the background gas to be lower than some 10^{-3} Pa. Thus coevaporation can take place only under high vacuum conditions in vacuum chambers.

The metal vapour condenses on the substrate if it is much cooler than the metal sources. The atoms first are adsorbed weakly ('physisorbed') keeping enough surface mobility to diffuse on the substrate or film surface until they get strongly bound ('chemisorbed') to the film by oxidation. The higher the substrate is heated, the further the adsorbants diffuse and get a chance to find their site in the lattice structure of YBCO. To allow for epitaxial YBCO growth, in TCE deposition processes substrates generally have to be heated to 650–680°C. Instead of gluing the substrates to a heater block, most of the TCE substrate heater concepts today rely on the transfer of heat by radiation from hot heater wires or quartz lamps which are not in contact with the substrate.

Each metal source is monitored separately by rate monitors like quartz crystal monitors (QCM). A proper collimation scheme ensures that each quartz is coated by one boat only and crosstalk between the QCMs is minimized. The measured evaporation rates are electronically fed back to the high current supplies of the individual boats to precisely control and lock the evaporation rate. To grow high quality epitaxial YBCO layers, the metals Y, Ba and Cu have to be evaporated in a stoichiometric ratio. Typical YBCO growth rates of 25 nm min^{-1} are easily achieved.

A standard TCE setup for small samples is sketched in figure B4.2.1.11. In the next paragraph some specific aspects of this technique will be discussed in detail.

B4.2.1.4.2 Substrate considerations

It is well known that grain boundaries degrade the critical current densities in YBCO thin films. Thus single crystalline YBCO films have to be grown epitaxially on single-crystal or highly textured substrates. A basic prerequisite for substrate materials is that crystal structure and lattice constant match those of YBCO. Single crystalline perovskites like $SrTiO_3$, $LaAlO_3$, or $NdGaO_3$ and the cubic MgO are commonly used. Materials which do not match the YBCO lattice properly can be buffered by a variety of appropriate buffer layers that may both transfer the crystalline order of the substrate while forming a proper template for the YBCO, and act as a diffusion barrier to species of the substrate material that tend to diffuse into the growing YBCO film. Using buffer layers, YBCO can be grown by TCE on semiconductors like Si and GaAs as well as on sapphire [62] or various metal alloys.

B4.1

Especially for the TCE deposition method, YSZ substrate surfaces react with condensing Ba metal atoms:

$$2Ba + 2ZrO_2 + 3O_2 \rightarrow 2BaZrO_3$$

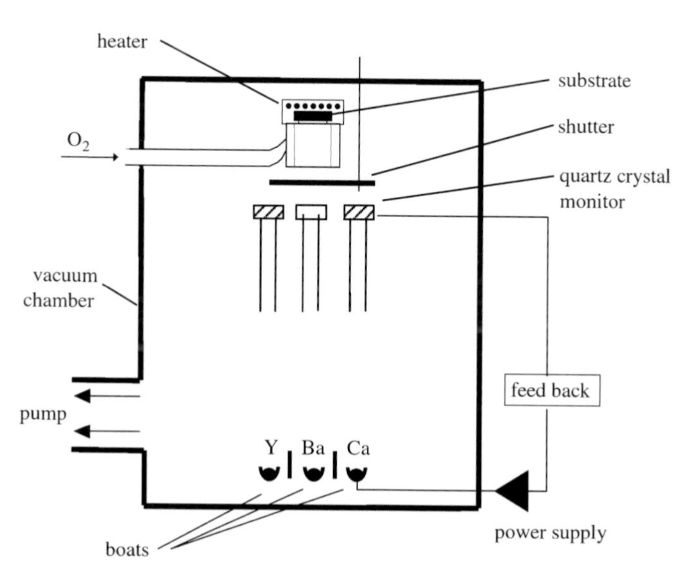

Figure B4.2.1.11. Basic setup of a TCE deposition system.

Table B4.2.1.2. Substrate materials and buffer layers for YBCO deposition

Substrate	$SrTiO_3$	MgO	$LaAlO_3$	Sapphire	YSZ	Si	GaAs	Ni
Buffer layers	–	–	–	CeO_2	Y_2O_3	YSZ/Y_2O_3	MgO	CeO_2

D1.2 As TEM studies reveal, an amorphous $BaZrO_3$ layer forms at the interface that blocks the epitaxial growth of subsequent YBCO. In addition, due to the loss of Ba the YBCO is nonstoichiometric and precipitates may form during the crucial first monolayers. The problem can be overcome by growing a thin passivation layer of CeO_2 or Y_2O_3 on the YSZ which does not act as a diffusion barrier but as a chemical reaction barrier during YBCO growth.

Due to substrate temperatures low in comparison to other YBCO coating methods, TCE is especially suitable for deposition on oxidation sensitive metal substrates like Hastelloy or nickel alloys which are currently discussed for long length coated conductors in combination with texturing B4.1 technologies like IBAD [1, 2] or RABiTS [4].

A selection of well established substrates and buffer layer combinations are listed in Table B4.2.1.2.

B4.2.1.4.3 Evaporation rate control and film composition

To grow high quality YBCO by TCE, it is crucial to offer the metals Y, Ba and Cu in a strictly stoichiometric ratio. Films with the best figures of merit like flat surface morphology, critical current D2.7 density $J_c(77\,K) > 2\,MA\,cm^{-2}$ or surface resistance $R_s < 0.5\,m\Omega$ at 10 GHz with good power handling capability are achieved only in a narrow region of $< 2\,\%$ deviation from the ideal composition. For larger deviations, the film morphology gets rougher and precipitates of CuO or Y_2O_3 form. While little excess Y or Cu practically does not affect the YBCO quality, excess Ba drastically reduces T_c and degrades all superconducting properties. Compositions more than 5% off stoichiometry result in very A2.3 poor superconducting properties or a complete loss of superconductivity [63].

As a consequence, the performance of a TCE deposition system is critically influenced by the ability to strictly control and tune the metal evaporation rates. A standard method to measure rates uses QCMs which are commercially available and well established. Its high precision and low noise level allows one to reproducibly keep the film composition within a 2% deviation from stoichiometry margin.

A more recent rate control system is based on atomic absorption spectrometry (AAS). Spectral light from sources (e.g. hollow cathode lamps) containing the evaporation materials is passed through the metal vapour in the chamber. From the detected amount of resonantly scattered photons of each light source the density of metal atoms can be calculated. While the evaporation rate R depends exponentially on boat temperature T_B, the mean velocity of the vapour $\langle v \rangle$ is only logarithmically dependent on T_B and thus can be approximated to be temperature independent. The evaporation rate

$$R = \langle v(T_B) \rangle n(T_B) \propto n$$

is roughly proportional to the vapour density. Deposition systems using AAS rate monitoring are currently developed in laboratories in the US and Germany [64].

B4.2.1.4.4 In situ oxidation approaches

Growing high quality YBCO films by TCE is not only a matter of stoichiometric metal fluxes. In addition one has to introduce a sufficient amount of oxygen to oxidize the metals *in situ* during the growth process. At substrate temperatures of 650–680°C the YBCO lattice forms if the oxygen partial pressure at the growing film exceeds 0.5 Pa. In contrast, as we learned earlier, no evaporation process can take place at this pressure. Therefore some special oxidation methods had to be developed.

For substrates of 1 cm^2 in size the conflict can be circumvented by directing an oxygen nozzle onto the substrate while heavily pumping the chamber (figure B4.2.1.11). A pressure of some 0.1 Pa can be achieved in the vicinity of the substrate while the remaining vacuum chamber is pumped to some 10^{-3} Pa. The metal atoms can propagate unscattered from the sources to the substrate and have to pass only a thin layer of high oxygen pressure directly in front of the film.

For large area substrates, a high oxygen pressure cannot be confined to the substrate but spreads easily all over the vacuum chamber. This limitation is overcome by locally separating the high vacuum metal evaporation from a high oxygen pressure oxidation zone (figure 4.2.1.12). For that purpose the substrate is mounted on a rotating disc heated radiatively from the rear. Metal atoms are evaporated onto the substrate in one part of the heater which is open to the sources below. The remaining part of the heater consists of a chamber adjusted very closely to the substrate. On introduction of oxygen, the low flow conductance of the small gap of typically 0.3 mm confines the gas efficiently to the oxygen pocket. A pressure drop of 1:500 from the oxidation zone to the vacuum chamber can easily be maintained. An oxygen pressure of 1 Pa in the oxygen pocket can be achieved while keeping the base pressure in the vacuum chamber below 2×10^{-3} Pa as required for the evaporation process. The periodic alternation of metal condensation and oxidation results in high quality YBCO films as long as the rotation frequency allows the oxidation to take place a couple of times per YBCO unit cell growth [65].

The rotating disc holder heater concept has been scaled up to a deposition area 9″ in diameter that can be covered by various combinations of wafers 1″ – 8″ in size. It is even possible to grow double sided YBCO films by simply flipping the wafers after growth of the first film and repeating the deposition process. Figure B4.2.1.13 shows critical current density scans of both sides of double-sided YBCO on a 2″ $LaAlO_3$ wafer. For substrate extensions larger than 8″ the rotating disc technique has been reversed. In deposition systems currently under development instead of moving the disc holder with respect to the oxygen zone the oxygen pocket itself is reciprocating below the substrate which is fixed (figure B4.2.1.14). This setup enhances the deposition flexibility in terms of both substrate sizes and shapes [64]. B4.1

B4.2.1.4.5 Barium fluoride based TCE

An alternative method of barium fluoride process (BFP) [66, 67] is based on a vacuum deposition of Y, Cu and BaF_2 precursors and a subsequent conversion to YBCO; so called *ex situ* process.

In the BFP, a congruent film deposition employing the TCE is performed for Y, Cu and BaF_2 in vacuum conditions of $\sim 10^{-6}$ mbar, and then, during an *ex situ* post-annealing (conversion) step, a BaF_2 decomposing and a hydroepitaxial growth of YBCO which starts from the buffer are provided. PLD and sputtering techniques can be also exploited for the deposition of the precursors. At the conversion step,

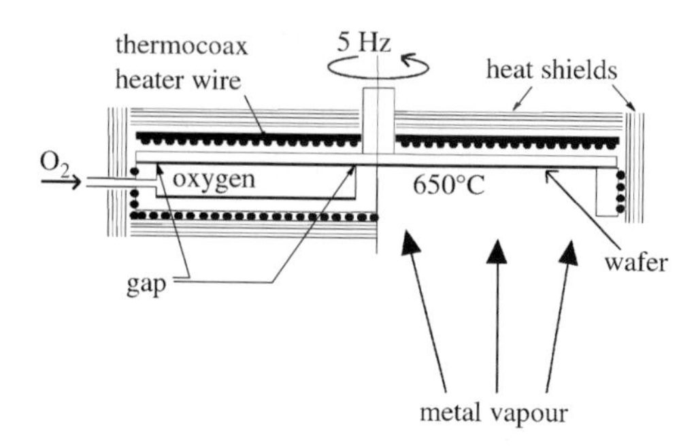

Figure B4.2.1.12. Schematic drawing of a rotating disc holder substrate heater.

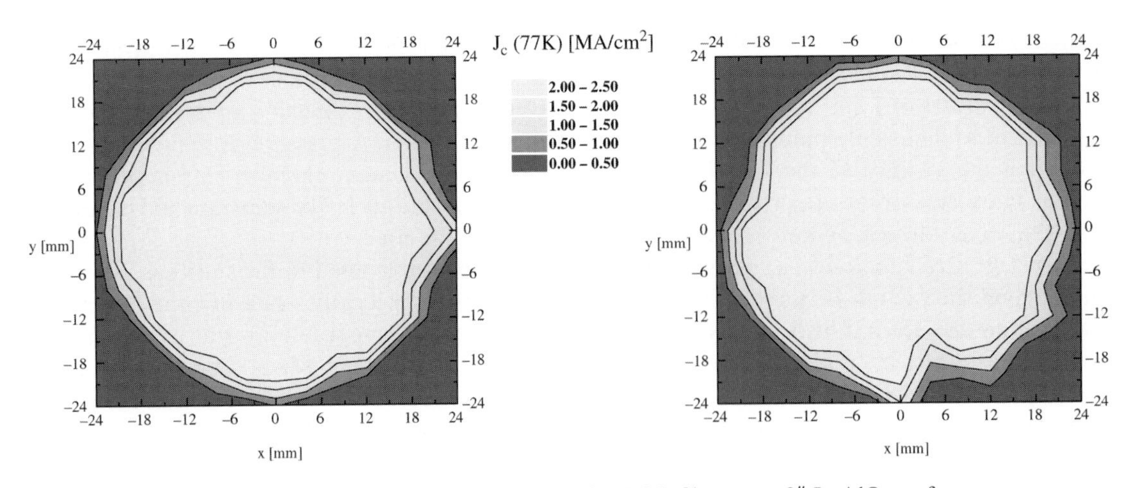

Figure B4.2.1.13. J_c maps of double-sided YBCO films on a 2″ $LaAlO_3$ wafer.

the temperatures of 700–800°C and the ambient pressures of 100–300 mbar for oxygen, and 20–150 mbar for water vapour were employed to convert the BaF_2 to BaO with a release of HF and oxidation of Y and Cu with subsequent formation of YBCO. Epitaxial growth proceeds as uniform advancement of the YBCO-precursor film interface from the substrate towards the surface of the film with the rate of 0.1–0.2 nm s^{-1}, while the deposition rate at the TCE step can be kept at > 10 nm s^{-1}. C1 YBCO film with J_c up to 3 MA cm^{-2} (0 T, 77 K) were manufactured by such a technique. An advantage of the BFP is a capability of large-scale film growth e.g., using a batch process and reel-to-reel tape transportation. Nevertheless, certain difficulties in removal of HF (reaction product of interaction of BaF_2 with H_2O) have been observed when the processing was upscaled.

B4.2.1.4.6 Conclusions

Thermal coevaporation allows growth of high quality YBCO films at reasonable rates and temperatures of 650°C on large areas. Both accurate control of the individual metal fluxes and suitable *in situ* oxidation conditions turn out to be crucial for successful depositions. YBCO films have been produced up to 8″ in diameter and can be manufactured double sided using a rotating disc holder heater concept. Thus TCE has been demonstrated to be a cost efficient YBCO coating technology with great potential

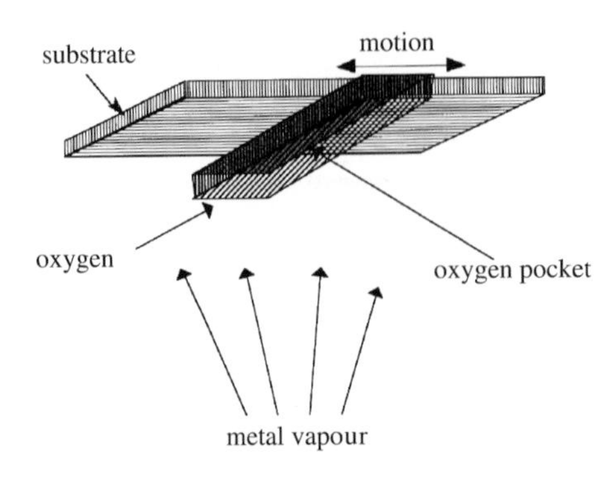

Figure B4.2.1.14. Oxidation scheme for TCE deposition with fixed substrate and reciprocating oxygen pocket.

for industrial production. BFP can be considered as a technique which considerably increases the efficiency of the film processing.

B4.2.1.5 Liquid phase epitaxy

Thin and thick films of (RE)$Ba_2Cu_3O_x$ (RE)BCO system and Bi–Sr–Ca–Cu–O system can be grown by the LPE process on some single crystalline substrates [67–86]. Different from other film formation methods, LPE process of oxide superconductors consists of the deposition from a high temperature solution similar to a single-crystal growth by a top seeded solution growth [87, 88]. Since the solution contains a denser concentration of solutes than the vapour phase, the growth of superconductor films can be one or two orders of magnitude faster than that from typical vapour growth. Growth rates of 10 $\mu m\ min^{-1}$ were obtained for NdBCO film formation [85] and, therefore, thick films exceeding 10 μm in thickness can be obtained in a short period. These are advantageous points to apply this process to the formation of a large, current carrying conductor since critical current density also reaches as high a value as about 1 $MA\ cm^{-2}$ [89, 90]. The LPE films grow under very low supersaturation, and the interstep distance becomes 10–100 times longer than that by other vapour phase techniques. This flatness may contribute to low surface resistance, which is satisfactory for microwave device applications [91]. The near equilibrium growth from a solution has the possibility to lead to a very flat surface over 10 μm by controlling the supersaturation [74, 79].

C1, C2

Three ideally independent processes govern the growth of LPE films, i.e., supersaturation, mass transfer to the growing film and heat transfer from the film-solution interface. They influence growth conditions, quality and surface morphology of the film. Supersaturation depends on the difference between real and equilibrium temperature at the film-solution interface. Mass transfer strongly depends on the convective flux of the solution, which is caused by natural convection and forced convection by substrate rotation in a crucible. Heat transfer from the interface mainly depends on the temperature gradient above the substrate. They are all involved in temperature and heat flow control. Therefore, a furnace structure that provides suitable heat flow circumstance is vitally important to conduct actual

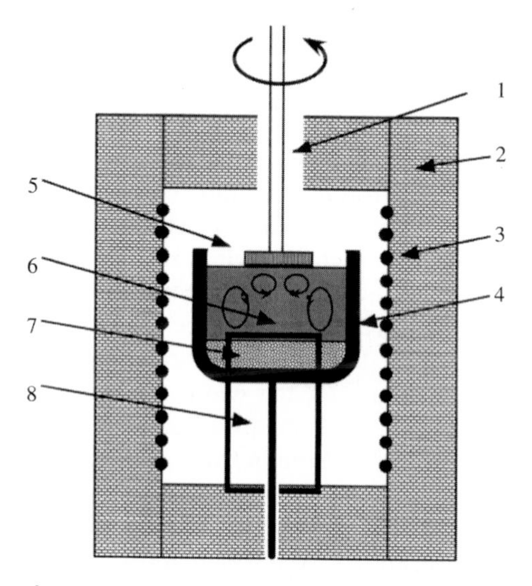

Figure B4.2.1.15. Configuration of LPE furnace and crucible; alumina rod (1), fire brick (2), heater element (3), yttria crucible (4), substrate (5), Ba–Cu–O solvent (6), nutrient (7) and thermocouple (8).

LPE growth. Figure B4.2.1.15 shows an example of LPE growth illustrating the furnace structure and arrangement of the crucible.

A proper arrangement of heating elements and thermal insulators is needed to stabilize the surface of the solution. Strong convection of the atmosphere above the crucible should be avoided since it may nucleate particles at the surface and, more severely, solidify the surface. In this case the $BaCuO_2$ phase is often formed as a metastable phase. An opaque solution of the Ba–Cu–O, which is a typical solvent for C1 (RE)BCO, can also be one of the origins to solidify the surface because radiation from the surface can cause steep temperature gradients in the vicinity of the surface. Accordingly, a thermal insulator above the crucible is often effective to overcome the problems.

Ba–Cu–O melt can react severely with the materials used for crucibles. The solvents wet the crucibles and often climb up the crucible wall and are lost. Large temperature gradients seem to enhance this tendency. Dissolved elements from the crucible materials cause a degradation of superconductivity of the film because of their incorporation into the crystal lattice. Many crucibles made of yttria, stabilized zirconia, barium zirconate, magnesia, alumina, platinum, gold etc. have been tried to contain such a highly reactive solvent. Zirconia and barium zirconate crucibles are often used in solution growth because they can keep the solution relatively clean for a long period. Yttria crucible for YBCO film formation (or neodymium oxide crucible for NdBCO film etc.) is also often used for contamination free films. The employment of double crucibles is sometimes useful to increase the solution life-time.

The solvent is selected to have sufficient solubility of the solute and be inert in superconducting films. For (RE)BCO crystal growth, $BaCuO_2$–CuO and $BaCuO_2$–CuO–BaF_2 [92] are successfully used as solvents, and alkaline halide is mainly used for bismuth system. The $BaCuO_2$–CuO system has a eutectic point at the composition of BaO : CuO = 3 : 7 at about 900°C under ambient pressure. The usage of this composition reduces the primary crystallization temperature to about 930°C. A solvent with the composition of BaO : CuO = 3 : 5 is used in combination with Y_2BaCuO_5 as a nutrient because A1.3 the solvent composition is maintained constant according to the following reaction:

B2.2.3

$$Y_2BaCuO_5 + Ba_3Cu_5O_8 \rightarrow Y123.$$

A low oxygen partial pressure and a BaF_2 addition decreases the growth temperature further to 860°C. Growth temperature, solvent composition and oxygen partial pressure are also important to B2.3.4 control crystal composition when the substitution of rare earth element and Ba occurs. $Nd_{1+x}Ba_{2-x}Cu_3O_{7-d}$ system, for example, contains Nd–Ba substitution and x varies from 0 to 1 by changing growth conditions, strongly affecting pinning properties.

Supersaturation can be driven by abrupt change of the solution temperature (step cooling), gradual cooling of the solution during growth (ramp cooling), contacting a cooler substrate to a higher temperature solution (transient mode) or cooling the part where the substrate is contacted (temperature gradient). All methods are based on the principle that a small deviation from equilibrium promotes crystallization. The deviation corresponds to the supersaturation of the order of $\sigma = 0.1$ or undercooling of the order of $\Delta T = 0.1$–1 K [78] which were determined from the solubility curve of the solute [71, 80, 93]. This small deviation from equilibrium results in a very flat surface due to the relationship between supersaturation and interstep width for spiral growth. Typical vapour growth causes a one to three orders of magnitude higher supersaturation.

Solute in the solution is provided by diffusion in front of the substrate surface and is promoted by convective flow in the solution. The growth rate of the film is considered to be restricted primarily by the mass transportation rate and secondarily by the surface growth kinetics in the solution growth of YBCO [94]. Therefore, diffusion and concentration of the solute and viscosity of the solution almost entirely B2.3.2 determine the growth rate. Measured solubility (about 0.5 at% of Y, which corresponds to the YBCO B2.3.3 fraction of 1/50 in volume of the solution), typical diffusion coefficient of 1×10^{-6} cm^2 s^{-1} in high temperature solution and a solute boundary layer of 50 μm in thickness estimated at moderate convection yield growth rates of the order of 1 μm min^{-1}.

Many available substrates were examined to determine whether they were suitable for the LPE growth of YBCO. They can be classified in two categories; one is a group of substrates that need to form a seed layer as a previous treatment and another group of them that can be used without seed layer. The former group consists of MgO, $SrTiO_3$, $LaAlO_3$, stabilized ZrO_2, Ag etc. They exhibit moderate lattice matching with YBCO, however, dissolve into the solution or react with it and form other compounds on the substrate surface. The seed layer, which is deposited by other processing involving higher supersaturation such as pulsed laser deposition in the thickness of about 100 nm, acts as a protective layer against the attack of the solution and provides nucleation sites for the growth [78]. The latter group contains $NdGaO_3$, $LaGaO_3$, $PrGaO_3$ etc, which are mostly rare earth gallates of perovskite-like structure. An orientation change for films grown on the hetroepitaxial substrates is reported. Small undercooling induces *c*-axis orientation whereas *a*-axis orientation occurs as undercooling increases [80, 81]. Temperature and solvent composition also seem to be the controlling factors for crystal orientation [76]. B4.1

References

[1] Iijima Y, Onabe K, Futaki N, Tanabe N, Sadakata N, Kohno O and Ikeno Y 1993 Structural and transport properties of biaxially aligned $YBa_2Cu_3O_{7-x}$ films on polycrystalline Ni-based alloy with ion-beam-modified buffer layers *J. Appl. Phys.* **74** 1905
[2] Iijima Y, Tanabe N, Kohno O and Ikeno Y 1991 *Physica* C **185** 1961
[3] Hasegawa K, Fujino K, Konishi M, Ohmatsu K, Hayashi K, Sato K, Honjo S, Sato Y, Ishii H and Iwata Y 1998 Development of YBCO film tapes by ISD method *The 1998 International Workshop on Superconductivity* (Toyko: ISTEC) p 73
[4] Goyal A, Norton D P, Budai J D, Paranthaman M, Specht E D, Kroeger D M, Christen D K, Qing H E, Saffian B and Sikka V K 1996 High critical current density superconducting tapes by epitaxial deposition of YBa2Cu3Ox thick films on biaxially textured metals *Appl. Phys. Lett.* **69** 1795
[5] Garcia-Moreno F, Usoskin A, Freyhardt H C, Issaev A, Wiesmann J, Hoffmann J, Heinemann K, Sievers S and Dzick J 1999 Laser deposition of YBCO on long-length technical substrates *IEEE Trans. Appl. Supercond.* **9** 2260
[6] Herbolts N, Appelton B R, Pennycook S J, Noggle T S and Zuhr R A 1986 *Mater. Res. Soc. Symp. Proc.* **51** 369
[7] Menzinger M and Wahlin L 1969 *Rev. Sci. Instrum.* **40** 102
[8] Freeman J H, Temple, Beanland D and Gard G A 1976 *AERE Report* 8287
[9] Miyake K, Ohashi K and Komuro M 1993 *Mater. Res. Soc. Symp. Proc.* **279** 787
[10] Matsuo T and Miyake K 1995 *J. Vac. Sci. Technol.* A **13** 2138
[11] Pruymboom A, *PhD Thesis* University of Leiden, The Netherlands
[12] Zaitsev AG Ockenfuss G and Wördenweber R 1997 *Appl. Superconductivity 1997a* eds H Rogalla and D Blank (*Inst. Phys. Conf. Ser.* 158) p. 25
[13] Miyake K, Yagi K and Tokuyama T 1982 *Nucl. Instrum. Methods* **198** 535
[14] Glocker D A and Shah S I (eds) 1995 *Handbook of Thin Film Process Technology* (Bristol: Institute of Physics Publishing) Ch A3
[15] Gilbert L R, Messier R and Krishnaswamy S V 1980 *J. Vac. Sci. Technol.* **17** 389
[16] Rossnagel S M and Cuomo J J 1988 Thin film processing and characterization of high-temperature superconductors *Conf. Proc.* **165** 106
[17] Xi X X, Linker G, Meyer O, Nold E, Obst B, Ratzel F, Smithley R, Strehlau B, Weschenfelder F and Geerk J 1989 *Z. Phys.* B **74** 13
[18] Newman N, Cole B F, Garrison S M, Char K and Taber R C 1991 *IEEE Trans. Magn.* **27** 1276
[19] Eom C B, Sun J Z, Yamamoto K, Marshall A F, Luther K E, Geballe T H and Laderman S S 1989 *Appl. Phys. Lett.* **55** 595
[20] Poppe U, Schubert J, Arons R R, Evers W, Freiburg C R, Reichert W, Schmidt K, Sybertz W and Urban K 1988 *Solid State Commun.* **66** 661
[21] Krüger U, Kutzner R and Wördenweber R *IEEE Trans. Appl. Supercond.* Submitted
[22] Wördenweber R, Einfeld J, Kutzner R, Zaitsev A G, Hein M A, Kaiser T and Müller G 1999 Large-Area YBCO Films on Sapphire for Microwave Applications *Proc. ASC 98* (Palm Springs)
[23] Schneider J, Einfeld J, Lahl P, Königs T, Kutzner R, Wördenweber R 1997 *Applied Superconductivity* ed Rogalla H and Blank D (*Inst. Phys. Conf. Ser.* 158) p. 221
[24] Ockenfuss G, Wördenweber R, Scherer T A, Unger R and Jutzi W 1995 *Physica* C **243** 24
[25] Zaitsev A G, Ockenfuss G, Guggi D, Wördenweber R and Krüger U 1997 *J. Appl. Phys.* **81** 3069

[26] Dam B, Koeman N J, Rector J H, Stäuble-Pümpin B, Poppe U and Griessen R 1996 *Physica* C **261** 1
[27] Wördenweber R *et al* unpublished
[28] Suzuki Y, Triscone J M, Eom C B, Beasley M R and Geballe T H 1994 *Phys. Rev. Lett.* **73** 328
[29] Fischer Ø, Triscone J M, Fivat P, Anderson M and Decroux M 1994 *Proc. SPIE* **2157** 134
[30] Wagner P, et al, 1994 *J. Supercond.* **7** 217
[31] Satoh T, Adachi H, Ichikawa Y, Setsune K and Wasa K 1994 *J. Mater. Res.* **9** 1961
[32] Jakob G, Moshchalkov V V and Bruynserade Y 1995 *Appl. Phys. Lett.* **66** 2564
[33] Foltin R S, Dye R C, Ott K C, Peterson E, Hubbard K M, Hutchinson W, Muenchausen R E, Estler R C and Wu X D 1991 Target modification in the excimer laser deposition of $YBa_2Cu_3O_{7-x}$ thin films *Appl. Phys. Lett.* **59** 594
[34] Venkatesan T, Wu X D, Muenchausen R and Pique A 1992 Pulsed laser deposition: future directions *MRS Bulletin* **17** 54
[35] Hass G and Thun R E (ed) 1966. *Physics of thin films* **V 3**, (New york: Academic) V3
[36] Izumi H, Ohata K, Sawada T, Morishta T and Tanaka S 1991 Direct observation of ions in the laser plume onto the substrate *Appl. Phys. Lett.* **59** 597
[37] Singh R K and Narayan J 1990 Pulsed-laser evaporation technique for deposition of thin films: Physics and theoretical model *Phys. Rev.* B **41** 8843
[38] Izumi H, Ohata K, Sawada T, Morishita T and Tanaka S 1991 Deposition pressure effects on laser plume of $YBa_2Cu_3O_{7-\delta}$ *Japan. J. Appl. Phys.* **30** 1956
[39] Witanachchi S, Ahmed K, Sakthivel P and Mukherjee P 1995 Dual-laser ablation for particulate-free film growth *Appl. Phys. Lett.* **66** 1469
[40] Koren G, Baseman R J, Gupta A, Lutwyche M I and Laibowitz R B 1990 Particulates reduction in laser-ablated $YBa_2Cu_3O_{7-\delta}$ thin films by laser-induced plume heating *Appl. Phys. Lett.* **56** 2144
[41] Holzapfel B, Kämmer K and Schultz L 1996 PLD plume diagnostics by high-speed ICCD photography *Lambda Physik Scientific Report* **7** p 1
[42] Misra D S and Palmer S B 1991 Laser ablated thin films of $Y_1Ba_2Cu_3O_{7-\delta}$: the nature and origin of the particulates *Physica* C **176** 43
[43] Craciun V, Craciun D, Bunescu M C, Boulmer-Leborgne C and Hermann J 1998 Subsurface boiling during pulsed laser ablation of Ge *Phys. Rev.* B **58** 6787
[44] Kelli R and Rothenberg J E 1985 Laser sputtering *Nucl. Instrum. Methods Phys. Res.* B **7/8** 755
[45] Nagaishi T, Itozaki H 1996 Method and apparatus for depositing superconducting layers onto the substrate surface via off-axis laser deposition *European Patent* EP 07022416 A1 *EP Bulletin No. 12*
[46] Barr W P 1969 *J. Plasma Phys. E* **2** 2
[47] Singh R K, Bhattacharya D and Narayan 1992 *J. Appl. Phys. Lett.* **61** 483
[48] Chiba O, Murakami K, Eryu O, Shihoyama K, Mochizuki T and Masuda K 1991 Laser excitation effects on laser ablated particles in fabrication of high T_c superconducting thin films 1997 *Japan. J. Appl. Phys.* **30** L732
[49] Cohen A, Allenspacher P, Bieger M M, Jeuck I and Opower H B e a m 1991 Target interaction during growth of $YBa_2Cu_3O_{7-x}$ by the laser ablation technique *Appl. Phys. Lett.* **59** 2186
[50] Dam B, Rector J, Chang M F, Kars S, de Groot D G and Griessen R 1994 Laser ablation threshold of $YBa_2Cu_3O_{6+x}$ *Appl. Phys. Lett.* **65** 1581
[51] Usoskin A, Freyhardt H C and Krebs H U 1999 Influence of light scattering on the development of laser-induced ridge-cone structures on target surfaces *Appl. Phys.* A **69** S823
[52] Usoskin A, Garcìa-Moreno F, Freyhardt H C, Knoke J, Sievers S, Gorkhover L, Hofmann A and Pink F 1999 Variable-azimuth laser ablation: principles and application for film deposition on long tubes and tapes *Appl. Phys.* A **69** S423
[53] Garcìa-Moreno F, Usoskin A, Freyhardt H C, Wiesmann J, Dzick J, Hoffmann J, Heinemann K and Issaev A 1997 High critical current densities in YBCO films on technical substrates *Appl. Supercond. (The Netherlands, 30 June-3 July 1997) (Inst. Phys. Conf. Ser.)* **158** p. 909
[54] Freyhardt H C, Hoffmann J, Wiesmann J, Dzick J, Heinemann K, Isaev A, Garcia-Moreno F, Sievers S and Usoskin A I 1997 YBaCuO thick films on planar and curved technical substrates *IEEE Trans. Appl. Supercond.* **7** 1426
[55] Usoskin A I, Freyhardt H C, García-Moreno F, Sievers S, Popova O, Heinemann K, Hoffmann J, Wiesmann J and Isaev A, 1995 Growth of HTSC films with high critical currents on polycrystalline technical substrates *Appl. Supercond. (Edinburgh 1995) (Inst. Phys. Conf. Ser.)* **148** p. 499
[56] Doughty C, Findikoglu A T and Venkatesan T 1995 Steady state pulsed laser deposition target scanning for improvement plume stability and reduced particle density *Appl. Phys. Lett.* **66** 1276
[57] Usoskin A, Knoke J, Garcia-Moreno F, Isaev A, Dzick J, Sievers S and Freyhardt H C YBaCuO thick films on planar and curved technical *ASC 2000: Appl. Supercond. Conf. Substrates Book of Abstracts* **3ME07** 202
Usoskin A, Knoke J, Garcia-Moreno F, Isaev A, Dzick J, Sievers S and Freyhardt HC *IEEE Trans. Appl. Supercond.* Submitted
[58] Usoskin A, Freyhardt H C and Chukanova I 1993 Possibility of striations in HTSL films caused by self-destabilization of the growth temperature *J. Alloys Comp.* **195** 117
[59] Usoskin A I, Freyhardt H C, Neuhaus W and Damaske M 1994 Thermal equilibrium YBaCuO film growth for high critical current conductors *Critical Currents in Superconductors*, ed H W Weber (Singapore: World Scientific) p 383

[60] Freyhardt HC, Usoskin A and Neuhaus W 1992 Verfahren und Vorrichtung zum Beschichten eines Substrats unter Verwendung einer gepulsten Quelle *German Patent* 42 28 573.9–45
[61] Usoskin A, García-Moreno F, Sievers S, Knoke J, Freyhardt HC and Dzick J 1999 Large-area HTS conductors obtained by novel PLD technique on metallic substrates *EUCAS 99 4th Eur. Conf. Appl. Superconductivity* (*Inst. Phys. Conf. Ser.* 167), vol 1 p 447
[62] Prusseit W, Utz B, Berberich P and Kinder H 1994 *Journal of Supercond.* **7** 231
[63] Baudenbacher F, Hirata K, Berberich P, Kinder H and Assmann W 1992 *High T_C Superconductor Thin Films*, ed L Correra (Amsterdam: Elsevier) p 365
[64] Kinder H, Berberich P, Prusseit W, Rieder-Zecha S and Utz B 1997 YBCO film deposition on very large areas up to $20\times20\,cm^2$ *Physica* C **282–287**
[65] Berberich P, Utz B, Prusseit W and Kinder H 1994 Homogeneous high quality $YBa_2Cu_3O_{7-x}$ films on 3′ and 4′ substrates *Physica* C **219** 497
[66] Mankiewich P M, Scofield J H, Skocpol W J, Howard R E, Dayem A H and Good E 1987 Reproducible technique for fabrication of thin films of high transition temperature superconductors *Appl. Phys. Lett.* **51** 1753
[67] Solovyov V F, Wiesmann H J, Wu L, Suenaga M and Feenstra R 1999 High rate deposition of 5 μm thick $YBa_2Cu_3O_7$ films using the BaF_2 ex-situ post annealing process *IEEE Trans. Appl. Supercond.* **9** 1467
[68] Belt R F, Ings J and Diercks G 1990 *Appl. Phys. Lett.* **56** 1805
[69] Scheel H J, Berkowsky M and Chabot B 1991 *J. Cryst. Growth* **115** 19
[70] Dubs C, Ficher K and Görnert P 1992 *J. Cryst. Growth* **123** 611
[71] Klementz C and Scheel H J 1993 *J. Cryst. Growth* **129** 421
[72] Görnert P, Ficher K and Dubs C 1993 *J. Cryst. Growth* **128** 751
[73] Scheel H. J. 1994 *MRS Bullitine/September* 26
[74] Scheel H J, Klementz C, Reinhatrt F K, Lang H P and Güntherodt H J 1994 *Appl. Phys. Lett.* **65** 901
[75] Kitamura T, Yoshida M, Yamada Y, Shiohara Y, Hirabayashi I and Tanaka S 1995 *Appl. Phys. Lett.* **66** 1421
[76] Kitamura T, Taniguti S, Shiohara Y, Hirabayashi I, Tanaka S, Sugawara Y and Ikuhara Y 1996 *J. Cryst. Growth* **158** 61
[77] Yamada Y, Kawashima J, Niiori Y and Hirabayashi I 1996 *Appl. Supercond.* **4** 479
[78] Yamada Y, Niiori Y, Yoshida Y, Hirabayashi I and Tanaka S 1996 *J. Cryst. Growth* **167** 566
[79] Klementz C and Scheel H J 1996 *Physica* C **265** 126
[80] Dubs C, Bornmann S, Schemelz M, Shüler T, Sandiumenge F, Bruchlos G and Görnert P 1996 *J. Cryst. Growth* **166** 836
[81] Aichele T, Bornmann S, Dubs C and Görnert P 1997 *Cryst. Res. Tecnol.* **32** 1145
[82] Utke I, Klementz C, Scheel H J, Sasaura M and Miyazawa S 1997 *J. Cryst. Growth* **74** 806
[83] Takagi A, Kitamura T, Taniguti S, Yamada Y, Shiohara Y, Hirabayashi I, Tanaka S and Mizutani U 1997 *IEEE Trans. Appl. Supercond.* **7** 1388
[84] Niiori Y, Yamada Y and Hirabayashi I 1998 *Physica* C **296** 65
[85] Takagi A, Wen J-G, Hirabayashi I and Mizutani U 1998 *J. Cryst. Growth* **193** 71
[86] Schneemeyer L F, van Dover R B, Glarum S H, Sunshine S A, Fleming R M, Batlogg B, Siegrist T, Marchall J H, Waszczak J V and Rupp L W 1988 *Nature* **332** 422
[87] Yamada Y and Shiohara H 1993 *Physica* C **217**
[88] Zhokhov A A and Emel'chenko G A 1993 *J. Cryst. Growth* **129** 786
[89] Yoshida M, Nakamoto T, Kitamura T, Hyun O-B, Hirabayashi I, Tanaka S, Tsuzuki A, Sugawara Y and Ikuhara Y 1994 *Appl. Phys. Lett.* **65** 1714
[90] Miura S, Hashimoto K, Wang F, Enomoto Y and Morishita T 1997 *Physica* C **278** 201
[91] Miura S, Wen J - G, Suzuki K, Morishita T, Yoshitake T, Fujii G and Suzuki S 1998 *Proc. 1998 Applied Superconductivity Conference Submitted*
[92] Yamada Y, Niiori Y, Yoshida Y, Hirabayashi I and Tanaka S 1996 *J. Cryst. Growth* **167** 566
[93] Krauns C, Sumida M, Tagami M, Yamada Y and Shiohara Y 1994 *Z. Phys.* B **96** 207
[94] Yamada Y, Krauns C, Nakamura M, Tagami M and Shiohara Y 1995 *J. Mater. Res.* **10** 1601

B4.2.2
Chemical vapour thin film deposition techniques

G Wahl, F Weiss, and O Stadel

B4.2.2.1 Introduction

Chemical vapour deposition (CVD) processes have been investigated in the past for the thin film synthesis of different low temperature superconductors (LTS): Nb_3Sn, Nb_3Ge, $NbN_{1-y}C_y$, Nb_3Si, V_3Si, Nb_3Ga, $TiC_{1-y}N_y$, $W_{1-y}Ge_y$ [1, 2]. These thin films were synthesized for the purpose of microelectronic applications and also for large-scale applications. CVD was then used because the corresponding phases could not be formed with good superconducting properties with the use of conventional metallurgical processing techniques. A1.3

In the case of high temperature superconductors (HTS), chemical deposition (CD) processes have been widely used, because they are simple and closely related to the ceramic route used for the synthesis of bulk materials. After the first success in thin film deposition of HTS ($YBa_2Cu_3O_7$) by CVD [3, 4], the interest in CD and CVD processes for oxide film synthesis grew constantly and many review papers appeared [5–11].

In the present paper, we discuss the chemical engineering of different CD processes used for the synthesis of thin and thick superconducting oxide layers.

The most investigated superconductor is $Y_1Ba_2Cu_3O_7$ (often abbreviated as YBCO) because of its high critical temperature and high critical current at high magnetic field [12]. C1

In the present review, the basic discussion concerning film preparation and film characterization will be concentrated on this compound. The other families of superconducting oxides such as Bi-, Tl-, and Hg-based superconductors have also been investigated and the main characteristics of these films will also be presented. C2, C3, C4

B.4.2.2.2 Chemical deposition: basic principles

CD processes used for the formation of thin and thick superconducting films are based on the decomposition or on chemical reactions between inorganic or organic metal containing species leading to the formation of a thin layer on an appropriate substrate. B4.2.1

These CD processes can be classified in two main groups.

Two-step processes, which are based on the deposition of an unreacted precursor film, followed by a recrystallization step where solid state reactions induce the formation of the desired superconducting phase.

In the case of HTS materials, the main methods which have been used are:

- screen printing (doctor blade method, YBCO powder is mixed with an organic binder which is later decomposed at high temperature during the second step heat treatment); B4.3 B2.1

B4.3
- metal-organic deposition (MOD) (sol–gel, dipping and spinning techniques, spray pyrolysis, where unreacted metal-organic metal mixtures and suitable organic solvent and binders are deposited as gels on a substrate before the final heat treatment).

All these processes are in general simple and inexpensive. However, most of them are mainly devoted to obtaining thick layers. They do not allow careful control of the thickness and homogeneity of the layers as well as the crystalline quality. It is difficult to obtain a good texturing, which is necessary to get high current carrying capabilities in the oxide superconductors.

One-step processes, using volatile chemical precursors, where the superconducting oxide phase is formed directly on the substrate (as in physical deposition processes) after chemical reactions in the gas phase or on the surface of the deposited material. These processes are: CVD or metal-organic chemical vapour deposition (MOCVD) (depending on the nature of chemical precursors used).

CVD is an important process because of its large throwing power [13]; deposition on complicated forms, e.g., for microwave resonators or on cables is also possible. The use of higher working pressures means a cost performance improvement in comparison to other vacuum methods [14].

B.4.2.2.2.1 Spray pyrolysis of HTS

Of all the two-step processes, research activity has mainly concentrated on one of them, the spray pyrolysis technique. Experimental procedure involves pyrolysis on a heated substrate of droplets produced by an aerosol of non-volatile liquid source precursors. To some extent, this procedure allows us to have reasonable control over the film thickness and the homogeneity.

In general, the standard deposition route for the preparation of HTS materials involves a deposition stage of a not completely reacted film (precursor film), followed by one or several annealing treatments allowing control of the film recrystallization and a final oxidation step to improve the superconducting properties.

Deposition modes

B4.2.1 In spray pyrolysis processes, the aerosol produced by pulverization is transported at nearly room temperature from the source to the reaction zone and to the deposition surface by convection. On the way to the surface, the fate of the droplet can be different depending on the flow conditions. With variation of the temperature in the reaction zone and of the substrate, different deposition modes are possible (figure B4.2.2.1).

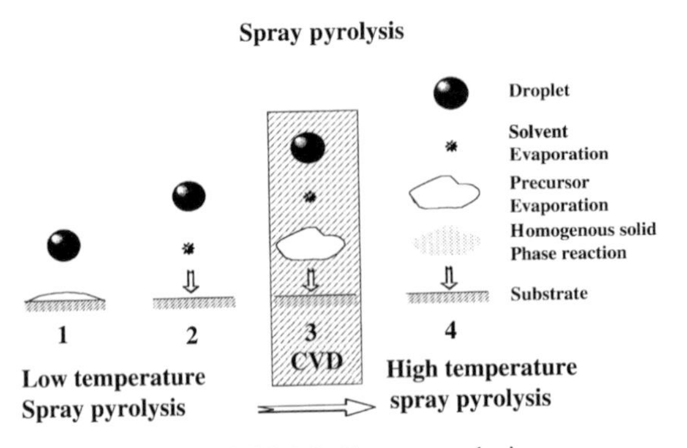

Figure B4.2.2.1. Spray pyrolysis.

(1) At low temperature, the aerosol droplet reaches the substrate in the liquid state. The solvent can evaporate at the surface leaving either dry precipitates or a gel, which can then be processed in the sol–gel route.

B4.3 B2.1

(2) At slightly higher temperatures, the solvent is evaporated before droplet reaches the substrate. Dry particles are precipitated on the substrate where they decompose and homogenous solid phase reactions take place (real spray pyrolysis).

B2.1

(3) After solvent evaporation, the precursor material is evaporated at a characteristic distance d_d from the surface and forms a reaction gas. At the substrate surface, heterogeneous solid–vapour phase reactions take place basically as in CVD - described in more detail in section B.4.2.2.2.2. (droplet-derived CVD).

(4) At higher temperatures, homogeneous solid phase reactions occur during the transport of the aerosol. The reaction product is deposited on the substrate in a pre-reacted form.

B2.1

In general, the experimental parameters (temperature, gas flow, etc) are adjusted to proceed in one of these four modes. The advantage of spray pyrolysis over conventional CVD techniques is the simple transport mode of the precursor material. In spray techniques, the aerosol droplets remain cold and thermally stable avoiding recondensation problems or premature reactions in the gas phase.

The droplet size is a crucial parameter in these processes. The evaporation rate, which depends directly on the droplet size, can induce the shift from one decomposition mode into another. In order to get good quality films, a sharp distribution in size for the droplets within the aerosol is therefore suitable and ultrasonic aerosol generation seems to be an appropriate technique for this purpose.

Precursor solution used for spray pyrolysis

The source solution for aerosol production must be a liquid. If the source compound based on the elements which are to be deposited does not exist in the liquid state, the solid source will be dissolved in an appropriate solvent. This solution must then be chemically stable during the nebulization time.

B2.1

The choice of the source compound is very important because it affects the deposition conditions. Two families of source compounds can be distinguished:

- mineral compounds;
- organo-metallic compounds.

Mineral compounds are generally non volatile:

- Chlorides and nitrates are solid at room temperature and their decomposition occurs at relatively elevated temperatures.
- Fluorides are also solid at room temperature. They have a good solubility mainly in alcohols and have a general tendency to hydrolyse in aqueous solutions.

Organo-metallic compounds can be classified into three groups:

(1) Carboxylates (acetates, propionates, citrates, ethylhexanoates, benzolates): these compounds are soluble in water and in some alcohols (methanol, ethanol); they have a low fusion point.

(2) Alcoholates (methoxydes, ethoxydes, buthoxydes, etc): they are generally liquid at room temperature.

(3) Diketonates (pentadionates = acetylacetonates = acac, tetramethylheptadions = tmhd compounds and fluorinated derivates): this group contains the largest amount of volatile precursors and essentially the β-diketonates, which are used for CVD.

These precursors are described in more detail in the next chapter.

The solvent used in spray pyrolysis must be chemically compatible with all the compounds, in order to have a stable single source solution. In addition, the solvent must be able to dissolve large amounts of these precursors. The minimal concentration (below which the deposition rate will be too low) is in the range 10^{-2} mol l^{-1}. The maximal concentration (above which the solution will be too dense) is generally around 0.5 mol l^{-1}.

C1, C2, C3, C4 For the spray pyrolysis of HTS materials (YBCO, BSCCO (= $Bi_2Sr_2Ca_2Cu_3O_x$ or $Bi_2Sr_2Ca_1Cu_2O_x$), Tl- and Hg-based superconductors), aqueous nitrate solutions have been used in most cases but sometimes also carboxylates. Nitrate solutions are generally preferred, especially in the case of Hg-based B2.1 superconductors, because they allow circumvention of problems of carbon contamination in the layer.

Aqueous nitrate solutions are prepared either by dissolving $Y(NO_3)_3{\cdot}3H_2O$, $Ba(NO_3)$, and $Cu(NO_3){\cdot}6H_2O$, $Bi(NO_3)_3{\cdot}5H_2O$, $Pb(NO_3)_2$, $Sr(NO_3)_2$ or $Ca(NO_3)_2{\cdot}4H_2O$ in highly distilled and deionized water to give the appropriate cationic ratio for deposition, or from single oxides or carbonates by dissolution in nitric acid. This last procedure gives a better control of stoichiometry because the water content of nitrates is not always exactly known.

B2.1 In the case of carboxylates, experiments have been made on YBCO and BSCCO with acetates dissolved in water or with YBCO or single oxides powders dissolved in 90% propionic acid and 10% deionized water.

Engineering of spray pyrolysis deposition

(1) *Production of the aerosol*: The transformation of a liquid source into a fine mist of small droplets (aerosol) can be achieved by two different methods: Pneumatic nebulization and ultrasonic nebulization.

In pneumatic nebulization, a carrier gas under pressure induces the explosion in fine droplets of the liquid which is pushed through a fine capillary tube (figure B4.2.2.2).

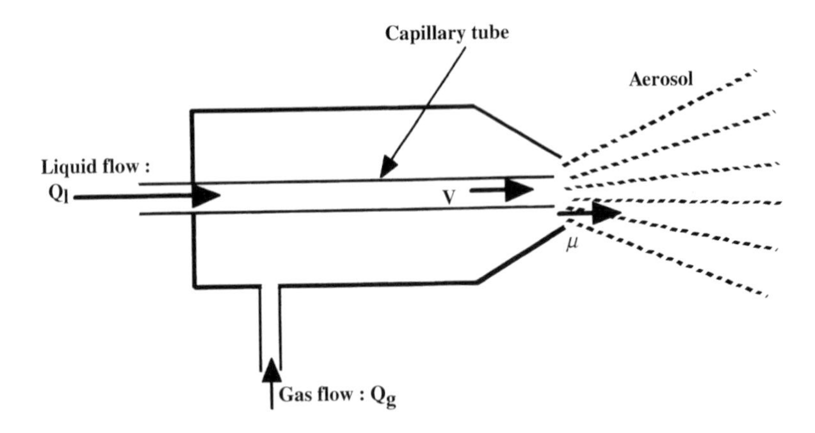

Figure B4.2.2.2. Pneumatic nebulization.

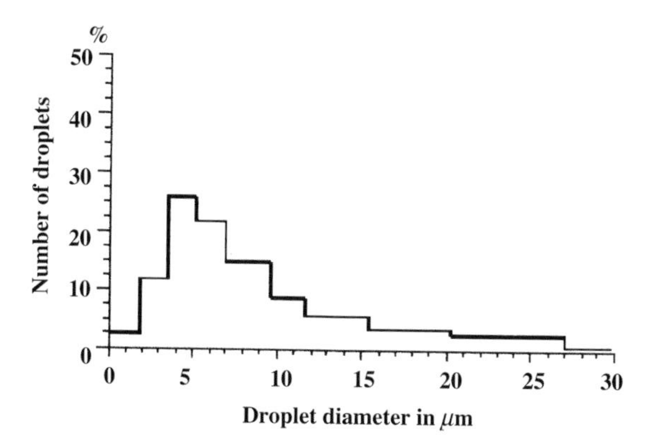

Figure B4.2.2.3. Diameter distribution of the droplets formed by pneumatic nebulization.

The droplet diameter d_p can be related empirically to different experimental parameters [15] such, as liquid density, the dynamic viscosity, the surface tension, the liquid velocity, the total liquid flow, the gas velocity and the gas flow.

Figure B4.2.2.3 shows the typical distribution in size of the droplets produced by pneumatic spray in the case of water. B2.1

In the ultrasonic nebulization method, an ultrasonic wave is focused on the liquid–gas interface and induces a deformation of this surface and the production of a geyser with a height related to the acoustic intensity. This geyser produces an aerosol. Figure B4.2.2.4 shows a typical deposition system used for the synthesis of thin films through ultrasonic spray pyrolysis.

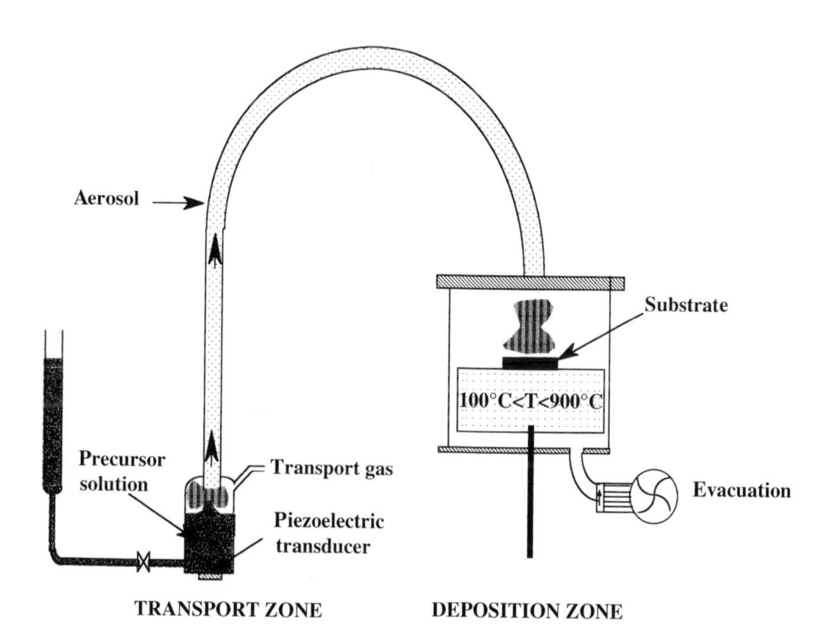

Figure B4.2.2.4. Ultrasonic nebulization.

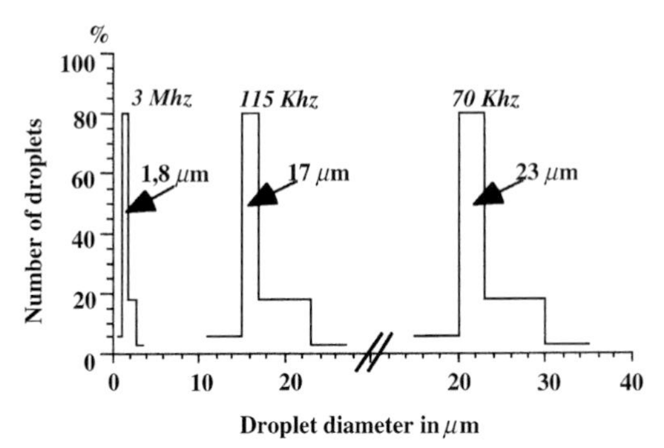

Figure B4.2.2.5. Diameter distribution of the droplets formed by ultrasonic nebulization.

The diameter of the droplets depends on the frequency ν of the ultrasonic wave and is given by equation (B4.2.2.1) [16]:

$$d_\mathrm{p} = \frac{\pi}{2}\sqrt[3]{\frac{\sigma}{\rho_0 4\pi^2 \nu^2}}. \tag{B4.2.2.1}$$

For water, with surface tension of $\sigma = 79\,\mathrm{dyn\,cm^{-1}}$, liquid density $\rho_0 = 1\,\mathrm{g\,cm^{-3}}$ and a typical ultrasonic frequency of $\nu = 1000\,\mathrm{kHz}$, the diameter is $d_\mathrm{p} = 2\,\mu\mathrm{m}$.

Figure B4.2.2.5 shows the typical narrow repartition in size of the droplets, which can be obtained by ultrasonic excitations in the case of water.

Figures B4.2.2.3 and B4.2.2.5 show that the diameter of droplets can be smaller by about an order of magnitude for droplets produced ultrasonically than in the case of droplets generated pneumatically.

The aerosol flow is dependent on the specific physical properties of the liquid, namely its vapour pressure and inversely on its viscosity and surface tension [17].

C1, C2, C3, C4

(2) *Experimental deposition procedure*: Thin and thick HTS films in the YBCO-, BSCCO-, Tl- and Hg-based oxide systems have been prepared in the three deposition modes involving 'real spray pyrolysis' (mode 1, 2, 4). Mode 3 is a CVD process and will therefore be described in the section B4.2.2.2.1.

In the two first pyrolysis modes, the film preparation is made in a two-step process.

Mode 1. In the so-called sol–gel route the precursors for the YBCO formation were carboxylates dissolved in propionic acid [18, 19]. The aerosol droplets were transported by an air flow onto the substrates heated to 100°C. At this temperature, the aerosol polymerized at the substrate surface leading to the deposition of a thin rigid and uniform amorphous gel. The deposition rate was simply controlled by the ultrasonic power level and deposition time. If the deposited films were calcined at 500°C in air in order to release the remaining water or carbon dioxide molecules, a volume shrinkage of about 90% of the layer was generated and cracks were formed in the film. Only films with a final thickness less than 1 μm appeared free of cracks. For this reason thicker films were prepared by repeated sequences of deposition and calcination with a maximum thickness of 0.2–1 μm per sequence.

YBCO films are finally obtained after a final annealing treatment in oxygen at temperatures ranging from 850 to 1000°C and after slow cooling to room temperature.

B2.1

Mode 2. In aerosol deposition from the nitrates, film formation also generally consists of two steps. The film deposition is made as previously in a few cycles (5–10), which are performed in order to avoid crack formation. The precursor film is deposited at a substrate temperature of 200–400°C. Amorphous films are formed. Deposition is then interrupted and the substrate temperature is increased to higher temperatures for recrystallization. During this annealing, a partial decomposition of nitrates takes place. This procedure is repeated until the desired thickness is reached.

In the case of nitrate decomposition, thermogravimetric measurements have been intensively performed to study the reaction paths for individual and mixed components.

The decomposition temperature of the various nitrates used for HTS film formation is very different. Bi and Cu nitrates decompose at around 200°C, $Y(NO_3)_3 \sim 400°C$, $Ca(NO_3)_2 \sim 600°C$ and Ba and Sr nitrates decompose at temperatures higher than 800°C. As a consequence, for a temperature range higher than 200°C, Cu components are preferentially deposited and the stoichiometry of the starting aerosol solution has to be adjusted in order to keep the correct cationic composition in the deposited layer [19]. B2.1

Mode 4. This is particularly true for deposition made at very high temperature (mode 4), where it is intended to get layers from pre-reacted nitrates. The starting Y: Ba: Cu stoichiometry must be considerably enriched in Y and Ba (Y: Ba: Cu = 1:2:0.75, for example, see [20]).

Achievements of films: (a) *YBCO*. YBCO films have been sprayed over single crystalline substrates, MgO, $SrTiO_3$ and on C fibres and wires [21] (sometimes also with Ag additions [22–24]). B4.1

After the deposition, films are in an amorphous state and therefore an annealing step at elevated temperature is necessary to build the proper structure. This annealing is performed either in an inert atmosphere or in a mixture of an inert atmosphere with O_2, with a final post-annealing under O_2 atmosphere.

The surface morphology of as-deposited films depends on the nebulization system (better with ultrasonic spray) and strongly on the deposition temperature. At temperatures below 500°C film surfaces are rough with pinholes and cracks. However, repeated spray-annealing cycles for films deposited in the sol–gel mode may fill in or bridge over the cracks to finally form thick films with a good connectivity.

At temperatures higher than 600°C, these surfaces are smoother and with a mirror-like aspect. This surface quality is further improved with increasing temperatures and with the addition of Ag. The enormous improvement in surface morphology reflects a large change in the deposition–decomposition process and opens the way to resolve partially the granularity problem of low temperature deposited films [25].

Nevertheless, these films can have a good *c*-axis texture, but they do not have a biaxial texture (in correlation with the nucleation and growth processes induced).

The highest reported T_c values are around 90 K with ΔT_c between 1.5 K and a few kelvin. Films prepared by melt textured growth during the second step annealing can reach critical current densities up to $J_c = 4800\ \mathrm{Acm^{-2}}$ at 77 K and a magnetic field of $B = 0$ T [26, 27]. C1

Recent studies have shown that biaxial texturing of YBCO can be reached using nitrate deposition at elevated temperatures on biaxially textured Ag substrates, in an *in situ* process [28, 29]. J_c values in the range 10^4–10^5 A $\mathrm{cm^{-2}}$ were first reported [30], but recently, in association with a perfect control of the growth and annealing parameters, critical current densities over 10^6 A $\mathrm{cm^{-2}}$ have been obtained on $SrTiO_3$ single crystalline substrates [31]. These values are very promising and open the way for a large scale, environmentally friendly and economic production process of YBCO thick films, mainly for coated conductor applications. C1

(b) *BSCCO*. BSCCO films have been produced basically in the same way as YBCO films [32–36] on single crystalline substrates (MgO, $SrTiO_3$) and on Ag foils [37]. Deposition occurred at temperatures between 150 and 600°C typically. Amorphous films were annealed in a second step in air at temperatures between 840 and 850°C. Usually a mixture of the different (2201, 2212, 2223) phases was obtained. Films annealed in a PbO atmosphere produced the highest $T_c = 110$ K for the Bi superconductor [38]. C2

Critical current densities in films densified by melt quenching reach $4 \times 10^3\ \mathrm{A\,cm^{-2}}$ at 77 K, $B = 0$ T [39].

(c) *Tl(Hg)–BCCO*. Tl, (and later Hg) based oxide superconductors have been widely deposited by spray pyrolysis techniques. C3, C4

The synthesis procedure is split into two distinct steps:

B4.1
- deposition of a Ba–Ca–Cu nitrate precursor layers on single crystalline or Ag substrates.

C3, C4
- Tl or Hg introduction into the pre-annealed precursor, using a Tl or Hg source from a Tl(Hg)BCCO pellet or from a Tl_2O_3 ($HgCu_2$) powder. This second step annealing is carried out at high temperatures (typically 800–860°C) with a careful control of Tl (Hg) partial pressures.

C3 Tl-based films may be prepared with very good superconducting properties. The most interesting Tl compound, Tl-1223, can be obtained as a relatively pure phase [40–48]. The best superconducting properties obtained so far are $T_c = 120$ K, $J_c = 10^6$ A cm^{-2} (77 K, 0 T) [49].

C4 Hg-1223 films present actually the most promising superconducting properties. Films with a thickness of 1–2 μm and $T_c \sim 130$ K have been epitaxially grown by spray pyrolysis on $SrTiO_3$ substrates [50–54]. The critical current density can be as high as: $J_c \sim 4.4 \times 10^5$ A cm^{-2} (77 K, 0 T) [54, 55].

B.4.2.2.2.2 Chemical vapour deposition of HTS

Chemical deposition principles

In a CVD process, the gaseous precursor molecules, which contain the elements of the coating, are transported through the gas phase to the deposition surface and they undergo a chemical reaction to obtain the required material. The total pressure in the gas phase is typically 1 hPa to 1000 hPa. In this pressure range, there are many collisions in the gas phase and the transport is controlled by convection and diffusion. Diffusion processes are mainly located near the deposition surface. Homogeneous reactions in the gas phase during the transport can modify the deposition process.

In chemical beam epitaxy (CBE) the precursor molecules flow through a reactor in which the pressure is much smaller than in CVD and therefore there are no collisions in the gas phase. At the surface, the precursor molecules decompose, react and form the layer. In this case, the kinetic gas theory can be used to describe the transport process (see review [6]). This technique is also called MBE (molecular beam epitaxy) or, if metal-organic molecules (MO) are used: MOMBE. By using a pulsed system, it is possible to realize atomic layer epitaxy (ALE: further literature about ALE: [9, 56, 57]).

The characteristic number that separates the CVD and the CBE modes is the Knudsen number: $Kn = \lambda/d$, where λ is the mean free path of the molecules in the gas phase and d is a characteristic length in the deposition process (e.g., distance between the source and the substrate). This number determines the number of collisions of the molecules between the source and the substrate and therefore the transport mechanism between source and substrate. The mean free path of air molecules at 10^5 Pa is approximately $\lambda = 0.1$ μm. This means that at $d = 1$ cm, the Knudsen number (Kn) is 10^{-5} and at 100 Pa, approximately $Kn = 10^{-2}$. In figure B4.2.2.6, the different chemical deposition processes (CVD and CBE) and the different theories used to describe the transport of the molecules are given [58]. The order parameter is the Knudsen number. The following abbreviations are additionally used: ACVD = CVD at atmospheric pressure; LP CVD = CVD at low pressure, mostly in the range 1–100 hPa ($Kn = 10^{-2}$ until 10^{-4} at $d = 1$ cm). According to the decreasing number of collisions with increasing Kn, the role of homogeneous reactions decreases with increasing Kn too. Therefore, the Kn expresses the role of possible homogeneous reactions in the gas phase. If such homogeneous processes which can produce pre-reacted powder must be avoided, the CVD processes must be carried out at large Kn.

The droplet-derived CVD process (Step 3 in figure B4.2.2.1) is slightly different from these processes. In this process, a liquid precursor or a precursor solution is used. Kn can be estimated with d_d: $Kn = \lambda/d_d$, where d_d—as defined in section B4.2.2.2.1— is the distance from the deposition surface

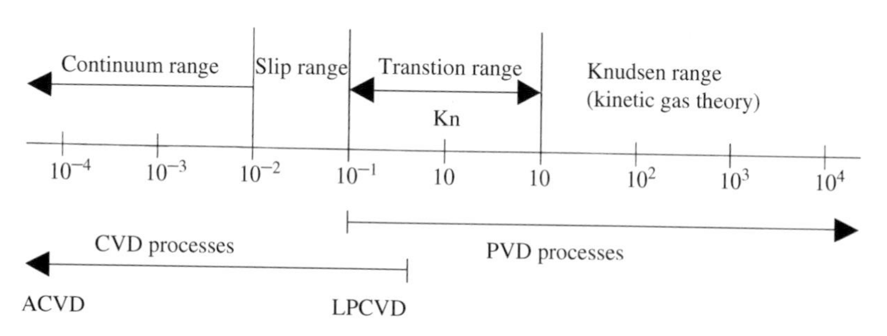

Figure B4.2.2.6. Pressure range for the different deposition processes and the models used for the transport calculations in the gas phase.

where the droplet is completely evaporated. If d_d is small, this number can be large at high pressures also, these considerations show an important advantage of droplet-derived CVD: The role of homogeneous reactions in the gas phase can be strongly decreased by droplet CVD. A further advantage is that in one droplet all precursors can be dissolved at the same time (= single source CVD as described later).

To achieve the aforementioned chemical deposition principles, the CVD equipments always consist of three parts: a unit for gas production, a deposition unit and an exhaust unit.

Precursors for CVD

Precursors for CVD should in general meet the following properties:

(1) The precursor should have an evaporation rate which is high enough, stable and constant with time.
(2) The precursor molecule should be chemically and thermally stable during the transport through the gas phase to the surface.
(3) The precursor should relatively easily be synthesized.
(4) The precursor should not be dangerous and should not produce dangerous side products.

In the case of droplet-derived deposition processes, the following condition must also be fulfilled:

(5) The precursor must be soluble and stable in a suitable solvent without the formation of precipitates.

For CVD processes, halogenides, hydrides or alcoholates are normally used [59, 60]. All these precursors, however, are not suitable for all the elements of group 2 of the periodic system, which are present in high temperature superconductors. These elements are very electropositive and therefore have a tendency to form compounds with ionic bonds with low vapour pressures. To obtain precursors with large vapour pressure, it is necessary to use molecules with small intermolecular interaction energies. Since the work of Yamane [3], chelate compounds are the main candidates for such precursors. The basic compound of the chelates is the β-diketone, which exists in two tautomer forms, the diketone and the enol form:

$$\underset{\text{(diketone)}}{\mathrm{R'{-}C(=O){-}CH2{-}C(=O)}{-}R} \leftrightarrow \underset{\text{(enol)}}{\mathrm{R'{-}C({-}OH) = CH{-}C(=O)}{-}R}$$

R and R′ are different organic ligands. The anion of the enol forms complexes with metals. Figure B4.2.2.7(a) shows the different β-diketones whose complexes with metals are used in CVD of superconductors [2, 6]. Figure B4.2.2.7(b) shows these complexes.

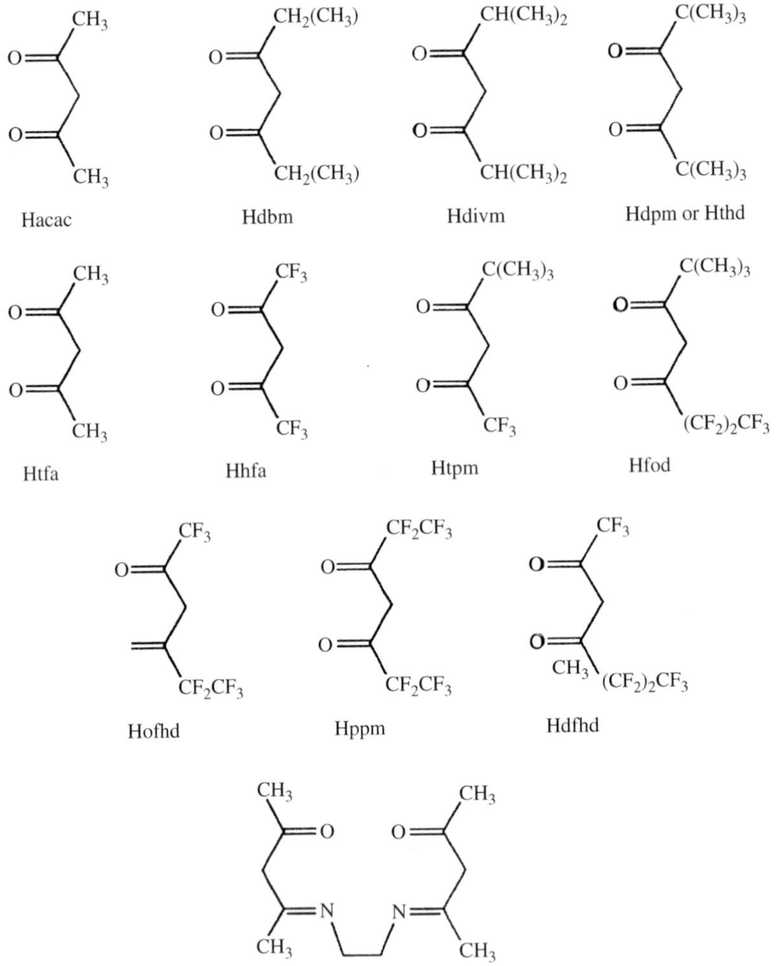

Hacac   Hdbm   Hdivm   Hdpm or Hthd

Htfa   Hhfa   Htpm   Hfod

Hofhd   Hppm   Hdfhd

Hcn-acac

(a)

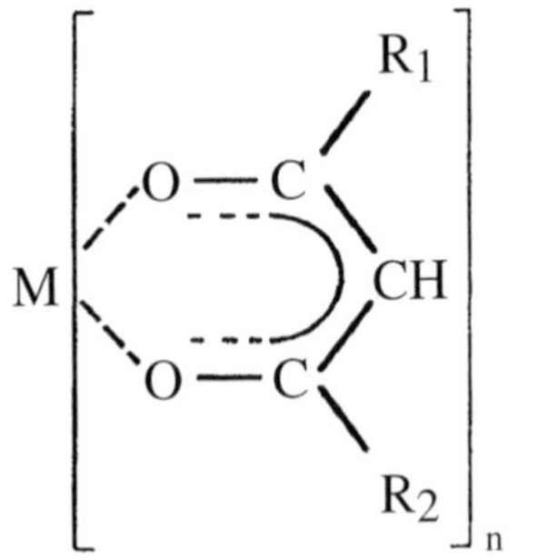

R_1	R_2	Short Name
CH_3	CH_3	$M(acac)_n$
t-C_4H_9	t-C_4H_9	$M(thd)_n$ $M(dpm)_n$
CF_3	CF_3	$M(hfac)_n$
n-C_3F_7	t-C_4H_9	$M(fod)_n$

(b)

Figure B4.2.2.7. Compounds used for precursor preparation (*a*) and some chelate compounds for MOCVD (*b*).

The evaporation enthalpy ΔH is approximately equal to the binding energy between the precursor molecules in the liquid or solid. ΔH can be estimated by the formula (B4.2.2.2) [61]

$$\Delta H = -\frac{3kT}{(4\pi\varepsilon_0)^2 r^6}\left[\frac{\mu^2}{3kT} + \alpha_0\right]^2 - \frac{3\alpha_0^2 I}{4(4\pi\varepsilon_0)^2 r^6}, \qquad \text{(B4.2.2.2)}$$

where k is the Boltzmann constant, r the distance between the molecules, α_0 the polarizability, I the ionization energy, μ the dipole of the molecule and T the temperature.

This formula explains qualitatively the following effects:

(1) The absolute value of ΔH increases and the vapour pressure decreases when the dipolar moment between metal and oxygen increases (when their difference in electronegativity increases). In the case of $Mg(tmhd)_2$ and $Cu(tmhd)_2$ or $Ca(tmhd)_2$ and $Pb(tmhd)_2$ [62], we therefore have:

$$p(Mg(tmhd)_2 < p(Cu(tmhd)_2,$$

$$p(Ca(tmhd)_2) < p(Pb(tmhd)_2).$$

(2) The ionic radius of the metal atom determines its distance to oxygen and the value of the dipolar moment $\mu = qd$ (q is the charge of the atoms, d the distance between O and the metal atom). In the case of lanthanide complexes, the electronegativity differences are similar and the vapour pressure, therefore, decreases with increasing radius [63]:

$$p(Yb(tmhd)_2) > p(Tm(tmhd)_2) > p(Er(tmhd)_2) > p(Ho(tmhd)_2) > p(Dy(tmhd)_2) > p(Tb(tmhd)_2)$$

$$> p(Gd(tmhd)_2) > p(Eu(tmhd)_2) > p(Sm(tmhd)_2) > p(Nd(tmhd)_2).$$

A similar relation is found for the vapour pressures of tmhd compounds with elements from group 2 and for the acac compounds of groups 9 and 10 where the radius increases from Mg to Ba, from Co to Ir and from Pt to Ni:

$$p(Mg(tmhd)_2) > p(Ca(tmhd)_2) > p(Sr(tmhd)_2) > p(Ba(tmhd)_2)\ [64],$$

$$p(Co(acac)_3 > p(Rh(acac)_3 > p(Ir(acac)_3,$$

$$p(Ni(acac)_2 > p(Pd(acac)_2 > p(Pt(acac)_2\ [65].$$

The vapour pressure of chelates with large metal atoms (like Ba) can be modified by association effects. This is particulary true for $Ba(tmhd)_2$. Because of the large ionic radius of Ba and the large coordination number c ($c = 6$) there is enough room between the ligands for coordination with other molecules. The formation of tetramers $Ba_4(tmhd)_8$ is therefore possible.

A further problem in the case of barium compounds is that the type of coordination depends on the synthesis method. If non-water-free synthesis methods are used the following complexes can be generated: $Ba_5(thd)_9(H_2O)_3(OH)$, $Ba_5(tmhd)_9(ClH_2O)_7$ [66, 67], $Ba_2(tmhd)_4(H_2O)$, $Ba_6(tmhd)_{12}(H_2O)_{13}$ [68] or $Ba_2(tmhd)_4)_3(H_2O)$ [69]. By sublimation or by water-free methods, $Ba_4(tmhd)_8$ is formed. All these Ba compounds have a great tendency to hydrolyse and to oligomerize in the presence of water.

The association effects can be minimized by the complexation of the molecules, by the addition of further ligands, e.g., by the formation of $Ba(tmhd)_2$ tetraglyme, $Ba(tmhd)_2Phen_2$ [70].

All these Ba precursors have different vapour pressures. Evaporation measurements of the precursors are therefore absolutely necessary before the deposition experiments in the case of conventional CVD processes. The evaporation process of the Ba precursors can, in addition, change with time due to hydrolysis or oligomerization [71]. The Ba precursor evaporates as oligomers [72].

(3) The size and the number of ligands determine the vapour pressure. Large ligands can induce large London forces [last term in equation (B4.2.2.2)]. On the other hand, large molecules can produce a larger screening of the Coulomb forces.

Al or Cr chelates illustrate the first case, where the vapour pressure decreases with the length of the ligand [73]: $p(\mathrm{M(acac)_3}) > p(\mathrm{M(tmhd)_3})$, M = Al, Cr; but screening is probably more important in the case of Pb compounds [74]:

$$p(\mathrm{Pb(acac)_3}) < p(\mathrm{Pb(tmhd)_3}).$$

(4) The substitution of H by F, with a smaller polarizability and therefore small dispersion forces, increases the vapour pressure [75]. This is shown in the following row:

$$p(\mathrm{Zr(acac)_4}) < p(\mathrm{Zr(tfacac)_4}) < p(\mathrm{Zr(hfacac)_4})\ \ [76],$$

$$p(\mathrm{Pb(acac)_2}) < p(\mathrm{Pb(tfacac)_2}) < p(\mathrm{Pb(hfacac)_2})\ \ [77],$$

$$p(\mathrm{M(acac)_3}) < p(\mathrm{M(fod)_3}) < p(\mathrm{M(tfacac)_3}) < p(\mathrm{M(hfacac)_3})\mathrm{M} = \mathrm{Al, Cr}\ \ [78].$$

Engineering of CVD deposition

Evaporation behaviour: The most commonly applied compounds for the synthesis of HTS are the tmhd compounds. They are free of fluorine and have vapour pressures which are high enough for CVD experiments. Good results have also been obtained with fluorine containing precursors [79]. Table B4.2.2.1 shows important physical properties necessary for the transport of these molecules in the gas phase. The Lennard Jones length σ and the Lennard Jones energy ε for the single molecules were taken from [80] and if data were not available, they were estimated with the Le Bas rule or from the boiling point T_b. For the Ba compound different molecules were calculated, because the nature of the evaporating molecules is not completely known. Only very few attempts have been made to use liquid precursors [81].

The most important properties of the precursors are their evaporation rate and their stability over time. A suitable method to measure these properties under CVD conditions is the microbalance method where the mass change of a container with the evaporating precursor is measured [82]. From these

Table B4.2.2.1. Gas properties

	σ (Å)	ε (K)	M (g mol^{-1})
O_2	3.467	106.7	32
N_2	3.798	71.4	28
$Y(thd)_3$	10.52	601	658
$Cu(thd)_2$	9.20	590	450
$Ba(thd)_2$			
$Ba(thd)_2$	8.20	932	503
$Ba_2(thd)_4$	11.59	932	1006
$Ba_4(thd)_8$	14.60	932	2012
$Ba_6(thd)_{12}(H_2O)_{13}$	16.99	932	3252

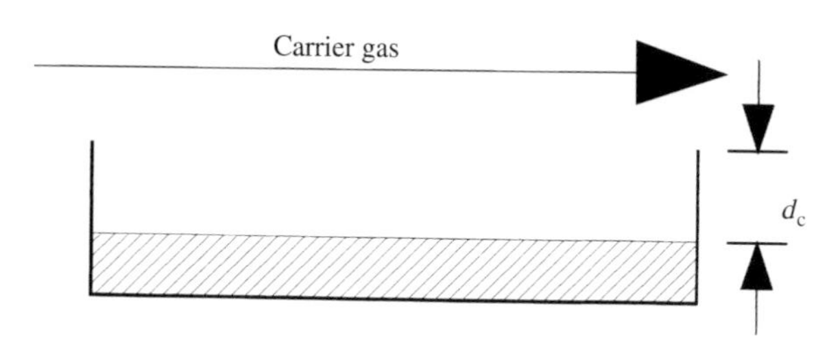

Figure B4.2.2.8. Crucible for the precursor evaporation.

evaporation rates the vapour pressure of the material can be estimated, if it is assumed that the vapour pressure is formed near the powder surface. This vapour pressure can then be used to calculate evaporation rates in different evaporators with different gas atmospheres and different flow conditions. A typical container is shown in figure B4.2.2.8 [6]. In the container a stagnant layer with thickness d_c is formed. The relation between molecular evaporation rate I and the equilibrium pressure p_{prs} near the precursor surface is given by

$$I = \frac{p_{prs} - p_{prg}}{kTDd_c} A, \qquad (B4.2.2.3)$$

where p_{prg} is the precursor partial pressure in the flowing gas outside of the container, p_{prs} the precursor partial pressure at the precursor surface ($p_{prg} \ll p_{prs}$), k the Boltzmann constant and A = evaporator surface (d_c: see figure B4.2.2.8).

The binary diffusion coefficient D of the precursor in the gas phase is calculated by (B4.2.2.4) [58]

$$D = \frac{0.00263 T^{3/2}}{p_{tot}\sigma^2\sqrt{2M}\Omega_D}, \qquad (B4.2.2.4)$$

with the total pressure p_{tot} (bar), the diffusion coefficient D (cm^2 s^{-1}), the absolute temperature T (K), the Lennard Jones collision length σ_{pc} (Å) for the collisions between the precursor and the carrier gas, the reduced molar mass M (g mol^{-1}) of the precursor mass and the mass of the carrier gas molecules, and the collision integral Ω_D which depends on the ratio ε_{pc}/k (ε = Lennard Jones interaction energy between the precursor molecules and the carrier gas molecules). The length σ_{pc} and the energy ε_{pc} must be calculated from the values σ and ε for the single molecules by the arithmetic, respectively, the geometric mean value [83]. The evaporation calculation with the help of equation (B4.2.2.3) gives only approximate values because the thickness d_c of the stagnant layer is not so clearly defined. Exact gas flow calculations give slightly different values (±30 %) [84].

All vapour pressures p_e can be described by the formula:

$$p_e = p_0 \exp(-E/RT). \qquad (B4.2.2.5)$$

Table B4.2.2.2 shows the pre-exponential factors p_0 and the evaporation energies E for precursors used for the deposition of YBaCuO. According to table B4.2.2.2 the effect of changing E is partly compensated by changing p_{eo} (compensation effect). An exception appears for the Ba compound. The reason might be that, in this case, we did not have Ba(tmhd$_2$), but a more complex compound, as mentioned above. The compensation effect is frequently found, e.g., for the evaporation of other metal-organic compounds [85], in the catalysis [86] and at the chemical adsorption of metals on tungsten [87]. The reason for this effect is still not clear.

Generation of the reaction gas mixture: The first CVD equipment for the synthesis of superconductors used different sources for the precursors (figure B4.2.2.9 [88]). The disadvantage of these

Table B4.2.2.2. Pre-exponential factor and activation energy for the vapour pressure of the precursors

Compound	$\mathrm{Log}(P_0)$ (P_a)	E (kJ mol^{-1})
$Zr(thd)_4$	14	117
$Ba(thd)_2$	13.19	129
$Y(thd)_3$	13.03	98
$Cu(thd)_2$	12.94	94

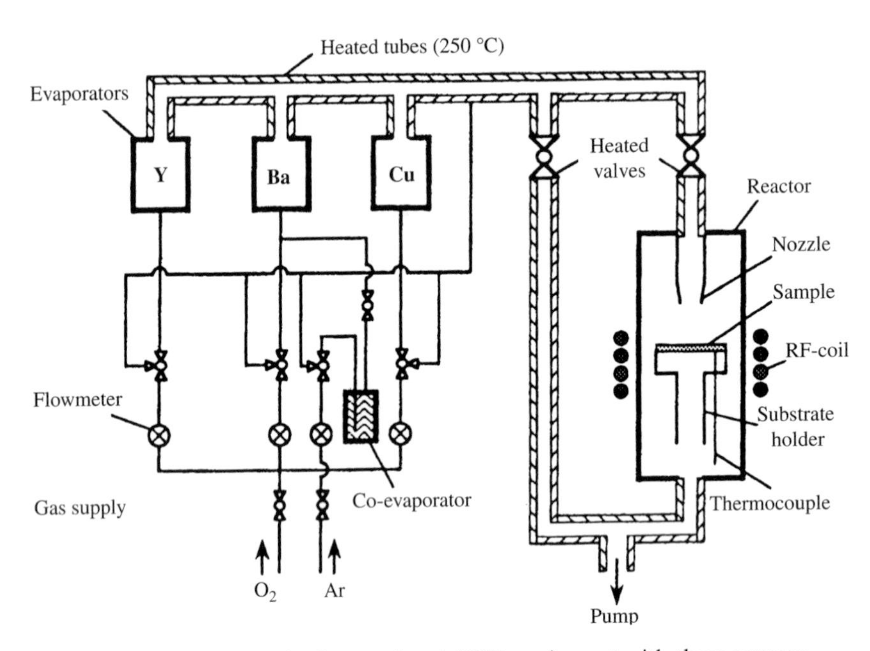

Figure B4.2.2.9. Conventional CVD equipment with three sources.

multi-source setup is the non-steady evaporation of $Ba(tmhd)_2$ and the difficult temperature control of the different sources.

Therefore, single-source equipment was constructed. Hiskes *et al* used a mixture of all precursor powders moving in a tube where the powder was flash evaporated (figure B4.2.2.10) [89].

Samoylenko [90] used vibration equipment to transport the precursors into a heated zone where again the powder was flash evaporated (figure B4.2.2.11).

A new class of single-source CVD equipments is based on liquid sources. There are different types of liquid source systems:

B2.1 (1) Droplets of the precursor containing solution are generated by ultrasonic nebulization and conducted through a heated zone where the droplets evaporate completely. The CVD process is then carried out in mode 3 of figure B4.2.2.1 [91] (figure B4.2.2.12).

(2) The solution is injected by an injector valve as is typically used in the car industry [92] (figure B4.2.2.13).

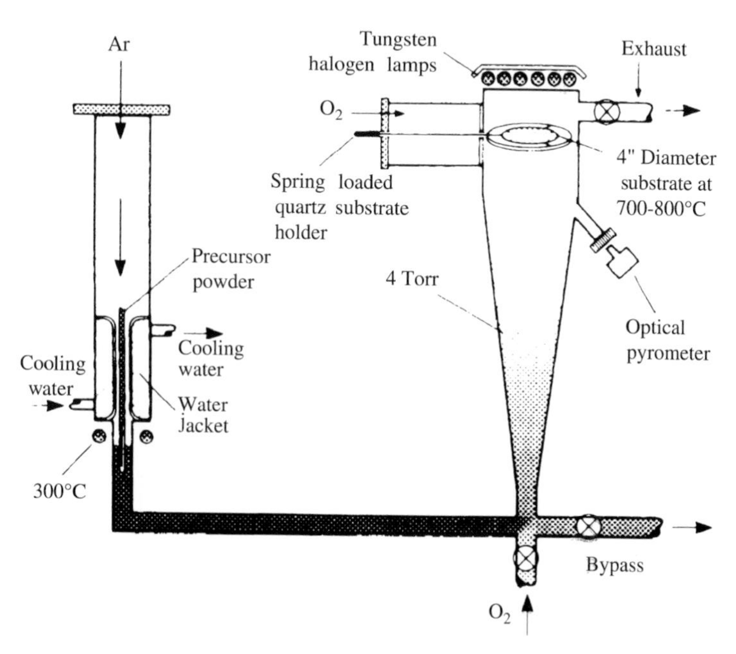

Figure B4.2.2.10. Single source evaporation system.

(3) A tape is wetted with the solution and dried. This tape is then moved through a hot zone where the precursor mixture is completely evaporated and transported into the reactor. By this method, it is possible to produce a solvent-free gas mixture. The tape can also be 'printed' with different precursors and alternating gas concentrations and multilayer coatings can be produced [93] (figure B4.2.2.14.).

(4) An evaporator that separates the solvent molecules from the precursor molecules and also guarantees long-term operation is the band evaporator as shown in figure B4.2.2.15 [94]. An infinite fibre band is continuously fed by the precursor solution. The condensate on the moving band is first

B4.2.1

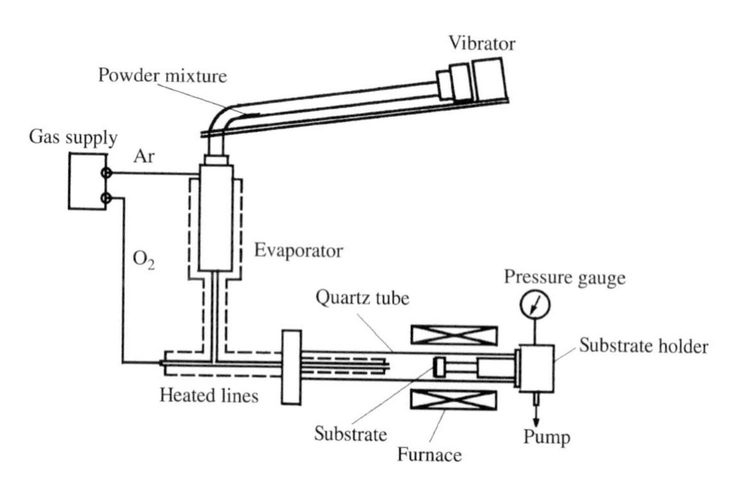

Figure B4.2.2.11. Vibrator evaporation system.

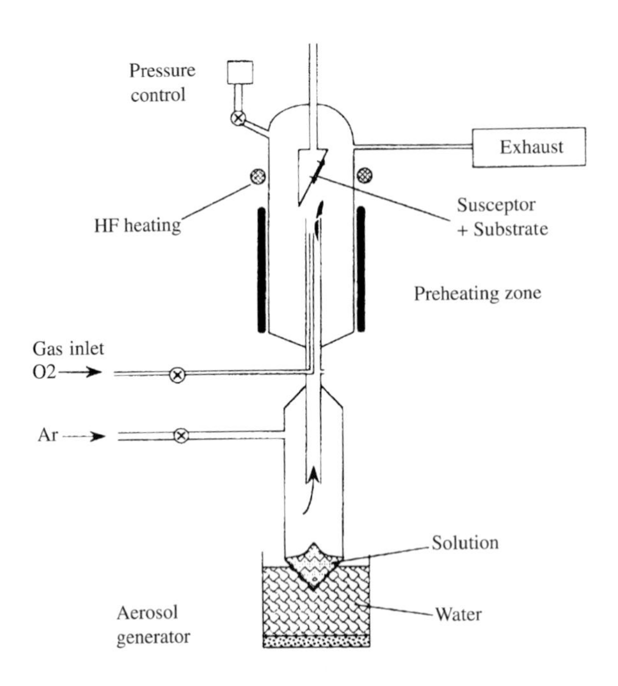

Figure B4.2.2.12. Ultrasonic nebulization connected with preheating zone where the droplets are evaporated.

transported through a hot zone with $T = 60°C$, where the solvent is evaporated and transported into the cooling trap. The solvent can be reused later. The band with the dry precursor then moves into the hot zone where at 350°C the precursors evaporate completely. In Figure B4.2.2.15 the band evaporator is connected to a tape deposition reactor which is described later. Mostly diglyme or monoglyme is used as a solvent.

(3) *Thermodynamic calculations*: The deposition processes of HTS are carried out at high temperature ($T > 700°C$). At these high temperatures, thermodynamic equilibrium calculations give information about the experimental conditions under which a deposition of superconducting YBaCuO can be expected. Such calculations are especially important, if gas mixtures with solvent molecules are used. These solvent molecules contain carbon and therefore they support the formation of carbonates instead of the superconducting phase (described in other parts of this book).

C1

B2.1

For the CVD of YBaCuO a gas mixture containing C, H, O, Y, Ba, Cu, Ar (carrier gas) atoms is mostly used. The following parameters determine the thermodynamic equilibrium: total pressure p_{tot}, temperature T and six relations which determine the composition of the gas mixture before the deposition process, e.g., the ratios $\alpha(C) = n(C)/n(Ar)$, $\alpha(H) = n(H)/n(Ar)$, $\alpha(O) = n(O)/n(Ar)$, $\alpha(Y) = n(Y)/n(Ar)$, $\alpha(Ba) = n(Ba)/n(Ar)$, $\alpha(Cu) = n(Cu)/n(Ar)$, where $n(\,)$ is the number of atoms independent of the bond state of the atoms. The carbon, e.g., can be bonded to the precursor molecule or to the solvent molecule. In the case of YBaCuO deposition from a solution, these parameters can be reduced to the values p_{tot}, T, and the molar fractions $x(\,)$ of the following compounds in the gas phase: $x(\text{solvent})$, $x(O_2)$, $x(Y(tmhd)_3)$, $x(Ba(tmhd)_2)$, $x(Cu(tmhd)_2)$. The molar fraction of the carrier gas (e.g. argon) is given by:

B2.1

$$x(Ar) = 1 - x(\text{solvent}) - x(O_2) - x(Y(tmhd)_3) - x(Ba(tmhd)_2) - x(Cu(tmhd)_2).$$

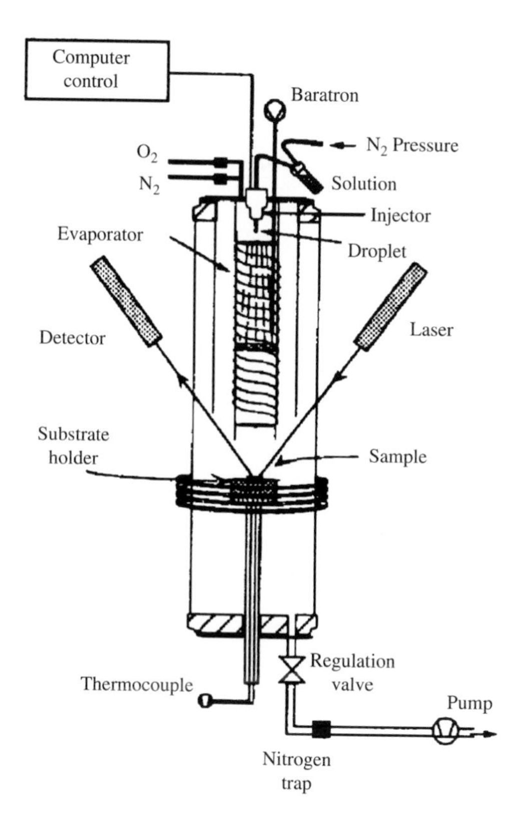

Figure B4.2.2.13. Single source injector system.

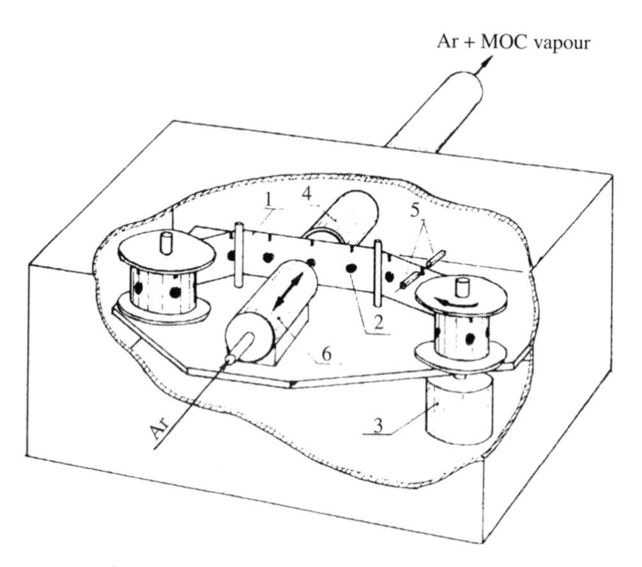

Figure B4.2.2.14. Tape evaporator for the deposition of multilayers—1: tape, 2: patches with different precursor precipitates, 3: rotating cylinder for the transport of the tape, 4: connection tube to the deposition reactor, 5: transport control, 6: heating system to heat the patches, MOC: metal-organic compound.

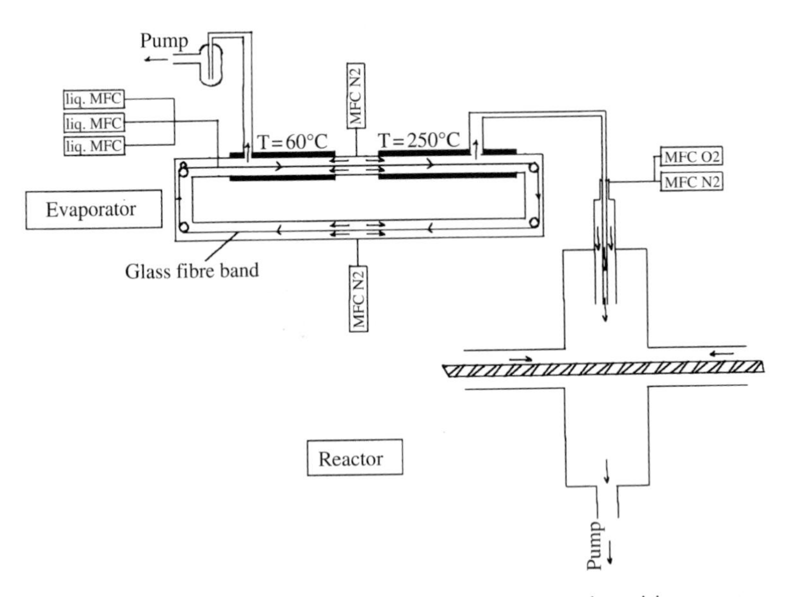

Figure B4.2.2.15. Band evaporator connected to a tape deposition reactor.

The molar fractions of the precursors can be calculated from the molar fractions x (precursor in solution) in the liquid solution:

$$x(\text{precursor}) = x(\text{solvent}) \times (\text{precursor in solution}).$$

Thermodynamic calculations (figure B4.2.2.16) were carried out in a model reactor for the following conditions [96]: $p_{tot} = 13.3\,\text{hPa}$, $T = 760$–$860°C$, dilution gas: argon, gas flow $I(Ar + O_2) = 27\,l_n\,h^{-1}$ ($= 1.2\,\text{mol}\,h^{-1}$), molar concentrations of precursors in diglyme: x(sum of all precursors in diglyme) = 0.0027 (= 0.02 mol l^{-1}), $x(Y(tmhd)_3)x(Ba(tmhd)_2)x(Cu(tmhd)_2)$ =1:2:3.

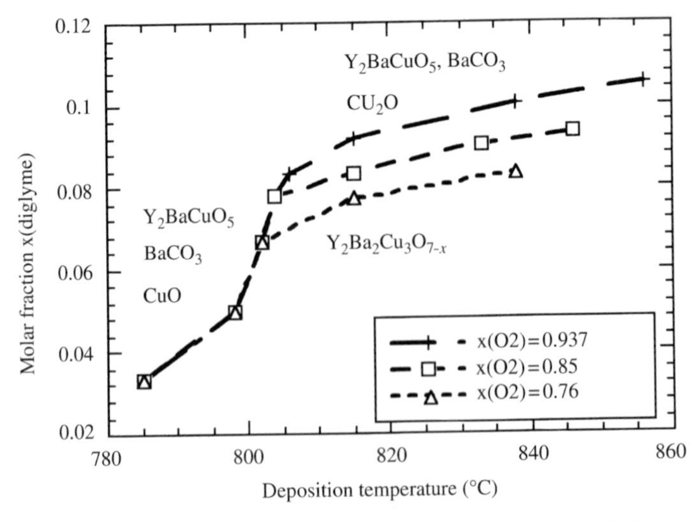

Figure B4.2.2.16. Equilibrium calculations for the phases formed in an Ar–O_2–diglyme–precursor system. The calculated points are indicated at the curves. Deposition conditions are given in the text.

The curves in figure B4.2.2.16 are calculated with different O_2 flows:

upper curve: $I(O_2) = 25.4\,l_n\,h^{-1}(= 1.13\,mol\,h^{-1})$,

middle curve: $I(O_2) = 23\,l_n\,h^{-1}(= 1\,mol/h)$,

lower curve: $I(O_2) = 20.5\,l_n\,h^{-1}(= 0.91\,mol\,h^{-1})$.

On the left axis the feeding rate of the diglyme solution is given and on the right axis the corresponding molar flow I(diglyme) of the solution.

Figure B4.2.2.16 shows that, at high feeding rates of the solution, carbonates can be formed. The boundary between carbonate and YBaCuO formation depends, at high temperatures only, weakly on the temperature. In order to understand this boundary, the oxidation is simplified and C and H in the chelates are neglected because of their small molar fractions in the gas phase. The complete burning of diglyme gives:

$$C_6H_{14}O_3 + 8O_2 \rightarrow 6CO_2 + 7H_2O.$$

A complete burning of diglyme is only possible if there is enough oxygen in the gas phase. The boundary value of the diglyme concentration is related to the O_2 molar fraction by:

$$x(\text{diglyme}) < \alpha x(O_2)) \quad \text{with} \quad \alpha = 1/8.$$

Only in this case is superconductor deposition possible. At higher diglyme concentrations, carbonates are formed. The last equation can be formed into gas flows which is more convenient for CVD reactors.

$$I(\text{diglyme}) < \alpha\ I(O_2).$$

A superconducting layer can only be deposited if this condition is fulfilled.

The critical gas flow I_{cr}(diglyme) where the composition of the layer switches from an oxide to an oxide carbonate is given by

$$I_{cr}(\text{diglyme}) = \alpha I(O_2).$$

With the oxygen gas flows used for the calculations for figure B4.2.2 16, the critical diglyme currents I_{cr} can be calculated with the help of the last equation:

upper curve: $I_{cr}(\text{diglyme}) = 0.14\,mol\,h^{-1}$,

middle curve: $I_{cr}(\text{diglyme}) = 0.125\,mol\,h^{-1}$,

lower curve: $I_{cr}(\text{diglyme}) = 0.113\,mol\,h^{-1}$.

These values agree quite well at high temperatures with the calculated boundaries, as can be seen in figure B4.2.2.16. At low temperatures, the exact thermodynamics is more complicated than the simple assumption of complete burning.

The boundary between carbonate and YBaCuO formation limits the maximum deposition rates of the superconductor because of the low solubility of the precursor in the diglyme. This limitation can be avoided, if the diglyme is separated from the precursor (band evaporation systems).

Mass transport to the surface: Most CVD experiments for superconductor deposition are conducted made in a stagnation flow geometry in a cold wall reactor. In this geometry, the gas flows perpendicularly onto the deposition surface, the gas is then heated up to the deposition temperature and deposition takes place. In the following, only this deposition geometry is discussed.

The CVD process consists of different steps:

(1) transport through the gas phase;

(2) chemical reaction on the surface, nucleation, film formation.

For the CVD of YBCO (figure B4.2.2.17, [84]), the mass deposition rate is plotted vs the reciprocal absolute temperature. This figure shows a temperature behaviour typical for the deposition of oxides from the compounds in a cold wall stagnation flow reactor. Similar curves were found for the ZrO_2 deposition [96]. All experiments shown in figure B4.2.2.17 were made at a precursor molar fraction $x(\text{precursor}) = 2 \times 10^{-4}$ at a total pressure of $p = 1000$ Pa and a total gas flow of $J = 15\,\text{l}_n\,\text{h}^{-1}$. The mass change was measured by a microbalance. The deposition process can be described by the following deposition model. The molecules diffuse through a stagnant boundary layer with the thickness δ and then react on the surface with a reaction constant k in a monomolecular reaction to form the oxide. This model gives the equation [97]:

$$J = \frac{n_{\text{pr}}}{\delta/D + 1/k} \tag{B4.2.2.6}$$

where D is the diffusion constant and n_{pr} the precursor concentration in the gas phase. The reaction constant k can be written as:

$$k = k_0 \ \exp\left(-\frac{E}{RT}\right) \tag{B4.2.2.7}$$

According to equation (B4.2.2.7), deposition at high temperatures is determined by diffusion in the gas phase

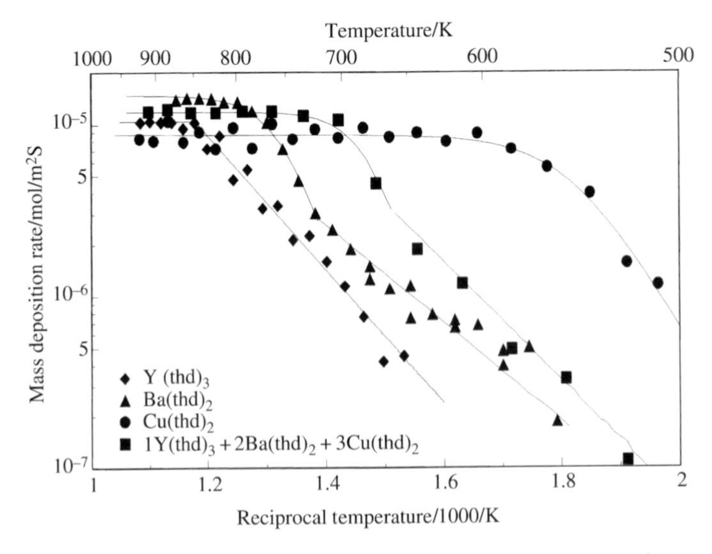

Figure B4.2.2.17. Mass transport calculations for the deposition of oxides. Comparison with calculations (large signs).

$$J_D = \frac{D}{\delta} n_{pr} \tag{B4.2.2.8}$$

and at low temperature by the kinetics of the surface reaction

$$J_k = k n_{pr}. \tag{B.4.2.2.9}$$

Equation (B4.2.2.8) can also be written with the help of equations. (B4.2.2.8) and (B4.2.2.9):

$$J = \frac{1}{1/J_D + 1/J_k}. \tag{B4.2.2.10}$$

Figure B4.2.2.17 shows that the deposition of YBaCuO is not the sum of the single curves. The deposition of Cu is possible at much lower temperatures in the case of pure Cu oxide than in the case of mixed oxide.

The deposition of superconductors is always carried out in the diffusion-controlled range. In this range, the deposition on the samples can be calculated with hydrodynamic calculations, whereby the diffusion data were calculated with the Lennard Jones parameters given in table B4.2.2.1. Thermodiffusion effects are included according to the formulae given in [58].

Table B4.2.2.3 shows a good accordance between the calculated deposition rates J_D (calc) and the experimental values J_D (exp) shown in figure B4.2.2.17. A good agreement was reached with the assumption that the Ba precursor in the gas phase is a tetramer [84]. The use of monomers generates deviations of 20%. In addition, in table B4.2.2.3 the pre-exponential factors and the activation energies for the deposition in the kinetically controlled range are given.

The deposition in the stagnation point in the diffusion-controlled range can also be calculated by the empirical formula [97]

$$J_D = D \frac{x(Pr\ \ ec)n}{d} Sh, \tag{B4.2.2.11}$$

where d is a characteristic length (nozzle–substrate distance) in the reactor. The Sherwood number Sh is given by

$$Sh = \alpha\ Re^{1/2}\ Sc^{1/3}, \tag{B4.2.2.12}$$

where the Reynolds number $Re = \rho v d/\eta$ (η = dynamic viscosity, v = characteristic velocity, ρ = mass density) and the Schmidt number, $Sc = \eta/(\rho D)$. α is a factor in the range of 1 [98].

According to the different diffusion coefficients for the precursor molecules, the composition of the cations in the gas phase is different to the composition of the film. This effect is described by the

Table B4.2.2.3. Experimental and calculated deposition rates in the diffusion-controlled range and activation energy and preexponential factor in the kinetically controlled range

	j_D(exp.) (μmol $m^{-2}s^{-1}$)	j_D(calc.) (μmol $m^{-2}s^{-1}$)	E (kJ mol^{-1})	K_0 (mol m^{-2} s^{-1})
$Y_1O_{1.5}$	10.0	11.1	75	9.8×10^3
CuO	8.7	9.4	120	4.3×10^9
BaO	14.3	13.8	207	2.9×10^{13}
$Y_1Ba_2Cu_3O_7$	11.8	11.7		

enrichment factor, which compares the deposition of the cation i with the cation J:

$$k_{ij} = \frac{x(i, film)x(J, gas)}{x(i, gas)x(J, film)}, \qquad \text{(B4.2.2.13)}$$

where x are the molar fractions of the cations in the gas outside of the concentration boundary layer and in the film. From equation (B4.2.2.10), equation (B4.2.2.14) is the k_D for the diffusion-controlled range given by:

$$k_{ij} = \left[\frac{D_i}{D_J}\right]^{2/3}. \qquad \text{(B4.2.2.14)}$$

In order to reach a stoichiometric $Y_1Ba_2Cu_3O_7$ composition, the gas composition must have the relation:

$$\text{Y Ba Cu} = 1:2k_{Y,Ba}:3k_{Y,Cu} = 1:2.9:2.3.$$

These values are calculated with the Lennard Jones values given in table B4.2.2.2 and the assumption that the Ba compound in the gas phase is a tetramer.

Thermodiffusion effects, which are neglected here, have only a small effect (approximately 10%) on these distribution factors [83]. In reality, these factors are often different because of side effects in the reactor, e.g., deposition on the walls.

(5) *Formation of the crystallized film*: CVD films are mostly used as current carrying element (e.g. cables). Then these films must have a high critical current density parallel to the surface (10^6 A cm^{-2}, depending on the application even higher). Because of the anisotropy of the properties of YBaCuO, the c axis must be perpendicular to the surface and the crystallites must be a–b oriented on the surface too as shown in figure B4.2.2.18, if possible with $\Delta\Phi = 0$. A growth of a–b oriented coatings with the c axis perpendicular to the surface can be easily produced on $SrTiO_3$ cut along the principal planes. On (110) planes, however, a mixture of (110) grains with the c axis parallel to the surface and (103)/(130) grains exhibiting a c axis aligned 45° out of the surface plane are formed [99]. Important parameters for the formation of the preferential orientation on common substrates are [100]:

(1) the c axis in plane orientation is favoured by highly matched substrates;

(2) the ratio of the content of (100) and (001) orientation decreases with increasing temperature;

(3) this ratio increases with increasing oxygen pressure;

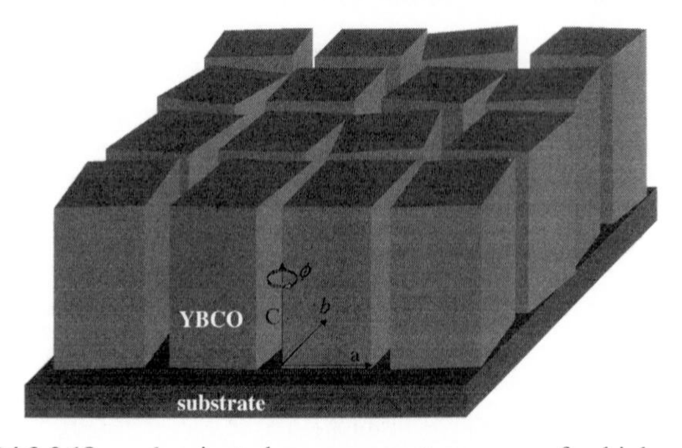

Figure B4.2.2.18. a–b-oriented arrangement necessary for high critical currents.

(4) the ratio also increases with increasing deposition rate.

Granozzio *et al* [100] have found, by thermodynamic Stranski–Kossel-like thermodynamic calculations, that the orientation of the grains on a (001) YBCO surface depends on the difference of the supersaturation. At small supersaturations nuclei with c axis perpendicular to the surface and at high supersaturation nuclei with c axis parallel to the surface are formed. This explains the experimental dependencies cited earlier.

(6) *Achievement of films*: HTS materials are fabricated by MOCVD on many single-crystal metal oxide substrates, such as $SrTiO_3$, $NdGaO_3$, MgO, Al_2O_3 and yttrium-stabilized zirconia. Two-dimensional, flat substrates, with a well-defined surface structure, and substrate sizes in the range of several inches are commercially available. For device applications, these substrates are important. However, for power applications, long length and flexible substrates are required. In this case two solutions are discussed, which both apply physical vapour deposition methods (see contribution in this book about physical deposition techniques): A1.3

(1) Metal tapes are coated with a 100 oriented Y stabilized ZrO_2 layer. This buffer layer is produced by an ion beam assisted deposition process and is in-plane oriented. On such substrates in-plane oriented YBCO layers with high critical currents $> 10^6$ A cm^{-2} were deposited by physical deposition methods [101]. B4.1

(2) The second method is the deposition on RABiTS (rolling assisted biaxially textured substrates) which have an interlayer (e.g. CeO_2–YSZ). On these substrates also a–b- oriented layers were deposited by physical deposition methods [102] with high current densities. First experiments to deposit YBCO by CVD with CVD interlayers (YSZ) on RABiTS are described in [103]. Ignatiev has deposited YBCO on CeO_2, YSZ on Ni tapes with a current density 5×10^5 A cm^{-2} [104]. B4.1 B4.2.1

Actually, instead of systems using thermal evaporation of solid or liquid precursors, systems that contain the dissolved precursor in an organic solvent (liquid delivery systems) are mainly used for the deposition of YBCO thin films. The precursor solution is kept at room temperature under an inert gas atmosphere. Both the low storage temperature of the precursor solution and the fast evaporation step reduce the possibility of precursor decomposition.

For the MOCVD of YBCO, a solution of $Y(thd)_3$, $Ba(thd)_2$ and $Cu(thd)_2$ dissolved in a tetrahydrofuran (THF) and mixtures with isopropanol and tetraglyme [105, 106] or diethylene glycol dimethylether (diglyme) [107–111] was used. The film composition is controlled by varying the Y: Ba: Cu molar ratio of the complexes in solution. THF or diglyme is chosen as a solvent for their solvating power and compatibility with the precursors. Isopropanol and tetraglyme are added to enhance the stability of the solution and of the $Ba(tmhd)_2$ during storage. Using aerosol-assisted MOCVD (AACVD) or injection CVD (ICVD), YBCO has been grown on single crystal substrates like $SrTiO_3$, $LaAlO_3$, MgO, and YSZ [106–109]; heterostructures on YSZ buffer layers on polycrystalline metallic substrates [110] and on sapphire substrates with a CeO_2 buffer layer [111, 112] have also been realized.

The resulting YBCO film on $LaAlO_3$ shows a $T_c = 91$ K and $J_c = 6 \times 10^6$ A cm^{-2} (77 K) [113].

YBCO films have been grown on a Hastelloy C-276 substrate with an in-plane biaxially aligned YSZ buffer layer. The YBCO film shows $J_c = 1.3 \times 10^6$ A cm^{-2} (77 K,0 T) and $J_c = 5.7 \times 10^5$ A cm^{-2} (77 K, 1 T) [114, 115].

YBCO was also grown with AACVD on polycrystalline Ag substrate. $J_c = 2.2 \times 10^5$ A cm^{-2} and $I_c = 3.58$ A (78 K, 0 T) with a $T_c = 89$–91 K were obtained [116]. C1

YBCO films have been deposited on moving (velocity = 4 m h^{-1}) polycrystalline YSZ tapes. Such superconducting films reached a maximum $J_c = 2.5 \times 10^6$ A cm^{-2} at 77 K [117]. Recently, a

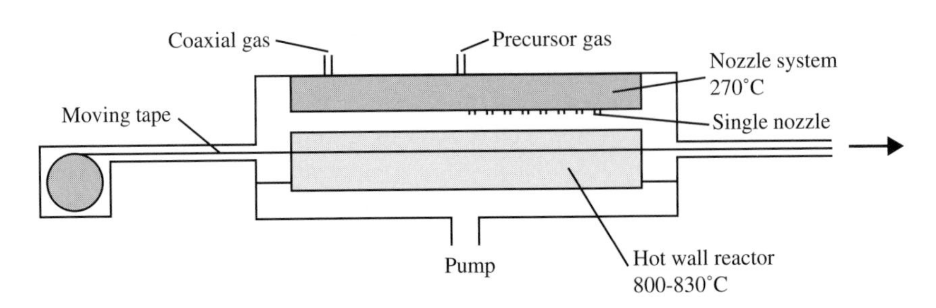

Figure B4.2.2.19. Multiple source system for the tape deposition.

B4.1 $CeO_2/YSZ/CeO_2$ buffered RABiTS coated under similar conditions in the same MOCVD system resulted in a YBCO-film with a $J_c = 2.5 \times 10^5\,A\,cm^{-2}$ (77 K) [118].

The growth rate, which should be as high as possible for industrial fabrication of YBCO films, can be considerably improved in MOCVD processes. YBCO thick films were grown on $LaAlO_3$ (100), in photo-assisted MOCVD, with an exceptionally high growth rate of about 0.5–0.8 $\mu m\,min^{-1}$. The results, $T_c > 89\,K$, and $J_c > 10^6\,A\,cm^{-2}$ even in 4.5 μm thick films [119, 120], are impressive and show that further optimization should be possible.

CVD methods are especially useful for industrial applications as mentioned in section B4.2.2.1. Figure B4.2.2.19 [117] shows as an example a scheme of a tape deposition chamber for the production of cables. The precursor evaporation was carried out with a band evaporator and the deposition unit is a stagnation flow arrangement of a multiple nozzle system with a heated rectangular box. The rectangular box have the dimensions $540 \times 60 \times 50\,mm^3$. The 1 cm wide tape is continuously transported with a velocity of 2–4 $m\,h^{-1}$ through this rectangular box, which had an approximately 220 mm long deposition zone [117]. Figure B4.2.2.20 shows the calculated flow lines in the transverse cross-section of this deposition chamber.

E3 For the application of high T_c superconductors in microwave devices, YBaCuO films can be deposited on both sides of a substrate. Ito *et al* [121] developed a parallel flow reactor and Kaul [122] constructed a reactor with a moving substrate.

C3 *Other superconductors*: Epitaxial $Tl_2Ba_2CaCu_2O_y$ films were grown on $LaAlO_3$ in a two-step process [123]. In the first step Ba, Ca, Cu oxides are deposited and then these layers were annealed in a Tl_2O atmosphere. The maximum current densities at 77 K were $1.2 \times 10^5\,A\,cm^{-2}$.

C2 $Bi_2Sr_2CaCu_2O_x$ films were deposited by Fuflyigin *et al* [124]. As precursors tmhd compounds for Sr, Cu and triphenylbismuth for Ca were used. Maximum current density on $SrTiO_3$ was 10^4 A/cm^2.

C4 In another two-step process, a 1 μm thick (Hg, Pb)-a1223 ($Hg_{1-x}Pb_xSr_{1.5}Ba_{0.5}Ca_2Cu_3O_z$, $x = 0.2$–0.3) film was grown on $LaAlO_3$. The J_c at 77 K was $2 \times 10^6\,A\,cm^{-2}$ with $T_c = 118\,K$ [125].

B4.2.2.3 Conclusions

It is shown in this paper that CVD is a very flexible technique and it can also be used for cable production. The first advantage of MOCVD is that, in comparison to physical vapour deposition techniques, an expensive deposition equipment can be avoided and therefore the process can be very economical. In principle, a CVD process can also work at atmospheric pressure if suitable precursors are developed and large areas can be coated with high rates.

B2.1 These requirements are not so far from the achievements of YBCO films by using spray pyrolysis. Biaxial texturing of YBCO coatings can be achieved by a careful control of the organic or metal-organic

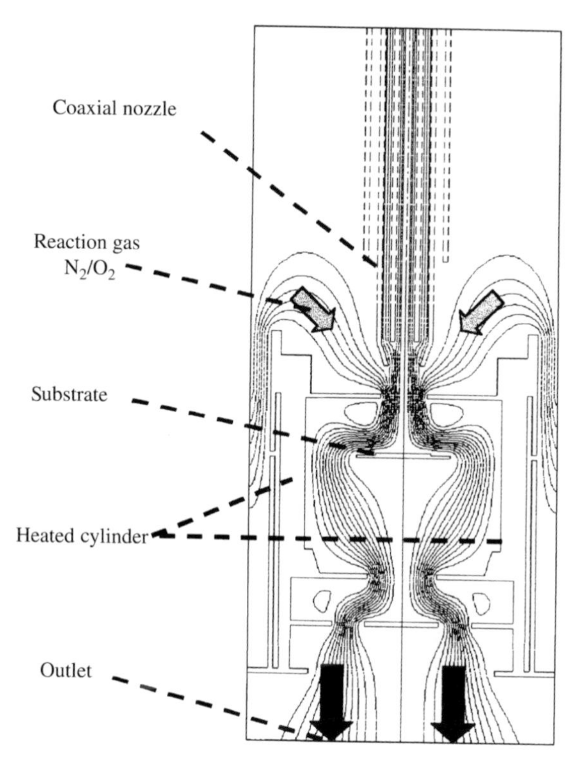

Figure B4.2.2.20. Gas flow in the reactor in the tape deposition equipment with one source. The tape is transported perpendicular to the paper surface.

precursors and their optimal annealing. Promising results are actually obtained by metal-organic decomposition (MOD) at atmospheric pressure of fluorinated precursors (trifluoroacetates) [126]. Chemical deposition techniques in this way become the most powerful synthesis routes for the development of a low cost 'coated conductor' technology.

B4.1

References

[1] Wahl G and Schmaderer F 1989 Review CVD of superconductors *J. Mater. Sci.* **24** 1141

[2] Schulz L and Marks T J 1196 Superconducting materials *CVD of Nonmetals*, ed W S Rees (Weinheim, Germany: VCH) p 39

[3] Yamane A, Kurosawa H and Hirai T 1988 *Chem. Lett.* 939

[4] Berry A D, Gaskill D K, Holm R T, Cukauskas E J, Kaplan R and Henry R L 1988 *Appl. Phys. Lett.* **52** 1743

[5] Schieber M 1991 *J. Cryst. Growth* **109** 401

[6] Dahmen K H and Gerfin T 1993 *Prog. Cryst. Growth Charact.* **27** 117

[7] Schulz D L and Marks T J 1994 *Adv. Mater.* 10 **6** 719

[8] Jergel M 1995 *Supercond. Sci. Technol.* **8** 67

[9] Leskela M, Molsa H and Niinisto L 1993 *Supercond. Sci. Technol.* **6** 627–656

[10] Watson I M 1997 *Chem. Vap. Dep.* **3** 9

[11] Bech M 1996 *Appl. Supercond.* 10–11 **V4** 465

[12] Komarek P 1995 *Hochstromanwendung der Supraleitung* (Stuttgart: BG Teubner)

[13] 1993 *Chemical Vapour Deposition*, eds M L Hitchman and K F Jensen (London: Academic Press)

[14] Malozemoff A P, Carter W, Fleshler S, Fritzemeier L, Li O, Masur L, Miles P, Parker D, Porrella R, Podtburg E, Riley G N, Rupich M, Scudiere J, Zhang W 1998 ASC Conf.

[15] Nukiyama S and Tanasiwa Y 1939 *Trans. Soc. Mech. Eng. Japan* **V5** 68

[16] Rayleigh J W S 1945 *The Theory of Sound* **vol 2**, (New York: Dover) p 344

[17] Gershenson E K and Eknadiosyantis O K 1964 *Sov. Phys. Acoust.* V10
[18] Langlet M, Senet E, Deschanvres J L, Weiss F, Joubert J C, Thomas O and Senateur J P 1989 *Physica* C **162–164** 137
[19] Langlet M, Senet E, Deschanvres J L, Delabouglise G, Weiss F and Joubert J C 1989 *Thin Solid Films* **174** 263
[20] Blanchet G and Fincher C R 1991 *Supercond. Sci. Technol.* **4** 69
[21] Wang J G and Yang R T 1990 *J. Appl. Phys.* 4 **67** 2160–2161
[22] Cukauskas E J, Allen L H, Newman H S, Henry R L and Damme P K V 1990 *J. Appl. Phys.* 11 **67** 6946–52
[23] Kumari S, Singh A K and Srivastava O N 1996 *Supercond. Sci. Technol.* **9** 405–411
[24] Kumari S and Srivastava O N 1998 *Physica C* 1&2 **297** 166–170
[25] Walia D K and Gupta A K 1991 *Supercond. Sci. Technol.* **4** 650–653
[26] Ban E, Matsuoka Y and Ogawa H 1990 *J. Appl. Phys. Pt. 1* 9–1 **67** 4367–4369
[27] Jergel M, Chromik S, Strbik V, Smatko V, Hanic F, Plesch G, Buchta S and Valtyniova S 1992 *Supercond. Sci. Technol.* **5** 225–230
[28] MacManus-Driscoll J L, Ferreri A, Wells J J and Nelstrop J G A 2001 *Supercond. Sci. Technol vol* **14** 1
[29] Schields T C, Kawano K, Button T W and Abell J S 2002 *Supercond. Sci. Technol.* **15 (1)** 99–103
[30] Blanchet G B and Fincher C R Jr 1992 *Supercond. Sci. Technol.* **5** 69
[31] Supardi Z, Sin A, Phok S, Odier P, Weiss F, Chaudouët P and Perkin G 2001 (Physique IV), Proceedings of the 2nd Thin Film Deposition of Oxide Multilayers (TFDOM). Industrial Scale Processing (Autrans, France, Oct. 18–19) (Patent pending)
[32] Nobumasa H, Shimizu K, Kitano Y and Kawai T 1988 *Jpn. J. Appl. Phys. Pt. 2* 9 **27** L1669–L1671
[33] Vaslow D F, Dieckmann G H, Elli D D, Ellis A B, Holmes D S, Lefkow A, MacGregor M, Nordman J E, Petras M F and Yang Y 1988 *Appl. Phys. Lett.* 4 **53** 324–326
[34] Walia D K, Gupta A K, Reddy G S N, Tomar V S, Kataria N D, Ojha V N and Khare N 1989 *Solid State Commun.* 11 **71** 987–990
[35] Walia D K, Gupta A K, Reddy G S N, Kataria N D, Khare N, Ojha V N and Tomar V S 1990 *Mod. Phys. Lett.* B 6 **4** 393–398
[36] Tomizawa T, Matsunaga H, Fujishiro M and Kakegawa K 1990 *J. Solid State Chem.* 1 **89** 212–214
[37] De Rochemont L P, Maroni V A, Klugerman M, Andrews R J and Kelliher W C 1994 *Appl. Supercond.* **2** 281
[38] Hur N H, Oh W K, Park Y K, Moon Y S and Park J C 1990 *Solid State Commun.* 2 **76** 149–152
[39] Hsu H M, Yee I, DeLuca J, Hilbert C, Miracky R F and Smith L N 1989 *Appl. Phys. Lett.* 10 **54** 957–959
[40] DeLuca J A, Garbauskas M F, Bolon R B, McMullen J G, Balz W E and Karas P L 1991 *J. Mater. Res.* 7 **6** 1415–1424
[41] Verma K K, Sexena A K, Singh A K, Tiwari R S and Srivastava O N 1992 *Supercond. Sci. Technol.* **5** 163–167
[42] Tkaczyk J E, DeLuca J A, Karas P L, Bednarczyk P J, Garbauskas M F, Arendt R H, Lay K W and Moodera J S 1992 *Appl. Phys. Lett.* 5 **61** 610–612
[43] DeLuca J A, Karas P L, Tkaczyk J E, Bednarczyk P J, Garbauskas M F, Briant C L and Sorensen D B 1993 *Physica* C 1&2 **205** 21–31
[44] Doi T J, Yuasa T, Ozawa T and Higashiyama K 1994 *Jpn. J. Appl. Phys. Pt. 1* 10 **33** 5692–5696
[45] Goyal A, Specht E D and Kroeger D M 1995 *Appl. Phys. Lett.* 17 **67** 2563–2565
[46] Specht E D, Goyal A, Kroeger D M, Mogro-Campero A, Bednarczyk P J, Tkaczyk J E and DeLuca J A 1996 *Physica* C 1&2 **270** 91–96
[47] Paranthaman M, List F A, Goyal A, Specht E D, Vallet C E, Kroeger D M and Christen D K 1997 *J. Mater. Res.* 3 **12** 619–623
[48] Phok S, Galez P, Jorda J L, Supardi Z, de Barros D, Sin A, Weiss F and Odier P Physica C - Proceedings of the 5th European Conference of Applied Superconductivity (EUCAS) (Copenhagen, Denmark, Aug. 2001 26–30) (at press)
[49] Phok S, Sin A, Supardi Z, Galez P, Jorda J L and Weiss F Proceedings of the 2nd Thin Film Deposition of Oxide Multilayers (TFDOM) (Industrial Scale Processing, Autrans France, Oct. 2001 18–19) *J. Phys. IV* **11** 157–161
[50] Sin A, Supardi Z, Odier P, Weiss F and Nunez-Regueiro M 2000 *Supercond. Sci. Technol.* **13** 617–621
[51] Sin A, Supardi Z, Odier P, Weiss F, Ortega L and Nunez-Regueiro M 2001 *IEEE trans. Supercond.* 1 **11** 2877–2880
[52] Sin A, Supardi Z, Rousseau B, de Barros D, Weiss F and Odier P J. Cryst. Solids, Proceeding of Sol–Gel 2001 (Padoue, Sept. 2001) submitted
[53] Sin A, Odier P, Supardi Z, Weiss F, Ortega L, Sulpice A and Nunez-Regueiro M 2001 *Thin Solid Films* 1–2 **388** 251–255
[54] Moriwaki Y, Sugano T, Gasser C, Fukuoka A, Nakanishi K, Adachi S and Tanabe K 1996 *Appl. Phys. Lett.* 22 **69** 3423–3425
[55] Kumari S, Singh A K and Srivastava O N 1997 *Supercond. Sci. Technol.* **10** 227–234
[56] Duray S J, Buchholz D B, Song S N, Richeson D S, Ketterson J B, Marks T J and Chang R P H 1991 *Appl. Phys. Lett.* **59** 1503
[57] Oda S, Yamamoto S, Wang Z, Tobisaka H and Nagata K 1999 High Temperature Superconductors and Novel Inorganic Materials, ed V Tendeloo, E V Antipov and S N Putilin (Dordrecht: Kluwer Academic Press) p 75
[58] Hirschfelder J O, Curtiss C h F and Bird R B 1964 *Molecular Theory of Gases and Liquids* (New York: Wiley)
[59] 1996 *CVD of Nonmetals*, ed W S Rees (Weinheim: VCH)
[60] Kodas T and Hampden-Smith M 1994 *The Chemistry of Metal CVD* (Weinheim: VCH)
[61] Israelachvili J N 1997 *Intermolecular and Surface Forces* 2nd edn (London: Academic Press)
[62] Strem Inc. Catalog No. 17 Newburyport, MA 1997
[63] Eisentraut K J and Sievers R E 1965 *J. Am. Chem. Soc.* **87** 5254

[64] Schwarberg J E, Sievers R E and Moshier R W 1970 *Anal. Chem.* **42** 1828
[65] Dechter J J and Kowalewski J 1984 *J. Magn. Reson.* **59** 146
[66] Turnipseed S B, Barkley R M and Sievers R E 1991 *Inorg. Chem.* **30** 1164
[67] Drozdov A A, Troyanov S, Pisrevski A and Struchkov Y 1995 *J. Physique IV Coll. C5, Suppl. J. Physique II* C **5** 503
[68] Drozdov A, Troyanov S, Pisrevski A and Struchkov Y 1994 *Polyhedron* **13** 2459
[69] Gorbenko, O 1997 *Inorg. Chemistry* Lomonossov Univ. of Moscow, Private communication
[70] Drozdov A A, Troyanov S, Pisrevski A and Struchkov Y 1995 *J. Physique IV Coll. C5, Suppl. J. Physique II* C **5** 503
[71] Fitzer E, Oetzmann H, Schmaderer F and Wahl G 1991 *J. Physique IV Coll. C2, Suppl. J. Phys. II* **1** 713
[72] Drozdov A A, Troyanov S I, Kuzmina N P, Martynenko L I, Alikhanyan A S and Malkerova L E 1993 *J. Physique IV Coll. C3, Suppl. J. Physique II* C **2** 379
[73] Eisentraut K J and Sievers R E 1967 *J. Inorg. Nucl. Chem.* **29** 1931
[74] Kisyuk V V, Turgambaeva A E and Igumenov I K 1998 *Chem. Vap. Dep.* **4** 43
[75] Sievers R E, Ponder B W, Morris M L and Moshier R W 1963 *Inorg. Chem.* **2** 693
[76] Balog M, Schieber M, Michman M and Patar S 1977 *Thin Solid Films* **47** 109
[77] Kriyuk V V, Turgambaeva A E and Igumenov I K 1998 *Chem. Vap. Dep.* **4** 43
[78] Eisentraut K J and Sievers R E 1967 *J. Inorg. Nucl. Chem.* **29** 1931
[79] Richards B C, Cook S L, Pinch D L, Andrews G W, Lengeling G, Schulte B, Jurgensen H, Shen Y Q, Vase P, Frelthoft T, Spee C I M A, Linden J L, Hitchman M L, Shamlian S H and Brown A 1995 *Physica* C **252** 229
[80] Reid R C, Prausnitz J M and Poling B E 1987 *The Properties of Gases and Liquids* (New York: Wiley)
[81] Yoshida Y, Ito Y, Yamada Y, Nagai H, Takai Y, Hirabayashi I and Tanaka Sh 1997 *Electrochem. Soc. Proc.* **97–25** 998
[82] Arndt J, Klippe L, Stolle R and Wahl G 1995 *J. Physique IV Coll. C5, Suppl. J Physique II* **5** C119
[83] Schmaderer F, Huber R, Oetzmann H and Wahl G 1990 *Proc. XI Int. CVD Conf.*, ed K E Spear and G W Cullen (Pennington, NJ: Electrochem. Soc.) p 211
[84] Klippe, L Entwicklung eines MOCVD Verfahrens zur Abscheidung Supraleitender YBaCuO Schichten auf bewegten Substraten Thesis TU Braunschweig
[85] Pulver, M Chemische Gasphasenabscheidung von Zirconiumdioxid, Yttriumoxid und Aluminiumoxid aus β Diketonaten und Alkoholaten Thesis Technical University, Braunschweig
[86] Cremer E 1955 Compensation effect in heterogeneous catalysis *Adv. Catal.* **7** 67
[87] Wahl, G Über Chemisorption von reinen Metallen auf reinen Polycrystallinen Wolframoberflächen Thesis Univ. Marburg
[88] Schmaderer F, Huber R, Oetzmann H and Wahl G 1991 *J. de Phys. IV, Coll. C2, Suppl. J. Phys. II* **1** 539
[89] Hiskes R, DiCarolis S A, Young J L, Laderman S S, Jacowitz R D and Taber R C 1991 *Appl. Phys. Lett.* **59** 606
[90] Samoylenko S V, Gorbenko O Y, Graboy I E, Kaul A R and Tretyakov Y D 1996 *J. Mare. Chem.* **6** 623
[91] Weiss F, Fröhlich K, Haase R, Labeau M, Selbmann D, Senateur J P and Thomas O 1993 *J. Physique IV,* C3 **3** 321
[92] Felten F, Senateur J P, Weiss F, Madar R and Abrutis A 1995 *J. Physique IV Coll. C5, Suppl. J. Phys. II* **5** 1079
[93] Kaul R, Seleznev B V 1993 *Proc. IX Europ. Conf. CVD* (Tampere, Finland)
[94] Wahl G, Pulver M, Decker W and Klippe L 1998 *Surf. Coat. Technol.* **100–101** 132
[95] Pisch, A Etude Thermodynamique et Depot Chimique en Phase Supraconductrice: YBaCuO These de Doctorat Inst. National Polytechnique de Grenoble
[96] Pulver M, Nemetz W and Wahl G 2000 *Surf. Coat. Technol.* **125** 400
[97] Wahl G 1993 *Chemical Vapour Deposition*, eds L H Hitchman and K F Jensen (London: Academic Press) p 376
[98] Chin D T and Tsang C H 1978 *J. Electrochem. Soc.* **125** 1461
[99] Poelder S, Auer R, Linker G, Smithley R and Schneider R 1995 *Physica,* C **247** 309
[100] Miletto Granozio F and Scotti di Uccia U 1997 *J. Cryst. Growth* **174** 409
[101] Willis J O, Arendt P N, Foltyn S R, Jia Q X, Groves J R, DePaula R F, Dowden P C, Peterson E J, Lolesinger T G, Coulter J Y, Ma M, Maley M P and Peterson D E 2000 *Physica* C **335** 73–77
[102] Goyal A, Lee D F, List F A, Specht E D, Feenstra R, Paranthaman M, Cui X, Lu S W, Martin P M, Kroeger D M, Christen D K, Kang B W, Norton D P, Park C, Verbelyi D T, Thomson J R, Williams R K, Aytug T and Cantoni C 2001 *Physica* C **357–360** 903–913
[103] Krellmann, M Chemische Gasphasenabscheidung oxidischer Schichten Thesis TU Braunschweig
[104] Ignatiev A, Chou P C, Zhong Q, Zhang X and Chen Y M 1998 *Int. J. Mod. Phys.* B **12** 3162
[105] Zhang J, Gardiner R A, Kirlin P S, Boerstler R W and Steinbeck J 1992 *Appl. Phys. Lett.* **61** 2884
[106] Matsuno S, Uchikawa F, Utsunomiya S and Nakabayashi S 1992 *Appl. Phys. Lett.* **60** 2427
[107] Gorbenko O Y, Fuflyigin V N, Erohkin Y Y, Graboy I E, Kaul A R, Tretyakov Y D, Wahl G and Klippe L 1994 *J. Mater. Chem.* **4** 1585
[108] Weiss F, Schmatz U, Pisch A, Felten F, Pignard S, Senateur J P, Abrutis A, Fröhlich K, Selbmann D and Klippe L 1997 *J. Alloys Compounds* **251** 264
[109] Abrutis A, Sénateur J P, Weiss F, Kubilius V, Bigelytė V, Altytė Z, Vengalis B and Jukna A 1997 *Supercond. Sci.Technol.* **10** 959
[110] Krellmann M, Selbmann D, Schmatz U and Weiss F 1997 *J. Alloys Compounds* **251** 307

[111] Fröhlich K, Souc J, Rosova A, Machajdik D, Graboy I E, Svetchnikov V L, Figueras A and Weiss F 1997 *Supercond. Sci. Technol.* **10** 657
[112] Abrutis A, Kubilius V, Bigelyte V, Teiserskis A, Saltyte Z, Senateur J P and Weiss F 1997 *Mater. Lett.* **31** 201
[113] Abrutis A, Senateur J P, Weiss F, Bigelyte V, Teiserskis A, Kubilius V, Galindo V and Balevicius S 1998 *J. Cryst. Growth* **191** 79–83
[114] Selvamanickam V, Carota G, Funk M, Vo N, Haldar P, Balachandran U, Chudzik M, Arendt P N, Groves J R, DePaula R F, Newnam B E and Peterson D E 2001 *IEEE Trans. Appl. Supercond.* 1 **11** 3379
[115] Selvamanickam V, Galinski G B, Carota G, DeFrank J, Trautwein C, Haldar FP, Balachandran U, Chudzik M, Coulter J Y, Arendt P N, Groves J R, DePaula R F, Newnam B E and Peterson D E 2000 *Physica* C **333** 155–162
[116] Yuan F, Xie Y, Chen J, Yang G and Cheng B 1996 *Supercond. Sci. Technol.* **9** 991
[117] Stadel O, Schmidt OJ, Markov N V, Samoylenkov S V, Wahl G, Jimenez C, Weiss F, Selbmann D, Eickemyer J, Gorbenko O Yu, Kaul A R, and Abrutis A, 2001 *J. Phys. IV* **11** 233–237
[118] Stadel O 2001 Private communication, TU Braunschweig
[119] Chou P C, Zhong Q, Li Q L, Abazajian Li, Ignatiev A, Whang C Y, Deal E E and Chen J E 1995 *Physica* C **2S4** 93
[120] Ignatiev A, Zhong Q, Chou P C, Zhang X, Riu J R and Chu W K 1997 *Appl. Phys. Lett.* **70** 1474
[121] Ito Y, Yoshida Y, Iwata M, Takai Y and Hirabayashi I 1997 *Physica* C **288** 178
[122] Wahl G, Stadel O, Klippe L, Kaul A R, Gorbenko O Y and Samoylenko S V 1999 High Temperature Superconductors and Novel Inorganic Materials, eds G V Tendeloo, E V Antipov and S N Putilin (Dordrecht: Kluwer) p 79
[123] Zhang X F, Sung Y S, Miller D J, Hinds B J, McNeely R J, Studebaker D L and Marks T J 1997 *Physica* C **275** 146
[124] Fuflyigin V N, Kaul A R, Pozigun S A, Tretyakov Yu D, Wahl G, Nürnberg A, Klippe L, and Stolle R 1993 Applied Superconductivity, ed H.C. Freyhardt, DGM Oberursel 587
[125] Samoilenkov S V, Gorbenko O Yu, Kiryakov N P, Emelianov D A, Liashenko A V, Lee S R, Kaul A R and Andranov D G 2001 MRS Proc. Vol. 659, High-Temperature Superconductors—Crystal Chemistry, Processing and Properties, eds U. Balachandran, H. C. Freyhardt, T. Izumi, D. C. Larbalestier, Paper II, 11.10
[126] Araki T, Kurosaki H, Yamada Y, Hirabayashi I, Shibata J and Hirayama T 2001 *Supercond. Sci. Technol.* **14** 783–786

B4.3
Thick film deposition techniques

N McN Alford, S J Penn and T W Button

B4.3.1 Introduction

The use of thick films in the European Electronics industry comprises a business of some $1.42 billion (1994). The hybrid market is around $2.9 billion (1994) and for comparison the market for thin films in Europe was $265 million in 1994. The development of superconducting thick films is an essential component in the growth of the superconducting electronics industry. This chapter of the handbook will review the materials, the deposition and manufacturing techniques associated with HTS thick films [1].

B4.3.1.1 Definitions

It is important at the outset of this section to define precisely what is meant by the term thick film in order to avoid confusion.

Thick film

The term 'thick film' relates not so much to the thickness of films but to the mode of deposition. In the electronics industry a thick film conductor of silver, copper or gold for example, is usually prepared by taking a powder of the metal, mixing the powder with a vehicle to make an ink, which is then deposited onto a suitable substrate. A thermal treatment is then needed in order to burn off the organic components, densify the film and optimize the properties. Alternative methods of deposition, including solution and plasma routes, are also classed under the general heading of thick film deposition techniques, and these are also discussed below.

Bulk materials

Bulk materials are classified as materials made by powder routes and may be freestanding objects such as discs, tubes or rods or may be produced by encapsulating the powder in a substrate such as silver and drawing into a tape or wire. The commonest method for producing HTS conductors is by the powder in tube (PIT) method in which a superconductor powder is placed inside a silver tube and swaged and/or rolled into the desired conductor shape. However, another technique which is also used is the deposition of a superconductor thick film paste or slurry onto a metallic (silver) substrate. In a variant, the superconductor may be deposited onto a barrier layer such as zirconia which itself is on a metallic substrate, e.g., a high temperature nickel alloy. This is classed as a thick film and conductors made by these methods are referred to as having been prepared by 'open' methods as opposed to 'closed' methods

B2.1
B2
B4.1
B4.2.1
B4.2.2

B3.1 such as PIT. Such conductors have been shown to possess excellent properties which will be discussed below.

Thin films

Thin film deposition techniques such as sputtering, laser ablation, chemical vapour deposition, evaporation etc differ from thick film methods in three very important respects.

- B4.1 first, thin film deposition usually requires the use of single crystal substrates
- B4.2.1 second, thin films are usually deposited with the intention of achieving a degree of epitaxy with respect to the crystallographic orientation of the substrate
- B4.2.2 third, thin films usually require the use of vacuum technology.

The most important distinction, however, is the deposition method.

In the literature the term thick films is unfortunately used somewhat loosely to refer to thin film deposition methods which have built up many layers of a compound so that effectively a 'thick' thin film results. In this review this is classed as a thin film. A good example of this is the use of thin film deposition techniques to produce films up to several micrometers thick for current carrying conductors. B4.1 For example, Wu *et al* [2] have produced thin films on flexible nickel substrates with YSZ buffer layers by ion beam assisted deposition (pulsed laser deposition) and have achieved very high critical currents using this process. This is discussed at the end of this section.

Vehicle and ink

The vehicle is a mixture of polymers and solvents which are thoroughly mixed with the chosen powder, usually on a three roll mill, until a homogeneous mixture is formed, known as an ink. The ink is then deposited onto a suitable substrate, for example polycrystalline alumina or polymer circuit board, by B2.2.1 screen printing and then dried to form a green (unsintered) coating. In the case of ceramic substrates, the coating is then fired at high temperature in order to sinter the particles in the ink so that a homogeneous electrical circuit is obtained.

B4.3.2 Thick film deposition—methods

C1, C2, C3 The most commonly reported materials are thick films of $YBa_2Cu_3O_x$ (YBCO,123) [3, 4], TlBaCaCuO (TBCCO 2223, 2212) [5, 6] and Bi(Pb)SrCaCuO (BSCCO 2223, 2212) [7]. Although relatively little work has been carried out on thick films of the mercury system ($HgBa_2Ca_2Cu_3O_{8+\delta}$) these are also included. Adjustments to the stoichiometry of these three main systems have also been reported. In general conductors are made with films of BSCCO or TBCCO taking advantage of the anisotropy in crystal E3 morphology and current carrying capacity which is greatest in the a–b plane. Radio frequency and microwave devices are more commonly made from YBCO thick films as these have been shown to C2, C3, possess a consistently lower surface resistance (R_s) [3] than films made with BSCCO or TBCCO.

D2.7 There is a wide range of techniques that fall within our broad definition for the deposition of polycrystalline thick films and which have all been used with varying degrees of success for the fabrication of HTS thick films. Although their use in this capacity was largely replaced by thin film methods producing films with superior properties, there has recently been renewed interest in many of the thick film techniques as candidates for scaleable, non-vacuum processes for the fabrication B4.1 of oriented buffer and YBCO layers on textured metallic substrates by the RABiTS technique.

The following sections outline the various thick film deposition techniques that have been used to deposit HTS materials and these are discussed mainly in terms of YBCO, as this is the composition on which most of the work has been carried out. Specific sections are devoted to the other HTS materials and to the deposition methods employed for BSCCO conductors, although these are also discussed in more detail in this paper. C1 C2

B4.3.2.1 Preparation of a powder based thick film

As an example we shall examine the preparation of YBCO thick films. In common with many thick film preparations, the starting point is the superconductor powder which is conveniently made by a mixed oxide route. The raw materials usually used, yttrium oxide, barium carbonate and copper oxide, are weighed out into the appropriate stoichiometric proportions. These are mixed in a ball mill or a vibro-energy mill in alcohol for 16 h and dried in a rotary evaporator. The mixed powder is then calcined in an air furnace. The process of calcination takes place below the melting temperature of the constituents and, because only a few point contacts exist in a powder bed, the process is often repeated until a homogeneous compound is attained. The resultant material is a friable 'crumble' that must be ground to a fine powder in alcohol in a ball mill or vibro-energy mill using ceramic (e.g. zirconia) milling media. This is again dried in a rotary evaporator. The resulting powder is then made into an ink by three roll milling using polymers and solvents. The ink is a viscous liquid which can be screen printed or 'doctor-bladed' onto a suitable substrate. If a spin or dip coating method is employed as a deposition technique then the viscosity must be adjusted accordingly. The temperatures of calcination are typically 900°C for YBCO, and closer to 800°C for BSCCO and TBCCO. TBCCO is unusual in that the thallium is often incorporated in a second stage process, e.g. as thallium oxide vapour, in order to minimize toxicity problems and difficulties associated with loss of thallium due to vaporization. B2.1 C2, C3

B4.3.2.2 Electrophoretic deposition

The phenomenon of electrophoresis (the motion of charged particles in a suspension under the influence of an electric field) can be utilized as a deposition technique. Electrophoretic deposition (EPD) is an example of colloidal processing whereby green films are shaped directly from a stable colloid suspension by a dc electric field which causes the charged particles to move towards, and deposit on, the oppositely charged electrode. In general, the technique requires the use of submicron powders in order to obtain a colloidally stable suspension (usually non-aqueous for HTS materials), is limited to the use of electrically conducting substrates and requires a subsequent heat treatment step in order to densify the film. Woolf *et al* [8] have used the concept for the coating of silver wires and Bhattacharya *et al* [9] have used electrophoretic deposition to deposit films of YBCO on silver and have observed extensive grain growth and recrystallization. They also observed a thickness dependence for the critical current density (J_c) with a J_c of 450 A cm^{-2} (77 K, $H = 0$) for films 65 μm thick and $J_c > 4000$ A cm^{-2} for films 3 μm thick. C1

B4.3.2.3 Sol–gel preparation

Strictly speaking sol–gel processing must involve the preparation of a sol (often described as a suspension of small but discrete particles remaining dispersed in a liquid phase) which undergoes a transition to a gel characterized by an infinite three-dimensional network structure spreading throughout the liquid medium. In practice this definition is not strictly adhered to and sol–gel preparation of HTS materials usually encompasses routes based on hydrolysis–condensation of metal alkoxides, the concentration of aqueous solutions involving metal-chelates, and the organic polymeric gel route. Following formation of the gel a subsequent thermal treatment or pyrolysis step is required in B2.1

order to develop the crystallinity and properties of the superconducting phase. There is a wide range of different approaches that have been reported and the reader is referred to a comprehensive review article on the sol–gel preparation of HTS materials for more detailed information and analysis of the various techniques [10]. In principle, sol–gel routes offer many potential advantages over conventional solid-state reaction routes including lower processing temperatures, higher homogeneity and purity, and the ability to fabrication films or fibres as well as submicron powders. The technique lends itself to easy deposition of films by spin-on and dip-coating methods, although many repetitive dip / pyrolysis cycles can be required to build up films with thicknesses $>1\,\mu m$.

Masuda *et al* [11] found that, using sols prepared from alkoxides, J_c values of up to $1.18\times 10^4\,A\,cm^{-2}$ were obtained at 77 K in zero magnetic field. They found also that melt processing in the peritectic reaction regime was beneficial. More recent work has concentrated on the deposition of films on single crystal and textured substrates in an effort to develop a scalable, non-vacuum processing technology, and the results are quite encouraging. For example, Sathyamurthy and Salama [12] have used metal-trifluoroacetate precursors to deposit YBCO films on (100) $SrTiO_3$ and $LaAlO_3$ substrates. A two stage heat treatment was used to yield $0.5\,\mu m$ thick films with a high degree of alignment and J_c values $>5\times 10^5\,A\,cm^{-2}$ (77 K). Yamagiwa and Hirabayashi [13] have demonstrated biaxial alignment and J_c values of approximately $10^5\,A\,cm^{-2}$ for $YbBa_2Cu_3O_x$ films produced using metal naphthenates on (100) $SrTiO_3$ substrates. Other workers are also investigating the technique for the deposition of the buffer layers necessary when depositing films on textured Ni substrates [14]. However, one of the problems which has hindered progress of sol–gel synthesis of HTS oxides to date is the rather poor understanding of the chemistry relevant to the necessary precursors in these complicated multi-cation systems.

B4.2.2

B4.1

B4.3.2.4 Plasma spraying

Plasma spraying is a well known technique for the deposition of thick layers of ceramics and metals and, as high rates of deposition can be achieved, large areas can be coated. Essentially, a plasma torch (20–35 kW) is used to melt a powder of the appropriate stoichiometry directly onto the chosen substrate, and dense films ($<5\%$ porosity) can be produced with thicknesses in the range 100–200 μm. A post-deposition heat treatment and oxygenation step is usually required in order to obtain the superconducting phase and optimize the film properties. The technical implementation of the plasma spray process and the subsequent thermal treatment for the production of the HTS layers is relatively straightforward. However, the deposition process, the complex phase relationships and incongruent melting behaviour of the superconductor compositions, together with interactions with the substrate material make the whole process very complicated. Two variations of the plasma spray technique have been used for the deposition of YBCO coatings—atmospheric plasma spraying (APS) where the deposition takes place in air, and low pressure plasma spraying (LPPS) where the deposition is carried out at reduced pressure in an appropriate atmosphere. Ezura *et al* [15] used LPPS to deposit YBCO thick films onto silver or nickel buffered copper substrates and measured the surface resistance (R_s) at 3 GHz. R_s was found to be less than that of copper at 77 K but inferior to results obtained on zirconia substrates as a consequence of HTS-substrate reaction. Hemmes *et al* [16] used both APS and LPPS to deposit YBCO and buffer layer films on a range of metallic substrates—pure nickel, nimonic alloys and non-magnetic stainless steels for use in magnetic shielding applications. The best results ($J_c = 425\,A\,cm^{-2}$ at 60 K) were obtained for YBCO deposited by APS directly on a NiCrAlY bond layer on a 316 stainless steel substrate tube, but again the results are inferior to those on zirconia, and substrate reaction and film cracking due to the thermal expansion mismatch remain severe problems. APS has also been used to deposit YSZ barrier layers onto 96% alumina substrates [17]. The films were subsequently heat treated at 1450°C for 1 h in air, followed by deposition of YBCO layers by screen printing and melt processing. The resulting films had R_s values

B4.1

D2.7

measured at 15 GHz which were inferior to those of YBCO films on YSZ substrates, but better than those of comparable films on alumina substrates with screen-printed YSZ barrier layers. B4.1

B4.3.2.5 Spray pyrolysis

The preparation of films by spray pyrolysis can take a number of forms; four modes of deposition and decomposition may be distinguished, all of which start with the production of aerosol droplets of a solution (usually nitrates) of the appropriate constituents [18]. In the first mode (1) the aerosol droplets splash on to the surface of the heated substrate, the solvent vaporizes and then decomposition of the dry precipitate occurs directly on the substrate. Mode 2 is similar except that the aerosol droplets dry on their way to the substrate so that only dry precipitates come into contact with the substrate surface. These two modes are designated as 'true pyrolysis' since thermal decomposition takes place on the heated substrate only. A third mode (3) can be considered as 'true CVD' (chemical vapour deposition) in which thermal decomposition of the dry precipitates starts on their way to the substrate. In this case mainly vapour arrives at the substrate. In the fourth mode (4) the full thermal decomposition takes place prior to arrival at the substrate surface and deposition of finely divided product occurs. The actual working mode in any deposition system thus depends not only on the substrate temperature, but also on other apparatus variables such as the temperature of the surroundings, the rate of mass transport and the nozzle-substrate distance. It is therefore important to know the details of the aerosol deposition conditions as well as the details of the chemical thermal decomposition processes. In common with the sol–gel method, it is usually necessary to carry out a final high temperature annealing step in order to fully develop the crystalline structure and superconducting properties, and a number of intermediate heat treatments are required in order to deposit thicker films of good quality. The spray pyrolysis technique has been used successfully to deposit films of all the major HTS materials, YBCO, BSCCO and TBCCO [18] (the latter two are discussed in more detail in later sections), and it is again receiving renewed study as a candidate for a scalable processing technology for texture films and buffer layers. B4.2.2 C1, C2, C3

Nickel alloys such as Inconel 600 and Inconel X have been examined by Schulz *et al* [19]. They sprayed pyrolysed YBCO onto the substrates and found that pre-reacted YBCO was preferable to aqueous metal nitrate solutions. They experienced severe difficulties due to substrate interactions. Vanolo *et al* [20] deposited YBCO using nickel based buffer layers on stainless steel containing chromium and found that the shielding effectiveness was comparable to shields made by depositing YBCO on silver. The main problem was oxygenation of the YBCO. Oskina *et al* [21] have deposited BSCCO thick films on nickel substrates with silver or gold interlayers. They did not observe any appreciable reaction between the interlayer and the BSCCO. BSCCO 2223 attained a transition temperature T_c ($R = 0$) of 92, 101 and 99 K on gold, silver and nickel with a silver buffer layer, respectively. The use of thin film deposition methods to produce conductors on nickel alloy substrates is discussed in more detail in section B4.3.3. B2

B4.3.2.6 Liquid phase epitaxy

Liquid phase epitaxial (LPE) growth of YBCO (see also chapter B4.2.1) utilizes epitaxial deposition onto a single crystal substrate of YBCO from a liquid phase comprising a molten oxide solvent or flux and YBCO solute. As such, the technique does not strictly fall within our definition of thick films but is included here as films with large thicknesses can be produced rapidly in non-vacuum conditions. The flux is contained in a crucible situated in a temperature gradient, the bottom being a few degrees higher than that of the surface. Conventional LPE uses a self-flux of BaO–CuO and, compared to other deposition processes, has a very high growth rate—typically 1–10 μm min^{-1}. YBCO grows as the primary phase under ambient atmospheres without any contamination from the solution, and J_c values $> 10^6$ A cm^{-2} have been obtained for 10 μm thick films [22] grown by this technique. However, the use of a BaO–CuO B4.2.1 C1 solvent requires a relatively high processing temperature of 1000°C which can result in reaction of the substrate with the solution, and limits the choice of substrate material (e.g. $NdGaO_3$ or YBCO). The use of a solvent modified with BaF_2 has been reported by Yamada *et al* [23] as decreasing the primary crystallization temperature to 920°C enabling a wider range of substrates to be used, including MgO, YSZ, $SrTiO_3$ and Ag, and growth on single crystal fibres has also been demonstrated. For growth on Ag substrates, Ag is also usually added to the flux. In order to inhibit reaction between the substrate and flux and to promote the desired crystal growth it is also usual to first deposit a thin seed or buffer layer of YBCO onto the substrate by some thin film technique, e.g. pulsed laser deposition, prior to LPE. Choice of crucible material is also important in order to minimize corrosion by the flux resulting in contamination, and YSZ and yttria (or other rare earth oxides for RE123 compositions) have proved successful. The high growth rates and resulting thick films lead to high strains being developed during cooling due to the thermal expansion mismatch between the film and substrate. This can lead to severe cracking problems in films grown by LPE [24].

B4.1

B4.3.3 Thick film deposition—HTS materials

B4.3.3.1 YBCO

C1 A1.3

With reference to published phase diagrams [25–27], it can be seen that for YBCO in air, a peritectic reaction occurs at around 1000°C. In such a reaction two phases, one of them a liquid, react to produce a new phase on cooling. 123 is not formed in equilibrium from the melt without the prior formation of 211. This is useful in two major respects. First, the liquid encourages grain growth as was explained above and second, the presence of small quantities of 211 encourages the formation of dislocations (caused by differential strain with respect to a 123 matrix) which aid pinning of the magnetic flux. The liquid phase has a dramatic effect on the grain growth as shown in figures B4.3.1(*a*) (sintered above the peritectic temperature) and (*b*) (sintered below the peritectic temperature) and it also enhances the performance (J_c, R_s) of the films [1, 3, 4]. In HTS materials of all three systems discussed in this handbook, the presence of a liquid phase is generally beneficial but great care must be taken in the time temperature profile of the sintering step. In BSCCO 2223 for example, although a very small amount of liquid is beneficial, if the sintering temperature is too high the 2223 phase will be lost and BSCCO 2212 is the resultant phase. In fact, BSCCO 2212 is the preferred phase for some manufacturers of HTS wire because there is a wider processing window and because the properties [$J_c(H)$] of BSCCO 2212 at liquid helium temperatures (and up to about 30 K) are preferred for current carrying applications.

C2

Pt doping has been used successfully in bulk materials to increase the critical current and has also been shown to improve the properties of YBCO thick films made by conventional powder routes. Langhorn *et al* [28, 29] showed that doping YBCO thick films with 0.1 wt% platinum powder caused an increase in the current carrying capacity of the material from $J_c = 1800$ to $5000\ \mathrm{A\,cm^{-2}}$. When 0.4 wt% $Ba_4Cu_{1-x}Pt_{2-x}O_{9-\delta}$ (0412) powder was added to the thick film ink the properties improved reaching $J_c > 7 \times 10^3\ \mathrm{A\,cm^{-2}}$. The presence of the 0412 powder, as with the Pt powder addition, caused a refinement (grain growth inhibition) of the 211 precipitates. This is believed to be beneficial due to the increased surface curvature of the 211 and its increased specific surface area leading to an enhanced dislocation density. TEM micrographs indicated increased dislocation densities at the 123/211 interface and it is believed that the dislocations themselves can act as flux pinning sites.

A1.3, B2.3.3 D1.2 A4.3

B4.3.3.2 BSCCO

C2

Holesinger *et al* [30] have used silver substrates to process BSCCO 2212 using an isothermal processing technique. In this process films are heated in an argon atmosphere at $10°\mathrm{C\,min^{-1}}$ to between 760 and

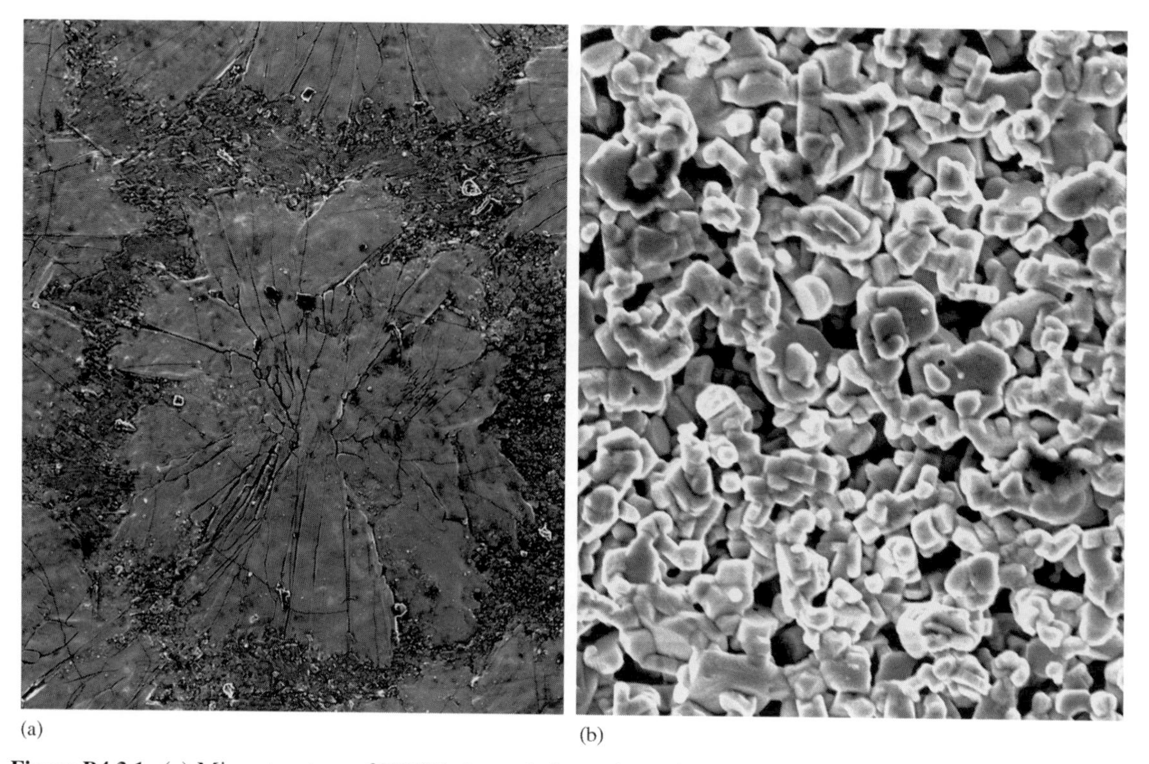

(a) (b)

Figure B4.3.1. (*a*) Microstructure of YBCO sintered above the peritectic temperature (width of view is 1060 μm). (*b*) Microstructure of YBCO sintered below the peritectic temperature (width of view is 17 μm).

860°C and held for 1 h. The atmosphere is then changed to 10% O_2–90% Ar for 15 h. The sample is then cooled at 5°C min^{-1} to room temperature. At 4 K the transport properties indicate that the samples processed at the higher temperatures, i.e. 860°C, are less susceptible to an applied field of 1 T (parallel to *a–b* planes) dropping by only 20% of their self field value, i.e. from $J_c = 2.87 \times 10^4$ to 2.26×10^4 A cm^{-2}. The highest J_c value was seen in samples processed at 780°C, 1.2×10^5 A cm^{-2} but this dropped by around 40% to 7.38×10^4 A cm^{-2} in 1 T. The microstructure of these films consisted mainly of a series of grain colonies separated by low angle grain boundaries. At 780°C there was very little evidence of 2201 needles.

C2

In a similar study Dimesso *et al* [31] examined the effects of annealing BSCCO 2212 on silver in an atmosphere of flowing nitrogen at temperatures between 300 and 700°C for up to 6 h. The nitrogen atmosphere significantly reduces the melting temperature of the BSCCO 2212. At an annealing temperature of 500°C the T_c was found to be 85 K and the J_c at 77 K was found to be 1600 A cm^{-2}. Unfortunately no transport data were available at 4 K.

B3.2.1

BSCCO 2223 thick films pose problems in that the achievement of the three layer 110 K 2223 phase is not straightforward and, in particular, there can be de-densification on annealing. Extra Ca and Cu have been found to be beneficial [32]. Pb substitution for Bi lowers the melting temperature and favours the diffusion of species for the growth of the high T_c phase. Finally, it is now well known that the addition of Ag influences the oxidation and annealing conditions, aids sintering, enhances the J_c and encourages the formation of the high T_c phase. The conditions which favour the formation of the high T_c phase have been analysed by several authors. The starting composition is ideally copper rich i.e. a 2224

B3.2.2

composition, then it appears that a short time i.e. less than 1 h at a temperature between 870 and 900°C followed by an extensive heat treatment of up to 100 h at around 860°C is required for the formation of the 2223 phase.

B4.3.3.3 TBCCO

C3 B2.1 B4.1

TBCCO thick films which show significant promise have been prepared by spray pyrolysis of solution, ink spraying, electro-deposition and sol–gel. In this review we include such techniques for producing thick films since they satisfy the criteria of not requiring high vacuum systems, the precursors are powders of the appropriate oxides and the substrates do not need to be single crystals, although excellent results are reported on small samples grown on single crystals.

D2.4

Schultz *et al* [33] have prepared TBCCO1223 films by spray pyrolysis. These were prepared by taking a precursor powder ($Pb_{0.46}Ba_{0.4}Sr_{1.52}Ca_{1.86}Cu_3O_x$) of 4–6 μm in size, mixed with a vehicle binder and diluted with alcohol. This mixture was sprayed onto a heated (80–100°C) substrate with an airbrush. The sample was then oxygen annealed at 700–800°C for 60–75 min to remove the organics. The sample was sintered in a two zone furnace at 770°C for the separate thallium oxide source and 931°C for the precursor for 2 h. The authors rank a series of substrates in order, using morphology as a criterion, $LaAlO_3 > NdGaO_3 > SrTiO_3 > MgO > YSZ$ is approximately the ranking in lattice mismatch. Using magnetization measurements the J_c was calculated to be 9×10^6 A cm^{-2} at 0 T, 5 K and 2×10^6 A cm^{-2} at 4.75 T, 5 K.

D1.2 A1.3

The films were found to have a high degree of *c*-axis texture but no macroscopic in-plane texture was found. The TEM of the grain boundaries revealed no second phase. There appears to be clustering of grain colonies which are locally aligned and where no large angle grain boundaries exist. The authors suggest that there is long range current transfer through a percolative network of small angle colonies. This biaxial texture reduces the effect of weak links and may provide an effective means of using thick film technology to prepare practical lengths of conductor in TBCCO. The properties of thallium cuprate films for conductor applications are discussed in more detail in section B4.3.3.4.

B4.3.3.4 Hg-1223

C4 D2.4

Tsabba and Reich [34] have prepared thick films of the Hg1223 compound. They used aqueous (and glycerine) solutions into which were dissolved the nitrates of the Cu, Ca and Ba. After some h a sol was formed to which was added Hg nitrate in an aqueous glycerine solution. The gel was spread onto a YSZ substrate and calcined at 900°C for 10 min. These films were then sealed in helium purged quartz ampoules with source pellets (HgO, BaO, CaO, CuO of composition 1223). The ampoule was baked at around 870°C for 2–3 h, annealed at 300°C for 3 h and cooled in oxygen. The films obtained were 5–10 μm thick with no epitaxy and the T_c was 110–117 K before oxygenation and 135 K after oxygenation. The resistive transition is broad but is centred around 135 K. Using magnetization measurements J_c was calculated to be 10^5 A cm^{-2} at 10 K and 500 A cm^{-2} at 40 K at a grain size of 5 μm.

B4.3.3.5 Conductors using thick films

C2 B3.2.1

Dip coated conductors using silver tapes on which the superconductor (BSCCO usually) is deposited have impressive properties. Tiefel and Jin [35], using $Bi_{1.6}Pb_{0.3}Sb_{0.1}Sr_2Ca_2Cu_3O_x$, show that transport J_c of 2.3×10^5 A cm^{-2} at 4.2 K and 8 T (H $\perp$ *ab*) is achievable over a 2 cm length. Yang *et al* [36] show that, in BSCCO 2212, electrophoretic deposition can be used to deposit material up to a few hundred microns, although optimal results were seen at 100 μm. A J_c of 180 000 A cm^{-2} at 4.2 K in zero field was measured over 10 cm and this reduced to 120 000 A cm^{-2} in a 2 T field applied parallel to the face of the

Table B4.3.1. J_c and I_c of BSCCO tapes using thick films (SSDIP and DIP)

		J_c A cm^{-2} × 10^5 in ceramic core	J_c A cm^{-2} × 10^3 full cross-section	I_c (A)
PIT	4.2 K, 0 T	2.0	67.0	101
PIT	4.2 K, 10 T	0.1	4.0	6.6
SSDIP	4.2 K, 0 T	9.1	5.4	328
SSDIP	4.2 K, 10 T	1.0	6.0	36
DIP	4.2 K, 0 T	2.2	90.0	439
DIP	4.2 K, 10 T	0.7	29.0	142

tape. Oxford Instruments (Cowey *et al* [37]) have evaluated the use of BSCCO thick films in dip coated silver tape (DIP) and in dip coated tape which is subsequently sheathed in silver (SSDIP). A 100 μm slurry is deposited onto a silver tape and wound and reacted. The sintered BSCCO thickness was around 30 μm. In the case of DIP tape the silver:superconductor ratio is 1:1 and in the case of SSDIP the ratio is 3:1. Short sample (4 cm) results are shown in table B4.3.1.

Cowey *et al* point out that there are severe difficulties in overcoming the length problem, i.e. the observed reduction in the I_c and J_c as longer lengths of conductor are produced. Recent results from Oxford Instruments on dip coated tapes indicate that $J_c = 4 \times 10^5$ are possible in lengths up to 20 cm and $J_c = 1 \times 10^5$ are achieved in lengths between 10–100 m.

Short samples of doctor bladed BSCCO 2212 have been tested by Kase *et al* [38]. The BSCCO powder was mixed with an organic and a tape was placed onto a silver sheet. Various heating and cooling schedules were investigated but the composite heated to 890°C and cooled to 870°C at 5°C h^{-1}, held for 1–24 h and then slowly cooled to room temperature displayed the best results. These were $J_c = 3.1 \times 10^5$ A cm^{-2} in 2 T and 4.2 K, 1.7×10^5 A cm^{-2} 10 T, 4.2 K and 1.4×10^5 A cm^{-2} at 4.2 K, 25 T. The field was applied perpendicular to the current and parallel to the tape surface. B3.2.1

Thallium based thick films have recently shown significant promise. For detailed information on such conductors the reader is referred to a comprehensive review by Jergel *et al* [39] in which methods such as electro-deposition, sol–gel, aerosol deposition and screen printing are described. Table B4.3.2 is adapted from Jergel *et al* [39] and shows Tl tapes in which thick films have been deposited by so called 'open deposition' methods i.e. not PIT. C3

The tapes described in [40] were prepared by electro-deposition. The nitrate precursors without Tl were dissolved in dimethyl sulfoxide and deposited on 125 μm silver coil. A two zone furnace was used with oxygen as a carrier gas for thallium oxide vapour. The zero field value for J_c is 4.42×10^5 A cm^{-2} but high field performance is impressive at $J_c = 8.2 \times 10^3$ A cm^{-2} at 5.5 T.

B4.3.4 Materials for substrates

The substrate performs an important function in supporting the film. The key properties of a substrate for thick films are the following. B4.1

- First, it should not react deleteriously with the superconductor film.
- Second, it should maintain shape at the processing temperature of the superconductor. In patterned films where the substrate is exposed to a radio frequency or microwave field there are other

Table B4.3.2. Results for thallium tapes prepared by open deposition methods

Compound TBCCO	Method	Substrate	Film thickness μm	J_c transport A cm^{-2} × 10^4 77 K, 0 T	J_c transport in field A cm^{-2} × 10^4 77 K	Tesla
1223	solution spray	YSZ	~ 3	2–11	0.7	1
T(B,S)CCO 1223	solution spray	SrTiO3/Ag	< 1	28	5	1
1223	solution spray	YSZ	3	10	5	0.1
1223	solution spray	Ag tape	1	9	0.7	1
(T,P)(B,S)CCO1223	Ink spray	LaAlO3	5–20	2.9	> 1	1
2223	electro-deposition	Ag	1.3	3.2	1	1
1223	electro-deposition	Ag	1	4.42	0.82	5.5
1223	electro-deposition	Ag	1	7	1	5
1223	sol–gel	Ag	10–30	2.5	0.1	1

Adapted from [39 and 1].

E3 important considerations, the most important of which is that the dielectric loss should be low which generally means that the dielectric constant is moderate at 30 or below.

- Finally the issue of cost is important.

B4.1 Table B4.3.3 shows some ceramic materials which have been used for thick film substrates. Some of the commonly used thin film substrate materials such as $LaAlO_3$ and buffered Al_2O_3 are generally unsuitable as substrates for thick film due to the higher processing temperatures required for the thick

Table B4.3.3. Properties of candidate materials for HTS thick film substrates (single crystal values added for information)

Substrate material	Dielectric constant ε'	Dielectric Loss @ 300 K tan δ	Frequency f	Adverse reaction with thick film HTS > 1000°C
Y-stabilized zirconia	25	10^{-3}	7.5 GHz	no
MgO	9	3×10^{-5}	10	yes
Al_2O_3	9.5	1.7×10^{-5}	10 GHz	yes
YBa_2NbO_6	29	1.4×10^{-4}	4.9 GHz	no
$NdBa_2NbO_6$	13.3	3×10^{-4}	8.6 GHz	no
$BaZrO_3$ at 90% of full density	12	2.5×10^{-4}	8 GHz	no
$LaAlO_3$, at 96% of full density	21	4×10^{-4}	8 GHz	yes
Single Crystals				
Al_2O_3	9.5	10^{-5}	8.5 GHz	
MgO	9	10^{-5}	7.35 GHz	
$LaAlO_3$	22.6	2.3×10^{-5}	7.34 GHz	
TiO_2	85	1.4×10^{-4}	8.6 GHz	

film process (in the range 900–1100°C for YBCO, 830–900°C for BSCCO and TBCCO) which causes rapid diffusion of species from the substrate to the superconductor with an adverse effect on superconducting properties. The presence of a liquid phase during sintering and a certain degree of reaction with the substrate is beneficial as it promotes good adhesion with the substrate. However, there is a balance between the degree of reaction needed to provide adhesion and that which causes a deleterious reduction in the superconducting properties. Other substrate materials such as $Sr(Al_{0.5}Ta_{0.5})O_3$ and $Sr(Al_{0.5}Nb_{0.5})$ are under consideration as thin film substrates and $LaSrGaO_4$ has also been tested as a possible candidate for thin films. They have not so far been assessed for thick film applications. Recently, Koshy *et al* [41] have made significant advances in the development of materials for substrates, some of which are presented in table B4.3.2. Measurements of tan δ and dielectric constant were measured at South Bank University in an oxygen free copper cylindrical cavity with a vertically adjustable top plate. The dielectric pucks were placed on a low loss, low permittivity quartz spacer. The transmission measurements were performed using a HP8719 vector network analyser with 1 Hz resolution.

C1, C2, C3

D2.7
E3.1

At present, the preferred substrate for microwave applications for availability, cost and properties is yttria stabilized zirconia. Materials for thick film HTS are preferably polycrystalline ceramics which can be made as large area planar or curved substrates at low cost. In general, the dielectric constant is not greatly affected by reducing the temperature of measurement but the dielectric loss is often reduced considerably at low temperatures. This is the subject of further investigation in our laboratories.

For conductors in particular there has been a good deal of interest in using metals as a substrate for thick films. The superconducting materials used for such conductors are either BSCCO or TBCCO which may be processed at a lower temperature than YBCO and hence the constraints on the choice of metal are not so severe. Silver (and some silver alloys) does not display a deleterious reaction with any of the three systems and has been used extensively.

C2, C3
C1

B4.3.5 Sintering of HTS thick films

Most of the deposition methods described above involve a final thermal treatment step to densify the film and optimize the properties. Each of the main material systems sinters in a different manner but the principles are the same. Once a film has been deposited onto the substrate the particles of superconductor powder must be encouraged to coalesce to form a uniform thick film. The driving force for sintering is the reduction of surface energy by the reduction of surface area and this is achieved by the reaction of the particles at high temperature. The reduction of surface area is achieved by material transport. If we imagine two spherical superconducting particles of radii r, then the contact between the particles will be characterized by a contact diameter x. In the initial stages of sintering the surface curvature at the contact between the two particles is directly related to the vapour pressure, which is far higher in areas of high surface curvature. Initial stage sintering is by evaporation and condensation.

B2.2.1

However, in the sintering of HTS materials two main processes can be distinguished. The first is solid-state diffusion which occurs by vacancy migration, described in equation (B4.3.1) [42].

$$\frac{x}{r} = \left[\frac{40\gamma a^3 D^* t}{kT}\right]^{\frac{1}{5}} r^{\frac{3}{5}} \qquad \text{(B4.3.1)}$$

where γ is the surface energy, a^3 is the volume of the lattice vacancy, D^* is the diffusion coefficient and t is time. In thick films of all systems however, it has been observed that sintering in the presence of a liquid phase is beneficial, thus liquid phase sintering is the second important process. In YBCO in particular the surface resistance and the current density are greatly improved if sintering takes place above the peritectic temperature (around 1030°C in oxygen and 1020°C in air) where there is both liquid and solid present. Under these conditions equation (B4.3.2) [43, 44] relating to viscous flow, applies.

$$\frac{x}{r} = \left[\frac{3\gamma t}{2\nu r}\right]^{\frac{1}{2}} \tag{B4.3.2}$$

where ν is the viscosity of the liquid phase. In YBCO thick films spherulitic grain growth is observed and grains with characteristic lengths of the order of millimetres are observed in samples sintered in the presence of a liquid phase. The kinetics of grain growth has been described by an isothermal grain growth model where the grain size l_g is related to a power law in time [45].

$$l_g = At^n \tag{B4.3.3}$$

where l_g is the grain size and A is a constant with an Arrhenius type temperature dependence. The activation energy for YBCO is 125 kJ mol^{-1}. The exponent n has a value of 1/5 for a sintering temperature of say 930°C with no liquid phase, and of 1/3 when sintered above the peritectic temperature in the presence of a liquid phase.

B4.3.6 Applications of thick films

A1.3 Perhaps the most significant aspect of the thick films described here is their ability to be deposited on curved substrates. For a more complete description of these applications the reader is referred to [1]. Some of the applications in this reference have now been replaced by thin film technologies or competing E3.1 technologies so that for example the thick film resonators described with Q factors in excess of 750 000 at 5 GHz [46] may be more conveniently replaced by thin film HTS resonators or dielectric resonators. However, the cylindrical cavities described provide a very useful means of shielding. The noise properties of YBCO–$SrTiO_3$–YBCO thin film multiloop magnetometers enclosed within YBCO thick film shields has been measured by Ludwig *et al* [47] and found to be very low at 37 fT Hz$^{-1/2}$ at 1 Hz and 18 fT Hz$^{-1/2}$ at 1 kHz. The use of thick film HTS in magnetic resonance imaging is interesting as they E2.3 have the potential for large area coverage and can be three-dimensional. The use of thick YBCO films was first reported by Penn *et al* [48] where it was found that HTS thick film coils showed improvements over a copper mimic at 77 K by a factor of around 3. This is highly significant in achieving better image resolution and reducing scan times.

B4.3.7 Conclusions

Thick film deposition of the sort described above is a fairly straightforward technique that is well understood in the electronics industry. We used the strict definition of a thick film where a powder is mixed with a vehicle and deposited onto a substrate either by screen printing or doctor blading. We noted that there were variants such as spray pyrolysis and sol–gel deposition that similarly did not B4.2.1 require vacuum deposition techniques. Since the melt processed technique was developed the properties B4.2.2 of thick films have been considerably surpassed by those of thin film methods involving vacuum deposition etc. However, there are certain applications where thick films will find use and these are, for E2.3 example, applications involving large area coated shields or large area MRI search coils of planar or cylindrical design.

B4.3.8 Troubleshooting and safety

Symptom	Remedy
YBCO Film has too much green 211 phase	Need thicker film to avoid substrate interaction
Film not superconducting	Check oxygenation, check stoichiometry
J_c too low	Check processing temperature — need to melt process at around 1050°C in oxygen for YBCO
Need *far* higher critical current — 10^4–10^5	Use BSCCO or Tl system and solution spray or consider electrophoretic deposition or reduce operating temperature
Need lower surface resistance	Use YBCO melt processed on YSZ substrate or reduce operating temperature
Need better microwave power handling	Use three-dimensional geometry, avoid stripline, avoid sharp edges and/or reduce operating temperature
Substrate is poisoning film	Use silver or YSZ substrates. Otherwise must use buffer layers on e.g. alumina or nickel alloys etc.

B4.3.8.1 Safety

Problem	Solution
Finely divided powders	
Are a potential explosion hazard	Need to avoid static build up
Can cause respiration problems	Wear dust mask or respirator helmet
Can be a skin irritant	Wear protective gloves
High temperature furnaces	
Static build up	Earth properly
Temperature overrun	Failsafe controller e.g. Eurotherm
Burn hazard	Wear protective gloves, use tongs for removal, use ceramic tiled benches, use extract
Thick film polymers/ solvents	
Storage issues	Use fireproof cabinets
Fume hazard	Use respirator helmets or fume cupboards for preparation
Toxicity issues	
Chemicals	Main problems are barium, lead and thallium in increasing toxicity order, and with some of the chemical components. With thallium in particular stringent measures must be taken including regular medical checks. Reference to handling of toxic chemical reference texts essential. If possible, avoid use of thallium unless absolutely necessary. Consult Health and Safety Executive before use. Consult colleagues experienced in handling before use.

References

[1] Alford N McN, Penn S J and Button T W 1997 *Supercond. Sci. Technol.* **10** 169–185

[2] Wu X U, Foltyn S R, Arendt P, Townsend J, Adams C, Campbell I H, Tiwari P, Coulter Y and Peterson D 1994 *Appl. Phys. Lett.* **65** 1961–1963

[3] Alford N McN, Button T W, Adams M J, Hedges S, Nicholson B and Phillips W A 1991 *Nature* **349** 680–683

[4] Button T W, Alford N McN, Wellhofer F, Shields T C, Abell J S and Day M 1991 *IEEE Trans Magn.* **27** 1434–1437

[5] He Q, Christen D K, Klabunde C E, Traczyk J E, Lay K W, Paranthaman M, Thompson J R, Goyal A, Padraza A J and Kroeger D M 1995 *Appl. Phys. Lett.* **67** 294–296

[6] Su L Y, Grovenor C R M, Goringe M J, Dewhurst C D, Cardwell D A, Jenkins R and Jones H 1994 *Physica* C **229** 70–78

[7] Holesinger T G, Phillips D S, Coulter J Y, Willis J O and Peterson D E 1995 *Physica* C **243** 93–102

[8] Woolf L D, *et al*, 1991 *Appl. Phys. Lett.* **58** 534–536

[9] Bhattacharya D, Roy S N, Basu R N, Dassharma A and Maiti H S 1993 *Mater. Lett.* **16** 337–341

[10] Kakihana M 1996 *J. Sol-Gel Sci. Technol.* **6** 7–55

[11] Masuda Y, Matsubara K, Ogawa R and Kawate Y 1992 *Japan J. Phys. Part 1:Regular Papers Short Notes and Review Papers* **31** 2709–2715

[12] Sathyamurthy S and Salama K 1998 *J. Supercond.* **11** 545–553

[13] Yamagiwa K and Hirabayashi I 1998 *Physica* C **304** 12–20

[14] Sheth A, Lasrado V, White M and Paranthaman M 1999 *IEEE Trans. Appl. Supercon.* **9** 1514–1518

[15] Ezura E, Asano K, Hayano H, Hosoyama K, Inagaki S, Isagawa S, Kabasawa M, Kojima Y, Kosuge S, Mitsunobu S, Momose T, Nakada K, Nakanishi H, Shimbo Y, Shishido T, Tachikawa K, Takahashi T and Yoshihara K 1993 *Japan J. Appl. Phys. Part 1:Regular Papers Short Notes and Review Papers* **32** 3435–3441

[16] Hemmes H, Rogalla H, Jager D, Smithers M, Vanderveer J and Stover D 1993 *Cryogenics* **33** 302–307

[17] Shields T C, Langhorn J B, Watcham S C, Abell J S and Button T W 1997 *IEEE Trans. Appl. Supercond.* **7** 1478–1481

[18] Jergel M 1995 *Supercond. Sci. Technol.* **8** 67–78

[19] Schulz D L, Parilla P A, Gopalswamy H, Swartzlande A, Duda A, Blaugher R D and Ginley D S 1995 *Mater. Res. Bull.* **30** 689–697

[20] Vanolo M, Pavese F, Giraudi D and Bianco M 1994 *Nuovo Cimento della societa Italiana di Fisica D: Condensed Matter* **16** 2119–2126

[21] Os'kina T E, Kazin P E and Tretyakov Yu D 1991 *Supercond. Sci. Technol.* **4** 301–305

[22] Miura S, Hashimoto K, Wang F, Enemoto Y and Morishita T 1997 *Physica* C **278** 201–206

[23] Yamada Y, Kawashima J, Niiori Y and Hirabayashi I 1996 *Appl. Supercond.* **4** 497–506

[24] Aichele T, Gornert P, Uecker R and Muhlberg M 1999 *IEEE Trans. Appl. Supercon.* **9** 1510–1513

[25] Oka K, Nakane K, Ito M, Saito M and Unoki H 1988 *Japan J. Appl. Phys. Lett.* **27** L1065–L1067

[26] Maeda M, Kadoi M and Ikeda T 1989 *Japan J. Appl. Phys.* **28** 1417–1420

[27] Bourdillon A and Tan Bourdillon N X 1994 *High Temperature Superconductors: Processing and Science* (Boston, SanDiego, NewYork, London, Sydney, Tokyo, Toronto: Academic Press, Harcourt Brace and Jovanovich)

[28] Langhorn J, Bi Y J and Abell J S 1996 *Physica* C **271** 164–170

[29] Langhorn J B 1996 *Platinum Metals Rev.* **40** 64–69

[30] Holesinger T G, Phillips D S, Coulter J Y, Willis J O and Peterson D E 1995 *Physica* C **243** 93–102

[31] Dimesso L, Masini R, Cavinato M L, Fiorani D, Testa A M and Aurisicchio C 1992 *Physica* C **203** 403–410

[32] Suyiama A, Yoshimoti T, Endo H, Tsuchiya J, Kijima N, Mizuno M and Oguri Y 1988 *Japan J. Appl. Phys.* **27** L542–544

[33] Schultz D L, Parilla P A, Ginley D S, Voight J A, Roth E P and Venturini E L 1995 *IEEE Trans. Appl. Supercon.* **5** 1962–1965

[34] Tsabba Y and Reich S 1995 *Physica* C **254** 21–25

[35] Tiefel T H and Jin S 1991 *J. Appl. Phys.* **70** 6510–6512

[36] Yang M, Goringe M J, Grovenor C R M, Jenkins R and Jones H 1994 *Supercond. Sci. Technol.* **7** 378–388

[37] Cowey L, Timms K, McDougall I, Marken K, Dai W and Hong S 1994 *Cryogenics* **34** 813–816

[38] Kase J L, Togano K, Kumakura H, Dietderich D R, Irisawa N, Morimito T and Maeda H 1990 *Japan Appl. Phys.* **29** L1096–L1099

[39] Jergel M, Conde Gallardo C, Falcony Guajardo C and Strbik V 1996 *Supercond. Sci. Technol.* **9** 427–446

[40] Bhattacharya R H, Duda A, Ginlet D S, DeLuca J A, Ren Z F, Wang C A and Wang J H 1994 *Physica* C **229** 145

[41] Koshy J, Kumar K S, Kurian J, Yadava Y P and Damodaran A D 1994 *Bull. Mater. Sci.* **17** 577–584 see also ref 2 for further references

[42] Kuczynski G C 1949 *J. Metals* **1** 169–173

[43] Frenkel J 1945 *J. Phys* **9** 385–391

[44] Exner H E and Petzow G 1973 *Sintering and Catalysis*, ed G C Kuczynski (New York: Plenum) pp 279–293

[45] Shin M W, Hare T M, Kingon A I and Koch C C 1991 *J. Mater. Res.* **6** 2026

[46] Button T W and Alford N McN 1992 High Q $YBa_2Cu_3O_x$ Cavities *Appl. Phys. Lett.* **60** 1378–1380

[47] Ludwig F, Dantsker E, Kleiner R, Koelle D, Clarke J, Knappe S, Drung D, Koch H, Alford N McN and Button T W 1995 *Appl. Phys. Lett.* **66** 1418–1420

[48] Penn S J, Alford N McN, Hall A S, Button T W, Johnstone R, Zammattio S J and Young I R 1995 *Appl. Supercond.* **1** 1995–1861

B4.4.1
High temperature superconductor films: processing techniques

P J Hirst and R G Humphreys

B4.4.1.1 Introduction

In this chapter, we describe the processes used in making circuits, starting with photolithography and then the methods by which a pattern is transferred from resist to the superconductor. Ion milling is the mainstream technology for HTS processing and will be discussed in some detail. Finally we will discuss some complete processes. These technologies were developed for other materials systems and have adapted so readily to high temperature superconductors that rather little systematic work has been published on the subject. We often have to refer to the literature of other materials to provide the background.

We distinguish two types of HTS circuits. The simplest are passive microwave devices, which are made by patterning a single superconducting layer and adding normal metal contacts. This technology is in industrial production. Although it might benefit from refinement, a functioning technology exists. Active circuits incorporating Josephson junctions [e.g. SQUIDs and single flux quantum (SFQ) logic circuits] require much more complex processing. Successive superconducting layers need to be grown and patterned, and then interleaved with insulating layers, while maintaining epitaxy and oxygenation in all the layers. This multilayer circuit technology, particularly all aspects of Josephson junctions, is much less advanced, and a deeper understanding of the processing mechanisms is needed if progress is to be made. We shall say little here about processes specifically aimed at making Josephson junctions, which are the subject of another chapter in this handbook.

E3
E4
E4.2, E4.5
B4.4.2
B4.4.3

B4.4.1.2 Lithography

Lithography is the process of transferring geometric shapes to the surface of a wafer according to a specified pattern. A more detailed explanation and extensive references for this topic can be found in [1]. First, we describe a generic lithographic process, and then we examine the individual steps in more detail.

The basic types of lithography processes are outlined schematically in figure B4.4.1.1. A radiation sensitive material, called resist, is coated onto the wafer to record the pattern. Regions of the resist are exposed to radiation according to the desired pattern (in the photolithography example shown, the radiation is UV light, and the pattern is transferred from a mask). The radiation interacts with the resist to change its chemical or physical properties. Following exposure, the resist is developed to remove the desired regions of resist. Positive resist remains on the wafer after development if it has not been exposed; negative resist remains on the wafer if it has been exposed.

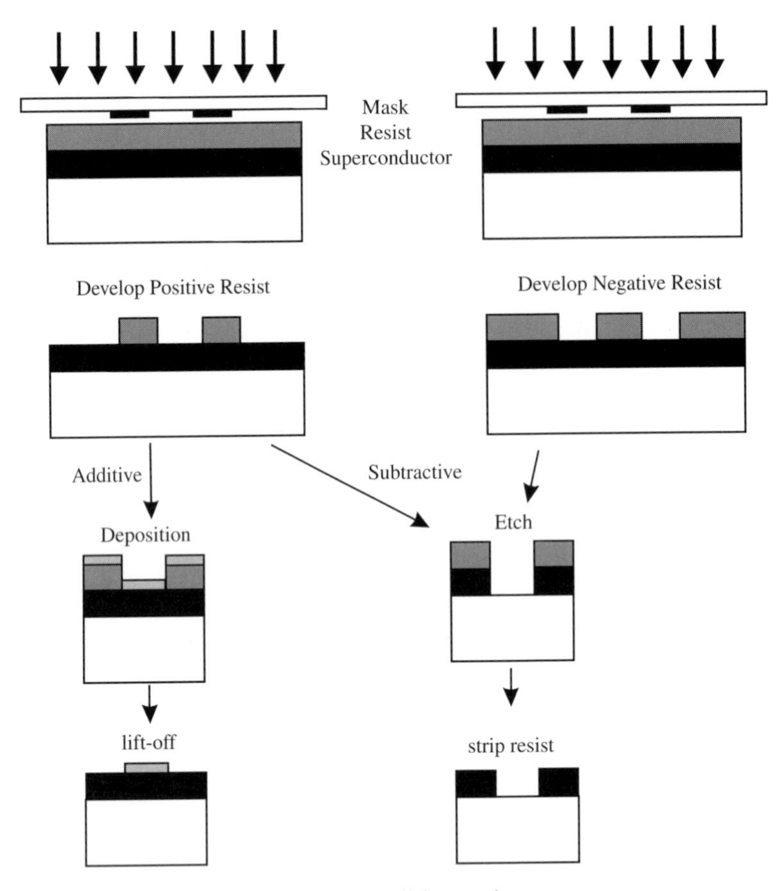

Figure B4.4.1.1. Photolithography processes.

Once the resist pattern is formed this can be used in two ways. (a) The resist acts as a mask where the exposed superconductor is removed—subtractive lithography or (b) a subsequent layer (e.g. a metal layer for contacting) is deposited through the resist mask. The resist is then dissolved away by a solvent, detaching the metal layer on top of it. This leaves an inverse of the resist pattern on the wafer. This is called additive lithography or 'lift off'. Lift-off can only be used with positive resists, as the negative varieties are very difficult to strip with liquid based processes (see resist stripping).

The most widely used exposure system is light, usually in the UV region of the spectrum. This has the attraction that the process is a parallel one (i.e. a whole wafer, or at least a large fraction of it, is exposed simultaneously) resulting in a high throughput. However, there are higher resolution lithographic techniques which use beams of electrons or ions. These write the pattern serially (i.e. a point at a time) and are comparatively slow.

B4.4.1.2.1 Resist materials and processes

In general, a resist material should satisfy the following criteria:

- high exposure speed (for high throughput);
- high resolution;

- etch resistance;
- very low defect density;
- wide process margins.

The resist systems in widespread use are commercially available and are the result of a very large investment by the semiconductor industry. The manufacturer's literature usually describes approximately how they should be used, but every process will need to be optimized to account for local variations in laboratory conditions. Typically, the primary parameters of this optimization are the exposure dose (intensity × time) and the development time, but each of these is linked to many subsidiary parameters.

Table B4.4.1.1 summarizes the most common resist systems used in HTS processing. Resists used for optical lithography are multi component, consisting of a resin, a photoactive component (PAC) and a casting solvent.

The phenolic novolak resin in positive diazonaphthoquinone novolak (DQN) resist is rendered insoluble in alkaline solutions by the addition of the diazonaphthoquinone (DQ) photoactive component. Under the action of UV radiation the DQ is broken down and the novalak resin can then be dissolved in alkaline solutions.

The negative, rubber based resists undergo cross-linking under the action of UV radiation to make the exposed regions insoluble. This is facilitated by a biazide sensitizer, which decomposes under the action of UV radiation and reacts with the rubber, causing dense cross-linking.

Since the mechanism for high energy sensitive electron, ion beam or x-ray resists capture the radiation by mass absorption, they do not require a photoactive component. The absorption of the high energy radiation causes polymer chain scission into lower molecular weight fractions which can be dissolved in organic solvents (e.g. polymethylmethacrylate (PMMA) to n-MMA).

The casting solvent does not take part in the exposure mechanism but simply allows the delivery of the resist to the wafer surface as a liquid for wafer coating. The wafer must be coated evenly with resist and this is carried out with a technique called spin coating. This involves applying a small volume of liquid resist onto the surface of the wafer and spinning the wafer at very high speeds. This ejects any excess resist, leaving an even coating on the wafer. Different thicknesses of resist can be achieved by changing the solvent content and finer adjustments can be made by changing the spin speed.

After coating, the excess solvent in the resist must be driven out prior to exposure to convert the liquid film into a solid polymer, a pre-bake. This process also promotes adhesion between the resist and wafer. The pre-bake is best carried out above the glass transition temperature (T_g) for the resin in the resist but below the decomposition temperature of the PAC. This is usually carried out on a hotplate or in a convection oven. Hotplates tend to be quicker and provide a better driving force for drying processes.

A typical resist film used in HTS processing is 1–2 μm thick, determined by the viscosity of the resist and the spin speed. Usually, the resist should be as thin as possible, but thick enough that some

Table B.4.4.1.1. Composition and properties of typical resist systems

Use	UV positive	UV negative	Positive electron, ion, x-ray resist
Resin	Novolak	Cyclized rubber	Polymethylmethacrylate
Sensitiser	Diazonaphthoquinone	Bisazide	
Casting Solvent	2-Ethoxyethylacetate	Xylene	Chlorobenzene
Prebake	75–95 °C	70–90 °C	175 °C
Developer	0.25 N KOH	Xylene	Methyl isobutyl ketone

resist remains (with some substantial safety margin) when a superconducting film has been completely etched through. Commercial suppliers carefully control the quality of their resist to ensure that it is free from particles much smaller than the resist thickness. However, so-called 'slugs' (high viscosity inhomogeneities) can form in storage and can make high resolution difficult to achieve. In some small scale research laboratories, resist is applied manually to the wafer. In our experience, these processes are difficult to control. Both these problems can be solved by using a resist dispenser that delivers a controlled volume of resist, reproducibly synchronised to the spin speed of the wafer, through a 'point of use' filter. We find it hard to believe that a high yield multilayer process could be achieved without automatic resist dispensing.

The resist is then exposed to the required dose of radiation. During optical exposure standing waves can be set up between the resist surface and the wafer, causing the modulation in exposure intensity through the depth of the resist layer. Hence, the edge of the developed resist is rippled parallel to the wafer. This can be avoided by using mixtures of wavelengths during exposure. If short wavelengths are being used, this is often not possible, but a post exposure bake (before development) can be used to allow the diffusion of PAC or monomer fractions to regain a smooth resist profile.

The resist is then developed, by dissolving the exposed part (positive resist) in a suitable developer solution. Following development it may also be desirable to carry out a post development bake to stabilise the resist and further promote adhesion of the resist to the wafer, particularly when using wet etching for pattern transfer. If the resist profile is to be maintained this will be carried out below the T_g (glass transition temperature) for the resist resin. However, for some processes (see subsection B4.4.1.3.3) it may be advantageous to slump (or reflow) the resist by post baking above T_g. Also, a post development bake may be carried out above the decomposition temperature of the PAC rendering the resist impervious to any further patterning steps. This can be useful in multi-layer resist processes, which are discussed below.

Early on in the development of the subject, there were fears that the wet processing used in lithography would not be compatible with HTS materials, but the organic or alkaline solutions used cause little harm to good quality material. Detailed studies are rare, however. Sheats *et al* [2] found that processing could increase both microwave surface resistance and J_c, and these changes were compounded rather than reversed on annealing. It would be good to see more systematic work in this area.

D2.7

By control of the exposure and development conditions the resist profile can be engineered for the particular process being used. This is summarized in table B4.4.1.2.

Table B.4.4.1.2. Profiles, expose-develop criteria and uses of positive resists

Profile	Exposure Dose	Developer Influence	Uses
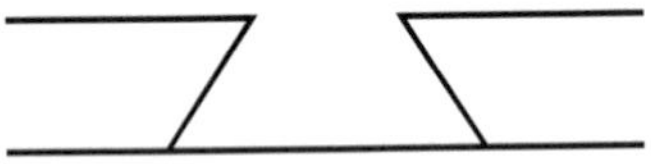	High	Low	Lift-off
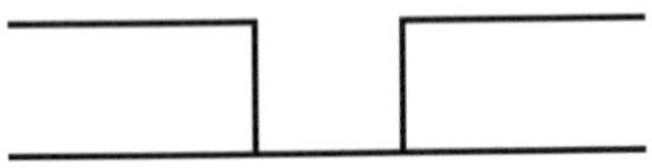	Medium	Moderate	Dry etch Lift-off Wet etch
	Low	High	Wet etch (N.B. a similar profile can also be achieved with a reflow process)

High resolution lithography is carried out in clean conditions to avoid particles of dust reducing the yield. Extreme cleanliness, such as is used in integrated circuit manufacture, is very expensive, and most HTS processing is carried out in less carefully controlled conditions. Production laboratories also have sophisticated temperature and humidity control. For the relatively simple circuits currently being made, quite simple precautions are all that is required. We have a simple temperature control system and laminar flow hoods. In our experience, yield is limited by factors other than particulate contamination.

B4.4.1.2.2 Optical exposure

By far the most popular exposure system used in both semiconductor and superconductor patterning is optical lithography, also called photolithography, using radiation in the range from blue to UV. In optical lithography, it is necessary to exclude all wavelengths to which the resist is sensitive from the environment. This is done by using yellow filters on lights (including microscope illuminators) and windows.

There are two main methods of optical lithography exposure.

(a) Contact and proximity printing

(b) Projection printing

In both cases, the fundamental limit to the resolution is set by diffraction: shorter wavelength equates to higher resolution. However, the shorter the wavelength becomes, the harder it is to find convenient, bright sources, so exposure times tend to increase. In addition, fewer materials transmit well at short wavelengths, and the optics become more expensive. In practice, the choice of wavelength is a trade-off between performance and convenience/cost. The sources used for UV exposure are summarized in table B4.4.1.3. For the majority of superconducting applications at the present state of the art, a minimum feature size of 1 μm is sufficient, and near UV exposure is most widely used.

Contact and proximity printing

The basic process is to flood a resist layer with a parallel beam of light through a mask with the required pattern on it. The attraction of this method is that it is a parallel process with a high throughput of wafers. Economies of scale can be achieved by increasing the wafer size. The resolution achievable in

Table B.4.4.1.3. UV exposure regions and the sources used to generate the exposure

Region	λ(nm)	Sources
Deep UV	150–350	Excimer laser — e.g. F_2 157 nm, ArF 193 nm, Cd lamp 240 nm
Mid UV	300–350	Laser Hg lamp — 313 and 334 nm lines
Near UV	350–500	Hg lamp — 365, 405 and 436 nm lines

optical exposure is limited by diffraction effects, given by [1]:

$$W \approx 0.7\sqrt{\lambda d}$$

where λ is the exposure wavelength, d is the wafer to mask distance. The shorter the wavelength and the closer the mask is to the wafer the better the resolution. Ideally, the gap between the wafer and mask should be zero to minimize diffraction effects. Thus, for high resolution, the resist must be as thin as possible.

During contact exposure, the mask is pressed against the resist coated wafer, and a 1:1 copy of the mask is made. The mask substrate must be transparent to the wavelength of radiation used and the pattern must be opaque. As most HTS exposure is carried out in the near UV, the most common mask system is chrome on crown glass. If deeper UV is used, then the mask must be made of quartz so that it is transparent.

With this technique, feature sizes well into the sub-micron region can be achieved (using large, flat substrates, thin resist and considerable care) with comparatively cheap equipment. In small scale research lithography, non-uniformity in the substrate or film can limit the contact achievable. A single tall outgrowth on a wafer can guarantee poor lithography.

B4.3 The spin coating technique for resist application will always leave a region at the edge of the wafer which is considerably thicker than the bulk of the resist—called a resist bead. This effect is amplified with HTS processing, where the substrate is often small and square; the resist at the corners is many times thicker than that at the edge. For high resolution lithography it is necessary to remove the bead prior to pattern exposure. This is done by exposing the edges of the wafer to a high exposure dose and developing away the resist bead prior to contact exposure. Microwave filter production is currently standardized on 2″ diameter circular wafers. For these the resist bead is much smaller.

To achieve good contact, a vacuum is pulled between the wafer and the mask. This will increase mask wear and eventually generates defects in the pattern, but this is only really an issue in volume production.

Present day HTS circuits are often made on small ($10 \times 10\,\mathrm{mm}^2$ or $25 \times 25\,\mathrm{mm}^2$) substrates. If substrates are polished on a chuck, first one side and then the other, they can often be curved. This can make it difficult to get a close spacing between wafer and mask. Introducing a thin polymer layer underneath such a substrate will allow it to deform enough that good contact can be achieved. The quality of contact can be assessed from the interference fringes seen in the mask aligner. Fringes are formed between the top of the resist and the bottom of the mask. Contact between the two is characterized by a dark fringe.

Projection printing

This technique, as the name suggests, projects a focused and reduced image of the reticle (mask) onto the wafer using a refractive or reflective exposure system. It is the mainstream technology for manufacture of large scale semiconductor integrated circuits. Typical reduction ratios are 5:1 or 2:1 between reticle and wafer. Typical exposure areas are between 0.5 and $3\,\mathrm{cm}^2$ and therefore the wafer is usually stepped between exposures to cover the whole wafer (hence the exposure tools are called steppers). The advantage of this technique is the relative ease of mask manufacture, because the mask image is reduced on the wafer. Also mask defects and imperfections are reduced in size and hence become less severe.

With projection systems the key parameters governing the resolution (R) are given by the Rayleigh equation, from [1]:

$$R = k_1 \frac{\lambda}{NA}$$

where λ is the exposure wavelength, NA the numerical aperture of the projection lens and k_1 a constant (≥ 0.5). A typical NA of 0.6 yields a resolution of 0.16 μm for deep UV (193 nm) exposure.

Present state of the art projection systems using excimer laser exposure at 193 nm are used to produce 0.18 μm line widths in production. The main disadvantages of these systems are their immense cost and limited area of exposure. If large scale production of HTS active circuits becomes a reality, this is likely to be the technique used.

E4

B4.4.1.2.3 Electron beam exposure

Until now we have only discussed exposure of resist using light. Particle beams such as electrons can also be used. The mask pattern is held in software and the beam is scanned across the wafer to pattern areas selectively. The fact that the pattern is held in software is an advantage when developing processes, as it can be changed much more easily than a chrome on glass mask. Older electron beam systems used a raster scan to cover the whole field of view and simply switched off the beam to select the areas for exposure. Modern systems use a vector scanning technique where each object is written separately. This is much faster, as time is not wasted on areas that are not to be patterned. Nevertheless, e-beam lithography is much slower than optical lithography, and it is mainly a research technique, except for manufacturing optical masks.

To compare the different methods, the De Broglie wavelength of the electron can be approximately calculated from $\lambda \approx 1.23/\sqrt{V}$, where λ is the wavelength in nm and V is accelerating voltage in volts. This gives, for 10–100 keV, wavelengths in the range of 0.0123–0.0039 nm. This is 1–2 orders of magnitude lower than for UV exposure, and comparable to the wavelength of x-rays. However, this is not the real resolution attainable in e-beam exposure due to electron – matter interactions. While beam diameters of 1 nm are achievable, as soon as the beam hits the solid it splays out, due to scattering within the solid. This can be analysed by considering the point spread function of the electrons, which is typically composed of two parts; a relatively narrow forward scattered component (*ca* 10 nm diameter at 100 keV) and a broad background caused by backscatter and second order effects. This means that areas adjacent to those being written are partially exposed (called the proximity effect). The proximity effect is worse for thicker resist films and electron beam resists tend to be quite thin; around 500 nm is typical. The backscatter effect is particularly large in the HTS systems due to the large atomic number of elements present in the materials (e.g. Ba, Z = 138). In fact, with some very sensitive chemically amplified resists the backscattered electrons are capable of exposing the underside of the resist adjacent to the feature being exposed, as shown in figure B4.4.1.2. This can be avoided by choosing the less sensitive PMMA resist systems. Another problem of the HTS systems is that the substrate is insulating. This means that particular care is needed in earthing of the HTS material to prevent charging, which will cause beam deflection. In spite of these problems, 50 nm features have been produced by electron beam exposure in HTS thin films [3].

B4.2.1, B4.2.2, B4.3

B4.4.1.2.4 Multi-layer resist processes

We have seen that achieving high lithographic resolution often demands the use of a thin resist layer. If the resist etches too quickly, it may be necessary to transfer the pattern to a more resilient or thicker material. This is achieved by using a multi-layer resist technique, an example of which is shown in figure B4.4.1.3. Here, the low ion milling rate material, carbon, is to be used as an etch mask [4, 5]. Carbon is a particularly convenient hard mask material that can be ashed away using an oxygen plasma without damaging the underlying superconductor [4]. Carbon has the lowest milling rate of any material, about half the rate of YBCO (see section B4.4.1.3).

B4.1

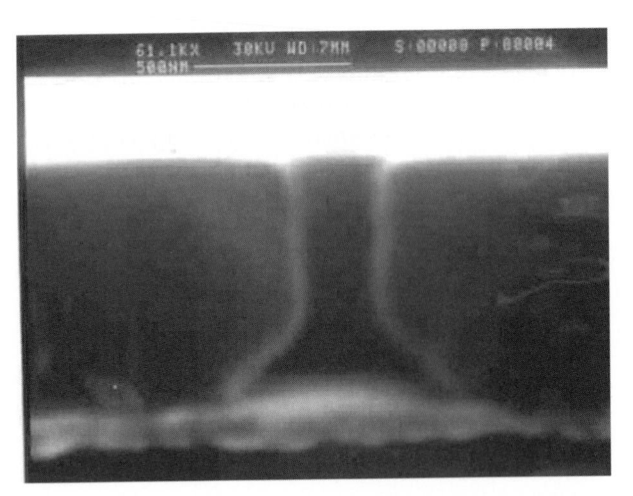

Figure B4.4.1.2. The effect of backscattered electrons upon the resist profile produced in a very sensitive chemically amplified resist, dose $= 10\,\mu$C. The sub-micron hole has an undercut next to the substrate caused by backscattered electrons.

In order to pattern the carbon a thin gold layer is deposited on top of it, followed by a layer of PMMA e-beam resist. The PMMA is exposed by the electron beam and then developed. Next, the pattern is transferred to the gold layer by ion beam milling. This can be done using thin resist because the gold layer is thin, and it mills faster than resist. The PMMA is then stripped and the gold now acts

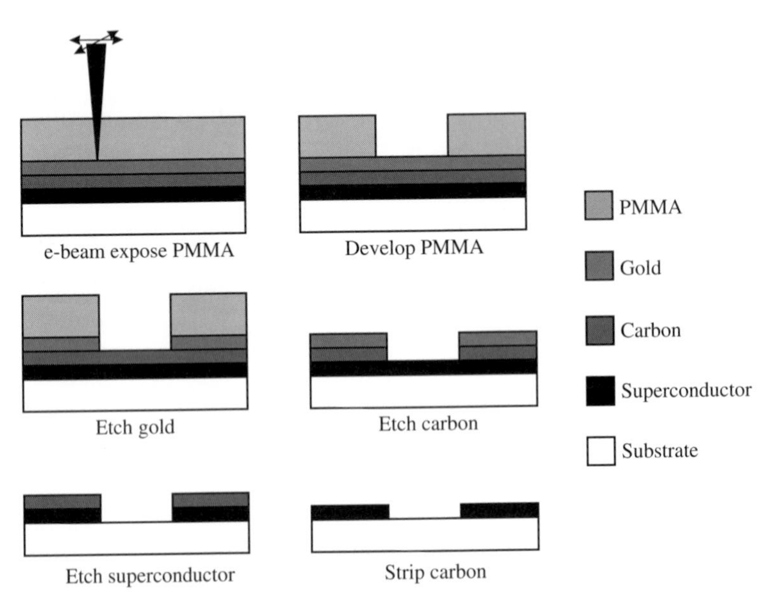

Figure B4.4.1.3. An example of a multilayer resist process. A trilayer resist process for the patterning of sub-micron features in superconductors. The e-beam resist, PMMA would not be sufficiently thick for the ion beam milling of the superconductor due to its low etch resistance. The pattern is transferred from the PMMA to gold and then to carbon, as the final etch resist.

as the resist for oxygen reactive ion etching of the carbon. With the right plasma conditions, the etch rate of gold is minimal and a near vertical wall is produced in the carbon. The gold is then stripped leaving a carbon mask through which to ion mill the superconductor.

B4.4.1.2.5 Lift-off patterning

Lift-off processes (described as additive in figure B4.4.1.1) are sometimes used in HTS systems in metallization and multi-layer resist processes. They have the attraction of requiring one fewer steps than subtractive patterning, but tend to have narrower process margins. Metal contacts are often made in this way. Lift-off depends on dissolving away the resist material after the metal deposition and allowing the unwanted parts of the metal layer to float away. For this to work well, the metal layer must not be continuous (figure B4.4.1.4). This is achieved by producing an undercut resist profile and depositing the metal by evaporation, which is directional, and makes full use of the shadowing of the undercut resist to break up the metal layer. Sputtering is more isotropic and is more likely to produce a continuous metal layer. B4.2.1

The main use of metallisation layers is as a contacting layer. Here it is necessary to ensure low contact resistance between the metal (usually gold or silver) and the superconductor. This has been studied by Huh *et al* [6], who found that the gold deposited directly onto grown surfaces has a lower contact resistance than if deposited onto a patterned surface as used in lift-off. They attributed this to a loss in surface oxygen in the superconductor during water immersion, but other interface contaminants are also possibilities. Many groups anneal metal contacts at around 300–500°C [7] to improve adhesion by metal diffusion into the film and reduce the contact resistance. This is usually carried out in an oxygen atmosphere to prevent oxygen loss from the superconductor, but it is worth noting that, while Ag is permeable to oxygen, Au is not. Ag is not compatible with annealing in active oxygen (atomic or ozone) as it forms silver pentoxide. The issue of minimization of contact resistance has become even more important recently where highly linear low value ($\sim$ mΩ) resistors are required for SFQ applications such as A-D converters. Typically, such resistors are dominated by the metal/superconductor contact resistance [8]. E4.5 E4.6

Lift-off processes for patterning a high temperature superconducting layer cannot use a photoresist mask, as the temperatures for HTS growth ($\sim$700°C) are far too high. Some attempts at lift off using a multi-layer resist process have been reported [9]. In this approach, amorphous CaO was deposited at room temperature over a photoresist pattern and the pattern lifted off with acetone. Then superconductor films could be grown through the CaO pattern and the unwanted superconductor lifted off by dissolving the CaO in water. Processes like this seem to have been abandoned because their resolution is poor.

B4.4.1.2.6 Resist stripping

After etching the resist must be stripped to leave a clean suface. This is particularly important if further epitaxial growth is to be carried out on the patterned layer.

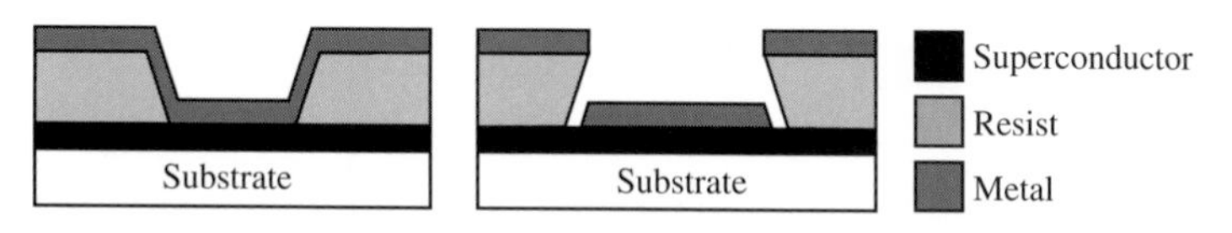

Figure B4.4.1.4. (*a*) An undesirable resist profile for lift-off patterning. (*b*) An undercut combined with directional deposition ensures that the metal layer is not continuous.

The positive DQN and e-beam resists can easily be stripped by chlorinated solvents (e.g. acetone) and alcohols but the cross-linked negative resists can not. These resists require either very aggressive solvents (e.g. dimethyl sulphoxide), which pose a safety risk, or strongly acidic/alkaline strip. These are incompatible with the HTS type materials which could be etched by such stripping agents. Another class of processes, which are used for all types of resist, are those plasma processes involving activated oxygen. The activated oxygen breaks the polymer backbone and the volatile species generated are pumped away.

The solvent based stripping processes tend to leave residues on the surface but the activated oxygen processes do not. To ensure epitaxy of subsequent layers we always surface clean the patterned wafer in atomic oxygen *in situ*, prior to epitaxial growth.

B4.4.1.3 Etching

The most common form of lithography used for semiconductor and superconductor patterning is subtractive patterning or etching. This is the most crucial stage in patterning because an error at this stage cannot be undone, unlike the resist pattern.

The ideal process will etch the film uniformly at a controllable rate. The etchant should exhibit a high degree of selectivity i.e. it should etch the film but not the resist, any underlying film or substrate. It will lead to the edge slope that the process designer wants. For example, the highest resolution demands vertical edges, but this is not compatible with the need for epitaxy in multilayer circuits. Edge slope is controlled by the anisotropy of the etch process, (see figure B4.4.1.5). If the etch is isotropic i.e. the etch rate is the same in all directions, an undercut of the resist mask will result, leading to a loss in pattern resolution. With an etch process that etches in depth faster than laterally the replication of the pattern is better and the edge slope increases. The methods used for achieving the desired edge profiles are discussed in more detail in section B4.4.1.4.

Essentially, etching mechanisms are either chemical, physical or a combination of the two. The main techniques are:

(a) wet etching;

(b) reactive ion etching;

(c) ion milling.

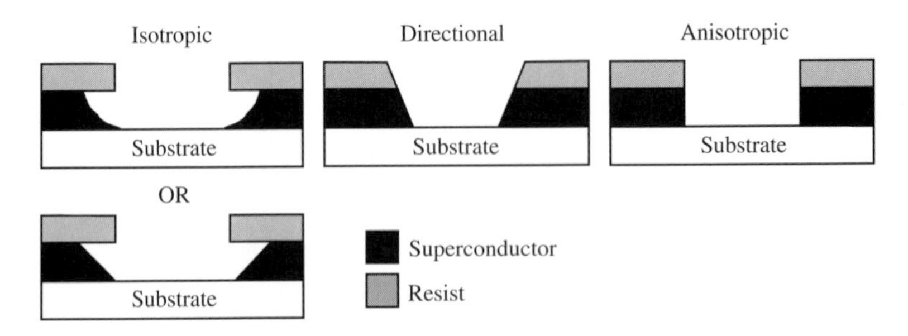

Figure B4.4.1.5. Edge profiles obtained with isotropic to anisotropic etches. The more anisotropic then the better the dimensional reproduction of the mask but vertical edges cannot be reproducibly grown over in the HTS cuprates (see 'Growth issues').

B4.4.1.3.1 Wet etching

The practicalities of wet etching are very simple with no need for expensive processing equipment. The superconductor is merely immersed in an etching solution for a predetermined length of time. In general, the cuprate superconductors are basic oxides and are etched by acid–base displacement reactions. This means that good selectivity to resist is easy to achieve, as polymeric materials tend to be impervious to aqueous acids.

The first etches used for HTS were acids such as phosphoric, nitric and hydrochloric acids in aqueous solution. These showed quite high etch rates ($>0.5\,\mu\text{m min}^{-1}$) even at low concentrations, making a controllable process difficult to achieve. It has been shown that the etch rate of sulphuric acid > nitric acid > hydrochloric acid > phosphoric acid [10]. Heavy undercuts have been observed when etching with all of these acids, limiting their applicability to minimum feature sizes of *ca* 3 μm. These etches have also been shown to degrade the surface of the superconductor, as monitored by surface resistance measurements [11]. These effects were attributed to the production of non-stoichiometric surface layers during the etch. This effect was greatly reduced by the purging of the etching solution of CO_2, which effects the dissolution kinetics of the surface reaction.

Another class of acid etches are those which contain carboxylic acids. These materials act to complex or chelate with the cations at the surface. This means that, once the acid has reacted with a surface cation, it can wrap around the cation and sequester it away from any further reaction. These etches have moderate and therefore more controllable etching rates. The most common chelating agent for metallic cations is ethylene diamine tetra acetic acid (EDTA). This has been successfully used for etching YBCO and TlBCCO with little effect upon the surface resistance [12]. Careful control of the pH C1, C3 of the solution is needed. A number of other di- and tricarboxylic acids have been studied by Ginley *et al* [13]. Of those examined, it was found that 0.1 M solutions of adipic or citric acid give a uniform etch rate with little effect upon the surface resistance after etching. It was found necessary to nitrogen purge the etchant of CO_2 in order to keep the surface resistance rise to below 10%. Also agitation or stirring was necessary to give uniform edge profiles. Some of these chemicals contain small amounts of sulphate ions as impurities. If they are used for etching Ba containing compounds, $BaSO_4$ can precipitate out at track edges. This can be solved by adding a small amount of a soluble Ba compound (e.g. $BaCl_2$) to the etch solution, to precipitate out the impurity.

The final class of wet etchants which have been studied are the halides in alcohols. This etch system avoids potential adverse reactions with aqueous solutions. However, conventional photoresists are soluble in lower alcohols; ethylene glycol is a possible solvent for use with resists. The chlorides of Ba and Sr and the fluorides of Sr are not soluble in alcohols and can therefore be discounted for etching the cuprate superconductors but the iodides and bromides are possible candidates. These have the possible advantage of not being an aqueous etch; the presence of water has been suggested as the source of the degradation of the surface properties of the superconductor [14]. In fact, some evidence shows that these etches may reduce the surface resistance of thin films.

Many conflicting results appear in the early literature as the etching properties have been found to be extremely dependent upon the structural properties of the films [2]. Wet etching is a straightforward, cheap patterning technique, but it is material dependent and cannot be used reliably for multilayer circuits.

B4.4.1.3.2 Plasma etching

We distinguish here two types of etching using ions or a plasma. One is chemical (reactive ion etching, RIE) and the other is mechanical (ion milling). In RIE, chemically reactive species are created in a plasma and react with the layer being etched to form volatile compounds which are pumped away. In ion milling, inert ions are accelerated by an electric field, and their kinetic energy is used to knock atoms or molecules from the layer being etched. In a general plasma process, the effect is often a mixture of the two. RIE processes are highly developed for semiconductor circuits, and are used for low temperature superconductors too. They are very attractive, as they cause minimal damage, can be designed to be anisotropic, and leave the surface clean, as all the reaction products are volatile.

B4.2.2 Although there were reports in the early HTS literature of RIE processes, we believe that all were actually ion milling. The problem with RIE of HTS is finding volatile species to transport the barium and (to a lesser extent) the Y or rare earth.

B4.4.1.3.3 Ion milling

Ion milling has been entirely replaced for semiconductor processing by reactive ion etching, but no such processes have yet been found for HTS. The basic work on the ion milling process that was done 20–30 years ago has been largely forgotten in the HTS literature and we will try to give some pointers to it here. A good starting point is [15], which gives numerous references, although the papers themselves are rather theoretical in nature. Recent work in this field has been rather specialized to secondary ion mass spectrometry (SIMS) with emphasis on optimizing depth resolution and quantitative analysis of thin films [16].

Ion milling was abandoned for semiconductors because it introduces damage, differential etching rates between materials are limited, and high aspect ratios were difficult to achieve. Damage is much less critical in superconductors, and much of it can be annealed out. High aspect ratios are not usually required in HTS circuits, but the absence of good etch stops in milling is a problem for the HTS process designer. Thus, these objections to milling are less strong for superconductors than for semiconductors.

In ion milling, an ion gun (Kauffman source) points at the wafer and the ions (usually argon) strike the wafer at an angle (see figure B4.4.1.6). Electrostatic effects are minimized by supplying electrons by

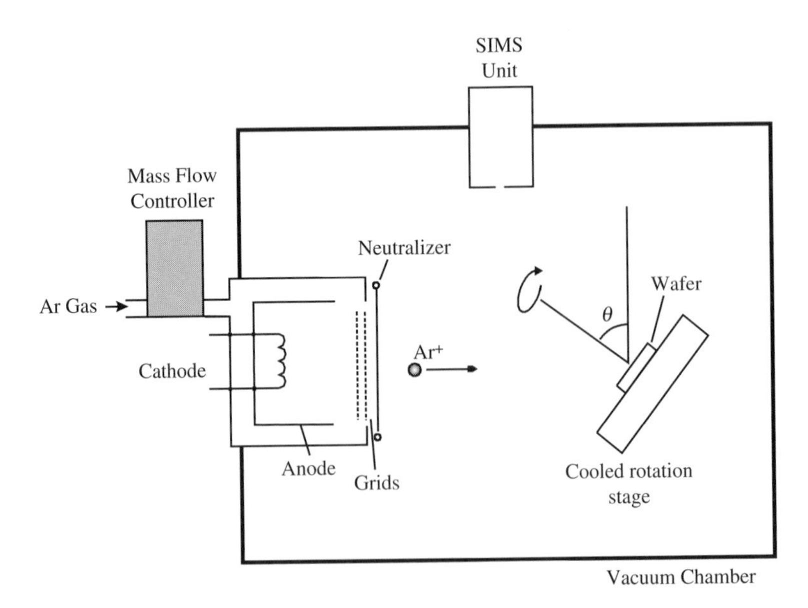

Figure B4.4.1.6. A typical ion milling setup. The Kauffman source is shown on the left. Argon ions are generated within the source and accelerated by a set of grids. Electrostatic effects are minimized by the use of a neutralizer that supplies electrons by thermionic emmision. The wafer is attached to a rotating cooled stage which can be set to the desired milling angle (θ). The SIMS unit contains a mass spectrometer and deflection plates to direct the emitted ions to the mass spectrometer.

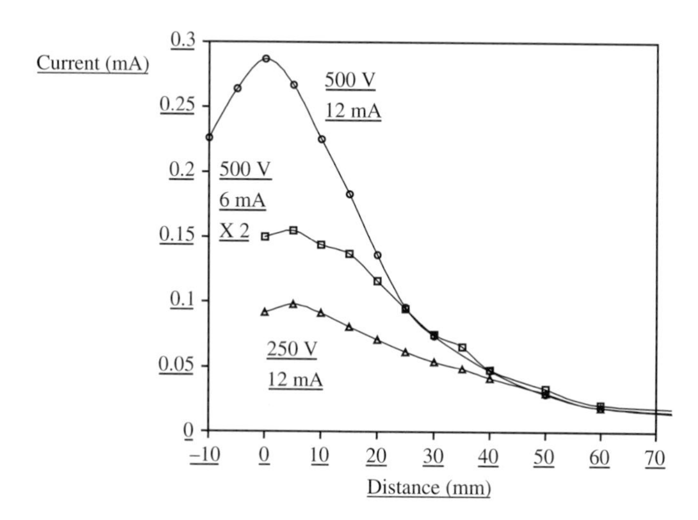

Figure B4.4.1.7. Ion current collected by a probe biased at −3 V ~15 cm from a 3 cm ion gun fed with 2 sccm (standard centimetres cubed per minute) argon. The probe was moved perpendicular to the beam. The accelerating voltage and total beam voltage are indicated. The current measured at 500 V and 6 mA has been multiplied by 2. The results have been shifted slightly so that the peaks approximately coincide.

thermionic emission from a 'neutralizer' filament, with an emission current slightly larger than the beam current. In most cases the sample is rotated during milling for improved uniformity. In research a small 3 cm diameter ion gun is often used. Our experience is that such a gun will often point 10° or so away from its apparent direction, so a mounting that allows the gun to be tilted is desirable. The ion beam direction can change with time, and needs to be checked regularly using a probe or Faraday cup. For large area processing, larger guns are used and these issues become less critical.

Figure B4.4.1.7 shows the profile of ions produced by a 3 cm gun for different sets of conditions. This figure shows that the gun produces a narrower beam at higher current and higher voltage. Thus, doubling the beam current from 6 to 12 mA will produce a much larger increase in the milling rate. Reducing the voltage from 500 to 250 V will reduce the milling rate by much more than just the reduction in milling yield. To some extent these effects will be less marked when a larger diameter gun is used. Finally, the angular spread of the ion beam is much greater at low voltages and currents. This is important in milling sharp edges, and is not ameliorated by using a large gun.

Temperature control during milling

The ions produced by an ion gun are usually accelerated by several hundred volts and carry considerable kinetic energy. It is easy to show experimentally or theoretically that any radiated heat input from the hot cathodes is negligible by comparison. If the sample is not heat sunk, its temperature will rise during milling leading to deformation or charring of the resist and possible deoxygenation of the superconductor. If the sample is to be kept cool during milling, it must be heat sunk. The substrate holder must either be massive enough that the temperature rise is small or it must be thermally linked to a cooling medium. A number of rotating sample holders are commercially available, although the design is quite tricky. So what can go wrong? The simplest mistake is to assume that laying the sample on top of a cold metal block implies that the sample is cold. In fact, such a sample will be supported on only three points and thermal conduction between the two will be extremely poor—hardly better than vacuum. Let us look at the problem more quantitatively.

Consider a substrate of thermal conductivity σ_1 and thickness d_1 connected to a heat sink through a layer of thermal conductivity σ_2 and thickness d_2. The power density input at the front surface is neV, where n is the ion flux/unit area, e the electronic charge and V the accelerating voltage. The temperature differences (ΔT_i) across the layers are given by

$$\Delta T_i = neV\frac{d_i}{\sigma_i}$$

Typical milling conditions used for superconductors are n $\sim 10^{15}$ ions/cm^2/s and $V = 500$ V, so the input power is $\sim 0.1\,\mathrm{W\,cm^{-2}}$. The substrate thickness is typically ~ 0.5 mm, and its thermal conductivity $0.1\,\mathrm{W\,cm^{-1}\,K^{-1}}$. It is immediately apparent that ΔT_1, the temperature difference across the substrate, is negligible, < 0.1 K for the above parameters. There are still occasional references in the literature to 'substrate surface temperature' in milling processes. Clearly, the substrate is nearly isothermal and the temperature of the surface is essentially that of the substrate.

Now suppose the heat sinking layer has $\sigma_2 = 2.5 \times 10^{-3}\,\mathrm{W\,cm^{-1}\,K^{-1}}$ (this is about the lowest thermal conductivity available in room temperature solids, characteristic of polymers) and a thickness of 0.25 mm. Then its conductance is $G = 0.1\,\mathrm{W\,cm^{-2}\,K^{-1}}$ and $\Delta T_2 = 1$ K, which is again negligible; any thermal contact through a solid (or liquid) is good enough to ensure that the sample is well heat sunk.

Thus, if the substrate temperature rises it is because of poor thermal contact. If no heat sinking layer is used, then the conductance between the wafer and its environment is the radiation conductance, for small temperature differences this is given by

$$G = 8\eta_s\sigma_{SB}T^3$$

where σ_{SB} is the Stephan–Boltzmann constant ($5.67 \times 10^{-12}\,\mathrm{W\,cm^{-2}\,K^{-4}}$) and η_s is the sample emissivity. The factor is 8 because the sample has two surfaces. Even if the emissivity is high, $G \sim 10^{-3}\,\mathrm{W\,cm^{-2}\,K^{-1}}$, about two orders of magnitude lower than the worst solid in the geometry considered above. If the sample is not heat sunk at all, its temperature rises by over 100°C (we have confirmed this experimentally). If we can arrange a thin, heat sinking layer that is in good thermal contact everywhere, then the substrate temperature rise can be kept small. Good thermal contact is much more important than thermal conductivity. We use PTFE tape, as used for water fittings, for this purpose. It has very poor thermal conductivity, but is thin and deforms well to give a good thermal contact.

Some groups use pulsed milling to limit the temperature rise [17]. As we have just seen, this is only helpful if the sample is poorly heat sunk. For a given average milling rate, the peak temperature of the substrate will be slightly larger if pulsed milling is used than if it is not. The only reason to use pulsed milling might be that the beam profile from the ion gun is better at a higher current. This is indeed the case, as we have seen, but is not the reason that is usually given for using pulsed milling.

B4.4.1.3.4 The ion milling process

Although little detailed work on ion milling has been reported in the HTS literature, there is a large literature on milling other materials. In this section we look at the fundamentals. The milling process is quite complicated [2, 18]. The milling rate depends on the following:

(a) the mass and energy of the projectile ion;

(b) the angle of incidence of the projectile ion on the surface being milled;

(c) the sample temperature;

(d) the rate of arrival of other species on the surface being milled;

(e) the mass, chemical environment and physical environment of the layer being milled.

Of these, the process designer has full control of the first three, and some control of the fourth.

The milling rate for a given species vanishes [19] below some energy, of the order of a few eV. Milling yields [20] at such energies are very low ($\sim 10^{-3}$ atoms/ion). At milling energies that are used in practice, the milling rate rises approximately linearly with energy, on a curve which extrapolates to zero milling rate between 50 and 100 eV [20], with milling yields ~ 1 atom/incident ion at around 500–1000 eV.

In thinking about milling HTS, comprising a number of different elements, we need to qualify the list of factors determining the milling rate. Evidently, a sample only has one milling rate at a point, but each element in the surface will sputter at a different rate. The obvious answer is that, initially, some elements mill faster than others but, quite quickly, an equilibrium is established in which the surface being milled has a different composition from the bulk material. The thickness of this surface layer is of the order of the stopping distance of the projectile ion, say about 1 nm for 400 eV Ar^+ ions [21]. It is well established that oxides tend to lose oxygen preferentially [22] and, once the oxygen is lost, milling proceeds much more rapidly. Kelly [23] has sought to correlate those oxides that dissociate under milling with their dissociation vapour pressure at high temperatures, although it seems that the correlation was not perfect. An extra idea that needs to be factored in is that the residual gas in the milling chamber is important. In [22], it was shown that the milling rate of SiO_2 (and the differential milling rate compared to photoresist) is greatest for low ($< 10^{-6}$ mbar) residual pressures of oxygen. It seems likely that water vapour (usually the dominant residual gas in vacuum systems) will play a similar role.

It is not enough to consider milling a plane surface; we have a pattern to transfer through a mask (usually photoresist), and the shape of the surface evolves with time, often in a non-intuitive way. The surprising shapes that can result are due to the angle dependence of milling: the milling rate for superconductors and photoresist peaks at an angle of incidence of around 45°. Facets develop and there is a direct analogy with Frank's theory of crystal dissolution [24, 25]. There is a tendency for a curved surface to turn into two plane surfaces separated by a sharp cusp [26], in much the same way as facets develop on dissolving crystals. On a convex surface, the fastest milling facets develop, and on a concave surface the slowest milling ones. Under some conditions, a smooth surface can spontaneously roughen under grazing incidence milling [27], conditions that might intuitively be expected to polish it.

While the theories can explain some of the features seen in experiments, many ignore second order effects. Wilson *et al* [28] identified three second order effects. These are ion reflection at grazing incidence, forward peaked sputtering at grazing incidence and redeposition of sputtered atoms. They also cite numerous further complications, but so far as we know, their three second order effects suffice to explain what is observed in patterning HTS. These effects are hard to model because they are non-local. Ion reflection and forward peaked sputtering are both potential contributors to the way morphology develops near a steep slope, and redeposition is observed almost anywhere that one looks for it.

The redeposition of milled material onto adjacent surfaces is of considerable practical importance. For example, suppose we are milling a straight vertical edge on an otherwise featureless sample, and that sputtered atoms are emitted at an angle θ from the surface normal with a $\cos\theta$ distribution. Then the redeposition rate on the vertical edge is half the milling rate—hardly a negligible effect. Thus the net milling rate of a slope will often be significantly different from that measured on a planar sample at the same milling angle. In addition, both the redeposition and milling process have a random element; they comprise atoms or clusters rather than a continuum. Surfaces subjected to milling in the presence of a flux of redeposited atoms tend to roughen. However, the presence of resputtered material can have a positive effect by reducing the retreat rate of a resist pattern as shown in figure B4.4.1.11. The effect here is amplified, as the sample was not rotated and the edge shown was in constant shadow. This effect is particularly important in HTS systems where the HTS to resist milling rate selectivity is poor.

In most milling processes the sample is rotated for improved uniformity and to get an isotropic process. This can make it more difficult to visualize what is happening; some parts of the evolving surface may be temporarily shadowed during rotation. During this period they may accumulate redeposited material. When the surface emerges from shadow, the redeposited material must be milled off before any net material removal takes place on that part of the surface.

The second non-local effect is due to the reflection of high energy ions from surfaces on which they are incident at near-grazing incidence. Little energy is lost in such a collision and the reflected ion is still able to sputter. The result is that a shallow trench is formed at the base of a vertical wall as the reflected ions enhance the milling rate near the wall. This effect is known as trenching [29] and is usually suppressed by milling away from normal incidence. A competing effect is the tendency near grazing incidence for material to be milled off in the forward direction, leading to a reduced net milling rate [30], and a foot forms near the wall.

B4.4.1.3.5 Milling high temperature superconductors

Little systematic information has been published that might allow a comparison between the relatively simple materials used in the studies cited above, and HTS. There are many papers that quote HTS milling rates for a specific set of conditions. These are broadly consistent with the above: milling yields are about 1 atom/ion at around 500 eV, typical of slow milling material [19]. Photoresist mills somewhat faster than HTS. The secondary ions and energies produced in milling with 4 KeV Ar^+ ions have been studied by Chenakin [31]. He found an initial transient in the milling rate consistent with the formation of a surface layer of changed stoichiometry and that secondary ion energies are ~ 10 eV. Detailed SIMS spectra were also reported by Gauzzi *et al* [32] using 5 KeV Ar^+ ions and by Ilyinsky *et al* [33] for lower energy ions. Ilyinsky *et al* found that the addition of oxygen to the milling gas resulted in significant changes in the secondary ions produced.

E3 Figure B4.4.1.8 shows the dependence of the milling rate on the angle of incidence of the Ar^+ ions for YBCO and other materials used in HTS circuits. Other data in the literature are broadly consistent with this picture, although there is considerable variation in detail.

Because ion milling has only small differentials between materials, etch stop layers are not available and it is very useful to be able to monitor the composition of the material being milled (end point detection). The most general method is secondary ion mass spectrometry (SIMS), which can easily give a depth resolution of 4 nm in routine use [35]. This paper also showed that the surface of the sample is fully hydrated for water vapour pressures in the milling chamber $> 10^{-6}$ mbar. Below this pressure, the secondary ions detected contain fewer species like $(YOH)^+$. For the milling conditions used, the threshold for dehydration of the surface corresponds to a water vapour incidence rate comparable to the rate at which material is removed from the milled surface. Any effect of the water vapour pressure on the milling rate of the sample was slight, however.

We might expect only the yttrium (or rare earth) oxygen bond to be strong enough to hold onto oxygen, with the result that the surface becomes significantly yttrium rich. This is observed in practice. Figure B4.4.1.9 shows the SIMS traces from milling through a layer of YBCO grown on a second YBCO film, where the surface of the first film has been prepared by ion milling *in situ* before growing the second layer. It is clear that the interface layer contains excess Y and is deficient in Ba and Cu. If the first film is not milled before the growth of the second, no interface peak is observed (figure 4.4.1.9(*b*)). It is difficult to make such data quantitative, but the Y excess at the interface corresponds to ~ 0.5 nm of Y_2O_3. A similar change in surface stoichiometry has also been observed by Wen *et al* [36]. Reducing

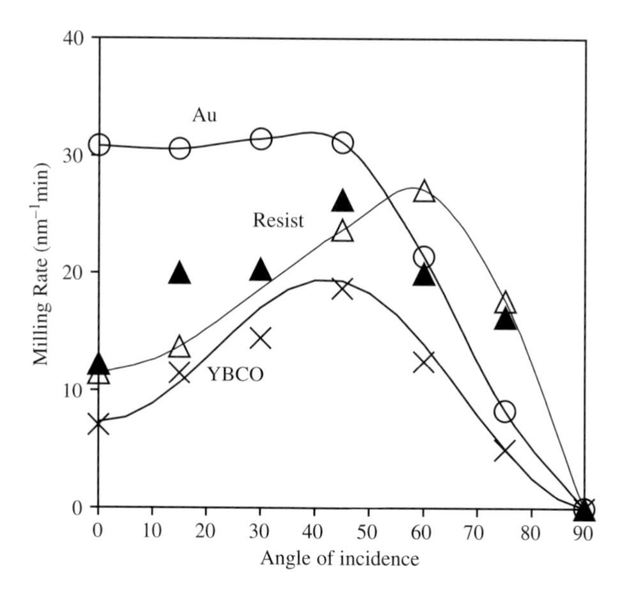

Figure B4.4.1.8. Angle dependence of the milling rate of YBCO, and AZ1518 resist (solid triangles) from Goodyear [34] for 500 eV Ar ions at a beam current density of $\sim 0.3\,\mathrm{mA\,cm^{-2}}$. These are compared to data for AZ1350 resist (open triangles) and Au from Johnson [30]. The relative vertical scales between data sets taken in different laboratories are our best guesses.

the milling energy is expected to reduce the depth of damage, but so far we have not identified a trend in the SIMS data. Low energy milling is quite widely used for preparing the surfaces of edge barrier junctions [37, 38], and there are many reports that this is necessary to prepare high J_c connections between layers, although the experience is not universal.

It is well known in the preparation of specimens for transmission electron microscopy that reducing the sample temperature during milling is advantageous for minimizing damage, and some groups have cooled their samples in HTS patterning. For example Schneidewind *et al* [39] have reported that

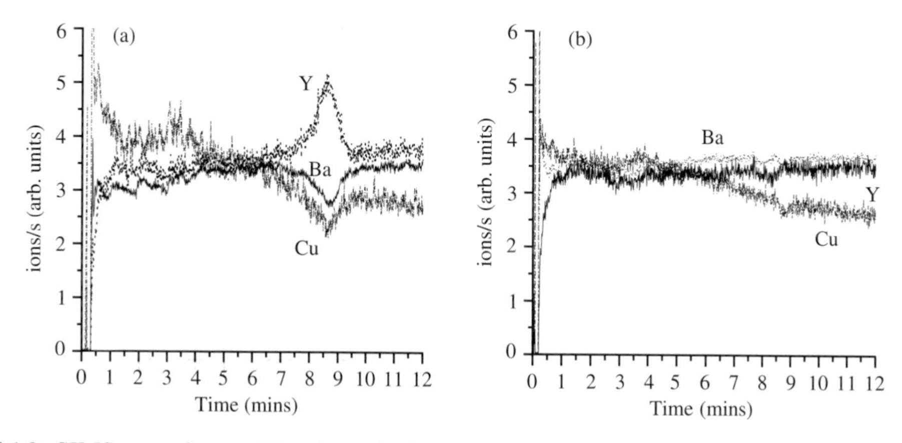

Figure B4.4.1.9. SIMS traces from milling through a layer of 0.1 μm thick grown onto another YBCO layer. (*a*) the surface was ion cleaned before the growth of the second layer for 2 min at 500 V followed by 2 min at 300 V. (*b*) No ion clean as all. The figure shows that ion milling leaves the surface Y rich.

the critical currents of microbridges milled at low temperature are higher than those milled at room temperature. A significant improvement in J_c and T_c was found even for 20 μm wide bridges. Bearing in mind that only the edges of the bridge are actually milled, this is hard to understand in terms of the expected chemical diffusion coefficient of oxygen [40]. Degradation on this scale is apparently implied by a number of transport measurements on narrow lines, e.g. [41]. However, in nearly all cases, the lines were long compared to their width and it is difficult to distinguish effects of film non-uniformity from deoxygenation. An exception is the work of de Nivelle *et al* [42] who made very short bridges. Their T_c degraded from 87 K in the original film to 83 K for a bridge width of 0.175 μm. This suggests that, at least in some materials, deoxygenation is only significant on a length scale $\sim$0.1 μm.

B4.4.3

Ban *et al* [43] have reported that introducing oxygen into the milling gas improves the quality of microwave resonators. Our attempts to look for similar effects [44] failed to identify any difference in properties between milled and wet etched resonators.

In our opinion, this area requires further study. What underlies the wide range of conclusions about damage during milling? Certainly some of the results will have been obtained on poor quality material or under poorly controlled conditions, and can be discounted. Nevertheless, there does seem to be a body of evidence that suggests that some materials can lose oxygen over a longer range than the known diffusion data would allow, but there are few studies in the literature of lateral diffusion in thin films. It is worth pointing out that there is a large spread in results of chemical diffusion measurements in HTS [45]. There are several reasons for this. First of all, the anisotropy is very large, and there is clearly a strong possibility that diffusion, particularly in the *c*-axis direction, is dominated by defects. This is believed to be the mechanism by which thin films oxygenate [46]. The second point is that if diffusion is the limiting mechanism, detailed balance requires that the rate of oxygenation should equal the rate of deoxygenation at sufficiently small deviations from equilibrium. This is seen in some sets of results (e.g. [47]) but by no means in all. Where the two rates differ, either the driving force is too large for a simple analysis or the rate limiting step is not diffusion but at the surface of the sample [48]. Finally, Yamamoto *et al* [18] have shown that oxygenation rates are much larger for atomic oxygen than molecular. These experiments were in the large signal limit, but nevertheless the authors argue fairly convincingly that the most likely explanation is a chemical potential gradient due to lattice strain. We have not studied deoxygenation of milled edges in detail, but we have concluded that re-oxygenation of all circuits after patterning is the safest route.

B4.4.1.4 HTS processes using ion milling

Having discussed the methods and materials issues of the processing of superconductors, now we try to illustrate how these disparate processes are stitched together to make circuits. It is worth noting at the outset that making a complete circuit involves a large number of process steps. The yield of the whole process will be the product of the yields of the individual steps. A complex process will only give a useful yield if all of the individual steps are very reliable.

E3

The only HTS thin film circuits currently in production are passive microwave circuits. These only require the patterning of a single layer of superconductor. The processing of multilayer circuits like SQUIDs, flux transformers and single flux quantum logic circuits, is much more complex than the single layer circuits used for passive microwave devices.

E4.2, E4.5

B4.4.1.4.1 Single layer circuits

E3

Passive microwave circuits require a single layer of superconductor to be grown on each side of a substrate, to a thickness >0.5 μm. Only one of these layers is patterned. Normal metal (Au or Ag) contacts to both layers are required. The processes used in production are proprietary, and the following is only an outline of how patterning might proceed.

Typically, the patterning is done using a photoresist mask, produced as described in section B4.4.1.2. Positive resist is used as it has better resolution than negative, is easier to strip, and is very reliable, with good process margins.

The wafer is then mounted in the ion miller with appropriate precautions to heat sink the sample. The wafer is then milled, rotating for improved uniformity, with a low angle of incidence ($\leq 30°$) to give a steep side wall. Dimensions can be controlled to sub-micron accuracy.

Normal metal contacts can be added either by lift off or by subtractive patterning. E5.1

Our standard process uses an anneal to re-oxygenate edges after patterning, although whether this is necessary for passive microwave devices is not clear.

An altogether more sophisticated process, but which still involves only a single superconducting layer, is required to make high sensitivity bolometers for infra-red detection. In this application, a superconducting meander line is maintained close to T_c and the heating of the meander line by the radiation is used as the detection mechanism. It is central to achieving high performance that the thermal conductance between the detector element and the substrate be minimized. This has led to some innovative HTS processes. Lee *et al* [49] made use of the fact that their thin film $SrTiO_3$ was soluble in HF, while their $YBa_2Cu_3O_7$ was not. This allowed them to etch away a $SrTiO_3$ layer from beneath an $YBa_2Cu_3O_7$ film, resulting in an air bridge. From the results in their paper, this process appears to have worked very well. C1

However, it is worth remarking that bulk $SrTiO_3$ does not dissolve in HF: their thin film $SrTiO_3$ must have done so because of defects. When we have attempted similar experiments with our $YBa_2Cu_3O_7$, it dissolved in the HF, so evidently the defects in our $YBa_2Cu_3O_7$ were different from those in the material used by Lee *et al*. This makes a very important point: processing is very dependent on material quality. A process may depend on being able to produce 'poor' quality material. Processes rarely transfer between laboratories without amendment.

A second generation process from the same laboratory [50] used a sacrificial YBCO layer, dissolved by dilute HNO_3 instead of the $SrTiO_3$. The circuit was separated from the sacrificial layer by ZrO_2:Y. This is obviously a rather difficult process, as any defects in either the resist or the ZrO_2 layer will lead to failure.

A more conservative solution to the same problem was presented by Mechin *et al* [51] They deposited YBCO films on Si substrates, using a CeO_2/ZrO_2 buffer layer. They were able to make these films free-standing using RIE etching of the underlying silicon. A similar process was also adopted by de Nivelle *et al* [52]. B4.2.1 B4.2.2

B4.4.1.4.2 Multilayer circuits

A minimum circuit with two superconducting layers, one insulator layer and a contact layer involves at least three growth processes, one metallisation and four patterning steps. Making circuits containing junctions is usually considerably more complex. This means that the yield at each step has to be very high if a circuit is to be completed successfully with a high probability. B4.4.2 B4.4.3

For multilayer circuits, the substrate must be dimensionally stable so that subsequent mask layers align properly with the first. This rules out substrates like $LaAlO_3$ that undergo a structural phase transition below the growth temperature, and change shape between one growth run and the next as a result. The most popular substrate is MgO, although $SrTiO_3$ is acceptable for some circuits. B4.1

Wet etching has only been used rather rarely [53] for making multilayer circuits; nearly all are made by ion beam milling. In milling, we believe end point detection is essential if errors are not to build up in processing, while for single layer circuits it is merely convenient.

In multilayer circuits, it is important to minimize parasitic inductances [54]. This demands steep edges. It is also vital to avoid introducing unintended grain boundaries, which suggests using gentle slopes. Thus it is important to produce well controlled slope angles on patterned edges.The most common way of doing this is to reflow the resist pattern after it has been exposed and developed by heating the wafer above the T_g of the resist. As shown in figure B4.4.1.10 this generates a sloped profile to the resist which when milled in rotation produces a sloped edge superconductor profile. Angles of the slope usually range from 15 to 45°. The slopes produced tend to be slightly concave but, for successful epitaxial growth over an edge, this is not a problem.

Another process, which is used mainly for ramp junctions, uses a uni-directional mill (no sample rotation) as shown schematically in figure B4.4.1.11. This process can produce reproducible edge angles up to 60° which are very straight. This is because the edge is in constant shadow allowing a thick redeposited layer to build up in the shadow region, halting the resist edge retreat. This process has the drawback that the slope only faces one direction. This may be acceptable for technology development but can be a design limitation for complex circuits.

Growth issues

Once the first layer of a circuit has been patterned, subsequent layers must be grown epitaxially everywhere, including over the patterned steps. It is well known that growth over steep steps in the substrate material can result in a-oriented growth on the step [55], so that grain boundaries form at the top and bottom of the step. The problem is more serious on MgO, where growth on a misoriented

B4.1

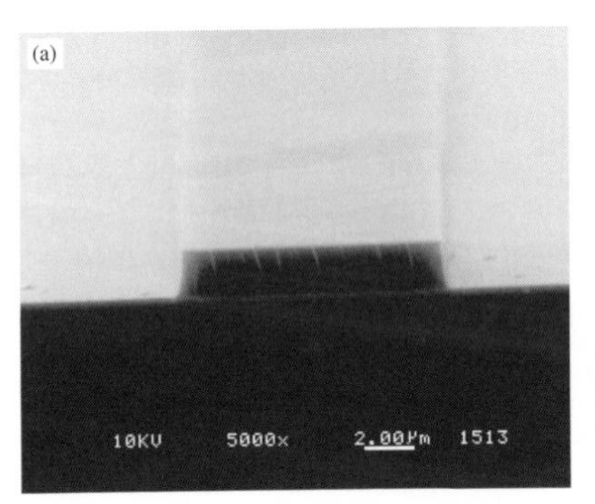

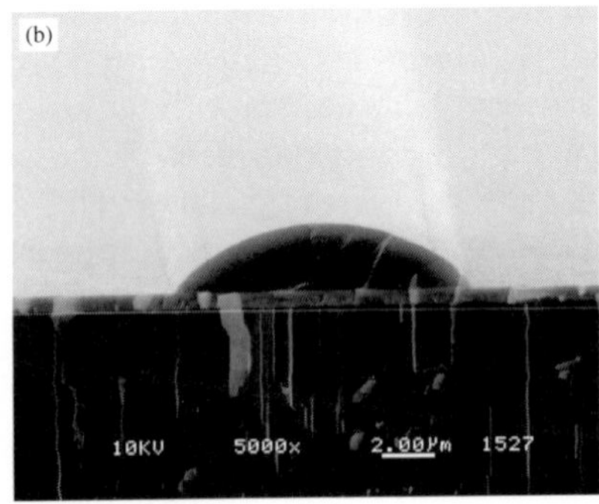

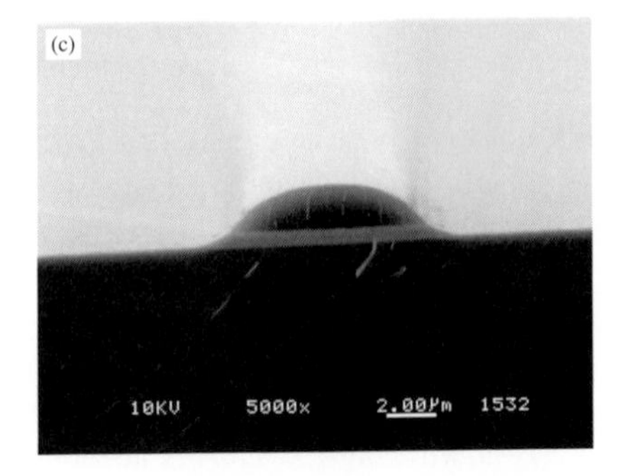

Figure B4.4.1.10. (*a*) As patterned resist profile with near vertical edges. (*b*) Resist profile of patterned and post development baked (reflow) at 120°C. (*c*) Profile of patterned, 120°C reflowed and ion milled 450 nm into a YBCO film. Note the retreat of the resist and the gentle concave edge slope of the superconductor track.

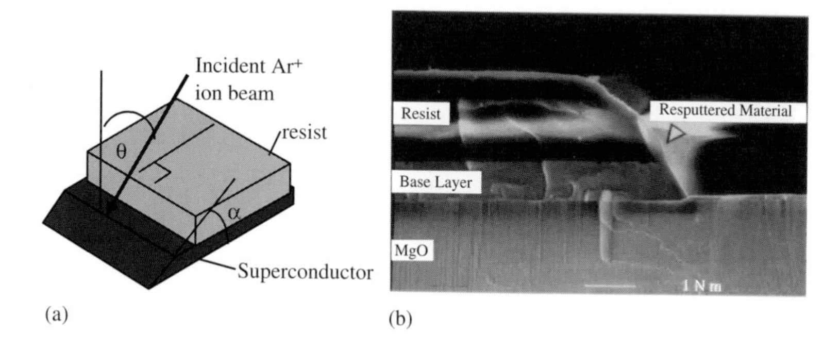

Figure B4.4.1.11. (*a*) Schematic view of the uni-directional milling process. The edge is milled with no rotation with the ion beam incident at some angle θ from normal and perpendicular to the edge, leaving this edge in constant shadow. (*b*) Cross section of a 800 nm thick YBCO film after milling using this process with the resist still attached, showing clearly the resputtered material. Note the retreat of the corner of the resist and that this is halted by the resputtered material.

substrate results in a film whose c-axis is tilted with respect to the substrate lattice. Even a very small step height, say 20 nm, is enough to result in grain boundaries in growth on a MgO step, and such a step is inevitable if patterning is to be taken to completion. It has been conjectured [56] that it may be necessary to use a planar technology to avoid grain boundaries when films are grown over steps, a very substantial complication, possibly enough to rule out HTS SFQ altogether. However, there is a simple solution, adopted by a number of groups, which is to grow a layer of $PrBa_2Cu_3O_7$ underneath the first superconducting layer. With this isostructural buffer layer *c*-axis orientation is assured even over steep slopes.

For successful growth over steps, it is important that the top of the step be free of redeposited material. A thick redeposited layer is usually associated with shadowing and, in most cases, the problem can be avoided by milling sufficiently close to normal incidence (say $\leq 30°$) that a thick redeposited layer does not form. In some cases even a thin layer will continue to adhere (this is a particular problem in making small vias, where a ring of redeposited material forms and has considerable strength because it forms a complete circle). Then the only recourse is to modify the process sequence or conditions to find a process that removes the redeposited material.

Even when these precautions have been taken, we find that growth over steps can lead to a rough surface. The roughness takes the form of multiple deep depressions that form above precipitates within the film [57]. Provided the edge is free of grain boundaries, however, it is not of crucial importance for interconnects. The problem has been studied more closely for ramp junctions, discussed in more detail in chapter B4.4.3. Several groups have found that a low energy ion clean is essential if a milled ramp is not to roughen when annealed under the growth conditions [58]. The underlying mechanism of this has not been clarified. It is tempting to suggest that the roughening is due to recrystallization of the Y-rich surface layer, and that low energy milling results in this layer being thinner. However, we can find no evidence that the amount of excess Y in this layer depends significantly on milling energy. We suggest, rather tentatively, that the roughening could be associated with recrystallization of the underlying YBCO as the milling damage anneals out.

B4.4.3

Another problem that can arise in the growth on patterned base layers is a change in orientation of growth on the substrate. This can be a particular problem on MgO, but other substrates can also show the effect. Early on, we used this effect for making a kind of bi-epitaxial junction [59]. We now believe that this was related to the presence of a very thin contaminant layer on the substrate surface due to redeposition from surrounding parts of the milling system. The effect of monolayer levels of contamination on film orientation has been graphically demonstrated by Garrison *et al* [60].

A number of groups have found that the quality of epitaxy and electrical properties are compromised in the growth of higher superconducting layers, although we believe that for most this problem has now been overcome. In our experience an *in situ* ion clean prior to growth is essential for reproducible results, although growth sometimes works satisfactorily without it. Nucleation at a reduced temperature can also be required to encourage epitaxy, followed by an increase in temperature once nucleation is complete.

Insulators

A multilayer circuit process requires not only the epitaxial growth of superconductor, but also an insulator to separate the layers. Two types of insulator can be considered. The most obvious are materials like Y_2O_3, $LaAlO_3$ and $SrTiO_3$ that support mutual epitaxy in both directions with superconducting films. These insulators have very low leakage resistance, although $SrTiO_3$ has a high dielectric constant and is thus often unsuitable. There are two problems with these insulators. They are usually cubic, so when the next superconducting layer is grown over a step, there is a risk that the film will not grow with the *c*-axis normal to the substrate, as is usually required. This problem can often be avoided by making step angles small, but this is undesirable in many circuits because it introduces extra inductance. The second problem with these insulators is that they are impermeable to oxygen if they are pinhole-free, and if they are not they may allow shorts between layers. In most growth processes, although perhaps not all [61], films are grown with a low oxygen content and must be oxygenated afterwards, usually during cooling after growth. Rowell [56] highlighted this as a key issue in processing HTS circuits. It may be possible to get round this problem by including holes in the insulator to allow oxygen through [62], but this is undesirable as it means compromising the circuit layout. In our experience, it is also difficult to achieve full oxygenation in this way. A possible solution that has not been attempted to our knowledge is to grow the insulator at a temperature where the superconductor can be kept oxygenated, and hope to lock the oxygen in the lower layers during growth of the higher ones. These very ionic oxides can be grown at very low temperatures provided they can be made to nucleate: MgO can be grown at 100 K (not C!), for example [63] but will not nucleate epitaxially on YBCO in our experience. A third problem, at least with Y_2O_3 [64], is that it can be grown very satisfactorily on YBCO, but cracks appear when it is then raised in temperature to grow the next superconducting layer. The cracks allow the base layer to oxygenate, but can result in shorts between layers.

E3, E4

C1

All these difficulties have led many groups to compromise on the insulator resistivity and use $PrBa_2Cu_3O_7$, which is isostructural with YBCO but not superconducting, as the 'insulator'. This immediately solves the problem of grain boundaries at steps, oxygen diffusion and cracking and replaces them with a problem of leakage resistance. This can be a problem in flux transformers, where leakage resistance causes Johnson noise. The dielectric constant of PBCO is also rather high (~ 80). The leakage resistance problem can be solved if necessary by doping e.g. with gallium [65]

Planarisation

For complex circuits, it is sometimes necessary to planarise. The similarity in milling rates at low angles of incidence between YBCO and photoresist can be used to planarize film surfaces. Resist forms a smooth surface over a rough film or fine patterning features and, if the resist is then milled away, the resulting film surface is much smoother, see figure B4.4.1.12. This can be used to open contact holes [66] or produce planar multilayer circuits [67]. Another approach to planarization is grazing incidence

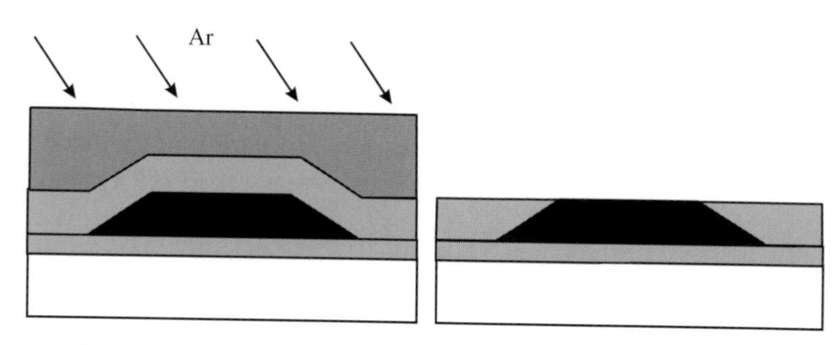

Figure B4.4.1.12. The etch back planarization process. The process here is being used to open up contact to a lower superconductor track.

Table B.4.4.1.4. A summary of some of the key technology choices made for multilayer processing for a number of research groups

Group	Substrate	Buffer	Insulator	S/C layers	EPD	Ref.
Jülich	STO		STO	2		[71]
NEC	STO		LSAT	3		[72]
Northrop-Grumman	$NdGaO_3$		STO, CeO_2	3	SIMS	[8]
QinetiQ	MgO	PBCO	PBCO	2	SIMS	[3]
TRW	$NdGaO_3$	STO	STO	3	SIMS	[70]
Twente	MgO	STO	PBCO	3	Optical[a]	[58]

[a] When using DyBCO as the superconductor a green plasma glow is given off during ion milling, this can be used as a crude endpoint detection system.

milling [68]. Milling using 1 kV Xe^+ ions at an angle of incidence of 80° was found to reduce the surface roughness of post-annealed films. Planarization can also be achieved by mechanical polishing [69].

It is difficult to summarize the current state of the art of multilayer HTS circuits. Each laboratory in the field has a suite of processes that are evolving continuously. A recent paper from TRW reviewed the historical development of their processes [70] and illustrates this evolution rather well. All the groups involved in making multilayer single flux quantum logic circuits use postbaked resist to round off the resist profile and hence produce controlled slopes, although a remarkable range of different baking conditions has been reported. Most groups use *in situ* ion cleaning of a patterned wafer before the next layer is grown. All use YBCO (or a rare earth 1–2–3 compound) as the superconductor. Table B4.4.1.4 gives a list of some of the key characteristics of processes used by groups working on HTS SFQ.

E3, E4
E4.5

B4.4.1.5 Conclusion

Solutions have been found to nearly all of the problems of processing HTS circuits. We believe that the biggest unsolved problem is making reproducible junctions, which is discussed in another chapter. However, there is much that we do not yet understand about how these processes work. If HTS circuits are to become a production technology, work will be needed to establish a more detailed understanding of what is happening in processing so that optimum processes can be found. In particular, we identify the oxygen diffusion problem, and what happens to the surface in milling as demanding further study.

E3, E4

References

[1] Moreau W M 1988 *Semiconductor Lithography. Principles, Practices and Materials* (New York: Plenum Press)
[2] Sheats J R, Newman N, Taber R C and Merchant P 1994 *J. Vac. Sci. Technol.* **A12** 388
[3] Larsson P, Nilsson B, Yi HR and Ivanov ZG. 1995 Proc. Appl. Supercond. 1995, Inst. Of Physics series 148 p 935
[4] Hirst P J, Henrici T G, Atkin I L, Satchell J S, Moxey J, Exon N J, Wooliscroft M, Horton T and Humphreys R G 1999 *IEEE Trans. Appl. Supercond.* **9** 3833 Part 3
[5] Yi H R, Winkler D, Ivanov Z G and Claeson T 1995 *IEEE Trans. Appl. Supercond.* **5** 2778
[6] Huh Y, Kim J-T, Hwang Y, Park Y, So J, Kim I S, Lee S-G, Park G, Park Y-K and Park J-C 1998 *Japan. J. Appl. Phys.* **37** p 2478
[7] Neiman R L, Giapintzaskis J, Ginsberg D M and Molech J M 1995 *J. Supercond.* **8** 383
[8] Forrester MG, Hunt BD, Miller DL, Talvacchio J and Young RM 1999 Proc. 7th Int. Superconductive Electronics Conference (ISEC) (Berkeley, USA, June 1999) p 29
[9] Roas B 1991 *Appl. Phys. Lett.* **59** 2594
[10] Shih I and Qiu C X 1988 *Appl. Phys. Lett.* **52** 1523–4
[11] Martens J S, Zipperian T E, Ginley D S, Hietala V M, Tigges C P and Plut T A 1991 *J. Appl. Phys.* **69** 8261
[12] Ashby C I H, Martens J, Plut T A, Ginley D S and Phillips J M 1992 *Appl. Phys. Lett.* **60** 2147
[13] Ginley D S, Barr L, Ashby C I H, Martens J, Plut T A, Urea D and Siegal M P 1994 *J. Mat. Res.* **9** 1126
[14] Vasquez R P, Hunt B D and Foote M C 1988 *Appl. Phys. Lett.* **53** 2692–4
[15] Auciello O Kelly R ed. 1984 *Ion Bombardment Modification of Surfaces; Fundamentals and Applications*, (Amsterdam: Elsevier)
[16] Wittmaak K 1994 *J. Vac. Sci. Technol.* B **12** 258
[17] Gao, Boguslavskÿ Y M, Klopman B B G, Terpstra D, Wÿbrans R, Gerritsma G J and Rogalla H 1992 *J. Appl. Phys.* **72** 573
[18] Yamamoto K, Lairson B M, Bravman J C and Geballe T H 1991 *J. Appl. Phys.* **69** 7189
[19] Spencer E G and Schmidt P H 1971 *J. Vac. Sci. Technol.* **8** S52
[20] Morgulis N D and Tischenko V D 1957 *Bull. Acad. Sci. USSR* **20** 1082
[21] Pivin J-C 1983 *J. Mat. Sci.* **18** 1267
[22] Mader L and Hoepfner J 1976 *J. Electrochem. Soc.* **123** 1893
[23] Kelly R 1978 *Nucl. Instrum. Methods* **149** 553
[24] Barber D J, Frank F C, Moss M, Steeds J W and Tsong I S T 1973 *J. Mat. Sci.* **8** 1030
[25] Carter G, Colligon J S and Nobes M J 1973 *J. Mat. Sci.* **8** 1473
[26] Nobes M J, Katardjiev I V, Carter G and Smith R 1987 *J Phys.* D **20** 870
[27] Sigmund P 1973 *J. Mat. Sci.* **8** 1545
[28] Wilson I H, Belson J and Auciello O 1984 *Ion Bombardment Modification of Surfaces; Fundamentals and Applications*, eds O Auciello and R Kelly (Amsterdam: Elsevier) p 225
[29] Ducommun J P, Cantagrel M and Moulin M 1975 *J. Mat. Sci.* **10** 52
[30] Johnson L F 1984 *Ion Beam Modification of Surfaces*, eds O Auciello and R Kelly (Amsterdam: Elsevier) p 361
[31] Chenakin S P 1990 *Vacuum* **42** 139
[32] Gauzzi A, Mathieu H J, James J H and Kellett B 1990 *Vacuum* **41** 870
[33] Ilyinsky L S, Emmoth B, Hollmann E K, Zaitsev A G and Lavrentyev A A 1995 *J. Phys.* D **28** 996
[34] Goodyear S W, PhD Thesis Birmingham University
[35] Humphreys R G, Chew N G, Morgan S F, Satchell J S, Cullis A G and Smith P W 1992 *Appl. Phys. Lett.* **61** 228
[36] Wen J G, Satoh T, Hidaka M, Tahara S and Koshizuka N 1999 *Appl. Phys. Lett.* **75** 2470
[37] Verhoeven M A J, Gerristsma G J and Rogalla H 1995 *Appl. Supercond.* **1+2 148** 1395
[38] Horstmann C, Leinenbach P, Engelhardt A, Gerber R, Jia J L, Dittman R, Hartmann U and Braginski A I 1998 *Physica* C **302** p 176
[39] Schneidewind H, Schmidl F, Linzen L and Seidel P 1995 *Physica* C **250** 191
[40] chemical diffusion ref.
[41] Sato H, Akoh H, Nishihara K, Aoyagi M and Takada S 1992 *Japan. J Appl. Phys.* **31** L1044
[42] Ban M, Takenaka T, Hayashi K, Suzuki K and Enomoto Y 1996 *Japan. J Appl. Phys.* **35** 4318
[43] de Nivelle M J M, Gerritsma G J and Rogalla H 1993 *Phys. Rev. Lett.* **70** 1525
[44] Porch A, Lancaster M J and Humphreys R G 1995 *IEEE Trans. Microwave Theory Tech.* **43** 306–14
[45] Conder K and Krüger C 1996 *Physica* C **269** 92
[46] Aarnink W A M, Ijsselsteijn R P J, Gao J, van Silfhout A and Rogalla H 1992 *Phys. Rev.* B **45** 13002
[47] Dediu V and Macotta F C 1996-II *Phys. Rev. B* **54** 16259
[48] Kittelberger S, Bolz U, Huebener R P, Hozapfel B and Mex L 1998 *Physica* C **302** 93
[49] Lee L P, Burns M J and Char K 1992 *Appl. Phys. Lett.* **61** 2706
[50] Berkowitz S J, Hirahara A S, Char K and Grossman E N 1996 *Appl. Phys. Lett.* **69** 2125
[51] Mechin L, Villegier J-C and Bloyet D 1997 *IEEE Trans. Appl. Supercond.* **7** 2382
[52] de Nivelle M J M, Bruijn M P, de Korte P A J, Sanchez S, Elenspoek M, Heidenblut T, Schwierzi B, Michalke W and Steinbeiss E 1999 *IEEE Trans. Appl. Supercond.* 3350

[53] Faley M I, Poppe U, Jia C L, Dahne U, Gondanov U, Klein N and Urban K 1995 *IEEE Trans. Supercond.* 2 **5** 2608
[54] Atkin I L, Satchell J S, Hirst P J and Humphreys R G 2000 *Appl. Phys. Lett.* **77** 1366
[55] Jia C L, Kabius B, Urban K, Herrman K, Jui G J, Shubert J, Zander W, Braginski Q A I and Heiden C 1991 *Physica* C **175** p. 545
[56] Rowell J M 1999 *IEEE Trans. Appl. Supercond.* **9** 2843
[57] Hirst P J, Barnett M A, Chew N G, Abell J S, Aindow M and Humphreys R G 1994 *Inst. Phys. Conf. Ser.* **158** 185
[58] Verhoeven M A J, Moerma R, Bijlsma M E, Rijnders A J H M and Blank D H A 1996 *Appl. Phys. Lett.* **68** 1276–8
[59] Chew N G, Goodyear S W, Humphreys R G, Satchell J S, Edwards J A and Keene M N 1991 *Appl. Phys. Lett.* **60** 1516
[60] Fork D K, Garrison S, Hawley M and Geballe T H 1992 *J. Mat. Res.* **7** 1641
[61] Teshima H, Shimada H, Imafuku M and Tanaka K 1993 *Physica* C **206** 103
[62] Talvacchio J, Young R M, Forrester M G and Hunt B D 1999 *IEEE Trans. Appl. Supercond.* **9** 1990
[63] Yardavalli S, Yang M H and Flynn C P 1990-I *Phs. Rev.* B **41** 7961
[64] Chew N G, PhD Thesis University of Birmingham
[65] Verhoeven M A J, Gerritsma G J, Rogalla H and Gulabov A A 1996 *Appl. Phys. Lett.* **69** 848
[66] Goodyear S W, Chew N G, Humphreys R G, Satchell J S and Lander K 1995 *IEEE Trans. Appl. Supercond.* **5** 3143
[67] Marathe A P, Ludwig F, van Duzer T and Lee L 1995 *IEEE Trans. Appl. Supercond.* **5** 3135
[68] Hebard A F, Fleming R M, Short K T, White A E, Rice C E, Levi A F J and Eick R H 1989 *Appl. Phys. Lett.* **55** 1915
[69] Nilsson P A, Brorson G, Orliaguet J M, Olsson E, Olin H, Gustafsson M and Claeson T 1995 *IEEE Trans. Appl. Supercond.* **5** 1653
[70] Pettiette-Hall C L, Murduck J, Burch J F, Sergant M, Hu R, Cordromp J and Aquilino H 1999 *IEEE Trans. Appl. Supercond.* **9** 1998
[71] Hansohm J K, Hadfield R H and Dittman R 1999 *Physica* C **326–327**
[72] Satoh H, Wen J G, Hidaka M, Koshizuka N and Tanaka S 2000 *Supercond. Sci. Technol.* **13** 88–92

B4.4.2
Processing and manufacture of Josephson junctions: low T_c

Jürgen Niemeyer and Hisao Hayakawa

B4.4.2.1 Introduction

The development of low-temperature superconductivity (LTS) junction and circuit technology was stimulated by all applications, which make use of flux quantization and which, therefore, are without competition by semiconductor circuits: SQUID circuits, Josephson array voltage standards and single flux quantum (SFQ) digital circuits. These applications set the following requirements for junction technology.

B4.4.3 A2.6 E4.2, A2.7 E4.5

- SQUIDs require one or two small, overdamped junctions (shunted, low junction capacitance) with moderate critical current densities in the range from 10^2 to $10^3\,\mathrm{A\,cm^{-2}}$ and large I_cR_n of 1 mV or more. Together with the signal coupling circuit, a multi-layer technology with moderate linewidths and complexity is required. E4.2 A2.7
- Conventional voltage standards make use of a series array with a large number of underdamped (high junction capacitance, no shunt) Josephson tunnel junctions with low critical current densities between 10 and $10^2\,\mathrm{A\,cm^{-2}}$. In principle, I_cR_n may be relatively small but, as unshunted tunnel junctions are used, it is found to be $I_cR_n = (\pi/4)V_g$, i.e. about 2 mV for a symmetric Nb tunnel junction. V_g is the gap voltage of the junction. Series array voltage standards require multi-layer circuits with moderate linewidths and integrated microwave circuit elements. No junction failure is permitted in general. A2.7
- Digital circuits (SFQ) and programmable voltage standards require overdamped junctions with critical current densities and I_cR_n products as large as possible ($J_c > 10^3\,\mathrm{A\,cm^{-2}}$, $I_cR_n > 0.5\,\mathrm{mV}$) to achieve high frequency operation. These requirements finally will lead to a complex small linewidth technology for circuits. E4.5

Moreover, these applications of LTS electronic circuits call for a junction technology which provides high stability to room temperature storage and thermal cycling as well as a very small parameter spread in multi-junction circuits. Internal shunting is advantageous for junction damping in high-speed circuits at the level of high integration. A2.7

B4.4.2.2 Junction technology

B4.4.2.2.1 Underdamped refractory metal tunnel junctions, circuits

Nb/Al$_2$O$_3$/Nb

To achieve the described objectives, it was suggested at the beginning of the 1980s using Al oxide barriers between Nb electrodes [1, 2]. Based on this invention, mainly in the 10 years that followed, a very reliable complete wafer trilayer sandwich technology was developed, most intensively in Japan [3–6]. This development became useful for all of the above-mentioned applications as well as for SIS mixer applications in radio astronomy; for an overview, see [7, 8]. The latter applications were important for the development of small area tunnel junctions with extremely low subgap leakage currents. A typical dc characteristic of a series array of nearly 10 000 Nb/Al$_2$O$_3$/Nb (SIS) tunnel junctions with the typical hysteretic switching behaviour is shown in figure B4.4.2.1. The figure shows that the junction parameter spread is very low [9] because the dc characteristic is similar to the characteristic of a single junction.

E4.7

Critical current densities of up to more than 10^5 A cm^{-2} can be reached [10]. High-quality Nb films with London penetration depths of 40 nm and Al films are deposited by dc magnetron sputtering processes. (In the case of a 6 inch sputtering target, typical process data for Nb are: film thickness, 150 nm; rate, 150 nm min^{-1}; dc voltage, 330 V; Ar pressure, 0.75 Pa; for Al: film thickness, 12 nm; rate, 8 nm min^{-1}; dc voltage, 270 V; Ar pressure, 1.2 Pa.) The Al$_2$O$_3$ barrier is formed by a temperature-, pressure-, and time-controlled oxidation process, for example, 250 mbar O$_2$ at 25°C for 24 h leads to a critical current density of 15 A cm^{-2} [11]. The activity of the oxide atmosphere can be enhanced by ultraviolet radiation during the oxidation process [12]. In most cases, the junction sandwich is prepared without interrupting the process by patterning, and it covers the substrate (typically an Si wafer) completely. After this process, the junction area and the base electrode are structured by a reactive ion etching process with CF$_4$ or by ion beam etching with Ar ions. Then an insulating window layer is applied (typically 250 nm SiO$_2$ free of stress [13]), which defines the openings on the surface of the junction top

A2.7

B4.2.1

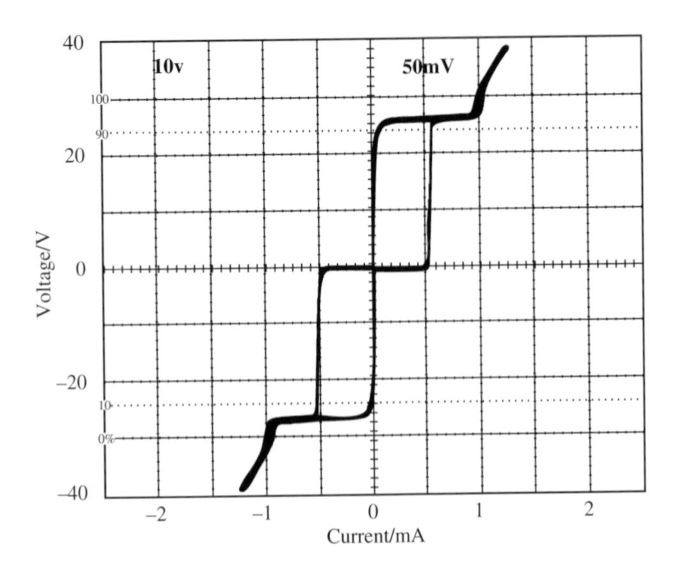

Figure B4.4.2.1. dc characteristic of a series array of 10 000 SIS junctions. Critical current: $I_c = 0.5$ mA, junction area: $A = 24\,\mu$m^2.

electrodes for the wiring connections and insulates the edges of the base electrode. After an Ar ion cleaning process of the top electrode surface inside the window area, an Nb wiring layer is sputtered onto the complete wafer and patterned afterwards, again by reactive ion etching. Figure B4.4.2.2 describes the sequence of fabrication steps for SIS circuits.

E4.7

In principle, this type of manufacturing process is used for patterning all refractory metal junctions, which consist of a trilayer sandwich. High-speed SFQ circuits based on Nb/Al oxide have reached

B4.4.1

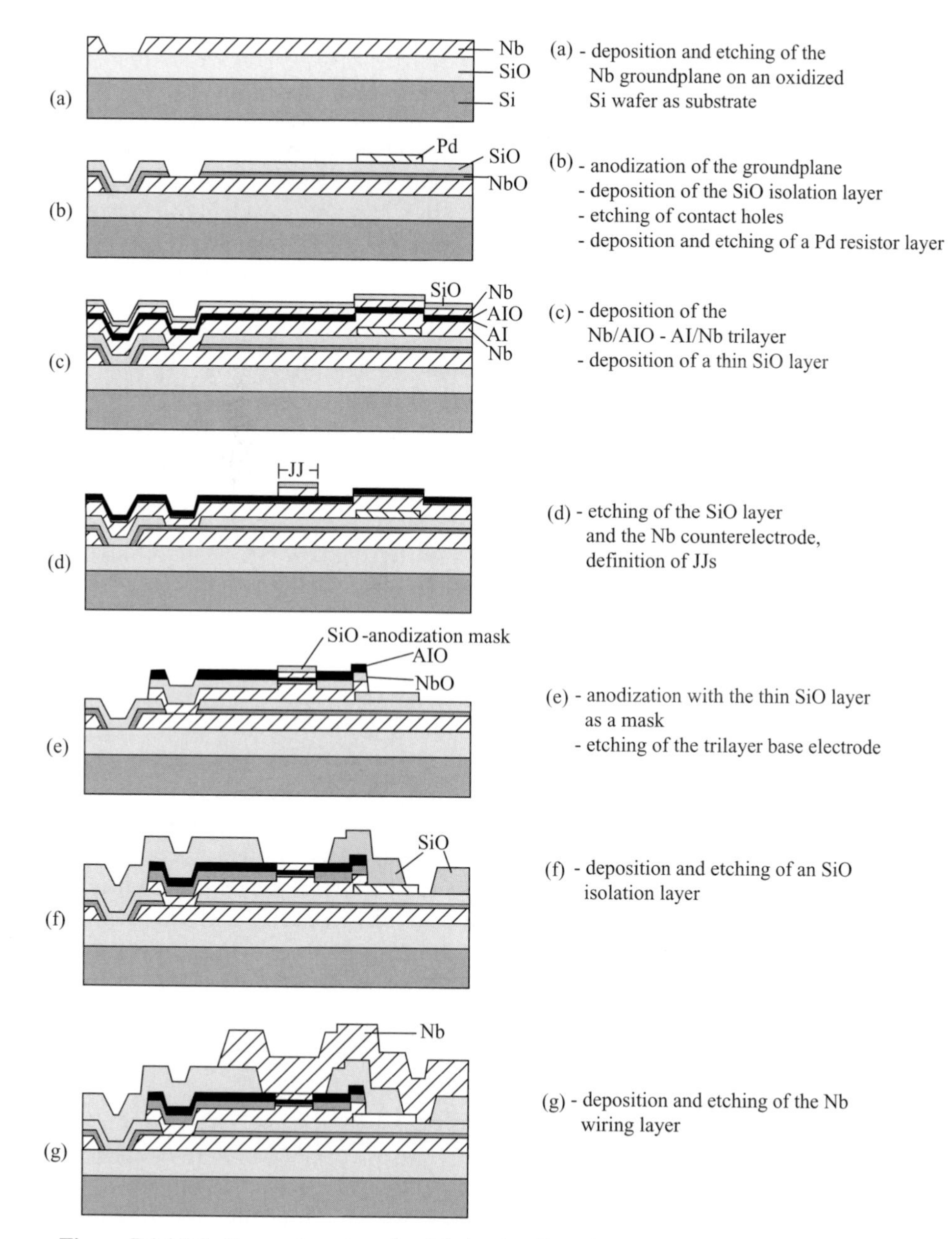

(a)
Nb
SiO
Si
(a) - deposition and etching of the Nb groundplane on an oxidized Si wafer as substrate
(b)
Pd
SiO
NbO
(b) - anodization of the groundplane
- deposition of the SiO isolation layer
- etching of contact holes
- deposition and etching of a Pd resistor layer
(c)
SiO
Nb
AlO
Al
Nb
(c) - deposition of the Nb/AlO - Al/Nb trilayer
- deposition of a thin SiO layer
(d)
JJ
(d) - etching of the SiO layer and the Nb counterelectrode, definition of JJs
(e)
SiO -anodization mask
AlO
NbO
(e) - anodization with the thin SiO layer as a mask
- etching of the trilayer base electrode
(f)
SiO
(f) - deposition and etching of an SiO isolation layer
(g)
Nb
(g) - deposition and etching of the Nb wiring layer

Figure B4.4.2.2. Processing steps for fabricating Josephson junctions in SIS technology.

a complexity of more than 40 000 junctions [14]. Junction damping is realized by external shunting with normal metal thin film resistors, e.g. Pd (figure B4.4.2.2). Five foundries are available for Nb/Al oxide consumer-designed integrated superconducting circuits [15–19]. The technology described is applicable to linewidths down to at least 3 μm. Standard optical lithography is sufficient for the structuring processes (for an overview, see [20]).

B4.4.1

NbN/MgO/NbN

In view of the applications, another important SIS tunnel junction type consists of NbN or NbC_xN_{1-x} electrodes with MgO barriers [21–23]. The NbN electrodes are deposited by reactive dc magnetron sputtering in a gas mixture of Ar and N_2 or of N_2, Ar and C_2H_2. In the first case, the sputtering rate is 50 nm min^{-1}, the total pressure of Ar and N_2 is about 8 mbar with a partial N_2 pressure of 4%. The MgO barrier is formed by a pure Ar (10 mbar) sputtering process. Due to film thickness variations, the barrier deposition process produces a junction parameter spread across the wafer higher than that of the natural oxidation process in the case of Al oxide barriers [11]. Besides its extreme stability to thermal cycling and room temperature storage, the main advantages of this junction type are higher operating temperatures of more than 10 K and the gap voltage of about twice the Nb value and, as a result of this, the larger I_cR_n product and higher maximum operating frequency. The main disadvantages are the very great London penetration depth of about 200 nm and the small coherence length of less than 10 nm for polycrystalline material. This makes the junction type difficult to apply in Josephson array voltage standards. In addition, circuits with ultrasmall linewidths might not work properly. To overcome the difficulties, the NbN can be backed by Nb, but the advantage of a higher operating temperature will then be lost. The use of monocrystalline NbN films [24, 25] also reduces the size of the London penetration depths. Both methods introduce the drawback of a more complex fabrication process. The junction patterning process for NbN/MgO junctions is very similar to that for Nb/Al_2O_3 junctions. The MgO barrier may be replaced by an AlN film which is deposited by reactive sputtering [26, 27].

B4.2.1 E4.1 A2.7 E5.3 A1.2 B4.4.1

One foundry exists for NbN/MgO consumer-designed circuits [16].

B4.4.2.2.2 Overdamped refractory metal junctions, circuits

Besides the SIS mixer and the conventional Josephson array voltage standard, all applications require overdamped junctions with a non-hysteretic dc characteristic. In the case of SIS tunnel junctions, overdamping can be achieved only by external shunting which, on the one hand, allows the normal state resistance to be chosen independent of the critical current up to a certain degree; on the other hand, the shunt provides parasitic inductances and capacitances and adds another fabrication step. Especially the contacting between the shunt and the superconducting layers and the shunt itself increases the parameter spread. Moreover, the shunt needs space. These disadvantages are of special importance in high-speed circuits with a very large number of junctions, such as SFQ circuits and programmable voltage standard/D–A converters. Intrinsically shunted refractory metal junctions have, therefore, been developed recently, which allow the patterning processes to be used which are successfully applied in the case of SIS junctions.

E4.7 E4.5 E4.6

SINIS

Double-barrier junctions consisting of $Nb/Al_2O_3/Al/Al_2O_3/Nb$ [27–30] or $Nb/Al_2O_3/Cu/Al_2O_3/Nb$ [32] are very attractive in many applications. They allow the junction parameters to be tuned from high to low damping by suitably choosing the transparency of the barriers and the thickness of the N layer. In particular, the Nb/Al-oxide version allows the well-established circuit technology to be applied to these materials. Figure B4.4.2.3 shows the dc characteristic of a series array of 7000 $Nb/Al_2O_3/Al/Al_2O_3/Nb$

E4.7

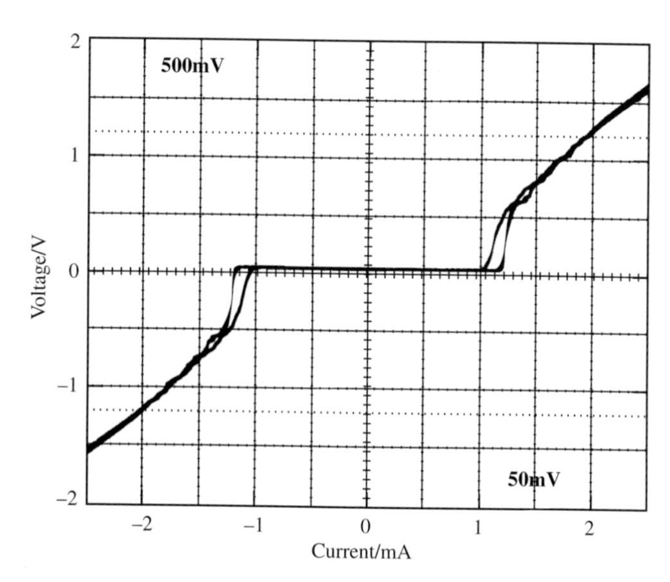

Figure B4.4.2.3. Series array of 7000 strongly damped SINIS junctions. Critical current: $I_c = 1.1$ mA, junction area: $A = 1000\ \mu m^2$.

junctions with the typical non-hysteretic switching behaviour of highly damped junctions. The junction exhibits a nearly hysteresis-free dc characteristic. The magnetic field dependence of the critical current shows a perfect Fraunhofer pattern. The structure at voltages higher than 0.5 mV is caused by resonances of the microwave stripline which is formed by the array. Due to the well-developed technology, the junction parameter spread is very low (2%) so that constant voltage steps of up to 1 V could be generated by 70 GHz microwave radiation. Critical current densities of up to 1 kA cm^{-2} and products of $I_c R_n = 160\ \mu V$ ($R_n = 0.8\ \Omega$) were achieved. Higher values seem to be possible.

A2.7

E4.1

In contrast to single junction preparation, the fabrication of integrated superconducting circuits is a more complex multilayer process which includes the deposition and structuring of additional normally conducting and dielectric layers for circuit resistances, inductances and microwave circuit elements. The following section describes the fabrication of $Nb/Al_2O_3/Al/Al_2O_3/Nb$ (SINIS) circuits. The fabrication process is representative of the simpler SIS circuits based on underdamped junctions. SINIS circuits can be used for all applications described. Figure B4.4.2.4 shows the SINIS fabrication steps which are based on the fabrication of shunted SIS junctions in 4 μm Nb/Al_2O_3–Al/Nb trilayer technology [7]. It has been modified and adapted to the special requirements of intrinsically shunted SINIS junctions [31].

E4.7

Thermally oxidized 3 inch silicon wafers are used as substrates. First, a Nb groundplane is deposited by dc magnetron sputtering and patterned by reactive ion etching (RIE) in a CF_4/O_2 plasma (figure B4.4.2.4(*a*)). To isolate other parts of the circuits, wet anodization of the groundplane is used up to a voltage of 27 V (figure B4.4.2.4(*b*)). The isolation is reinforced by an rf-sputtered SiO_2 layer. In order to connect the groundplane to other metal layers, holes are etched into the double dielectric layer in a CHF_3/O_2 RIE process. The next step (not included in figure B4.4.2.4) is the deposition of a Cr/Pt/Cr layer which is etched by Ar ion milling to form bias resistors [33]. After this, the $Nb/Al_2O_3/Al/Al_2O_3/Nb$ multilayer is deposited (see section B4.4.2.2.1) and covered with a thin SiO_2 layer which serves as a mask during anodization and eliminates the undercut of simple photoresist masks (figures B4.4.2.4(*c*) and (*d*)). The areas of the junctions are protected by the photoresist and SiO_2 layer combination, and the Nb counterelectrode is subsequently exterminated outside by an RIE process in CHF_3/O_2 and CF_4/O_2 plasmas with the Al_2O_5–Al layer acting as an etch stop (figure B4.4.2.4(*e*)). Then, the sample is anodized

B4.4.1

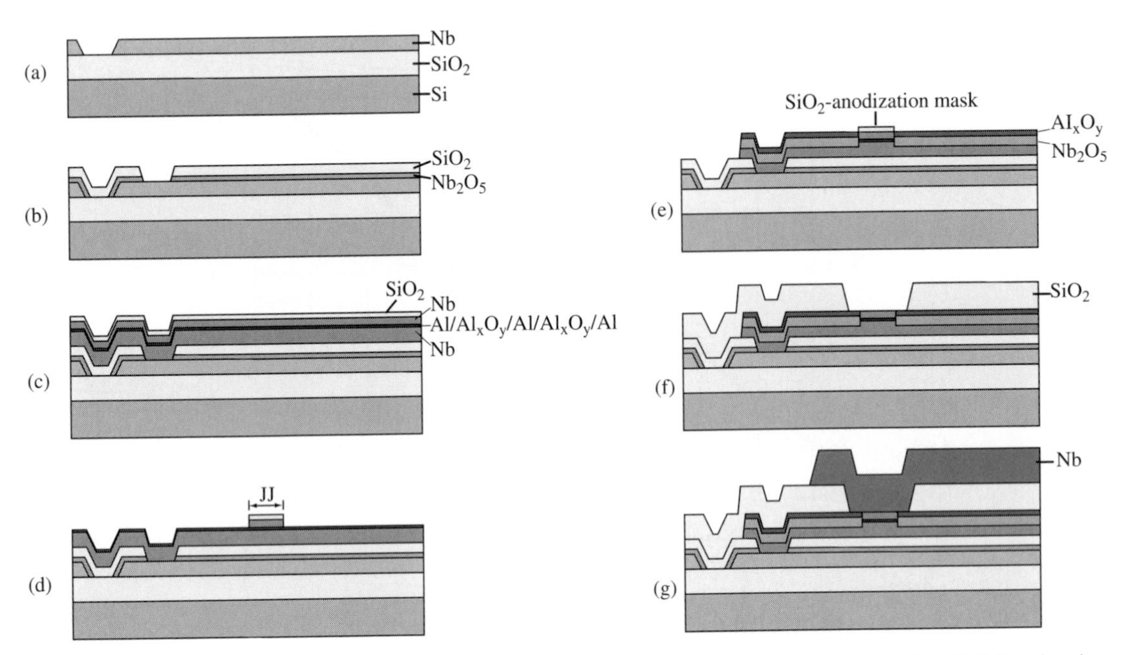

Figure B4.4.2.4. Processing steps for fabricating intrinsically shunted Josephson junctions in SINIS technology.

up to a voltage of 45 V. The following photomask is used to pattern the Nb base electrode. First, the Al_2O_3 is etched by Ar ion milling and then the Nb_2O_5 and Nb layers are etched in a CF_4/O_2 RIE process. An SiO_2 layer is sputtered onto the sample to insulate the edges of the base electrode and strengthen the anodic oxide (figure B4.4.2.4(*f*)). By etching contact holes into this SiO_2 layer, the thin SiO_2 anodization mask is removed from the junction surface. The circuits are completed by sputtering of an Nb wiring layer and by structuring this layer in a CF_4/O_2 RIE process (figure B4.4.2.4(*g*)). The main parameters of the technology process are summarized in table B4.4.2.1.

B4.4.1

Different thicknesses of the Al_2O_3 insulation layers within the multilayer junctions have been obtained by varying the oxidation times of the Al surfaces exposed to oxygen during the single processes of wafer production. They have been chosen to be between 1 and 4 min, in each case at a constant oxygen pressure $p = 0.4$ Pa. This leads to values of J_c between 416 and 1000 A cm^{-2} and of I_cR_n between 160 and 230 μV.

E4.1

SNS, SSeS

If the interfaces are clean and if the barrier consists of a more or less clean metal, the main disadvantage of SNS junctions is their relatively low I_cR_n product. Figure B4.4.2.5 shows the dc characteristic of a 3 μm linewidth series array of 1000 Nb/PdAu/Nb junctions, which demonstrates the low I_cR_n product but also a very low parameter spread.

E4.7 E4.1

Other normal metal barriers such as Ti, Cu or Al may be used instead of PdAu [34]. Due to the progress in small linewidth patterning technology, this junction type may become very attractive in the future for very fast highly integrated digital circuits because the critical current density of SNS junctions may be very high (at least 10^6 A cm^{-2}). This provides a critical current still high enough for applications even at a junction size which is so small that I_cR_n may become large enough for high frequency digital applications. Moreover, recent experiments have shown that it is possible to increase the junction

E4

Table B4.4.2.1. SINIS technology parameters

Step	Function	Material	SINIS code	Thickness (nm)
	Substrate	Si, SiO_2		
(a)	Ground	Nb		200
(b)	1. Groundpl. isol.	Nb_2O_5		50
(b)	2. Groundpl. isol.	SiO_2		200
	Resistors	Cr/Pt/Cr		15/90/15
(c,d)	Base electrode	Nb	S	200
(c,d)	Normal metal	Al		10
(c,d)	Tunnel barrier	Al_xO_y	I	1
(c,d)	Normal metal	Al	N	10
(c,d)	Tunnel barrier	Al_xO_y	I	1
(c,d)	Normal metal	Al		10
(c,d)	Counter electrode	Nb	S	100
(e)	1. Multilayer isol.	Al_xO_y		45
(e)	2. Multilayer isol.	Nb_2O_5		75
(f)	3. Multilayer isol.	SiO_2		280
(g)	Wiring	Nb		350
	Contact pads	Pd		75

resistance arbitrarily by using a strongly disturbed N layer such as TiN [35] or metallic Ta oxide [36] as a barrier. NbC_xN_{1-x} junctions with TiN_x or MgO/TiN_x barriers [37] and NbN junctions with Mg barriers [38] or α-NbN_x barriers [39] are interesting representatives of this junction type. By properly adjusting the partial N_2 pressure during the α-NbN_x barrier deposition, overdamped and underdamped junctions of this type can be prepared because the barrier changes from a metallic conductor to semiconductor. This leads to another attractive junction type (SSeS) with semiconducting 2DEG barriers instead of

B4.4.1

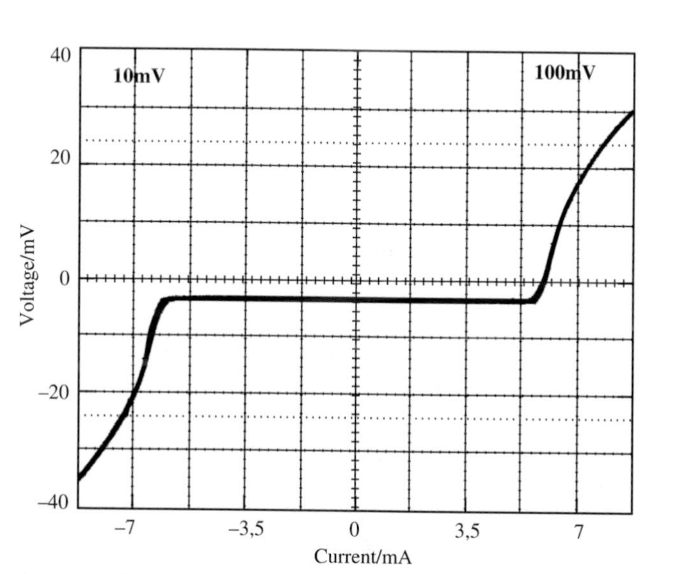

Figure B4.4.2.5. Series array of 1000 SNS junctions. Critical current: $I_c = 5.5\,\mathrm{mA}$, junction area: $A = 6.25\,\mu\mathrm{m}^2$.

a normal metal. Such a system would allow low-temperature semiconducting circuits to be directly integrated into superconducting SFQ circuits. In particular, InAs 2DEGs develop a very small Schottky barrier to Nb [40–43]. It could be shown that the critical current in such a junction can be tuned by light radiation [44, 45].

E4.5

Large circuits, so far made only of Nb/PdAu/Nb SNS junctions, show a remarkably low parameter spread.

A very prominent example of a complex circuit is the 30 000 junction series array in the form of a binary sequence. This array is integrated into a coplanar microwave transmission line and operates as programmable voltage standard with output voltages of up to 1 V [34, 46].

E3.2 E5.3

B4.4.2.3 Outlook

Superconductive circuit technology is being developed to higher integration and speed, resulting in smaller linewidths. Moreover, smaller junction circuits decrease the total active junction area per circuit, which is most sensitive to failures by shorts across the barrier. The reduction of such failures will be extremely important for circuits with more than 100 000 junctions. Size reduction will be supported by the development of optical lithography for nanometre structures in the area of CMOS technology. Smaller structures will require planarization techniques to guarantee a safe edge insulation and continuous edge crossing of ultrasmall lines [47–51]. An $Nb/Al_2O_3/Nb$ SFQ circuit with 0.5 μm linewidths was successfully operated at 750 GHz [52]. It has been shown that $Nb/Al_2O_3/Nb$ junctions with linewidths down to 0.2 μm can be prepared by planar technology, still showing a high quality quasiparticle characteristic [48, 53–55].

E4.1 B4.4.1

However, to achieve critical currents high enough for applications (200 μA at 4 K), very small junctions must provide extremely high critical current densities of up to 10^6 A cm^{-2}, which are difficult to reach with pure oxide barriers in SIS and SINIS contacts. It is, therefore, expected that the SNS or SSeS junctions will gain importance for very complex and fast circuits. This is supported by the development of HTS junctions which are now available only as SNS type, at least for circuit applications (see chapter B4.3.3).

E4.7 B4.3.3

References

[1] Rowell J M, Gurvitch M and Geerk J 1981 *Phys. Rev.* B **24** 2278
[2] Gurvitch M, Washington M A, Huggins H A and Rowell J M 1983 *IEEE Trans. Magn.* **19** 791
[3] Shoji A, Kosaka S, Shinoki F, Aoyagi M and Hayakawa H 1983 All refractory Josephson tunnel junctions fabricated by reactive ion etching *IEEE Trans. Magn.* **19** 827
[4] Hasuo S 1992 Toward the realization of a Josephson computer *Science* **255** 301
[5] Haykawa H and Kotani S 1993 Josephson digital devices *Compound and Josephson High-Speed Devices*, ed T Isugi and A Shibatomi (New York: Plenum) p 255
[6] Niemeyer J, Sakamoto Y, Vollmer E, Hinken J H, Shoji A, Nakagawa H, Takada S and Kosaka S 1986 Nb/Al-oxide/Nb and NbN/MgO/NbN tunnel junctions in large series arrays for voltage standards *Japan. J. Appl. Phys.* **25** L343
[7] Gundlach K H 1989 Principles of direct and heterodyne detection with SIS mixers *Superconducting Electronics*, eds W Weinstock and M Nisenhoff (Berlin: Springer) p 259
[8] Koshelets V P and Shitov S V 2000 Integrated superconducting receivers *Supercond. Sci. Technol.* **13** R53
[9] Dolata R, Khabipov M I, Buchholz F-Im, Kessel W and Niemeyer J 1995 Nb/Al2O3Al/Nb process development for the fabrication of fast-switching circuits in RSFQ logic *Applied Superconductivity Vol 2 Inst. Phys. Conf. Ser. 148* (Bristol: Institute of Physics Publishing) p 1709
[10] Kleinsasser A W, Miller R E and Mallison W H 1995 Dependence of critical current density on oxygen exposure in Nb–AlO_x–Nb tunnel junctions *IEEE Trans. Appl. Supercond.* **5** 26
[11] Müller F, Behr R, Kohlmann J, Pöpel R, Niemeyer J, Wende G, Fritzsch L, Thrum F, Meyer H-G and Krasnopolin I Y 1997 *Optimized 10-V Josephson Series Arrays: Fabrication and Properties ISEC'97 Extended Abstracts: Vol 1 Plenary Contributions*, eds H Koch and S Knappe (Berlin: PTB) p 95

[12] Fritsch L, Köhler H, Thrum F, Wende G and Meyer H-G 1998 Preparation of Nb/Al–AlO_x/Nb Josephson junctions with critical current densities down to 0.1–1 A cm^{-2} *Physica* B **296** 319
[13] Meng X, Bhat A and van Duzer T 1999 Very low critical current spreads in Nb/AlO_x/Nb integrated circuits using low temperature and low stress ECR PECVD silicon oxide films ASC'98 *IEEE Trans. Appl. Supercond.* **9** 3208
[14] Nagasawa S, Numata H, Hashimoto Y and Tahara S 1997 *High-Frequency Clock Operation of Josephson RAMs ISEC'97 Extended Abstracts: Vol 2 Junctions and Digital Applications*, eds H Koch and S Knappe (Berlin: PTB) p 29
[15] http://www.hypres.com/
[16] http://www.trw.com/superconductivity/asc98 kerber.pdf
[17] NEC Corporation, Tsukuba-shi, 305-8501 Japan; E-mail: tahara@qwave.cl.nec.co.jp
[18] http://www.ipht-jena.de/abt31/foundry.html
[19] http://www.starcryo.com/foundry.htm
[20] Abelson L A 1997 *Superconductive Electronics Process Technologies 1997 ISEC'97 Extended Abstracts: Vol 1 Plenary Contributions*, eds H Koch and S Knappe (Berlin: PTB) p 1
[21] Shoji A, Kosaka S, Shinoki F, Aoyagi M and Hayakawa H 1985 Niobium nitride Josephson tunnel junctions with magnesium oxide barriers *Appl. Phys. Lett.* **46** 1098
[22] Hunt B D, LeDuc H G, Cypher S R and Stern J A 1989 NbN/MgO/NbN edge-geometry tunnel junctions *Appl. Phys. Lett.* **55** 81
[23] Robertazzi R P and Buhrman R A 1989 Josephson terahertz local oscillator *IEEE Trans. Magn.* **25** 1384
[24] Shooji A 1991 Fabrication of all-NbN Josephson tunnel junctions using single crystal NbN films for the base electrodes *IEEE Trans. Magn.* **27** 3184
[25] Shoji A, Kiryu S, Kashiwaya S, Kohjiro S, Kosaka S and Koyanagi M 1992 Preparation and characteristics of full epitaxial NbC_xN_{1-x}/MgO/NbC_xN_{1-x} Josephson tunnel junctions *Superconducting Devices and their Applications: Proceedings in Physics 64* (Berlin: Springer) p 208
[26] Wang Z, Kawakami A and Uzawa Y 1996 NbN/AlN/NbN tunnel junctions with high current density up to 54 kA/cm^2 *Appl. Phys. Lett.* **70** 114
[27] Villegier J-C, Delaet B, Larrey V, Salez M, Delorme Y and Munier J-M 1997 *Fabrication of NbN/AlN/NbN Junctions with Al Embedding Circuits on Si Membrane for 1.5 THz SIS Mixers ISEC'97 Extended Abstracts: Vol 3 Microwave Devices, SQUIDs and Other Applications*, eds H Koch and S Knappe (Berlin: PTB) p 111
[28] Maezawa M and Shoji A 1997 Overdamped Josephson junctions with Nb/Al/Al_xO_y/Al/Al_xO_y/Nb structure for integrated circuit application *Appl. Phys. Lett.* **70** 3603
[29] Sugiyama H, Yanada A, Ota M, Fujimaki A and Hayakawa H 1997 Characteristics of Nb/Al/Al_xO_y/Al/Al_xO_y/Nb junctions based on the proximity effect *Japan. J. Appl. Phys.* **36** L1157
[30] Schulze H, Behr R, Müller F and Niemeyer J 1998 Nb/Al/Al_xO_y/Al/Al_xO_y/Nb-Josephson junctions for programmable voltage standards *Appl. Phys. Lett.* **73** 996
[31] Balashov D, Buchholz F-Im, Schulze H, Khabipov M, Kessel W and Niemeyer J 1998 Superconductor–insulator–normalconductor–insulator–superconductor (Nb/Al_xO_y/Al/Al_xO_y/Nb) process development for integrated circuit applications *Supercond. Sci. Technol.* **11** 1401
[32] Leoni R, Castellano M G, Torrioli G, Carelli P, Gerardino A and Melchiorri F 1997 Study of superconductor–insulator normal–insulator–superconductor (SINIS) junctions *Applied Superconductivity 1997: Proceedings of EUCAS'97 Vol 1* eds H Rogalla and D H A Blank (Bristol: Institute of Physics Publishing) p 809
[33] Dolata R and Balashov D 1998 Platinum thin film resistors with Cr under- and overlayers for the Nb/AlO_x/Nb technology *Physica* C **295** 247
[34] Burroughs C J, Benz S P, Hamilton C A and Harvey T E 1999 1 Volt dc Programmable voltage standard (ASC'98) *IEEE Trans. Appl. Supercond.* **9** 4145
[35] Harvey T E, Smith M J, Smit A F and Benz S P 1998 Superconductor–normal–superconductor junctions with titanium nitride barriers (ASC'98) *Plenary and Electronic Abstracts ETB-02* p 143
[36] Laquaniti V, Gonzini S, Maggi S, Monticone E, Steni R and Andreone D 1999 Nb-based SNS Junctions with Al and TaOx barriers for a programmable Josephson voltage standard (ASC'98) *IEEE Trans. Appl. Supercond.* **9** 4245
[37] Wang Q, Katsuta J, Kikuchi T, Sasaki H and Shoji A 1997 *Properties of Nonhysteretic NbC_xN_{1-x} Josephson Junctions with MgO/TiN_x Multilayered Barriers ISEC'97 Extended Abstracts: Vol 2 Junctions and Digital Applications*, ed H Koch and S Knappe (Berlin: PTB) p 173
[38] Radpavar M and Hunt R 1998 Niobium nitride-based superconductor–normal–superconductor junctions (ASC'98) *Plenary and Electronic Abstracts EFB-09* p 86
[39] Yamamori H, Yoshizawa K, Fujimaki A and Hayakawa H 1996 Fabrication of all NbN Josephson junctions using semiconductive amorphous NbN_x barriers *Supercond. Sci. Technol.* **9** A30
[40] Kawakami T and Takayanagi H 1985 Single-crystal n-InAs coupled Josephson junction *Appl. Phys. Lett.* **46** 92
[41] Akazaki T, Nitta J, Takayanagi H and Arai K 1995 Superconducting junctions using a 2DEG in a strained InAs quantum well inserted into an InAlAs/InGaAs MD structure *IEEE Trans. Appl. Supercond.* **5** 2887
[42] Shäpers T, Kaluza A, Neurohr K, Malindretos J, Grecelius G, Van der Hart A, Hardtdegen H and Lüth A I 1997 Josephson effect in Nb/two-dimensional electron gas structures using a pseudomorphic $In_xGa_{1-x}As$/InP heterostructure

[43] Bastian G, Göbel E O, Zorin A B, Schulze H, Niemeyer J, Weimann T, Bennett M R and Singer K E 1998 Quasiparticle interference effects in a ballistic superconductor–semiconductor–superconductor Josephson junction *Phys Rev. Lett.* **81** 1686

[44] Takaoka S, Nakao Y, Kousai Haruta K, Oto K, Murase K and Gamo K 1995 Investigation of photoresponse in photoconducting semiconductor-superconductor microstructure *Japan. J. Appl. Phys.* **34** 5585

[45] Bastian G, Göbel E O, Schmitz J, Walther M and Wagner J 1999 Optically induced flip-flop in a superconductor–semiconductor–superconductor Josephson junction *Appl. Phys. Lett.* **75** 94
Bastian G, Göbel E O, Schmitz J, Waither M and Wagner J 1997 Deutsches Patent AZ 197 05 239.8 (1997)

[46] Benz S P and Burroughs C J 1997 Constant-voltage steps in arrays of Nb–PdAu–Nb Josephson junctions *IEEE Trans. Appl. Supercond.* **7** 2434

[47] Ketchen M B, Pearson D, Kleinsasser A W, Hu C K, Smyth M, Logan J, Stawiasz K, Baran E, Jaso M, Ross T, Petrillo K, Manny M, Basavaiah S, Brodsky S, Kaplan S B, Gallagher W J and Bushan M 1991 Sub-μm, planarized, Nb–AlO_x–Nb Josephson process for 125 mm wafers developed in partnership with Si-technology *Appl. Phys. Lett.* **59** 2609

[48] Bao Z, Bushan M, Han S and Lukens J E 1995 Fabrication of high quality, deep-submicron Nb/AlO_x/Nb Josephson junctions using chemical mechanical polishing *IEEE Trans. Appl. Supercond.* **5** 2731

[49] Numata H, Nagasawa S, Tanaka M and Tahara S 1999 Fabrication technology for high-density Josephson integrated circuits using mechanical polishing planarisation (ASC'98) *IEEE Trans. Appl. Supercond.* **9** 3198

[50] Kohjiro S, Yamamori H and Shoji A 1999 Fabrication of niobium-carbonitride Josephson junctions on magnesium-oxide substrates using chemical–mechanical polishing (ASC'98) *IEEE Trans. Appl. Supercond.* **9** 4464

[51] Berggren K K, Macedo E M, Feld D A and Sage J P 1999 Low T_c superconductive circuits fabricated on 150-mm-diameter wafers using a doubly planarized Nb/AlO_x/Nb process (ASC'98) *IEEE Trans. Appl. Supercond.* **9** 3271

[52] Chen W, Rylakov A V, Patel V, Lukens J E and Likharev K K 1998 Superconductor digital frequency divider operating up to 750 GHz *Appl. Phys. Lett.* **73** 2817

[53] Dolata R, Weimann T, Scherer H and Niemeyer J 1999 Sub μm Nb/AlO_x/Nb Josephson junctions prepared by anodization techniques (ASC'98) *IEEE Trans. Appl. Supercond.* **9** 3255

[54] Flees, D J Experimental studies of band-structure properties in Bloch transistors *Thesis* State University of New York at Stony Brook

[55] Pavolotsky A B, Weimann T, Scherer H, Krupenin V A, Niemeyer J and Zorin A B 1999 Multilayer technique for fabricating niobium junction circuits exhibiting charging effects *J. Vac. Sci. Technol.* B **17** 2030

B4.4.3
Processing and manufacture of Josephson junctions: high T_c

Regina Dittmann and Alex I Braginski

B4.4.3.1 Introduction

As noted in chapter B4.4.2, Josephson junctions are the key nonlinear elements in all applications which make use of flux quantization, such as SQUIDs, voltage standards or digital circuits. In this chapter, we provide the description of most common fabrication processes for HTS thin-film Josephson junctions. As in the case of LTS junctions, these processes involve film deposition and patterning techniques, which, however, may differ significantly from those employed in conventional superconductor processing.

B4.4.2
A2.6, E4.2
B4.2.1
B4.4.1

All HTS cuprate materials have anisotropic layered structures with an extremely short superconducting coherence length, especially in the c-axis direction ($\xi_{ab} = 20$–30 Å, $\xi_c = 2$–3 Å). Due to this short coherence length, and d-wave pairing mechanism, transport properties of all such materials are direction-dependent and extremely sensitive to defects on atomic length scales. Coupling across higher-angle grain boundaries is weakened enough to exhibit Josephson effects. Hence, the use of epitaxial films on lattice-matched single crystal substrates is a basic requirement for fabrication of controllable HTS Josephson junctions.

C
A2.3
A2.7

The highly anisotropic BSCCO and TlBCCO crystals can be regarded as stacks of superconducting and insulating layers, where the current carrier transport in c-axis direction occurs via tunneling. These, so-called, intrinsic Josephson junction stacks exhibit Josephson effects without any additional barrier layer [1–3]. Generally, the shorter ξ_c of HTS is the reason why in almost all practical HTS junctions with additional barriers, the Josephson coupling between ab-planes is used to attain higher I_c and better reproducibility.

C2, C3
E4.1

The key task in fabricating all Josephson junctions is to produce a sufficiently thin uniform insulating barrier or normal conducting interlayer between two superconducting electrodes to ensure controllable weak exchange of Cooper pairs and quasiparticles.

E2.4

Since the critical current, I_c, decreases exponentially with increasing barrier thickness, insufficient thickness uniformity leads to a wide spread of junction parameters. In HTS junctions, structural complexity of material interfaces, directionality of material properties, and presence of atomic scale defects are additional causes of wide parameter spreads. In the case of high, uncontrollable interface resistivities, the characteristics of the junction may be determined by the interface rather than by the barrier or interlayer.

In most applications, the essential junction parameters are I_c, the normal resistance R_N and their product I_cR_N. These parameters strongly depend on the used junction type and the details of

E4.1

the fabrication process. In almost all applications, a high I_cR_N product is favourable, because it results in a high upper limit of the device or circuit working frequency.

B4.4.3.2 Overview of HTS junctions

In this section, we describe processes used to fabricate more established HTS junction types, shown in figure B4.4.1. The complexity of various junctions and of their fabrication process is compared in table B4.4.1. We emphasize junction fabrication by optical lithography. Submicron size junctions usually require e-beam lithography. Close control of their dimension tolerances and properties is much more difficult. For simplicity, we restrict the overview to the most widely used superconductor YBCO. However, for most of the junction types, the fabrication process is the same, irrespective of the HTS material used.

B4.4.1

Most of the HTS junctions are intrinsically-shunted, and can in principle be described by the RSJ model [4]. If the junctions are dominated by a few microshorts in the barrier layer, the shape of the current–voltage characteristics may strongly differ from the RSJ model, because the voltage drop could be caused by the dissipation of moving fluxons. Therefore, the close similarity of the current–voltage characteristics to that of the RSJ model is usually an indicator of quality of a HTS junction. Furthermore, the similarity of the experimental I_c versus H pattern to the Fraunhofer pattern, expected for a homogeneous current distribution, is an important check for the homogeneity of the junctions [5]. Some deviations from the RSJ model, e.g. excess currents and conductance peaks, may be intrinsically

A2.7

B4.4.2

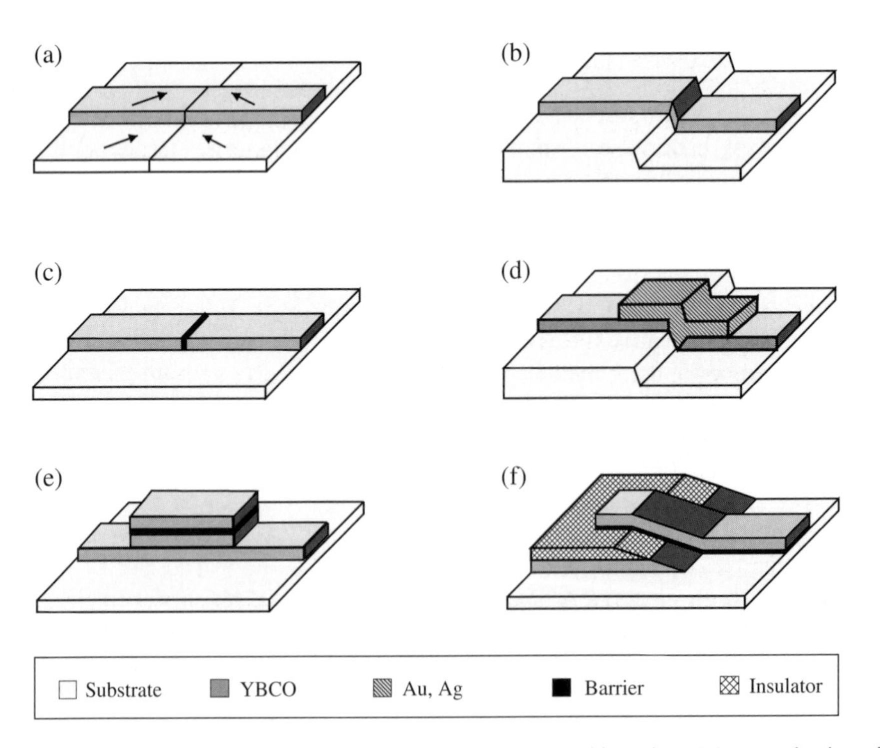

Figure B4.4.3.1. Schematic view of different junction types: (*a*) bicrystal junction, (*b*) step-edge junction, (*c*) locally damaged junction (FEBI, Ion-beam damaged junction), (*d*) SNS step-edge junction (*e*) planar junction, (*f*) edge junction.

Table B4.4.3.1. Needed equipment and opportunities of the different junction fabrication processes. For all fabrication processes, thin film deposition, photolithography and ion beam etching equipment is needed

Junction type	Fabrication process complexity	Special equipment	Obtainable circuit complexity
Bicrystal junctions	Low	–	Low
Step-edge junctions	Medium	–	Low
FEBI junctions	Medium	e-beam writing gun (energy ≥ 100 keV)	Medium
Ion-beam damaged junctions	High	e-beam lithography, ion implanter	High
Step-edge SNS junctions	Medium	–	Low
Planar junctions	High	–	High
Edge junctions	High	–	High

correlated with the mechanism of the current transport across the interface between superconductor and barrier [6–8]. Properties of HTS Josephson junctions are described in more detail in chapter E4.1. E4.1

B4.4.3.2.1 Grain boundary junctions

The simplest way to fabricate a thin barrier in an HTS thin film is to nucleate a grain boundary. Due to the local perturbation of the crystal structure, the current density of all HTS cuprates is strongly reduced at grain boundaries. In consequence, a microbridge across a grain boundary exhibits Josephson effects. A2.7 The critical current density of grain boundary junctions decreases roughly exponentially with increasing misorientation angle up to 45° [9,10].

Bicrystal junctions

The most convenient way to fabricate a grain boundary junction with controlled misorientation angle is to use bicrystal substrates. Bicrystal substrates consist of two single crystals, which are fused together with a certain misorientation angle. The HTS thin films grow according to the substrate crystal orientation and the grain boundary is replicated with the same misorientation angle (see figure B4.4.3.1(*a*)). Bicrystal substrates of a variety of substrate materials (e.g. $SrTiO_3$, MgO) are commercially available with several misorientation angles (e.g. 24, 30, 36.8°), symmetrical and asymmetrical.

The fabrication of bicrystal junctions consists of the growth of a superconducting thin film and the patterning of microbridges (usually 2–10 μm wide) across the grain boundary(ies). Patterning by B4.4.1 photolithography and ion beam etching is most widely used.

Junction parameters can be adjusted by controlling the oxygen deficiency of the grain boundary. When increasing that deficiency by annealing in vacuum, the critical current can be decreased and the normal resistance increased following the scaling law: $I_c R_N \sim \sqrt{J_c}$. The opposite effect can be obtained by annealing in oxygen or oxygen plasma. Values of $I_c R_N$ product up to 10 mV at 4.2 K can be achieved, E4.1 depending on the substrate material and the degree of oxygenation [9].

Another possibility to influence the current density of bicrystal junctions is to locally dope the grain boundary by using Ca doped YBCO films [11]. The current density at 4.2 K can be increased by one C1 order of magnitude by substituting 30% of Y by Ca [12]. To obtain an enhancement of I_c at 77 K, multilayers of undoped and Ca doped YBCO have to be used to compensate the supression of T_c in the superconducting electrodes due to Ca doping [11]. The I_cR_N product is reduced by Ca doping since the normal resistance is reduced more than the I_c is increased by this method.

Although the use of bicrystals is the most convenient way to fabricate RSJ-like HTS junctions with an I_c-spread of $\pm 15\%$[13], these junctions are rather unsuitable for use in complex electronic circuits because their positions are restricted to bicrystal lines. An ingeneous attempt to overcome this limitation was the concept of biepitaxial junctions [14]. However, their low I_cR_N product prevented their practical implementation. Recently, an attempt to revive this approach was published [15,16].

Step-edge junctions

Another method of producing grain boundaries in a thin film is to grow HTS films on steps in the substrate (see figure B4.4.3.1(*b*)) [17]. High-resolution transmission electron microscope investigations showed that the growth of a YBCO thin film over such steps depends strongly on the step angle [18]. On steep steps (angle $\geq 45°$), the film on the step grows also with *c*-axis orientation, thus resulting in two 90° grain boundaries. Near 45°, two growth orientations are equally probable, and that causes the formation of multiple 90° grain boundaries. In contrast, on low angle steps (angle $\leq 45°$) in well-lattice-matched substrates such as $SrTiO_3$ or $LaAlO_3$, YBCO films can grow without any perturbation of the crystalline orientation. On cubic substrates with a large lattice mismatch, such as MgO, the microstructure can be different. Grain boundaries in YBCO on MgO can occur even on very shallow steps [19].

B4.2.1
B4.2.2

The most important junction processing step is the fabrication of the step in the substrate. In principle, ion beam etching offers the possibility of producing steep steps, but strongly rounded photoresist mask profiles may result in shallow step angles. Therefore, metal [20] or carbon [21] masks have also been employed, and step angles of 70–80° achieved.

Film thickness of 1/2–2/3 of the step height is needed for obtaining controllable step-edge junctions. The properties of such junctions strongly depend on the film morphology and microstructure in the step area. After the thin film growth, patterning into microbridges (usually 2–10 μm wide) occurs by photolithography and ion beam etching.

As in bicrystal junctions, step-edge junction parameters can be adjusted by annealing in suitable atmosphere, which controls oxygen content at the grain boundary [22]. Typically, an I_cR_N product of 1 mV at 4.2 K and 0.1 mV at 77 K can be achieved.

Step-edge junctions have a higher design flexibility than bicrystal junctions, but the fabrication process is less reliable and parameter spreads are rather large. Hence, step-edge junctions are not suitable for the fabrication of complex circuits.

B4.4.3.2.2 Locally damaged junctions

A2.7

Due to the sensitivity of the superconducting state to local lattice perturbations, point defects created by various means (e.g., electron and ion irradiation) can result in partial or even complete suppression of high-temperature superconductivity. If a sufficiently thin barrier of the material with reduced T_c ($T_{c,N}$) is created in a HTS film, proximity coupling can occur and a Josephson junction is formed. To ensure a sufficiently thin barrier, the irradiation should be carried out using a focused beam of electrons or ions. Alternately, the HTS film can be damaged via a sufficiently narrow slit in a suitable masking layer.

B4.4.1

The fabrication of junctions starts with the patterning of microbridges from YBCO thin films, usually by ion beam etching. The next, key fabrication step is the local damage of the YBCO microbridges. The two most important approaches are described below.

Electron beam damaged junctions

A comprehensive overview of this subject can be found in [23]. There exists a threshold for the energy of the electrons below which no displacement of oxygen in YBCO is possible. The displacement energy for chain oxygen and in-plane oxygen is about 2 and 8 eV, respectively. This corresponds to electron beam energies of 20 and 60 keV. Usually, energies of the order of 100 keV are used for the fabrication of focused electron beam irradiated (FEBI) junctions [23–25]. The $T_{c,N}$ decreases with increasing irradiation dose. B4.4.1

The lateral extent of the damaged zone is limited by the incident beam diameter, rather than by lateral scattering within the film. Therefore, a zone as narrow as 10 nm can be fabricated, and this makes proximity coupling possible. Junctions with RSJ-like I–V curves and Fraunhofer patterns indicating a homogeneous current distribution have been fabricated. The critical current of these junctions could be adjusted by the irradiation dose, because the ξ_N of the normal-conducting zone decreases with increasing dose. Furthermore, the junction parameters can be subsequently adjusted by annealing above room temperature to partially reorder the oxygen sublattice. B4.4.2 A2.3

The low activation energy of oxygen defects introduced by FEBI means that the properties of junctions stored at room temperature will change significantly over a few months. However, quasistable FEBI junctions can be produced by first overdamaging and then annealing the junctions to remove defects with lower activation energies. This technique results in junctions with a satisfactory reliability and small on-chip I_c spreads 1σ below 10% [26]. The I_cR_N products which can be obtained are lower than those of grain boundary junctions (up to 1 mV at 4.2 K), and it is difficult to achieve normal resistance values above 1 Ω without patterning of submicron microbridge width [26]. An obstacle to use FEBI junctions in complex circuits is the long e-beam writing time resulting from sequential writing.

Ion beam damaged junctions

In contrast to electron beam irradiation, ion implantation at energies of the order of 100 keV causes considerable damage of the cation lattice, leading to a more stable structural change. Most often, oxygen implantation process for irradiating YBCO was used [27,28] but He_2^+ [29] and Ne_2^+ [30] ions have also been employed. B4.4.1

Since wide-beam ion sources are readily available, sequential writing is not necessary. The whole chip or wafer can be irradiated through a slit mask to produce junctions. However, fabrication of masks with nanoscale slits represents a considerable difficulty. The mask should completely stop the ions, so that they are prevented from penetrating the YBCO thin film, while the slit has to be of the order of the normal conducting coherence length of YBCO, ideally about 10 nm. For this purpose, electron beam lithography and a complicated multilayer mask is needed. For example, a multilayer of carbon, conventional photoresist, Ti and e-beam resist was used [31].

As in the case of electron irradiation, the $T_{c,N}$ and ξ_N are reduced with increasing ion dose. However, the lateral defect distribution also increases, and so does the thickness of the normal conducting zone. This makes it difficult to adjust the junction parameters, because lengths of the normal conducting path and of coherence cannot be influenced independently. Furthermore, the length of the normal-conducting (lower $T_{c,N}$) path is mainly determined by the lateral ion damage (ca. 200 nm for 200 keV ion energy). Hence, it cannot be significantly influenced by the slit width. Employing a 50 nm slit mask, RSJ-like Josephson junctions with Fraunhofer patterns indicating a homogeneous current distribution were fabricated by oxygen implantation [28]. Due to the longer length of the normal zone, current densities were lower than in the case of e-beam damaged junctions and I_cR_N-products of the order of only 100 μV at 4.2 K could be attained. In junctions fabricated by Ne^+ irradiation with identical parameters, I_c spreads of 12–40% were obtained between 50 and 65 K [30].

B4.4.3.2.3 *Josephson junctions with artificial barriers*

B4.2.1 B4.2.2
Josephson junctions with artificial barriers or proximity interlayers, which are fabricated by thin film deposition, offer, in principle, the highest degree of parameter flexibility. The ranges of I_c, R_N and I_cR_N can be set by the choice of an appropriate barrier material suitable for the intended application. Thus far, noble metals Ag and Au, and a wide variety of different perovskites have been used. The most widely used perovskites have been the nonsuperconducting $PrBa_2Cu_3O_7$, also with some possible substitutions (Ga [32,33], Y [34]), and the Co-substituted YBCO [35,36] which is a superconductor with reduced T_c. Typical barriers or interlayers have been 10–40 nm thick.

Thickness inhomogeneity of the barrier or interlayer strongly influences the spread of junction parameters, since the I_c is exponentially determined by this thickness. Consequently, epitaxial growth of the barrier should be controlled on the atomic scale. One approach to such a control is the use of atomic B4.2.2 layer-by-layer (ALL) MBE technique. Although the possibility of growing multilayers with defect- and pinhole-free interlayers of the order of one unit cell was demonstrated[37], this complicated technique has not been adopted widely. Another possibility to overcome the problem of controlling epitaxial growth of a thin film barrier on an atomic scale is to produce a barrier layer by treatment of the bottom electrode surface. Within the last few years, these so-called interface-engineered junctions became of E4.1 great interest [38]. Although planar interface-engineered junctions were demonstrated [39,40], interface-engineered junctions are usually fabricated using the edge geometry. Therefore, we shall describe this method in the context of the fabrication of edge junctions.

SNS step edge junctions

E4.7
For Josephson junctions with noble metal proximity interlayer, a special geometry is needed because no heteroepitaxial growth of the metal on YBCO is possible. The most established geometry is the step-edge SNS junction which can be seen in figure B4.4.3.1(*d*). A *c*-axis YBCO film is deposited across a step in the substrate such that film continuity is disrupted and the *ab*-plane terminations in both electrodes are exposed. The metal is then deposited to fill the spacing between electrodes and form the interlayer. This approach requires directional deposition of YBCO and of the metal. Off-axis sputtering as well as PLD under 45° have been used for this purpose [41,42]. The interlayer thickness can partly be influenced by the relation of step height to the YBCO film thickness. Nevertheless, junction parameters are mainly E4.1 determined by the Au–YBCO interface resistance and the microstructure in the step area. Although I_cR_N products of the order of 10 mV, and normal resistivities of the order of 10 were achieved at 4.2 K [43], these junctions are difficult to control and not suitable for the fabrication of complex circuits.

Planar junctions

Planar junctions are the most widely used approach to fabricate Josephson junctions of conventional superconductors. A trilayer of superconductor, barrier and superconductor is fabricated *in situ* and patterned into small mesa structures (see figure B4.4.3.1(*e*)). The advantage of such trilayers is that no degradation of the superconductor–barrier interface occurs due to *ex situ* processing [45].

This fabrication process is used also for all types of HTS junctions where vertical current transport E4.1 is needed. Examples are the intrinsic Josephson junctions [43,44] or BSCCO tunnel junctions [37] fabricated from *c*-axis films. In YBCO trilayers with artificial barrier layers, *a*-axis oriented films are needed to ensure Josephson coupling via the *ab*-planes.

The fabrication process is shown in figure B4.4.3.2. The first step (*a*) is the *in situ* deposition of a superconductor–barrier–superconductor trilayer. To protect it, and especially the YBCO surface during the following fabrication steps, a thin Au layer (20–50 nm) is often deposited *in situ* prior to B4.4.1 photolithographic processing. In step (*b*), the pattern for the later bottom electrode has to be patterned

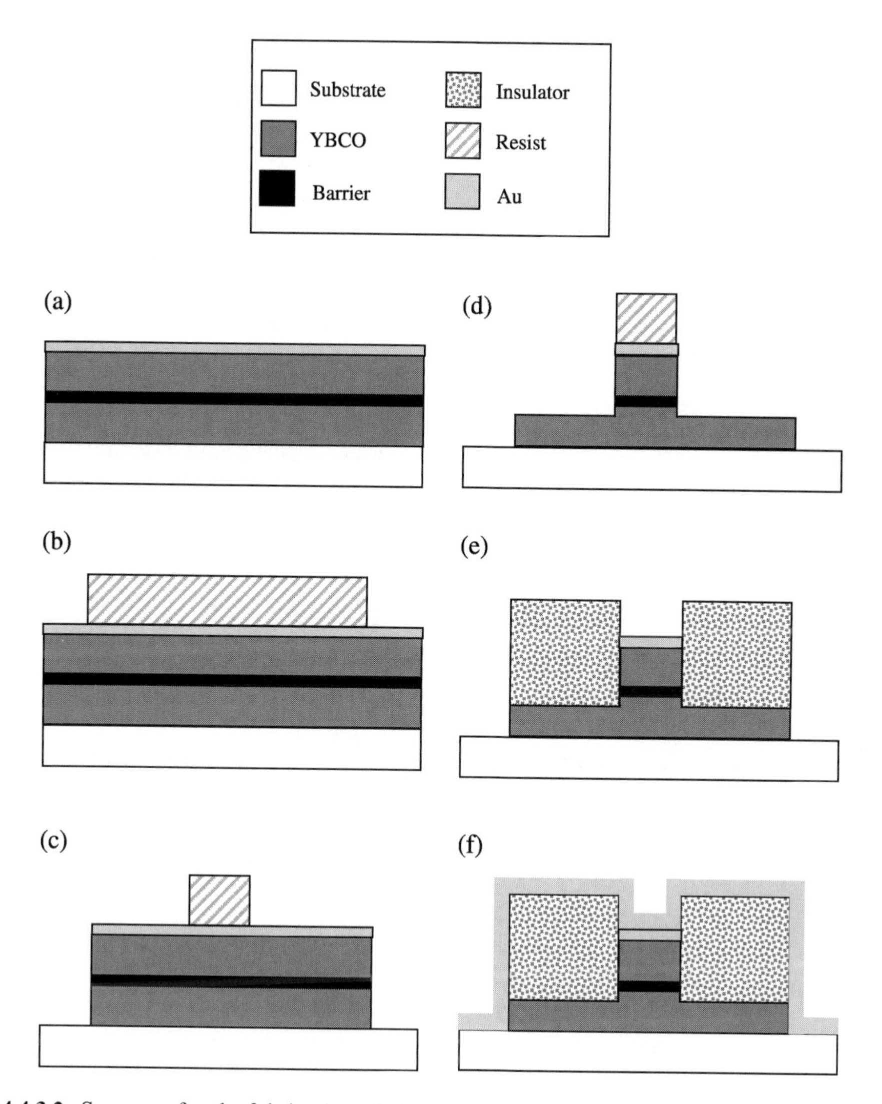

Figure B4.4.3.2. Sequence for the fabrication of planar junctions: (*a*) deposition of the trilayer and a Au protection layer, (*b*) patterning of the bottom electrode, (*c*, *d*) patterning of the mesa structure, (*e*) patterning of the insulation layer, (*f*) deposition and patterning of the Au connection to the top electrode.

by photolithography and ion beam etching. Afterwards, the mesas have to be defined by photoresist (figure B4.4.3.2(*c*)). Outside the mesa area, the trilayer will be etched down to the bottom YBCO layer (figure B4.4.3.2(*d*)). The photoresist of this etching step can be used as lift-off mask for the patterning of an subsequently deposited insulating layer (figure B4.4.3.2(*e*)). The most established insulator is amorphous SiO, but any other insulating material, which can be deposited at room temperature, may be used as well.

To connect the top electrode for junction characterization, a metal film is deposited and then patterned by lift off or by ion-beam etching (figure B4.4.3.2(*f*)). Two gold leads are separated on

B5.1

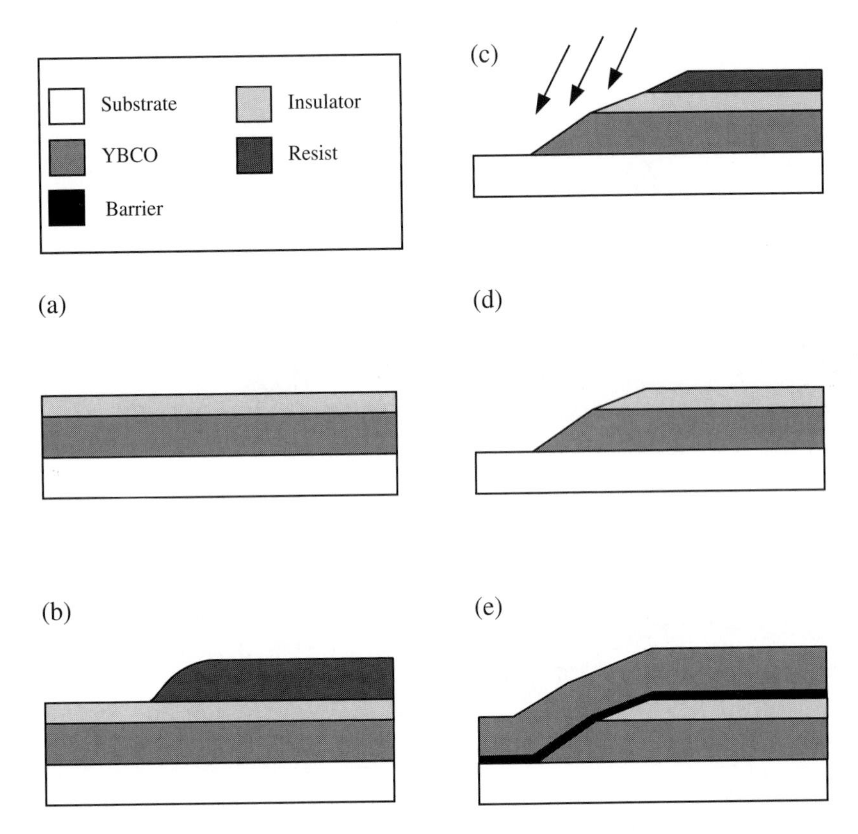

Figure B4.4.3.3. Sequence for the fabrication of edge junctions: (*a*) deposition of a YBCO-insulator bilayer, (*b–d*) preparation of the edge mask and ion beam etching of the ramp, (*e*) deposition of barrier layer and top electrode.

top of the mesa by photolithography and ion beam etching, if four-probe measurement of the mesa structure is desired. Using this method, RSJ-like junctions with an I_c spread $I\sigma \approx 34\%$ were fabricated [46]. However, there are several disadvantages of this geometry which limits their application.

E4.1

- The *a*-axis oriented YBCO films and multilayers are more difficult to fabricate and to pattern than *c*-axis oriented films.
- The SN-interface has to be atomically smooth within the relatively large junction area, which can be fabricated by conventional photolithography(typically $2 \times 2\,\mu$m). Therefore, the surface and interface roughness of the thin films is more detrimental than in junctions of other geometries.
- The fabrication process shown in figure. B4.4.3.2 ensures only a metallic, nonsuperconducting connection to the top electrode. This is an obstacle for most application. Superconducting connections to the top electrode require an even more complicated multilayer deposition and patterning process including planarization like that used for the fabrication of *c*-axis microbridges (CAM-junctions) [47].

B5.1

Edge junctions

In edge or ramp HTS junctions, the superconducting electrodes are weakly coupled along *ab*-planes via an epitaxial barrier at a shallow edge in a *c*-axis oriented YBCO film (see figure B4.4.3.1(*f*)). The advantage of this geometry is that small contact areas can be obtained without submicron patterning, and, at the same time, *ab*-coupling between *c*-axis-oriented films is possible.

C1

The first fabrication step is the deposition of a YBCO-insulator double layer (figure B4.4.3.3(*a*)). The surface morphology of this YBCO-insulator double layer influences the roughness of the edge, and thus the homogeneity of the barrier layer. The crucial task is the fabrication of a shallow edge (ramp) in the double layer (figures B4.4.3.3(*b*)–(*d*)). Its angle has to be less than 45° to prevent the formation of grain boundaries in the barrier and in the top YBCO layer.

B4.4.1

The most common way to produce shallow ramps in the YBCO-insulator double layer is by ion beam etching [48]. A sufficiently shallow ramp can be achieved by employing a shallow resist mask which was produced by re-flowing of the resist at 120–150°C [49]. Ion beam etching offers a high design flexibility, but may result in the formation of defects at the interface. Furthermore, degradation of YBCO in the ramp area may occur due to contact with humidity and chemical solvents. To minimize these problems, one includes some kind of cleaning procedures in the fabrication process. One possibility is the *in situ* use of low-energy ion beam cleaning step, which can remove adsorbates without producing new defects by ion damage [50] Another approach is to remove amorphous, nonstochiometric material from the ion-beam-etched surface by a short dip in bromine ethanol prior to the deposition of the barrier and the top electrode [35,49].

B4.4.1

In interface engineered junctions, the ion-beam damaged and degraded edge surface layer is not removed but employed as a barrier [51–53]. The edge surface is intentionally damaged by plasma treatments [38] or by ion beam etching at high energies (1200 V) [54]. During the subsequent annealing procedure, a barrier is sufficiently re-crystallized to be lattice-matched to YBCO. Many different process conditions have been used for fabrication of interface engineered junctions [38,51–53,55]. Depending on the process parameters, current densities between 10^6 and 10^3 A cm^{-2} could be achieved at 4.2 K [53]. Junction properties reported by various authors strongly differ, and no consistent picture of the barrier crystal structure emerged. A cubic YBCO phase [56] as well as a nonstoichiometric Ba-rich phase [57] were reported.

E4.1

Although the fabrication process for interface engineered junctions is not understood well enough, reported junction properties were much more reproducible than those of junctions with epitaxially grown barrier, which were fabricated in the same laboratories. For 12 junctions a 1σ of 5% at 4.2 K was achieved and 10% for an array of 1000 junctions [45]. For edge junctions with epiaxially grown barrier, the best result was 1σ of 10% for 10–20 junctions with Co-doped YBCO interlayers [35,36] and PrBaCuO barriers [52].

Although fabrication of edge junctions is quite complicated and consists of many steps, it holds the promise of a reliable process with excellent circuit design flexibility.

B4.4.3.3 Conclusion

Fabrication of highly reproducible HTS junctions with narrow spreads of the parameters, comparable to spreads attained in LTS junctions, is the biggest unsolved technology problem. Hence, the fabrication of functional multilayered HTS integrated circuits with high junction numbers remains impractical. At this juncture, interface engineered junctions appear rather promising. Nevertheless, better understanding of fabrication processes at microscopic scale, resulting in new concepts and technological approaches, may be needed to solve the problem.

B4.4.2

References

[1] Kleiner R, Steinmeyer F, Kunkel G and Müller P 1992 Intrinsic Josephson effects in BSCCO single crystals *Phys. Rev. Lett.* **68** 2394

[2] Odagawa A, Sakai M, Adachi H, Setsune K, Hirao T and Yoshida K 1997 Observation of intrinsic josephson junction properties on B(Pb)SCCO thin films Japan. *J. Appl. Phys.* **36** L21

[3] Kleiner R and Müller P 1997 Intrinsic Josephson effects in layered superconductors *Physica* C **293** 156

[4] Likharev K K 1986 *Dynamics of Josephson Junctions and Circuits* (New York: Gordon and Breach)

[5] Barone A and Paterno G 1982 *Physics and Applications of the Josephson Effect* (New York: Wiley)

[6] Gross R, Alff L, Beck A, Fröhlich O, Koelle D and Marx A 1997 Physics and technology of high temperature superconducting Josephson junctions *IEEE Trans. Appl. Supercond.* **7** 2929

[7] Engelhardt A, Dittmann R and Braginski A I 1999 Subgap conductance features of YBCO edge Josephson junctions *Phys. Rev B* **59** 3815

[8] Hirst P J, Henrici T G, Atkin I L, Satchell J S, Moxey J, Exon N J, Wooliscoft M J, Horton T J and Humphreys R G 1999 *C*-axis microbridges for rapid single flux quantum logic *IEEE Trans. Supercond.* **9** 3833

[9] Gross R and Meyer B 1991 Transport properties and noise in YBCO grain boundary junctions *Physica* C **180** 235

[10] Gross R 1994 Grain-boundary Josephson junctions in the high-temperature superconductors *Interfaces in High-T_c Superconducting Systems* ed S L Shinde and D Rudman (New York: Springer) p 176

[11] Hammerl G, Schmehl A, Schulz R R, Goetz B, Bielefeldt H, Schneider C W, Hilgenkamp H and Mannhardt J 2000 Enhanced supercurrent density in polycrystalline YBCO at 77 K from calcium doping of grain boundaries *Nature* **407** 162

[12] Schmehl A, Goetz B, Schulz R R, Schneider C W, Bielefeldt H, Hilgenkamp H and Mannhardt J 1999 Doping induced enhancement of the critical currents of grain boundaries in YBCO *Europhys. Lett.* **47** 110

[13] Beuven S, Darula M, Schubert J, Zander W, Siegel M and Seidel P 1995 Phase locking of HTS Josephson junctions closed into a superconducting loop *IEEE Trans. Appl. Supercond.* **5** 3288

[14] Char K, Colclough M S, Garrison S M, Newmann N and Zaharchuk G 1991 Biepitaxial grain-boundary junctions by lattice engineering *Appl. Phys. Lett.* **59** 733

[15] Tafuri F, Miletto Granozio F, Carillo F, Di Chiara A, Verbist K and Van Tedeloo G 1999 Microstructure and Josephson phenomenology in 45° tilt and twist YBCO artificial grain boundaries *Phys. Rev.* B **59** 11523

[16] Testa G, Sarnelli E, Carillo F and Tafuri F A 1999 *A*-axis tilt grain boundaries for YBCO superconducting quantum interference devices *Appl. Phys. Lett.* **75** 3542

[17] Simon R W, et al 1990 Progress towards a YBCO circuit progress *Science and Technology of Thin Film Superconductors*, ed R D McConnell and R Noufi (New York: Plenum Press) p 549

[18] Jia C L, Kabius B, Urban K, Herrmann K, Cui G J, Schubert J, Zander W, Braginski A I and Heiden C Microstructure of epitaxial YBCO films on step-edge $SrTiO_3$ substrates *Physica* C **175** 545 Jia C L, Kabius B, Urban K, Herrmann K, Schubert J, Zander W, and Braginski A I The microstructure of epitaxial YBCO films on steep steps in $LaAlO_3$ substrates. *Physica* C 196 1992 211–226

[19] Edwards J A, Satchell J S, Chew N G, Humphreys R G, Keene M N and Dosser O D 1992 YBCO thin-film step junctions on MgO substrates *Appl. Phys. Lett.* **60** 2433–2435

[20] Herrmann K, Kunkel G, Siegel M, Schubert J, Zander W, Braginski A I, Jia C L, Kabius B and Urban K 1995 Correlation of YBCO step-edge junction characteristics with microstructure *J. Appl. Phys.* **78** 1131–1139

[21] Yi H R, Ivanov Z G, Winkler D, Zhang Y M, Olin H, Larson P and Claeson T 1994 Improved step edges on $LaAlO_3$ substrates by using amorphous carbon etch masks *Appl. Phys. Lett.* **65** 1177

[22] Dillmann F, Glyantsev V N and Siegel M 1996 Performance of YBCO direct current SQUIDS with high-resistance step-edge junctions *Appl. Phys. Lett.* **69** 1948

[23] Pauza A J, Booij W E, Herrmann K, Moore D F, Blamire M G, Rudman D A and Vale L R 1997 Electron-beam damaged high-temperature superconductor Josephson junctions *J. Appl. Phys.* **82** 5612

[24] Tolpygo S K, Shokhor S, Nadgorny B, Lin J Y, Gurvitch M, Bourdillon A, Hou S Y and Phillips J M 1993 High quality YBCO Josephson junctions made by direct electron beam writing *Appl. Phys. Lett.* **63** 1696

[25] Davidson B A, Nordman J E, Hinaus B M, Rzchowski M S, Siangchaev K and Libera M 1996 Superconductor-normal-superconductor behavior of Josephson junctions scribed in YBCO by a high brightness electron source *Appl. Phys. Lett.* **68** 3811

[26] Pauza A J, Moore D F, Campbell A M, Broers A N and Char K 1995 Electron beam damaged high-T_c junctions–stability, reproducibility and scaling laws *IEEE Trans. Appl. Supercond.* **5** 3410

[27] Tinchev S S 1995 Low-frequency noise in high-T_c rf superconducting quantum interference devices made by oxygen-ion irradiation *J. Appl. Phys.* **77** 3563

[28] Kahlmann F, Engelhardt A, Schubert J, Zander W, Buchal C h and Hollkott J 1998 Superconductor-normal-superconductor Josephson junctions fabricated by oxygen implantation into YBCO *Appl. Phys. Lett.* **73** 2354

[29] Booij W E, Elwell C A, Tarte E J, McBrien P F, Kahlmann F, Moore D F, Blamire M G, Peng N H and Jeynes C 1999 Electrical properties of electron and ion beam irradiated YBCO *IEEE Trans. Appl. Supercond.* **9** 2886

[30] Katz A S, Woods S I and Dynes R C 2000 Transport properties of high-T_c planar Josephson junctions fabricated by nanolithography and ion implantation *J. Appl. Phys.* **87** 2978

[31] Hollkott J, Hu S, Becker C, Auge J, Spangenberg B and Kurz H 1996 Combined method of electron-beam lithography and ion implantation techniques for the fabrication of high-temperature superconductor Josephson junctions *J. Vac. Sci. Technol.* B **14** 4100

[32] Verhoeven M A J, Gerritsma G J and Rogalla H 1994 Charge transport through $PrBa_2Cu_{3-x}Ga_xO_{7-\delta}$ *Physica* C **235–240** 3261

[33] Horibe M, Kawai K, Fujimaki A and Hayakawa H 1998 Ramp-edge Josephson junctions using barriers of various resistivities *IEICE Trans. Electron.* **81-C** 1526

[34] Stölzel C, Siegel M, Adrian G, Krimmer C, Söllner J, Wilkens W, Schulz G and Adrian H 1993 Transport properties of $YBa_2Cu_3O_{7-\delta}/Y_{0.3}Pr_{0.7}Ba_2Cu_3O_{7-\delta}/YBa_2Cu_3O_{7-\delta}$ Josephson junctions *Appl. Phys. Lett.* **63** 2970

[35] Hunt B D, Forrester M G, Talvacchio J, Young R M and McCambridge J D 1997 High-T_c SNS Edge junctions with integrated YBCO groundplanes *IEEE Trans. Appl. Supercond.* **7** 2936

[36] Mallison W H, Berkowitz S J, Hirahara A S, Neal M J and Char K 1996 A multilayer YBCO Josephson junction process for digital circuit applications *Appl. Phys. Lett.* **68** 3808

[37] Bozovic I, Eckstein J N, Virshup G F, Chaiken A, Wall M, Howell R and Fluss M 1994 Atomic-layer engineering of cuprate superconductors *J. Supercond.* **7** 187

[38] Moeckly B H and Char K 1997 Properties of interface-engineered high T_c Josephson junctions *Appl. Phys. Lett.* **71** 2526

[39] Moeckly B H 2001 All YBCO *c*-axis trilayer interface-engineered Josephson junctions *Appl. Phys. Lett.* **78** 791

[40] Eckstein J N, Bozovic I, Virshup Ono R H and Benz S P 1995 Stacked series arrays of high-T_c trilayer Josephson junctions *IEEE Trans. Supercond.* **5** 3284

[41] DiIorio M S, Yoshizumi S, Yang K Y, Zhang J and Maung M 1991 Practical high-T_c junctions and SQUIDs operating above 85 K *Appl. Phys. Lett.* **58** 2552

[42] Ono R H, Beall J A, Cromar R W, Harvey T E, Johansson M E, Reintsema C D and Rudman D 1991 High-T_c SNS Josephson microbridges with high-resistance normal metal links *Appl. Phys. Lett.* **59** 1126

[43] Rosenthal P A, Grossmann E N, Ono R H and Vale L R 1993 High temperature superconductor-normal metal-superconductor Josephson junctions with high characteristic voltages *Appl. Phys. Lett.* **63** 1984

[44] Xiao Y G, Dömel R, Jia C L, Osthöver C and Kohlstedt H 1996 Fabrication of stacked intrinsic Josephson junctions from $Bi^2Sr^2CaCuO_{8+x}$ thin films *Supercond. Sci. Technol.* **9** A22

[45] Hashimoto T, Sagoi M, Mizutani Y, Yoshida J and Mizushima K 1992 Josephson characteristics in *a*-axis oriented YBCO/PrBCO/YBCO junctions *Appl. Phys. Lett.* **60** 1756

[46] Sato H, Nakamura N, Gjoen S R and Hiroshi A 1996 Improvement in parameter spreads of YBCO/PrBCO/YBCO trilayer junctions Japan *J. Appl. Phys.* **35** L1411

[47] Goodyear S W, Chew N G, Humphreys R G, Satchell J S and Lander K 1995 Vertical *c*-axis microbridge junctions in YBCO/PrBCO thin films *IEEE Trans. Supercond.* **5** 3143

[48] Gao J, Aarnink A M, Gerritsma G J and Rogalla H 1990 Controlled preparation of all high-T_c SNS-type edge junctions and DC SQUIDs *Physica* C **171** 126

[49] Horstmann C, Leinenbach P, Engelhardt A, Gerber R, Jia C L, Dittmann R, Memmert U, Hartmann U and Braginski A I 1998 Influence of ramp shape and morphology on the properties of YBCO-ramp-type junctions *Physica* C **302** 176

[50] Verhoeven M A J, Gerritsma G J, Rogalla H and Golubov A A 1995 Ramp-type junctions with very thin PBCO barriers *IEEE Trans. Appl. Supercond.* **5** 2095

[51] Satoh T, Wen J G, Hidaka M, Tahara S, Koshizuka N and Tanaka S 2001 High-temperature superconducting edge-type Josephson junctions with modified interface barriers *IEEE Trans. Appl. Supercond.* **11** 770

[52] Soutome Y, Hanson R, Fukazawa T, Saaitoh K, Tsukamoto A, Tarutani Y and Takagi K 2001 Investigation of Ramp-type Josephson junctions with surface-modified barriers *IEEE Trans. Appl. Supercond.* **11** 163

[53] Horibe M, Takuma I, Inagaki Y, Matsuda G, Fujimaki A and Hayakawa H 2001 Preparation of Ramp-edge interface modified junctions for HTS SFQ circuits *IEEE Trans. Appl. Supercond.* **11** 159

[54] Heinsohn J K, Hadfield R H and Dittmann R 1999 Effects of process parameters on the fabrication of edge-type YBCO Josephson junctions by interface treatments *Physica* C **326–327** 157

[55] Heinsohn J K, Dittmann R, Rodrguez Contreras J, Scherbel J, Klushin A, Siegel M, Jia C L, Golubov S and Kupryanov M Y u 2001 Current transport in Ramp-type junctions with engineered interface *J. Appl. Phys.* **89** 3852

[56] Huang Y, Merkle K L, Moeckly B H and Char K 1999 The effect of microstructure on the electrical properties of YBCO interface-engineered Josephson junctions *Physica* C **314** 36

[57] Wen J G, Koshizuka N, Tanaka S, Satoh T, Hidaka M and Tahara S 1999 Atomic structure and composition of the barrier in the modified interface high-T_c Josephson junction studied by transmission electron microscopy *Appl. Phys. Lett.* **75** 2470

[58] Satoh T, Hidaka M and Tahara S 1998 Fabrication process for high-T_c superconducting integrated circuits based on edge-type Josephson junctions *IEICE Trans. Electron.* **81-C** 1532

Further Reading

Braginski A I 1993 Thin film structures *The New Superconducting Electronics* ed H Weinstock and R W Ralston (Bristol: Institute of Physics Publishing)

More details about the patterning and the properties of thin film structures and Josephson junctions.

Gross R, Alff L, Beck A, Fröhlich O, Koelle D, Marx A 1997 Physics and technology of high temperature superconducting Josephson junctions *IEEE Trans. Appl. Supercond.* **7** 2929

Details about the physics and technology of HTS junctions.

B5
Introduction to section B5: Superconductor contacts

Harry Jones

Throughout all electrical/electronic engineering, electrical contacts are a subject of immense importance, although probably not regarded as particularly exciting, and the popular view is that the technology is well established and few problems remain. Whether or not this is true, generally speaking, it certainly is not the case in superconducting technology.

Consider first the traditional low temperature superconductors (LTS). These have to operate in the liquid helium regime and this gives rise immediately to at least two problems. Firstly, at liquid helium temperatures, the specific heat of most materials is very low. The energy equivalent of dropping a pin 10 cm will raise the temperature of NbTi, for instance, by 10 K. Secondly, if heat transfer exceeds $1\ \mathrm{W\,cm^{-2}}$, helium goes from the nucleate to film boiling regime and the concomitant, 'superheat' can be at least 10 K. Hence dissipative contacts are a real problem.

High temperature superconductors (HTS) are popularly thought of as operating at 77 K, or thereabouts, and consequently the problems related above disappear, or considerably lessen. However, many applications and measurements are at low temperatures still and so the problems still affect best practice. Furthermore, if one reflects that the received wisdom for soldering, say, to conventional conductors is, 'first remove the oxide layer', these materials, in their bulk, unclad form, *are* oxides! So, the techniques for making contacts to HTS are considerably more complex. As an example of the way in which good contacts are crucial, take critical current measurements at high magnetic fields. Here, if one has bad/unbalanced contacts in limited length samples, a variety of thermoelectric and magnetothermogalvanic effects can completely distort the measurements [1].

This important chapter is written from the standpoint of considerable hands on experience and therefore fits well with the tutorial aspirations of this handbook. After B5.1.1 (the introduction) there follow four sections. B5.1.2 introduces the concept of specific contact resistivity, effectively the figure of merit for contacts. In B5.1.3 the topic of contacts for high current applications is thoroughly reviewed. Contacts to thin films are a topic in themselves and these are dealt with in B5.1.4. Again with the tutorial slant in mind, B5.1.5 presents worked examples of minimum contact areas for the important subject of critical current measurements mentioned above. The last section, B5.1.6, is quite special and discusses how the limited thickness of contact pads can lead to an unexpectedly small contact area in practice.

References

[1] Jones H, Cowey L and Dew-Hughes D 1989 Contact problems in J_c measurements on high temperature superconductors *Cryogenics* **29** 795–9

B5.1
Superconductor contacts

Jack Ekin

B5.1.1 Introduction

Electrical contacts are the Achilles heel of most superconductor transport measurements. Poor contacts have led to more 'unusual' effects and scuttled more critical current measurements than any other known factor. At best, Joule heating at the contacts makes it difficult to control and measure sample temperature; at worst, it triggers thermal runaway and sample destruction. For many materials, it is fairly straightforward to make good contacts to samples, but not for the oxide-based high T_c oxide superconductors (HTS). They have a unique interfacial chemistry that is counter intuitive. C

In this chapter, we focus mainly on contact techniques for the oxide superconductors. However, many of the topics are widely applicable to other materials as well. Such general topics include: contact-soldering techniques, measuring specific-interface resistivity, calculating contact areas needed to prevent sample heating, and determining spreading-resistance effects.

After the introduction, the chapter is divided into two major parts: the first describes techniques for contacting *high current* ($\gg 1$ A) superconductors; the second focuses on contacting low current *thin film* superconductors. Thick film, high current conductors, such as $YBa_2Cu_3O_7$ (YBCO) 'coated superconductors', are hybrids; they are treated mainly in the high current section. At the end of the chapter, several calculational examples are given which illustrate methods for determining the minimum contact *area* needed to limit sample heating to an acceptable level. D2.2 B4.1, B4.2.1

This monograph is an excerpt from a forthcoming book on experimental techniques, which contains extensive additional information about heat transfer, construction techniques, cryogenic wiring, sample holder design and properties of solids at low temperatures (see 'Further Reading' at the end of the chapter).

B5.1.2 Definition of specific contact resistivity—values for practical applications

Before we launch into contact techniques and the unique interfacial chemistry of high T_c superconductors, first we need a quantity to describe the resistance of a contact in a way that is independent of contact area—the specific contact resistivity ρ_c

$$\rho_c \equiv R_c A_c \qquad \text{(B5.1.1)}$$

Here R_c is the contact resistance and A_c is the contact area. If, for example, we increase the area of a contact A_c, the measured contact resistance R_c would of course get proportionately smaller, but the $R_c A_c$ product remains the same. Thus, the $R_c A_c$ product, ρ_c, gives us a means to quantify the specific resistivity of the contact interface independent of its area. Note that the *contact interface* resistivity ρ_c is measured in $\Omega\,\mathrm{cm}^2$ (that is, *square* centimeters), not the usual $\Omega\,\mathrm{cm}$ used for the *bulk* resistivity of materials ρ. (The two quantities could be mathematically related by $\rho_c \equiv \rho t$, where t is the thickness of the interface; however, for practical purposes the 'thickness' of an interface cannot be defined and so the bulk resistivity of an interface is not a useful quantity.)

We can get a feeling for the contact resistivity requirements in different applications by referring to the information below the line in figure B5.1.1. Superconductor applications such as magnets and transmission lines require a ρ_c less than 10^{-5}–$10^{-4}\,\Omega\,\mathrm{cm}^2$ to prevent unacceptably high heat and voltage generation at current connections. Values of ρ_c for transport critical current *measurements* usually also fall in this range. Sometimes with short-sample measurements, pulsed current methods can be used to minimize heating and we can get away with marginally high ρ_c values (but such techniques do not work in actual applications, of course, where current is usually steady-state). For applications involving thin film package interconnects, we see from figure B5.1.1 that nominal values of ρ_c are needed in the range of 10^{-8}–$10^{-7}\,\Omega\,\mathrm{cm}^2$. The most stringent requirements are for on-chip interconnects, where there are many contacts in series and the strip line width is narrow; in this case an extremely low contact resistivity in the 10^{-10}–$10^{-9}\,\Omega\,\mathrm{cm}^2$ range is necessary.

E1.1, E1.3.2

D3.1

As a benchmark, the specific contact resistivity of two pieces of copper soldered together under moderate pressure with common eutectic 63Sn–37Pb solder is quite low, about $4\times10^{-9}\,\Omega\,\mathrm{cm}^2$ [8]. This is near the extreme right end of figure B5.1.1 and, thus, for conductors sheathed with copper or silver, the usual soft-soldering techniques adequately serve a wide range of magnet and transmission-line applications. However, when contacts are formed to bare high T_c superconductors (bulk, film, or thick film conductors) it is another story. Here the range of ρ_c values is enormous, spanning eight orders of magnitude, as indicated by the information above the line figure B5.1.1. This is because high T_c oxide superconductors have a unique interfacial chemistry that must be accommodated; otherwise a semiconducting interface is formed (described later) that results in highly resistive contacts.

B

An overview of ρ_c values for the entire range of contact techniques for high T_c superconductors, as well as comments on their common usage, is given in table B5.1.1. The techniques giving the lowest contact resistivity are listed first, with a soldered Cu contact shown for reference at the top of the table.

In the next section, techniques for contacting *high current* samples are presented, with particular emphasis on the problematic case of bare oxide superconductors. High current 'coated conductors' are

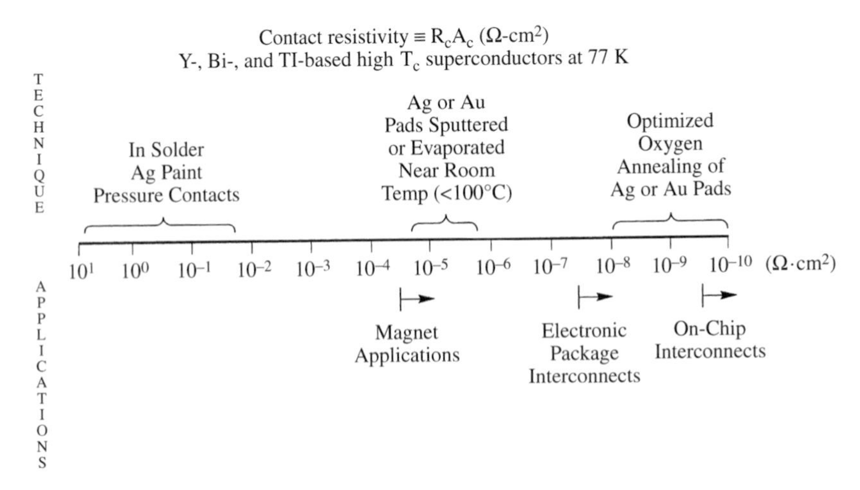

Figure B5.1.1. Representative values of contact resistivity ($\rho_c \equiv R_cA_c$) for high T_c superconductors obtained using various contact fabrication techniques. The lower half of the figure shows typical current-contact requirements for applications.

included in this section. *Thin film* contacts are treated in a later section; although the principles for making thin film contacts of high quality are the same as those discussed for bulk samples, the techniques are sometimes unique.

B5.1.3 Contact techniques for high current superconductors

B5.1.3.1 Overview

For low T_c superconductors with a copper sheath, or bismuth-based high T_c superconductor having a relatively thick (>20 μm) silver-sheath, common Pb–Sn soft solder works fine. On the other hand, oxide

Table B5.1.1. Overview of representative contact resistivity ρ_c values for various high-temperature superconductor connections

Contact type	Specific contact resistivity ρ_c ($\Omega\,cm^2$)	Common usage/comments
Cu/63Sn–37Pb/Cu[a] (soldered under hand-applied pressure)	4×10^{-9}	Normal copper conductor contacts
Silver sheath/BSCCO[b]	$\ll 10^{-8}$	High current BSCCO[c] tape contacts
Silver or gold deposited on YBCO[d]		
In situ[e] deposited; no anneal[f]	10^{-9}–10^{-7}	Contacts to oxide superconductor films, both for electronics and high current transmission
Ex situ[e] deposited; oxygen annealed[g]	10^{-9}–10^{-6}	Contacts to bare oxide superconductor surfaces (e.g. HTS[h] bulk current leads, YBCO thick film 'coated conductors')
Ex situ deposited; no oxygen anneal[i]	10^{-5}–10^{-1}	Contacts where annealing is not possible (Could be improved several orders of magnitude by oxygen annealing, as seen in previous line.) Lowest (best) values of ρ_c obtained when surface is well cleaned
Indium-soldered silver or gold pad on YBCO	10^{-1}–10^{-6}	Lowest (best) values of ρ_c obtained when gold or silver pad thickness is at least 7–10 μm thick (see text 'Soldering to noble metal contact pads' section B5.1.3)
In/YBCO[j]	10^{-2}–10^{-1}	Soldered voltage contacts for bulk oxide superconductors

[a] [8].
[b] [2].
[c] BSCCO ≡ Bi–Sr–Ca–Cu–O oxide superconductors.
[d] YBCO ≡ Y–Ba–Cu–O oxide superconductors.
[e] '*Ex situ*' and '*in situ*' contacts refer to whether the superconductor surface is exposed to air before the noble metal contact pad is deposited, as described in section B5.4 on thin film contact techniques.
[f] [6, 12].
[g] [5].
[h] HTS ≡ high T_c superconductors.
[i] [4].
[j] Ekin, unpublished.

superconductors that are bare or that have only a thin silver coating are more challenging. This includes materials such as YBCO 'coated conductors' or high T_c superconductors that are unsheathed. Table B5.1.2 outlines some preferences for contacts to these materials. Fabrication techniques for each of the methods listed in table B5.1.2 are given in the subsections that follow.

B4.1, B4.2.1

The contact techniques shown in table B5.1.2 are listed according to whether the methods are appropriate for *voltage* or *current* connections, because the requirements for the two types of contacts are very different. Contacts for current leads usually need to carry large currents and so must have a much lower resistivity than contacts for attaching voltage leads. [When a four-terminal technique is used (two contacts for current leads and two for voltage taps), the ratio of current through the *voltage* contacts to that through the *current* contacts is given approximately by the ratio of sample resistance to voltmeter-input impedance. Ratio values are typically $<10^{-6}$ and so the resulting current in the voltage leads is quite small, usually only a fraction of a microampere.]

D2.2

B5.1.3.2 Voltage contacts

Since voltage contacts do not need to carry much current, the contact resistivity can be relatively high, opening up the use of simple soldering techniques. Wires used as voltage leads should be small in diameter—much less than the sample cross-sectional dimensions, if possible, so the voltage contact point is well defined. Small voltage leads also reduce the chance of damaging the sample from handling, which can occur if the voltage lead is so large that its strength becomes comparable to that of the sample. Using a small size for the voltage lead also minimizes the risk that a voltage lead will pull off from differential thermal contraction during cooling.

Soldered voltage contacts

Soldered contacts are robust and work well for copper-sheathed samples, silver-sheathed oxide superconductors and, with the right solder, even for bare oxide superconductors. Some samples, however, would be damaged by soldering (either chemically or thermally). Also, at times, soldering is inconvenient, such as when many similar samples need to be tested in rapid succession. In that case, spring-pressure voltage contacts are worth making, as described later in this section.

Eutectic Pb–63%Sn solder is fairly strong and works well with copper- and silver-sheathed superconductors. I generally give the voltage lead a gentle pull with a pair of tweezers after making the connection to make sure the connection is mechanically sound. A few times I have detected bad solder joints this way; in each case, the extra ten seconds invested saved me a couple of hours of testing time.

When selecting a soldering material for voltage taps, a few precautions are in order. Depending on the particular solder chosen, the heat from soldering can cause oxygen loss in YBCO above about 150°C and in bismuth-based superconductors above 250–300°C (solder melting temperatures are given in table B5.1.4 below). The relatively high melting temperature of some soldering materials can also lead to thermal shock in the case of bismuth conductors if the superconducting material is fragile and porous. The soldering temperature also needs to be kept below the melting temperature of any soldering materials used in the fabrication of some multilayer composites.

C1

For bare oxide superconductors, Pb–Sn-based solders will not wet the oxide superconductor surface, unlike indium-based solders which will. Pure indium (T_{melt} = 157°C) is extremely soft and weak, however, so indium solder alloys, such as eutectic In–48%Sn (T_{melt} = 118°C) or eutectic In–3%Ag (T_{melt} = 143°C), are better choices. These alloys are stronger than pure indium, their soldering temperatures lower, and they work well for voltage contacts.

Table B5.1.2. Contacts for voltage and current connections to bare YBCO superconductors (listed within each category by contact resistivity). Any of the current contact methods can also be used for voltage contacts, but are more complicated to fabricate than the simple techniques listed for voltage contacts. Contact materials that do not work well with the oxide superconductors are included at the end of the table for pedagogical reasons (see text)

Contact Method	Procedure	ρ_c ($\Omega\,cm^2$)	Comments
Voltage contacts			
In–3wt.%Ag solder	To make solder bond, lightly scratch sample surface under molten solder with soldering iron, or use ultrasonic soldering iron; see text 'Soldered voltage contacts'	10^{-2}–10^{-1}	$T_{melt} = 143°C$; eutectic
In–48wt.%Sn solder	Same procedure as for previous material.	10^{-2}–10^{-1}	$T_{melt} = 118°C$; eutectic; dissolves thin Ag or Au films
Spring contacts	Beryllium copper or other conducting spring stock is used to contact the sample; see text 'Pressure contacts'		May require silver or gold pads deposited on the test sample for good contact
Silver paint		10^{-1}–1	Weak connection, but sometimes needed for delicate samples
Current contacts			
In situ gold or silver pad deposited on superconductor, no oxygen anneal	Descriptions of *in situ* versus *ex situ* deposition techniques are given in section B5.1.4	10^{-9}–10^{-7}	The lowest ρ_c
			In situ contacts are mainly amenable to HTS film fabrication techniques
			Gold is more expensive than silver contact pads, but does not tarnish as readily
Ex situ gold or silver pad deposited on superconductor, with oxygen annealing	See text 'Oxygen annealing' under 'Current contacts/Fab. procedure…'	10^{-9}–10^{-6}	

Table B5.1.2. (*Continued*)

Contact Method	Procedure	ρ_c ($\Omega\,cm^2$)	Comments
Ex situ gold or silver pad deposited on superconductor, no oxygen annealing	See text 'Noble metal deposition' under 'Current contacts/Fab. procedure…,	10^{-5}–10^{-1}	Used for applications where oxygen annealing is not possible, or where very low ρ_c s are not needed. ρ_c depends on how well the surface is cleaned
Indium-soldered connection to silver or gold pad	Make silver or gold pad thickness at least 7–10 μm. See text 'Soldering to noble metal contact pads' under 'Current contacts/Fab. procedure…,	10^{-1}–10^{-6}	Used for high-current contacts between the sample-contact pad and Cu wire or bus bar. ρ_c depends strongly on soldering technique used (see text)
Poor contacts			
Copper pad deposited on superconductor	Sputter deposited	10^{-2}	ρ_c comparable to that of In solder, but with a lot more work, so why use it?
Au/Cr pad deposited on superconductor	Sputter deposited	10^{-1}	Contact commonly used for semiconductors, but terrible for superconductors
Pb–Sn solder		no bond	

Soldering procedures for current contacts are a bit more involved if the solder joint is made to thin (<25 μm) noble metal contact pads (as in the case of the YBCO coated conductors). Soldering techniques for this case are described in the section on current contacts below.

Wetting the oxides

Solder flux is death for most bare, sintered (porous) oxide-superconductor materials; the flux wicks into the superconductor along grain boundaries and deteriorates the transport properties of the superconductor. On the other hand, pure indium without any solder flux will wet almost anything, including glass and ceramics. Just make sure the sample surfaces are clean—very clean. Either pure indium or (stronger) indium alloys can be made to wet bare oxide superconductors by lightly scratching the sample surface with a soldering iron *under* the molten indium (again, with no solder flux). To avoid overheating the sample, one suggestion is to keep the temperature of the soldering iron less than ~30–50°C above the solder's melting temperature by using a variable temperature soldering iron or by plugging the soldering iron into variable power source.

An even more sure method of wetting bare oxide superconductors is to use an ultrasonic soldering iron. Operate the ultrasonic iron at very low ultrasonic power to avoid fracturing the sample. Also, for ultrasonic soldering to work well, the sample needs to be placed on a mechanically rigid surface. A drop of molten solder is applied to the hot tip of the ultrasonic soldering iron, and the vibration of the tip chips away at the brittle superconductor's surface beneath the molten solder drop, exposing a fresh surface that the solder readily wets. For fragile samples, however, the ultrasonic power can fracture the specimen. In this case it is best to patiently scratch the oxide surface by hand, under a drop of molten indium-based solder, as mentioned in the preceding paragraph.

D.3.3

While ultrasonic soldering and scratching work well with brittle materials, these abrasion techniques are surprisingly useless with copper leads, because copper is ductile and the tough copper-oxide layer on the surface does not chip away. A mild ZnCl flux is needed to chemically cut through the copper-oxide coating on the surface and allow the indium-based solders to wet the copper surface. Thus, flux is essential for 'tinning' copper leads with indium before soldering to high T_c superconductors. After tinning, however, be sure to wipe all the excess flux off the copper leads before attempting to attach them to the oxide superconductors, or once again it is 'death by flux'.

Pressure contacts

For quick sample changing or when many samples need to be tested, spring pressure contacts are convenient. They are a little more work to fabricate initially, but can be used to make voltage contacts to almost any sample material. A 'springy' material such as beryllium–copper–alloy spring stock is used to make contact with the sample by forming the alloy stock into a spring or clip that presses on the sample surface with an arrangement like that shown in figure B5.1.2, for example. Contact pressure needs to be kept light ($<$ approximately 0.5 N for small, brittle samples) so the sample will not be damaged; the sample also needs to be supported with a solid sample-holder base, as shown in figure B5.1.2. If the contact resistance between the spring clip and the sample is too high, try coating the surface of the spring (where it will touch the sample) with indium solder. Again, a mild flux is needed to get indium to wet copper-based materials such as beryllium and copper; but be sure to clean the residual solder flux off the indium after it solidifies, before making a pressure contact to the superconductor. If, on the other hand, silver or gold contact pads have been deposited on the superconductor for current connections (see below), an extra set of voltage contact pads can readily be deposited on the superconductor at the same time. Such pads work extremely well to lower the resistance of the spring-clip contact where it presses on the sample.

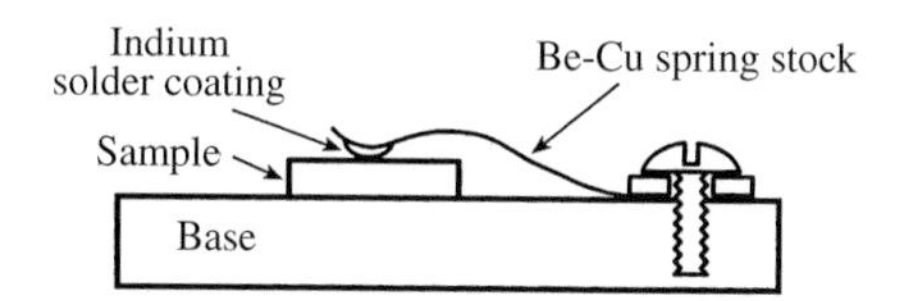

Figure B5.1.2. Spring contact for convenient, quick changing of voltage contacts when testing many samples in succession.

Silver paint, paste and epoxy

Lower reliability contact methods are silver paint silver paste and silver-based epoxy. The joint is extremely weak (for paint and paste) and the voltage contact usually ends up as a sizable blob. Annealing the silver paint or paste in oxygen (using the oxygen annealing procedure described below) can greatly improve the contact resistivity. The organic carrier used in compounding the silver paint or paste can wick in along pores and grain boundaries, however, degrading the sample. Nevertheless, they do make contact and are sometimes needed for delicate samples.

B5.1.3.3 Current contacts for oxide superconductors

These are the challenging connections. Because current contacts carry much higher current than voltage contacts, the requirements on ρ_c are orders of magnitude more stringent. For silver-sheathed conductors, excellent low-resistivity current contacts can be made simply by soldering to the thick silver sheath, as long as the sheath is more than about 20 μm thick. But if the silver layer is only a few micrometres thick, or if the contact must be made directly to a bare oxide sample, then soldering does not work for current contacts—the resulting ρ_c is far too high.

The bottom of table B5.1.2 lists my early failures at soldering or sputtering contact materials onto the oxide superconductors. Failures are instructive, however. Cu or AuCr (a great semiconductor-contact material) sputter deposited onto a clean oxide-superconductor surface is terrible. Pb–Sn solder does not form a bond at all. Most of the voltage contact methods at the top of table B5.1.2 would also fail as current contacts. Silver paint adheres, but has a high interface resistivity unless oxygen annealed at high temperatures (>approximately 400–500°C). Indium and indium-alloy solders, even when applied to a freshly exposed superconductor surface under molten solder, produce contacts with ρ_cs orders of magnitude too high for use in most current connections.

Interfacial chemistry

The problem is that all these 'bad' contact materials have a high affinity for oxygen. The top of figure B5.1.3 shows an Auger electron spectrograph (AES) depth profile of the interface between YBCO and a typical indium-alloy solder (In–2%Ag). The AES signal strength along the vertical axis indicates the relative amounts of different elements at the interface; the sputter time along the horizontal axis indicates the relative position across the interface as the probe sputters through pure indium on the left side of the figure, into pure YBCO on the right side. The AES depth profile shows that indium does not end abruptly at the YBCO surface, but instead diffuses far into the YBCO, reacting with the oxygen in the YBCO. This oxygen reaction causes two major problems: first, the YBCO superconductor is depleted of its oxygen and ceases to be superconducting at the surface, and second, the oxygen gets bound into an indium-oxide compound that forms a resistive barrier at the interface. The compound layer (indicated by the nearly constant oxygen and indium signals from about 50 to 150 min sputter time in the top panel of figure B5.1.3), is probably In_2O_3, which is semiconducting. It has a band gap of 3.5 eV and a resistivity at liquid nitrogen temperature that is orders of magnitude higher than that of most metals. An interfacial In_2O_3 layer would thus explain the poor contact ρ_c observed for In/YBCO contacts. Similarly, Pb, Sn, Cr, and Cu all readily form oxide compounds, which would explain the poor contact resistivity for these materials as well (see table B5.1.2). For silver paint without oxygen annealing, ρ_c is probably high because of a similar degradation of the superconductor surface by the organic carrier in the paint.

The solution is to use contact pads made from materials with a low affinity for oxygen—the *noble* metals. Verification of this is shown by the AES depth profile of an Ag/YBCO interface in the middle panel of figure B5.1.3. The oxygen signal decays very quickly into the silver, and there is no telltale

D1.3
C1

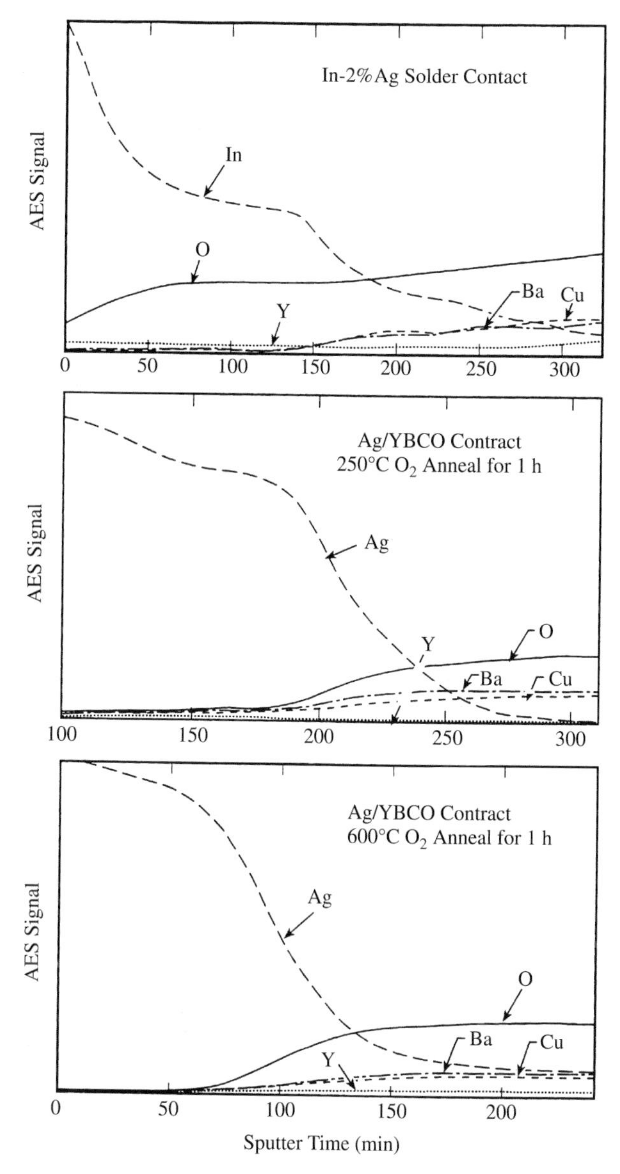

Figure B5.1.3. Auger electron spectrometry (AES) depth profile for three contacts to YBCO: top—indium-solder contact; middle—Ag/YBCO contact [given an oxygen anneal at a very low temperature (250°C)]; and bottom—Ag/YBCO contact fully oxygen annealed at 600°C for 1 h [5]. Auger electron intensity is plotted in arbitrary units, as a function of ion sputter time (equivalent to position across the interface).

section of nearly constant silver and oxygen signal levels, which would indicate the formation of an oxide-barrier compound, as in the In/YBCO depth profile in the top panel of figure B5.1.3. The ρ_c values for sputter-deposited silver and gold contacts to YBCO are shown in the current contact section of table B5.1.2. These values are four to six orders of magnitude lower than those for the high oxygen affinity materials presented in the top or bottom sections of table B5.1.2.

C1

Fabrication procedures for HTS current contacts

Thus, the trick for fabricating high quality HTS contacts is to avoid formation of a resistive oxide-compound layer at the contact interface. A generic method that consistently yields contacts of high quality and low ρ_c, consists of the following three basic processing steps:

(a) *Clean* the superconductor surface.

(b) Use a *noble metal*, such as silver or gold, for the contact material. If current leads are to be *soldered* to the noble metal pads, make sure the noble metal pads are at least 7–10 μm thick (see next section).

(c) *Oxygen anneal* the noble metal/superconductor interface (this step is not needed if ρ_cs over $10^{-5}\,\Omega\,\mathrm{cm}^2$ are acceptable.

(a) Cleaning

The surface of YBCO reacts with water vapour and carbon dioxide in air to produce a resistive layer at the superconductor surface. The effect of this barrier on contact resistivity is shown in figure B5.1.4, where contacts were made to YBCO films that had been exposed for different times to air, CO_2, N_2, O_2, and vacuum *prior* to depositing a silver contact pad. As shown in figure B5.1.4, the ρ_c of such contacts rises rapidly for air exposure times greater than approximately 1000 min. So, for air-exposed

D3.1

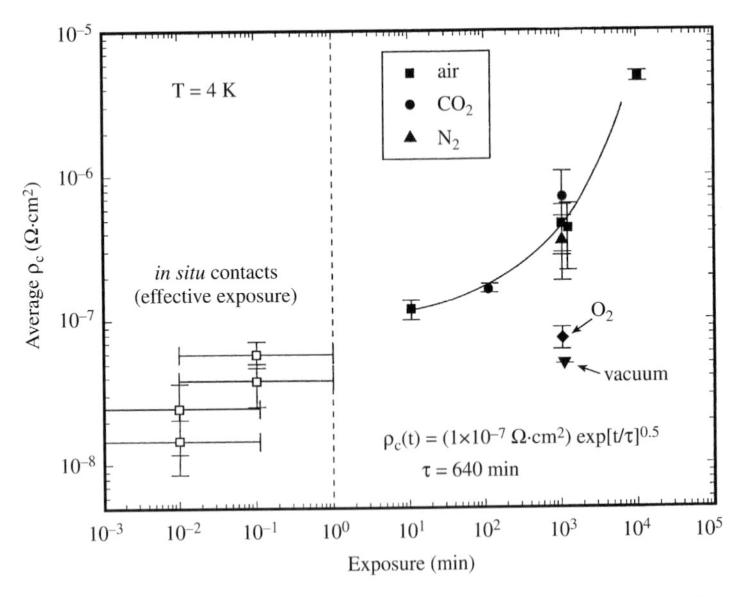

Figure B5.1.4. Contact resistivity of Ag/YBCO interfaces at 4 K as a function of exposure time to air, CO_2, N_2, O_2 and vacuum [16]. The term '*in situ* contacts,' designating the data to the left of the vertical dashed line, refers to noble metal contact pads deposited *in situ* on freshly fabricated superconductor surfaces not exposed to air. The 'effective' exposure times for these *in situ* contacts result from the small residual partial pressure of water vapour in the vacuum chamber, which reacts with the sample surface during the period between the end of superconductor deposition and the start of noble metal deposition.

superconductor surfaces, the first step before depositing a noble metal contact pad is to remove the degraded superconductor surface. This cleaning step is helpful, but not essential, if the contacts are later oxygen annealed (described below).

To clean bulk samples, the surface can be sanded lightly where the contact pads will be made and then use low energy sputter etching (a physical etching process carried out in a low-pressure argon atmosphere, see 'Further reading') or ion milling (carried out in a vacuum, typically with argon ions accelerated from a source). Use sputtering or milling to clean away the remaining degraded layer on the oxide superconductor surface just before the noble metal contact pads are deposited. It is best to carry out the sputtering or milling in the same vacuum system where the noble metal pads will be deposited so the samples are not exposed to air between cleaning and deposition. (For thin film samples, or course, no sanding is done and just the ion milling is performed.)

It is important in this step to keep the ion milling or sputter energy low to avoid destroying the weak crystal structure of the superconductor. However, if the energy is too low, the damaged surface material is not removed. An optimum reduction in ρ_c is achieved by ion milling or sputter etching with argon at about 200 V, which, at a typical ion current density of 0.5 mA cm^{-2}, will remove about 16 nm in about 2 min (further information is given in figure B5.1.10 in the 'Cleaning etch' section for films in section B5.1.4 below).

(b) Noble metal deposition

Immediately after the superconductor surface is sputter cleaned or ion milled, the noble metal contact layer needs to be deposited onto the superconductor surface. Preferably, this should be done without breaking vacuum between the cleaning and deposition operations, or at least limit the intervening period to less than a few hours (see figure B5.1.4).

Silver or gold can be either thermally evaporated or sputtered onto the surface. The relatively higher energies of sputtering compared with evaporation can damage the superconductor's crystal structure at the surface. However, if the first few tens of nanometers of sputter deposition are carried out at a relatively low cathode voltage (e.g. $<$ approximately 750 V) in argon at a reasonably high gas pressure (1–2 Pa) to reduce the ion energy, sputtering produces a layer of noble metal strongly bonded to the superconductor surface with minimal surface damage. After a few tens of nanometers of noble metal are deposited, the sputter deposition energy (and rate) can be increased, since this initial layer then protects the superconductor surface.

Even with this variable-energy sputtering technique, the overall difference in deposition rate between sputtering and evaporation becomes important when depositing *thick* contact pads. The pad thickness needs to be about 7–10 μm thick when copper wires or bus bars are soldered to the pads. This keeps the solder from completely alloying through the pad and coming into contact with the sensitive oxide superconductor surface (see the next section describing soldering techniques). Such a thick pad can take literally all day to deposit when sputtering, since sputter deposition rates are usually about 0.3–0.5 nm s^{-1}, compared with evaporation rates, which are typically 1–2 nm s^{-1}. So, for speed and simplicity, it is usually best to deposit the noble metal contact pad by thermal evaporation, typically from an inexpensive, resistively heated source. This usually produces ρ_c values of about $10^{-4}\,\Omega\,cm^2$ before oxygen annealing.

(c) Oxygen annealing

It is possible to obtain another three to four orders of magnitude of improvement in ρ_c, depending on how well the superconductor's surface has been cleaned prior to depositing the noble metal contact pad. This is done by annealing the contact pad in flowing oxygen after it is deposited. The resulting ρ_c is

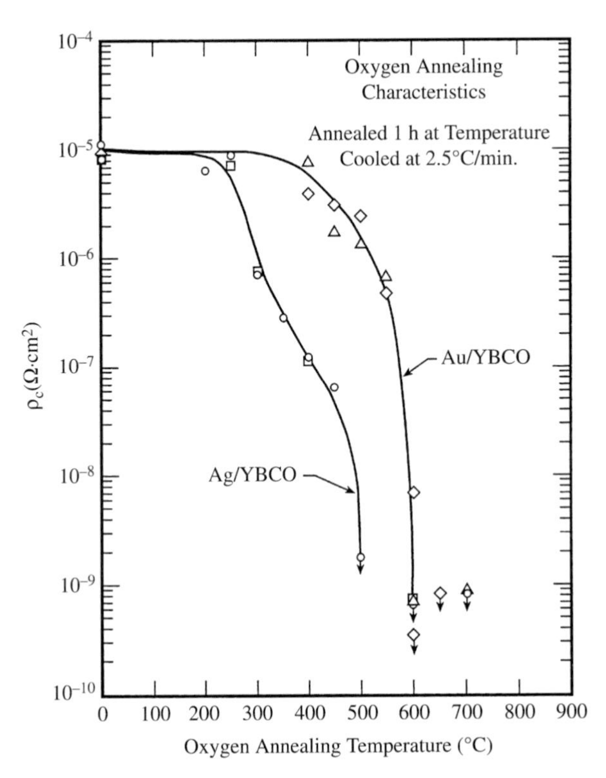

Figure B5.1.5. Oxygen annealing characteristic for silver and gold contacts on sintered bulk YBCO; annealing time was 1 h and oxygen was supplied at atmospheric pressure, flowing at a rate of about $2 \times 10^{-6}\,m^3\,s^{-1}$ ($\sim$0.3 scfh, standard cubic feet per hour) [5].

D1.3 usually in the range of 10^{-7}–$10^{-8}\,\Omega\,cm^2$. A comparison of the middle and bottom AES depth profiles in figure B5.1.3 shows that, after oxygen annealing at 600°C, silver is diffused much farther into the superconductor (probably mainly along grain boundaries) providing better contact with the C1 superconductor. Oxygen concentration near the YBCO surface is also enhanced.

Figure B5.1.5 shows the reduction of ρ_c obtained for Au/YBCO and Ag/YBCO contacts after annealing in oxygen at various temperatures. The annealing was carried out in flowing oxygen at atmospheric pressure for 1 h at each temperature. As seen in figure B5.1.5, a rapid reduction in ρ_c for gold contacts occurs at an annealing temperature about 100°C higher than for silver, probably because higher temperatures are required for diffusion of gold compared with silver. Another difference between the two noble metals is that oxygen readily diffuses through silver, whereas gold is impervious to oxygen. The data indicate that the optimum condition for annealing Ag/YBCO contacts is about 30–60 min at about 500°C in flowing oxygen, whereas the optimum annealing temperature for Au/YBCO contacts is slightly higher, about 600°C. It is important not to anneal silver contacts at too high a temperature because the silver contact pad can completely diffuse into the YBCO and disappear. Annealing for longer than 1 h does not significantly improve ρ_c. Optimum annealing conditions are summarized for B2.2.2 bulk sintered HTS materials in table B5.1.3. (For film superconductors, annealing times and temperatures are slightly lower, as described in section B5.1.4.)

Oxygen annealing curves for silver contact pads on bulk sintered bismuth-based superconductors are given in figure B5.1.6. Between 200 and 400°C, the contact interface resistivity drops four orders of magnitude to the 10^{-9}–$10^{-8}\,\Omega\,cm^2$ range. The optimum range of annealing temperature for silver

Table B5.1.3. Optimum annealing conditions for silver and gold contacts to Y-, Bi-, and Tl-based high T_c superconductors. Annealing times are about 30–60 min for bulk sintered contacts, 30 min or less (at temperature) for thin film contacts

Contacts to sintered high T_c superconductors	Annealing temperature
Ag/YBCO[a]	500°C in O_2
Au/YBCO[a]	600°C in O_2
Ag/BiPbSrCaCuO	~400°C in O_2
Ag/TlCaBaCuO	500°C in O_2
Contacts to YBCO film superconductors	
Ag(<1 μm)/YBCO film[b]	400°C in O_2
Au(<1 μm)/YBCO[b] film[c]	450–500°C in O_2

Annealing carried out in oxygen at atmospheric pressure, flowing at a rate of about $2 \times 10^{-6}\,m^3\,s^{-1}$ (~0.3 scfh, standard cubic feet per hour) using a furnace such as shown in figure B5.1.8.
For YBCO, the contacts were cooled in oxygen in the annealing furnace at a slow rate (~2.5°C min^{-1} for the bulk superconductors used in these tests; rates for thin films should also be kept low, although rates up to 50°C min^{-1} have been used successfully. This allows time for the crystal structure to take up oxygen as it cools and minimize oxygen disorder, [13]).
For silver-contact pads that are thin (<1 μm), the pad will 'ball up' at oxygen annealing temperatures higher than about 400°C (see figure B5.1.12). For thick silver-contact pads (≫ 1 μm), the optimum annealing temperature can be slightly higher [15].
[a] Data from [5].
[b] Data from [7].
[c] Data from [20].

contacts with bulk sintered bismuth-based superconductors is lower than that for YBCO, about 350–400°C.

For silver contact pads on bulk sintered Tl-based superconductors, the optimum annealing temperature range is about 500°C, as shown in figure B5.1.7. If too high a temperature is used, thallium loss results in a degraded superconductor [17]. C3

Figure B5.1.8 shows an example of a typical tube furnace arrangement for performing oxygen annealing. The sample (with contact pads already deposited on its surface) is placed in the tube furnace, and oxygen is fed into one end of the tube (shown in figure B5.1.8 on the left). A valve is used to control the flow at a low rate of about $2 \times 10^{-6}\,m^3\,s^{-1}$ (~0.3 scfh). The oxygen flows down a small quartz tube inserted into a larger quartz tube holding the sample. In this way, the oxygen is warmed to furnace temperature before it flows back across the sample. Eventually, it vents through a flexible plastic tube with a porous plug (cotton or a paper tissue stuffed into its end) to prevent back-flow of air into the furnace tube. At such a low flow rate (~$2 \times 10^{-6}\,m^3\,s^{-1}$) no special precautions are usually needed for venting pure O_2 into a reasonably large laboratory space with normal room ventilation. Air is purged from the sample tube by starting the oxygen flow at a high rate (about $2\text{–}5 \times 10^{-5}\,m^3\,s^{-1}$) approximately 15 min before turning on the furnace, and then reducing the rate to the low value for the actual anneal (to save oxygen). After they are annealed, YBCO samples need to be slowly cooled (typically 2.5°C min^{-1}) in flowing oxygen to allow the crystal structure to take up its full complement of oxygen. C1

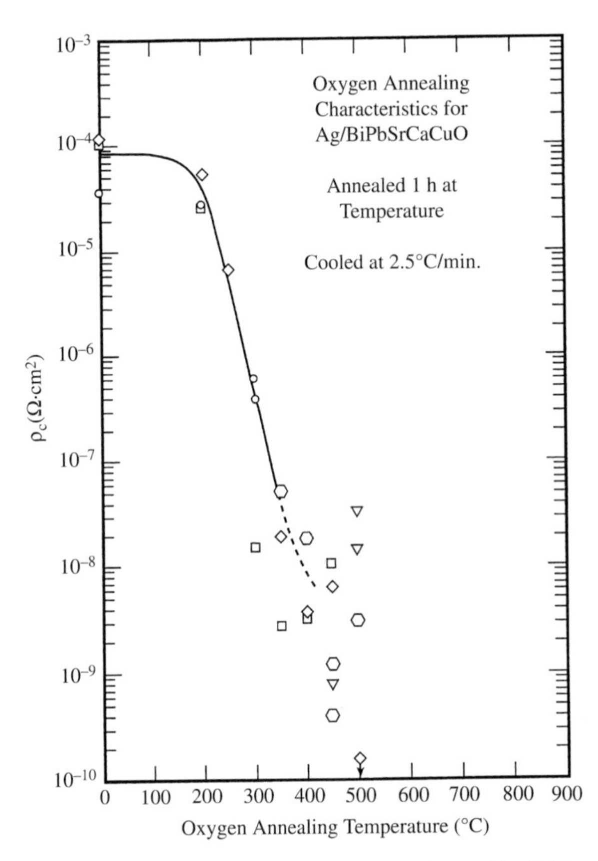

Figure B5.1.6. Oxygen annealing characteristic for silver contacts on sintered bulk bismuth-based superconductors; annealing time was 1 h and oxygen was supplied at atmospheric pressure, flowing at a rate of about $2 \times 10^{-6}\ m^3\ s^{-1}$ ($\sim$0.3 scfh) (from Ekin, unpublished data).

Soldering to noble metal contact pads

Soldering to noble metal contact pads is tricky; if not done right, the contact pad will be useless as a buffer. Typically, the solder (usually indium-based) will partially alloy through the noble metal contact pad, bringing oxygen-hungry indium into direct contact with the oxide superconductor, which would form the infamous indium-oxide barrier (top of figure B5.1.3). In the worst situation, the molten indium can completely alloy and dissolve thin noble metal pads. The result is degradation of the interface resistivity of a perfectly good contact, from 10^{-7} or $10^{-8}\ \Omega\ cm^2$ before soldering to typically 10^{-1}–$10^{-3}\ \Omega\ cm^2$ after soldering.

Destruction of the noble metal interface layer is avoided by using the following soldering procedure:

(i) Deposit a noble metal contact pad at least 7–10 μm thick (thinner pads only $\sim$0.2 μm thick are fine for attaching wire bonds to thin films, but not for soldering).

(ii) Coat the noble metal contact pad with eutectic In–3%Ag solder, without using solder flux. Adjust the soldering iron temperature so it is only about 30–50°C above the solder melting temperature (143°C) to minimize dissolving any more of the pad than necessary. Avoid

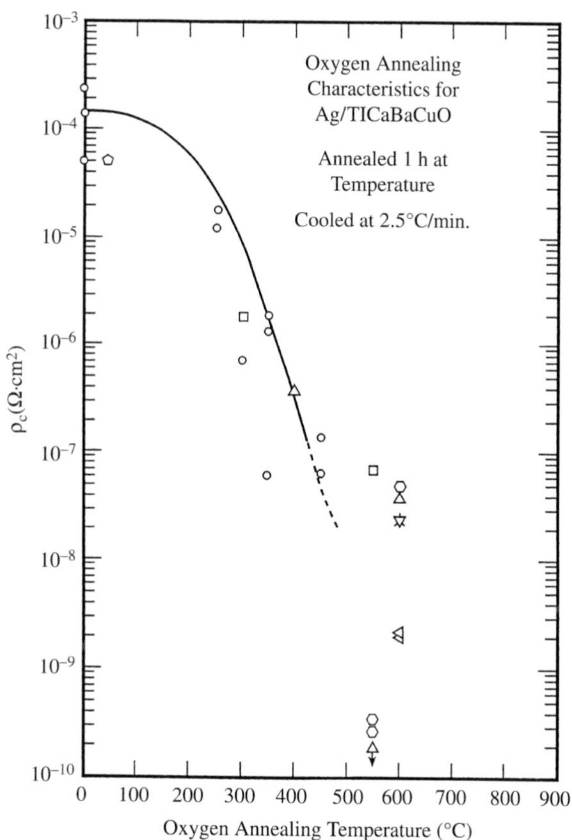

Figure B5.1.7. Oxygen annealing characteristic for silver contacts on sintered bulk Tl-based superconductors; annealing time was 1 h and oxygen was supplied at atmospheric pressure, flowing at a rate of about 2×10^{-6} m^3 s^{-1} ($\sim$0.3 scfh) (from Ekin, unpublished data).

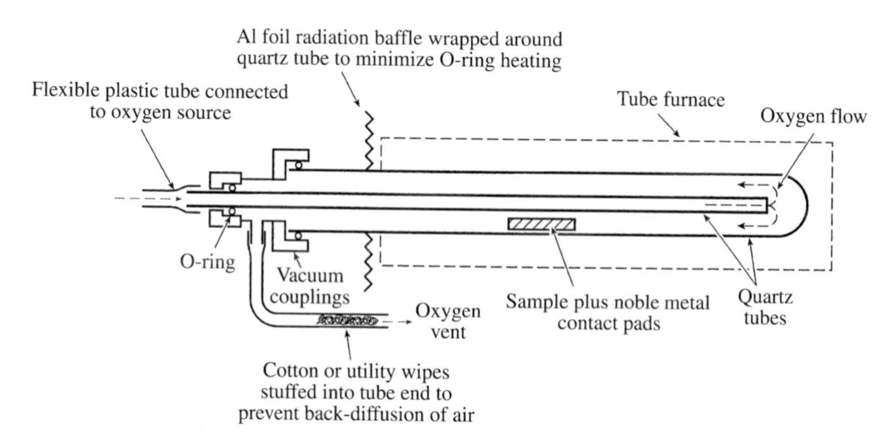

Figure B5.1.8. Furnace arrangement for counter-flow oxygen annealing of high T_c superconductor contacts, in which the oxygen is warmed to the temperature of the furnace before it flows back across the sample.

scratching the contact pad with the soldering iron. I have found In–3%Ag to be generally the best because it avoids the use of Sn, which is a strong leaching agent for silver. Furthermore, since In–3%Ag solder already has exactly the eutectic concentration of silver in it, minimal silver uptake should occur if the soldering temperature is kept close to the eutectic melting temperature.

(iii) Wet the copper wire or bus bar that is to be attached to the noble metal pad with a coating of In–3%Ag solder, using a mild flux such as zinc chloride. Clean away the flux with water or alcohol before attaching the copper conductor to the noble metal contact pad.

(iv) Fuse the solder contact that is on the copper conductor to the solder-coated contact pad using no flux, with minimal heat (by using the moment at which the solder melts as an indicator of when the temperature reaches 143°C). Avoid scratching the contact pad.

Even using this solder method, some indium degradation of the contact interface usually occurs. About the best ρ_c we can obtain when soldering is $10^{-6}\,\Omega\,cm^2$.

Silver-sheathed HTS materials

B3.1 For a sample with a Ag or Ag–Mg sheath that is co-processed with the superconductor (such as a bismuth-based superconductor tape) the resistivity of the Ag/HTS interface is typically very good, less than $10^{-8}\,\Omega\,cm^2$. Because the silver sheath of these samples is so thick (typically hundreds of micrometers), no particular care about leaching silver or diffusion of In is required when soldering. A solder should be chosen, however, that has not too high a melting temperature, such as In–48%Sn (118°C), In–3%Ag (143°C), or standard eutectic Pb–63%Sn solder (183°C), so that heating does not result in oxygen loss from the superconductor crystal structure. The critical temperature and current of bismuth-based superconductors starts to degrade from oxygen loss at temperatures above 250–300°C, so eutectic Pb-Sn solder is fine. A similar degradation from oxygen loss in YBCO materials heated in air starts to occur at lower temperatures of only about 150°C. High melting temperature solder also can damage structurally weak porous materials and delaminate some composite samples fabricated with solder.

B5.1.3.4 Measuring contact resistivity

Measuring the contact resistance of high current superconductor samples is relatively easy, as shown in figure B5.1.9. It consists of attaching an extra voltage lead to the current contact pads being tested, in order to monitor the potential difference between the contact pad and any voltage tap on the superconductor. Such a scheme readily works, since at measuring current below I_c, the superconductor

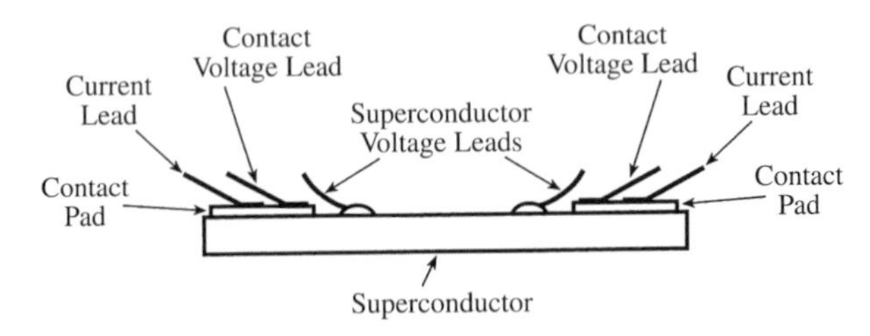

Figure B5.1.9. Diagram of contact resistance measurement, showing an extra voltage lead attached to the current pad close to the current lead. The potential difference between the extra voltage lead and a voltage tap anywhere on the superconductor can be used to monitor the voltage across the pad/superconductor interface.

has negligible resistance compared with the contact-pad interface resistance being measured. Thus, the superconductor is effectively an equipotential and a voltage lead attached *anywhere* on its surface will monitor the potential at the superconductor surface under the contact pads.

The *total* resistance of both contacts and the sample can be monitored during an experiment by simply attaching a voltage monitoring lead to each contact (see figure B5.1.9) and measuring the potential difference across the entire ensemble. This is a good measurement practice, which acts as a warning signal if significant heating is occurring at either of the current contacts.

Keep in mind that the spreading resistance of the contact pad can limit the effective contact area, as discussed below in section B5.1.6. So it is important to attach the extra voltage lead close to where the current lead attaches to the contact pad, especially if the contact pad is thin and the interface resistivity between the pad and superconductor is expected to be low. If the tap is *not* located well within the effective spreading distan/ce ξ given by equation (B5.1.8) below, then the tap will be connected to a part of the contact pad where current is not flowing and a contact resistance will be measured that is falsely low.

B5.1.4 Contact techniques for films

B5.1.4.1 Overview

In this section, we focus on techniques for making effective electrical interfaces between superconductor thin films and contact pads; external connections between the contact pads and the outside world using wire bonds [9] or pogo pins (spring-loaded contact pins) are described in the references listed in the suggested reading section at the end of this chapter.

Before describing contact techniques for oxide superconducting films in particular, we make some brief observations about techniques that apply to film samples in general.

- Contact pads for thin films are usually much thinner (usually 50–200 nm thick) compared with high current samples, because connections to the outside world can be made using low-current wire bonds and pogo pins rather than solder. For high current thick film YBCO 'coated conductors,' on the other hand, noble metal contact pads need to be 7–10 μm thick if soldered (see discussion in the subtopic on soldering in section B5.1.3.3). B4.1,
- Gold is a popular film contact pad material because it maintains a clean contact surface and does not tarnish like silver; however, the cost of gold becomes significant when depositing thick contact pads on high current coated conductors, for example, where silver is a more economical choice.
- Before depositing a contact pad, the surface of the film is usually given a light cleaning etch with a low-voltage ion mill; then the contact pad is immediately deposited on the film. B4.4.1
- When larger area (>1 × 1 mm) contact pads are used, it is not necessary to use photolithographic techniques (photosensitive films deposited on the film surface and used for patterning); in this case processing time can be saved by evaporating the contact layer through a shadow mask to define the contact pads.
- When small area contact pads (<1 × 1 mm) are required, the contact layer is usually deposited over the entire test film surface and then patterned using photolithographic techniques (an introduction to such patterning techniques if given, for example, by [1]).

Thin film electronic applications can place fairly severe demands on the specific contact resistivity between the pad and film, as indicated on the chart in figure B5.1.1 at the beginning of this chapter. This is illustrated, for example, by the stringent requirement for on-chip interconnects, where thin film E4

E3.2 transmission lines are of the order of a micrometer wide, which results in very small contact areas requiring contact-interface resistivities in the range of 10^{-9}–$10^{-10}\,\Omega\,cm^2$.

We now look at techniques for contacting oxide-superconductor films that can produce such quality contacts. For the most part, the same contact techniques and annealing procedures apply either E1 to thin films used in electronic devices or to thick film YBCO coated conductors used for power applications.

B5.1.4.2 Contacts for high T_c oxide superconductor films

To achieve high quality contacts to oxide-film superconductors, the same interfacial chemistry problems have to be faced as discussed above (section B5.1.3—Interfacial chemistry) for high current HTS materials. The contact pads must be made from a material with a low oxygen affinity, like gold or silver. As described below, the main difference in technique between the bulk and film cases is that the optimum oxygen annealing temperatures are slightly lower and times are shorter for films. Another important distinction for film samples (versus bulk samples) is that it is possible in some cases to use an *in situ* technique to deposit contact pads.

In situ versus ex situ contacts

For some film applications, it is possible to deposit the contact noble metal in the same vacuum chamber used to make the superconducting film, thereby not exposing the superconductor surface to air before the contact pad is deposited. This so-called *in situ* technique results in the lowest values of ρ_c, usually in the range from 10^{-7} to $10^{-9}\,\Omega\,cm^2$, and requires no oxygen annealing step [6, 12].

Often the *in situ* process is incompatible with other film processing steps that commonly need to be performed on the superconductor film before contact pads are deposited. Then the *ex situ* contact technique is used, where the contact film is deposited after an intervening exposure of the B4.4.1 superconductor surface to air or photoresist processing chemicals. When exposed to air, some degradation of the YBCO surface results from reaction with water vapour and carbon dioxide, as shown in figure B5.1.4 above. However, the degradation of the resulting ρ_c is gradual until 100–1000 min of exposure to air, after which it increases rapidly.

Ex situ contacts usually have contact resistivities in the range from 10^{-5}–$10^{-1}\,\Omega\,cm^2$, depending on the quality of cleaning before deposition of the contact pad. However, oxygen annealing can be used to reduce ρ_c of *ex situ* contacts to the range from 10^{-6}–$10^{-8}\,\Omega\,cm^2$, which is almost as good as *in situ* contacts. (Oxygen annealing is not effective in further reducing the ρ_c of *in situ* contacts.)

Cleaning etch

Before depositing a noble metal contact pad on a superconductor surface that has been exposed to air (when forming an *ex situ* contact), it is best first to clean the superconductor film surface using a gentle low-voltage ion mill. (For *in situ* contacts, no such cleaning is required, but because of the small residual water-vapour partial pressure in the vacuum chamber and impurities in process gases, it is best to minimize the time between fabrication of the superconductor and deposition of the contact pad, preferably not exceeding 100–1000 min as noted above.) As with bulk high T_c superconductors, care should be taken to keep the ion energy low to avoid destroying the delicate crystal structure of the superconductor; too low an energy, however, results in no milling at all. Figure B5.1.10 shows a typical B4.4.1 voltage dependence for the sputter yield, which is proportional to the milling rate. (The results are shown for argon ions incident on silicon, but the relationship is similar for most physical sputtering B4.2.1 applications such as ion milling and sputter etching.) For cleaning YBCO with argon ions, an energy of about 200 eV is optimum.

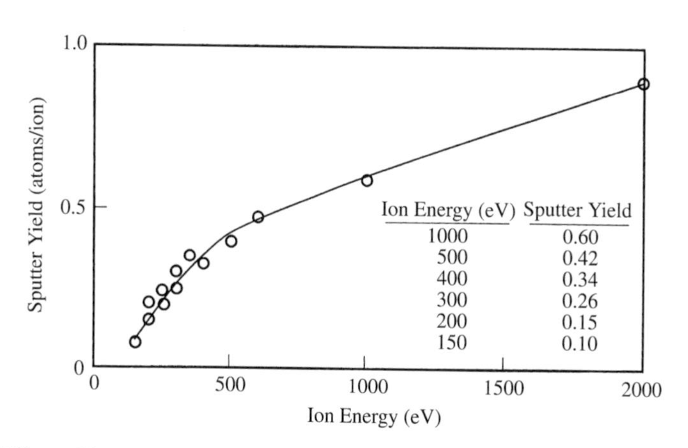

Figure B5.1.10. Effect of ion energy on sputter yield (number of sputtered atoms per incident ion) [10].

The amount of material removed needs to be sufficient to clean the surface, but not so great that film samples are significantly thinned in the process. Milling away approximately 10 nm of a YBCO film surface is usually enough to remove the degraded surface layer.

Thickness monitors, such as resonant quartz-crystal monitors, are commercially available for measuring film thickness when *depositing* material, but such real-time measurements are not generally available when *removing* material (unless the quartz crystal is coated with the same material, which can occur when deposition and milling are carried out in the same chamber). Ion milling rates are usually experimentally determined for any particular system using test films. For example, photoresist droplets can be placed on a test film and the film's surface ion milled for a time long enough to produce an easily measurable step. After rinsing away the photoresist with acetone, the step height (at the edge of where the droplets used to be) can then be measured with a sensitive depth gauge.

An approximate milling rate can also be determined using handbook values. Milling rates using argon ions at 200 and 500 eV, for example, are given by Vossen and Kern [19] for many elements and a few common compounds. Rates at voltages other than those tabulated can be estimated from the sputter-yield curve shown in figure B5.1.10. This yield curve will vary for different materials, especially at low voltage, but can at least give a first approximation for how the milling rate will change with ion energy (accelerating voltage). B4.2.1

Sometimes the superconductor's surface needs to be ion cleaned *after* photoresist processing, as with the 'lift-off' patterning technique (see, e.g. [1]). Especially after developing and rinsing small areas, a residual film of photoresist can remain on the 'cleared' areas where the contact pads are supposed to adhere. Unfortunately, it does not take much residual photoresist on the surface to give erroneously small milling depths in these areas. A trick that helps is to add about 20% oxygen to the argon gas used for milling in order to ion mill *organic* material reactively. This preferentially increases the milling rate for any residual photoresist without a commensurate rise in etch rate for the superconductor film surface.

Noble metal deposition and thickness

Techniques for depositing gold or silver pads on films are similar to those described above for bulk HTS contacts. Thermal evaporation is the easiest method. Typically we use 99.99% ('four nines pure') silver or gold evaporated from a resistively heated tungsten boat at a deposition rate of about $1\,\mathrm{nm\,s^{-1}}$. Alternatively, e-beam evaporation can be used, but this requires a greater initial outlay of funds for

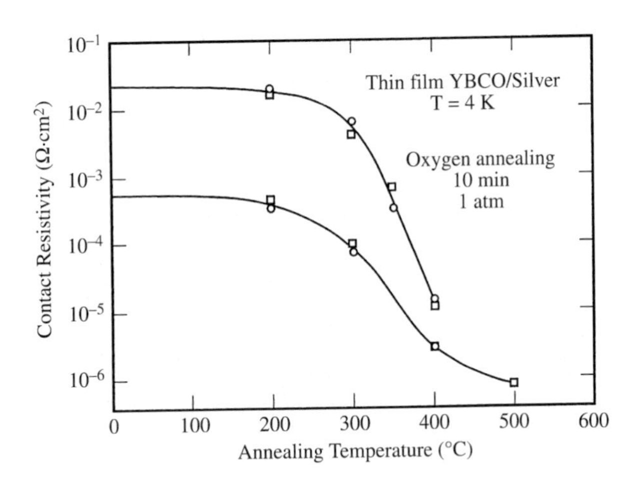

Figure B5.1.11. Contact resistivity ρ_c as a function of annealing temperature for two silver thin film contacts (500 nm thick) deposited *ex situ* on YBCO films [7]. Annealing was carried out at atmospheric pressure in flowing oxygen for 10 min (after reaching the annealing temperature). The contact resistivity starts to drop at an annealing temperature of about 250°C, and shows marked improvement, to less than $10^{-5}\,\Omega\,\mathrm{cm}^2$, after annealing at approximately 400°C. At higher annealing temperatures, thin silver films agglomerate and electrical continuity is lost. (The initial contact resistivity depends mostly on the exposure time of YBCO to air, as well as on the cleaning procedure before coating with silver).

the deposition equipment. The pads do not have to be very thick, about 50–200 nm for ribbon-bond lead attachment or for pogo pin contacts.

Film contact annealing

For *ex situ* contacts, the contact resistivity can be improved by three to four orders of magnitude by annealing the contact pads in oxygen. As shown in the Auger depth profile in the bottom of figure B5.1.3, oxygen annealing diffuses silver into the superconductor, probably mainly along grain boundaries, thereby increasing the effective contact area. The surface layers of YBCO are also reoxygenated.

Oxygen annealing schedules for YBCO thin films are summarized along with those for bulk samples in table B5.1.3. Figure B5.1.11 shows the reduction in contact resistivity that results from annealing C1 500 nm thick silver contacts to YBCO films in flowing oxygen at atmospheric pressure. With *ex situ* deposited films, the ultimate contact resistivities reached after annealing appear to be not quite as low as for bulk-sintered samples, probably because for sintered superconductors there is more grain-boundary B2.2.2 surface area over which silver can diffuse. ('Melt textured' bulk superconductors are expected to be an intermediate case because they are less porous than bulk sintered samples.) Annealing times are usually somewhat shorter for films than for bulk samples (in our case, 10 min at full temperature for films versus ~1 h for bulk samples). As seen in figure B5.1.11, the optimum annealing temperature for thin ($<$ approximately 1 μm) silver pads on YBCO films is about 400°C, which is about 100°C lower than for bulk sintered YBCO. The slightly reduced temperature is particularly important for thin films of silver, because the silver film starts to agglomerate or 'ball up' at annealing temperatures above approximately 400°C. Figure B5.1.12 shows a silver film after 10 min oxygen annealing at 500°C; there is very little contact left. As the thickness of the silver film increases, the temperature limit to avoid agglomeration will also increase somewhat [15].

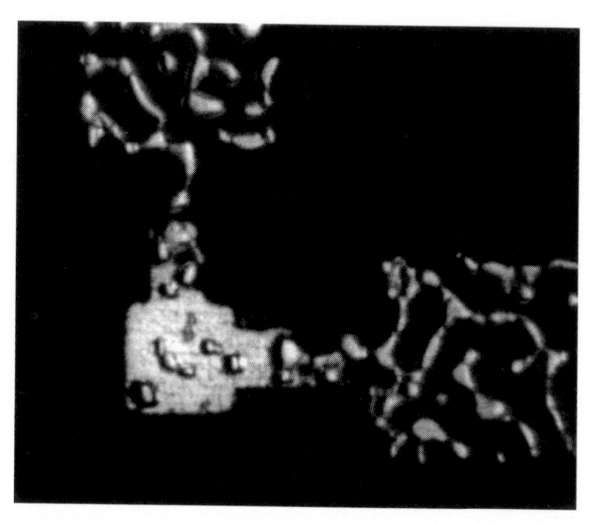

Figure B5.1.12. Optical micrograph of a 500 nm thick silver film on a YBCO film after annealing at 400 and 500°C in oxygen for 10 min and a 50°C min^{-1} cool down rate, which shows extensive agglomeration of silver that results in the loss of electrical continuity along the contact pad [7].

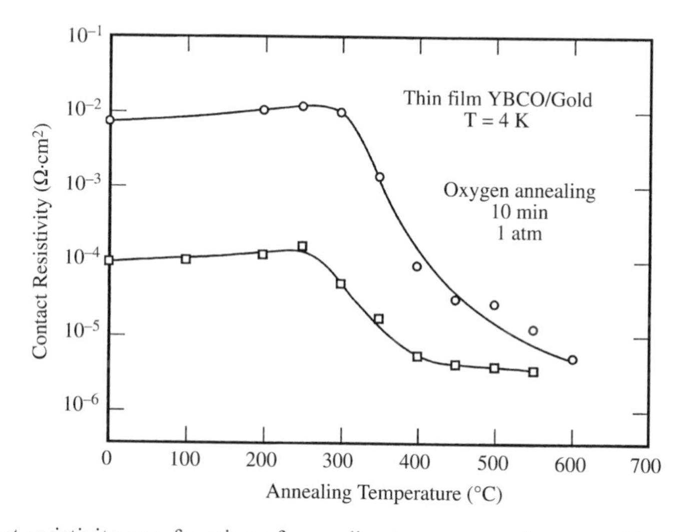

Figure B5.1.13. Contact resistivity as a function of annealing temperature for two gold thin film (350 nm thick) *ex situ* contacts on YBCO films, with different initial contact resistivities [20]. Annealing was carried out at atmospheric pressure in flowing oxygen for 10 min (after reaching the annealing temperature). The optimum annealing temperature is about 50–100°C higher than for silver contacts, see figure B5.1.11. (Similar to silver contacts, the initial contact resistivity of each contact depends mostly on the YBCO air exposure time and cleaning procedure before coating with gold.)

For gold pads, on the other hand, agglomeration is not a problem (presumably because of the lower surface energy and higher melting temperature of gold). Figure B5.1.13 shows the reduction in contact resistivity that results from annealing 350 nm thick gold contacts to YBCO films in flowing oxygen at atmospheric pressure. The optimum oxygen annealing temperature for thin (< approximately 1 μm) gold contact pads on YBCO films is in the range of about 450–500°C, for 10–30 min.

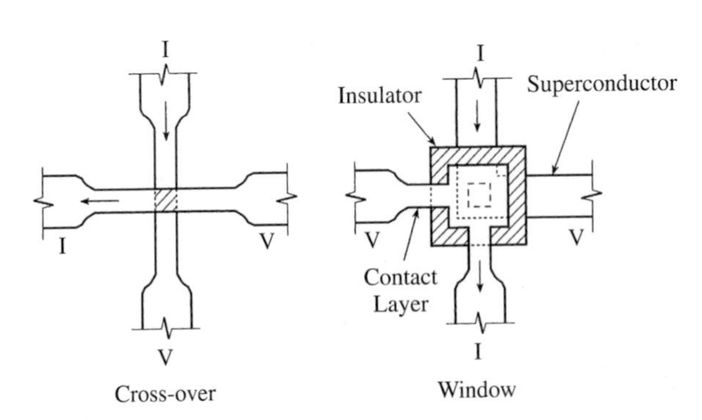

Figure B5.1.14. Geometries for measuring film interface resistivity: (left) cross-over geometry, for simplicity in patterning; and (right) window geometry, to minimize sheet resistance effects in the normal-metal overlayer [3].

The oxygen furnace described in Section B5.1.3 'Fabrication procedures' for bulk samples works perfectly well for annealing film samples as well. Rapid thermal annealing can also be performed using radiant heat lamps, but the initial temperature calibration has to be done carefully by attaching a thermocouple to a dummy film–substrate combination that has an emissivity and mass similar to those of the actual sample chip. When annealing YBCO, the film is cooled slowly in oxygen, no faster than approximately 50°C min^{-1}, to avoid loss of critical current from oxygen disorder [13].

C1

B5.1.4.3 Measuring film contact resistivity

When initially checking the performance of a new technique for depositing contact pads, the interface resistivity between the contact pad and superconductor film surface can be measured with one of the test patterns illustrated in figure B5.1.14. The simplest arrangement is a cross-strip geometry, shown on the left side of figure B5.1.14. The superconductor is patterned into a strip and the contact material deposited on top. The contact layer is then patterned to form the cross-strip. One arm of the superconductor strip and one arm of the contact material strip are used to supply current through the interface formed where the two films cross (the arms labelled 'I' in figure B5.1.14). The other two arms are used to detect the potential difference across the contact interface (labelled 'V' in figure B5.1.14). When measuring very low interface *resistivity*, the crossover area needs to be kept small to obtain a contact resistance large enough to measure. This usually requires photolithographic patterning techniques.

B4.4.1

Another arrangement that is effective in minimizing potential differences due to sheet resistance in the normal-metal layer is shown on the right side of figure B5.1.14. With this arrangement, current is not forced to make a 90° bend at the contact, as in the cross geometry on the left side of figure B5.1.14. Instead, the current continues to flow in a straight, nearly symmetrical path, from the top of the diagram to the bottom, passing vertically through the contact hole in the insulator layer at the middle of the pattern [3]. The symmetrical pattern of current flow and perpendicular voltage detection cancels to first order the effect of potential differences generated in the normal-metal film.

Wire bond or ribbon bond connections, such as these shown in the top part of figure B5.1.15, present a different set of measurement problems. To measure the quality of such connections, a second ribbon can be bonded right on top of the first ribbon bond. This double-stacking arrangement avoids the severe limitations on voltage tap placement imposed by spreading resistance effects, as calculated in section B5.1.6. A measurement of the overall resistance of the ribbon bond contact is obtained by introducing current into the first ribbon bond that is attached directly to the sample (the ribbon bond being measured) and detecting voltage using the second ribbon. The return voltage tap can be located anywhere on the superconductor surface because the superconductor is effectively an equipotential surface (as long as the measuring current is well below the critical current).

B5.1.5 Example calculations of minimum contact area needed for critical current measurements

In this section and the next, several examples are given to illustrate the calculation of the contact area needed to prevent overheating at sample contacts for critical current measurements. Knowing the required contact area is especially relevant to measuring critical current (the maximum current a superconductor can carry without losing superconductivity), since the current densities are usually high and the space for contacts can be severely limited by the confined space of the test apparatus. Although examples are given for critical current measurements, the same calculation procedure applies equally well to other types of transport measurements.

The first step in the calculation is to determine the maximum contact resistance that can be tolerated to keep joule heating below an acceptable level. The second step is to calculate the minimum contact area needed to obtain this resistance value. We illustrate this first for *low* T_c superconductors such as NbTi and Nb_3Sn operating in liquid helium. A brief discussion is also given of how to calculate the minimum contact area when measurements are carried out in a gas or vacuum environment (rather than in liquid helium), such as on the cooled stage of a cryocooler. This is followed by an example for *high* T_c superconductors operating in liquid nitrogen. In the last section, we show how these minimum area calculations are modified by the effects of *spreading resistance* when the contact pads are thin.

B3.3.1

B5.1.5.1 NbTi at 4 K

Example: Suppose we are designing a sample holder for measuring the critical current of NbTi in liquid helium where currents up to 600 A are expected. We assume that the transport current connections to the sample are made by using the common Pb–Sn eutectic solder to attach copper wires or bus bars to the conductor's outer copper sheath. The design question is: what is the minimum contact area we will need to keep the contact joule heating at 600 A from raising the temperature of the sample contacts by no more than 100 mK? This temperature rise corresponds, for example, to a decrease of about 2% in the critical current of NbTi at 4.2 K and 0 T, or to a decrease of about 5% in a magnetic field of 7 T.

The contact heat generation, $\dot{q}$, that would give rise to a temperature difference ΔT of 100 mK between the test sample and liquid helium is determined by the solid/liquid heat transfer rate for liquid helium, which at low heat flux ($\lesssim 10^4\,\mathrm{W\,m^{-2}}$ or $\Delta T \lesssim 0.4\,\mathrm{K}$) is approximately [11, 18]

$$\dot{q} \cong 6\times10^4 \Delta T^{2.5} (\mathrm{W\,m^{-2}}) \times A_{\text{total cooled surface}} \tag{B5.1.2}$$

where $A_{\text{total cooled surface}}$ is the total cooled surface area. We make the worst case approximation that the heat generated at the contact is dissipated into the liquid helium bath only over the contact area, and ignore the cooling of the contact that occurs from solid conduction along the current bus bar. So, if we make the conservative assumption that the total cooled surface area $A_{\text{total cooled surface}}$ is only the contact area A_c, we find from equation (B5.1.2) that

$$\dot{q} \cong 6\times10^4 \Delta T^{2.5} (\mathrm{W\,m^{-2}}) \times A_c = [6\times10^4 (10^{-1}\,\mathrm{K})^{2.5}\ \mathrm{W\,m^{-2}}]\,A_c \cong (1.9\times10^2\ \mathrm{W\,m^{-2}})\,A_c.$$

Since the design critical current in our example is 600 A, then from Ohm's law, this would require a contact resistance

$$R_c \le \dot{q}\, I^{-2} = (1.9\times10^2\ \mathrm{W\,m^{-2}})\,A_c\,(600\,\mathrm{A})^{-2} = (5\times10^{-4}\ \Omega\,\mathrm{m^{-2}})\,A_c. \tag{B5.1.3}$$

Next we determine the minimum contact area from equation (B5.1.1), using for ρ_c the interfacial contact resistivity given at the top of table B5.1.1 for two pieces of copper soldered together under moderate pressure with the eutectic Pb–Sn solder:

$$A_c \geq \rho_c \times R_c^{-1} \cong (4 \times 10^{-9}\,\Omega\,\mathrm{cm}^2)[(5 \times 10^{-8}\,\Omega\,\mathrm{cm}^{-2})\,A_c]^{-1}. \quad \text{(B5.1.4)}$$

Thus

$$A_c^2 \geq 8 \times 10^{-2}\,\mathrm{cm}^4$$

or

$$A_c \geq 0.28\,\mathrm{cm}^2.$$

Because of the worst case assumptions used in the calculation, a smaller contact area would probably meet the temperature difference requirement (100 mK in this example). Note that the requirements on contact area can become more stringent to achieve a smaller temperature difference ΔT between the sample contact and helium bath because the nucleate-boiling heat transfer rate for liquid helium scales as $\Delta T^{2.5}$ in equation (B5.1.2).

When measurements are carried out with the sample mounted on the cooled stage of a cryocooler, the sample is in a *gas or vacuum* environment (rather than in a liquid cryogen). In this case, the contacts will be cooled mainly by solid conduction along the current leads and through the sample holder. The calculation of the minimum contact area is carried out in the same way as above, except that the heat transfer rate for solid conduction to the cryocooler head is used, instead of using equation (B5.1.2) for heat conduction across a solid/liquid interface. Usually, the main factor limiting heat flux is the thermal-boundary resistance at joints. Solder joints generally have the highest thermal conductance, followed by varnish, grease and pressure joints (thermal conductance data at cryogenic temperatures for a wide range of solid/solid joints are given, for example, in Ekin, 2004). Such measurements in a vacuum or gas environment typically require a larger contact area than for cooling directly in liquid helium, because the rate of heat conduction through the joints and various sample mounting members to a refrigerated stage is usually lower than the heat transfer rate for direct immersion in liquid helium.

B5.1.5.2 Nb_3Sn at 4 K: resistive matrix contribution

Another factor in calculating minimum contact areas for practical multifilamentary conductors is the heat generated in the resistive matrix surrounding the superconducting filaments. This can lead to an effective contact resistivity $\rho_{c\ \mathrm{effective}}$ much larger than that given in table B5.1.1. In general, $\rho_{c\ \mathrm{effective}}$ is given by

$$\rho_{c\ \mathrm{effective}} \cong \rho_c + \rho_{\mathrm{matrix}}\,t \quad \text{(B5.1.5)}$$

where ρ_c is the *interface* resistivity of the contact (having units of $\Omega\,\mathrm{cm}^2$), ρ_{matrix} is the *bulk* resistivity of the superconductor's matrix material (having units of $\Omega\,\mathrm{cm}$), and t is the thickness of the matrix material between the sample surface and the superconducting filaments. Heating in the matrix material is usually not much of an issue for NbTi superconductors, which have a copper matrix with a low electrical resistivity, but it becomes a significant factor when the matrix is more resistive, such as the bronze matrix in Nb_3Sn superconductors or the silver-*alloy*-sheath material in BSCCO superconductors.

B3.3.2
B3.3.3, C2

Example: Suppose the test sample in the first example above is a Nb_3Sn multifilamentary composite conductor (rather than a NbTi conductor) with a soldered contact length at least one twist-pitch long (so that all filaments come into close proximity to the contact). Furthermore, let us assume that the outermost Nb_3Sn filaments in the composite wire are separated from the high-conductivity copper stabilizer surface by a Cu–Sn bronze matrix layer that has an average thickness of about 0.1 mm. What is the contact area requirement in this case?

Cu–Sn bronze typically has a $\rho_{\text{matrix}} \cong 2 \times 10^{-6}\,\Omega\,\text{cm}$, as shown in table B5.1.4. Thus, we see from equation (B5.1.5) that the effective contact resistivity becomes much higher than that of just soldered Cu-to-Cu interfaces ($4 \times 10^{-9}\,\Omega\,\text{cm}^2$), namely

$$\rho_{\text{c effective}} = 4 \times 10^{-9}\,\Omega\,\text{cm}^2 + (2 \times 10^{-6}\,\Omega\,\text{cm})(0.01\,\text{cm}) = 24 \times 10^{-9}\,\Omega\,\text{cm}^2.$$

For the same contact resistance requirement as in the previous example, this necessitates a larger contact area

$$A_c \geq \rho_c R_c^{-1} \cong (24 \times 10^{-9}\,\Omega\,\text{cm}^2)[(5 \times 10^{-8}\,\Omega\,\text{cm}^{-2})\,A_c]^{-1}.$$

Solving for A_c, we obtain

$$A_c \geq 0.69\,\text{cm}^2,$$

which is about 2.5 times larger than the required contact area calculated for NbTi in the first example. (Note that the presence of an outer *pure* copper sheath does not significantly affect the contact resistivity in either case, since, at 4 K, copper typically has a ρ_{matrix} much smaller than that of bronze ($1.7 \times 10^{-8}\,\Omega\,\text{cm}^2$ versus $2 \times 10^{-6}\,\Omega\,\text{cm}$, see table B5.1.4). B3.3.3

B5.1.5.3 High T_c superconductors at 77 K

We now calculate how large to make the noble metal contact pads when measuring the critical current of high T_c superconductors operating in *liquid nitrogen*. As in the earlier examples, the focus will be on critical current testing, since it usually poses the most demanding requirements on contacts. A similar calculational procedure, however, would apply to current contacts for just about any kind of transport measurement. Because the contact pads deposited on YBCO conductors are usually thin, *spreading resistance* within the contact pad area can also be a problem, so the same example will be continued in section B5.1.6 to illustrate the calculation of spreading resistance effects. C1

Example: Suppose we want to size the current connections to a YBCO coated conductor cooled by liquid-nitrogen. Assume we are making a critical current measurement where the maximum current is expected to be 150 A, and that we want to limit the sample's temperature rise from contact heating at 150 A to less than 0.5 K. (This temperature rise corresponds to about a 5% decrease in the critical current of YBCO at 77 K in self field, for example). To obtain low contact resistance we deposit thick (about 7–10 m) silver contact pads on the sample ends and use In–3%Ag eutectic solder to attach current wires or a copper bus bar to the contact pads. Let us further assume that the total liquid nitrogen cooled surface area of the contact and current bus bar is about $10\,\text{cm}^2$. After determining the minimum contact area, we also want to check that the resistivity of the contact pad after soldering (through its thickness from top to bottom) will not limit the overall contact resistance. B4.1, B4.2.1

Proceeding as in the previous examples, the 0.5 K temperature rise places a limit on power generation q, which is given by the solid/liquid heat transfer rate for liquid nitrogen. Similar

Table B5.1.4. Bulk resistivity of common solders, contact pads and matrix materials for estimating total contact resistivity using equation (B5.1.5)

Material	ρ(295 K) ($\mu\Omega$ cm)	ρ(77 K) ($\mu\Omega$ cm)	ρ(4 K) ($\mu\Omega$ cm)
Solder (numerical values in wt%)			
52In–48Sn[a] (eutectic) ($T_{melt} = 118°C$)	18.8		SC
97In–3Ag[a] (eutectic) ($T_{melt} = 143°C$)	9.7	1.8	0.02
90In–10Ag[a]	9.1	1.8	0.03
In[a] ($T_{melt} = 157°C$)	8.8	1.6	0.002
63Sn–37Pb[a] (standard eutectic soft solder) ($T_{melt} = 183°C$)	15	3.0	SC
91Sn–9Zn[a] ($T_{melt} = 199°C$)	12.2	2.3	0.07
Contact pad material			
Silver (pure: evaporated, sputtered, or plasma sprayed)	1.6	0.27	
Gold (pure: evaporated or sputtered)	2.2	0.43	
Low T_c superconductor matrix materials			
Copper	1.7	~0.2	~0.017
Bronze (Cu–13wt%Sn)		~2	~2
High T_c superconductor matrix materials			
Silver	1.6	0.27	
Silver dispersion strengthened with 1 at%Mn[b]	4.0	2.7	2.2
Silver dispersion strengthened with 2 at%Mn[b]	~6.0	~4.6	~4.1

SC: Superconducting.
[a] Data from C. Clickner (1999), NIST, unpublished.
[b] [14].

to equation (B5.1.2), this is given approximately at low heat flux ($< 2 \times 10^5$ W m^{-2} or $\Delta T \lesssim 10$ K) by [11]

$$\dot{q} \cong 4 \times 10^2 \Delta T^{2.5} (\mathrm{W\,m^{-2}}) \times A_{\text{total cooled surface}}. \tag{B5.1.6}$$

Comparing equation (B5.1.6) with equation (B5.1.2), we find that the solid/liquid heat transfer rate for *liquid nitrogen* is about two orders of magnitude *smaller* than for liquid helium. So, unlike the previous examples for liquid helium temperature, where we assumed liquid helium cooling of the contact area alone, we also consider in this case the liquid nitrogen cooling of the *current bus bar* in addition to the contact area (assumed to total 10 cm^2). This is an optimistic assumption and so it must be treated carefully. In many cases, however, it is a reasonable assumption if we compare the relatively high heat conduction along a typical ETP or oxygen-free copper bus bar with the solid/liquid heat transfer rate into liquid nitrogen. [That is, if we assume the dimensions of the contact are of the order of 1 cm^2 and that ΔT in equation (B5.1.6) is in the range of 1 K, then the heat transfer rate into liquid nitrogen is $\dot{q} \cong 4 \times 10^{-2}$ W, which can be significantly smaller than the solid heat conduction along an ETP or oxygen-free copper bus bar on a centimeter-size scale, since the thermal conductivity at 77 K is about $5\,\mathrm{W\,(cm\,K)^{-1}}$.] Of course, much more sophisticated heat transfer calculations can be done, but using the combined cooled surface area of the contact *and* bus bar for $A_{\text{total cooled surface}}$ in equation (B5.1.6) gives us a quick, simple method of obtaining a rough estimate of the contact cooling rate,

D3.2.2

$$\dot{q} \cong [4 \times 10^2\ (0.5\,\mathrm{K})^{2.5}\ \mathrm{W\,m^{-2}}](10 \times 10^{-4}\,\mathrm{m^2}) \cong 71\,\mathrm{mW}.$$

For a 150 A test current, this limits the contact resistance to about

$$R_c \leq \dot{q}I^{-2} = (71\,\mathrm{mW})(150\,\mathrm{A})^{-2} = 3.2 \times 10^{-6}\,\Omega.$$

Next, to determine the minimum contact area we will need to obtain this resistance by first estimating the total effective contact resistivity using equation (B5.1.5)

$$\rho_{c\ \text{effective}} \cong \rho_c + \rho_{\text{pad}}\, t_{\text{pad}}. \qquad \text{(B5.1.7)}$$

Here, ρ_c is the interface contact resistivity, and ρ_{pad} and t_{pad} represent the bulk resistivity and thickness of the soldered noble metal contact pad, respectively. Let us assume that we are careful about our soldering technique (section B5.1.3) so that we end up with $\rho_c \cong 10^{-6}\,\Omega\,\mathrm{cm^2}$. For ρ_{pad}, we assume the worst case and take ρ_{pad} to be dominated by the In–3%Ag solder, since the contact pad will partially alloy with the In–3%Ag solder. That is, we assume $\rho_{\text{pad}} = 1.8\,\mu\Omega\,\mathrm{cm}$, from table B5.1.4. (If we were to attach a wire bond to the pad instead, then we would use the bulk resistivity of pure silver from table B5.1.4.) So, assuming we have made the contact pad 10 μm thick (to avoid having solder alloying completely destroy the interface), we find from equation (B5.1.7) that

$$\rho_{c\,\text{effective}} = 10^{-6}\,\Omega\,\mathrm{cm^2} + (1.8 \times 10^{-6}\,\Omega\,\mathrm{cm})(10^{-3}\,\mathrm{cm}) \cong 10^{-6}\,\Omega\,\mathrm{cm^2}.$$

That is, the effective contact resistivity is totally dominated by the interface and not by the contact-pad thickness, even after soldering. (For Bi-based superconductors made with high resistivity sheaths of dispersion strengthened silver, such as Ag–Mg, the added resistivity of the matrix will need to be taken into account using equation (B5.1.5) and the data in table B5.1.4. The effect of the dispersion strengthened sheath on $\rho_{c\,\text{effective}}$ will be very similar to that calculated above for the bronze matrix in Nb_3Sn superconductors.)

Substituting this value for $\rho_{c\ \text{effective}}$ in equation (B5.1.4), we find that we need a contact area of at least

$$A_c \geq \rho_{c\,\text{effective}}\, R_c^{-1} = (10^{-6}\,\Omega\,\mathrm{cm^2})(3.2 \times 10^{-6}\,\Omega)^{-1} = 0.31\,\mathrm{cm^2}.$$

Thus, if the 150 A conductor in this example were a high current YBCO coated conductor having a width of 3 mm, this would require a contact length of about 1 cm.

Wider tapes carrying more current will require a larger contact area. The required contact area scales quadratically with current if the total cooled surface area of the bus bar does not increase. Thus, for example, a YBCO coated conductor 1 cm wide, having a critical current of 500 A (bus bar cooled surface area unchanged at $10\,\mathrm{cm^2}$), would require a sample contact area of about $3.5\,\mathrm{cm^2}$ (and a corresponding contact length of about 3.5 cm).

B4.1

If we want to reduce the temperature difference between the sample and liquid nitrogen bath to less than 0.5 K, the heat transfer rate decreases as $\Delta T^{2.5}$ in equation (B5.1.6) and so the required contact area will increase greatly. Alternatively, the surface area cooled by liquid nitrogen [$A_{\text{total cooled surface}}$ in equation (B5.1.6)] could be increased by using a larger copper current bus bar.

The example above is for a YBCO superconductor; however, the calculation to size contacts to a Bi-2223 superconductor in liquid nitrogen would proceed in the same way, except the interface

B3.2.2

contact resistivity ρ_c to use in equation (B5.1.7) would be the silver-sheath/BSCCO value in table B5.1.1, instead of ρ_c for silver contact pads deposited on YBCO.

When high T_c superconductor contacts will be utilized in a *vacuum* environment (rather than liquid nitrogen), the contacts will be cooled mainly by solid conduction along the current leads and sample holder to the refrigerator. The calculation is carried out in the same way as above, except that, instead of equation (B5.1.6), we substitute the heat transfer rate for solid conduction along the current bus bar and sample holder. The main factor usually limiting heat flux is the thermal-boundary resistance across joints. At liquid nitrogen temperature, the highest thermal *conductivity* is obtained across solder joints [about 1–40 $W(cm^2 K)^{-1}$], followed by varnish joints [~0.5–2 $W(cm^2 K)^{-1}$] and grease joints (~0.1–1 $W(cm^2 K)^{-1}$]. Pressure joints typically have a lower thermal conductance (0.01–3 $W K^{-1}$), varying with the force applied to the joint, and the surface condition of the parts, gold plating being the best (a more detailed discussion is given in Ekin, 2004). Thus, the size of the electrical contacts in vacuum environments is determined primarily by the heat conductance between the sample and cold head, as well as the refrigeration power. Thermal conductance along the sample will also be a significant factor with measurements in vacuum, and usually requires thermally anchoring the sample to a high thermal conductivity sample holder with varnish or grease to help stabilize the temperature of the superconductor.

D3.2.2

B5.1.6 Spreading resistance of the contact pad

There is still another caveat to the high T_c superconductor example above. We cannot assume that noble metal contact pads will distribute the current uniformly over their entire area because the contact pads are usually quite thin. If the current is injected locally into the middle of a pad by a wire, the spreading resistance of the pad itself can limit the effective area of the contact pad to a distance ξ around the wire, as illustrated in figure B5.1.15. The spreading distance ξ is determined by the sheet resistance of the contact pad, R_s, and the contact interface resistivity ρ_c between the pad and superconductor. It is given approximately by [6]

$$\xi \cong (\rho_c/R_s)^{0.5}, \quad \text{(B5.1.8)}$$

where the sheet resistance R_s is the quotient of the film's *bulk* resistivity ρ and thickness t (assumed uniform)

$$R_s \equiv \rho/t\,[\Omega/\square]. \quad \text{(B5.1.9)}$$

The sheet resistance is expressed in units of 'ohms per square' ($\Omega/\square$)because, physically, it corresponds to the in-plane resistance of a square-shaped piece of the film: 1 μm × 1 μm, 1 m × 1 m, or whatever, just so long as it is *square*.

B5.1.6.1 Example of spreading resistance effect on YBCO coated conductor contacts

Example: In the above example for YBCO coated conductors, if we inject current locally at the middle of the contact pad, the effective contact area may still be inadequate because of spreading resistance. Let us determine the effective spreading distance ξ assuming we inject the current into the contact pad via a wire soldered to the middle of the contact pad, using In–3%Ag solder.

Assume that the solder coating the contact pad is about 1 mm thick and spread uniformly over the entire contact pad. Then the solder layer will be much thicker than the pad itself and provides the main path for distributing the current over the contact pad area. We determine the approximate sheet resistance of the solder layer from equation (B5.1.9) and substitute the bulk resistivity of In–

3%Ag solder from table B5.1.4 to obtain

$$R_{\text{s solder layer}} \equiv \rho/t = 1.8\,\mu\Omega\,\text{cm}/0.1\,\text{cm} = 1.8\times10^{-5}\,\Omega/\Box.$$

As in the previous example we assume that the contact resistivity between the pad and superconductor is that of a relatively good soldered contact having $\rho_c = 10^{-6}\,\Omega\,\text{cm}^2$. Then, from equation (B5.1.8) we have, for the effective spreading distance

$$\xi \cong (\rho_c/R_s)^{0.5} = [(10^{-6}\,\Omega\,\text{cm}^2)/(1.8\times10^{-5}\,\Omega/\Box)]^{0.5} = 0.24\,\text{cm}.$$

So, regardless of how big a silver contact pad we deposit, if we inject current at a point at its middle, it will function like a contact pad with a radius of only 2.4 mm, significantly less than the 31 mm^2 contact area we needed in the previous example. Thus, to obtain an effective contact area of about 31 mm^2, we would need to use a copper bus bar like that shown at the bottom of figure B5.1.15 to spread the current uniformly over the entire area of the contact pad.

B5.1.6.2 Example of spreading resistance effect on thin film contacts

For thin film devices where wire bond, ribbon bond, or pogo pin connections can be used, the contact interface resistivity is usually much lower than for soldered contact pads, usually of the order of $10^{-8}\,\Omega\,\text{cm}^2$ or less. When the interface resistivity is so low, the spreading resistance of the contact pad usually limits the effective contact area to only the very small area just under the ribbon bond or pogo pin, and essentially none of the rest of the contact pad area is utilized. This is illustrated in the following example.

Example—Spreading resistance for wire or ribbon bond contact pads: A typical wire bond or ribbon bond arrangement is illustrated in figure B5.1.15. We calculate the effective current spreading distance ξ around the bond for an oxygen-annealed gold contact pad at 77 K, which we assume is 200 nm thick.The sheet resistance R_s of the gold contact pad will be, from equation (B5.1.9),

$$R_s \equiv \rho/t = (0.43\,\mu\Omega\,\text{cm})/(2\times10^{-5}\,\text{cm}) = 2.1\times10^{-2}\Omega/\Box,$$

where we have used a bulk resistivity of 0.43 $\mu\Omega$ cm at 77 K for evaporated gold, from table B5.1.4. If the gold contact pad is a high quality *in situ* deposited contact, it will have a low specific contact resistivity between the pad and superconductor of about $\rho_c = 10^{-8}\,\Omega\,\text{cm}^2$, from table B5.1.2. Thus, from equation (B5.1.8), the effective current spreading distance ξ around the ribbon bond will be only

$$\xi \cong (\rho_c/R_s)^{0.5} = [(10^{-8}\,\Omega\,\text{cm}^2)/(2.1\times10^{-2}\,\Omega/\Box)]^{0.5} = 7\,\mu\text{m}.$$

This is small compared to the contact size of a typical wire bond or ribbon bond (usually in the range 25–100 μm across); so for contact pads with a relatively low ρ_c, the contact area is effectively that of the ribbon bond area alone. If the contact interface resistivity ρ_c is much higher than the $10^{-8}\,\Omega\,\text{cm}^2$ value assumed in this example, ξ will become correspondingly greater and the contact pad will then assist in improving the contact resistance by providing a larger effective pad area.

Use caution when *measuring* the interface resistivity of thin film contacts, however. Current spreading effects can lead to significant error in the apparent value of ρ_c in this situation, because contact areas are usually kept very small (of the order of several micrometers) in order to develop a detectable voltage across the contact. With such a small contact area, ξ can become a significant fraction of the contact dimension. Furthermore, the voltage on the normal-metal side of the contact must be

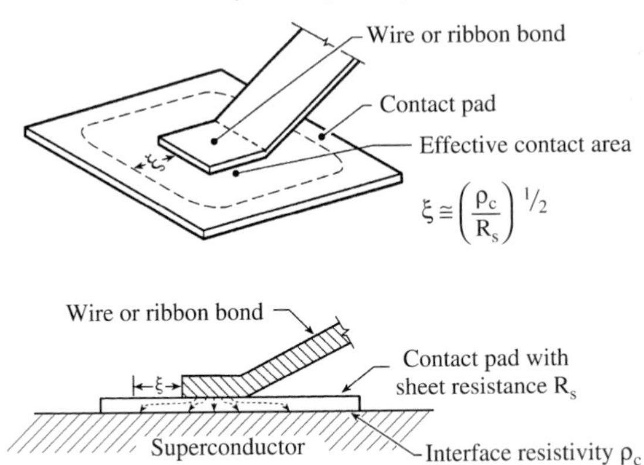

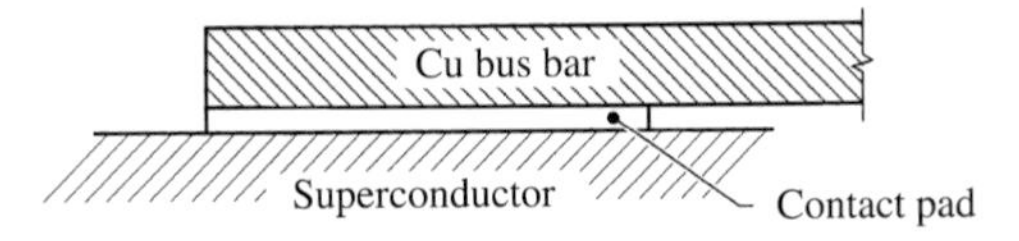

Figure B5.1.15. Spreading resistance of a contact pad. When current is injected into the middle of a thin contact pad (using a soldered wire, wire bond or ribbon bond, for example) the spreading resistance of the pad limits the effective contact area to a distance ξ beyond the injection perimeter. Effects of spreading resistance are not a problem when current is injected uniformly over the entire contact pad area by a thick copper bus bar, for example, as shown in the lower part of the figure.

detected within ξ of the contact pad, or else the 'measured' potential difference across the contact will appear erroneously low. Measurement geometries like those shown in figure B5.1.14 are needed to ensure both accurate area determination and voltage detection.

References

[1] Anner G E 1990 *Planar Processing Primer* (New York: Van Nostrand-Rienhold)

[2] Cha Y S, Lanagan M T, Gray K E, Jankus V Z and Fang Y 1994 Analysis and interpretation of critical current experiments for bismuth-based high-temperature superconductors made by powder-in-tube processing *Appl. Super. Cond.* **2** 47

[3] Cohen S S and Gildenblat G S 1986 *VLSI Electronics 13* Metal-Semiconductor Contacts and Devices (New York: Academic) p 87

[4] Ekin J W, Panson A J and Blankenship B A 1988 *Appl. Phys. Lett.* **52** 331

[5] Ekin J W, Larson T M, Bergren N F, Nelson A J, Swartzlander A B, Kazmerski L L, Panson A J and Blankenship B A 1988 High T_c superconductor/noble metal contacts with surface resistivities in the $10^{-10}\,\Omega\,\mathrm{cm}^2$ range *Appl. Phys. Lett.* **52** 1819

[6] Ekin J W, Russek S E, Clickner C C and Jeanneret B 1993 *In situ* noble metal $YBa_2Cu_2O_7$ thin film contacts *Appl. Phys. Lett.* **62** 369

[7] Ekin J W, Clickner C C, Russek S E and Sanders S C 1995 Oxygen annealing of *ex situ* YBCO/Ag thin film interfaces *IEEE Trans. Appl. Supercond.* **5** 2400

[8] Goodrich L F and Ekin J W 1981 Lap joint resistance and intrinsic critical current measurements on a NbTi superconducting wire *IEEE Trans. Magn.* **17** 69

[9] Harman G G 1997 *Wire Bonding in Microelectronics: Materials, Processes, Reliability, and Yield* (New York: McGraw-Hill)

[10] Kaufman H R and Robinson R S 1989 *Vacuum* **39** 1175

[11] Kutateladze S S 1952 Statistical science and technical publications of literature on machinery *Atomic Energy Commission Translation 3770* (Oak Ridge, TN: Technical Information Services)
[12] Lee M, Lew D, Eom C B, Geballe T H and Beasley M R 1990 *Appl. Phys. Lett.* **57** 1152
[13] Moeckly B H, Lathrop D K and Buhrman R A 1993 *Phys. Rev.* **B47** 400
[14] Putti M, Ferdeghini C, Grasso G, Manca A and Goldacker W 2000 *Proc. M2S-HTSC-IV Conf.* (Houston Texas).
[15] Roshko A, Ono R H, Beall J A, Moreland J, Nelson A J and Asher S E 1991 *IEEE Trans. Magn.* **27** 1616
[16] Russek S E, Sanders S C, Roshko A and Ekin J W 1994 Surface degradation of superconducting YBCO thin films *Appl. Phys. Lett.* **64** 3649
[17] Siegal M P, Overmyer D L, Venturini E L, Padilla R R and Provencio P N 1999 Stability of Tl–Ba–Ca–Cu–O superconducting thin films *J. Mater. Res.* **14** 4482
[18] van Sciver S W 1986 *Helium Cryogenics* (New York: Plenum) p 213
[19] Vossen J L and Kern W 1991 *Thin Film Processes II* (San Diego: Academic)
[20] Xu Y, Ekin J W, Clickner C C and Fiske R L 1998 Oxygen annealing of YBCO/gold thin film contacts *Adv.Cryo. Eng.* **44** 381

Further Reading

Talvacchio J Electrical Contact to Superconductors 1989 *IEEE Trans. Components, Hybrids, Manuf. Tech.* **12** 21; and Preparation of low-resistivity contacts for high T_c superconductors Ekin J W 1992 *Processing and Properties of High-T_c Superconductors, Vol. 1 Bulk Materials,* ed Sungho Jin (Singapore: World Scientific) p. 371.

Reviews of the early literature on contacts for high T_c superconductors are described.

Vossen J L and Kern W 1991 *Thin Film Processes II*, (San Diego:Academic) Anner G E *Planar Processing Primer* (New York: Van Nostrand-Rienhold) 1990

A description of thin film etching, deposition, and patterning techniques is given.

Harman G G 1997 *Wire Bonding in Microelectronics: Materials, Processes, Reliability, and Yield*, (New York:McGraw-Hill) Wire bonds are treated in detail.

Cohen S S and Gildenblat G S 1986 *VLSI Electronics 13*, Metal-Semiconductor Contacts and Devices, (New York:Academic Press) p 87

Thin film contact measurement techniques (used in the semiconductor industry, but adaptable to superconductor testing) are described in more detail.

Ekin J W 2004 *Experimental Techniques for Low-Temperature Measurements* (Oxford: Oxford University Press)

This monograph is an excerpt from the above forthcoming introductory textbook. The book also includes detailed information on related topics: cryogenic heat-transfer data, sample-holder construction, techniques for heat-sinking measurement leads, optimal sizing of high current leads, techniques for minimizing noise from thermoelectric voltages, vacuum connectors that survive cryogenic thermal cycling, thermometer selection and temperature control, techniques for critical current testing (along with corrections for temperature, current-transfer length and strain effects) extensive tables and appendices on the physical, electrical, thermal, magnetic and mechanical properties of many technical materials used in construction of cryogenic apparatus.

PART C

HIGH TEMPERATURE SUPERCONDUCTORS

C
Introduction to section C: High temperature superconductors

David Shaw

The term 'high temperature superconductors' (HTS) refers to a group of materials with critical temperatures (T_cs) higher than 23 K, the record T_c for the Nb_3Ge intermetallic superconductors. Until recently, all of these materials were perovskite-like ceramic oxides containing layers of Cu–O planes. The first HTS materials were discovered in 1986 by researchers at the IBM Research Laboratory in Rüschlikon, Switzerland, who created a brittle $(La,Ba)_2CuO_4$ ceramic compound that superconducted at approximately 30 K [1]. What made the discovery so remarkable was that—unlike the conventional low temperature superconductors (LTS), which are mostly intermetallic—ceramic oxide compounds are normally insulators and were not expected to be suitable candidates for superconductivity.

Bednorz and Müller's paper triggered a flurry of interest in the search for new HTS materials. Within a year, groups at the University of Alabama and the University of Houston jointly announced the discovery of superconductivity at 90 K—well above the boiling point of liquid nitrogen (77 K)—in a Y–Ba–Cu–O (YBCO) system [5]. Subsequently, groups at the National Research Institute for Metals in Tsukuba, Japan (Maeda *et al* 88) and the University of Arkansas [4] announced the discovery of the Bi–Sr–Ca–Cu–O (BSCCO) system with critical temperatures up to 115 K and the Tl–Ba–Ca–Cu–O (TBCCO) system with critical temperatures up to 125 K. To date, researchers have observed superconductivity in more than 50 cuprate compounds by substituting exotic and sometimes toxic, elements into the basic Y-, Bi-, and Tl-based HTS systems. At the present time, the ceramic cuprate superconductor with the highest T_c (138 K) is the thallium-doped Hg–Ba–Ca–Cu–O (HBCCO) system reported by Schilling *et al* [3] of Zurich, Switzerland.

Following Schilling's discovery, no new HTS materials were reported until 2001, when a group led by Akimitsu at Aoyna-Gakuin University in Tokyo, Japan discovered superconductivity in MgB_2 at approximately 40 K [2]. Since the new superconductor was announced in January 2001, over 260 reports have appeared on this superconducting intermetallic compound. Rapid progress in understanding and manufacturing the new MgB_2 materials into useful superconductors has been made, due in part to the experience accumulated in research on cuprate HTS systems. A comprehensive discussion of MgB_2 is included in chapter C1.4.

References

[1] Bednorz G J and Müller K A 1986 *Phys. B* **64** 189
[2] Nagamatsu J, Nakagawa N, Muranaka T, Zenitani Y and Akimitsu J 2002 *Nature* **410** 63

[3] Schilling A, Cantoni M, Guo J D and Ott H R 1993 *Nature* **363** 56
[4] Sheng Z Z and Hermann Z M 1988 *Nature* **332** 138
[5] Wu M K, Asburn R, Torng J, Hor P H, Meng L, Huang J, Wang Y Q and Chu C W 1987 *Phys. Rev. Lett.* **58** 908

C1
YBCO

A Koblischka-Veneva, N Sakai, S Tajima and Masato Murakami

C1.1 Introduction

The most extensively studied high-temperature superconductor (HTSC) oxide is $YBa_2Cu_3O_{7-x}$, or abbreviated YBCO-123; Y-123 ($T_c = 93$ K). This compound was discovered in late 1986 by Wu *et al* [1] and is the first compound for which T_c exceeded the boiling point of liquid nitrogen (77 K at 1 atm). It seems fitting that the first material found to be superconducting above liquid nitrogen temperature, exhibits one of the most interesting and complex relationships between chemistry, crystal structure and physical properties of any ceramic material ever studied.

The oxygenation procedure of YBCO samples turned out to be a very crucial step. In a typical preparation procedure, YBCO is formed with an oxygen content close to six and then has to undergo the phase transition from a tetragonal to orthorhombic material, including the characteristic formation of twin boundaries. In principle, it is extremely difficult to reach the $x = 0$ state within reasonable time frames [2–4]. B2.2.2

This high sensitivity of YBCO concerning the oxygen content leads to a variety of observations, which are caused by inhomogeneous oxygenation, including the so-called fishtail effect in the magnetization loops [5, 6]. B2.3.3

YBCO is—despite the other superconducting compounds with higher critical temperatures—still the most promising material for applications aimed to operate at 77 K. Soon after the discovery of YBCO, it was found that the resulting bulk materials are severely suffering from the presence of high-angle grain boundaries, which made the fabrication of silver-cladded YBCO wires [7, 8] practically impossible. Therefore, different preparation techniques—taking into account the ceramic nature of YBCO—had to be found. At the present time, the superconducting material chosen for the second generation tapes is again YBCO. These types of YBCO tapes on Ni or similar substrates reach values of the critical current density, J_c, of 10^6 A/cm^2 at 77 K (self-field) and lengths of about a metre [9]. Further research is required to enable the production of longer length conductors required for applications. A1.3 B2.2.2 B4.1

Melt texturing was found to be a promising way to produce bulk YBCO samples, thereby eliminating most of the high-angle grain boundaries. Currently, bulk samples of YBCO with diameters up to 10 cm and J_c of 10^6 A/cm^2 at 77 K (self-field) can be produced [10]. B2.3.3

Most of the research in the beginning of the high T_c period was devoted to YBCO; therefore, many basic properties of the perovskite superconductors were studied on YBCO. Doping effects were studied in detail in a quest to increase T_c even further. Another important aspect of doping is to obtain a denser ceramic body approaching the theoretical density and a microstructure with an optimal grain size and pinning site density for high mechanical strength, high J_c and high T_c. One can classify the effects of dopants into four groups following [11]: dopants in the first category dissolve and substitute in G5 the copper sublattice; therefore, there is a significant effect on the superconducting transition temperature. Examples for these dopants are Zn, Mg, Fe and Pr. Dopants in the second category dissolve and substitute in the yttrium and barium sublattice, but these dopants are not likely to have a potent effect on the superconductivity of YBCO. Examples here are SrO and other rare earth oxides. Dopants in the third category have a limited solubility in YBCO, but have a strong tendency to decompose YBCO. T_c of the undecomposed YBCO, however, remains unaffected. Examples for this are SiO_2 and Al_2O_3. The dopants of the fourth category also have a very limited solubility in YBCO. However, these are virtually non-reactive with YBCO, so they are present as a second phase. Typical examples of these dopants are silver and gold. Nevertheless, the doping of YBCO with small amounts of T_c-depressing elements like Pr or Zn was found to be beneficial for the flux pinning properties, which could be improved considerably [10]. The use of $BaZrO_3$ crucibles [12–14] enabled ultra-pure YBCO single crystals to be prepared, which allowed a variety of measurements until then hampered by the presence of impurities.

B2.3.3

In this chapter, the structural, thermal, mechanical, chemical, optical, normal and superconducting properties of YBCO will be described.

D3, D2

C1.2 Structural properties of Y-123, Y-124 and Y-247

C1.2.1 Crystal structure

D1.1.1

At the time that the crystal structure of $YBa_2Cu_3O_{7-x}$, was first determined, it was a unique and highly unusual atomic arrangement among known complex oxides. It is now known that YBCO-123 is one member of a family of structurally related HTSC YBCO compounds. Other members of the family $YBa_2Cu_4O_8$ (YBCO-124; Y-124) and $Y_2Ba_4Cu_7O_{15}$ (YBCO-247; Y-247), require more exotic synthetic conditions or chemistry and therefore were discovered and identified later [15–27].

The YBCO superconductors have layered perovskite-like and highly anisotropic crystal structure. Figure C1.1 shows the crystal structure of YBCO-123, which is an oxygen deficient perovskite with ordered vacancies, consisting of three unit cells, with four different layers stacked sequentially as BaO–CuO–BaO–CuO_2–Y–CuO_2–BaO–CuO–BaO [28]. In the CuO layer, each Cu atom is coordinated by four oxygen atoms, which is different from the five oxygen atoms surrounding the Cu atom in the CuO_2 layer. The $YBa_2Cu_3O_{7-x}$ compound has both orthorhombic ($x < 0.5$) and tetragonal ($x \geq 0.5$)

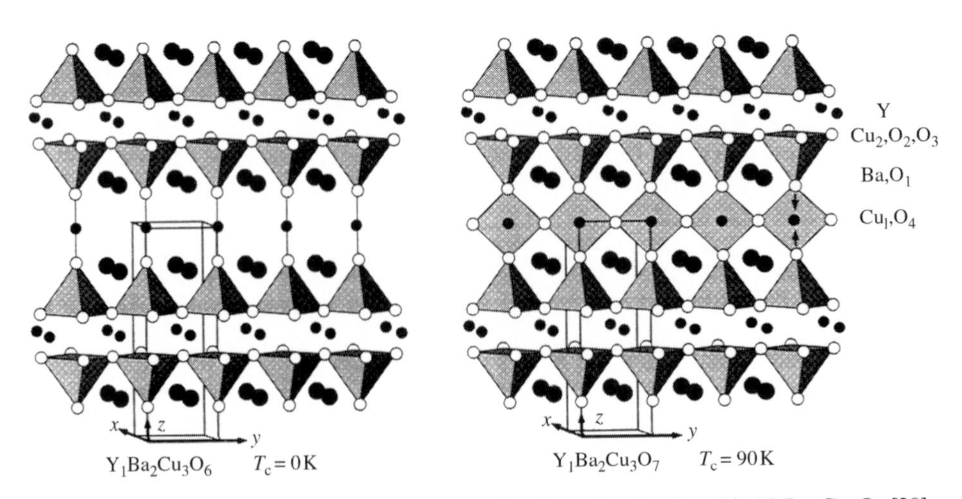

Figure C1.1. Structures of tetragonal $Y_1Ba_2Cu_3O_6$ and orthorhombic $Y_1Ba_2Cu_3O_7$ [28].

structures. The superconducting phase, $YBa_2Cu_3O_{7-x}$ has an orthorhombic structure. The relation between parameters of crystal structure and oxygen content in Y-123 has been investigated [29–33]. Data for the lattice parameters and bond lengths are given in table C1.1 .

A slight difference between a and b lattice constants in the orthorhombic superconducting phase is caused by oxygen vacancy ordering in the CuO layer (chains) sandwiched between BaO layers. D1.1.1

The typical X-ray diffraction patterns of Y-123 with various oxygen content are presented in figure C1.2 [37], and the position of the main diffraction lines of $YBa_2Cu_3O_7$ sample, their intensity and $(h\,k\,l)$ indices, are shown in table C1.2 [35].

The structures of Y-124 and Y-247 compounds are similar to YBCO-123. These crystal structures are shown schematically in figure C1.3(a)–(b) [28]. Data for the lattice parameters and bond lengths are given in table C1.3 .

The superconducting phase $YBa_2Cu_4O_8$ (YBCO-124, $T_c = 80$ K) was originally discovered as a lattice defect in $YBa_2Cu_3O_7$ [38]. The structure of YBCO-124 is closely related to YBCO-123, but with one additional CuO chain layer in the unit cell. Because the position of Cu in adjacent Cu–O chains differs by $b/2$ along the b-axis [38], c is doubled (see table C1.3). Two distinct Cu sites exist in the crystal structure: Cu(1), lying in the Cu–O chains, with four-fold square planar coordination of oxygen and Cu(2) with a five-fold pyramidal coordination of oxygen in the CuO_2 planes. Unlike YBCO-123, this compound is stable with respect to the oxygen stoichiometry. However, this HTSC has to be synthesized in high oxygen pressure at a high temperature to stabilize the structure. When the oxygen content is not stoichiometric the solid decomposes into YBCO-123 and CuO, or $Y_2Ba_4Cu_7O_{15}$ (YBCO-247) and CuO [38]. The Y-247 superconductor was first observed by Karpinski *et al* [39] as an impurity phase in the investigation of the pressure–temperature–composition phase diagram in the system of $YBa_2Cu_3O_{6+x}O_2$. The structure in the c-direction consists of alternating blocks with CuO single-chains (123-units) and with CuO double-chains (124-units) [40].

The typical X-ray diffraction pattern of Y-124 and Dy-247 are presented in figure C1.4(a)–(c) [23, 41, 42], and the position of the main diffraction lines of a Y-124 and Dy-247 samples, their intensity and $h\,k\,l$ indices, are shown in tables C1.4 [43] and C1.5 [44]. D1.1.2

Table C1.1. Lattice parameters, space group and bond lengths for YBCO-123 HTSC materials

Abbreviation	Compound	Space group	Lattice parameters (Å)			Ref.
YBCO-123	$YBa_2Cu_3O_{7-x}$	*Pmmm*	$a = 3.8227$	$b = 3.8872$	$c = 11.6802$	[34]
YBCO-123	$YBa_2Cu_3O_7$	*Pmmm*	$a = 3.8185(4)$	$b = 3.8856(3)$	$c = 11.6804(7)$	[35]
YBCO-123	$YBa_2Cu_3O_{6.8}$	*Pmmm*	$a = 3.8214(7)$	$b = 3.8877(7)$	$c = 11.693(2)$	[35]
YBCO-123	$YBa_2Cu_3O_{6.56}$	*Pmmm*	$a = 3.8336(4)$	$b = 3.8807(4)$	$c = 11.7355(10)$	[35]

		Bond lengths (Å)			
		Ref. [31]	Ref. [32]	Ref. [36]	Ref. [30]
YBCO-123	Cu1–O1	1.836	1.876	1.848(4)	1.857
	Cu1–O4	1.944	1.921	1.926	1.944
	Cu2–O1	2.306	2.293	2.302(5)	2.296
	Cu2–O2/3	1.946	1.931	1.944(1)	1.946
	Ba–O1	2.744	2.735	2.744(1)	2.742
	Ba–O2/3	2.972	3.168	2.951(5)	2.976
	Ba–O4	2.879	2.835	2.908(3)	2.879

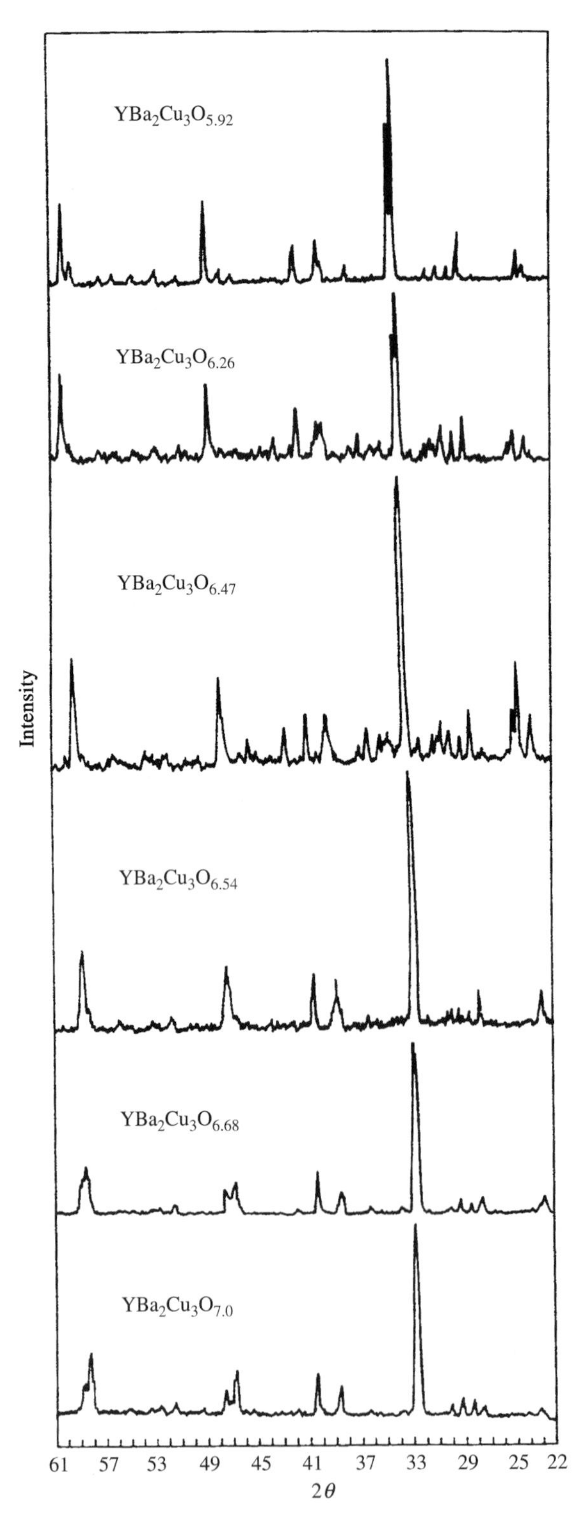

Figure C1.2. X-ray diffraction patterns of YBCO samples with different oxygen contents [37].

Table C1.2. $YBa_2Cu_3O_7$ x-ray diffraction lines

2-θ	Int.	*h k l*	2-θ	Int.	*h k l*	2-θ	Int.	*h k l*
7.557	<1	0 1 0	53.400	2	0 3 2	77.480	4	1 9 0
15.169	4	0 2 0	54.997	2	0 7 0	77.654	6	0 9 1
22.835	10	0 3 0	55.313	1	2 2 1	77.827	5	3 0 1
23.274	4	0 0 1	58.207	26	1 6 1	79.088	5	0 3 3
27.554	3	1 2 0	58.826	13	1 3 2	79.749	3	2 7 1
25.893	5	0 2 1	60.309	<1	1 7 0	81.141	1	1 2 3
30.618	<1	0 4 0	60.494	1	0 7 1	81.812	2	2 5 2
32.538	55	1 3 0	62.080	2	2 5 0	82.335	<1	3 3 1
32.842	100	0 3 1	62.262	2	2 4 1	82.498	1	0 10 0
33.757	2	1 1 1	62.809	3	0 5 2	83.650	1	1 3 3
36.370	3	1 2 1	65.571	2	1 7 1	87.028	3	1 10 0
38.512	13	0 5 0	68.134	5	2 6 0	87.287	6	2 8 1
38.799	5	0 4 1	68.618	5	1 8 0	87.748	4	1 8 2
40.384	14	1 3 1	68.797	13	0 8 1	90.351	1	3 5 1
45.524	2	1 4 1	68.889	12	2 0 2	91.089	1	2 9 0
46.633	22	0 6 0	72.820	1	0 9 0	91.772	1	0 9 2
46.725	21	2 0 0	72.995	<1	3 0 0	93.027	1	0 11 0
47.580	12	0 0 2	73.561	2	1 6 2	93.786	1	2 7 2
51.485	4	1 5 1	74.997	<1	2 7 0	95.853	4	3 6 1
52.526	3	1 6 0	75.615	1	0 7 2	96.394	4	3 3 2
52.733	4	0 6 1	77.247	<1	2 4 2	97.145	4	1 6 3

Radiation: Cu Kα1, $\lambda = 1.540598$ Å [35].

C1.2.2 Defects in YBCO

All of YBCO HTSC compounds have highly defective structures. In such crystal structures several types of defects are easily incorporated: oxygen vacancies, cation vacancies, twin boundaries, intergrowth, dislocations, stacking faults, interstitials, Schotky defects, anti-site defects and Frenkel defects [45–50]. D1.1.2

In the Y-123, the defects are oxygen atoms that can occupy, in a random or ordered way, available lattice sites in the chain region of the structure. The range of defect concentrations in this compound is unusually large, allowing the properties to be varied from insulation to superconduction [46]. The influence of oxygen stoichiometry on properties of YBCO HTSC, especially with respect to the tetragonal (T) to orthorhombic (O) phase transformation, has been the subject of much research. The simplest phase diagram for Y-123 as a function of temperature and x has a T field at high temperature and high x, an O field at lower temperature and low x, and a T + O two-phase field that separates the two at low temperature (figure C1.5). The tetragonal phase has the space group $P4/mmm$; at $YBa_2Cu_3O_{6.5}$ stoichiometry, the z = O basal has half as many oxygen atoms as oxygen sites. At the composition, $YBa_2Cu_3O_7$, the orthorhombic phase has a perfect ordering of oxygen atoms and vacancies in orthogonal sets of chains. At the composition $YBa_2Cu_3O_6$ the basal planes are completely depleted of oxygen, and therefore the tetragonal crystal structure is observed. D1.1.1

The most frequently observed planar stacking fault in YBCO is a $(CuO)_2$ double layer formed by insertion of an extra CuO layer at the CuO layer between the two BaO layers, leading to a local composition of Y-124 in Y-123 or $YBa_2Cu_5O_9$ in Y-124 [48, 49]. Figure C1.6 shows a typical example for planar defects [47].

In YBCO, the twinning is associated with the tetragonal to orthorhombic order–disorder phase transformation. Typically, these twins exhibit a 90° (1 0 0) rotation of the lattice at the boundary. Across the boundary of these twins the *c* axis of the matrix continues as an *a* or *b* axis within the twin, as can be seen from figure C1.7 [48]. High-resolution transmission electron microscopy (HRTEM) studies have led to the suggestion that some twin boundaries are oxygen deficient [45].

D1.2

Three types of dislocations have been observed in YBCO. The first type is an edge dislocation running along (1 0 0) with a Burgers vector of **a** [0 1 0]; the second—with a Burgers vector of **a** [0 0 1], and third—with a Burgers vector of $\mathbf{a}/2 + \mathbf{b}/2$ [0 0 1]. The last two types of dislocations are observed more often in materials containing many (1 0 0) 90° rotation twins. The lattice image of the Y-124 in the region indicated by arrow C in figure C1.8, [49] shows the presence of an edge dislocation with a Burgers vector of either $b = \langle 0\,1\,0\rangle$ or $\langle 1\,0\,0\rangle$ on [0 0 1].

D1.1.2

C1.2.3 Anisotropy

Besides high transition temperatures, the HTSCs are characterized by their pronounced anisotropy. Such a high anisotropy originates from the fact that the superconductivity occurs mainly in the CuO_2 planes and there are weak interlayer couplings between them. The anisotropy is usually expressed as a dimensionless parameter γ, which is defined as $\gamma = \xi_{ab}/\xi_c$, where ξ_{ab} and ξ_c are the superconducting coherence length parallel and perpendicular to the *a–b* plane, respectively. In the case of the Y-123 compounds a further anisotropy has been seen. The dependence of several physical properties, measured in the *a–b* plane, on parameters such as temperature and frequency of the applied electromagnetic wave, has been found to have a directional aspect. It is generally believed that this further anisotropy is a result of the orthorhombic crystal structure and existence of CuO chains along *b* axis [51–53].

A2.3

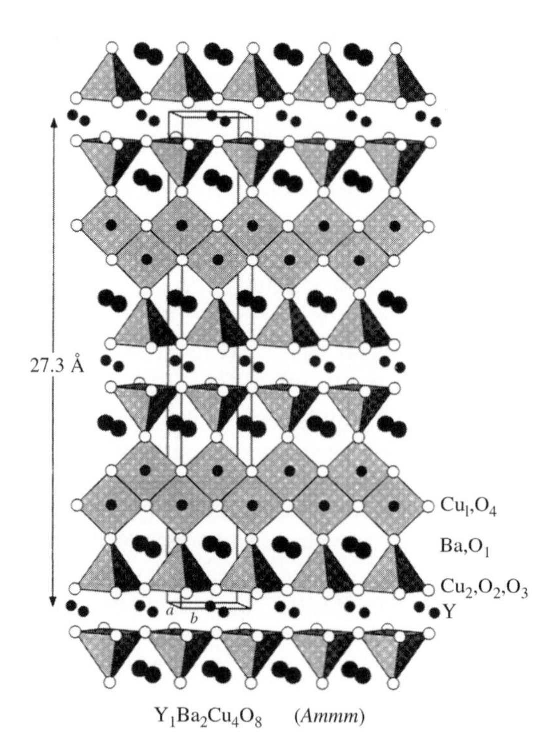

Figure C1.3. (*a*) Structures of $Y_1Ba_2Cu_4O_8$ (Y-124) and (*b*) $Y_2Ba_4Cu_7O_{15}$ (Y-247) [28].

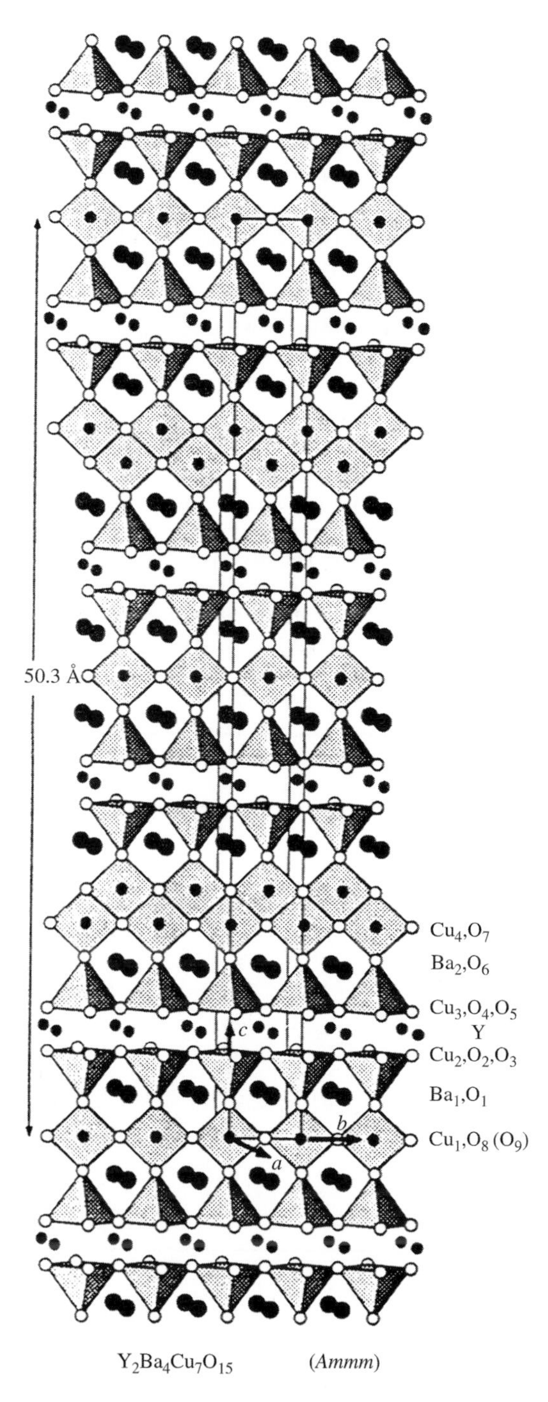

Figure C1.3. (Continued).

C1.3 Thermal properties

C1.3.1 Specific heat of YBCO

A2.4, A3.2
D3.2.1

There are some peculiar features in the specific heat of YBCO compared with those of conventional BCS superconductors. They are the finite $\gamma^* T$ term in the superconducting state, the upturn of C_p/T and anomalous peak shape around T_c and at $\sim$220 K, for example, double peak structure and fluctuation phenomena [54–59]. These anomalies have been paid considerable attention, because they may provide

Table C1.3. Lattice parameters, space group and bond lengths for $YBa_2Cu_4O_8$ and $Y_2Ba_4Cu_7O_{15}$ HTSC materials

Abbreviation	Compound	Space group	Lattice parameters (Å)			Ref.
YBCO-124	$YBa_2Cu_4O_8$	*Ammm*	$a = 3.838$	$b = 3.868$	$c = 27.20$	[43]
YBCO-247	$Y_2Ba_4Cu_7O_{15}$	*Ammm*	$a = 3.833$	$b = 3.878$	$c = 50.59$	[44]
		Bond lengths (Å) for Y-124, [19]				
Y–O2*x*4	Y–O3*x*4	Ba–O1*x*4	Ba–O2*x*2	Ba–O3*x*2	Ba–O4*x*2	
2.400(2)	2.396(2)	2.740(0)	2.962(3)	2.939(3)	2.980(2)	
Cu1–O1	Cu1–O4*x*2	Cu1–O4	Cu2–O1	Cu2–O2*x*2	Cu2–O3*x*2	
1.830(4)	1.943(0)	1.876(4)	2.276(4)	1.935(0)	1.950(0)	

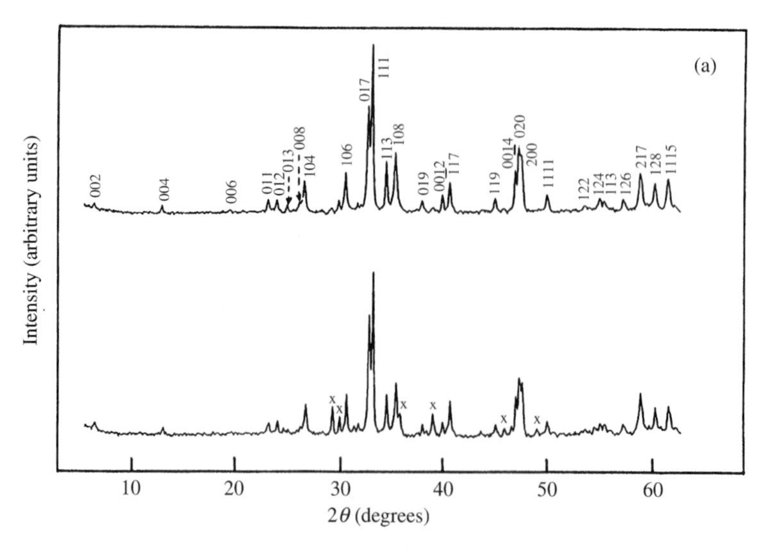

Figure C1.4. (*a*) X-ray diffraction patterns of ground Y-124 pellets. The Y-124 peaks are indexed and impurity peaks are marked with crosses. The upper plot is a starting composition of Y/Ba/Cu = 1/2/4; the lower plot a starting composition of Y/Ba/Cu = 1/2/3. (*b*) X-ray diffraction pattern for Y-124 starting material. Appropriate peaks are marked with their *h k l* reflection indices. The lower panel shows the x-ray diffraction pattern after partial conversion into Y-123 + CuO. The asterisks (*) identify the Y-123 peaks [23, 41]. (*c*) X-ray powder diffraction patterns for the Y-124 and Y-247 phases in the Dy–Ba–Cu–O system. Each pattern shows a single phase, except for additional peaks identified as CuO [42].

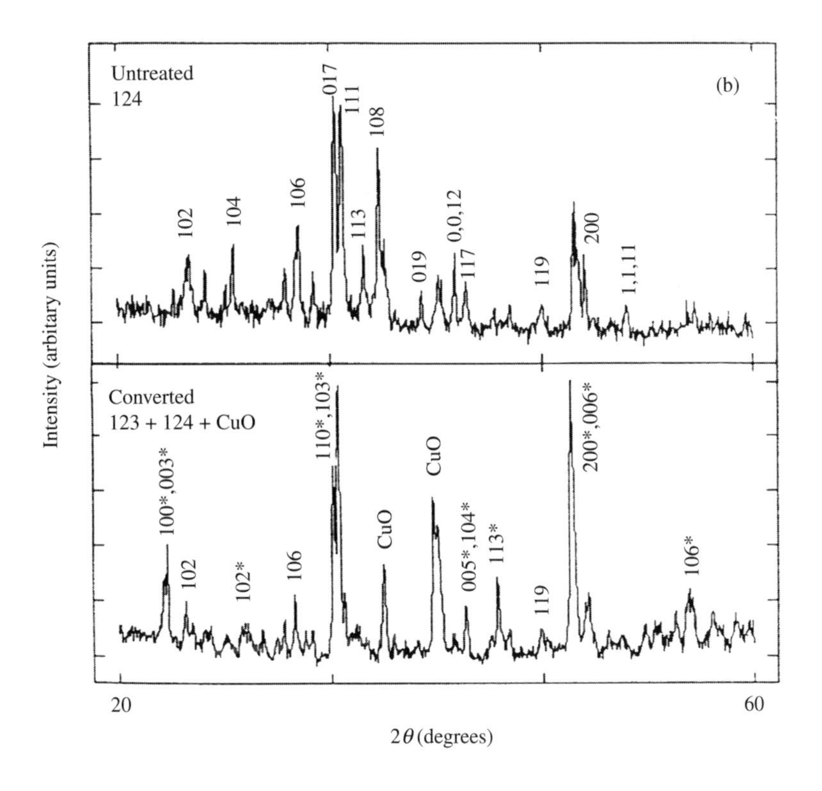

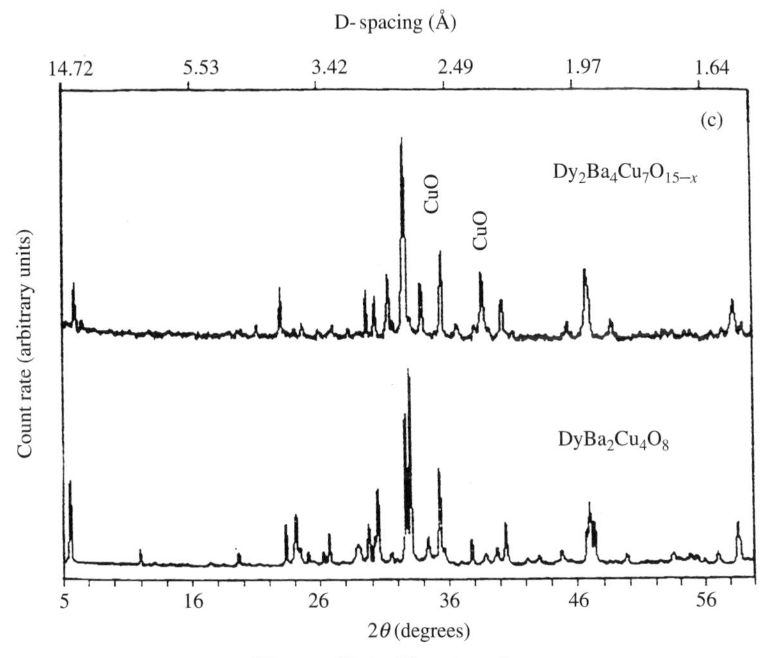

Figure C1.4. (Continued).

Table C1.4. $YBa_2Cu_4O_8$ X-ray diffraction lines

2-θ	Int.	$h\,k\,l$	2-θ	Int.	$h\,k\,l$	2-θ	Int.	$h\,k\,l$
6.494	275	0 0 2	52.963	4	1 2 0	74.215	3	0 3 3
13.00	88	0 0 4	53.348	25	2 1 1	74.411	3	3 0 2
19.566	20	0 0 6	53.422	27	1 2 2	75.045	4	2 2 8
23.203	111	1 0 0	54.258	11	2 1 3	75.552	10	3 0 4
23.203	111	0 1 1	54.534	7	0 2 8	75.737	6	0 3 5
24.070	139	1 0 2	54.783	65	1 2 4	76.250	1	2 1 15
25.003	46	0 1 3	54.887	37	2 0 8	77.443	10	3 0 6
26.188	21	0 0 8	55.311	5	1 1 13	77.678	6	0 1 21
26.645	305	1 0 4	56.050	5	2 1 5	78.002	29	0 3 7
28.283	21	0 1 5	57.003	58	1 2 6	78.211	68	1 3 1
30.490	217	1 0 6	58.670	147	2 1 7	78.444	32	2 2 10
32.635	490	0 1 7	58.838	79	2 0 10	78.798	57	3 1 1
33.010	999*	1 1 1	59.412	3	1 0 16	78.961	33	1 3 3
34.336	254	1 1 3	60.021	83	1 2 8	79.547	13	3 1 3
35.233	242	1 0 8	61.360	8	1 1 15	80.067	15	3 0 8
36.862	4	1 1 5	62.053	8	2 1 9	80.994	2	0 3 9
37.739	22	0 1 9	62.794	1	0 1 17	80.994	2	3 1 5
39.733	12	0 0 12	63.134	4	0 2 12	81.768	1	1 0 22
40.396	69	1 1 7	63.456	4	2 0 12	82.239	1	2 1 17
43.394	1	0 1 11	63.772	1	1 2 10	82.545	3	2 2 12
44.744	29	1 1 9	66.141	1	2 1 11	82.685	6	1 3 7
46.489	6	1 0 12	66.435	1	1 0 18	83.265	6	3 1 7
46.715	24	0 0 14	67.874	1	1 1 17	84.699	1	0 3 11
46.932	339	0 2 0	68.200	2	1 2 12	85.259	1	1 2 18
47.324	325	2 0 0	68.377	8	0 2 14	85.641	3	0 0 24
47.433	183	0 2 2	68.856	159	2 2 0	85.641	3	1 3 9
47.822	7	2 0 2	68.996	83	0 0 20	85.857	2	0 1 23
48.913	6	0 2 4	70.019	1	0 1 19	86.219	3	3 1 9
49.294	6	2 0 4	70.424	4	2 2 4	87.343	5	2 2 14
49.484	3	0 1 13	70.883	1	2 1 13	87.633	2	0 2 20
49.751	19	1 1 11	72.364	2	2 2 6	88.879	1	2 1 19
51.308	3	0 2 6	73.449	4	0 3 1	89.321	2	1 3 11
51.676	3	2 0 6	74.029	1	3 0 0	89.897	2	3 1 11
52.764	1	1 0 14	74.215	3	0 2 16			

Radiation: Cu Kα1, $\lambda = 1.54060$ Å [43].

key information to the understanding of the high T_c mechanism. In order to clarify the origins and nature of these anomalies, researchers performed both low and high temperature specific heat studies on YBCO HTSC [60–66]. In a temperature range of 1.4–7 K, the data are well approximated by [63]:

$$C_p(T) = \gamma^* T + \beta_3 T^3 + \beta_5 T^5 \quad \text{(C1.1)}$$

where $\beta^3 = (12/5)\pi^4 R\theta^{-3}(0)$; and at temperatures above 8 K by [64]:

$$C_p(T) = \gamma^* T + \beta T^3 \quad \text{(C1.2)}$$

Table C1.5. $Y_2Ba_4Cu_7O_{15}$ X-ray diffraction lines

2-θ	Int.	$h\,k\,l$	2-θ	Int.	$h\,k\,l$	2-θ	Int.	$h\,k\,l$
6.986	11	0 0 4	34.061	45	1 1 5	51.862	1	0 2 12
10.645	3	0 0 6	34.061	45	1 0 14	52.422	1	2 0 12
13.997	8	0 0 8	35.164	4	1 1 7	52.705	1	1 0 26
21.055	4	0 0 12	35.451	2	0 0 20	52.855	1	1 2 0
22.980	10	0 1 1	36.836	6	1 0 16	53.314	2	2 1 1
23.186	5	1 0 0	38.083	7	0 1 17	53.411	2	1 2 4
23.447	4	1 0 2	38.370	3	1 1 11	53.590	2	0 2 14
24.238	11	1 0 4	39.133	13	0 0 22	54.055	2	1 2 6
24.599	7	0 1 5	40.358	24	1 1 13	54.346	4	0 1 27
24.599	7	0 0 14	42.611	2	1 1 15	54.346	4	0 0 30
25.479	3	1 0 6	42.801	2	1 0 20	54.981	3	1 2 8
26.064	1	0 1 7	44.232	2	0 1 21	56.595	5	1 1 25
27.181	13	1 0 8	45.067	3	1 1 17	57.541	6	1 2 12
27.920	4	0 1 9	46.640	33	0 0 26	58.234	6	2 0 18
29.198	1	1 0 10	46.793	36	0 2 0	58.335	7	0 0 32
31.520	25	1 0 12	47.391	24	2 0 0	58.689	22	2 1 13
31.830	7	0 0 18	47.663	4	1 1 19	59.169	11	1 2 14
32.569	100	0 1 13	49.115	1	0 2 8	59.853	13	1 1 27
32.864	100	1 1 1	49.684	1	2 0 8	59.853	13	1 0 30
33.254	11	1 1 3	50.517	5	1 1 21			

Radiation: Cu Kα1, $\lambda = 1.54056$ Å [44].

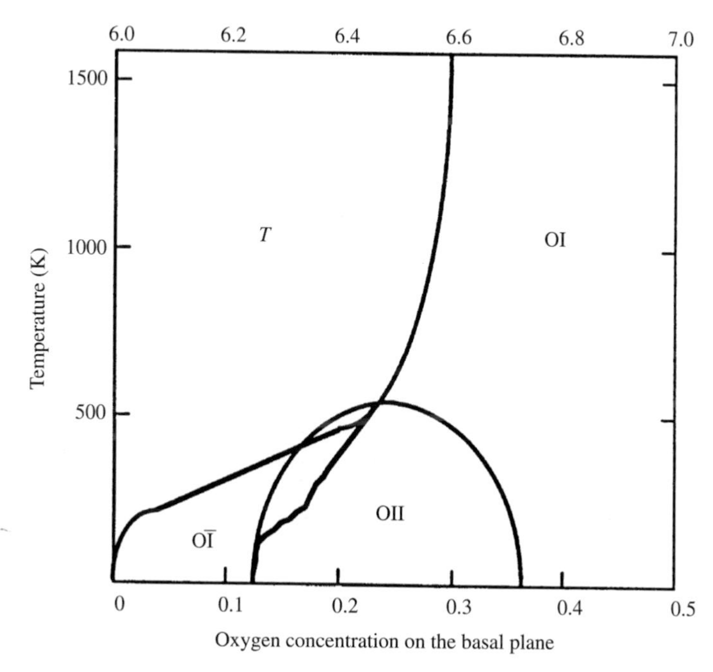

Figure C1.5. Phase diagram for Y-123 as a function of oxygen stoichiometry.

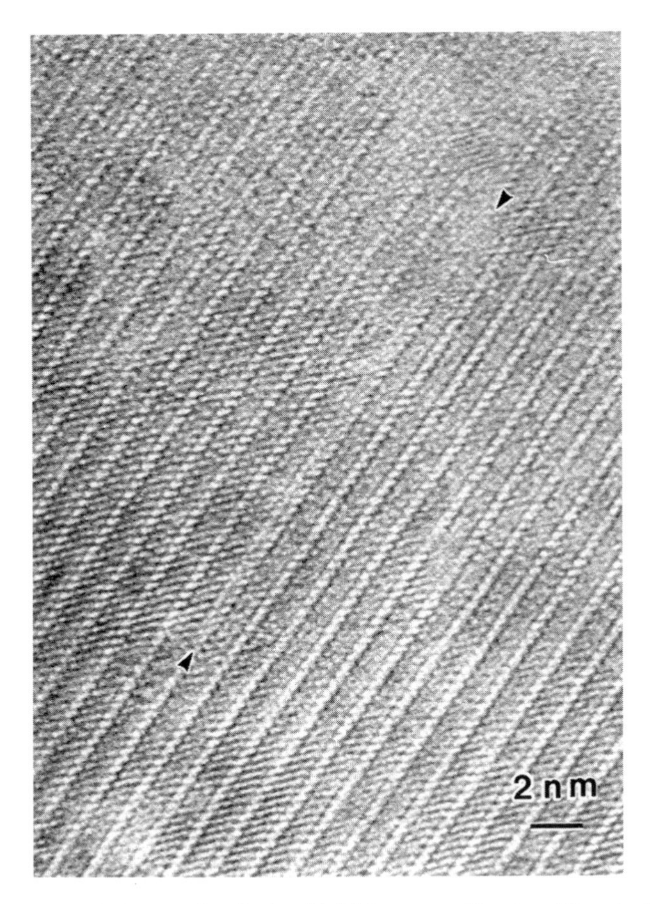

Figure C1.6. A TEM picture of the stacking fault with higher magnification. The stacking faults have a structure with an extra CuO plane inserted to the chain site of the 123 structure [47].

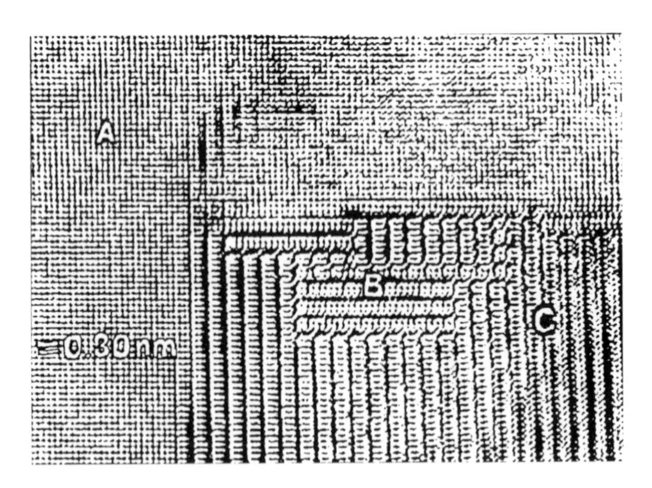

Figure C1.7. Micrograph of $Y_1Ba_2Cu_3O_{7-x}$ showing [1 0 0] 90° rotation twin boundaries [48].

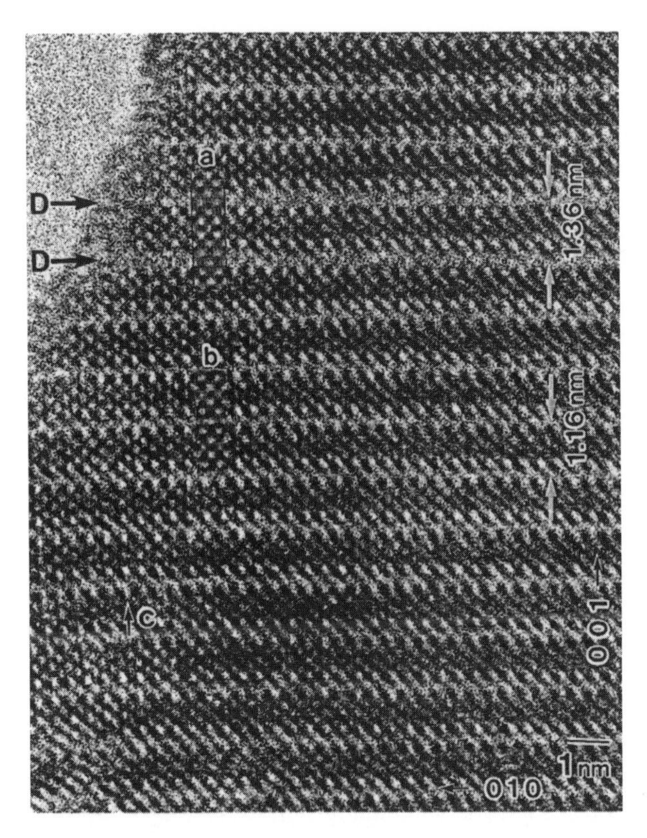

Figure C1.8. A HRTEM micrograph of YBCO crystal observed along the [1 0 0] direction. Arrow D indicates thick and bright bands parallel to the basal plane. A simulated image of Y-124 in a [1 0 0] projection is inserted at (*a*) and a simulated image of Y-123 in the same projection is inserted at (*b*). Arrow C indicates an edge dislocation in Y-123 with a Burgers vector $\mathbf{b} = \mathbf{a}\langle 1\ 0\ 0\rangle$ or $\mathbf{b} = \mathbf{b}\langle 0\ 1\ 0\rangle$ [49].

where $\gamma^* = 0$ and $\beta = 0.60\,\mathrm{mJ/(mol\,K^4)}$. Figure C1.9 shows typical examples of the temperature dependencies of specific heat for the Y-123 and Y-124 systems [64]. The anomaly at ~220 K is visible, but the anomaly at the superconducting transition (90–80 K) is not easily seen on this scale. To study the specific heat jump at the superconducting transition, it is customary to plot the data as C_p/T *versus* T. D3.2.1 Figure C1.10 shows an example of such a plot for Y-123 and Y-124 [64]. Anomalies at T_c are observed for all samples, but they are considerably smaller for Y-124 than for Y-123. A very faint peak can be noted in the specific heat of all samples at $T = 86$ K. Its origin has been ascribed to the presence of impurity phases (as $BaCuO_{2+\delta}$). The anomaly at ~220 K in YBCO has been interpreted as being of structurally related origin [62]. The effect of oxygen content on the transition temperature is well known, and the connection between the 90 and 220 K anomalies suggests that ordering in the oxygen may be responsible for the 220 K anomaly.

A systematic study of the effects of Zn, Ni and Al on the specific heat of Y-123 has been reported [36]. The results of low temperature specific heat experiments suggest that the finite T linear (γ^*) term remaining in the superconducting state and the upturn of C_p/T at the lowest temperatures are related to disorder in the CuO_2 plane and CuO chain, respectively. There have been reports on investigation of Y-123 doped with Fe [66] and Zn [67]. The low temperature linear term in the specific heat reflects a growing non-superconducting fraction. The absence of any definite specific heat jump at T_c in the sample

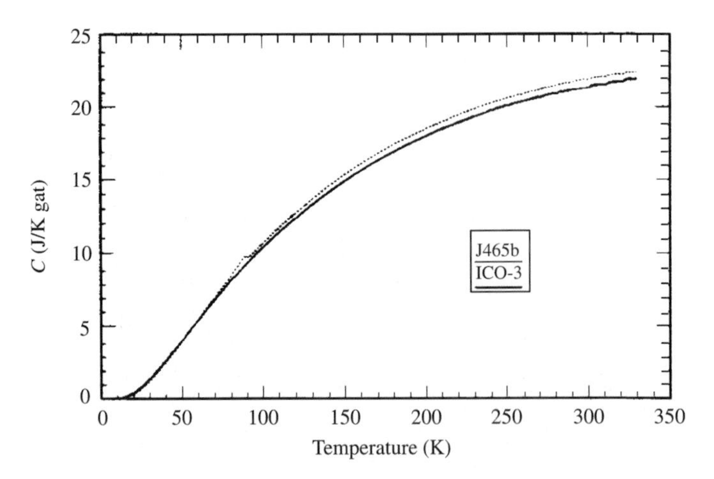

Figure C1.9. Specific heat of Y-123 (sample J465b) and Y-124 (sample JCO-3) up to room temperature [64].

$YBa_2(Cu_{1-x}Fe_x)O_{7-\delta}$ with $x = 4\%$ is understood as an effect of the short coherence length in the presence of repulsive defects [66].

C1.3.2 *Thermal conductivity of YBCO*

D3.2.2 Studies of thermal conductivity κ give valuable information on the interaction between charge carriers and phonons and on the scattering of both defects and impurities. Since κ is also unusual in having a nonzero value in both normal and superconducting states, there has recently been a surge of interest in data for κ in HTSC. One basic characteristic of copper oxide superconductors is their poor thermal conductivity. Experimental data from thermal conductivity studies of YBCO have been reviewed in several papers [68–76]. It is worth noting that results for $\kappa(T)$ of the Y-123 HTSC compounds at $T > 10$ K differ by almost one order of magnitude depending on sample microstructure, oxygen

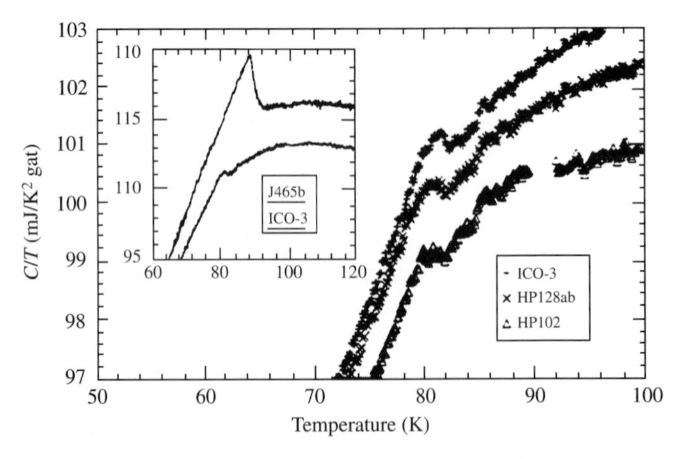

Figure C1.10. Specific heat divided by the temperature versus temperature for three Y-124 samples. For clarity, the curve of sample HP128AB is shifted by $-1\,\mathrm{mJ/Kg^2}$ at. and that for sample HP102 is shifted by $-2\,\mathrm{mJ/Kg^2}$ at. The transition of Y-123 (sample J465b) and Y-124 (sample JCO-3) are shown for comparison in the inset [64].

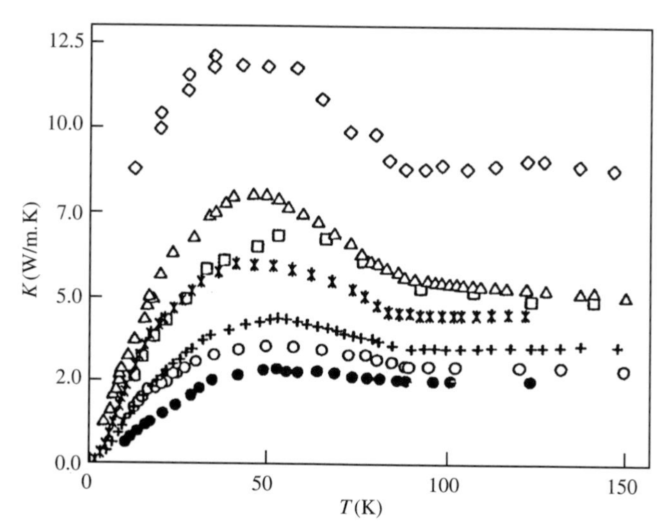

Figure C1.11. Thermal conductivity of different samples of $Y_1Ba_2Cu_3O_{7-x}$: single crystal (◇) and polycrystalline samples [70].

content, admixture and crystallite borders. Temperature dependencies of κ obtained for YBCO are all qualitatively similar, in that κ shows the maximum in the temperature range 40–90 K and a step change of a slope at T_c. The maximum temperature depends on the oxygen content. Examples of the temperature dependence of thermal conductivity for Y-123 can be found in figure C1.11 [70]. All the results show a dominant phonon contribution above T_c. For the YBCO system in the 90–240 K temperature range, the contribution of the electron part is from 1 to 20% depending on the sample. The electrons contribute 60% of the total thermal conductivity at 300 K in single YBCO crystals [74, 75].

D3.2.2

The data for κ in Y-124 differ to a surprisingly large extent from those for Y-123, despite their structural similarity, and κ is dominated by phonon–phonon interactions [77–80]. At all temperatures, phonon–phonon interactions give rise to a much larger thermal resistivity in Y-124 than do electron–phonon interactions, while point-defect scattering was found to be negligible, probably because of the stable oxygen stoichiometry of Y-124. In contrast, point-defect scattering dominates κ completely in Y-123, even for single crystals [78, 81, 82]. Both phonon–phonon and phonon–electron interaction strengths are similar to those in Y-123, and therefore different behaviour is ascribed to the near absence of phonon-defect scattering in Y-124 [77].

Few experimental results have been published concerning thermal conductivity in doped YBCO [83–86]. The thermal conductivity of $YBa_2(Cu_{1-x}Zn_x)_3O_{7-\delta}$ has been investigated in [83, 86]. Figure C1.12 shows the thermal conductivity as a function of temperature for the 90 and 60 K phase $YBa_2(Cu_{1-x}Zn_x)_3O_{7-\delta}$ [83]. The characteristic enhancement of κ just below T_c in pure Y-123 was suppressed by the Zn substitution. On the basis of the phonon heat conduction model, the disappearance of the κ enhancement is attributed to the depressed phonon–electron scattering caused by the Zn substitution. Investigations of the effect of Fe substitution at the Cu site of YBCO on the thermal conductivity have been reported in [84, 85]. The thermal conductivity of all Fe-doped Y-123 samples exhibit a minimum in the vicinity of the critical temperature T_c and maximum near $T_c/2$. The results are interpreted in terms of an electronic model based on the idea that the main contribution to the thermal conductivity below T_c is due to electron scattering in the CuO_2 planes.

D3.2.2

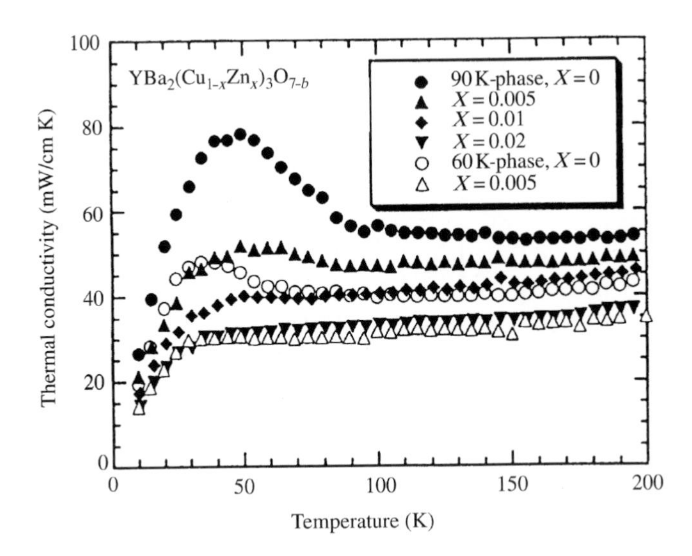

Figure C1.12. Temperature dependence of thermal conductivity κ of the 90 K phase and 60 K phase $YBa_2(Cu_{1-x}Zn_x)O_{7-x}$ samples with various Zn concentrations [83].

C1.4 Mechanical properties of RE123

D3.3 For practical applications, it is important to improve mechanical properties, since the superconductors experience various kinds of forces such as thermal stress due to the heat cycles, and the electromagnetic force during superconductivity operation. High temperature superconductors are brittle and sometimes fracture, therefore mechanical properties are important for engineering applications.

In this section, mechanical properties and related physical parameters of $REBa_2Cu_3O_{7-\delta}$ (RE123, RE; rare earth elements) will be reviewed along with the methods to measure mechanical properties.

C1.4.1 Thermal expansion coefficient

D3.2.3 Cracking in ceramics is generally caused by the difference in thermal expansion coefficients between different phases in contact.

The thermal expansion coefficient is defined as the change of length or volume per degree of temperature. The coefficient of the line thermal expansion (α) is defined by:

$$\alpha = \mathrm{d}l/l\,\mathrm{d}T \qquad \text{(C1.3)}$$

and the volume thermal expansion (γ) is defined by:

$$\gamma = \mathrm{d}v/v\,\mathrm{d}T \qquad \text{(C1.4)}$$

where l is the length, v is the volume and T is temperature. Although these values are temperature dependent, the mean value averaged over a certain temperature range is used as coefficient value.

With increasing temperature, the amplitude of atomic vibration at the equilibrium position is increased, leading to an increase in the bond length and, thus, lattice expansion. The volume change with the lattice vibration will cause an increase in the total free energy of the lattice. Therefore, the D3.2.1 temperature dependence of the thermal expansion coefficient is similar to that of the specific heat. At low temperatures, the thermal expansion coefficient increases abruptly with temperature, and becomes almost constant above Debye temperature Θ_D on the assumption that the lattice simply expands. However, for most materials, it can increase even above Θ_D through the formation of lattice defects. Then the concentration of the defects is directly related to the thermal expansion.

Since the absolute value of the thermal expansion coefficient is closely connected with the crystal structure and the strength of the chemical bond, that of RE123 shows the anisotropy reflecting its layered structure like other high temperature superconductors.

The thermal expansion coefficient can be determined from the temperature dependence of lattice parameters using X-ray or neutron diffraction [87–91]. It is also common to measure the coefficient directly with dilatometers [92–99].

Figure C1.13 shows the thermal expansion coefficients of the principal crystalline axes of untwinned Y-123 single crystal at low temperatures [92]. The maximum thermal expansion is the largest along the *c* axis and the smallest along the *b* axis. Temperature dependence of the thermal expansion coefficient of Y-123 follows the general behaviour and rapidly increases with increasing temperature up to Θ_D of Y-123, which is reported to be around 370–450 K [90, 93], and thereafter gradually increases.

D3.2.3

Table C1.6 shows the typical reported values of the thermal expansion coefficients for RE123 compounds [87–90, 98–100]. Relatively large scatter in the literature data probably arises from the differences in sample quality such as the density of porosity, oxygen content and mismatch in the crystal orientation.

C1.4.2 Elastic modulus

Elastic modulus is defined as the slope of an initial linear portion of the stress–strain curve, in which the stress is proportional to the strain up to the proportional limit. It is represented by the following equation:

$$\sigma = E\varepsilon \tag{C1.5}$$

where σ is the uniaxial stress, E is Young's modulus and ε is the strain. Shear stress τ is proportional to the shear strain (γ):

$$\tau = G\gamma \tag{C1.6}$$

where G is uniaxial modulus of rigidity, which is the shear elastic modulus. The thickness decreases with stretching the sample. Poisson ratio (ν) is defined by:

$$\nu = (\Delta d/d)/(\Delta l/l) \tag{C1.7}$$

where $\Delta d/d$ is the change in the sample length and $\Delta l/l$ is the change in its thickness. In plastic deformation and creep, the volume is constant, then ν is 0.5. Generally, in elastic deformation, the Poisson ratio varies from 0.2 to 0.3. The Poisson ratio is given by using the Young's modulus (E) and the shear elastic modulus (G) as follows:

$$\nu = E/(2G - 1) \tag{C1.8}$$

This formula can only be used for isotropic materials in which the elastic modulus is independent of the measurement angle, and will be a good approximation for polycrystalline ceramics.

Velocity measurement of ultrasonic wave (pulse echo technique [101–110], vibrating reed [111–114]) is usually used for determining the elastic modulus values.

Here, the shear modulus (G) and the bulk modulus (B) are represented by using longitudinal sound velocity (v_L), transverse sound velocity (v_S) and the density (ρ) as follows:

$$G = \rho v_S^2; \quad B = (3\rho v_L^2 - 4G)/3 \tag{C1.9}$$

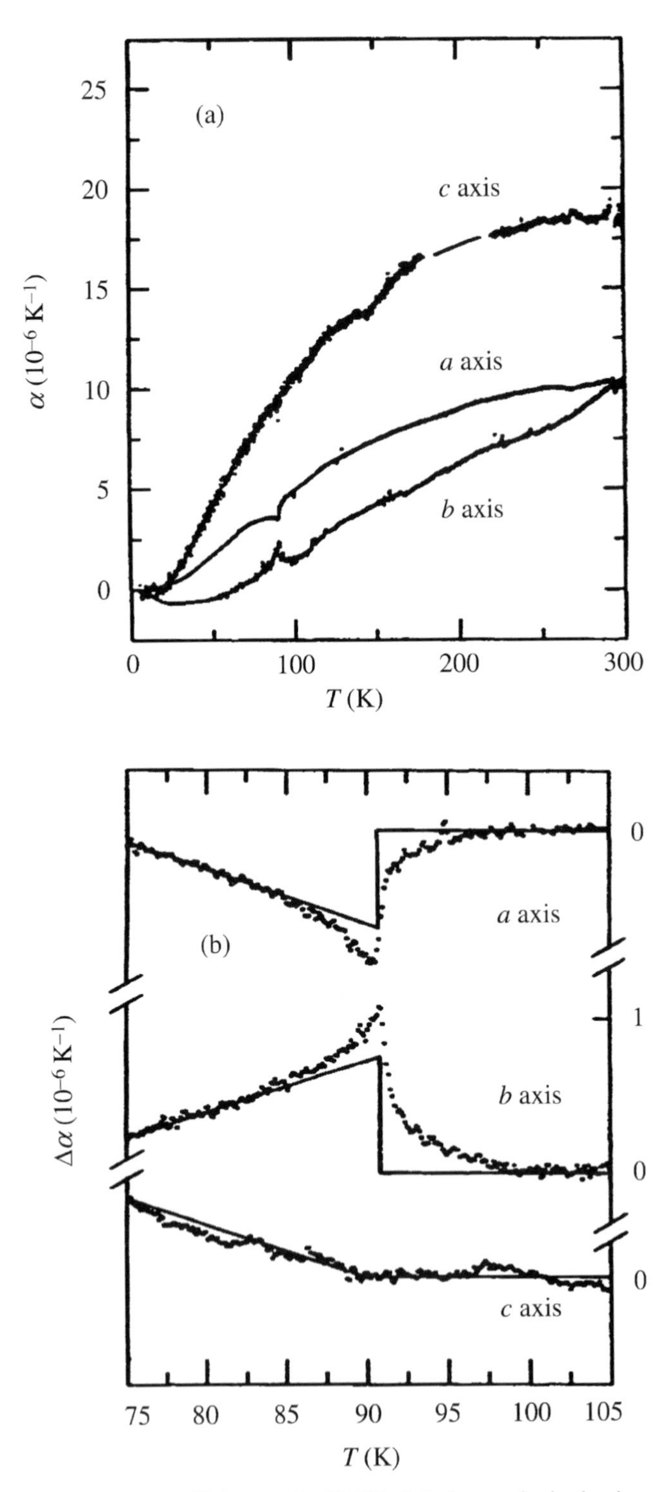

Figure C1.13. Line thermal expansion coefficients α for Y-123: (*a*) along principal axis: and (*b*) magnified views of $\Delta\alpha$ near T_c [92].

Table C1.6. Thermal expansion coefficients, α_a, α_b, α_c, for RE123 compounds

Sample	α_a ($10^{-5}K^{-1}$)	α_b ($10^{-5}K^{-1}$)	α_c ($10^{-5}K^{-1}$)	Temp. range (K)	Method	Ref.
Y-123	1.1	1.1	2.5	298–873	XRD	[87]
Y-123	0.7	0.7	1.5	298–1200	XRD	[88]
Y-123 (O = 6.95)	0.52	0.92	1.57	150–300	XRD	[89]
Y-123	1	–	–	250	Strain Gauge	[99]
Y-123	1.69	–	–	303–1173	Dilatometer	[98]
Y-123	1.67	–	–	–	–	[100]
Y-123/15 vol% Ag	1.72	–	–	–	–	[100]
Nd123	1.3	–	1.9	273–1073	Neutron	[91]
Sm123	1.16	–	–	250	Strain Gauge	[99]
Yb123	1.5	–	–	303–1173	Dilatometer	[97]

When the material is isotropic:

$$E = 9BG/(3B + G) \tag{C1.10}$$

Table C1.7 shows reported data of various elastic moduli E, G and B and Poisson ratio for RE123 [100–103]. Elastic moduli and Poisson ratio are strongly sensitive to the oxygen content, because the orthorhombic phase is significantly stiffer than the tetragonal phase [97, 107]. Several kinds of abnormalities are observed in the temperature dependencies of the elastic moduli for RE123 superconductors, which is in part explained in terms of the phase transition [114].

D1.1.2

B2.3.2

C1.4.3 Fracture toughness

Fracture toughness reveals the resistance against crack propagation. Vickers indentation method (ID) [11, 57–121] and single-edge notch beam techniques (SENB) [100, 122–127] are generally used to characterize the fracture toughness for a brittle material like RE123.

D3.3

In the indentation method, the threshold stress intensity factor K_{th} (= K_c^{air}), which is calculated from an impression half-diagonal (a), a radial/median crack length (c) and indentation peak load (P) as follows [121]:

Table C1.7. Elastic moduli E, B, G and Poisson ratio ν for various RE123

Sample	Density (g/cm^3)	Porosity (%)	x	E [GPa]	B [GPa]	G [GPa]	ν	Ref.
YBCO[a]	5.952			95.89	41.79	45.3	0.14	[101]
Y-123	6.07	4.2		104		51.6	0.163	[103]
Y-123/15 vol% Ag	6.58	5.4		119		91	0.282	[103]
Y-123	6.171		6.2	91.3	38.5	48.6	0.187	[102]
Y-123	6.381		6.91	126.2	50.5	84.5	0.25	[102]
Y-123				116			0.26	[100]
Y-123/15 vol% Ag				105			0.285	[100]

[a] Melt processed Y-123/Y211.

$$K_{th} = x(E/H_v)^{1/2} \times P/c^{3/2} \tag{C1.11}$$

$$H_v = 1.854P/(2a)^2 \tag{C1.12}$$

where x is 0.016 and a material independent constant, E is Young's modulus, H_v is Vickers hardness. Cooks *et al* [120] chose the value of $E/H = 40$ for ceramics with small elastic recovery, and $K_{th} = 0.74$ [MPa m$^{1/2}$] was estimated for Y-123.

D3.3

Table C1.8 shows the reported fracture toughness and hardness obtained by the ID method and also by the SENB method. Hardness is unaffected by the presence of twin boundaries and moisture, whereas the fracture toughness is enhanced by twin boundaries, and the moisture in the air promotes crack growth, thereby degrading K_c [117]. Compared with single crystals, relatively low H_v values in ceramic materials are attributed to the presence of a large amount of voids [119]. It is reported that an addition of Ag and excess amount of RE211 phase is effective in increasing the fracture toughness of the RE123 [97, 100].

A1.3

C1.4.4 Fracture strength

D3.3

Unlike metals, the absence of multiple active slip systems makes ceramic RE123 materials very brittle and they fracture at the elastic limit, which is defined as the fracture strength. When stressed in tension, the stress at which the sample breaks is called the tensile strength. When stressed in bending or in

Table C1.8. Hardness H, fracture toughness K_c, and Young's modulus E for RE123

Sample	H [GPa]	K_c^{air} [MPa m]	E [GPa]	Method	Ref.
Y-123[a] twinned (1 0 0)/(0 1 0) on {0 0 1}	8.7	0.74	($E/H = 40$)	ID	[120]
Y-123[a] detwinned (1 0 0) on {0 0 1}	9.4	0.59	157	ID	[117]
Y-123[a] detwinned (0 1 0) on {0 0 1}	9.4	0.47	157	ID	[117]
Y-123[a] twinned (1 0 0)/(0 1 0) on {0 0 1}	9.5	0.66	157	ID	[117]
Y-123[a] twinned (0 0 1) on {1 0 0}/{0 1 0}	9.8	0.8	157	ID	[117]
Y-123[a] twinned (1 0 0) on {1 0 0}/{0 1 0}	9.8	0.32	89	ID	[117]
Y-123[b] (1 0 0)/(0 1 0) on {0 0 1}	6.7	0.67	182	ID	[118]
Y-123[b] (1 0 0)/(0 1 0) on {0 0 1}	7.35	0.84	($E/H = 40$)	ID	[115]
YBCO[c] (1 0 0)/(0 1 0) on {0 0 1}		1.48	($E/H = 40$)	ID	[117]
Nd123[a] (1 0 0)/(0 1 0) on {0 0 1}	7.79	0.703	158	ID	[116]
Y-123		1.8	116	SENB	[100]
Y-123/15 wt% Ag		3.6	105	SENB	[100]
Y-123/19 wt% Ag		3.6		SENB	[125]
Y-123 ($n = 18.4\%$)	2.1	0.71	66.8	ID	[119]
Y-123 ($n = 11.8\%$)	4.4	0.87	76.5	ID	[119]
Y-123/0.1 mol% Ag_2O ($n = 9.0\%$)	3.9	1.37	80.6	ID	[119]
Y-123 ($n = 33\%$)		1.05		SENB	[126]
Y-123 ($n = 13\%$)		1.4		SENB	[126]
Y-123/30 wt% Ag ($n = 13\%$)		2.4		SENB	[126]

ID: indentation; SENB: single edge notch beam; n: porosity.
[a] Single crystal.
[b] Melt processed sample.
[c] Melt processed Y-123 with a large amount of Y211.

Table C1.9. Fracture strength σ_f of RE123

Sample	σ_f (MPa)	E (Gpa)[a]	Method	Ref.
Y-123	40	81	3-Point	[131]
Y-123/15 wt% Ag	60	96	3-Point	[131]
YBCO/15 wt% Ag[b]	70	100	3-Point	[131]
Y-123	87	–	4-Point	[129]
Y-123/10 vol% Ag	116	–	4-Point	[129]
Y-123/30 vol% Ag	136	–	4-point	[129]
Y-123	38	–	3-Point	[132]
Y-123/10 vol% Ag + 0.8 mol% Zr	280	–	3-Point	[132]
YBCO[b]	28	–	3-Point	[130]
YBCO[b] unannealed	110	130	3-Point	[128]
YBCO[b] annealed	77	128	3-Point	[128]

[a] Calculated from stress–strain curves.
[b] Melt processed sample. (Y123 + Y211).

compression, the stresses are called the flexural strength and the compression strength, respectively. Generally, the compression strength of conventional ceramics is about ten times larger than the tensile strength, and the flexural strength is slightly larger than the tensile strength, which depends on the amount of defects. A bending test is usually employed to estimate the fracture strength for brittle materials using three-point or four-point bending test techniques [127–132]. Flexural strength (σ_f) is given by [131]: D3.3

$$\sigma_f = 14.72LP/Bt^2 \tag{C1.13}$$

where L, P, B, and t are span, load, width and thickness of the bar shaped sample, respectively. Table C1.9 shows reported data of flexural strengths for RE123 compounds. Here, E is calculated from the slope of a linear portion of the load–displacement curve for bent bars.

The fracture strength of Y-123 is reported to range from 40 to 200 MPa, depending on the sample quality. The scatter in the data is attributed to the variation in the porosity density, the oxygen content and the amount of cracking [128–132].

An addition of the second phase is effective in increasing the fracture strength of RE123 matrix. For instance, the flexural strength is increased with Ag addition [100, 131]. When metal inclusions like Ag or Au are dispersed in the matrix, crack tips can be blunted since they are ductile materials [129, 133]. It is also probable that Ag particles induce the compressive stress to resist crack propagation at the crack tip [129]. An addition of RE211 phase or ZrO_2 has also been reported to be effective in improving mechanical properties [132]. A1.3 A2.3.3

C1.4.5 Plastic deformation

Oxide superconductors are so brittle that plastic deformation such as drawing or rolling is impossible at room temperature. However, they are known to deform plastically at temperatures above 800°C, since the yield stress markedly decreases with increasing temperature. Y-123 deforms more than 50% in compressive strain at temperatures above 840°C as shown in figure C1.14 [134]. The flow stress is A1.3

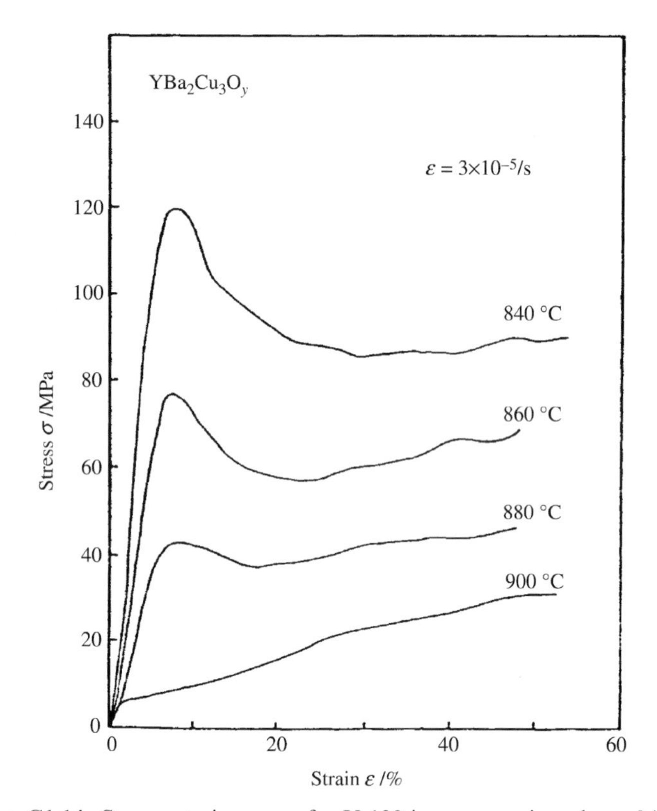

Figure C1.14. Stress–strain curves for Y-123 in compression above 840°C [134].

gradually decreased as the deformation proceeds after yielding in a temperature range between 840 and 900°C. This seems to be a kind of yield–drop phenomenon and the density of mobile dislocations is increasing during the deformation.

D1.1.2

In order to deform polycrystals, it is necessary for five independent slip systems to be activated, as pointed out by von Mises [135]. Since the major slip system of ⟨0 1 0⟩(0 0 1) has only two independent slip systems, secondary slip systems must operate to satisfy the von Mises condition. As suggested by Suzuki *et al* [136], a possible secondary system is the ⟨0 $\underline{3}$ 1⟩{1 1 3}, which is active above 817°C. Above this temperature, plastic deformation occurs with the motion of dislocations in these two slip systems.

C1.5 Chemical properties

C1.5.1 Phase diagrams for YBCO materials

B2.3.2

For the past few years, the phase diagram of the YBCO system has been of interest to many laboratories, as its knowledge is of great importance for preparation procedure [137–148]. The most familiar diagrams published are top–down projections of the three-component plane. Figure C1.15 is typical and shows the phase equilibrium relations in the ternary system $YO_{1.5}$–BaO–CuO [146]. It is an isotherm at 900°C (in air) and it includes the tie lines connecting the most important phases. The stable compounds in the composition triangle at 900°C have been revealed to be Y-123, Y_2BaCuO_5, $YBa_3Cu_2O_x$, $BaCuO_2$ and $Y_2Ba_2O_5$ [146]. The YBCO is a quaternary system, which makes its investigations rather

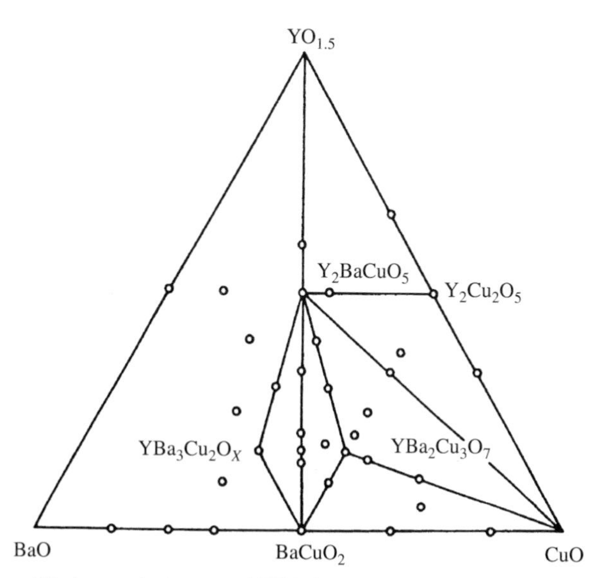

Figure C1.15. The phase equilibrium relations at 900°C in the ternary system YO1.5–BaO–CuO. Open circles indicate the compositions at which the phase identification experiments were performed. The tie-like triangles are shown by solid lines in the composition triangle [146].

complicated. In order to make phase relationships more clear, the sections of the ternary Y_2BaCuO_5 (2 1 1), $Ba_2Cu_3O_5$ (0 2 3), CuO (0 0 1) system along the 123-CuO tie line have been presented. Figure C1.16 shows a part of the ternary system with vertical T (temperature) axes at $P(O_2) = 1$ bar [145]. All three superconducting YBCO phases, Y-123, Y-124 and Y-247, are stable at $P(O_2) = 1$ bar for a limited temperature range. In reality, the phase diagram changes depending on whether the YBCO is being made in air or in oxygen of various pressures. B2.3.2

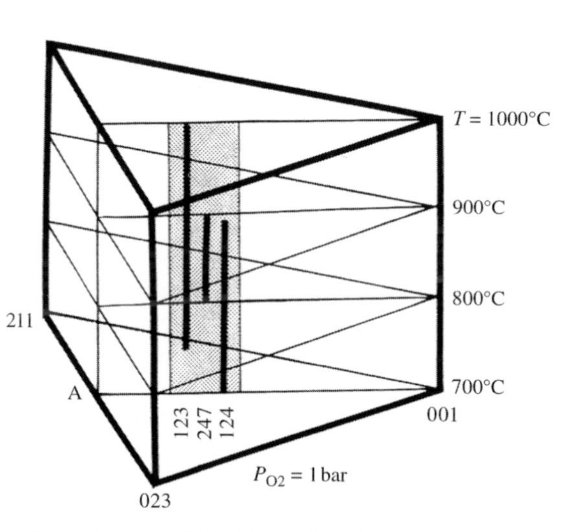

Figure C1.16. T–x section of ternary Y_2BaCuO_5–$Ba_2Cu_3O_5$–CuO system at $P(O_2) = 1$ bar [145].

C1.5.2 Melting point versus O_2 partial pressure; phase changes versus temperature: reaction temperature/atmospheres and stability

B2.4 In order to synthesize ceramic samples of high quality and to grow single crystals of YBCO the knowledge of phase relations and phase stability as a function of temperature and oxygen partial pressure is of considerable importance. In particular, the occurrence of melts is either a restriction or a precondition to guarantee optimum manufacturing conditions. The peritectic melting point of Y-123 has been reported to occur at many different temperatures between 900 and 1050°C [149–152]. The partial pressure of oxygen has been identified as an important variable. Therefore, the melting point (T_m) B2.3.2 of HTSC has been investigated as a function of oxygen partial pressure. It has been concluded that the melting point is a kind of reduction reaction in the form [151]:

$$\text{Solid(I)} = \text{Solid(II)} + \text{Liquid} + O_2 \qquad \text{(C1.14)}$$

Since the equilibrium is shifted to the left side when the oxygen partial pressure is increased, the melting point becomes higher as the oxygen partial pressure is increased. The melting point of Y-123 for different oxygen partial pressure: $P_{O_2} = 1$ atm, $T_m = 1055$°C; $P_{O_2} = 0.1$ atm, $T_m = 1017$°C; $P_{O_2} = 0.01$ atm, $T_m = 984$°C [151]. However, different values have been reported [149–152] for the peritectic melting point of Y-123 using the same atmosphere. The difference in the reported values is probably due to B2.1 experimental factors and powder characteristics including impurity and CO_2-content. The phase diagrams of YBCO HTSC with compositions of Y-123, Y-124, and Y-247 have been investigated as a function of temperature and partial pressure of oxygen over the range 700–1000°C and $0.5 \leq P(O_2) \leq 10$ atm, respectively [139, 141, 145, 147, 149]. Figure C1.17 shows temperature–composition (T–C) diagrams in the YBCO system for different values of $P(O_2)$ [147]. Two regions can be discerned (in terms of number of stable phases and the temperature and pressure regions of the stable phase): from the starting composition of 123, the 247 phase is not formed at any temperature, except for the coexistence point (247 and 124), while from starting compositions of 247, 124 and 125, the 247 phase is formed at intermediate temperature between the 123 and 124 stability regions. From figure C1.17, under this oxygen potential $P(O_2) = 10$ atm, 123 is stable only above 880°C. It is also clear from figures C1.17(*a*) and (*b*), that under $P(O_2) = 1$ atm, the phase boundary temperatures of 123/247 and 247/124 are about 870 and 817°C, respectively, and 123/124 is 760°C.

Figure C1.18 shows the pressure–temperature (P–T) phase diagrams ($\log P(O_2)$ *versus* $1/T$) for Y-123 and Y-124 [147]. As shown in this figure, the 123 phase is stable only at high temperature and low $P(O_2)$, while 124 is stable at high $P(O_2)$ and low temperature. The 247 phase is stable at intermediate temperatures and oxygen pressures between 123 and 124 stability regions. The tetragonal–orthorhombic phase boundary of 123 lies to the right of the 123/124 phase boundary within the 124 phase. This result is evidence that 123 is stable only in its tetragonal phase (semiconducting) and the B2.3.2 orthorhombic phase (superconducting) is metastable at all temperatures. The 124 phase is more stable than the 123 orthorhombic phase.

The relationship between phase stability of YBCO type superconductors, sintering temperature and oxygen partial pressure has been reported in the form of a phase diagram by Karpinski *et al* [15]. The phase diagram showed that Y-124 was stable at high $P(O_2)$ and high T, Y-123 phase at low $P(O_2)$ and high T, and Y-247 at high $P(O_2)$ and high T. Their samples were canned in an encapsule containing oxygen gas as a pressure medium for sintering, therefore the total was not distinguished. The three-dimensional $\log P(O_2)$–$\log P_{tot}$–T phase diagram of an YBCO type superconductor has been determined [153]. This phase diagram shows that the phase transformation temperature from 124 phase to 247 phase with CuO at 10 MPa of oxygen partial pressure increases as the total gas pressure is raised and, at a total gas pressure of 200 MPa, 123 phase with CuO transforms to 124 phase at 7 MPa oxygen partial pressure for the temperature range 1170 – 1370 K.

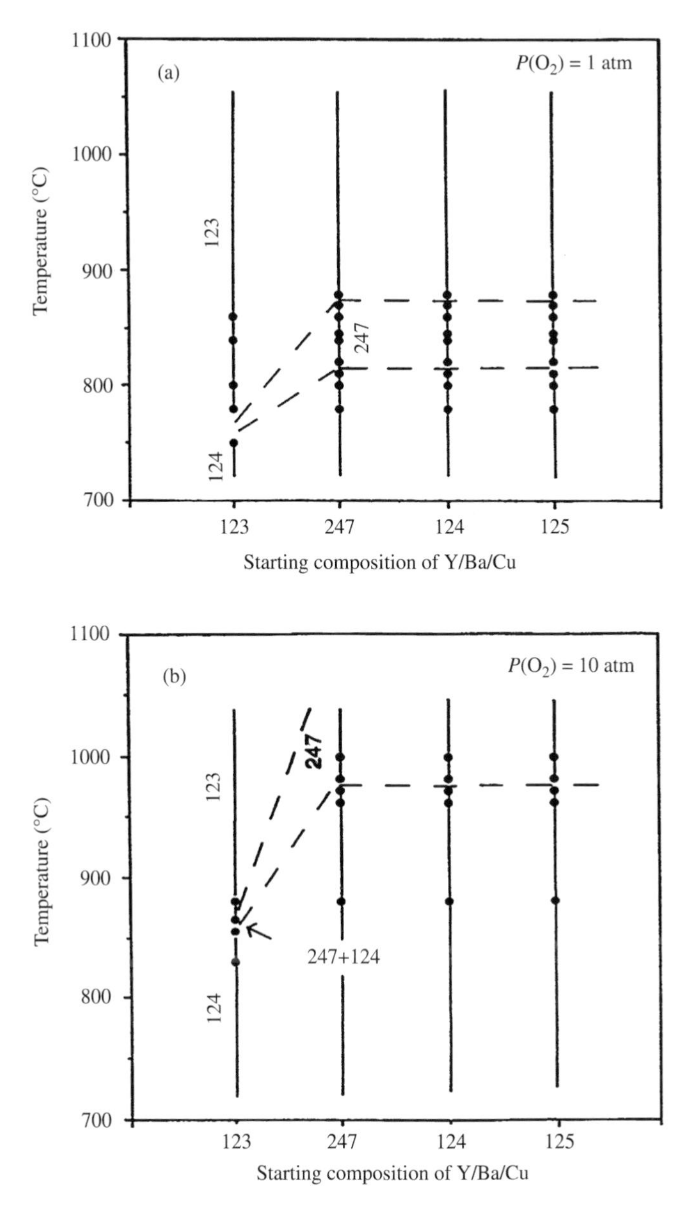

Figure C1.17. Temperature ranges for stability of Y-123, Y-124 and, Y-247 phases at: (*a*) $P(O_2) = 10\,\mathrm{atm}$ and (*b*) $P(O_2) = 1\,\mathrm{atm}$ [147].

C1.5.3 Phase diagram for Y-123 materials versus oxygen: chemical compatibility

The influence of oxygen stoichiometry on the properties of YBCO HTSC, especially with respect to the tetragonal (T) to orthorhombic (O) phase transformation, has been the subject of much research. The relationship between phase stability of Y-123 type superconductors and oxygen content for various values of $P(O_2)$ in the form of a phase diagram has been reported [15, 145, 154–158]. For example, the temperature–oxygen content diagram for Y-123 is shown in figure C1.19 [158]. Long-dashed isobars in

B2.3.2

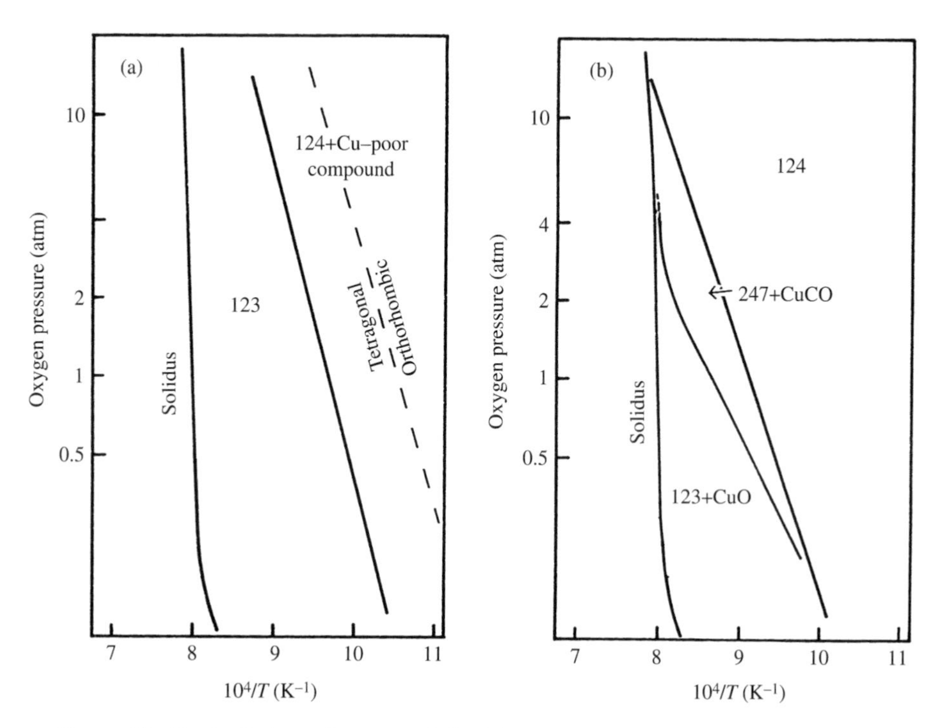

Figure C1.18. P–T phase diagrams of the (*a*) Y-123 and (*b*) Y-124 composition [147].

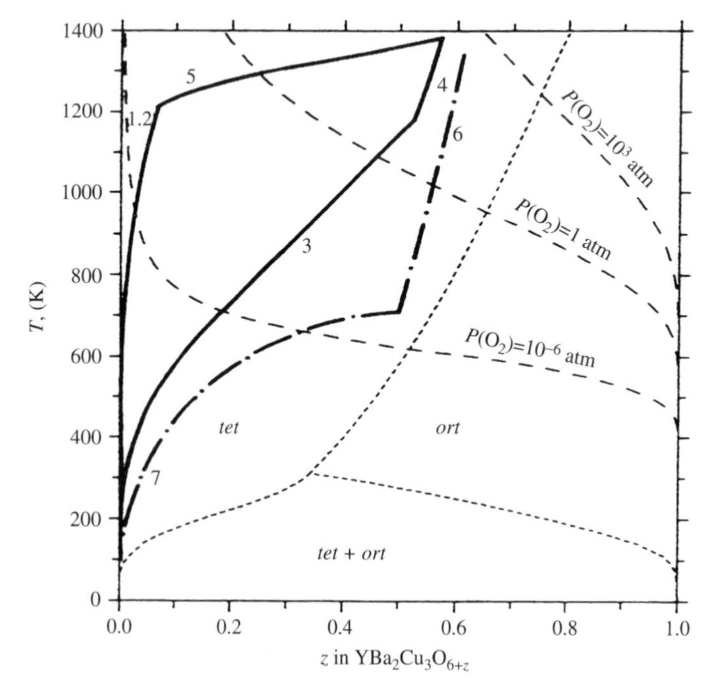

Figure C1.19. Stability field of the Y-123 phase. This phase is thermodynamically stable between the bold lines 1–5. The lines 1–7 represent decomposition reactions to equations (C1.15)–(C1.21) [158].

this figure represent the temperature dependence of equilibrium of the solid solution, and short dashed lines represent metastable equilibrium of the orthorhombic and tetragonal phases as well as a metastable miscibility gap of solid solution. Bold lines show the equilibrium conditions for several decomposition reactions of 123, which outline its thermodynamic stability range. These reactions are [158]:

B2.3.2

$$9YBa_2Cu_3O_{6+z} = 4Y_2BaCuO_5 + YBa_4Cu_3O_{8.5+q} + 10BaCu_2O_2 + [(9z + 5.5 - q)/2]O_2 \quad (C1.15)$$

$$2YBa_2Cu_3O_{6+z} = Y_2BaCuO_5 + BaCuO_2 + 2BaCu_2O_{2+[(2z+1)/2]}O_2 \quad (C1.16)$$

$$6YBa_2Cu_3O_{6+z} = Y_2BaCuO_5 + 3BaCuO_2 + 2Y_2Ba_4Cu_7O_{14+w} + [(6z - 2w - 3)/2]O_2 \quad (C1.17)$$

$$6YBa_2Cu_3O_{6+z} = 2Y_2BaCuO_5 + 3Ba_2Cu_3O_{5+y} + Y_2Ba_4Cu_7O_{14+w} + [(6z - 3 - 3y - w)/2]O_2 \quad (C1.18)$$

$$YBa_2Cu_3O_{6+z} \rightarrow Y_2BaCuO_5 + \text{Liquid} + O_2 \quad (C1.19)$$

$$4YBa_2Cu_3O_{6+z} = 2Y_2BaCuO_5 + 3Ba_2Cu_3O_{5+y} + CuO + [(4z - 2 - 3y)/2]O_2 \quad (C1.20)$$

$$2YBa_2Cu_3O_{6+z} = Y_2BaCuO_5 + 3BaCuO_2 + 2CuO + [(2z - 1)/2]O_2. \quad (C1.21)$$

C1.6 Optical properties

As is commonly observed in all high temperature superconductors, a reflectivity spectrum $R(\omega)$ and, thus, a conductivity spectrum $\sigma_1(\omega)$ changes dramatically with hole doping for all polarization directions. Figure C1.20 shows the doping dependence (namely, oxygen content dependence) of reflectivity spectrum of twinned and untwinned $YBa_2Cu_3O_{6+x}$ for $E//a$ and $E//b$ [159]. For the low oxygen content ($x = 0.1$), reflectivity is very low, and the spectrum is flat below 1 eV except for the phonon absorption peaks below 0.1 eV, which are typical in an insulator. High ω spectrum is characterized by a charge transfer gap excitation at around 1.7 eV and some interband excitations at higher energies. With increasing oxygen content, reflectivity increases below 1 eV for $E//a$ and below 2 eV for $E//b$, forming a Drude like metallic spectrum, in compensation for a reduction in a charge transfer absorption. Similar growth of low-ω spectrum is also seen for $E//c$ (see figure C1.21) [160]. As the oxygen content increases, the insulator-like spectrum at $x \sim 0.1$, dominated by the far-infrared phonon peaks, changes into a Drude-like spectrum at $x \sim 0.9$. All these changes indicate that the material varies from an insulator to a metal.

When temperature is lowered, far-infrared reflectivity and/or conductivity increase in highly oxygenated YBCO with the metallic resistivity $\rho(T)$. In the superconducting state, reflectivity is enhanced and conductivity is suppressed below 2Δ (Δ, the maximum gap amplitude). The T-dependencies of the far-infrared conductivity spectra are shown in figure C1.22 for $E//a$ [161] and in figure C1.23 for $E//c$ [160].

D3.1

In high doping at $x \sim 0.9$, conductivity suppression just sets in at T_c both for $E//a$ and $E//c$, which indicates that the observed suppression is related to the opening of a superconducting gap. Although there is still an argument that the conductivity suppression for $E//a$ is a result of Drude narrowing of quasiparticles below T_c and therefore, the absorption edge around 500 cm^{-1} is not related to a gap, one can roughly estimate a maximum gap energy to be 600 cm^{-1} ($2\Delta \sim 74$ meV) from the c-axis spectrum. A long tail below 2Δ suggests an anisotropic gap like a d-wave one.

A2.4

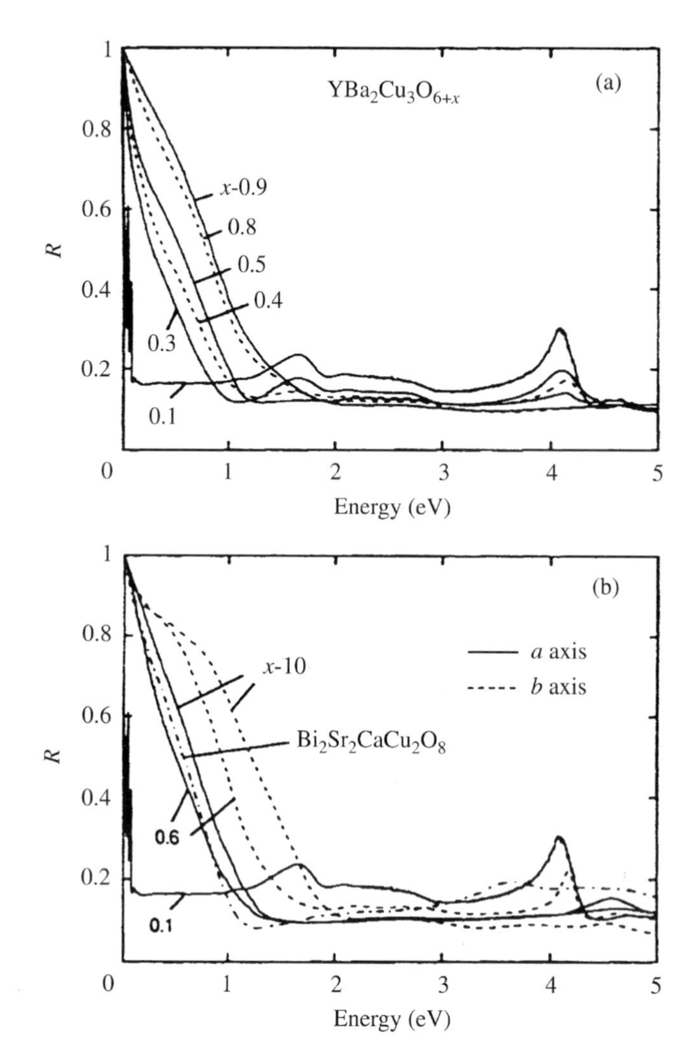

Figure C1.20. Room temperature reflectivity spectra of twinned (*a*) and untwinned $YBa_2Cu_3O_{6+x}$ crystals (*b*) with $E//a$ and $E//b$ for various x [159].

For under-doped crystals with $T_c \sim 50-70$ K, the suppression of conductivity begins well above T_c. The critical temperature for this phenomenon to take place is different between $E//a$ and $E//c$. Concerning the in-plane spectrum with $E//a$, the suppression of σ_1 is considered as a psuedogap or a spin-gap related feature. The origin of this normal-state gap attracts worldwide attention. For $E//c$, conductivity suppression is closely correlated with a dc resistivity behaviour, which exhibits an upturn at a certain temperature much higher than the in-plane pseudo-gap temperature. Such upturn of $\rho_\chi(T)$ represents the anomalous c-axis conduction mechanism in high temperature superconductors.

A2.4
D3.1

C1.7 Normal state properties

D3.1

When the oxygen content y is close to six in $YBa_2Cu_3O_y$, the material is an antiferromagnetic insulator due to a strong electron–electron correlation within the CuO_2 plane. As holes are doped into the CuO_2

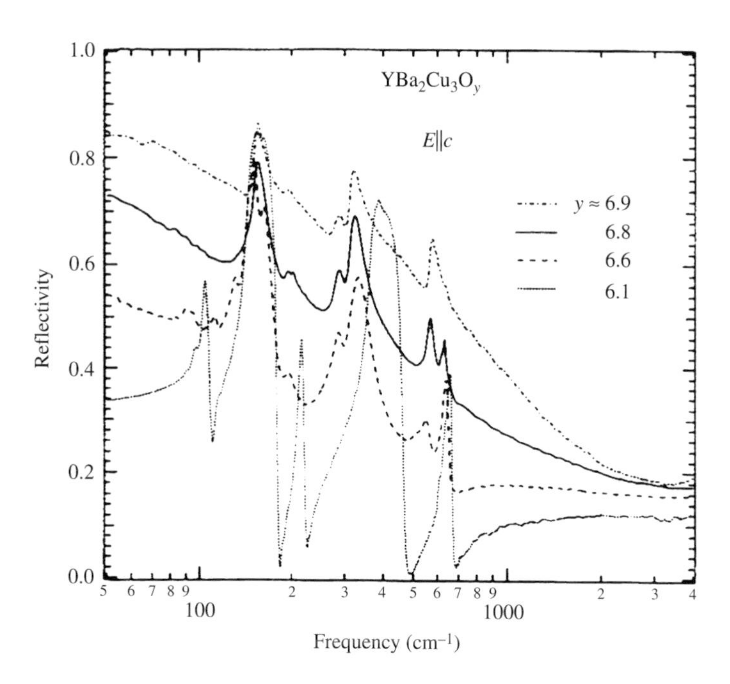

Figure C1.21. Room temperature reflectivity spectra of $YBa_2Cu_3O_y$ with $E//c$ for various oxygen contents [160].

planes with increasing oxygen content, the material changes from an insulator to a metal. However, this electronic change caused by the hole doping cannot be described within a rigid band picture for a doped Mott-insulator, but suggests more radical change in the electronic structure. All anomalies in the normal state properties originate from this unclear electronic state in the crossover region, which is predominantly characterized by the so-called 'pseudo-gap' existing well above T_c. A3.2

With increasing oxygen content, resistivity decreases in all directions as shown in figures C1.24 and C1.25 [162, 163]. For optimal doping around $y \sim 6.88$, the in-plane resistivity $\rho_\alpha(T)$ is almost linearly temperature dependent, and $\rho_\beta(T)$ contains a small amount of a T^2-term owing to a chain conductivity, while the c-axis resistivity $\rho_\chi(T)$ has a large residual component and shows a slight upturn just above T_c. The effect of pseudo-gap manifests itself in the downward bending of $\rho_\alpha(T)$ (see figure C1.25), which is due to a reduction in carrier scattering caused by a spin fluctuation [162]. The onset temperature of this phenomenon increases with decreasing oxygen content. In a low doping level, $\rho_\chi(T)$ shows a more pronounced upturn like a semiconductor, whereas $\rho_\alpha(T)$ is still metallic. This anisotropy in the temperature dependence of resistivity is regarded as one of the most anomalous properties of high temperature superconductors, which cannot be understood in the framework of a conventional metal. The strange anisotropy can be seen in the ratio ρ_c/ρ_a as well. First, the ratio is much larger than the effective mass ratio estimated from the band calculation even in the over-doped crystals. Second, it changes dramatically with temperature and hole doping. This implies that the effective mass model does not hold in this compound. D3.1 D1.1.2

In a conventional metal, the Hall coefficient is almost constant for all temperatures. But, as is shown in figure C1.26, the Hall coefficient of YBCO increases with decreasing temperature, forming a bump at an intermediate temperature [164]. On the other hand, the Seebeck coefficient, which should change linearly with temperature in a conventional metal, is only weakly temperature dependent (see D3.1

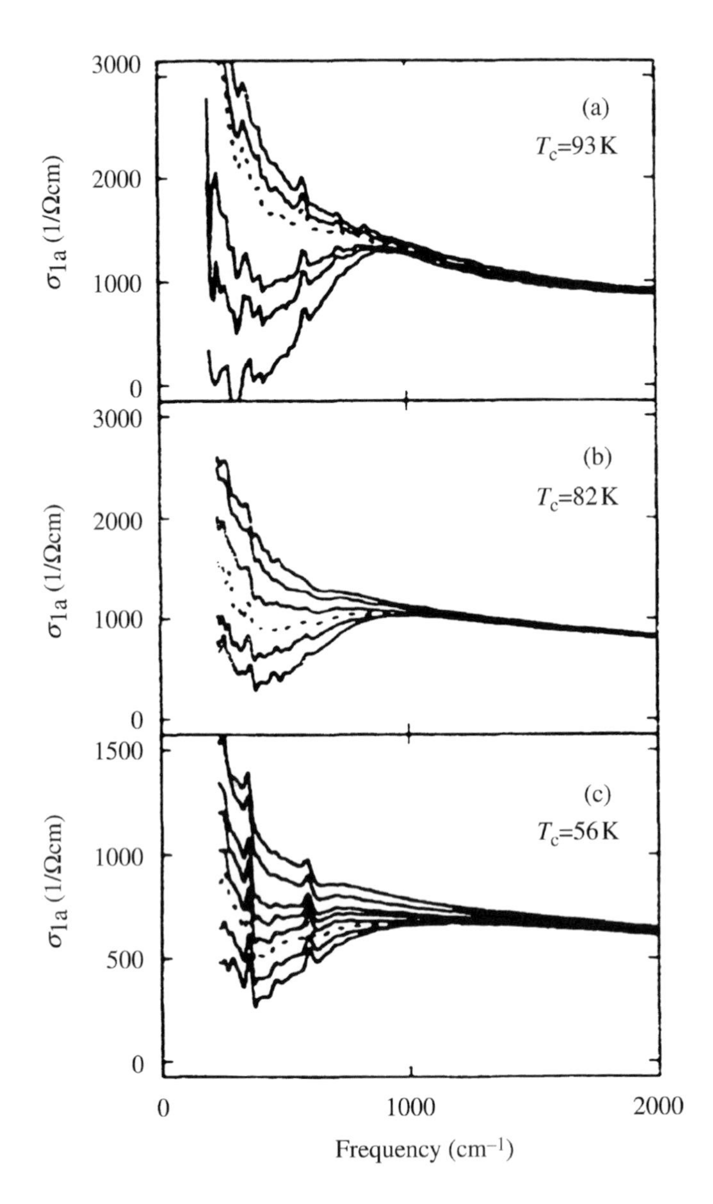

Figure C1.22. Temperature dependence of conductivity spectrum of $YBa_2Cu_3O_y$ with $E//a$. (*a*) Optimally doped crystal with $T_c = 93$ K for $T = 120$, 100, 90 (dashed), 80, 70 and 30 K (from top to bottom). (*b*) Underdoped crystal with $T_c = 82$ K for $T = 150$, 120, 90, 80 (dashed), 70 and 20 K (from top to bottom). (*c*) Underdoped crystal with $T_c = 56$ K for $T = 200$, 150, 120, 100, 80, 60 (dashed), 50 and 20 K (from top to bottom) [161].

figure C1.27) [163]. Both quantities strongly depend on oxygen content. Anomalous T-dependencies are typically observed in the under-doped crystals with low oxygen contents. These non-monotonous behaviours are attributed to the opening of a pseudo-gap or a spin gap at a certain temperature T^*, which increases with reducing oxygen content. Note that, at the optimal doping, T^* is nearly equal to T_c.

A3.2

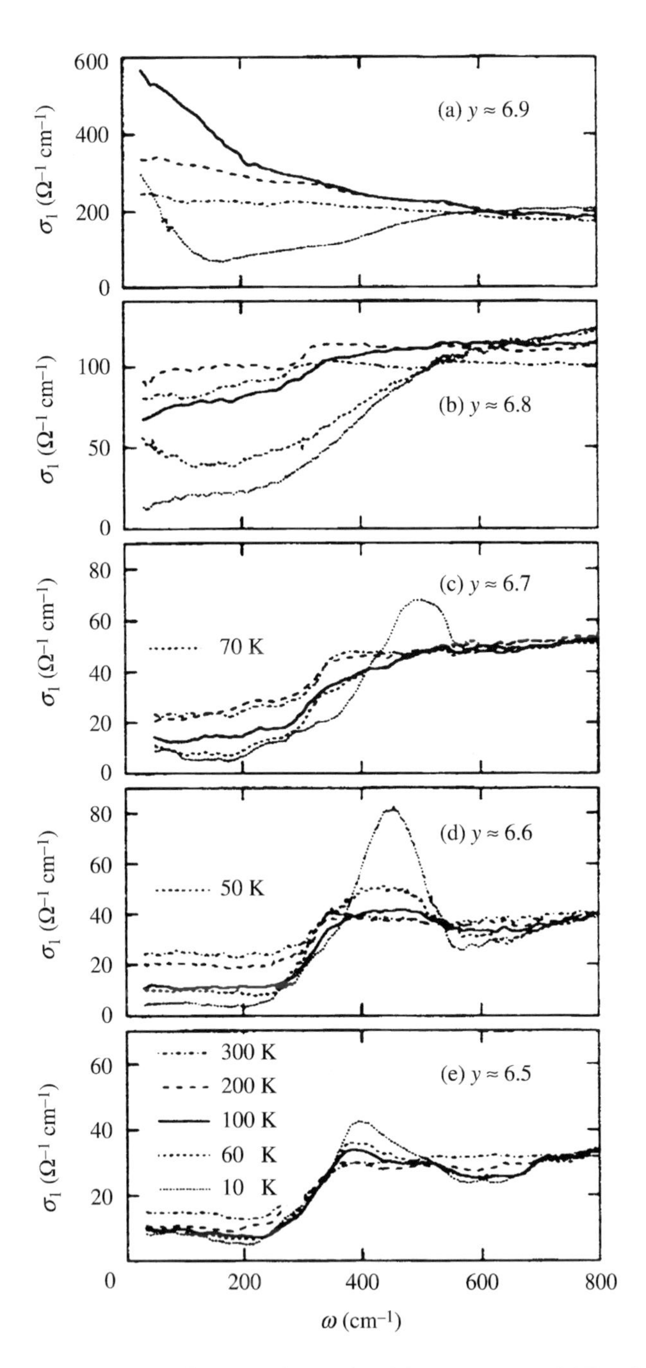

Figure C1.23. Temperature dependence of the c-axis conductivity spectrum of $YBa_2Cu_3O_y$ for various oxygen contents. Here, all phonon contributions are subtracted using the Lorentz fitting [160].

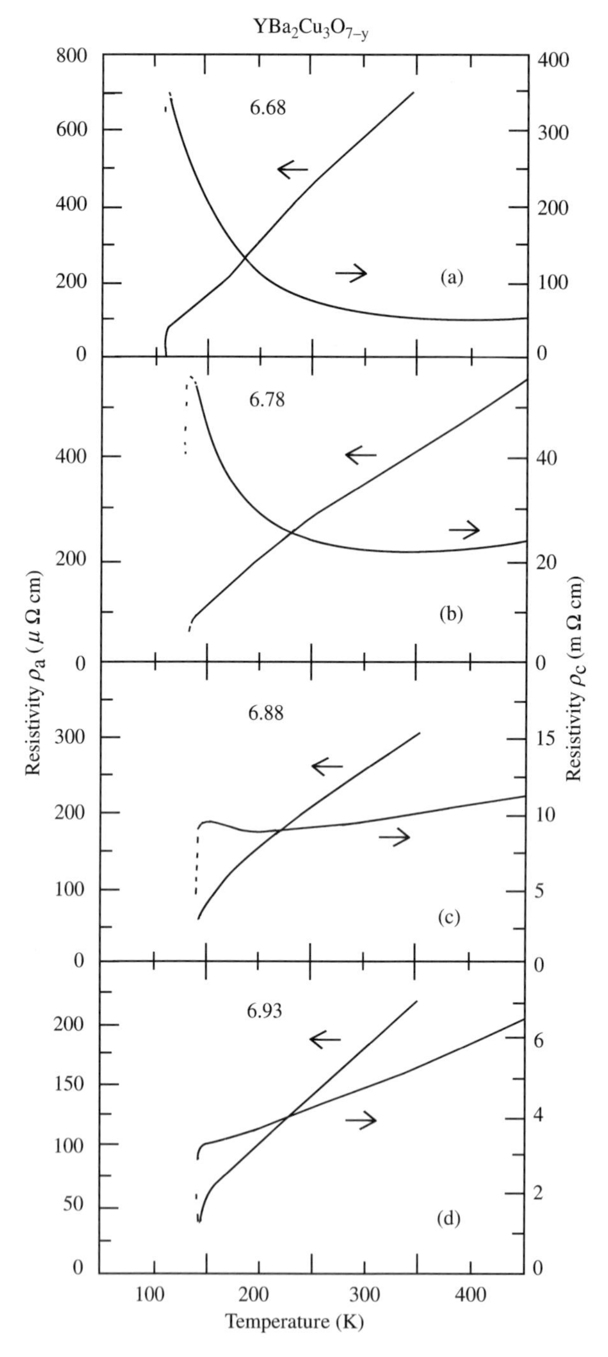

Figure C1.24. Temperature dependence of in-plane (ρ_α) and out-of-plane (ρ_β) resistivity of untwinned $YBa_2Cu_3O_{7-y}$ for various oxygen contents [162].

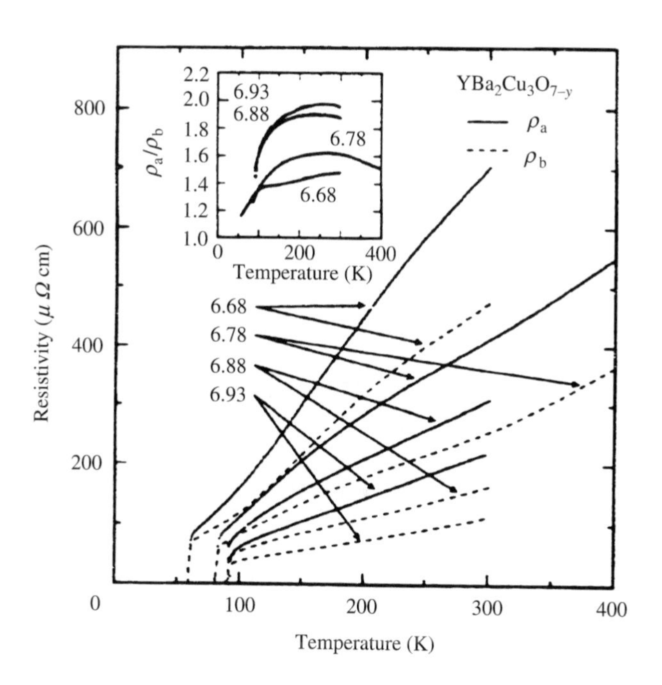

Figure C1.25. Temperature dependence of two in-plane resistivity components in untwinned $YBa_2Cu_3O_{7-y}$ for various oxygen contents [162].

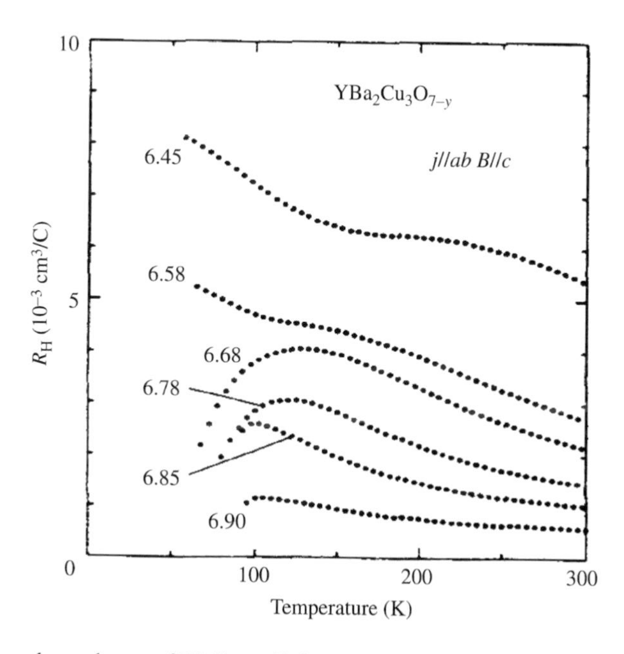

Figure C1.26. Temperature dependence of Hall coefficient of two in-plane resistivity components in untwinned $YBa_2Cu_3O_{7-y}$ single crystals for various oxygen contents [164].

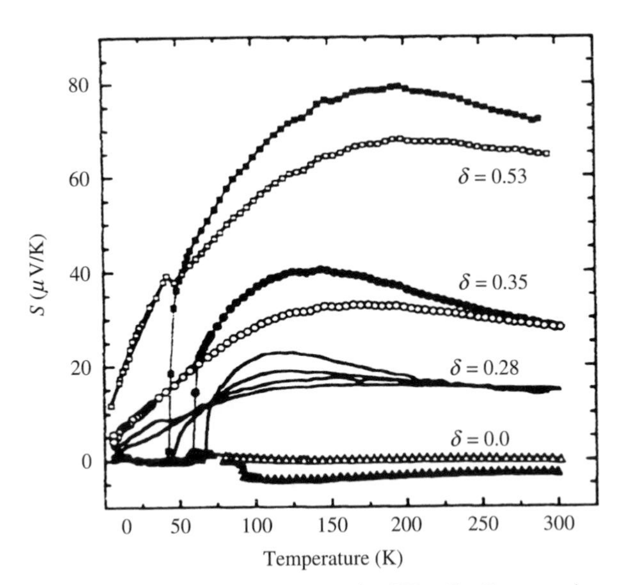

Figure C1.27. Temperature dependence of thermopower in $YBa_2Cu_3O_{7-\delta}$ polycrystals for various oxygen contents. Solid symbols denote non-substituted and open symbols denote 3.0% Zn-substituted YBCO. The family of solid curves for $\delta = 0.28$ in the downward order denote 0, 1.0, 3.0 and 5.0% Zn-substitution [163].

C1.8 Superconducting properties

C1.8.1 Transition temperature (T_c), coherence length (ξ), penetration depth (λ) and Ginzburg–Landau coefficient (κ)

D1.1.2 The strong anisotropy in YBCO structures has an important effect on the electrical transport properties and the equilibrium flux line lattice properties, and hence the flux pinning behaviour. Since the transition temperature strongly depends on carrier concentration, proper doping must be provided to tailor the charge distribution (electrons and holes) between the conducting planes and along the direction out of the planes. For different practical applications of YBCO HTSC it is necessary that they have a complex of critical characteristics: high critical transition temperature (T_c), high critical current density (J_c) and D2.7 low microwave surface resistance (R_s). The penetration depth (λ) and coherence length (ξ) are important parameters in a superconductor. For example, the penetration depth, which measures the length over which magnetic fields are attenuated near the surface of a superconductor, is related to the effective mass A2.3 and density of superconducting pairs. The coherence length measures the size of Cooper pairs, or, equivalently, the size of the vortex flux quantum core. They are referred to in the literature as the characteristic lengths of superconductors [165]. The temperature dependence of coherence length $\xi(T)$ in the critical region is described by a simple power law [166]:

$$\xi(T) = \xi(0)[1 - (T/T_c)]^{-1/2} \tag{C1.22}$$

A3.1 The conventional Ginzburg–Landau (GL) theory is very successful in describing the temperature dependence of several thermodynamic properties of superconductors, such as the critical fields (H_c) and other characteristic material parameters [167]. For isotropic material, the GL parameter is defined as:

$$\kappa = \lambda/\xi \tag{C1.23}$$

and readers are referred to some newly developed theories [168–170]. For anisotropic superconductors, $\xi = (\xi_a \xi_b \xi_c)^{1/3}$ and $\lambda = (\lambda_a \lambda_b \lambda_c)^{1/3}$ [165]. Since $\xi_a \cong \xi_b$, $\lambda_a \cong \lambda_b$ for YBCO, the following definitions are introduced: D1.1.2

$$\kappa_{\parallel} = \gamma \kappa_{\perp} \tag{C1.24}$$

$$\xi_{\parallel} = \gamma \xi_{\perp} \tag{C1.25}$$

$$\lambda_{\parallel} = \lambda_{\perp}/\gamma \tag{C1.26}$$

where γ is the anisotropy factor. YBCO is an anisotropic material, and both the ξ and λ are anisotropic.

Table C1.10 summarizes data for the most important parameters of YBCO HTSC: transition temperature (T_c), coherence length (ξ), penetration depth (λ), GL coefficient (κ) and lower (H_{c1}) and upper (H_{c2}) critical field. Note that, typically, a–b anisotropy in orthorhombic crystal is not detectable in measurements of ξ_i and λ_i, but flux lattice decoration experiments have been able to resolve anisotropy. A2.3, A3.1 A1.1, A2.6

C1.8.2 Pinning energies

Critical currents in type II superconductors are due to vortex pinning induced by small regions in the material where the superconductivity is weakened [176]. If a vortex core passes through one of these A4.2, A4.3

Table C1.10. Transition temperature (T_c), coherence length (ξ), penetration depth (λ), GL coefficient (κ), and lower (H_{c1}) and upper (H_{c2}) critical field for YBCO

Parameter	Y-123 (Ref.)
T_c (K)	92 [1]
Coherence length (nm)	
$\xi_0^{\parallel}$	0.6 [171]
	0.4 [172]
$\xi_0^{\perp}$	2.7 [171]
	3.1 [172]
Penetration depth (nm)	
$\lambda_L^{\parallel}$	26 [173]
	90 [172]
	130 [174]
$\lambda_L^{\perp}$	125 [173]
	800 [172]
	450 [174]
GL coefficient	
$\kappa^{\parallel}$	7.6
$\kappa^{\perp}$	37
Lower critical field, H_{c1} (T)	
$H_{c1}^{\parallel}$	53 [175]
$H_{c1}^{\perp}$	520 [175]
Upper critical field, H_{c2} (T)	
$H_{c2}^{\parallel}$	140 [173]
$H_{c2}^{\perp}$	650 [173]

defects, the system gains an amount of pinning energy equal to some fraction of the condensation energy in the volume ν of the vortex core that is pinned;

$$U_0 = (H_c^2/8\pi)\nu \qquad \text{(C1.27)}$$

where H_c is the thermodynamic critical field. In the case of Y-123, the pinning energy at zero temperature due to an optimum pinning site volume is $U_0 \cong 9\,\text{meV}$ (in units of temperature, $U_0/k_B \cong 110\,\text{K}$) [177, 178]. In order to determine the flux pinning energies, different methods were used, such as magnetic relaxation measurements [179], dc-resistance measurements [180], ac-resistance measurements [181], noise measurements [182] and 90° rotating sample magnetic measurements [183]. Zeldov *et al* [184] have studied the resistive transition in Y-123 and reported pinning energies as high as 6 eV in a magnetic field of 0.5 T. On the other hand, Hagen and Griessen [185] deduced a much lower value of the activation energy in YBCO from flux creep measurements. They found a distribution of activation energies with a peak near 0.06 eV. Ferrari *et al* [182] used the temperature and frequency dependence of the noise to determine the pinning energies of individual flux vortices in thermal equilibrium. They found a distribution of pinning energies with two peaks: a low energy peak, below 0.1 eV and a higher energy peak near 0.35 eV. The higher energy peak has been associated with grain boundaries or *a*-axis grains, while the lower energy peak may represent an intrinsic, intragranular pinning energy. Generally, an average pinning energy was obtained, which depends on temperature and field in a complicated way [186]. Since there are various defects in a material, a single value of pinning energy cannot describe the interaction between flux lines and the defects completely. There should be a distribution of pinning energies in the YBCO material [185, 187].

D3.1 A4.3 A4.3 D1.7.2

C1.8.3 *Hysteresis curves, J_c versus field and temperature*

Plots of measured magnetic moment *versus* magnetic field show 'hysteresis' [188–191]. The Anderson 'flux pinning' model [192] is usually used to explain hysteresis; in this model, bundles of magnetic flux are caught or pinned on crystal disorder. When the bundles are depinned, flux creeps from the sample, and the sample loses magnetization. Typical magnetic hysteresis loops for YBCO at 77 K are shown in figure C1.28 [193].

A4.3 A1.3

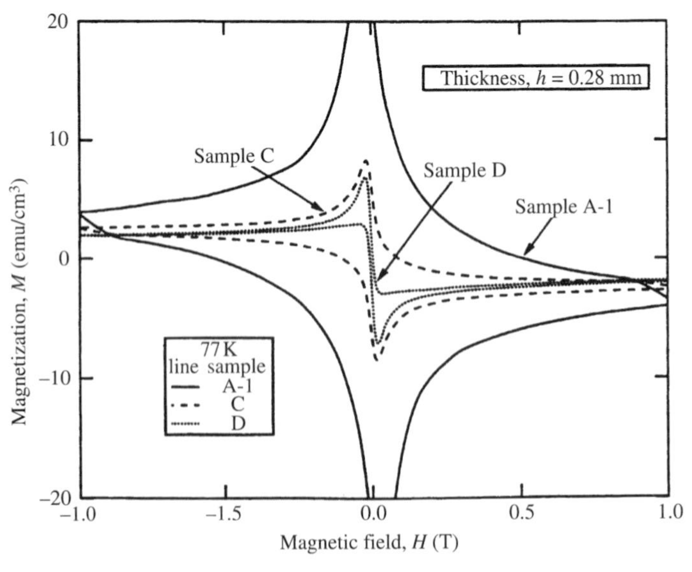

Figure C1.28. Magnetic hysteresis loops of YBCO samples measured at 77 K [193].

The low-field magnetization behaviour of the sintered material can be understood by the presence of a weakly coupled region between grains.

The measurement of magnetic hysteresis plays a vital role in providing the information regarding the temperature and field dependence of critical current density, $J_c(H,T)$ in HTSC. The relationship between critical current density and isothermal magnetic hysteresis has been established by Bean [194, 195]. The variation of critical current density of YBCO with magnetic field and temperature has been investigated [196]. The results are analyzed on the basis of Bean's critical state model [194] and its various extensions. According to the critical state model, the current density $J_c(H)$ is proportional to ΔM. Figures C1.29 (*a*) and (*b*) shows the field and temperature dependence of ΔM for Y-124 [196]. From this figure, it is evident that $J_c(T)$ follows a power law in the field region between field of full penetration and field where the remnant magnetization saturates. $J_c(T)$ varies exponentially in a limited temperature and field region. Three methods are applied to determine the critical current density of YBCO: the usual transport method, the calculation from magnetization measurements, and the derivation from torque

B2.3.3
D2.4

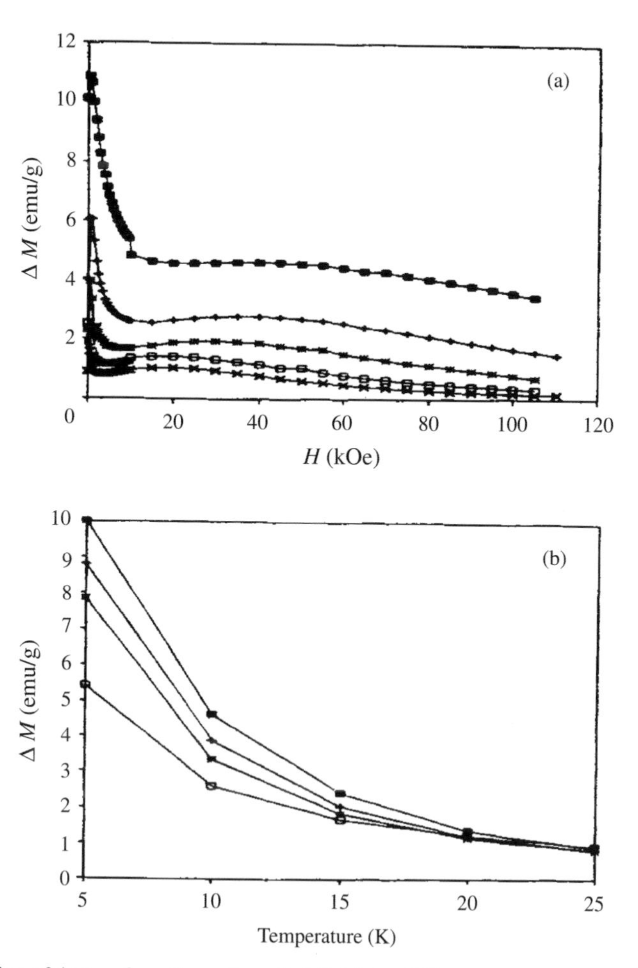

Figure C1.29. (*a*) The plot of ΔM as function of applied field for different temperatures (filled box at 5 K, plus sign at 10 K, star at 15 K, open box at 20 K and cross at 25 K) obtained from hysteresis curves; (*b*) variation of ΔM *versus* temperature for different values of applied fields (filled box at 2 kOe, plus sign at 3 kOe, star at 4 kOe and open box at 10 kOe) for Y-124 specimen [196].

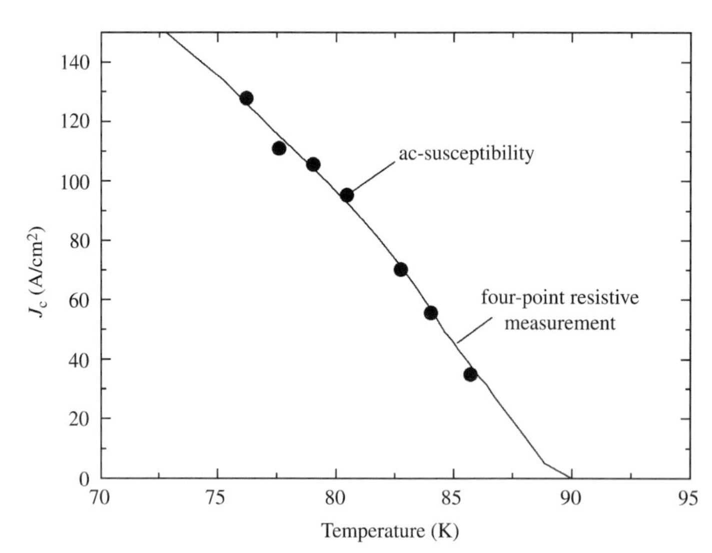

Figure C1.30. The comparison between J_c-data obtained from ac-susceptibility measurements (markers) and data from standard four-point measurements (solid line). Both measurements were done using sample geometry ($2a = 0.45$ mm, $l = 8$ mm) [197].

D2.4 measurements. Comparison between J_c-data obtained from ac-susceptibility measurements and data from standard four-point measurements of Y-123 is shown in figure C1.30 [197].

C1.8.4 Irreversibility data

A1.3 The measurement of ac complex susceptibility remains one of the most powerful methods to obtain important information on dissipation mechanisms in polycrystalline YBCO HTSC. It is commonly B2.2.2 accepted that bulk sintered superconductors generally consist of two components: anisotropic grains with a quite large lower critical field $H_{c1,g}$ and their coupling matrix having negligibly low $H_{c1,m}$ if existing [197]. Two peaks in the temperature dependence of the imaginary part of the complex ac-D2.5, D2.6 susceptibility reflecting the intra- and intergranular losses can be distinguished at high ac-fields. Generally speaking, a maximum in $\chi''(T)$ appears when the supercurrents penetrate just to the center of their circumferential paths, which may be the center of the superconducting grain, or the coupling matrix.

A1.3 An important feature of the HTSC is the existence of the so-called 'irreversibility line' (IL) or 'quasi de Almeda–Thouless line', where magnetic irreversibility sets in [198–200]. Brandt [201] has listed eight D2.5 experiments for monitoring the 'irreversibility line'. ac-susceptibility is pointed out as one of the most sensitive methods among them. By this method, the IL is defined either as the relation between the temperature, at which the maximum in the imaginary component appears, and the dc-field H_{DC} ($H_{DC} >> H_{AC}$), at which the maximum is measured, or as a relation between T_m and the ac-field amplitude ($H_{DC} = 0$) [197, 202–205]. The difference between $T_m(H_{AC})$ for $H_{DC} = 0$ and $T_m(H_{DC})$ for $H_{DC} >> H_{AC}$, i.e., between the two cases mentioned above, has been studied by Muller [206]. For both the inter- and intragranular χ'' peak temperatures a linear dependence of T_m and H_{AC} has been found in agreement with the experimental data reported in [197]. Figure C1.31(a) and (b) shows T_m and T_g as a function of the ac-field, H_m, at which the χ'' peak temperature from the matrix and the grains, respectively, is observed for YBCO samples with different thickness [197]. The onset of irreversible behaviour at the 'irreversibility line' T_{irr}/H is described, in most experiments, by a power law [207, 208]:

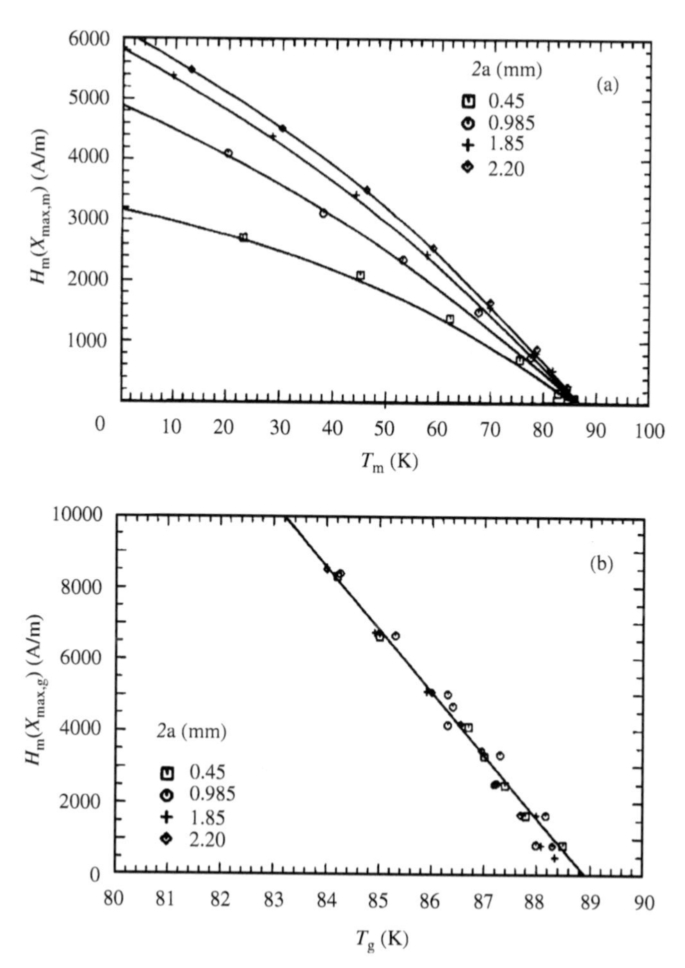

Figure C1.31. (*a*) Intergranular χ'' peak temperature T_m *versus* intergranular ac-field amplitude $H_m(\chi_{max,m})$ for samples with different thicknesses; (*b*) intragranular χ'' peak temperature T_g *versus* intragranular ac-field amplitude $H_m(\chi_{max,m})$ for samples with different thicknesses [197].

$$1 - T/T_c = AH^n, \tag{C1.28}$$

where A is a frequency-dependent constant and the power n (in dc measurements) is approximately 2/3. Some experimental investigations have shown that, in many cases, the exponent n is different from the theoretical values of 2/3 or 3/4 and a change in the value of n is observed in different regions of applied magnetic field strength [209, 210].

Investigations of Y-124 have shown that the IL line obeys a power-law behaviour similar to that of Y-123 (equation C1 23), with $n \sim 1.51$ [205].

C1.8.5 Grain boundaries in YBCO

With the discovery of YBCO, it seemed that the vision of superconducting power cables operating at liquid nitrogen temperature was close to realization. The critical current density, J_c, however, is in E1.3.2

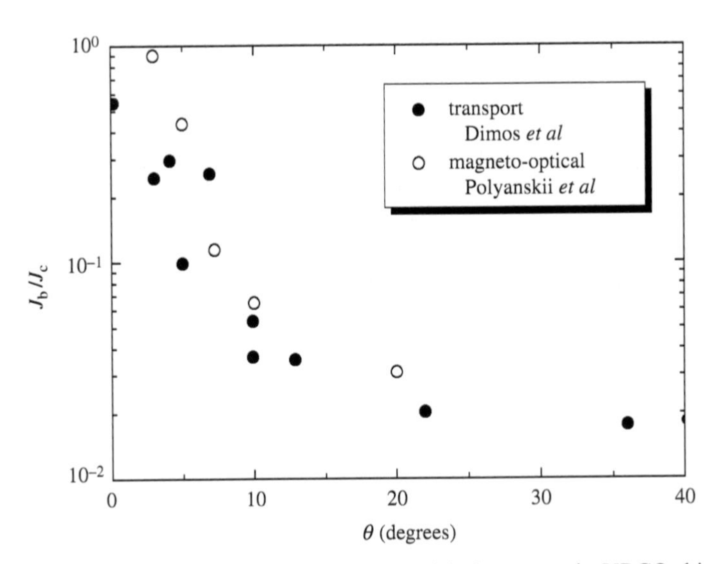

Figure C1.32. Plot of the ratio intergrain J_b and intragrain J_c critical currents in YBCO thin film bicrystals *versus* the misorientation tilt angle θ. The measurements [215, 218, 229] were taken at $T = 6\,K$.

general suppressed at grain boundaries by phenomena such as interface charging and bending of the electronic band structure [211, 212] as is also known from other perovskite ceramics [213, 214]. For YBCO, the ever-present high-angle grain boundaries are the most severe problem for applications. In experiments on bicrystals and bicrystalline thin films [215–219], the critical current density across a grain boundary has been shown to be strongly dependent on the misorientation angle as indicated in figure C1.32 . A rapid decrease in the critical current density of the boundary (J_b), normalized in these cases to the intragranular current density (J_c) of the adjacent grains, with increasing angle of crystallographic misorientation is evident.

D1.3 In figure C1.33 , a SEM photograph of a typical polycrystalline YBCO sample is shown; figure C1.34 presents an electron backscatter diffraction (EBSD) mapping of a similar YBCO sample [220],

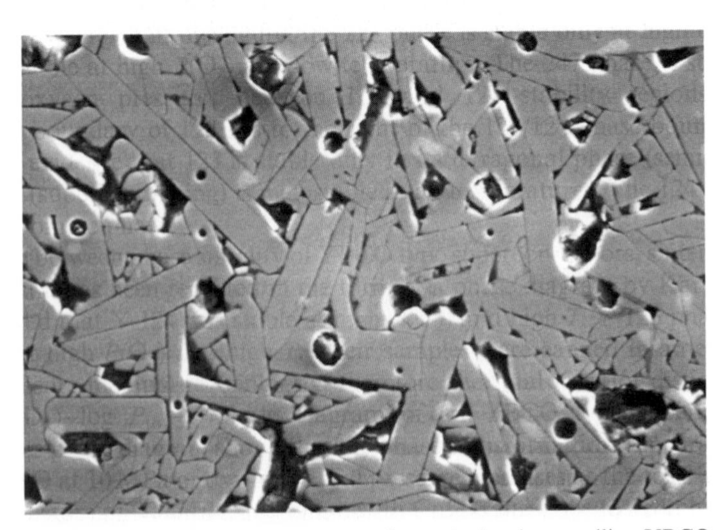

Figure C1.33. SEM image of the grain structure of a typical polycrystalline YBCO sample [221].

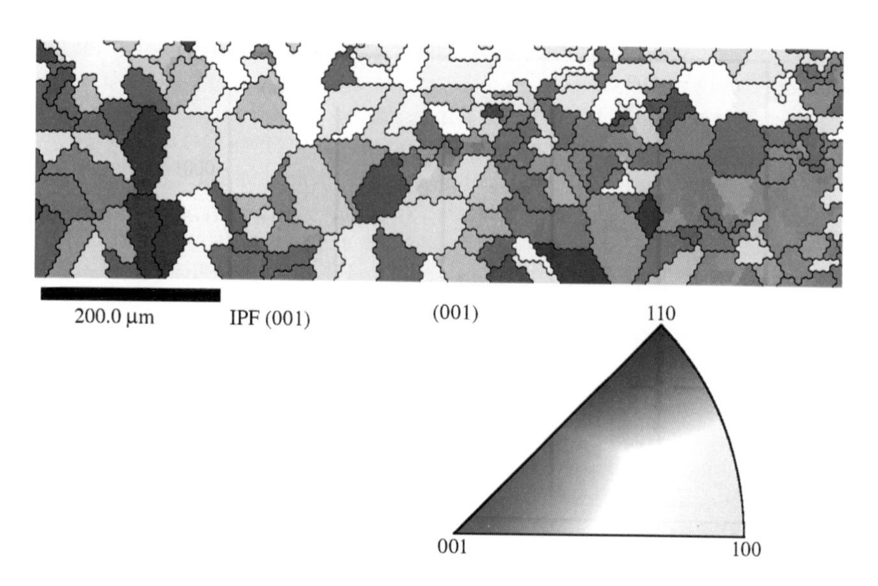

Figure C1.34. EBSD mapping of a polycrystalline YBCO sample with Rb-doping [220], presenting the crystal orientations normal [0 0 1] to the sample surface. The individual crystallographic directions are indicated in the colour-coded triangle.

indicating the crystallographic directions of the individual grains. As a result, the flux penetration into such a material occurs firstly *via* the grain boundary regions and, as soon as the lower critical field of the grains is reached, the flux enters in the grains (figure C1.35). This granular flux penetration behaviour

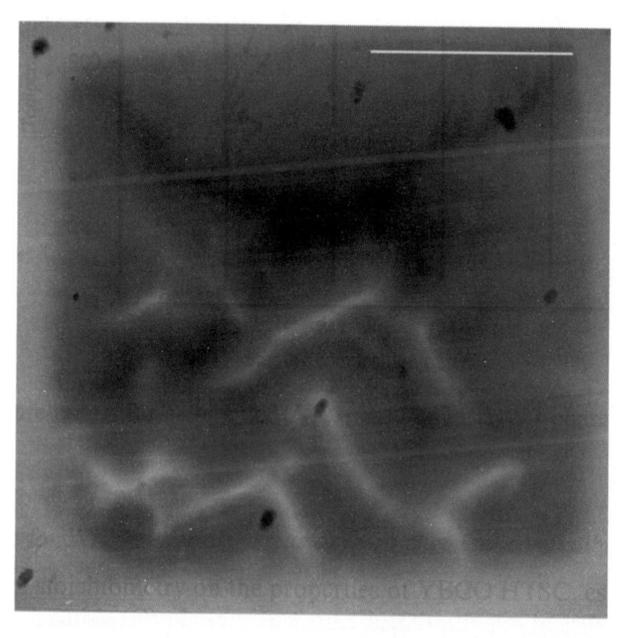

Figure C1.35. MO flux patterns of a $KClO_3$-doped YBCO sample at $T = 50$ K and an applied field of 100 mT. The magnetic field is imaged as bright areas; the Meissner phase remains dark. Only in the lower part of the image, flux entering along grain boundaries can be observed. The marker is 500 μm long [221].

D3.4 was discussed in detail by Koblischka *et al* [221, 222], based on magneto-optical imaging (see section D3.4). The high-angle grain boundaries basically act as weak links or Josephson junctions; thus a transport current across these grain boundaries ('intergranular' current density) is severly reduced as compared to the current density inside the grains ('intragranular' current density). By means of a local measurement technique, these two current densities can be measured directly [221]. The effect of the D2.4 granularity is also seen in the magnetization loops (figure C1.28) or in ac-susceptibility measurements. In transport current measurements, it was observed that the current density is rapidly decreasing when applying small magnetic fields [223]. In some experiments, however, the grain boundaries seem not to exhibit these detrimental effects [224, 225], e.g. high-angle grain boundaries of YBCO with a misorientation relationship near 90° [0 1 0] that are not parallel to the (0 0 1) plane of one crystal do not show weak-link behaviour.

Recent progress in the understanding of the behaviour of the grain boundaries was achieved by Mannhart *et al* [226–229]. The charging of the grain boundaries was shown to be removed by means of overdoping YBCO by adding Ca. The results of these measurements are illustrated in figure C1.36 [228]. This doping of the YBCO material may be beneficial for the fabrication of YBCO coated conductors or B4.1 even wires.

C1.8.6 Microwave surface resistance (R_s)

D2.7 Since the discovery of superconductivity in the YBCO system, there have been numerous efforts to measure the microwave surface resistance (R_s) of these compounds in order to identify the conduction mechanism and the nature of the superconducting state [230–234]. It has been estimated that the grain boundary in the polycrystalline sample significantly affected microwave surface resistance in a magnetic field [230, 235–240]. In most of the studies, surface resistance has been determined by measuring

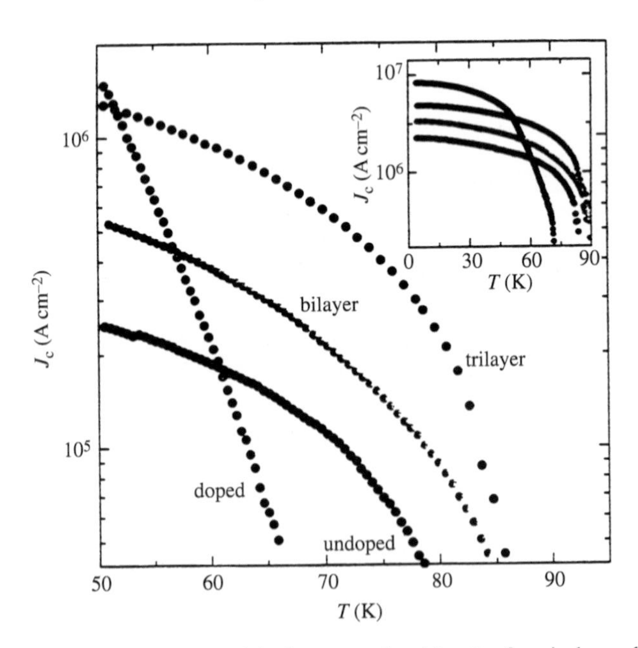

Figure C1.36. Temperature dependence of the critical current densities J_c of grain boundaries with various doping configurations, indicating the strong enhancement of J_c in doping heterostructures. Displayed are the $J_c(T)$ dependencies of symmetric 24° [0 0 1] tilt grain boundaries in various bicrystalline samples: a YBCO film (undoped), a YBCO film (Ca-doped), a bilayer (YBCO doped/undoped) and a trilayer (doped/undoped/doped) [228].

Table C1.11. Microwave surface resistance (R_s) data for YBCO

Composition	R_s(mΩ)	f (GHz)	T (K)	Ref.
Y-123	51	12	77	[240]
Y-123 Na0.40	50	12	77	[240]
Y-Rb	42	14.3	77	[243]
Y-K	58	14	77	[241]
YBCO, a–b plane-oriented	12.8	9.45	6	[239]
YBCO, a–c plane-oriented	26.2	9.45	6	[239]

the unloaded quality factor Q of a cylindrical copper cavity in which the sample replaces one of the end plates [230–240]. Wosik *et al* [235] have measured the magnetic field and temperature dependence of R_s on a–b plane and a–c plane-oriented YBCO sample surfaces. The measurements have shown that R_s depends exponentially on magnetic field H for fields below H_{c1} and linearly for fields higher than H_{c1}. This behaviour has been explained by a model with an array of Josephson junctions and vortex motion in the presence of a magnetic field. There have been reports on measurements of surface resistance for grain-aligned Y-123 bulk material with surfaces oriented perpendicular and parallel to the c axis of the grains [235]. For the parallel configuration, the R_s at 77 K and 80 GHz is near 100 mΩ; for the perpendicular configuration, the R_s is less than 50 mΩ, at 88 K. The effect of microstructure on R_s has been discussed. There have been reports on investigation of alkali metal additions on R_s [240–243]. The presence of alkali metals during the YBCO synthesis affects the intergranular medium of the end product and R_s. Table C1.11 summarizes some data for R_s of YBCO.

A2.7

D2.7

C1.8.7 Radiation effects: types of ions employed and changes in pinning and pinning energies

B2.6, A4.3

A large number of results on the effects of particle irradiation on the flux pinning of HTSC are now available [244–268]. The general conclusion from those studies is that defects introduced by irradiation can act as effective pinning sites in HTSC materials. Electron, proton and light ion irradiations generate mostly point defects, which are effective pinning sites [244, 262–268]. The influence of proton and Au ion irradiations on the flux pinning of YBCO have been investigated [244]. Proton irradiation leads only to small changes of the pinning force. After Au ion irradiation changes of the pinning force are clearly visible, indicating that the nature of new defects is different in comparison with those existing in the pre-irradiated state. Neutrons and some heavier ions produce bigger defects, like cascades or clusters of few tens of angstrom in diameter [249–260]. For example, notable changes in the microstructure and critical current density of Y-123 after irradiation with fast neutrons have been reported [249]. Authors proposed that the observed defect structure in the irradiated material is responsible for increased pinning and consequently higher J_c. Irradiation with very heavy ions can create columnar defects, which have more effective pinning centers at high temperatures and fields. Information regarding the pinning energy (U) of defects can be obtained from measurements of the thermal relaxation of the magnetization. Maley *et al* [258] use a method, which allows one to build up the function $U(J)$ over an extended range of J values by matching small portions of the curve obtained from relaxation measurements at various temperature. Using this type of analysis, Lessure *et al* [245] have obtained $U(J)$ for Y-123 ceramics before and after irradiation with fast neutrons (to a fluence of 2.1×10^{18} neutrons cm^{-2}). The increase in U after neutron irradiation is apparent, as is the deviation from linearity and the diverging behaviour of U at low currents.

C1.9 Summary

In summary, YBCO exhibits a wide range of properties according to the oxygen content, so the oxygenation procedure of 123-superconductors is a very crucial one. YBCO is also the material with the smallest anisotropy among the high T_c compounds and the one with the highest critical current densities at elevated temperatures (i.e. 77 K). Therefore, YBCO and its derivatives of the 123 family (NdBCO, SmBCO, GdBCO and binary and ternary rare earth compounds) are still the best suited high T_c superconducting materials for a large variety of applications including all thin film applications and many bulk applications. Concerning the basic research, YBCO has served as the model system to explore all pecularities of high T_c compounds concerning structural, chemical, mechanical, optical, normal and superconducting properties; therefore, a vast literature is available (see the reference list). The research on YBCO ceramics provided also benefits for the understanding of different ceramic materials, not only for the other high T_c compounds.

B2.3.4
D1, D2, D3

References

[1] Wu M K, Ashburn J R, Torng C J, Hor P H, Meng R L, Huang Z J, Wang Y Q and Chu C W 1987 Superconductuvity at 93 K in a new mixed-phase Y–Ba–Cu–O compound system at ambient pressure *Phys. Rev. Lett.* **58** 908

[2] Pinol S, Gomis V, Gou A, Martinez B, Fontcuberta J and Obradors X 1994 Oxygenation and aging processes in melt textured $YBa_2Cu_3O_{7-\delta}$ *Physica* C **235–240** 3045

[3] Shi D, Qu D and Tent B A 1997 Effect of oxygenation on levitation force in seeded melt grown single-domain $YBa_2Cu_3O_x$ *Physica* C **291** 181

[4] Chen T G, Li S, Gao W, Xianyu Z, Liu H K and Dou S X 1998 The oxygenation kinetics of $YBa_2Cu_3O_{7-\delta}$–(0–30 percent) Ag superconductors *Supercond. Sci. Technol.* **11** 1193

[5] Erb A, Manuel A A, Dhalle M, Marti F, Genoud J Y, Revaz B, Junod A, Vasumathi D, Ishibashi S, Shukla A, Walker E, Fischer O, Flukiger R, Pozzi R, Mali M and Brinkmann D 1999 Experimental evidence for fast cluster formation of chain oxygen vacancies in $YBa_2Cu_3O_{7-\delta}$ as the origin of the fishtail anomaly *Solid State Commun.* **112** 245

[6] Koblischka M R and Murakami M 2000 Pinning mechanisms in bulk high T_c superconductors *Supercond. Sci. Technol.* **13** 738

[7] Okada M, Okayama A, Matsumoto T, Aihara K, Matsuda S, Ozawa K, Morii Y and Funahashi S 1988 Neutron diffraction study on preferred orientation of Ag-sheathed YBCO superconductor tape with $J_c = 1000–3000\ A/cm^2$ *Japan. J. Appl. Phys.* **27** L1715

[8] Okada M, Okayama A, Morimoto T, Matsumoto T, Aihara K and Matsuda S 1988 Fabrication of Ag-sheathed YBCO superconductor tapes *Japan. J. Appl. Phys.* **27** L185

[9] Goyal A, Norton D P, Kroeger D M, Christen D K, Paranthaman M, Specht E D, Budai J D, He Q, Saffian B, List F A, Lee D F, Hartfield E, Martin P M, Klabunde R E, Mathis J and Park C 1997 Conductors with controlled grain boundaries: An approach to the next generation, high temperature superconducting wire *J. Mater. Res.* **12** 2924

[10] Krabbes G, Fuchs G, Schätzle P, Gruss S, Park J W, Hardinghaus F, Stover G, Hayn R, Drechsler S L and Fahr T 2000 Zn doping of $YBa_2Cu_3O_7$ in melt textured materials: peak effect and high trapped fields *Physica* C **330** 181

[11] Yan M F, Rhodes W W and Gallagher P K 1988 Dopant effects on the superconductivity of YBCO ceramics *J. Appl. Phys.* **63** 821

[12] Erb A, Walker E and Flükiger R 1995 $BaZrO_3$: the solution for the crucible corrosion problem during the single crystal growth of high- T_c superconductors $REBa_2Cu_3O_{7-\delta}$; RE = Y, Pr *Physica* C **245** 245

[13] Erb A, Walker E and Flükiger R 1996 The use of $BaZrO_3$ crucibles in crystal growth of the high- T_c superconductors; progress in crystal growth as well as in sample quality *Physica* C **258** 9

[14] Erb A, Walker E, Genoud J-Y and Flükiger R 1997 10 years of crystal growth of the 123- and 124-high T_c superconductors: from Al_2O_3 to $BaZrO_3$. Progress in crystal growth and sample quality and its impact on physics *Physica* C **282–287** 459

[15] Karpinski J, Kaldis E, Jilek E, Rusiecki S and Bucher B 1988 Bulk synthesis of the 81-K superconductor $YBa_2Cu_4O_8$ at high oxygen pressure *Nature* **336** 660

[16] Bordet P, Chaillout C, Chenavas J, Hodeau J L, Marezio M, Karpinski J and Kaldis E 1988 Structure determination of the new high-temperature superconductor $Y_2Ba_4Cu_7O_{14+x}$ *Nature* **334** 596

[17] Fisher P, Karpinski J, Kaldis E, Jilek E and Rusiecki S 1989 High pressure preparation and neutron-diffraction study of the high T_c superconductor $YBa_2Cu_4O_{8+x}$ *Solid State Commun.* **69** 531

[18] Bucher B, Karpinski J, Kaldis E and Wachter P 1989 Strong pressure dependence of T_c of the new 80 K phase $YBa_2Cu_4O_{8+x}$ *Physica* C **157** 478

[19] Kaldis E, Fisher P, Hewat A W, Karpinski J and Rusiecki S 1989 Low temperature anomalies and pressure effects on the structure and T_c of the superconductor $YBa_2Cu_4O_8$ ($T_c = 80$ K) *Physica* C **159** 668

[20] van Eenige E N, Griessen R and Wijngaarden 1990 Superconductivity at 108 K in $YBa_2Cu_4O_8$ at pressure up to 12 GPa *Physica* C **168** 482

[21] Yamada Y, Yata M, Kaeda Y, Irie H and Matsumoto T 1989 The phase decomposition of $YBa_2Cu_3O_{6.9}$ induced by HIP *Japan. J. Appl. Phys.* **28** L797

[22] Kourtakis K, Robbins M, Gallagher K P and Tiefel T 1989 Synthesis of $Ba_2YCu_4O_8$ by anionic oxidation-reduction *J. Mater. Res.* **4** 1289

[23] Murakami H, Yaegashi S, Nishino J, Shiohara Y and Tanaka S 1990 Synthesis of $YBa_2Cu_4O_8$ powders by sol–gel method under ambient presure *Japan. J. Appl. Phys.* **29** L445

[24] Wada T, Suzuki N, Yamaguchi K, Ichinose A, Yaegashi Y, Yamauchi H, Koshizuka N and Tanaka S 1991 Superconductive $(Y_{1-x}Ca_x)Ba_2Cu_4O_8$ ($x = 0.0$ and 0.05) ceramics prepared by low and high oxygen partial pressure techniques *J. Mater. Res.* **6** 18

[25] Chen T-M, Ho S J, Liu S R and Koo S H 1993 Synthesis of a 90 K $Y_2Ba_4Cu_7O_{15-x}$ Superconductor under ambient pressure by triethylammonimoxalate co-precipitation *Physica* C **215** 435

[26] Schwer H, Kaldis E, Karpinski J and Rossel C 1993 Effect of structural changes on the transition temperatures in $Y_2Ba_4Cu_7O_{14+x}$ single crystals *Physica* C **211** 165

[27] Matic M V 1993 Low-temperature behaviour of oxygen ordering in $YBa_2Cu_3O_{6+c}$: calculation of length of O(1)-Cu(1)-O(1) chains at $T < 100$ K *Physica* C **211** 217

[28] Hewat A W 1994 Neutron powder diffraction on the ILL high flux reactor and high T_c superconductors *Materials and Crystallographic Aspects of HT$_c$-Superconductivity* NATO ASI Series E: Applied Sciences **263**, ed E Kaldis (Dordrecht, The Netherlands: Kluwer) p 17

[29] Capponi J J, Chaillout C, Hewat A W, Lejay P, Marezio M, Nguyen N, Raveau B, Soubeyroux J L, Tholence J L and Tournier R 1987 Structure of the 100 K superconductor $Ba_2YCu_3O_7$ between $(5 \div 300)$ K by neutron powder diffraction *Europhys. Lett.* **3** 1301

[30] Jorgensen J D, Veal B W, Paulikas A P, Nowicki L J, Crabtree G W, Claus H and Kwok W 1990 Structural properties of oxygen-deficient $YBa_2Cu_3O_{7-\delta}$ *Phys. Rev.* B **41** 1863

[31] Cava R J, Hewat A W, Hewat E A, Batlogg B, Marezio M, Rabe K M, Krajevski J J, Peck W F Jr and Rupp L W 1990 *Physica* C **165** 419

[32] Brown I D 1991 The influence of internal strain on the charge distribution and superconducting transition temperature in $Ba_2YCu_3O_x$ *J. Solid State Chem.* **90** 155

[33] Kruger Ch, Conder K, Schwer H and Kaldis E 1997 The dependence of the lattice parameters on oxygen content in orthorhombic $YBa_2Cu_3O_{6+x}$: high precision reinvestigation of near equilibrium samples *J. Solid State Chem.* **134** 356

[34] Shaked H, Keane P M, Rodriguez J C, Owen F F, Hitterman R L and Jorgensen J D 1994 *Crystal Structures of the High-T$_c$ Superconducting Copper-oxides, Physica* C (Amsterdam, The Netherlands: Elsevier) p 44

[35] *PCPDFWIN, Version 2.0* 1998 *JCPDS-ICDD, PDF-2 Data Base (Sets 1–48 plus 70–85), File no. 38-1433, 39-486, 39-1434* JCPDS—International Centre for Diffraction Data.

[36] Shibata H, Kinoshita K and Yamada T 1990 Crystal structure of low-oxygen-defect tetragonal $Ba_2YCu_3O_{6.75}$ *Japan. J. Appl. Phys.* **29** L423

[37] Dharwadkar S R, Jakkai V S, Yakhmi J V, Gopalakrishnan I K and Iyer R M 1987 X-ray diffraction coupled thermogravimetric investigations of $YBa_2Cu_3O_{7-x}$ *Solid State Commun.* **64** 1429

[38] Zandbergen W H, Gronsky R, Wang K and Thomas G 1988 Structure of $(CuO)_2$ double layers in superconducting $YBa_2Cu_3O_7$ *Nature* **331** 596

[39] Karpinski J, Beeli C, Kaldis E, Wisard A and Jilek E 1988 Crystallization of YBCO from nearly stoichiometric melts, under oxygen pressures up to 2800 bar *Physica* C **153–155** 830

[40] Schwer H, Kaldis E, Karpinski J and Rossel C 1993 Effect of structural changes on the transition temperature in $Y_2Ba_4Cu_7O_{14+x}$ single crystals *Physica* C **211** 165

[41] Morris D E, Markelz A G, Fayn B and Nickel J H 1990 Conversion of 124 into 123 + CuO and 124, 123 and 247 phase regions in the Y–Ba–Cu–O system *Physica* C **168** 153

[42] Morris D E, Asmar N G, Nickel J H, Sid R L, Wie J Y T and Post J E 1989 Stability of 124, 123, and 247 superconductors *Physica* C **159** 287

[43] *PCPDFWIN, Version 2.0* 1998 *JCPDS-ICDD, PDF-2 Data Base (Sets 1–48 plus 70–85), File no. 84-2460* JCPDS—International Centre for Diffraction Data.

[44] *PCPDFWIN, Version 2.0* 1998 *JCPDS-ICDD, PDF-2 Data Base (Sets 1–48 plus 70–85), File no. 43-0410* JCPDS—International Centre for Diffraction Data.

[45] McHenry M E and Sutton R A 1994 Flux pinning and Dissipation in High Temperature Oxide Superconductors *Progress in Materials Science* Vol 38, eds J W Christian and T B Massalski (Oxford: Pergamon) p 160

[46] Jorgensen J D 1991 Defects and superconductivity in the copper oxides *Phys. Today* **44** 34

[47] Murakami M 1992 *MeltProcessed High-temperature Superconductors* (Singapore: World Scientific) p 87

[48] Zandbergen H W and Tendeloo G 1991 Microstructures in High Temperature Superconductors *High-temperature Superconductors—Materials Aspects* Vol 2 eds H C Freyhardt, R Flükiger and M Peuckert (Verlag: Informationsgesellschaft) p 544

[49] Hashimoto K, Akiyoshi M, Wisniewski A, Jenkins M L, Toda Y and Yano T 1996 A high-resolution electron microscopy study of structural defects in $YBa_2Cu_4O_8$ superconductor *Physica* C **269** 139

[50] Domenges B, Hervieu M, Raveau B, Karpinski J, Kaldis E and Rusiecki S 1991 High-resolution electron microscopy study of defects in 'high oxygen pressure' $YBa_2Cu_4O_8$ *J. Solid State Chem.* **93** 316

[51] Buan J, Zhou B, Huang C C, Liu J Z and Shelton R N 1994 Anisotropy of the thermodynamic response along the *a* and *b* axes of the 1:2:3 compounds *Phys. Rev.* B **49** 12220

[52] Schlesinger Z, Collins R T, Holtzberg F, Feild C, Blanton S H, Welp U, Crabtree G W, Fang Y and Liu J Z 1990 Superconducting energy gap and normal state conductivity of a single-domain YBCO crystal *Phys. Rev. Lett.* **65** 801

[53] Zibold A, Widder K, Geserich H P, Scherer T, Marienhoff P, Neuhaus M, Jutzi W, Erb A and Müller-Vogt G 1992 Optical anisotropy of $YBa_2Cu_3O_{7-\delta}$ films on $NaGaO_3$ (001) substrates: a comparison with single crystals *Appl. Phys. Lett.* **61** 345

[54] Nakazawa Y, Takeya J and Ishikava M 1992 Specific heat study of $Ba_2YCu_3O_y$ ($6 < y < 7$) and $Ba_2Y(Cu_{1-x}M_x)_3O_y$ (M = Zn, Ni and Al) in *The Physics and chemistry of oxide Superconductors, Springer Proceedings in Physics* Vol 60, eds Y Iye and H Yasuoka (Berlin: Springer) p 283

[55] Slaski M, Laegried T, Nes O M and Fossheim K 1989 On the specific heat anomalies in the temperature range 200 K to 240 K in CuO and $YBa_2Cu_3O_{7-\delta}$ (YBCO) *Mod. Phys. Lett.* B **3** 585

[56] Shehata L N 1990 Specific heat and critical fluctuations in anisotropic high-T_c oxide superconductors *Solid State Commun.* **73** 827

[57] Gordon J E, Fisher R A, Kamin S and Phillips N E 1991 Specific heat evidence for strong coupling in $YBa_2Cu_3O_7$ *Supercond. Sci. Technol.* **4** S280

[58] Liang W Y, Loram J W, Mirza K A, Atanassopoulou L and Cooper J R 1996 Specific heat and susceptibility determination of the pseudogap in YBCO *Physica* C **263** 277

[59] Junod A, Roulin M, Genoud J-Y, Revaz B, Erb A and Walker E 1997 Specific heat peaks observed up to 16 T on the melting line of the vortex lattice in $YBa_2Cu_3O_7$ *Physica* C **275** 245

[60] Zhaojia C, Hongshun Y, Dongin Z, Baimei W, Kegin W, Yong Z, Zuyao C, Yitai Q and Qirui Z 1987 Specific heat study on Y–Ba–Cu–O system *Int. J. Mod. Phys.* **1** 419

[61] Zhaojia C, Yong Z, Hongshun Y, Zuyao C, Dongin Z, Yitai Q, Baimei W and Qirui Z 1987 Specific heat anomaly of single phase $YBa_2Cu_3O_{7-\delta}$ at superconducting transition temperature *Solid State Commun.* **64** 685

[62] Laegreid T, Fossheim K, Sandvold E and Julsurd S 1987 Specific heat anomaly at 220 K connected with superconductivity at 90 K in ceramic $YBa_2Cu_3O_{7-\delta}$ *Nature* **330** 637

[63] Schilling A, Bernasconi A, Ott H R and Hulliger F 1990 Specific heat, resistivity and magnetization study on polycrystalline $YBa_2Cu_4O_8$ *Physica* C **169** 237

[64] Junod A, Eckert D, Graf T, Kaldis E, Karpinski E, Rusiecki S, Sanchez D, Triscone G and Muller J 1990 Specific heat of the superconductor $YBa_2Cu_4O_8$ from 1.5 to 300 K *Physica* C **168** 47

[65] Ota S B 1991 Specific heat in the mixed state of superconducting $YBa_2Cu_3O_{7-x}$ *Phys. Rev.* B **43** 1237

[66] Junod A, Bezinge A, Eckert D, Graf T and Muller J 1988 Specific heat, magnetic susceptibility and superconductivity of $YBa_2Cu_3O_{7-\delta}$ doped with iron *Physica* C **152** 495

[67] Loram J W, Mirza K A, Cooper J R and Liang Y W 1993 Electronic specific heat of $YBa_2Cu_3O_{7-x}$ from 1.8 to 300 K *Phys. Rev. Lett.* **71** 1740

[68] Morelli D T, Heremans J and Swets D E 1987 Thermal conductivity of superconductive Y–Ba–Cu–O *Phys. Rev.* B **36** 3917

[69] Uher C and Kaiser A B 1987 Thermal transport properties $YBa_2Cu_3O_7$ superconductors *Phys. Rev.* B **36** 5680

[70] Reguerio M D N and Castello D 1991 Thermal conductivity of high temperature superconductors *Int. J. Mod. Phys.* B **5** 2003

[71] Heremans J, Morelli D T, Smith G W and Strite S C III 1988 Thermal and electronic properties of rare- earth $Ba_2Cu_3O_{7-x}$ superconductors *Phys. Rev.* B **37** 1604

[72] Zhu D M, Anderson A C, Friedmann T A and Ginsberg D M 1990 Thermal conductivity of ploycrystalline $YBa_2Cu_3O_{7-\delta}$ in a magnetic field *Phys. Rev.* B **41** 6605

[73] Delap M R and Bernhoet F 1992 A model for the thermal conductivity of $YBa_2Cu_3O_{7-\delta}$ *Physica* C **195** 301

[74] Terzijska B M, Wawryk R, Dimitrov D A, Marucha C, Kovachev V T and Rafalowic Z 1992 Thermal conductivity of YBCO and thermal conductance at YBCO/ruby boundary Part 1: joint experimental set-up for simultaneous measurements and experimental study in the temperature range 10–260 K *Cryogenics* **32** 53

[75] Terzijska B M, Wawryk R, Dimitrov D A, Marucha C, Kovachev V T and Rafalowic Z 1992 Thermal conductivity of YBCO and thermal conductance at YBCO/ruby boundary Part 2: hysteresis behaviour between 40 and 230 K *Cryogenics* **32** 60

[76] Williams W S 1993 Thermal conductivity peaks of old and new ceramic superconductors *Solid State Commun.* **87** 355

[77] Andersson B M and Sundqvist B 1993 Thermal conductivity of $YBa_2Cu_3O_8$ dominated by phonon–phonon interactions *Phys. Rev.* B **48** 3575

[78] Andersson B M and Sundqvist B 1994 Thermal conductivity of polycrystalline $YBa_2Cu_3O_8$ *Phys. Rev.* B **49** 4189

[79] Williams R K, Scarbrough J O, Schmitz J M and Thompson J R 1998 Thermal conductivity of polycrystalline $YBa_2Cu_3O_8$ from 10 to 300 K *Phys. Rev.* B **57** 10923

[80] Sundqvist B and Andersson B M 1994 Thermal conductivity of polycrystalline $YBa_2Cu_3O_8$ and the phonon transport model *Physica* C **235–240** 1377

[81] Cohn J L, Wolf S A, Vanderah T A, Selvamanickam V and Salama K 1992 Lattice thermal conductivity of $YBa_2Cu_3O_{7-\delta}$ *Physica* C **192** 435

[82] Peacor S D, Richardson R A, Nori F and Uher C 1991 Theoretical analysis of the thermal conductivity of $YBa_2Cu_3O_{7-\delta}$ single crystals *Phys. Rev.* B **44** 9508

[83] Fujishiro H, Ikebe M, Nakasato K and Noto K 1996 Influence of Cu site impurities on the thermal conductivity of $YBa_2Cu_3O_{7-\delta}$ *Physica* C **263** 305

[84] Bougrine H, Sergeenkov S, Ausloos M and Mehbod M 1993 Thermal conductivity of twinned $YBa_2Cu_3O_{7-x}$ and tweeded $YBa_2(Cu_{0.95}Fe_{0.05})_3O_{7-x}$ in a magnetic field: evidence for intrinsic proximity effect *Solid State Commun.* **86** 513

[85] Houssa M, Bougrine H, Ausloos M, Grandjeant I and Mehbod M 1994 Thermal conductivity of pure or iron-doped $YBa_2Cu_3O_{7-\delta}$ with or without an excess of CuO *J. Phys. Condens. Matter* **6** 6305

[86] Ravindran T R, Sankaranaryanan V and Srinivasan R 1994 Thermal conductivity of $YBa_2Cu_{3-x}Zn_xO_{7-y}$ *Physica* C **235–240** 1381

[87] Yukino K, Sato T, Ooba S, Ohta M, Okamura F P and Ono A 1987 Studies on the thermal behaviour of $Ba_2YCu_3O_{7-x}$ by X-ray powder diffraction method *Japan. J. Appl. Phys.* **26** L869

[88] Momin A C, Mathews M D, Jakkal V S, Gopalakrishnan L K, Yakhmi J V and Iyer R M 1987 High temperature X-ray powder diffractometric studies of the superconducting compound $YBa_2Cu_3O_{7-x}$ from room temperature to 1300 K in air *Solid state commun.* **64** 329

[89] Usami K, Kobayashi N and Doi T 1990 Temperature dependence of lattice parameters of $YBa_2Cu_3O_x$ superconductor at low temperature *Japan. J. Appl. Phys.* **30** L96

[90] You H, Axe J D, Kan X B, Hashimoto S, Moss S C, Liu J Z, Crabtree G W and Lam D J 1988 Phase constitution and thermal expansion of $YBa_2Cu_3O_{7-\delta}$ single crystals *Phys. Rev.* B **38** 9213

[91] Marti W, Altorfer F and Fischer P 1993 Thermal expansion coefficients of $NdBa_2Cu_3O_{7-\delta}$ *Physica* C **206** 158

[92] Meingast C, Kraut O, Wolf T and Wühl 1991 Large a–b anisotropy of the expansivity anormaly at T_c in untwinned $YBa_2Cu_3O_{7-\delta}$ *Phys. Rev. Lett.* **67** 1634

[93] Haetinger C, Castillo I A, Kunzler J V, Ghivelder L, Pureur P and Reich S 1996 Thermal expansion and specific heat of non-random YBCO/Ag composites *Supercond. Sci. Technol.* **9** 639

[94] Kund M and Andres K 1993 Anisotropic stress dependence of T_c in $YBa_2Cu_3O_{7-\delta}$ single crystals deduced from thermal expansion, measured with a capacitive quartz-dilatometer *Physica* C **205** 32

[95] Schnelle W, Braun E, Broicher H, Domel R, Ruppel S, Braunisch W, Harnischmacher J and Wohlleben D 1990 Fluctuation specific heat and thermal expansion of YBaCuO and DyBaCuO *Physica* C **168** 465

[96] Ruan Y Z, Li L P, Hu X L, Peng D K, Hu J B and Zhang Y H 1989 Thermal expansion coefficients of Y with orthorhombic and tetragonal phases *Mod. Phys. Lett.* B **3** 325

[97] Jericho M H, Simpson A M, Tarascon J M, Green L H, McKinnon R and Hall G 1988 Thermal expansion and velosity of ultrasonic waves in single phase $YBa_2Cu_3O_{7-x}$ *Solid State Commun.* **65** 978

[98] Hashimoto T, Fueki K, Kishi A, Azumi T and Koinuma H 1988 Thermal expansion coefficients of high T_c superconductors *Japan. J. Appl. Phys.* **27** L214

[99] Moral A D, Ibarra M R, Algarabel P A and Arnaudas J I 1989 Magnetostriction and thermal expansion of high- T_c magnetic superconductors $REBa_2Cu_3O_{7-x}$ (RE = Sm, Eu, Gd, Dy, Ho, Er, Tm and Y) *Physica* C **161** 48

[100] Singh J P, Joo J, Singh D, Warzynski T and Poeppel R B 1993 Effects of silver additions in resistance to thermal shock and delayed failure of $YBa_2Cu_3O_{7-\delta}$ superconductors *J. Mater. Res.* **8** 1226

[101] Reddy R R, Murakami M, Tanaka S and Reddy P V 1996 Elastic behaviour of a Y–Ba–Cu–O sample prepared by MPMG method *Physica* C **257** 137

[102] Ledbetter H 1992 Elastic constants of polycrystalline $Y_1Ba_2Cu_3O_x$ *J. Mater. Res.* **7** 2905

[103] Cankurtaran M and Saunders G A 1992 Ultrasonic determination of the elastic moduli and their temperature and pressure dependences in $YBa_2Cu_3O_{7-x}$/Ag(15 vol.%) composite *Supercond. Sci. Technol.* **5** 210

[104] Holcomb D J and Mayo M J 1990 Effect of microcracking on the measured moduli of bulk $YBa_2Cu_3O_x$ *J. Mater. Res.* **5** 1827

[105] Ledbetter H M, Austin M W, Kim S A and Lei M 1987 Elastic constants and Debye temperature of polycrystalline $Y_1Ba_2Cu_3O_x$ *J. Mater. Res.* **2** 786

[106] Shindo Y, Ledbetter H and Nozaki H 1995 Elastic constants and microcracks in $YBa_2Cu_3O_7$ *J. Mater. Res.* **10** 7

[107] Suasmoro S, Smith D S, Lejeune M, Huger M and Gault C 1992 High temperature ultrasonic characterization of intrinsic and microstructural changes in ceramic $YBa_2Cu_3O_{7-x}$ *J. Mater. Res.* **7** 1629

[108] Almond D P, Wang Q, Freestone J, Lambson E F, Chapman B and Saunders G A 1989 An ultrasonic study of superconducting and non-superconducting $GdBa_2Cu_3O_{7-x}$ *J. Phys: Condens. Matter.* **1** 6853

[109] Cankurtaran M, Saunders G A, Willis J R, Kheffaji A A and Almond D P 1989 Bulk modulus and its pressure derivative of $YBa_2Cu_3O_{7-x}$ *Phys. Rev.* B **39** 2872

[110] Cankurtaran M, Saunders G A, Goretta K C and Poeppel R B 1992 Ultrasonic determination of the elastic properties and their pressure and temperature dependences in very dense $YBa_2Cu_3O_{7-x}$ *Phys. Rev.* B **46** 1157

[111] Bonetti E, Campari E G, Manfredini T and Mantovani S 1991 Orthorhombic to tetragonal phase transition in $YBa_2Cu_3O_{7-x}$ observed by dynamic Young's modulus measurements *Physica* C **179** 381

[112] Kusz B, Barczynski R, Gazda M, Sadowski W, Murawski L, Ozowski O, Davoli I and Stizza S 1990 Elastic constant and internal friction in $YBa_2Cu_3O_x$ single crystal *Solid State Commun.* **76** 357

[113] Shi X D, Yu R C, Wang Z Z, Ong N P and Chaikin P M 1989 Sound velocity and attenuation in single-crystal $YBa_2Cu_3O_{7-\delta}$ *Phys. Rev.* B **39** 827

[114] Anderson A R, Murakami M, Nagashima K and Russell G J 1998 Evidence for a structural change in TSMG Y123 at 225 K *Physica* C **306** 15

[115] Fujimoto H, Murakami M, Oyama T, Shiohara Y, Koshizuka N and Tanaka S 1990 Fracture toughness of YBaCuO prepared by MPMG process *Japan. J. Appl. Phys.* **29** L1793

[116] Murayama T, Sakai N, Yoo S I and Murakami M 1996 Mechanical properties of OCMG-processed Nd–Ba–Cu–O bulk superconductors *Proceedings of International Symposium on Advances in Superconductivity: New Materials, Critical Currents and Devices (ASMCCD'96) Mumbai, India* p 321

[117] Fujimoto H, Murakami M and Koshizuka N 1992 Effect of Y_2BaCuO_5 on fracture toughness of YBCO prepared by a MPMG process *Physica* C **203** 103

[118] Leeders A, Ullrich M and Freyhardt H C 1997 Influence of thermal cycling on the mechanical properties of VGF melt-textured YBCO *Physica* C **279** 173

[119] Ochiai S, Osamura K and Takayama T 1988 Fracture toughness measurements of $Ba_2YCu_3O_{7-x}$ superconducting oxide by means of indentation technique *Japan. J. Appl. Phys.* **27** L1101

[120] Cook R F, Dinger T R and Clarke D R 1987 Fracture toughness measurements of $YBa_2Cu_3O_x$ single crystals *Appl. Phys. Lett.* **51** 454

[121] Anstis G R, Chantikul P, Lawn B R and Marshall D B 1981 *J. Am. Ceram. Soc.* **64** 532

[122] Joo J, Singh J P, Warzynski T, Grow A and Poeppel R B 1994 Role of silver addition in mechanical and superconducting properties of high T_c superconductors *Appl. Supercond.* **2** 401

[123] Oka T, Itoh Y, Yanagi Y, Tanaka H, Takashima S and Mizutani U 1992 Metallurgical reactions and their relationships to enhanced mechanical strength in Zr-bearing YBCO composite superconductors *Japan. J. Appl. Phys.* **31** 1760

[124] Xu J A 1994 Elastic properties of high T_c superconductors: elastic systematics in $YBa_2Cu_3O_{7-x}$/Ag composites *Supercond. Sci. Technol.* **7** 1

[125] Yeou L S and White K W 1992 The development of high fracture toughness $YBa_2Cu_3O_{7-x}$/Ag composites *J. Mater. Res.* **7** 1

[126] Yeh F and White K W 1991 Fracture toughness behaviour of the $YBa_2Cu_3O_{7-x}$ superconducting ceramic with silver oxide additions *J. Appl. Phys.* **70** 4989

[127] Singh J P, Leu H J, Poeppel R B, Voorhees E V, Goudey G T and Winsley K 1989 Effect of silver and silver oxide additions on the mechanical and superconducting properties of $YBa_2Cu_3O_{7-\delta}$ superconductors *J. Appl. Phys.* **66** 3154

[128] Yu F, White K W and Meng R 1997 Mechanical characterization of top-seeded melt-textured $YBa_2Cu_3O_{7-\delta}$ single crystal *Physica* C **276** 295

[129] Joo J, Kim J G and Nah W 1998 Improvement of mechanical properties of YBCO-Ag composite superconductors made by mixing with metallic Ag powder and $AgNO_3$ solution *Supercond. Sci. Technol.* **11** 645

[130] Oka T, Itoh Y, Yanagi Y, Tanaka H, Takashima S, Yamada Y and Mizutani U 1992 Critical current density and mechanical strength of $YBa_2Cu_3O_{7-\delta}$ superconducting composites containing Zr, Ag and Y_2BaCuO_5 dispersions by melt-processing *Physica* C **200** 55

[131] Lee D and Salama K 1990 Enhancements in current density and mechanical properties of Y–Ba–Cu–O/Ag composites *Japan. J. Appl. Phys.* **29** L2017

[132] Oka T, Ogasawara F, Itoh Y, Suganuma M and Mizutani U 1990 Mechanical and superconducting properties of Ag/YBCO composite superconductors reinforced by the addition of Zr *Japan. J. Appl. Phys.* **29** 1924

[133] Wu N L, Hsu C C and Lee C H 1998 Reduced cracking in melt-grown $RBa_2Cu_3O_7$ (R = Nd, Sm)/gold composites *Japan. J. Appl. Phys.* **37** L438

[134] Higashida K and Narita N 1991 High temperature deformation and textures in oxide superconductors *Advances in Superconductivity III Proceedings of the 3rd ISS'90, Sendai* (Springer: Tokyo) p 805

[135] von Mises R 1928 *Z. Angew. Math. Mech.* **8** 161

[136] Suzuki T and Takeuchi S 1989 *JJAP Series 2, Lattice Defects in Ceramics* (Tokyo: Publication Office, JJAP) pp 9–15

[137] Guangcan C, Jingkui L, Wei C, Sishen X, Yude Y, Qiansheng Y, Yuengming N, Guirueng L and Genguha C 1987 Study on phase diagram of $BaO-Y_2O_3-CuO$ ternary system *Int. J. Mod. Phys.* B **1** 363

[138] Hinks D G, Soderholm L, Cappone I D W, Jorgensen J D, Schuller I K, Segre C U, Zhang K and Grace J D 1987 Phase diagram and superconductivity in the Y–Ba–Cu–O system *Appl. Phys. Lett.* **50** 1688

[139] Karpinski J and Kaldis E 1988 Equilibrium pressures of oxygen above $YBa_2Cu_3O_{7-x}$ up to 2000 bar *Nature* **331** 242

[140] de Leeuw D M, Mutsares A H A C, Langereis C, Smoorenburg H C A and Rommers P J 1988 Compounds and phase compatibilities in the system $Y_2O_3-BaO-CuO$ at 950°C *Physica* C **152** 39

[141] Karpinski J, Rusiecki S, Kaldis E, Bucher B and Jilek E 1989 Phase diagrams of $YBa_2Cu_4O_8$ and $YBa_2Cu_{3.5}O_{7.5}$ in the pressure range 1 bar $\leq PO_2 \leq$ 3000 bar *Physica* C **160** 449
[142] Chandrachood M R, Morris D E and Sinha A P B 1990 Phase diagram and new phases in the Y–Ba–Cu–O system *Physica* C **171** 187
[143] Sestak J 1992 Phase diagrams in CuO_x based superconductors *Pure Appl. Chem.* **64** 125
[144] Krabbes G, Bieger W, Wiesner U, Ritschel M and Teresiak A 1993 Isothermal sections and primary crystallization in the quasiternary $YO_{1.5}$–BaO–CuO_x system at $P(O_2) = 0.21 \times 10^5$ Pa *J. Solid State Chem.* **103** 420
[145] Karpinski J, Conder K, Kruger Ch, Schwer H, Mangeschots I, Jilek E and Kaldis E 1994 Phase diagram, synthesis and crystal growth of YBaCuO phases at high oxygen pressures $PO_2 < 3000$ bar in *Materials and Crystallographic Aspects of HT_c-Superconductivity* NATO ASI Series E: Applied Sciences **263**, ed E Kaldis (Dordrecht, The Netherlands: Kluwer) p 555
[146] Maeda M, Kadoi M and Ikeda T 1989 The phase diagram of the $YO_{1.5}$–Bao–CuO ternary system and growth of $YBa_2Cu_3O_7$ single crystals *Japan. J. Appl. Phys.* **28** 1417
[147] Murakami H, Suga T, Noda T, Shiohara Y and Tanaka S 1990 Phase diagram of $YBa_2Cu_3O_{7-x}$, $Y_2Ba_4Cu_7O_{15-x}$ and $YBa_2Cu_4O_8$ superconductors *Japan. J. Appl. Phys.* **29** 2720
[148] Moiseev K G, Vatolin A N, Zaizeva I S, Ilyinych I N, Tsagareishvily S D, Gvelesiani B I and Sestak J 1992 Calculation of thermodynamic properties of the phases in the Y–Ba–Cu–O system *Thermochim. Acta* **198** 267
[149] Barus A M M and Taylor J A T 1994 The effect of particle size distribution on the phase composition in $YBa_2Cu_3O_{7-x}$ as detemined by DTA *Physica* C **225** 374
[150] Rodriguez M A, Snyder R L, Chen B J, Matheis D P, Misture S T and Frechette K 1993 The high-temperature reactions of $YBa_2Cu_3O_{7-\delta}$ *Physica* C **206** 43
[151] Idemoto Y and Fueki K 1990 Melting point of superconducting oxides as a function of oxygen partial pressure *Japan. J. Appl. Phys.* **29** 2729
[152] Ono A and Tanaka T 1987 Preparation of single crystals of the superconductor $Ba_2YCu_3O_{6.5+x}$ *Japan. J. Appl. Phys.* **26** 1825
[153] Sawai Y, Ishizaki K and Takata M 1991 Stability of $YBa_2Cu_3O_7$, $Y_2Ba_4Cu_7O_{15}$ and $YBa_2Cu_4O_8$ superconductors under varying oxygen partial pressure, total gas and pressure and temperature *Physica* C **176** 147
[154] Hohlwein D 1994 Superstructures in 123 compounds. X-ray and neutron diffraction *Materials and Crystallographic Aspects of HT_c-Superconductivity* NATO ASI Series E: Applied Sciences **263**, ed E Kaldis (Dordrecht, The Netherlands: Kluwer) p 65
[155] Lindemer T B, Washburn F A, MacDougall C S, Feenstra R and Cavin O B 1991 Decomposition of $YBa_2Cu_3O_{7-x}$ and $YBa_2Cu_4O_8$ for $p_{O2} \leq 0.1$ MPa *Physica* C **178** 93
[156] Ceder G, Asta M and Fontaine D 1991 Computation of the OI–OII–OIII phase diagram and local oxygen configuration for $YBa_2Cu_3O_z$ with z between 6.5 and 7 *Physica* C **177** 106
[157] Morris D E, Marathe A P and Sinha A P B 1990 Destabilisation of 124 into 247 phases of Y–Ba–Cu–O by Fe substitution at elevated oxygen pressures *Physica* C **169** 386
[158] Voronin G F 1994 Thermodynamic stability of superconductors in the Y–Ba–Cu–O systen *Materials and Crystallographic Aspects of HT_c-Superconductivity* NATO ASI Series E: Applied Sciences **263**, ed E Kaldis (Dordrecht, The Netherlands: Kluwer) p 585
[159] Cooper S L, Reznik D, Kotz A, Karlow M A, Liu R, Klein M V, Lee W C, Giapintzakis J and Ginsberg D M 1993 Optical studies of the a-, b-, and c-axis charge dynamics in YBCO *Phys. Rev.* B **47** 8233
[160] Tajima S, Schützmann J, Miyamoto S, Teraski I, Sato Y and Hauff R 1997 Optical study of c-axis charge dynamics in YBCO: carrier self-confinement in the normal and the superconducting states *Phys. Rev.* B **55** 6051
[161] Rotter L D, Schlesinger Z, Collins R T, Holtzberg F, Feild C, Welp U, Crabtree G W, Liu J Z, Fang Y, Vandervoort K G and Fleshler S 1991 Dependence of the infrared properties of single-domain YBCO on oxygen content *Phys. Rev. Lett.* **67** 2741
[162] Takenaka K, Mizuhashi K, Takagi H and Uchida S 1994 Interplane charge transport in YBCO: spin-gap effect on in-plane and out-of-plane resistivity *Phys. Rev. Lett.* **50** 6534
[163] Tallon J L, Cooper J R, de Silva P S I P N, Williams G V M and Loram J W 1995 Thermoelectric power: a simple instructive probe of high- T_c superconductors *Phys. Rev. Lett.* **75** 4114
[164] Ito T, Takenaka K and Uchida S 1993 Systematic deviation from T-linear behaviour in the in-plane resistivity of $YBa_2Cu_3O_{7-y}$: evidence for dominant spin scattering *Phys. Rev. Lett.* **70** 3995
[165] Jiang H, Yuan T, How H, Widom A, Vittoria C and Drehman A 1993 Measurements of anisotropic characteristic lengths in YBCO films at microwave frequencies *J. Appl. Phys.* **73** 5865
[166] Rose-Innes A C and Rhoderick E H 1978 *Introduction to Superconductivity* (Oxford: Pergamon) p 67
[167] Ginzburg V L and Landau D 1950 *Zh. Eksp. Teor. Fiz.* **20** 1064 English translation in Landau LD 1965 *Men of Physics* vol 1, ed D Ter Haar (New York: Pergamon Press)
[168] Coffey M W and Clem J R 1992 Theory of high-frequency linear response of isotropic type-II superconductors in the mixed state *Phys. Rev.* B **46** 11757
[169] Coffey M W and Clem J R 1992 Theory of rf magnetic permeability of isotropic type-II superconductors in a parallel field *Phys. Rev.* B **45** 9872
[170] Coffey M W and Clem J R 1991 Unified theory of effects of vortex pinning and flux creep upon the rf surface impedance of type-II superconductors *Phys. Rev. Lett.* **67** 386

[171] Hikita M, Tajima Y, Katsui A, Hidaka Y, Iwata S and Tsurumi S 1987 Electrical properties of high T_c superconducting single-crystal $Eu_1Ba_2Cu_3O_y$ *Phys. Rev.* B **36** 7199
[172] Gallagher W J 1988 Studies at IBM on anisotropy in single crystals of the high-temperature oxide superconductor $Y_1Ba_2Cu_3O_{7-x}$ *J. Appl. Phys.* **63** 4216
[173] Worthington T K, Gallagher W J and Dinger T R 1987 Anisotropic nature of high-temperature superconductivity in single-crystal $Y_1Ba_2Cu_3O_{7-x}$ *Phys. Rev. Lett.* **59** 1160
[174] Feinberg D and Villard C 1990 Intrinsic pinning and lock-in transition of flux lines in layered type-II superconductors *Phys. Rev. Lett.* **65** 919
[175] Dinger T R, Worthington T K, Gallagher W J and Sandstrom R L 1987 Direct observation of electronic anisotropy in single-crystal $Y_1Ba_2Cu_3O_{7-x}$ *Phys. Rev. Lett.* **58** 2687
[176] Campbell A M and Evetts J E 1972 Flux vortices and transport current in type-II superconductors *Adv. Phys.* **21** 199
[177] Hao Z, Clem J R, McElfresh M W, Civale L, Malozemoff A P and Holtzberg F 1991 Model for the reversible magnetization of high-κ type-II superconductors: application to high- T_c superconductors *Phys. Rev.* B **43** 2844
[178] Welp U, Kwok W K, Crabtree G W, Vandervoort K G and Liu J Z 1989 Magnetic measurements of the field of $YBa_2Cu_3O_{7-\delta}$ single crystal *Phys. Rev. Lett.* **62** 1908
[179] Yeshurun Y and Malozemoff A P 1988 Direct measurement of the temperature-dependent magnetic penetration depth in Y–Ba–Cu–O crystals *Phys. Rev. Lett.* **60** 2202
[180] Palstra T T M, Batlogg B, van Dover R B, Schneemeyer L F and Waszczak J V 1989 Critical currents and thermally activated flux motion in high-temperature superconductors *Appl. Phys. Lett.* **54** 763
[181] Kes P H, Berguis P, Guo S Q, Dam B and Stallman G M 1989 *J. Less-Common. Met.* **151** 325
[182] Ferrari M J, Johnson M, Wellstood F C, Clarke J, Mitzi D, Rosenthal P A, Eom C B, Geballe T H, Kapitulnik A and Beasley M R 1990 Distribution of flux-pinning energies in $YBa_2Cu_3O_{7-y}$ and $Bi_2Sr_2CaCu_2O_{8+y}$ from flux noise *Phys. Rev. Lett.* **64** 72
[183] Li G, Sun Y, Liu S, Liu G, Yan S, Xiao L and Fu X 1993 Distribution of flux pinning energies in powder-melt-textured-grown $YBa_2Cu_3O_{7-\delta}$ *Solid State Commun.* **88** 451
[184] Zeldov E, Amer M N, Koren G, Gupta A, Gambino R J and McElfresh M W 1989 Optical and electrical enhancement of flux creep in $YBa_2Cu_3O_{7-\delta}$ epitaxial films *Phys. Rev. Lett.* **62** 3093
[185] Hagen C W and Griessen R 1989 Distribution of activation energies for thermally activated flux motion in high T_c superconductors: an inversion scheme *Phys. Rev. Lett.* **62** 2857
[186] Sun Y R, Thompson J R, Christen D K, Holtzberg F, Marwick A D and Ossandon J G 1992 *Physica* C **194** 403
[187] Yan S, Liang S, Ma H, Feng Q, Sun Y, Gao Y and Zhang H 1989 *Solid State Commun.* **70** 553
[188] Parish J L 1994 Note on magnetic hysteresis in HTSC *Physica* C **226** 325
[189] Andrä W, Bruchlos H, Eick T, Hergt R, Michalke W, Schuppel W and Steenbeck K 1991 Critical current density and flux pinning determined by different methods *Physica* C **180** 184
[190] Nojima T and Fujita T 1991 Universality of magnetization curves in superconducting Bi–Sr–Ca–Cu–O and Y–Ba–Cu–O films *Physica* C **178** 140
[191] Schlenker C, Liu C J, Buder R, Schubert J and Stritzker B 1991 Magnetic properties and critical currents in $YBa_2Cu_3O_7$ thin films *Physica* C **180** 148
[192] Anderson P W and Kim Y B 1964 Hard superconductors: theory of the motion of Abrikosov flux lines *Rev. Mod. Phys.* **36** 39
[193] Kamiya H, Kondo A, Yokoyama T, Naito M, Jimbo G, Nagaya S, Miyajima M and Hirabayashi I 1994 Effect of Y_2BaCuO_5 particle size on the properties of $YBa_2Cu_3O_{7-\delta}$ superconductor *Adv. Powder Technol.* **5** 339
[194] Bean C P 1962 Magnetisation of hard superconductors *Phys. Rev. Lett.* **9** 309
[195] Bean C P 1964 Magnetisation of high-field superconductors *Rev. Mod. Phys.* **36** 31
[196] Kumar R, Walia R, Oussena M, de Groot P A J, Lanchester P C, Currie D B and Weller M T 1994 High field magnetization study of $YBa_2Cu_4O_8$ *Solid State Commun.* **9** 783
[197] Skumryev V, Koblischka M R and Kronmüller H 1991 Sample size dependence of the AC-susceptibility of sintered $YBa_2Cu_3O_{7-\delta}$ superconductors *Physica* C **184** 332
[198] Müller K A, Takashige M and Bednorz J G 1987 Flux trapping and superconductive glass state in La_2CuO_{4-y}: Ba *Phys. Rev. Lett.* **58** 1143
[199] Malozemoff A P, Worthington T K, Yeshurun Y, Holtzberg F and Kes P H 1988 Frequency dependence of the ac susceptibility in an YBaCuO crystal: a reinterpretation of H_{c2} *Phys. Rev.* B **38** 7203
[200] Brandt E H 1989 Thermal fluctuation and melting of the vortex lattice in oxide superconductors *Phys. Rev. Lett.* **63** 1106
[201] Brandt E H 1991 Thermal depinning and melting of the flux-line lattice in high T_c superconductors *Int. J. Mod. Phys.* B **5** 751
[202] Yeshurun Y, Malozemoff A P and Shaulov A 1996 Magnetic relaxation in high-temperature superconductors *Rev. Mod. Phys.* **68** 911
[203] Suryanarayanan R, Leelaprute S and Niarchos D 1993 Irreversibility line of $Y_{1-x}Ca_xSrBaCu_{2.9}Co_{0.1}O_{6+z}$ $(0 < x < 0.25)$. A comparative study of DC magnetization and complex AC susceptibility *Physica* C **214** 277
[204] Zheng D N, Campbell A M, Johnson J D, Cooper J R, Blunt F J, Porch A and Freeman P A 1994 Magnetic susceptibilities, critical fields, and critical currents of Co- and Zn-doped $YBa_2Cu_3O_7$ *Phys. Rev.* B **49** 1417

[205] Lee W C and Ginsberg D M 1992 Magnetic measurements of the upper critical field, irreversibility line, anisotropy, and magnetic penetration depth of grain-aligned $YBa_2Cu_4O_8$ *Phys. Rev.* B **45** 7402

[206] Müller K H 1989 AC susceptibility of high temperature superconductors in a critical state model *Physica* C **159** 717

[207] Yacoby E R, Shaulov A, Yeshurun Y, Konczykowski M and Rullier-Albenque F 1992 Irreversibility line in $YBa_2Cu_3O_7$ samples. A comparison between experimental techniques and effect of electron irradiation *Physica* C **199** 15

[208] El-Abbar A A, King P J, Maxwell K J, Owers-Bradley J R and Roys W B 1992 The irreversibility line in polycrystalline $YBa_2Cu_3O_7$ superconductors with Y_2BaCuO_5 inclusions *Physica* C **198** 81

[209] Almasan C C, Seaman C L, Dalichaouch Y and Maple M B 1991 Irreversibility line and magnetic relaxation in a $Sm_{1.85}Ce_{0.15}CuO_{4-y}$ single crystal *Physica* C **174** 93

[210] Sagdahl L T, Laegreid T, Fossheim K, Murakami M, Fujimoto H, Gotoh S, Yamaguchi K, Yamauchi H, Koshizuka N and Tanaka S 1990 Restricted reversible region and strongly enhanced pinning in MPMG $YBa_2Cu_3O_7$ with Y_2BaCuO_5 inclusions *Physica* C **172** 495

[211] Mannhart J and Hilgenkamp H 1998 Possible influence of band bending on the normal state properties of grain boundaries in high- T_c superconductors *Mater. Sci. Eng.* B **56** 77

[212] Gurevich A and Pashitskii E A 1998 Current transport through low-angle grain boundaries in high-temperature superconductors *Phys. Rev.* B **57** 13878

[213] Buessem W R, Cross L E and Goswami A K 1966 Phenomenological theory of high permittivity in fine-grained $BaTiO_3$ *J. Am. Ceram. Soc.* **49** 33

[214] Waser R 1995 Electronic properties of grain boundaries in $SrTiO_3$ and $BaTiO_3$ ceramics *Solid State Ionics* **75** 89

[215] Dimos J, Chaudhari P, Mannhart J and LeGoues F K 1988 Orientation dependence of grain boundary critical currents in YBCO bicrystals *Phys. Rev. Lett.* **61** 219

[216] Dimos J, Chaudhari P and Mannhart J 1990 Superconducting transport properties of grain boundaries in YBCO bicrystals *Phys. Rev.* B **41** 4038

[217] Babcock S E, Cai X Y, Kaiser D L and Larbalestier D C 1990 Weak-link free behaviour of high-angle YBCO grain boundaries in high magnetic fields *Nature* **347** 167

[218] Heinig N F, Redwing R D, Tsu I F, Gurevich A, Nordman J E, Babcock S E and Larbalestier D C 1996 Evidence for channel conduction in low misorientation angle [0 0 1] tilt YBCO bicrystal films *Appl. Phys. Lett.* **69** 577

[219] Polyanskii A A, Gurevich A, Pashitski A E, Heinig N F, Redwing R D, Nordman J E and Larbalestier D C 1996 Magneto-optical study of flux penetration and critical current densities in [0 0 1] tilt YBCO thin-film bicrystals *Phys. Rev.* B **53** 8687

[220] Koblischka-Veneva A and Koblischka M R 2002 EBSD on high T_c superconductors in: *Studies of high temperature superconductors* Vol 41, ed A Narlikar (Commack, NY: Nova Science Publishers) p 1

[221] Koblischka M R, Schuster T and Kronmüller H 1994 Flux penetration in granular YBCO samples: a magneto-optical study *Physica* C **219** 205

[222] Koblischka M R, van Dalen A J J and Ravikumar G 1994 Influence of melt-processing on flux behaviour and critical current densities *Proc. 7th IWCC Conference, 24–27 Jan. 1994, Alpbach, Austria*, ed H W Weber (Singapore: World Scientific) p. 399

[223] Schuster T, Koblischka M R, Reininger T, Ludescher B, Henes R and Kronmüller H 1992 Influence of low magnetic fields on the transport properties of sintered YBCO with different grain sizes *Supercond. Sci. Technol.* **5** 614

[224] Salama K, Mironova M, Stolbov S and Sathyamurthy S 2000 Grain boundaries in bulk YBCO *Physica* C **341–348** 1401

[225] Eom C B, Marshall A F, Suzuki Y, Boyer B, Pease R F W and Geballe T H 1991 Absence of weak-link behaviour in YBCO grains connected by 90° (0 1 0) twist boundaries *Nature* **353** 544

[226] Mannhart J, Bielefeldt H, Goetz B, Hilgenkamp H, Schmehl A, Schneider C W and Schulz R R 2000 Grain boundaries in high- T_c superconductors: insights and improvements *Phil. Mag.* B **80** 827

[227] Mannhart J, Bielefeldt H, Goetz B, Hilgenkamp H, Schmehl A, Schneider C W and Schulz R R 2000 Doping induced enhancement of the critical currents of grain boundaries in high T_c superconductors *Physica* C **341–348** 1393

[228] Hammerl G, Schmehl A, Schulz R R, Goetz B, Bielefeldt H, Schneider C W, Hilgenkamp H and Mannhart J 2000 Enhanced supercurrent density in polycrystalline YBCO at 77 K from calcium doping of grain boundaries *Nature* **407** 162

[229] Babcock S E 1999 Roles for electron microscopy in establishing structure-property relationships for high T_c superconductor grain boundaries *Micron* **30** 449

[230] Sankawa I, Sato M, Konaka T, Kobayashi M and Ishihara K 1988 Microwave surface resistance studies of $YBa_2Cu_3O_{7-\delta}$ single crystal *Japan. J. Appl. Phys.* **27** L1637

[231] Cohen L, Gray I R, Porch A and Waldram J R 1987 Surface impedance measurements of superconducting $YBa_2Cu_3O_{6+x}$ *J. Phys.* F **17** L179

[232] Martens J S, Beyer J B and Ginley D S 1988 Microwave surface resistance of $YBa_2Cu_3O_{6.9}$ superconducting films *Appl. Phys. Lett.* **52** 1822

[233] Wu D-H, Shiffman C A and Sridhar S 1988 Field variation of the penetration depth in ceramic $Y_1Ba_2Cu_3O_y$ *Phys. Rev.* B **38** 9311

[234] Srikanth H, Zhai Z, Sridhar S, Erb A and Walker E 1998 Systematic of two-component superconductivity in $YBa_2Cu_3O_{6.95}$ from microwave measurements of high-quality single crystals *Phys. Rev.* B **57** 7986

[235] Wosik J, Kranenburg R A, Wolfe J C, Selvamanickam V and Salama K 1991 Millimeter wave surface resistance of grain-aligned $Y_1Ba_2Cu_3O_x$ bulk material *J. Appl. Phys.* **69** 874

[236] Awasthi A M, Carini J P, Alavi B and Gruner G 1988 Millimeter-wave surface impedance measurements of $Y_1Ba_2Cu_3O_{7-\delta}$ ceramic superconductors *Solid State Commun.* **67** 373

[237] Klein N, Müller G, Piel H, Roas B, Schultz L, Klein U and Peiniger M 1989 Millimeter wave surface resistance of epitaxially grown $YBa_2Cu_3O_{7-x}$ thin films *Appl. Phys. Lett.* **54** 757

[238] Cooke D W, Gray E R, Javadi H H S, Houlton R J, Rusnak B, Meyer E A, Arendt P N, Klein N, Muller G, Orbach S, Piel H, Drabeck L, Gruner G, Josefowicz J Y, Rensch D B and Krajenbrink F 1990 Frequency dependence of the surface resistance in high-temperature superconductors *Solid State Commun.* **73** 297

[239] Wosik J, Xie L M, Chau R, Samaan A, Wolfe J C, Selvamanickam V and Salama K 1993 Effect of DC magnetic fields on the microwave properties of grain-aligned $YBa_2Cu_3O_x$ bulk material *Physica* C **180** 532

[240] Nurgaliev T, Miteva S, Nedkov I, Veneva A and Taslakov M 1994 Na-doping effect on the magnetic properties of the YBCO ceramics *J. Appl. Phys.* **76** 7118

[241] Nedkov I and Veneva A 1994 Alkali metal impurity influence on magnetic and electric properties of YBCO *J. Appl. Phys.* **75** 6726

[242] Nedkov I, Veneva A, Miteva S, Nurgaliev T and Lovchinov V 1994 Magnetic properties of Rb-doped YBaCuO *IEEE Trans. Magn.* **30** 1187

Nurgaliev T, Miteva S, Nedkov I, Veneva A 1995 Magnetic characteristics of K, Rb, Na-doped YBCO ceramics *J. Magn. Magn. Mater.* **140–144** 1305.

[243] Nedkov I, Veneva A, Miteva S, Nurgaliev T and Lovchinov V 1994 Microwave surface resistance of Rb-doped YBCO ceramics *Proc. of the 6th Workshop on RF Superconductivity* Vol 1, ed R Sundelin (VA: CEBAF-Newport News) p 556

[244] Wisniewski A, Baran M, Przyslupski P, Szymczak H, Pajaczkowska A, Pytel B and Pytel K 1988 Magnetization studies of $YBa_2Cu_3O_{7-x}$ irradiated by fast neutrons *Solid State Commun.* **65** 577

[245] Lessure H S, Simizu S, Sankar S G, McHenry M E, Cost J R and Maley M P 1991 Critical current density and flux pinning dominated by neutron irradiation induced defects in $YBa_2Cu_3O_{7-x}$ *J. Appl. Phys.* **70** 6513

[246] Sickafus K E, Willis J O, Kung P J, Wilson W B, Parkin D M, Maley M P, Clinard F W Jr, Salgado C J, Dye R P and Hubbard K M 1992 Neutron-radiation-induced flux pinning in Gd-doped $YBa_2Cu_3O_{7-x}$ and $GdBa_2Cu_3O_{7-x}$ *Phys. Rev.* B **46** 11862

[247] Bulaevskii L N, Vinokur V M and Maley M P 1996 Reversible magnetization of irradiated high T_c superconductors *Phys. Rev. Lett.* **77** 936

[248] Ghigo G, Gerbaldo R, Gozzelino L, Mezzetti E, Minetti B and Wisniewski A 1997 Influence of irradiation-induced correlated and random disorder on flux pinning in bulk YBCO melt-textured samples *J. Supercond.* **10** 541

[249] Lee J-W, Lessure H S, Laughlin D E, McHenry M E, Sankar S G, Willis J O, Cost J R and Maley M P 1990 Observation of proposed flux pinning sites in neutron-irradiated $YBa_2Cu_3O_{7-x}$ *Appl. Phys. Lett.* **57** 2150

[250] Cost J R, Willis J O, Thompson J D and Peterson D E 1988 Fast-neutron irradiation of YBa_2Cu_3O x *Phys. Rev.* B **37** 1563

[251] Sauerzopf F M, Wiesinger H P, Weber H W and Crabtree G W 1995 Analysis of pinning effects in $YBa_2Cu_3O_{7-\delta}$ single crystals after fast neutron irradiation *Phys. Rev.* B **51** 6002

[252] Sauerzopf F M, Wiesinger H P, Kritscha W, Weber H W, Crabtree G W and Liu J Z 1991 Neutron-irradiation effects on critical current densities in single-crystalline $YBa_2Cu_3O_{7-y}$ *Phys. Rev.* B **43** 3091

[253] Wisniewski A, Czurda C, Weber H W, Baran M, Reissner M, Steiner W, Zhang P X and Zhou L 1996 Influence of oxygen deficiency and of neutron-induced defects on flux pinning in melt textured bulk $YBa_2Cu_3O_{7-x}$ samples *Physica* C **266** 309

[254] Werner M, Sauerzopf F M, Weber H W, Veal B D, Licci F, Winzer K and Koblischka M R 1994 Fishtails in 123-superconductors *Physica* C **235–240** 2833

[255] Frischherz M C, Kirk M A, Farmer J, Greenwood L R and Weber H W 1994 Defect cascades produced by neutron irradiation in $YBa_2Cu_3O_{7-\delta}$ *Physica* C **232** 309

[256] Wisniewski A, Schalk R M, Weber H W, Reissner M and Steiner W 1992 Comparison of neutron irradiation effects in the 90 K and 60 K phases of YBCO ceramics *Physica* C **197** 365

[257] Wisniewski A, Brandstätter G, Czurda C, Weber H W, Morawski A and Lada T 1994 Comparison of fast-neutron irradiation effects in $YBa_2Cu_3O_{7-x}$ (123) and $YBa_2Cu_4O_8$ (124) ceramics *Physica* C **220** 181

[258] Maley M P, Willis J O, Lessure H and McHenry M E 1990 Dependence of flux-creep activation energy upon current density in grain-aligned $YBa_2Cu_3O_{7-x}$ *Phys. Rev.* B **42** 2639

[259] Fleischer R L, Hart H R Jr, Lay K W and Luborsky F E 1989 Increased flux pinning upon thermal-neutron irradiation of uranium-doped $YBa_2Cu_3O_7$ *Phys. Rev.* B **40** 2163

[260] Luborsky F E, Arendt R H, Fleischer R L, Hart H R Jr, Lay K W, Tkaczyk J E and Orsini D 1991 Critical currents after thermal neutron irradiation of uranium doped superconductors *J. Mater. Res.* **6** 28

[261] Wheeler R, Kirk M A, Marwick A D, Civale L and Holtzberg F H 1993 Columnar defects in $YBa_2Cu_3O_{7-\delta}$ induced by irradiation with high energy heavy ions *Appl. Phys. Lett.* **63** 1573

[262] Vichery H, Rullier-Albenque F, Pascard H, Konczykowski M, Kormann R, Favrot D and Colling G 1989 Effects of low temperature electron irradiation on single crystal and sintered $YBa_2Cu_3O_{7-\delta}$ *Physica* C **159** 697

[263] Konczykowski M and Gilchrist J 1990 AC screening measurement for the characterization of ceramic superconductors. II: electron irradiation effect on $YBa_2Cu_3O_7$ *Physica* C **168** 131

[264] Konczykowski M, Rullier-Albenque F, Yeshurun Y, Yacoby E R and Shaulov A 1991 Effects of electron irradiation on irreversibility line in superconducting $YBa_2Cu_3O_7$ *Supercond. Sci. Technol.* **4** S445

[265] Yacoby E R, Shaulov A, Yeshurun Y, Konczykowski M and Rullier-Albenque F 1992 Irreversibility line in $YBa_2Cu_3O_7$ samples. A comparison between experimental techniques and effect of electron irradiation *Physica* C **199** 15

[266] Civale L, Marwick A D, McElfresh M W, Worthington T K, Malozemoff A P, Holtzberg F H, Thompson J R and Kirk M A 1990 Defect independence of the irreversibility line in proton-irradiated Y–Ba–Cu–O crystals *Phys. Rev. Lett.* **65** 1164

[267] Thompson J R, Sun Y R, Civale L, Malozemoff A P, McElfresh M W, Marwick A D and Holtzberg F 1993 Effect of flux creep on the temperature dependence of the current density in Y–Ba–Cu–O crystals *Phys. Rev.* B **47** 14440

[268] Civale L, McElfresh M W, Marwick A D, Holtzberg F, Feild C, Thompson J R and Christen D K 1991 Scaling of the hysteretic magnetic behaviour in $YBa_2Cu_3O_7$ single crystals *Phys. Rev.* B **43** 13732

C2
BSCCO

P N Mikheenko, K K Uprety and S X Dou

C2.1 Introduction

Superconductors in the Bi–Sr–Ca–Cu–O system have been attracting great attention due to peculiar relationships between the structure and the superconductivity characteristics. A series of $Bi_2Sr_2Ca_{n-1}Cu_nO_{2n+4+y}$ compounds exists in the Bi–Sr–Ca–Cu–O system. The best known are the compounds with $n = 1$–3, namely $Bi_2Sr_2CuO_{6+y}$ (Bi2201), $Bi_2Sr_2CaCu_2O_{8+y}$ (Bi2212) and $Bi_2Sr_2Ca_2Cu_3O_{10+y}$ (Bi2223). Superconductivity in Bi2201 system was discovered by Michel *et al* [1] at a relatively low critical temperature T_c. Addition of Ca to this Bi2201 system led to the report of a superconducting transition between 80 and 110 K by Maeda *et al* (Bi2212 and Bi2223 phases) [2]. The T_c in the Bi–Sr–Ca–Cu–O system is not only a function of doped carrier density (holes) in every CuO_2 plane but also depends on the number of CuO_2 layers in a unit cell. D1.1.2

Bi compounds are of special importance for HTS applications, having apparently strongly coupled bulk critical currents [3]. Bi2212 silver sheathed tapes show a critical current density higher than 2×10^5 A cm^{-2} at 4.2 K and 10 T, and there is a plateau in $J_c(H)$ at low temperatures, extending beyond 20 T. This is well above the critical current density in the conventional superconductors NbTi and Nb_3Sn [3]. Bi2223 tapes show an excellent performance at high temperatures. For instance, a maximum critical current of 1.8×10^5 A cm^{-2} is reported in short filaments extracted from the Ag sheath tapes at 77 K ($H = 0$) [4]. This critical current density is above the commercial range for most applications (3–6×10^4 A cm^{-2}), and the main problem is to maintain it over the whole cross-section and over long lengths of tape. Already, tens of thousands of metres of the tapes are produced monthly, with the length of separate tapes up to 1000 m. This opens wide possibilities for BSCCO prototypes such as motors, power transmission lines, transformers, current limiters and magnetic separators. B3.1 B3.3.2, B3.3.3 E1.4.2 E1.3.2, E1.3.

Besides the technical applications, Bi–Sr–Ca–Cu–O compounds are very interesting in the study of the chemistry and physics of these materials. The structure of the Bi-compounds can be viewed as a stack of CuO_2 layers interleaved with Ca layers and BiO layers and sandwiched between SrO layers. The three layered phases ($n = 1$–3) have an energy of formation that is very close, preventing creation of a single phase (free of stacking faults) in the Bi–Sr–Ca–Cu–O bulk materials. For example, the $n = 3$ phase is always contaminated by $n = 2$ and 1 phases. A study of the chemical substitution or chemical doping in the system is important to understand the origin of the structural modulation in the Bi-phases, to synthesize new phases and to understand the superconductivity mechanism. Significant improvement in the stability of the $n = 3$ phase is observed on the partial substitution of Bi by Pb. Chemical doping in the $n = 2$ and 3 compounds is also relevant to study the flux pinning mechanism. Heavy Pb doping in the Bi2212 ($n = 2$) single crystal is reported to increase flux pinning. It should be noted that the Pb in E1.3.3, E2.2 D1.1.2

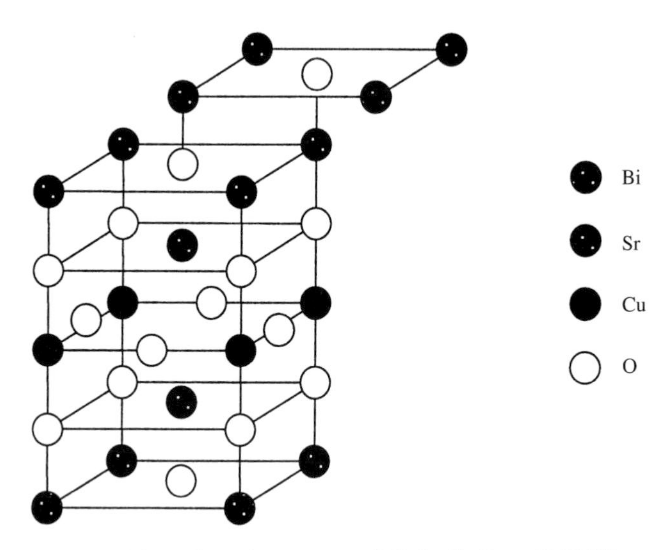

Figure C2.1. Crystal structure of $Bi_2Sr_2CuO_{6+y}$ (Bi2201).

A2.7 the Bi2212 crystal replaces Bi and increases *c*-axis conductivity (Josephson current). The enhanced pinning is due to strong Josephson coupling of the two-dimensional pancake vortices.

D1, D2, D3 In this chapter, the structural, thermal, mechanical, chemical, optical, normal and superconducting properties of BSCCO will be described.

C2.2 Structural properties

D1.1.2 The crystal structures of $Bi_2Sr_2Ca_{n-1}Cu_nO_{2n+4+y}$ ($n = 1$–3) are shown in figures C2.1, C2.3 and C2.4 [5].

Bi2201 phase (figure C2.1) has a pseudotetragonal symmetry with lattice parameters $a \approx b \approx 5.4$ Å and $c \approx 24.4$ Å [5]. The unit cell of Bi2201 contains four formula units and is a stack of atomic planes in the sequence $(BiO)_2/SrO/CuO_2/SrO/(BiO)_2/SrO/CuO_2/SrO/(BiO)_2$.

The typical atomic positions in the Bi2201 unit-cell are given in table C2.1 [5].

D1.1.2 The typical x-ray diffraction pattern of Bi2201 is presented in figure C2.2 ($n = 1$) [6], and the position of the main diffraction lines of a Bi2201 sample, their intensity and *h*, *k*, *l* indexes, are shown in table C2.2 [7].

Table C2.1. Bi2201 structure: symmetry and positions of ions

Atoms	Site	Symmetry	*x*	*y*	*z*
Bi	8l	m	0	0.2758(5)	0.0660(2)
Sr	8l	m	1/2	0.2479(9)	0.1790(4)
Cu	4e	2/m	1/2	3/4	1/4
O1	8g	2	3/4	1/2	0.246(2)
O2	8l	m	0	0.226(12)	0.145(4)
O3	8l	m	1/2	0.334(15)	0.064(5)

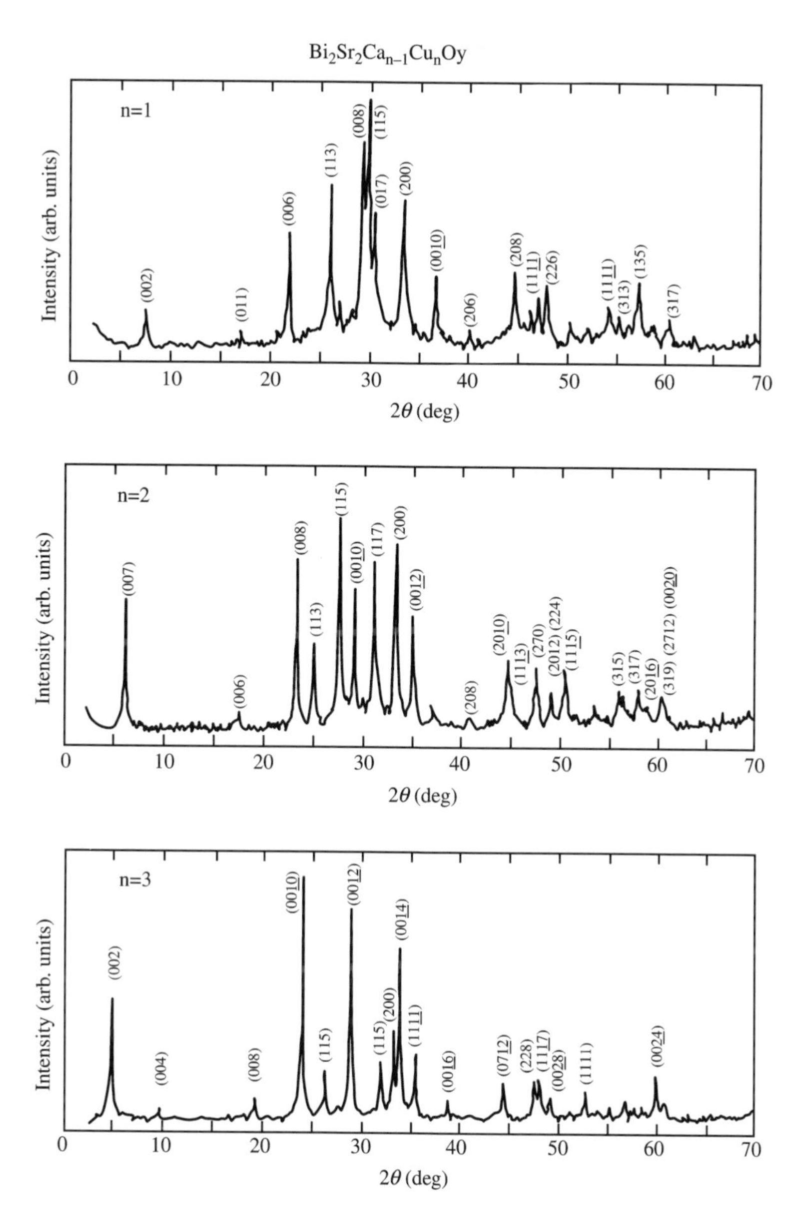

Figure C2.2. Cu Kα powder x-ray diffraction pattern for typical single-phase samples of the $n = 1$ (2:2:0:1), $n = 2$ (2:2:1:2), and $n = 3$ (2:2:2:3) samples. Nominal compositions are $Bi_{2.1}Sr_{1.9}CuO_y$, $Bi_2Sr_{1.8}Ca_{1.2}Cu_2O_y$ and $Bi_{1.85}Pb_{0.35}Sr_2Ca_2Cu_{3.1}O_y$, respectively. Reproduced by permission of Maeda [6].

Bi2212 phase (figure C2.3) has a pseudo-tetragonal structure with lattice parameters $a \approx b \approx 5.4$ Å and $c \approx 30.8$ Å [5]. The unit-cell of Bi2212 contains four formula units and is a stack of atomic planes in the sequence $(BiO)_2/SrO/CuO_2/Ca/CuO_2/SrO/(BiO)_2/SrO/CuO_2/Ca/CuO_2/SrO$.

The atomic positions in the Bi2212 unit-cell are given in table C2.3 [5].

A typical x-ray diffraction pattern of Bi2212 is presented in figure C2.2 ($n = 2$), and the position of the main diffraction lines of a sample, their intensity and h, k, l indexes, are shown in table C2.4 [8]. D1.1.2

Table C2.2. Bi2201 x-ray diffraction lines. Radiation: CuKα1, $\lambda = 1.54056$ Å [7]

2-θ	Int.	$h\ k\ l$	2-θ	Int.	$h\ k\ l$	2-θ	Int.	$h\ k\ l$
4.776	< 1		30.396	27	1 0 7	50.662	6	3 0 0
7.196	3	0 0 2	31.683	9	1 1 6	51.601	38	0 1 13
9.196	18		32.794	1	0 0 9	51.795	1	3 0 3
12.337	13		33.416	48	2 0 1	52.226	10	0 3 3
13.386	1		34.148	4	0 2 2	53.099	10	2 2 6
14.485	4	0 0 4	34.889	5	1 1 7	53.823	2	1 3 0
16.834	3	0 1 1	35.716	19		54.565	10	0 3 5
18.512	33	0 0 5	36.278	3	2 0 4	55.183	4	1 3 3
18.990	21		37.471	2	1 2 1	56.054	2	1 3 4
20.337	2		38.139	10	0 2 5	56.798	16	2 2 8
21.276	5		38.926	24	2 1 3	67.406	2	1 3 5
21.769	7	0 0 6	40.309	2	1 2 4	58.490	8	3 1 6
22.251	2	0 1 4	41.113	5	1 1 9	59.170	1	2 2 9
24.139	30	1 1 2	41.617	14	2 1 5	60.021	4	0 2 13
24.758	3	0 1 5	42.258	17	2 0 7	60.710	5	0 0 16
25.854	55	1 1 3	43.692	4	1 2 6	61.844	4	3 0 9
26.711	12		44.944	11	0 2 8	62.697	2	1 2 13
28.011	100	1 1 4	45.861	7	1 2 7	63.576	2	0 2 14
28.581	3		47.734	26	0 1 12	65.593	2	1 1 16
28.933	43	0 0 8	48.952	16	2 2 3	66.536	4	3 2 6
29.282	15		49.250	6		67.426	4	0 1 17
29.704	35	1 1 5	49.756	4	2 2 4	68.466	9	2 3 7

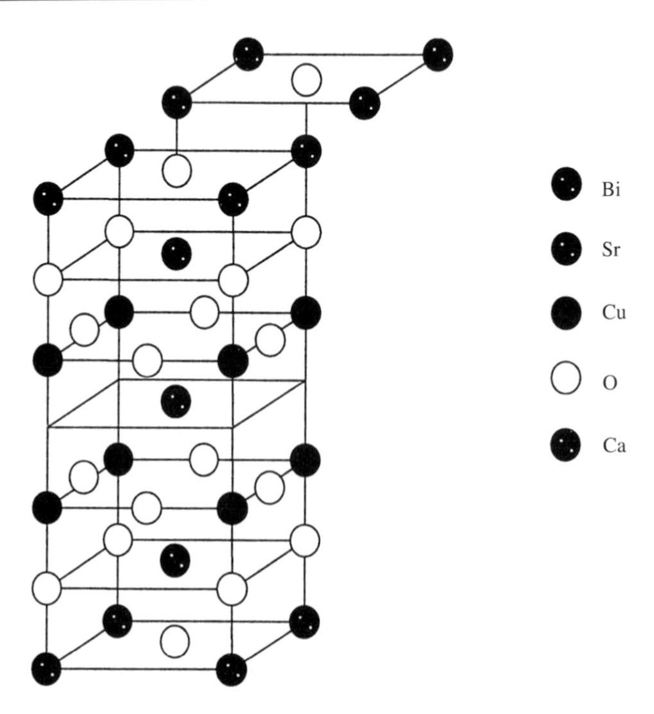

Figure C2.3. Crystal structure of $Bi_2Sr_2CaCu_2O_{8+y}$ (Bi2212).

Table C2.3. Bi2212 structure: symmetry and positions of ions

Atoms	Site	Symmetry	x	y	z
Bi	8l[a]	m	1/2	0.23	0.05
Sr	8l	m	0	0.25	0.14
Ca	4e	2/m	0	1/4	1/4
Cu	8l	m	1/2	0.25	0.20
O1	8h	2	1/4	0	0.19
O2	8h	2	1/4	1/2	0.20
O3	8l[a]	m	1/2	0.31	0.12
O4	8l[a]	m	0	0.20	0.05

[a] Atoms Bi, O3 and O4 can be modelled as split atoms in the 16m site.

Bi2223 phase (figure C2.4) has a pseudotetragonal symmetry with lattice parameters $a \approx b \approx 5.4$ Å B3.2.1 and $c \approx 37$ Å. The unit-cell of Bi2223 contains four formula units and is a stack of atomic planes in the sequence $(BiO)_2/SrO/CuO_2/Ca/CuO_2/Ca/CuO_2/SrO/(BiO)_2/SrO/CuO_2/Ca/CuO_2/SrO$. The atomic positions in the Bi2223 unit-cell are given in table C2.5 [5].

The typical x-ray diffraction pattern of the Pb-doped Bi2223 compound $Bi_{1.85}Pb_{0.35}Sr_2Ca_2Cu_{3.1}O_y$, D1.1.2 important for the manufacture of Bi2223/Ag tapes, is presented in figure C2.2 ($n = 3$). The position of the x-ray diffraction lines for Bi2223, their intensity and h, k, l indexes, are shown in table C2.6 [9].

The Bi2201, Bi2212 and Bi2223 structures are similar, and the difference between them is in the number of Ca/CuO_2 block layers. The block layers containing Sr, Bi, Ca, Cu and O, e.g., $(BiO)_2/SrO/CuO_2/Ca/CuO_2/SrO/(BiO)_2/SrO/CuO_2/Ca/CuO_2/SrO$ in Bi2212, have a perovskite type structure, and they are responsible for the transport of supercurrent. The $SrO/(BiO)_2/SrO$ block layer has the same structure as NaCl, and is a charge reservoir. The modulated structures arise due to the mismatch between NaCl-type layers and the perovskite-type layers [10].

Table C2.4. Bi2212 x-ray diffraction lines. Radiation: CuKα1, $\lambda = 1.5409$ Å [8]

2-θ	Int.	$h\,k\,l$	2-θ	Int.	$h\,k\,l$	2-θ	Int.	$h\,k\,l$
5.697	14	0 0 2	31.310	4	1 0 7	53.424	3	2 2 8
16.587	3	0 1 1	33.072	74	0 2 0, 2 0 0	54.311	1	3 1 3, 1 3 3
17.170	1	0 0 6	33.590	8	0 2 2, 2 0 2	55.691	15	3 1 5, 1 3 5
23.016	26	0 0 8	34.869	25	0 0 12	56.576	5	2 2 10
23.384	3	1 1 1	35.100	20	1 1 9	57.770	17	3 1 7, 1 3 7
27.214	5		36.809	6		58.620	6	0 2 16, 2 0 16
27.394	100	1 1 5	40.670	2	2 0 8, 0 2 8	60.357	15	2 2 12, 1 3 9
28.335	5	1 0 8	44.539	28	2 0 10, 0 2 10	67.449	2	1 3 13, 3 1 13
28.914	25	0 0 10	45.013	9	1 1 13	69.405	4	0 4 0, 4 0 0
29.639	6		47.520	26	2 2 0	70.194	1	0 2 20, 2 0 20
30.902	77		48.859	7	2 0 12, 0 2 12	71.714	1	1 3 15, 3 1 15
31.232	4	0 0 6	50.329	20	1 1 15			

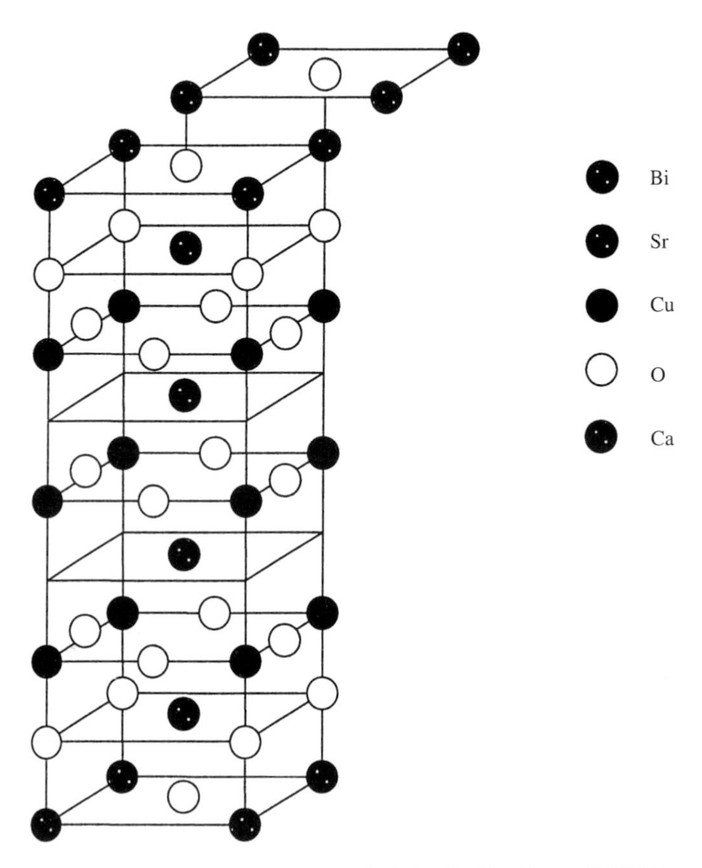

Figure C2.4. Crystal structure of $Bi_2Sr_2Ca_2Cu_3O_{10+y}$ (Bi2223).

D1.1.2 Bismuth compounds are susceptible to structural defects such as vacancies, intersite substitutions, interstitial oxygen, intergrowths and stacking faults. Despite this, the overall oxygen content of BSCCO C1 is stable. A notable difference of BSCCO from RE-123 HTS compounds is the absence of CuO chains

Table C2.5. Bi2223 structure: symmetry and positions of ions

Atoms	Site	Symmetry	x	y	z
Bi	4e	4 mm	0	0	0.21
Sr	4e	4 mm	1/2	1/2	0.16
Ca	4e	4 mm	1/2	1/2	0.05
Cu1	2a	4/mmm	0	0	0
Cu2	4e	4 mm	0	0	0.11
O1	4c	mmm	0	1/2	0
O2	8 g	mm	0	1/2	0.11
O3	4e	4 mm	0	0	0.16
O4	4e	4 mm	1/2	1/2	0.21

Table C2.6. Bi2223 x-ray diffraction lines. Radiation: CuKα1, $\lambda = 1.5406$ Å

2-θ	Int.	$h\,k\,l$	2-θ	Int.	$h\,k\,l$	2-θ	Int.	$h\,k\,l$
4.694	4	0 0 2	33.306	27	0 0 2	46.428	10	
16.524	5	0 1 1	33.634	24	1 1 6	47.399	37	2 0 2
23.405	4	1 1 1, 0 1 7	32.794	1	0 0 14	47.714	7	1 1 17
24.251	5	1 1 3	34.151	3		47.971	7	2 0 14
24. 792	5		34.476	2	2 0 4	48.656	2	2 2 4
25.597	19	0 0 4	35.060	5		50.219	11	
26.092	13	1 1 5	35.323	11	0 1 13	51.185	4	0 3 3
26.383	5		36.186	2	2 0 6	51.373	1	
27.397	8		36.719	8		52.701	3	2 1 15
27.738	6		41.094	5	0 2 10	55.010	2	3 1 5
28.701	10	1 1 7, 0 0 12	42.453	4		56.432	4	3 1 7
29.656	17		42.791	5		57.166	4	1 1 21
29.916	8		44.415	12	2 0 12	57.510	8	
31.595	7		44.634	5		60.448	6	1 3 11
31.932	41		44.876	6		60.659	5	0 3 13
32.995	100	0 2 0	45.864	1				

and a tetragonal–orthorhombic transition. As a consequence, (110) twins are absent in BSCCO. The compositions 2201, 2212 and 2223 are actually idealized, and represent formula numbers that are not exactly integers. D1.1.2

All of the Bi based superconductors exhibit structural modulations which are often incommensurate with the basic structure. The relationship of the structural modulation to the element substitution or chemical doping and their role with respect to the superconducting properties are of considerable interest and have been discussed in various papers [11–14]. The structural modulation in this superconductor can be obtained (a) from the decrease of Sr content in the SrO layers, resulting in the shift of the apical oxygen in the CuO_6 octahedron [11] (b) from ordering of Bi-depleted and Bi-concentrated zones and site exchange of cations [12] (c) from the extra oxygen in the Bi_2O_2 layers [13] (d) from the presence of Sr vacancies, leading to the distortion of the Sr–O bond [14]. Figure C2.5(*a*) shows an incommensurate structural modulation in a selected area electron diffraction patterns of Bi2212 single crystal. The incident beam is parallel to the [100] direction. The superlattice modulation has an incommensurate periodicity 4.7b along the *b*-axis. The stronger intensities of satellites as shown in the figure were observed due to multiple scattering along the *c*-axis. The incommensurate superlattice structure in the b^*–c^* plane of reciprocal lattice space is shown in the figure C2.5(*b*). Figures C2.5(*c*) and (*d*) are superlattice structure in the b^*–c^* plane but with 45 and 90° from *c*-axis, respectively. In [12], a study on the structural modulation of $Bi_2(Sr_{0.9}La_{0.1})_2Cu_{1-x}Co_xO_y$ crystal with $x = 0.2$, 0.6 and 1 using electron diffraction patterns have been presented. The authors have mentioned that the structural modulation with $x = 0.2$ and 0.6 is incommensurate, but with $x = 1.0$ is commensurate. Figure C2.6 shows the [001] high-resolution electron microscope image of the Bi2212 crystal. The Bi-concentrated bands, indicated by the wavy dotted lines, and the Bi deficient bands in the figure show the lattice distortion associated with the incommensurate modulation. D1.5

A structural modulation in Bi2201 leads to a monoclinic superlattice: $A = a = 5.4$ Å, $B = 5b = 26$ Å, $C = c/\cos(\alpha - 90°) = 28$ Å with average $\alpha = 120°$. The modulation wave vector has two incommensurate components: $q = \beta b^* + \gamma c^*$. Here β and γ are coefficients larger than zero. The value of α changes site by site between 110 and 130°. The structural modulation in Bi2212 leads to

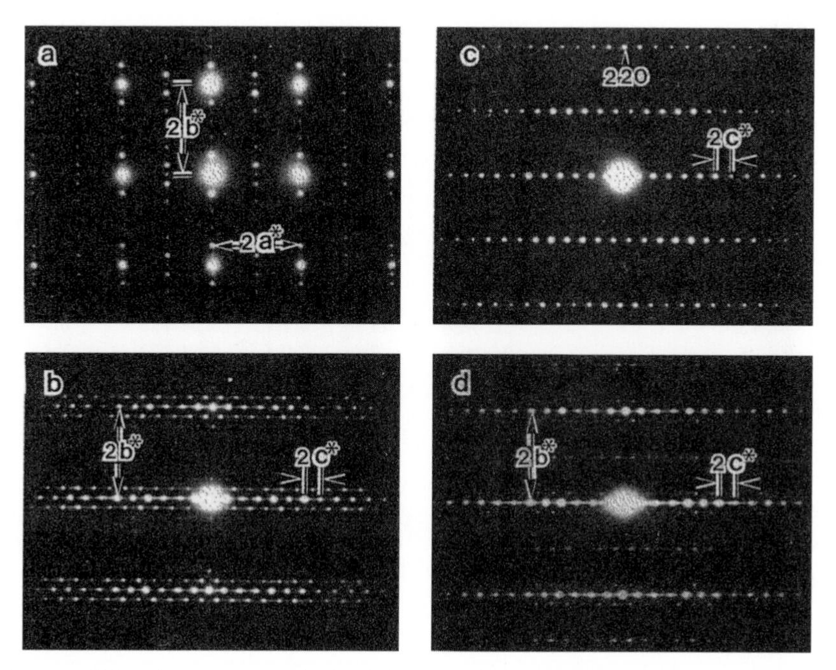

Figure C2.5. Electron diffraction patterns obtained from Bi2212 crystal projected along the [100] direction. A strong structural modulation is seen in the figure [17].

an orthorhombic supercell: $A = a = 5.4\,\text{Å}$, $B = 5b = 27\,\text{Å}$, $C = c = 30.8\,\text{Å}$. A transformation from the orthorhombic to a monoclinic structure has also been reported [15, 16]. Similarly, the supercell parameters in modulated Bi2223 are $A = a = 5.4\,\text{Å}$, $B = 5b = 26\,\text{Å}$, $C = c = 36.5\,\text{Å}$.

A clear understanding of the grain boundaries in Bi based superconductors is very important for technological applications. It is reported that textured Bi–Sr–Ca–O samples that are well aligned with the c-axis perpendicular to the direction of current flow, have a relatively higher critical current density than in more randomly oriented samples [19]. This suggests that the current-carrying capacity is a

Figure C2.6. [100] projected high resolution electron microscope image showing the incommensurate modulation in Bi2212 crystal [18].

function of the grain structure. Twin boundaries are absent in the Bi based superconductors. However, other types of grain boundaries such as twist, tilt and mixed boundaries are commonly observed. Figure C2.7 shows a schematic illustration of these three types of boundaries.

Twist boundaries are formed by one grain being rotated a certain angle with respect to the other about the common c-axis. Twist boundaries are best illustrated by a high resolution electron microscope (HREM). Figure C2.8 shows HREM images of 37.45° twist boundary in the Bi2212 bicrystals (crystal A and crystal B). In the figure, the boundary is located in the middle of BiO double layers. At the twist boundary a lamella of $n = 3$ phase is seen in the $n = 2$ matrix directions (see figure C2.8). The dark wavy line represents the BiO layer. Tilt boundaries are obtained by rotating one part of a crystal by a certain angle, ψ, with respect to the other around a common rotation axis within the grain boundary plane (see figure D.1.2.7). Similarly, mixed boundaries are obtained by both twisting and tilting operations. Detailed studies of the twist, tilt and mixed boundaries are given in [19].

D1.2

Lead doping in the Bi based superconductors plays a very important role in the phase stability mechanism. Bulk materials with a single phase do not exist in the Br–Sr–Ca–Cu–O system. For example, in the microstructure of the $n = 3$ phase, generally single (c) or half unit cells (c/2) of $n = 1$ and 2 phases are observed [19]. This is because the energies of formation of the $n = 1$, 2 and 3 phases are similar. The phase contamination leads to a formation of intergrowth defects. The intergrowth defects can be considered as stacking defects of the perovskite layers along the c-axis. These defects are also observable near the grain boundaries. The intergrowth defects result in the appearance of steps in the resistivity plot of the materials. With Pb doping, intergrowth defects are reduced and within a grain, almost no intergrowth defects are present. This increases electrical conductivity between the grains, and as a result of which a sharp zero resistance temperature is observed. It is mentioned that Pb doping leads to a structural modulation of Bi-based superconductors, since Pb in the superstructure modifies the incommensurate superlattice modulation present in the undoped materials. Recently, it has been reported that heavy Pb doping in Bi2212 single crystal improves flux pinning [20, 21]. There is now great interest in studying the physics behind the improved flux pinning.

B3.2.1
B2.3.5
A4.3

As in the case of figure C2.2 ($n = 3$), the Bi2223 phase is usually obtained with partial substitution of Pb for Bi. This substitution promotes the formation and stabilization of the Bi2223 phase [6, 19]. The mechanism for the stabilization is well explained by the structure stability and adjustable oxygen stoichiometry due to the variable valence of Pb (Pb^{4+}/Pb^{2+}). The x-ray diffraction patterns for Bi2212

D1.1.2

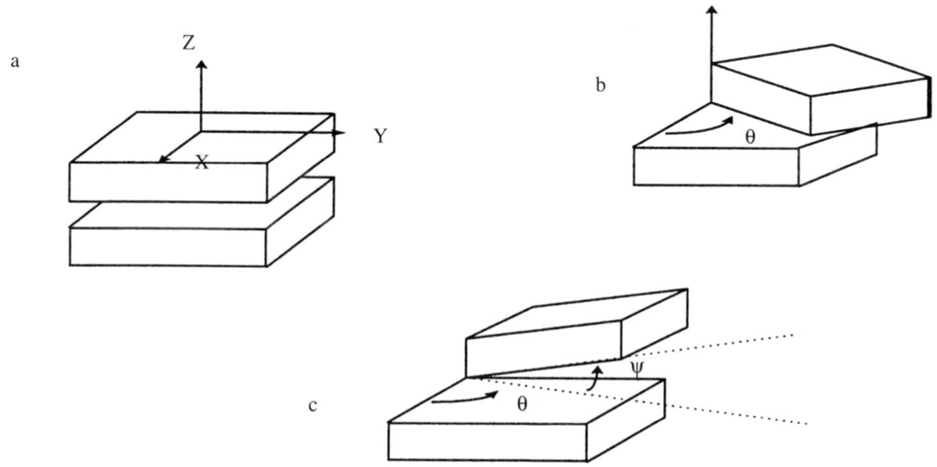

Figure C2.7. Schematic diagrams showing (a) a twist boundary (b) a tilt boundary and (c) a mixed boundary.

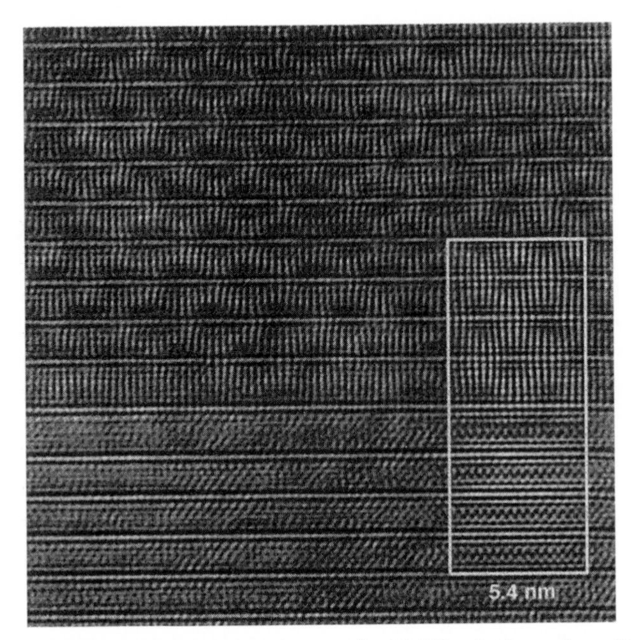

Figure C2.8. HREM of Bi2212 bicrystals showing an image of a 37.45° twist boundary. The embedded image is the calculated [001] twist boundary. The boundary is seen in the middle of BiO double layers of the embedded image [18].

samples with up to 40% of substituted Bi ions are basically the same as those from undoped material, except that peaks are systematically shifted to higher angles with increasing Pb substitution [22].

C2.3 Thermal properties

C2.3.1 Specific heat

In BSCCO, the temperature dependence of specific heat (C_p) is different in the low-T and high-T ranges. Unlike YBCO, the specific heat of the BSCCO superconductors does not have a linear term at low temperature [23–26]. It is suggested that the linear term in Y123 arises due to the concentration of the Cu^{2+} moments in the Y123 lattice, which locally suppresses the superconductivity and that normal conductivity region contributes to the linear term in the specific heat [27]. The absence of the linear term at low temperature in the BSCCO system is described in terms of much lower concentration of Cu^{2+} in its lattice [24]. A detailed work on the specific heat at low temperature has been done in Bi2212 samples doped with Co, Fe and Zn by Yu *et al* [24, 25]. They observed a linear term in Co or Fe doped Bi2212 samples. No such linear behaviour is observed in Zn-doped Bi2212 samples. The linear term in the specific heat at low temperature arises from the magnetic impurities in BSCCO lattice because the Co or Fe substitutes magnetically in BISCO, whereas the Zn substitutes non-magnetically. In the low-T range, the specific heat of BSCCO system is described by the expression $C_p(T) = AT^{-2} + \gamma^* T + \beta T^3$, where AT^{-2} is the low temperature upturn due to Schottky anomalies, $\gamma^* T$ is a linear term and βT^3 is the phonon term. In Bi2201, the value of γ^* is low, and varies for different samples from 3 to 10 mJ K^{-2} mol. The Debye temperature is in the range of 218–232 K [28]. In Bi2212, $\gamma^* \approx 2$ mJ K^{-2} mol. In Bi2223, γ^* varies from 0 to 11 mJ K^{-2} mol. The value $\gamma^* \approx 2$ mJ K^{-2} mol can be considered as intrinsic, and it is a useful measure of the sample quality.

A2.4, D3.2.1

C1

Important information about the nature of superconductivity is provided by the specific-heat jump at the critical temperature. Despite many attempts, no specific-heat anomaly has been observed in Bi2201. A T_c anomaly in Bi2212 and Bi2223 is present, though it is not the specific-heat jump, but rather a small peak formed by the critical fluctuations. Far enough from T_c, the specific heat shows features of two-dimensional behaviour. A2.4

C2.3.2 *Thermal conductivity*

The low value or absence of the γ^* linear term in the specific heat is correlated with the absence of linear T dependence in the thermal conductivity (k) of Bi22(1-n)n. This absence is in contrast to $k(T)$ behaviour of most other HTS. The low-temperature double logarithmic plot of $k(T)$ for hot-pressed and single-crystalline Bi2212 specimens is shown in figure C2.9 [29]. The T^2 instead of the T behaviour is evident. A2.4 D3.2.2

The temperature dependence of thermal conductivity of Bi2212 crystals along the *ab*-plane and *c*-axis are shown in figures C2.10 and C2.11, respectively [30, 31]. Figure C2.10 shows phonon and electron components of the thermal conductivity. An anomaly at the critical temperature is evident in the *ab*-plane conductivity. The fast drop of k below T_c is connected with the formation of Cooper pairs that do not carry entropy (loss of the electron thermal conductivity). The k increases at intermediate temperatures because of the growth in the mean free path of phonons that are not scattered by the Cooper pairs. No substantial anomaly at T_c is found in the *c*-axis thermal conductivity. Figure C2.12 demonstrates the growth of the anisotropy ratio for the thermal conductivity below T_c [31].

The influence of magnetic field on the *ab*-plane thermal conductivity of Bi2212 is shown in figure C2.13 [32]. The decrease of thermal conductivity below T_c is the result of the additional phonon scattering on magnetic flux vortices. This effect is absent at very low temperatures, because the phonon A4.2

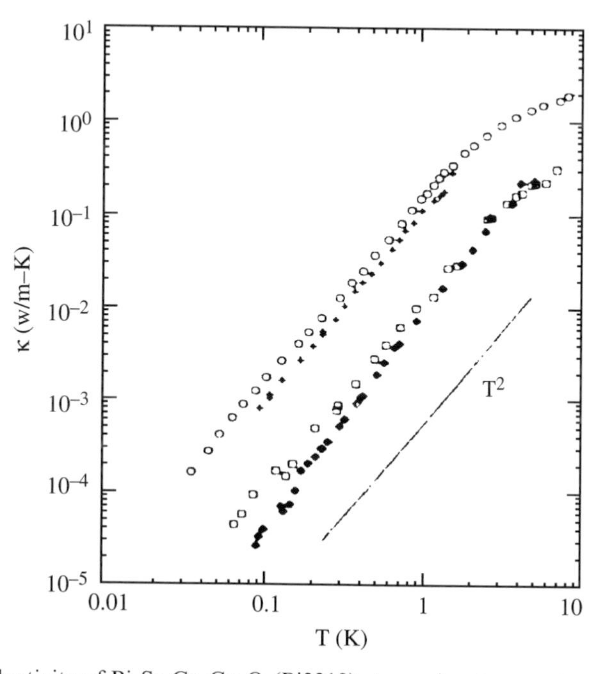

Figure C2.9. Thermal conductivity of $Bi_2Sr_2Ca_1Cu_2O_8$(Bi2212) at very low temperatures. ◇ and □ are hot-pressed materials; △ and ○ are single crystals. Reproduced by permission of Uher [29].

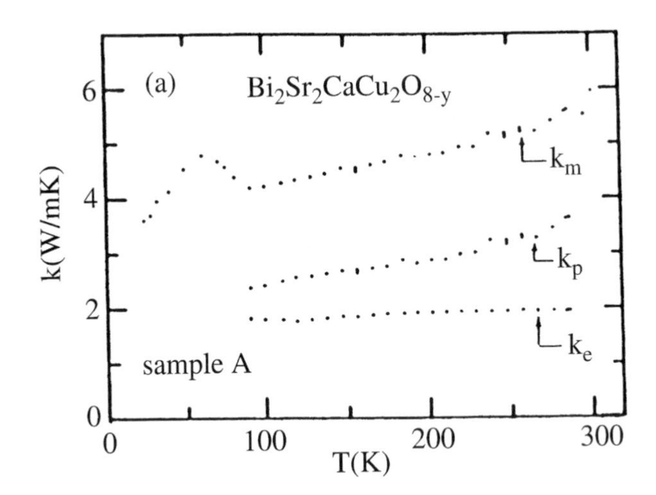

Figure C2.10. Temperature dependence of *ab*-plane thermal conductivity (k_m) of $Bi_2Sr_2CaCu_2O_{8-y}$ (Bi2212). Phonon (k_p) and electron (k_e) thermal conductivities are indicated. Reproduced by permission of Crommie [30].

wave length becomes larger than the vortex core size. Comparable to 1-2-3 compounds, the influence of magnetic field on thermal properties is less pronounced in BSCCO.

D3.2.2 The thermal conductivity of highly c-oriented $(BiPb)_2Sr_2Ca_2Cu_3O_{10}$ tapes has been investigated by Castellazzi *et al* in the range 20–220 K [33]. In the tapes, a maximum thermal conductivity has been observed just below the critical temperature. As with Bi2212 crystals, a fast drop of thermal conductivity A3.2 below T_c for Bi2223 is related to the formation of the Cooper pair. Furthermore, the authors have employed three different hypotheses: phonon approach, electron approach; both phonon and electron approach to analyse the experimental data. The best fit function with the experimental data was reported to be obtained only from the electron phonon approach, supporting electron–phonon coupling.

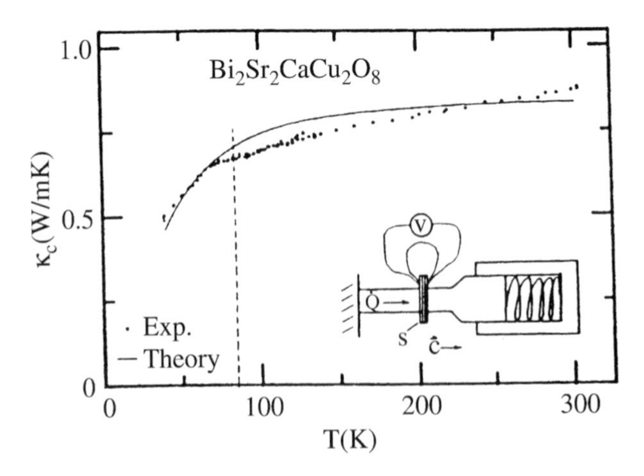

Figure C2.11. *c*-axis thermal conductivity (k_m) for $Bi_2Sr_2CaCu_2O_8$ (Bi2212). T_c is indicated by a dashed vertical line. The solid line is a fit to theory. Reproduced by permission of Crommie [31].

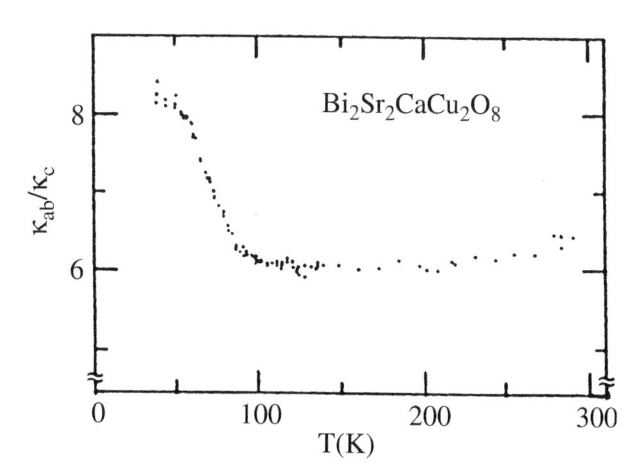

Figure C2.12. Thermal conductivity anisotropy ratio (k_{ab}/k_c) for $Bi_2Sr_2CaCu_2O_8$ (Bi2212). Reproduced by permission of Crommie [31].

C2.3.3 Thermoelectric power

The thermoelectric power (TEP), e.g. voltage that appears across the sample at the application of a temperature gradient, in BSCCO is similar to TEP of other HTS. The temperature dependence of TEP for Bi2212 sintered rods at different hole concentrations is shown in figure C2.14 [34]. The TEP changes sign near the maximum critical temperature, and it is negative in the overdoped region. The TEP across the two superconducting–non-superconducting boundaries changes continuously with doping. There is

D3.1

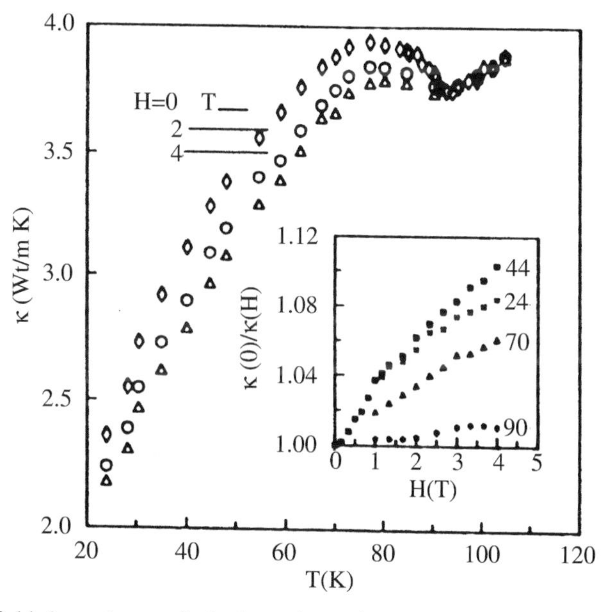

Figure C2.13. Magnetic field dependence of *ab*-plane thermal conductivity for $Bi_2Sr_2Ca_1Cu_2O_x$. Reproduced by permission of Zavaritsky [32].

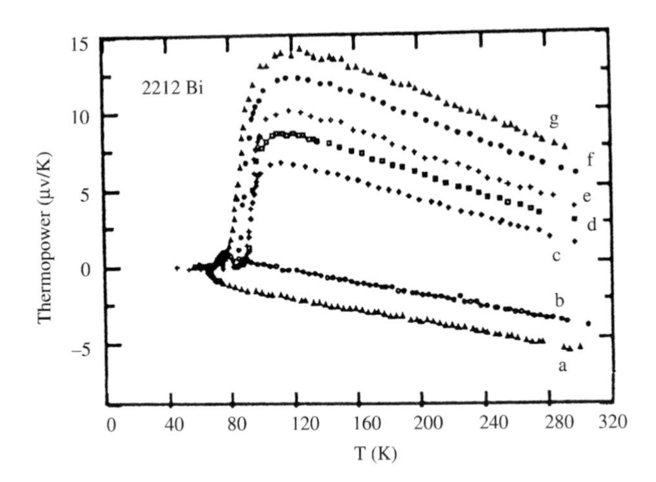

Figure C2.14. Thermoelectric power *versus* temperature for $Bi_2Sr_2Ca_1Cu_2O_{8+x}$. The relative values of x are 0.077, 0.051, 0, 0.01, -0.07, -0.025 and -0.035 for curves a, b, c, d, e, f, and g, respectively. Reproduced by permission of Obertelli [34].

a universal linear correspondence of the room temperature TEP with the doping level for nearly all HTS. This reflects the dominant contribution to TEP from the CuO_2 planes [34]. The temperature D3.1 dependencies of TEP for Bi2212 and Bi2223 single crystals and sintered ceramics are similar to each other [34].

C2.4 Mechanical properties

D3.3 The maximum density (ρ) of Bi2201, Bi2212 and Bi2223 ceramics is around 7.05, 6.51 and 6.19 g cm^{-3}, respectively. The specific mechanical properties of BSCCO reflect weak bonds between BiO layers and B2.2.3 the micaceous structure of polycrystals that allow the plastic deformation. Due to the weak bonds between Bi layers, Bi2212 crystals are easily cleaved and can be peeled off by a gentle rubbing of the surface. This produces perfectly glossy surfaces with an intact oxygen content. This procedure erodes the sample slowly and can be repeated many times [35].

The deviation of the crystal structure in BSCCO from the ideal perovskite structure is larger than in most other HTS. Because of this, the compressibility of BSCCO is nearly twice as high as in other HTS. The change in the Bi2212 unit-cell volume at 10 GPa is as large as 10.6%. The corresponding changes of the a, b and c lattice parameters are 2.2, 2.2 and 6.2%, respectively.

Important mechanical parameters of HTS are the Young's modulus, rigidity modulus, flexural stress, fracture toughness, critical crack length and hardness. In polycrystalline and highly textured Bi2212 and Bi2223 samples, all parameters are strongly dependent on the porosity. The elastic and mechanical properties of Bi2212 and Bi2223 are very close to each other (within a difference of 2%). The elastic stiffness in BSCCO is inferior to what is seen in other HTS because of the weak bonds between the Bi layers.

The fraction porosity dependence of Young's modulus for Bi2223 rods is shown in figure C2.15 [36]. The Young's modulus decreases with the porosity. The extrapolation of Young's modulus to zero porosity yields 127 GPa. This is higher than what is found from longitudinal shear wave velocity C1 measurements in Bi(Pb)2212 (about 82 GPa) [37], close to the data in $YBa_2Cu_3O_x$ and $La_{1.8}Sr_{0.2}CuO_4$ [38] and about half the value from atomistic calculations (230 GPa) [39].

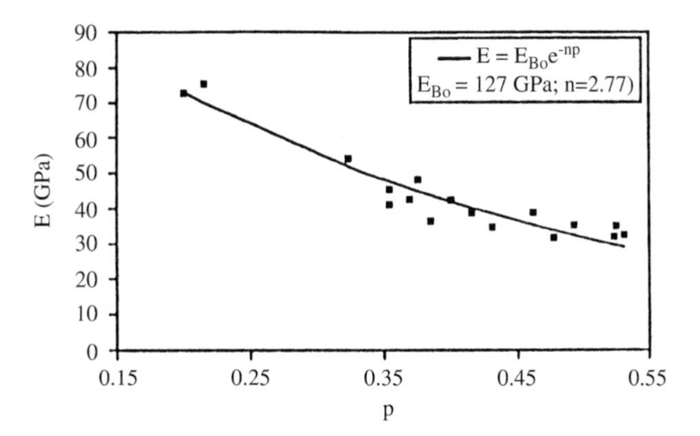

Figure C2.15. Variation of Young's modulus, *E*, for Bi2212 and Bi2223 rods with volume fraction porosity, *p*. Reproduced by permission of Oduleye [36].

The variation of bending strength with porosity is shown in figure C2.16. Usually, the bending strength is 2–3 orders of magnitude less than the Young's modulus. The reduction in the bending strength is not the direct result of porosity, but rather a reflection of the reduction in fracture energy and elastic modulus. The zero porosity fracture toughness in Bi2223 is 1.1 MPa $m^{1/2}$, and the critical crack length is 20–100 μm [36]. The mechanical parameters of Bi2223 rods are very close to the mechanical parameters of Bi2223/Ag tapes, and this is responsible for the strain tolerance of Bi2223/Ag tapes around 0.2%. The rigidity modulus in BSCCO is usually 2–3 times less than its Young's modulus [37].

D3.3
B3.2.1
B3.2.2

An important mechanical parameter is the Vickers microhardness (HV). In BSCCO, it is a strong function of porosity or fractional density (see figure C2.17) [36]. Data similar to what is shown in figure C2.17, are also found for Bi2223/Ag tapes, with an important exclusion for the case of hot hydrostatic extrusion where HV is as high as 330 [40].

In most works on the micro hardness of HTS, a large applied force of the order of 1 N was used. An experiment on Bi2212 single crystals shows, however, that cracks can be induced at an applied force as low as 20 mN [41]. Using low force piezo-electric measurements, the microhardness in Bi2212 crystals was measured and found to be 374 and 72 HV perpendicular to and along the *ab*-plane, respectively,

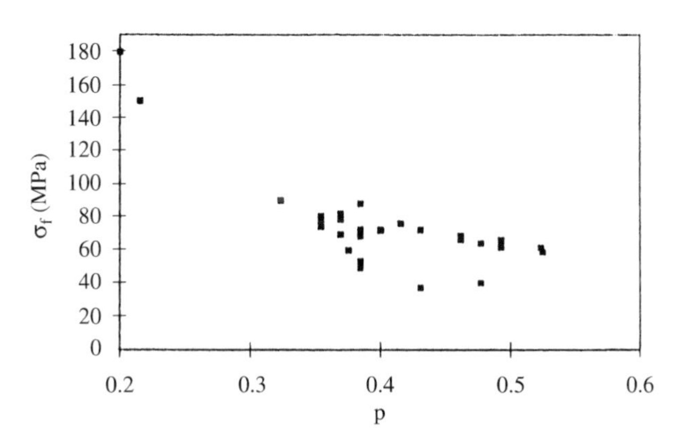

Figure C2.16. Variation of bending strength, σ_f,for Bi2212 and Bi2223 rods with volume fraction porosity, *p*. Reproduced by permission of Oduleye [36].

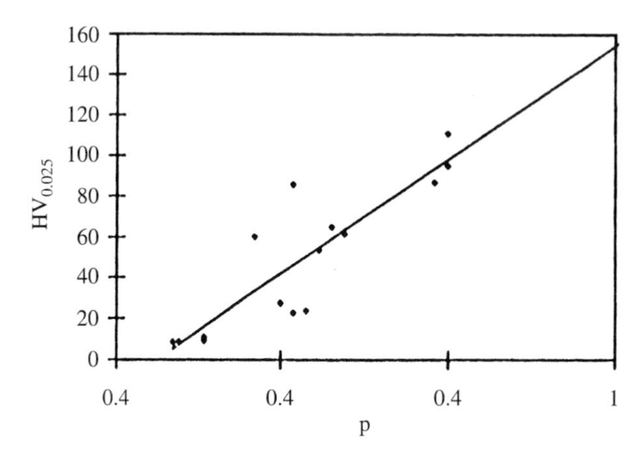

Figure C2.17. Variation of microhardness, HV, for Bi2212 and Bi2223 extruded rods with fractional density, ρ. Reproduced by permission of Oduleye [36].

both at a maximum load force of 10 mN. This five times anisotropy is also evident in the Young's modulus and the compression strength. The anisotropy in the elastic recovery rate and irrecoverable energy losses during the load–unload cycle were also determined [42]. The single crystal data are consistent with the data of the best polycrystal samples.

B2.4

D3.3

There is a clear relationship between mechanical properties and the velocity (V) of the ultrasonic waves. Thus, measurements of the velocity allow us to investigate the anisotropic mechanical properties of BSCCO. The anisotropic ultrasonic parameters of highly textured Bi2212 ceramics are shown in table C2.7 [42]. The velocity of the longitudinal wave along the *ab*-plane is 4370 ms^{-1}, nearly twice as high as along the *c*-axis (2670 ms^{-1}). This means that elastic stiffness in the *ab*-plane is large compared to the *c*-axis stiffness defined by the weak interlayer forces. In figure C2.18, the temperature dependence of both the *ab*-plane and *c*-axis components of the wave velocity is shown. Both components grow as the temperature decreases, showing no anomaly at the critical temperature. The pressure dependence of the extracted elastic modulus (ρV^2) for both components is shown in figure C2.19 [42]. The pressure dependence of the elastic modulus along the *c*-direction is substantially higher than the pressure dependence of the *ab*-plane elastic modulus.

C2.5 Chemical properties

B3.2.1

B2.3.5

The Bi2201, Bi2212 and Bi2223 phases are homologous phases and often intergrow. Due to different reactivity and some volatility of constituents, it is difficult to prepare single-phase specimens. Being formed in a conventional solid-state reaction, the phases are frequently accompanied by other compounds.

B2.1

Bi2201 is usually prepared from high purity $SrCO_3$, Bi_2O_3 and CuO. The preparation includes several 16–20 h calcinations at 600–850°C with intermediate grindings and quenching to room or liquid N_2 temperature. The $SrCO_3$–Bi_2O_3–CuO phase diagrams for Bi2201are presented in figure C2.20 for the treatment at 875–925°C [43, 44]. The Bi2201 phase shown in the diagrams as 2:2:1 is formed in a narrow area and can be accompanied by $Sr_8Bi_4Cu_5O_{19+x}$ (8:4:5), $Sr_3Bi_2Cu_2O_8$ (3:2:2), $Sr_{14}Cu_{24}O_{41}$ and Raveau solid solution $Sr_{1.8-x}Bi_{2.2+x}Cu_{1\pm x/2}O_z$ (R_{ss}). Figure C2.20(*b*) is an enlargement of the dotted region of (*a*). In this figure, two distinct phases are observed. The phase 2:2:1indicated by an arrow in (*b*)

Table C2.7. Measured velocity, V, modulus, ρV^2 and hydrostatic pressure derivative, $d(\rho V^2)/dP$, for $f = 10$ MHz ultrasonic waves propagated in highly textured Bi2212 at 290 K

Direction of Propagation	Mode	V (ms^{-1}) ($\pm$10)	ρV^2 (GPa)($\pm$1)	$d(\rho V^2)/dP$
Along *c*	Longitudinal	2670	44	22
Along *c*	Shear	1750	19	3.6
Along *ab*-plane	Longitudinal	4370	118	12.6
Along *ab*-plane	Fast shear[a]	2460	38	3.4
Along *ab*-plane	Fast shear[b]	1740	19	2.9
45° *ab*-plane	Longitudinal	3430	73	
45° *ab*-plane	Shear	2150	29	

Reproduced by permission of Chang [42].
[a] Polarized in *ab*-plane.
[b] Polarized along *c*-axis.

is a semiconductor. This phase is very close to the ideal composition $Bi_2Sr_2CuO_6$. Superconductivity is observed only in the phase $Sr_{1.8-x}Bi_{2.2+x}Cu_{1\pm x/2}O_z$ (Raveau solid solution, R_{ss}).

There are a number of different techniques for the growth of Bi2212 and Bi2223. One of them is a solid-state reaction. High purity powders of SrO, CaO, Bi_2O_3 and CuO are used in the preparation of Bi2212 and Bi2223. SrO and CaO are usually obtained by calcining $SrCO_3$ and $CaCO_3$ at 1200°C. The B2.1 B2.2.3

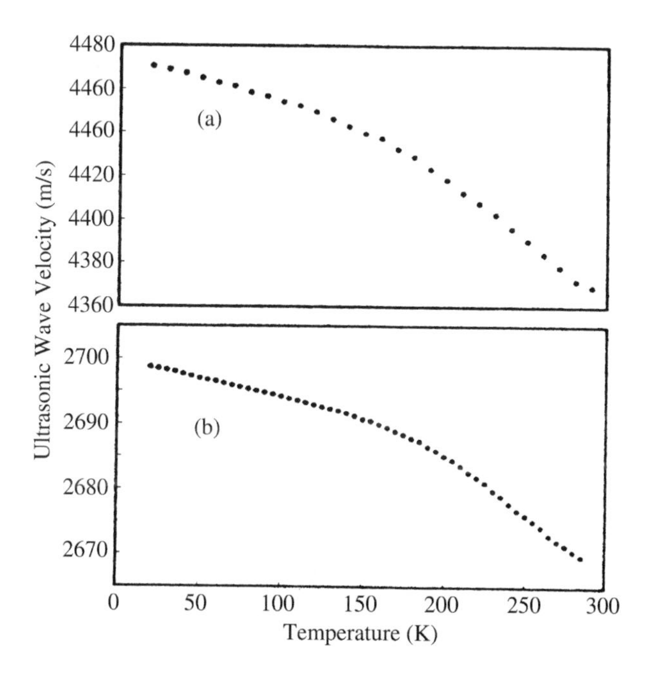

Figure C2.18. Temperature dependence obtained during cooling of the velocity of longitudinal ultrasonic waves propagating (*a*) in the *ab*-plane and (*b*) along the *c*-direction of textured Bi2212. Reproduced by permission of Chang [42].

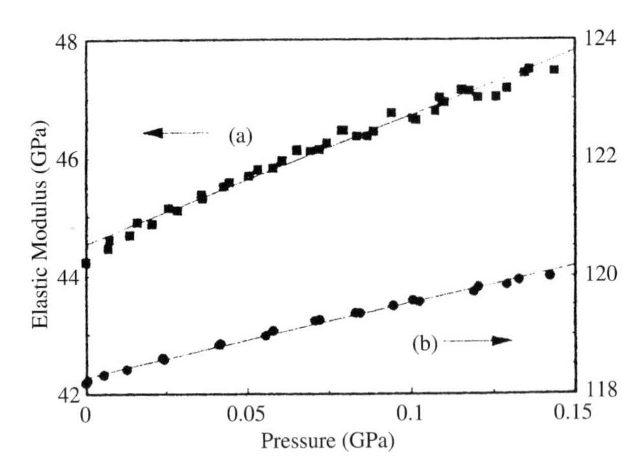

Figure C2.19. Hydrostatic pressure dependence of elastic stiffness for longitudinal waves propagating (*a*) along the *c*-direction and (*b*) in the *ab*-plane of textured Bi2212. Reproduced by permission of Chang [42].

preparation includes several calcinations of SrO, CaO, Bi_2O_3 and CuO at 700–800°C with intermediate grindings and quenching to room or liquid N_2 temperature. The typical quaternary sections of Bi2212 and Bi2223 phase diagrams for a treatment temperature of 850°C are presented in figures C2.21 and C2.22, respectively [45]. For the composition (2SrO + CaO)–Bi_2O_3–CuO, only Bi2212 is formed.

B3.2.1 B2.3.5

The compounds in the Bi–Sr–Ca–Cu–O system have a complicated series of binary, ternary and quaternary phases. Discussions of the different phase equilibria at various temperatures in the binary, ternary and quaternary systems have been made by Majewski [46]. The phase equilibria in the quaternary system are represented by the systems: Bi_2O_3–SrO–CaO–CuO. However, the phase equilibria in binary system are constructed by choosing two metal oxides at a time, for example Bi_2O_3–SrO or SrO–CaO or others. In the ternary system, the phase equilibria are constructed from the systems: SrO–CaO–CuO or Bi_2O_3–SrO–CuO or Bi_2O_3–CaO–CuO or Bi_2O_3–SrO–CaO. Majewski has reported 16 four phase equilibria of Bi2212 at 830°C in air and 6 four phase equilibria of Bi2223 at 850° C [46]. In (Bi, Pb)–2223, the 11 four-phase equilibria has been reported [46]. Wong-Ng *et al* have described 29 five phase equilibria of the (Bi, Pb)–Sr–Ca–Cu–O at different temperatures in 7.5% O_2 and16 five phase equilibria containing 2223 + 2212 phases [47, 48]. The projection of the single phases Bi2212 and Bi2223 onto the ternary system at 830°C and Bi2201 at 850°C is shown in figure C2.23 .

B3.2.2

To facilitate the formation of Bi2223, PbO is frequently added. The stabilizing and promoting effect of Pb was mentioned above [22]. An effect of Pb on the Bi2223 phase evolution and microstructure development has also been described by Luo *et al* [49]. The authors suggested that the Pb addition promotes the formation of the liquid phase and this liquid phase converts non-superconducting phase to Bi2212 grains and finally to Bi2223 phase. A typical phase diagram for $Sr_xCa_yBi_{1.84}Pb_{0.34}Cu_2O_w$ is shown in figure C2.24 [50]. Bi2223 phase is formed in the area labelled ‘C’. A slight excess of Ca and Cu is necessary to form Bi2223. In Pb-doped samples, the Ca_2PbO_4 is responsible for the accelerated formation rate of Bi2223. A pseudo phase diagram for the $(Bi_{1.6}Pb_{0.4})_2Sr_2CaCuO_2$ system has been described by Strobel *et al* using the data of differential thermal analyses (DTA) and x-ray powder diffraction (XRD) on quenched samples [51, 52]. This diagram has many transformation lines and therefore is difficult to study. Recently, Suzuki *et al* have presented a simpler pseudo-binary phase diagram for the $(Bi_{0.8}Pb_{0.2})_2Sr_2CaCuO_2$ system using a newly developed high-temperature XRD apparatus with the simultaneous use of an optical microscope [53]. This phase diagram is shown in figure C2.25 . In this plot, vertically aligned dots indicate the measurements on heating by 1°C per min, open

B2.1 D1.1.2

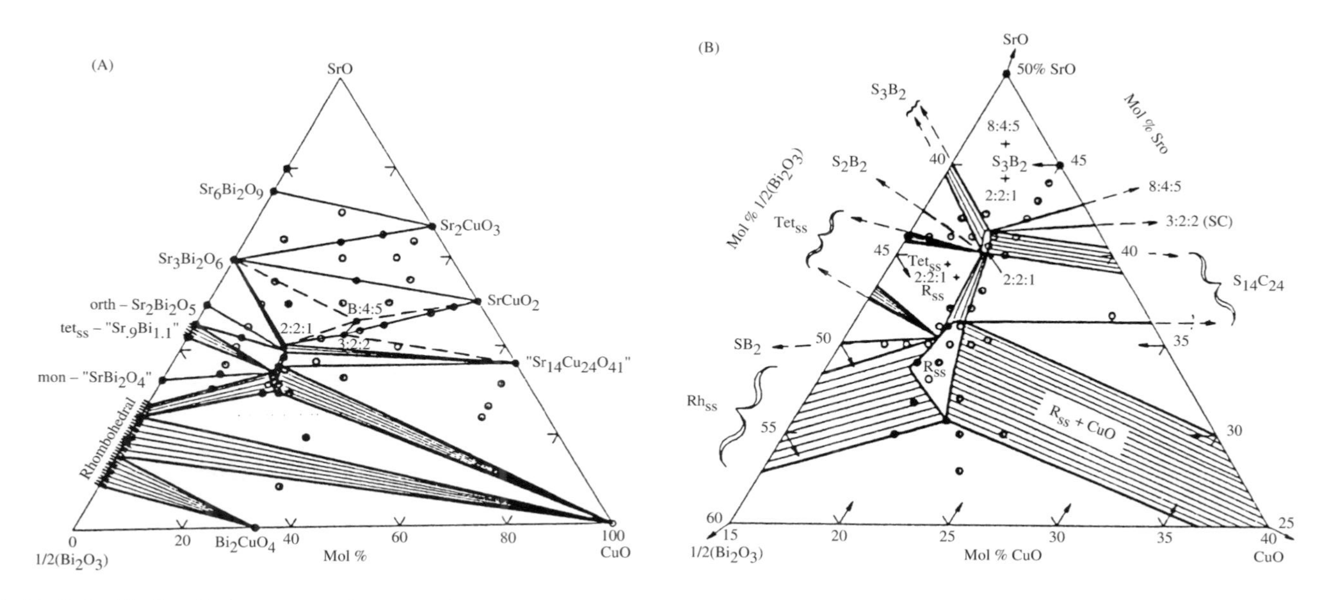

Figure C2.20. SrO–Bi_2O_3–CuO system at 875 – 925°C. (*a*) Melting occurs below 875°C below the CuO-rhombohedral solid solution join; subsolidus relations are shown. (*b*) Enlargement of dotted region of figure (*a*). 8:4:5 = $Sr_8Bi_4Cu_5O_{19+x}$; 3:2:2 = $Sr_3Bi_2Cu_2O_8$; 2:2:1 = $Sr_2Bi_2CuO_6$; R_{ss} = Raveau solid solution $Sr_{1.8-x}Bi_{2.2+x}Cu_{1\pm x/2}O_z$. Reproduced by permission of Roth [43, 44].

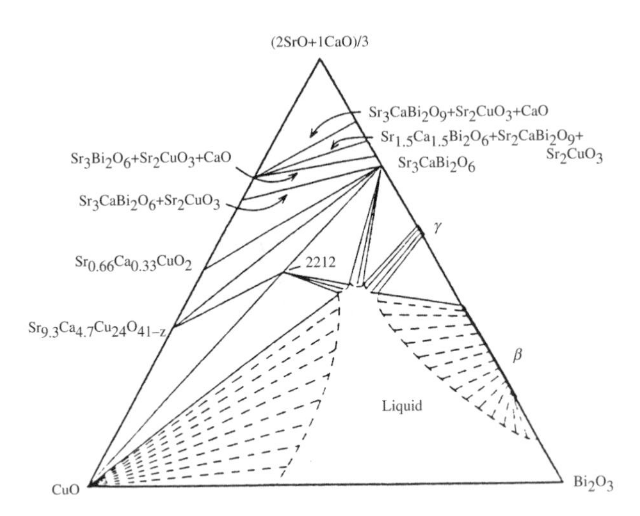

Figure C2.21. Quaternary section (2SrO + CaO)/3 − 1/2Bi_2O_3 − CuO at 850°C in air. Compositions expressed in terms of SrO:CaO:1/2Bi_2O_3:CuO ratios. Reproduced by permission of Schulze [45].

circles mean holding points for the measurements at fixed temperatures, the horizontal dotted line between 2212 and 2223 indicates the decomposition line, the mark 'am' means amorphous phase and 'L' is a liquid phase. The phase diagram shows melting of 2223 phase above 874°C and the presence of the amorphous phase or low crystallinity solid state between 896 and 925°C. At 925°C, the amorphous phase disappears and mixed phases, (Sr,Ca)O, $(Sr,Ca)_2CuO_3$ and 'L' appear. The diagram also shows that 2212 phase begins to melt above 881°C and the 2201 phase melts above 891°C.

Chemical substitutions in the Bi-based superconductors are considered as a powerful tool for understanding the origin of structural modulation in the Bi-phases, for synthesizing new phases and for understanding the superconductivity mechanism. Most of the studies on substitution are done in $n = 1$

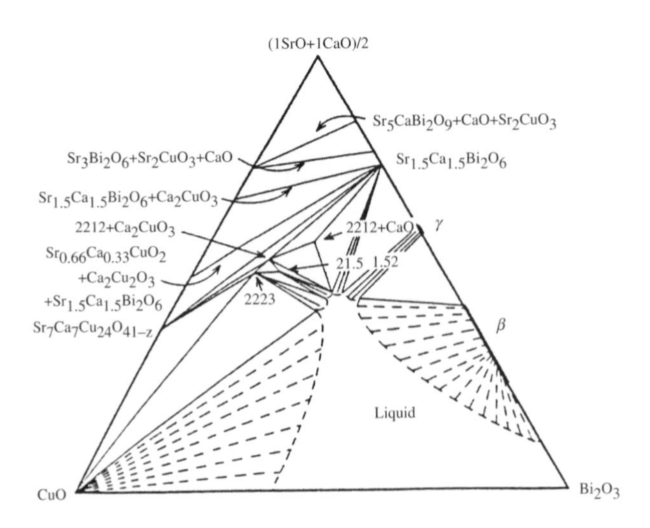

Figure C2.22. Quaternary section (SrO + CaO)/3 − 1/2Bi_2O_3 − CuO at 850°C in air. Compositions expressed in terms of SrO:CaO:1/2Bi_2O_3:CuO ratios. Reproduced by permission of Schulze [45].

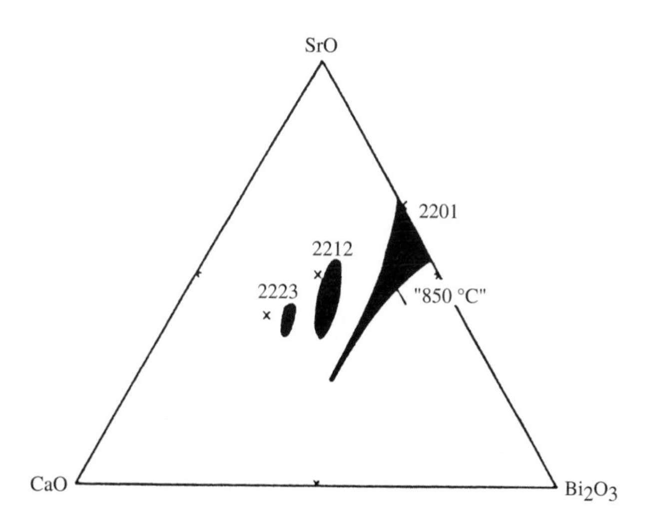

Figure C2.23. Projection of single phase regions onto $SrO-CaO-1/2Bi_2O_3$ face. Compositions expressed in terms of $SrO:CaO:1/2Bi_2O_3:CuO$ ratios. Reproduced by permission of Schulze [45].

and 2 phases. This is because of the difficulty of obtaining a homogenous single phase in $n = 3$ material. It is well established that the CuO_2 planes are the centre of superconductivity in all Bi based superconductors. Superconductivity is directly affected if the substitution is made in the CuO_2 plane itself or indirectly affected if the substitution is made in Bi–O or Sr–O planes (because of a change in bond length, affecting a charge transfer to or from the CuO_2 plane). Another important factor which helps in understanding superconductivity is the oxygen state of Cu–O. The valence of Cu is always

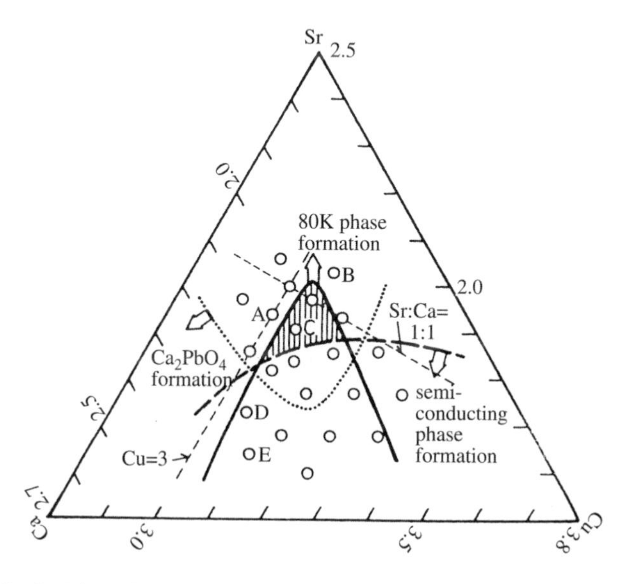

Figure C2.24. System $Sr_xCa_yBi_{1.84}Pb_{0.34}Cu_2O_w$. Effect of variable Sr, Ca, and Cu with Bi:Pb = 1.84:0.34. Reproduced by permission of Koyama [50].

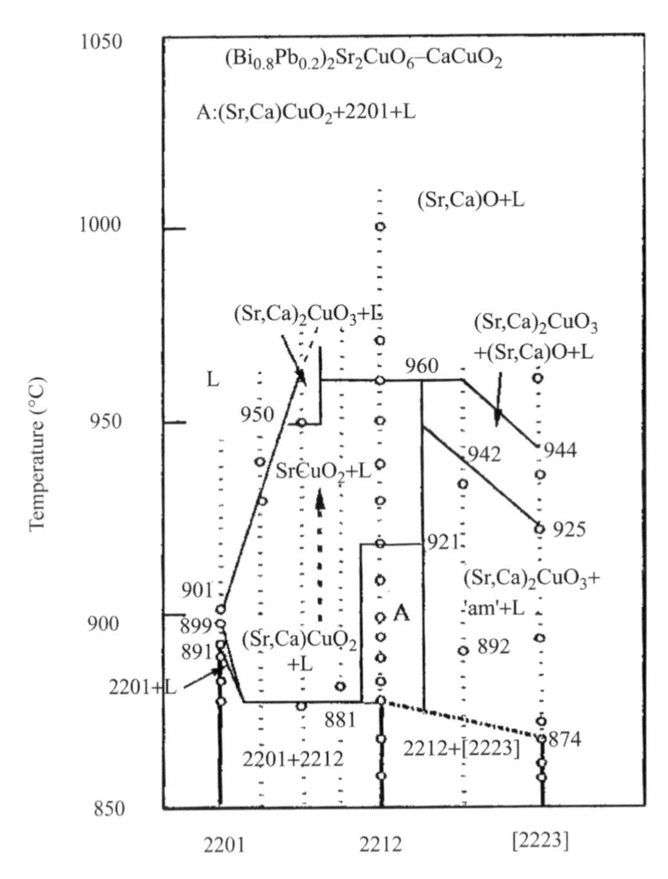

Figure C2.25. Pseudo-binary phase diagram for the $(Bi_{0.8}Pb_{0.2})_2Sr_2CaCuO_2$ system. The vertically aligned dots indicate the measurements on heating of 1°C per min, open circles mean holding points for the measurements at fixed temperatures, the horizontal dotted line between 2212 and 2223 phases indicates decomposition line, the mark 'am' means amorphous phase and 'L' is a liquid phase.

greater than two for p type materials and less than two for n type materials. In BSCCO, the valence of Cu is around 2.1 [54]. The substitutions have direct or indirect effects on the CuO_2 planes. Some substitutions have little or no effect on the planes. The direct effect on the CuO_2 planes (on the superconducting properties) is observed by substituting Cu with 3d elements, such as Zn, Co, Ni and Fe. These dopants have been used to study the superconducting–insulating transition and also in the flux pinning mechanism [55–57]. A strong suppression of T_c has been observed with 3d element substitution. The T_c suppression is interpreted as due to the scattering of charge carriers by impurities or the breaking of the Cooper pairs [55, 57]. With rare earth ion (Pr) substitution at the Ca site of $n = 1$ phase $Bi_2Sr_2Ca_{1-x}Pr_xCu_2O_y$ single crystal, the a and b axis lengths were reported to increase with increasing Pr concentration ($x = 0$–0.78),while the c-axis length and the period modulation were decreased [13]. This is due to the increased oxygen content in the Bi–O layer and a lattice mismatch due to Pr doping. The superconducting temperature was increased slightly at first, then dropped gradually with the increase in Pr concentration and was suppressed completely when the Pr content reached 0.60 [13]. Strong lattice distortion or a change in the lattice parameters has been observed with La doping at the Ca site of $n = 1$ phase single crystal [58]. La has the largest ionic radius in the rare earth family. The suppression of T_c is

A1.3

very severe here. Lattice modulations are also reported with Y and Pb doping [59, 60]. The substitution sites of Pb and Y in $n = 2$ phase single crystal are, Bi and Ca, respectively. Decreasing Sr content in the Sr–O layers affects superconductivity dramatically [11]. It induces a shift of the apical oxygen of the CuO_6 octahedron and also decreases the incommensurate modulation wavelength in $n = 1$ phase crystal [11]. D1.1.2

In [61, 62],the noble metals Au, Ag, Pt, Re, Rh, Ru and Pd have been used for doping in the Bi22($n-1$)n. It is reported that Au and Pt-group metals significantly suppress or eliminate superconducting transition. Only Ag is found to be non-toxic and this is responsible for the current large scale production of Ag sheathed Bi22(n-1)n tapes. In some cases, the Ag addition sharpens the superconducting transition of polycrystalline materials. Doping with alkali elements Li, Na and K in $n = 1$ and 2 materials has been discussed. In $n = 3$ materials, small Li substitutions (0.1–0.2 at. %) led to an increase in the I_c value by 10%. Alkali elements are reported to be excellent sintering aids for $n = 2$ materials due to their fluxing action (formation of a liquid that provides fast diffusion of Ca, Sr and Cu) [63]. Since they have a $+1$ valence states, their substitution for any elements in the Bi based superconductors would reduce oxygen in the structure or create holes in the Cu–O-planes due to charge balance requirements [63].

The reactivity with different oxides is of great importance for Bi–Pb–Sr–Ca–Cu–O because of the use of these materials for melt processing, single crystal growth, preparation of films and for creation of pinning centres. It was found that Bi-compounds show higher tolerance to oxide additions than $YBa_2Cu_3O_x$. Figure C2.26 shows the influence of SiO_2, ZrO_2, Al_2O_3 and MgO on resistivity and critical temperature of $Bi_{1.6}Pb_{0.4}Sr_{1.6}Ca_2Cu_3A_xO_y$, where A = Mg, Al, Zr or Si and $x = 0$–5 [64]. C1

The room temperature resistivity increases rapidly, and the T_c decreases rapidly with SiO_2 and ZrO_2 additions. The Al_2O_3 additions have less deleterious effects on T_c, and the MgO hardly any influence on either ρ or T_c. The non-poisoning effect of MgO is important in the preparation of substrates for Bi22(n-1)n and also for the introduction of fine MgO particles to create pinning centres. Another material important for the creation of pinning centres is Ca_2CuO_3 oxide combined from the elements already present in Bi–Pb–Sr–Ca–Cu–O. A significant increase in critical current density was achieved by introduction of Ca_2CuO_3 into Bi–Pb–Sr–Ca–Cu–O [65]. D3.1 B4.1

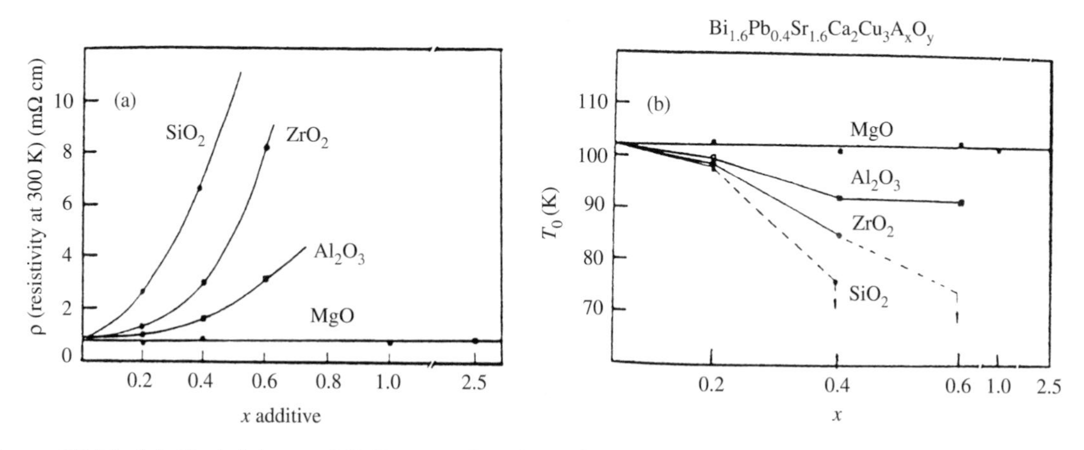

Figure C2.26. (a) Resistivity at 300 K as a function of the amount of oxide additive in BPSCCO. (b) Superconducting transition temperature (T_0) as a function of the amount of the additive level (x). Reproduced by permission of Dou [64].

The stability of Bi-compounds in different solutions is a matter of interest because of the possible corrosion of liquid-immersed samples, especially near the contact pads. In [66], the Bi2212 powder was immersed at room temperature into 2 mol l^{-1} HCl, 2 mol l^{-1} NaCl, distilled water, CH_3OH or 2 mol l^{-1} NaOH. In HCl, O_2 evolved with the formation of a brown precipitate that further dissolved after 2 days. No reaction was observed in NaCl and CH_3OH, very slow evolution of gas was found in H_2O, and a strong reaction with NaOH was found with the formation of blue solution and brown precipitate. The leaching process for Bi2212 pellets was analysed, and it was found that the rate of reaction with HCl is two times smaller in Bi2212 compared to $YBa_2Cu_3O_x$ [66]. The resistance increase was monitored for electrochemical cells comprised of Bi2212 and $YBa_2Cu_3O_x$ immersed to distilled water [66]. For unprotected Bi2212, the 0.3 mm diameter contact pads were completely corroded in 220 h, while for $YBa_2Cu_3O_x$, only 55 h were necessary to complete the corrosion. Using epoxy resin significantly decreased the rate of corrosion.

C1

B2.3.4

NH_4OH/H_2O solution reacts actively with Ag for protection but is an inert substance for Bi2212. Using this, the extraction of 5–10 μm thick filaments was achieved from the multifilamentary Bi2223/Ag tapes [67].

A1.3

Probably the most important feature of BSCCO is the degree of sensitivity to environmental conditions. A detailed investigation that involved measurements of resistance, critical current, x-ray diffraction patterns, x-ray photoemission spectra and energy dispersive x-ray spectra during 1 year of aging, showed that both pure Bi2223 and Bi2212 samples are structurally very stable in ambient atmosphere. Only a small decrease in T_c was evident [68]. Bi2223 is also stable in a humid atmosphere, whereas Bi2212 decomposes into Bi_2CuO_4, $SrCO_3$, $CaCO_3$ and CuO. In contrast to this, multiphase BSCCO is unstable. In ambient atmosphere, Bi2223 transforms partially into Bi2212, and the degradation is more pronounced with lower Bi2223 content. In its turn, the Bi2212 phase decomposes into the compounds mentioned for humid atmosphere. The extent of the degradation can be greatly reduced by the addition of Pb [68].

D1.1.2

C2.5.1 Raman scattering

D1.6

Raman scattering allows us to investigate the properties at low frequencies of several hundreds cm^{-1}. In BSCCO, Raman spectra contain both electronic and phonon components. The phonon component forms the peaks of the scattering intensity, whereas the electron component forms a background. The Raman spectra of a Bi2212 crystal in the *ab*-plane at temperatures below and above the critical temperature are shown in figure C2.27. The *c*-axis spectrum is much stronger than the *ab*-plane, which indicates that Bi2212 is a quasi-two-dimensional metal. The comparison of Raman phonon spectra at different polarizations shows the orthorhombicity of the crystal, while the electronic spectra are well described within tetragonal symmetry. The latter is an indication that only electrons in the CuO_2 planes are responsible for the electron Raman component [69]. The difference between electron Raman spectra below and above T_c gives directly the energy gap peak in the *ab*-plane (figure C2.28). This yields $2\Delta/kT_c$ ratios of 5.5, 6.3 and 8.2 for A_{1g}, B_{1g} and B_{2g} symmetries, respectively, that are the possible consequences of the $d_{x^2-y^2}$ pairing state [70].

Raman scattering is also used to determine the location of doped holes in Bi based superconductors. Several articles in relation to the effect of cation substitution and oxygen content on the Raman scattering have been published [71–73]. The determination of the location of holes with a change of the local structure (due to cation substitution) is important for determining the mechanism of superconductivity. It should be noted that, in the Raman scattering experiments, the shift of the energy and the scattering intensity of the oxygen atom on each site, and the frequencies of some phonon modes are sensitive to the relevant bond length (which is changed by cation substitution) and are used in the determination of the location [73].

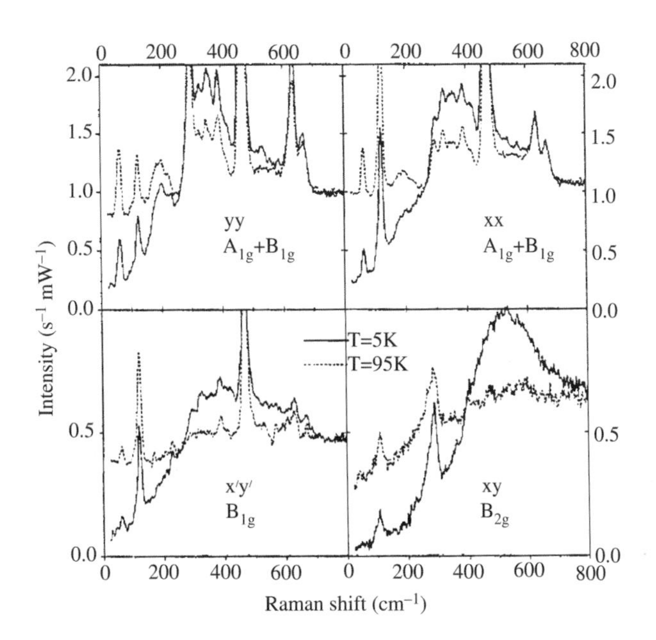

Figure C2.27. Raman spectra of the optimally doped Bi2212 crystal taken with four different polarizations at temperatures above ($T = 95$ K, dotted lines) and below ($T = 5$ K, solid lines) the temperature of the superconducting phase transition $T_c = 91$ K. Reproduced by permission of Misochko [69].

C2.6 Optical properties

C2.6.1 Infrared optical spectra

The optical reflectance of BSCCO in the range of frequencies 4000–25 000 cm^{-1} is shown in figure C2.29 [6]. The reflectance drops in nearly linear fashion throughout the infrared region. The reflectance minimum is seen to shift upward with the number of CuO_2 planes. The calculated Drude plasma frequencies are shown by arrows, although in general BSCCO, like other HTS, shows non-Drude behaviour. One of the non-Drude features is the linear (instead of ω^2) mid-infrared transmittance of the Bi2212 crystals (figure C2.30) [74]. D1.6

The optical conductivity can be derived from the transmittance of single crystals by Kramers–Kronig analysis [75]. The main features of the optical conductivity are similar in Bi2212 and Bi2201. The typical frequency dependence of the optical conductivity for Bi2212 is presented in figure C2.31 [76]. In the normal state, the low frequency conductivity approaches DC conductivity and decreases with frequency as expected for the Drude response: $\sigma \propto \omega^{-2}$. Above 300 cm^{-1}, the decrease becomes non-Drude, closer to ω^{-1}. Below T_c, σ has a broad maximum around 1000 cm^{-1}. In Bi2212, in addition to the large *c*-axis–*ab*-plane anisotropy, there is an *ab*-plane anisotropy, i.e., the *a*-axis is more transparent (less conductive) than the *b*-axis.

C2.6.2 High resolution photoemission

High resolution photoemission tests unoccupied states just above the Fermi level (E_F). Bi2212 crystals are among the rare HTS that give a well-defined Fermi edge in inverse photoemission spectra. A typical photoemission spectrum for Bi2212 is shown in figure C2.32 [76], together with the soft x-ray absorption A2.1 B2.4 D1.1.2

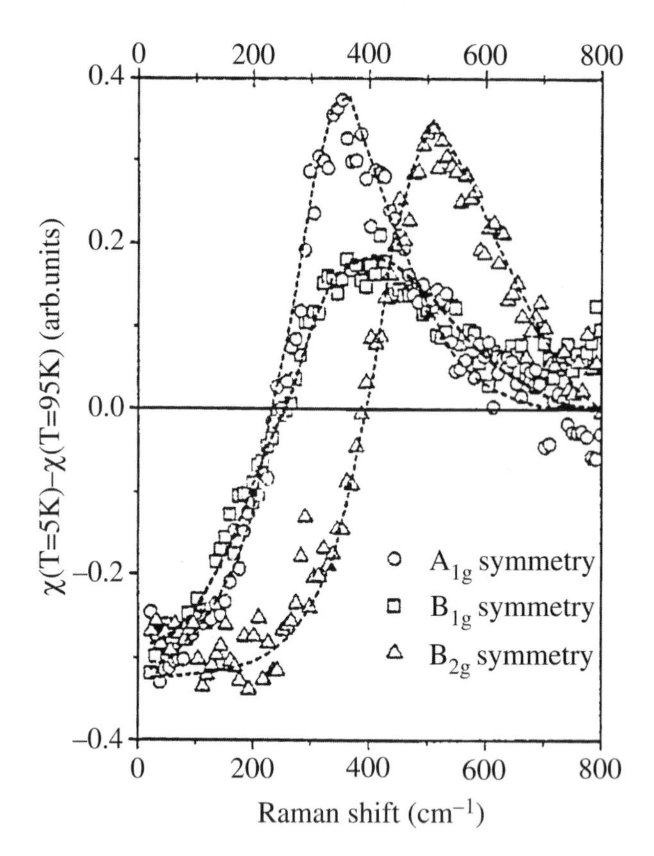

Figure C2.28. Comparison of the different symmetry components representing the superconducting condensate response. Each component is obtained by subtracting the pure symmetry spectra corrected for the Bose factor at temperatures above T_c from that at $T = 5$ K. The dashed lines are just a guide to the eye. Reproduced by permission of Misochko [69].

data. The analysis of this spectrum makes it evident that a metallic band of partial O 2p character extends continuously through the Fermi level. The decrease of the photon intensity at 1.5 eV above E_F delineates the upper edge of the metallic band around the Fermi level. A peak at $E_F + 2.9$ eV can be assigned to unoccupied Cu 3d states (if they are localized) or to Bi 6p states. Figure C2.33 shows the change in the photoemission spectra at the transition to the superconducting state. The appearance of a peak is in accordance with simple BCS model calculations (shown in the bottom of figure C2.33) and allows us to determine directly the energy gap (Δ) of superconductor. According to figure C2.33, $2\Delta/kT_c = 8 \pm 1.4$.

A2.1
A3.2

C2.7 Normal state properties

C2.7.1 Magnetic properties

D3.1

The BSCCO compounds can be in normal or superconducting states depending on temperature and hole density p. The schematic p–T diagram is given in figure C2.34 [78], and the superconducting state boundaries for Bi2201 and Bi2212 are given in figure C2.35 [79]. The normal state can be metallic,

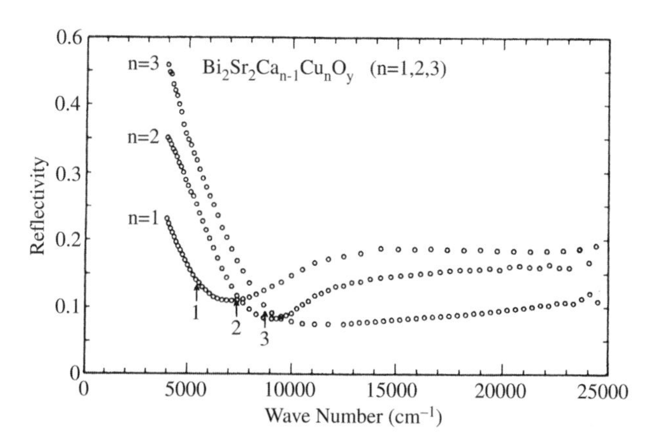

Figure C2.29. Reflectance of ceramic samples of $Bi_{2.1}Sr_{1.9}CuO_y$ (2201), $Bi_2Sr_{1.8}Ca_{1.2}Cu_2O_y$ (2212) and $Bi_{1.85}Pb_{0.35}Sr_2Ca_{1.2}Cu_{3.1}O_y$ (2223). The arrows show the results for the plasma frequency from fits to the Drude formula. Reproduced by permission of Maeda [6].

insulating and antiferromagnetic (see figure C2.34). At low doping, three-dimensional antiferromagnetic order appears. Above the Neel temperature, the two-dimensional Heisenberg antiferromagnetic short order appears as a result of exchange coupling between the Cu moments in the CuO_2 planes.

The magnetic susceptibility of BSCCO as a function of temperature shows a peak with pronounced negative slope at high temperatures (figure C2.36) [80]. Near the optimum hole doping (curves at $T_c = 94.4$ and 89.4 K in figure C2.36), $d\chi/dT$ vanishes, and the system changes over to Pauli-like behaviour. Further increases in the hole doping lead to Curie–Weiss behaviour due to localized D2.5

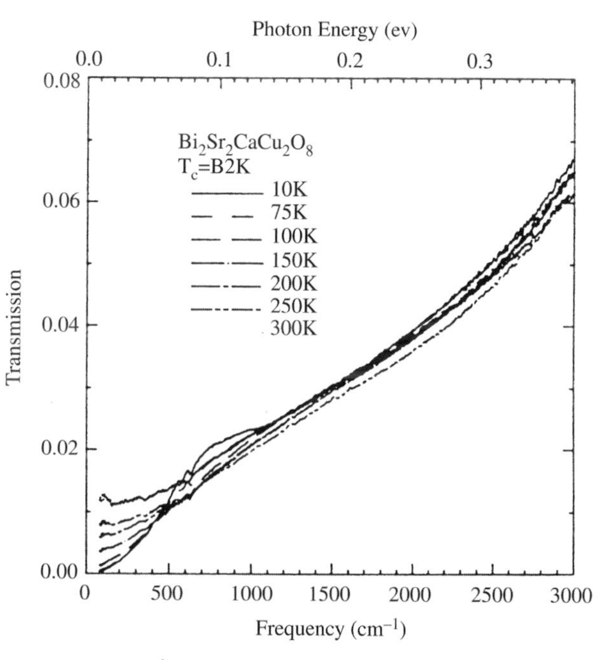

Figure C2.30. Transmittance of a 1350 Å crystal of $Bi_2Sr_2CaCu_2O_8$. Reproduced by permission of Romero [74].

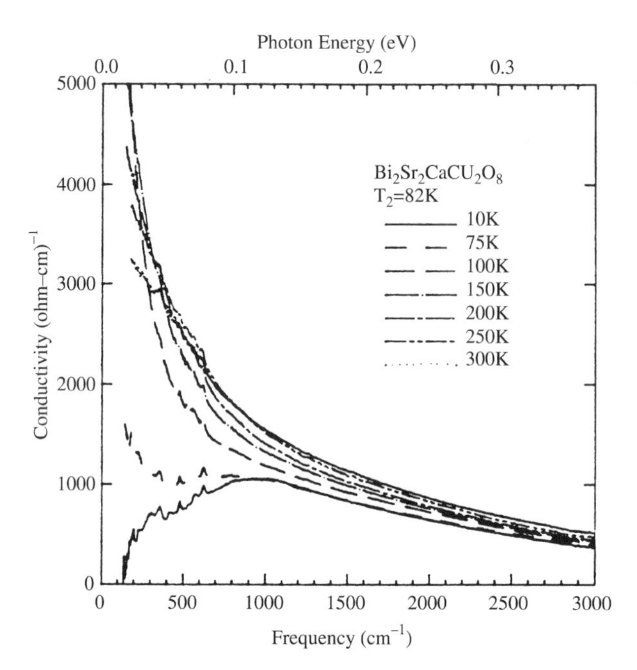

Figure C2.31. Frequency-dependent conductivity of $Bi_2Sr_2CaCu_2O_8$ between 20 and 300 K. Reproduced by permission of Romero [75].

moments. A small temperature-independent anisotropic component exists in all three major BSCCO phases due to the van Vleck contribution.

C2.7.2 Resistivity

D3.1 The normal state resistivity of BSCCO is different along different directions and at different oxygen content, as shown in figure C2.37 [81]. An important feature of in-plane resistivity $\rho_a(T)$ is T-linear $\rho(T)$ dependence. For the underdoped single crystal, the in plane resistivity is reported to deviate from high temperature T-linear behaviour due to the presence of a pseudogap. The interpretation of the linear $\rho_a(T)$ dependence is an important test for different HTS models. T-linear behaviour is natural for the scattering on phonons at $T \gg$ the 'transport' Debye temperature (Θ). One can expect the transition to a higher power of T at $T < 0.25\ \Theta$ (usually T^5), but it could be hidden by the superconducting transition. The T-linear dependence for Bi2201 down to 10 K indicates that not only electron–phonon scattering is involved in this process, because fitting to the Bloch–Grüneisen formula would demand a lower than physically reasonable value of $\Theta \approx 35$ K. Several models, e.g. Fermi-liquid interpretations, involve the renormalized electron–electron interactions. T-linear dependence is an intrinsic feature of the resonance valence bond (RVB) model [82].

D3.1 The c-axis resistivity (ρ_c) of Bi compounds is strongly influenced by the oxygen content in the crystal. The figure (C2.37) shows semiconducting behaviour for an oxygen underdoped crystal and a metallic behaviour for an oxygen overdoped crystal [83, 84]. In [85], ρ_c is reported to decrease with B2.4 increasing Pb content in $Bi_{2.1-x}Pb_xSr_{1.8}CaCu_2O_y$ single crystal. The ρ_c in Bi22(1-n)n is the largest among the HTS (shown in figure C2.38 [86]). This is connected with the insulating properties of the layers between CuO_2 planes. Being close to the insulating phase shown in figure C2.34, different substitutions

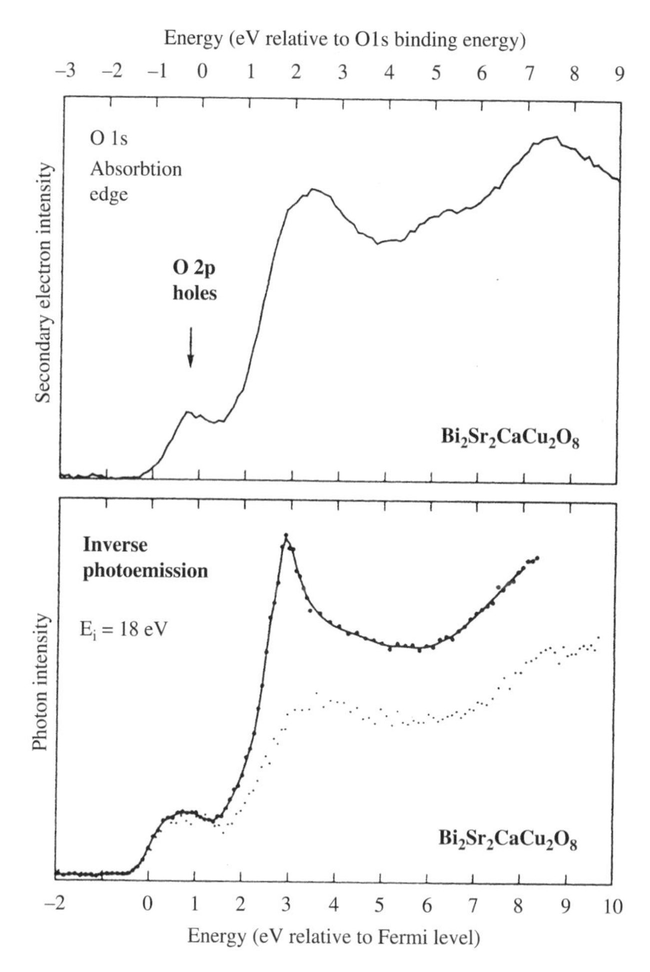

Figure C2.32. Unoccupied states of $Bi_2Sr_2CaCu_2O_8$ measured by inverse photoemission (bottom) and by soft x-ray absorption (top, from [77]). The two inverse photoemission spectra span a range of different sample preparations. Reproduced by permission of Drube [76].

transform metallic behaviour into an insulator-like one and suppress the superconducting transition. Such an influence is shown for $Bi_2Sr_2Ca_{1-x}Y_xCu_2O_{8+y}$ in figure C2.39 [87].

C2.7.3 Hall effect

The Hall effect provides information on the carrier density and the sign of the charge carriers. According to the Hall effect, the charge carriers in BSCCO are holes. In HTS, the Hall coefficient (R_H) is large and strongly temperature dependent. The Hall density of charge carriers n_H can be extracted as $1/R_He$, and the carrier density per Cu ion p_H can be defined as $p_H = n_HV/N$, where V is the unit-cell volume, and N is the number of Cu ions per unit-cell. D3.1

The temperature dependence of R_H, and the charge carrier density for $Bi_2Sr_2Ca_{1-x}Y_xCu_2O_{8+y}$ determined by the Hall effect and iodometric titration technique, are shown in figures C2.40 and C2.41, respectively [87].

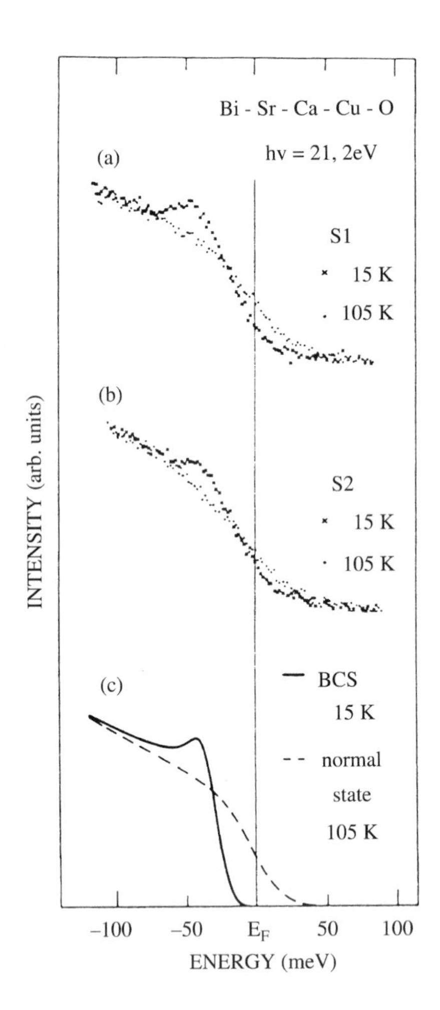

Figure C2.33. (*a*) and (*b*) Comparison between the experimental photoemission spectra and (*c*) a model calculation of the density of states. Reproduced by permission of Imer [35].

D3.1 The value of the Hall coefficient is smaller in Bi2201 than in Bi2212 [88]. In BSCCO, the Hall coefficient is strongly anisotropic. R_H is negative for $H \perp c$ in Bi2201 and, at low temperatures, in Bi2212 [88]. The in-plane Hall coefficient drops near the superconducting transition. There is an anti-correlation in Hall coefficient and magnetic susceptibility behaviour: at a T_c-decreasing doping R_H becomes less temperature dependent, whereas the temperature dependence of χ becomes more pronounced. The negative value of R_H is connected with non-metallic charge transport along the *c*-direction due to a small overlap between the relevant electron orbitals of the CuO_2 planes. The value of p_H is 0.4 in Bi2212 and about 1.0 in Bi2201.

C2.7.4 Pseudogap

A specific property of HTS, among them Bi22$(n-1)n$ compounds, is the presence of a normal state gaplike structure in the electronic excitation spectra. Considerable evidence for this pseudogap has been accumulated in angle resolved photo emission spectroscopy, nuclear magnetic resonance spectra,

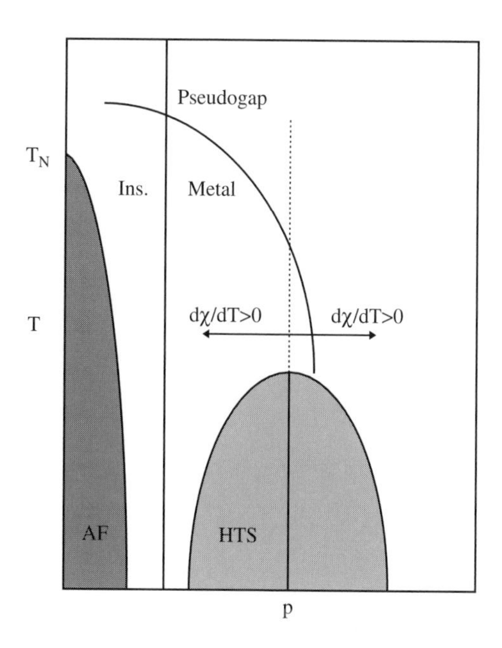

Figure C2.34. Schematic hole density–temperature diagram. Representation of the dependence of T_c and the Neel temperature T_N on the hole density p.

infrared conductivity, neutron scattering, transport properties, specific heat, thermoelectric power, spin susceptibility and Raman spectra [54, 89, 90]. The position of the pseudogap is shown schematically in figure C2.34. It is mainly a property of underdoped HTS, but the gaplike structure is also present in the

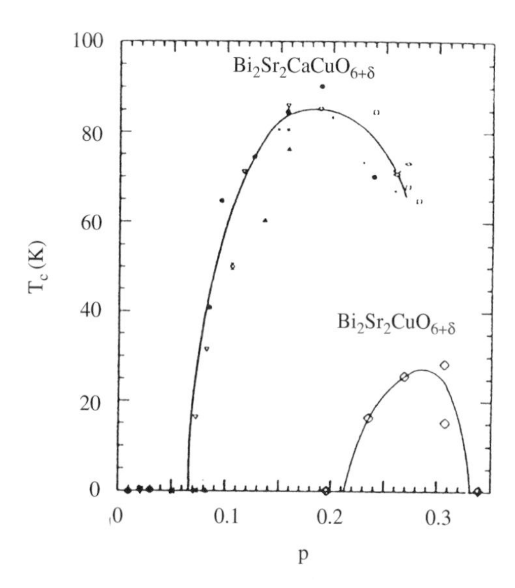

Figure C2.35. Critical temperature versus hole density for the Bi2201 and Bi2212 phases. Reproduced by permission of Groen [79].

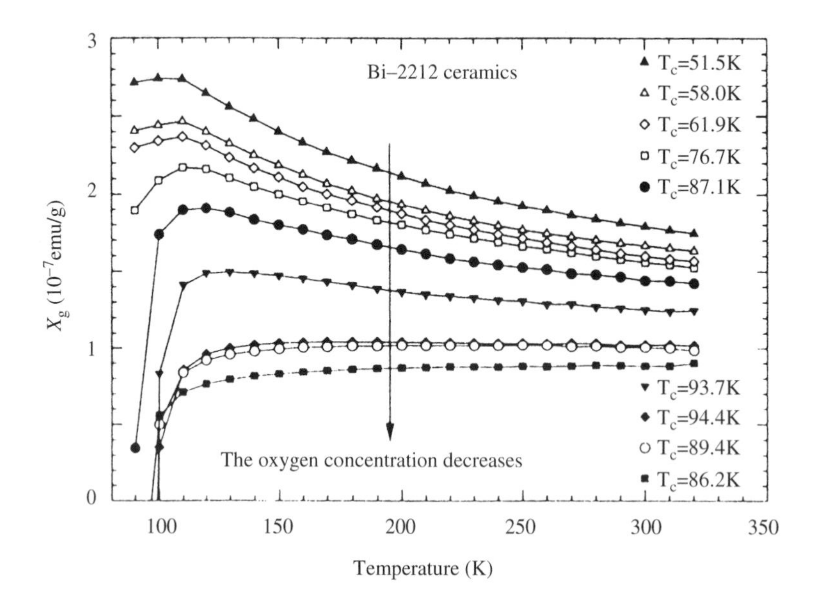

Figure C2.36. Normal-state susceptibility *versus* temperature for a batch of $Bi_{2.0}Sr_{2.0}Ca_{1.0}Cu_{2.0}O_x$ ceramics with different oxygen concentrations. The oxygen concentration decreases from top to bottom. The maximum $T_c \sim 95$ K is obtained for a sample that has a constant susceptibility. Reproduced by permission of Triscone [80].

normal state of overdoped superconductors [91]. In underdoped superconductors, it persists to temperatures higher than 300 K. Recent tunnel spectrometry investigations show that the pseudogap is essentially temperature independent and transforms smoothly to the superconducting gap at the T_c [54].

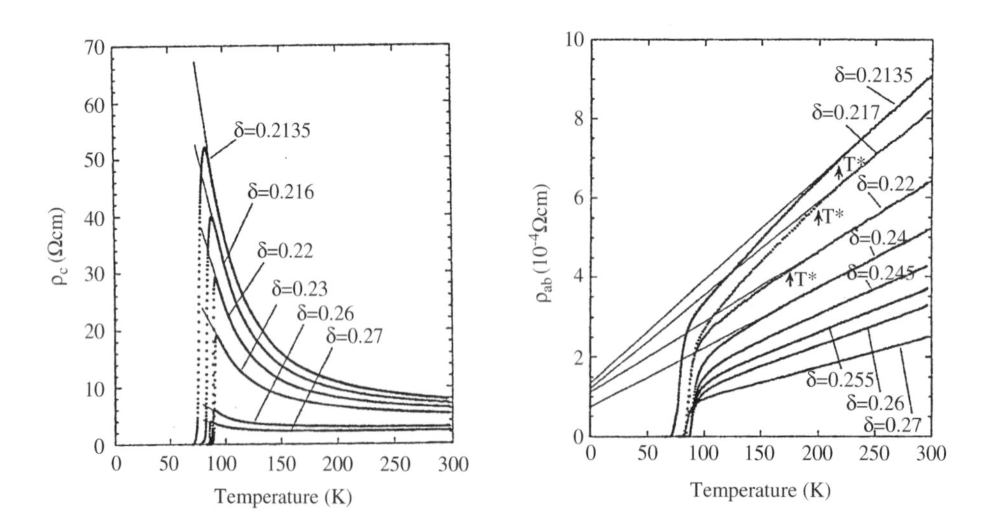

Figure C2.37. Temperature dependence of the in-plane ρ_a and out plane ρ_c resistivity for $Bi_2Sr_2CaCu_2O_{8+\delta}$ crystals for various oxygen contents (δ) [83]. The temperature T^* at which the ρ_a deviates from T-linear behaviour are shown by arrows for the underdoped crystals.

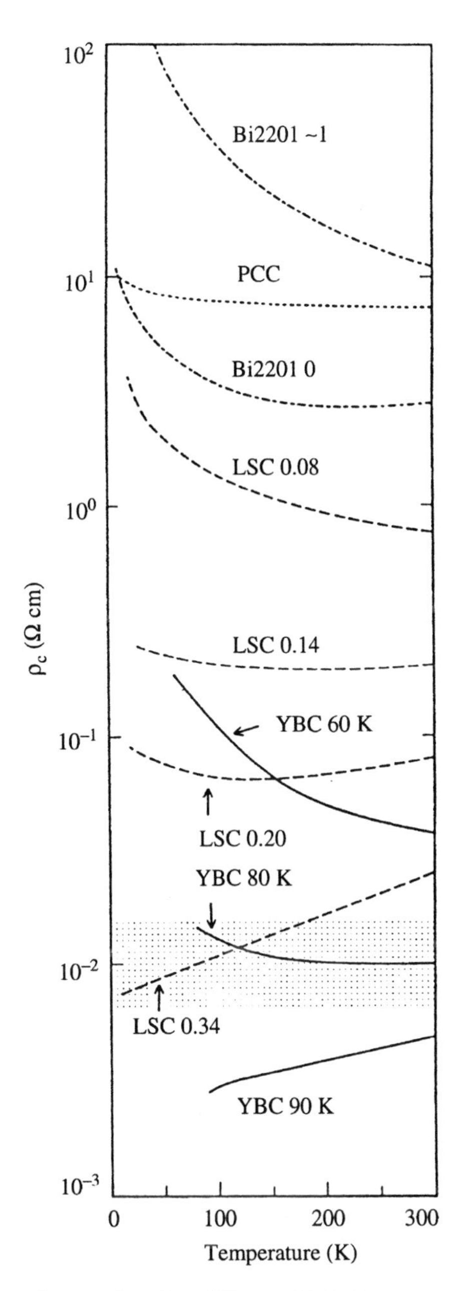

Figure C2.38. Temperature dependence of ρ_c for different high T_c cuprates. The hatched region indicates the resistivity range roughly corresponding to the Ioffe-Regel limit. $\rho_c(T)$ bifurcates between metallic and semiconducting behaviour depending on whether the resistivity is below or above this range. Reproduced by permission of Ito [86].

The absolute value of the pseudogap (Δ_p) is equal to the absolute value of the superconducting gap ($2\Delta_p/kT_c = 12.3$–8.7 in Bi2212 [54]), and the pseudogap reveals d-wave symmetry identical to that of the superconducting gap [92, 93]. A simple explanation for this could be that the pseudogap is

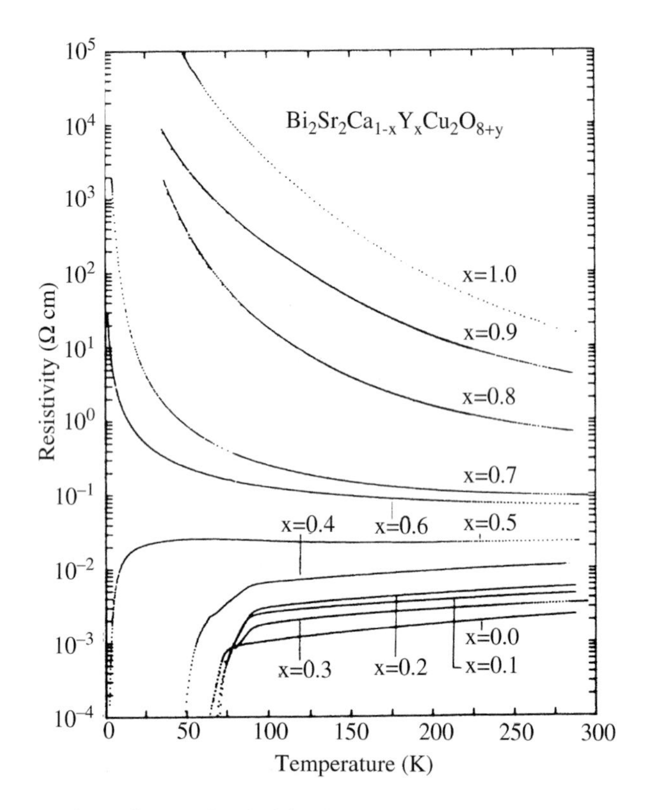

Figure C2.39. Temperature dependence of resistivity in $Bi_2Sr_2Ca_{1-x}Y_xCu_2O_{8+y}$ for different x. Reproduced by permission of Tamegai [87].

the precursor of superconducting gap, and that the pairing amplitude develops at much higher temperatures than the temperature (T_c) at which phase coherence sets in.

C2.8 Superconducting properties

The typical values of superconducting parameters for BSCCO are shown in table C2.8 . Little is known about Bi2201 because many high quality Bi2201 crystals are not superconducting.

B2.4

C2.8.1 Critical temperature

Only non-stoichiometric Bi2201 shows superconductivity. The critical temperature of Bi2201 is ~ 10 K [94] but strongly varies because of the difficulty in obtaining pure single phase composition. Many high quality Bi2201 crystals show no superconductivity above 5 K. However, substitutions of Sr for La, Pr, Nd and Sm in Bi2201 increase the critical temperature up to 33 K [95]. The critical temperature of Bi2212 is near 85 K [94], but some crystals show T_c up to 96 K [96, 97]. This is a result of different oxygen content and cation intersubstitution [96, 97]. $T_c(y)$ for $Bi_2Sr_2CaCu_2O_{8+y}$ is shown in figure C2.42 [98]. The critical temperature of Bi2223 is close to 110 K [94].

B3.2.2

In figure C2.42, the pressure dependence coefficient, dT_c/dP, measured at pressures up to 0.6 GPa is shown. Several experimental groups have found a change of dT_c/dP to negative values at a pressure

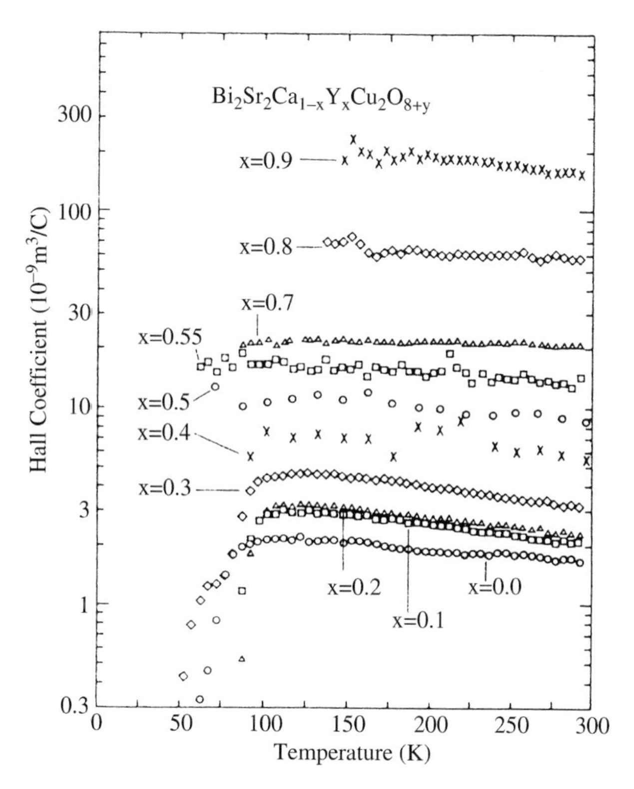

Figure C2.40. Temperature dependencies of R_H for the system $Bi_2Sr_2Ca_{1-x}Y_xCu_2O_{8+y}$ measured in a field of 5 T. Reproduced by permission of Tamegai [87].

of 2–4 GPa. The dT_c/dP in Bi2223 is positive up to 8 GPa, and T_c increases monotonically from 110 K (normal pressure) to 119 K (8 GPa) [99].

C2.8.2 Critical fields and characteristic lengths

The important physical parameters are the first and second anisotropic critical fields at zero temperature: $H_{c1}^{c}(0)$, $H_{c1}^{ab}(0)$, $H_{c2}^{c}(0)$ and $H_{c2}^{ab}(0)$, respectively; coherence length [$\xi_{ab}(0)$, $\xi_c(0)$]; penetration depth [$\lambda_{ab}(0)$, $\lambda_c(0)$]; Ginzburg–Landau parameter $k = \lambda_{ab}/\xi_{ab}$ and anisotropy parameter $\gamma = \xi_{ab}/\xi_c = \lambda_c/\lambda_{ab} = H_{c2}^{ab}/H_{c2}^{c} = (\rho_c/\rho_{ab})^{0.5}$. There are simple relations between some of these parameters. Their temperature dependence is mostly tabulated and not very different from conventional superconductors. A2.3 A3.1

The anisotropic second critical field and coherence length can be obtained from magneto-resistive data.In Bi2212 this was done shortly after the first synthesis of this compound, whereas single crystal whiskers of Bi2223 have become available only recently.The values of these parameters for Bi2212 and Bi2223 are shown in table C2.8 [100, 101]. D1.1.2

The first and second critical fields can also be obtained from the reversible magnetization recorded as a function of temperature and field if the Josephson-coupled layered structure and thermal fluctuations are correctly taken into account. Measuring Bi2212 crystals by SQUID magnetometer, Schilling *et al* obtained $H_{c2}^{c}(0) = 90$ T, $\xi_{ab} = 19$ Å and $\lambda_{ab} = 0.21\ \mu$m resulting in the temperature-independent $k = 113$, and they derived $H_{c1}^{c} = 19$ mT [101]. The first critical field and penetration depth A4.3 D2.4

Table C2.8. Main superconducting parameters for Bi2201, Bi2212 and Bi2223

Parameter	Bi2201	Bi2212	Bi2223
T_c(K)	10 [94]; 8–25 [6]; 33 [95]	85 [6, 94]; 92 [98]	110 [6, 94]
$H_{c1}^{c}(0)$ (mT)		22 [136]; 19 [86]	13.5 [87]
$H_{c1}^{ab}(0)$ (mT)		0.1 [112]	0.04 [87][a]
$H_{c2}^{c}(0)$ (T)		57 [100]; 90 [101]	39 [100]
$H_{c2}^{ab}(0)$ (T)		3403 [100]	1210 [100]
$\xi_{ab}(0)$(Å)		27 [106]; 19 [101]	29 [100]
$\xi_c(0)$(Å)		0.45 [107]	0.93 [100]
$\lambda_{ab}(0)(\mu m)$		0.18–0.25 [107]; 0.21 [101]	0.245 [102]
$\lambda_c(0)(\mu m)$		100 [103]	7.6 [102][a]
γ		60 [106]; 150–1000 [110]; > 150 [111]	31 [100]
k		150 [106]; 113 [101]	65 [109]; 84 [102]
$J_c^{ab}(0)(A\,cm^{-2})$		1.3×10^6–3.5×10^6 [107]	$\sim 10^6$ [89, 92, 93]
$J_c^{c}(0)(A\,cm^{-2})$		5×10^3 [94]; 0.1–1.5×10^3 [98]	$\sim 10^3$ [95]
$2\Delta(0)/kT_c$	6–7 [104]	6–7 [104]	6–7 [104]
$\rho_{ab}(300\,K)(\mu\Omega\,cm)$	500–2000 [110]	150 [103]	380 [100]
$\rho_c(300\,K)(\Omega\,cm)$		7 [103]; 4.3–28 [110]	0.37 [100][a]

[n][a] — evaluated using data of [n].

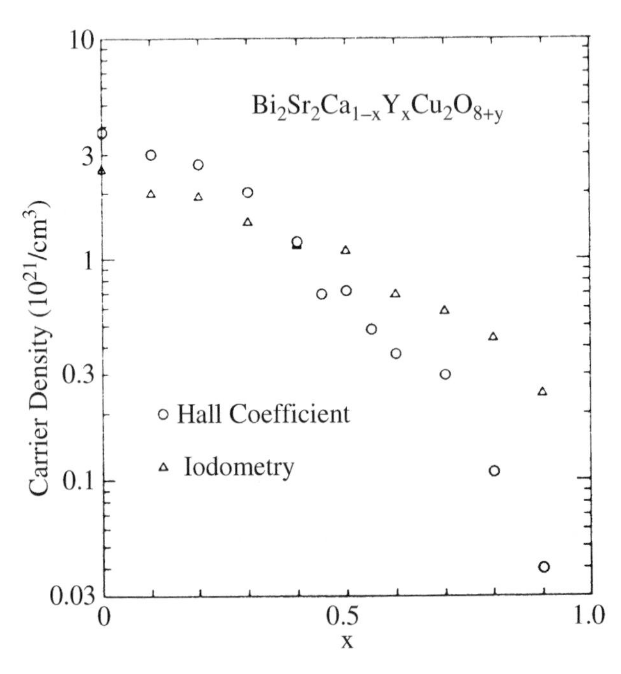

Figure C2.41. Variation of the carrier concentration with x in $Bi_2Sr_2Ca_{1-x}Y_xCu_2O_{8+y}$ determined by the Hall coefficient and by the iodometric titration technique. Reproduced by permission of Tamegai [87].

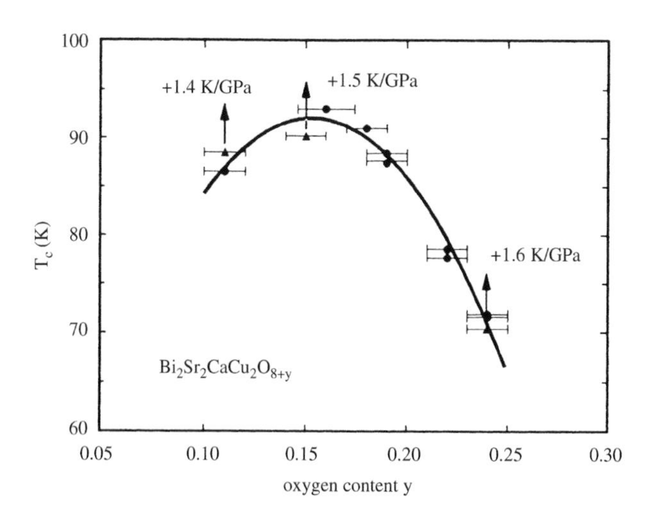

Figure C2.42. Dependence of T_c on oxygen content y for $Bi_2Sr_2CaCu_2O_{8+y}$ crystals at three values of $y = 0.11, 0.15$, and 0.24. T_c increases with pressure. Reproduced by permission of Sieburger [98].

from reversible magnetization in a Bi2223 whisker was found by Matsubara *et al* [102]. The large *ab*-plane penetration depth in Bi2212 single crystals and $H_{c1} = 0.1$ mT were found by Cooper *et al* [103]. The critical parameters determined in other works are close to these values [104–111].

C2.8.3 Critical current and irreversibility line

The critical current density in Bi based superconductors is a primary parameter in technological development for applications. The critical current density along the *c*-axis (J_c^c) in this highly anisotropic Bi compound is much lower than the critical current density along the *ab*- plane (J_c^{ab}). For example, single crystalline Bi2212 and Bi2223 have their critical current density above 10^6 A cm^{-2} along the *ab* plane [107, 112–114] and 10^2–10^3 A cm^{-2} along the *c*-axis [110, 115–117]. The low J_c along the *c*-axis is the result of a weak Josephson interaction between every pair of CuO_2 double or triple layers [110]. The critical current of the best textured BSCCO samples is one or two orders of magnitude lower than J_c^{ab} of single crystals. Bi2212 superlattices show pure two-dimensional behaviour, and their critical current does not depend on the magnetic field parallel to the *ab*-plane up to the value of 20 T [118].

D1.1.2

The irreversibility line (IL) in high T_c superconductors is an important phase boundary since it separates the hysteresis response of the materials from the reversible magnetic response. The physical interpretation or the origin of this phase boundary in Bi2201, Bi2212 and Bi2223 is still the subject of extensive research work [119–121]. In [119], the authors have interpreted the irreversible field of Bi2201 single crystal as a melting of the flux line. Majer *et al* [120] using Hall probe measurements, argued that the irreversibility line and the melting line have completely different physical origins. They observed the irreversibility line of Bi2212 single crystal to be lying in the vortex-solid region significantly below the vortex melting line. Farrell *et al* [121] reported that H_{irr} in Bi2212 single crystal coincides with the melting field line.

A1.3
B3.2.1
B2.3.5

C2.8.4 Energy gap

The 'effective' double energy gap that does not include the anisotropy is in the range of 6–7 kT_c in optimally doped Bi2201, Bi2212 and Bi2223 [122]. The gap anisotropy is, however, strong in BSCCO both in the ab-plane and between the ab-plane and the c-axis [123]. According to point contact measurements in a Bi2212 single crystal, $2\Delta(0)^{ab}/kT_c = 6.2 \pm 0.3$, whereas $2\Delta(0)^{c}/kT_c = 3.3 \pm 0.3$ [124]. Boekholt *et al* [125] found $2\Delta(0)^{ab}/kT_c = 5.8$ and $2\Delta(0)^{c}/kT_c = 3.6$ in accordance with results obtained by polarized Raman spectrometry [$2\Delta(0)^{XX}/kT_c = 5.7 \pm 0.2$ and $2\Delta(0)^{YY}/kT_c = 3.5 \pm 0.1$]. Ekino and Akimitsu [126] performed point contact measurements on the face and edge of c-axis oriented films, and they obtained $2\Delta(0)^{ab}/kT_c = 10-12$ and $2\Delta(0)^{c}/kT_c = 6$ for Bi2212, and $2\Delta(0)^{ab}/kT_c = 9-11$ and $2\Delta(0)^{c}/kT_c = 5.5-6.3$ for Bi2223. Although in the last work the absolute values were larger than in other works, the anisotropy ratio is about the same in all works: 1.7–1.8 for Bi2212 and 1.5–1.8 for Bi2223.

The gap anisotropy in the ab-plane reflects the $d_{x^2-y^2}$ symmetry: $\Delta(0)^{ab} = \Delta_0\cos(2\theta)$, where θ is the angle in the ab-plane [127]. Tunnelling or Raman spectra usually give average values of $\Delta(0)^{ab}$. The registered difference in $2\Delta(0)/kT_c$ arise mainly from differences in the doping level. This is clarified by the oxygen doping level (p) dependence of T_c and Δ_0 as shown in figure C2.43 . According to this figure, $T_c(p)$ has a maximum at $p = 0.17$, whereas $\Delta(p)$ increases monotonically with decreasing p [127]. A systematic increase of $2\Delta(0)/kT_c$ with decreasing T_c (by changing oxygen concentration) has also been reported in [128, 129]. Their interpretation was that inelastic scattering processes were responsible for the reduction in T_c. In [130], a very small effect on the in-plane energy gap in Pd doped Bi2212 single crystal is mentioned. This leads to an increase in $2\Delta(0)/kT_c$ due to a reduced T_c with Pb doping. However, a strong effect on the out of plane energy gap is observed with Pb doping. The reduction of T_c in this crystal is therefore due to out of plane substitution of Pb for Bi, which changes the carrier density in the copper oxide planes.

B2.4

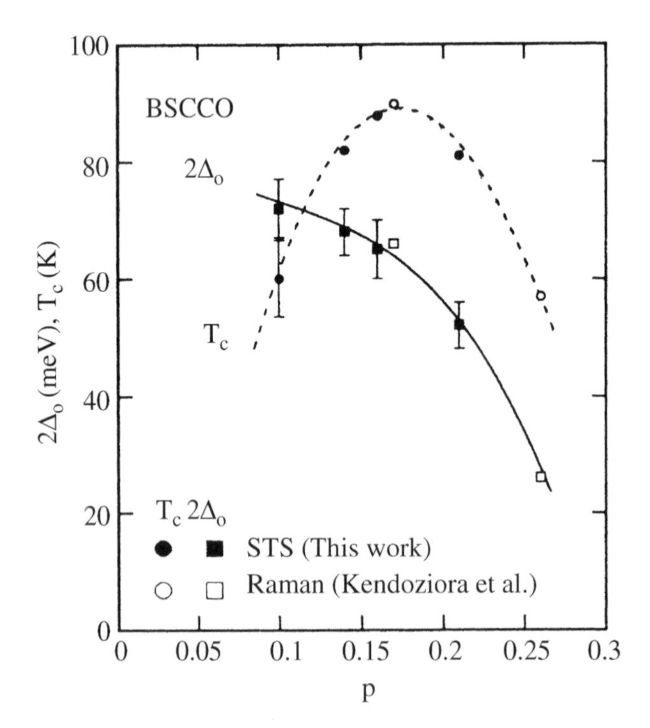

Figure C2.43. Doping dependence of superconducting gap amplitude $2\Delta_0$. The $2\Delta_0$ values determined from Raman spectra [131] are also shown. Reproduced by permission of Oda [127].

C2.8.5 Layered structure, intrinsic Josephson behaviour, two-dimensional effects

There is strong evidence that superconductivity arises in the CuO_2 planes. The characteristics of BSCCO include a pronounced layered structure and the weak coupling between CuO_2 layers. This gives a low value of c-axis coherence length, and a high value of the parameter γ (see table C2.8). According to recent data, in BSCCO, Ca–O and Sr–O layers are insulating, whereas the Bi–O plane is semiconducting with a gap ~ 100 meV [132]. Due to the weak Josephson coupling between CuO_2 layers, Bi2212 can be regarded as sequence of intrinsic SIS Josephson contacts along the c-axis with the size of half unit cell (15 Å). The Josephson nature of intrinsic junctions in Bi2212 crystals was proved by the direct observation of a dc and ac Josephson effect by Kleiner and Muller [110]. This included observation of multiple-branch current–voltage characteristics, collective gap, dc critical current with the specific Ambegaokar–Baratoff dependence on the temperature, Fraunhofer patterns in the magnetic field dependence of critical current and ac Shapiro steps. The microwave emission was directly detected from Bi2212 crystals [110]. It was shown that every pair of CuO_2 double layers forms a working Josephson contact.

A4.3

A2.7

The confinement of superconducting electrons in CuO_2 planes defines the two-dimensional features that are especially pronounced in highly anisotropic BSCCO. The description of electrical and magnetic properties can be done using the pancake vortices introduced by Clem [133, 134]. Actually, the behaviour of HTS is intermediate between three-dimensional and two-dimensional. Although the pancake vortices are two-dimensional objects regarded as situated in zero-thickness CuO_2 planes, magnetic and Josephson interactions can easily align them into 'three-dimensional' line vortices. Even if the Josephson interaction is weak (large γ), the magnetic interaction can still align vortices in different planes, although for Bi2212 it is an unusual alignment. Whether the field is tilted towards the c-axis or not, the alignment is always along the c-axis, i.e., Bi2212 crystals are transparent to the ab-plane component of the magnetic field [135].

A4.3

Another important phenomenon caused by the Josephson interlayer interaction is the collective Josephson plasma resonance, now a powerful method in HTS physics. Josephson oscillations of supercurrent along the c-axis usually take place at a frequency of 10^{10}–10^{12} Hz. Josephson oscillations can propagate both along the c-axis (longitudinal plasma) or along the ab-plane (transverse plasma) [136]. The Josephson plasma oscillations are evident in far infrared reflectivity measurements [137] and microwave absorption [138]. Since the Josephson plasma mode lies at an energy well below the superconducting gap, the plasma damping processes are mainly prohibited, and this mode is very stable [139].

A2.7

Since the Josephson plasma directly reflects the interlayer interaction, the excitation of the Josephson plasma is a powerful method in the investigation of the vortex state. Using the Josephson plasma resonance, the nature of the first order melting transition in $Bi_2Sr_2Ca_1Cu_2O_{8+\delta}$ was investigated, and it was shown that vortex lines are almost decoupled above the melting line, i.e. the vortex solid undergoes a transition to a vortex gas rather than to a vortex line liquid (sublimation transition) [138, 139]. Another important result obtained using the collective Josephson mode is that columnar defects transform the vortex gas into a liquid of vortex lines [139].

A4.2

C2.8.6 Vortex phase diagram

BSCCO compounds are extreme type II superconductors (see table C2.8) and, except in very small fields, their electromagnetic properties are defined by vortices. The behaviour of vortices is different at different temperatures and magnetic fields. A vortex phase diagram for Bi2212 is shown schematically in figure C2.44 [140, 141]. Some lines of this diagram, for instance, the melting line T_m and the pinning induced second peak line T_{sp} are generally accepted, whereas other lines are still controversial and need further clarification.

A4.2

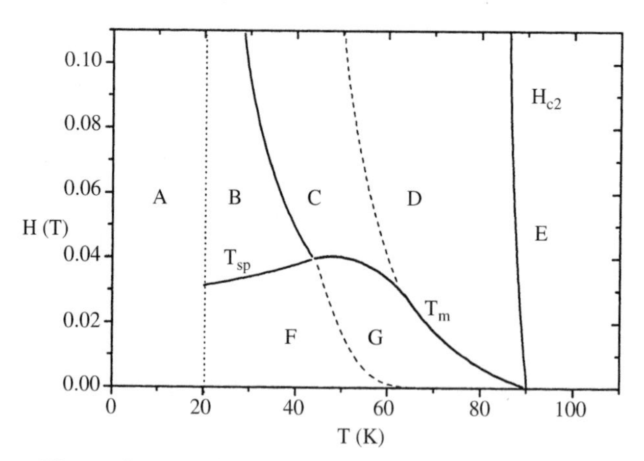

Figure C2.44. Schematic vortex phase diagram for Bi2212.

At low temperatures ($T < 20$ K, area A in figure C2.44), point disorder is sufficiently strong to induce zero-dimensional pinning and a strong critical current [140]. A vortex lattice is easily observed at low temperatures. It is a metastable vortex lattice frozen from higher temperatures where it is stable. The zero-dimensional pinning only slightly deforms the vortex lattice, and keeps it intact in the flow of transport or magnetization current.

D1.7.2

D2.2, D2.4

At $T > 20$ K, a vortex lattice or Bragg glass is stable at low fields (areas F and G). The areas F and G are restricted from above by the T_m and T_{sp} lines that meet in a critical point. The melting line T_m is the first order phase transition line, and a local jump in reversible magnetization can be found on it as shown in figure C2.45 [142, 143]. The magnetization jump results in a local paramagnetic peak in the ac susceptibility (shown in figure C2.46) [144]. On the second peak line T_{sp} that is situated approximately at $20 < T < 40$ K, a well-defined jump in irreversible magnetization is present as shown in figure C2.47 (second peak or 'fishtail' anomaly). This jump corresponds to the pinning induced order–disorder

A4.3

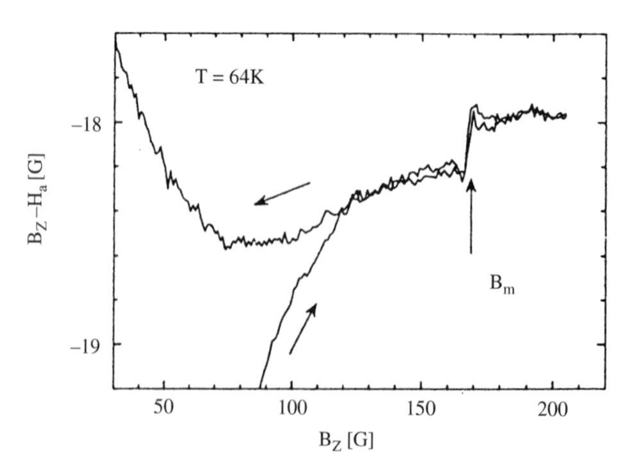

Figure C2.45. Local magnetization loop at elevated temperature showing the first order transition manifested by the sharp step in equilibrium magnetization at $B_m(T)$. Reproduced by permission of Doyle [143].

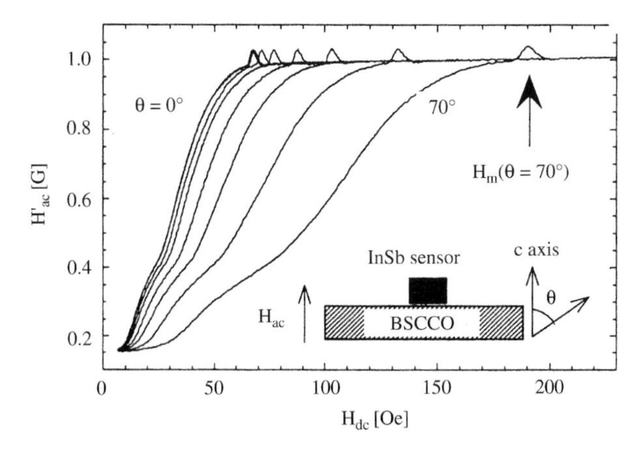

Figure C2.46. Transmitted ac field as a function of applied dc field for different angles θ between the c-axis and the dc field direction, $\theta = 0$–$70°$ in $10°$ intervals. $T = 80\,\mathrm{K}$, $H_{ac} = 1\,\mathrm{Oe}$, $f = 7.75\,\mathrm{Hz}$. The inset gives a schematic description of the setup and the orientations of the applied field. Reproduced by permission of Schmidt [144].

transition in the vortex solid [145]. Recently, sensitive measurements of the current distribution and the height of the Bean–Livingston barrier [146] indicated a depinning transition line that divides areas F and G [147].

At high temperatures, the first-order phase transition on the melting line T_m is a sublimation transition, i.e., simultaneous melting and decoupling (the line divides areas G and D in figure C2.44). The presence of extended defects, e.g., column tracks, can prevent decoupling, and the transition could be from a vortex lattice to a vortex liquid (melting transition) [148]. The Bean–Livingston barrier sensitive measurements revealed the line that divides areas C and D on which a vortex gas transforms to a vortex liquid or specific vortex solid [147]. The extension of the depinning line that divides areas F and G to higher fields gives a depinning line of the vortex glass. On this line, the bulk pinning drops below the

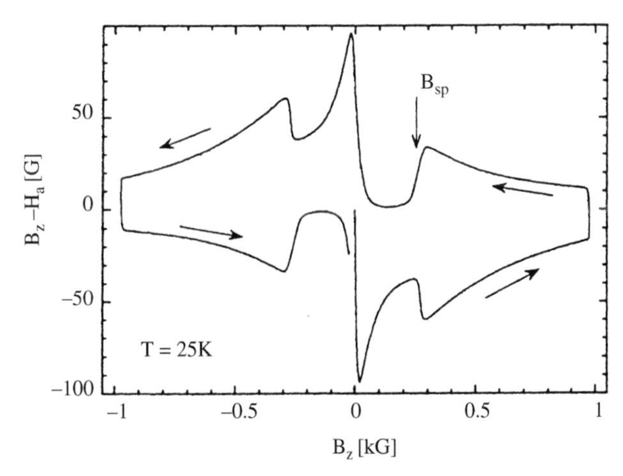

Figure C2.47. Local magnetization loop in the vicinity of the second magnetization peak B_{sp}. Reproduced by permission of Doyle [143].

detectable level. Mixed vortex solid–vortex liquid and vortex liquid–vortex gas areas are also possible in the crystals.

References

[1] Michel C, Hervieu M, Borel M M, Grandin A, Deslandes F, Provost J and Raveau B 1987 Superconductivity in Bi–Sr–Cu–O system *Z. Phys.* B **68** 421

[2] Maeda H, Tanaka Y, Fukutomi M and Asano T 1998 A new high-T_c oxide superconductor without a rare earth element *Japan. J. Appl. Phys.* **27** L209

[3] Malozemoff A P, Li Q and Fleshler S 1998 Progress in BSCCO-2223 tape technology *Physica* C **282–287** 424

[4] Polyanskii A, Beilin V M, Yashchin E, Goldgirsh A, Roth M and Larbalestier D 2001 Fast healing of deformation-induced damage in Ag/Bi-2223 tapes *IEEE Trans. Appl. Supercond.* **11** 3736

[5] Hazen R M 1990 Crystal structures of high-temperature superconductors *Physical Properties of High Temperature Superconductors II*, ed D M Ginsberg (Singapore: World Scientific) p 121

[6] Maeda A, Hase M, Tsukada I, Noda K, Takebayashi S and Uchinokura K 1990 Physical properties of $Bi_2Sr_2Ca_{n-1}Cu_nO_y$ ($n = 1, 2, 3$) *Phys. Rev.* B **41** 6418

[7] Set 43-42 1995 *ICDD powder diffraction file PDF-2 database sets 1-45 CDRS1309170*, PDF-2_45 International Centre for Diffraction Data, Release A6, Dataware Technologies

[8] Set 41-317 1995 *ICDD powder diffraction file PDF-2 database sets 1-45 CDRS1309170*, PDF-2_45 International Centre for Diffraction Data, Release A6, Dataware Technologies

[9] Set 42-743 1995 *ICDD powder diffraction file PDF-2 database sets 1-45 CDRS1309170*, PDF-2_45 International Centre for Diffraction Data, Release A6, Dataware Technologies

[10] Horiuchi S and Takayama-Muromachi E 1996 Crystal structure *Bithmuth-Based High-Temperature Superconductors*, eds H Maeda and K Togano (New York: Dekker) p 7

[11] Yan H, Mao Z, Xu G and Zhang Y 1999 Shift of the apical oxygen of the CuO6 octahedron and the superconductivity in $Bi_{1.8}Pb_{0.2}Sr_{2-x}CuO_y$ system *Phys. Rev.* B **59** 8459

[12] Jiang H, Li F H, Liu W, Zhang Y and Mao Z Q 1998 Structural modulation in $Bi_2(Sr_{0.9}La_{0.1})_2Cu_{1-x}Co_xO_y$ ($x = 0.2, 0.6, 1.0$) *Physica* C **305** 202

[13] Sun X, Zhao X, Wu W, Fan X, Li X G and Ku H C 1998 Pr-doping effect on the structure and superconductivity of the $Bi_2Sr_2Ca_{1-x}Pr_xCu_2O_y$ single crystal *Physica* C **307** 67

[14] Yan H, Mao Z, Xu G, Shi L, Tian M and Zhang Y 1998 Distortion of the microstructure in $Bi_{1.8}Pb_{0.2}Sr_{2-x}CuO_y$ system *Physica* C **309** 263

[15] Gladyshevskii R E and Flukiger R 1996 Modulated structure $Bi_2Sr_2CaCu_2O_{8+\delta}$, a high T_c superconductor with monoclinic symmetry *Acta Crystallogr.* **52** 38

[16] Zhao X, Wu W, Sun X and Li X G 1999 New experimental evidence of the structural modification in single crystal $Bi_2Sr_2CaCu_2O_y$ by x-ray diffraction observation *Physica* C **320** 225

[17] Ginsberg DM 1990 *Physical Properties of High Temperature Superconductors (II)* (Singapore: World Scientific)

[18] Cai Z X and Zhu Y 1998 *Microstructure and Structural Defects in High-Temperature Superconductors* (Singapore: World Scientific)

[19] Yan Y, Kirk M A and Evetts J E 1997 Structure of grain boundaries: correlation to supercurrent transport in textured $Bi_2Sr_2Ca_{n-1}Cu_nO_x$ bulk material *J. Mater. Res.* **12** 3009

[20] Nishiyama M, Ogawa K, Chong I, Hiroi Z and Takano M 1999 Scanning tunnelling microscope studies on the atomic structures in $Bi_2Sr_2CaCu_2O_{8+\delta}$ highly doped with Pb *Physica* C **314** 299

[21] Uprety K K, Horvat J, Wang X L, Ionescu M, Liu H K and Dou S X 2001 Enhancement of vortex pinning by Josephson coupling of two-dimensional pancake vortices in heavily Pb-doped $Bi_{2-x}Pb_xSr_2CaCu_2O_{8+\delta}$ *Supercond. Sci. Technol.* **14** 479

[22] Dou S X, Liu H K, Bourdillion A J, Kviz M, Tan N X and Sorrell C C 1989 Stability of superconducting phases in Bi–Sr–Ca–Cu–O and the role of Pb doping *Phys. Rev.* B **40** 5266

[23] Fisher R A, Kim S, Lacy S E, Phillips N E, Morris D E, Markelz A G, Wei J Y T and Ginley D S 1988 Specific-heat measurements on superconducting Bi–Ca–Sr–Cu and Tl–Ca–Ba–Cu oxides: absence of a linear term in the specific heat of Bi–Ca–Sr–Cu oxides *Phys. Rev.* B **38** 11942

[24] Yu M K and Franck J P 1993 Specific heat of Zn- and Co-substituted $Bi_{1.8}Pb_{0.2}Sr_2Ca(Cu_{1-x}M_x)_2O_y$ *Phys. Rev.* B **48** 13939

[25] Yu M K and Franck J P 1996 Comparison of the low-temperature specific heat of Fe- and Co-substituted $Bi_{1.8}Pb_{0.2}Sr_2Ca(Cu_{1-x}M_x)_2O_y$ (M = Fe or Co): anomolously enhanced electronic contribution due to Fe doping *Phys. Rev.* B **53** 8651

[26] Yu M K and Franck J P 1994 The specific heat of the 2201-BISCO high-T(C) superconductor *Physica* C **223** 57

[27] Phillips N E, Fisher R A, Gordon J E, Kim S, Stacy A M, Crawford M K and McCarron E M 1990 Specific Heat of YBa2Cu3O7: origin of the linear term and volume fraction of superconductivity *Phys. Rev. Lett.* **65** 357

[28] Collocott S J, Driver R and Andrikis C 1991 Specific heat of the ceramic superconductor $Bi_2Sr_2CuO_6$ from 0.4 to 20 K *Physica* C **173** 117

[29] Uher C 1992 Thermal conductivity of high-temperature superconductors *Physical Properties of High Temperature Superconductors III*, ed D M Ginsberg (Singapore: World Scientific) p 254
[30] Crommie M F and Zettl A 1990 Thermal conductivity of single-crystal Bi–Sr–Ca–Cu–O *Phys. Rev.* B **41** 10978
[31] Crommie M F and Zettl A 1991 Thermal conductivity anisotropy of single-crystal $Bi_2Sr_2CaCu_2O_8$ *Phys. Rev.* B **43** 408
[32] Zavaritsky N V, Samoilov A V and Yurgens A A 1991 Mixed state properties of $Bi_2Sr_2CaCu_2O_x$ single crystals with $T_c \approx 95$ K *Physica* C **185–189** 1817
[33] Castellazzi S, Cimberle M R, Ferdeghini C, Giannini E, Grasso G, Marre D, Putti M and Siri A S 1997 Thermal conductivity of a BSCCO(2223) c-oriented tape—a discussion on the origin of the peak *Physica* C **273** 314
[34] Obertelli S D, Cooper J R and Tallon J L 1992 Systematics in the thermoelectric power of high-T_c oxides *Phys. Rev.* B **46** 14928
[35] Imer J-M, Patthey F, Dardel B, Schneider W-D, Baer Y, Perroff Y and Zettl A 1989 High-resolution photoemission study of the *low-energy* excitations reflecting the superconducting state of Bi–Sr–Ca–Cu–O single crystals *Phys. Rev. Lett.* **62** 336
[36] Oduleye O O, Penn S J and Alford N Mc N 1998 The mechanical properties of (Bi–Pb)SrCaCuO *Supercond. Sci. Technol.* **11** 859
[37] Reddy R R, Muralidhar M, Baru V H and Reddy P V 1995 The relationship between the porosity and elastic moduli of the Bi-Pb-2212 high-T_c superconductor *Supercond. Sci. Technol.* **8** 101
[38] Alford N M cN, Birchall J D, Clegg W J, Harmer M A, Kendall K, Eaglesham D J, Humphreys C J and Jones D R 1988 *Br. Ceram. Proc.* **40** 149
[39] Mackrodt W C 1989 Calculated lattice structure, stability and properties of the series $Bi_2X_2CuO_6$ (X = Ca, Sr, Ba), $Bi_2X_2YCu_2O_8$ (X = Ca, Sr, Ba; Y = Mg, Ca, Sr, Ba) and $Bi_2X_2Y_2Cu_3O_{10}$ (X = Ca, Sr, Ba; Y = Ba, Ca, Sr, Mg) *Supercond. Sci. Technol.* **2** 343
[40] Husek I, Kovac P and Pachla W 1995 Microhardness profiles in BSCCO/Ag composites made by various technological steps *Supercond. Sci. Technol.* **8** 617
[41] Ionescu M, Zeimetz B and Dou S X 1998 Microhardness anisotropy of Bi-2212 crystals *Physica* C **306** 213
[42] Chang F, Ford P J, Saunders G A, Jiaqiang L, Almond D P, Chapman B, Cankurtaran M, Poeppel R B and Goretta K C 1993 Anisotropic elastic and nonlinear acoustic properties of very dense textured $Bi_2Sr_2CaCu_2O_{8+y}$ *Supercond. Sci. Technol.* **6** 484
[43] Roth R S, Rawn C J, Burton B P and Beech F 1990 Phase equilibria and crystal chemistry in portions of the system SrO–CaO–Bi_2O_3–CuO, part II. The system SrO–Bi_2O_3–CuO *J. Res. Natl. Inst. Stand. Technol.* **95** 291
[44] Roth R S, Rawn C J and Bendersky L A 1990 Crystal chemistry of the compound $Sr_2Bi_2CuO_6$ *J. Mat. Res.* **5** 46
[45] Schulze K, Majewski P, Hettich B and Petzov G 1990 Phase equilibria in the system Bi_2O_3–SrO–CaO–CuO with emphasis on the high-T_c superconducting compounds *Z. Metallkd.* **81** 836
[46] Majewski P 2000 Materials aspects of the high-temperature superconductors in the system Bi_2O_3–SrO–CaO–CuO *J. Mater. Res.* **15** 854
[47] Wong-Ng W, Cook L P, Jiang F, Greenwood W, Balachandran U and Lanagan M 1997 Subsolidus phase equilibria of coexisting high-T_c Pb-2223 and 2212 superconductors in the (Bi,Pb)–Sr–Ca–Cu–O system under 7.5% O_2 *J. Mater. Res.* **12** 2855
[48] Wong-Ng W, Cook L P, Greenwood W and Kearsley A 2000 Effect of Ag on the primary field of the high–T_c (Bi,Pb)-2223 superconductor *J. Mater. Res.* **15** 296
[49] Luo J S, Dorris S E, Fischer A K, LeBoy J S, Maroni V A, Feng Y and Larbalestier D C 1996 Mode of lead addition and its effects on the phase formation and microstructure development in Ag/$(Bi,Pb)_2Sr_2Ca_2Cu_3O_x$ composite conductors *Supercond. Sci. Technol.* **9** 412
[50] Koyama S, Endo U and Kawai T 1988 Preparation of single 110 K phase of the Bi–Pb–Sr–Ca–Cu–O superconductor *Japan. J. Appl. Phys.* **27** L1861
[51] Strobel P and Fournier T 1990 Phase diagram studies in the Bi(Pb)–Sr–Ca–Cu–O system *J. Less Common Met.* **164–165** 519
[52] Strobel P, Toledano J C, Morin D, Schneck J, Vacquier G, Monnereau O, Primot J and Fournier T 1992 Phase diagram of the system $Bi_{1.6}Pb_{0.4}Sr_2CuO_6$–$CaCuO_2$ between 825°C and 1100°C *Physica* C **201** 27
[53] Suzuki T, Yumoto K, Mamiya M, Haegawa M and Takei H 1998 Phase relation studies on the $(Bi_{0.8}Pb_{0.2})_2Sr_2CuO_6$–$CaCuO_2$ between 850 and 1020°C *Physica* C **307** 1
[54] Renner Ch, Revaz B, Genoud J-Y, Kadowaki K and Fisher O 1998 Pseudogap precursor of the superconducting gap in under- and overdoped $Bi_2Sr_2Ca_1Cu_2O_{8+\delta}$ *Phys. Rev. Lett.* **80** 149
[55] vom Hedt B, Lisseck W, Westerholt K and Bach H 1994 Superconductivity in $Bi_2Sr_2CaCu_2O_{8+\delta}$ single crystals doped with Fe, Ni, and Zn *Phys. Rev.* B **49** 9898
[56] Chen X H, Li S Y, Ruan K Q, Sun Z, Cao Q and Cao L Z 1999 Effect of doping on the anisotropic resistivity in single-crystal $Bi_2Sr_2CaCu_2O_{8+\delta}$ *Phys. Rev.* B **60** 15339
[57] Kuo Y K, Schneider C W, Verebelyi D T, Nevitt M V, Skove M J, Tessema G X, Li He and Pond J M 1999 The effect of Co substitution for Cu in $Bi_2Sr_2CaCu_2O_{8-\delta}$ *Physica* C **319** 1
[58] Jin H and Kotzler J 1999 Effect of La doping on growth and superconductivity of Bi-2212 crystals *Physica* C **325** 153
[59] Pillo T h, Hayoz J, Schwaller P, Berger H, Aepi A and Schlapbach L 1999 Substitution sites of Pb and Y in $Bi_2Sr_2CaCu_2O_{8+\delta}$: x-ray photoelectron diffraction as fingerprinting tool *Appl. Phys. Lett.* **75** 1550

[60] Manifacier L, Collin G and Blanchard N 1999 Correlation between crystallographic and physical properties in (Bi, Pb)$_2$Sr$_2$(Ca,Y)Cu$_2$O$_{8+\delta}$ superconductors *Physica* B **259–261** 562
[61] Jin S, Sherwood R C, Tiefel T H, Kammlott G W, Fastnacht R A, Davis M E and Zahurak S M 1988 Superconductivity in the Bi–Sr–Ca–Cu–O compounds with noble metal additions *Appl. Phys. Lett.* **52** 1628
[62] Dou S X, Liu H K, Sorrell C C, Song K-H, Apperley M H, Guo S J, Easterling K E and Jones W K 1990 Chemistry related problems in the processing of Bi-based cuprate superconductors *Mater. Forum* **14** 92
[63] Liu H K, Dou S X, Savvides N, Zhou J P, Tan N X, Bourdillion A J, Kviz M and Sorrell C C 1989 Stabilisation of 110 K superconducting phase in Bi–Sr–Ca–Cu–O Pb substitution *Physica* C **157** 93
[64] Dou S X, Liu H K, Guo S J, Easterling K E and Mikael J 1989 Superconductivity in the Bi–Pb–Sr–Ca–Cu–O system with oxide additions *Supercond. Sci. Technol.* **2** 274
[65] Dou S X, Guo S J, Liu H K and Easterling K E 1989 Enhancement of critical current density in the Bi–Pb–Sr–Ca–Cu–O system by addition of Ca_2CuO_3 *Supercond. Sci. Technol.* **2** 308
[66] Liu H K, Dou S X, Bourdillion A J and Sorrell C C 1988 A comparison of the stability of $Bi_2Sr_2CaCu_2O_{8+y}$ and $YBa_2Cu_3O_{6.5+y}$ in various solutions *Supercond. Sci. Technol.* **1** 194
[67] Cai X Y, Polyanskii A, Li Q, Riley G N Jr and Larbalestier D C 1998 Current-limiting mechanisms in individual filaments extracted from the superconducting tapes *Nature* **392** 906
[68] Gogia B, Kashyap S C, Pandya D K and Chopra K L 1993 Possible degradation mechanism of BSCCO superconductors *Supercond. Sci. Technol.* **6** 497
[69] Misochko O V and Gu G 1997 Study of electronic Raman continua in single crystal $Bi_2Sr_2CaCu_2O_{8+x}$ *Physica* C **288** 115
[70] Devereaux T P, Einzel D, Stadlober B, Hackl R, Leach D H and Neumeier J J 1994 Electronic Raman scattering in high-T_c superconductors: a probe of $d_{x^2-y^2}$ pairing *Phys. Rev. Lett.* **72** 396
[71] Qian G G, Chen X H, Ruan K Q, Li S Y, Cao Q, Wang C Y, Cao L Z and Yu M 1999 Raman-active phonons in $Bi_2Sr_{2-x}La_xCaCu_2O_y$: effect of the oxygen content induced by La doping *Physica* C **312** 3
[72] Li, P., Yang, W., Tan, P., Wen, H., and Zhao, Z. (2000) Raman-forbidden and oxygen ordering in $Bi_2Sr_2La_xCuO_{6+\delta}$. *Phys. Rev.* **61** 11324
[73] Chen X H, Ruan K Q, Qian G G, Li S Y, Cao L Z, Zou J and Xu C Y 1998 Effects of doping on phonon Raman scattering in Bi-based 2212 system *Phys. Rev.* B **58** 5868
[74] Romero D B, Carr G L, Tanner D B, Forro L, Mandrus D, Mihály L and Williams G P 1991 $12k_BT_c$ optical signature of superconductivity in single-domain $Bi_2Sr_2CaCu_2O_8$ *Phys. Rev.* B **44** 2818
[75] Romero D B, Porter C D, Tanner D B, Forro L, Mandrus D, Mihály L, Carr G L and Williams G P 1992 On the phenomenology of the infrared properties of the copper-oxide superconductors *Solid State Commun.* **82** 183
[76] Drube W, Himpsel F J, Chandrashekhar G V and Shafer M W 1989 Empty states near the Fermi level in $Bi_2Sr_2CaCu_2O_8$ *Phys. Rev.* B **39** 7328
[77] Moog E R, Bader S D, Arko A J and Flandermeyer B K 1987 Photoemission search for the superconducting energy gap of high-T_c $YBa_2Cu_3O_7$ *Phys. Rev.* B **36** 5583
[78] Triscone G and Jenod A 1996 Thermal and magnetic properties *Bismuth-Based High-Temperature Superconductors*, eds H Maeda and K Togano (New York: Dekker) p 33
[79] Groen W A, Leeuw D M and Feiner L F 1990 Hole concentration and T_c in $Bi_2Sr_2CaCu_2O_{8+\delta}$ *Physica* C **165** 55
[80] Triscone G, Jenoud J-Y, Graf T and Muller J 1991 Variation of the superconducting properties of $Bi_2Sr_2CaCu_2O_{8+x}$ with oxygen content *Physica* C **176** 247
[81] Forro L, Ilakovac V and Keszey B 1990 High pressure study of $Bi_2Sr_2CaCu_2O_8$ single crystals *Phys. Rev.* B **41** 9551
[82] Anderson P W and Zou Z 1988 Normal tunnelling and normal transport: diagnostics for the resonating-valence-bond state *Phys. Rev. Lett.* **60** 132
[83] Watanabe T, Fujii T and Matsuda A 1997 Anisotropy resistivity of orecisely oxygen controlled single-crystal $Bi_2Sr_2CaCu_2O_{8+\delta}$: systematic study on spin gap effect *Phys. Rev. Lett.* **79** 2113
[84] Chen X H, Yu M, Ruan K Q, Li S Y, Gui Z, Zhang G C and Cao L Z 1998 Anisotropic resistivity of single crystal $Bi_2Sr_2CaCu_2O_{8+\delta}$ with different oxygen content *Phys. Rev.* B **58** 14219
[85] Motohashi T, Nakayama Y, Fujita T, Kitazawa K, Shimoyama J and Kishio K 1999 Systematic decrease of resistivity anisotropy in $Bi_2Sr_2CaCu_2O_y$ by Pb doping *Phys. Rev.* B **59** 14080
[86] Ito T, Takagi H, Ishibashi S, Ido T and Uchida S 1991 Normal-state conductivity between CuO_2 planes in copper oxide superconductors *Nature* **350** 596
[87] Tamegai T, Koga K, Suzuki K, Ichihara M, Sakai F and Iye Y 1989 Metal-insulator transition in the $Bi_2Sr_2Ca_{1-x}Y_xCu_2O_{8+y}$ system *Japan. J. Appl. Phys.* **28** L112
[88] Forro L, Mandrus D, Kendzioa C, Mihaly L and Reeder R 1990 Hall-effect measurements on superconducting and nonsuperconducting copper-oxide-based metals *Phys. Rev.* B **42** 8704
[89] Ishida K, Yoshida K, Mito T, Tokunaga Y, Kitaoka Y, Asayama K, Nakayama Y, Shimoyama J and Kishio K 1998 Pseudogap behaviour in single-crystal $Bi_2Sr_2Ca_1Cu_2O_{8+\delta}$ probed by Cu NMR *Phys. Rev.* B **58** 5960
[90] Opel M, Nemmetescek F, Venturini R, Hackl R, Erb A, Walker E, Berger H and Forro L 1999 Raman spectroscopy in $Yba_2Cu_3O_{6+x}$ and $Bi_2Sr_2(Ca_xY_{1-x})Cu_2O_{8+\delta}$: pseudogap and superconducting gap *Phys. Stat. Sol.* **215** 471

[91] Vobornik I, Berger H, Grioni M, Margaritondo G, Forro L and Ruller-Albenque F 2000 Alternative scenario: spectroscopic analogies between underdoped and disordered $Bi_2Sr_2Ca_1Cu_2O_{8+\delta}$ *Phys. Rev.* B **61** 12248

[92] Ding H, Yokoya T, Campuzano J C, Takahashi T, Randeria M, Norman M R, Mochiku T, Kadowaki K and Giapintzakis J 1996 Spectroscopic evidence for a pseudogap in the normal state of underdoped high-T_c superconductors *Nature* **382** 51

[93] Engelbrecht J R, Nazarenko A, Randeria M and Dagotto E 1998 Pseudogap above T_c in a model with $d_{x^2-y^2}$ pairing *Phys. Rev.* B **57** 13406

[94] Tarascon J M, McKinnon W R, Barboux P, Hwang D M, Bagley B G, Greene L H, Hull G W, LePage Y, Stoffel N and Giroud M 1988 Preparation, structure, and properties of the superconducting compound series $Bi_2Sr_2Ca_{n-1}Cu_nO_y$ with $n = 1, 2$, and 3 *Phys. Rev.* B **38** 8885

[95] Lan Y C, Che G C, Jia S L, Wu F, Dong C, Chen H and Zhao Z X 1996 The effect of composition, synthesis conditions, oxygen content and F doping on superconductivity and structure for R-substituted Bi-2201 *Supercond. Sci. Technol.* **9** 297

[96] Xenikos D G and Strobel P 1995 Critical-temperature optimisation in $Bi_2Sr_2CaCu_2O_{8+\delta}$ crystal *Physica* C **248** 343

[97] Ionescu M, Murashov V and Dou S X 1997 Crystallisation in Bi_2O_3–SrO–CaO–CuO system, and the influence of stoichiometry, and oxygen content on the T_c of Bi-2212 crystals *Physica* C **282–287** 437

[98] Sieburger R, Müller P and Schilling J S 1991 Pressure dependence of the superconducting transition temperature in $Bi_2Sr_2CaCu_2O_{8+y}$ as a function of oxygen content *Physica* C **181** 335

[99] Schilling J S and Klotz S 1992 The influence of high pressure on the superconducting and normal properties of high temperature superconductors *Physical Properties of High Temperature Superconductors III*, ed D M Ginsberg (Singapore: World Scientific) p 100

[100] Matsubara I, Tanigawa H, Ogura T, Yamashita H, Kinoshita M and Kawai T 1992 Upper critical field and anisotropy of the high-T_c $Bi_2Sr_2Ca_2Cu_3O_x$ phase *Phys. Rev.* B **45** 7414

[101] Schilling A, Jin R, Guo J D and Ott H R 1994 Critical field and characteristic lengths of monocrystalline superconducting $Bi_2Sr_2Ca_1Cu_2O_8$ derived from magnetization measurements in the mixed state *Physica* B **194–196** 2185

[102] Matsubara I, Funahashi R, Ueno K, Yamashita H and Kawai T 1996 Lower critical field and reversible magnetization of $(Bi,Pb)_2Sr_2Ca_2Cu_3O_x$ superconducting whiskers *Physica* C **256** 33

[103] Cooper J R, Forro L and Keszei B 1990 Direct evidence for a very large penetration depth in superconducting $Bi_2Sr_2CaCu_2O_8$ *Nature* **343** 444

[104] Bohmer C, Brandstatter G and Weber H W 1997 The lower critical field of high temperature superconductors *Supercond. Sci. Technol.* **10** A1

[105] Hedt Bvom and Westerholt K 1995 Flux penetration in Bi(2212) single crystals at low magnetic fields *Physica* C **243** 389

[106] Palstra T T M, Batlogg B, Schneemeyer L F and Cava R J 1988 Angular dependence of the upper critical field of $Bi_{2.2}Sr_2Ca_{0.8}Cu_2O_{8+\delta}$ *Phys. Rev.* B **38** 5102

[107] Li T W, Menovsky A A, Franse J J M and Kes P H 1996 Flux pinning in Bi-2212 single crystals with various oxygen contents *Physica* C **257** 179

[108] Kogan V G, Ledvij M, Simonov A Y, Cho J H and Johnston D C 1993 Role of vortex fluctuations in determining superconducting parameters from magnetization data for layered superconductors *Phys.Rev.Lett.* **70** 1870

[109] Pekala M, Bougrine H, Lada T, Morawski A and Ausloos M 1995 Anisotropy in the mixed state in textured Pb-free Bi-2223 superconductor: electrical resistivity, Seebeck and Nernst effects *Supercond. Sci. Technol.* **8** 726

[110] Kleiner R and Muller P 1994 Intrinsic Josephson effect in high-T_c superconductors *Phys.Rev.* B **49** 1327

[111] Kossler W J, Dai Y, Petzinger K G, Greer A J, Williams D L i, Koster E, Harshman D R and Mitzi D B 1998 Transparency of the ab planes of $Bi_2Sr_2CaCu_2O_{8+\delta}$ to magnetic fields *Phys. Rev. Lett.* **80** 592

[112] Muller K-H, Andrikidis C, Liu K H and Dou C X 1994 Intergranular and intragranular critical currents in silver-sheathed Bi–Sr–Ca–Cu–O tapes *Phys. Rev.* B **50** 10218

[113] Hakuraku Y and Mori Z 1993 Rapid annealing effect in the superconducting 2223 Bi(Pb)SrCaCuO thin films prepared by sputtering *J. Appl. Phys.* **73** 309

[114] Wagner P, Frey U, Hillmer E and Adrian H 1995 Evidence for a vortex-liquid vortex-glass transition in epitaxial $Bi_2Sr_2Ca_2Cu_3O_{10}$ thin-films *Phys. Rev.* B **51** 1206

[115] Kung P J, McHenry M E, Maley M P, Kes P H, Laughlin D E and Mullins W W 1995 Critical-current anisotropy, intergranular coupling, and effective pinning energy in $Bi_2Sr_2Ca_1Cu_2O_{8+\delta}$ single crystals and Ag sheathed $(Bi,Pb)_2Sr_2Ca_2Cu_3O_{10+\delta}$ tapes *Physica* C **249** 53

[116] Hensel B, Grasso G, Grindatto D P, Nissen H-U and Flukiger R 1995 Critical current density normal to the tape plane and current-transfer length of $(Bi,Pb)_2Sr_2Ca_2Cu_3O_{10}$ silver-sheathed tapes *Physica* C **249** 247

[117] Mros N, Krasnov V M, Yurgens A, Winkler D and Claeson T 1998 Multiple-valued *c*-axis critical current and phase locking in $Bi_2Sr_2CaCu_2O_{8+\delta}$ single crystals *Phys. Rev.* B **57** R8135

[118] Labdi S, Kim S F, Li Z Z, Megtert S, Raffy H, Laborde O and Monceau P 1997 Magnetic field dependence of the critical current in $Bi_2Sr_2Ca_1Cu_2O_8/Bi_2Sr_2CuO_6$ multilayers: an approach to an ideal two dimensional superconductor *Phys. Rev. Lett.* **79** 1381

[119] Morello A, Jansen A G M, Gonnelli R S and Vedeneev S I 2000 Irreversibility line of overdoped $Bi_{2+x}Sr_{2-(x+y)}Cu_{1+y}O_{6\pm\delta}$ at ultralow temperature and high magnetic field *Phys. Rev.* B **61** 9113

[120] Majer D, Zeldov E and Konczykowski M 1995 Separation of the irreversibility and melting lines in $Bi_2Sr_2CaCu_2O_8$ crystals *Phys. Rev. Lett.* **75** 1166

[121] Farrell D E, Johnston-Halperin E, Klein L, Fournier P, Rae A I M, Li T W, Trawick M L, Sasik R and Gerland J C 1996 Magnetisation jumps and irreversibility in $Bi_2Sr_2CaCu_2O_8$ *Phys. Rev.* B **53** 11807

[122] Hudakova N, Samuely P, Szabo P, Plechacek V, Knizek K and Sedmidubsky D 1995 Scaling of the superconducting order parameter in Bi cuprates with T_c *Physica* C **246** 163

[123] Kane J and Ng K W 1996 Angular dependence of the in-plane energy gap of $Bi_2Sr_2CaCu_2O_8$ by tunnelling spectroscopy *Phys. Rev.* B **53** 2819

[124] Briceno G and Zettl A 1989 Tunnelling spectrometry in $Bi_2Sr_2CaCu_2O_8$: is the energy gap anisotropic? *Solid State Commun.* **70** 1055

[125] Boekholt M, Hoffmann M and Guntherodt G 1991 Detection of an anisotropy of the superconducting gap in $Bi_2Sr_2Ca_1Cu_2O_{8+\delta}$ single crystals by Raman and tunnelling spectroscopy *Physica* C **175** 127

[126] Ekino T and Akimitsu J 1989 Energy gaps in Bi–Sr–Ca–Cu–O and Bi–Sr–Cu–O systems by electron tunnelling *Phys. Rev.* B **40** 6902

[127] Oda M, Hoya K, Kubota R, Manabe C, Momono N, Nakano T and Ido M 1997 STM/STS studies for doping effects on the symmetry and magnitude of superconducting gap in $Bi_2Sr_2Ca_1Cu_2O_{8+\delta}$ *Physica* C **282–287** 1499

[128] Harris J M, Shen Z-X, White P J, Marshall D S, Schabel M C, Eckstein J N and Bozovic I 1996 Anomalous superconducting state gap size versus T_c behaviour in underdoped $Bi_2Sr_2Ca_{1-x}Dy_xCu_2O_{8+\delta}$ *Phys. Rev.* B **54** R15665

[129] Li Y, Liu J and Leiber C 1993 Dependence of the energy gap on T_c: absence of scaling in the copper oxide superconductors *Phys. Rev. Lett.* **70** 3494

[130] Kane J, Ng K W and Moecher D 1998 Effects of lead doping on T_c and energy gap of $Bi_2Sr_2CaCu_2O_8$ by tunnelling spectrometry *Physica* C **294** 176

[131] Kendziora C, Kelley R J and Onellion M 1996 Superconducting gap anisotropy vs doping level in high-T_c cuprates *Phys. Rev. Lett.* **77** 727

[132] Oda M, Manabe C and Ido M 1996 STM images of a superconducting Cu–O plane and the corresponding tunnelling spectrum in $Bi_2Sr_2Ca_1Cu_2O_{8+\delta}$ *Phys. Rev.* B **53** 2253

[133] Clem J R 1991 Two-dimensional vortices in a stack of thin superconducting films: a model for high-temperature superconducting multilayers *Phys. Rev.* B **43** 7837

[134] Clem J R 1998 Anisotropy and two-dimensional behaviour in the high-temperature superconductors *Supercond. Sci. Technol.* **11** 909

[135] Martinez J C, Brongersma S H, Koshelev A, Ivlev B, Kes P H, Griessen R P, de Groot D G, Tarnavski Z and Menovsky A A 1992 Magnetic anisotropy of a $Bi_2Sr_2Ca_1Cu_2O_x$ single crystal *Phys. Rev. Lett.* **69** 2276

[136] Tachiki M 1997 Josephson plasma in cuprate high T_c superconductors *Physica* C **282–287** 383

[137] Tamasaku K, Nakamura Y and Uchida S 1992 Charge dynamics across the CuO_2 planes in $La_{2-x}Sr_xCuO_4$ *Phys. Rev. Lett.* **69** 1455

[138] Matsuda Y, Gaifullin M B, Kumagai K, Kadowaki K and Mochiku T 1995 Collective Josephson plasma resonance in the vortex state of $Bi_2Sr_2Ca_1Cu_2O_{8+\delta}$ *Phys. Rev. Lett.* **75** 4512

[139] Matsuda Y, Kosugi M, Gaifullin M B, Kumagai K, Hirata K, Chikumoto N, Konczykowski M, Watauchi S, Shimoyama J and Kishio K 1997 Interlayer phase coherence in the vortex state of $Bi_2Sr_2Ca_1Cu_2O_{8+\delta}$ proved by Josephson plasma resonance *Physica* C **282–287** 282

[140] Goffman M F, Herbsommer J A, de la Cruz F, Li T W and Kes P H 1998 Vortex phase diagram of $Bi_2Sr_2Ca_1Cu_2O_{8+\delta}$: c-axis superconducting correlation in the different vortex phases *Phys. Rev.* B **57** 3663

[141] Khaykovich B, Konczykowski M, Zeldov E, Doyle R A, Majer D, Kes P H and Li T W 1997 Vortex-matter phase transitions in $Bi_2Sr_2Ca_1Cu_2O_8$: effects of weak disorder *Phys. Rev.* B **56** 517

[142] Zeldov E, Majer D, Konczykowski M, Geshkenbein V B, Vinokur V M and Shtrikman H 1995 Thermodynamic observation of first-order vortex-lattice melting transition in $Bi_2Sr_2Ca_1Cu_2O_8$ *Nature* **375** 373

[143] Doyle R A, Khaykovich B, Konczykowski M, Zeldov E, Morozov N, Majer D, Kes P H and Vinokur V M 1997 Vortex-matter phase transitions in $Bi_2Sr_2Ca_1Cu_2O_8$ *Physica* C **282–287** 323

[144] Schmidt B, Konczykowski M, Morozov N and Zeldov E 1997 Angular dependence of the first-order vortex-lattice phase transition in $Bi_2Sr_2Ca_1Cu_2O_8$ *Phys. Rev.* B **55** 8705

[145] Koshelev A E and Vinokur V M 1998 Pinning induced transition to disordered vortex phase in layered superconductors *Phys. Rev.* B **57** 8026

[146] Fuchs D T, Zeldov E, Rappaport M, Tamegai T, Ooi S and Shtrikman H 1998 Transport properties governed by surface barriers in $Bi_2Sr_2Ca_1Cu_2O_8$ *Nature* **391** 373

[147] Fuchs D T, Zeldov E, Tamegai T, Ooi S, Rappaport M and Shtrikman H 1998 Possible new vortex-matter phases in $Bi_2Sr_2Ca_1Cu_2O_8$: effects of weak disorder *Phys. Rev. Lett.* **80** 4971

[148] Khaykovich B, Konczykowski M, Teitelbaum K, Zeldov E, Shtrikman H and Rappaport M 1998 Effect of columnar defects on the vortex-solid melting transition in $Bi_2Sr_2Ca_1Cu_2O_8$ *Phys. Rev.* B **57** 14088

C3
TIBCCO

Emilio Bellingeri and René Flükiger

C3.1 Introduction

Thallium based superconductors form one of the largest family of high temperature superconductors and, up to the discovery of mercury based cuprates, Tl2223 held the record of the highest critical temperature. C4

These compounds are very interesting both for application and for fundamental studies. In particular the Tl1223 compound presents a high irreversibility field allowing, in principle, the transport of high current in high magnetic field. Unfortunately up to now, the transport current is dramatically limited by the weak-link phenomenon. D2.2

Other compounds, e.g. Tl1212 and Tl2212, are used in fabrication of devices operating in the field of microwaves exploiting their low surface resistance at frequencies up to 10 GHz. D2.7

Tl1212 is also a interesting compound for fundamental studies: when Ca is partially substituted by Y the phase transforms continuously, from a superconductor with T_c as high as 110 K to an antiferromagnetic insulator exhibiting a metal–insulator transition, without any structural modification, offering an exceptional opportunity of investigation. Studies of the pairing symmetry were performed on thin films of Tl2201. D1.1.2

In contrast to Bi based superconductors, these phases are very 'tolerant' to substitution and deviations from stoichiometry. This fact, from one side, makes an organic and complete description of their properties difficult but offers, on the other side, opportunities for improving both the properties and the synthesis. C2

The toxicity and volatility of Tl is often reported as an obstacle to the preparation of this class of compounds. For this reason some precaution must be adopted during the handling, but the preparation of Tl based compounds can now be considered as safe. B2.2.4

C3.2 Structural properties [1]

The first Tl based superconducting material was reported in the Tl–Ba–Cu–O system by Kondoh *et al* [2]. A critical temperature of 19 K was measured for a sample of nominal composition $Tl_{1.2}Ba_{0.8}CuO$. Independently, Shen and Herman [3] reported critical temperatures up to 90 K for the nominal compositions $Tl_2Ba_2Cu_3O_{8+\delta}$ $TlBaCu_3O_{5.5+\delta}$ and $Tl_{1.5}Ba_2Cu_3O_{7.3+\delta}$: the superconducting compound was later identified as Tl2201. A short time later the same authors [4] succeeded in preparing superconductors with $T_c = 120$ K for the nominal composition $Tl_2BaCa_{1.15}Cu_3O_{8.5+\delta}$ and $Tl_{1.86}BaCaCu3O_{7.8+\delta}$. D1.1.2 D1.1.1

At present Tl based cuprates constitute one of the largest chemical families of high temperature superconductors, forming two distinct structure series with the general formulas Tl(Ba or Sr)$_2$Ca$_{n-1}$-Cu$_n$O$_{2n+3}$ and Tl$_2$Ba$_2$Ca$_{n-1}$Cu$_n$O$_{2n+4}$. The two families can be both described by the general formula Tl$_m$Ba$_2$Ca$_{n-1}$Cu$_n$O$_{2n+m+2}$ where m can be 1 or 2 and $n = 1, 2, 3$ or 4 for bulk samples prepared by solid state chemistry and arrive up to 5 and 6 for samples synthesized under very high pressure or for thin films. From n and m, it is also possible to generate the common adopted abbreviated name for these phases (m 2 $n-1$ n): for example Tl1201 or Tl2234.

B2.2.4

These series of compounds include one or more two dimensional CuO$_2$ layers (*conducting layers*), if the number, n in the general formula, of this layer is > 1 they are separated by $n-1$ Ca layers (*separating layers*). Together these layers form the conducting block; on both sides of it are always present two, BaO (SrO) layers (*bridging layers*), one on each side. The structure is completed by m (1 or 2) TlO layers (*additional layers*). All the layers consist in an approximately square mesh:

D1.1.2

- in the CuO$_2$ layers the Cu atom lies on the centre of the square and the O atoms on the centre of the square edges;
- the Ca atom in the separating layer occupies the corner of the square;
- in the bridging layer the Ba (Sr) atom is positioned in the corner of the square and the O atom in the centre, on the top (or bottom) of underlying Cu atom;
- the TlO layer is identical to the bridging layer but is shifted by 1/2 1/2 in the plane direction.

The structures are obtained by stacking the layers one on top of each other so that the cations of two consecutive layer are always shifted by 1/2 1/2; as a consequence of this systematic shifts the translation period in the stacking direction must contain an even number of layers. It should be noted that in single Tl layer compounds ($m = 1$) the conventional cell of the structure contains one stacking unit, i.e. one formula unit ($Z = 1$), whereas in double Tl layers compounds ($m = 2$) it contains two stacking unit ($Z = 2$). In the first case the tetragonal cell of the structure is primitive, in the second case it is body-centred because of the shift by 1/2 1/2 observed within the plane when two additional layers are present. See for example figures C3.1 and C3.7 where the crystal structure of Tl1201 and Tl2201 are drawn.

The stacking of the layers forms for the $n = 1$ compounds an octahedral CuO$_6$ block (e.g. figure C3.1), the $n = 2$ compounds present two pyramidal CuO$_5$ blocks (e.g. figure C3.3)and the $n = 3$ compounds have two pyramidal and one square CuO$_4$ blocks (e.g. figure C3.5).

In the single layer compound it is possible to substitute partially or completely the Ba by Sr in the bridging layer. For both structure series, the Ca sites are mainly occupied by Ca with small amounts of Tl. A common feature of the structures of Tl based superconductors is the displacement of the Tl and O sites in the Tl layer from the ideal position on four-fold rotation axes. The resulting tetrahedral coordination of Tl atoms is typical for Tl^{3+}. Vacancies may occur on the O sites in the Tl layer.

C3.2.1 The Tl1201 phase

D1.1.2

The structure of stoichiometric TlBa$_2$CuO$_5$ is tetragonal (space group P4/mmm) [5] whereas that of the analogous TlSr$_2$CuO$_5$ is reported to be either tetragonal [6] or orthorhombic (Pmmm) [7] depending on the oxygen content. Superconductivity may be induced by a partial substitution of Ba^{2+} or Sr^{2+} by La^{3+} or rare-earth elements [8, 9] or by preparing samples under reducing atmosphere [10]. Increasing the Pb content in Tl$_{1-x}$Pb$_x$Sr$_2$CuO$_5$, a transformation from orthorhombic to tetragonal is observed for $x = 0.12$. A superstructure with doubling of the b parameter and an ordered arrangement of oxygen

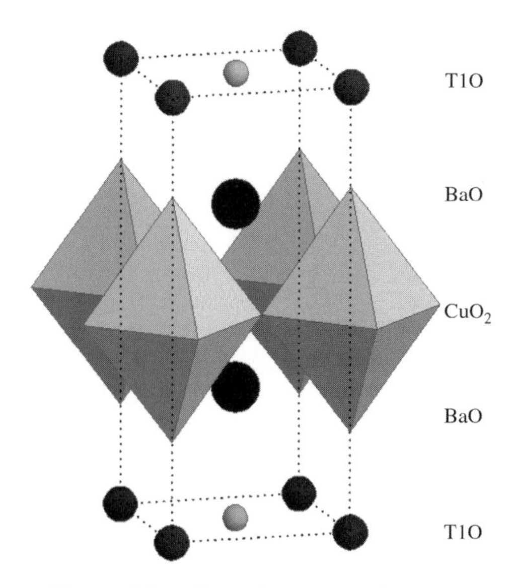

Figure C3.1. Crystal structure of Tl1201.

vacancies in the CuO_2 layer is observed in $TlSr_2CuO_{4.515}$ [11]. A schematic drawing of the structure is represented in figure C3.1; calculated peaks position and intensity in a XRD diagram are plotted in figure C3.2 (table C3.1). D1.1.2

C3.2.2 The Tl1212 phase

The structure of $TlBa_2CaCu_2O_7$ is tetragonal (P4/mmm) with $a = 3.8472$ and $c = 12.721$ Å (figure C3.3, table C3.2); the O site in the Tl layer was found to be only occupied for 3/4 and Tl and Ba atoms were

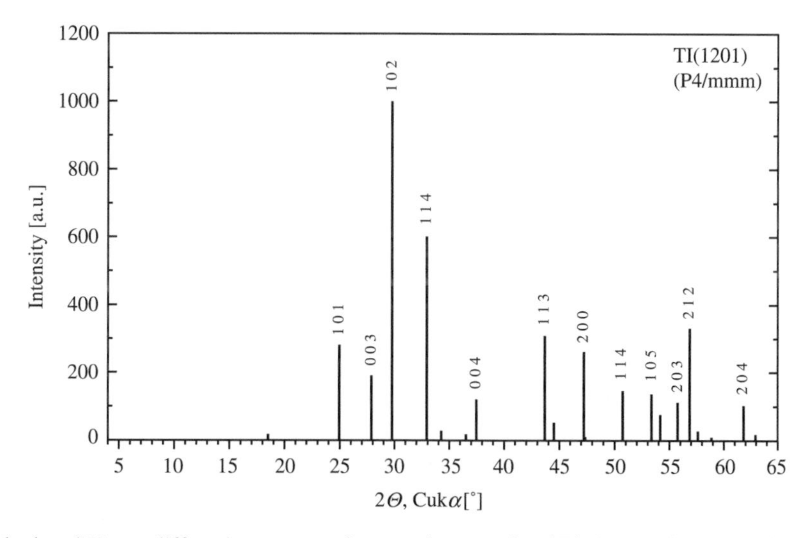

Figure C3.2. Calculated X-ray diffraction pattern for powder sample of Tl1201. Indexes are given for the stronger reflections.

Table C3.1. Atoms coordinates of $Tl_{0.92}Ba_{1.2}La_{0.8}CuO_{4.864}$ P4/mmm $a = 3.8479$, $c = 9.0909$ Å, $Z = 1$ [12]

Atom	Site	x	y	z	Occ
Tl	4 (l)	0.0801	0	0	0.230
Ba(La)	2 (h)	1/2	1/2	0.2942	
Cu	1 (b)	0	0	1/2	
O1	4 (n)	0.4281	1/2	0	
O2	2 (g)	0	0	0.2250	
O3	2 (e)	0	1/2	1/2	

found to be displaced towards the oxygen vacancies [13]. In the analogous family $(Tl_{0.5}M_{0.5})Sr_2CaCuO$, M = Pb, Bi both O and Tl site in Tl layer are displaced from their ideal positions resulting in a tetrahedral coordination of the Tl atoms. The calculated XRD diagram is shown in figure C3.4.

D1.1.2

C3.2.3 The Tl1223 phase

Superconductivity at 110 K was first identified for the composition $TlBa_2Ca_2Cu_3O_9$ with a structure in the group P4/mmm $a = 3.8429$, $c = 15.871$ [15] (figures C3.5 and C3.6). The Tl site was found to

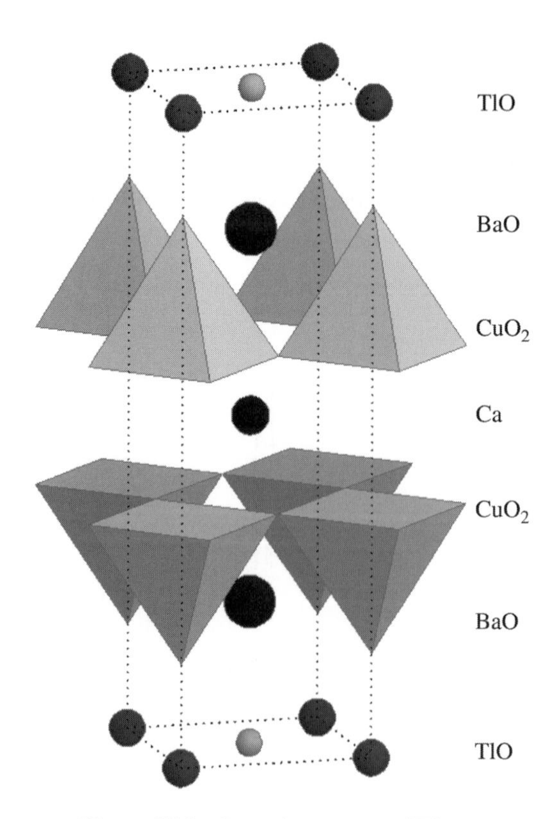

Figure C3.3. Crystal structure of Tl1212.

Table C3.2. Atoms coordinates of $Tl_{1.13}Ba_2Ca_{0.87}Cu_2O_7$ P4/mmm $a = 3.8472$, $c = 12.721$ Å, $Z = 1$ [14]

Atom	Site	x	y	z	Occ
Tl	4 (l)	0.0877	0	0	0.250
Ba	2 (h)	1/2	1/2	0.2155	
Cu	2 (g)	0	0	0.3740	
Ca	1 (d)	1/2	1/2	1/2	
O1	1 (c)	1/2	1/2	0	
O2	2 (g)	0	0	0.1582	
O3	4 (i)	0	1/2	0.3797	

be displaced from the ideal position along the short translation vectors by about 0.35 Å. Vacancies were observed on Ba sites (occupancy 0.94) and a small amount of Tl was observed on Ca site (5 at%). For $TlBa_2Ca_2Cu_3O_{8.62}$ two partly occupied sites, one in the ideal position (0,0,0) and the other one, displaced in (0.1138,0,0) were considered in [16] and associated with the oxidation states Tl^{+} and Tl^{+3}, respectively. The O site in the Tl layer was found to be only partly occupied (occupancy 0.62). In the Sr analogous superconductor, with half Tl replaced by Pb [17] Tl and O sites in Tl layer and O site in Sr layer were found to be off centred in the [1 1 0] direction of about 0.24, 0.29 and 0.27 Å, respectively. The partial substitution of Ca by Tl was also observed in this material (table C3.3). D1.1.2

C3.2.4 *The Tl2201 phase*

Tl-2201 crystallizes with two structural modifications, one orthorhombic, and the other tetragonal, and the latter being superconducting with T_c up to 90 K for an optimal oxygen content. In tetragonal D1.1.2

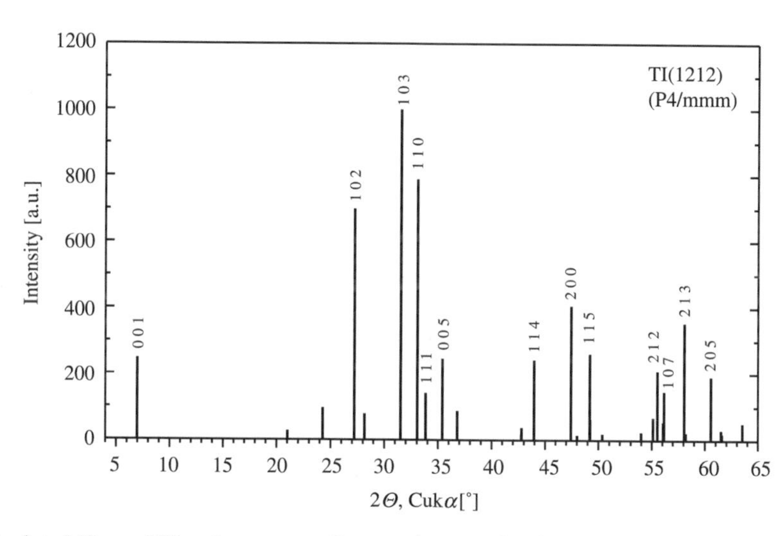

Figure C3.4. Calculated X-ray diffraction pattern for powder sample of Tl1212. Indexes are given for the stronger reflections.

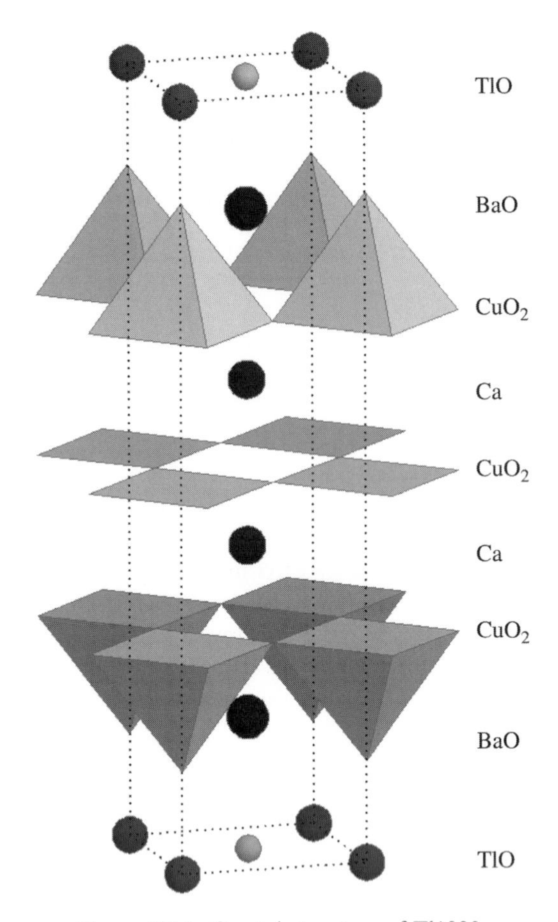

Figure C3.5. Crystal structure of Tl1223.

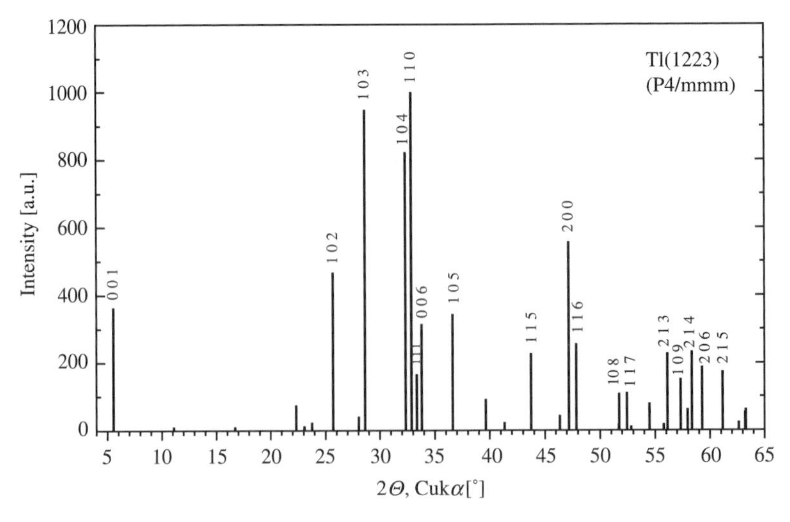

Figure C3.6. Calculated X-ray diffraction pattern for powder sample of Tl1223. Indexes are given for the stronger reflections.

Table C3.3. Atoms coordinates of $Tl_{0.56}Pb_{0.56}Sr_2Ca1_{1.88}Cu_3O_9$ P4/mmm $a = 3.808$, $c = 15.232$ Å, $Z = 1$ [17]

Atom	Site	x	y	z	Occ
Tl	4 (l)	0.067	0	0	0.250
Sr	2 (h)	1/2	1/2	0.1709	
Cu1	2 (g)	0	0	0.2868	
Ca	1 (h)	1/2	1/2	0.3928	
Cu2	1 (b)	0	0	1/2	
O1	4 (n)	0.4	1/2	0	
O2	2 (g)	0	0	0.1582	0.250
O3	4 (i)	0	1/2	0.3797	
O4	2 (e)	0	1/2	1/2	

Table C3.4. Atoms coordinates of $Tl_2Ba_2CuO_6$, $T_c = 90$ K, *I4/mmm*, $a = 3.866$, $c = 23.239$ Å, $Z = 2$ [29]

Atom	Site	x	y	z	Occ
Tl	4 (e)	0	0	0.2974	
Ba	4 (e)	1/2	1/2	0.4170	
Cu	2 (b)	0	0	1/2	
O(1)	16 (n)	0.405	1/2	0.2829	0.25
O(2)	4 (e)	0	0	0.3832	
O(3)	4 (c)	0	1/2	1/2	

$Tl_2Ba_2CuO_6$ (space group I4/mmm $a = 3.866$ $c = 23.239$ Å) (table C3.4) the Tl site was found to be displaced from its ideal position and a partial substitution of up to 7 at% Tl by Cu was observed [18, 19]. The presence of an extra oxygen with very low occupation (from 0.0005 to 0.028) located between the two Tl layers was suggested in [20]. Non-superconducting Tl-2201 phase with orthorhombic structure ($a = 5.4451$, $b = 5.4961$, $c = 23.153$ Å) was first reported and the space group Fmmm was proposed in [21]. A schematic drawing of the structure is presented in figure C3.7, while figure C3.8 shows calculated XRD diagrams of tetragonal (a) and orthorhombic (b) Tl2201. As shown in figure C3.9 in the space group Fmmm the a and b parameters are rotated by 45° with respect to those in the other space groups considered here. The crystal structure was also refined in subgroups of Fmmm, in order to account for positional disorder in additional TlO layers (A2aa [22], Abma [23, 24]). It is generally agreed that the formation of tetragonal phase is favoured by Tl-deficient composition, and the formation of the orthorhombic phase by high Tl content [25]. To a minor extent, the tetragonal–orthorhombic transition depends also on oxygen content, a higher oxygen content favouring the orthorhombic modification [24, 26]. Thallium stoichiometric tetragonal and orthorhombic phases were obtained by high-pressure synthesis [27, 28] (table C3.5).

D1.1.2

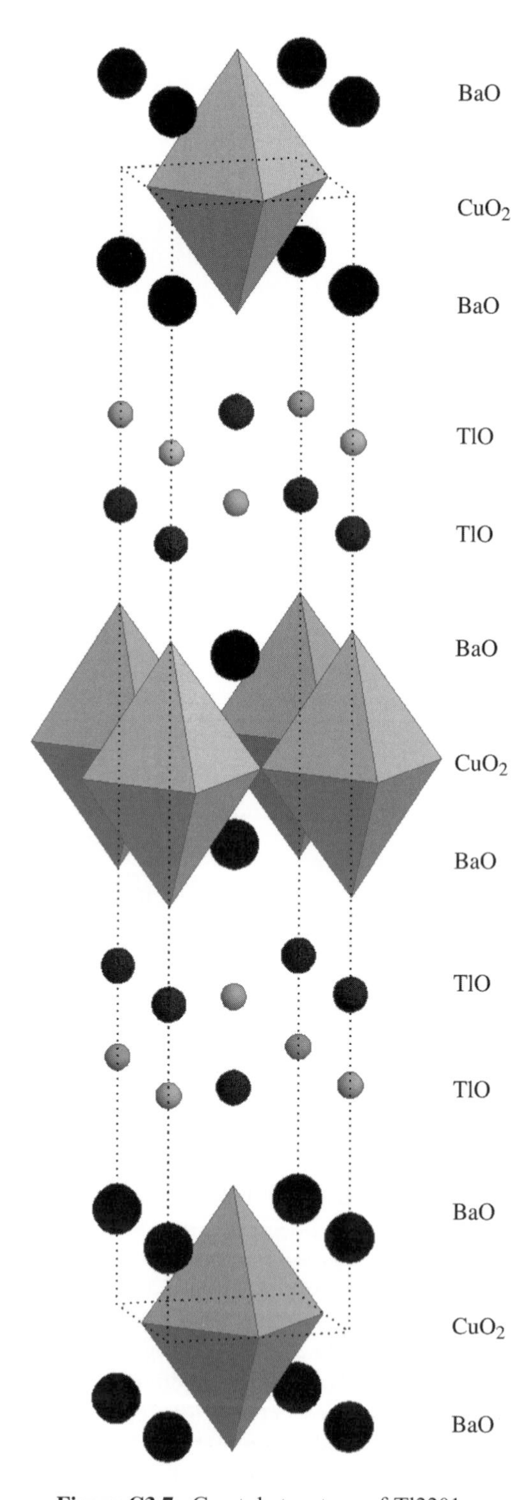

Figure C3.7. Crystal structure of Tl2201.

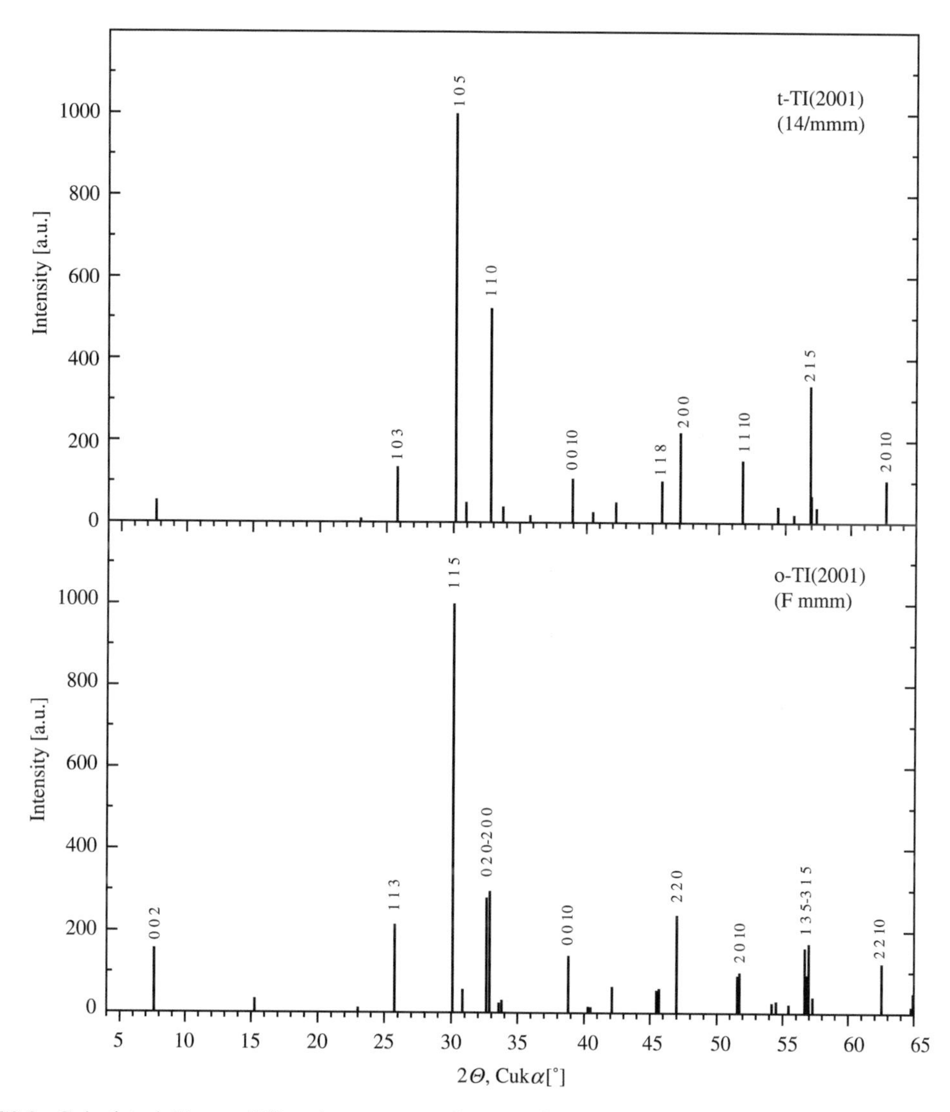

Figure C3.8. Calculated X-ray diffraction pattern for powder sample of tetragonal (*a*) and orthorhombic (*b*) Tl2201. Indexes are given for the stronger reflections.

C3.2.5 *The Tl2212 phase*

Tl2212 phase, $Tl_2Ba_2CaCu_2O_{8+\delta}$, was first identified by Hazen *et al* in [31] and the structural refinement carried out by Subramanian in [32]. In the majority of structural studies, the Ca site was found to be partly occupied by Tl (12–28 at%), whereas reduced scattering density on Tl site was attributed to either partial substitution of Tl by Ca (10–11 at%) [33–35] or by Cu (9 at%) [36], or simply Tl deficiency (6–13%) [37–39]. Vacancies on O site in *additional* layers were detected in [38] (5–6%), [35] (12–16%). Displacement of Tl site from four-fold rotation axis, split into 32(o) 0.013 0.041 0.28635 was reported

D1.1.2

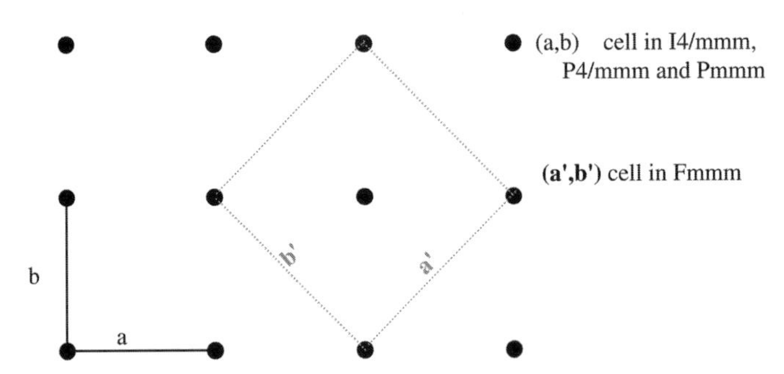

Figure C3.9. In-plane crystallographic axis in I4/mmm, P4/mmm, Pmmm (a, b) and in Fmmm (a′, b′).

in [39]. Structure and calculated XRD diagram are plotted in figures C3.10 and C3.11, respectively (table C3.6).

C3.2.6 The Tl2223 phase

The Tl2223 phase was identified by Politis *et al* in [40] and first the structural refinement carried out by Torardi *et al* in [41] with partial disorder between the Tl and Ca site. The Ca site was found to be partly occupied by Tl atoms (3–7 at%), also in [42–44]. In [42, 43] [45], vacancies on the Tl site (6–12%) were detected, whereas in [44] partial substitution of Tl by Cu (14 at%) was reported. O site in *additional* layers was found to be split from the ideal position into 16(n) 0.6112 1/2 0.2753, with occupancy 0.234 [43, 45]. In [44] the Tl and O sites in *additional* layers were refined both in 16(m) 0.0276 0.0276 0.27921, with occupancy 0.215 for Tl and 0.5819 0.5819 0.2756, with occupancy 0.25 for O. The structure and the calculated XRD spectra are presented in figures C3.12 and C3.13 (table C3.7).

D1.1.2

C3.2.7 Exotic structures

The structure of less common superconducting phases of the family are reported in this paragraph: Tl1222 and the two four copper layer compound Tl1234 and Tl2234.

Tl1222 phases were first reported in [46] for compositions $Tl(Ba_{0.84}Tl_{0.16})_2Pr_2Cu_2O_9$ and $Tl_{0.9}(Sr_{0.8}Tl_{0.2})_2Pr_2Cu_2O_9$. Structural refinements were carried out in space group *I4/mmm* with $a = 3.900$, $c = 30.273$ Å and $a = 3.8635$, $c = 29.535$ Å, respectively. In both structures the Tl site,

Table C3.5. Atoms coordinates of $Tl_2Ba_2CuO_{6.10}$, *Fmmm*, $a = 5.4604$, $b = 5.4848$, $c = 23.2038$ Å, $Z = 4$ [30]

Atom	Site	x	y	z	Occ
Tl	16 (m)	0	− 0.025	0.2976	0.5
Ba	8 (i)	1/2	0	0.4172	
Cu	4 (b)	0	0	1/2	
O(1)	8 (f)	1/4	1/4	1/4	0.07
O(2)	16 (m)	1/2	− 0.059	0.2895	0.49
O(3)	8 (i)	0	0	0.3837	
O(4)	8 (e)	1/4	1/4	1/2	

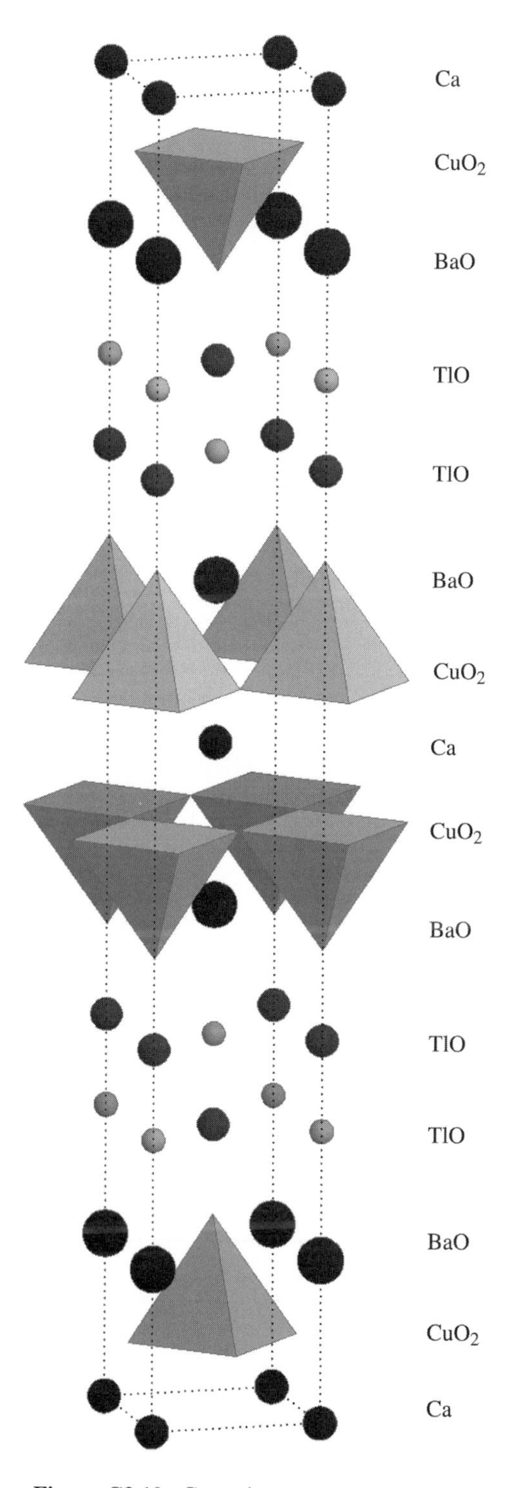

Figure C3.10. Crystal structure of Tl2212.

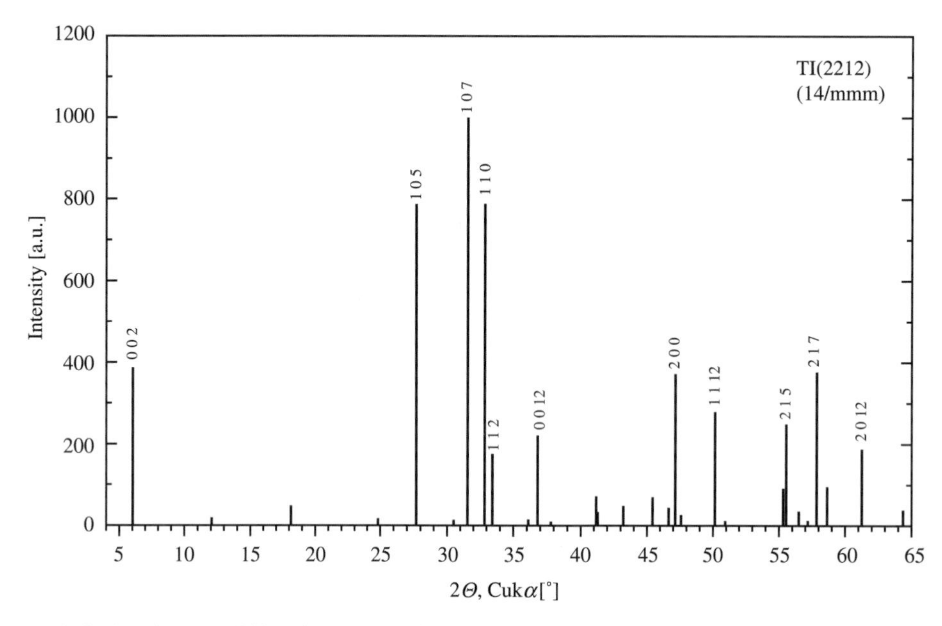

Figure C3.11. Calculated X-ray diffraction pattern for powder sample of Tl2212. Indexes are given for the stronger reflections.

deficient in the structure of Sr-containing phase, was found to be located on four-fold rotation axis in 2(*a*) 0 0 0. Superconductivity ($T_c = 40$ K) was first reported by Iqbal *et al* in [47] for composition $(Tl_{0.5}Pb_{0.5})Sr_2(Eu_{0.9}Ce_{0.1})_2Cu_2O_9$ and subsequently by Liu *et al* [48] for Pb-free $TlBa_2(Eu_{0.75}Ce_{0.25})_2$-$Cu_2O_9$ (table C3.8).

The superconducting ($T_c = 122$ K) Tl1234 phase with tetragonal structure ($a = 3.85$ and $c = 19.1$ Å) was first reported in [49] for the composition $TlBa_2Ca_3Cu_4O_{11}$ (table C3.9).

The superconducting ($T_c = 114$ K) Tl2234 phase was first reported by Hervieu *et al* in [51] for composition $Tl_2Ba_2Ca_3Cu_4O_{12}$ and a structural model was proposed in space group *I*4/*mmm* ($a = 3.852$, $c = 42.00$ Å). In [52] disorder between the Tl and Ca site was reported: 40 at% of Ca on the Tl site and 10–15 at% of Tl on the Ca sites. In addition, vacancies on Tl (6–7%) and O (0–5%) sites in TlO layers were detected, the latter depending on the annealing conditions (table C3.10).

Table C3.6. Atoms coordinates of $Tl_2Ba_2CaCu_2O_8$, *I*4/*mmm*, $a = 3.8550$, $c = 29.318$ Å, $Z = 2$ [32]

Atom	Site	x	y	z	Occ
Tl	4 (e)	0	0	0.2864	
Ba	4 (e)	1/2	1/2	0.3782	
Cu	4 (e)	0	0	0.4460	
Ca	2 (a)	1/2	1/2	1/2	
O(1)	16 (n)	0.396	1/2	0.2815	0.25
O(2)	4 (e)	0	0	0.3539	
O(3)	8 (g)	0	1/2	0.4469	

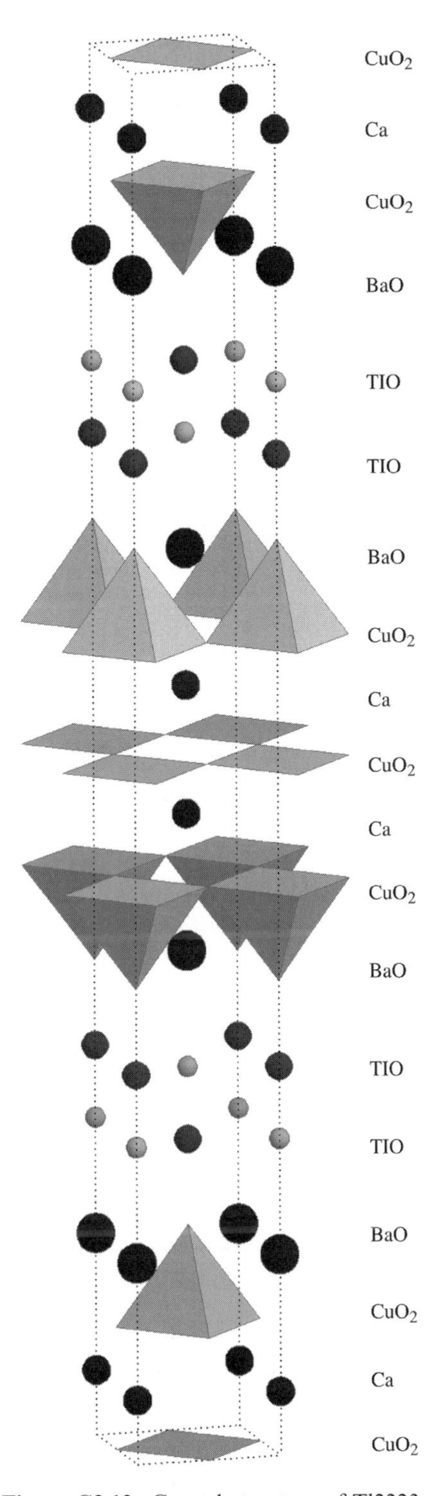

Figure C3.12. Crystal structure of Tl2223.

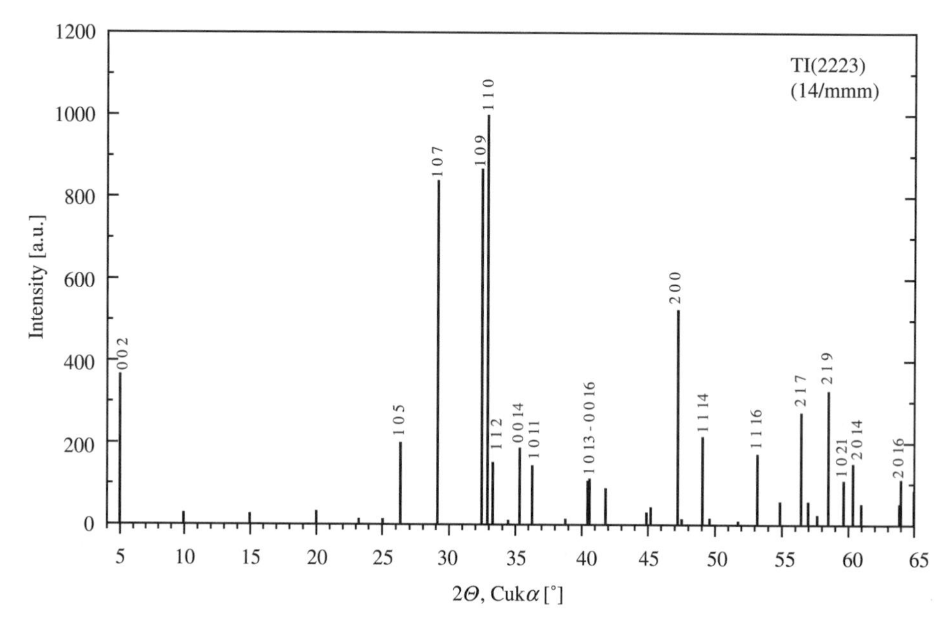

Figure C3.13. Calculated X-ray diffraction pattern for powder sample of Tl2223. Indexes are given for the stronger reflections.

C3.3 Chemical properties [54]

B2.2.4 One of the more serious difficulties in processing Tl based superconductors is the high volatility of Tl and Tl oxides at the temperatures required for the synthesis and the sintering of the superconducting phases. Furthermore, Tl is highly toxic adding significant safety problems for the operator.

There are basically three ways of synthesis of Tl based superconductors: in a sealed (or well closed) crucible (as an example see [55]), in a two zone furnace (e.g.[56]), providing a Tl oxide source, or in a (moderately) high pressure furnace preventing Tl oxides evaporation (e.g. [57]). A precise knowledge of the vapour solid equilibrium of thallium oxide is necessary for the first two methods the reaction being

Table C3.7. Atoms coordinates of $Tl_{1.94}Ba_2Ca_{2.06}Cu_3O_{10}$, $I4/mmm$, $a = 3.8503$, $c = 35.88$ Å, $Z = 2$ [41]

Atom	Site	x	y	z	Occ
Tl[a]	4 (e)	0	0	0.2799	
Ba	4 (e)	1/2	1/2	0.3552	
Cu(1)	4 (e)	0	0	0.4104	
Ca[b]	4 (e)	1/2	1/2	0.4537	
Cu(2)	2 (b)	0	0	1/2	
O(1)	4 (e)	1/2	1/2	0.2719	
O(2)	4 (e)	0	0	0.3412	
O(3)	8 (g)	0	1/2	0.4125	
O(4)	4 (c)	0	1/2	1/2	

Table C3.8. Atoms coordinates of $TlBa_2Eu_{1.5}Ce_{0.5}Cu_2O_9$, *I4/mmm*, $a = 3.8782$, $c = 30.423$ Å, $Z = 2$ [48]

Atom	Site	*x*	*y*	*z*	Occ
Tl	8(i)	0.082	0	0	0.25
Ba	4(e)	1/2	1/2	0.0872	
Cu	4(e)	0	0	0.1520	
Eu[a]	4(e)	1/2	1/2	0.2087	
O(1)	2(b)	1/2	1/2	0	
O(2)	4(e)	0	0	0.0831	
O(3)	8(g)	0	1/2	0.1554	
O(4)	4(d)	0	1/2	1/4	

[a] Eu = $Eu_{0.75}Ce_{0.25}$.

Table C3.9. Atoms coordinates of $Tl_{0.996}Ba_2Ca_{2.96}Cu_4O_{11}$, *P4/mmm*, $a = 3.84809$, $c = 19.0005$ Å, $Z = 1$ [50]

Atom	Site	*x*	*y*	*z*	Occ
Tl	4 (l)	0.086	0	0	0.249
Ba	2 (h)	1/2	1/2	0.1437	
Cu(1)	2 (g)	0	0	0.2487	
Ca(1)	2 (h)	1/2	1/2	0.3292	0.98
Cu(2)	2 (g)	0	0	0.4158	
Ca(2)	1 (d)	1/2	1/2	1/2	
O(1)	1 (c)	1/2	1/2	0	
O(2)	2 (g)	0	0	0.1099	
O(3)	4 (i)	0	1/2	0.2515	
O(4)	4 (i)	0	1/2	0.4156	

Table C3.10. Atoms coordinates of $Tl_{1.64}Ba_2Ca_3Cu_4O_{12}$, *I4/mmm*, $a = 3.84877$, $c = 42.0494$ Å, $Z = 2$ [53]

Atom	Site	*x*	*y*	*z*	Occ
Tl	4 (e)	0	0	0.2757	0.82
Ba	4 (e)	1/2	1/2	0.3387	
Cu(1)	4 (e)	0	0	0.3866	
Ca(1)	4 (e)	1/2	1/2	0.4236	
Cu(2)	4 (e)	0	0	0.4636	
Ca(2)	2 (a)	1/2	1/2	1/2	
O(1)	16 (n)	0.394	1/2	0.2664	0.25
O(2)	4 (e)	0	0	0.3234	
O(3)	8 (g)	0	1/2	0.3874	
O(4)	8 (g)	0	1/2	0.4626	

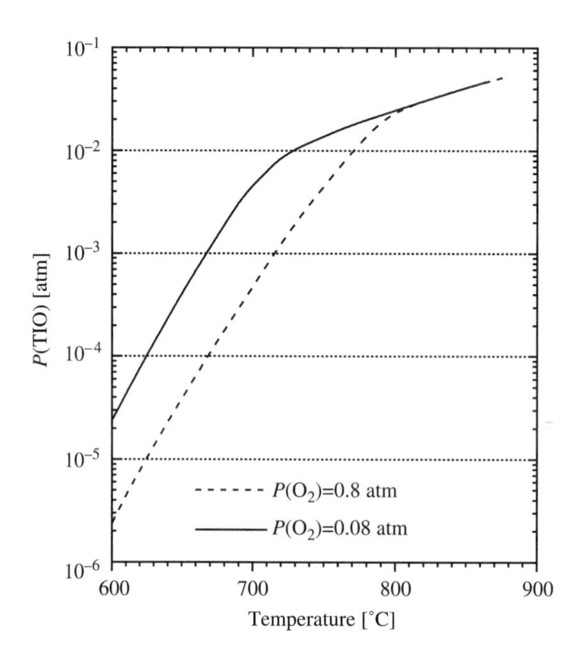

Figure C3.14. Vapour pressure of TlO over a solid (or liquid) source of thallium oxides *versus* temperature for different partial pressures of oxygen [58].

mainly accomplished between Tl oxide vapour and a solid precursor. The vapour pressure of thallous oxide over condensed thallium oxides rises rapidly with increasing temperature. At high temperature the condensed thallium oxide phase may be Tl_2O_3, Tl_4O_3, Tl_2O or a solid or liquid solution of two, depending on temperature and O partial pressure whereas the vapour phase is Tl_2O. In figure C3.14 the vapour pressure of TlO is plotted versus temperature [54, 58].

The representation of a phase diagram for these superconductors requires at least a three dimensional space, the minimum constituents of the phase (Tl, Ba, Ca, Cu oxides) being 4. Since all the nominal composition lies in a single plane, we can represent the phase diagram in a pseudo-quaternary tetrahedron with components $TlO_{1.5}$, Ba_2CuO_3 and $CaCuO_2$ (figure C3.15). The nominal composition of each quaternary superconductor lies at the intersection of one of the two lines originating at the component $CaCuO_2$ and terminating at the single copper layer phases 1201 and 2201, and one of the three lines originated in the corner $TlO_{1.5}$ and terminating at the Tl free composition 0212, 0223, 0234. Moving on any of these latter lines correspond to adding or removing thallium layers, varying the amount of Tl, in principle progresses up or down any line of this set. Travelling along any of the first two lines means the addition (or removal) of a CuO_2 layer with its spacing Ca ion.

Some experiments [59] shows that this diagram is too simplified, i.e. it is not possible to find an equilibrium between Tl2212, Tl2223 and Tl2234 with the Tl oxide; other Tl rich phases form in excess of Tl. The diagrams in figure C3.16 [60] show the stability of different Tl containing phases at different $p(O_2)$ and $p(TlO)$ are shown. A composition of Ba: Ca: Cu = 2:1:2 and 2:2:3 is fixed for diagrams (*a*) and (*b*), respectively, and the diagrams are drawn for temperatures of a few degrees lower than the melting point. The melting temperatures depend on $p(O_2)$ so the diagrams do not represent section at $T =$ constant.

In both the diagrams, i.e. for all the compositions, the Tl2212 results as the most stable phase at high values of $p(TlO)$ before the decomposition of any layered structure.

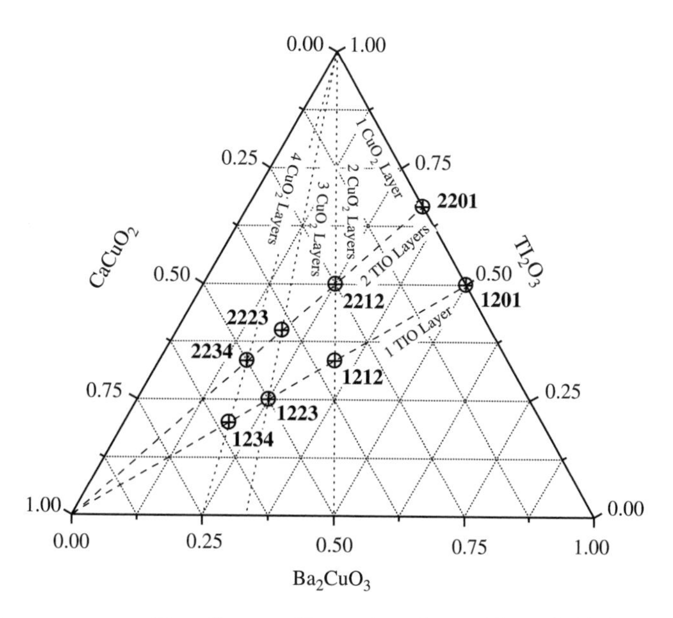

Figure C3.15. Pseudo-ternary composition diagram illustrating the position of all the superconducting phases of the Tl system [54].

Figure C3.16(a) shows also that Tl1212 is stable only if $p(O2) > 0.1$ atm and for low p(TlO); as the latter is increased Tl2212 immediately starts to form. With a complete occupancy of each cation and oxygen site in Tl1212 phase, the formal valence of Cu is calculated to be 2.5. Due to this high formal oxidation state Tl1212 should become less stable with decreasing $p(O_2)$, confirming the results presented

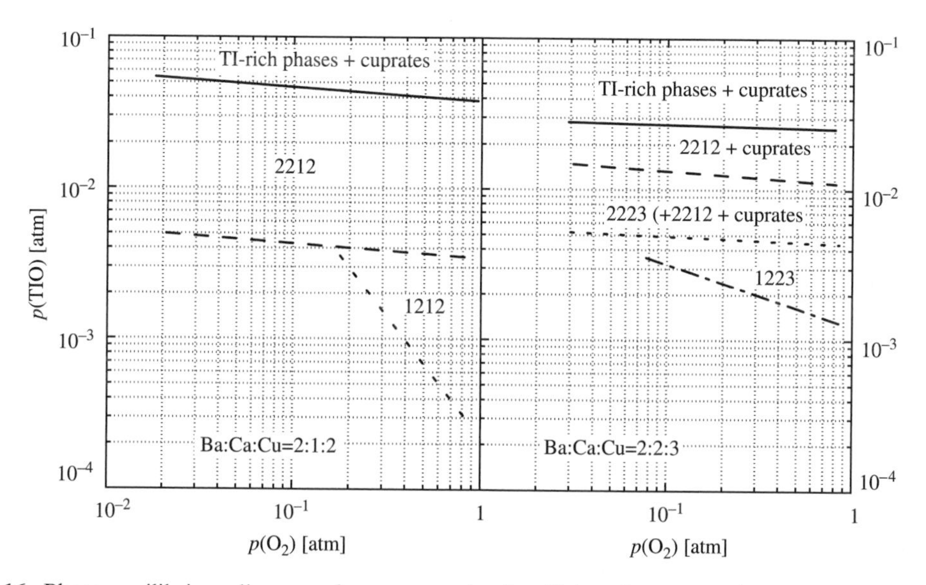

Figure C3.16. Phase equilibrium diagrams for superconducting Tl based compounds at temperature just below melting (diagrams are not sections at constant temperature) in p(O_2), p(TlO) plane. Precursor composition: (a) Ba:Ca:Cu = 2:1:2, (b) Ba:Ca:Cu = 2:2:3. Non-superconducting cuprates are not indicated [60].

in diagram (*a*). A similar behaviour of single TlO layer phase Tl1223 is observed in figure C3.16(*b*) where Tl1223 transforms into Tl2223 with increasing *p*(TlO). The region of stability of this latter phase is very narrow, a Tl2212 is found for a further increase in the *p*(TlO).

All these transformations are reversible and thus at thermodynamical equilibrium, a remarkable difference in the kinetics of formation is observed: the 3 CuO_2 layer phases (Tl1223 and Tl2223) have a slower growth kinetics than the double CuO_2 layer compounds (Tl1212 and Tl2212).

In the case of the Sr homologues, the diagrams are different, the double TlO layer member of the family being absent. In this case the phase with the larger stability domain turns out to be the Tl1212 through which it is necessary to pass for the formation of the Tl1223. The thermodynamic stability of the Tl1223 phase can be improved by appropriate cationic and anionic substitutions. The compound $Tl_{0.5}Pb_{0.5}Sr_{1.6}Ba_{0.4}Ca_2Cu_3O_x$ has a large stability domain that can be further enlarged by the partial substitution of O by F [61].

In the case of synthesis by means of the two zone furnace method, *p*(TlO) is a parameter that can be controlled by setting the temperature of the thallium oxide source, while in a sealed crucible it is determined (at fixed reaction temperature) by the ratio between the volume of the Tl_2O_3 and the volume of the crucible.

B2.2.4

If the synthesis is performed in open or quasi-closed systems, the starting composition is usually in excess of thallium oxide to compensate for the Tl losses during the treatment. The reaction is performed moving down, with the time, in the diagrams in figure C3.16 through the different phases and the final product can be controlled by accurate setting of the duration of the thermal treatment.

The synthesis path is completely different if the reaction is performed under high isostatic pressure. Experiments show that a moderate high pressure of 50 atm is already sufficient to prevent evaporation of Tl oxides up to 1100°C. In this case the phase formation is mainly realized through a solid state diffusion reaction [62]. In figure C3.17, the stability region of a Tl1223 phase is represented in a temperature–oxygen partial pressure diagram. Tl1212 turns out to be the more stable phase at temperature both lower and higher than the Tl1223 stability region. Since the kinetics of formation of the double CuO layer phase is very high, in normal conditions, Tl1223 is formed through Tl1212. More details on the high-pressure reaction method are presented in the chapter on the Tl1223 tapes.

C3.4 Mechanical properties

D3.3

Very few experiments are reported on mechanical properties of Tl based superconductors. Our unpublished results on room temperature Vickers microhardness (HV) measured on the core of monofilamentary Ag sheathed tapes, show that the single layer compounds Tl1212 and Tl1223 show high hardness values (order of 250 HV: g mm^{-2}). Comparable but slightly lower values are observed for the double layer compounds Tl2223. The microhardness decreases if the Tl is partially substituted by Bi (about 150 HV for Tl:Bi = 1:1 in Tl1223), approaching the values measured on similar samples of Bi2223 ($\sim$ 70–120 HV) (see also chapter D3.3 of this Handbook). This effect can be explained if the stereochemistry of Tl and Bi is considered: Bi has three short bonds with O mutually perpendicular, two are in the additional layer so that only one strong bond with the oxygen of the bridging layer is present. In contrast to Bi, Tl forms strong bonds both towards the underlying bridging and the upper bridging or additional layer, resulting in a more robust structure.

D3.3

The different hardness of the Tl- and Bi-based phases can also partially explain the failure of the powder-in-tube (PIT) technique when applied to prepare high critical current tapes of Tl-based superconductors. In the deformation process of the tape manufacturing, the grains of these phases are broken and not, as in Bi-based superconductors, partially deformed and oriented.

B3.1

Some deformation studies at high temperature demonstrated that extensive plasticity in Tl1223 is possible [63]. Stress *versus* strain curves show that the deformation is controlled by viscous flow for low

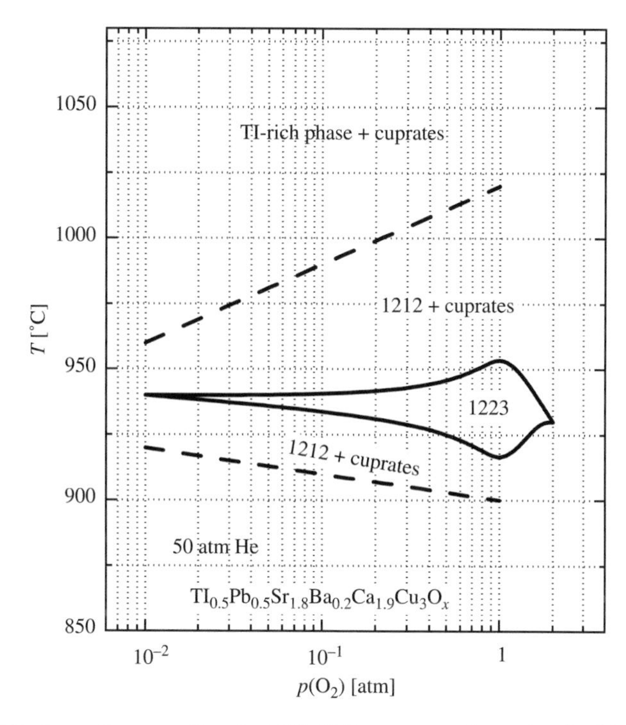

Figure C3.17. Phase equilibrium diagrams for the TlSrCCO system where the double Tl layer compounds are absent. The diagram in the p(O_2), T plane is drawn for a pressure of 50 bar of He.

stresses and high temperature (~ 850°C), whereas at lower temperature (~ 750°C) and higher stress, dislocation gliding or climbing plays a significant role in the deformation process. However, the adoption of a high temperature deformation process for the preparation of superconducting tapes is strongly complicated by the difficulty in manipulating these materials at high temperature, mainly because of evaporation of toxic Tl.

D1.2

C3.5 Optical properties: IR reflectivity [64]

D1.6

For infrared studies, it is important to have well defined materials and furthermore high quality sample surface, the penetration depth of the infrared radiation being of the order of 100 nm.

Interesting information can be extracted from the reflectivity measurements by a Kramers–Kronig analysis that permits the calculation of the dynamic conductivity and by the use of the lattice dynamical results makes it possible to characterize infrared active phonon modes [65–67].

Figures C3.18–C3.21 show far infrared reflectivity spectra of Tl-based compounds at different temperatures. Common to all phases is a high reflectivity at small frequencies, a decrease towards large frequencies and a pronounced phonon structure. The smooth background is mainly due to electronic excitations in the (a, b) plane while the phonon structure is due to the infrared active phonons with displacements in the c direction.

Anomalous behaviour of the oscillation strength is also observed in the reflectivity spectra of Tl1212 (figure C3.18). Comparing the spectra of superconducting ($Tl_{0.55}Pb_{0.45}Sr_2Ca_1Cu_2O_7$) and non-superconducting ($Tl_{0.8}Pb_{0.2}Sr_2Ca_1Cu_3O_7$) it is found that the oscillator strength increases strongly with decreasing temperature for all the infrared active phonons in the superconducting phase, while it remains almost constant for the normal conducting compound [68].

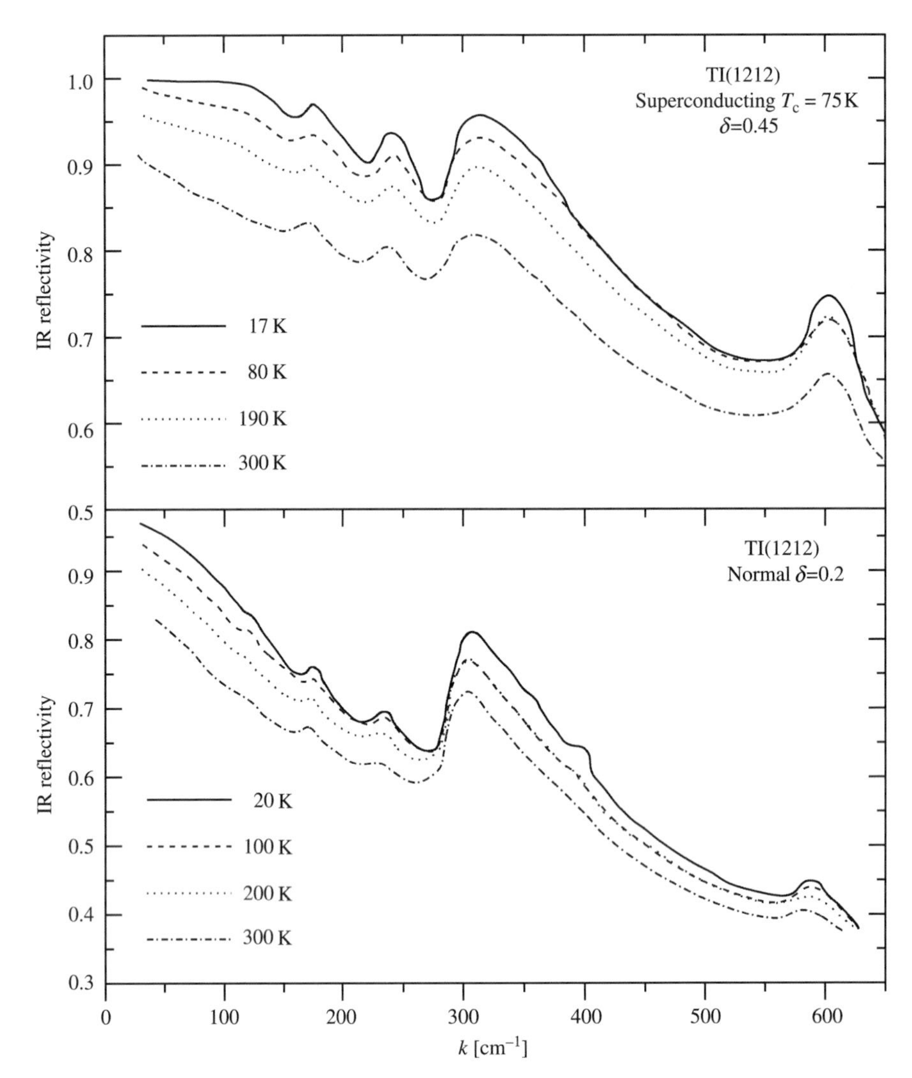

Figure C3.18. Infrared reflectivity of superconducting ($\delta = 0.45$) (*a*) and normal-conducting ($\delta = 0.2$) (*b*) $Tl_{\delta-1}Pb_{\delta}Sr_2Ca_1Cu_2O_z$ [64].

Common to all double Tl layer phases are the two resonance like reflection minima near 80 and 140 cm^{-1} that correspond mainly to Ba and Cu vibrations and to reststrahlen-like maxima near 580 cm^{-1} that can be attributed mainly to vibration of oxygen in the BaO layer against oxygen in the TlO plane.

D1.6 The reflectivity of both the superconducting and non-superconducting 2201 does not change much below 100 K (figure C3.19) while strong changes are seen for the 2212 phase (figure C3.20). An infrared active phonon that corresponds mainly to a vibration of the bridging oxygen atoms against the Ca atoms softens in Tl2212 by 3% of the resonance frequency (311 cm^{-1}) and in Tl2223 (figure C3.21) by 7% of the resonance frequency (305 cm^{-1}) when the temperature is decreased below T_c. The oscillator strengths of most of the phonons show strong temperature dependence, with a strong increase of oscillator strength below T_c, including the phonon with the anomalous shift in the 2212 phase, while the phonon with the

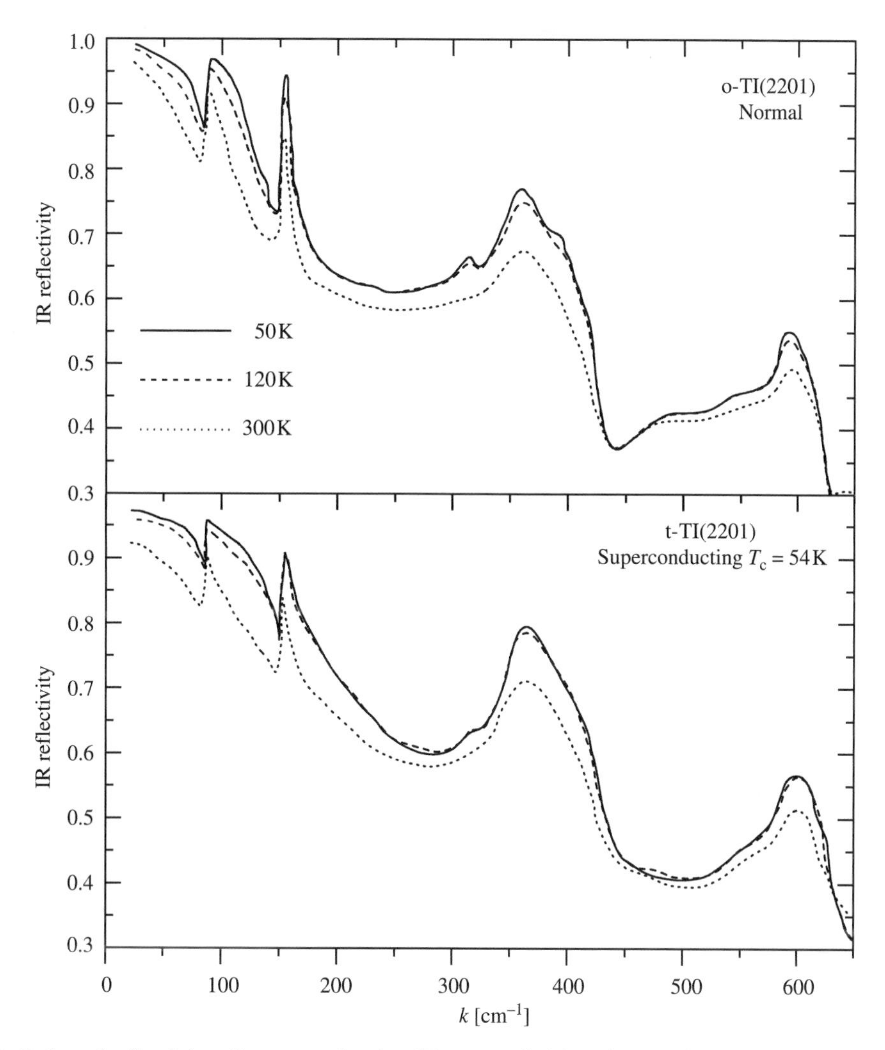

Figure C3.19. Infrared reflectivity of superconducting (Tetragonal) (*a*) and normal-conducting (Orthorhombic) (*b*) Tl2201 [64].

anomalous shift in the 2223 phase shows a strong decrease of strength; also, another phonon at 580 cm^{-1} of the 2223 phase that corresponds mainly to vibrations of the bridging oxygen atom against the oxygen atom in the TlO layer is decreased in strength [69].

C3.6 Normal state properties

C3.6.1 Electrical and thermal transport properties

A complete and accurate discussion of the transport properties is complicated by the existence of many different stoichiometric compositions for each phase, sometimes with very different properties. Some well-established characteristics can, however, be found. D3.1

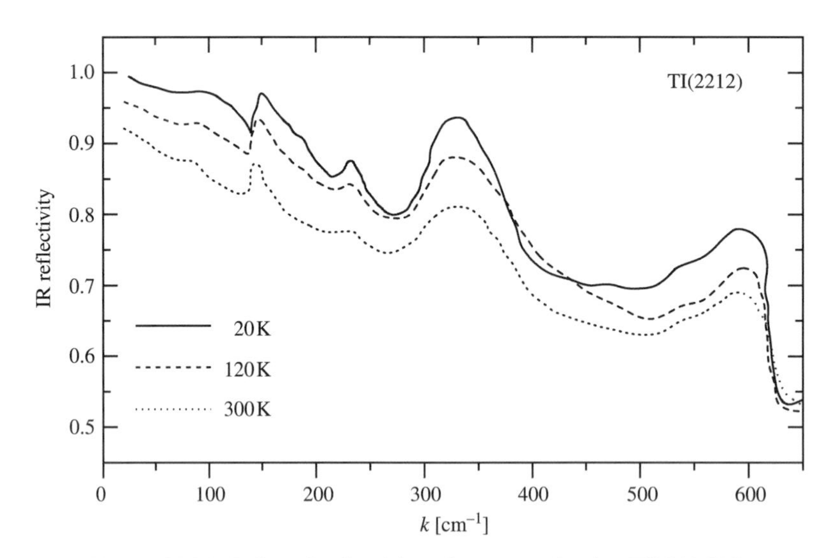

Figure C3.20. Infrared reflectivity of superconducting Tl2212 [64].

Experimentally, the resistivity for optimally doped and overdoped samples can be described by a power law $\rho = \rho_0 + \beta T^n$. For an optimally doped sample a linear dependence is observed ($n = 1$) whereas in the overdoped case the resistivity shows a quadratic dependence ($n \approx 2$) (figure C3.22). For underdoped samples the resistivity is semiconductor-like as clearly shown in figure C3.23 [70, 78]. For single crystal samples ρ_{ab} and ρ_c show strong similarities in temperature dependence but ρ_c is two or three order of magnitude greater than ρ_{ab}. The qualitative similarity of ρ_c and ρ_{ab}, as in some other cuprate superconductors, suggests metal-like conduction in the out-of-plane direction as well as in CuO_2 planes.

The Hall coefficient in a single band model is $R_H^{-1} = \pm n_H e$ where n_H is the number of carriers of charge $\pm e$; the sign of R_H is positive for holes conductors and negative for electrons conductors.

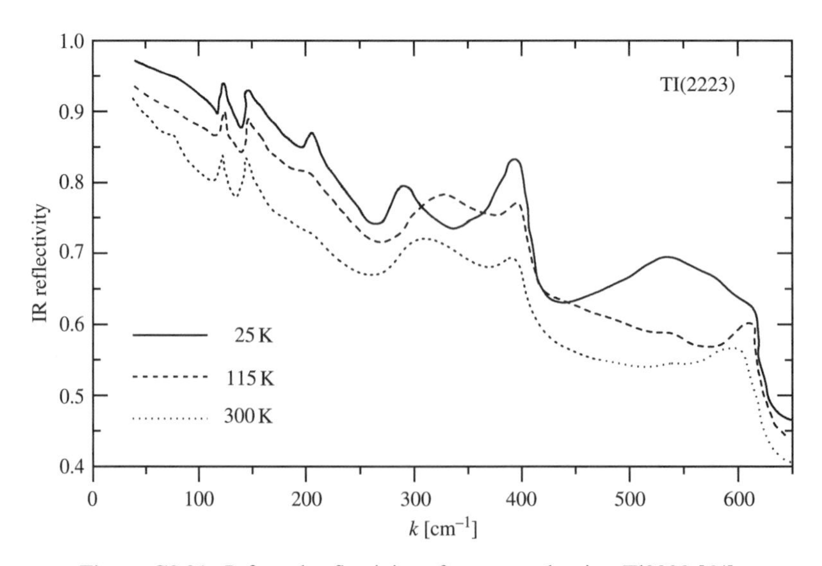

Figure C3.21. Infrared reflectivity of superconducting Tl2223 [64].

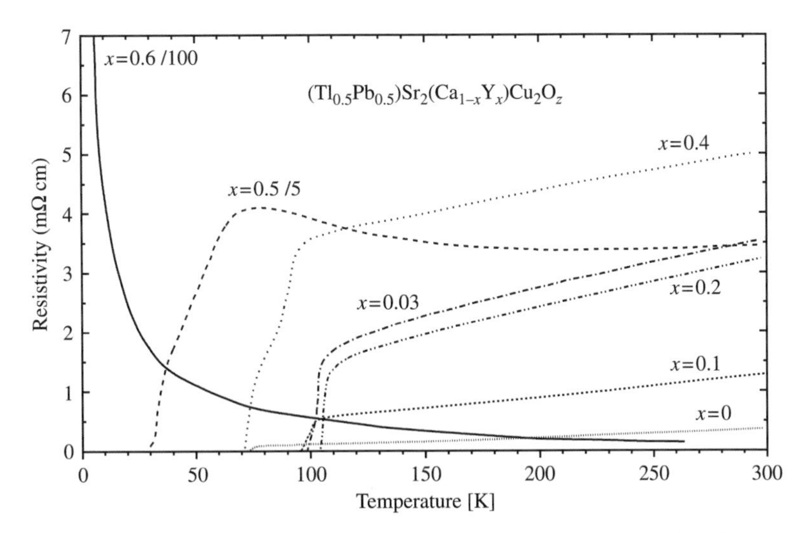

Figure C3.22. Electrical resistivity of Tl1212 doped with Y *versus* temperature. Varying the Y content the material changes from a superconductor with T_c up to 105 K to a semiconductor. The resistivity for an Y content of 0.05 and 0.06 per formula unit are divided by 5 and 100, respectively to fit in the scale [78].

Great caution should be exercised anyway in equating the Hall number from this simple expression with the actual carrier density n in cases where the carriers are not homogeneous and more than one conducting band exist, since cancellation effects may occur. In the Tl based superconductor R_H is positive for all of the phases (figure C3.24): the underdoped and optimally doped samples generally exhibit a linear temperature dependence for R_H^{-1}, $R_H^{-1} = A + BT$ with $A > 0$ (figure C3.25) [71]; in the overdoped case the R_H^{-1} temperature dependence is much weaker and shows a minimum (figure C3.26) [70]. A temperature independent behaviour as expected for good metallic samples is not observed. The positive sign of the Hall coefficient indicates that unless some unusual effect reverses its sign, the dominant carriers are hole-type rather than electron-type. D3.1

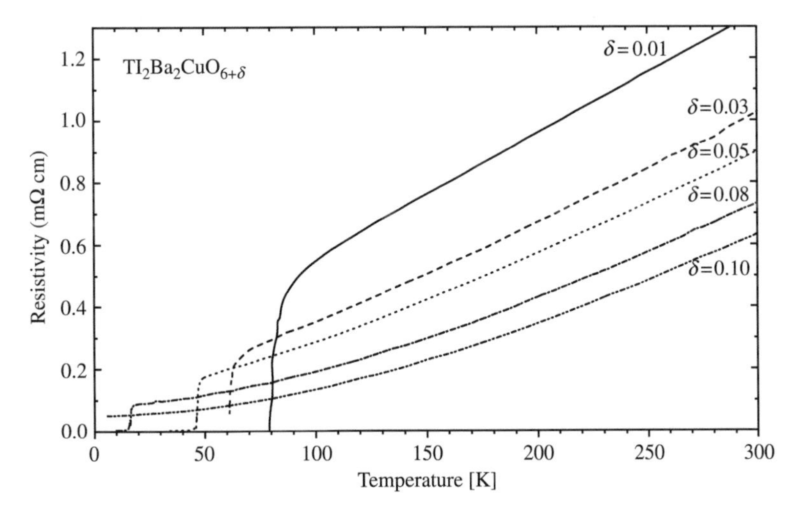

Figure C3.23. Electrical resistivity of Tl2201 for different oxygen contents (overdopings) [70].

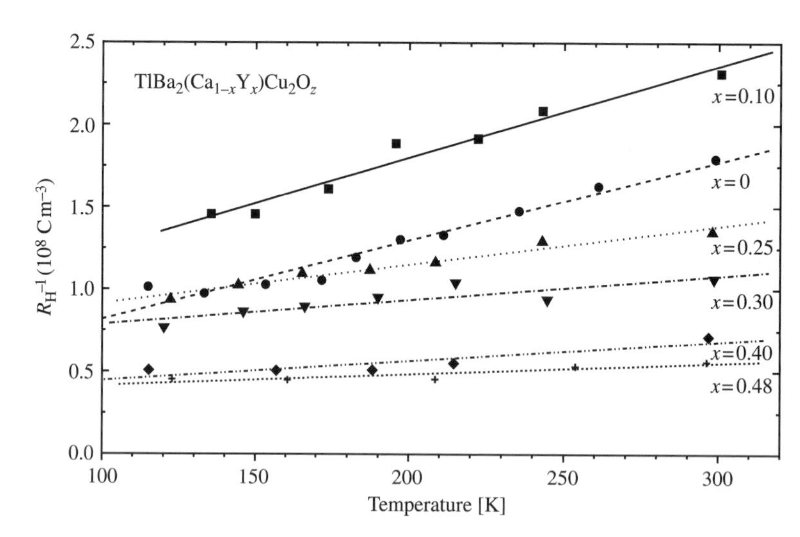

Figure C3.24. Hall coefficient of Tl1212 doped with Y *versus* temperature for different Y contents [89].

D3.1 The thermopower S given by $S = \Delta V/\Delta T$, at temperature above T_c, can be described approximately by a linear equation: $S = A' - B'T$ where A' and B' are always positive. The constant term turns out to be very small for overdoped samples but strongly increased in underdoped samples, whereas the temperature dependence increases with the doping (figure C3.27). It follows from the linearity of the thermopower that phonon drag does not play a major role in producing the shift of thermopower away from $S \propto T$ that characterizes the superconducting samples. In the underdoped limit where T_c is decreased below its maximum value, the thermopower becomes large and resembles a hopping-type thermopower consistent with the larger resistivity that acquires a semiconductor-like temperature dependence in this limit. In the overdoped limit the thermopower approaches the traditional metallic

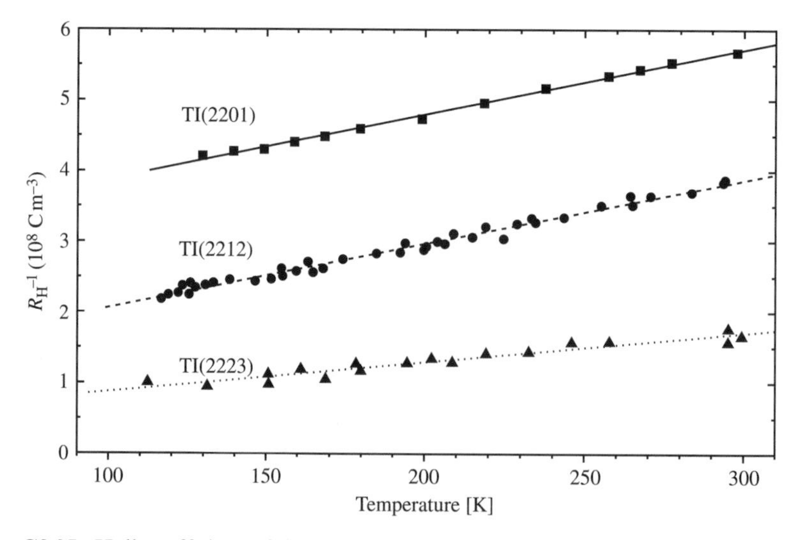

Figure C3.25. Hall coefficient of double Tl layer superconductors *versus* temperature [71].

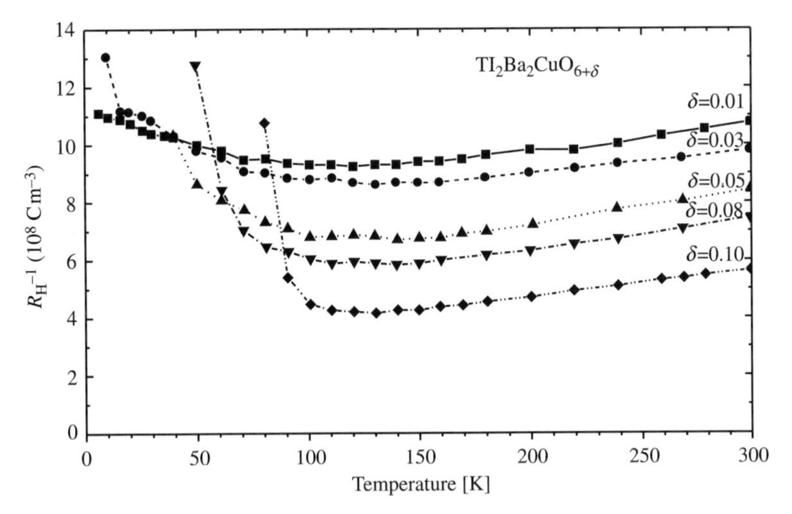

Figure C3.26. Hall coefficient of Tl2201 for different oxygen contents (overdopings) [70].

behaviour described by the Mott formula: $S = \frac{\pi^2 K^2 T}{3e}\left[\frac{\partial \ln \sigma(\varepsilon)}{\partial \varepsilon}\right]_{\varepsilon_F}$ where K is Boltzmann's constant and $\sigma(\varepsilon)$ is a conductivity-like function of electron energy ε [71–73].

C3.6.2 *Magnetic properties*

A characteristic property of any conductor such as a normal metal or a superconducting solid above T_c, is the Pauli paramagnetism (χ_P). In the paramagnetic state there is no magnetic order; an external magnetic field can enforce order of the spin of the conduction electrons and produce a net magnetization parallel to the applied field. This contribution to the magnetic susceptibility is small and temperature independent since only those electrons that are thermally excitable above the Fermi surface can line up with the field.

D2.5

A2.1

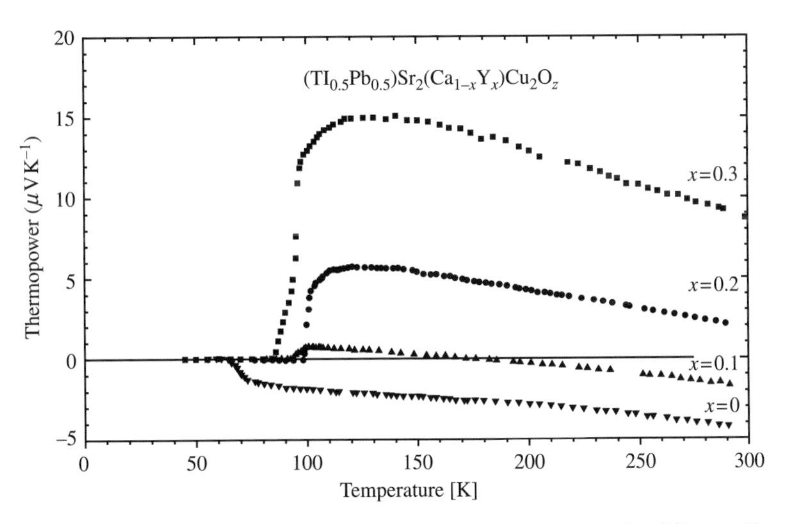

Figure C3.27. Thermopower of Tl1212 doped with Y *versus* temperature for different Y contents [78].

A stronger and temperature dependent magnetization is originated by localized moments well described by a Curie–Weiss model:

$$\chi_{C\text{-}W} = \frac{C}{T - \Theta_C}$$

where Θ_C is the Curie Temperature.

In this kind of material only the two terms (χ_P and $\chi_{C\text{-}W}$) contribute significantly to the total susceptibility, while the other terms like Landau-Peierls, core electron diamagnetism and van Vleck type account for only a few percent of the already small Pauli contribution [74].

Accurate measurements of $\chi(T)$ in the normal state well above T_c allow calculating the localized moments and their interaction starting from C and Θ_C, respectively.

The density of state at the Fermi surface $D(E_F)$ can be calculated from χ_P by the expression $D(E_F) = \frac{3.1\times10^{-4}}{f.u.(Cu)}\chi_P$, where $D(E_F)$ is expressed in *states per* eV and χ_p in emu mol^{-1}; *f.u.*(Cu) is the number of Cu in a formula unit [e.g. *f.u.*(Cu) = 3 for Tl1223].

C3.7 Superconducting properties

C3.7.1 Transition temperatures

Critical temperatures and their variations for all the Tl based HTCS are plotted *versus* the number of CuO_2 layers in figure C3.28. Oxygen deficiency seems to be at the basis of superconductivity in the single Tl layer TBCCO compounds. In the simplest material, the $TlBa_2CuO_6$ (1201) phase, all the copper should be nominally Cu^{3+}, resulting in a strong overdoping of the CuO_2 layers. Superconductivity is observed only after annealing in reducing atmosphere. The progressive reduction of the oxygen content produces at first the semiconductor–metal transition and then a spectacular increase in T_c from 0 to more than 70 K. The T_c value of $TlBa_2CaCu_2O_7$ (1212) is very variable and also depends on the conditions utilized for the synthesis. As with the 1201 phase, 1212 is overdoped and has a formal copper valence of +2.5. Transition temperatures of as-prepared samples normally range from 80 to 90 K, but

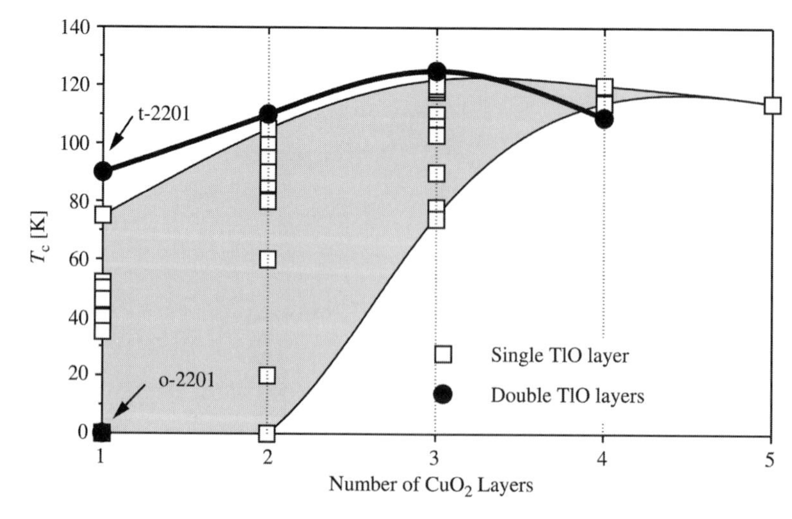

Figure C3.28. Critical temperature and its variation with chemical substitution in Tl based superconductors, plotted *versus* the number of CuO_2 layers.

combinations of oxygen deficiency and/or thallium substitution on the Ca site can decrease the copper valence to about +2.2, for which T_c exceeding 110 K was reported (see below). The maximum T_c increases at about 120 K in 1223, where the formal valence of copper in the fully oxygenated sample is +2.33 and therefore near the optimum value. The variation of T_c with the annealing conditions in these phases is consequently more limited. The superconductivity in the $m = 2$ phases of the Tl and Bi system, for which the stoichiometric compositions lead to a nominal copper valence of +2, is still related to oxygen non-stoichiometry and cationic substitution that occurs spontaneously in the samples. On the contrary to the $m = 1$ phases, the oxygen non-stoichiometry consists of an excess rather than a deficiency of oxygen. The insertion of extra oxygen occurs in the Tl compounds on an interstitial site between the double Tl_2O_2 layers, whereas in the Bi phases the extra oxygen is intercalated within the BiO layers along the chains of short Bi–O bonds running along the direction of the modulation wave vector. For a long time, it has been believed that the oxygen excess would represent alone, at the same time, the origin of superconductivity in these systems and the origin of the structural modulation in Bi system. However, it was demonstrated [75] that stoichiometric Bi2212, prepared in argon, shows an almost commensurate structural modulation and superconductivity at 86 K. For this compound an increase in the oxygen content led to a shortening of the modulation period and to a decrease in the transition temperature. The presence of holes in this compound, as well as in stoichiometric Tl2201 phase where superconductivity was observed, could be explained by the overlap (hybridization) of the TI and Bi 6 s with Cu $3d_{x^2-y^2}$ bands at the Fermi level as schematically shown in figure C3.29 [76]. This would result in the formation of holes in the CuO_2 layers and in a semi-metallic character of the BiO and TlO layers. This does not happen in the corresponding $m = 1$ Tl compounds, where the 6 s bands lie above the Fermi level. The chemical control of the superconductivity in these compounds is further complicated by the occurrence of spontaneous cationic substitutions during the synthesis, which cannot be easily avoided inspite of accurate control of the reaction atmosphere and of the preparative procedure.

The Tl 2201 compound $Tl_2Ba_2Cu0_{6+\delta}$, which ranges from a normal metal (as-prepared) to a superconductor with T_c as high as 90 K (after annealing in argon), is very sensitive to small variations of the oxygen stoichiometry. The compound is often Tl deficient and copper rich. In this case Cu^{2+} was

C2
C2
A2.1
D3.1

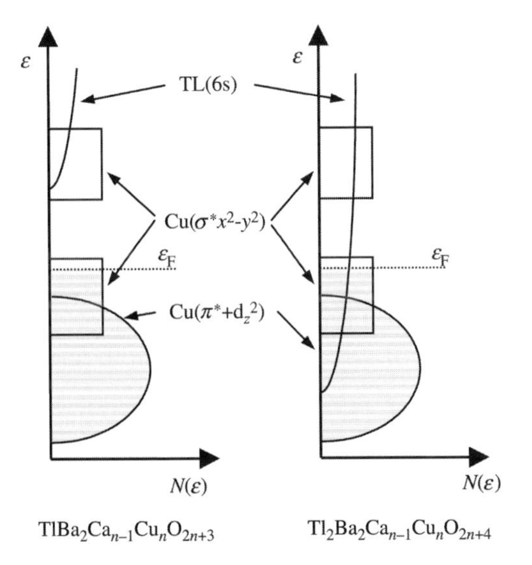

Figure C3.29. Schematic diagram of band structures for single (a) and double (b) Tl layer compounds with two copper layers [76].

found to substitute Tl^{3+} (about 5%) producing an overdoped material for which the maximum T_c can be achieved only by suppressing all the oxygen excess by argon reduction. The corresponding $n = 2$ phase $Tl_2Ba_2CaCu_2O_{8+d}$, also has a critical temperature that is dependent on the oxygen content. Even in this case an annealing under argon of the as-prepared material, hence a reduction of the hole concentration, is able to increase T_c to 112 K. Site disorder was also observed in this phase; in particular the TI site contains calcium or copper, whereas the Ca site may be occupied by Tl. The 2223 phase possesses the highest T_c of the family, 128 K, which is rather less dependent on the oxygen content and on the thermal history since only a variation of a few degrees K can be produced. The reduction, by increasing n of the range in which it is possible to vary T_c, can be understood in terms of the same holes coming from the TlO bilayers, that can produce different overdopings as a function of the number of involved CuO_2 planes.

C3.7.2 Chemical control of superconductivity

Thallium as well as its neighbours in the periodic table, lead and bismuth, presents a further stable valency that is lower by two units from the group valency. In addition to the normal group oxidation states Tl(III), Pb(IV) and Bi(V), also Tl(I), Pb(II) and Bi(III) are stable oxidation states presenting the outer $6s^2$ electronic configuration. These two electrons ionize or participate in covalent bond formation with difficulty because the outer s and p states are separated by a large difference in energy, and are thus called an *inert (lone) electron pair*. On the contrary, the pair often participates in bonding resulting in asymmetrical stereochemistries its volume being comparable to that of an anion in solid [77].

There is a structural size mismatch between the TlO and the CuO_2 layers, which produces a stretching of the TlO layers resulting in displacements of the atoms of the rocksalt block from their ideal positions. In thallium compounds the displacements are in most cases statistically distributed, giving rise to an average tetragonal symmetry, but a lowering of the symmetry from tetragonal ($a = a_p$) to orthorhombic ($a \approx b \approx a_p = \sqrt{2}$) is observed in some cases as a result of ordering as previously discussed in section C3.1 for Tl2201. In bismuth compounds, (see also chapter C1.2.2 of this handbook) due to the lone pair character of Bi^{3+} which produces a remarkable distortion of the coordination (there are three short and three long bonds for the six-fold coordinated Bi), the mismatch is more pronounced than in thallium compounds and gives rise in all cases to an incommensurate unidimensional structural modulation with a modulation period $\sim 5a$ of the orthorhombic ($a \approx b \approx a_p = \sqrt{2}$) fundamental cell. The presence of the lone pair on Bi^{3+} results in another important structural difference between Tl based and Bi based phases: Tl based phases are much more three-dimensional in character than the corresponding Bi based phases as a result of the fact that the three short Bi–O bonds are mutually perpendicular, so that only one strong bond in the direction perpendicular to the BiO layer (the one with the oxygen of the SrO layer) can be formed. Adjacent BiO layers are consequently weakly bound with a spacing of about 3.2 Å whereas the TlO layers are much more strongly bonded and have an interlayer spacing of about 2.0 Å. This characteristic of the BiO layer is moreover responsible for the impossibility of producing Bi based phases with $m = 1$. In spite of the similitude of the structural sequences and of the formal valence of Bi and Tl in the structures, it is impossible to find a simple unified mechanism to describe the chemical control of superconductivity in these systems. Oxygen non-stoichiometry and cationic substitutions play an important role in defining the superconducting properties but their effects often overlap and in some cases they seem not to be sufficient for justifying the observed superconductivity. A remarkable example of chemical control of the electronic properties is the compound $Tl_{1-y}Pb_ySr_2Ca_{1-x}Y_xCu_2O_z$ in the family of Tl1212 materials. The compound $TlSr_2CaCu_2O_z$ is itself a metal but no superconducting transition is observed down to 4 K; the nominal Cu valency of this compound is +2.5 which indicates an excess of hole carriers in the CuO_2 layers (overdoped state). To induce the appearance of superconductivity it is possible to reduce the overdoping by replacing

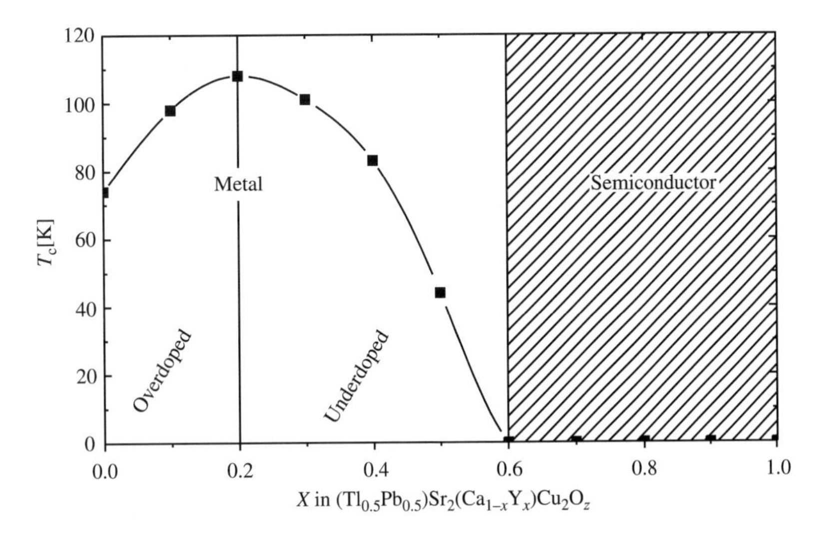

Figure C3.30. Critical temperature *versus* Y content in $Tl_{0.5}Pb_{0.5}Sr_2Ca_{1-x}Y_xCu_2O_z$ [78].

the Tl^{3+} by Pb^{4+} and/or Ca^{2+} by Y^{3+}. This substitution is proved to introduce an excess of electrons in the conducting CuO_2 planes, therefore reducing the net hole number (see figure C3.25) [78].

For the phase without Pb ($y = 0$), varying the Y content (x) it is possible to transform this material from a normal metal $TlSr_2CaCu_2O_z$ ($x = 0$) to a superconductor (with a maximum of $T_c = 80$ K for $x = 0.6$) to an insulator $TlSr_2YCu_2O_z$. The phase with the Tl site half substituted by Pb ($y = 0.5$), as shown in figure C3.30, is a superconductor with a T_c of ~ 80 K for $x = 0$; reaches the highest T_c of the family (108 K) for the composition $Tl_{0.5}Pb_{0.5}Sr_2Ca_{0.8}Y_{0.2}Cu_2O_z$ ($x = 0.2$) and finally becomes an antiferromagnetic insulator for $x = 1$ [78].

C3.7.3 Magnetic properties

Diamagnetic results of single crystals of Tl based superconductors demonstrate perfect bulk superconductivity with very sharp transition and a complete diamagnetic state is reached. In contrast to the conventional conductors the Sommerfeld constant of these compounds is rather small, especially for such a high T_c. Temperature dependence of magnetic penetration depth indicates a low symmetry non s-wave ground state; a result recently confirmed by measurements of 2201 thin films using tricrystal ring magnetometry [79, 80] and angle dependent torque magnetometry [81]. Possibly there are some regions on the Fermi surface with no gap or pseudogap. The London penetration depth is very large ($\lambda_{ab} \sim 200$ nm) and the coherence length is very small ($\zeta_{ab} \sim 2$ nm), so all these compounds are definitively type II superconductors in the clean limit. As all high T_c superconductors they present a large effective mass anisotropy ($m_c^*/m_{ab}^* \sim 100$); and also a quite pronounced anisotropy in the penetration depth is observed ($\lambda_c/\lambda_{ab} \sim 15$) (table C3.11).

B2.4

A2.1, A1.2

Consistent with this anisotropy, diamagnetic fluctuation indicates the existence of two-dimensional superconductivity. The anisotropy turns out to be more pronounced in double Tl layer compounds than in the single Tl layer ones. Softness of the flux line lattice is evidenced in the broadening of the transition and the reduced irreversibility at the high field-high temperature region of the phase space. Figure C3.31 shows the temperature dependence of the irreversibility field determined from the 'pinch-off' field in magnetization hysteresis loops for a variety of thallium HTSC [82]. Except for Tl2234, data are reported for samples where T_c has been approximately maximized by atomic substitution or by appropriate

D1.7.2

Table C3.11. Typical superconductive intrinsic parameter of Tl based compound reported in literature

Compound	T_c	λ_{ab}(nm)	λ_c (nm)	ζ_{ab} (nm)	ζ_c (nm)
Tl1201	52	–	–	–	–
Tl1212	80	210	–	2.0	–
Tl1223	120	200	–	1.8	–
Tl2201	90	170	–	5.2	0.3
Tl2212	110	215	–	2.2	0.5
Tl2223	125	205	480	1.3	–

A1.3 oxygen annealing to obtain the optimum hole doping. As shown in figure C3.31 the irreversibility line for Tl1223 exceeds that of any Tl based HTSC. The Tl1223 irreversibility line is higher than the Bi2223. C2, C1 Below approximately 1 T it is also higher than the one of YBCO (see also C2 and C1).

Beyond this, it is possible to observe some systematic trends: (a) the irreversibility field for the double-layered TI compounds (2201, 2212, 2223 and 2234) displays a similar temperature dependence although the number of CuO layers and T_c(max) differs between compositions. There is a slow steepening of $H_{irr}(T)$ as n decreases. (b) the single-layered Tl-based materials (1212 and 1223) are quite different. Both compositions display an irreversibility field that increases more rapidly with decreasing temperature than any of the double-layered compositions. Thus $H_{irr}(T)$ is strongly dependent on m and only weakly dependent on n. These data are consistent with the model that pinning of three-dimensional vortex lines is much more effective than that of decoupled two-dimensional pancake vortices in the CuO_2 planes. The anisotropy (more two-dimensional-like behaviour) increases with the number of Tl layers due to the fact that the distance between the CuO_2 layer is increased and consequently the coupling between the superconducting sheets decreased.

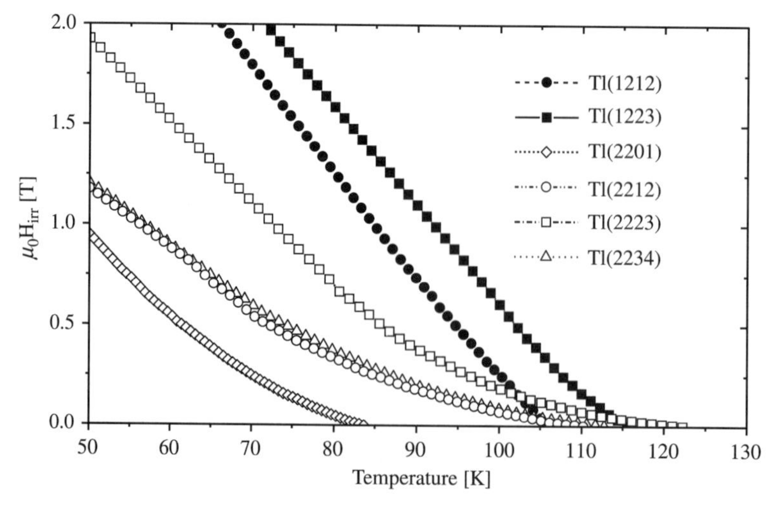

Figure C3.31. Irreversibility line from magnetization measurements for Tl based superconductors [82].

For power applications like transmission lines and various magnet systems two Tl-based compounds may be of special interest, namely the double layer Tl2223 and particularly the single layer Tl1223, which has a higher irreversibility line in the $H_{irr}(T)$ representation.

A1.3

C3.7.4 Transport properties

The transport critical current is strongly dependent on the preparation technique utilized for the sample preparation. Two basic processing procedures of Tl-based conductor preparation are used: the so-called closed approach represented by the PIT method and the more convenient and successful open approach represented by various relatively simple but efficient non-vacuum methods such as aerosol deposition, electro-deposition, sol–gel and screening or painting. For a more exhaustive discussion on the superconducting transport properties of different kinds of Tl-based conductors see also chapter B3.2.3 of this Handbook.

B3.2.3

The closed approach PIT process is able to produce conductors with the required core thickness, however, from the point of view of achieving high J_c values, its development has so far stopped at values around 2×10^4 A cm^{-2} at 77 K/0 T [83, 84]. Small increase in J_c may be achieved with improved contacts between the grains that can be obtained by increasing the material density [85] or by the formation of a transient liquid during the sintering by appropriate substitutions [61]. Anyway J_c decreases dramatically in very low magnetic fields of only 0.1 T due to the weak-link behaviour caused by the lack of the necessary grain alignment. The weak link phenomenon is well evidenced by the fact that the transport current in a magnetic field of 0.2 T is decreased by a factor $\sim$ 15 of the zero field value and also that a strong hysteresis is present when the magnetic field is cycled (figure C3.32). However, it is interesting to note that J_c does not decrease further increasing the magnetic field up to almost 10 T, meaning that some 'strong' current path has formed (figure C3.33).

B3.1

The open system would appear to be a substantially more hopeful approach where more successful methods exist, the results of which have already overwhelmed those obtained by the PIT procedure. Of the open approach methods the most successful in this respect are aerosol deposition from solution and the electro-deposition [86] on Ag substrates, with $J_c > 10^5$ A cm^{-2} at 77 K/0 T, and current densities

B3.2.3

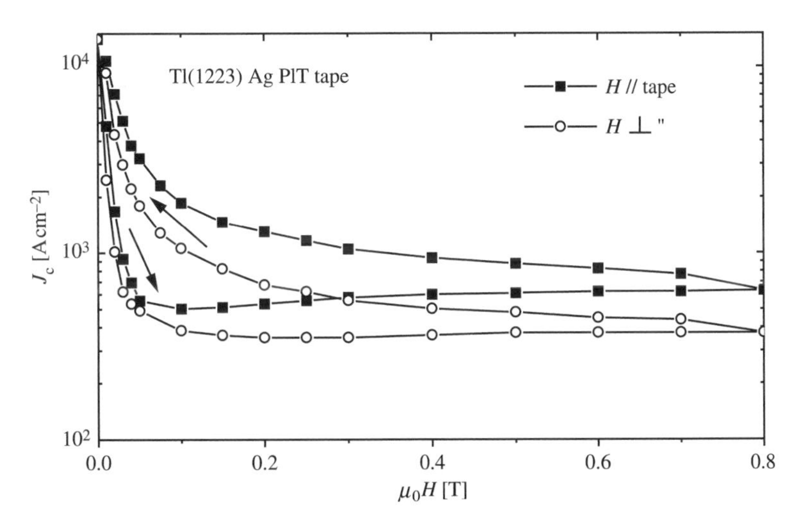

Figure C3.32. Transport critical current densities *versus* magnetic field at 77 K of a Tl1223/Ag PIT tape with the surface of the tape parallel and perpendicular to the magnetic field. The arrows indicate if the measure was taken with increasing or decreasing magnetic field.

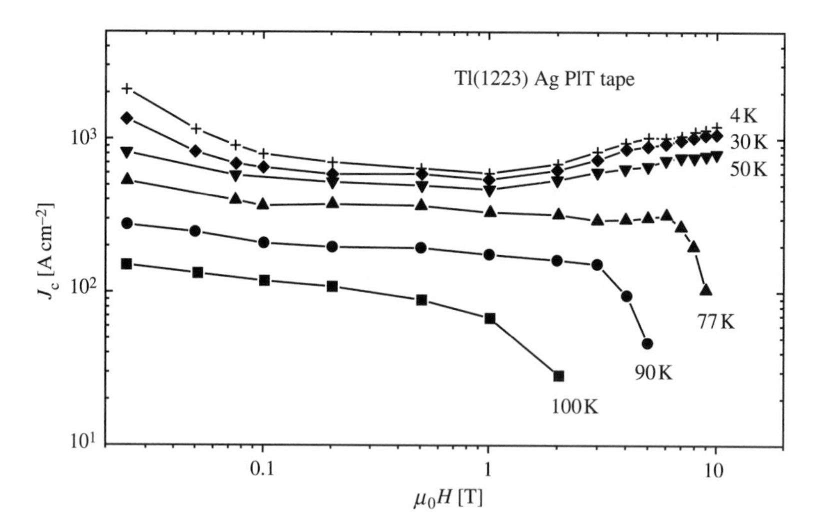

Figure C3.33. Transport critical current densities *versus* magnetic field of a Tl1223/Ag PIT tape for different temperature.

higher than 10^4 A cm^{-2} are sustained in high magnetic fields (5 T) at LN_2 temperatures. As shown in figure C3.34, depositions on single crystalline substrates [87] present critical current densities higher than 10^6 A cm^{-2} at 77 K/0 T. The reason for such a tremendous improvement over the PIT method, lies in the essential grain alignment of the polycrystalline films, a fundamental condition to obtain high J_c values in magnetic fields. The grains should be both *c*- and *a*-axis aligned in such a way that the microstructure approaches that of a single crystal. In this respect an important discovery has been made demonstrating that the bi-axially textured polycrystalline Ag substrate helps to align the superconducting grains grown on such a type of substrate. For power applications, high I_c values are also required, which corresponds to grow HTCS films with thicknesses of the order of 10 μm or more. From this point of view, spray deposition from ink, sol–gel and screening or printing and electrodeposition are of considerable interest.

B4.1
E1
B4.3

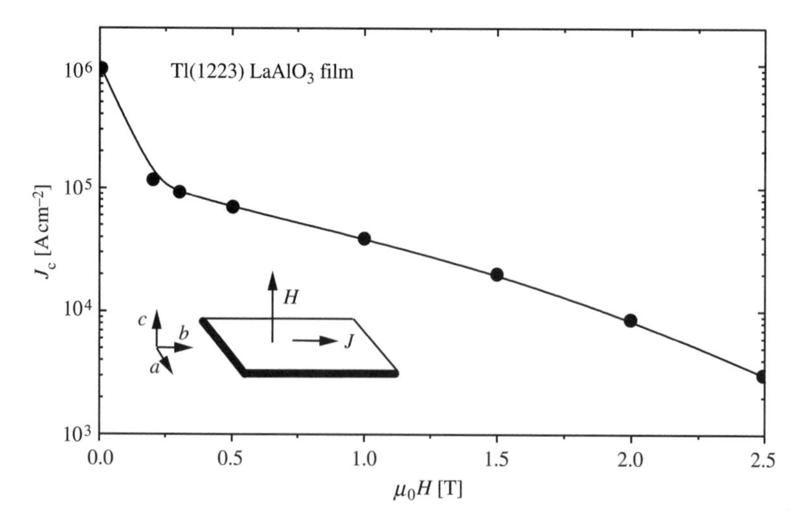

Figure C3.34. Transport critical current densities as a function of magnetic field applied parallel to the *c*-axis. The sample carried 51 A at zero field [87].

The great challenge associated with the problem of thick-film growing involves developing and sustaining the substrate texturing up to the very surface of the thick film.

B4.3

C3.8 Thallium safety [88]

Thallium and most of its compounds are potentially toxic. Hundreds of deaths have resulted from accidental, as well as homicidal and suicidal ingestion of Tl compounds. Another problem arises from the fact that soluble Tl compound (e.g. $TlNO_3$) can penetrate the unbroken skin adding another dimension to their hazard potential. The lowest dose that will cause death is estimated to be about 15 mg of Tl per kg of body weight regardless of the route of administration making Tl as lethal as arsenic. Unlike Pb and Hg that the human body is not able to eliminate, Tl is excreted, primarily in the urine but also in the faeces, considerably reducing the risk of long-term accumulation and its dramatic effects. The biological half-life (i.e. the time required to excrete half of the remaining body burden of Tl) is estimated to be around 10 days. Exposure to Tl can cause fatigue, weakness, poor appetite, insomnia, confusion and mood changes. Higher exposures can damage the nervous system causing numbness, pain and ‘pins and needles’ in arms and legs and may affect the liver and kidneys.

The recommended airborne exposure limit is 0.1 mg m^{-3} averaged over an 8 h workshift. The above exposure limits are for air levels only. When skin contact also occurs, you may be overexposed, even though air levels are less than the limit listed above.

Engineering controls are the most effective way of reducing exposure. The best protection is to enclose operations and/or provide local exhaust ventilation at the site of chemical release. Isolating operations can also reduce exposure. Using respirators or protective equipment is less effective than the controls mentioned above, but is sometimes necessary. Good work practices can help to reduce hazardous exposures. The following work practices are recommended:

- On skin contact with Tl, immediately wash or shower to remove the chemical. At the end of the workshift, wash any areas of the body that may have contacted it, whether or not known skin contact has occurred.
- Do not eat, smoke, or drink where Tl is handled, processed, or stored, since the chemical can be swallowed. Wash hands carefully before eating or smoking.
- Use a vacuum or a wet method to reduce dust during clean-up. Do not dry sweep.

Workplace controls are better than personal protective equipment. However, for some jobs or in addition, personal protective equipment may be appropriate.

- Avoid skin contact with Tl. Wear protective gloves and clothing. Safety equipment suppliers and manufacturers can provide recommendations on the most protective glove and clothing material for your operation.
- Wear dust-proof goggles and face shield when working with powders or dust, unless full facepiece respiratory protection is worn.
- Use respiratory protection. Be sure to consider all potential exposures in your workplace. You may need a combination of filters, prefilters, cartridges, or canisters to protect against different forms of a chemical (such as vapour and mist) or against a mixture of chemicals.
- Exposure to 15 mg m^{-3} is immediately dangerous to life and health. If the possibility of exposure above 15 mg m^{-3} exists, use a self-contained breathing apparatus with a full facepiece operated in continuous flow or other positive pressure mode.

References

[1] Gladyshevskii R E and Galez P 1999 *Handbook of Superconductivity* ed C P Poole (San Diego: Academic)
[2] Kondoh S, Ando Y, Onoda M, Sato M and Akimitsu J 1988 *Solid State Commun.* **65** 132
[3] Sheng Z Z and Hermann A M 1988 *Nature* **332** 55
[4] Sheng Z Z and Hermann A M 1988 *Nature* **332** 138
[5] Parkin S S P, Lee V Y, Nazzal A I, Savoy R, Huang T C, Gorman G and Beyers R 1988 *Phys. Rev.* B **38** 653
[6] Kim J S, Swinnea J S and Steinfink H 1989 *J. Less-Common Met.* **156** 347
[7] Ganguli A K and Subramanian M A 1991 *J. Solid State Chem.* **93** 250
[8] Manako T, Shimakawa Y, Kubo Y, Satoh T and Igarashi H 1989 *Physica* C **158** 143
[9] Subramanian M A 1990 *Mater. Res. Bull.* **25** 191
[10] Gopalakrishnan I K, Yakhmi J V and Iyer R M 1991 *Physica* C **175** 183
[11] Chshima E, Kikuchi M, Izumi F, Hiraga K, Oku T, Nakajima S, Ohnishi N, Morii Y, Funahashi S and Syono Y 1994 *Physica* C **221** 261
[12] Subramanian M A, Kwei G H, Parise J B, Goldstone J A and Von Dreele R B 1990 *Physica* C **166** 19
[13] Morosin B, Ginley D S, Hlava P F, Carr M J, Baughman R J, Schirber J E, Venturini E L and Kwak J F 1988 *Physica* C **152** 413
[14] Kolesnikov N N, Korotkov V E, Kulakov M P, Lagvenov G A, Molchanov V N, Muradyan L A, Simonov V I, Tanazyan R A, Shibaeva R P and Shchegolev I F 1989 *Physica* C **162–164** 1663
[15] Parkin S S P, Lee V Y, Nazzal A I, Savoy R, Beyers R and La Placa S J 1988 *Phys. Rev. Lett.* **61** 750
[16] Morosin B, Venturini E L and Ginley D S 1991 *Physica* C **183** 90
[17] Subramanian M A, Torardi C C, Gopalakrishnan J, Gai P L, Calabrese J C, Askew T R, Flippen R B and Sleight A W 1988 *Science* **242** 249
[18] Liu R S, Hughes S D, Angel R J, Hackwell T P, Mackenzie A P and Edwards P P 1992 *Physica* C **198** 203
[19] Opagiste C, Couach M, Khoder A F, Abraham R, Jondo T K, Jorda J-L, Cohen-Adad M Th, Junod A, Triscone G and Muller J 1993 *J. Alloys Compd.* **195** 47
[20] Shimakawa Y, Kubo Y, Manako T, Igarashi H, Izumi F and Asano H 1990 *Phys. Rev.* B **42** 10165
[21] Huang T C, Lee V Y, Karimi R, Beyers R and Parkin S S P 1988 *Mater. Res. Bull.* **23** 1307
[22] Hewat A W, Bordet P, Capponi J J, Chaillout C, Chenavas J, Godinho M, Hewat E A, Hodeau J L and Marezio M 1988 *Physica* C **156** 369
[23] Parise J B, Gopalakrishnan J, Subramanian M A and Sleight A W 1988 *J. Solid State Chem.* **76** 432
[24] Ström C, Eriksson S-G, Johansson L-G, Simon A, Mattausch H J and Kremer R K 1994 *J. Solid State Chem.* **109** 321
[25] Shimakawa Y 1993 *Physica* C **204** 247
[26] Jorda J L, Jondo T K, Abraham R, Cohen-Adad M Th, Opagiste C, Couach M, Khoder A and Sibieude F 1993 *Physica* C **205** 177
[27] Opagiste C, Triscone G, Couach M, Jondo T K, Jorda J-L, Junod A, Khoder A F and Muller J 1993 *Physica* C **213** 17
[28] Jorda J L, Jondo T K, Abraham R, Cohen-Adad M Th, Opagiste C, Couach M, Khoder A F and Triscone G 1994 *J. Alloys Compd.* **215** 135
[29] Torardi C C, Subramanian M A, Calabrese J C, Gopalakrishnan J, McCarron E M, Morrissey K J, Askew T R, Flippen R B, Chowdhry U and Sleight A W 1988 *Phys. Rev.* B **38** 225
[30] Parise J B, Torardi C C, Subramanian M A, Gopalakrishnan J, Sleight A W and Prince E 1989 *Physica* C **159** 239
[31] Hazen R M, Finger L W, Angel R J, Prewitt C T, Ross N L, Hadidiacos C G, Heaney P J, Veblen D R, Sheng Z Z, El Ali A and Hermann A M 1988 *Phys. Rev. Lett.* **60** 1657
[32] Subramanian M A, Calabrese J C, Torardi C C, Gopalakrishnan J, Askew T R, Flippen R B, Morrissey K J, Chowdhry U and Sleight A W 1988 *Nature* **332** 420
[33] Maignan A, Michel C, Hervieu M, Martin C, Groult D and Raveau B 1988 *Mod. Phys. Lett.* B **2** 681
[34] Kikuchi M, Kajitani T, Suzuki T, Nakajima S, Hiraga K, Kobayashi N, Iwasaki H, Syono Y and Muto Y 1989 *Japan. J. Appl. Phys.* **28** L382
[35] Johansson L-G, Ström C, Eriksson S and Bryntse I 1994 *Physica* C **220** 295
[36] Onoda M, Kondoh S, Fukuda K and Sato M 1988 *Japan. J. Appl. Phys.* **27** L1234
[37] Morosin B, Ginley D S, Venturini E L, Baughman R J and Tigges C P 1991 *Physica* C **172** 413
[38] Ogborne D M, Weller M T and Lanchester P C 1992 *Physica* C **200** 207
[39] V Molchanov N, Tamazyan R A, Simonov V I, Blomberg M K, Merisalo M J and Mironov V S 1994 *Physica* C **229** 331
[40] Politis C and Luo H L 1988 *Mod. Phys. Lett.* B **2** 793
[41] Torardi C C, Subramanian M A, Calabrese J C, Gopalakrishnan J, Morrissey K J, Askew T R, Flippen R B, Chowdhry U and Sleight A W 1988 *Science* **240** 631
[42] Hervieu M, Michel C, Maignan A, Martin C and Raveau B 1988 *J. Solid State Chem.* **74** 428
[43] Morosin B, Venturini E L and Ginley D S 1991 *Physica* C **175** 241
[44] Sinclair D C, Aranda M A G, Attfield P and Rodríguez-Carvajal J 1994 *Physica* C **225** 307
[45] Ogborne D M, Weller M T and Lanchester P C 1992 *Physica* C **200** 167
[46] Martin C, Bourgault D, Hervieu M, Michel C, Provost J and Raveau B 1989 *Mod. Phys. Lett.* B **3** 993

[47] Iqbal Z, Sinha A P B, D Morris E, Barry J C, Auchterlonie G J and Ramakrishna B L 1991 *J. Appl. Phys.* **70** 2234
[48] Liu R S, Hervieu M, Michel C, Maignan A, Martin C, Raveau B and Edwards P P 1992 *Physica* C **197** 131
[49] Ihara H, Sugise R, Hirabayashi M, Terada N, Jo M, Hayashi K, Negishi A, Tokumoto M, Kimura Y and Shimomura T 1988 *Nature* **334** 510
[50] Ogborne D M and Weller M T 1994 *Physica* C **230** 153
[51] Hervieu M, Maignan A, Martin C, Michel C, Provost J and Raveau B 1988 *Mod. Phys. Lett.* B **2** 1103
[52] Ogborne D M and Weller M T 1994 *Physica* C **223** 283
[53] Ogborne D M and Weller M T 1992 *Physica* C **201** 53
[54] Siegal M P, Venturini E L, Morosin B and Aselage T L 1997 *J. Mater. Res.* **12** 2825
[55] Ruckenstein E and Wu N L 1994 *Thallium-Based High-Temperature Superconductors* ed A M Hermann and J V Yakhmi (New York: Dekker) p 119
[56] DeLuca J A, Garbauskas M F, Bolon R B, McMullin J G, Balz W E and Karas P L 1991 *J. Mater. Res.* **6** 1415
[57] Flükiger R, Gladyshevkii R E and Bellingeri E 1998 *J. Supercond.* **11** 23
[58] Holstein W L 1993 *J. Phys. Chem.* **97** 4224
[59] Aselage T L, Voigt J A and Keefer K D 1990 *J. Am. Ceram. Soc.* **73** 3345
[60] Aselage T L, Venturini E L and Van Deusen S B 1993 *J. Appl. Phys* **75** 1023
[61] Bellingeri E, Gladyshevskii R, Marti F, Dhallè M and Flükiger R 1998 *Supercond. Sci. Technol.* **11** 810
[62] Gladyshevkii R E, Bellingeri E and Flükiger R 1998 *J. Supercond.* **11** 109
[63] Routbort J L, Miller D J, Zamirowski E J and Gorretta K C 1993 *Supercond. Sci. Technol.* **6** 337
[64] Renk K F 1994 *Thallium-Based High-Temperature Superconductors* ed A M Hermann and J V Yakhmi (New York: Dekker) p 477
[65] Webb B C and Sievers A J 1986 *Phys. Rev. Lett* **i** 1951
[66] Thomas G A, Orensein J, Rapkine D H, Capizzi M, Millis A J, Bhatt R N, Schneemeyer L F and Waszczak J 1988 *Phys. Rev. Lett.* **61** 1313
[67] Tinkham M 1970 *Far-Infrared Properties of Solid* ed S S Mitra and S Nudelman (New York: Plenum) p 223
[68] Zetter T, Franz M, Schutzmann J, Ose W, Otto H H and Renk K F 1990 *Solid State Commun.* **75** 325
[69] Zetter T, Franz M, Schutzmann J, Ose W, Otto H H and Renk K F 1990 *Phys. Rev.* B **41** 9499
[70] Kubo Y, Shimakawa Y, Manako T and Igarashi H 1991 *Phys. Rev.* B **43** 7875
[71] Naugle D G and Kaiser A B 1994 *Thallium-Based High-Temperature Superconductors* ed A M Hermann and J V Yakhmi (New York: Dekker) p 543
[72] Siri S 1996 *High Temperature Superconductivity, Models and Measurements* ed M Acquarone (Singapore: World Scientific) p 445
[73] Mott N F and Davis E A 1979 *Electronic Process in Non-Crystalline Materials* (Oxford: Clarendon)
[74] Datta T 1994 *Thallium-Based High-Temperature Superconductors* ed A M Hermann and J V Yakhmi (New York: Dekker) p 407
[75] Pham A Q, Maigna A, Hervieu M, Michel C, Provost J and Raveu B 1992 *Physica* C **191** 77
[76] Liu R S and Edwards P P 1994 *Thallium-Based High-Temperature Superconductors* ed A M Hermann and J V Yakhmi (New York: Dekker) p 325
[77] Hulliger F 1976 *Structural Chemistry of Layer-Type Phases* ed D Reidel (Kluwer: Dordrecht)
[78] Liu R S and Edwards P P 1993 *Mater. Sci. Forum* **130–132** 435
[79] Tsuei C C, Kirtley J R, Rupp M, Sun J Z, Gupta A, Ketchen M B, Wang C A, Ren Z F, Wang J H and Bushan M 1996 *Science* **271** 326
[80] Tsuei C C, Kirtley J R, Ren Z F, Wang J H, Raffy H and Li Z Z 1997 *Nature* **387** 481
[81] Rossel C, Willemin M, Hofer J, Keller H, Ren Z F and Wang J H 1997 *Physica* C **282–287** 136
[82] Presland M R, Tallon J L, Flower N E, Buckley R G, Mawdsley A, Staines M P and Fee M G 1993 *Cryogenics* **33** 502
[83] Ren Z F and Wang J H 1993 *Physica* C **216** 199
[84] Bellingeri E, Gladyshevkii R E and Flükiger R 1997 *Il Nuovo Cimento* D **19** 1117
[85] Jeong D Y, Kim H K and Kim Y C 1999 *Physica* C **314** 139
[86] Bhattacharya R N, Blaugher R D, Ren Z F, Li W, Wang J H, Paranthaman M, Verebelyi D T and Christen D K 1998 *Physica* C **304** 55
[87] Li W, Wang D Z, Lao J Y, Ren Z F, Wang J H, Paranthaman M, Verebelyi D T and Christen D K 1999 *Supercond. Sci. Technol.* **12** L1
[88] New Jersey Department of Health and Senior Services 1998 *Hazardous substance fact sheet*
[89] Poddar A, Mandal P, Das A N, Ghosh B and Choudhury P 1991 *Phys. Rev* B **44** 2757

C4
HgBCCO

J Schwartz and P V P S S Sastry

C4.1 Introduction

The series of discoveries of superconductivity in ceramic oxides containing Cu–O layers that began in 1986 with the discovery of the La–Sr/Ba–Cu–O system culminated in the mid-1990s when the highest confirmed critical temperature reached 164 K. This was an increase of over 130 K from the previous record held by intermetallic A15 compounds. In the years since that first discovery, much progress has been made in the physical understanding of the phenomenon of superconductivity and in the technological development of thin and thick films, wires, cables, and their applications. This chapter focuses on progress since 1993 on the Hg–Ba–Ca–Cu–O system, which holds the record for the highest T_c of any known compound. After a brief history of the compounds, their structural properties are discussed. Then, synthesis and processing techniques for bulk material, thin and thick films and tapes are discussed, including discussion on the effects of dopants on the processing parameters. Then, their superconducting properties and the effects of pressure and irradiation on the superconducting behaviour are summarized. Lastly, prospects for applications are considered.

B2.5

C4.2 History of Hg–Ba–Ca–Cu–O

The first report of superconductivity in the Hg–Ba–Ca–Cu–O system was authored by Putilin *et al* [1] and appeared in *Nature* on March 18, 1993. In this letter, the authors reported a transition temperature of 94 K in $HgBa_2CuO_{4+\delta}$ (Hg1201). Interestingly, the same first author had reported in 1991 that $HgBa_2RCu_2O_{6+\delta}$ was not superconducting [2]. Subsequently, $HgBa_2CaCu_2O_{6+\delta}$ (Hg1212) and $HgBa_2Ca_2Cu_3O_{8+\delta}$ (Hg1223) were reported in *Nature* by Schilling *et al* [3] and in *Physica* C by Putilin *et al* [4] with critical temperatures of 125 and 135 K, respectively [3, 4]. Today, the $n = 1 \ldots 5$ members of the homologous series $HgBa_2Ca_{n-1}Cu_nO_{2n+2+\delta}$ have been shown to be superconducting, with the $n = 3$ member having the highest T_c. Note that this series strongly resembles the homologous series of single-layer Tl-based superconducting compounds [5–7]. It is also noteworthy that the Hg–Sr–Ca–Cu–O system has also been shown to be superconducting. This series, however, is less stable and has lower T_c than the Hg–Ba–... series, and can be considered a substitutional variation of the Hg–Ba–... series [8–12]. The Hg–Sr–... series will not be discussed further in this chapter.

D1.1.1
D1.1.2

C4.3 Crystal structure of $HgBa_2Ca_{n-1}Cu_nO_{2n+2}$ superconductors

The crystal structures of the members of homologous series, $Hg_1Ba_2Ca_{n-1}Cu_nO_{2n+2}$, are similar to the other high temperature superconducting cuprate systems [13]. The structures of Hg-based superconductors are isomorphous to the single layer Tl-based superconductors $Tl_1Ba_2Ca_{n-1}Cu_nO_{2n+2}$ [5–7].

C3 The main difference between the Tl and Hg analogues is that the oxygen occupancy in the Hg layers is only partial, with Hg being only 2-coordinated, whereas in the Tl compounds, Tl is 6-coordinated. The significant oxygen deficiency results in the commonly observed synthesis difficulties and the relative instability of Hg-based superconducting materials. Partial substitution of Hg by any of several dopants such as Re, Pb, Bi and Cr improves the stability by altering the defect structure [14–18]. The dopants, being in a higher valance state than Hg, pull additional oxygen atoms into the Hg/M layer, stabilizing the structure [14–18].

D1.1.1 All the superconducting phases of the $Hg_1Ba_2Ca_{n-1}Cu_nO_{2n+2}$ system crystallize with a tetragonal cell having the symmetry of space group P4/mmm. The a-parameter is ~ 3.5 Å and the c-parameters of the various phases follow the formula $c \sim 9.5 + 3.2(n-1)$ Å, n being the number of Cu–O planes in the structure. The crystal structures are based on the layer sequence:

$$\ldots\{(BaO)\ (HgO)\ (BaO)\ (CuO_2)\ [(n-1)\ (Ca)\ (CuO_2)]\}\ (BaO)\ldots$$

The blocks (BaO) (HgO) (BaO) have the rock-salt structure with a thickness of about 5.5 Å and alternate with blocks $(CuO_2)\,[(n-1)\,(Ca)\,(CuO_2)]$ having a perovskite-like structure with an approximate thickness of $[4.0 + 3.16(n-1)]$ Å. The crystal structures for $n = 1, 2, 3$ and 4 are shown in figure C4.1 [19].

D1.1.2 The detailed crystal structures have been studied by x-ray diffraction, neutron diffraction and high-resolution electron microscopy [14–30]. The oxygen deficiency in HgO layers varies with n, the number of CuO_2 planes in the structure. Partial substitution of Hg by several dopants has been shown to alter the oxygen defect structure of the host Hg–O layer [14–18]. Chmaissem *et al* [15] reported structural features of Cr-doped Hg1201, in which Cr substitutes at the Hg site and is tetrahedrally coordinated to four oxygen atoms. Villard *et al* [16] showed that partial substitution of Hg by V in Hg1201 shortens the c-axis parameter due to the substitution of VO_4 tetrahedra for HgO_2 dumbbells. Schwer *et al* [17] observed similar contraction of the Hg1201 lattice by Pb-doping. Pelloquin *et al* [18] reported structural features of single crystals of Hg1201 with 16% of Hg substituted by Cu. Chmaissem *et al* [14] reported the effect of Re-substitution on the defect structure of $(Hg,Re)_1Ba_2Ca_{n-1}Cu_nO_{2n+2}$ ($n = 2, 3$, and 4). Re atoms substitute at the Hg site and alter the defect structure of the host layer by pulling in four oxygen atoms to form an octahedron around Re.

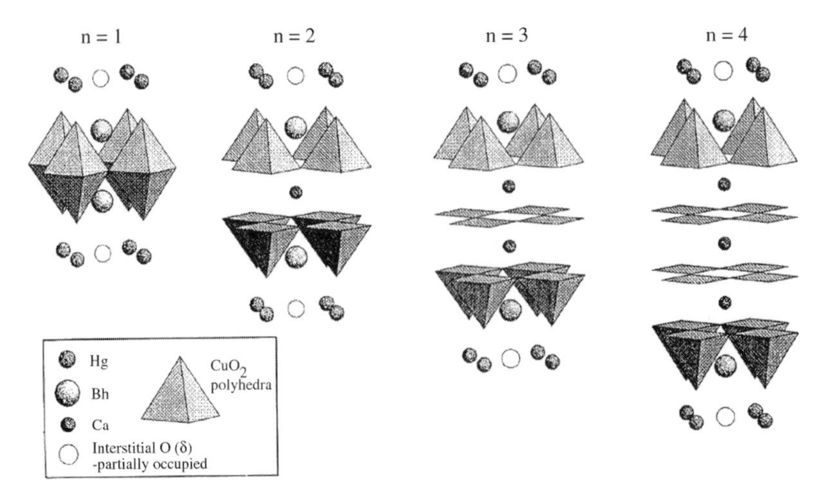

Figure C4.1. Crystal structures for $HgBa_2Ca_{n-1}Cu_nO_{2n+2+\delta}$ ($n = 1, 2, 3$ and 4).

There have been a few studies on the crystal structures of the Hg superconducting phases at high pressures. These studies have been aimed at understanding the structural features of the Hg superconducting phases that are responsible for the observed high dependence of the superconducting transition temperature, T_c, with applied pressure [12, 31, 32]. Gonzalez *et al* [31] conducted an *in situ* high pressure study up to 5 GPa to measure the anisotropic compressibility of polycrystalline Hg1201 superconductor. Aksenov *et al* [22] used neutron diffraction to investigate the structure of Hg1201 under external pressure up to 5 GPa. Hunter *et al* [32] compared the pressure induced structural changes in $Hg_1Ba_2Ca_{n-1}Cu_nO_{2n+2}$ ($n = 1$, 2 and 3) compounds using neutron diffraction. The compressibility along the c-axis is nearly the same for the three phases and up to two times larger than the compressibility along the a-axis. Hg1201 shows the largest a-axis compressibility, while Hg1223 shows the smallest a-axis compressibility [32].

D1.1.1

B2.2.4

C4.4 Synthesis of $Hg_1Ba_2Ca_{n-1}Cu_nO_{2n+2}$ superconductors

The synthesis of the superconducting phases of the homologous series $Hg_1Ba_2Ca_{n-1}Cu_nO_{2n+2}$ has been more challenging than most of the other high T_c cuprate superconductors. The high volatility of mercury necessitates heating the samples in sealed reaction. The resulting lack of independent control of mercury and oxygen partial pressures during the reaction makes it difficult to synthesize phase-pure samples of the individual superconducting phases. The synthesis is also sensitive to trace quantities of moisture and carbon dioxide and demands freshly prepared oxide precursors. Partial substitution of Hg by several dopants significantly improves the ease of formation and enhances the stability of the superconducting phases without affecting the T_c [33–44]. The control of mercury and oxygen partial pressures during the synthesis, however, needs intelligent approaches in the reaction configuration and synthesis pathways. Over the past five years, there has been enormous progress in understanding the reaction sequence and the relative phase stability of the superconducting and the intermediate phases that form during the synthesis process. Phase pure materials of the first three ($n = 1$, 2 and 3) superconducting phases have been synthesized by several research groups in both the undoped and doped compositions. The stability and superconducting properties of the doped materials are, in general, superior to the pure materials. Many research groups, however, synthesize and study the undoped materials to understand the role of dopants in enhancing the phase stability and superconducting properties. The following sections describe the synthesis and processing conditions for each of the $n = 1$, 2 and 3 phases. The synthesis of the phases with $n \geq 4$ is commonly carried out under high pressures and high temperatures. A brief description of the synthesis of $n \geq 4$ phases is given at the end of the synthesis section.

B2.2.4

B2.1

The most common synthesis procedure of $Hg_1Ba_2Ca_{n-1}Cu_nO_{2n+2}$ consists of two steps. In the first step, Ba–Ca–Cu–O precursor powders are prepared by reacting stoichiometric mixtures of simple oxides, nitrates, and/or carbonates of Ba, Ca, and Cu. Since the presence of even small amounts of carbon is detrimental to the formation of the superconducting phases, it is better to avoid using carbonates. Barium carbonate is particularly problematic to decompose completely. It is easier to complete the formation of Ba–Ca–Cu–O precursors starting from nitrates. The nitrates, however, are hygroscopic and may result in off-stoichiometric reaction mixtures. Stoichiometric mixtures of barium peroxide, calcium carbonate, and cupric oxide generally lead to completely reacted Ba–Ca–Cu–O precursor powders. The reaction mixture is heated at 900–920°C for 3–4 days with intermittent grindings. If nitrates are used, the initial decomposition of nitrates is carried out at 600°C for 10 h before increasing the reaction temperature to 900°C. The precursor $Ba_2CuO_{3+\delta}$, used for the synthesis of the $n = 1$ phase (Hg1201), can be made single phase, whereas the precursors with nominal compositions $Ba_2Ca_1Cu_2O_{5+\delta}$ and $Ba_2Ca_2Cu_3O_{7+\delta}$ used for the synthesis of Hg1212 and Hg1223 superconductors, respectively, are mixtures of $Ba_2CuO_{3+\delta}$, $BaCuO_2$ and Ca_2CuO_3.

B2.1

In the second step, Ba–Ca–Cu–O precursor powders are reacted with HgO in an evacuated quartz tube. The Ba–Ca–Cu–O precursor could be intimately mixed with the required amount of yellow HgO to obtain the reaction mixture. Small sections of compacted pellets of Ba–Ca–Cu–O + HgO are inserted in an alumina liner and sealed in a quartz tube. The purpose of the alumina liner is to avoid direct contact of the reaction mixture with quartz. This type of reaction configuration is termed the internal mercury source method. The reaction with HgO can also be carried out by placing the mercury source in a separate alumina liner, which is sealed in the quartz tube along with the Ba–Ca–Cu–O precursor. Mercury vapours react with Ba–Ca–Cu–O precursor by vapour phase transport during the high temperature reaction. The materials used as mercury sources include HgO, Ba–Ca–Cu–O + HgO, and $CaHgO_2$. This reaction configuration is termed the external mercury source method as the source of Hg–O is physically separated from the Ba–Ca–Cu–O precursors.

B2.1 In the synthesis of doped Hg–Ba–Ca–Cu–O superconductors, stoichiometric amounts of the dopant oxide are mixed with the Ba–Ca–Cu–O precursor powder and reacted with Hg–O using one of the above methods.

B2.2.4 The phase purity of $Hg_1Ba_2Ca_{n-1}Cu_nO_{2n+2}$ depends on several synthesis parameters such as reaction temperature, and the partial pressures of mercury and oxygen during the synthesis. For a given mass of the reaction mixture, the mercury and oxygen partial pressures depend on the volume of the sealed reaction tube. Hence, it is necessary to keep the volume of the reaction tube fixed while synthesizing a series of samples with varying composition or temperature to optimize reaction conditions. The mercury and oxygen partial pressures during the reaction also depend on the chemical nature of the mercury source used. For example, decomposition of HgO starts at around 450°C, whereas some binary oxides such as $CaHgO_2$ decompose at 650°C or higher. Hence, the equilibrium mercury and oxygen partial pressures in the sealed reaction tubes depend on the stoichiometry and structure of the mercury source. Oxygen partial pressure can also be controlled using redox couples such as Co_3O_4/CoO, CuO/Cu_2O, or Mn_2O_3/Mn_3O_4. Changing the temperature of the redox couple in a multi-temperature zone furnace controls the oxygen partial pressure.

C4.4.1 Synthesis of $Hg_1Ba_2CuO_{4+\delta}$ (n = 1 phase, Hg1201)

D1.1.1 In the original discovery of superconductivity in $HgBa_2CuO_{4+\delta}$ at 94 K, Putilin *et al* [1] synthesized the material by solid state reaction between stoichiometric mixtures of $Ba_2CuO_{3+\delta}$ and yellow HgO. The reaction mixture was placed in silica tubes that were then evacuated and heated slowly to about 800°C in 5 h. The samples were cooled in the furnace after 10 h at 800°C. All mixing and grinding operations were performed in a dry box. Most reports on the synthesis of $HgBa_2CuO_{4+\delta}$ used the same basic synthesis protocol described by Putilin *et al* [1]. El-Sayed *et al* [45] reported synthesis of $Hg_1Ba_2CuO_{4+\delta}$ at 600°C with a rapid heating rate of 900°C h^{-1} to reach the reaction temperature. Some researchers used an excess of 10–20% HgO in the reaction mixtures [46].

The reported T_c of as-prepared $HgBa_2CuO_{4+\delta}$ samples is around 95 K. The T_c of $Hg_1Ba_2CuO_{4+\delta}$, however, can be changed reversibly between 20 and 97 K by changing the oxygen stoichiometry (δ) by post-annealing in either oxygen or an inert atmosphere at 200–300°C [47–50]. Control of the oxygen partial pressure during the synthesis was achieved using transition metal oxide redox couples discussed previously. Control of mercury, and oxygen partial pressures can also be achieved by using a compound of mercury, such as $CaHgO_2$ or $HgBa_2CuO_{4+\delta}/Ba_2CuO_{3+\delta}$, as the mercury source instead of HgO [47]. Peacock *et al* [47] reported the synthesis of highly underdoped $HgBa_2CuO_{4+\delta}$ with a T_c of 35 K using elemental Hg instead of HgO in the reaction. Post-annealing of the sample in oxygen at 300°C increased T_c to 92–96 K.

Doped $Hg_{1-x}M_xBa_2CuO_{4+\delta}$ has been reported with M = Bi, Cr, Mo, Pb, Re, Ti, Tl, V and W (see table C4.1). The doped Hg1201 materials are synthesized by heating the mixtures of HgO, MO and

Table C4.1. Literature references for various dopants

	(HgM)1201	(HgM)1212	(HgM)1223
Au			[69]
Bi	[51]		[30, 36, 40, 42, 43, 51, 85, 86]
Cr	[10, 15]	[10]	[10]
Cu	[18]		
In		[57]	
Li			[84]
Mo	[10]	[10]	[10]
Na			[44]
Pb	[17, 52]	[59, 80]	[33–35, 40, 41, 44, 70, 79–83]
Re	[43]	[14, 38, 53, 60, 105]	[14, 35–40, 44, 60, 75–78, 121]
Sn			[71–74]
Ti	[10]	[10]	[10]
Tl	[24]		[87–92]
V	[16, 52, 54]	[54]	[10]
W	[10]	[10]	

$Ba_2CuO_{3+\delta}$ in sealed quartz tubes similar to the methods used for undoped materials. The solubility limits are around 20% and vary depending on the dopant. In general, all the dopants enhance the stability of $Hg_1Ba_2CuO_{4+\delta}$ phase but lower T_c to 80–95 K. All the dopants, due to their higher valance state, introduce additional oxygen in the (Hg,M)–O layers of the crystal structures. The *c*-axis length of all the doped materials is smaller than that of the undoped material. The reported $Hg_{1-x}M_xBa_2CuO_{4+\delta}$ materials and the corresponding literature references are listed in table C4.1.

Single crystals of $Hg_1Ba_2CuO_{4+\delta}$ have been grown from mixtures of HgO and Ba_2CuO_3 in sealed quartz tubes. Pissas *et al* [26] reported growth of $Hg_1Ba_2CuO_{4+\delta}$ single crystals. Pelloquin *et al* [18] reported single crystals of $(Hg,Cu)_1Ba_2CuO_{4+\delta}$. Schwer *et al* [17] reported growth of single crystals of $(Hg,Pb)_1Ba_2CuO_{4+\delta}$. B2.4

C4.4.2 Synthesis of $Hg_1Ba_2Ca_1Cu_2O_{6+\delta}$ (n = 2 phase, Hg1212)

The synthesis of $HgBa_2CaCu_2O_{6+\delta}$ is usually carried out starting from a mixture of $Ba_2CaCu_2O_{5+\delta}$ precursor and HgO. The reacted precursor with nominal composition $Ba_2CaCu_2O_{5+\delta}$ is typically a mixture of $Ba_2CuO_{3+\delta}$, $BaCuO_2$ and Ca_2CuO_3. The mixture of $Ba_2Ca_1Cu_2O_{5+\delta}$ and HgO is compacted into pellets and heated in sealed quartz tubes at around 850°C. The as-prepared samples of Hg1212 show a T_c of 115 K. Post-annealing at 300°C in an oxygen atmosphere for about 10 h after synthesis increases T_c to 128 K. The T_c of Hg1212 can be reversibly changed between 98 and 128 K by annealing the samples in oxygen or an inert atmosphere at 300°C [55, 56]. B2.2.4

Doped $Hg_{1-x}M_xBa_2Ca_1Cu_2O_{6+\delta}$ (HgM1212) has been synthesized with M = Cr, In, Mo, Pb, Re, Ti and W (see table C4.1). The doped materials are synthesized by reacting mixtures of $Ba_2Ca_1Cu_2O_{5+\delta}$ and MO with HgO. The doped materials exhibit optimized T_c values in the range of 110–125 K [54–59]. Many of the doped materials are easier to synthesize phase pure than the undoped materials.

Among the HgM1212 materials reported, HgPb1212 and HgRe1212 have been investigated for improving the microstructural and superconducting properties [58–60]. HgPb1212 and HgRe1212 with high phase purity and good microstructural and superconducting properties were synthesized from commercial Ba–Ca–Cu–O precursor powders using $CaHgO_2$ as the external mercury source [60]. It has been shown that the phase stability of HgM1212 depends on both the reaction temperature and starting composition. Temperatures below 750°C typically result in the formation of Hg1201 and $CaHgO_2$ phases. Temperatures above 850°C result in the formation of Hg1223 phase [60].

C4.4.3 *Synthesis of* $Hg_1Ba_2Ca_2Cu_3O_{8+\delta}$ *(n = 3 phase, Hg1223)*

B2.2.4 $Hg_1Ba_2Ca_2Cu_3O_{8+\delta}$ exhibits the highest T_c of all the known superconducting materials. In some of the initial studies, Hg1223 samples were made from mixtures of component oxides: HgO, BaO, CaO and CuO. Most of the recent syntheses of Hg1223 reacted precursor powders with nominal composition $Ba_2Ca_2Cu_3O_{7+\delta}$ with HgO using either an internal or an external mercury source. The source material used in most studies was either HgO or a mixture of HgO and $Ba_2Ca_2Cu_3O_{7+\delta}$. Recently, Sastry *et al* reported synthesis of doped Hg1223 superconductors using $CaHgO_2$ as the external mercury source [39–43]. When the HgO and $Ba_2Ca_2Cu_3O_{7+\delta}$ mixture is used as the mercury source, excess HgO is usually required for completing the reaction because some HgO reacts with the Ba–Ca–Cu–O in the precursor [63]. The amount of excess HgO required depends on the volume of the sealed reaction tube and the amount of Ba–Ca–Cu–O buffer in the source. The sealed reaction tubes are usually heated for 10 h at 850–900°C followed by furnace cooling to room temperature. The resulting Hg1223 samples exhibit T_c in the range of 110–125 K. The samples are post-annealed in oxygen at 300°C to increase T_c to 132–135 K [61–63].

Addition of Hg-halogenide and In_2O_3 during synthesis has been reported to promote the formation of Hg1223 [64–66]. The mechanism of promotion has, however, not been investigated. HgF_2 and $HgCl_2$ decompose during the high temperature reaction, releasing mercury and Cl_2 or F_2 and changing the partial pressures of mercury and oxygen. The presence of BaF_2 is also known to alter the melting characteristics, which affects the grain growth process in other high T_c cuprate systems [67–68].

Several HgM1223 materials have been synthesized with compositions $Hg_{1-x}M_xBa_2Ca_2Cu_3O_{8+\delta}$: M = Bi, Cr, In, Mo, Pb, Re, Ti, Tl and V, $0.0 < x < 0.2$. Dopant concentrations higher than 20% have been achieved with Tl, and it should be noted that Tl forms an analogous $Tl_1Ba_2Ca_2Cu_3O_{8+\delta}$ superconductor [5–7].

D1.1.1 Due to the improved stability of HgM1223 relative to Hg1223, most recent synthesis and processing studies focused on HgM1223 compositions. Among the several dopants, Bi, Pb and Re have been shown to enhance the properties of Hg1223 without reducing the T_c [32–44]. Detailed studies showed that the phase stability and microstructural characteristics of HgM1223 materials vary significantly with dopant. Recent studies of Sastry and Schwartz reported a novel synthesis technique for HgM1223 using $CaHgO_2$ as the external mercury source. They have also reported that a reduced temperature annealing stage after the high temperature reaction improves grain growth and microstructural characteristics of HgM1223 [41–43]. Using their technique, HgBi1223, HgPb1223 and HgRe1223 were synthesized from commercial $Ba_2Ca_2Cu_3O_{7+\delta}$ precursor powders under similar conditions. These studies showed that the phase stability and microstructural characteristics of HgM1223 materials vary significantly depending on the dopant, M (see figure C4.2). HgM1223 phase forms over a wide range of temperatures, 750–950°C, 750–880°C and 840–880°C for M = Re, Pb and Bi, respectively. At T < 750°C, HgM1212 phase forms for M = Re and Pb. Interestingly, the HgBi1212 phase does not form. The existence of a partial melting stage at T > 880°C for M = Bi and Pb allows the synthesis of dense materials with

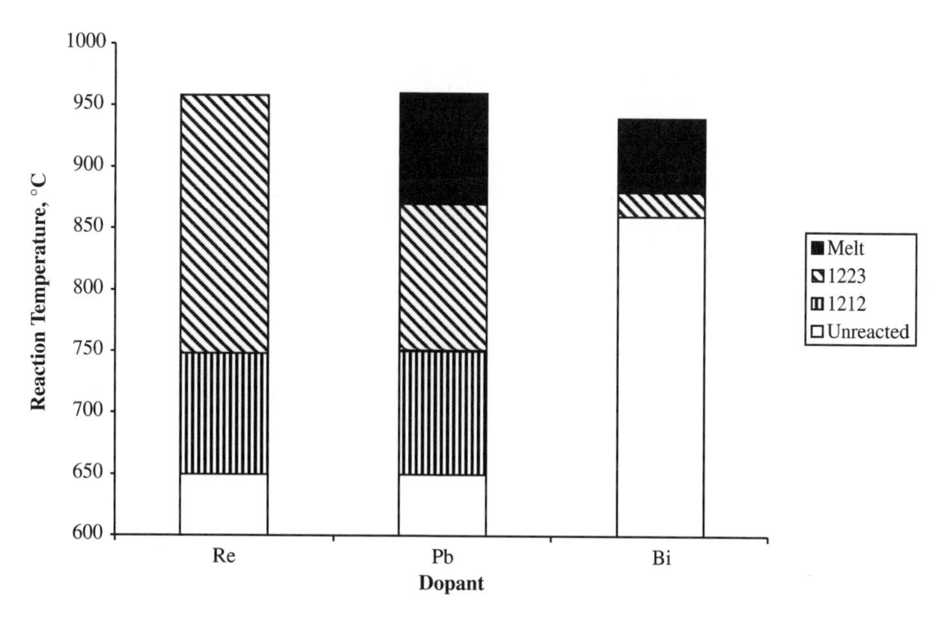

Figure C4.2. Phase stability versus temperature and dopant M (M = Re, Pb, Bi) for a $HgMBa_2Ca_2Cu_3O_{8+\delta}$ stoichiometry.

enhanced grain growth and texture [see figures C4.3 (*a*) and (*b*)]. The absence of a partial-melt stage for $T < 950°C$ in Re-doped compositions is reflected in the randomly oriented small grains typically seen in HgRe1223 materials [see figure C4.3(*c*)]. The phase stability and microstructure of HgM1223 materials also depend on the starting composition. Figure C4.4 shows the phase stability of Re-doped materials as a function of composition. HgRe1223 phase forms both in the $Hg_{0.8}Re_{0.2}Ba_2Ca_1Cu_2O_{6+\delta}$ and $Hg_{0.8}Re_{0.2}Ba_2Ca_2Cu_3O_{8+\delta}$ composition depending on the reaction temperature. The microstructure of the resulting HgRe1223 phase, however, varies significantly depending on the starting composition. The HgRe1223 formed from $Hg_{0.8}Re_{0.2}Ba_2Ca_1Cu_2O_{6+\delta}$, which has excess Ba and Cu relative to the stoichiometric composition, results in significantly larger HgRe1223 grains [60]. Similar observations were made on HgPb1223 materials synthesized from compositions containing an excess of Ba and Cu [83]. The presence of excess Ba and Cu lowers the melting temperature, enhancing the grain growth. The off-stoichiometric compositions, however, result in materials with some non-superconducting impurity phases. Careful choice of dopant, composition and heat treatment is necessary to achieve phase purity, grain growth and texture in HgM1223 materials.

C4.4.4 Synthesis of $Hg_1Ba_2Ca_{n-1}Cu_nO_{2n+2}$, $n \geq 4$ phases

Synthesis of the higher members ($n > 3$) of the homologous series $Hg_1Ba_2Ca_{n-1}Cu_nO_{2n+2}$ requires high pressures and high temperatures [93–95]. Synthesis of these phases is carried out from stoichiometric mixtures of individual oxides. The compacted powders are packed in a gold capsule and heated to about 900°C under pressure up to 20 kbar for 1–20 h using a belt-type apparatus or a Boyd and England type piston/cylinder [93–94]. The pressure is released after cooling the sample to room temperature. The quantity of superconductor that can be synthesized using high pressure, high temperature cells is usually less than a gram due to the limited reaction cell volume. D1.1.1

Scott *et al* [93] synthesized $Hg12\{n-1\}n$ phases with n up to 9 using a Boyd and England type piston/cylinder under a pressure of 1.8 GPa in the temperature range of 950–1050°C [93]. They used

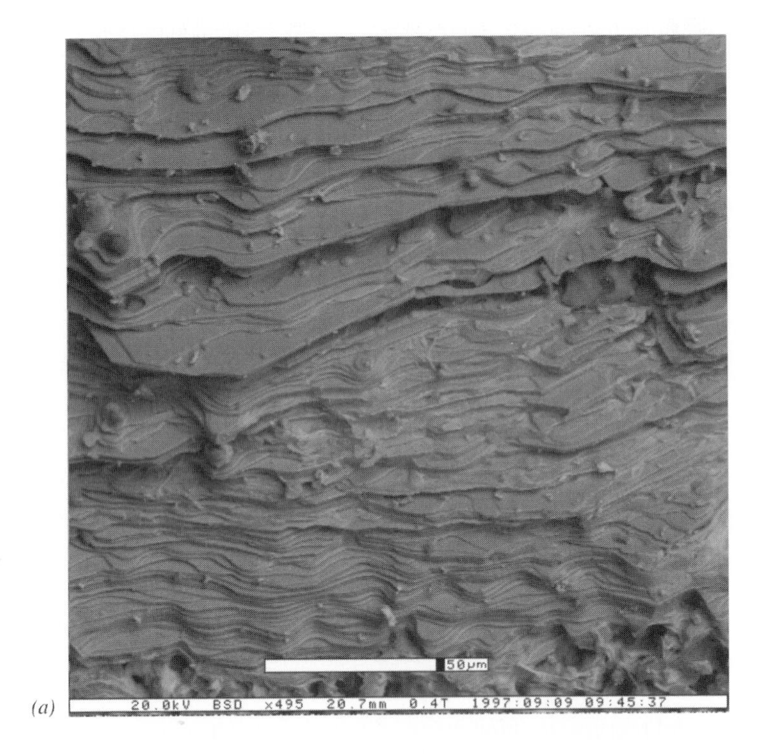

(a)

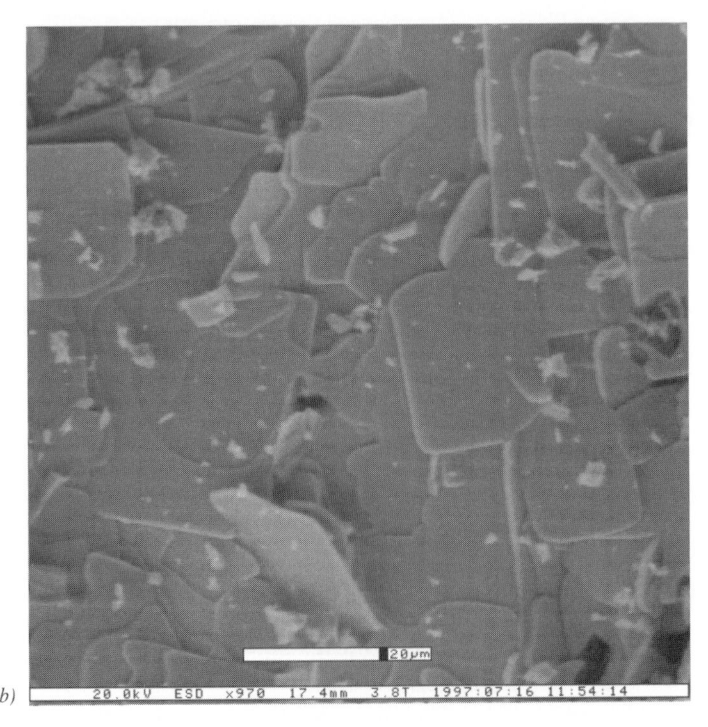

(b)

Figure C4.3. Scanning electron micrographs of $HgMBa_2Ca_2Cu_3O_{8+\delta}$ (*a*) M = Bi, (*b*) M = Pb and (*c*) M = Re.

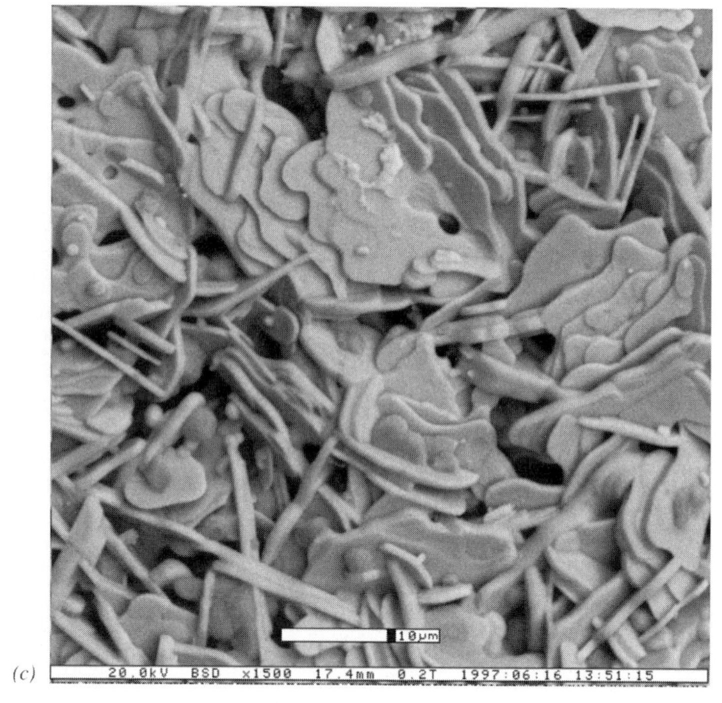

(c)

Figure C4.3. (Continued)

a high purity Al_2O_3 liner and crushable Al_2O_3 discs as the reaction capsule. The samples reacted at 950°C resulted in a mixture of two or more superconducting phases with $n = 4$–7, depending on the starting composition. The samples with a starting composition of $n = 9$ and reacted at 1050°C contained mainly the $n = 8$ superconducting phase. The T_c value of the $n > 3$ phases decreases gradually with increasing n,

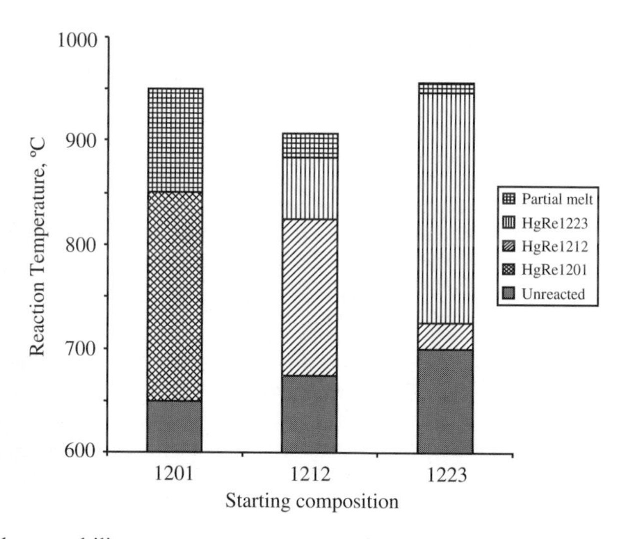

Figure C4.4. Phase stability versus temperature and composition for HgM–Ba–Ca–Cu–O.

reaching 90 K for the $n = 7$ phase. Lokshin *et al* [95] showed that the T_c values of each of the $n > 3$ phases vary within a range of 10–15 K depending on the oxygen content.

C4.5 Thin films

B4 Attempts to synthesize thin films of $Hg_1Ba_2Ca_{n-1}Cu_nO_{2n+2}$ began shortly after their discovery. Early results on $n = 2$ films from IBM were very encouraging for the prospects of $Hg_1Ba_2Ca_{n-1}Cu_nO_{2n+2}$ applications in general [96]. The discussion of $Hg_1Ba_2Ca_{n-1}Cu_nO_{2n+2}$ thin films will focus on a number of key variables, including the phase to be formed (n), substrate selection, deposition method, *in situ* versus *ex situ* growth and the chemistry of the Hg-source. Thin films of $n = 2$ and 3 have been formed B4.1 with varying degrees of film quality. Most films have been grown on $SrTiO_3$ (STO) and $LaAlO_3$ (LAO) substrates, but MgO, $NdGaO_3$ and $Y_{0.15}Zr_{0.85}O_2$ (YSZ) have also been used successfully.

An early attempt to grow $Hg_1Ba_2Ca_{n-1}Cu_nO_{2n+2}$ was reported in late 1993 by Adachi *et al* [97]. In this report, rf magnetron sputtering and subsequent Hg-free annealing was used to grow films on a STO substrate. Although the attempt was to grow the $n = 2$ or 3 phase, only $n = 1$ with $T_c = 85$ K was obtained. This may be the only published report of a Hg1201 thin film.

C4.5.1 $Hg_1Ba_2Ca_1Cu_2O_{6+\delta}$ ($n = 2$ phase, Hg1212) thin films

Shortly after the aforementioned report by Adachi *et al*, four papers appeared that demonstrated the potential for $n = 2$ thin films. Two from IBM reported that *c*-axis oriented epitaxial films could be grown on STO by pulsed laser deposition (PLD) [98, 99] with a zero-resistance T_c as high as 124 K. In the IBM process, stoichiometric Hg1212 is deposited on the STO substrate with atomic-level mixing by using two targets. The film is subsequently covered with a 'cap layer' of HgO, MgO or STO to protect the film from potential CO_2 and H_2O contamination problems and annealed. Only small amounts of D1.1.2, impurity phases appeared by x-ray diffraction. Grain boundary junctions and SQUIDs were sub- E4.2 sequently reported with high critical current densities (10^6 A cm^{-2} at 100 K), albeit with significant weak links. Although the use of a cap layer is not a practical approach for thin film applications, this early report of successful film growth with high J_c was significant for demonstrating the potential of $Hg_1Ba_2Ca_1Cu_2O_{6+\delta}$ films.

Around the same time as the two IBM papers were two reports of Hg1212 films from Mitsubishi [100, 101]. The first, by Miyashita *et al*, reported Hg1212 films on STO by laser ablation with an onset T_c of 123 K and a zero-resistance T_c of 97 K. As with the IBM films, stoichiometric Hg1212 was deposited on the substrate. Here, the film was sealed in a quartz tube with a stoichiometric Hg1212 pellet and annealed to form the superconducting film. When the pellet was not sealed in the quartz tube with the film, no superconductor was formed. Two weeks later, a report from the same group by Higuma *et al* B4.2.1 followed the same process of laser ablation onto STO and post-annealing, but with Pb-doping in the Hg1212 film. In this case, the zero-resistance T_c was increased to 117 K and the required annealing temperature was reduced from 850 to 760°C.

Encouraged by the early success of these two groups, many others began studying the synthesis of Hg1212 films. A number of key new processing features have been introduced, including: the use of dopants, primarily to enhance grain growth while reducing sensitivity to air contamination; the use of an external source of Hg, thus using vapour transport during the growth reaction; and the use of substrates B4.1 other than STO. For example, Brazdeikis *et al* [102] reported that small amounts of Tl_2O_3 promote the growth of Hg1212 on LAO and STO. In this report, the film was deposited by laser ablation and the Hg was within the as-deposited film.

Two of the most active groups in the processing of Hg1212 films are in the Physics Departments at the University of Kansas and Peking University. These groups were among the first to use an external source of Hg for the growth reaction. Wu *et al* [103] reported epitaxial growth of Hg1212 on STO using rf magnetron sputtering of a Hg-free precursor. The Hg was contained within a stoichiometric (but unreacted) Hg1212 pellet that was encapsulated with the precursor film. Despite some reaction between Ba and the STO substrate, a high quality film was grown. They reported that increasing the ramp rate to the growth temperature minimized $CaHgO_2$ formation in the film [103]. Subsequently, they improved resistance to air contamination by including small amounts of Cl or F in the Hg source pellet. These films were grown on LAO [104]. This is in contrast to other studies, including one report from ISTEC, in which chemical stability was improved by including Re in the deposited precursor [105]. These films were also deposited on LAO with the growth reaction occurring in the presence of an external Hg vapour. B4.1

The Peking University group has developed processes that react an external source of Hg vapour with precursor films deposited by laser ablation [106]. In one report, they show that Bi can be successfully added into Hg1212 films. The Bi-doped films showed a broad transition with an onset T_c of 130 K, indicating that Bi may enhance the formation of Hg1223 rather than Hg1212 [107]. This is consistent with the bulk synthesis results discussed previously. Lastly, this group did an extensive study of substrate material options, investigating growth on STO, $NdGaO_3$ (NGO), LAO and YSZ [108]. Good T_c values were obtained on all substrates. The authors reported, however, that while the films grown on STO and NGO were shiny, those on LAO and YSZ were not smooth. This is consistent with the T_c data. It is also noteworthy that the (110) NGO has the closest lattice match with Hg1212 and is a low rf-loss substrate.

Recently, a new approach to the formation of Hg1212 films was demonstrated by Wu *et al* [109]. In this approach, Tl1212 films are grown and subsequently annealed in the presence of a Hg vapour. The annealing removes the Tl from the film and replaces it with Hg. Using this cation exchange process, the authors obtained films with J_cs close to 1 MA cm^{-2} at 110 K. At 5 K, the J_c of the Hg1212 film is about D2.2 twice that of the original Tl1212 film. At 77 K, the Hg1212 film has a J_c about two orders of magnitude greater than the Tl1212 counterpart (note that the T_c of the Tl1212 is about 90–95 K, while that of the D2.4 Hg1212 is about 115 K).

Lastly, Hitachi and ISTEC have reported high-performance Josephson junctions of HgRe1212 B4.4.3 grown on STO using PLD [110, 111]. These films showed transport J_c (77 K, self-field) $= 5 \times 10^6$ A cm^{-2}, J_c (100 K, self-field) $= 1.5 \times 10^6$ A cm^{-2} and J_c (77 K, 1 T) $= 4 \times 10^5$ A cm^{-2}. In this same report, the authors show that Mo-doped films also have excellent superconducting properties. These films exhibited a supercurrent at temperatures close to the film T_c. Dc SQUIDs made of these films showed field-induced E4.2 periodic voltages up to 111 K.

C4.5.2 $Hg_1Ba_2Ca_2Cu_3O_{8+\delta}$ (*n* = 3 phase, Hg1223) thin films

The development of Hg1223 thin films has closely followed and benefited greatly from the development of Hg1212 thin films. Typically, obtaining phase purity in Hg1223 is more difficult than with Hg1212, requiring higher processing temperatures. In general, all groups grow films by first depositing the precursor on the substrate and subsequently reacting with Hg vapour in a sealed quartz tube. In some cases, ion implantation has been used to introduce the Hg rather than a vapour phase transport [112, 113]. STO is the most-used substrate, however some success has been obtained on LAO. Typically, B4.1 precursors are deposited by rf sputtering, PLD, or laser ablation. Some success has been obtained using a sol–gel technique [66, 114, 115] and spray pyrolysis [116]. Dopants used include Re, In and Li.

The group at the University of Kansas first reported the fabrication of *c*-axis oriented Hg1223 films in 1995 [117]. The film precursor was deposited by rf sputtering onto STO with no dopant. A HgO cap layer was deposited atop the film that was subsequently annealed to grow the Hg1223 phase. The resulting onset T_c was between 130–132 K. Less than one year later, the same group reported significantly improved films that were grown without the cap layer but with unreacted Hg1223 pellets as the Hg source for the growth reaction [118]. As before, these films were deposited by rf sputtering on STO with no dopant. In this case, oxygen annealing was introduced after the growth reaction. High magnetization J_c was reported (2.3×10^7 A cm^{-2} at 5 K, zero-field; 5×10^5 A cm^{-2} at 100 K). As with their Hg1212 films, the Kansas group also used a fast temperature ramping technique to grow Hg1223 films on LAO with success [119]. Most recently, this group has shown that doping with Li significantly reduces the required sintering temperature for growth of the Hg1223 phase, while maintaining good phase purity (73%) [84].

B2.2.4

Recent reports of high quality Hg1223 films have come from ISTEC and TCSUH [110, 120, 121]. Both report films deposited by PLD onto STO with Re-doping that show improved grain morphology and film properties. ISTEC, which used a HgO cap over the precursor layer, reported a zero-resistivity T_c of 127.5 K and a *transport* J_c (77 K) of 1.5×10^6 A cm^{-2} in self-field at 2×10^5 A cm^{-2} in a 1 T background. Note that these values are actually slightly lower than those obtained with HgRe1212 [110]. TCSUH reported a zero-resistance T_c of 131 K with a narrow (1.5 K) transition and a magnetization J_c of 1.1×10^7 A cm^{-2} at 10 K and 1.2×10^5 A cm^{-2} at 120 K (both at zero-field). Subsequently, TCSUH reported that oxygen annealing increased T_c to 133 K and J_c to 1.9×10^7 A cm^{-2} at 10 K and 2.6×10^5 A cm^{-2} at 120 K [122].

D2.4

C4.6 Thick films and tapes

B4.3

There have been efforts to fabricate thick films of Hg–Ba–Ca–Cu–O superconductors on metallic and ceramic substrates. The choice of metals is limited due to amalgamation and subsequent melting of most metals under the processing conditions required for synthesis. The solubility of mercury in a given metal varies with temperature, making it difficult to optimize the processing conditions. Among the metals investigated, platinum and palladium were found to react severely with Hg–Ba–Ca–Cu–O [123]. Nickel was found to be stable in the presence of mercury, but Meng *et al* [35] observed that the adherence of Ba–Ca–Cu–O precursor materials on nickel was poor. Silver and gold were found to be compatible with Hg–Ba–Ca–Cu–O materials even though both metals absorb significant quantities of mercury [123–125].

Lechter *et al* [124] showed that gold was compatible with Hg–Ba–Ca–Cu–O superconductors with little reaction with the components of the superconducting system. They synthesized Hg1223 phase encapsulated in gold foil by hot isostatic pressing at 900°C. This process resulted in samples with densities as high as 97% containing majority Hg1223 phase and small fractions of Hg1212 and $BaCuO_2$ phases. Amm *et al* [123] showed that the gold interface enhances the grain growth and alignment of HgBi1223 by altering the mercury partial pressure at the gold-oxide interface. Large colonies of aligned grains of HgBi1223 were synthesized by reacting a coating of the mixture of Bi_2CuO_4 and Ba–Ca–Cu–O on gold at 845°C using $CaHgO_2$ as the mercury source [123].

Meng *et al* [35] fabricated HgRe1223 films on Ni substrates with a Cr buffer layer of 300–500 Å. They also used a 50 Å layer of silver to take advantage of the lower melting temperature of the superconductor on silver. The coatings were made by spraying a 10–40 μm thick layer of the mixture of ReO_2 and Ba–Ca–Cu–O in alcohol. The reaction was carried out at 850–870°C. The reacted tapes had ~40 μm thick *c*-oriented Hg1223 film with a self-field J_c of ~ 2.5×10^4 Aμcm^{-2} at 77 K [35].

Recently Sastry and Schwartz fabricated HgPb1223 on silver at 780°C using $CaHgO_2$ as the external mercury source. The precursor coatings were made by dip-coating the mixture of $Ba_2Ca_2Cu_3O_7 + 0.2\,PbO$ on silver foils. The precursor coatings were heated at 700°C in flowing oxygen for 10 h followed by cooling to room temperature at a rate of 2°C per minute. Precursor films were vacuum-sealed in quartz tubes along with 0.5 g $Ba_2Ca_2Cu_3O_7$ and 0.5 g $CaHgO_2$ and were reacted at 780°C for 10 h. The films had large regions of aligned grains of HgPb1223 and showed a resistive transition at 133 K.

B3.1

Two groups have attempted to fabricate Hg1201 wires/tapes using the powder-in-tube technique [125, 126]. Both studies used silver for the sheath material and reported high volume fraction of Hg1201 in the composite. Schwartz *et al* [125] used a two-zone technique to control the mercury pressure inside the sealed quartz reaction tubes. The Ag-Hg1201 composite wires were reacted at 800–850°C at a mercury pressure of 15 bar achieved by controlling the temperature of the mercury reservoir at 450°C. Peacock *et al* [126] achieved high volume fraction of Hg1201 phase from a reaction mixture that contains 50% excess of HgO. They utilized a progressive thermal cycling process that involves heat treatment for 1 h at each of four different temperatures between 600 and 800°C. The final tapes had > 90% Hg1201 phase as analysed by XRD and magnetization measurements. The tapes, however, were porous with poor intergrain coupling and did not support any transport current [126]. These studies clearly demonstrate that silver is compatible with the Hg–Ba–Ca–Cu–O system. More studies are essential to improve the processing to achieve density and intergrain connectivity.

D1.1.2

Thick films of Hg1223 were fabricated on YSZ, MgO and ZrO_2 ceramic substrates [114, 115, 127]. Tsabba and Reich [114, 115] used sol–gel techniques to deposit thick Hg1223 films on single crystal YSZ substrates. They used minute quantities of In_2O_3 additions to promote the formation and growth of Hg1223. They obtained films with a T_c of 132 K and magnetic J_c of 10^7 A cm^{-2} at 10 K and 10^5 A cm^{-2} at 100 K. Yoo *et al* [127] also made thick films of Hg1223 on polycrystalline MgO using a sol–gel technique. The as-synthesized films showed a T_c of 118 K, which was enhanced to 130 K after post-annealing in oxygen at 350–400°C. Singh *et al* [89] used spray pyrolysis to make Ag doped HgTl1223 films on single crystal MgO and ZrO_2. The films showed spiral like growth of HgTl1223 grains. The transport critical current density of the films was $\sim 7 \times 10^4$ A cm^{-2} at 20 K.

B4.3
D2.4
D2.2

C4.7 Key superconducting properties

The superconducting properties of the $Hg_1Ba_2Ca_{n-1}Cu_nO_{2n+2}$ system are important for both fundamental and technological reasons. The fundamental superconducting properties and the magnetic properties are important in the efforts to elucidate the mechanism for superconductivity in the oxides and, because the $n = 3$ compound has the highest T_c of all compounds, it provides insight into approaches for discovering compounds with even higher T_c. Furthermore, from the technological perspective, T_c only gives a preliminary 'necessary but not sufficient' indication of the potential utility of a superconducting material. Ultimately, J_c as a function of temperature and magnetic field is the primary determining factor, and this is governed first by the magnetic behaviour of the superconductor.

A1.3

C4.7.1 Fundamental properties

The fundamental properties of the $n = 1$, 2 and 3 phases have been investigated by a few authors. Early studies on Hg1201 showed that the Fermi surface is similar to that of the infinite layer compounds, a potentially important insight for understanding the superconductivity mechanism [128]. Another early study on Hg1201 estimates the energy gap $\Delta \sim 5$–20 meV, with values around 15 meV for the best samples [129]. Magnetization studies of Hg1212 [130] and specific heat measurements of Hg1223 [131] both indicated that thermal fluctuations near T_c play an important role in the behaviour of these compounds. For Hg1212, superconducting parameters were determined: $H_{c2}(0) = 53$ T, $\kappa = 60$, $\xi_{ab}(0) \sim 2.5$ nm and $\lambda_{ab}(0) \sim 150$ nm. For $n = 3$, the most thorough study is found in [132, 133]. Here, for HgBi1223, the following values, which are consistent with other authors for undoped and Pb-doped [134, 135] samples, were obtained:

A3.1

$$\text{(clean limit) } 158 \pm 58\,\text{T} \leq H_{c2}(0\,\text{K}) \leq 166 \pm 74\,\text{T} \text{ (dirty limit)}$$

$$\text{(clean limit) } 1.41\,\text{nm} \leq \xi_{ab}(0\,\text{K}) \leq 1.44\,\text{nm} \text{ (dirty limit)}$$

$$H_{c2}(60\,\mathrm{K}) = 88 \pm 8\,\mathrm{T}$$

$$H_{c2}(110\,\mathrm{K}) = 60 \pm 10\,\mathrm{T}$$

$$E_F = 0.67\,\mathrm{eV}$$

$$\Delta(0) = 53.5 \pm 0.2\,\mathrm{meV}.$$

C4.7.2 Magnetic properties

A4.3 The magnetic studies of $Hg_1Ba_2Ca_{n-1}Cu_nO_{2n+2}$ have focused on the nature of flux pinning in these materials and on the irreversibility behaviour. One early paper on grain-aligned Hg1212 powder was key in confirming collective pinning theory for anisotropic materials [136]. This paper also showed that Hg1212 had a relatively low anisotropy ratio $\epsilon \sim 1/5$. This was an early indication that Hg_1Ba_2-$Ca_{n-1}Cu_nO_{2n+2}$ may have technologically interesting magnetic behaviour. Furthermore, these samples were used to show that the activation barrier for magnetic relaxation is well described by $U(j) = U_c\ln(j_c/j)$, and to demonstrate the existence of quantum creep [137].

The magnetic properties of $Hg_1Ba_2Ca_{n-1}Cu_nO_{2n+2}$ have also been studied extensively in the context of surface versus bulk pinning. It has been shown that both Hg1212 and Hg1223, in the as-grown condition, can be dominated by surface pinning [138–140]. When bulk defects are subsequently B2.6 introduced by neutron irradiation, the bulk contribution to flux pinning improves, while the surface-pinning contribution is destroyed.

The irreversibility line of undoped and doped $Hg_1Ba_2Ca_{n-1}Cu_nO_{2n+2}$ shows the highest values of irreversibility field at high temperature of any known superconductor. Interestingly, it is strongly dependent upon dopants, which is important from both fundamental and technological points-of-view. A1.3 Consistent with the model that the irreversibility field as a function of normalized temperature is inversely proportional to the separation between the CuO_2 layers, the H_{irr} versus T/T_c was shown to be above that of the Bi–Sr–Ca–Cu–O double layer compounds, but below that for Y–Ba–Cu–O [141]. Owing to the record high T_c, however, the absolute value of $H_{irr}(T)$ is as high as Y–Ba–Cu–O around 77 K, and above it for higher temperatures [142]. These early reports estimated H_{irr} (77 K) = 2 T for Hg1212 and Hg1223 and about 0.5 T for Hg1201 [141–143]. Similar conclusions were obtained in Pb-doped Hg1223 [33].

Efforts to improve the irreversibility field of $Hg_1Ba_2Ca_{n-1}Cu_nO_{2n+2}$ have been both chemical and nuclear. An early report from the University of Tokyo showed that the magnetization hysteresis of HgRe1223 remained open up to 12 T at 77 K [144]. Although these values have not been confirmed, significant improvements in the critical current density, depinning activation energy and irreversibility line in Hg1212 and Hg1223 thin films by Re and Mo doping have been demonstrated [110]. These films showed 77 K irreversibility fields around 6 T as obtained by transport measurements. As transport is often limited not by magnetic behaviour but by phase purity and intergranular connectivity, the actual A1.3 irreversibility field may be higher. It has been proposed that improvements from the dopants may be inherent and related to metallization of the blocking layers [140].

B2.6 Irradiation is an effective method for improving the flux pinning in high temperature superconductors. The first report of high critical current density in Hg1201 was a result of neutron irradiation, which increased the 77 K irreversibility field from about 0.25 to 2 T [145]. Large increases with neutron irradiation were subsequently demonstrated in Hg1212 and Hg1223 [139]. Interestingly, an inversion of the magnetic anisotropy was also observed [146]. Lastly, heavy ion irradiation that results in fission of the Hg cations and columnar defects in the crystal structure has been shown to greatly improve the magnetic behaviour of Hg1212 and Hg1223 [147, 148]. B2.6

C4.7.3 *Pressure effects*

The initial discoveries of superconductivity in $Hg_1Ba_2Ca_{n-1}Cu_nO_{2n+2}$ sparked a flurry of activity to increase T_c to new record-high values. A successful approach that significantly increased T_c of Hg1201, Hg1212, and Hg1223 was the application of hydrostatic pressure during the T_c measurement [32, 149–153]. Ultimately, T_cs of Hg1201, Hg1212 and Hg1223 reached 118, 154, and 164 K, respectively. Although pressure is not practical for applications, it demonstrates the great potential of $Hg_1Ba_2Ca_{n-1}Cu_nO_{2n+2}$ and high temperature superconductors in general.

C4.8 Future directions for $Hg_1Ba_2Ca_{n-1}Cu_nO_{2n+2}$

The great potential in the $Hg_1Ba_2Ca_{n-1}Cu_nO_{2n+2}$ superconductors lies in their record high critical temperatures. Despite the significant progress in the processing of bulk, thin film and thick film/tape materials, quite a bit of work remains to be done before these materials will be seen in applications. One area where application may be viable in the short term, however, is in current leads. Because current leads do not require the long lengths of continuous superconductor that are required for most applications, the limits associated with batch processing in sealed quartz chambers are not a significant issue. Furthermore, current leads can take full advantage of the expanded temperature/magnetic field operating space afforded by the Hg1212 and Hg1223 compounds. Applications of thin films are also of great potential, again because of the relatively small batch size requirement. As cooling technologies like cryocoolers and pulse tubes develop, the ability to take advantage of the broad temperature window also becomes attractive. Large-scale applications based upon long-length wires or tapes of $Hg_1Ba_2Ca_{n-1}Cu_nO_{2n+2}$ appear to be the furthest away. No one has demonstrated significant, reproducible transport J_c in a length of conductor. Although breakthroughs in coated Y–Ba–Cu–O conductor technology may be applicable to the $Hg_1Ba_2Ca_{n-1}Cu_nO_{2n+2}$ compounds, it is difficult to envision applications before short-sample conductor properties are demonstrated.

B2.2.4 B3.2.5 F A1.3 C1

References

[1] Putilin S N, Antipov E V, Chmaissem O and Marezio M 1993 Superconductivity at 94 K in $HgBa_2CuO_{4+\delta}$ *Nature* **362** 226
[2] Putilin S N, Bryntse I and Antipov E V 1991 *Mat. Res. Bull.* **26** 1299
[3] Schilling A, Cantoni M, Guo J D and Ott H R 1993 Superconductivity above 130 K in the Hg–Ba–Ca–Cu–O system *Nature* **363** 56
[4] Putilin S N, Antipov E V and Marezio M 1993 Superconductivity above 120 K in $HgBa_2CaCu_2O_{6+\delta}$ *Physica* C **212** 266
[5] Parkin S S P, Lee V Y, Nazzal A I, Savoy R, Bayers R and La Placa S J 1988 $Tl_1Ca_{n-1}Cu_nO_{2n+3}$ ($n = 1, 2, 3$): a new class of crystal structures exhibiting volume superconductivity at up to ≈100 K *Phys. Rev. Lett.* **61** 750
[6] Sugise R, Hirabayashi M, Terada N, Jo M, Shimonura T and Ihara H 1988 The formation process of new high T_c superconductors with single layer thallium-oxide $TlBa_2Ca_2Cu_3O_y$ and $TlBa_2Ca_3Cu_4O_y$ *Japan. J. Appl. Phys.* **27** L1709
[7] Morosin A, Baughman R J, Ginley D S, Schirber J E and Venturini E L 1990 Structure studies on Tl–Ca–Ca–Cu–O high T_c superconductors: effects of cation disorder and oxygen vacancies *Physica* C **161** 115
[8] Shimoyama J, Hahakura S, Kitazawa K, Yamafuji K and Kishio K 1994 A new mercury-based superconductor: (Hg, Cr)Sr_2CuO_y *Physica* C **224** 1
[9] Singh K K, Kirtikar V, Sinha A P B and Morris D E 1994 $HgSr_2CuO_{4+\delta}$: a new 78 K superconductor by Mo substitution *Physica* C **231** 9
[10] Maignan A, Pelloquin D, Malo S, Michel C, Hervieu M and Raveau B 1995 The great ability of mercury-based cuprates to accommodate transition elements *Physica* C **243** 233
[11] Shi J B 1996 Superconductivity and magnetism of $(Hg_{0.7}Bi_{0.3})Sr_2(Ca_{1-x}R_x)Cu_2O_{6+\delta}$ system (R = Gd, Nd and Y) *Physica* C **270** 97

[12] Chmaissem O, Wessels L and Sheng Z Z 1994 Synthesis and characterization of (Hg,Bi)-based 1212-type cuprate superconductor $(Hg_{0.67}Bi_{0.33})Sr_2(Y_{0.67}Ca_{0.33})Cu_2O_{6+\delta}$ ($\delta = 0.68$) *Physica* C **228** 190

[13] Aranda M A C 1994 Crystal structures of copper-based high T_c superconductors *Adv. Mater.* **6** 905

[14] Chmaissem O, Guptasarma P, Welp U, Hinks D G and Jorgensen J D 1997 Effect of Re substitution on the defect structure and superconducting properties of $(Hg_{1-x}Re_x)$ $Ba_2Ca_{n-1}Cu_nO_{2n+2+\delta}$ ($n = 2, 3, 4$) *Physica* C **292** 305

[15] Chmaissem O, Jorgensen J D, Hinks D G, Storey B G, Dabrowski B, Zhang H and Marks L D 1997 Structure and superconductivity in Cr-substituted $HgBa_2CuO_{4+\delta}$ *Physica* C **279** 1

[16] Villard G, Pelloquin D and Maignan A 1998 Structure of new $Hg_{0.75}V_{0.25}Ba_2CuO_{4+\delta}$ superconducting single crystals: effect of overdoping on the magnetization second peak *Physica* C **307** 128

[17] Schwer H, Kopnin E, Molinski R, Jun J, Meijer G I, Conder K, Rossel C and Karpinski J 1997 Effect of Pb doping on the structure of $HgBa_2CuO_{4+\delta}$ single crystals *Physica* C **276** 281

[18] Pelloquin D, Hardy H, Maignan A and Raveau B 1997 Single crystals of the superconductor $(Hg,Cu)Ba_2CuO_{4+\delta}$: growth, structure and magnetism *Physica* C **273** 205

[19] Edwards P P, Peacock G B, Hodges J P, Asab A and Gameson I 1997 Mercurocuprates: the highest transition-temperature superconductors *High-T_c Superconductivity: Ten Years after the Discovery*, ed E Kaldis, E Liarokapais and K A Muller (Dordrecht: Kluwer) p 135

[20] Tokiwa-Yamamoto A, Isawa K, Itoh M, Adachi S and Yamauchi H 1993 Composition, crystal structure and superconducting properties of Hg–Ba–Cu–O and Hg–Ba–Ca–Cu–O superconductors *Physica* C **216** 250

[21] Eggert J H, Hu J Z, Mao H K, Beauvais L, Meng R L and Chu C W 1994 Compressibility of the $HgBa_2Ca_{n-1}Cu_nO_{2n+2+\delta}$ ($n = 1, 2, 3$) high-temperature superconductors *Phys. Rev.* B **49** 15299

[22] Aksenov V L, Balagurov A M, Savenko B N, Sheptyakov D V, Glazkov V P, Somenkov V A, Shilshtein S S, Antipov E V and Putilin S N 1997 Investigation of the $HgBa_2CuO_{4+\delta}$ structure under external pressures up to 5 GPa by neutron powder diffraction *Physica* C **275** 87

[23] Finger L W, Hazen R M, Meng R L and Chu C W 1994 Crystal chemistry of $HgBa_2CaCu_2O_{6+\delta}$: single-crystal x-ray diffraction results *Physica* C **226** 216

[24] Bandyopadhyay B, Mandal J B and Ghosh B 1998 Effect of Tl doping on the structural and physical properties of $HgBa_2CuO_{4+\delta}$ superconductor *Physica* C **298** 95

[25] Chmaissem O, Huang Q, Antipov E V, Putilin S N, Marezio M, Loureiro S M, Capponi J J, Tholence J L and Santoro A 1993 Neutron powder diffraction study at room temperature and at 10 K of the crystal structure of the 133 K superconductor $HgBa_2Ca_2Cu_3O_{8+\delta}$ *Physica* C **217** 265

[26] Pissas M, Billon B, Charalambous M, Chaussy J, LeFloch S, Bordet P and Capponi J J 1997 Single-crystal growth and characterization of the superconductor $HgBa_2CuO_{4+\delta}$ *Supercond. Sci. Technol.* **10** 598

[27] Antipov E V, Capponi J J, Challout C, Chmaissem O, Loureiro S M, Marezio M, Putilin S N, Santoro A and Tholence J L 1993 Synthesis and neutron powder diffraction study of the superconductor $HgBa_2CaCu_2O_{6+\delta}$ before and after heat treatment *Physica* C **218** 348

[28] Loureiro S M, Antipov E V, Tholence J L, Capponi J J, Chmaissem O, Huang Q and Marezio M 1993 Synthesis and structural characterization of the 127 K $HgBa_2CaCu_2O_{6.22}$ superconductor *Physica* C **217** 253

[29] van Tendeloo G, Chaillout C, Capponi J J, Marezio M and Antipov E V 1994 Atomic structure and defect structure of the superconducting $HgBa_2Ca_{n-1}Cu_nO_{2n+2+\delta}$ homologous series *Physica* C **223** 219

[30] Pelloquin D, Hardy V and Maignan A 1996 Synthesis and characterization of single crystals of the superconductors $Hg_{0.8}Bi_{0.2}Ba_2Ca_{n-1}Cu_nO_{2n+2+\delta}$ ($n = 2, 3$) *Phys. Rev.* B **54** 16246

[31] Gonzalez E J, Wong-Ng W, Piermarini G J, Wolters C and Schwartz J 1997 X-ray diffraction study of $HgBa_2CuO_{4+\delta}$ at high pressures *Powder Diffraction* **12** 106

[32] Hunter B A, Jorgensen J D, Wagner J L, Radaelli P G, Hinks P G, Shaked H, Hitterman R L and Von Dreele R B 1994 Pressure-induced structural changes in superconducting $HgBa_2Ca_{n-1}Cu_nO_{2n+2+\delta}$ ($n = 1, 2, 3$) compounds *Physica* C **221** 1

[33] Isawa K, Higuchi T, Machi T, Tokiwa-Yamamoto A, Adachi S, Murakami M and Yamauchi H 1994 Irreversibility line for a Pb-doped Hg–Ba–Ca–Cu–O superconductor *Appl. Phys. Lett.* **64** 1301

[34] Isawa K, Tokiwa-Yamamoto A, Itoh M, Adachi S and Yamauchi H 1993 The effect of Pb doping in $HgBa_2Ca_2Cu_3O_{8+\delta}$ superconductor *Physica* C **217** 11

[35] Meng R L, Hickey B R, Wang Y Q, Sun Y Y, Gao L, Xue Y Y and Chu C W 1996 Processing of highly oriented $(Hg_{1-x}Re_x)Ba_2Cu_3O_{8+\delta}$ tape with $x \sim 0.1$ *Appl. Phys. Lett.* **68** 3177

[36] Reder M, Krelaus J, Schmidt L, Heinemann K and Freyhardt H C 1998 Effects of Re-doping on microstructure and superconductivity of $HgBa_2Ca_2Cu_3O_{8+\delta}$ *Physica* C **306** 289

[37] Puzniak R, Karpinski J, Wisniewski A, Szymczak R, Angst M, Schwer H, Molinski R and Kopnin E M 1998 Influence of Re substitution on the flux pinning in $(Hg,Re)Ba_2Ca_2Cu_3O_{8+\delta}$ single crystals *Physica* C **309** 161

[38] Wolters C H, Amm K M, Sun Y R and Schwartz J 1996 Synthesis of $(Hg,Re)Ba_2Ca_{n-1}Cu_nO_y$ superconductors *Physica* C **267** 164

[39] Sastry P V P S S, Amm K M, Knoll D C, Peterson S C, Wolters C H and Schwartz J 1998 Synthesis of $(HgX)Ba_2Ca_2Cu_3O_x$ superconductors. *J. Supercond.* **11** 49

[40] Amm K M, Sastry P V P S S, Knoll D C, Peterson S C and Schwartz J 1998 Effects of an Au interface on (HgBi)$Ba_2Ca_2Cu_3O_x$ superconductor *J. Supercond.* **11** 75

[41] Sastry P V P S S, Amm K M, Knoll D C, Peterson S C and Schwartz J 1998 Synthesis and processing of $(Hg,Pb)_1Ba_2Ca_2Cu_3O_y$ superconductors *Physica* C **297** 223

[42] Sastry P V P S S, Amm K M, Knoll D C, Peterson S C and Schwartz J 1998 Synthesis and processing of Bi-doped $Hg_1Ba_2Ca_2Cu_3O_y$ superconductors *Physica* C **300** 125

[43] Sastry P V P S S and Schwartz J 1998 Synthesis and processing of doped $Hg_1Ba_2Ca_2Cu_3O_y$ superconductor *J. Supercond.* **11** 595

[44] Aytug T, Gapud A A, Yoo S H, Kang B W, Gapud S D and Wu J Z 1999 Effect of sodium doping on the oxygen distribution of Hg-1223 superconductors *Physica* C **313** 121

[45] El-Sayed A H, Bellingeri E, Calzona V, Cimberle M R, Eggenhöffner R, Ferdeghini C, Grasso G, Marrè D, Putti M, Siri A S, Costa G A, Kaiser E and Masini R 1994 Synthesis and properties of superconducting $HgBa_2CuO_{4+x}$ from a single-step low-temperature solid state reaction *Supercond. Sci. Technol.* **7** 36

[46] Du Z L, Fung W, Chow J C L, Luo Y Y and Li Q Y 1996 Fabrication of $HgBa_2CuO_x$ superconductor under atmospheric pressure *J. Supercond.* **9** 43

[47] Peacock G B, Fletcher A, Gameson I and Edwards P P 1998 Synthesis and superconductivity of highly underdoped $HgBa_2CuO_{4+\delta}$ *Physica* C **301** 1

[48] Xiong Q, Cao Y, Chen F, Xue Y Y and Chu C W 1994 Annealing temperature and O_2 partial pressure dependence of T_c in $HgBa_2CuO_{4+\delta}$ *J. Appl. Phys.* **76** 7127

[49] Xiong Q, Xue Y Y, Chen F, Cao Y, Sun Y Y, Liu L M, Jacobson A J and Chu C W 1994 Oxygen annealing and the related thermodynamics in $HgBa_2CuO_{4+\delta}$ *Physica* C **231** 233

[50] Alyoshin V A, Mikhailova D A and Antipov E V 1996 Synthesis of $HgBa_2CuO_{4+\delta}$ under controlled mercury and oxygen pressures *Physica* C **271** 197

[51] Michel C, Hervieu M, Maignan A, Pelloquin D, Badri V and Raveau B 1995 Stabilization of mercury cuprates by bismuth superconductors $Hg_{1-x}Bi_xBa_2Ca_{m-1}Cu_mO_{2m+2+\delta}$ *Physica* C **241** 1

[52] Tampieri A, Calestani G, Celotti G, Micheletti C and Rinaldi D 1998 Preparation of Hg1201 superconductor by hot isostatic pressing *Physica* C **298** 10

[53] Su J H, Sastry P V P S S and Schwartz J 2001 Synthesis and characterization of $(Hg_{0.8}Re_{0.2})$ $Ba_2CaCu_2O_{6+\delta}$ thick films on Ag obtained by a two-step dip-coating/rolling method *Physica* C **361** 292

[54] Maignan A, Pelloquin D, Hervieu M, Michel C and Raveau B 1995 New superconducting vanadomercury cuprates $Hg_{0.8}V_{0.2}Ba_2Ca_{m-1}Cu_mO_{2m+2+\delta}$ *Physica* C **243** 214

[55] Loureiro S M, Antipov E V, Tholence J L, Capponi J J, Chmaissem O, Huang Q and Marezio M 1993 Synthesis and structural characterization of the 127 K $HgBa_2CaCu_2O_{6.22}$ superconductor *Physica* C **217** 253

[56] Xu Q, Tang T B and Chen Z 1994 The synthesis and thermal stability of superconducting $HgBa_2CaCu_2O_{6+\delta}$ *Supercond. Sci. Technol.* **7** 828

[57] Li J Q, Lam C C, Feng J and Hung K C 1998 Effects of in doping in $Hg_{1-x}In_xBa_2CaCu_2O_{6+\delta}$ *Supercond. Sci. Technol.* **11** 217

[58] Gasser C, Moriwaki Y, Sugano T, Nakanishi K, Wu X J, Adachi S and Tanabe K 1998 Orientation control of ex situ $(Hg_{1-x}Re_x)$ $Ba_2CaCu_2O_y$ ($x \approx 0.1$) thin films on $LaAlO_3$ *Appl. Phys. Lett.* **72** 972

[59] Higuma H, Miyashita S and Uchikawa F 1994 Synthesis of superconducting Pb-doped $HgBa_2CaCu_2O_y$ films by laser ablation and post-annealing *Appl. Phys. Lett.* **65** 743

[60] Sastry P V P S S and Schwartz J 1999 Synthesis and stability of HgRe1212 and HgRe1223 superconductors *IEEE Trans. Appl. Supercond.* **9** 1684

[61] Paranthaman M 1994 Single step synthesis of bulk $HgBa_2Ca_2Cu_3O_{8+\delta}$ superconductors *Physica* C **222** 7

[62] Huang Z J, Meng R L, Qiu X D, Sun Y Y, Kulik J, Xue Y Y and Chu C W 1993 Superconductivity, structure and resistivity in $HgBa_2Ca_2Cu_3O_{8+\delta}$ *Physica* C **217** 1

[63] Meng R L, Beauvais L, Zhang X N, Huang Z J, Sun Y Y, Xue Y Y and Chu C W 1993 Synthesis of the high-temperature superconductors $HgBa_2Ca_2Cu_2O_{6+\delta}$ *Physica* C **216** 21

[64] Meng R L, Hickey B R, Sun Y Y, Cao Y, Kinalidis C, Meen J, Xue Y Y and Chu C W 1996 Formation of $HgBa_2Ca_2Cu_3O_{8+\delta}$ with additives under ambient conditions *Physica* C **260** 1

[65] Goto T, Watanabe K and Awaji S 1996 Field dependence of J_c for F-doped Hg1223 filament *Japan. J. Appl. Phys.* **35** L1404

[66] Tsabba Y and Reich S 1995 Superconducting Hg-1223 films obtained by a sol–gel process *Physica* C **254** 21

[67] Tachikawa K, Kikuchi A, Kinoschita T and Komiya S 1995 Critical current in Tl-base high T_c oxides synthesized through a diffusion process *IEEE Trans. Appl. Supercond.* **5** 2019

[68] Hamdan N M, Ziq Kh A, Al-Harthi A S and Shirokoff J 1998 The effect of fluorine on the phase formation and properties of Tl-based superconductors *J. Supercond.* **11** 95

[69] Colson D, Viallet V, Forget A, Poissonnet S, Schmirgeld-Mignot L, Marucco J F and Bertinotti A 1998 Gold substitution in $HgBa_2Ca_2Cu_3O_{8+\delta}$ single crystals *Physica* C **295** 186

[70] Hung K C, Jin X, Lam C C, Geng J F, Chen W M and Shao H M 1997 Thermally assisted flux flow in $Hg_{0.69}Pb_{0.31}Ba_2Ca_3Cu_4O_{10+\delta}$ superconductor *Supercond. Sci. Technol.* **10** 562

[71] Li J Q, Lam C C, Feng J and Hung K C 1997 Effects of Sn doping in $Hg_{1-x}Sn_xBa_2Ca_2Cu_3O_{8+\delta}$ superconductors *Physica* C **292** 295

[72] Li J Q, Lam C C, Fu E C L and Feng J 1998 Thermal stabilization of impurity $HgCaO_2$ and superconducting phase $Hg(Sn)Ba_2Ca_2Cu_3O_{8+\delta}$ in $Hg_{0.9}Sn_{0.1}Ba_2Ca_2Cu_3O_{8+\delta}$ superconductor *Supercond. Sci. Technol.* **11** 603

[73] Balchev N, Van Allemeersch F, Persyn F, Schroeder J, Deltour R and Hoste S 1997 The effect of Sn substitution in the $(Hg_{1-x}Sn_x)Ba_2Ca_2Cu_3O_y$ superconducting system *Supercond. Sci. Technol.* **10** 65

[74] Li J Q, Lam C C, Hung K C and Shen L J 1998 Enhancement of critical current density in $HgBa_2Ca_2Cu_3O_{8+\delta}$ superconductor doped by Sb *Physica* C **304** 133

[75] Higuma H, Miyashita S and Wakata M 1997 High pressure synthesis and magnetic properties for $Hg_{1-x}Re_xBa_2Ca_2Cu_3O_y$ *Physica* C **291** 302

[76] Lin C T, Yan Y, Peters K, Schönherr E and Cardona M 1998 Flux growth of $Hg_{1-x}Re_xBa_2Ca_{n-1}Cu_nO_{2n+2+\delta}$ single crystals by self-atmosphere *Physica* C **300** 141

[77] Sin A, Cunha A G, Calleja A, Orlando M T D, Emmerich F G, Baggio-Saitovich E, Piñol S, Chimenos J M and Obradors X 1998 Formation and stability of $HgCaO_2$, a competing phase in the synthesis of $Hg_{1-x}Re_xBa_2Ca_2Cu_3O_{8+\delta}$ superconductor *Physica* C **306** 34

[78] Yamasaki H, Nakagawa Y, Mawatari Y and Cao B 1997 Preparation and magnetic properties of Re-doped $HgBa_2Ca_2Cu_3O_{8+\delta}$ superconductors *Physica* C **274** 213

[79] Isawa K, Tokiwa-Yamamoto A, Itoh M, Adachi S and Yamauchi H 1994 Pb-doping effect on irreversibility fields of $HgBa_2Ca_2Cu_3O_{8+\delta}$ superconductors *Appl. Phys. Lett.* **65** 2105

[80] Plesch G, Chromik S, Strbik V, Mair M, Gritzner G, Benacka S, Sargankova I and Buckuliakova A 1998 Thin (Hg, Pb)$Ba_2CaCu_2O_y$ films prepared from thermally evaporated precursors by post annealing in Hg-atmosphere *Physica* C **307** 74

[81] Kumari S, Singh A K and Srivastava O N 1997 On the synthesis and characterization of superconducting $Hg_{1-x}Pb_xBa_2Ca_2Cu_3O_{8+\delta}$ films prepared through spray pyrolysis *Supercond. Sci. Technol.* **10** 235

[82] Sargánkova I, Diko P, Kavecansky V, Kovác J, Konig W, Mair M, Gritzner G and Longauer S 1997 Influence of the variation of the Hg and Pb stoichiometry on the microstructure and T_c of $Hg_xPb_yBa_2Ca_2Cu_3O_{8+\delta}$ *Superlattices and Microstruct.* **21** 1

[83] Li Y, Sastry P V P S S, Knoll D C, Peterson S C and Schwartz J 1999 Synthesis of HgPb1223 superconductor *IEEE Trans. Appl. Supercond.* **9** 1767

[84] Gapud A A, Aytug T, Yoo S H, Xie Y Y, Kang B W, Gapud S D, Wu J Z, Wu S W, Liang W Y, Cui X T, Liu J R and Chu W K 1998 Lithium-doping-assisted growth of $HgBa_2Ca_2Cu_3O_{8+\delta}$ superconducting phase in bulks and thin films *Physica* C **308** 264

[85] Amm K M, Sastry P V P S S, Knoll D C, Peterson S C and Schwartz J 1998 Effects of an Au interface on (HgBi)$Ba_2Ca_2Cu_3O_x$ superconductor *J. Supercond.* **11** 75

[86] Amm K M, Sastry P V P S S, Knoll D C, Peterson S C and Schwartz J 1998 The influence of metallic interfaces on the properties of (Hg,Bi)$Ba_2Ca_2Cu_3O_y$ superconductors *Supercond. Sci. Technol.* **11** 793

[87] Xu Q L, Foong F, Liou S H, Cao L Z and Zhang Y H 1997 Synthesis and thermal stabilization of nearly single-phase superconductor $HgBa_2Ca_2Cu_3O_{8+\delta}$ by Tl substitution *Supercond. Sci. Technol.* **10** 218

[88] Tatsuki T, Tokiwa-Yamamoto A, Tamura T, Adachi S and Tanabe K 1997 Annealing study on $(Hg, Tl)_2Ba_2Ca_{n-1}Cu_nO_y$ (n = 1–5) superconductors *Physica* C **278** 160

[89] Singh H K, Saxena A K and Srivastava O N 1997 Effect of Ag doping on the transition temperature and critical current density of $Hg_{1-x}Tl_xBa_2Ca_2Cu_3O_{8+\delta}$ films fabricated through spray pyrolysis *Physica* C **273** 181

[90] Asthana A and Srivastava O N 1998 Fabrication and characterization of Hg(Tl) HTSC tapes *Supercond. Sci. Technol.* **11** 244

[91] Pandey A K, Verma G D and Srivastava O N 1998 Investigations on the Tl-doped Hg–Ba–Ca–Cu–O high temperature superconductors in regard to hole doping and microstructural characteristics *Physica* C **306** 47

[92] Titova S, Bryntse I, Irvine J, Mitchell B and Balakirev V 1998 Structural anomalies of 1223 Hg(Tl)–Ba–Ca–Cu–O superconductors in the temperature range 100–300 K *J. Supercond.* **11** 471

[93] Scott B A, Suard E Y, Tsuei C C, Mitzi D B, McGuire T R, Chen B H and Walker D 1994 Layer dependence of the superconducting transition temperature of $HgBa_2Ca_{n-1}Cu_nO_{2n+2+\delta}$ *Physica* C **230** 239

[94] Abílio C C, Loureiro S M, Capponi J J and Godinho M 1995 Magnetic properties of the $HgBa_2Ca_3Cu_4O_{10+\delta}$ superconducting phase *Physica* C **245** 1

[95] Lokshin K A, Pavlov D A, Kovba M L, Antipov E V, Kuzemskaya I G, Kulikova L F, Davydov V V, Morozov I V and Itskevich E S 1998 Synthesis and characterization of overdoped Hg-1234 and Hg-1245 phases; the universal behaviour of T_c variation in the $HgBa_2Ca_{n-1}Cu_nO_{2n+2+\delta}$ series *Physica* C **300** 71

[96] Krusin-Elbaum L, Tsuei C C and Gupta A 1995 High current densities above 100 K in the high-temperature superconductor $HgBa_2CaCu_2O_{6+\delta}$ *Nature* **373** 679

[97] Adachi H, Satoh T and Setsune K 1993 Highly oriented Hg–Ba–Ca–Cu–O superconducting thin films *Appl. Phys. Lett.* **63** 3628

[98] Tsuei C C, Gupta A, Trafas G and Mitzi D 1994 Superconducting mercury-based cuprate films with a zero-resistance transition temperature of 124 kelvin *Science* **263** 1259

[99] Gupta A, Sun J Z and Tsuei C C 1994 Mercury-based cuprate high-transition temperature grain-boundary junctions and SQUIDS operating above 110 kelvin *Science* **265** 1075

[100] Miyashita S, Higuma H and Uchikawa F 1994 Structure and superconducting properties of $HgBa_2CaCu_2O_y$ films prepared by laser ablation *Japan. J. Appl. Phys.* **33** 931

[101] Higuma H, Miyashita S and Uchikawa F 1994 Synthesis of superconducting Pb-doped $HgBa_2CaCu_2O_y$ films by laser ablation and post-annealing *Appl. Phys. Lett.* **65** 743

[102] Brazdeikis A, Flodstrom A S and Bryntse I 1996 Effect of thallium oxide, Tl_2O_3 on the formation of superconducting HgBaCaCuO films *Physica* C **265** 1

[103] Wu J Z, Yun S H, Gapud A, Kang B W, Kang W N, Tidrow S C, Monahan T P, Cui X T and Chu W K 1997 Epitaxial growth of $HgBa_2CaCu_2O_{6+\delta}$ thin films on $SrTiO_3$ substrates *Physica* C **277** 219

[104] Kang B W, Gapud A A, Fei X, Aytug T and Wu J X 1998 Minimization of detrimental effect of air in $HgBa_2CaCu_2O_{6+\delta}$ thin film processing *Appl. Phys. Lett.* **72** 1766

[105] Gasser C, Moriwaki Y, Sugano T, Nakanishi K, Wu X J, Adachi S and Tanabe K 1998 Orientation control of *ex situ* $(Hg_{1-x}Re_x)$ $Ba_2CaCu_2O_y$ $(x \approx 0.1)$ thin films on $LaAlO_3$ *Appl. Phys. Lett.* **72** 972

[106] Guo J D, Xiong G C, Yu D P, Feng Q R, Xu X L, Lian G J and Hu Z H 1997 Preparation of superconducting $HgBa_2CaCu_2O_x$ films with a zero-resistance transition temperature of 121 K *Physica* C **276** 277

[107] Guo J D, Xiong G C, Yu D P, Feng Q R, Xu X L, Lian G J, Xiu K and Hu Z H 1997 Preparation of superconducting $HgBa_2CaCu_2O_x$ and $Hg_{0.8}Bi_{0.2}Ba_2CaCu_2O_x$ films by means of annealing of mercury-free precursor films *Physica* C **282–287** 645

[108] Sun Y, Guo J D, Xu X L, Lian G J, Wang Y Z and Xiong G C 1999 Superconducting $HgBa_2CaCu_2O_y$ thin films growth on $NdGaO_3$, $SrTiO_2$, $LaAlO_3$ and Y-ZrO_2 substrates *Physica* C **312** 197

[109] Wu J Z, Yan S L and Xie Y Y 1999 Cation exchange: a scheme for synthesis of mercury-based high-temperature superconducting epitaxial thin films *Appl. Phys. Lett.* **74** 1469

[110] Moriwaki Y, Sugano T, Adachi S and Tanabe K 1998 Transport properties of (Hg, M)-12($n-1$) n (M = Re, Mo; n = 2,3) superconducting thin films *IEEE Trans. Appl. Supercond.* **9** 2390

[111] Tsukamoto A, Takagi K, Moriwaki Y, Sugano T, Adachi S and Tanabe K 1998 High-performance (Hg, Re)$Ba_2CaCu_2O_y$ grain-boundary Josephson junctions and dc superconducting quantum interference devices *Appl. Phys. Lett.* **73** 990

[112] Wu X S, Shao H M, Yao X X, Jiang S S, Wang D W, Wu Z H, Cai Y M, Chen L J and Wu Z 1996 Synthesis of the superconducting thin film of $HgBa_2Ca_2Cu_3O_{8+\delta}$ *Appl. Phys. Lett.* **68** 1723

[113] Wu X S, Shao H M, Jiang S S and Yao X X 1996 Synthesis and properties of the Hg1223 superconducting thin films prepared by ion implantation *Solid State Commun.* 10 **99** 733

[114] Tsabba Y and Reich S 1996 Giant mass anisotropy and high critical current in Hg1223 superconducting films *Physica* C **269** 1

[115] Reich S and Tsabba Y 1997 Growth mode of $HgBa_2Ca_2Cu_3O_{8+\delta}$ superconducting films prepared by a sol–gel method on a $Y_{0.15}Zr_{0.85}O_{1.93}$ substrate *Adv. Mater.* **9** 329

[116] Moriwaki Y, Sugano T, Gasser C, Nakanishi K, Adachi S and Tanabe K 1997 Highly c-axis oriented $HgBa_2Ca_2Cu_3O_y$ thick and thin films with $T_c > 130$ K *Physica* C **282–287** 643

[117] Yun S H, Wu J Z, Kang B W, Ray A N, Gapud A, Yang Y, Farr R, Sun G F, Yoo S H, Xin Y and He W S 1995 Fabrication of c-oriented $HgBa_2Ca_2Cu_3O_{8+\delta}$ superconducting thin films *Appl. Phys. Lett.* **67** 2866

[118] Yun S H and Wu J Z 1996 Superconductivity above 130 K in high-quality mercury-based cuprate thin films *Appl. Phys. Lett.* **68** 862

[119] Yun S H, Wu J Z, Tidrow S C and Eckart D W 1996 Growth of $HgBa_2Ca_2Cu_3O_{8+\delta}$ thin films on $LaAlO_3$ substrates using fast temperature ramping Hg-vapour annealing *Appl. Phys. Lett.* **68** 2565

[120] Moriwaki Y, Sugano T, Tsukamoto A, Gasser C, Nakanishi K, Adachi S and Tanabe K 1998 Fabrication and properties of c-axis Hg-1223 superconducting thin films *Physica* C **303** 65

[121] Kang W N, Meng R L and Chu C W 1998 Growth of $HgBa_2Ca_2Cu_3O_8$ thin films using stable $Re_{0.1}Ba_2Ca_2Cu_3O_x$ precursor by pulsed laser deposition *Appl. Phys. Lett.* **73** 381

[122] Kang W N, Lee S and Chu C W 1999 Oxygen annealing and superconductivity of $HgBa_2Ca_2Cu_3O_{8+y}$ thin films *Physica* C **315** 223

[123] Amm K M, Wolters C h, Knoll D C, Peterson S C and Schwartz J 1997 Growth of $Hg_{0.9}Re_{0.1}Ba_2Ca_2Cu_3O_{8+x}$ on a metallic substrate *IEEE Trans. Appl. Supercond.* **7** 1973

[124] Lechter W, Toth L, Osofsky M, Skelton E, Soulen R J Jr, Qadri S, Schwartz J, Wolters C h and Kessler J 1995 One step reaction and consolidation of Hg based high temperature superconductors by hot isostatic pressing *Physica* C **249** 213

[125] Schwartz J, Amm K M, Sun Y R and Wolters Ch 1996 HgBaCaCuO superconductors: processing, properties and potential *Physica* B **216** 261

[126] Peacock G B, Gameson I, Edwards P P, Khaliq M, Yang G, Shields T C and Abell J S 1997 Fabrication of high-temperature superconducting $HgBa_2CuO_{4+\delta}$ within silver-sheathed tapes *Physica* C **273** 193

[127] Yoo S H, Wong K W and Xin Y 1997 Thick film of $HgBa_2Ca_2Cu_3O_{8+\delta}$ via the sol–gel technique *Physica* C **273** 189

[128] Novikov D L and Freeman A J 1993 Electronic structure and Fermi surface of the $HgBa_2CuO_{4+\delta}$ superconductor *Physica* C **212** 233

[129] Chen J, Zasadzinski J F, Gray K E, Wagner J L and Hinks D G 1994 Point-contact tunneling study of $HgBa_2CuO_{4+\delta}$: BCS-like gap structure *Phys. Rev.* B **49** 3683

[130] Huang Z J, Xue Y Y, Meng R L, Qiu X D, Hao Z D and Chu C W 1994 Thermal fluctuation in high-temperature superconductor $HgBa_2CaCu_2O_{6+\delta}$ *Physica* C **228** 211

[131] Jeandupeux O, Schilling A and Ott H R 1993 Specific heat of superconducting $HgBa_2Ca_2Cu_3O_8$ between 65 and 200 K *Physica* C **216** 17

[132] Nakamae S, Crow J and Schwartz J 1999 Neutron irradiation effect on magnetization and thermal conductivity of $(Hg_{1-x}Bix)Ba_2Ca_2Cu_3O_y$ superconductor *IEEE Trans. Appl. Supercond.* **9** 2300

[133] Nakamae, S Study of high magnetic field and neutron irradiation effects on the thermal conductivity of high-T_c superconductors *PhD Thesis* Florida State University

[134] Zhuo Y, Choi J H, Kim M S, Lee S and Lee S I 1997 Equilibrium magnetic properties and vortex fluctuation behaviour of grain-aligned $Hg_{0.8}Pb_{0.2}Ba_{1.5}Sr_{0.5}Ca_2Cu_3O_y$ *Physica* C **282–287** 2003

[135] Kim Y C, Thompson J R, Ossandon J G, Christen D K and Paranthaman M 1995 Equilibrium superconducting properties of grain-aligned $HgBa_2Ca_2Cu_3O_{8+\delta}$ *Phys. Rev.* B **51** 11767

[136] Sun Y R and Schwartz J 1996 Anisotropy studies on aligned $HgBa_2CaCu_2O_{6+\delta}$ powder: confirmation of the collective pinning theory for anisotropic materials *Phys. Rev.* B **53** 5830

[137] Gjolmesli S, Fossheim K, Sun Y R and Schwartz J 1995 Logarithmic current density dependence on the activation barrier in superconducting $HgBa_2CaCu_2O_{6+x}$ *Phys. Rev.* B **52** 10447

[138] Sun Y R, Thompson J R, Schwartz J, Christen D K, Kim Y C and Paranthaman M 1995 Surface barrier in Hg-based polycrystalline superconductors *Phys. Rev.* B **51** 581

[139] Sun Y R, Amm K M and Schwartz J 1995 Flux pinning and magnetic anisotropy in neutron irradiated Hg–Ba–Ca–Cu–O *IEEE Trans. Appl. Supercond.* **5** 1870

[140] Krelaus J, Reder M, Hoffmann J and Freyhardt H C 1999 Magnetization and relaxation in Hg-1223: bulk vs. surface irreversibility, anisotropy and the influence of Re-doping *Physica* C **314** 81

[141] Welp U, Crabtree U W, Wagner J L and Hinks D G 1993 Flux pinning and the irreversibility lines in the $HgBa_2CuO_{4+\delta}$, $HgBa_2CaCu_2O_{6+\delta}$ and $HgBa_2Ca_2Cu_3O_{8+\delta}$ compounds *Physica* C **218** 373

[142] Huang Z J, Xue Y Y, Meng R L and Chu C W 1994 Irreversibility line of the $HgBa_2CaCu_2O_{6+\delta}$ high-temperature superconductors *Phys. Rev.* B **49** 4218

[143] Schilling A, Jeandupeux O, Guo J D and Ott H R 1993 Magnetization and resistivity study on the 130 K superconductor in the Hg–Ba–Ca–Cu–O system *Physica* C **216** 6

[144] Shimoyama J, Kishio K, Hahakura S, Kitazawa K, Yamaura K, Hiroi Z and Takano M 1995 Chemical stabilization and irreversible magnetic behaviour of $HgM_2Ca_{n-1}Cu_nO_y$ with M = Sr(Ba) and $n = 1$ to 3 *Advances in Superconductivity* **vol. 7**, ed K Yamafuji and T Morishita (Tokyo: Springer-Verlag) p 287

[145] Schwartz J, Nakamae S, Raban G W Jr, Heuer J K, Wu S, Wagner J L and Hinks D G 1993 Large critical current density in neutron-irradiated polycrystalline $HgBa_2CuO_{4+\delta}$ *Phys. Rev.* B **48** 9932

[146] Amm K M and Schwartz J 1995 Enhanced flux pinning in $HgBa_2CuO_{4+x}$ by neutron irradiation and its relationship to magnetic anisotropy *J. Appl. Phys.* **78** 2575

[147] Thompson J R, Ossandon J G, Krusin-Elbaum L, Song K J, Christen D K, Paranthaman M, Wu J Z and Ullmann J L 1997 J_c and vortex pinning enhancements in Bi-, Tl-, and Hg-based cuprate superconductors via GeV proton irradiation *Proc. 8th US–Japan Workshop on High-T_c Superconductors*, ed J Schwartz (National High Magnetic Field Laboratory, Tallahassee, FL) pp 146–152

[148] Krusin-Elbaum L, Petrov D K, Lopez D, Thompson J R, Wheeler R, Ullmann J, Chu C W and Lin Q M 1997 Enhanced supercurrents above 100 K in mercury cuprates via fission of mercury *Proc. 8th US–Japan Workshop on High-T_c Superconductors*, ed J Schwartz (National High Magnetic Field Laboratory, Tallahassee, FL) pp 158–165

[149] Chu C W, Gao L, Chen F, Huang Z J, Meng R L and Xue Y Y 1993 Superconductivity above 150 K in $HgBa_2Ca_2Cu_3O_{8+\delta}$ at high pressures *Nature* **365** 323

[150] Nuñez-Regueiro M, Tholence J L, Antipov E V, Capponi J J and Marezio M 1993 Pressure-induced enhancement of T_c above 150 K in Hg1223 *Science* **262** 97

[151] Takahashi H, Tokiwa-Yamamoto A, Môri N, Adachi S, Yamauchi H and Tanaka S 1993 Large enhancement of T_c in the 134 K superconductor $HgBa_2Ca_2Cu_3O_y$ under high pressure *Physica* C **218** 1

[152] Gao L, Chen F, Meng R L, Xue Y Y and Chu C W 1993 Superconductivity up to 147 K in $HgBa_2CaCu_2O_{6+\delta}$ under quasi-hydrostatic pressure *Phil. Mag. Lett.* **68** 345

[153] Gao L, Xue Y Y, Chen F, Xiong Q, Meng R L, Ramirez D, Chu C W, Eggert J H and Mao H K 1994 Superconductivity up to 164 K in $HgBa_2Ca_{m-1}Cu_mO_{2m+2+\delta}$ ($m = 1$, 2 and 3) under quasihydrostatic pressures *Phys. Rev.* B *Rapid Commun.* **50** 4260

C5
Magnesium diboride

Cristina Buzea and Tsutomu Yamashita

C5.1 Introduction

MgB_2 is an 'old' material, known since the early 1950s, but only recently (January 2001) discovered to be superconducting at a remarkably high critical temperature — about 40 K — for its simple hexagonal structure.

In the framework of BCS theory, the low mass elements result in higher frequency phonon modes that may lead to enhanced transition temperatures. The highest superconducting temperature is predicted for the lightest element, hydrogen, under high pressure. In 1986, investigations of the electrical resistance of Li under pressure up to 410 kbar showed a sudden electrical resistance drop at around 7 K between 220 and 230 kbar, suggesting a possible superconducting transition. Extremely pure beryllium superconducts at ordinary pressure with a T_c of 0.026 K. Its critical temperature can be increased to about 9–10 K for amorphous films. The recent discovery of superconductivity in MgB_2 confirms the predictions of higher T_c in compounds containing light elements, where it is believed that the metallic B layers play a crucial role in the superconductivity of MgB_2. A3.2

The discovery of superconductivity in MgB_2 certainly revived interest in the field of superconductivity, especially in the study of non-oxides, and initiated a search for superconductivity in related boron compounds. Because of its high critical temperature, higher T_c can be obtained in simple compounds. The superconductivity in MgB_2 was the catalyst for the discovery of several superconductors, some related to magnesium diboride, TaB_2 ($T_c = 9.5$ K), $BeB_{2.75}$ ($T_c = 0.7$ K), and graphite–sulphur composites ($T_c = 35$ K), and another not related, but 'inspired' by it, $MgCNi_3$ ($T_c = 8$ K). Probably the most impressive is the recent report of superconductivity under pressure in B, which has a very high critical temperature (11.2 K) for a simple element.

The critical temperature of MgB_2 (about 40 K) is close to or above the theoretical value predicted from the BCS theory. This may be a strong argument to consider MgB_2 as a non-conventional superconductor. A3.2

Since Akimitsu's group reported the superconductivity of MgB_2 [1], more than 260 studies on this superconductor have appeared. These studies have covered a wide array of subjects, including: preparation, the effect of substitution of various elements on T_c, isotope and Hall effect measurements, thermodynamic properties, critical current and field dependencies, and microwave and tunnelling properties. D3.1 A2.5, A2.7

Much effort has been expended on understanding the origin of superconductivity in this compound. Several theories have already been proposed. However, the superconductivity mechanism in MgB_2 is still to be decided. Recent calculations try to theoretically forecast the electronic properties of this material and similar compounds.

Why such a large interest in MgB_2 from the physics community? After all its critical temperature is only 40 K, more than three times lower than 134 K attained by the mercury-based high-T_c superconducting (HTSC) cuprates. Wires made of high-T_c copper oxides already operate above liquid nitrogen temperature (77 K). One important reason is the cost — HTSC wires are 70% silver, therefore, expensive. Unlike the cuprates, MgB_2 has lower anisotropy, larger coherence length, and transparency of the grain boundaries to current flow, which makes it a good candidate for applications. MgB_2 promises a higher operating temperature and higher device speed than the present electronics based on Nb. Moreover, high critical current densities (J_c) can be achieved in magnetic fields by oxygen alloying, and irradiation shows an increase of J_c values.

A1.3 E3 B2.6

According to initial findings, MgB_2 seemed to be a low-T_c superconductor with a remarkably high critical temperature, its properties resembling those of conventional superconductors rather than of high-T_c cuprates. These properties include isotope effect, a linear temperature-dependence of the upper critical field with a positive curvature near T_c (similar to borocarbides), and a shift to lower temperatures of both T_c (onset) and T_c (end) at increasing magnetic fields as observed in resistivity $R(T)$ measurements.

A3.1 D3.1

On the other hand, the quadratic T-dependence of the penetration depth, $\lambda(T)$ as well as the sign reversal of the Hall coefficient near T_c indicate unconventional superconductivity similar to cuprates. More attention should be paid to the layered structure of MgB_2, which may be the key to a higher T_c, as in cuprates and borocarbides.

D1.1.2

C5.2 MgB_2 and other diborides

MgB_2 possesses the simple hexagonal AlB_2-type structure (space group P6/mmm) [1, 2], which is common among borides. The MgB_2 structure is shown in figure C5.1. It contains graphite-type boron layers that are separated by hexagonal close-packed layers of magnesium. The magnesium atoms are located at the centre of hexagons formed by borons and donate their electrons to the boron planes. Similar to graphite, MgB_2 exhibits a strong anisotropy in the B–B lengths: the distance between the boron planes is significantly longer than in-plane B–B distance. Its transition temperature is almost twice as high as the highest T_c found in binary superconductors (Nb_3Ge with a $T_c = 23$ K).

D1.1.1

In comparison to other types of superconductors (figure C5.2), MgB_2 may be the 'ultimate' low-T_c superconductor with the highest critical temperature.

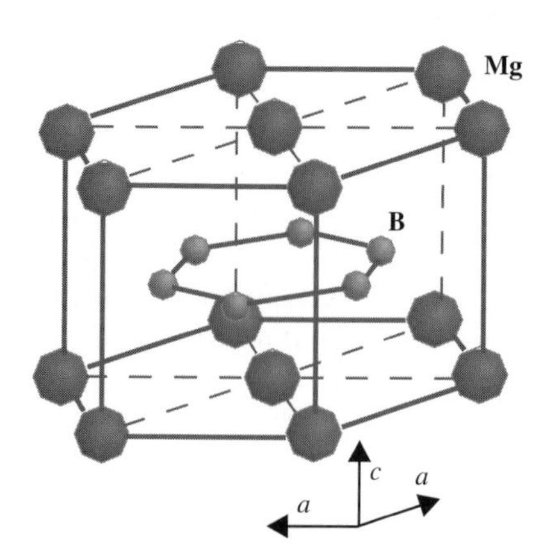

Figure C5.1. The structure of MgB_2 containing graphite-type B layers separated by hexagonal close-packed layers of Mg.

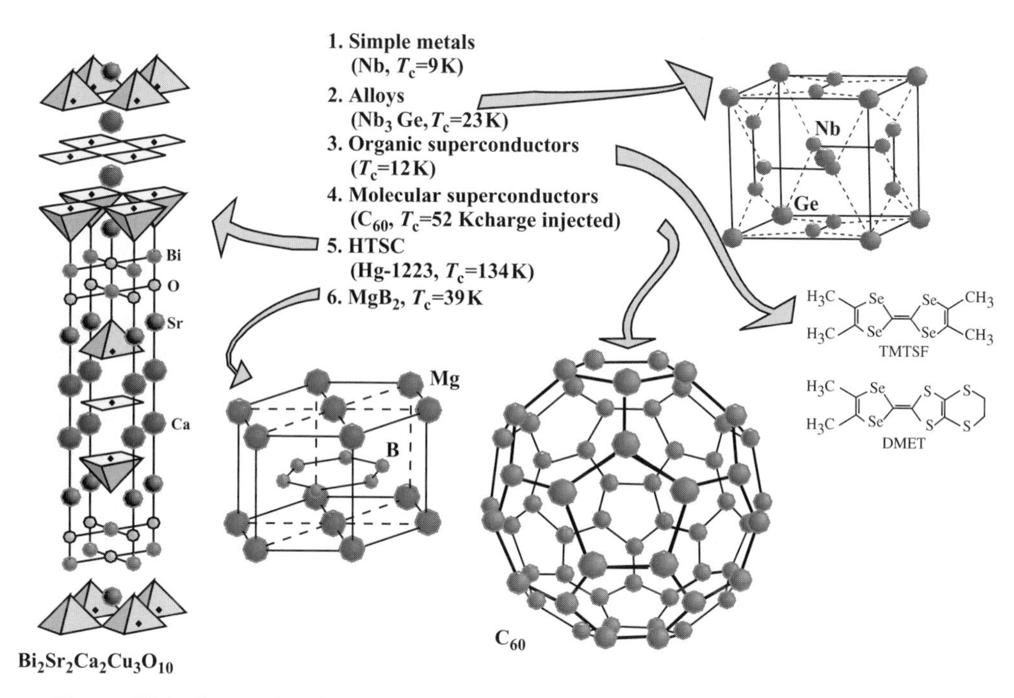

Figure C5.2. Comparison between the structures of different classes of superconductors.

After the announcement of MgB_2 superconductivity, it was hoped that this would be the first of a series of diborides with much higher T_c. However, to date MgB_2 has the highest T_c among borides, as can be seen in table C5.1 [3].

The search for superconductivity in borides dates from 1949 when Kiessling found a T_c of 4 K in TaB [4]. In the 1970s Cooper *et al* [5] and Leyarovska and Leyarovski [6] looked for superconductivity in various borides.

Since the discovery of superconductivity in MgB_2 [2], there have been several theoretical studies to search for potential high T_c binary and ternary borides in isoelectronic systems (BeB_2 and CaB_2), transition metal (TM) diborides (TMB_2), hole doped systems ($Mg_{1-x}Li_xB_2$, $Mg_{1-x}Na_xB_2$, and $Mg_{1-x}Cu_xB_2$), noble metal diborides (AgB_2 and AuB_2), CuB_2 and related compounds [7–14].

Also, there were further attempts to prepare new superconducting borides. The reports are still controversial, some authors reporting superconductivity in one compound and other authors finding the same material normal. This is the case of TaB_2, which was reported non-superconductive in earlier experiments [6] and was recently discovered to have a transition temperature of 9.5 K [15]. Similar situations apply for ZrB_2 (found non-superconductive by Kaczorowski *et al* [15] and superconducting at 5.5 K by Gasparov *et al* [16]) as well as BeB_2 (found non-superconductive in stoichiometric form [17] but superconductive at 0.7 K for the composition $BeB_{2.75}$ [18]).

The fact that some borides have been found superconducting by some authors while other authors found no traces of superconductivity in the same materials, suggests that non-stoichiometry may be an important factor in the superconductivity of this family. Extrapolating, in the case of MgB_2, it is also possible that the composition for which the critical temperature is a maximum to be slightly non-stoichiometric. The non-stoichiometry requirement for best superconducting properties is frequently seen in low-T_c as well as in high-T_c superconductors.

Table C5.1. List of diborides, their critical temperature (T_c) and structure type [3]

Compound	T_c(K)	Structure
MgB_2	40	AlB_2
NbB_2	0.62	AlB_2
$NbB_{2.5}$	6.4	AlB_2
$Nb_{0.95}Y_{0.05}B_{2.5}$	9.3	AlB_2
$Nb_{0.9}Th_{0.1}B_{2.5}$	7	AlB_2
MoB_2	–	AlB_2
$MoB_{2.5}$	8.1	AlB_2
$Mo_{0.9}Sc_{0.1}B_{2.5}$	9	AlB_2
$Mo_{0.95}Y_{0.05}B_{2.5}$	8.6	AlB_2
$Mo_{0.85}Zr_{0.15}B_{2.5}$	11.2	AlB_2
$Mo_{0.9}Hf_{0.1}B_{2.5}$	8.7	AlB_2
$Mo_{0.85}Nb_{0.15}B_{2.5}$	8.5	AlB_2
TaB_2	9.5	
BeB_2	–	AlB_2
$BeB_{2.75}$	0.7	not AlB_2
ZrB_2	5.5	
$ReB_{1.8-2}$	4.5–6.3	
TiB_2	–	
HfB_2	–	
VB_2	–	
CrB_2	–	

C5.3 Preparation

One of the advantages of MgB_2 fabrication is that magnesium diboride is already available from chemical suppliers, as it has been synthesized since the early 1950s. However, sometimes the quality of the MgB_2 powder, commercially available, is not as high as desirable. For example, MgB_2 commercial powders have a wider transition in the superconductive state and slightly lower T_c than the materials prepared in the laboratory from stoichiometric Mg and B powders.

B2.1

Figure C5.3 is a schematic picture of the fabrication methods used to date for MgB_2 thin films, powders, single crystals, wires and tapes.

B4.2.1

Typical methods of film fabrication used to date are: pulsed laser deposition (PLD), co-evaporation, deposition from suspension, Mg diffusion, and magnetron sputtering. Please note that some authors refer to their preparation method as PLD, but they use in fact the Mg diffusion method for B films prepared by PLD. Different substrates have already been used for the deposition of MgB_2 thin films: SiC, Si, $LaAlO_3$, $SrTiO_3$, MgO, Al_2O_3, and stainless steel (SS).

B2.2.4

In the case of film fabrication, Mg volatility reflects the need for unheated substrates and Mg enriched targets. Due to magnesium volatility, an essential problem is to establish the minimum deposition and growth temperature at which the film crystallizes into the hexagonal structure, but at which Mg is not lost from the film. A recent report has used thermodynamics to predict the conditions under which MgB_2 synthesis would be possible under vacuum conditions [19]. Important information on the thermal stability of MgB_2 can be found in an experimental study that measures the MgB_2 decomposition rate [20].

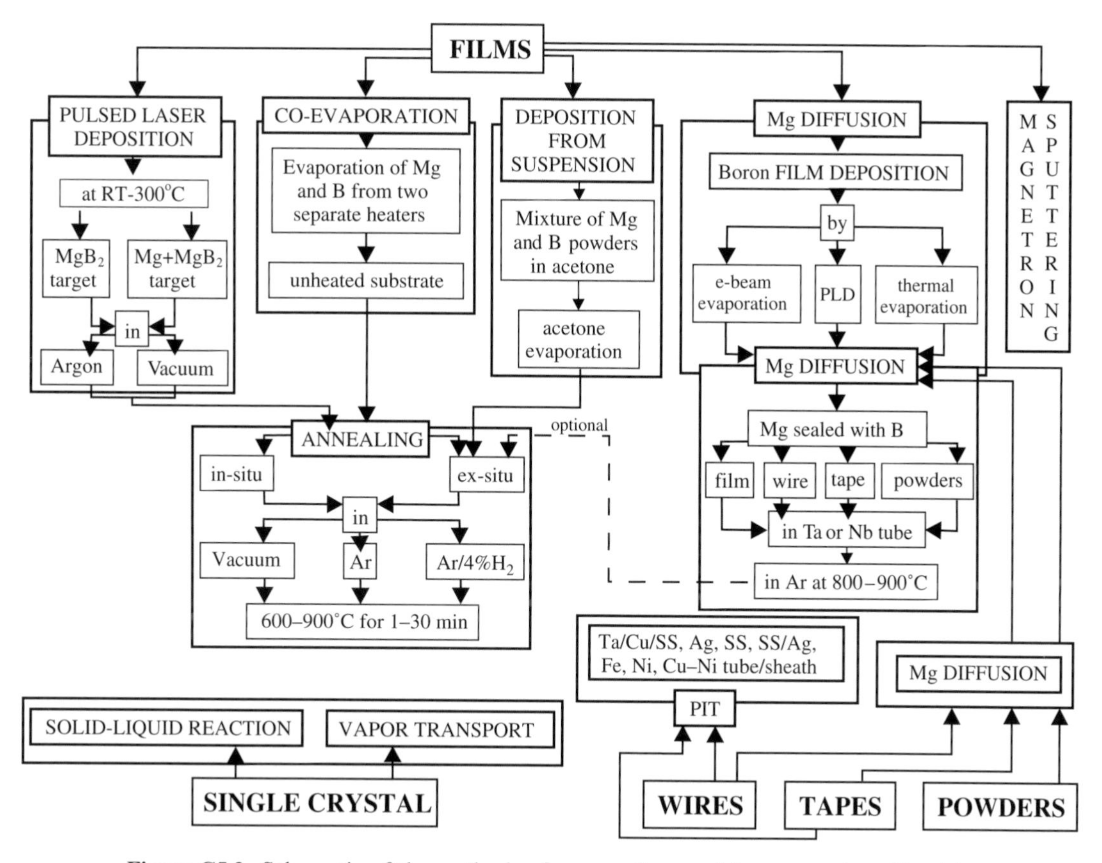

Figure C5.3. Schematic of the methods of preparation used for magnesium diboride.

Figure C5.4 shows the critical temperature of films prepared by different methods on different substrates: Al_2O_3, $SrTiO_3$, Si, SiC, MgO and SS. The reports using sapphire show the highest T_c and sharpest transitions by Mg diffusion method [21–26].

For the same substrate, Al_2O_3, the thin films prepared by PLD have lower T_c and usually wider transitions [21, 27–29] than the films prepared using a Mg diffusion method. In order to prepare better quality films by PLD, the fabrication procedure must be optimized. A recent report on PLD shows that the temperature of the film varies during PLD, the variation depending on the deposition parameters: substrate temperature, pressure in the ablation chamber, and deposition rate [30]. This may be an important factor because of Mg volatility. B4.2.1

In addition to sapphire, good quality films can be prepared on $SrTiO_3$ [31], Si [25], and SS [32]. However, the films prepared on SS have poor adhesion on the substrate [32]. B4.1

Figure C5.4 shows that the most important factor is the deposition method and not the type of substrate. The best method for MgB_2 thin film fabrication has proven to be Mg diffusion. The type of substrate is less important probably because the hexagonal structure of MgB_2 can accommodate substrates with different lattice parameters. However, we expect that further experiments will show a dependence of the critical temperature of MgB_2 on the type of substrate, as the critical temperature varies with the B–B bond length.

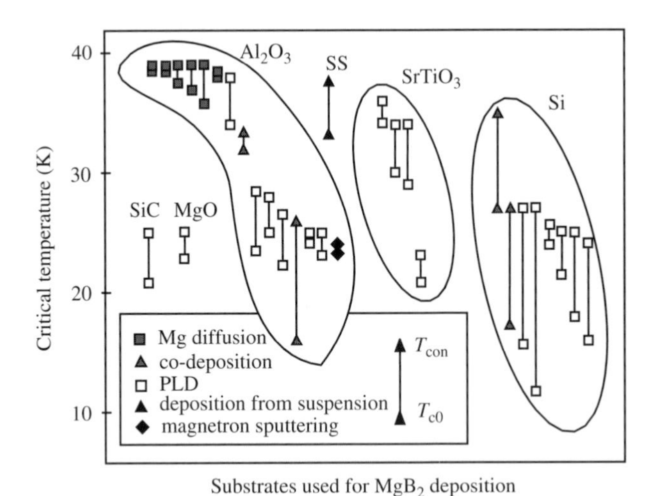

Figure C5.4. Critical temperature and critical temperature width for MgB_2 films deposited on different substrates.

For electronic applications, it is desirable that films with a high T_c of 39 K be made by a single step *in situ* process. Usually, the magnesium diboride films with high superconducting temperatures made to date are fabricated in a two-step process, film deposition followed by annealing.

B4

It is interesting to note that the Mg diffusion method is used not only for the fabrication of thin films, but also for the fabrication of bulk, powders, wires, and tapes. This method consists in Mg diffusion into B with different geometries. Due to the fact that Mg is highly volatile, Mg together with B are sealed in Nb or Ta tubes and heated up to 800–900°C. During this procedure, magnesium diffuses into the boron, increasing the size of the final reacted material.

B2.2.5

E1.1, E1.3.2

For practical applications of MgB_2 (such as magnets and cables), it is necessary to develop tapes and wires.

B3.3.6

Various research groups have already reported the fabrication of tapes and wires. Several critical issues relevant for practical fabrication of bulk wires remain unresolved. One of them is that MgB_2 is mechanically hard and brittle, therefore the drawing into fine-wire geometry is not possible. The wire and tape fabrication is achieved by two methods: Mg diffusion and powder-in-tube (PIT) method.

B2.2.5

Mg diffusion into B wires is a relatively easy method, which can rapidly convert already commercially existing B wires into superconductive MgB_2 wires [33, 34]. Attempts have been made to use magnesium diffusion for fabricating tapes [35].

B3.1

However, the PIT method is the most popular for achieving good quality wires [36–39] and tapes [40–45]. The PIT approach has been used to fabricate metal- clad MgB_2 wires/ribbons using various metals, such as: SS, Cu, Ag, Ag/SS, Ni, Cu–Ni, Nb, Ta/Cu/SS, and Fe. Usually, the PIT method consists of the following procedure. MgB_2 reacted powder or a mixture of Mg and B powders with stoichiometric composition is packed in various metal tubes or sheaths. These tubes are drawn into wires and cold-worked into ribbons. These steps are followed by an optional heat treatment at 900–1000°C.

For fabricating metal-clad MgB_2 wires/ribbons, hard but ductile and malleable metals are essential. These metals have to act as diffusion barriers for the volatile and reactive Mg. Also, it is important to find a suitable sheath material that does not degrade the superconductivity. Mg and MgB_2 tend to react and combine with many metals, such as Cu and Ag, forming solid solutions or intermetallics with low melting points, which render the metal cladding useless during sintering of MgB_2 at 900–1000°C. One can see that there are only a small number of metals that are not soluble or do not form intermetallic compounds with Mg [38]. These are Fe, Mo, Nb, V, Ta, Hf, and W. Of these, the refractory metals (Mo,

Nb, V, Ta, Hf, and W) have inferior ductility compared to Fe, which makes iron the best candidate material as a practical cladding metal or diffusion barrier for MgB_2 wire and tape fabrication which includes annealing.

B3.3.6

If the annealing process is skipped, more metals could be used as sheaths, their reactivity with Mg being on a secondary plane. A fabrication process with no heat treatment would also reduce the fabrication costs.

In order to improve the superconducting properties of bulk MgB_2, two methods have been used: hot deformation [46–49] and high-pressure sintering [50–53].

Single crystals are currently obtained by solid–liquid reaction method from Mg-rich precursor [54], under high pressure in Mg–B–N system [55], and by vapour transport method [56].

B2.4

C5.4 Hall coefficient

The existing reports on Hall effect [57–59] agree with the fact that the normal state Hall coefficient (R_H) is positive. The charge carriers in magnesium diboride are holes with a density at 300 K of between 1.7 and 2.8×10^{23} holes cm^{-3}, about two orders of magnitude higher than the charge carrier density for Nb_3Sn and YBCO.

D3.1

C1

C5.5 Pressure dependent properties

C5.5.1 Critical temperature versus pressure

The response of MgB_2 crystal structure to pressure is important for testing the predictions of competing theoretical models, but it also might give valuable clues for chemical substitutions. For example, in simple metal BSC-like superconductors, like aluminium, critical temperature (T_c) decreases under pressure due to the reduced electron–phonon coupling energy from lattice stiffening [60]. Also, a large magnitude of the pressure derivative dT_c/dP is a good indication that higher values of T_c may be obtained through chemical means.

A3.2

The pressure effect on the superconducting transition of MgB_2 is negative to the highest pressure studied. Figure C5.5 shows the evolution of the critical temperature with pressure from several references [61–70]. All reports agree with the fact that the critical temperature of MgB_2 is shifted to lower values, giving different rates of decrease $-dT_c/dP$.

T_c follows a quadratic or linear dependence on applied pressure, decreasing monotonically. Despite the fact that $T_c(P)$ data from different authors differ considerably in figure C5.5, one can notice a pattern. Samples with lower T_c at zero pressure have a much steeper $T_c(P)$ dependence than the samples with higher T_c. More exactly, the initial slope rate of the samples with lower T_c is about $-2\,K\,GPa^{-1}$, while that of samples with higher T_c is about $-0.2\,K\,GPa^{-1}$, as can be seen in figure C5.5 inset.

The initial rate of the derivative $-dT_c/dP$ is inversely proportional to pressure, most of the data falling in the quadratic dependence as shown in figure C5.5 inset in the shadowed region. Several data do not fit this dependence [66], but taking into account the solid pressure medium (steatite) they used, the quasi-hydrostatic nature of their experiment makes this explainable.

Also, from figure C5.5, it can be seen that samples with higher T_c have a negative curvature of $T_c(P)$ dependence, changing to positive for samples with lower T_c.

Taking into account the strong compressibility anisotropy, which will be described in the next paragraph, it is likely that shear stress of sufficient magnitude will cause important changes in the $T_c(P)$ dependence.

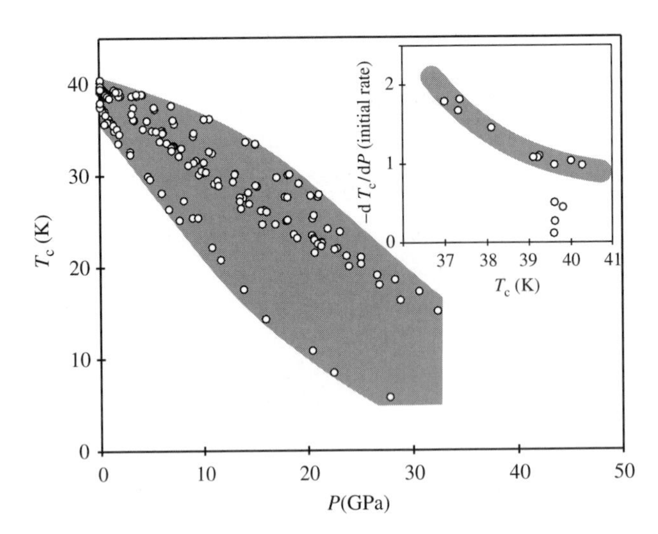

Figure C5.5. Critical temperature of MgB_2 versus applied pressure. Inset shows the initial rate of variation of the derivative *versus* T_c at zero pressure.

D3.3 Large shear stresses are generated by changing the pressure on a solid medium, such as steatite [70]. The shear stresses generated in cooling Fluorinert or other liquids with similar melting curves are much smaller, depending on the experimental procedure (cooling rate, change in applied force).

One report shows the existence of a cusp in the $T_c(P)$ dependence at about 9 GPa, the authors attributing it to a pressure-induced electronic transition [69]. However, the data from other reports do not show any cusp.

The discrepancy in the T_c dependence by various groups may arise partially due to different pressure transmitting media used in experiments, He, Fluorinert, methanol–ethanol, steatite, N, and NaF. Considering its anisotropic structure, MgB_2 may be sensitive to non-hydrostatic pressure components, which would explain the spread of dT_c/dP values reported in the literature.

More interestingly, several authors report different $T_c(P)$ dependencies for different MgB_2 samples measured in the same experimental set up [61, 66], which points towards the Mg non-stoichiometry as an important factor in determining the pressure dependent behaviour of the critical temperature.

The reduction of T_c under pressure is consistent with a BCS-type pairing interaction mediated by high-frequency boron phonon modes. This indicates that the reduction of the density of states at the Fermi energy, due to the contraction of B–B and B–Mg bonds, dominates the hardening phonon frequencies which could cause an increase in T_c as an external pressure is applied.

A hole-based theoretical scenario for explaining the superconductivity of MgB_2 predicted a positive pressure coefficient on T_c, as a result of decreasing in-plane B–B distance with increasing pressure [71–73]. This contradicts all experimental data. However, the situation is more complex if pressure also affects the charge transfer between Mg–B, resulting in different responses of the system in the underdoped and overdoped regimes.

C5.5.2 Anisotropic compressibility

D1.1.2 Diffraction studies at room temperature under pressure have been performed by a series of authors. Most of the reports study the lattice compression up to 6 GPa [63, 68, 74, 75], while there is a report which goes up to 30 GPa [61].

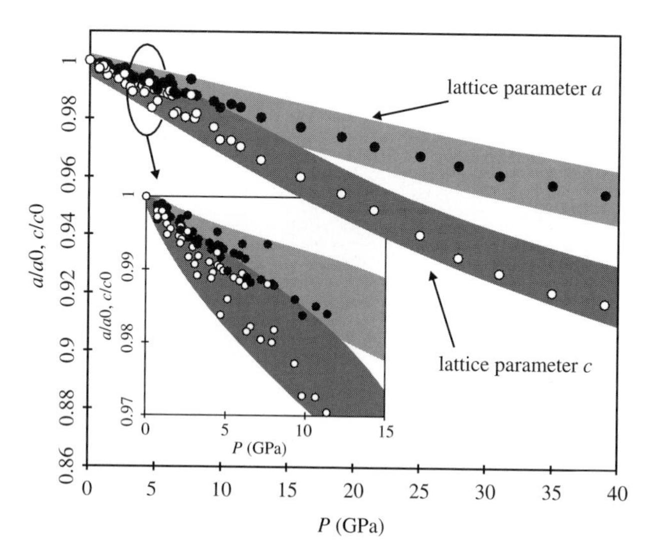

Figure C5.6. The normalized lattice parameters to the zero pressure value *versus* applied pressure. Inset shows the same data at lower pressures at an enlarged scale.

MgB_2 remains strictly hexagonal until the highest pressure, with no sign of a structural transition being seen. This is illustrated in the pressure variation of the normalized hexagonal lattice constants *a* and *c* from figure C5.6. There is a clear anisotropy in the bonding of the MgB_2 structure. All reports show that the lattice parameter along the *c*-axis decreases faster with pressure than along *a*-axis (figure C5.6), demonstrating that the out-of-plane Mg–B bonds are much weaker than in-plane Mg–Mg bonds. This fact is also emphasized by the lattice parameters variation *versus* temperature. The difference in compressibility values obtained in different reports may arise from the fact of using different pressure-transmitting media.

The compressibility anisotropy decreases linearly with pressure, as shown in figure C5.7.

From the critical temperature dependence on applied pressure corroborated with the data of compressibility, we calculated the dependence of T_c on the unit cell volume. We plotted $T_c(V)$ in figure C5.7 inset [61, 63, 68]. The large value of critical temperature variation with small modification in the unit cell volume demonstrates that Mg–B and B–B bonding distances are crucial in the superconductivity of MgB_2 at such a high T_c compared to other materials. The reduction of critical temperature by 1 K is achieved by lowering the unit cell volume only 0.17 Å^3. This implies a very sensitive dependence of the superconducting properties on the interatomic distances. D1.1.1

C5.6 Thermal expansion

Thermal expansion, analogous to compressibility, exhibits a pronounced anisotropy, with the *c*-axis responses substantially higher than *a*-axis, as shown in figure C5.8. The lattice parameter along the *c*-axis increases twice compared to the lattice parameter along *a*-axis at the same temperature [76]. This fact demonstrates that the out-of-plane Mg–B bonds are much weaker than in-plane Mg–Mg bonds. D3.2.3

Band structure calculations clearly reveal that, while strong B–B covalent bonding is retained, Mg is ionized and its two electrons are fully donated to the B-derived conduction band [77]. Then it may be assumed that the superconductivity in MgB_2 is essentially due to the metallic nature of the two-dimensional sheets of boron and the high vibrational frequencies of the light boron atoms which leads to the high T_c of this compound.

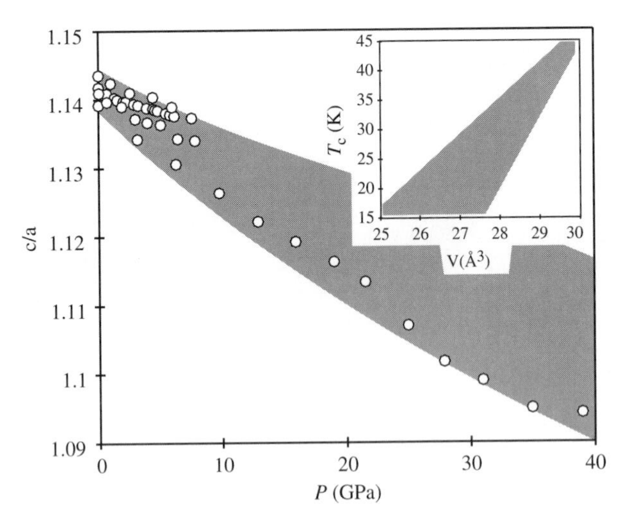

Figure C5.7. The ratio between the lattice parameters along c- and a-axes *versus* pressure. Inset shows the critical temperature of MgB_2 *versus* the volume of the unit cell.

C5.7 Effect of substitutions on critical temperature

The substitutions are important from several points of view. First, it may increase the critical temperature of one compound. Secondly, it may suggest the existence of a related compound with higher T_c. And last but not least, the doped elements that do not lower the T_c considerably may act as pinning centers and increase the critical current density.

A4.3

In the case of MgB_2, several substitutions have been tried up to date: carbon [12, 78–81]; aluminium [82–89]; lithium, silicon [83, 90]; beryllium [12, 17]; zinc [91, 92]; copper [12, 91]; manganese [88, 92]; niobium, titanium [88]; iron, cobalt, and nickel [92]. The critical temperature decreases at various rates for different substitutions, as can be seen in figure C5.9. The largest reduction is given by Mn [92],

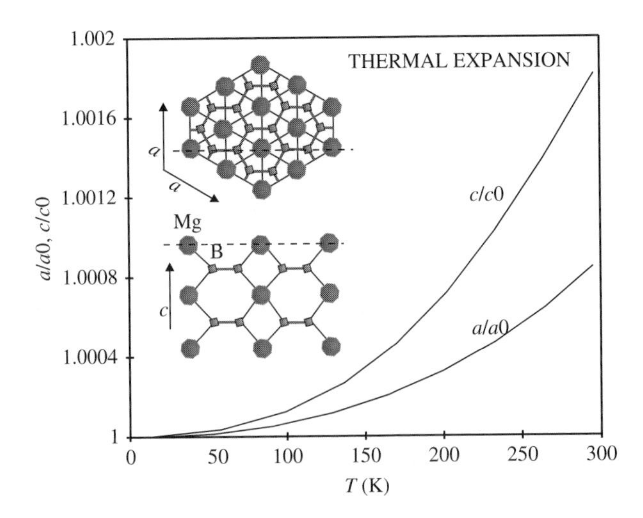

Figure C5.8. The normalized thermal expansion along a and c-axes. Inset shows the boron–boron and magnesium–boron bonds.

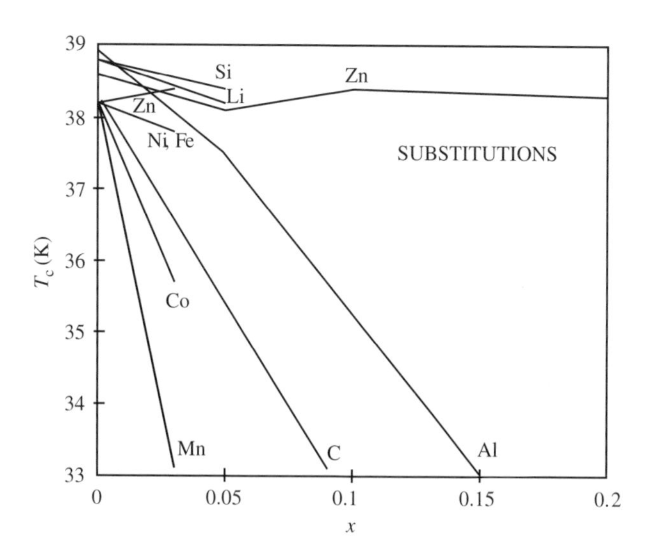

Figure C5.9. Critical temperature dependence on doping content x for substitutions with Zn, Si, Li, Ni, Fe, Al, C, Co, and Mn.

followed by Co [92], C [80], Al [84], and Ni and Fe [92]. The elements that do not reduce the critical temperature of MgB_2 considerably are Si and Li [83].

To date, all the substitutions alter the critical temperature of MgB_2 with an exception: Zn, which increases T_c slightly, by less than one degree [91, 92]. There are only two reports regarding Zn doping. Both agree with the fact that at a certain doping level T_c increases, but disagree with the doping level at which this occurs. This may be due to the incorporation of a smaller amount of Zn than the doping content. Anyway, Zn doping deserves further attention.

However, in order to have a clear picture about the effect of substitutions on MgB_2, more data on a wider range of doping levels are necessary.

C5.8 Total isotope effect

Figure C5.10 shows the critical temperature of MgB_2 with isotopic substitutions of Mg and B. The large value of the partial boron isotope exponent (α_B) of 0.26 [93] and 0.3 [94] shows that phonons associated with B vibration play a significant role in MgB_2 superconductivity. On the other hand, the magnesium isotope effect (α_{Mg}) is very small, 0.02 [94], as can be seen in figure C5.10 inset. This means that the vibrational frequencies of Mg have a low contribution to T_c. The B isotope substitution shifts T_c about 1 K, while the Mg isotope substitution changes T_c 10 times less. Overall, the presence of an isotope effect clearly indicates a phonon coupling contribution to T_c. The difference between the value of the total isotope effect $\alpha_T = \alpha_B + \alpha_{Mg} \approx 0.3$ in MgB_2 and the 0.5 BCS value may be related to the high T_c of this material. D2.4

C5.9 Testardi correlation between T_c and RR

Another proof in favour of a dominant phonon mechanism in MgB_2 superconductivity is the correlation between T_c and the ratio of resistivity at room temperature and near T_c, $\mathrm{RR} = R(300\,\mathrm{K})/R(T_c)$, also known as the Testardi correlation [95–98]. In 1975, Testardi showed that disorder decreases both the McMillan electron–phonon coupling constant (λ) and phonon-limited resistivity of normal transport D3.1

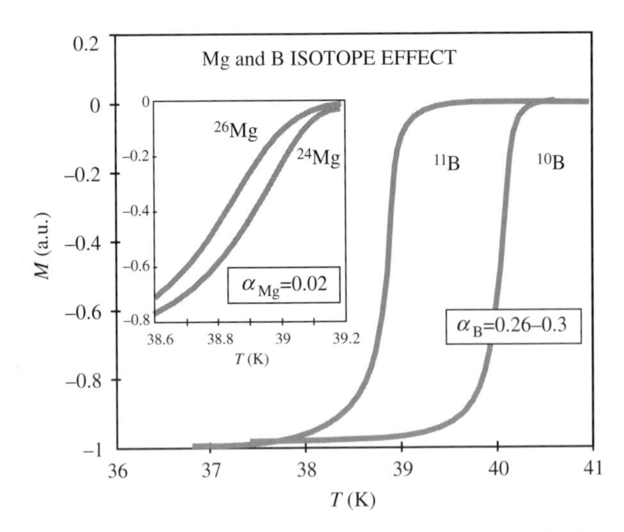

Figure C5.10. The relative magnetization *versus* temperature for B isotopically substituted samples. Inset shows Mg isotope effect.

phenomena (λ_{tr}), leading to the universal correlation between T_c and RR [95]. Decreasing T_c, no matter how it is achieved, is accompanied by the loss of thermal resistivity (electron–phonon interaction) [96]. D3.1 The Testardi correlation means that: samples with metallic behaviour will have higher T_c than samples with higher resistivity near T_c.

Figure C5.11 shows the critical temperature of zero resistivity normalized to the onset critical temperature versus the ratio of resistance at 300 K to the resistance near T_c, i.e., the Testardi correlation, for MgB_2 [3] and A15 compounds [96]. From figure C5.11, one notices that MgB_2 shows Testardi correlation between the critical temperature and the resistivity ratio in normal state and near T_c, being one more proof in favour of a phonon-mediated mechanism in the superconductivity of this compound.

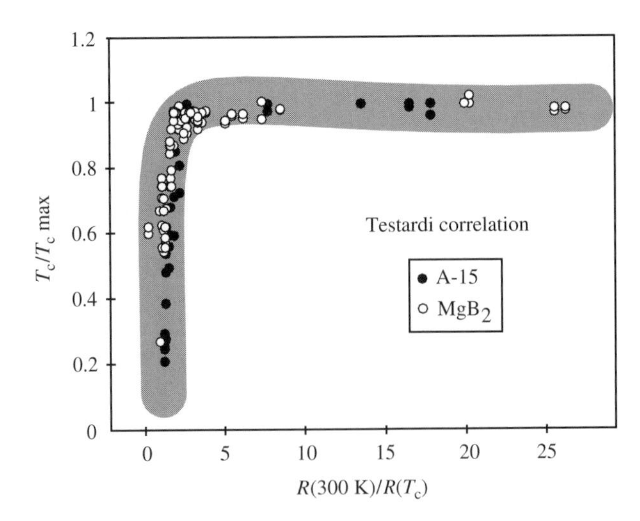

Figure C5.11. The correlation between the critical temperature of zero resistivity normalized to the onset critical temperature *versus* the ratio of resistance at 300 K to the resistance near T_c for MgB_2 and A15 compounds.

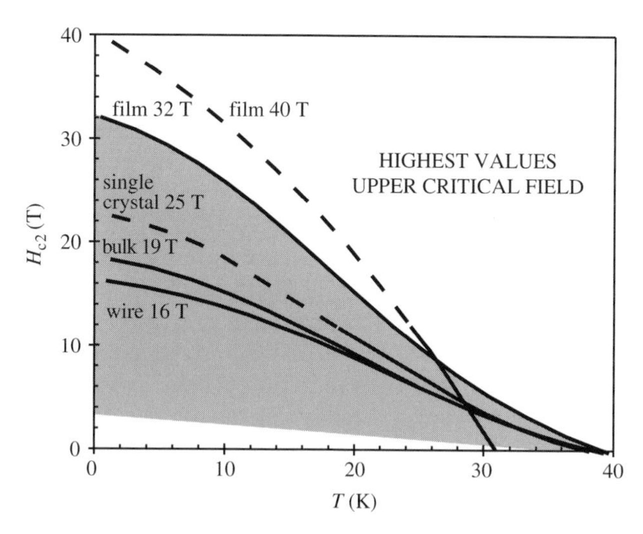

Figure C5.12. Highest values of $H_{c2}(T)$ for MgB_2 in different geometries (bulk, single crystals, wires, and films).

C5.10 Critical fields

C5.10.1 $H_{c2}(T)$ highest values

Measurements of the upper critical field in temperature show a wide range of values for the $H_{c2}(0)$, from 2.5 up to 32 T, as shown in figure C5.12. However, even higher upper critical fields (40 T) may be obtained for films with oxygen incorporated [99]. Unfortunately, due to oxygen alloying, these films have a lower T_c, of about 31 K. Although, shortening the coherence length of MgB_2 is the basis of improving high field performances, the ability to maintain high ξ is very advantageous for electronic applications. Understanding and controlling the superconducting properties of MgB_2 by alloying will be crucial in the future applications of this material. A3.1

Figure C5.12 shows the curves $H_{c2}(T)$ with the highest values at low temperatures for MgB_2 in different configurations. The highest values of the upper critical field are achieved for films. The films with the usual critical temperature of 39 K have upper critical fields of $H_{c2}(0) = 32$ T [100]. However, films with lower T_c can reach higher upper critical fields up to 40 T [99]. The second best values for the upper critical fields are attained by single crystals with $H_{c2}(0) = 25$ T [56], followed by bulk with $H_{c2}(0) = 19$ T [51, 101] and wires 16 T [102].

The $H_{c2}(T)$ dependence is linear in a large T range, saturating at low temperatures. A particular feature of $H_{c2}(T)$ curve for MgB_2 is the pronounced positive curvature near T_c, similar to the one observed in borocarbides YNi_2B_2C and $LuNi_2B_2C$, which are considered superconductors in the clean limit [103]. A3.1

C5.10.2 $H_{c2}(T)$ anisotropy

Anisotropy is very important both for basic understanding of this material and for practical applications, strongly affecting the pinning and critical currents. The degree of anisotropy in MgB_2 is still unresolved, reports giving values ranging between 1.1 and 9. A3.1

For textured bulk and partially oriented crystallites, the anisotropy ratio $\gamma = H_{c2}//ab/H_{c2}//c$ is reported to be between 1.1 and 1.7 [46, 104, 105]; for c-axis oriented films 1.2–2 [99, 100, 106]. Single

B2.4 crystals have slightly larger values than aligned powders or films, between 1.7–2.7 [53, 55, 56], and powders have unexpectedly large values, ranging from 5 to 9 [107, 108].

Generally, the anisotropy of one material can be estimated on aligned powders, epitaxial films, or/and single crystals. The method using aligned powders consists of mixing the superconducting powders with epoxy, followed by the alignment in magnetic fields made permanent by curing the epoxy. In order to give reliable results, the powders must consist of single crystalline grains with a considerable normal state magnetic anisotropy. This method regularly gives underestimates of anisotropy coefficient (γ) due to uncertainties in the degree of alignment. The *c*-axis oriented films may also have a certain degree of misorientation, therefore, the anisotropy coefficient will be smaller than the real value. Usually, the most reliable values are for single crystals.

Recently, Bud'ko *et al* [107] proposed a method of extracting the anisotropy parameter $\gamma = H_{c2}^{max}/H_{c2}^{min}$ from the magnetization $M(H, T)$ of randomly oriented powders. Their method is based on two features in $(\partial M/\partial T)_H$. The maximum upper critical field (H_{c2}^{max}) is associated with the onset of diamagnetism at T_c^{max} and the minimum upper critical field (H_{c2}^{min}) is associated with a kink $\partial M/\partial T$ at lower temperatures, T_c^{min}. In order to prove that this method is reliable, the anisotropy coefficient for $LuNi_2B_2C$ and $YNBi_2B_2C$ powders was measured. The data they obtained are in agreement with the previous values reported in the literature. For MgB_2 powders, a very large anisotropy factor $\gamma \approx 6$–7 was obtained [107].

Still, even higher H_{c2} anisotropy $\gamma = 6$–9 was inferred from conduction electron spin resonance measurements on high purity and high residual resistance samples [108].

A3.1 Figure C5.13 shows the anisotropic upper critical fields measured for single crystals [55, 56], and powders and wires [107] together with the data for bulk materials [46, 51, 101, 108–112]. The values for bulk are situated between the anisotropic upper critical field curves for *H*//*ab* and *H*//*c*. The anisotropic upper critical field value $H_{c2}//ab$ for both single crystals and powders is close to highest values for bulk, suggesting the upper limit of H_{c2} determined from anisotropy measurements may be close to the real value for MgB_2.

On the other side, the upper critical field for fields parallel to *c*-axis $H_{c2}//c$ inferred from non-aligned powders by the new method of Bud'ko *et al* [107] is much lower than the lowest values obtained for bulk

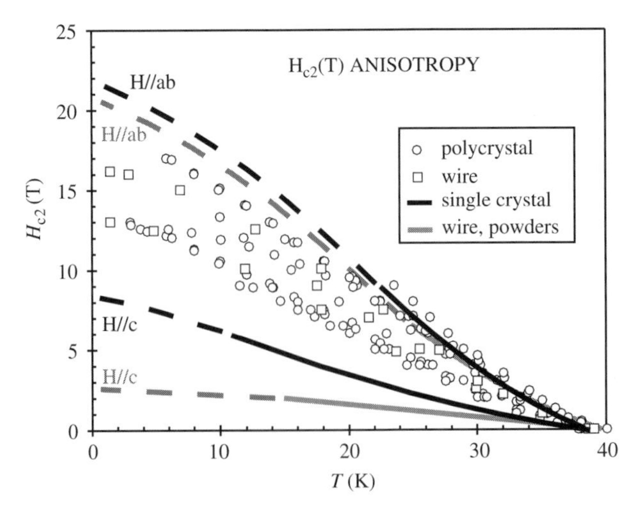

Figure C5.13. Upper critical field anisotropy *versus* temperature for MgB_2 single crystals, wires, and powders. Note that the $H_{c2}(T)$ data for MgB_2 bulk fall between the anisotropic dependencies of $H_{c2}(T)$ for *H*//*c* and *H*//*ab*.

Table C5.2. Anisotropy of the upper critical field and coherence lengths inferred from experiments on aligned powders, thin films, single crystals, and randomly aligned powders

Form	Reference	$H_{c2}//ab(0)$ [T]	$H_{c2}//c(0)$ [T]	ξ_{ab} (0) [nm]	$\xi_c(0)$ [nm]	γ
Textured bulk	[46]	12	11	5.5	5.0	1.1
Aligned crystallites	[104]	11	6.5	7.0	4.1	1.7
	[105]	12.5	7.8	6.5	4.0	1.6
Films	[100]	30	24	3.7	3.0	1.25
	[106]	26.4	14.6	4.7	2.6	1.8
	[99]	22.5	12.5	5.0	2.8	1.8
	[99]	24.1	12.7	5.0	2.6	1.9
	[99]	39	19.5	4.0	2.0	2
Single crystals	[53]	14.5	8.6	6.1	3.7	1.7
	[56]	25.5	9.2	6.5	2.5	2.6
	[55]					2.7
Powders	[107]	20	2.5	11.4	1.7	5–8
	[108]	16	2	12.8	1.6	6–9

and single crystals, implying an underestimation of $H_{c2}//c$. In order to determine with certainty the anisotropy of MgB_2, more experiments on larger single crystals are necessary. B2.4

C5.10.3 Coherence lengths

A comparison between the values of the coherence lengths, the anisotropy parameter γ, and upper critical field determined from experiments performed on aligned powders, thin films, single crystals, and randomly aligned powders can be seen in table C5.2. In order to deduce the values of the anisotropic coherence lengths from the upper critical fields, we used the anisotropic Ginzburg–Landau theory equations: for the magnetic field applied along the c-axis $H_{c2}//c = \phi_0/2\pi\xi_{ab}^2$, and for the magnetic field applied in the ab-plane $H_{c2}//ab = \phi_0/2\pi\xi_{ab}\xi_c$, where ϕ_0 is the flux quantum, and ξ_{ab} and ξ_c are the coherence lengths along the ab plane and the c-axis. The previous formulae are in CGS system. Overall, the coherence lengths values along the ab-plane range between $\xi_{ab}(0) = 3.7$ and 12.8 nm and along c-axis between $\xi_c(0) = 1.6$ and 5.0 nm. A2.3 A3.1

Probably, the most reliable data are for single crystals with $\xi_{ab}(0) = 6.1$–6.5 nm and $\xi_c(0) = 2.5$–3.7 nm.

C5.10.4 Lower critical field $H_{c1}(T)$

The lower critical field data *versus* temperature is shown in figure C5.14. Most of the values are situated between 25 and 48 mT. The data of anisotropic $H_{c1}//ab$ and $H_{c1}//c$ measured using single crystals [56] do not encompass the values for bulk [51, 113–115], suggesting that the data for single crystals are not accurate. The values of the penetration depth deduced from the lower critical field data range between 85 and 203 nm. A3.1

C5.10.5 Irreversibility field $H_{irr}(T)$

The knowledge of the irreversibility line is important in potential applications, as non-zero critical currents are confined to magnetic fields below this line. The irreversibility fields extrapolated at zero A1.3

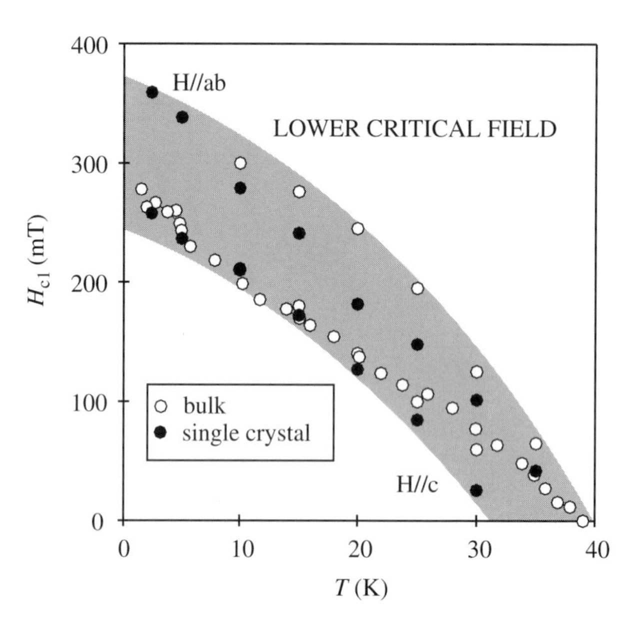

Figure C5.14. Lower critical field *versus* temperature.

temperature range between 6 and 12 T for MgB_2 bulk, films, wires, tapes and powders. A substantial enhancement of the irreversibility line accompanied by a significantly large J_c between 10^6 and 10^7 A cm^{-2} at 4.2 K and 1 T and have been reported in MgB_2 thin films with lower T_c [31, 99, 106]. These results give further encouragement to the development of MgB_2 for high current applications.

E1

C5.11 Critical current density *versus* applied magnetic field $J_c(H)$

C5.11.1 $J_c(H)$ in bulk

Many groups have measured the critical current density and its temperature and magnetic field dependence for different geometrical configurations of MgB_2, including powders [109, 116], bulk [47, 51, 109, 110, 113, 117–119], films [22, 23, 31, 32, 120–122], tapes [35, 40, 41, 43–45], and wires [33, 36–39].

B2.1

The consensus that seems to emerge is that, unlike in high temperature superconductors (HTSC), $J_c(T, H)$ in MgB_2 is determined by its pinning properties and not by weak link effects. These pinning properties are strongly field dependent, becoming rather poor in modest magnetic fields. The inductive measurements indicate that in dense bulk samples, the microscopic current density is practically identical to the intragranular J_c measured in dispersed powders [109], therefore, the current is not limited by grain boundaries [123].

In figure C5.15, a shadowed region indicates critical current *versus* applied magnetic field, $J_c(H)$, for bulk MgB_2 samples, taken between 5 and 30 K. For comparison, the $J_c(H)$ data for Nb–Ti [124] and Nb_3Sn [125] at 4.2 K are shown. In self fields bulk MgB_2 achieves moderate critical current densities, up to 10^6 A cm^{-2}. In applied magnetic fields of 6 T, J_c remains above 10^4 A cm^{-2}, while in a 10 T field, J_c is about 10^2 A cm^{-2}.

D2.4

C5.11.2 $J_c(H)$ in powders

Figure C5.16 shows the critical current density *versus* field for MgB_2 powders [51, 109, 116]. Very high current densities can be achieved in low fields, of up to 3×10^6 A cm^{-2}. However, magnetic fields of 7 T

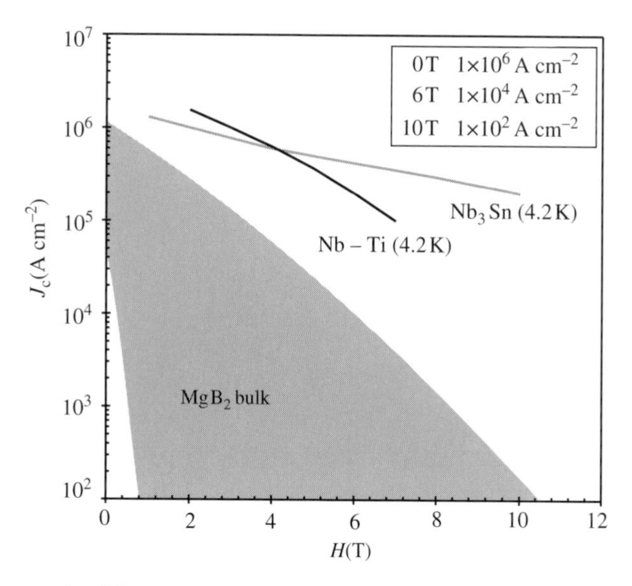

Figure C5.15. Critical current densities *versus* magnetic field for MgB_2 bulk samples. The data for Nb–Ti and Nb_3Sn at 4.2 K are shown for comparison.

quench the current density to low values: 10^2 A cm^{-2}, indicating that for powders $J_c(H)$ has a steeper dependence on field than bulk MgB_2.

C5.11.3 $J_c(H)$ in wires and tapes

Figure C5.17 shows the critical current density dependence in magnetic field for MgB_2 wires and tapes [33, 35–39, 43–45, 126]. Compared to bulk and powder MgB_2, the wires and tapes have lower values of J_c B3.3.6

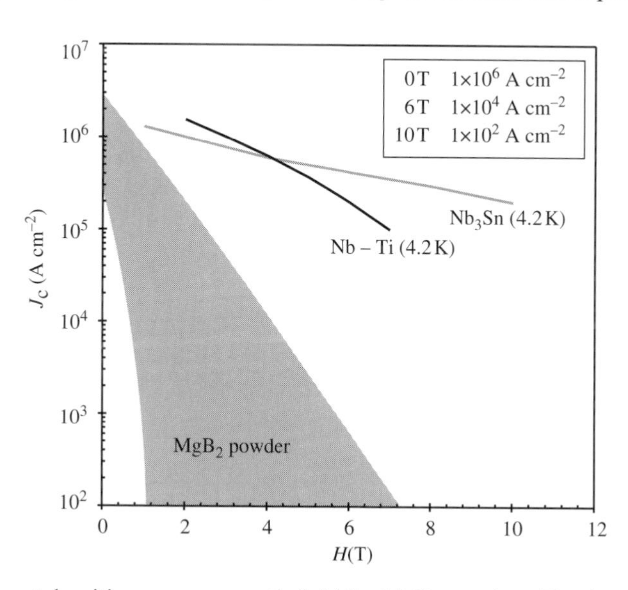

Figure C5.16. Critical current densities *versus* magnetic field for MgB_2 powders. The data for Nb–Ti and Nb_3Sn at 4.2 K are shown for comparison.

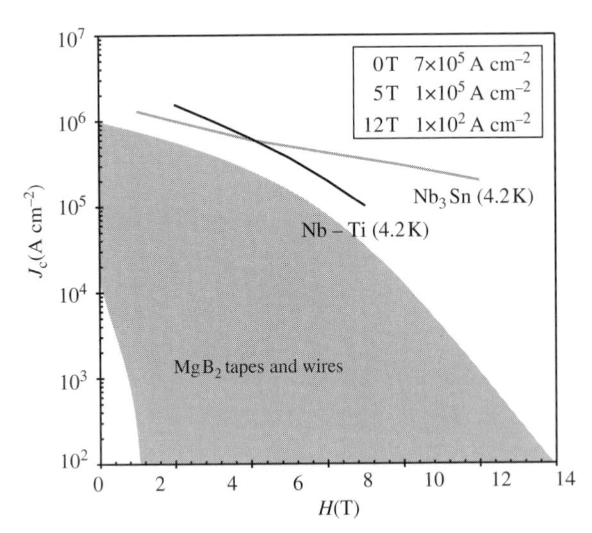

Figure C5.17. Critical current densities *versus* magnetic field for MgB_2 wires and tapes. The data for Nb–Ti and Nb_3Sn at 4.2 K are shown for comparison.

in low fields, about 6×10^5 A cm^{-2}. However, the $J_c(H)$ dependence becomes more gradual in field, allowing larger current density values in higher fields, $J_c(5\,\mathrm{T}) > 10^5$ A cm^{-2}. Due to geometrical shielding properties, the tapes can achieve superior currents in relatively high magnetic fields than the wires.

Suo *et al* [126] found that annealing of the tapes increases core density and sharpens the superconducting transition, raising J_c by more than a factor of 10 [126].

Wang *et al* [39] studied the effect of sintering time on the critical current density of MgB_2 wires. They found that there is no need for prolonged heat treatment in the fabrication of Fe-clad wires. Several minutes sintering gives the same performances as longer sintering time. Therefore, these findings substantially simplify the fabrication process and reduce the cost for large-scale production of MgB_2 wires.

Jin *et al* [38] showed that alloying MgB_2 with Ti, Ag, Cu, Mo, and Y has an important effect on J_c, despite the fact that T_c remains unaffected or slightly reduced by these elements. Iron addition seems to be least damaging, whereas Cu addition causes J_c to be significantly reduced by 2–3 orders of magnitude.

Iron is also beneficial as metal cladding, as it shields the core from external fields, the shielding being less effective for fields parallel to the tape plane [45]. When there is no external field, the transport current will generate a self-field surrounding the tape. Since Fe is ferromagnetic, the flux lines will suck into the Fe sheath, particularly at the edges of the tape. Therefore, the sheath will reduce the effect of self-field on I_c. When external fields are applied, the Fe sheath acts as a shield, reducing the effect of external field. Therefore, using Fe clad tapes may be beneficial for power transmission lines. B3.3.6

In order to increase J_c in wires and tapes, the fabrication process must be optimized by using finer starting powders or by incorporating nanoscale chemically inert particles that inhibit the grain growth. B2.1

C5.11.4 $J_c(H)$ in thin films

Figure C5.18 shows the values of critical current density versus magnetic field in MgB_2 films [22, 23, 31]. Surprisingly, the data for thin films show that the performances of MgB_2 can rival and perhaps eventually exceed the performances of existing superconducting wires. Figure C5.18 shows that in low fields, the current density in MgB_2 is higher [22, 31] than the current in Nb_3Sn films [125] and Nb–Ti [124]. In larger magnetic fields J_c in MgB_2 decreases faster than for Nb–Sn and Nb–Ti superconductors. B3.3.2 B3.3.3

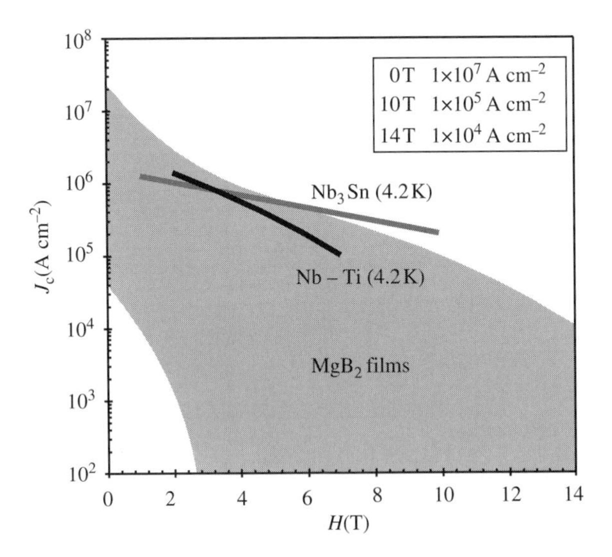

Figure C5.18. Critical current densities *versus* magnetic field for MgB_2 films. The data for Nb–Ti and Nb_3Sn at 4.2 K are shown for comparison.

However, a J_c of 10^4 A cm^{-2} can be attained in 14 T for films with oxygen and MgO incorporated [31]. These high current densities, exceeding 1 MA cm^{-2}, measured in films [22, 31], demonstrate the potential for further improving the current carrying capabilities of wires and tapes.

D2.2
B3.3.6

C5.11.5 Highest $J_c(H)$ at different temperatures

MgB_2 has a great potential for high-current and high-field applications as well as for microelectronics. MgB_2 Josephson junctions may be much easier to fabricate than those made from HTSC, performing like conventional superconductors (Nb, NbN) but operating at much higher temperatures. In particular, as shown in figure C5.19(*a*), MgB_2 has critical current density in low temperatures similar to the best existing superconductors.

To date several authors have succeeded in improving the J_c of MgB_2 by oxygen alloying [31] and proton irradiation [116], while others have studied the influence of doping [38] or sample preparation [109] on J_c.

B2.6

To take advantage of the relatively high T_c (39 K) of MgB_2, it is important to have high J_c values at temperatures above 20 K. The boiling point of H at atmospheric pressure is 20.13 K, so that is possible to use liquid hydrogen as cryogen for cooling MgB_2. Figure C5.19(*b*) shows the best values of $J_c(H)$ for temperatures of 25–30 K, respectively. For applications above 20 K it will be necessary to improve the flux-pinning properties through structural and microstructural modifications (e.g. chemical doping, introduction of precipitates, and atomic-scale control of defects such as vacancies, dislocations, grain boundaries).

A4.3

C5.11.6 Absence of weak links

Many magnetization and transport measurements show that MgB_2 does not exhibit weak-link electromagnetic behaviour at grain boundaries [111] or fast flux creep [127], phenomena which limit the performances of high-T_c superconducting cuprates.

A4.3

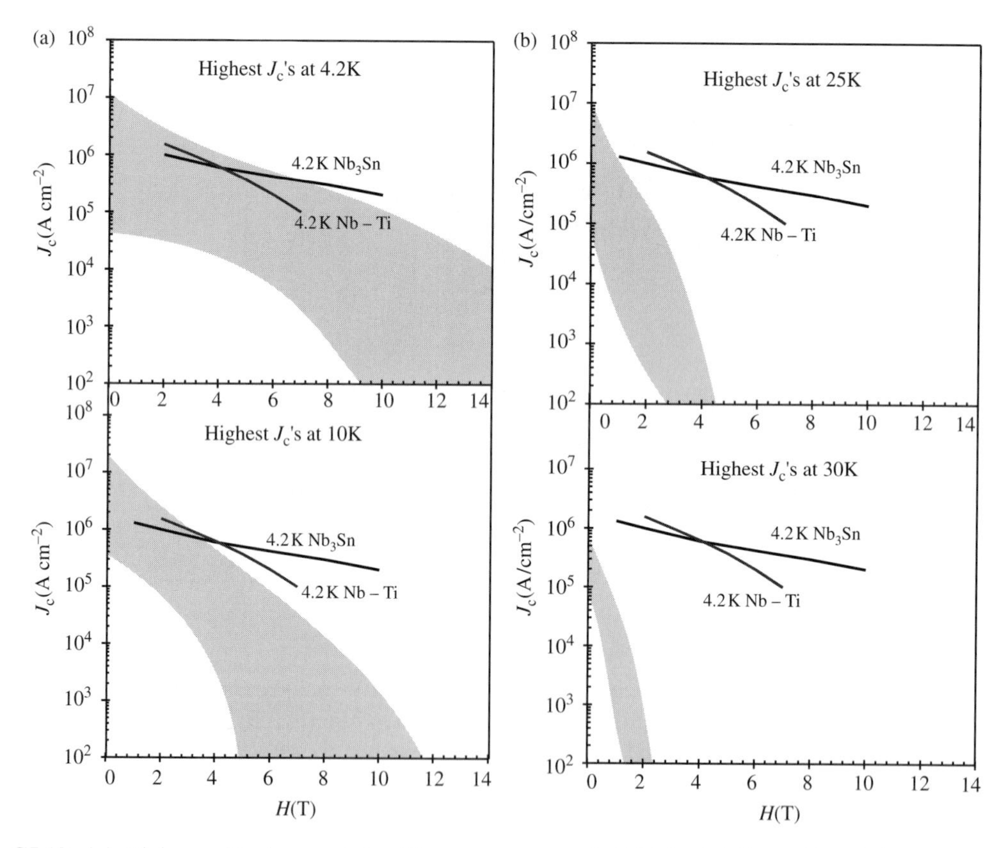

Figure C5.19. (*a*) Highest critical current densities *versus* magnetic field for MgB_2 at 4.2 and 10 K. The data for Nb–Ti and Nb_3Sn at 4.2 K are shown for comparison. (*b*) Highest critical current densities versus magnetic field for MgB_2 at 25 and 30 K. The data for Nb–Ti and Nb_3Sn at 4.2 K are shown for comparison.

As stated previously, high critical current densities have been observed in bulk samples, regardless of the degree of grain alignment [121, 126]. This would be an advantage for making wires or tapes with no degradation of J_c, in contrast to the degradation due to grain boundary induced weak-links, which is a common and serious problem in cuprate HTS.

A1.3

The transport measurements in high magnetic fields of dense MgB_2 bulk samples yield very similar J_c values as the inductive measurements [109, 121]. This confirms that the inductive current flows coherently throughout the sample, unaffected by grain boundaries. Therefore, the flux motion will determine J_c dependence in field and temperature.

A4.2

C5.12 Energy gap

There is no consensus yet about the gap values in MgB_2 and whether or not this material has a single anisotropic gap or a double gap, as shown in figure C5.20.

Energy gap values have been inferred by using tunneling spectroscopy [115, 127, 129–133], point contact tunnelling [134–138], specific heat studies [139–145], high-resolution photoemission spectroscopy (HRPS) [146, 147], far-infrared transmission studies (FIRT) [148–150], Raman spectroscopy [151, 152], and tunnelling junctions [153]. Energy gaps in superconductors are usually

D1.6

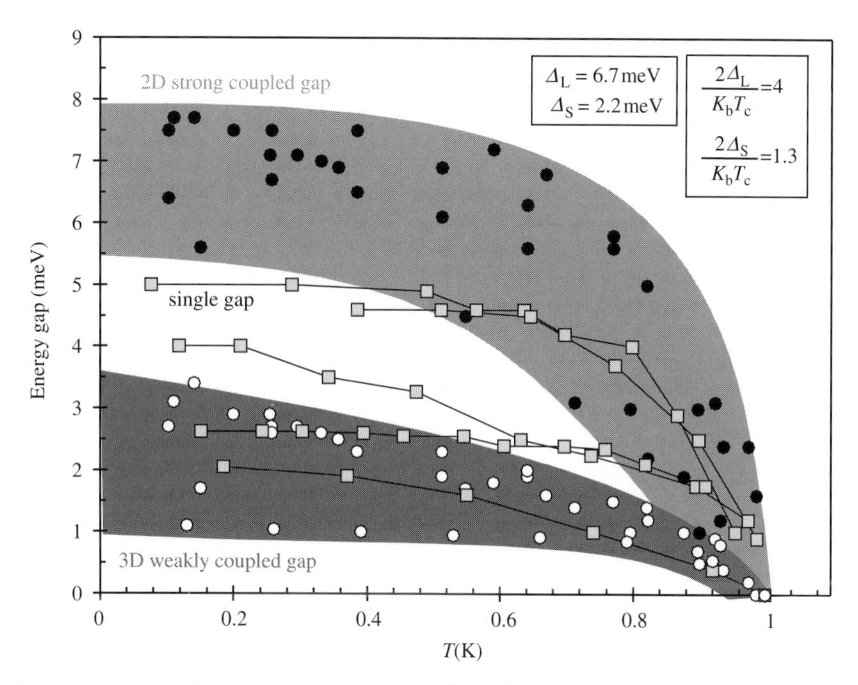

Figure C5.20. Energy gap dependence on temperature obtained from point contact spectroscopy, HRPS, scanning tunnelling spectroscopy, tunnelling, FIRT, and Raman spectroscopy experiments.

investigated by spectroscopic techniques, which are subjected to errors associated with surface impurities or non-uniformity. In the case of MgB_2, the gap structure is so pronounced that specific heat measurements can be used to infer its values. D3.2.1

As shown in figure C5.20, several experiments measured a single gap, with values between 2.5 and 5 meV, while latest experiments claim to have brought some clarification about the gap features in MgB_2. According to tunneling spectroscopy, point contact spectroscopy, and Raman scattering, there is evidence, suggested earlier by Liu *et al* [154], of two distinct gaps associated with the two separate segments of the Fermi surface [155]. The width values of these two gaps were determined to be between 1.8 and 3 meV for the small three-dimensional weakly coupled gap, and between 5.8 and 7.7 meV for the large strongly coupled gap. D1.6

Specific heat measurements show that it is necessary to involve either two gaps or a single anisotropic gap to describe the data.

Microwave measurement results can be explained by the existence of an anisotropic superconducting gap or the presence of a secondary phase, with lower gap width, in some of the MgB_2 samples. A2.5

C5.13 Conclusions

To summarize, MgB_2 has an unusual high critical temperature of about 40 K among binary compounds, with an AlB_2-type structure with graphite-type boron layers separated by hexagonal close-packed layers of Mg. The presence of the light boron as well as its layered structure may be important factors that contribute to superconductivity at such a high temperature for a binary compound.

According to initial findings, MgB_2 seemed to be a low-T_c superconductor with a remarkably high critical temperature. Its properties resemble those of conventional superconductors rather than of

Table C5.3. List of superconducting parameters of MgB_2

Parameter	Values
Critical temperature	$T_c = 39$–40 K
Hexagonal lattice parameters	$a = 0.3086$ nm, $b = 0.3524$ nm
Theoretical density	$\rho = 2.55$ g cm^{-3}
Pressure coefficient	$dT_c/dP = -1.1$–2 K Gpa^{-1}
Carrier density	$n_s = 1.7$–2.8×10^{23} holes cm^{-3}
Isotope effect	$\alpha_T = \alpha_B + \alpha_{Mg} = 0.3 + 0.02$
Resistivity near T_c	$\rho(40\,K) = 0.4$–$16\,\mu\Omega$cm
Resistivity ratio	$RR = \rho(40\,K)/\rho(300\,K) = 1$–$27$
Upper critical field	$H_{c2}//ab(0) = 14$–39 T, $H_{c2}//c(0) = 2$–24 T
Lower critical field	$H_{c1}(0) = 27$–48 mT
Coherence lengths	$\xi_{ab}(0) = 3.7$–12 nm, $\xi_c(0) = 1.6$–3.6 nm
Penetration depths	$\lambda(0) = 85$–180 nm
Energy gap	$\Delta(0) = 1.8$–7.5 meV
Critical current densities	$J_c(4.2\,K, 0\,T) > 10^7$ A cm^{-2}
	$J_c(4.2\,K, 4\,T) = 10^6$ A cm^{-2}
	$J_c(4.2\,K, 10\,T) > 10^5$ A cm^{-2}
	$J_c(25\,K, 0\,T) > 5 \times 10^6$ A cm^{-2}
	$J_c(25\,K, 2\,T) > 10^5$ A cm^{-2}

high-T_c cuprates. These properties include isotope effect, a linear T-dependence of the upper critical field with a positive curvature near T_c (similar to borocarbides), and a shift to lower temperatures of both D3.1 T_c(onset) and T_c(end) at increasing magnetic fields as observed in resistivity $R(T)$ measurements. On the other hand, the quadratic T-dependence of the penetration depth ($\lambda(T)$), as well as the sign reversal of the Hall coefficient near T_c, indicates unconventional superconductivity similar to cuprates.

Table C5.3 presents a list with the most important parameters of MgB_2.

Altogether, its relatively low fabrication cost, high critical current and field, large coherence length, F high critical temperature (39 K), and absence of weak links, make MgB_2 a promising material for applications at above 20.13 K, the temperature of boiling hydrogen at normal pressure.

References

[1] Akimitsu J 2001 *Symposium on Transition Metal Oxides* Sendai 10 January 2001
[2] Nagamatsu J, Nakagawa N, Muranaka T, Zenitani Y and Akimitsu J 2001 Superconductivity at 39 K in magnesium diboride *Nature* **410** 63
[3] Buzea C and Yamashita T 2001 Review of superconducting properties of MgB_2 *Supercond. Sci. Technol.* **14** R115
[4] Kiessling R 1949 *Acta Chem. Scand.* **3** 603
[5] Cooper A S, Corenzwit E, Longinotti L D, Matthias B T and Zachariasen W H 1970 *Proc. Natl Acad. Sci. USA* **67** 313
[6] Leyarovska L and Leyarovski E 1979 *J. Less-Common Met.* **67** 249
[7] Satta G, Profeta G, Bernardini F, Continenza A and Massidda S 2001 Electronic and structural properties of superconducting MgB_2, $CaSi_2$ and related compounds *Phys. Rev.* B **64** 104507
[8] Neaton B and Perali A 2001 On the possibility of superconductivity at higher temperatures in sp-valent diborides *Preprint* cond-mat/0104098
[9] Medvedeva N I, Ivanovskii A L, Medvedeva J E and Freeman A J 2001 Band structure of superconducting MgB_2 compound and modeling of related ternary systems *JETP Lett.* **73** 336
[10] Medvedeva N I, Ivanovskii A L, Medvedeva J E and Freeman A J 2001 Electronic structure of superconducting MgB_2 and related binary and ternary borides *Phys. Rev.* B **64** 020502R

[11] Medvedeva N I, Ivanovskii A L, Medvedeva J E, Freeman A J and Novikov D L 2001 Electronic structure and electric field gradient in MgB_2 and related s-, p- and d-metal diborides: possible correlation with superconductivity *Phys. Rev.* B **65** 052501

[12] Mehl M J, Papaconstantopoulos D A and Singh D J 2001 Effects of C, Cu and Be substitutions in superconducting MgB_2 *Phys. Rev.* B **64** 140509

[13] Kwon S K, Youn S J, Kim K S and Min B I 2001 New high temperature diboride superconductors: AgB_2 and AuB_2 *Preprint* cond-mat/0106483

[14] Ravindran P, Vajeeston P, Vidya R, Kjekshus A and Fjellvig H 2001 Detailed electronic structure studies on superconducting MgB_2 and related compound *Phys. Rev.* B **64** 224509

[15] Kaczorowski D, Zaleski A J, Zogal O J and Klamut J 2001 Incipient superconductivity in TaB_2 *Preprint* cond-mat/0103571

[16] Gasparov V A, Sidorov N S, Zver'kova I I and Kulakov M P 2001 Electron transport in diborides: observation of superconductivity in ZrB_2 *JETP Lett.* **73** 532

[17] Felner I 2001 Absence of superconductivity in BeB_2 *Physica* C **353** 11

[18] Young D P, Adams P W, Chan J Y and Fronczek F R 2001 Structure and superconducting properties of BeB2 *Preprint* cond-mat/0104063

[19] Liu Z K, Scholm D G, Li Q and Xi X X 2001 Thermodynamics of the Mg–B system: implications for the deposition of MgB_2 thin films *Appl. Phys. Lett.* **78** 3678

[20] Fan Z Y, Hinks D G, Newman N and Rowell J M 2001 Experimental study of MgB_2 decomposition *Appl. Phys. Lett.* **79** 87

[21] Zhai H Y, Christen H M, Zhang L, Paranthaman M, Cantoni C, Sales B C, Fleming P H, Christen D K and Lowndes D H 2001 Growth mechanism of superconducting MgB_2 films prepared by various methods *Mater. Res.* **16** 2759

[22] Kim H J, Kang W N, Choi E M, Kim M S, Kim K H P and Lee S I 2001 High current-carrying capability in *c*-axis-oriented superconducting MgB_2 thin films *Phys. Rev. Lett.* **87** 087002

[23] Paranthaman M, Cantoni C, Zhai H Y, Christen H M, Aytug T, Sathyamurthy S, Specht E D, Thompson J R, Lowndes D H, Kerchner H R and Christen D K 2001 Superconducting MgB_2 films via precursor post-processing approach *Appl. Phys. Lett.* **78** 3669

[24] Kang W N, Kim H J, Choi E M, Jung C U and Lee S I 2001 MgB_2 superconducting thin films with a transition temperature of 39 Kelvin *Science* **292** 1521

[25] Plecenik A, Satrapinsky L, Kus P, Gazi S, Benacka S, Vavra I and Kostic I 2001 MgB_2 superconductor thin films on Si and Al_2O_3 substrates *Physica* C **363** 224

[26] Wang S F, Dai S Y, Zhou Y L, Chen Z H, Cui D F, Yu J D, He M, Lu H B and Yang G Z 2001 Superconducting MgB_2 thin films with T_c of about 39 K grown by pulsed laser deposition *Chin. Phys. Lett.* **18** 967

[27] Zeng X H, Sukiasyan A, Xi X X, Hu Y F, Wertz E, Li Q, Tian W, Sun H P, Pan X Q, Lettieri J, Schlom D G, Brubaker C O, Liu Z K and Li Q 2001 Superconducting properties of nanocrystalline MgB_2 thin films made by an in situ annealing process *Appl. Phys. Lett.* **79** 1840

[28] Grassano G, Ramadan W, Ferrando V, Bellingeri E, Marre D, Ferdeghini C, Grasso G, Putti M, Siri A S, Manfrinetti P, Palenzona A and Chincarini A 2001 In-situ magnesium diboride superconducting thin films grown by pulsed laser deposition *Supercond. Sci. Technol.* **14** 762

[29] Christen H M, Zhai H Y, Cantoni C, Paranthaman M, Sales B C, Rouleau C, Norton D P, Christen D K and Lowndes D H 2001 Superconducting magnesium diboride films with $T_c = 24$ K grown by pulsed laser deposition with in situ anneal *Physica* C **353** 157

[30] Buzea C, Wang H B, Nakajima K, Kim S J and Yamashita T 1999 Comprehensive study of the film surface temperature and plasma thermo-kinetics during $La_{1.85}Sr_{0.15}CuO_4$ deposition by laser ablation *J. Appl. Phys.* **86** 2856

[31] Eom C B *et al* 2001 Thin film magnesium boride superconductor with very high critical current density and enhanced irreversibility field *Nature* **411** 558

[32] Li A H, Wang X L, Ionescu M, Soltonian S, Horvat J, Silver T, Liu H K and Dou S X 2001 Fast formation and superconductivity of MgB_2 thick films grown on stainless steel substrate *Physica* C **361** 73

[33] Canfield P C, Finnemore D K, Bud'ko S L, Ostenson J E, Lapertot G, Cunningham C E and Petrovic C 2001 Superconductivity in dense MgB_2 wires *Phys. Rev. Lett.* **86** 2423

[34] Cunningham C E, Petrovic C, Lapertot G, Bud'ko S L, Laabs F, Strazheim W, Finnemore D K and Canfield P C 2001 Synthesis and processing of MgB_2 powders and wires *Physica* C **353** 5

[35] Che G C, Li S L, Ren Z A, Li L, Jia S L, Ni Y M, Chen H, Dong C, Li J Q, Wen H H and Zhao Z X 2001 Preparation and superconductivity of a MgB_2 superconducting tape *Preprint* cond-mat/0105215

[36] Glowacki B A, Majoros M, Vickers M, Evvets J E, Shi Y and McDougall I 2001 Superconductivity of powder-in-tube MgB_2 wires *Supercond. Sci. Technol.* **14** 193

[37] Goldacker W, Schlachter S I, Zimmer S and Reiner H 2001 High transport currents in mechanically reinforced MgB_2 wires *Supercond. Sci. Technol.* **14** 787

[38] Jin S, Mavoori H and van Dover R B 2001 High critical currents in iron-clad superconducting MgB_2 wires *Nature* **411** 563

[39] Wang X L, Soltanian S, Horvat J, Qin M J, Liu H K and Dou S X 2001 Very fast formation of superconducting MgB_2/Fe wires with high J_c *Physica* C **361** 149

[40] Grasso G, Malagoli A, Ferdeghini C, Roncallo S, Braccini V, Cimberle M R and Siri A S 2001 Large transport critical currents in unsintered MgB_2 superconducting tapes *Appl. Phys. Lett.* **79** 230

[41] Sumption M D, Peng X, Lee E, Tomsic M and Collings E W 2001 Transport current in MgB_2 based superconducting strand at 4.2 K and self-field *Preprint* cond-mat/0102441

[42] Liu C F, Du S J, Yan G, Feng Y, Wu X, Wang J R, Liu X H, Zhang P X, Wu X Z, Zhou L, Cao L Z, Ruan K Q, Wang C Y, Li X G, Zhou G E and Zhang Y H 2001 Preparation of 18-filament Cu/NbZr/MgB_2 tape with high transport critical current density *Preprint* cond-mat/0106061

[43] Song K J, Lee N J, Jang H M, Ha H S, Ha D W, Oh S S, Sohn M H, Kwon Y K and Ryu K S 2001 Single-filament composite MgB_2/SUS ribbons by powder-in-tube process *Preprint* cond-mat/0106124

[44] Kumakura H, Matsumoto A, Fujii H and Togano K 2001 High transport critical current density obtained for powder-in-tube-processed MgB_2 tapes and wires using stainless steel and Cu–Ni tubes *Physica* C **363** 179

[45] Soltanian S, Wang X L, Kusevic I, Babic E, Li A H, Liu H K, Collings E W and Dou S X 2001 High transport critical current density above 30 K in pure Fe-clad MgB_2 tape *Physica* C **361** 84

[46] Handstein A, Hinz D, Fuchs G, Muller K H, Nenkov K, Gutfleisch O, Narozhnyi V N and Schultz L 2001 Fully dense MgB_2 superconductor textured by hot deformation *Preprint* cond-mat/0103408

[47] Frederick N A, Li S, Maple M B, Nesterenko V F and Indrakanti S S 2001 Improved superconducting properties of MgB_2 *Physica* C **363** 1

[48] Indrakanti S S, Nesterenko V F, Maple M B, Frederick N A, Yuhasz W M and Li S 2001 Hot isostatic pressing of bulk magnesium diboride: superconducting properties *Preprint* cond-mat/0105485

[49] Shields T C, Kawano K, Holdom D and Abell J S 2001 Microstructure and superconducting properties of hot isostatically pressed MgB_2 *Appl. Phys. Lett.* **79** 227

[50] Jung C U, Park M S, Kang W N, Kim M S, Kim K H P, Lee S Y and Lee S I 2001 Effect of sintering temperature under high pressure in the superconductivity for MgB_2 *Appl. Phys. Lett.* **78** 4157

[51] Takano Y, Takeya H, Fujii H, Kumakura H, Hatano T, Togano K, Kito H and Ihara H 2001 Superconducting properties of MgB_2 bulk materials prepared by high pressure sintering *Appl. Phys. Lett.* **78** 2914

[52] Tsvyashchenko A V, Fomicheva L N, Magnitskaya M V, Shirani E N, Brudanin V B, Filossofov D V, Kochetov O I, Lebedev N A, Novgorodov A F, Salamatin A V, Korolev N A, Velichkov A I, Timkin V V, Menushenkov A P, Kuznetsov A V, Shabanov V M and Akselrod Z Z 2001 Electric field gradients in MgB_2 synthesized at high pressure: 111Cd TDPAC study and ab initio calculation *Preprint* cond-mat/0104560

[53] Jung C U, Park M S, Kang W N, Kim M S, Lee S Y and Lee S I 2001 Temperature- and magnetic-field-dependences of normal state resistivity of MgB_2 prepared at high temperature and high pressure condition *Physica* C **353** 162

[54] Kim K H P, Choi J H, Jung C U, Chowdhury P, Park M S, Kim H J, Kim J Y, Du Z, Choi E M, Kim M S, Kang W N, Lee S Y, Sung G Y and Lee J Y 2001 Superconducting properties of well-shaped MgB_2 single crystal *Preprint* cond-mat/0105330

[55] Lee S, Mori H, Masui T, Eltsev Yu, Yamamoto A and Tajima S 2001 Growth, structure analysis and anisotropic superconducting properties of MgB_2 single crystals *Phys.Soc. Japan* **70** 2255

[56] Xu M, Kitazawa H, Takano Y, Ye J, Nishida K, Abe H, Matsushita A and Kido G 2001 Single crystal MgB_2 with anisotropic superconducting properties *Appl. Phys. Lett.* **79** 2779

[57] Kang W N, Jung C U, Kim K H P, Park M S, Lee S Y, Kim H J, Choi E M, Kim K H, Kim M S and Lee S I 2001 Hole carrier in MgB_2 characterized by Hall measurements *Appl. Phys. Lett.* **79** 982

[58] Kang W N, Ki K H P, Kim H J, Choi E M, Park M S, Kim M S, Du Z, Jung C U, Kim K H, Lee S I and Mun M O 2001 Fluctuation magnetoconductance in MgB_2 *Preprint* cond-mat/0103161

[59] Jin R, Paranthaman M, Zhai H Y, Christen H M, Christen D K and Mandrus D 2001 Unusual Hall effect in superconducting MgB_2 films: analogy to high- T_c cuprates *Phys. Rev.* B **64** 220506

[60] Gubser D U and Webb A W 1975 *Phys. Rev. Lett.* **35** 104

[61] Bordet P, Mezouar M, Nunez-Regueiro M, Monteverde M, Nunez-Regueiro M D, Rogado N, Regan K A, Hayward M A, He T, Loureiro S M and Cava R J 2001 Absence of a structural transition up to 40 Gpa in MgB_2 and the relevance of magnesium non-stoichiometry *Phys. Rev.* B **64** 172502

[62] Deemyad S, Schilling J S, Jorgensen J D and Hinks D G unpublished

[63] Goncharov A F, Struzhkin V V, Gregoryanz E, Mao H K, Hemley R J, Lapertot G, Bud'ko S L, Canfield P C and Mazin I I 2001 Pressure dependence of the Raman spectrum, lattice parameters and superconducting critical temperature of MgB_2 *Phys. Rev.* B **64** 100509

[64] Lorenz B, Meng R L and Chu C W 2001 High pressure study on MgB_2 *Phys. Rev.* B **64** 012507

[65] Lorenz B, Meng R L and Chu C W 2001 Hydrostatic pressure effect on the superconducting transition temperature of MgB_2 *Preprint* cond-mat/0104303

[66] Monteverde M, Nunez-Regueiro M, Rogado N, Regan K A, Hayward M A, He T, Loureiro S M and Cava R J 2001 Pressure dependence of the superconducting transition temperature of magnesium diboride *Science* **292** 75

[67] Saito E, Takenobu T, Ito T, Iwasa Y, Prassides K and Arima T 2001 Pressure dependence of T_c in the MgB_2 superconductor as probed by resistivity measurements *J. Phys. Condens. Matter* **13** L267

[68] Schlachter S I, Fietz W H, Grube K and Goldacker W 2001 High pressure studies of T_c and lattice parameters of MgB_2 *Preprint* cond-mat/0107205

[69] Tissen V G, Nefedova M V, Kolesnikov N N and Kulakov M P 2001 Effect of pressure on the superconducting T_c of MgB_2 *Physica* C **363** 194
[70] Tomita T, Hamlin J J, Schilling J S, Hinks D G and Jorgensen J D 2001 Dependence of T_c on hydrostatic pressure in superconducting MgB_2 *Phys. Rev.* B **64** 092505
[71] Hirsch J E 2001 Hole superconductivity in MgB_2: a high T_c cuprate without Cu *Phys. Lett.* A **282** 392
[72] Hirsch J E 2001 Hole superconductivity in MgB_2, cuprates, and other materials *Preprint* cond-mat/0106310
[73] Hirsch J E and Marsiglio F 2001 Electron–phonon or hole superconductivity in MgB_2? *Phys. Rev.* B **64** 144523
[74] Prassides K, Iwasa Y, Ito T, Chi D H, Uehara K, Nishibori E, Takata M, Sakata S, Ohishi Y, Shimomura O, Muranaka T and Akimitsu J 2001 Compressibility of the MgB_2 Superconductor *Phys. Rev.* B **64** 012509
[75] Vogt T, Schneider G, Hriljac J A, Yang G and Abell J S 2001 Compressibility and electronic structure of MgB_2 up to 8 GPa *Phys. Rev.* B **63** 220505
[76] Jorgensen J D, Hinks D G and Short S 2001 Lattice properties of MgB_2 versus temperature and pressure *Phys. Rev.* B **63** 224522
[77] Kortus J, Mazin I I, Belashchenko K D, Antropov V P and Boyer L L 2001 Superconductivity of metallic boron in MgB_2 *Phys. Rev. Lett.* **86** 4656
[78] Ahn J S and Choi E J 2001 Carbon substitution effect in MgB_2 *Preprint* cond-mat/0103169
[79] Paranthaman M, Thompson J R and Christen D K 2001 Effect of carbon-doping in bulk superconducting MgB_2 samples *Physica* C **355** 5
[80] Takenobu T, Itoh T, Chi D H, Prassides K and Iwasa Y 2001 Interlayer carbon substitution in the MgB_2 superconductor *Phys. Rev.* B **64** 134513
[81] Zhang S Y, Zhang J, Zhao T Y, Rong C B, Shen B G and Cheng Z H 2001 Structure and superconductivity of $Mg(B_{1-x}C_x)_2$ compounds *Preprint* cond-mat/0103203
[82] Bianconi A, Di Castro D, Agrestini S, Campi G, Saini N L, Saccone A, De Negri S and Giovannini M 2001 A superconductor made by a metal heterostructure at the atomic limit tuned at the 'shape resonance': MgB_2 *J. Phys. Condens. Matter* **13** 7383
[83] Cimberle M R, Novak M, Manfrinetti P and Palenzona A 2002 Magnetic characterization of sintered MgB_2 samples: effect of the substitution or doping with Li, Al and Si *Supercond. Sci. Technol.* **15** 43
[84] Li J Q, Li L, Liu F M, Dong C, Xiang J Y and Zhao Z X 2001 Superconductivity and aluminum ordering in $Mg_{1-x}Al_xB_2$ *Preprint* cond-mat/0104320
[85] Lorenz B, Meng R L, Xue Y Y and Chu C W 2001 Thermoelectric power and transport properties of pure and Al-doped MgB_2 *Phys. Rev.* B **64** 052513
[86] Slusky J S, Rogado N, Regan K A, Hayward M A, Khalifah P, He T, Inumaru K, Loureiro S M, Haas M K, Zandbergen H W and Cava R J 2001 Loss of superconductivity with the addition of Al to MgB_2 and a structural transition in $M_{1-x}Al_xB_2$ *Nature* **410** 343
[87] Xiang J Y *et al* 2001 Study of superconducting properties and observation of *c*-axis superstructure in $Mg_{1-x}Al_xB_2$ *Preprint* cond-mat/0104366
[88] Ogita N, Kariya T, Hiraoka K, Nagamatsu J, Muranaka T, Takagiwa H, Akimitsu J and Udagawa M 2001 Micro-Raman scattering investigation of MgB_2 and RB_2 (R = Al, Mn, Nb and Ti) *Preprint* cond-mat/0106147
[89] Postorino P, Congeduti A, Dore P, Nucara A, Bianconi A, Di Castro D, De Negri S and Saccone A 2002 Effect of Al doping on the optical phonon spectrum in $Mg_{1-x}Al_xB_2$ *Phys. Rev.* B **65** 020507
[90] Zhao Y G, Zhang X P, Qiao P T, Zhang H T, Jia S L, Cao B S, Zhu M H, Han Z H, Wang X L and Gu B L 2001 Effect of Li doping on structure and superconducting transition temperature of $Mg_{1-x}Li_xB_2$ *Physica* C **361** 91
[91] Kazakov S M, Angst M and Karpinski J 2001 Substitution effect of Zn and Cu in MgB_2 on T_c and structure *Preprint* cond-mat/0103350
[92] Moritomo Y and Xu S 2001 Effects of transition metal doping in MgB_2 superconductor *Preprint* cond-mat/0104568
[93] Bud'ko S L, Lapertot G, Petrovic C, Cunningham C E, Anderson N and Canfield P C 2001 Boron isotope effect in superconducting MgB_2 *Phys. Rev. Lett.* **86** 1877
[94] Hinks D G, Claus H and Jorgensen J D 2001 The complex nature of superconductivity in MgB_2 as revealed by the reduced total isotope effect *Nature* **411** 457
[95] Testardi R L, Meek R L, Poate J M, Royer W A, Storm A R and Wernick J H 1975 Preparation and analysis of superconducting Nb–Ge films *Phys. Rev.* B **11** 4303
[96] Testardi R L, Poate J M and Levinstein H L 1977 Anomalous electrical resistivity and defects in A-15 compounds *Phys. Rev.* B **15** 2570
[97] Poate J M, Testardi R L, Storm A R and Augustyniak W M 1975 4He-Induced damage in superconducting Nb–Ge films *Phys. Rev. Lett.* **35** 1290
[98] Park M A, Savran K and Kim Y J 2001 A new method of probing the phonon mechanism in superconductors, including MgB_2 *Supercond. Sci. Technol.* **14** L31
[99] Patnaik S, Cooley L D, Gurevich A, Polyanskii A A, Jiang J, Cai X Y, Squitieri A A, Naus M T, Lee M K, Choi J H, Belenky L, Bu S D, Letteri J, Song X, Schlom D G, Babcock S E, Eom C B, Hellstrom E E and Larbalestier D C 2001 Electronic

anisotropy, magnetic field-temperature phase diagram and their dependence on resistivity in *c*-axis oriented MgB_2 thin films *Supercond. Sci. Technol.* **14** 315

[100] Jung M H, Jaime M, Lacerda A H, Boebinger G S, Kang W N, Kim H J, Choi E M and Lee S I 2001 Anisotropic superconductivity in epitaxial MgB_2 films *Chem. Phys. Lett.* **343** 447

[101] Fuchs G, Muller K H, Handstein A, Nenkov K, Narozhnyi V N, Eckert D, Wolf M and Schultz L 2001 Upper critical field and irreversibility line in superconducting MgB_2 *Solid State Commun.* **118** 497

[102] Bud'ko S L, Petrovic C, Lapertot G, Cunningham C E, Canfield P C, Jung M H and Lacerda A H 2001 Magnetoresistivity and $Hc_2(T)$ in MgB_2 *Phys. Rev.* B **63** 220503

[103] Shulga S V, Drechsler S L, Fucks G, Muller K H, Winzer K, Heinecke M and Krug K 1998 *Phys. Rev. Lett.* **80** 1730

[104] de Lima O F, Ribeiro R A, Avila M A, Cardoso C A and Coelho A A 2001 Anisotropic superconducting properties of aligned MgB_2 crystallites *Phys. Rev. Lett.* **86** 5974

[105] de Lima O F, Cardoso C A, Ribeiro R A, Avila M A and Coelho A A 2001 Angular dependence of the bulk nucleation field H_{c2} of aligned MgB_2 crystallites *Phys. Rev.* B **64** 144517

[106] Ferdeghini C *et al* 2001 Growth of *c*-oriented MgB_2 thin films by pulsed laser deposition: structural characterization and electronic anisotropy *Supercond. Sci. Technol.* **14** 952

[107] Bud'ko S L, Kogan V G and Canfield P C 2001 Determination of superconducting anisotropy from magnetization data on random powders as applied to $LuNi_2B_2C$, YNi_2B_2C and MgB_2 *Phys. Rev.* B **64** 180501

[108] Simon F *et al* 2001 Anisotropy of superconducting MgB_2 as seen in electron spin resonance and magnetization data *Phys. Rev. Lett.* **87** 047002

[109] Dhalle M, Toulemonde P, Beneduce C, Musolino N, Decroux M and Flukiger R 2001 Transport and inductive critical current densities in superconducting MgB_2 *Physica* C **363** 155

[110] Finnemore D K, Ostenson J E, Bud'ko S L, Lapertot G and Canfield P C 2001 Thermodynamic and transport properties of superconducting MgB_2 *Phys. Rev. Lett.* **86** 2420

[111] Larbalestier D C *et al* 2001 Strongly linked current flow in polycrystalline forms of the superconductor MgB_2 *Nature* **410** 186

[112] Muller K H, Fuchs G, Handstein A, Nenkov K, Narozhnyi V N and Eckert D 2001 The upper critical field in superconducting MgB_2 *J. Alloys Compd.* **322** L10

[113] Joshi A G, Pillai C G S, Raj P and Malik S K 2001 Magnetization studies on superconducting MgB_2 – lower and upper critical fields and critical current density *Solid State Commun.* **118** 445

[114] Li S L, Wen H H, Zhao Z W, Ni Y M, Ren Z A, Che G C, Yang H P, Liu Z Y and Zhao Z X 2001 Lower critical field at odds with a s-wave superconductivity in the new superconductor MgB_2 *Phys. Rev.* B **64** 094522

[115] Sharoni A, Felner I and Millo O 2001 Tunneling spectroscopy measurement of the superconducting properties of MgB_2 *Phys. Rev.* B **63** 220508R

[116] Bugoslavsky Y, Cohen L F, Perkins G K, Polichetti M, Tate T J, Gwilliam R and Caplin A D 2001 Enhancement of the high-field critical current density of superconducting MgB_2 by proton irradiation *Nature* **411** 561

[117] Bugoslavsky Y, Perkins G K, Qi X, Cohen L F and Caplin A D 2001 Critical currents and vortex dynamics in superconducting MgB_2 *Preprint* cond-mat/0102353

[118] Kambara M, Hari Babu N, Sadki E S, Cooper J R, Minami H, Cardwell D A, Campbell A M and Inoue I H 2001 High intergranular critical currents in metallic MgB_2 superconductor *Supercond. Sci. Technol.* **14** L5

[119] Wen H H, Li S L, Zhao Z W, Ni Y M, Ren Z A, Che G C and Zhao Z X 2001 Magnetic relaxation and critical current density of MgB_2 films *Phys. Rev.* B **64** 134505

[120] Johansen T H, Baziljevich M, Shantsev D V, Goa P E, Galperin Y M, Kang W N, Kim H J, Choi E M, Kim M S and Lee S I 2001 Complex flux dynamics in MgB_2 films *Preprint* cond-mat/0104113

[121] Kim K H P, Kang W N, Kim M S, Jung C U, Kim H J, Choi E M, Park M S and Lee S I 2001 Origin of the high dc transport critical current density for the MgB_2 superconductor *Preprint* cond-mat/0103176

[122] Moon S H, Yun J H, Lee H N, Kye J I, Kim H G, Chung W and Oh B 2001 High critical current densities in superconducting MgB_2 thin films *Appl. Phys. Lett.* **79** 2429

[123] Kawano K, Abell J S, Kambara M, Hari Babu N and Cardwell D A 2001 Evidence for high inter-granular current flow in single-phase polycrystalline MgB_2 superconductor *Appl. Phys. Lett.* **79** 2216

[124] Heussner R W, Marquardt J D, Lee P J and Larbalestier D C 1997 Increased critical current density in Nb–Ti wires having Nb artificial pinning centers *Appl. Phys. Lett.* **70** 17

[125] Kim Y B and Stephen M J 1969 Flux flow and irreversible effects. In *Superconductivity* ed Parks R D (New York: Dekker) **2** p. 1107

[126] Suo H L, Beneduce C, Dhalle M, Musolino N, Genoud J Y and Flukiger R 2001 Large transport critical currents in dense Fe- and Ni-clad MgB_2 superconducting tapes *Appl. Phys. Lett.* **79** 3116

[127] Thompson J R, Paranthaman M, Christen D K, Sorge K D, Kim H J and Ossandon J G 2001 High temporal stability of supercurrents in MgB_2 materials *Supercond. Sci. Technol.* **14** L17

[128] Karapetrov G, Iavarone M, Kwok W K, Crabtree G W and Hinks D G 2001 Scanning tunneling spectroscopy in MgB_2 *Phys. Rev. Lett.* **86** 4374

[129] Sharoni A, Millo O, Leitus G and Reich S 2001 Spatial variations of the superconductor gap structure in MgB_2/Al composite *J. Phys. Condens. Matter.* **13** L503

[130] Chen C T, Seneor P, Yeh N C, Vasquez R P, Jung C U, Park M S, Kim H J, Kang W N and Lee S I 2001 Spectroscopic evidence for anisotropic S-wave pairing symmetry in MgB_2 *Preprint* cond-mat/0104285
[131] Giubileo F, Roditchev D, Sacks W, Lamy R and Klein J 2001 Strong coupling and double gap density of states in superconducting MgB_2 *Preprint* cond-mat/0105146
[132] Giubileo F, Roditchev D, Sacks W, Lamy R, Thanh D X, Klein J, Miraglia S, Fruchart D and Monod Ph 2001 Two gap state density in MgB_2: a true bulk property or a proximity effect? *Phys. Rev. Lett.* **87** 177008
[133] Rubio-Bollinger G, Suderow H and Vieira S 2001 Tunneling spectroscopy in small grains of superconducting MgB_2 *Phys. Rev. Lett.* **86** 5582
[134] Schmidt H, Zasadzinski J F, Gray K E and Hinks D G 2001 Energy gap from tunneling and metallic sharvin contacts onto MgB_2: evidence for a weakened surface layer *Phys. Rev.* B **63** 220504
[135] Szabo P, Samuely P, Kacmarcik J, Klein Th, Marcus J, Fruchart D, Miraglia S, Marcenat C and Jansen A G M 2001 Evidence for two superconducting energy gaps in MgB_2 by point-contact spectroscopy *Phys. Rev. Lett.* **87** 137005
[136] Laube F, Goll G, Hagel, Luhneysen H, Ernst D and Wolf T 2001 Superconducting energy gap distribution of MgB_2 investigated by point-contact spectroscopy *Preprint* cond-mat/0106407
[137] Zhang Y, Kinion D, Chen J, Hinks D G, Crabtree G W and Clarke J 2001 MgB_2 tunnel junctions and 19 K low-noise dc superconducting quantum interference devices *Appl. Phys. Lett.* **79** 3995
[138] Gonnelli R S, Calzolari A, Daghero D, Ummarino G A, Stepanov V A, Fino P, Giunchi G, Ceresara S and Ripamonti G 2001 Temperature and junction-type dependency of Andreev reflection in MgB_2 *Preprint* cond-mat/0107239
[139] Kremer R K, Gibson B J and Ahn K 2001 Heat capacity of MgB_2: evidence for moderately strong coupling behaviour *Preprint* cond-mat/0102432
[140] Walti Ch, Felder E, Dengen C, Wigger G, Monnier R, Delley B and Ott H R 2001 Strong electron-phonon coupling in superconducting MgB_2: a specific heat study *Phys. Rev.* B **64** 172515
[141] Wang Y, Plackowski T and Junod A 2001 Specific heat in the superconducting and normal state (2–300 K, 0–16 Teslas), and magnetic susceptibility of the 38-K superconductor MgB_2: evidence for a multicomponent gap *Physica* C **355** 179
[142] Bauer E, Paul Ch, Berger St, Majumdar S, Michor H, Giovannini M, Saccone A and Bianconi A 2001 Thermal conductivity of superconducting MgB_2 *J. Phys. Condens. Matter.* **13** L487
[143] Junod A, Wang Y, Bouquet F and Toulemonde P 2001 Specific heat of the 38-K superconductor MgB_2 in the normal and superconducting state: bulk evidence for a double gap *Preprint* cond-mat/0106394
[144] Fisher R A, Bouquet F, Phillips N E, Hinks D G and Jorgensen J D 2001 Identification and characterization of two energy gaps in superconducting MgB_2 by specific-heat measurements *Preprint* cond-mat/0107072
[145] Bouquet F, Wang Y, Fisher R A, Hinks D G, Jorgensen J D, Junod A and Phillips N E 2001 Phenomenological two-gap model for the specific heat of MgB_2 *Euro. Phys. Lett.* **56** 856
[146] Takahashi T, Sato T, Souma S, Muranaka T and Akimitsu J 2001 High-resolution photoemission study of MgB_2 *Phys. Rev. Lett.* **86** 4915
[147] Tsuda S, Yokoya T, Kiss T, Takano Y, Togano K, Kitou H, Ihara H and Shin S 2001 Direct evidence for a multiple superconducting gap in MgB_2 from high-resolution photoemission spectroscopy *Phys. Rev. Lett.* **87** 177006
[148] Gorshunov G, Kuntscher C A, Haas P, Dressel M, Mena F P, Kuz'menko A B, van Marel D, Muranaka T and Akimitsu J 2001 Optical measurements of the superconducting gap in MgB_2 *Preprint* cond-mat/0103164
[149] Jung J H, Kim K W, Lee H J, Kim M W, Noh T W, Kang W N, Kim H J, Choi E M, Jung C U and Lee S I 2002 Far-infrared transmission studies of *c*-axis oriented superconducting MgB_2 thin film *Phys. Rev.* B **65** 052413
[150] Kaindl R A, Carnahan M A, Orenstein J, Chemla D S, Christen H M, Zhai H, Paranthaman M and Lowndes D H 2001 Far-infrared optical conductivity gap in superconducting MgB_2 films *Phys. Rev. Lett.* **88** 027003
[151] Chen X K, Konstantinovic M J, Irwin J C, Lawrie D D and Franck J P 2001 Investigation of the superconducting gap in MgB_2 by Raman spectroscopy *Phys. Rev. Lett.* **87** 157002
[152] Quilty J W, Lee S, Yamamoto A and Tajima S 2001 The superconducting gap in MgB_2: electronic Raman scattering measurements of single crystals *Phys. Rev. Lett.* **88** 087001
[153] Plecenik A, Benacka S, Kus P and Grajcar M 2002 Superconducting gap parameters of MgB_2 obtained on MgB_2/Ag and MgB_2/In junctions *Physica* C **368** 251
[154] Liu A Y, Mazin I I and Kortus J 2001 Beyond Eliashberg superconductivity in MgB_2: anharmonicity, two-phonon scattering, and multiple gaps *Phys. Rev. Lett.* **87** 087005
[155] Belashchenko K D, Antropov V P and Rashkeev S N 2001 Anisotropy of p states and ^{11}B nuclear spin-lattice relaxation in $(Mg,Al)B_2$ *Phys. Rev.* B **64** 132506